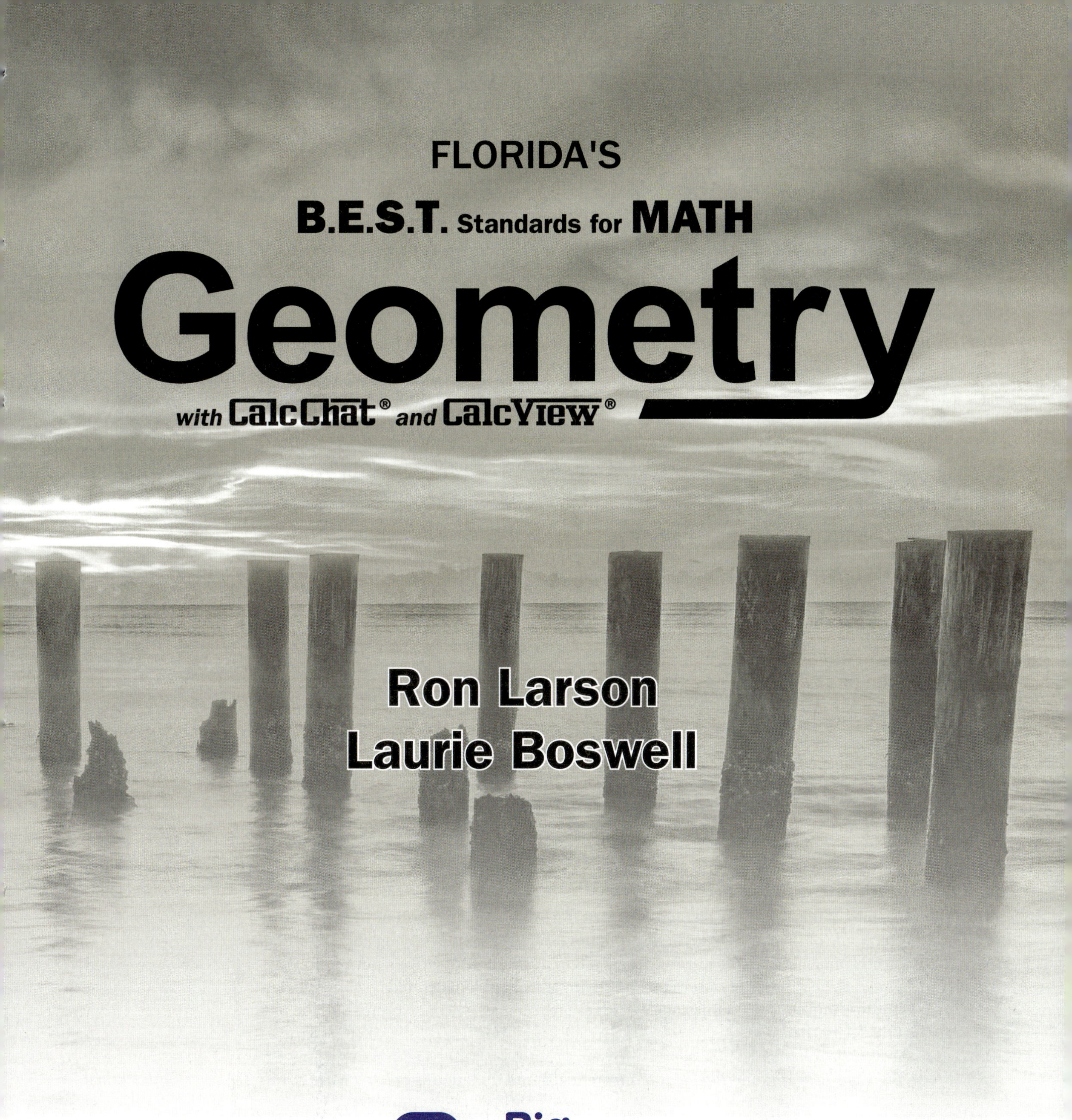

FLORIDA'S

B.E.S.T. Standards for **MATH**

Geometry

with CalcChat® *and* CalcView®

Ron Larson

Laurie Boswell

Big Ideas Learning™

Erie, Pennsylvania

BigIdeasLearning.com

Big Ideas Learning, LLC
1762 Norcross Road
Erie, PA 16510-3838
USA

For product information and customer support, contact Big Ideas Learning at **1-877-552-7766** or visit us at ***BigIdeasLearning.com***.

Cover:
finepic/Shutterstock.com

Front Matter:
bgblue/DigitalVision Vectors/Getty Images

Printed in the U.S.A.

ISBN 13: 978-1-63736-234-1

2 3 4 5 6 7 8 9—25 24 23 22

A Single Authorship Team

Written by renowned authors, Dr. Ron Larson and Dr. Laurie Boswell, *Florida's B.E.S.T. Standards for MATH* offers a seamless math pedagogy from Kindergarten through Algebra 2. Together, Ron and Laurie provide a consistent voice that encourages students to make connections through cohesive progressions and clear instruction. Since 1992, Ron and Laurie have authored over 50 mathematics programs.

"Each time Laurie and I start working on a new program, we spend time putting ourselves in the position of the reader. How old is the reader? What is the reader's experience with mathematics? The answers to these questions become our writing guides. Our goal is to make the learning targets understandable and to develop these targets in a clear path that leads to student success."

Ron Larson

Ron Larson, Ph.D., is well known as lead author of a comprehensive and widely used mathematics program that ranges from elementary school through college. He holds the distinction of Professor Emeritus from Penn State Erie, The Behrend College, where he taught for nearly 40 years. He received his Ph.D. in mathematics from the University of Colorado. Dr. Larson engages in the latest research and advancements in mathematics education and consistently incorporates key pedagogical elements to ensure focus, coherence, rigor, and student self-reflection.

"My passion and goal in writing is to provide an essential resource for exploring and making sense of mathematics. Our program is guided by research around the learning and teaching of mathematics in the hopes of improving the achievement of all students. May this be a successful year for you!"

Laurie Boswell

Laurie Boswell, Ed.D., is the former Head of School at Riverside School in Lyndonville, Vermont. In addition to authoring textbooks, she provides mathematics consulting and embedded coaching sessions. Dr. Boswell received her Ed.D. from the University of Vermont in 2010. She is a recipient of the Presidential Award for Excellence in Mathematics Teaching and later served as president of CPAM. Laurie has taught math to students at all levels, elementary through college. In addition, Laurie has served on the NCTM Board of Directors and as a Regional Director for NCSM. Along with Ron, Laurie has co-authored numerous math programs and has become a popular national speaker.

A Program Built for Florida

Florida Landmarks

Cocoa Beach Pier, Cocoa Beach, Florida

Pier 60, Clearwater, Florida

Old Pier, Naples, Florida

Opportunities for Deeper Thinking

Course Project

One of the course projects suggested at the end of this book is related to the Florida landmark on the cover. You can choose to study the numerous structures of the Old Pier, or perhaps an environmental issue related to the coastal region, or any other topic of interest to you. Several topics are suggested. Work on your project throughout the course as you investigate, explore, and analyze the world around you.

The Old Pier

The Old Pier is one of the best-known landmarks of Naples, Florida. Originally constructed in 1888, the pier has had an eventful life. Four hurricanes, a fire, and worms (yes—worms) have all damaged or destroyed the pier over the years, requiring a series of reconstructions and repairs.

Piers are surprisingly complex structures and contain numerous geometric and mathematical features. The length, width, and layout of the pier; the height and shape of the pilings; the dimensions of the building materials used, their number, and how they are assembled; and the depth of the water and contours of the ocean floor can all be described using geometry and other types of mathematics.

Research the Old Pier and its history. Pay particular attention to the specifications of the various reconstructions, repairs, and remodels the pier has undergone. Use what you learn to choose a topic and generate a driving question that you will seek to answer. Use factual information or invent fictitious stories and situations based on real-life information to create a presentation that answers your question and explains how the history of the Old Pier relates to what you are learning in Geometry.

Embedded Florida Honors Content

2.6 Solving Absolute Value Inequalities (H)

Learning Target: Write and solve inequalities involving absolute value.

Success Criteria:
- I can write a compound inequality related to a given absolute value inequality.
- I can solve absolute value inequalities.
- I can use absolute value inequalities to solve real-life problems.

EXPLORE IT! Solving an Absolute Value Inequality

Work with a partner. Consider the absolute value inequality

$|x + 2| \le 3.$

a. Explain what you think this inequality means.

b. Can you find a number that makes the inequality true? If so, what is the number?

c. Do you think there are other numbers that make the inequality true? If so, find several of them. Compare your answers with your classmates'.

d. On the real number line below, locate the point for which the expression $|x + 2|$ is equal to 0.

−10 −9 −8 −7 −6 −5 −4 −3 −2 −1 0 1 2 3 4 5 6 7 8 9 10

Then locate the numbers you found in parts (b) and (c) on the real number line. What do you notice?

e. Can you write two linear inequalities that use the expression $x + 2$ to represent the solutions of $|x + 2| \le 3$? Explain.

f. Repeat parts (b)–(e) for the inequality $|x + 2| \ge 3$. Compare your results with those for the inequality $|x + 2| \le 3$.

g. Describe how to find the solutions of the absolute value inequalities algebraically. Then find the solutions.

i. $|x - 4| \le 2$

ii. $|x - 4| \ge 2$

x_1	$\lvert x_1 - 4\rvert$
−6	10
−5	9
−4	
−3	
−2	
−1	
0	
1	
2	

5 MTR USE STRUCTURE How can you change one of the absolute value inequalities shown so that it has no solution?

3 MTR h. CHOOSE A METHOD Solve the absolute value inequalities in part (g) in a different way. Explain your method.

Algebraic Reasoning
(H) **MA.912.AR.4.2** Given a mathematical or real-world context, write and solve one-variable absolute value inequalities. Represent solutions algebraically or graphically.

GO DIGITAL

2.6 Solving Absolute Value Inequalities 85

Throughout this program, look for (H) to denote honors content. This icon may appear at the beginning of a section or within a section.

Dig Deeper

Go deeper in your thinking on every topic by accessing *Dig Deeper* problems online. These problems allow you to work with more complex and interconnected ideas to achieve deeper understanding of conceptual themes throughout your course. These additional higher order thinking problems are available for every section.

Dig Deeper

1. You and your friend run at constant rates. The linear function $d = 8t$ represents the distance d (in feet) that you run in t seconds. The table shows the distance that your friend runs over time.

Time (seconds), t	Distance (feet), d
20	170
40	340
60	510

a. Does the table re

b. Who is running a

Dig Deeper

1. Write a quadratic function represented by the table, if possible. If not, explain why.

x	−3	4	10	15
y	6	0	0	6

Research, Contributors, and Reviewers

Research

Ron Larson and Laurie Boswell used the latest in educational research, along with the body of knowledge collected from expert mathematics educators, to develop the *Florida's B.E.S.T. Standards for MATH* series. The pedagogical approach used in this program follows the best practices outlined in the most prominent and widely accepted educational research, including:

- *B.E.S.T. Standards for Mathematics*, Florida Department of Education ©2020
- *Visible Learning*, John Hattie ©2009
- *Visible Learning for Mathematics* John Hattie ©2017
- *Visible Learning Feedback* John Hattie ©2018
- *Teaching Mathematics in the Visible Learning Classroom, High School* John Almarode, Douglas Fisher, Joseph Assof, John Hattie, and Nancy Frey ©2018
- *The Teacher Clarity Playbook, Grades K–12* Douglas Fisher, Nancy Frey, Olivia Amador, and John Hattie ©2018
- *The Distance Learning Playbook, Grades K–12* Douglas Fisher, Nancy Frey, and John Hattie ©2020
- *Principles to Actions: Ensuring Mathematical Success for All* NCTM ©2014
- *Adding It Up: Helping Children Learn Mathematics* National Research Council ©2001
- *Mathematical Mindsets: Unleashing Students' Potential through Creative Math, Inspiring Messages and Innovative Teaching* Jo Boaler ©2015
- *What Works in Schools: Translating Research into Action* Robert Marzano ©2003
- *Classroom Instruction That Works: Research-Based Strategies for Increasing Student Achievement* Marzano, Pickering, and Pollock ©2001
- *Principles and Standards for School Mathematics* NCTM ©2000
- *Rigorous PBL by Design: Three Shifts for Developing Confident and Competent Learners* Michael McDowell ©2017
- *Universal Design for Learning Guidelines* CAST ©2011
- Rigor/Relevance Framework® International Center for Leadership in Education
- *Understanding by Design* Grant Wiggins and Jay McTighe ©2005
- Achieve, ACT, and The College Board
- *Elementary and Middle School Mathematics: Teaching Developmentally* John A. Van de Walle and Karen S. Karp ©2015
- *Evaluating the Quality of Learning: The SOLO Taxonomy* John B. Biggs & Kevin F. Collis ©1982
- *Unlocking Formative Assessment: Practical Strategies for Enhancing Students' Learning in the Primary and Intermediate Classroom* Shirley Clarke, Helen Timperley, and John Hattie ©2004
- *Formative Assessment in the Secondary Classroom* Shirley Clarke ©2005
- *Improving Student Achievement: A Practical Guide to Assessment for Learning* Toni Glasson ©2009

Contributing Specialists and Reviewers

Big Ideas Learning would like to express our gratitude to the mathematics education and instruction experts from Florida who served as our advisory panel, in addition to all the contributing specialists and reviewers who played a key role during the writing of *Florida's B.E.S.T. Standards for MATH*. Their input was an invaluable asset during the development of this program.

- **Sophie Murphy**, Ph.D. Candidate, Melbourne School of Education, Melbourne, Australia
  Learning Targets and Success Criteria Specialist and Visible Learning Reviewer
- **Judy Hickman**, Florida Education Consultant Northern Pan Handle, FL
  Content Reviewer and Assessment Specialist
- **India White**, Ph.D., National Education Consultant Team, Tampa, FL
  Content Reviewer
- **Dakeyan Chá Dré Graham**, Ph.D., Executive Director, Independent Education and Parental Choice, Florida Department of Education, Tallahassee, FL
  Equity Specialist
- **Noelle Camel**, Special Education Teacher, Osceola, FL
  Advisory Panel
- **Elisa Greco**, Math Teacher, Osceola, FL
  Advisory Panel
- **Benjamin Hicks**, Math Teacher, Alachua, FL
  Advisory Panel
- **Jake Kahn**, Math Teacher, Duval, FL
  Advisory Panel
- **Kéto Nord**, Math Department Head, Hillsborough, FL
  Advisory Panel
- **Denise Williams**, Secondary Math Coordinator, Leon, FL
  Advisory Panel
- **Jason Berkholz**, Mathematics Department Chair, Metropolitan School District of Washington Township, Indianapolis, IN
  Content Reviewer
- **Matthew L. Beyranevand**, Ed.D, K–12 Mathematics Coordinator, Chelmsford, MA
  Content Reviewer
- **Elizabeth Caccavella**, Ed.D, Supervisor of Mathematics, Paterson, NJ
  Content Reviewer
- **Larry Dorf**, Secondary Mathematics Teacher, Harrisburg, PA
  Content Reviewer
- **Linda Hall**, Mathematics Educational Consultant, Edmond, OK
  Teaching Edition Contributor and Advisory Panel
- **Jill Kalb**, Secondary Math Content Specialist, Arvada, CO
  Content Reviewer
- **Michael McDowell**, Ed.D., Superintendent, Ross, CA
  Project-Based Learning Specialist
- **Dr. Connie S. Schrock**, Professor of Mathematics, Emporia, KS
  Teaching Edition Contributor
- **Nancy Siddens**, Independent Language Teaching Consultant, Las Cruces, NM
  English Language Learner Specialist
- **Beverly Stitzel**, Secondary Mathematics Teacher, Oxford, MI
  Content Reviewer

Focus, Coherence, and Rigor

Instructional Design

A single authorship team from Kindergarten through Algebra 2 results in a logical progression of focused topics with thoughtful coherence and rigor throughout the curriculum.

FOCUS

Focused on Florida's *B.E.S.T. Standards for Mathematics*, each section displays the precise language of Florida benchmarks, making the expectations clear. Learning Targets and Success Criteria are aligned to those expectations.

Florida benchmarks are shown at the beginning of each section, with related **Learning Targets** and **Success Criteria** to guide your learning.

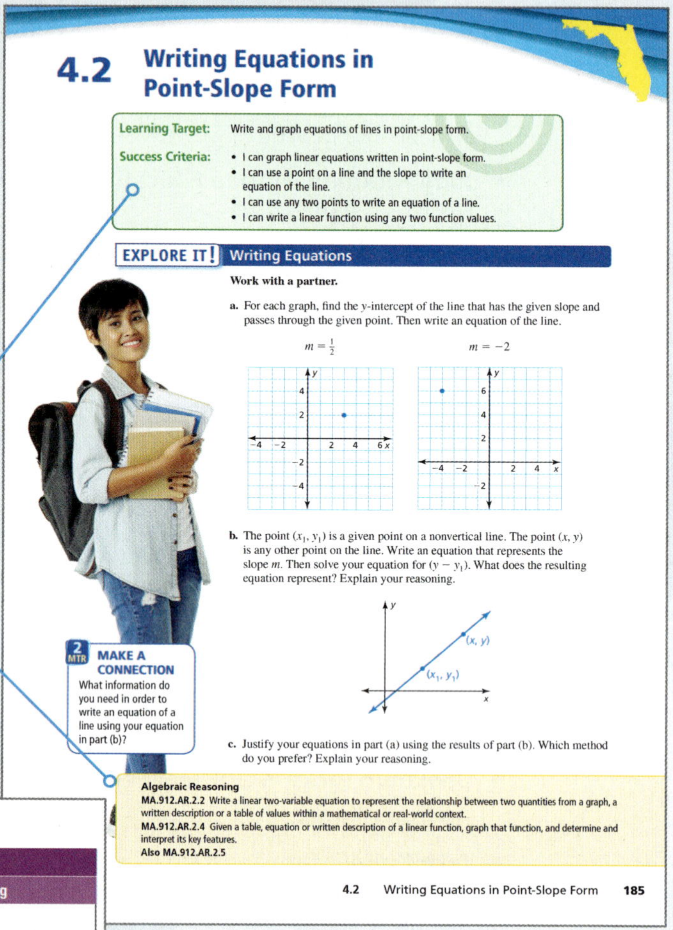

4.2 Writing Equations in Point-Slope Form

Learning Target: Write and graph equations of lines in point-slope form.

Success Criteria:
- I can graph linear equations written in point-slope form.
- I can use a point on a line and the slope to write an equation of the line.
- I can use any two points to write an equation of a line.
- I can write a linear function using any two function values.

EXPLORE IT! Writing Equations

Work with a partner.

a. For each graph, find the y-intercept of the line that has the given slope and passes through the given point. Then write an equation of the line.

$m = \frac{1}{2}$ $\qquad$ $m = -2$

b. The point (x_1, y_1) is a given point on a nonvertical line. The point (x, y) is any other point on the line. Write an equation that represents the slope m. Then solve your equation for $(y - y_1)$. What does the resulting equation represent? Explain your reasoning.

c. Justify your equations in part (a) using the results of part (b). Which method do you prefer? Explain your reasoning.

2 MTR MAKE A CONNECTION
What information do you need in order to write an equation of a line using your equation in part (b)?

Algebraic Reasoning
MA.912.AR.2.2 Write a linear two-variable equation to represent the relationship between two quantities from a graph, a written description or a table of values within a mathematical or real-world context.
MA.912.AR.2.4 Given a table, equation or written description of a linear function, graph that function, and determine and interpret its key features.
Also MA.912.AR.2.5

4.2 Writing Equations in Point-Slope Form 185

Progressions

COHERENCE Through the Grades		
Prior Learning	**Current Learning**	**Future Learning**
Middle School • **MA.8.GR.2.3** Describe and apply translations, reflections, rotations, and dilations using coordinates and the coordinate plane. • **MA.8.AR.3.4** Graph linear equations. • **MA.8.AR.3.2, MA.8.AR.3.5** Find and interpret the slope and y-intercept of a line. • **MA.8.F.1.1** Determine whether a relation is a function. Find the domain and range of a relation. • **MA.8.F.1.2** Recognize linear functions represented as tables, equations, and graphs. • **MA.8.F.1.3** Describe relationships between quantities in graphs. Sketch graphs given	**Chapter 3** • **MA.912.AR.2.5** Understand the concept of a function. Find the domain and range of a function. • **MA.912.AR.2.4, MA.912.AR.2.5, MA.912.AR.3.7, MA.912.AR.3.8** Describe characteristics of functions. • **MA.912.AR.2.4** Sketch a graph of a function from a verbal description. • **MA.912.F.1.6** Compare key features of linear and nonlinear functions. • **MA.912.AR.2.4, MA.912.AR.2.5** Graph linear functions and interpret key features of the graphs. • **MA.912.F.1.1** Identify linear functions using graphs, tables, and equations.	**Algebra 1** • **MA.912.AR.2.2** Write equations of lines in slope-intercept, point-slope, a[…] form. • **MA.912.AR.9.1** Write and sol[…] linear equations by graphing[…] • **MA.912.AR.2.1** Solve linear e[…] graphing. • **MA.912.AR.2.8** Graph the sol[…] linear inequality in two varia[…] • **MA.912.AR.9.4** Graph the sol[…] system of linear inequalities[…] • **MA.912.F.1.1** Distinguish bet[…] exponential functions. • **MA.912.AR.5.6** Graph expon[…] and show key features of the[…]

Every chapter is written to follow the progressions built into the Florida benchmarks. The **Progressions** detailed in the Teaching Edition show how content progresses from grade to grade. These Algebra 1 standards, on functions, build on the foundational work with equations and graphs from middle school.

COHERENCE

The sequence of topics, from Kindergarten to Algebra 2, follows the benchmarks and clarifications for each grade and progresses meaningfully within each grade and between grade levels.

from a Single Authorship Team

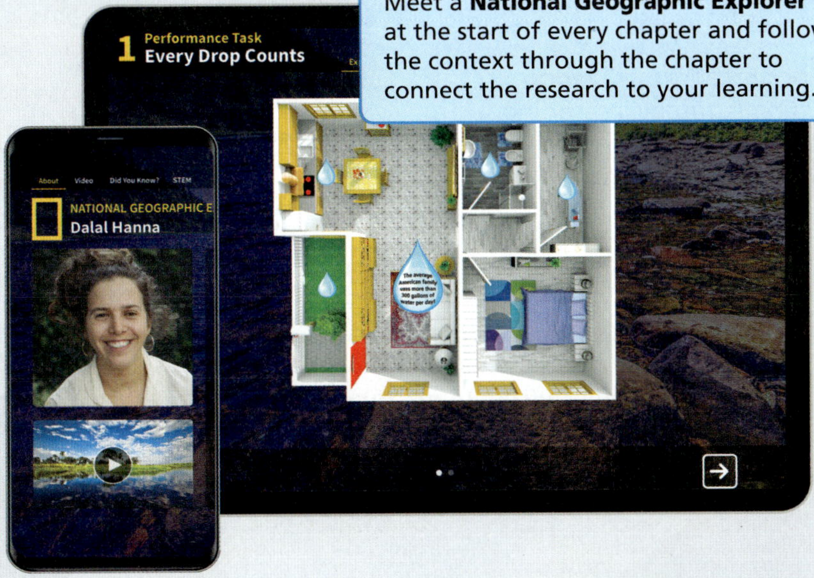

Meet a **National Geographic Explorer** at the start of every chapter and follow the context through the chapter to connect the research to your learning.

RIGOR

In every chapter, you have opportunities to explore, discover, and solidify conceptual understanding, then apply and transfer that learning. This program weaves together the three important building blocks of rigor:

- **Conceptual Understanding**
  Discovering why
- **Procedural Fluency**
  Learning how
- **Application**
  Knowing when to apply

EXPLORE IT! Modeling a Real-Life Problem

Work with a partner. The Okavango Delta is the largest freshwater wetland in southern Africa and is the main source of water for one million people.

1 MTR ANALYZE A PROBLEM
How can you use the graph to determine the quantities involved and the relationship between the quantities?

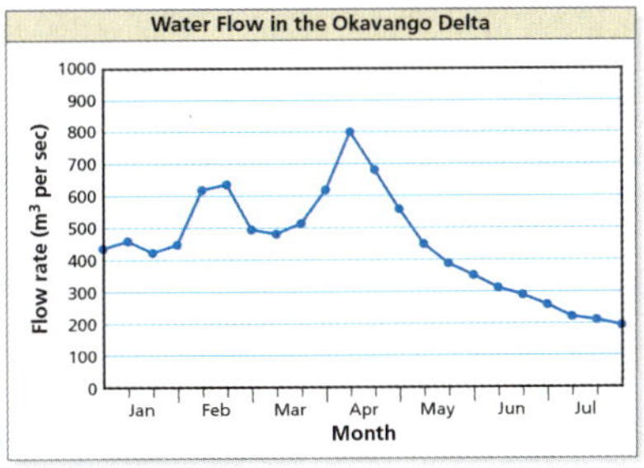

Develop foundational concepts with discovery in **Explore It!**

Apply your knowledge with **Modeling Real Life**, **Dig Deeper**, and other non-routine problems to achieve deep levels of learning. Apply your knowledge to different contexts, see connections between ideas, and justify your thinking.

7 MTR **33. MODELING REAL LIFE** A city's commuter system has three zones. Zone 1 serves people living within 3 miles of the city's center. Zone 2 serves those between 3 and 7 miles from the center. Zone 3 serves those more than 7 miles from the center. *(See Example 6.)*

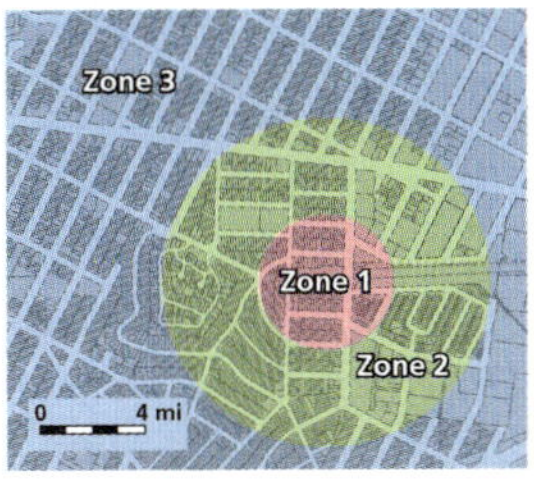

a. Graph this situation in a coordinate plane where each unit corresponds to 1 mile. Locate the city's center at the origin.

b. Determine which zone serves people whose homes are represented by the points (3, 4), (6, 5), (1, 2), (0, 3), and (1, 6).

Build procedural fluency with clear **Key Ideas** and **Examples**, followed by abundant opportunities to practice skills in **Self-Assessment**, **Practice**, and **Review & Refresh**.

EXAMPLE 2 Combining Like Terms to Solve an Equation (WATCH)

Solve $-12 = 9x - 6x + 15$. Check your solution.

SOLUTION

$-12 = 9x - 6x + 15$	Write the equation.
$-12 = 3x + 15$	Combine like terms.
Undo the addition. → $\underline{-15} \quad \underline{-15}$	Subtraction Property of Equality
$-27 = 3x$	Simplify.
Undo the multiplication. → $\frac{-27}{3} = \frac{3x}{3}$	Division Property of Equality
$-9 = x$	Simplify.

▶ The solution is $x = -9$.

Check
$-12 = 9x - 6x + 15$
$-12 \stackrel{?}{=} 9(-9) - 6(-9) + 15$
$-12 = -12$ ✓

A Program Geared Toward Fluency

What is Fluency?

Fluency is more than the memorization of facts or procedures. Fluency builds on a foundation of conceptual understanding, strategic reasoning, and problem-solving to achieve automaticity. You connect your conceptual understanding with strategies and methods that make sense to you.

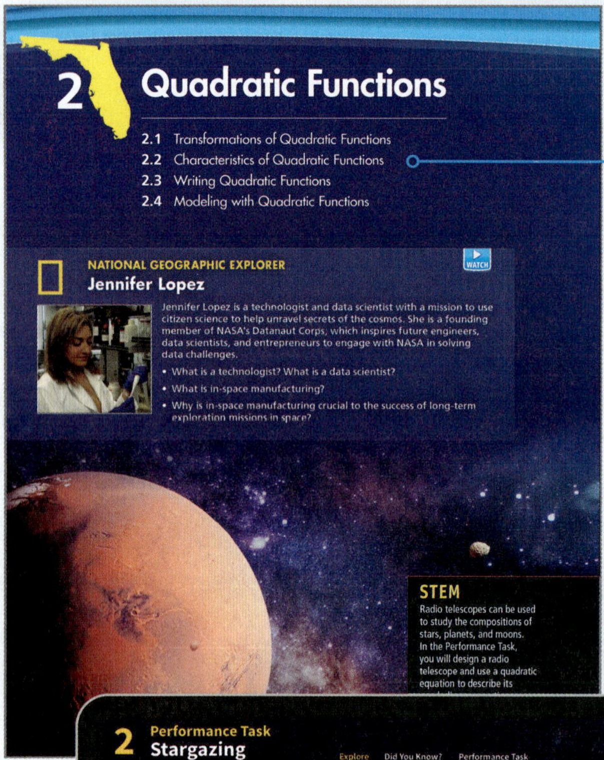

Begin every chapter discovering the research from a National Geographic Explorer and thinking about the world around you. Apply what you learn in the chapter with a related **Performance Task**.

Why Fluency Matters

By building fluency in arithmetic, you can efficiently use foundational skills to solve deeper, more meaningful problems about the world around you. Fluency will contribute to your success not only in school, but also in your daily life.

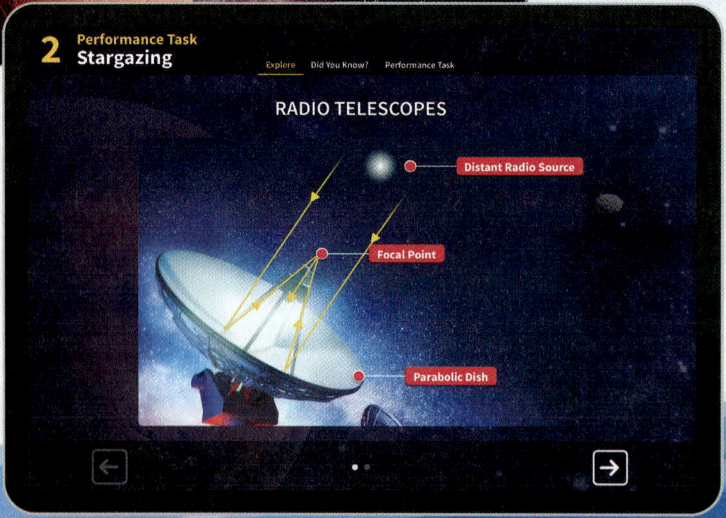

Modeling Real Life, Dig Deeper, and other non-routine problems help you apply and deepen your learning.

69. DIG DEEPER The function $t(r) = -35 \ln\left(1 - \frac{r}{100}\right)$ represents the time (in minutes) it takes to recharge a tablet battery from 0% to r% of its full charge.

a. Use technology to graph t for $0 < r < 100$ and for $0 < t < 350$.

b. How long will it take to recharge the battery to 40% of its full charge?

c. Describe what happens to $t(r)$ as r increases in this situation.

with B.E.S.T. Mathematics

Procedural Fluency

In previous grades, you solidified your understanding of arithmetic operations, found reliable methods for solving, and worked toward becoming independently fluent to solve problems. This program provides ample opportunities for you to practice and achieve fluency in any skills from prior grades, with support resources such as the **Skills Trainer**.

Use procedures and standard algorithms to solve math problems and justify your thinking.

EXAMPLE 1 Finding Side Lengths in 45°-45°-90° Triangles WATCH

Find the value of x. Write your answer in simplest form.

a.

b.

SOLUTION

a. By the Triangle Sum Theorem, the measure of the third angle must be 45°, so the triangle is a 45°-45°-90° triangle.

hypotenuse = leg • $\sqrt{2}$	45°-45°-90° Triangle Theorem
$x = 8 \cdot \sqrt{2}$	Substitute.
$x = 8\sqrt{2}$	Simplify.

▶ The value of x is $8\sqrt{2}$.

b. By the Base Angles Theorem and the Corollary to the Triangle Sum Theorem, the triangle is a 45°-45°-90° triangle.

hypotenuse = leg • $\sqrt{2}$	45°-45°-90° Triangle Theorem
$5\sqrt{2} = x \cdot \sqrt{2}$	Substitute.
$\frac{5\sqrt{2}}{\sqrt{2}} = \frac{x\sqrt{2}}{\sqrt{2}}$	Division Property of Equality
$5 = x$	Simplify.

▶ The value of x is 5.

SELF-ASSESSMENT 1 I don't understand yet. 2 I can do it with help. 3 I can do it on my own. 4 I can teach

Find the missing side length(s). Write your answer(s) in simplest form.

1.

2.

3.

4.

GO DIGITAL

Practice using standard algorithms to solve exercises in **Self-Assessment**, **Practice**, and **Review & Refresh**.

This cumulative practice feature gives you an opportunity to independently practice using accurate and efficient procedures.

9 Chapter Review WITH CalcChat®

Chapter Learning Target: Understand right triangles and trigonometry.

Chapter Success Criteria:
- ◆ I can use the Pythagorean Theorem to solve problems.
- ◆ I can find side lengths in special right triangles.
- ■ I can explain how similar triangles are used with trigonometric ratios.
- ■ I can use trigonometric ratios to solve problems.

◆ Surface ■ Deep

SELF-ASSESSMENT 1 I don't understand yet. 2 I can do it with help. 3 I can do it on my own. 4 I can teach someone else.

9.1 The Pythagorean Theorem *(pp. 453–460)* WATCH

Learning Target: Understand and apply the Pythagorean Theorem.

Vocabulary
pictorial proof
Pythagorean triple

Find the value of x. Then tell whether the side lengths form a Pythagorean triple.

1.

2.

3.

4.

Tell whether the triangle is a right triangle.

5.

6.

Embedded Mathematical Thinking

Encouraging Mathematical Mindsets

Developing proficiency in the **Mathematical Thinking and Reasoning (MTR) Standards** is about becoming a mathematical thinker. Actively learn to ask why, and to reason and communicate with others as you learn. Use this guide to develop proficiency with each standard.

1

Actively Participate in Effortful Learning by staying engaged and maintaining a positive mindset when working to solve tasks. Ask questions and analyze the problem in a way that makes sense, persevering and modifying as needed, and support each other during challenging tasks or when attempting a new method or approach.

Look for labels such as:

- Analyze a Problem
- Ask a Question
- Persevere
- Stay Positive
- Help a Classmate

EXPLORE IT! Graphing a Linear Inequality in Two Variables

Work with a partner. You have \$60 to spend on sand and gravel to make a pen for your dog.

a. Use an inequality to represent the situation.

b. Identify several solu... the inequality. Plot y... in the coordinate pla...

1 MTR **ANALYZE A PROBLEM** If you used only sand, about how much could you purchase? If you used only gravel, about how much could you purchase? What are the advantages and disadvantages of each for a dog pen?

1 MTR **20. HELP A CLASSMATE** Explain to a classmate how to find $m\angle A$.

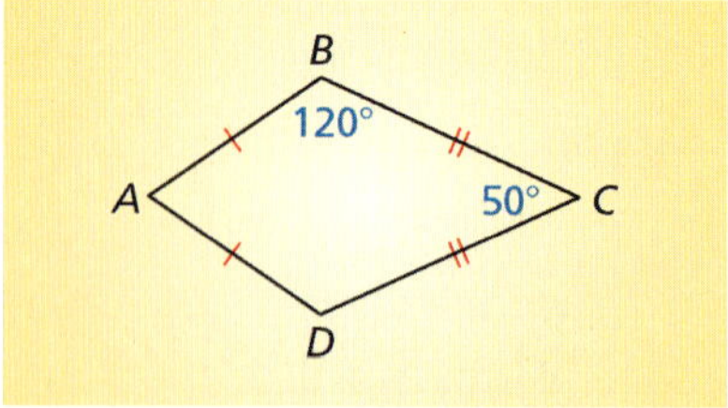

BUILDING TO FULL UNDERSTANDING

Throughout each course, you have opportunities to demonstrate specific aspects of the Mathematical Thinking and Reasoning Standards. Labels throughout the book indicate gateways to those aspects. Collectively, these opportunities will lead students to a full understanding of each standard. Developing these mindsets and habits will give meaning to the mathematics you learn.

and Reasoning Standards

2

Demonstrate Understanding by Representing Problems in Multiple Ways through modeling and by using objects, drawings, tables, and graphs to represent solutions. Progress from choosing representations to using algorithms and equations to connect concepts with models.

Look for labels such as:

- Model a Problem
- Use Another Method
- Multiple Representations
- Choose a Representation
- Make a Connection

EXAMPLE 2 Using the ASA Congruence Theorem WATCH

Write a proof.

Given $\overline{AD} \parallel \overline{EC}$, $\overline{BD} \cong \overline{BC}$

Prove $\triangle ABD \cong \triangle EBC$

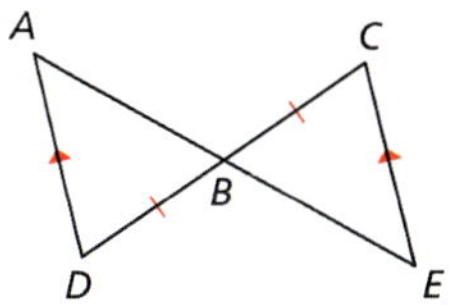

SOLUTION

	STATEMENTS	REASONS
	1. $\overline{AD} \parallel \overline{EC}$	**1.** Given
A	**2.** $\angle D \cong \angle C$	**2.** Alternate Interior Angles Theorem
S	**3.** $\overline{BD} \cong \overline{BC}$	**3.** Given
A	**4.** $\angle ABD \cong \angle EBC$	**4.** Vertical Angles Congruence Theorem
	5. $\triangle ABD \cong \triangle EBC$	**5.**

2 MTR **USE ANOTHER METHOD**
Use the AAS Congruence Theorem to prove that $\triangle ABD \cong \triangle EBC$.

3 MTR **51. ADAPT A PROCEDURE** You and your friend plan to meet each other at the water fountain in a park. Write a system of equations that represents a possible route you take to the fountain and a possible route your friend takes to the fountain. Is the solution of your system the fountain? Justify your answer.

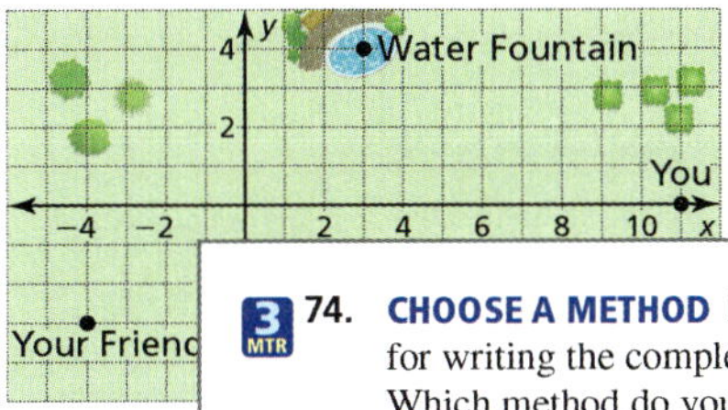

When you **Complete Tasks with Mathematical Fluency**, you select efficient methods to complete tasks accurately and with confidence. You stay flexible, using feedback to improve efficiency and adapting procedures to new concepts.

3

3 MTR **74. CHOOSE A METHOD** Describe the methods shown for writing the complex expression in standard form. Which method do you prefer? Explain.

Method 1

$$4i(2-3i)+4i(1-2i)=8i-12i^2+4i-8i^2$$
$$=8i-12(-1)+4i-8(-1)$$
$$=20+12i$$

Method 2

$$4i(2-3i)+4i(1-2i)=4i[(2-3i)+(1-2i)]$$
$$=4i[3-5i]$$
$$=12i-20i^2$$
$$=12i-20(-1)$$
$$=20+12i$$

Look for labels such as:

- Choose a Method
- Select Methods
- Maintain Accuracy
- Adapt a Procedure
- Reflect on Your Method

Embedded Mathematical Thinking

4

When you **Engage in Discussions that Reflect on the Mathematical Thinking of Self and Others**, you analyze and compare your own mathematical ideas and thinking together with your peers. By recognizing errors and justifying results, you can construct possible arguments based on evidence.

Look for labels such as:

- Communicate Clearly
- Discuss Mathematical Thinking
- Error Analysis
- Compare Methods
- Construct an Argument
- Making an Argument
- Justify a Result
- Which One Doesn't Belong?
- Different Words, Same Question

EXPLORE IT! Displaying Data

Work with a partner. Analyze the data and then create a display that best represents the data. Explain your choice of data display.

a. The Montana Department of Fish, Wildlife & Parks reported the following numbers of claims made to retrieve elk killed on roadways.

adult male: 69 adult female: 178 calf: 52

b. The data below show the numbers of deer killed on roads in one region of Colorado from 2007 to 2018.

OBSERVED DEER FATALITIES

2007: 352	2011: 315	2015: 159
2008: 348	2012: 275	2016: 301
2009: 264	2013: 139	2017: 291
2010: 336	2014: 116	2018: 220

c. A yearlong study by volun... ...s of animals killed by motor...

4 MTR COMPARE METHODS

Compare your data displays in parts (a)–(c) with other students. Can more than one display be appropriate for a set of data?

5

Use Patterns and Structure to Help Understand and Connect Concepts by focusing on details, finding logical order, or breaking down a problem into smaller parts. You often look for similarities between a new concept and something you learned before.

Look for labels such as:

- Make a Plan
- Relate Concepts
- Connecting Concepts
- Use a Similar Problem
- Decompose a Problem
- Patterns
- Structure

EXAMPLE 5 Finding the Surface Area of a Similar Solid WATCH

Prism A and prism B are similar. Find the surface area of prism B.

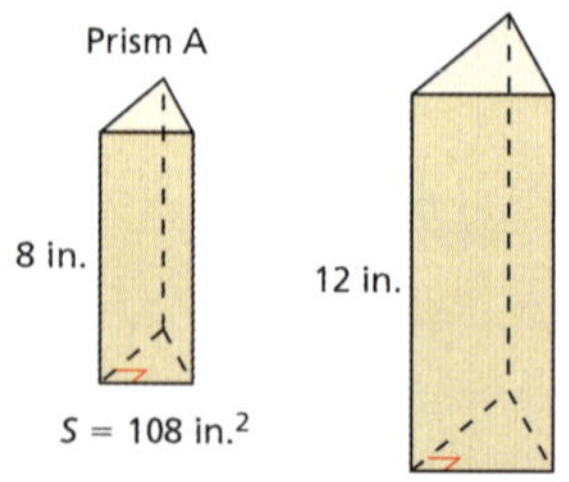

SOLUTION

The scale factor is $k = \frac{\text{Height of prism B}}{\text{Height of prism A}}$

$= \frac{12}{8}$

$= \frac{3}{2}$

Use the scale factor to find the surface area of prism B.

$\frac{\text{Surface area of prism B}}{\text{Surface area of prism A}} = k^2$ The ratio of the surface areas is k^2.

5 MTR RELATE CONCEPTS

Explain how your previous understanding of dilations and scale factors in two dimensions is related to finding surface areas of similar solids.

6

When you **Assess the Reasonableness of Solutions**, you are developing a habit of checking your calculations when solving problems. Estimate to determine possible solutions and use benchmarks to determine if a solution makes sense.

Look for labels such as:
- Assess Reasonableness
- Justifying Steps

EXPLORE IT! Transforming Graphs of Cubic and Quartic Functions

Work with a partner. The graphs of the parent cubic function $f(x) = x^3$ and the parent quartic function $g(x) = x^4$ are shown.

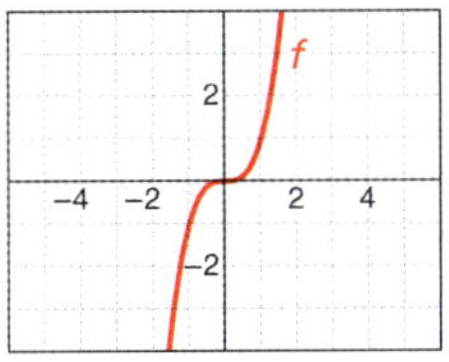

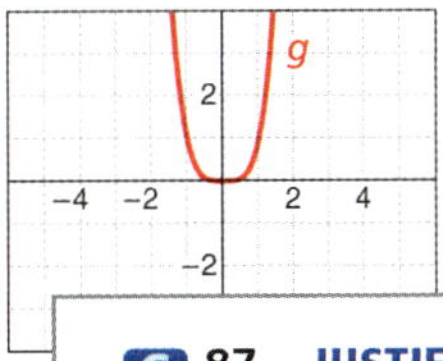

6 MTR **ASSESS REASONABLENESS**
Explain why it is reasonable that the range of *f* includes negative numbers, but the range of *g* does not.

6 MTR **87. JUSTIFYING STEPS** Justify each step in the simplification of i^2.

Algebraic Step	Justification
$i^2 = (\sqrt{-1})^2$	
$= -1$	

7

When you **Apply Mathematics to Real-World Contexts**, you connect concepts to everyday experiences and use models and methods to understand, represent, and solve problems.

Look for labels such as:
- Apply Mathematics
- Modeling Real Life
- Investigate
- Performance Task

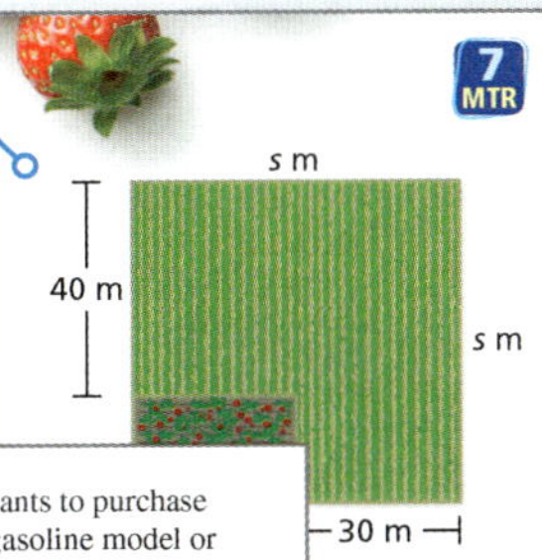

7 MTR **EXAMPLE 4** Modeling Real Life WATCH

A farmer plants a rectangular strawberry patch in a corner of a square plot of land. The area of the strawberry patch is 600 square meters. What is the area of the square plot of land?

SOLUTION

1. **Understand the Problem** You are given the area of a strawberry patch and a diagram showing dimensions of a plot of land that contains the strawberry patch. You are asked to find the total area of the plot of land.
2. **Make a Plan** The length of the strawberry patch is $(s - 30)$ meters, and the width is $(s - 40)$ meters. Write and solve an equation to find the side length s. Then use the solution to find the area of the square plot of land.
3. **Solve and Check** Use the formula for the area of a rectangle to write an equation. Then solve to find the side length s of the square plot of land.

$600 = (s - 30)(s - 40)$ Write an equation.

$600 = s^2 - 70s + 1200$ Multiply.

ws that
more than

7 MTR **25. PERFORMANCE TASK** Your family wants to purchase a new vehicle that comes in either a gasoline model or an electric model.

a. Using the information shown, the approximate number of miles your family drives per year, and gas and electricity prices in your area, determine which vehicle is a better buy. Use linear equations to support your answer.

b. Research other factors that affect the cost of vehicle ownership. How might these factors support or change your answer?

Gasoline
Price: $17,950
Fuel Economy:
26 mi/gal city
38 mi/gal highway

Electric
Price: $29,120
Fuel Economy:
31 kWh/100 mi

Visible Learning Through Learning Targets,

Making Learning Visible

Knowing the learning intention of a chapter or section helps you focus on the purpose of an activity, rather than simply completing it in isolation. This program supports visible learning through the consistent use of Learning Targets and Success Criteria to ensure positive outcomes for all students.

Every chapter and section shows a **Learning Target** and related **Success Criteria**. These are purposefully integrated into each carefully written lesson.

Chapter Learning Target: Understand reasoning and proofs.

Chapter Success Criteria:
- ◆ I can use inductive and deductive reasoning.
- ◆ I can justify steps using algebraic reasoning.
- ■ I can explain postulates using diagrams.
- ■ I can prove geometric relationships.

2.3 Postulates and Diagrams

Learning Target: Interpret and sketch diagrams.

Success Criteria:
- I can identify postulates represented by diagrams.
- I can sketch a diagram given a verbal description.
- I can interpret a diagram.

SELF-ASSESSMENT 1 I don't understand yet. 2 I can do it with help. 3 I can do it on my own. 4 I can teach someone else.

2.1 Conditional Statements *(pp. 65–72)* WATCH

Learning Target: Understand and write conditional statements.

Write the if-then form, the converse, the inverse, the contrapositive, and the biconditional of the conditional statement.

1. Two lines intersect in a point.
2. $4x + 9 = 21$ because $x = 3$.
3. Supplementary angles sum to 180°.
4. Right angles are 90°.

Decide whether the statement about the diagram is true. Explain your answer using the definitions you have learned.

5. S is the midpoint of $\overline{EF}$.

Vocabulary AZ VOCAB
conditional statement
if-then form
hypothesis
conclusion
negation
converse
inverse
contrapositive
equivalent statements

The **Self-Assessment** and **Chapter Review** remind you to rate your understanding of the Learning Targets. In the Chapter Review, you can review each section with a reminder of that section's Learning Target.

QUESTIONS FOR LEARNING

As you progress through a section, you should be able to answer the following questions.

- What am I learning?
- Why am I learning this?
- Where am I in my learning?
- How will I know when I have learned it?
- Where am I going next?

Success Criteria, and Self-Assessment

Chapter 4 Writing Linear Functions		
Learning Target: Understand writing linear functions.	1 2 3 4	
I can determine the slope of a line given ordered pairs, a graph, a table, or a context.	1 2 3 4	
I can write the equation of a line in different forms.	1 2 3 4	
I can write equations to model and solve real-life problems.	1 2 3 4	
…et scatter plots and analyze lines of fit.	1 2 3 4	

Use a 4-point scale to rate your understanding of each success criterion. Keep track of your learning on paper or online.

1 I don't understand yet.
2 I can do it with help.
3 I can do it on my own.
4 I can teach someone else.

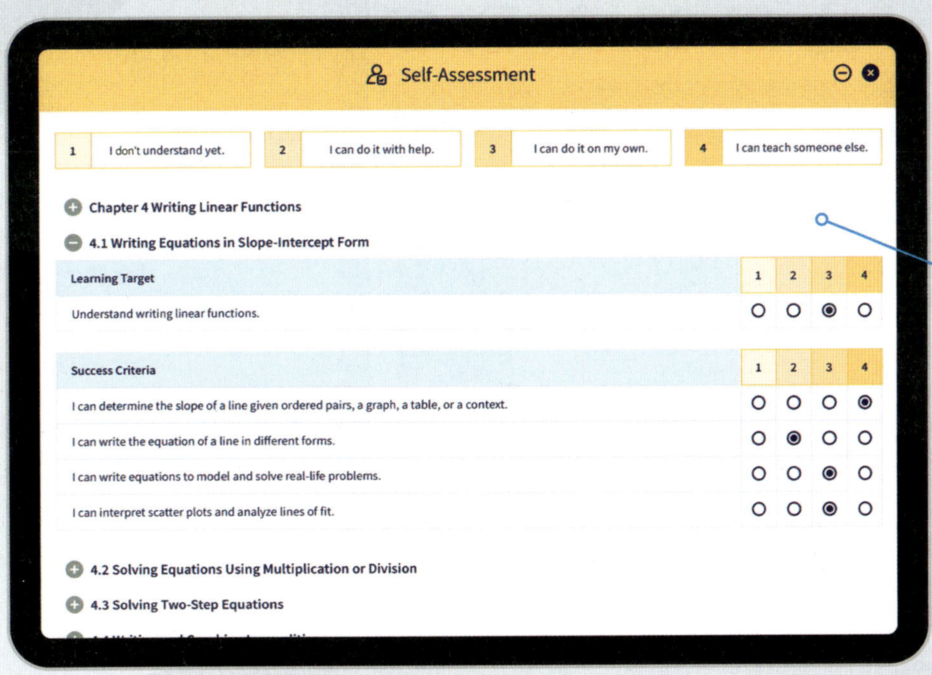

Self-Assessments are included throughout every section, and in the **Chapter Review,** to help you take ownership of your learning and think about where to go next.

Ensuring Positive Outcomes

John Hattie's *Visible Learning* research consistently shows that using Learning Targets and Success Criteria can result in two years' growth in one year, ensuring positive outcomes for your learning and achievement.

Sophie Murphy, M.Ed., wrote the chapter-level Learning Targets and Success Criteria for this program. Sophie is currently completing her Ph.D. at the University of Melbourne in Australia with Professor John Hattie as her leading supervisor. Sophie completed her Master's thesis with Professor John Hattie in 2015. Sophie has over 20 years of experience as a teacher and school leader in private and public school settings in Australia.

Strategic Support for Online Learning

Get the Support You Need, When You Need It

There will be times throughout this course when you may need help. Whether you missed a section, did not understand the content, or just want to review, take advantage of the resources provided in the *Dynamic Student Edition*.

Use the **Self-Assessment** tool to keep track of your understanding of the section's Learning Target and Success Criteria.

Take notes throughout the section using the **My Notes** function. These notes will be organized by chapter and lesson.

Check your answers to selected exercises as you work through the lesson. Use the **Help** option to view the Digital Example and Extra Example videos.

Use **tools** to help clearly show your work and emphasize your math knowledge.

Use **CalcChat**® to view worked-out solutions for select exercises. You can also chat with a live tutor.

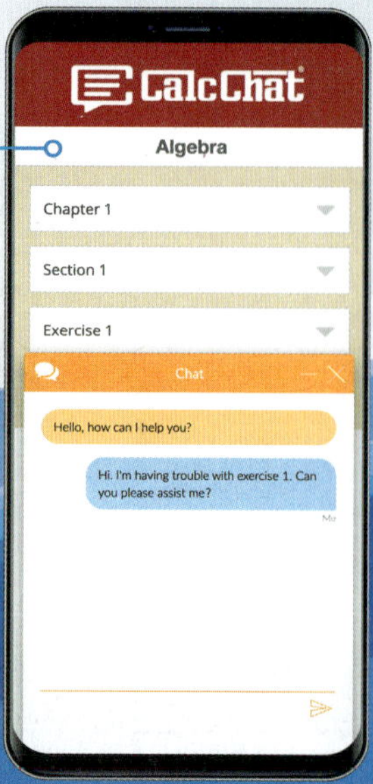

Use **CalcView**® to watch a video of a worked-out solution for any exercise with a blue triangle. A teacher will explain, step-by-step, how to solve the problem.

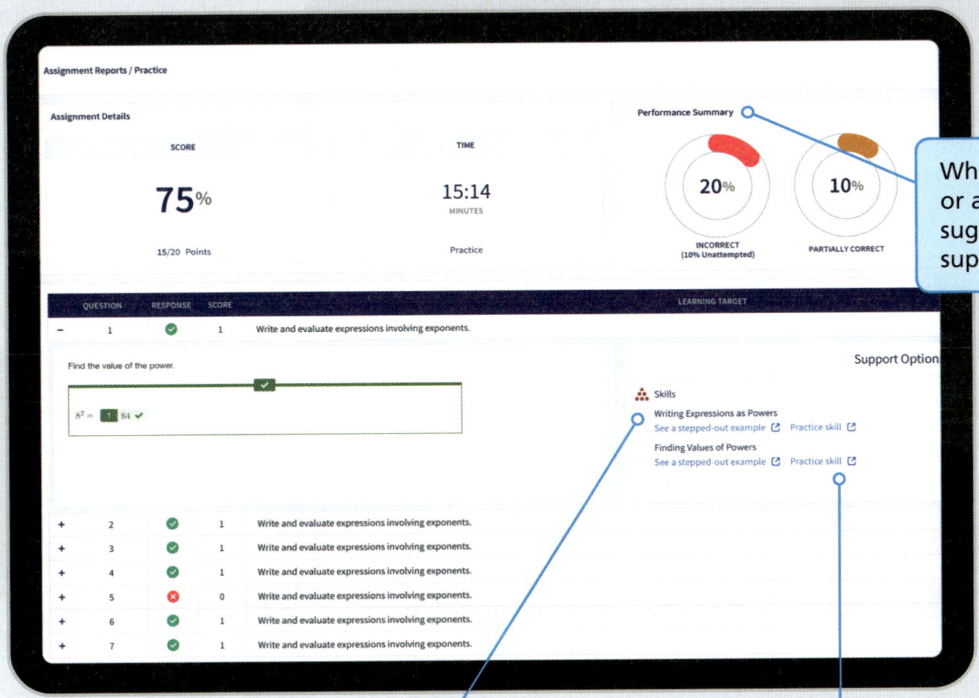

When you submit an assignment or assessment digitally, use the suggested links to get additional support where you need it.

Choose a skill to review and watch a video to see a stepped-out example of that skill. Whether you get a problem incorrect, or want a second explanation, these videos can provide additional help with homework.

Choose a skill and launch the **Skills Trainer** for additional practice on that skill. Practicing similar problems with instant feedback can help build confidence when solving problems.

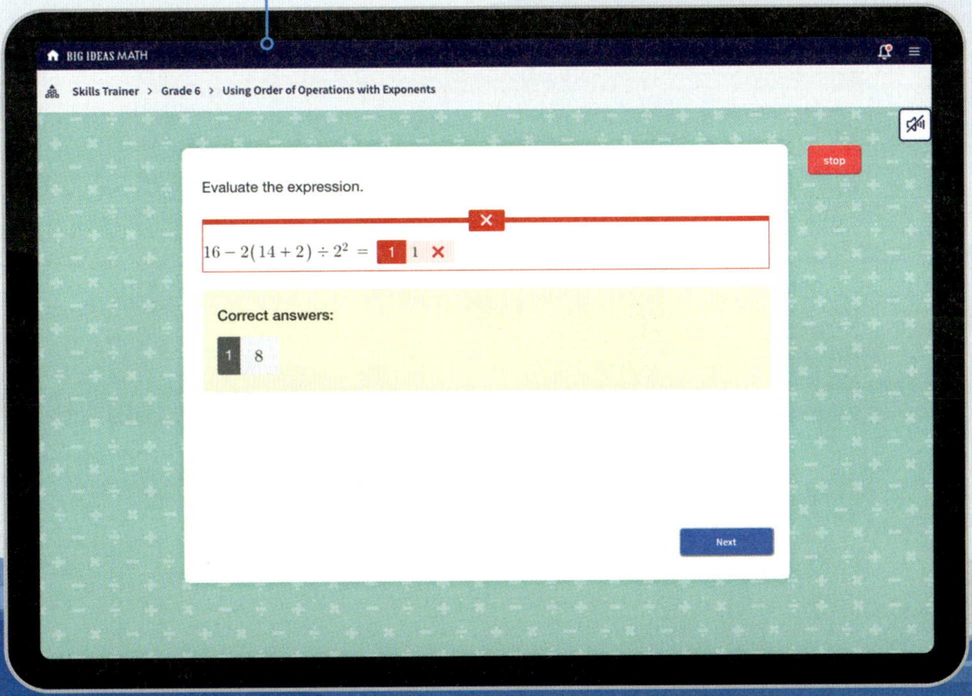

1 Basics of Geometry

Ecology

Design a wildlife reservation to provide a protected habitat for a tiger population.

Reasoning and Proofs

2

Climate Change

See how greenhouse gases warm the planet. Research some of the effects of climate change and write conditional statements based on your research.

3 Parallel and Perpendicular Lines

Geothermal Science

Find a location for a new power plant that will provide electricity to several cities.

Transformations

4

Entomology

Sketch a butterfly species and show how to construct a dilation of your sketch.

H means the entire section is honors.

H means that the section contains honors content.

5 Congruent Triangles

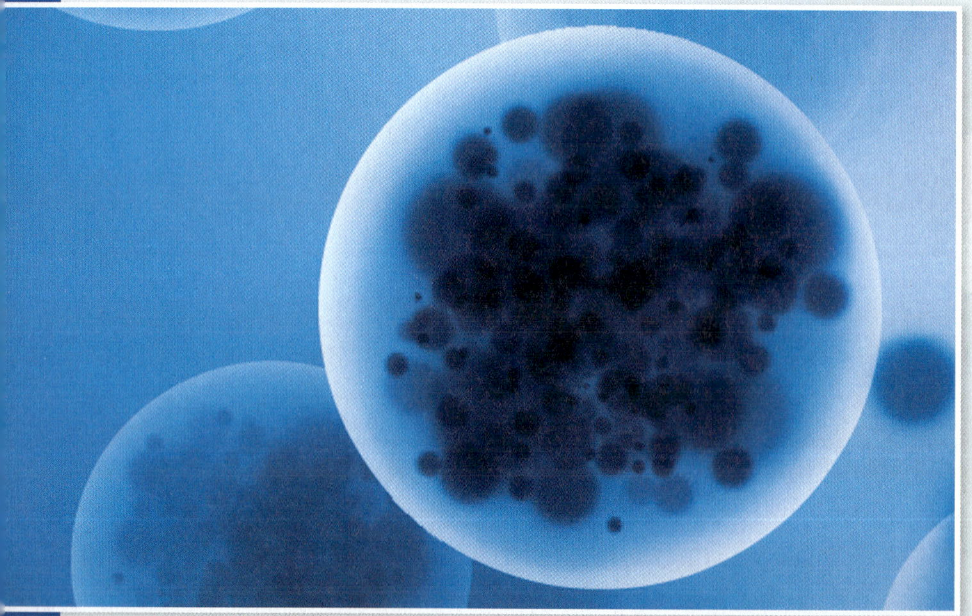

Diagnostic Technologies

Analyze a drawing of a virus known as a *bacteriophage*.

H means the entire section is honors.

H means that the section contains honors content.

Relationships Within Triangles

6

Archaeology
Use a fragment to find the diameter of an ancient plate.

H means the entire section is honors.

H means that the section contains honors content.

7 Quadrilaterals and Other Polygons

Astrobiology

Use angle measures and side lengths to investigate polygons found in several different constellations.

Similarity

8

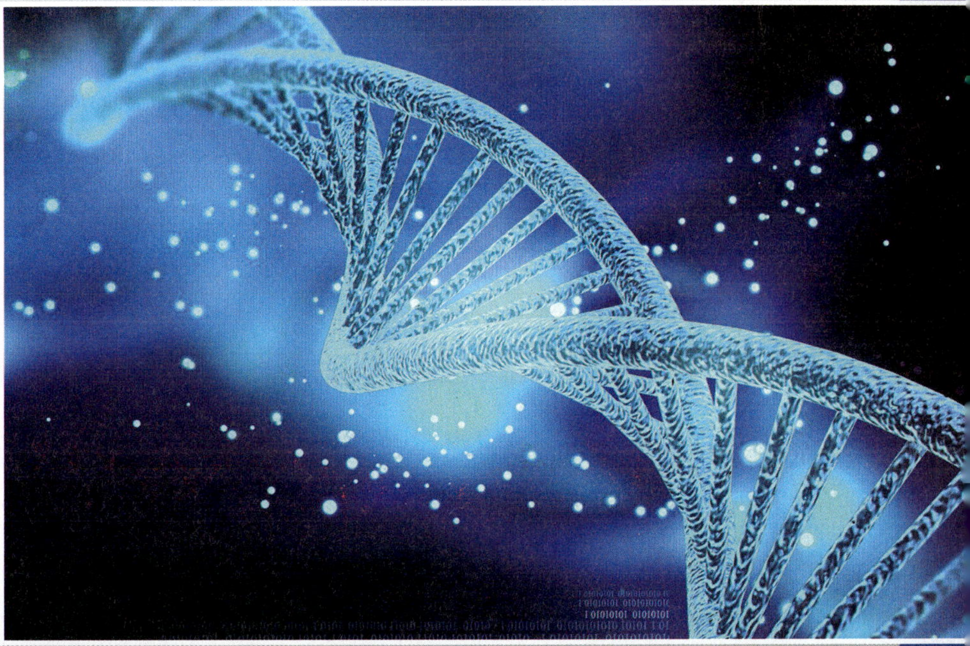

DNA and Genomes

Create a brochure for the African Burial Ground National Monument that includes a scale drawing of the site.

H means the entire section is honors.

H means that the section contains honors content.

9 Right Triangles and Trigonometry

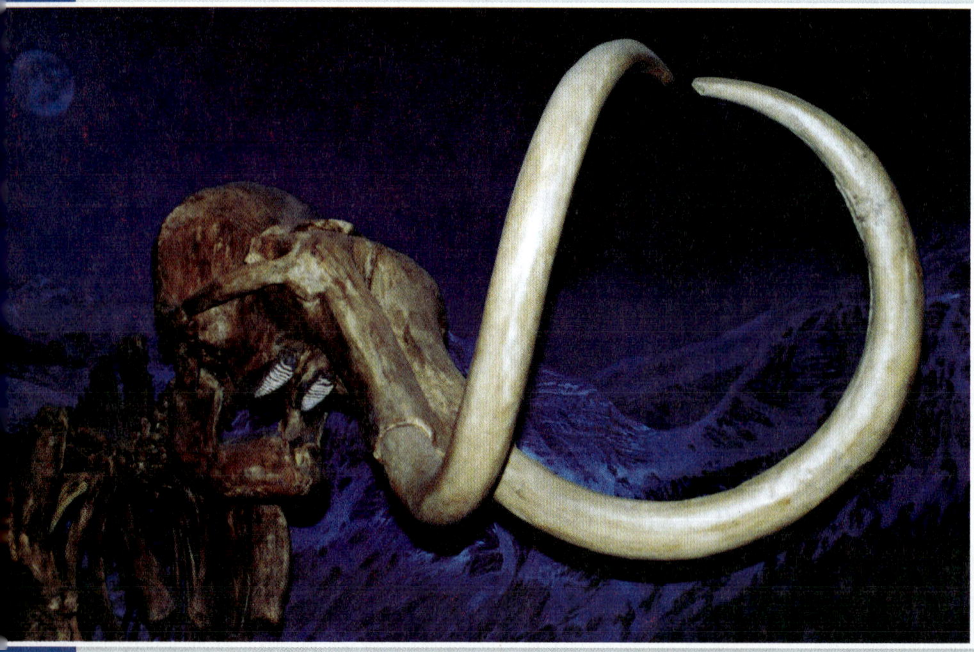

Extinct Species
Work on a new woolly mammoth exhibit at a museum.

H means the entire section is honors.

H means that the section contains honors content.

Circles

10

Bioarchaeology

Find geometric relationships in Stonehenge and analyze their possible significance.

H means the entire section is honors.

H means that the section contains honors content.

11 Circumference and Area

Bioresource Engineering

Design your own center-pivot irrigation system.

Surface Area and Volume

12

Additional Resources

Conservation Biology

Design an artificial bat cave and estimate the number of hibernating bats that it can accommodate.

1 Basics of Geometry

NATIONAL GEOGRAPHIC EXPLORER

WATCH

Rae Wynn-Grant

Dr. Rae Wynn-Grant is an ecologist who uses statistical modeling to investigate how anthropogenic factors can influence the spatial patterns of carnivore behavior and ecology. She studies the ecological and social drivers of human-carnivore conflict.

- What is a carnivore? Name several large carnivores that live in North America.
- Ecology is the branch of biology that deals with relationships among animals. Give several examples of predator-prey relationships in North America.

STEM

When a carnivore's habitat is diminished, the likelihood of human-carnivore conflict increases. In the Performance Task, you will design a wildlife reservation to provide a protected habitat for a tiger population.

GO DIGITAL

Ecology

Preparing for Chapter 1

Chapter Learning Target: Understand basics of geometry.

Chapter Success Criteria:
- ◆ I can name points, lines, and planes.
- ◆ I can measure segments and angles.
- ■ I can use formulas in the coordinate plane.
- ■ I can construct segments and angles.

◆ Surface
■ Deep

AZ VOCAB

Chapter Vocabulary

Work with a partner. Discuss each of the vocabulary terms.

point
line
plane
line segment
angle
acute angle
right angle
obtuse angle
complementary angles
supplementary angles

Mathematical Thinking and Reasoning

1 MTR

ACTIVELY PARTICIPATE IN EFFORTFUL LEARNING BOTH INDIVIDUALLY AND COLLECTIVELY

Mathematicians who actively participate in effortful learning both individually and with others ask questions that will help with solving tasks.

Work with a partner. The figure shown represents a polar bear enclosure at a zoo, where 1 centimeter represents 25 feet.

1. What information do you need to find the perimeter of the enclosure? Explain how you can find this information. Then find the perimeter.
2. What information do you need to find the area of the enclosure? Explain how you can find this information. Then find the area.

GO DIGITAL

1 Prepare WITH CalcChat®

Finding Absolute Value

WATCH

Example 1 **Simplify $|-7 - 1|$.**

$$|-7 - 1| = |-7 + (-1)| \quad \text{Add the opposite of 1.}$$
$$= |-8| \quad \text{Add.}$$
$$= 8 \quad \text{Find the absolute value.}$$

▶ $|-7 - 1| = 8$

Simplify the expression.

1. $|8 - 12|$
2. $|-6 - 5|$
3. $|4 + (-9)|$
4. $|13 + (-4)|$
5. $|6 - (-2)|$
6. $|5 - (-1)|$
7. $|-8 - (-7)|$
8. $|8 - 13|$
9. $|-14 - 3|$

Finding the Area of a Triangle

WATCH

Example 2 **Find the area of the triangle.**

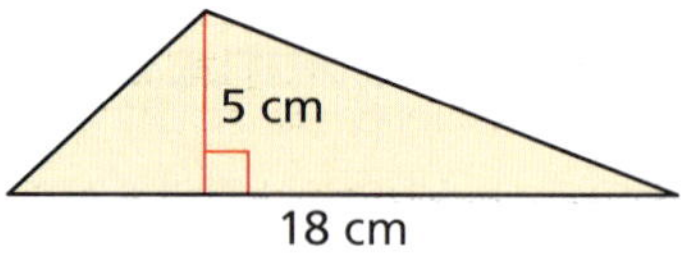

$$A = \frac{1}{2}bh \quad \text{Write the formula for area of a triangle.}$$
$$= \frac{1}{2}(18)(5) \quad \text{Substitute 18 for } b \text{ and 5 for } h.$$
$$= \frac{1}{2}(90) \quad \text{Multiply 18 and 5.}$$
$$= 45 \quad \text{Multiply } \tfrac{1}{2} \text{ and 90.}$$

▶ The area of the triangle is 45 square centimeters.

Find the area of the triangle.

10.

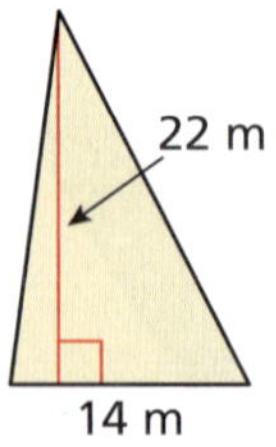

11.

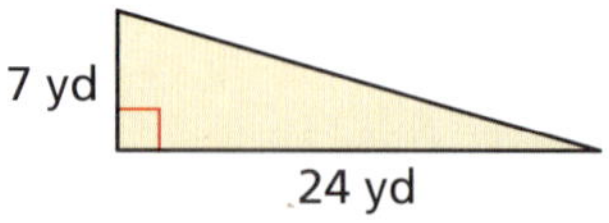

12.

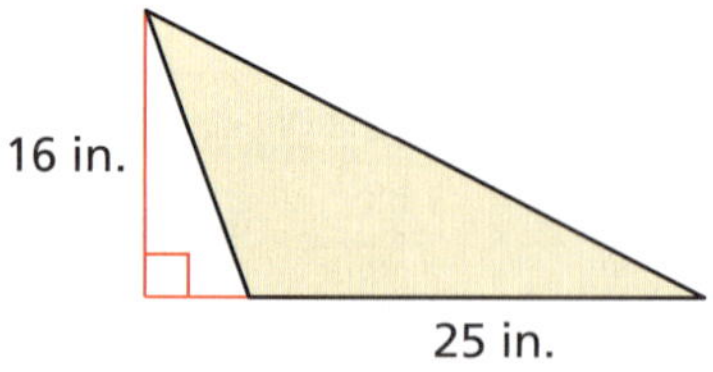

1 MTR **13.** **ANALYZE A PROBLEM** Describe the possible values for x and y when $|x - y| > 0$. What does it mean when $|x - y| = 0$? Can $|x - y| < 0$? Explain your reasoning.

GO DIGITAL

1.1 Points, Lines, and Planes

Learning Target: Use defined terms and undefined terms.

Success Criteria:
- I can describe a point, a line, and a plane.
- I can define and name segments and rays.
- I can sketch intersections of lines and planes.

EXPLORE IT! Using Technology

Work with a partner.

a. Use technology to draw several points. Also, draw some lines, line segments, and rays.

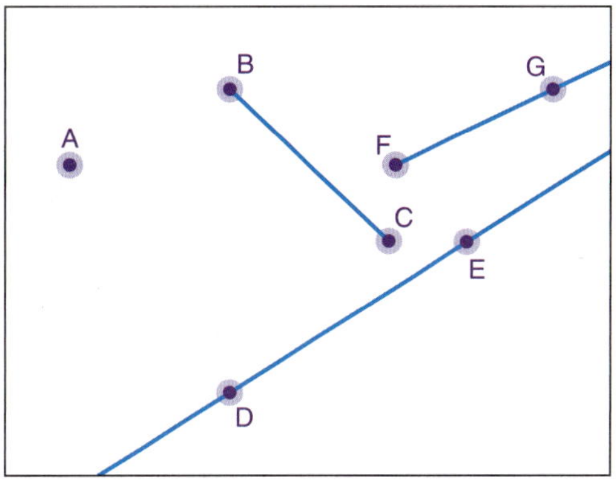

b. How would you describe a line? a point?

c. What is the difference between a line and a line segment? a line and a ray?

d. Write your own definitions for a line segment and a ray, based on how they relate to a line.

e. The diagram shows plane P and plane Q intersecting. How would you describe a plane?

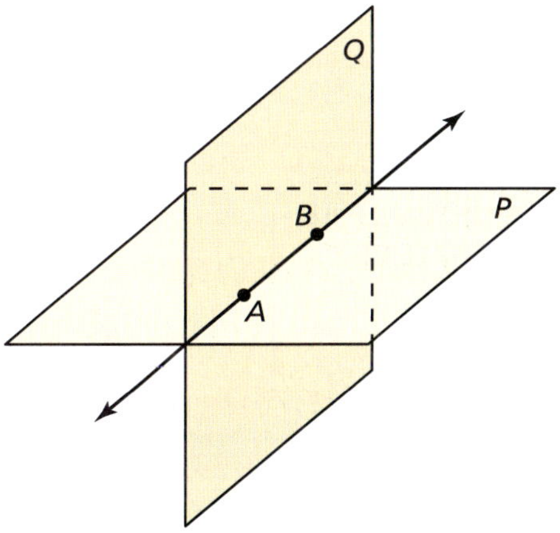

f. Describe the ways in which each of the following can intersect and not intersect. Provide a sketch of each type of intersection.

i. two lines

ii. a line and a plane

iii. two planes

2 MTR MODEL A PROBLEM

Use a real-life object to model each type of intersection in part (f).

Geometric Reasoning

preparing for MA.912.GR.1.1 Prove relationships and theorems about lines and angles. Solve mathematical and real-world problems involving postulates, relationships and theorems of lines and angles.

GO DIGITAL

Vocabulary

undefined terms, *p. 4*
point, *p. 4*
line, *p. 4*
plane, *p. 4*
collinear points, *p. 4*
coplanar points, *p. 4*
defined terms, *p. 5*
line segment, or segment, *p. 5*
endpoints, *p. 5*
ray, *p. 5*
opposite rays, *p. 5*
intersection, *p. 6*

Using Undefined Terms

In geometry, the words *point*, *line*, and *plane* are **undefined terms**. These words do not have formal definitions, but there is agreement about what they mean.

KEY IDEAS

Undefined Terms: Point, Line, and Plane

Point A **point** has no dimension. A dot represents a point.

point *A*

Line A **line** has one dimension. It is represented by a line with two arrowheads, but it extends without end.

Through any two points, there is exactly one line. You can use any two points on a line to name it.

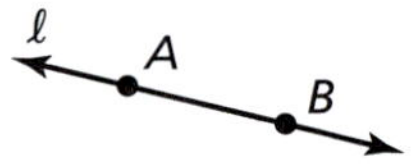

line ℓ, line *AB* ($\overleftrightarrow{AB}$), or line *BA* ($\overleftrightarrow{BA}$)

Plane A **plane** has two dimensions. It is represented by a shape that looks like a floor or a wall, but it extends without end.

Through any three points not on the same line, there is exactly one plane. You can use three points that are not all on the same line to name a plane.

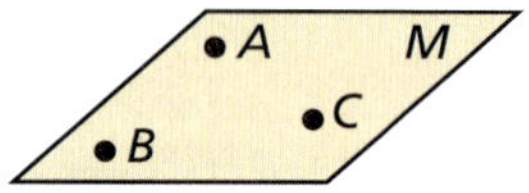

plane *M*, or plane *ABC*

Collinear points are points that lie on the same line. **Coplanar points** are points that lie in the same plane.

EXAMPLE 1 Naming Points, Lines, and Planes

WATCH

a. Give two other names for $\overleftrightarrow{PQ}$ and plane *R*.

b. Name three points that are collinear. Name four points that are coplanar.

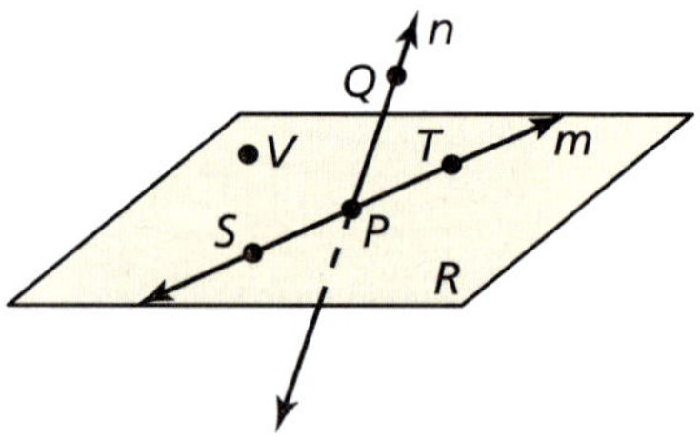

SOLUTION

a. Other names for $\overleftrightarrow{PQ}$ are $\overleftrightarrow{QP}$ and line *n*. Other names for plane *R* are plane *SVT* and plane *PTV*.

b. Points *S*, *P*, and *T* lie on the same line, so they are collinear. Points *S*, *P*, *T*, and *V* lie in the same plane, so they are coplanar.

4 MTR CONSTRUCT AN ARGUMENT
Is there a plane that contains *Q*, *T*, *S*, and *V*? Explain.

SELF-ASSESSMENT 1 I don't understand yet. 2 I can do it with help. 3 I can do it on my own. 4 I can teach someone else.

1. Use the diagram in Example 1. Give two other names for $\overleftrightarrow{ST}$. Name a point that is *not* coplanar with points *Q*, *S*, and *T*.

2. WRITING Compare collinear points and coplanar points.

GO DIGITAL

Using Defined Terms

In geometry, terms that can be described using known words such as *point* or *line* are called **defined terms**.

KEY IDEAS

Defined Terms: Segment and Ray

The diagrams below use the points A and B and parts of the line AB.

Segment A **line segment**, or **segment**, is a part of a line that consists of two **endpoints** and all points on the line between the endpoints.

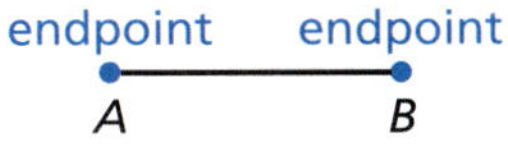

segment AB ($\overline{AB}$), or segment BA ($\overline{BA}$)

Ray A **ray** is a part of a line that consists of an endpoint and all points on the line on one side of the endpoint.

ray AB ($\overrightarrow{AB}$)

Opposite Rays Two rays that have the same endpoint and form a line are **opposite rays**.

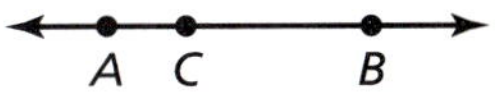

$\overrightarrow{CA}$ and $\overrightarrow{CB}$ are opposite rays.

STUDY TIP

Note that $\overrightarrow{AB}$ and $\overrightarrow{BA}$ are different rays.

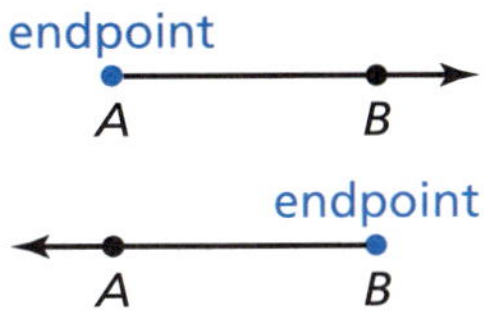

Segments and rays are collinear when they lie on the same line. So, opposite rays are collinear. Lines, segments, and rays are coplanar when they lie in the same plane.

EXAMPLE 2 Naming Segments, Rays, and Opposite Rays

WATCH

a. Give another name for $\overline{GH}$.

b. Name all rays with endpoint J. Which of these rays are opposite rays?

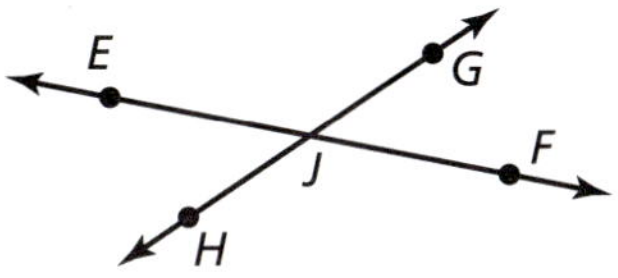

COMMON ERROR

In Example 2, $\overrightarrow{JG}$ and $\overrightarrow{JF}$ have a common endpoint, but they are not collinear. So, they are *not* opposite rays.

SOLUTION

a. Another name for $\overline{GH}$ is $\overline{HG}$.

b. The rays with endpoint J are $\overrightarrow{JE}$, $\overrightarrow{JG}$, $\overrightarrow{JF}$, and $\overrightarrow{JH}$. The pairs of opposite rays with endpoint J are $\overrightarrow{JE}$ and $\overrightarrow{JF}$, and $\overrightarrow{JG}$ and $\overrightarrow{JH}$.

SELF-ASSESSMENT

1 I don't understand yet. 2 I can do it with help. 3 I can do it on my own. 4 I can teach someone else.

Use the diagram.

3. Give another name for $\overline{KL}$.

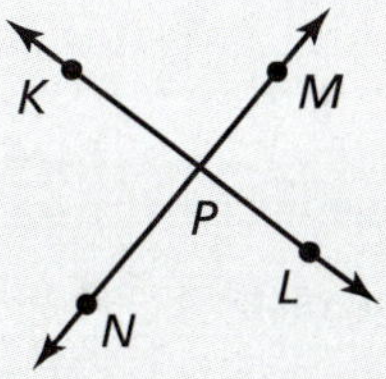

4 MTR **4. MAKING AN ARGUMENT** Are $\overrightarrow{KP}$ and $\overrightarrow{PK}$ the same ray? Are $\overrightarrow{NP}$ and $\overrightarrow{NM}$ the same ray? Explain.

GO DIGITAL

Sketching Intersections

Two or more geometric figures *intersect* when they have one or more points in common. The **intersection** of the figures is the set of points the figures have in common. Some examples of intersections are shown below.

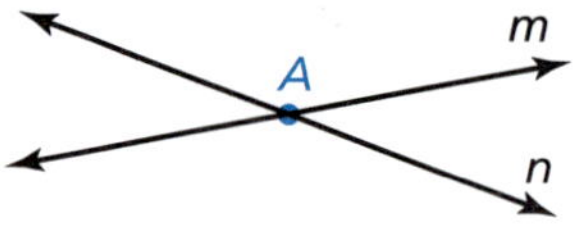

The intersection of these two lines is a point.

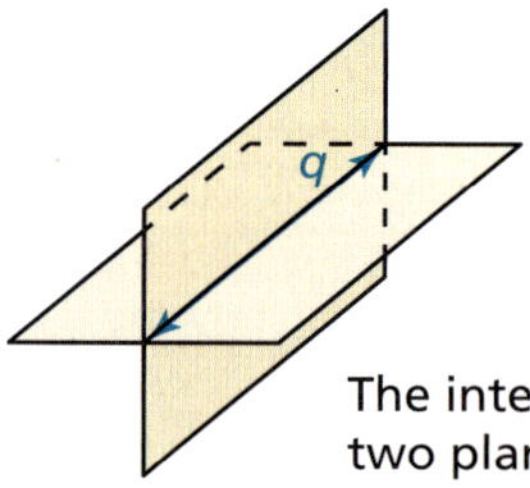

The intersection of these two planes is a line.

EXAMPLE 3 **Sketching Intersections of Lines and Planes** WATCH

a. Sketch a plane and a line that is in the plane.

b. Sketch a plane and a line that does not intersect the plane.

c. Sketch a plane and a line that intersects the plane at a point.

SOLUTION

a.

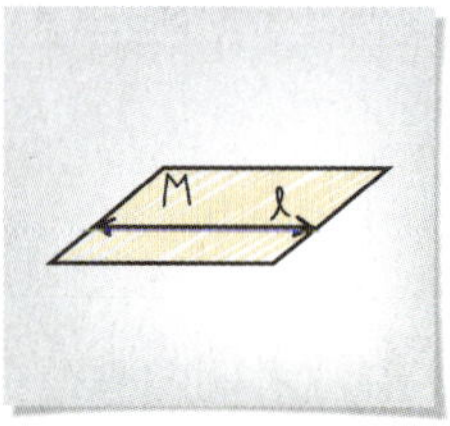

b.

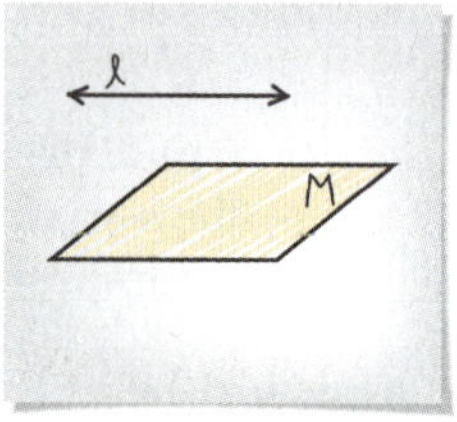

c.

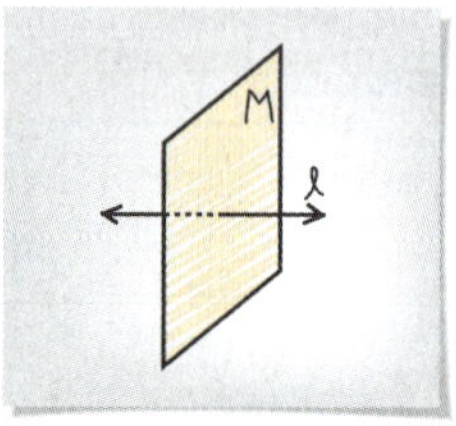

EXAMPLE 4 **Sketching an Intersection of Planes** WATCH

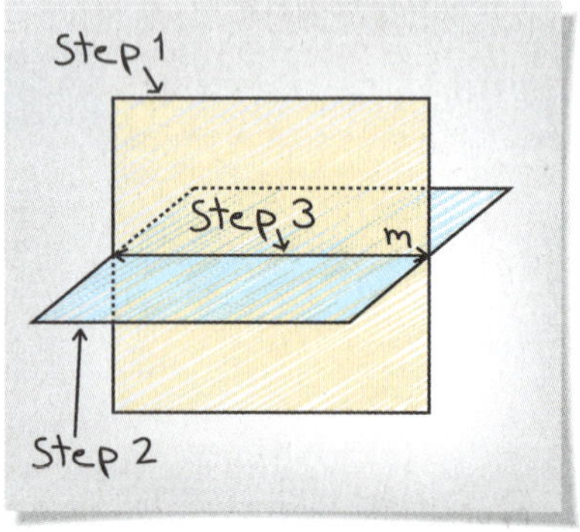

Sketch two planes that intersect in a line.

SOLUTION

Step 1 Draw a vertical plane. Shade the plane.

Step 2 Draw a second plane that is horizontal. Shade this plane a different color. Use dashed lines to show where planes are hidden.

Step 3 Draw the line of intersection.

SELF-ASSESSMENT

1 I don't understand yet. 2 I can do it with help. 3 I can do it on my own. 4 I can teach someone else.

5. Sketch two different lines that intersect a plane at the same point.

6. Sketch two planes that do not intersect.

Use the diagram.

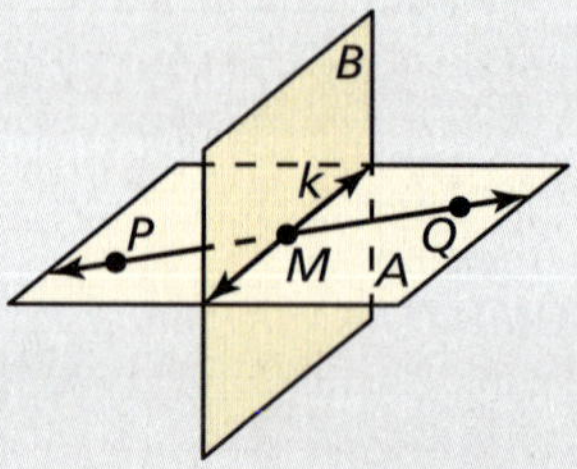

7. Name the intersection of $\overleftrightarrow{PQ}$ and line k.

8. Name the intersection of plane A and plane B.

9. Name the intersection of line k and plane A.

GO DIGITAL

Solving Real-Life Problems

Electric utilities use sulfur hexafluoride as an insulator. Leaks in electrical equipment contribute to the release of sulfur hexafluoride into the atmosphere.

7 MTR

EXAMPLE 5 Modeling Real Life

WATCH

The diagram shows a model of a molecule of sulfur hexafluoride, the most potent greenhouse gas in the world. Name two different planes that contain line *r*.

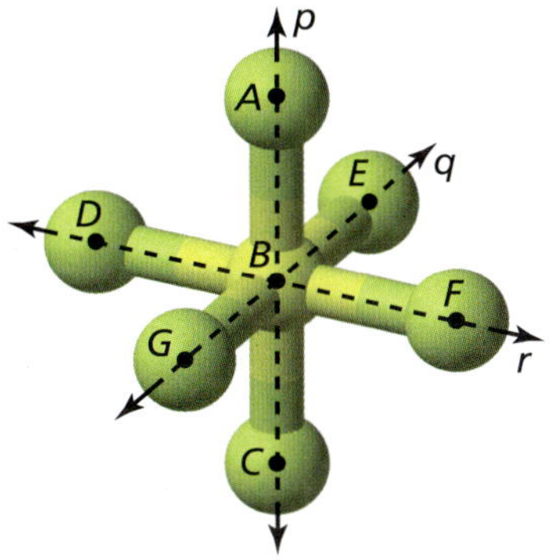

SOLUTION

To name a plane that contains line *r*, use two points on line *r* and one point not on line *r*. Points *D* and *F* lie on line *r*. Points *C* and *E* do not lie on line *r*.

▶ So, plane *DEF* and plane *CDF* both contain line *r*.

COMMON ERROR

Because point *B* also lies on line *r*, you cannot use points *D*, *B*, and *F* to name a single plane. There are infinitely many planes that pass through these points.

Check The question asks for two *different* planes. Check whether plane *DEF* and plane *CDF* are two unique planes or the same plane named differently. Because point *C* does not lie in plane *DEF*, plane *DEF* and plane *CDF* are different planes.

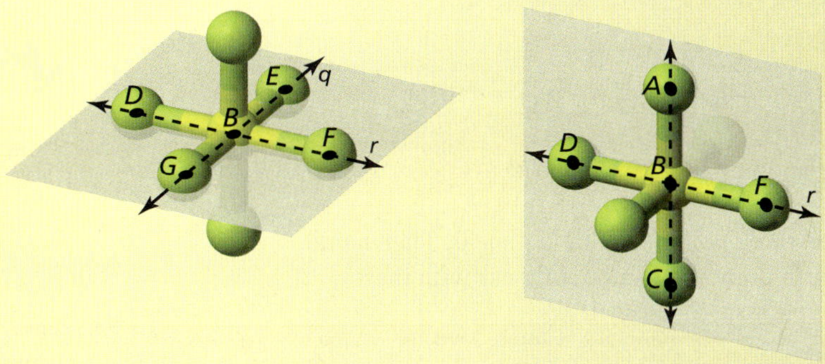

SELF-ASSESSMENT 1 I don't understand yet. 2 I can do it with help. 3 I can do it on my own. 4 I can teach someone else.

Use the diagram that shows a model of a molecule of phosphorus pentachloride.

10. Name two different planes that contain line *s*.

11. Name three different planes that contain point *K*.

12. Name two different planes that contain $\overrightarrow{HJ}$.

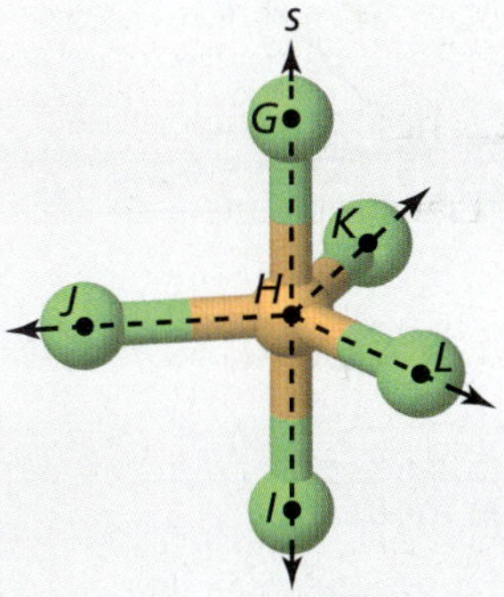

GO DIGITAL

1.1 Practice WITH CalcChat® AND CalcView®

In Exercises 1–4, use the diagram.

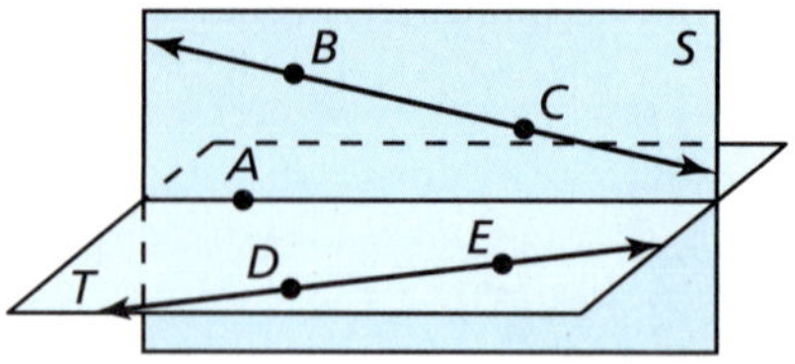

1. Name four points.

2. Name two lines.

3. Name the plane that contains points A, B, and C.

4. Name the plane that contains points A, D, and E.

In Exercises 5–8, use the diagram. *(See Example 1.)*

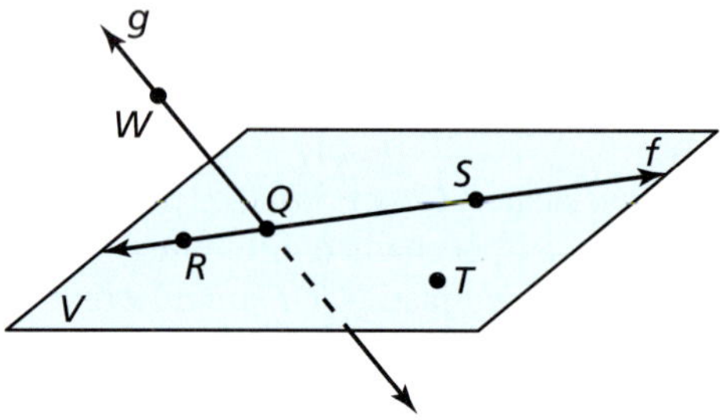

▶ **5.** Give two other names for $\overleftrightarrow{WQ}$.

6. Give another name for plane V.

7. Name three points that are collinear. Then name a fourth point that is not collinear with these three points.

8. Name a point that is not coplanar with R, S, and T.

In Exercises 9–14, use the diagram. *(See Example 2.)*

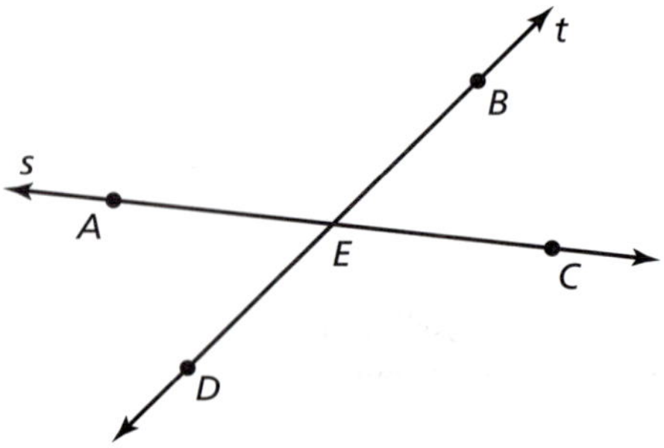

▶ **9.** What is another name for $\overline{BD}$?

10. What is another name for $\overline{AC}$?

11. What is another name for $\overrightarrow{AE}$?

12. Name all rays with endpoint E.

13. Name two pairs of opposite rays.

14. Name one pair of rays that are not opposite rays.

4 MTR **ERROR ANALYSIS In Exercises 15 and 16, describe and correct the error in naming opposite rays in the diagram.**

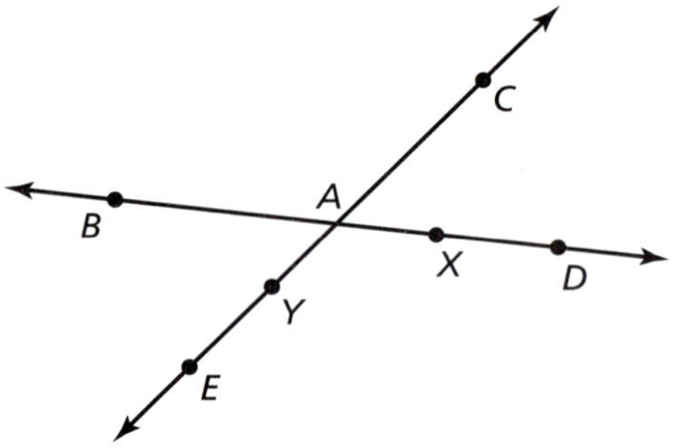

15. ✗ $\overrightarrow{AD}$ and $\overrightarrow{AC}$ are opposite rays.

16. ✗ $\overline{YC}$ and $\overline{YE}$ are opposite rays.

In Exercises 17–24, sketch the figure described. *(See Examples 3 and 4.)*

▶ **17.** plane P and line ℓ intersecting at one point

18. plane K and line m intersecting at all points on line m

19. $\overrightarrow{AB}$ and $\overleftrightarrow{AC}$

20. $\overrightarrow{MN}$ and $\overrightarrow{NX}$

21. plane M and $\overrightarrow{NB}$ intersecting at point B

22. plane M and $\overrightarrow{NB}$ intersecting at point A

23. plane A and plane B not intersecting

24. plane C and plane D intersecting at $\overleftrightarrow{XY}$

GO DIGITAL

In Exercises 25–32, use the diagram.

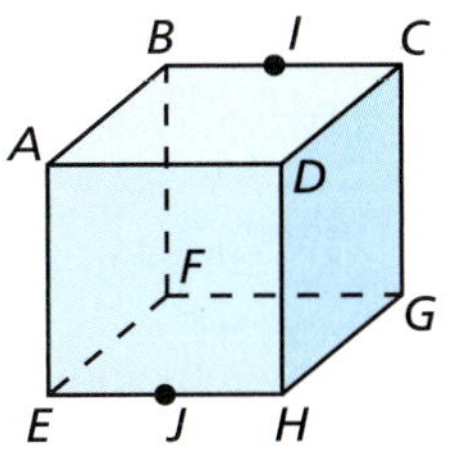

25. Name a point that is collinear with points E and H.

26. Name a point that is collinear with points B and I.

27. Name a point that is not collinear with points E and H.

28. Name a point that is not collinear with points B and I.

29. Name a point that is coplanar with points D, A, and B.

30. Name a point that is coplanar with points C, G, and F.

31. Name the intersection of plane AEH and plane FBE.

32. Name the intersection of plane BGF and plane HDG.

7 MTR **MODELING REAL LIFE** **In Exercises 33 and 34, use the diagram.** *(See Example 5.)*

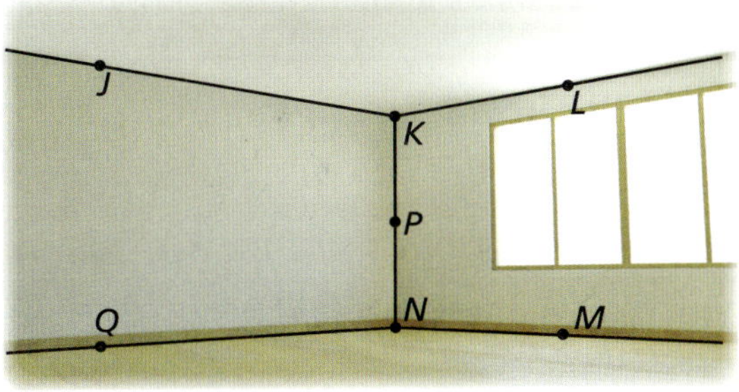

33. Name two points that are collinear with P.

34. Name two planes that contain J.

7 MTR **35. MODELING REAL LIFE** When two trucks traveling in different directions approach an intersection at the same time, one of the trucks must change its speed or direction to avoid a collision. Two airplanes, however, can travel in different directions and cross paths without colliding. Explain how this is possible.

36. REASONING Given two points on a line and a third point not on the line, is it possible to draw a plane that includes the line and the third point? Explain your reasoning.

GO DIGITAL

In Exercises 37–40, name the geometric term modeled by the part of the object indicated with an arrow.

37.

38.

39.

40.

In Exercises 41–44, use the diagram to name all the points that are *not* coplanar with the given points.

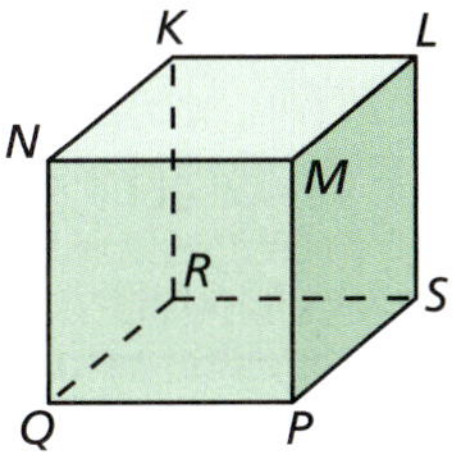

41. N, K, and L

42. P, Q, and N

43. P, Q, and R

44. R, K, and N

45. REASONING Is it possible to draw two planes that intersect at one point? Explain your reasoning.

46. HOW DO YOU SEE IT?
You and your friend walk in opposite directions, forming opposite rays. You were originally on the corner of Apple Avenue and Cherry Court.

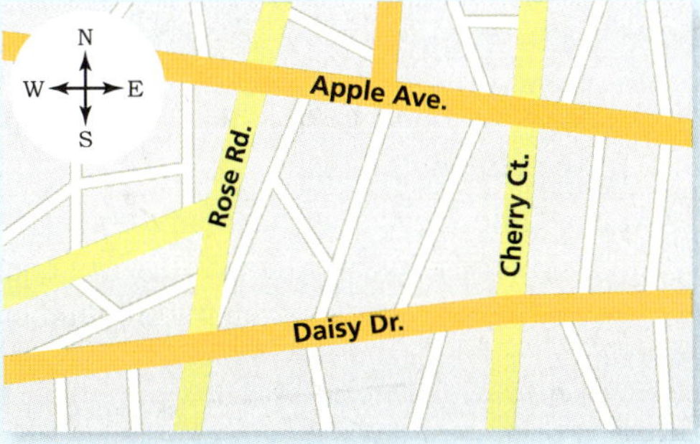

a. Name two possibilities of the road and direction you and your friend may have traveled.

b. Your friend claims he went north on Cherry Court, and you went east on Apple Avenue. Make an argument for why you know this could not have happened.

47. **REASONING** Explain why a four-legged chair may rock from side to side even if the floor is level. Would a three-legged chair on the same level floor rock from side to side? Why or why not?

7 MTR **48.** **MODELING REAL LIFE** You are designing a living room. Counting the floor, walls, and ceiling, you want the design to contain at least eight different planes. Draw a diagram of your design. Label each plane in your design.

5 MTR **CONNECTING CONCEPTS** **In Exercises 49 and 50, graph the inequality on a number line. Tell whether the graph is a *segment*, a *ray* or *rays*, a *point*, or a *line*.**

49. $x \leq 3$

50. $-7 \leq x \leq 4$

4 MTR **DISCUSS MATHEMATICAL THINKING** **In Exercises 51–58, complete the statement with *always*, *sometimes*, or *never*. Explain your reasoning.**

51. A line ____________ has endpoints.

52. A line and a point ____________ intersect.

53. A plane and a point ____________ intersect.

54. Two planes ____________ intersect in a line.

55. Two points ____________ determine a line.

56. Any three points ____________ determine a plane.

57. Any three points not on the same line ____________ determine a plane.

58. Two lines that are not parallel ____________ intersect.

5 MTR **59.** **STRUCTURE** Two coplanar intersecting lines will always intersect at one point. What is the greatest number of intersection points that exist if you draw four coplanar lines? Explain.

60. **THOUGHT PROVOKING**
Is it possible for three planes to never intersect? to intersect in one line? to intersect in one point? Sketch the possible situations.

REVIEW & REFRESH

WATCH

In Exercises 61 and 62, determine which of the lines, if any, are parallel or perpendicular. Explain.

61. Line a passes through (1, 3) and (−2, −3).
Line b passes through (−1, −5) and (0, −3).
Line c passes through (3, 2) and (1, 0).

62. Line a: $y + 4 = \frac{1}{2}x$

Line b: $2y = -4x + 6$

Line c: $y = 2x - 1$

In Exercises 63 and 64, solve the equation.

63. $18 + x = 43$

64. $x - 23 = 19$

7 MTR **65.** **MODELING REAL LIFE** You bike at a constant speed of 10 miles per hour. You plan to bike 30 miles, plus or minus 5 miles. Write and solve an equation to find the minimum and maximum numbers of hours you bike.

In Exercises 66 and 67, evaluate the expression.

66. $\sqrt[3]{8^5}$

67. $36^{1/2}$

68. Graph $f(x) = -\frac{1}{3}x + 5$ and $g(x) = f(x - 4)$. Describe the transformation from the graph of f to the graph of g.

In Exercises 69–75, use the diagram.

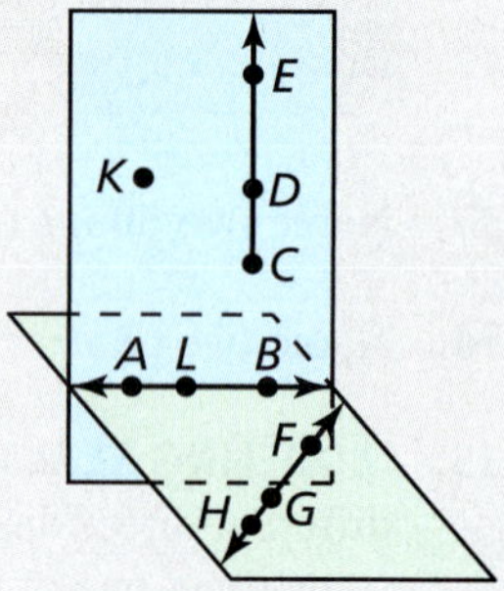

69. Name four points.

70. Name two lines.

71. Name three rays.

72. Name three collinear points.

73. Name three coplanar points.

74. Give two names for the plane shaded blue.

75. Name three line segments.

In Exercises 76–79, solve the equation.

76. $2x(x - 5)(x + 8) = 0$

77. $4x^3 - 64x = 0$

78. $3x^3 + 3x^2 - 6x = 0$

79. $-x(x + 1)(x - 7) = 0$

In Exercises 80 and 81, make a box-and-whisker plot that represents the data.

80. Scores on a test: 76, 90, 84, 97, 82, 100, 92, 90, 88

81. Minutes spent at the gym: 60, 45, 50, 45, 65, 50, 55, 60, 60, 50

GO DIGITAL

1.2 Measuring and Constructing Segments

Learning Target: Measure and construct line segments.

Success Criteria:
- I can measure a line segment.
- I can copy a line segment.
- I can explain and use the Segment Addition Postulate.

EXPLORE IT! Measuring and Copying Line Segments

Work with a partner. A *straightedge* is a tool that you can use to draw a straight line. An example of a straightedge is a ruler. A *compass* is a tool that you can use to draw circles and arcs, and copy segments. Choose from these tools to complete the following tasks.

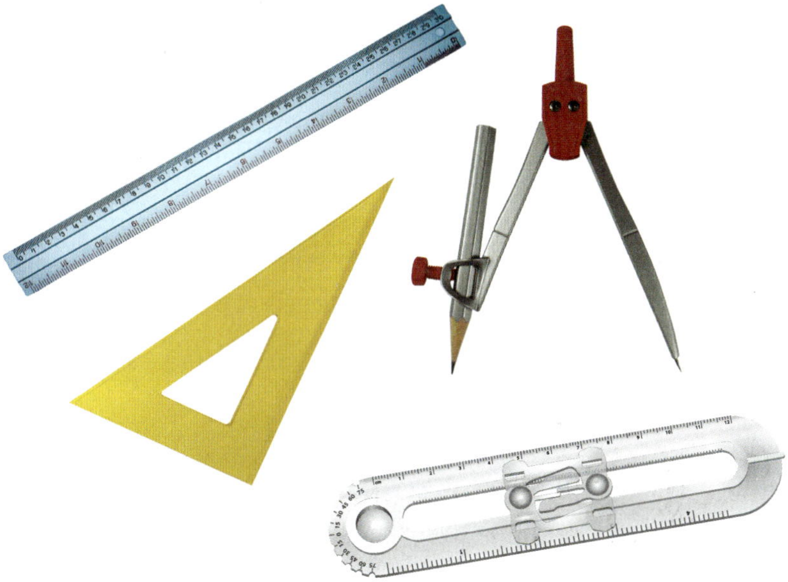

a. Find the length of the line segment.

b. Make a copy of the line segment in part (a). Explain your process.

c. Draw a different line segment that has a length between 4 centimeters and 10 centimeters.

d. Make a copy of the line segment in part (c) using a different method than you used in part (b). Explain your process.

e. Find the lengths x, y, and z. What do you notice?

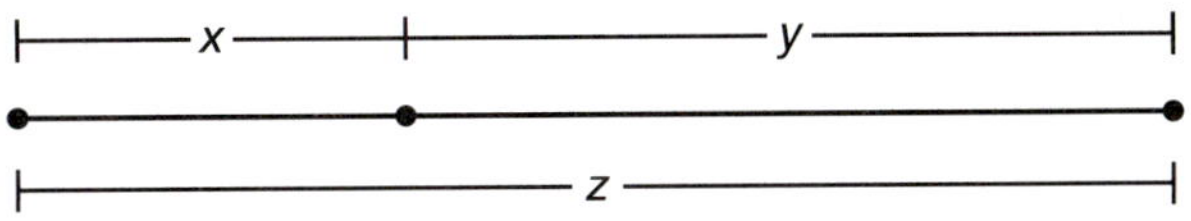

5 MTR MAKE A PLAN

How can you use a paper clip to compare the lengths of two different line segments?

Geometric Reasoning
MA.912.GR.5.1 Construct a copy of a segment or an angle.

GO DIGITAL

Using the Ruler Postulate

Vocabulary AZ VOCAB

postulate, *p. 12*
axiom, *p. 12*
coordinate, *p. 12*
distance between two points, *p. 12*
construction, *p. 13*
congruent segments, *p. 13*
between, *p. 14*

In geometry, a rule that is accepted without proof is called a **postulate** or an **axiom**. A rule that can be proved is called a *theorem*, as you will see later. Postulate 1.1 shows how to find the distance between two points on a line.

POSTULATE

1.1 Ruler Postulate

The points on a line can be matched one to one with the real numbers. The real number that corresponds to a point is the **coordinate** of the point.

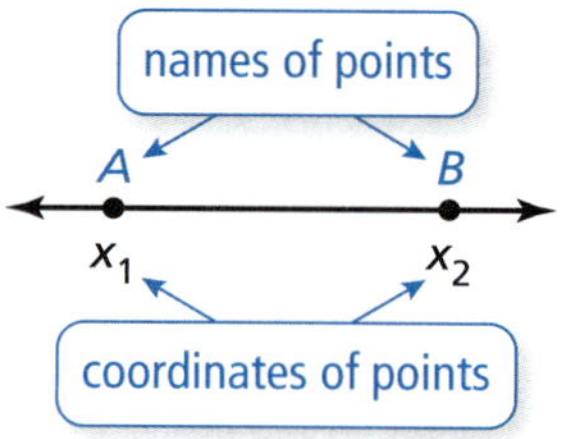

The **distance** between points A and B, written as AB, is the absolute value of the difference of the coordinates of A and B.

$AB = |x_2 - x_1|$

EXAMPLE 1 Using the Ruler Postulate

WATCH

Measure the length of $\overline{ST}$ to the nearest tenth of a centimeter.

SOLUTION

Align one mark of a metric ruler with S. Then estimate the coordinate of T. For example, when you align S with 2, T appears to align with 5.4.

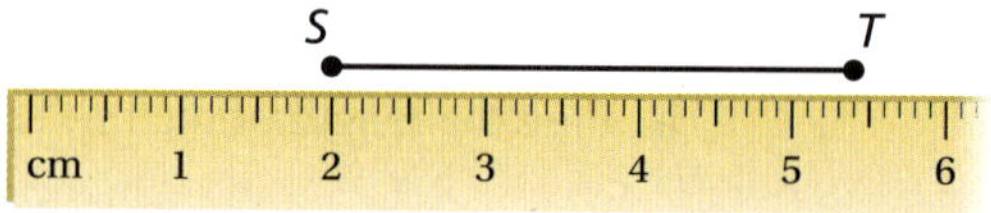

$ST = |5.4 - 2| = 3.4$ Ruler Postulate

▶ So, the length of $\overline{ST}$ is about 3.4 centimeters.

SELF-ASSESSMENT 1 I don't understand yet. 2 I can do it with help. 3 I can do it on my own. 4 I can teach someone else.

Use a ruler to measure the length of the segment to the nearest $\frac{1}{8}$ inch.

1. 

2. 

3. 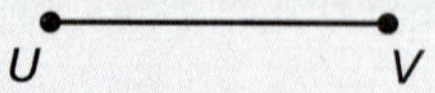

4. 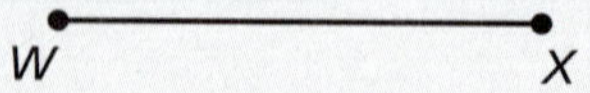

5. WRITING Explain how $\overline{XY}$ and XY are different.

GO DIGITAL

Constructing and Comparing Congruent Segments

A **construction** is a geometric drawing that uses a limited set of tools, usually a *compass* and *straightedge*.

CONSTRUCTION **Copying a Segment** WATCH

Use a compass and straightedge to construct a line segment that has the same length as $\overline{AB}$.

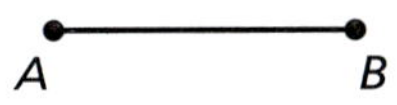

SOLUTION

Step 1

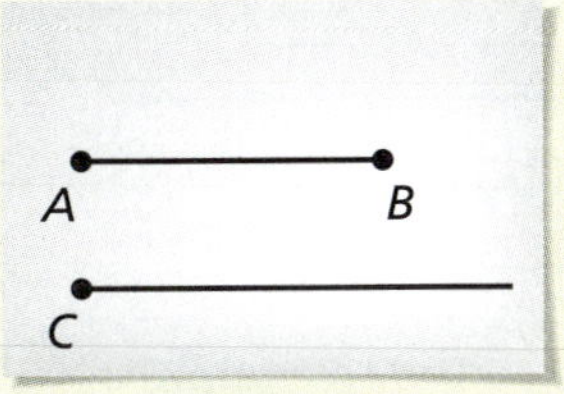

Draw a segment Use a straightedge to draw a segment longer than AB. Label point C on the new segment.

Step 2

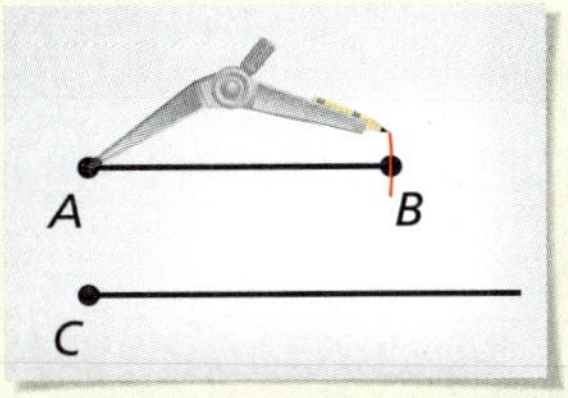

Measure length Set your compass at the length of $\overline{AB}$.

Step 3

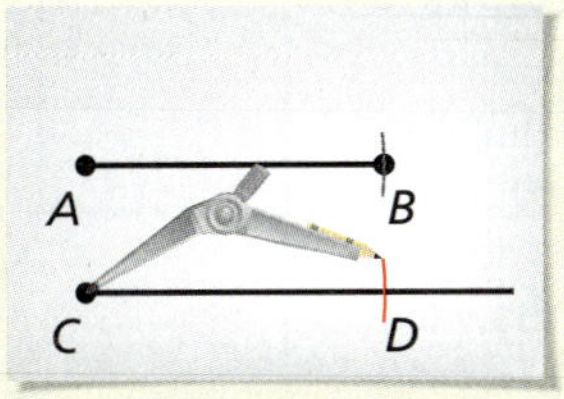

Copy length Place the compass at C. Mark point D on the new segment. So, $\overline{CD}$ has the same length as $\overline{AB}$.

KEY IDEA

Congruent Segments

Line segments that have the same length are called **congruent segments**. You can say "the length of $\overline{AB}$ is equal to the length of $\overline{CD}$," or you can say "$\overline{AB}$ *is congruent to* $\overline{CD}$." The symbol $\cong$ means "is congruent to."

Lengths are equal.
$AB = CD$
"is equal to"

Segments are congruent.
$\overline{AB} \cong \overline{CD}$
"is congruent to"

READING

In the diagram, the red tick marks indicate $\overline{AB} \cong \overline{CD}$. When there is more than one pair of congruent segments, use multiple tick marks.

EXAMPLE 2 **Comparing Segments for Congruence** WATCH

Plot $J(-3, 4)$, $K(2, 4)$, $L(1, 3)$, and $M(1, -2)$ in a coordinate plane. Then determine whether $\overline{JK}$ and $\overline{LM}$ are congruent.

SOLUTION

Plot the points, as shown. To find the length of a horizontal segment, find the absolute value of the difference of the x-coordinates of the endpoints.

$JK = |2 - (-3)| = 5$ Ruler Postulate

To find the length of a vertical segment, find the absolute value of the difference of the y-coordinates of the endpoints.

$LM = |-2 - 3| = 5$ Ruler Postulate

▶ $\overline{JK}$ and $\overline{LM}$ have the same length. So, $\overline{JK} \cong \overline{LM}$.

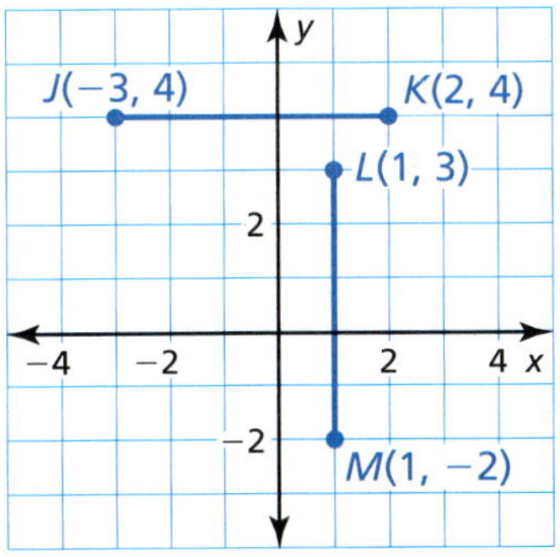

GO DIGITAL

Using the Segment Addition Postulate

When three points are collinear, you can say that one point is **between** the other two.

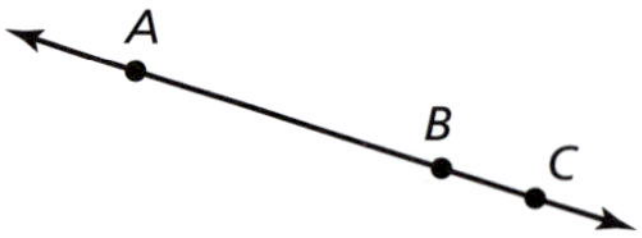

Point *B* is between points *A* and *C*.

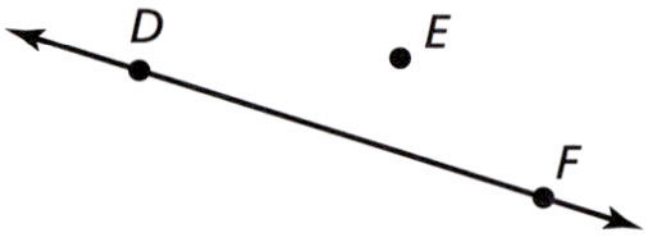

Point *E* is not between points *D* and *F*.

4 MTR CONSTRUCT AN ARGUMENT
Consider a fourth point *D* such that *C* is between *B* and *D*. Use the Segment Addition Postulate to show that $AB + BC + CD = AD$.

POSTULATE

1.2 Segment Addition Postulate

If *B* is between *A* and *C*, then $AB + BC = AC$.

If $AB + BC = AC$, then *B* is between *A* and *C*.

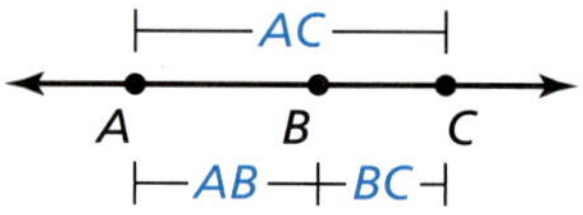

EXAMPLE 3 Using the Segment Addition Postulate

a. Find *DF*.

b. Find *GH*.

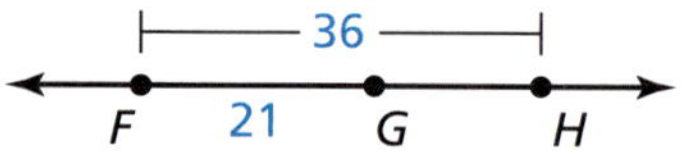

SOLUTION

a. Use the Segment Addition Postulate to write an equation. Then solve the equation to find *DF*.

$DF = DE + EF$	Segment Addition Postulate
$DF = 23 + 35$	Substitute 23 for *DE* and 35 for *EF*.
$DF = 58$	Add.

b. Use the Segment Addition Postulate to write an equation. Then solve the equation to find *GH*.

$FH = FG + GH$	Segment Addition Postulate
$36 = 21 + GH$	Substitute 36 for *FH* and 21 for *FG*.
$15 = GH$	Subtract 21 from each side.

SELF-ASSESSMENT

1 I don't understand yet. 2 I can do it with help. 3 I can do it on my own. 4 I can teach someone else.

6. Plot $A(-2, 4)$, $B(3, 4)$, $C(0, 2)$, and $D(0, -2)$ in a coordinate plane. Then determine whether $\overline{AB}$ and $\overline{CD}$ are congruent.

4 MTR **7. DIFFERENT WORDS, SAME QUESTION** Which is different? Find "both" answers.

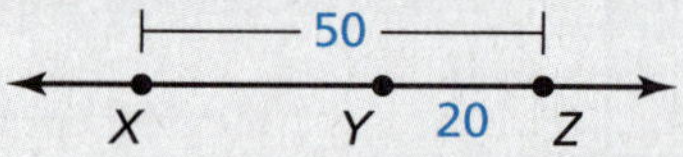

Find $XZ - YZ$.

Find $XZ - XY$.

Find XY.

Find the length of $\overline{XY}$.

GO DIGITAL

7 MTR **EXAMPLE 4** **Modeling Real Life** WATCH

The cities shown on the map lie approximately in a straight line. Find the distance from Orlando to Lake City.

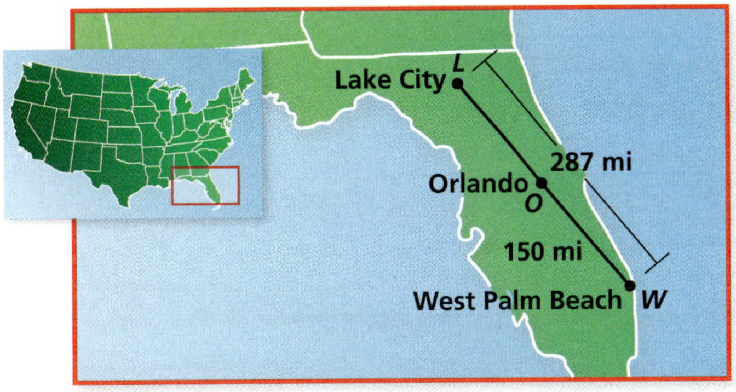

SOLUTION

1. **Understand the Problem** You know that the three cities are approximately collinear. The map shows the distances from West Palm Beach to Lake City and from West Palm Beach to Orlando. You need to find the distance from Orlando to Lake City.
2. **Make a Plan** Use the Segment Addition Postulate to find the distance from Orlando to Lake City.
3. **Solve and Check** Use the Segment Addition Postulate to write an equation. Then solve the equation to find OL.

$WL = WO + OL$ Segment Addition Postulate

$287 = 150 + OL$ Substitute 287 for *WL* and 150 for *WO*.

$137 = OL$ Subtract 150 from each side.

▶ So, the distance from Orlando to Lake City is about 137 miles.

Check The distance from West Palm Beach to Lake City is 287 miles. By the Segment Addition Postulate, the distance from West Palm Beach to Orlando plus the distance from Orlando to Lake City should equal 287 miles.

$150 + 137 = 287$ ✓

SELF-ASSESSMENT

1 I don't understand yet. 2 I can do it with help. 3 I can do it on my own. 4 I can teach someone else.

8. The cities shown on the map lie approximately in a straight line. Find the distance from Albuquerque, New Mexico, to Provo, Utah.

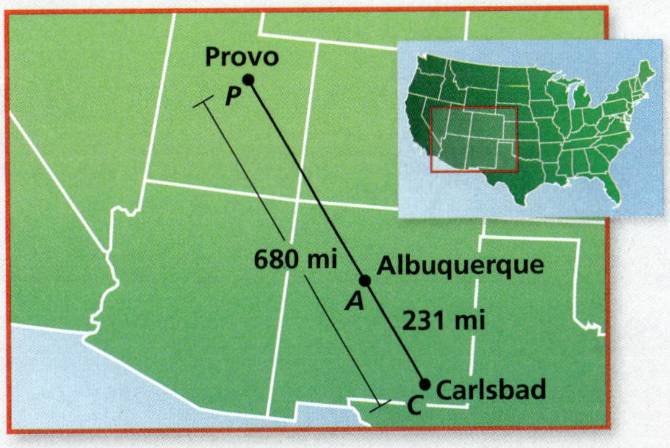

GO DIGITAL

1.2 Practice WITH CalcChat® AND CalcView®

In Exercises 1–4, use a ruler to measure the length of the segment to the nearest tenth of a centimeter. *(See Example 1.)*

▶ **1.**

2. 

3.

4. 

CONSTRUCTION **In Exercises 5 and 6, use a compass and straightedge to construct a copy of the segment.**

5. Copy the segment in Exercise 3.

6. Copy the segment in Exercise 4.

In Exercises 7–12, plot the points in a coordinate plane. Then determine whether $\overline{AB}$ and $\overline{CD}$ are congruent. *(See Example 2.)*

▶ **7.** $A(-4, 5)$, $B(-4, 8)$, $C(2, -3)$, $D(2, 0)$

8. $A(6, -1)$, $B(1, -1)$, $C(2, -3)$, $D(4, -3)$

9. $A(8, 3)$, $B(-1, 3)$, $C(5, 10)$, $D(5, 3)$

10. $A(6, -8)$, $B(6, 1)$, $C(7, -2)$, $D(-2, -2)$

11. $A(-5, 6)$, $B(-5, -1)$, $C(-4, 3)$, $D(3, 3)$

12. $A(10, -4)$, $B(3, -4)$, $C(-1, 2)$, $D(-1, 5)$

In Exercises 13–20, find FH. *(See Example 3.)*

▶ **13.**

14.

15.

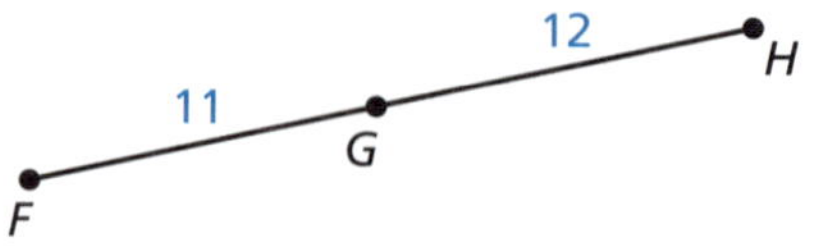

16.

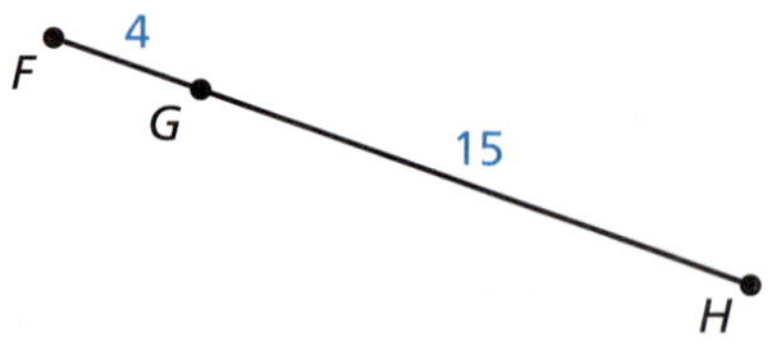

17.

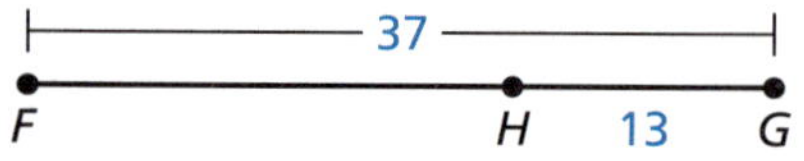

18.

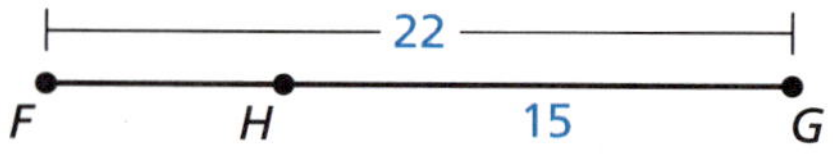

19.

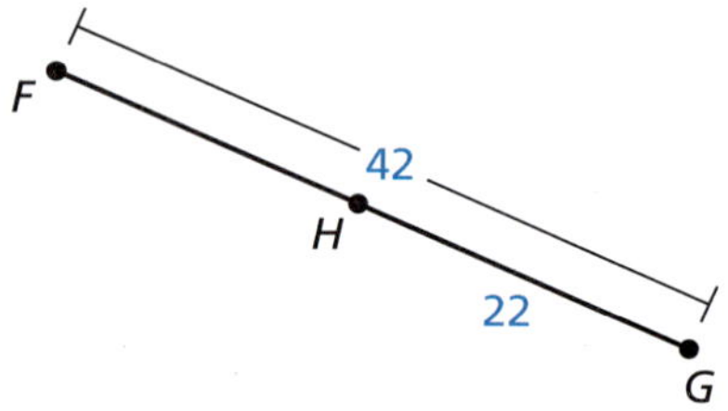

20.

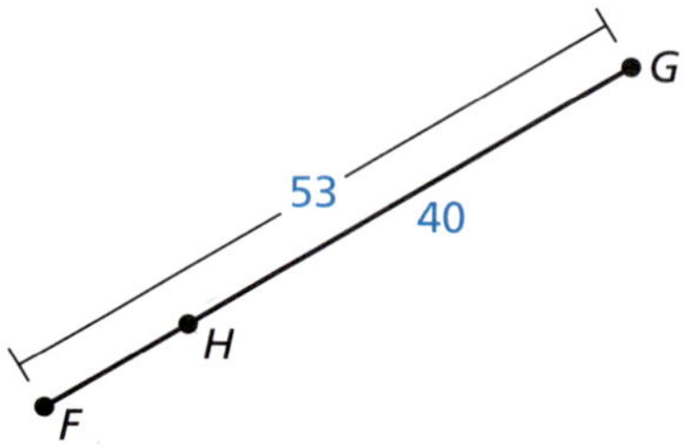

4 MTR **ERROR ANALYSIS** **In Exercises 21 and 22, describe and correct the error in finding the length of $\overline{AB}$.**

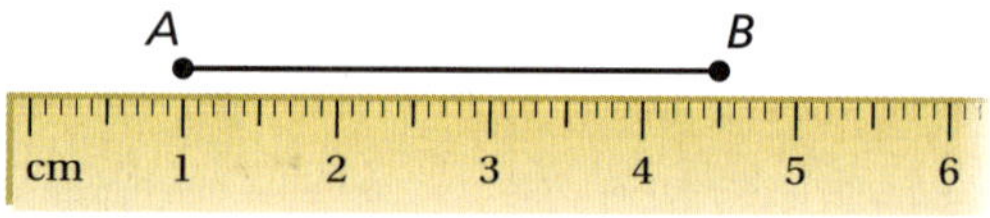

21. ✗ $AB = 1 - 4.5 = -3.5$

22. ✗ $AB = |1 + 4.5| = 5.5$

23. **COLLEGE PREP** Which expression does *not* equal 10?

Ⓐ $AC + CB$

Ⓑ $BA - CA$

Ⓒ AB

Ⓓ $CA + BC$

GO DIGITAL

3 MTR **24. MAINTAIN ACCURACY** The diagram shows an insect called a walking stick. Use the ruler to estimate the length of the abdomen and the length of the thorax to the nearest $\frac{1}{4}$ inch. How much longer is the walking stick's abdomen than its thorax? How many times longer is its abdomen than its thorax?

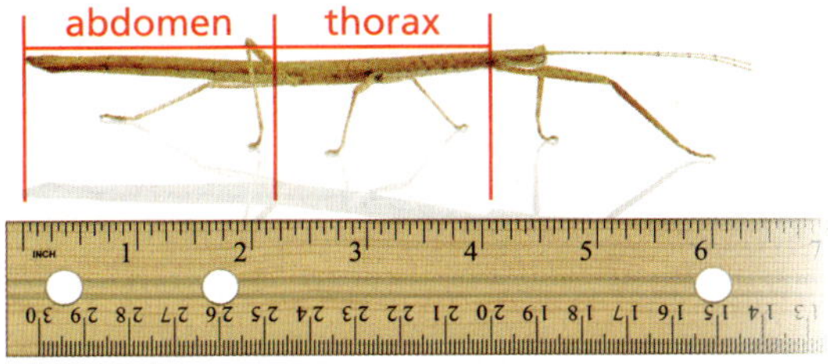

7 MTR **25. MODELING REAL LIFE** In 2003, a remote-controlled model airplane became the first ever to fly nonstop across the Atlantic Ocean. The map shows the airplane's position at three different points during its flight. Point A represents Cape Spear, Newfoundland, point B represents the approximate position after one day, and point C represents Mannin Bay, Ireland. The airplane left from Cape Spear and landed in Mannin Bay. *(See Example 4.)*

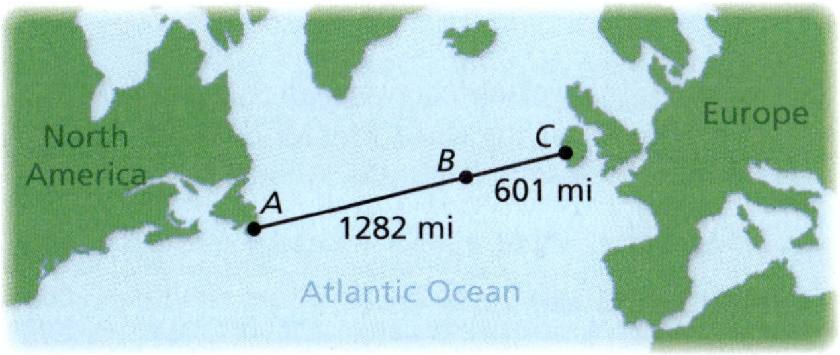

a. Find the total distance the model airplane flew.

b. The flight lasted nearly 38 hours. Estimate the airplane's average speed in miles per hour.

7 MTR **26. MODELING REAL LIFE** You walk in a straight line from Room 103 to Room 117 at a speed of 4.4 feet per second.

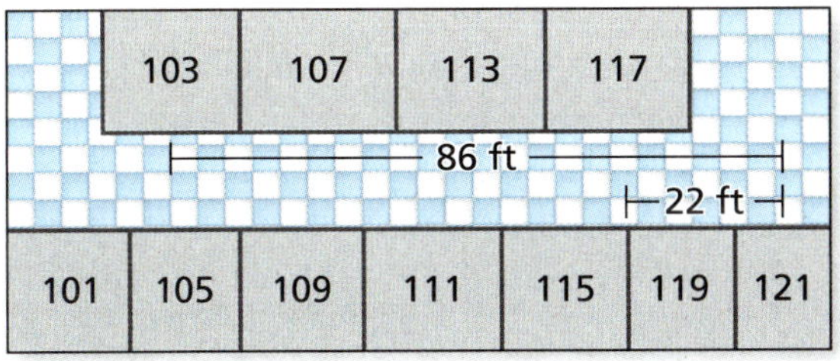

a. How far do you walk?

b. How long does it take you to get to Room 117?

c. Why might it actually take you longer than the time in part (b)?

GO DIGITAL

5 MTR **27. STRUCTURE** Determine whether each statement is *true* or *false*. Explain your reasoning.

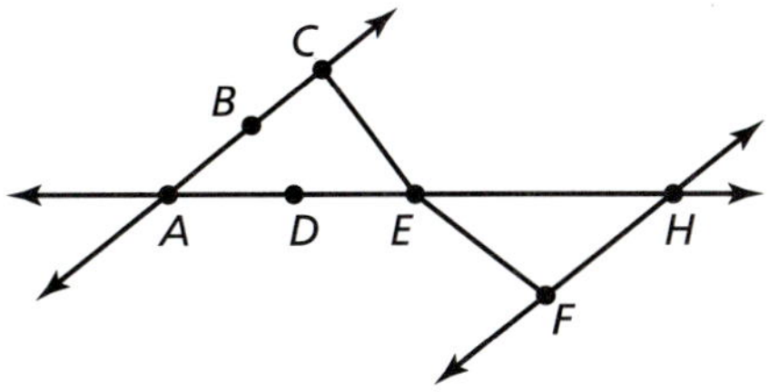

a. B is between A and C.

b. C is between B and E.

c. D is between A and H.

d. E is between C and F.

5 MTR **28. CONNECTING CONCEPTS** Point S is between points R and T on $\overline{RT}$. Use the information to write an equation in terms of x. Then solve the equation and find RS, ST, and RT.

a. $RS = 2x + 10$
$ST = x - 4$
$RT = 21$

b. $RS = 4x - 9$
$ST = 19$
$RT = 8x - 14$

4 MTR **29. MAKING AN ARGUMENT** Your friend says that when measuring with a ruler, you must always line up objects at the zero on the ruler. Is your friend correct? Explain your reasoning.

30. HOW DO YOU SEE IT?
The bar graph shows the win-loss record for a lacrosse team over a period of three years. Explain how you can apply the Ruler Postulate and the Segment Addition Postulate when interpreting a stacked bar graph like the one shown.

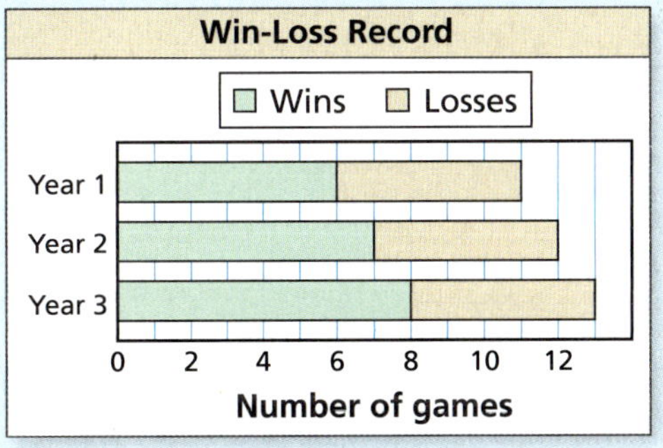

31. REASONING The round-trip distance between City X and City Y is 647 miles. A national park is between City X and City Y, and is 27 miles from City X. Find the round-trip distance between the national park and City Y. Justify your answer.

32. **REASONING** The points (a, b) and (c, b) form a segment, and the points (d, e) and (d, f) form a segment. The segments are congruent. Write an equation that represents the relationship among the variables. Are any of the variables *not* used in the equation? Explain.

33. **CONNECTING CONCEPTS** In the diagram, $\overline{AB} \cong \overline{BC}$, $\overline{AC} \cong \overline{CD}$, and $AD = 12$. Find the lengths of all segments in the diagram. You choose one of the segments at random. What is the probability that the length of the segment is greater than 3? Explain your reasoning.

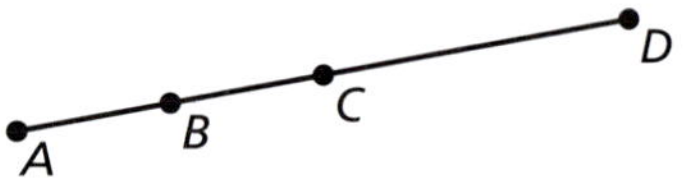

34. **REASONING** Points A, B, and C lie on a line where $AB = 35$ and $AC = 93$. What are the possible values of BC?

35. **DIG DEEPER** Is it possible to use the Segment Addition Postulate to show that $FB > CB$? that $AC > DB$? Explain your reasoning.

36. **THOUGHT PROVOKING**
Is it possible to design a table where no two legs have the same length? Assume that the endpoints of the legs (that are not attached to the table) must all lie in the same plane. Include a diagram with your answer.

REVIEW & REFRESH

WATCH

In Exercises 37–40, solve the equation.

37. $3 + y = 12$

38. $-5x = 10$

39. $5x + 7 = 9x - 17$

40. $\dfrac{-5 + x}{2} = -9$

41. Sketch plane P and $\overleftrightarrow{YZ}$ intersecting at point Z.

42. Write an inequality that represents the graph.

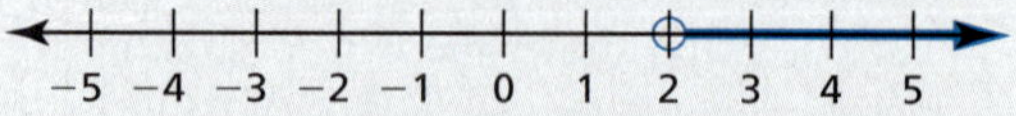

In Exercises 43 and 44, use intercepts to graph the linear equation. Label the points corresponding to the intercepts.

43. $4x + 3y = 24$

44. $-2x + 4y = -16$

45. Determine whether the relation is a function. Explain.

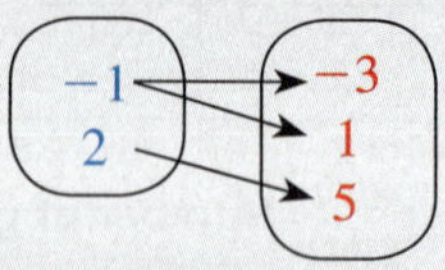

In Exercises 46–49, solve the inequality. Graph the solution.

46. $x - 6 \leq 13$

47. $-3t > 15$

48. $5 - \dfrac{c}{3} < 12$

49. $6 - v < 8$ *or* $-4v \geq 40$

In Exercises 50 and 51, graph the function. Identify the asymptote. Find the domain and range of *f*.

50. $f(x) = 2^x - 3$

51. $f(x) = 3(0.5)^x$

52. Is there a correlation between amusement park attendance and the wait times for rides? If so, is there a causal relationship? Explain your reasoning.

In Exercises 53 and 54, find *AC*.

53.

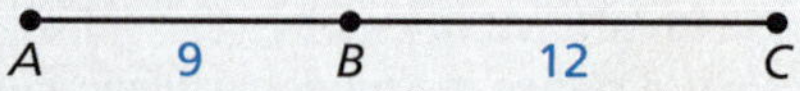

54.

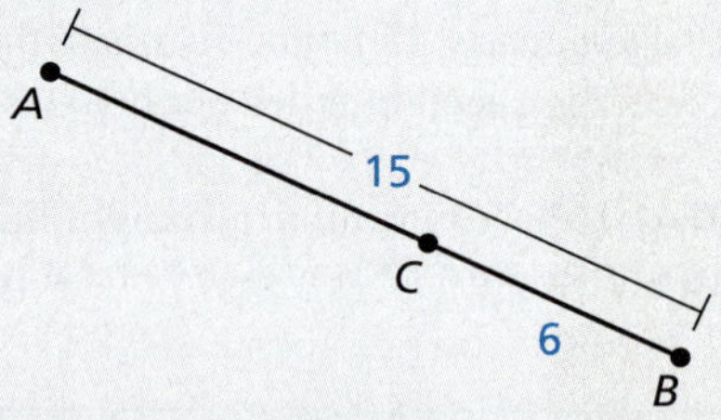

55. **MODELING REAL LIFE** A football team scores a total of 7 touchdowns and field goals in a game. The team scores an extra point with each touchdown, so each touchdown is worth 7 points. Each field goal is worth 3 points. The team scores a total of 41 points. How many touchdowns does the team score? How many field goals?

In Exercises 56 and 57, write an equation in slope-intercept form of the line that passes through the given points.

56. $(0, 3), \left(\frac{1}{2}, 0\right)$

57. $(-8, -8), (12, -3)$

GO DIGITAL

1.3 Using Midpoint and Distance Formulas

Learning Target: Find midpoints and lengths of segments.

Success Criteria:
- I can find lengths of segments.
- I can construct a segment bisector.
- I can find the weighted average of two or more points on a number line.
- I can find the midpoint of a segment.

EXPLORE IT! Finding Midpoints of Line Segments

1 MTR ANALYZE A PROBLEM
Draw a right triangle with hypoteneuse $\overline{AB}$. How can you find the length of $\overline{AB}$?

Work with a partner.

a. Plot any two points A and B. Then graph $\overline{AB}$. Identify the point M on $\overline{AB}$ that is halfway between points A and B, called the *midpoint* of $\overline{AB}$. Explain how you found the midpoint.

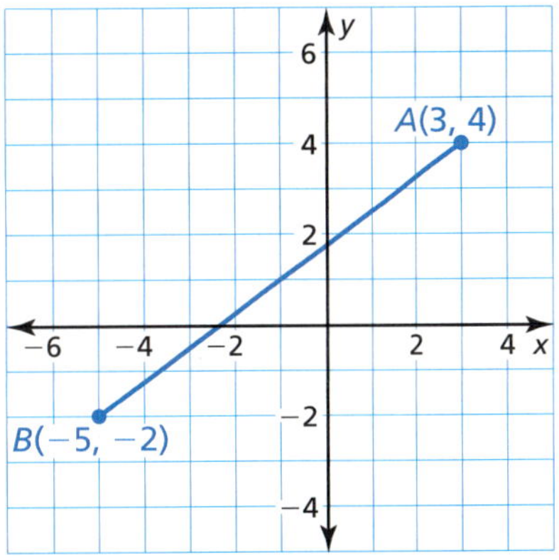

b. Repeat part (a) five times and complete the table.

Coordinates of A	Coordinates of B	Coordinates of M

c. Compare the x-coordinates of A, B, and M. Compare the y-coordinates of A, B, and M. How are the coordinates of the midpoint M related to the coordinates of A and B?

Geometric Reasoning
MA.912.GR.3.1 Determine the weighted average of two or more points on a line.
MA.912.GR.3.3 Use coordinate geometry to solve mathematical and real-world geometric problems involving lines, circles, triangles and quadrilaterals.
Also MA.912.GR.5.2

GO DIGITAL

Midpoints and Segment Bisectors

Vocabulary

midpoint, *p. 20*
segment bisector, *p. 20*
weighted average, *p. 22*

READING

The word *bisect* means "to cut into two equal parts."

KEY IDEAS

Midpoints and Segment Bisectors

The **midpoint** of a segment is the point that divides the segment into two congruent segments.

M is the midpoint of $\overline{AB}$.
So, $\overline{AM} \cong \overline{MB}$ and $AM = MB$.

A **segment bisector** is a point, ray, line, line segment, or plane that intersects the segment at its midpoint. A midpoint or a segment bisector *bisects* a segment.

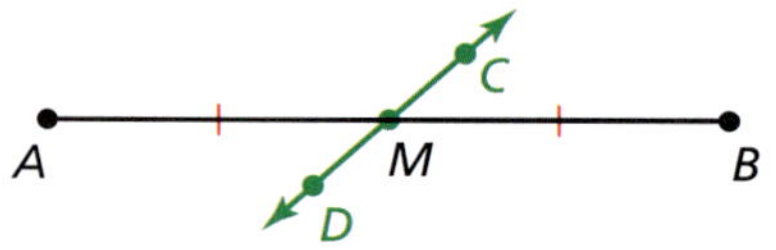

$\overleftrightarrow{CD}$ is a segment bisector of $\overline{AB}$.
So, $\overline{AM} \cong \overline{MB}$ and $AM = MB$.

EXAMPLE 1 Finding Segment Lengths

WATCH

In the skateboard design, $XT = 39.9$ cm. Identify the segment bisector of $\overline{XY}$. Then find XY.

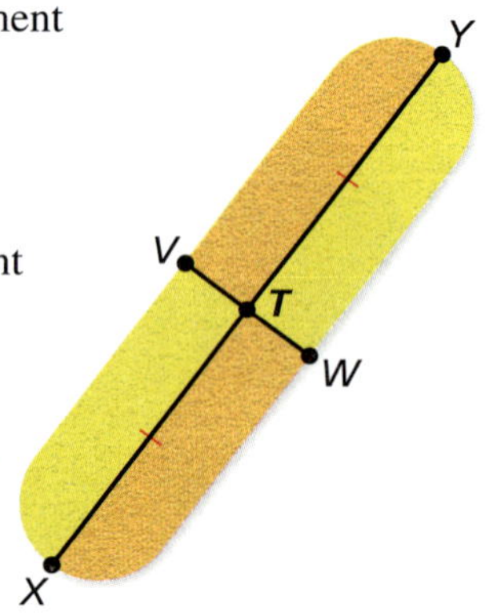

SOLUTION

The design shows that $\overline{XT} \cong \overline{TY}$. So, point T is the midpoint of $\overline{XY}$, and $XT = TY = 39.9$ cm. Because $\overline{VW}$ intersects $\overline{XY}$ at its midpoint T, $\overline{VW}$ bisects $\overline{XY}$. Find XY.

$XY = XT + TY$	Segment Addition Postulate
$= 39.9 + 39.9$	Substitute.
$= 79.8$	Add.

▶ $\overline{VW}$ is the segment bisector of $\overline{XY}$, and XY is 79.8 centimeters.

SELF-ASSESSMENT 1 I don't understand yet. 2 I can do it with help. 3 I can do it on my own. 4 I can teach someone else.

Identify the segment bisector of $\overline{PQ}$. Then find PQ.

1.

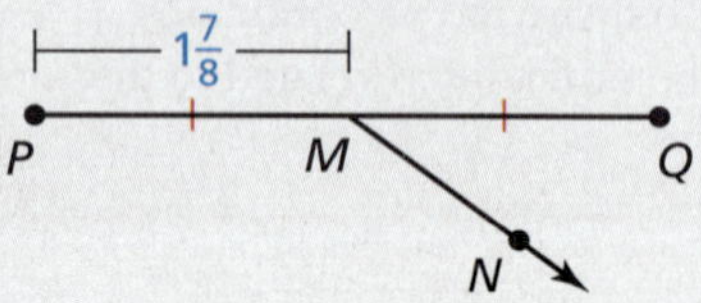

2.

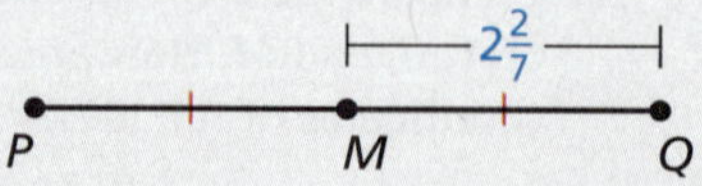

3. **VOCABULARY** If a point, ray, line, line segment, or plane intersects a segment at its midpoint, then what does it do to the segment?

GO DIGITAL

EXAMPLE 2 Using Algebra with Segment Lengths

WATCH

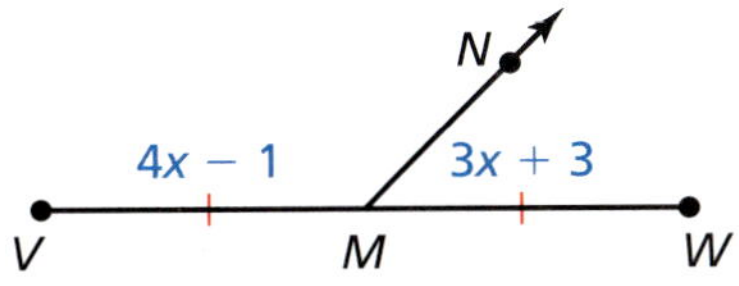

Identify the segment bisector of $\overline{VW}$. Then find VM.

SOLUTION

The figure shows that $\overline{VM} \cong \overline{MW}$. So, point M is the midpoint of $\overline{VW}$, and $VM = MW$. Because $\overrightarrow{MN}$ intersects $\overline{VW}$ at its midpoint M, $\overrightarrow{MN}$ bisects $\overline{VW}$. Find VM.

Step 1 Write and solve an equation to find VM.

$VM = MW$	Write the equation.
$4x - 1 = 3x + 3$	Substitute.
$x - 1 = 3$	Subtract $3x$ from each side.
$x = 4$	Add 1 to each side.

Step 2 Evaluate the expression for VM when $x = 4$.

$$VM = 4x - 1 = 4(4) - 1 = 15$$

▶ $\overrightarrow{MN}$ is the segment bisector of $\overline{VW}$, and VM is 15.

Check Because $VM = MW$, the length of $\overline{MW}$ should be 15.

$$MW = 3x + 3$$
$$= 3(4) + 3$$
$$= 15 \checkmark$$

SELF-ASSESSMENT

1 I don't understand yet. 2 I can do it with help. 3 I can do it on my own. 4 I can teach someone else.

4. Identify the segment bisector of $\overline{PQ}$. Then find MQ.

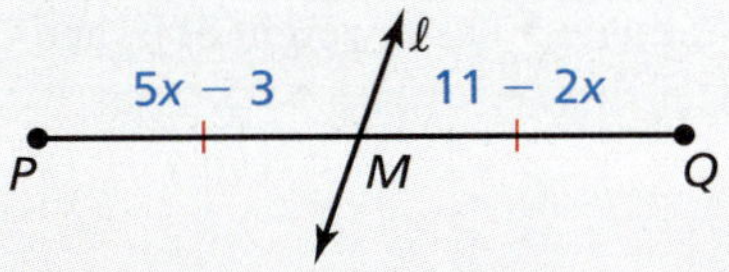

5. Identify the segment bisector of $\overline{RS}$. Then find RS.

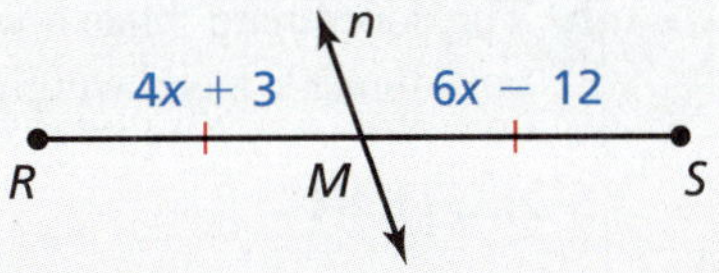

CONSTRUCTION Bisecting a Segment

WATCH

Construct a segment bisector of $\overline{AB}$ by paper folding. Then label the midpoint M of $\overline{AB}$.

SOLUTION

Step 1

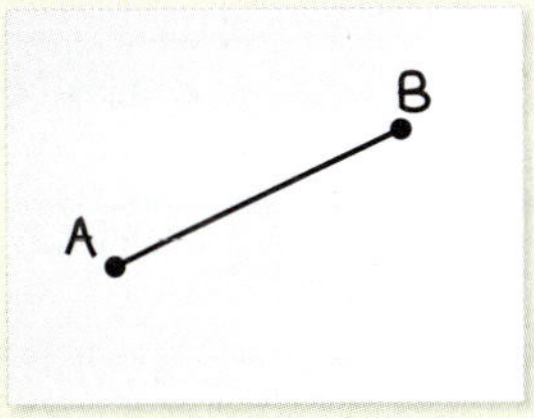

Draw a segment
Use a straightedge to draw $\overline{AB}$ on a piece of paper.

Step 2

Fold the paper
Fold the paper so that B is on top of A.

Step 3

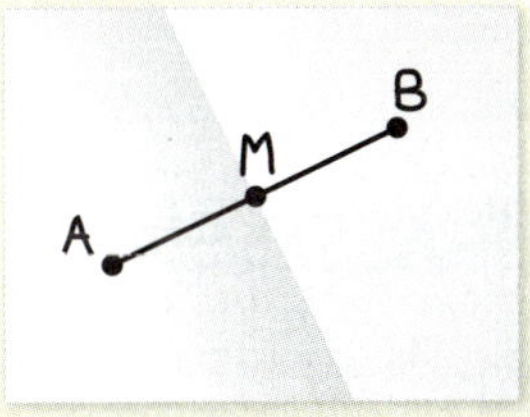

Label the midpoint
Label point M. Compare AM, MB, and AB.

$AM = MB = \frac{1}{2}AB$

GO DIGITAL

Finding Weighted Averages on a Number Line

When two distinct points A and B are weighted equally, the average is the midpoint of $\overline{AB}$. When points are weighted unequally, the average is a *weighted average*.

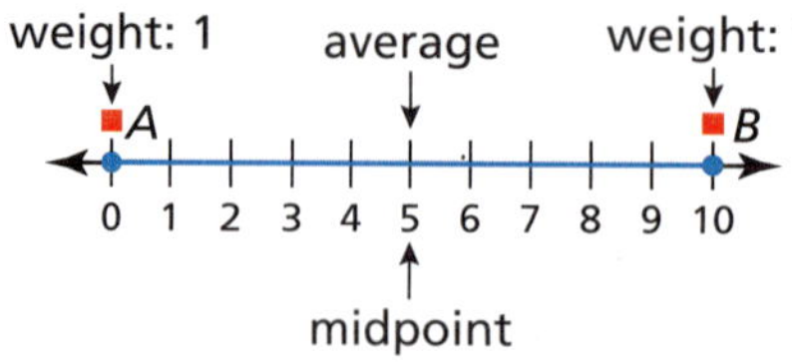

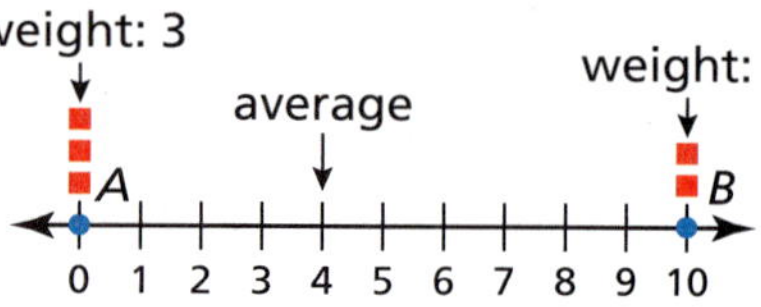

KEY IDEA

Weighted Averages

To find the **weighted average** of points on a number line, multiply the coordinate of each point by its weight, and divide the sum of the weighted values by the sum of the weights.

$$W = \frac{\text{sum of the weighted values}}{\text{sum of the weights}}$$

EXAMPLE 3 **Finding Weighted Averages** WATCH

4 MTR CONSTRUCT AN ARGUMENT
How does increasing or decreasing the weight on a point affect the weighted average? Explain.

Find each weighted average.

a. The coordinate 2 has a weight of 1, and the coordinate 8 has a weight of 2.

b. The coordinate 3 has a weight of 1, the coordinate 5 has a weight of 3, and the coordinate 9 has a weight of 2.

SOLUTION

a. Multiply each coordinate by its weight. Divide the sum of the weighted values by the sum of the weights.

$$\frac{\text{sum of the weighted values}}{\text{sum of the weights}} = \frac{1(2) + 2(8)}{1 + 2} = 6$$

▶ The weighted average is 6.

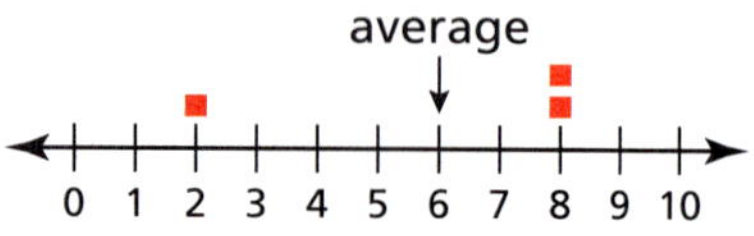

b. Multiply each coordinate by its weight. Divide the sum of the weighted values by the sum of the weights.

$$\frac{\text{sum of the weighted values}}{\text{sum of the weights}} = \frac{1(3) + 3(5) + 2(9)}{1 + 3 + 2} = 6$$

▶ The weighted average is 6.

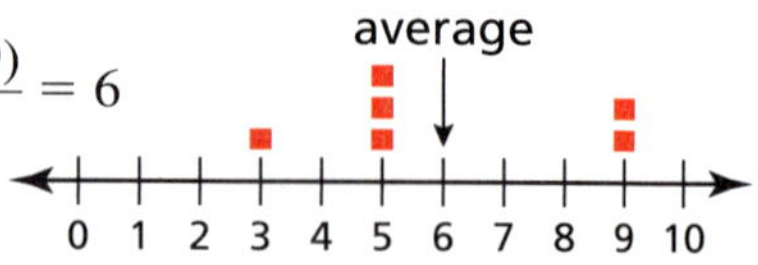

SELF-ASSESSMENT

1 I don't understand yet. 2 I can do it with help. 3 I can do it on my own. 4 I can teach someone else.

6. The coordinate 3 has a weight of 2, and the coordinate 9 has a weight of 1. Find the weighted average.

7. The coordinate 2 has a weight of 2, the coordinate 3 has a weight of 2, and the coordinate 10 has a weight of 1. Find the weighted average.

8. Your final grade in a class is based on your score on two tests and the final exam. You scored 86% and 94% on the tests and 92% on the final exam. Find your final grade when the final exam has twice the weight of each test.

GO DIGITAL

Using the Midpoint Formula

You can use the coordinates of the endpoints of a segment in the coordinate plane to find the coordinates of the midpoint.

KEY IDEA

The Midpoint Formula

The coordinates of the midpoint of a segment are the averages of the x-coordinates and of the y-coordinates of the endpoints.

If $A(x_1, y_1)$ and $B(x_2, y_2)$ are points in a coordinate plane, then the midpoint M of $\overline{AB}$ has coordinates

$$\left(\frac{x_1 + x_2}{2}, \frac{y_1 + y_2}{2}\right).$$

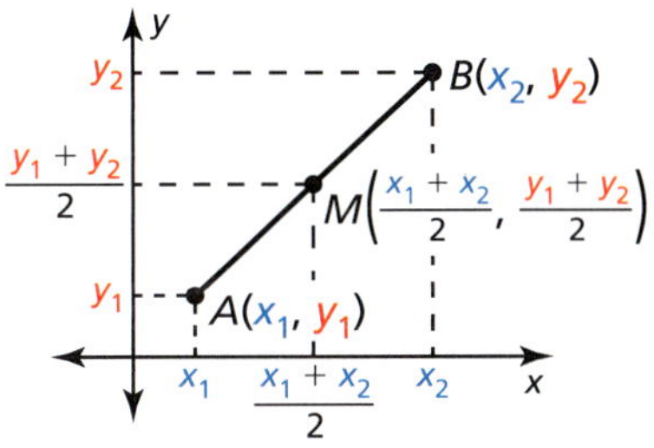

EXAMPLE 4 Using the Midpoint Formula

WATCH

a. The endpoints of $\overline{RS}$ are $R(1, -3)$ and $S(4, 2)$. Find the coordinates of the midpoint M.

b. The midpoint of $\overline{JK}$ is $M(2, 1)$. One endpoint is $J(1, 4)$. Find the coordinates of endpoint K.

SOLUTION

a. Use the Midpoint Formula.

$$M\left(\frac{1 + 4}{2}, \frac{-3 + 2}{2}\right) = M\left(\frac{5}{2}, -\frac{1}{2}\right)$$

▶ The coordinates of the midpoint M are $\left(\frac{5}{2}, -\frac{1}{2}\right)$.

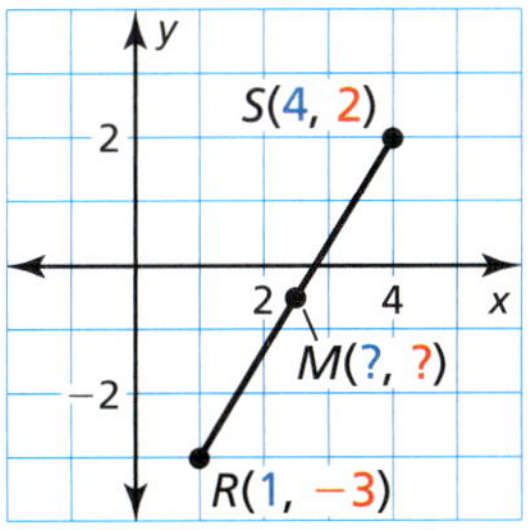

b. Let (x, y) be the coordinates of endpoint K. Use the Midpoint Formula.

Step 1 Find x.

$$\frac{1 + x}{2} = 2$$

$$1 + x = 4$$

$$x = 3$$

Step 2 Find y.

$$\frac{4 + y}{2} = 1$$

$$4 + y = 2$$

$$y = -2$$

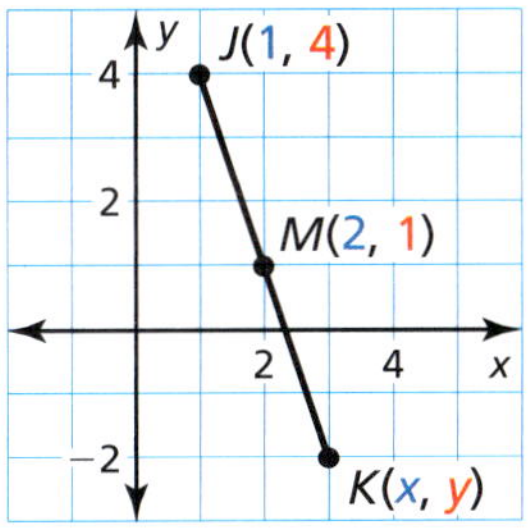

▶ The coordinates of endpoint K are $(3, -2)$.

SELF-ASSESSMENT

1 I don't understand yet. 2 I can do it with help. 3 I can do it on my own. 4 I can teach someone else.

The endpoints of $\overline{AB}$ are given. Find the coordinates of the midpoint M.

9. $A(1, 2)$ and $B(7, 8)$

10. $A(-4, 3)$ and $B(-6, 5)$

The midpoint M and one endpoint of $\overline{TU}$ are given. Find the coordinates of the other endpoint.

11. $T(1, 1)$ and $M(2, 4)$

12. $U(4, 4)$ and $M(-1, -2)$

GO DIGITAL

Using the Distance Formula

You can use the Distance Formula to find the distance between two points in a coordinate plane. You can derive the Distance Formula from the *Pythagorean Theorem*, which you will see again when you work with right triangles.

Pythagorean Theorem

$c^2 = a^2 + b^2$

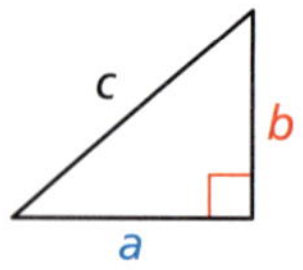

Distance Formula

$(AB)^2 = (x_2 - x_1)^2 + (y_2 - y_1)^2$

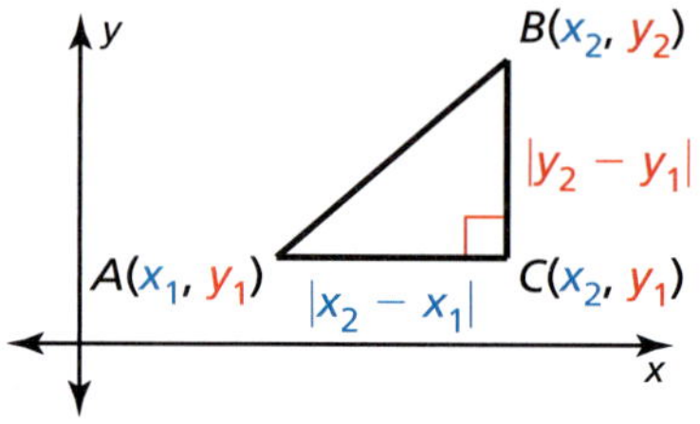

READING

The red mark at the corner of the triangle that makes a right angle indicates a right triangle.

KEY IDEA

The Distance Formula

If $A(x_1, y_1)$ and $B(x_2, y_2)$ are points in a coordinate plane, then the distance between A and B is

$$AB = \sqrt{(x_2 - x_1)^2 + (y_2 - y_1)^2}.$$

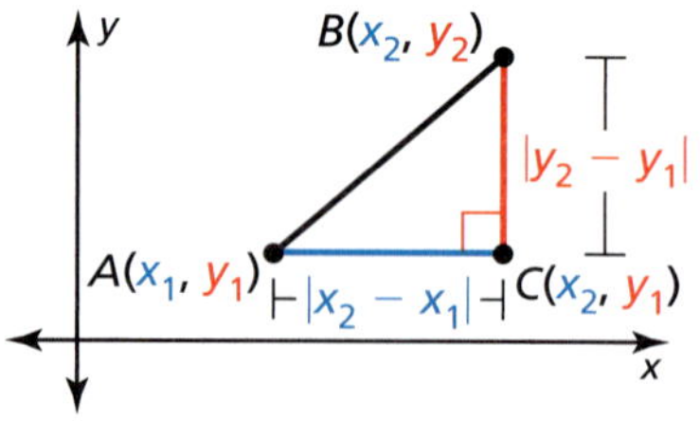

7 MTR

EXAMPLE 5 **Modeling Real Life** WATCH

Your school is 4 miles east and 1 mile south of your apartment. A recycling center, where your class is going on a field trip, is 2 miles east and 3 miles north of your apartment. Estimate the distance between the recycling center and your school.

SOLUTION

You can model the situation using a coordinate plane with your apartment at the origin (0, 0). The coordinates of the recycling center and the school are $R(2, 3)$ and $S(4, -1)$, respectively. Use the Distance Formula. Let $(x_1, y_1) = (2, 3)$ and $(x_2, y_2) = (4, -1)$.

$RS = \sqrt{(x_2 - x_1)^2 + (y_2 - y_1)^2}$	Distance Formula
$= \sqrt{(4 - 2)^2 + (-1 - 3)^2}$	Substitute.
$= \sqrt{2^2 + (-4)^2}$	Subtract.
$= \sqrt{4 + 16}$	Evaluate powers.
$= \sqrt{20}$	Add.
≈ 4.5	Use technology.

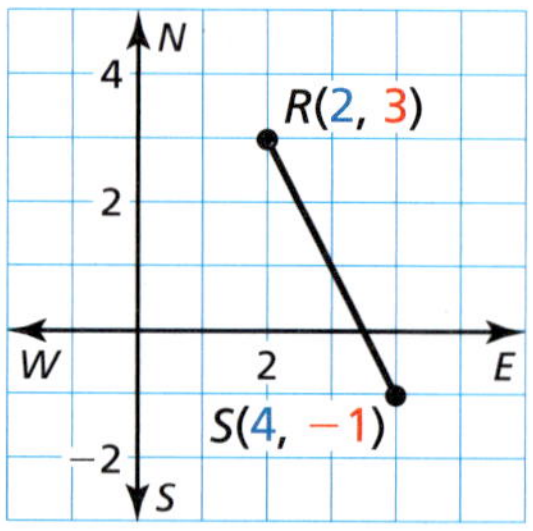

READING

The symbol ≈ means "is approximately equal to."

▶ So, the distance between the recycling center and your school is about 4.5 miles.

SELF-ASSESSMENT 1 I don't understand yet. 2 I can do it with help. 3 I can do it on my own. 4 I can teach someone else.

13. In Example 5, a park is 3 miles east and 4 miles south of your apartment. Estimate the distance between the park and your school.

GO DIGITAL

1.3 Practice WITH CalcChat® AND CalcView®

In Exercises 1–4, identify the segment bisector of $\overline{RS}$. Then find RS. *(See Example 1.)*

1.

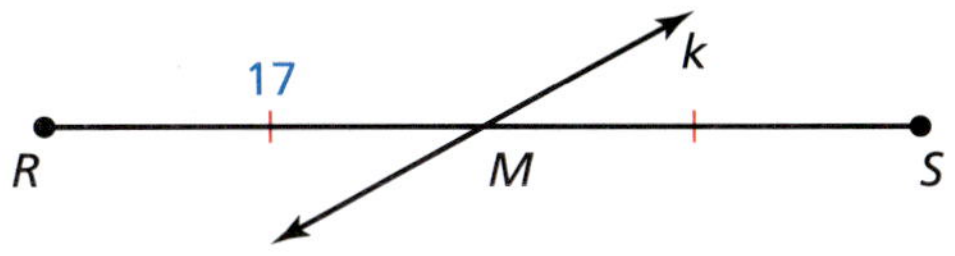

2.

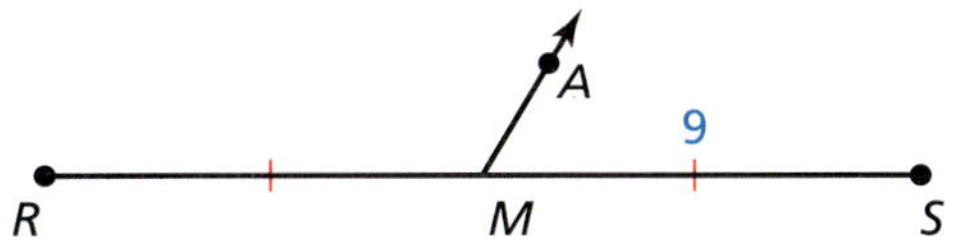

3.

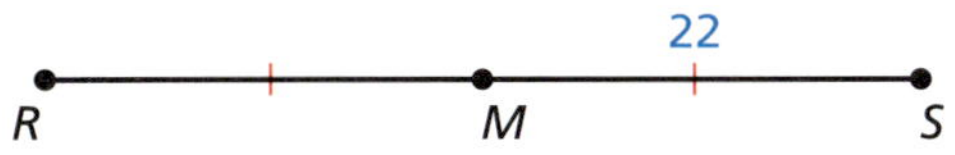

4.

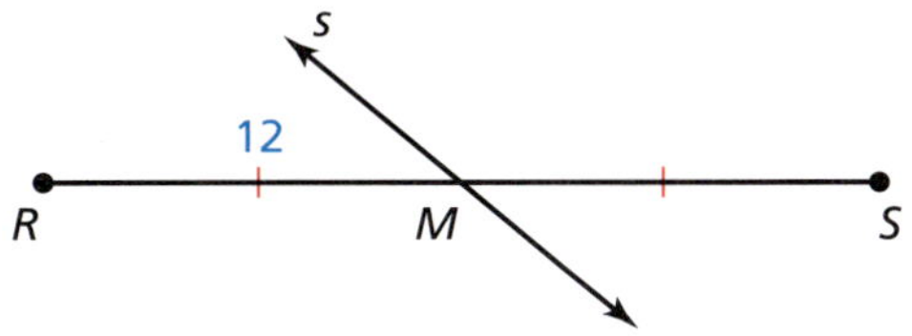

In Exercises 5 and 6, identify the segment bisector of $\overline{JK}$. Then find JM. *(See Example 2.)*

5.

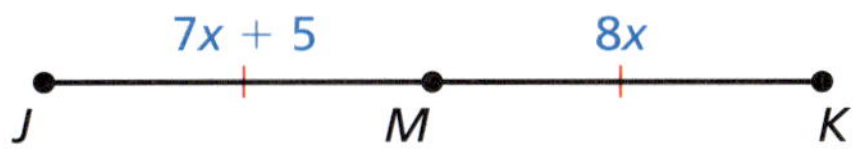

6.

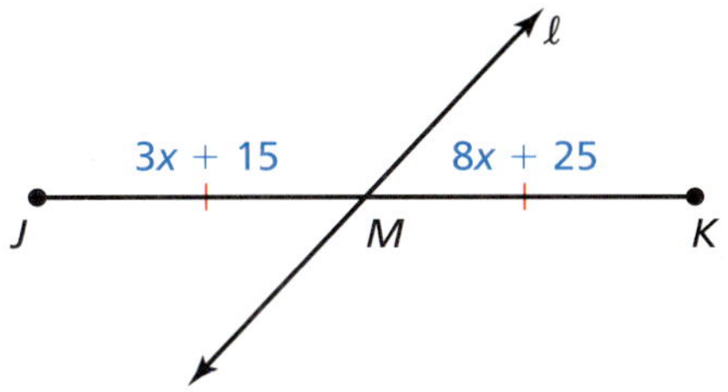

In Exercises 7 and 8, identify the segment bisector of $\overline{XY}$. Then find XY. *(See Example 2.)*

7.

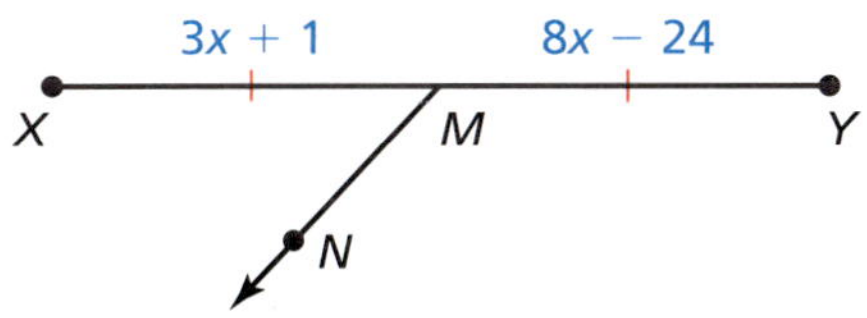

8.

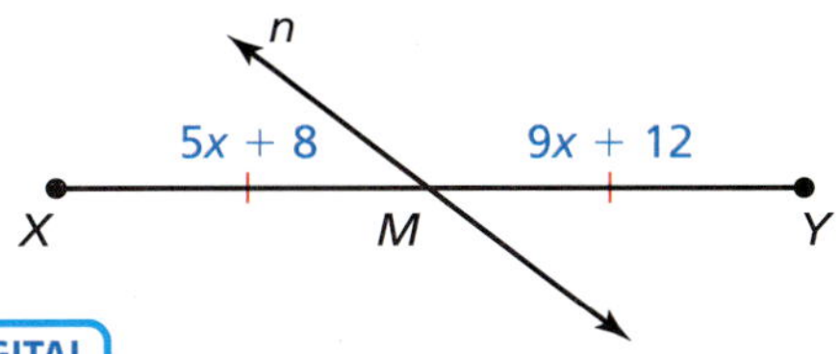

GO DIGITAL

CONSTRUCTION In Exercises 9–12, copy the segment and construct a segment bisector by paper folding. Then label the midpoint M.

9.

10.

11.

12.

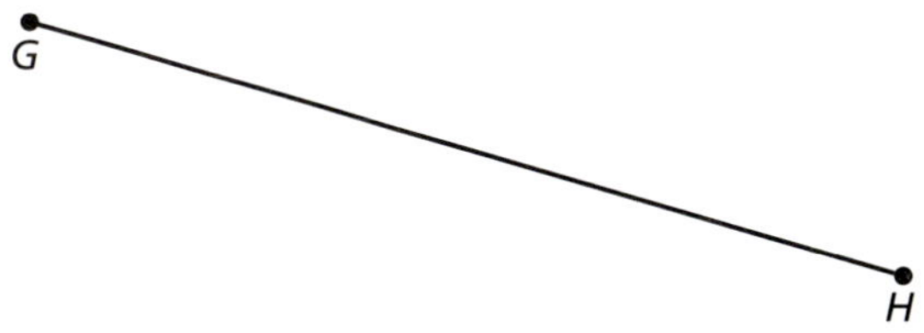

In Exercises 13–20, find the weighted average. *(See Example 3.)*

13. The coordinate 3 has a weight of 2, and the coordinate 8 has a weight of 3.

14. The coordinate −6 has a weight of 3, and the coordinate 2 has a weight of 1.

15. The coordinate −3 has a weight of 2, and the coordinate 4 has a weight of 5.

16. The coordinate 1 has a weight of 4, and the coordinate 4 has a weight of 2.

17. The coordinate −1 has a weight of 1, the coordinate 4 has a weight of 2, and the coordinate 9 has a weight of 2.

18. The coordinate 2 has a weight of 2, the coordinate 5 has a weight of 1, and the coordinate 7 has a weight of 3.

19. The coordinate −2 has a weight of 1, the coordinate 0 has a weight of 1, and the coordinate 3 has a weight of 2.

20. The coordinate 6 has a weight of 1, the coordinate 8 has a weight of 3, and the coordinate 10 has a weight of 1.

In Exercises 21–26, the endpoints of $\overline{CD}$ are given. Find the coordinates of the midpoint M. *(See Example 4.)*

▶ **21.** $C(3, -5)$ and $D(7, 9)$

22. $C(-4, 7)$ and $D(0, -3)$

23. $C(-2, 0)$ and $D(4, 9)$

24. $C(-8, -6)$ and $D(-4, 10)$

25. $C(-3, 5)$ and $D(4, -2)$

26. $C(-7, -3)$ and $D(2, 1)$

In Exercises 27–32, the midpoint M and one endpoint of $\overline{GH}$ are given. Find the coordinates of the other endpoint. *(See Example 4.)*

▶ **27.** $G(5, -6)$ and $M(4, 3)$

28. $H(-3, 7)$ and $M(-2, 5)$

29. $G(-7, 2)$ and $M(-1, 3)$

30. $H(4, -4)$ and $M(-2, 0)$

31. $H(-2, 9)$ and $M(8, 0)$

32. $G(-4, 1)$ and $M\left(-\frac{13}{2}, -6\right)$

In Exercises 33–40, find the distance between the two points. *(See Example 5.)*

33. $A(13, 2)$ and $B(7, 10)$

34. $C(-6, 5)$ and $D(-3, 1)$

▶ **35.** $E(3, 7)$ and $F(6, 5)$

36. $G(-5, 4)$ and $H(2, 6)$

37. $J(-8, 0)$ and $K(1, 4)$

38. $L(7, -1)$ and $M(-2, 4)$

39. $R(0, 1)$ and $S(6, 3.5)$

40. $T(13, 1.6)$ and $V(5.4, 3.7)$

ERROR ANALYSIS In Exercises 41 and 42, describe and correct the error in finding the distance between $A(6, 2)$ and $B(1, -4)$.

41.

$$AB = (6-1)^2 + [2-(-4)]^2$$
$$= 5^2 + 6^2$$
$$= 25 + 36$$
$$= 61$$

42.

$$AB = \sqrt{(6-2)^2 + [1-(-4)]^2}$$
$$= \sqrt{4^2 + 5^2}$$
$$= \sqrt{16 + 25}$$
$$= \sqrt{41}$$

In Exercises 43–46, the endpoints of two segments are given. Find the length of each segment. Tell whether the segments are congruent. If they are not congruent, tell which segment is longer.

43. $\overline{AB}$: $A(0, 2)$, $B(-3, 8)$ and $\overline{CD}$: $C(-2, 2)$, $D(0, -4)$

44. $\overline{EF}$: $E(1, 4)$, $F(5, 1)$ and $\overline{GH}$: $G(-3, 1)$, $H(1, 6)$

45. $\overline{WX}$: $W(-3, -1)$, $X(3, 4)$ and $\overline{YZ}$: $Y(2, -6)$, $Z(7, -1)$

46. $\overline{LM}$: $L(-5, 1)$, $M(2, -2)$ and $\overline{NP}$: $N(-1, -3)$, $P(2, 4)$

47. MODELING REAL LIFE In baseball, the strike zone is the region a baseball needs to pass through for the umpire to declare it a strike when the batter does not swing. The bottom of the strike zone is a horizontal plane passing through a point just below the kneecap. The top of the strike zone is a horizontal plane passing through the midpoint of the top of the batter's shoulders and the top of the uniform pants when the player is in a batting stance. Find the height of T.

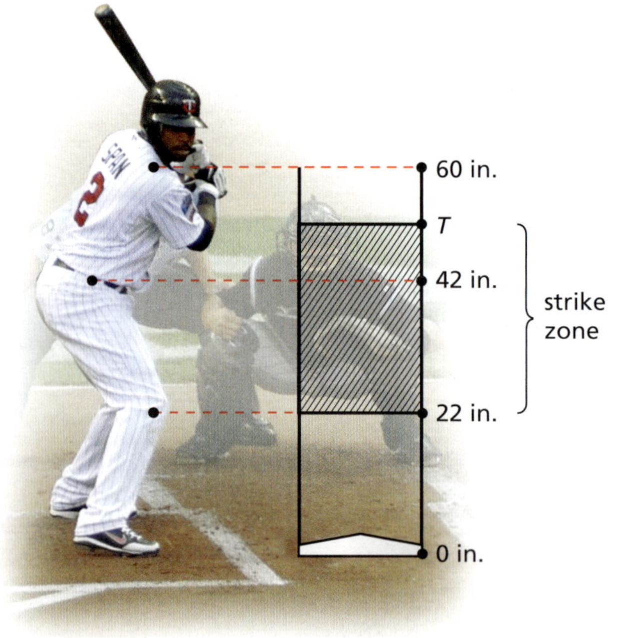

GO DIGITAL

7 MTR **48.** **MODELING REAL LIFE** Two wolves spot a deer in a field. The positions of the animals are shown. Which wolf is closer to the deer?

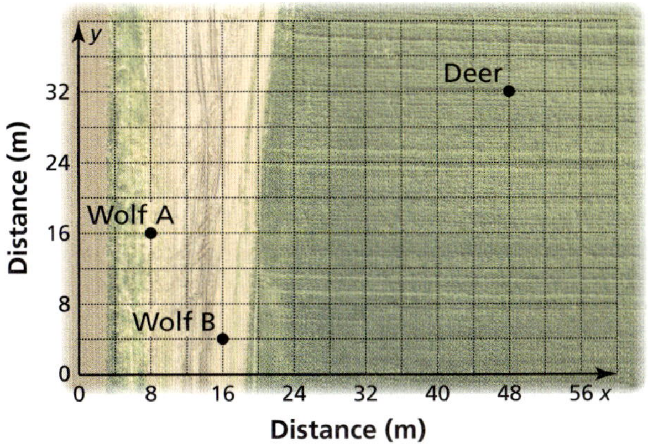

7 MTR **49.** **MODELING REAL LIFE** A theater is 3 miles east and 1 mile north of a bus stop. A museum is 4 miles west and 3 miles south of the bus stop. Estimate the distance between the theater and the museum.

7 MTR **50.** **MODELING REAL LIFE** Your school is 20 blocks east and 12 blocks south of your home. The mall, where you plan to go after school, is 7 blocks west and 10 blocks north of your home. One block is 0.1 mile. Estimate the distance in miles between your school and the mall.

4 MTR **51.** **MAKING AN ARGUMENT** Your friend claims there is an easier way to find the length of a segment than using the Distance Formula when the x-coordinates of the endpoints are equal. He claims all you have to do is subtract the y-coordinates. Do you agree with his statement? Explain your reasoning.

5 MTR **52.** **STRUCTURE** The endpoints of a segment are located at (a, c) and (b, c). Find the coordinates of the midpoint and the length of the segment in terms of a, b, and c.

7 MTR **53.** **MODELING REAL LIFE** The chart shows the weights of the assignments in your math class and your corresponding scores.

Assignment	Percent of grade	Score
Homework	10%	95%
Quizzes	20%	75%
Midterm	30%	85%
Final Exam	40%	90%

a. Find your final grade.

b. Your friend scores 90% on homework, 70% on quizzes, and 80% on the midterm. Is it possible for your friend to obtain a higher final grade than you? Explain.

GO DIGITAL

6 MTR **54.** **ASSESS REASONABLENESS** Panama City is about a 375-mile drive from Tampa. Your friend claims that the midpoint between Panama City and Tampa is about 187.5 miles between the two cities. Is your friend's claim reasonable? Explain.

5 MTR **55.** **CONNECTING CONCEPTS** Triangle LMN is shown.

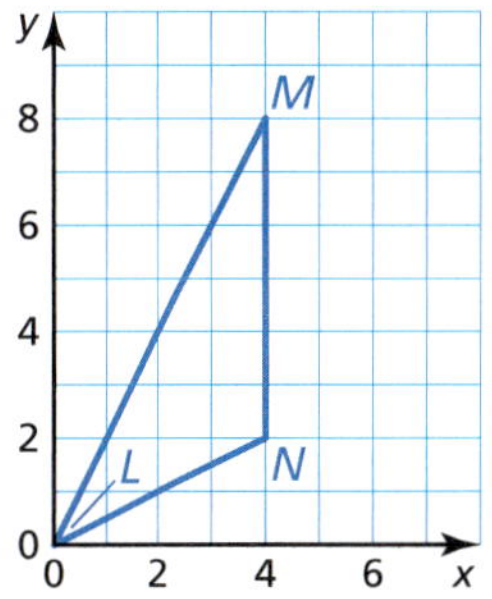

a. Label point M' as the midpoint of segment LM.

b. Label point N' as the midpoint of segment LN.

c. Is $LM'N'$ a dilation of LMN? Justify your answer.

56. **HOW DO YOU SEE IT?**
$\overline{AB}$ contains midpoint M and points C and D, as shown. Compare the lengths. If you cannot draw a conclusion, write *impossible to tell*. Explain your reasoning.

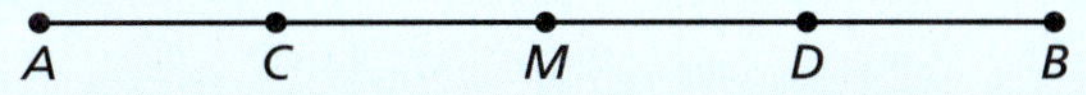

a. AM and MB

b. AC and MB

c. CM and MD

d. MB and DB

57. **PROBLEM SOLVING** A new bridge is constructed in the triangular park shown. The bridge spans from point Q to the midpoint M of $\overline{PR}$. A person jogs from P to Q to M to R to Q and back to P at an average speed of 150 yards per minute. About how many minutes does it take? Explain your reasoning.

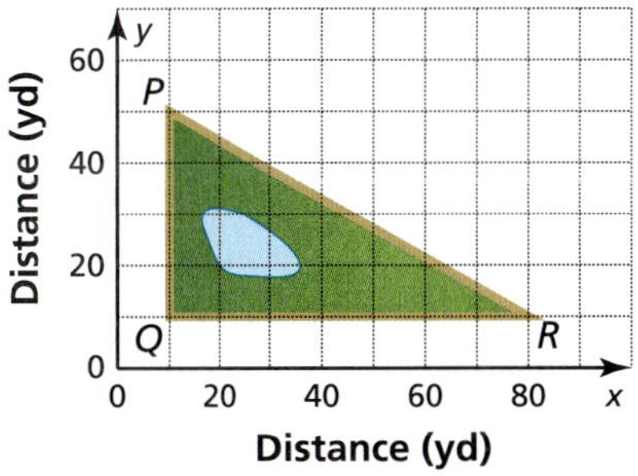

5 MTR **58.** **STRUCTURE** The length of $\overline{XY}$ is 24 centimeters. The midpoint of $\overline{XY}$ is M, and point C lies on $\overline{XM}$ so that XC is $\frac{2}{3}$ of XM. Point D lies on $\overline{MY}$ so that MD is $\frac{3}{4}$ of MY. What is the length of $\overline{CD}$?

59. **DIG DEEPER** The endpoints of $\overline{AB}$ are $A(2x, y - 1)$ and $B(y + 3, 3x + 1)$. The midpoint of $\overline{AB}$ is $M\left(-\frac{7}{2}, -8\right)$. What is the length of $\overline{AB}$?

60. **THOUGHT PROVOKING**
The distance between $K(1, -5)$ and a point L with integer coordinates is $\sqrt{58}$ units. Find all the possible coordinates of point L.

REVIEW & REFRESH

WATCH

In Exercises 61–64, find the perimeter and area of the figure.

61.

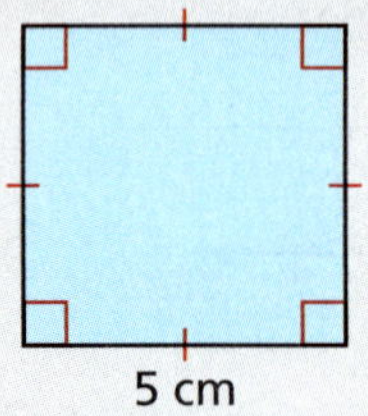

62.

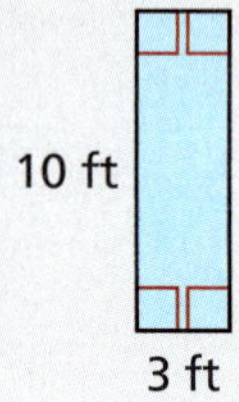

63.

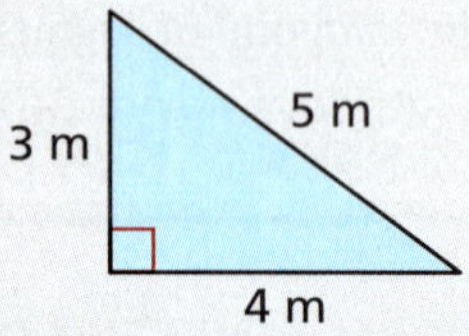

64.

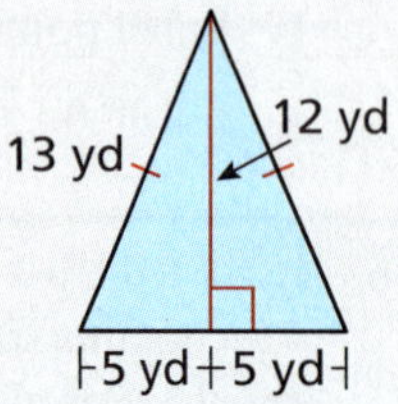

In Exercises 65–68, solve the inequality. Graph the solution.

65. $a + 18 < 7$

66. $y - 5 \geq 8$

67. $-3x > 24$

68. $\frac{z}{4} \leq 12$

69. The endpoints of $\overline{YZ}$ are $Y(1, -6)$ and $Z(-2, 8)$. Find the coordinates of the midpoint M. Then find YZ.

70. Solve the literal equation $5x + 15y = -30$ for y.

71. Find the average rate of change of $f(x) = 3^x$ from $x = 1$ to $x = 3$.

In Exercises 72–75, factor the polynomial.

72. $3x^2 - 36x$

73. $n^2 + 3n - 70$

74. $121p^2 - 100$

75. $15y^2 + 4y - 4$

76. Name two pairs of opposite rays in the diagram.

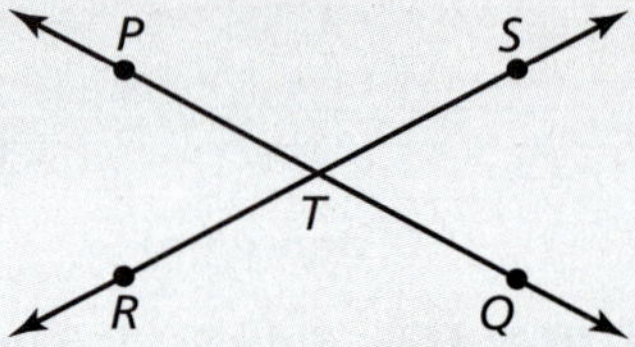

In Exercises 77 and 78, simplify the expression. Write your answer using only positive exponents.

77. $\dfrac{b^4 \cdot b^{-2}}{b^{10}}$

78. $\left(\dfrac{2}{5t^4}\right)^{-3}$

79. Plot $A(-3, 3)$, $B(1, 3)$, $C(3, 2)$, and $D(3, -2)$ in a coordinate plane. Then determine whether $\overline{AB}$ and $\overline{CD}$ are congruent.

7 MTR **80.** **MODELING REAL LIFE** The function $p(x) = 80 - 2x$ represents the number of points earned on a test with x incorrect answers.

a. How many points are earned with 2 incorrect answers?

b. How many incorrect answers are there when 68 points are earned?

81. Convert 320 fluid ounces to gallons.

82. Determine when the function $y = -|x| - 3$ is positive, negative, increasing, or decreasing. Then describe the end behavior of the function.

83. Write an equation of the line in slope-intercept form.

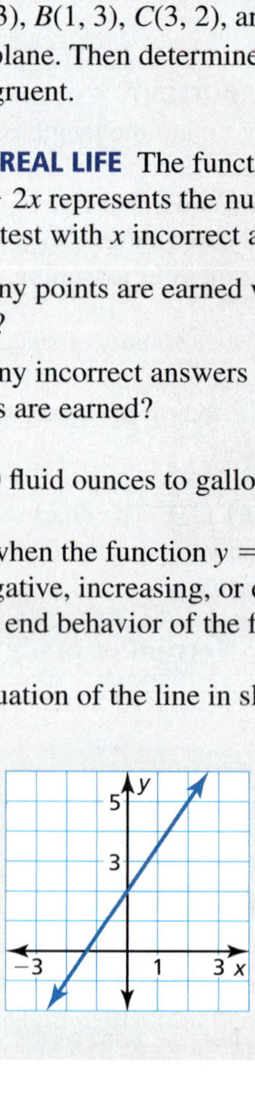

GO DIGITAL

1.4 Perimeter and Area in the Coordinate Plane

Learning Target: Find perimeters and areas of polygons in the coordinate plane.

Success Criteria:
- I can classify and describe polygons.
- I can find perimeters of polygons in the coordinate plane.
- I can find areas of polygons in the coordinate plane.

EXPLORE IT! Finding the Perimeter and Area of a Quadrilateral

Work with a partner.

a. Use a piece of graph paper to draw a quadrilateral $ABCD$ in a coordinate plane. At most, two sides of your quadrilateral can be horizontal or vertical. Plot and label the vertices of $ABCD$.

b. Make several observations about quadrilateral $ABCD$. Can you use any other names to classify your quadrilateral? Explain.

c. Explain how you can find the perimeter of quadrilateral $ABCD$. Then find the perimeter. Compare your method with those of your classmates.

d. Explain how you can find the area of quadrilateral $ABCD$. Then find the area. Compare your method with those of your classmates.

e. Use the methods from parts (c) and (d) to find the perimeter and area of the polygon below. Explain your reasoning.

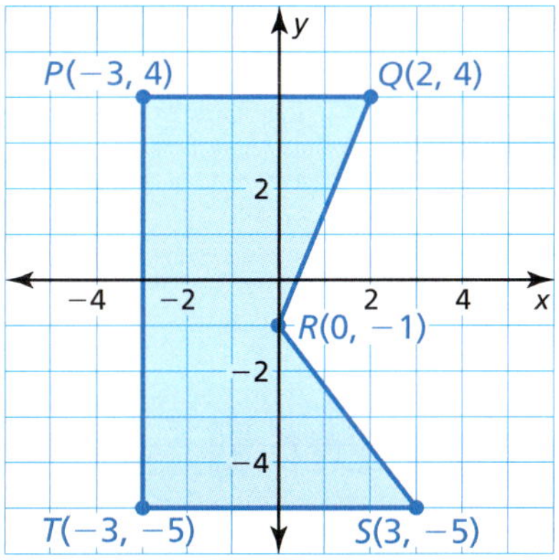

5 MTR DECOMPOSE A PROBLEM
In part (d), how might it be helpful to visualize a polygon as being composed of one or more smaller polygons?

Geometric Reasoning
MA.912.GR.3.3 Use coordinate geometry to solve mathematical and real-world geometric problems involving lines, circles, triangles and quadrilaterals.
MA.912.GR.3.4 Use coordinate geometry to solve mathematical and real-world problems on the coordinate plane involving perimeter or area of polygons.
Also MA.912.GR.3.2, MA.912.GR.4.4

GO DIGITAL

Classifying Polygons

KEY IDEA

Polygons

In geometry, a figure that lies in a plane is called a plane figure. Recall that a *polygon* is a closed plane figure formed by three or more line segments called *sides*. Each side intersects exactly two sides, one at each *vertex*, so that no two sides with a common vertex are collinear.

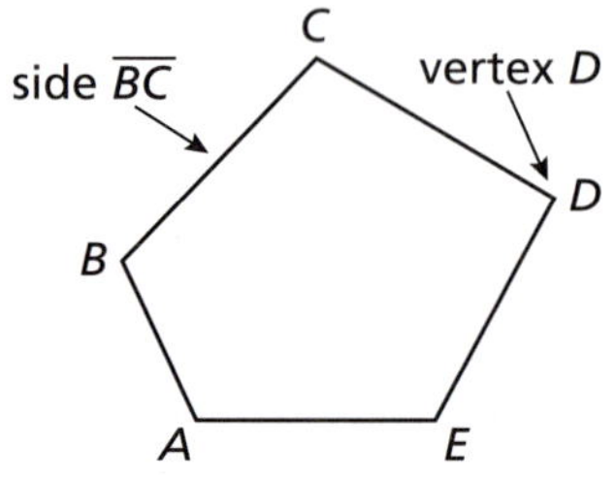

polygon *ABCDE*

READING

You can name a polygon by listing the vertices in consecutive order.

Number of sides	Type of polygon
3	Triangle
4	Quadrilateral
5	Pentagon
6	Hexagon
7	Heptagon
8	Octagon
9	Nonagon
10	Decagon
12	Dodecagon
n	n-gon

The number of sides determines the type of polygon, as shown in the table. You can also name a polygon using the term n-gon, where n is the number of sides. For instance, a 14-gon is a polygon with 14 sides.

A polygon is *convex* when no line that contains a side of the polygon contains a point in the interior of the polygon. A polygon that is not convex is *concave*.

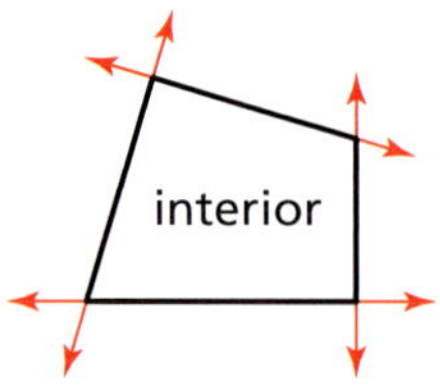

convex polygon

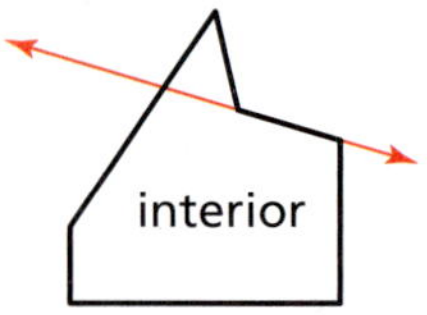

concave polygon

EXAMPLE 1 Classifying Polygons

WATCH

Classify each polygon by the number of sides. Tell whether it is *convex* or *concave*.

a.

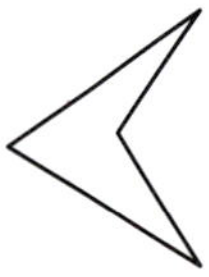

b.

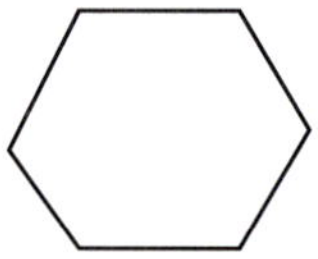

SOLUTION

a. The polygon has four sides. So, it is a quadrilateral. The polygon is concave.

b. The polygon has six sides. So, it is a hexagon. The polygon is convex.

SELF-ASSESSMENT

1 I don't understand yet. 2 I can do it with help. 3 I can do it on my own. 4 I can teach someone else.

Classify the polygon by the number of sides. Tell whether it is *convex* or *concave*.

1.

2.

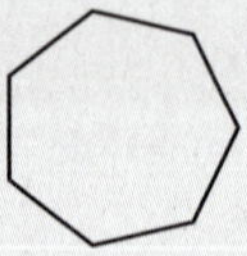

3.

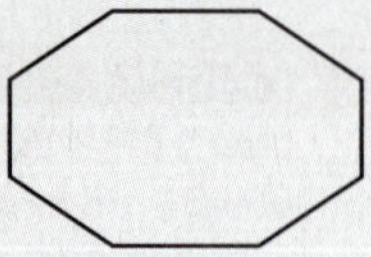

4. REASONING Can you draw a concave triangle? If so, draw one. If not, explain why not.

GO DIGITAL

Finding Perimeter and Area in the Coordinate Plane

You can use the formulas below and the Distance Formula to find perimeters and areas of polygons in the coordinate plane.

Perimeter and Area

Triangle

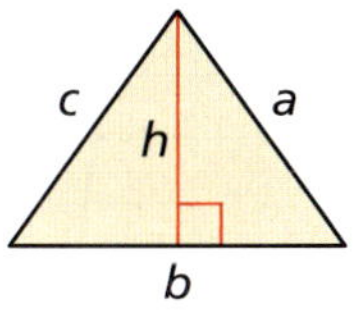

$P = a + b + c$

$A = \frac{1}{2}bh$

Square

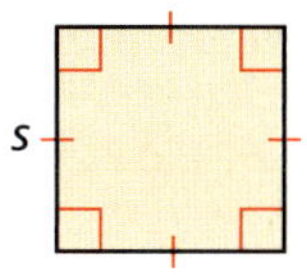

$P = 4s$

$A = s^2$

Rectangle

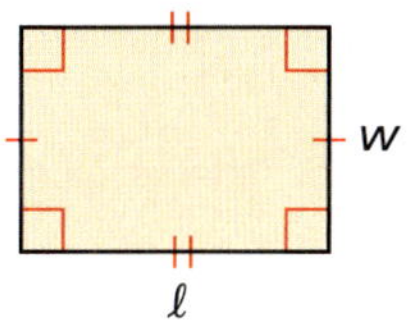

$P = 2\ell + 2w$

$A = \ell w$

Parallelogram

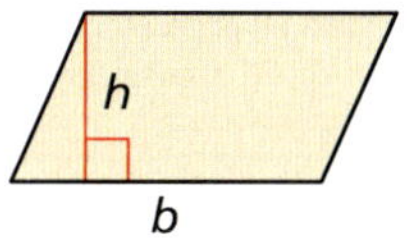

$A = bh$

READING

You can read the notation $\triangle DEF$ as "triangle *D E F*."

EXAMPLE 2 Finding Perimeter in the Coordinate Plane

WATCH

Find the perimeter of $\triangle DEF$ with vertices $D(1, 3)$, $E(4, -3)$, and $F(-4, -3)$.

SOLUTION

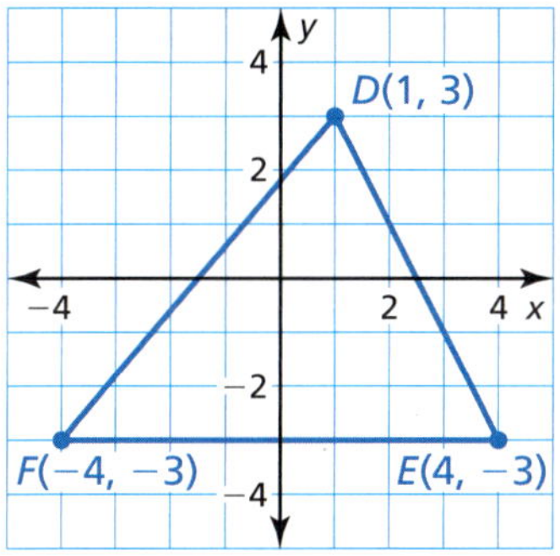

Step 1 Draw the triangle in a coordinate plane by plotting the vertices and connecting them.

Step 2 Find the length of each side.

$\overline{DE}$ Let $(x_1, y_1) = (1, 3)$ and $(x_2, y_2) = (4, -3)$.

$DE = \sqrt{(x_2 - x_1)^2 + (y_2 - y_1)^2}$ Distance Formula

$= \sqrt{(4 - 1)^2 + (-3 - 3)^2}$ Substitute.

$= \sqrt{3^2 + (-6)^2}$ Subtract.

$= \sqrt{45}$ Simplify.

$\overline{EF}$ $EF = |-4 - 4| = |-8| = 8$ Ruler Postulate

$\overline{FD}$ Let $(x_1, y_1) = (-4, -3)$ and $(x_2, y_2) = (1, 3)$.

$FD = \sqrt{(x_2 - x_1)^2 + (y_2 - y_1)^2}$ Distance Formula

$= \sqrt{[1 - (-4)]^2 + [3 - (-3)]^2}$ Substitute.

$= \sqrt{5^2 + 6^2}$ Subtract.

$= \sqrt{61}$ Simplify.

Step 3 Find the sum of the side lengths.

$$DE + EF + FD = \sqrt{45} + 8 + \sqrt{61} \approx 22.52 \text{ units}$$

▶ So, the perimeter of $\triangle DEF$ is about 22.52 units.

REMEMBER

Perimeter has linear units, such as feet or meters. Area has square units, such as square feet or square meters.

SELF-ASSESSMENT

1 I don't understand yet. 2 I can do it with help. 3 I can do it on my own. 4 I can teach someone else.

Find the perimeter of the polygon with the given vertices.

5. $G(-3, 2)$, $H(2, 2)$, $J(-1, -3)$

6. $Q(-4, -1)$, $R(1, 4)$, $S(4, 1)$, $T(-1, -4)$

GO DIGITAL

EXAMPLE 3 Finding Area in the Coordinate Plane

WATCH

Find the area of $\square JKLM$ with vertices $J(-3, 5)$, $K(1, 5)$, $L(2, -1)$, and $M(-2, -1)$.

READING

You can read the notation $\square JKLM$ as "parallelogram J K L M."

SOLUTION

Step 1 Draw the parallelogram in a coordinate plane by plotting the vertices and connecting them.

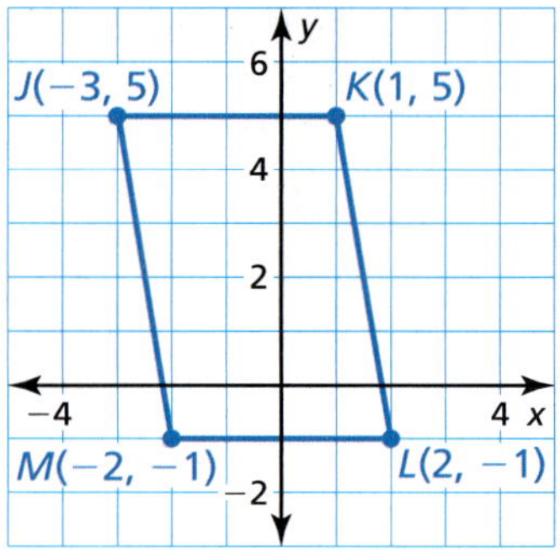

Step 2 Find the length of the base and the height.

Base

Let $\overline{JK}$ be the base. Use the Ruler Postulate to find the length of $\overline{JK}$.

$JK = |1 - (-3)| = |4| = 4$ Ruler Postulate

So, the length of the base is 4 units.

Height

Let the height be the distance from point M to $\overline{JK}$. By counting grid lines, you can determine that the height is 6 units.

ANOTHER WAY

You can also find the area of $\square JKLM$ by decomposing the parallelogram into a rectangle and two triangles, then finding the sum of their areas.

Step 3 Substitute the values for the base and height into the formula for the area of a parallelogram.

$A = bh$ Write the formula for area of a parallelogram.

$= 4(6)$ Substitute.

$= 24$ Multiply.

▶ So, the area of $\square JKLM$ is 24 square units.

SELF-ASSESSMENT

1 I don't understand yet. 2 I can do it with help. 3 I can do it on my own. 4 I can teach someone else.

Find the area of the polygon with the given vertices.

7.

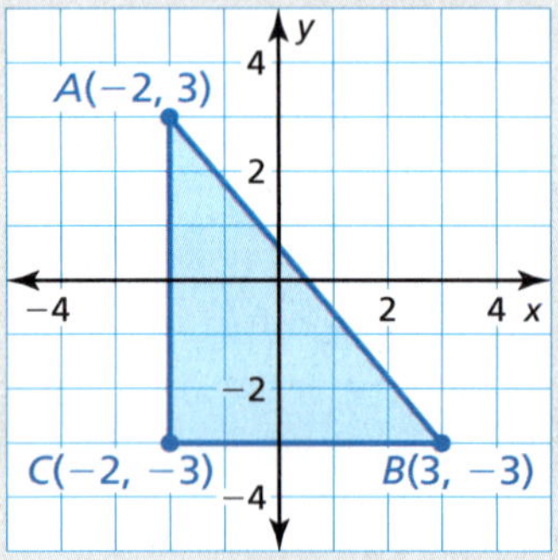

8.

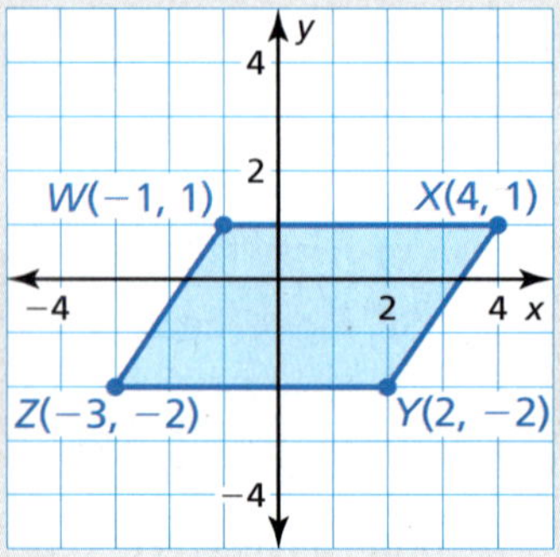

9. $N(-1, 1), P(2, 1), Q(2, -2), R(-1, -2)$

10. $K(-3, 3), L(3, 3), M(3, -1), N(-3, -1)$

GO DIGITAL

7 MTR EXAMPLE 4 Modeling Real Life WATCH

You are building a shed in your backyard. The diagram shows the four vertices of the shed floor. Each unit in the coordinate plane represents 1 foot. Find the perimeter and the area of the floor of the shed.

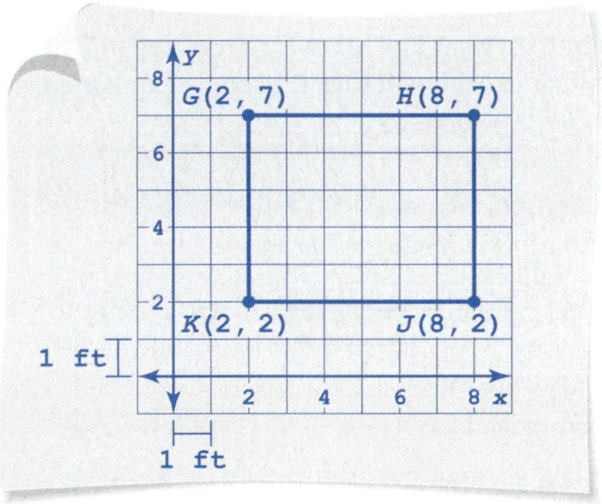

SOLUTION

1. **Understand the Problem** You are given the coordinates of the vertices of the shed floor. You need to find the perimeter and the area of the floor.
2. **Make a Plan** The floor of the shed is rectangular, so use the coordinates of the vertices to find the length and the width. Then use formulas to find the perimeter and area.
3. **Solve and Check**

Step 1 Find the length and the width.

Length $GH = |8 - 2| = 6$ Ruler Postulate

Width $KG = |7 - 2| = 5$ Ruler Postulate

The shed has a length of 6 feet and a width of 5 feet.

Step 2 Substitute the values for the length and width into the formulas for the perimeter P and area A of a rectangle.

$P = 2\ell + 2w$	Write formulas.	$A = \ell w$
$= 2(6) + 2(5)$	Substitute.	$= 6(5)$
$= 22$	Evaluate.	$= 30$

▶ The perimeter of the floor of the shed is 22 feet, and the area is 30 square feet.

Check To check the perimeter, count the grid lines around the floor of the shed. There are 22 grid lines. To check the area, count the number of grid squares that make up the floor. There are 30 grid squares. ✓

SELF-ASSESSMENT

1 I don't understand yet. 2 I can do it with help. 3 I can do it on my own. 4 I can teach someone else.

11. You are building a patio in your school's courtyard. The diagram shows the four vertices of the patio. Each unit in the coordinate plane represents 1 yard. Find the perimeter and the area of the patio.

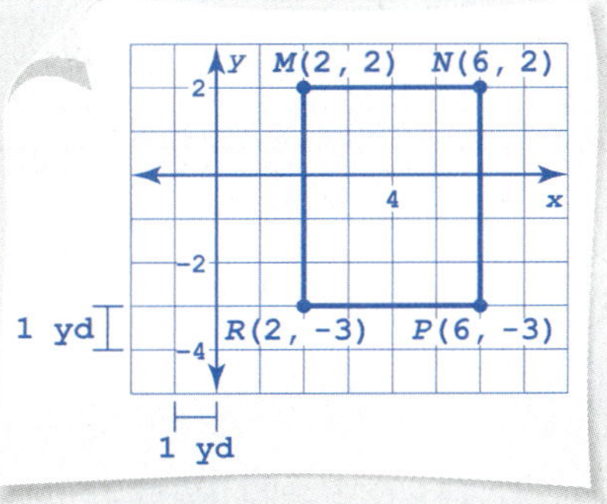

12. You are building a doghouse with four corners. The first corner is 30 inches west of the second corner. The second corner is 35 inches south of the third corner. The third corner is 30 inches east of the fourth corner. Draw the doghouse in a coordinate plane. Then find the perimeter and area of the doghouse.

GO DIGITAL

1.4 Practice WITH CalcChat® AND CalcView®

In Exercises 1–4, classify the polygon by the number of sides. Tell whether it is *convex* or *concave*. *(See Example 1.)*

1. 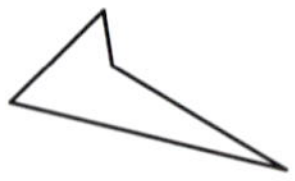

2. 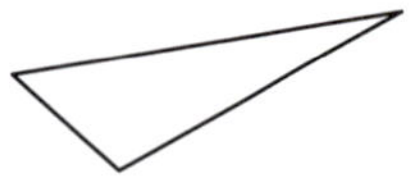

3. 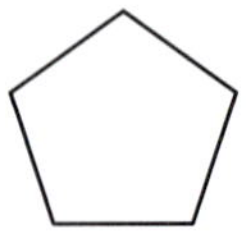

4. 

In Exercises 5–10, find the perimeter of the polygon with the given vertices. *(See Example 2.)*

5. $G(2, 4), H(2, -3), J(-2, -3), K(-2, 4)$

6. $Q(-3, 2), R(1, 2), S(1, -2), T(-3, -2)$

7. $U(-2, 4), V(3, 4), W(3, -4)$

8. $X(-1, 3), Y(3, 0), Z(-1, -2)$

9.

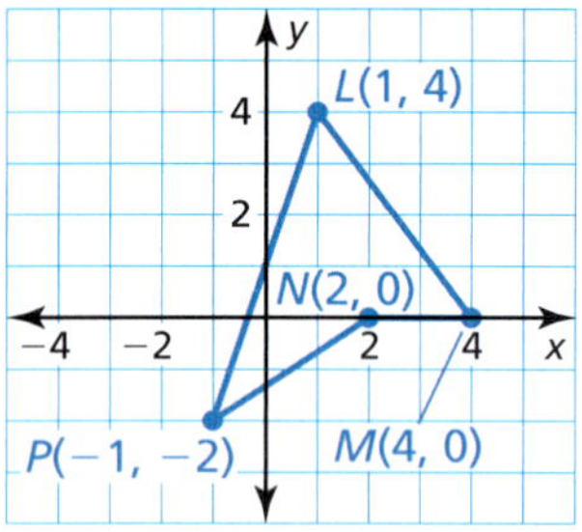

10.

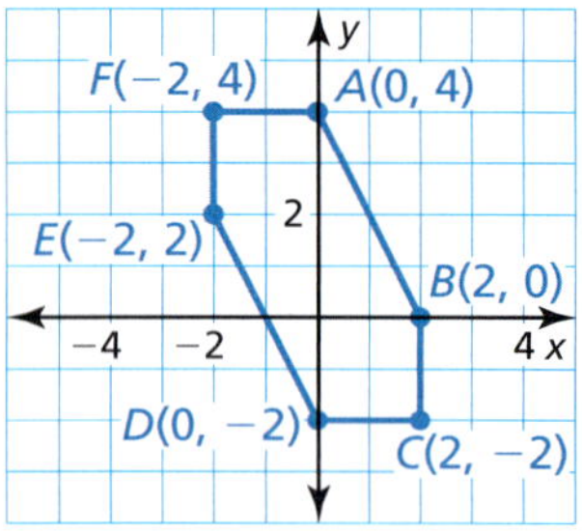

In Exercises 11–14, find the area of the polygon with the given vertices. *(See Example 3.)*

11. $E(3, 1), F(3, -2), G(-2, -2)$

12. $J(-3, 4), K(4, 4), L(3, -3)$

13. $W(0, 0), X(0, 3), Y(-3, 3), Z(-3, 0)$

14. $N(-4, 1), P(1, 1), Q(3, -1), R(-2, -1)$

In Exercises 15–18, use the diagram to find the perimeter and the area of the polygon.

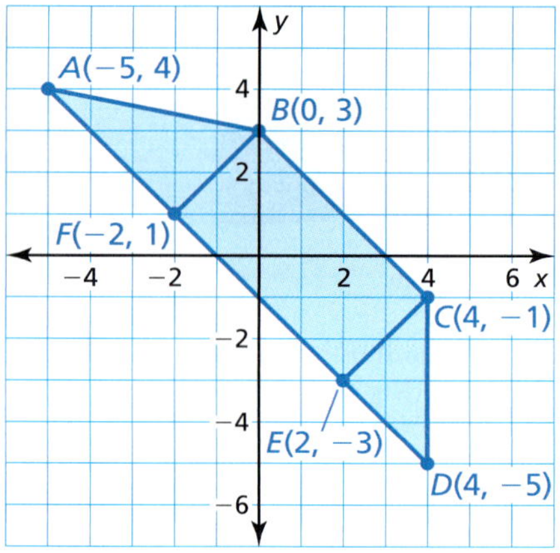

15. $\triangle CDE$

16. $\triangle ABF$

17. rectangle $BCEF$

18. quadrilateral $ABCD$

4 MTR **19. ERROR ANALYSIS** Describe and correct the error in finding the area of the triangle.

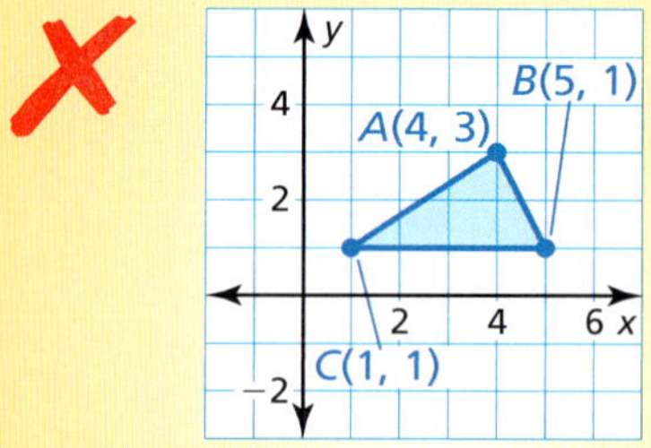

$b = |5 - 1| = 4$

$h = \sqrt{(5 - 4)^2 + (1 - 3)^2} = \sqrt{5}$

$A = \frac{1}{2}bh = \frac{1}{2}(4)(\sqrt{5}) = 2\sqrt{5}$

The area is $2\sqrt{5}$ square units.

20. REASONING Use the diagram.

a. Find the perimeter and area of each square.

b. What happens to the area of a square when its perimeter increases by a factor of n?

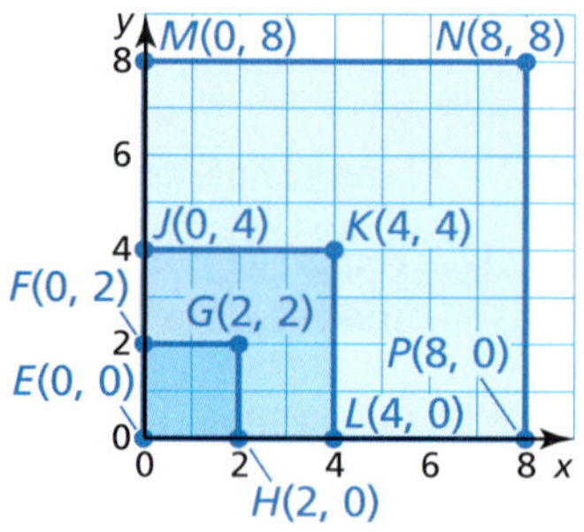

GO DIGITAL

COLLEGE PREP **In Exercises 21 and 22, use the diagram.**

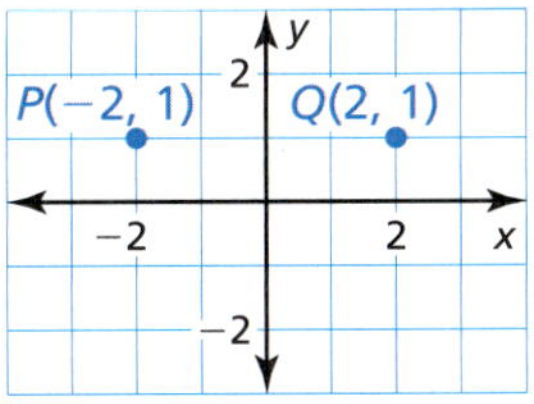

21. Determine which point is the remaining vertex of a triangle with an area of 4 square units.

Ⓐ $R(2, 0)$ Ⓒ $T(-1, 0)$

Ⓑ $S(-2, -1)$ Ⓓ $U(2, -2)$

22. Determine which points are the remaining vertices of a rectangle with a perimeter of 14 units.

Ⓐ $A(2, -1)$ and $B(-2, -1)$

Ⓑ $C(-1, -2)$ and $D(1, -2)$

Ⓒ $E(-2, -2)$ and $F(2, -2)$

Ⓓ $G(2, 0)$ and $H(-2, 0)$

7 MTR **23. MODELING REAL LIFE** You are building a school garden. The diagram shows the four vertices of the garden. Each unit in the coordinate plane represents 1 foot. Find the perimeter and the area of the garden. *(See Example 4.)*

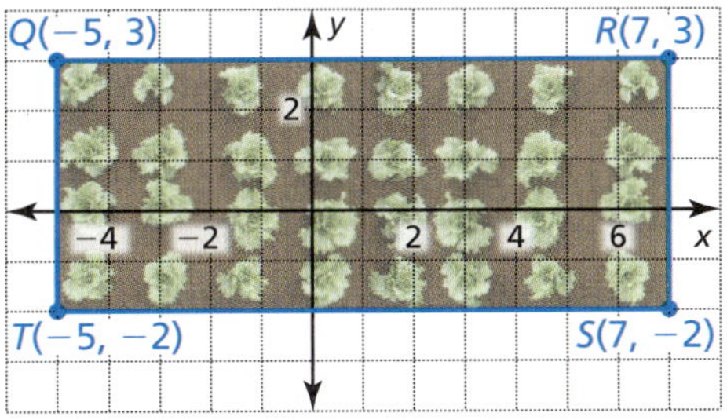

7 MTR **24. MODELING REAL LIFE** The diagram shows the vertices of a lion sanctuary. Each unit in the coordinate plane represents 100 feet. Find the perimeter and the area of the sanctuary.

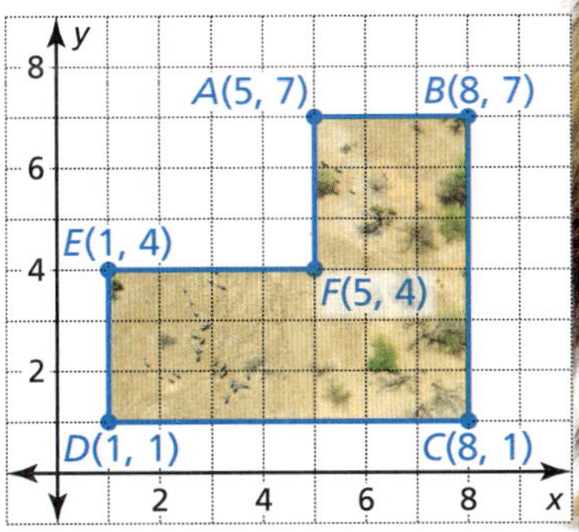

7 MTR **25. MODELING REAL LIFE** You and your friend hike to a waterfall that is 4 miles east of where you left your bikes. You then hike to a lookout point that is 2 miles north of your bikes. From the lookout point, you return to your bikes.

a. About how far do you hike? Assume you travel along straight paths.

b. From the waterfall, your friend hikes to a wishing well before going to the lookout point and returning to the bikes. The wishing well is 3 miles north and 2 miles west of the lookout point. About how far does your friend hike?

26. HOW DO YOU SEE IT?
Without performing any calculations, determine whether the triangle or the rectangle has a greater area. Which polygon has a greater perimeter? Explain your reasoning.

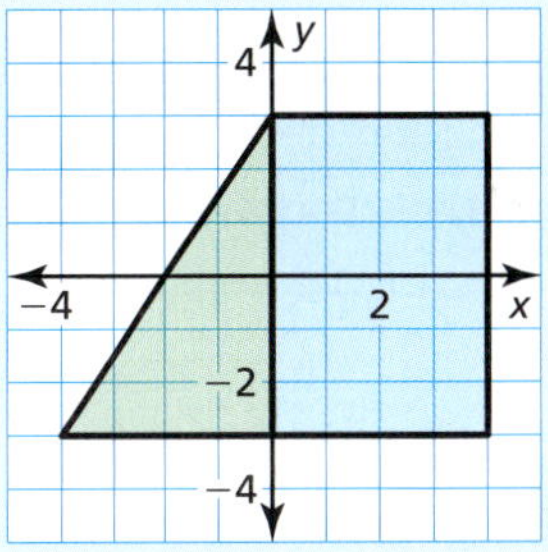

27. REASONING Use the diagram.

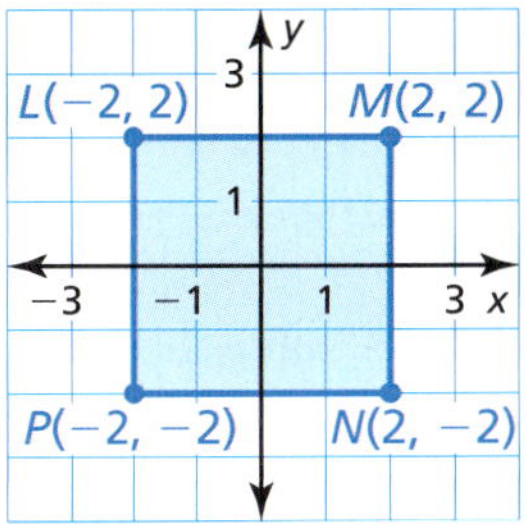

a. Find the perimeter and the area of the square.

b. Connect the midpoints of the sides of the given square to make a quadrilateral. Is this quadrilateral a square? Explain your reasoning.

c. Find the perimeter and the area of the quadrilateral you made in part (b). Compare this area to the area of the square you found in part (a).

GO DIGITAL

28. **CONNECTING CONCEPTS** The lines $y_1 = 2x - 6$, $y_2 = -3x + 4$, and $y_3 = -\frac{1}{2}x + 4$ intersect to form the sides of a right triangle. Find the perimeter and the area of the triangle.

29. **MAKING AN ARGUMENT** Will a rectangle that has the same perimeter as $\triangle QRS$ have the same area as the triangle? Explain your reasoning.

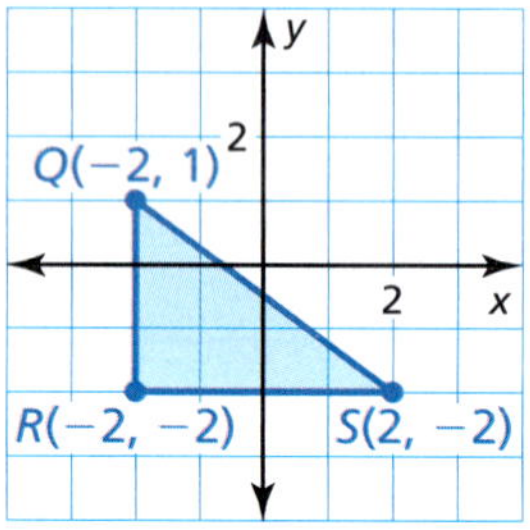

30. THOUGHT PROVOKING
A café that has an area of 350 square feet is being expanded to occupy an adjacent space that has an area of 150 square feet. Draw a diagram of the remodeled café in a coordinate plane.

31. **REASONING** Triangle ABC has a perimeter of 12 units. The vertices of the triangle are $A(x, 2)$, $B(2, -2)$, and $C(-1, 2)$. Find the value of x.

32. **PERFORMANCE TASK** As a graphic designer, your job is to create a company logo that includes at least two different polygons and has an area of at least 50 square units. Draw your logo in a coordinate plane, and record its perimeter and area. Describe the company, and create a proposal explaining how your logo relates to the company.

33. **DIG DEEPER** Find the area of $\triangle XYZ$. (*Hint:* Draw a rectangle whose sides contain points X, Y, and Z.)

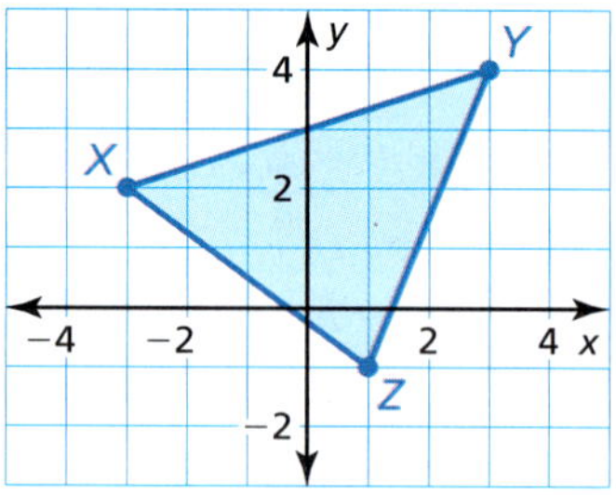

REVIEW & REFRESH

WATCH

34. Does the table represent a *linear* or *nonlinear* function? Explain.

x	−1	0	1	2	3
y	−9	−7	−5	−3	−1

In Exercises 35–38, solve the equation.

35. $3x - 7 = 2$

36. $4 = 9 + 5x$

37. $x + 4 = x - 12$

38. $\frac{x + 1}{2} = 4x - 3$

In Exercises 39 and 40, use the diagram.

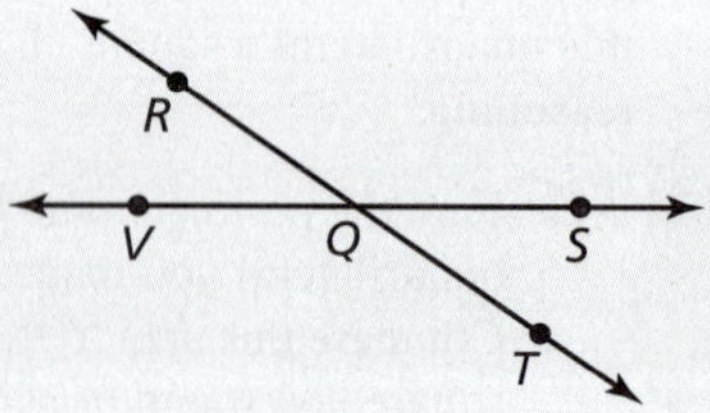

39. Give another name for $\overline{RT}$.

40. Name two pairs of opposite rays.

In Exercises 41 and 42, the endpoints of a segment are given. Find the coordinates of the midpoint *M* and the length of the segment.

41. $J(4, 3)$ and $K(2, -3)$

42. $L(-4, 5)$ and $N(5, -3)$

43. Use a compass and straightedge to construct a copy of the line segment.

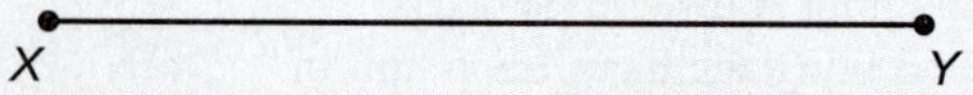

44. **MODELING REAL LIFE** You deposit \$200 into a savings account that earns 5% annual interest compounded quarterly. Write a function that represents the balance y (in dollars) after t years.

In Exercises 45 and 46, graph the function. Then describe the transformations from the graph of $f(x) = |x|$ to the graph of the function.

45. $g(x) = |x| + 5$

46. $h(x) = -3|x|$

47. Find the perimeter and the area of $\square ABCD$ with vertices $A(3, 5)$, $B(6, 5)$, $C(4, -1)$, and $D(1, -1)$.

GO DIGITAL

1.5 Measuring and Constructing Angles

Learning Target: Measure, construct, and describe angles.

Success Criteria:
- I can measure and classify angles.
- I can construct congruent angles.
- I can find angle measures.
- I can construct an angle bisector.

EXPLORE IT! Analyzing a Geometric Figure

7 MTR APPLY MATHEMATICS
Give some real-life situations that can be represented by these types of angles. Research different ways you can measure these types of angles in real life.

Work with a partner.

a. Identify the figure shown at the right. Then define it in your own words.

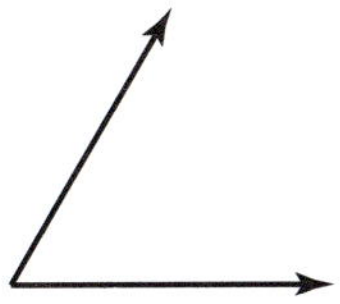

b. Label and name the figure. Then compare your results with those of your classmates.

c. How can you *measure* the figure?

d. Describe each angle below. How would you group these angles? Explain.

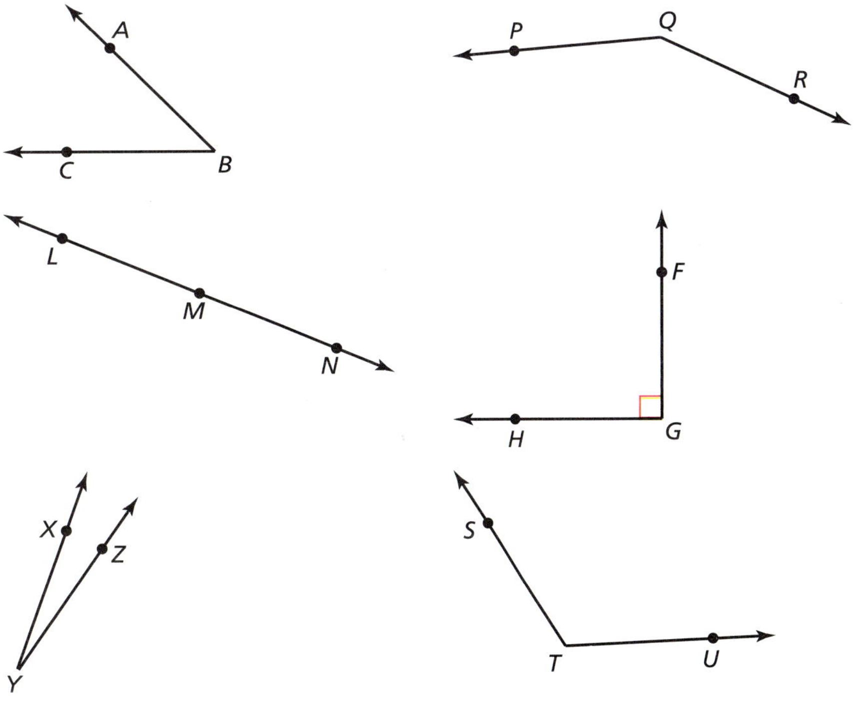

e. Construct a copy of an angle from part (d). Explain your method.

f. Construct each of the following. Explain your method.

i. An angle that is twice the measure of the angle in part (a)

ii. Separate the angle in part (a) into two angles with the same measure.

Geometric Reasoning
MA.912.GR.5.1 Construct a copy of a segment or an angle.
MA.912.GR.5.2 Construct the bisector of a segment or an angle, including the perpendicular bisector of a line segment.

GO DIGITAL

Vocabulary

angle, *p. 38*
vertex, *p. 38*
sides of an angle, *p. 38*
interior of an angle, *p. 38*
exterior of an angle, *p. 38*
measure of an angle, *p. 38*
acute angle, *p. 39*
right angle, *p. 39*
obtuse angle, *p. 39*
straight angle, *p. 39*
congruent angles, *p. 40*
angle bisector, *p. 42*

Naming Angles

An **angle** is a set of points consisting of two different rays that have the same endpoint, called the **vertex**. The rays are the **sides** of the angle.

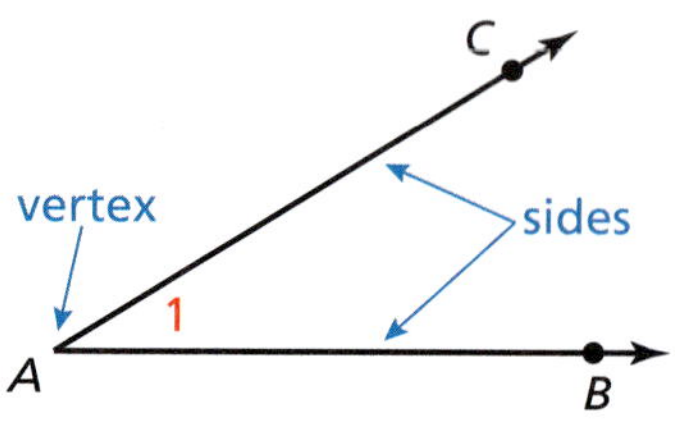

You can name an angle in several different ways. The symbol ∠ represents an angle.

- Use its vertex, such as ∠*A*.
- Use a point on each ray and the vertex, such as ∠*BAC* or ∠*CAB*.
  Make sure the vertex is the middle letter.
- Use a number, such as ∠1.

The region that contains all the points between the sides of the angle is the **interior of the angle**. The region that contains all the points outside the angle is the **exterior of the angle**.

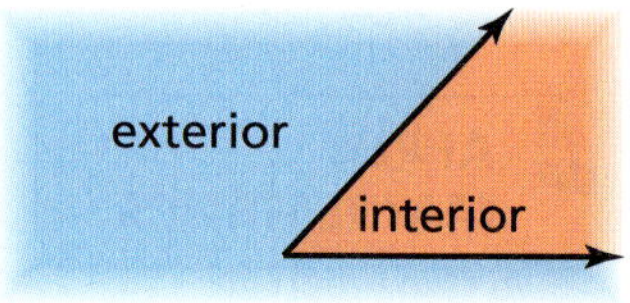

EXAMPLE 1 Naming Angles

A lighthouse keeper measures the angles formed by the lighthouse at point *M* and three boats. Name three angles shown in the diagram.

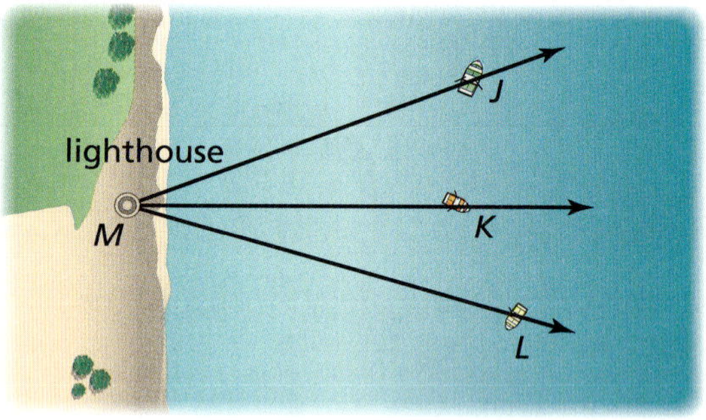

SOLUTION

∠*JMK* or ∠*KMJ*

∠*KML* or ∠*LMK*

∠*JML* or ∠*LMJ*

COMMON ERROR
When a point is the vertex of more than one angle, you cannot use the vertex alone to name the angle.

Measuring and Classifying Angles

A protractor helps you approximate the *measure* of an angle. The measure is usually given in *degrees*.

POSTULATE

1.3 Protractor Postulate

Consider $\overleftrightarrow{OB}$ and a point *A* on one side of $\overleftrightarrow{OB}$. The rays of the form $\overrightarrow{OA}$ can be matched one to one with the real numbers from 0 to 180.

The **measure** of ∠*AOB*, which can be written as *m*∠*AOB*, is equal to the absolute value of the difference between the real numbers matched with $\overrightarrow{OA}$ and $\overrightarrow{OB}$ on a protractor.

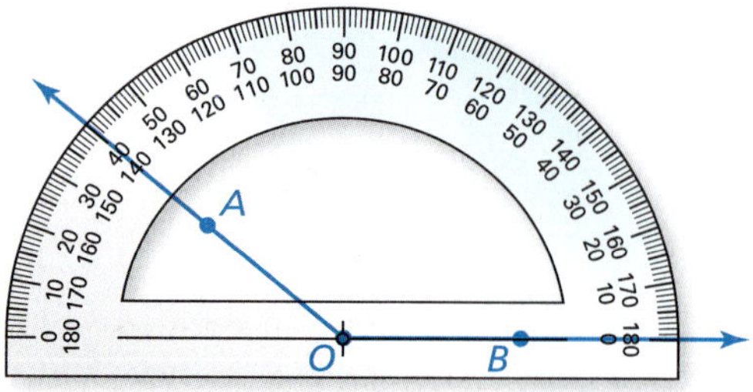

GO DIGITAL

You can classify angles according to their measures.

KEY IDEA

Types of Angles

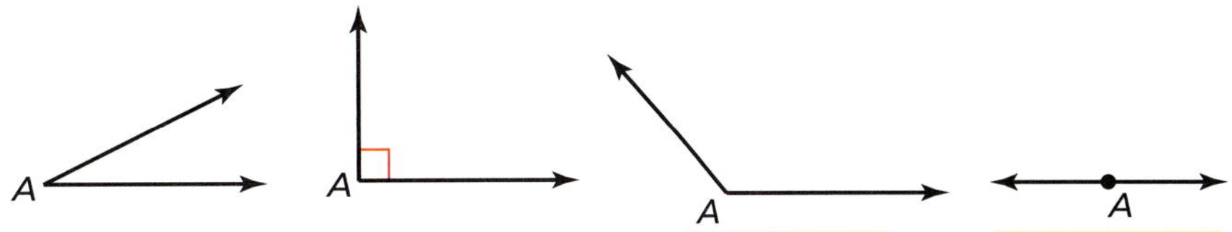

acute angle

Measures greater than 0° and less than 90°

right angle

Measures 90°

obtuse angle

Measures greater than 90° and less than 180°

straight angle

Measures 180°

EXAMPLE 2 Measuring and Classifying Angles

WATCH

COMMON ERROR

Most protractors have an inner and an outer scale. When measuring, make sure you are using the correct scale.

Find the measure of each angle. Then classify the angle.

a. $\angle GHK$

b. $\angle JHL$

c. $\angle LHK$

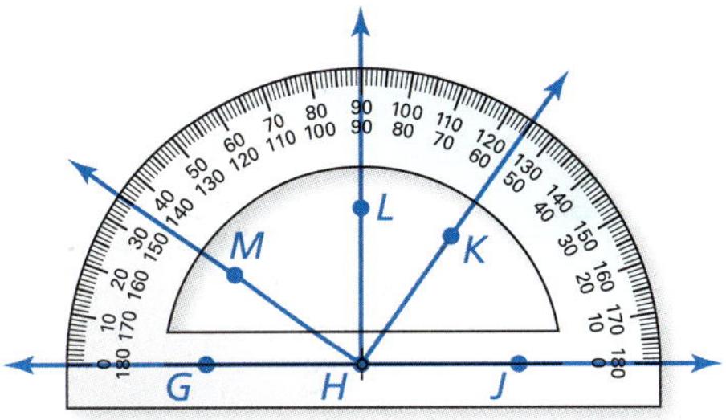

SOLUTION

a. $\overrightarrow{HG}$ lines up with 0° on the outer scale of the protractor. $\overrightarrow{HK}$ passes through 125° on the outer scale. So, $m\angle GHK = 125°$. It is an *obtuse* angle.

b. $\overrightarrow{HJ}$ lines up with 0° on the inner scale of the protractor. $\overrightarrow{HL}$ passes through 90°. So, $m\angle JHL = 90°$. It is a *right* angle.

c. $\overrightarrow{HL}$ passes through 90°. $\overrightarrow{HK}$ passes through 55° on the inner scale. So, $m\angle LHK = |90 - 55| = 35°$. It is an *acute* angle.

SELF-ASSESSMENT

1 I don't understand yet. 2 I can do it with help. 3 I can do it on my own. 4 I can teach someone else.

Write three names for the angle.

1.

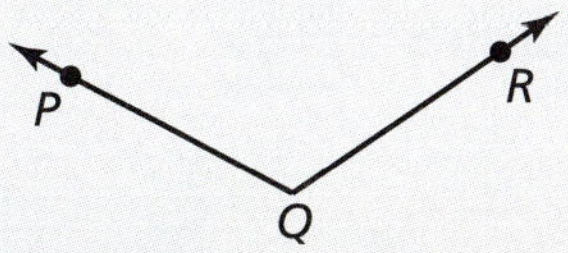

2.

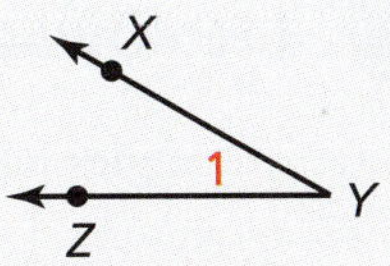

3.

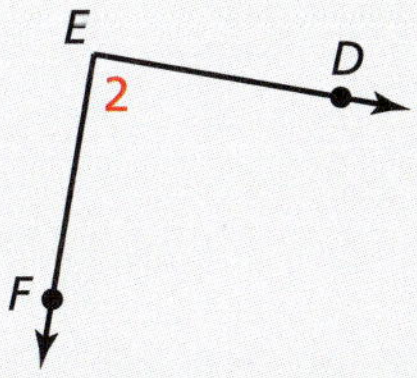

4 MTR **4. WHICH ONE DOESN'T BELONG?** Which angle name does *not* belong with the other three? Explain your reasoning.

$\angle BCA$	$\angle BAC$	$\angle 1$	$\angle CAB$

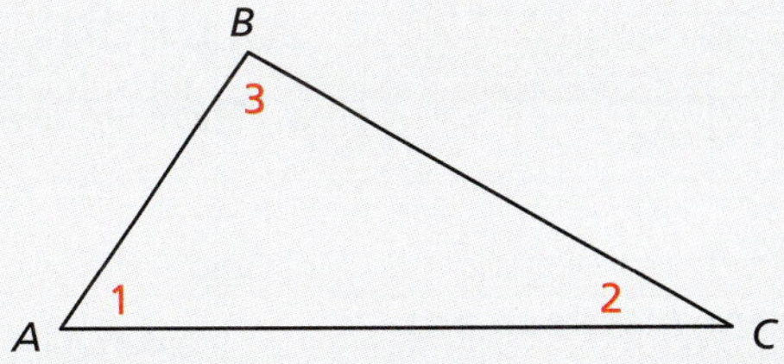

Use the diagram in Example 2 to find the measure of the angle. Then classify the angle.

5. $\angle JHM$

6. $\angle MHK$

7. $\angle MHL$

GO DIGITAL

Identifying Congruent Angles

You can use a compass and straightedge to construct an angle that has the same measure as a given angle.

CONSTRUCTION Copying an Angle

WATCH

Use a compass and straightedge to construct an angle that has the same measure as $\angle A$. In this construction, the *center* of an arc is the point where the compass point rests. The *radius* of an arc is the distance from the center of the arc to a point on the arc drawn by the compass.

SOLUTION

Step 1

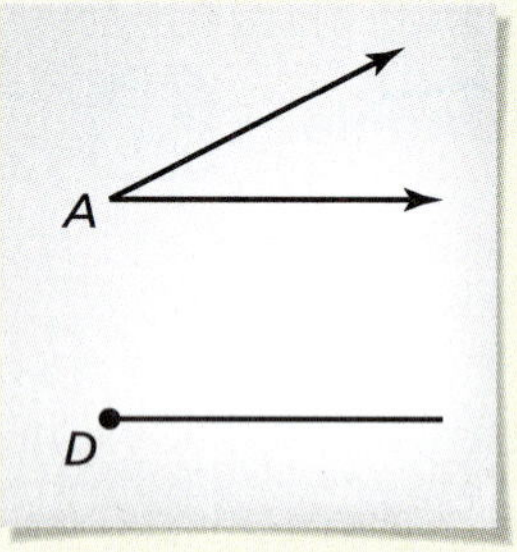

Draw a segment Draw an angle such as $\angle A$, as shown. Then draw a segment. Label point D on the segment.

Step 2

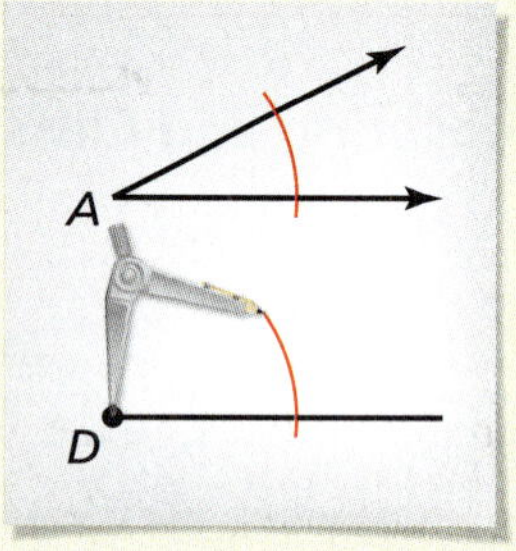

Draw arcs Draw an arc with center A. Using the same radius, draw an arc with center D.

Step 3

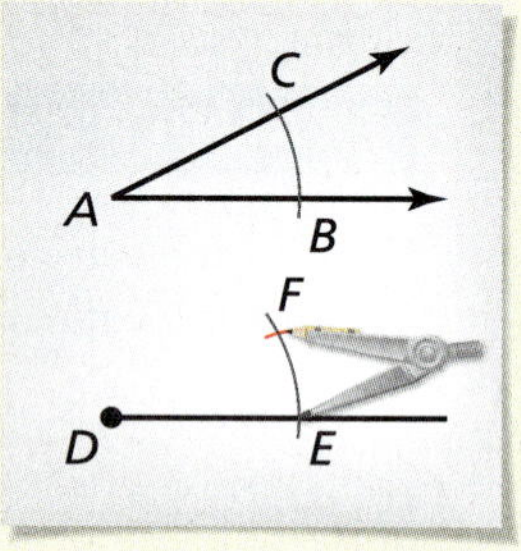

Draw an arc Label B, C, and E. Draw an arc with radius BC and center E. Label the intersection F.

Step 4

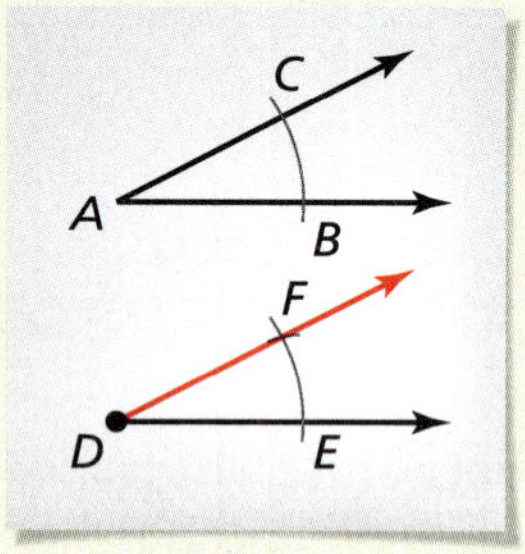

Draw a ray Draw $\overrightarrow{DF}$. $\angle D$ has the same measure as $\angle A$.

Two angles are **congruent angles** when they have the same measure. In the construction above, $\angle A$ and $\angle D$ are congruent angles. So,

$m\angle A = m\angle D$ The measure of angle *A* is *equal to* the measure of angle *D*.

and

$\angle A \cong \angle D$. Angle *A* is *congruent to* angle *D*.

EXAMPLE 3 Identifying Congruent Angles

WATCH

a. Identify the congruent angles labeled in the quilt design.

b. $m\angle ADC = 140°$. What is $m\angle EFG$?

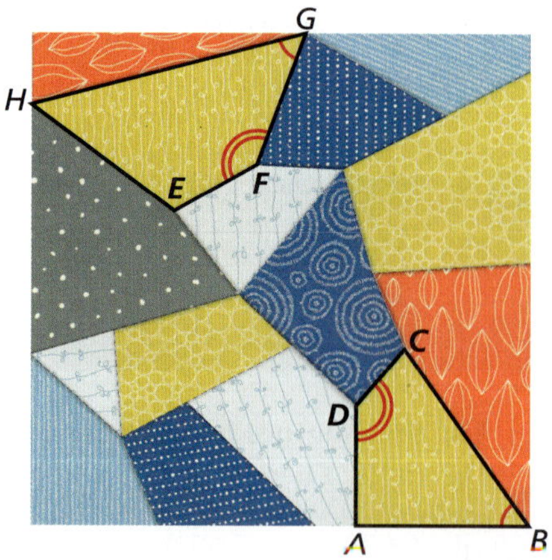

SOLUTION

a. There are two pairs of congruent angles:

$$\angle ABC \cong \angle FGH$$

and

$$\angle ADC \cong \angle EFG.$$

b. Because $\angle ADC \cong \angle EFG$, $m\angle ADC = m\angle EFG$.

▶ So, $m\angle EFG = 140°$.

READING

In a diagram, matching arcs indicate congruent angles. When there is more than one pair of congruent angles, use multiple arcs.

GO DIGITAL

Using the Angle Addition Postulate

POSTULATE

1.4 Angle Addition Postulate

Words If P is in the interior of $\angle RST$, then the measure of $\angle RST$ is equal to the sum of the measures of $\angle RSP$ and $\angle PST$.

Symbols If P is in the interior of $\angle RST$, then

$$m\angle RST = m\angle RSP + m\angle PST.$$

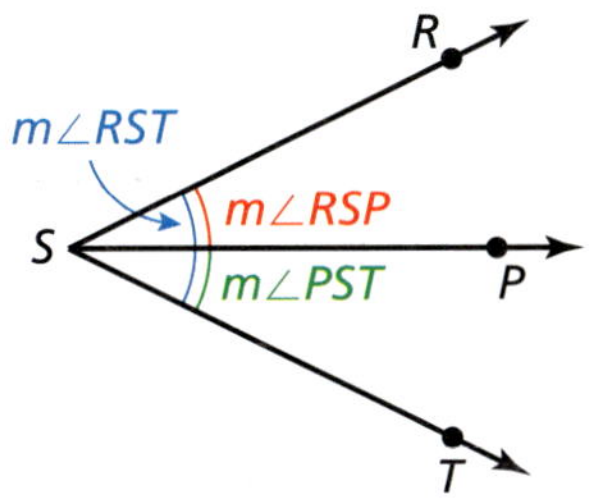

EXAMPLE 4 Finding Angle Measures

WATCH

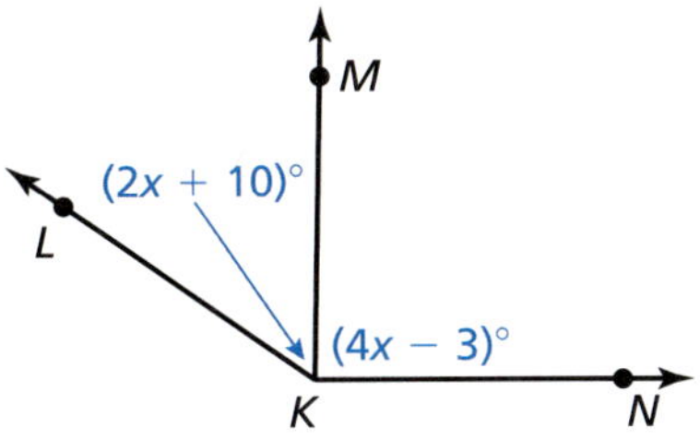

Given that $m\angle LKN = 145°$, find $m\angle LKM$ and $m\angle MKN$.

SOLUTION

Step 1 Write and solve an equation to find the value of x.

$m\angle LKN = m\angle LKM + m\angle MKN$	Angle Addition Postulate
$145° = (2x + 10)° + (4x - 3)°$	Substitute angle measures.
$145 = 6x + 7$	Combine like terms.
$138 = 6x$	Subtract 7 from each side.
$23 = x$	Divide each side by 6.

Step 2 Evaluate the given expressions when $x = 23$.

$$m\angle LKM = (2x + 10)° = (2 \cdot 23 + 10)° = 56°$$

$$m\angle MKN = (4x - 3)° = (4 \cdot 23 - 3)° = 89°$$

▶ So, $m\angle LKM = 56°$ and $m\angle MKN = 89°$.

SELF-ASSESSMENT

1 I don't understand yet. 2 I can do it with help. 3 I can do it on my own. 4 I can teach someone else.

6 MTR **8. ASSESS REASONABLENESS** Without measuring, determine whether $\angle DAB$ and $\angle FEH$ in Example 3 appear to be congruent. Explain your reasoning. Use a protractor to verify your answer.

Find the indicated angle measures.

9. Given that $\angle KLM$ is a straight angle, find $m\angle KLN$ and $m\angle NLM$.

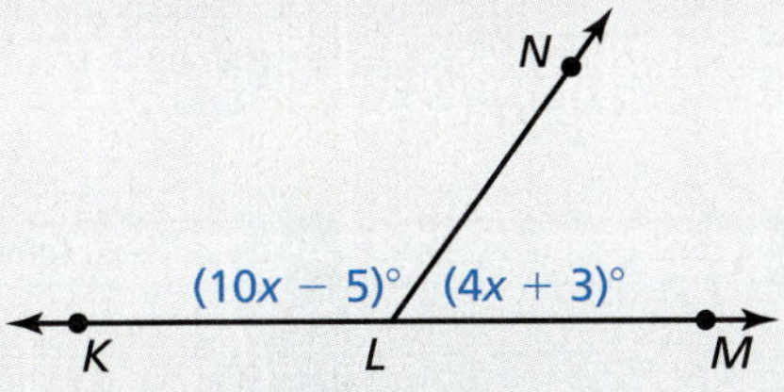

10. Given that $\angle EFG$ is a right angle, find $m\angle EFH$ and $m\angle HFG$.

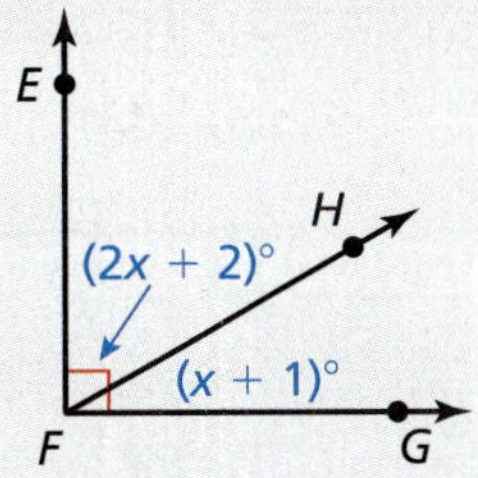

GO DIGITAL

Bisecting Angles

An **angle bisector** is a ray that divides an angle into two angles that are congruent. In the figure, $\overrightarrow{YW}$ bisects $\angle XYZ$, so $\angle XYW \cong \angle ZYW$.

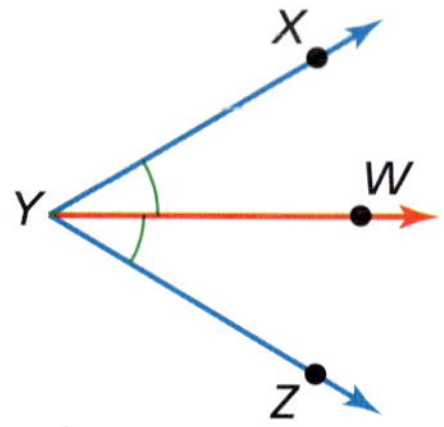

You can use a compass and straightedge to bisect an angle.

CONSTRUCTION **Bisecting an Angle** WATCH

Construct an angle bisector of $\angle A$ with a compass and straightedge.

SOLUTION

Step 1

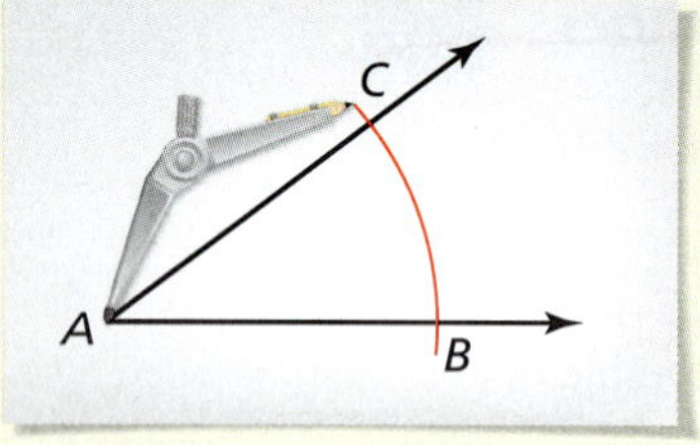

Draw an arc Draw an angle such as $\angle A$, as shown. Place the compass at A. Draw an arc with center A that intersects both sides of the angle. Label the intersections B and C.

Step 2

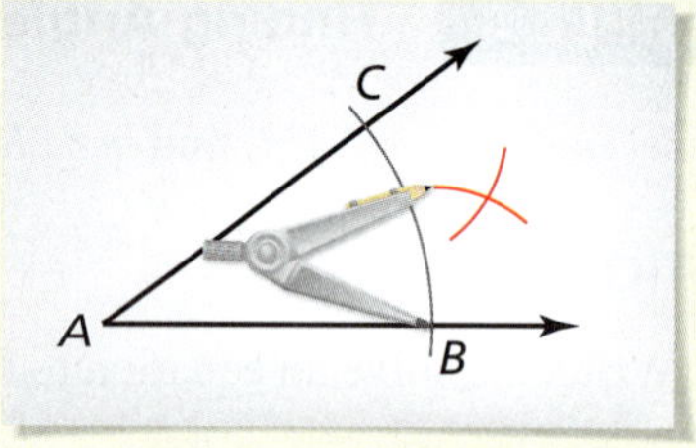

Draw arcs Draw an arc with center C. Using the same radius, draw an arc with center B.

Step 3

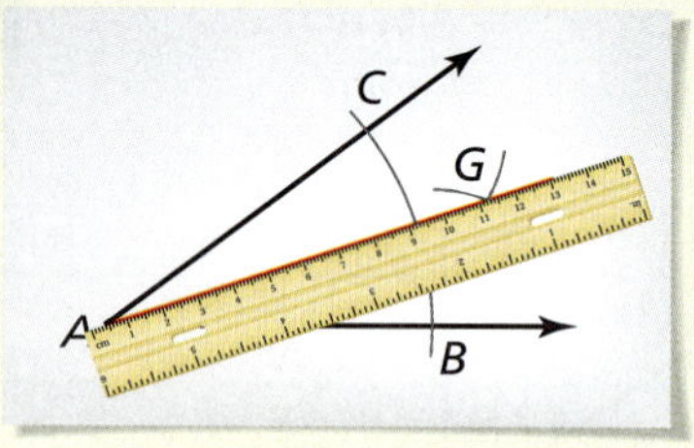

Draw a ray Label the intersection G. Use a straightedge to draw $\overrightarrow{AG}$. $\overrightarrow{AG}$ bisects $\angle A$.

EXAMPLE 5 **Using a Bisector to Find Angle Measures** WATCH

$\overrightarrow{QS}$ bisects $\angle PQR$, and $m\angle PQS = 24°$. Find $m\angle PQR$.

SOLUTION

Step 1 Draw a diagram.

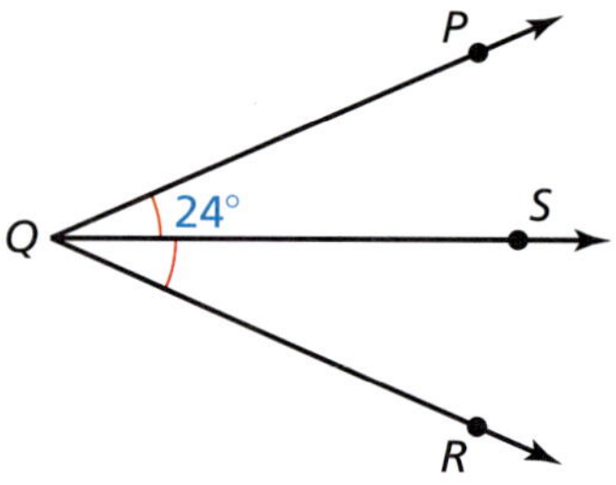

Step 2 Because $\overrightarrow{QS}$ bisects $\angle PQR$, $m\angle PQS = m\angle RQS$. So, $m\angle RQS = 24°$. Use the Angle Addition Postulate to find $m\angle PQR$.

$m\angle PQR = m\angle PQS + m\angle RQS$ — Angle Addition Postulate

$= 24° + 24°$ — Substitute angle measures.

$= 48°$ — Add.

▶ So, $m\angle PQR = 48°$.

SELF-ASSESSMENT 1 I don't understand yet. 2 I can do it with help. 3 I can do it on my own. 4 I can teach someone else.

11. Angle MNP is a straight angle, and $\overrightarrow{NQ}$ bisects $\angle MNP$. Draw $\angle MNP$ and $\overrightarrow{NQ}$. Use matching arcs to indicate congruent angles in your diagram. Find the angle measures of these congruent angles.

GO DIGITAL

1.5 Practice WITH CalcChat® AND CalcView®

In Exercises 1–4, write three names for the angle.

1. 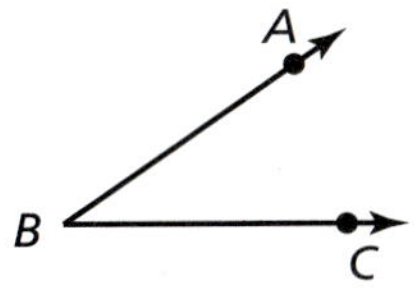

2. 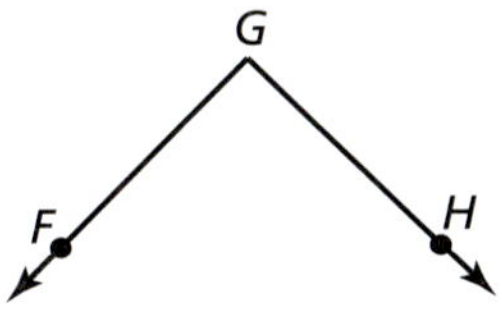

3. 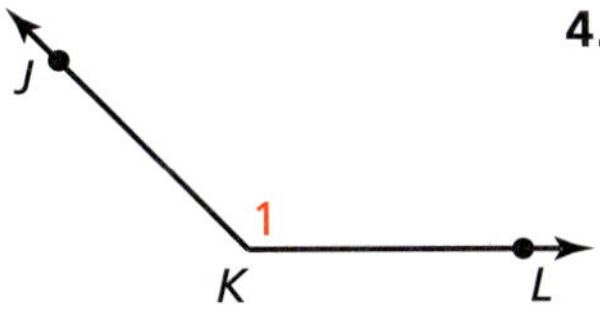

4. 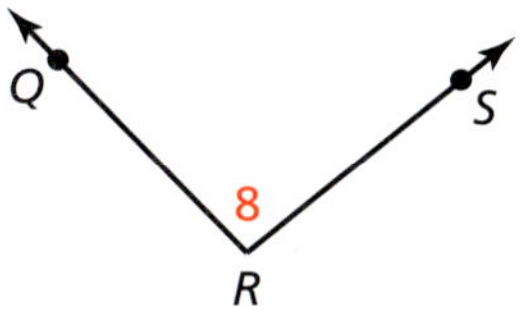

In Exercises 5 and 6, name three different angles in the diagram. *(See Example 1.)*

▶ 5. 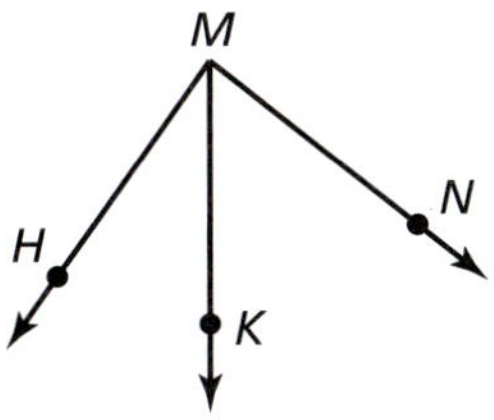

6. 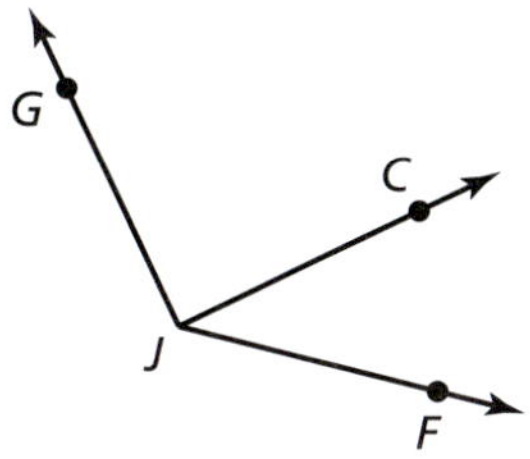

In Exercises 7–10, find the angle measure. Then classify the angle. *(See Example 2.)*

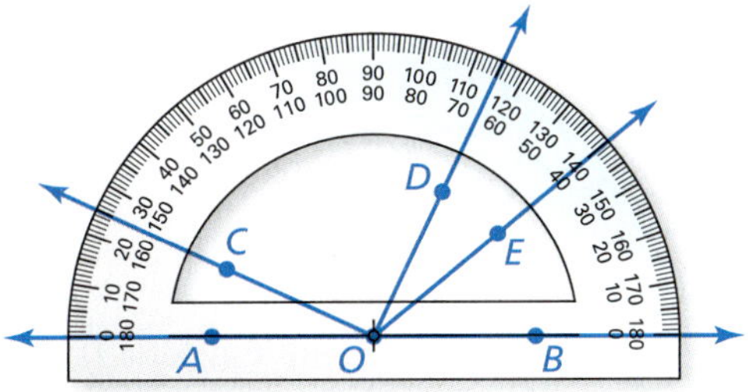

7. $m\angle BOD$

8. $m\angle AOE$

▶ 9. $m\angle COE$

10. $m\angle COD$

4 MTR **ERROR ANALYSIS** **In Exercises 11 and 12, describe and correct the error in finding the angle measure. Use the diagram from Exercises 7–10.**

11. ✗ $m\angle BOC = 25°$

12. ✗ $m\angle DOE = 65°$

GO DIGITAL

CONSTRUCTION **In Exercises 13 and 14, use a compass and straightedge to copy the angle.**

13. 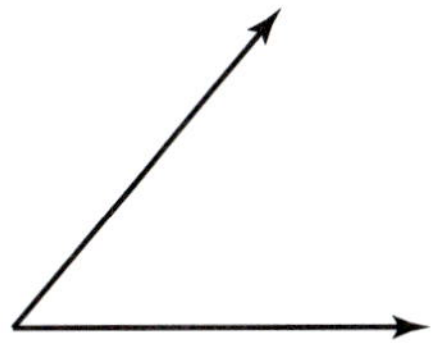

14. 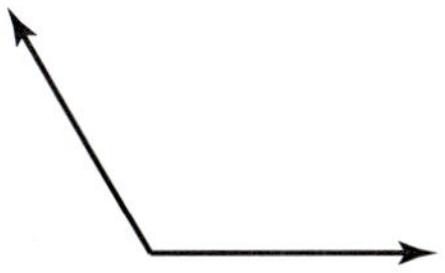

In Exercises 15–18, the design of a nomadic tribal rug is shown, where $m\angle AED = 34°$ and $m\angle EAD = 112°$. *(See Example 3.)*

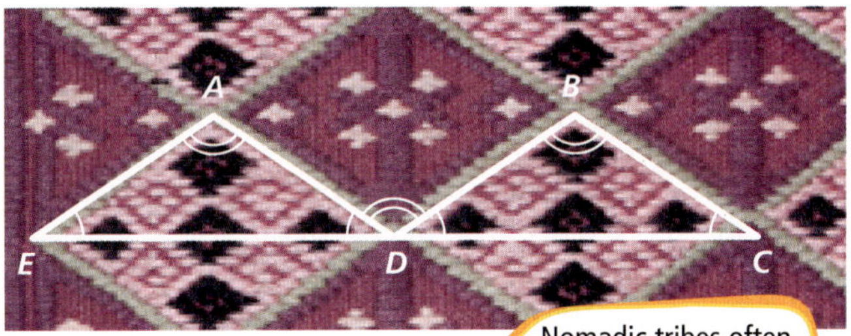

Nomadic tribes often used triangular motifs to represent water.

15. Identify the angles congruent to $\angle AED$.

16. Identify the angles congruent to $\angle EAD$.

▶ 17. Find $m\angle BDC$.

18. Find $m\angle ADB$.

In Exercises 19–22, find the indicated angle measure.

19. Find $m\angle ABC$.

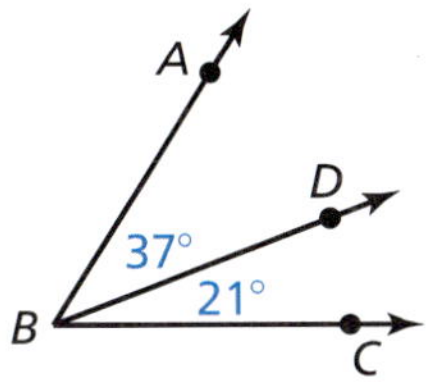

20. Find $m\angle LMN$.

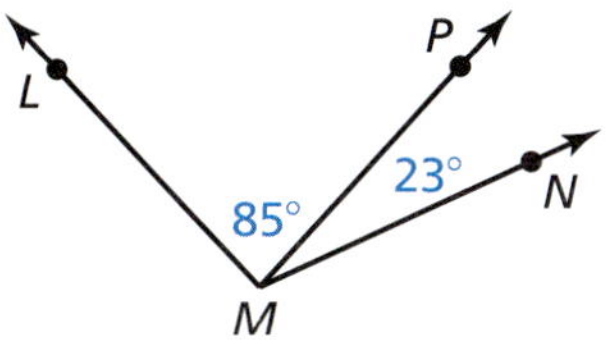

21. $m\angle RST = 114°$. Find $m\angle RSV$.

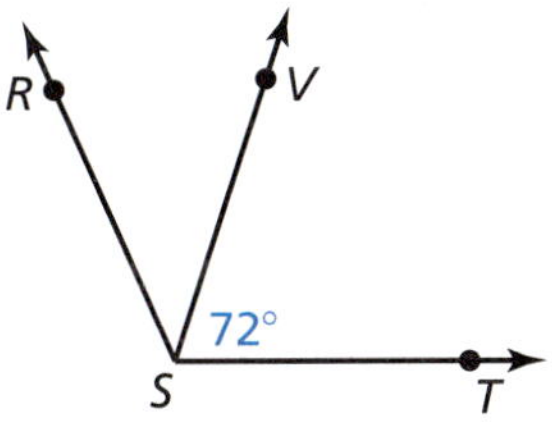

22. $\angle GHK$ is a straight angle. Find $m\angle LHK$.

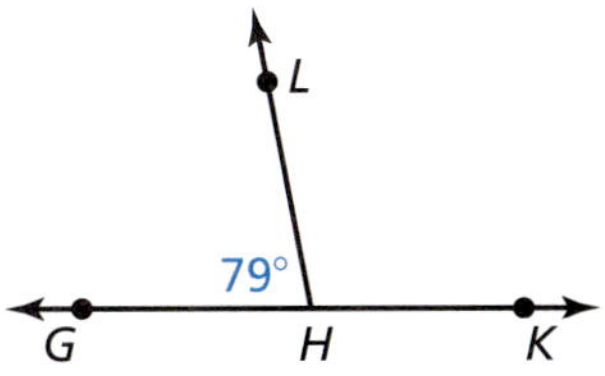

In Exercises 23–28, find the indicated angle measures. *(See Example 4.)*

23. $m\angle ABC = 95^\circ$. Find $m\angle ABD$ and $m\angle DBC$.

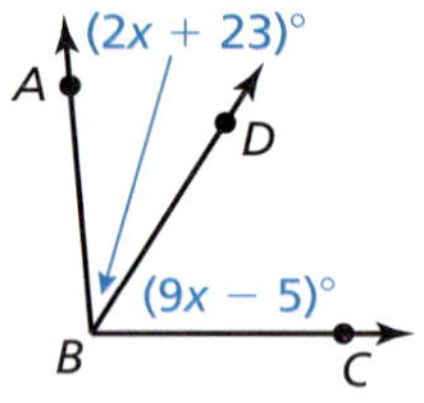

24. $m\angle XYZ = 117^\circ$. Find $m\angle XYW$ and $m\angle WYZ$.

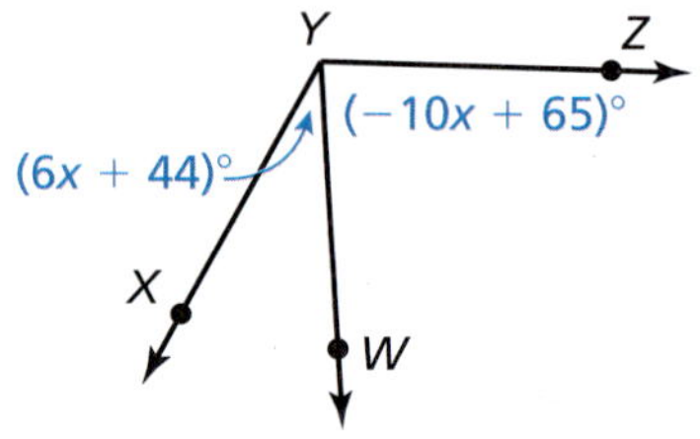

25. $\angle LMN$ is a straight angle. Find $m\angle LMP$ and $m\angle NMP$.

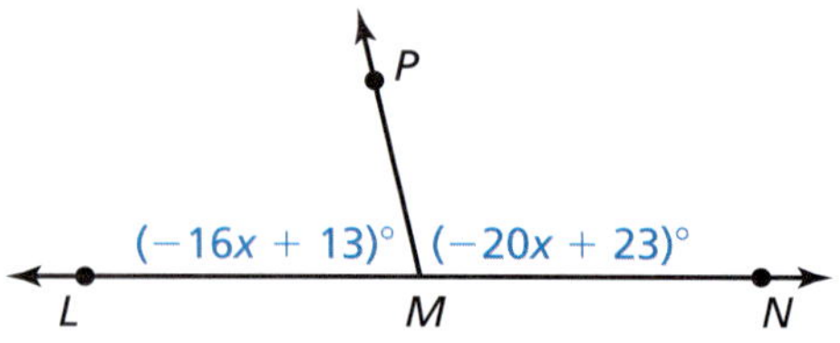

26. $\angle ABC$ is a straight angle. Find $m\angle ABX$ and $m\angle CBX$.

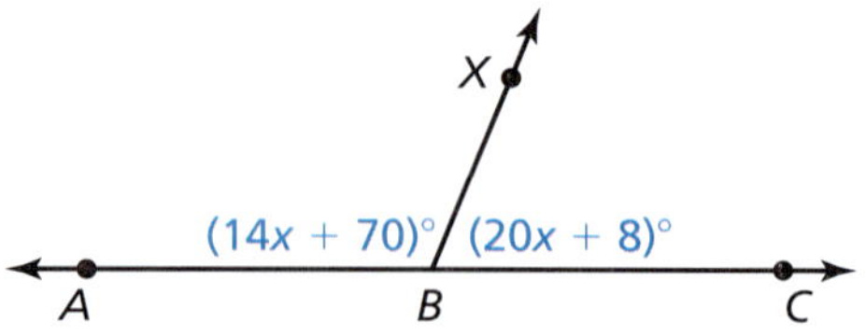

27. Find $m\angle RSQ$ and $m\angle TSQ$.

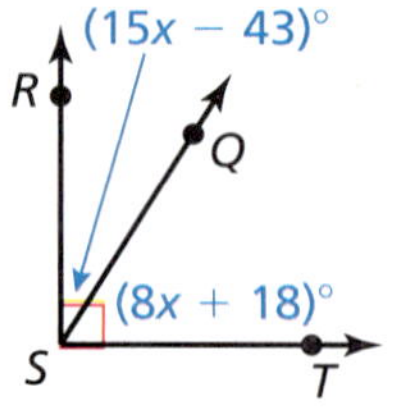

28. Find $m\angle DEH$ and $m\angle FEH$.

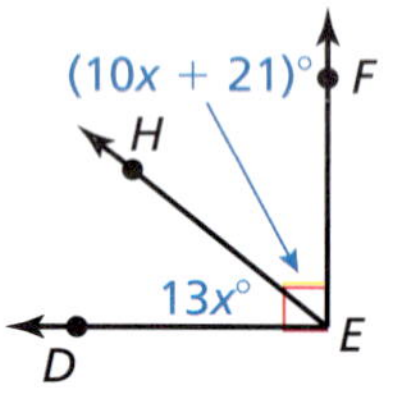

CONSTRUCTION In Exercises 29 and 30, copy the angle. Then construct the angle bisector with a compass and straightedge.

29.

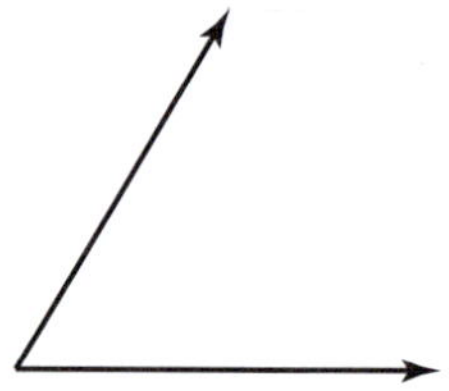

30.

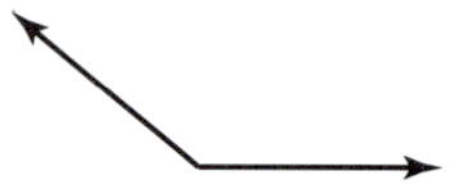

In Exercises 31–34, $\overrightarrow{FH}$ bisects $\angle EFG$. Find the indicated angle measures. *(See Example 5.)*

31. $m\angle EFH = 63^\circ$. Find $m\angle GFH$ and $m\angle EFG$.

32. $m\angle GFH = 71^\circ$. Find $m\angle EFH$ and $m\angle EFG$.

33. $m\angle EFG = 124^\circ$. Find $m\angle EFH$ and $m\angle GFH$.

34. $m\angle EFG = 119^\circ$. Find $m\angle EFH$ and $m\angle GFH$.

In Exercises 35–38, $\overrightarrow{BD}$ bisects $\angle ABC$. Find $m\angle ABD$, $m\angle CBD$, and $m\angle ABC$.

35.

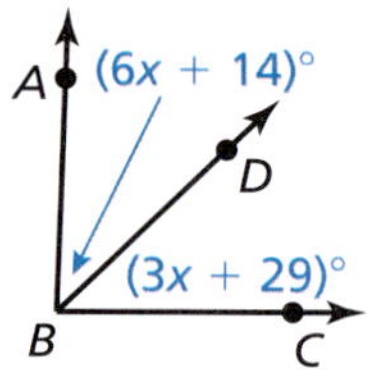

36.

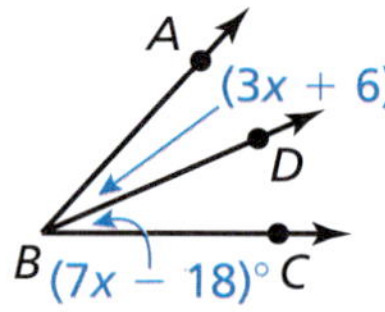

37. $m\angle ABC = (2 - 16x)^\circ$

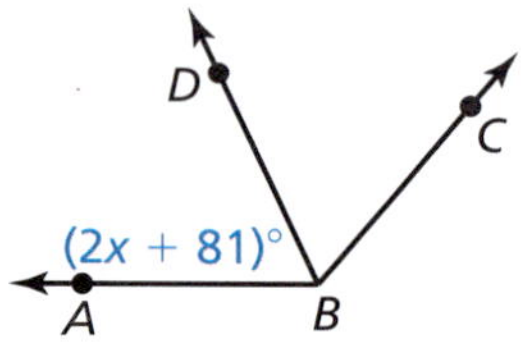

38. $m\angle ABC = (25x + 34)^\circ$

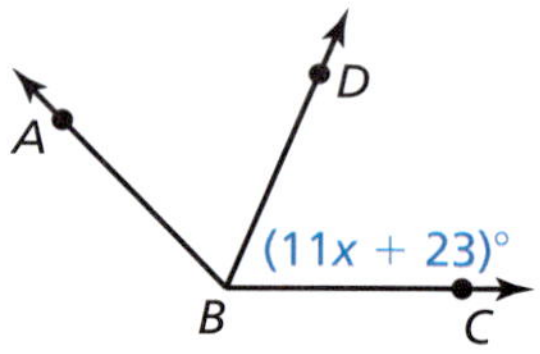

7 MTR **39.** **MODELING REAL LIFE** The map shows the intersections of three roads. Malcom Way intersects Sydney Street at an angle of 162°. Park Road intersects Sydney Street at an angle of 87°. Find the angle at which Malcom Way intersects Park Road.

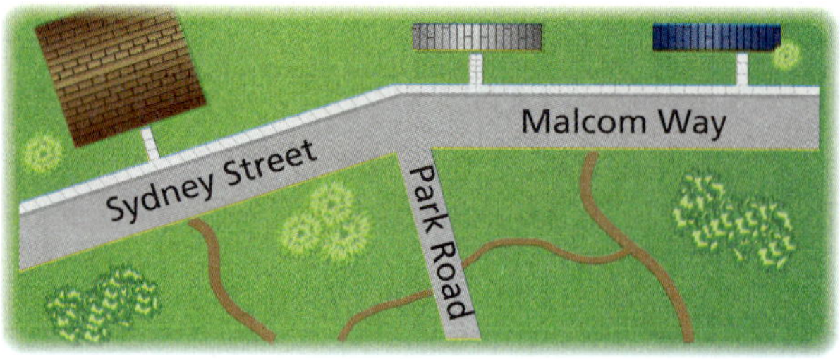

GO DIGITAL

7 MTR **40. MODELING REAL LIFE** In the sculpture shown, the measure of $\angle LMN$ is 76°, and the measure of $\angle PMN$ is 36°. What is the measure of $\angle LMP$?

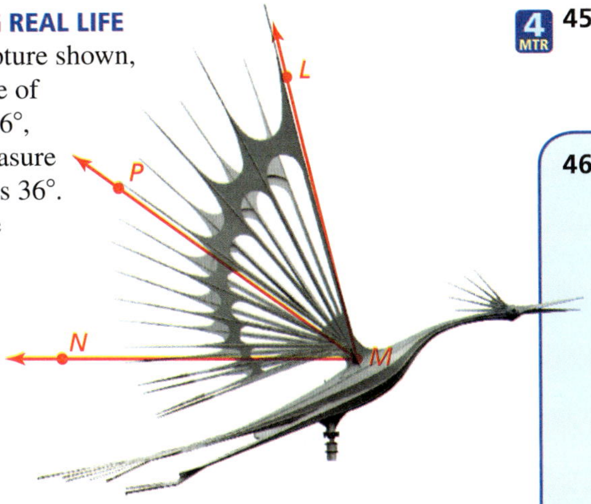

7 MTR **MODELING REAL LIFE In Exercises 41 and 42, use the diagram of the roof truss.**

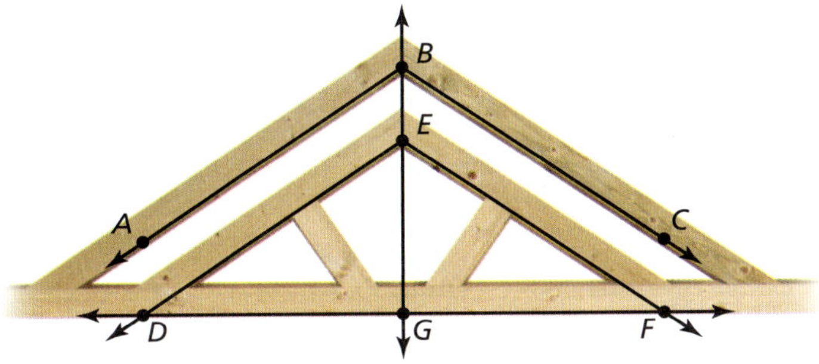

41. $\overrightarrow{BG}$ bisects $\angle ABC$ and $\angle DEF$, $m\angle ABC = 112°$, and $\angle ABC \cong \angle DEF$. Find the measure of each angle.

a. $\angle DEF$

b. $\angle ABG$

c. $\angle CBG$

d. $\angle DEG$

42. $\angle DGF$ is a straight angle, and $\overrightarrow{GB}$ bisects $\angle DGF$. Find $m\angle DGE$ and $m\angle FGE$.

43. NUMBER SENSE Given $\angle ABC$, X is in the interior of the angle, $m\angle ABX$ is 12° more than 4 times $m\angle CBX$, and $m\angle ABC = 92°$. Find $m\angle ABX$ and $m\angle CBX$.

5 MTR **44. STRUCTURE** In a coordinate plane, the ray from the origin through (4, 0) forms one side of an angle. Use the numbers below as x- and y-coordinates to create each type of angle.

−2 −1 0 1 2

a. acute angle

b. right angle

c. obtuse angle

d. straight angle

GO DIGITAL

4 MTR **45. MAKING AN ARGUMENT** Is it possible for a straight angle to consist of two obtuse angles? Explain your reasoning.

46. HOW DO YOU SEE IT?
Is it possible for $\angle XYZ$ to be a straight angle? Explain your reasoning. If it is not possible, what can you change in the diagram so that $\angle XYZ$ is a straight angle?

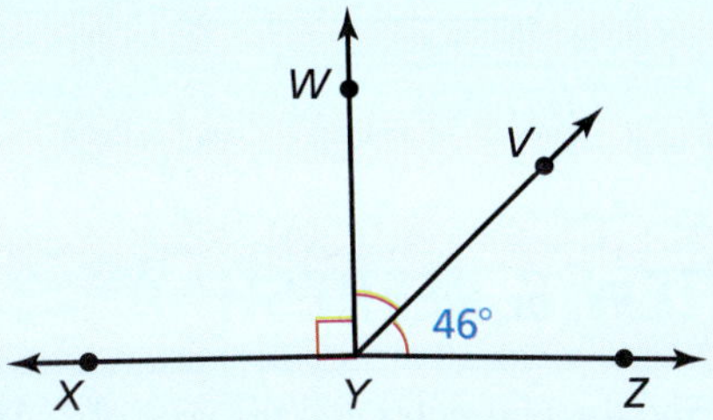

47. REASONING Classify the angles that result from bisecting each type of angle.

a. acute angle

b. right angle

c. obtuse angle

d. straight angle

48. REASONING Classify the angles that result from drawing a ray from the vertex through a point in the interior of each type of angle. Include all possibilities, and explain your reasoning.

a. acute angle

b. right angle

c. obtuse angle

d. straight angle

49. COLLEGE PREP In the diagram, $m\angle AGC = 38°$, $m\angle CGD = 71°$, and $m\angle FGC = 147°$. Which of the following statements are true? Select all that apply.

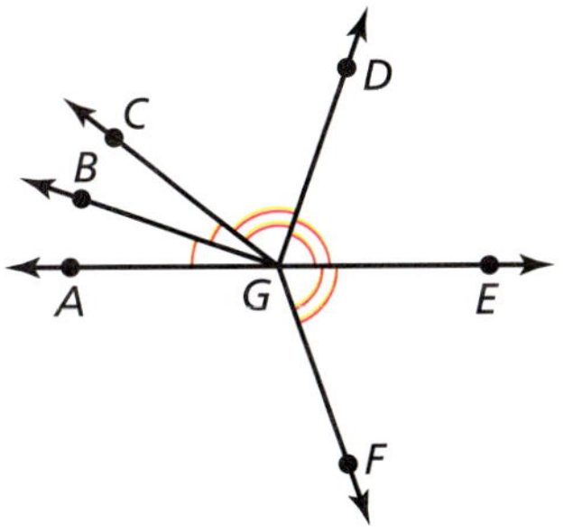

Ⓐ $m\angle AGB = 19°$

Ⓑ $m\angle DGF = 142°$

Ⓒ $m\angle AGF = 128°$

Ⓓ $\angle BGD$ is a right angle.

50. REASONING Copy the angle. Then construct an angle with a measure that is $\frac{1}{4}$ the measure of the given angle. Explain your reasoning.

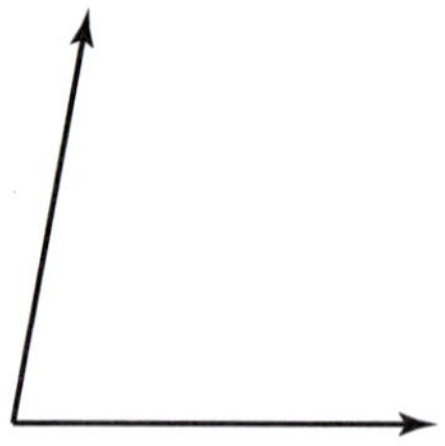

51. PROBLEM SOLVING $\overrightarrow{SQ}$ bisects $\angle RST$, $\overrightarrow{SP}$ bisects $\angle RSQ$, and $\overrightarrow{SV}$ bisects $\angle RSP$. The measure of $\angle VSP$ is 17°. Find $m\angle TSQ$. Explain.

52. THOUGHT PROVOKING
How many times between 12 A.M. and 12 P.M. do the minute hand and hour hand of a clock form a right angle? (Be sure to consider how the hour hand moves, in addition to how the minute hand moves.)

REVIEW & REFRESH

WATCH

53. Find the perimeter and the area of $\triangle ABC$ with vertices $A(-1, 1)$, $B(2, 1)$, and $C(1, -2)$.

In Exercises 54–56, solve the equation.

54. $3x + 15 + 4x - 9 = 90$

55. $\frac{1}{2}(4x + 6) - 11 = 5x + 7$

56. $3(6 - 8x) = 2(-12x + 9)$

In Exercises 57–60, simplify the expression.

57. $\sqrt{160}$

58. $\sqrt[3]{135}$

59. $\sqrt{\frac{21}{100}}$

60. $\frac{\sqrt{11}}{\sqrt{5}}$

7 MTR **61. MODELING REAL LIFE** The positions of three players during part of a water polo match are shown. Player A throws the ball to Player B, who then throws the ball to Player C.

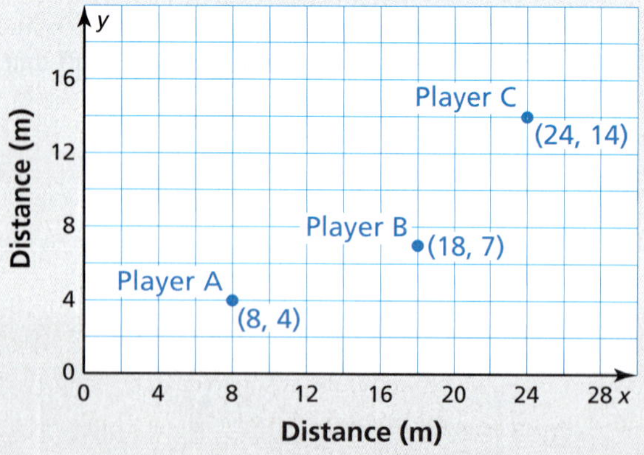

a. Who throws the ball farther, Player A or B?

b. About how far would Player A have to throw the ball to throw it directly to Player C?

In Exercises 62 and 63, graph the inequality in a coordinate plane.

62. $x \geq -2.5$

63. $y < -\frac{1}{3}x + 2$

In Exercises 64 and 65, find the indicated angle measures.

64. $\overrightarrow{KM}$ bisects $\angle JKL$. Find $m\angle JKM$ and $m\angle JKL$.

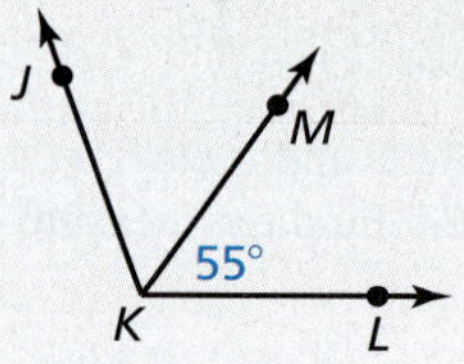

65. $m\angle DEF = 76°$. Find $m\angle DEG$ and $m\angle GEF$.

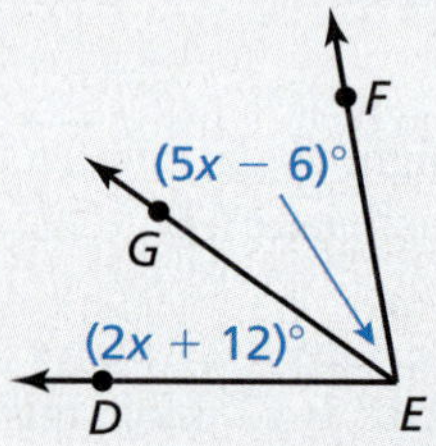

In Exercises 66 and 67, solve the system using any method. Explain your choice of method.

66. $2x + 3y = 3$
$x = y - 11$

67. $3x - 4y = 24$
$-5x + 2y = -26$

68. Plot $A(3, 2)$, $B(3, 5)$, $C(-4, -1)$, $D(2, -1)$ in a coordinate plane. Then determine whether $\overline{AB}$ and $\overline{CD}$ are congruent.

69. Point Y is between points X and Z on $\overline{XZ}$. $XY = 27$ and $YZ = 8$. Find XZ.

GO DIGITAL

1.6 Describing Pairs of Angles

Learning Target: Identify and use pairs of angles.

Success Criteria:
- I can identify complementary and supplementary angles.
- I can identify linear pairs and vertical angles.
- I can find angle measures in pairs of angles.

EXPLORE IT! Identifying Pairs of Angles

Work with a partner. The Blackfriars Street Bridge in London, Ontario, Canada, is a bowstring arch-truss bridge. Use the diagram to complete parts (a)–(c).

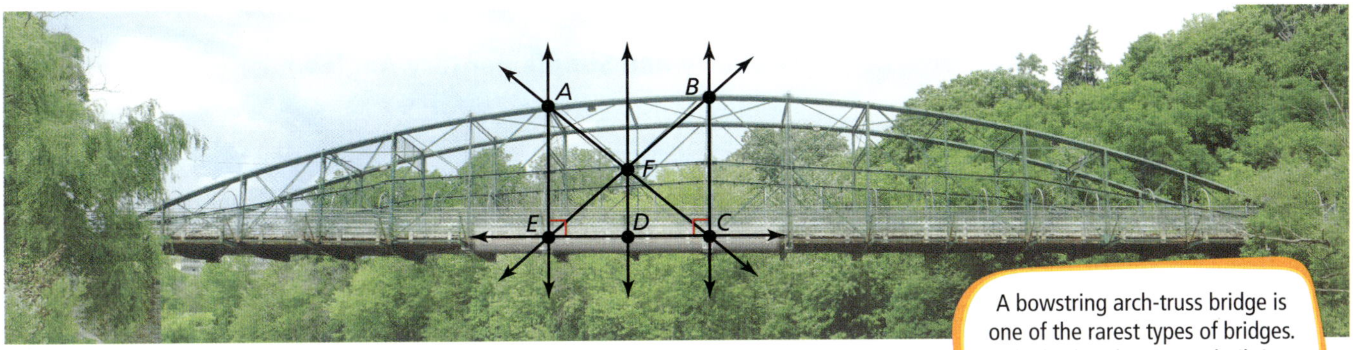

A bowstring arch-truss bridge is one of the rarest types of bridges. The bridge shown was built in 1875. Few bridges of this type remain today.

a. Identify a pair of the indicated angles. Do not use the same pair of angles twice.

  i. complementary angles

  ii. supplementary angles

  iii. adjacent angles

  iv. vertical angles

b. Suppose $\angle EDF$ and $\angle CDF$ are congruent. What can you conclude about $\overleftrightarrow{DF}$ and $\overleftrightarrow{EC}$? Explain.

c. What does it mean for two angles to form a *linear pair*? Identify a linear pair.

d. Research different bridge designs. Make sketches of the designs, and identify pairs of complementary, supplementary, adjacent, and vertical angles. Why are these types of angles used when building bridges?

4 MTR COMMUNICATE CLEARLY
What does it mean to be *nonadjacent*? Identify a pair of nonadjacent angles in the diagram.

Geometric Reasoning
preparing for MA.912.GR.1.1 Prove relationships and theorems about lines and angles. Solve mathematical and real-world problems involving postulates, relationships and theorems of lines and angles.

GO DIGITAL

Using Complementary and Supplementary Angles

Pairs of angles can have special relationships. The measurements of the angles or the positions of the angles in the pair determine the relationship.

Vocabulary

adjacent angles, *p. 48*
complementary angles, *p. 48*
supplementary angles, *p. 48*
linear pair, *p. 50*
vertical angles, *p. 50*

KEY IDEAS

Adjacent Angles

Adjacent angles are two angles that share a common vertex and side, but have no common interior points.

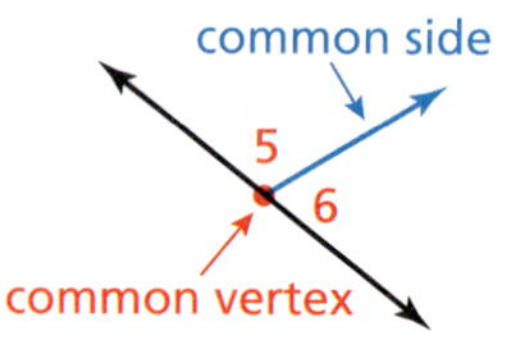

$\angle 5$ and $\angle 6$ are adjacent angles.

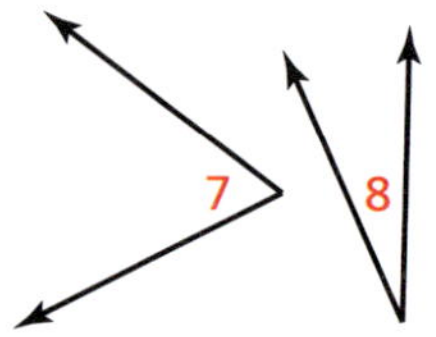

$\angle 7$ and $\angle 8$ are *nonadjacent* angles.

Complementary and Supplementary Angles

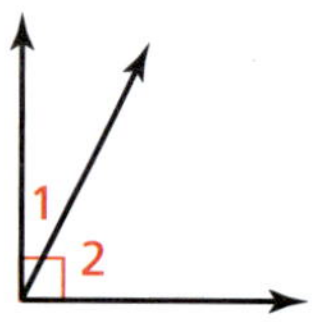

$\angle 1$ and $\angle 2$

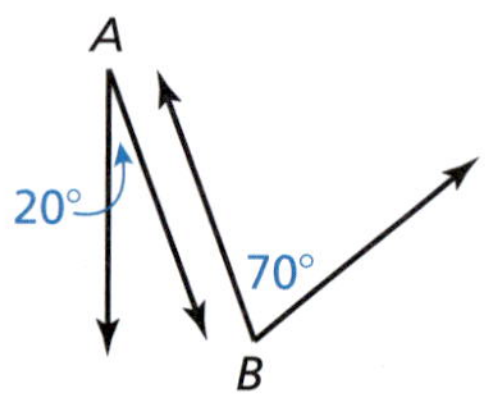

$\angle A$ and $\angle B$

complementary angles

Complementary angles are two positive angles whose measures have a sum of 90°. Each angle is the *complement* of the other.

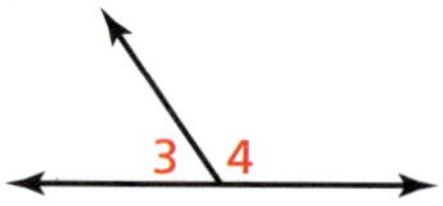

$\angle 3$ and $\angle 4$

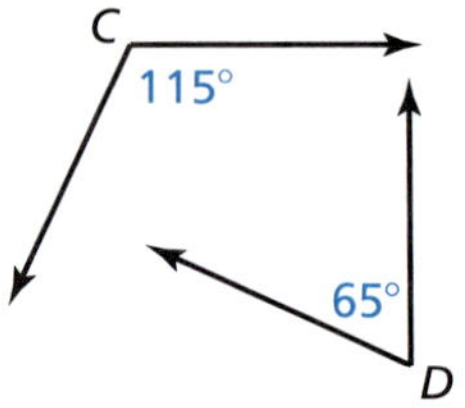

$\angle C$ and $\angle D$

supplementary angles

Supplementary angles are two positive angles whose measures have a sum of 180°. Each angle is the *supplement* of the other.

STUDY TIP

Complementary angles and supplementary angles can be adjacent or nonadjacent.

EXAMPLE 1 Identifying Pairs of Angles

WATCH

In the diagram, name a pair of adjacent angles, a pair of complementary angles, and a pair of supplementary angles.

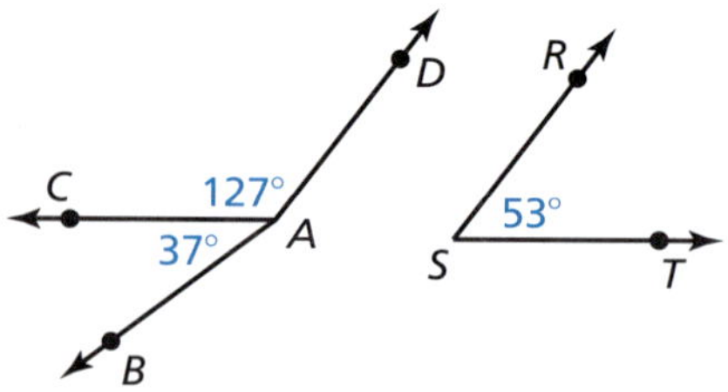

SOLUTION

$\angle BAC$ and $\angle CAD$ share a common vertex and side, but have no common interior points. So, they are adjacent angles.

Because $37° + 53° = 90°$, $\angle BAC$ and $\angle RST$ are complementary angles.

Because $127° + 53° = 180°$, $\angle CAD$ and $\angle RST$ are supplementary angles.

COMMON ERROR

In Example 1, $\angle DAC$ and $\angle DAB$ share a common vertex and a common side, but they also share common interior points. So, they are *not* adjacent angles.

GO DIGITAL

EXAMPLE 2 Finding Angle Measures

WATCH

a. $\angle 1$ is a complement of $\angle 2$, and $m\angle 1 = 62°$. Find $m\angle 2$.

b. $\angle 3$ is a supplement of $\angle 4$, and $m\angle 4 = 47°$. Find $m\angle 3$.

COMMON ERROR

Do not confuse angle names with angle measures.

SOLUTION

a. Draw a diagram with complementary adjacent angles to illustrate the relationship.

$$m\angle 2 = 90° - m\angle 1 = 90° - 62° = 28°$$

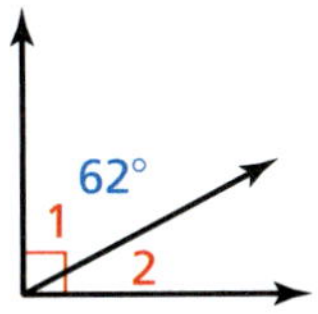

b. Draw a diagram with supplementary adjacent angles to illustrate the relationship.

$$m\angle 3 = 180° - m\angle 4 = 180° - 47° = 133°$$

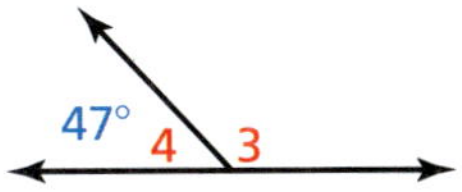

7 MTR

EXAMPLE 3 Modeling Real Life

WATCH

When viewed from the side, the frame of a ball-return net forms a pair of supplementary angles with the ground. Find $m\angle BCE$ and $m\angle ECD$.

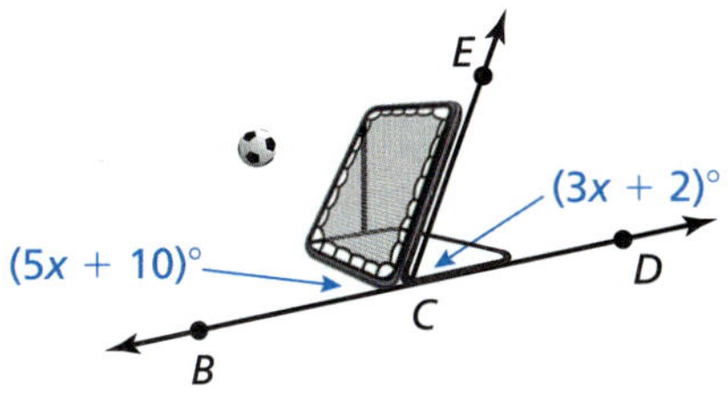

SOLUTION

Step 1 Use the fact that the sum of the measures of supplementary angles is 180°.

$m\angle BCE + m\angle ECD = 180°$	Write an equation.
$(5x + 10)° + (3x + 2)° = 180°$	Substitute angle measures.
$8x + 12 = 180$	Combine like terms.
$x = 21$	Solve for x.

Step 2 Evaluate the given expressions when $x = 21$.

$$m\angle BCE = (5x + 10)° = (5 \cdot 21 + 10)° = 115°$$

$$m\angle ECD = (3x + 2)° = (3 \cdot 21 + 2)° = 65°$$

▶ So, $m\angle BCE = 115°$ and $m\angle ECD = 65°$.

SELF-ASSESSMENT

1 I don't understand yet. 2 I can do it with help. 3 I can do it on my own. 4 I can teach someone else.

In Exercises 1 and 2, use the diagram.

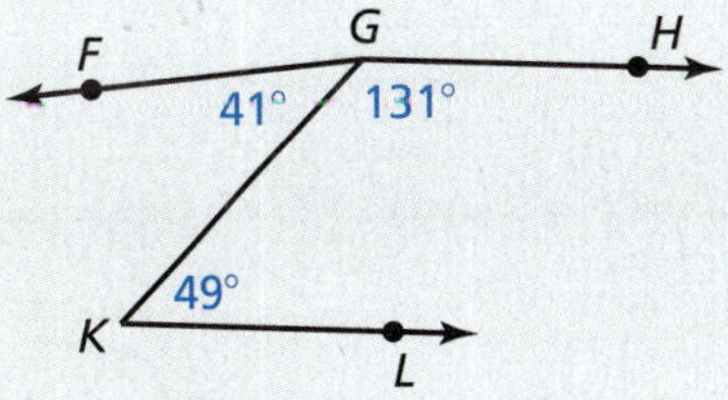

1. Name a pair of adjacent angles, a pair of complementary angles, and a pair of supplementary angles.

2. Are $\angle KGH$ and $\angle LKG$ adjacent angles? Are $\angle FGK$ and $\angle FGH$ adjacent angles? Explain.

3. $\angle 1$ is a complement of $\angle 2$, and $m\angle 2 = 5°$. Find $m\angle 1$.

4. $\angle 3$ is a supplement of $\angle 4$, and $m\angle 3 = 148°$. Find $m\angle 4$.

5. $\angle LMN$ and $\angle PQR$ are complementary angles. Find the measures of the angles when $m\angle LMN = (4x - 2)°$ and $m\angle PQR = (9x + 1)°$.

GO DIGITAL

Using Other Angle Pairs

KEY IDEAS

Linear Pairs and Vertical Angles

Two adjacent angles are a **linear pair** when their noncommon sides are opposite rays. The angles in a linear pair are supplementary angles.

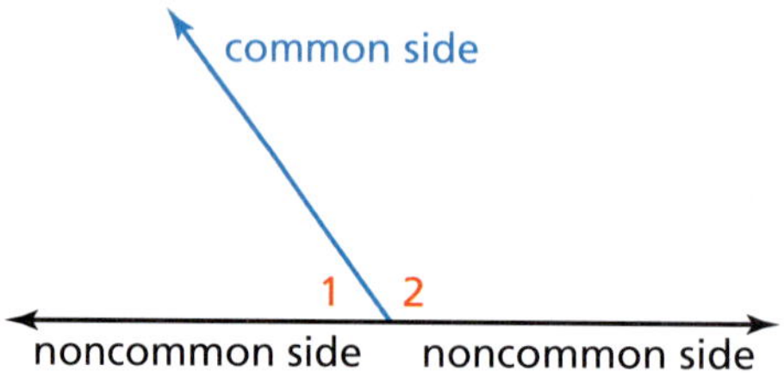

∠1 and ∠2 are a linear pair.

Two angles are **vertical angles** when their sides form two pairs of opposite rays.

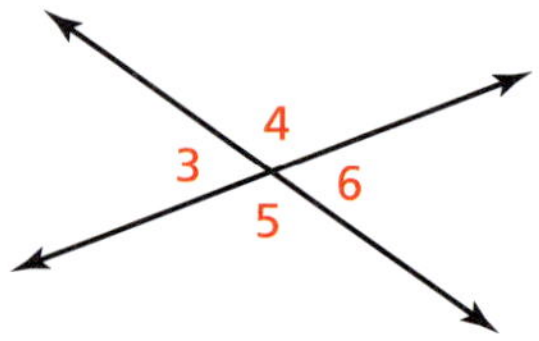

∠3 and ∠6 are vertical angles.
∠4 and ∠5 are vertical angles.

EXAMPLE 4 Identifying Angle Pairs

Identify all the linear pairs and all the vertical angles in the diagram.

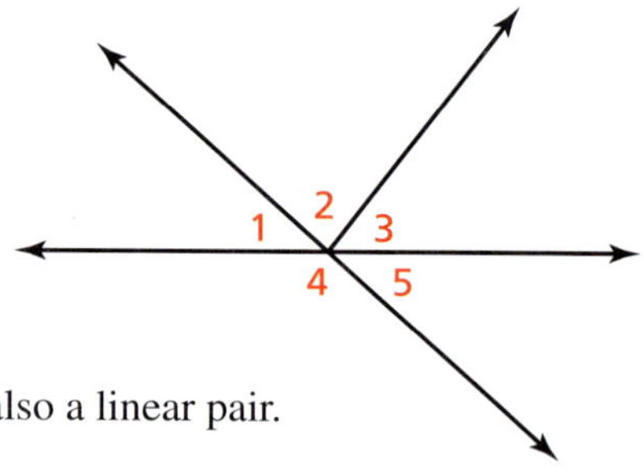

SOLUTION

To find linear pairs, look for adjacent angles whose noncommon sides are opposite rays.

▶ ∠1 and ∠4 are a linear pair. ∠4 and ∠5 are also a linear pair.

To find vertical angles, look for pairs of opposite rays.

▶ ∠1 and ∠5 are vertical angles.

> **COMMON ERROR**
> In Example 4, one side of ∠1 and one side of ∠3 are opposite rays. However, the angles are not a linear pair because they are *nonadjacent*.

EXAMPLE 5 Finding Angle Measures in a Linear Pair

Two angles form a linear pair. The measure of one angle is five times the measure of the other angle. Find the measure of each angle.

SOLUTION

Step 1 Draw a diagram. Let $x°$ be the measure of one angle. The measure of the other angle is $5x°$.

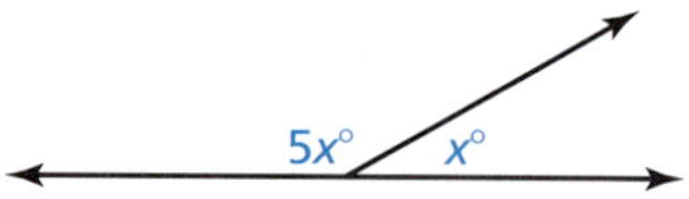

Step 2 Use the fact that the angles of a linear pair are supplementary to write an equation.

$$x° + 5x° = 180°$$ Write an equation.

$$6x = 180$$ Combine like terms.

$$x = 30$$ Divide each side by 6.

▶ The measures of the angles are 30° and 5(30°) = 150°.

GO DIGITAL

SELF-ASSESSMENT
1 I don't understand yet. **2** I can do it with help. **3** I can do it on my own. **4** I can teach someone else.

6. WRITING Explain the difference between adjacent angles and vertical angles.

4 MTR **7. WHICH ONE DOESN'T BELONG?** Which one does *not* belong with the other three? Explain your reasoning.

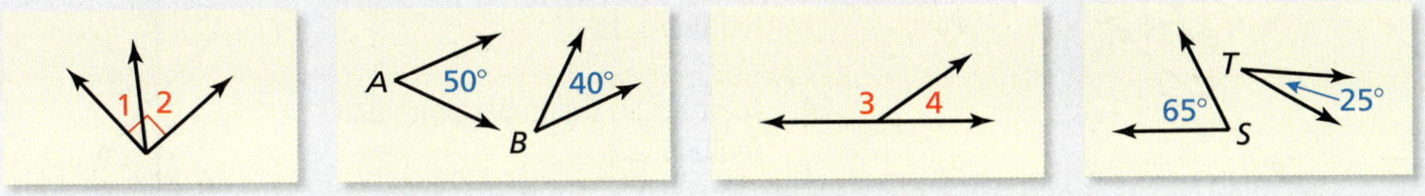

8. Do any of the numbered angles in the diagram form a linear pair? Which angles are vertical angles? Explain your reasoning.

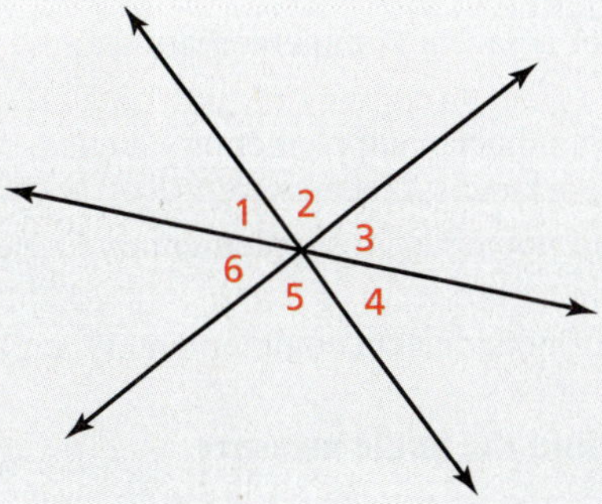

9. The measure of an angle is twice the measure of its complement. Find the measure of each angle.

10. Two angles form a linear pair. The measure of one angle is $1\frac{1}{2}$ times the measure of the other angle. Find the measure of each angle.

CONCEPT SUMMARY

Interpreting a Diagram

There are some things you can conclude from a diagram, and some you cannot. For example, here are some things you ***can* conclude** from the diagram.

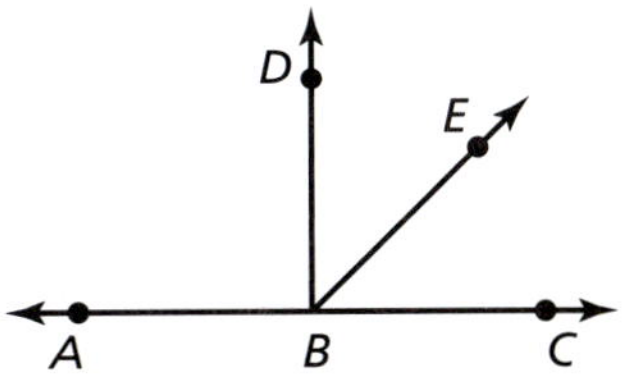

YOU CAN CONCLUDE

- All points shown are coplanar.
- Points A, B, and C are collinear, and B is between A and C.
- $\overleftrightarrow{AC}$, $\overrightarrow{BD}$, and $\overrightarrow{BE}$ intersect at point B.
- $\angle DBE$ and $\angle EBC$ are adjacent angles, and $\angle ABC$ is a straight angle.
- Point E lies in the interior of $\angle DBC$.

Here are some things you ***cannot* conclude** from the diagram above.

YOU CANNOT CONCLUDE

- $\overline{AB} \cong \overline{BC}$
- $\angle DBE \cong \angle EBC$
- $\angle ABD$ is a right angle.

To make such conclusions, the information in the diagram at the right must be given.

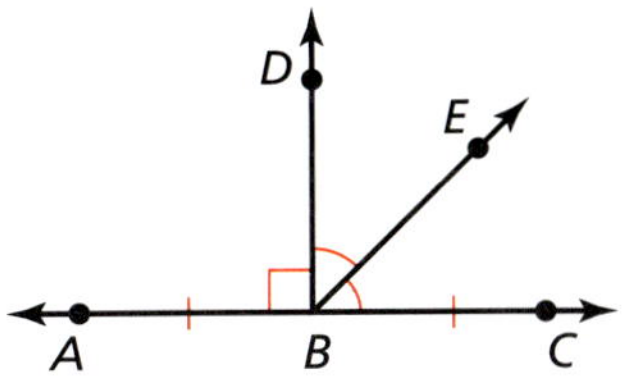

GO DIGITAL

1.6 Practice WITH CalcChat® AND CalcView®

In Exercises 1–4, use the diagrams. *(See Example 1.)*

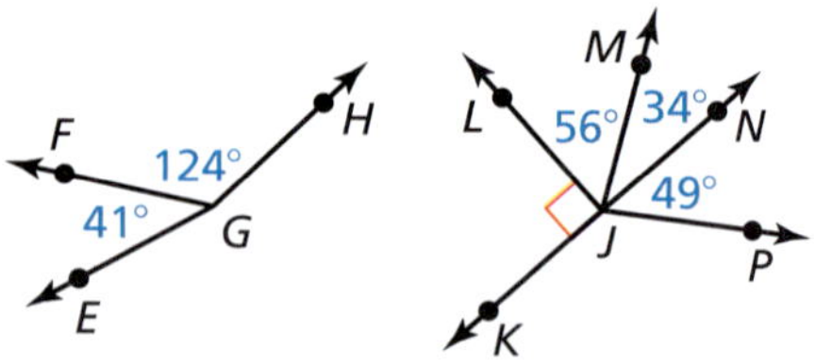

1. Name a pair of adjacent complementary angles.

2. Name a pair of adjacent supplementary angles.

3. Name a pair of nonadjacent supplementary angles.

4. Name a pair of nonadjacent complementary angles.

In Exercises 5–8, find the angle measure.
(See Example 2.)

5. $\angle 1$ is a complement of $\angle 2$, and $m\angle 1 = 23°$. Find $m\angle 2$.

6. $\angle 3$ is a complement of $\angle 4$, and $m\angle 3 = 46°$. Find $m\angle 4$.

7. $\angle 5$ is a supplement of $\angle 6$, and $m\angle 5 = 78°$. Find $m\angle 6$.

8. $\angle 7$ is a supplement of $\angle 8$, and $m\angle 7 = 109°$. Find $m\angle 8$.

In Exercises 9–12, find the measure of each angle.
(See Example 3.)

9.

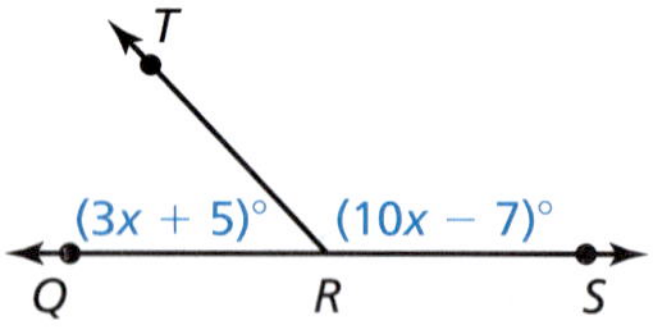

10.

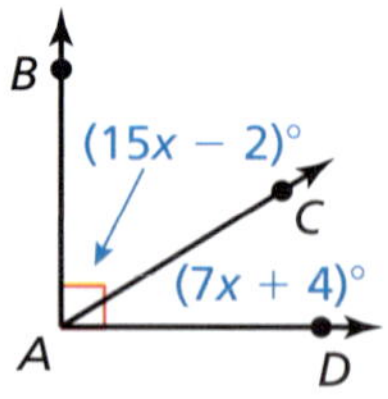

11. $\angle UVW$ and $\angle XYZ$ are complementary angles, $m\angle UVW = (x - 10)°$, and $m\angle XYZ = (4x - 10)°$.

12. $\angle EFG$ and $\angle LMN$ are supplementary angles, $m\angle EFG = (3x + 17)°$, and $m\angle LMN = \left(\frac{1}{2}x - 5\right)°$.

In Exercises 13–16, use the diagram. *(See Example 4.)*

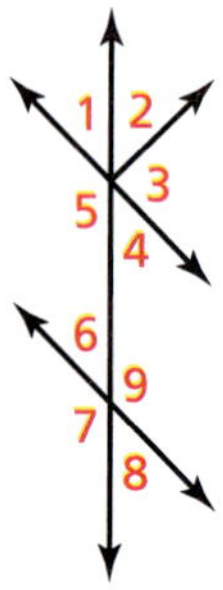

13. Identify all the linear pairs that include $\angle 1$.

14. Identify all the linear pairs that include $\angle 7$.

15. Are $\angle 6$ and $\angle 8$ vertical angles? Explain your reasoning.

16. Are $\angle 2$ and $\angle 5$ vertical angles? Explain your reasoning.

4 MTR **ERROR ANALYSIS In Exercises 17 and 18, describe and correct the error in identifying pairs of angles in the diagram.**

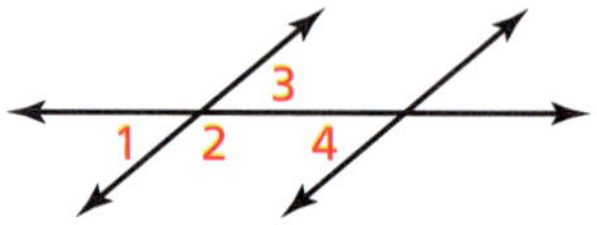

17. ✗ $\angle 2$ and $\angle 4$ are adjacent angles.

18. ✗ $\angle 1$ and $\angle 3$ are a linear pair.

In Exercises 19–24, find the measure of each angle.
(See Example 5.)

19. Two angles form a linear pair. The measure of one angle is twice the measure of the other angle.

20. Two angles form a linear pair. The measure of one angle is $\frac{1}{3}$ the measure of the other angle.

21. The measure of an angle is $\frac{1}{4}$ the measure of its complement.

22. The measure of an angle is nine times the measure of its complement.

23. The ratio of the measure of an angle to the measure of its complement is 4 : 5.

24. The ratio of the measure of an angle to the measure of its complement is 2 : 7.

GO DIGITAL

MODELING REAL LIFE In Exercises 25 and 26, the picture shows the Alamillo Bridge in Seville, Spain. In the picture, $m\angle 1 = 58°$ and $m\angle 2 = 24°$.

25. Find the measure of the supplement of $\angle 1$.

26. Find the measure of the supplement of $\angle 2$.

27. **MODELING REAL LIFE** The foul lines of a baseball field intersect at home plate to form a right angle. A batter hits a fair ball such that the path of the baseball forms an angle of 27° with the third-base foul line. What is the measure of the angle between the first-base foul line and the path of the baseball?

28. **COLLEGE PREP** The arm of a crossing gate moves 42° from a vertical position. How many more degrees does the arm have to move so that it is horizontal?

Ⓐ 42° Ⓒ 48°
Ⓑ 138° Ⓓ 90°

29. **CONSTRUCTION** Construct a linear pair where one angle measure is 115°.

30. **CONSTRUCTION** Construct a pair of adjacent angles that have angle measures of 45° and 97°.

CONNECTING CONCEPTS In Exercises 31–34, write and solve an algebraic equation to find the measure of each angle described.

31. The measure of an angle is 6° less than the measure of its complement.

32. The measure of an angle is 12° more than twice the measure of its complement.

GO DIGITAL

33. The measure of an angle is 3° more than $\frac{1}{2}$ the measure of its supplement.

34. Two angles form a linear pair. The measure of one angle is 15° less than $\frac{2}{3}$ the measure of the other angle.

35. **COLLEGE PREP** $m\angle U = 2x°$, and $m\angle V = 4(m\angle U)$. Which value of x makes $\angle U$ and $\angle V$ complements of each other?

Ⓐ 25 Ⓑ 9 Ⓒ 36 Ⓓ 18

36. **HOW DO YOU SEE IT?**
Determine whether you can conclude each statement from the diagram. Explain your reasoning.

a. $\overline{CA} \cong \overline{AF}$

b. Points C, A, and F are collinear.

c. $\angle CAD \cong \angle EAF$

d. $\overline{BA} \cong \overline{AE}$

e. $\angle DAE$ is a right angle.

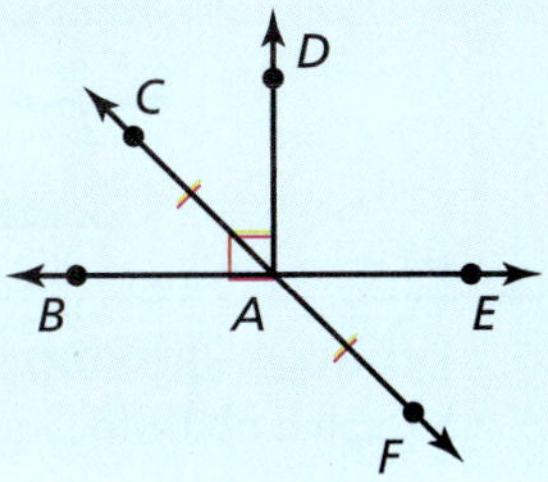

DISCUSS MATHEMATICAL THINKING In Exercises 37–42, tell whether the statement is *always*, *sometimes*, or *never* true. Explain your reasoning.

37. Complementary angles are adjacent.

38. Angles in a linear pair are supplements of each other.

39. Vertical angles are adjacent.

40. Vertical angles are supplements of each other.

41. If an angle is acute, then its complement is greater than its supplement.

42. If two complementary angles are congruent, then the measure of each angle is 45°.

43. **CONNECTING CONCEPTS** Use the diagram. You write the measures of $\angle TWU$, $\angle TWX$, $\angle UWV$, and $\angle VWX$ on separate pieces of paper and place the pieces of paper in a box. You choose two pieces of paper out of the box at random. Find the probability that the angle measures you choose represent supplementary angles. Explain.

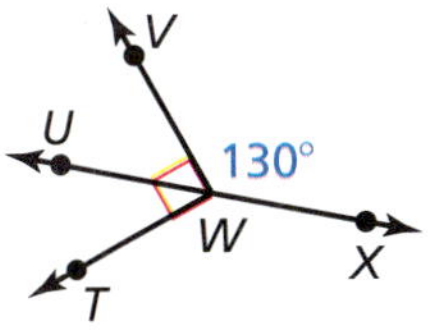

44. REASONING $\angle KJL$ and $\angle LJM$ are complements, and $\angle MJN$ and $\angle LJM$ are complements. Can you show that $\angle KJL \cong \angle MJN$? Explain your reasoning.

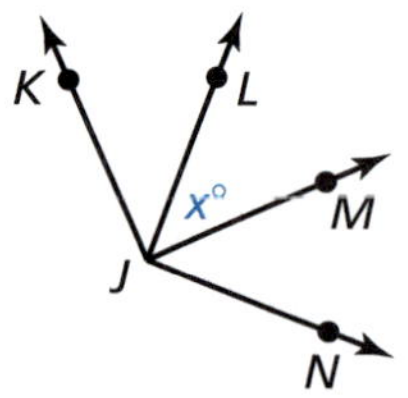

45. MAKING AN ARGUMENT Light from a flashlight strikes a mirror and is reflected so that the angle of reflection is congruent to the angle of incidence. Your classmate claims that $\angle QPR$ is congruent to $\angle TPU$ regardless of the measure of $\angle RPS$. Is your classmate correct? Explain your reasoning.

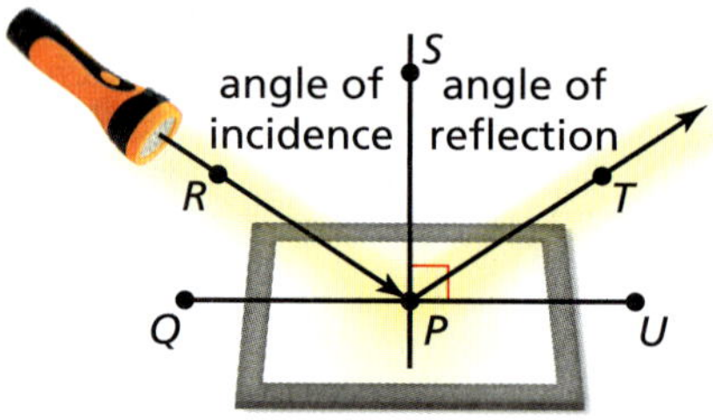

46. THOUGHT PROVOKING
Sketch a real-life situation that shows supplementary, complementary, and vertical angles.

47. STRUCTURE Use the diagram.

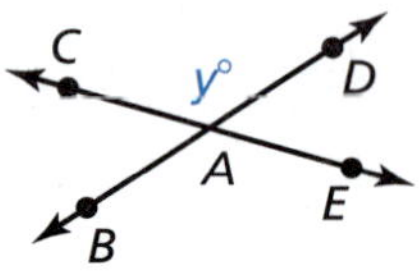

a. Write expressions for the measures of $\angle BAE$, $\angle DAE$, and $\angle CAB$.

b. What do you notice about the measures of vertical angles? Explain your reasoning.

48. CONNECTING CONCEPTS Let $m\angle 1 = x°$, $m\angle 2 = y_1°$, and $m\angle 3 = y_2°$. $\angle 2$ is the complement of $\angle 1$, and $\angle 3$ is the supplement of $\angle 1$.

a. Write equations for y_1 as a function of x and y_2 as a function of x. What is the domain of each function? Explain.

b. Graph each function and find its range.

49. CONNECTING CONCEPTS The sum of the measures of two complementary angles is 74° greater than the difference of their measures. Find the measure of each angle. Explain how you found the angle measures.

REVIEW & REFRESH

In Exercises 50 and 51, find the area of the polygon with the given vertices.

50. $K(-3, 4)$, $L(1, 4)$, $M(-4, -2)$, $N(0, -2)$

51. $X(-1, 2)$, $Y(-1, -3)$, $Z(4, -3)$

52. The midpoint of $\overline{JK}$ is $M(0, 1)$. One endpoint is $J(-6, 3)$. Find the coordinates of endpoint K.

53. Identify the segment bisector of $\overline{RS}$. Then find RS.

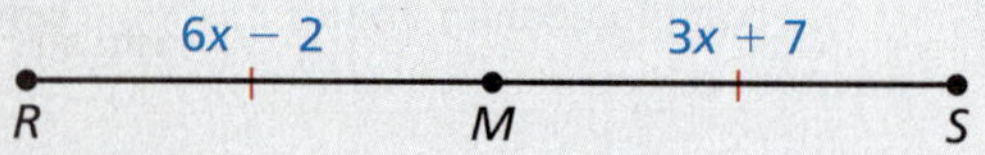

In Exercises 54–57, solve the equation. Graph the solution(s), if possible.

54. $|t + 5| = 3$

55. $\left|\frac{1}{4}d - 1\right| + 2 = 5$

56. $-4|7 + 2n| = 12$

57. $|-1.6q| = 7.2$

In Exercises 58 and 59, find the product.

58. $(8x^2 - 16 + 3x^3)(-4x^5)$

59. $(4s - 3)(7s + 5)$

60. MODELING REAL LIFE The total cost (in dollars) of renting a cabin for n days is represented by the function $f(n) = 75n + 200$. The daily rate is doubled. The new total cost is represented by the function $g(n) = f(2n)$. Describe the transformation from the graph of f to the graph of g.

In Exercises 61 and 62, find the slope and the y-intercept of the graph of the linear equation.

61. $y = -5x + 2$

62. $y + 7 = \frac{3}{2}x$

63. Given that $m\angle EFG = 126°$, find $m\angle EFH$ and $m\angle HFG$.

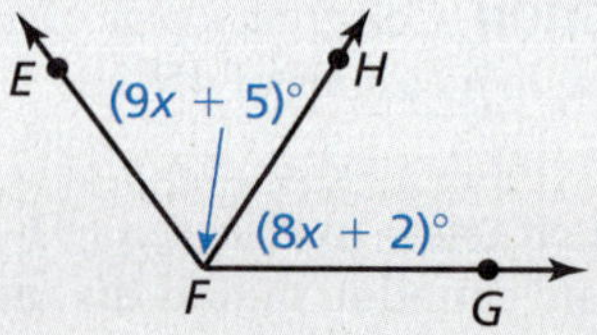

In Exercises 64 and 65, find the angle measure.

64. $\angle 1$ is a supplement of $\angle 2$, and $m\angle 1 = 57°$. Find $m\angle 2$.

65. $\angle 3$ is a complement of $\angle 4$, and $m\angle 4 = 34°$. Find $m\angle 3$.

1 Chapter Review WITH CalcChat®

Chapter Learning Target: Understand basics of geometry.

Chapter Success Criteria:
- ◆ I can name points, lines, and planes.
- ◆ I can measure segments and angles.
- ■ I can use formulas in the coordinate plane.
- ■ I can construct segments and angles.

◆ Surface
■ Deep

SELF-ASSESSMENT 1 I don't understand yet. 2 I can do it with help. 3 I can do it on my own. 4 I can teach someone else.

1.1 Points, Lines, and Planes (pp. 3–10) WATCH

Learning Target: Use defined terms and undefined terms.

Vocabulary AZ VOCAB
undefined terms
point
line
plane
collinear points
coplanar points
defined terms
line segment, or segment
endpoints
ray
opposite rays
intersection

Use the diagram.

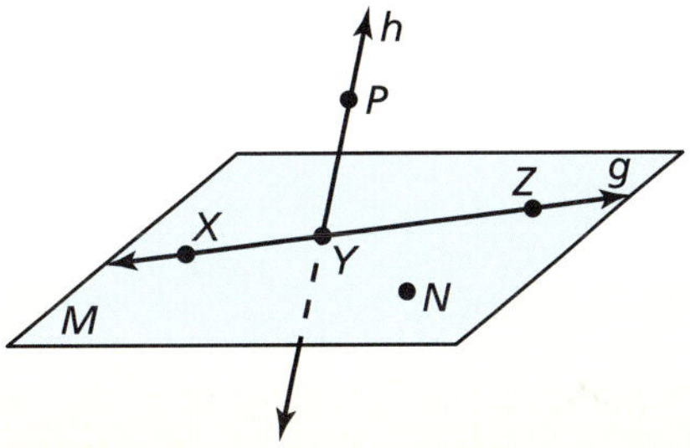

1. Give another name for plane M.
2. Name a line in plane M.
3. Name a line intersecting plane M.
4. Name two rays.
5. Name a pair of opposite rays.
6. Name a point not in plane M.
7. Is it possible for the intersection of two planes to be a segment? a line? a ray? Sketch the possible situations.

1.2 Measuring and Constructing Segments (pp. 11–18) WATCH

Learning Target: Measure and construct line segments.

Vocabulary AZ VOCAB
postulate
axiom
coordinate
distance between two points
construction
congruent segments
between

Find XZ.

8. 

9. 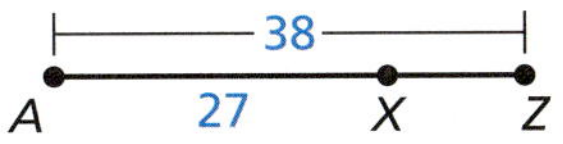

10. Plot $A(8, -4)$, $B(3, -4)$, $C(7, 1)$, and $D(7, -3)$ in a coordinate plane. Then determine whether $\overline{AB}$ and $\overline{CD}$ are congruent.

11. You pass by school and the library on a walk from home to the bookstore, as shown below. How far from school is the library? How long does it take you to walk from home to the bookstore at an average speed of 68 meters per minute?

GO DIGITAL

1.3 Using Midpoint and Distance Formulas *(pp. 19–28)* WATCH

Learning Target: Find midpoints and lengths of segments.

Vocabulary AZ VOCAB
midpoint
segment bisector
weighted average

The endpoints of $\overline{ST}$ are given. Find the coordinates of the midpoint M. Then find the length of $\overline{ST}$.

12. $S(-2, 4)$ and $T(3, 9)$

13. $S(6, -3)$ and $T(7, -2)$

14. The midpoint of $\overline{JK}$ is $M(6, 3)$. One endpoint is $J(14, 9)$. Find the coordinates of endpoint K.

15. Point M is the midpoint of $\overline{AB}$, where $AM = 3x + 8$ and $MB = 6x - 4$. Find AB.

16. The coordinate 3 has a weight of 2, and the coordinate 9 has a weight of 1. Find the weighted average.

17. The coordinate 4 has a weight of 1, the coordinate 5 has a weight of 3, and the coordinate 11 has a weight of 1. Find the weighted average.

18. The coordinate plane shows distances (in feet) on a baseball infield. The pitcher's plate is about 3 feet closer to home plate than the midpoint between home plate and second base is to home plate. Estimate the distance between home plate and the pitcher's plate. Explain how you found your answer.

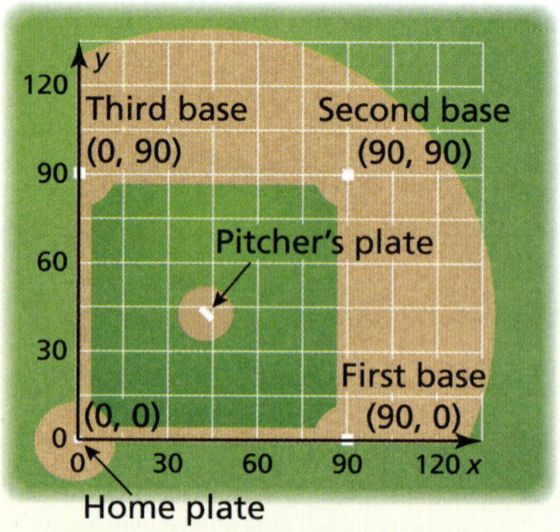

1.4 Perimeter and Area in the Coordinate Plane *(pp. 29–36)* WATCH

Learning Target: Find perimeters and areas of polygons in the coordinate plane.

Classify the polygon by the number of sides. Tell whether it is *convex* or *concave*.

19.

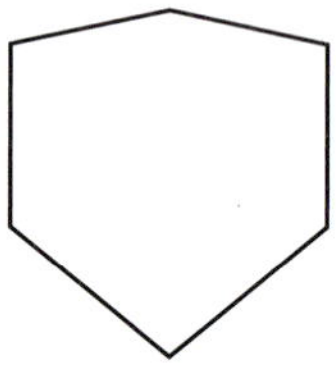

20.

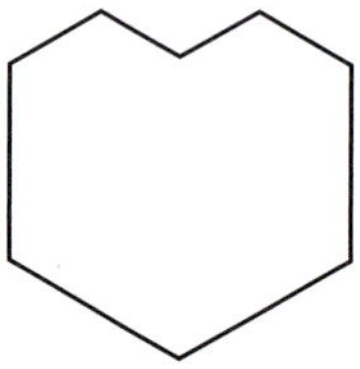

Find the perimeter and area of the polygon with the given vertices.

21. $W(5, 6)$, $X(5, -1)$, $Y(2, -1)$, $Z(2, 6)$

22. $E(6, -2)$, $F(6, 5)$, $G(-1, 5)$

23. Two polygons are shown. Compare the area and perimeter of the concave polygon with the area and perimeter of the convex polygon.

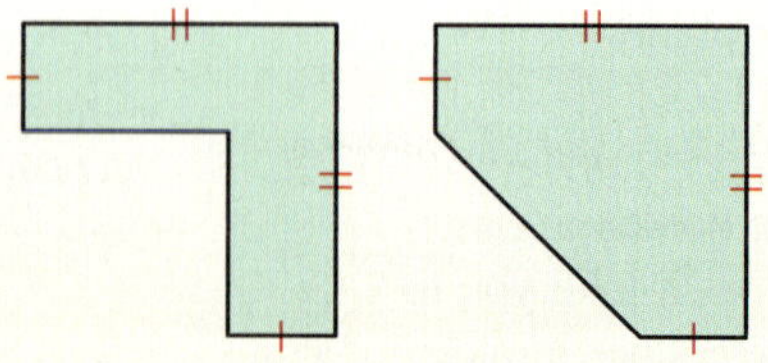

24. Find the perimeter of quadrilateral $RSTU$ in the coordinate plane at the right.

25. Find the area of quadrilateral $RSTU$.

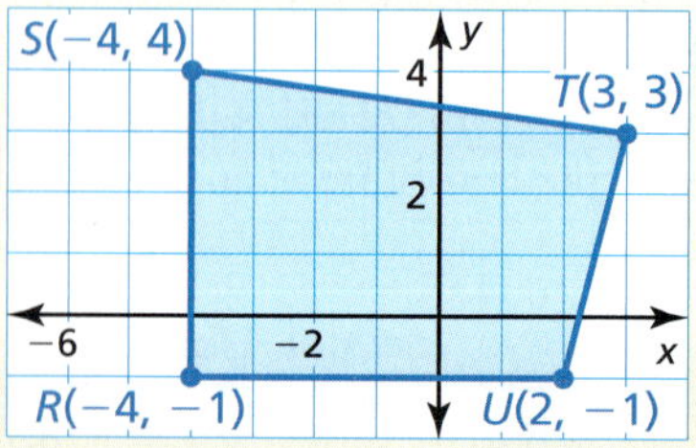

GO DIGITAL

1.5 Measuring and Constructing Angles (pp. 37–46) WATCH

Learning Target: Measure, construct, and describe angles.

Find $m\angle ABD$ and $m\angle CBD$.

26. $m\angle ABC = 77°$

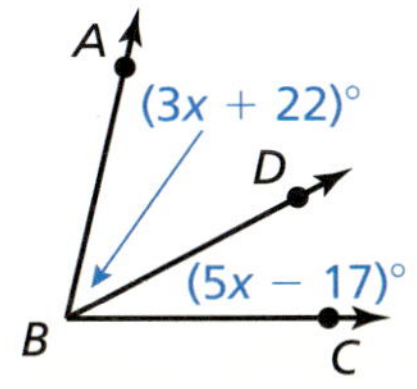

27. $m\angle ABC = 111°$

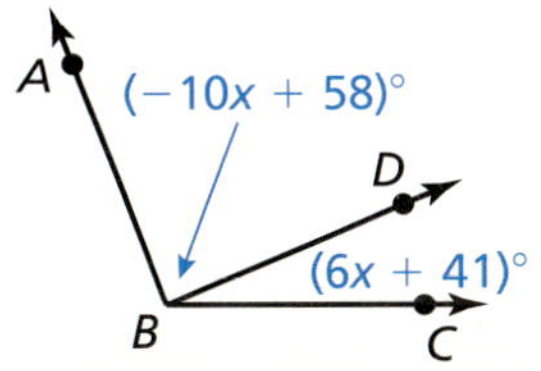

28. Find the measure of the angle using a protractor.

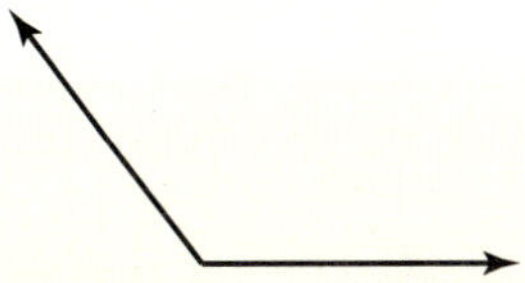

29. Given that P is in the interior of $\angle ABC$, $m\angle PBC = 30°$, $\angle DBC$ is a right angle, and $m\angle ABD = 26°$, what are the possible measures of $\angle ABP$?

Vocabulary (AZ VOCAB)

angle
vertex
sides of an angle
interior of an angle
exterior of an angle
measure of an angle
acute angle
right angle
obtuse angle
straight angle
congruent angles
angle bisector

1.6 Describing Pairs of Angles (pp. 47–54) WATCH

Learning Target: Identify and use pairs of angles.

$\angle 1$ and $\angle 2$ are complementary angles. Given $m\angle 1$, find $m\angle 2$.

30. $m\angle 1 = 12°$

31. $m\angle 1 = 83°$

$\angle 3$ and $\angle 4$ are supplementary angles. Given $m\angle 3$, find $m\angle 4$.

32. $m\angle 3 = 116°$

33. $m\angle 3 = 56°$

34. Construct a linear pair, where one angle measure is 35°. Label the measures of both angles.

35. The measure of an angle is 4 times the measure of its supplement. Find the measures of both angles.

36. Find the measures of $\angle ADB$ and $\angle BDC$ formed between the escalators shown at the right.

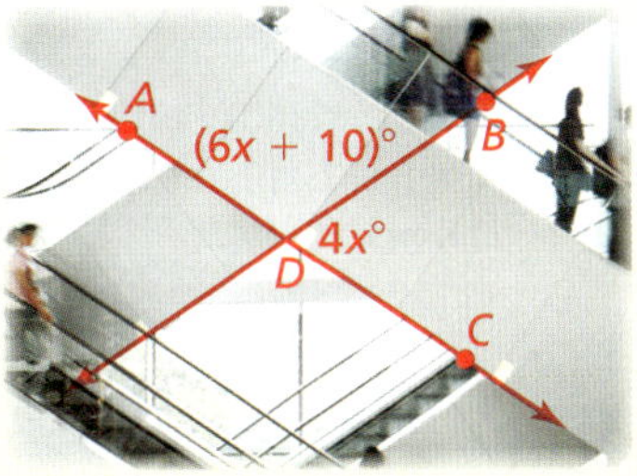

Vocabulary (AZ VOCAB)

adjacent angles
complementary angles
supplementary angles
linear pair
vertical angles

Mathematical Thinking and Reasoning

MTR 1 **ACTIVELY PARTICIPATE IN EFFORTFUL LEARNING BOTH INDIVIDUALLY AND COLLECTIVELY**

Mathematicians who actively participate in effortful learning both individually and with others ask questions that will help with solving tasks.

1. In Exercise 48 on page 27, what steps did you take before solving the problem?
2. A garden is the shape of a semicircle and a rectangle as shown. What information do you need to find the area of the garden?

GO DIGITAL

1 Practice Test WITH CalcChat®

Find QS. Explain how you found your answer.

1. 

2. 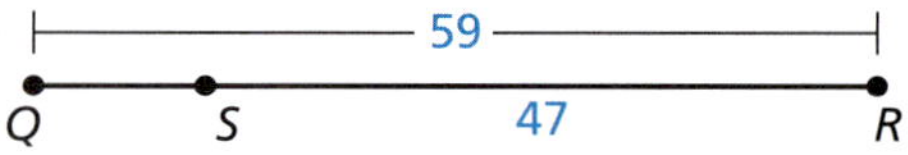

3. The endpoints of $\overline{AB}$ are $A(-4, -8)$ and $B(-1, 4)$. Find the coordinates of the midpoint M. Then find the length of $\overline{AB}$.

4. The midpoint of $\overline{EF}$ is $M(1, -1)$. One endpoint is $E(-3, 2)$. Find the coordinates of endpoint F.

Use the diagram to decide whether the statement is true or false.

5. Points A, R, and B are collinear.

6. $\overleftrightarrow{BW}$ and $\overleftrightarrow{AT}$ are lines.

7. $\overrightarrow{RB}$ and $\overrightarrow{RT}$ are opposite rays.

8. Plane D could also be named plane ART.

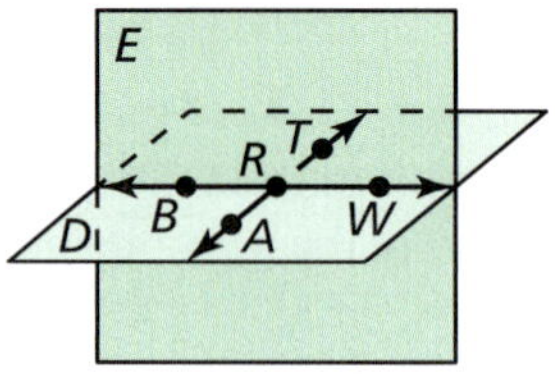

Find the perimeter and area of the polygon with the given vertices.

9. $P(-3, 4)$, $Q(1, 4)$, $R(3, -2)$, $S(-1, -2)$

10. $J(-1, 3)$, $K(5, 3)$, $L(2, -2)$

11. The coordinate 6 has a weight of 2, the coordinate 9 has a weight of 2, and the coordinate 15 has a weight of 1. Find the weighted average.

12. $\overrightarrow{BX}$ bisects $\angle ABC$ to form two congruent acute angles. What type of angle can $\angle ABC$ be?

13. Given $\angle RST$, U is in the interior of the angle, $m\angle TSU$ is 6° less than 5 times $m\angle RSU$, and $m\angle RST = 48°$. Write and solve a system of equations to find $m\angle RSU$ and $m\angle TSU$.

14. In the diagram at the right, identify all supplementary and complementary angles. Explain. Then find $m\angle DFE$, $m\angle BFC$, and $m\angle BFE$.

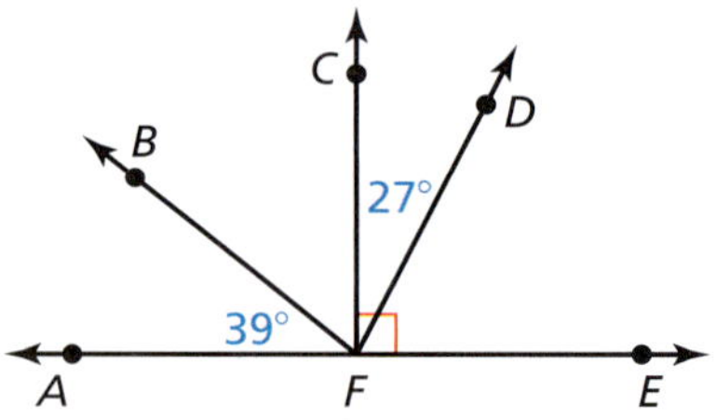

15. Sketch a figure that contains a plane and two lines that all intersect at one point.

16. Your community decides to install a rectangular swimming pool in a park. There is a 48-foot by 81-foot rectangular area available. There must be at least a 3-foot border around the pool. Draw a diagram of this situation in a coordinate plane. Based on the constraints, find the perimeter and the area of the largest swimming pool possible.

17. Four wildland firefighting crews are approaching a small, growing fire, as shown. The crews are at the positions labeled A, B, C, and D. Which position is closest to the fire?

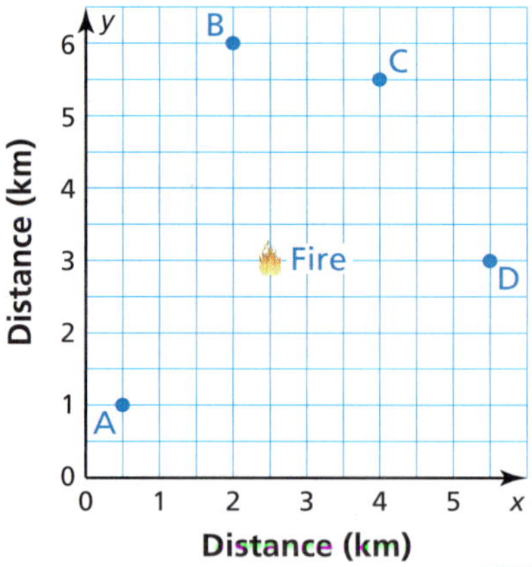

GO DIGITAL

1 Performance Task
Eye of the Tiger

Conflict between humans and wild animals is a major threat to both humans and animals.

Causes of human-tiger conflict include the following:

- **HABITAT AVAILABILITY**
  Deforestation, habitat degradation, and increasing human populations force tigers and humans into closer proximity.
- **SOCIOECONOMIC FACTORS**
  Attitudes, perceptions, beliefs, education, and economic situations affect views on how to interact with tigers.
- **WILD PREY AVAILABILITY**
  Prey species are diminished by overexploitation and competition with livestock. A low density of wild prey increases the chance of human-tiger conflict.
- **IMPROPER LIVESTOCK MANAGEMENT**
  Herding practices and locations of grazing pastures can leave livestock susceptible to attacks.
- **HUMAN BEHAVIOR**
  Baiting or hunting tigers, and sleeping in exposed locations increase the risk of an attack.

Estimated Wild Tiger Populations

The wild tiger population has decreased about 96% since 1900.

GO DIGITAL

WILDLIFE RESERVATION

You propose a new wildlife reservation in an attempt to limit human-tiger conflict. Use points and line segments to sketch the outline of your reservation in a coordinate plane. Name each point and line segment in your sketch.

A local government requires several details before considering your proposal. Provide the following information:

- the length of each side of the reservation
- the area of the reservation
- the measures of the angles formed by the sides of the reservation
- the coordinates of at least three gates, located at midpoints of the sides of the reservation

1 Test Prep WITH CalcChat®
Cumulative Practice

Tutorial videos are available for each exercise.

1. Which inequality is represented by the graph?

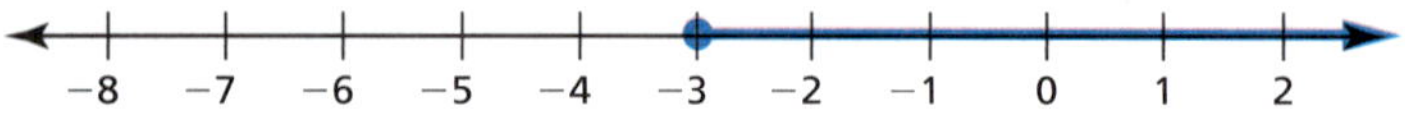

Ⓐ $x \leq -3$

Ⓑ $x < -3$

Ⓒ $x \geq -3$

Ⓓ $x > -3$

2. Order the terms so that each consecutive term builds off the previous term.

plane	segment	line	point	ray

3. The endpoints of a line segment are $(-6, 13)$ and $(11, 5)$. Which of the following is the midpoint and the length of the segment?

Ⓐ $\left(\frac{5}{2}, 4\right)$; $\sqrt{353}$ units

Ⓑ $\left(\frac{5}{2}, 9\right)$; $\sqrt{353}$ units

Ⓒ $\left(\frac{5}{2}, 4\right)$; $\sqrt{89}$ units

Ⓓ $\left(\frac{5}{2}, 9\right)$; $\sqrt{89}$ units

4. Solve the system of linear equations.

$$7x + 4y = 0$$

$$y = \frac{1}{4}x - 8$$

5. Which of the following is the perimeter and area of the figure shown?

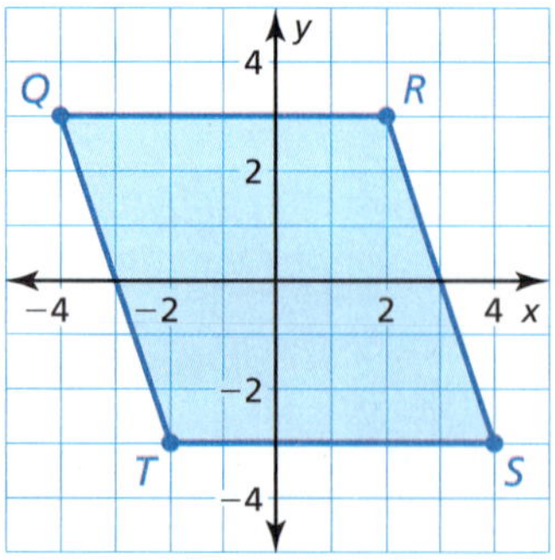

Ⓐ $6 + 2\sqrt{10}$ units; 36 square units

Ⓑ $6 + 2\sqrt{10}$ units; $12\sqrt{10}$ square units

Ⓒ $12 + 4\sqrt{10}$ units; 36 square units

Ⓓ $12 + 4\sqrt{10}$ units; $12\sqrt{10}$ square units

GO DIGITAL

6. Three roads come to an intersection point that the people in your town call Five Corners, as shown in the figure.

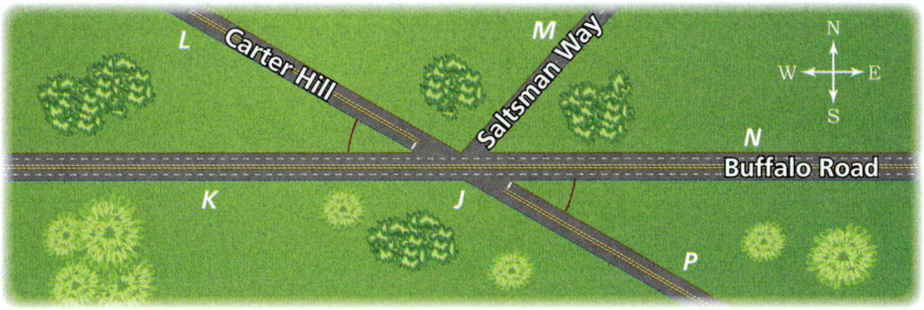

Answer parts (a) through (c) using the angles given below.

a. You are traveling west on Buffalo Road and turn left onto Carter Hill. What is the name of the angle through which you turn?

b. Identify all the vertical angles.

c. Identify all the linear pairs.

$\angle KJL$	$\angle KJM$	$\angle KJN$	$\angle KJP$	$\angle LJM$
$\angle LJN$	$\angle LJP$	$\angle MJN$	$\angle MJP$	$\angle NJP$

7. Use the steps in the construction to explain how you know that $\overrightarrow{PS}$ is the angle bisector of $\angle RPQ$.

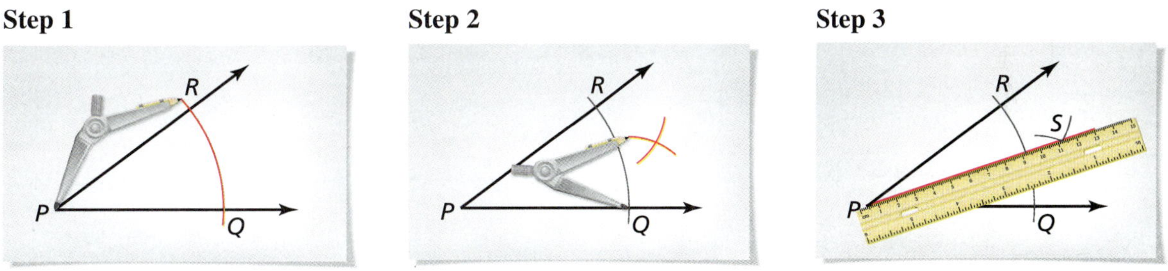

8. Which factorization can be used to find the zeros of the function $f(x) = 15x^2 + 14x - 8$?

Ⓐ $f(x) = x(15x + 14) - 8$

Ⓑ $f(x) = 15x^2 + 2(7x - 4)$

Ⓒ $f(x) = (5x + 8)(3x - 1)$

Ⓓ $f(x) = (5x - 2)(3x + 4)$

9. Plot the points $W(-1, 1)$, $X(5, 1)$, $Y(5, -2)$, and $Z(-1, -2)$ in a coordinate plane. What type of polygon do the points form? Your friend claims that you could use this polygon to represent a basketball court with an area of 4050 square feet and a perimeter of 270 feet. Do you support your friend's claim? Explain.

GO DIGITAL

2 Reasoning and Proofs

NATIONAL GEOGRAPHIC EXPLORER

Caroline Quanbeck

WATCH

Caroline Quanbeck is a geologist whose research aims to reconstruct the evolution of sea level rise in Western Australia during the Last Interglacial, a warm period 125,000 years ago when temperatures were similar to today but global average sea level was 20 to 30 feet higher.

- What causes rising sea levels?
- Explain some of the effects of rising sea levels on coastal communities.

STEM

Greenhouse gases are major contributors to climate change. In the Performance Task, you will see how greenhouse gases warm the planet. Then you will research some of the effects of climate change and write conditional statements based on your research.

GO DIGITAL

Climate Change

Preparing for Chapter 2

Chapter Learning Target: Understand reasoning and proofs.

Chapter Success Criteria:
- ◆ I can use inductive and deductive reasoning.
- ◆ I can justify steps using algebraic reasoning.
- ■ I can explain postulates using diagrams.
- ■ I can prove geometric relationships.

◆ Surface
■ Deep

AZ VOCAB

Chapter Vocabulary

Work with a partner. Discuss each of the vocabulary terms.

conditional statement
hypothesis
conclusion
equivalent statements
biconditional statement
conjecture
inductive reasoning
proof
theorem

Mathematical Thinking and Reasoning

4 MTR

ENGAGE IN DISCUSSIONS THAT REFLECT ON THE MATHEMATICAL THINKING OF SELF AND OTHERS

Mathematicians who engage in discussions that reflect on the mathematical thinking of self and others recognize errors and suggest how to correctly solve tasks.

Work with a partner. When you use *deductive reasoning*, you start with two or more true statements and *deduce* or *infer* the truth of another statement. Here is an example.

Premise: If a vehicle uses a gasoline-powered engine, then it produces carbon dioxide.

Premise: Your school bus uses a gasoline-powered engine.

Conclusion: Therefore, your school bus produces carbon dioxide.

1. This pattern for deductive reasoning is called a *syllogism*. For each syllogism below, determine whether the conclusion is valid. Explain your reasoning.

a. Polar bears reside in the Arctic.
You see a polar bear.
Therefore, you must be in the Arctic.

b. If the ice caps melt, then sea levels will rise.
If sea levels rise, then coasts will flood.
Therefore, if the ice caps melt, coasts will flood.

2. Use deductive reasoning to write another syllogism. Exchange syllogisms with your partner and determine whether your partner's conclusion is valid. Explain your reasoning.

GO DIGITAL

2 Prepare WITH CalcChat®

Solving Equations with Variables on Both Sides

WATCH

Example 1 **Solve $2(2x - 5) = 3x + 4$. Check your solution.**

$2(2x - 5) = 3x + 4$	Write the equation.
$4x - 10 = 3x + 4$	Distributive Property
$\underline{+10} \quad \underline{+10}$	Addition Property of Equality.
$4x = 3x + 14$	Simplify.
$\underline{-3x} \quad \underline{-3x}$	Subtraction Property of Equality
$x = 14$	Simplify.

▶ The solution is $x = 14$.

Solve the equation. Check your solution.

1. $5x + 9 = 2x$

2. $7x - 3 = 5 + 3x$

3. $\frac{1}{6}x + 4 = \frac{1}{2}x$

4. $\frac{1}{5}(x + 6) = 3x$

5. $4(3x - 2) = 2(x - 1)$

6. $3.2x = 3.6(2x + 3)$

Rewriting Literal Equations

WATCH

Example 2 **Solve the literal equation $3x + 6y = 24$ for y.**

$3x + 6y = 24$	Write the equation.
$3x - 3x + 6y = 24 - 3x$	Subtraction Property of Equality
$6y = 24 - 3x$	Simplify.
$\frac{6y}{6} = \frac{24 - 3x}{6}$	Division Property of Equality
$y = 4 - \frac{1}{2}x$	Simplify.

▶ The rewritten literal equation is $y = 4 - \frac{1}{2}x$.

Solve the literal equation for x.

7. $2y - 2x = 10$

8. $20y + 5x = 15$

9. $4y - 5 = 4x + 7$

10. $y = 8x - x$

11. $y = 4x + zx + 6$

12. $z = 2x + 6xy$

4 MTR **13.** **MAKING AN ARGUMENT** Determine whether the statement below is *true* or *false*. Explain.

If two equations are equivalent, then you can use properties of equality to obtain one equation from the other.

GO DIGITAL

2.1 Conditional Statements

Learning Target: Understand and write conditional statements.

Success Criteria:
- I can identify the hypothesis and conclusion of a statement.
- I can write conditional statements and their related conditional statements.
- I can write biconditional statements.

A *conditional statement*, symbolized by $p \rightarrow q$, can be written as an "if-then statement" that contains a *hypothesis* p and a *conclusion* q. Here is an example.

If a polygon is a triangle, then the sum of its angle measures is 180°.

hypothesis, p (If a polygon is a triangle) — conclusion, q (then the sum of its angle measures is 180°)

EXPLORE IT! Determining Whether Statements Are True or False

Work with a partner. A hypothesis can be either true or false. The same is true of a conclusion. When a conditional statement is true, the hypothesis and conclusion do not necessarily both have to be true.

a. Determine whether each conditional statement is true or false. Justify your answer.

i. If yesterday was Wednesday, then today is Thursday.

ii. If an angle is acute, then it has a measure of 30°.

S	M	T	W	T	F	S
1	2	3	4	5	6	7
8	9	10	11	12	13	14
15	16	17	18	19	20	21
22	23	24	25	26	27	28
29	30					

iii. If a month has 30 days, then it is June.

iv. If $\triangle ADC$ is a right triangle, then the Pythagorean Theorem is valid for $\triangle ADC$.

v. If a polygon is a quadrilateral, then the sum of its angle measures is 180°.

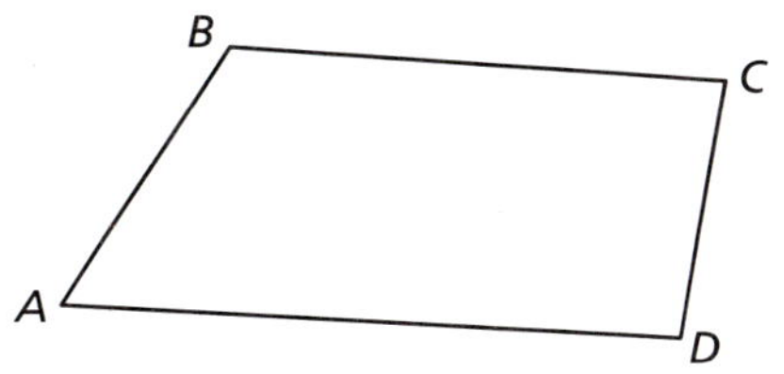

vi. If points A, B, and C are collinear, then they lie on the same line.

b. Write one true conditional statement and one false conditional statement that are different from those given in part (a). Justify your answer.

c. Conditional statements do not have to be written in if-then form. Determine whether each conditional statement is true or false. Justify your answer.

i. Two angles are complementary if the sum of their measures is 90°.

ii. The product of two numbers is negative when both numbers are negative.

5 MTR DECOMPOSE A PROBLEM
Which parts of the statements in part (c) are the hypotheses? the conclusions?

Logic and Discrete Theory
MA.912.LT.4.3 Identify and accurately interpret "if . . . then," "if and only if," "all" and "not" statements. Find the converse, inverse and contrapositive of a statement.

GO DIGITAL

Writing Conditional Statements

Vocabulary AZ VOCAB

conditional statement, *p. 66*
if-then form, *p. 66*
hypothesis, *p. 66*
conclusion, *p. 66*
negation, *p. 66*
converse, *p. 67*
inverse, *p. 67*
contrapositive, *p. 67*
equivalent statements, *p. 67*
perpendicular lines, *p. 68*
biconditional statement, *p. 69*

KEY IDEA

Conditional Statement

A **conditional statement** is a logical statement that has two parts, a *hypothesis* p and a *conclusion* q. When a conditional statement is written in **if-then form**, the "if" part contains the **hypothesis** and the "then" part contains the **conclusion**.

Words If p, then q. **Symbols** $p \rightarrow q$ (read as "p implies q")

EXAMPLE 1 Rewriting a Statement in If-Then Form

WATCH

Identify the hypothesis and the conclusion. Then rewrite the conditional statement in if-then form.

a. All birds have feathers. **b.** You are in Florida if you are in Miami.

SOLUTION

a. All birds have feathers.

All birds: hypothesis
feathers: conclusion

▶ If an animal is a bird, then it has feathers.

b. You are in Florida if you are in Miami.

You are in Florida: conclusion
you are in Miami: hypothesis

▶ If you are in Miami, then you are in Florida.

KEY IDEA

Negation

The **negation** of a statement is the *opposite* of the original statement. To write the negation of a statement p, you write the symbol for negation ($\sim$) before the letter.

Words not p **Symbols** $\sim p$ (read as "not p")

EXAMPLE 2 Writing a Negation

WATCH

Write the negation of each statement.

a. The ball is red. **b.** The cat is not black.

SOLUTION

a. The ball is not red. **b.** The cat is black.

SELF-ASSESSMENT

1 I don't understand yet. 2 I can do it with help. 3 I can do it on my own. 4 I can teach someone else.

Identify the hypothesis and the conclusion. Then rewrite the conditional statement in if-then form.

1. All 30° angles are acute angles.

2. $2x + 7 = 1$, because $x = -3$.

Write the negation of the statement.

3. The shirt is green.

4. The shoes are not red.

GO DIGITAL

KEY IDEA

Related Conditionals

Consider the conditional statement below.

Words If p, then q. **Symbols** $p \rightarrow q$

Converse To write the **converse** of a conditional statement, exchange the hypothesis and the conclusion.

Words If q, then p. **Symbols** $q \rightarrow p$

Inverse To write the **inverse** of a conditional statement, negate both the hypothesis and the conclusion.

Words If not p, then not q. **Symbols** $\sim p \rightarrow \sim q$

Contrapositive To write the **contrapositive** of a conditional statement, first write the converse. Then negate both the hypothesis and the conclusion.

Words If not q, then not p. **Symbols** $\sim q \rightarrow \sim p$

A conditional statement and its contrapositive are either both true or both false. Similarly, the converse and inverse of a conditional statement are either both true or both false. In general, when two statements are both true or both false, they are called **equivalent statements**.

COMMON ERROR

Just because a conditional statement and its contrapositive are both true does not mean that its converse and inverse are both false. The converse and inverse can also both be true.

EXAMPLE 3 Writing Related Conditional Statements

WATCH

Let p be "you are a guitar player" and let q be "you are a musician." Write each statement in words. Then decide whether it is *true* or *false*.

a. the conditional statement $p \rightarrow q$

b. the converse $q \rightarrow p$

c. the inverse $\sim p \rightarrow \sim q$

d. the contrapositive $\sim q \rightarrow \sim p$

SOLUTION

a. Conditional: If you are a guitar player, then you are a musician.
true; Guitar players are musicians.

b. Converse: If you are a musician, then you are a guitar player.
false; Not all musicians play the guitar.

c. Inverse: If you are not a guitar player, then you are not a musician.
false; Even if you do not play the guitar, you can still be a musician.

d. Contrapositive: If you are not a musician, then you are not a guitar player.
true; A person who is not a musician cannot be a guitar player.

SELF-ASSESSMENT

1 I don't understand yet. 2 I can do it with help. 3 I can do it on my own. 4 I can teach someone else.

5. Repeat Example 3 when p is "the stars are visible" and q is "it is night."

4 MTR **6.** **WHICH ONE DOESN'T BELONG?** Which statement does *not* belong with the other three? Explain your reasoning.

If today is Tuesday, then tomorrow is Wednesday.	If it is Independence Day, then it is July.
If an angle is acute, then its measure is less than 90°.	If you are an athlete, then you play soccer.

GO DIGITAL

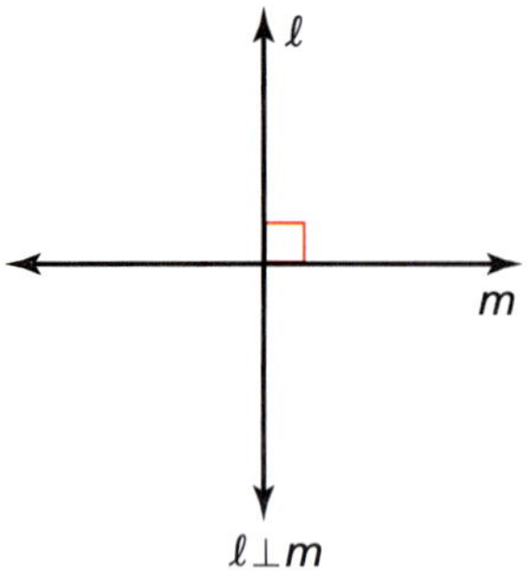

You can write a definition as a conditional statement in if-then form or as its converse. Both the conditional statement and its converse are true for definitions. For example, consider the definition of *perpendicular lines*.

If two lines intersect to form a right angle, then they are **perpendicular lines**.

You can also write the definition using the converse: If two lines are perpendicular lines, then they intersect to form a right angle.

You can write "line ℓ is perpendicular to line m" as $\ell \perp m$.

EXAMPLE 4 Using Definitions

WATCH

Decide whether each statement about the diagram is true. Explain your answer using the definitions you have learned.

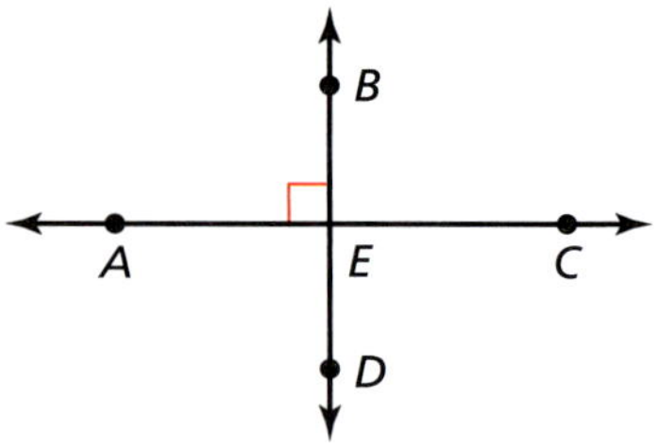

a. $\overleftrightarrow{AC} \perp \overleftrightarrow{BD}$

b. $\angle AEB$ and $\angle CEB$ are a linear pair.

c. $\overrightarrow{EA}$ and $\overrightarrow{EB}$ are opposite rays.

SOLUTION

a. This statement is *true*. The right-angle symbol in the diagram indicates that the lines intersect to form a right angle. So, you can say the lines are perpendicular.

b. This statement is *true*. By definition, if the noncommon sides of adjacent angles are opposite rays, then the angles are a linear pair. Because $\overrightarrow{EA}$ and $\overrightarrow{EC}$ are opposite rays, $\angle AEB$ and $\angle CEB$ are a linear pair.

c. This statement is *false*. The rays have the same endpoint, but they do not form a line. So, the rays are not opposite rays.

SELF-ASSESSMENT

1 I don't understand yet. 2 I can do it with help. 3 I can do it on my own. 4 I can teach someone else.

Use the diagram. Decide whether the statement is true. Explain your answer using the definitions you have learned.

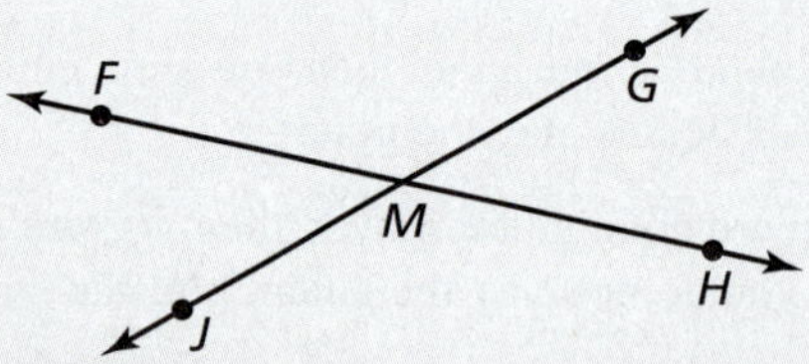

7. $\angle JMF$ and $\angle FMG$ are supplementary.

8. Point M is the midpoint of $\overline{FH}$.

9. $\angle JMF$ and $\angle HMG$ are vertical angles.

10. $\overleftrightarrow{FH} \perp \overleftrightarrow{JG}$

GO DIGITAL

Writing Biconditional Statements

KEY IDEA

Biconditional Statement

When a conditional statement and its converse are both true, you can write them as a single *biconditional statement*. A **biconditional statement** is a statement that contains the phrase "if and only if."

Words p if and only if q **Symbols** $p \leftrightarrow q$

Any definition can be written as a biconditional statement.

EXAMPLE 5 Writing a Biconditional Statement

Rewrite the definition of perpendicular lines as a biconditional statement.

Definition If two lines intersect to form a right angle, then they are perpendicular lines.

SOLUTION

Let p be "two lines intersect to form a right angle" and let q be "they are perpendicular lines."
Use red to identify p and blue to identify q.
Write the definition $p \rightarrow q$.

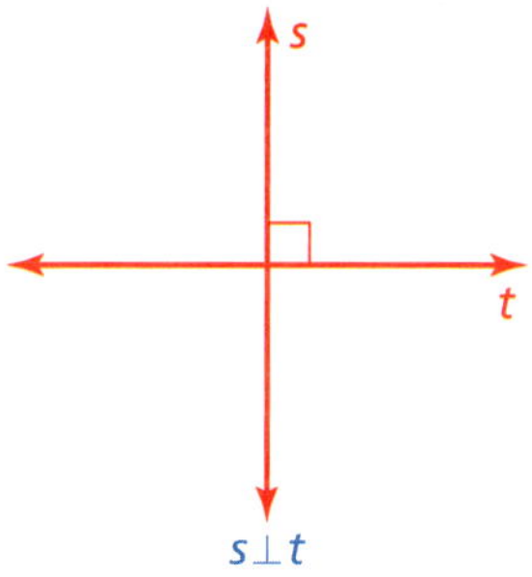

Definition If two lines intersect to form a right angle, then they are perpendicular lines.

Write the converse $q \rightarrow p$.

Converse If two lines are perpendicular lines, then they intersect to form a right angle.

Use the definition and its converse to write the biconditional statement $p \leftrightarrow q$.

▶ **Biconditional** Two lines intersect to form a right angle if and only if they are perpendicular lines.

SELF-ASSESSMENT 1 I don't understand yet. 2 I can do it with help. 3 I can do it on my own. 4 I can teach someone else.

11. Rewrite the definition of a right angle as a single biconditional statement.

Definition If an angle is a right angle, then its measure is 90°.

12. Rewrite the definition of congruent segments as a single biconditional statement.

Definition If two line segments have the same length, then they are congruent segments.

Rewrite the statements as a biconditional statement.

13. If Mary is taking theater class, then she will be in the fall play. If Mary is in the fall play, then she must be taking theater class.

14. If you can run for president, then you are at least 35 years old. If you are at least 35 years old, then you can run for president.

GO DIGITAL

2.1 Practice WITH CalcChat® AND CalcView®

In Exercises 1–4, identify the hypothesis and the conclusion.

1. If a polygon is a pentagon, then it has five sides.

2. If two lines form vertical angles, then they intersect.

3. If you run, then you are fast.

4. If you like math, then you like science.

In Exercises 5–10, rewrite the conditional statement in if-then form. *(See Example 1.)*

5. $9x + 5 = 23$, because $x = 2$.

6. Today is Friday, and tomorrow is the weekend.

▶ **7.** When a glacier melts, the sea level rises.

8. Two right angles are supplementary angles.

9. Only people who are registered are allowed to vote.

10. The measures of complementary angles sum to 90°.

In Exercises 11–14, write the negation of the statement. *(See Example 2.)*

▶ **11.** The sky is blue.

12. The lake is cold.

13. The ball is not pink.

14. The dog is not a Labrador retriever.

In Exercises 15–22, write the conditional statement $p \to q$, the converse $q \to p$, the inverse $\sim p \to \sim q$, and the contrapositive $\sim q \to \sim p$ in words. Then decide whether each statement is *true* or *false*. *(See Example 3.)*

▶ **15.** Let p be "two angles are supplementary" and let q be "the measures of the angles sum to 180°."

16. Let p be "you are in math class" and let q be "you are in Geometry."

17. Let p be "you do your math homework" and let q be "you will do well on the test."

18. Let p be "you are not an only child" and let q be "you have a sibling."

19. Let p be "an instrument is a *cuatro*" and let q be "the instrument is a guitar."

20. Let p be "the Sun is out" and let q be "it is daytime."

21. Let p be "$3x - 7 = 20$" and let q be "$x = 9$."

22. Let p be "it is Pascua Florida Day" and let q be "it is April."

In Exercises 23–26, decide whether the statement about the diagram is true. Explain your answer using the definitions you have learned. *(See Example 4.)*

23. $m\angle ABC = 90°$

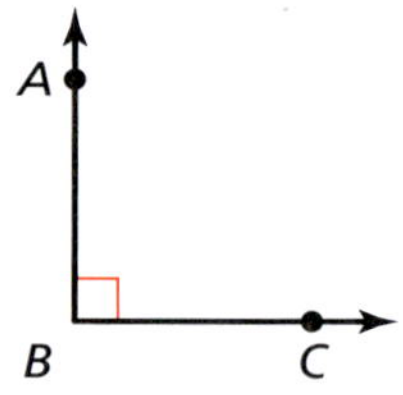

24. $\angle S \cong \angle T$

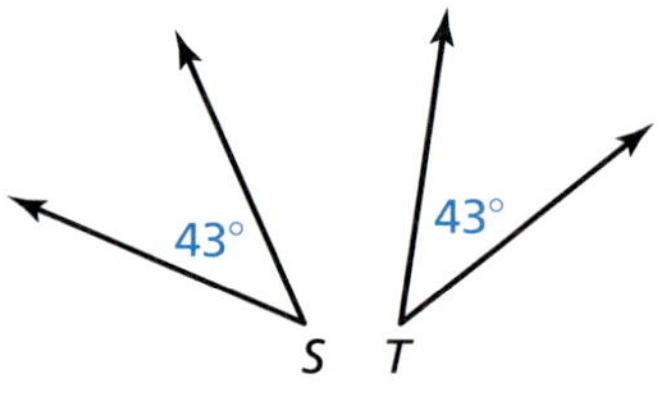

▶ **25.** $m\angle 2 + m\angle 3 = 180°$

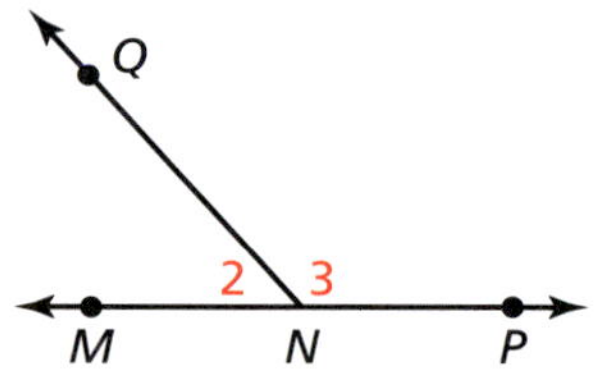

26. M is the midpoint of $\overline{AB}$.

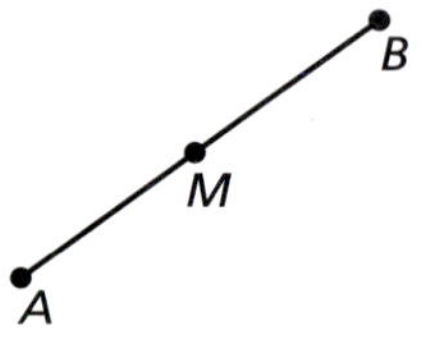

In Exercises 27–30, rewrite the definition as a biconditional statement. *(See Example 5.)*

▶ **27.** The *midpoint* of a segment is the point that divides the segment into two congruent segments.

28. Two angles are *vertical angles* when their sides form two pairs of opposite rays.

29. *Adjacent angles* are two angles that share a common vertex and side but have no common interior points.

30. Two angles are *supplementary angles* when the sum of their measures is 180°.

GO DIGITAL

In Exercises 31–34, rewrite the statements as a biconditional statement.

31. If a polygon has three sides, then it is a triangle.
If a polygon is a triangle, then it has three sides.

32. If a polygon has four sides, then it is a quadrilateral.
If a polygon is a quadrilateral, then it has four sides.

33. If an angle is a right angle, then it measures 90°.
If an angle measures 90°, then it is a right angle.

34. If an angle is obtuse, then it has a measure between 90° and 180°.
If an angle has a measure between 90° and 180°, then it is obtuse.

4 MTR **35. ERROR ANALYSIS** Describe and correct the error in writing the converse of the conditional statement.

> ✗ *Conditional statement*
> If it is raining, then I will bring an umbrella.
>
> *Converse*
> If it is not raining, then I will not bring an umbrella.

4 MTR **36. ERROR ANALYSIS** Describe and correct the error in determining the truth value of the statement.

> ✗ *Conditional statement*
> If a triangle is concave, then a square is a quadrilateral.
>
> The hypothesis is false and the conclusion is true, so the conditional statement is false.

37. REASONING You know that the contrapositive of a statement is true. Does that help you determine whether the statement can be rewritten as a true biconditional statement? Explain your reasoning.

4 MTR **38. MAKING AN ARGUMENT** Can the statement "If I bought a shirt, then I went to the mall" be rewritten as a true biconditional statement? Explain your reasoning.

5 MTR **39. CONNECTING CONCEPTS** Can the statement "If $x = 4$, then $x^2 - 10 = x + 2$" be combined with its converse to form a true biconditional statement? Explain your reasoning.

GO DIGITAL

5 MTR **40. STRUCTURE** Use the conditional statement to identify the if-then statement as the converse, inverse, or contrapositive of the conditional statement. Then use the symbols to represent both statements.

Conditional statement

If I rode my bike to school, then I did not walk to school.

If-then statement

If I did not ride my bike to school, then I walked to school.

p	q	$\sim$	$\rightarrow$	$\leftrightarrow$

41. COLLEGE PREP The given statement is true. Which of the following statements must be true? Select all that apply. Explain your reasoning.

Given statement

If I go to the movie theater, then I will eat popcorn.

Ⓐ If I do not eat popcorn, then I did not go to the movie theater.

Ⓑ I will go to the movie theater if and only if I eat popcorn.

Ⓒ If I eat popcorn, then I went to the movie theater.

Ⓓ If I do not go to the movie theater, then I will not eat popcorn.

42. HOW DO YOU SEE IT?
The Venn diagram represents all the musicians at a high school. Write three conditional statements in if-then form describing the relationships between the various groups of musicians.

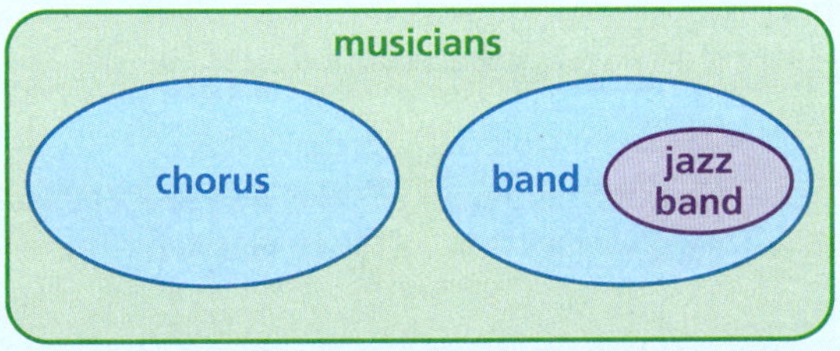

43. OPEN-ENDED Advertising slogans such as "Buy these shoes! They will make you a better athlete!" often imply conditional statements. Find an advertisement or write your own slogan. Then write it as a conditional statement.

2 MTR **44. MULTIPLE REPRESENTATIONS** Create a Venn diagram representing each conditional statement. Write the converse of each conditional statement. Then determine whether each conditional statement and its converse are true or false. Explain your reasoning.

a. If you go to the zoo to see a lion, then you will see a cat.

b. If you play a sport, then you wear a helmet.

c. If this month has 31 days, then it is not February.

45. OPEN-ENDED Write a conditional statement that is true, but its converse is false.

46. THOUGHT PROVOKING
Write three conditional statements, where one is always true, one is always false, and one depends on the person interpreting the statement.

47. REASONING Write a series of if-then statements that allow you to find the measure of each angle, given that $m\angle 1 = 90°$.

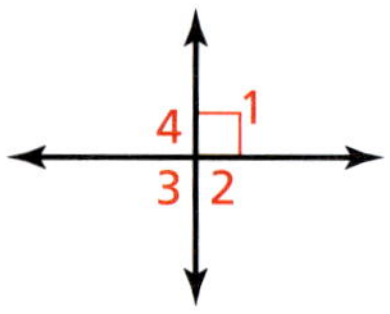

REVIEW & REFRESH

WATCH

48. Determine whether the graph represents a function. Explain.

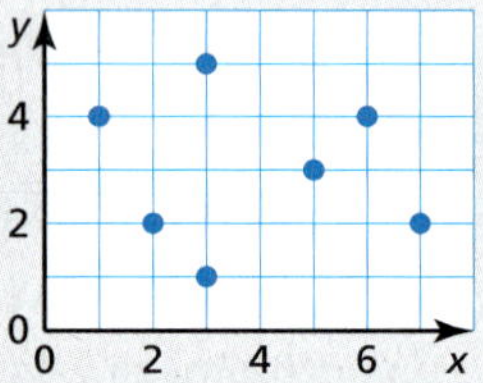

In Exercises 49 and 50, solve the equation.

49. $x^2 - 2x - 7 = 0$

50. $2x^2 + 3x - 5 = 0$

In Exercises 51 and 52, use the graphs of *f* and *g* to describe the transformation from the graph of *f* to the graph of *g*.

51. $f(x) = 2x + 1, g(x) = f(x) + 5$

52. $f(x) = \frac{1}{3}x - 6, g(x) = 3f(x)$

53. Find the measure of each angle.

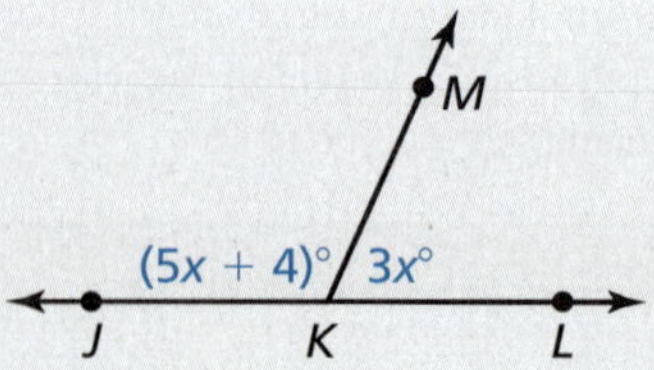

54. Find the perimeter and the area of $\triangle QRS$ with vertices $Q(-3, 4)$, $R(5, 4)$, and $S(1, -2)$.

7 MTR **55. MODELING REAL LIFE** The average distance from Earth to the moon is 3.844×10^5 kilometers. Write this number in standard form.

56. In the diagram, $\overrightarrow{WY}$ bisects $\angle XWZ$, and $m\angle YWZ = 49°$. Find $m\angle XWY$ and $m\angle XWZ$.

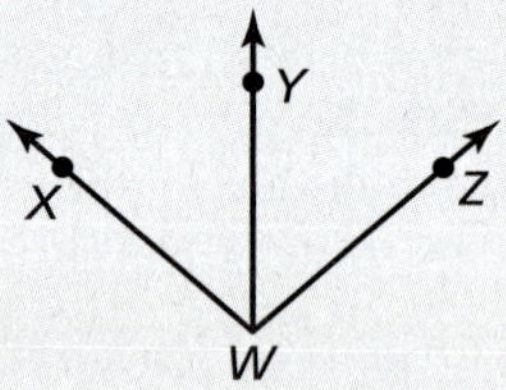

In Exercises 57–60, perform the operation.

57. $3x^2(-x + 7)$

58. $(z - 1)(z + 8)$

59. $(5b^2 - 6b + 3) - (4b - 2)$

60. $(-4n^3 - n^2 + 8) + (6n^2 + 5n - 9)$

61. Write an inequality that represents the graph.

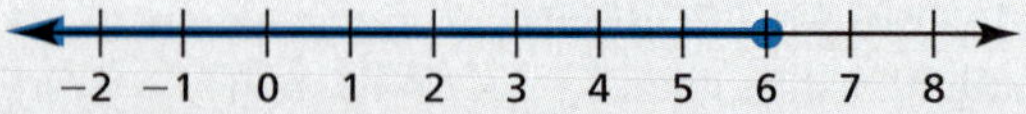

62. Let *p* be "you play a video game" and let *q* be "you beat the video game." Write the conditional statement $p \rightarrow q$, the converse $q \rightarrow p$, the inverse $\sim p \rightarrow \sim q$, and the contrapositive $\sim q \rightarrow \sim p$ in words. Then decide whether each statement is *true* or *false*.

GO DIGITAL

2.2 Inductive and Deductive Reasoning

Learning Target: Use inductive and deductive reasoning.

Success Criteria:
- I can use inductive reasoning to make conjectures.
- I can use deductive reasoning to verify conjectures.
- I can distinguish between inductive and deductive reasoning.

EXPLORE IT! Using Inductive and Deductive Reasoning

Work with a partner.

a. A *conjecture* is an unproven statement that is based on observations.

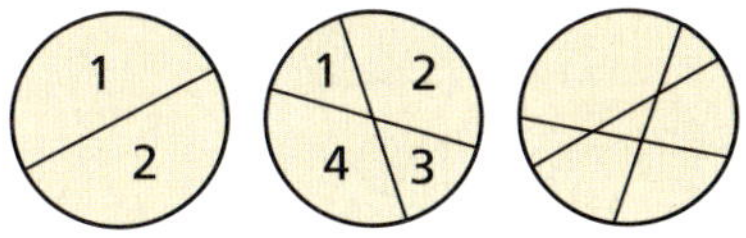

i. You can use line segments to divide a circle into regions. Use technology to complete the table. Then make a conjecture about the maximum number of regions into which a circle can be divided using n line segments. Explain your reasoning.

Number of Segments, n	1	2	3	4	5
Maximum Number of Regions	2	4			

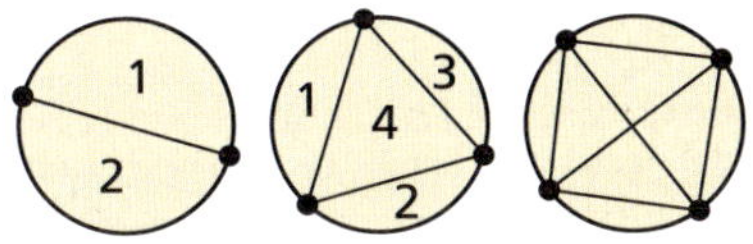

ii. You can connect points on a circle to divide the circle into regions. Use technology to complete the table. Then make a conjecture about the maximum number of regions into which a circle can be divided by connecting n points on the circle. Explain your reasoning.

Number of Points, n	2	3	4	5	6
Maximum Number of Regions	2	4			

1 MTR ANALYZE A PROBLEM
Why might a person incorrectly predict the maximum number of regions for 6 points? What does this tell you about reasoning from examples?

b. The Venn diagram shows the relationships among different types of quadrilaterals. Use the Venn diagram to determine whether each statement is true or false. Justify your answer.

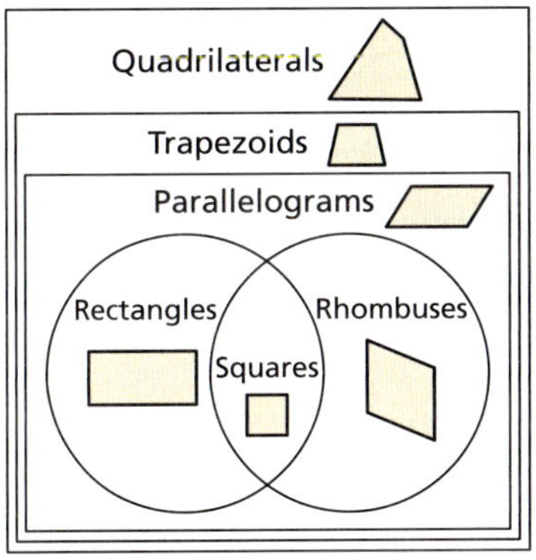

i. If a quadrilateral is a square, then it is a rectangle.

ii. If a quadrilateral is a rhombus, then it is a square.

iii. If a quadrilateral is a rectangle, then it is a parallelogram.

c. *Deductive reasoning* uses facts, definitions, accepted properties, and the laws of logic to form a logical argument. This is different from *inductive reasoning*, which uses specific examples and patterns to form a conjecture. In this Explore It!, how did you use inductive reasoning? deductive reasoning?

Logic and Discrete Theory
MA.912.LT.4.10 Judge the validity of arguments and give counterexamples to disprove statements.

GO DIGITAL

Using Inductive Reasoning

Vocabulary

conjecture, *p. 74*
inductive reasoning, *p. 74*
counterexample, *p. 75*
deductive reasoning, *p. 76*

KEY IDEA

Inductive Reasoning

A **conjecture** is an unproven statement that is based on observations. You use **inductive reasoning** when you find a pattern in specific cases and then write a conjecture for the general case.

EXAMPLE 1 Describing a Visual Pattern

Describe how to sketch the fourth figure in the pattern. Then sketch the fourth figure.

Figure 1 Figure 2 Figure 3

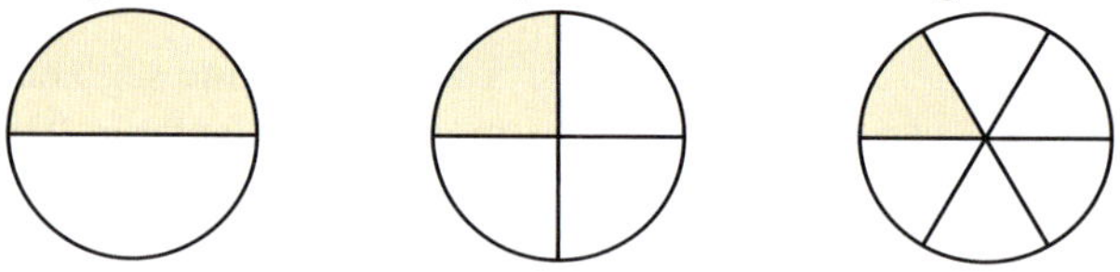

5 MTR USE A PATTERN
Describe the *n*th figure in the pattern.

SOLUTION

Each circle is divided into twice as many equal regions as the figure number. Sketch the fourth figure by dividing a circle into eighths. Shade the section just above the horizontal segment at the left.

Figure 4

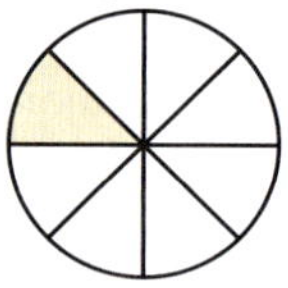

SELF-ASSESSMENT

1 I don't understand yet. 2 I can do it with help. 3 I can do it on my own. 4 I can teach someone else.

1. Sketch the fifth figure in the pattern in Example 1.

Describe the pattern. Then write or draw the next two numbers, letters, or figures.

2. S, R, Q, P, . . .

3. 1, 2, 6, 24, . . .

4.

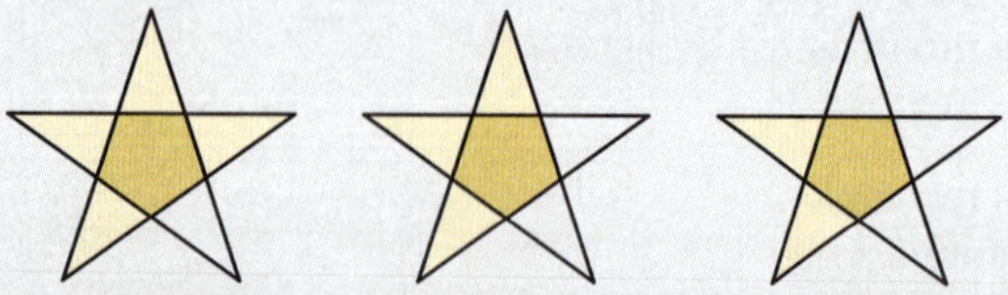

5.

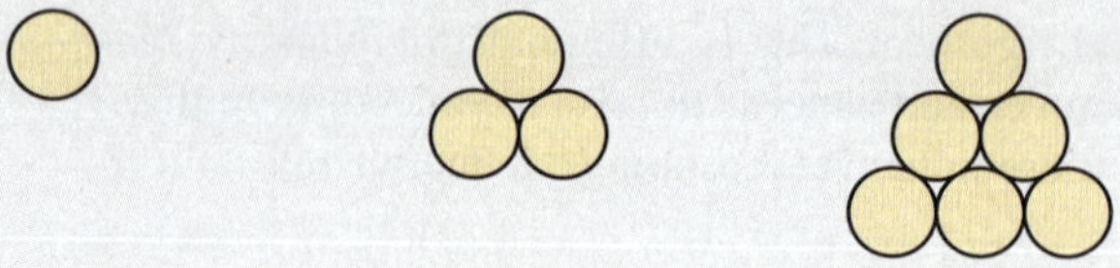

GO DIGITAL

EXAMPLE 2 Making and Testing a Conjecture

WATCH

Consecutive integers follow each other in order, such as 3, 4, and 5. Make and test a conjecture about the sum of any three consecutive integers.

SOLUTION

Step 1 Find a pattern using a few groups of small numbers.

$3 + 4 + 5 = 12 = 4 \cdot 3$ $\qquad$ $7 + 8 + 9 = 24 = 8 \cdot 3$

$10 + 11 + 12 = 33 = 11 \cdot 3$ $\qquad$ $16 + 17 + 18 = 51 = 17 \cdot 3$

Step 2 Make a conjecture.

Conjecture The sum of any three consecutive integers is three times the second number.

Step 3 Test your conjecture using other numbers. For example, test that it works with the groups −1, 0, 1 and 100, 101, 102.

$-1 + 0 + 1 = 0 = 0 \cdot 3$ ✓

$100 + 101 + 102 = 303 = 101 \cdot 3$ ✓

5 MTR USE STRUCTURE

Explain how you can use the greatest common factor to find the pattern in Step 1.

KEY IDEA

Counterexample

To show that a conjecture is true, you must show that it is true for all cases. You can show that a conjecture is false, however, by finding just one *counterexample*. A **counterexample** is a specific case for which the conjecture is false.

WORDS AND MATH

When you counter something, you are opposing it. A counterexample opposes a statement by demonstrating that it is false.

EXAMPLE 3 Finding a Counterexample

WATCH

A student makes the following conjecture about the sum of two numbers. Find a counterexample to disprove the student's conjecture.

Conjecture The sum of two numbers is always more than the greater number.

SOLUTION

To find a counterexample, you need to find a sum that is less than the greater number.

$-2 + (-3) = -5$

$-5 \not> -2$

▶ Because a counterexample exists, the conjecture is false.

SELF-ASSESSMENT

1 I don't understand yet. 2 I can do it with help. 3 I can do it on my own. 4 I can teach someone else.

6. Make and test a conjecture about the sign of the product of any three negative integers.

7. Make and test a conjecture about the sum of any five consecutive integers.

Find a counterexample to show that the conjecture is false.

8. The value of x^2 is always greater than the value of x.

9. The sum of two numbers is always greater than their difference.

GO DIGITAL

Using Deductive Reasoning

KEY IDEAS

Deductive Reasoning

Deductive reasoning uses facts, definitions, accepted properties, and the laws of logic to form a logical argument. This is different from *inductive reasoning*, which uses specific examples and patterns to form a conjecture.

Laws of Logic

Law of Detachment

If the hypothesis of a true conditional statement is true, then the conclusion is also true.

Law of Syllogism

If hypothesis p, then conclusion q.
If hypothesis q, then conclusion r. } If these statements are true,

If hypothesis p, then conclusion r. ← then this statement is true.

READING

A true conditional statement means that $p \rightarrow q$ is true.

EXAMPLE 4 Using the Law of Detachment

If two segments have the same length, then they are congruent. You know that $BC = XY$. Using the Law of Detachment, what statement can you make?

SOLUTION

Because $BC = XY$ satisfies the hypothesis of a true conditional statement, the conclusion is also true.

▶ So, $\overline{BC} \cong \overline{XY}$.

EXAMPLE 5 Using the Law of Syllogism

If possible, use the Law of Syllogism to write a new conditional statement that follows from the pair of true statements.

a. If $x^2 > 25$, then $x^2 > 20$.

If $x > 5$, then $x^2 > 25$.

b. If a polygon is regular, then all angles in the interior of the polygon are congruent.

If a polygon is regular, then all its sides are congruent.

SOLUTION

a. Notice that the conclusion of the second statement is the hypothesis of the first statement. The order in which the statements are given does not affect whether you can use the Law of Syllogism. So, you can write the following new statement.

▶ If $x > 5$, then $x^2 > 20$.

b. Neither statement's conclusion is the same as the other statement's hypothesis.

▶ You cannot use the Law of Syllogism to write a new conditional statement.

GO DIGITAL

EXAMPLE 6 Using Inductive and Deductive Reasoning

WATCH

What conclusion can you make about the product of an even integer and any other integer?

SOLUTION

Step 1 Look for a pattern in several examples. Use inductive reasoning to make a conjecture.

$(-2)(2) = -4$	$(-1)(2) = -2$	$2(2) = 4$	$3(2) = 6$
$(-2)(-4) = 8$	$(-1)(-4) = 4$	$2(-4) = -8$	$3(-4) = -12$

Conjecture Even integer • Any integer = Even integer

Step 2 Let n and m each be any integer. Use deductive reasoning to show that the conjecture is true.

$2n$ is an even integer because any integer multiplied by 2 is even.

$2nm$ represents the product of an even integer $2n$ and any integer m.

$2nm$ is the product of 2 and an integer nm. So, $2nm$ is an even integer.

▶ The product of an even integer and any integer is an even integer.

4 MTR CONSTRUCT AN ARGUMENT

Which type of reasoning helps you make a conjecture? Which type helps you justify a conjecture? How do you know when to use each type?

EXAMPLE 7 Comparing Inductive and Deductive Reasoning

WATCH

Decide whether inductive reasoning or deductive reasoning is used to reach each conclusion. Explain your reasoning.

a. Each time Monica kicks a ball up in the air, it returns to the ground. So, the next time Monica kicks a ball up in the air, it will return to the ground.

b. All reptiles are cold-blooded. Parrots are not cold-blooded. Sue's pet parrot is not a reptile.

SOLUTION

a. Inductive reasoning, because a pattern is used to reach the conclusion.

b. Deductive reasoning, because facts about animals and the laws of logic are used to reach the conclusion.

SELF-ASSESSMENT

1 I don't understand yet. 2 I can do it with help. 3 I can do it on my own. 4 I can teach someone else.

10. If $90° < m\angle R < 180°$, then $\angle R$ is obtuse. The measure of $\angle R$ is 155°. Using the Law of Detachment, what statement can you make?

11. Use the Law of Syllogism to write a new conditional statement that follows from the pair of true statements.

If you get an A on your math test, then you can go to the movies.
If you go to the movies, then you can watch your favorite actor.

12. Use inductive reasoning to make a conjecture about the sum of a number and itself. Then use deductive reasoning to show that the conjecture is true.

13. Decide whether inductive reasoning or deductive reasoning is used to reach the conclusion. Explain your reasoning.

All multiples of 8 are divisible by 4.
64 is a multiple of 8.
So, 64 is divisible by 4.

GO DIGITAL

2.2 Practice WITH CalcChat® AND CalcView®

In Exercises 1–6, describe the pattern. Then write or draw the next two numbers, letters, or figures. *(See Example 1.)*

1. 1, −2, 3, −4, 5, . . .
2. 0, 2, 6, 12, 20, . . .
3. Z, Y, X, W, V, . . .
4. A, D, G, J, M, . . .
5. 
6. 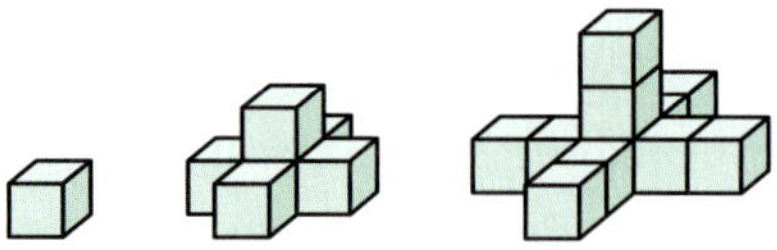

In Exercises 7–10, make and test a conjecture about the given quantity. *(See Example 2.)*

7. the sum of an even integer and an odd integer
8. the product of any two even integers
9. the quotient of a number and its reciprocal
10. the quotient of two negative integers

In Exercises 11–14, find a counterexample to show that the conjecture is false. *(See Example 3.)*

11. The product of two positive numbers is always greater than either number.
12. If n is a nonzero integer, then $\frac{n+1}{n}$ is always greater than 1.
13. If two angles are supplements of each other, then one of the angles must be acute.
14. A line s divides $\overline{MN}$ into two line segments. So, the line s is a segment bisector of $\overline{MN}$.

In Exercises 15–18, use the Law of Detachment to determine what you can conclude from the given information, if possible. *(See Example 4.)*

15. If you download a GIF, then your device crashes. You download a GIF.
16. If your cousin lets you borrow a car, then you will go to the mountains with your friend. You will go to the mountains with your friend.
17. If a quadrilateral is a square, then it has four right angles. Quadrilateral $QRST$ has four right angles.
18. If a point divides a line segment into two congruent line segments, then the point is a midpoint. Point P divides $\overline{LH}$ into two congruent line segments.

In Exercises 19–22, use the Law of Syllogism to write a new conditional statement that follows from the pair of true statements, if possible. *(See Example 5.)*

19. If $x < -2$, then $|x| > 2$. If $x > 2$, then $|x| > 2$.
20. If $a = 3$, then $5a = 15$. If $\frac{1}{2}a = 1\frac{1}{2}$, then $a = 3$.
21. If a figure is a rhombus, then the figure is a parallelogram. If a figure is a parallelogram, then the figure has two pairs of opposite sides that are parallel.
22. If a figure is a square, then the figure has four congruent sides. If a figure is a square, then the figure has four right angles.

In Exercises 23–26, state the law of logic that is illustrated.

23. If you do your homework, then you can watch TV. If you watch TV, then you can watch your favorite show.

    If you do your homework, then you can watch your favorite show.

24. If you miss practice the day before a game, then you will not be a starting player in the game.

    You miss practice on Tuesday. You will not start the game Wednesday.

25. If $x > 12$, then $x + 9 > 20$. The value of x is 14.

    So, $x + 9 > 20$.

26. If $\angle 1$ and $\angle 2$ are vertical angles, then $\angle 1 \cong \angle 2$. If $\angle 1 \cong \angle 2$, then $m\angle 1 = m\angle 2$.

    If $\angle 1$ and $\angle 2$ are vertical angles, then $m\angle 1 = m\angle 2$.

In Exercises 27 and 28, use inductive reasoning to make a conjecture about the given quantity. Then use deductive reasoning to show that the conjecture is true. *(See Example 6.)*

27. the sum of two odd integers
28. the product of two odd integers

GO DIGITAL

In Exercises 29–32, decide whether inductive reasoning or deductive reasoning is used to reach the conclusion. Explain your reasoning. *(See Example 7.)*

29. Each time you go to bed, you charge your electronic devices. So, the next time you go to bed, you will charge your electronic devices.

30. Even numbers are divisible by 2. Odd numbers are not divisible by 2. So, 4 is an even number.

31. All photosynthetic organisms produce oxygen. Phytoplankton are photosynthetic organisms. So, phytoplankton produce oxygen.

32. Each time you clean your room, you are allowed to go out with your friends. So, the next time you clean your room, you will be allowed to go out with your friends.

4 MTR **ERROR ANALYSIS** **In Exercises 33 and 34, describe and correct the error in interpreting the statement.**

33. If a figure is a rectangle, then the figure has four sides. A trapezoid has four sides.

> ✗ Using the Law of Detachment, you can conclude that a trapezoid is a rectangle.

34. Each day, you get to school before your friend.

> ✗ Using deductive reasoning, you can conclude that you will arrive at school before your friend tomorrow.

7 MTR **35. MODELING REAL LIFE** The table shows the average weights of several subspecies of tigers. What conjecture can you make about the relation between the weights of female tigers and the weights of male tigers? Explain your reasoning.

STEM

	Weight of female (pounds)	Weight of male (pounds)
Siberian	370	660
Bengal	300	480
South China	240	330
Sumatran	200	270
Indo-Chinese	250	400

GO DIGITAL

36. HOW DO YOU SEE IT?
The graph shows the total mass loss of the Greenland ice sheet since 2007. Write a conjecture using the graph.

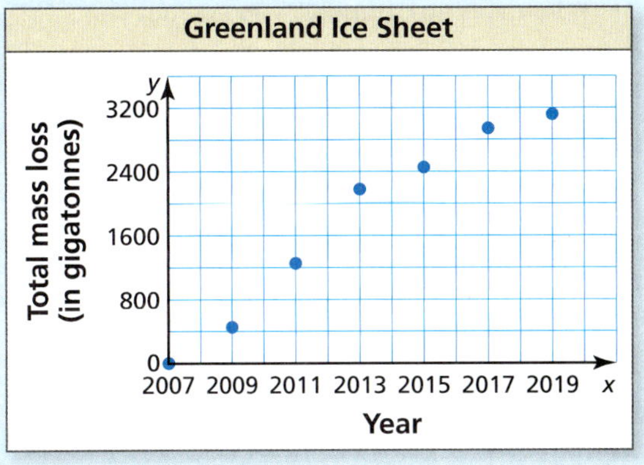

5 MTR **37. CONNECTING CONCEPTS** Use the table to make a conjecture about the relationship between x and y. Then write an equation for y in terms of x. Use the equation to test your conjecture for other values of x.

x	0	1	2	3	4
y	2	5	8	11	14

38. THOUGHT PROVOKING
The first two terms of a sequence are $\frac{1}{4}$ and $\frac{1}{2}$. Describe three different possible patterns for the sequence. List the first five terms for each sequence.

5 MTR **39. PATTERNS** The following are the first nine *Fibonacci numbers*.

1, 1, 2, 3, 5, 8, 13, 21, 34, . . .

a. Make a conjecture about each of the Fibonacci numbers after the first two.

b. Write the next three numbers in the pattern.

c. Research to find and explain a real-world example of this pattern.

4 MTR **40. MAKING AN ARGUMENT** Which argument is correct? Explain your reasoning.

Argument 1: If two angles measure 30° and 60°, then the angles are complementary. $\angle 1$ and $\angle 2$ are complementary. So, $m\angle 1 = 30°$ and $m\angle 2 = 60°$.

Argument 2: If two angles measure 30° and 60°, then the angles are complementary. The measure of $\angle 1$ is 30°, and the measure of $\angle 2$ is 60°. So, $\angle 1$ and $\angle 2$ are complementary.

41. REASONING Decide whether each conclusion is valid. Explain your reasoning.

- You and your friend went camping at Silver Springs State Park.
- When you go camping, you go canoeing.
- If you go on a hike, your friend goes with you.
- You go on a hike.
- There is a trail near your campsite.

a. Your friend went canoeing.

b. You and your friend went on a hike on the trail near your campsite.

5 MTR **42. PATTERNS** Each figure is made of squares that are 1 unit by 1 unit.

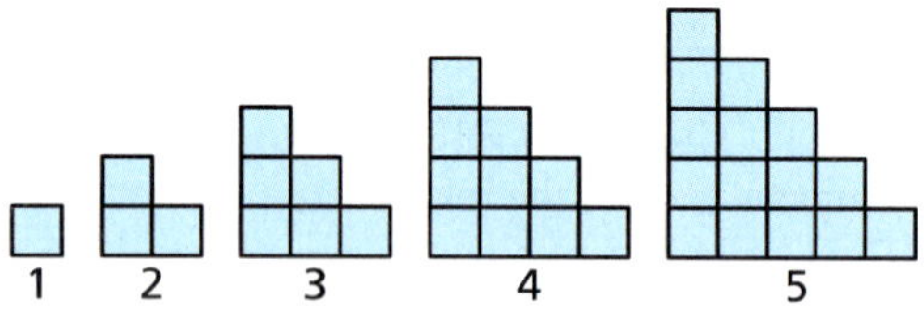

Predict the perimeter of the 20th figure.

43. REASONING Geologists use the Mohs scale to determine a mineral's hardness. Using the scale, a mineral with a higher rating will leave a scratch on a mineral with a lower rating. Testing a mineral's hardness can help identify the mineral.

Mineral	 Talc	 Gypsum	 Calcite	 Fluorite
Mohs rating	1	2	3	4

a. The four minerals are randomly labeled A, B, C, and D. Mineral A is scratched by Mineral B. Mineral C is scratched by all three of the other minerals. What can you conclude? Explain.

b. How can you identify *all* the minerals in part (a)?

5 MTR **44. CONNECTING CONCEPTS** Use inductive reasoning to write a formula for the sum of the first n positive even integers.

REVIEW & REFRESH

WATCH

45. Identify the hypothesis and the conclusion. Then rewrite the conditional statement in if-then form.

Storm surge causes the erosion of coastline.

7 MTR **46. MODELING REAL LIFE** Classify the polygon by the number of sides. Tell whether it is *convex* or *concave*.

47. Write an equation of the line that passes through the points in the table.

x	0	1	2	3
y	8	6	4	2

48. $\angle 3$ is a complement of $\angle 4$, and $m\angle 3 = 19°$. Find $m\angle 4$.

49. Determine whether the equation $y = x^2 + 5$ represents a *linear* or *nonlinear* function.

50. Solve the equation $-6x + 5 = x - 9$. Check your solution.

51. Graph $g(x) = 2x^2$. Compare the graph to the graph of $f(x) = x^2$.

52. Approximate $\sqrt{70}$ to the nearest (a) integer and (b) tenth.

53. REASONING Determine which postulate is illustrated by the statement $m\angle DAC = m\angle DAE + m\angle EAB$.

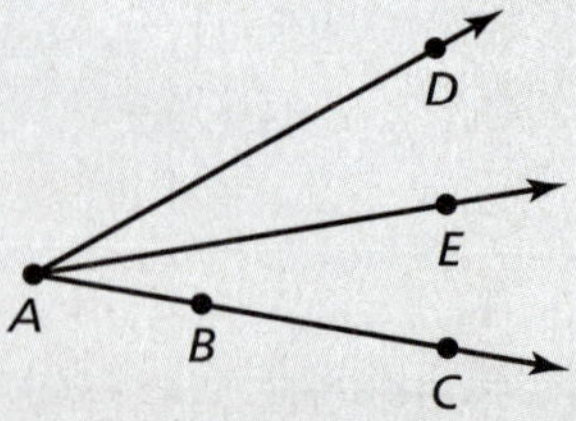

54. Find $(2n^2 - n - 6) + (-5n^2 - 4n + 8)$.

55. Find the next two numbers in the pattern $2, -2, 3, -3, 4, \ldots$.

56. Point M is the midpoint of $\overline{AB}$, where $AM = 2x + 9$ and $MB = 4x - 3$. Find AB.

GO DIGITAL

2.3 Postulates and Diagrams

Learning Target: Interpret and sketch diagrams.

Success Criteria:
- I can identify postulates represented by diagrams.
- I can sketch a diagram given a verbal description.
- I can interpret a diagram.

EXPLORE IT! Interpreting Diagrams

Work with a partner.

a. On a piece of paper, draw two perpendicular lines. Label them $\overleftrightarrow{AB}$ and $\overleftrightarrow{CD}$. Look at the diagram from different angles. Do the lines appear perpendicular regardless of the angle at which you look at them? Describe *all* the angles at which you can look at the lines and have them appear perpendicular.

4 MTR COMMUNICATE CLEARLY
From the diagram, can you assume that the lines intersect? Without the right-angle symbol, can you assume the lines are perpendicular?

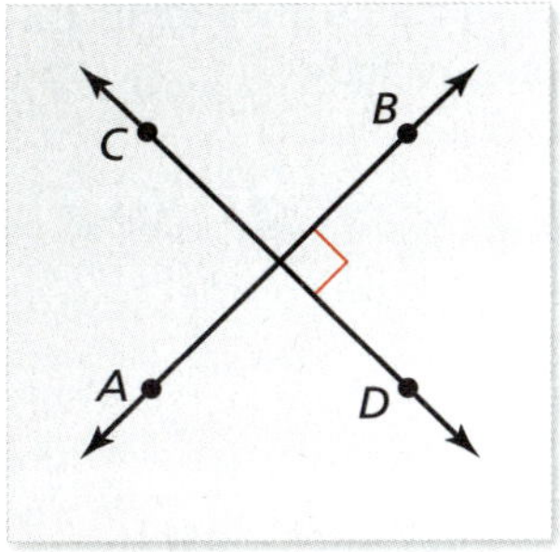

view from above

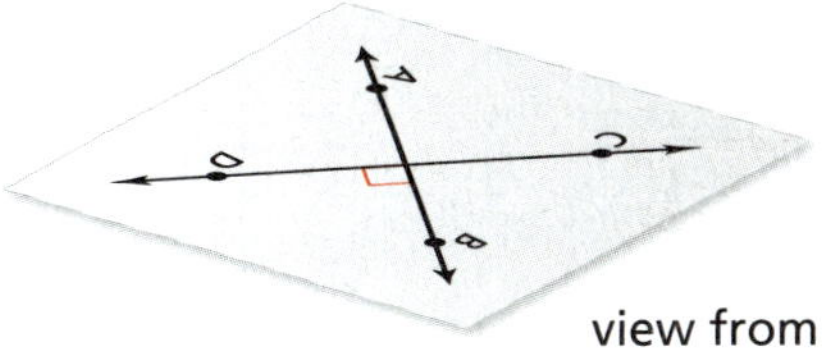

view from upper right

b. When you draw a diagram, you are communicating with others. It is important that you include sufficient information in the diagram. Use the diagram below to determine which of the following statements you can assume to be true. Explain your reasoning.

i. Points D, G, and I are collinear.

ii. Points A, C, and H are collinear.

iii. $\overleftrightarrow{EG}$ and $\overleftrightarrow{AH}$ are perpendicular.

iv. $\overleftrightarrow{AF}$ and $\overleftrightarrow{BD}$ are perpendicular.

v. $\overleftrightarrow{AF}$ and $\overleftrightarrow{BD}$ are coplanar.

vi. $\overleftrightarrow{EG}$ and $\overleftrightarrow{BD}$ do not intersect.

vii. $\overleftrightarrow{AF}$ and $\overleftrightarrow{BD}$ intersect.

viii. $\overleftrightarrow{AC}$ and $\overleftrightarrow{FH}$ are the same line.

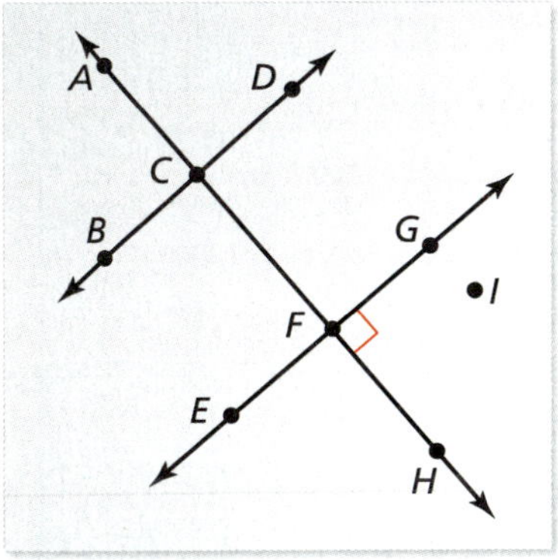

c. Use the diagram in part (b) to write two statements you can assume to be true and two statements you cannot assume to be true. Your statements should be different from those given in part (b). Explain your reasoning.

Geometric Reasoning
preparing for MA.912.GR.1.1 Prove relationships and theorems about lines and angles. Solve mathematical and real-world problems involving postulates, relationships and theorems of lines and angles.

GO DIGITAL

Identifying Postulates

Here are seven more postulates involving points, lines, and planes.

Vocabulary

line perpendicular to a plane, *p. 84*

POSTULATES

Point, Line, and Plane Postulates

Postulate | **Example**

2.1 Two Point Postulate

Through any two points, there exists exactly one line.

2.2 Line-Point Postulate

A line contains at least two points.

Through points A and B, there is exactly one line ℓ. Line ℓ contains at least two points.

2.3 Line Intersection Postulate

If two lines intersect, then their intersection is exactly one point.

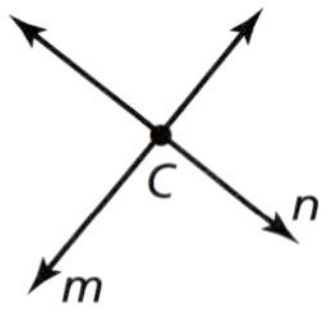

The intersection of line m and line n is point C.

2.4 Three Point Postulate

Through any three noncollinear points, there exists exactly one plane.

2.5 Plane-Point Postulate

A plane contains at least three noncollinear points.

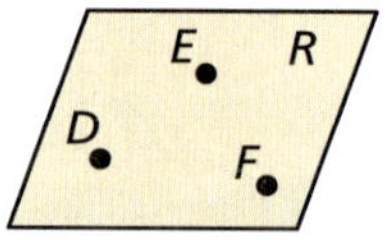

Through points D, E, and F, there is exactly one plane, plane R. Plane R contains at least three noncollinear points.

2.6 Plane-Line Postulate

If two points lie in a plane, then the line containing them lies in the plane.

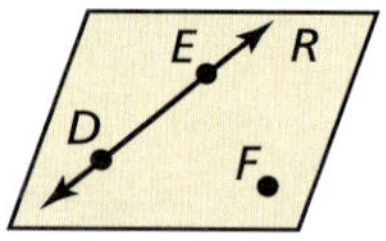

Points D and E lie in plane R, so $\overleftrightarrow{DE}$ lies in plane R.

2.7 Plane Intersection Postulate

If two planes intersect, then their intersection is a line.

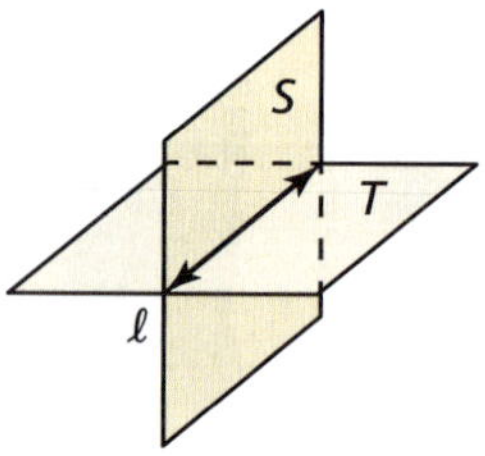

The intersection of plane S and plane T is line ℓ.

GO DIGITAL

EXAMPLE 1 Identifying a Postulate Using a Diagram

State the postulate illustrated by the diagram.

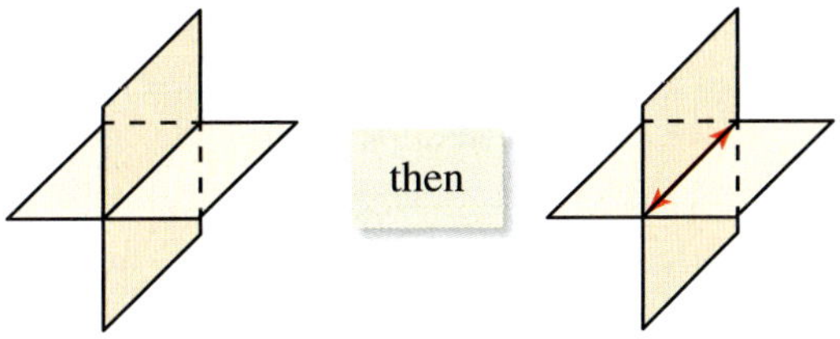

SOLUTION

a. Line Intersection Postulate If two lines intersect, then their intersection is exactly one point.

b. Plane Intersection Postulate If two planes intersect, then their intersection is a line.

EXAMPLE 2 Identifying Postulates from a Diagram

Use the diagram to write examples of the Plane-Point Postulate and the Plane-Line Postulate.

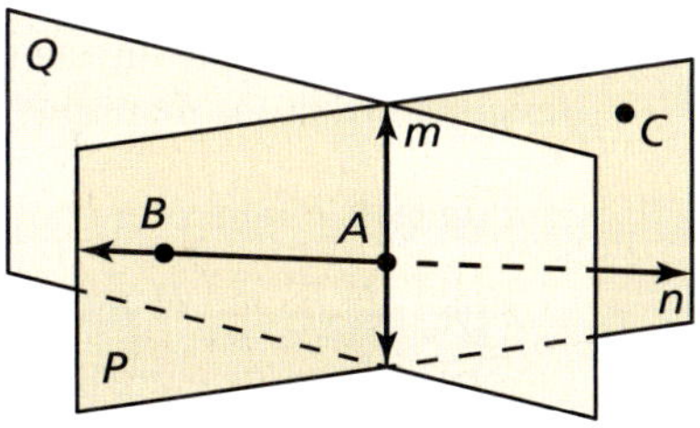

SOLUTION

Plane-Point Postulate Plane P contains at least three noncollinear points, A, B, and C.

Plane-Line Postulate Point A and point B lie in plane P. So, line n containing points A and B also lies in plane P.

SELF-ASSESSMENT

1 I don't understand yet. 2 I can do it with help. 3 I can do it on my own. 4 I can teach someone else.

1. In the diagram in Example 2, which postulate allows you to say that the intersection of plane P and plane Q is a line?

2. Use the diagram in Example 2 to write an example of each postulate.
   a. Two Point Postulate
   b. Line-Point Postulate
   c. Line Intersection Postulate

3. **WRITING** Explain why at least three noncollinear points are needed to determine a plane.

GO DIGITAL

Sketching and Interpreting Diagrams

EXAMPLE 3 Sketching a Diagram

Sketch a diagram showing $\overleftrightarrow{TV}$ intersecting $\overline{PQ}$ at point W, so that $\overline{TW} \cong \overline{WV}$.

SOLUTION

Step 1 Draw $\overleftrightarrow{TV}$ and label points T and V.

Step 2 Draw point W at the midpoint of $\overline{TV}$. Mark the congruent segments.

Step 3 Draw $\overline{PQ}$ through W.

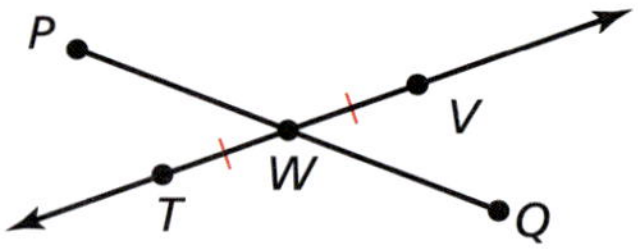

ANOTHER WAY

In Example 3, there are many ways you can sketch the diagram. Another way is shown below.

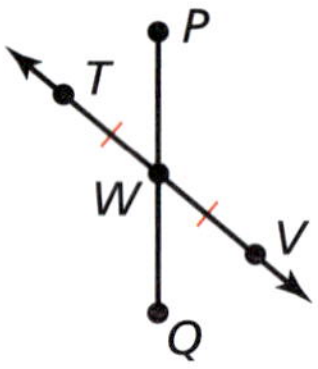

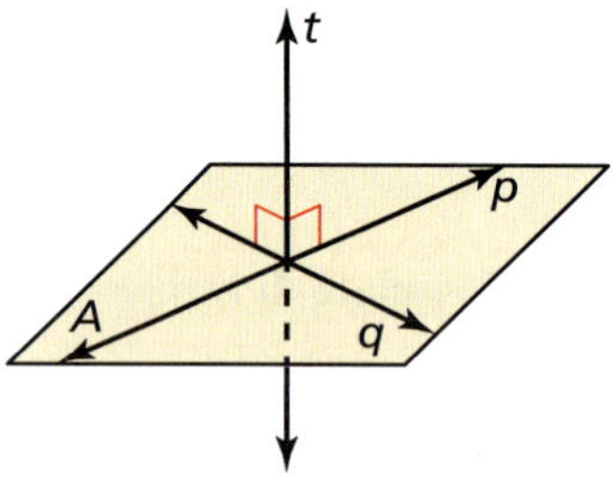

A line is a **line perpendicular to a plane** if and only if the line intersects the plane in a point and is perpendicular to every line in the plane that intersects it at that point.

In a diagram, a line perpendicular to a plane must be marked with a right-angle symbol, as shown.

EXAMPLE 4 Interpreting a Diagram

Which of the following statements *cannot* be assumed from the diagram?

- Points A, B, and F are collinear.
- Points E, B, and D are collinear.
- $\overleftrightarrow{AB} \perp$ plane S
- $\overleftrightarrow{CD} \perp$ plane T
- $\overleftrightarrow{AF}$ intersects $\overleftrightarrow{BC}$ at point B.

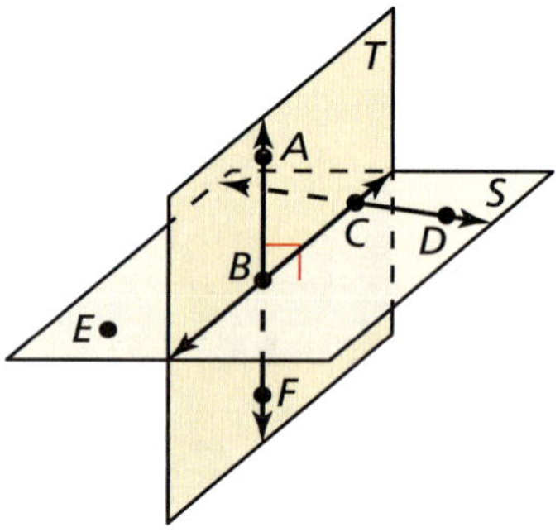

SOLUTION

No drawn line connects points E, B, and D. So, you cannot assume they are collinear. With no right angle marked, you cannot assume $\overleftrightarrow{CD} \perp$ plane T.

SELF-ASSESSMENT

1 I don't understand yet. 2 I can do it with help. 3 I can do it on my own. 4 I can teach someone else.

Use the diagram in Example 3.

4. If it is given that $\overline{PW}$ and $\overline{QW}$ are congruent, how can you indicate this in the diagram?

5. Name a pair of supplementary angles in the diagram. Explain.

Use the diagram in Example 4.

6. Can you assume that plane S intersects plane T at $\overleftrightarrow{BC}$? Explain.

7. Explain how you know that $\overleftrightarrow{AB} \perp \overleftrightarrow{BC}$.

GO DIGITAL

2.3 Practice WITH CalcChat® AND CalcView®

In Exercises 1 and 2, state the postulate illustrated by the diagram. *(See Example 1.)*

▶ **1.** 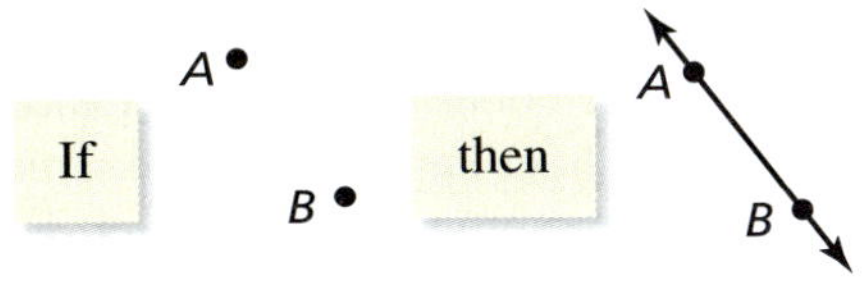

2. 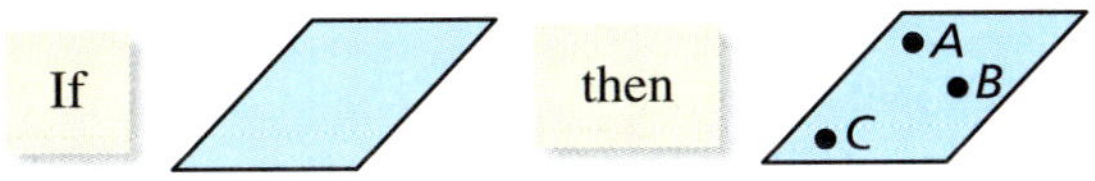

In Exercises 3–6, use the diagram to write an example of the postulate. *(See Example 2.)*

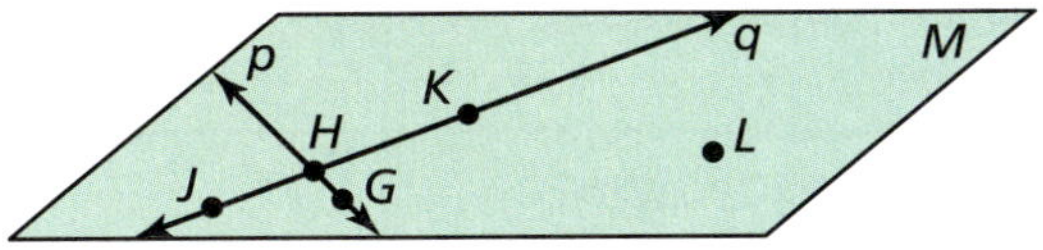

▶ **3.** Line-Point Postulate

4. Line Intersection Postulate

5. Three Point Postulate

6. Plane-Line Postulate

In Exercises 7–10, sketch a diagram of the description. *(See Example 3.)*

▶ **7.** plane P and line m perpendicular to plane P

8. $\overline{XY}$ in plane P, $\overline{XY}$ bisected by point A, and point C not on $\overline{XY}$

9. $\overline{XY}$ intersecting $\overline{WV}$ at point A, so that $XA = VA$

10. $\overline{AB}$, $\overline{CD}$, and $\overline{EF}$ are all in plane P, and point X is the midpoint of all three segments.

In Exercises 11–18, use the diagram to determine whether you can assume the statement. *(See Example 4.)*

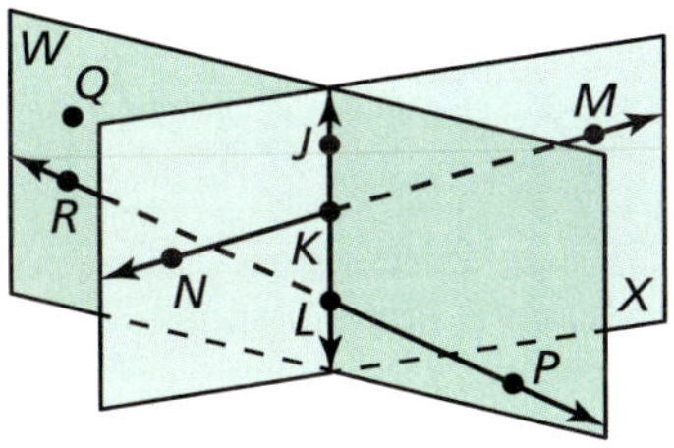

▶ **11.** Planes W and X intersect at $\overleftrightarrow{KL}$.

12. Points K, L, M, and N are coplanar.

13. Points Q, J, and M are collinear.

14. $\overleftrightarrow{MN}$ and $\overleftrightarrow{RP}$ intersect.

15. $\overleftrightarrow{JK}$ lies in plane X.

16. $\angle PLK$ is a right angle.

17. $\angle NKL$ and $\angle JKM$ are vertical angles.

18. $\angle NKJ$ and $\angle JKM$ are supplementary angles.

4 MTR **ERROR ANALYSIS In Exercises 19 and 20, describe and correct the error in the statement made about the diagram.**

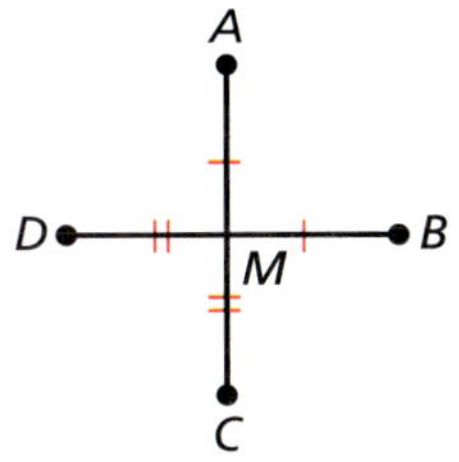

19. ✗ M is the midpoint of $\overline{AC}$ and $\overline{BD}$.

20. ✗ $\overline{AC}$ intersects $\overline{BD}$ at a 90° angle, so $\overline{AC} \perp \overline{BD}$.

21. COLLEGE PREP Select all the statements about the diagram that you *cannot* conclude.

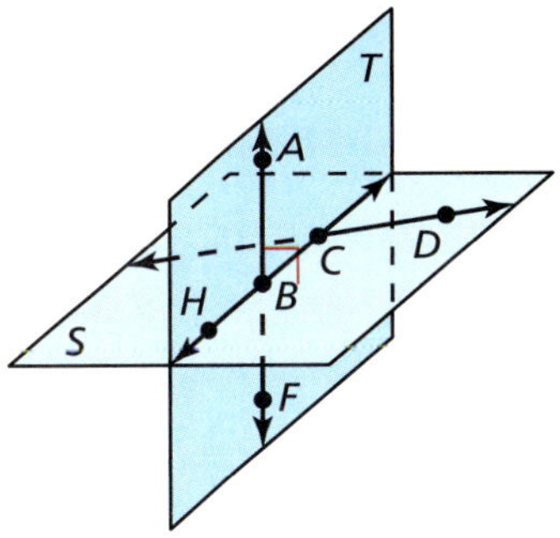

Ⓐ A, B, and C are coplanar.

Ⓑ Plane T intersects plane S in $\overleftrightarrow{BC}$.

Ⓒ $\overleftrightarrow{AB}$ intersects $\overleftrightarrow{CD}$.

Ⓓ $\overleftrightarrow{HC} \perp \overleftrightarrow{CD}$

Ⓔ Plane $T \perp$ plane S

Ⓕ Point B bisects $\overline{HC}$.

Ⓖ $\angle ABH$ and $\angle HBF$ are a linear pair.

Ⓗ $\overleftrightarrow{AF} \perp \overleftrightarrow{CD}$

GO DIGITAL

22. HOW DO YOU SEE IT?
Use the diagram of line m and point C. Make a conjecture about how many planes can be drawn so that line m and point C lie in the same plane. Use postulates to justify your conjecture.

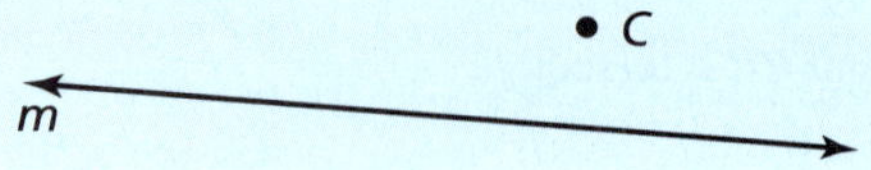

5 MTR **23. CONNECTING CONCEPTS** One way to graph a linear equation is to plot two points whose coordinates satisfy the equation and then connect them with a line. Which postulate guarantees this process works for any linear equation?

5 MTR **24. CONNECTING CONCEPTS** The graph of a system of two linear equations consists of two different lines that intersect. A solution of the system is a point of intersection of the graphs of the equations. Which postulate guarantees that the system has exactly one solution?

In Exercises 25 and 26, rewrite the postulate in if-then form.

25. Two Point Postulate

26. Plane-Point Postulate

4 MTR **27. MAKING AN ARGUMENT** Your friend claims that by the Plane Intersection Postulate, any two planes intersect in a line. Is your friend's interpretation of the Plane Intersection Postulate correct? Explain your reasoning.

28. DIG DEEPER If two lines intersect, then they intersect in exactly one point by the Line Intersection Postulate. Do the two lines have to be in the same plane? Draw a picture to support your answer. Then explain your reasoning.

29. REASONING Points E, F, and G all lie in plane P and in plane Q. What must be true about points E, F, and G so that planes P and Q are different planes? What must be true about points E, F, and G to force planes P and Q to be the same plane? Make sketches to support your answers.

30. THOUGHT PROVOKING
The postulates in this book represent Euclidean geometry. In spherical geometry, all points are points on the surface of a sphere. A line is a circle on the sphere whose diameter is equal to the diameter of the sphere. A plane is the surface of the sphere. Find a postulate on page 82 that is not true in spherical geometry. Explain your reasoning.

REVIEW & REFRESH

WATCH

31. Find a counterexample to show that the conjecture is false.

If a figure has four sides, then it is a rectangle.

In Exercises 32–35, solve the equation. Justify each step.

32. $t - 6 = -4$

33. $3x = 21$

34. $9 + x = 13$

35. $5 = \frac{x}{7}$

36. $\angle 1$ is a supplement of $\angle 2$, and $m\angle 2 = 27°$. Find $m\angle 1$.

7 MTR **37. MODELING REAL LIFE** A locker in the shape of a rectangular prism has a width of 12 inches. Its height is four times its depth. The volume of the locker is 10,800 cubic inches. Find the height and depth of the locker.

38. Write an equation in point-slope form of the line that passes through $(-8, 2)$ and has a slope of $\frac{3}{4}$.

39. Rewrite the statements as a single biconditional statement.

If you can vote, then you are at least 18 years old.
If you are at least 18 years old, then you can vote.

In Exercises 40–43, use the diagram to determine whether you can assume the statement.

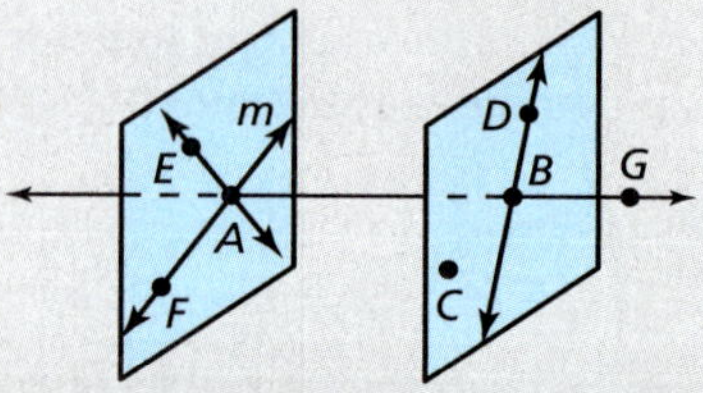

40. Points D, B, and C are coplanar.

41. $m\angle DBG = 90°$

42. Line m intersects $\overleftrightarrow{AB}$ at point A.

43. $\overleftrightarrow{DC}$ lies in plane DBC.

GO DIGITAL

2.4 Algebraic Reasoning

Learning Target: Use properties of equality to solve problems.

Success Criteria:
- I can identify algebraic properties of equality.
- I can use algebraic properties of equality to solve equations.
- I can use properties of equality to solve for geometric measures.

EXPLORE IT! Justifying Steps in a Solution

3 MTR REFLECT ON YOUR METHOD
Why is it helpful to justify each step of a solution?

Work with a partner. In previous courses, you studied different properties, such as the properties of equality and the Distributive, Commutative, and Associative Properties. You can use these properties to help solve equations.

a. A student solves the equation $\frac{5}{2}(x + 4) - 6 = 5x + 9$ as shown. Justify each step in the solution.

Algebraic Step	Justification
$\frac{5}{2}(x+4) - 6 = 5x + 9$	Write given equation.
$\frac{5}{2}x + 10 - 6 = 5x + 9$	______________
$\frac{5}{2}x + 4 = 5x + 9$	______________
$\frac{5}{2}x - \frac{5}{2}x + 4 = 5x - \frac{5}{2}x + 9$	______________
$4 = \frac{5}{2}x + 9$	______________
$4 - 9 = \frac{5}{2}x + 9 - 9$	______________
$-5 = \frac{5}{2}x$	______________
$\frac{2}{5} \cdot (-5) = \frac{2}{5} \cdot \frac{5}{2}x$	______________
$-2 = x$	______________

b. Identify another method you can use to solve the equation in part (a). Use this method to solve and justify each step. Then compare your method with the one in part (a) and the methods used by your classmates.

c. Solve $3(x + 1) - 1 = -13$ using two different methods. Justify each step for each method. Compare your answers with those of your classmates.

d. How can algebraic properties help you solve an equation?

Geometric Reasoning
preparing for MA.912.GR.1.1 Prove relationships and theorems about lines and angles. Solve mathematical and real-world problems involving postulates, relationships and theorems of lines and angles.

GO DIGITAL

Using Algebraic Properties

When you *solve an equation*, you use properties of real numbers. Segment lengths and angle measures are real numbers, so you can also use these properties to write logical arguments about geometric figures.

KEY IDEA

Algebraic Properties

Let a, b, and c be real numbers.

Addition Property of Equality	If $a = b$, then $a + c = b + c$.
Subtraction Property of Equality	If $a = b$, then $a - c = b - c$.
Multiplication Property of Equality	If $a = b$, then $a \cdot c = b \cdot c, c \neq 0$.
Division Property of Equality	If $a = b$, then $\frac{a}{c} = \frac{b}{c}, c \neq 0$.
Substitution Property of Equality	If $a = b$, then a can be substituted for b (or b for a) in any equation or expression.
Distributive Property	**Sum** $a(b + c) = ab + ac$ **Difference** $a(b - c) = ab - ac$

EXAMPLE 1 Justifying Steps

WATCH

Solve $3x + 2 = 23 - 4x$. Justify each step.

SOLUTION

Equation	Explanation	Reason
$3x + 2 = 23 - 4x$	Write the equation.	Given
$3x + 2 + 4x = 23 - 4x + 4x$	Add $4x$ to each side.	Addition Property of Equality
$7x + 2 = 23$	Combine like terms.	Simplify.
$7x + 2 - 2 = 23 - 2$	Subtract 2 from each side.	Subtraction Property of Equality
$7x = 21$	Combine constant terms.	Simplify.
$x = 3$	Divide each side by 7.	Division Property of Equality

REMEMBER
Inverse operations "undo" each other. Addition and subtraction are inverse operations. Multiplication and division are inverse operations.

▶ The solution is $x = 3$.

REMEMBER
Be sure to always check your solutions.

Check

$3x + 2 = 23 - 4x$

$3(3) + 2 \stackrel{?}{=} 23 - 4(3)$

$9 + 2 \stackrel{?}{=} 23 - 12$

$11 = 11$ ✓

GO DIGITAL

EXAMPLE 2 Justifying Steps

WATCH

Solve $-5(7w + 8) = 30$. Justify each step.

SOLUTION

Equation	Explanation	Reason
$-5(7w + 8) = 30$	Write the equation.	Given
$-35w - 40 = 30$	Multiply.	Distributive Property
$-35w = 70$	Add 40 to each side.	Addition Property of Equality
$w = -2$	Divide each side by -35.	Division Property of Equality

▶ The solution is $w = -2$.

7 MTR

EXAMPLE 3 Modeling Real Life

WATCH

You get a raise at your part-time job. To write your raise as a percent, use the formula $p(r + 1) = n$, where p is your previous wage, r is the percent increase (as a decimal), and n is your new wage. Solve the formula for r. What is your raise written as a percent when your hourly wage increases from \$15.25 to \$15.86 per hour?

SOLUTION

Step 1 Solve for r in the formula $p(r + 1) = n$.

Equation	Explanation	Reason
$p(r + 1) = n$	Write the equation.	Given
$pr + p = n$	Multiply.	Distributive Property
$pr = n - p$	Subtract p from each side.	Subtraction Property of Equality
$r = \dfrac{n - p}{p}$	Divide each side by p.	Division Property of Equality

Step 2 Evaluate $r = \dfrac{n - p}{p}$ when $n = 15.86$ and $p = 15.25$.

$$r = \frac{n - p}{p} = \frac{15.86 - 15.25}{15.25} = \frac{0.61}{15.25} = 0.04$$

REMEMBER

When evaluating expressions, use the order of operations.

▶ Your raise is 4%.

SELF-ASSESSMENT

1 I don't understand yet. 2 I can do it with help. 3 I can do it on my own. 4 I can teach someone else.

Solve the equation. Justify each step.

1. $6x - 11 = -35$

2. $-2p - 9 = 10p - 17$

3. $39 - 5z = -1 + 5z$

4. $3(3x + 14) = -3$

5. $4 = -10b + 6(2 - b)$

6. $-3(2r - 5) = 2(9 - 4r)$

7. Solve the formula $A = \frac{1}{2}bh$ for b. Justify each step. Then find the base of a triangle whose area is 952 square feet and whose height is 56 feet.

GO DIGITAL

Using Other Properties of Equality

The following properties of equality are true for all real numbers. Segment lengths and angle measures are real numbers, so these properties of equality are true for all segment lengths and angle measures.

KEY IDEAS

Reflexive, Symmetric, and Transitive Properties of Equality

	Real Numbers	Segment Lengths	Angle Measures
Reflexive Property	$a = a$	$AB = AB$	$m\angle A = m\angle A$
Symmetric Property	If $a = b$, then $b = a$.	If $AB = CD$, then $CD = AB$.	If $m\angle A = m\angle B$, then $m\angle B = m\angle A$.
Transitive Property	If $a = b$ and $b = c$, then $a = c$.	If $AB = CD$ and $CD = EF$, then $AB = EF$.	If $m\angle A = m\angle B$ and $m\angle B = m\angle C$, then $m\angle A = m\angle C$.

EXAMPLE 4 Using Properties of Equality with Angle Measures

You reflect the beam of a spotlight off a mirror lying flat on a stage, as shown. Determine whether $m\angle DBA = m\angle EBC$.

WATCH

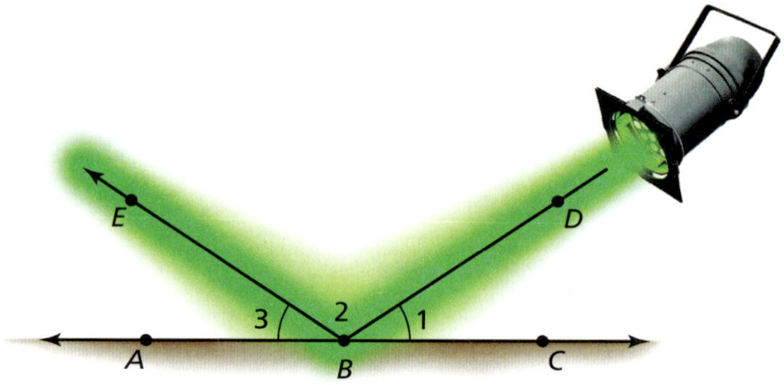

SOLUTION

Equation	Explanation	Reason
$m\angle 1 = m\angle 3$	Marked in diagram.	Given
$m\angle DBA = m\angle 3 + m\angle 2$	Add measures of adjacent angles.	Angle Addition Postulate
$m\angle DBA = m\angle 1 + m\angle 2$	Substitute $m\angle 1$ for $m\angle 3$.	Substitution Property of Equality
$m\angle 1 + m\angle 2 = m\angle EBC$	Add measures of adjacent angles.	Angle Addition Postulate
$m\angle DBA = m\angle EBC$	Both measures are equal to the sum $m\angle 1 + m\angle 2$.	Transitive Property of Equality

GO DIGITAL

7 MTR

EXAMPLE 5 Modeling Real Life

WATCH

A park, a shoe store, a pizza shop, and a movie theater are located, in that order, on a city street. The distance between the park and the shoe store is the same as the distance between the pizza shop and the movie theater. Show that the distance between the park and the pizza shop is the same as the distance between the shoe store and the movie theater.

SOLUTION

1. **Understand the Problem** You know that the locations lie in order and that the distance between two of the locations (park and shoe store) is the same as the distance between the other two locations (pizza shop and movie theater). You need to show that two of the other distances are the same.

2. **Make a Plan** Draw and label a diagram to represent the situation.

Modify your diagram by letting the points P, S, Z, and M represent the park, the shoe store, the pizza shop, and the movie theater, respectively. Show any mathematical relationships.

Use the Segment Addition Postulate to show that $PZ = SM$.

3. **Solve and Check**

Equation	Explanation	Reason
$PS = ZM$	Marked in diagram.	Given
$PZ = PS + SZ$	Add lengths of adjacent segments.	Segment Addition Postulate
$SM = SZ + ZM$	Add lengths of adjacent segments.	Segment Addition Postulate
$PS + SZ = ZM + SZ$	Add SZ to each side of $PS = ZM$.	Addition Property of Equality
$PZ = SM$	Substitute PZ for $PS + SZ$ and SM for $ZM + SZ$.	Substitution Property of Equality

Look Back Reread the problem. Make sure your diagram is drawn precisely using the given information. Check the steps in your solution.

SELF-ASSESSMENT

1 I don't understand yet. 2 I can do it with help. 3 I can do it on my own. 4 I can teach someone else.

Name the property of equality that the statement illustrates.

8. If $m\angle 6 = m\angle 7$, then $m\angle 7 = m\angle 6$.

9. If $JK = KL$ and $KL = 16$, then $JK = 16$.

10. $m\angle 1 = m\angle 2$ and $m\angle 2 = m\angle 5$. So, $m\angle 1 = m\angle 5$.

11. $ZY = ZY$

12. In Example 5, a hot dog stand is located halfway between the shoe store and the pizza shop, at point H. Show that $PH = HM$.

GO DIGITAL

2.4 Practice WITH CalcChat® AND CalcView®

6 MTR **JUSTIFYING STEPS** **In Exercises 1 and 2, write the property that justifies each step.**

1.

$3x - 12 = 7x + 8$	Given
$-4x - 12 = 8$	________
$-4x = 20$	________
$x = -5$	________

2.

$5(x - 1) = 4x + 13$	Given
$5x - 5 = 4x + 13$	________
$x - 5 = 13$	________
$x = 18$	________

In Exercises 3–12, solve the equation. Justify each step. *(See Examples 1 and 2.)*

3. $5x - 10 = -40$

4. $6x + 17 = -7$

▶**5.** $2x - 8 = 6x - 20$

6. $4x + 9 = 16 - 3x$

7. $5(3x - 20) = -10$

8. $3(2x + 11) = 9$

9. $2(-x - 5) = 12$

10. $44 - 2(3x + 4) = -18x$

11. $4(5x - 9) = -2(x + 7)$

12. $3(4x + 7) = 5(3x + 3)$

4 MTR **ERROR ANALYSIS** **In Exercises 13 and 14, describe and correct the error in solving the equation and justifying each step.**

13.

$7x = x + 24$	Given
$8x = 24$	Addition Property of Equality
$x = 3$	Division Property of Equality

14.

$6x + 14 = 32$	Given
$6x = 18$	Division Property of Equality
$x = 3$	Simplify.

In Exercises 15–20, solve the equation for *y*. Justify each step.

15. $5x + y = 18$

16. $-4x + 2y = 8$

17. $2y + 0.5x = 16$

18. $\frac{1}{2}x - \frac{3}{4}y = -2$

19. $12 - 3y = 30x + 6$

20. $3x + 7 = -7 + 9y$

In Exercises 21–24, solve the formula for the given variable. Justify each step.

21. $C = 2\pi r$; r

22. $I = Prt$; P

23. $S = 180(n - 2)$; n

24. $S = 2\pi r^2 + 2\pi rh$; h

25. **PROBLEM SOLVING** The formula for the perimeter P of a rectangle is

$$P = 2\ell + 2w$$

where ℓ is the length and w is the width. Solve the formula for ℓ. Justify each step. Then find the length of a rectangular lawn with a perimeter of 32 meters and a width of 5 meters. *(See Example 3.)*

26. **PROBLEM SOLVING** The formula for the area A of a trapezoid is

$$A = \frac{1}{2}h(b_1 + b_2)$$

where h is the height and b_1 and b_2 are the lengths of the two bases. Solve the formula for b_1. Justify each step. Then find the length of one of the bases of the trapezoid when the area of the trapezoid is 91 square meters, the height is 7 meters, and the length of the other base is 20 meters.

In Exercises 27–34, name the property of equality that the statement illustrates.

27. If $x = y$, then $3x = 3y$.

28. If $AM = MB$, then $AM + 5 = MB + 5$.

29. $x = x$

30. If $x = y$, then $y = x$.

31. If $m\angle A = 29°$ and $m\angle B = 29°$, then $m\angle A = m\angle B$.

32. $m\angle Z = m\angle Z$

33. If $AB = LM$, then $LM = AB$.

34. If $BC = XY$ and $XY = 8$, then $BC = 8$.

GO DIGITAL

In Exercises 35–38, use the property to complete the statement.

35. Substitution Property of Equality:
If $AB = 20$, then $AB + CD =$ ____.

36. Symmetric Property of Equality:
If $m\angle 1 = m\angle 2$, then ____.

37. Transitive Property of Equality:
If $m\angle 1 = m\angle 2$ and $m\angle 2 = m\angle 3$, then ____.

38. Reflexive Property of Equality:
$m\angle ABC =$ ____.

39. **REASONING** In the diagram, $m\angle ABD = m\angle CBE$. Show that $m\angle 1 = m\angle 3$. *(See Example 4.)*

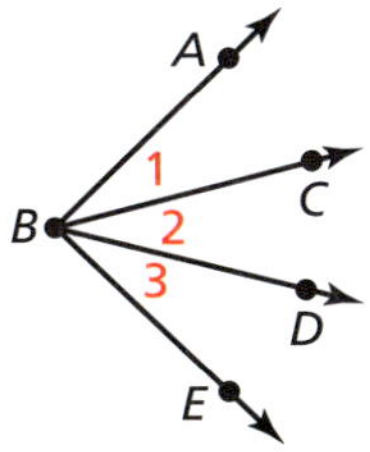

40. **REASONING** Show that $m\angle 2 = m\angle 3$.

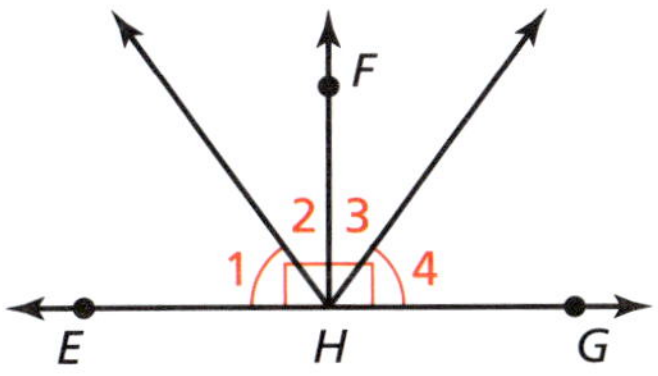

41. **MODELING REAL LIFE** You run a road race with two water stops. The distance between the starting line and the second water stop is the same as the distance between the first water stop and the finish line. Show that the distance between the starting line and the first water stop is the same as the distance between the second water stop and the finish line. *(See Example 5.)*

GO DIGITAL

42. **MODELING REAL LIFE** You cut half of a circular pizza into four pieces as shown. The two end pieces are cut at congruent angles. Show that if you eat any three consecutive pieces, you are eating the same amount of pizza.

43. **REASONING** At least how many segment lengths or angle measures are needed to demonstrate the Reflexive Property? the Symmetric Property? the Transitive Property? Explain your reasoning.

44. **HOW DO YOU SEE IT?**
In the diagram, $\overline{AD} \cong \overline{EC}$. Give an example of the Reflexive, Symmetric, and Transitive Properties of Equality.

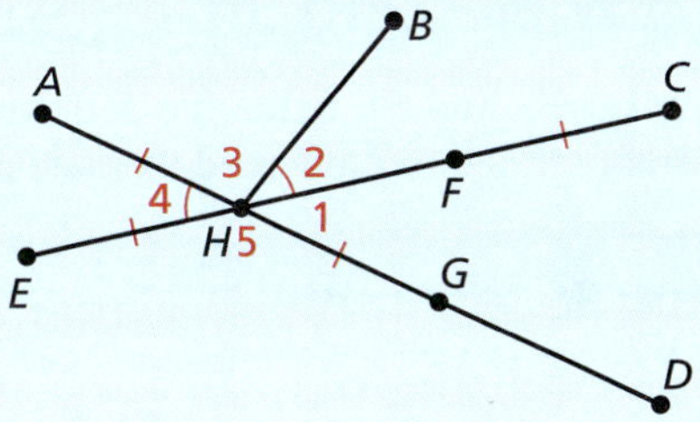

REASONING **In Exercises 45 and 46, show that the perimeter of $\triangle ABC$ is equal to the perimeter of $\triangle ADC$.**

45.

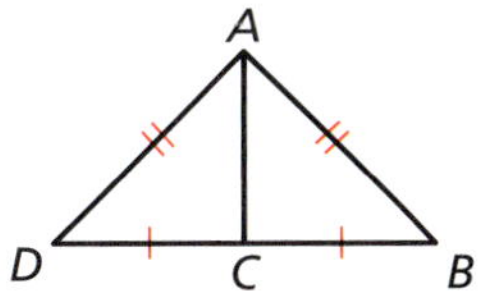

46.

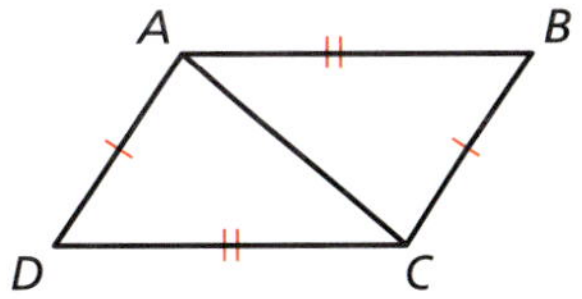

47. **WRITING** Compare the Reflexive Property of Equality with the Symmetric Property of Equality. How are the properties similar? How are they different?

48. **CONNECTING CONCEPTS** In the figure, $\overline{ZY} \cong \overline{XW}$, $ZX = 5x + 17$, $YW = 10 - 2x$, and $YX = 3$. Find ZY and XW.

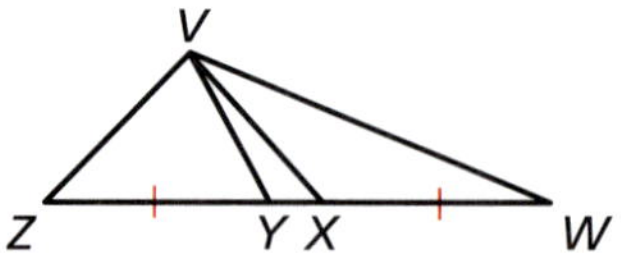

49. **MULTIPLE REPRESENTATIONS** The formula to convert a temperature in degrees Fahrenheit (°F) to degrees Celsius (°C) is $C = \frac{5}{9}(F - 32)$.

a. Solve the formula for F. Justify each step.

b. Make a table that shows the conversion to Fahrenheit for each temperature: 0°C, 20°C, 32°C, and 41°C.

c. Use your table to graph the temperature in degrees Fahrenheit as a function of the temperature in degrees Celsius. Is this a linear function?

50. **THOUGHT PROVOKING**
Write examples from your everyday life to help you remember the Reflexive, Symmetric, and Transitive Properties of Equality. Justify your answers.

51. **COLLEGE PREP** Which of the following statements illustrate the Symmetric Property of Equality? Select all that apply.

Ⓐ If $AC = RS$, then $RS = AC$.

Ⓑ If $x = 9$, then $9 = x$.

Ⓒ If $AD = BC$, then $DA = CB$.

Ⓓ $AB = BA$

Ⓔ If $AB = LM$ and $LM = RT$, then $AB = RT$.

Ⓕ If $XY = EF$, then $FE = XY$.

52. **REASONING** Select all of the properties that also apply to inequalities. Explain your reasoning.

Ⓐ Addition Property

Ⓑ Subtraction Property

Ⓒ Substitution Property

Ⓓ Reflexive Property

Ⓔ Symmetric Property

Ⓕ Transitive Property

REVIEW & REFRESH

WATCH

In Exercises 53 and 54, name the definition, property, or postulate that is represented by each diagram.

53.

$XY + YZ = XZ$

54.

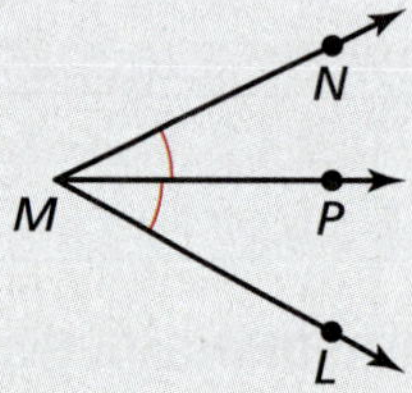

55. Solve $6x - 2y = 12$ for y. Justify each step.

In Exercises 56–58, solve the inequality. Graph the solution, if possible.

56. $|d - 3| > 7$

57. $16 < 2|5w + 4|$

58. $-3|1 - 4t| + 4.3 < -3.5$

59. Sketch a diagram of $\overleftrightarrow{XY}$ intersecting $\overline{WV}$ at point Z, so that $\overline{XZ} \cong \overline{ZY}$.

In Exercises 60 and 61, rewrite the conditional statement in if-then form.

60. When it storms, soccer practice is canceled.

61. Only people taller than 4 feet are allowed to ride the roller coaster.

62. **MODELING REAL LIFE** You work between 20 and 40 hours per week at a coffee shop. You earn $16.75 per hour plus tips. Write and graph a system that represents the situation.

63. Use inductive reasoning to make a conjecture about the difference of two even integers. Then use deductive reasoning to show that the conjecture is true.

In Exercises 64 and 65, find the product.

64. $(x + 8)(x + 4)$

65. $(y - 2)(y + 9)$

66. Approximate when the function is positive, negative, increasing, or decreasing. Then describe the end behavior of the function.

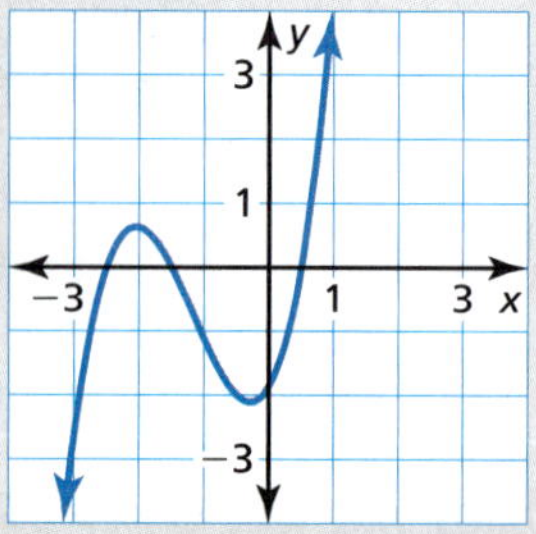

GO DIGITAL

2.5 Proving Statements about Segments and Angles

Learning Target: Prove statements about segments and angles.

Success Criteria:
- I can explain the structure of a two-column proof.
- I can write a two-column proof.
- I can identify properties of congruence.

A **proof** is a logical argument that uses deductive reasoning to show that a statement is true.

EXPLORE IT! Completing Proofs

Work with a partner.

a. Complete the statements to prove that $AB = BC$.

Given $AC = AB + AB$

Prove $AB = BC$

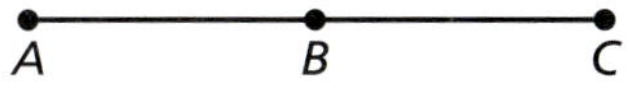

You are given that $AC =$ _____ . By the _____ , $AB + BC = AC$.

$AB + BC = AB + AB$ by the _____ . Then by the _____ , $AB = BC$.

b. Seven steps of a proof are shown. Complete the statements to prove that $\overleftrightarrow{JM} \perp \overleftrightarrow{LN}$.

Given $\angle JKL \cong \angle MKL$

Prove $\overleftrightarrow{JM} \perp \overleftrightarrow{LN}$

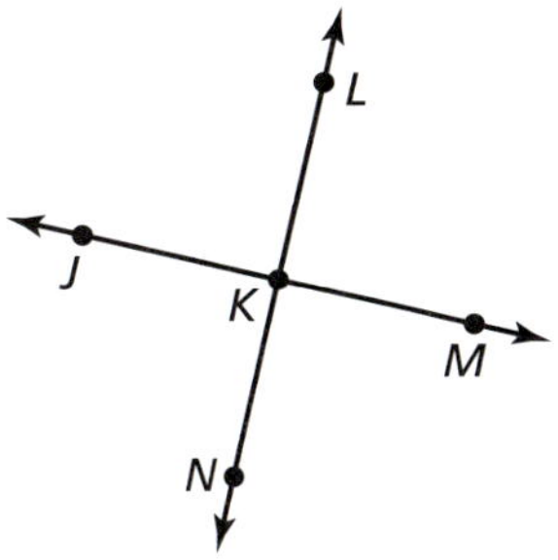

STATEMENTS	REASONS
1. $\angle JKL \cong \angle MKL$	1. Given
2. $m\angle JKL = m\angle MKL$	2. __________
3. $m\angle JKL + m\angle MKL = 180°$	3. __________
4. $m\angle JKL +$ _______ $= 180°$	4. Substitution Property of Equality
5. $2(m\angle JKL) = 180°$	5. __________
6. __________	6. Division Property of Equality
7. $\overleftrightarrow{JM} \perp \overleftrightarrow{LN}$	7. __________

4 MTR CONSTRUCT AN ARGUMENT

Why does the order of the statements and reasons in a proof matter?

Geometric Reasoning

MA.912.GR.1.1 Prove relationships and theorems about lines and angles. Solve mathematical and real-world problems involving postulates, relationships and theorems of lines and angles.

GO DIGITAL

Writing Two-Column Proofs

Vocabulary

proof, *p. 96*
two-column proof, *p. 96*
theorem, *p. 97*

A **proof** is a logical argument that uses deductive reasoning to show that a statement is true. There are several formats for proofs. A **two-column proof** has numbered statements and corresponding reasons that show an argument in a logical order.

In a two-column proof, each statement in the left-hand column is either given information or the result of applying a known property or fact to statements already made. Each reason in the right-hand column is an explanation for the corresponding statement.

EXAMPLE 1 Writing a Two-Column Proof

Write a two-column proof for the situation in Example 4 of Section 2.4.

Given $m\angle 1 = m\angle 3$

Prove $m\angle DBA = m\angle EBC$

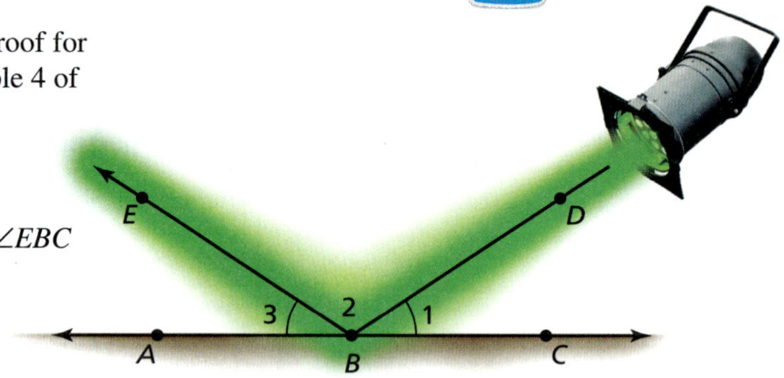

STATEMENTS	REASONS
1. $m\angle 1 = m\angle 3$	**1.** Given
2. $m\angle DBA = m\angle 3 + m\angle 2$	**2.** Angle Addition Postulate
3. $m\angle DBA = m\angle 1 + m\angle 2$	**3.** Substitution Property of Equality
4. $m\angle 1 + m\angle 2 = m\angle EBC$	**4.** Angle Addition Postulate
5. $m\angle DBA = m\angle EBC$	**5.** Transitive Property of Equality

SELF-ASSESSMENT

1 I don't understand yet. 2 I can do it with help. 3 I can do it on my own. 4 I can teach someone else.

1. Complete the proof.

Given T is the midpoint of $\overline{SU}$.

Prove $x = 5$

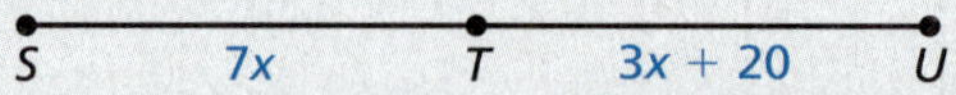

STATEMENTS	REASONS
1. T is the midpoint of $\overline{SU}$.	**1.** ______________
2. $\overline{ST} \cong \overline{TU}$	**2.** Definition of midpoint
3. $ST = TU$	**3.** Definition of congruent segments
4. $7x = 3x + 20$	**4.** ______________
5. ______________	**5.** Subtraction Property of Equality
6. $x = 5$	**6.** ______________

GO DIGITAL

Using Properties of Congruence

> **STUDY TIP**
> In this lesson, most of the proofs involve showing that congruence and equality are equivalent. These may seem basic, but it is important to practice writing these proofs to help you prepare for writing more complicated proofs in later chapters.

The reasons used in a proof can include definitions, properties, postulates, and *theorems*. A **theorem** is a statement that can be proven. Once you have proven a theorem, you can use the theorem as a reason in other proofs.

THEOREMS

2.1 Properties of Segment Congruence

Segment congruence is reflexive, symmetric, and transitive.

Reflexive For any segment AB, $\overline{AB} \cong \overline{AB}$.

Symmetric If $\overline{AB} \cong \overline{CD}$, then $\overline{CD} \cong \overline{AB}$.

Transitive If $\overline{AB} \cong \overline{CD}$ and $\overline{CD} \cong \overline{EF}$, then $\overline{AB} \cong \overline{EF}$.

Prove this Theorem Exercises 7 and 9, page 99

2.2 Properties of Angle Congruence

Angle congruence is reflexive, symmetric, and transitive.

Reflexive For any angle A, $\angle A \cong \angle A$.

Symmetric If $\angle A \cong \angle B$, then $\angle B \cong \angle A$.

Transitive If $\angle A \cong \angle B$ and $\angle B \cong \angle C$, then $\angle A \cong \angle C$.

Proof Concept Summary, page 98

Prove this Theorem Exercise 8, page 99; Exercise 35, page 113

EXAMPLE 2 Naming Properties of Congruence

WATCH

Name the property that each statement illustrates.

a. If $\angle T \cong \angle V$ and $\angle V \cong \angle R$, then $\angle T \cong \angle R$.

b. If $\overline{JL} \cong \overline{YZ}$, then $\overline{YZ} \cong \overline{JL}$.

SOLUTION

a. Transitive Property of Angle Congruence

b. Symmetric Property of Segment Congruence

EXAMPLE 3 Proving a Symmetric Property of Congruence

WATCH

> **STUDY TIP**
> When writing a proof, organize your reasoning by copying or drawing a diagram for the situation described. Then identify the **Given** and **Prove** statements.

Write a two-column proof for the Symmetric Property of Segment Congruence.

Given $\overline{LM} \cong \overline{NP}$

Prove $\overline{NP} \cong \overline{LM}$

STATEMENTS	REASONS
1. $\overline{LM} \cong \overline{NP}$	**1.** Given
2. $LM = NP$	**2.** Definition of congruent segments
3. $NP = LM$	**3.** Symmetric Property of Equality
4. $\overline{NP} \cong \overline{LM}$	**4.** Definition of congruent segments

GO DIGITAL

EXAMPLE 4 Writing a Two-Column Proof

WATCH

Prove this property of midpoints: If you know that M is the midpoint of $\overline{AB}$, prove that AB is two times AM and AM is one-half AB.

Given M is the midpoint of $\overline{AB}$.

Prove $AB = 2AM$, $AM = \frac{1}{2}AB$

STATEMENTS	REASONS
1. M is the midpoint of $\overline{AB}$.	1. Given
2. $\overline{AM} \cong \overline{MB}$	2. Definition of midpoint
3. $AM = MB$	3. Definition of congruent segments
4. $AM + MB = AB$	4. Segment Addition Postulate
5. $AM + AM = AB$	5. Substitution Property of Equality
6. $2AM = AB$	6. Distributive Property
7. $AM = \frac{1}{2}AB$	7. Division Property of Equality

SELF-ASSESSMENT

1 I don't understand yet. 2 I can do it with help. 3 I can do it on my own. 4 I can teach someone else.

2. **WRITING** How is a theorem different from a postulate?

Name the property that the statement illustrates.

3. $\overline{GH} \cong \overline{GH}$

4. If $\angle K \cong \angle P$, then $\angle P \cong \angle K$.

5. **WHAT IF?** In Example 4, you want to prove that $AB = 2MB$ and that $MB = \frac{1}{2}AB$ instead. How would the proof be different?

CONCEPT SUMMARY

Writing a Two-Column Proof

In a proof, you make one statement at a time until you reach the conclusion. Because you make statements based on facts, you are using deductive reasoning. Usually the first statement-and-reason pair you write is given information.

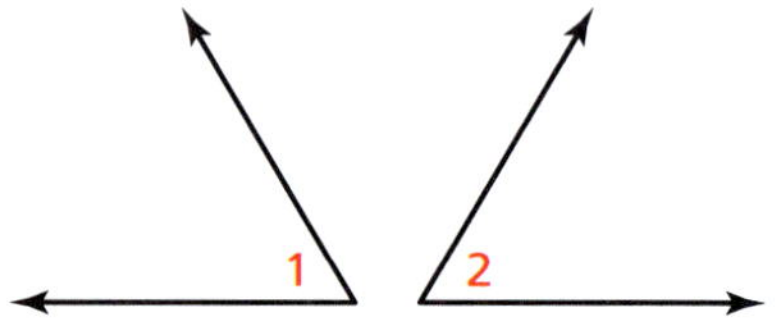

Copy or draw diagrams and label given information to help develop proofs. Do not mark or label the information from the Prove statement.

Proof of the Symmetric Property of Angle Congruence

Given $\angle 1 \cong \angle 2$

Prove $\angle 2 \cong \angle 1$

STATEMENTS	REASONS
1. $\angle 1 \cong \angle 2$	1. Given
2. $m\angle 1 = m\angle 2$	2. Definition of congruent angles
3. $m\angle 2 = m\angle 1$	3. Symmetric Property of Equality
4. $\angle 2 \cong \angle 1$	4. Definition of congruent angles

statements based on facts that you know or on conclusions from deductive reasoning

definitions, postulates, or proven theorems that allow you to state the corresponding statement

The number of statements will vary.

Remember to give a reason for the last statement.

GO DIGITAL

2.5 Practice WITH CalcChat® AND CalcView®

In Exercises 1 and 2, complete the proof. *(See Example 1.)*

1. Given $PQ = RS$

Prove $PR = QS$

STATEMENTS	REASONS
1. $PQ = RS$	**1.** _______________
2. $PQ + QR = RS + QR$	**2.** _______________
3. _______________	**3.** Segment Addition Postulate
4. $RS + QR = QS$	**4.** Segment Addition Postulate
5. $PR = QS$	**5.** _______________

2. Given $\angle 1$ is a complement of $\angle 2$.
$\angle 2 \cong \angle 3$

Prove $\angle 1$ is a complement of $\angle 3$.

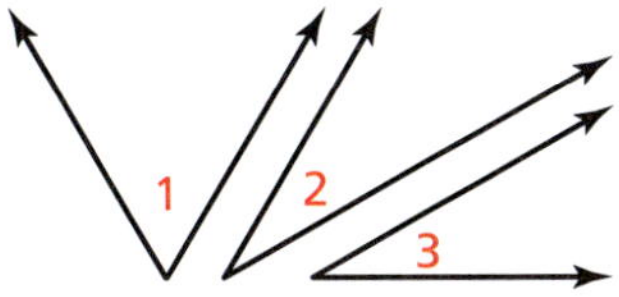

STATEMENTS	REASONS
1. $\angle 1$ is a complement of $\angle 2$.	**1.** Given
2. $\angle 2 \cong \angle 3$	**2.** _______________
3. $m\angle 1 + m\angle 2 = 90°$	**3.** _______________
4. $m\angle 2 = m\angle 3$	**4.** Definition of congruent angles
5. _______________	**5.** Substitution Property of Equality
6. $\angle 1$ is a complement of $\angle 3$.	**6.** _______________

In Exercises 3–6, name the property that the statement illustrates. *(See Example 2.)*

3. If $\overline{PQ} \cong \overline{ST}$ and $\overline{ST} \cong \overline{UV}$, then $\overline{PQ} \cong \overline{UV}$.

4. $\angle F \cong \angle F$

5. If $XY \cong UV$, then $UV \cong XY$.

6. If $\angle L \cong \angle M$ and $\angle M \cong \angle N$, then $\angle L \cong \angle N$.

PROVING A THEOREM In Exercises 7–9, write a two-column proof for the property. *(See Example 3.)*

7. Reflexive Property of Segment Congruence

8. Transitive Property of Angle Congruence

9. Transitive Property of Segment Congruence

10. ERROR ANALYSIS In the diagram, $\overline{MN} \cong \overline{LQ}$ and $\overline{LQ} \cong \overline{PN}$. Describe and correct the error in the reasoning.

✗ *Because $\overline{MN} \cong \overline{LQ}$ and $\overline{LQ} \cong \overline{PN}$, then $\overline{MN} \cong \overline{PN}$ by the Reflexive Property of Segment Congruence.*

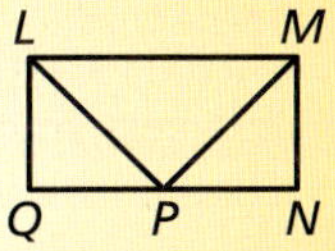

PROOF In Exercises 11 and 12, write a two-column proof. *(See Example 4.)*

11. Given $\angle GFH \cong \angle GHF$

Prove $\angle EFG$ and $\angle GHF$ are supplementary.

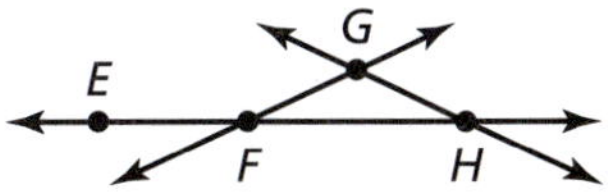

12. Given $\overline{AB} \cong \overline{FG}$, $\overleftrightarrow{BF}$ bisects $\overline{AC}$ and $\overline{DG}$.

Prove $\overline{BC} \cong \overline{DF}$

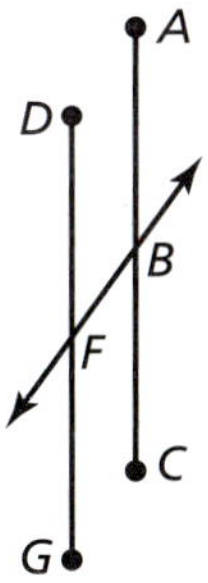

13. MAKING AN ARGUMENT In the figure, $\overline{SR} \cong \overline{CB}$ and $\overline{AC} \cong \overline{QR}$. Your friend claims that, because of this, $\overline{CB} \cong \overline{AC}$ by the Transitive Property of Segment Congruence. Is your friend correct? Explain your reasoning.

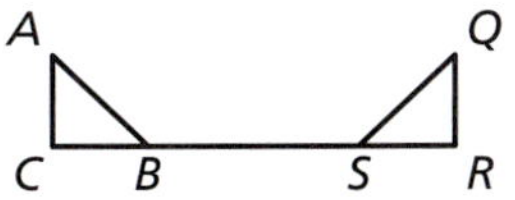

GO DIGITAL

14. REASONING In the sign, $\angle 1 \cong \angle 2$ and $\angle 2 \cong \angle 3$. Classify the triangle and justify your answer.

7 MTR **15. MODELING REAL LIFE** Use the diagram and the steps below to prove that the distance from the restaurant to the movie theater is the same as the distance from the café to the dry cleaners.

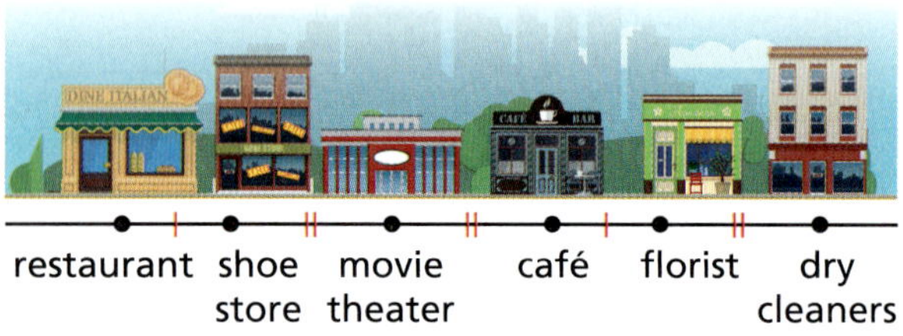

a. State what is given and what is to be proven for the situation.

b. Write a two-column proof.

16. THOUGHT PROVOKING
The distance from Springfield to Lake City is equal to the distance from Springfield to Bettsville. Janisburg is 50 miles farther from Springfield than Bettsville is. Moon Valley is 50 miles farther from Springfield than Lake City is. Use line segments to draw a diagram that represents this situation. Then prove that the distance from Springfield to Janisburg is equal to the distance from Moon Valley to Springfield.

5 MTR **17. CONNECTING CONCEPTS** Write a two-column proof. Then solve for x. Justify each step.

Given $\overline{QR} \cong \overline{PQ}$, $\overline{RS} \cong \overline{PQ}$

Prove $\overline{QR} \cong \overline{RS}$

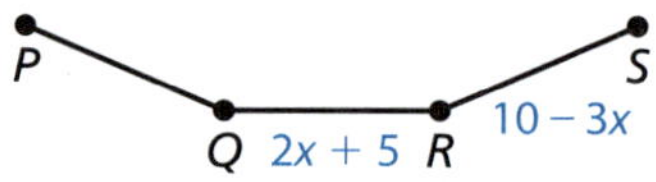

18. HOW DO YOU SEE IT?
Use the figure to write Given and Prove statements for the conclusion "A segment connecting the midpoints of two sides of a triangle is half as long as the third side."

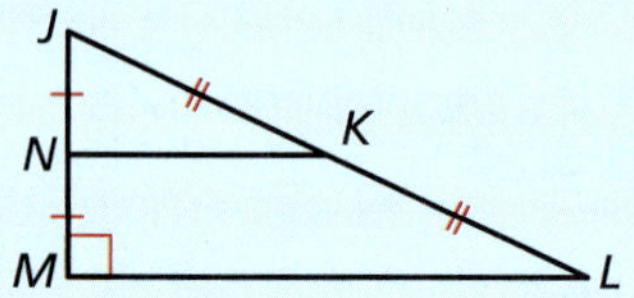

19. DIG DEEPER Fold two corners of a piece of paper so their edges match, as shown.

a. What do you notice about the angle formed at the top of the page where the folds meet?

b. Write a two-column proof to show that the angle measure is always the same no matter how you make the folds.

REVIEW & REFRESH

WATCH

In Exercises 20 and 21, solve the equation using any method. Explain your choice of method.

20. $4x^2 - 87 = 109$

21. $3x^2 - 2x - 7 = 0$

22. Does the table represent a *linear* or *nonlinear* function? Explain.

x	2	4	6	8	10
y	$\frac{1}{2}$	1	2	4	8

In Exercises 23 and 24, find the angle measure.

23. $\angle 1$ is a complement of $\angle 4$, and $m\angle 1 = 33°$. Find $m\angle 4$.

24. $\angle 3$ is a supplement of $\angle 2$, and $m\angle 2 = 147°$. Find $m\angle 3$.

25. Use inductive reasoning to make a conjecture about the sum of two negative integers. Then use deductive reasoning to show that the conjecture is true.

26. Solve $-3(6x - 1) = 6x - 9$. Justify each step.

In Exercises 27 and 28, name the property that the statement illustrates.

27. $\overline{JK} \cong \overline{JK}$

28. $\angle C \cong \angle C$

7 MTR **29. MODELING REAL LIFE** A fitness center charges members an initial fee of \$10 and a monthly fee of \$21.99. Find the total cost of 1 year of membership.

30. Sketch a diagram showing $\overline{AB}$ intersecting $\overleftrightarrow{CD}$ at point K, so that $\overline{AK} \cong \overline{KB}$ and $\overline{CK} \cong \overline{KD}$.

GO DIGITAL

2.6 Proving Geometric Relationships

Learning Target: Prove geometric relationships.

Success Criteria:
- I can prove geometric relationships by writing flowchart proofs.
- I can prove geometric relationships by writing paragraph proofs.

EXPLORE IT! Completing Flowchart Proofs

Work with a partner.

a. Complete the flowchart to prove that $AB = BC$.

Given $AC = AB + AB$

Prove $AB = BC$

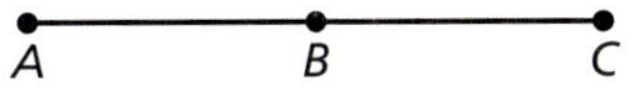

$AC = AB + AB$ →

$AB + BC = AC$ → $AB + BC = AB + AB$ → $AB = BC$

b. Complete the flowchart to prove that $\overleftrightarrow{JM} \perp \overleftrightarrow{LN}$.

Given $\angle JKL \cong \angle MKL$

Prove $\overleftrightarrow{JM} \perp \overleftrightarrow{LN}$

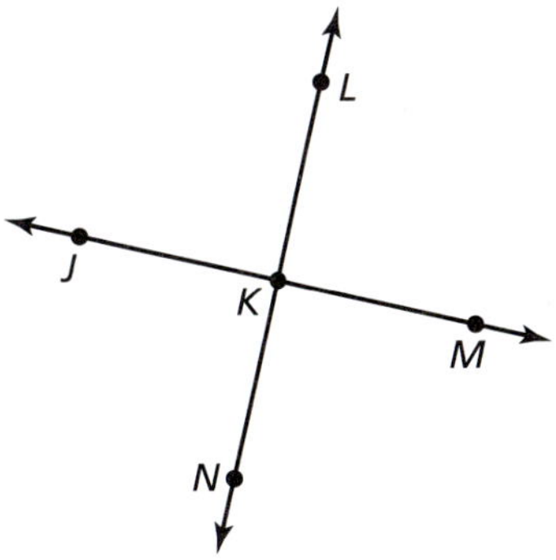

$\angle JKL \cong \angle MKL$

↓

$m\angle JKL = m\angle MKL$

↓

$m\angle JKL + m\angle MKL = 180°$ → $m\angle JKL + m\angle JKL = 180°$ → $2(m\angle JKL) = 180°$

↓

$m\angle JKL = 90°$

↓

$\overleftrightarrow{JM} \perp \overleftrightarrow{LN}$

c. How can you use a flowchart to prove a mathematical statement?

d. Compare the flowchart proofs above with the proofs in the 2.5 Explore It! Explain the advantages and disadvantages of each.

3 MTR REFLECT ON YOUR METHOD
Why is it helpful to sketch a diagram when writing certain proofs?

Geometric Reasoning
MA.912.GR.1.1 Prove relationships and theorems about lines and angles. Solve mathematical and real-world problems involving postulates, relationships and theorems of lines and angles.

GO DIGITAL

Writing Flowchart Proofs

Vocabulary (AZ VOCAB)

flowchart proof, or flow proof, *p. 102*
paragraph proof or narrative proof, *p. 104*

Another proof format is a **flowchart proof**, or **flow proof**, which uses boxes and arrows to show the flow of a logical argument. Each reason is below the statement it justifies. A flowchart proof of the *Right Angles Congruence Theorem* is shown in Example 1.

THEOREM

2.3 Right Angles Congruence Theorem

All right angles are congruent.

EXAMPLE 1 Proving the Right Angles Congruence Theorem

Use the given flowchart proof to write a two-column proof of the Right Angles Congruence Theorem.

WATCH

STUDY TIP

When you prove a theorem, write the hypothesis of the theorem as the **Given** statement. The conclusion is what you must **Prove**.

Given $\angle 1$ and $\angle 2$ are right angles.

Prove $\angle 1 \cong \angle 2$

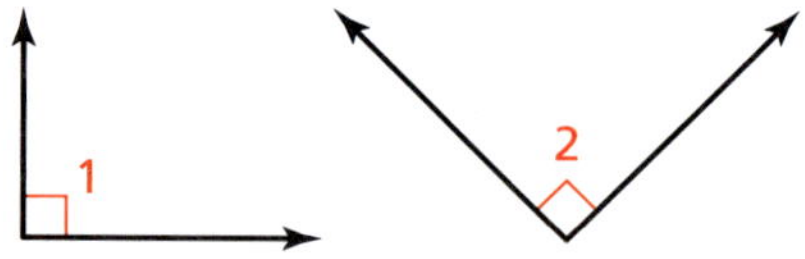

Flowchart Proof

$\angle 1$ and $\angle 2$ are right angles.
Given
↓
$m\angle 1 = 90°, m\angle 2 = 90°$ → $m\angle 1 = m\angle 2$ → $\angle 1 \cong \angle 2$
Definition of right angle | Transitive Property of Equality | Definition of congruent angles

Two-Column Proof

STATEMENTS	REASONS
1. $\angle 1$ and $\angle 2$ are right angles.	**1.** Given
2. $m\angle 1 = 90°, m\angle 2 = 90°$	**2.** Definition of right angle
3. $m\angle 1 = m\angle 2$	**3.** Transitive Property of Equality
4. $\angle 1 \cong \angle 2$	**4.** Definition of congruent angles

3 MTR SELECT METHODS

Which type of proof is easier to write when there are many statements?

SELF-ASSESSMENT 1 I don't understand yet. 2 I can do it with help. 3 I can do it on my own. 4 I can teach someone else.

1. Complete the flowchart proof. Then write a two-column proof.

Given $\overline{AB} \perp \overline{BC}, \overline{DC} \perp \overline{BC}$

Prove $\angle B \cong \angle C$

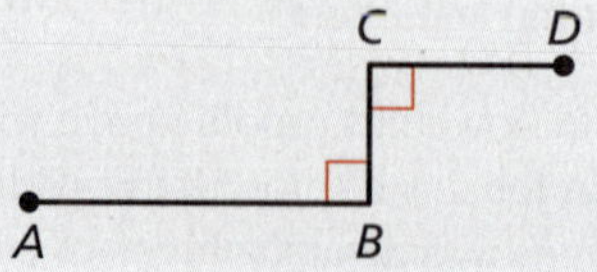

$\overline{AB} \perp \overline{BC}, \overline{DC} \perp \overline{BC}$ → [ ] → $\angle B \cong \angle C$
Given | Definition of ⊥ lines | [ ]

GO DIGITAL

THEOREMS

2.4 Congruent Supplements Theorem

If two angles are supplementary to the same angle (or to congruent angles), then they are congruent.

If $\angle 1$ and $\angle 2$ are supplementary and $\angle 3$ and $\angle 2$ are supplementary, then $\angle 1 \cong \angle 3$.

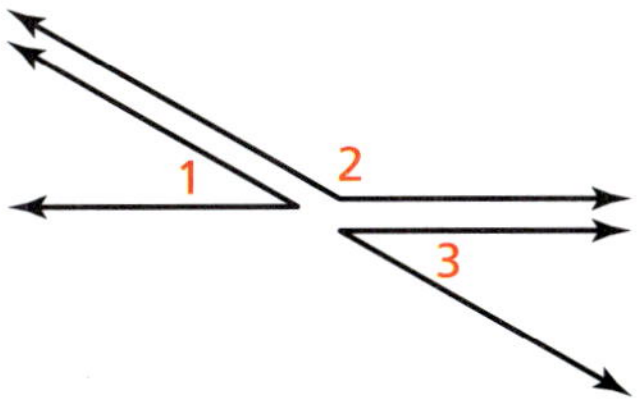

Prove this Theorem Exercise 20 (case 2), page 109

2.5 Congruent Complements Theorem

If two angles are complementary to the same angle (or to congruent angles), then they are congruent.

If $\angle 4$ and $\angle 5$ are complementary and $\angle 6$ and $\angle 5$ are complementary, then $\angle 4 \cong \angle 6$.

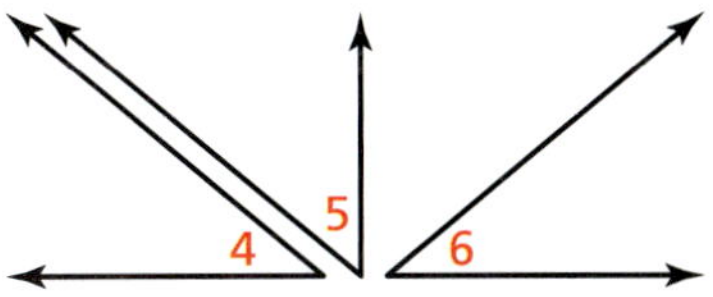

Prove this Theorem Exercise 19 (case 1), page 108; Exercise 24 (case 2), page 110

STUDY TIP

To prove the Congruent Supplements Theorem, you must prove two cases: one with angles supplementary to the same angle and one with angles supplementary to congruent angles. The proof of the Congruent Complements Theorem also requires two cases.

EXAMPLE 2 Proving a Case of the Congruent Supplements Theorem

WATCH

Use the given two-column proof to write a flowchart proof that proves that two angles supplementary to the same angle are congruent.

Given $\angle 1$ and $\angle 2$ are supplementary.
$\angle 3$ and $\angle 2$ are supplementary.

Prove $\angle 1 \cong \angle 3$

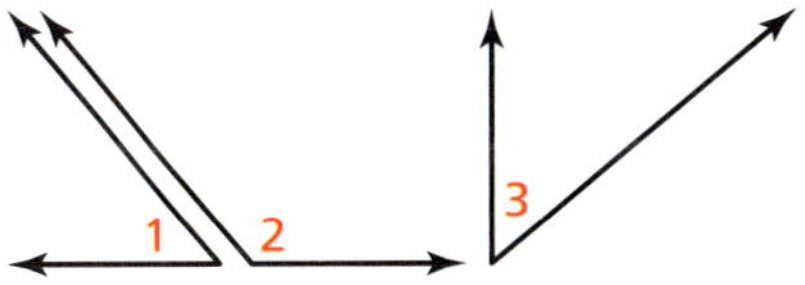

Two-Column Proof

STATEMENTS	REASONS
1. $\angle 1$ and $\angle 2$ are supplementary. $\angle 3$ and $\angle 2$ are supplementary.	**1.** Given
2. $m\angle 1 + m\angle 2 = 180°$, $m\angle 3 + m\angle 2 = 180°$	**2.** Definition of supplementary angles
3. $m\angle 1 + m\angle 2 = m\angle 3 + m\angle 2$	**3.** Transitive Property of Equality
4. $m\angle 1 = m\angle 3$	**4.** Subtraction Property of Equality
5. $\angle 1 \cong \angle 3$	**5.** Definition of congruent angles

Flowchart Proof

$\angle 1$ and $\angle 2$ are supplementary. → $m\angle 1 + m\angle 2 = 180°$ → $m\angle 1 + m\angle 2 = m\angle 3 + m\angle 2$

Given

Definition of supplementary angles

Transitive Property of Equality

$m\angle 1 = m\angle 3$

Subtraction Property of Equality

$\angle 1 \cong \angle 3$

Definition of congruent angles

$\angle 3$ and $\angle 2$ are supplementary. → $m\angle 3 + m\angle 2 = 180°$

Given

Definition of supplementary angles

GO DIGITAL

Writing Paragraph Proofs

Another proof format is a **paragraph proof**, or **narrative proof**, which presents the statements and reasons of a proof as sentences in a paragraph. It uses words to explain the logical flow of the argument.

Two intersecting lines form pairs of vertical angles and linear pairs. The *Linear Pair Postulate* formally states the relationship between angles that form linear pairs. You can use this postulate to prove the *Vertical Angles Congruence Theorem*.

POSTULATE AND THEOREM

2.8 Linear Pair Postulate

If two angles form a linear pair, then they are supplementary.

$\angle 1$ and $\angle 2$ form a linear pair, so $\angle 1$ and $\angle 2$ are supplementary and $m\angle 1 + m\angle 2 = 180°$.

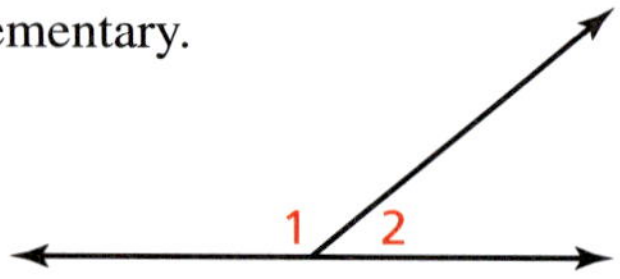

2.6 Vertical Angles Congruence Theorem

Vertical angles are congruent.

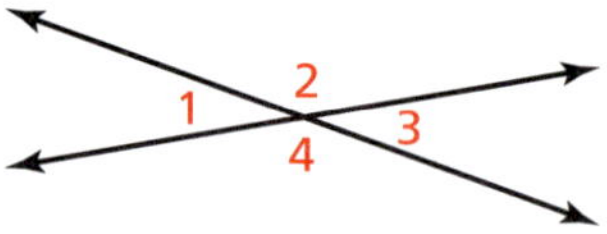

$\angle 1 \cong \angle 3$, $\angle 2 \cong \angle 4$

EXAMPLE 3 Proving the Vertical Angles Congruence Theorem

WATCH

Use the given paragraph proof to write a two-column proof of the Vertical Angles Congruence Theorem.

Given $\angle 5$ and $\angle 7$ are vertical angles.

Prove $\angle 5 \cong \angle 7$

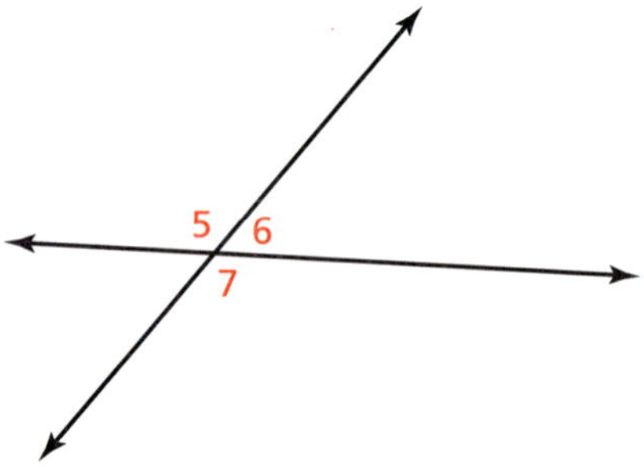

STUDY TIP

In paragraph proofs, *transitional words* such as *so*, *then*, and *therefore* help make the logic clear.

Paragraph Proof

$\angle 5$ and $\angle 7$ are vertical angles formed by intersecting lines. As shown in the diagram, $\angle 5$ and $\angle 6$ are a linear pair, and $\angle 6$ and $\angle 7$ are a linear pair. Then, by the Linear Pair Postulate, $\angle 5$ and $\angle 6$ are supplementary and $\angle 6$ and $\angle 7$ are supplementary. So, by the Congruent Supplements Theorem, $\angle 5 \cong \angle 7$.

Two-Column Proof

STATEMENTS	REASONS
1. $\angle 5$ and $\angle 7$ are vertical angles.	**1.** Given
2. $\angle 5$ and $\angle 6$ are a linear pair. $\angle 6$ and $\angle 7$ are a linear pair.	**2.** Definition of linear pair, as shown in the diagram
3. $\angle 5$ and $\angle 6$ are supplementary. $\angle 6$ and $\angle 7$ are supplementary.	**3.** Linear Pair Postulate
4. $\angle 5 \cong \angle 7$	**4.** Congruent Supplements Theorem

STUDY TIP

Your proof can use information that is labeled in a diagram.

GO DIGITAL

EXAMPLE 4 Using Angle Relationships

WATCH

Find the value of x.

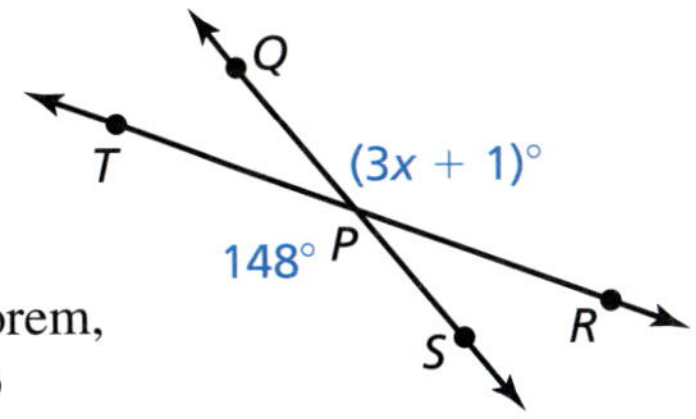

SOLUTION

$\angle TPS$ and $\angle QPR$ are vertical angles. By the Vertical Angles Congruence Theorem, the angles are congruent. Use this fact to write and solve an equation.

$m\angle TPS = m\angle QPR$	Definition of congruent angles
$148° = (3x + 1)°$	Substitute angle measures.
$147 = 3x$	Subtract 1 from each side.
$49 = x$	Divide each side by 3.

▶ So, the value of x is 49.

SELF-ASSESSMENT

1 I don't understand yet. 2 I can do it with help. 3 I can do it on my own. 4 I can teach someone else.

2. Complete the two-column proof. Then write a flowchart proof.

Given $AB = DE$, $BC = CD$

Prove $\overline{AC} \cong \overline{CE}$

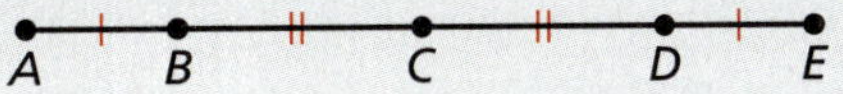

STATEMENTS	REASONS
1. $AB = DE$, $BC = CD$	**1.** Given
2. $AB + BC = BC + DE$	**2.** Addition Property of Equality
3. ______________	**3.** Substitution Property of Equality
4. $AB + BC = AC$, $CD + DE = CE$	**4.** ______________
5. ______________	**5.** Substitution Property of Equality
6. $\overline{AC} \cong \overline{CE}$	**6.** ______________

3. Write a two-column proof of the Vertical Angle Congruence Theorem without using the Congruent Supplements Theorem. Compare your proof with the proof in Example 3.

Use the diagram and the given angle measure to find the other three angle measures.

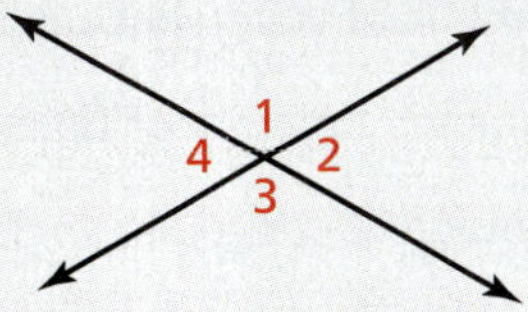

4. $m\angle 1 = 117°$

5. $m\angle 2 = 59°$

6. $m\angle 4 = 88°$

7. Find the value of w.

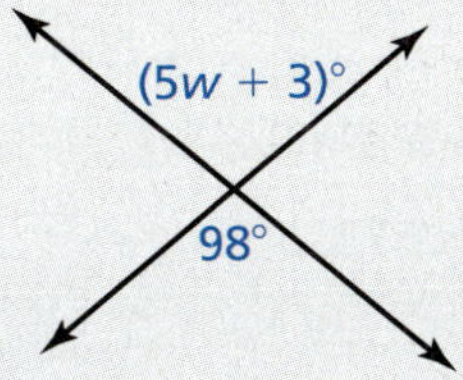

8. Find the values of a and b.

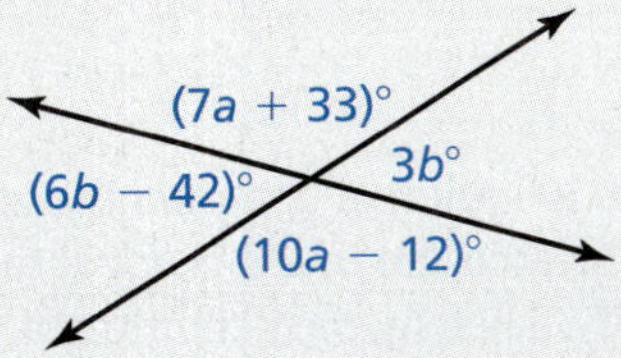

GO DIGITAL

EXAMPLE 5 Using the Vertical Angles Congruence Theorem

WATCH

Write a paragraph proof.

Given $\angle 1 \cong \angle 4$

Prove $\angle 2 \cong \angle 3$

Paragraph Proof

$\angle 1$ and $\angle 4$ are congruent. By the Vertical Angles Congruence Theorem, $\angle 1 \cong \angle 2$ and $\angle 3 \cong \angle 4$. By the Transitive Property of Angle Congruence, $\angle 2 \cong \angle 4$. Using the Transitive Property of Angle Congruence once more, $\angle 2 \cong \angle 3$.

4 MTR COMPARE METHODS

Rewrite the proof as a flowchart proof and a two-column proof. Compare the proofs. Which format do you prefer? Explain.

SELF-ASSESSMENT

1 I don't understand yet. 2 I can do it with help. 3 I can do it on my own. 4 I can teach someone else.

9. Write a paragraph proof.

Given $\angle 1$ is a right angle.

Prove $\angle 2$ is a right angle.

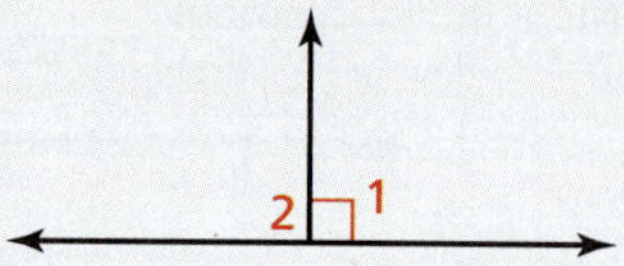

CONCEPT SUMMARY

Types of Proofs for the Symmetric Property of Angle Congruence

Given $\angle 1 \cong \angle 2$

Prove $\angle 2 \cong \angle 1$

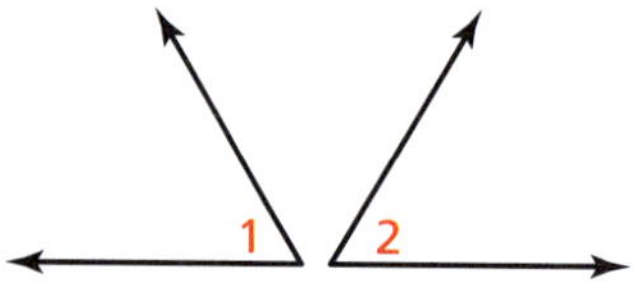

Two-Column Proof

STATEMENTS	REASONS
1. $\angle 1 \cong \angle 2$	**1.** Given
2. $m\angle 1 = m\angle 2$	**2.** Definition of congruent angles
3. $m\angle 2 = m\angle 1$	**3.** Symmetric Property of Equality
4. $\angle 2 \cong \angle 1$	**4.** Definition of congruent angles

Flowchart Proof

$\angle 1 \cong \angle 2$ (Given) → $m\angle 1 = m\angle 2$ (Definition of congruent angles) → $m\angle 2 = m\angle 1$ (Symmetric Property of Equality) → $\angle 2 \cong \angle 1$ (Definition of congruent angles)

Paragraph Proof

$\angle 1$ is congruent to $\angle 2$. By the definition of congruent angles, the measure of $\angle 1$ is equal to the measure of $\angle 2$. The measure of $\angle 2$ is equal to the measure of $\angle 1$ by the Symmetric Property of Equality. Then by the definition of congruent angles, $\angle 2$ is congruent to $\angle 1$.

GO DIGITAL

2.6 Practice WITH CalcChat® AND CalcView®

In Exercises 1–4, identify the pair(s) of congruent angles in the figures. Explain how you know they are congruent.

▶ **1.**

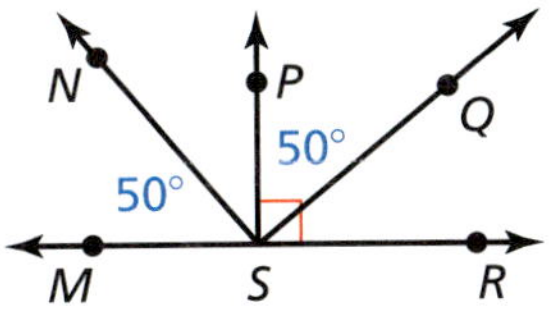

2.

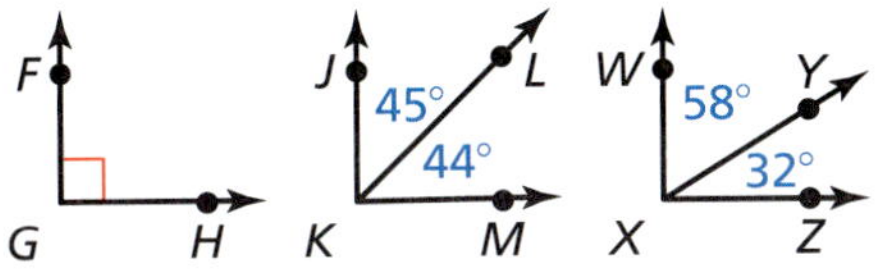

3.

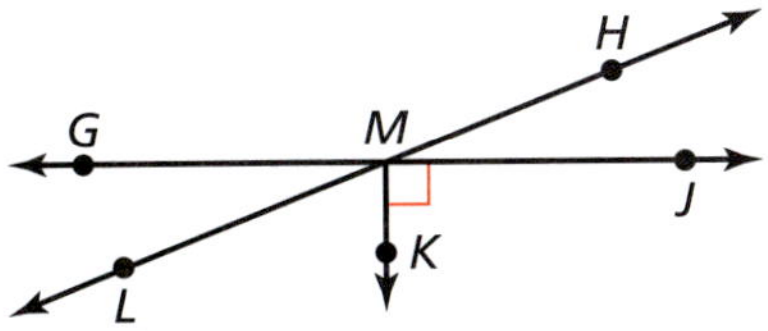

4. $\angle ABC$ is supplementary to $\angle CBD$.
$\angle CBD$ is supplementary to $\angle DEF$.

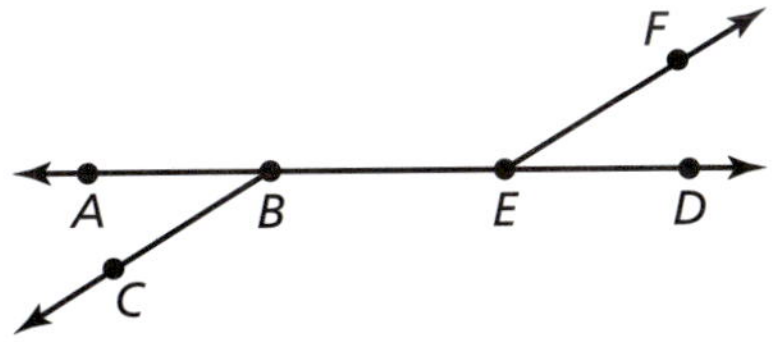

In Exercises 5–8, use the diagram and the given angle measure to find the other three measures.

▶ **5.** $m\angle 1 = 143°$

6. $m\angle 3 = 159°$

7. $m\angle 2 = 34°$

8. $m\angle 4 = 29°$

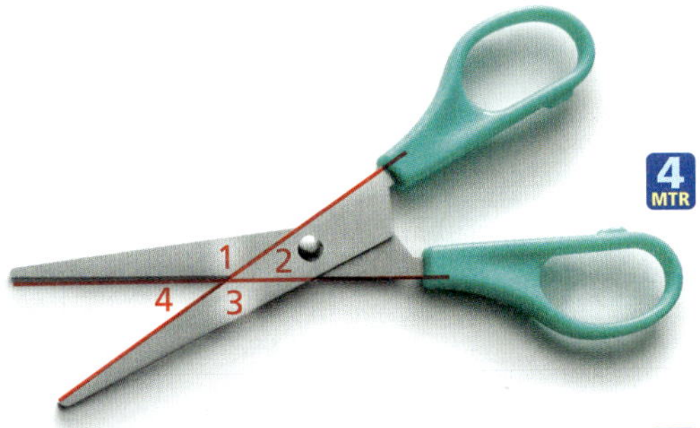

In Exercises 9–12, find the value of each variable.
(See Example 4.)

▶ **9.**

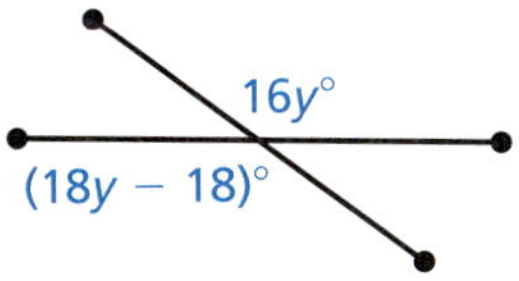

10.

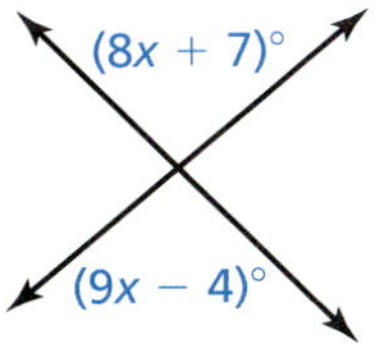

11.

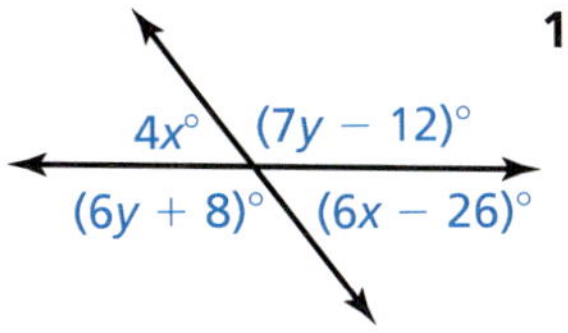

12.

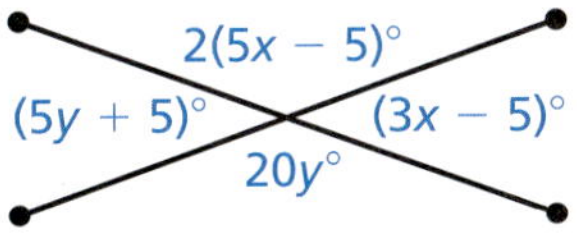

4 MTR **13. ERROR ANALYSIS** Describe and correct the error in using the diagram to find the value of x.

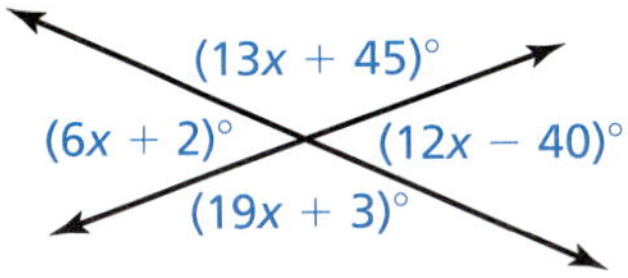

✗
$$(13x + 45)° + (19x + 3)° = 180°$$
$$32x + 48 = 180$$
$$32x = 132$$
$$x = 4.125$$

14. COLLEGE PREP Which statements can you conclude from the diagram? Select all that apply.

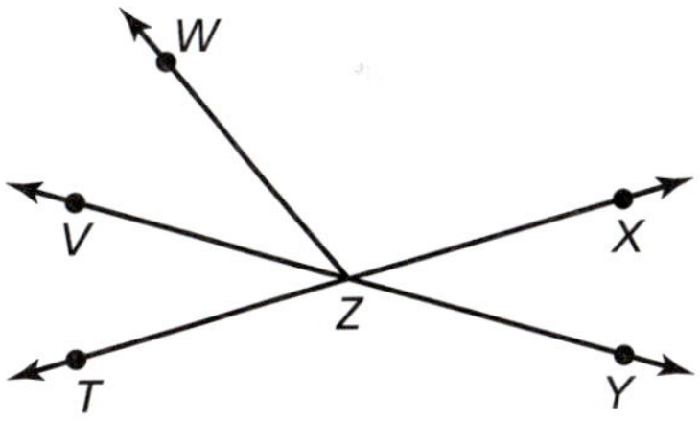

Ⓐ $\angle VZT \cong \angle XZY$ Ⓒ $\angle VZX \cong \angle TZY$

Ⓑ $\angle VZT \cong \angle VZW$ Ⓓ $\angle VZW \cong \angle XZY$

4 MTR **15. MAKING AN ARGUMENT** Your friend claims that $\angle 1 \cong \angle 4$ because they are vertical angles. Is your friend correct? Support your answer with definitions or theorems.

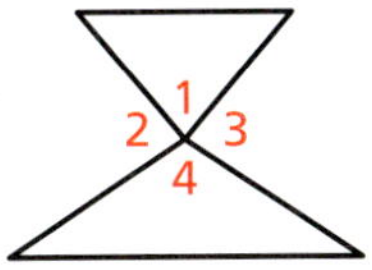

5 MTR **16. STRUCTURE** Find the measure of each angle in the diagram.

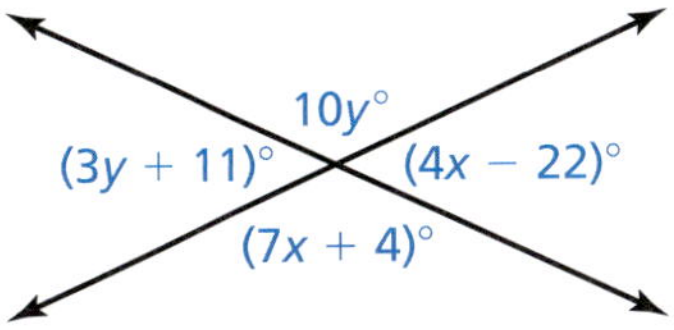

GO DIGITAL

17. PROOF Complete the flowchart proof. Then write a two-column proof. *(See Example 1.)*

Given ∠1 ≅ ∠3

Prove ∠2 ≅ ∠4

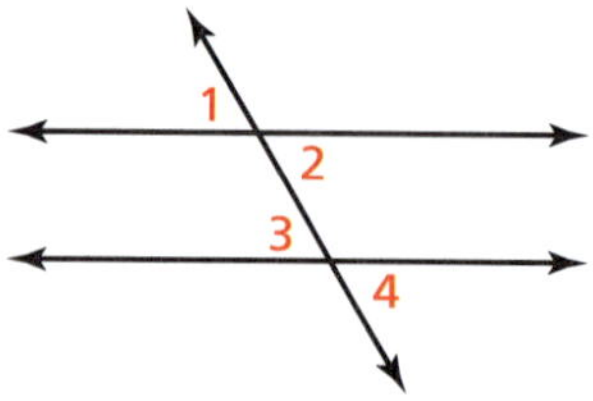

∠1 ≅ ∠3 → ∠1 ≅ ∠2, ∠3 ≅ ∠4 → ∠2 ≅ ∠3 → ∠2 ≅ ∠4

∠1 ≅ ∠3	∠1 ≅ ∠2, ∠3 ≅ ∠4	∠2 ≅ ∠3	∠2 ≅ ∠4
Given	Vertical Angles Congruence Theorem	_____	_____

18. PROOF Complete the two-column proof. Then write a flowchart proof. *(See Example 2.)*

Given ∠*ABD* is a right angle.
∠*CBE* is a right angle.

Prove ∠*ABC* ≅ ∠*DBE*

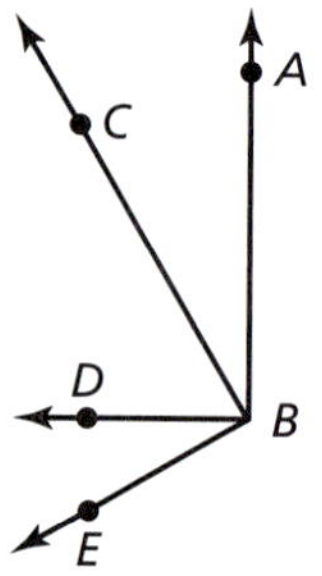

STATEMENTS	REASONS
1. ∠*ABD* is a right angle. ∠*CBE* is a right angle.	**1.** _______________
2. ∠*ABC* and ∠*CBD* are complementary.	**2.** Definition of complementary angles
3. ∠*DBE* and ∠*CBD* are complementary.	**3.** _______________
4. ∠*ABC* ≅ ∠*DBE*	**4.** _______________

19. PROVING A THEOREM Complete the paragraph proof for the Congruent Complements Theorem. Then write a two-column proof. *(See Example 3.)*

Given ∠1 and ∠2 are complementary.
∠1 and ∠3 are complementary.

Prove ∠2 ≅ ∠3

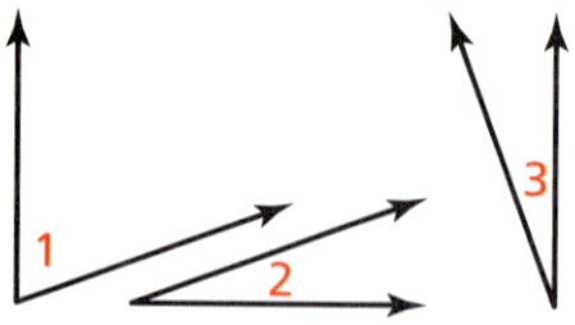

∠1 and ∠2 are complementary, and ∠1 and ∠3 are complementary. By the definition of ____________ angles, $m\angle 1 + m\angle 2 = 90°$ and ____________ $= 90°$. By the ________________________, $m\angle 1 + m\angle 2 = m\angle 1 + m\angle 3$. By the Subtraction Property of Equality, ________________. So, ∠2 ≅ ∠3 by the definition of ____________.

GO DIGITAL

20. PROVING A THEOREM Complete the two-column proof for the Congruent Supplements Theorem. Then write a paragraph proof. *(See Example 5.)*

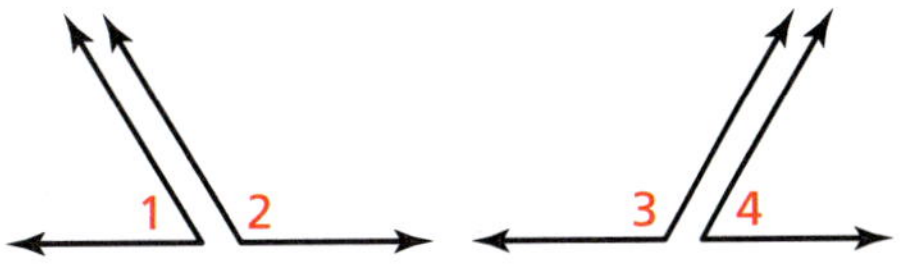

Given ∠1 and ∠2 are supplementary.
∠3 and ∠4 are supplementary.
∠1 ≅ ∠4

Prove ∠2 ≅ ∠3

STATEMENTS	REASONS
1. ∠1 and ∠2 are supplementary. ∠3 and ∠4 are supplementary. ∠1 ≅ ∠4	1. Given
2. $m\angle 1 + m\angle 2 = 180°$, $m\angle 3 + m\angle 4 = 180°$	2. ________
3. ________ $= m\angle 3 + m\angle 4$	3. Transitive Property of Equality
4. $m\angle 1 = m\angle 4$	4. Definition of congruent angles
5. $m\angle 1 + m\angle 2 =$ ________	5. Substitution Property of Equality
6. $m\angle 2 = m\angle 3$	6. ________
7. ________	7. ________

21. WRITING Explain why you do not use inductive reasoning when writing a proof.

22. HOW DO YOU SEE IT?
Consider the two-column proof. What is the writer trying to prove?

Given ∠1 ≅ ∠2
∠1 and ∠2 are supplementary.

Prove ________

STATEMENTS	REASONS
1. ∠1 ≅ ∠2 ∠1 and ∠2 are supplementary.	1. Given
2. $m\angle 1 = m\angle 2$	2. Definition of congruent angles
3. $m\angle 1 + m\angle 2 = 180°$	3. Definition of supplementary angles
4. $m\angle 1 + m\angle 1 = 180°$	4. Substitution Property of Equality
5. $2(m\angle 1) = 180°$	5. Simplify.
6. $m\angle 1 = 90°$	6. Division Property of Equality
7. $m\angle 2 = 90°$	7. Transitive Property of Equality
8. ________	8. ________

GO DIGITAL

3 MTR **CHOOSE A METHOD** **In Exercises 23–26, write a proof using any format. Explain your choice.**

23. **Given** $\angle QRS$ and $\angle PSR$ are supplementary.

Prove $\angle QRL \cong \angle PSR$

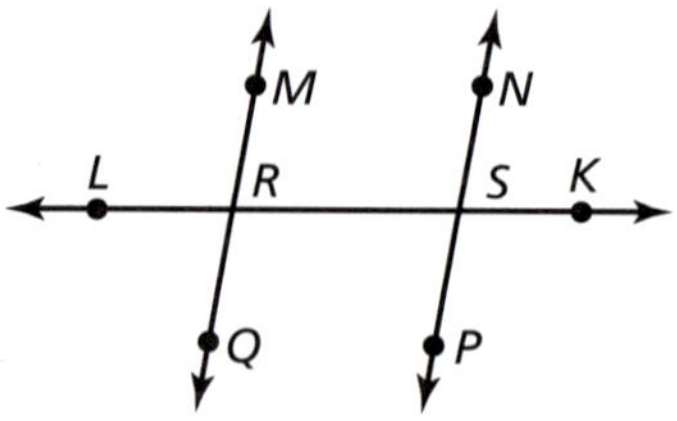

24. **Given** $\angle 1$ and $\angle 3$ are complementary.
$\angle 2$ and $\angle 4$ are complementary.

Prove $\angle 1 \cong \angle 4$

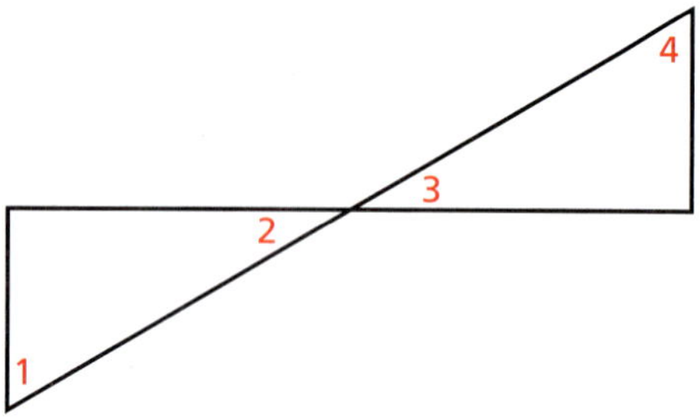

25. **Given** $\overline{JK} \perp \overline{JM}$, $\overline{KL} \perp \overline{ML}$,
$\angle J \cong \angle M$, $\angle K \cong \angle L$

Prove $\overline{JM} \perp \overline{ML}$ and $\overline{JK} \perp \overline{KL}$

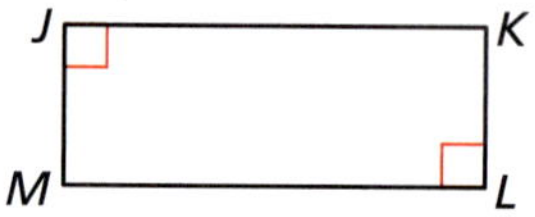

26. **Given** $\angle AEB \cong \angle DEC$

Prove $\angle AEC \cong \angle DEB$

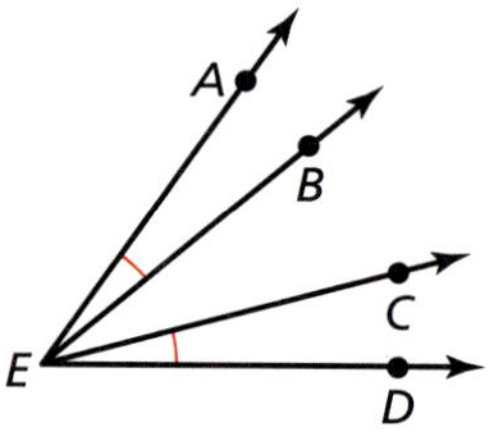

27. **REASONING** Is the converse of the Linear Pair Postulate true? If so, write a biconditional statement. If not, explain why not.

28. **THOUGHT PROVOKING**
Draw three lines all intersecting at the same point. Label two of the angle measures so that you can find the remaining four angle measures. Explain how you chose which angle measures to label.

REVIEW & REFRESH

WATCH

29. Complete the statement. Name the property you use.

If $\overline{RS} \cong \overline{TU}$ and $\overline{TU} \cong \overline{VW}$, then ____ $\cong$ ____.

In Exercises 30–32, use the cube.

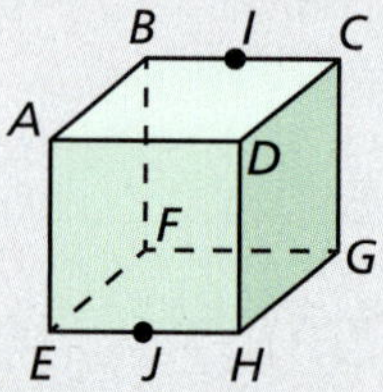

30. Name three collinear points.

31. Write an example of the Three Point Postulate.

32. Name two planes containing $\overline{BC}$.

7 MTR **33.** **MODELING REAL LIFE** The final velocity v_f of an object is given by the formula $v_f = v_i + at$, where v_i is the initial velocity, a is the acceleration, and t is the time.

a. Solve the formula for t. Justify each step.

b. A car with an initial velocity of 14 meters per second accelerates at a constant rate of 2.5 meters per second squared. How many seconds does it take the car to reach a final velocity of 29 meters per second?

34. Complete the square for $x^2 - 14x$. Then factor the trinomial.

35. Complete the two-column proof. Then write a flowchart proof.

Given $\angle 1 \cong \angle 2$

Prove $\angle 1 \cong \angle 3$

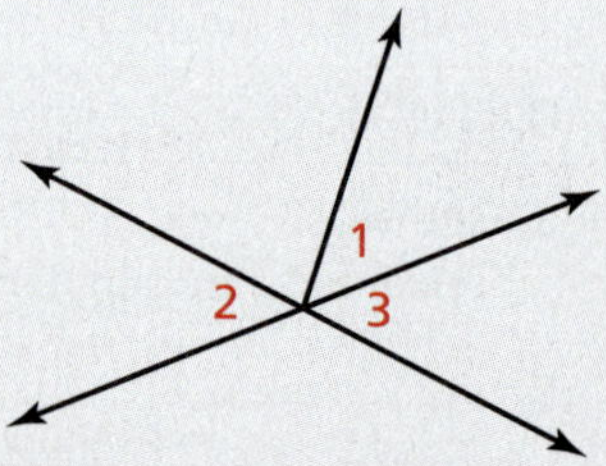

STATEMENTS	REASONS
1. $\angle 1 \cong \angle 2$	**1.** Given
2. $\angle 2 \cong \angle 3$	**2.** ____________
3. ________	**3.** Transitive Property of Angle Congruence

In Exercises 36–39, graph the function.

36. $f(x) = -x^2 - 6$

37. $g(x) = 2x^2 - 2x + 3$

38. $y = (x + 2)(x - 4)$

39. $y = -3(x - 1)^2 + 4$

GO DIGITAL

2 Chapter Review WITH CalcChat®

Chapter Learning Target: Understand reasoning and proofs.

Chapter Success Criteria:
- ◆ I can use inductive and deductive reasoning.
- ◆ I can justify steps using algebraic reasoning.
- ■ I can explain postulates using diagrams.
- ■ I can prove geometric relationships.

◆ Surface
■ Deep

SELF-ASSESSMENT 1 I don't understand yet. 2 I can do it with help. 3 I can do it on my own. 4 I can teach someone else.

2.1 Conditional Statements (pp. 65–72)

WATCH

Learning Target: Understand and write conditional statements.

Vocabulary (VOCAB)
conditional statement
if-then form
hypothesis
conclusion
negation
converse
inverse
contrapositive
equivalent statements
perpendicular lines
biconditional statement

Write the if-then form, the converse, the inverse, the contrapositive, and the biconditional of the conditional statement.

1. Two lines intersect in a point.
2. $4x + 9 = 21$ because $x = 3$.
3. Supplementary angles sum to 180°.
4. Right angles are 90°.

Decide whether the statement about the diagram is true. Explain your answer using the definitions you have learned.

5. S is the midpoint of $\overline{EF}$.
6. $\overline{ES} \cong \overline{ST}$
7. $\overleftrightarrow{ST}$ is a segment bisector of $\overline{EF}$.

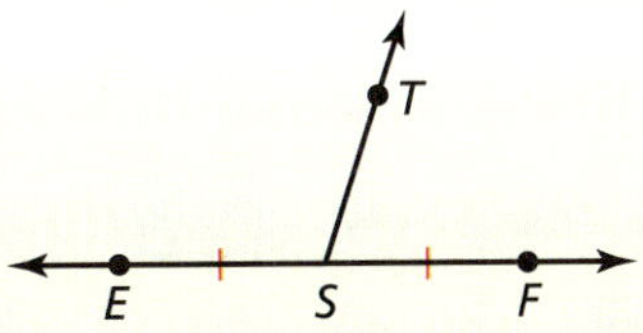

2.2 Inductive and Deductive Reasoning (pp. 73–80)

WATCH

Learning Target: Use inductive and deductive reasoning.

Vocabulary (VOCAB)
conjecture
inductive reasoning
counterexample
deductive reasoning

8. Make and test a conjecture about the difference of any two odd integers.
9. Make and test a conjecture about the product of an even and an odd integer.
10. If an angle is a right angle, then the angle measures 90°. $\angle B$ is a right angle. Using the Law of Detachment, what statement can you make?
11. Use the Law of Syllogism to write a new conditional statement that follows from the pair of true statements: If $x = 3$, then $2x = 6$. If $4x = 12$, then $x = 3$.

Decide whether inductive reasoning or deductive reasoning is used to reach the conclusion. Explain your reasoning.

12. The wolf population in a park has increased each year for the last 10 years. So, the wolf population will increase again next year.
13. The dew point is the warmest temperature at which the relative humidity reaches 100 percent. On a given night, the relative humidity reaches 100 percent at the moment the temperature drops to 72°. So, the dew point is 72°.

GO DIGITAL

2.3 Postulates and Diagrams (pp. 81–86) WATCH

Learning Target: Interpret and sketch diagrams.

Vocabulary AZ VOCAB
line perpendicular to a plane

14. State the postulate illustrated by the diagram.

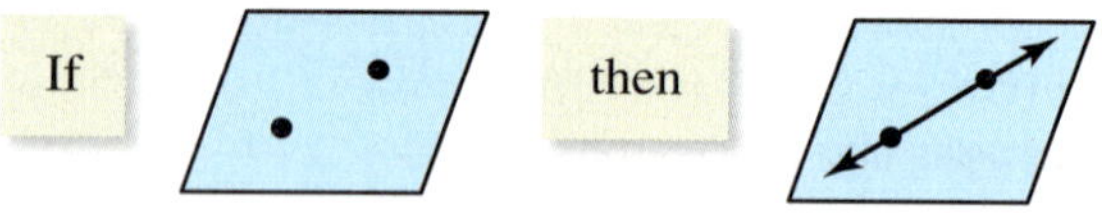

Sketch a diagram of the description.

15. $\angle ABC$, an acute angle, is bisected by $\overrightarrow{BE}$.

16. $\angle CDE$, a straight angle, is bisected by $\overleftrightarrow{DK}$.

17. Plane P intersects plane R at $\overleftrightarrow{XY}$. $\overline{ZW}$ lies in plane P. Plane $P \perp$ plane R.

Use the diagram at the right to determine whether you can assume the statement.

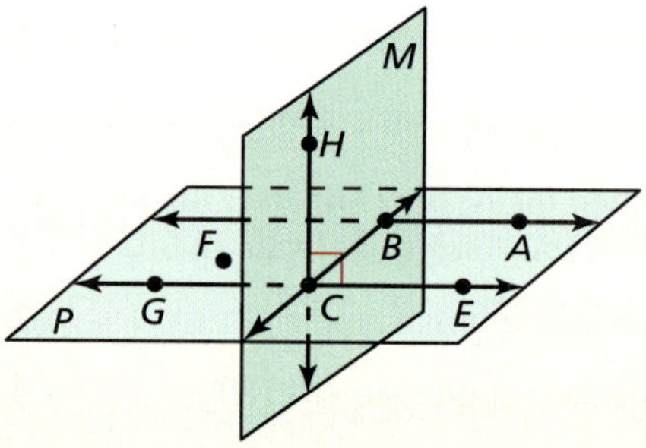

18. Points A, B, C, and E are coplanar.

19. $\overleftrightarrow{HC} \perp \overleftrightarrow{GE}$

20. Points F, B, and G are collinear.

21. $\overleftrightarrow{AB} \parallel \overleftrightarrow{GE}$

22. $\overleftrightarrow{AB}$ lies in plane S. Points A, B, and C lie in plane R. Point C does not lie in S. Describe the intersection of plane R and plane S. Make a sketch to support your answer. State three postulates that support your answer.

2.4 Algebraic Reasoning (pp. 87–94) WATCH

Learning Target: Use properties of equality to solve problems.

Solve the equation. Justify each step.

23. $-9x - 21 = -20x - 87$

24. $15x + 22 = 7x + 62$

25. $3(2x + 9) = 30$

26. $5x + 2(2x - 23) = -154$

Name the property of equality that the statement illustrates.

27. If $LM = RS$ and $RS = 25$, then $LM = 25$.

28. $AM = AM$

29. If $3XY = 36$, then $XY = 12$.

30. If $m\angle A = m\angle B$, then $m\angle B = m\angle A$.

31. A formula for the volume V (in gallons) of a cylinder with a diameter of 22.5 inches is

$$V = \frac{11.25^2 \pi h}{231}$$

where h is the height of the cylinder in inches. Solve the formula for h. Justify each step. Then find the height of the 55-gallon steel drum.

GO DIGITAL

2.5 Proving Statements about Segments and Angles (pp. 95–100)

WATCH

Learning Target: Prove statements about segments and angles.

Vocabulary

proof
two-column proof
theorem

Name the property that the statement illustrates.

32. If $\angle DEF \cong \angle JKL$, then $\angle JKL \cong \angle DEF$.

33. $\angle C \cong \angle C$

34. If $MN = PQ$ and $PQ = RS$, then $MN = RS$.

35. Write a two-column proof for the Reflexive Property of Angle Congruence.

36. A ramp at a skate park is constructed with $\angle BAD \cong \angle CDA$, as pictured. Prove that $\angle EAB \cong \angle FDC$.

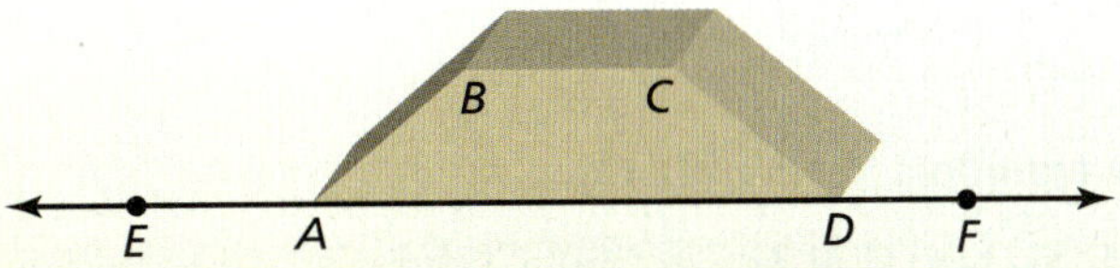

2.6 Proving Geometric Relationships (pp. 101–110)

WATCH

Learning Target: Prove geometric relationships.

Vocabulary

flowchart proof, or flow proof
paragraph proof, or narrative proof

37. Complete the flowchart proof. Then write a two-column proof.

Given $\angle 3$ and $\angle 2$ are complementary.
$m\angle 1 + m\angle 2 = 90°$

Prove $\angle 3 \cong \angle 1$

$m\angle 1 + m\angle 2 = 90°$ → $\angle 1$ and $\angle 2$ are complementary. → $\angle 3$ and $\angle 2$ are complementary. → $\angle 3 \cong \angle 1$

Given | Definition of complementary angles | ______ | ______

38. Write a proof using any format. Explain your choice.

Given $m\angle ABC = 48°$, $m\angle ABD = 24°$

Prove $\overrightarrow{BD}$ bisects $\angle ABC$.

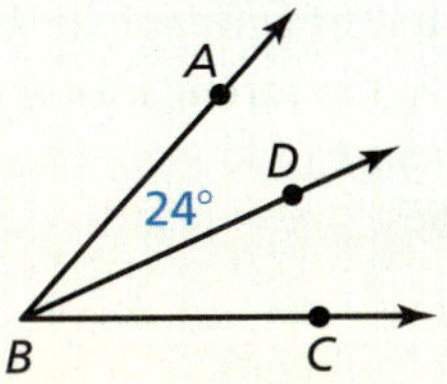

Mathematical Thinking and Reasoning

4 MTR **ENGAGE IN DISCUSSIONS THAT REFLECT ON THE MATHEMATICAL THINKING OF SELF AND OTHERS**

Mathematicians who engage in discussions that reflect on the mathematical thinking of self and others recognize errors and suggest how to correctly solve tasks.

1. In Exercise 15 on page 107, write an argument your friend might use to support the claim that $\angle 1 \cong \angle 4$. Is this reasoning based on the Law of Detachment or the Law of Syllogism? Explain why the argument fails.

2. Write the inverse of the Right Angles Congruence Theorem. Does this statement provide a valid way to show that two given acute angles are not congruent? Explain your reasoning.

GO DIGITAL

2 Practice Test WITH CalcChat®

Use the diagram to determine whether you can assume the statement. Explain your reasoning.

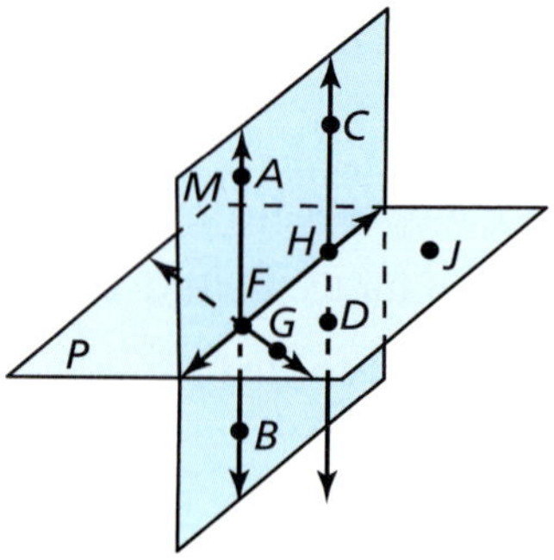

1. $\overleftrightarrow{CD} \perp$ plane P
2. Points A, B, and C are coplanar.
3. Planes M and P intersect at $\overleftrightarrow{FH}$.
4. $\overleftrightarrow{DJ}$ intersects $\overleftrightarrow{FG}$ at point G.

Solve the equation. Justify each step.

5. $9x + 31 = -23 + 3x$
6. $26 + 2(3x + 11) = -18$

Write the if-then form, the converse, the inverse, the contrapositive, and the biconditional of the conditional statement. Then decide whether each statement is *true* or *false*.

7. Two planes intersect at a line.
8. A relation that pairs each input with exactly one output is a function.

9. Use inductive reasoning to make a conjecture about the sum of three odd integers. Then use deductive reasoning to show that the conjecture is true.

10. Both statements below are true. Can you apply the Law of Detachment to conclude something about the situation? Explain.

    If a figure is a rectangle, then it has four sides.
    Quadrilateral $ABCD$ has four sides.

11. The formula for the area A of a triangle is $A = \frac{1}{2}bh$, where b is the base and h is the height. Solve the formula for h and justify each step. Then find the height of a standard yield sign when the area is 558 square inches and each side is 36 inches.

12. You visit the zoo and notice the following.

    - The elephants, giraffes, lions, tigers, and zebras are located along a straight walkway.
    - The giraffes are halfway between the elephants and the lions.
    - The tigers are halfway between the lions and the zebras.
    - The lions are halfway between the giraffes and the tigers.

    Draw and label a diagram that represents this information. Then prove that the distance between the elephants and the giraffes is equal to the distance between the tigers and the zebras. Use any proof format. Explain your choice.

13. Write a proof using any format. Explain your choice.

    Given $\angle 2 \cong \angle 3$

    $\overrightarrow{TV}$ bisects $\angle UTW$.

    Prove $\angle 1 \cong \angle 3$

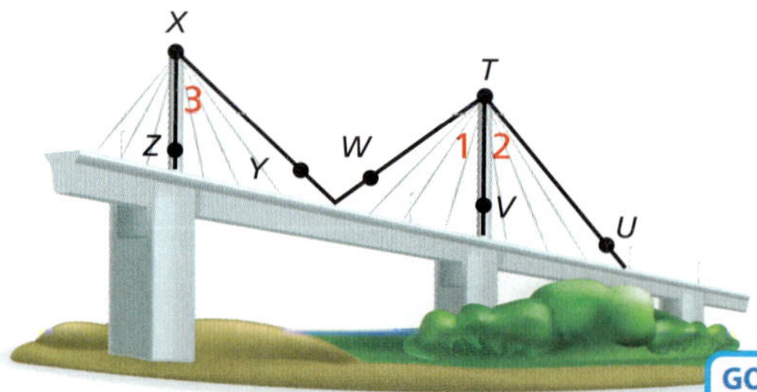

GO DIGITAL

2 Performance Task

The Greenhouse Effect

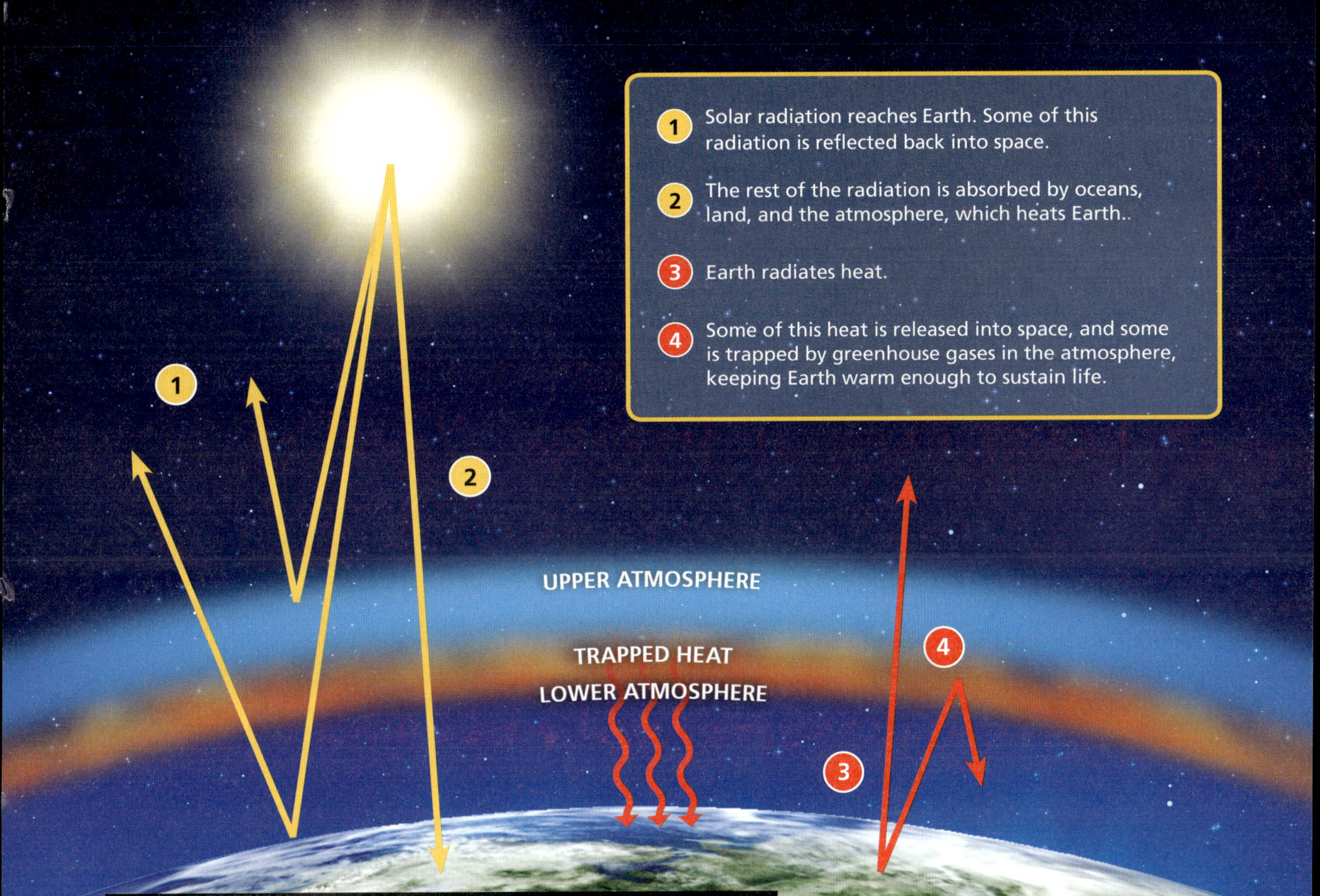

1. Solar radiation reaches Earth. Some of this radiation is reflected back into space.
2. The rest of the radiation is absorbed by oceans, land, and the atmosphere, which heats Earth..
3. Earth radiates heat.
4. Some of this heat is released into space, and some is trapped by greenhouse gases in the atmosphere, keeping Earth warm enough to sustain life.

CLIMATE CHANGE CONDITIONALS

Greenhouse gases are a major contributor to climate change. Use the Internet or other resources to research the effects of climate change.

- Write three conditional statements based on your research.
- Write the converse, inverse, and contrapositive of each statement. Then explain whether each statement is *true* or *false*.
- Find a conjecture about climate change, and a counterexample that shows that the conjecture is false.

Greenhouses Gases Include:

- Carbon Dioxide (CO_2)
- Methane (CH_4)
- Nitrous Oxide (N_2O)
- Hydrofluorocarbons (HFCs)
- Sulfur Hexafluoride (SF_6)
- Nitrogen Trifluoride (NF_3)

GO DIGITAL

2 Test Prep WITH CalcChat®
Cumulative Practice

WATCH Tutorial videos are available for each exercise.

1. Use the diagram to write an example of each postulate.

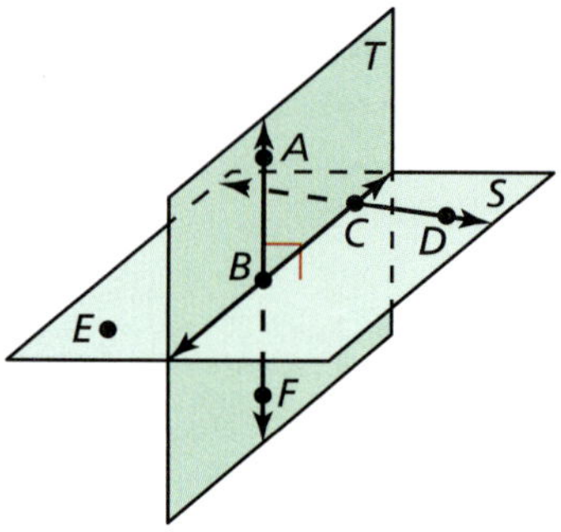

a. **Two Point Postulate** Through any two points, there exists exactly one line.

b. **Line Intersection Postulate** If two lines intersect, then their intersection is exactly one point.

c. **Plane-Line Postulate** If two points lie in a plane, then the line containing them lies in the plane.

d. **Plane Intersection Postulate** If two planes intersect, then their intersection is a line.

2. Complete the two-column proof.

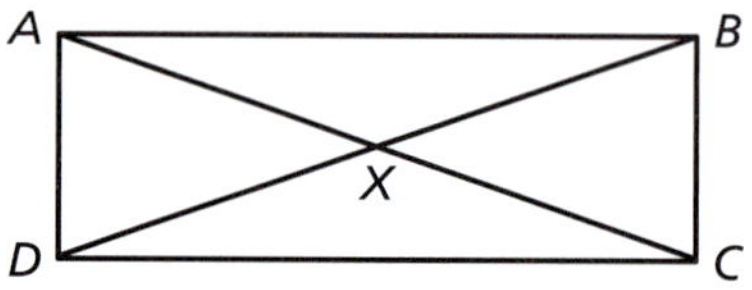

Given $\overline{AX} \cong \overline{DX}, \overline{XB} \cong \overline{XC}$

Prove $\overline{AC} \cong \overline{BD}$

STATEMENTS	REASONS
1. $\overline{AX} \cong \overline{DX}$	**1.** Given
2. $AX = DX$	**2.** ________________
3. $\overline{XB} \cong \overline{XC}$	**3.** Given
4. $XB = XC$	**4.** ________________
5. $AX + XC = AC$	**5.** ________________
6. $DX + XB = DB$	**6.** ________________
7. $AC = DX + XB$	**7.** ________________
8. $AC = BD$	**8.** ________________
9. $\overline{AC} \cong \overline{BD}$	**9.** ________________

3. Which lines are segment bisectors when $x = 3$?

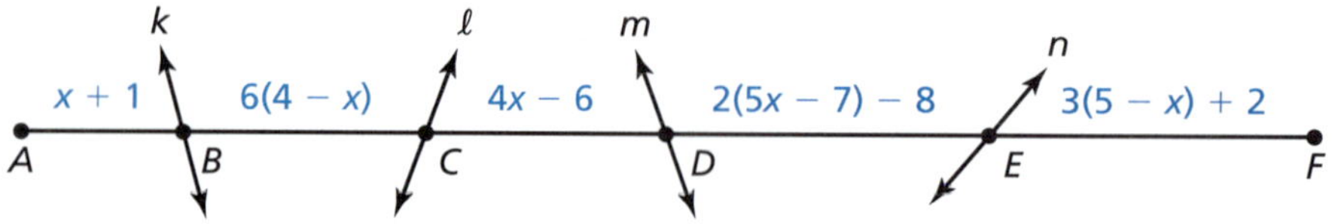

Ⓐ line n

Ⓑ line ℓ and line n

Ⓒ line k, line ℓ, and line n

Ⓓ line ℓ, line m, and line n

GO DIGITAL

4. Classify each related conditional statement, based on the conditional statement "If I study, then I will pass the final exam."

a. I will pass the final exam if and only if I study.

b. If I do not study, then I will not pass the final exam.

c. If I pass the final exam, then I studied.

d. If I do not pass the final exam, then I did not study.

5. Consider the partial construction of $\angle S$ with the same measure as $\angle P$. What is the next step in the construction?

Ⓐ Draw a ray through S that intersects the arc.

Ⓑ Draw a ray through T that intersects the arc.

Ⓒ Draw an arc with radius QR and center T.

Ⓓ Draw an arc with radius PQ and center T.

6. Which pair of points are the endpoints of the longest line segment?

Ⓐ $A(-6, 1), B(-1, 6)$

Ⓑ $E(2, 7), F(4, -2)$

Ⓒ $J(-4, -2), K(1, -5)$

Ⓓ $L(3, -8), M(7, -5)$

7. Which statement *cannot* be concluded from the diagram?

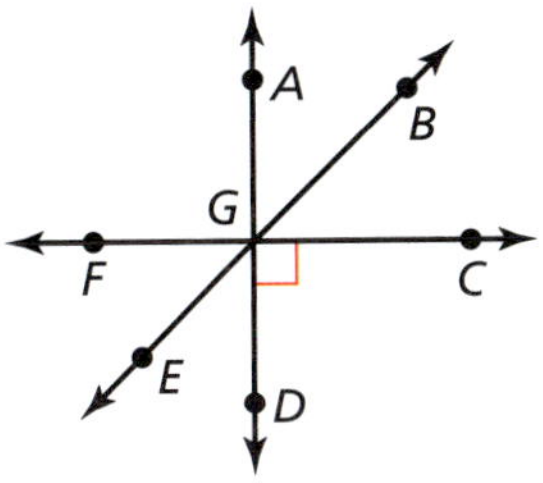

Ⓐ $\angle AGB \cong \angle DGE$

Ⓑ $\angle AGC$ is a right angle.

Ⓒ $\angle AGB$ and $\angle DGB$ form a linear pair.

Ⓓ $\angle AGE$ and $\angle BGC$ are supplementary angles.

8. Your teacher assigns your class a homework problem that asks you to prove the Vertical Angles Congruence Theorem using the diagram and information given.

Given $\angle 1$ and $\angle 3$ are vertical angles.

Prove $\angle 1 \cong \angle 3$

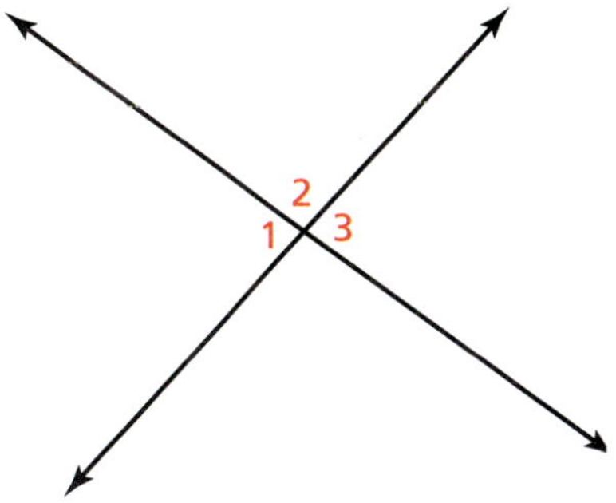

Which property must you use in your proof?

Ⓐ Line Intersection Postulate

Ⓑ Linear Pair Postulate

Ⓒ Congruent Complements Theorem

Ⓓ Reflexive Property of Angle Congruence

GO DIGITAL

3 Parallel and Perpendicular Lines

NATIONAL GEOGRAPHIC EXPLORER

Andrés Ruzo

WATCH

Andrés Ruzo is a geothermal scientist known for his work at the world's largest documented thermal river, the Boiling River of the Amazon. He is the founder and director of the Boiling River Project, conducting scientific research and conservation work in the Boiling River area. Andrés is also heavily involved in education. He serves on the boards of a high school and a university in Costa Rica, and is a Student Independent Research Teacher at schools in the United States.

- What is geothermal science?
- How can geothermal energy be used to generate electricity? What percent of the electricity used in the United States is geothermal?
- What is the typical temperature of Earth's crust at a depth of 1 mile? 2 miles?

STEM

Geothermal power plants harness the heat within the earth and use it to generate electricity. In the Performance Task, you will find a location for a new power plant that will provide electricity to several cities.

GO DIGITAL

Geothermal Science

Preparing for Chapter 3

Chapter Learning Target: Understand parallel and perpendicular lines.

Chapter Success Criteria:

- ◆ I can describe angle relationships formed by parallel lines and a transversal.
- ◆ I can construct parallel and perpendicular lines.
- ■ I can prove theorems involving parallel and perpendicular lines.
- ■ I can write equations of parallel and perpendicular lines.

◆ Surface
■ Deep

AZ VOCAB

Chapter Vocabulary

Work with a partner. Discuss each of the vocabulary terms.

parallel lines
parallel planes
transversal
corresponding angles
alternate interior angles
alternate exterior angles
consecutive interior angles
perpendicular bisector

Mathematical Thinking and Reasoning

3 MTR

COMPLETE TASKS WITH MATHEMATICAL FLUENCY
Mathematicians who complete tasks with mathematical fluency select efficient and appropriate methods for solving problems within the given context.

Work with a partner. The diagram shows two *parallel lines* that are cut by a *transversal.*

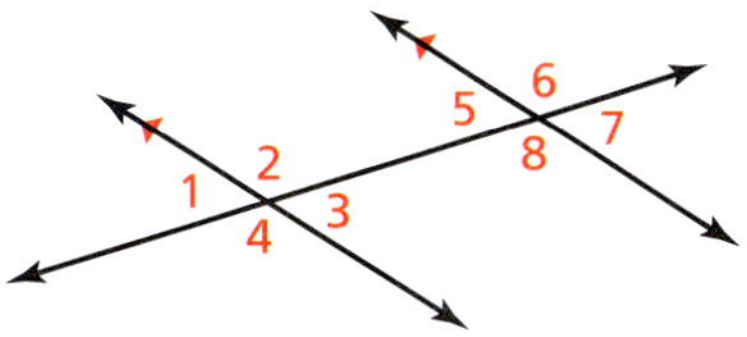

1. Draw two parallel lines cut by a transversal. Describe your method, and then make several observations about the angle measures.
2. The measure of $\angle 1$ in the diagram shown above is 50°. Use your observations in Exercise 1 to find the measures of the other angles. Explain your reasoning.
3. Two parallel lines cut by a transversal form eight angles. Is it possible for all of the angles to be acute? Explain how you found your answer.

GO DIGITAL

3 Prepare WITH CalcChat®

Finding the Slope of a Line

WATCH

Example 1 **Find the slope of the line.**

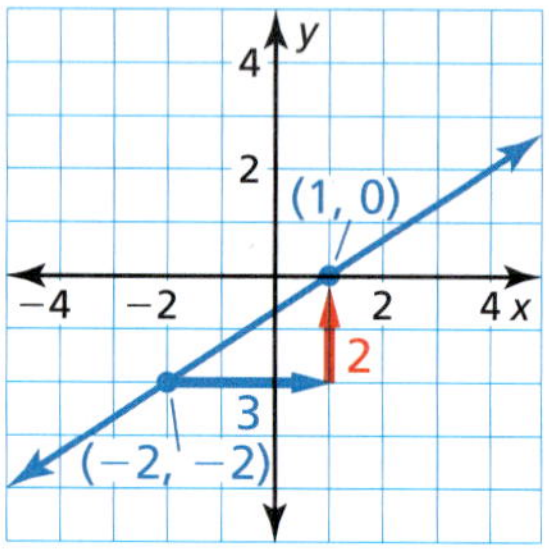

Let $(x_1, y_1) = (-2, -2)$ and $(x_2, y_2) = (1, 0)$.

$$\text{slope} = \frac{y_2 - y_1}{x_2 - x_1}$$ Write formula for slope.

$$= \frac{0 - (-2)}{1 - (-2)}$$ Substitute.

$$= \frac{2}{3}$$ Simplify.

Find the slope of the line.

1.

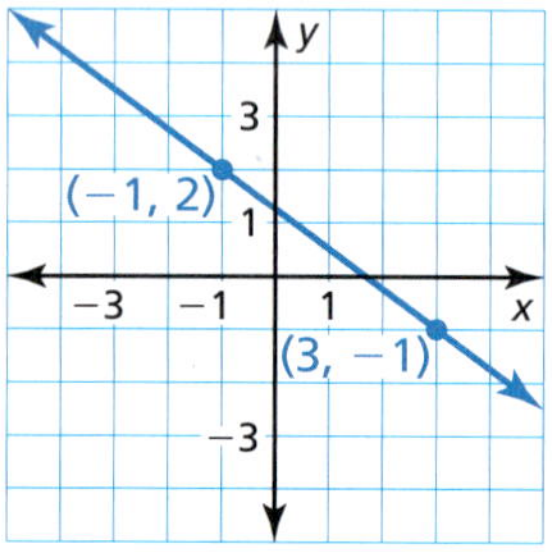

2.

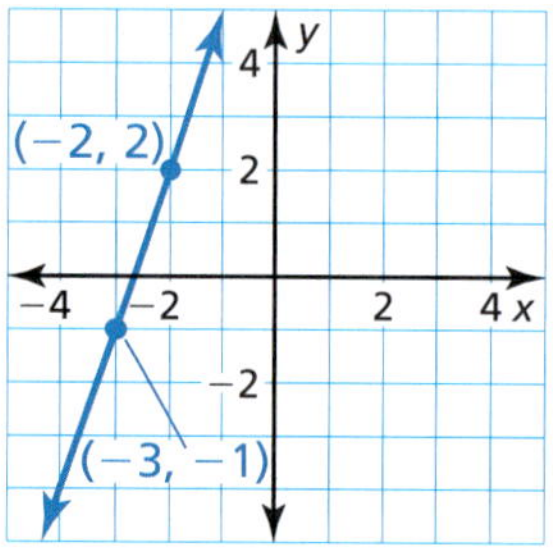

3.

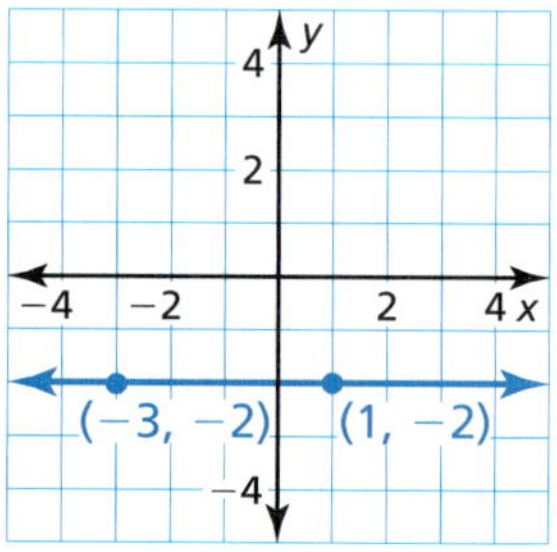

Writing Equations of Lines

WATCH

Example 2 **Write an equation of the line that passes through the point (−4, 5) and has a slope of $\frac{3}{4}$.**

$y = mx + b$	Write the slope-intercept form.
$5 = \frac{3}{4}(-4) + b$	Substitute $\frac{3}{4}$ for m, −4 for x, and 5 for y.
$5 = -3 + b$	Simplify.
$8 = b$	Solve for b.

▶ So, an equation is $y = \frac{3}{4}x + 8$.

Write an equation of the line that passes through the given point and has the given slope.

4. $(6, 1)$; $m = -3$

5. $(-3, 8)$; $m = -2$

6. $(-1, 5)$; $m = 4$

7. $(2, -4)$; $m = \frac{1}{2}$

8. $(-8, -5)$; $m = -\frac{1}{4}$

9. $(0, 9)$; $m = \frac{2}{3}$

4 MTR **10.** **DISCUSS MATHEMATICAL THINKING** Why does a horizontal line have a slope of 0, but a vertical line has an undefined slope?

GO DIGITAL

3.1 Pairs of Lines and Angles

Learning Target: Understand lines, planes, and pairs of angles.

Success Criteria:
- I can identify lines and planes.
- I can identify parallel and perpendicular lines.
- I can identify pairs of angles formed by transversals.

EXPLORE IT! Defining Parallel Lines

4 MTR CONSTRUCT AN ARGUMENT
How can you use counterexamples to identify flaws in each definition?

Work with a partner. Use a straightedge to draw a pair of *parallel lines*.

a. Three students write the definition of parallel lines in their own words. Analyze the definitions and explain how to improve them, if necessary.

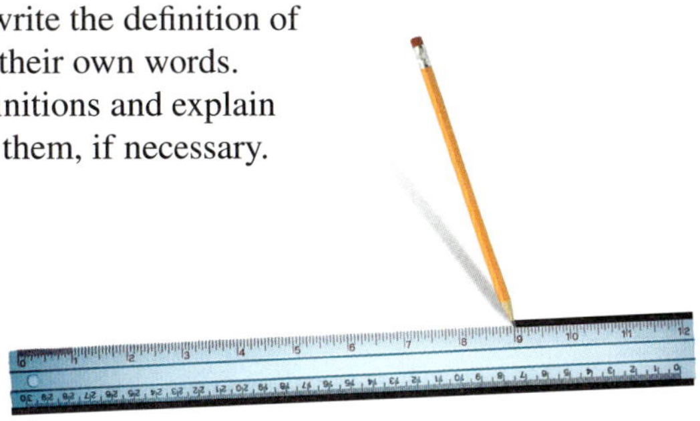

Student A
Two lines are parallel if they don't intersect.

Student B
Two lines are parallel if they have the same slope.

Student C
Two lines are parallel if the distance between them is always the same.

b. Write your own definition of parallel lines.

c. Write your own definition of perpendicular lines.

Geometric Reasoning
preparing for MA.912.GR.1.1 Prove relationships and theorems about lines and angles. Solve mathematical and real-world problems involving postulates, relationships and theorems of lines and angles.

GO DIGITAL

Vocabulary

parallel lines, *p. 122*
skew lines, *p. 122*
parallel planes, *p. 122*
transversal, *p. 124*
corresponding angles, *p. 124*
alternate interior angles, *p. 124*
alternate exterior angles, *p. 124*
consecutive interior angles, *p. 124*

KEY IDEAS

Parallel Lines, Skew Lines, and Parallel Planes

Two lines that do not intersect are either *parallel lines* or *skew lines*. Two lines are **parallel lines** when they do not intersect and are coplanar. Two lines are **skew lines** when they do not intersect and are not coplanar. Two planes that do not intersect are **parallel planes**.

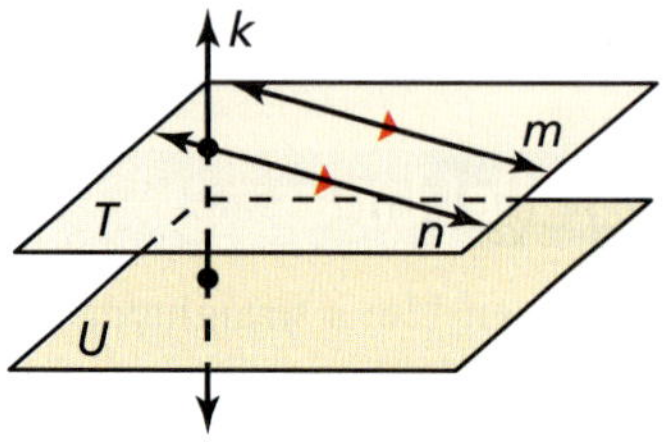

Lines m and n are parallel lines ($m \parallel n$).

Lines m and k are skew lines.

Planes T and U are parallel planes ($T \parallel U$).

Lines k and n are intersecting lines, and there is a plane (not shown) containing them.

Small directed arrows, as shown in red on lines m and n above, are used to show that lines are parallel. The symbol $\parallel$ means "is parallel to," as in $m \parallel n$.

Segments and rays are parallel when they lie in parallel lines. A line is parallel to a plane when the line is in a plane parallel to the given plane. In the diagram above, line n is parallel to plane U.

EXAMPLE 1 Identifying Lines and Planes

WATCH

REMEMBER

Recall that if two lines intersect to form a right angle, then they are perpendicular lines.

Consider the lines that contain the segments in the figure and the planes that contain the faces of the figure. Which line(s) or plane(s) appear to fit each description?

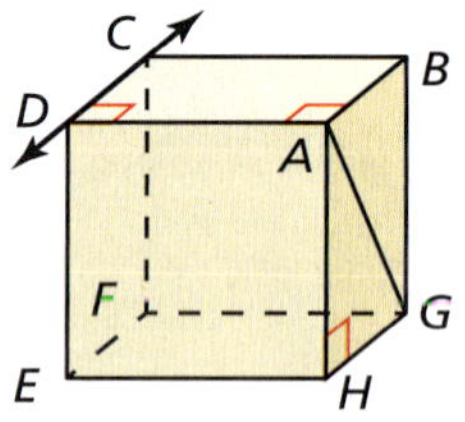

a. line(s) parallel to $\overleftrightarrow{CD}$ and containing point A

b. line(s) skew to $\overleftrightarrow{CD}$ and containing point A

c. line(s) perpendicular to $\overleftrightarrow{CD}$ and containing point A

d. plane(s) parallel to plane EFG and containing point A

SOLUTION

a. $\overleftrightarrow{AB}$, $\overleftrightarrow{HG}$, and $\overleftrightarrow{EF}$ all appear parallel to $\overleftrightarrow{CD}$, but only $\overleftrightarrow{AB}$ contains point A.

b. Both $\overleftrightarrow{AG}$ and $\overleftrightarrow{AH}$ appear skew to $\overleftrightarrow{CD}$ and contain point A.

c. $\overleftrightarrow{BC}$, $\overleftrightarrow{AD}$, $\overleftrightarrow{DE}$, and $\overleftrightarrow{FC}$ all appear perpendicular to $\overleftrightarrow{CD}$, but only $\overleftrightarrow{AD}$ contains point A.

d. Plane ABC appears parallel to plane EFG and contains point A.

SELF-ASSESSMENT 1 I don't understand yet. 2 I can do it with help. 3 I can do it on my own. 4 I can teach someone else.

Use the diagram in Example 1.

1. Name the line(s) through point F that appear skew to $\overleftrightarrow{EH}$.

2. Name the line(s) through point A that appear perpendicular to $\overleftrightarrow{GH}$.

GO DIGITAL

Identifying Parallel and Perpendicular Lines

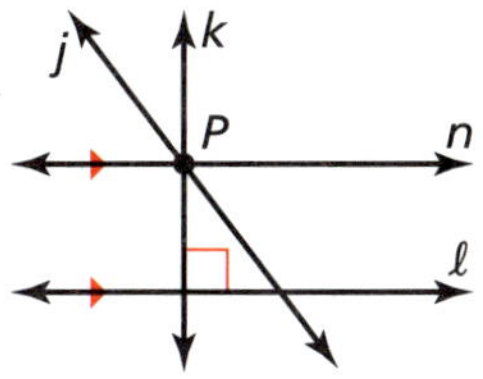

Two lines in the same plane are either parallel, like line ℓ and line n, or intersect in a point, like line j and line n.

Through a point not on a given line, there are infinitely many lines. Exactly one of these lines is parallel to the given line, and exactly one of them is perpendicular to the given line. For example, line k is the line through point P perpendicular to line ℓ, and line n is the line through point P parallel to line ℓ.

POSTULATES

3.1 Parallel Postulate

If there is a line and a point not on the line, then there is exactly one line through the point parallel to the given line.

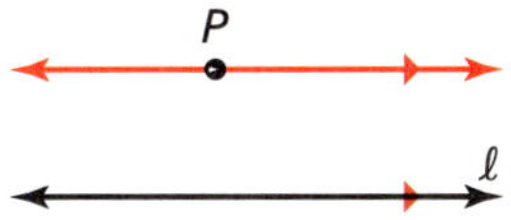

There is exactly one line through P parallel to ℓ.

3.2 Perpendicular Postulate

If there is a line and a point not on the line, then there is exactly one line through the point perpendicular to the given line.

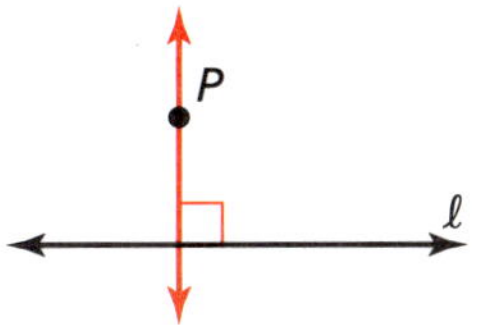

There is exactly one line through P perpendicular to ℓ.

EXAMPLE 2 Identifying Parallel and Perpendicular Lines

WATCH

The map shows how the roads in a town are related to one another.

a. Name a pair of parallel lines.

b. Name a pair of perpendicular lines.

c. Is $\overleftrightarrow{FE} \parallel \overleftrightarrow{AC}$? Explain.

SOLUTION

a. $\overleftrightarrow{MD} \parallel \overleftrightarrow{FE}$

b. $\overleftrightarrow{MD} \perp \overleftrightarrow{BF}$

c. $\overleftrightarrow{FE}$ is not parallel to $\overleftrightarrow{AC}$, because $\overleftrightarrow{MD}$ is parallel to $\overleftrightarrow{FE}$, and by the Parallel Postulate, there is exactly one line parallel to $\overleftrightarrow{FE}$ through M.

SELF-ASSESSMENT 1 I don't understand yet. 2 I can do it with help. 3 I can do it on my own. 4 I can teach someone else.

3. In Example 2, can you use the Perpendicular Postulate to show that $\overleftrightarrow{CE}$ is *not* perpendicular to $\overleftrightarrow{BF}$? Explain why or why not.

GO DIGITAL

Identifying Pairs of Angles

A **transversal** is a line that intersects two or more coplanar lines at different points.

WORDS AND MATH

The prefix *trans-* means "across." A transversal is a line that intersects, or lies across, two or more other lines.

KEY IDEA

Angles Formed by Transversals

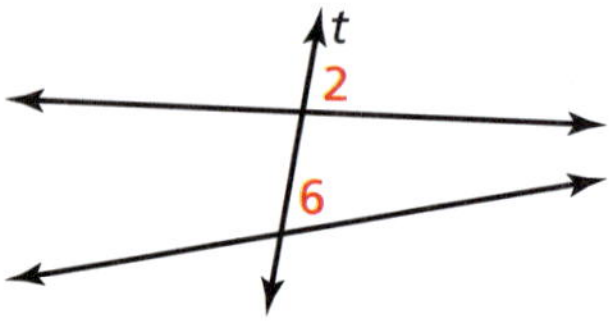

Two angles are **corresponding angles** when they have corresponding positions. For example, ∠2 and ∠6 are above the lines and to the right of the transversal *t*.

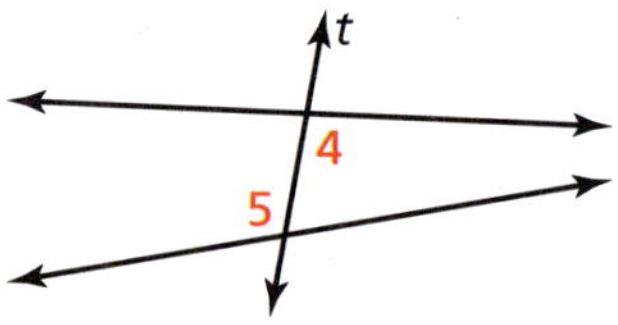

Two angles are **alternate interior angles** when they lie between the two lines and on opposite sides of the transversal *t*.

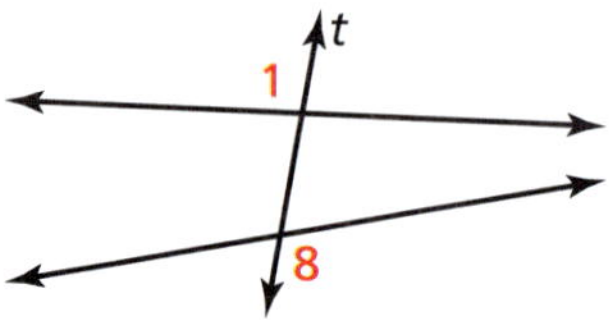

Two angles are **alternate exterior angles** when they lie outside the two lines and on opposite sides of the transversal *t*.

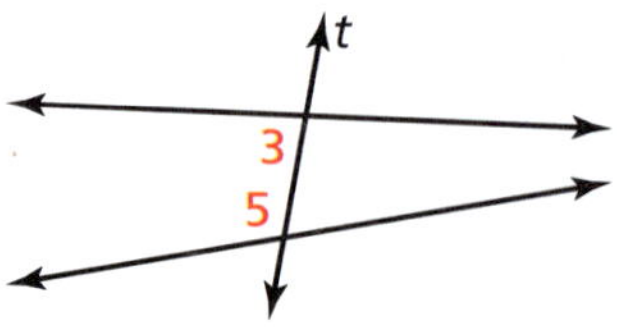

Two angles are **consecutive interior angles** when they lie between the two lines and on the same side of the transversal *t*.

EXAMPLE 3 Identifying Pairs of Angles

WATCH

Identify all pairs of angles of the given type.

a. corresponding

b. alternate interior

c. alternate exterior

d. consecutive interior

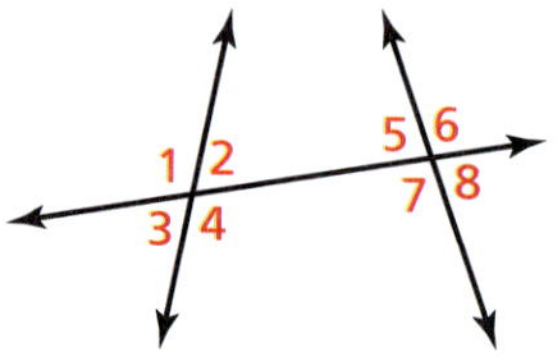

SOLUTION

a. ∠1 and ∠5
∠2 and ∠6
∠3 and ∠7
∠4 and ∠8

b. ∠2 and ∠7
∠4 and ∠5

c. ∠1 and ∠8
∠3 and ∠6

d. ∠2 and ∠5
∠4 and ∠7

SELF-ASSESSMENT 1 I don't understand yet. 2 I can do it with help. 3 I can do it on my own. 4 I can teach someone else.

Classify the pair of numbered angles.

4.

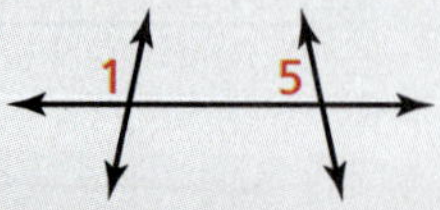

5.

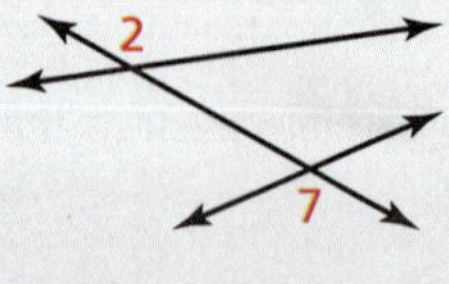

6.

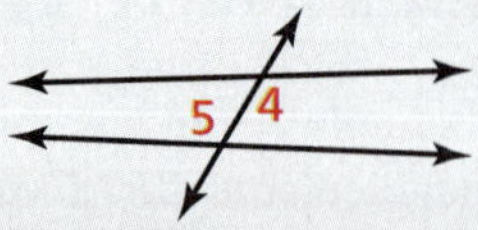

GO DIGITAL

3.1 Practice WITH CalcChat® AND CalcView®

In Exercises 1–4, consider the lines that contain the segments in the figure and the planes that contain the faces of the figure. All angles are right angles. Which line(s) or plane(s) contain point *B* and appear to fit the description? *(See Example 1.)*

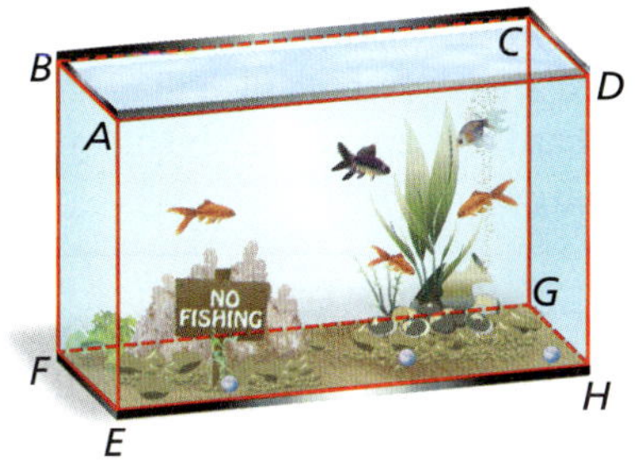

▶ **1.** line(s) parallel to $\overleftrightarrow{CD}$

2. line(s) perpendicular to $\overleftrightarrow{CD}$

3. line(s) skew to $\overleftrightarrow{CD}$

4. plane(s) parallel to plane *CDH*

In Exercises 5–8, use the diagram. *(See Example 2.)*

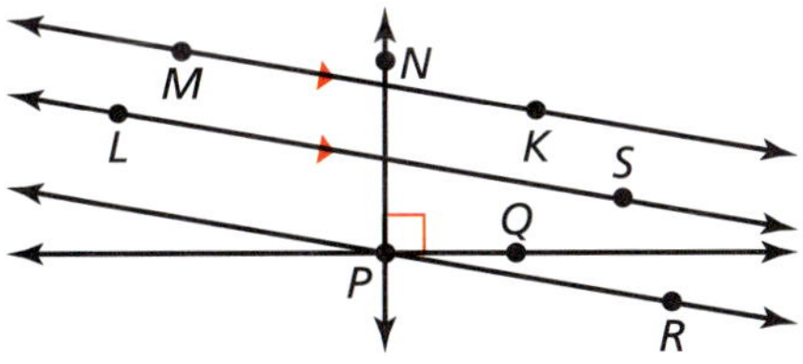

5. Name a pair of parallel lines.

6. Name a pair of perpendicular lines.

▶ **7.** Is $\overleftrightarrow{PN} \parallel \overleftrightarrow{KM}$? Explain.

8. Is $\overleftrightarrow{PR} \perp \overleftrightarrow{NP}$? Explain.

In Exercises 9–12, identify all pairs of angles of the given type. *(See Example 3.)*

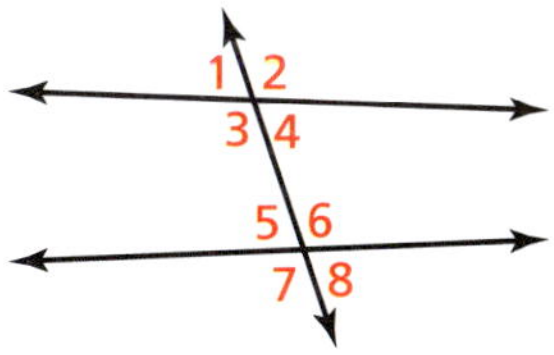

▶ **9.** corresponding

10. alternate interior

11. alternate exterior

12. consecutive interior

GO DIGITAL

5 MTR **STRUCTURE** **In Exercises 13–16, classify the angle pair as *corresponding*, *alternate interior*, *alternate exterior*, or *consecutive interior* angles.**

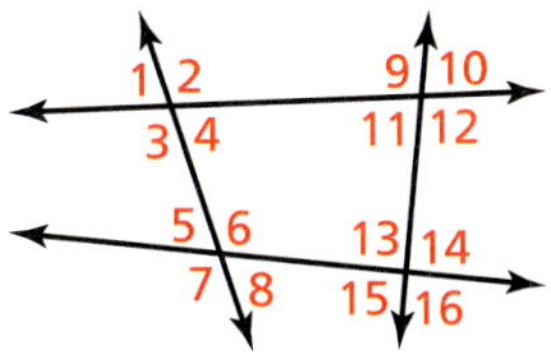

▶ **13.** ∠5 and ∠1

14. ∠11 and ∠13

15. ∠6 and ∠13

16. ∠2 and ∠11

4 MTR **ERROR ANALYSIS** **In Exercises 17 and 18, describe and correct the error in the conditional statement.**

17.

✗ If two lines do not intersect, then they are parallel.

18.

✗ If there is a line and a point not on the line, then there is exactly one line through the point that intersects the given line.

7 MTR **19. MODELING REAL LIFE** Use the image to determine whether the line containing the *main vent* is skew to the line containing the *secondary vent*. Explain.

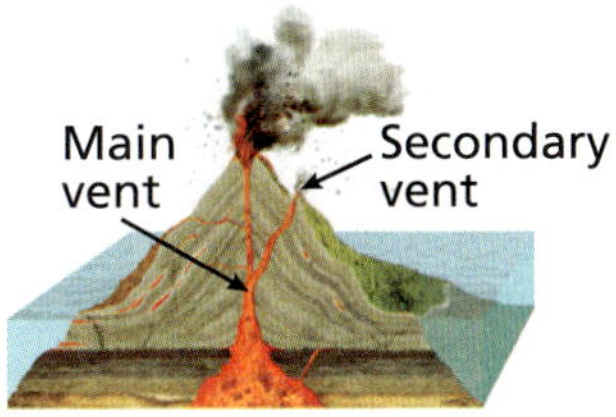

7 MTR **20. MODELING REAL LIFE** Use the photo to decide whether the statement is true or false. Explain.

STEM

a. The plane containing the floor of the tree house is parallel to the ground.

b. All the lines containing the balusters, such as $\overleftrightarrow{CD}$, are perpendicular to the plane containing the floor of the tree house.

4 MTR **21. MAKING AN ARGUMENT** Your friend claims the uneven parallel bars in gymnastics are not really parallel. She says one is higher than the other, so they cannot be in the same plane. Is she correct? Explain.

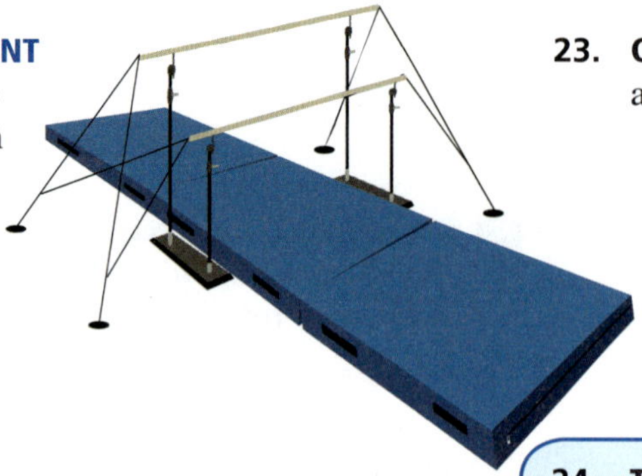

22. HOW DO YOU SEE IT?
Consider the lines that contain the segments in the figure and the planes that contain the faces of the figure. All angles are right angles.

a. Which lines are parallel to $\overleftrightarrow{NQ}$?

b. Which lines intersect $\overleftrightarrow{NQ}$?

c. Which lines are skew to $\overleftrightarrow{NQ}$?

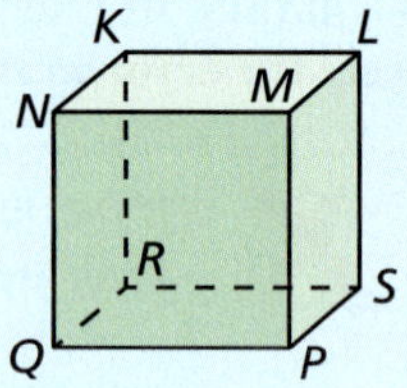

d. Should you have named all the lines on the cube in parts (a)–(c) except $\overleftrightarrow{NQ}$? Explain.

23. OPEN-ENDED Draw a transversal $\overleftrightarrow{AG}$ so that $\angle BCG$ and $\angle HFB$ are corresponding angles.

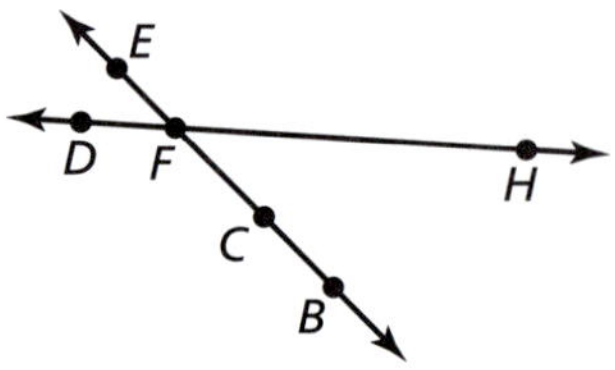

24. THOUGHT PROVOKING
If two lines are intersected by a third line, is the third line necessarily a transversal? Justify your answer with a diagram.

25. DIG DEEPER Create a rule for the maximum number of points of intersection for n lines. An example for $n = 4$ is shown.

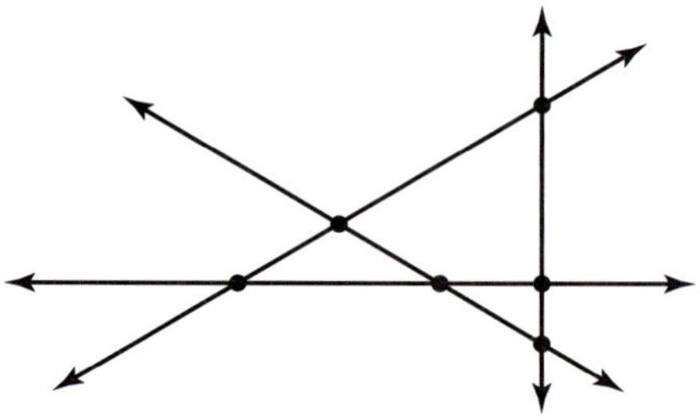

REVIEW & REFRESH

WATCH

26. Copy the segment and construct a segment bisector by paper folding. Then label the midpoint M.

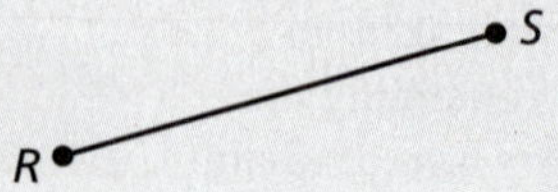

27. Solve the inequality $t - (-3) \geq 7$. Graph the solution.

28. Name the property that the statement $\overline{FG} \cong \overline{FG}$ illustrates.

29. Classify the pair of numbered angles.

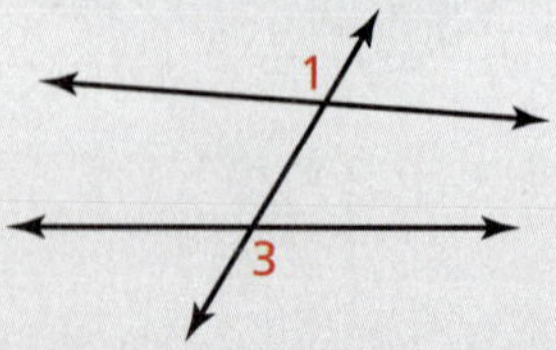

In Exercises 30 and 31, solve the system.

30. $y = \frac{1}{6}x + 1$
$y = \frac{1}{3}x + 2$

31. $-0.5x - 1.5y = -7$
$3.5x + 1.5y = 4$

32. Use the Transitive Property of Equality to complete the statement. If $m\angle 3 = m\angle 5$ and $m\angle 5 = m\angle 7$, then ________.

3 MTR **33. CHOOSE A METHOD** Write a proof using any format. Explain your choice.

Given $\angle 1$ and $\angle 3$ are complementary.
$\angle 2$ and $\angle 4$ are complementary.

Prove $\angle 1 \cong \angle 4$

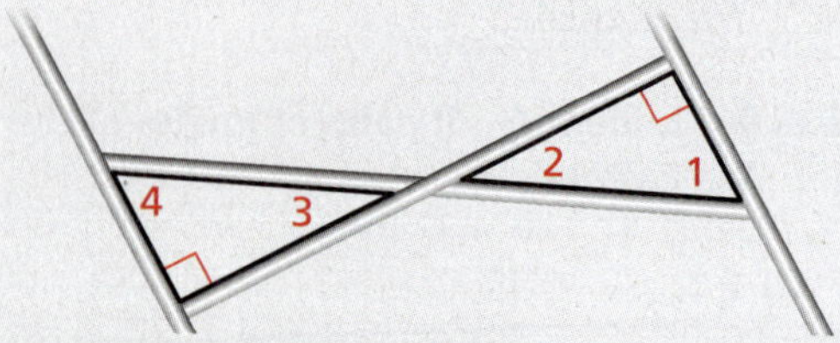

34. Write an equation of the line that passes through the point $(1, 3)$ and has a slope of -2.

35. Evaluate $\sqrt[3]{-125}$.

36. Find the volume of a cylinder with a radius of 5 inches and a height of 2 inches. Round your answer to the nearest tenth.

GO DIGITAL

3.2 Parallel Lines and Transversals

Learning Target: Prove and use theorems about parallel lines.

Success Criteria:
- I can use properties of parallel lines to find angle measures.
- I can prove theorems about parallel lines.

EXPLORE IT! Making Conjectures about Parallel Lines

Work with a partner. Draw two parallel lines. Draw a transversal that intersects both parallel lines.

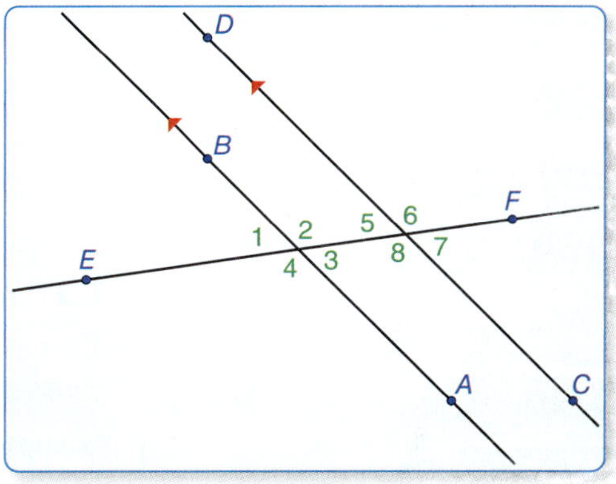

a. Find the measures of the eight angles that are formed. What do you notice?

b. Adjust the parallel lines and transversal so they intersect at different angles. Repeat part (a). How do your results compare to part (a)?

c. Write conjectures about each pair of angles formed by two parallel lines and a transversal.

i. corresponding angles

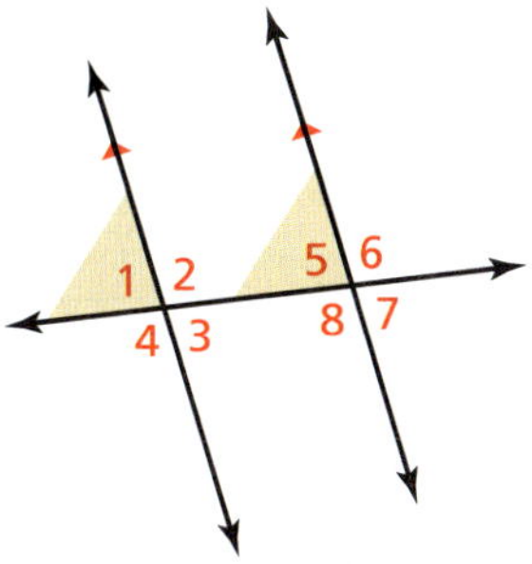

ii. alternate interior angles

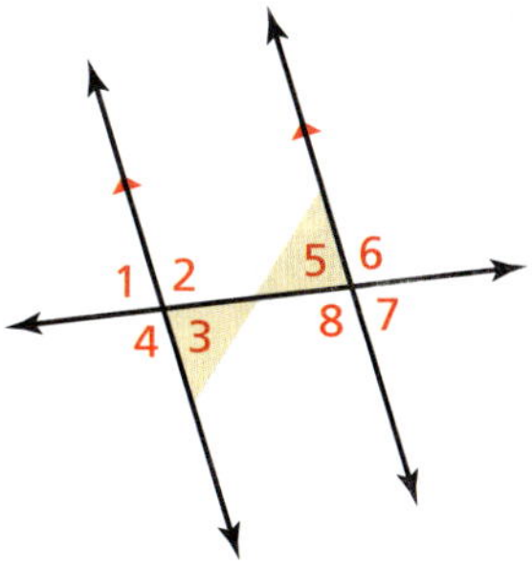

iii. consecutive interior angles

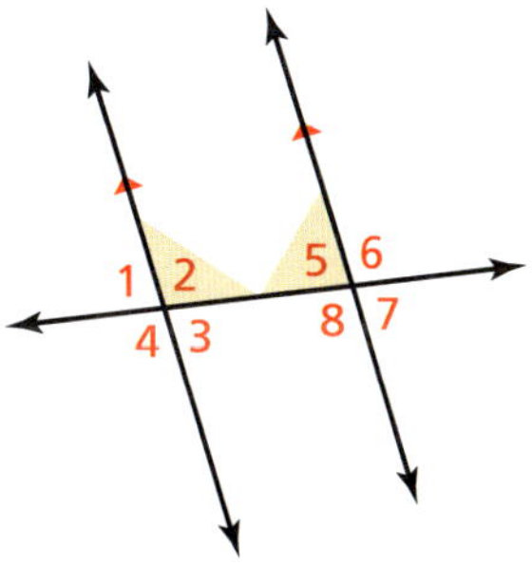

iv. alternate exterior angles

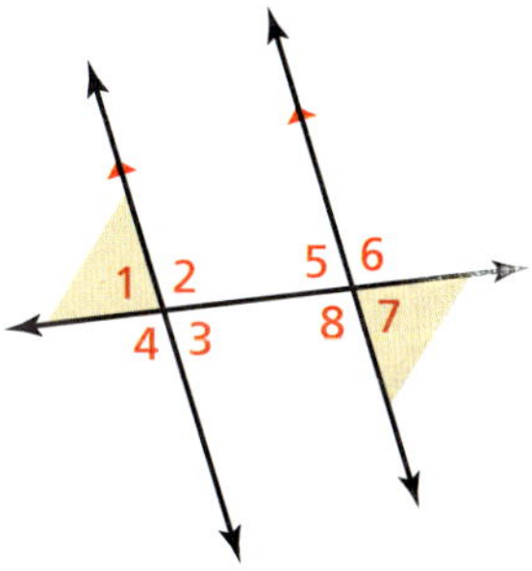

1 MTR ANALYZE A PROBLEM
What is the sum of the measures of a pair of consecutive interior angles? Does it stay the same or change when you adjust the lines?

Geometric Reasoning
MA.912.GR.1.1 Prove relationships and theorems about lines and angles. Solve mathematical and real-world problems involving postulates, relationships and theorems of lines and angles.

GO DIGITAL

Using Properties of Parallel Lines

THEOREMS

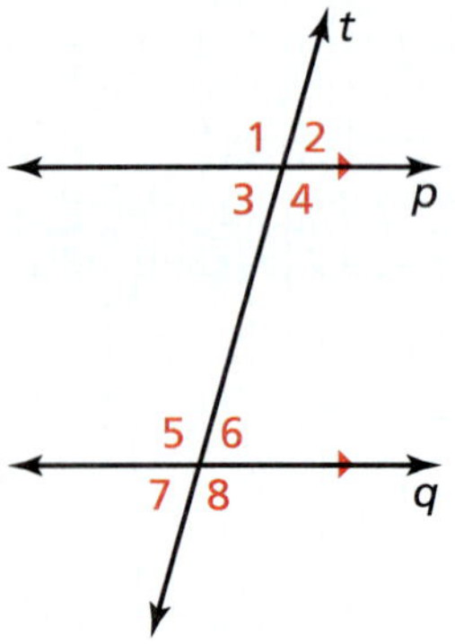

3.1 Corresponding Angles Theorem

If two parallel lines are cut by a transversal, then the pairs of corresponding angles are congruent.

Examples In the diagram at the left, $\angle 1 \cong \angle 5$, $\angle 2 \cong \angle 6$, $\angle 3 \cong \angle 7$, and $\angle 4 \cong \angle 8$.

Prove this Theorem Exercise 35, page 174

3.2 Alternate Interior Angles Theorem

If two parallel lines are cut by a transversal, then the pairs of alternate interior angles are congruent.

Examples In the diagram at the left, $\angle 3 \cong \angle 6$ and $\angle 4 \cong \angle 5$.

Prove this Theorem Exercise 17, page 131

3.3 Alternate Exterior Angles Theorem

If two parallel lines are cut by a transversal, then the pairs of alternate exterior angles are congruent.

Examples In the diagram at the left, $\angle 1 \cong \angle 8$ and $\angle 2 \cong \angle 7$.

Proof Example 4, page 130

3.4 Consecutive Interior Angles Theorem

If two parallel lines are cut by a transversal, then the pairs of consecutive interior angles are supplementary.

Examples In the diagram at the left, $\angle 3$ and $\angle 5$ are supplementary, and $\angle 4$ and $\angle 6$ are supplementary.

Prove this Theorem Exercise 18, page 131

EXAMPLE 1 Identifying Angles

WATCH

The measures of three of the numbered angles are 120°. Identify the angles. Explain your reasoning.

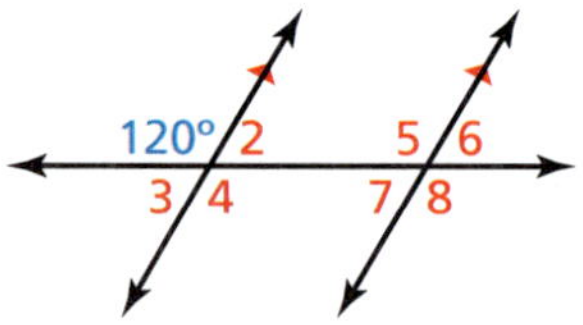

SOLUTION

Using the Alternate Exterior Angles Theorem, $m\angle 8 = 120°$.

$\angle 5$ and $\angle 8$ are vertical angles. Using the Vertical Angles Congruence Theorem, $m\angle 5 = 120°$.

$\angle 5$ and $\angle 4$ are alternate interior angles. Using the Alternate Interior Angles Theorem, $m\angle 4 = 120°$.

▶ So, the three angles that each have a measure of 120° are $\angle 4$, $\angle 5$, and $\angle 8$.

2 MTR USE ANOTHER METHOD

Solve Example 1 another way by finding the measure of a different angle first.

GO DIGITAL

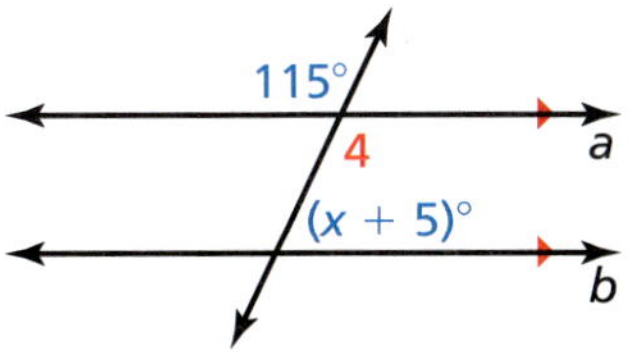

Check

$115° + (x + 5)° = 180°$

$115 + (60 + 5) \stackrel{?}{=} 180$

$180 = 180$ ✓

EXAMPLE 2 Using Properties of Parallel Lines

WATCH

Find the value of x.

SOLUTION

By the Vertical Angles Congruence Theorem, $m\angle 4 = 115°$. Lines a and b are parallel, so you can use theorems about parallel lines.

$m\angle 4 + (x + 5)° = 180°$	Consecutive Interior Angles Theorem
$115° + (x + 5)° = 180°$	Substitute 115° for $m\angle 4$.
$x + 120 = 180$	Combine like terms.
$x = 60$	Subtract 120 from each side.

▶ So, the value of x is 60.

EXAMPLE 3 Using Properties of Parallel Lines

WATCH

Find the value of x.

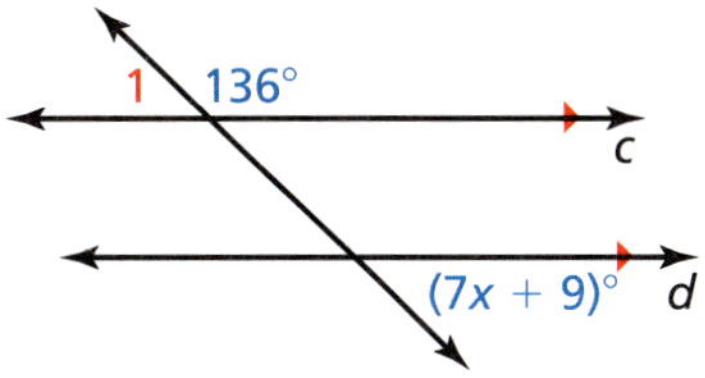

SOLUTION

By the Linear Pair Postulate, $m\angle 1 = 180° - 136° = 44°$. Lines c and d are parallel, so you can use theorems about parallel lines.

$m\angle 1 = (7x + 9)°$	Alternate Exterior Angles Theorem
$44° = (7x + 9)°$	Substitute 44° for $m\angle 1$.
$35 = 7x$	Subtract 9 from each side.
$5 = x$	Divide each side by 7.

▶ So, the value of x is 5.

Check

$44° = (7x + 9)°$

$44 \stackrel{?}{=} 7(5) + 9$

$44 = 44$ ✓

SELF-ASSESSMENT

1 I don't understand yet. 2 I can do it with help. 3 I can do it on my own. 4 I can teach someone else.

Use the diagram.

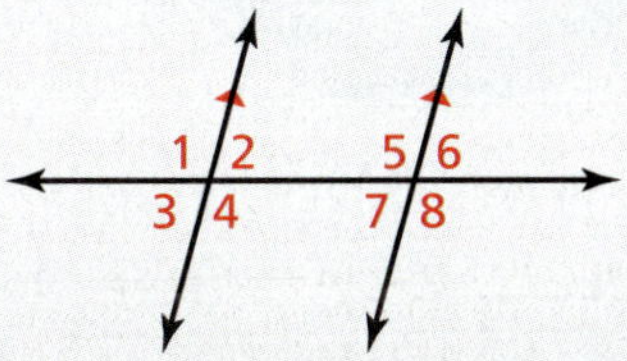

1. Given $m\angle 1 = 105°$, find $m\angle 4$, $m\angle 5$, and $m\angle 8$. Tell which theorem you use in each case.

2. Given $m\angle 3 = 68°$ and $m\angle 8 = (2x + 4)°$, what is the value of x? Show your steps.

4 MTR 3. **WHICH ONE DOESN'T BELONG?** Which pair of angle measures does *not* belong with the other three? Explain.

$m\angle 1$ and $m\angle 3$ | $m\angle 2$ and $m\angle 4$

$m\angle 2$ and $m\angle 3$ | $m\angle 1$ and $m\angle 5$

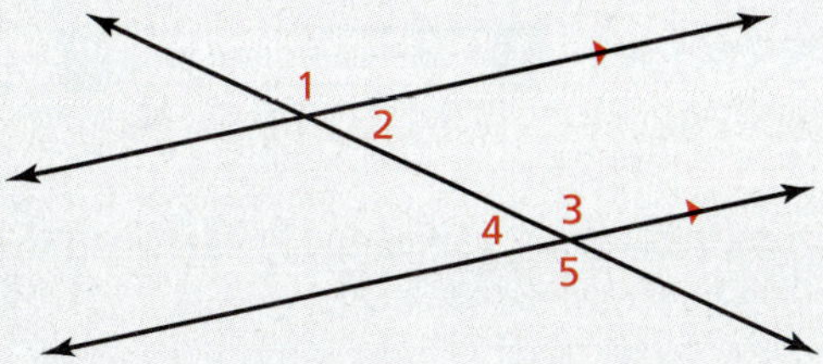

GO DIGITAL

Proving Theorems about Parallel Lines

EXAMPLE 4 **Proving the Alternate Exterior Angles Theorem** WATCH

4 MTR **COMPARE METHODS**
Rewrite the proof in Example 4 as a flowchart proof. Explain which version of the proof you prefer and why.

Prove that if two parallel lines are cut by a transversal, then the pairs of alternate exterior angles are congruent.

SOLUTION

Draw a diagram. Label a pair of alternate exterior angles as $\angle 1$ and $\angle 2$. You are looking for an angle that is related to both $\angle 1$ and $\angle 2$. Notice that one angle is a vertical angle with $\angle 2$ and a corresponding angle with $\angle 1$. Label it $\angle 3$.

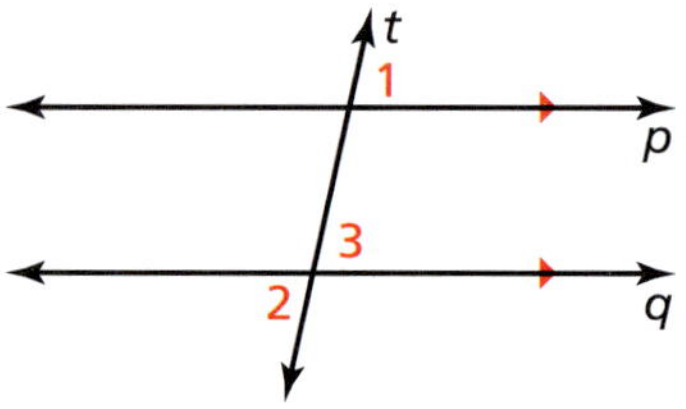

REMEMBER
Before you write a proof, identify the **Given** and **Prove** statements for the situation described or for any diagram you draw.

Given $p \parallel q$

Prove $\angle 1 \cong \angle 2$

STATEMENTS	REASONS
1. $p \parallel q$	**1.** Given
2. $\angle 1 \cong \angle 3$	**2.** Corresponding Angles Theorem
3. $\angle 3 \cong \angle 2$	**3.** Vertical Angles Congruence Theorem
4. $\angle 1 \cong \angle 2$	**4.** Transitive Property of Angle Congruence

Solving Real-Life Problems

7 MTR **EXAMPLE 5** **Modeling Real Life** WATCH

When sunlight enters a drop of rain, different colors of light leave the drop at different angles. This process is what makes a rainbow. For violet light, $m\angle 2 = 40^\circ$. What is $m\angle 1$? How do you know?

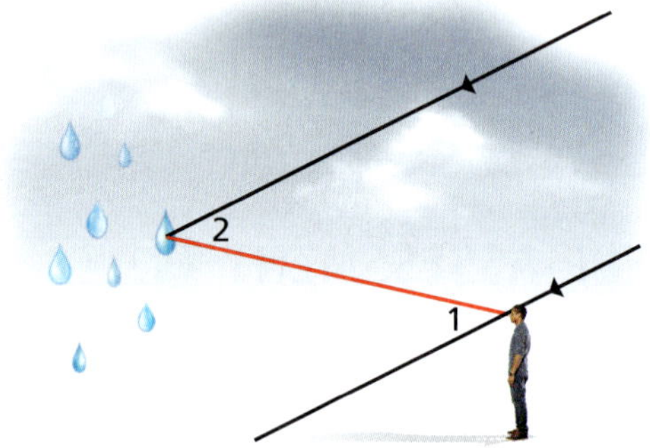

SOLUTION

The Sun's rays are parallel, and $\angle 1$ and $\angle 2$ are alternate interior angles. By the Alternate Interior Angles Theorem, $\angle 1 \cong \angle 2$.

▶ So, by the definition of congruent angles, $m\angle 1 = m\angle 2 = 40^\circ$.

SELF-ASSESSMENT 1 I don't understand yet. 2 I can do it with help. 3 I can do it on my own. 4 I can teach someone else.

4. Write an alternative proof for the Alternate Exterior Angles Theorem.

5. **WHAT IF?** In Example 5, yellow light leaves a drop at an angle of $m\angle 2 = 41^\circ$. What is $m\angle 1$? How do you know?

GO DIGITAL

3.2 Practice WITH CalcChat® AND CalcView®

In Exercises 1–4, find $m\angle 1$ and $m\angle 2$. Tell which theorem you use in each case. *(See Example 1.)*

▶ **1.**

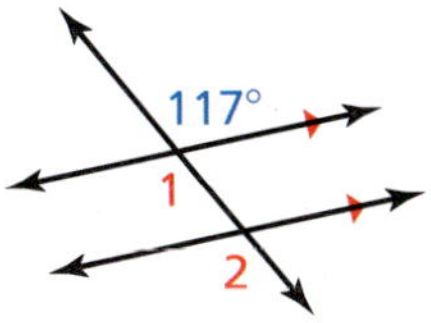

2.

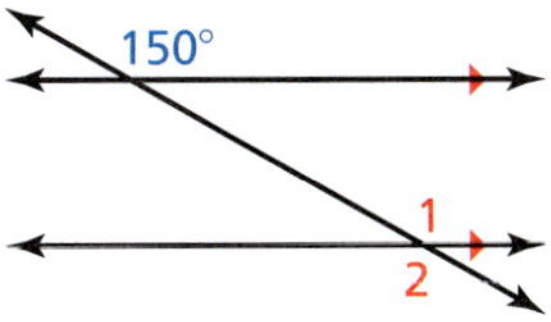

3.

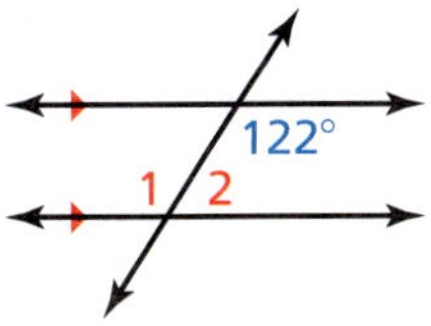

4.

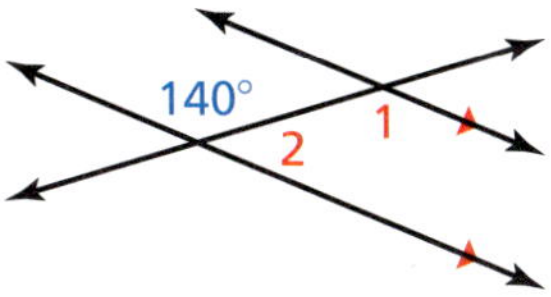

In Exercises 5–10, find the value of x. Show your steps. *(See Examples 2 and 3.)*

5.

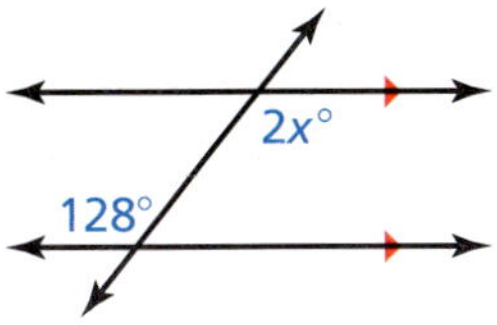

6.

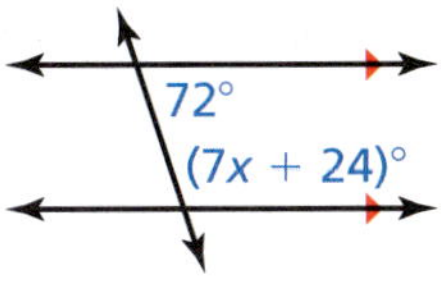

▶ **7.**

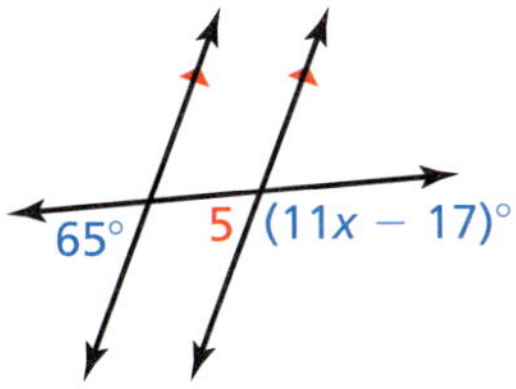

8.

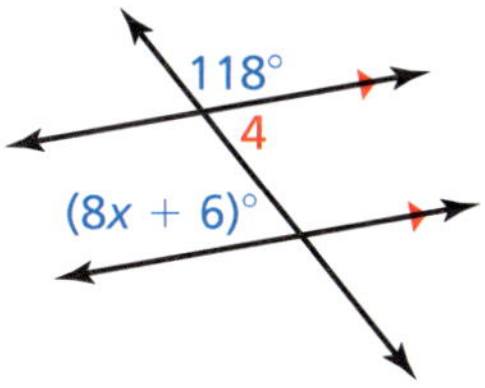

9.

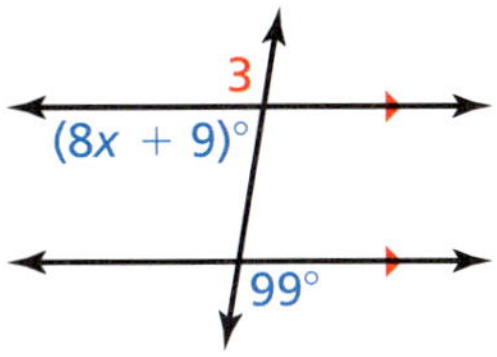

10.

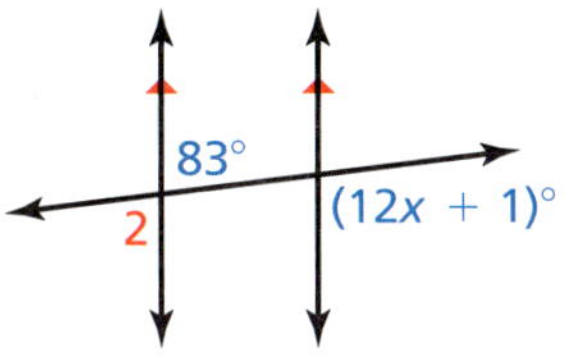

In Exercises 11–14, find $m\angle 1$, $m\angle 2$, and $m\angle 3$. Explain your reasoning.

11.

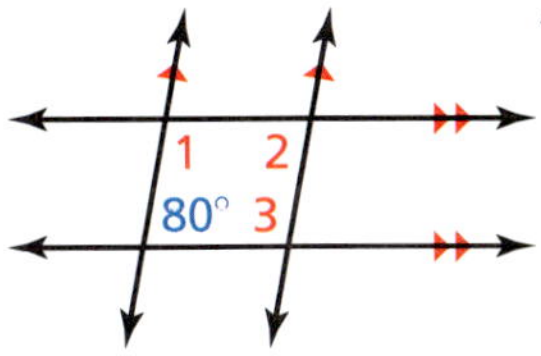

12.

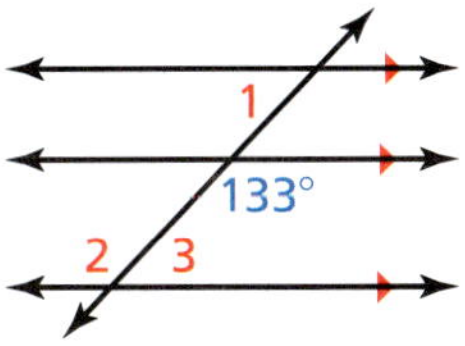

13.

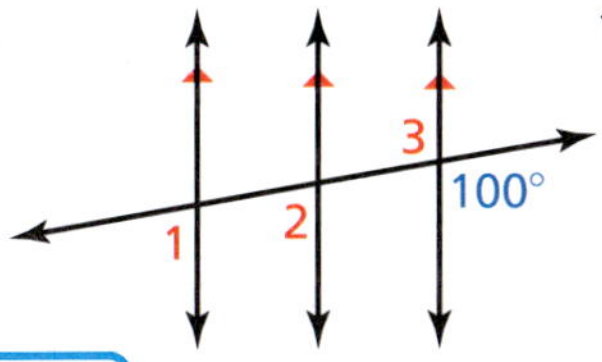

14.

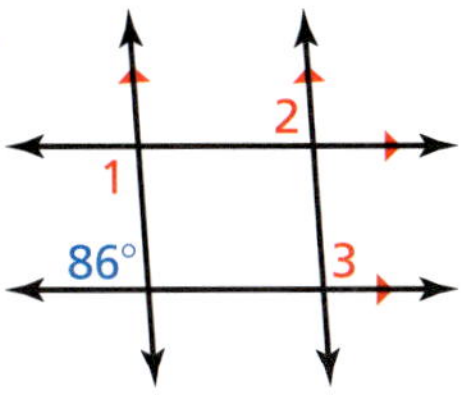

GO DIGITAL

4 MTR ERROR ANALYSIS In Exercises 15 and 16, describe and correct the error in the student's reasoning.

15.

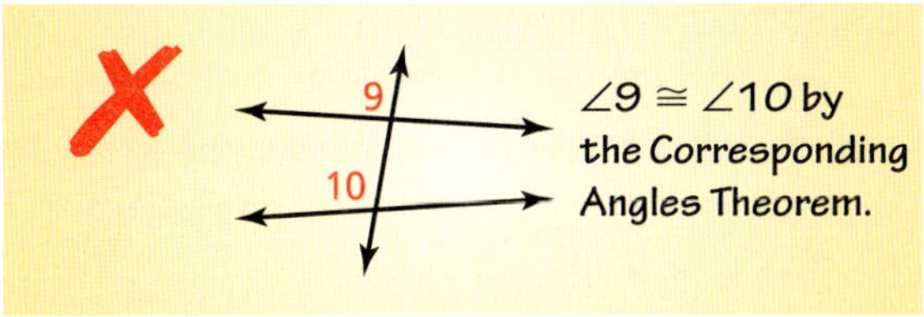

16.

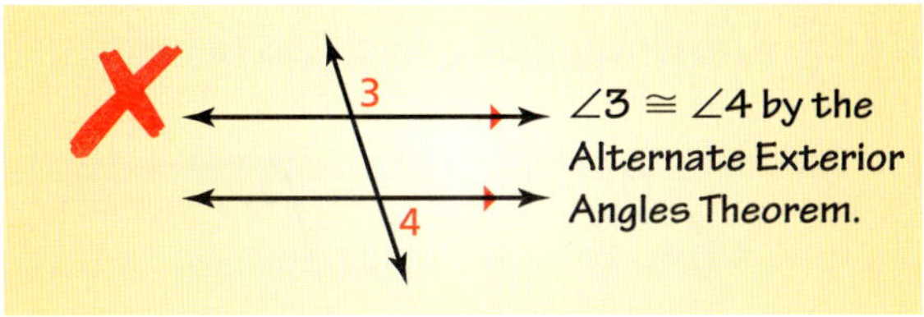

PROVING A THEOREM In Exercises 17 and 18, prove the theorem. *(See Example 4.)*

▶ **17.** Alternate Interior Angles Theorem

18. Consecutive Interior Angles Theorem

7 MTR 19. MODELING REAL LIFE A group of campers at Ocala National Forest tie up their food in a bear bag between two parallel trees, as shown. The rope is pulled taut, forming a straight line. Find $m\angle 2$. Explain your reasoning. *(See Example 5.)*

7 MTR 20. MODELING REAL LIFE You are designing a box like the one shown.

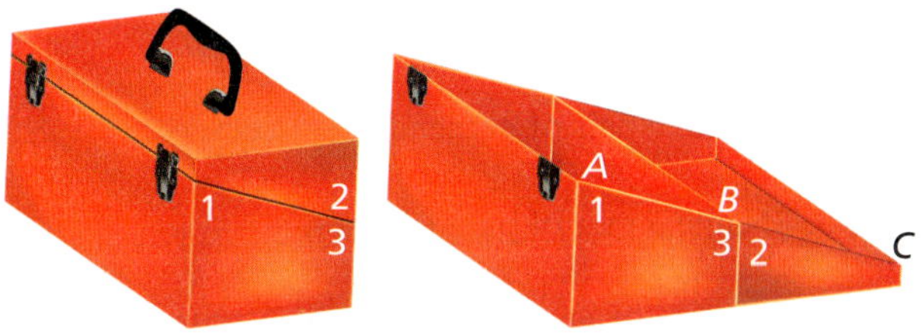

a. The measure of $\angle 1$ is 70°. Find $m\angle 2$ and $m\angle 3$.

b. Explain why $\angle ABC$ is a straight angle.

c. If $m\angle 1$ is 60°, will $\angle ABC$ still be a straight angle? Will the opening of the box be *more steep* or *less steep*? Explain.

21. REASONING Is it possible for consecutive interior angles to be congruent? Explain.

22. HOW DO YOU SEE IT?
Use the diagram.

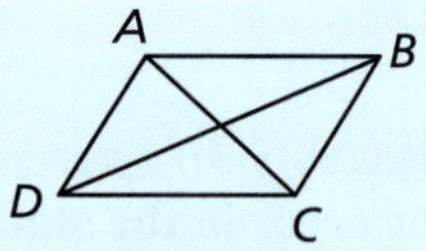

a. Name two pairs of congruent angles when $\overline{AD}$ and $\overline{BC}$ are parallel. Explain your reasoning.

b. Name two pairs of supplementary angles when $\overline{AB}$ and $\overline{DC}$ are parallel. Explain your reasoning.

5 MTR **CONNECTING CONCEPTS** **In Exercises 23 and 24, write and solve a system of linear equations to find the values of *x* and *y*.**

23.

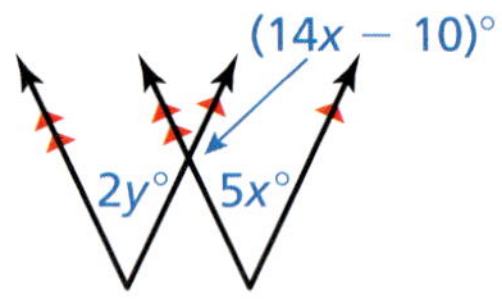

24.

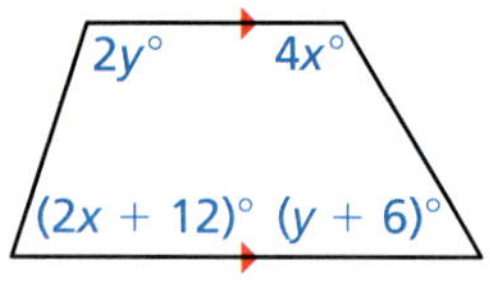

25. DIG DEEPER In the diagram, $\angle 4 \cong \angle 5$ and $\overline{SE}$ bisects $\angle RSF$. Find $m\angle 1$. Explain your reasoning.

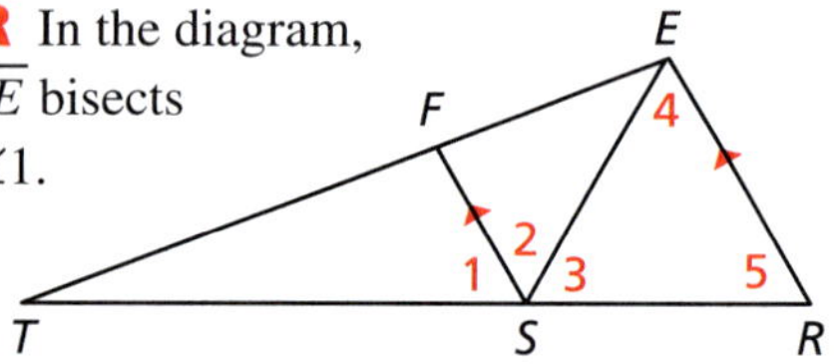

4 MTR **26. MAKING AN ARGUMENT** During a game of pool, your friend claims to be able to make the shot shown in the diagram by hitting the cue ball so that $m\angle 1 = 25°$. Is your friend correct? Explain your reasoning.

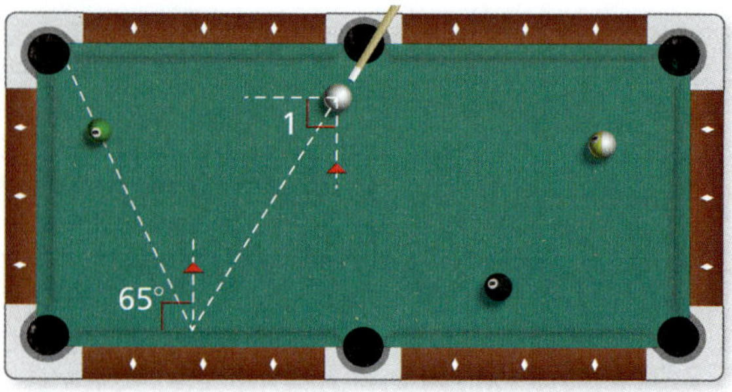

27. OPEN-ENDED Draw a real-life situation modeled by parallel lines and transversals. Describe the relationships between the angles.

28. THOUGHT PROVOKING
The postulates and theorems in this book represent Euclidean geometry. In spherical geometry, all points are points on the surface of a sphere. A line is a circle on the sphere whose diameter is equal to the diameter of the sphere. In spherical geometry, is it possible that a transversal intersects two parallel lines? Explain your reasoning.

REVIEW & REFRESH

WATCH

In Exercises 29–31, use the diagram.

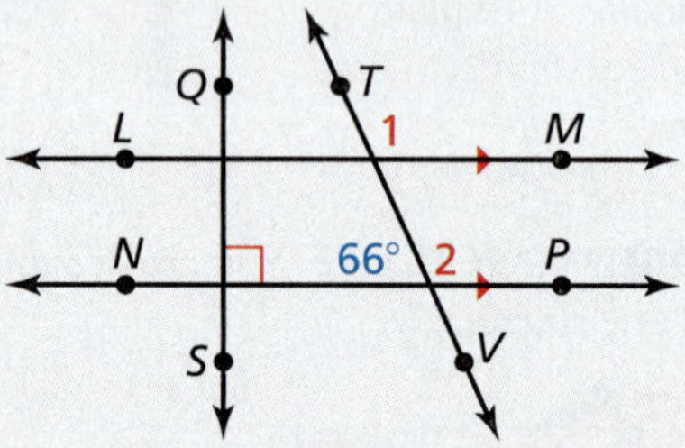

29. Name a pair of perpendicular lines.

30. Name a pair of parallel lines.

31. Find $m\angle 1$ and $m\angle 2$. Tell which postulates or theorems you used.

In Exercises 32 and 33, name the property that the statement illustrates.

32. If $\angle F \cong \angle G$ and $\angle G \cong \angle H$, then $\angle F \cong \angle H$.

33. If $\overline{WX} \cong \overline{YZ}$, then $\overline{YZ} \cong \overline{WX}$.

In Exercises 34 and 35, factor the polynomial completely.

34. $t^3 - 5t^2 + 3t - 15$ **35.** $4x^4 - 36x^2$

36. Find the *x*- and *y*-intercepts of the graph of $7x - 4y = 28$.

7 MTR **37. MODELING REAL LIFE** A square painting is surrounded by a frame with uniform width. The painting has a side length of $(x - 3)$ inches. The side length of the frame is $(x + 2)$ inches. Write an expression for the area of the square frame. Then find the area of the frame when $x = 12$.

38. Find the value of *x* in the diagram.

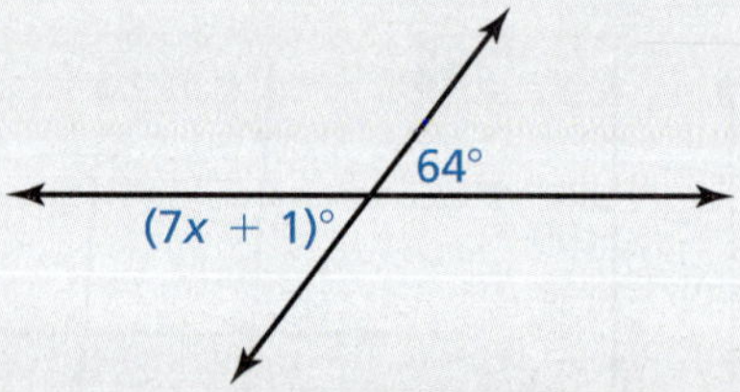

GO DIGITAL

3.3 Proofs with Parallel Lines

Learning Target: Prove and use theorems about identifying parallel lines.

Success Criteria:
- I can use theorems to identify parallel lines.
- I can construct parallel lines.
- I can prove theorems about identifying parallel lines.

EXPLORE IT! Determining Whether Converses Are True

4 MTR CONSTRUCT AN ARGUMENT
What can you conclude about the inverse of each conditional statement? Explain.

Work with a partner. Write the converse of each conditional statement. Determine whether the converse is true. Justify your conclusion.

a. Corresponding Angles Theorem
If two parallel lines are cut by a transversal, then the pairs of corresponding angles are congruent.

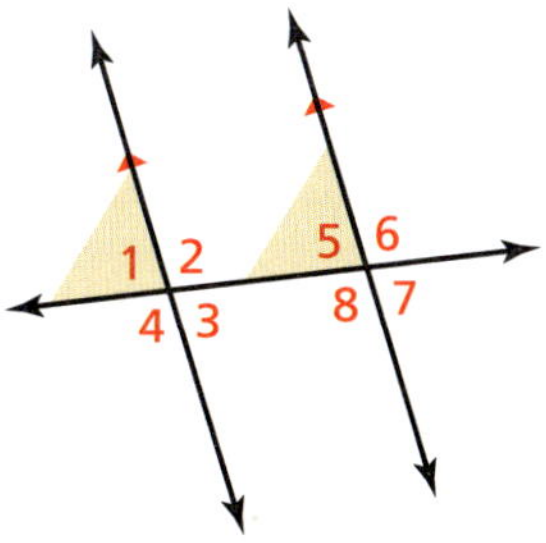

b. Alternate Interior Angles Theorem
If two parallel lines are cut by a transversal, then the pairs of alternate interior angles are congruent.

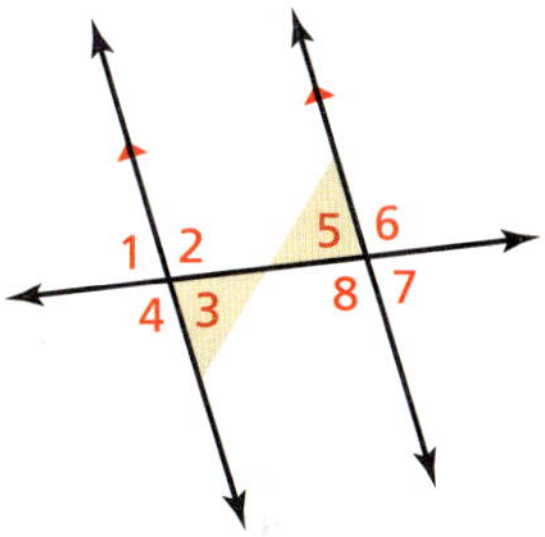

c. Alternate Exterior Angles Theorem
If two parallel lines are cut by a transversal, then the pairs of alternate exterior angles are congruent.

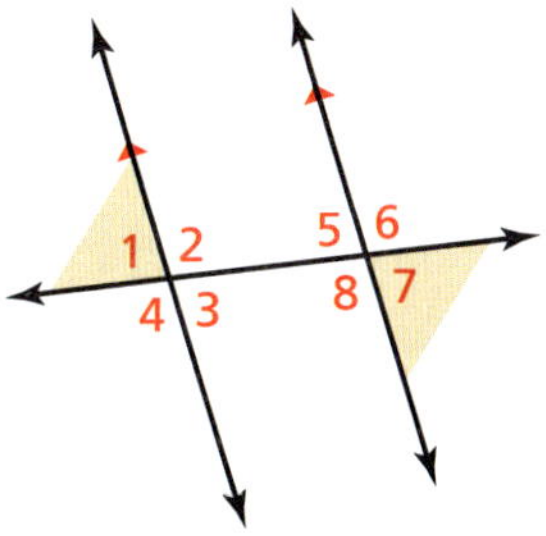

d. Consecutive Interior Angles Theorem
If two parallel lines are cut by a transversal, then the pairs of consecutive interior angles are supplementary.

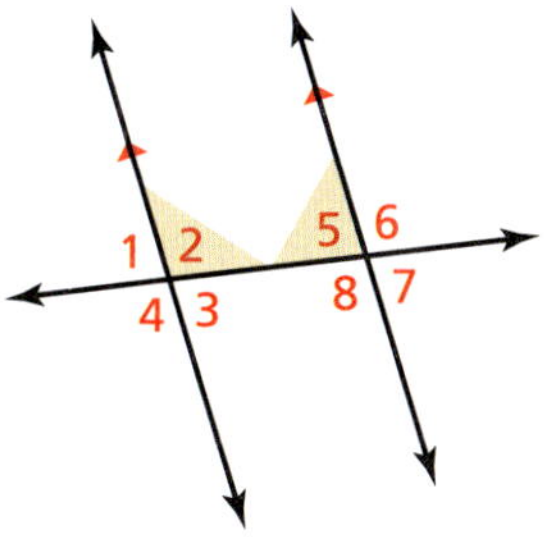

Geometric Reasoning
MA.912.GR.1.1 Prove relationships and theorems about lines and angles. Solve mathematical and real-world problems involving postulates, relationships and theorems of lines and angles.

GO DIGITAL

Using the Corresponding Angles Converse

Theorem 3.5 below is the converse of the Corresponding Angles Theorem. Similarly, the other theorems about angles formed when parallel lines are cut by a transversal have true converses. Remember that the converse of a true conditional statement is not necessarily true, so you must prove each converse of a theorem.

THEOREM

3.5 Corresponding Angles Converse

If two lines are cut by a transversal so the corresponding angles are congruent, then the lines are parallel.

Prove this Theorem Exercise 35, page 174

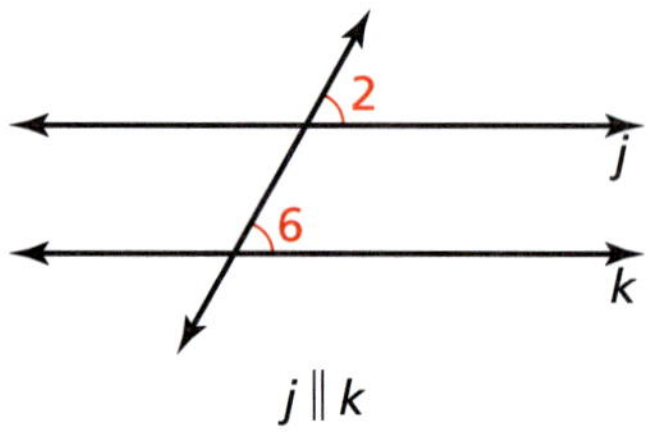

EXAMPLE 1 Using the Corresponding Angles Converse

WATCH

Find the value of x that makes $m \parallel n$.

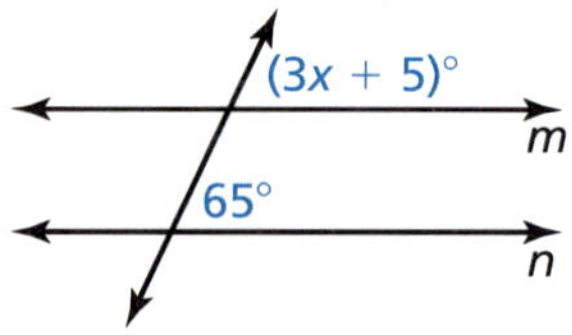

SOLUTION

Lines m and n are parallel when the marked corresponding angles are congruent.

$(3x + 5)° = 65°$	Use the Corresponding Angles Converse to write an equation.
$3x = 60$	Subtract 5 from each side.
$x = 20$	Divide each side by 3.

▶ So, lines m and n are parallel when $x = 20$.

SELF-ASSESSMENT

1 I don't understand yet. 2 I can do it with help. 3 I can do it on my own. 4 I can teach someone else.

1. Find the value of x that makes $m \parallel n$.

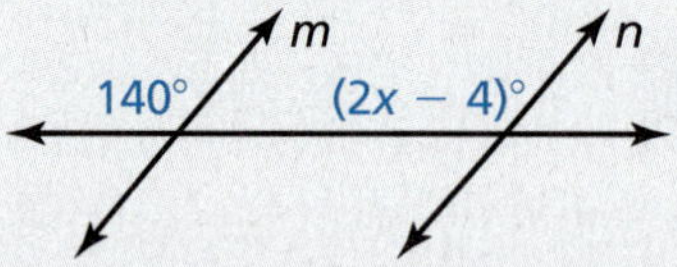

2. **REASONING** Is there enough information in the diagram to conclude that $m \parallel n$? Explain.

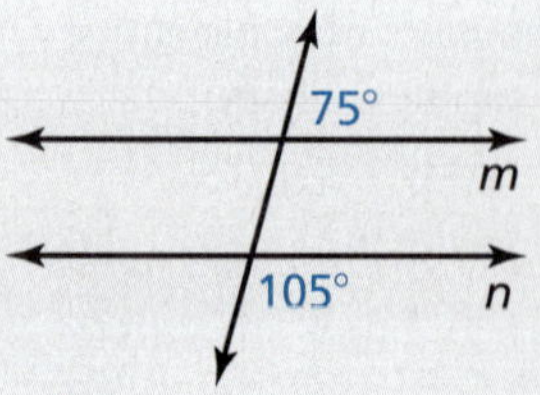

3. Explain why the Corresponding Angles Converse is the converse of the Corresponding Angles Theorem.

GO DIGITAL

Constructing Parallel Lines

The Corresponding Angles Converse justifies the construction of parallel lines, as shown below.

CONSTRUCTION **Constructing Parallel Lines** WATCH

Use a compass and straightedge to construct a line through point P that is parallel to line m.

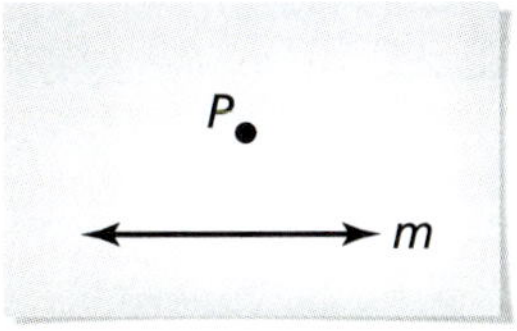

SOLUTION

Step 1

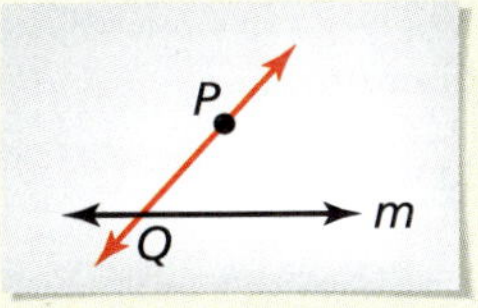

Draw a point and line Start by drawing point P and line m. Choose a point Q anywhere on line m and draw $\overleftrightarrow{QP}$.

Step 2

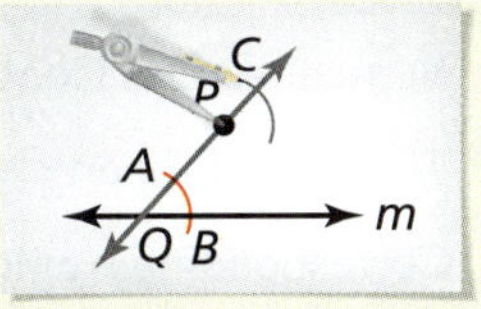

Draw arcs Draw an arc with center Q that crosses $\overleftrightarrow{QP}$ and line m. Label points A and B. Using the same compass setting, draw an arc with center P. Label point C.

Step 3

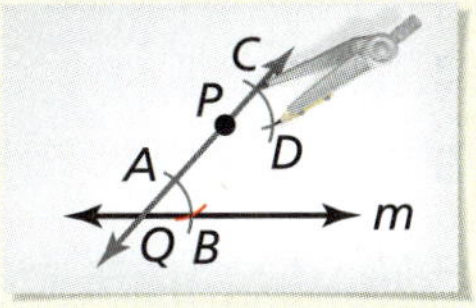

Copy angle Draw an arc with radius AB and center A. Using the same compass setting, draw an arc with center C. Label the intersection D.

Step 4

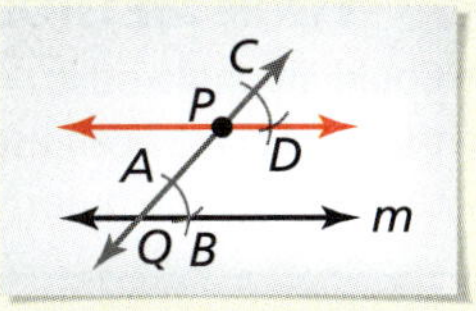

Draw parallel lines Draw $\overleftrightarrow{PD}$. This line is parallel to line m.

THEOREMS

3.6 Alternate Interior Angles Converse

If two lines are cut by a transversal so the alternate interior angles are congruent, then the lines are parallel.

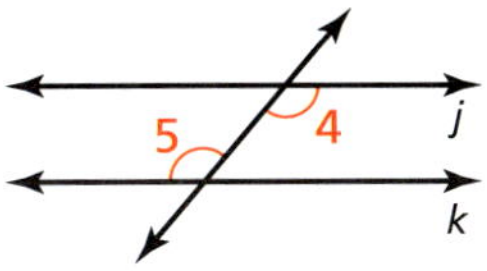

$j \parallel k$

Prove this Theorem Example 2, page 136

3.7 Alternate Exterior Angles Converse

If two lines are cut by a transversal so the alternate exterior angles are congruent, then the lines are parallel.

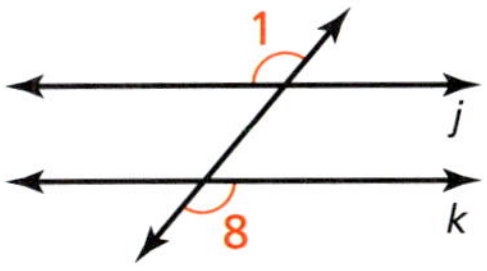

$j \parallel k$

Prove this Theorem Exercise 9, page 138

3.8 Consecutive Interior Angles Converse

If two lines are cut by a transversal so the consecutive interior angles are supplementary, then the lines are parallel.

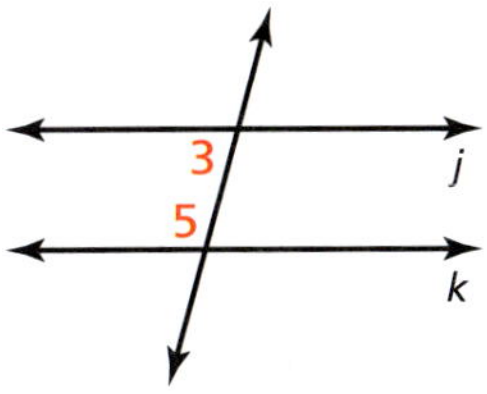

If $\angle 3$ and $\angle 5$ are supplementary, then $j \parallel k$.

Prove this Theorem Exercise 10, page 138

GO DIGITAL

Proving Theorems about Parallel Lines

EXAMPLE 2 Proving the Alternate Interior Angles Converse

WATCH

Prove that if two lines are cut by a transversal so the alternate interior angles are congruent, then the lines are parallel.

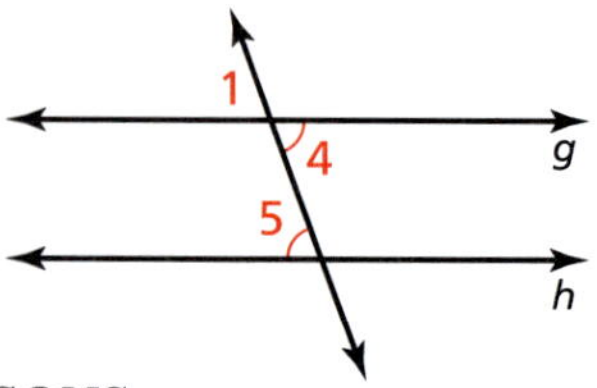

SOLUTION

Given $\angle 4 \cong \angle 5$

Prove $g \parallel h$

STATEMENTS	REASONS
1. $\angle 4 \cong \angle 5$	**1.** Given
2. $\angle 1 \cong \angle 4$	**2.** Vertical Angles Congruence Theorem
3. $\angle 1 \cong \angle 5$	**3.** Transitive Property of Angle Congruence
4. $g \parallel h$	**4.** Corresponding Angles Converse

3 MTR REFLECT ON YOUR METHOD
Could an angle other than $\angle 1$ have been used to complete the proof?

EXAMPLE 3 Determining Whether Lines Are Parallel

WATCH

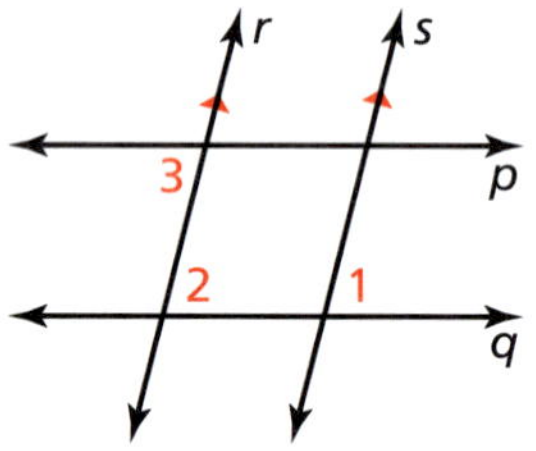

In the diagram, $r \parallel s$ and $\angle 1$ is congruent to $\angle 3$. Prove $p \parallel q$.

SOLUTION

Look at the diagram to make a plan. The diagram suggests that you look at angles 1, 2, and 3. Also, you may find it helpful to focus on one pair of lines and one transversal at a time.

Plan for Proof **a.** Look at $\angle 1$ and $\angle 2$. $\angle 1 \cong \angle 2$ because $r \parallel s$.

b. Look at $\angle 2$ and $\angle 3$. If $\angle 2 \cong \angle 3$, then $p \parallel q$.

Plan in Action **a.** It is given that $r \parallel s$, so by the Corresponding Angles Theorem, $\angle 1 \cong \angle 2$.

b. It is also given that $\angle 1 \cong \angle 3$. Then $\angle 2 \cong \angle 3$ by the Transitive Property of Angle Congruence.

▶ So, by the Alternate Interior Angles Converse, $p \parallel q$.

SELF-ASSESSMENT

1 I don't understand yet. **2** I can do it with help. **3** I can do it on my own. **4** I can teach someone else.

4. Complete the following paragraph proof of the Alternate Interior Angles Converse using the diagram in Example 2.

It is given that $\angle 4 \cong \angle 5$. By the ______, $\angle 1 \cong \angle 4$. Then by the Transitive Property of Angle Congruence, ______. So, by the ______, $g \parallel h$.

5. In the diagram, $j \parallel k$ and $\angle 1$ is congruent to $\angle 3$. Prove $m \parallel n$.

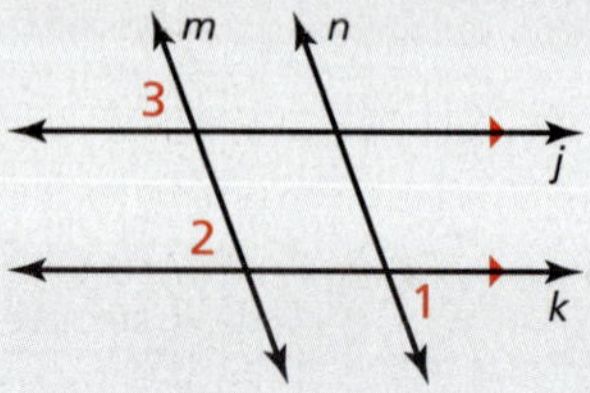

GO DIGITAL

Using the Transitive Property of Parallel Lines

THEOREM

3.9 Transitive Property of Parallel Lines

If two lines are parallel to the same line, then they are parallel to each other.

Prove this Theorem Exercise 38, page 140; Exercise 48, page 156

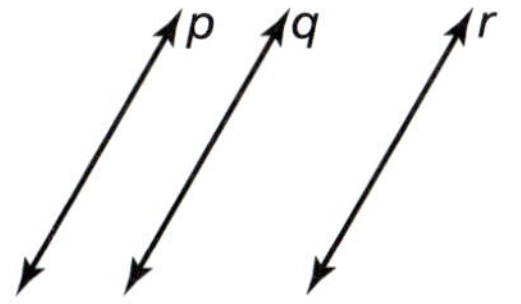

If $p \parallel q$ and $q \parallel r$, then $p \parallel r$.

EXAMPLE 4 **Using the Transitive Property of Parallel Lines**

WATCH

The flag of Puerto Rico has five alternating red and white stripes. Each stripe is parallel to the stripe immediately below it. Explain why the top stripe is parallel to the bottom stripe.

SOLUTION

You can name the stripes from top to bottom as s_1, s_2, s_3, s_4, and s_5. Each stripe is parallel to the one immediately below it, so $s_1 \parallel s_2$, $s_2 \parallel s_3$, and so on. Then $s_1 \parallel s_3$ by the Transitive Property of Parallel Lines. Similarly, because $s_3 \parallel s_4$, it follows that $s_1 \parallel s_4$. By continuing this reasoning, $s_1 \parallel s_5$.

▶ So, the top stripe is parallel to the bottom stripe by the Transitive Property of Parallel Lines.

SELF-ASSESSMENT 1 I don't understand yet. 2 I can do it with help. 3 I can do it on my own. 4 I can teach someone else.

6. Each step is parallel to the step immediately above it. The bottom step is parallel to the ground. Explain why the top step is parallel to the ground.

7. In the diagram below, $p \parallel q$ and $q \parallel r$. Find $m\angle 8$. Explain your reasoning.

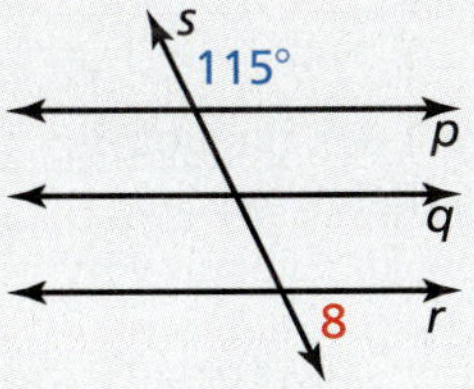

GO DIGITAL

3.3 Practice WITH CalcChat® AND CalcView®

In Exercises 1–6, find the value of x that makes $m \parallel n$. Explain your reasoning. *(See Example 1.)*

▶ **1.**
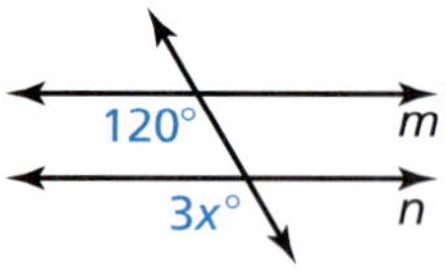

2.
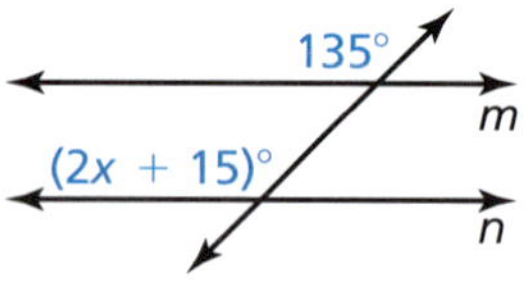

3.
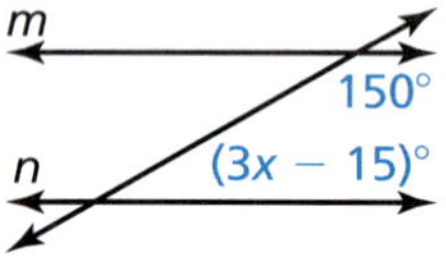

4.
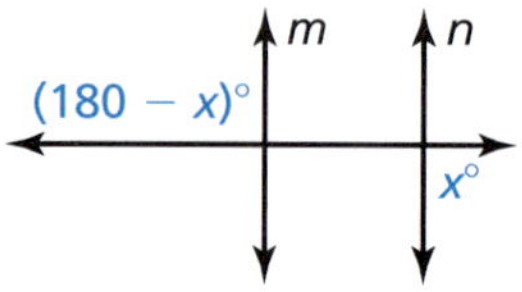

5.
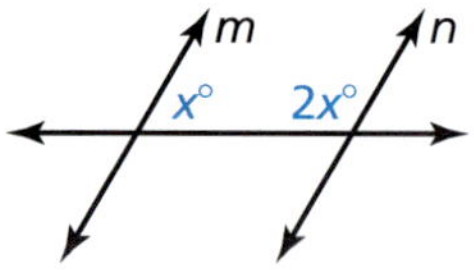

6.
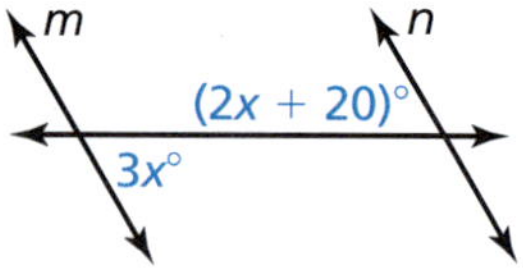

CONSTRUCTION **In Exercises 7 and 8, trace line m and point P. Then use a compass and straightedge to construct a line through point P that is parallel to line m.**

7.
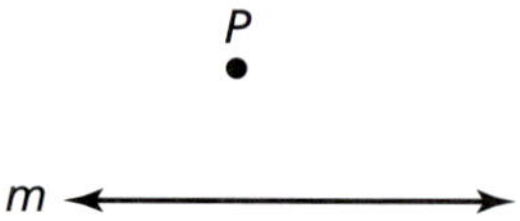

8.
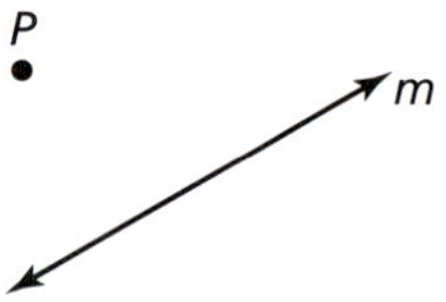

PROVING A THEOREM **In Exercises 9 and 10, prove the theorem.** *(See Example 2.)*

▶ **9.** Alternate Exterior Angles Converse

10. Consecutive Interior Angles Converse

In Exercises 11–16, decide whether there is enough information to prove that $m \parallel n$. If so, state the theorem you can use. *(See Example 3.)*

▶ **11.**
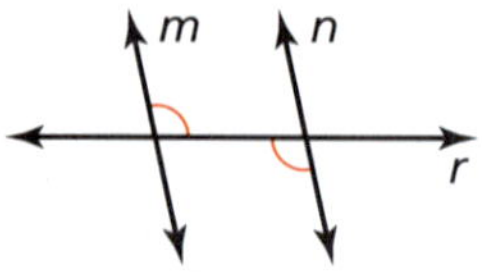

12.
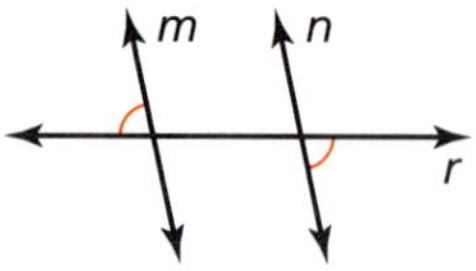

13.
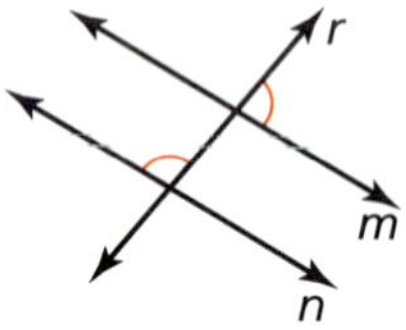

14.
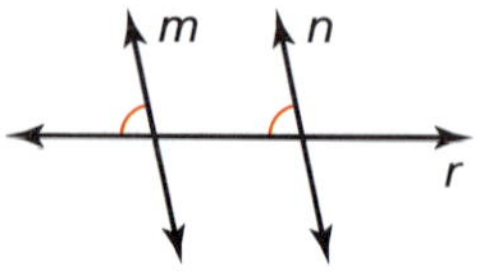

15.
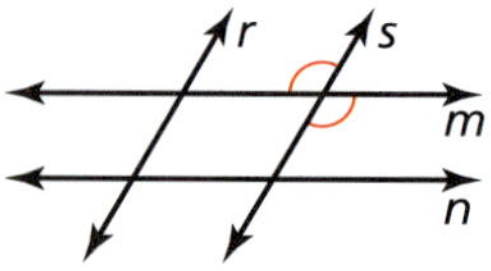

16.
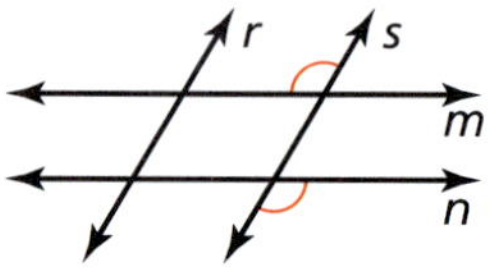

4 MTR **ERROR ANALYSIS** **In Exercises 17 and 18, describe and correct the error in the reasoning.**

17.
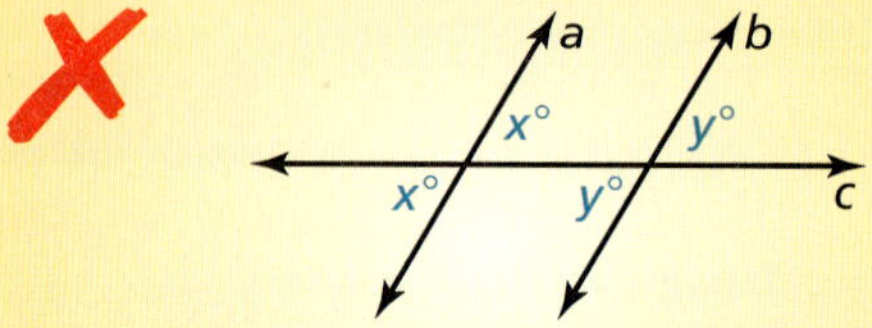

$a \parallel b$ by the Vertical Angles Congruence Theorem.

18.
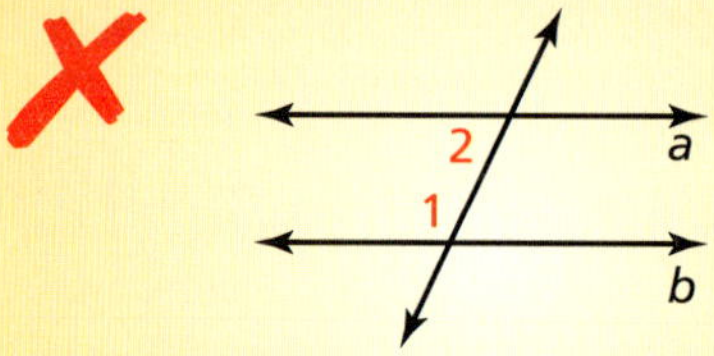

$a \parallel b$ by the Consecutive Interior Angles Converse.

In Exercises 19–22, are $\overleftrightarrow{AC}$ and $\overleftrightarrow{DF}$ parallel? Explain your reasoning.

19.
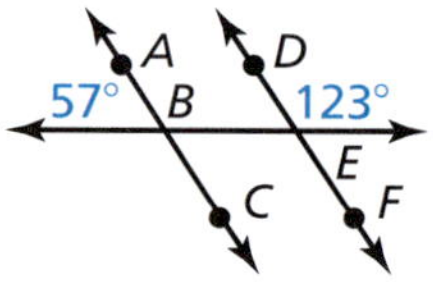

20.
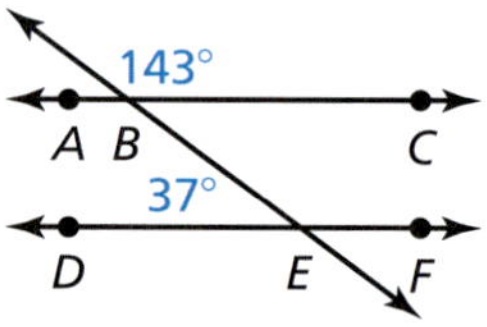

21.
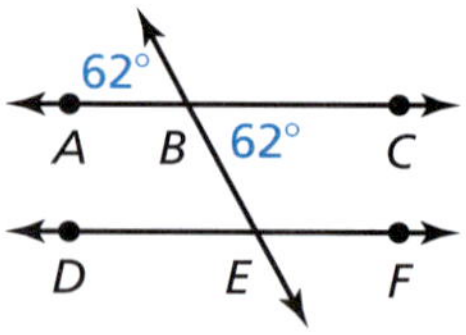

22.
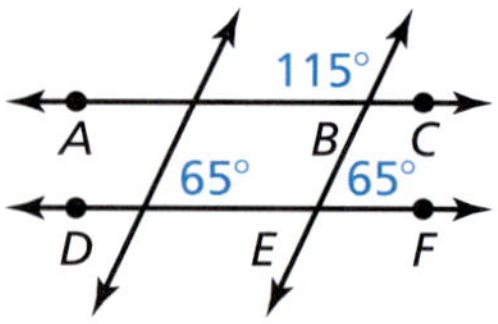

▶ **23.** **REASONING** Each rung of the ladder is parallel to the rung directly above it. Explain why the top rung is parallel to the bottom rung. *(See Example 4.)*

GO DIGITAL

24. REASONING The map shows part of Little Havana in Miami. Use the markings on the map. Are SW 4th Street, SW 5th Street, SW 6th Street, and SW 7th Street parallel to one another? Explain.

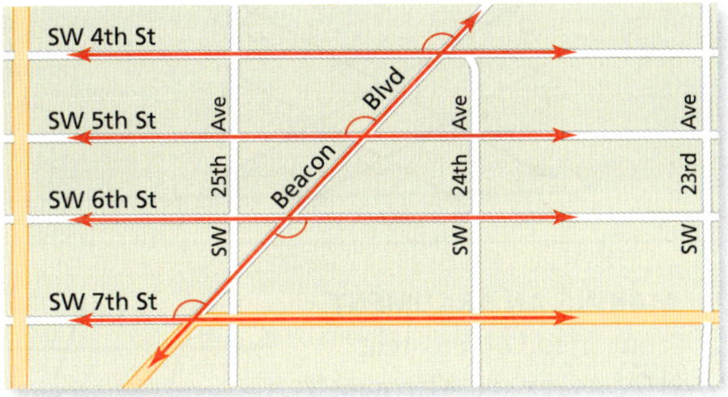

25. MODELING REAL LIFE The diagram of the control bar of the kite shows the angles formed between the control bar and the kite lines. How do you know that n is parallel to m?

26. MODELING REAL LIFE One way to build stairs is to attach triangular blocks to an angled support, as shown. The sides of the angled support are parallel. If the support makes a 32° angle with the floor, what must $m\angle 1$ be so the top of the step will be parallel to the floor? Explain your reasoning.

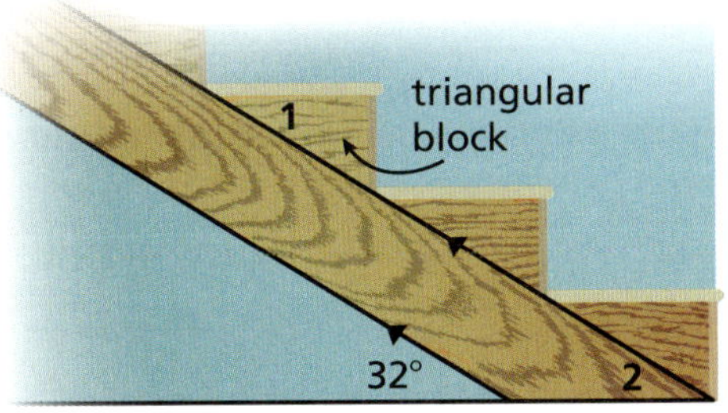

27. REASONING Which rays are parallel? Which rays are not parallel? Explain your reasoning.

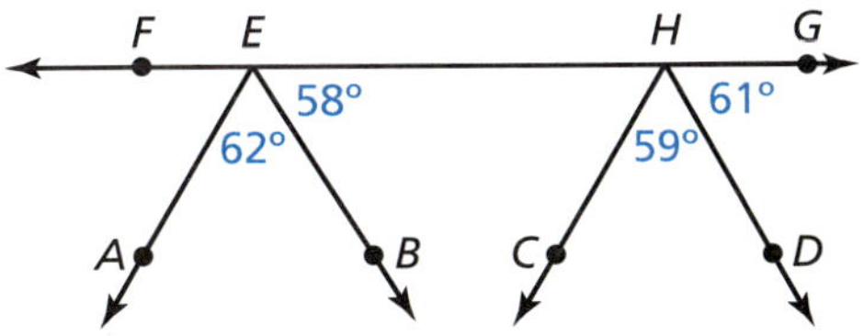

28. COLLEGE PREP Which theorems allow you to conclude that $m \parallel n$? Select all that apply.

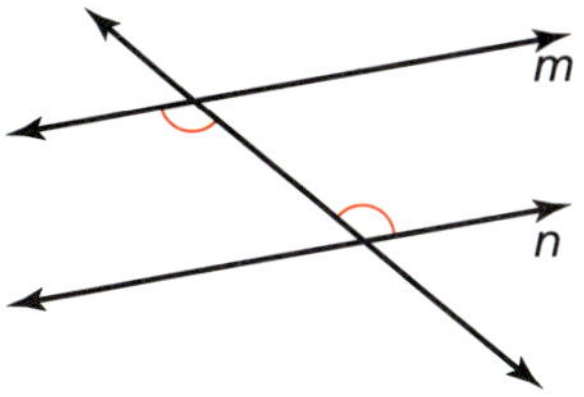

Ⓐ Corresponding Angles Converse

Ⓑ Alternate Interior Angles Converse

Ⓒ Alternate Exterior Angles Converse

Ⓓ Consecutive Interior Angles Converse

PROOF In Exercises 29–32, write a proof.

29. Given $m\angle 1 = 115°$, $m\angle 2 = 65°$

Prove $m \parallel n$

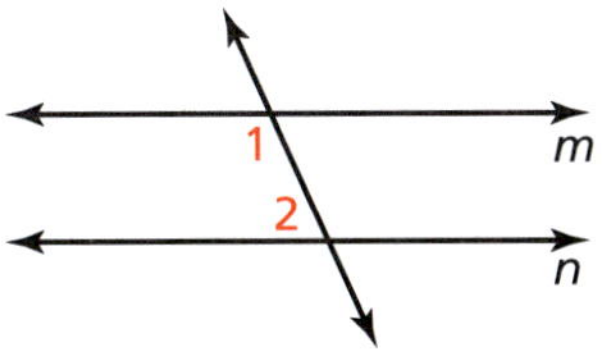

30. Given $\angle 1$ and $\angle 3$ are supplementary.

Prove $m \parallel n$

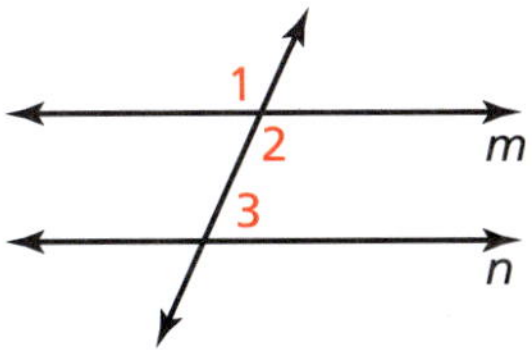

31. Given $\angle 1 \cong \angle 2$, $\angle 3 \cong \angle 4$

Prove $\overline{AB} \parallel \overline{CD}$

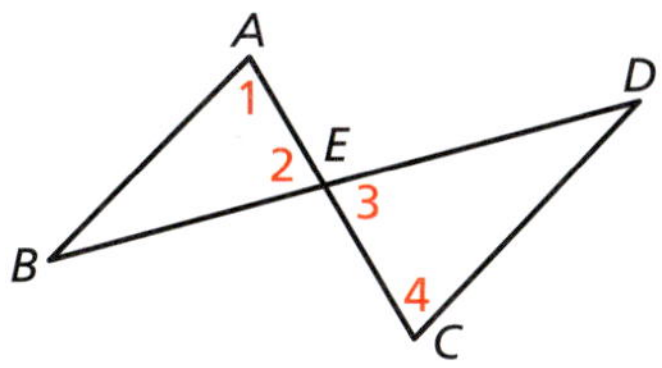

32. Given $a \parallel b$, $\angle 2 \cong \angle 3$

Prove $c \parallel d$

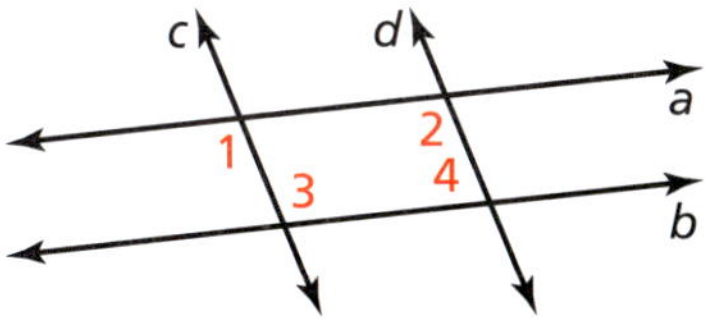

33. REASONING How many angle measures must be given to determine whether $j \parallel k$? Give four examples that would allow you to conclude that $j \parallel k$ using the theorems from this lesson.

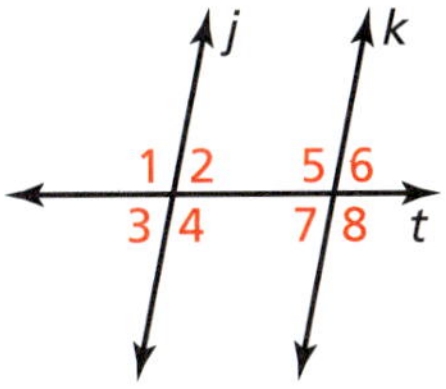

GO DIGITAL

34. THOUGHT PROVOKING Draw a diagram of at least two lines cut by at least one transversal. Mark your diagram so that it cannot be proven that any lines are parallel. Then explain how your diagram needs to change to prove that lines are parallel.

36. HOW DO YOU SEE IT? Are the markings on the diagram enough to conclude that any lines are parallel? If so, which ones? If not, what information is needed?

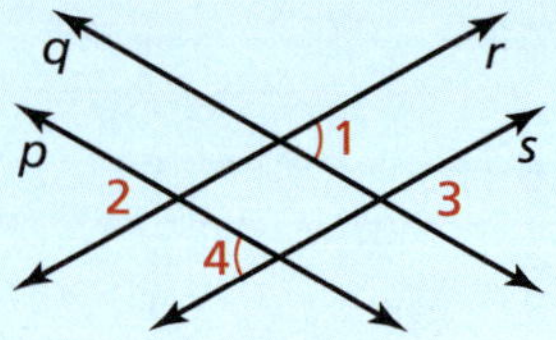

5 MTR **35. CONNECTING CONCEPTS** Use the diagram.

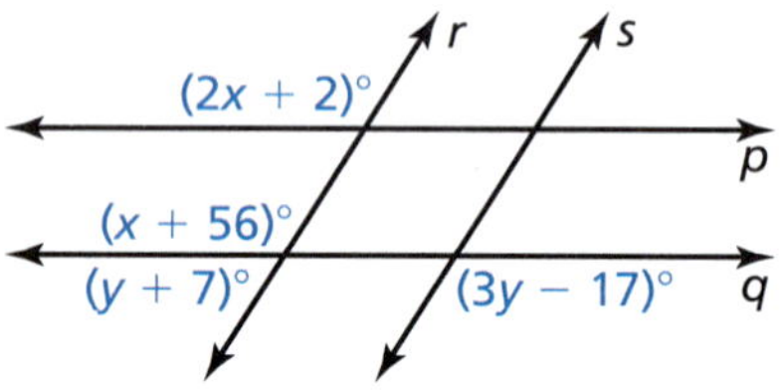

a. Find the value of x that makes $p \parallel q$.

b. Find the value of y that makes $r \parallel s$.

c. Using the values from parts (a) and (b), can r be parallel to s and can p be parallel to q at the same time? Explain your reasoning.

4 MTR **37. MAKING AN ARGUMENT** Your friend claims that $\overleftrightarrow{AD} \parallel \overleftrightarrow{BC}$ by the Alternate Interior Angles Converse. Is your friend correct? Explain.

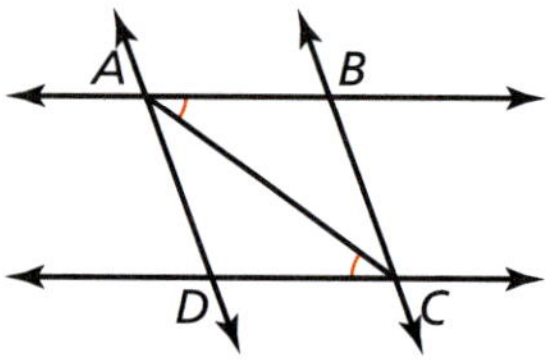

38. PROVING A THEOREM Prove the Transitive Property of Parallel Lines Theorem.

REVIEW & REFRESH

WATCH

In Exercises 39 and 40, find the distance between the two points.

39. (5, −4) and (0, 8) **40.** (13, 1) and (9, −4)

41. Find the value of x.

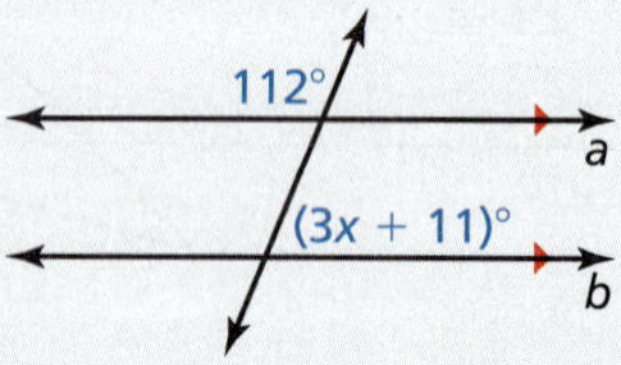

7 MTR **42. MODELING REAL LIFE** The height (in feet) of a T-shirt t seconds after it is launched into a crowd can be represented by $h(t) = -16t^2 + 96t + 4$. Estimate and interpret the maximum value of the function.

43. Find the value of x that makes $m \parallel n$. Explain your reasoning.

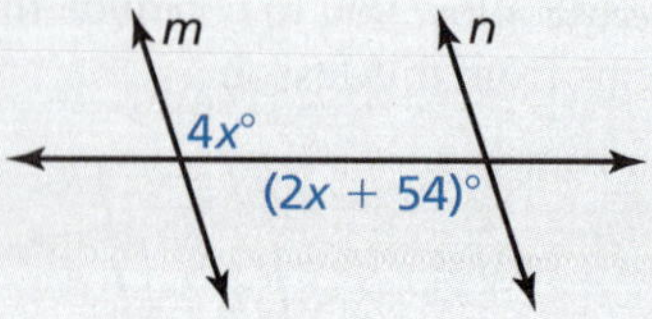

In Exercises 44 and 45, use the diagram.

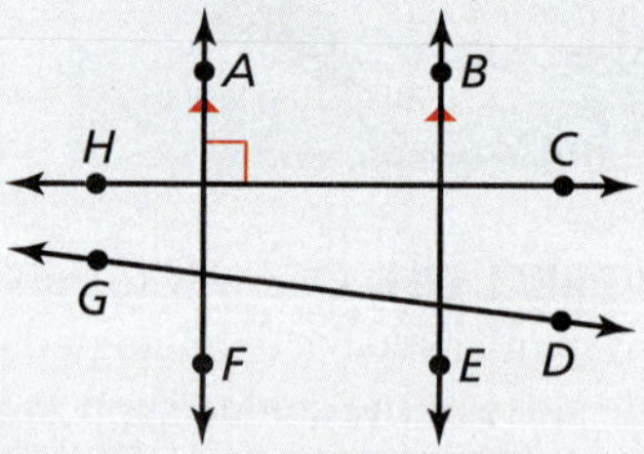

44. Name a pair of perpendicular lines.

45. Is $\overleftrightarrow{HC} \parallel \overleftrightarrow{GD}$? Explain.

3 MTR **CHOOSE A METHOD In Exercises 46 and 47, solve the system using any method. Explain your choice of method.**

46. $x = 2y + 5$
$3y - x = -9$

47. $-2x + 3y = 20$
$4x - 2y = -16$

48. Evaluate $f(x) = -3x + 5$ when $x = -2$, 3, and 5.

3 MTR **49. CHOOSE A METHOD** Write a proof using any format. Explain your choice.

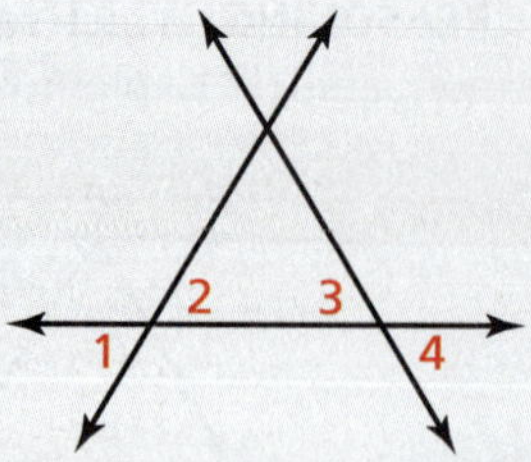

Given $\angle 1 \cong \angle 3$

Prove $\angle 2 \cong \angle 4$

GO DIGITAL

3.4 Proofs with Perpendicular Lines

Learning Target: Prove and use theorems about perpendicular lines.

Success Criteria:

- I can find the distance from a point to a line.
- I can construct perpendicular lines and perpendicular bisectors.
- I can prove theorems about perpendicular lines.

EXPLORE IT! Constructing Perpendicular Lines

Work with a partner.

a. Use a piece of paper.

i. Fold and crease the piece of paper, as shown. Label the ends of the crease as A and B.

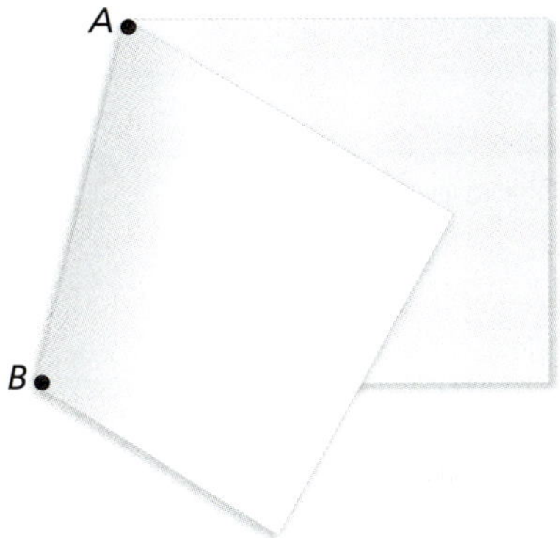

ii. Fold the paper again so that point A coincides with point B. Crease the paper on that fold.

iii. Unfold the paper and examine the four angles formed by the two creases. What can you conclude about the four angles?

b. Use a new piece of paper and repeat the first step of part (a).

i. Unfold the paper and draw a point not on the crease, as shown. Label the point C.

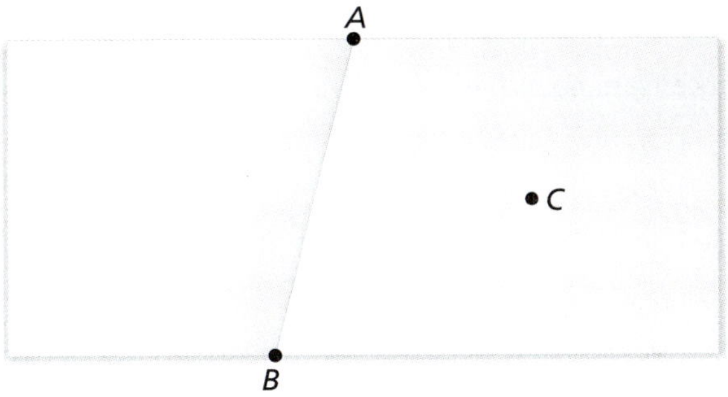

ii. Fold the paper so that the existing crease lies on top of itself and point C lies on the new fold. Crease the paper on the new fold.

iii. Unfold the paper and examine the four angles formed by the two creases. What can you conclude about the four angles?

iv. Can you find a line segment that connects $\overline{AB}$ and C that is shorter than the one on the new fold? Explain.

2 MTR USE ANOTHER METHOD

What other tools can you use to perform the construction in part (b)?

Geometric Reasoning

MA.912.GR.1.1 Prove relationships and theorems about lines and angles. Solve mathematical and real-world problems involving postulates, relationships and theorems of lines and angles.

MA.912.GR.5.2 Construct the bisector of a segment or an angle, including the perpendicular bisector of a line segment.

Also MA.912.GR.3.3

GO DIGITAL

Finding the Distance from a Point to a Line

Vocabulary AZ VOCAB

distance from a point to a line, *p. 142*
perpendicular bisector, *p. 143*

The **distance from a point to a line** is the length of the perpendicular segment from the point to the line. The length of this segment is the shortest distance between the point and the line. For example, the distance between point A and line k is AB.

distance from a point to a line

EXAMPLE 1 **Finding the Distance from a Point to a Line** WATCH

Find the distance from point A to $\overleftrightarrow{BD}$.

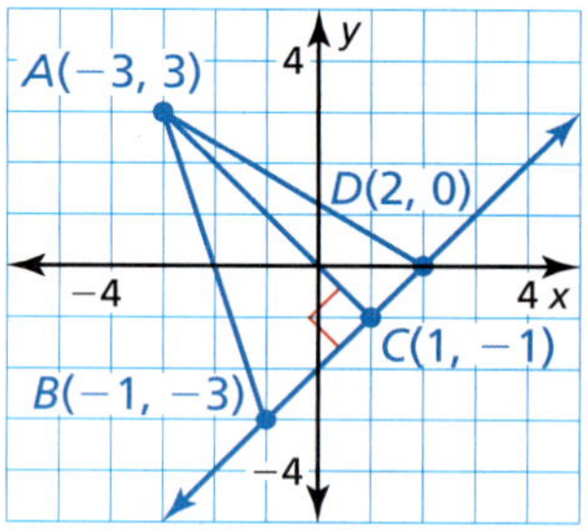

REMEMBER

Recall that the distance between $A(x_1, y_1)$ and $C(x_2, y_2)$ is

$AC = \sqrt{(x_2 - x_1)^2 + (y_2 - y_1)^2}$.

SOLUTION

Because $\overline{AC} \perp \overleftrightarrow{BD}$, the distance from point A to $\overleftrightarrow{BD}$ is AC. Use the Distance Formula.

$$AC = \sqrt{[1 - (-3)]^2 + (-1 - 3)^2} = \sqrt{4^2 + (-4)^2} = \sqrt{32} \approx 5.7$$

▶ So, the distance from point A to $\overleftrightarrow{BD}$ is about 5.7 units.

SELF-ASSESSMENT

1 I don't understand yet. 2 I can do it with help. 3 I can do it on my own. 4 I can teach someone else.

1. Find the distance from point E to $\overleftrightarrow{FH}$.

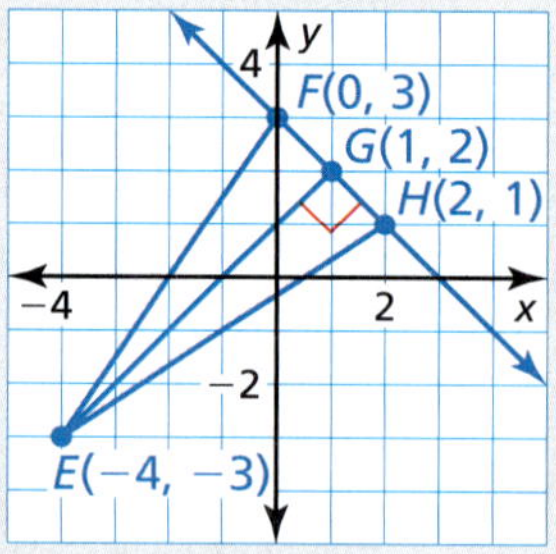

4 MTR **2. DIFFERENT WORDS, SAME QUESTION** Which is different? Find "both" answers.

Find the distance from point X to line $\overleftrightarrow{WZ}$.

Find XZ.

Find the length of $\overline{XY}$.

Find the distance from line ℓ to point X.

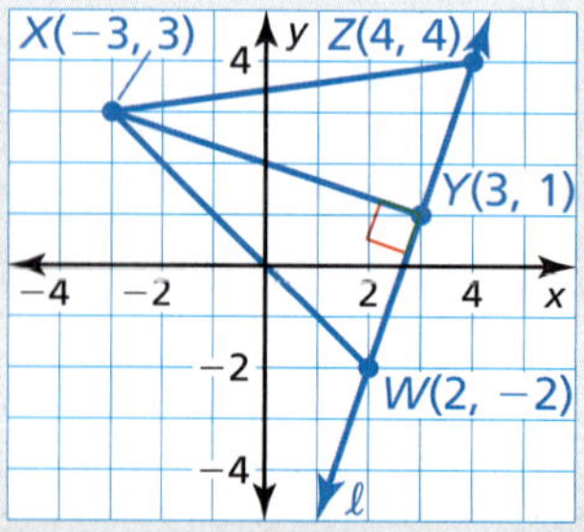

GO DIGITAL

Constructing Perpendicular Lines

CONSTRUCTION Constructing a Perpendicular Line

Use a compass and straightedge to construct a line perpendicular to line m through point P, which is not on line m.

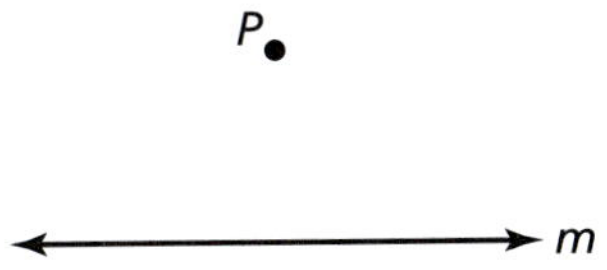

SOLUTION

Step 1

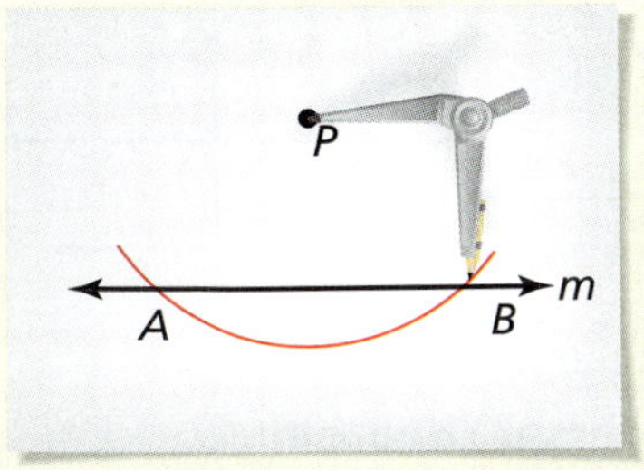

Draw arc with center P Place the compass at point P and draw an arc that intersects the line twice. Label the intersections A and B.

Step 2

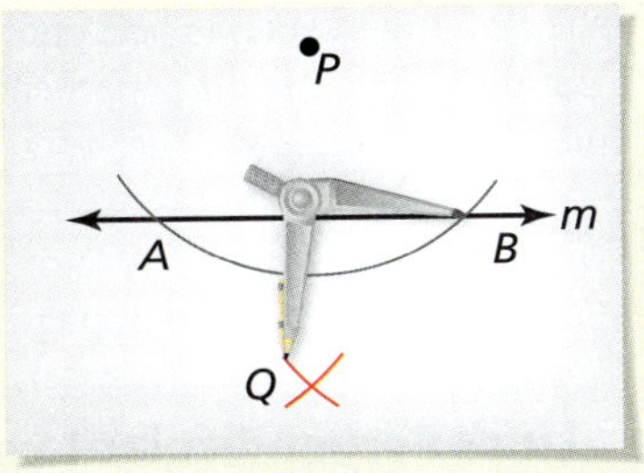

Draw intersecting arcs Draw an arc with center A. Using the same radius, draw an arc with center B. Label the intersection of the arcs Q.

Step 3

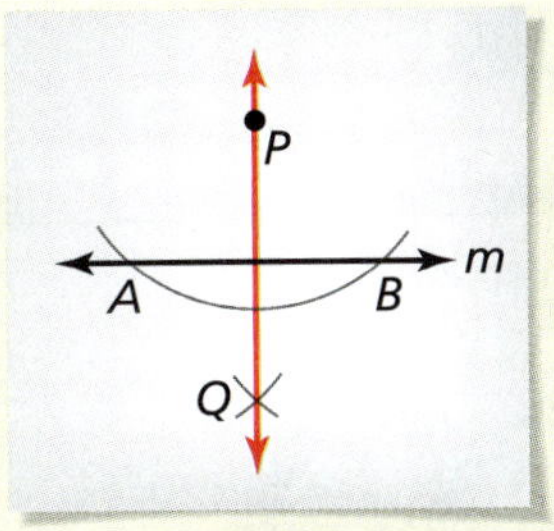

Draw perpendicular line Draw $\overleftrightarrow{PQ}$. This line is perpendicular to line m.

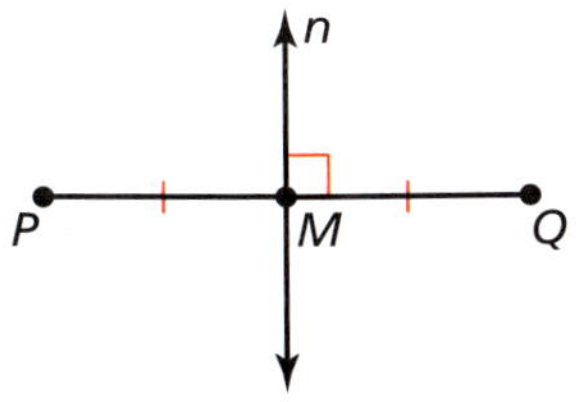

The **perpendicular bisector** of a line segment $\overline{PQ}$ is the line n with the following two properties.

- $n \perp \overline{PQ}$
- n passes through the midpoint M of $\overline{PQ}$.

CONSTRUCTION Constructing a Perpendicular Bisector

Use a compass and straightedge to construct the perpendicular bisector of $\overline{AB}$.

SOLUTION

Step 1

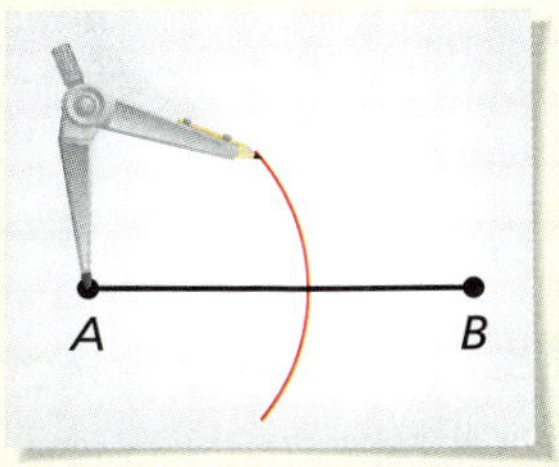

Draw an arc Place the compass at A. Use a compass setting that is greater than half the length of $\overline{AB}$. Draw an arc.

Step 2

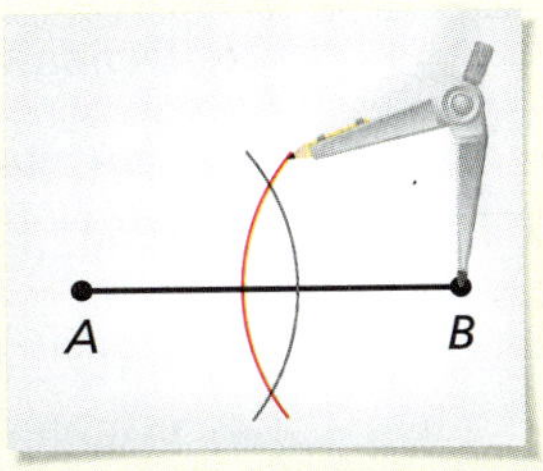

Draw a second arc Keep the same compass setting. Place the compass at B. Draw an arc. It should intersect the other arc at two points.

Step 3

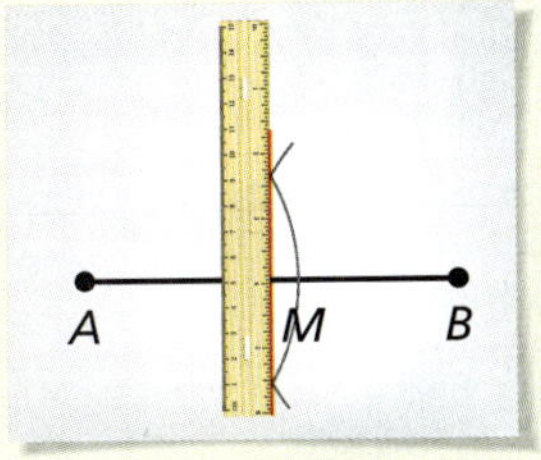

Bisect segment Draw a line through the two points of intersection. This line is the perpendicular bisector of $\overline{AB}$. It passes through M, the midpoint of $\overline{AB}$. So, $AM = MB$.

GO DIGITAL

Proving Theorems about Perpendicular Lines

THEOREMS

3.10 Linear Pair Perpendicular Theorem

If two lines intersect to form a linear pair of congruent angles, then the lines are perpendicular.

If $\angle 1 \cong \angle 2$, then $g \perp h$.

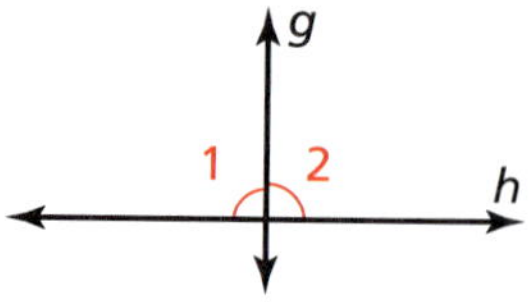

Prove this Theorem Exercise 9, page 146

3.11 Perpendicular Transversal Theorem

In a plane, if a transversal is perpendicular to one of two parallel lines, then it is perpendicular to the other line.

If $h \parallel k$ and $j \perp h$, then $j \perp k$.

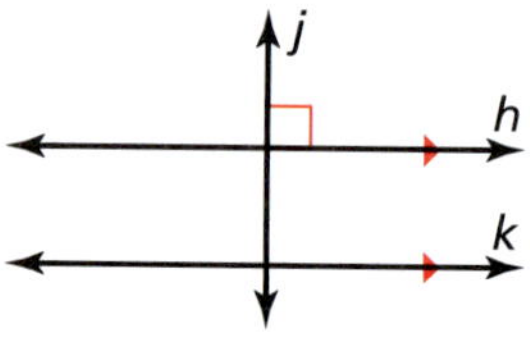

Prove this Theorem Exercise 3, page 144

3.12 Lines Perpendicular to a Transversal Theorem

In a plane, if two lines are perpendicular to the same line, then they are parallel to each other.

If $m \perp p$ and $n \perp p$, then $m \parallel n$.

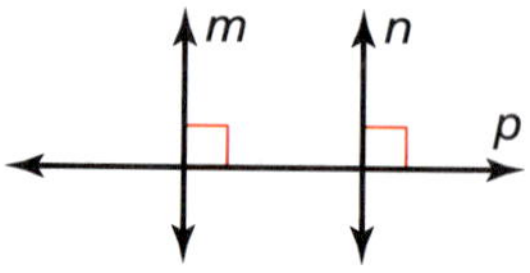

Prove this Theorem Exercise 10, page 146;
Exercise 47, page 156

EXAMPLE 2 Proving the Perpendicular Transversal Theorem

Use the diagram to prove the Perpendicular Transversal Theorem.

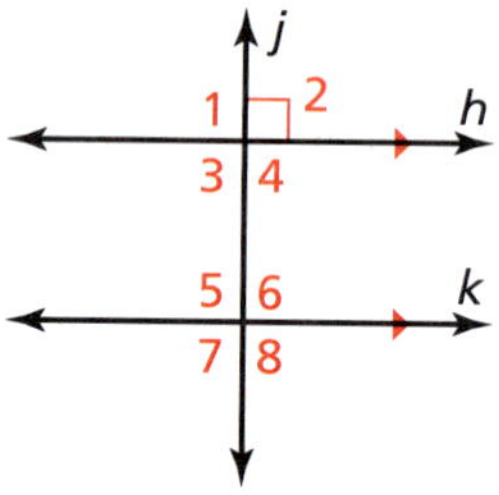

WATCH

SOLUTION

Given $h \parallel k, j \perp h$

Prove $j \perp k$

STATEMENTS	REASONS
1. $h \parallel k, j \perp h$	**1.** Given
2. $m\angle 2 = 90°$	**2.** Definition of perpendicular lines
3. $\angle 2 \cong \angle 6$	**3.** Corresponding Angles Theorem
4. $m\angle 2 = m\angle 6$	**4.** Definition of congruent angles
5. $m\angle 6 = 90°$	**5.** Transitive Property of Equality
6. $j \perp k$	**6.** Definition of perpendicular lines

SELF-ASSESSMENT 1 I don't understand yet. 2 I can do it with help. 3 I can do it on my own. 4 I can teach someone else.

3. Prove the Perpendicular Transversal Theorem using the diagram in Example 2 and the Alternate Exterior Angles Theorem.

GO DIGITAL

Solving Real-Life Problems

7 MTR **EXAMPLE 3** **Modeling Real Life** WATCH

The photo shows the layout of a neighborhood. Determine which lines, if any, must be parallel in the diagram. Explain your reasoning.

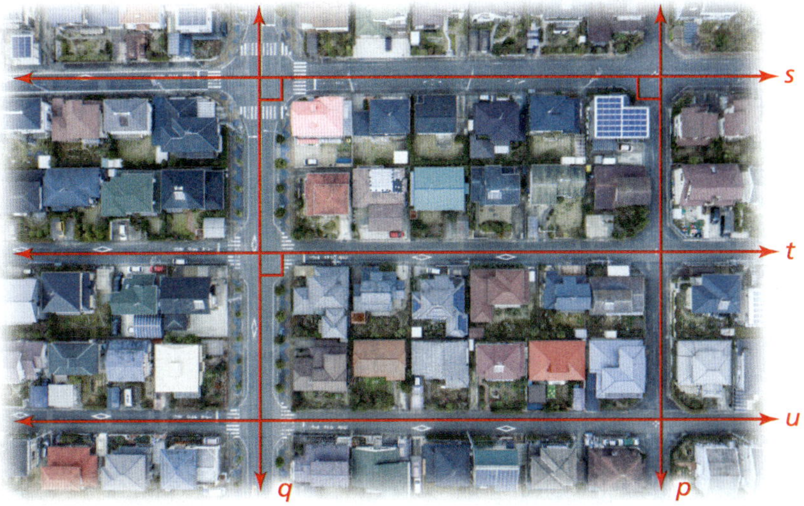

SOLUTION

Lines p and q are both perpendicular to s, so by the Lines Perpendicular to a Transversal Theorem, $p \parallel q$. Also, lines s and t are both perpendicular to q, so by the Lines Perpendicular to a Transversal Theorem, $s \parallel t$.

▶ So, from the diagram you can conclude $p \parallel q$ and $s \parallel t$.

SELF-ASSESSMENT

1 I don't understand yet. 2 I can do it with help. 3 I can do it on my own. 4 I can teach someone else.

Use the lines marked in the photo.

4. Is $b \parallel a$? Explain your reasoning.

5. Is $b \perp c$? Explain your reasoning.

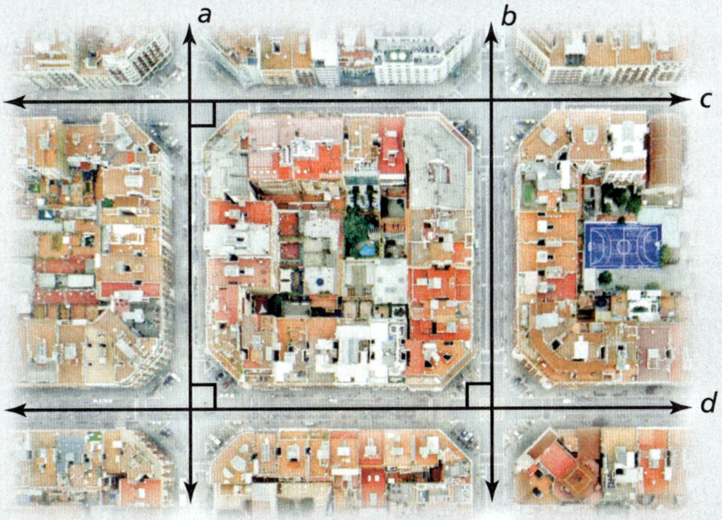

GO DIGITAL

3.4 Practice WITH CalcChat® AND CalcView®

In Exercises 1 and 2, find the distance from point A to $\overleftrightarrow{XZ}$. *(See Example 1.)*

1.

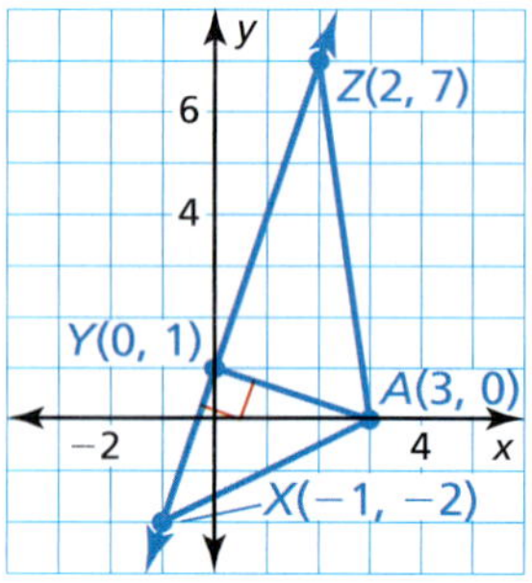

2.

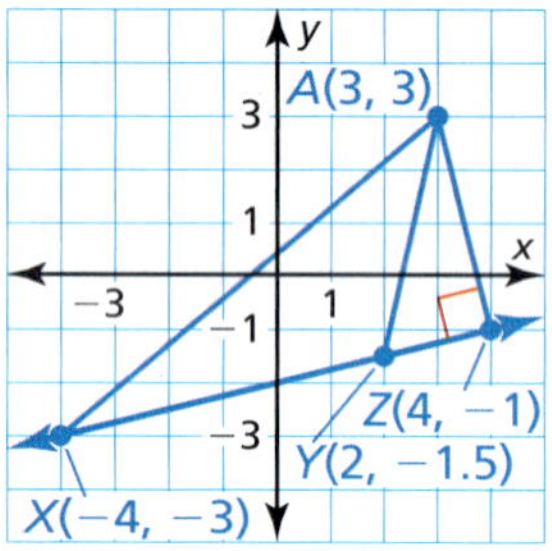

CONSTRUCTION **In Exercises 3–6, trace line m and point P. Then use a compass and straightedge to construct a line perpendicular to line m through point P.**

3.

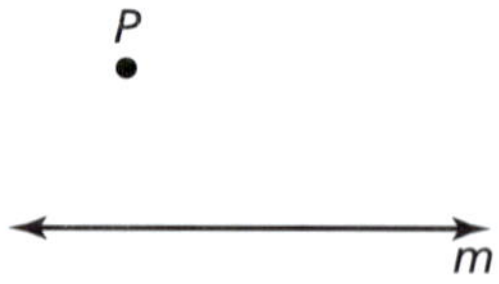

4.

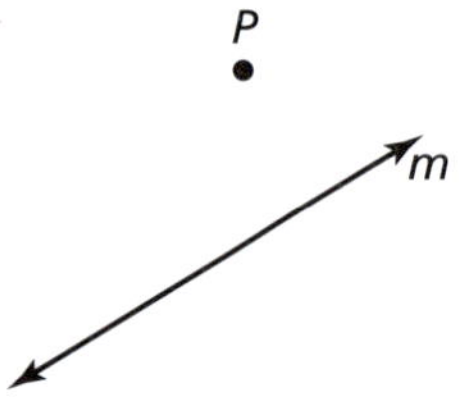

5.

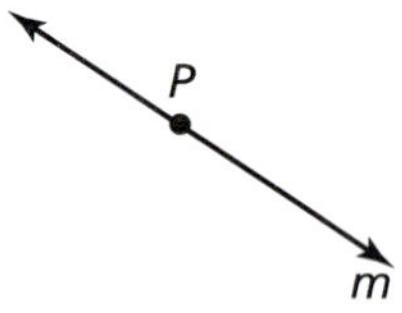

6.

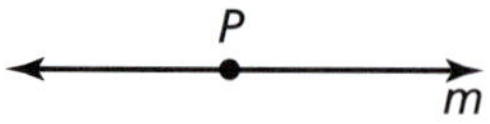

CONSTRUCTION **In Exercises 7 and 8, trace $\overline{AB}$. Then use a compass and straightedge to construct the perpendicular bisector of $\overline{AB}$.**

7.

8.

PROVING A THEOREM **In Exercises 9 and 10, prove the theorem.** *(See Example 2.)*

9. Linear Pair Perpendicular Theorem

10. Lines Perpendicular to a Transversal Theorem

In Exercises 11–16, determine which lines, if any, must be parallel. Explain your reasoning. *(See Example 3.)*

11.

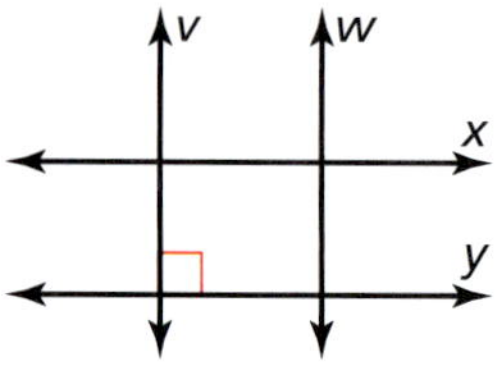

12.

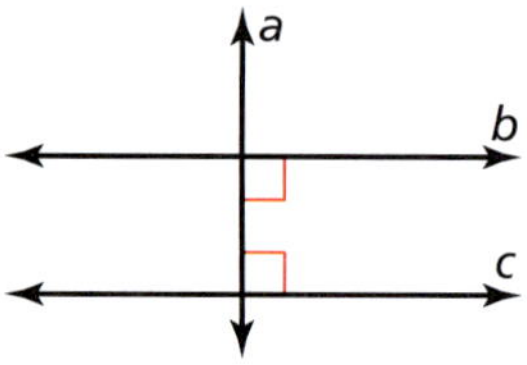

13.

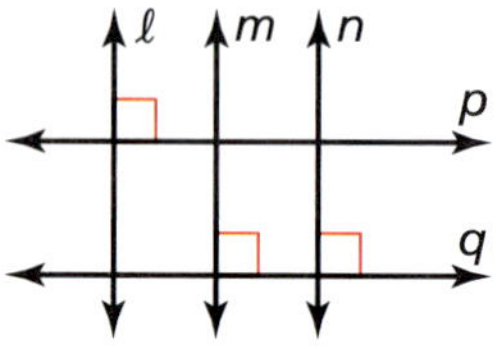

14.

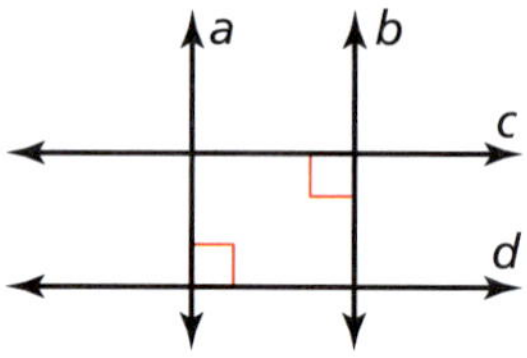

15.

16.

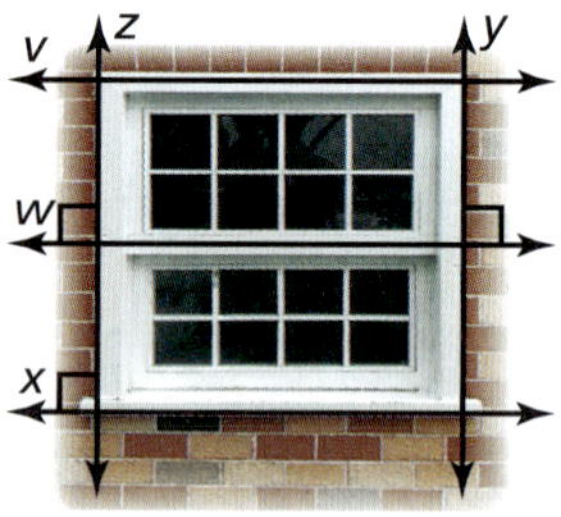

ERROR ANALYSIS **In Exercises 17 and 18, describe and correct the error in the statement about the diagram.**

17.

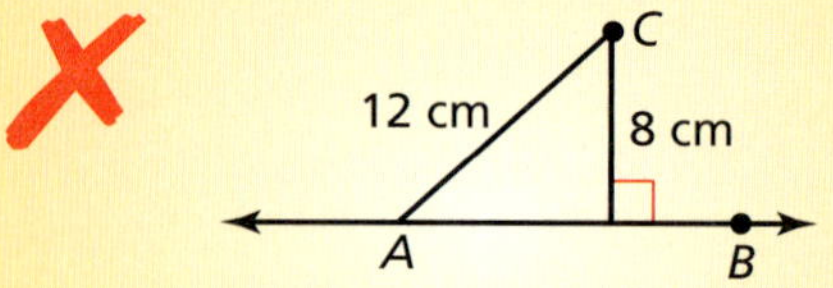

The distance from point C to $\overleftrightarrow{AB}$ is 12 centimeters.

18.

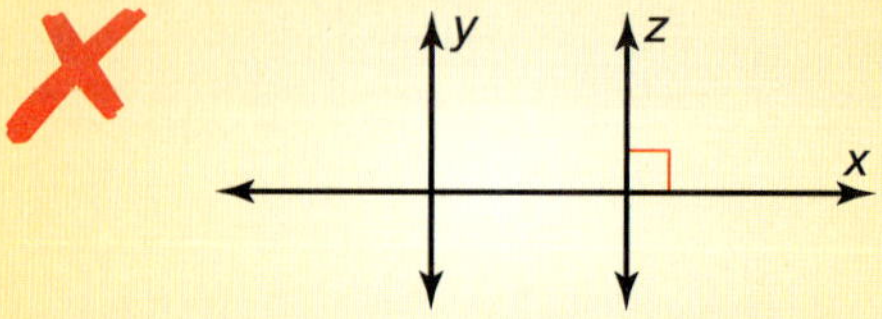

Lines y and z are parallel.

GO DIGITAL

PROOF In Exercises 19 and 20, use the diagram to write a proof of the statement.

19. If two intersecting lines are perpendicular, then they intersect to form four right angles.

Given $a \perp b$

Prove $\angle 1$, $\angle 2$, $\angle 3$, and $\angle 4$ are right angles.

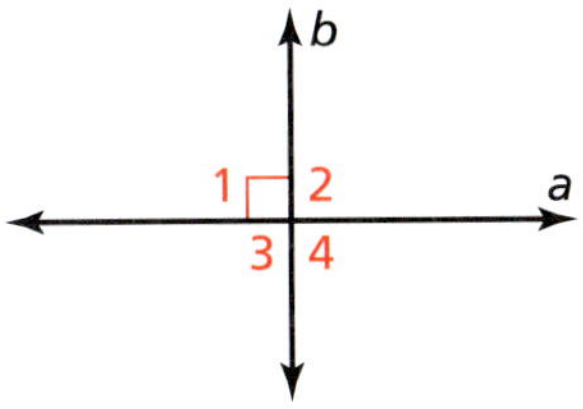

20. If two sides of two adjacent acute angles are perpendicular, then the angles are complementary.

Given $\overrightarrow{BA} \perp \overrightarrow{BC}$

Prove $\angle 1$ and $\angle 2$ are complementary.

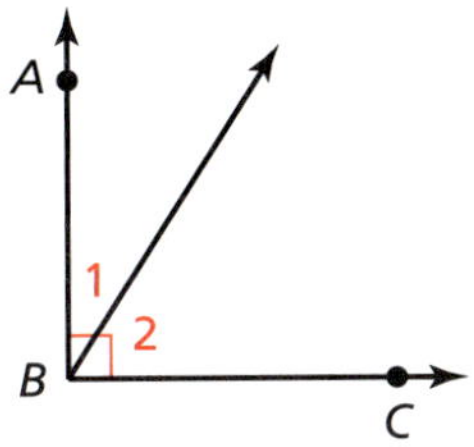

6 MTR **21.** **JUSTIFYING STEPS** Find all the unknown angle measures in the diagram. Justify your answer for each angle measure.

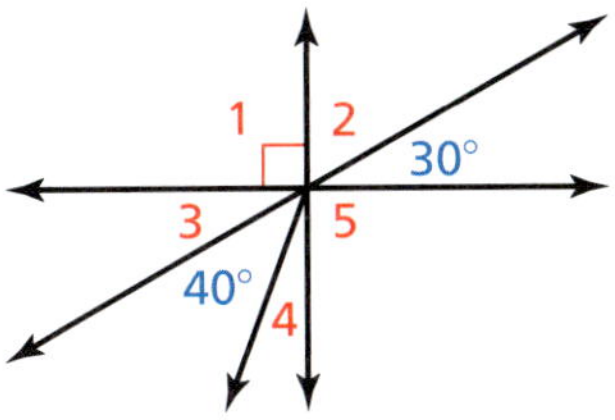

4 MTR **22.** **MAKING AN ARGUMENT** Your friend claims that because you can find the distance from a point to a line, you should be able to find the distance between any two lines. Is your friend correct? Explain your reasoning.

5 MTR **23.** **STRUCTURE** In the diagram, $a \perp b$. Find the value of x that makes $b \parallel c$.

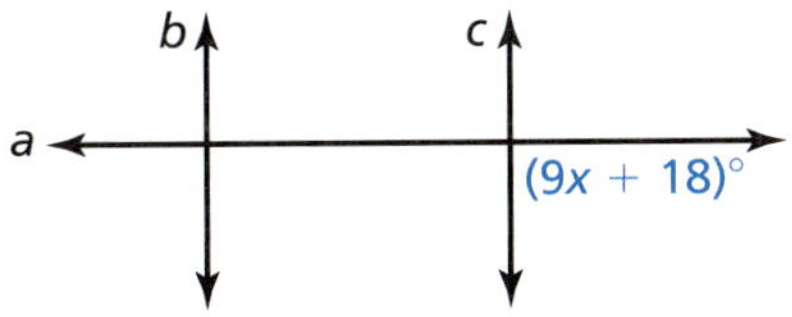

24. HOW DO YOU SEE IT?

You are trying to cross a stream from point A. Which point should you jump to in order to jump the shortest distance? Explain your reasoning.

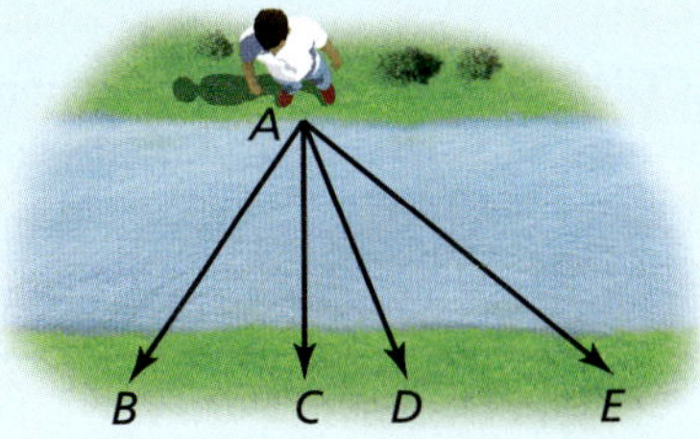

7 MTR **25.** **MODELING REAL LIFE** Park officials want to build a boardwalk that passes a geyser. A proposed design is shown. The boardwalk must be at least 100 feet from the center of the geyser at point G. Does the design meet this requirement? Explain.

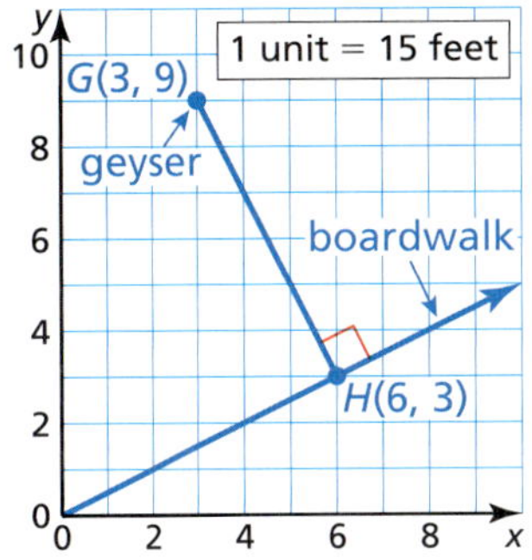

7 MTR **26.** **MODELING REAL LIFE** The painted line segments that form the path of a crosswalk can be painted as shown, or they can be perpendicular to the two parallel lines of the crosswalk. Which type of pattern requires less paint if both types use lines of equal thickness? Explain your reasoning.

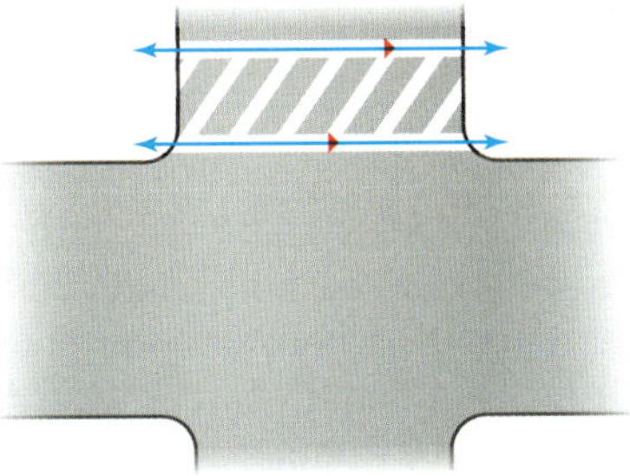

27. **CONSTRUCTION** Construct a square of side length AB.

GO DIGITAL

28. **REASONING** Two lines, a and b, are perpendicular to line c. Line d is parallel to line c. The distance between lines a and b is x meters. The distance between lines c and d is y meters. What shape is formed by the intersections of the four lines?

5 MTR 29. **CONNECTING CONCEPTS** Let C be a point on line n. Does the area of $\triangle ABC$ depend on the location of C? Explain your reasoning.

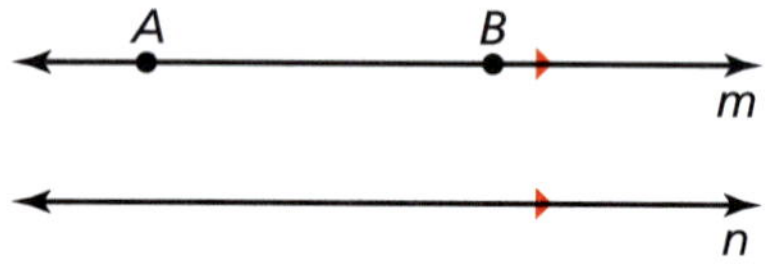

30. **THOUGHT PROVOKING**
The postulates and theorems in this book represent Euclidean geometry. In spherical geometry, all points are points on the surface of a sphere. A line is a circle on the sphere whose diameter is equal to the diameter of the sphere. In spherical geometry, how many right angles are formed by two perpendicular lines? Justify your answer.

31. **REASONING** Describe how you can find the distance from a point to a plane. Can you find the distance from a line to a plane? Explain your reasoning.

REVIEW & REFRESH

WATCH

In Exercises 32 and 33, find the slope and the y-intercept of the graph of the linear equation.

32. $y = \frac{1}{6}x - 8$

33. $-3x + y = 9$

34. Two angles form a linear pair. The measure of one angle is 77°. Find the measure of the other angle.

35. Find the distance from point W to $\overleftrightarrow{YZ}$.

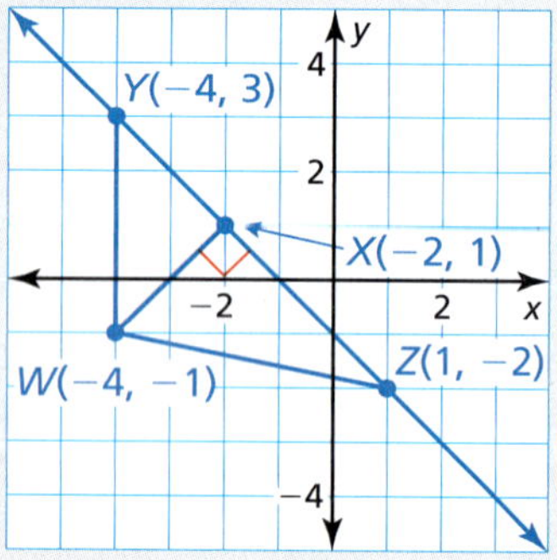

36. Find the slope of the line that passes through $(-4, 3)$ and $(6, 8)$.

7 MTR 37. **MODELING REAL LIFE** The post office and the grocery store are both on the same straight road between the school and your house. The distance from the school to the post office is 376 yards, the distance from the post office to your house is 929 yards, and the distance from the grocery store to your house is 513 yards.

a. What is the distance from the post office to the grocery store?

b. What is the distance from the school to your house?

3 MTR 38. **CHOOSE A METHOD** Solve the system using any method. Explain your choice of method.

$$2x + y = 5$$
$$3x + 2y = 6$$

In Exercises 39–42, consider the lines that contain the segments in the figure and the planes that contain the faces of the figure. Which line(s) or plane(s) contain point G and appear to fit the description?

39. line(s) parallel to $\overleftrightarrow{EF}$

40. line(s) perpendicular to $\overleftrightarrow{EF}$

41. line(s) skew to $\overleftrightarrow{EF}$

42. plane(s) parallel to plane ADE

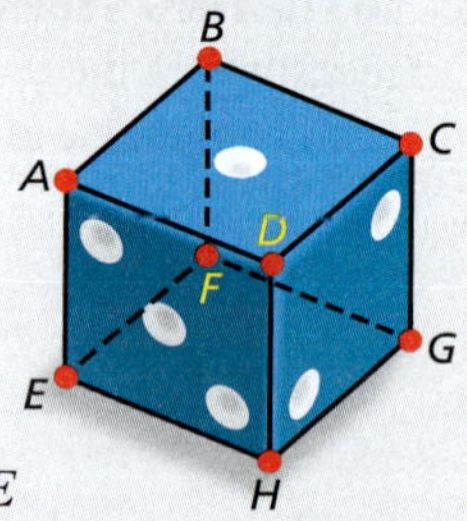

In Exercises 43 and 44, graph the function. Compare the graph to the graph of $f(x) = |x|$. Describe the domain and range.

43. $g(x) = |x| + 9$

44. $h(x) = -\frac{3}{2}|x|$

In Exercises 45 and 46, use the diagram.

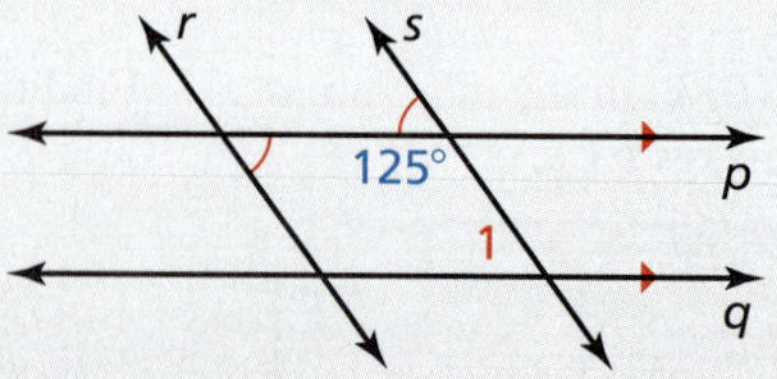

45. Find $m\angle 1$. Tell which theorem you use.

46. Is there enough information to prove that $r \parallel s$? If so, state the theorem you can use.

GO DIGITAL

3.5 Equations of Parallel and Perpendicular Lines

Learning Target: Partition a directed line segment and understand slopes of parallel and perpendicular lines.

Success Criteria:

- I can partition directed line segments using slope.
- I can use slopes to identify parallel and perpendicular lines.
- I can write equations of parallel and perpendicular lines.
- I can find the distance from a point to a line.

EXPLORE IT! Drawing Parallel and Perpendicular Lines

Work with a partner.

a. Plot any two points. Draw a line through the points. Translate the line vertically k units. Draw the image.

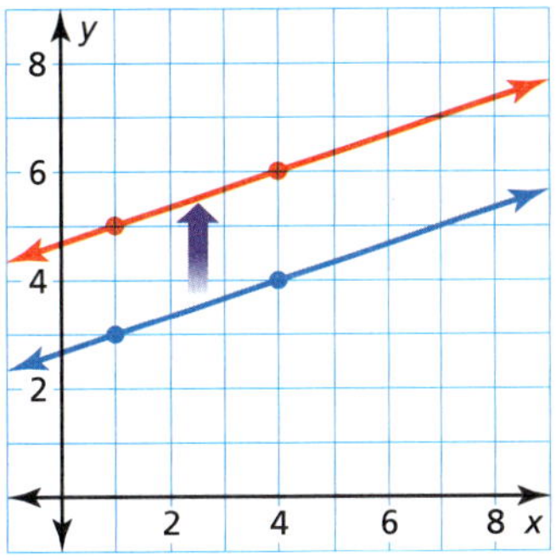

i. What points does the translated line pass through?

ii. Do the lines intersect? Explain.

iii. Find the slope of each line. What do you notice? Make a conjecture based on your observations.

b. Plot any two points. Draw a line through the points. Rotate the line 90° counterclockwise about the origin. Draw the image.

i. Describe the intersection of the lines.

ii. Find the slope of each line. What do you notice? Make a conjecture based on your observations.

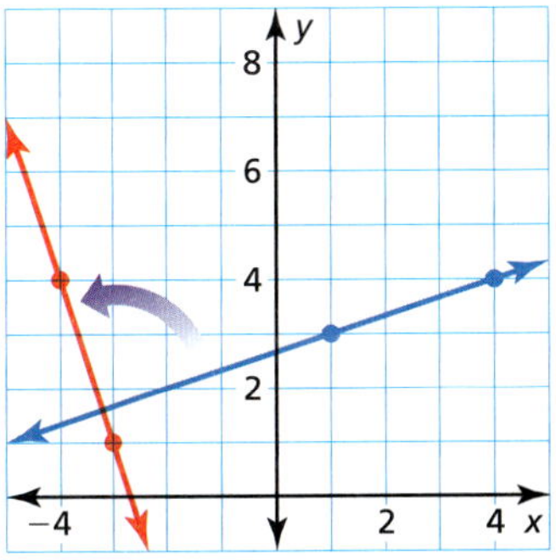

5 MTR RELATE CONCEPTS

Recall that a rotation of 90° counterclockwise about the origin maps a point (x, y) to $(-y, x)$. How can you find the slope of the image without graphing?

Geometric Reasoning

MA.912.GR.3.2 Given a mathematical or real-world context, use coordinate geometry to classify or justify definitions, properties and theorems involving circles, triangles or quadrilaterals.

MA.912.GR.3.3 Use coordinate geometry to solve mathematical and real-world geometric problems involving lines, circles, triangles and quadrilaterals.

GO DIGITAL

Vocabulary

directed line segment, *p. 150*

REMEMBER

Recall that the slope of a line or line segment through two points, (x_1, y_1) and (x_2, y_2), is defined as follows.

$$m = \frac{y_2 - y_1}{x_2 - x_1} = \frac{\text{change in } y}{\text{change in } x} = \frac{\text{rise}}{\text{run}}$$

You can choose either of the two points to be (x_1, y_1).

Partitioning a Directed Line Segment

A **directed line segment** AB is a segment that represents moving from point A to point B. The following example shows how to use slope to find a point on a directed line segment that partitions the segment in a given ratio.

EXAMPLE 1 Partitioning a Directed Line Segment

Find the coordinates of point P along the directed line segment AB so that the ratio of AP to PB is 3 to 2.

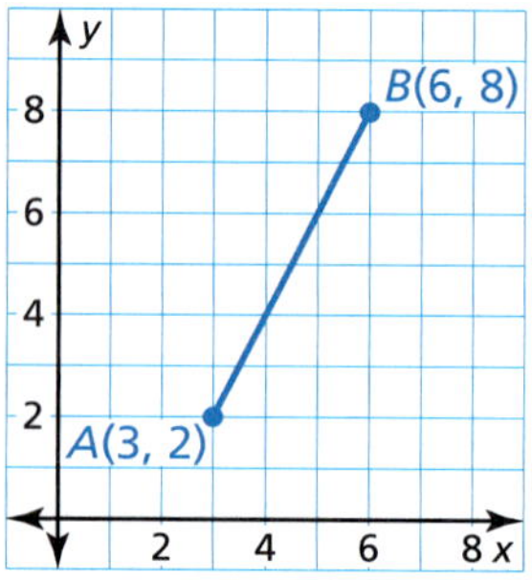

SOLUTION

To divide the segment in the ratio 3 to 2, think of dividing, or *partitioning*, the segment into 3 + 2, or 5 congruent pieces.

Point P is the point that is $\frac{3}{5}$ of the way from point A to point B.

Find the rise and run from point A to point B. Leave the slope in terms of rise and run, and do not simplify.

$$\text{slope of } \overline{AB}: \; m = \frac{8 - 2}{6 - 3} = \frac{6}{3} = \frac{\text{rise}}{\text{run}}$$

To find the coordinates of point P, add $\frac{3}{5}$ of the run to the x-coordinate of A, and add $\frac{3}{5}$ of the rise to the y-coordinate of A.

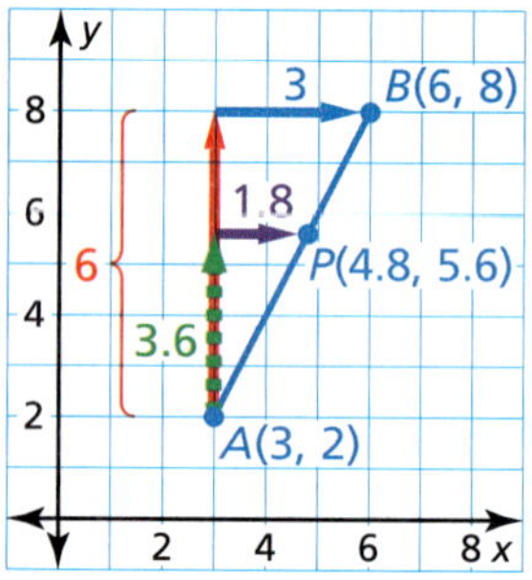

run: $\frac{3}{5}$ of $3 = \frac{3}{5} \cdot 3 = 1.8$

rise: $\frac{3}{5}$ of $6 = \frac{3}{5} \cdot 6 = 3.6$

▶ So, the coordinates of P are $(3 + 1.8, 2 + 3.6) = (4.8, 5.6)$.

Check

$AP = \sqrt{(4.8 - 3)^2 + (5.6 - 2)^2} = \sqrt{16.2}$

$PB = \sqrt{(6 - 4.8)^2 + (8 - 5.6)^2} = \sqrt{7.2}$

$\frac{AP}{PB} = \sqrt{\frac{16.2}{7.2}} = 1.5 = \frac{3}{2}$ ✓

SELF-ASSESSMENT

1 I don't understand yet. 2 I can do it with help. 3 I can do it on my own. 4 I can teach someone else.

Find the coordinates of point *P* along the directed line segment *AB* so that *AP* to *PB* is the given ratio.

1. 4 to 1

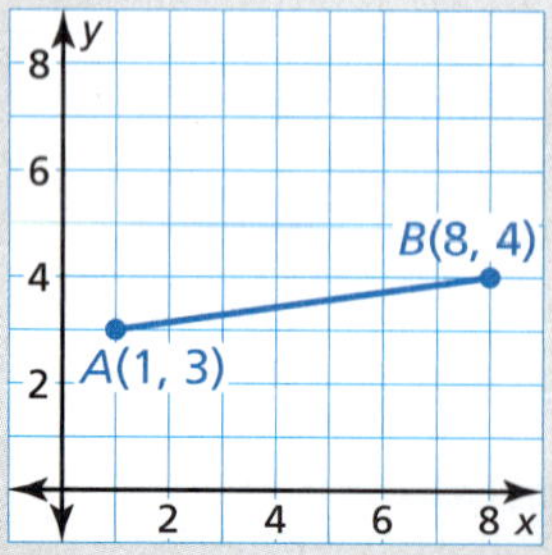

2. 3 to 7

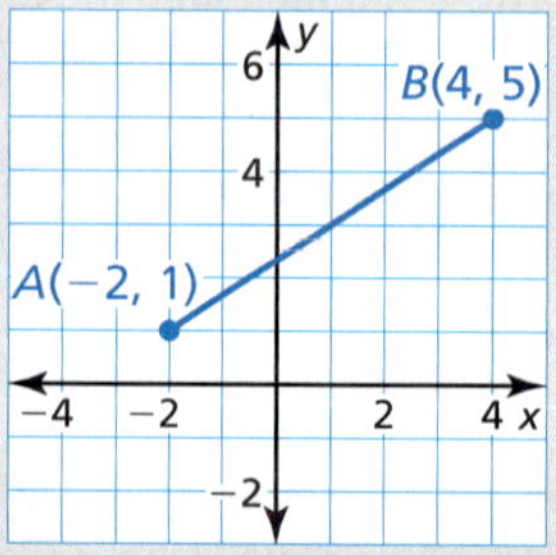

GO DIGITAL

Identifying Parallel and Perpendicular Lines

In the coordinate plane, the x-axis and the y-axis are perpendicular. Horizontal lines are parallel to the x-axis, and vertical lines are parallel to the y-axis.

THEOREMS

3.13 Slopes of Parallel Lines

In a coordinate plane, two nonvertical lines are parallel if and only if they have the same slope.

Any two vertical lines are parallel.

Proof page 423
Prove this Theorem Exercise 29, page 427

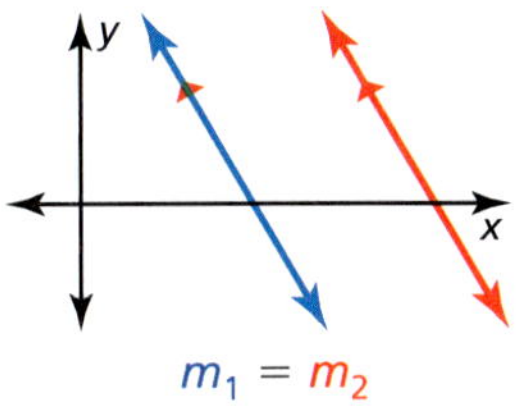

3.14 Slopes of Perpendicular Lines

In a coordinate plane, two nonvertical lines are perpendicular if and only if the product of their slopes is -1.

Horizontal lines are perpendicular to vertical lines.

Proof page 424
Prove this Theorem Exercise 30, page 427

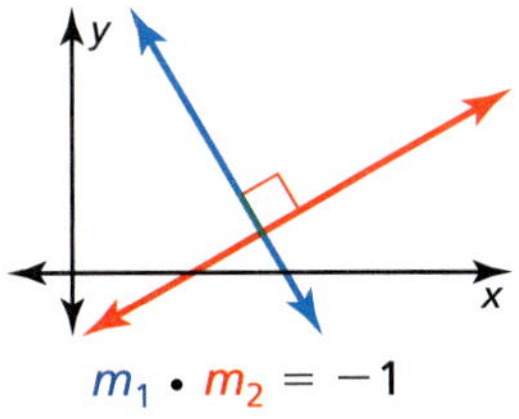

READING

If the product of two numbers is -1, then the numbers are called *negative reciprocals*.

EXAMPLE 2 Identifying Parallel and Perpendicular Lines

WATCH

Determine which lines are parallel and which lines are perpendicular.

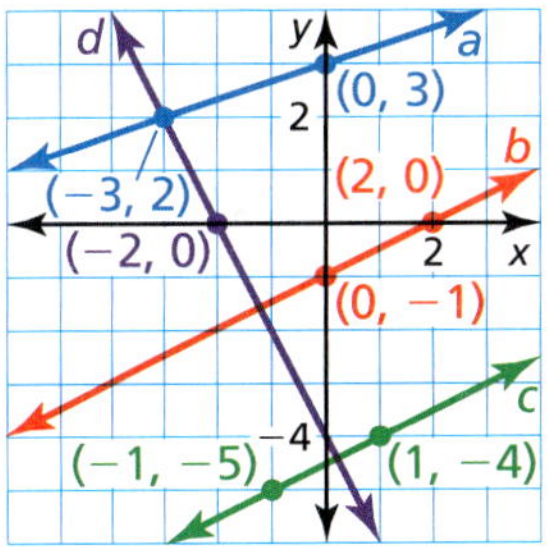

SOLUTION

Find the slope of each line.

Line a: $m = \dfrac{3 - 2}{0 - (-3)} = \dfrac{1}{3}$

Line b: $m = \dfrac{0 - (-1)}{2 - 0} = \dfrac{1}{2}$

Line c: $m = \dfrac{-4 - (-5)}{1 - (-1)} = \dfrac{1}{2}$ Line d: $m = \dfrac{2 - 0}{-3 - (-2)} = -2$

▶ Because lines b and c have the same slope, lines b and c are parallel. Because $\frac{1}{2}(-2) = -1$, lines b and d are perpendicular and lines c and d are perpendicular.

SELF-ASSESSMENT

1 I don't understand yet. 2 I can do it with help. 3 I can do it on my own. 4 I can teach someone else.

3. Determine which lines are parallel and which lines are perpendicular.

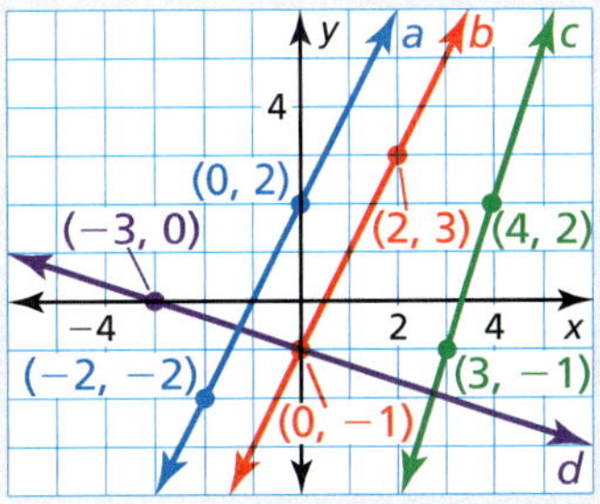

GO DIGITAL

Writing Equations of Parallel and Perpendicular Lines

REMEMBER
The linear equation $y = 2x - 3$ is written in slope-intercept form $y = mx + b$, where m is the slope and b is the y-intercept.

You can apply the Slopes of Parallel Lines Theorem and the Slopes of Perpendicular Lines Theorem to write equations of parallel and perpendicular lines.

EXAMPLE 3 Writing an Equation of a Parallel Line

WATCH

Write an equation of the line passing through the point $(-1, 1)$ that is parallel to the line $y = 2x - 3$.

SOLUTION

Step 1 Find the slope m of the parallel line. The line $y = 2x - 3$ has a slope of 2. By the Slopes of Parallel Lines Theorem, a line parallel to this line also has a slope of 2. So, $m = 2$.

Step 2 Find the y-intercept b by using $m = 2$ and $(x, y) = (-1, 1)$.

$y = mx + b$ — Use slope-intercept form.

$1 = 2(-1) + b$ — Substitute for m, x, and y.

$3 = b$ — Solve for b.

▶ Because $m = 2$ and $b = 3$, an equation of the line is $y = 2x + 3$.

Check
Use a graph.

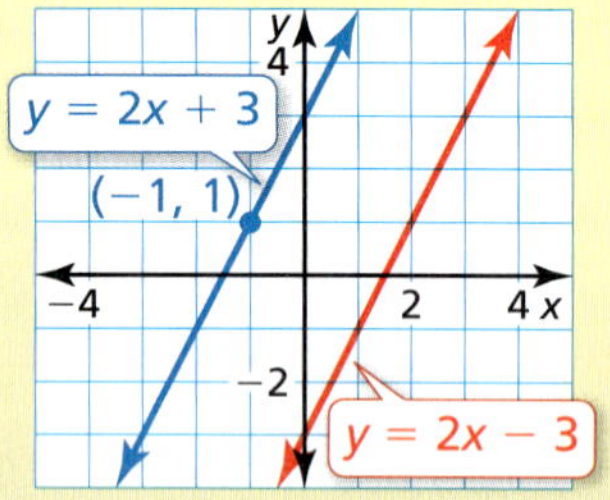

EXAMPLE 4 Writing an Equation of a Perpendicular Line

WATCH

Write an equation of the line passing through the point $(2, 3)$ that is perpendicular to the line $2x + y = 2$.

SOLUTION

Step 1 Find the slope m of the perpendicular line. The line $2x + y = 2$, or $y = -2x + 2$, has a slope of -2. Use the Slopes of Perpendicular Lines Theorem.

$-2 \cdot m = -1$ — The product of the slopes of $\perp$ lines is -1.

$m = \frac{1}{2}$ — Divide each side by -2.

Step 2 Find the y-intercept b by using $m = \frac{1}{2}$ and $(x, y) = (2, 3)$.

$y = mx + b$ — Use slope-intercept form.

$3 = \frac{1}{2}(2) + b$ — Substitute for m, x, and y.

$2 = b$ — Solve for b.

▶ Because $m = \frac{1}{2}$ and $b = 2$, an equation of the line is $y = \frac{1}{2}x + 2$.

Check
Use a graph.

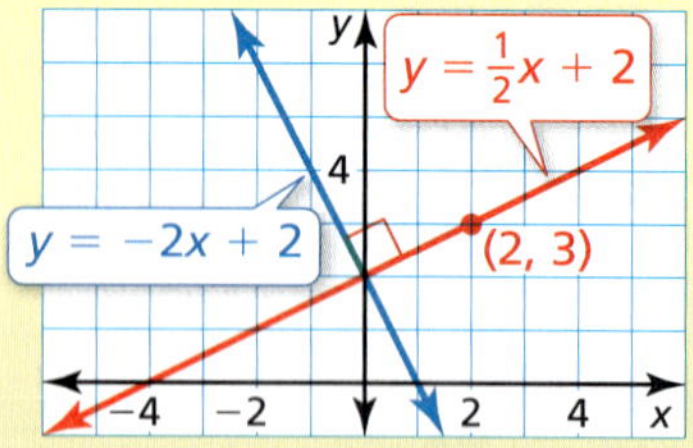

SELF-ASSESSMENT 1 I don't understand yet. 2 I can do it with help. 3 I can do it on my own. 4 I can teach someone else.

4. Write an equation of the line that passes through the point $(1, 5)$ and is (a) parallel to the line $y = 3x - 5$ and (b) perpendicular to the line $y = 3x - 5$.

5. REASONING How do you know that the lines $x = 4$ and $y = 2$ are perpendicular?

GO DIGITAL

Finding the Distance from a Point to a Line

Recall that the distance from a point to a line is the length of the perpendicular segment from the point to the line.

EXAMPLE 5 Finding the Distance from a Point to a Line

WATCH

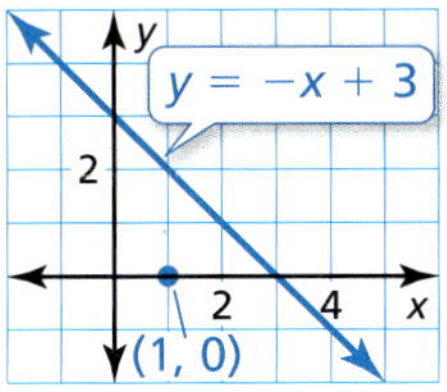

Find the distance from the point (1, 0) to the line $y = -x + 3$.

SOLUTION

Step 1 Find an equation of the line perpendicular to the line $y = -x + 3$ that passes through the point (1, 0).

First, find the slope m of the perpendicular line. The line $y = -x + 3$ has a slope of -1. Use the Slopes of Perpendicular Lines Theorem.

$-1 \cdot m = -1$	The product of the slopes of ⊥ lines is -1.
$m = 1$	Divide each side by -1.

Then find the y-intercept b by using $m = 1$ and $(x, y) = (1, 0)$.

$y = mx + b$	Use slope-intercept form.
$0 = 1(1) + b$	Substitute for m, x, and y.
$-1 = b$	Solve for b.

Because $m = 1$ and $b = -1$, an equation of the line is $y = x - 1$.

REMEMBER

Recall that the solution of a system of two linear equations in two variables is the ordered pair that represents the intersection of the graphs of the equations.

There are also two special cases to consider.

- When the equations represent parallel lines, the system has *no solution*.
- When the equations represent the same line, the system has *infinitely many solutions*.

Step 2 Use the two equations to write and solve a system of equations to find the point where the two lines intersect.

$y = -x + 3$	Equation 1
$y = x - 1$	Equation 2

Substitute $-x + 3$ for y in Equation 2.

$y = x - 1$	Equation 2
$-x + 3 = x - 1$	Substitute $-x + 3$ for y.
$x = 2$	Solve for x.

Substitute 2 for x in Equation 1 and solve for y.

$y = -x + 3$	Equation 1
$y = -2 + 3$	Substitute 2 for x.
$y = 1$	Simplify.

So, the perpendicular lines intersect at (2, 1).

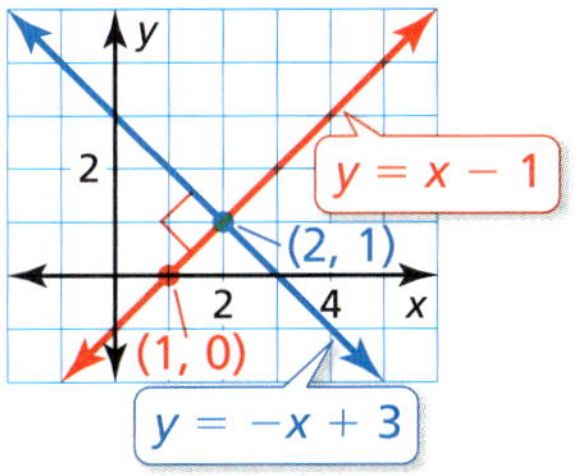

Step 3 Use the Distance Formula to find the distance from (1, 0) to (2, 1).

$$\text{distance} = \sqrt{(2 - 1)^2 + (1 - 0)^2} = \sqrt{1^2 + 1^2} = \sqrt{2} \approx 1.4$$

▶ So, the distance from the point (1, 0) to the line $y = -x + 3$ is about 1.4 units.

SELF-ASSESSMENT

1 I don't understand yet. 2 I can do it with help. 3 I can do it on my own. 4 I can teach someone else.

6. Find the distance from the point (6, 4) to the line $y = x + 4$.

7. Find the distance from the point $(-1, 6)$ to the line $y = -2x$.

GO DIGITAL

3.5 Practice WITH CalcChat® AND CalcView®

In Exercises 1–4, find the coordinates of point P along the directed line segment AB so that AP to PB is the given ratio. *(See Example 1.)*

1. $A(8, 0)$, $B(3, -2)$; 1 to 4

2. $A(-2, -4)$, $B(6, 1)$; 3 to 2

▶ 3. $A(1, 6)$, $B(-2, -3)$; 5 to 1

4. $A(-3, 2)$, $B(5, -4)$; 2 to 6

In Exercises 5 and 6, determine which of the lines are parallel and which of the lines are perpendicular. *(See Example 2.)*

▶ 5.

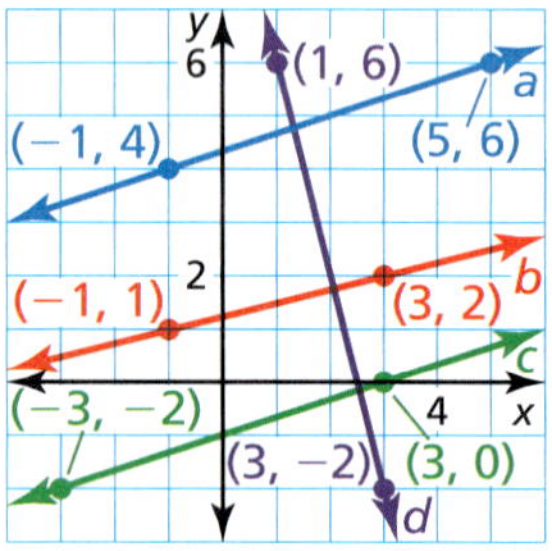

6.

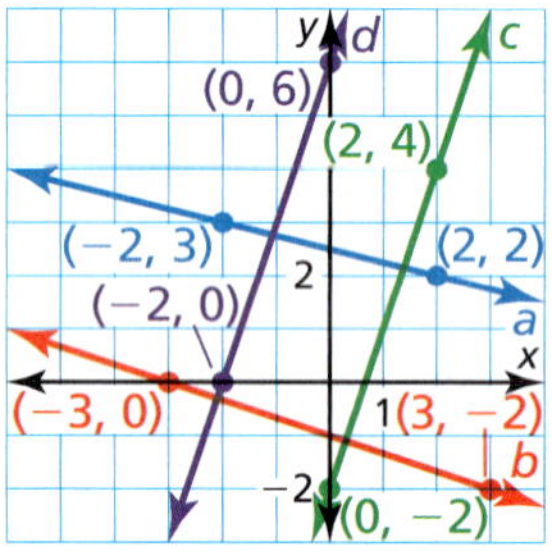

In Exercises 7–10, tell whether the lines through the given points are *parallel*, *perpendicular*, or *neither*. Justify your answer.

7. **Line 1:** $(1, 0)$, $(7, 4)$
   Line 2: $(7, 0)$, $(3, 6)$

8. **Line 1:** $(-3, 1)$, $(-7, -2)$
   Line 2: $(2, -1)$, $(8, 4)$

9. **Line 1:** $(-9, 3)$, $(-5, 7)$
   Line 2: $(-11, 6)$, $(-7, 2)$

10. **Line 1:** $(10, 5)$, $(-8, 9)$
    Line 2: $(2, -4)$, $(11, -6)$

In Exercises 11–14, write an equation of the line passing through point P that is parallel to the given line. Graph the equations of the lines to check that they are parallel. *(See Example 3.)*

▶ 11. $P(0, 1)$, $y = -2x + 3$

12. $P(3, 8)$, $y = \frac{1}{5}(x + 4)$

13. $P(-2, 6)$, $x = -5$

14. $P(4, 0)$, $-x + 2y = 12$

In Exercises 15–18, write an equation of the line passing through point P that is perpendicular to the given line. Graph the equations of the lines to check that they are perpendicular. *(See Example 4.)*

▶ 15. $P(0, 0)$, $y = -9x - 1$

16. $P(4, -6)$, $y = -3$

17. $P(2, 3)$, $y + 2 = -2x$

18. $P(-8, 0)$, $3x - 5y = 6$

In Exercises 19–22, find the distance from point A to the given line. *(See Example 5.)*

▶ 19. $A(-1, 7)$, $y = 3x$

20. $A(-9, -3)$, $y = x - 6$

21. $A(15, -21)$, $5x + 2y = 4$

22. $A\left(-\frac{1}{4}, 5\right)$, $-x + 2y = 14$

4 MTR 23. **ERROR ANALYSIS** Describe and correct the error in determining whether the lines are parallel, perpendicular, or neither.

> Line 1: $(3, -5)$, $(2, -1)$
> Line 2: $(0, 3)$, $(1, 7)$
> $$m_1 = \frac{-1 - (-5)}{2 - 3} = -4$$
> $$m_2 = \frac{7 - 3}{1 - 0} = 4$$
> Lines 1 and 2 are perpendicular.

4 MTR 24. **ERROR ANALYSIS** Describe and correct the error in writing an equation of the line that passes through the point $(3, 4)$ and is parallel to the line $y = 2x + 1$.

> $y = 2x + 1$, $(3, 4)$
> $4 = m(3) + 1$
> $1 = m$
> The line $y = x + 1$ is parallel to the line $y = 2x + 1$.

GO DIGITAL

In Exercises 25–28, find the midpoint of $\overline{PQ}$. Then write an equation of the line that passes through the midpoint and is perpendicular to $\overline{PQ}$.

25. $P(-4, 3), Q(4, -1)$

26. $P(-5, -5), Q(3, 3)$

27. $P(0, 2), Q(6, -2)$

28. $P(-7, 0), Q(1, 8)$

In Exercises 29–32, find the distance between the parallel lines.

29.

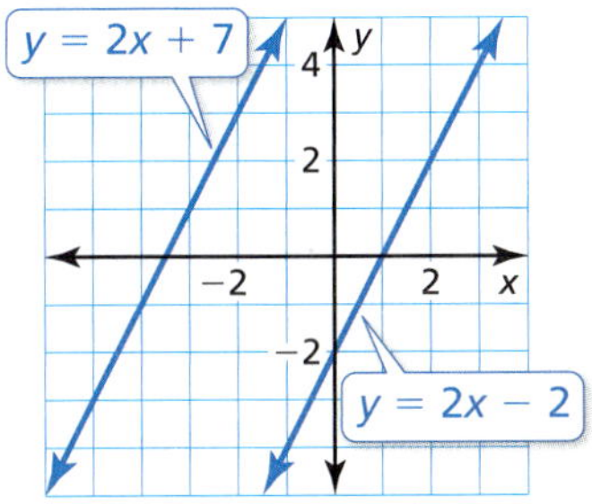

30.

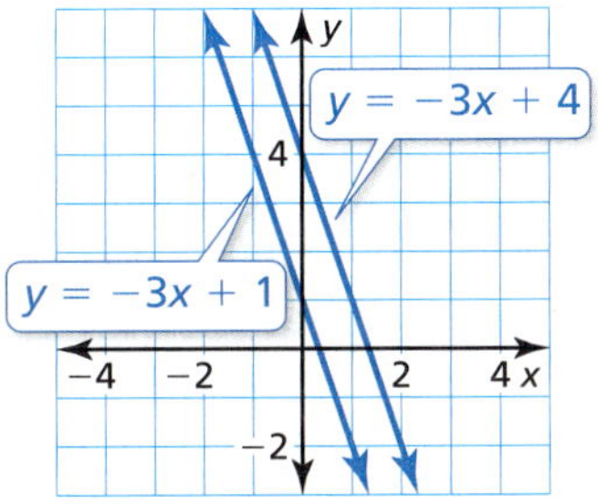

31. $y = -\frac{2}{3}x + 8$, parallel line that passes through $(0, 0)$

32. $y = -\frac{5}{6}x - 1$, parallel line that passes through $(6, -4)$

33. **STRUCTURE** Point C lies on $\overline{AB}$, where $A(-4, -2)$ and $B(5, 2)$. The distance from A to C is one-fourth of the distance from C to B. What are the coordinates of C?

34. **STRUCTURE** Suppose point P divides the directed line segment XY so that the ratio of XP to PY is 3 to 5. Describe the point that divides the directed line segment YX so that the ratio of YP to PX is 5 to 3.

35. **REASONING** Is triangle LMN a right triangle? Explain your reasoning.

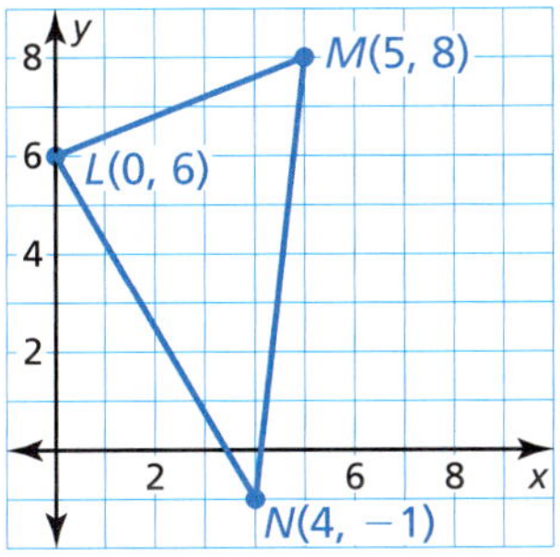

36. **REASONING** Is quadrilateral $QRST$ a parallelogram? Explain your reasoning.

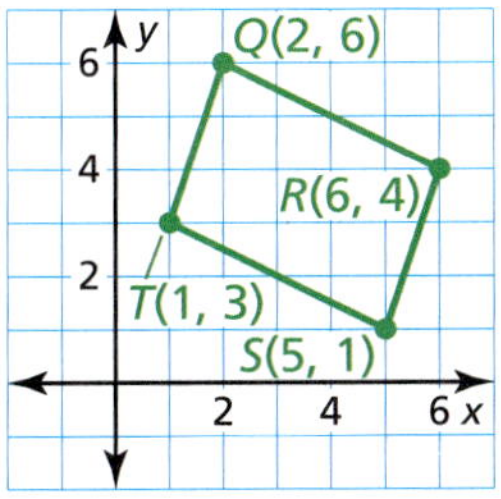

37. **PROBLEM SOLVING** A bike path is constructed perpendicular to Washington Boulevard starting at point $P(2, 2)$. An equation of the line representing Washington Boulevard is $y = -\frac{2}{3}x$. Each unit in the coordinate plane corresponds to 10 feet. Approximately how far is the starting point from Washington Boulevard?

38. **PROBLEM SOLVING** You go on a hayride to take photographs of landmarks. The map shows your path and two landmarks. Each unit in the coordinate plane corresponds to 10 yards. Approximate your minimum distance from the giant pumpkin.

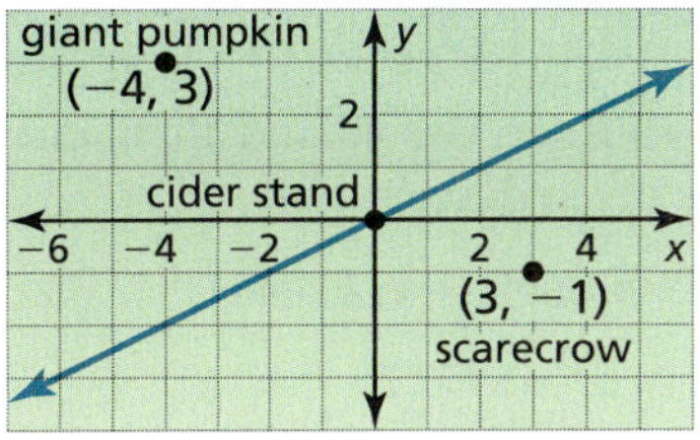

39. **STRUCTURE** The slope of line ℓ is greater than 0 and less than 1. Write an inequality for the slope of a line perpendicular to ℓ.

40. **HOW DO YOU SEE IT?**
Determine whether quadrilateral $JKLM$ is a square. Explain your reasoning.

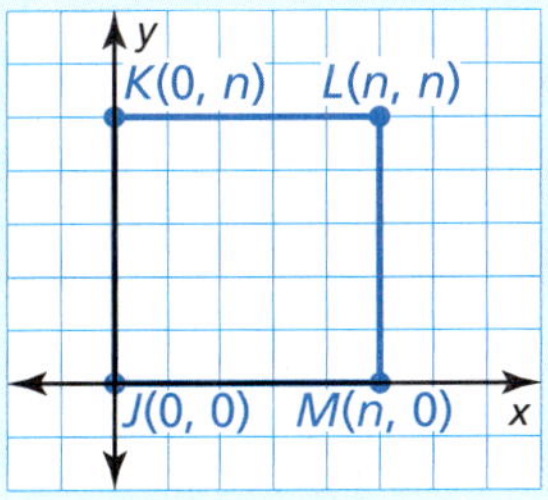

41. **PERFORMANCE TASK** Utility mapping involves determining the positions of underground pipes and cables.

a. A gas pipe runs along a street with multiple residences. Sewer pipes intersect the gas pipe at right angles in front of each residence. Create a map on a coordinate plane that represents residences and the pipes. Provide a scale.

b. Lampposts must be installed at least 5 feet from the nearest gas or sewer pipe. Add lampposts to your diagram in part (a). Verify that the locations of the lampposts meet the requirement.

42. **MAKING AN ARGUMENT** Your classmate claims that no two nonvertical parallel lines can have the same y-intercept. Is your classmate correct? Explain.

GO DIGITAL

43. **CONNECTING CONCEPTS** Solve each system of equations algebraically. Make a conjecture about what the solution(s) tell you about the graph of the system.

a. $y = 4x + 9$
$4x - y = 1$

b. $3y + 4x = 16$
$2x - y = 18$

c. $y = -5x + 6$
$10x + 2y = 12$

44. THOUGHT PROVOKING
Find a formula for the distance from the point (x_0, y_0) to the line $ax + by = 0$. Verify your formula using a point and a line.

STRUCTURE In Exercises 45 and 46, find a value for k based on the given description.

45. The line through $(-1, k)$ and $(-7, -2)$ is parallel to the line $y = x + 1$.

46. The line through $(k, 2)$ and $(7, 0)$ is perpendicular to the line $y = x - \frac{28}{5}$.

PROVING A THEOREM In Exercises 47 and 48, use the slopes of lines to write a paragraph proof of the theorem.

47. Lines Perpendicular to a Transversal Theorem: In a plane, if two lines are perpendicular to the same line, then they are parallel to each other.

48. Transitive Property of Parallel Lines Theorem: If two lines are parallel to the same line, then they are parallel to each other.

PROOF In Exercises 49 and 50, prove the statement.

49. If two lines are vertical, then they are parallel.

50. If two lines are horizontal, then they are parallel.

51. **CONNECTING CONCEPTS** Rewrite each statement from Exercises 49 and 50 as a contrapositive statement. Then prove each contrapositive statement.

52. **PROOF** Prove that horizontal lines are perpendicular to vertical lines.

REVIEW & REFRESH

53. Find the value of x that makes $m \parallel n$. Explain your reasoning.

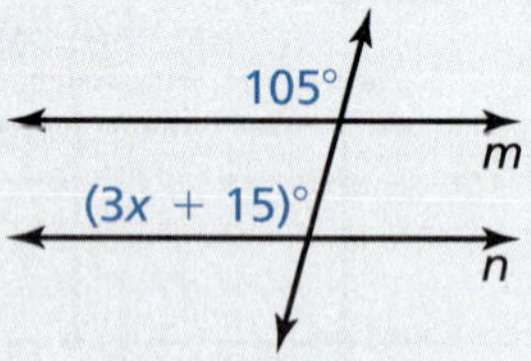

54. Make and test a conjecture about the product of three consecutive odd numbers.

55. Find the perimeter of the triangle with the vertices (0, 1), (3, 6), and (4, 1).

56. Determine which lines, if any, must be parallel. Explain your reasoning.

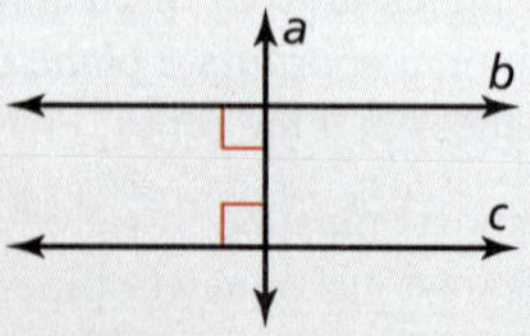

57. Factor $4h^2 + 8h + 3$.

58. Write an equation of the line passing through point $P(2, 5)$ that is parallel to $y = 4x + 1$.

59. Solve $x^2 - 9x = 0$.

60. **MODELING REAL LIFE** A chute forms a line between two parallel supports, as shown. Find $m\angle 2$. Explain your reasoning.

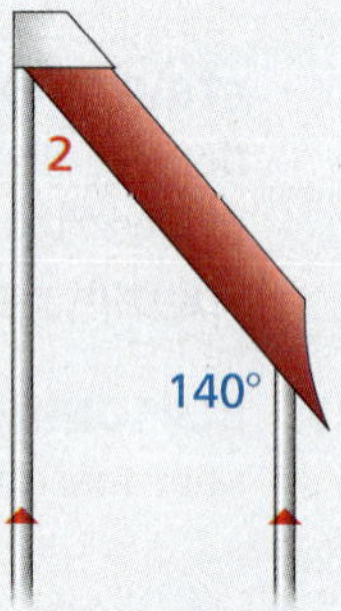

61. Solve the inequality $2w - 3 > 2w + w - 7$. Graph the solution.

CHOOSE A METHOD In Exercises 62 and 63, solve the system using any method. Explain your choice of method.

62. $6x + 3y = 24$
$2x - 3y = -16$

63. $-y = 3$
$2x - 3y = 11$

64. Write a linear function f with the values $f(3) = -4$ and $f(1) = 2$.

65. Graph the quadratic function $y = (x - 1)(x + 2)$. Label the vertex, axis of symmetry, and x-intercepts. Find the domain and range of the function.

GO DIGITAL

3 Chapter Review WITH CalcChat®

Chapter Learning Target: Understand parallel and perpendicular lines.

Chapter Success Criteria:
- ◆ I can describe angle relationships formed by parallel lines and a transversal.
- ◆ I can construct parallel and perpendicular lines.
- ■ I can prove theorems involving parallel and perpendicular lines.
- ■ I can write equations of parallel and perpendicular lines.

◆ Surface
■ Deep

SELF-ASSESSMENT 1 I don't understand yet. 2 I can do it with help. 3 I can do it on my own. 4 I can teach someone else.

3.1 Pairs of Lines and Angles (pp. 121–126) WATCH

Learning Target: Understand lines, planes, and pairs of angles.

Vocabulary
- parallel lines
- skew lines
- parallel planes
- transversal
- corresponding angles
- alternate interior angles
- alternate exterior angles
- consecutive interior angles

Consider the lines that contain the segments in the figure and the planes that contain the faces of the figure. All angles are right angles. Which line(s) or plane(s) contain point N and appear to fit the description?

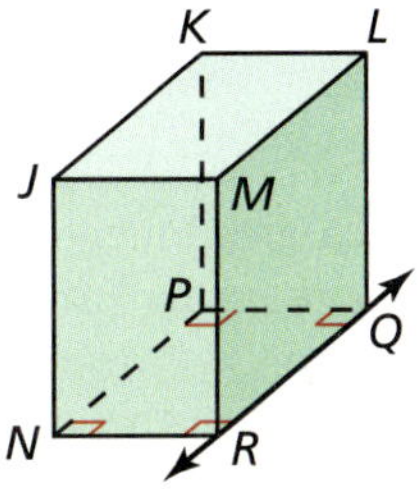

1. line(s) perpendicular to $\overleftrightarrow{QR}$
2. line(s) parallel to $\overleftrightarrow{QR}$
3. line(s) skew to $\overleftrightarrow{QR}$
4. plane(s) parallel to plane LMQ

Identify all pairs of angles of the given type.

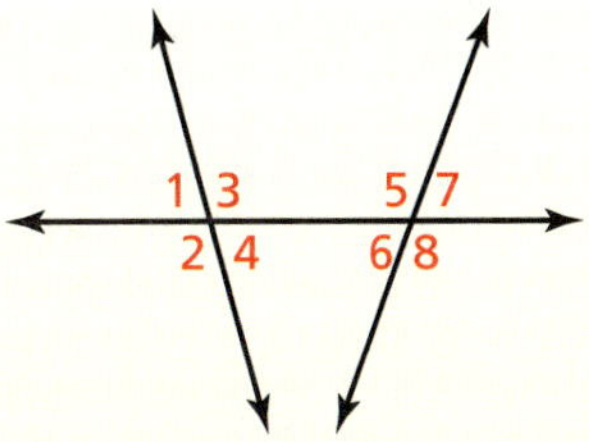

5. consecutive interior
6. alternate interior
7. corresponding
8. alternate exterior

3.2 Parallel Lines and Transversals (pp. 127–132) WATCH

Learning Target: Prove and use theorems about parallel lines.

Find the values of x and y. Show your steps.

9.

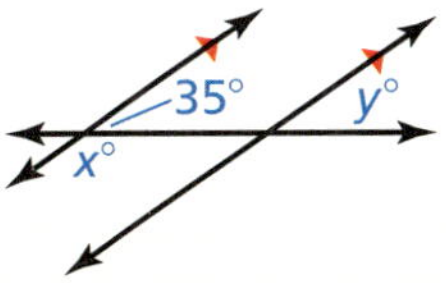

10.

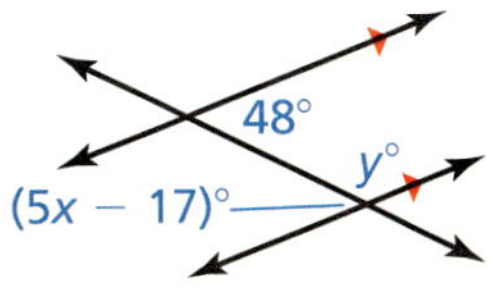

11.

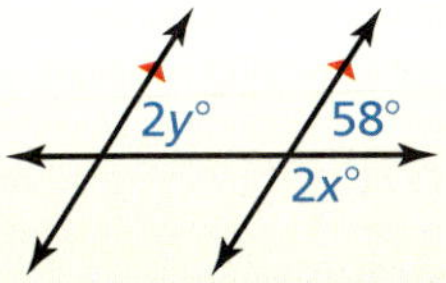

12.

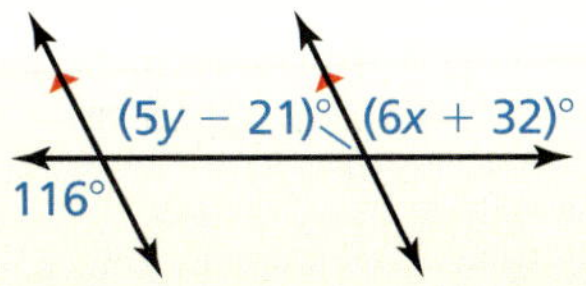

GO DIGITAL

3.3 Proofs with Parallel Lines (pp. 133–140) WATCH

Learning Target: Prove and use theorems about identifying parallel lines.

Find the value of *x* that makes *m* ∥ *n*. Explain your reasoning.

13.

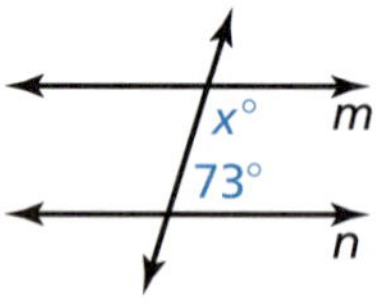

14.

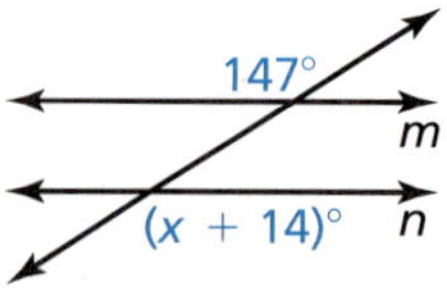

15.

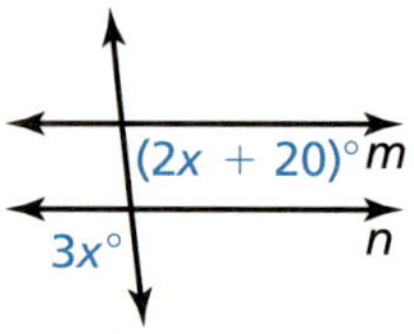

16.

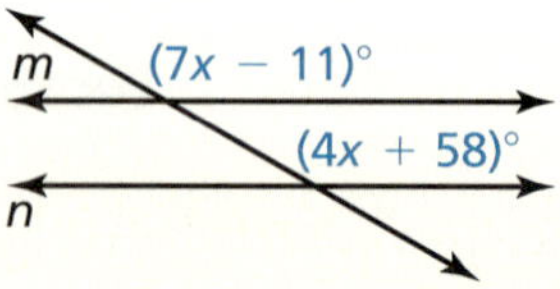

17. The strings of a musical instrument called an *oud* are shown. Are the outer strings of the oud parallel? Explain.

3.4 Proofs with Perpendicular Lines (pp. 141–148) WATCH

Learning Target: Prove and use theorems about perpendicular lines.

Vocabulary VOCAB
distance from a point to a line
perpendicular bisector

Determine which lines, if any, must be parallel. Explain your reasoning.

18.

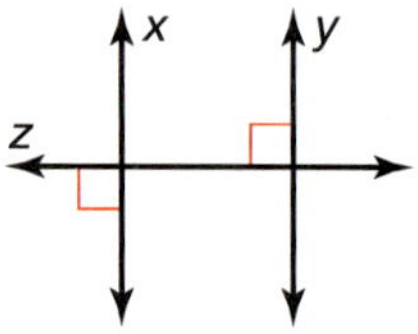

19.

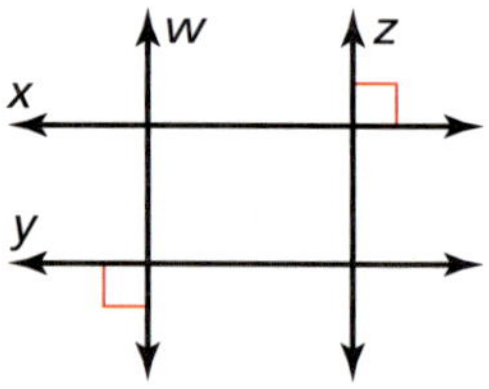

20.

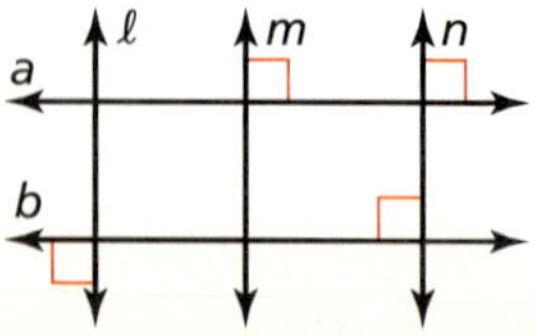

21.

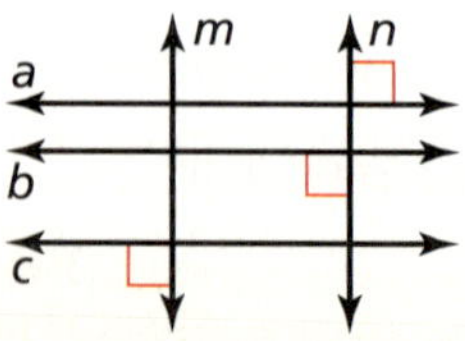

22. Use the diagram to write a proof.

Given $\angle 1 \cong \angle 2$, $h \perp k$

Prove $g \parallel h$

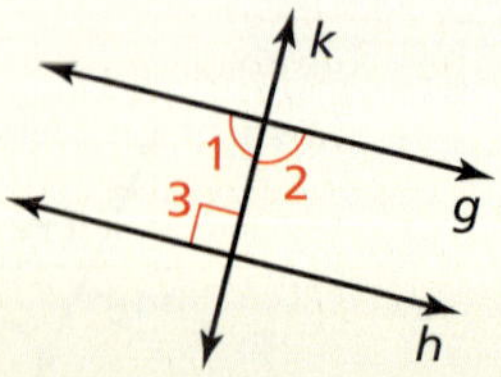

GO DIGITAL

3.5 Equations of Parallel and Perpendicular Lines (pp. 149–156) WATCH

Learning Target: Partition a directed line segment and understand slopes of parallel and perpendicular lines.

Vocabulary AZ VOCAB
directed line segment

Find the coordinates of point P along the directed line segment AB so that AP to PB is the given ratio.

23. $A(-3, 2), B(5, 5)$; 1 to 3

24. $A(-2, 4), B(4, -3)$; 3 to 2

25. Determine which of the lines are parallel and which of the lines are perpendicular.

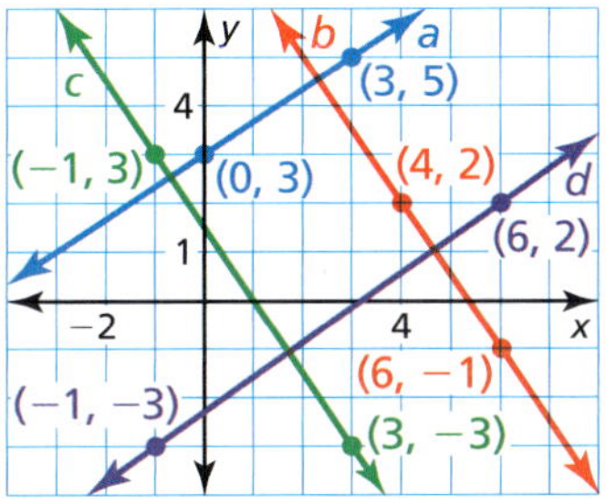

Write an equation of the line passing through point A that is parallel to the given line.

26. $A(3, -4), y = -x + 8$

27. $A(-6, 5), y = \frac{1}{2}x - 7$

Write an equation of the line passing through point A that is perpendicular to the given line.

28. $A(6, -1), y = -2x + 8$

29. $A(-1, 5), y = \frac{1}{7}x + 4$

Find the distance from point A to the given line.

30. $A(2, -1), y = -x + 4$

31. $A(-2, 3), y = \frac{1}{2}x + 1$

32. A coordinate plane is superimposed on a property map. Each unit in the coordinate plane represents 1 foot. The location of a wellhead on the property is given by the point (12, 6). How close is the wellhead to a property line modeled by the line $y = \frac{1}{4}x - 14$?

Mathematical Thinking and Reasoning

3 MTR COMPLETE TASKS WITH MATHEMATICAL FLUENCY
Mathematicians who complete tasks with mathematical fluency select efficient and appropriate methods for solving problems within the given context.

1. Look back at 3.5 Explore It! part (b) on page 149. Did you use technology, or did you perform the steps by hand? Complete part (b) using the other method. Describe both processes. Then describe the advantages and disadvantages of each.

2. Explore the inverse of the Corresponding Angles Theorem: “If two lines that are not parallel are cut by a transversal, then the pairs of corresponding angles are not congruent.” Tell what tools you use, and describe your results.

GO DIGITAL

3 Practice Test WITH CalcChat®

Find the values of *x* and *y*. State which theorem(s) you used.

1.

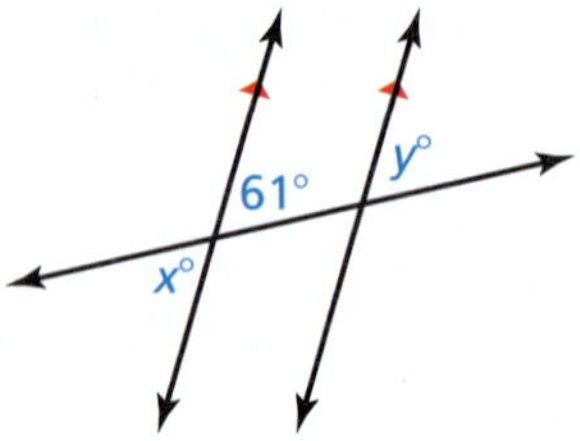

2.

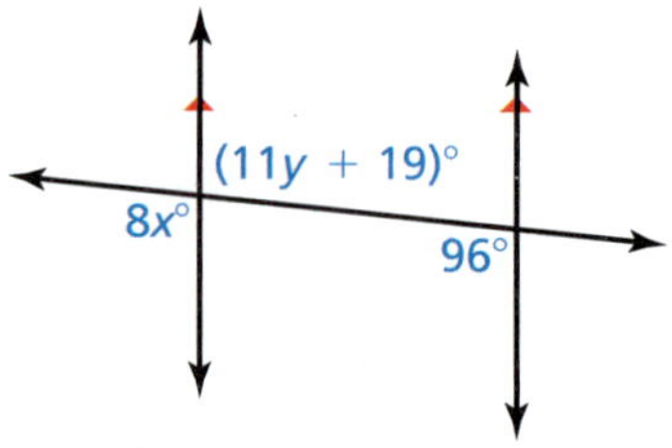

3.

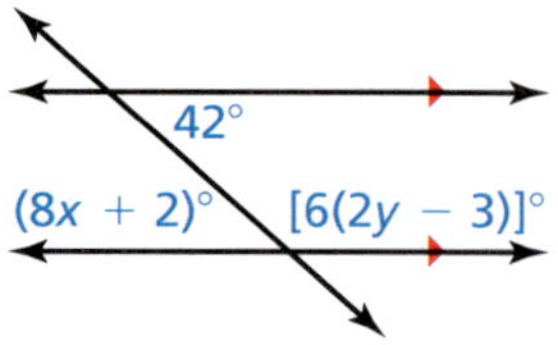

Find the value of *x* that makes $m \parallel n$.

4.

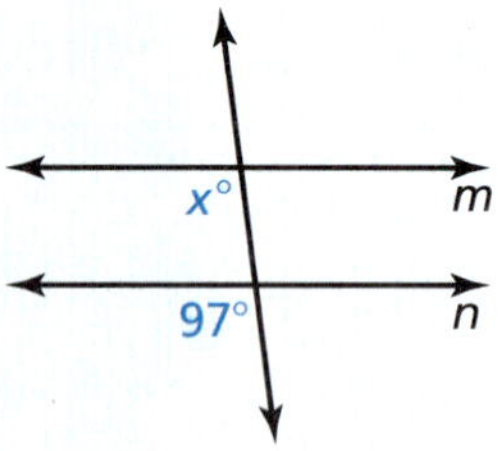

5.

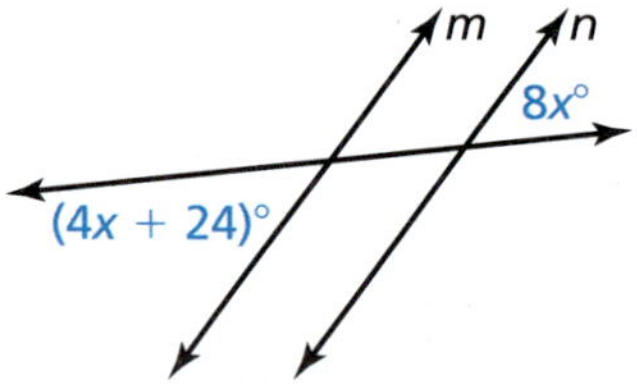

6.

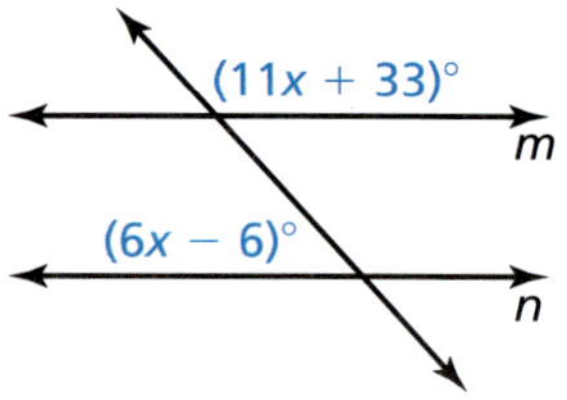

7. A student says, "Because $j \perp k$, $j \perp \ell$." What missing information is the student assuming from the diagram at the right? Which theorem is the student trying to use?

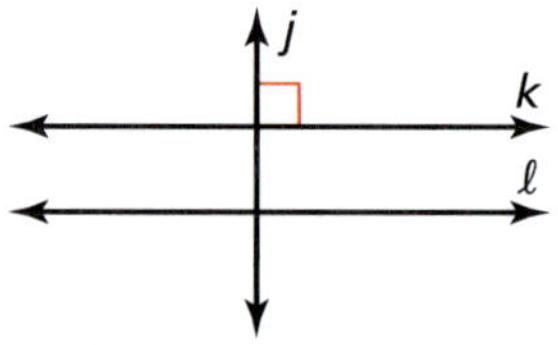

8. Find the distance from point $A(-3, 7)$ to the line $y = \frac{1}{3}x - 2$.

9. Write and solve a system of equations to find the values of *x* and *y* in the diagram at the right.

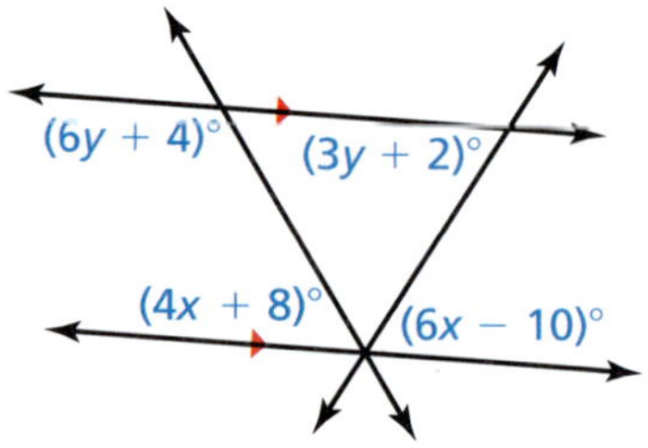

10. Write an equation of the line that passes through the point $(-5, 2)$ and is (a) parallel and (b) perpendicular to the line $y = 2x - 3$.

11. Your school lies directly between your home and the movie theater. The distance from your home to the school is one-fourth of the distance from the school to the movie theater. What point on the graph represents your school?

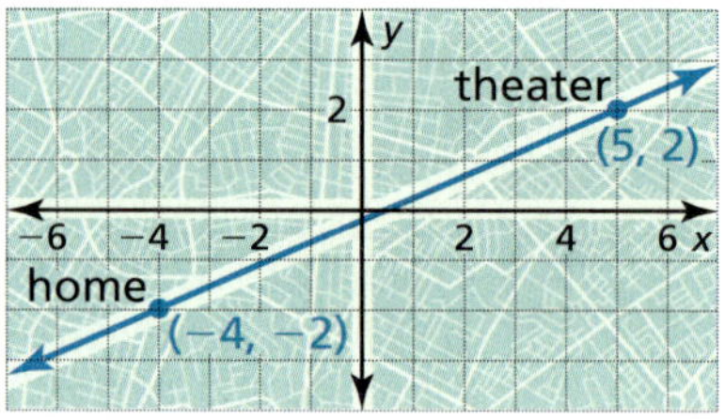

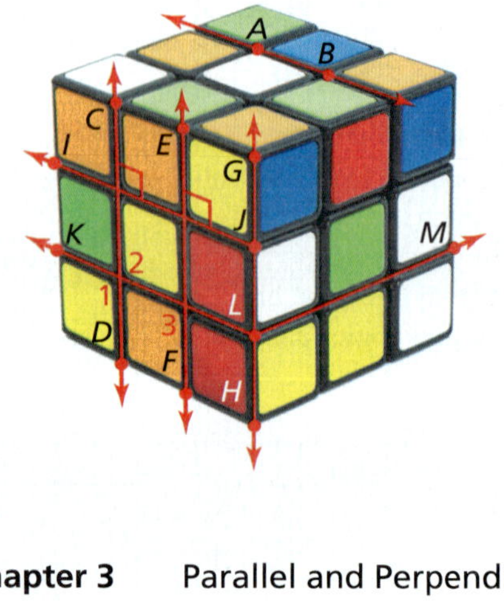

12. Identify an example on the puzzle cube of each description. Explain your reasoning.

a. a pair of skew lines

b. a pair of perpendicular lines

c. a pair of parallel lines

d. a pair of congruent corresponding angles

e. a pair of congruent alternate interior angles

GO DIGITAL

3 Performance Task
Geothermal Energy

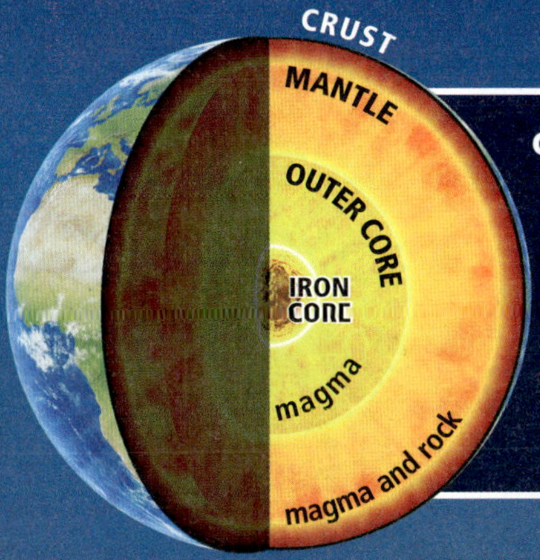

Geothermal Energy: Energy produced by Earth's internal heat

- Heat is produced when radioactive particles slowly decay in Earth's core.
- Magma comes close to the surface in areas where the crust is thin, faulted, or fractured by plate tectonics.
- The heat from this magma is transferred to underground water, which produces geothermal energy.

Temperatures at a Depth of 10 Kilometers

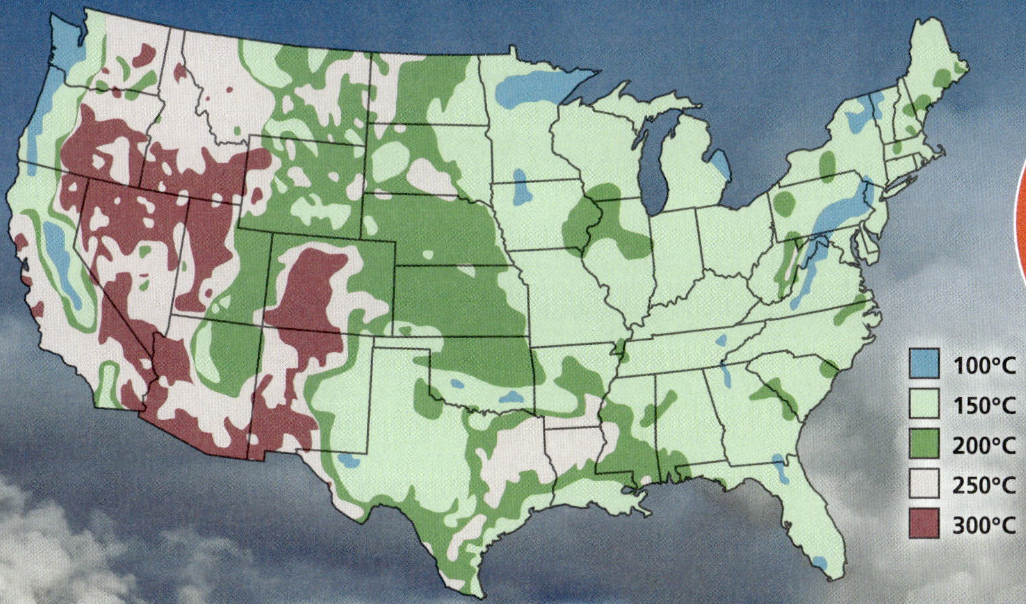

Western states are generally more favorable locations for geothermal power plants, where heat from magma is usually closer to the surface.

NEW POWER PLANT

You work for an energy company that wants to build a new geothermal power plant.

- Use the Internet or another resource to research and explain how geothermal energy is harnessed. Then choose a state for the power plant. Explain your reasoning.
- Plot the approximate locations of two cities in the state in a coordinate plane, and connect them with a line segment. Then construct the perpendicular bisector of the segment. Where could the power plant be located so that it is equidistant from the two cities? Explain.
- Plot the location of a third city in the coordinate plane. Show how to find a location for the power plant that is equidistant from all three cities.

GO DIGITAL

3 Test Prep WITH CalcChat® Cumulative Practice

WATCH Tutorial videos are available for each exercise.

1. Which postulate is *not* correctly demonstrated by the given example from the diagram?

Ⓐ Plane Intersection Postulate; Planes X and Y intersect at $\overleftrightarrow{CD}$.

Ⓑ Plane-Line Postulate; C, D, and $\overleftrightarrow{CD}$ all lie in plane X.

Ⓒ Plane-Point Postulate; Plane Y contains point E.

Ⓓ Line Intersection Postulate; Line n and line m intersect at point G.

2. The equation of a line is $x + 2y = 10$.

a. Use the numbers and symbols to create the equation of a line in slope-intercept form that passes through the point $(4, -5)$ and is parallel to the given line.

b. Use the numbers and symbols to create the equation of a line in slope-intercept form that passes through the point $(2, -1)$ and is perpendicular to the given line.

y	x	$=$	$+$	$-$	-9	-2	-1
$-\frac{1}{2}$	$\frac{1}{2}$	1	$\frac{3}{2}$	2	3	4	5

3. Classify each pair of angles whose measurements are given. Select all that apply.

a.

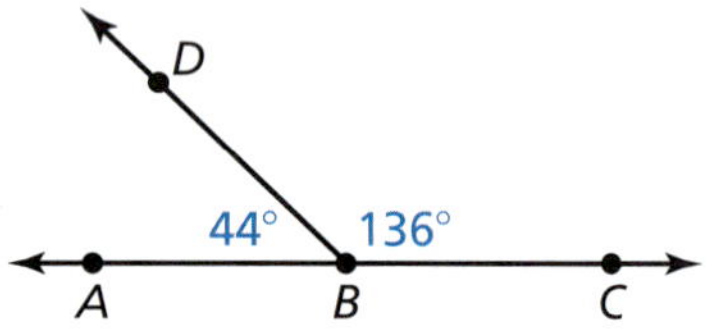

b.

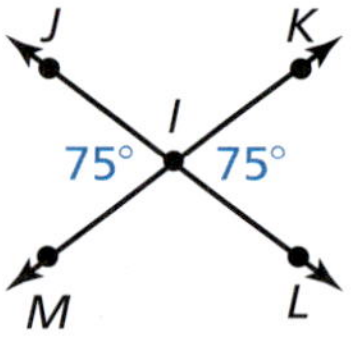

c.

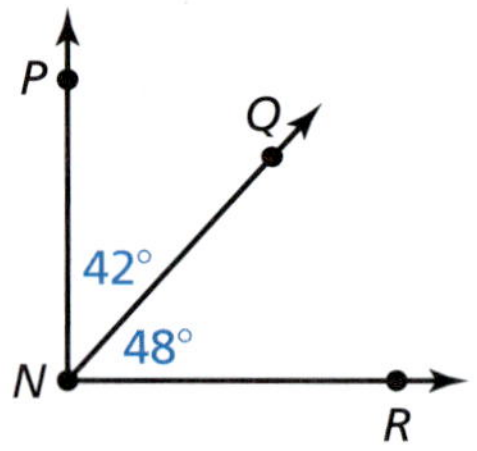

Supplementary angles	Adjacent angles	Vertical angles	Complementary angles	Linear pair

4. Your school is installing new turf on the soccer field with rubber edging around the perimeter of the field. Each unit in the coordinate plane represents 10 yards. How many yards of rubber edging are needed to completely surround the field?

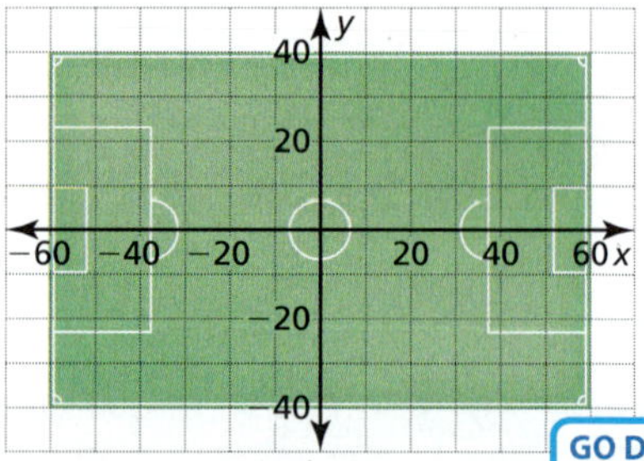

Ⓐ 200 yd

Ⓑ 400 yd

Ⓒ 960 ft

Ⓓ 9600 yd

GO DIGITAL

5. Complete the two-column proof.

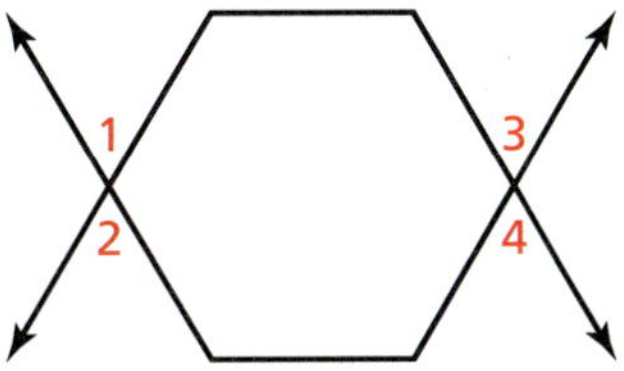

Given $\angle 1 \cong \angle 3$

Prove $\angle 2 \cong \angle 4$

STATEMENTS	REASONS
1. $\angle 1 \cong \angle 3$	**1.** Given
2. $\angle 1 \cong \angle 2$	**2.** ______________________
3. $\angle 2 \cong \angle 3$	**3.** ______________________
4. ______________	**4.** Vertical Angles Congruence Theorem
5. $\angle 2 \cong \angle 4$	**5.** ______________________

6. Which of the following is true when $\overleftrightarrow{AB}$ and $\overleftrightarrow{CD}$ are skew?

Ⓐ $\overleftrightarrow{AB}$ and $\overleftrightarrow{CD}$ are parallel.

Ⓑ $\overleftrightarrow{AB}$ and $\overleftrightarrow{CD}$ intersect.

Ⓒ $\overleftrightarrow{AB}$ and $\overleftrightarrow{CD}$ are perpendicular.

Ⓓ A, B, and C are noncollinear.

7. Select the angle that makes the statement true.

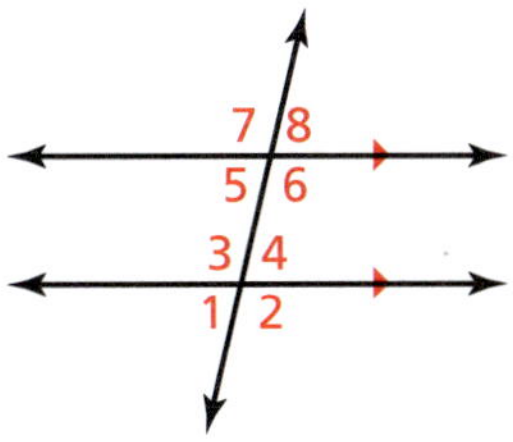

a. $\angle 4 \cong$ ___ by the Alternate Interior Angles Theorem.

b. $\angle 2 \cong$ ___ by the Corresponding Angles Theorem.

c. $\angle 1 \cong$ ___ by the Alternate Exterior Angles Theorem.

d. $m\angle 6 + m$ _____ $= 180°$ by the Consecutive Interior Angles Theorem.

$\angle 1$ $\angle 2$ $\angle 3$ $\angle 4$ $\angle 5$ $\angle 6$ $\angle 7$ $\angle 8$

8. You and your friend meet at the halfway point between your homes and then walk to school together from there. Each unit in the coordinate plane corresponds to 50 yards. What is the distance that the two of you walk together?

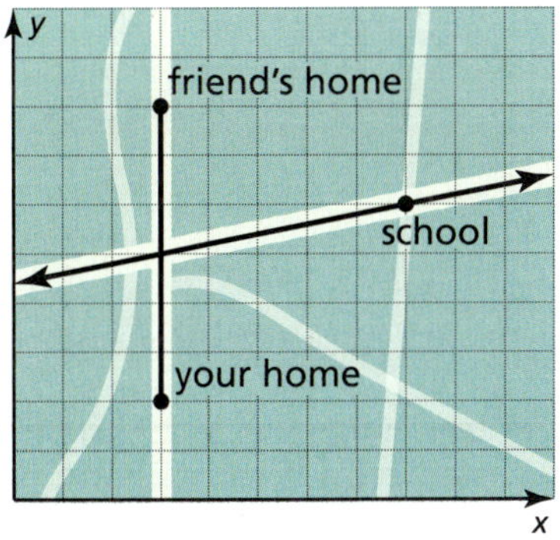

Ⓐ 250 yd

Ⓑ 255 yd

Ⓒ 320 yd

Ⓓ 405 yd

GO DIGITAL

4 Transformations

NATIONAL GEOGRAPHIC EXPLORER

Aaron Pomerantz

WATCH

Aaron Pomerantz is an entomologist whose research aims to discover the genetic basis of color and wing transparency in butterflies. He has also applied novel technology such as origami-based portable microscopes and handheld gene sequencers to conduct fieldwork in remote tropical rain forests.

- What is entomology? What is lepidopterology?
- How many wings do all butterflies have?
- What is metamorphosis? What are the stages of a butterfly's life?

STEM

Entomologists often sketch insects that they find in nature. Sketches can be dilated to enlarge certain physical features. In the Performance Task, you will sketch a butterfly species and show how to construct a dilation of your sketch.

GO DIGITAL

Entomology

Preparing for Chapter 4

Chapter Learning Target: Understand transformations.

Chapter Success Criteria:
- ◆ I can identify transformations.
- ◆ I can perform translations, reflections, rotations, and dilations.
- ■ I can describe congruence and similarity transformations.
- ■ I can solve problems involving transformations.

◆ Surface
■ Deep

Chapter Vocabulary

Work with a partner. Discuss each of the vocabulary terms.

transformation	glide reflection	congruent figures
translation	line symmetry	dilation
rigid motion	rotation	scale factor
reflection	rotational symmetry	similar figures

Mathematical Thinking and Reasoning

4 MTR
ENGAGE IN DISCUSSIONS THAT REFLECT ON THE MATHEMATICAL THINKING OF SELF AND OTHERS
Mathematicians who engage in discussions that reflect on the mathematical thinking of self and others communicate mathematical ideas, vocabulary, and methods efficiently.

Work with a partner. An entomology club wants to create a new logo. The logo must satisfy the following requirements.

- The logo must resemble the shape of an insect.
- The logo must be drawn using only line segments.
- The logo must be symmetric about both the x-axis and the y-axis in a coordinate plane.

1. Draw a logo that meets the requirements.
2. Explain how someone can replicate your logo given only the portion in the first quadrant.
3. The members of the club like your logo, but want to increase the size. Explain how to draw the logo so that it has the same shape, but is a larger size. Then draw the enlarged logo in a coordinate plane.

GO DIGITAL

4 Prepare WITH CalcChat®

Identifying Transformations

WATCH

Example 1 **Tell whether the red figure is a *translation*, *reflection*, *rotation*, or *dilation* of the blue figure.**

a.

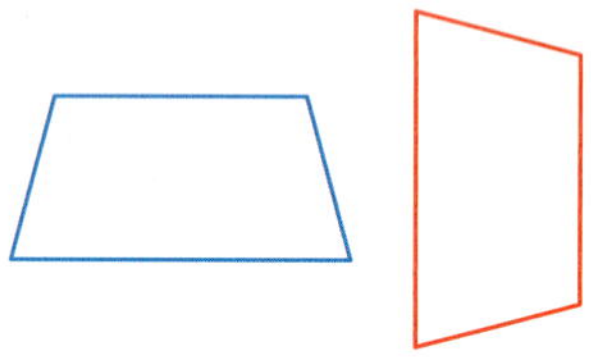

The blue figure turns to form the red figure, so it is a rotation.

b.

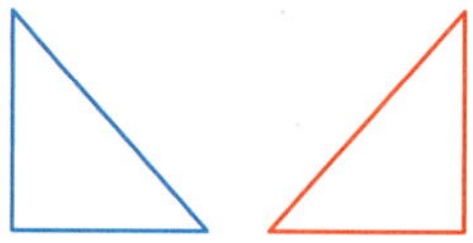

The red figure is a mirror image of the blue figure, so it is a reflection.

Tell whether the red figure is a *translation*, *reflection*, *rotation*, or *dilation* of the blue figure.

1.

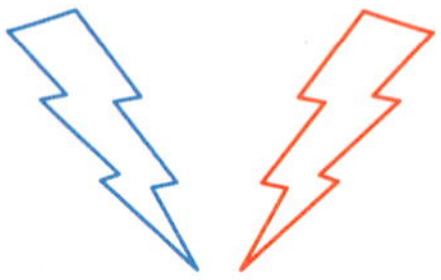

2.

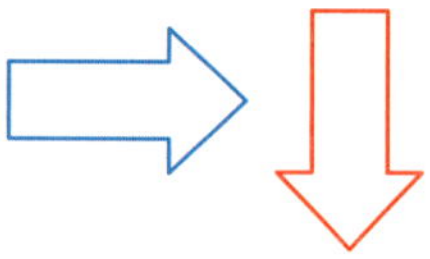

3.

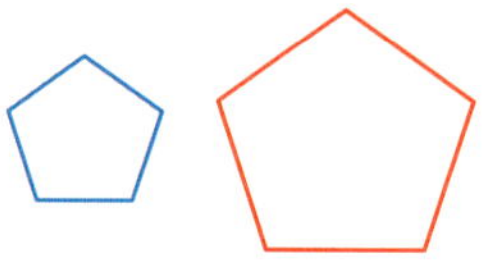

4.

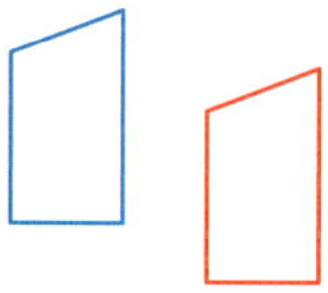

Translating Figures in the Coordinate Plane

WATCH

Example 2 **A triangle has vertices $A(-1, 0)$, $B(0, 3)$, and $C(2, 2)$. Draw the triangle and its image after a translation 2 units left and 4 units down. What are the coordinates of the image?**

Draw the triangle. Then move each vertex 2 units left and 4 units down.

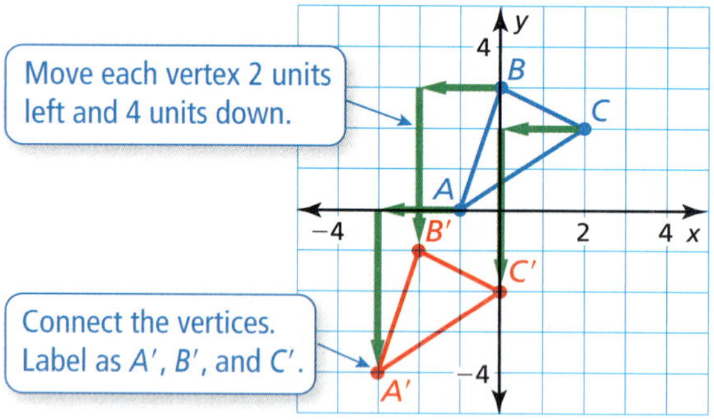

▶ The coordinates of the image are $A'(-3, -4)$, $B'(-2, -1)$, and $C'(0, -2)$.

Draw the figure and its image after the given translation. What are the coordinates of the image?

5. $A(-3, -3)$, $B(0, 0)$, $C(1, -2)$; 3 units right and 3 units up

6. $A(-1, -4)$, $B(-3, -3)$, $C(3, 0)$, $D(4, -2)$; 1 unit left and 6 units up

1 MTR **7.** **ANALYZE A PROBLEM** Describe two different ways to translate a figure that result in the same image. Justify your answer.

GO DIGITAL

4.1 Translations

Learning Target: Understand translations of figures.

Success Criteria:
- I can translate figures.
- I can write a translation rule for a given translation.
- I can explain what a rigid motion is.
- I can perform a composition of translations on a figure.

EXPLORE IT! Translating a Polygon

Work with a partner.

a. The diagram shows the *translation* of $\triangle DEF$ to $\triangle D'E'F'$ (read as "triangle D prime, E prime, F prime"). How would you define a translation?

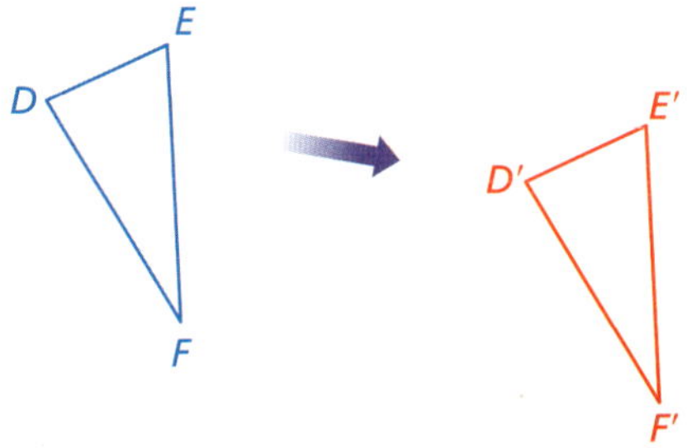

b. Use technology to draw any triangle and label it $\triangle ABC$.

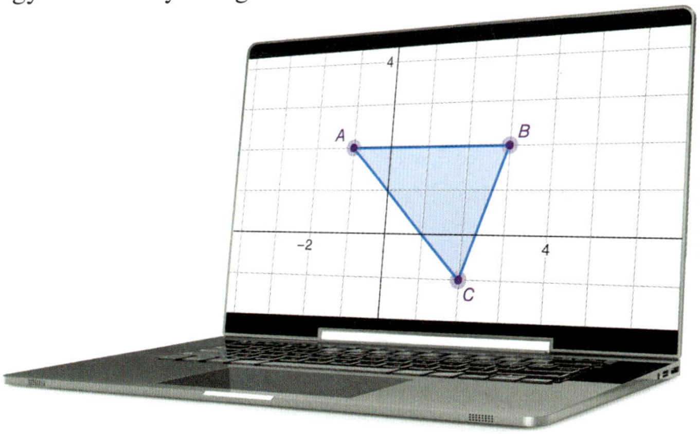

c. Translate the triangle to form a new triangle, called $\triangle A'B'C'$. What do you observe about the side lengths, angle measures, and coordinates of the vertices of the two triangles?

d. The point (x, y) is translated a units horizontally and b units vertically. Write a rule to determine the coordinates of the image of (x, y).

$$(x, y) \rightarrow (\quad, \quad)$$

e. Based on your results in parts (b)–(d), is there anything you would like to change or include in your definition in part (a)? Explain.

4 MTR CONSTRUCT AN ARGUMENT
Do you expect your observations about side lengths and angle measures in part (c) to be true for any translation? Explain.

Geometric Reasoning
MA.912.GR.2.1 Given a preimage and image, describe the transformation and represent the transformation algebraically using coordinates.
MA.912.GR.2.5 Given a geometric figure and a sequence of transformations, draw the transformed figure on a coordinate plane.
Also MA.912.GR.2.2

GO DIGITAL

Performing Translations

A **vector** is a quantity that has both direction and *magnitude*, or size, and is represented in the coordinate plane by an arrow drawn from one point to another.

Vocabulary

vector, *p. 168*
initial point, *p. 168*
terminal point, *p. 168*
horizontal component, *p. 168*
vertical component, *p. 168*
component form, *p. 168*
transformation, *p. 168*
image, *p. 168*
preimage, *p. 168*
translation, *p. 168*
rigid motion, *p. 170*
composition of transformations, *p. 170*

KEY IDEA

Vectors

The diagram shows a vector. The **initial point**, or starting point, of the vector is P, and the **terminal point**, or ending point, is Q. The vector is named $\overrightarrow{PQ}$, which is read as "vector PQ." The **horizontal component** of $\overrightarrow{PQ}$ is 5, and the **vertical component** is 3. The **component form** of a vector combines the horizontal and vertical components. So, the component form of $\overrightarrow{PQ}$ is $\langle 5, 3 \rangle$.

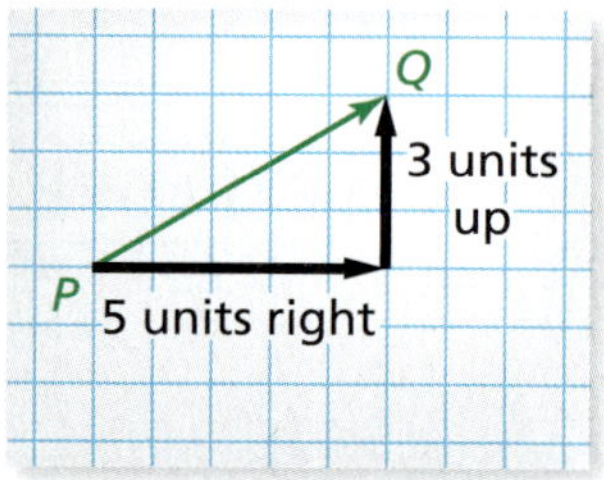

EXAMPLE 1 Identifying Vector Components

Name the vector and write its component form.

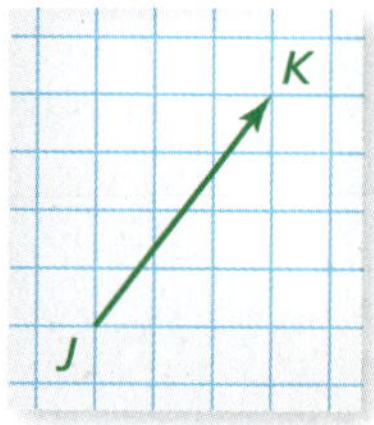

SOLUTION

The vector is $\overrightarrow{JK}$. To move from the initial point J to the terminal point K, move 3 units right and 4 units up. So, the component form is $\langle 3, 4 \rangle$.

1 MTR HELP A CLASSMATE

Explain to a classmate why $\overrightarrow{KJ}$ is not a valid name for the vector.

A **transformation** is a function that moves or changes a figure in some way to produce a new figure called an **image**. Another name for the original figure is the **preimage**. The points on the preimage are the inputs for the transformation, and the points on the image are the outputs.

KEY IDEA

Translations

A **translation** moves every point of a figure the same distance in the same direction. More specifically, a translation *maps*, or moves, the points P and Q of a plane figure along a vector $\langle a, b \rangle$ to the points P' and Q', so that one of the following statements is true.

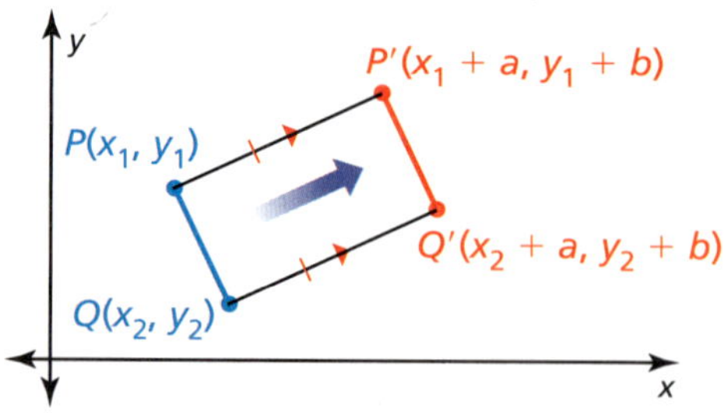

- $PP' = QQ'$ and $\overline{PP'} \parallel \overline{QQ'}$ or
- $PP' = QQ'$, and $\overline{PP'}$ and $\overline{QQ'}$ are collinear.

STUDY TIP

You can use *prime notation* to name an image. For example, if the preimage is point P, then its image is point P', read as "point P *prime*."

Translations map lines to parallel lines and segments to parallel segments. For instance, in the figure above, $\overline{PQ} \parallel \overline{P'Q'}$.

GO DIGITAL

EXAMPLE 2 Translating a Figure Using a Vector

The vertices of $\triangle ABC$ are $A(0, 3)$, $B(2, 4)$, and $C(1, 0)$. Translate $\triangle ABC$ using the vector $\langle 5, -1 \rangle$.

STUDY TIP

Notice that the vectors drawn from preimage vertices to image vertices are parallel.

SOLUTION

First, graph $\triangle ABC$. Use $\langle 5, -1 \rangle$ to move each vertex 5 units right and 1 unit down. Label the image vertices and draw $\triangle A'B'C'$.

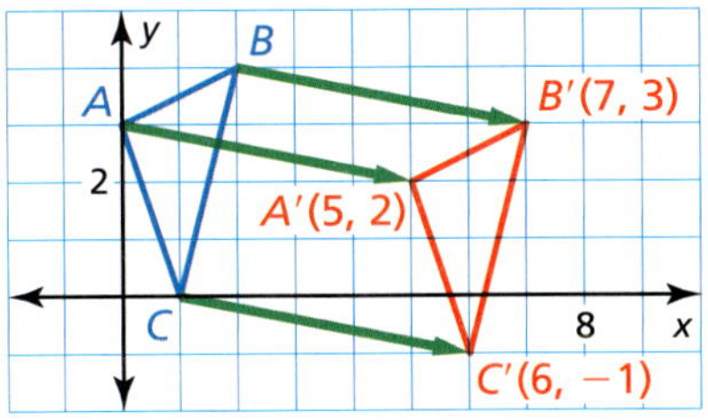

You can also express a translation along the vector $\langle a, b \rangle$ using a rule, which has the notation $(x, y) \rightarrow (x + a, y + b)$.

EXAMPLE 3 Writing a Translation Rule

Write a rule for the translation of $\triangle PQR$ to $\triangle P'Q'R'$.

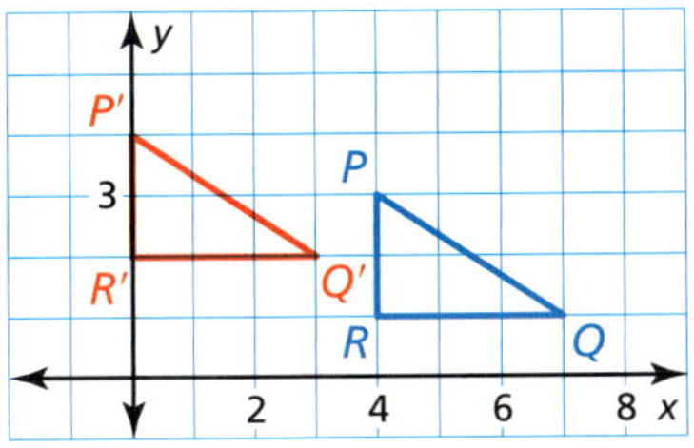

SOLUTION

To go from P to P', move 4 units left and 1 unit up, along the vector $\langle -4, 1 \rangle$.

▶ So, a rule for the translation is $(x, y) \rightarrow (x - 4, y + 1)$.

EXAMPLE 4 Translating a Figure in the Coordinate Plane

Graph quadrilateral $ABCD$ with vertices $A(-1, 2)$, $B(-1, 5)$, $C(4, 6)$, and $D(4, 2)$ and its image after the translation $(x, y) \rightarrow (x + 3, y - 1)$.

SOLUTION

Graph quadrilateral $ABCD$. To find the coordinates of the vertices of the image, add 3 to the x-coordinates and subtract 1 from the y-coordinates of the vertices of the preimage. Then graph the image, as shown at the left.

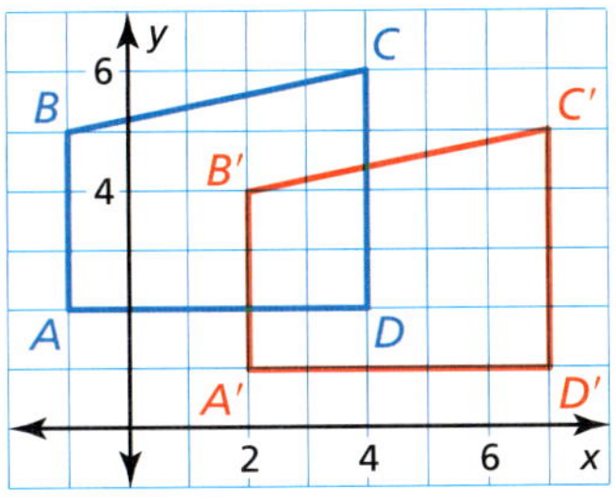

$$(x, y) \rightarrow (x + 3, y - 1)$$

$$A(-1, 2) \rightarrow A'(2, 1)$$

$$B(-1, 5) \rightarrow B'(2, 4)$$

$$C(4, 6) \rightarrow C'(7, 5)$$

$$D(4, 2) \rightarrow D'(7, 1)$$

SELF-ASSESSMENT

1 I don't understand yet. 2 I can do it with help. 3 I can do it on my own. 4 I can teach someone else.

1. Name the vector and write its component form.

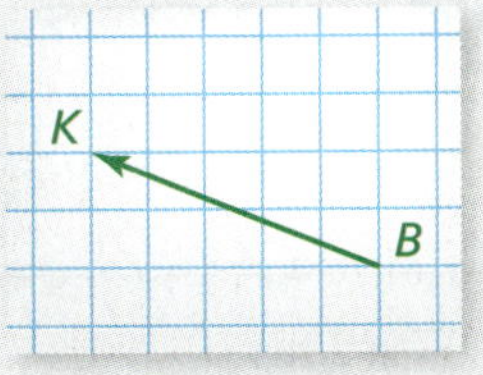

2. The vertices of $\triangle LMN$ are $L(2, 2)$, $M(5, 3)$, and $N(9, 1)$. Translate $\triangle LMN$ using the vector $\langle -2, 6 \rangle$.
3. In Example 3, write a rule to translate $\triangle P'Q'R'$ back to $\triangle PQR$.
4. Graph $\triangle RST$ with vertices $R(2, 2)$, $S(5, 2)$, and $T(3, 5)$ and its image after the translation $(x, y) \rightarrow (x + 1, y + 2)$.

GO DIGITAL

Performing Compositions

A **rigid motion** is a transformation that preserves length and angle measure. Another name for a rigid motion is an *isometry*. A rigid motion maps lines to lines, rays to rays, and segments to segments.

POSTULATE

4.1 Translation Postulate

A translation is a rigid motion.

Because a translation is a rigid motion, and a rigid motion preserves length and angle measure, the following statements are true for the translation shown.

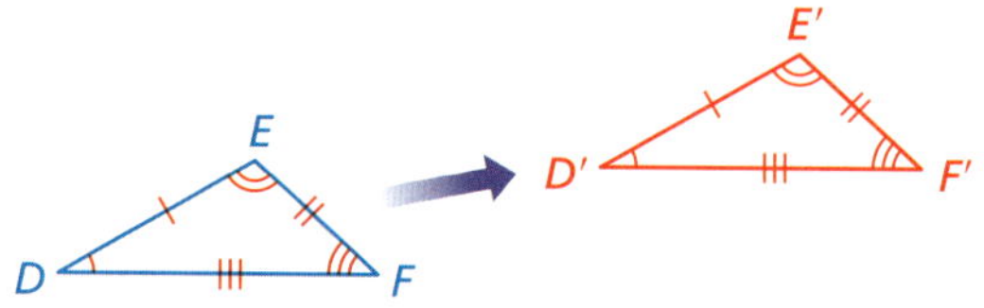

- $DE = D'E'$ $\quad EF = E'F'$ $\quad FD = F'D'$
- $m\angle D = m\angle D'$ $\quad m\angle E = m\angle E'$ $\quad m\angle F = m\angle F'$

When two or more transformations are combined to form a single transformation, the result is a **composition of transformations**.

THEOREM

4.1 Composition Theorem

The composition of two (or more) rigid motions is a rigid motion.

Proof Exercise 34, page 174

The theorem above is important because it states that no matter how many rigid motions you perform, lengths and angle measures will be preserved in the final image. For instance, the composition of two or more translations is a translation.

EXAMPLE 5 Performing a Composition

WATCH

Graph $\overline{RS}$ with endpoints $R(-8, 5)$ and $S(-6, 8)$ and its image after the composition.

Translation: $(x, y) \rightarrow (x + 5, y - 2)$

Translation: $(x, y) \rightarrow (x - 4, y - 2)$

SOLUTION

Step 1 Graph $\overline{RS}$.

Step 2 Translate $\overline{RS}$ 5 units right and 2 units down. $\overline{R'S'}$ has endpoints $R'(-3, 3)$ and $S'(-1, 6)$.

Step 3 Translate $\overline{R'S'}$ 4 units left and 2 units down. $\overline{R''S''}$ has endpoints $R''(-7, 1)$ and $S''(-5, 4)$.

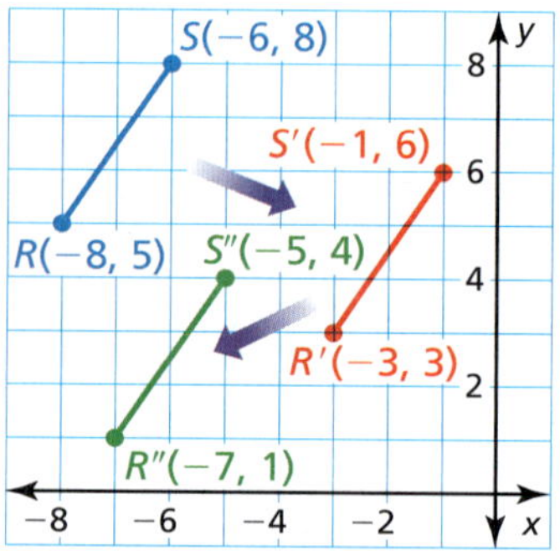

GO DIGITAL

Solving Real-Life Problems

7 MTR **EXAMPLE 6** **Modeling Real Life** WATCH

You use floor tiles to help you learn a new dance move. You move 1 tile right and 3 tiles down. Then you move 2 tiles left and 1 tile up. Rewrite the composition as a single translation.

SOLUTION

1. **Understand the Problem** You are given two translations. You need to rewrite the result of the composition of the two translations as a single translation.

2. **Make a Plan** Let your starting position be an arbitrary point $A(x, y)$ on a coordinate plane where each grid square represents 1 tile. Determine the horizontal and vertical shifts in the coordinates of the point after both translations. This tells you how much you need to shift each coordinate to map the starting position to the final position.

3. **Solve and Check** After the first translation, the coordinates of your position are

$$A'(x + 1, y - 3).$$

The second translation maps $A'(x + 1, y - 3)$ to

$$A''(x + 1 - 2, y - 3 + 1) = A''(x - 1, y - 2).$$

▶ The single translation rule for the composition is $(x, y) \rightarrow (x - 1, y - 2)$. So, your final position is 1 tile left and 2 tiles down from your starting position.

Check Test a point to check that the rule is correct. For instance, let $A(0, 0)$ be your starting position.

Apply the two translations to $A(0, 0)$.

$$A(0, 0) \rightarrow A'(0 + 1, 0 - 3) = A'(1, -3)$$

$$A'(1, -3) \rightarrow A''(1 - 2, -3 + 1) = A''(-1, -2)$$

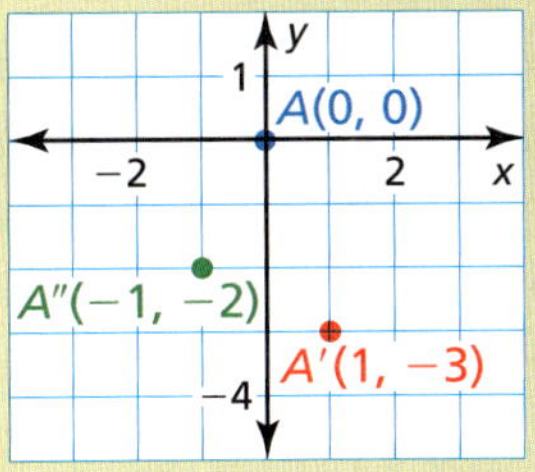

Check that the rule you found in Step 3 gives the same result.

$$A(0, 0) \rightarrow A''(0 - 1, 0 - 2) = A''(-1, -2) \checkmark$$

SELF-ASSESSMENT

1 I don't understand yet. 2 I can do it with help. 3 I can do it on my own. 4 I can teach someone else.

5. Graph $\overline{TU}$ with endpoints $T(1, 2)$ and $U(4, 6)$ and its image after the composition.

   Translation: $(x, y) \rightarrow (x - 2, y - 3)$

   Translation: $(x, y) \rightarrow (x - 4, y + 5)$

6. Graph $\overline{VW}$ with endpoints $V(-6, -4)$ and $W(-3, 1)$ and its image after the composition.

   Translation: $(x, y) \rightarrow (x + 3, y + 1)$

   Translation: $(x, y) \rightarrow (x - 6, y - 4)$

7. In Example 6, you move 2 tiles right and 3 tiles up. Then you move 1 tile left and 1 tile down. Rewrite the composition as a single translation.

GO DIGITAL

4.1 Practice WITH CalcChat® AND CalcView®

In Exercises 1 and 2, name the vector and write its component form. *(See Example 1.)*

1.

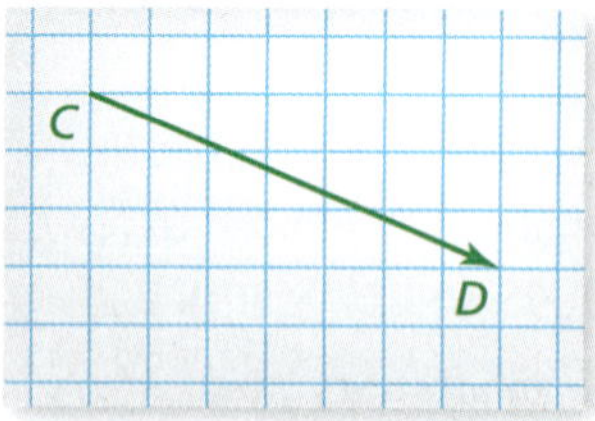

2.

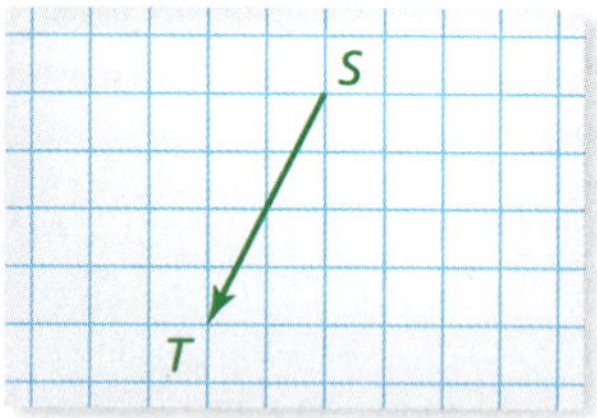

In Exercises 3–6, the vertices of $\triangle DEF$ are $D(2, 5)$, $E(6, 3)$, and $F(4, 0)$. Translate $\triangle DEF$ using the given vector. Graph $\triangle DEF$ and its image. *(See Example 2.)*

3. $\langle 6, 0 \rangle$ **4.** $\langle 5, -1 \rangle$

5. $\langle -3, -7 \rangle$ **6.** $\langle -2, -4 \rangle$

In Exercises 7 and 8, find the component form of the vector that translates $P(-3, 6)$ to P'.

7. $P'(0, 1)$ **8.** $P'(-4, 8)$

In Exercises 9 and 10, write a rule for the translation of $\triangle LMN$ to $\triangle L'M'N'$. *(See Example 3.)*

9.

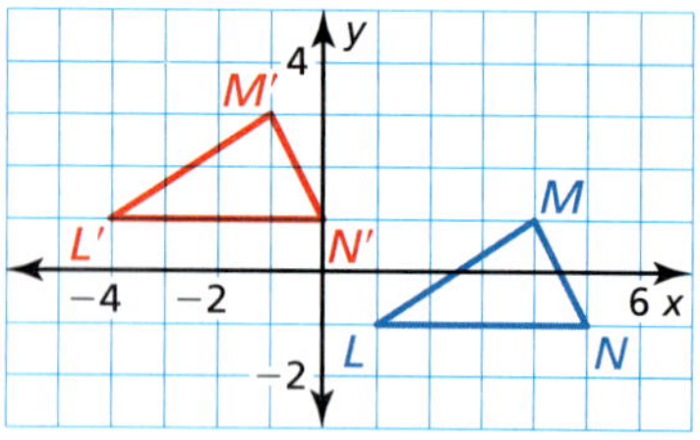

10.

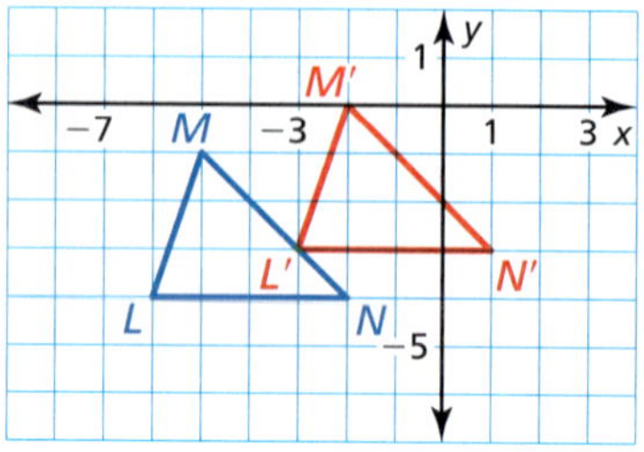

In Exercises 11–14, use the translation.

$$(x, y) \to (x - 8, y + 4)$$

11. What is the image of $A(2, 6)$?

12. What is the image of $B(-1, 5)$?

13. What is the preimage of $C'(-3, -10)$?

14. What is the preimage of $D'(4, -3)$?

In Exercises 15–18, graph quadrilateral $PQRS$ with vertices $P(-2, 3)$, $Q(1, 2)$, $R(3, -1)$, and $S(-2, -1)$ and its image after the translation. *(See Example 4.)*

15. $(x, y) \to (x + 4, y + 6)$ **16.** $(x, y) \to (x + 9, y - 2)$

17. $(x, y) \to (x - 2, y - 5)$ **18.** $(x, y) \to (x - 1, y + 3)$

19. **ERROR ANALYSIS** Describe and correct the error in graphing the image of quadrilateral $EFGH$ after the translation $(x, y) \to (x - 1, y - 2)$.

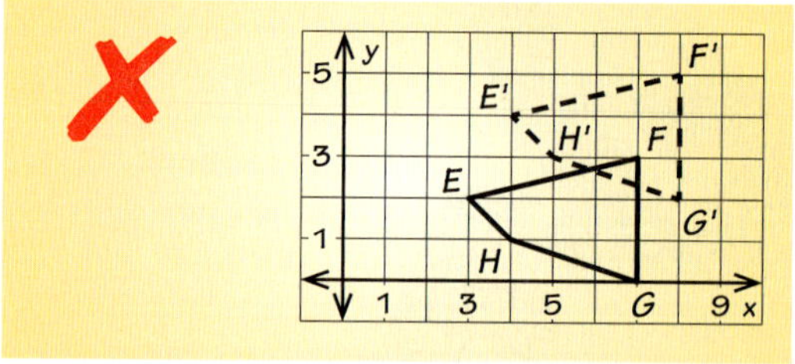

20. **ERROR ANALYSIS** Describe and correct the error in graphing the image of $\triangle ABC$ after the translation $(x, y) \to (x - 3, y + 2)$.

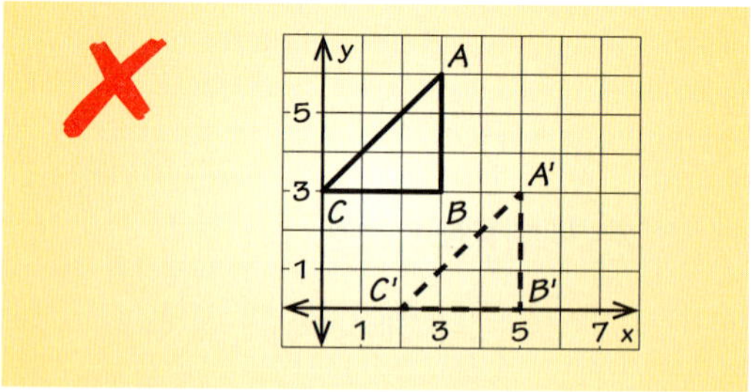

In Exercises 21 and 22, graph $\triangle XYZ$ with vertices $X(2, 4)$, $Y(6, 0)$, and $Z(7, 2)$ and its image after the composition. *(See Example 5.)*

21. **Translation:** $(x, y) \to (x + 12, y + 4)$
Translation: $(x, y) \to (x - 5, y - 9)$

22. **Translation:** $(x, y) \to (x - 6, y)$
Translation: $(x, y) \to (x + 2, y + 7)$

GO DIGITAL

In Exercises 23 and 24, write rules for the composition of translations.

23.

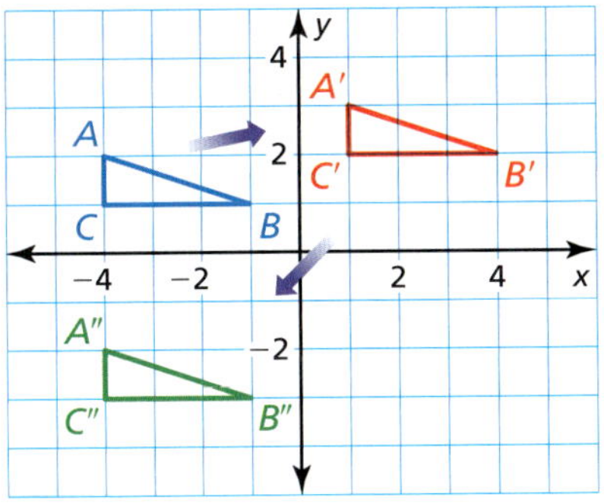

24.

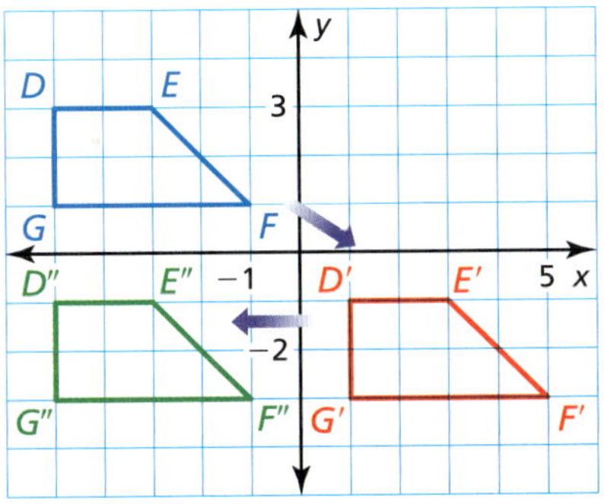

7 MTR **25. MODELING REAL LIFE** In chess, the knight (the piece shaped like a horse) moves in an L pattern. The board shows two consecutive moves of a black knight during a game. Write a composition of translations for the moves. Then rewrite the composition as a single translation that moves the knight from its original position to its ending position. *(See Example 6.)*

7 MTR **26. MODELING REAL LIFE** You are studying an amoeba through a microscope. The amoeba moves on a grid-indexed microscope slide in a straight line from square B3 to square G7.

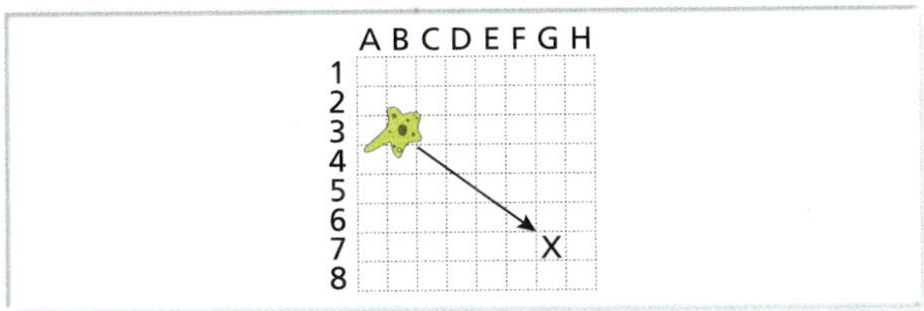

a. Describe the translation.

b. The side length of each grid square is 2 millimeters. How far does the amoeba travel?

c. The amoeba moves from square B3 to square G7 in 24.5 seconds. What is its speed in millimeters per second?

GO DIGITAL

5 MTR **CONNECTING CONCEPTS In Exercises 27 and 28, a translation maps the blue figure to the red figure. Find the value of each variable.**

27.

28.

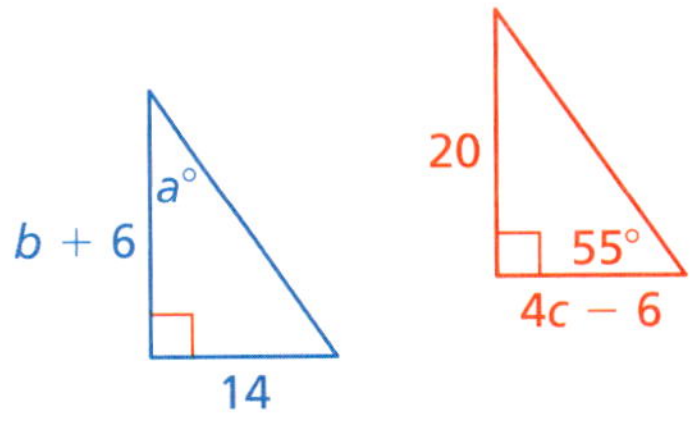

5 MTR **29. CONNECTING CONCEPTS** Translation A maps (x, y) to $(x + n, y + t)$. Translation B maps (x, y) to $(x + s, y + m)$.

a. Translate a point using Translation A, followed by Translation B. Write a rule for the final image of the point after this composition.

b. Translate a point using Translation B, followed by Translation A. Write a rule for the final image of the point after this composition.

c. Compare the rules you wrote for parts (a) and (b). Does it matter which translation is first? Explain your reasoning.

30. HOW DO YOU SEE IT?
Which two figures represent a translation? Describe the translation.

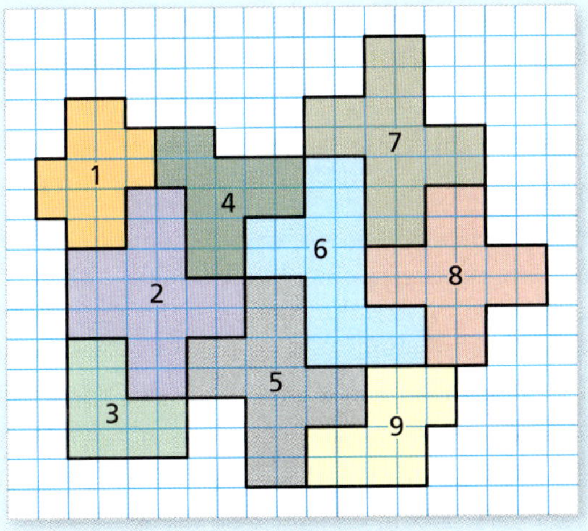

5 MTR **31. STRUCTURE** Quadrilateral $DEFG$ has vertices $D(-1, 2)$, $E(-2, 0)$, $F(-1, -1)$, and $G(1, 3)$. A translation maps quadrilateral $DEFG$ to quadrilateral $D'E'F'G'$. The image of D is $D'(-2, -2)$. What are the coordinates of E', F', and G'?

32. **REASONING** The translation $(x, y) \rightarrow (x + m, y + n)$ maps $\overline{PQ}$ to $\overline{P'Q'}$. Write a rule for the translation of $\overline{P'Q'}$ to $\overline{PQ}$. Explain your reasoning.

5 MTR 33. **CONNECTING CONCEPTS** The vertices of a rectangle are $Q(2, -3)$, $R(2, 4)$, $S(5, 4)$, and $T(5, -3)$.

a. Translate rectangle $QRST$ 3 units left and 3 units down to produce rectangle $Q'R'S'T'$. Find the area of rectangle $QRST$ and the area of rectangle $Q'R'S'T'$.

b. Compare the areas. Make a conjecture about the areas of a preimage and its image after a translation. Justify your answer.

34. **PROVING A THEOREM** Prove the Composition Theorem.

35. **PROVING A THEOREM** Use properties of translations to prove each theorem.

a. Corresponding Angles Theorem

b. Corresponding Angles Converse

36. **OPEN-ENDED** Draw a line segment in Quadrant II. Identify a composition of translations that maps the line segment entirely to Quadrant IV.

4 MTR 37. **MAKING AN ARGUMENT** A translation maps $\overline{GH}$ to $\overline{G'H'}$. Your friend claims that if you draw segments connecting G to G' and H to H', then the resulting quadrilateral is a parallelogram. Is your friend correct? Explain your reasoning.

5 MTR 38. **CONNECTING CONCEPTS** The vector $\langle 4, 1 \rangle$ describes the translation of $A(-1, w)$ to $A'(2x + 1, 4)$ and $B(8y - 1, 1)$ to $B'(3, 3z)$. Find the values of w, x, y, and z.

39. **REASONING** The vertices of $\triangle ABC$ are $A(2, 2)$, $B(4, 2)$, and $C(3, 4)$. Graph the image of $\triangle ABC$ after the transformation $(x, y) \rightarrow (x + y, y)$. Is this transformation a translation? Explain your reasoning.

40. **THOUGHT PROVOKING**
You are a graphic designer for a company that manufactures floor tiles. Design a floor tile in a coordinate plane. Then use translations to show how the tiles cover a floor. Describe the translations that map the original tile to four other tiles.

41. **PROOF** $\overline{MN}$ is perpendicular to line ℓ. $\overline{M'N'}$ is the translation of $\overline{MN}$ 2 units left. Prove that $\overline{M'N'}$ is perpendicular to ℓ.

REVIEW & REFRESH

WATCH

42. Decide whether there is enough information to prove that $m \parallel n$. If so, state the theorem you can use.

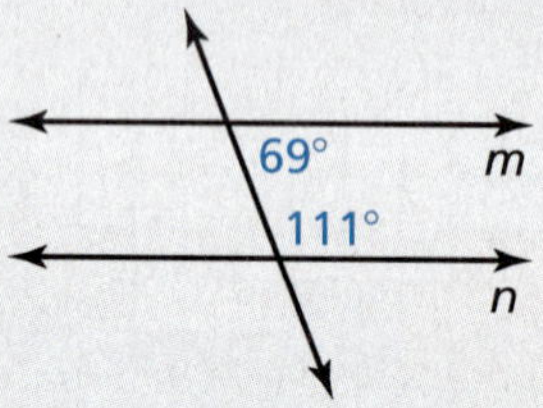

In Exercises 43 and 44, write an equation of the line passing through point P that is perpendicular to the given line. Graph the equations to check that the lines are perpendicular.

43. $P(4, -1)$, $y - 3 = -4(x + 7)$

44. $P(-3, 5)$, $5x - 2y = 8$

45. Graph quadrilateral $ABCD$ with vertices $A(-4, 1)$, $B(-3, 3)$, $C(0, 1)$, and $D(-2, 0)$ and its image after the translation $(x, y) \rightarrow (x + 4, y - 2)$.

In Exercises 46 and 47, the endpoints of $\overline{GH}$ are given. Find the coordinates of the midpoint M.

46. $G(-6, 2)$ and $H(4, 8)$

47. $G(3, -1)$ and $H(9, 7)$

In Exercises 48 and 49, solve the equation.

48. $6x^3 - 12x^2 = 0$

49. $-32y = 8y^2$

In Exercises 50 and 51, graph the function. Compare the graph to the graph of $f(x) = x^2$.

50. $g(x) = 1.6x^2$

51. $p(x) = -\frac{7}{2}x^2$

52. Determine which lines, if any, must be parallel. Explain your reasoning.

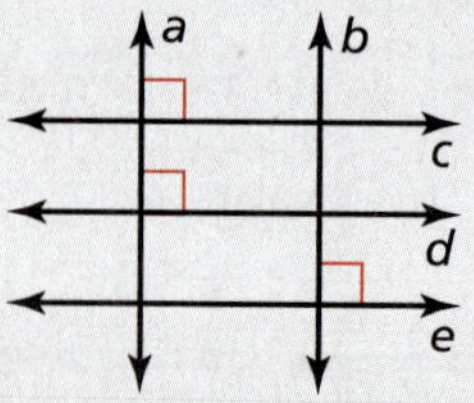

7 MTR 53. **MODELING REAL LIFE** The function $h(x) = -16x^2 + 48x + 6$ models the height h (in feet) of a ball ejected from a ball launcher after x seconds. When does the ball reach a height of 30 feet?

54. Find the perimeter and the area of $\triangle DEF$ with vertices $D(-2, 2)$, $E(4, 2)$, and $F(2, 5)$.

GO DIGITAL

4.2 Reflections

Learning Target: Understand reflections of figures.

Success Criteria:
- I can reflect figures.
- I can perform compositions with reflections.
- I can identify line symmetry in polygons.

EXPLORE IT! Reflecting a Polygon

Work with a partner.

a. The diagram shows the *reflection* of $\triangle DEF$ to $\triangle D'E'F'$. How would you define a reflection?

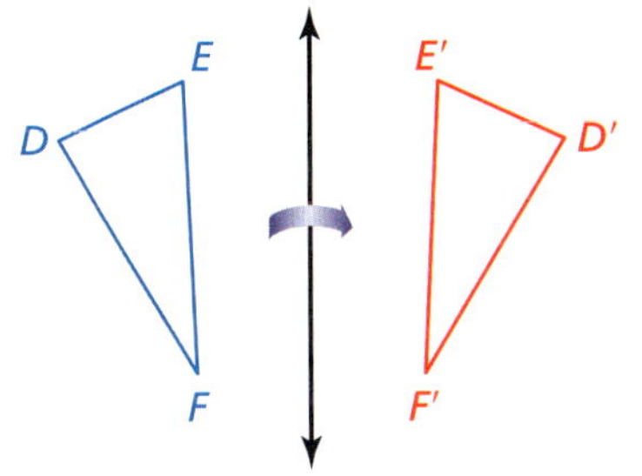

b. Use technology to draw any triangle and label it $\triangle ABC$.

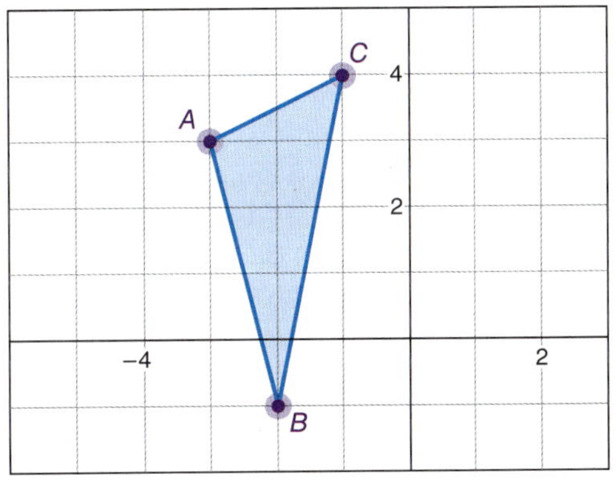

3 MTR SELECT METHODS
What other tools can you use to perform reflections of figures?

c. Reflect the triangle in the y-axis to form $\triangle A'B'C'$. What do you observe about the side lengths, angle measures, and coordinates of the vertices of the two triangles?

d. How do your observations in part (c) change when you reflect $\triangle ABC$ in the x-axis?

e. Write rules to determine the coordinates of the image of (x, y) when it is reflected in the x-axis or the y-axis.

Reflection in x-axis:	**Reflection in y-axis:**
$(x, y) \rightarrow (\square, \square)$	$(x, y) \rightarrow (\square, \square)$

f. Based on your results in parts (b)–(e), is there anything you would like to change or include in your definition in part (a)? Explain.

g. Is a reflection a rigid motion? Explain your reasoning.

Geometric Reasoning
MA.912.GR.2.1 Given a preimage and image, describe the transformation and represent the transformation algebraically using coordinates.
MA.912.GR.2.5 Given a geometric figure and a sequence of transformations, draw the transformed figure on a coordinate plane.
Also MA.912.GR.2.2, MA.912.GR.2.3, H MA.912.GR.2.4

GO DIGITAL

Performing Reflections

Vocabulary AZ VOCAB

reflection, *p. 176*
line of reflection, *p. 176*
glide reflection, *p. 178*
line symmetry, *p. 179*
line of symmetry, *p. 179*

KEY IDEA

Reflections

A **reflection** is a transformation that uses a line like a mirror to reflect a figure. This line is called the **line of reflection**.

A reflection in a line m maps every point P in the plane to a point P', so that for each point, one of the following properties is true.

- If P is not on m, then m is the perpendicular bisector of $\overline{PP'}$, or
- If P is on m, then $P = P'$.

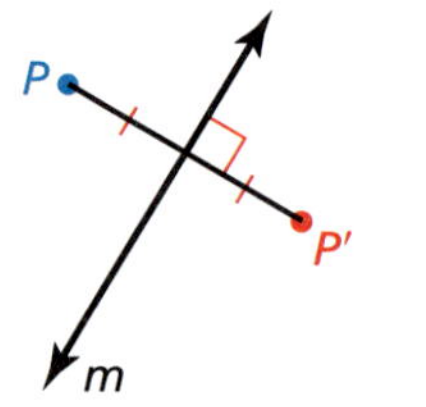

point P not on m

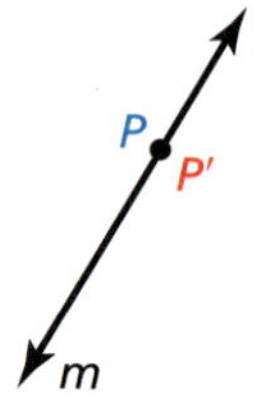

point P on m

EXAMPLE 1 Reflecting in Horizontal and Vertical Lines WATCH

Graph $\triangle ABC$ with vertices $A(1, 3)$, $B(5, 2)$, and $C(2, 1)$ and its image after the reflection described.

a. In the line m: $y = 1$

b. In the line n: $x = 3$

SOLUTION

a. Point A is 2 units above line m, so A' is 2 units below line m at $(1, -1)$. Also, B' is 1 unit below line m at $(5, 0)$. Because point C is on line m, you know that $C = C'$.

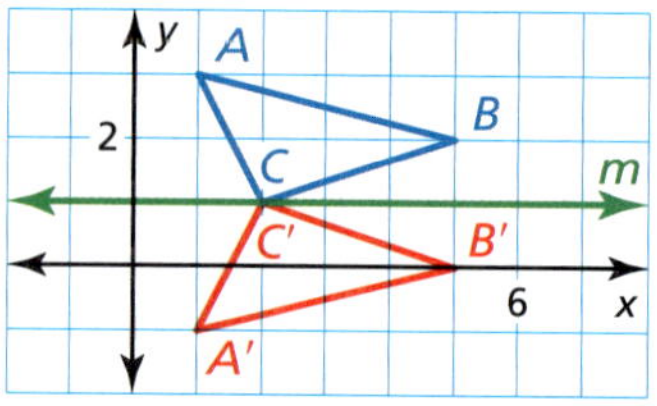

b. Point A is 2 units left of line n, so its reflection A' is 2 units right of line n at $(5, 3)$. Also, B' is 2 units left of line n at $(1, 2)$, and C' is 1 unit right of line n at $(4, 1)$.

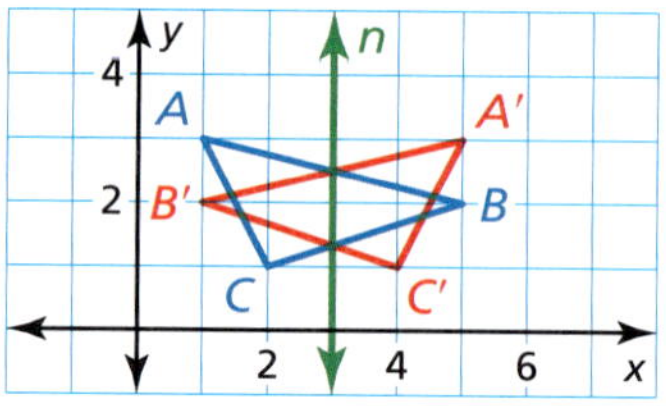

STUDY TIP

In part (a), $C = C'$ because C is on the line of reflection. The line of reflection is the perpendicular bisector of $\overline{AA'}$ and $\overline{BB'}$.

SELF-ASSESSMENT 1 I don't understand yet. 2 I can do it with help. 3 I can do it on my own. 4 I can teach someone else.

Graph $\triangle ABC$ from Example 1 and its image after a reflection in the given line.

1. $x = 4$ **2.** $x = -3$ **3.** $y = 2$ **4.** $y = -1$

4 MTR **5. WHICH ONE DOESN'T BELONG?** Which transformation does *not* belong with the other three? Explain your reasoning.

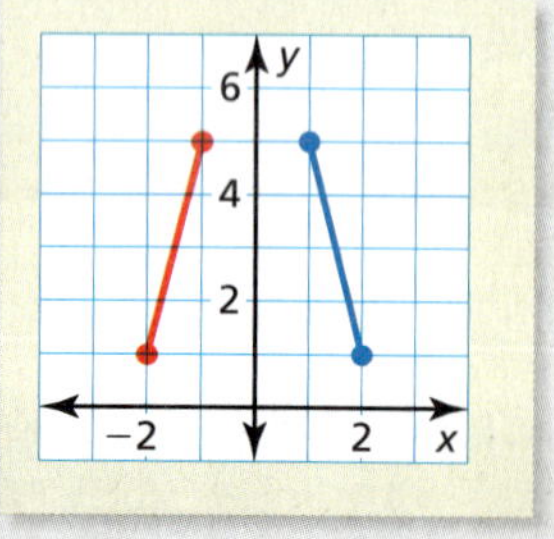

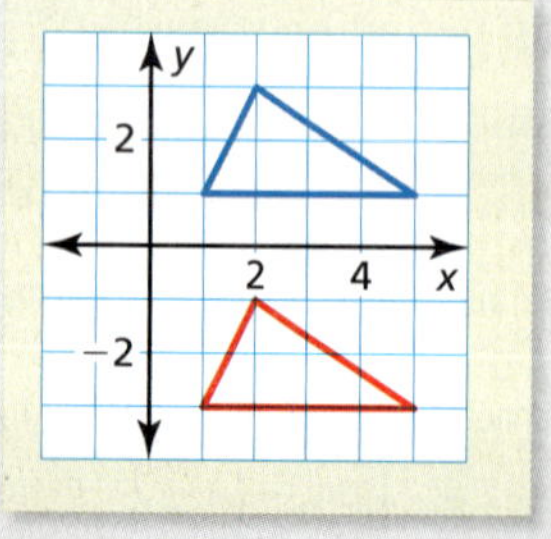

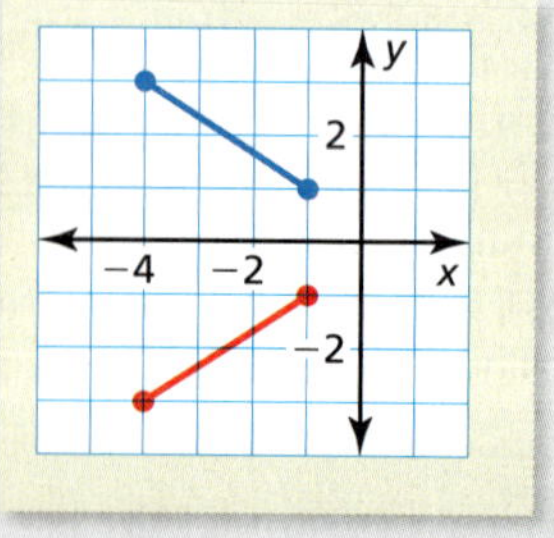

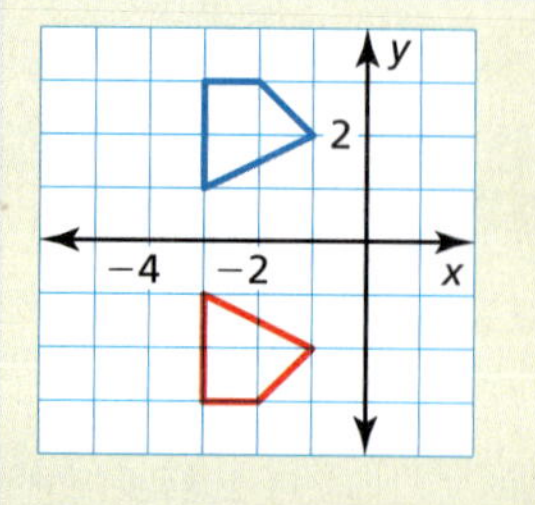

GO DIGITAL

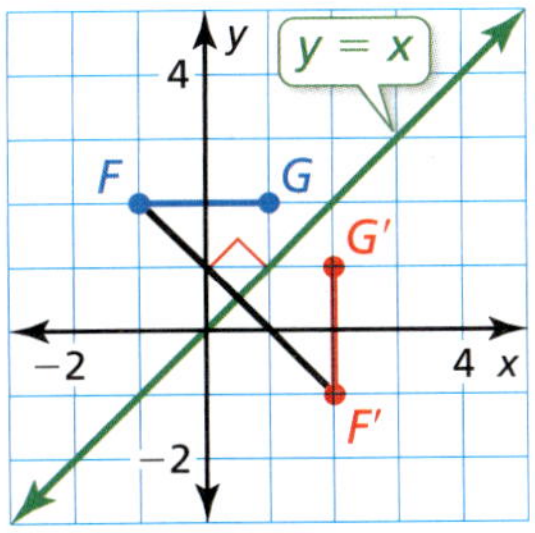

EXAMPLE 2 Reflecting in the Line $y = x$

WATCH

Graph $\overline{FG}$ with endpoints $F(-1, 2)$ and $G(1, 2)$ and its image after a reflection in the line $y = x$.

SOLUTION

Graph $\overline{FG}$ and the line $y = x$. The slope of $y = x$ is 1. The segment from F to its image F' is perpendicular to the line of reflection $y = x$, so the slope of $\overline{FF'}$ is -1. Notice that the intersection point of $\overline{FF'}$ and $y = x$ is 1.5 units right and 1.5 units down from F. So, from this intersection point, move 1.5 units right and 1.5 units down to locate $F'(2, -1)$.

The slope of $\overline{GG'}$ is also -1. The intersection point of $\overline{GG'}$ and $y = x$ is 0.5 unit right and 0.5 unit down from G. So, from this intersection point, move 0.5 unit right and 0.5 unit down to locate $G'(2, 1)$. Graph $\overline{F'G'}$.

REMEMBER

The product of the slopes of perpendicular lines is -1. Because the slope of $y = x$ is 1 and $1(-1) = -1$, the slope of $\overline{FF'}$ is -1.

You can use coordinate rules to find the images of points reflected in four special lines.

KEY IDEA

Coordinate Rules for Reflections

- If (a, b) is reflected in the x-axis, then its image is the point $(a, -b)$.
- If (a, b) is reflected in the y-axis, then its image is the point $(-a, b)$.
- If (a, b) is reflected in the line $y = x$, then its image is the point (b, a).
- If (a, b) is reflected in the line $y = -x$, then its image is the point $(-b, -a)$.

EXAMPLE 3 Reflecting in the Line $y = -x$

WATCH

Graph $\overline{FG}$ from Example 2 and its image after a reflection in the line $y = -x$.

SOLUTION

Graph $\overline{FG}$ and the line $y = -x$. Use the coordinate rule for reflecting in the line $y = -x$ to find the coordinates of the endpoints of the image. Then graph the image.

$$(a, b) \rightarrow (-b, -a)$$
$$F(-1, 2) \rightarrow F'(-2, 1)$$
$$G(1, 2) \rightarrow G'(-2, -1)$$

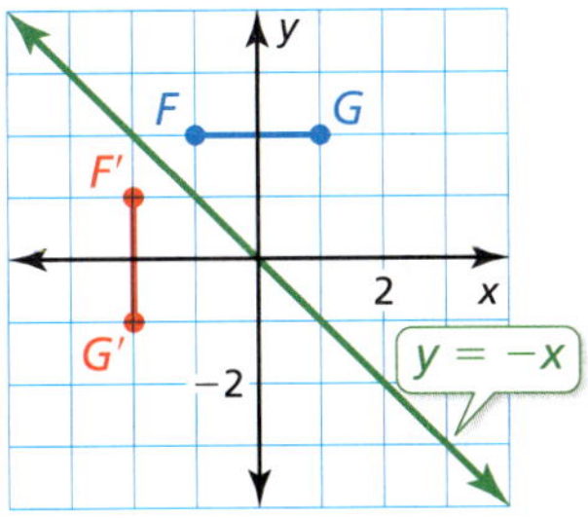

SELF-ASSESSMENT

1 I don't understand yet. 2 I can do it with help. 3 I can do it on my own. 4 I can teach someone else.

The vertices of $\triangle JKL$ are $J(1, 3)$, $K(4, 4)$, and $L(3, 1)$. Graph $\triangle JKL$ and its image after a reflection in the given line.

6. x-axis

7. y-axis

8. $y = x$

9. $y = -x$

10. In Example 3, verify that $\overline{FF'}$ is perpendicular to $y = -x$.

GO DIGITAL

Performing Glide Reflections

POSTULATE

4.2 Reflection Postulate

A reflection is a rigid motion.

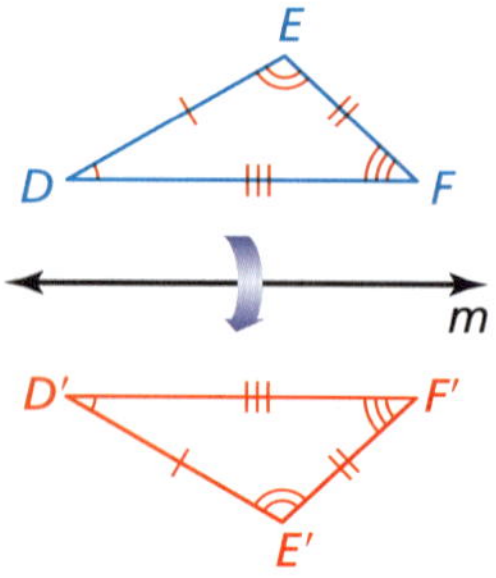

Because a reflection is a rigid motion, and a rigid motion preserves length and angle measure, the following statements are true for the reflection shown.

- $DE = D'E'$ $\quad$ $EF = E'F'$ $\quad$ $FD = F'D'$
- $m\angle D = m\angle D'$ $\quad$ $m\angle E = m\angle E'$ $\quad$ $m\angle F = m\angle F'$

Because a reflection is a rigid motion, the Composition Theorem guarantees that any composition of reflections and translations is a rigid motion.

STUDY TIP

The line of reflection must be parallel to the direction of the translation to be a glide reflection.

A **glide reflection** is a transformation involving a translation followed by a reflection in which every point P is mapped to a point P'' by the following steps.

Step 1 First, a translation maps P to P'.

Step 2 Then a reflection in a line k parallel to the direction of the translation maps P' to P''.

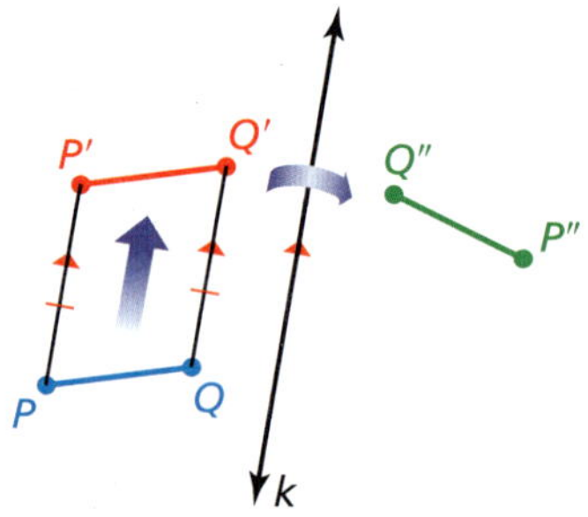

EXAMPLE 4 Performing a Glide Reflection

WATCH

2 MTR USE ANOTHER METHOD

Rewrite the glide reflection as a single transformation. Use your rule to graph the image after the glide reflection.

Graph $\triangle ABC$ with vertices $A(3, 2)$, $B(6, 3)$, and $C(7, 1)$ and its image after the glide reflection.

Translation: $(x, y) \rightarrow (x - 12, y)$

Reflection: in the x-axis

SOLUTION

Begin by graphing $\triangle ABC$. Then graph $\triangle A'B'C'$ after a translation 12 units left. Finally, graph $\triangle A''B''C''$ after a reflection in the x-axis.

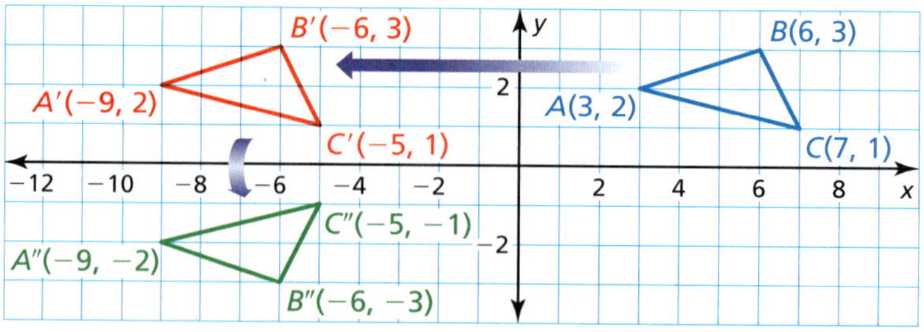

SELF-ASSESSMENT 1 I don't understand yet. 2 I can do it with help. 3 I can do it on my own. 4 I can teach someone else.

11. WHAT IF? In Example 4, $\triangle ABC$ is translated 4 units down and then reflected in the y-axis. Graph $\triangle ABC$ and its image after the glide reflection.

12. In Example 4, describe a glide reflection from $\triangle A''B''C''$ to $\triangle ABC$.

GO DIGITAL

Identifying Lines of Symmetry

A figure in the plane has **line symmetry** when the figure can be mapped onto itself by a reflection in a line. This line of reflection is a **line of symmetry**, such as line m at the left. A figure can have more than one line of symmetry.

Butterflies almost always have symmetrical wing patterns because of the genes responsible for the pattern. Cells on both wings have the same genetic code. The zebra longwing, pictured above, is the state butterfly of Florida.

EXAMPLE 5 Identifying Line Symmetry

WATCH

Determine whether each polygon has line symmetry. If so, draw the line(s) of symmetry and describe any reflections that map the polygon onto itself.

a. trapezoid

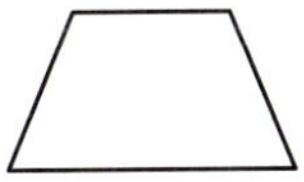

b. regular hexagon

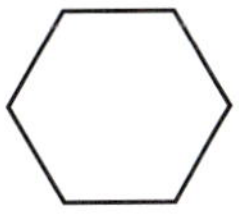

c. parallelogram

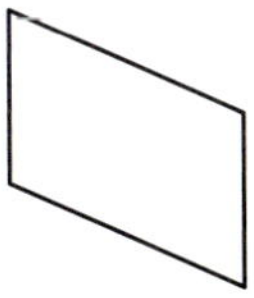

SOLUTION

a. The trapezoid has line symmetry. The line of symmetry is shown. A reflection in the line of symmetry maps the trapezoid onto itself.

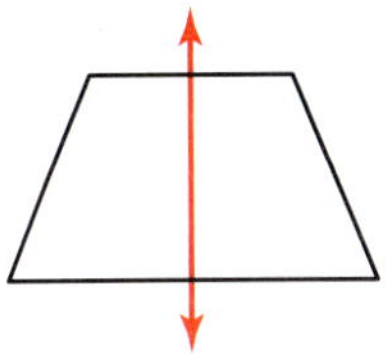

b. The regular hexagon has line symmetry. The 6 lines of symmetry are shown. A reflection in any of the lines of symmetry maps the hexagon onto itself.

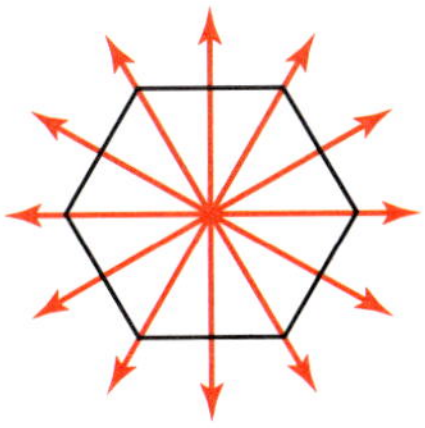

c. The parallelogram does not have line symmetry because the figure cannot be mapped onto itself by a reflection in a line.

SELF-ASSESSMENT

1 I don't understand yet. 2 I can do it with help. 3 I can do it on my own. 4 I can teach someone else.

Determine whether the polygon has line symmetry. If so, draw the line(s) of symmetry and describe any reflections that map the polygon onto itself.

13. rectangle

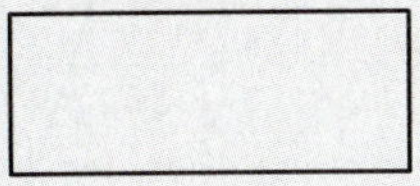

14. regular pentagon

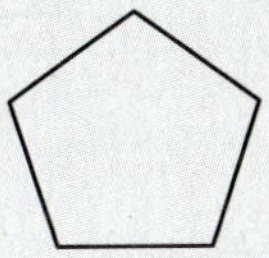

15. isosceles triangle

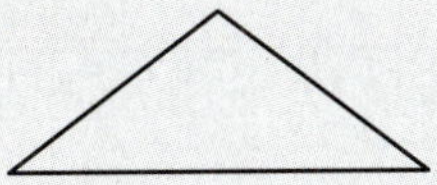

16. OPEN-ENDED Draw a polygon with no lines of symmetry.

17. REASONING Is it true that all parallelograms have no line of symmetry? Explain your reasoning.

GO DIGITAL

4.2 Practice WITH CalcChat® AND CalcView®

In Exercises 1–8, graph the polygon with the given vertices and its image after a reflection in the given line. *(See Example 1.)*

▶ **1.** $J(2, -4)$, $K(3, 7)$, $L(6, -1)$; x-axis

2. $J(5, 3)$, $K(1, -2)$, $L(-3, 4)$; y-axis

3. $J(2, -1)$, $K(4, -5)$, $L(3, 1)$; $x = -1$

4. $J(1, -1)$, $K(3, 0)$, $L(0, -4)$; $x = 2$

5. $J(2, 4)$, $K(-4, -2)$, $L(-1, 0)$; $y = 1$

6. $J(3, -5)$, $K(4, -1)$, $L(0, -3)$; $y = -3$

7. $J(2, 1)$, $K(3, 5)$, $L(6, 5)$, $M(5, 1)$; $x = 1$

8. $J(-3, 1)$, $K(0, 3)$, $L(4, -1)$, $M(1, -3)$; $y = -2$

In Exercises 9–12, write a rule for the reflection.

9.

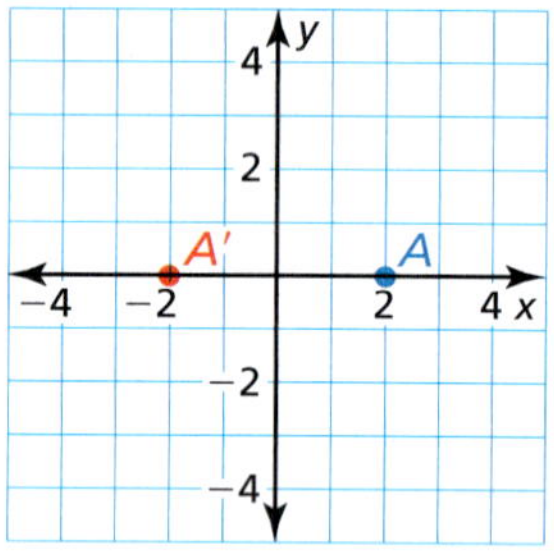

10.

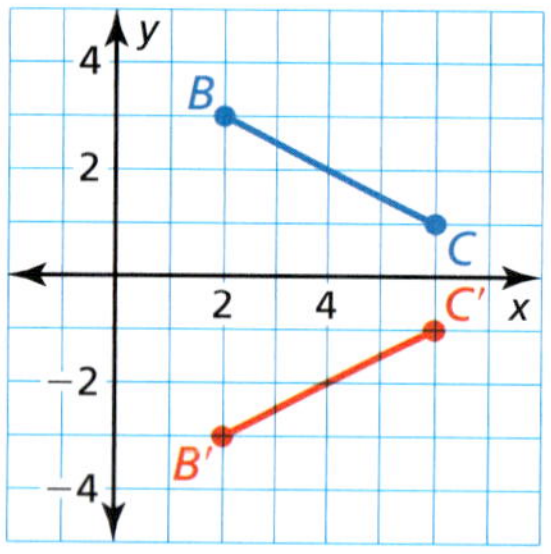

11.

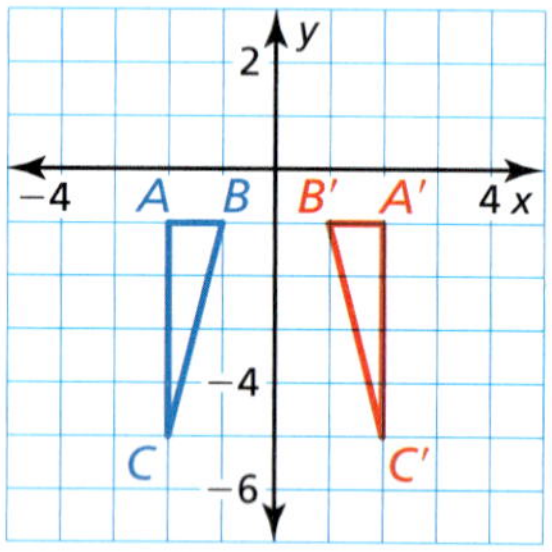

12.

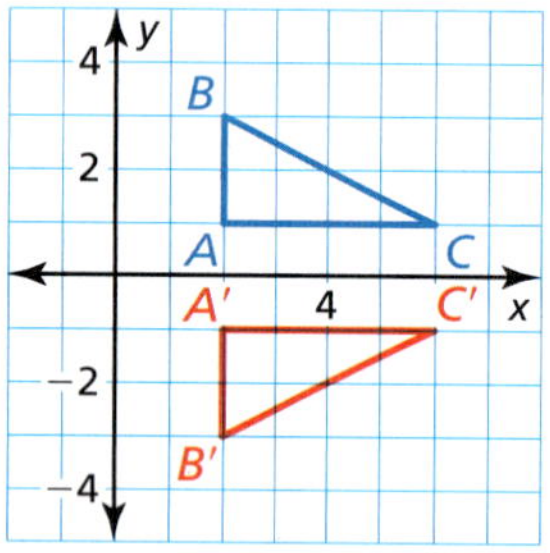

In Exercises 13–16, graph the polygon with the given vertices and its image after a reflection in the line $y = x$. *(See Example 2.)*

▶ **13.** $A(6, -3)$, $B(1, -2)$, $C(4, 1)$

14. $A(0, -3)$, $B(2, 2)$, $C(5, 0)$

15. $A(-2, 4)$, $B(1, 4)$, $C(1, 2)$, $D(-2, 2)$

16. $A(2, -1)$, $B(-1, 2)$, $C(2, 3)$, $D(4, 2)$

In Exercises 17–20, graph the polygon with the given vertices and its image after a reflection in the line $y = -x$. *(See Example 3.)*

▶ **17.** $A(1, 2)$, $B(4, 2)$, $C(3, -2)$

18. $A(-2, -3)$, $B(-2, 0)$, $C(0, 1)$

19. $A(-3, 2)$, $B(1, -1)$, $C(-2, -2)$, $D(-4, -1)$

20. $A(2, 0)$, $B(3, 4)$, $C(6, 4)$, $D(5, 0)$

In Exercises 21 and 22, graph $\triangle RST$ with vertices $R(4, 1)$, $S(7, 3)$, and $T(6, 4)$ and its image after the glide reflection. *(See Example 4.)*

▶ **21.** **Translation:** $(x, y) \rightarrow (x, y - 1)$
Reflection: in the y-axis

22. **Translation:** $(x, y) \rightarrow (x - 3, y)$
Reflection: in the line $y = -1$

In Exercises 23 and 24, graph quadrilateral $QRST$ with vertices $Q(1, -1)$, $R(1, 2)$, $S(4, -1)$, and $T(4, -4)$ and its image after the glide reflection.

23. **Translation:** $(x, y) \rightarrow (x + 2, y)$
Reflection: in the x-axis

24. **Translation:** $(x, y) \rightarrow (x, y - 2)$
Reflection: in the line $x = 2$

In Exercises 25–30, determine whether the figure has line symmetry. If so, draw the line(s) of symmetry and describe any reflections that map the figure onto itself. *(See Example 5.)*

▶ **25.**

26.

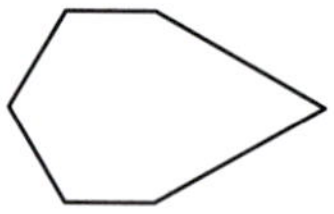

27.

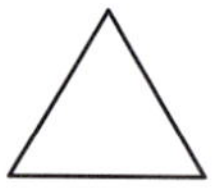

28.

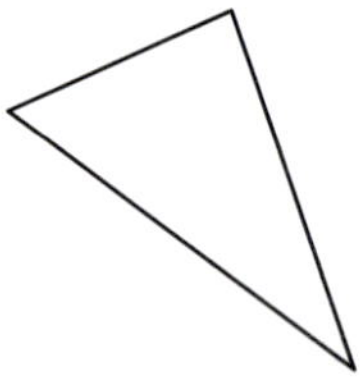

29.

30.

GO DIGITAL

31. STRUCTURE Describe the line symmetry (if any) of each word.

a. LOOK **b.** OX

32. ERROR ANALYSIS Identify and correct the error in describing the transformation.

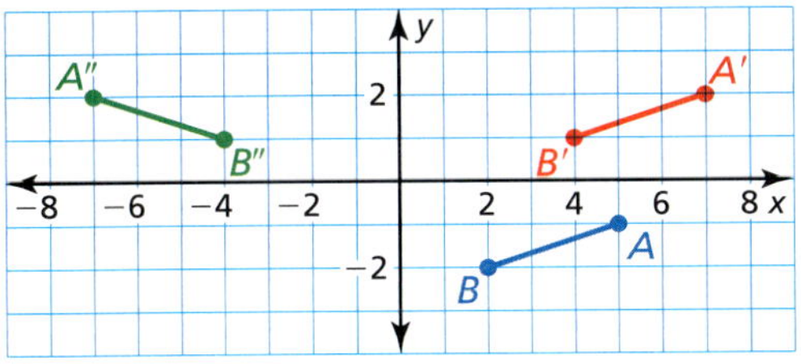

✗ $\overline{AB}$ to $\overline{A''B''}$ is a glide reflection.

33. REASONING Use the numbers, variables, and symbols to create the glide reflection resulting in the image shown.

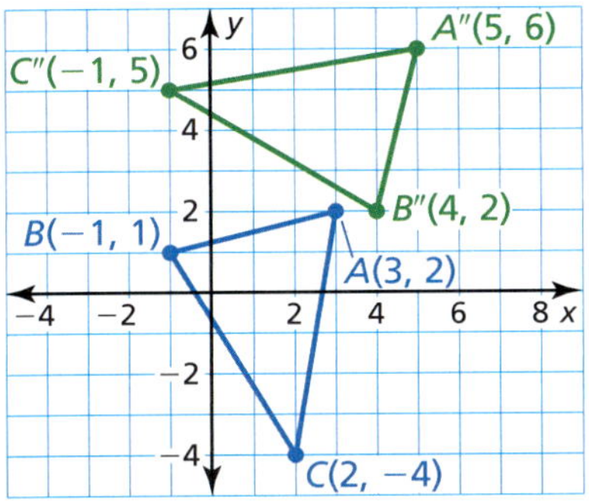

Translation: $(x, y) \rightarrow ($ ▢, ▢ $)$

Reflection: in the line $y = x$

1 2 3

x y $+$ $-$

34. COLLEGE PREP Which transformations do *not* result in mapping $(-k, k)$ to $(k, -k)$?

Ⓐ reflection in the line $y = x$

Ⓑ reflection in the line $y = -x$

Ⓒ reflection in the y-axis, followed by a reflection in the x-axis

Ⓓ translation k units right, followed by a reflection in the y-axis

GO DIGITAL

35. CONNECTING CONCEPTS The line $y = 3x + 2$ is reflected in the line $y = -1$. What is the equation of the image?

36. HOW DO YOU SEE IT?
Use Figure A.

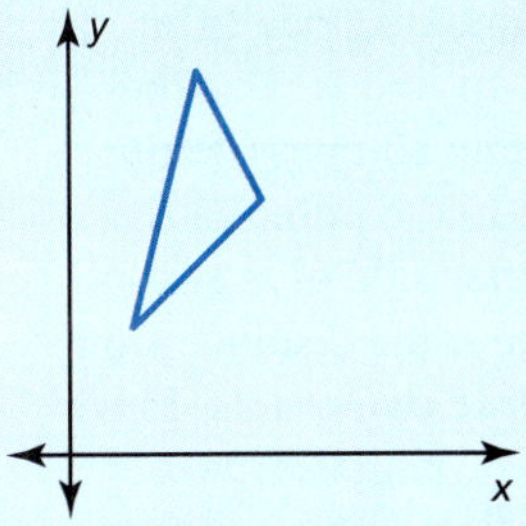

Figure A

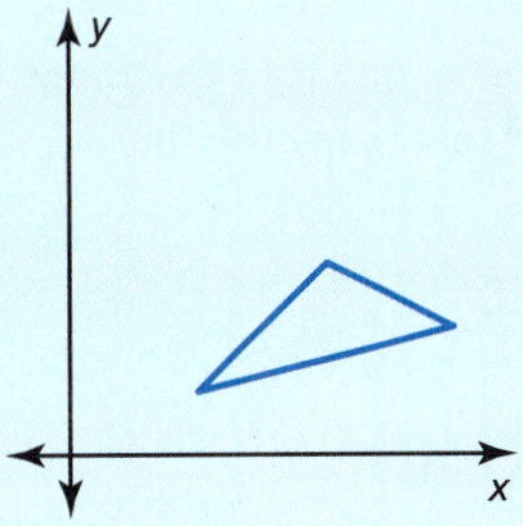

Figure 1

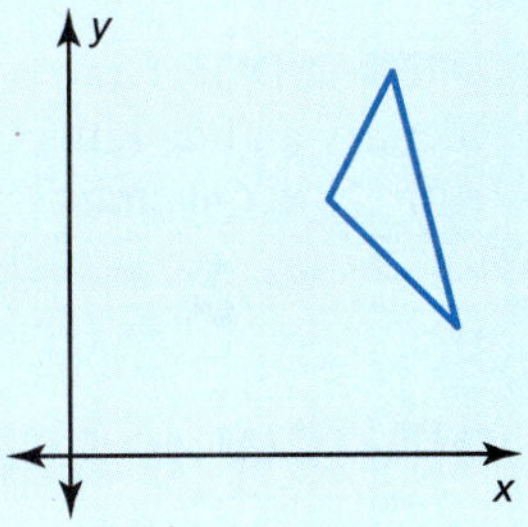

Figure 2

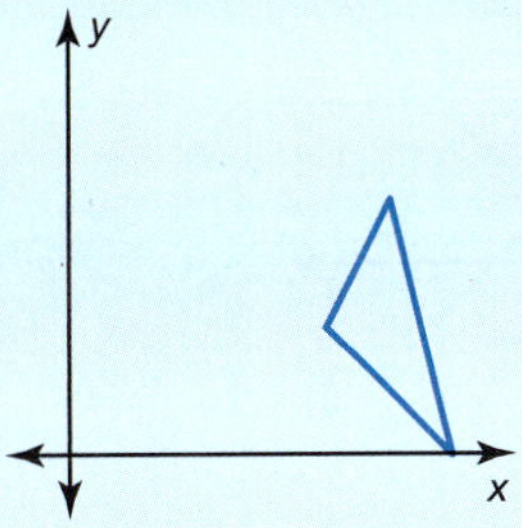

Figure 3

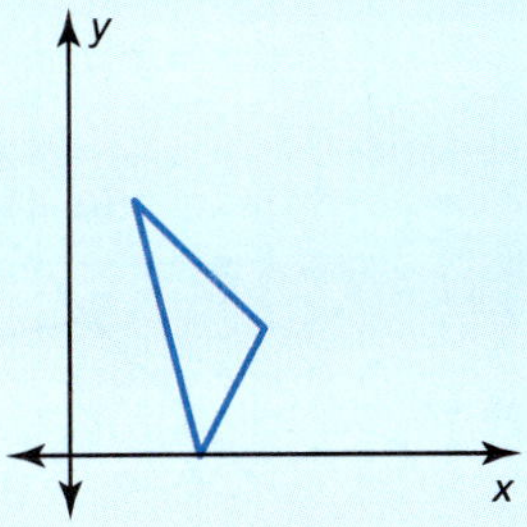

Figure 4

a. Which figure is a reflection of Figure A in the line $x = a$? Explain.

b. Which figure is a reflection of Figure A in the line $y = b$? Explain.

c. Which figure is a reflection of Figure A in the line $y = x$? Explain.

d. Is there a figure that represents a glide reflection? Explain your reasoning.

In Exercises 37 and 38, point B' is the image of point B after a reflection in a line. Write an equation of the line.

37. $B(0, 3), B'(0, -1)$ **38.** $B(2, -4), B'(6, -4)$

39. CONSTRUCTION Follow the steps below to construct a reflection of $\triangle ABC$ in the line m. Use a compass and straightedge.

Step 1 Draw $\triangle ABC$ and line m.

Step 2 Use one compass setting to find two points that are the same distance from A on line m. Use the same compass setting to find a point on the other side of m that is the same distance from these two points. Label that point as A'.

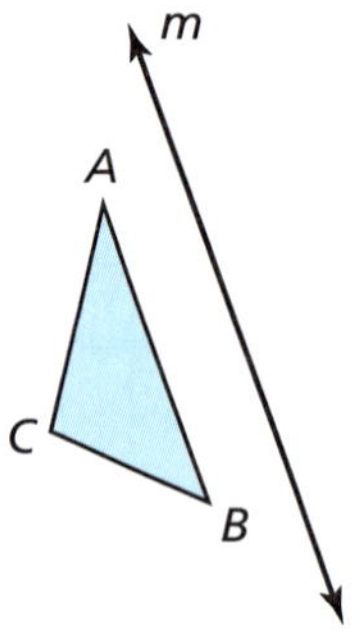

Step 3 Repeat Step 2 to find points B' and C'. Draw $\triangle A'B'C'$.

40. OPEN-ENDED Draw a line segment in Quadrant III. Identify a glide reflection that maps the line segment entirely to Quadrant I.

CONSTRUCTION In Exercises 41–46, use a reflective device to perform the construction.

41. copying a segment

42. copying an angle

43. constructing a segment bisector

44. constructing an angle bisector

45. constructing a perpendicular bisector

46. constructing the reflection of a figure over a line

47. DIG DEEPER Point $B'(1, 4)$ is the image of $B(3, 2)$ after a reflection in line c. Write an equation of line c.

48. THOUGHT PROVOKING
Is the composition of a translation and a reflection commutative? (In other words, do you obtain the same image regardless of the order in which you perform the transformations?) Justify your answer.

REVIEW & REFRESH

WATCH

49. Find the distance from point A to $\overleftrightarrow{XZ}$.

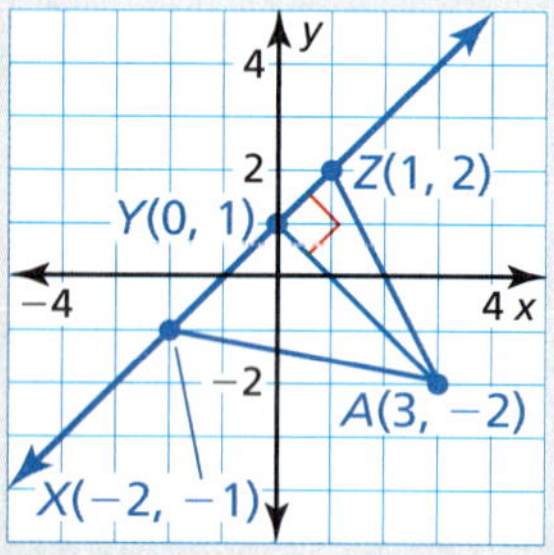

In Exercises 50 and 51, solve the equation. Check your solution.

50. $|2x| = |x + 6|$

51. $5x - 7 = 2(x + 1)$

52. Graph $\triangle ABC$ with vertices $A(0, 2)$, $B(3, 2)$, and $C(2, 1)$ and its image after a reflection in the x-axis.

53. Name the property that "If $\angle A \cong \angle B$ and $\angle B \cong \angle C$, then $\angle A \cong \angle C$" illustrates.

54. Make a scatter plot of the data. Then describe the relationship between the data.

x	0.3	0.6	0.8	1.2	1.4	1.9	2.0	2.1
y	1.1	1.4	1.7	2.3	2.6	3.0	3.1	3.3

7 MTR **55. MODELING REAL LIFE** A subway train travels at a speed of 30 miles per hour. What is the speed of the subway train in feet per second?

56. Find the distance from the point $(-3, 6)$ to the line $y = -\frac{3}{2}x - 5$.

57. Evaluate $h(x) = -3x + 7$ when $x = -1$.

58. Solve $A = \frac{1}{2}h(b_1 + b_2)$ for b_2.

59. Use the translation $(x, y) \rightarrow (x, y + 4)$ to find the image of $A(0, -4)$.

60. Factor $-2t^2 + 9t - 7$.

61. Find the sum of 4×10^6 and 1.7×10^6.

62. Estimate the intercepts of the graph of the function.

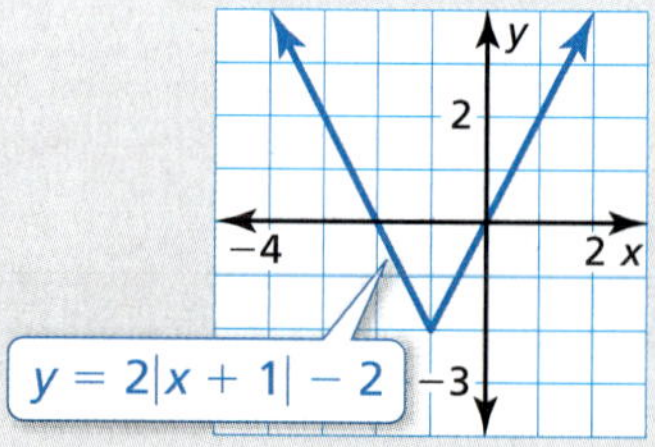

63. Find the product of $x + 5$ and $x + 2$.

GO DIGITAL

4.3 Rotations

Learning Target: Understand rotations of figures.

Success Criteria:
- I can rotate figures.
- I can perform compositions with rotations.
- I can identify rotational symmetry in polygons.

EXPLORE IT! Rotating a Polygon

Work with a partner.

a. The diagram shows a *rotation* of $\triangle DEF$ to $\triangle D'E'F'$. How would you define a rotation?

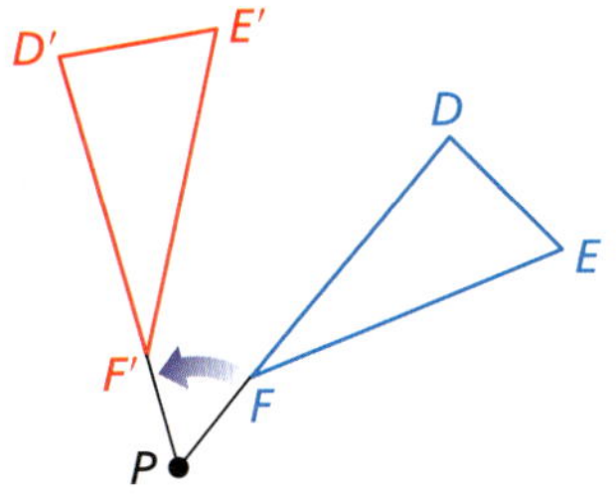

b. Use technology to draw any triangle and label it $\triangle ABC$.

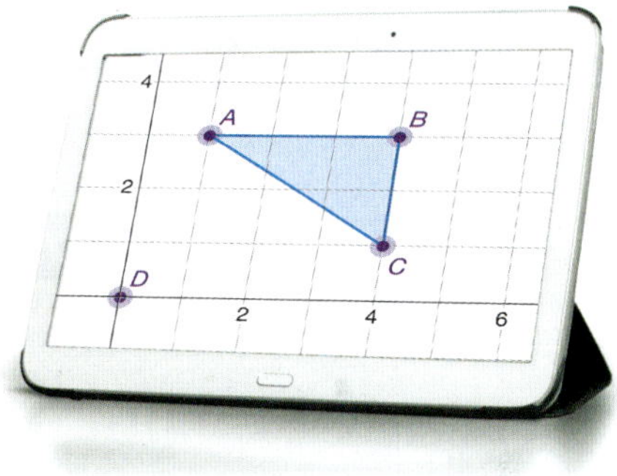

c. Rotate the triangle 90° counterclockwise about the origin to form $\triangle A'B'C'$. What do you observe about the side lengths, angle measures, and coordinates of the vertices of the two triangles?

d. Write rules to determine the coordinates of the image of (x, y) when it is rotated 90°, 180°, and 270° counterclockwise about the origin.

90° Rotation:	**180° Rotation:**	**270° Rotation:**
$(x, y) \rightarrow (\ \_\_\ ,\ \_\_\ )$	$(x, y) \rightarrow (\ \_\_\ ,\ \_\_\ )$	$(x, y) \rightarrow (\ \_\_\ ,\ \_\_\ )$

e. Based on your results in parts (b)–(d), is there anything you would like to change or include in your definition in part (a)? Explain.

f. Is a rotation a rigid motion? Explain your reasoning.

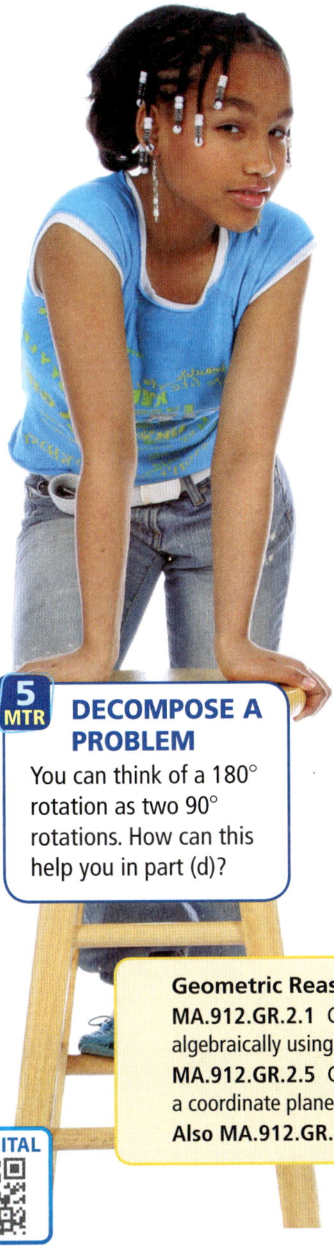

5 MTR DECOMPOSE A PROBLEM
You can think of a 180° rotation as two 90° rotations. How can this help you in part (d)?

Geometric Reasoning
MA.912.GR.2.1 Given a preimage and image, describe the transformation and represent the transformation algebraically using coordinates.
MA.912.GR.2.5 Given a geometric figure and a sequence of transformations, draw the transformed figure on a coordinate plane.
Also MA.912.GR.2.2, MA.912.GR.2.3, H MA.912.GR.2.4

GO DIGITAL

Performing Rotations

Vocabulary

rotation, *p. 184*
center of rotation, *p. 184*
angle of rotation, *p. 184*
rotational symmetry, *p. 188*
center of symmetry, *p. 188*
order of rotational symmetry, *p. 188*

KEY IDEA

Rotations

A **rotation** is a transformation in which a figure is turned about a fixed point called the **center of rotation**. Rays drawn from the center of rotation to a point and its image form the **angle of rotation**.

A rotation about a point P through an angle of $x°$ maps every point Q in the plane to a point Q' so that one of the following properties is true.

- If Q is not the center of rotation P, then $QP = Q'P$ and $m\angle QPQ' = x°$, or
- If Q is the center of rotation P, then $Q = Q'$.

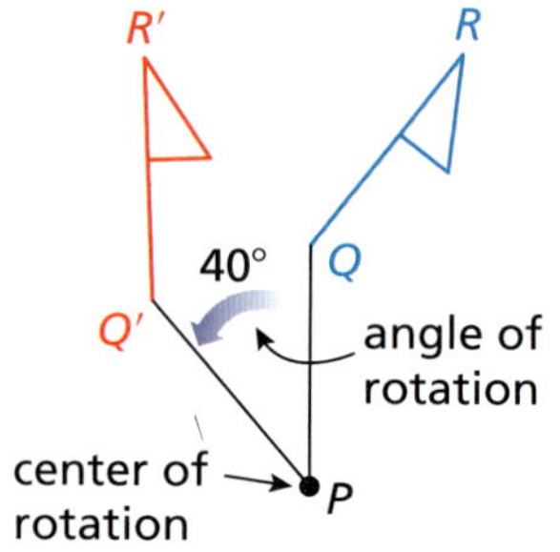

The figure above shows a 40° counterclockwise rotation. Rotations can be *clockwise* or *counterclockwise*.

Direction of rotation

EXAMPLE 1 Drawing a Rotation

Draw a 120° counterclockwise rotation of $\triangle ABC$ about point P.

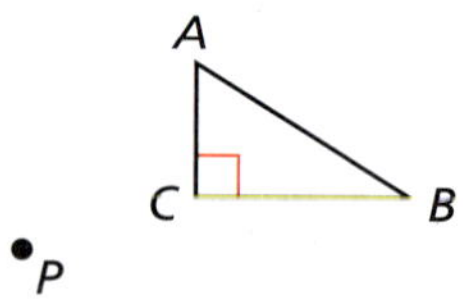

SOLUTION

Step 1 Draw a segment from P to A.

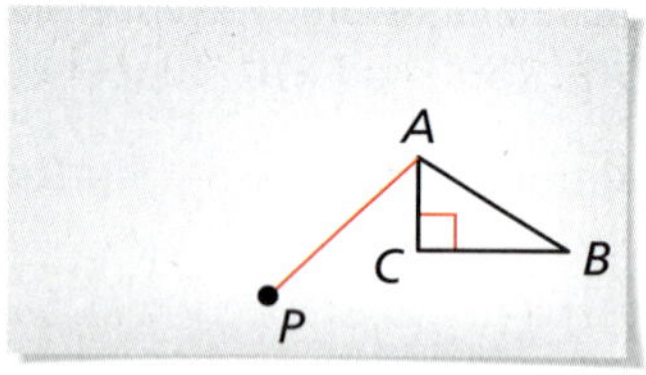

Step 2 Draw a ray to form a 120° angle with $\overline{PA}$.

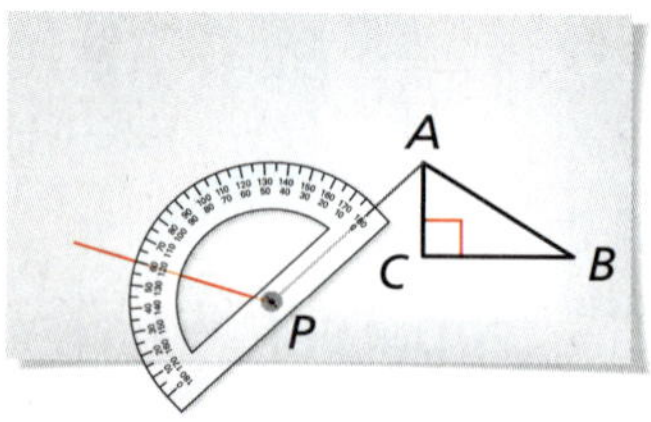

Step 3 Draw A' so that $PA' = PA$.

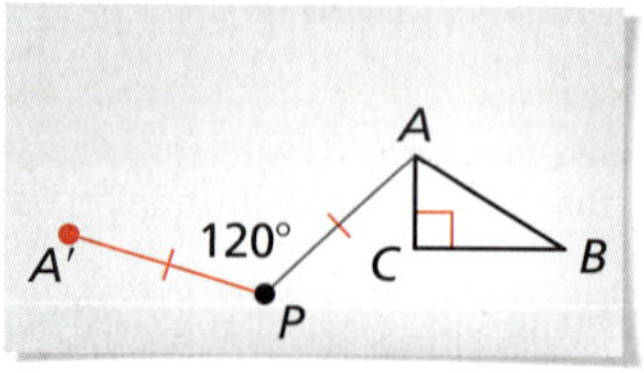

Step 4 Repeat Steps 1–3 for each remaining vertex. Draw $\triangle A'B'C'$.

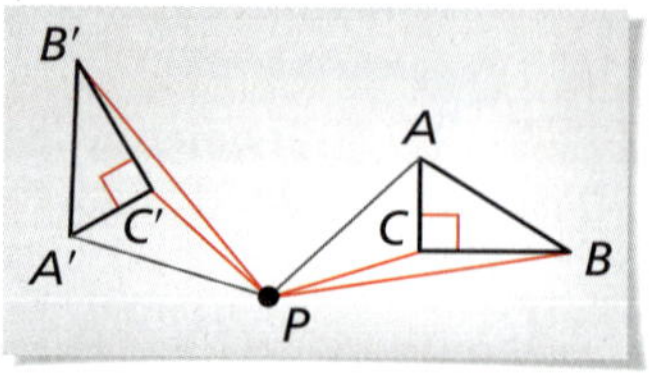

GO DIGITAL

You can rotate a figure more than 180°. The diagram shows counterclockwise rotations of point A 130°, 220°, and 310° about the origin. Notice that point A and its images all lie on the same circle. A rotation of 360° maps a figure onto itself.

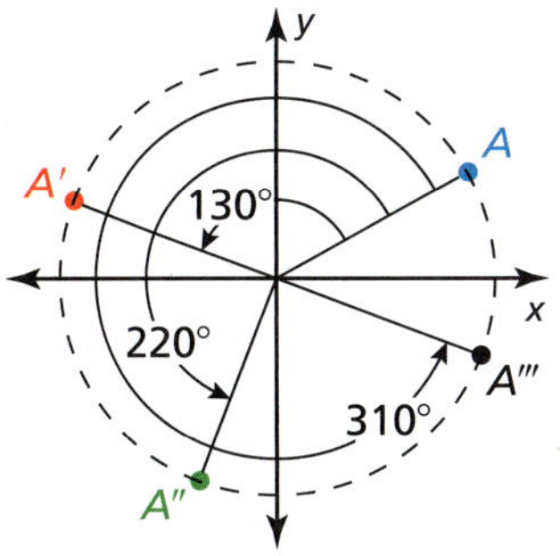

You can use coordinate rules to find the coordinates of a point after a rotation of 90°, 180°, or 270° about the origin.

STUDY TIP

You can rotate a figure more than 360°. The effect, however, is the same as rotating the figure by the angle minus a multiple of 360°.

KEY IDEA

Coordinate Rules for Rotations about the Origin

When a point (a, b) is rotated counterclockwise about the origin, the following are true.

- For a rotation of 90°, $(a, b) \rightarrow (-b, a)$.
- For a rotation of 180°, $(a, b) \rightarrow (-a, -b)$.
- For a rotation of 270°, $(a, b) \rightarrow (b, -a)$.

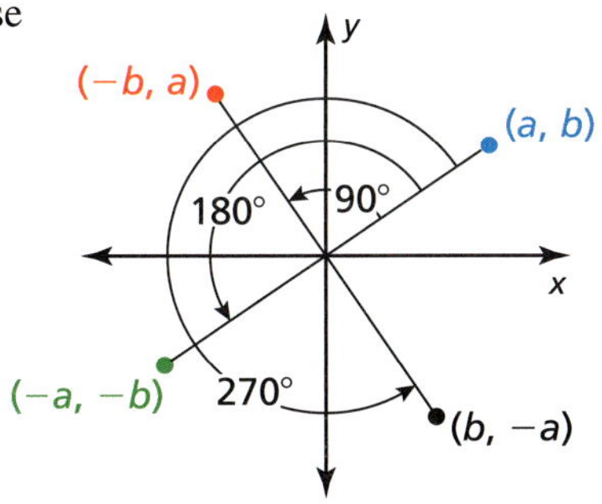

A counterclockwise rotation of $n°$ is the same as a clockwise rotation of $(360 - n)°$. Similarly, a clockwise rotation of $n°$ is the same as a counterclockwise rotation of $(360 - n)°$.

EXAMPLE 2 Rotating a Figure in the Coordinate Plane

WATCH

Graph quadrilateral $RSTU$ with vertices $R(3, 1)$, $S(5, 1)$, $T(5, -3)$, and $U(2, -1)$ and its image after a 270° counterclockwise rotation about the origin.

SOLUTION

Use the coordinate rule for a 270° rotation to find the coordinates of the vertices of the image. Then graph quadrilateral $RSTU$ and its image.

$$(a, b) \rightarrow (b, -a)$$

$$R(3, 1) \rightarrow R'(1, -3)$$

$$S(5, 1) \rightarrow S'(1, -5)$$

$$T(5, -3) \rightarrow T'(-3, -5)$$

$$U(2, -1) \rightarrow U'(-1, -2)$$

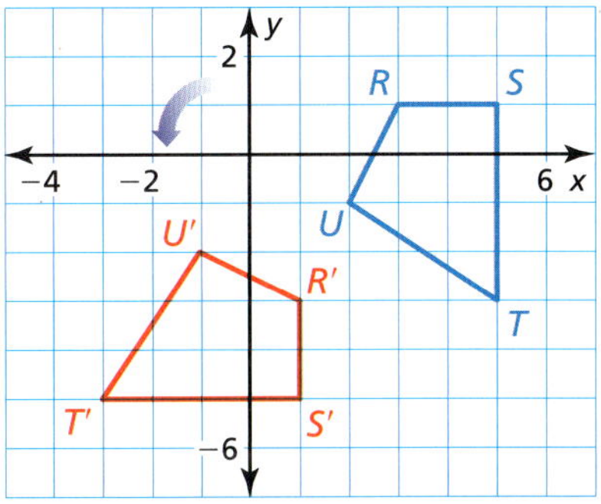

SELF-ASSESSMENT 1 I don't understand yet. 2 I can do it with help. 3 I can do it on my own. 4 I can teach someone else.

1. Trace $\triangle DEF$ and point P. Then draw a 50° clockwise rotation of $\triangle DEF$ about point P.

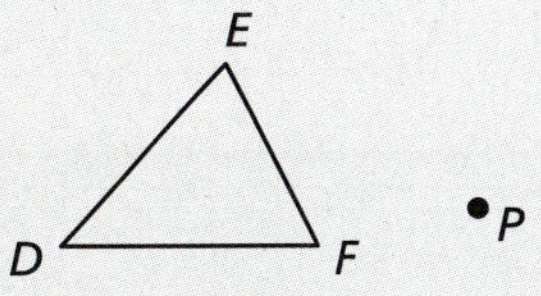

2. Graph $\triangle JKL$ with vertices $J(3, 0)$, $K(4, 3)$, and $L(6, 0)$ and its image after a 90° clockwise rotation about the origin.

GO DIGITAL

Performing Compositions with Rotations

POSTULATE

4.3 Rotation Postulate

A rotation is a rigid motion.

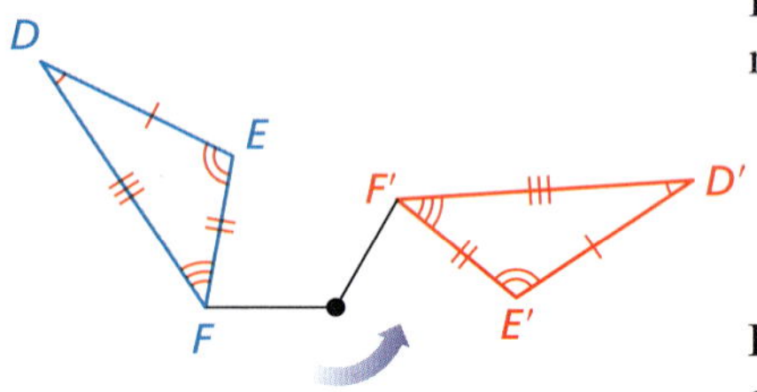

Because a rotation is a rigid motion, and a rigid motion preserves length and angle measure, the following statements are true for the rotation shown.

- $DE = D'E'$ $\quad EF = E'F'$ $\quad FD = F'D'$
- $m\angle D = m\angle D'$ $\quad m\angle E = m\angle E'$ $\quad m\angle F = m\angle F'$

Because a rotation is a rigid motion, the Composition Theorem guarantees that compositions of rotations and other rigid motions, such as translations and reflections, are rigid motions.

EXAMPLE 3 Performing a Composition

WATCH

a. Graph $\overline{RS}$ with endpoints $R(-1, -3)$ and $S(-2, -6)$ and its image after the composition.

Reflection: in the y-axis

Rotation: 90° clockwise about the origin

b. Write a coordinate rule for the composition.

COMMON ERROR

Unless you are told otherwise, perform the transformations in the order given.

SOLUTION

a. Step 1 Graph $\overline{RS}$.

Step 2 Reflect $\overline{RS}$ in the y-axis. $\overline{R'S'}$ has endpoints $R'(1, -3)$ and $S'(2, -6)$.

Step 3 Rotate $\overline{R'S'}$ 90° clockwise or $(360 - 90)° = 270°$ counterclockwise about the origin. $\overline{R''S''}$ has endpoints $R''(-3, -1)$ and $S''(-6, -2)$.

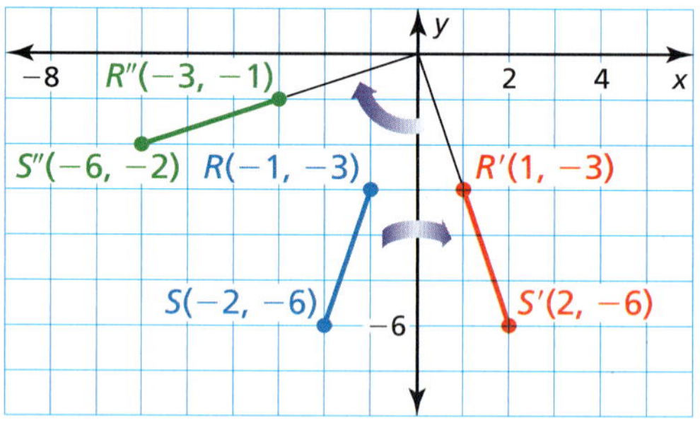

b. The first transformation is a reflection in the y-axis, so $(a, b) \rightarrow (-a, b)$. The second transformation is a rotation 90° clockwise, or 270° counterclockwise, about the origin, so $(-a, b) \rightarrow (b - (-a)) = (b, a)$.

▶ So, a coordinate rule for the composition is $(a, b) \rightarrow (b, a)$.

Check Test the endpoints of $\overline{RS}$ to check that the rule is correct.

$$(a, b) \rightarrow (b, a)$$

$$(-1, -3) \rightarrow (-3, -1)$$

$$(-2, -6) \rightarrow (-6, -2) \checkmark$$

GO DIGITAL

You can use a composition to rotate a figure when the center of rotation is a point on the figure.

KEY IDEA

Rotating When the Center Is a Point on the Figure

To rotate a figure when the center of rotation is a point P on the figure, translate the figure along $\overrightarrow{PO}$, so the center is mapped to the origin O. Then use the appropriate coordinate rule for a rotation about the origin. Finally, translate the center of rotation back to P along $\overrightarrow{OP}$.

EXAMPLE 4 **Rotating When the Center is on the Figure**

WATCH

Graph $\triangle ABC$ with vertices $A(1, 2)$, $B(2, 4)$, and $C(3, 3)$ and its image after a 180° rotation about point A.

SOLUTION

Step 1 Translate the figure along $\overrightarrow{AO}$.

$$(a, b) \rightarrow (a - 1, b - 2)$$

$$A(1, 2) \rightarrow A'(1 - 1, 2 - 2) = A'(0, 0)$$

$$B(2, 4) \rightarrow B'(2 - 1, 4 - 2) = B'(1, 2)$$

$$C(3, 3) \rightarrow C'(3 - 1, 3 - 2) = C'(2, 1)$$

Step 2 Use the coordinate rule for a 180° rotation about the origin.

$$(a, b) \rightarrow (-a, -b)$$

$$A'(0, 0) \rightarrow A''(0, 0)$$

$$B'(1, 2) \rightarrow B''(-1, -2)$$

$$C'(2, 1) \rightarrow C''(-2, -1)$$

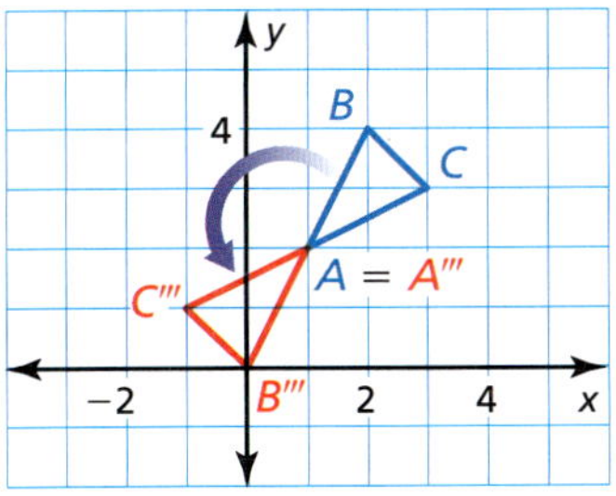

Step 3 Translate back along $\overrightarrow{OA}$.

$$(a, b) \rightarrow (a + 1, b + 2)$$

$$A''(0, 0) \rightarrow A'''(0 + 1, 0 + 2) = A'''(1, 2)$$

$$B''(-1, -2) \rightarrow B'''(-1 + 1, -2 + 2) = B'''(0, 0)$$

$$C''(-2, -1) \rightarrow C'''(-2 + 1, -1 + 2) = C'''(-1, 1)$$

SELF-ASSESSMENT

1 I don't understand yet. 2 I can do it with help. 3 I can do it on my own. 4 I can teach someone else.

3. Graph $\overline{RS}$ from Example 3. Perform the rotation, followed by the reflection. Does the order of the transformations matter? Explain.

4. Graph $\overline{AB}$ with endpoints $A(-4, 4)$ and $B(-1, 7)$ and its image after the composition. Write a coordinate rule for the composition.

Translation: $(x, y) \rightarrow (x - 2, y - 1)$ **Rotation:** 90° counterclockwise about the origin

4 MTR **5.** **DIFFERENT WORDS, SAME QUESTION** Which is different? Find "both" answers.

Rotate $A(2, 3)$ 90° counterclockwise about the origin.	Rotate $A(2, 3)$ 270° clockwise about the origin.
Rotate $A(2, 3)$ 180° about the origin and then 90° counterclockwise about the origin.	Rotate $A(2, 3)$ 180° about the origin and then 90° clockwise about the origin.

6. Graph $\triangle ABC$ with vertices $A(-4, -3)$, $B(0, -2)$, and $C(2, -2)$ and its image after a 90° clockwise rotation about point C.

GO DIGITAL

Identifying Rotational Symmetry

A figure in the plane has **rotational symmetry** when the figure can be mapped onto itself by a rotation of 180° or less about the center of the figure. This point is the **center of symmetry**. Note that the rotation can be either clockwise or counterclockwise.

STUDY TIP

This figure also has *point symmetry*, which is 180° rotational symmetry.

For example, the figure below has rotational symmetry, because a rotation of either 90° or 180° maps the figure onto itself (although a rotation of 45° does not).

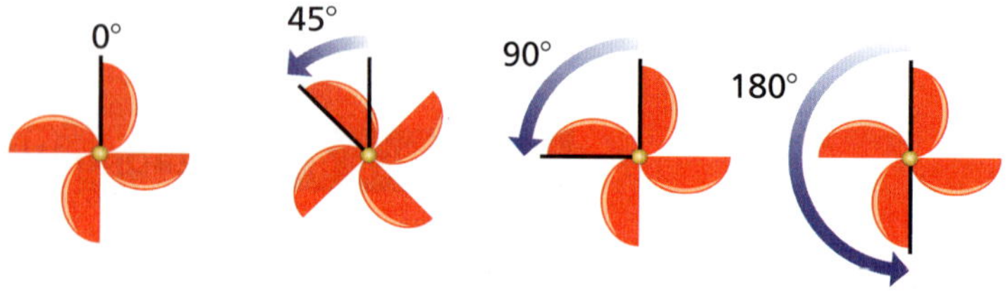

For a figure with rotational symmetry, the **order of rotational symmetry** is the number of times a figure can be mapped onto itself in one 360° rotation about the center of the figure.

EXAMPLE 5 **Identifying Rotational Symmetry** WATCH

Determine whether each polygon has rotational symmetry. If so, identify the order of rotational symmetry.

a. regular hexagon

b. trapezoid

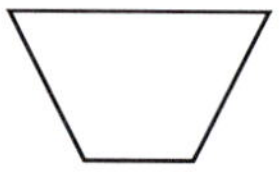

SOLUTION

a. The regular hexagon has rotational symmetry because a 60° rotation about its center maps the hexagon onto itself. The center is the intersection of the diagonals. There are 6 rotations that map the hexagon onto itself, so the order of rotational symmetry is 6.

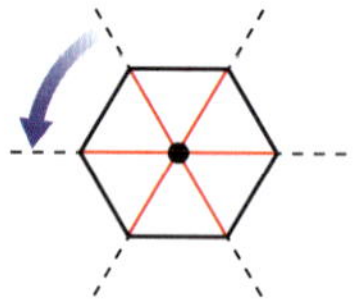

b. The trapezoid does not have rotational symmetry because no rotation of 180° or less maps the trapezoid onto itself.

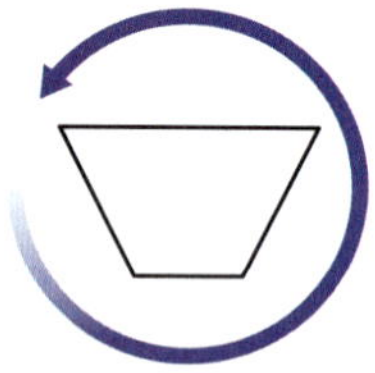

1 MTR HELP A CLASSMATE

Help a classmate understand why there are 6 rotations that map the hexagon onto itself.

SELF-ASSESSMENT 1 I don't understand yet. 2 I can do it with help. 3 I can do it on my own. 4 I can teach someone else.

Determine whether the polygon has rotational symmetry. If so, describe any rotations that map the polygon onto itself and identify the order of rotational symmetry.

7. rhombus

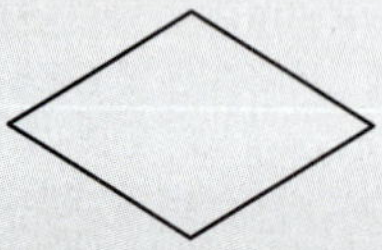

8. octagon

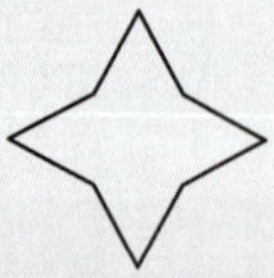

9. right triangle

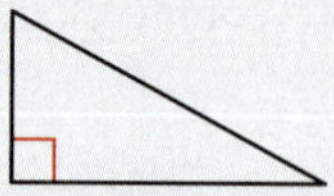

GO DIGITAL

4.3 Practice WITH CalcChat® AND CalcView®

In Exercises 1–4, trace the polygon and point *P*. Then draw the given rotation of the polygon about point *P*. *(See Example 1.)*

1. 30° counterclockwise

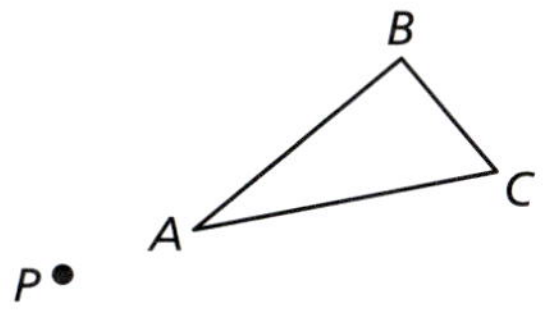

2. 90° clockwise

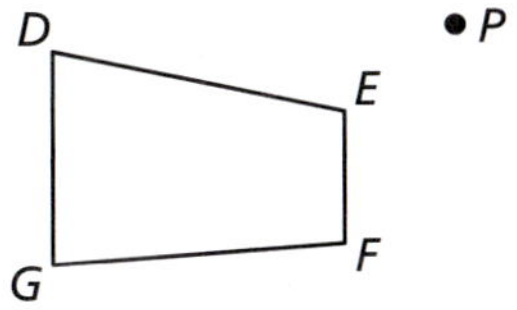

▶ 3. 180°

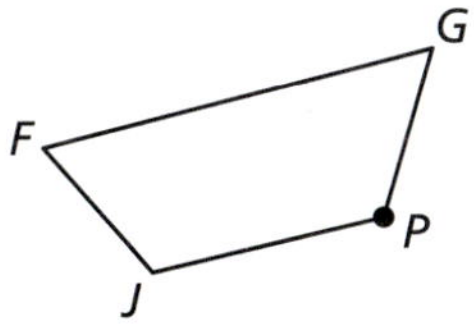

4. 130° counterclockwise

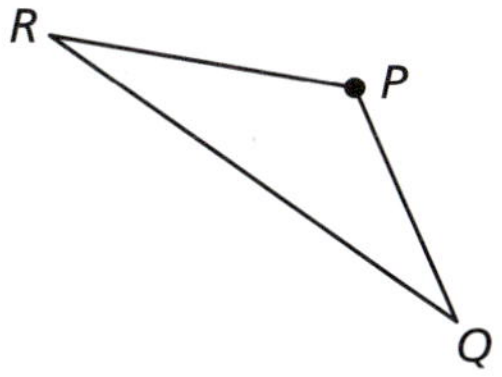

In Exercises 5–8, graph the polygon with the given vertices and its image after the given rotation about the origin. *(See Example 2.)*

▶ 5. $A(-3, 2)$, $B(2, 4)$, $C(3, 1)$; 90° counterclockwise

6. $D(-3, -1)$, $E(-1, 2)$, $F(4, -2)$; 180°

7. $J(1, 4)$, $K(5, 5)$, $L(7, 2)$, $M(2, 2)$; 180°

8. $Q(-6, -3)$, $R(-5, 0)$, $S(-3, 0)$, $T(-1, -3)$; 270° clockwise

In Exercises 9–12, describe the rotation as a (a) counterclockwise rotation and (b) clockwise rotation. Then write a rule for the rotation.

9.

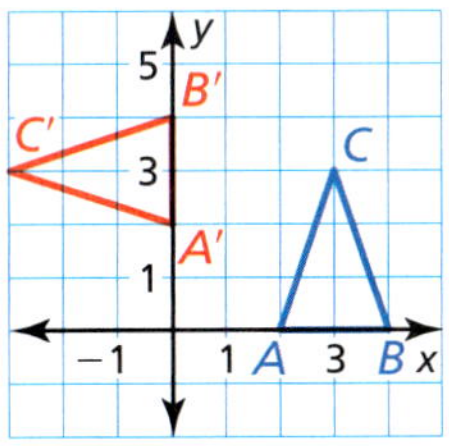

10.

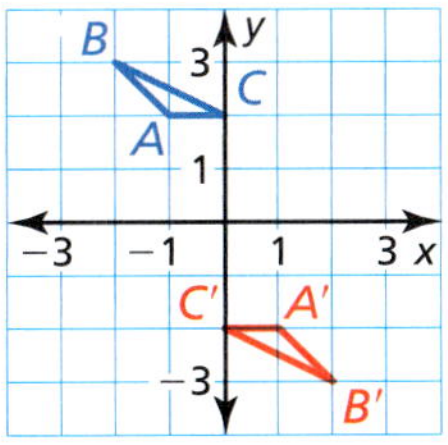

11.

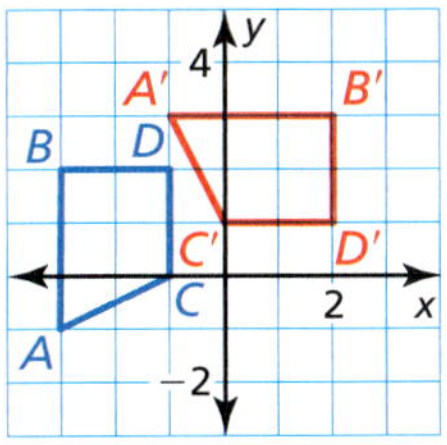

12.

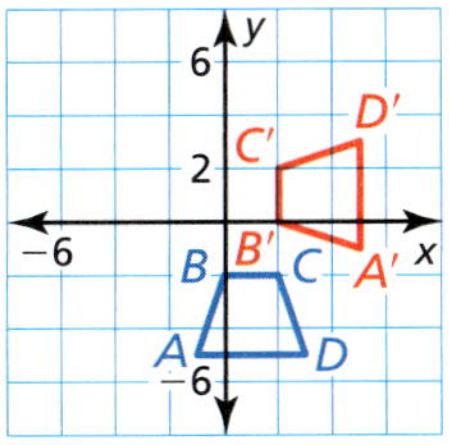

ERROR ANALYSIS **In Exercises 13 and 14, the endpoints of $\overline{CD}$ are $C(-1, 1)$ and $D(2, 3)$. Describe and correct the error in finding the coordinates of the endpoints of the image after a rotation of 270° counterclockwise about the origin.**

13.

✗ $C(-1, 1) \rightarrow C'(-1, -1)$
$D(2, 3) \rightarrow D'(2, -3)$

14.

✗ $C(-1, 1) \rightarrow C'(1, -1)$
$D(2, 3) \rightarrow D'(3, 2)$

In Exercises 15–18, graph $\overline{XY}$ with endpoints $X(-3, 1)$ and $Y(4, -5)$ and its image after the composition. Write a coordinate rule for the composition. *(See Example 3.)*

▶ 15. **Translation:** $(x, y) \rightarrow (x, y + 2)$
Rotation: 90° counterclockwise about the origin

16. **Rotation:** 180° about the origin
Translation: $(x, y) \rightarrow (x - 1, y + 1)$

17. **Rotation:** 270° counterclockwise about the origin
Reflection: in the y-axis

18. **Reflection:** in the line $y = x$
Rotation: 90° clockwise about the origin

In Exercises 19–22, graph $\triangle LMN$ with vertices $L(1, 6)$, $M(-2, 4)$, and $N(3, 2)$ and its image after the composition.

19. **Rotation:** 90° clockwise about the origin
Translation: $(x, y) \rightarrow (x - 3, y + 2)$

20. **Reflection:** in the x-axis
Rotation: 270° counterclockwise about the origin

21. **Translation:** $(x, y) \rightarrow (x + 1, y + 4)$
Rotation: 180° about the origin

22. **Rotation:** 90° counterclockwise about the origin
Reflection: in the y-axis

23. **REASONING** Use the coordinate rules for counterclockwise rotations about the origin to write coordinate rules for clockwise rotations of 90°, 180°, and 270° about the origin.

24. **REASONING** You transform an arrow using a rotation, a translation, and a second rotation. What do you know about the rotations when the image is pointing the same direction as the original figure?

GO DIGITAL

In Exercises 25–28, graph the polygon with the given vertices and its image after the given rotation about point A. *(See Example 4.)*

▶ **25.** $A(-6, 4)$, $B(5, 2)$, $C(2, 1)$; 90° clockwise

26. $A(-2, -3)$, $B(-3, 2)$, $C(3, -1)$; 180°

27. $A(1, 4)$, $B(6, 5)$, $C(8, 2)$, $D(3, 2)$; 90° counterclockwise

28. $A(-2, -5)$, $B(-7, 3)$, $C(-4, 3)$, $D(-1, -3)$; 270° counterclockwise

In Exercises 29–34, determine whether the figure has rotational symmetry. If so identify the order of rotational symmetry of the figure. *(See Example 5.)*

▶ **29.**

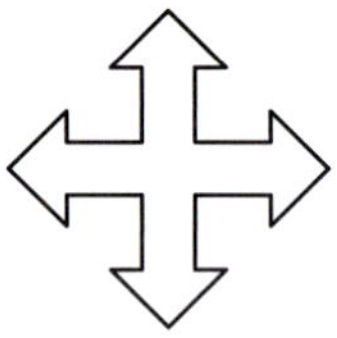

30.

31.

32.

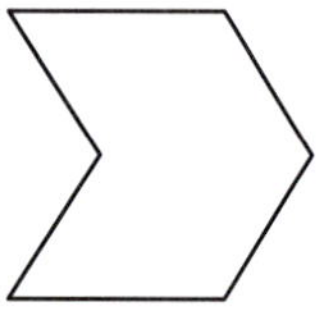

33.

34.

COLLEGE PREP In Exercises 35–38, select all angles of counterclockwise rotation about the center of the regular polygon that map the polygon onto itself.

Ⓐ 30° Ⓒ 60° Ⓔ 90° Ⓖ 144°

Ⓑ 45° Ⓓ 72° Ⓕ 120° Ⓗ 180°

35.

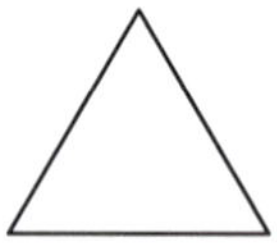

36.

37.

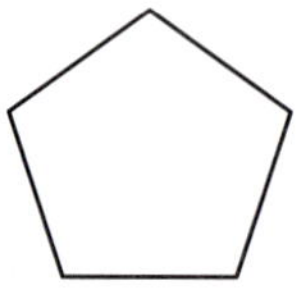

38.

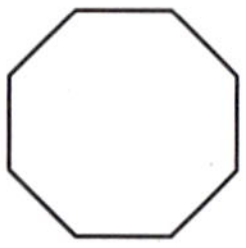

39. COLLEGE PREP What are the vertices of the image of quadrilateral $WXYZ$ after a rotation of 270° clockwise about the origin, followed by a translation 4 units right?

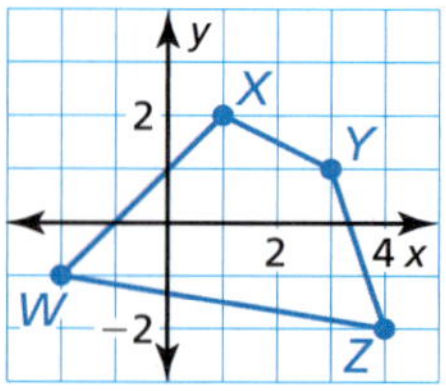

Ⓐ $W''(3, 2)$, $X''(6, -1)$, $Y''(5, -3)$, $Z''(2, -4)$

Ⓑ $W''(5, -2)$, $X''(2, 1)$, $Y''(3, 3)$, $Z''(6, 4)$

Ⓒ $W''(-3, -2)$, $X''(-6, 1)$, $Y''(-5, 3)$, $Z''(-2, 4)$

Ⓓ $W''(1, 2)$, $X''(-2, 5)$, $Y''(-1, 7)$, $Z''(2, 8)$

40. HOW DO YOU SEE IT?
You are finishing the puzzle. The remaining two pieces have rotational symmetry.

a. Describe the rotational symmetry of Piece 1 and of Piece 2.

b. Without considering the picture on Piece 1, how many different ways can it fit into the puzzle?

c. Before putting Piece 1 into the puzzle, you connect it to Piece 2. Now how many ways can it fit into the puzzle? Explain.

41. CONSTRUCTION Follow these steps to construct a counterclockwise rotation of $\triangle ABC$ by angle D about a point O. Use a compass and straightedge.

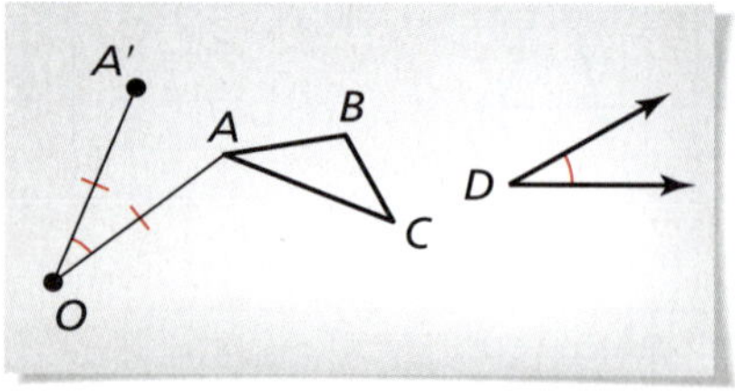

Step 1 Draw $\triangle ABC$, $\angle D$, and O (the center of rotation).

Step 2 Draw $\overline{OA}$. Use the construction for copying an angle to copy $\angle D$ at O, as shown. Then draw A' so that $OA' = OA$.

Step 3 Repeat Step 2 to find points B' and C'. Draw $\triangle A'B'C'$.

GO DIGITAL

42. REASONING You enter a revolving door.

STEM

a. You rotate the door 180°. What does this mean in the context of the situation? Explain.

b. You rotate the door 360°. What does this mean in the context of the situation? Explain.

5 MTR **43. CONNECTING CONCEPTS** Use the graph of $y = 2x - 3$.

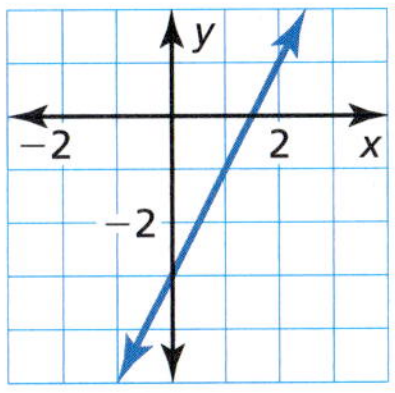

a. Rotate the line 90°, 180°, and 270° counterclockwise about the origin. Write the equation of the line for each image. Describe the relationship between the equation of the preimage and the equation of each image.

b. Are the relationships you described in part (a) true for any line that is not vertical or horizontal? Explain your reasoning.

7 MTR **44. MODELING REAL LIFE** Polar bearings use a polar coordinate system. A polar coordinate system locates a point in a plane by its distance from the origin O and by the measure of an angle with its vertex at the origin. For example, the point $A(2, 30°)$ is 2 units from the origin and $m\angle XOA = 30°$. What are the polar coordinates of the image of point A after a 90° clockwise rotation? a 180° rotation? a 90° counterclockwise rotation? Explain.

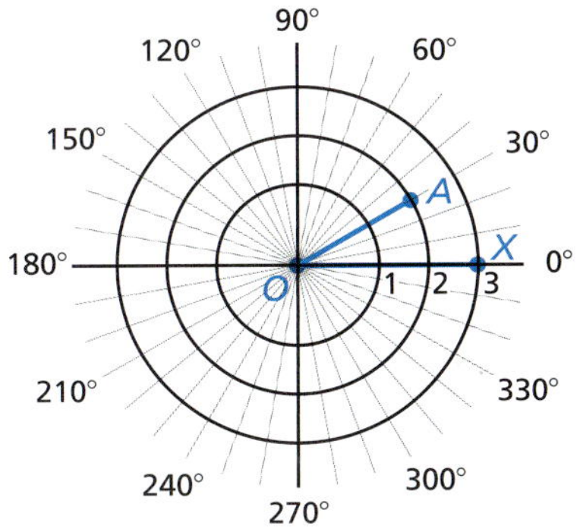

45. NUMBER SENSE The endpoints of $\overline{JK}$ are $J(-5, -4)$ and $K(-3, -1)$. What are the coordinates of the endpoints of the image after a rotation of 630° about the origin? 900° about the origin?

GO DIGITAL

6 MTR **46. ASSESS REASONABLENESS** You take a picture of a malachite butterfly during a class trip to the Florida Museum of Natural History's Butterfly Rainforest. Your friend says that the butterfly's wing markings have rotational symmetry because a 180° rotation of the markings around the butterfly maps the markings onto one another. Is your friend's claim reasonable? Explain.

4 MTR **47. DISCUSS MATHEMATICAL THINKING** Graph $\triangle ABC$ with the vertices $A(-2, 1)$, $B(3, 4)$, and $C(-1, 3)$ and its image after a 270° clockwise rotation around the given center.

a. the origin

b. point A

c. point B

d. point C

e. Write coordinate rules for parts (a)–(d). How do the rules differ? Explain your reasoning.

7 MTR **48. PERFORMANCE TASK** From the Northern Hemisphere, constellations appear to rotate counterclockwise about the North Celestial Pole, approximated by Polaris. A 360° rotation is completed about once every 24 hours.

a. Research *circumpolar constellations* and choose one that is visible in the Northern Hemisphere. Use the Internet or another reference to sketch the current orientation of the constellation (when you are facing north).

b. You are planning a stargazing event later this evening. Choose a time for the event. Then create a stargazing guide for your constellation that includes at least two facts about the constellation and a drawing of its orientation in the sky at that time. Use geometric tools to construct the drawing.

49. REASONING A figure has only point symmetry. How many rotations that map the figure onto itself can be performed before it is back where it started?

4 MTR **50. DISCUSS MATHEMATICAL THINKING** Is it possible for a figure to have 90° rotational symmetry but not 180° rotational symmetry? Explain your reasoning.

51. DIG DEEPER $\triangle XYZ$ has vertices $X(2, 5)$, $Y(3, 1)$, and $Z(0, 2)$. Rotate $\triangle XYZ$ 90° counterclockwise about the point $P(-2, -1)$.

52. THOUGHT PROVOKING
Can rotations of 90°, 180°, 270°, and 360° be written as the compositions of two reflections? Justify your answers.

53. REASONING The vertices of a quadrilateral after a reflection in the line $y = -x$, followed by a rotation of 90° counterclockwise about the origin, are $D''(-1, -2)$, $E''(4, -1)$, $F''(3, 2)$, and $G''(1, 1)$. What are the vertices of quadrilateral $DEFG$?

REVIEW & REFRESH

WATCH

In Exercises 54 and 55, $\overrightarrow{DF}$ bisects $\angle CDE$. Find the indicated angle measures.

54. $m\angle CDF = 43°$. Find $m\angle EDF$ and $m\angle CDE$.

55. $m\angle CDE = 102°$. Find $m\angle CDF$ and $m\angle EDF$.

56. The figures are congruent. Name the corresponding angles and the corresponding sides.

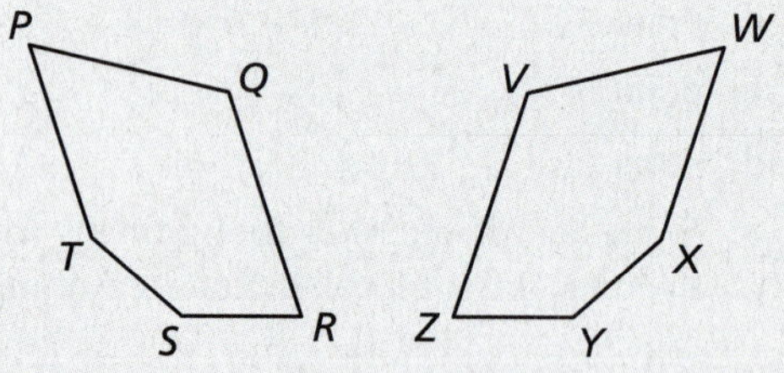

57. The endpoints of the directed line segment AB are $A(1, 2)$ and $B(10, 5)$. Find the coordinates of point P along segment AB so that the ratio of AP to PB is 1 to 2.

58. Graph the system. Identify a solution.

$$y < x + 3$$
$$y \geq \tfrac{1}{2}x + 1$$

59. Determine whether the table represents a *linear* or an *exponential* function. Explain.

x	−1	0	1	2	3
y	0.5	2	8	32	128

60. Classify the polygon by the number of sides. Tell whether it is *convex* or *concave*.

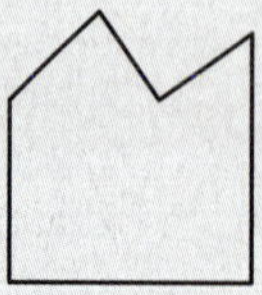

In Exercises 61–63, graph the polygon with the given vertices and its image after the indicated transformation.

61. $A(-4, 1)$, $B(-3, 3)$, $C(-1, 2)$
Reflection: in the x-axis

62. $J(-2, -3)$, $K(-1, 1)$, $L(2, 0)$, $M(1, -4)$
Translation: $(x, y) \rightarrow (x + 3, y - 2)$

63. $X(2, -1)$, $Y(5, 1)$, $Z(4, -4)$
Rotation: 180° about the origin

In Exercises 64 and 65, decide whether inductive reasoning or deductive reasoning is used to reach the conclusion. Explain your reasoning.

64. Each time you ride your bike, you take a bottle of water. So, the next time you ride your bike, you will take a bottle of water.

65. All birds have feathers. A robin is a type of bird. So, a robin has feathers.

66. Identify the population and sample in the study below. Describe the sample.

In the United Kingdom, a survey of 1467 high school students found that 723 of them planned to attend a four-year college.

7 MTR **67. MODELING REAL LIFE** Cell phones use bars like the ones shown to indicate how much signal strength a phone receives from the nearest service tower. Each bar is parallel to the bar directly next to it. Explain why the tallest bar is parallel to the shortest bar.

GO DIGITAL

4.4 Congruence and Transformations

Learning Target: Understand congruence transformations.

Success Criteria:

- I can identify congruent figures.
- I can describe congruence transformations.
- I can identify the types of symmetry in a tessellation.
- I can use congruence transformations to solve problems.

EXPLORE IT! Reflecting Figures in Lines

Work with a partner. Use technology to draw any scalene triangle and label it $\triangle ABC$. Draw any line, $\overleftrightarrow{DE}$, and another line that is parallel to $\overleftrightarrow{DE}$.

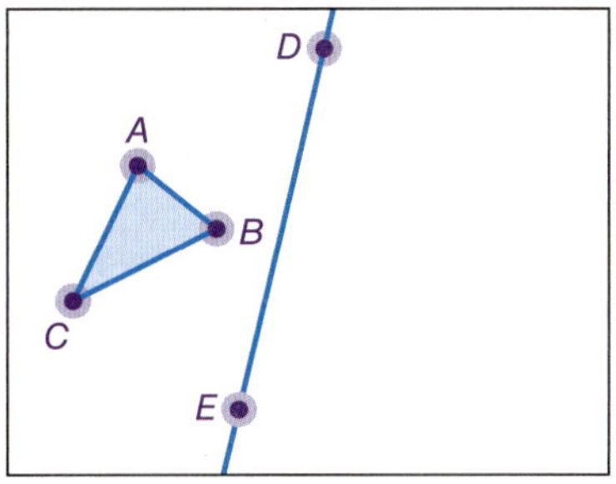

a. Reflect $\triangle ABC$ in $\overleftrightarrow{DE}$, followed by a reflection in the other line to form $\triangle A''B''C''$. What do you notice? Make several observations.

b. Is there a single transformation that maps $\triangle ABC$ to $\triangle A''B''C''$? Explain.

c. Repeat parts (a) and (b) with other figures. What do you notice?

d. Using the same triangle and line $\overleftrightarrow{DE}$, draw line $\overleftrightarrow{DF}$ that intersects $\overleftrightarrow{DE}$ at point D so that $\angle EDF$ is an acute or right angle. Then reflect $\triangle ABC$ in $\overleftrightarrow{DE}$, followed by a reflection in $\overleftrightarrow{DF}$ to form $\triangle A''B''C''$. What do you notice? Make several observations.

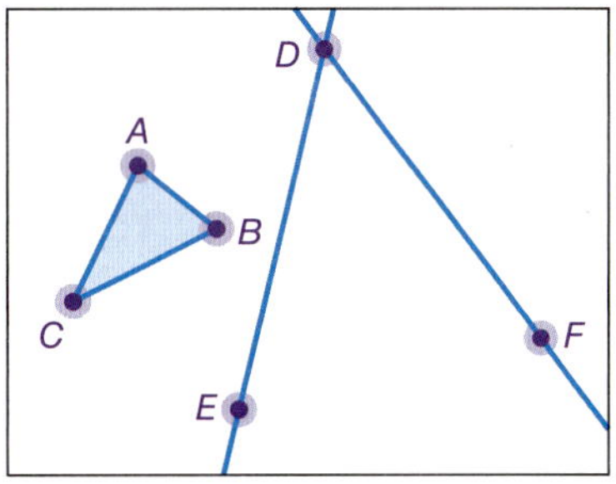

e. Is there a single transformation that maps $\triangle ABC$ to $\triangle A''B''C''$? Explain.

f. Repeat parts (d) and (e) with other figures. What do you notice?

4 MTR CONSTRUCT AN ARGUMENT
What conjectures can you make based on your results? How can you test your conjectures?

Geometric Reasoning
MA.912.GR.2.1 Given a preimage and image, describe the transformation and represent the transformation algebraically using coordinates.
MA.912.GR.2.6 Apply rigid transformations to map one figure onto another to justify that the two figures are congruent.
Also MA.912.GR.1.6, MA.912.GR.2.3, H MA.912.GR.2.4

GO DIGITAL

Identifying Congruent Figures

Vocabulary

congruent figures, *p. 194*
congruence transformation, *p. 195*
tessellation, *p. 196*
translational symmetry, *p. 196*

Two geometric figures are **congruent figures** if and only if there is a rigid motion or a composition of rigid motions that maps one of the figures to the other. Congruent figures have the same size and same shape.

Congruent

same size and same shape

Not congruent

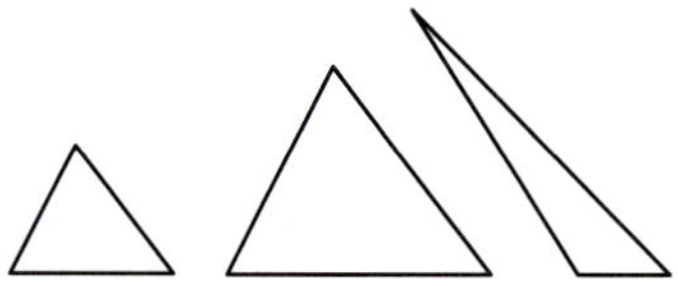

different sizes or shapes

You can identify congruent figures in the coordinate plane by identifying the rigid motion or composition of rigid motions that maps one of the figures to the other. Recall from Postulates 4.1–4.3 and Theorem 4.1 that translations, reflections, rotations, and compositions of these transformations are rigid motions.

5 MTR MAKE A PLAN

How can you determine which figures may be congruent before identifying a transformation?

EXAMPLE 1 Identifying Congruent Figures

Identify any congruent figures in the coordinate plane. Explain.

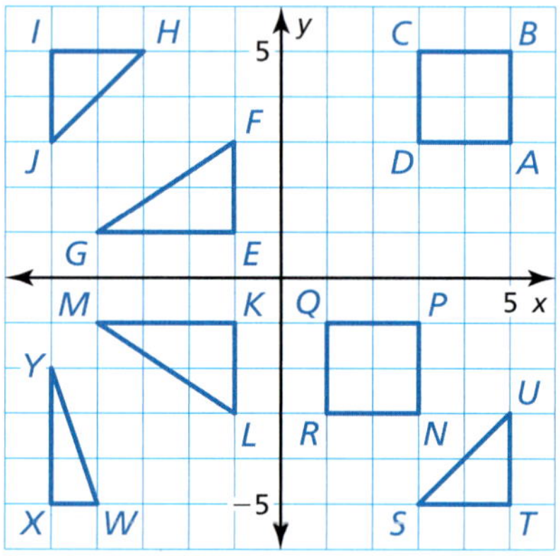

SOLUTION

Square $NPQR$ is a translation of square $ABCD$ 2 units left and 6 units down. So, square $ABCD$ and square $NPQR$ are congruent.

$\triangle KLM$ is a reflection of $\triangle EFG$ in the x-axis. So, $\triangle EFG$ and $\triangle KLM$ are congruent.

$\triangle STU$ is a 180° rotation of $\triangle HIJ$. So, $\triangle HIJ$ and $\triangle STU$ are congruent.

SELF-ASSESSMENT

1 I don't understand yet. 2 I can do it with help. 3 I can do it on my own. 4 I can teach someone else.

1. Identify any congruent figures in the coordinate plane. Explain.

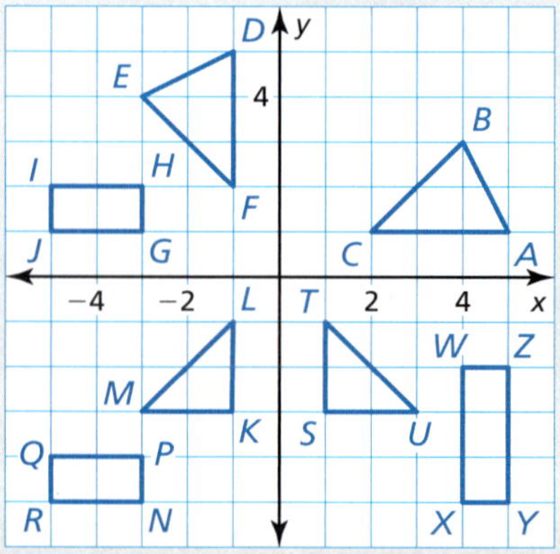

GO DIGITAL

Congruence Transformations

Another name for a rigid motion or a combination of rigid motions is a **congruence transformation** because the preimage and image are congruent. The terms *rigid motion* and *congruence transformation* are interchangeable.

EXAMPLE 2 Describing a Congruence Transformation

WATCH

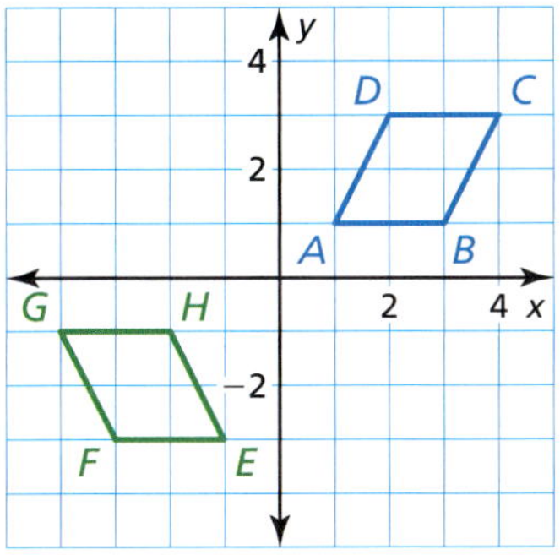

Describe a congruence transformation that maps ▱*ABCD* to ▱*EFGH*.

SOLUTION

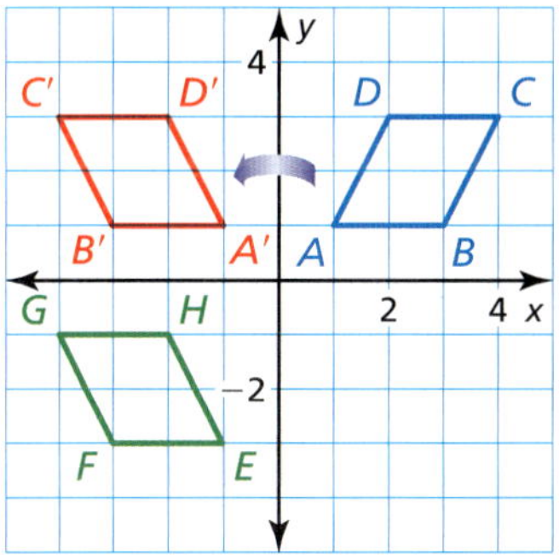

Two sides of ▱*ABCD* rise from left to right, and the corresponding sides of ▱*EFGH* fall from left to right. If you reflect ▱*ABCD* in the *y*-axis, as shown, then the image, ▱*A′B′C′D′*, will have the same orientation as ▱*EFGH*.

Then you can map ▱*A′B′C′D′* to ▱*EFGH* using a translation 4 units down.

▶ So, a congruence transformation that maps ▱*ABCD* to ▱*EFGH* is a reflection in the *y*-axis, followed by a translation 4 units down.

Check Verify that the corresponding sides and the corresponding angles are congruent.

$AB = EF = 2$ and $CD = GH = 2$, so $\overline{AB} \cong \overline{EF}$ and $\overline{CD} \cong \overline{GH}$.

Using the Distance Formula, $BC = FG = \sqrt{5}$ and $AD = EH = \sqrt{5}$, so $\overline{BC} \cong \overline{FG}$ and $\overline{AD} \cong \overline{EH}$.

Using a protractor to estimate the angle measures, you can see that the corresponding angles appear to be congruent. ✓

SELF-ASSESSMENT

1 I don't understand yet. 2 I can do it with help. 3 I can do it on my own. 4 I can teach someone else.

2. **REASONING** A composition of rigid motions maps one figure to another figure. Is the image at each step of the composition congruent to the preimage and the final image? Explain.

3. In Example 2, describe another congruence transformation that maps ▱*ABCD* to ▱*EFGH*.

4. Describe a congruence transformation that maps △*JKL* to △*MNP*.

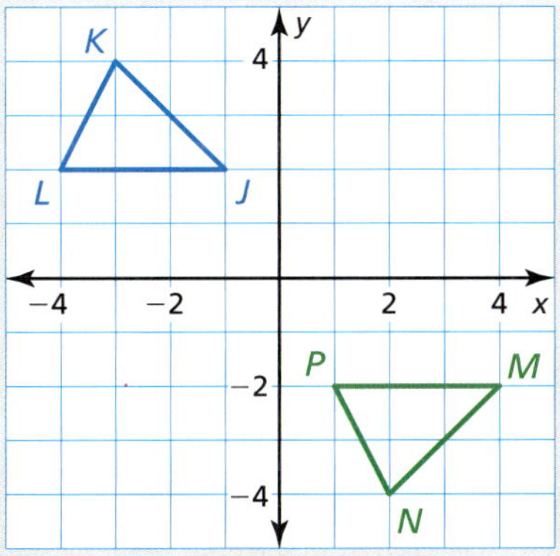

GO DIGITAL

Tessellations

A **tessellation** is the covering of a plane with one or more congruent figures so that there are no gaps or overlaps. A tessellation has **translational symmetry** when the tessellation can be mapped onto itself by a translation.

EXAMPLE 3 Identifying Transitional Symmetry

WATCH H

Determine whether the tessellation has translational symmetry. If so, identify two translations that map the tessellation onto itself.

SOLUTION

Because a translation can map the tessellation onto itself, it has translational symmetry. Two such translations are shown below.

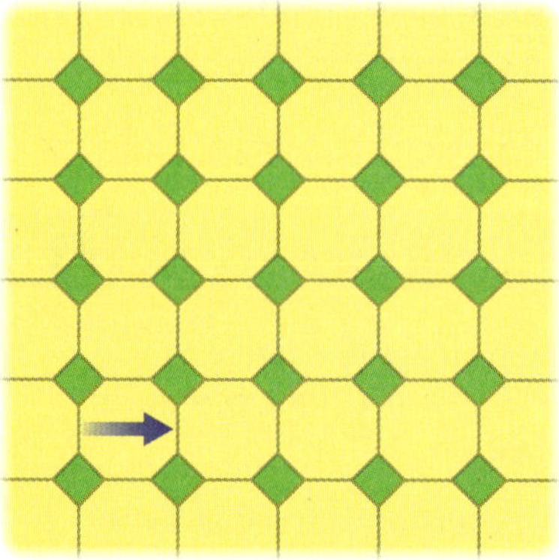

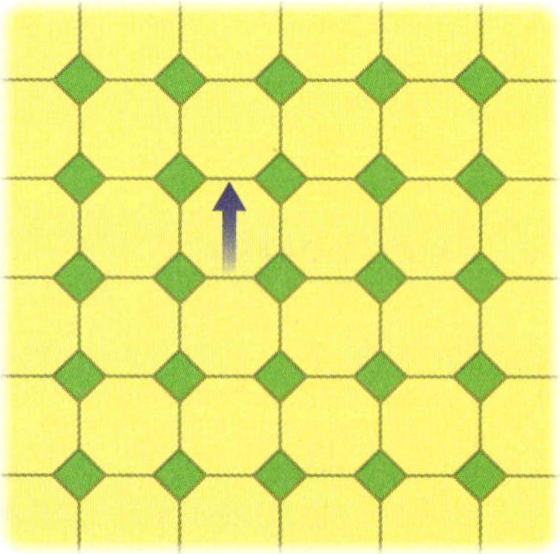

SELF-ASSESSMENT 1 I don't understand yet. 2 I can do it with help. 3 I can do it on my own. 4 I can teach someone else.

5. Determine whether the tessellation has translational symmetry. If so, identify two translations that map the tessellation onto itself.

GO DIGITAL

Using Theorems about Congruence Transformations

Compositions of two reflections result in either a translation or a rotation. A composition of two reflections in parallel lines results in a translation, as described in the following theorem.

THEOREM

4.2 Reflections in Parallel Lines Theorem

If lines k and m are parallel, then a reflection in line k followed by a reflection in line m is the same as a translation.

If A'' is the image of A, then

1. $\overline{AA''}$ is perpendicular to k and m, and
2. $AA'' = 2d$, where d is the distance between k and m.

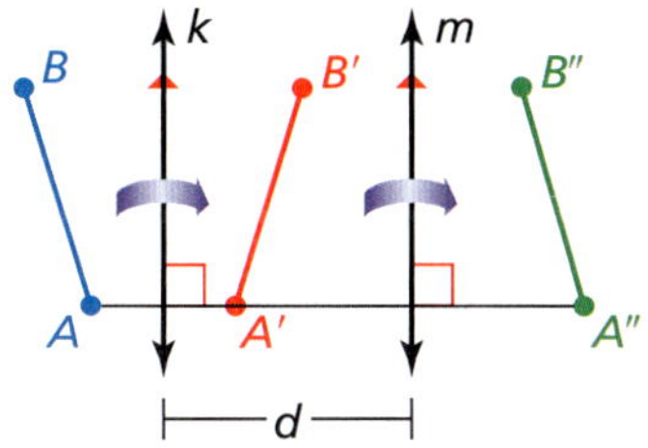

Proof Exercise 40, page 201

EXAMPLE 4 Using the Reflections in Parallel Lines Theorem

In the diagram, a reflection in line k maps $\overline{GH}$ to $\overline{G'H'}$. A reflection in line m maps $\overline{G'H'}$ to $\overline{G''H''}$. Also, $HB = 9$ and $H''D = 4$.

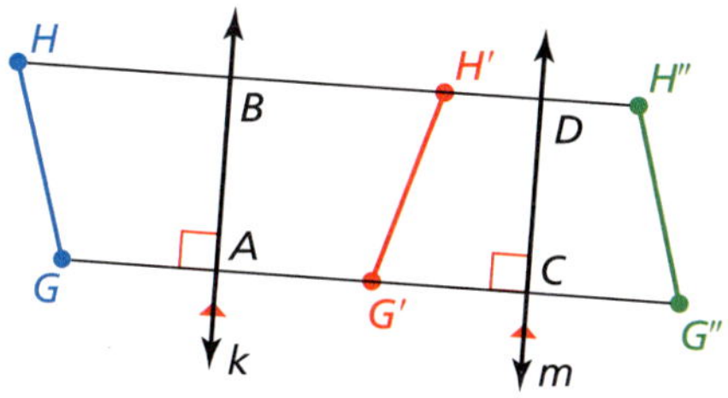

a. Name any segments congruent to each segment: $\overline{GH}$, $\overline{HB}$, and $\overline{GA}$.

b. Does $AC = BD$? Explain.

c. What is the length of $\overline{GG''}$?

SOLUTION

a. $\overline{GH} \cong \overline{G'H'}$, and $\overline{GH} \cong \overline{G''H''}$. $\overline{HB} \cong \overline{H'B}$. $\overline{GA} \cong \overline{G'A}$.

b. Yes, $\overline{GG''}$ and $\overline{HH''}$ are perpendicular to both k and m. So, $ABDC$ is a rectangle because it has four right angles. $\overline{AC}$ and $\overline{BD}$ are opposite sides of rectangle $ABDC$, so $AC = BD$.

c. By the properties of reflections, $H'B = 9$ and $H'D = 4$. The Reflections in Parallel Lines Theorem implies that $GG'' = HH'' = 2 \cdot BD$, so the length of $\overline{GG''}$ is $2(9 + 4) = 26$ units.

SELF-ASSESSMENT

1 I don't understand yet. 2 I can do it with help. 3 I can do it on my own. 4 I can teach someone else.

Use the figure. The distance between line k and line m is 1.6 centimeters.

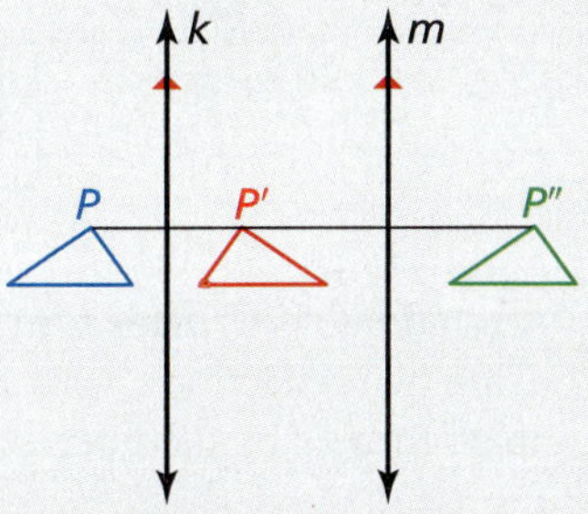

6. The preimage is reflected in line k, then in line m. Describe a single transformation that maps the blue figure to the green figure.
7. What is the relationship between $\overline{PP'}$ and line k? Explain.
8. What is the distance between P and P''?

GO DIGITAL

A composition of two reflections in intersecting lines results in a rotation, as described in the following theorem.

THEOREM

4.3 Reflections in Intersecting Lines Theorem

If lines k and m intersect at point P, then a reflection in line k followed by a reflection in line m is the same as a rotation about point P.

The angle of rotation is $2x°$, where $x°$ is the measure of the acute or right angle formed by lines k and m.

Proof Exercise 22, page 248

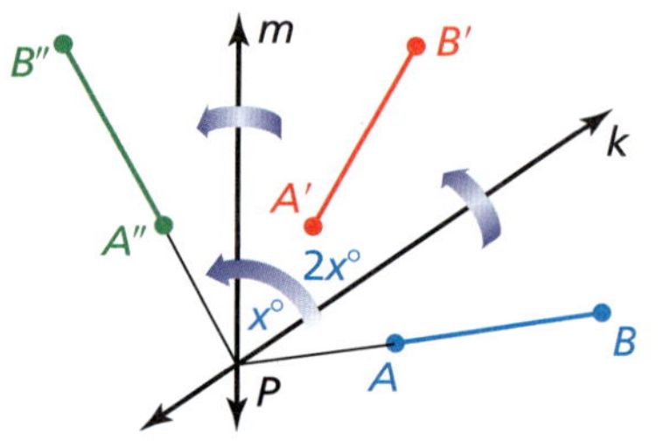

$m\angle BPB'' = 2x°$

EXAMPLE 5 Using the Reflections in Intersecting Lines Theorem

WATCH

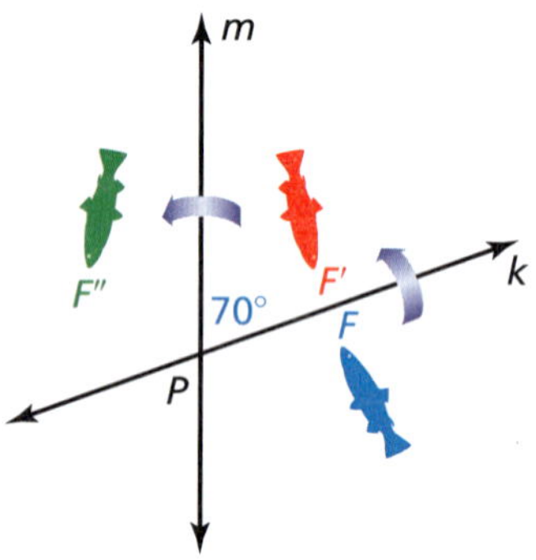

In the diagram, the figure is reflected in line k. The image is then reflected in line m. Describe a single transformation that maps F to F''.

SOLUTION

By the Reflections in Intersecting Lines Theorem, a reflection in line k followed by a reflection in line m is the same as a rotation about point P. The measure of the acute angle formed by lines k and m is 70°. So, by the Reflections in Intersecting Lines Theorem, the angle of rotation is $2(70°) = 140°$.

▶ A single transformation that maps F to F'' is a 140° rotation about point P. You can check that this is correct by tracing lines k and m and point F, then rotating the point 140°.

SELF-ASSESSMENT

1 I don't understand yet. 2 I can do it with help. 3 I can do it on my own. 4 I can teach someone else.

9. In the diagram, the preimage is reflected in line k, then in line m. Describe a single transformation that maps the blue figure to the green figure.

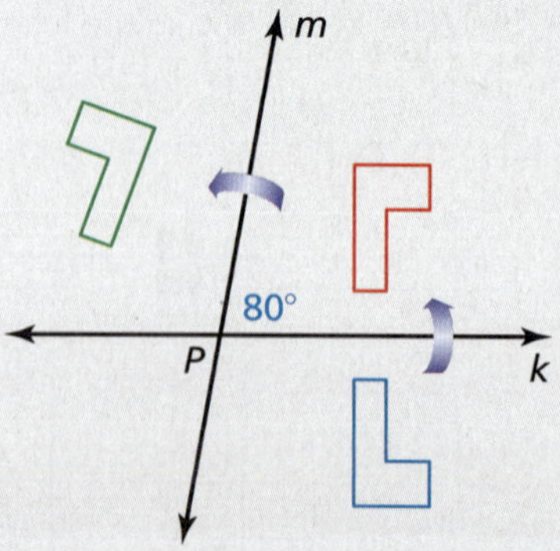

10. A rotation of 76° maps C to C'. To map C to C' using two reflections, what is the measure of the angle formed by the intersecting lines of reflection?

GO DIGITAL

4.4 Practice WITH CalcChat® AND CalcView®

In Exercises 1 and 2, identify any congruent figures in the coordinate plane. Explain. *(See Example 1.)*

1. 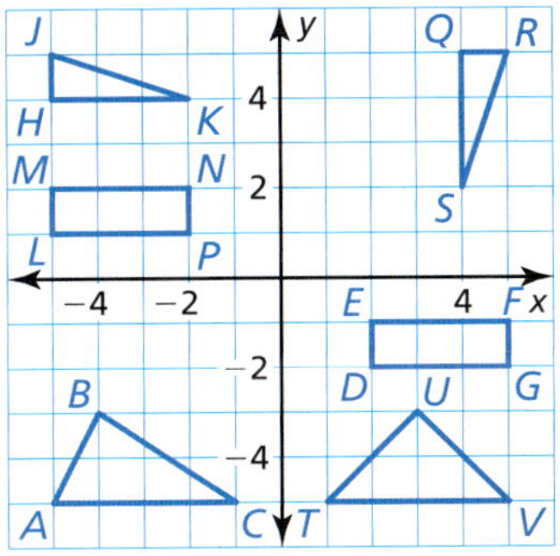

2. 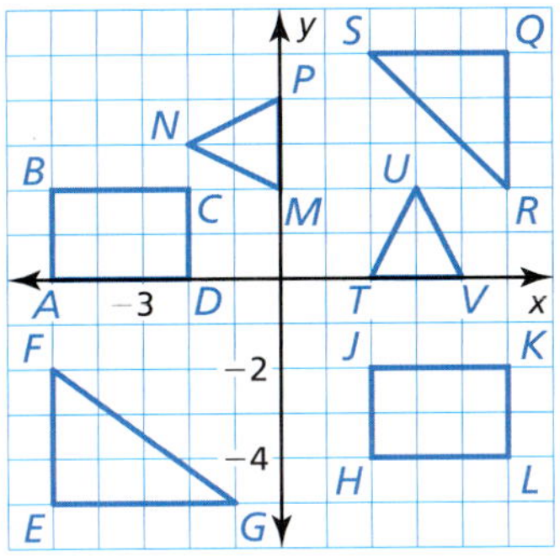

In Exercises 3 and 4, describe a congruence transformation that maps the blue preimage to the green image. Verify that corresponding sides and corresponding angles are congruent. *(See Example 2.)*

3. 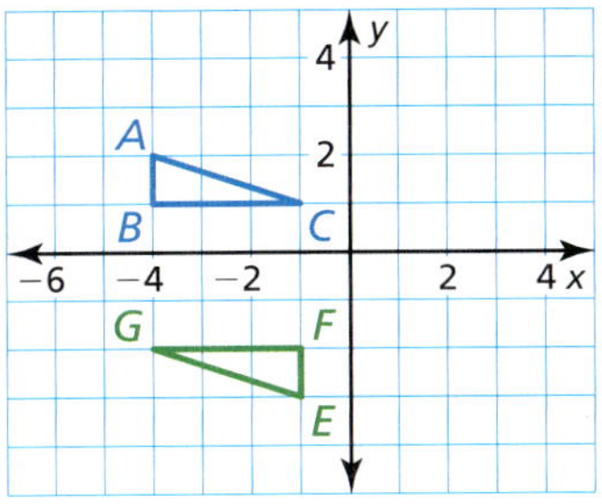

4. 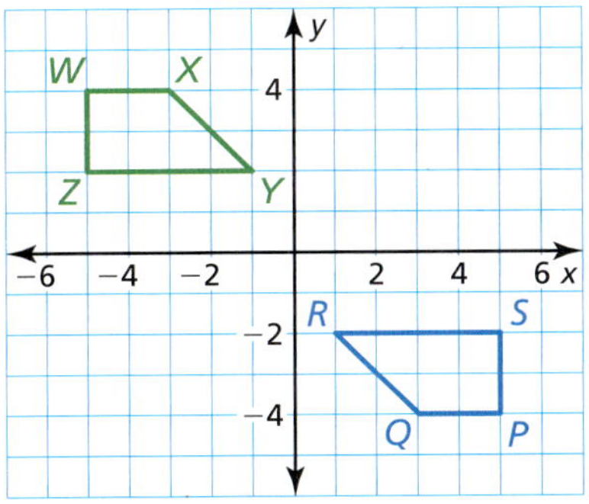

In Exercises 5–8, determine whether the polygons with the given vertices are congruent. Use transformations to explain your reasoning.

5. $Q(2, 4)$, $R(5, 4)$, $S(4, 1)$ and $T(6, 4)$, $U(9, 4)$, $V(8, 1)$

6. $W(-3, 1)$, $X(2, 1)$, $Y(4, -4)$, $Z(-5, -4)$ and $C(-1, -3)$, $D(-1, 2)$, $E(4, 4)$, $F(4, -5)$

7. $J(1, 1)$, $K(3, 2)$, $L(4, 1)$ and $M(6, 1)$, $N(5, 2)$, $P(2, 1)$

8. $A(0, 0)$, $B(1, 2)$, $C(4, 2)$, $D(3, 0)$ and $E(0, -5)$, $F(-1, -3)$, $G(-4, -3)$, $H(-3, -5)$

In Exercises 9–12, determine whether the tessellation has translational symmetry. If so, identify two translations that map the tessellation onto itself. *(See Example 3.)*

9. 

10. 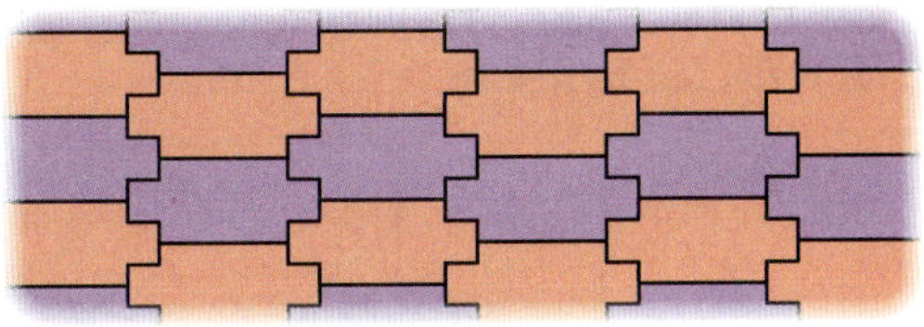

11. 

12. 

GO DIGITAL

13. MODELING REAL LIFE The Alhambra is a palace in Granada, Spain. It is known for its geometric patterns, especially symmetric patterns. The image depicts a portion of decorative tile wall in the Alhambra. Determine whether the tessellation has translational symmetry. If so, identify two translations that map the tessellation onto itself.

14. MODELING REAL LIFE You use a tessellation to model a honeycomb. Identify two different translations that map the tessellation onto itself.

In Exercises 15–18, $\triangle ABC$ is reflected in line k, and $\triangle A'B'C'$ is reflected in line m. *(See Example 4.)*

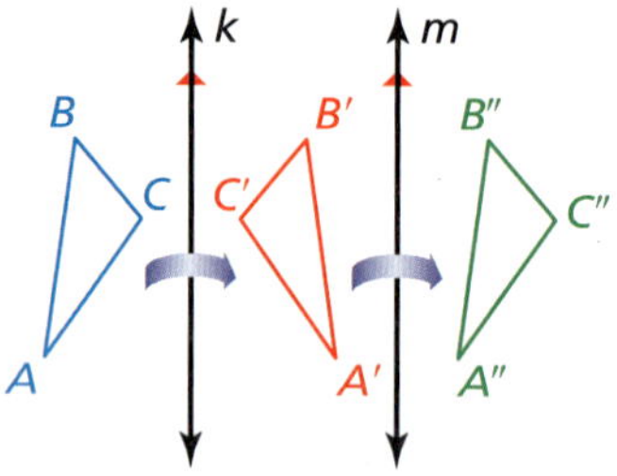

15. A translation maps $\triangle ABC$ to which triangle?

16. Which lines are perpendicular to $\overline{AA''}$?

17. If the distance between k and m is 2.6 inches, what is the length of $\overline{CC''}$?

18. Is the distance from B' to m the same as the distance from B'' to m? Explain.

In Exercises 19 and 20, describe a single transformation that maps A to A''. *(See Example 5.)*

19.

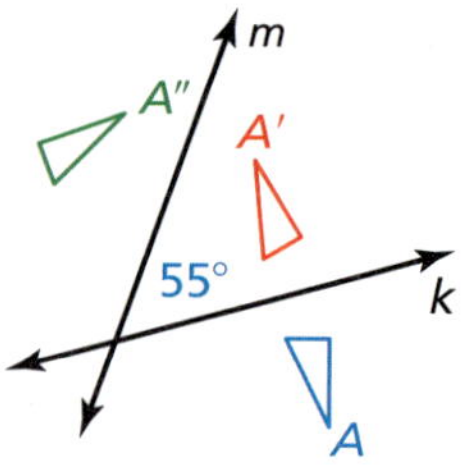

20.

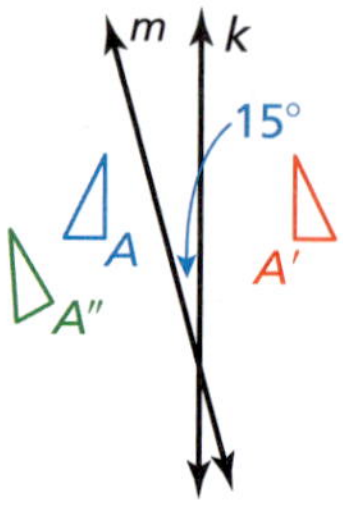

21. ERROR ANALYSIS Describe and correct the error in using the Reflections in Intersecting Lines Theorem.

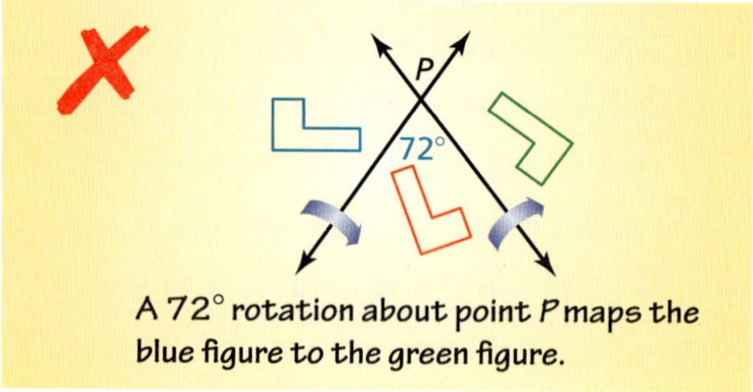

22. ERROR ANALYSIS Identify and correct the error in describing the congruence transformation.

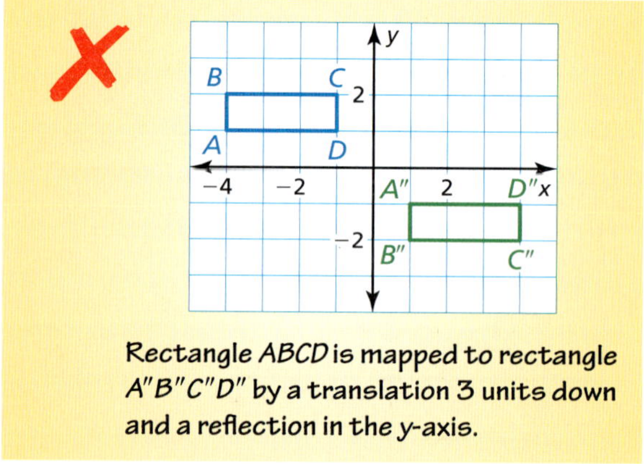

In Exercises 23–26, find the measure of the acute or right angle formed by intersecting lines so that C can be mapped to C' using two reflections.

23. A rotation of 84° maps C to C'.

24. A rotation of 24° maps C to C'.

25. The rotation $(x, y) \rightarrow (-x, -y)$ maps C to C'.

26. The rotation $(x, y) \rightarrow (y, -x)$ maps C to C'.

GO DIGITAL

27. **REASONING** Use the Reflections in Parallel Lines Theorem to explain how you can make a glide reflection using three reflections. How are the lines of reflection related?

5 MTR 28. **STRUCTURE** $\triangle ABC$ with vertices $A(1, 4)$, $B(4, 5)$, and $C(5, 1)$ is translated 6 units left and 2 units up, then rotated 180° about the origin. Determine which quadrant the image is located in without performing the transformations. Explain.

4 MTR **DISCUSS MATHEMATICAL THINKING** **In Exercises 29–32, tell whether the statement is *always*, *sometimes*, or *never* true. Explain your reasoning.**

29. A congruence transformation changes the size of a figure.

30. If two figures are congruent, then there is a rigid motion or a composition of rigid motions that maps one figure to the other.

31. The composition of two reflections results in a rotation.

32. The composition of two reflections results in a translation.

4 MTR 33. **MAKING AN ARGUMENT** $\overline{PQ}$ with endpoints $P(1, 3)$ and $Q(3, 2)$ is reflected in the y-axis. The image $\overline{P'Q'}$ is then reflected in the x-axis to produce the image $\overline{P''Q''}$. Your friend says that $\overline{PQ}$ is mapped to $\overline{P''Q''}$ by a 180° rotation about the origin. Is your friend correct? Explain your reasoning.

34. **HOW DO YOU SEE IT?**
What type of congruence transformation verifies each statement about the stained-glass window?

a. Triangle 5 is congruent to Triangle 8.
b. Triangle 1 is congruent to Triangle 4.
c. Triangle 2 is congruent to Triangle 7.
d. Pentagon 3 is congruent to Pentagon 6.

35. **REASONING** Can any polygon be used to construct a tessellation? Can any regular polygon be used to construct a tessellation? Explain.

GO DIGITAL

5 MTR 36. **STRUCTURE** Use the figure below to construct a tessellation in the coordinate plane.

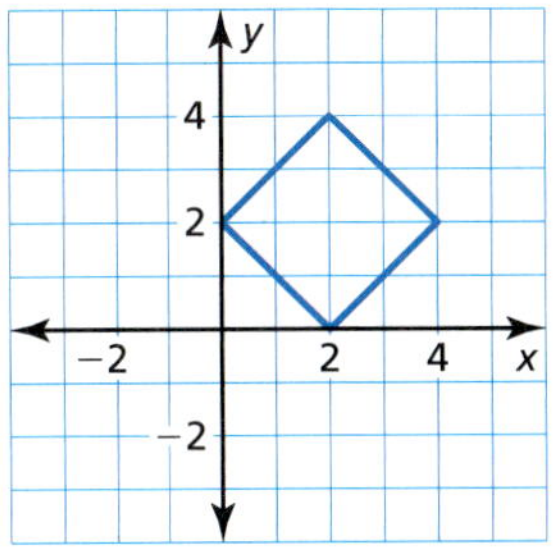

CONSTRUCTION **In Exercises 37 and 38, use a compass and straightedge to construct two lines of reflection that produce a composition of reflections resulting in the given transformation.**

37. **Translation:** $\triangle ABC \rightarrow \triangle A''B''C''$

38. **Rotation about** ***P*****:** $\triangle XYZ \rightarrow \triangle X''Y''Z''$

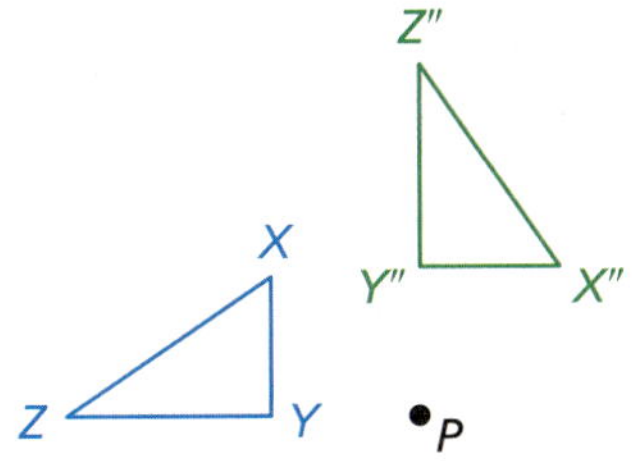

39. **REASONING** Use the figures below to construct a tessellation.

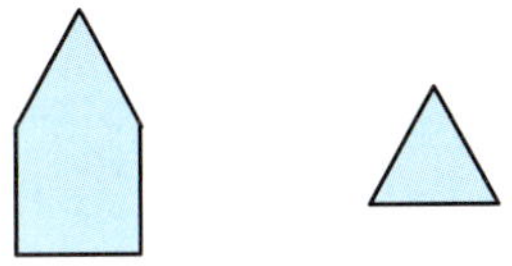

40. **PROVING A THEOREM** Prove the Reflections in Parallel Lines Theorem.

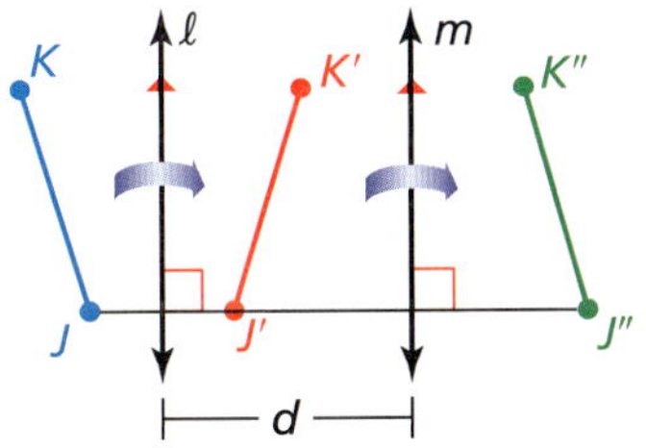

Given A reflection in line ℓ maps $\overline{JK}$ to $\overline{J'K'}$, a reflection in line m maps $\overline{J'K'}$ to $\overline{J''K''}$, and $\ell \parallel m$.

Prove **a.** $\overline{JJ''}$ is perpendicular to ℓ and m.

b. $JJ'' = 2d$, where d is the distance between ℓ and m.

41. **DISCUSS MATHEMATICAL THINKING** You reflect a figure in each of two parallel lines. Is the order of the reflections important? Justify your answer.

42. **OPEN-ENDED** Describe three examples of the use of tessellations in architecture.

43. **DIG DEEPER** Are any two rays congruent? If so, describe a congruence transformation that maps a ray to any other ray. If not, explain why not.

44. **THOUGHT PROVOKING**
Describe a sequence of two or more different transformations where the order of the transformations does not affect the final image.

REVIEW & REFRESH

In Exercises 45–47, solve the equation.

45. $12 + 6m = 2m$

46. $-2(8 - y) = -6y$

47. $7(2n + 1) = \frac{1}{3}(6n - 15)$

48. **MODELING REAL LIFE** Last month, a charity received \$500 in donations. This month, the charity received \$625 in donations. What is the percent of change?

49. Describe a congruence transformation that maps $\triangle JKL$ to $\triangle XYZ$.

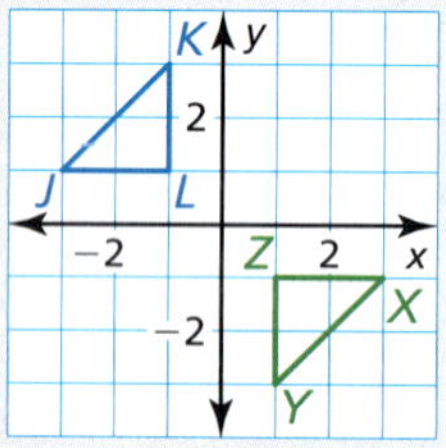

In Exercises 50–52, graph quadrilateral $QRST$ with vertices $Q(2, -1)$, $R(5, -2)$, $S(5, -4)$, and $T(2, -4)$ and its image after the composition.

50. **Translation:** $(x, y) \rightarrow (x - 5, y + 3)$
Reflection: in the line $y = -3$

51. **Reflection:** in the x-axis
Rotation: $90°$ counterclockwise about the origin

52. **Translation:** $(x, y) \rightarrow (x + 2, y - 1)$
Translation: $(x, y) \rightarrow (x + 3, y + 4)$

In Exercises 53 and 54, graph the linear equation. Identify the x-intercept.

53. $y = -\frac{3}{4}x + 2$ **54.** $3x + y = -5$

55. Write an inequality that represents the graph.

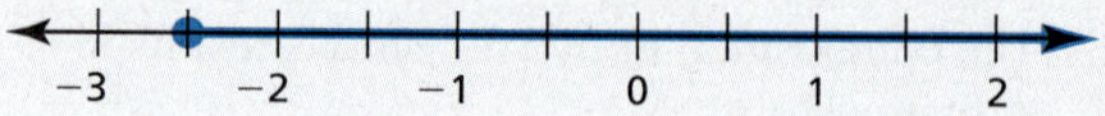

56. Let p be "you ride a roller coaster" and let q be "you go to an amusement park." Write the conditional statement $p \rightarrow q$, the converse $q \rightarrow p$, the inverse $\sim p \rightarrow \sim q$, and the contrapositive $\sim q \rightarrow \sim p$ in words. Then decide whether each statement is *true* or *false*.

In Exercises 57 and 58, find the value of x. Show your steps.

57.

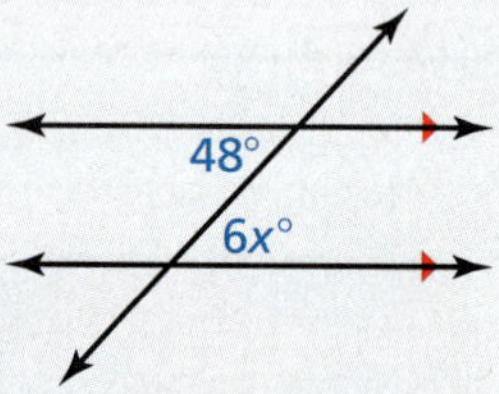

58.

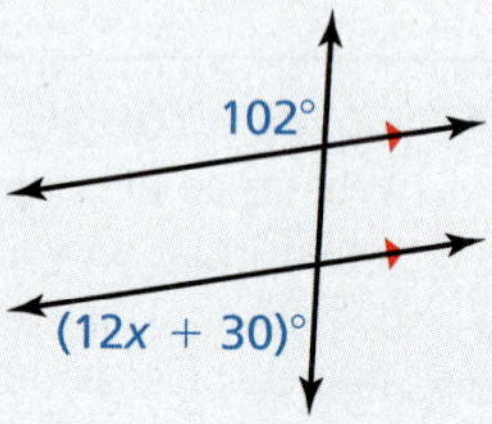

In Exercises 59 and 60, find the area of the polygon with the given vertices.

59.

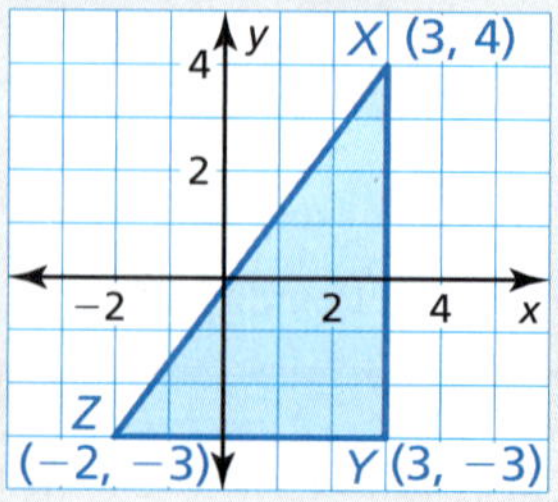

60.

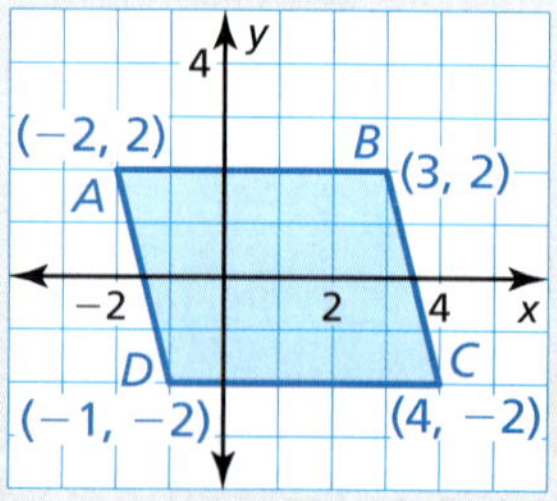

GO DIGITAL

4.5 Dilations

Learning Target: Understand dilations of figures.

Success Criteria:
- I can identify dilations.
- I can dilate figures.
- I can solve real-life problems involving scale factors and dilations.

EXPLORE IT! Dilating Figures

Work with a partner.

a. The diagram shows a *dilation* of $\triangle DEF$ to $\triangle D'E'F'$. How would you define a dilation?

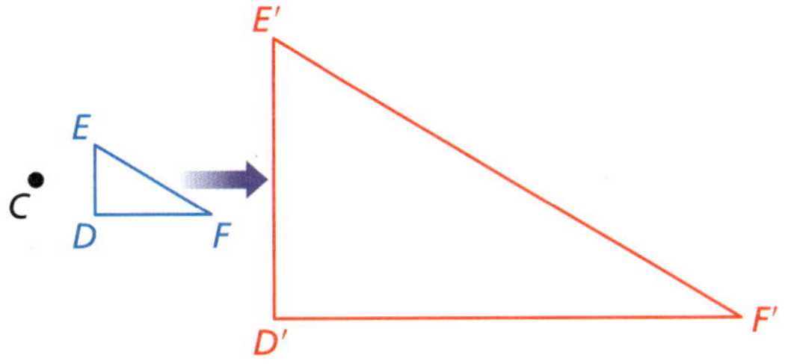

b. Use technology to draw any line segment, $\overline{PQ}$, and a point C not on the line segment.

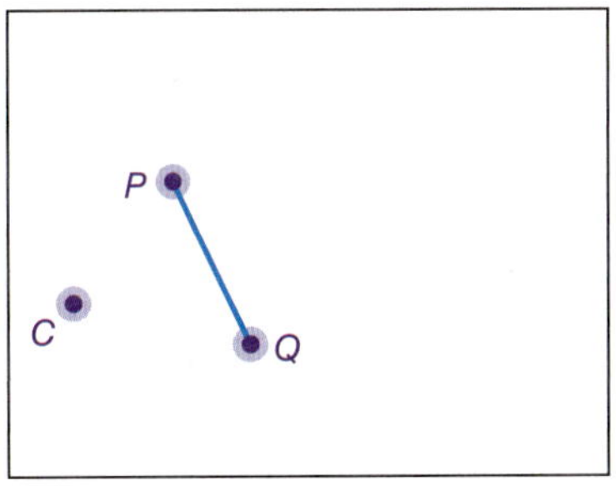

i. *Dilate* $\overline{PQ}$ using a *scale factor* of 2 and the *center of dilation* C to form $\overline{P'Q'}$. What do you notice? Make several observations.

ii. Choose two other scale factors, where one scale factor is greater than 1 and the other scale factor is between 0 and 1. What conclusions can you make?

iii. What scale factor results in $\overline{PQ}$ and $\overline{P'Q'}$ being congruent?

c. Use technology to draw any $\triangle PQR$ and a point C not in $\triangle PQR$.

i. Dilate $\triangle PQR$ using the center of dilation C and several different scale factors to form $\triangle P'Q'R'$. Compare $\triangle PQR$ and $\triangle P'Q'R'$.

ii. Make a conjecture about the side lengths and angle measures of the image of $\triangle PQR$ after a dilation with a scale factor of k.

d. Based on your results in parts (b) and (c), is there anything you would like to change or include in your definition in part (a)? Explain.

e. Is a dilation a rigid motion? Explain your reasoning.

4 MTR CONSTRUCT AN ARGUMENT
What happens to a figure when a scale factor is less than 0?

Geometric Reasoning
MA.912.GR.2.1 Given a preimage and image, describe the transformation and represent the transformation algebraically using coordinates.
MA.912.GR.2.5 Given a geometric figure and a sequence of transformations, draw the transformed figure on a coordinate plane.
Also MA.912.GR.2.2

GO DIGITAL

Identifying and Performing Dilations

Vocabulary

dilation, *p. 204*
center of dilation, *p. 204*
scale factor, *p. 204*
enlargement, *p. 204*
reduction, *p. 204*

KEY IDEA

Dilations

A **dilation** is a transformation in which a figure is enlarged or reduced with respect to a fixed point C called the **center of dilation** and a **scale factor** k, which is the ratio of the lengths of the corresponding sides of the image and the preimage.

A dilation with center of dilation C and scale factor k maps every point P in a figure to a point P' so that the following are true.

- If P is the center of dilation C, then $P = P'$.
- If P is not the center of dilation C, then the image point P' lies on $\overrightarrow{CP}$. The scale factor k is a positive number such that $k = \dfrac{CP'}{CP}$.
- Angle measures are preserved.

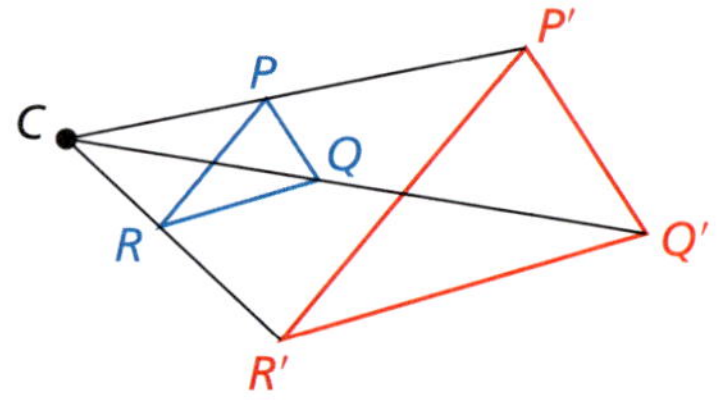

A dilation does not change any line that passes through the center of dilation. A dilation maps a line that does not pass through the center of dilation to a parallel line. In the figure above, $\overleftrightarrow{PR} \parallel \overleftrightarrow{P'R'}$, $\overleftrightarrow{PQ} \parallel \overleftrightarrow{P'Q'}$, and $\overleftrightarrow{QR} \parallel \overleftrightarrow{Q'R'}$.

When the scale factor $k > 1$, a dilation is an **enlargement**. When $0 < k < 1$, a dilation is a **reduction**.

READING

The scale factor of a dilation can be written as a fraction, decimal, or percent.

EXAMPLE 1 Identifying Dilations

WATCH

Find the scale factor of the dilation. Then tell whether the dilation is a *reduction* or an *enlargement*.

a.

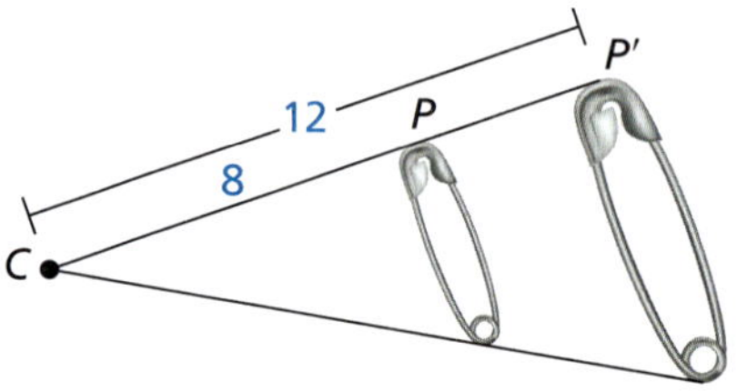

b.

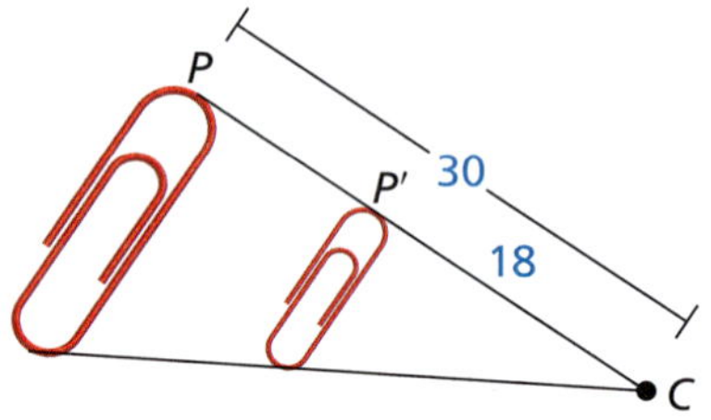

SOLUTION

a. Because $\dfrac{CP'}{CP} = \dfrac{12}{8}$, the scale factor is $k = \dfrac{3}{2}$. So, the dilation is an enlargement.

b. Because $\dfrac{CP'}{CP} = \dfrac{18}{30}$, the scale factor is $k = \dfrac{3}{5}$. So, the dilation is a reduction.

SELF-ASSESSMENT 1 I don't understand yet. 2 I can do it with help. 3 I can do it on my own. 4 I can teach someone else.

1. In a dilation, $CP' = 3$ and $CP = 12$. Find the scale factor. Then tell whether the dilation is a *reduction* or an *enlargement*.

4 MTR 2. **WHICH ONE DOESN'T BELONG?** Which scale factor does *not* belong with the other three? Explain your reasoning.

$\frac{5}{4}$ | 60% | 115% | 2

GO DIGITAL

KEY IDEA

Coordinate Rule for Dilations

If $P(x, y)$ is the preimage of a point, then its image after a dilation centered at the origin (0, 0) with scale factor k is the point $P'(kx, ky)$.

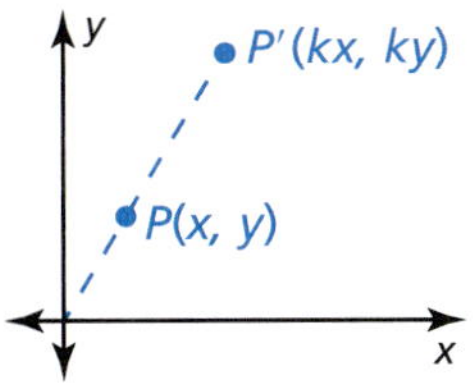

$(x, y) \rightarrow (kx, ky)$

STUDY TIP

In this chapter, for all dilations in the coordinate plane, the center of dilation is the origin unless otherwise noted.

EXAMPLE 2 Dilating a Figure in the Coordinate Plane

WATCH

Graph $\triangle ABC$ with vertices $A(2, 1)$, $B(4, 1)$, and $C(4, -1)$ and its image after a dilation with a scale factor of 2.

SOLUTION

Use the coordinate rule for a dilation with $k = 2$ to find the coordinates of the vertices of the image. Then graph $\triangle ABC$ and its image.

$$(x, y) \rightarrow (2x, 2y)$$

$A(2, 1) \rightarrow A'(4, 2)$

$B(4, 1) \rightarrow B'(8, 2)$

$C(4, -1) \rightarrow C'(8, -2)$

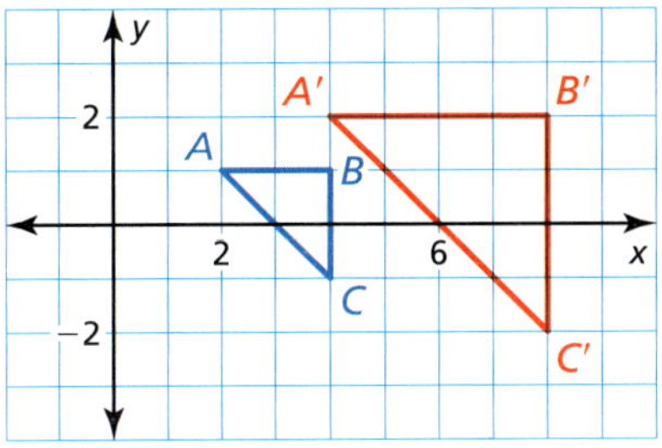

Notice the relationships between the lengths and slopes of the sides of the triangles in Example 2. Each side length of $\triangle A'B'C'$ is longer than its corresponding side by the scale factor. The corresponding sides are parallel because their slopes are the same.

EXAMPLE 3 Dilating a Figure in the Coordinate Plane

WATCH

Graph quadrilateral $KLMN$ with vertices $K(-3, 6)$, $L(0, 6)$, $M(3, 3)$, and $N(-3, -3)$ and its image after a dilation with a scale factor of $\frac{1}{3}$.

SOLUTION

Use the coordinate rule for a dilation with $k = \frac{1}{3}$ to find the coordinates of the vertices of the image. Then graph quadrilateral $KLMN$ and its image.

$$(x, y) \rightarrow \left(\tfrac{1}{3}x, \tfrac{1}{3}y\right)$$

$K(-3, 6) \rightarrow K'(-1, 2)$

$L(0, 6) \rightarrow L'(0, 2)$

$M(3, 3) \rightarrow M'(1, 1)$

$N(-3, -3) \rightarrow N'(-1, -1)$

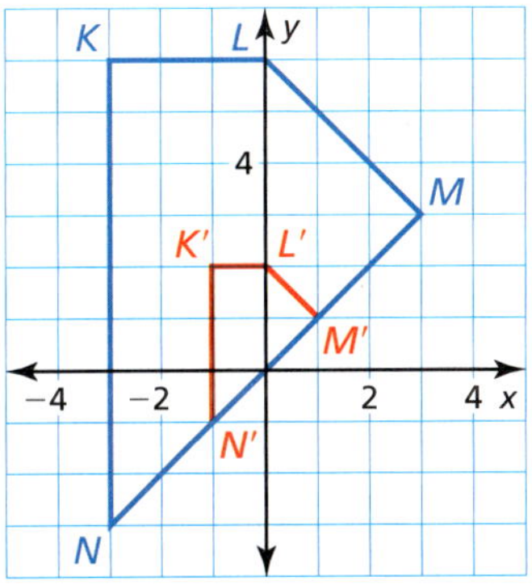

SELF-ASSESSMENT 1 I don't understand yet. 2 I can do it with help. 3 I can do it on my own. 4 I can teach someone else.

Graph $\triangle PQR$ and its image after a dilation with scale factor k.

3. $P(-2, -1)$, $Q(-1, 0)$, $R(0, -1)$; $k = 4$

4. $P(5, -5)$, $Q(10, -5)$, $R(10, 5)$; $k = 0.4$

GO DIGITAL

CONSTRUCTION Constructing a Dilation

Use a compass and straightedge to construct a dilation of $\triangle PQR$ with a scale factor of 2. Use a point C outside the triangle as the center of dilation.

SOLUTION

Step 1

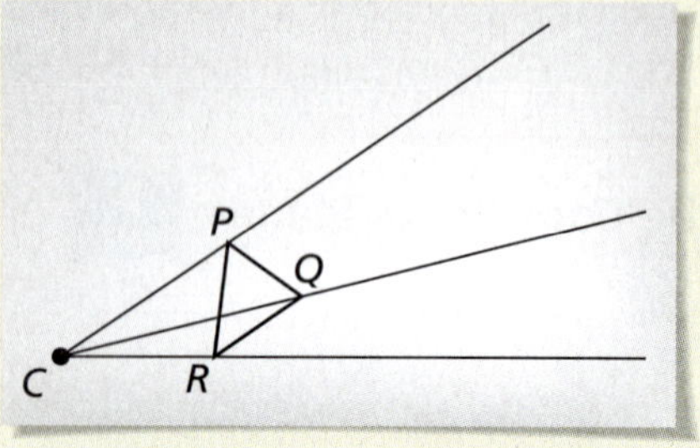

Draw a triangle Draw $\triangle PQR$ and choose the center of the dilation C outside the triangle. Draw rays from C through the vertices of the triangle.

Step 2

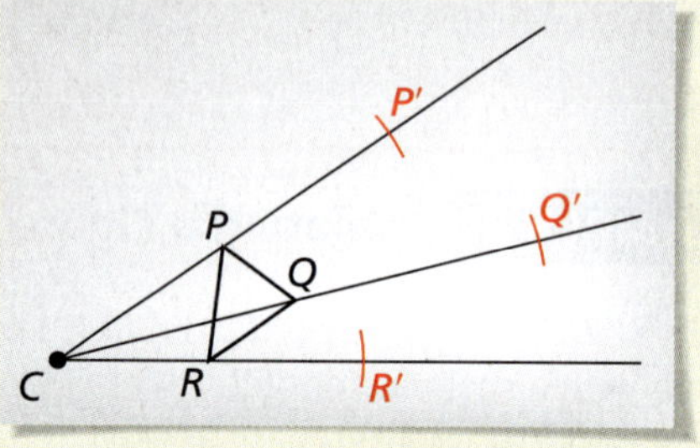

Use a compass Use a compass to locate P' on $\overrightarrow{CP}$ so that $CP' = 2(CP)$. Locate Q' and R' using the same method.

Step 3

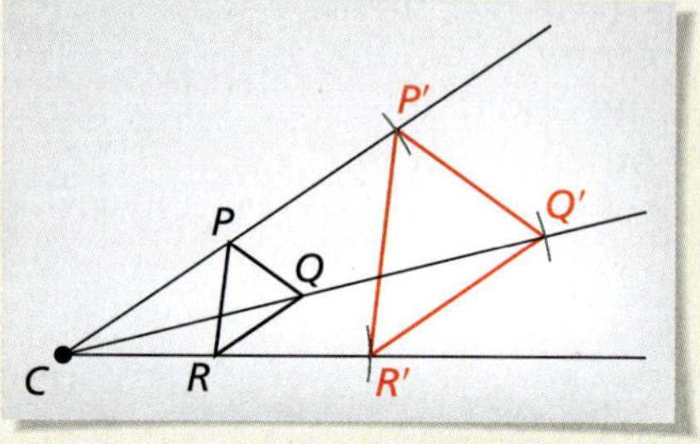

Connect points Connect points P', Q', and R' to form $\triangle P'Q'R'$.

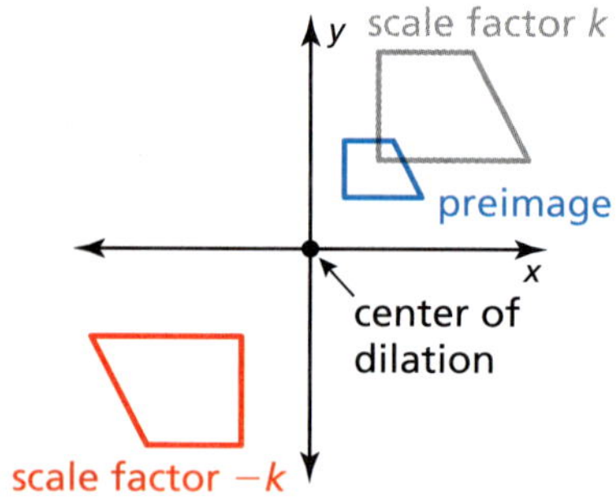

Scale factors can be negative numbers. When this occurs, the figure rotates 180°. So, when $k > 0$, a dilation with a scale factor of $-k$ is the same as the composition of a dilation with a scale factor of k followed by a rotation of 180° about the center of dilation. Using the coordinate rules for a dilation and a rotation of 180°, you can think of the notation as

$$(x, y) \rightarrow (kx, ky) \rightarrow (-kx, -ky).$$

EXAMPLE 4 Using a Negative Scale Factor

Graph $\triangle FGH$ with vertices $F(-4, -2)$, $G(-2, 4)$, and $H(-2, -2)$ and its image after a dilation with a scale factor of $-\frac{1}{2}$.

SOLUTION

Use the coordinate rule for a dilation with $k = -\frac{1}{2}$ to find the coordinates of the vertices of the image. Then graph $\triangle FGH$ and its image.

$$(x, y) \rightarrow \left(-\tfrac{1}{2}x, -\tfrac{1}{2}y\right)$$

$F(-4, -2) \rightarrow F'(2, 1)$

$G(-2, 4) \rightarrow G'(1, -2)$

$H(-2, -2) \rightarrow H'(1, 1)$

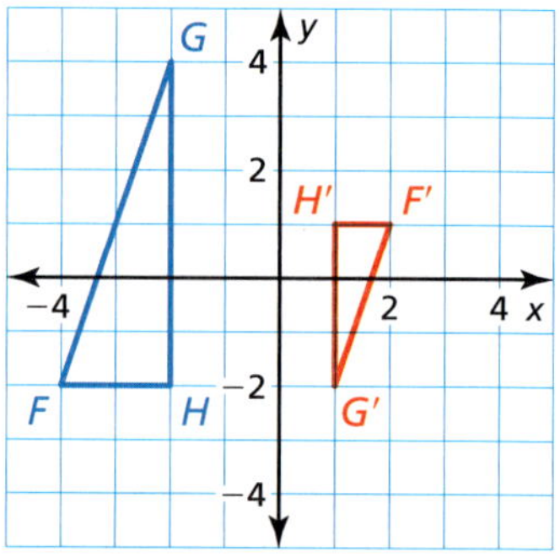

SELF-ASSESSMENT

1 I don't understand yet. 2 I can do it with help. 3 I can do it on my own. 4 I can teach someone else.

5. Graph $\triangle PQR$ with vertices $P(1, 2)$, $Q(3, 1)$, and $R(1, -3)$ and its image after a dilation with a scale factor of -2.

6. **REASONING** A polygon with a vertex at the origin is dilated. Explain why the corresponding vertex of the image is also at the origin.

GO DIGITAL

Solving Real-Life Problems

EXAMPLE 5 Finding a Scale Factor

WATCH

You are making your own photo stickers. Your photo is 4 inches by 4 inches. The image on the stickers is 1.1 inches by 1.1 inches. What is the scale factor of this dilation?

READING

A scale factor is written so that the units in the numerator and denominator divide out.

SOLUTION

The scale factor is the ratio of a side length of the sticker image to a side length of the original photo, or $\frac{1.1 \text{ in.}}{4 \text{ in.}}$.

▶ So, in simplest form, the scale factor is $\frac{11}{40}$.

EXAMPLE 6 Finding the Length of an Image

WATCH

You are using a magnifying glass that shows the image of an object as six times the object's actual size. Determine the length of the image of the spider seen through the magnifying glass.

SOLUTION

cm → image length / cm → actual length

$\frac{\text{image length}}{\text{actual length}} = k$ Write ratio of corresponding lengths.

$\frac{x}{1.5} = 6$ Substitute values.

$x = 9$ Multiply each side by 1.5.

▶ So, the image length seen through the magnifying glass is 9 centimeters.

SELF-ASSESSMENT 1 I don't understand yet. 2 I can do it with help. 3 I can do it on my own. 4 I can teach someone else.

7. An optometrist dilates the pupils of a patient's eyes to get a better look at the back of the eyes. A pupil dilates from 4.5 millimeters to 8 millimeters. What is the scale factor of this dilation?

8. The image of another spider seen through the magnifying glass in Example 6 is shown at the right. Find the actual length of the spider.

When a transformation, such as a dilation, changes the shape or size of a figure, the transformation is *nonrigid*. In addition to dilations, there are many possible nonrigid transformations. Two examples are shown. It is important to pay close attention to whether a nonrigid transformation preserves lengths and angle measures.

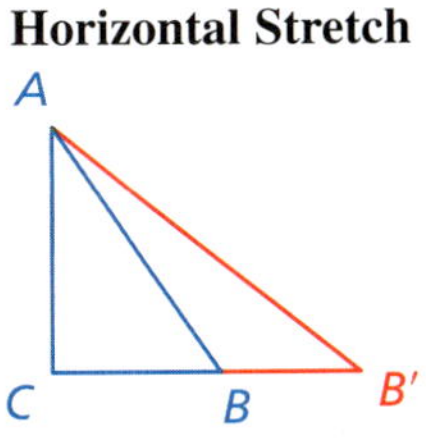

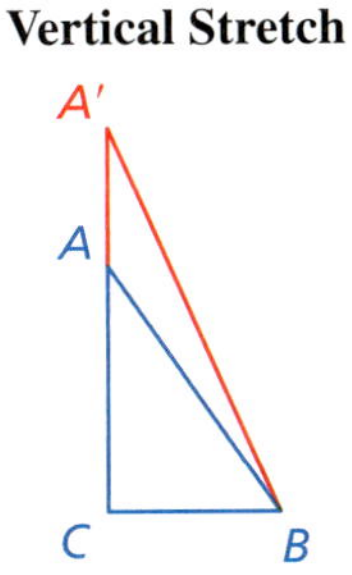

GO DIGITAL

4.5 Practice WITH CalcChat® AND CalcView®

In Exercises 1–4, find the scale factor of the dilation. Then tell whether the dilation is a *reduction* or an *enlargement*. (See Example 1.)

1\.
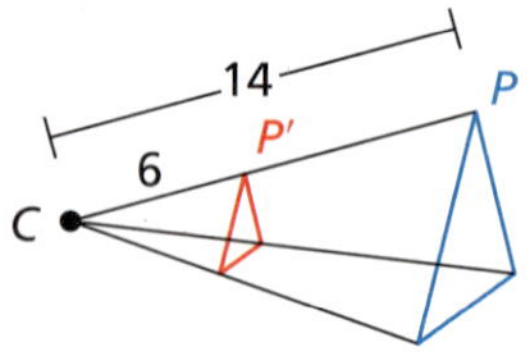

2\.
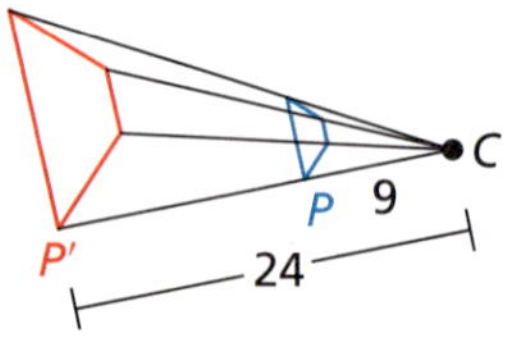

3\.
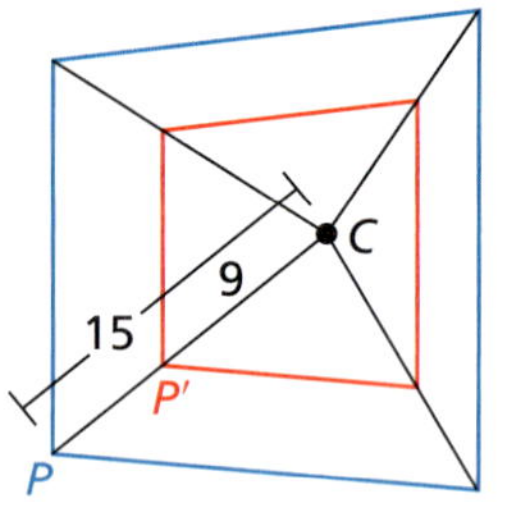

4\.
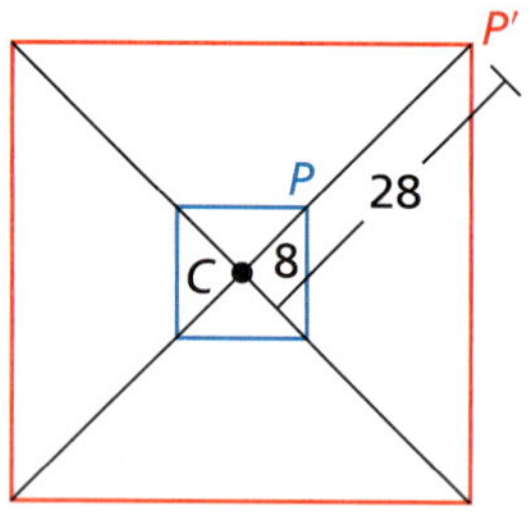

CONSTRUCTION In Exercises 5–8, copy the diagram. Then use a compass and straightedge to construct a dilation of $\triangle LMN$ with the given center and scale factor k.

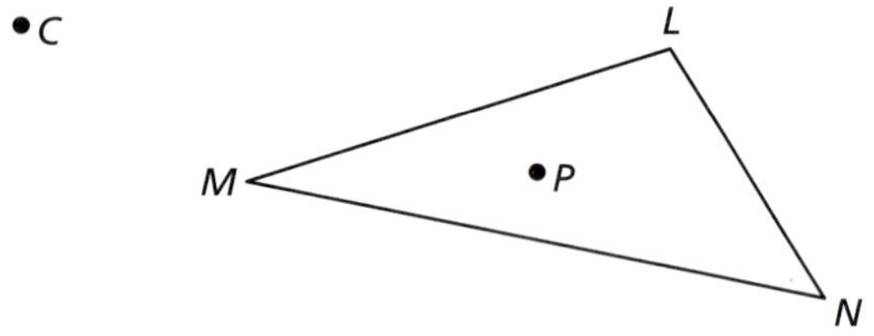

5\. Center C, $k = 2$

6\. Center P, $k = 3$

7\. Center M, $k = \frac{1}{2}$

8\. Center C, $k = 25\%$

CONSTRUCTION In Exercises 9–12, copy the diagram. Then use a compass and straightedge to construct a dilation of quadrilateral $RSTU$ with the given center and scale factor k.

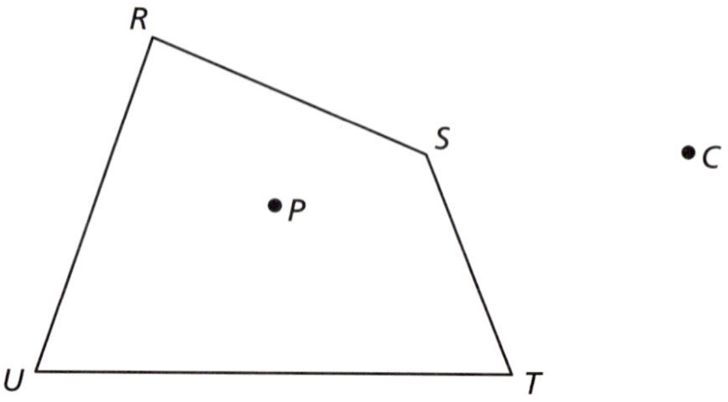

9\. Center P, $k = 2$

10\. Center C, $k = 3$

11\. Center C, $k = 75\%$

12\. Center R, $k = 0.25$

In Exercises 13–18, graph the polygon with the given vertices and its image after a dilation with scale factor k. (See Examples 2 and 3.)

13\. $X(6, -1)$, $Y(-2, -4)$, $Z(1, 2)$; $k = 3$

14\. $Q(-3, -2)$, $R(1, 2)$, $S(4, 1)$; $k = 4$

15\. $A(0, 5)$, $B(-10, -5)$, $C(5, -5)$; $k = 120\%$

16\. $D(-6, 4)$, $E(-4, -6)$, $F(4, 8)$; $k = 50\%$

17\. $J(4, 0)$, $K(-8, 4)$, $L(0, -4)$, $M(12, -8)$; $k = 0.25$

18\. $T(9, -3)$, $U(6, 0)$, $V(3, 9)$, $W(0, 0)$; $k = \frac{2}{3}$

In Exercises 19 and 20, write a coordinate rule for the dilation.

19\.
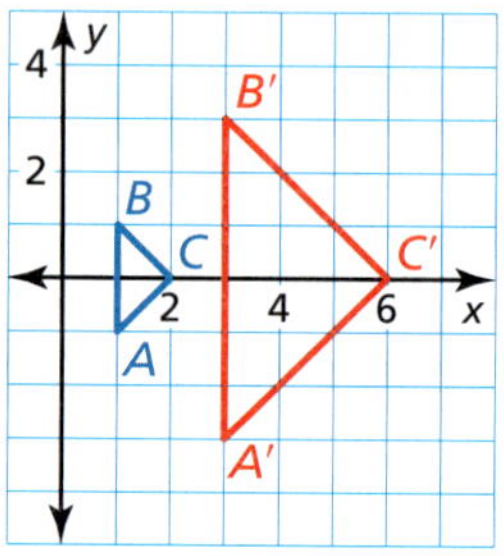

20\.
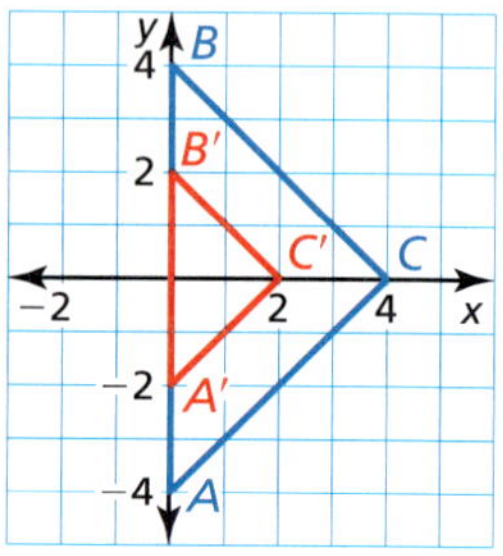

In Exercises 21–24, graph the polygon with the given vertices and its image after a dilation with scale factor k. (See Example 4.)

21\. $B(-5, -10)$, $C(-10, 15)$, $D(0, 5)$; $k = -\frac{1}{5}$

22\. $L(0, 0)$, $M(-4, 1)$, $N(-3, -6)$; $k = -3$

23\. $R(-7, -1)$, $S(2, 5)$, $T(-2, -3)$, $U(-3, -3)$; $k = -4$

24\. $W(8, -2)$, $X(6, 0)$, $Y(-6, 4)$, $Z(-2, 2)$; $k = -0.5$

25\. **FINDING A SCALE FACTOR** You receive wallet-sized photos of your school picture. The photo is 2.5 inches by 3.5 inches. You ask the photographer to dilate the photo to 5 inches by 7 inches. What is the scale factor of this dilation? (See Example 5.)

26\. **FINDING A SCALE FACTOR** Your friend asks you to enlarge your notes to study because your writing is small. Your writing covers 7.5 inches by 10 inches on a piece of paper. The writing on the enlarged copy has a smaller side with a length of 9 inches. What is the scale factor of this dilation?

GO DIGITAL

4 MTR **ERROR ANALYSIS** **In Exercises 27 and 28, describe and correct the error in finding the scale factor of the dilation.**

27.

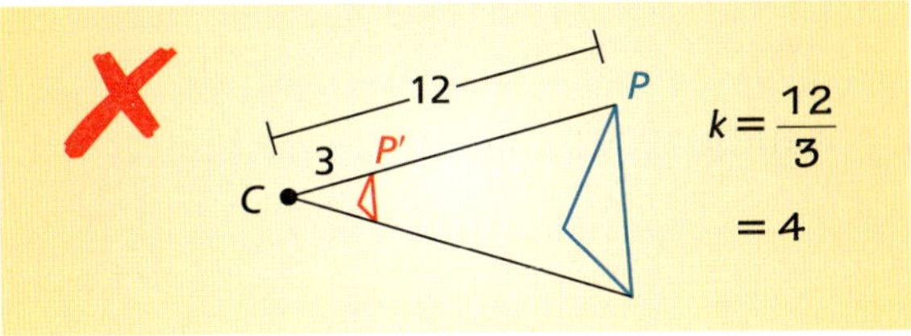

28.

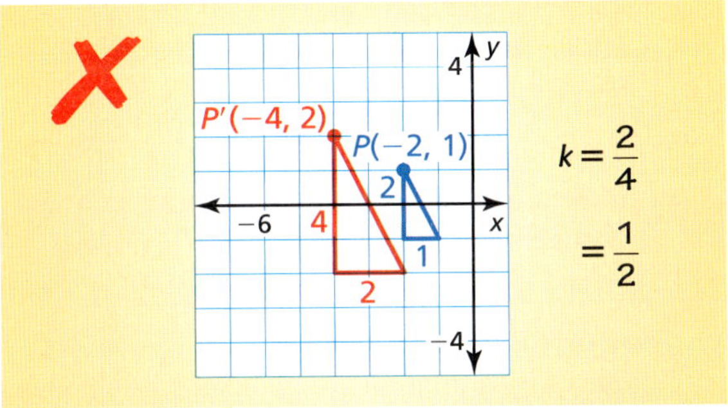

In Exercises 29–32, determine whether the dilated figure or the original figure is closer to the center of dilation. Use the given location of the center of dilation and the scale factor k.

29. Center of dilation: inside the figure; $k = 3$

30. Center of dilation: inside the figure; $k = \frac{1}{2}$

31. Center of dilation: outside the figure; $k = 0.1$

32. Center of dilation: outside the figure; $k = 120\%$

In Exercises 33–36, you are using a magnifying glass. Use the actual length of the insect and the magnification level to determine the length of the image of the insect seen through the magnifying glass. *(See Example 6.)*

▶ **33.** emperor moth
Magnification: 5×

34. ladybug
Magnification: 10×

35. dragonfly
Magnification: 20×

36. carpenter ant
Magnification: 15×

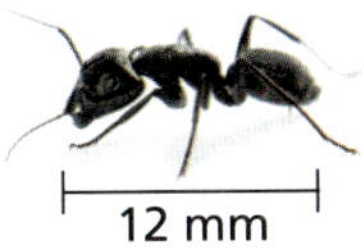

GO DIGITAL

37. **REASONING** Use the given actual and magnified lengths to determine which of the following insects were seen using the same magnification level. Explain your reasoning.

grasshopper
Actual: 2 in.
Magnified: 15 in.

black beetle
Actual: 0.6 in.
Magnified: 4.2 in.

honeybee
Actual: $\frac{5}{8}$ in.
Magnified: $\frac{75}{16}$ in.

monarch butterfly
Actual: 3.9 in.
Magnified: 29.25 in.

38. **HOW DO YOU SEE IT?**
Point C is the center of dilation of the figures. The scale factor is $\frac{1}{3}$. Which figure is the preimage? Which figure is the image? Explain your reasoning.

39. **REASONING** You have a 4-inch by 6-inch photo from the school dance. You have an 8-inch by 10-inch frame. Can you enlarge the photo without cropping to fit the frame? Explain your reasoning.

40. **WRITING** Is a scale factor of 2 the same as a scale factor of 200%? Explain your reasoning.

4 MTR **41.** **DISCUSS MATHEMATICAL THINKING** Dilate the line through $O(0, 0)$ and $A(1, 2)$ using a scale factor of 2.

a. What do you notice about the lengths of $\overline{O'A'}$ and $\overline{OA}$?

b. What do you notice about $\overleftrightarrow{O'A'}$ and $\overleftrightarrow{OA}$?

5 MTR **42. CONNECTING CONCEPTS** The larger triangle is a dilation of the smaller triangle. Find the values of x and y.

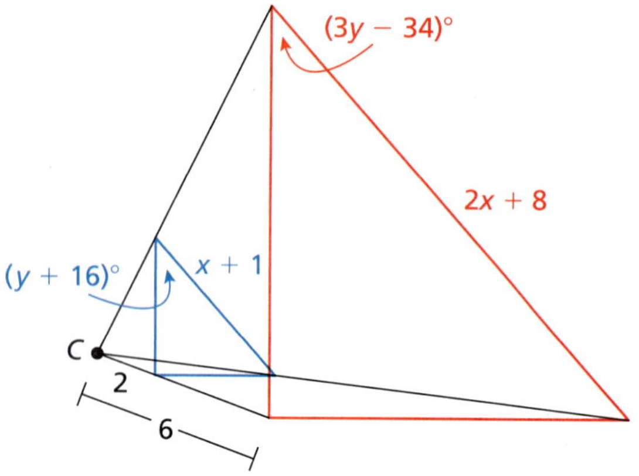

4 MTR **DISCUSS MATHEMATICAL THINKING In Exercises 43 and 44, graph $\triangle ABC$ with vertices $A(-3, 4)$, $B(-4, 2)$, and $C(0, -4)$ and its image after the transformation. Then determine whether the transformation is a dilation. Explain your reasoning.**

43. $(x, y) \rightarrow (2x, y)$

44. $(x, y) \rightarrow (x, 3y)$

45. REASONING You put a reduction of a rectangle on the original rectangle. Explain why there is a point that is in the same location on both rectangles.

4 MTR **46. MAKING AN ARGUMENT** Your friend claims that dilating a figure by 1 is the same as dilating the figure by -1 because the original figure will not be enlarged or reduced. Is your friend correct? Explain your reasoning.

5 MTR **47. STRUCTURE** Rectangle $WXYZ$ has vertices $W(-3, -1)$, $X(-3, 3)$, $Y(5, 3)$, and $Z(5, -1)$.

a. Find the perimeter and the area of the rectangle.

b. Dilate the rectangle using a scale factor of 3. Find the perimeter and the area of the image. Compare the perimeters and the areas of the rectangles. What do you notice?

c. Repeat part (b) using a scale factor of $\frac{1}{4}$.

d. Make conjectures for how the perimeter and area change when a figure is dilated.

48. THOUGHT PROVOKING
Explain why a dilation with a negative scale factor results in a rotation.

49. DIG DEEPER $\triangle ABC$ has vertices $A(4, 2)$, $B(4, 6)$, and $C(7, 2)$. Find the coordinates of the vertices of the image after a dilation with center $(4, 0)$ and scale factor 2.

REVIEW & REFRESH

WATCH

In Exercises 50–53, graph the polygon with the given vertices and its image after the indicated transformation.

50. $A(2, -1)$, $B(0, 4)$, $C(-3, 5)$
Translation: $(x, y) \rightarrow (x - 1, y + 3)$

51. $A(-5, 6)$, $B(-7, 8)$, $C(-3, 11)$
Reflection: in the x-axis

52. $D(-3, 2)$, $E(-1, 4)$, $F(1, 2)$, $G(1, -1)$
Rotation: 270° counterclockwise about the origin

53. $J(0, 4)$, $K(4, 4)$, $L(8, 0)$, $M(4, 0)$
Dilation: scale factor $k = 75\%$

54. Simplify $\dfrac{4^{-2}b^{-3}}{2^{-1}a^0b^{-4}}$.

7 MTR **55. MODELING REAL LIFE** You are painting a rectangular canvas that is 30 inches wide and 40 inches long. Your friend is painting a rectangular canvas, where the width and length are each x inches shorter. When $x = 3$, what is the area of your friend's canvas?

56. Describe a congruence transformation that maps the blue preimage to the green image.

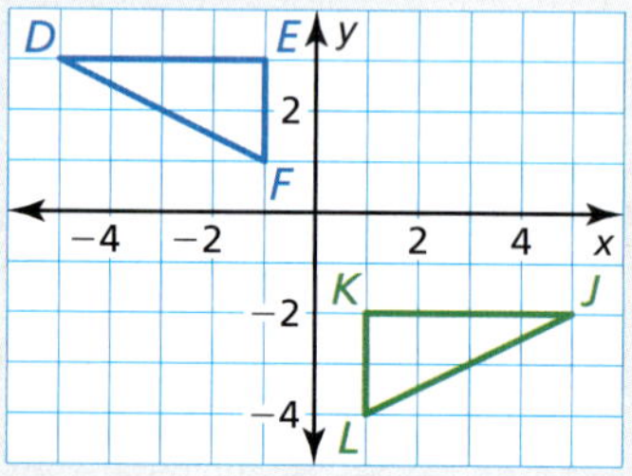

57. Graph $g(x) = 3(x - 1)^2 + 7$. Compare the graph to the graph of $f(x) = x^2$.

In Exercises 58 and 59, find the product.

58. $(3x - 4)^2$

59. $(w - 5)(6 + 2w)$

3 MTR **CHOOSE A METHOD In Exercises 60 and 61, solve the system using any method. Explain your choice of method.**

60. $y = -3x + 5$
$2x + 4y = 15$

61. $0.6y + 0.5x = 1$
$0.25x = -0.5y + 2$

GO DIGITAL

4.6 Similarity and Transformations

Learning Target: Understand similarity transformations.

Success Criteria:
- I can perform similarity transformations.
- I can describe similarity transformations.
- I can prove that figures are similar.

Two figures are *similar figures* when they have the same shape but not necessarily the same size.

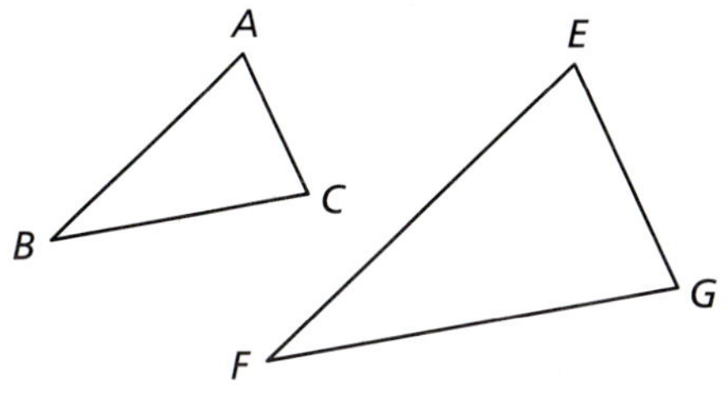

Similar Triangles

EXPLORE IT! Transforming Figures and Determining Similarity

Work with a partner. Use technology to draw any triangle and label it $\triangle ABC$.

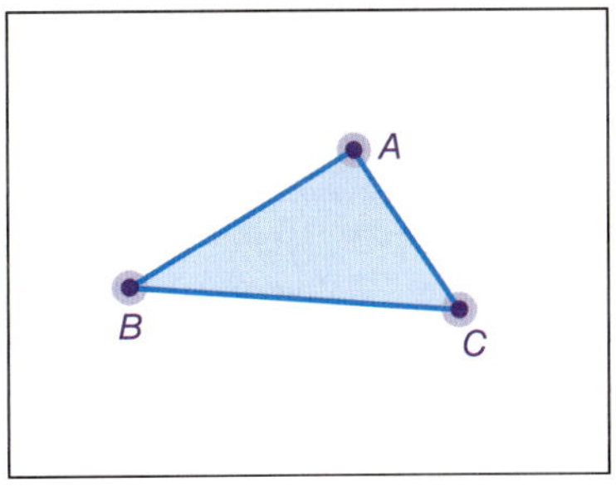

a. Translate $\triangle ABC$ several times using different translations. Are the images similar to the preimage? Justify your answer.

b. Reflect $\triangle ABC$ several times using different lines of reflection. Are the images similar to the preimage? Justify your answer.

c. Rotate $\triangle ABC$ several times using different centers of rotation and angles of rotation. Are the images similar to the preimage? Justify your answer.

d. Dilate $\triangle ABC$ several times using different scale factors and centers of dilation. Are the images similar to the preimage? Justify your answer.

e. A figure undergoes a composition of transformations, which includes translations, reflections, rotations, and dilations. Is the image similar to the preimage? Explain your reasoning.

f. Explain the difference between congruent figures and similar figures.

4 MTR COMMUNICATE CLEARLY

What are some examples of nonrigid transformations that result in an image *not* being similar to the preimage?

Geometric Reasoning

MA.912.GR.2.1 Given a preimage and image, describe the transformation and represent the transformation algebraically using coordinates.

MA.912.GR.2.8 Apply an appropriate transformation to map one figure onto another to justify that the two figures are similar.

Also MA.912.GR.1.6, MA.912.GR.2.3, MA.912.GR.2.5

GO DIGITAL

Performing Similarity Transformations

Vocabulary

similarity transformation, *p. 212*
similar figures, *p. 212*

Dilations are nonrigid motions because they preserve shape but not necessarily size. A **similarity transformation** is a dilation or a composition of rigid motions and dilations. Two geometric figures are **similar figures** if and only if there is a similarity transformation that maps one of the figures to the other.

Unlike congruence transformations that preserve length and angle measure, similarity transformations preserve angle measure only, unless the dilations have scale factors of 1 or −1. So, similar figures have the same shape but not necessarily the same size.

EXAMPLE 1 Performing a Similarity Transformation

Graph $\triangle ABC$ with vertices $A(-4, 1)$, $B(-2, 2)$, and $C(-2, 1)$ and its image after the similarity transformation.

Translation: $(x, y) \rightarrow (x + 5, y + 1)$

Dilation: $(x, y) \rightarrow (2x, 2y)$

SOLUTION

Step 1 Graph $\triangle ABC$.

Step 2 Translate $\triangle ABC$ 5 units right and 1 unit up. $\triangle A'B'C'$ has vertices $A'(1, 2)$, $B'(3, 3)$, and $C'(3, 2)$.

Step 3 Dilate $\triangle A'B'C'$ using a scale factor of 2. $\triangle A''B''C''$ has vertices $A''(2, 4)$, $B''(6, 6)$, and $C''(6, 4)$.

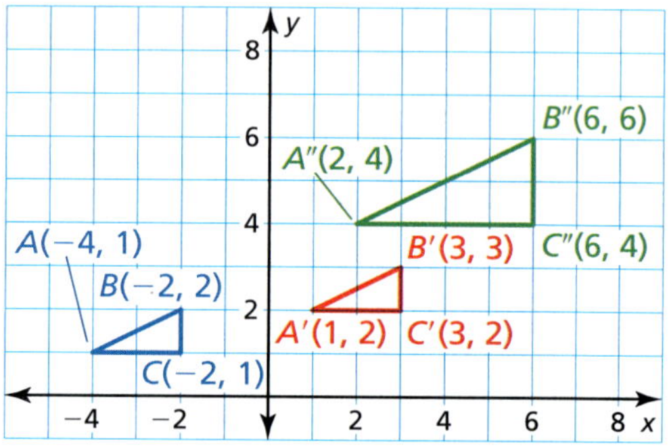

SELF-ASSESSMENT

1 I don't understand yet. 2 I can do it with help. 3 I can do it on my own. 4 I can teach someone else.

1. **WRITING** Explain the difference between each pair of vocabulary terms.
   a. congruent figures and similar figures
   b. congruence transformations and similarity transformations

2. **VOCABULARY** Explain why a similarity transformation includes at least one dilation.

3. Graph $\overline{CD}$ with endpoints $C(-2, 2)$ and $D(2, 2)$ and its image after the similarity transformation.

   Rotation: 90° clockwise about the origin

   Dilation: $(x, y) \rightarrow \left(\frac{1}{2}x, \frac{1}{2}y\right)$

4. Graph $\triangle FGH$ with vertices $F(1, 2)$, $G(4, 4)$, and $H(2, 0)$ and its image after the similarity transformation.

   Reflection: in the x-axis

   Dilation: $(x, y) \rightarrow (1.5x, 1.5y)$

GO DIGITAL

Describing Similarity Transformations

EXAMPLE 2 Describing a Similarity Transformation

WATCH

Describe a similarity transformation that maps trapezoid *PQRS* to trapezoid *WXYZ*.

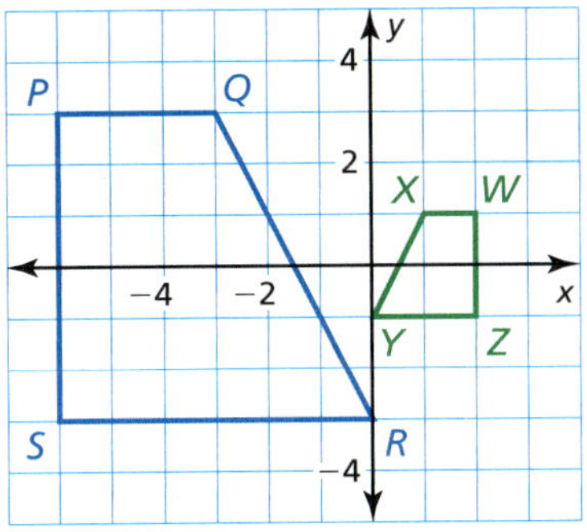

SOLUTION

$\overline{QR}$ falls from left to right, and $\overline{XY}$ rises from left to right. If you reflect trapezoid *PQRS* in the *y*-axis as shown, then the image, trapezoid $P'Q'R'S'$, will have the same orientation as trapezoid *WXYZ*.

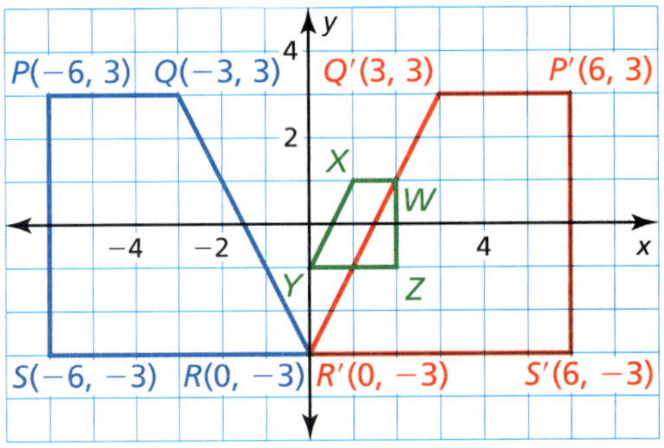

Trapezoid *WXYZ* appears to be about one-third as large as trapezoid $P'Q'R'S'$. Dilate trapezoid $P'Q'R'S'$ using a scale factor of $\frac{1}{3}$.

$$(x, y) \rightarrow \left(\tfrac{1}{3}x, \tfrac{1}{3}y\right)$$

$$P'(6, 3) \rightarrow P''(2, 1)$$

$$Q'(3, 3) \rightarrow Q''(1, 1)$$

$$R'(0, -3) \rightarrow R''(0, -1)$$

$$S'(6, -3) \rightarrow S''(2, -1)$$

The vertices of trapezoid $P''Q''R''S''$ match the vertices of trapezoid *WXYZ*.

▶ So, a similarity transformation that maps trapezoid *PQRS* to trapezoid *WXYZ* is a reflection in the *y*-axis, followed by a dilation with a scale factor of $\frac{1}{3}$.

Check Verify that the corresponding sides are proportional and the corresponding angles are congruent.

$$\frac{WX}{PQ} = \frac{XY}{QR} = \frac{YZ}{RS} = \frac{ZW}{SP} = \frac{1}{3}$$

$\angle P$ and $\angle W$ are right angles, so they are congruent. $\angle S$ and $\angle Z$ are right angles, so they are congruent. Using a protractor to estimate the other angle measures, you can see that the other corresponding angles appear to be congruent. ✓

SELF-ASSESSMENT

1 I don't understand yet. 2 I can do it with help. 3 I can do it on my own. 4 I can teach someone else.

5. In Example 2, describe another similarity transformation that maps trapezoid *PQRS* to trapezoid *WXYZ*.

6. Describe a similarity transformation that maps quadrilateral *DEFG* to quadrilateral *STUV*.

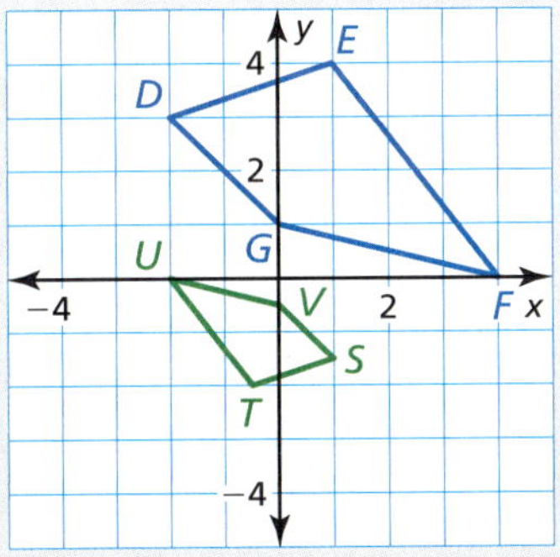

GO DIGITAL

EXAMPLE 3 Writing a Coordinate Rule for a Similarity Transformation

WATCH

Write a coordinate rule for the composition in Example 2.

SOLUTION

The first transformation is a reflection in the y-axis, so $(a, b) \rightarrow (-a, b)$. The second transformation is a dilation with a scale factor of $\frac{1}{3}$, so $(-a, b) \rightarrow \left(-\frac{1}{3}a, \frac{1}{3}b\right)$.

▶ So, a coordinate rule for the composition is $(a, b) \rightarrow \left(-\frac{1}{3}a, \frac{1}{3}b\right)$.

You can use a composition to dilate a figure when the center of dilation is a point on the figure.

KEY IDEA

Dilating When the Center Is a Point on the Figure

To dilate a figure when the center of dilation is a point P on the figure, translate the figure along $\overrightarrow{PO}$, so the center is mapped to the origin O. Then use the appropriate coordinate rule for a dilation centered at the origin. Finally, translate the center of dilation back to P along $\overrightarrow{OP}$.

EXAMPLE 4 Dilating When the Center Is on the Figure

WATCH

Graph $\triangle ABC$ with vertices $A(1, 1)$, $B(3, 2)$, and $C(4, 1)$ and its image after a dilation centered at point A with a scale factor of 2.

SOLUTION

Step 1 Translate the figure along $\overrightarrow{AO}$.

$$(a, b) \rightarrow (a - 1, b - 1)$$

$A(1, 1) \rightarrow A'(1 - 1, 1 - 1) = A'(0, 0)$

$B(3, 2) \rightarrow B'(3 - 1, 2 - 1) = B'(2, 1)$

$C(4, 1) \rightarrow C'(4 - 1, 1 - 1) = C'(3, 0)$

Step 2 Use the coordinate rule for a dilation centered at the origin with a scale factor of 2.

$$(a, b) \rightarrow (2a, 2b)$$

$A'(0, 0) \rightarrow A''(2(0), 2(0)) = A''(0, 0)$

$B'(2, 1) \rightarrow B''(2(2), 2(1)) = B''(4, 2)$

$C'(3, 0) \rightarrow C''(2(3), 2(0)) = C''(6, 0)$

Step 3 Translate back along $\overrightarrow{OA}$.

$$(a, b) \rightarrow (a + 1, b + 1)$$

$A''(0, 0) \rightarrow A'''(0 + 1, 0 + 1) = A'''(1, 1)$

$B''(4, 2) \rightarrow B'''(4 + 1, 2 + 1) = B'''(5, 3)$

$C''(6, 0) \rightarrow C'''(6 + 1, 0 + 1) = C'''(7, 1)$

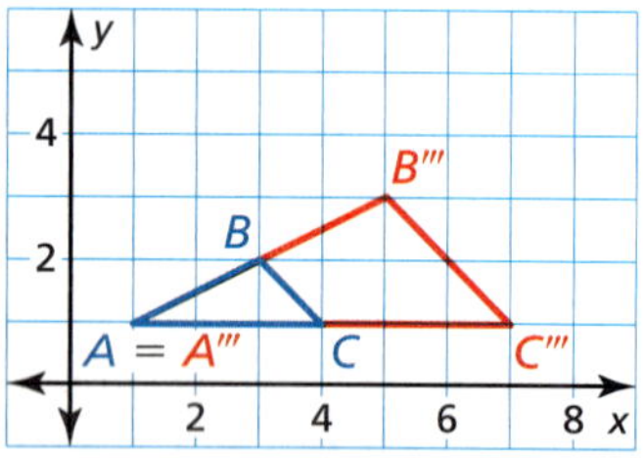

SELF-ASSESSMENT

1 I don't understand yet. 2 I can do it with help. 3 I can do it on my own. 4 I can teach someone else.

7. Write a coordinate rule for the composition in Exercise 6.

8. Graph ABC with vertices $A(-5, -2)$, $B(-5, -6)$, and $C(-1, -2)$ and its image after a dilation centered at point C with a scale factor of $\frac{1}{2}$.

GO DIGITAL

Proving Figures Are Similar

To prove that two figures are similar, you must prove that a similarity transformation maps one of the figures to the other.

EXAMPLE 5 Proving that Two Squares Are Similar

WATCH

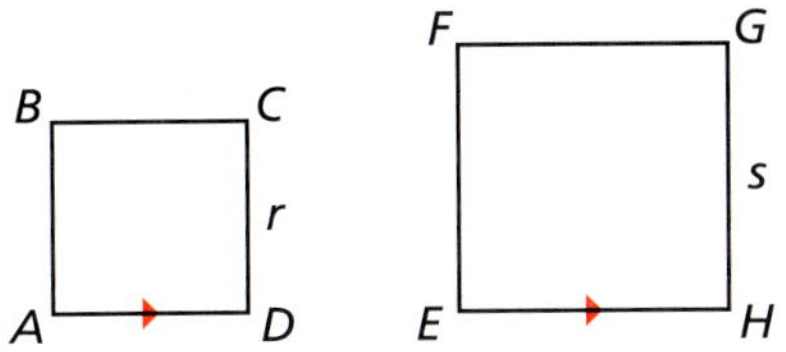

Prove that square $ABCD$ is similar to square $EFGH$.

Given Square $ABCD$ with side length r, square $EFGH$ with side length s, $\overline{AD} \parallel \overline{EH}$

Prove Square $ABCD$ is similar to square $EFGH$.

SOLUTION

Translate square $ABCD$ so that point A maps to point E. Because translations map segments to parallel segments and $\overline{AD} \parallel \overline{EH}$, the image of $\overline{AD}$ lies on $\overline{EH}$.

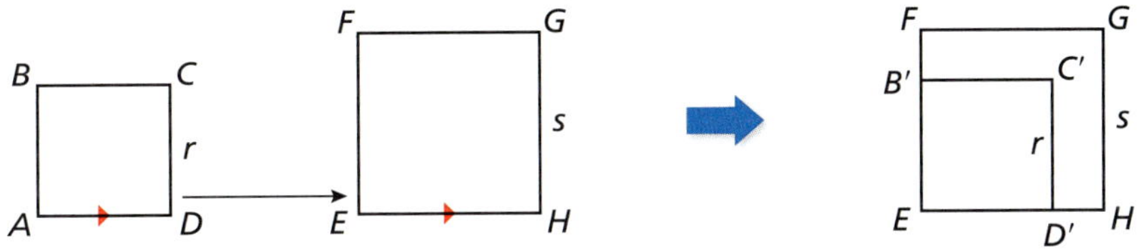

Because translations preserve length and angle measure, the image of square $ABCD$, $EB'C'D'$, is a square with side length r. Because all the interior angles of a square are right angles, $\angle B'ED' \cong \angle FEH$. When $\overrightarrow{ED'}$ coincides with $\overrightarrow{EH}$, $\overrightarrow{EB'}$ coincides with $\overrightarrow{EF}$. So, $\overline{EB'}$ lies on $\overline{EF}$. Next, dilate square $EB'C'D'$ using the center of dilation E. Choose the scale factor to be the ratio of the side lengths of $EFGH$ and $EB'C'D'$, which is $\frac{s}{r}$.

This dilation maps $\overline{ED'}$ to $\overline{EH}$ and $\overline{EB'}$ to $\overline{EF}$ because the images of $\overline{ED'}$ and $\overline{EB'}$ have side lengths $\frac{s}{r}(r) = s$ and the segments $\overline{ED'}$ and $\overline{EB'}$ lie on lines passing through the center of dilation. So, the dilation maps B' to F and D' to H. The image of C' lies $\frac{s}{r}(r) = s$ units to the right of the image of B' and $\frac{s}{r}(r) = s$ units above the image of D'. So, the image of C' is G.

▶ A similarity transformation maps square $ABCD$ to square $EFGH$. So, square $ABCD$ is similar to square $EFGH$.

SELF-ASSESSMENT

1 I don't understand yet. 2 I can do it with help. 3 I can do it on my own. 4 I can teach someone else.

9. Prove that $\triangle JKL$ is similar to $\triangle MNP$.

Given Right isosceles $\triangle JKL$ with leg length t, right isosceles $\triangle MNP$ with leg length v, $\overline{LJ} \parallel \overline{PM}$

Prove $\triangle JKL$ is similar to $\triangle MNP$.

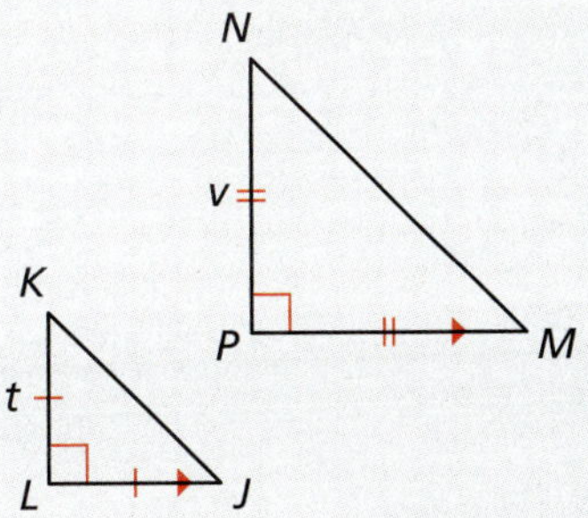

GO DIGITAL

4.6 Practice WITH CalcChat® AND CalcView®

In Exercises 1–4, graph $\triangle FGH$ with vertices $F(-2, 2)$, $G(-2, -4)$, and $H(-4, -4)$ and its image after the similarity transformation. *(See Example 1.)*

1. **Translation:** $(x, y) \rightarrow (x + 3, y + 1)$
   Dilation: $(x, y) \rightarrow (2x, 2y)$

2. **Dilation:** $(x, y) \rightarrow \left(\frac{1}{2}x, \frac{1}{2}y\right)$
   Reflection: in the y-axis

▶ 3. **Dilation:** $(x, y) \rightarrow \left(\frac{3}{4}x, \frac{3}{4}y\right)$
   Reflection: in the x-axis

4. **Rotation:** 90° counterclockwise about the origin
   Dilation: $(x, y) \rightarrow (3x, 3y)$

In Exercises 5–8, describe a similarity transformation that maps the blue preimage to the green image. Verify that the corresponding sides are proportional and the corresponding angles are congruent. *(See Example 2.)*

▶ 5.
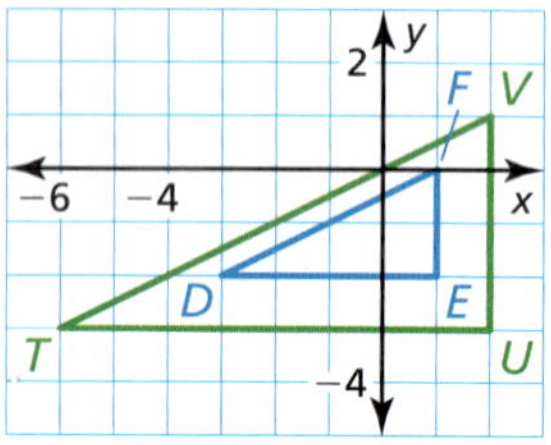

6.
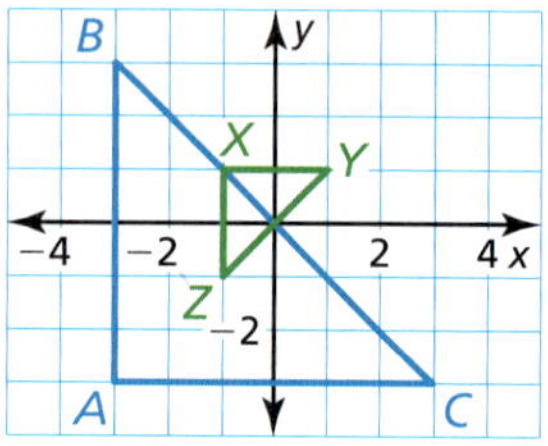

7.
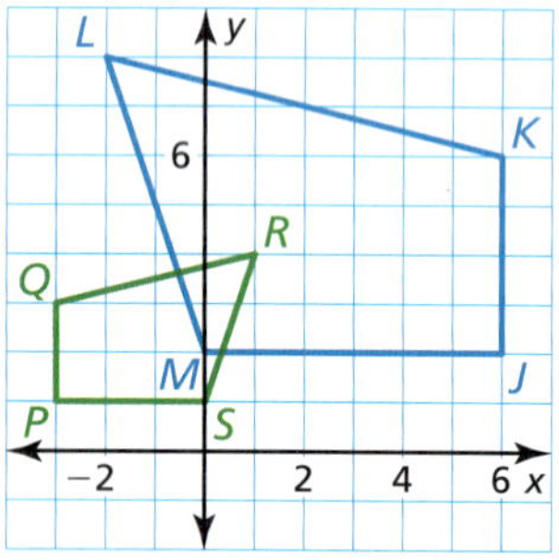

8.
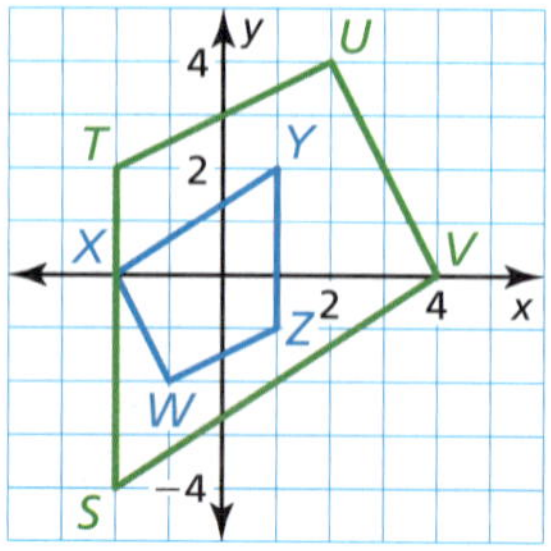

In Exercises 9 and 10, write a coordinate rule for the composition. *(See Example 3.)*

▶ 9.
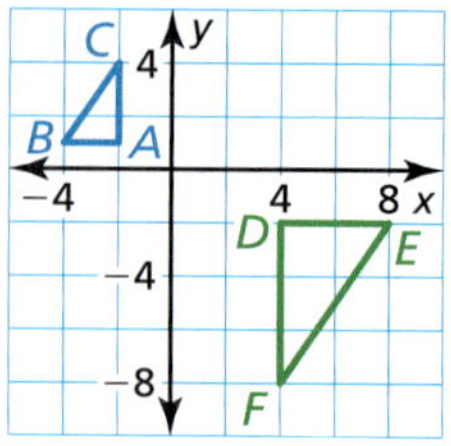

10.
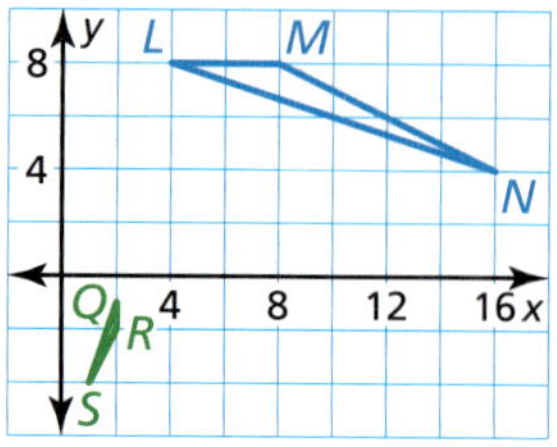

11. Write a coordinate rule for the composition in Exercise 5.

12. Write a coordinate rule for the composition in Exercise 6.

13. Write a coordinate rule for the composition in Exercise 7.

14. Write a coordinate rule for the composition in Exercise 8.

In Exercises 15–18, graph the figure and its image after a dilation centered at the given point using the given scale factor. *(See Example 4.)*

▶ 15. $A(-1, 2)$, $B(-1, -1)$, $C(2, -1)$; dilation centered at point B with a scale factor of 2

16. $D(0, -4)$, $E(4, 0)$, $F(4, -4)$; dilation centered at point D with a scale factor of $\frac{1}{4}$

17. $H(-2, -1)$, $I(-2, 1)$, $J(1, 2)$, $K(1, 0)$; dilation centered at point J with a scale factor of 3

18. $L(-3, -5)$, $M(-3, 1)$, $N(1, 1)$, $P(3, -1)$; dilation centered at point P with a scale factor of $\frac{1}{2}$

In Exercises 19–22, determine whether the polygons with the given vertices are similar. Use transformations to explain your reasoning.

19. $A(6, 0)$, $B(9, 6)$, $C(12, 6)$ and $D(0, 3)$, $E(1, 5)$, $F(2, 5)$

20. $Q(-1, 0)$, $R(-2, 2)$, $S(1, 3)$, $T(2, 1)$ and $W(0, 2)$, $X(4, 4)$, $Y(6, -2)$, $Z(2, -4)$

21. $G(-2, 3)$, $H(4, 3)$, $I(4, 0)$ and $J(1, 0)$, $K(6, -2)$, $L(1, -2)$

22. $D(-4, 3)$, $E(-2, 3)$, $F(-1, 1)$, $G(-4, 1)$ and $L(1, -1)$, $M(3, -1)$, $N(6, -3)$, $P(1, -3)$

GO DIGITAL

In Exercises 23 and 24, prove that the figures are similar. *(See Example 5.)*

23. Given Right isosceles $\triangle ABC$ with leg length j, right isosceles $\triangle RST$ with leg length k, $\overline{CA} \parallel \overline{RT}$

Prove $\triangle ABC$ is similar to $\triangle RST$.

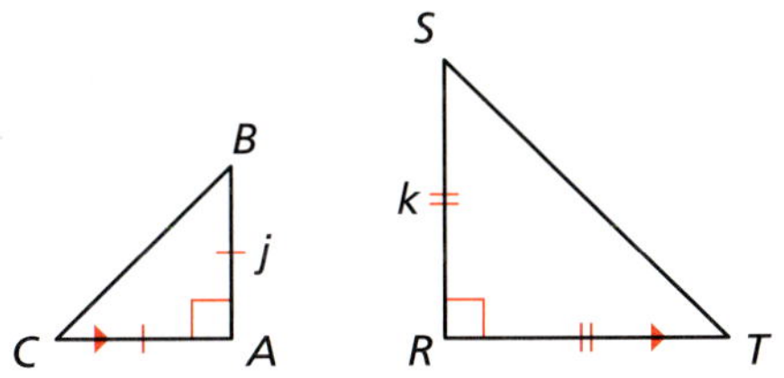

24. Given Rectangle $JKLM$ with side lengths x and y, rectangle $QRST$ with side lengths $2x$ and $2y$

Prove Rectangle $JKLM$ is similar to rectangle $QRST$.

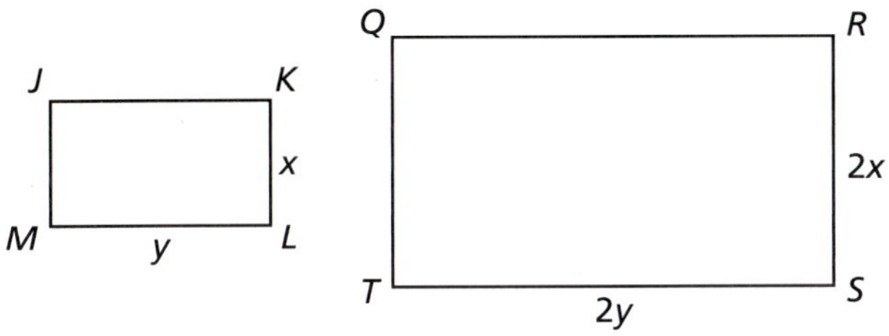

25. ERROR ANALYSIS Describe and correct the error in determining whether the figures are similar.

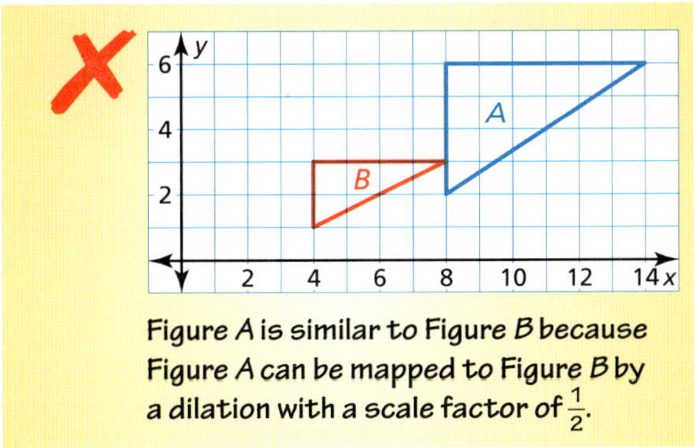

26. MODELING REAL LIFE The stop sign and the stop sign sticker are regular octagons. Determine whether they are similar. Explain your reasoning.

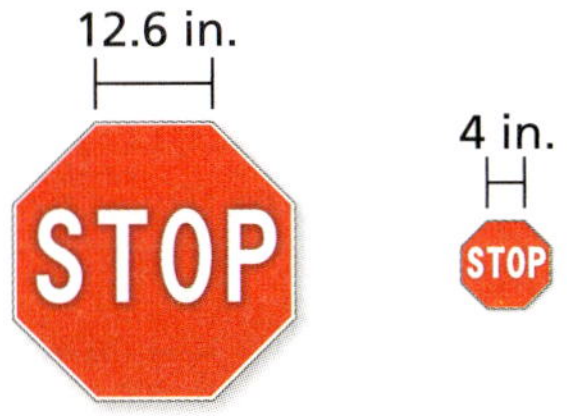

27. STRUCTURE Quadrilateral $JKLM$ is mapped to quadrilateral $J'K'L'M'$ using the dilation $(x, y) \rightarrow \left(\frac{3}{2}x, \frac{3}{2}y\right)$. Then quadrilateral $J'K'L'M'$ is mapped to quadrilateral $J''K''L''M''$ using the translation $(x, y) \rightarrow (x + 3, y - 4)$. The vertices of quadrilateral $J'K'L'M'$ are $J'(-12, 0)$, $K'(-12, 18)$, $L'(-6, 18)$, and $M'(-6, 0)$. Find the coordinates of the vertices of quadrilateral $JKLM$ and quadrilateral $J''K''L''M''$. Are quadrilateral $JKLM$ and quadrilateral $J''K''L''M''$ similar? Explain.

28. HOW DO YOU SEE IT?
Determine whether each pair of figures is similar. Explain your reasoning.

a.

b.

29. DISCUSS MATHEMATICAL THINKING $\triangle ABC$ can be mapped to $\triangle DEF$ by a translation 3 units right and 2 units up, followed by a dilation with a scale factor of $\frac{1}{2}$. Describe a similarity transformation that maps $\triangle DEF$ to $\triangle ABC$. Justify your answer.

30. STRUCTURE A similarity transformation maps $JKLM$ to its image $PQRS$. Find the coordinates of point S. Then write a coordinate rule for the composition.

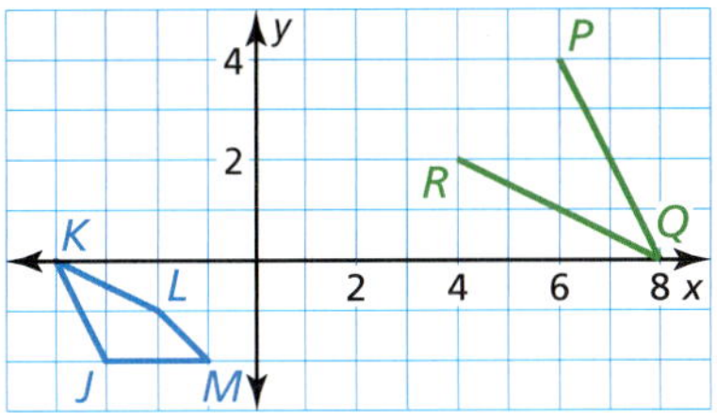

31. REASONING $\triangle QRS$ has vertices $Q(1, 1)$, $R(1, 5)$, and $S(7, 1)$.

a. Graph $\triangle QRS$. Then connect the midpoints of the sides of $\triangle QRS$ to make another triangle. Are the triangles similar? Justify your answer.

b. Repeat part (a) for several other triangles. What conjecture can you make?

32. DISCUSS MATHEMATICAL THINKING Consider ABC and its image after a dilation centered at point A with a scale factor of 2. Is there a dilation you can perform to the image that will map it to the original figure? Explain why or why not.

GO DIGITAL

33. DIG DEEPER A rectangle has vertices $A(0, 0)$, $B(0, 6)$, $C(9, 6)$, and $D(9, 0)$. Explain how to dilate the rectangle to produce an image whose area is twice the area of the original rectangle. Make a conjecture about how to dilate any polygon to produce an image whose area is n times the area of the original polygon.

34. THOUGHT PROVOKING
Is the composition of a rotation and a dilation commutative? (In other words, do you obtain the same image regardless of the order in which you perform the transformations?) Justify your answer.

5 MTR **35. STRUCTURE** Describe as many similarity transformations as you can that map square $QRST$ to square $WXYZ$.

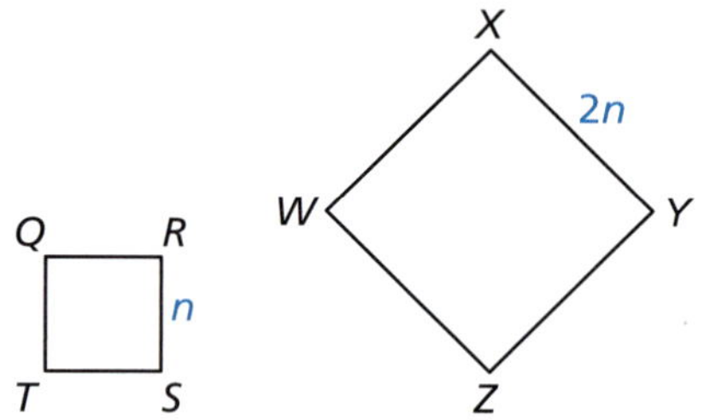

REVIEW & REFRESH

WATCH

In Exercises 36–39, classify the angle.

36.

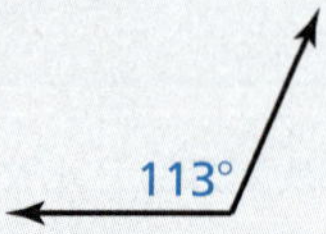

37.

38.

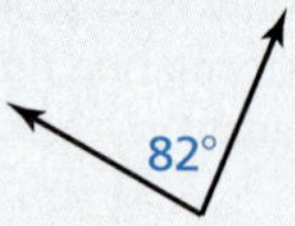

39.

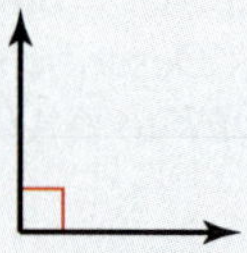

40. Graph $\overline{PQ}$ with endpoints $P(-6, 2)$ and $Q(-2, 2)$ and its image after the similarity transformation.

Dilation: $(x, y) \rightarrow \left(\frac{3}{2}x, \frac{3}{2}y\right)$

Translation: $(x, y) \rightarrow (x + 6, y - 5)$

In Exercises 41 and 42, graph $\triangle DEF$ with vertices $D(4, -6)$, $E(4, -2)$, and $F(8, -2)$ and its image after the transformation.

41. a dilation with a scale factor of 2.5

42. a 270° clockwise rotation about the origin

43. Write an equation of the line passing through $(2, 3)$ that is perpendicular to the line $y = -\frac{1}{4}x + 1$.

44. Solve $9x - 2 = 13 + 6x$. Justify each step.

7 MTR **45. MODELING REAL LIFE** The linear function $m = 60 - 5c$ represents the amount of time m (in minutes) that you have left to escape a room in a game after receiving c clues.

a. Find the domain of the function. Is the domain discrete or continuous? Explain.

b. Graph the function using its domain.

In Exercises 46 and 47, solve the inequality. Graph the solution.

46. $4m - 1 > 9 + 2m$

47. $-1 < 4z - 5 < 19$

48. Determine whether the quadrilaterals with vertices $J(-4, -2)$, $K(-1, -1)$, $L(-1, -7)$, $M(-4, -5)$ and $W(6, 2)$, $X(3, 3)$, $Y(3, -3)$, $Z(6, -1)$ are congruent. Use transformations to explain your reasoning.

In Exercises 49–52, simplify the expression.

49. $(3x^2 + x + 7) + (-x^2 - x - 4)$

50. $(5x^2 - 8x - 13) - (2x^2 - x + 9)$

51. $(x - 2)(x^2 + 3x - 5)$

52. $\dfrac{3x^4 - 6x^2 + 9x}{3x}$

53. Describe the transformation from the graph of f to the graph of g.

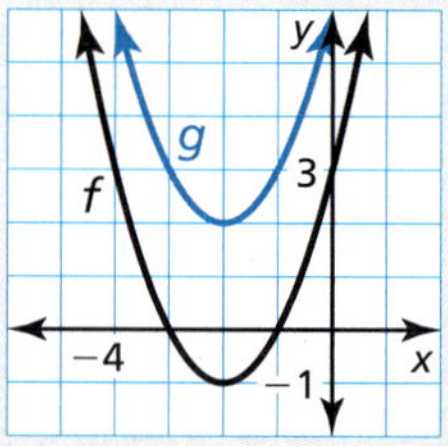

In Exercises 54 and 55, factor the expression.

54. $x^2 - 9x + 8$

55. $4x^2y + 32xy + 60y$

GO DIGITAL

4 Chapter Review WITH CalcChat®

Chapter Learning Target: Understand transformations.

Chapter Success Criteria:
- ◆ I can identify transformations.
- ◆ I can perform translations, reflections, rotations, and dilations.
- ■ I can describe congruence and similarity transformations.
- ■ I can solve problems involving transformations.

◆ Surface
■ Deep

SELF-ASSESSMENT 1 I don't understand yet. 2 I can do it with help. 3 I can do it on my own. 4 I can teach someone else.

4.1 Translations (pp. 167–174) WATCH

Learning Target: Understand translations of figures.

Vocabulary AZ VOCAB
vector
initial point
terminal point
horizontal component
vertical component
component form
transformation
image
preimage
translation
rigid motion
composition of transformations

The vertices of $\triangle XYZ$ are $X(2, 3)$, $Y(-3, 2)$, and $Z(-4, -3)$. Translate $\triangle XYZ$ using the given vector or rule. Graph $\triangle XYZ$ and its image.

1. $\langle 0, 2\rangle$
2. $\langle -3, 4\rangle$
3. $(x, y) \rightarrow (x + 3, y - 1)$
4. $(x, y) \rightarrow (x + 4, y + 1)$

Graph $\triangle PQR$ with vertices $P(0, -4)$, $Q(1, 3)$, and $R(2, -5)$ and its image after the composition.

5. **Translation:** $(x, y) \rightarrow (x + 1, y + 2)$
   Translation: $(x, y) \rightarrow (x - 4, y + 1)$
6. **Translation:** $(x, y) \rightarrow (x, y + 3)$
   Translation: $(x, y) \rightarrow (x - 1, y + 1)$
7. A translation maps $\triangle ABC$ to $\triangle A'B'C'$ using the vector $\langle -3, 5\rangle$. A second translation maps $\triangle A'B'C'$ to $\triangle A''B''C''$ using the vector $\langle 4, -2\rangle$. Write a rule for translating $\triangle ABC$ to $\triangle A''B''C''$.

4.2 Reflections (pp. 175–182) WATCH

Learning Target: Understand reflections of figures.

Vocabulary AZ VOCAB
reflection
line of reflection
glide reflection
line symmetry
line of symmetry

8. Graph the polygon with vertices $J(1, 2)$, $K(3, 4)$, and $L(5, 1)$ and its image after a reflection in the line $x = 4$.

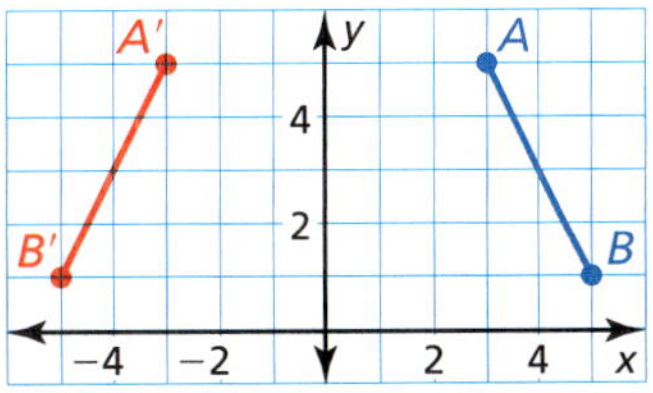

9. Write a coordinate rule for the reflection shown.
10. Graph $\overline{RS}$ with endpoints $R(2, 3)$ and $S(4, -1)$ and its image after the glide reflection.

    Translation: $(x, y) \rightarrow (x, y + 3)$
    Reflection: in the y-axis

11. Determine whether the figure has line symmetry. If so, draw the line(s) of symmetry and describe any reflections that map the figure onto itself.

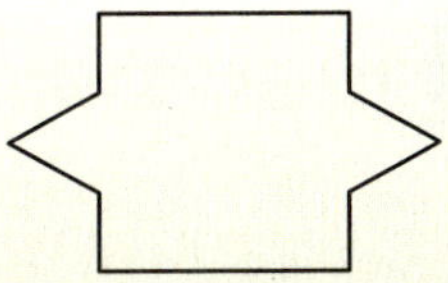

GO DIGITAL

4.3 Rotations (pp. 183–192) WATCH

Learning Target: Understand rotations of figures.

Vocabulary AZ VOCAB
rotation
center of rotation
angle of rotation
rotational symmetry
center of symmetry
order of rotational symmetry

12. Graph the polygon with vertices $A(-3, -1)$, $B(2, 2)$, $C(3, -3)$ and its image after a rotation of 180° about the origin.

13. Graph $\overline{XY}$ with endpoints $X(5, -2)$ and $Y(3, -3)$ and its image after the composition. Write a coordinate rule for the composition.

Reflection: in the x-axis
Rotation: 270° counterclockwise about the origin

Determine whether the figure has rotational symmetry. If so, identify the order of rotational symmetry of the figure.

14.

15.

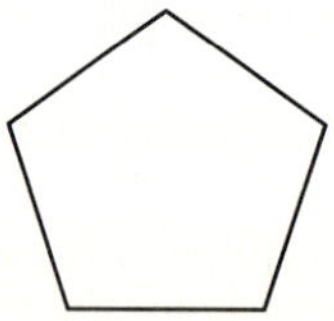

16. The diagram shows a game in which you use the pieces at the top to form solid rows at the bottom. Using only translations and rotations, describe the transformations of the pieces at the top that will form two solid rows at the bottom.

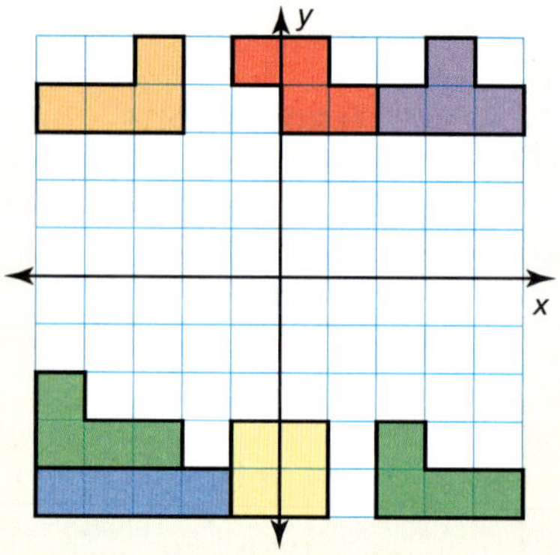

4.4 Congruence and Transformations (pp. 193–202) WATCH

Learning Target: Understand congruence transformations.

Vocabulary AZ VOCAB
congruent figures
congruence transformation
tessellation
transitional symmetry

17. Identify any congruent figures in the coordinate plane at the right. Explain.

18. Refer to Exercise 17. Describe the congruence transformations that map the congruent figures onto one another. Verify that the corresponding sides and the corresponding angles are congruent.

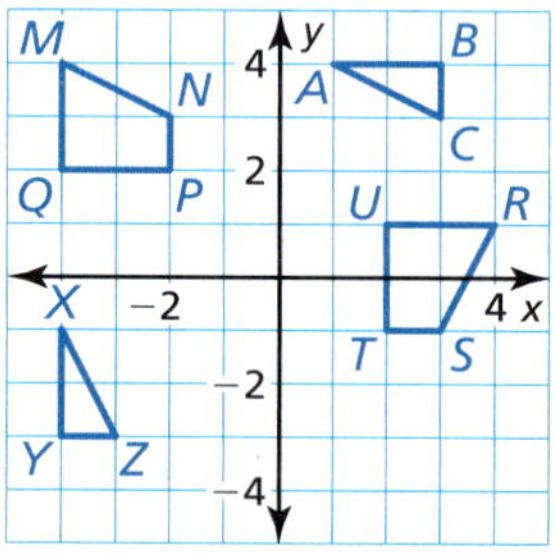

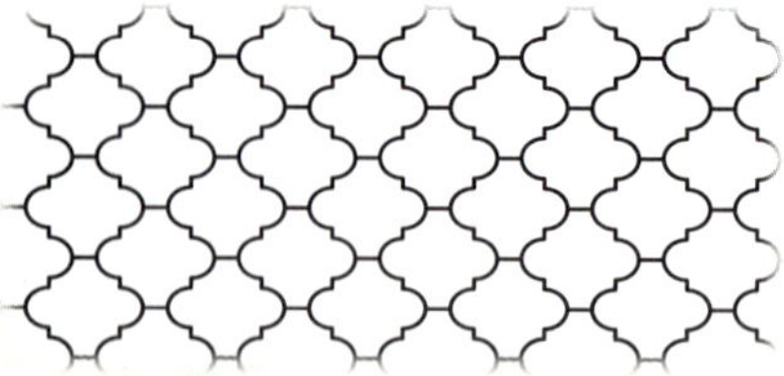

19. Determine whether the tessellation has translational symmetry. If so, identify two translations that map the tessellation onto itself.

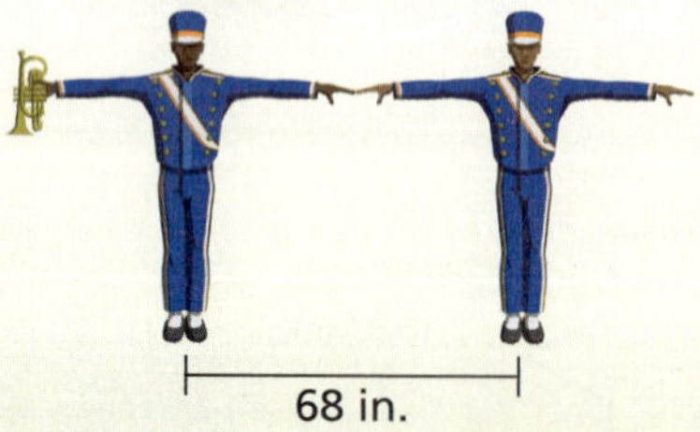

20. In a marching band maneuver, the marcher at the left spins 180° in place with arms spread, and hands the horn to the marcher at the right, who spins 180° with arms spread, and stops.

a. How far from the initial position is the horn?

b. What composition maps the horn from the initial position to the final position?

GO DIGITAL

4.5 Dilations (pp. 203–210) WATCH

Learning Target: Understand dilations of figures.

21. Find the scale factor of the dilation. Then tell whether the dilation is a *reduction* or an *enlargement*.

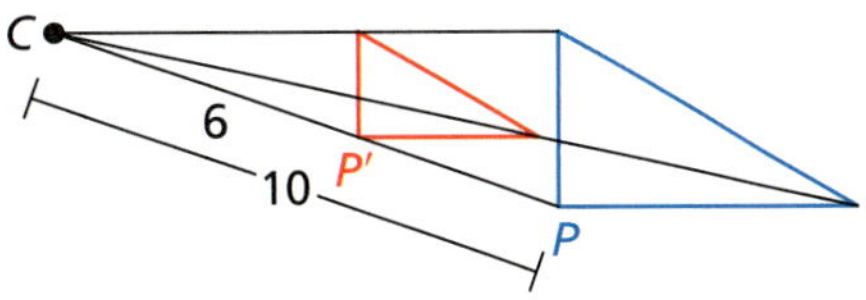

Graph the triangle with the given vertices and its image after a dilation with scale factor k.

22. $P(2, 2)$, $Q(4, 4)$, $R(8, 2)$; $k = \frac{1}{2}$

23. $X(-3, 2)$, $Y(2, 3)$, $Z(1, -1)$; $k = -3$

24. Write a coordinate rule for the dilation.

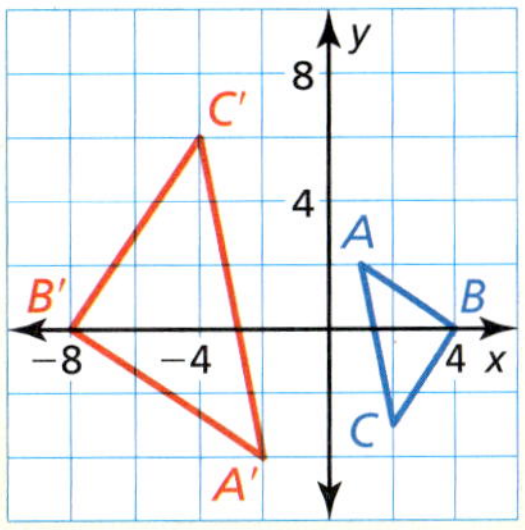

25. You are using a magnifying glass that shows the image of an object as four times the object's actual size. The image has a length of 15.2 centimeters. Find the actual length of the object.

Vocabulary (AZ VOCAB)

dilation
center of dilation
scale factor
enlargement
reduction

4.6 Similarity and Transformations (pp. 211–218) WATCH

Learning Target: Understand similarity transformations.

26. Graph $\triangle DEF$ with vertices $D(-3, 4)$, $E(3, 4)$, and $F(1, 2)$ and its image after the similarity transformation.

Translation: $(x, y) \rightarrow (x + 1, y - 2)$
Dilation: $(x, y) \rightarrow \left(\frac{3}{2}x, \frac{3}{2}y\right)$

Describe a similarity transformation that maps $\triangle ABC$ to $\triangle RST$.

27. $A(1, 0)$, $B(-2, -1)$, $C(-1, -2)$ and $R(-3, 0)$, $S(6, -3)$, $T(3, -6)$

28. $A(6, 4)$, $B(-2, 0)$, $C(-4, 2)$ and $R(2, 3)$, $S(0, -1)$, $T(1, -2)$

29. $A(3, -2)$, $B(0, 4)$, $C(-1, -3)$ and $R(-4, -6)$, $S(8, 0)$, $T(-6, 2)$

30. Write a coordinate rule for the composition in Exercise 29.

Vocabulary (AZ VOCAB)

similarity transformation
similar figures

Mathematical Thinking and Reasoning

4 MTR **ENGAGE IN DISCUSSIONS THAT REFLECT ON THE MATHEMATICAL THINKING OF SELF AND OTHERS**

Mathematicians who engage in discussions that reflect on the mathematical thinking of self and others communicate mathematical ideas, vocabulary, and methods effectively.

1. Is it possible for a triangle to have more than one line of symmetry? Explain your reasoning.

2. In Example 5 on page 215, one of the given statements is $\overline{AD} \parallel \overline{EH}$. Explain how this statement is used in proving that square $ABCD$ is similar to square $EFGH$. Describe the additional step(s) needed in the proof if this statement is not given.

GO DIGITAL

4 Practice Test WITH CalcChat®

1. Graph $\triangle RST$ with vertices $R(-4, 1)$, $S(-2, 2)$, and $T(3, -2)$ and its image after the translation $(x, y) \rightarrow (x - 4, y + 1)$.

Graph the polygon with the given vertices and its image after a rotation of the given number of degrees about the origin.

2. $D(-1, -1)$, $E(-3, 2)$, $F(1, 4)$; 270° clockwise

3. $J(-1, 1)$, $K(3, 3)$, $L(4, -3)$, $M(0, -2)$; 180°

Determine whether the polygons with the given vertices are congruent or similar. Use transformations to explain your reasoning.

4. $Q(2, 4)$, $R(5, 4)$, $S(6, 2)$, $T(1, 2)$ and $W(6, -12)$, $X(15, -12)$, $Y(18, -6)$, $Z(3, -6)$

5. $A(-6, 6)$, $B(-6, 2)$, $C(-2, -4)$ and $D(9, 7)$, $E(5, 7)$, $F(-1, 3)$

Determine whether the figure has line symmetry. If so, draw the line(s) of symmetry and describe any reflections that map the figure onto itself. Then determine whether the figure has rotational symmetry. If so, identify the order of rotational symmetry of the figure.

6. 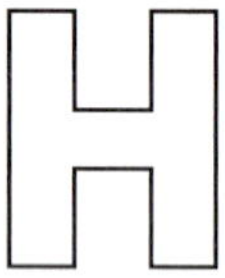

7. 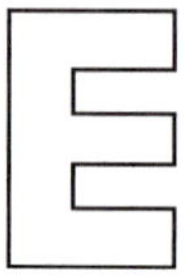

8. 

9. In the coordinate plane, you dilate $\triangle ABC$ with a scale factor of 3. The image is $\triangle A'B'C'$. What scale factor will map $\triangle A'B'C'$ back to $\triangle ABC$?

10. Write a composition of transformations that maps $\triangle ABC$ to $\triangle CDB$ in the tessellation shown. Is the composition a congruence transformation? Explain your reasoning.

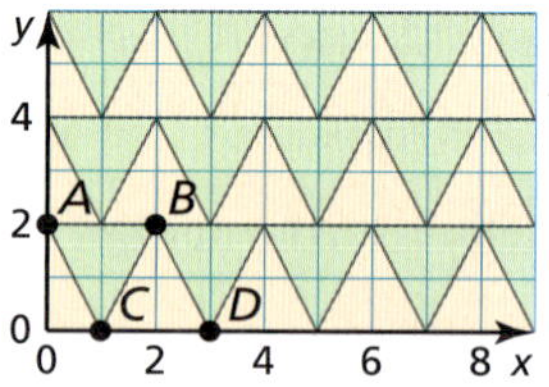

11. There is one slice of a large pizza and one slice of a small pizza in the box. Each unit in the coordinate plane corresponds to three centimeters.

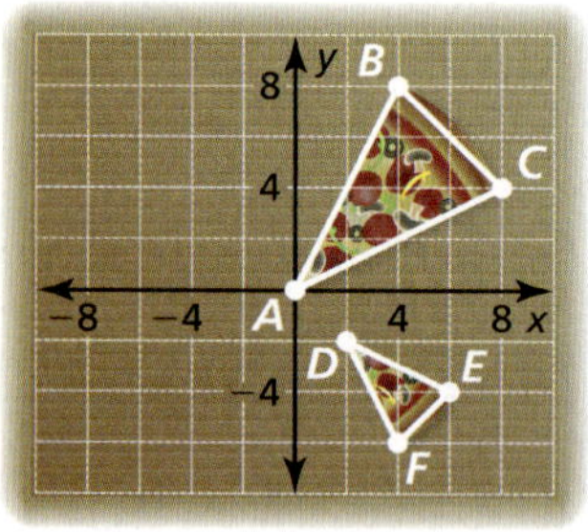

a. Describe a similarity transformation that maps $\triangle ABC$ to $\triangle DEF$.

b. What is one possible scale factor for a slice of a medium pizza?

c. Write a coordinate rule for the composition.

d. Approximate and compare the areas of the large and medium pizza slices.

12. The original photograph shown is 4 inches by 6 inches.

a. What transformations produce the new photograph?

b. The scale factor of the dilation involved in producing the new photograph is $\frac{1}{2}$. What are the dimensions of the new photograph?

c. You have a frame that holds photos that are 8.5 inches by 11 inches. Can you enlarge the original photograph without cropping to fit the frame? Explain your reasoning.

new

original

GO DIGITAL

4 Performance Task

Butterfly Identification

CITRUS SWALLOWTAIL

$3\frac{1}{2}$–$4\frac{3}{8}$ in.

ZEBRA SWALLOWTAIL

$2\frac{1}{2}$–4 in.

MEXICAN YELLOW

$1\frac{3}{4}$–$2\frac{1}{2}$ in.

DESERT ORANGETIP

1–$1\frac{1}{2}$ in.

GREEN-UNDERSIDE BLUE

1–$1\frac{1}{2}$ in.

ARCTIC SKIPPER

1–$1\frac{1}{4}$ in.

EUROPEAN PEACOCK

2–$2\frac{1}{2}$ in.

EMERALD SWALLOWTAIL

3–4 in.

SPICEBUSH SWALLOWTAIL

3–4 in.

THE APOLLO

$2\frac{1}{2}$–$3\frac{3}{4}$ in.

EASTERN TIGER SWALLOWTAIL

$2\frac{1}{2}$–$4\frac{1}{2}$ in.

MONARCH

$3\frac{3}{8}$–$4\frac{7}{8}$ in.

Out of nearly 20,000 species of butterflies, about 750 can be found in the United States.

GO DIGITAL

BUTTERFLY RESEARCH

Entomologists often create illustrations of the insects they are studying. Research one of the butterfly species shown. Create a life-size illustration of the species and show any lines of symmetry. Then construct a dilation that shows how the species would look under a magnifying glass. State the magnification level that you used. Finally, give information about how to identify the species, its habitat, and its diet.

4 Test Prep WITH CalcChat® Cumulative Practice

WATCH Tutorial videos are available for each exercise.

1. Which compositions of transformations map $\triangle ABC$ to $\triangle DEF$? Select all that apply.

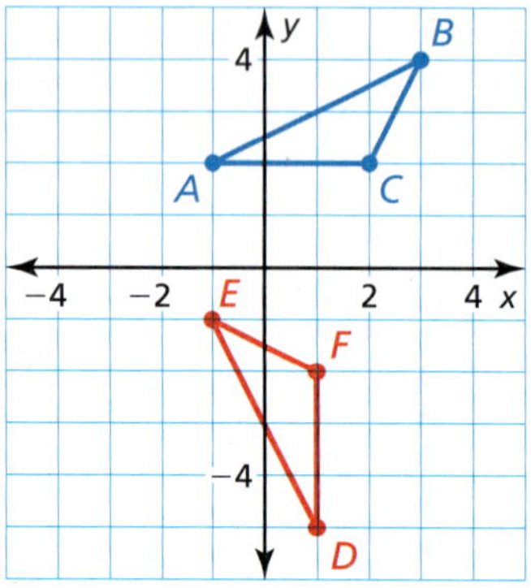

Ⓐ **Rotation:** 90° counterclockwise about the origin
Translation: $(x, y) \rightarrow (x + 3, y - 4)$

Ⓑ **Translation:** $(x, y) \rightarrow (x - 4, y - 3)$
Rotation: 90° counterclockwise about the origin

Ⓒ **Translation:** $(x, y) \rightarrow (x - 4, y - 3)$
Rotation: 270° counterclockwise about the origin

Ⓓ **Rotation:** 270° counterclockwise about the origin
Translation: $(x, y) \rightarrow (x + 3, y - 4)$

2. Describe each step in constructing a line perpendicular to line m through point P, which is not on line m.

Step 1

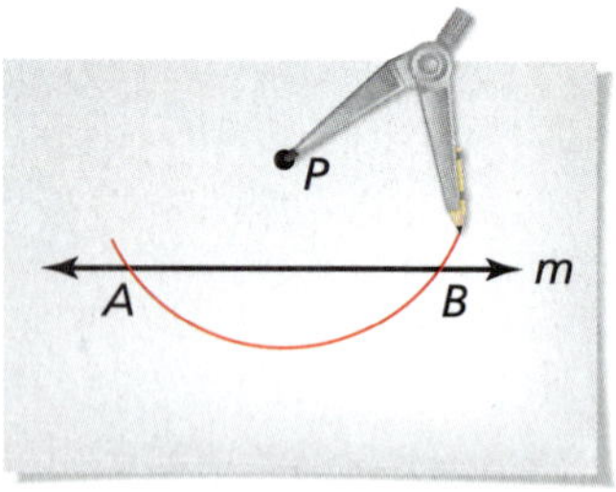

Step 2

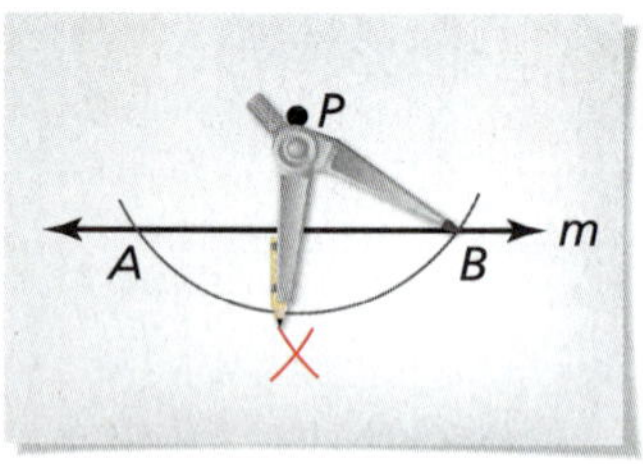

Step 3

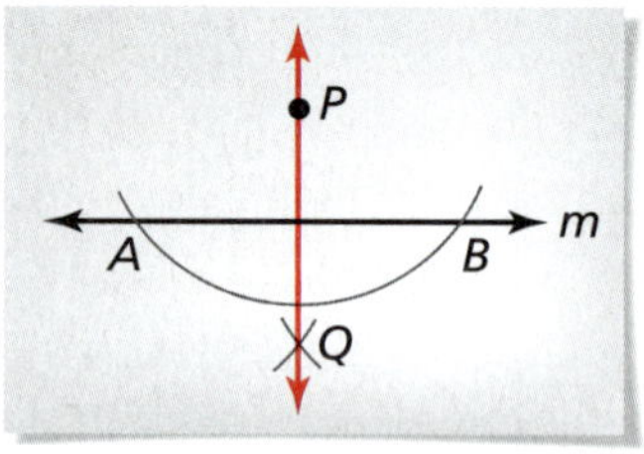

3. Your friend claims that it is possible to find the perimeter of the school crossing sign without using the Distance Formula. Do you support your friend's claim? Explain your reasoning.

GO DIGITAL

4. The endpoints of the directed line segment ST are $S(-3, -2)$ and $T(4, 5)$. Which are the coordinates of point P along segment ST so that the ratio of SP to PT is 3 to 4?

Ⓐ $\left(-1\frac{1}{4}, -\frac{1}{4}\right)$

Ⓑ $\left(2\frac{1}{4}, 3\frac{1}{4}\right)$

Ⓒ $(0, 1)$

Ⓓ $(1, 2)$

5. Which statements can you assume from the diagram? Select all that apply.

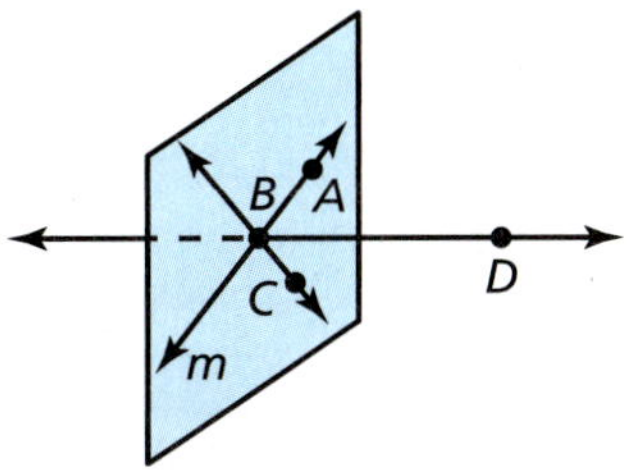

Ⓐ $\overleftrightarrow{BC} \perp$ line m

Ⓑ Points A, B, and C are coplanar.

Ⓒ Line m lies in plane ABC.

Ⓓ $\overleftrightarrow{BD} \perp$ plane ABC

6. Which equation represents the line passing through the point $(-6, 3)$ that is parallel to the line $y = -\frac{1}{3}x - 5$?

Ⓐ $y = 3x + 21$

Ⓑ $y = -\frac{1}{3}x + 5$

Ⓒ $y = 3x - 15$

Ⓓ $y = -\frac{1}{3}x + 1$

7. Which scale factor(s) would create a dilation of $\overline{AB}$ that is shorter than $\overline{AB}$? Select all that apply.

$\frac{1}{3}$ $\frac{7}{2}$ $\frac{3}{2}$ 3 $\frac{3}{4}$ 2 1 $\frac{1}{2}$

8. List one possible set of coordinates of the vertices of quadrilateral $ABCD$ for each description.

a. A reflection in the y-axis maps quadrilateral $ABCD$ onto itself.

b. A reflection in the x-axis maps quadrilateral $ABCD$ onto itself.

c. A rotation of 90° counterclockwise about the origin maps quadrilateral $ABCD$ onto itself.

d. A rotation of 180° about the origin maps quadrilateral $ABCD$ onto itself.

9. Two angles form a linear pair. The measure of one angle is 6° less than twice the measure of the other angle. What is the measure of the smaller angle?

Ⓐ 32°

Ⓑ 54°

Ⓒ 58°

Ⓓ 62°

GO DIGITAL

5 Congruent Triangles

NATIONAL GEOGRAPHIC EXPLORER

WATCH

Aydogan Ozcan

Dr. Aydogan Ozcan's research focuses on the use of computation to create new optical microscopy, sensing, and diagnostic technologies. He pioneered the use of smartphone biomedical tools such as diagnostic test readers, bacteria sensors, blood analyzers, and allergen detectors, all integrated with mobile phones using compact interfaces.

- What is microscopy?
- What is an allergen detector? How might you use your smartphone to determine whether you are allergic to cats?
- Name some types of bacteria that are beneficial to humans. Name some types that are harmful to humans.

STEM

Scientists draw diagrams of bacteria and viruses to document their features. In the Performance Task, you will analyze a drawing of a virus known as a *bacteriophage*.

GO DIGITAL

Diagnostic Technologies

Preparing for Chapter 5

Chapter Learning Target: Understand congruent triangles.

Chapter Success Criteria:
- ◆ I can classify triangles by sides and angles.
- ◆ I can solve problems involving congruent triangles.
- ■ I can use different theorems to prove two triangles are congruent.
- ■ I can write a coordinate proof.

◆ Surface
■ Deep

AZ VOCAB Chapter Vocabulary

Work with a partner. Discuss each of the vocabulary terms.

interior angles
exterior angles
corresponding parts
base
base angles
hypotenuse
coordinate proof

Mathematical Thinking and Reasoning

4 MTR ENGAGE IN DISCUSSIONS THAT REFLECT ON THE MATHEMATICAL THINKING OF SELF AND OTHERS

Mathematicians who engage in discussions that reflect on the mathematical thinking of self and others justify results by explaining methods and processes.

Work with a partner. Two triangles that have the same size and the same shape are *congruent* triangles.

1. Consider a triangle with vertices A, B, and C. Determine whether each set of measurements describes *one* or *many* triangles. Explain your reasoning.

a. $m\angle A = 60°, AB = 3 \text{ cm}, m\angle B = 45°$

b. $m\angle A = 30°, m\angle B = 50°, m\angle C = 100°$

c. $m\angle B = 50°, m\angle C = 40°, CA = 7 \text{ cm}$

d. $AB = 4 \text{ cm}, m\angle B = 40°, BC = 2 \text{ cm}$

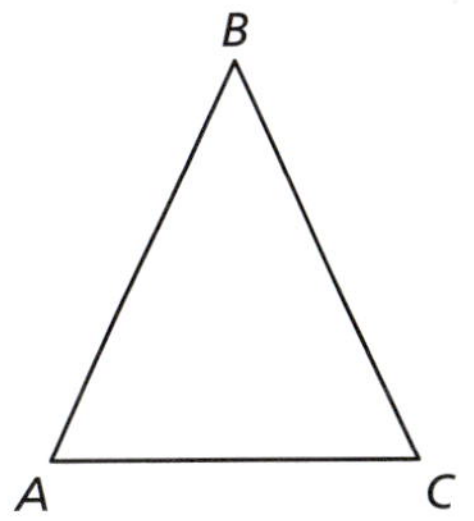

2. Your friend says that if three angles of one triangle have the same measures as three angles of another triangle, then the triangles are congruent. Is your friend correct? Explain.

3. Using your results in Exercise 1, determine several ways to conclude that two triangles are congruent. Explain.

GO DIGITAL

5 Prepare WITH CalcChat®

Solving Equations with Variables on Both Sides

WATCH

Example 1 **Solve $2 - 5x = -3x$.**

$2 - 5x = -3x$	Write the equation.
$\underline{+5x} \quad \underline{+5x}$	Addition Property of Equality
$2 = 2x$	Simplify.
$\dfrac{2}{2} = \dfrac{2x}{2}$	Division Property of Equality
$1 = x$	Simplify.

▶ The solution is $x = 1$.

Solve the equation.

1. $7x + 12 = 3x$

2. $5p + 10 = 8p + 1$

3. $z - 2 = 4 + 9z$

Using the Midpoint and Distance Formulas

WATCH

Example 2 **The endpoints of $\overline{AB}$ are $A(-2, 3)$ and $B(4, 7)$. Find the coordinates of the midpoint M.**

Use the Midpoint Formula.

$$M\left(\frac{-2+4}{2}, \frac{3+7}{2}\right) = M\left(\frac{2}{2}, \frac{10}{2}\right)$$
$$= M(1, 5)$$

▶ The coordinates of the midpoint M are $(1, 5)$.

WATCH

Example 3 **Find the distance between $C(0, -5)$ and $D(3, 2)$.**

$CD = \sqrt{(x_2 - x_1)^2 + (y_2 - y_1)^2}$	Distance Formula
$= \sqrt{(3 - 0)^2 + [2 - (-5)]^2}$	Substitute.
$= \sqrt{3^2 + 7^2}$	Subtract.
$= \sqrt{9 + 49}$	Evaluate powers.
$= \sqrt{58}$	Add.
≈ 7.6	Use technology.

▶ The distance between $C(0, -5)$ and $D(3, 2)$ is about 7.6.

Find the coordinates of the midpoint M of the segment with the given endpoints. Then find the distance between the endpoints.

4. $P(-4, 1)$ and $Q(0, 7)$

5. $G(3, 6)$ and $H(9, -2)$

6. $U(-1, -2)$ and $V(8, 0)$

4 MTR **7.** **DISCUSS MATHEMATICAL THINKING** Explain why you can use the Pythagorean Theorem to find the distance between two points in a coordinate plane.

GO DIGITAL

5.1 Angles of Triangles

Learning Target: Prove and use theorems about angles of triangles.

Success Criteria:
- I can classify triangles by sides and by angles.
- I can prove theorems about angles of triangles.
- I can find interior and exterior angle measures of triangles.

When you demonstrate a concept without formally justifying each step, you are constructing an *informal proof*.

EXPLORE IT! Analyzing Angle Measures of Triangles

Work with a partner.

a. Use technology to draw a triangle. Find the measures of the interior angles of the triangle.

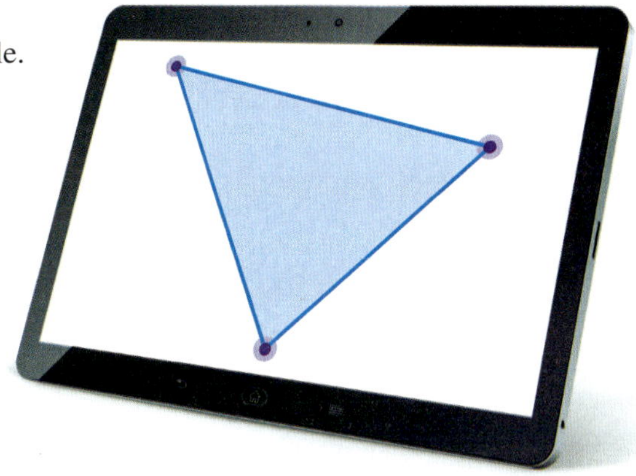

b. Repeat part (a) for several triangles. What do you noticc about the sum of the interior angle measures? Make a conjecture.

4 MTR COMMUNICATE CLEARLY
An informal proof is a useful stepping-stone toward a formal proof. Explain why your informal proofs are insufficient to prove your conjectures.

c. Construct and cut out a triangle. Tear off two of its angles and arrange them to construct an informal proof of your conjecture in part (b).

d. Use technology to draw a triangle and an exterior angle. Find the measures of the interior angles and exterior angle.

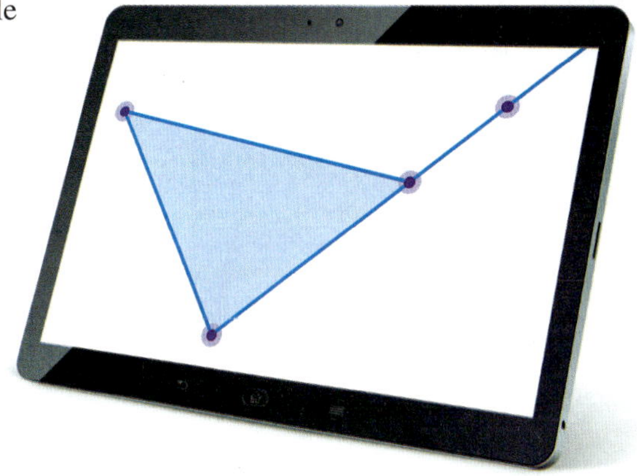

e. Repeat part (d) for several triangles. What do you notice about the measures of the interior angles and exterior angles'? Make a conjecture.

f. Construct and cut out a triangle. Trace the triangle on a piece of paper and extend one side to form an exterior angle. Tear off the angles of the triangle that are not adjacent to the exterior angle. Arrange the pieces to construct an informal proof of your conjecture in part (e).

Geometric Reasoning
MA.912.GR.1.3 Prove relationships and theorems about triangles. Solve mathematical and real-world problems involving postulates, relationships and theorems of triangles.
MA.912.GR.3.2 Given a mathematical context, use coordinate geometry to classify or justify definitions, properties and theorems involving circles, triangles or quadrilaterals.

GO DIGITAL

Classifying Triangles by Sides and by Angles

Vocabulary AZ VOCAB

interior angles, *p. 231*
exterior angles, *p. 231*
corollary to a theorem, *p. 233*

Recall that a *triangle* is a polygon with three sides. You can classify triangles by sides and by angles, as shown below.

KEY IDEAS

Classifying Triangles by Sides

Scalene Triangle

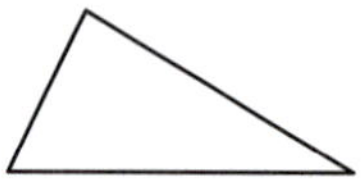

no congruent sides

Isosceles Triangle

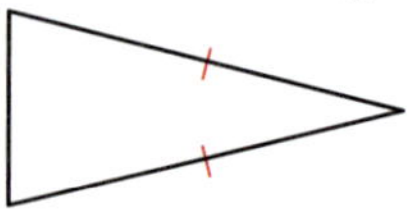

at least 2 congruent sides

Equilateral Triangle

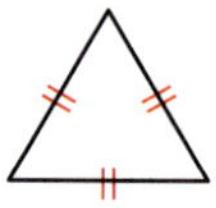

3 congruent sides

Classifying Triangles by Angles

Acute Triangle

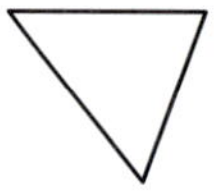

3 acute angles

Right Triangle

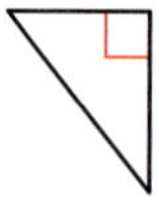

1 right angle

Obtuse Triangle

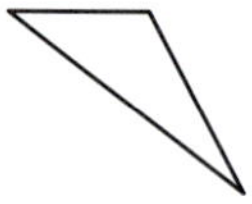

1 obtuse angle

Equiangular Triangle

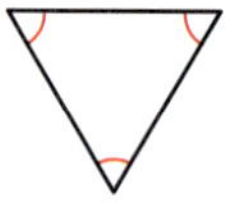

3 congruent angles

STUDY TIP

Notice that an equilateral triangle is also isosceles. An equiangular triangle is also acute.

EXAMPLE 1 Classifying Triangles by Sides and by Angles

WATCH

Classify the triangular tile on Spaceship Earth at Epcot Center in Disney World by its sides and by measuring its angles.

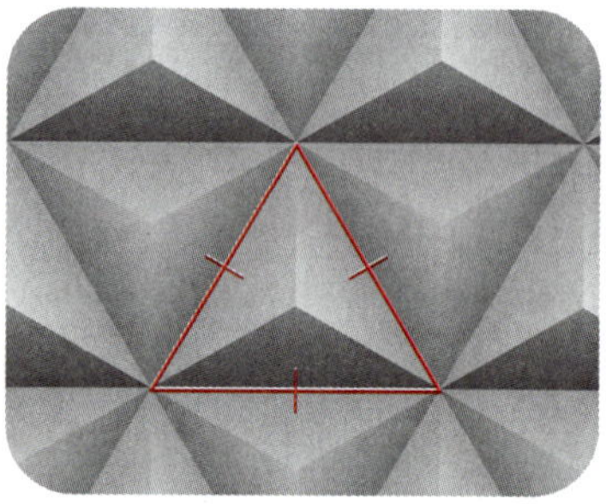

SOLUTION

All the sides are congruent, so it is equilateral. By measuring, you can determine that all the angles are 60°.

▶ So, it is an equiangular equilateral triangle.

SELF-ASSESSMENT

1 I don't understand yet. 2 I can do it with help. 3 I can do it on my own. 4 I can teach someone else.

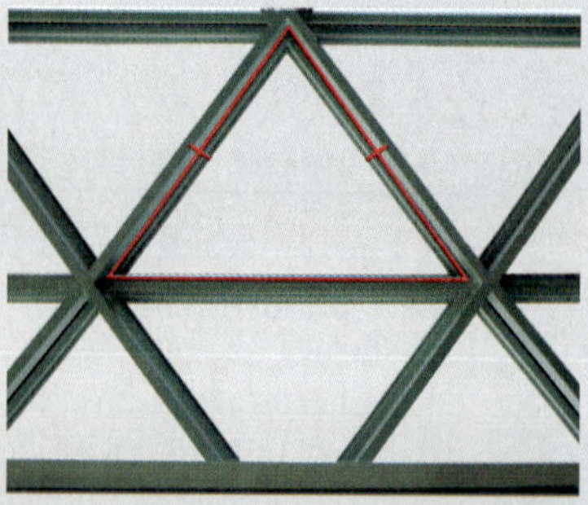

1. Classify the triangle by its sides and by measuring its angles.

2. Draw an obtuse isosceles triangle and an acute scalene triangle.

3. **WRITING** Can a right triangle also be obtuse? Explain your reasoning.

GO DIGITAL

EXAMPLE 2 Classifying a Triangle in the Coordinate Plane

WATCH

Classify $\triangle OPQ$ by its sides. Then determine whether it is a right triangle.

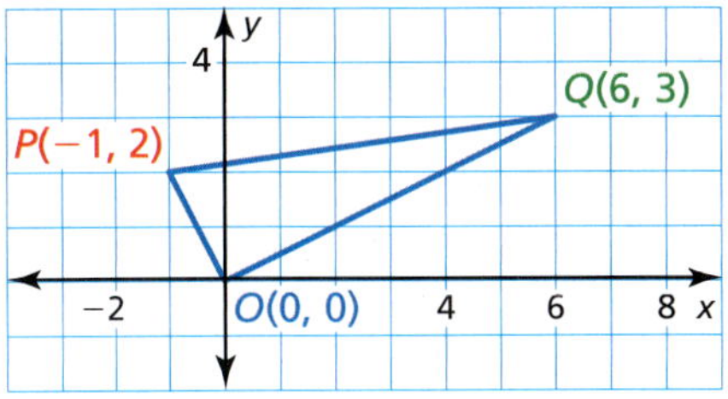

SOLUTION

Step 1 Use the Distance Formula to find the side lengths.

$$OP = \sqrt{(x_2 - x_1)^2 + (y_2 - y_1)^2} = \sqrt{(-1 - 0)^2 + (2 - 0)^2} = \sqrt{5} \approx 2.2$$

$$OQ = \sqrt{(x_2 - x_1)^2 + (y_2 - y_1)^2} = \sqrt{(6 - 0)^2 + (3 - 0)^2} = \sqrt{45} \approx 6.7$$

$$PQ = \sqrt{(x_2 - x_1)^2 + (y_2 - y_1)^2} = \sqrt{[6 - (-1)]^2 + (3 - 2)^2} = \sqrt{50} \approx 7.1$$

Because no sides are congruent, $\triangle OPQ$ is a scalene triangle.

Step 2 Check for right angles. The slope of $\overline{OP}$ is $\frac{2-0}{-1-0} = -2$. The slope of $\overline{OQ}$ is $\frac{3-0}{6-0} = \frac{1}{2}$. The product of the slopes is $-2\left(\frac{1}{2}\right) = -1$. So, $\overline{OP} \perp \overline{OQ}$ and $\angle POQ$ is a right angle.

▶ So, $\triangle OPQ$ is a right scalene triangle.

SELF-ASSESSMENT

1 I don't understand yet. 2 I can do it with help. 3 I can do it on my own. 4 I can teach someone else.

4. $\triangle ABC$ has vertices $A(0, 0)$, $B(3, 3)$, and $C(-3, 3)$. Classify the triangle by its sides. Then determine whether it is a right triangle.

Finding Angle Measures of Triangles

When the sides of a polygon are extended, other angles are formed. The original angles are the **interior angles**. The angles that form linear pairs with the interior angles are the **exterior angles**.

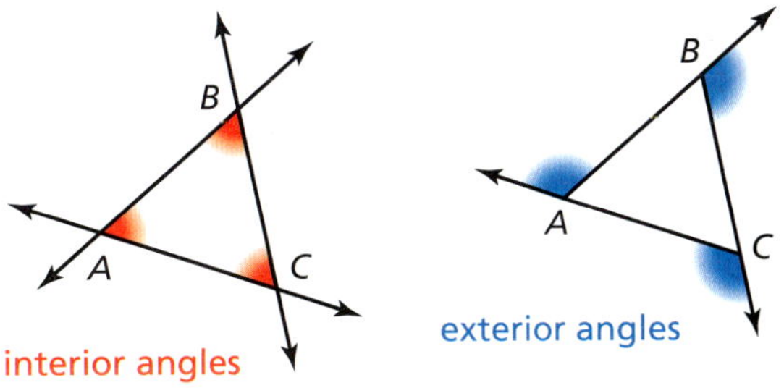

THEOREM

5.1 Triangle Sum Theorem

The sum of the measures of the interior angles of a triangle is 180°.

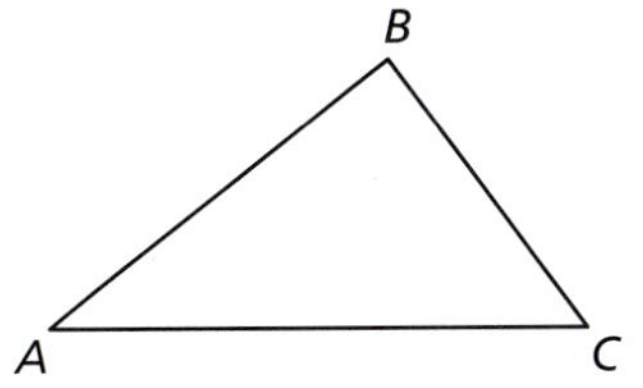

$m\angle A + m\angle B + m\angle C = 180°$

Proof page 232

Prove this Theorem Exercise 50, page 236

GO DIGITAL

To prove certain theorems, you may need to add a line, a segment, or a ray to a given diagram. An *auxiliary* line is used in the proof of the Triangle Sum Theorem.

PROOF Triangle Sum Theorem

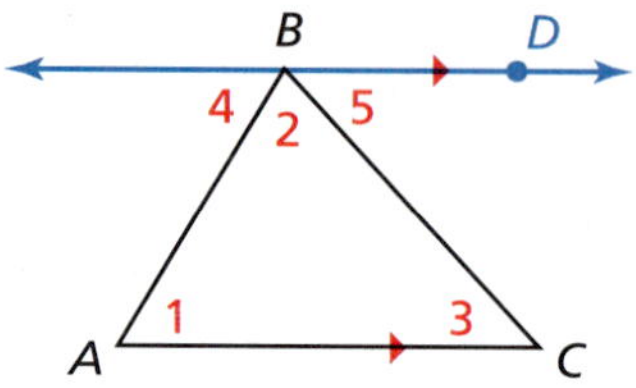

Given $\triangle ABC$

Prove $m\angle 1 + m\angle 2 + m\angle 3 = 180^\circ$

Plan for Proof

a. Draw an auxiliary line through B that is parallel to $\overline{AC}$.

b. Show that $m\angle 4 + m\angle 2 + m\angle 5 = 180^\circ$, $\angle 1 \cong \angle 4$, and $\angle 3 \cong \angle 5$.

c. By substitution, $m\angle 1 + m\angle 2 + m\angle 3 = 180^\circ$.

Plan in Action	STATEMENTS	REASONS
a.	**1.** Draw $\overleftrightarrow{BD}$ parallel to $\overline{AC}$.	**1.** Parallel Postulate
b.	**2.** $m\angle 4 + m\angle 2 + m\angle 5 = 180^\circ$	**2.** Angle Addition Postulate and definition of straight angle
	3. $\angle 1 \cong \angle 4$, $\angle 3 \cong \angle 5$	**3.** Alternate Interior Angles Theorem
	4. $m\angle 1 = m\angle 4$, $m\angle 3 = m\angle 5$	**4.** Definition of congruent angles
c.	**5.** $m\angle 1 + m\angle 2 + m\angle 3 = 180^\circ$	**5.** Substitution Property of Equality

THEOREM

5.2 Exterior Angle Theorem

The measure of an exterior angle of a triangle is equal to the sum of the measures of the two nonadjacent interior angles.

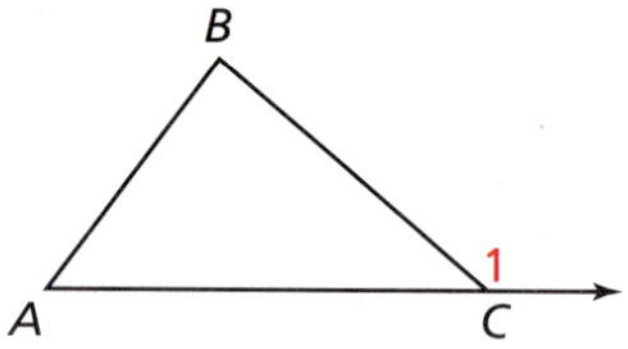

$m\angle 1 = m\angle A + m\angle B$

Prove this Theorem Exercise 42, page 235

EXAMPLE 3 Finding an Angle Measure

WATCH

Find $m\angle JKM$.

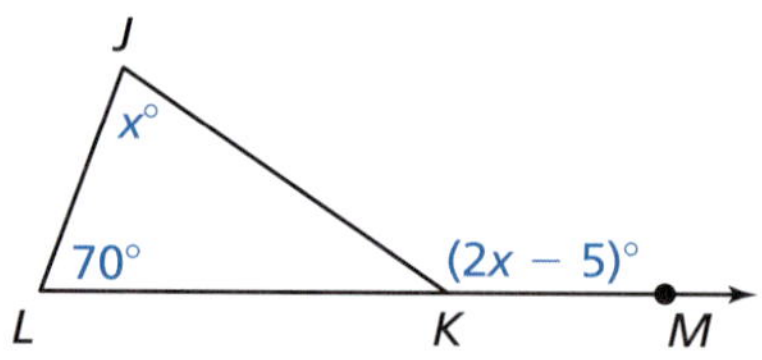

SOLUTION

Step 1 Write and solve an equation to find the value of x.

$(2x - 5)^\circ = 70^\circ + x^\circ$ Apply the Exterior Angle Theorem.

$x = 75$ Solve for x.

Step 2 Substitute 75 for x in $2x - 5$ to find $m\angle JKM$.

$2x - 5 = 2 \cdot 75 - 5 = 145$

▶ So, the measure of $\angle JKM$ is 145°.

GO DIGITAL

WORDS AND MATH

A *corollary* is something that accompanies, or follows, something else. In mathematics, a corollary to a theorem is a statement that follows from the theorem.

A **corollary to a theorem** is a statement that can be proved easily using the theorem. The corollary below follows from the Triangle Sum Theorem.

COROLLARY

5.1 Corollary to the Triangle Sum Theorem

The acute angles of a right triangle are complementary.

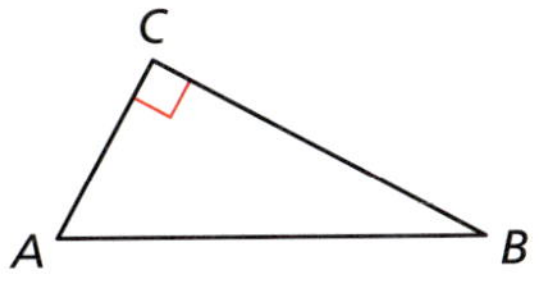

Prove this Corollary Exercise 41, page 235

$m\angle A + m\angle B = 90^\circ$

7 MTR

EXAMPLE 4 Modeling Real Life

WATCH

You are designing doors for a new building. The doors are right triangles. For each door, the measure of one acute angle in the triangle is twice the measure of the other. Find the measure of each acute angle for one door.

SOLUTION

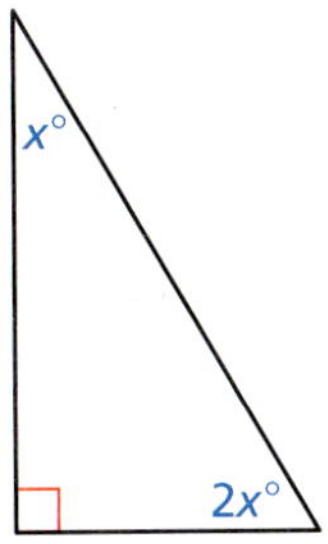

1. **Understand the Problem** You are given the relationship between the two acute angles in a door shaped like a right triangle. You need to find the measure of each acute angle.

2. **Make a Plan** Sketch a diagram of the situation. You can use the Corollary to the Triangle Sum Theorem and the given relationship between the two acute angles to write and solve an equation to find the measure of each acute angle.

3. **Solve and Check** Let the measure of the smaller acute angle be x°. Then the measure of the larger acute angle is $2x^\circ$. The Corollary to the Triangle Sum Theorem states that the acute angles of a right triangle are complementary.

   Use the corollary to set up and solve an equation.

   $x^\circ + 2x^\circ = 90^\circ$ Corollary to the Triangle Sum Theorem

   $x = 30$ Solve for x.

▶ So, the measures of the acute angles are 30° and 2(30°) = 60°.

Check Add the two angle measures and check that their sum satisfies the Corollary to the Triangle Sum Theorem.

$30^\circ + 60^\circ = 90^\circ$ ✓

SELF-ASSESSMENT

1 I don't understand yet. 2 I can do it with help. 3 I can do it on my own. 4 I can teach someone else.

5. Find $m\angle 1$.

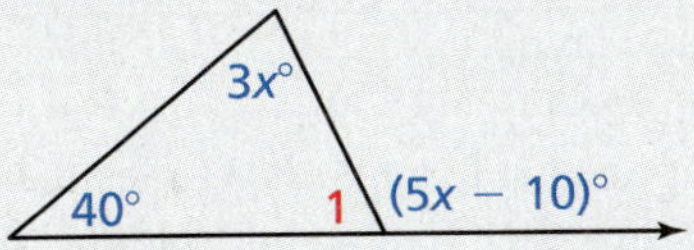

6. Find the measure of each acute angle.

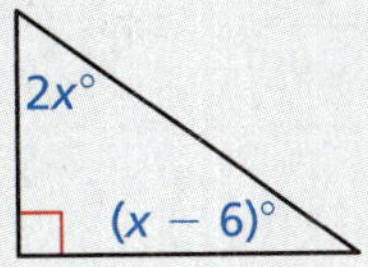

7. **WHAT IF?** In Example 4, find the measure of each acute angle when the measure of one acute angle in the triangle is three times the measure of the other.

GO DIGITAL

5.1 Practice WITH CalcChat® AND CalcView®

In Exercises 1–4, classify the triangle by its sides and by measuring its angles. *(See Example 1.)*

1.

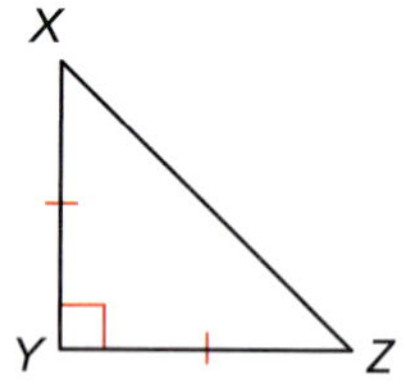

2.

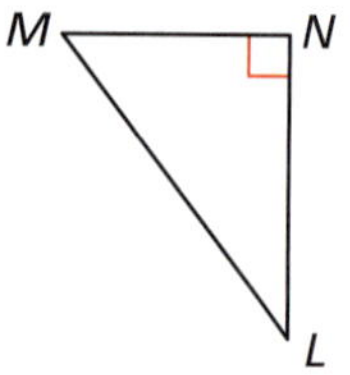

▶ 3.

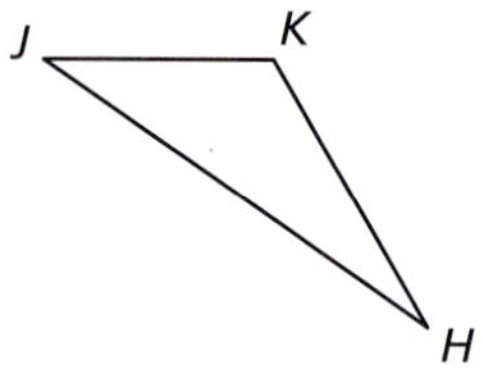

4.

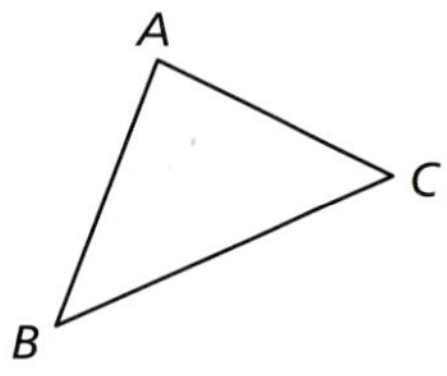

In Exercises 5–8, classify $\triangle ABC$ by its sides. Then determine whether it is a right triangle. *(See Example 2.)*

▶ 5. $A(2, 3)$, $B(6, 3)$, $C(2, 7)$

6. $A(3, 3)$, $B(6, 9)$, $C(6, -3)$

7. $A(1, 9)$, $B(4, 8)$, $C(2, 5)$

8. $A(-2, 3)$, $B(0, -3)$, $C(3, -2)$

In Exercises 9–12, find $m\angle 1$. Then classify the triangle by its angles.

9.

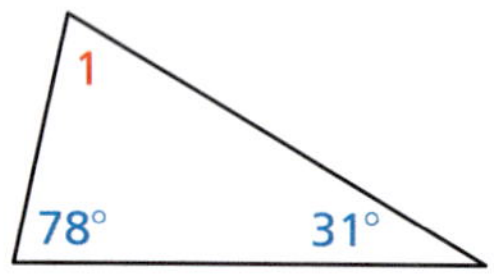

10.

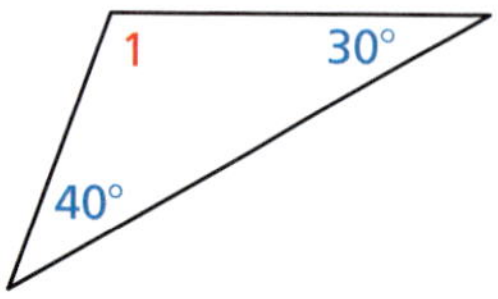

11.

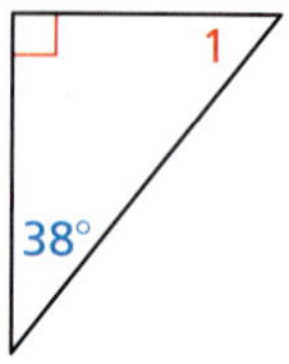

12.

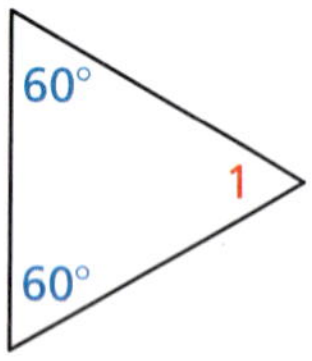

In Exercises 13–16, find the measure of the exterior angle. *(See Example 3.)*

▶ 13.

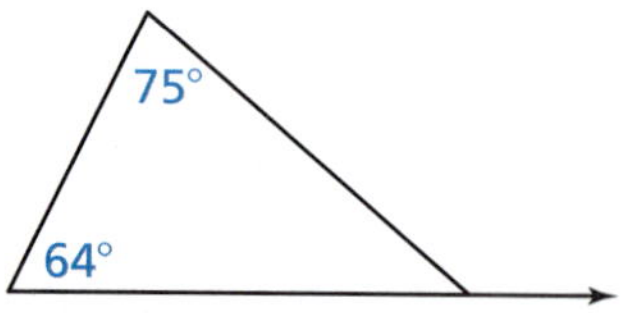

14.

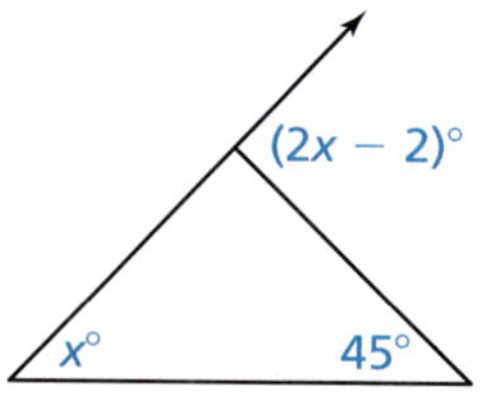

15.

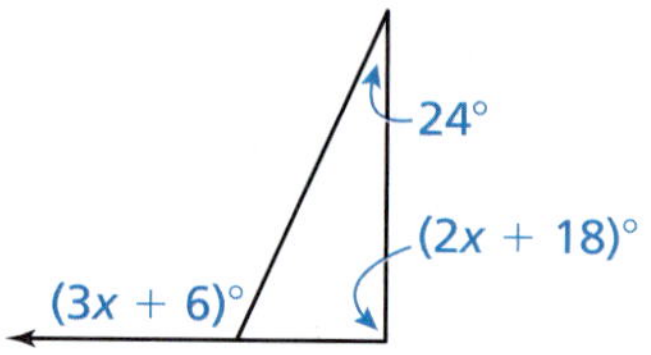

16.

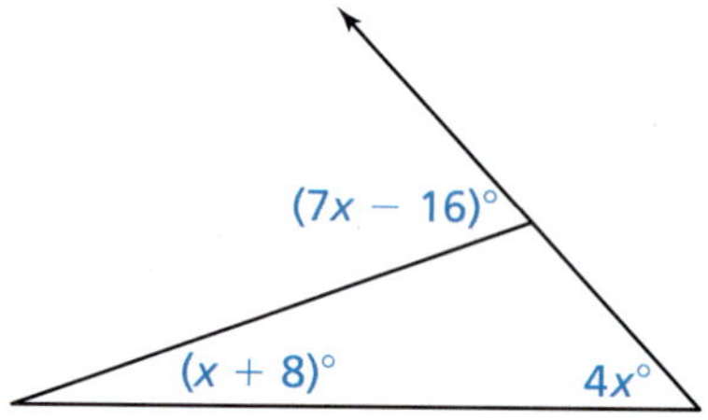

In Exercises 17–20, find the measure of each acute angle.

17.

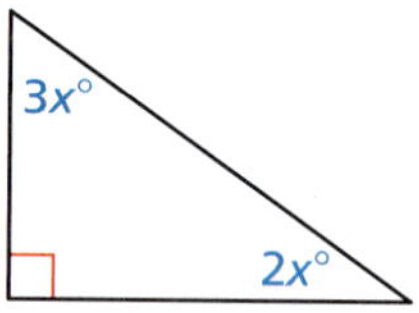

18.

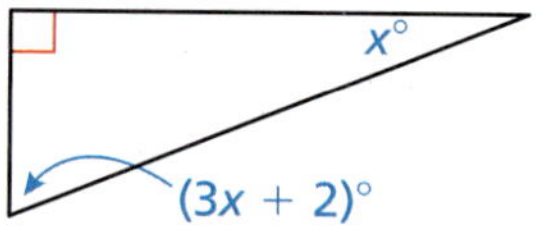

19.

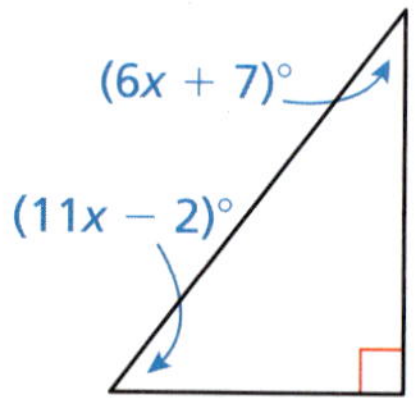

20.

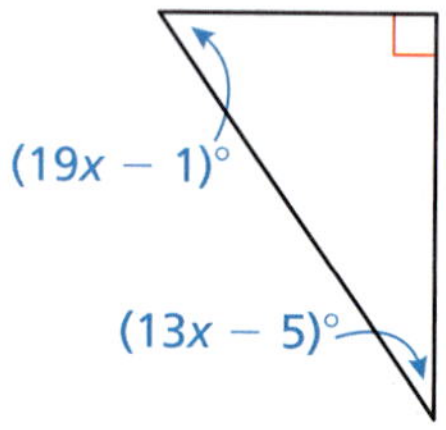

In Exercises 21–24, find the measure of each acute angle in the right triangle. *(See Example 4.)*

▶ 21. The measure of one acute angle is 5 times the measure of the other acute angle.

22. The measure of one acute angle is 8 times the measure of the other acute angle.

23. The measure of one acute angle is 3 times the sum of the measure of the other acute angle and 8.

24. The measure of one acute angle is twice the difference of the measure of the other acute angle and 12.

GO DIGITAL

ERROR ANALYSIS **In Exercises 25 and 26, describe and correct the error in finding $m\angle 1$.**

25.

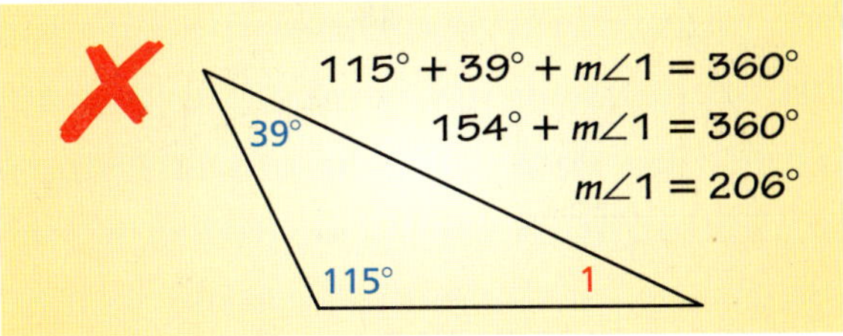

$$115^\circ + 39^\circ + m\angle 1 = 360^\circ$$
$$154^\circ + m\angle 1 = 360^\circ$$
$$m\angle 1 = 206^\circ$$

26.

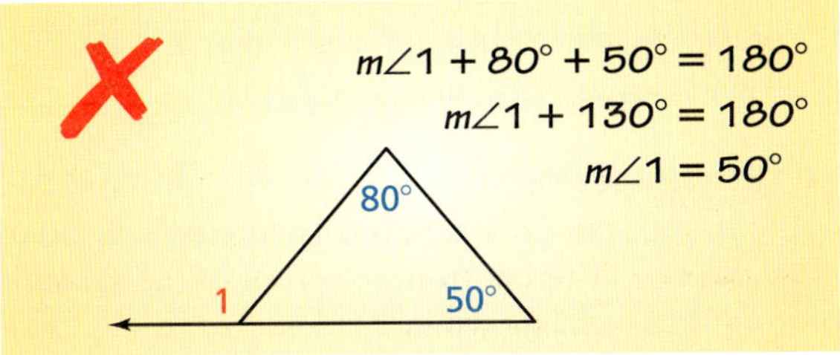

$$m\angle 1 + 80^\circ + 50^\circ = 180^\circ$$
$$m\angle 1 + 130^\circ = 180^\circ$$
$$m\angle 1 = 50^\circ$$

In Exercises 27–34, find the measure of the numbered angle.

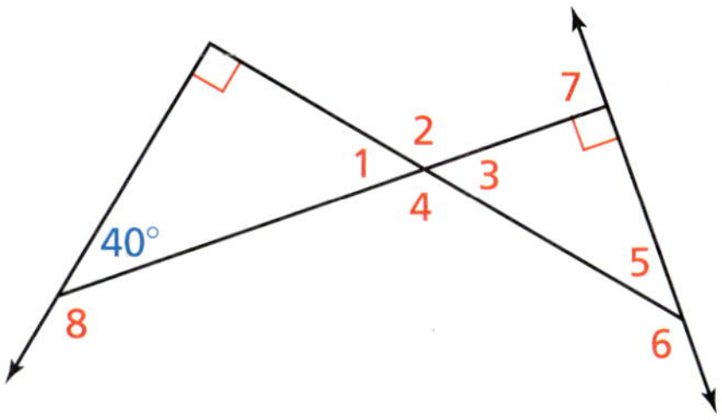

27. $\angle 1$

28. $\angle 2$

29. $\angle 3$

30. $\angle 4$

31. $\angle 5$

32. $\angle 6$

33. $\angle 7$

34. $\angle 8$

35. OPEN-ENDED Find and draw an object (or part of an object) that can be modeled by a triangle and an exterior angle.

36. OPEN-ENDED Construct a triangle with an exterior angle measure of 110°.

37. MODELING REAL LIFE A face of a bike ramp is shown. Classify the triangle by its sides and by its angles.

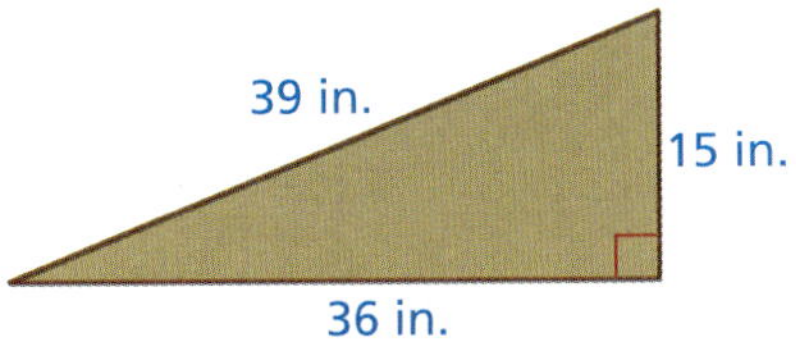

GO DIGITAL

38. MAKING AN ARGUMENT The Golden Triangle in Florida is formed by the cities of Eustis, Mount Dora, and Tavares. Your friend says the triangle can be classified as an equilateral and equiangular triangle. Is your friend's claim reasonable? Explain.

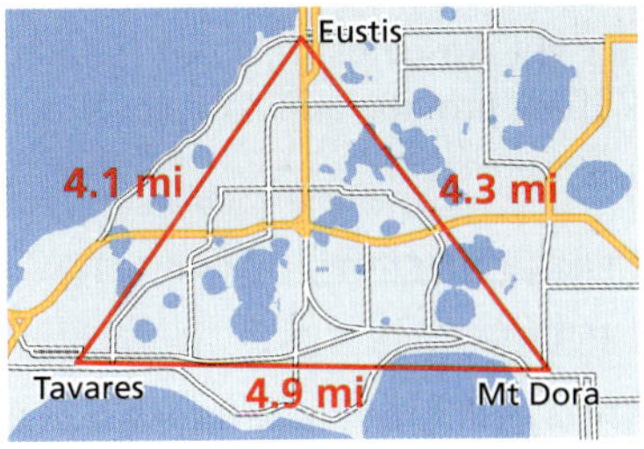

39. REASONING You are bending a strip of metal into an isosceles triangle for a sculpture. The strip of metal is 20 inches long. The first bend is made 6 inches from one end. Describe two ways you could complete the triangle.

40. HOW DO YOU SEE IT?
In as many ways as possible, classify each triangle by its appearance.

a.

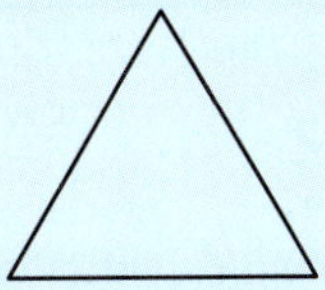

b.

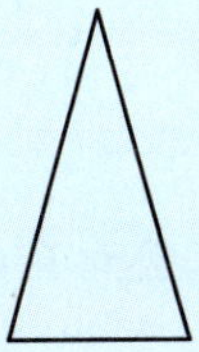

c.

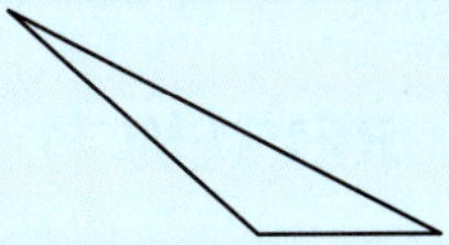

d.

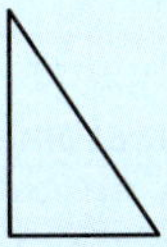

41. PROVING A COROLLARY Prove the Corollary to the Triangle Sum Theorem.

Given $\triangle ABC$ is a right triangle.

Prove $\angle A$ and $\angle B$ are complementary.

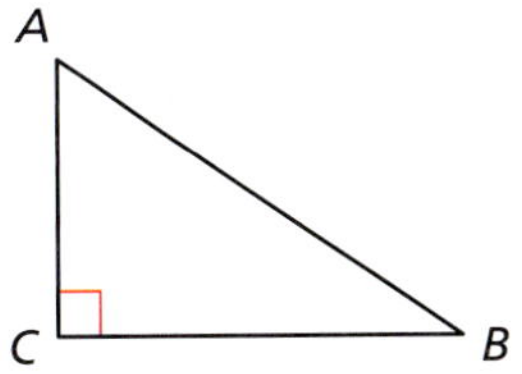

42. PROVING A THEOREM Prove the Exterior Angle Theorem.

Given $\triangle ABC$, exterior $\angle BCD$

Prove $m\angle A + m\angle B = m\angle BCD$

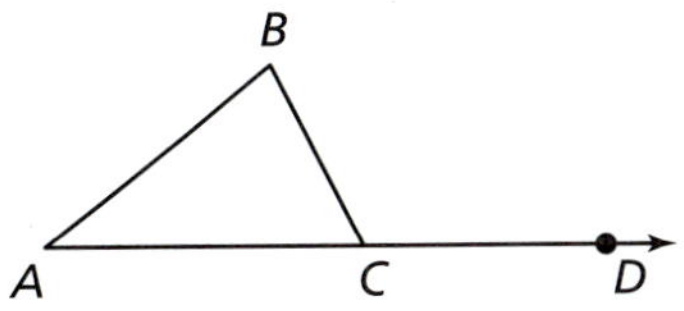

4 MTR **43. DISCUSS MATHEMATICAL THINKING** Is it possible to draw an obtuse isosceles triangle? obtuse equilateral triangle? If so, provide an example. If not, explain why it is not possible.

4 MTR **44. DISCUSS MATHEMATICAL THINKING** Is it possible to draw a right isosceles triangle? right equilateral triangle? If so, provide an example. If not, explain why it is not possible.

5 MTR **45. CONNECTING CONCEPTS** $\triangle ABC$ is isosceles, $AB = x$, and $BC = 2x - 4$.

a. Find two possible values for x when the perimeter of $\triangle ABC$ is 32.

b. How many possible values are there for x when the perimeter of $\triangle ABC$ is 12?

46. THOUGHT PROVOKING Let the measures of the exterior angles of a triangle, one angle at each vertex, be $x°$, $y°$, and $z°$, as shown. What can you prove?

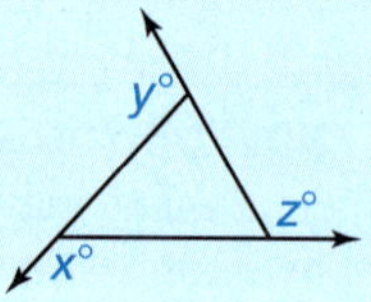

5 MTR **CONNECTING CONCEPTS In Exercises 47 and 48, find the values of x and y.**

47.

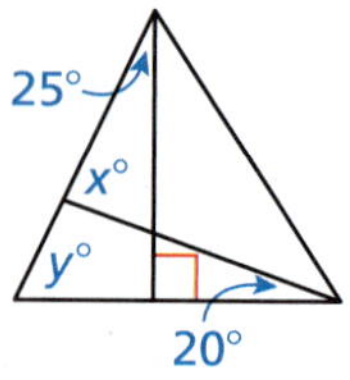

48.

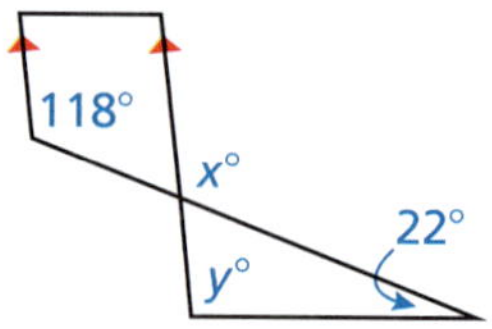

4 MTR **49. MAKING AN ARGUMENT** Your friend claims the measure of an exterior angle will always be greater than each interior angle measure. Is your friend correct? Explain.

50. PROVING A THEOREM Use the diagram to write a proof of the Triangle Sum Theorem. Your proof should be different from the proof of the Triangle Sum Theorem shown in this lesson.

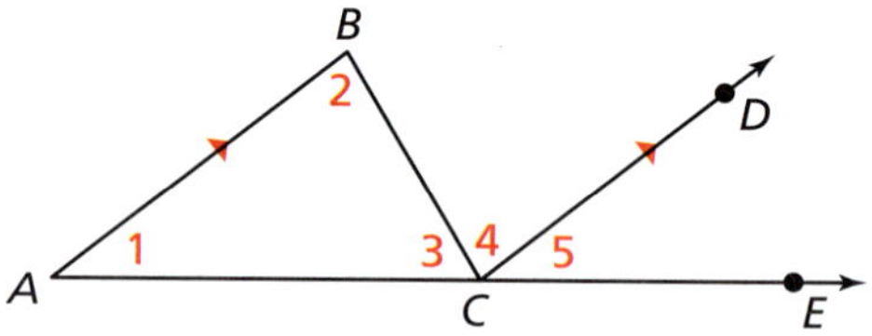

REVIEW & REFRESH

WATCH

51. Determine whether the triangles with the vertices $Q(1, 1)$, $R(2, 4)$, $S(5, 1)$ and $T(3, 0)$, $U(4, 5)$, $V(7, 0)$ are congruent. Use transformations to explain your reasoning.

52. Find the measure of each acute angle in a right triangle in which the measure of one acute angle is 11 times the measure of the other acute angle.

In Exercises 53 and 54, solve the equation.

53. $|2b - 5| = 9$ **54.** $x + 6 = -2x$

55. Find the scale factor of the dilation. Then tell whether the dilation is a *reduction* or an *enlargement*.

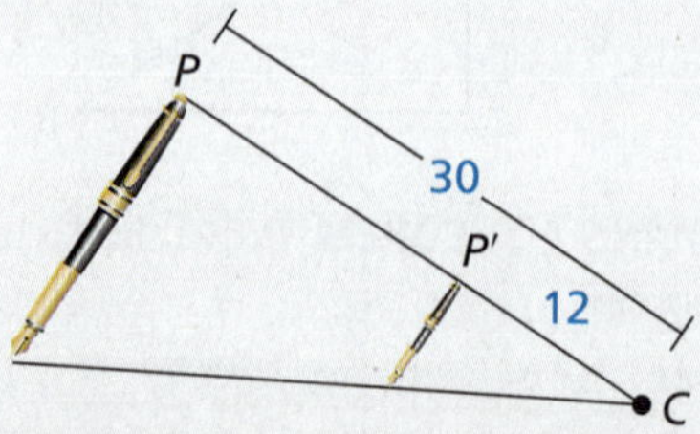

56. Determine whether $y = 2(0.5)^t$ represents *exponential growth* or *exponential decay*. Identify the percent rate of change.

57. Describe a similarity transformation that maps $\triangle DEF$ to $\triangle TUV$. Verify that the corresponding sides are proportional and the corresponding angles are congruent.

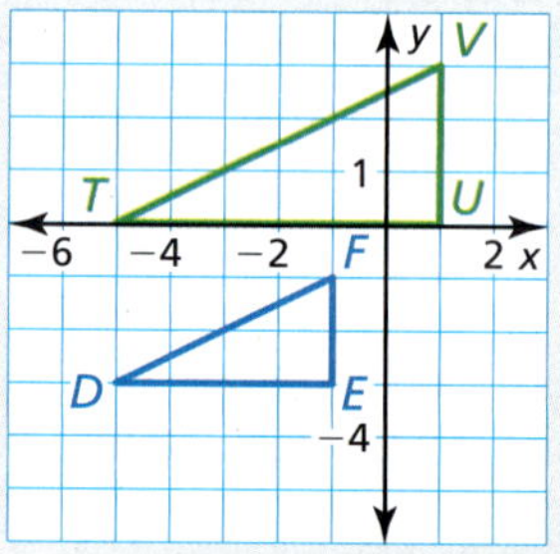

In Exercises 58 and 59, use the diagram.

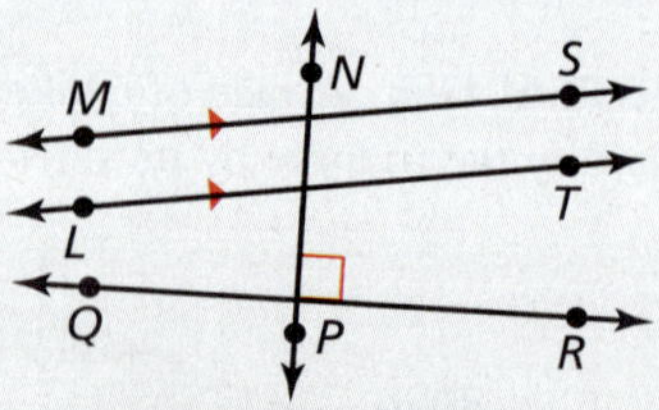

58. Name a pair of parallel lines.

59. Name a pair of perpendicular lines.

GO DIGITAL

5.2 Congruent Polygons

Learning Target: Understand congruence in terms of rigid motions.

Success Criteria:
- I can use rigid motions to show that two triangles are congruent.
- I can identify corresponding parts of congruent polygons.
- I can use congruent polygons to solve problems.

EXPLORE IT! Describing Rigid Motions

4 MTR COMMUNICATE CLEARLY

When the corresponding sides and corresponding angles of $\triangle ABC$ and $\triangle DEF$ are congruent, can you conclude that a rigid motion or a composition of rigid motions maps $\triangle ABC$ to $\triangle FED$? Explain.

Work with a partner.

a. Consider $\triangle ABC$ and $\triangle DEF$.

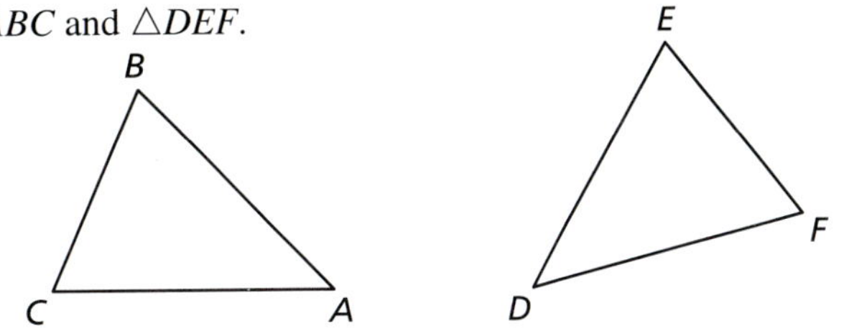

i. Explain why the statement is true.

If a rigid motion or a composition of rigid motions maps ΔABC to ΔDEF, then the corresponding sides and corresponding angles of ΔABC and ΔDEF are congruent.

ii. Show why the statement is true.

If the corresponding sides and corresponding angles of ΔABC and ΔDEF are congruent, then a rigid motion or a composition of rigid motions maps ΔABC to ΔDEF.

b. For each pair of triangles, describe a rigid motion or a composition of rigid motions that maps one of the triangles to the other. Then explain what you can conclude about each pair of triangles.

i. $\triangle ABC$ and $\triangle BAD$

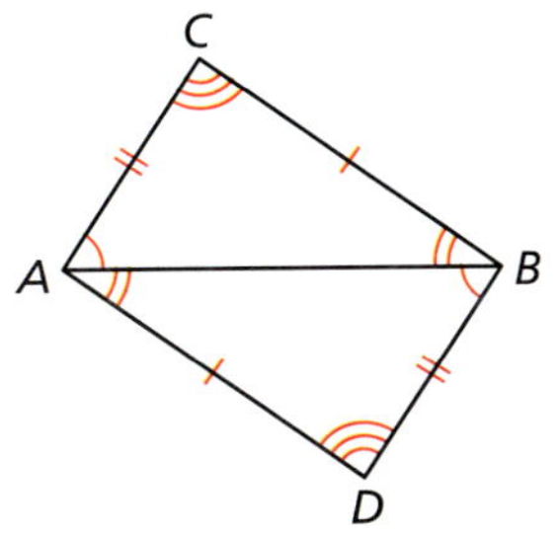

ii. $\triangle ABC$ and $\triangle ADE$

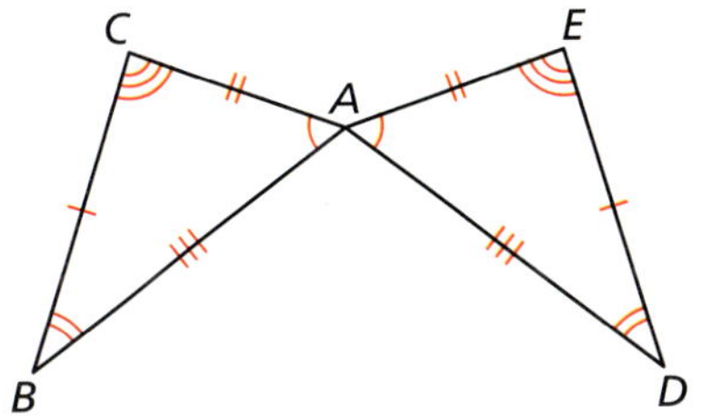

Geometric Reasoning

MA.912.GR.1.6 Solve mathematical and real-world problems involving congruence or similarity in two-dimensional figures.

MA.912.GR.2.6 Apply rigid transformations to map one figure onto another to justify that the two figures are congruent.

H **Also MA.912.GR.2.7**

GO DIGITAL

Identifying and Using Corresponding Parts

Vocabulary AZ VOCAB

corresponding parts, *p. 238*

Recall that two geometric figures are congruent if and only if a rigid motion or a composition of rigid motions maps one of the figures onto the other. A rigid motion maps each part of a figure to a **corresponding part** of its image. Because rigid motions preserve length and angle measure, corresponding parts of congruent figures are congruent. In congruent polygons, this means that the *corresponding sides* and the *corresponding angles* are congruent.

When $\triangle DEF$ is the image of $\triangle ABC$ after a rigid motion or a composition of rigid motions, you can write congruence statements for the corresponding angles and corresponding sides.

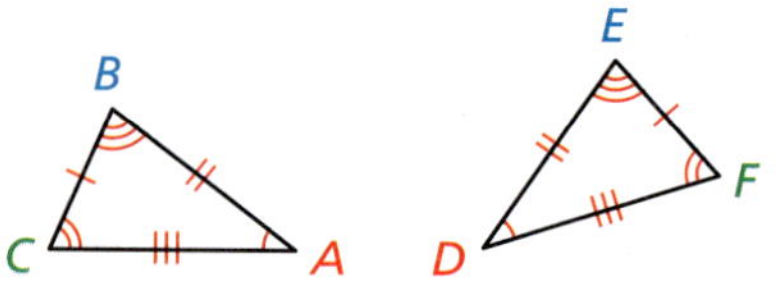

Corresponding angles	Corresponding sides
$\angle A \cong \angle D$, $\angle B \cong \angle E$, $\angle C \cong \angle F$	$\overline{AB} \cong \overline{DE}$, $\overline{BC} \cong \overline{EF}$, $\overline{AC} \cong \overline{DF}$

STUDY TIP

Notice that both of the following statements are true.

1. If two triangles are congruent, then all their corresponding parts are congruent.
2. If all the corresponding parts of two triangles are congruent, then the triangles are congruent.

When you write a congruence statement for two polygons, always list the corresponding vertices in the same order. You can write congruence statements in more than one way. Two possible congruence statements for the triangles above are $\triangle ABC \cong \triangle DEF$ or $\triangle BCA \cong \triangle EFD$.

When all the corresponding parts of two triangles are congruent, you can show that the triangles are congruent. Using the triangles above, first translate $\triangle ABC$ so that point A maps to point D. This translation maps $\triangle ABC$ to $\triangle DB'C'$. Next, rotate $\triangle DB'C'$ counterclockwise through $\angle C'DF$ so that the image of $\overrightarrow{DC'}$ coincides with $\overrightarrow{DF}$. Because $\overline{DC'} \cong \overline{DF}$, the rotation maps point C' to point F. So, this rotation maps $\triangle DB'C'$ to $\triangle DB''F$.

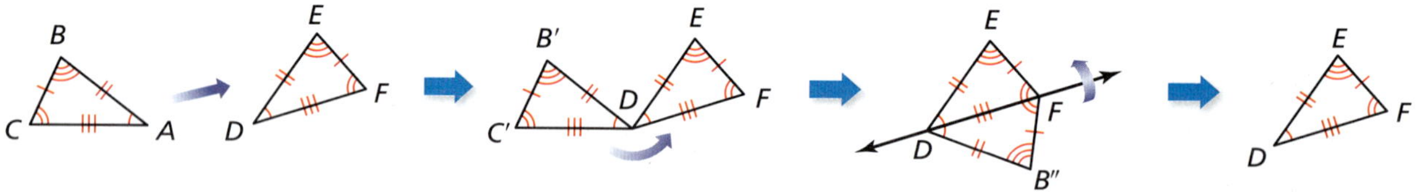

Now, reflect $\triangle DB''F$ in the line through points D and F. This reflection maps the sides and angles of $\triangle DB''F$ to the corresponding sides and corresponding angles of $\triangle DEF$, so $\triangle ABC \cong \triangle DEF$.

So, to show that two triangles are congruent, it is sufficient to show that their corresponding parts are congruent. In general, this is true for all polygons.

STUDY TIP

To help you identify corresponding parts, rotate $\triangle TSR$.

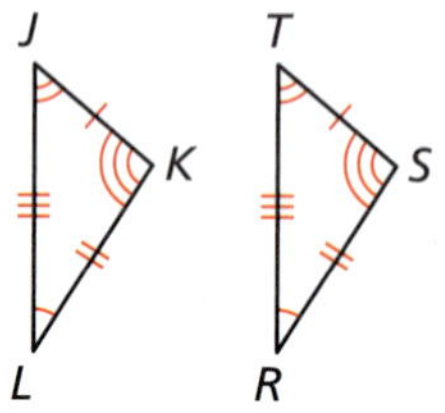

EXAMPLE 1 Identifying Corresponding Parts

WATCH

Write a congruence statement for the triangles. Identify all pairs of congruent corresponding parts.

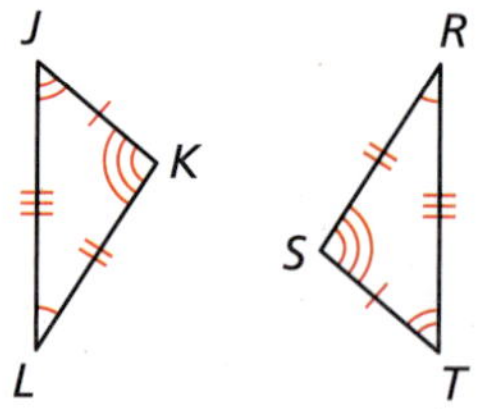

SOLUTION

The diagram indicates that $\triangle JKL \cong \triangle TSR$.

Corresponding angles $\angle J \cong \angle T$, $\angle K \cong \angle S$, $\angle L \cong \angle R$

Corresponding sides $\overline{JK} \cong \overline{TS}$, $\overline{KL} \cong \overline{SR}$, $\overline{LJ} \cong \overline{RT}$

GO DIGITAL

EXAMPLE 2 Using Properties of Congruent Figures

WATCH

In the diagram, $DEFG \cong SPQR$.

a. Find the value of x.

b. Find the value of y.

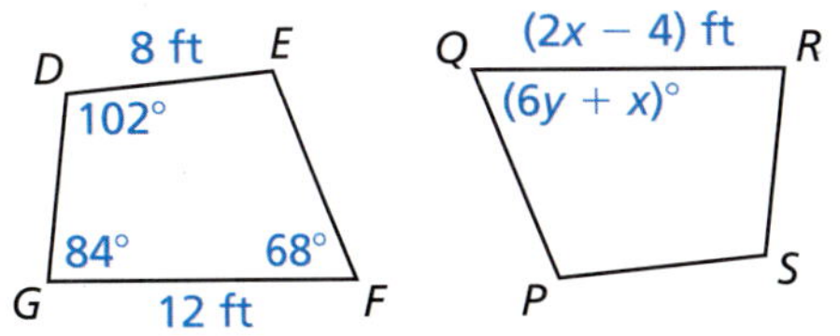

SOLUTION

a. You know that $\overline{FG} \cong \overline{QR}$.

$FG = QR$

$12 = 2x - 4$

$16 = 2x$

$8 = x$

b. You know that $\angle F \cong \angle Q$.

$m\angle F = m\angle Q$

$68^\circ = (6y + x)^\circ$

$68 = 6y + 8$

$10 = y$

EXAMPLE 3 Showing that Figures Are Congruent

WATCH

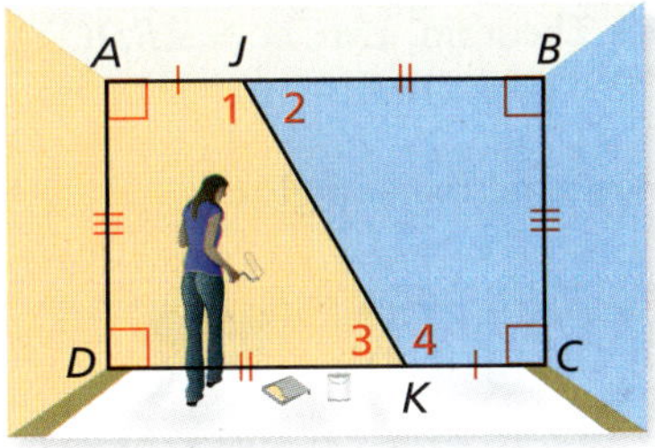

You divide a wall into orange and blue sections along $\overline{JK}$. Will the sections of the wall be the same size and shape? Explain.

SOLUTION

From the diagram, $\angle A \cong \angle C$ and $\angle D \cong \angle B$ because all right angles are congruent. Also, by the Lines Perpendicular to a Transversal Theorem, $\overline{AB} \parallel \overline{DC}$. Then $\angle 1 \cong \angle 4$ and $\angle 2 \cong \angle 3$ by the Alternate Interior Angles Theorem. So, all pairs of corresponding angles are congruent. The diagram shows $\overline{AJ} \cong \overline{CK}$, $\overline{KD} \cong \overline{JB}$, and $\overline{DA} \cong \overline{BC}$. By the Reflexive Property of Segment Congruence, $\overline{JK} \cong \overline{KJ}$. So, all pairs of corresponding sides are congruent. Because all corresponding parts are congruent, $AJKD \cong CKJB$.

▶ Yes, the two sections will be the same size and shape.

THEOREM

5.3 Properties of Triangle Congruence

Triangle congruence is reflexive, symmetric, and transitive.

Reflexive For any triangle $\triangle ABC$, $\triangle ABC \cong \triangle ABC$.

Symmetric If $\triangle ABC \cong \triangle DEF$, then $\triangle DEF \cong \triangle ABC$.

Transitive If $\triangle ABC \cong \triangle DEF$ and $\triangle DEF \cong \triangle JKL$, then $\triangle ABC \cong \triangle JKL$.

Proof *BigIdeasMath.com*

STUDY TIP

The properties of congruence that are true for segments and angles are also true for triangles.

SELF-ASSESSMENT

1 I don't understand yet. 2 I can do it with help. 3 I can do it on my own. 4 I can teach someone else.

In the diagram, $ABGH \cong CDEF$.

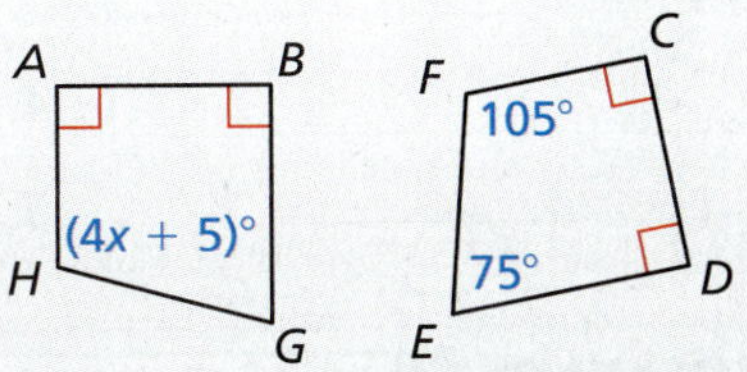

1. Identify all pairs of congruent corresponding parts.

2. Find the value of x.

4 MTR

3. **DIFFERENT WORDS, SAME QUESTION** Which is different? Find *both* answers.

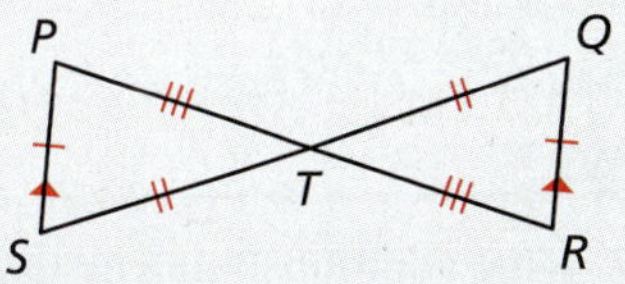

Is $\triangle PTS \cong \triangle RTQ$?	Is $\triangle TSP \cong \triangle TQR$?
Is $\triangle SPT \cong \triangle QRT$?	Is $\triangle TPS \cong \triangle TQR$?

GO DIGITAL

Using the Third Angles Theorem

THEOREM

5.4 Third Angles Theorem

If two angles of one triangle are congruent to two angles of another triangle, then the third angles are also congruent.

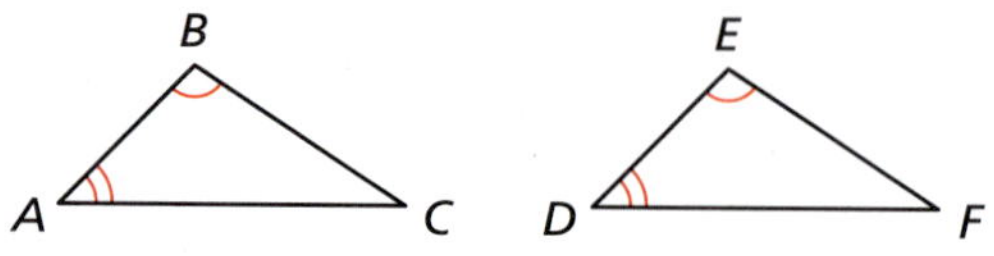

If $\angle A \cong \angle D$ and $\angle B \cong \angle E$, then $\angle C \cong \angle F$.

Prove this Theorem Exercise 21, page 242

EXAMPLE 4 Using the Third Angles Theorem

WATCH

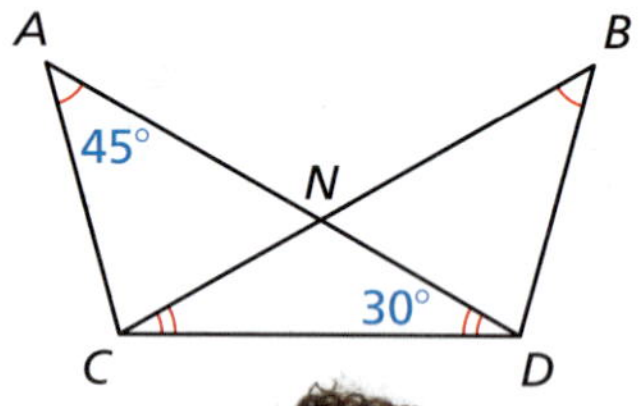

Find $m\angle BDC$.

SOLUTION

$\angle A \cong \angle B$ and $\angle ADC \cong \angle BCD$, so by the Third Angles Theorem, $\angle ACD \cong \angle BDC$. By the Triangle Sum Theorem, $m\angle ACD = 180° - 45° - 30° = 105°$.

▶ So, $m\angle BDC = m\angle ACD = 105°$ by the definition of congruent angles.

EXAMPLE 5 Proving that Triangles Are Congruent

WATCH

Use the information in the figure to prove that $\triangle ACD \cong \triangle CAB$.

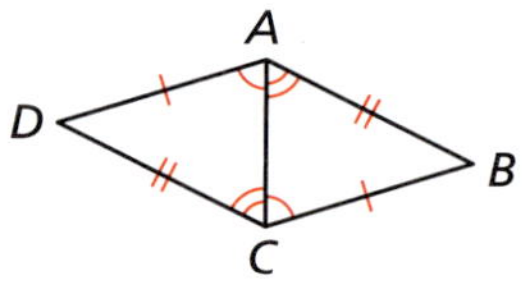

SOLUTION

Given $\overline{AD} \cong \overline{CB}, \overline{DC} \cong \overline{BA}$, $\angle ACD \cong \angle CAB$, $\angle CAD \cong \angle ACB$

Prove $\triangle ACD \cong \triangle CAB$

Plan for Proof
a. Use the Reflexive Property of Segment Congruence to show that $\overline{AC} \cong \overline{CA}$.
b. Use the Third Angles Theorem to show that $\angle B \cong \angle D$.

Plan in Action

	STATEMENTS	REASONS
	1. $\overline{AD} \cong \overline{CB}, \overline{DC} \cong \overline{BA}$	**1.** Given
a.	**2.** $\overline{AC} \cong \overline{CA}$	**2.** Reflexive Property of Segment Congruence
	3. $\angle ACD \cong \angle CAB$, $\angle CAD \cong \angle ACB$	**3.** Given
b.	**4.** $\angle B \cong \angle D$	**4.** Third Angles Theorem
	5. $\triangle ACD \cong \triangle CAB$	**5.** All corresponding parts are congruent.

SELF-ASSESSMENT

1 I don't understand yet. 2 I can do it with help. 3 I can do it on my own. 4 I can teach someone else.

Use the diagram.

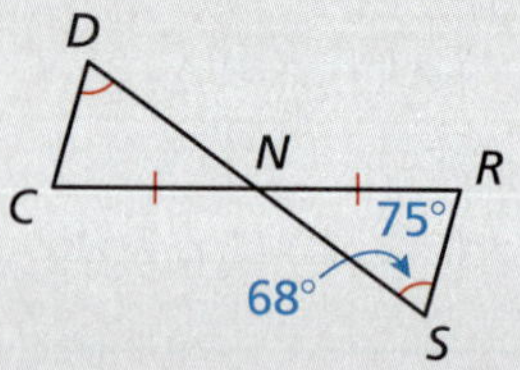

4. Find $m\angle DCN$.

5. REASONING What additional information is needed to conclude that $\triangle NDC \cong \triangle NSR$?

GO DIGITAL

5.2 Practice WITH CalcChat® AND CalcView®

In Exercises 1 and 2, write a congruence statement for the polygons. Identify all pairs of congruent corresponding parts. *(See Example 1.)*

▶ **1.**

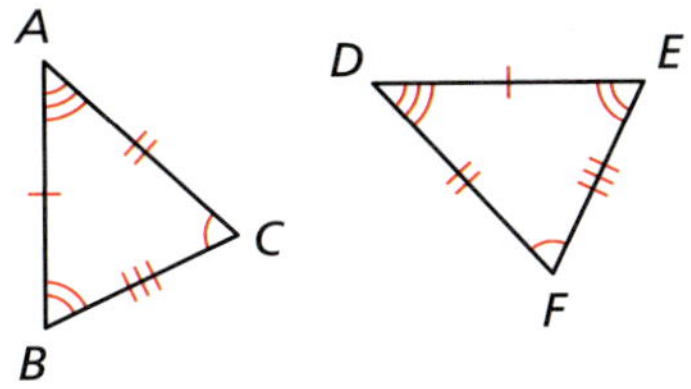

2.

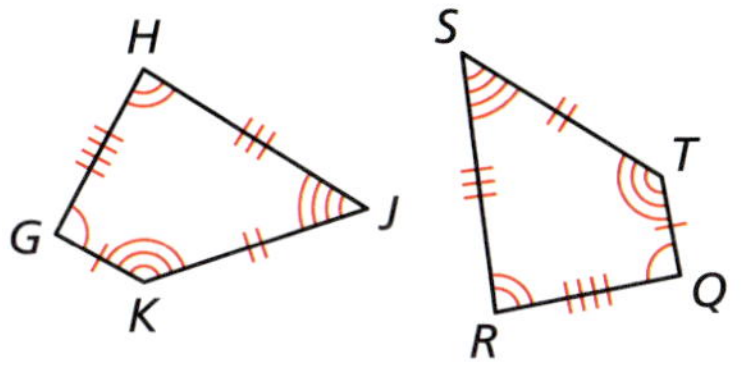

In Exercises 3–6, $\triangle XYZ \cong \triangle MNL$. Complete the statement.

3. $m\angle Y =$ ______

4. $m\angle M =$ ______

5. $m\angle Z =$ ______

6. $XY =$ ______

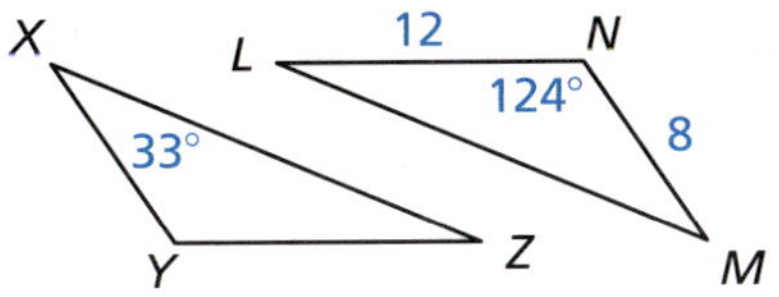

In Exercises 7 and 8, find the values of x and y. *(See Example 2.)*

▶ **7.** $ABCD \cong EFGH$

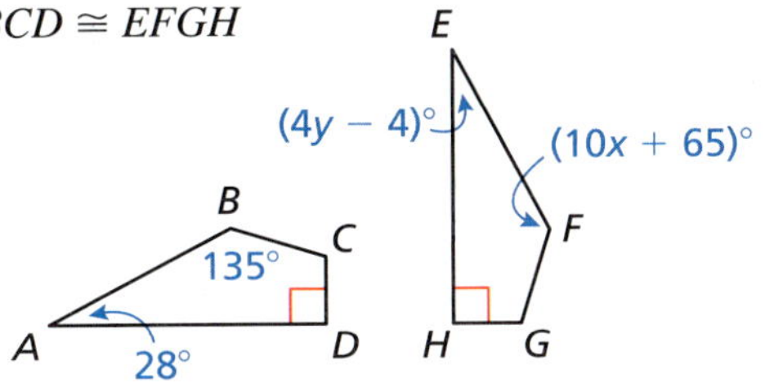

8. $\triangle MNP \cong \triangle TUS$

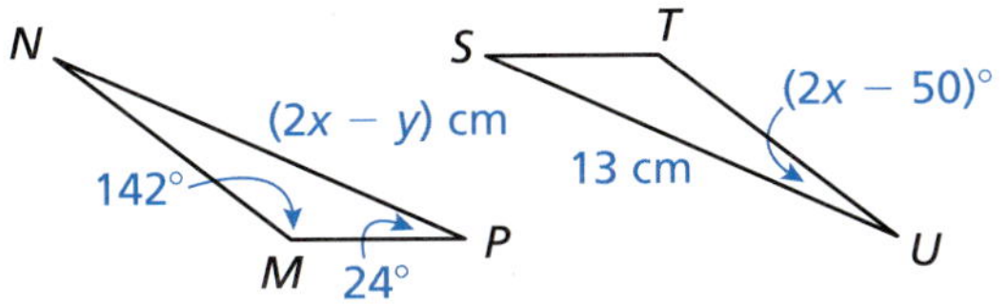

In Exercises 9 and 10, show that the polygons are congruent. Explain your reasoning. *(See Example 3.)*

▶ **9.**

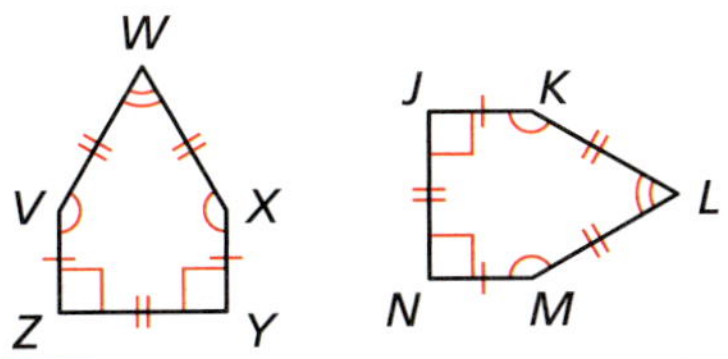

10.

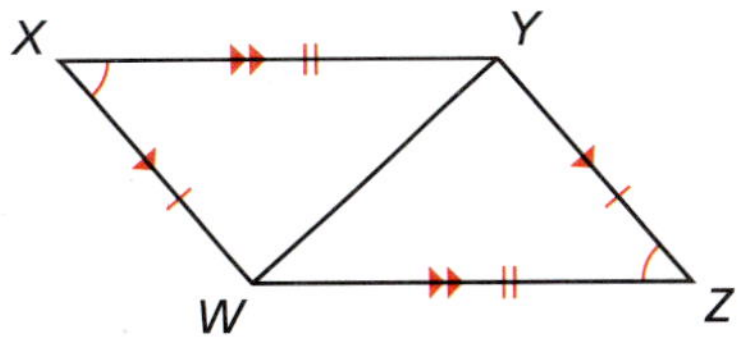

In Exercises 11 and 12, name the property that the statement illustrates.

11. If $\triangle QRS \cong \triangle TUV$ and $\triangle TUV \cong \triangle XYZ$, then $\triangle QRS \cong \triangle XYZ$.

12. If $\triangle EFG \cong \triangle JKL$, then $\triangle JKL \cong \triangle EFG$.

In Exercises 13 and 14, find $m\angle 1$. *(See Example 4.)*

▶ **13.**

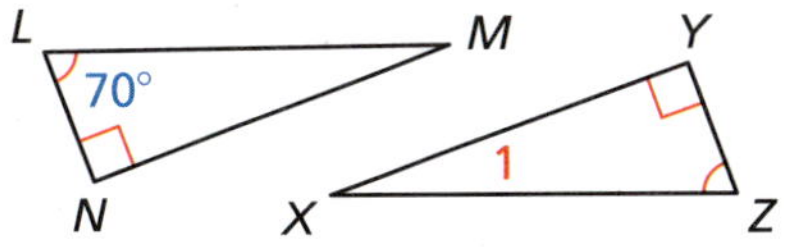

14.

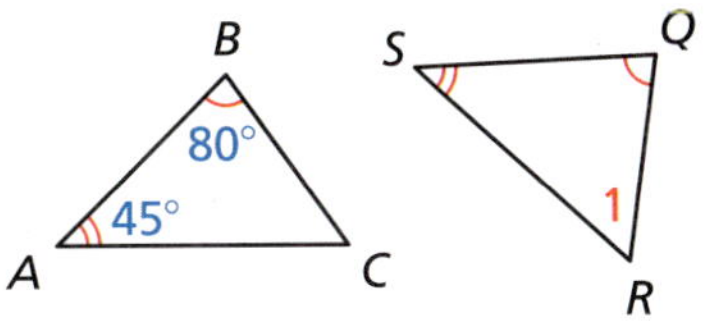

15. PROOF Triangular postage stamps, like the ones shown, are highly valued by stamp collectors. Prove that $\triangle AEB \cong \triangle CED$. *(See Example 5.)*

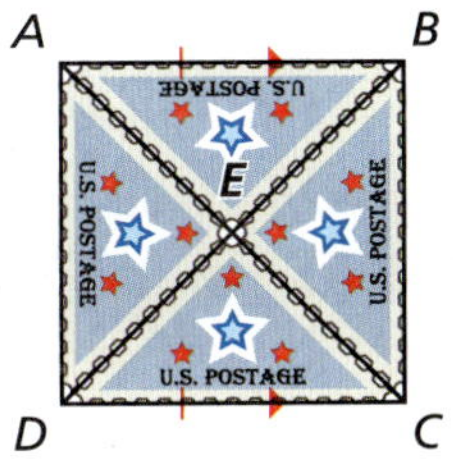

Given $\overline{AB} \parallel \overline{DC}$, $\overline{AB} \cong \overline{DC}$, E is the midpoint of $\overline{AC}$ and $\overline{BD}$.

Prove $\triangle AEB \cong \triangle CED$

16. PROOF Use the information in the figure to prove that $\triangle ABG \cong \triangle DCF$.

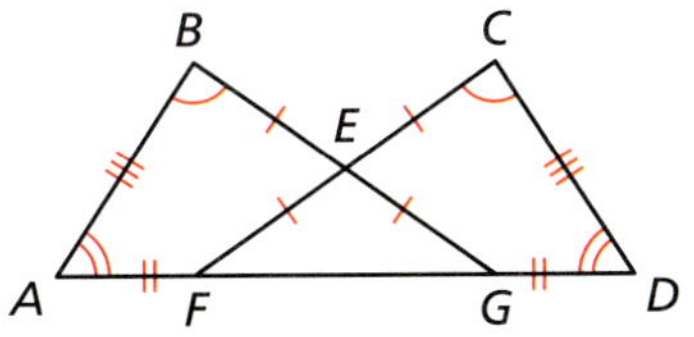

17. HELP A CLASSMATE Explain to a classmate how to find $m\angle R$.

Given $\triangle QRS \cong \triangle XZY$

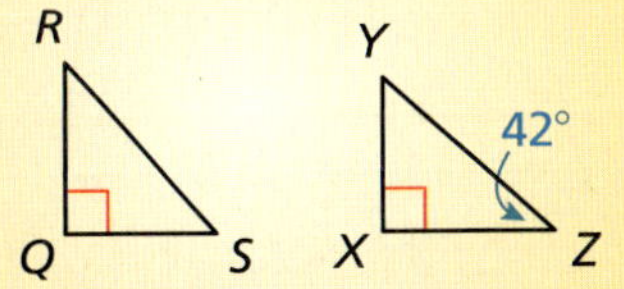

GO DIGITAL

18. HOW DO YOU SEE IT?
In the diagram, $ABEF \cong CDEF$.

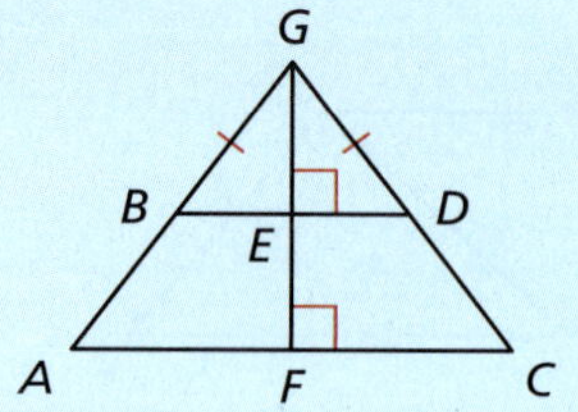

a. Explain how you know that $\overline{BE} \cong \overline{DE}$ and $\angle ABE \cong \angle CDE$.

b. Explain how you know that $\angle GBE \cong \angle GDE$.

c. Explain how you know that $\angle GEB \cong \angle GED$.

d. Do you have enough information to prove that $\triangle BEG \cong \triangle DEG$? Explain.

19. REASONING Which of the triangles appear congruent? Which of the triangles appear similar? Explain your reasoning. How can you justify your answer?

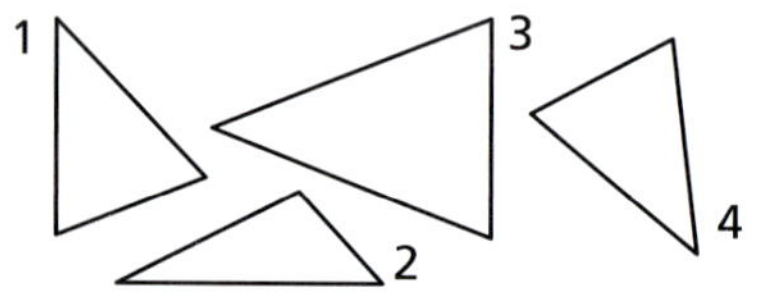

20. REASONING $\triangle JKL$ is congruent to $\triangle XYZ$. Identify all pairs of congruent corresponding parts.

21. PROVING A THEOREM Prove the Third Angles Theorem by using the Triangle Sum Theorem.

22. THOUGHT PROVOKING
Draw a triangle. Copy the triangle multiple times to create a rug design made of congruent triangles. Which property guarantees that all the triangles are congruent?

5 MTR **CONNECTING CONCEPTS** **In Exercises 23 and 24, use the given information to write and solve a system of linear equations to find the values of *x* and *y*.**

23. $\triangle LMN \cong \triangle PQR$, $m\angle L = 40°$, $m\angle M = 90°$, $m\angle P = (17x - y)°$, $m\angle R = (2x + 4y)°$

24. $\triangle STU \cong \triangle XYZ$, $m\angle T = 28°$, $m\angle U = (4x + y)°$, $m\angle X = 130°$, $m\angle Y = (8x - 6y)°$

25. PROOF Prove that the criteria for congruent triangles are equivalent to the definition of congruence in terms of rigid motions.

REVIEW & REFRESH

WATCH

26. Find the measure of the exterior angle.

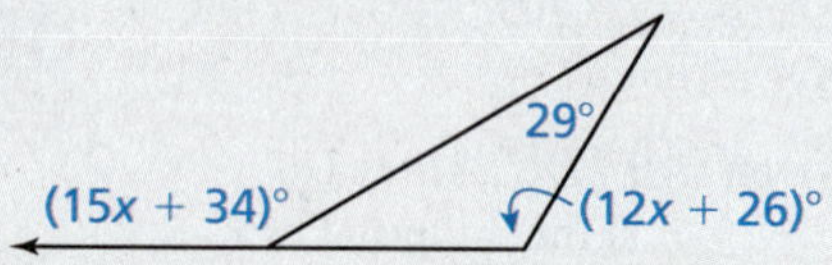

In Exercises 27 and 28, graph $\triangle FGH$ with vertices $F(-6, 3)$, $G(3, 0)$, and $H(3, -6)$ and its image after the similarity transformation.

27. **Translation:** $(x, y) \rightarrow (x + 2, y - 1)$
Dilation: $(x, y) \rightarrow (2x, 2y)$

28. **Dilation:** $(x, y) \rightarrow \left(\frac{1}{3}x, \frac{1}{3}y\right)$
Reflection: in the y-axis

29. Write a congruence statement for the quadrilaterals. Identify all pairs of congruent corresponding parts.

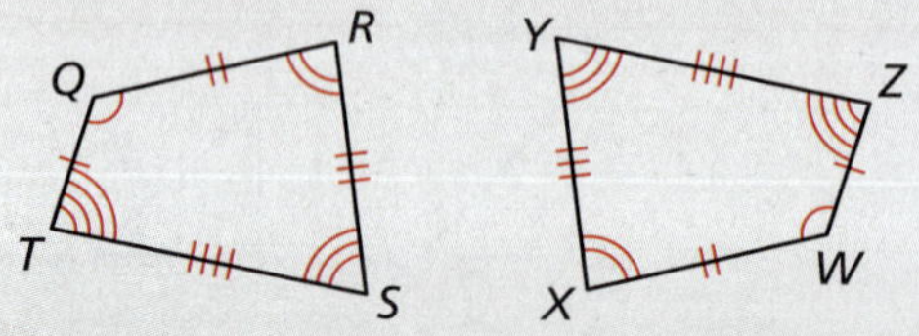

7 MTR **30. MODELING REAL LIFE** You design a logo for your soccer team. The logo is 3 inches by 5 inches. You decide to dilate the logo to 1.5 inches by 2.5 inches. What is the scale factor of this dilation?

In Exercises 31 and 32, factor the polynomial.

31. $t^2 + 7t + 10$

32. $2x^2 + 5x - 12$

In Exercises 33 and 34, using *f*, graph *g*. Describe the transformation from the graph of *f* to the graph of *g*.

33. $f(x) = 2x$; $g(x) = 4f(x)$

34. $f(x) = x^2 + 3$; $g(x) = f(x - 2)$

In Exercises 35 and 36, determine whether the equation represents a *linear* or *nonlinear* function. Explain.

35. $y = 5 - 2x$

36. $18 = -y + x^2$

GO DIGITAL

5.3 Proving Triangle Congruence by SAS

Learning Target: Prove and use the Side-Angle-Side Congruence Theorem.

Success Criteria:
- I can use rigid motions to prove the SAS Congruence Theorem.
- I can use the SAS Congruence Theorem.

EXPLORE IT! Reasoning about Triangles

Work with a partner.

a. Use technology to construct two circles with different radii and the same center. Draw an angle that has a measure less than 180° with its vertex at the center of the circles. Label the vertex A.

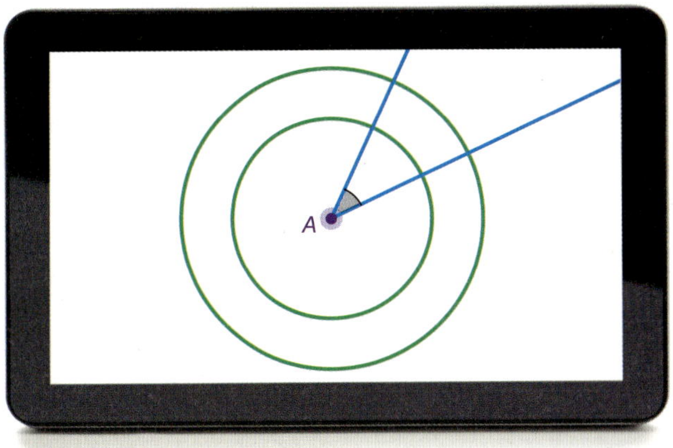

4 MTR CONSTRUCT AN ARGUMENT
What type of triangle is $\triangle ABC$ when you use one circle instead of two in parts (a) and (b)?

b. Locate the point where one side of $\angle A$ intersects the smaller circle and label this point B. Locate the point where the other side of $\angle A$ intersects the larger circle and label this point C. Then draw $\triangle ABC$.

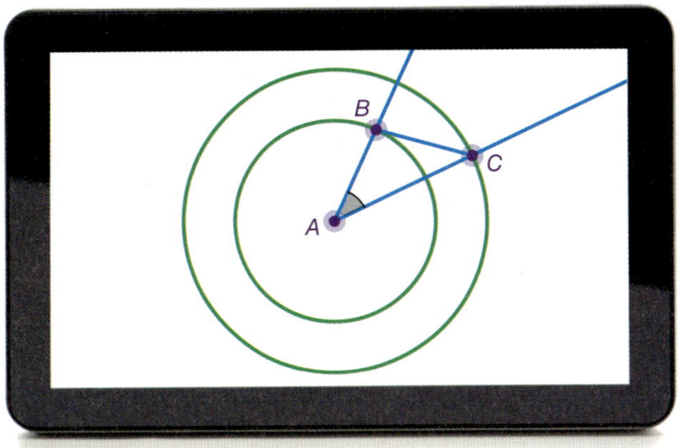

c. Find BC, $m\angle B$, and $m\angle C$.

d. Repeat parts (a)–(c) several times, redrawing $\angle A$ in different orientations. Use the same two radii and the same measure of $\angle A$ each time. What do you notice? Make a conjecture.

e. Can you prove your conjecture in part (d)? If so, write your proof.

Geometric Reasoning
MA.912.GR.1.2 Prove triangle congruence or similarity using Side-Side-Side, Side-Angle-Side, Angle-Side-Angle, Angle-Angle-Side, Angle-Angle and Hypotenuse-Leg.
MA.912.GR.1.6 Solve mathematical and real-world problems involving congruence or similarity in two-dimensional figures.
Also MA.912.GR.2.6, H MA.912.GR.2.7

GO DIGITAL

Using the Side-Angle-Side Congruence Theorem

THEOREM

5.5 Side-Angle-Side (SAS) Congruence Theorem

If two sides and the included angle of one triangle are congruent to two sides and the included angle of a second triangle, then the two triangles are congruent.

If $\overline{AB} \cong \overline{DE}$, $\angle A \cong \angle D$, and $\overline{AC} \cong \overline{DF}$, then $\triangle ABC \cong \triangle DEF$.

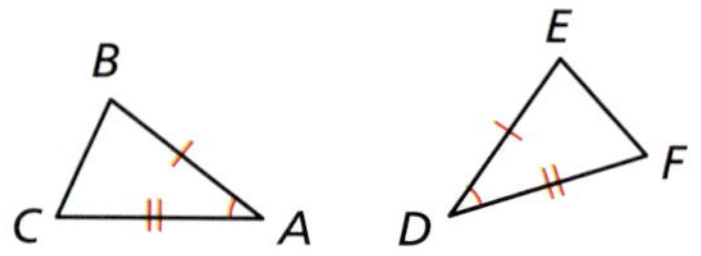

STUDY TIP

The *included angle* of two sides of a triangle is the angle formed by the two sides.

PROOF Side-Angle-Side (SAS) Congruence Theorem **H**

Given $\overline{AB} \cong \overline{DE}$, $\angle A \cong \angle D$, $\overline{AC} \cong \overline{DF}$

Prove $\triangle ABC \cong \triangle DEF$

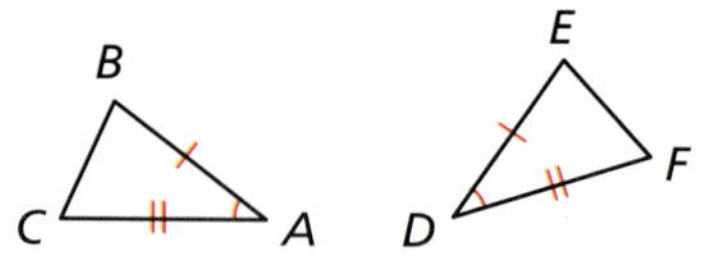

First, translate $\triangle ABC$ so that point A maps to point D, as shown below.

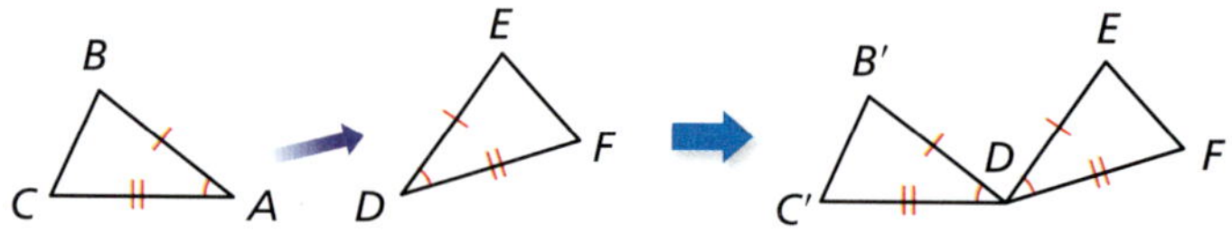

This translation maps $\triangle ABC$ to $\triangle DB'C'$. Next, rotate $\triangle DB'C'$ counterclockwise through $\angle C'DF$ so that the image of $\overrightarrow{DC'}$ coincides with $\overrightarrow{DF}$, as shown below.

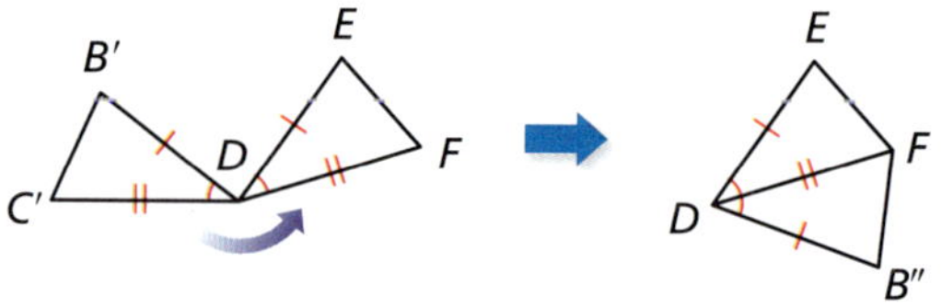

Because $\overline{DC'} \cong \overline{DF}$, the rotation maps point C' to point F. So, this rotation maps $\triangle DB'C'$ to $\triangle DB''F$. Now, reflect $\triangle DB''F$ in the line through points D and F, as shown below.

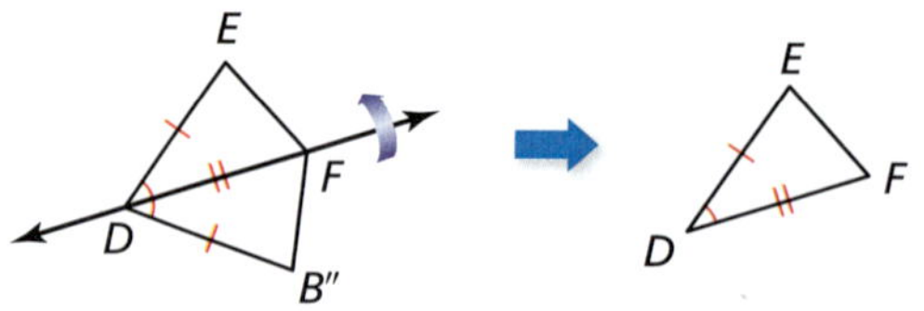

Because points D and F lie on $\overleftrightarrow{DF}$, this reflection maps them onto themselves. Because a reflection preserves angle measure and $\angle B''DF \cong \angle EDF$, the reflection maps $\overrightarrow{DB''}$ to $\overrightarrow{DE}$. Because $\overline{DB''} \cong \overline{DE}$, the reflection maps point B'' to point E. So, this reflection maps $\triangle DB''F$ to $\triangle DEF$.

Because you can map $\triangle ABC$ to $\triangle DEF$ using a composition of rigid motions, $\triangle ABC \cong \triangle DEF$.

GO DIGITAL

EXAMPLE 1 Using the SAS Congruence Theorem

WATCH

2 MTR USE ANOTHER METHOD

Use rigid motions to prove that $\triangle ABC \cong \triangle CDA$.

Write a proof.

Given $\overline{BC} \cong \overline{DA}$, $\overline{BC} \parallel \overline{AD}$

Prove $\triangle ABC \cong \triangle CDA$

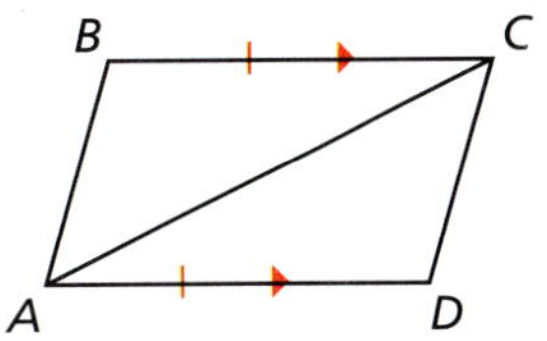

SOLUTION

	STATEMENTS	REASONS
S	**1.** $\overline{BC} \cong \overline{DA}$	**1.** Given
	2. $\overline{BC} \parallel \overline{AD}$	**2.** Given
A	**3.** $\angle BCA \cong \angle DAC$	**3.** Alternate Interior Angles Theorem
S	**4.** $\overline{AC} \cong \overline{CA}$	**4.** Reflexive Property of Segment Congruence
	5. $\triangle ABC \cong \triangle CDA$	**5.** SAS Congruence Theorem

STUDY TIP

Make your proof easier to read by identifying the steps where you show congruent sides (S) and angles (A).

EXAMPLE 2 Using SAS and Properties of Shapes

WATCH

In the diagram, $\overline{QS}$ and $\overline{RP}$ pass through the center M of the circle. What can you conclude about $\triangle MRS$ and $\triangle MPQ$?

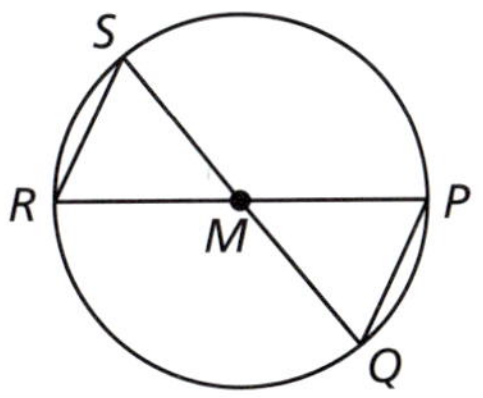

SOLUTION

Because they are vertical angles, $\angle PMQ \cong \angle RMS$. All points on a circle are the same distance from the center, so $\overline{MP}$, $\overline{MQ}$, $\overline{MR}$, and $\overline{MS}$ are all congruent.

▶ So, $\triangle MRS$ and $\triangle MPQ$ are congruent by the SAS Congruence Theorem.

SELF-ASSESSMENT

1 I don't understand yet. 2 I can do it with help. 3 I can do it on my own. 4 I can teach someone else.

1. Use rigid motions to prove that $\triangle JKL \cong \triangle MNP$.

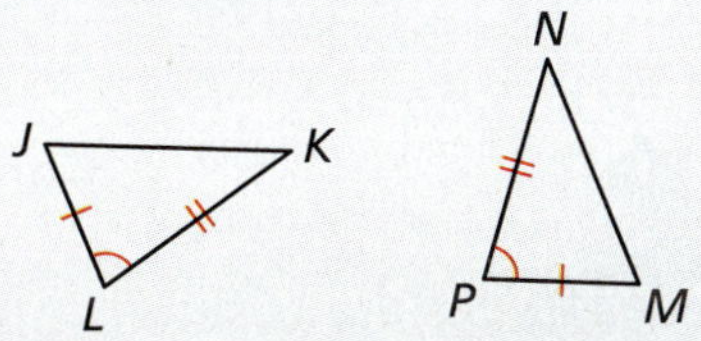

In the diagram, $ABCD$ is a square and R, S, T, and U are the midpoints of the sides of $ABCD$. Also, $\overline{RT} \perp \overline{SU}$ and $\overline{SV} \cong \overline{VU}$.

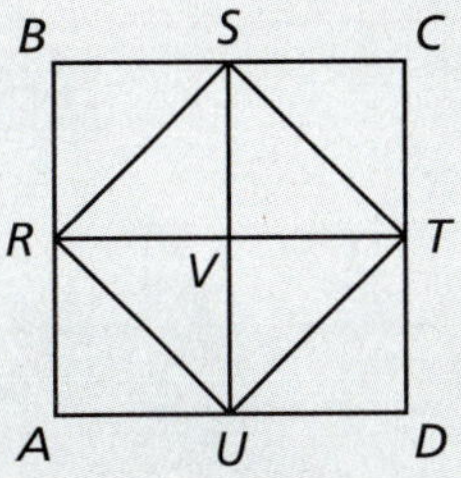

2. Prove that $\triangle SVR \cong \triangle UVR$.

3. Prove that $\triangle BSR \cong \triangle DUT$.

GO DIGITAL

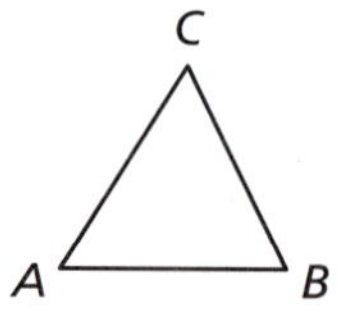

CONSTRUCTION Copying a Triangle Using SAS

Construct a triangle that is congruent to $\triangle ABC$ using the SAS Congruence Theorem. Use a compass and straightedge.

SOLUTION

Step 1

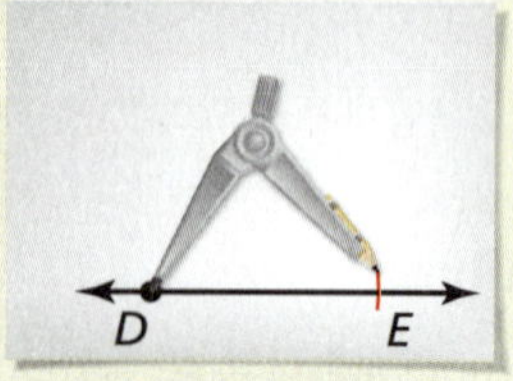

Construct a side
Construct $\overline{DE}$ so that it is congruent to $\overline{AB}$.

Step 2

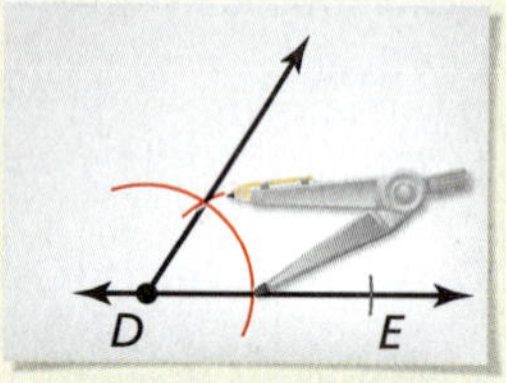

Construct an angle
Construct $\angle D$ with vertex D and side $\overrightarrow{DE}$ so that it is congruent to $\angle A$.

Step 3

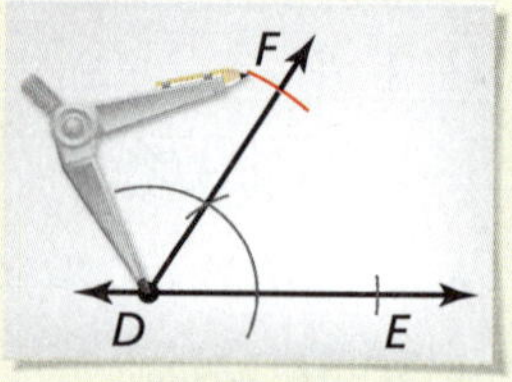

Construct a side
Construct $\overline{DF}$ so that it is congruent to $\overline{AC}$.

Step 4

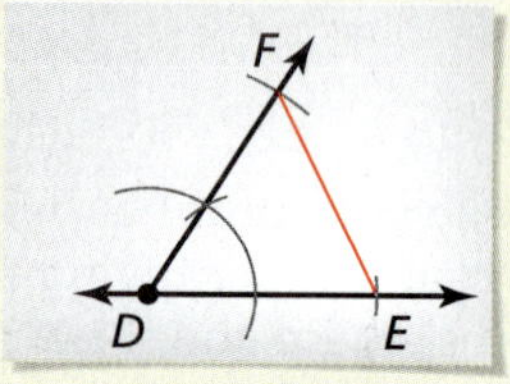

Draw a triangle
Draw $\triangle DEF$. By the SAS Congruence Theorem, $\triangle ABC \cong \triangle DEF$.

Solving Real-Life Problems

7 MTR

EXAMPLE 3 Modeling Real Life

You are making a canvas sign to hang on the triangular portion of the building shown in the photo. You think you can use two identical triangular sheets of canvas. You know that $\overline{RP} \perp \overline{QS}$ and $\overline{PQ} \cong \overline{PS}$. Use the SAS Congruence Theorem to prove that $\triangle PQR \cong \triangle PSR$.

SOLUTION

You are given that $\overline{PQ} \cong \overline{PS}$. By the Reflexive Property of Segment Congruence, $\overline{RP} \cong \overline{RP}$. By the definition of perpendicular lines, both $\angle RPQ$ and $\angle RPS$ are right angles, so they are congruent by the Right Angles Congruence Theorem. So, two pairs of sides and their included angles are congruent.

▶ $\triangle PQR$ and $\triangle PSR$ are congruent by the SAS Congruence Theorem.

SELF-ASSESSMENT

1 I don't understand yet. 2 I can do it with help. 3 I can do it on my own. 4 I can teach someone else.

4. You are designing the window shown. You want to make $\triangle DRA$ congruent to $\triangle DRG$. You design the window so that $\overline{DA} \cong \overline{DG}$ and $\angle ADR \cong \angle GDR$. Use the SAS Congruence Theorem to prove $\triangle DRA \cong \triangle DRG$.

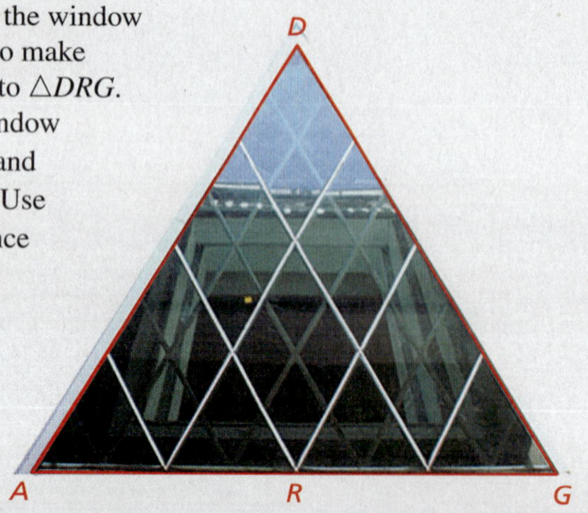

GO DIGITAL

5.3 Practice WITH CalcChat® AND CalcView®

In Exercises 1–4, decide whether enough information is given to prove that the triangles are congruent using the SAS Congruence Theorem. Explain.

1. $\triangle ABD$, $\triangle CDB$

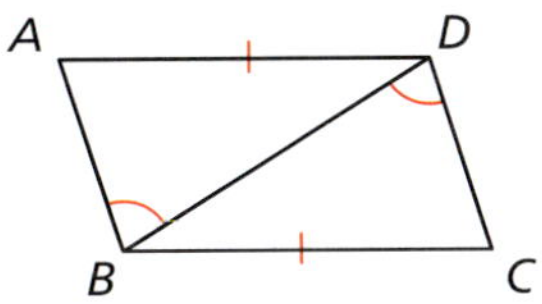

2. $\triangle LMN$, $\triangle NQP$

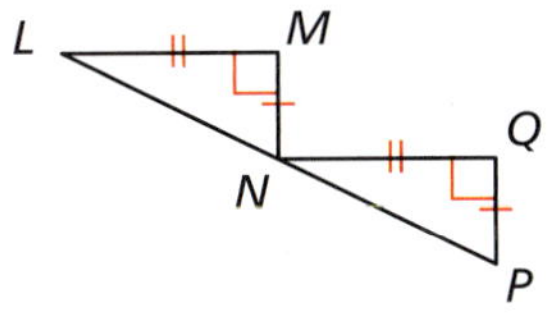

3. $\triangle EFH$, $\triangle GHF$

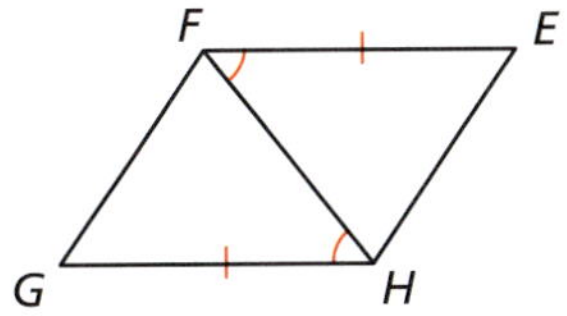

4. $\triangle KLM$, $\triangle MNK$

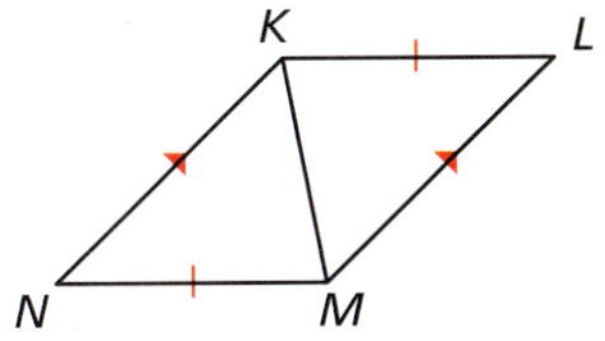

PROOF In Exercises 5–8, write a proof. *(See Example 1.)*

5. Given C is the midpoint of $\overline{AE}$ and $\overline{BD}$.

Prove $\triangle ABC \cong \triangle EDC$

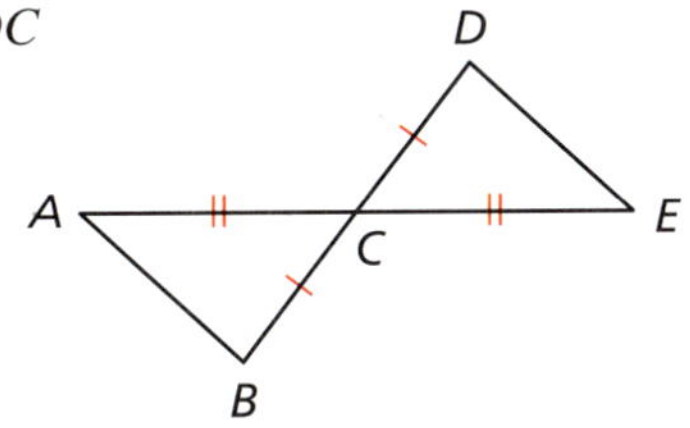

6. Given $\overline{PT} \cong \overline{RT}$, $\overline{QT} \cong \overline{ST}$

Prove $\triangle PQT \cong \triangle RST$

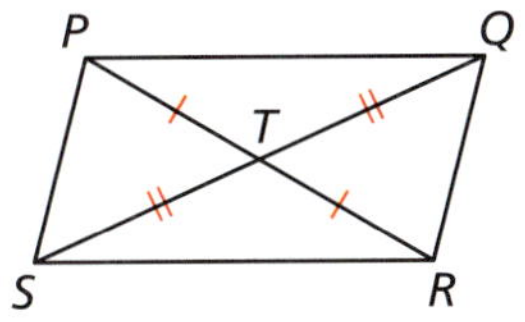

7. Given $\overline{PQ}$ bisects $\angle SPT$, $\overline{SP} \cong \overline{TP}$

Prove $\triangle SPQ \cong \triangle TPQ$

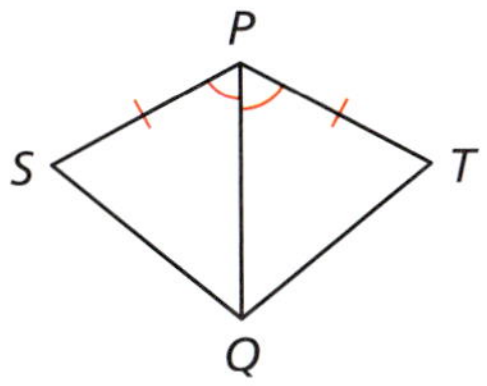

8. Given $\overline{AB} \cong \overline{CD}$, $\overline{AB} \parallel \overline{CD}$

Prove $\triangle ABC \cong \triangle CDA$

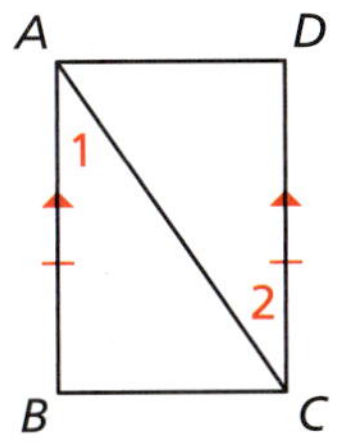

GO DIGITAL

In Exercises 9–12, use the given information to name two triangles that are congruent. Explain your reasoning. *(See Example 2.)*

9. $ABCD$ is a square.

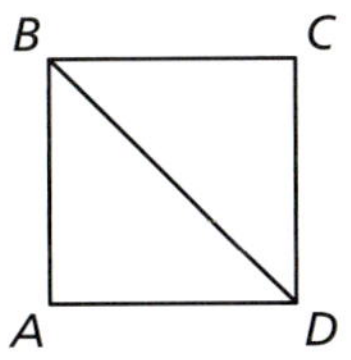

10. $RSTUV$ is a regular pentagon.

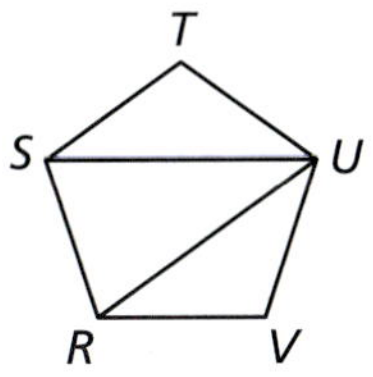

11. $\angle SRT \cong \angle URT$, and R is the center of the circle.

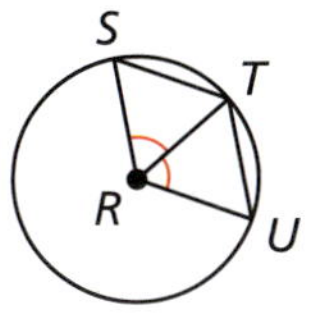

12. $\overline{MK} \perp \overline{MN}$, $\overline{KL} \perp \overline{NL}$, and M and L are centers of circles.

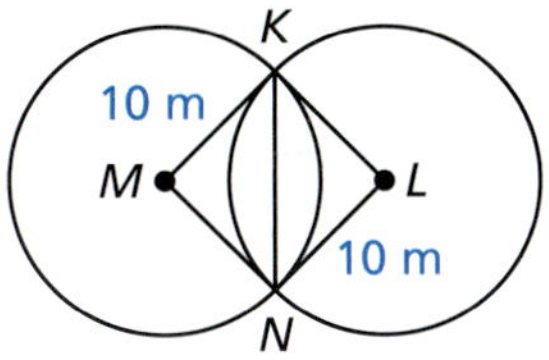

CONSTRUCTION In Exercises 13 and 14, construct a triangle that is congruent to $\triangle ABC$ using the SAS Congruence Theorem.

13.

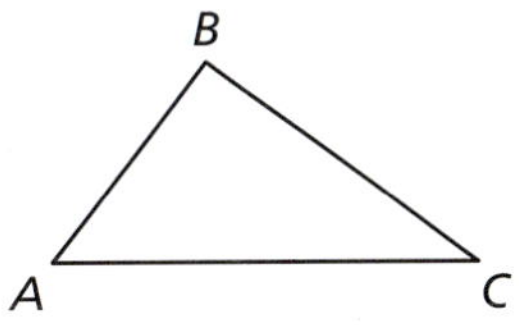

14.

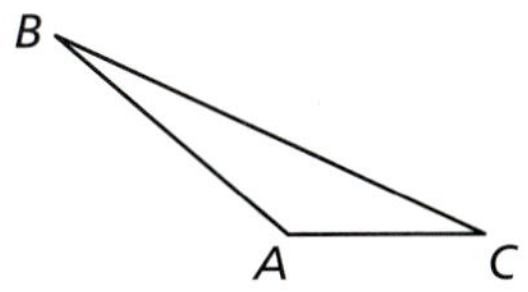

7 MTR **15. MODELING REAL LIFE** The *epidemiologic triangle* is a model used to understand what causes an infectious disease (agent), how the disease is transmitted (vector), who carries the disease (host), and where the disease is likely to be transmitted (environment). In the model, $\triangle AHE$ is equilateral and $\overline{HV}$ bisects $\angle AHE$. Use the SAS Congruence Theorem to prove that $\triangle AHV \cong \triangle EHV$. *(See Example 3.)*

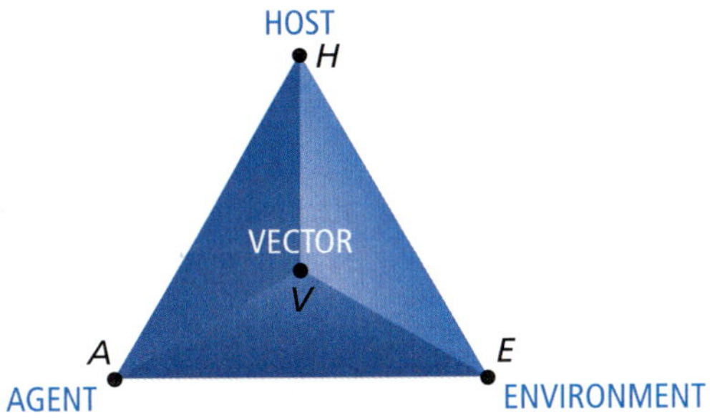

16. MODELING REAL LIFE The Navajo rug is made of isosceles triangles. You know $\angle B \cong \angle D$. Use the SAS Congruence Theorem to prove that $\triangle ABC \cong \triangle CDE$.

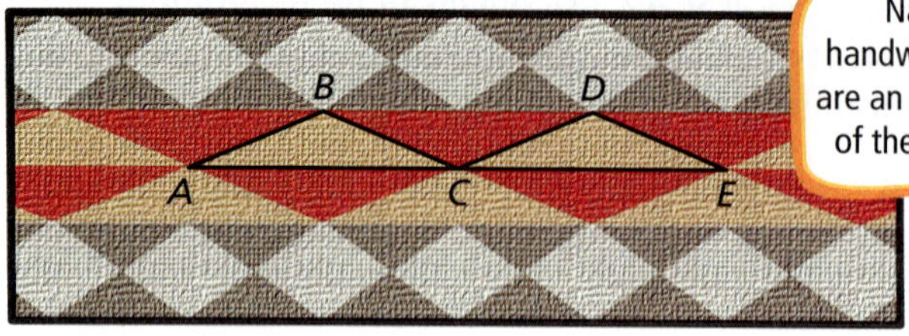

Navajo rugs are handwoven textiles that are an important element of the Navajo economy.

17. MAKING AN ARGUMENT Your friend says that $\triangle YXZ \cong \triangle WXZ$ by the SAS Congruence Theorem. Is your friend correct? Explain.

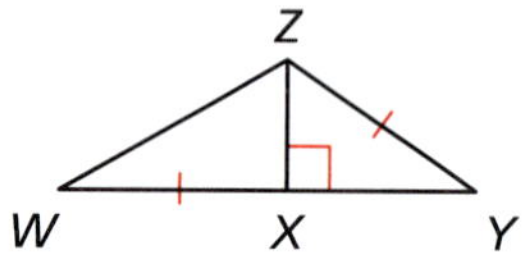

18. HOW DO YOU SEE IT?
What additional information do you need to prove that $\triangle ABC \cong \triangle DBC$?

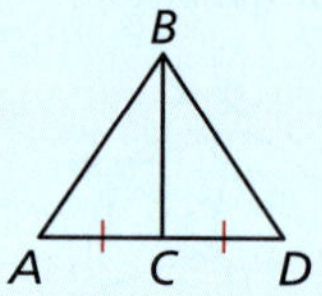

19. CONNECTING CONCEPTS Prove that $\triangle ABC \cong \triangle DEC$. Then find the values of x and y.

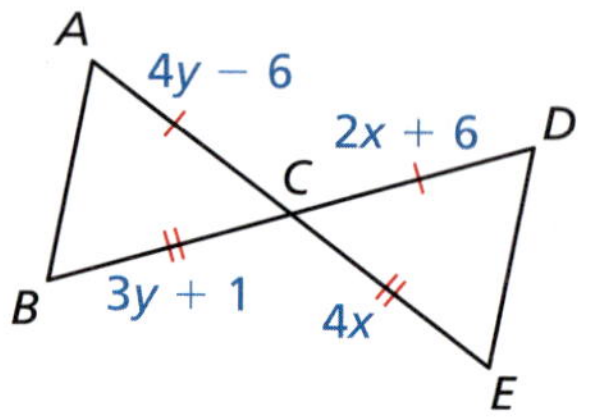

20. THOUGHT PROVOKING
Is it possible to prove that two triangles are congruent when you know that two corresponding sides and a non-included angle are congruent? (In other words, is SSA a valid congruence theorem?) If so, write a proof. If not, provide a counterexample.

21. DIG DEEPER The legs of two isosceles triangles are congruent. Must the triangles be congruent? Justify your answer.

22. PROVING A THEOREM Prove the Reflections in Intersecting Lines Theorem.

REVIEW & REFRESH

WATCH

In Exercises 23 and 24, classify the triangle by its sides and by measuring its angles.

23.

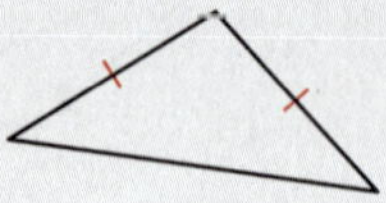

24.

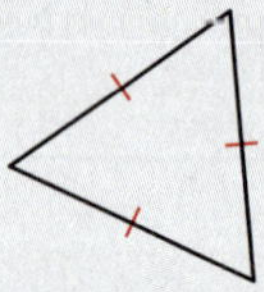

25. Graph $\triangle JKL$ with vertices $J(1, -2)$, $K(3, -1)$, and $L(4, -4)$ and its image after the similarity transformation.

Reflection: in the line $y = x$

Dilation: $(x, y) \to (2x, 2y)$

26. Decide whether enough information is given to prove that $\triangle QRV \cong \triangle TSU$ by the SAS Congruence Theorem. Explain.

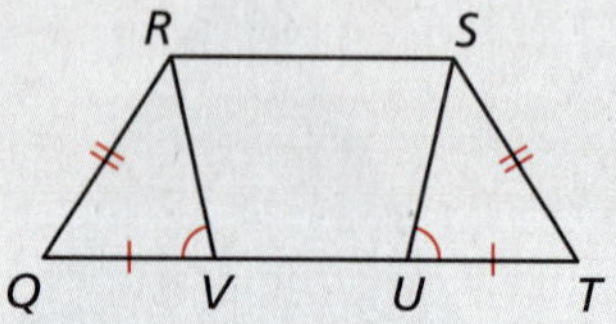

In Exercises 27–30, solve the inequality. Graph the solution.

27. $8 + y < -1$

28. $\frac{w}{6} \geq -4$

29. $2d + 7 \leq 25$

30. $|b + 9| > 4$

31. Find the values of x and y when $\triangle MNP \cong \triangle QRS$.

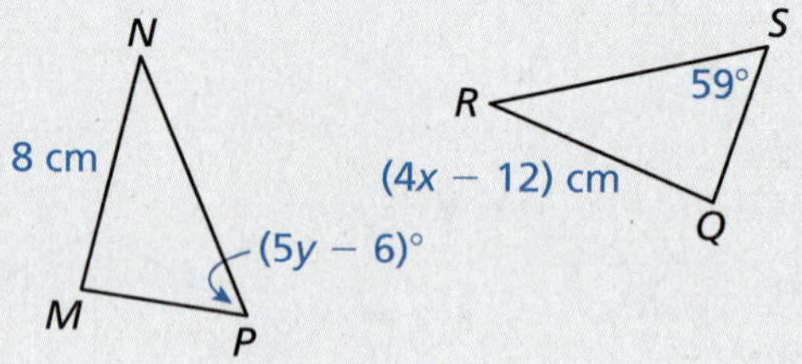

32. MODELING REAL LIFE You conduct a survey that asks 100 students in your school whether they like playing kickball or volleyball in gym class. Sixty-three of the students like playing volleyball, and sixteen of those students do not like playing kickball. Fourteen of the students do not like playing kickball or volleyball. Organize the results in a two-way table. Include the marginal frequencies.

33. Evaluate $g(x) = 14 - \frac{1}{2}x$ when $x = -4, 0$, and 8.

GO DIGITAL

5.4 Equilateral and Isosceles Triangles

Learning Target: Prove and use theorems about isosceles and equilateral triangles.

Success Criteria:
- I can prove and use theorems about isosceles triangles.
- I can prove and use theorems about equilateral triangles.

EXPLORE IT! Reasoning about Isosceles Triangles

Work with a partner.

a. Construct several circles. For each circle, draw a triangle with one vertex at the center of the circle and two vertices on the circle. Recall that a triangle is *isosceles* when it has at least two congruent sides. Explain why the triangles are isosceles. Then find the angle measures of each triangle. What do you notice? Make a conjecture about the angle measures of an isosceles triangle.

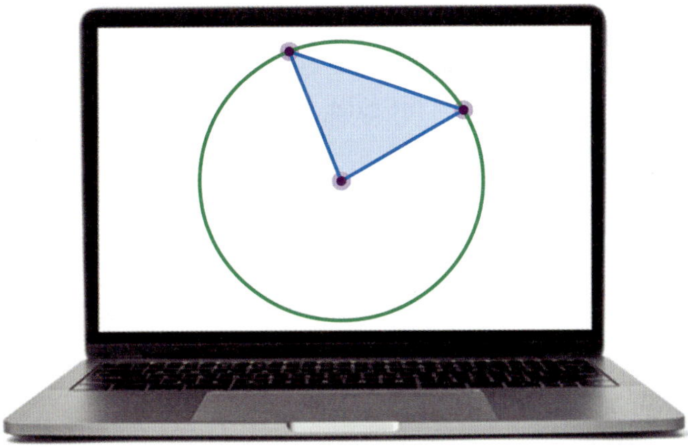

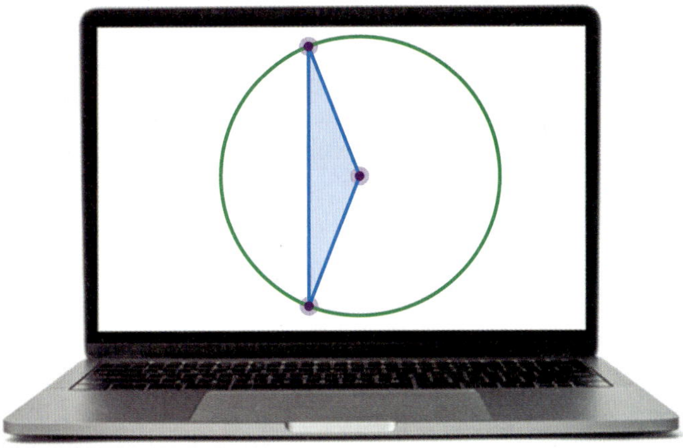

b. $\triangle ABC$ is an isosceles triangle. Given that $\overline{AB} \cong \overline{AC}$, show that $\angle B \cong \angle C$ when

i. $\overline{AD}$ bisects $\angle CAB$.

ii. $\overline{AD}$ is the perpendicular bisector of $\overline{BC}$.

c. Trace the isosceles triangle in part (b) and use paper folding to construct an informal proof showing that $\angle B \cong \angle C$.

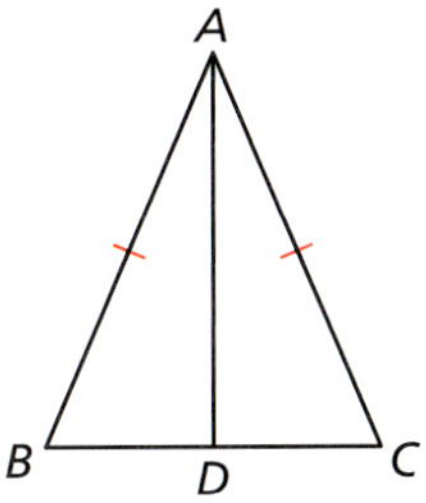

4 MTR CONSTRUCT AN ARGUMENT

How can you use the SAS Congruence Theorem to show that $\angle B$ and $\angle C$ are congruent without using $\overline{AD}$? Explain.

Geometric Reasoning

MA.912.GR.1.3 Prove relationships and theorems about triangles. Solve mathematical and real-world problems involving postulates, relationships and theorems of triangles.

GO DIGITAL

Using the Base Angles Theorem

Vocabulary AZ VOCAB

legs of an isosceles triangle, *p. 250*
vertex angle, *p. 250*
base, *p. 250*
base angles, *p. 250*

A triangle is isosceles when it has at least two congruent sides. When an isosceles triangle has exactly two congruent sides, these two sides are the **legs**. The angle formed by the legs is the **vertex angle**. The third side is the **base** of the isosceles triangle. The two angles adjacent to the base are called **base angles**.

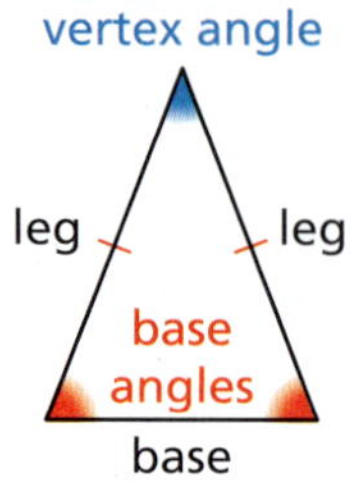

THEOREMS

5.6 Base Angles Theorem

If two sides of a triangle are congruent, then the angles opposite them are congruent.

If $\overline{AB} \cong \overline{AC}$, then $\angle B \cong \angle C$.

Prove this Theorem Explore It! part (b), page 249

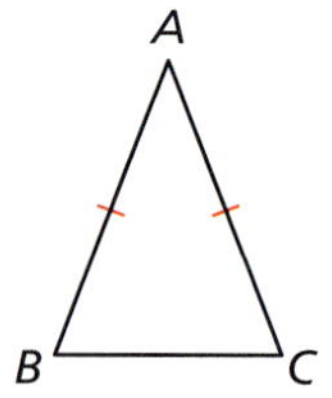

5.7 Converse of the Base Angles Theorem

If two angles of a triangle are congruent, then the sides opposite them are congruent.

If $\angle B \cong \angle C$, then $\overline{AB} \cong \overline{AC}$.

Prove this Theorem Exercise 20, page 271

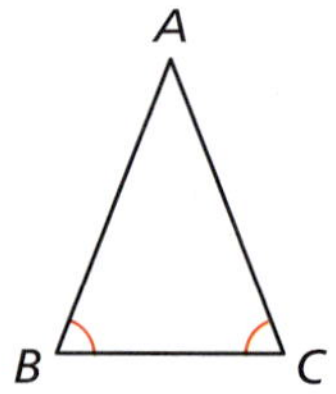

PROOF Base Angles Theorem

Given $\overline{AB} \cong \overline{AC}$

Prove $\angle B \cong \angle C$

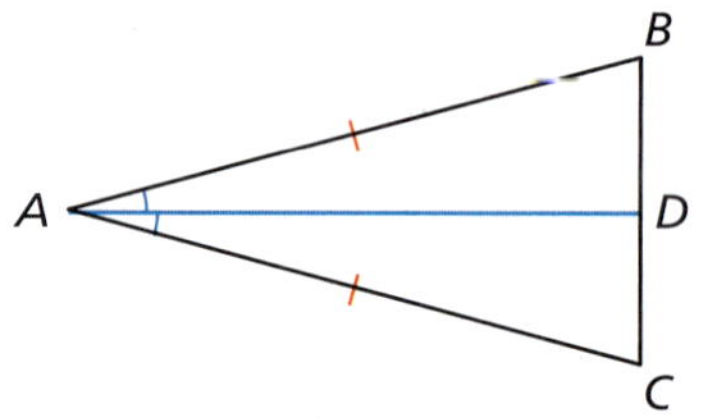

Plan for Proof

a. Draw $\overline{AD}$ so that it bisects $\angle CAB$.

b. Use the SAS Congruence Theorem to show that $\triangle ADB \cong \triangle ADC$.

c. Use properties of congruent triangles to show that $\angle B \cong \angle C$.

Plan in Action	STATEMENTS	REASONS
a.	**1.** Draw $\overline{AD}$, the angle bisector of $\angle CAB$.	**1.** Construction of angle bisector
	2. $\angle CAD \cong \angle BAD$	**2.** Definition of angle bisector
	3. $\overline{AB} \cong \overline{AC}$	**3.** Given
	4. $\overline{DA} \cong \overline{DA}$	**4.** Reflexive Property of Segment Congruence
b.	**5.** $\triangle ADB \cong \triangle ADC$	**5.** SAS Congruence Theorem
c.	**6.** $\angle B \cong \angle C$	**6.** Corresponding parts of congruent triangles are congruent.

GO DIGITAL

EXAMPLE 1 Using the Base Angles Theorem

In $\triangle DEF$, $\overline{DE} \cong \overline{DF}$. Name two congruent angles.

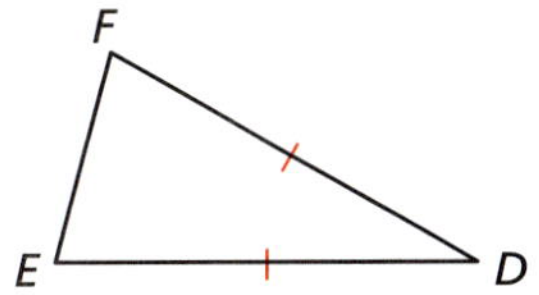

SOLUTION

▶ $\overline{DE} \cong \overline{DF}$, so by the Base Angles Theorem, $\angle E \cong \angle F$.

COROLLARIES

READING

The corollaries state that a triangle is *equilateral* if and only if it is *equiangular*.

5.2 Corollary to the Base Angles Theorem

If a triangle is equilateral, then it is equiangular.

Prove this Corollary Exercise 29, page 255; Exercise 10, page 346

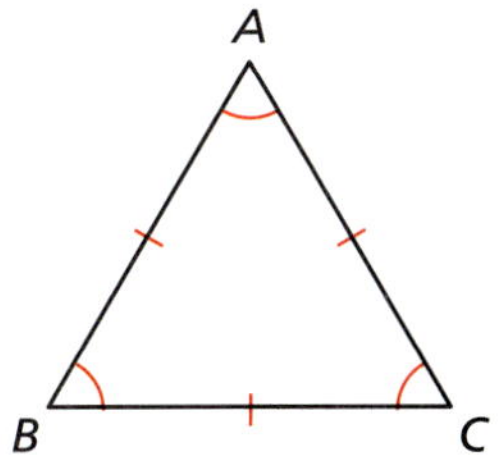

5.3 Corollary to the Converse of the Base Angles Theorem

If a triangle is equiangular, then it is equilateral.

Prove this Corollary Exercise 31, page 256

EXAMPLE 2 Finding Measures in a Triangle

Find the measures of $\angle P$, $\angle Q$, and $\angle R$.

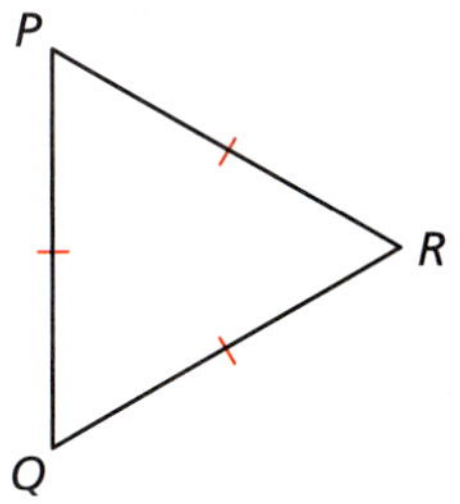

SOLUTION

The diagram shows that $\triangle PQR$ is equilateral. So, by the Corollary to the Base Angles Theorem, $\triangle PQR$ is equiangular. So, $m\angle P = m\angle Q = m\angle R$.

$3(m\angle P) = 180°$ Triangle Sum Theorem

$m\angle P = 60°$ Divide each side by 3.

▶ The measures of $\angle P$, $\angle Q$, and $\angle R$ are all 60°.

SELF-ASSESSMENT

1 I don't understand yet. 2 I can do it with help. 3 I can do it on my own. 4 I can teach someone else.

Complete the statement.

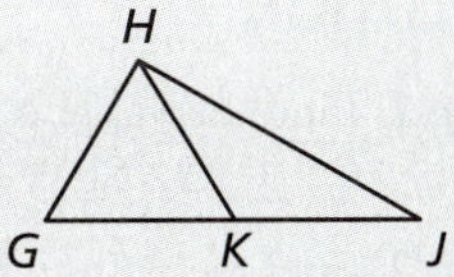

1. If $\overline{HG} \cong \overline{HK}$, then $\angle$_____ $\cong \angle$_____.
2. If $\angle KHJ \cong \angle KJH$, then _____ $\cong$ _____.
3. Find the length of $\overline{ST}$ for the triangle.

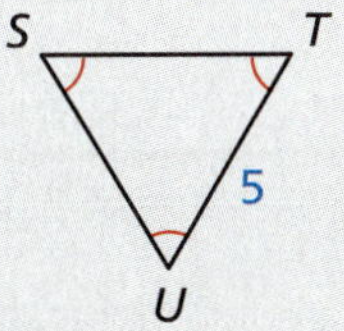

GO DIGITAL

Using Isosceles and Equilateral Triangles

CONSTRUCTION **Constructing an Equilateral Triangle** WATCH

Construct an equilateral triangle that has side lengths congruent to $\overline{AB}$. Use a compass and straightedge.

SOLUTION

Step 1

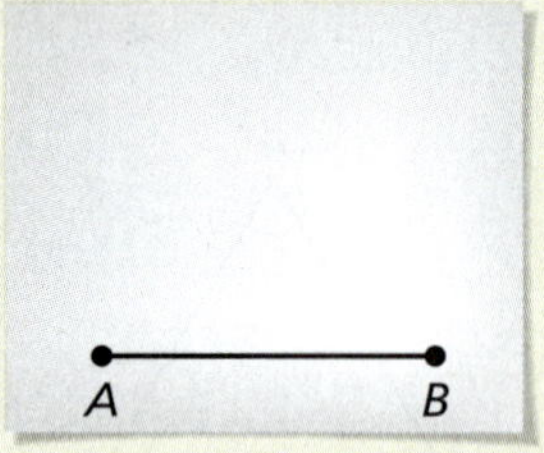

Copy a segment Copy $\overline{AB}$.

Step 2

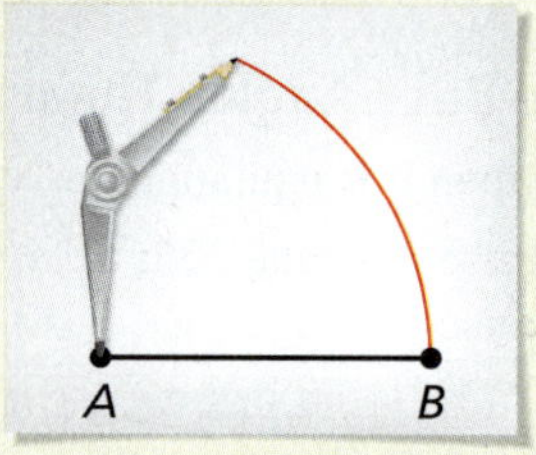

Draw an arc Draw an arc with center A and radius AB.

Step 3

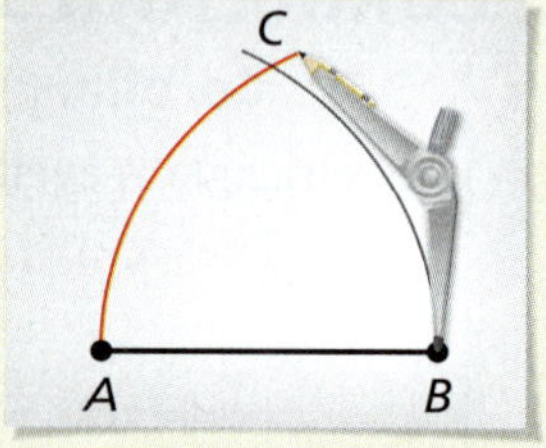

Draw an arc Draw an arc with center B and radius AB. Label the intersection of the arcs from Steps 2 and 3 as C.

Step 4

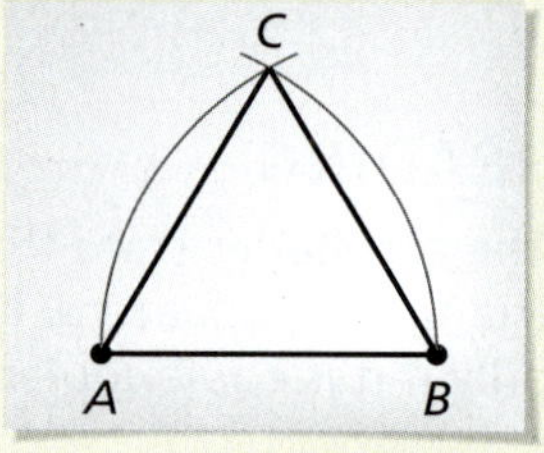

Draw a triangle Draw $\triangle ABC$. Because $\overline{AB}$ and $\overline{AC}$ are radii of the same circle, $\overline{AB} \cong \overline{AC}$. Because $\overline{AB}$ and $\overline{BC}$ are radii of the same circle, $\overline{AB} \cong \overline{BC}$. By the Transitive Property of Segment Congruence, $\overline{AC} \cong \overline{BC}$. So, $\triangle ABC$ is equilateral.

EXAMPLE 3 **Using Isosceles and Equilateral Triangles** WATCH

Find the values of x and y in the diagram.

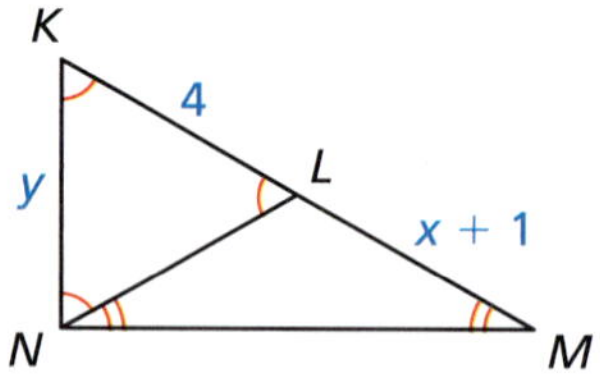

SOLUTION

Step 1 Find the value of y. Because $\triangle KLN$ is equiangular, it is also equilateral and $\overline{KN} \cong \overline{KL}$. So, $y = 4$.

Step 2 Find the value of x. Because $\angle LNM \cong \angle LMN$, $\overline{LN} \cong \overline{LM}$, and $\triangle LMN$ is isosceles. You also know that $LN = 4$ because $\triangle KLN$ is equilateral.

$LN = LM$	Definition of congruent segments
$4 = x + 1$	Substitute 4 for LN and $x + 1$ for LM.
$3 = x$	Subtract 1 from each side.

COMMON ERROR

You cannot use N to refer to $\angle LNM$ because three angles have N as their vertex.

GO DIGITAL

EXAMPLE 4 Modeling Real Life

In the lifeguard tower, $\overline{PS} \cong \overline{QR}$ and $\angle QPS \cong \angle PQR$.

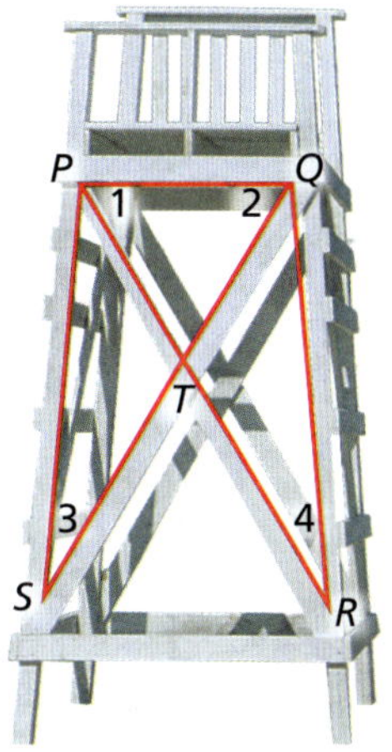

a. Explain how to prove that $\triangle QPS \cong \triangle PQR$.

b. Explain why $\triangle PQT$ is isosceles.

SOLUTION

a. Draw and label $\triangle QPS$ and $\triangle PQR$ so that they do not overlap. You can see that $\overline{PQ} \cong \overline{QP}$, $\overline{PS} \cong \overline{QR}$, and $\angle QPS \cong \angle PQR$. So, by the SAS Congruence Theorem, $\triangle QPS \cong \triangle PQR$.

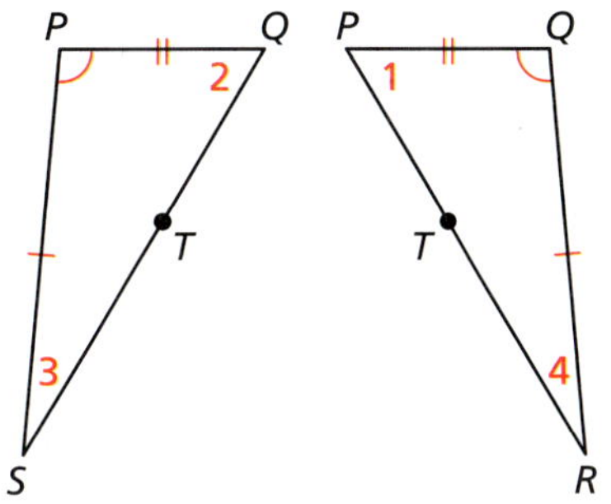

b. From part (a), you know that $\angle 1 \cong \angle 2$ because corresponding parts of congruent triangles are congruent. By the Converse of the Base Angles Theorem, $\overline{PT} \cong \overline{QT}$, and $\triangle PQT$ is isosceles.

COMMON ERROR

When you redraw the triangles so that they do not overlap, be careful to copy all given information and labels correctly.

SELF-ASSESSMENT

1 I don't understand yet. 2 I can do it with help. 3 I can do it on my own. 4 I can teach someone else.

Find the values of x and y in the diagram.

4.

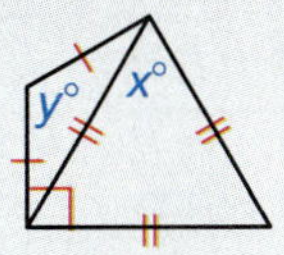

5.

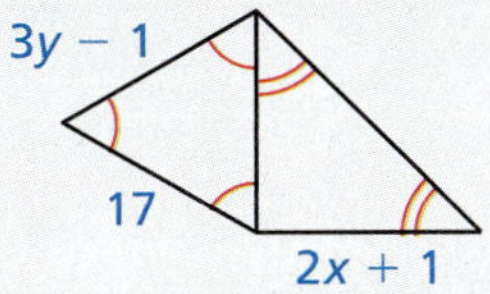

6.

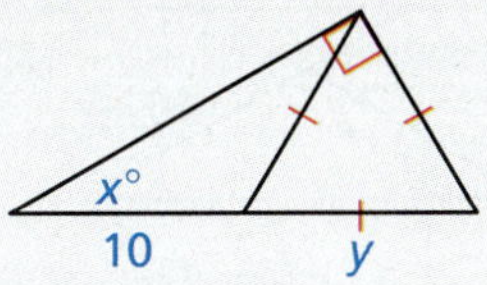

7. In Example 4, show that $\triangle PTS \cong \triangle QTR$.

GO DIGITAL

5.4 Practice WITH CalcChat® AND CalcView®

In Exercises 1–4, complete the statement. State which theorem you used. *(See Example 1.)*

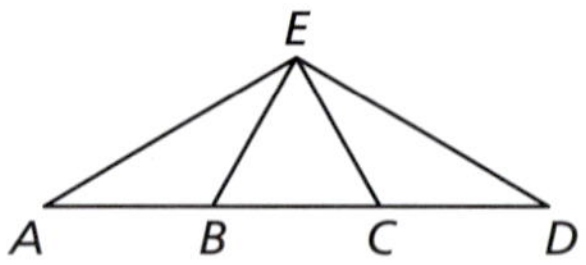

1. If $\overline{AE} \cong \overline{DE}$, then $\angle$___ $\cong \angle$___.

2. If $\overline{AB} \cong \overline{EB}$, then $\angle$___ $\cong \angle$___.

▶ 3. If $\angle D \cong \angle CED$, then ___ $\cong$ ___.

4. If $\angle EBC \cong \angle ECB$, then ___ $\cong$ ___.

In Exercises 5–8, find the value of *x*. *(See Example 2.)*

▶ 5.

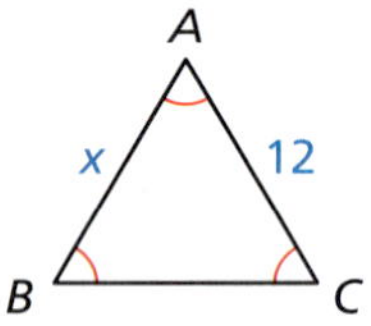

6.

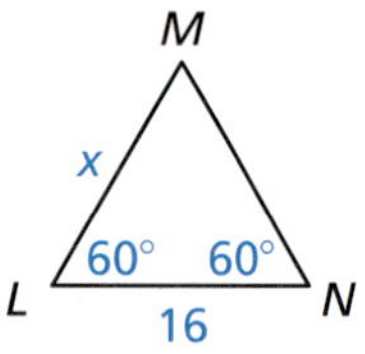

7.

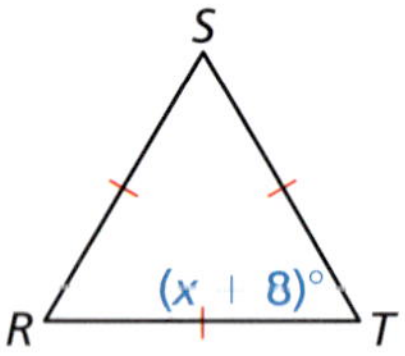

8.

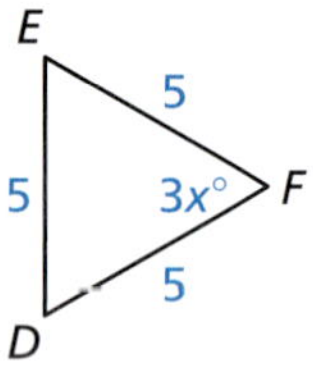

CONSTRUCTION In Exercises 9 and 10, construct an equilateral triangle whose sides are the given length.

9. 3 inches

10. 1.25 inches

In Exercises 11–14, find the values of *x* and *y*. *(See Example 3.)*

▶ 11.

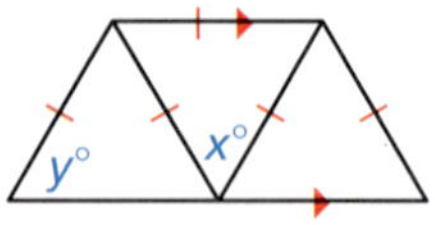

12.

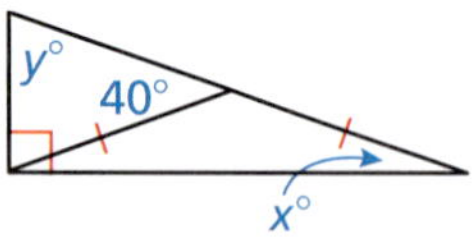

13.

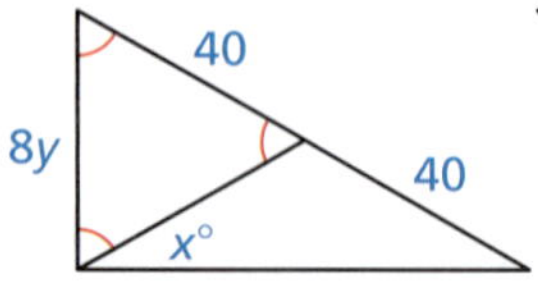

14.

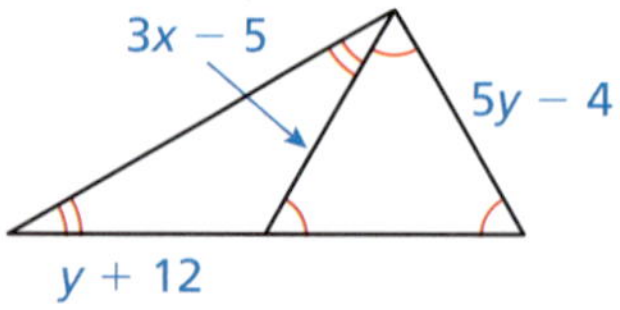

7 MTR 15. **MODELING REAL LIFE** Find the values of x and y in the diagram of the sports pennant.

7 MTR 16. **MODELING REAL LIFE** The image printed on a Chinese checkers board is shown. Without measuring, classify each of the six triangles surrounding the hexagon by sides and angle measures. Explain your reasoning.

4 MTR 17. **ERROR ANALYSIS** Describe and correct the error in finding the length of $\overline{BC}$.

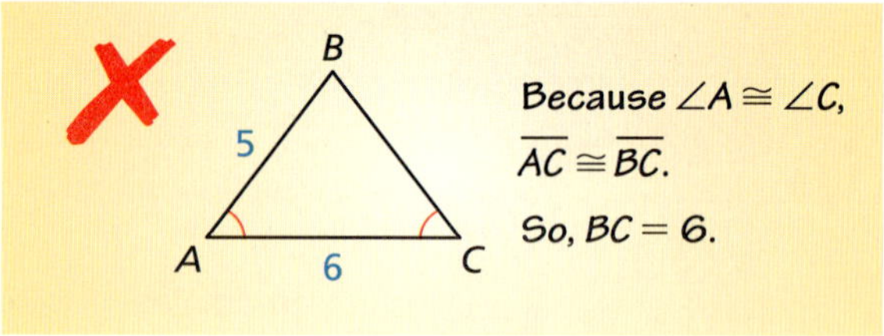

18. **COLLEGE PREP** The base of isosceles $\triangle XYZ$ is $\overline{YZ}$. What can you prove? Select all that apply.

Ⓐ $\overline{XY} \cong \overline{XZ}$

Ⓑ $\angle X \cong \angle Y$

Ⓒ $\angle Y \cong \angle Z$

Ⓓ $\overline{YZ} \cong \overline{ZX}$

7 MTR 19. **MODELING REAL LIFE** The diagram represents part of the exterior of the Bow Tower in Calgary, Alberta, Canada. In the diagram, $\triangle ABD$ and $\triangle CBD$ are congruent equilateral triangles. *(See Example 4.)*

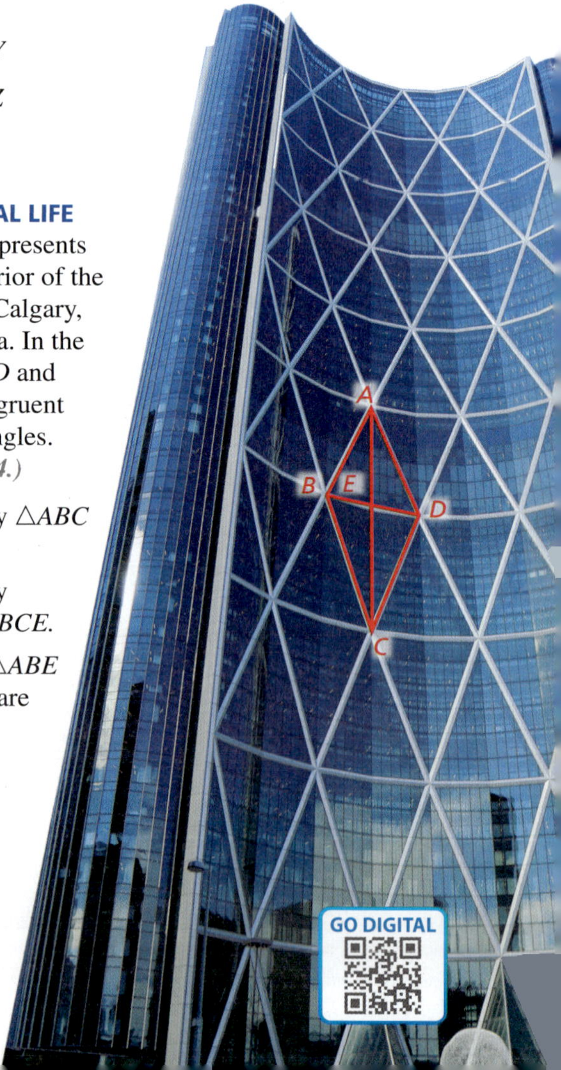

a. Explain why $\triangle ABC$ is isosceles.

b. Explain why $\angle BAE \cong \angle BCE$.

c. Show that $\triangle ABE$ and $\triangle CBE$ are congruent.

d. Find the measure of $\angle BAE$.

20. **MODELING REAL LIFE** The diagram is based on a color wheel. The 12 triangles in the diagram are isosceles triangles with congruent vertex angles.

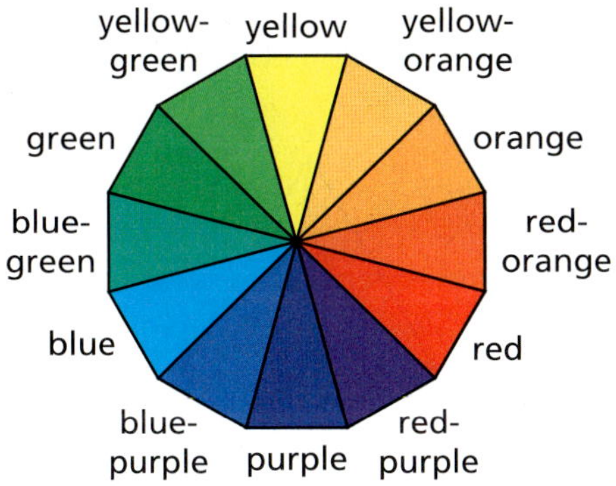

a. Complementary colors lie directly opposite each other on the color wheel. Explain how you know that the yellow triangle is congruent to the purple triangle.

b. The measure of the vertex angle of the yellow triangle is 30°. Find the measures of the base angles.

c. Trace the color wheel. Then form a triangle whose vertices are the midpoints of the bases of the red, yellow, and blue triangles. (These colors are the *primary colors*.) What type of triangle is this?

In Exercises 21 and 22, find the perimeter of the triangle.

21.

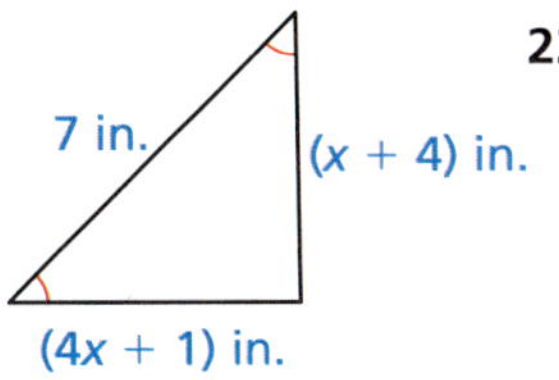

22.

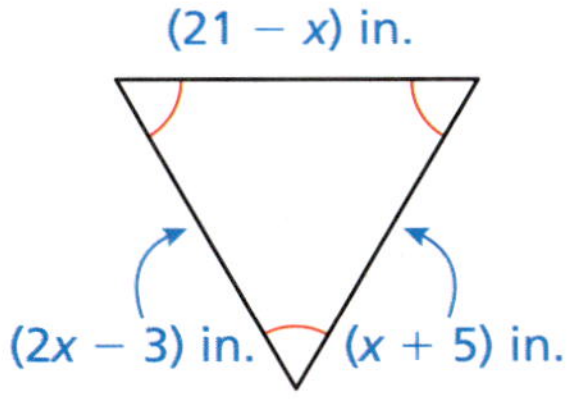

23. **CONNECTING CONCEPTS** The lengths of the sides of a triangle are $3t$, $5t - 12$, and $t + 20$. Find the values of t that make the triangle isosceles. Explain your reasoning.

24. **CONNECTING CONCEPTS** The measure of an exterior angle of an isosceles triangle is $x°$. Write expressions representing the possible angle measures of the triangle in terms of x.

25. **WRITING** Explain why the measure of the vertex angle of an isosceles triangle must be an even number of degrees when the measures of all the angles of the triangle are whole numbers.

26. **OPEN-ENDED** Find and draw an object (or part of an object) that can be modeled by an isosceles or equilateral triangle. Describe the relationship between the interior angles and sides of the triangle in terms of the object.

GO DIGITAL

27. **PROBLEM SOLVING** The triangular faces of the peaks on a roof are congruent isosceles triangles with vertex angles U and V.

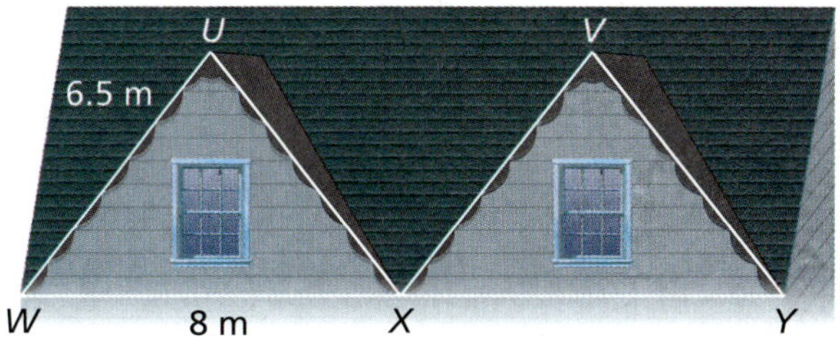

a. Name two angles congruent to $\angle WUX$. Explain your reasoning.

b. Find the distance between points U and V.

28. **REASONING** A boat is traveling parallel to the shore along $\overrightarrow{RT}$. When the boat is at point R, the captain measures the angle to the lighthouse L as 35°. After the boat has traveled 2.1 miles, the captain measures the angle to the lighthouse as 70°.

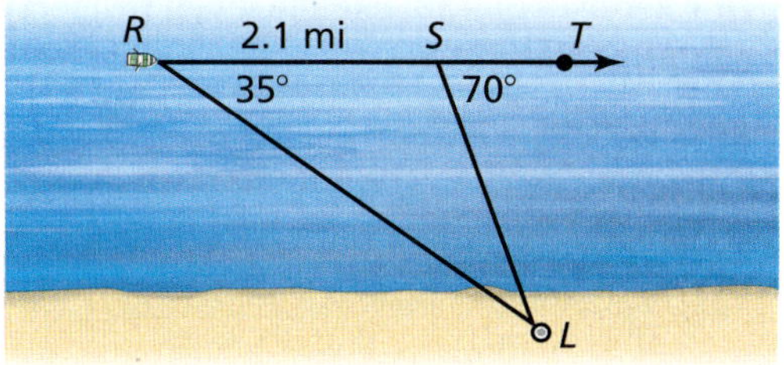

a. Find SL. Explain your reasoning.

b. Explain how the captain can use a similar process to find the distance between the boat and the shoreline.

29. **PROVING A COROLLARY** Prove that the Corollary to the Base Angles Theorem follows from the Base Angles Theorem.

30. **HOW DO YOU SEE IT?**
Use the image of the purse shown.

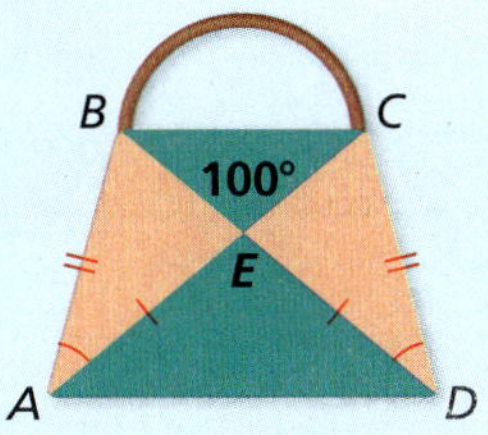

a. Explain why $\triangle ABE \cong \triangle DCE$.

b. Name the isosceles triangles in the purse.

c. Name three angles that are congruent to $\angle EAD$.

31. PROVING A COROLLARY Prove that the Corollary to the Converse of the Base Angles Theorem follows from the Converse of the Base Angles Theorem.

32. THOUGHT PROVOKING
The postulates and theorems in this book represent Euclidean geometry. In spherical geometry, all points are points on the surface of a sphere. A line is a circle on the sphere whose diameter is equal to the diameter of the sphere. In spherical geometry, do all equiangular triangles have the same angle measures? Justify your answer.

4 MTR **33. MAKING AN ARGUMENT** The coordinates of two points are $T(0, 6)$ and $U(6, 0)$. Will the points T, U, and V always be the vertices of an isosceles triangle when V is any point on the line $y = x$? Explain.

6 MTR **34. JUSTIFYING STEPS** Use the construction of an equilateral triangle and rotations to draw a hexagon whose sides are 2 inches long. Justify your steps.

35. PROOF Use the diagram to prove that $\triangle DEF$ is equilateral.

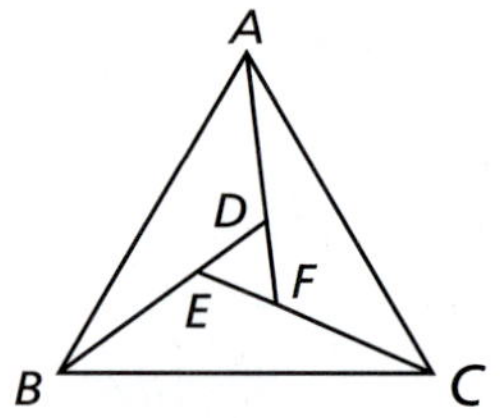

Given $\triangle ABC$ is equilateral.
$\angle CAD \cong \angle ABE \cong \angle BCF$

Prove $\triangle DEF$ is equilateral.

REVIEW & REFRESH

WATCH

In Exercises 36–38, use the given property to complete the statement.

36. Reflexive Property of Segment Congruence:
____ $\cong \overline{SE}$

37. Symmetric Property of Segment Congruence:
If ____ $\cong$ ____, then $\overline{RS} \cong \overline{JK}$.

38. Transitive Property of Segment Congruence:
If $\overline{EF} \cong \overline{PQ}$, and $\overline{PQ} \cong \overline{UV}$, then ____ $\cong$ ____.

7 MTR **39. MODELING REAL LIFE** The figure shows a stained-glass window. Is there enough information given to prove that $\triangle 1 \cong \triangle 2$? Explain.

40. Find $m\angle 1$.

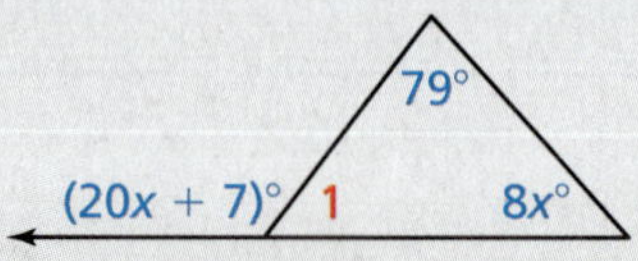

In Exercises 41 and 42, graph $\overline{YZ}$ with endpoints $Y(-4, 5)$ and $Z(-2, 1)$ and its image after the composition.

41. Rotation: 270° clockwise about the origin
Translation: $(x, y) \rightarrow (x - 3, y - 2)$

42. Reflection: in the line $y = -x$
Rotation: 180° about the origin

43. In the diagram, $ABCD \cong JKLM$. Find $m\angle L$ and JK.

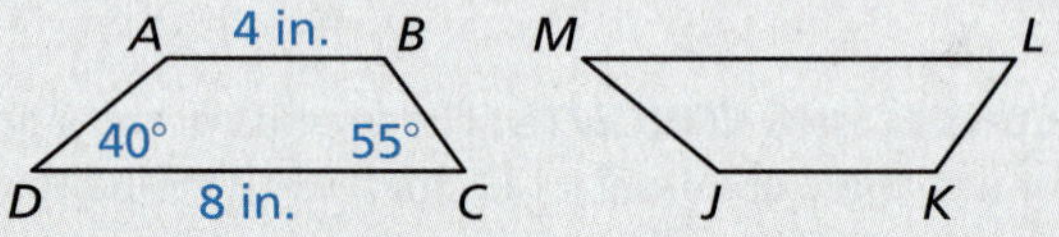

44. Find the distance from the point $(-4, -7)$ to the line $y = -\frac{1}{2}x - 4$.

45. Find the values of x and y.

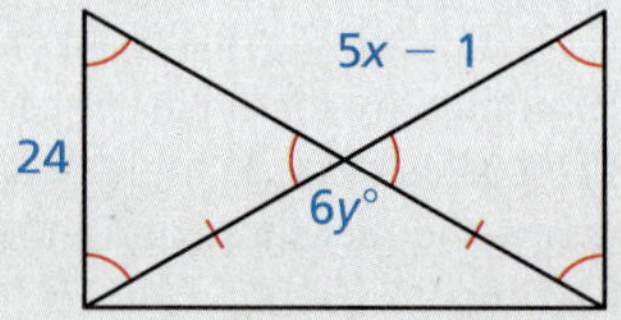

46. Find the mean, median, mode, range, and interquartile range of the data set.

13, 18, 17, 13, 15, 14, 20, 12, 13

GO DIGITAL

5.5 Proving Triangle Congruence by SSS

Learning Target: Prove and use the Side-Side-Side Congruence Theorem.

Success Criteria:
- I can use rigid motions to prove the SSS Congruence Theorem.
- I can use the SSS Congruence Theorem.
- I can use the Hypotenuse-Leg Congruence Theorem.

EXPLORE IT! Reasoning about Triangles

3 MTR REFLECT ON YOUR METHOD
Can you use any length for $\overline{BC}$ in part (b)? Explain.

Work with a partner.

a. Use technology to construct circles with center A and radii of 2 units and 3 units.

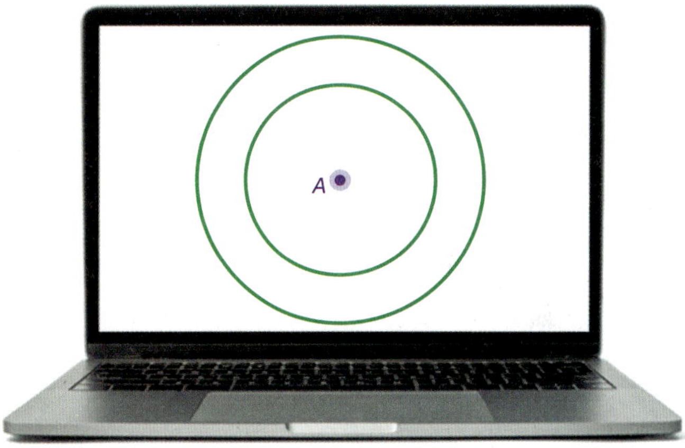

b. Draw $\overline{BC}$ so that $BC = 4$, B is on the smaller circle, and C is on the larger circle. Then draw $\triangle ABC$.

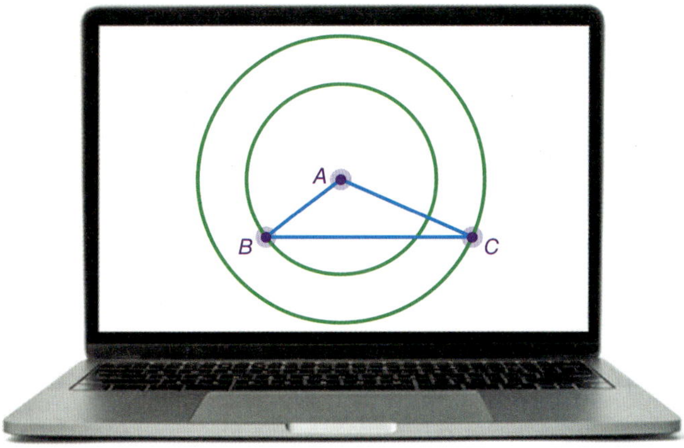

c. Explain why the side lengths of $\triangle ABC$ are 2, 3, and 4 units.

d. Find $m\angle A$, $m\angle B$, and $m\angle C$.

e. Repeat parts (b)–(d) several times, redrawing $\overline{BC}$ in different locations. What do you notice? Make a conjecture.

f. Can you prove your conjecture in part (e)? If so, write your proof.

Geometric Reasoning
MA.912.GR.1.2 Prove triangle congruence or similarity using Side-Side-Side, Side-Angle-Side, Angle-Side-Angle, Angle-Angle-Side, Angle-Angle and Hypotenuse-Leg.
MA.912.GR.1.6 Solve mathematical and real-world problems involving congruence or similarity in two-dimensional figures.
MA.912.GR.2.6, **H** **MA.912.GR.2.7**

GO DIGITAL

Using the Side-Side-Side Congruence Theorem

Vocabulary

legs of a right triangle, *p. 260*
hypotenuse, *p. 260*

THEOREM

5.8 Side-Side-Side (SSS) Congruence Theorem

If three sides of one triangle are congruent to three sides of a second triangle, then the two triangles are congruent.

If $\overline{AB} \cong \overline{DE}$, $\overline{BC} \cong \overline{EF}$, and $\overline{AC} \cong \overline{DF}$, then $\triangle ABC \cong \triangle DEF$.

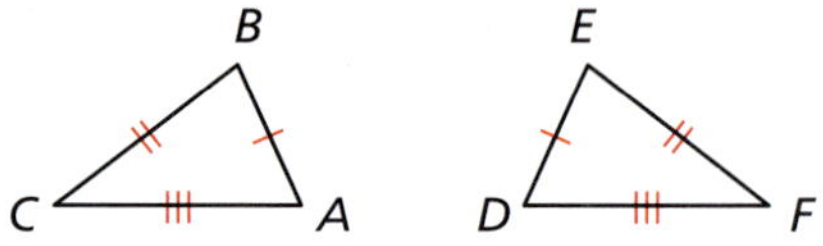

Prove this Theorem Exercise 1, page 259

PROOF Side-Side-Side (SSS) Congruence Theorem

Given $\overline{AB} \cong \overline{DE}$, $\overline{BC} \cong \overline{EF}$, $\overline{AC} \cong \overline{DF}$

Prove $\triangle ABC \cong \triangle DEF$

First, translate $\triangle ABC$ so that point A maps to point D, as shown below.

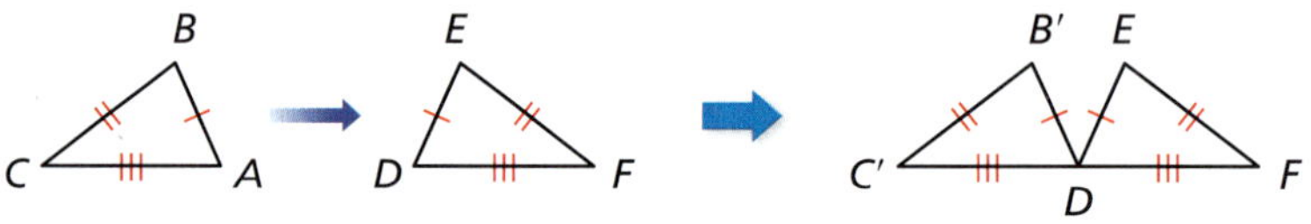

This translation maps $\triangle ABC$ to $\triangle DB'C'$. Next, rotate $\triangle DB'C'$ counterclockwise through $\angle C'DF$ so that the image of $\overrightarrow{DC'}$ coincides with $\overrightarrow{DF}$, as shown below.

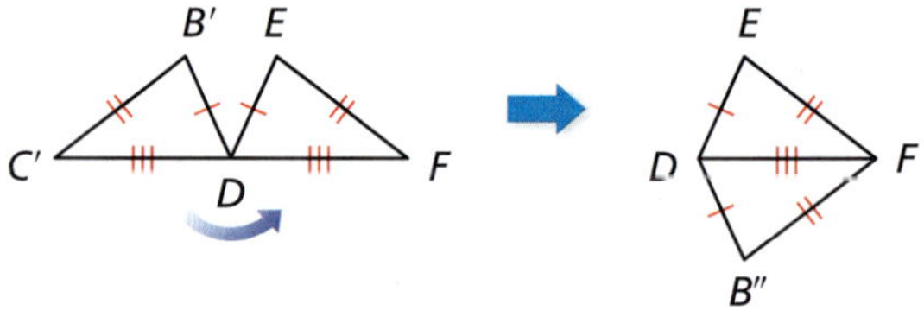

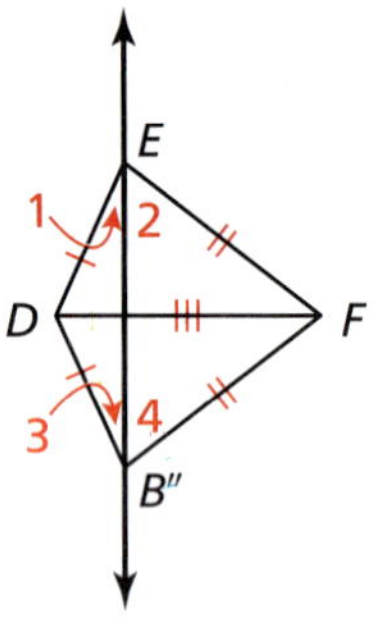

Because $\overline{DC'} \cong \overline{DF}$, the rotation maps point C' to point F. So, this rotation maps $\triangle DB'C'$ to $\triangle DB''F$. Draw an auxiliary line through points E and B''. This line creates $\angle 1$, $\angle 2$, $\angle 3$, and $\angle 4$, as shown at the left.

Because $\overline{DE} \cong \overline{DB''}$, $\triangle DEB''$ is an isosceles triangle. Because $\overline{FE} \cong \overline{FB''}$, $\triangle FEB''$ is an isosceles triangle. By the Base Angles Theorem, $\angle 1 \cong \angle 3$ and $\angle 2 \cong \angle 4$. By the definition of congruence, $m\angle 1 = m\angle 3$ and $m\angle 2 = m\angle 4$. By construction, $m\angle DEF = m\angle 1 + m\angle 2$ and $m\angle DB''F = m\angle 3 + m\angle 4$. You can now use the Substitution Property of Equality to show $m\angle DEF = m\angle DB''F$.

$m\angle DEF = m\angle 1 + m\angle 2$	Angle Addition Postulate
$= m\angle 3 + m\angle 4$	Substitute $m\angle 3$ for $m\angle 1$ and $m\angle 4$ for $m\angle 2$.
$= m\angle DB''F$	Angle Addition Postulate

By the definition of congruence, $\angle DEF \cong \angle DB''F$. So, two pairs of sides and their included angles are congruent. By the SAS Congruence Theorem, $\triangle DB''F \cong \triangle DEF$. So, a composition of rigid motions maps $\triangle DB''F$ to $\triangle DEF$. Because a composition of rigid motions maps $\triangle ABC$ to $\triangle DB''F$ and a composition of rigid motions maps $\triangle DB''F$ to $\triangle DEF$, a composition of rigid motions maps $\triangle ABC$ to $\triangle DEF$. So, $\triangle ABC \cong \triangle DEF$.

GO DIGITAL

EXAMPLE 1 Using the SSS Congruence Theorem

WATCH

Write a proof.

Given $\overline{KL} \cong \overline{NL}$, $\overline{KM} \cong \overline{NM}$

Prove $\triangle KLM \cong \triangle NLM$

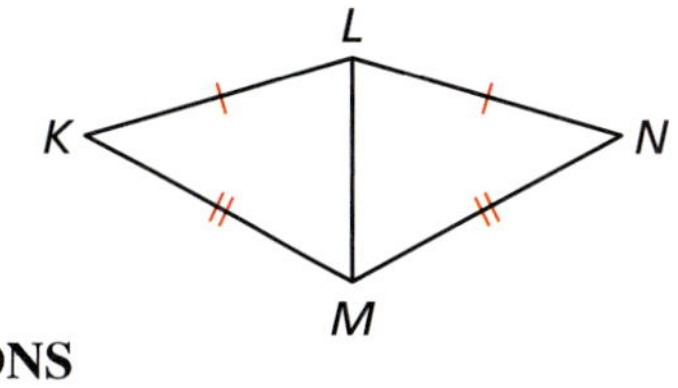

SOLUTION

STATEMENTS	REASONS
S **1.** $\overline{KL} \cong \overline{NL}$	**1.** Given
S **2.** $\overline{KM} \cong \overline{NM}$	**2.** Given
S **3.** $\overline{LM} \cong \overline{LM}$	**3.** Reflexive Property of Segment Congruence
4. $\triangle KLM \cong \triangle NLM$	**4.** SSS Congruence Theorem

2 MTR **USE ANOTHER METHOD**

Use rigid motions to prove that $\triangle KLM \cong \triangle NLM$.

7 MTR

EXAMPLE 2 Modeling Real LIfe

WATCH

Explain why the bench with the diagonal support is stable, whereas the one without the support can collapse.

SOLUTION

The bench with the diagonal support forms triangles with fixed side lengths. By the SSS Congruence Theorem, these triangles cannot change shape, so the bench is stable. The bench without the diagonal support is not stable because there are many possible quadrilaterals with the given side lengths.

SELF-ASSESSMENT

1 I don't understand yet. 2 I can do it with help. 3 I can do it on my own. 4 I can teach someone else.

1. Use rigid motions to prove that $\triangle DFG \cong \triangle HJK$.

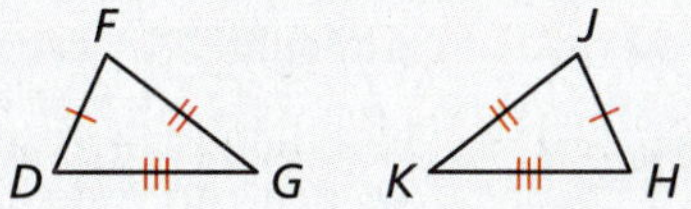

2. Write a proof.

Given $\overline{JK} \cong \overline{KL} \cong \overline{LM} \cong \overline{MJ}$

Prove $\triangle JKL \cong \triangle LMJ$

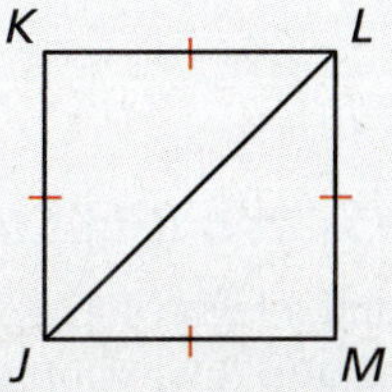

Determine whether the figure is stable. Explain your reasoning.

3.

4.

5.

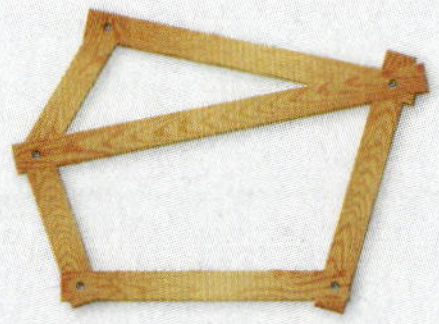

GO DIGITAL

CONSTRUCTION Copying a Triangle Using SSS

WATCH

Construct a triangle that is congruent to $\triangle ABC$ using the SSS Congruence Theorem. Use a compass and straightedge.

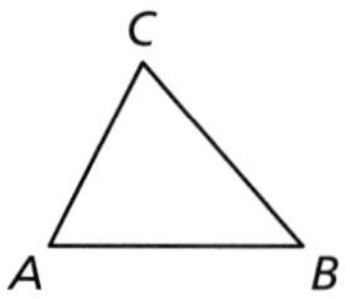

SOLUTION

Step 1

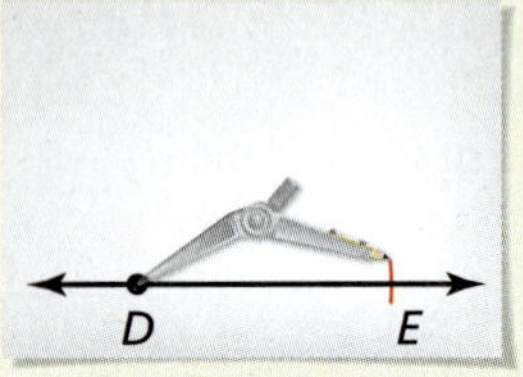

Construct a side Construct $\overline{DE}$ so that it is congruent to $\overline{AB}$.

Step 2

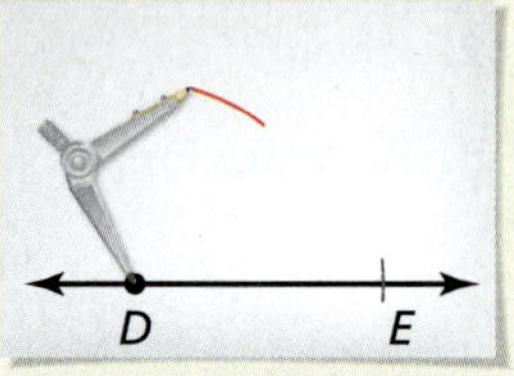

Draw an arc Open your compass to the length AC. Use this length to draw an arc with center D.

Step 3

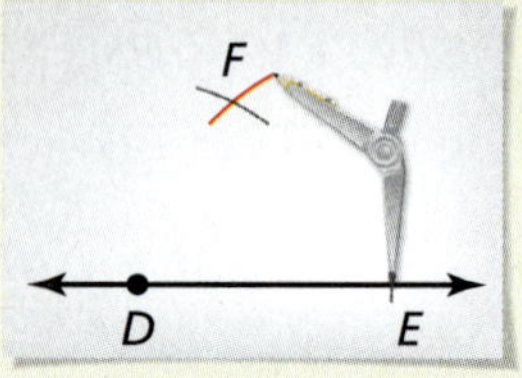

Draw an arc Draw an arc with radius BC and center E that intersects the arc from Step 2. Label the intersection point F.

Step 4

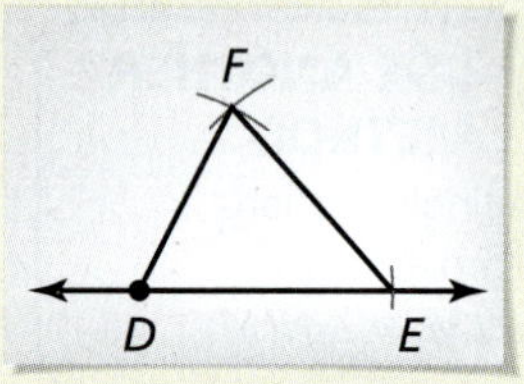

Draw a triangle Draw $\triangle DEF$. By the SSS Congruence Theorem, $\triangle ABC \cong \triangle DEF$.

Using the Hypotenuse-Leg Congruence Theorem

You know that SAS and SSS are valid methods for proving that triangles are congruent. What about SSA?

In general, SSA is *not* a valid method for proving that triangles are congruent. In the triangles below, two pairs of sides and a pair of angles not included between them are congruent, but the triangles are not congruent.

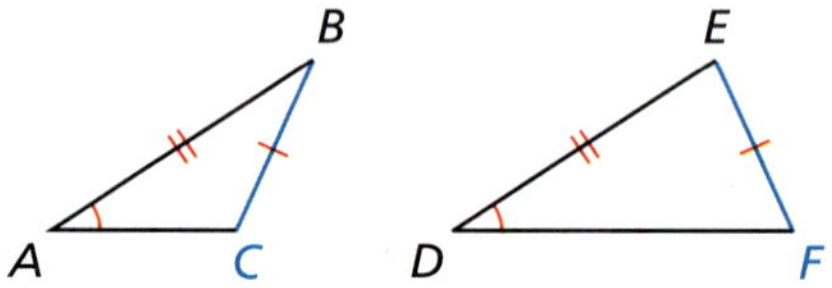

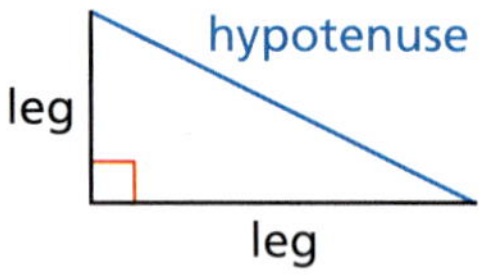

Although SSA is not valid in general, there is a special case for right triangles.

In a right triangle, the sides adjacent to the right angle are called the **legs**. The side opposite the right angle is called the **hypotenuse** of the right triangle.

THEOREM

5.9 Hypotenuse-Leg (HL) Congruence Theorem

If the hypotenuse and a leg of a right triangle are congruent to the hypotenuse and a leg of a second right triangle, then the two triangles are congruent.

If $\overline{AB} \cong \overline{DE}$, $\overline{AC} \cong \overline{DF}$, and $m\angle C = m\angle F = 90°$, then $\triangle ABC \cong \triangle DEF$.

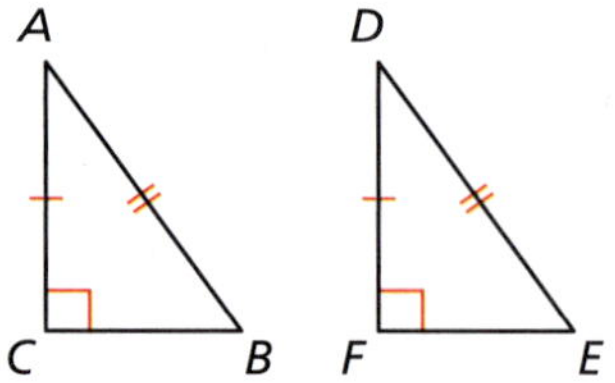

Prove this Theorem Exercise 30, page 459
Proof *BigIdeasMath.com*

GO DIGITAL

EXAMPLE 3 Using the Hypotenuse-Leg Congruence Theorem

WATCH

Write a proof.

Given $\overline{WY} \cong \overline{XZ}$, $\overline{WZ} \perp \overline{ZY}$, $\overline{XY} \perp \overline{ZY}$

Prove $\triangle WYZ \cong \triangle XZY$

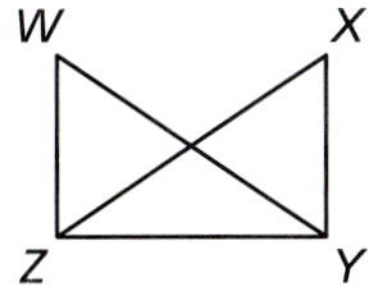

SOLUTION

Redraw the triangles so they are side by side with corresponding parts in the same position. Mark the given information in the diagram.

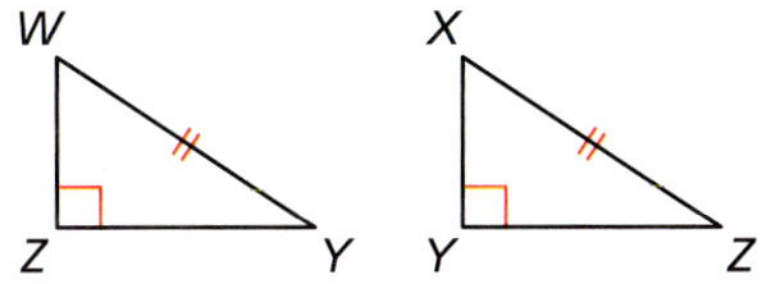

STUDY TIP

If you have trouble matching vertices to letters when you separate the overlapping triangles, leave the triangles in their original orientations.

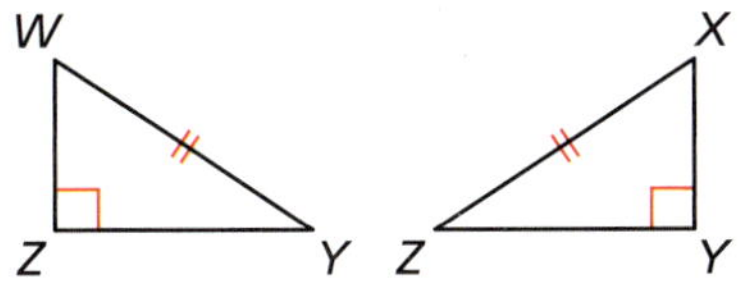

STATEMENTS	REASONS
H **1.** $\overline{WY} \cong \overline{XZ}$	**1.** Given
2. $\overline{WZ} \perp \overline{ZY}, \overline{XY} \perp \overline{ZY}$	**2.** Given
3. $\angle Z$ and $\angle Y$ are right angles.	**3.** Definition of $\perp$ lines
4. $\triangle WYZ$ and $\triangle XZY$ are right triangles.	**4.** Definition of a right triangle
L **5.** $\overline{ZY} \cong \overline{YZ}$	**5.** Reflexive Property of Segment Congruence
6. $\triangle WYZ \cong \triangle XZY$	**6.** HL Congruence Theorem

EXAMPLE 4 Using the Hypotenuse-Leg Congruence Theorem

WATCH

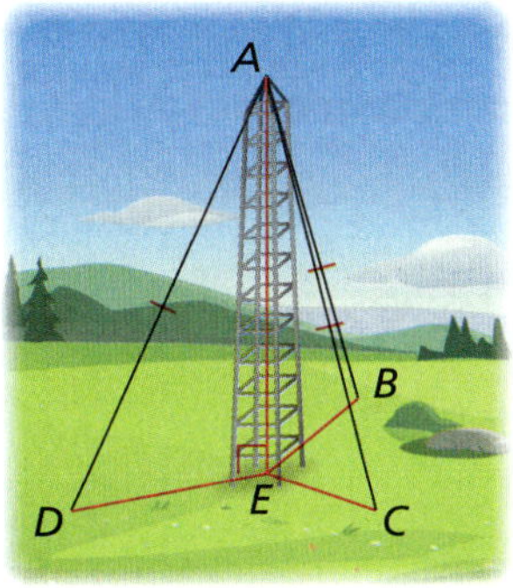

The television antenna is perpendicular to the plane containing points B, C, D, and E. Each of the cables running from the top of the antenna to B, C, and D has the same length. Prove that $\triangle AEB$, $\triangle AEC$, and $\triangle AED$ are congruent.

Given $\overline{AE} \perp \overline{EB}$, $\overline{AE} \perp \overline{EC}$, $\overline{AE} \perp \overline{ED}$, $\overline{AB} \cong \overline{AC} \cong \overline{AD}$

Prove $\triangle AEB \cong \triangle AEC \cong \triangle AED$

SOLUTION

You are given that $\overline{AE} \perp \overline{EB}$ and $\overline{AE} \perp \overline{EC}$. So, $\angle AEB$ and $\angle AEC$ are right angles by the definition of perpendicular lines. By definition, $\triangle AEB$ and $\triangle AEC$ are right triangles. You are given that the hypotenuses of these two triangles, $\overline{AB}$ and $\overline{AC}$, are congruent. Also, $\overline{AE}$ is a leg for both triangles, and $\overline{AE} \cong \overline{AE}$ by the Reflexive Property of Segment Congruence. So, by the Hypotenuse-Leg Congruence Theorem, $\triangle AEB \cong \triangle AEC$. You can use similar reasoning to prove that $\triangle AEC \cong \triangle AED$.

▶ So, by the Transitive Property of Triangle Congruence, $\triangle AEB \cong \triangle AEC \cong \triangle AED$.

SELF-ASSESSMENT

1 I don't understand yet. 2 I can do it with help. 3 I can do it on my own. 4 I can teach someone else.

Use the diagram.

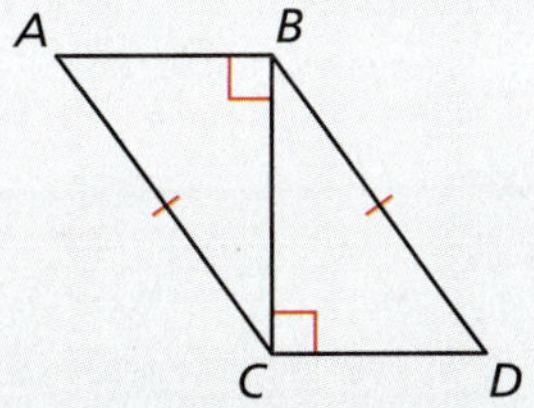

6. Redraw $\triangle ABC$ and $\triangle DCB$ side by side with corresponding parts in the same position.

7. Use the information in the diagram to prove that $\triangle ABC \cong \triangle DCB$.

GO DIGITAL

5.5 Practice WITH CalcChat® AND CalcView®

In Exercises 1 and 2, decide whether enough information is given to prove that the triangles are congruent using the SSS Congruence Theorem. Explain.

1. $\triangle ABC$, $\triangle DBE$

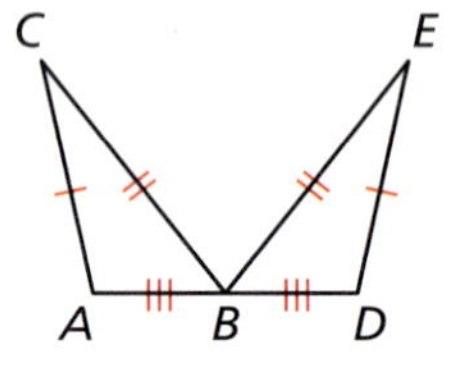

2. $\triangle PQS$, $\triangle RQS$

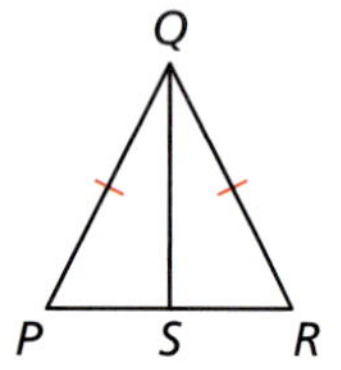

In Exercises 3 and 4, decide whether enough information is given to prove that the triangles are congruent using the HL Congruence Theorem. Explain.

3. $\triangle ABC$, $\triangle FED$

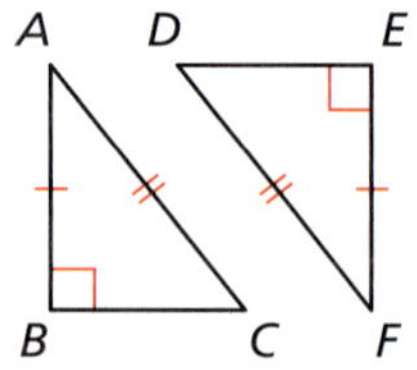

4. $\triangle PQT$, $\triangle SRT$

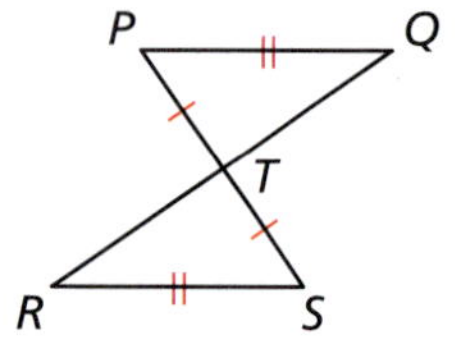

In Exercises 5 and 6, decide whether the congruence statement is true. Explain your reasoning.

5. $\triangle RST \cong \triangle TQP$

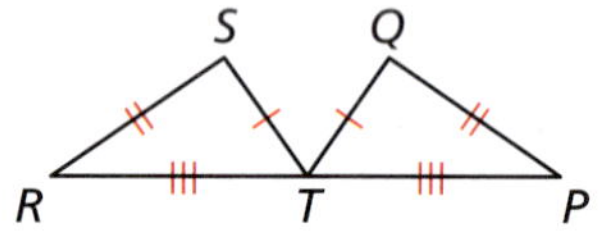

6. $\triangle ABD \cong \triangle CDB$

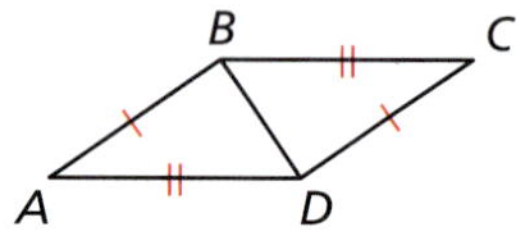

In Exercises 7 and 8, write a proof. *(See Example 1.)*

▶**7.** **Given** $\overline{LM} \cong \overline{JK}$, $\overline{MJ} \cong \overline{KL}$

Prove $\triangle LMJ \cong \triangle JKL$

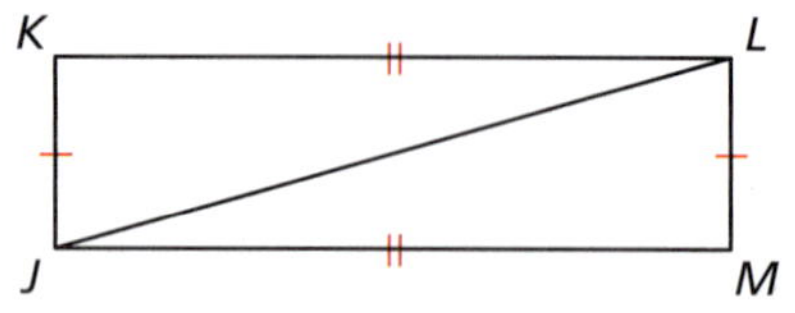

8. **Given** $\overline{WX} \cong \overline{VZ}$, $\overline{WY} \cong \overline{VY}$, $\overline{YZ} \cong \overline{YX}$

Prove $\triangle VWX \cong \triangle WVZ$

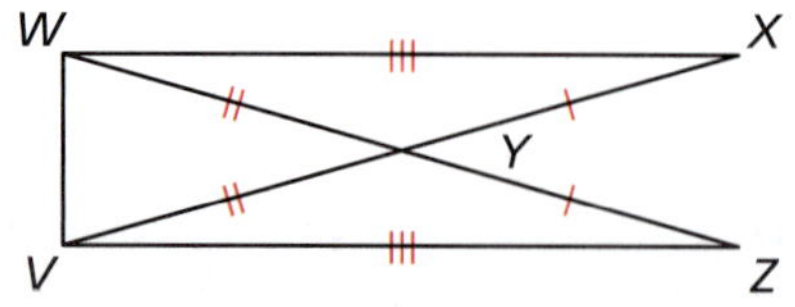

In Exercises 9 and 10, determine whether the figure is stable. Explain your reasoning. *(See Example 2.)*

▶**9.**

10.

In Exercises 11 and 12, redraw the triangles so they are side by side with corresponding parts in the same position. Then write a proof. *(See Example 3.)*

▶**11.** **Given** $\overline{AC} \cong \overline{BD}$, $\overline{AB} \perp \overline{AD}$, $\overline{CD} \perp \overline{AD}$

Prove $\triangle BAD \cong \triangle CDA$

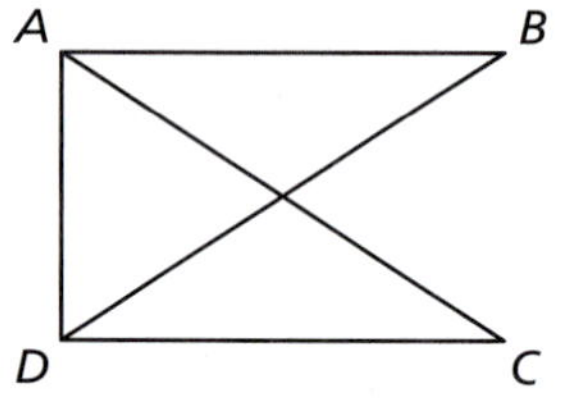

12. **Given** G is the midpoint of $\overline{EH}$, $\overline{FG} \cong \overline{GI}$, $\angle E$ and $\angle H$ are right angles.

Prove $\triangle EFG \cong \triangle HIG$

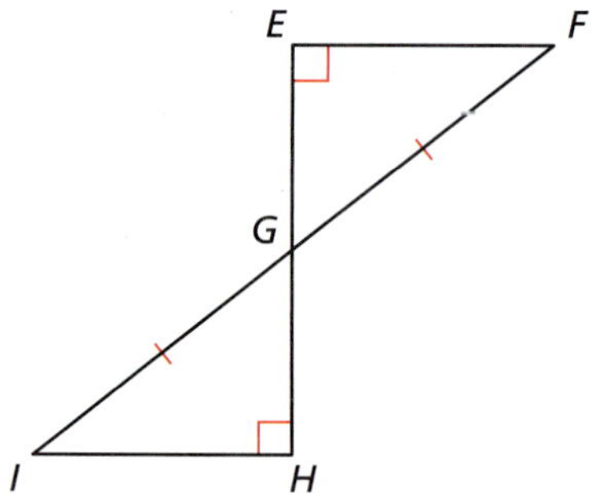

4 MTR **13. ERROR ANALYSIS** Describe and correct the error in identifying congruent triangles.

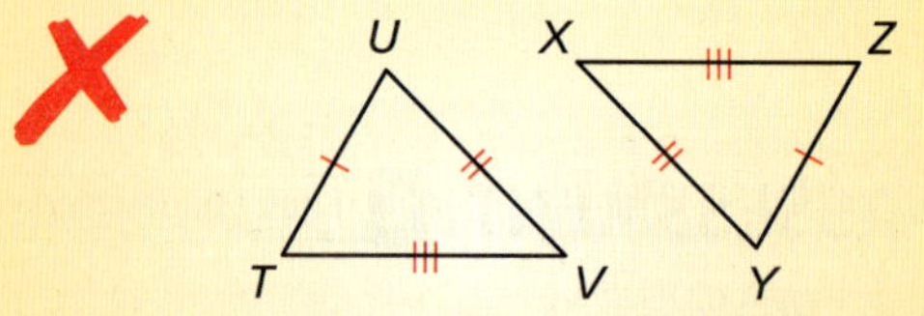

$\triangle TUV \cong \triangle XYZ$ by the SSS Congruence Theorem.

GO DIGITAL

4 MTR **14. ERROR ANALYSIS** Describe and correct the error in determining whether $\triangle JKL \cong \triangle LMJ$.

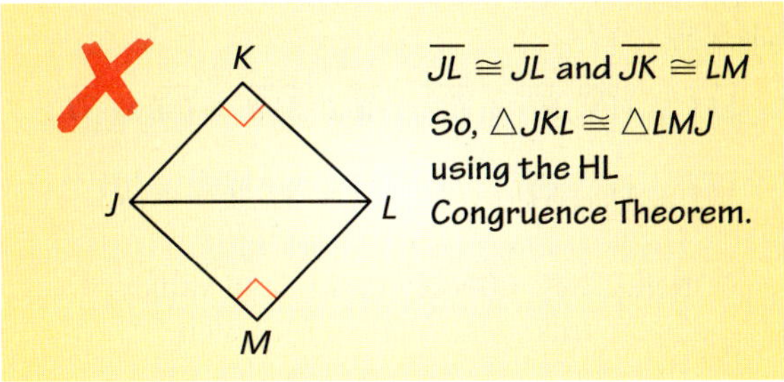

CONSTRUCTION In Exercises 15 and 16, construct a triangle that is congruent to $\triangle QRS$ using the SSS Congruence Theorem.

15.

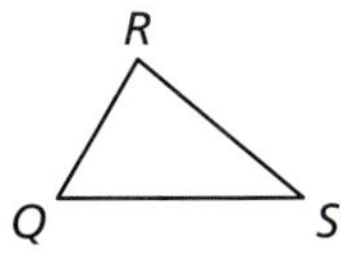

16.

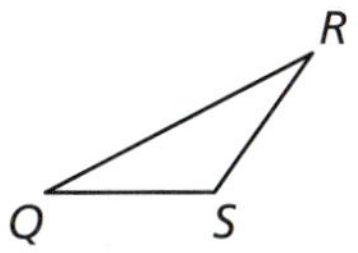

7 MTR **17. MODELING REAL LIFE** The distances between consecutive bases on a softball field are the same. The distance from home plate to second base is the same as the distance from first base to third base. The angles created at each base are 90°. Prove $\triangle HFS \cong \triangle FST \cong \triangle STH$. *(See Example 4.)*

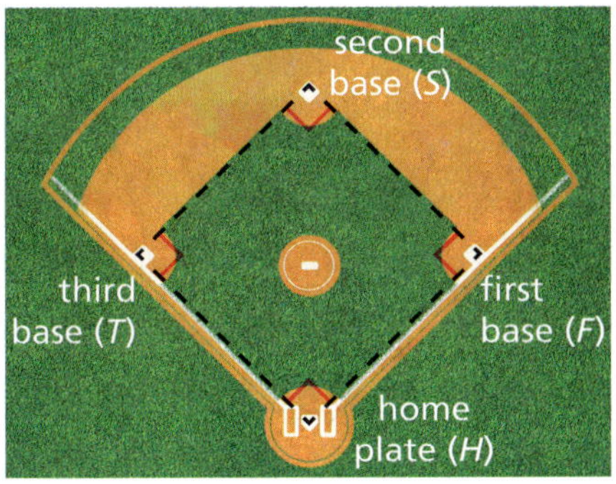

7 MTR **18. MODELING REAL LIFE** The diagram shows the light created by two spotlights. Both spotlights are the same distance from the stage. Prove $\triangle ABD \cong \triangle CBD$. Then determine whether all four triangles in the diagram are congruent. Explain your reasoning.

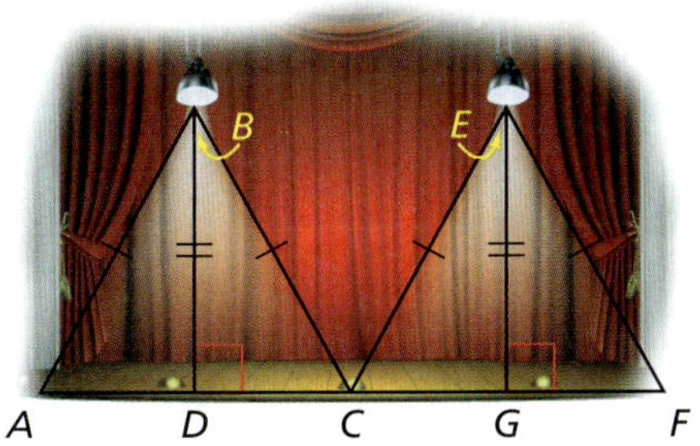

GO DIGITAL

19. REASONING To support a tree, you attach wires from the trunk of the tree to stakes in the ground, as shown in the diagram.

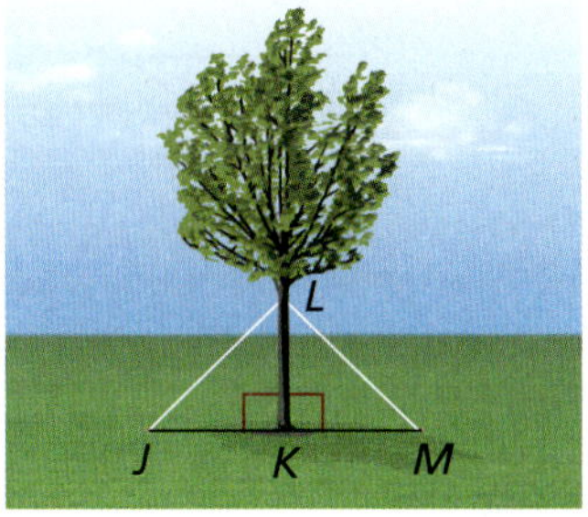

a. What additional information do you need to use the HL Congruence Theorem to prove that $\triangle JKL \cong \triangle MKL$?

b. Suppose K is the midpoint of $\overline{JM}$. Name a theorem you can use to prove that $\triangle JKL \cong \triangle MKL$. Explain your reasoning.

20. REASONING Use the photo of the phone case, where $\overline{AB} \cong \overline{DE}$ and $\overline{AC} \cong \overline{CE}$.

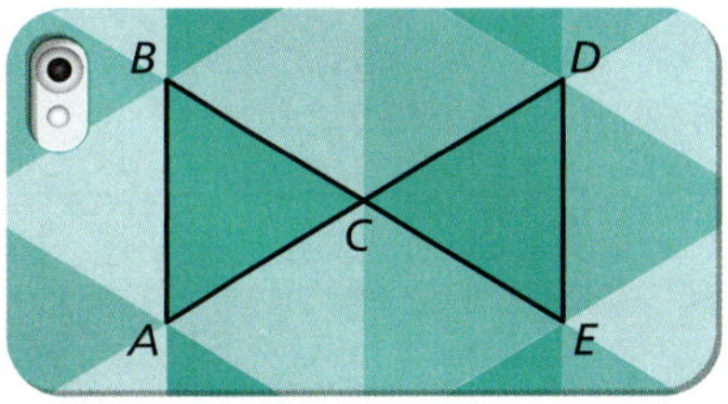

a. What additional information do you need to use the SSS Congruence Theorem to prove that $\triangle ABC \cong \triangle EDC$?

b. What additional information do you need to use the SAS Congruence Theorem to prove that $\triangle ABC \cong \triangle EDC$?

In Exercises 21 and 22, use the given coordinates to determine whether $\triangle ABC \cong \triangle DEF$.

21. $A(-2, -2)$, $B(4, -2)$, $C(4, 6)$, $D(5, 7)$, $E(5, 1)$, $F(13, 1)$

22. $A(-5, 7)$, $B(-5, 2)$, $C(0, 2)$, $D(0, 6)$, $E(0, 1)$, $F(4, 1)$

23. REASONING Describe how to determine whether two triangles are congruent using only a piece of string.

7 MTR **24. PERFORMANCE TASK** Research camping supplies such as chairs, tents, and grills. Describe aspects of the designs that make them stable. Then write a letter that you could send to the manufacturer with suggestions for improving the stability of one item.

4 MTR **25. MAKING AN ARGUMENT** Explain how you can use the HL Congruence Theorem to show that $\triangle PTQ \cong \triangle STR$.

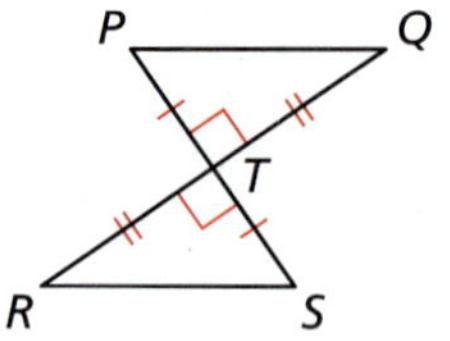

26. HOW DO YOU SEE IT?
Determine whether each statement is true. Explain your reasoning.

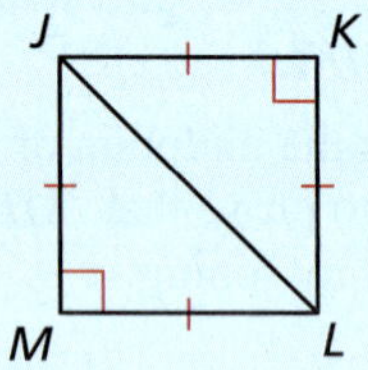

a. $\triangle JKL \cong \triangle LMJ$ using the SSS Congruence Theorem.

b. $\triangle JKL \cong \triangle LMJ$ using the HL Congruence Theorem.

USING TOOLS In Exercises 27 and 28, use the given information to sketch $\triangle LMN$ and $\triangle STU$. Mark the triangles with the given information. Then prove that $\triangle LMN \cong \triangle STU$.

27. $\overline{LM} \perp \overline{MN}$, $\overline{ST} \perp \overline{TU}$, $\overline{LM} \cong \overline{NM} \cong \overline{UT} \cong \overline{ST}$

28. $\overline{LM} \perp \overline{MN}$, $\overline{ST} \perp \overline{TU}$, $\overline{LM} \cong \overline{ST}$, $\overline{LN} \cong \overline{SU}$

5 MTR **29. CONNECTING CONCEPTS** Find all values of x that make the triangles congruent. Explain.

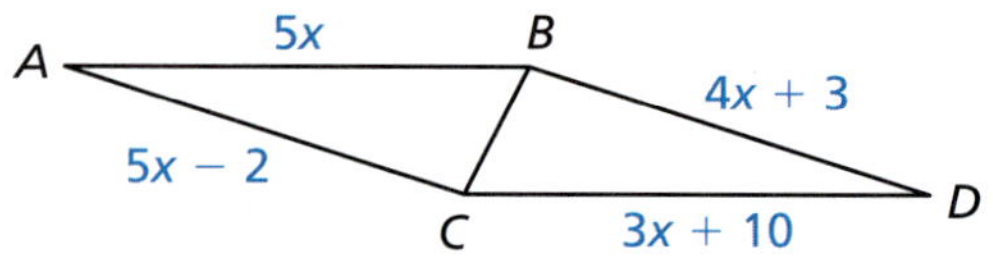

30. THOUGHT PROVOKING
The postulates and theorems in this book represent Euclidean geometry. In spherical geometry, all points are points on the surface of a sphere. A line is a circle on the sphere whose diameter is equal to the diameter of the sphere. In spherical geometry, are two triangles congruent if their corresponding sides are congruent? Explain.

REVIEW & REFRESH

WATCH

31. Are $\overleftrightarrow{AC}$ and $\overleftrightarrow{DF}$ parallel? Explain your reasoning.

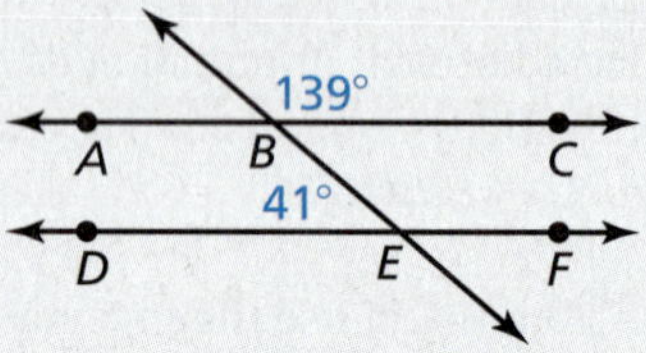

32. Find $m\angle 1$.

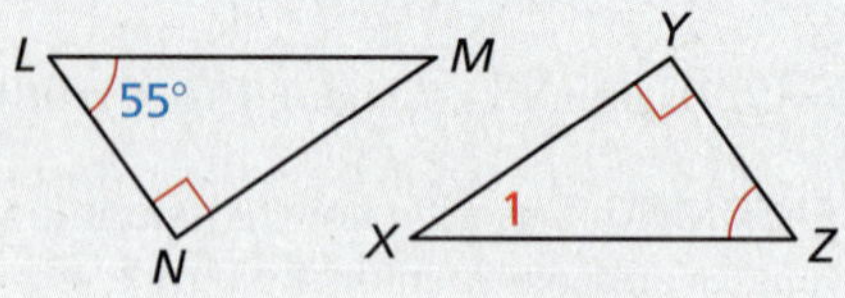

33. Write a linear function f with $f(1) = -1$ and $f(2) = 5$.

7 MTR **34. MODELING REAL LIFE** Find the values of x and y in the diagram of the magnet.

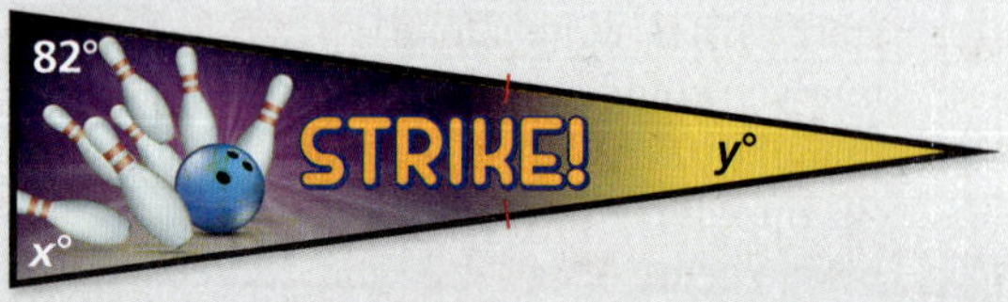

35. Write a proof.

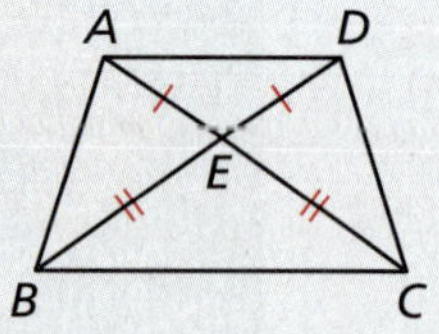

Given $\overline{AE} \cong \overline{DE}$, $\overline{BE} \cong \overline{CE}$, $\angle AEB$ and $\angle DEC$ are vertical angles.

Prove $\triangle AEB \cong \triangle DEC$

36. Graph the polygon given by $J(0, 3)$, $K(2, -2)$, $L(1, -4)$ and its image after a reflection in the y-axis.

37. Graph $y \leq 4 - x$ in a coordinate plane.

38. Decide whether the congruence statement $\triangle JKL \cong \triangle MNK$ is true. Explain your reasoning.

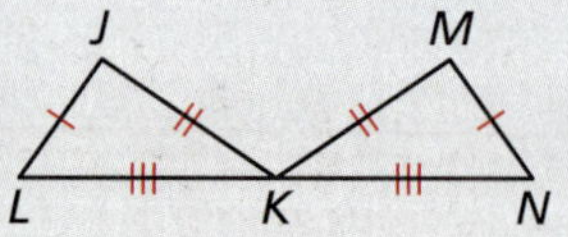

39. Tell whether the points $(-2, 8)$, $(-1, 3)$, $(0, 0)$, $(1, -1)$, and $(2, 0)$ appear to represent a *linear*, an *exponential*, or a *quadratic* function.

GO DIGITAL

5.6 Proving Triangle Congruence by ASA and AAS

Learning Target: Prove and use the Angle-Side-Angle Congruence Theorem and the Angle-Angle-Side Congruence Theorem.

Success Criteria:
- I can use rigid motions to prove the ASA Congruence Theorem.
- I can prove the AAS Congruence Theorem.
- I can use the ASA and AAS Congruence Theorems.

EXPLORE IT! Determining Valid Congruence Theorems

Work with a partner.

a. Use technology to construct a circle with center A.

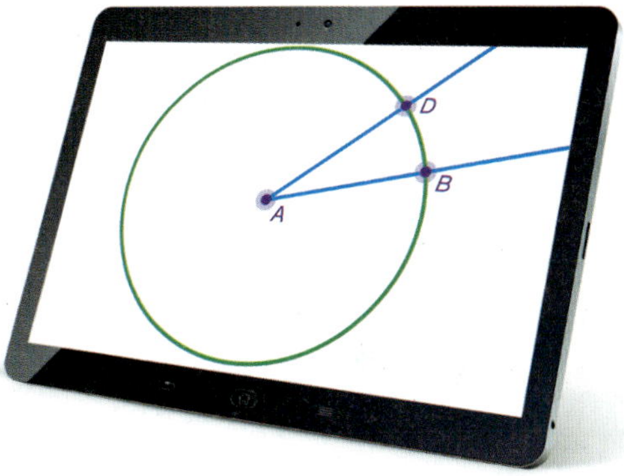

b. Draw an angle that has a measure less than 180° with vertex A. Label one of the points where a side of the angle intersects the circle as B. Label the other point of intersection as D.

c. Does the length of $\overline{AB}$ depend on the position of $\angle A$? Explain.

d. Draw an angle with vertex B and side $\overrightarrow{BA}$ so that the other side intersects $\overrightarrow{AD}$. Label the point where the other side of the angle intersects $\overrightarrow{AD}$ as C.

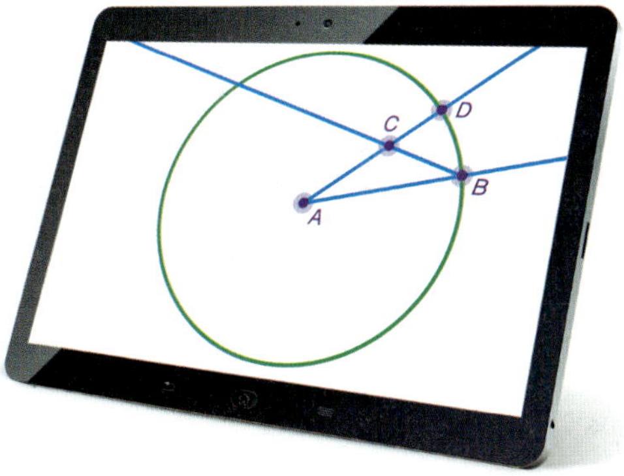

e. Find $m\angle ACB$, BC, and AC.

f. Repeat parts (a)–(e) several times, redrawing $\angle A$ in different orientations. Use the same angle measure for $\angle A$, the same radius for the circle, and the same angle measure for $\angle B$. What do you notice? Make a conjecture.

g. Determine whether each of the following is a valid triangle congruence theorem. Explain.

i. Angle-Angle-Side (AAS)

ii. Angle-Angle-Angle (AAA)

4 MTR CONSTRUCT AN ARGUMENT

When two triangles have four pairs of congruent corresponding parts, can you conclude that the triangles are congruent? Explain.

Geometric Reasoning

MA.912.GR.1.2 Prove triangle congruence or similarity using Side-Side-Side, Side-Angle-Side, Angle-Side-Angle, Angle-Angle-Side, Angle-Angle and Hypotenuse-Leg.

MA.912.GR.1.6 Solve mathematical and real-world problems involving congruence or similarity in two-dimensional figures.

MA.912.GR.2.6, H MA.912.GR.2.7

GO DIGITAL

Using the ASA and AAS Congruence Theorems

THEOREM

5.10 Angle-Side-Angle (ASA) Congruence Theorem

If two angles and the included side of one triangle are congruent to two angles and the included side of a second triangle, then the two triangles are congruent.

If $\angle A \cong \angle D$, $\overline{AC} \cong \overline{DF}$, and $\angle C \cong \angle F$, then $\triangle ABC \cong \triangle DEF$.

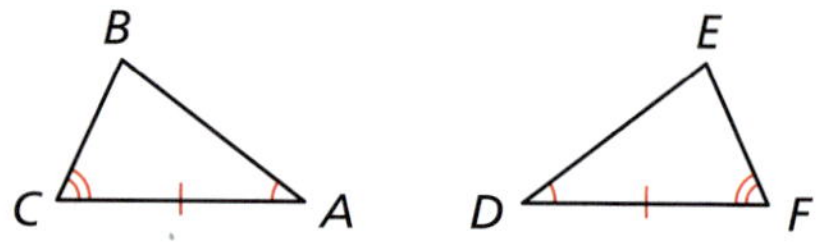

Prove this Theorem Exercise 1, page 267

PROOF Angle-Side-Angle (ASA) Congruence Theorem H

Given $\angle A \cong \angle D$, $\overline{AC} \cong \overline{DF}$, $\angle C \cong \angle F$

Prove $\triangle ABC \cong \triangle DEF$

First, translate $\triangle ABC$ so that point A maps to point D, as shown below.

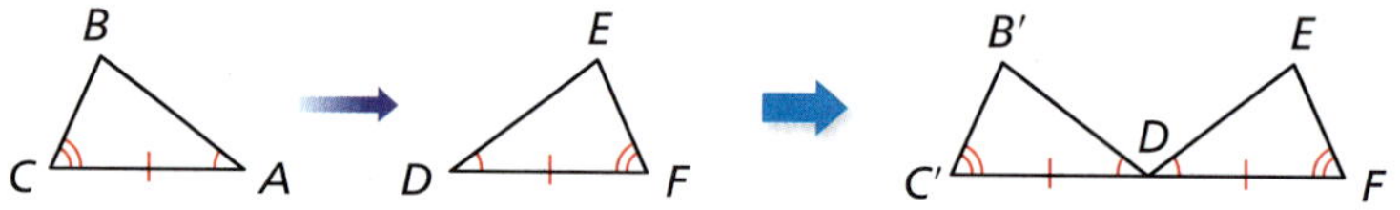

This translation maps $\triangle ABC$ to $\triangle DB'C'$. Next, rotate $\triangle DB'C'$ counterclockwise through $\angle C'DF$ so that the image of $\overrightarrow{DC'}$ coincides with $\overrightarrow{DF}$, as shown below.

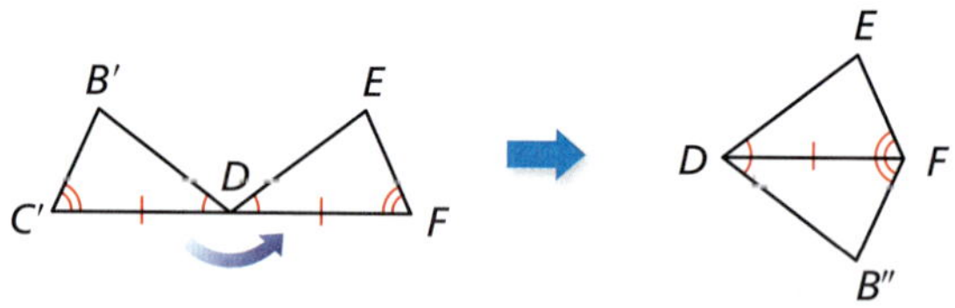

Because $\overline{DC'} \cong \overline{DF}$, the rotation maps point C' to point F. So, this rotation maps $\triangle DB'C'$ to $\triangle DB''F$. Now, reflect $\triangle DB''F$ in the line through points D and F, as shown below.

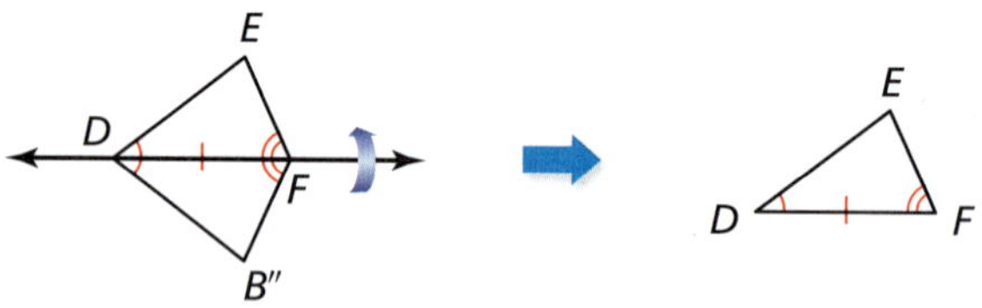

Because points D and F lie on $\overleftrightarrow{DF}$, this reflection maps them onto themselves. Because a reflection preserves angle measure and $\angle B''DF \cong \angle EDF$, the reflection maps $\overrightarrow{DB''}$ to $\overrightarrow{DE}$. Similarly, because $\angle B''FD \cong \angle EFD$, the reflection maps $\overrightarrow{FB''}$ to $\overrightarrow{FE}$. The image of B'' lies on $\overrightarrow{DE}$ and $\overrightarrow{FE}$. Because $\overrightarrow{DE}$ and $\overrightarrow{FE}$ have only point E in common, the image of B'' must be E. So, this reflection maps $\triangle DB''F$ to $\triangle DEF$.

Because you can map $\triangle ABC$ to $\triangle DEF$ using a composition of rigid motions, $\triangle ABC \cong \triangle DEF$.

GO DIGITAL

THEOREM

5.11 Angle-Angle-Side (AAS) Congruence Theorem

If two angles and a non-included side of one triangle are congruent to two angles and the corresponding non-included side of a second triangle, then the two triangles are congruent.

If $\angle A \cong \angle D$, $\angle C \cong \angle F$, and $\overline{BC} \cong \overline{EF}$, then $\triangle ABC \cong \triangle DEF$.

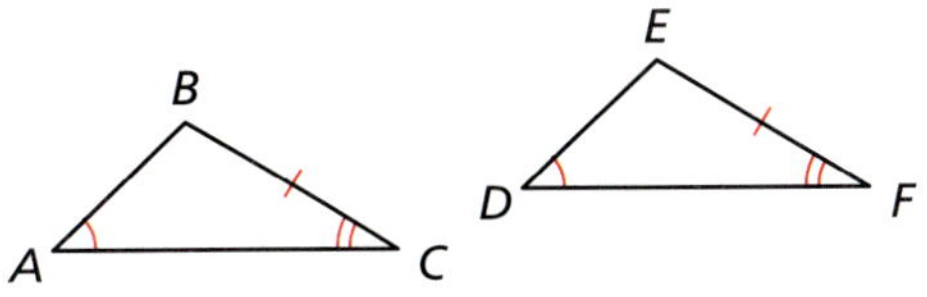

PROOF Angle-Angle-Side (AAS) Congruence Theorem

Given $\angle A \cong \angle D$, $\angle C \cong \angle F$, $\overline{BC} \cong \overline{EF}$

Prove $\triangle ABC \cong \triangle DEF$

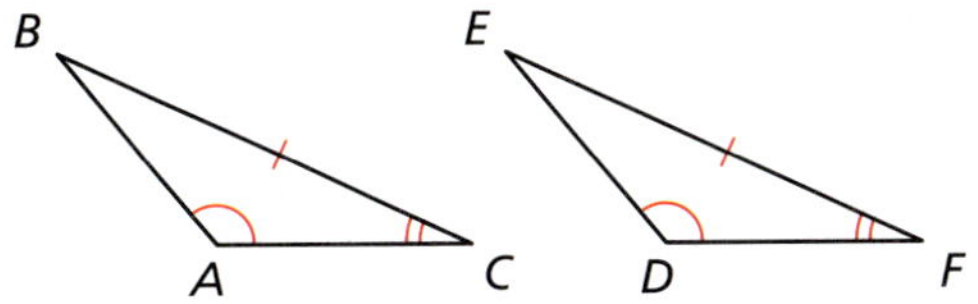

You are given $\angle A \cong \angle D$ and $\angle C \cong \angle F$. By the Third Angles Theorem, $\angle B \cong \angle E$. You are given $\overline{BC} \cong \overline{EF}$. So, two pairs of angles and their included sides are congruent. By the ASA Congruence Theorem, $\triangle ABC \cong \triangle DEF$.

EXAMPLE 1 Identifying Congruent Triangles

WATCH

Can the triangles be proven congruent with the information given in the diagram? If so, state the theorem you can use.

a. 

b. 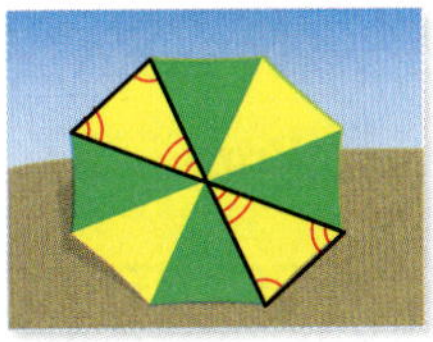

c. 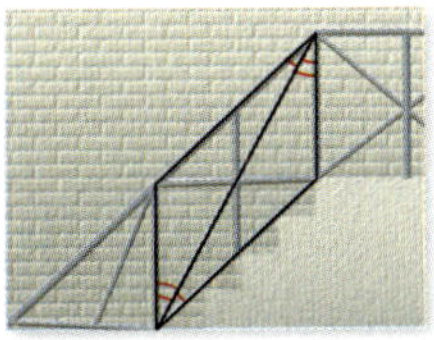

COMMON ERROR

You need at least one pair of congruent corresponding sides to prove two triangles are congruent.

SOLUTION

a. The vertical angles are congruent, so two pairs of angles and a pair of non-included sides are congruent. The triangles are congruent by the AAS Congruence Theorem.

b. There is not enough information to prove the triangles are congruent, because no sides are known to be congruent.

c. Two pairs of angles and their included sides are congruent. The triangles are congruent by the ASA Congruence Theorem.

SELF-ASSESSMENT

1 I don't understand yet. 2 I can do it with help. 3 I can do it on my own. 4 I can teach someone else.

1. Use rigid motions to prove that $\triangle DFG \cong \triangle HJK$.

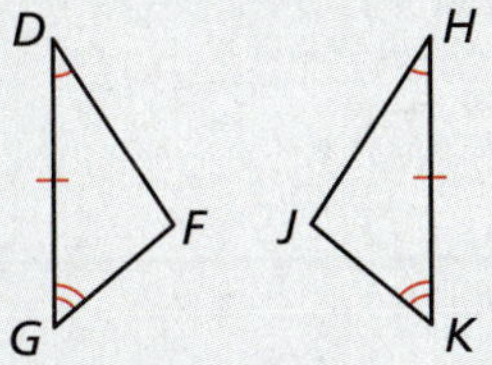

2. Can the triangles be proven congruent with the information given in the diagram? If so, state the theorem you can use.

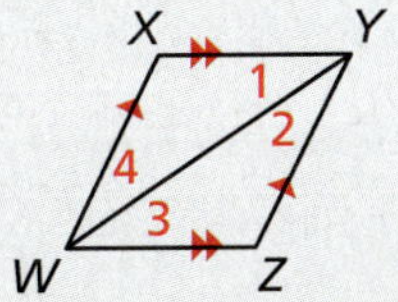

GO DIGITAL

CONSTRUCTION Copying a Triangle Using ASA

Construct a triangle that is congruent to $\triangle ABC$ using the ASA Congruence Theorem. Use a compass and straightedge.

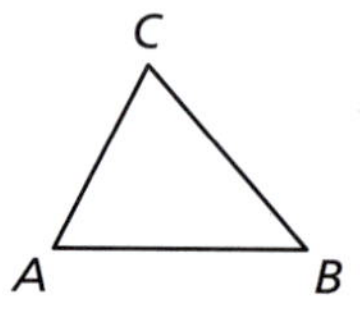

SOLUTION

Step 1

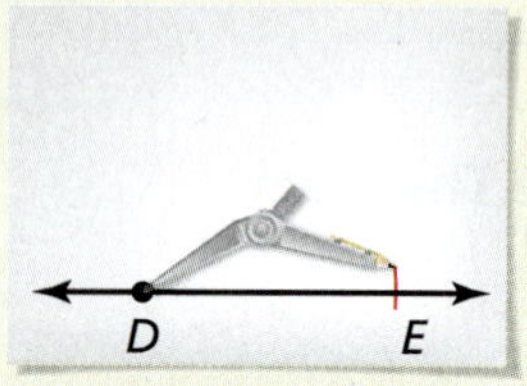

Construct a side
Construct $\overline{DE}$ so that it is congruent to $\overline{AB}$.

Step 2

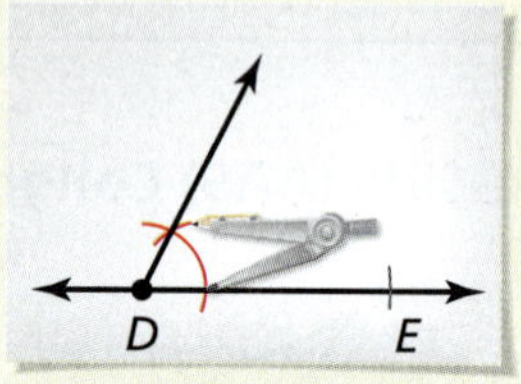

Construct an angle
Construct $\angle D$ with vertex D and side $\overrightarrow{DE}$ so that it is congruent to $\angle A$.

Step 3

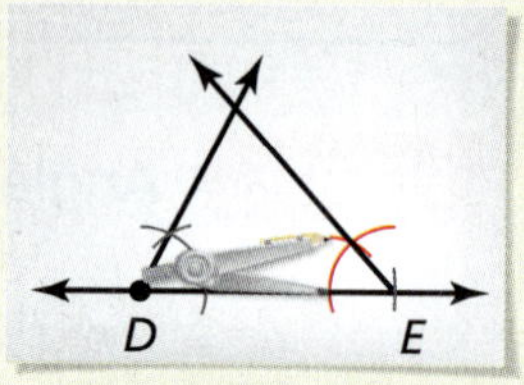

Construct an angle
Construct $\angle E$ with vertex E and side $\overrightarrow{ED}$ so that it is congruent to $\angle B$.

Step 4

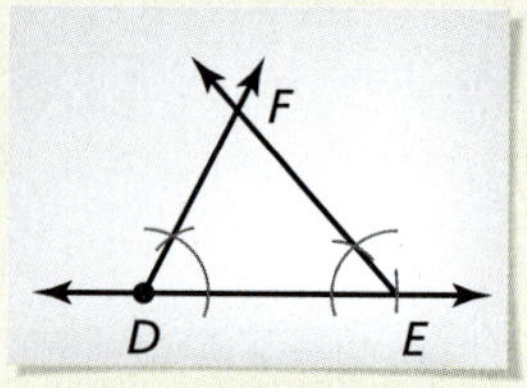

Label a point
Label the intersection of the sides of $\angle D$ and $\angle E$ that you constructed in Steps 2 and 3 as F. By the ASA Congruence Theorem, $\triangle ABC \cong \triangle DEF$.

EXAMPLE 2 Using the ASA Congruence Theorem

Write a proof.

Given $\overline{AD} \parallel \overline{EC}$, $\overline{BD} \cong \overline{BC}$

Prove $\triangle ABD \cong \triangle EBC$

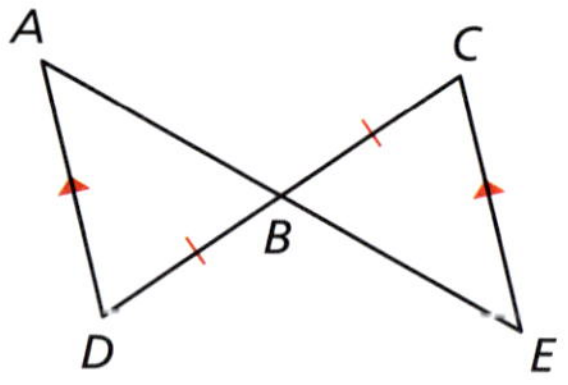

SOLUTION

STATEMENTS	REASONS
1. $\overline{AD} \parallel \overline{EC}$	1. Given
A 2. $\angle D \cong \angle C$	2. Alternate Interior Angles Theorem
S 3. $\overline{BD} \cong \overline{BC}$	3. Given
A 4. $\angle ABD \cong \angle EBC$	4. Vertical Angles Congruence Theorem
5. $\triangle ABD \cong \triangle EBC$	5. ASA Congruence Theorem

2 MTR USE ANOTHER METHOD
Use the AAS Congruence Theorem to prove that $\triangle ABD \cong \triangle EBC$.

SELF-ASSESSMENT

1 I don't understand yet. 2 I can do it with help. 3 I can do it on my own. 4 I can teach someone else.

3. In the diagram, $\overline{AB} \perp \overline{AD}$, $\overline{DE} \perp \overline{AD}$, and $\overline{AC} \cong \overline{DC}$. Prove $\triangle ABC \cong \triangle DEC$.

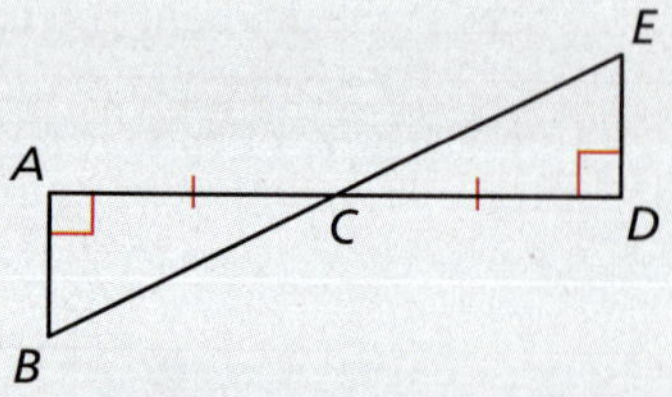

GO DIGITAL

EXAMPLE 3 Using the AAS Congruence Theorem

WATCH

Write a proof.

Given $\overline{HF} \parallel \overline{GK}$, $\angle F$ and $\angle K$ are right angles.

Prove $\triangle HFG \cong \triangle GKH$

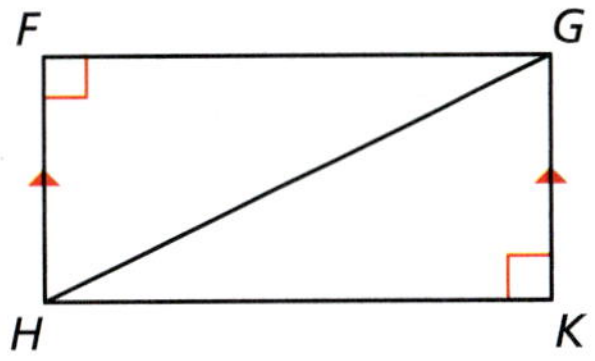

SOLUTION

STATEMENTS	REASONS
1. $\overline{HF} \parallel \overline{GK}$	**1.** Given
A **2.** $\angle GHF \cong \angle HGK$	**2.** Alternate Interior Angles Theorem
3. $\angle F$ and $\angle K$ are right angles.	**3.** Given
A **4.** $\angle F \cong \angle K$	**4.** Right Angles Congruence Theorem
S **5.** $\overline{HG} \cong \overline{GH}$	**5.** Reflexive Property of Segment Congruence
6. $\triangle HFG \cong \triangle GKH$	**6.** AAS Congruence Theorem

2 MTR USE ANOTHER METHOD

Use rigid motions to prove that $\triangle HFG \cong \triangle GKH$.

SELF-ASSESSMENT

1 I don't understand yet. 2 I can do it with help. 3 I can do it on my own. 4 I can teach someone else.

4. In the diagram, $\angle S \cong \angle U$ and $\overline{RS} \cong \overline{VU}$. Prove $\triangle RST \cong \triangle VUT$.

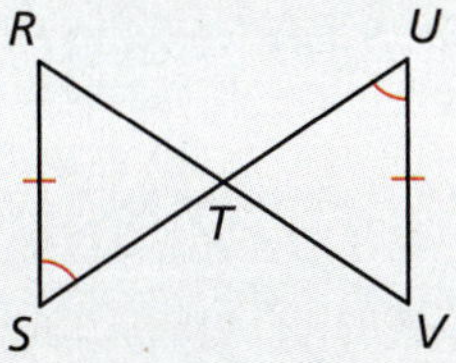

CONCEPT SUMMARY

Triangle Congruence Theorems

You have learned five methods for proving that triangles are congruent.

SAS	SSS	HL (right △ only)	ASA	AAS
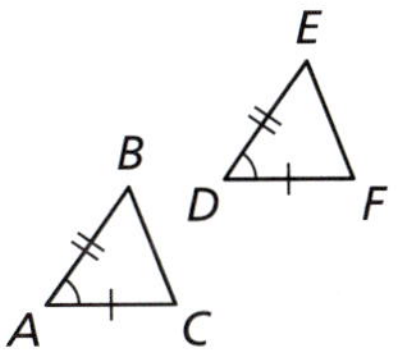	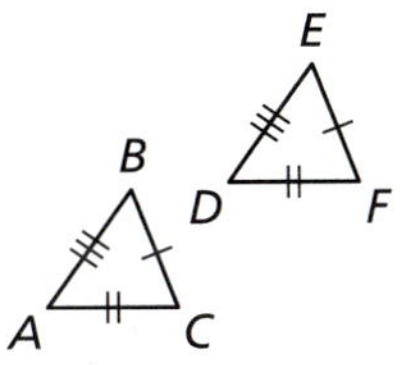	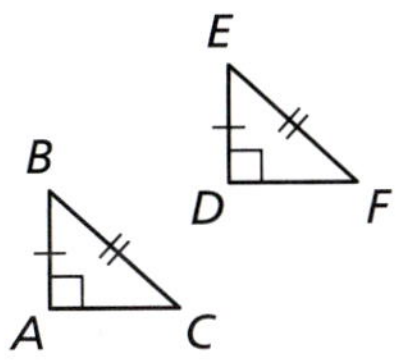	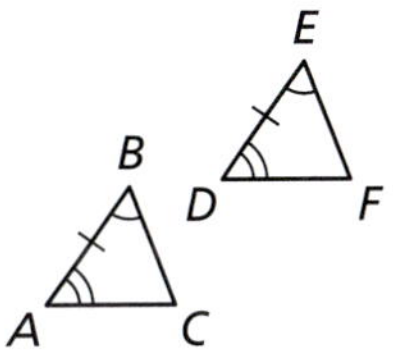	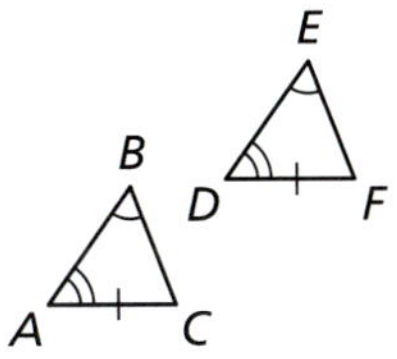
Two sides and the included angle are congruent.	All three sides are congruent.	The hypotenuse and one of the legs are congruent.	Two angles and the included side are congruent.	Two angles and a non-included side are congruent.

In the Exercises, you will prove three additional theorems about the congruence of right triangles: **H**ypotenuse-**A**ngle, **L**eg-**L**eg, and **A**ngle-**L**eg.

GO DIGITAL

5.6 Practice WITH CalcChat® AND CalcView®

In Exercises 1–4, decide whether enough information is given to prove that the triangles are congruent. If so, state the theorem you can use. *(See Example 1.)*

▶ 1. $\triangle ABC$, $\triangle QRS$

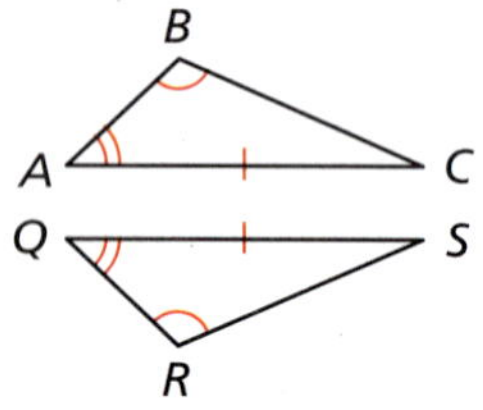

2. $\triangle ABC$, $\triangle DBC$

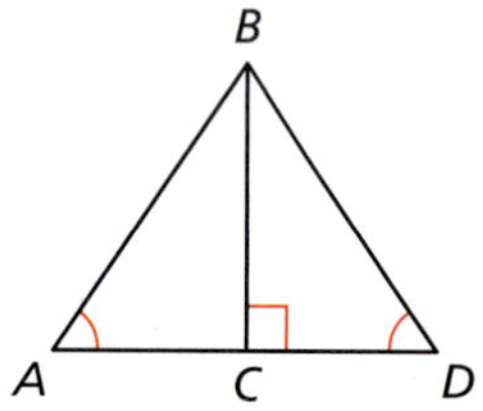

3. $\triangle XYZ$, $\triangle JKL$

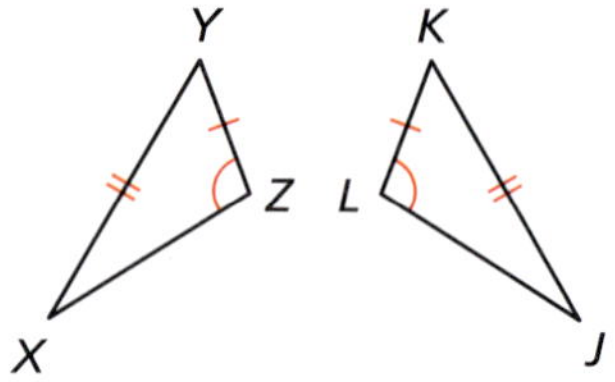

4. $\triangle RSV$, $\triangle UTV$

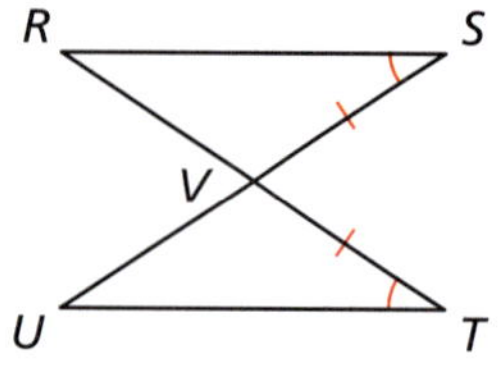

In Exercises 5–8, decide whether you can use the given information to prove that $\triangle ABC \cong \triangle DEF$. Explain your reasoning.

5. $\angle A \cong \angle D$, $\angle C \cong \angle F$, $\overline{AC} \cong \overline{DF}$

6. $\angle C \cong \angle F$, $\overline{AB} \cong \overline{DE}$, $\overline{BC} \cong \overline{EF}$

7. $\angle B \cong \angle E$, $\angle C \cong \angle F$, $\overline{AC} \cong \overline{DE}$

8. $\angle A \cong \angle D$, $\angle B \cong \angle E$, $\overline{BC} \cong \overline{EF}$

CONSTRUCTION In Exercises 9 and 10, construct a triangle that is congruent to the given triangle using the ASA Congruence Theorem. Use a compass and straightedge.

9. 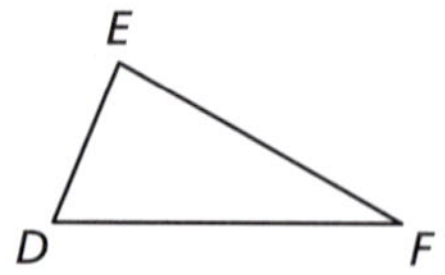

10. 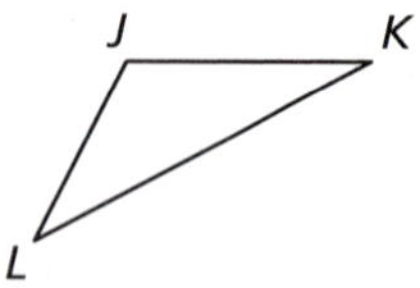

4 MTR ERROR ANALYSIS In Exercises 11 and 12, describe and correct the error.

11. 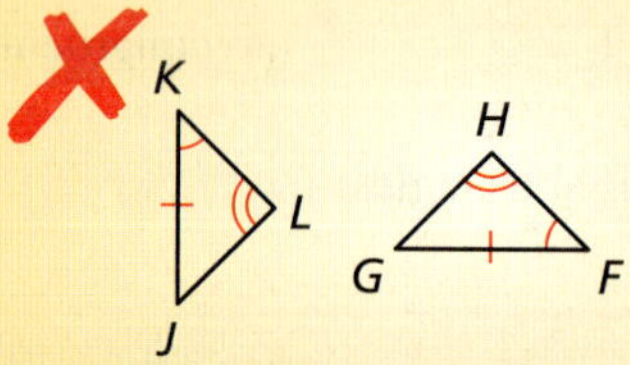

$\triangle JLK \cong \triangle FHG$ by the AAS Congruence Theorem.

12. 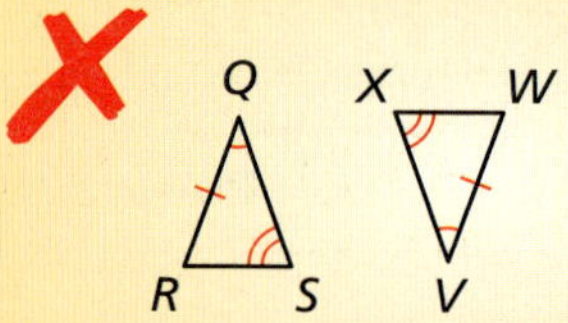

$\triangle QRS \cong \triangle VWX$ by the ASA Congruence Theorem.

PROOF In Exercises 13 and 14, prove that the triangles are congruent using the ASA Congruence Theorem. *(See Example 2.)*

▶ 13. **Given** M is the midpoint of $\overline{NL}$.
$\overline{NL} \perp \overline{NQ}$, $\overline{NL} \perp \overline{MP}$, $\overline{QM} \parallel \overline{PL}$

Prove $\triangle NQM \cong \triangle MPL$

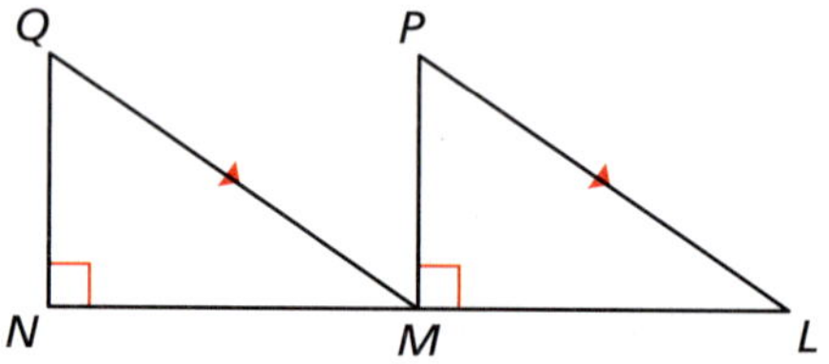

14. **Given** $\overline{AJ} \cong \overline{KC}$, $\angle BJK \cong \angle BKJ$, $\angle A \cong \angle C$

Prove $\triangle ABK \cong \triangle CBJ$

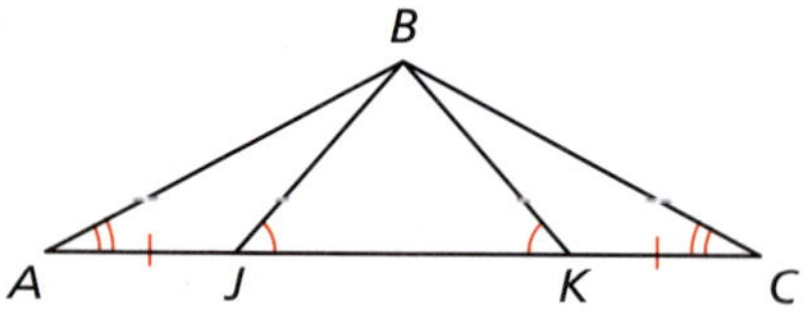

PROOF In Exercises 15 and 16, prove that the triangles are congruent using the AAS Congruence Theorem. *(See Example 3.)*

▶ 15. **Given** $\overline{VW} \cong \overline{UW}$, $\angle X \cong \angle Z$

Prove $\triangle XWV \cong \triangle ZWU$

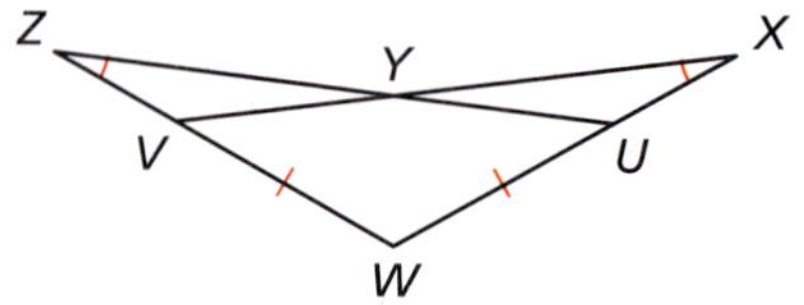

16. **Given** $\angle NKM \cong \angle LMK$, $\angle L \cong \angle N$

Prove $\triangle NMK \cong \triangle LKM$

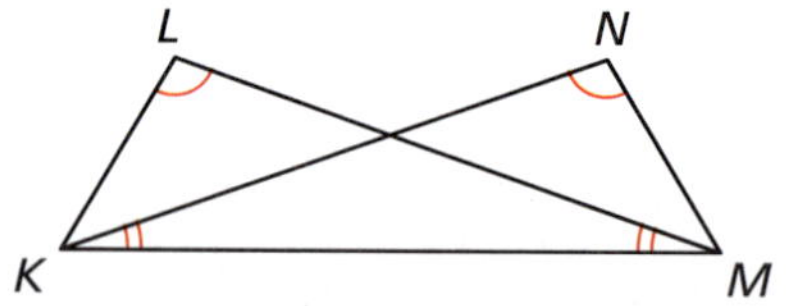

GO DIGITAL

PROVING A THEOREM **In Exercises 17–19, write a paragraph proof for the theorem about right triangles.**

17. **Hypotenuse-Angle (HA) Congruence Theorem** If an acute angle and the hypotenuse of a right triangle are congruent to an acute angle and the hypotenuse of a second right triangle, then the triangles are congruent.

18. **Leg-Leg (LL) Congruence Theorem** If the legs of a right triangle are congruent to the legs of a second right triangle, then the triangles are congruent.

19. **Angle-Leg (AL) Congruence Theorem** If an acute angle and a leg of a right triangle are congruent to an acute angle and the corresponding leg of a second right triangle, then the triangles are congruent.

20. **PROVING A THEOREM** Prove the Converse of the Base Angles Theorem. (*Hint:* Draw an auxiliary line inside the triangle.)

21. **COLLEGE PREP** What additional information do you need to prove $\triangle JKL \cong \triangle MNL$ by the ASA Congruence Theorem?

Ⓐ $\overline{KM} \cong \overline{NJ}$

Ⓑ $\overline{KH} \cong \overline{NH}$

Ⓒ $\angle M \cong \angle J$

Ⓓ $\angle LKJ \cong \angle LNM$

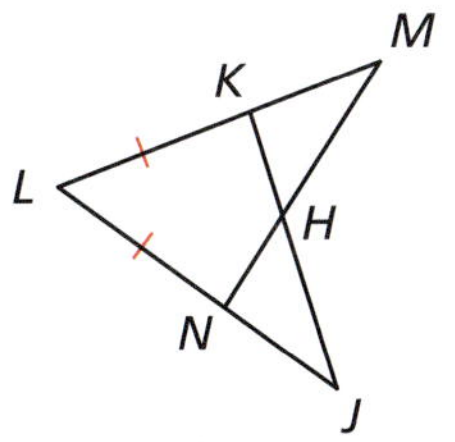

22. **COLLEGE PREP** Which of the following congruence statements are true? Select all that apply.

Ⓐ $\overline{TU} \cong \overline{UV}$

Ⓑ $\triangle STV \cong \triangle XVW$

Ⓒ $\triangle TVS \cong \triangle VWU$

Ⓓ $\triangle VST \cong \triangle VUW$

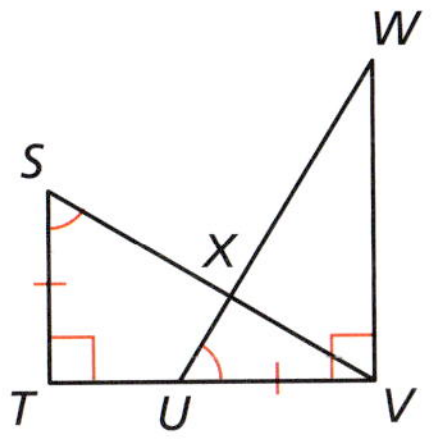

5 MTR **23.** **CONNECTING CONCEPTS** The toy contains $\triangle ABC$ and $\triangle DBC$. Can you conclude that $\triangle ABC \cong \triangle DBC$ from the given angle measures? Explain.

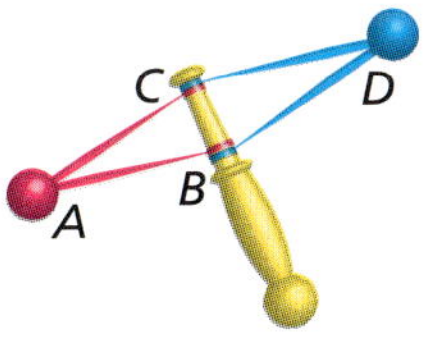

$m\angle ABC = (8x - 32)^\circ$ $\quad$ $m\angle DBC = (4y - 24)^\circ$

$m\angle BCA = (5x + 10)^\circ$ $\quad$ $m\angle BCD = (3y + 2)^\circ$

$m\angle CAB = (2x - 8)^\circ$ $\quad$ $m\angle CDB = (y - 6)^\circ$

GO DIGITAL

7 MTR **24.** **MODELING REAL LIFE** When a light ray from an object meets a mirror, it is reflected back to your eye. For example, in the diagram, a light ray from point C is reflected at point D and travels back to point A. The *law of reflection* states that the angle of incidence, $\angle CDB$, is congruent to the angle of reflection, $\angle ADB$.

a. Prove that $\triangle ABD$ is congruent to $\triangle CBD$.

Given $\angle CDB \cong \angle ADB$, $\overline{DB} \perp \overline{AC}$

Prove $\triangle ABD \cong \triangle CBD$

b. Verify that $\triangle ACD$ is isosceles.

c. Does moving away from the mirror affect the amount of his or her reflection a person sees? Explain.

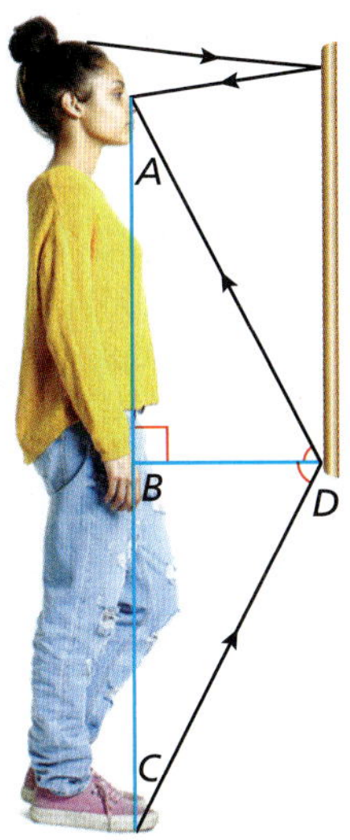

5 MTR **25.** **STRUCTURE** Which of the triangle congruence theorems in the Concept Summary on page 269 can be used to prove that $\triangle EFG \cong \triangle EHG$? Explain your reasoning.

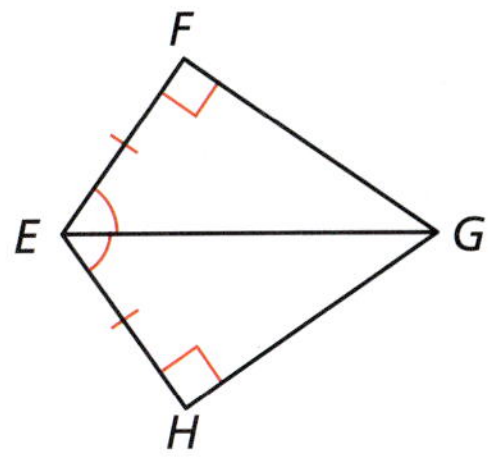

26. **HOW DO YOU SEE IT?**
Name as many pairs of congruent triangles as you can from the diagram. Explain how you know that each pair of triangles is congruent.

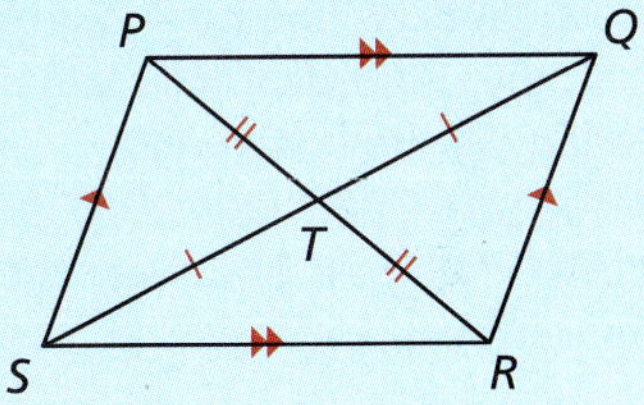

27. **REASONING** In $\triangle ABC$ and $\triangle DEF$, $m\angle A = 68^\circ$, $m\angle C = m\angle F = 59^\circ$, $m\angle E = 53^\circ$, and $AB = DE = 8$ units. Can you prove that $\triangle ABC \cong \triangle DEF$? If so, write a proof. If not, explain why not.

4 MTR **28.** **MAKING AN ARGUMENT** Is it possible to rewrite any proof that uses the AAS Congruence Theorem as a proof that uses the ASA Congruence Theorem? Explain your reasoning.

5 MTR **29. CONNECTING CONCEPTS** Six statements are given about $\triangle TUV$ and $\triangle XYZ$.

$\overline{TU} \cong \overline{XY}$ $\quad$ $\overline{UV} \cong \overline{YZ}$ $\quad$ $\overline{TV} \cong \overline{XZ}$

$\angle T \cong \angle X$ $\quad$ $\angle U \cong \angle Y$ $\quad$ $\angle V \cong \angle Z$

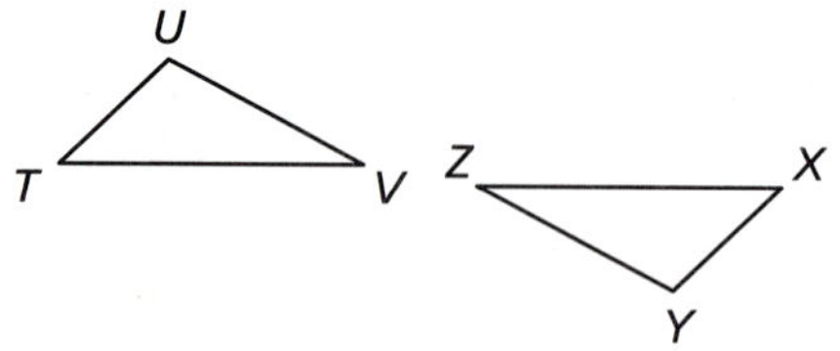

a. List all combinations of three given statements that provide enough information to prove that $\triangle TUV$ is congruent to $\triangle XYZ$.

b. You choose three statements at random. What is the probability that the statements you choose provide enough information to prove that the triangles are congruent?

30. THOUGHT PROVOKING
Graph theory is a branch of mathematics that studies vertices and the way they are connected. In graph theory, two polygons are *isomorphic* if there is a one-to-one mapping from one polygon's vertices to the other polygon's vertices that preserves adjacent vertices. In graph theory, are any two triangles isomorphic? Explain your reasoning.

31. DIG DEEPER Determine whether each is a valid method of showing that two quadrilaterals are congruent. If so, write a proof. If not, provide a counterexample.

a. SASA $\qquad$ **b.** SASAS

c. SSSS $\qquad$ **d.** SSSA

e. ASAAS $\qquad$ **f.** AASA

REVIEW & REFRESH

32. Find the coordinates of the midpoint of the line segment with endpoints $J(-2, 3)$ and $K(4, -1)$.

33. Copy the angle using a compass and straightedge.

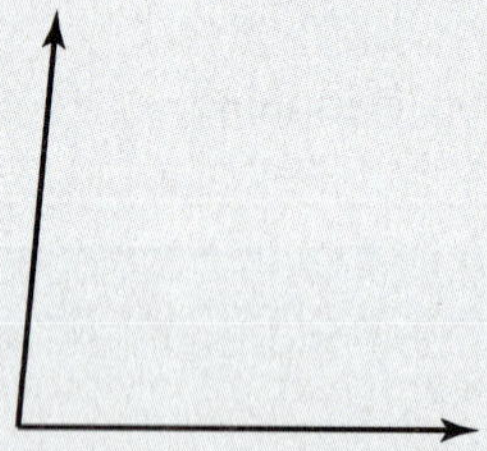

34. REASONING You know that a pair of triangles has two pairs of congruent corresponding angles. What other information do you need to show that the triangles are congruent?

In Exercises 35 and 36, complete the statement. State which theorem you use.

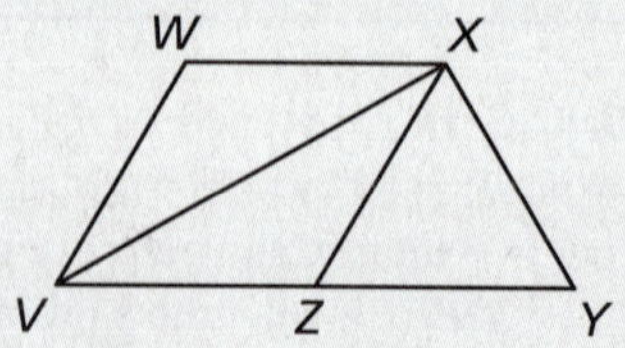

35. If $\overline{VW} \cong \overline{WX}$, then $\angle$___ $\cong \angle$___.

36. If $\angle XYZ \cong \angle ZXY$, then ___ $\cong$ ___.

7 MTR **37. MODELING REAL LIFE** You are using a microscope that shows the image of an object as 100 times the object's actual size. Use the actual length of the bacterium to determine the length of its image seen through the microscope.

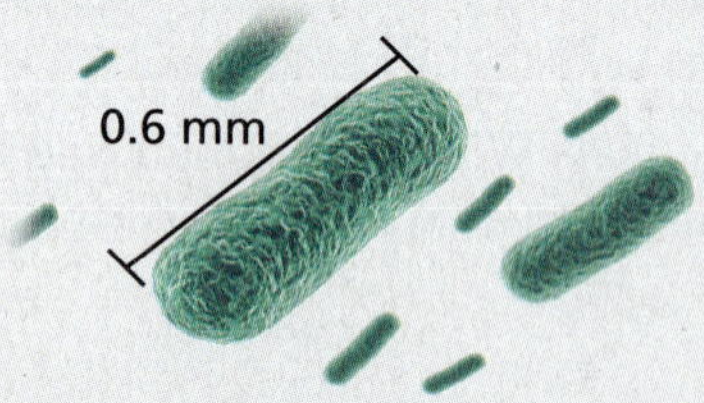

In Exercises 38–41, decide whether enough information is given to prove that the triangles are congruent. If so, state the theorem you can use.

38. $\triangle BAD$, $\triangle DCB$

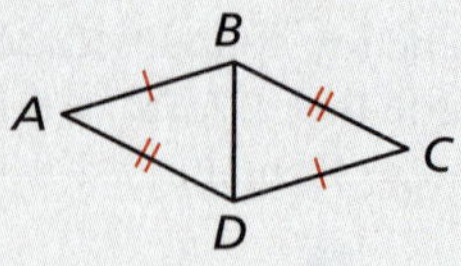

39. $\triangle EFH$, $\triangle GFH$

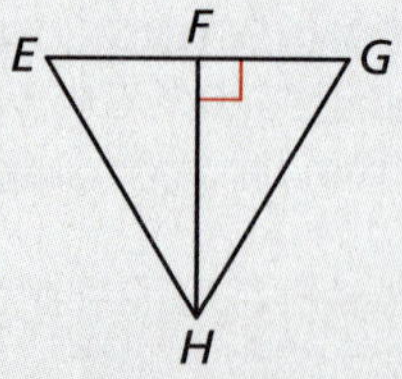

40. $\triangle JKM$, $\triangle LMK$

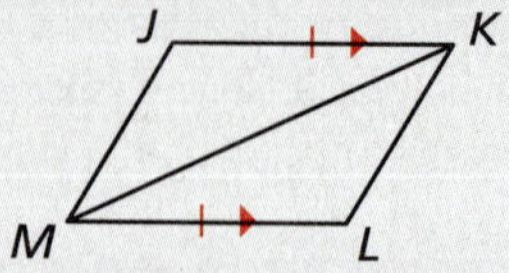

41. $\triangle QPT$, $\triangle QSR$

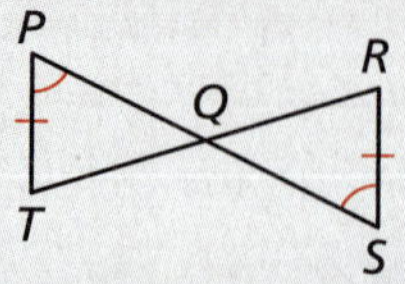

5.7 Using Congruent Triangles

Learning Target: Use congruent triangles in proofs and to measure distances.

Success Criteria:
- I can use congruent triangles to prove statements.
- I can use congruent triangles to solve real-life problems.
- I can use congruent triangles to prove constructions.

EXPLORE IT! Measuring the Width of a River

Work with a partner.

a. The figure shows how a surveyor can measure the width of a river by making measurements on only one side of the river. Study the figure.

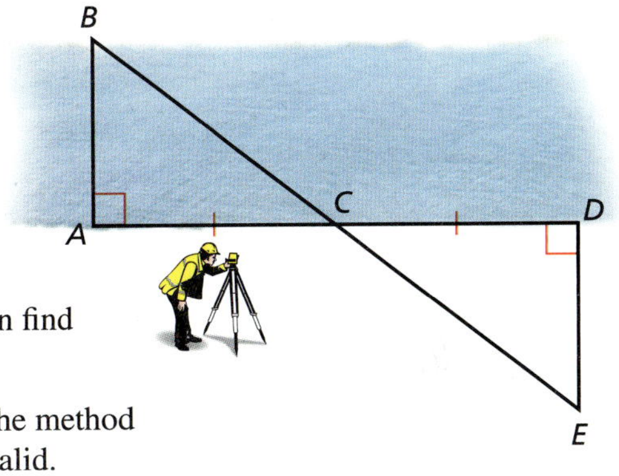

i. Explain how the surveyor can find the width of the river.

ii. Write a proof to verify that the method you described in part (a) is valid.

Given $\angle A$ is a right angle, $\angle D$ is a right angle, $\overline{AC} \cong \overline{CD}$

iii. Exchange proofs with another group and discuss the reasoning used.

4 MTR COMMUNICATE CLEARLY
Why do you think the types of measurements described in parts (a) and (b) are called *indirect measurements*?

b. It was reported that one of Napoleon's officers estimated the width of a river as follows. The officer stood on the bank of the river and lowered the visor on his cap until the farthest thing visible was the edge of the bank on the other side. He then turned and noted the point on his side that was in line with the tip of his visor and his eye. The officer then paced the distance to this point and concluded that distance was the width of the river. Study the figure.

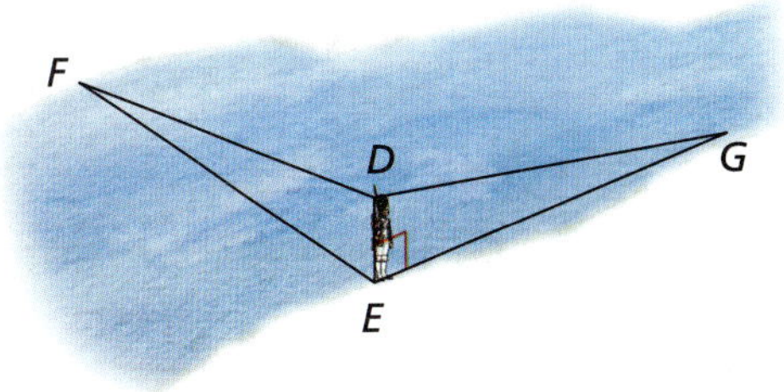

i. Explain how the officer concluded that the width of the river is EG.

ii. Write a proof to verify that the conclusion the officer made is correct.

Given $\angle DEG$ is a right angle, $\angle DEF$ is a right angle, $\angle EDG \cong \angle EDF$

iii. Exchange proofs with another group and discuss the reasoning used.

Geometric Reasoning
MA.912.GR.1.2 Prove triangle congruence or similarity using Side-Side-Side, Side-Angle-Side, Angle-Side-Angle, Angle-Angle-Side, Angle-Angle and Hypotenuse-Leg.
MA.912.GR.1.6 Solve mathematical and real-world problems involving congruence or similarity in two-dimensional figures.

GO DIGITAL

Using Congruent Triangles

Congruent triangles have congruent corresponding parts. So, if you can prove that two triangles are congruent, then you know that their corresponding parts must be congruent as well.

EXAMPLE 1 Using Congruent Triangles

WATCH STEM

Explain how you can use the given information to prove that the hang glider parts are congruent.

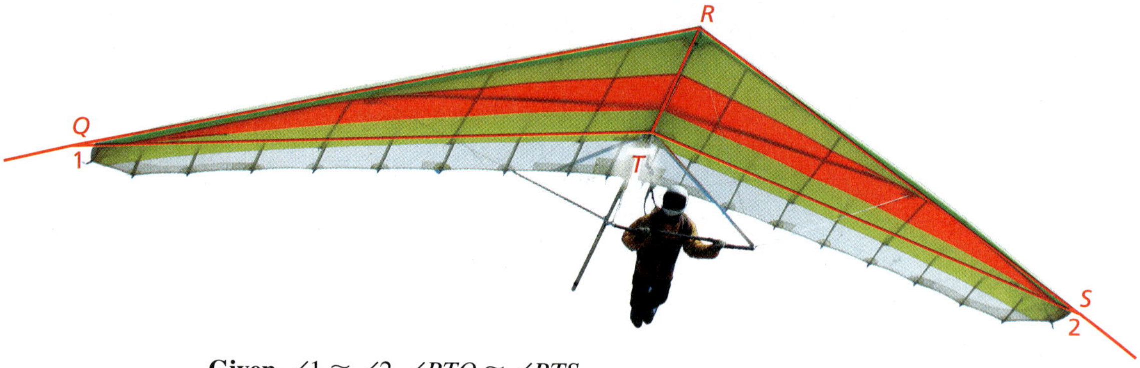

Given $\angle 1 \cong \angle 2$, $\angle RTQ \cong \angle RTS$

Prove $\overline{QT} \cong \overline{ST}$

SOLUTION

If you can show that $\triangle QRT \cong \triangle SRT$, then you will know that $\overline{QT} \cong \overline{ST}$. First, copy the diagram and mark the given information. Then mark the information that you can deduce. In this case, $\angle RQT$ and $\angle RST$ are supplementary to congruent angles, so $\angle RQT \cong \angle RST$. Also, $\overline{RT} \cong \overline{RT}$ by the Reflexive Property of Segment Congruence.

Mark given information.

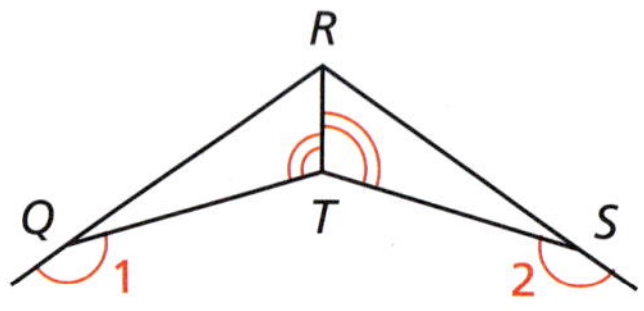

Mark deduced information.

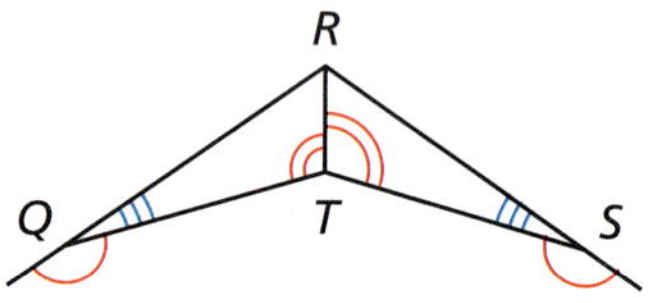

Two angle pairs and a non-included side are congruent, so by the AAS Congruence Theorem, $\triangle QRT \cong \triangle SRT$.

▶ Because corresponding parts of congruent triangles are congruent, $\overline{QT} \cong \overline{ST}$.

SELF-ASSESSMENT

1 I don't understand yet. 2 I can do it with help. 3 I can do it on my own. 4 I can teach someone else.

1. Explain how you can prove that $\angle A \cong \angle C$.

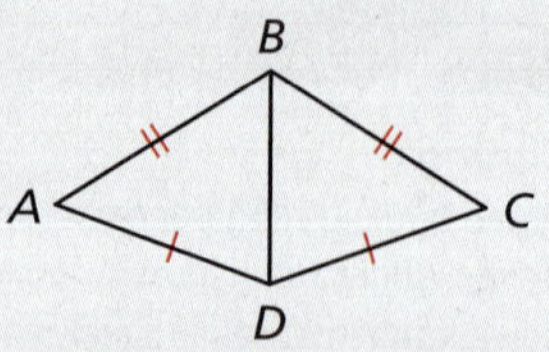

GO DIGITAL

EXAMPLE 2 Using Congruent Triangles for Measurement

Use the following method to find the distance across the Kissimmee River, from point N to point P.

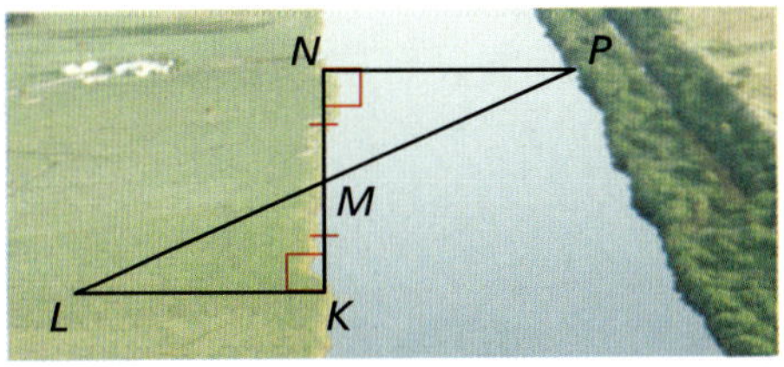

- Place a stake at K on the near side so that $\overline{NK} \perp \overline{NP}$.
- Find M, the midpoint of $\overline{NK}$.
- Locate the point L so that $\overline{NK} \perp \overline{KL}$ and L, P, and M are collinear.

Explain how this plan allows you to find the distance.

SOLUTION

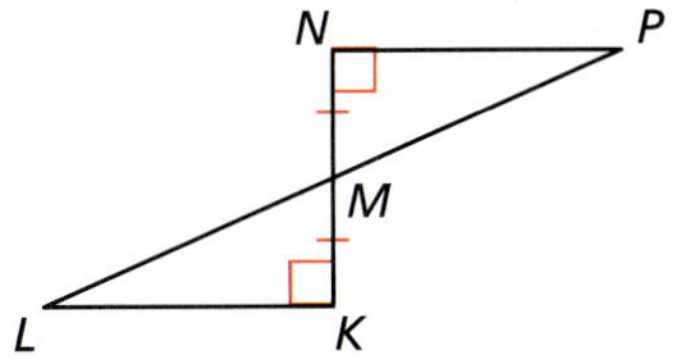

Because $\overline{NK} \perp \overline{NP}$ and $\overline{NK} \perp \overline{KL}$, $\angle N$ and $\angle K$ are congruent right angles. Because M is the midpoint of $\overline{NK}$, $\overline{NM} \cong \overline{KM}$. The vertical angles $\angle KML$ and $\angle NMP$ are congruent. So, $\triangle MLK \cong \triangle MPN$ by the ASA Congruence Theorem. Then because corresponding parts of congruent triangles are congruent, $\overline{KL} \cong \overline{NP}$. So, you can find the distance NP across the Kissimmee River by measuring KL.

EXAMPLE 3 Planning a Proof Involving Pairs of Triangles

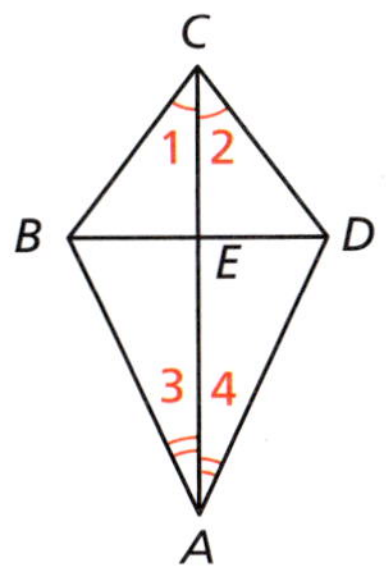

Use the given information to write a plan for proof.

Given $\angle 1 \cong \angle 2$, $\angle 3 \cong \angle 4$

Prove $\triangle BCE \cong \triangle DCE$

SOLUTION

In $\triangle BCE$ and $\triangle DCE$, you know that $\angle 1 \cong \angle 2$ and $\overline{CE} \cong \overline{CE}$. If you can show that $\overline{CB} \cong \overline{CD}$, then you can use the SAS Congruence Theorem.

To prove that $\overline{CB} \cong \overline{CD}$, you can first prove that $\triangle CBA \cong \triangle CDA$. You are given $\angle 1 \cong \angle 2$ and $\angle 3 \cong \angle 4$. $\overline{CA} \cong \overline{CA}$ by the Reflexive Property of Segment Congruence. You can use the ASA Congruence Theorem to prove that $\triangle CBA \cong \triangle CDA$.

▶ **Plan for Proof** Use the ASA Congruence Theorem to prove that $\triangle CBA \cong \triangle CDA$. Then state that $\overline{CB} \cong \overline{CD}$. Use the SAS Congruence Theorem to prove that $\triangle BCE \cong \triangle DCE$.

SELF-ASSESSMENT

1 I don't understand yet. 2 I can do it with help. 3 I can do it on my own. 4 I can teach someone else.

2. In Example 2, does it matter how far from point N you place a stake at point K? Explain.

3. Write a plan to prove that $\triangle PTU \cong \triangle UQP$.

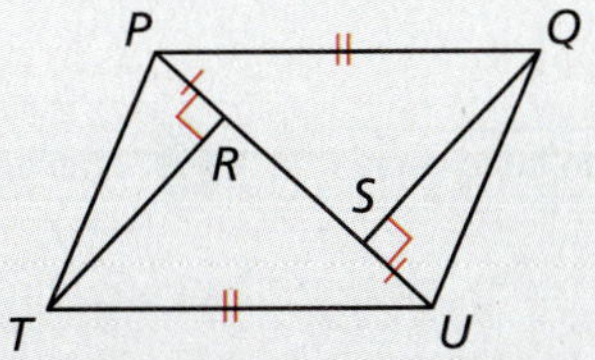

GO DIGITAL

Proving Constructions

Recall that you can use a compass and a straightedge to copy an angle. The construction is shown below. You can use congruent triangles to prove that this construction is valid.

Step 1

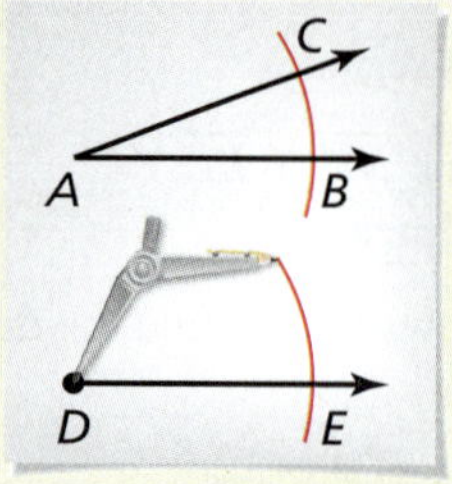

Draw a segment and arcs

To copy $\angle A$, draw a segment with initial point D. Draw an arc with center A. Using the same radius, draw an arc with center D. Label points B, C, and E.

Step 2

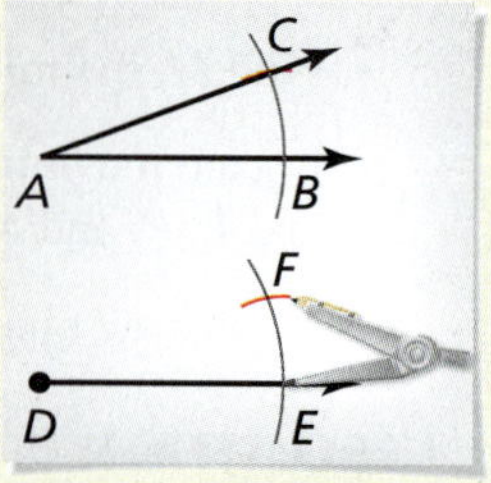

Draw an arc

Draw an arc with radius BC and center E. Label the intersection F.

Step 3

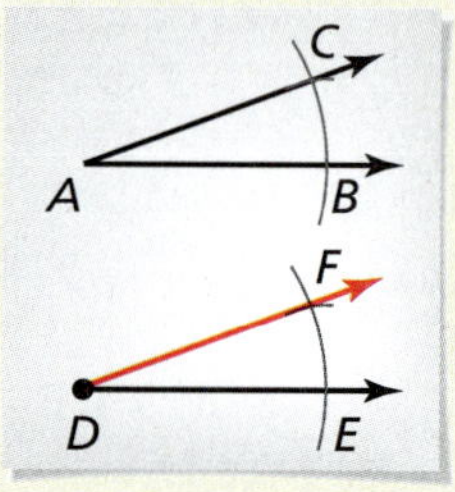

Draw a ray

Draw $\overrightarrow{DF}$. In Example 4, you will prove that $\angle D \cong \angle A$.

EXAMPLE 4 Proving a Construction

WATCH

Write a proof to verify that the construction for copying an angle is valid.

SOLUTION

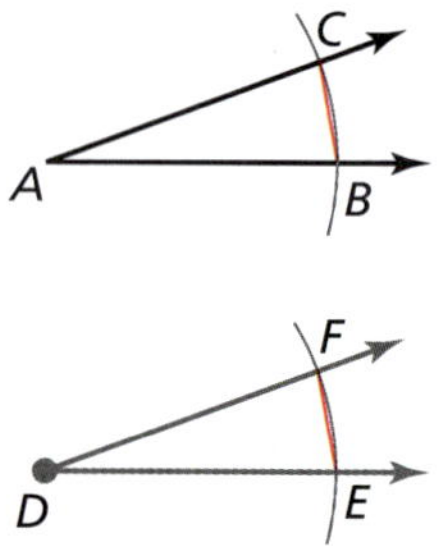

Add $\overline{BC}$ and $\overline{EF}$ to the diagram. In the construction, one compass setting determines $\overline{AB}$, $\overline{DE}$, $\overline{AC}$, and $\overline{DF}$, and another compass setting determines $\overline{BC}$ and $\overline{EF}$. So, you can assume the following as given statements.

Given $\overline{AB} \cong \overline{DE}$, $\overline{AC} \cong \overline{DF}$, $\overline{BC} \cong \overline{EF}$

Prove $\angle D \cong \angle A$

Plan for Proof Show that $\triangle DEF \cong \triangle ABC$, so you can conclude that the corresponding parts $\angle D$ and $\angle A$ are congruent.

Plan in Action

STATEMENTS	REASONS
1. $\overline{AB} \cong \overline{DE}$, $\overline{AC} \cong \overline{DF}$, $\overline{BC} \cong \overline{EF}$	1. Given
2. $\triangle DEF \cong \triangle ABC$	2. SSS Congruence Theorem
3. $\angle D \cong \angle A$	3. Corresponding parts of congruent triangles are congruent.

SELF-ASSESSMENT 1 I don't understand yet. 2 I can do it with help. 3 I can do it on my own. 4 I can teach someone else.

4. Write a proof to verify that the construction of an angle bisector on page 42 is valid.

GO DIGITAL

5.7 Practice WITH CalcChat® AND CalcView®

In Exercises 1–6, explain how to prove that the statement is true. *(See Example 1.)*

1. $\angle A \cong \angle D$

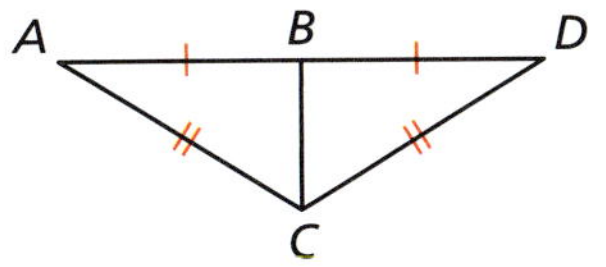

2. $\angle Q \cong \angle T$

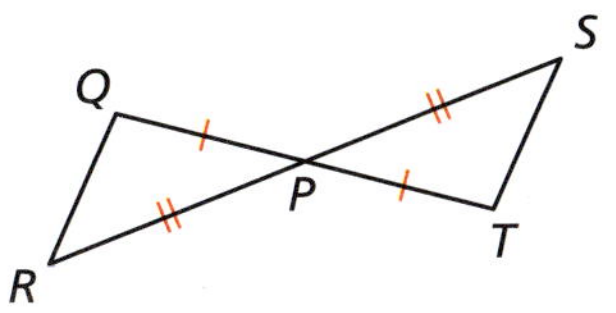

3. $\overline{JM} \cong \overline{LM}$

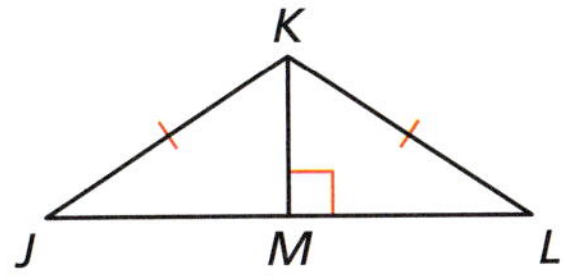

4. $\overline{AC} \cong \overline{DB}$

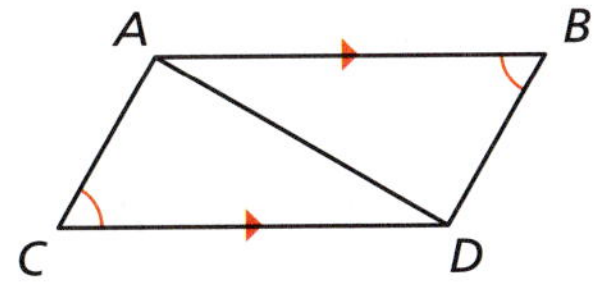

5. $\overline{GK} \cong \overline{HJ}$

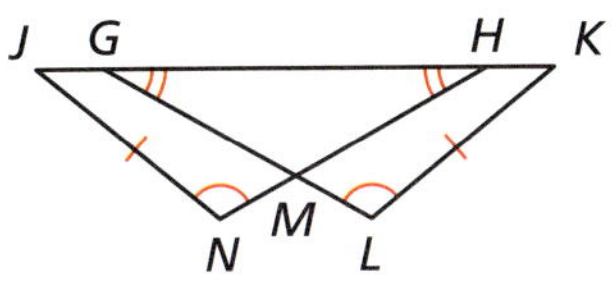

6. $\overline{QW} \cong \overline{VT}$

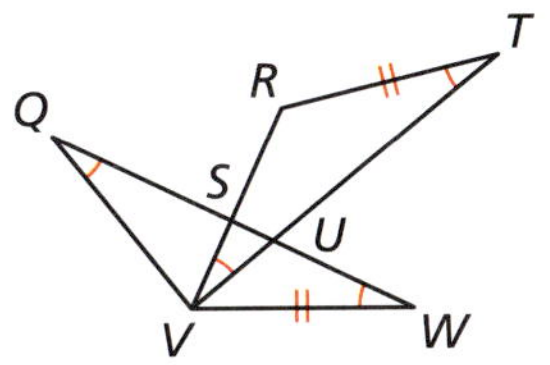

In Exercises 7–10, write a plan to prove that $\angle 1 \cong \angle 2$. *(See Example 3.)*

7.

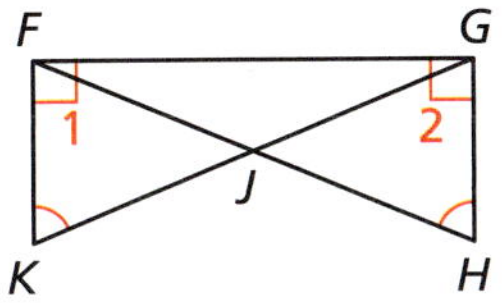

8.

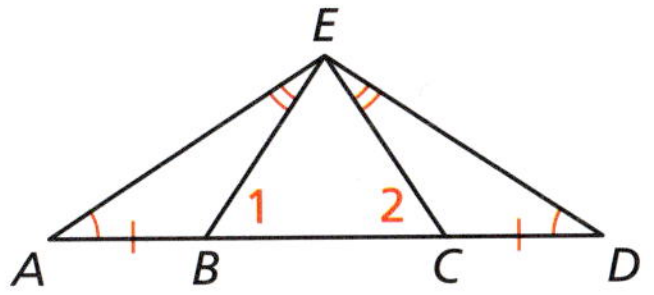

9.

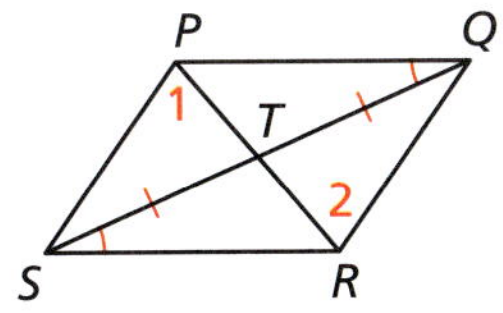

10.

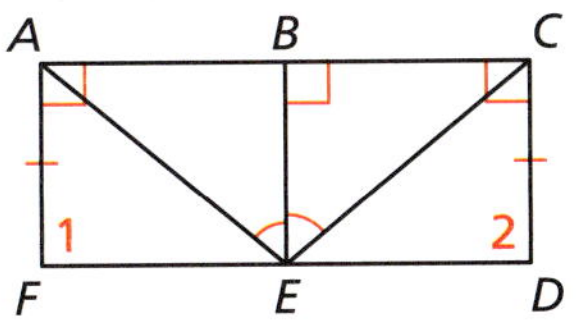

In Exercises 11 and 12, write a proof to verify that the construction is valid. *(See Example 4.)*

11. Line perpendicular to a line through a point not on the line (page 143)

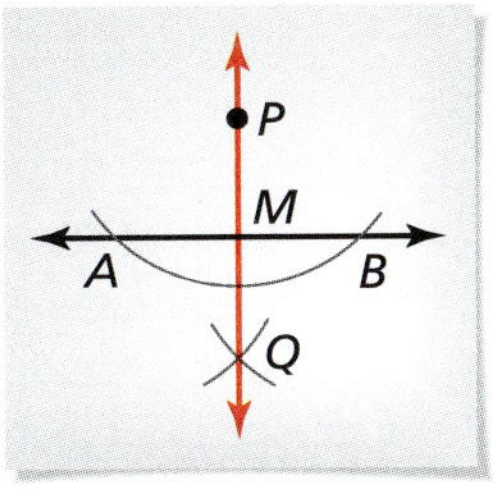

12. Line perpendicular to a line through a point on the line (Exercise 6, page 146)

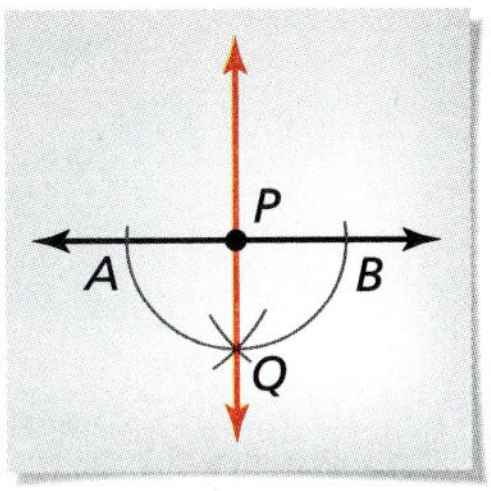

In Exercises 13 and 14, use the information given in the diagram to write a proof.

13. **Prove** $\overline{FL} \cong \overline{HN}$

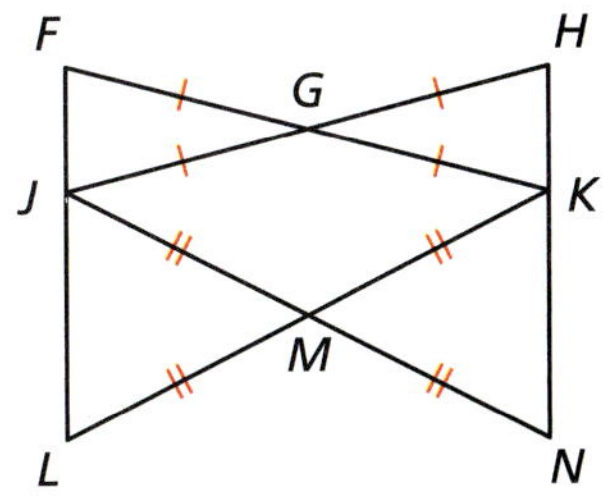

14. **Prove** $\triangle PUX \cong \triangle QSY$

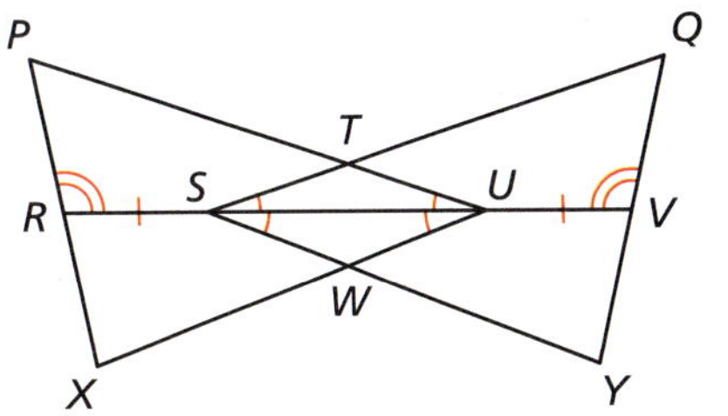

7 MTR **15.** **MODELING REAL LIFE** Explain how to find the distance across the canyon. *(See Example 2.)*

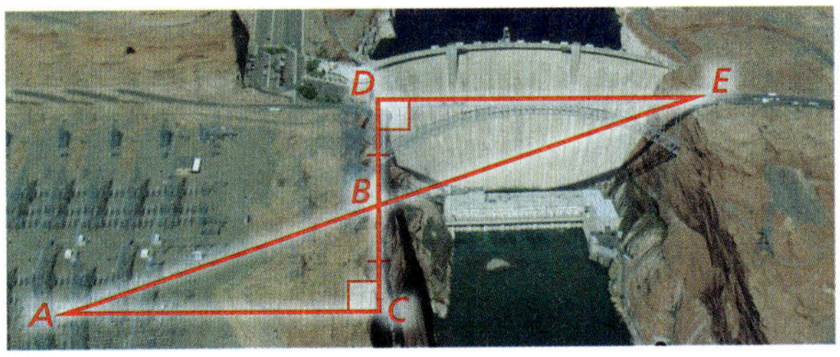

16. **HOW DO YOU SEE IT?**
Name two different ways to indirectly find the length of $\overline{AB}$.

GO DIGITAL

17. **PROOF** Prove that the green triangles in the Jamaican flag are congruent if $\overline{AD} \parallel \overline{BC}$ and E is the midpoint of $\overline{AC}$.

18. **THOUGHT PROVOKING**
The Bermuda Triangle is a region in the Atlantic Ocean. The vertices are Miami, San Juan, and Bermuda. Use the Internet or some other resource to find the side lengths (in miles), the perimeter, and the area of this triangle (in square miles). Then create a congruent triangle on land using cities as vertices.

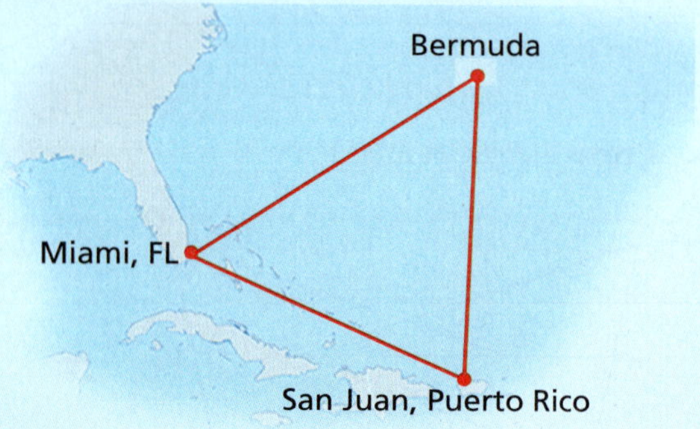

4 MTR **19.** **MAKING AN ARGUMENT** Your friend claims that $\triangle WZY$ can be proven congruent to $\triangle YXW$ using the HL Congruence Theorem. Is your friend correct? Explain your reasoning.

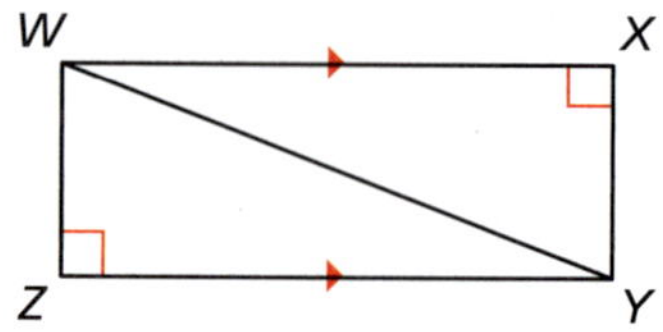

5 MTR **20.** **CONNECTING CONCEPTS** Determine whether each conditional statement is true or false. If the statement is false, rewrite it as a true statement using the converse, inverse, or contrapositive.

a. If two triangles have the same perimeter, then they are congruent.

b. If two triangles are congruent, then they have the same area.

3 MTR **21.** **MAINTAIN ACCURACY** Which triangles are congruent to $\triangle ABC$? Select all that apply.

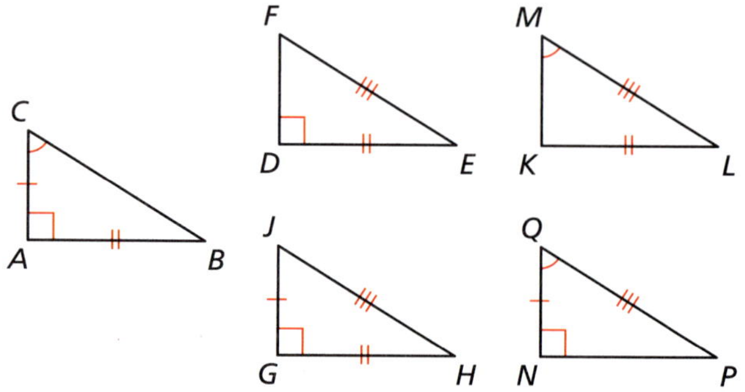

REVIEW & REFRESH

WATCH

22. Explain how you can prove that $\overline{EH} \cong \overline{GF}$ using a paragraph proof.

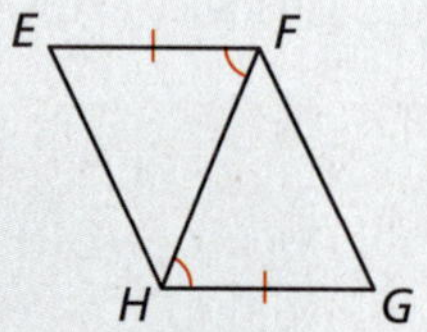

In Exercises 23 and 24, find the perimeter of the polygon with the given vertices.

23. $A(-1, 1), B(4, 1), C(4, -2), D(-1, -2)$

24. $J(-5, 3), K(-2, 1), L(3, 4)$

25. Find the value of x.

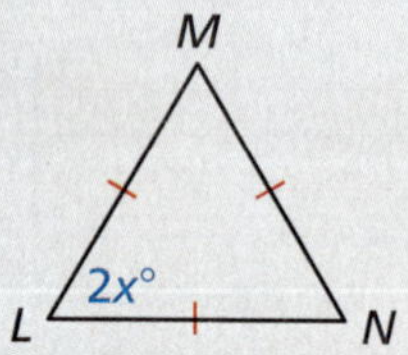

26. Simplify $(3y - 2) + (-y + 4)$.

In Exercises 27 and 28, decide whether enough information is given to prove that the triangles are congruent. If so, state the theorem you can use.

27. $\triangle TUV, \triangle XYZ$

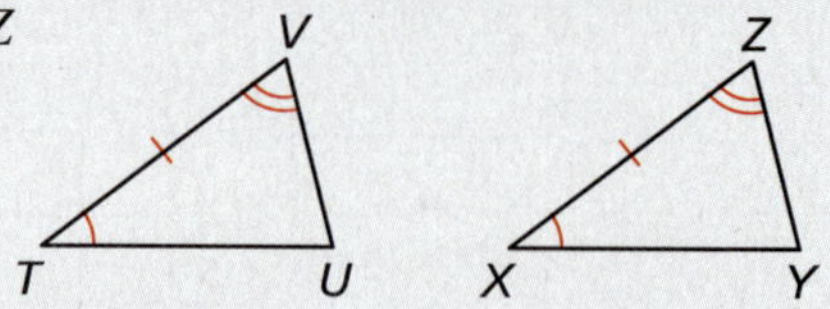

28. $\triangle JKM, \triangle LKM$

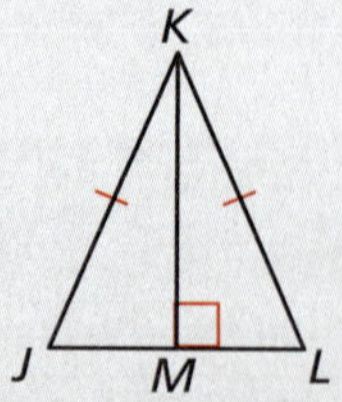

7 MTR **29.** **MODELING REAL LIFE** There are 380 students at your school. The number of students is increasing at an annual rate of 2%. Write and graph an exponential function to model the number of students at your school as a function of number of years.

GO DIGITAL

5.8 Coordinate Proofs

Learning Target: Use coordinates to write proofs.

Success Criteria:
- I can place figures in a coordinate plane.
- I can write plans for coordinate proofs.
- I can write coordinate proofs.

EXPLORE IT! Writing a Proof Using Coordinates

Work with a partner.

a. Draw $\overline{AB}$ with endpoints $A(-3, 0)$ and $B(3, 0)$. Then draw $\triangle ABC$ so that C lies on the y-axis.

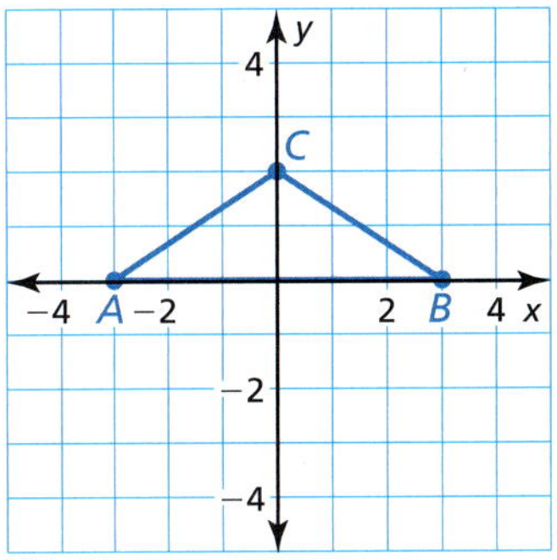

5 MTR USE STRUCTURE
How can you write coordinates that represent any point on the y-axis?

b. Classify $\triangle ABC$ by its sides.

c. Repeat parts (a) and (b), drawing C at different points on the y-axis. What do you notice?

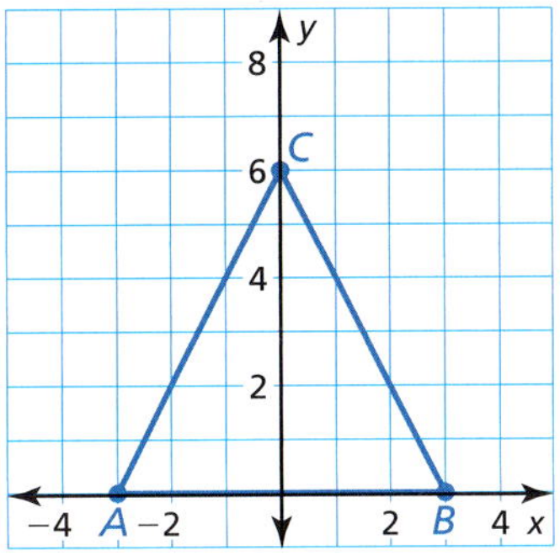

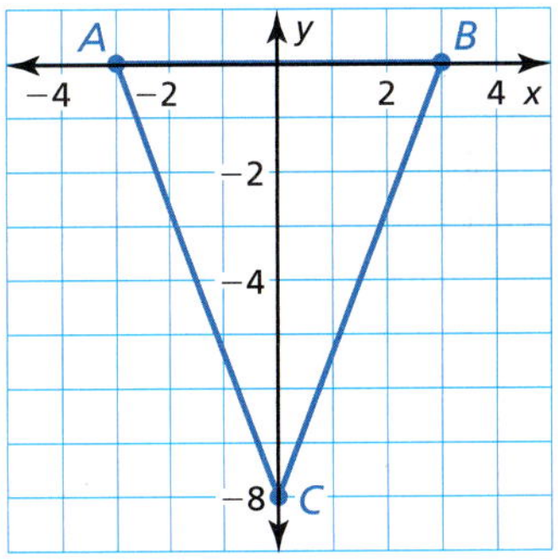

d. How can you prove that if C lies on the y-axis, then $\triangle ABC$ is an isosceles triangle?

e. What coordinates of C make $\triangle ABC$ an equilateral triangle?

Geometric Reasoning
MA.912.GR.3.2 Given a mathematical context, use coordinate geometry to classify or justify definitions, properties and theorems involving circles, triangles or quadrilaterals.

GO DIGITAL

Placing Figures in a Coordinate Plane

Vocabulary

coordinate proof, *p. 280*

A **coordinate proof** involves placing geometric figures in a coordinate plane. When you use variables to represent the coordinates of a figure in a coordinate proof, the results are true for all figures of that type.

EXAMPLE 1 Placing a Figure in a Coordinate Plane

Place each figure in a coordinate plane in a way that is convenient for finding side lengths. Assign coordinates to each vertex.

a. a rectangle **b.** a triangle

SOLUTION

It is easy to find lengths of horizontal and vertical segments and distances from (0, 0), so place one vertex at the origin and one or more sides on an axis.

a. Let h represent the length and k represent the width.

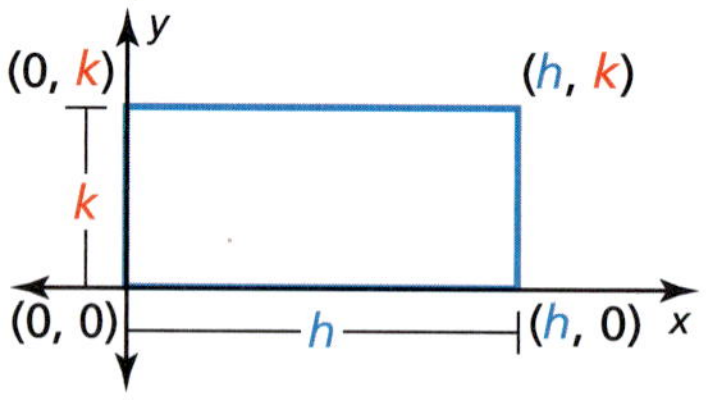

b. Notice that you need to use three different variables.

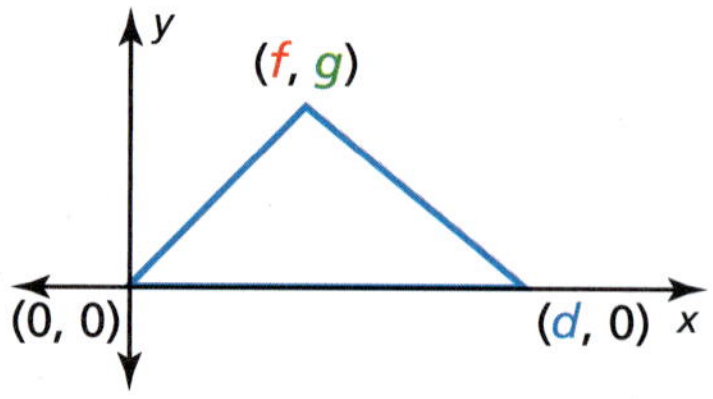

EXAMPLE 2 Applying Variable Coordinates

Place an isosceles right triangle in a coordinate plane. Then find the length of the hypotenuse and the coordinates of its midpoint M.

SOLUTION

Place $\triangle PQO$ with the right angle at the origin. Let the length of the legs be k. Then the vertices are located at $P(0, k)$, $Q(k, 0)$, and $O(0, 0)$.

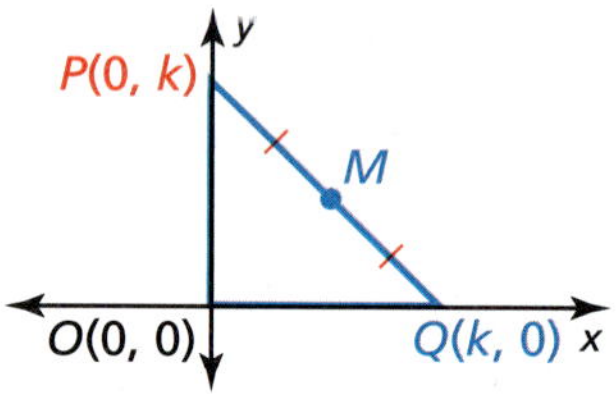

Use the Distance Formula to find PQ, the length of the hypotenuse.

$$\begin{aligned} PQ &= \sqrt{(k-0)^2 + (0-k)^2} \\ &= \sqrt{k^2 + (-k)^2} \\ &= \sqrt{k^2 + k^2} \\ &= \sqrt{2k^2} \\ &= k\sqrt{2} \end{aligned}$$

Use the Midpoint Formula to find the midpoint M of the hypotenuse.

$$M\left(\frac{0+k}{2}, \frac{k+0}{2}\right) = M\left(\frac{k}{2}, \frac{k}{2}\right)$$

▶ So, the length of the hypotenuse is $k\sqrt{2}$ and the midpoint of the hypotenuse is $\left(\frac{k}{2}, \frac{k}{2}\right)$.

2 MTR USE ANOTHER METHOD

Show how to solve Example 2 by drawing the hypotenuse on the *x*-axis and a vertex on the *y*-axis.

GO DIGITAL

EXAMPLE 3 Writing a Plan for a Coordinate Proof

WATCH

Write a plan to prove that $\overrightarrow{SO}$ bisects $\angle PSR$.

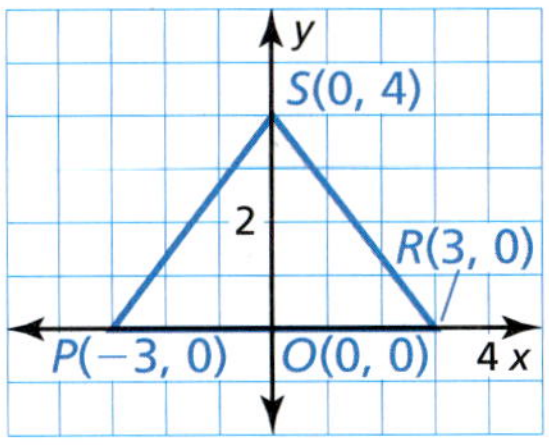

Given Coordinates of vertices of $\triangle POS$ and $\triangle ROS$

Prove $\overrightarrow{SO}$ bisects $\angle PSR$.

SOLUTION

Plan for Proof Use the Distance Formula to find the side lengths of $\triangle POS$ and $\triangle ROS$. Then use the SSS Congruence Theorem to show that $\triangle POS \cong \triangle ROS$. Finally, use the fact that corresponding parts of congruent triangles are congruent to conclude that $\angle PSO \cong \angle RSO$, which implies that $\overrightarrow{SO}$ bisects $\angle PSR$.

SELF-ASSESSMENT

1 I don't understand yet. 2 I can do it with help. 3 I can do it on my own. 4 I can teach someone else.

1. Show another way to place the rectangle in Example 1 part (a) that is convenient for finding side lengths. Assign new coordinates.

2. A square has vertices (0, 0), (m, 0), and (0, m). Find the fourth vertex.

3. Graph the points $O(0, 0)$, $H(m, n)$, and $J(m, 0)$. What kind of triangle is $\triangle OHJ$? Find the side lengths and the coordinates of the midpoint of each side.

4. Write a plan for the proof.

Given $\overrightarrow{GJ}$ bisects $\angle OGH$.

Prove $\triangle GJO \cong \triangle GJH$

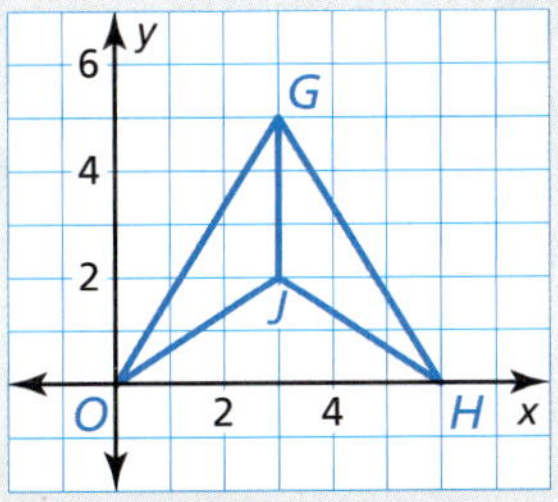

The coordinate proof in Example 3 applies to a specific triangle. When you want to prove a statement about a more general set of figures, it is helpful to use variables as coordinates.

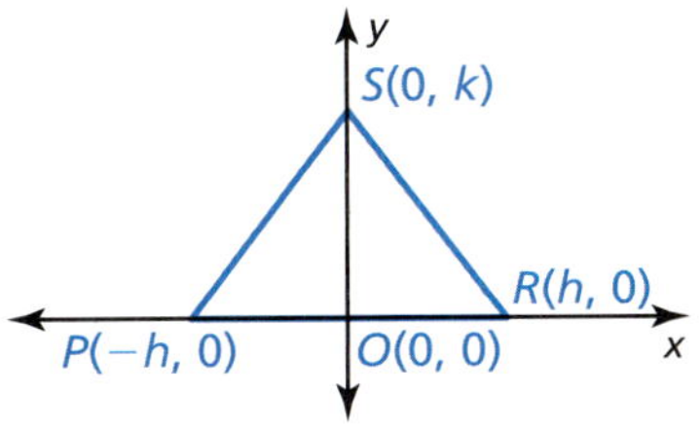

For instance, you can use the variable coordinates shown at the left to duplicate the proof in Example 3. Once this is done, you can conclude that $\overrightarrow{SO}$ bisects $\angle PSR$ for any triangle whose coordinates fit the given pattern.

When writing a coordinate proof, you may not be given the coordinates of a figure. You may have to place the figure in the coordinate plane. Once the figure is placed in a coordinate plane, you may be able to prove statements about the figure.

GO DIGITAL

Writing Coordinate Proofs

EXAMPLE 4 Writing a Coordinate Proof

WATCH

Write a coordinate proof.

Given Coordinates of vertices of quadrilateral $OTUV$

Prove $\triangle OTU \cong \triangle UVO$

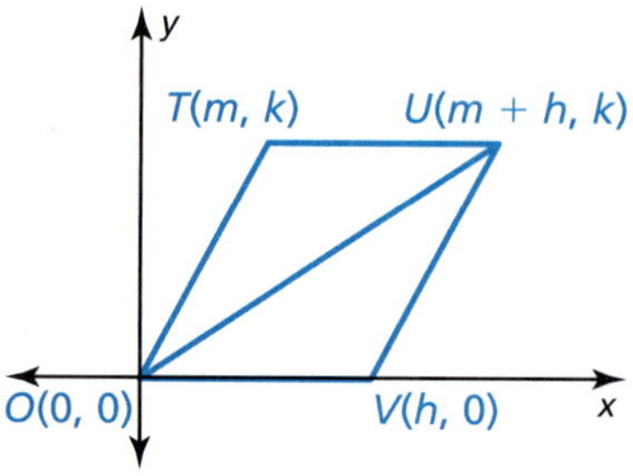

SOLUTION

Segments $\overline{OV}$ and $\overline{UT}$ have the same length.

$$OV = |h - 0| = h$$

$$UT = |(m + h) - m| = h$$

Horizontal segments $\overline{UT}$ and $\overline{OV}$ each have a slope of 0, which implies that they are parallel. Segment $\overline{OU}$ intersects $\overline{UT}$ and $\overline{OV}$ to form congruent alternate interior angles, $\angle TUO$ and $\angle VOU$. By the Reflexive Property of Segment Congruence, $\overline{OU} \cong \overline{OU}$.

▶ So, you can apply the SAS Congruence Theorem to conclude that $\triangle OTU \cong \triangle UVO$.

7 MTR

EXAMPLE 5 Modeling Real Life

WATCH

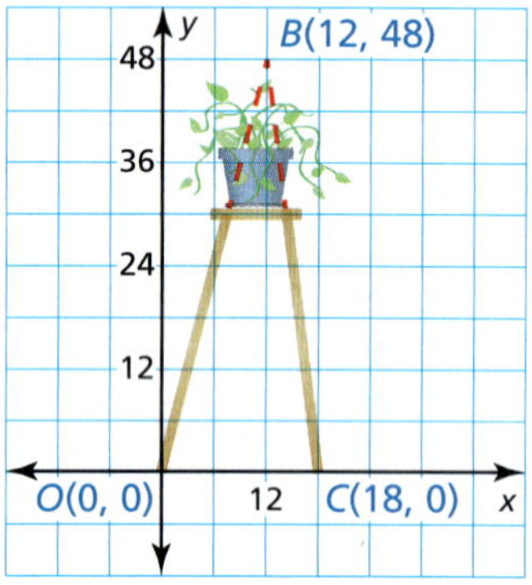

You buy a tall, four-legged plant stand. When you place a plant on the stand, the stand appears to be unstable under the weight of the plant. The diagram at the left shows a coordinate plane superimposed on one pair of the plant stand's legs. One unit in the coordinate plane represents one inch. The legs are extended to form $\triangle OBC$. Prove that $\triangle OBC$ is a scalene triangle. Explain why the plant stand may be unstable.

SOLUTION

First, find the side lengths of $\triangle OBC$.

$$OB = \sqrt{(12 - 0)^2 + (48 - 0)^2} = \sqrt{2448} \approx 49.5 \text{ in.}$$

$$BC = \sqrt{(18 - 12)^2 + (0 - 48)^2} = \sqrt{2340} \approx 48.4 \text{ in.}$$

$$OC = |18 - 0| = 18 \text{ in.}$$

▶ Because $\triangle OBC$ has no congruent sides, $\triangle OBC$ is a scalene triangle by definition. The plant stand may be unstable because $\overline{OB}$ is longer than $\overline{BC}$, so the plant stand is leaning to the right.

SELF-ASSESSMENT

1 I don't understand yet. 2 I can do it with help. 3 I can do it on my own. 4 I can teach someone else.

5. Write a coordinate proof.

Given Coordinates of vertices of $\triangle NPO$ and $\triangle NMO$

Prove $\triangle NPO \cong \triangle NMO$

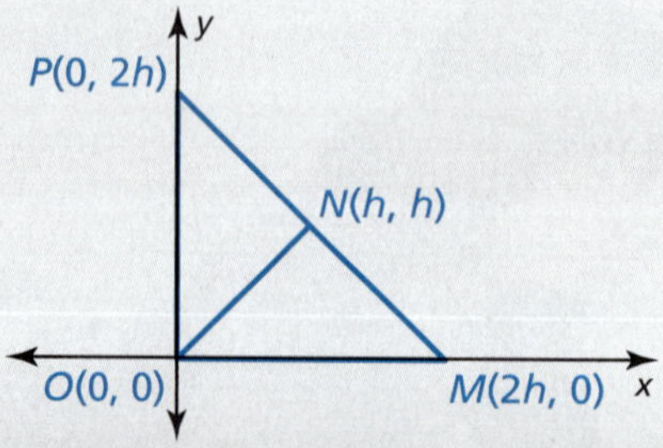

GO DIGITAL

5.8 Practice WITH CalcChat® AND CalcView®

In Exercises 1–4, place the figure in a coordinate plane in a convenient way. Assign coordinates to each vertex. Explain the advantages of your placement. *(See Example 1.)*

▶ **1.** a right triangle with leg lengths of 3 units and 2 units

2. a square with a side length of 3 units

3. an isosceles right triangle with leg length p

4. a scalene triangle with one side length of $2m$

In Exercises 5–8, place the figure in a coordinate plane and find the indicated length. *(See Example 2.)*

▶ **5.** a right triangle with leg lengths of 7 and 9 units; Find the length of the hypotenuse.

6. an isosceles triangle with a base length of 60 units and a height of 50 units; Find the length of one of the legs.

7. a rectangle with a length of 5 units and a width of 4 units; Find the length of the diagonal.

8. a square with side length n; Find the length of the diagonal.

In Exercises 9 and 10, graph the triangle with the given vertices. Find the length and the slope of each side of the triangle. Then find the coordinates of the midpoint of each side. Is the triangle a right triangle? isosceles? Explain. (Assume all variables are positive and $m \neq n$.)

▶ **9.** $A(0, 0)$, $B(h, h)$, $C(2h, 0)$

10. $D(0, n)$, $E(m, n)$, $F(m, 0)$

In Exercises 11 and 12, find the coordinates of any unlabeled vertices. Then find the indicated length(s).

11. Find ON and MN.

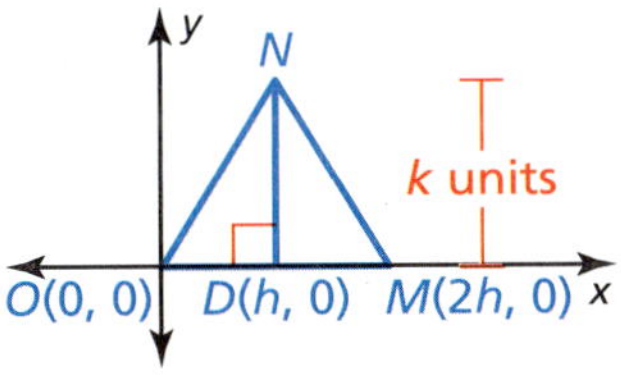

12. Find OT.

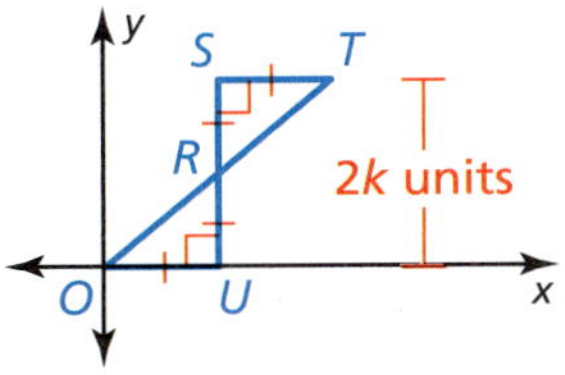

GO DIGITAL

In Exercises 13 and 14, write a plan for the proof. *(See Example 3.)*

▶ **13.** **Given** Coordinates of vertices of $\triangle OPM$ and $\triangle ONM$

Prove $\triangle OPM$ and $\triangle ONM$ are isosceles triangles.

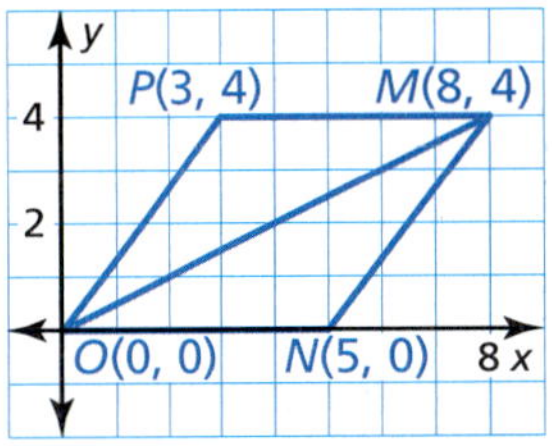

14. **Given** G is the midpoint of $\overline{HF}$.

Prove $\triangle GHJ \cong \triangle GFO$

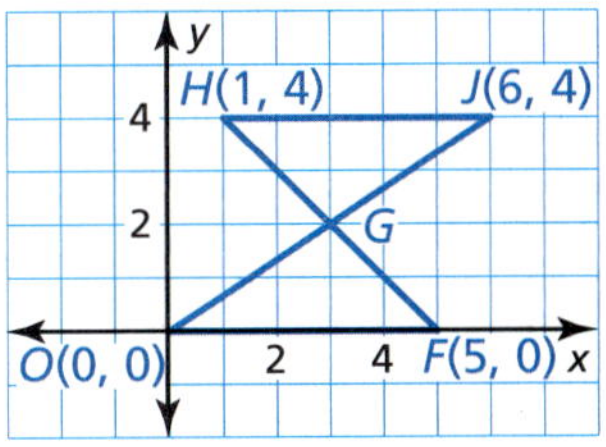

PROOF In Exercises 15 and 16, write a coordinate proof. *(See Example 4.)*

▶ **15.** **Given** Coordinates of vertices of $\triangle DEC$ and $\triangle BOC$

Prove $\triangle DEC \cong \triangle BOC$

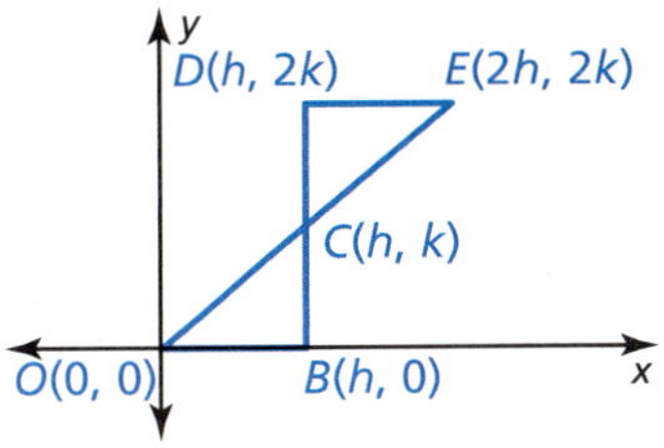

16. **Given** Coordinates of vertices of $\triangle DEA$, H is the midpoint of $\overline{DA}$, G is the midpoint of $\overline{EA}$.

Prove $\overline{DG} \cong \overline{EH}$

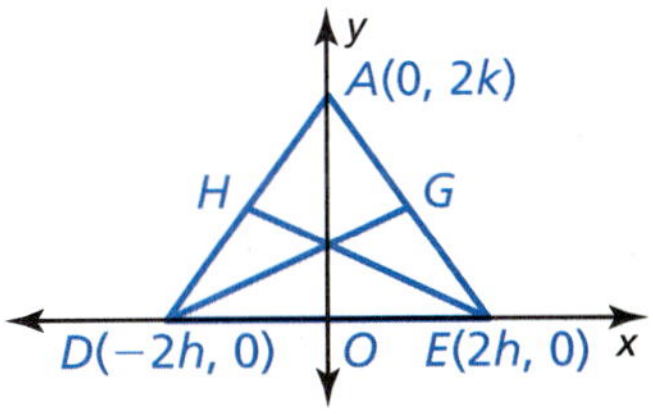

7 MTR **17.** **MODELING REAL LIFE** A manufacturer cuts a piece of metal for a microscope. The resulting piece of metal can be represented in a coordinate plane by a triangle with $A(0, 0)$, $B(5, 12)$, and $C(10, 0)$. One unit in the coordinate plane represents one millimeter. Prove that $\triangle ABC$ is isosceles. *(See Example 5.)*

7 MTR **18. MODELING REAL LIFE** You design the front of a phone tripod using a coordinate plane on a computer program. The coordinates of the vertices of the triangle at the front of the tripod are $A(0, 0)$, $B(12, 16)$, and $C(22, 0)$. One unit in the coordinate plane represents one inch. Prove that $\triangle ABC$ is a scalene triangle. Describe how to adjust point C to improve stability.

4 MTR **19. MAKING AN ARGUMENT** Your friend says that quadrilateral $PQRS$ with vertices $P(0, 2)$, $Q(3, -4)$, $R(1, -5)$, and $S(-2, 1)$ is a rectangle. Is your friend correct? Explain.

20. THOUGHT PROVOKING
Choose one of the theorems you have encountered that is easier to prove with a coordinate proof than with another type of proof. Explain. Then write a coordinate proof.

5 MTR **21. STRUCTURE** Write an algebraic expression for the coordinates of each endpoint of a line segment whose midpoint is the origin.

22. HOW DO YOU SEE IT?
Without performing any calculations, how do you know that the diagonals of square $TUVW$ are perpendicular to each other?

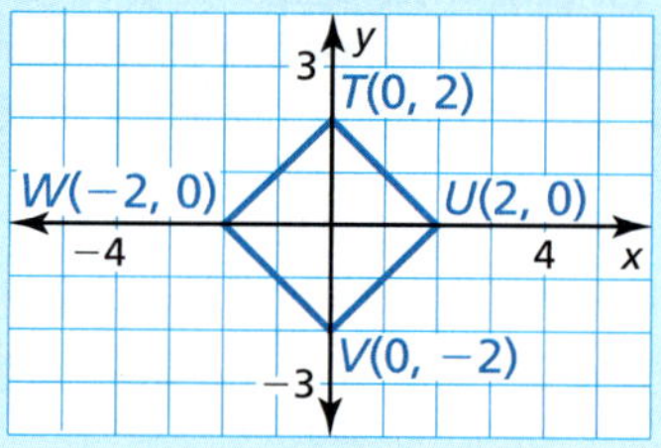

23. REASONING A rectangle has a length of ℓ inches and width of w inches. Represent this rectangle in the coordinate plane. Show that the diagonals of the rectangle are congruent.

24. PROOF Write a coordinate proof for each statement.

a. The midpoint of the hypotenuse of a right triangle is the same distance from each vertex of the triangle.

b. Any two congruent right isosceles triangles can be combined to form a single isosceles triangle.

REVIEW & REFRESH

WATCH

In Exercises 25 and 26, solve the equation. Justify each step.

25. $6x + 13 = -5$

26. $3(x - 1) = -(x + 10)$

27. Factor the polynomial $14a^2 + 23a + 3$.

28. Explain how to prove that $\angle J \cong \angle L$.

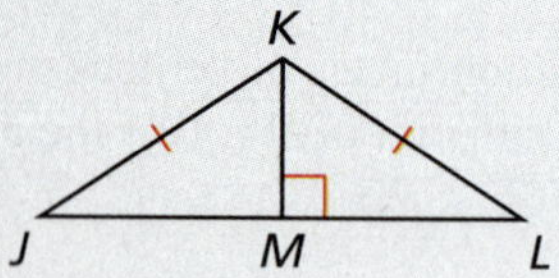

29. Decide whether enough information is given to prove that the $\triangle DEF$ and $\triangle JKL$ are congruent. If so, state the theorem you can use.

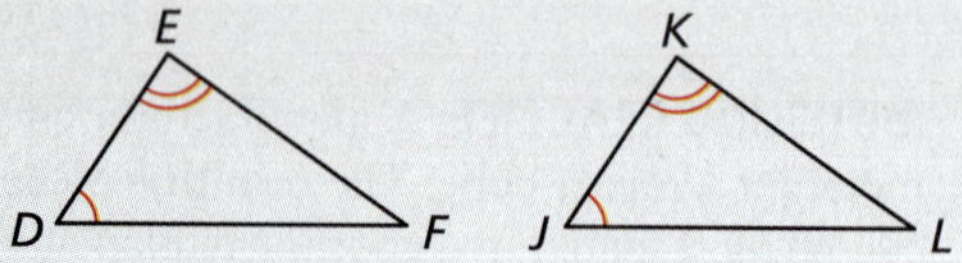

30. Write a proof.

Given $\overline{XY} \cong \overline{ZY}$, $\overline{WY} \cong \overline{VY}$, $\overline{VW} \cong \overline{ZX}$

Prove $\triangle VWZ \cong \triangle ZXV$

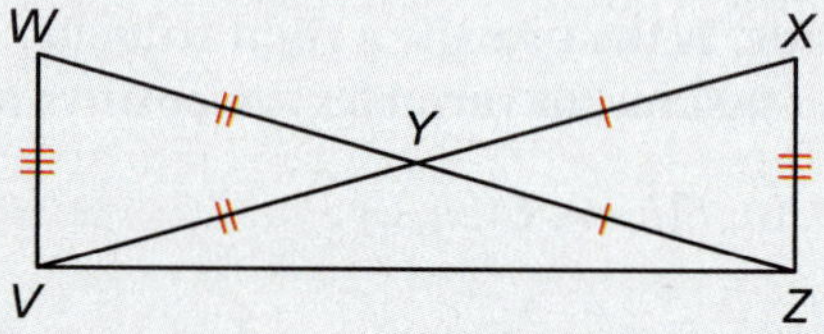

31. Place a square with a side length of m units in a coordinate plane in a convenient way for finding the length of the diagonal. Assign coordinates to each vertex.

In Exercises 32 and 33, find the value of x.

32.

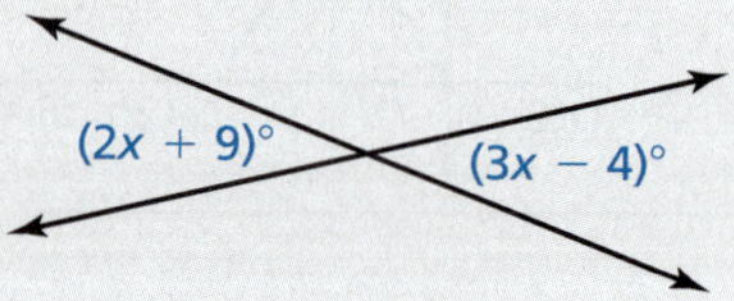

33.

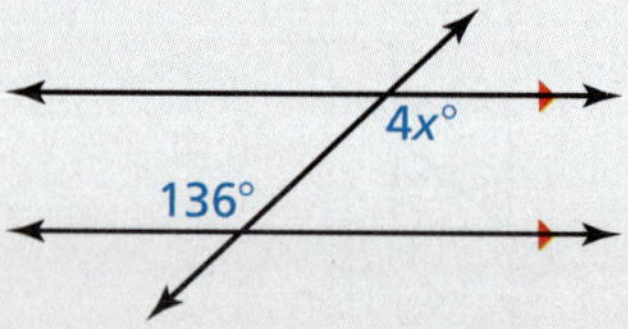

GO DIGITAL

5 Chapter Review WITH CalcChat®

Chapter Learning Target: Understand congruent triangles.

Chapter Success Criteria:
- ◆ I can classify triangles by sides and angles.
- ◆ I can solve problems involving congruent triangles.
- ■ I can use different theorems to prove two triangles are congruent.
- ■ I can write a coordinate proof.

◆ Surface
■ Deep

SELF-ASSESSMENT 1 I don't understand yet. 2 I can do it with help. 3 I can do it on my own. 4 I can teach someone else.

5.1 Angles of Triangles *(pp. 229–236)* WATCH

Learning Target: Prove and use theorems about angles of triangles.

Vocabulary VOCAB
- interior angles
- exterior angles
- corollary to a theorem

Classify the triangle by its sides and by measuring its angles.

1.

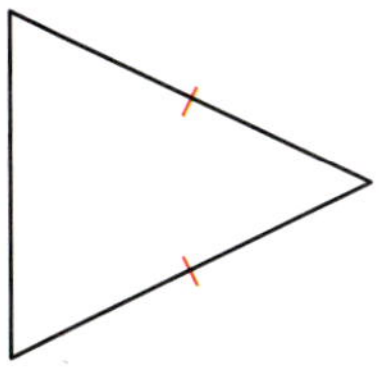

2.

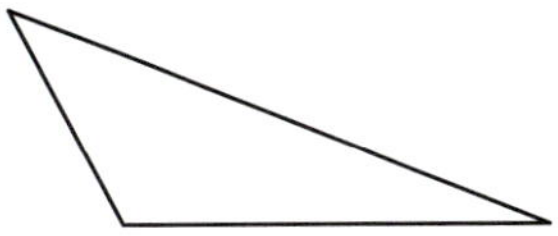

Classify $\triangle ABC$ by its sides. Then determine whether it is a right triangle.

3. $A(-2, 3)$, $B(3, 4)$, $C(1, -1)$

4. $A(2, 3)$, $B(6, 3)$, $C(2, 7)$

Find the measure of the exterior angle.

5.

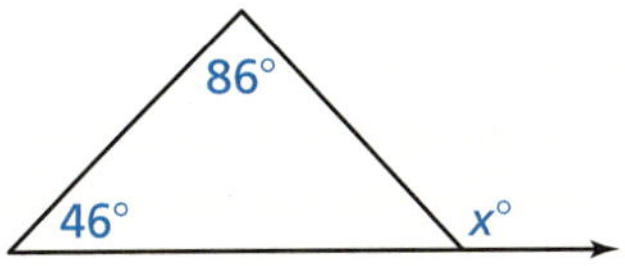

6.

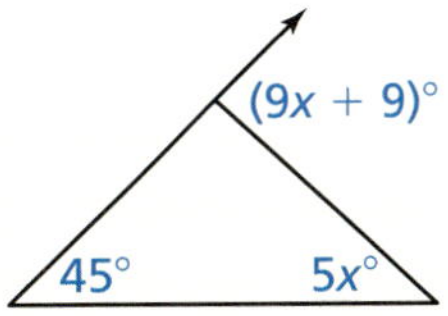

Find the measure of each acute angle.

7.

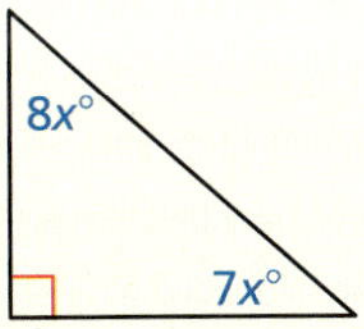

8.

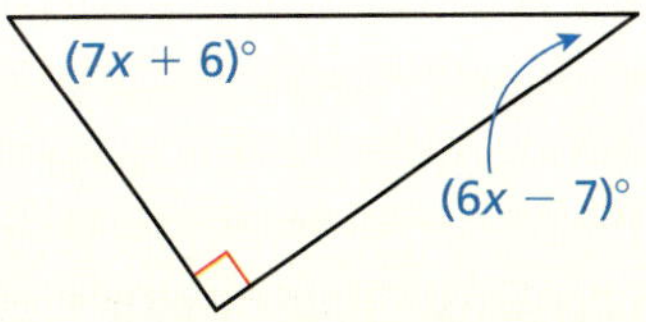

9. In a right triangle, the measure of one acute angle is 4 times the difference of the measure of the other acute angle and 5. Find the measure of each acute angle in the triangle.

GO DIGITAL

5.2 Congruent Polygons (pp. 237–242) WATCH

Learning Target: Understand congruence in terms of rigid motions.

Vocabulary VOCAB
corresponding parts

10. In the diagram, $GHJK \cong LMNP$. Identify all pairs of congruent corresponding parts. Then write another congruence statement for the quadrilaterals.

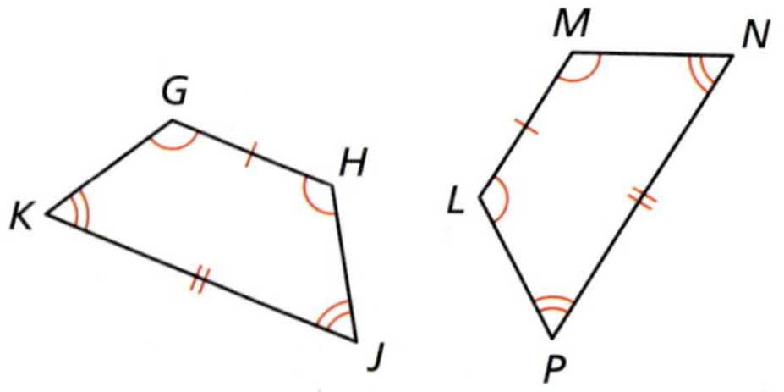

11. Find $m\angle V$.

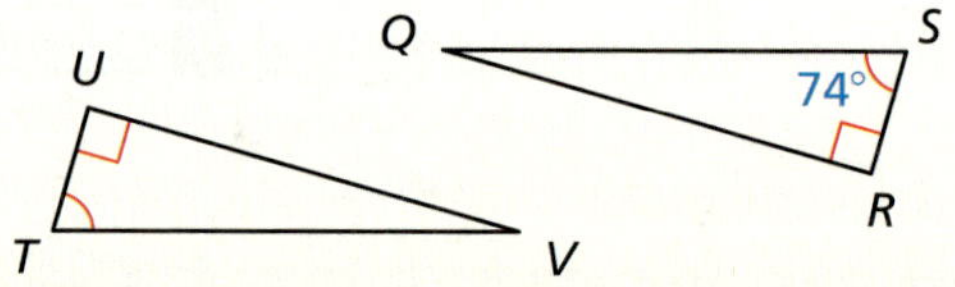

12. The figure shows the flag of the Czech Republic. Write a congruence statement for two of the polygons. Then show that those polygons are congruent.

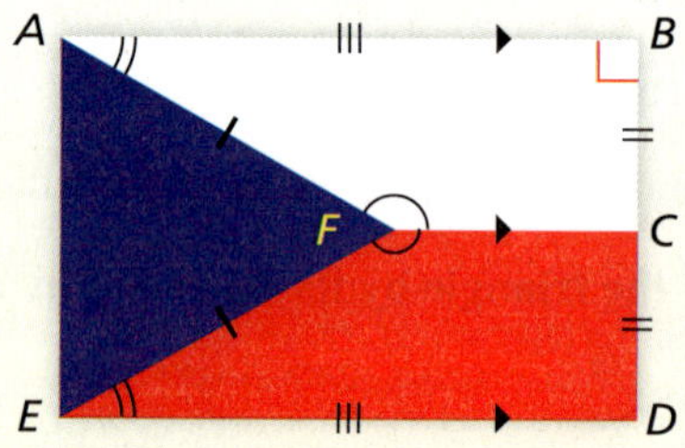

5.3 Proving Triangle Congruence by SAS (pp. 243–248) WATCH

Learning Target: Prove and use the Side-Angle-Side Congruence Theorem.

Decide whether enough information is given to prove that $\triangle WXZ \cong \triangle YZX$ using the SAS Congruence Theorem. If so, write a proof. If not, explain why.

13.

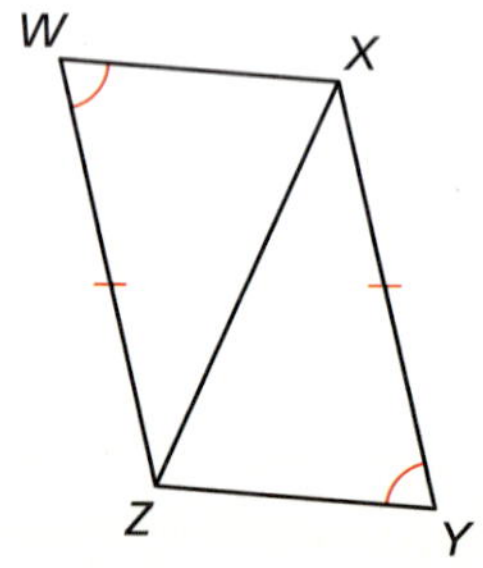

14.

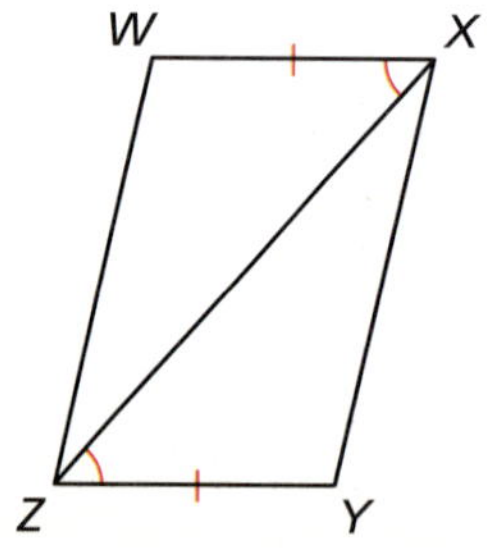

15. Construct a triangle that is congruent to $\triangle RST$ using the SAS Congruence Theorem.

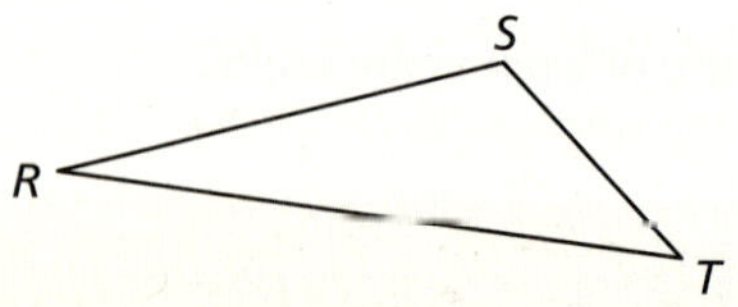

16. In the pyramid-shaped picture frame shown, $AD = CD$ and $\angle ADB \cong \angle CDB$. Use the SAS Congruence Theorem to prove that $\triangle ADB \cong \triangle CDB$.

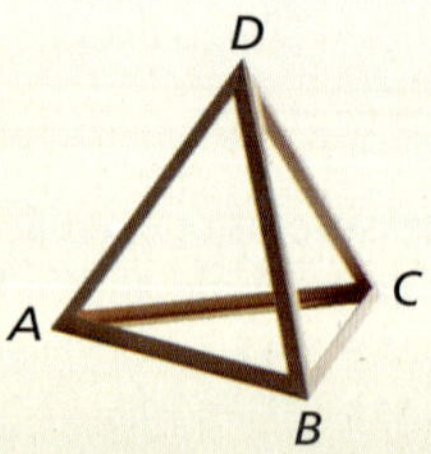

GO DIGITAL

5.4 Equilateral and Isosceles Triangles (pp. 249–256) WATCH

Learning Target: Prove and use theorems about isosceles and equilateral triangles.

Vocabulary AZ VOCAB
legs of an isosceles triangle
vertex angle
base
base angles

Complete the statement. State which theorem you used.

17. If $\overline{QP} \cong \overline{QR}$, then $\angle$__ $\cong \angle$__.

18. If $\angle TRV \cong \angle TVR$, then ___ $\cong$ ___.

19. If $\overline{RQ} \cong \overline{RS}$, then $\angle$__ $\cong \angle$__.

20. If $\angle SRV \cong \angle SVR$, then ___ $\cong$ ___.

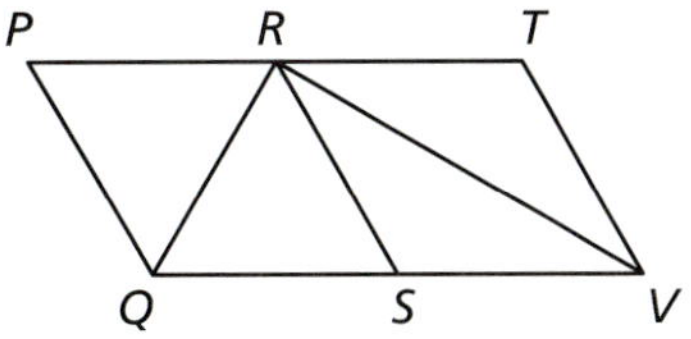

Find the values of *x* and *y*.

21.

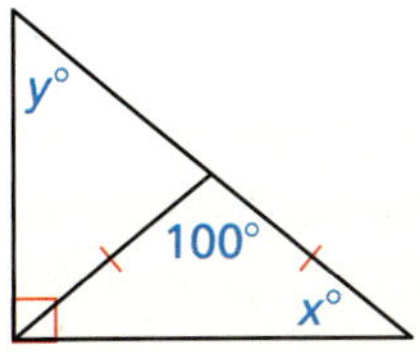

22.

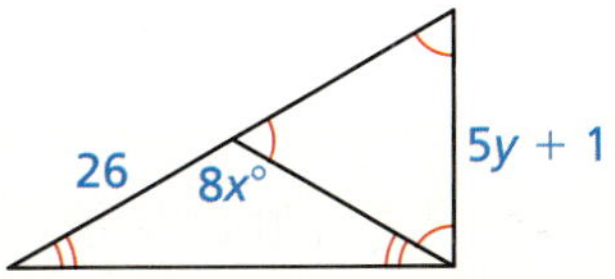

23. Find the perimeter of the triangular hedge.

5.5 Proving Triangle Congruence by SSS (pp. 257–264) WATCH

Learning Target: Prove and use the Side-Side-Side Congruence Theorem.

Vocabulary AZ VOCAB
legs of a right triangle
hypotenuse

24. Decide whether enough information is given to use the SSS Congruence Theorem to prove that $\triangle LMP \cong \triangle NPM$. If so, write a proof. If not, explain why.

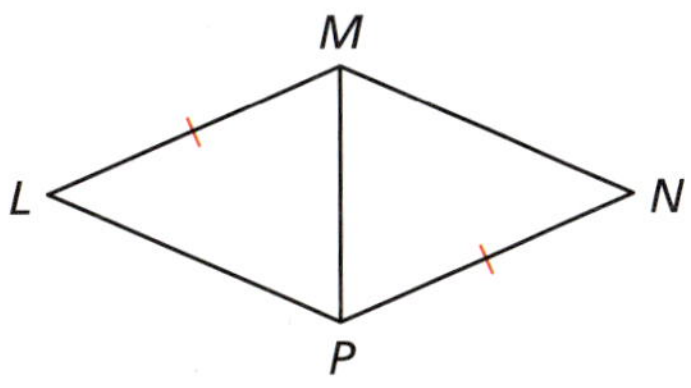

25. Decide whether enough information is given to use the HL Congruence Theorem to prove that $\triangle WXZ \cong \triangle YZX$. If so, write a proof. If not, explain why.

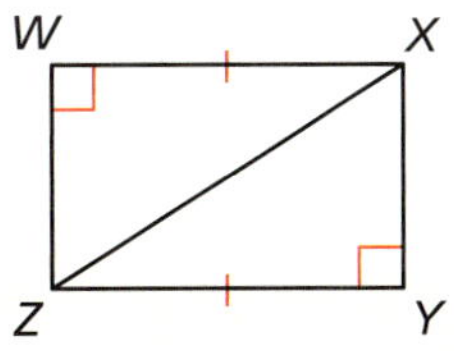

26. The photo shows two triangular windows.

a. What additional information do you need to prove that $\triangle ABD \cong \triangle CBD$ using the HL Congruence Theorem?

b. What additional information do you need to prove that $\triangle ABD \cong \triangle CBD$ using the SSS Congruence Theorem?

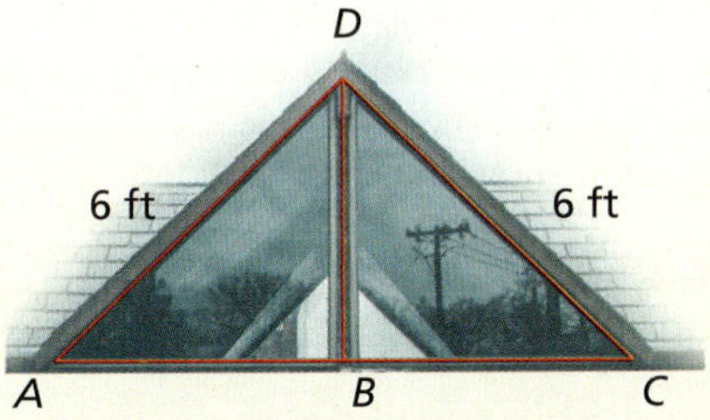

GO DIGITAL

5.6 Proving Triangle Congruence by ASA and AAS (pp. 265–272)

WATCH

Learning Target: Prove and use the Angle-Side-Angle Congruence Theorem and the Angle-Angle-Side Congruence Theorem.

Decide whether enough information is given to prove that the triangles are congruent using the AAS Congruence Theorem. If so, write a proof. If not, explain why.

27. $\triangle EFG$, $\triangle HJK$

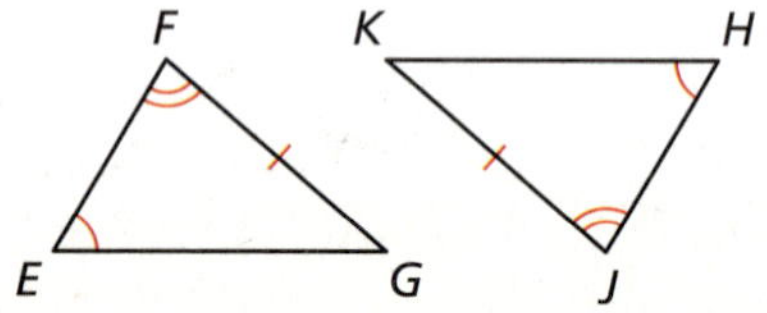

28. $\triangle TUV$, $\triangle QRS$

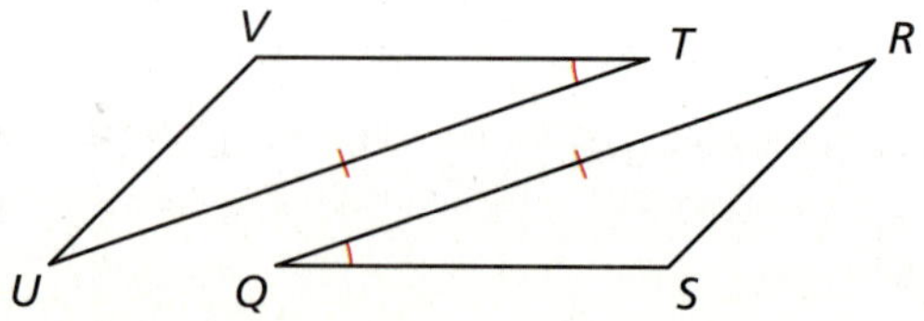

Decide whether enough information is given to prove that the triangles are congruent using the ASA Congruence Theorem. If so, write a proof. If not, explain why.

29. $\triangle LPN$, $\triangle LMN$

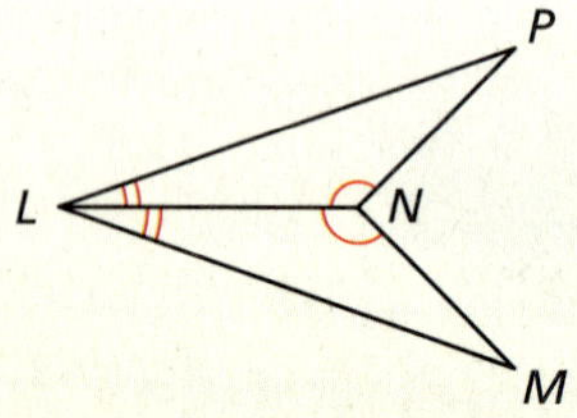

30. $\triangle WXZ$, $\triangle YZX$

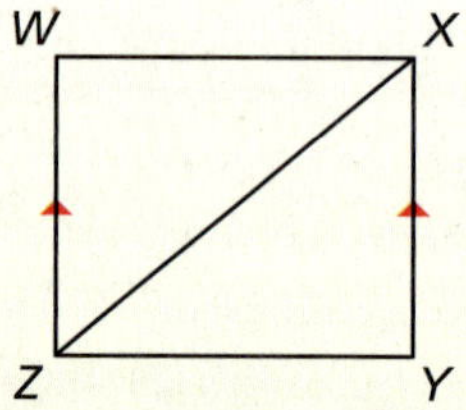

5.7 Using Congruent Triangles (pp. 273–278)

WATCH

Learning Target: Use congruent triangles in proofs and to measure distances.

Explain how to prove that the statement is true.

31. $\angle K \cong \angle N$

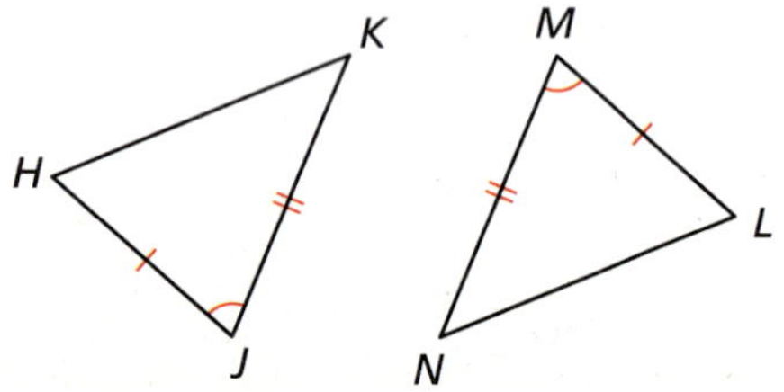

32. $\overline{AD} \cong \overline{CB}$

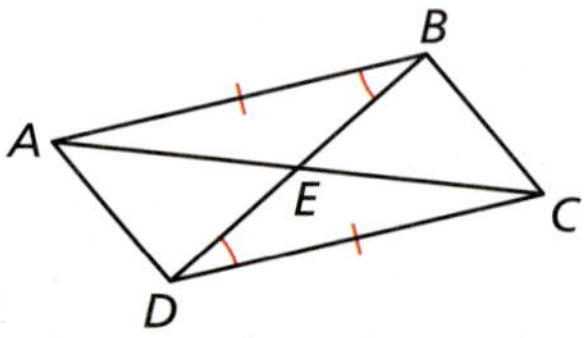

Write a plan to prove that $\angle 1 \cong \angle 2$.

33.

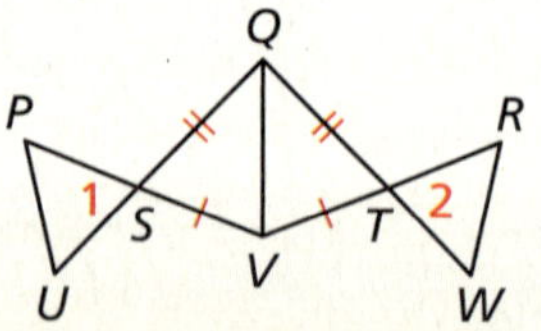

34.

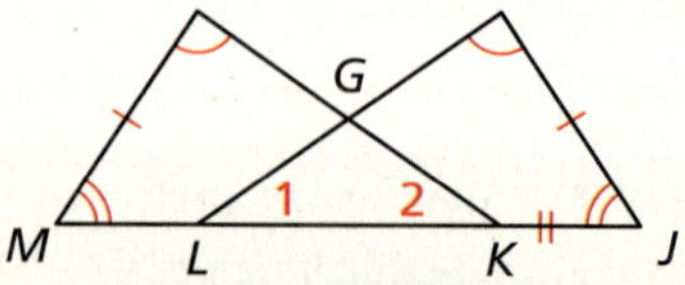

35. The diagram shows the shortest route around the buoys at points A, B, C, and D for a boat race. Use the information in the diagram to prove that $\angle D \cong \angle B$.

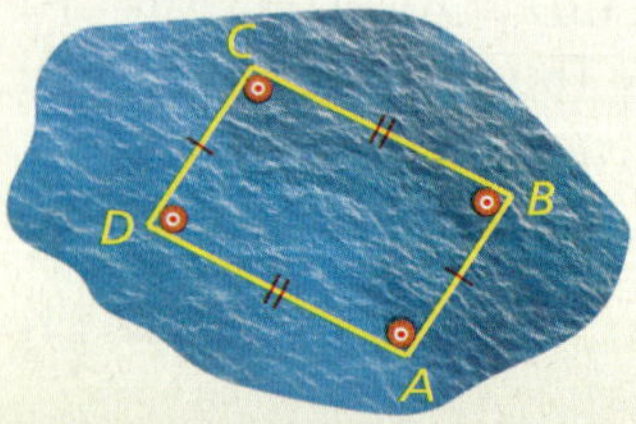

GO DIGITAL

5.8 Coordinate Proofs (pp. 279–284) WATCH

Learning Target: Use coordinates to write proofs.

Vocabulary VOCAB
coordinate proof

Place the figure in a coordinate plane in a convenient way. Assign coordinates to each vertex. Explain the advantages of your placement.

36. an isosceles triangle

37. a trapezoid with a pair of adjacent right angles

38. A rectangle has vertices $(0, 0)$, $(2k, 0)$, and $(0, k)$. Find the fourth vertex.

39. A square has vertices $(-k, 0)$, $(0, k)$, and $(k, 0)$. Find the fourth vertex.

40. Write a coordinate proof.

Given Coordinates of vertices of $\triangle ODB$ and $\triangle BDC$

Prove $\triangle ODB \cong \triangle BDC$

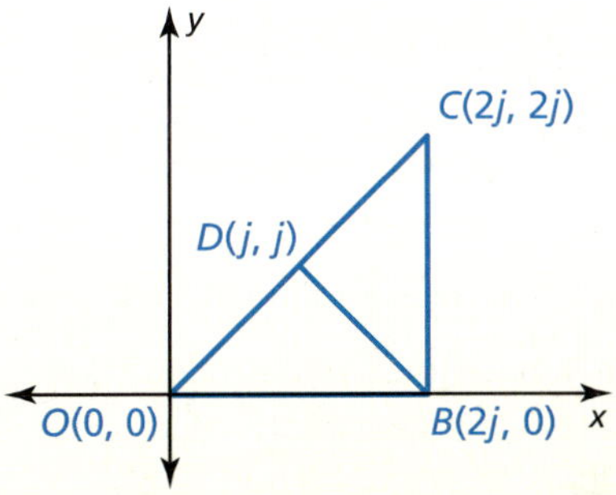

41. Write a coordinate proof.

Given Coordinates of vertices of quadrilateral $OPQR$

Prove $\triangle OPQ \cong \triangle QRO$

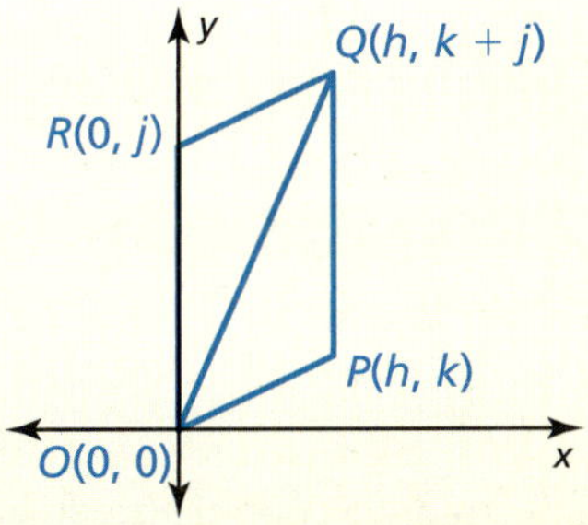

Mathematical Thinking and Reasoning

4 MTR **ENGAGE IN DISCUSSIONS THAT REFLECT ON THE MATHEMATICAL THINKING OF SELF AND OTHERS**

Mathematicians who engage in discussions that reflect on the mathematical thinking of self and others justify results by explaining methods and processes.

1. In 5.8 Explore It! on page 279, describe the different cases you explored. What did the different cases help you to notice about the situation? How might your conclusions have changed had you considered only a single case?

2. Your friend uses the figure to say that all the exterior angles of any isosceles triangle are congruent. Is your friend correct? Explain your reasoning.

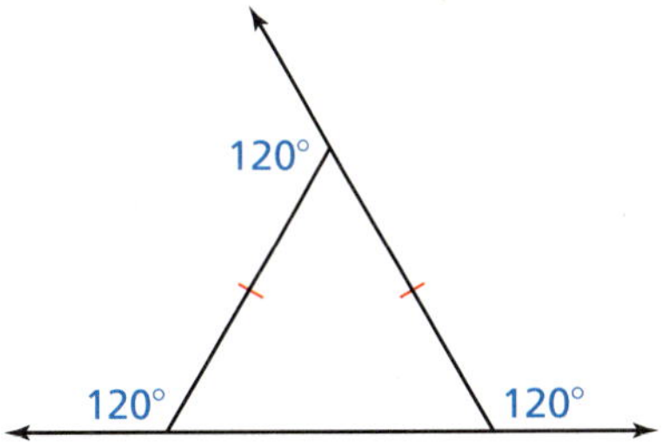

GO DIGITAL

5 Practice Test WITH CalcChat®

Write a proof.

1. **Given** $\overline{CA} \cong \overline{CB} \cong \overline{CD} \cong \overline{CE}$

   Prove $\triangle ABC \cong \triangle EDC$

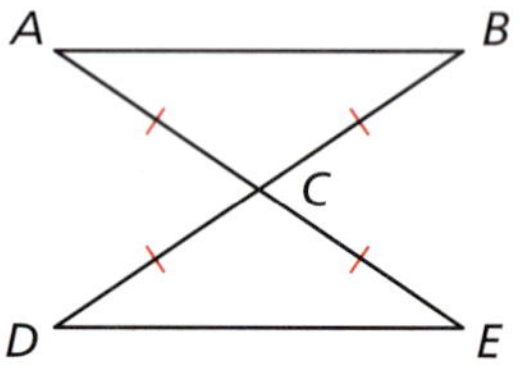

2. **Given** $\overline{JK} \parallel \overline{ML}, \overline{MJ} \parallel \overline{KL}$

   Prove $\triangle MJK \cong \triangle KLM$

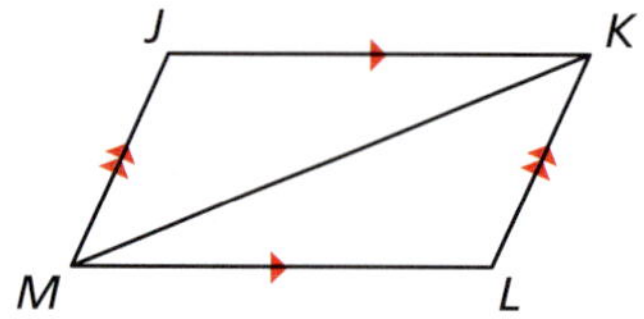

3. **Given** $\overline{QR} \cong \overline{RS}, \angle P \cong \angle T$

   Prove $\triangle SRP \cong \triangle QRT$

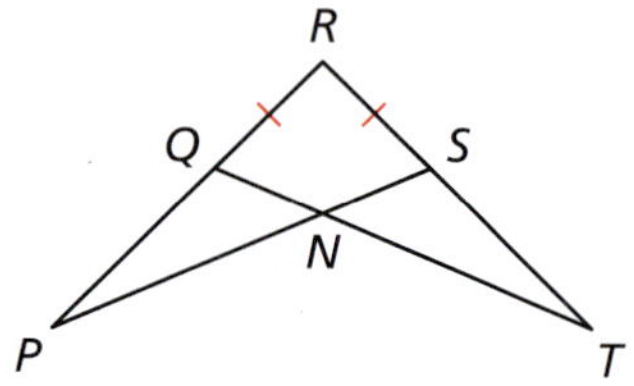

4. Find the measure of each acute angle in the figure at the right.
5. Is it possible to draw an equilateral triangle that is not equiangular? If so, provide an example. If not, explain why.
6. Can you use the Third Angles Theorem to prove that two triangles are congruent? Explain your reasoning.

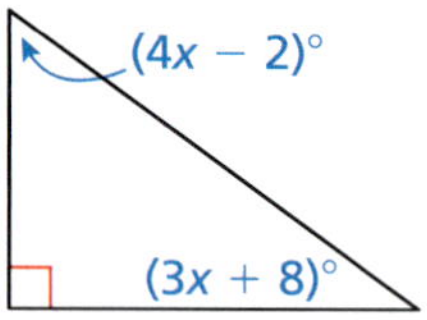

Write a plan to prove that $\angle 1 \cong \angle 2$.

7.

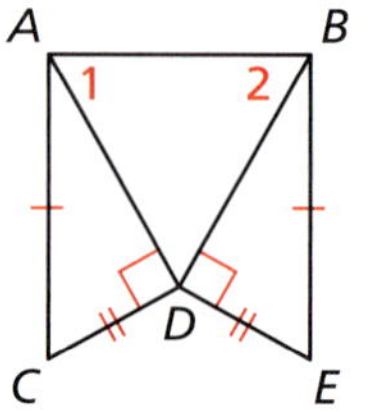

8.

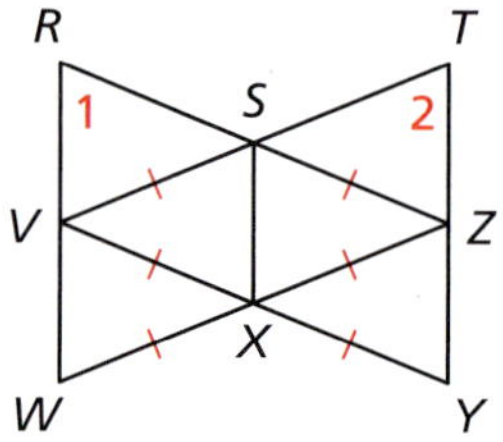

9. Name all the triangle congruence theorems that can be used to prove that $\triangle ABD \cong \triangle CDB$. Explain your reasoning.

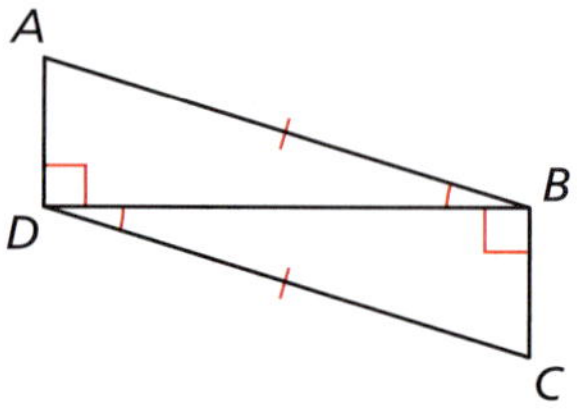

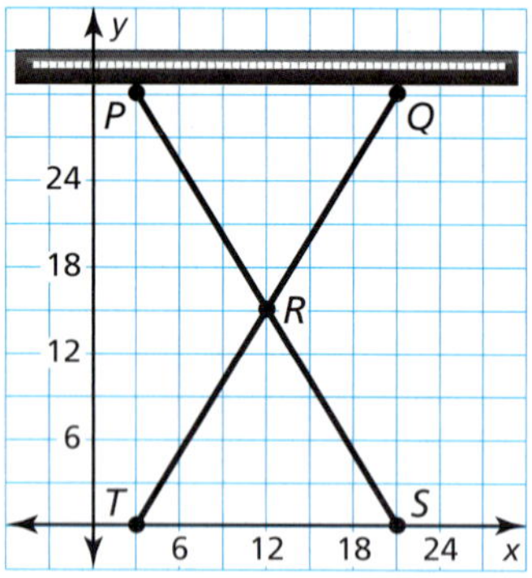

10. Write a coordinate proof to show that the triangles created by the keyboard stand at left are congruent.
11. An equiangular triangle has side lengths of $6x + 1$, $6y - 5$, and $7y - 2x$. Find the values of x and y.

12. The diagram shows a triangular face of the Pyramid of Cestius in Rome, Italy. The length of the base of the triangle is 30 meters. The lengths of the other two sides of the triangle are both 36 meters.

    a. Classify the triangle by its sides.

    b. The measure of $\angle 3$ is 40°. What are the measures of $\angle 1$ and $\angle 2$? Explain your reasoning.

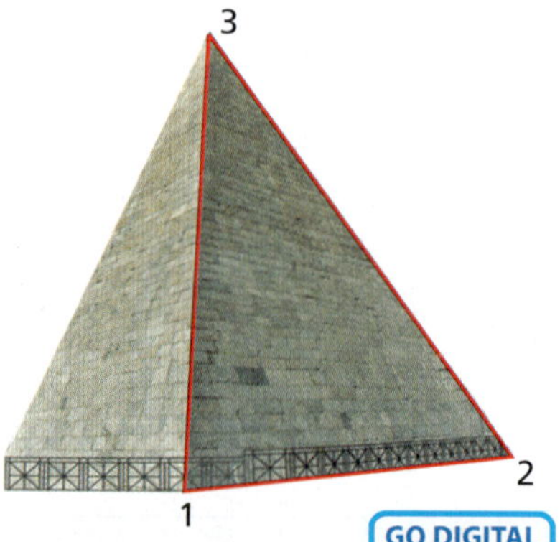

GO DIGITAL

5 Performance Task
Bacteriophages

Bacteriophages are viruses that infect bacteria.

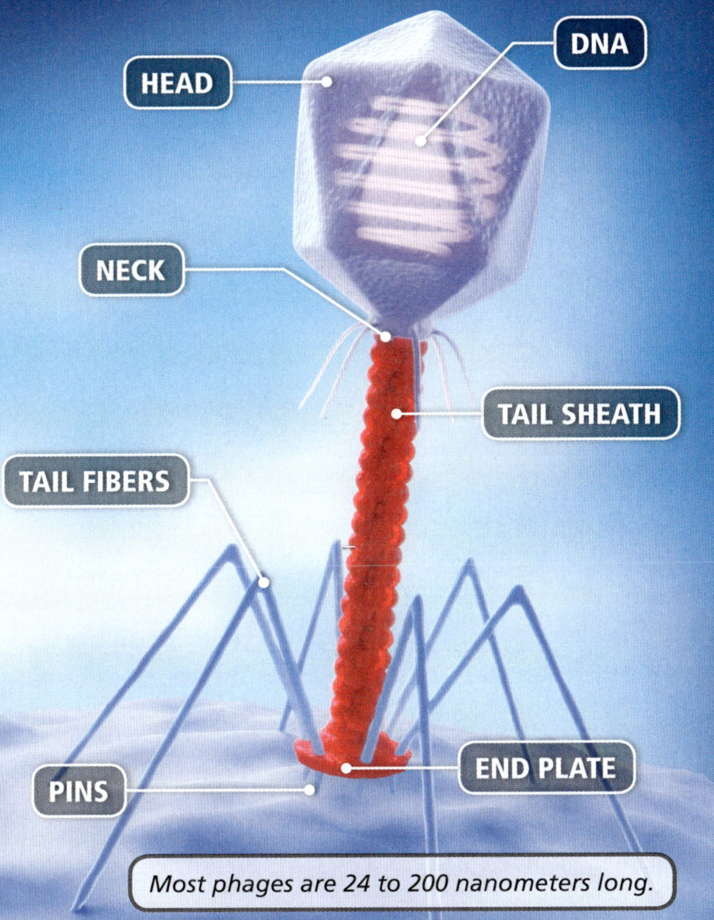

Most phages are 24 to 200 nanometers long.

1

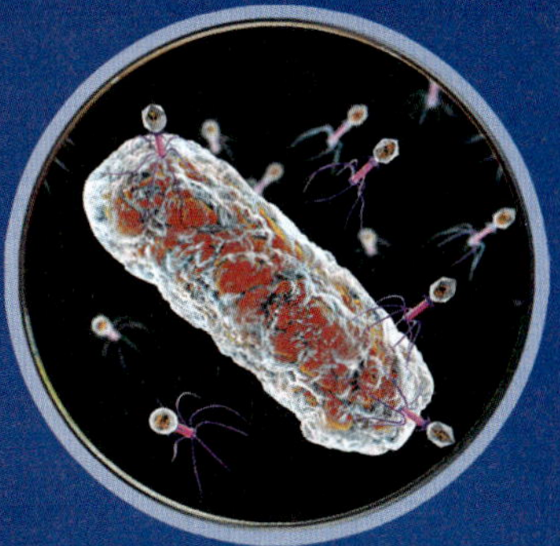

A phage lands on a bacterium.

2

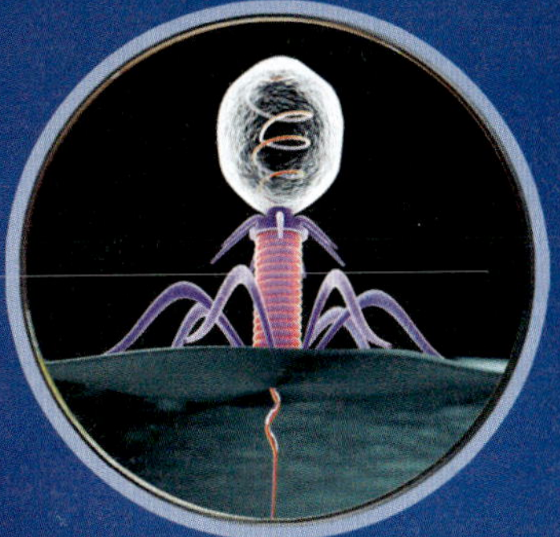

The phage uses its pins to inject its DNA into the bacterium.

3

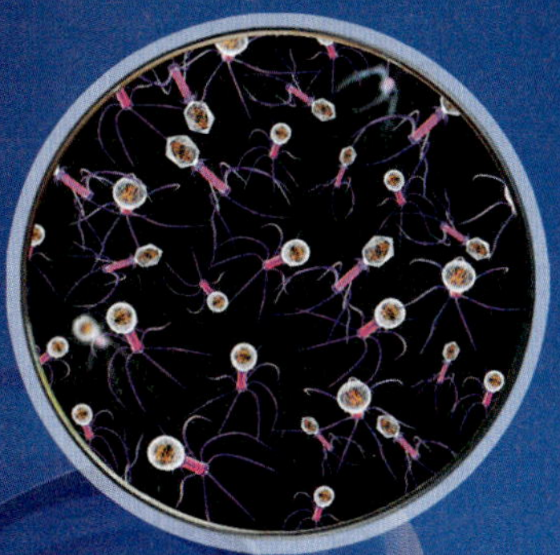

The DNA is copied, and new phages assemble inside the bacterium.

4

The new phages produce chemicals that rupture the bacterium, killing it in the process.

GOING VIRAL

A scientist views a bacteriophage under a microscope and draws a net of the head as shown. All line segments that appear parallel are parallel. Identify any triangles that appear to be congruent in the diagram. If possible, prove that they are congruent. If not possible, explain what additional information is needed.

GO DIGITAL

5 Test Prep WITH CalcChat® Cumulative Practice

WATCH Tutorial videos are available for each exercise.

1. Use the steps in the construction to explain how you know that the line through point P is parallel to line m.

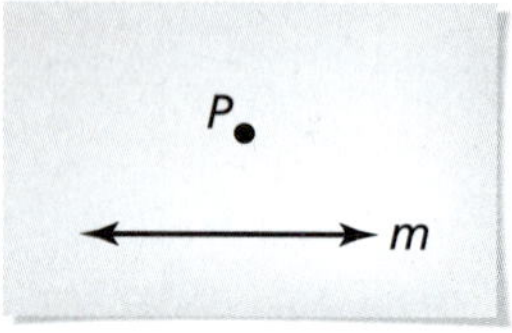

Step 1

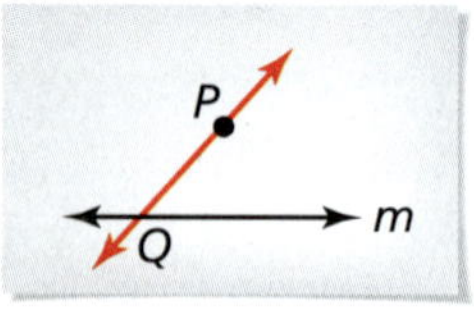

Step 2

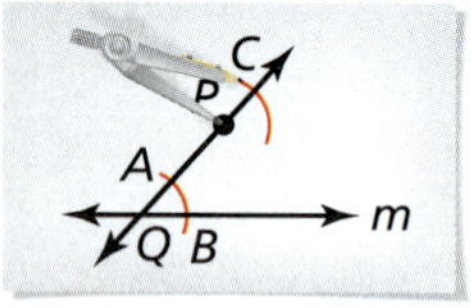

Step 3

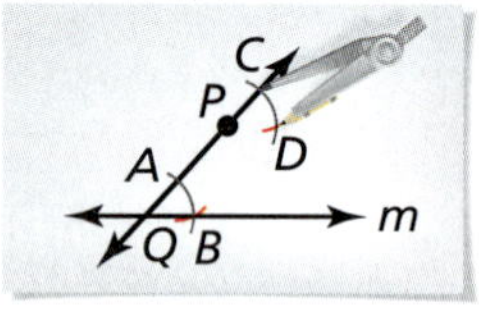

Step 4

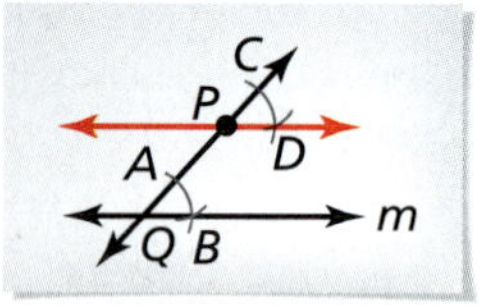

2. The coordinate plane shows $\triangle JKL$ and $\triangle XYZ$. Which composition of transformations maps $\triangle JKL$ to $\triangle XYZ$?

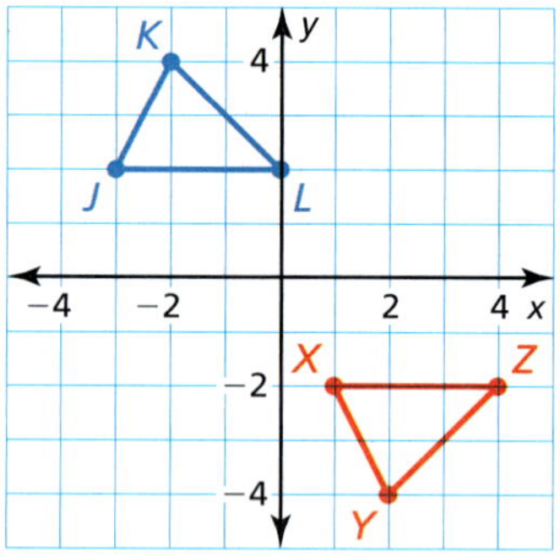

Ⓐ a reflection in the x-axis, followed by a reflection in the line $x = \frac{1}{2}$

Ⓑ a rotation of 180°, followed by a translation 1 unit right

Ⓒ a translation 1 unit left, followed by a rotation of 180°

Ⓓ a reflection in the x-axis, followed by a translation 4 units right

3. What are the coordinates of the image of $\triangle ABC$ after a dilation with a scale factor of 5?

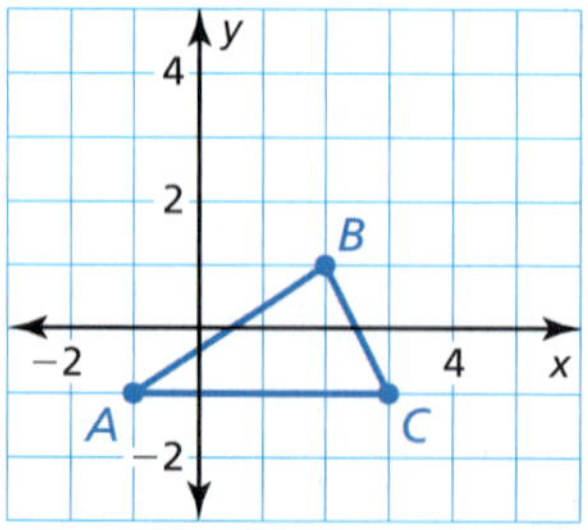

Ⓐ $A'(4, 4)$, $B'(7, 6)$, $C'(8, 4)$

Ⓑ $A'(-6, -6)$, $B'(-3, -4)$, $C'(-2, -6)$

Ⓒ $A'(-5, -5)$, $B'(10, 5)$, $C'(15, -5)$

Ⓓ $A'\left(-\frac{1}{5}, -\frac{1}{5}\right)$, $B'\left(\frac{2}{5}, \frac{1}{5}\right)$, $C'\left(\frac{3}{5}, -\frac{1}{5}\right)$

GO DIGITAL

4. The coordinate plane shows $\triangle ABC$ and $\triangle DEF$.

a. Prove $\triangle ABC \cong \triangle DEF$ using the given information.

b. Describe the composition of rigid motions that maps $\triangle ABC$ to $\triangle DEF$.

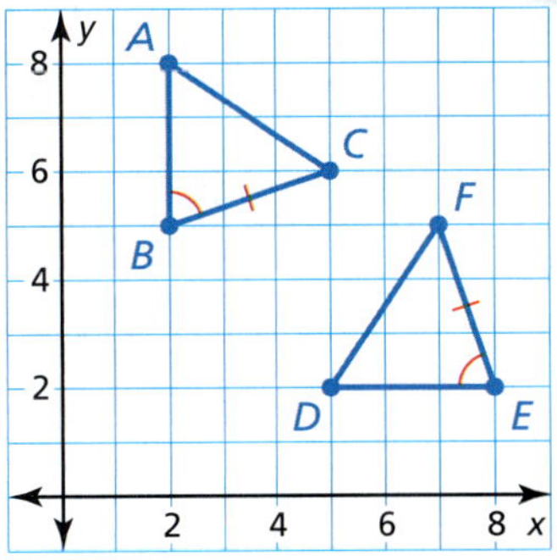

5. Which figure(s) have rotational symmetry? Select all that apply.

Ⓐ 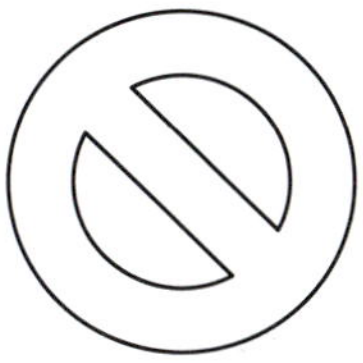

Ⓑ 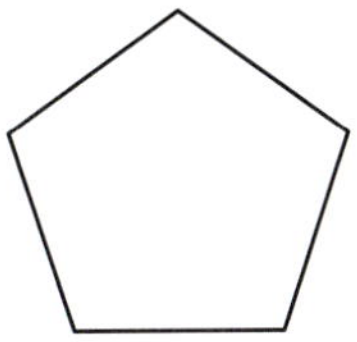

Ⓒ 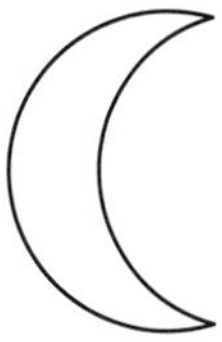

Ⓓ 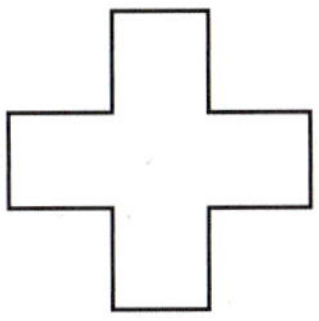

6. $m\angle ABC = 63°$. Find $m\angle CBD$.

Ⓐ 17°

Ⓑ 23°

Ⓒ 32°

Ⓓ 40°

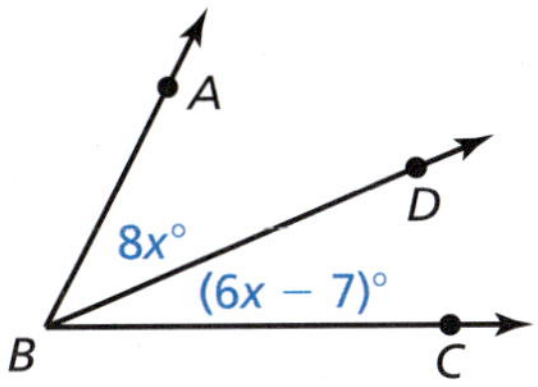

7. Write a coordinate proof.

Given Coordinates of vertices of quadrilateral *ABCD*

Prove Quadrilateral *ABCD* is a rectangle.

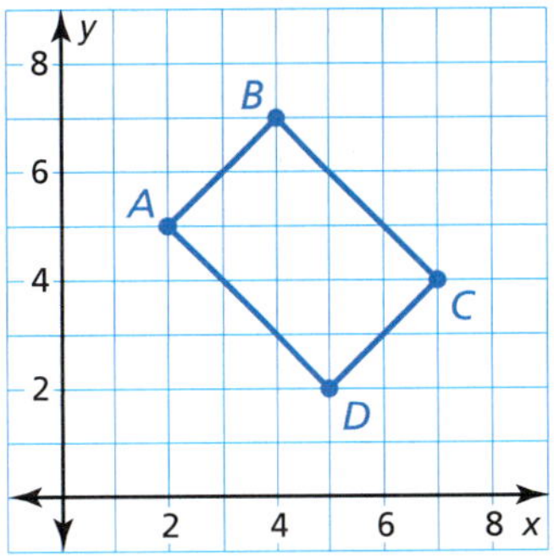

8. Write a proof to verify that the construction of the equilateral triangle shown below is valid.

Step 1

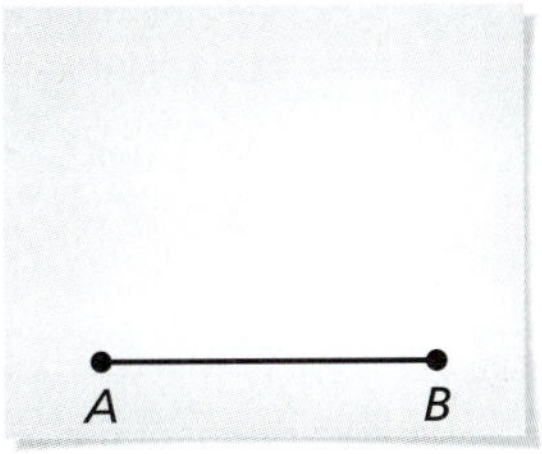

Draw $\overline{AB}$.

Step 2

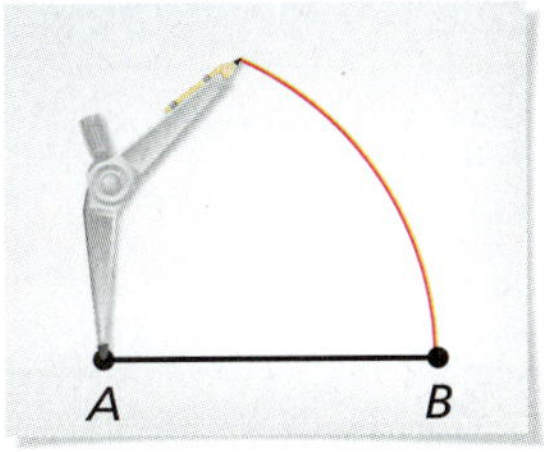

Draw an arc with center *A* and radius *AB*.

Step 3

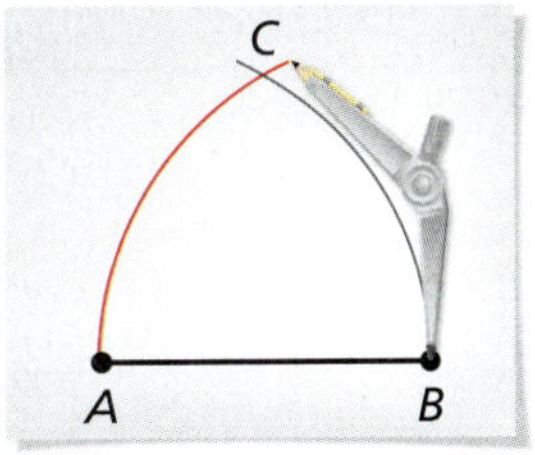

Draw an arc with center *B* and radius *AB*.

Step 4

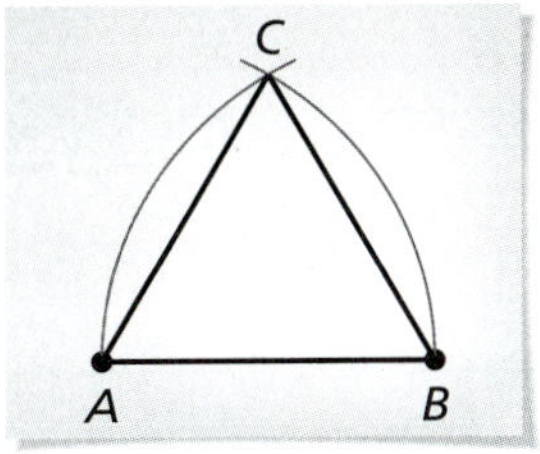

Draw $\triangle ABC$.

GO DIGITAL

6 Relationships Within Triangles

WATCH

NATIONAL GEOGRAPHIC EXPLORER

Amy E. Gusick

Dr. Amy E. Gusick is an associate curator of archaeology and chair of the Department of Anthropology at the Natural History Museum of Los Angeles. Her research is centered in Alta and Baja, California, home to some of the earliest-known human occupation in the New World. One of Amy's interests is human coastal migration and settlement of the late Pleistocene and early Holocene epochs. She uses both land and underwater methods in her research.

- What does an archaeologist do?
- What is the Pacific Rim?
- When did the Pleistocene epoch end? When did the Holocene epoch begin?
- Humans who lived in these eras are classified as hunter-gatherers. What is a hunter-gatherer?

STEM

When archaeologists discover artifact fragments, they try to piece them together to determine the size and shape of the original object. In the Performance Task, you will use a fragment to find the diameter of an ancient plate.

GO DIGITAL

Preparing for Chapter 6

Chapter Learning Target: Understand relationships within triangles.

Chapter Success Criteria:
- ◆ I can identify special segments in a triangle.
- ◆ I can use special segments in a triangle to solve problems.
- ■ I can prove statements using special segments in a triangle.
- ■ I can compare measures within triangles and between two triangles.

◆ Surface
■ Deep

Chapter Vocabulary

Work with a partner. Discuss each of the vocabulary terms.

equidistant
concurrent
altitude of a triangle
midsegment of a triangle
proof by contradiction

Mathematical Thinking and Reasoning

ACTIVELY PARTICIPATE IN EFFORTFUL LEARNING BOTH INDIVIDUALLY AND COLLECTIVELY

Mathematicians who actively participate in effortful learning both individually and with others analyze problems in a way that makes sense given the tasks.

Work with a partner. An archaeologist is studying the triangular region shown, where each vertex represents a village and each segment represents a road.

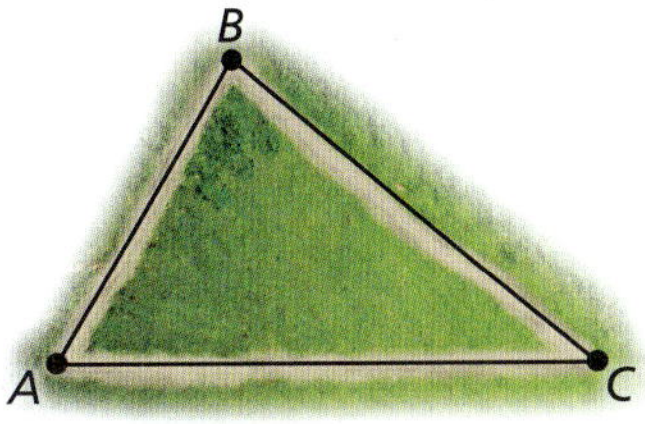

1. Each diagram below shows an intersection point of three line segments in the region. The archaeologist wants to establish a headquarters at either point P or point Q.

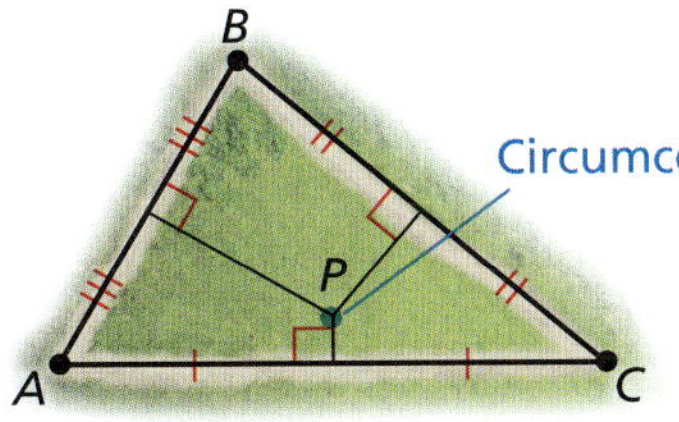

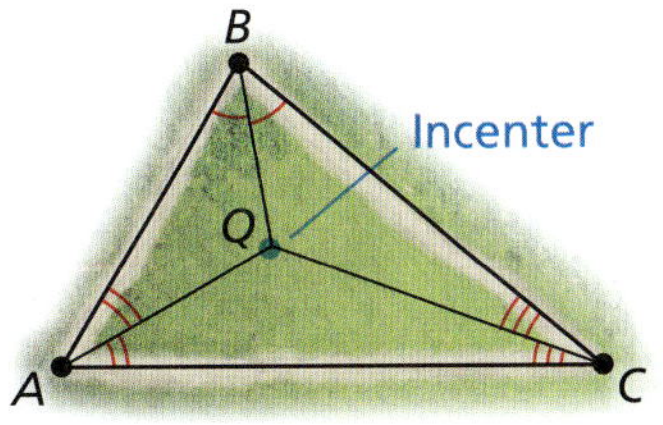

a. For each diagram, describe the given information.

b. Is point P closest to point A, B, or C? Explain.

c. Is point Q closest to $\overline{AB}$, $\overline{BC}$, or $\overline{AC}$? Explain.

2. Why might the archaeologist want to locate the headquarters at point P? at point Q?

GO DIGITAL

6 Prepare WITH CalcChat®

Writing an Equation of a Perpendicular Line

WATCH

Example 1 **Write an equation of the line passing through the point (−2, 0) that is perpendicular to the line $y = 2x + 8$.**

Step 1 Find the slope m of the perpendicular line. The line $y = 2x + 8$ has a slope of 2. Use the Slopes of Perpendicular Lines Theorem.

$2 \cdot m = -1$ — The product of the slopes of ⊥ lines is −1.

$m = -\frac{1}{2}$ — Divide each side by 2.

Step 2 Find the y-intercept b by using $m = -\frac{1}{2}$ and $(x, y) = (-2, 0)$.

$y = mx + b$ — Use slope-intercept form.

$0 = -\frac{1}{2}(-2) + b$ — Substitute for m, x, and y.

$-1 = b$ — Solve for b.

▶ Because $m = -\frac{1}{2}$ and $b = -1$, an equation of the line is $y = -\frac{1}{2}x - 1$.

Write an equation of the line passing through point P that is perpendicular to the given line.

1. $P(3, 1)$, $y = \frac{1}{3}x - 5$ **2.** $P(4, -3)$, $y = -x - 5$ **3.** $P(-1, -2)$, $y = -4x + 13$

Writing Compound Inequalities

WATCH

Example 2 **Write each sentence as an inequality.**

a. A number x is greater than or equal to −1 and less than 6.

A number x is greater than or equal to −1 and less than 6.

$x \geq -1$ *and* $x < 6$

▶ An inequality is $-1 \leq x < 6$.

b. A number y is at most 4 or at least 9.

A number y is at most 4 or at least 9.

$y \leq 4$ *or* $y \geq 9$

▶ An inequality is $y \leq 4$ *or* $y \geq 9$.

Write the sentence as an inequality.

4. A number w is at least −3 and no more than 8. **5.** A number d is fewer than −1 or no less than 5.

4 MTR **6.** **DISCUSS MATHEMATICAL THINKING** Is it possible for the solution of a compound inequality to be all real numbers? Explain your reasoning.

GO DIGITAL

6.1 Perpendicular and Angle Bisectors

Learning Target: Use theorems about perpendicular and angle bisectors.

Success Criteria:
- I can identify a perpendicular bisector and an angle bisector.
- I can use theorems about bisectors to find measures in figures.
- I can write equations of perpendicular bisectors.

EXPLORE IT! Drawing Perpendicular and Angle Bisectors

4 MTR COMPARE METHODS
What advantages does technology have over a compass and straightedge when making these conjectures?

Work with a partner.

a. Use technology to draw any segment and its perpendicular bisector. What do you notice about the distances between any point on the perpendicular bisector and the endpoints of the line segment? Explain why this is true.

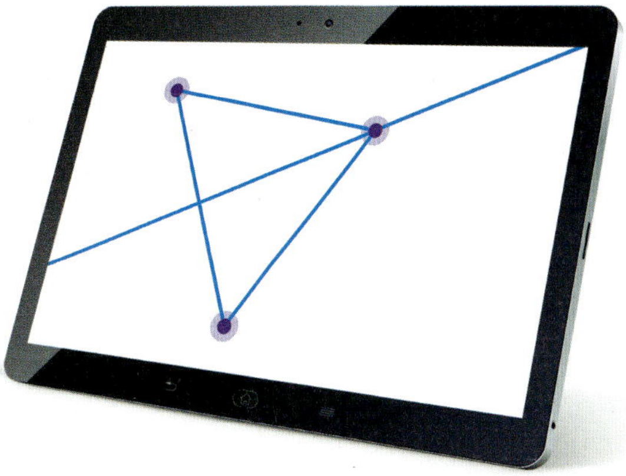

b. Use technology to draw any angle and its angle bisector. What do you notice about the distances between any point on the angle bisector and the sides of the angle? Explain why this is true.

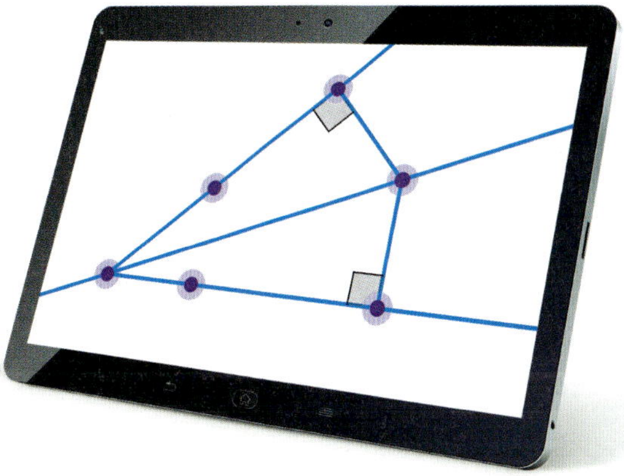

c. What conjectures can you make using your results in parts (a) and (b)? Write your conjectures as conditional statements written in if-then form.

Geometric Reasoning
MA.912.GR.1.1 Prove relationships and theorems about lines and angles. Solve mathematical and real-world problems involving postulates, relationships and theorems of lines and angles.

GO DIGITAL

Using Perpendicular Bisectors

Vocabulary AZ VOCAB

equidistant, *p. 298*

Recall that a *perpendicular bisector* of a line segment is the line that is perpendicular to the segment at its midpoint.

A point is **equidistant** from two figures when the point is the *same distance* from each figure.

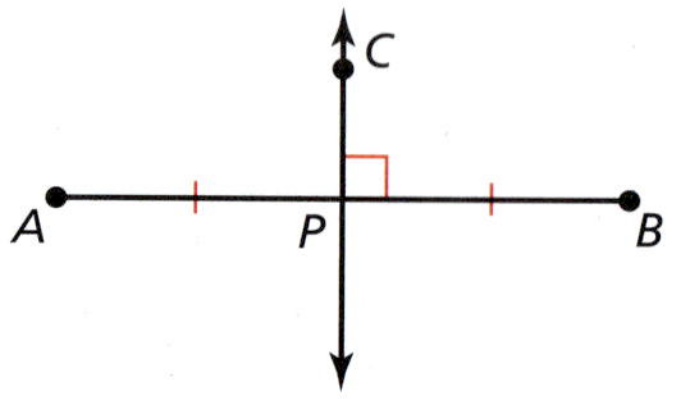

$\overleftrightarrow{CP}$ is a $\perp$ bisector of $\overline{AB}$.

STUDY TIP

A perpendicular bisector can be a segment, a ray, a line, or a plane.

THEOREMS

6.1 Perpendicular Bisector Theorem

In a plane, if a point lies on the perpendicular bisector of a segment, then it is equidistant from the endpoints of the segment.

If $\overleftrightarrow{CP}$ is the $\perp$ bisector of $\overline{AB}$, then $CA = CB$.

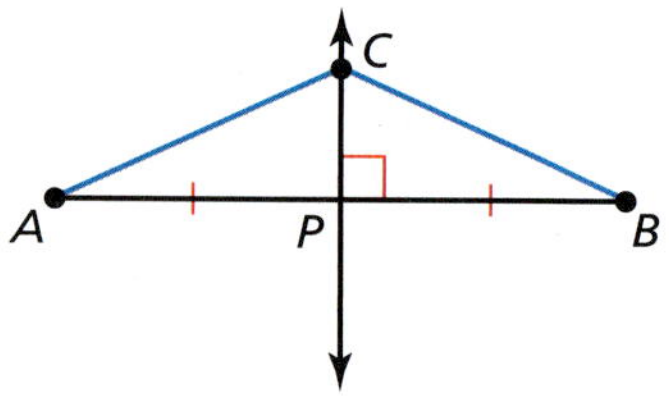

Prove this Theorem Exercise 1, page 299

6.2 Converse of the Perpendicular Bisector Theorem

In a plane, if a point is equidistant from the endpoints of a segment, then it lies on the perpendicular bisector of the segment.

If $DA = DB$, then point D lies on the $\perp$ bisector of $\overline{AB}$.

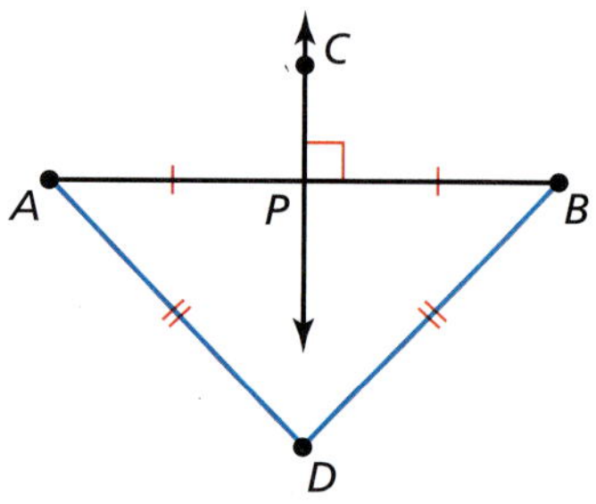

Prove this Theorem Exercise 30, page 303

PROOF Perpendicular Bisector Theorem

Given $\overleftrightarrow{CP}$ is the perpendicular bisector of $\overline{AB}$.

Prove $CA = CB$

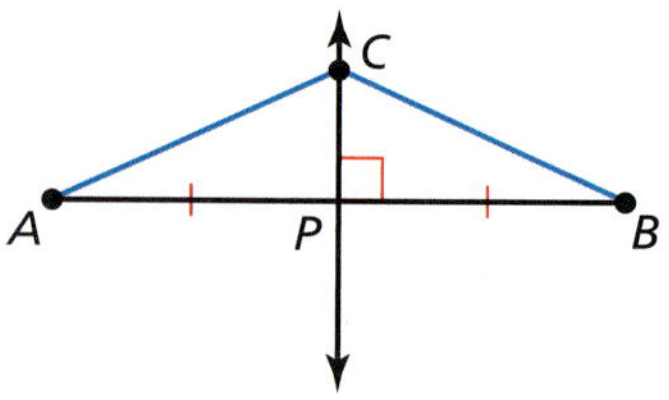

Paragraph Proof Because $\overleftrightarrow{CP}$ is the perpendicular bisector of $\overline{AB}$, $\overleftrightarrow{CP}$ is perpendicular to $\overline{AB}$ and point P is the midpoint of $\overline{AB}$. By the definition of midpoint, $AP = BP$, and by the definition of perpendicular lines, $m\angle CPA = m\angle CPB = 90°$. Then by the definition of segment congruence, $\overline{AP} \cong \overline{BP}$, and by the definition of angle congruence, $\angle CPA \cong \angle CPB$. By the Reflexive Property of Segment Congruence, $\overline{CP} \cong \overline{CP}$. So, $\triangle CPA \cong \triangle CPB$ by the SAS Congruence Theorem, and $\overline{CA} \cong \overline{CB}$ because corresponding parts of congruent triangles are congruent. So, $CA = CB$ by the definition of segment congruence.

4 MTR COMPARE METHODS

Compare the paragraph proof to other formats you could have used. Which format do you prefer?

GO DIGITAL

The Warren Pony Swing Bridge is a swing bridge in Florida that spans the Miami River Canal. A swing bridge is a bridge that can rotate to let boats pass.

EXAMPLE 1 Using a Diagram

Is there enough information in the diagram to conclude that point N lies on the perpendicular bisector of $\overline{KM}$?

SOLUTION

It is given that $\overline{KL} \cong \overline{ML}$. So, $\overline{LN}$ is a segment bisector of $\overline{KM}$. You do not know whether $\overline{LN}$ is perpendicular to $\overline{KM}$ because it is not indicated in the diagram.

▶ So, you cannot conclude that point N lies on the perpendicular bisector of $\overline{KM}$.

EXAMPLE 2 Using the Perpendicular Bisector Theorems

Find each measure.

a. RS

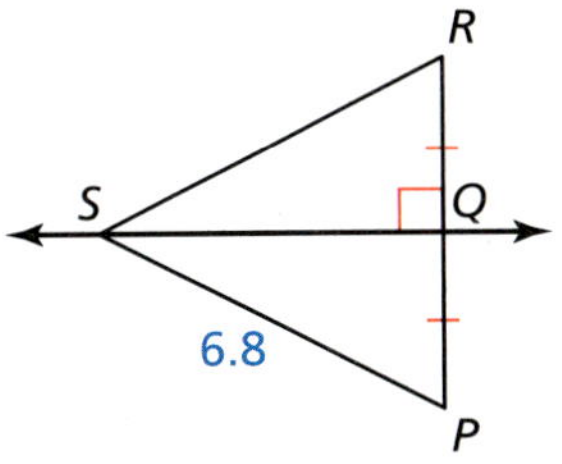

From the figure, $\overleftrightarrow{SQ}$ is the perpendicular bisector of $\overline{PR}$. By the Perpendicular Bisector Theorem, $PS = RS$.

▶ So, $RS = PS = 6.8$.

b. EG

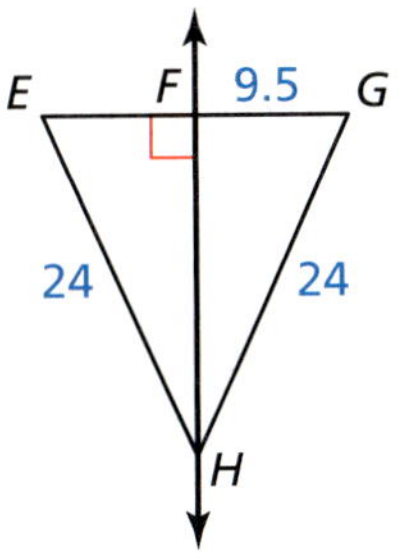

Because $EH = GH$ and $\overleftrightarrow{HF} \perp \overline{EG}$, $\overleftrightarrow{HF}$ is the perpendicular bisector of $\overline{EG}$ by the Converse of the Perpendicular Bisector Theorem. So, F is the midpoint of $\overline{EG}$, and $EF = GF$.

▶ So, $EG = EF + GF = 9.5 + 9.5 = 19$.

c. AD

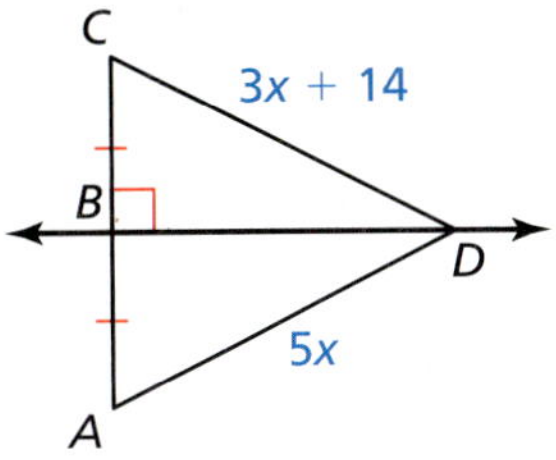

From the figure, $\overleftrightarrow{BD}$ is the perpendicular bisector of $\overline{AC}$.

$AD = CD$	Perpendicular Bisector Theorem
$5x = 3x + 14$	Substitute.
$x = 7$	Solve for x.

▶ So, $AD = 5x = 5(7) = 35$.

SELF-ASSESSMENT

1 I don't understand yet. 2 I can do it with help. 3 I can do it on my own. 4 I can teach someone else.

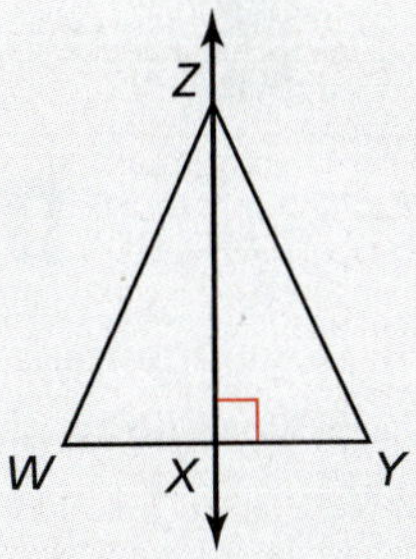

1. Point M is the midpoint of $\overline{PQ}$, and $\overleftrightarrow{LM}$ is the perpendicular bisector of $\overline{PQ}$. Write a two-column proof to show that $LP = LQ$.
2. Is there enough information in the figure to conclude that point Z lies on the perpendicular bisector of $\overline{WY}$? Explain your reasoning.

Use the figure and the given information to find the indicated measure.

3. $\overleftrightarrow{ZX}$ is the perpendicular bisector of $\overline{WY}$, and $YZ = 13.75$. Find WZ.
4. $\overleftrightarrow{ZX}$ is the perpendicular bisector of $\overline{WY}$, $WZ = 4n - 13$, and $YZ = n + 17$. Find YZ.
5. Find WX when $WZ = 20.5$, $WY = 14.8$, and $YZ = 20.5$.

GO DIGITAL

Using Angle Bisectors

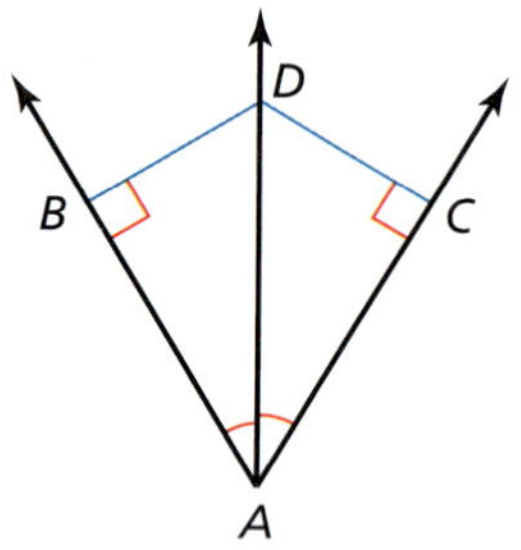

You know that an *angle bisector* is a ray that divides an angle into two congruent adjacent angles. You also know that the *distance from a point to a line* is the length of the perpendicular segment from the point to the line. So, in the figure, $\overrightarrow{AD}$ is the bisector of $\angle BAC$, and the distance from point D to $\overrightarrow{AB}$ is DB, where $\overline{DB} \perp \overrightarrow{AB}$.

THEOREMS

6.3 Angle Bisector Theorem

If a point lies on the bisector of an angle, then it is equidistant from the two sides of the angle.

If $\overrightarrow{AD}$ bisects $\angle BAC$ and $\overline{DB} \perp \overrightarrow{AB}$ and $\overline{DC} \perp \overrightarrow{AC}$, then $DB = DC$.

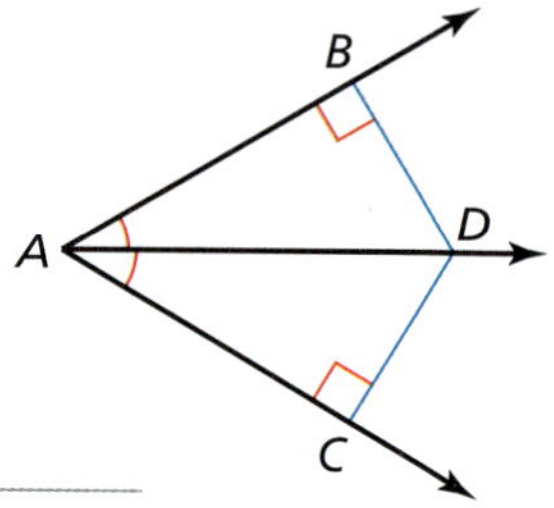

Prove this Theorem Exercise 31(a), page 303

6.4 Converse of the Angle Bisector Theorem

If a point is in the interior of an angle and is equidistant from the two sides of the angle, then it lies on the bisector of the angle.

If $\overline{DB} \perp \overrightarrow{AB}$ and $\overline{DC} \perp \overrightarrow{AC}$ and $DB = DC$, then $\overrightarrow{AD}$ bisects $\angle BAC$.

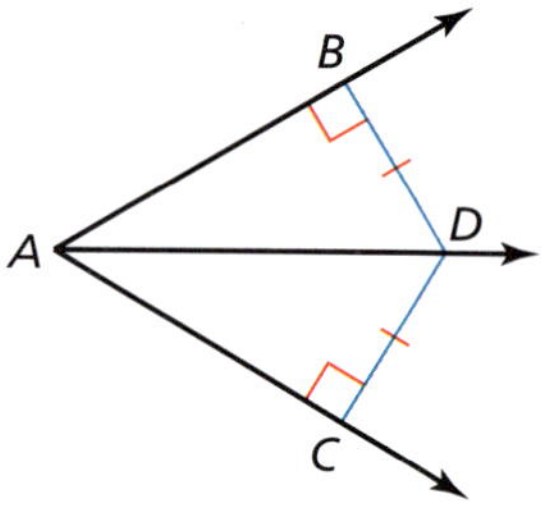

Prove this Theorem Exercise 31(b), page 303

EXAMPLE 3 Using a Diagram

WATCH

A soccer goalie stands in the interior of $\angle LBR$, which is formed by $\overrightarrow{BL}$ and $\overrightarrow{BR}$, the paths from the ball to the goalposts. Why might the goalie want to stand on the bisector of $\angle LBR$?

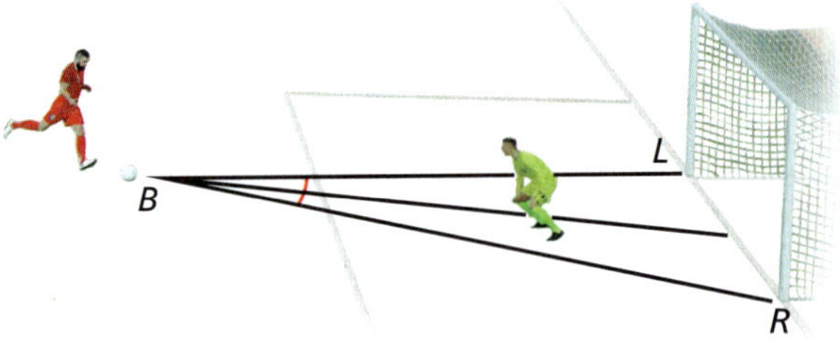

SOLUTION

By the Angle Bisector Theorem, if the goalie stands on the bisector of $\angle LBR$, then the goalie is equidistant from $\overrightarrow{BL}$ and $\overrightarrow{BR}$.

▶ So, the goalie might want to stand on the bisector of $\angle LBR$ in order to move the same distance to block a shot toward either goalpost.

SELF-ASSESSMENT

1 I don't understand yet. 2 I can do it with help. 3 I can do it on my own. 4 I can teach someone else.

6. Is there enough information in the figure to conclude that $\overrightarrow{QS}$ bisects $\angle PQR$? Explain.

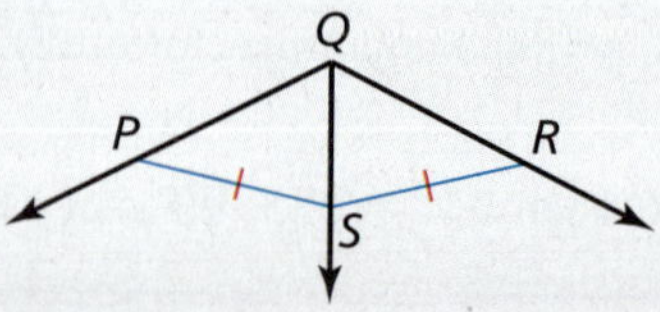

GO DIGITAL

EXAMPLE 4 Using the Angle Bisector Theorems

WATCH

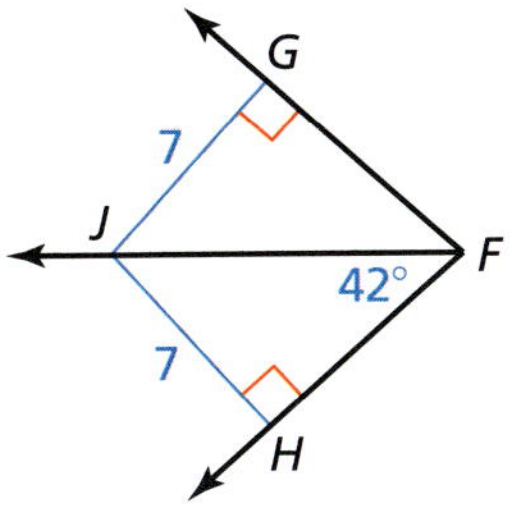

Find each measure.

a. $m\angle GFJ$

Because $\overline{JG} \perp \overrightarrow{FG}$ and $\overline{JH} \perp \overrightarrow{FH}$ and $JG = JH = 7$, $\overrightarrow{FJ}$ bisects $\angle GFH$ by the Converse of the Angle Bisector Theorem.

▶ So, $m\angle GFJ = m\angle HFJ = 42°$.

b. RS

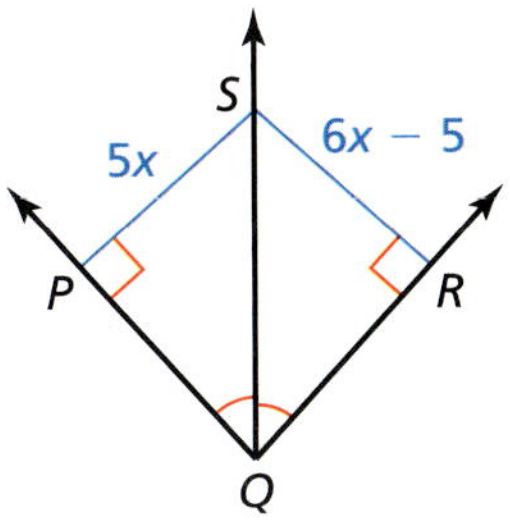

From the figure, $\overrightarrow{QS}$ is the angle bisector of $\angle PQR$.

$PS = RS$	Angle Bisector Theorem
$5x = 6x - 5$	Substitute.
$5 = x$	Solve for x.

▶ So, $RS = 6x - 5 = 6(5) - 5 = 25$.

Writing Equations of Perpendicular Bisectors

EXAMPLE 5 Writing an Equation of a Bisector

WATCH

Write an equation of the perpendicular bisector of the segment with endpoints $P(-2, 3)$ and $Q(4, 1)$.

SOLUTION

Step 1 Graph $\overline{PQ}$. By definition, the perpendicular bisector of $\overline{PQ}$ is perpendicular to $\overline{PQ}$ at its midpoint.

Step 2 Find the midpoint M of $\overline{PQ}$.

$$M\left(\frac{-2+4}{2}, \frac{3+1}{2}\right) = M\left(\frac{2}{2}, \frac{4}{2}\right) = M(1, 2)$$

Step 3 Find the slope of the perpendicular bisector.

$$\text{slope of } \overline{PQ} = \frac{1-3}{4-(-2)} = \frac{-2}{6} = -\frac{1}{3}$$

Because the slopes of perpendicular lines are negative reciprocals, the slope of the perpendicular bisector is 3.

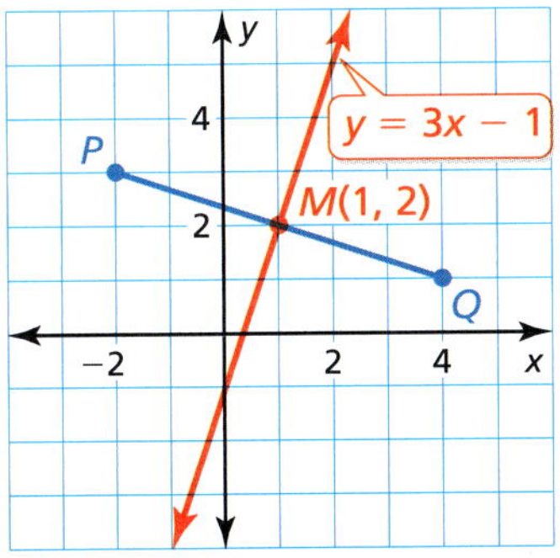

Step 4 Write an equation. The perpendicular bisector of $\overline{PQ}$ has slope 3 and passes through (1, 2).

$y = mx + b$	Use slope-intercept form.
$2 = 3(1) + b$	Substitute for m, x, and y.
$-1 = b$	Solve for b.

▶ So, an equation of the perpendicular bisector of $\overline{PQ}$ is $y = 3x - 1$.

SELF-ASSESSMENT

1 I don't understand yet. 2 I can do it with help. 3 I can do it on my own. 4 I can teach someone else.

Use the figure and the given information to find the indicated measure.

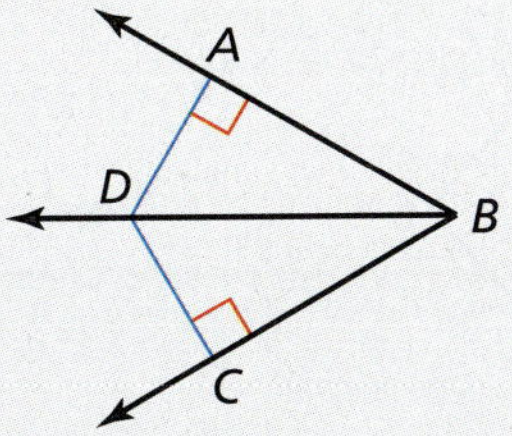

7. $\overrightarrow{BD}$ bisects $\angle ABC$, $AD = 3z + 7$, and $CD = 2z + 11$. Find CD.

8. Find $m\angle ABC$ when $AD = 3.2$, $CD = 3.2$, and $m\angle DBC = 39°$.

9. Write an equation of the perpendicular bisector of the segment with endpoints $(-1, -5)$ and $(3, -1)$.

GO DIGITAL

6.1 Practice WITH CalcChat® AND CalcView®

In Exercises 1–4, tell whether the information in the figure allows you to conclude that point P lies on the perpendicular bisector of $\overline{LM}$. Explain your reasoning. *(See Example 1.)*

▶ **1.**

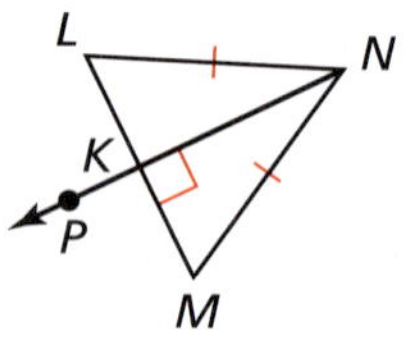

2.

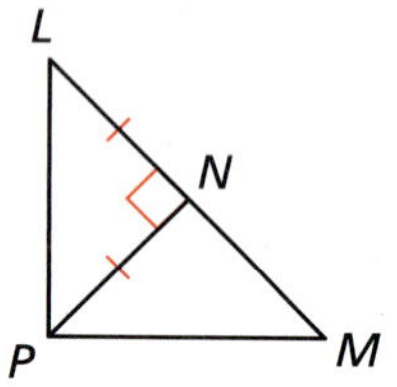

3.

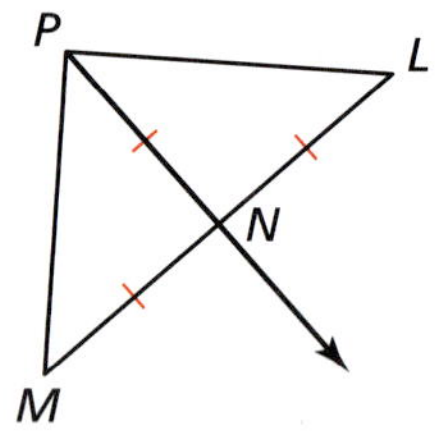

4.

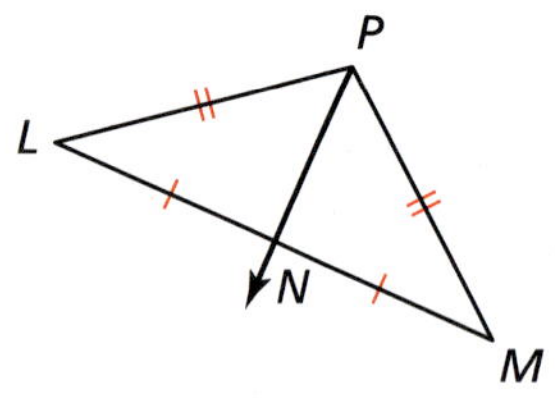

In Exercises 5–8, find the indicated measure. Explain your reasoning. *(See Example 2.)*

5. GH

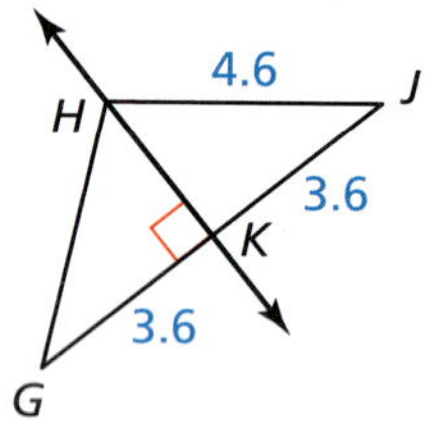

6. QR

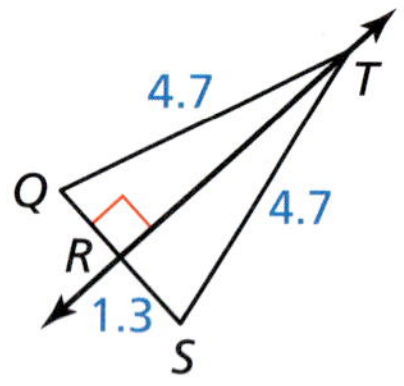

▶ **7.** AB

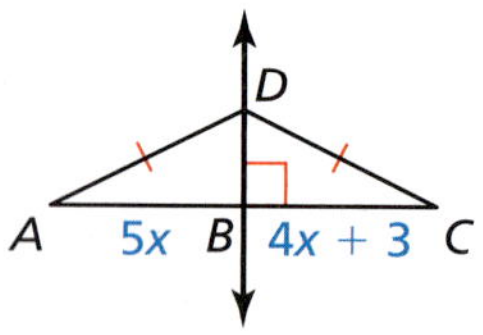

8. UW

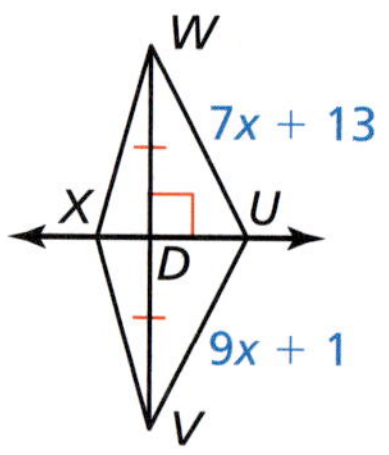

In Exercises 9 and 10, tell whether the information in the figure allows you to conclude that $\overrightarrow{EH}$ bisects $\angle FEG$. Explain your reasoning. *(See Example 3.)*

▶ **9.**

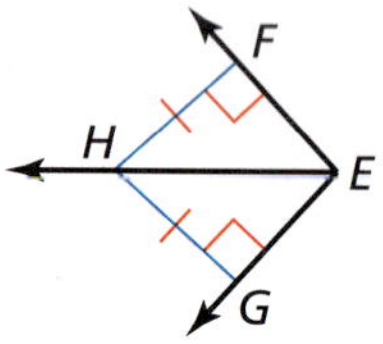

10.

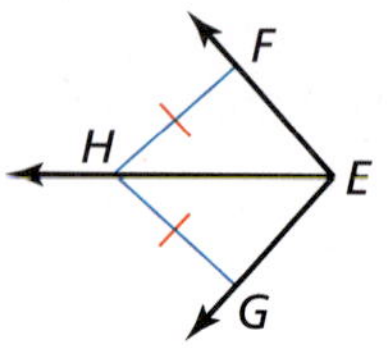

In Exercises 11 and 12, tell whether the information in the figure allows you to conclude that $DB = DC$. Explain your reasoning.

11.

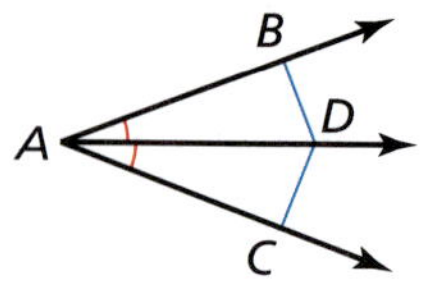

12.

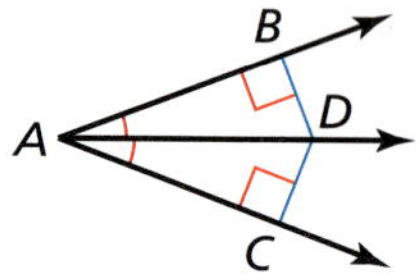

In Exercises 13–16, find the indicated measure. Explain your reasoning. *(See Example 4.)*

▶ **13.** $m\angle ABD$

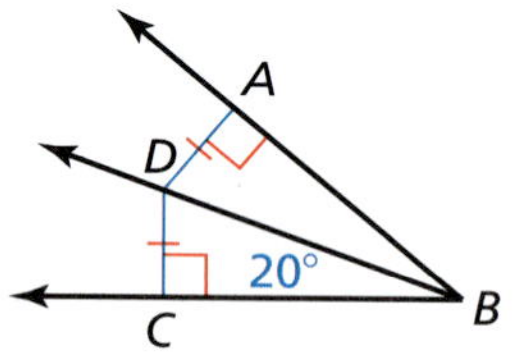

14. PS

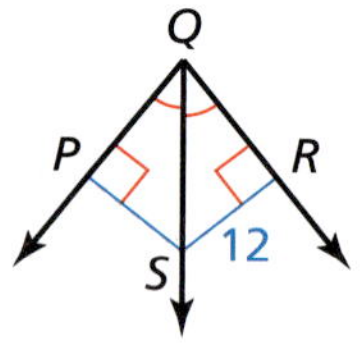

15. FG

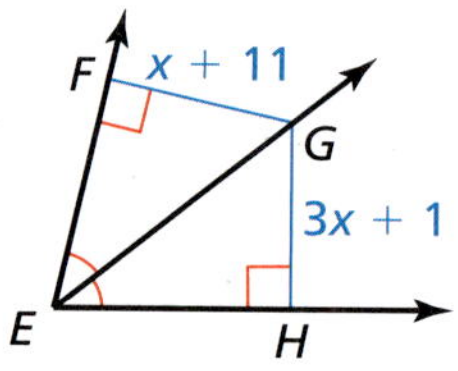

16. $m\angle KJL$

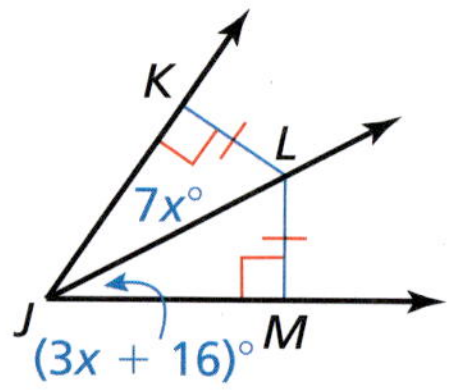

4 MTR **ERROR ANALYSIS** In Exercises 17 and 18, describe and correct the error in the student's reasoning.

17.

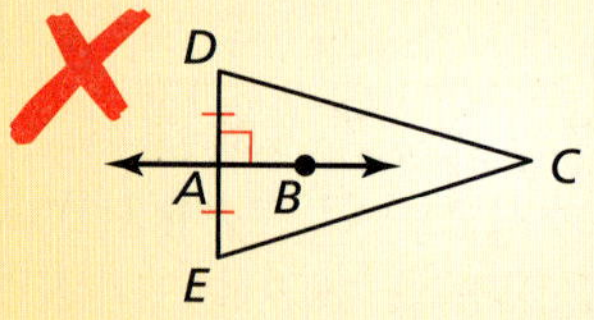

Because $AD = AE$, $\overrightarrow{AB}$ will pass through point C.

18.

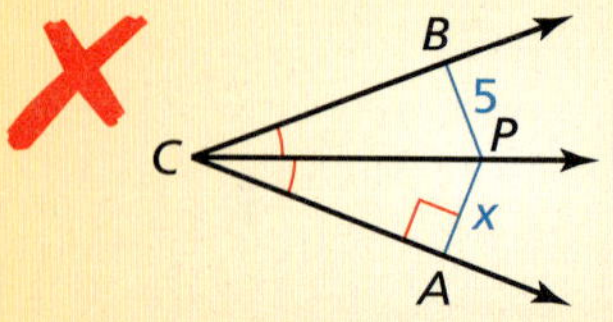

By the Angle Bisector Theorem, $x = 5$.

In Exercises 19–22, write an equation of the perpendicular bisector of the segment with the given endpoints. *(See Example 5.)*

▶ **19.** $M(1, 5)$, $N(7, -1)$

20. $Q(-2, 0)$, $R(6, 12)$

21. $U(-3, 4)$, $V(9, 8)$

22. $Y(10, -7)$, $Z(-4, 1)$

GO DIGITAL

7 MTR 23. **MODELING REAL LIFE** In the diagram, the road is perpendicular to the support beam and $\overline{AB} \cong \overline{CB}$. Can you conclude that $\overline{AD} \cong \overline{CD}$? Explain.

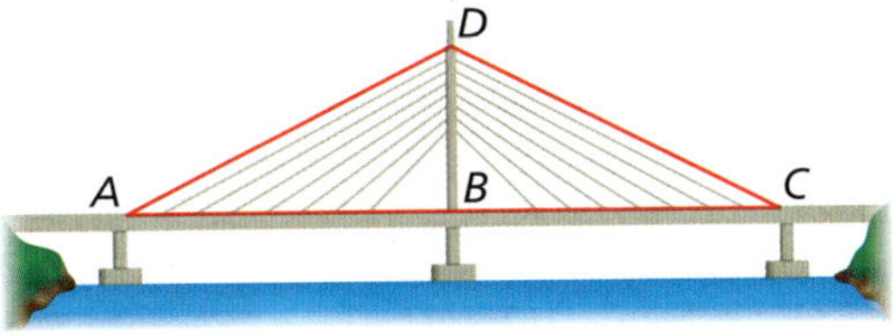

7 MTR 24. **MODELING REAL LIFE** The diagram shows the position of the goalie and the puck during a hockey game. The goalie is at point G, and the puck is at point P. $\overrightarrow{PA}$ and $\overrightarrow{PB}$ represent the paths from the puck to the goalposts.

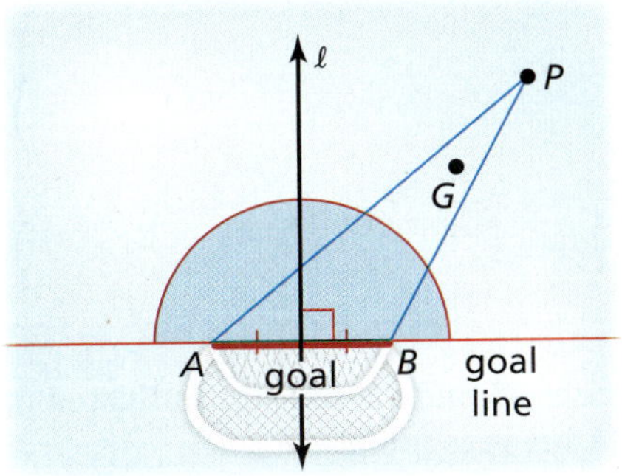

a. What should be the relationship between $\overrightarrow{PG}$ and $\angle APB$ so that the goalie is equidistant from $\overrightarrow{PA}$ and $\overrightarrow{PB}$?

b. How does $m\angle APB$ change as the puck gets closer to the goal? Does this change make it easier or more difficult for the goalie to defend the goal? Explain your reasoning.

25. **CONSTRUCTION** Use a compass and straightedge to construct a copy of $\overline{XY}$. Construct a perpendicular bisector and plot a point Z on the bisector so that the distance between point Z and $\overline{XY}$ is 3 centimeters. Measure $\overline{XZ}$ and $\overline{YZ}$. Which theorem does this construction demonstrate?

26. **WRITING** Explain how the Converse of the Perpendicular Bisector Theorem is related to the construction of a perpendicular bisector.

27. **COLLEGE PREP** What is the value of x in the figure?

Ⓐ 13

Ⓑ 18

Ⓒ 33

Ⓓ not enough information

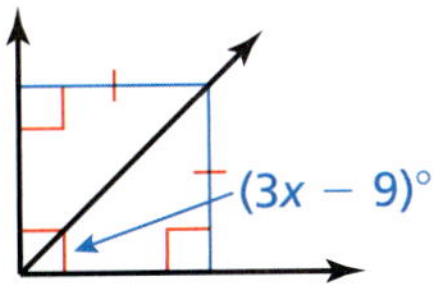

GO DIGITAL

28. **COLLEGE PREP** Which point lies on the perpendicular bisector of the segment with endpoints $M(7, 5)$ and $N(-1, 5)$?

Ⓐ (2, 0) Ⓒ (4, 1)

Ⓑ (3, 9) Ⓓ (1, 3)

4 MTR 29. **MAKING AN ARGUMENT** Is it possible for an angle bisector of a triangle to be the same line as the perpendicular bisector of the opposite side? Explain.

30. **PROVING A THEOREM** Prove the Converse of the Perpendicular Bisector Theorem. (*Hint:* Use an auxiliary line.)

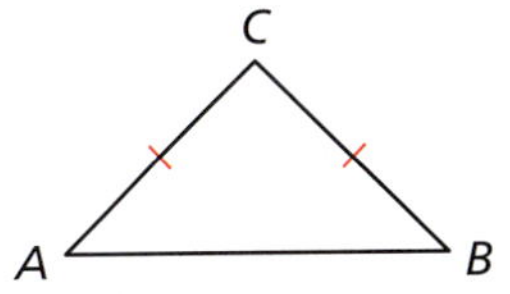

Given $CA = CB$

Prove Point C lies on the perpendicular bisector of $\overline{AB}$.

31. **PROVING A THEOREM** Use a congruence theorem to prove each theorem.

a. Angle Bisector Theorem

b. Converse of the Angle Bisector Theorem

32. **HOW DO YOU SEE IT?**
The map of a city is shown. The city is arranged so each block north to south is the same length and each block east to west is the same length. You are equidistant from the two hospitals. Describe your possible locations. Use a theorem to explain your reasoning.

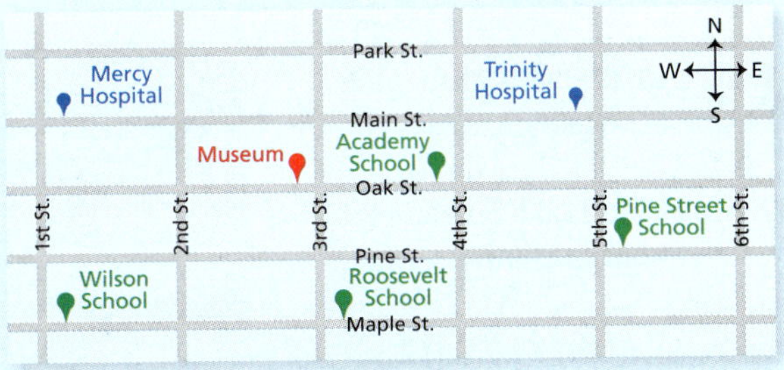

5 MTR 33. **CONNECTING CONCEPTS** Write an equation whose graph consists of all the points in the given quadrants that are equidistant from the x- and y-axes.

a. I and III

b. II and IV

c. I and II

d. III and IV

34. PROOF Use the information in the figure to prove that $\overline{AB} \cong \overline{CB}$ if and only if points D, E, and B are collinear.

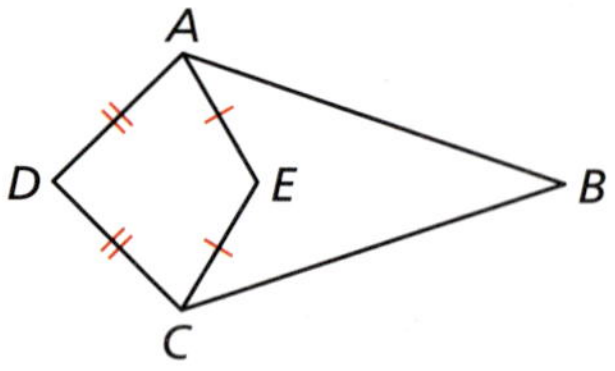

35. PROOF In the figure, plane P is a perpendicular bisector of $\overline{XZ}$ at point Y. Prove that $\angle VXW \cong \angle VZW$.

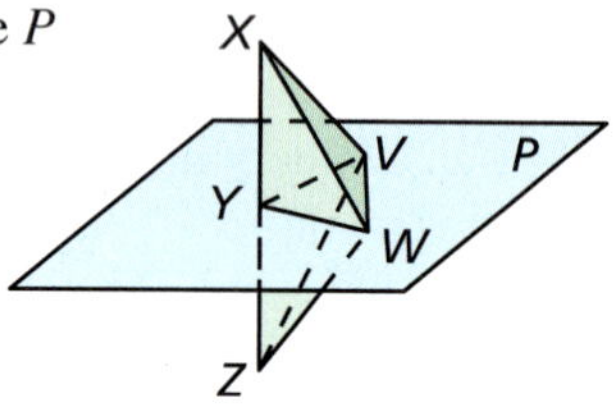

36. THOUGHT PROVOKING
The postulates and theorems in this book represent Euclidean geometry. In spherical geometry, all points are on the surface of a sphere. A line is a circle on the sphere whose diameter is equal to the diameter of the sphere. In spherical geometry, is it possible for two lines to be perpendicular but not bisect each other? Explain your reasoning.

37. DIG DEEPER $\triangle J'K'L'$ is a rotation of $\triangle JKL$. Explain how to find the center of rotation.

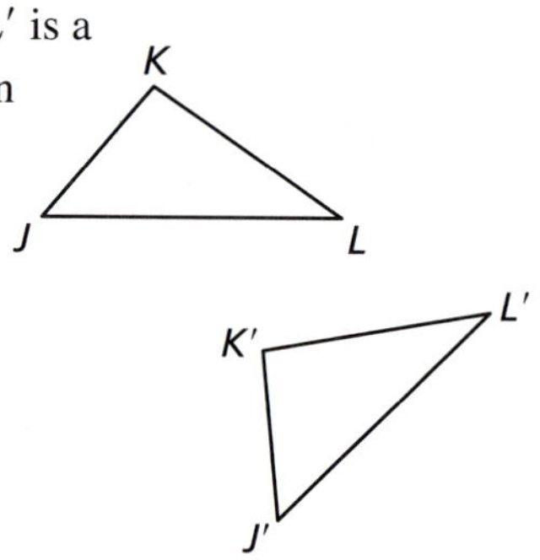

REVIEW & REFRESH

WATCH

In Exercises 38 and 39, classify the triangle by its sides.

38.

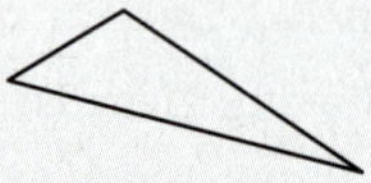

39.

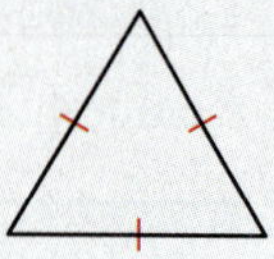

In Exercises 40 and 41, classify the triangle by its angles.

40.

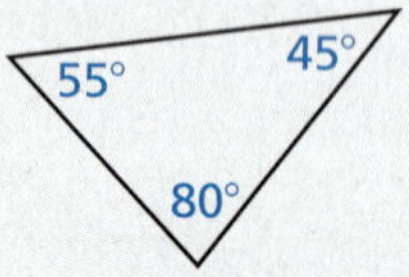

41.

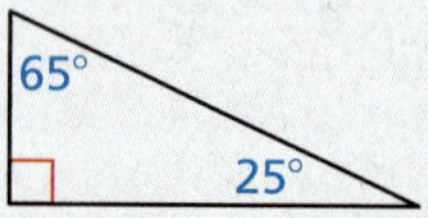

7 MTR **42. MODELING REAL LIFE** A wooden gate is designed as shown. Find $m\angle 2$. Explain your reasoning.

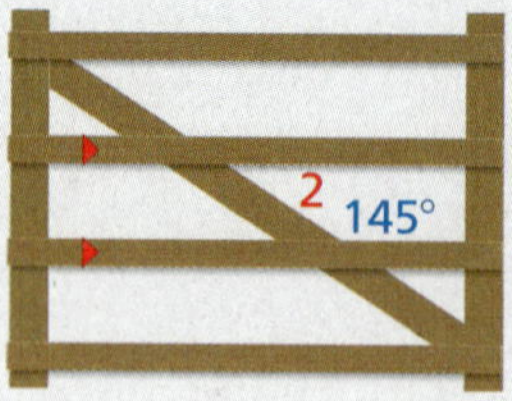

43. Use the given information to write a plan for proof.

Given $\overline{AD} \cong \overline{CD}$, $\overline{AE} \cong \overline{CE}$

Prove $\triangle BDA \cong \triangle BDC$

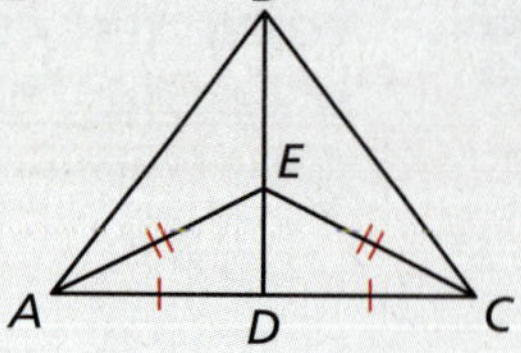

44. Find the product of $-2x^2$ and $3x^4 + 12x^3 - 14$.

In Exercises 45 and 46, find the indicated measure. Explain your reasoning.

45. QP

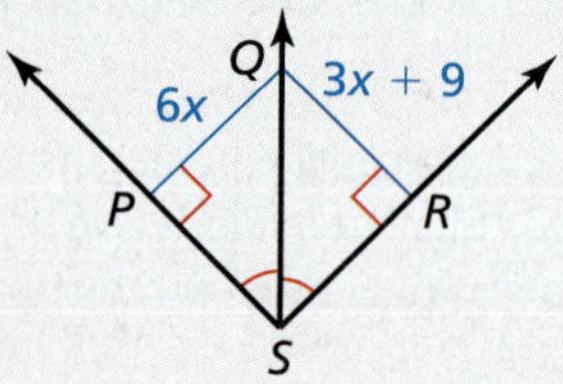

46. UW

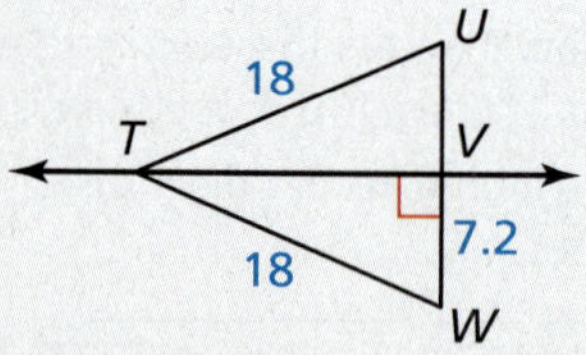

47. In $\triangle RST$ and $\triangle XYZ$, $\overline{RS} \cong \overline{XY}$ and $\angle R \cong \angle X$. What is the third congruence statement that is needed to prove that $\triangle RST \cong \triangle XYZ$ using the ASA Congruence Theorem? the AAS Congruence Theorem?

48. Graph the square with the vertices $A(0, 0)$, $B(0, k)$, $C(k, k)$, and $D(k, 0)$. Then find the coordinates of the midpoint of each side.

49. Evaluate $4096^{1/6}$.

GO DIGITAL

6.2 Bisectors of Triangles

Learning Target: Use bisectors of triangles.

Success Criteria:
- I can find the circumcenter and incenter of a triangle.
- I can circumscribe a circle about a triangle.
- I can inscribe a circle within a triangle.
- I can use points of concurrency to solve real-life problems.

EXPLORE IT! Analyzing Bisectors of Triangles

Work with a partner.

a. Use technology to draw any triangle and the perpendicular bisectors of all three sides of the triangle. What do you notice? What happens when you move the vertices of the triangle?

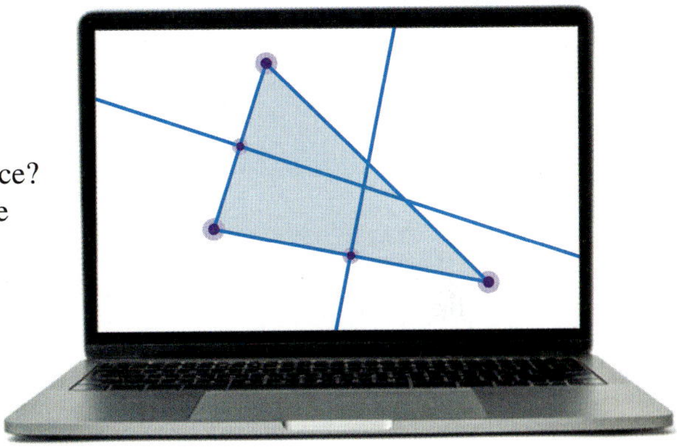

b. Draw the circle with its center at the intersection of the perpendicular bisectors that passes through a vertex of the triangle. What do you notice? What does this mean?

c. Use technology to draw a different triangle and the angle bisectors of all three angles of the triangle. What do you notice? What happens when you move the vertices of the triangle?

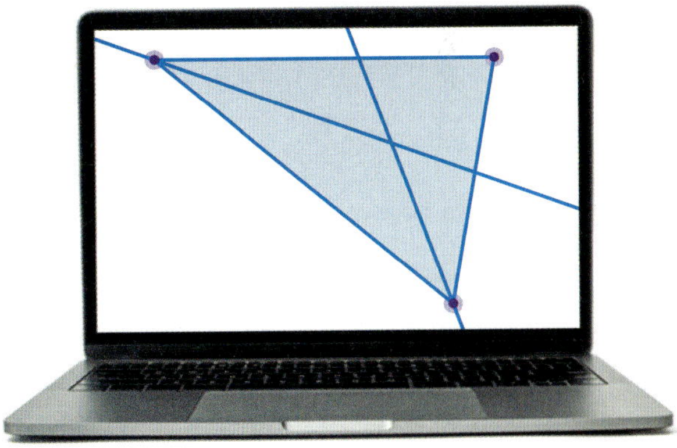

d. Find the distance r between the intersection of the angle bisectors and one of the sides of the triangle. Draw the circle with its center at the intersection of the angle bisectors and radius r. What do you notice? What does this mean?

e. What conjectures can you make using your results in parts (a)–(d)? Write your conjectures as conditional statements written in if-then form.

4 MTR CONSTRUCT AN ARGUMENT
In part (d), does it matter which side you choose when finding r? Explain your reasoning.

Geometric Reasoning
MA.912.GR.1.3 Prove relationships and theorems about triangles. Solve mathematical and real-world problems involving postulates, relationships and theorems of triangles.
MA.912.GR.5.3 Construct the inscribed and circumscribed circles of a triangle.

GO DIGITAL

Using the Circumcenter of a Triangle

Vocabulary (VOCAB)

concurrent, *p. 306*
point of concurrency, *p. 306*
circumcenter, *p. 306*
incenter, *p. 309*

When three or more lines, rays, or segments intersect in the same point, they are called **concurrent** lines, rays, or segments. The point of intersection of the lines, rays, or segments is called the **point of concurrency**.

In a triangle, the three perpendicular bisectors are concurrent. The point of concurrency is the **circumcenter** of the triangle.

THEOREM

6.5 Circumcenter Theorem

The circumcenter of a triangle is equidistant from the vertices of the triangle.

If $\overline{PD}$, $\overline{PE}$, and $\overline{PF}$ are perpendicular bisectors, then $PA = PB = PC$.

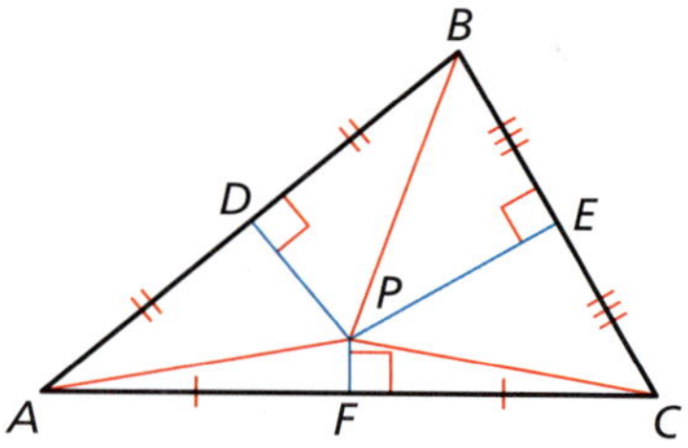

PROOF Circumcenter Theorem

Given $\triangle ABC$; the perpendicular bisectors of $\overline{AB}$, $\overline{BC}$, and $\overline{AC}$

Prove The perpendicular bisectors intersect in a point; that point is equidistant from A, B, and C.

Plan for Proof Show that P, the point of intersection of the perpendicular bisectors of $\overline{AB}$ and $\overline{BC}$, also lies on the perpendicular bisector of $\overline{AC}$. Then show that point P is equidistant from the vertices of the triangle.

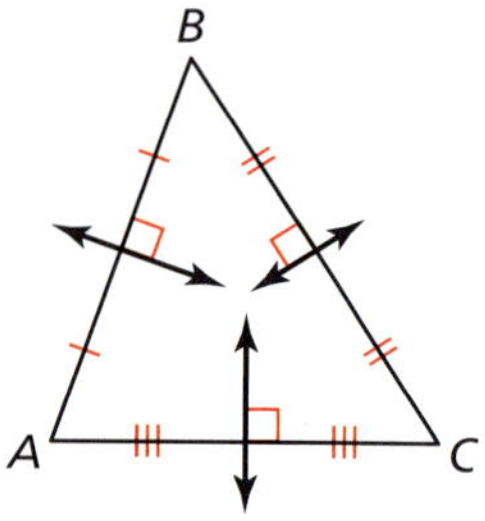

4 MTR COMPARE METHODS
Is a flowchart proof a reasonable format for this proof? Explain your reasoning.

Plan in Action

STATEMENTS	REASONS
1. $\triangle ABC$; the perpendicular bisectors of $\overline{AB}$, $\overline{BC}$, and $\overline{AC}$	**1.** Given
2. The perpendicular bisectors of $\overline{AB}$ and $\overline{BC}$ intersect at some point P.	**2.** Because the sides of a triangle cannot be parallel, these perpendicular bisectors must intersect in some point. Call it P.
3. Draw $\overline{PA}$, $\overline{PB}$, and $\overline{PC}$.	**3.** Two Point Postulate
4. $PA = PB$, $PB = PC$	**4.** Perpendicular Bisector Theorem
5. $PA = PC$	**5.** Transitive Property of Equality
6. P is on the perpendicular bisector of $\overline{AC}$.	**6.** Converse of the Perpendicular Bisector Theorem
7. $PA = PB = PC$. So, P is equidistant from the vertices of the triangle.	**7.** From the results of Steps 4 and 5, and the definition of equidistant

STUDY TIP
Use diagrams like the one below to help visualize your proof.

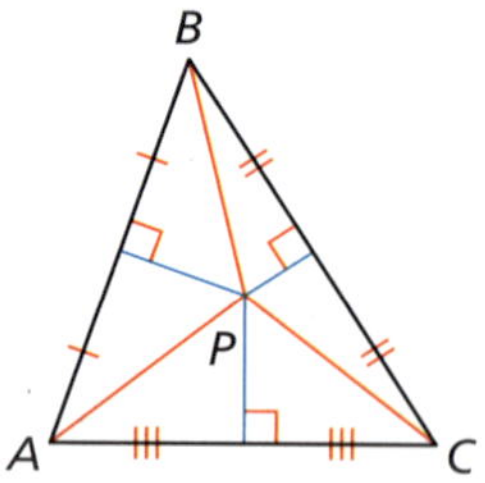

GO DIGITAL

EXAMPLE 1 Modeling Real Life

WATCH

Three snack carts sell frozen yogurt at points A, B, and C in a city. Each of the three carts is the same distance from the frozen yogurt distributor. Find the location of the distributor.

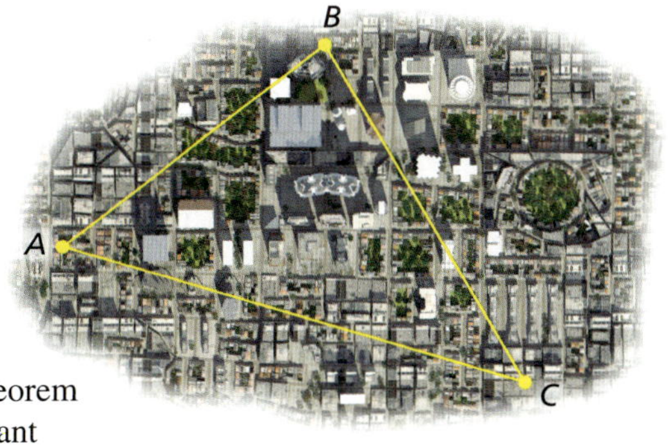

SOLUTION

The distributor is equidistant from the three snack carts. The Circumcenter Theorem shows that you can find a point equidistant from three points by using the perpendicular bisectors of the triangle formed by the three points.

Copy points A, B, and C and connect them to draw $\triangle ABC$. Then use a ruler and protractor to draw the three perpendicular bisectors of $\triangle ABC$. The circumcenter D is the location of the distributor.

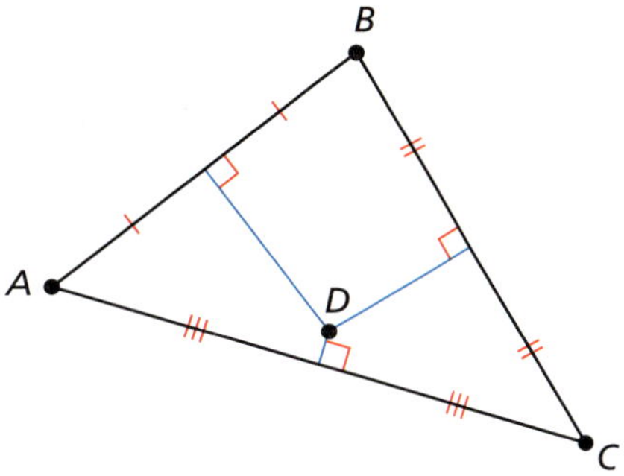

SELF-ASSESSMENT

1 I don't understand yet. 2 I can do it with help. 3 I can do it on my own. 4 I can teach someone else.

1. Three snack carts sell hot pretzels at points A, B, and E. Each of the three carts is the same distance from the pretzel distributor. Find the location of the distributor.

READING

The prefix *circum-* means "around" or "about," as in *circumference* (distance around a circle).

The circumcenter P is equidistant from the three vertices, so P is the center of a circle that passes through all three vertices. As shown below, the location of P depends on the type of triangle. The circle with center P is said to be *circumscribed* about the triangle.

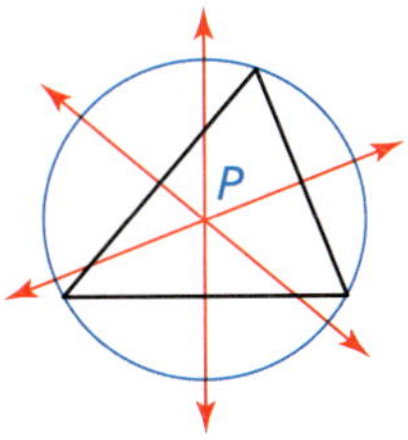

Acute triangle
P is inside triangle.

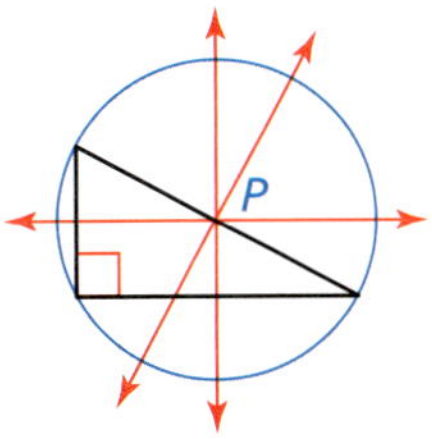

Right triangle
P is on triangle.

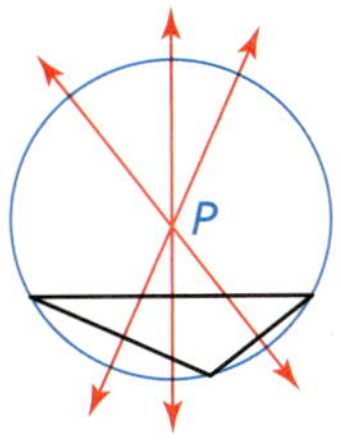

Obtuse triangle
P is outside triangle.

GO DIGITAL

CONSTRUCTION Circumscribing a Circle About a Triangle

WATCH

Use a compass and straightedge to construct a circle that is circumscribed about $\triangle ABC$.

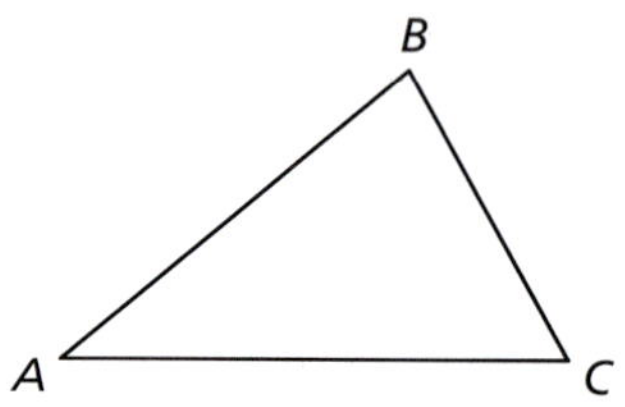

SOLUTION

Step 1

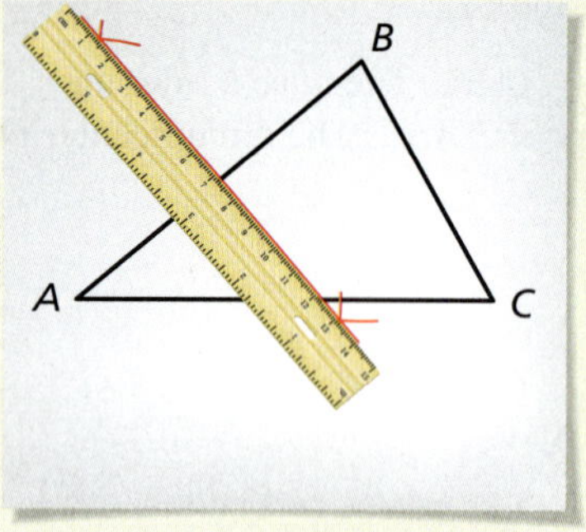

Draw a perpendicular bisector Draw the perpendicular bisector of $\overline{AB}$.

Step 2

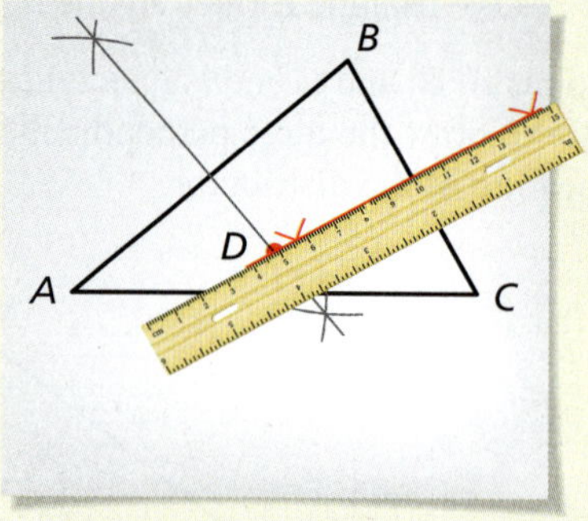

Draw a perpendicular bisector Draw the perpendicular bisector of $\overline{BC}$. Label the intersection of the bisectors D. This is the circumcenter.

Step 3

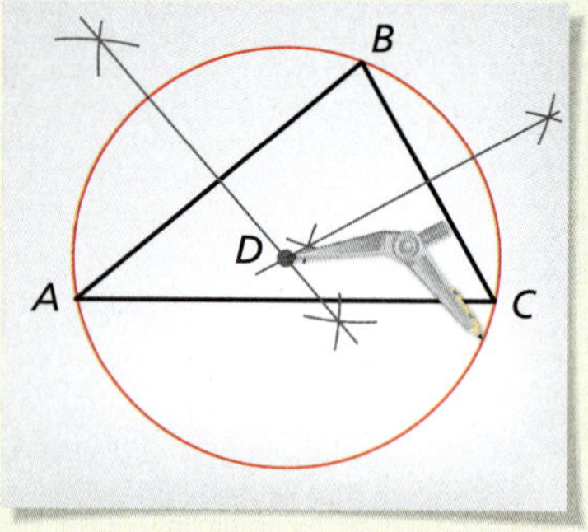

Draw a circle Place the compass at D. Set the width by using any vertex of the triangle. This is the radius of the *circumcircle*. Draw the circle. It should pass through all three vertices A, B, and C.

EXAMPLE 2 Finding the Circumcenter of a Triangle

WATCH

Find the coordinates of the circumcenter of $\triangle ABC$ with vertices $A(0, 3)$, $B(0, -1)$, and $C(6, -1)$.

SOLUTION

Step 1 Graph $\triangle ABC$.

Step 2 Find equations of two perpendicular bisectors. Use the Slopes of Perpendicular Lines Theorem, which states that horizontal lines are perpendicular to vertical lines.

The midpoint of $\overline{AB}$ is $(0, 1)$. The line through $(0, 1)$ that is perpendicular to $\overline{AB}$ is $y = 1$.

The midpoint of $\overline{BC}$ is $(3, -1)$. The line through $(3, -1)$ that is perpendicular to $\overline{BC}$ is $x = 3$.

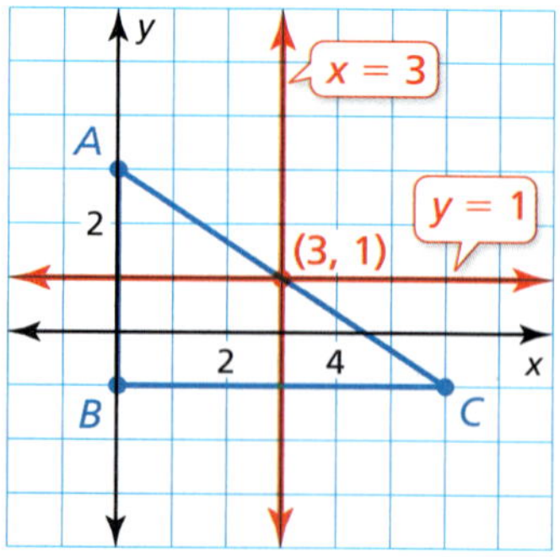

STUDY TIP

You need to find the equations of only *two* perpendicular bisectors. You can use the perpendicular bisector of the third side to verify your result.

Step 3 Find the point where $x = 3$ and $y = 1$ intersect. They intersect at $(3, 1)$.

▶ So, the coordinates of the circumcenter are $(3, 1)$.

GO DIGITAL

Using the Incenter of a Triangle

Just as a triangle has three perpendicular bisectors, it also has three angle bisectors. The angle bisectors of a triangle are also concurrent. This point of concurrency is the **incenter** of the triangle. For any triangle, the incenter always lies inside the triangle.

THEOREM

6.6 Incenter Theorem

The incenter of a triangle is equidistant from the sides of the triangle.

If $\overline{AP}$, $\overline{BP}$, and $\overline{CP}$ are angle bisectors of $\triangle ABC$, then $PD = PE = PF$.

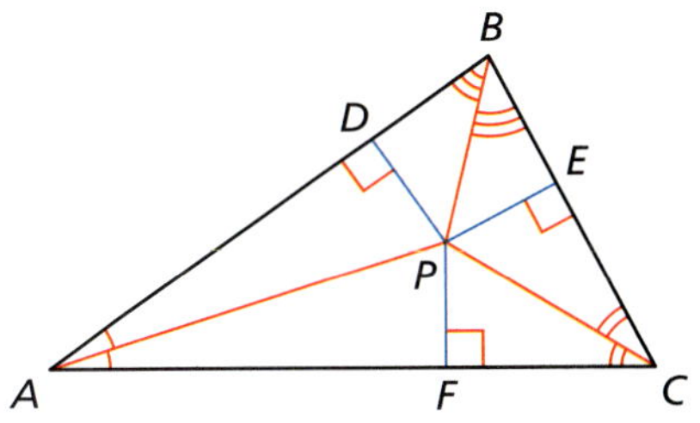

Prove this Theorem Exercise 38, page 313

EXAMPLE 3 Using the Incenter of a Triangle

WATCH

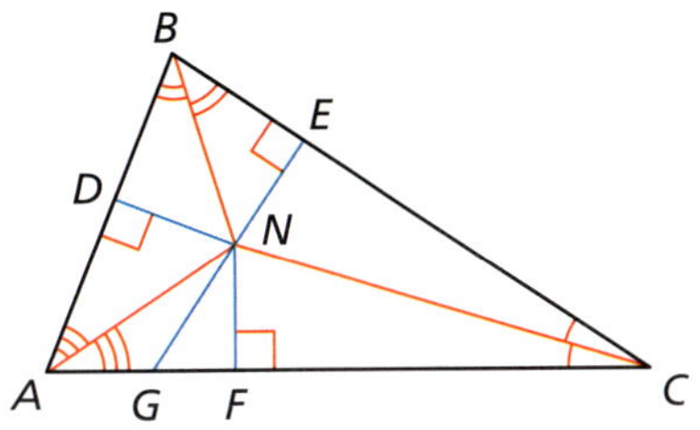

In the figure shown, $ND = 5x - 1$ and $NE = 2x + 11$.

a. Find NF.

b. Can NG be equal to 18? Explain your reasoning.

SOLUTION

a. Point N is the incenter of $\triangle ABC$ because it is the point of concurrency of the three angle bisectors. So, by the Incenter Theorem, $ND = NE = NF$.

Step 1 Solve for x.

$ND = NE$	Incenter Theorem
$5x - 1 = 2x + 11$	Substitute.
$x = 4$	Solve for x.

Step 2 Find ND (or NE).

$$ND = 5x - 1 = 5(4) - 1 = 19$$

▶ So, because $ND = NF$, $NF = 19$.

b. Recall that the shortest distance between a point and a line is the length of a perpendicular segment. In this case, the perpendicular segment is $\overline{NF}$, which has a length of 19. Because $18 < 19$, NG cannot be equal to 18.

SELF-ASSESSMENT

1 I don't understand yet. 2 I can do it with help. 3 I can do it on my own. 4 I can teach someone else.

Find the coordinates of the circumcenter of the triangle with the given vertices.

2. $R(-2, 5)$, $S(-6, 5)$, $T(-2, -1)$
3. $W(-1, 4)$, $X(1, 4)$, $Y(1, -6)$
4. In the figure shown, $QM = 3x + 8$ and $QN = 7x + 2$. Find QP.

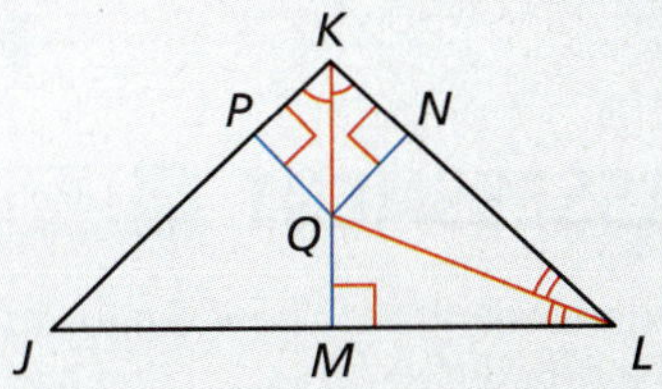

GO DIGITAL

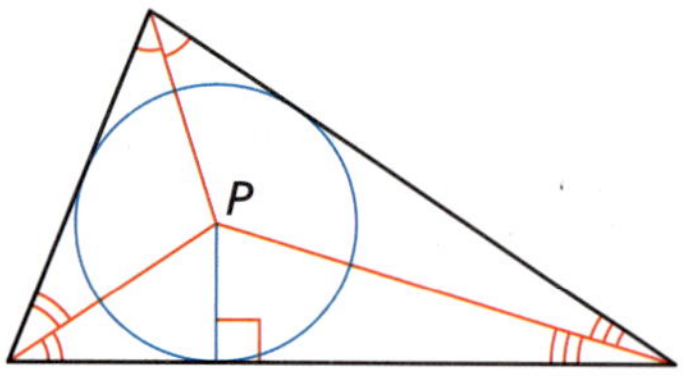

Because the incenter P is equidistant from the three sides of the triangle, a circle drawn using P as the center and the distance to one side of the triangle as the radius will just touch the other two sides of the triangle. The circle is said to be *inscribed* within the triangle.

CONSTRUCTION Inscribing a Circle Within a Triangle

WATCH

Use a compass and straightedge to construct a circle that is inscribed within $\triangle ABC$.

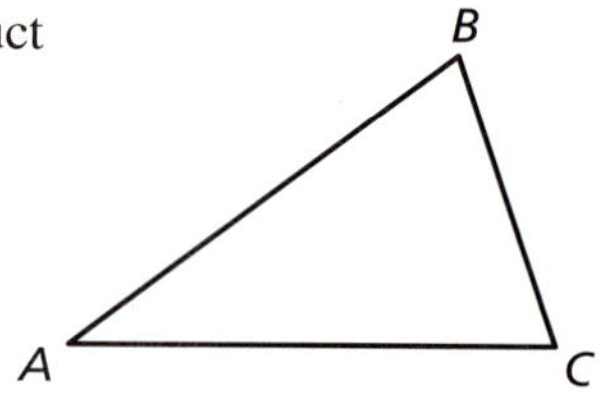

SOLUTION

Step 1

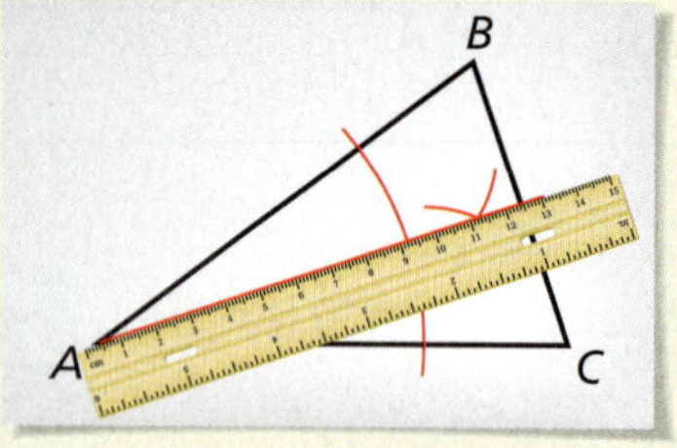

Draw an angle bisector Draw the angle bisector of $\angle A$.

Step 2

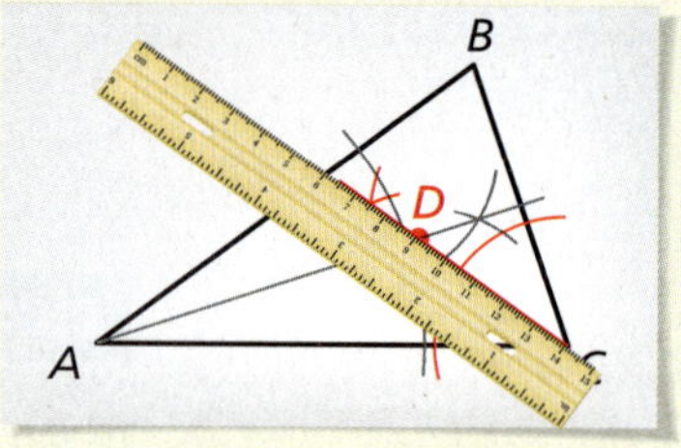

Draw an angle bisector Draw the angle bisector of $\angle C$. Label the intersection of the bisectors D. This is the incenter.

Step 3

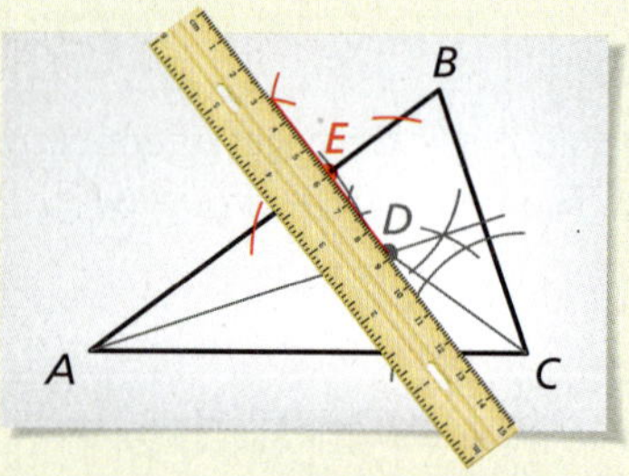

Draw a perpendicular line Draw the perpendicular line from D to $\overline{AB}$. Label the point where it intersects $\overline{AB}$ as E.

Step 4

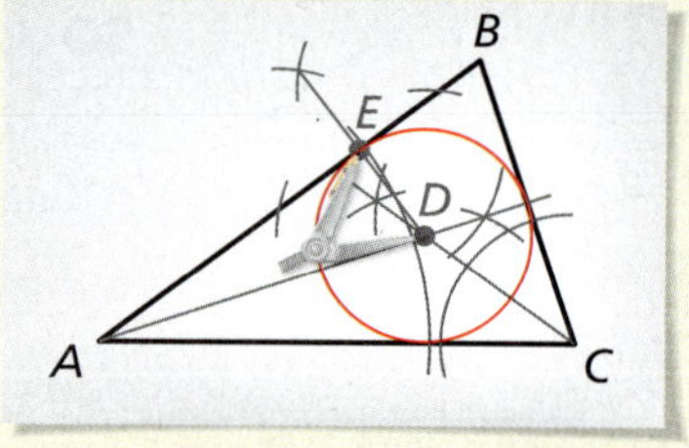

Draw a circle Place the compass at D. Set the width to E. This is the radius of the *incircle*. Draw the circle. It should touch each side of the triangle.

7 MTR

EXAMPLE 4 Modeling Real Life

WATCH

City officials want to place a lamppost near the streets shown so that the lamppost is the same distance from all three streets. Should the lamppost be at the *circumcenter* or *incenter* of the triangular piece of land? Explain.

SOLUTION

Because the shape of the land is an obtuse triangle, the circumcenter lies outside the triangle and is not equidistant from the sides of the triangle. By the Incenter Theorem, the incenter of the triangle is equidistant from the sides of the triangle.

▶ So, the lamppost should be at the incenter of the triangular piece of land.

SELF-ASSESSMENT

1 I don't understand yet. 2 I can do it with help. 3 I can do it on my own. 4 I can teach someone else.

5. Draw a sketch to show the location L of the lamppost in Example 4.

GO DIGITAL

6.2 Practice WITH CalcChat® AND CalcView®

In Exercises 1 and 2, the perpendicular bisectors of $\triangle ABC$ intersect at point G and are shown in blue. Find the indicated measure.

1. BG

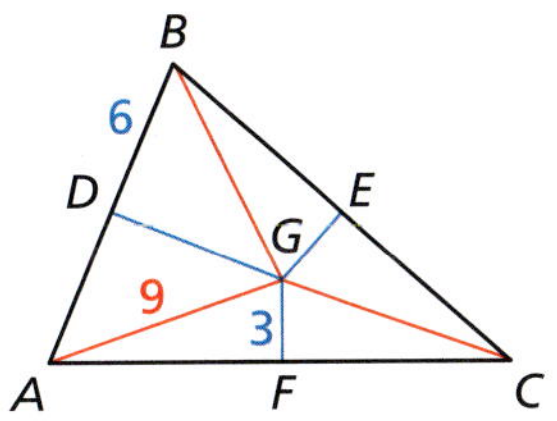

2. GA

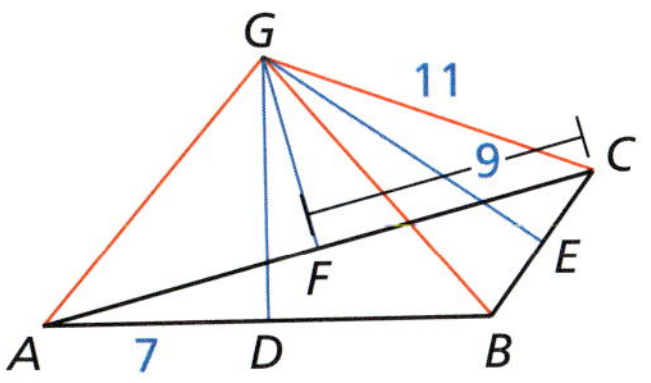

3. MODELING REAL LIFE You and two friends plan to walk your dogs together. You want the meeting place to be the same distance from each person's residence. Explain how you can use the diagram to locate the meeting place. *(See Example 1.)*

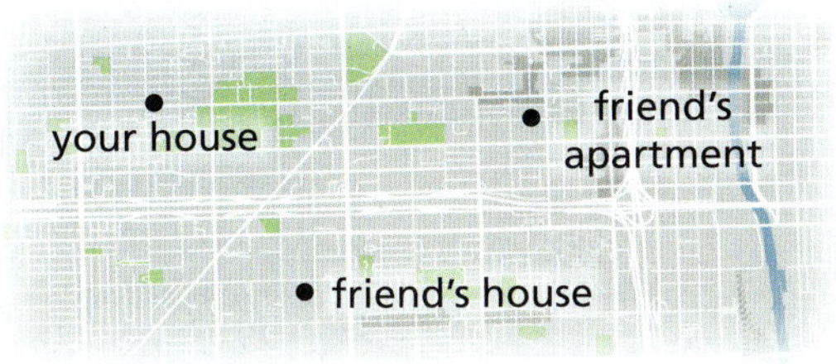

4. MODELING REAL LIFE You open a donation center the same distance from each of the three stores shown. Use the diagram to determine the location of the donation center.

In Exercises 5 and 6, the angle bisectors of $\triangle XYZ$ intersect at point P and are shown in red. Find the indicated measure.

5. PB

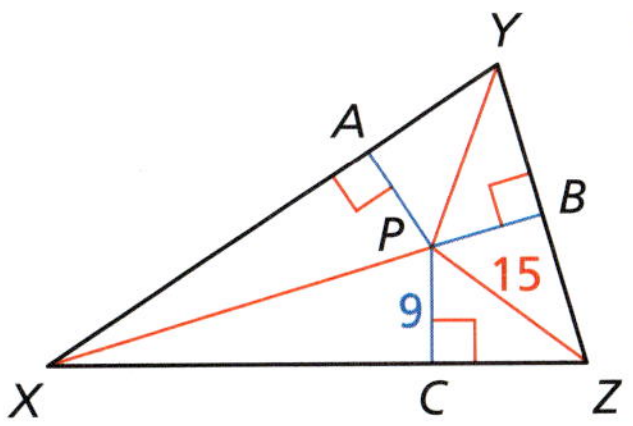

6. HP

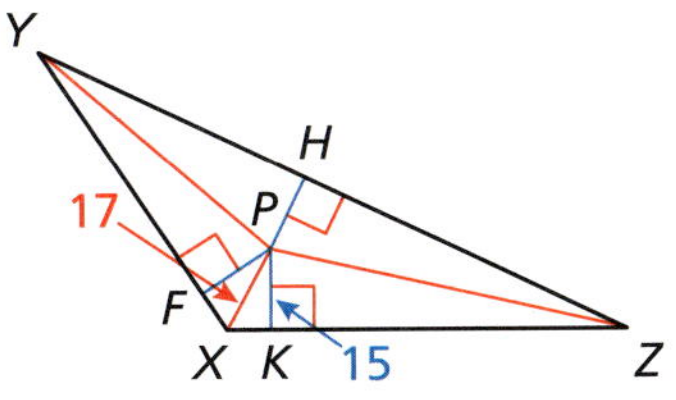

GO DIGITAL

In Exercises 7–10, find the coordinates of the circumcenter of the triangle with the given vertices. *(See Example 2.)*

7. $A(0, 0)$, $B(0, 8)$, $C(6, 0)$

8. $A(2, 2)$, $B(2, 4)$, $C(8, 4)$

9. $H(-10, 7)$, $J(-6, 3)$, $K(-2, 3)$

10. $L(3, -6)$, $M(5, -3)$, $N(8, -6)$

In Exercises 11–14, point N is the incenter of $\triangle ABC$. Use the given information to find the indicated measure. *(See Example 3.)*

11. $ND = 6x - 2$
$NE = 3x + 7$
Find NF.

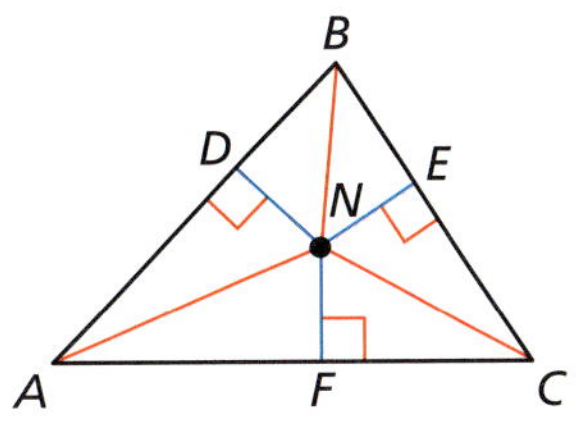

12. $NG = x + 3$
$NH = 2x - 3$
Find NJ.

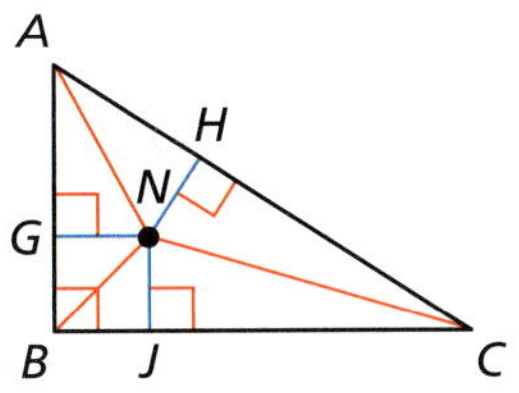

13. $NK = 2x - 2$
$NL = -x + 10$
Find NM.

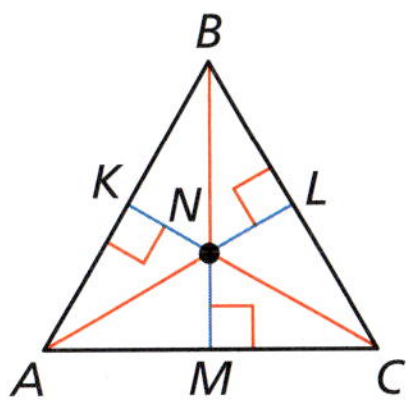

14. $NQ = 2x$
$NR = 3x - 2$
Find NS.

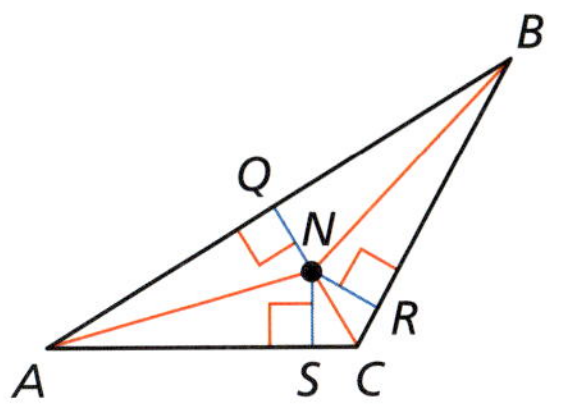

15. Point P is the circumcenter of $\triangle XYZ$. Use the given information to find PZ.

$PX = 3x + 2$
$PY = 4x - 8$

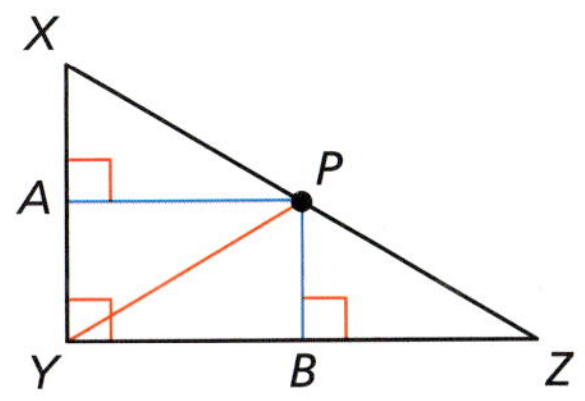

16. Point P is the circumcenter of $\triangle XYZ$. Use the given information to find PY.

$PX = 4x + 3$
$PZ = 6x - 11$

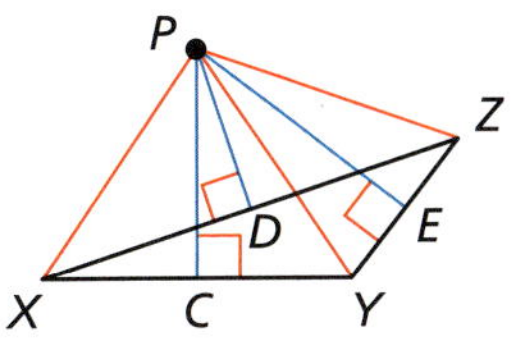

CONSTRUCTION In Exercises 17–20, draw a triangle of the given type. Find the circumcenter. Then construct the circumscribed circle.

17. right

18. obtuse

19. acute isosceles

20. equilateral

CONSTRUCTION In Exercises 21–24, copy the triangle with the given angle measures. Find the incenter. Then construct the inscribed circle.

21.

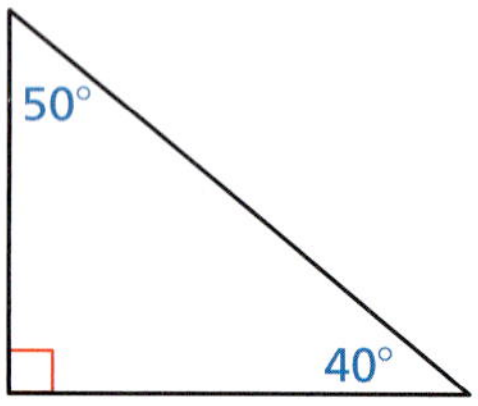

22.

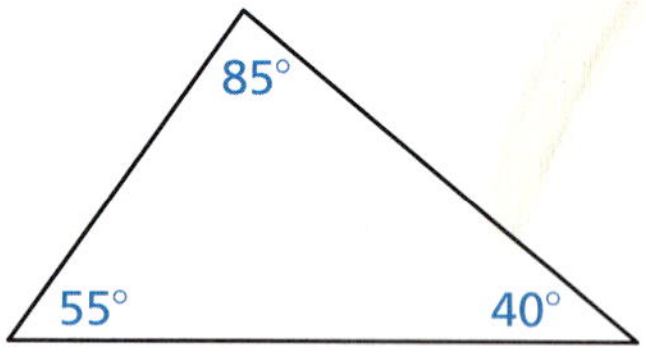

23.

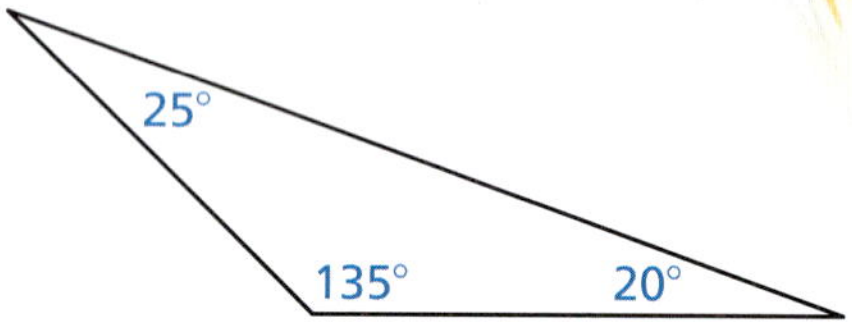

24.

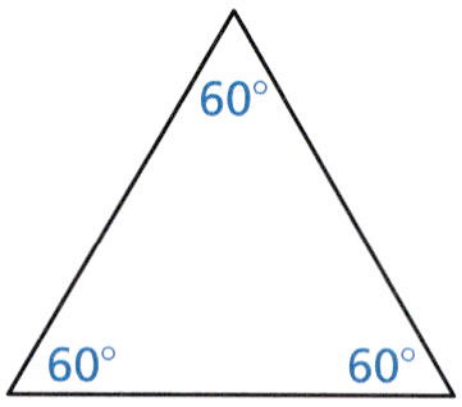

ERROR ANALYSIS In Exercises 25 and 26, describe and correct the error in identifying equal distances.

25.

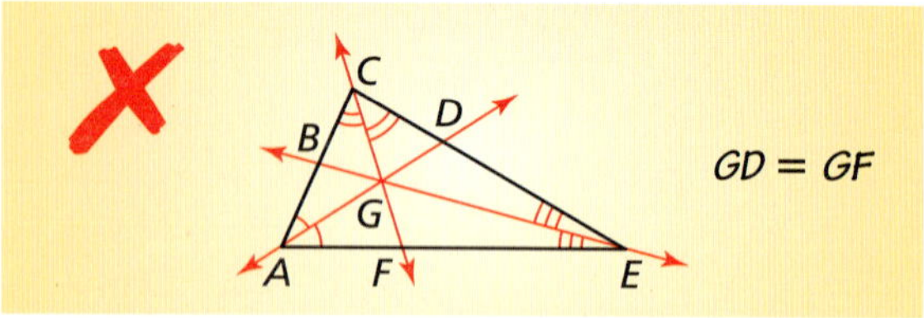

26.

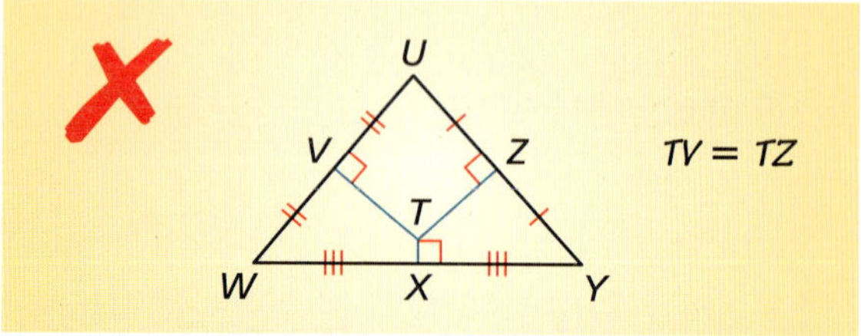

27. MODELING REAL LIFE You are placing a fountain in a triangular koi pond. You want the fountain to be the same distance from each side of the pond. Should the fountain be at the *circumcenter* or *incenter* of the triangular pond? Explain. *(See Example 4.)*

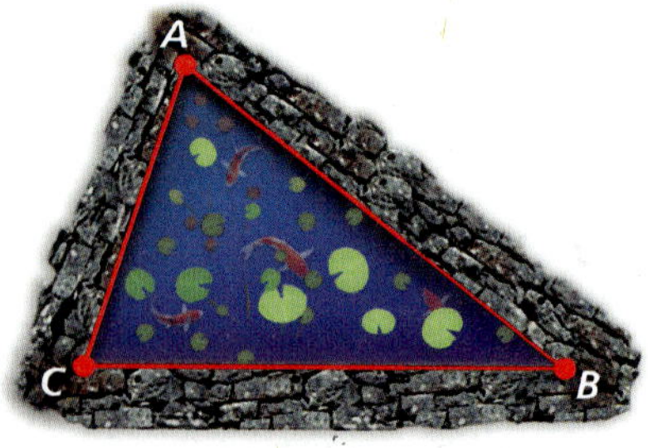

28. MODELING REAL LIFE A marching band director wants a soloist to be the same distance from each side of the triangular formation shown below. Should the soloist be at the *cirumcenter* or *incenter* of the triangular formation? Explain.

DISCUSS MATHEMATICAL THINKING In Exercises 29–32, complete the statement with *always*, *sometimes*, or *never*. Explain your reasoning.

29. The circumcenter of a scalene triangle is __________ inside the triangle.

30. If the perpendicular bisector of one side of a triangle intersects the opposite vertex, then the triangle is __________ isosceles.

31. The perpendicular bisectors of a triangle intersect at a point that is __________ equidistant from the midpoints of the sides of the triangle.

32. The angle bisectors of a triangle intersect at a point that is __________ equidistant from the sides of the triangle.

REASONING In Exercises 33 and 34, find the coordinates of the circumcenter of the triangle with the given vertices.

33. $A(2, 5)$, $B(6, 6)$, $C(12, 3)$

34. $D(-9, -5)$, $E(-5, -9)$, $F(-2, -2)$

GO DIGITAL

5 MTR **STRUCTURE** **In Exercises 35 and 36, find the value of x that makes point N the incenter of the triangle.**

35.

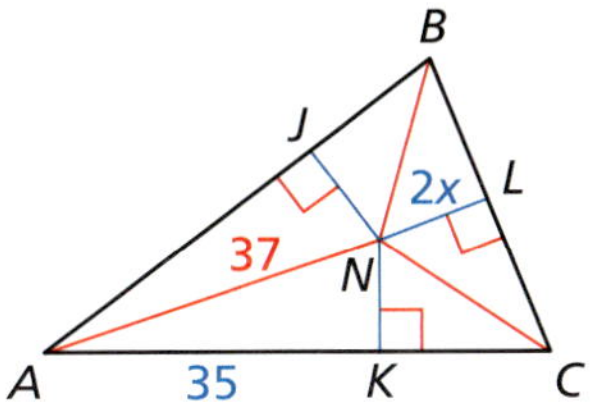

36.

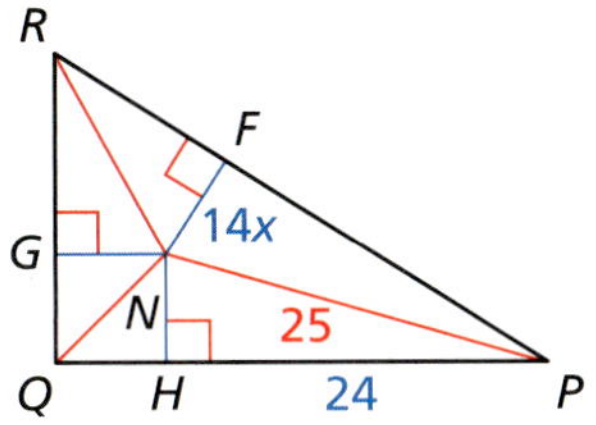

37. PROOF Where is the circumcenter located in any right triangle? Write a coordinate proof of this result.

38. PROVING A THEOREM Write a proof of the Incenter Theorem.

Given $\triangle ABC$, $\overline{AD}$ bisects $\angle CAB$, $\overline{BD}$ bisects $\angle CBA$, $\overline{DE} \perp \overline{AB}$, $\overline{DF} \perp \overline{BC}$, and $\overline{DG} \perp \overline{CA}$

Prove The angle bisectors intersect at D, which is equidistant from $\overline{AB}$, $\overline{BC}$, and $\overline{CA}$.

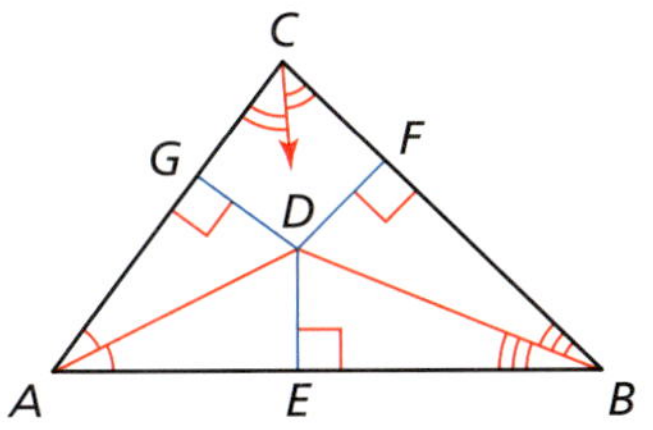

39. PROBLEM SOLVING Archaeologists find three stones. They believe that the stones were once part of a circle of stones with a community firepit at its center. They mark the locations of stones A, B, and C on a coordinate plane, where distances are measured in feet.

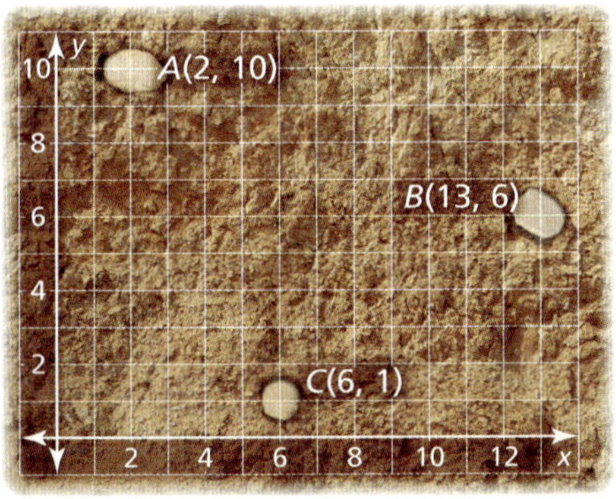

a. Explain how archaeologists can use a sketch to estimate the center of the circle of stones.

b. Copy the diagram and find the approximate coordinates of the point at which the archaeologists should look for the firepit.

7 MTR **40. PERFORMANCE TASK** You plan to install and furnish the largest circular patio possible inside a triangular yard with side lengths of 9 meters, 12 meters, and 15 meters.

a. Create a blueprint. Include the center and radius of the circle. Estimate the circumference and the area of the circle.

b. Research costs of materials and labor. Then prepare an itemized cost estimate for installing and furnishing the patio.

41. REASONING Is it possible for the incenter and the circumcenter of a triangle to be the same point? Use diagrams to support your reasoning.

42. HOW DO YOU SEE IT?
The arms of the windmill are the angle bisectors of the red triangle. What point of concurrency is the point that connects the three arms?

4 MTR **43. DISCUSS MATHEMATICAL THINKING** Explain why the incenter of a triangle is always located inside the triangle.

44. REASONING Use reflections to show that the three lines of symmetry of an equilateral triangle are perpendicular bisectors of the sides of the triangle.

45. USING TOOLS Cut the largest circle possible from an isosceles triangle made of paper whose sides are 8 inches, 12 inches, and 12 inches. Find the radius of the circle. State whether you used perpendicular bisectors or angle bisectors.

46. DIG DEEPER A high school is being built to accommodate the towns shown. Each town agrees that the school should be an equal distance from each of the four towns. Is there a single point where they could agree to build the school? If so, find it. If not, explain why not. Justify your answer with a diagram.

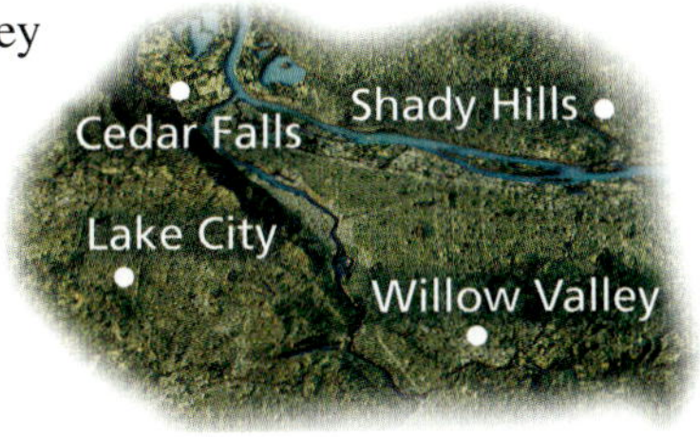

GO DIGITAL

47. REASONING Point D is the incenter of $\triangle ABC$. Write an expression for the length x in terms of the three side lengths AB, AC, and BC.

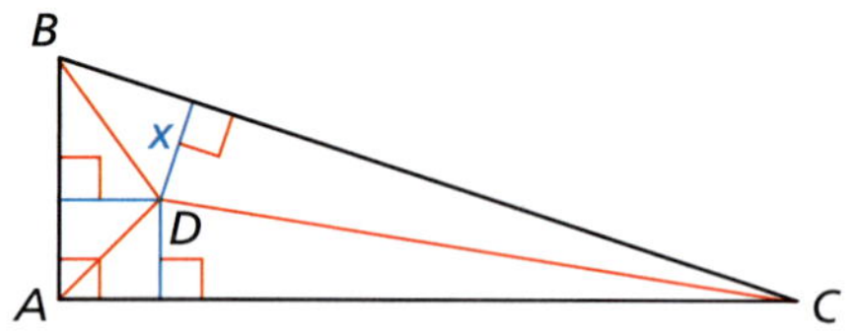

48. THOUGHT PROVOKING
You are asked to draw a triangle and its perpendicular bisectors and angle bisectors.

a. For which type of triangle would you need the fewest segments? What is the minimum number of segments you would need? Explain.

b. For which type of triangle would you need the most segments? What is the maximum number of segments you would need? Explain.

REVIEW & REFRESH

WATCH

49. Determine whether $\triangle QRS$ and $\triangle TUV$ with the given vertices are congruent. Use transformations to explain your reasoning.

$Q(1, 2)$, $R(3, 5)$, $S(5, 1)$
$T(1, 1)$, $U(3, 3)$, $V(5, 0)$

7 MTR **50. MODELING REAL LIFE** The largest concentrated solar power plant in the world is the Noor Complex, located in the Sahara Desert. It cost 3.9 billion dollars to construct. Use this information to write two true conditional statements.

51. Find $m\angle 1$. Then classify the triangle by its angles.

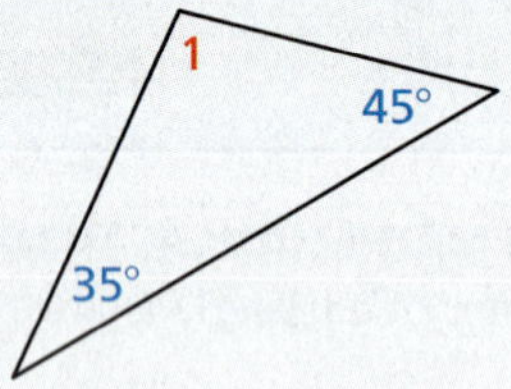

52. Copy the triangle with the given angle measures. Find the incenter. Then construct the inscribed circle.

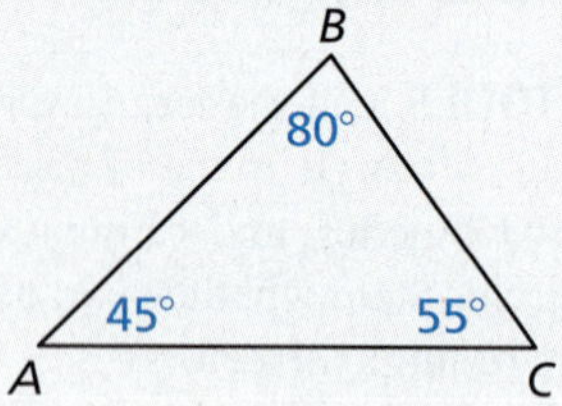

53. Explain how to prove that $\angle A \cong \angle D$.

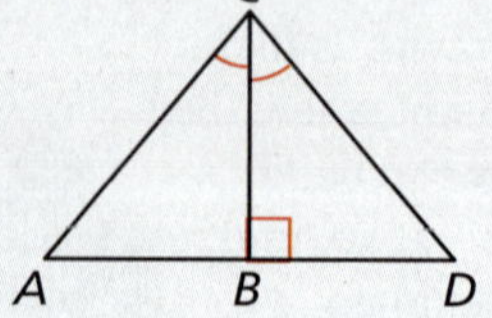

54. Factor $5n^5 + 25n^3$.

55. Find $m\angle ABD$.

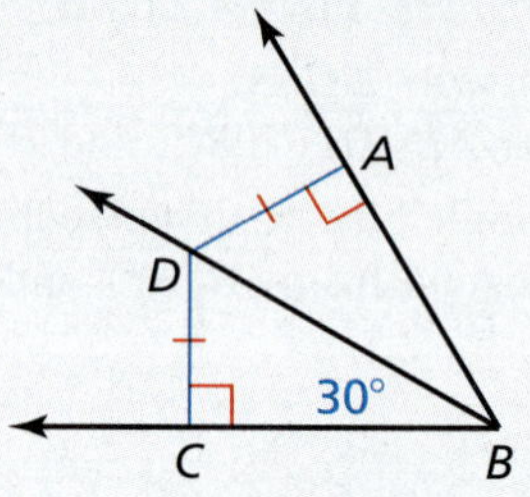

56. Find the distance from point A to $\overleftrightarrow{XZ}$.

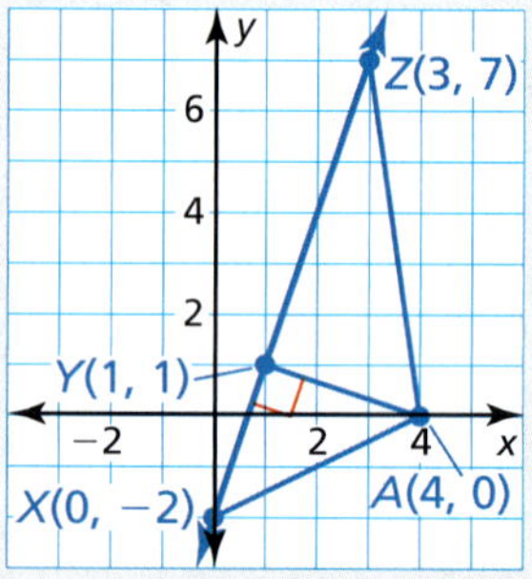

57. A triangle has vertices $A(0, 0)$, $B(8, 12)$, and $C(16, 0)$. Prove that $\triangle ABC$ is isosceles.

58. The endpoints of $\overline{AB}$ are $A(-3, 5)$ and $B(3, 5)$. Find the coordinates of the midpoint M. Then find AB.

59. For $h(x) = -7x$, find the value of x for which $h(x) = 42$.

60. Determine whether the table represents a *linear* or *nonlinear* function. Explain.

x	5	7	9	11
y	−4	−1	2	5

GO DIGITAL

6.3 Medians and Altitudes of Triangles

Learning Target: Use medians and altitudes of triangles.

Success Criteria:
- I can draw medians and altitudes of triangles.
- I can find the centroid of a triangle.
- I can find the orthocenter of a triangle.

EXPLORE IT! Drawing Medians and Altitudes of Triangles

Work with a partner. A *median* of a triangle is a segment from a vertex to the midpoint of the opposite side. An *altitude* of a triangle is the perpendicular segment from a vertex to the opposite side or to the line that contains the opposite side.

a. Use technology to draw any triangle. Construct the medians of the triangle. What do you notice? What happens when you move the vertices of the triangle?

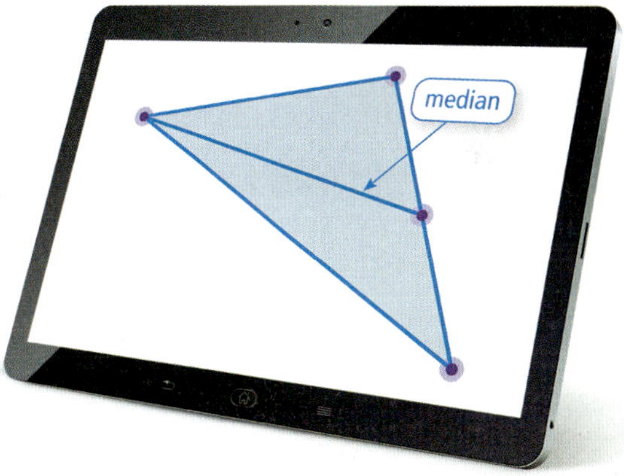

5 MTR MAKE A PLAN

Explain how you can prove your conjectures in part (d).

b. In part (a), the point of concurrency divides each median into two segments. Find a relationship between the lengths of the segments.

c. Use technology to draw any triangle. Construct the altitudes of the triangle. What do you notice? What happens when you move the vertices of the triangle?

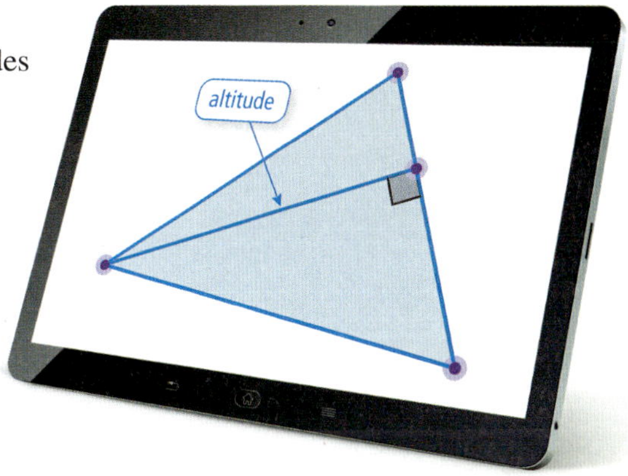

d. What conjectures can you make using your results in parts (a)–(c)?

Geometric Reasoning

MA.912.GR.1.3 Prove relationships and theorems about triangles. Solve mathematical and real-world problems involving postulates, relationships and theorems of triangles.

MA.912.GR.3.3 Use coordinate geometry to solve mathematical and real-world geometric problems involving lines, circles, triangles and quadrilaterals.

GO DIGITAL

Using the Median of a Triangle

Vocabulary

median of a triangle, *p. 316*
centroid, *p. 316*
altitude of a triangle, *p. 317*
orthocenter, *p. 317*

A **median of a triangle** is a segment from a vertex to the midpoint of the opposite side. The three medians of a triangle are concurrent. The point of concurrency, called the **centroid**, is inside the triangle.

THEOREM

6.7 Centroid Theorem

The centroid of a triangle is two-thirds of the distance from each vertex to the midpoint of the opposite side.

The medians of $\triangle ABC$ meet at point P, and $AP = \frac{2}{3}AE$, $BP = \frac{2}{3}BF$, and $CP = \frac{2}{3}CD$.

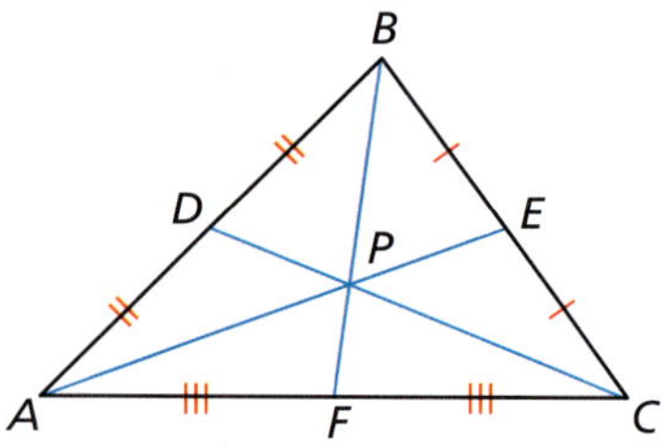

Proof *BigIdeasMath.com*

Prove this Theorem Exercise 51, page 322

CONSTRUCTION Finding the Centroid of a Triangle

Use a compass and straightedge to construct the medians of $\triangle ABC$.

SOLUTION

Step 1

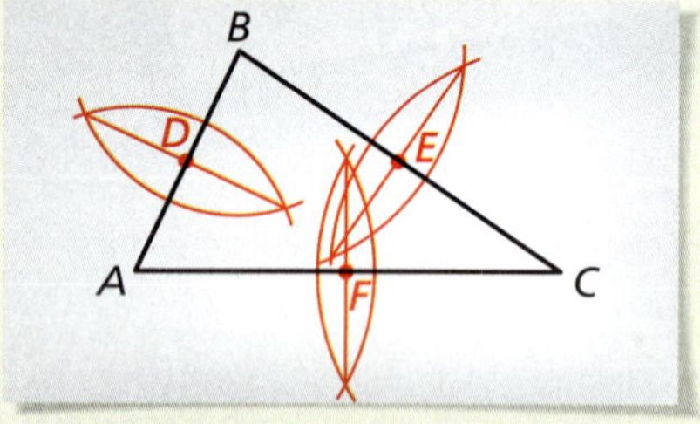

Find midpoints Draw $\triangle ABC$. Find the midpoints of $\overline{AB}$, $\overline{BC}$, and $\overline{AC}$. Label the midpoints of the sides D, E, and F, respectively.

Step 2

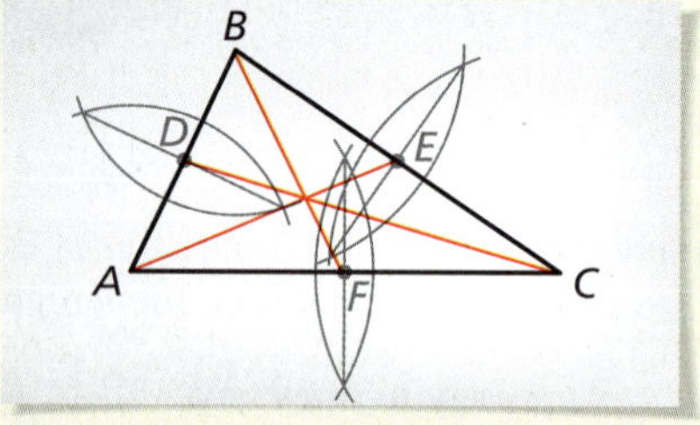

Draw medians Draw $\overline{AE}$, $\overline{BF}$, and $\overline{CD}$. These are the three medians of $\triangle ABC$.

Step 3

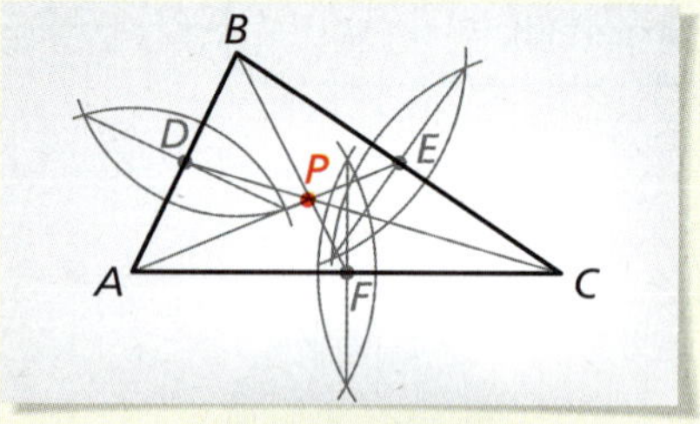

Label a point Label the point where $\overline{AE}$, $\overline{BF}$, and $\overline{CD}$ intersect as P. This is the centroid.

EXAMPLE 1 Using the Centroid of a Triangle

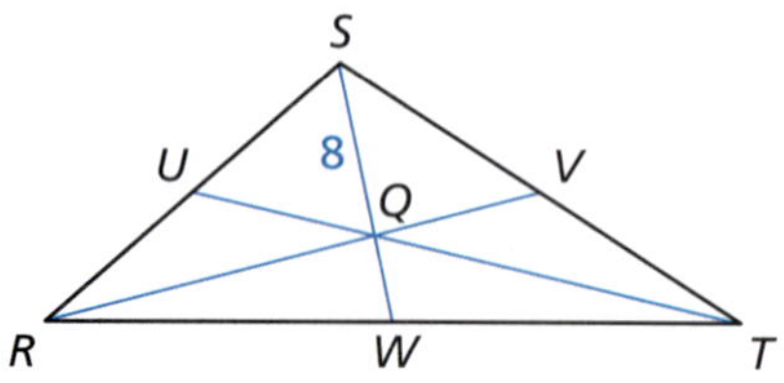

In $\triangle RST$, point Q is the centroid, and $SQ = 8$. Find QW and SW.

SOLUTION

$SQ = \frac{2}{3}SW$	Centroid Theorem
$8 = \frac{2}{3}SW$	Substitute 8 for SQ.
$12 = SW$	Multiply each side by the reciprocal, $\frac{3}{2}$.

Then $QW = SW - SQ = 12 - 8 = 4$.

▶ So, $QW = 4$ and $SW = 12$.

GO DIGITAL

EXAMPLE 2 Finding the Centroid of a Triangle

WATCH

Find the coordinates of the centroid of $\triangle RST$ with vertices $R(2, 1)$, $S(5, 8)$, and $T(8, 3)$.

3 MTR REFLECT ON YOUR METHOD

Why is it more convenient to use the midpoint of $\overline{RT}$ than the midpoint of $\overline{RS}$ or $\overline{ST}$ to solve this problem?

SOLUTION

Step 1 Graph $\triangle RST$.

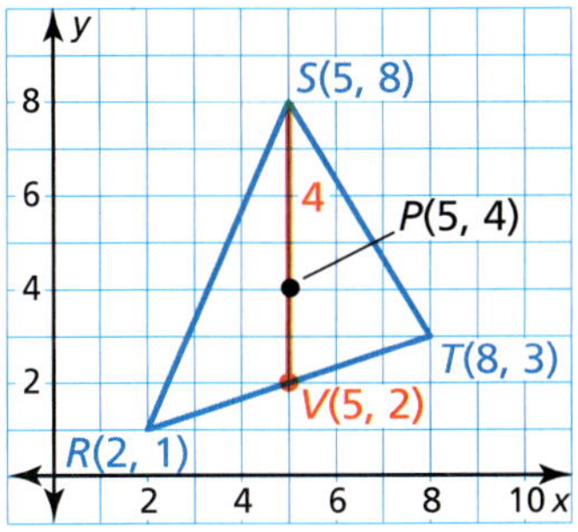

Step 2 Use the Midpoint Formula to find the midpoint V of $\overline{RT}$, and sketch median $\overline{SV}$.

$$V\left(\frac{2+8}{2}, \frac{1+3}{2}\right) = V(5, 2)$$

Step 3 Find the centroid. It is two-thirds of the distance from each vertex to the midpoint of the opposite side.

The distance from vertex $S(5, 8)$ to $V(5, 2)$ is $8 - 2 = 6$ units.
So, the centroid is $\frac{2}{3}(6) = 4$ units down from vertex S on $\overline{SV}$.

▶ So, the coordinates of the centroid P are $(5, 8 - 4)$, or $(5, 4)$.

SELF-ASSESSMENT

1 I don't understand yet. 2 I can do it with help. 3 I can do it on my own. 4 I can teach someone else.

There are three paths through a triangular park. Each path connects the midpoint of one side to the opposite corner. The paths meet at point *P*.

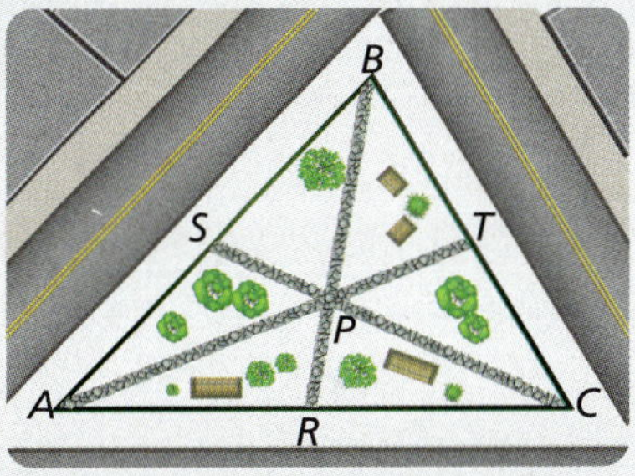

1. Find PS and PC when $SC = 2100$ feet.
2. Find TC and BC when $BT = 1000$ feet.
3. Find PA and TA when $PT = 800$ feet.

Find the coordinates of the centroid of the triangle with the given vertices.

4. $F(2, 5)$, $G(4, 9)$, $H(6, 1)$
5. $X(-3, 3)$, $Y(1, 5)$, $Z(-1, -2)$

Using the Altitude of a Triangle

An **altitude of a triangle** is the perpendicular segment from a vertex to the opposite side or to the line that contains the opposite side.

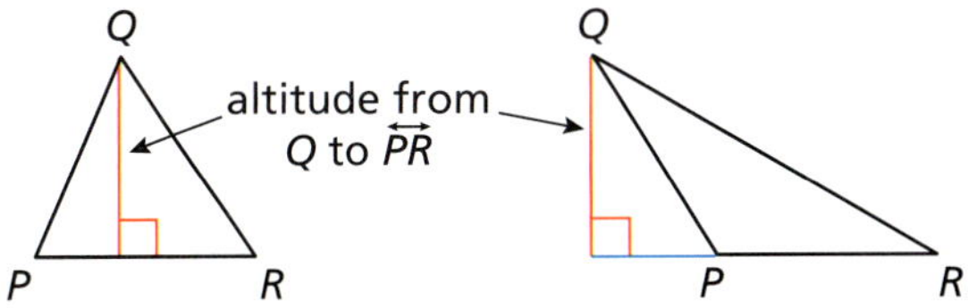

KEY IDEA

Orthocenter

The lines containing the altitudes of a triangle are concurrent. This point of concurrency is the **orthocenter** of the triangle.

The lines containing $\overline{AF}$, $\overline{BD}$, and $\overline{CE}$ meet at the orthocenter G of $\triangle ABC$.

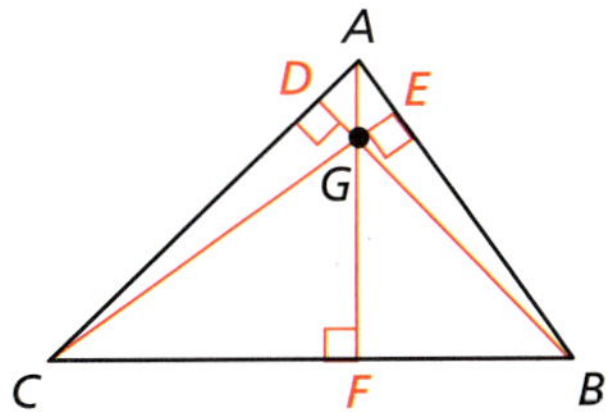

GO DIGITAL

As shown below, the location of the orthocenter P of a triangle depends on the type of triangle.

> **STUDY TIP**
> The altitudes are shown in red. Notice that in the right triangle, the legs are also altitudes. The altitudes of the obtuse triangle are extended to find the orthocenter.

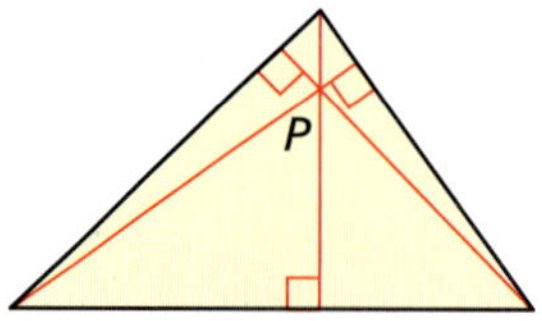

Acute triangle
P is inside triangle.

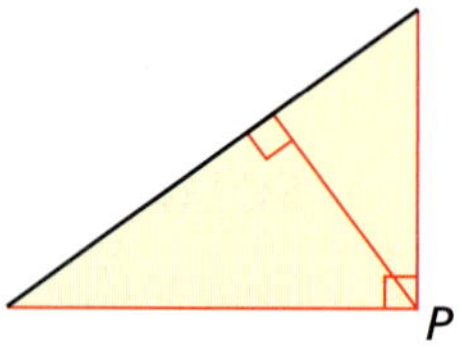

Right triangle
P is on triangle.

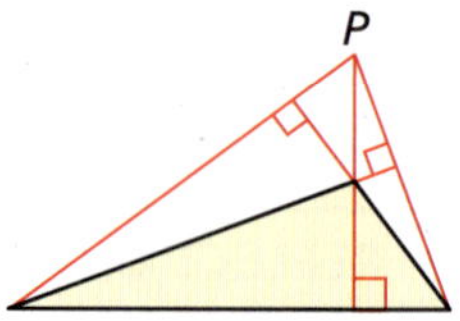

Obtuse triangle
P is outside triangle.

EXAMPLE 3 Finding the Orthocenter of a Triangle

WATCH

Find the coordinates of the orthocenter of $\triangle XYZ$ with vertices $X(-5, -1)$, $Y(-2, 4)$, and $Z(3, -1)$.

SOLUTION

Step 1 Graph $\triangle XYZ$.

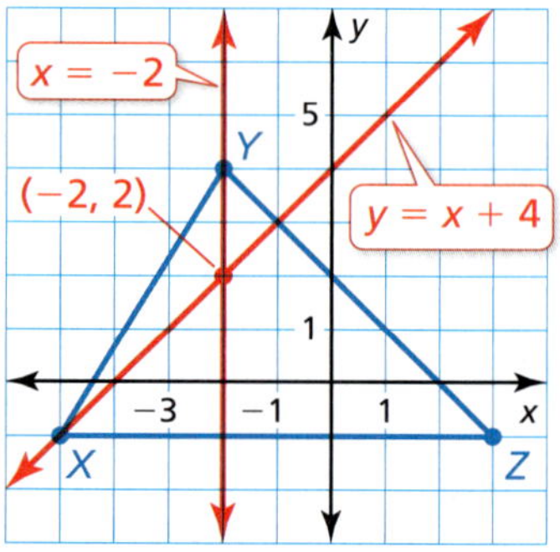

Step 2 Find an equation of the line that contains the altitude from Y to $\overline{XZ}$. Because $\overline{XZ}$ is horizontal, the altitude is vertical. The line that contains the altitude is $x = -2$ because it passes through $Y(-2, 4)$.

Step 3 Find an equation of the line that contains the altitude from X to $\overline{YZ}$.

$$\text{slope of } \overleftrightarrow{YZ} = \frac{-1 - 4}{3 - (-2)} = -1$$

Because the product of the slopes of two perpendicular lines is -1, the slope of a line perpendicular to $\overleftrightarrow{YZ}$ is 1. The line passes through $X(-5, -1)$.

$y = mx + b$	Use slope-intercept form.
$-1 = 1(-5) + b$	Substitute -1 for y, 1 for m, and -5 for x.
$4 = b$	Solve for b.

So, an equation of the line is $y = x + 4$.

Step 4 Find the point of intersection of the graphs of the equations $x = -2$ and $y = x + 4$.

Substitute -2 for x in the equation $y = x + 4$. Then solve for y.

$y = x + 4$	Write equation.
$y = -2 + 4$	Substitute -2 for x.
$y = 2$	Add.

▶ So, the coordinates of the orthocenter are $(-2, 2)$.

SELF-ASSESSMENT

1 I don't understand yet. 2 I can do it with help. 3 I can do it on my own. 4 I can teach someone else.

Tell whether the orthocenter of the triangle with the given vertices is *inside*, *on*, or *outside* the triangle. Then find the coordinates of the orthocenter.

6. $A(0, 3)$, $B(0, -2)$, $C(6, -3)$

7. $J(-3, -4)$, $K(-3, 4)$, $L(5, 4)$

GO DIGITAL

In an isosceles triangle, the perpendicular bisector, angle bisector, median, and altitude from the vertex angle to the base are all the same segment. In an equilateral triangle, this is true for any vertex.

EXAMPLE 4 Proving a Property of Isosceles Triangles

WATCH

Prove that the median from the vertex angle to the base of an isosceles triangle is an altitude.

SOLUTION

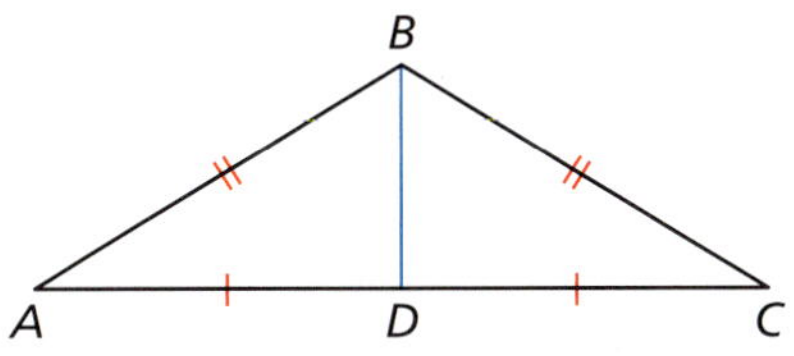

Given $\triangle ABC$ is isosceles, with base $\overline{AC}$. $\overline{BD}$ is the median to base $\overline{AC}$.

Prove $\overline{BD}$ is an altitude of $\triangle ABC$.

Paragraph Proof Legs $\overline{AB}$ and $\overline{BC}$ of isosceles $\triangle ABC$ are congruent. $\overline{CD} \cong \overline{AD}$ because $\overline{BD}$ is the median to $\overline{AC}$. Also, $\overline{BD} \cong \overline{BD}$ by the Reflexive Property of Segment Congruence. So, $\triangle ABD \cong \triangle CBD$ by the SSS Congruence Theorem. $\angle ADB \cong \angle CDB$ because corresponding parts of congruent triangles are congruent. Also, $\angle ADB$ and $\angle CDB$ are a linear pair. $\overline{BD}$ and $\overline{AC}$ intersect to form a linear pair of congruent angles, so $\overline{BD} \perp \overline{AC}$ by the Linear Pair Perpendicular Theorem and $\overline{BD}$ is an altitude of $\triangle ABC$.

SELF-ASSESSMENT

1 I don't understand yet. 2 I can do it with help. 3 I can do it on my own. 4 I can teach someone else.

8. **WHAT IF?** In Example 4, you want to show that median $\overline{BD}$ is also an angle bisector. How would your proof be different?

CONCEPT SUMMARY

Segments, Lines, Rays, and Points in Triangles

	Example	Point of Concurrency	Property	Example
perpendicular bisector	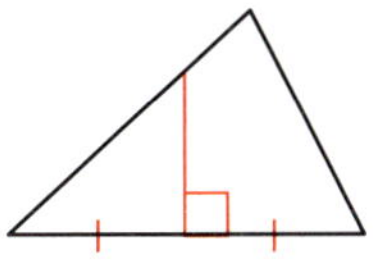	circumcenter	The circumcenter P of a triangle is equidistant from the vertices of the triangle.	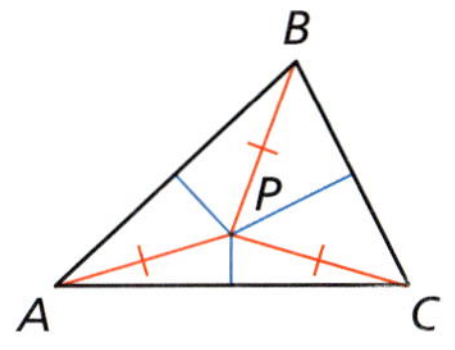
angle bisector	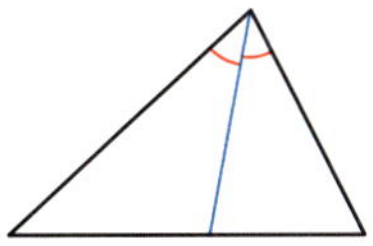	incenter	The incenter I of a triangle is equidistant from the sides of the triangle.	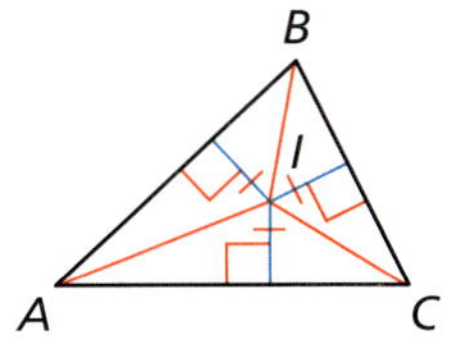
median	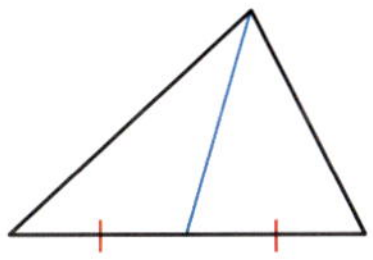	centroid	The centroid R of a triangle is two-thirds of the distance from each vertex to the midpoint of the opposite side.	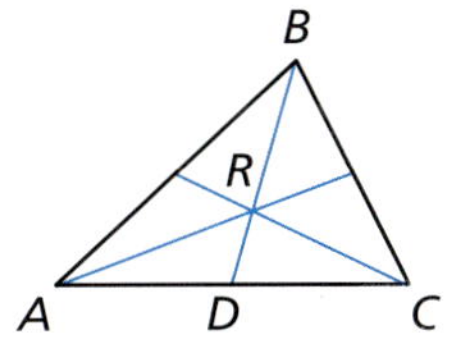
altitude	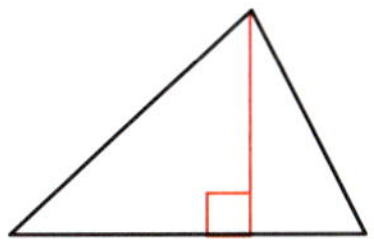	orthocenter	The lines containing the altitudes of a triangle are concurrent at the orthocenter O.	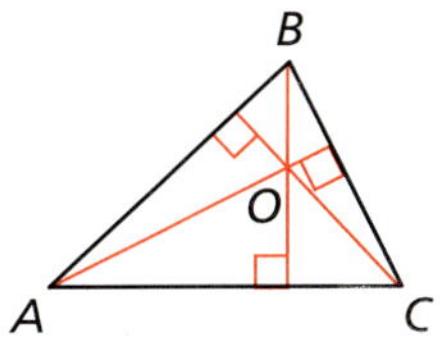

GO DIGITAL

6.3 Practice WITH CalcChat® AND CalcView®

In Exercises 1–4, point P is the centroid of $\triangle LMN$. Find PN and QP. *(See Example 1.)*

▶ **1.** $QN = 9$

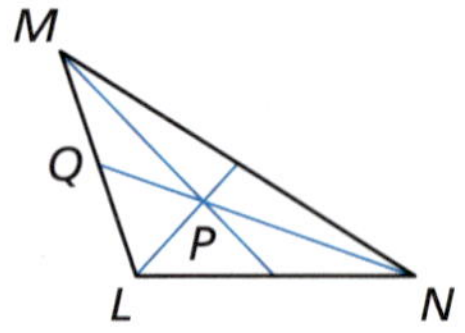

2. $QN = 21$

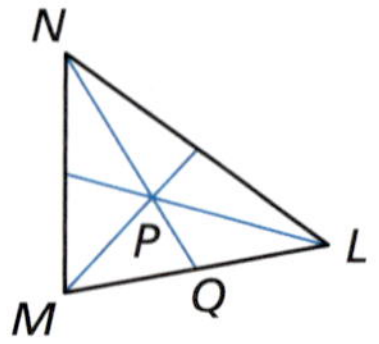

3. $QN = 30$

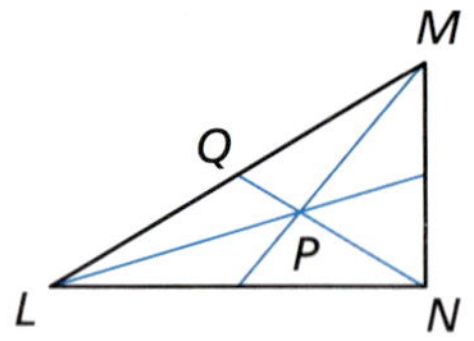

4. $QN = 42$

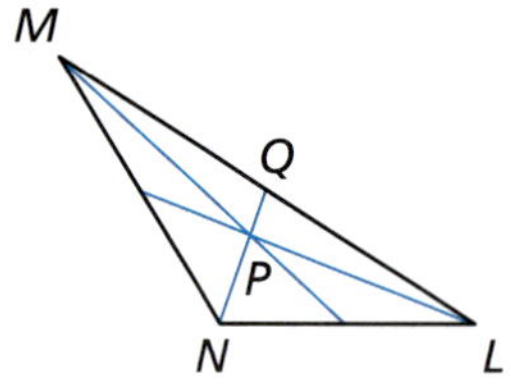

In Exercises 5–8, point D is the centroid of $\triangle ABC$. Find CD and CE.

5. $DE = 5$

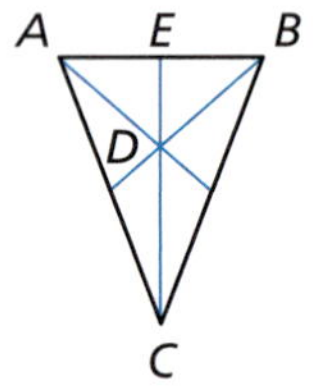

6. $DE = 11$

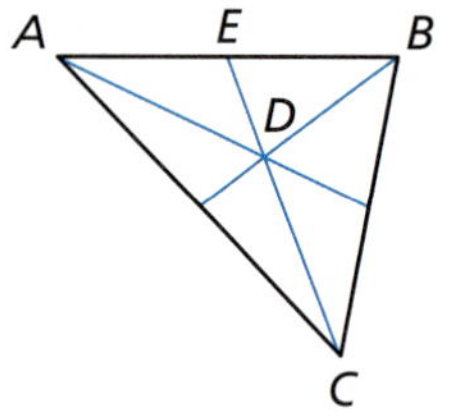

7. $DE = 9$

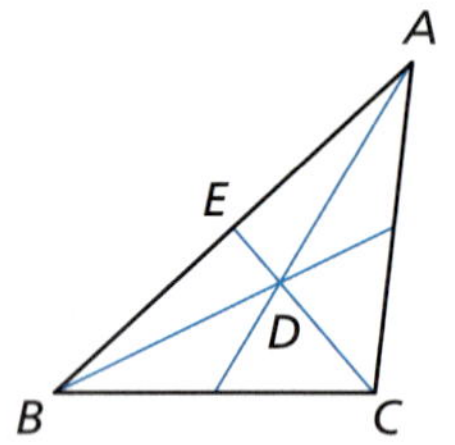

8. $DE = 15$

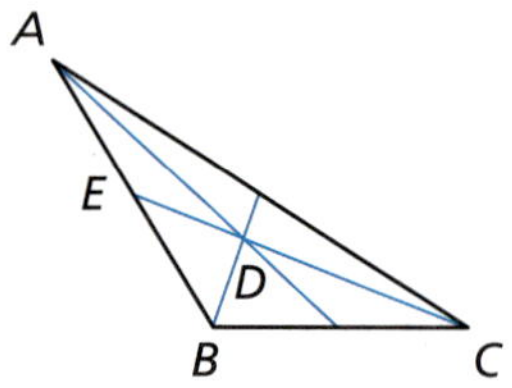

In Exercises 9–12, point G is the centroid of $\triangle ABC$. $BG = 6$, $AF = 12$, and $AE = 15$. Find the length of the segment.

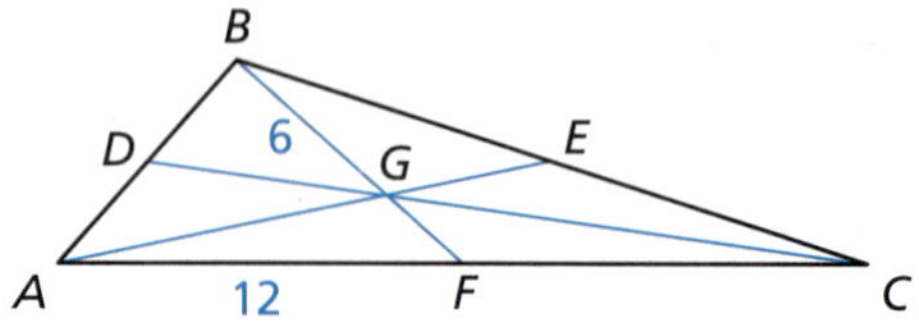

9. $\overline{FC}$

10. $\overline{BF}$

11. $\overline{AG}$

12. $\overline{GE}$

4 MTR **ERROR ANALYSIS** **In Exercises 13 and 14, describe and correct the error in finding DE. Point D is the centroid of $\triangle ABC$.**

13.

✗ $DE = \frac{2}{3}AE$

$DE = \frac{2}{3}(18)$

$DE = 12$

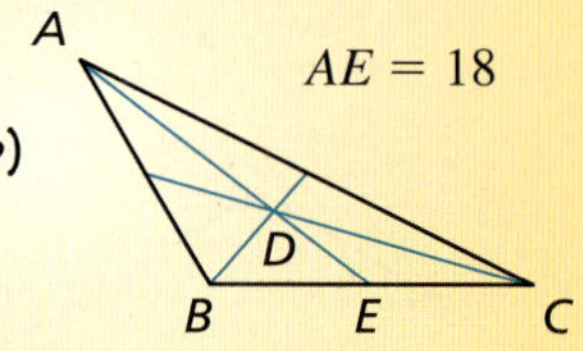

14.

✗ $DE = \frac{2}{3}AD$

$DE = \frac{2}{3}(24)$

$DE = 16$

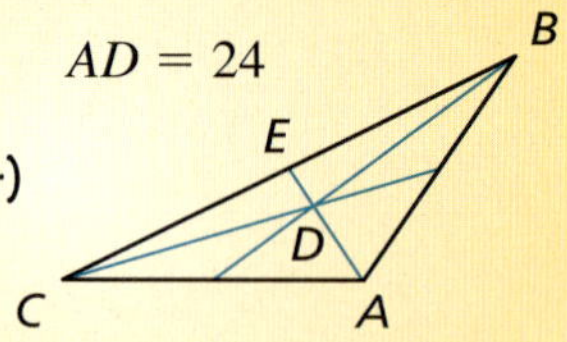

In Exercises 15 and 16, find the coordinates of the centroid of the triangle with the given vertices. *(See Example 2.)*

▶ **15.** $A(2, 3)$, $B(8, 1)$, $C(5, 7)$

16. $F(1, 5)$, $G(-2, 7)$, $H(-6, 3)$

In Exercises 17–20, tell whether the orthocenter of the triangle with the given vertices is *inside*, *on*, or *outside* the triangle. Then find the coordinates of the orthocenter. *(See Example 3.)*

▶ **17.** $L(0, 5)$, $M(3, 1)$, $N(8, 1)$

18. $X(-3, 2)$, $Y(5, 2)$, $Z(-3, 6)$

19. $A(-4, 0)$, $B(1, 0)$, $C(-1, 3)$

20. $T(-2, 1)$, $U(2, 1)$, $V(0, 4)$

CONSTRUCTION **In Exercises 21–24, draw the indicated triangle and find its centroid and orthocenter.**

21. right isosceles triangle

22. obtuse scalene triangle

23. right scalene triangle

24. acute isosceles triangle

PROOF **In Exercises 25 and 26, write a proof of the statement.** *(See Example 4.)*

▶ **25.** The angle bisector from the vertex angle to the base of an isosceles triangle is a median.

26. The altitude from the vertex angle to the base of an isosceles triangle is a perpendicular bisector.

GO DIGITAL

4 MTR **DISCUSS MATHEMATICAL THINKING** **In Exercises 27–32, complete the statement with *always*, *sometimes*, or *never*. Explain your reasoning.**

27. The centroid is __________ on the triangle.

28. The orthocenter is __________ outside the triangle.

29. A median is __________ the same line segment as a perpendicular bisector.

30. An altitude is __________ the same line segment as an angle bisector.

31. The centroid and orthocenter are __________ the same point.

32. The centroid is __________ formed by the intersection of the three medians.

33. **WRITING** Compare an altitude of a triangle with a perpendicular bisector of a triangle.

34. **WRITING** Compare a median, an altitude, and an angle bisector of a triangle.

7 MTR **35.** **MODELING REAL LIFE** Find the area of the triangular part of the paper airplane wing that is outlined in red. Which special segment of the triangle did you use?

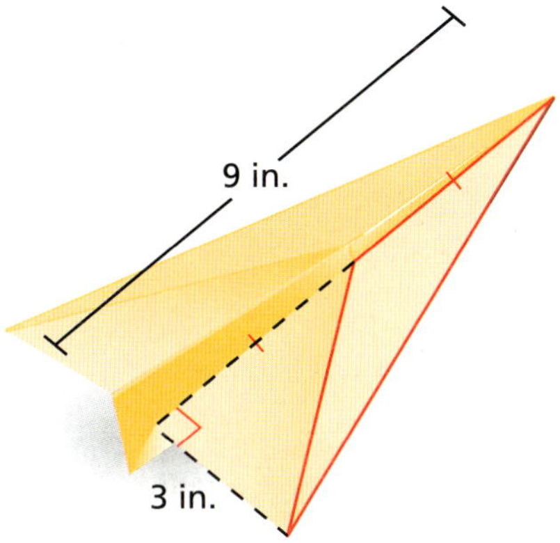

36. **REASONING** Complete the statement for $\triangle DEF$ with centroid K and medians $\overline{DH}$, $\overline{EJ}$, and $\overline{FG}$.

a. $EJ =$ _____ KJ **b.** $DK =$ _____ KH

c. $FG =$ _____ FK **d.** $KG =$ _____ FG

5 MTR **37.** **CONNECTING CONCEPTS** Graph the lines on the same coordinate plane. Find the centroid of the triangle formed by their intersections.

$y_1 = 3x - 4$

$y_2 = \frac{3}{4}x + 5$

$y_3 = -\frac{3}{2}x - 4$

GO DIGITAL

38. **PROBLEM SOLVING** In what type(s) of triangles can a vertex be one of the points of concurrency? Explain.

5 MTR **CONNECTING CONCEPTS** **In Exercises 39–42, point *D* is the centroid of $\triangle ABC$. Use the given information to find the value of *x*.**

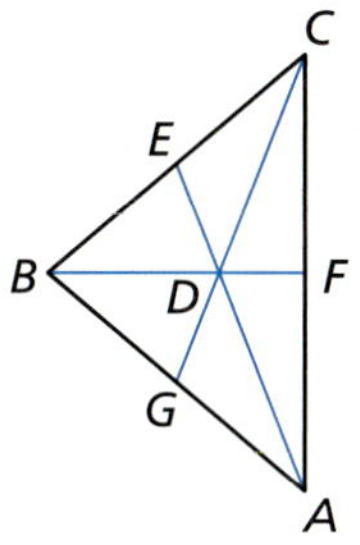

39. $BD = 4x + 5$ and $BF = 9x$

40. $DG = 2x - 8$ and $CG = 3x + 3$

41. $AD = 5x$ and $DE = 3x - 2$

42. $DF = 4x - 1$ and $BD = 6x + 4$

43. **NUMBER SENSE** Use the numbers, symbols, and segment lengths to write three different equations for PE.

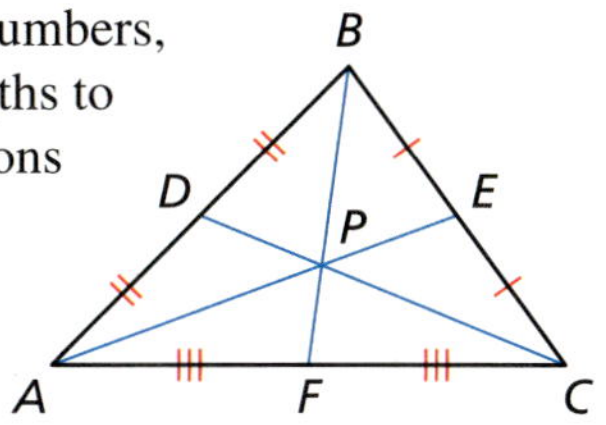

PE	AE	AP	$+$	$-$
$=$	$\frac{1}{4}$	$\frac{1}{3}$	$\frac{1}{2}$	$\frac{2}{3}$

44. **HOW DO YOU SEE IT?**
Use the figure.

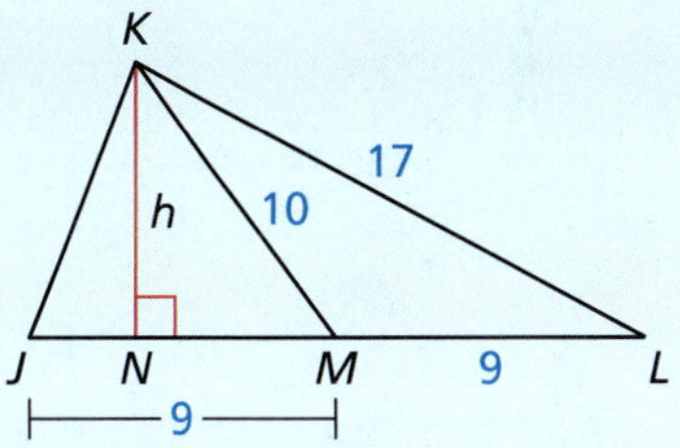

a. What type of segment is $\overline{KM}$? Which point of concurrency lies on $\overline{KM}$?

b. What type of segment is $\overline{KN}$? Which point of concurrency lies on $\overline{KN}$?

c. Compare the areas of $\triangle JKM$ and $\triangle KLM$. Do you think the areas of two triangles formed by the median of any triangle will always compare this way? Explain your reasoning.

4 MTR **45.** **MAKING AN ARGUMENT** Is it possible for the circumcenter, incenter, centroid, and orthocenter to all be the same point? Explain your reasoning.

46. CONSTRUCTION The center of gravity of a triangle, the point where a triangle can balance on the tip of a pencil, is one of the four points of concurrency. Draw and cut out a large scalene triangle on a piece of cardboard. Which of the four points of concurrency is the center of gravity? Explain.

47. PROOF Prove that a median of an equilateral triangle is an angle bisector, a perpendicular bisector, and an altitude.

48. CONSTRUCTION Follow the steps to construct a nine-point circle. Why is it called a nine-point circle?

Step 1 Draw a large acute scalene triangle.

Step 2 Find the orthocenter and circumcenter of the triangle.

Step 3 Find the midpoint between the orthocenter and circumcenter.

Step 4 Find the midpoint between each vertex and the orthocenter.

Step 5 Construct a circle. Use the midpoint in Step 3 as the center of the circle, and the distance from the center to the midpoint of a side of the triangle as the radius.

49. PROOF Prove the statements in parts (a)–(c).

Given $\overline{LP}$ and $\overline{MQ}$ are medians of scalene $\triangle LMN$. Point R is on $\overrightarrow{LP}$ such that $\overline{LP} \cong \overline{PR}$. Point S is on $\overrightarrow{MQ}$ such that $\overline{MQ} \cong \overline{QS}$.

Prove **a.** $\overline{NS} \cong \overline{NR}$

b. $\overline{NS}$ and $\overline{NR}$ are both parallel to $\overline{LM}$.

c. R, N, and S are collinear.

50. THOUGHT PROVOKING
Construct an acute scalene triangle. Find the orthocenter, centroid, and circumcenter. What can you conclude about the three points of concurrency?

51. PROOF In the figure, $\overline{AE}$, $\overline{BF}$, and $\overline{CD}$ are the medians of $\triangle ABC$. How can you show that all three medians intersect at point G? Prove that the three medians of any triangle are concurrent.

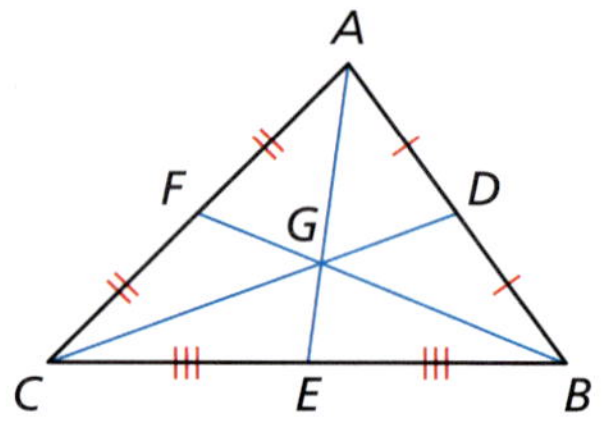

REVIEW & REFRESH

52. Find $m\angle GJK$. Explain your reasoning.

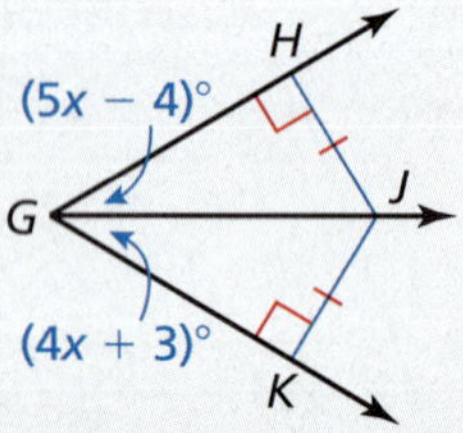

53. Find the coordinates of the circumcenter of the triangle with vertices $D(3, 5)$, $E(7, 9)$, and $F(11, 5)$.

54. Tell whether the orthocenter of $\triangle XYZ$ with vertices $X(-1, -4)$, $Y(7, -4)$, and $Z(7, 4)$ is *inside*, *on*, or *outside* the triangle. Then find its coordinates.

In Exercises 55 and 56, determine whether $\overline{AB}$ is parallel to $\overline{CD}$.

55. $A(5, 6)$, $B(-1, 3)$ and $C(-4, 9)$, $D(-16, 3)$

56. $A(-3, 6)$, $B(5, 4)$ and $C(-14, -10)$, $D(-2, -7)$

57. Graph $g(x) = x^2 + 7$. Compare the graph to the graph of $f(x) = x^2$.

In Exercises 58 and 59, solve the system. Check your solution.

58. $x - 3y = -6$
$5y = 2 + 3x$

59. $2x = 11 - 5y$
$0.5x + 7y = 20$

In Exercises 60 and 61, solve the equation.

60. $3x^2 + 48 = 0$

61. $x^2 - 8x = -4$

6 MTR **62. ASSESS REASONABLENESS** You and a friend are visiting the vendors at the Tampa Bay Black Heritage Festival. The layout of the vendors can be modeled in a coordinate plane where you are at $C(2, 3)$ and your friend is at $D(5, 7)$. Your friend estimates the midpoint for you to meet is $M(1.5, 2)$. Is your friend's estimate reasonable? Explain.

63. Write a plan for the proof.

Given $\overline{EG}$ bisects $\angle DEF$.

Prove $\triangle EGD \cong \triangle EGF$

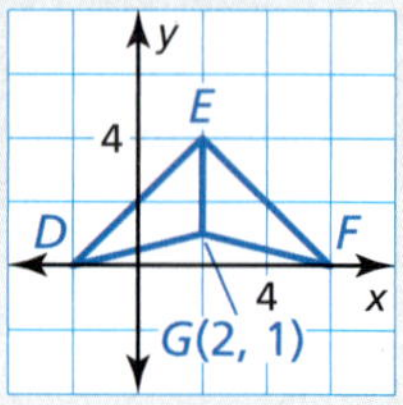

GO DIGITAL

6.4 The Triangle Midsegment Theorem

Learning Target: Find and use midsegments of triangles.

Success Criteria:
- I can use midsegments of triangles in the coordinate plane to solve problems.
- I can solve real-life problems involving midsegments.

EXPLORE IT! Drawing Midsegments of a Triangle

Work with a partner. Use technology.

a. Draw any $\triangle ABC$. Plot and connect the midpoints of two sides of $\triangle ABC$. Why is this called a *midsegment* of $\triangle ABC$?

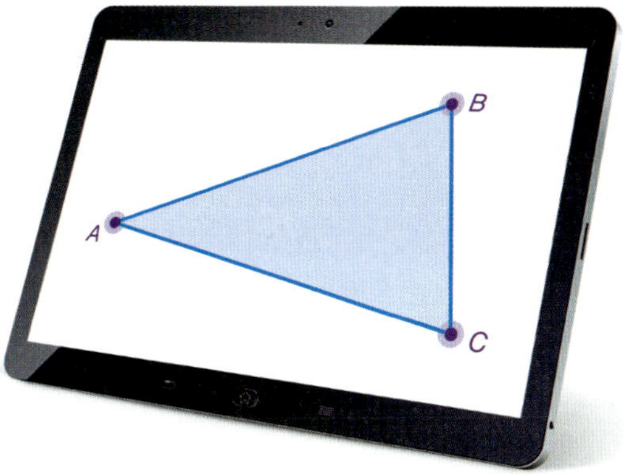

2 MTR USE ANOTHER METHOD

Draw a triangle. Use a compass and straightedge to construct a midsegment and verify your conjecture in part (c).

b. Analyze the midsegments and sides of $\triangle ABC$. What do you notice?

c. Move the vertices of $\triangle ABC$ to form different triangles. Then write a conjecture about the relationships between the midsegments and sides of a triangle.

d. Label the triangle formed by the midsegments of $\triangle ABC$ as shown. Analyze and compare $\triangle ABC$ and $\triangle DEF$. What do you notice? Make several observations.

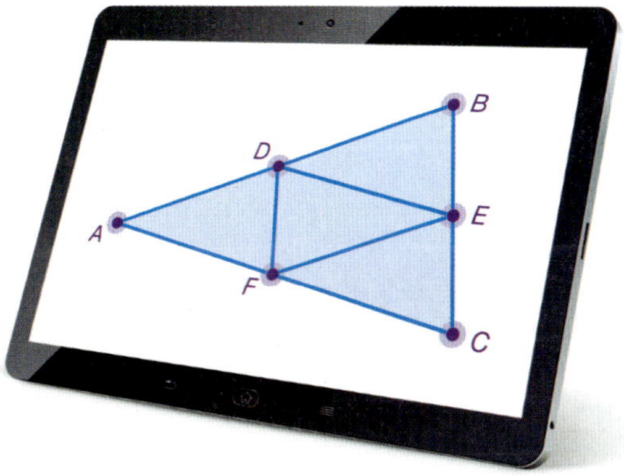

Geometric Reasoning

MA.912.GR.1.3 Prove relationships and theorems about triangles. Solve mathematical and real-world problems involving postulates, relationships and theorems of triangles.

GO DIGITAL

Using the Midsegment of a Triangle

Vocabulary

midsegment of a triangle, *p. 324*

A **midsegment of a triangle** is a segment that connects the midpoints of two sides of the triangle. Every triangle has three midsegments, which form the *midsegment triangle*.

The midsegments of $\triangle ABC$ at the right are $\overline{MP}$, $\overline{MN}$, and $\overline{NP}$. Midsegment $\overline{MP}$ can be called "the midsegment opposite $\overline{AC}$." The *midsegment triangle* is $\triangle MNP$.

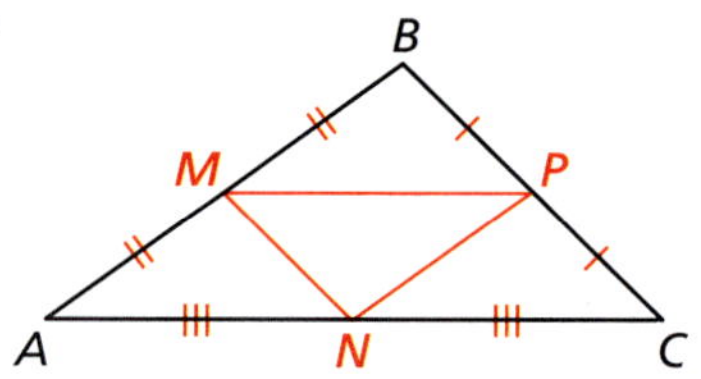

EXAMPLE 1 Using a Midsegment in the Coordinate Plane

WATCH

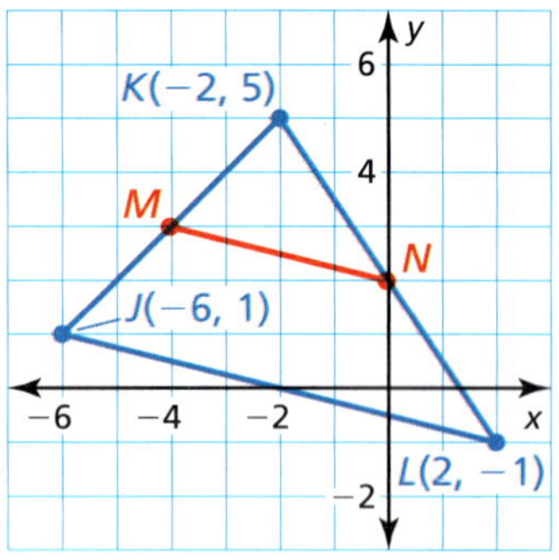

In $\triangle JKL$, show that midsegment $\overline{MN}$ is parallel to $\overline{JL}$ and that $MN = \frac{1}{2}JL$.

SOLUTION

Step 1 Find the coordinates of M and N by finding the midpoints of $\overline{JK}$ and $\overline{KL}$.

$$M\left(\frac{-6+(-2)}{2}, \frac{1+5}{2}\right) = M\left(\frac{-8}{2}, \frac{6}{2}\right) = M(-4, 3)$$

$$N\left(\frac{-2+2}{2}, \frac{5+(-1)}{2}\right) = N\left(\frac{0}{2}, \frac{4}{2}\right) = N(0, 2)$$

Step 2 Find and compare the slopes of $\overline{MN}$ and $\overline{JL}$.

$$\text{slope of } \overline{MN} = \frac{2-3}{0-(-4)} = -\frac{1}{4} \qquad \text{slope of } \overline{JL} = \frac{-1-1}{2-(-6)} = -\frac{2}{8} = -\frac{1}{4}$$

▶ Because the slopes are the same, $\overline{MN}$ is parallel to $\overline{JL}$.

Step 3 Find and compare the lengths of $\overline{MN}$ and $\overline{JL}$.

$$MN = \sqrt{[0-(-4)]^2 + (2-3)^2} = \sqrt{16+1} = \sqrt{17}$$

$$JL = \sqrt{[2-(-6)]^2 + (-1-1)^2} = \sqrt{64+4} = \sqrt{68} = 2\sqrt{17}$$

▶ Because $\sqrt{17} = \frac{1}{2}(2\sqrt{17})$, $MN = \frac{1}{2}JL$.

4 MTR JUSTIFY A RESULT

Show that $\triangle MKN$ is a dilation of $\triangle JKL$ centered at point K with scale factor $\frac{1}{2}$.

SELF-ASSESSMENT

1 I don't understand yet. 2 I can do it with help. 3 I can do it on my own. 4 I can teach someone else.

Use the graph of $\triangle ABC$.

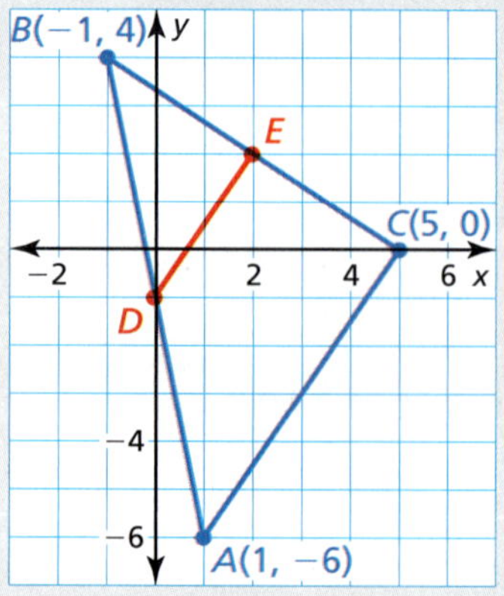

1. Show that midsegment $\overline{DE}$ is parallel to $\overline{AC}$ and that $DE = \frac{1}{2}AC$.

2. Find the coordinates of the endpoints of midsegment $\overline{EF}$, which is opposite $\overline{AB}$. Show that $\overline{EF} \parallel \overline{AB}$ and that $EF = \frac{1}{2}AB$.

4 MTR 3. **WHICH ONE DOESN'T BELONG?** Which one does *not* belong with the other three? Explain your reasoning.

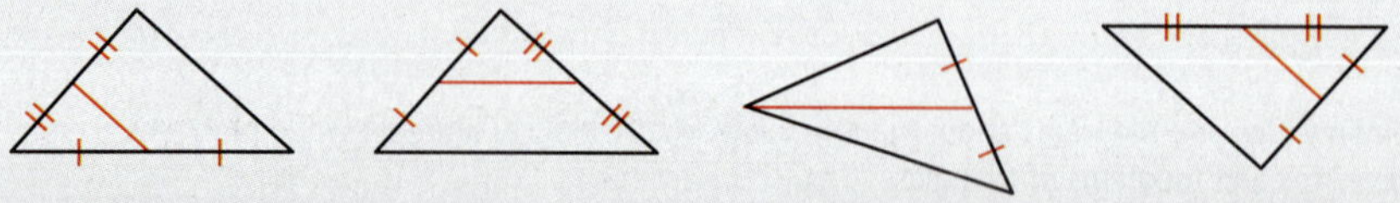

GO DIGITAL

Using the Triangle Midsegment Theorem

THEOREM

6.8 Triangle Midsegment Theorem

The segment connecting the midpoints of two sides of a triangle is parallel to the third side and is half as long as that side.

$\overline{DE}$ is a midsegment of $\triangle ABC$, $\overline{DE} \parallel \overline{AC}$, and $DE = \frac{1}{2}AC$.

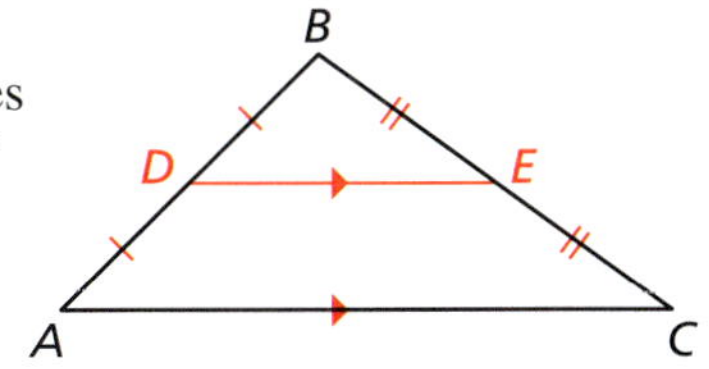

Prove this Theorem Exercise 4, page 326; Exercise 5, page 327

EXAMPLE 2 Proving the Triangle Midsegment Theorem

WATCH

Write a coordinate proof of the Triangle Midsegment Theorem for one midsegment.

Given $\overline{DE}$ is a midsegment of $\triangle OBC$.

Prove $\overline{DE} \parallel \overline{OC}$ and $DE = \frac{1}{2}OC$

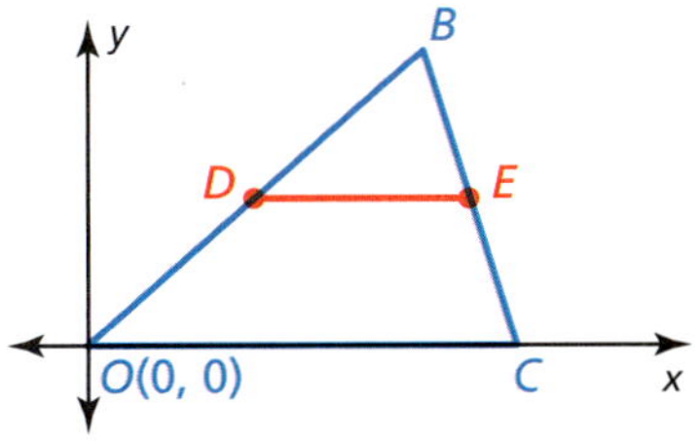

SOLUTION

STUDY TIP

When assigning coordinates, try to choose coordinates that make some of the computations easier. In Example 2, you can avoid fractions by using $2p$, $2q$, and $2r$.

Step 1 Assign coordinates to $\triangle OBC$ in the coordinate plane. Because you are finding midpoints, use $2p$, $2q$, and $2r$. Then find the coordinates of D and E.

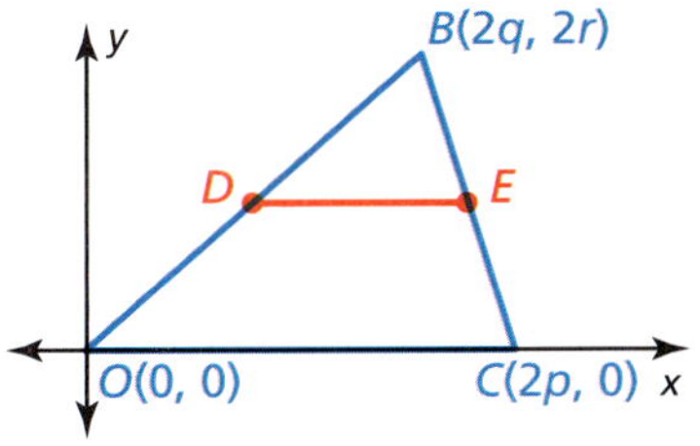

$$D\left(\frac{2q+0}{2}, \frac{2r+0}{2}\right) = D(q, r)$$

$$E\left(\frac{2q+2p}{2}, \frac{2r+0}{2}\right) = E(q+p, r)$$

Step 2 Prove $\overline{DE} \parallel \overline{OC}$. The y-coordinates of D and E are the same, so $\overline{DE}$ has a slope of 0. $\overline{OC}$ is on the x-axis, so its slope is 0.

▶ Because their slopes are the same, $\overline{DE} \parallel \overline{OC}$.

Step 3 Prove $DE = \frac{1}{2}OC$. Use the Ruler Postulate to find DE and OC.

$$DE = |(q+p) - q| = p \qquad OC = |2p - 0| = 2p$$

▶ Because $p = \frac{1}{2}(2p)$, $DE = \frac{1}{2}OC$.

EXAMPLE 3 Using the Triangle Midsegment Theorem

WATCH STEM

Triangles are used for strength in roof trusses. In the diagram, $\overline{UV}$ and $\overline{VW}$ are midsegments of $\triangle RST$. Find UV and RS.

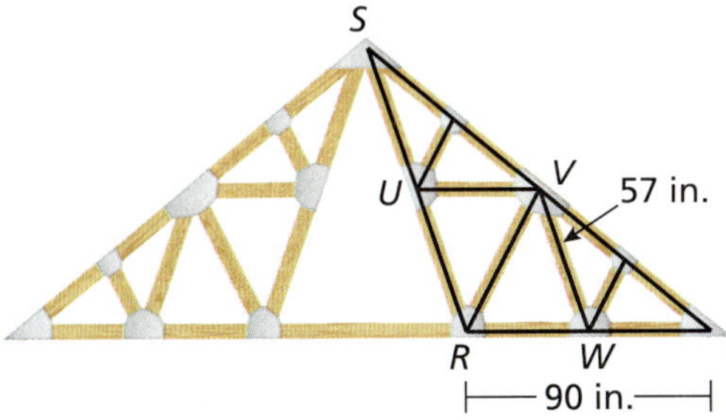

SOLUTION

$$UV = \frac{1}{2} \cdot RT = \frac{1}{2}(90 \text{ in.}) = 45 \text{ in.}$$

$$RS = 2 \cdot VW = 2(57 \text{ in.}) = 114 \text{ in.}$$

GO DIGITAL

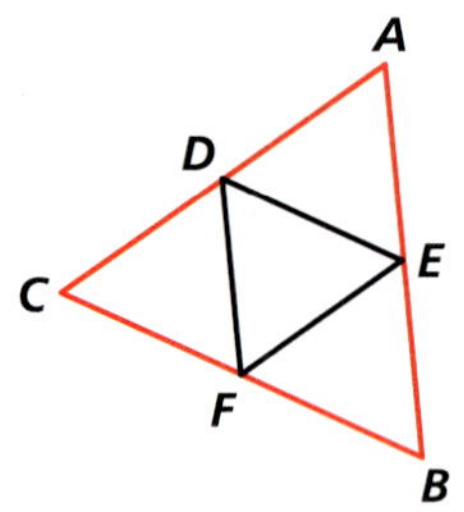

EXAMPLE 4 Using the Triangle Midsegment Theorem

In the diagram, $\overline{AE} \cong \overline{BE}$ and $\overline{AD} \cong \overline{CD}$. Show that $\overline{CB} \parallel \overline{DE}$.

SOLUTION

Because $\overline{AE} \cong \overline{BE}$ and $\overline{AD} \cong \overline{CD}$, E is the midpoint of $\overline{AB}$ and D is the midpoint of $\overline{AC}$ by definition. Then $\overline{DE}$ is a midsegment of $\triangle ABC$ by definition, and $\overline{CB} \parallel \overline{DE}$ by the Triangle Midsegment Theorem.

7 MTR

EXAMPLE 5 Modeling Real Life

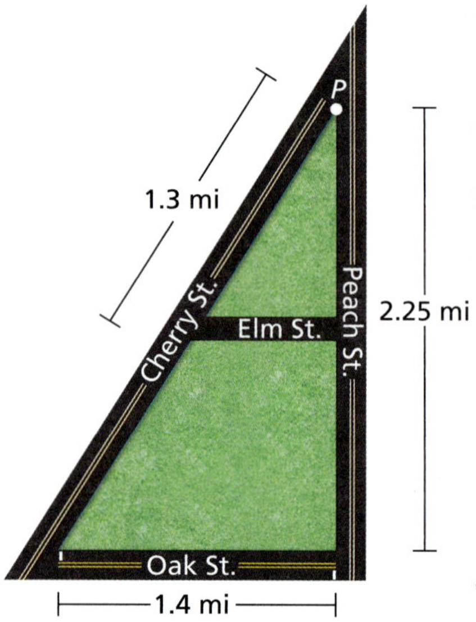

Elm Street intersects Cherry Street and Peach Street at their midpoints. Your home is at point P. You leave your home and jog down Cherry Street to Oak Street, over Oak Street to Peach Street, up Peach Street to Elm Street, over Elm Street to Cherry Street, and then back home up Cherry Street. About how many miles do you jog?

SOLUTION

1. **Understand the Problem** You know the distances from your home to Oak Street along Peach Street, from Peach Street to Cherry Street along Oak Street, and from Elm Street to your home along Cherry Street. You need to find the other distances along your route, then find the total number of miles you jog.

2. **Make a Plan** By definition, you know that Elm Street is a midsegment of the triangle formed by the other three streets. Use the Triangle Midsegment Theorem to find the length of Elm Street and the definition of midsegment to find the length of Cherry Street. Then add the distances along your route.

3. **Solve and Check**

length of Elm Street $= \frac{1}{2} \cdot$ (length of Oak St.) $= \frac{1}{2}(1.4 \text{ mi}) = 0.7$ mi

length of Cherry Street $= 2 \cdot$ (length from P to Elm St.) $= 2(1.3 \text{ mi}) = 2.6$ mi

distance along your route: $2.6 + 1.4 + \frac{1}{2}(2.25) + 0.7 + 1.3 = 7.125$

▶ So, you jog about 7 miles.

Check Reasonableness Use compatible numbers to check that your answer is reasonable.

Total distance:

$$2.6 + 1.4 + \tfrac{1}{2}(2.25) + 0.7 + 1.3 \approx 2.5 + 1.5 + 1 + 0.5 + 1.5 = 7 \checkmark$$

1 MTR **HELP A CLASSMATE**

Explain to a classmate how the compatible numbers were chosen.

SELF-ASSESSMENT

1 I don't understand yet. 2 I can do it with help. 3 I can do it on my own. 4 I can teach someone else.

4. Use the diagram from Example 2 to prove the Triangle Midsegment Theorem for midsegment $\overline{FE}$, where F is the midpoint of $\overline{OC}$.

5. Copy the diagram in Example 3. Draw and name the third midsegment. Then find the length of $\overline{VS}$, when the length of the third midsegment is 81 inches.

6. In Example 4, F is the midpoint of $\overline{CB}$. What do you know about $\overline{DF}$?

7. **WHAT IF?** In Example 5, you leave your home and jog down Peach Street to Oak Street, over Oak Street to Cherry Street, up Cherry Street to Elm Street, over Elm Street to Peach Street, and then back home up Peach Street. Do you jog more miles in Example 5? Explain.

GO DIGITAL

6.4 Practice WITH CalcChat® AND CalcView®

In Exercises 1–4, use the graph of $\triangle ABC$ with midsegments $\overline{DE}$, $\overline{EF}$, and $\overline{DF}$. *(See Example 1.)*

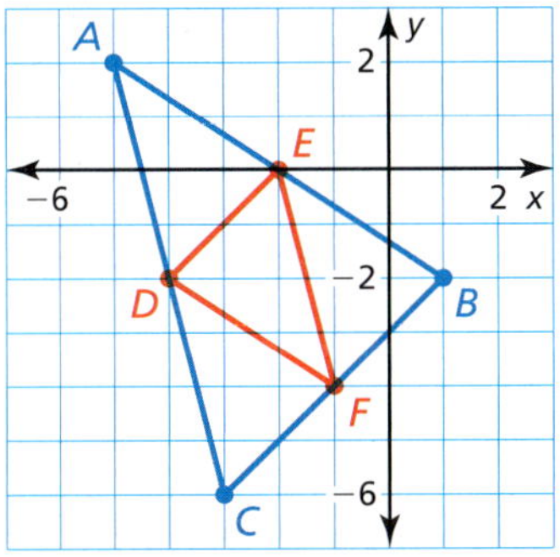

1. Find the coordinates of points D, E, and F.

2. Show that $\overline{DE} \parallel \overline{CB}$ and that $DE = \frac{1}{2}CB$.

▶ 3. Show that $\overline{EF} \parallel \overline{AC}$ and that $EF = \frac{1}{2}AC$.

4. Show that $\overline{DF} \parallel \overline{AB}$ and that $DF = \frac{1}{2}AB$.

▶ 5. **PROVING A THEOREM** Use the diagram from Example 2 to prove the Triangle Midsegment Theorem for midsegment $\overline{DF}$, where F is the midpoint of $\overline{OC}$. *(See Example 2.)*

6. **PROOF** Write a proof.

Given $\overline{MN}$, $\overline{MP}$, and $\overline{PN}$ are midsegments of $\triangle QRS$.

Prove $\triangle QMP \cong \triangle MRN$

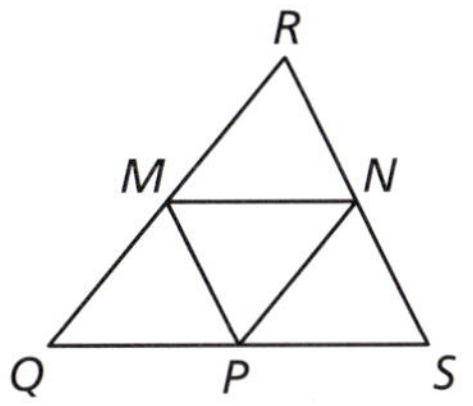

In Exercises 7–10, $\overline{DE}$ is a midsegment of $\triangle ABC$. Find the value of x. *(See Example 3.)*

▶ 7.

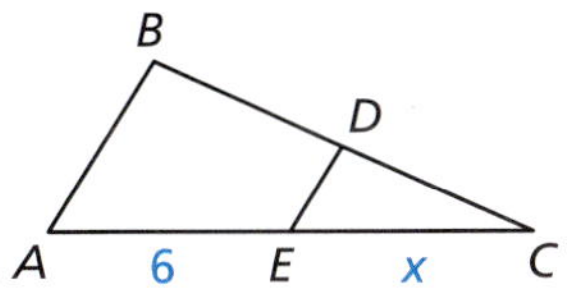

8.

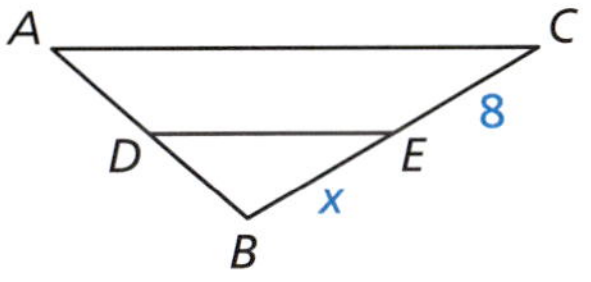

9.

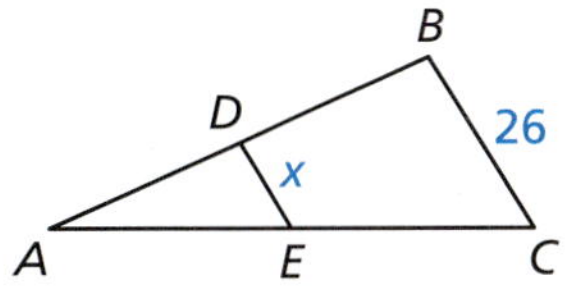

10.

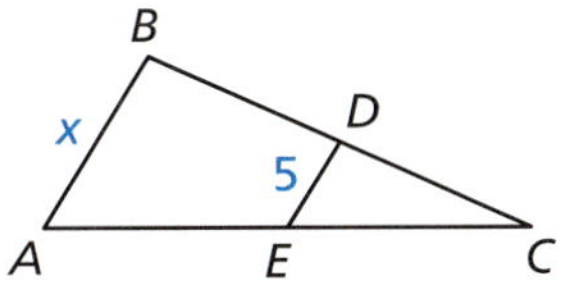

GO DIGITAL

In Exercises 11–16, $\overline{XJ} \cong \overline{JY}$, $\overline{YL} \cong \overline{LZ}$, and $\overline{XK} \cong \overline{KZ}$. Complete the statement. *(See Example 4.)*

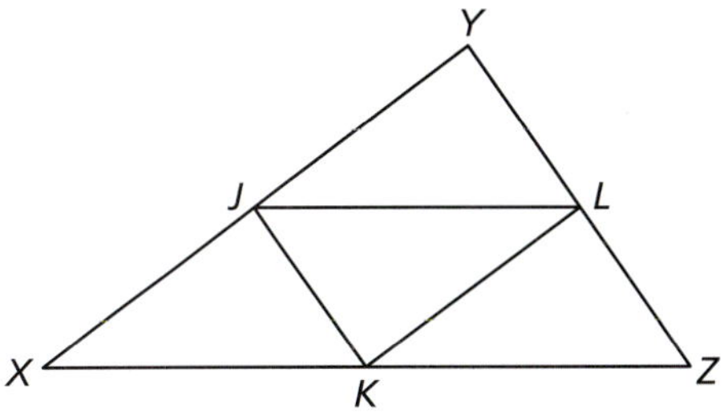

▶ 11. $\overline{JK} \parallel$ ___

12. $\overline{JL} \parallel$ ___

13. $\overline{XY} \parallel$ ___

14. $\overline{JY} \cong$ ___ $\cong$ ___

15. $\overline{JL} \cong$ ___ $\cong$ ___

16. $\overline{JK} \cong$ ___ $\cong$ ___

7 MTR ▶ 17. **MODELING REAL LIFE** The distance between consecutive bases on a baseball field is 90 feet. The pitcher fields a ball halfway between first base and third base. Find the distance between the shortstop and the pitcher. *(See Example 5.)*

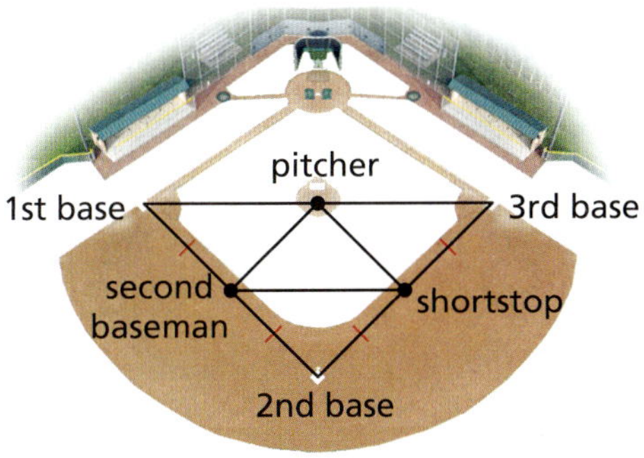

7 MTR 18. **MODELING REAL LIFE** Oak Street intersects Walnut Street and Maple Street at their midpoints. A parade float starts at point S, travels up Walnut Street to Oak Street, up Oak Street to Maple Street, over Maple Street to Spruce Street, and then down Spruce Street to the starting point. About how far does the float travel?

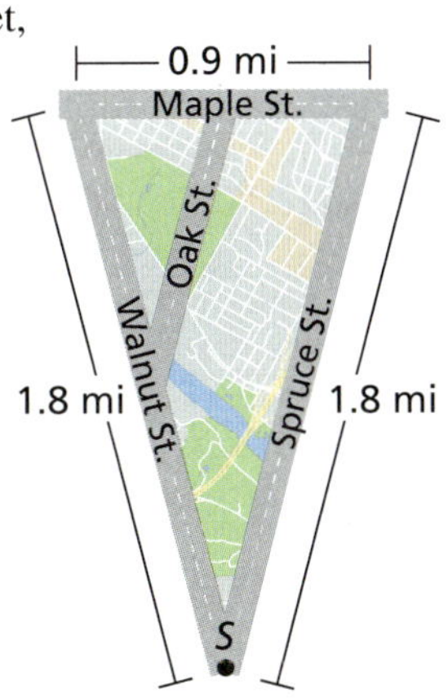

4 MTR **19. ERROR ANALYSIS** Describe and correct the error.

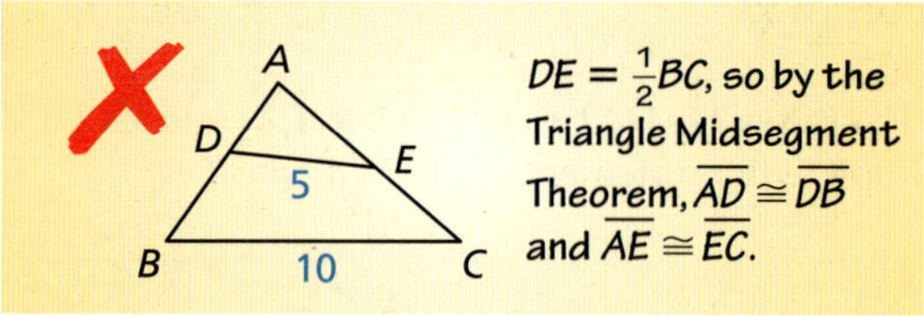

20. HOW DO YOU SEE IT?
Explain how you know that the yellow triangle is the midsegment triangle of the red triangle in the pattern of floor tiles.

5 MTR **21. CONNECTING CONCEPTS** In $\triangle GHJ$, A is the midpoint of $\overline{GH}$, $\overline{CB}$ is a midsegment, and $\overline{CB} \parallel \overline{GH}$. What is GA, when $GH = 7z - 1$ and $CB = 4z - 3$?

22. THOUGHT PROVOKING
$\overline{XY}$ is a midsegment of $\triangle LMN$. $\overline{DE}$ is called a "quarter segment" of $\triangle LMN$. What do you think an "eighth segment" would be? Make conjectures about the properties of a quarter segment and an eighth segment. Use variable coordinates to verify your conjectures.

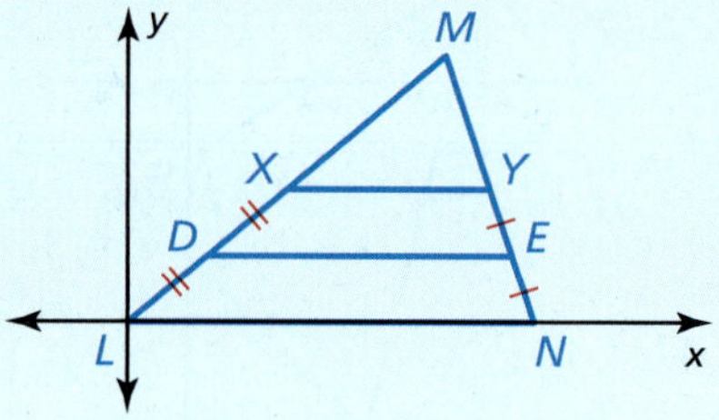

23. DIG DEEPER The midpoints of the sides of a triangle are given. Find the vertices of the original triangle. Explain your process and how you can check your answer.

a. $P(2, 1)$, $Q(4, 5)$, $R(7, 4)$

b. $T(4, 12)$, $U(5, 15)$, $V(6.4, 10.8)$

REVIEW & REFRESH

WATCH

24. Find a counterexample to show that the conjecture is false.

Conjecture The difference of two numbers is always less than the greater number.

25. Find UV. Explain your reasoning.

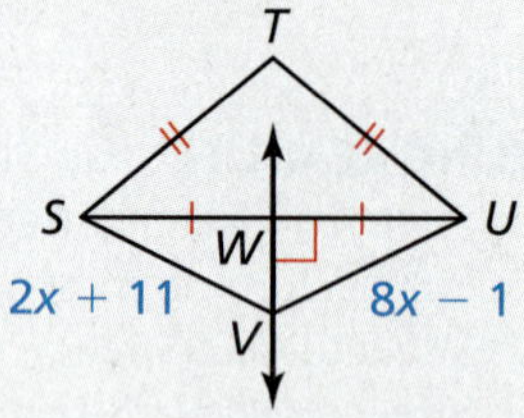

26. Find the coordinates of the centroid of $\triangle MNP$ with vertices $M(-8, -6)$, $N(-4, -2)$, and $P(0, -4)$.

27. The incenter of $\triangle ABC$ is point N. $NQ = 2x + 1$ and $NR = 4x - 9$. Find NS.

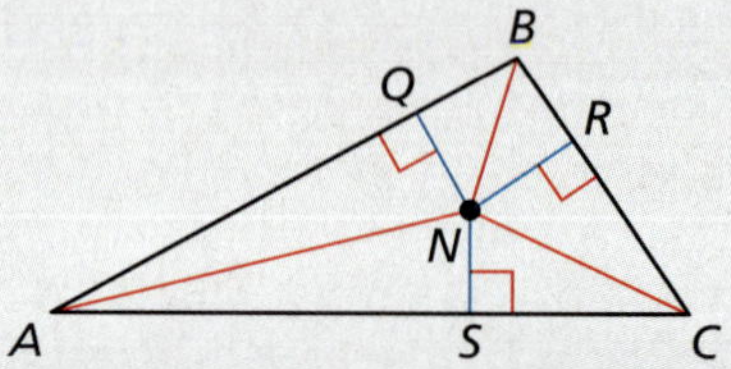

28. $\overline{JK}$ is a midsegment of $\triangle FGH$. Find the values of x and y.

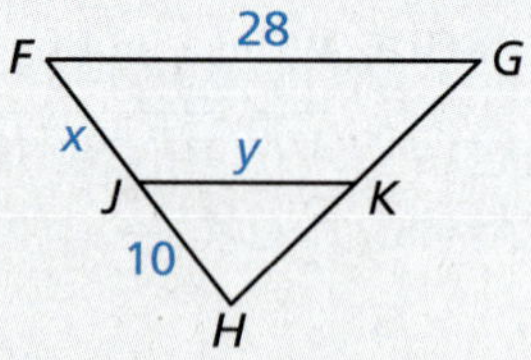

7 MTR **29. MODELING REAL LIFE** The two-way table shows the results of a survey that asked students whether they have been to each theme park. Find and interpret the marginal frequencies.

		Walt Disney World	
		Yes	No
Universal Studios	Yes	23	62
	No	71	49

In Exercises 30 and 31, determine whether the equation represents a *linear* or *nonlinear* function. Explain.

30. $y = -|x + 1|$

31. $y = \frac{1}{3}x + 2$

GO DIGITAL

6.5 Proof by Contradiction and Inequalities in One Triangle

Learning Target: Write proofs by contradiction and understand inequalities in a triangle.

Success Criteria:

- I can write proofs by contradiction. H
- I can order the angles of a triangle given the side lengths.
- I can order the side lengths of a triangle given the angle measures.
- I can determine possible side lengths of triangles.

EXPLORE IT! Analyzing Angle Measures and Side Lengths of Triangles

Work with a partner.

a. Use technology to draw any scalene $\triangle ABC$. Find the side lengths and angle measures of the triangle.

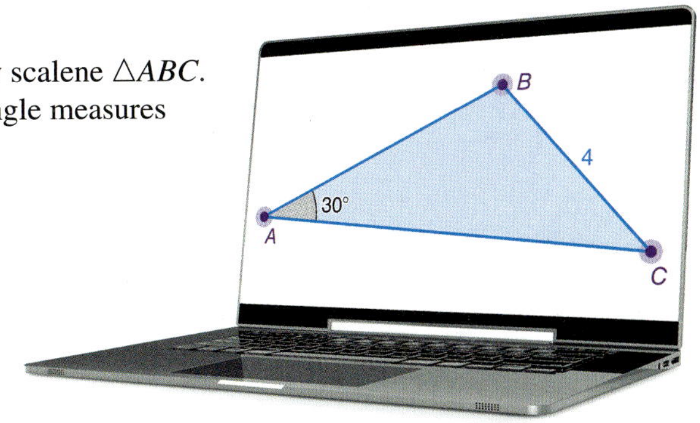

4 MTR CONSTRUCT AN ARGUMENT

In part (d), can *F* lie on any point of the circle to form a triangle?

b. Order the side lengths of $\triangle ABC$ from shortest to longest. Order the angle measures from smallest to largest. What do you notice? What happens when you move the vertices of the triangle?

c. Write a conjecture about the relationship between side lengths and angle measures of a triangle.

d. Use technology to draw a side $\overline{DE}$ of a triangle that has a length of 5 units. Then draw a circle with center *D* that has a radius of 2 units. Choose a point *F* that lies on the circle. Find the side lengths of $\triangle DEF$.

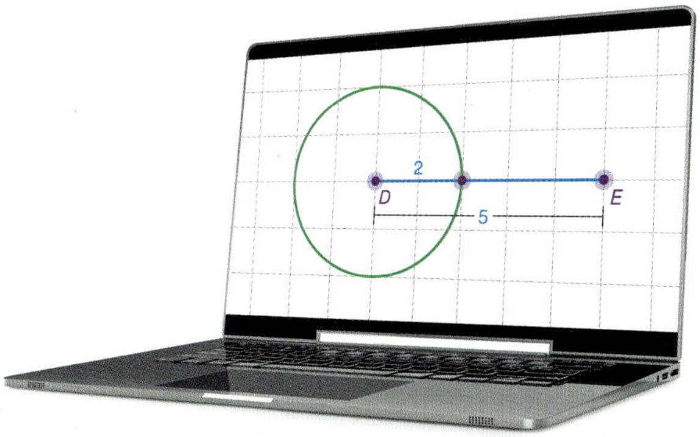

e. Move *F* around the circle to form new triangles. Describe the possible values of *EF*. Make observations about the side lengths of $\triangle DEF$.

Geometric Reasoning

MA.912.GR.1.3 Prove relationships and theorems about triangles. Solve mathematical and real-world problems involving postulates, relationships and theorems of triangles.

Logic and Discrete Theory

H **MA.912.LT.4.8** Construct proofs, including proofs by contradiction.

GO DIGITAL

Writing a Proof by Contradiciton

Vocabulary AZ VOCAB

proof by contradiction, p. 330

Suppose a student looks around the cafeteria, concludes that hamburgers are not being served, and explains as follows.

At first, I assumed that we are having hamburgers because today is Tuesday, and Tuesday is usually hamburger day.

There is always ketchup on the table when we have hamburgers, so I looked for the ketchup, but I didn't see any.

So, my assumption that we are having hamburgers must be false.

The student uses a *contradiction*. In a **proof by contradiction**, you start by making the temporary assumption that the desired conclusion is false. By then showing that this assumption leads to a logical impossibility, you prove the original statement true by contradiction. In the situation above, the temporary assumption is that hamburgers are being served and the desired conclusion is that hamburgers are not being served.

KEY IDEA

How to Write a Proof by Contradiction

Step 1 Identify the statement you want to prove. Assume temporarily that this statement is false by assuming that its opposite is true.

Step 2 Reason logically until you reach a contradiction.

Step 3 Point out that the desired conclusion must be true because the contradiction proves the temporary assumption false.

EXAMPLE 1 Writing a Proof by Contradiction

WATCH H

Write a proof by contradiction that in a given triangle, there can be at most one right angle.

Given $\triangle ABC$

Prove $\triangle ABC$ can have at most one right angle.

SOLUTION

Step 1 Assume temporarily that $\triangle ABC$ has two right angles. Then assume $\angle A$ and $\angle B$ are right angles.

Step 2 By the definition of right angle, $m\angle A = m\angle B = 90°$. By the Triangle Sum Theorem, $m\angle A + m\angle B + m\angle C = 180°$. Using the Substitution Property of Equality, $90° + 90° + m\angle C = 180°$. So, $m\angle C = 0°$ by the Subtraction Property of Equality. A triangle cannot have an angle measure of 0°. So, this contradicts the given information.

Step 3 So, the assumption that $\triangle ABC$ has two right angles must be false, which proves that $\triangle ABC$ can have at most one right angle.

STUDY TIP

You have reached a *contradiction* when you have two statements about the same concept that cannot both be true.

SELF-ASSESSMENT 1 I don't understand yet. 2 I can do it with help. 3 I can do it on my own. 4 I can teach someone else.

1. **WRITING** Explain the difference between a counterexample and a contradiction.
2. Write a proof by contradiction that a scalene triangle cannot have two congruent angles.

GO DIGITAL

Relating Sides and Angles of a Triangle

EXAMPLE 2 **Relating Side Length and Angle Measure** WATCH

Draw an obtuse scalene triangle. Label the largest angle and the longest side. Then label the smallest angle and the shortest side. What do you notice?

SOLUTION

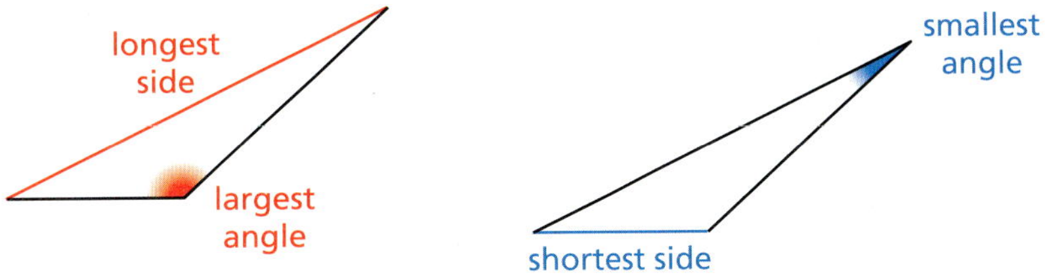

The longest side and largest angle are opposite each other.

The shortest side and smallest angle are opposite each other.

The relationships in Example 2 are true for all triangles, as stated in the two theorems below. These relationships can help you decide whether a particular arrangement of side lengths and angle measures in a triangle may be possible.

COMMON ERROR

Be careful not to confuse the symbol ∠ meaning "angle" with the symbol < meaning "is less than." Notice that the bottom edge of the angle symbol is horizontal.

THEOREMS

6.9 Triangle Longer Side Theorem

If one side of a triangle is longer than another side, then the angle opposite the longer side is larger than the angle opposite the shorter side.

Prove this Theorem Exercise 41, page 336

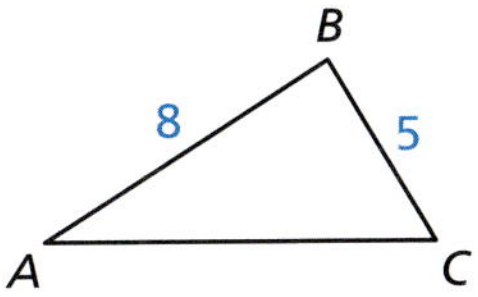

$AB > BC$, so $m\angle C > m\angle A$.

6.10 Triangle Larger Angle Theorem

If one angle of a triangle is larger than another angle, then the side opposite the larger angle is longer than the side opposite the smaller angle.

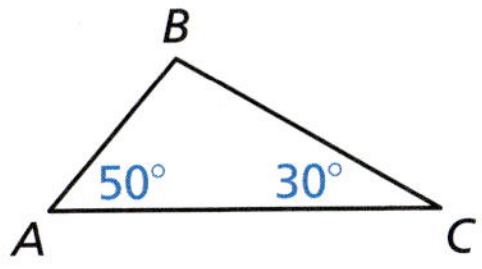

$m\angle A > m\angle C$, so $BC > AB$.

PROOF **Triangle Larger Angle Theorem** H

Given $m\angle A > m\angle C$

Prove $BC > AB$

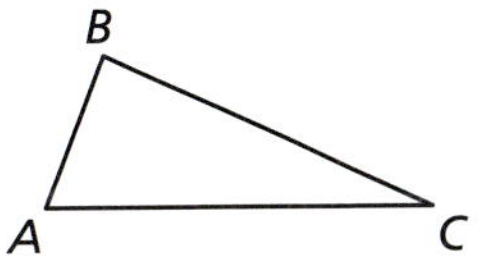

Proof by Contradiction

Step 1 Assume temporarily that $BC \not> AB$. Then it follows that either $BC < AB$ or $BC = AB$.

Step 2 If $BC < AB$, then $m\angle A < m\angle C$ by the Triangle Longer Side Theorem.

If $BC = AB$, then $m\angle A = m\angle C$ by the Base Angles Theorem.

Step 3 Both conclusions contradict the given statement that $m\angle A > m\angle C$. So, the temporary assumption that $BC \not> AB$ cannot be true. This proves that $BC > AB$.

COMMON ERROR

Be sure to consider all cases when assuming the opposite is true.

GO DIGITAL

EXAMPLE 3 Ordering Angle Measures of a Triangle

WATCH

An archaeologist finds artifacts at three different locations. The pottery is 82 meters away from the tools and 89 meters away from the drawings. The tools and drawings are 96 meters apart. List the angles of $\triangle TPD$ in order from smallest to largest.

SOLUTION

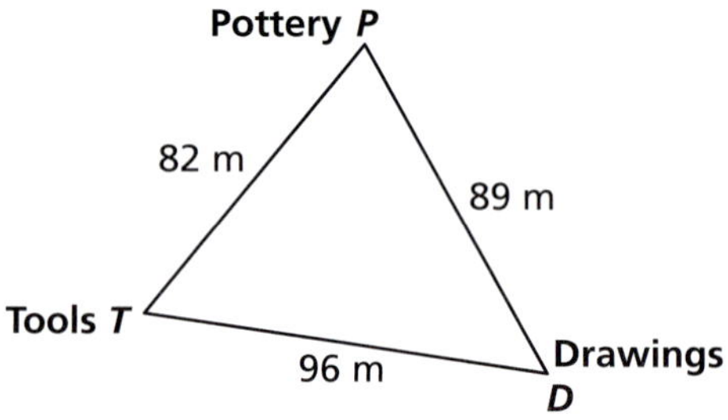

Draw the triangle that represents the situation. Label the side lengths.

The sides from shortest to longest are $\overline{TP}$, $\overline{PD}$, and $\overline{TD}$. The angles opposite these sides are $\angle D$, $\angle T$, and $\angle P$, respectively.

▶ So, by the Triangle Longer Side Theorem, the angles from smallest to largest are $\angle D$, $\angle T$, and $\angle P$.

EXAMPLE 4 Ordering Side Lengths of a Triangle

WATCH

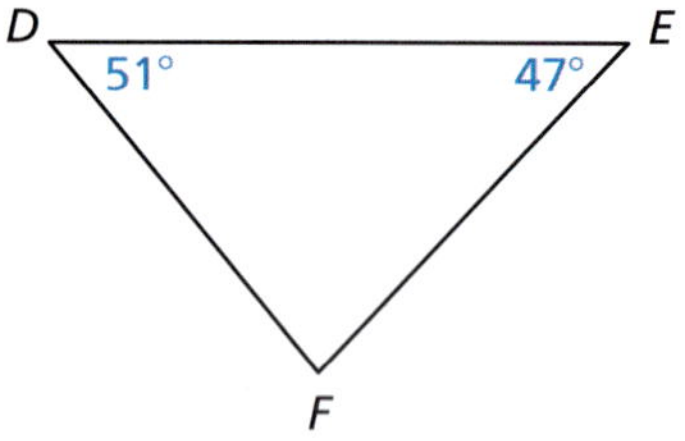

List the sides of $\triangle DEF$ in order from shortest to longest.

SOLUTION

Find $m\angle F$.

$m\angle D + m\angle E + m\angle F = 180°$	Triangle Sum Theorem
$51° + 47° + m\angle F = 180°$	Substitute angle measures.
$m\angle F = 82°$	Solve for $m\angle F$.

The angles from smallest to largest are $\angle E$, $\angle D$, and $\angle F$. The sides opposite these angles are $\overline{DF}$, $\overline{EF}$, and $\overline{DE}$, respectively.

▶ So, by the Triangle Larger Angle Theorem, the sides from shortest to longest are $\overline{DF}$, $\overline{EF}$, and $\overline{DE}$.

SELF-ASSESSMENT

1 I don't understand yet. 2 I can do it with help. 3 I can do it on my own. 4 I can teach someone else.

3. **WRITING** How can you tell which side of a triangle is the longest from the angle measures of the triangle? How can you tell which side is the shortest?

4. List the angles of $\triangle PQR$ in order from smallest to largest.

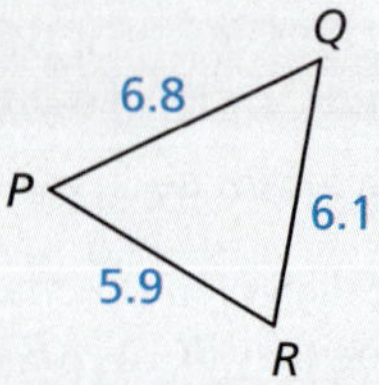

5. List the sides of $\triangle RST$ in order from shortest to longest.

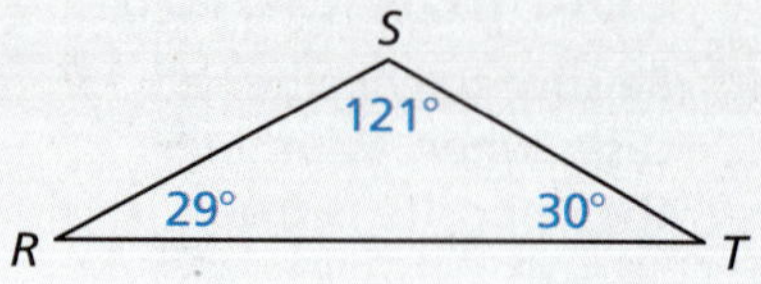

GO DIGITAL

Using the Triangle Inequality Theorem

Not every group of three segments can be used to form a triangle. The lengths of the segments must satisfy a certain relationship. For example, three attempted triangle constructions using segments with given lengths are shown below. Only the first group of segments forms a triangle.

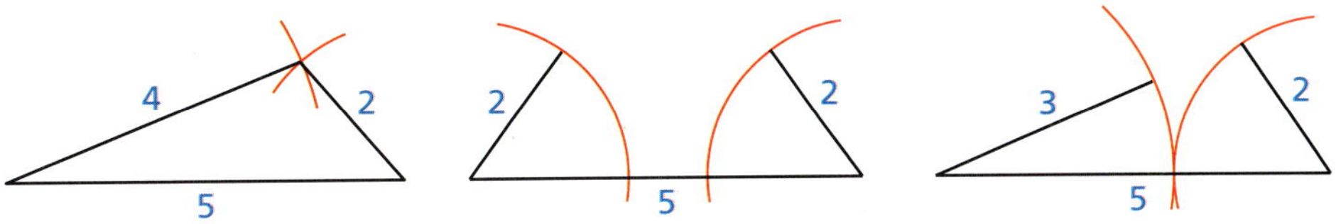

When you start with the longest side and attach the other two sides at its endpoints, you can see that the other two sides are not long enough to form a triangle in the second and third figures. This leads to the *Triangle Inequality Theorem.*

THEOREM

6.11 Triangle Inequality Theorem

The sum of the lengths of any two sides of a triangle is greater than the length of the third side.

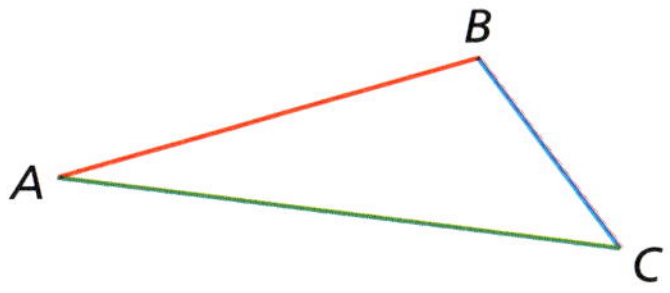

$AB + BC > AC$ $\quad$ $AC + BC > AB$ $\quad$ $AB + AC > BC$

Prove this Theorem Exercise 42, page 336

EXAMPLE 5 Finding Possible Side Lengths

WATCH

A triangle has one side of length 14 and another side of length 9. Describe the possible lengths of the third side.

SOLUTION

Let x represent the length of the third side. Draw diagrams to help visualize the small and large values of x. Then use the Triangle Inequality Theorem to write and solve inequalities.

READING

You can combine the two inequalities, $x > 5$ and $x < 23$, to write the compound inequality $5 < x < 23$. This can be read as *x is between 5 and 23*.

Small values of x

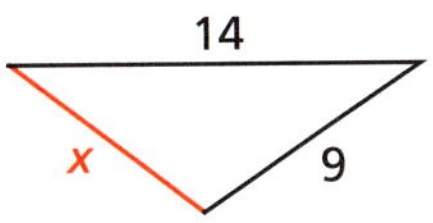

$x + 9 > 14$

$x > 5$

Large values of x

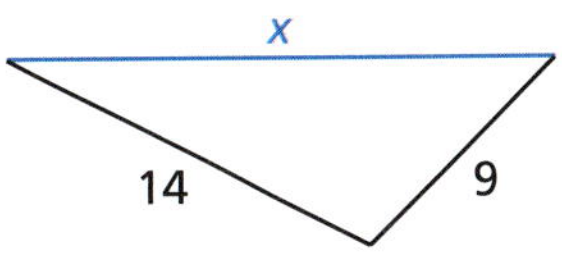

$9 + 14 > x$

$23 > x$, or $x < 23$

▶ The length of the third side must be greater than 5 and less than 23.

SELF-ASSESSMENT

1 I don't understand yet. 2 I can do it with help. 3 I can do it on my own. 4 I can teach someone else.

6. A triangle has one side of length 12 inches and another side of length 20 inches. Describe the possible lengths of the third side.

Decide whether it is possible to construct a triangle with the given side lengths. Explain your reasoning.

7. 4 ft, 9 ft, 10 ft

8. 8 m, 9 m, 18 m

9. 5 cm, 7 cm, 12 cm

GO DIGITAL

6.5 Practice WITH CalcChat® AND CalcView®

In Exercises 1–4, write the first step in a proof by contradiction of the statement.

1. If $WV + VU \neq 12$ inches and $VU = 5$ inches, then $WV \neq 7$ inches.

2. If x and y are odd integers, then xy is odd.

3. In $\triangle ABC$, if $m\angle A = 100°$, then $\angle B$ is not a right angle.

4. In $\triangle JKL$, if M is the midpoint of $\overline{KL}$, then $\overline{JM}$ is a median.

In Exercises 5 and 6, determine which two statements contradict each other. Explain your reasoning.

5. (A) $\triangle LMN$ is a right triangle.
   (B) $\angle L \cong \angle N$
   (C) $\triangle LMN$ is equilateral.

6. (A) Both $\angle X$ and $\angle Y$ have measures greater than 20°.
   (B) Both $\angle X$ and $\angle Y$ have measures less than 30°.
   (C) $m\angle X + m\angle Y = 62°$

▶ 7. **PROOF** Write a proof by contradiction that an odd number is not divisible by 4. *(See Example 1.)*

8. **PROOF** Write a proof by contradiction of the statement "In $\triangle QRS$, if $m\angle Q + m\angle R = 90°$, then $m\angle S = 90°$."

In Exercises 9 and 10, use a ruler and protractor to draw the given type of triangle. Label the largest angle and the longest side. Then label the smallest angle and the shortest side. What do you notice? *(See Example 2.)*

▶ 9. acute scalene

10. right scalene

In Exercises 11–14, list the angles of the triangle in order from smallest to largest. *(See Example 3.)*

▶ 11.
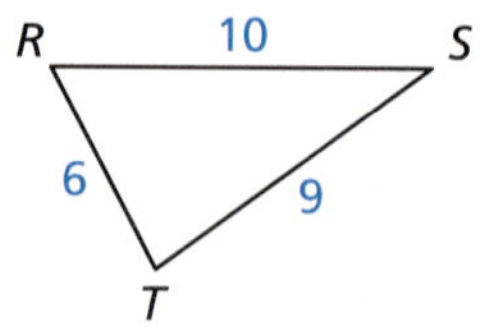

12.
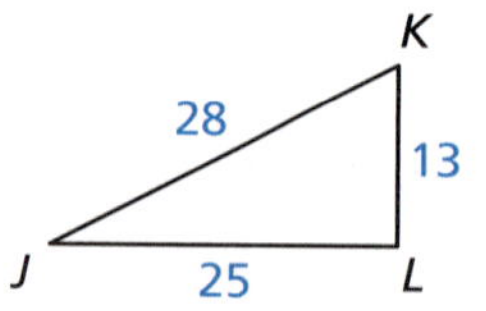

13.
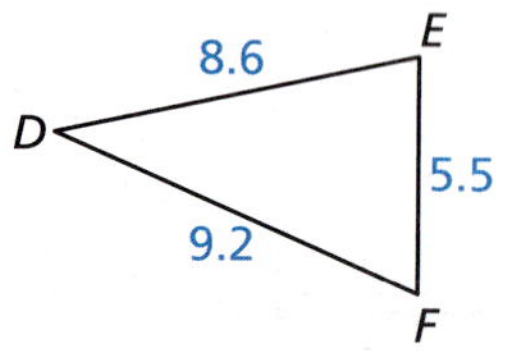

14.
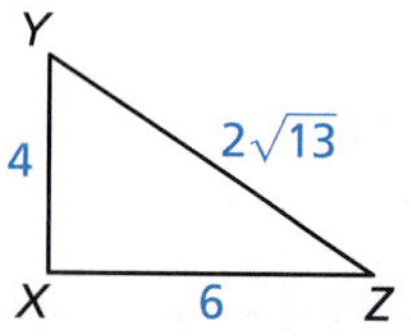

In Exercises 15–18, list the sides of the triangle in order from shortest to longest. *(See Example 4.)*

▶ 15.
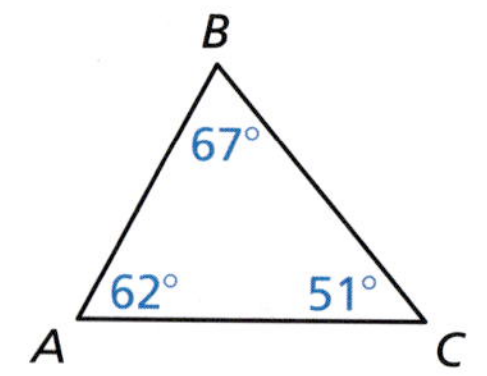

16.
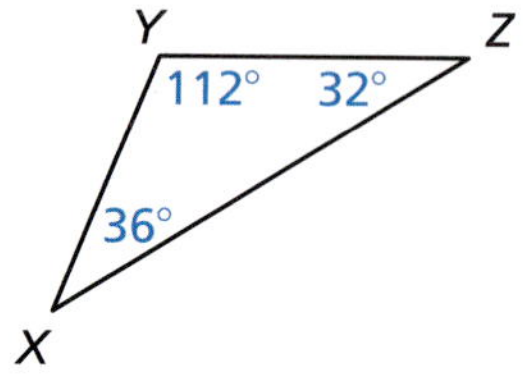

17.
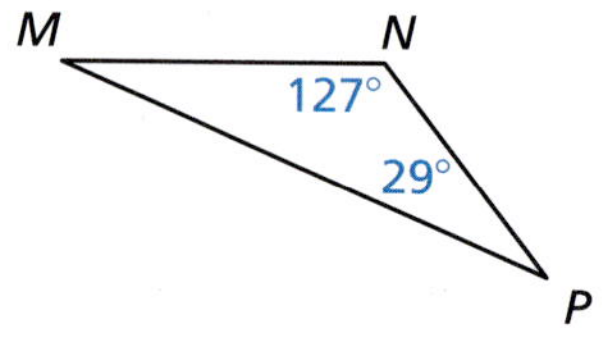

18.
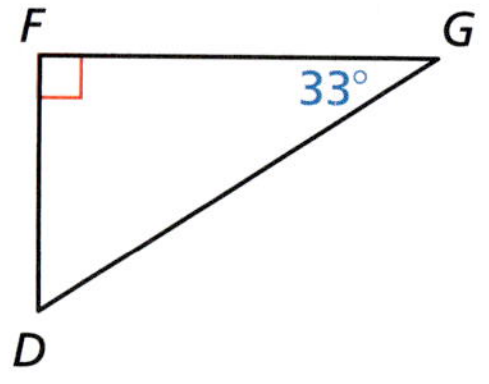

In Exercises 19–22, describe the possible lengths of the third side of the triangle given the lengths of the other two sides. *(See Example 5.)*

▶ 19. 5 inches, 12 inches

20. 12 feet, 18 feet

21. 2 feet, 40 inches

22. 25 meters, 25 meters

In Exercises 23–26, is it possible to construct a triangle with the given side lengths? If not, explain why not.

23. 6, 7, 11

24. 3, 6, 9

25. 28, 17, 46

26. 35, 120, 125

4 MTR 27. **ERROR ANALYSIS** Describe and correct the error in listing the angles of $\triangle ABC$ in order from smallest to largest.

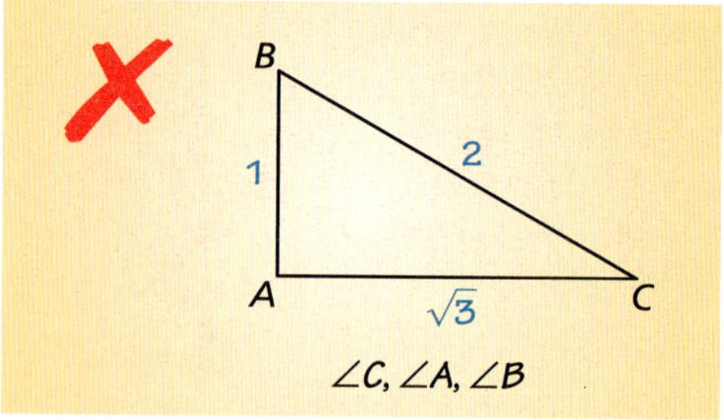

7 MTR 28. **MODELING REAL LIFE** You construct a stage prop of a triangular mountain. $\overline{JL}$ is about 32 feet long, $\overline{JK}$ is about 24 feet long, and $\overline{KL}$ is about 26 feet long. List the angles of $\triangle JKL$ in order from smallest to largest.

GO DIGITAL

29. **REASONING** You are a lawyer representing a client who has been accused of a crime. The crime took place in Los Angeles, California. Security footage shows your client in New York at the time of the crime. Explain how to prove by contradiction that your client is innocent.

30. **REASONING** Your class has fewer than 30 students. The teacher divides your class into two groups. The first group has 15 students. Use a proof by contradiction to show that the second group must have fewer than 15 students.

31. **COLLEGE PREP** Which statement about $\triangle TUV$ is false?

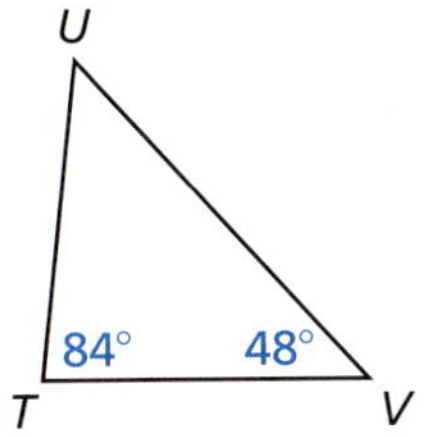

Ⓐ $UV > TU$

Ⓑ $UV + TV > TU$

Ⓒ $UV < TV$

Ⓓ $\triangle TUV$ is isosceles.

32. **COLLEGE PREP** In $\triangle RST$, which is a possible side length for ST? Select all that apply.

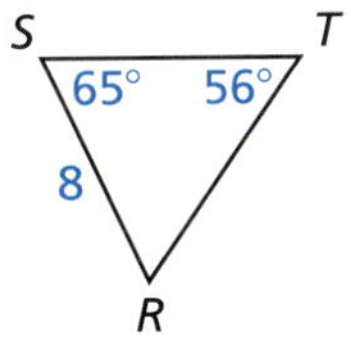

Ⓐ 7

Ⓑ 8

Ⓒ 9

Ⓓ 10

33. **WRITING** Explain why the hypotenuse of a right triangle must always be longer than either leg.

34. **REASONING** In the diagram, $\overline{XY}$ bisects $\angle WYZ$. List all six angles of $\triangle XYZ$ and $\triangle WXY$ in order from smallest to largest. Explain your reasoning.

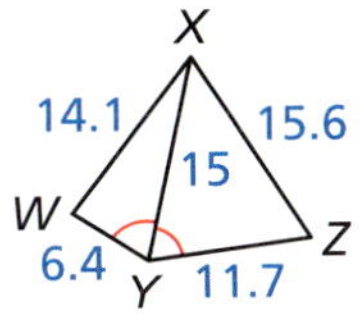

7 MTR 35. **MODELING REAL LIFE** You travel from Daytona Beach to Ocala and from Ocala to Orlando.

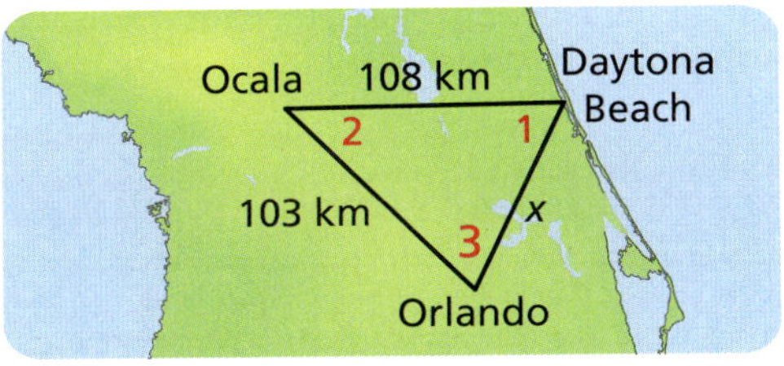

a. Describe the possible distances from Orlando back to Daytona Beach.

b. How is your answer to part (a) affected if you know that $m\angle 2 < m\angle 1$ and $m\angle 2 < m\angle 3$?

7 MTR 36. **MODELING REAL LIFE** You can estimate the width of the river from point A to the tree at point B by measuring the angle to the tree at several locations along the riverbank. The diagram shows the results for locations C and D.

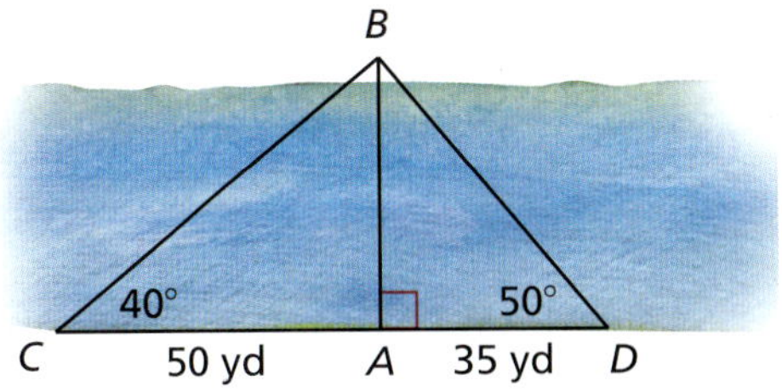

a. Determine the possible widths of the river. Explain your reasoning.

b. Explain what you can do differently to obtain a closer estimate.

5 MTR 37. **CONNECTING CONCEPTS** Describe the possible values of x.

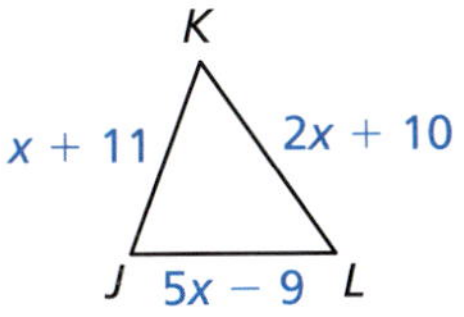

38. **HOW DO YOU SEE IT?**
Your home is on the corner of Hill Street and Eighth Street. The library is on the corner of View Street and Seventh Street. What is the shortest route from your home to the library? Explain.

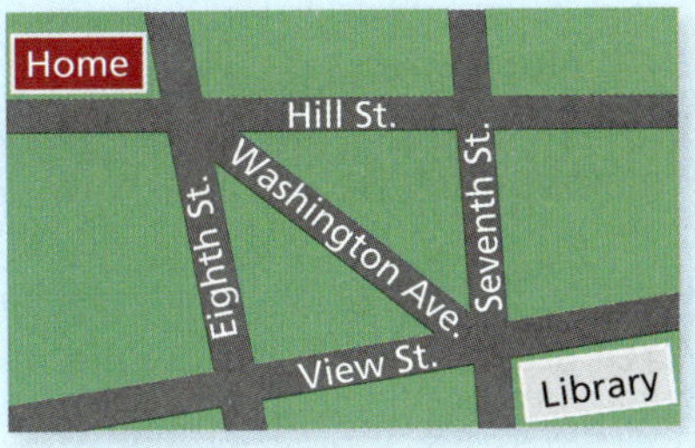

4 MTR 39. **DISCUSS MATHEMATICAL THINKING** Another triangle inequality relationship is given by the Exterior Angle Inequality Theorem. It states the following:

The measure of an exterior angle of a triangle is greater than the measure of either of the nonadjacent interior angles.

Explain how you know that $m\angle 1 > m\angle A$ and $m\angle 1 > m\angle B$ in $\triangle ABC$ with exterior angle $\angle 1$.

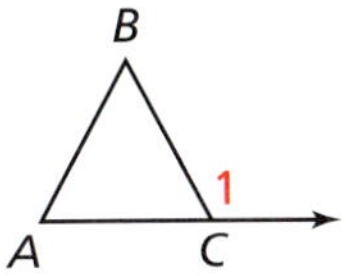

4 MTR 40. **MAKING AN ARGUMENT** Is it possible to draw a triangle with one side length of 13 inches and a perimeter of 2 feet? Explain your reasoning.

GO DIGITAL

41. PROVING A THEOREM Use the diagram to prove the Triangle Longer Side Theorem.

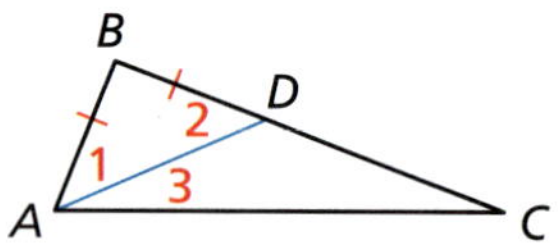

Given $BC > AB$, $BD = BA$

Prove $m\angle BAC > m\angle C$

42. PROVING A THEOREM Prove the Triangle Inequality Theorem.

Given $\triangle ABC$

Prove $AB + BC > AC$, $AC + BC > AB$, and $AB + AC > BC$

3 MTR **43. MAINTAIN ACCURACY** The perimeter of $\triangle HGF$ must be between what two measurements? Explain your reasoning.

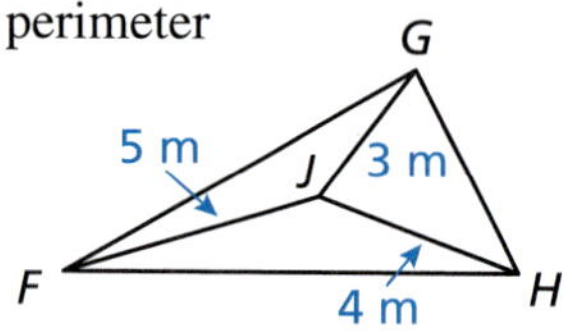

4 MTR **44. DISCUSS MATHEMATICAL THINKING** The length of the base of an isosceles triangle is ℓ. Describe the possible lengths of each leg. Explain your reasoning.

45. PROOF Write a proof by contradiction that a perpendicular segment is the shortest segment from a point to a plane.

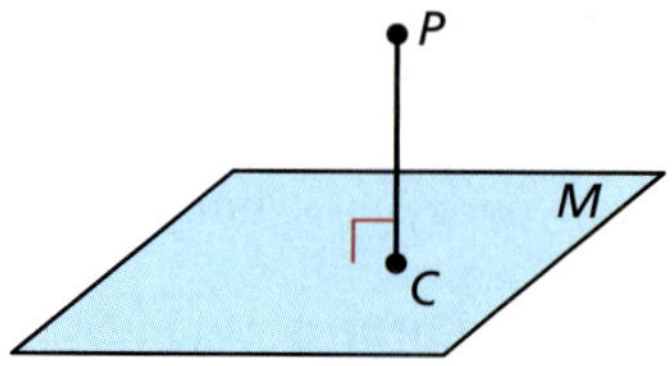

Given $\overline{PC} \perp$ plane M

Prove $\overline{PC}$ is the shortest segment from point P to plane M.

46. THOUGHT PROVOKING
The longest side of a triangle has length z. You randomly choose lesser positive numbers x and y. Use the Triangle Inequality Theorem and the graph to find the probability that x, y, and z represent the side lengths of a triangle. Explain.

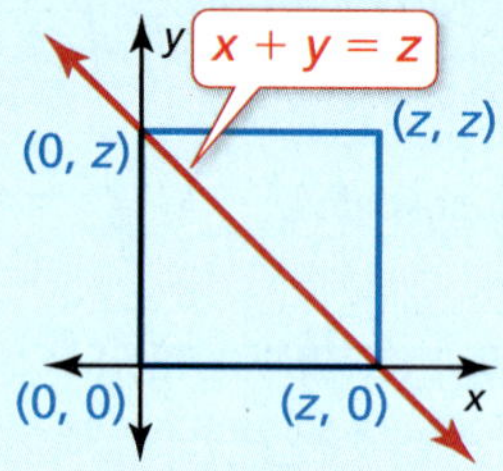

REVIEW & REFRESH

WATCH

47. $\triangle XYZ$ has vertices $X(-5, -2)$, $Y(-3, 4)$, and $Z(3, 0)$. Find the coordinates of the vertices of the midsegment triangle of $\triangle XYZ$.

7 MTR **48. MODELING REAL LIFE** You have a digital photograph that is 960 pixels by 720 pixels. You reduce the size of the photograph to 240 pixels by 180 pixels to post it on a website. What is the scale factor of this dilation?

49. The incenter of $\triangle ABC$ is point N. $NU = -3x + 6$ and $NV = -5x$. Find NT.

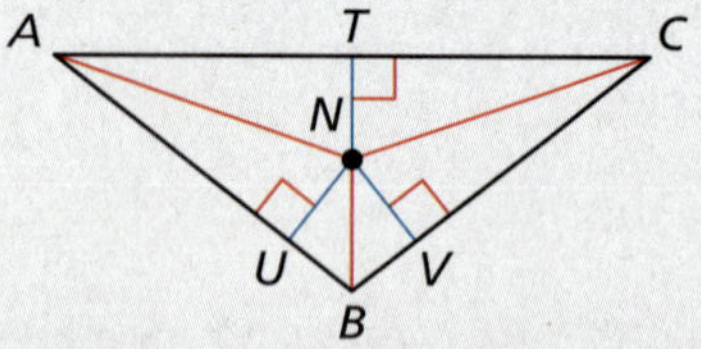

50. Tell whether the orthocenter of $\triangle TUV$ with vertices $T(-2, 5)$, $U(0, 1)$, and $V(2, 5)$ is *inside*, *on*, or *outside* the triangle. Then find its coordinates.

51. Graph $\overline{XY}$ with endpoints $X(-1, 3)$ and $Y(6, -2)$ and its image after a rotation of 270° counterclockwise about the origin, followed by a reflection in the x-axis.

In Exercises 52 and 53, decide whether enough information is given to prove that the triangles are congruent. If so, state the theorem you would use.

52. $\triangle FGJ$, $\triangle HGJ$

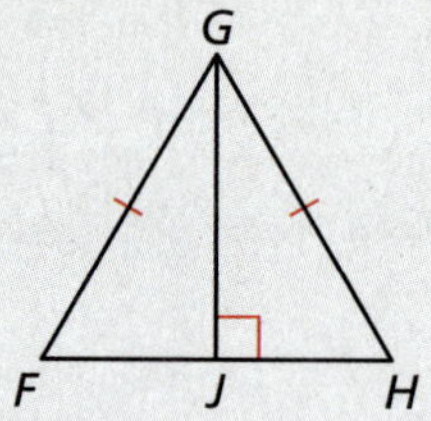

53. $\triangle KLM$, $\triangle KNM$

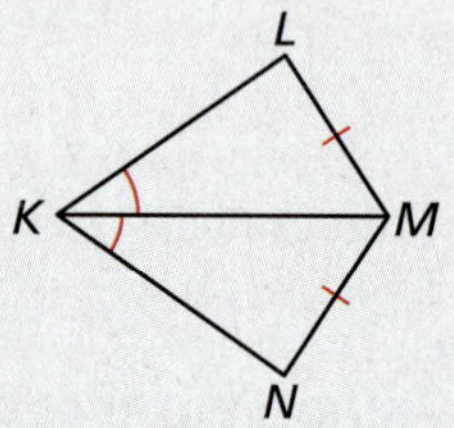

54. A triangle has one side length of 4 centimeters and another side length of 10 centimeters. Describe the possible lengths of the third side.

55. Find the value of x.

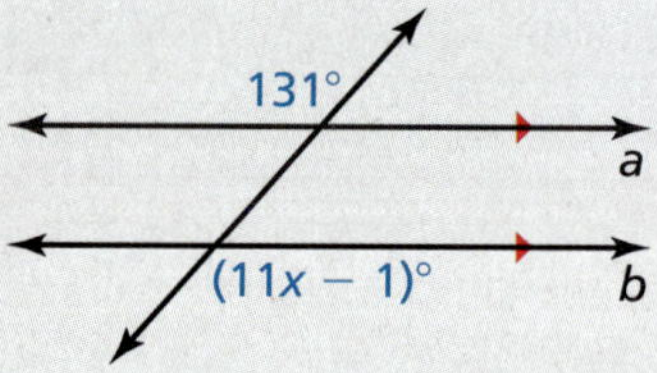

56. Solve $\frac{1}{4}x - \frac{3}{2}y = -1$ for y. Justify each step.

GO DIGITAL

6.6 Inequalities in Two Triangles

Learning Target: Understand inequalities in two triangles.

Success Criteria:
- I can explain the Hinge Theorem.
- I can compare measures in triangles.
- I can solve real-life problems using the Hinge Theorem.

EXPLORE IT! Comparing Measures in Triangles

Work with a partner. Use technology to draw any $\triangle ABC$. Then draw a circle with center B through vertex A.

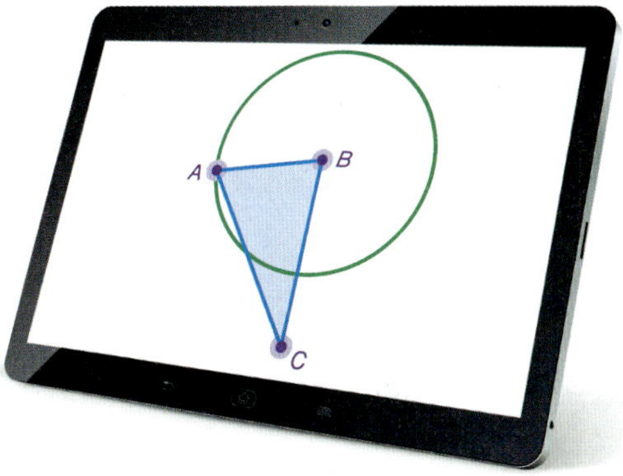

a. Draw $\triangle DBC$ so that D is a point on the circle. Then compare the side lengths of $\triangle ABC$ and $\triangle DBC$.

b. Compare the lengths of $\overline{AC}$ and $\overline{DC}$ and the measures of $\angle ABC$ and $\angle DBC$. What do you notice?

c. Move point D to several locations on the circle. At each location, repeat part (b). Record your results in the table. Then write a conjecture that summarizes your observations.

	AB	*BC*	*AC*	*DC*	$m\angle ABC$	$m\angle DBC$
1.						
2.						
3.						
4.						
5.						

d. If two sides of one triangle are congruent to two sides of another triangle, what can you say about the included angles and the third sides of the triangles?

2 MTR MODEL A PROBLEM

How does the hinge below model the concept you described in part (d)?

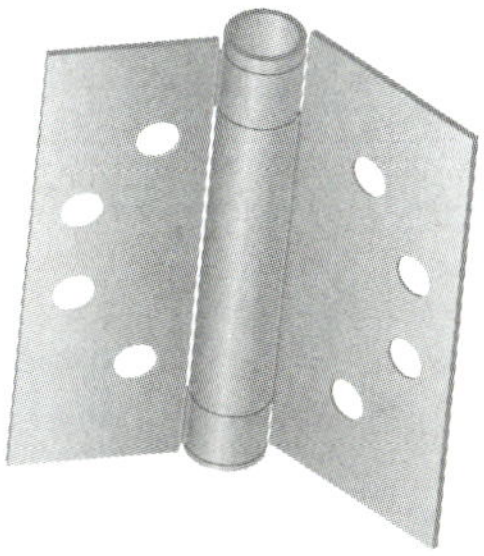

Geometric Reasoning

MA.912.GR.1.3 Prove relationships and theorems about triangles. Solve mathematical and real-world problems involving postulates, relationships and theorems of triangles.

Logic and Discrete Theory

H **MA.912.LT.4.8** Construct proofs, including proofs by contradiction.

GO DIGITAL

Comparing Measures in Triangles

Imagine a gate between fence posts A and B that has hinges at A and swings open at B.

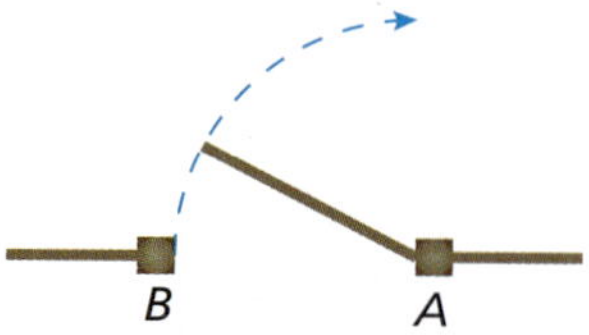

As the gate swings open, you can think of $\triangle ABC$, with side $\overline{AC}$ formed by the gate itself, side $\overline{AB}$ representing the distance between the fence posts, and side $\overline{BC}$ representing the opening between post B and the outer edge of the gate.

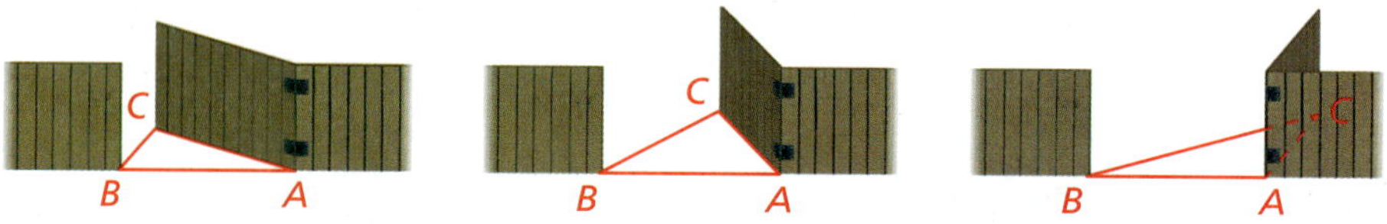

Notice that as the gate opens wider, both the measure of $\angle A$ and the distance BC increase. This suggests the *Hinge Theorem*.

THEOREMS

6.12 Hinge Theorem

If two sides of one triangle are congruent to two sides of another triangle, and the included angle of the first is larger than the included angle of the second, then the third side of the first is longer than the third side of the second.

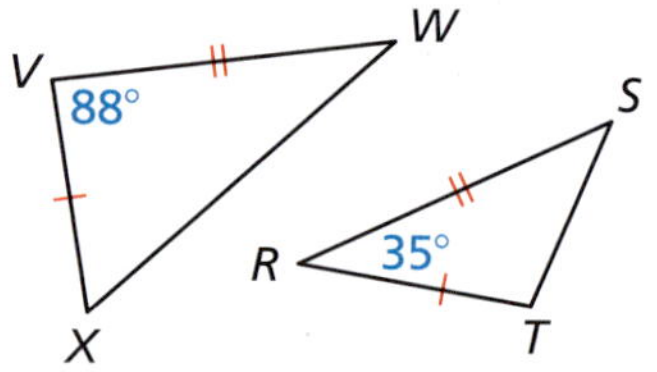

$WX > ST$

Prove this Theorem Exercise 19, page 342

6.13 Converse of the Hinge Theorem

If two sides of one triangle are congruent to two sides of another triangle, and the third side of the first is longer than the third side of the second, then the included angle of the first is larger than the included angle of the second.

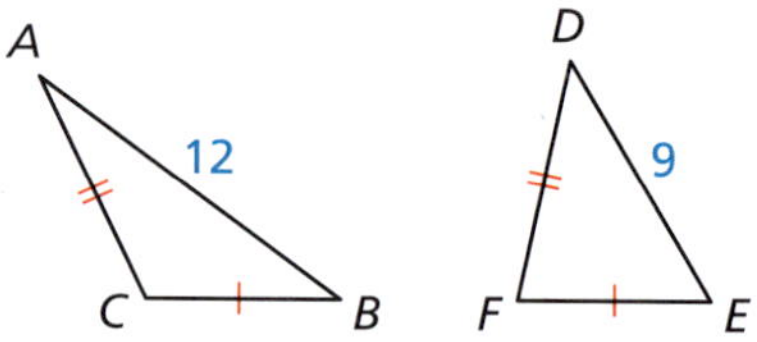

$m\angle C > m\angle F$

Proof page 339

EXAMPLE 1 Using the Converse of the Hinge Theorem

WATCH

Given that $\overline{ST} \cong \overline{PR}$, how does $m\angle PST$ compare to $m\angle SPR$?

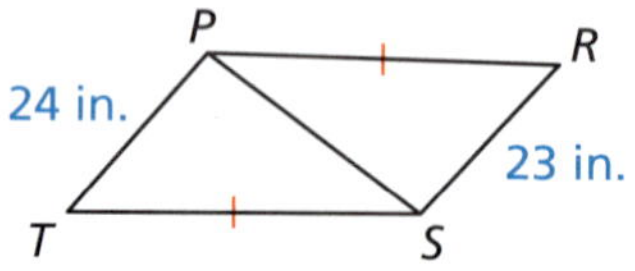

SOLUTION

You are given that $\overline{ST} \cong \overline{PR}$, and you know that $\overline{PS} \cong \overline{PS}$ by the Reflexive Property of Segment Congruence. Because 24 inches > 23 inches, $PT > SR$. So, two sides of $\triangle STP$ are congruent to two sides of $\triangle PRS$, and the third side of $\triangle STP$ is longer.

▶ By the Converse of the Hinge Theorem, $m\angle PST > m\angle SPR$.

GO DIGITAL

EXAMPLE 2 Using the Hinge Theorem

WATCH

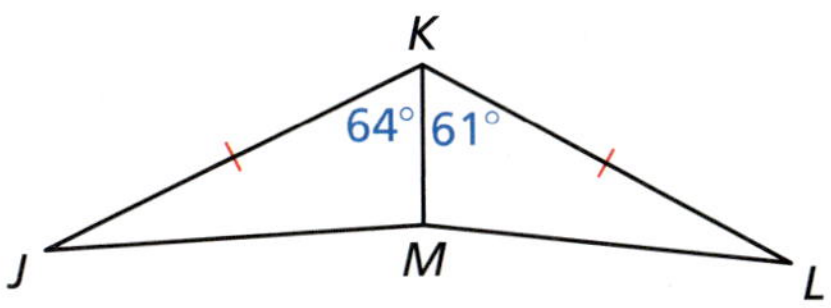

Given that $\overline{JK} \cong \overline{LK}$, how does JM compare to LM?

SOLUTION

You are given that $\overline{JK} \cong \overline{LK}$, and you know that $\overline{KM} \cong \overline{KM}$ by the Reflexive Property of Segment Congruence. Because $64° > 61°$, $m\angle JKM > m\angle LKM$. So, two sides of $\triangle JKM$ are congruent to two sides of $\triangle LKM$, and the included angle of $\triangle JKM$ is larger.

▶ By the Hinge Theorem, $JM > LM$.

PROOF Converse of the Hinge Theorem

H

Given $\overline{AB} \cong \overline{DE}$, $\overline{BC} \cong \overline{EF}$, $AC > DF$

Prove $m\angle B > m\angle E$

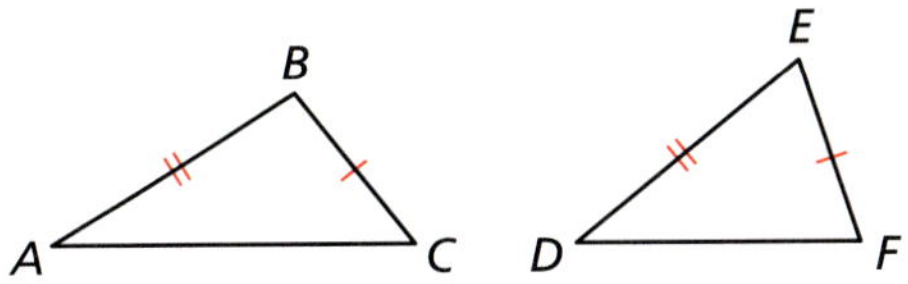

Proof by Contradiction

Step 1 Assume temporarily that $m\angle B \ngtr m\angle E$. Then it follows that either $m\angle B < m\angle E$ or $m\angle B = m\angle E$.

Step 2 If $m\angle B < m\angle E$, then $AC < DF$ by the Hinge Theorem.

If $m\angle B = m\angle E$, then $\angle B \cong \angle E$. So, $\triangle ABC \cong \triangle DEF$ by the SAS Congruence Theorem and $AC = DF$.

Step 3 Both conclusions contradict the given statement that $AC > DF$. So, the temporary assumption that $m\angle B \ngtr m\angle E$ cannot be true. This proves that $m\angle B > m\angle E$.

EXAMPLE 3 Proving Triangle Relationships

WATCH

Write a paragraph proof.

Given $\angle XWY \cong \angle XYW$, $WZ > YZ$

Prove $m\angle WXZ > m\angle YXZ$

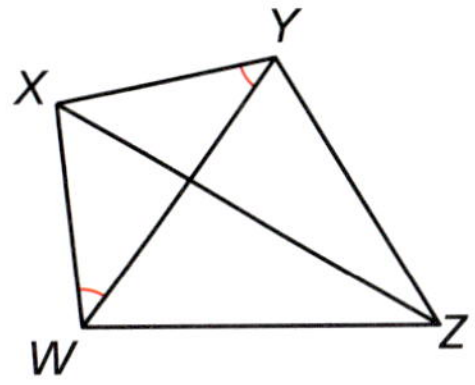

Paragraph Proof Because $\angle XWY \cong \angle XYW$, $\overline{XY} \cong \overline{XW}$ by the Converse of the Base Angles Theorem. By the Reflexive Property of Segment Congruence, $\overline{XZ} \cong \overline{XZ}$. Because $WZ > YZ$, $m\angle WXZ > m\angle YXZ$ by the Converse of the Hinge Theorem.

SELF-ASSESSMENT

1 I don't understand yet. 2 I can do it with help. 3 I can do it on my own. 4 I can teach someone else.

1. **WRITING** Explain why Theorem 6.12 is named the Hinge Theorem.

In Exercises 2 and 3, use the diagram.

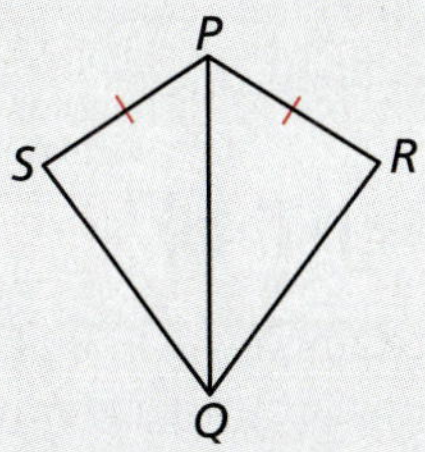

2. If $PR = PS$ and $m\angle QPR > m\angle QPS$, which is longer, $\overline{SQ}$ or $\overline{RQ}$?
3. If $PR = PS$ and $RQ < SQ$, which is larger, $\angle RPQ$ or $\angle SPQ$?
4. Write a temporary assumption you can make to prove the Hinge Theorem by contradiction indirectly. What two cases does that assumption lead to?

GO DIGITAL

Solving Real-Life Problems

7 MTR **EXAMPLE 4** **Modeling Real Life** WATCH

Two groups of bicyclists leave the same camp heading in opposite directions. Each group travels 2 miles, then changes direction and travels 1.2 miles. Group A starts due east and then turns 45° toward north. Group B starts due west and then turns 30° toward south. Which group is farther from camp? Explain your reasoning.

SOLUTION

1. **Understand the Problem** You know the distances and directions that the groups of bicyclists travel. You need to determine which group is farther from camp. You can interpret a turn of 45° toward north as shown.

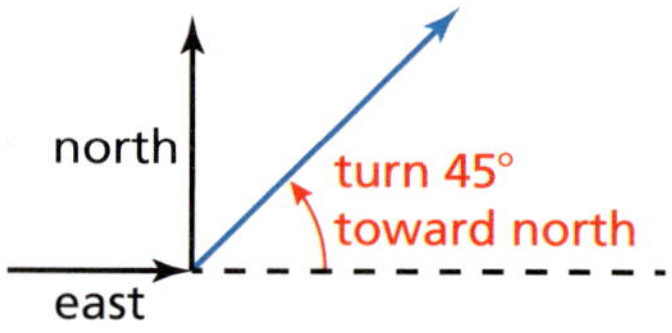

2. **Make a Plan** Draw a diagram that represents the situation and mark the given measurements. The distances that the groups bike and the distances back to camp form two triangles. The triangles each have two congruent sides with lengths of 2 miles and 1.2 miles.

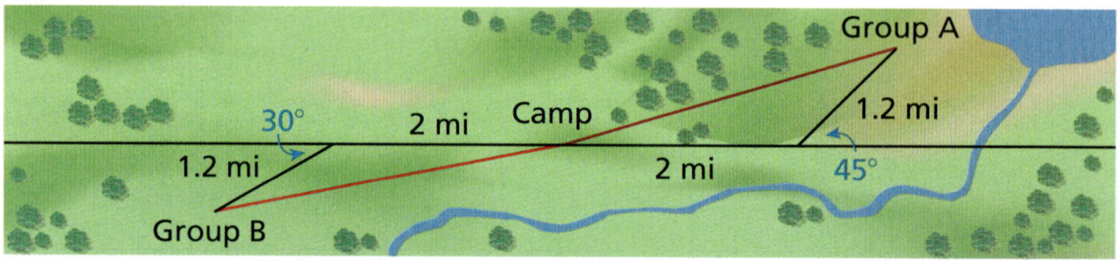

3. **Solve and Check** Use linear pairs to find the included angles formed by the paths that the groups take.

Group A: $180° - 45° = 135°$ **Group B:** $180° - 30° = 150°$

The included angles are 135° and 150°.

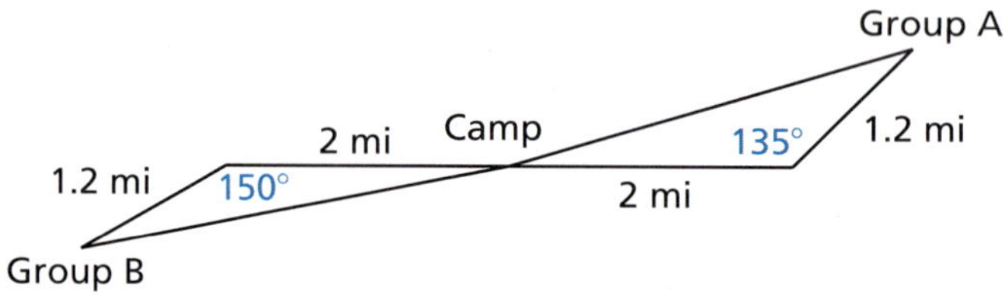

Because 150° > 135°, Group B's distance from camp is greater than Group A's distance from camp by the Hinge Theorem.

▶ So, Group B is farther from camp.

Look Back Because the turn toward north for Group A is greater than the turn toward south for Group B, you can reason that Group A would be closer to camp than Group B. So, Group B is farther from camp.

SELF-ASSESSMENT 1 I don't understand yet. 2 I can do it with help. 3 I can do it on my own. 4 I can teach someone else.

5. **WHAT IF?** In Example 4, Group C leaves camp and travels 2 miles due north, then turns 40° toward east and travels 1.2 miles. Compare the distances from camp for all three groups.

GO DIGITAL

6.6 Practice WITH CalcChat® AND CalcView®

In Exercises 1–4, complete the statement with < or >. Explain your reasoning. *(See Example 1.)*

1. $m\angle 1$ _____ $m\angle 2$

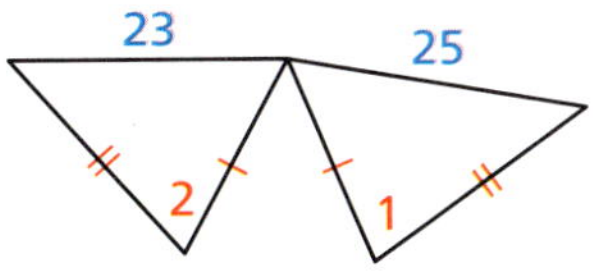

2. $m\angle 1$ _____ $m\angle 2$

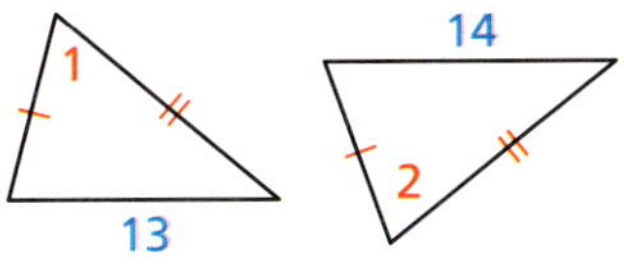

3. $m\angle 1$ _____ $m\angle 2$

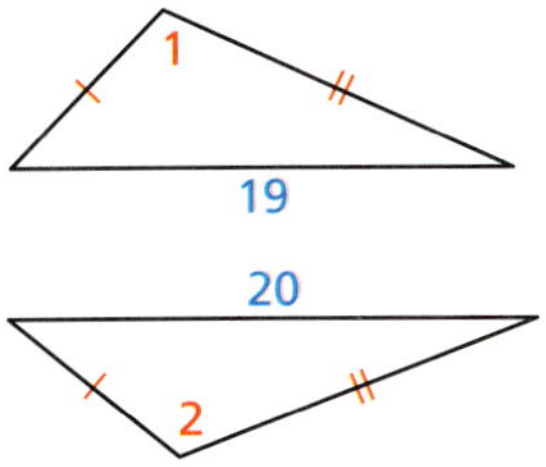

4. $m\angle 1$ _____ $m\angle 2$

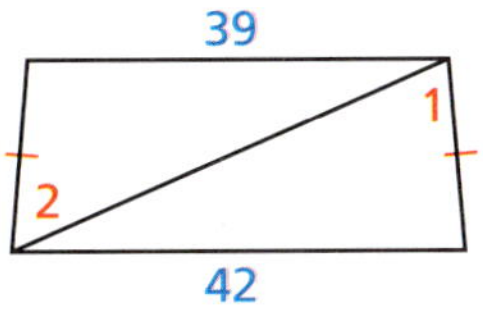

In Exercises 5–8, complete the statement with < or >. Explain your reasoning. *(See Example 2.)*

5. AC _____ DC

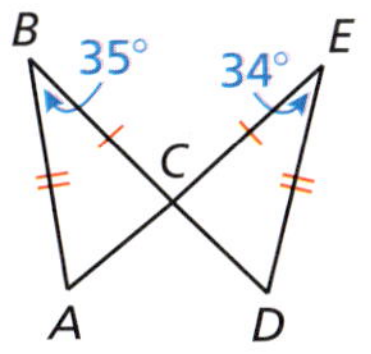

6. MN _____ LK

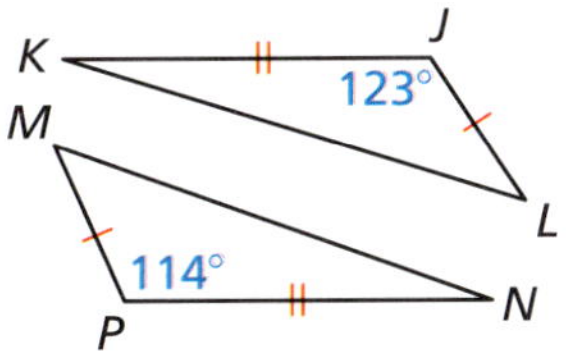

7. TR _____ UR

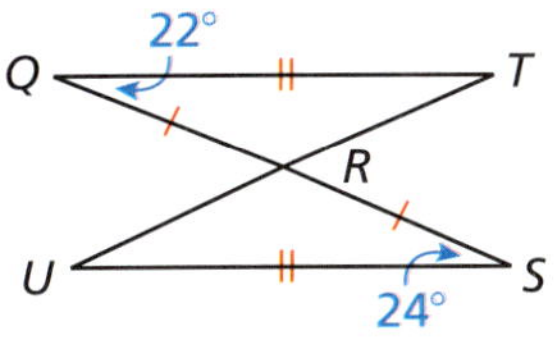

8. AD _____ CD

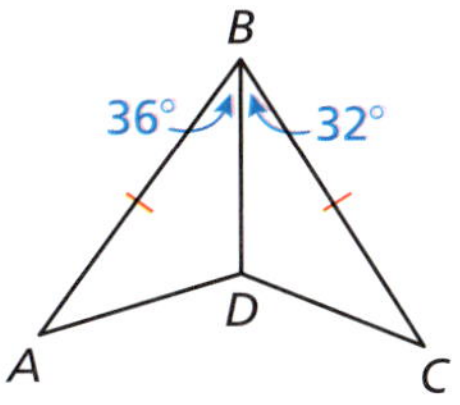

PROOF In Exercises 9 and 10, write a proof. *(See Example 3.)*

9. **Given** $\overline{XY} \cong \overline{ZY}$, $m\angle WYZ > m\angle WYX$

Prove $WZ > WX$

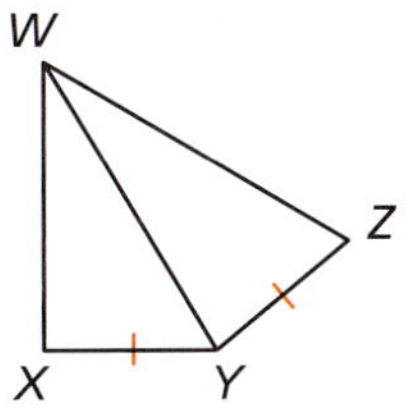

10. **Given** $\overline{BC} \cong \overline{DA}$, $DC < BA$

Prove $m\angle BCA > m\angle DAC$

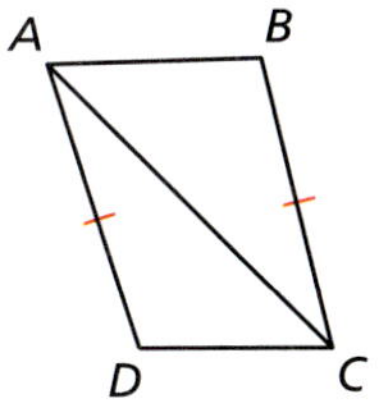

MODELING REAL LIFE In Exercises 11 and 12, two flights leave from the same airport. Determine which flight is farther from the airport. Explain your reasoning. *(See Example 4.)*

11. **Flight 1:** Flies 100 miles due west, then turns 20° toward north and flies 50 miles.

Flight 2: Flies 100 miles due north, then turns 30° toward east and flies 50 miles.

12. **Flight 1:** Flies 210 miles due south, then turns 70° toward west and flies 80 miles.

Flight 2: Flies 80 miles due north, then turns 50° toward east and flies 210 miles.

13. **ERROR ANALYSIS** Describe and correct the error in using the Hinge Theorem.

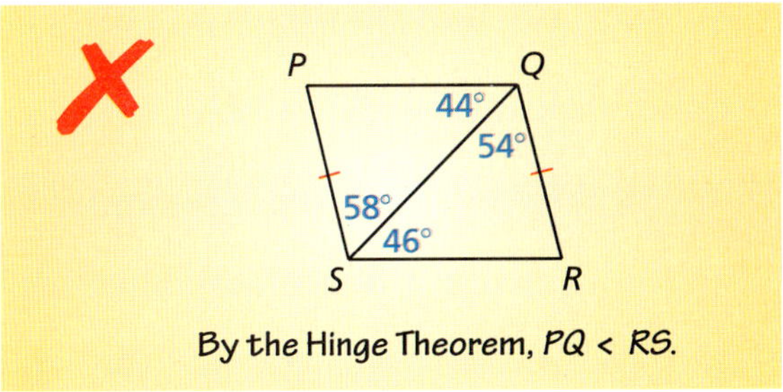

14. **COLLEGE PREP** Which is a possible measure of $\angle JKM$? Select all that apply.

Ⓐ 15°

Ⓑ 22°

Ⓒ 25°

Ⓓ 35°

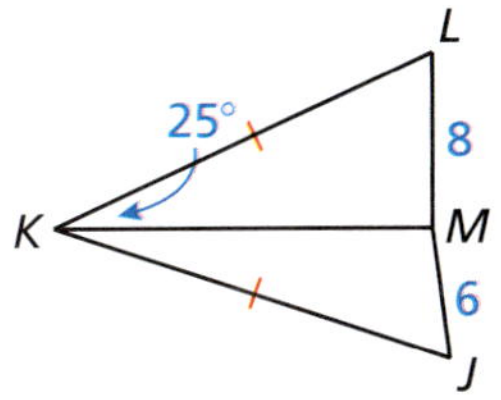

15. **CONNECTING CONCEPTS** Find the possible values of x in the figure. Explain your reasoning.

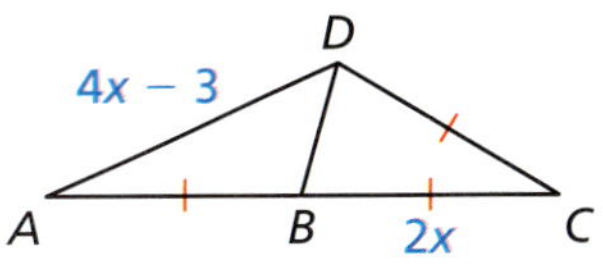

GO DIGITAL

16. HOW DO YOU SEE IT? In the diagram, triangles are formed by the locations of players on the basketball court. The dashed lines represent possible paths of the basketball as the players pass. How does $m\angle ACB$ compare with $m\angle ACD$? Explain your reasoning.

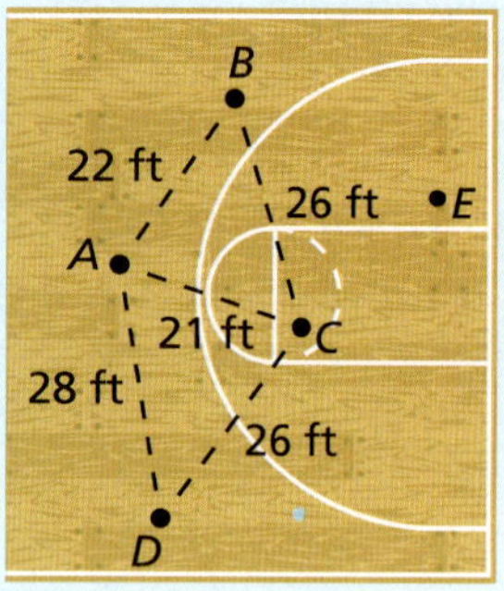

17. REASONING $\overline{NR}$ is a median of $\triangle NPQ$, and $NQ > NP$. Explain why $\angle NRQ$ is obtuse.

18. THOUGHT PROVOKING
The postulates and theorems in this book represent Euclidean geometry. In spherical geometry, all points are on the surface of a sphere. A line is a circle on the sphere whose diameter is equal to the diameter of the sphere. In spherical geometry, state an inequality involving the sum of the angles of a triangle. Find a formula for the area of a triangle in spherical geometry.

19. PROVING A THEOREM Use the Plan for Proof to prove the Hinge Theorem.

Given $\overline{AB} \cong \overline{DE}$, $\overline{BC} \cong \overline{EF}$, $m\angle ABC > m\angle DEF$

Prove $AC > DF$

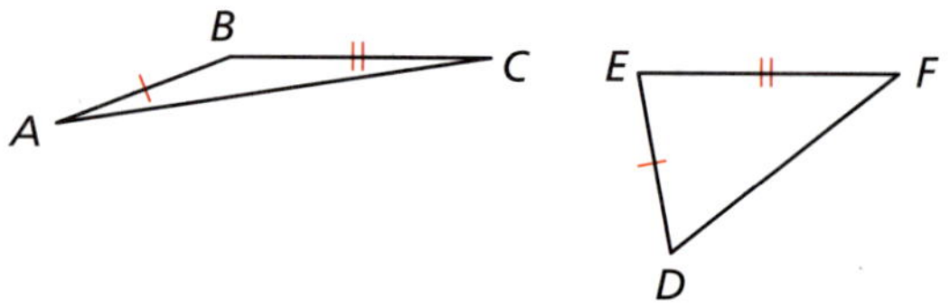

Plan for Proof

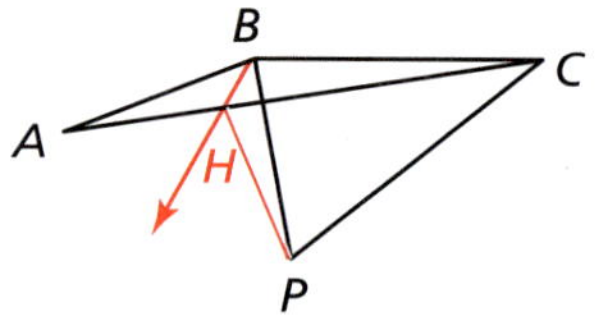

1. Because $m\angle ABC > m\angle DEF$, you can locate a point P in the interior of $\angle ABC$ so that $\angle CBP \cong \angle FED$ and $\overline{BP} \cong \overline{ED}$. Draw $\overline{BP}$ and show that $\triangle PBC \cong \triangle DEF$.
2. Locate a point H on $\overline{AC}$ so that $\overrightarrow{BH}$ bisects $\angle PBA$, and show that $\triangle ABH \cong \triangle PBH$.
3. Begin with the statement $AC = AH + HC$. Then show that $AC > DF$.

REVIEW & REFRESH

WATCH

In Exercises 20 and 21, find the value of *x*.

20.

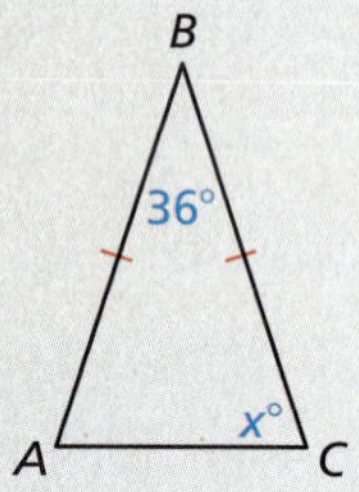

21. $\overline{ST}$ is a midsegment of $\triangle PQR$.

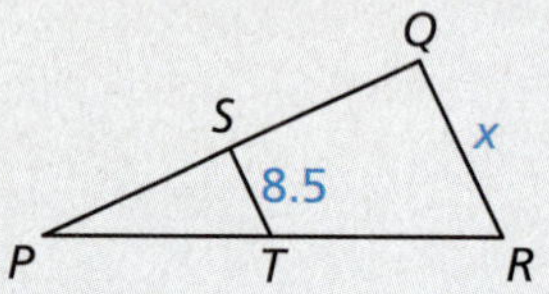

22. Graph quadrilateral $ABCD$ with vertices $A(-8, 2)$, $B(-6, 8)$, $C(-2, 6)$, and $D(-4, 2)$ and its image after the similarity transformation.

Reflection: in the x-axis

Dilation: $(x, y) \rightarrow (0.5x, 0.5y)$

23. Which is longer, $\overline{FG}$ or $\overline{KL}$? Explain your reasoning.

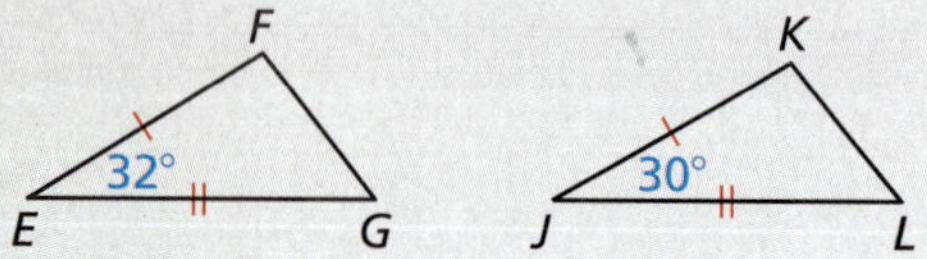

7 MTR **24. MODELING REAL LIFE** The Deer County Parks Committee plans to build a park. The committee sketches one possible location of the park represented by point P.

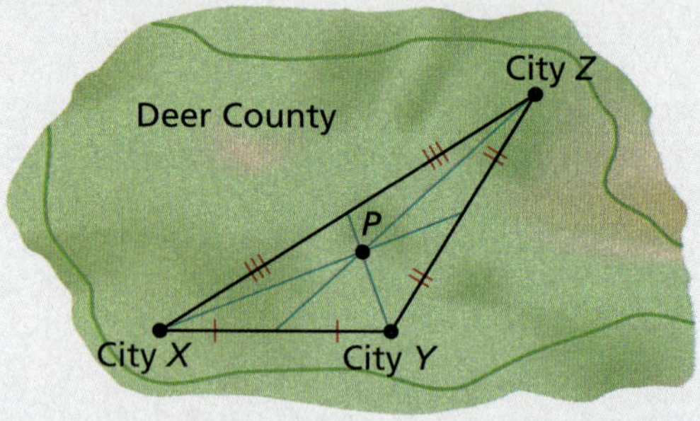

a. Which point of concurrency did the committee use in the sketch as the location of the park?

b. The committee wants the park to be equidistant from the three cities. Should the point of concurrency in the sketch be the location of the park? If not, which point of concurrency should the committee use? Explain.

25. Is it possible to construct a triangle with side lengths of 9 inches, 11 inches, and 21 inches? Explain your reasoning.

GO DIGITAL

6 Chapter Review WITH CalcChat®

Chapter Learning Target: Understand relationships within triangles.

Chapter Success Criteria:
- ◆ I can identify special segments in a triangle.
- ◆ I can use special segments in a triangle to solve problems.
- ■ I can prove statements using special segments in a triangle.
- ■ I can compare measures within triangles and between two triangles.

◆ Surface
■ Deep

SELF-ASSESSMENT 1 I don't understand yet. 2 I can do it with help. 3 I can do it on my own. 4 I can teach someone else.

6.1 Perpendicular and Angle Bisectors (pp. 297–304) WATCH

Learning Target: Use theorems about perpendicular and angle bisectors.

Vocabulary AZ VOCAB
equidistant

Find the indicated measure. Explain your reasoning.

1. DC

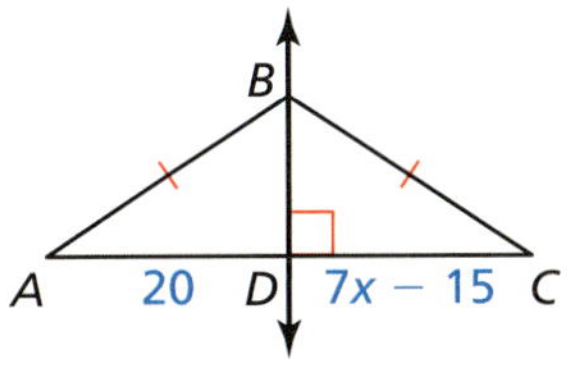

2. RS

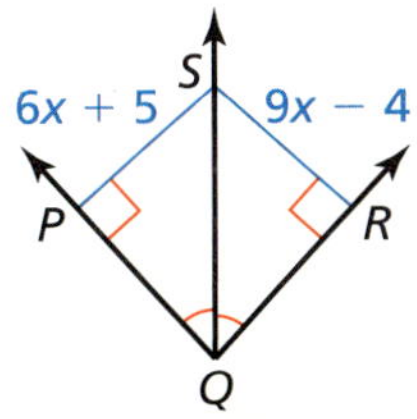

3. $m\angle JFH$

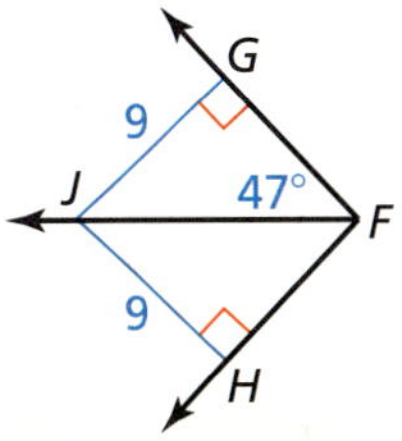

4. Is there enough information in the diagram to find ST? If so, find the length. If not, explain why not.

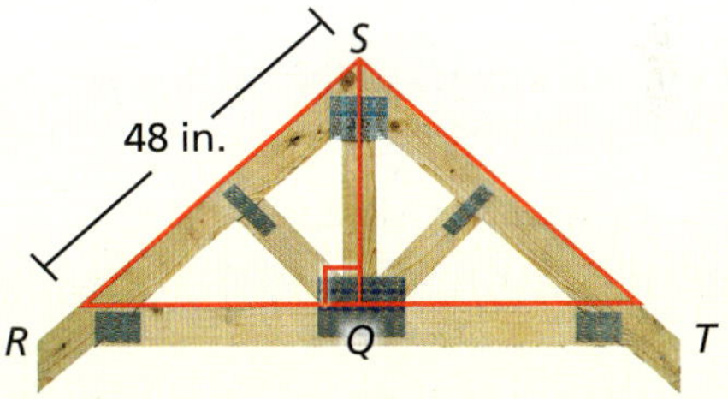

6.2 Bisectors of Triangles (pp. 305–314) WATCH

Learning Target: Use bisectors of triangles.

Vocabulary AZ VOCAB
concurrent
point of concurrency
circumcenter
incenter

Find the coordinates of the circumcenter of the triangle with the given vertices.

5. $T(-6, -5)$, $U(0, -1)$, $V(0, -5)$
6. $X(-2, 1)$, $Y(2, -3)$, $Z(6, -3)$
7. Point D is the incenter of $\triangle LMN$. Find the value of x.

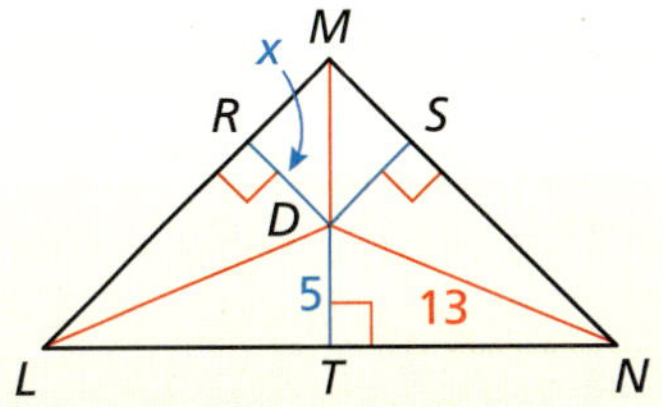

8. Draw an acute $\triangle ABC$. Construct the circumscribed circle about $\triangle ABC$.
9. Draw an obtuse $\triangle DEF$. Construct the inscribed circle within $\triangle DEF$.

GO DIGITAL

6.3 Medians and Altitudes of Triangles (pp. 315–322)

WATCH

Learning Target: Use medians and altitudes of triangles.

Vocabulary (VOCAB)

median of a triangle
centroid
altitude of a triangle
orthocenter

Point D is the centroid of $\triangle ABC$. Find ED and DC.

10. $EC = 18$

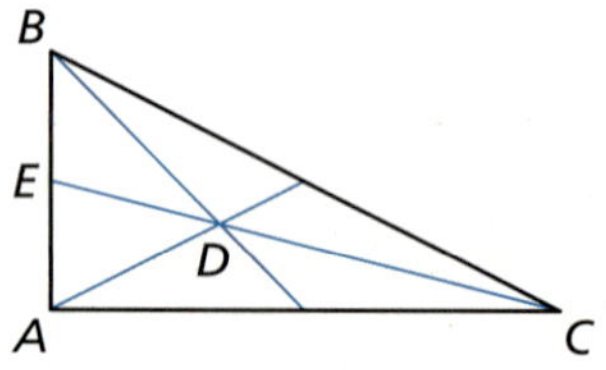

11. $EC = 27$

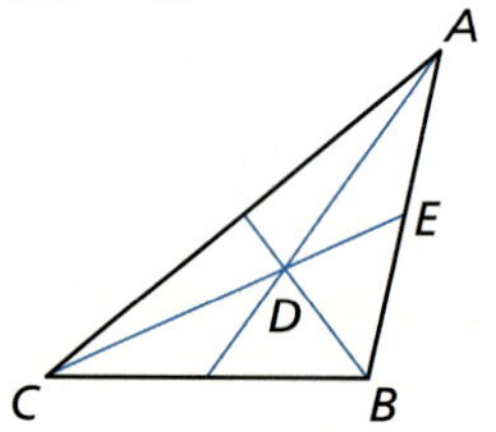

Find the coordinates of the centroid of the triangle with the given vertices.

12. $A(-10, 3)$, $B(-4, 5)$, $C(-4, 1)$

13. $D(2, -8)$, $E(2, -2)$, $F(8, -2)$

Tell whether the orthocenter of the triangle with the given vertices is *inside*, *on*, or *outside* the triangle. Then find the coordinates of the orthocenter.

14. $G(1, 6)$, $H(5, 6)$, $J(3, 1)$

15. $K(-8, 5)$, $L(-6, 3)$, $M(0, 5)$

16. The centroid of $\triangle ABC$ lies on one of the altitudes of $\triangle ABC$. What can you conclude about $\triangle ABC$?

17. Draw a triangle so that one of its vertices is the orthocenter. Explain how you drew the triangle to meet this condition.

6.4 The Triangle Midsegment Theorem (pp. 323–328)

WATCH

Learning Target: Find and use midsegments of triangles.

Vocabulary (VOCAB)

midsegment of a triangle

Find the coordinates of the vertices of the midsegment triangle for the triangle with the given vertices.

18. $A(-6, 8)$, $B(-6, 4)$, $C(0, 4)$

19. $D(-3, 1)$, $E(3, 5)$, $F(1, -5)$

$\overline{DE}$ is a midsegment of $\triangle ABC$. Find the value of x.

20.

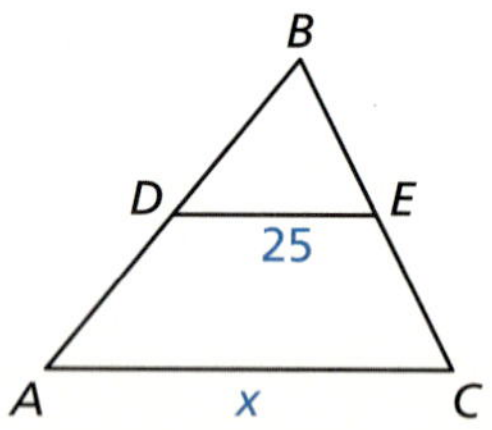

21.

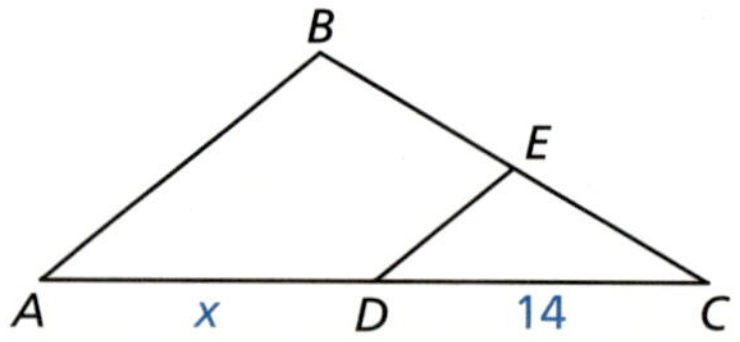

22. The diagram shows your first lap in mowing the triangular field. How far do you travel in this lap?

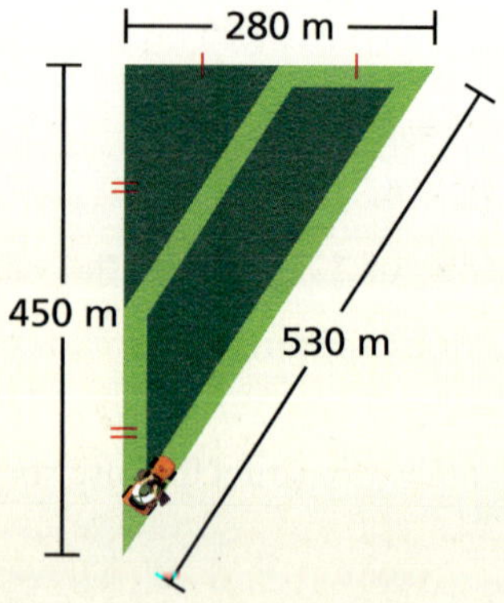

GO DIGITAL

6.5 Proof by Contradiction and Inequalities in One Triangle (pp. 329–336)

WATCH

Learning Target: Write proofs by contradictions and understand inequalities in a triangle.

Vocabulary

proof by contradiction

23. Write a proof by contradiction of the statement "In $\triangle XYZ$, if $XY = 4$ and $XZ = 8$, then $YZ > 4$."

List the sides of the triangle in order from shortest to longest.

24.

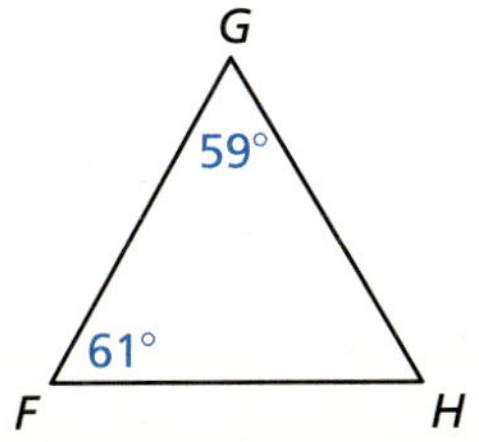

25.

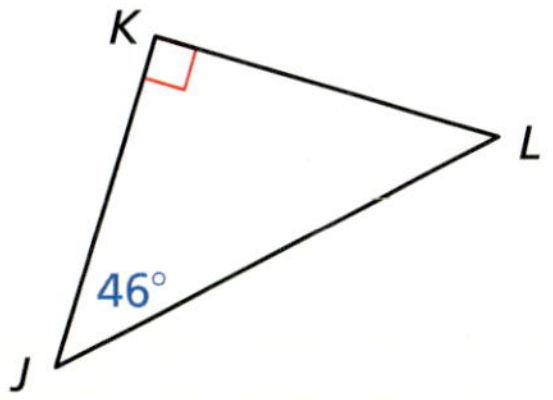

Describe the possible lengths of the third side of the triangle given the lengths of the other two sides.

26. 4 inches, 8 inches

27. 6 meters, 9 meters

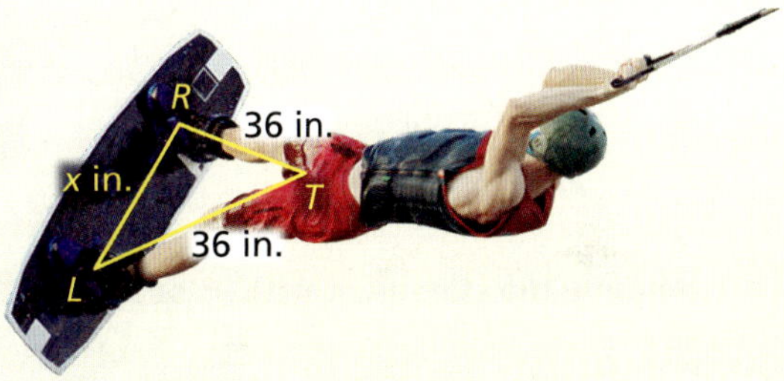

28. The widest adjustment possible for a wakeboarder's feet on the wakeboard is $x = 24$ inches. Write a proof of contradiction that there can be at most two congruent angles of the triangle formed by the wakeboarder's legs and the wakeboard.

6.6 Inequalities in Two Triangles (pp. 337–342)

WATCH

Learning Target: Understand inequalities in two triangles.

Use the diagram.

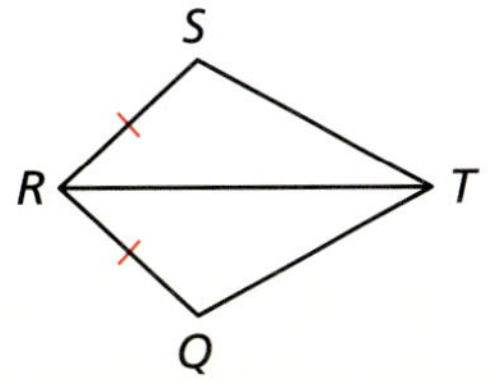

29. If $RQ = RS$ and $m\angle QRT > m\angle SRT$, then how does $\overline{QT}$ compare to $\overline{ST}$?

30. If $RQ = RS$ and $QT > ST$, then how does $\angle QRT$ compare to $\angle SRT$?

31. Two boats leave the same island. The first boat travels 20 miles due south, then turns 20° toward east and travels 15 miles. The second boat travels 15 miles due east, then turns 10° toward north and travels 20 miles. Which boat is farther from the island? Explain.

Mathematical Thinking and Reasoning

1 MTR ACTIVELY PARTICIPATE IN EFFORTFUL LEARNING BOTH INDIVIDUALLY AND COLLECTIVELY

Mathematicians who actively participate in effortful learning both individually and with others analyze problems in a way that makes sense given the tasks.

1. In Exercise 35 on page 313, what was the goal of the problem? What information was given? Describe the relationships that you used to obtain your answer.

2. Refer to Exercise 11 on page 341. Describe the first step you take when solving the problem. Does your answer change when Flight 2 turns west instead of east? Explain your reasoning.

GO DIGITAL

6 Practice Test WITH CalcChat®

In Exercises 1 and 2, $\overline{MN}$ is a midsegment of $\triangle JKL$. Find the value of x.

1.

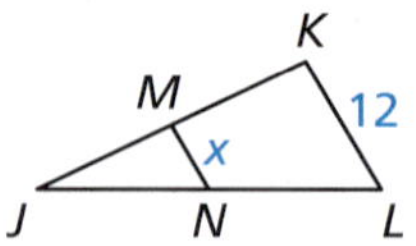

2.

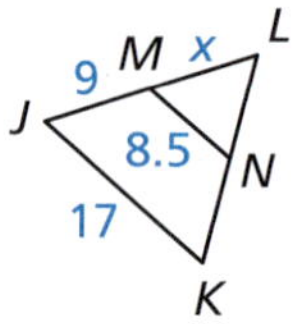

Find the indicated measure. Explain your reasoning.

3. ST

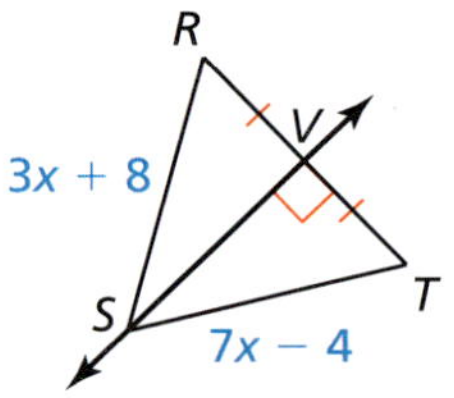

4. WY

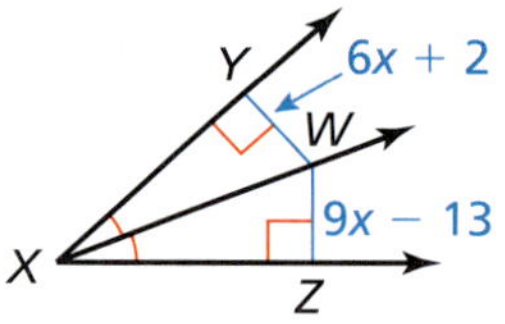

5. BW

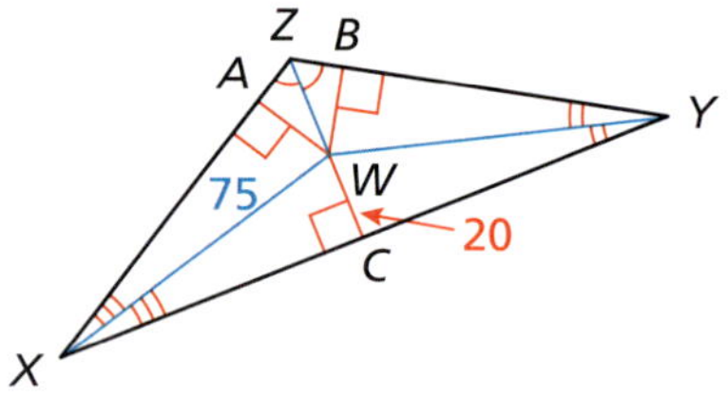

Complete the statement with < or >.

6. AB __ CB

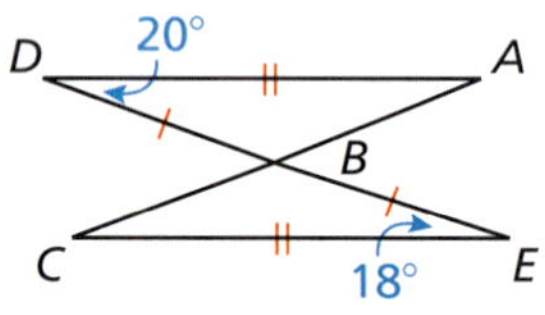

7. $m\angle 1$ ___ $m\angle 2$

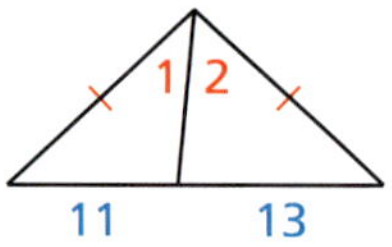

8. $m\angle MNP$ ___ $m\angle NPM$

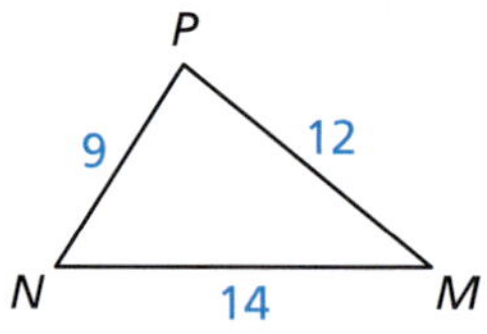

9. Find the coordinates of the circumcenter, orthocenter, and centroid of the triangle with vertices $A(0, -2)$, $B(4, -2)$, and $C(0, 6)$.

10. Write a proof by contradiction of the Corollary to the Base Angles Theorem: If $\triangle PQR$ is equilateral, then it is equiangular.

11. $\triangle DEF$ is a right triangle with area A and vertices $D(0, 0)$, $E(a, b)$, and $F(a, 0)$. Point G is the midpoint of $\overline{DE}$, and point H is the midpoint of $\overline{DF}$. Write an expression for the area of $\triangle DGH$. Justify your answer.

12. Given $A(0, a)$ and $B(b, 0)$, find coordinates of C so that the orthocenter of $\triangle ABC$ lies on the triangle.

13. Two hikers set out from camp. The first hikes 4 miles due west, then turns 40° toward south and hikes 1.8 miles. The second hikes 4 miles due east, then turns 52° toward north and hikes 1.8 miles. Which hiker is farther from camp? Explain.

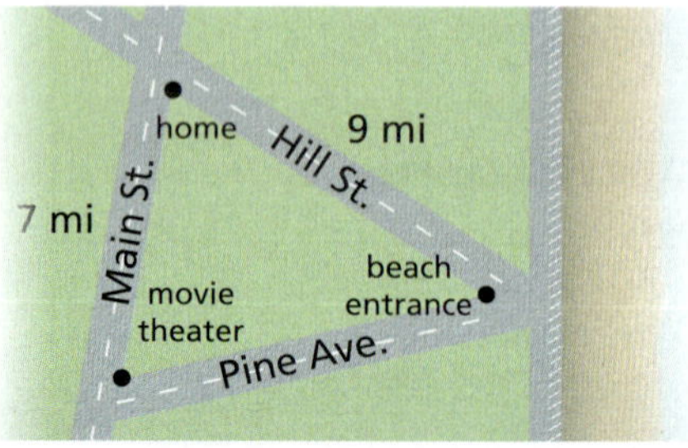

In Exercises 14–16, use the map.

14. Describe the possible lengths of Pine Avenue.

15. You bike along a trail that is the shortest distance from the beach entrance to Main Street. You end up halfway between home and the movie theater. How long is Pine Avenue? Explain.

16. A market is the same distance from home, the movie theater, and the beach entrance. Copy the map and locate the market. What point of concurrency does this location represent?

GO DIGITAL

6 Performance Task
Archaeology

Archaeology is the study of human past through material remains.

The Archaeological Process

Selecting an Excavation Site
A site is selected by consulting historic records, conducting field surveys, and sometimes by accidentally discovering artifacts.

Conducting Research
Archaeologists find reliable sources to learn about a specific people, place, and time period.

Excavating the Site
Archaeologists make a detailed map and establish a physical grid over the site. Each grid square is then excavated one layer at a time. As artifacts are discovered, their grid coordinates and layer numbers are recorded.

Cleaning and Cataloging Artifacts
Artifacts are cleaned carefully to avoid inflicting damage. Then they are sorted into categories, and fragments are pieced together to reveal the size and shape for photographs.

Reporting the Results
The final step is to analyze and interpret the findings, which are reported in publications, reports, lectures, and exhibitions.

- In archaeology, broken pieces of pottery are called *sherds*, and broken pieces of glass are called *shards*.
- No written records exist for about 99% of human history.
- The oldest stone tools ever discovered are estimated to be 3.3 million years old.

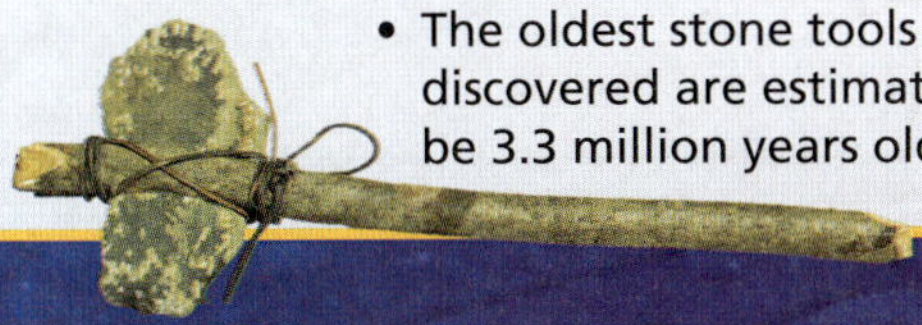

RECONSTRUCTING THE PAST

You are an archaeologist learning about the daily lives of an ancient civilization. You discover the fragments of a circular plate shown above. Use one of the fragments and the mathematical relationships in this chapter to find the diameter of the original plate. Explain each step so that other archaeologists can replicate your method.

GO DIGITAL

6 Test Prep WITH CalcChat®
Cumulative Practice

WATCH Tutorial videos are available for each exercise.

1. Write a two-column proof.

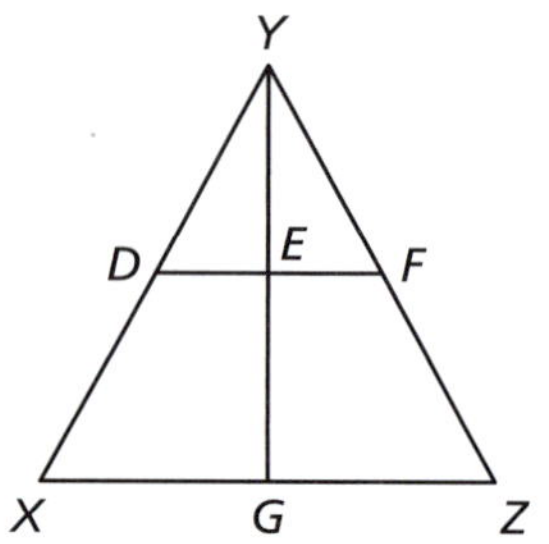

Given $\overline{YG}$ is the perpendicular bisector of $\overline{DF}$.

Prove $\triangle DEY \cong \triangle FEY$

2. Given the endpoint $G(-3, -4)$ and the midpoint $M(1, -1)$ of $\overline{GH}$, find the coordinates of endpoint H, and find GH.

Ⓐ $H(-7, -7);\ GH = 5$

Ⓑ $H(-1, -2.5);\ GH = 2.5$

Ⓒ $H(5, 2);\ GH = 5$

Ⓓ $H(5, 2);\ GH = 10$

3. A triangle has vertices $X(-2, 2)$, $Y(1, 4)$, and $Z(2, -2)$. Which are the coordinates of the image after a translation of $(x, y) \rightarrow (x + 2, y - 3)$, followed by a dilation with a scale factor of 3?

Ⓐ $X''(-4, 3), Y''(5, 9), Z''(8, -9)$

Ⓑ $X''(0, 12), Y''(-6, 3), Z''(6, -6)$

Ⓒ $X''(0, -3), Y''(9, 3), Z''(12, -15)$

Ⓓ $X''\left(0, -\frac{1}{3}\right), Y''\left(1, \frac{1}{3}\right), Z''\left(\frac{4}{3}, -\frac{5}{3}\right)$

4. Which is *not* a true statement about the construction shown?

Step 1

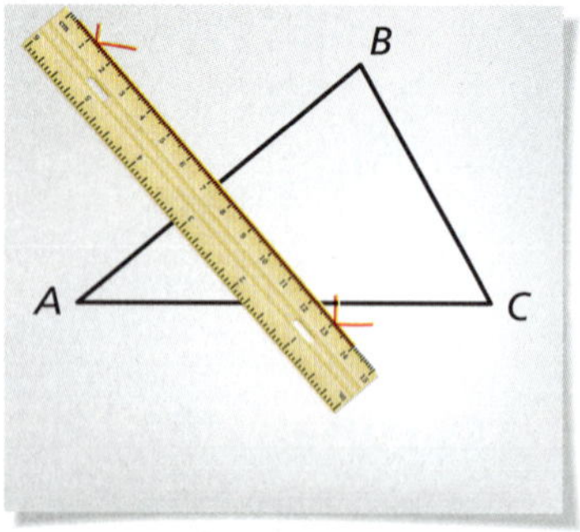

Step 2

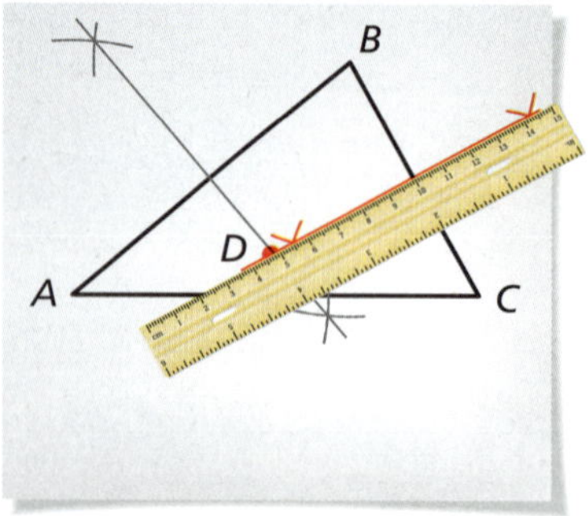

Step 3

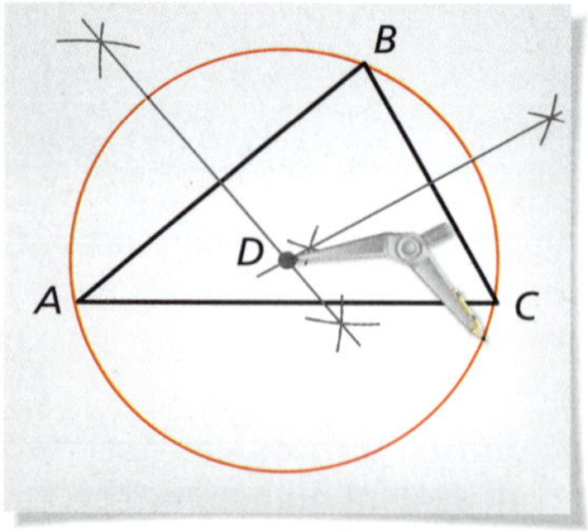

Ⓐ The perpendicular bisectors of $\overline{AB}$ and $\overline{BC}$ intersect at the circumcenter.

Ⓑ The incenter is equidistant from vertices A, B, and C.

Ⓒ The circumcenter is the center of the circle circumscribed about $\triangle ABC$.

Ⓓ The circumscribed circle passes through the vertices of $\triangle ABC$.

GO DIGITAL

5. Complete the proof of the Base Angles Theorem.

Given $\overline{AB} \cong \overline{AC}$

Prove $\angle B \cong \angle C$

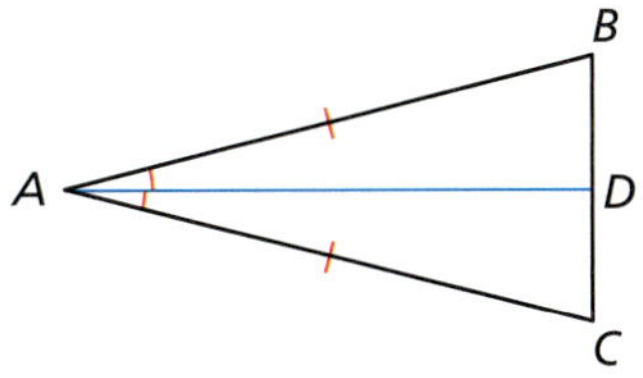

STATEMENTS	REASONS
1. Draw $\overline{AD}$, the angle bisector of $\angle CAB$.	**1.** Construction of angle bisector
2. $\angle CAD \cong \angle BAD$	**2.** ____________________
3. $\overline{AB} \cong \overline{AC}$	**3.** ____________________
4. $\overline{DA} \cong \overline{DA}$	**4.** ____________________
5. $\triangle ADB \cong \triangle ADC$	**5.** ____________________
6. $\angle B \cong \angle C$	**6.** ____________________

6. Use the graph of $\triangle QRS$.

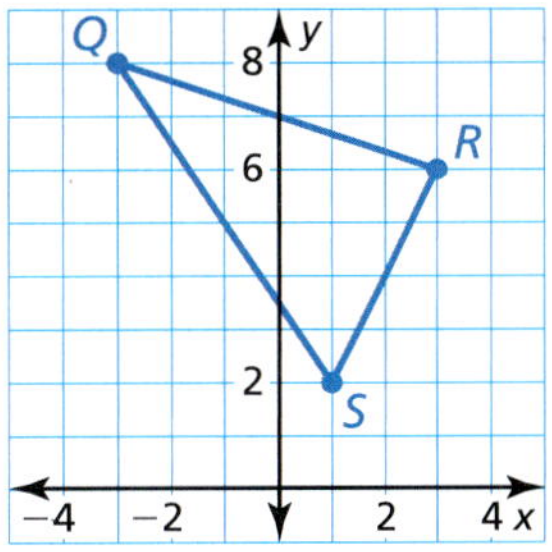

a. Find the coordinates of the vertices of the midsegment triangle. Label the vertices T, U, and V.

b. Show that each midsegment connecting the midpoints of two sides is parallel to the third side and is half as long as that side.

7. What are the coordinates of the centroid of $\triangle LMN$?

Ⓐ $\left(\frac{2}{3}, \frac{44}{9}\right)$

Ⓑ $(3, 5)$

Ⓒ $\left(3, \frac{13}{2}\right)$

Ⓓ $\left(\frac{25}{6}, \frac{91}{18}\right)$

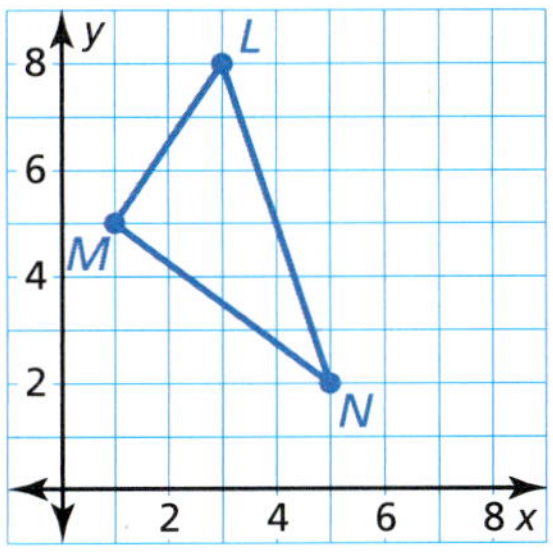

8. The graph shows a dilation of quadrilateral $ABCD$ by a scale factor of 2. Show that the line containing points B and D is parallel to the line containing points B' and D'.

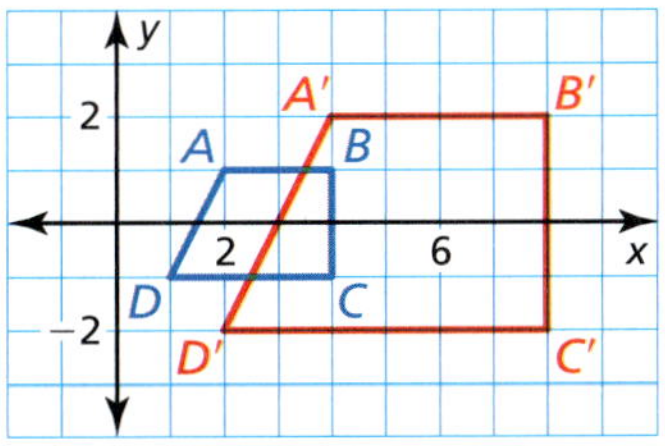

GO DIGITAL

7 Quadrilaterals and Other Polygons

NATIONAL GEOGRAPHIC EXPLORER

WATCH

Brendan Lawrence Mullan

Dr. Brendan Lawrence Mullan is an astrobiologist who believes that people are fascinated by the night sky, space missions, and photos of planets and galaxies. Dr. Mullan says that "human beings want to connect with some cosmic context greater than themselves. I want to give them that chance." He believes that inspiring a new generation of scientists is crucial to solving the 21st-century problems that humanity faces.

- What is astrobiology?
- What does *terrestrial* mean? What does *extraterrestrial* mean?
- Do you think there is intelligent life on planets in other parts of the universe? If so, do you think they have communicated with humans?
- What is a UFO? Do you think humans have had experiences with UFOs?

STEM

Humans have identified constellations in the night sky for millennia. In the Performance Task, you will use angle measures and side lengths to investigate polygons found in several different constellations.

GO DIGITAL

Astrobiology

Preparing for Chapter 7

Chapter Learning Target: Understand quadrilaterals and other polygons.

Chapter Success Criteria:
- ◆ I can find angles of a polygon.
- ◆ I can describe properties of parallelograms.
- ■ I can use properties of parallelograms to solve problems.
- ■ I can justify why a figure is a special quadrilateral.

◆ Surface
■ Deep

Chapter Vocabulary

Work with a partner. Discuss each of the vocabulary terms.

diagonal
equilateral polygon
equiangular polygon
parallelogram
rhombus
rectangle
square
trapezoid
base angles
isosceles trapezoid
midsegment of a trapezoid
kite

Mathematical Thinking and Reasoning

6 MTR

ASSESS THE REASONABLENESS OF SOLUTIONS
Mathematicians who assess the reasonableness of solutions verify possible solutions by explaining the methods used.

Work with a partner. When viewed from Earth, stars appear as if they lie in the same two-dimensional plane.

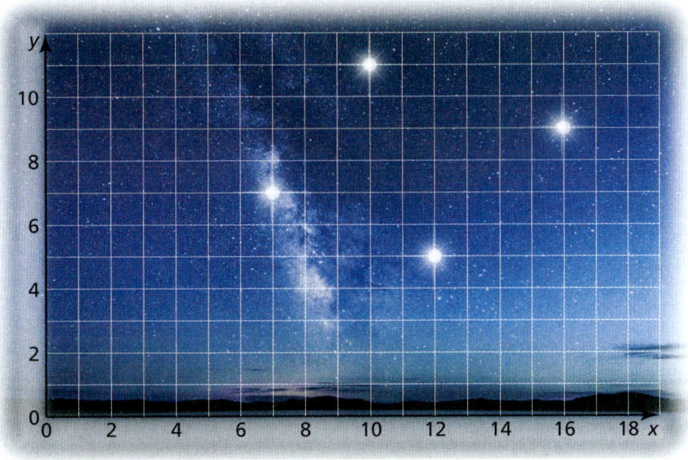

1. Your friend says that the stars shown can be represented by the vertices of a parallelogram. Is your friend's claim reasonable? Explain.
2. What are some ways you can determine if any four points represent the vertices of a parallelogram?

GO DIGITAL

7 Prepare WITH CalcChat®

Using Structure to Solve a Multi-Step Equation

WATCH

Example 1 **Solve $3(2 + x) = -9$.**

$3(2 + x) = -9$	Write the equation.
$\frac{3(2 + x)}{3} = \frac{-9}{3}$	Division Property of Equality
$2 + x = -3$	Simplify.
$\underline{-2} \quad \underline{-2}$	Subtraction Property of Equality
$x = -5$	Simplify.

Solve the equation. Check your solution.

1. $4(7 - x) = 16$ **2.** $7(1 - x) + 2 = -19$ **3.** $3(x - 5) + 8(x - 5) = 22$

Identifying Parallel and Perpendicular Lines

WATCH

Example 2 **Determine which lines are parallel and which are perpendicular.**

Find the slope of each line.

Line a: $m = \frac{3 - (-3)}{-4 - (-2)} = -3$

Line b: $m = \frac{-1 - (-4)}{1 - 2} = -3$

Line c: $m = \frac{2 - (-2)}{3 - 4} = -4$

Line d: $m = \frac{2 - 0}{2 - (-4)} = \frac{1}{3}$

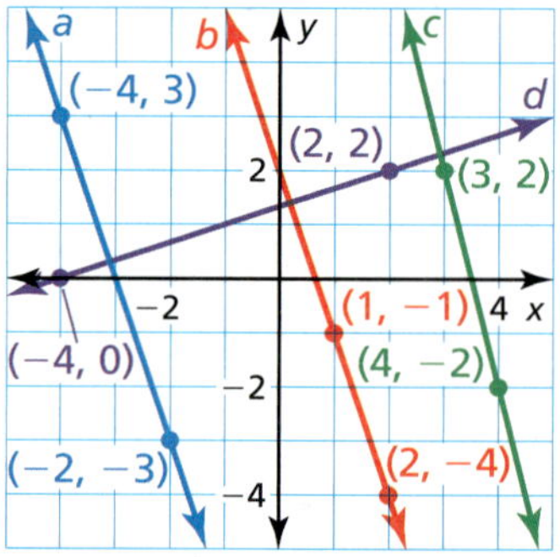

▶ Lines a and b have slopes of -3, so they are parallel. Line d has a slope of $\frac{1}{3}$, the negative reciprocal of -3, so it is perpendicular to lines a and b.

Determine which lines are parallel and which are perpendicular.

4.

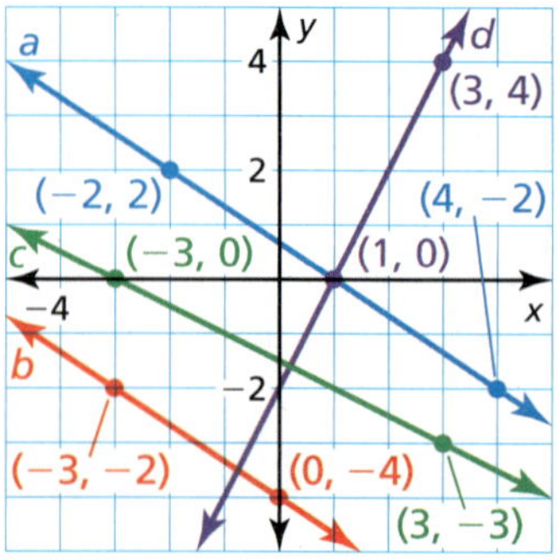

5.

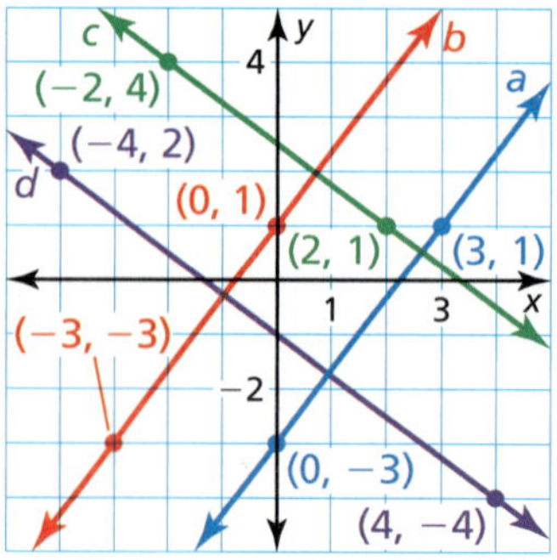

6.

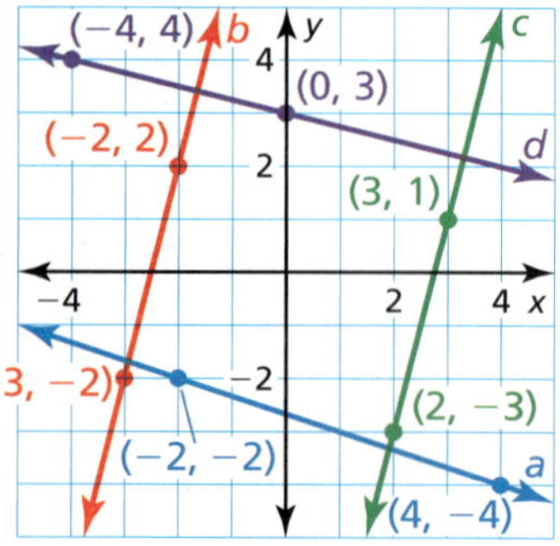

1 MTR **7.** **PERSEVERE** Line a is perpendicular to line b, line b is perpendicular to line c, line c is parallel to line d, and line d is perpendicular to line e. Describe the relationship between line a and line e.

GO DIGITAL

7.1 Angles of Polygons

Learning Target: Find angle measures of polygons.

Success Criteria:
- I can find the sum of the interior angle measures of a polygon.
- I can find interior angle measures of polygons.
- I can find exterior angle measures of polygons.

EXPLORE IT! Finding Sums of Interior Angle Measures of Polygons

Work with a partner.

a. Draw a quadrilateral and a pentagon. Use what you know about the interior angle measures of triangles to find the sum of the interior angle measures of each polygon. Explain your reasoning.

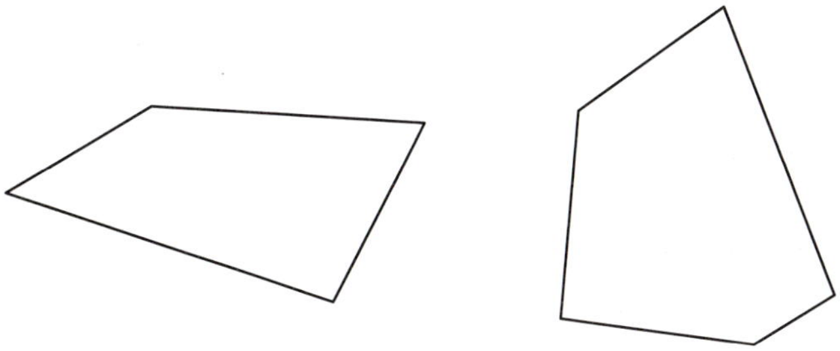

2 MTR MAKE A CONNECTION
What do you expect the graph of the functions in parts (c) and (d) to look like? Explain your reasoning.

b. Draw other convex polygons and find the sums of the measures of their interior angles using the method in part (a). Record your results in the table below.

Number of sides, n	3	4	5	6	7	8	9
Number of triangles							
Sum of angle measures, S							

c. Write an equation that represents S as a function of n. Explain what the function represents.

d. Use the function you found in part (c) to write a new function that represents the measure A of one interior angle in a polygon with n congruent sides and n congruent angles.

e. Does the function in part (c) apply to the polygon at the right? Explain your reasoning.

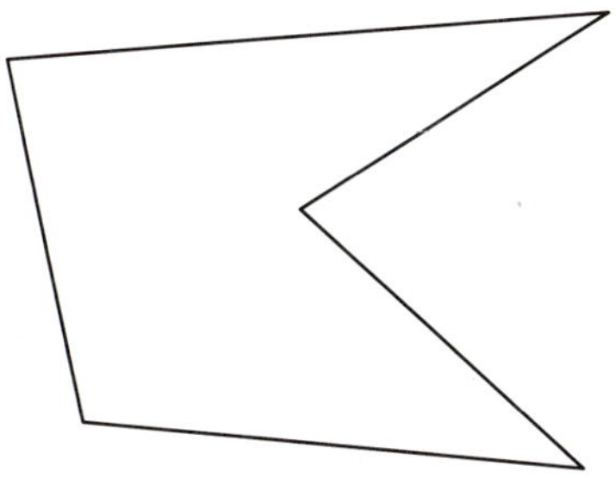

Geometric Reasoning
preparing for MA.912.GR.1.4 Prove relationships and theorems about parallelograms. Solve mathematical and real-world problems involving postulates, relationships and theorems of parallelograms.

GO DIGITAL

Using Interior Angle Measures of Polygons

Vocabulary

diagonal, *p. 354*
equilateral polygon, *p. 356*
equiangular polygon, *p. 356*
regular polygon, *p. 356*

In a polygon, two vertices that are endpoints of the same side are called *consecutive vertices*. A **diagonal** of a polygon is a segment that joins two *nonconsecutive vertices*.

Polygon *ABCDE*

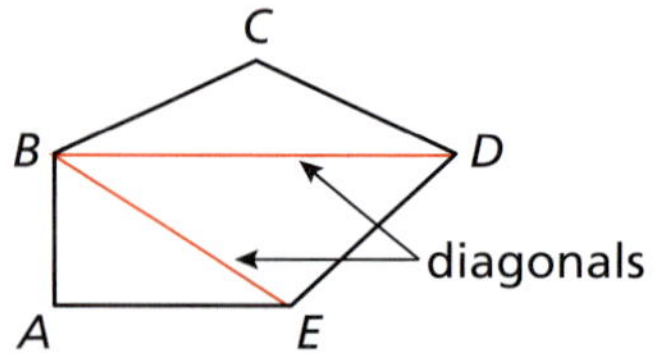

A and *B* are consecutive vertices.
Vertex *B* has two diagonals, $\overline{BD}$ and $\overline{BE}$.

As you can see, the diagonals from one vertex divide a polygon into triangles. Dividing a polygon with n sides into $(n - 2)$ triangles shows that the sum of the measures of the interior angles of a polygon is a multiple of 180°.

REMEMBER

A polygon is *convex* when no line that contains a side of the polygon contains a point in the interior of the polygon.

THEOREM

7.1 Polygon Interior Angles Theorem

The sum of the measures of the interior angles of a convex n-gon is $(n - 2) \cdot 180°$.

$$m\angle 1 + m\angle 2 + \cdots + m\angle n = (n - 2) \cdot 180°$$

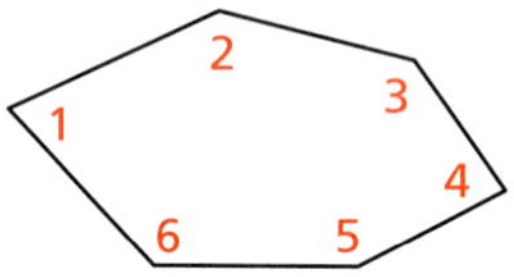

$n = 6$

Prove this Theorem Exercise 39 (for pentagons), page 359

EXAMPLE 1 Finding the Sum of Angle Measures in a Polygon

Find the sum of the measures of the interior angles of the figure.

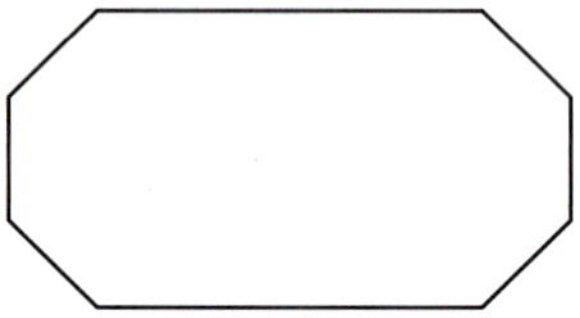

SOLUTION

The figure is a convex octagon. It has 8 sides. Use the Polygon Interior Angles Theorem.

$(n - 2) \cdot 180° = (8 - 2) \cdot 180°$ Substitute 8 for *n*.

$= 6 \cdot 180°$ Subtract.

$= 1080°$ Multiply.

▶ The sum of the measures of the interior angles of the figure is 1080°.

SELF-ASSESSMENT

1 I don't understand yet. 2 I can do it with help. 3 I can do it on my own. 4 I can teach someone else.

1. The shape on the coin is an 11-gon. Find the sum of the measures of the interior angles.

2. **REASONING** Why do vertices connected by a diagonal of a polygon have to be nonconsecutive?

GO DIGITAL

EXAMPLE 2 Finding the Number of Sides of a Polygon

WATCH

The sum of the measures of the interior angles of a convex polygon is 900°. Classify the polygon by the number of sides.

SOLUTION

Use the Polygon Interior Angles Theorem to write an equation involving the number of sides n. Then solve the equation to find the number of sides.

$(n-2) \cdot 180^\circ = 900^\circ$	Polygon Interior Angles Theorem
$n - 2 = 5$	Divide each side by 180°.
$n = 7$	Add 2 to each side.

▶ The polygon has 7 sides. It is a heptagon.

COROLLARY

7.1 Corollary to the Polygon Interior Angles Theorem

The sum of the measures of the interior angles of a quadrilateral is 360°.

Prove this Corollary Exercise 40, page 359

EXAMPLE 3 Finding an Unknown Interior Angle Measure

WATCH

Find the value of x in the diagram.

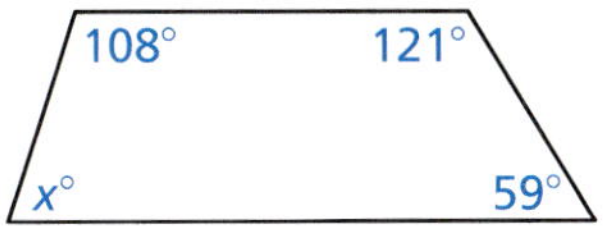

SOLUTION

The polygon is a quadrilateral. Use the Corollary to the Polygon Interior Angles Theorem to write an equation involving x. Then solve the equation.

$x^\circ + 108^\circ + 121^\circ + 59^\circ = 360^\circ$	Corollary to the Polygon Interior Angles Theorem
$x + 288 = 360$	Combine like terms.
$x = 72$	Subtract 288 from each side.

▶ The value of x is 72.

SELF-ASSESSMENT

1 I don't understand yet. 2 I can do it with help. 3 I can do it on my own. 4 I can teach someone else.

3. The sum of the measures of the interior angles of a convex polygon is 1440°. Classify the polygon by the number of sides.

4. Find the value of x in the diagram.

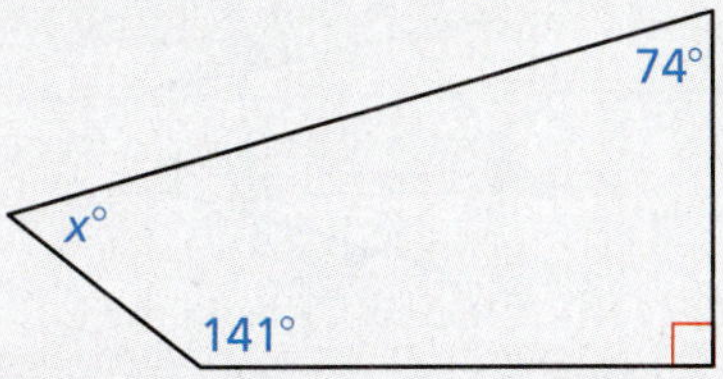

5. The measures of the interior angles of a quadrilateral are x°, $3x^\circ$, $5x^\circ$, and $7x^\circ$. Find the measures of all the interior angles.

GO DIGITAL

WORDS AND MATH

Equilateral and *equiangular* both share the prefix *equi-*, which means "equal." *Equilateral polygons* have sides with equal lengths. *Equiangular polygons* have angles with equal measures.

In an **equilateral polygon**, all sides are congruent.

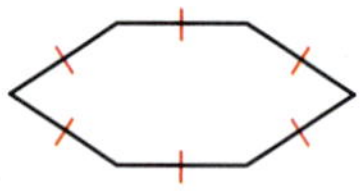

In an **equiangular polygon**, all angles in the interior of the polygon are congruent.

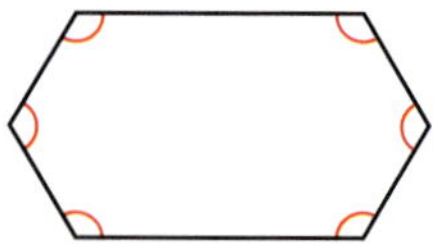

A **regular polygon** is a convex polygon that is both equilateral and equiangular.

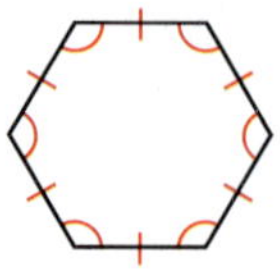

EXAMPLE 4 Finding Angle Measures in Polygons

WATCH

A home plate for a baseball field is shown.

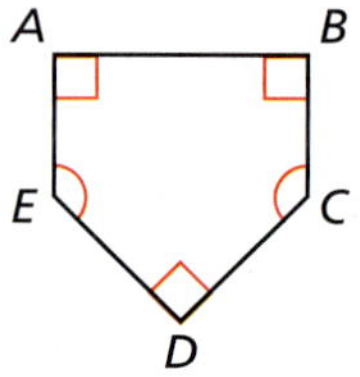

a. Is the polygon regular? Explain your reasoning.

b. Find the measures of $\angle C$ and $\angle E$.

SOLUTION

a. The polygon is not equilateral or equiangular. So, the polygon is not regular.

b. Find the sum of the measures of the interior angles.

$(n - 2) \cdot 180^\circ = (5 - 2) \cdot 180^\circ = 540^\circ$ — Polygon Interior Angles Theorem

Let $x^\circ = m\angle C = m\angle E$. Write an equation involving x and solve the equation.

$x^\circ + x^\circ + 90^\circ + 90^\circ + 90^\circ = 540^\circ$ — Write an equation.

$2x + 270 = 540$ — Combine like terms.

$x = 135$ — Solve for x.

▶ So, $m\angle C = m\angle E = 135^\circ$.

SELF-ASSESSMENT

1 I don't understand yet. 2 I can do it with help. 3 I can do it on my own. 4 I can teach someone else.

6. Find $m\angle S$ and $m\angle T$ in the diagram.

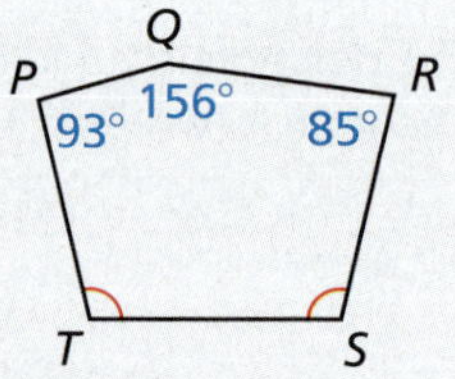

7. OPEN-ENDED Sketch a pentagon that is equilateral but not equiangular.

Using Exterior Angle Measures of Polygons

4 MTR JUSTIFY A RESULT

Show how you can use two straight angles to justify the conclusion.

Unlike the sum of the interior angle measures of a convex polygon, the sum of the exterior angle measures does *not* depend on the number of sides of the polygon. The diagrams suggest that the sum of the measures of the exterior angles, one angle at each vertex, of a pentagon is 360°. In general, this sum is 360° for any convex polygon.

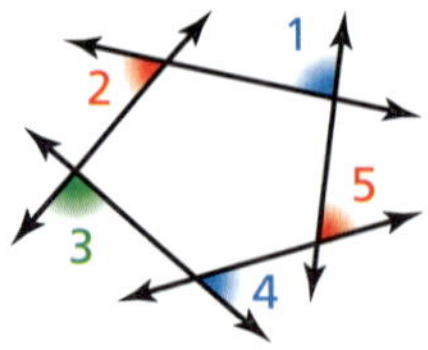

Step 1 Shade one exterior angle at each vertex.

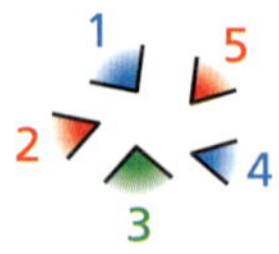

Step 2 Cut out the exterior angles.

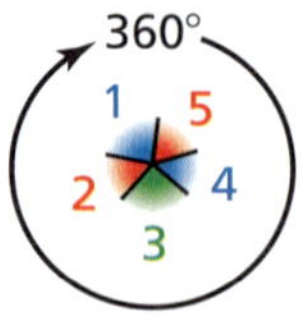

Step 3 Arrange the exterior angles to form 360°.

GO DIGITAL

THEOREM

7.2 Polygon Exterior Angles Theorem

The sum of the measures of the exterior angles of a convex polygon, one angle at each vertex, is 360°.

$$m\angle 1 + m\angle 2 + \cdots + m\angle n = 360^\circ$$

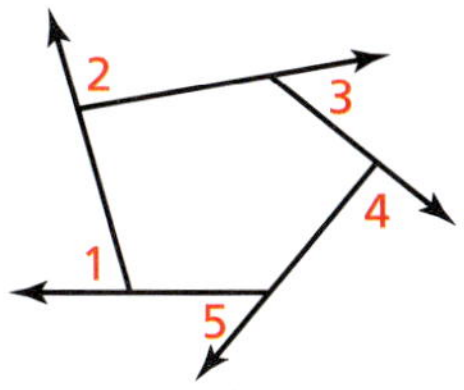

$n = 5$

Prove this Theorem Exercise 46, page 360

EXAMPLE 5 Finding an Unknown Exterior Angle Measure

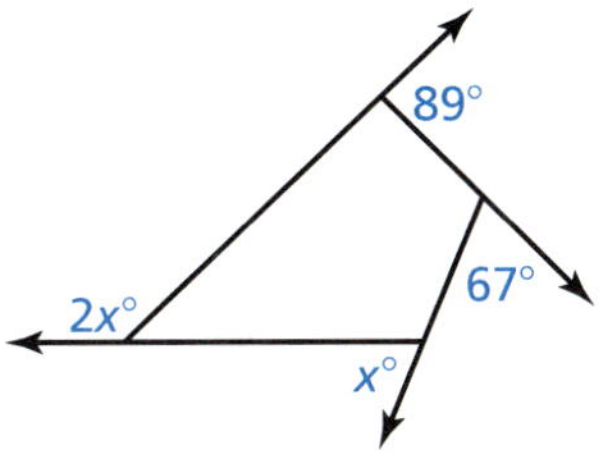

Find the value of x in the diagram.

SOLUTION

Use the Polygon Exterior Angles Theorem to write and solve an equation.

$x^\circ + 2x^\circ + 89^\circ + 67^\circ = 360^\circ$	Polygon Exterior Angles Theorem
$3x + 156 = 360$	Combine like terms.
$x = 68$	Solve for x.

▶ The value of x is 68.

EXAMPLE 6 Finding Angle Measures in Regular Polygons

REMEMBER
A *dodecagon* is a polygon with 12 sides and 12 vertices.

The trampoline is shaped like a regular dodecagon. Find the measure of each (a) interior angle and (b) exterior angle.

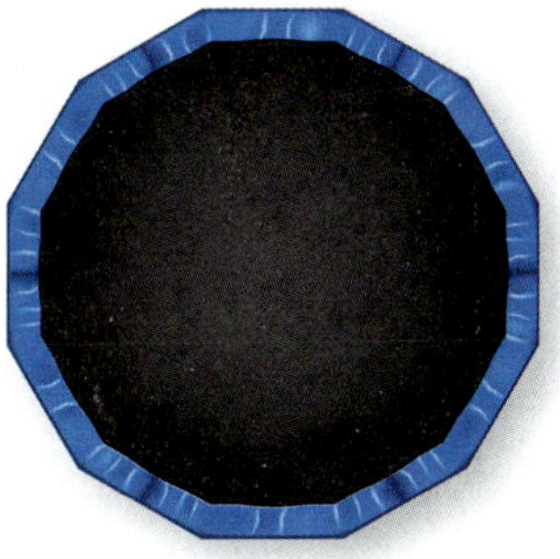

SOLUTION

a. Use the Polygon Interior Angles Theorem to find the sum of the measures of the interior angles.

$$(n - 2) \cdot 180^\circ = (12 - 2) \cdot 180^\circ$$
$$= 1800^\circ$$

Then find the measure of one interior angle. Because there are 12 congruent interior angles, divide 1800° by 12 to obtain 150°.

▶ The measure of each interior angle in the dodecagon is 150°.

b. By the Polygon Exterior Angles Theorem, the sum of the measures of the exterior angles, one angle at each vertex, is 360°. Because there are 12 congruent exterior angles, divide 360° by 12 to obtain 30°.

▶ The measure of each exterior angle in the dodecagon is 30°.

SELF-ASSESSMENT

1 I don't understand yet. 2 I can do it with help. 3 I can do it on my own. 4 I can teach someone else.

8. A convex hexagon has exterior angles with measures 34°, 49°, 58°, 67°, and 75°. What is the measure of an exterior angle at the sixth vertex?

9. **REASONING** An interior angle and an adjacent exterior angle of a polygon form a linear pair. How can you use this fact as another method to find the measure of each exterior angle in Example 6?

GO DIGITAL

7.1 Practice WITH CalcChat® AND CalcView®

In Exercises 1–4, find the sum of the measures of the interior angles of the indicated convex polygon. *(See Example 1.)*

▶ **1.** nonagon
2. 14-gon
3. 16-gon
4. 20-gon

In Exercises 5–8, the sum of the measures of the interior angles of a convex polygon is given. Classify the polygon by the number of sides. *(See Example 2.)*

▶ **5.** 720°
6. 1080°
7. 2520°
8. 3240°

In Exercises 9–14, find the value of *x*. *(See Example 3.)*

▶ **9.**

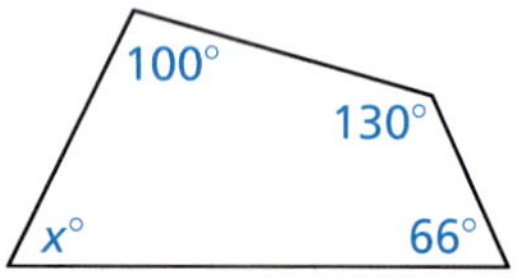

10.

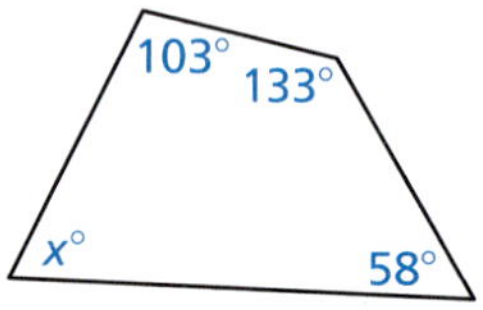

11.

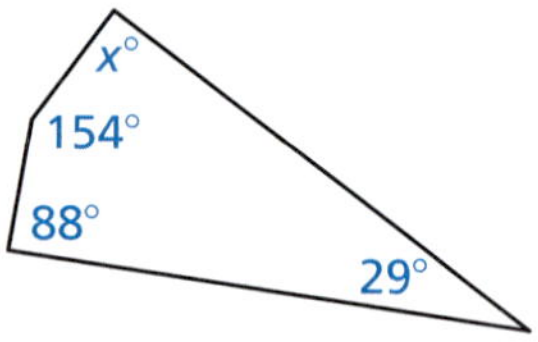

12.

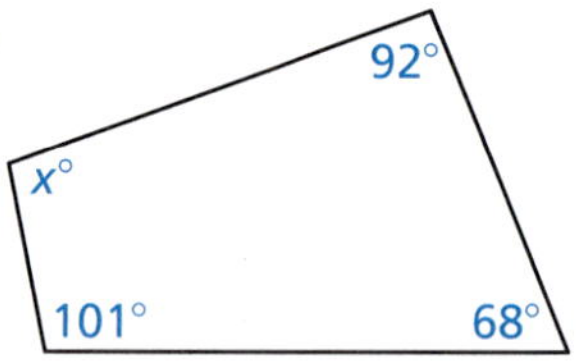

13.

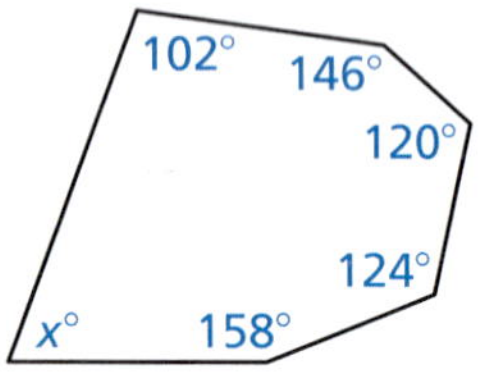

14.

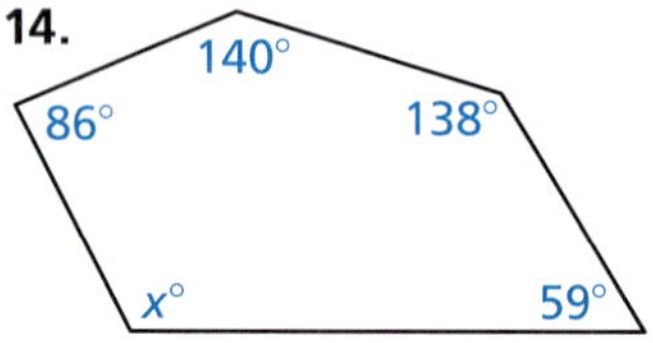

4 MTR **15. ERROR ANALYSIS** Describe and correct the error in finding the value of *x*.

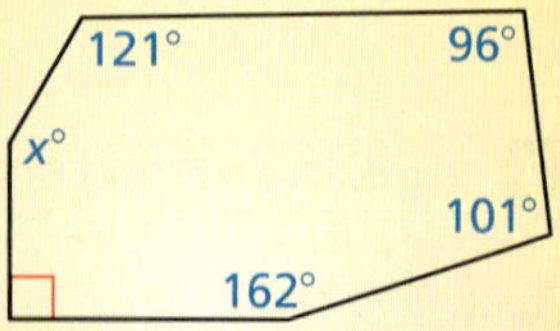

$$x° + 121° + 96° + 101° + 162° = 540°$$
$$x + 480 = 540$$
$$x = 60$$

4 MTR **16. ERROR ANALYSIS** Describe and correct the error in finding the sum of interior angles for an 18-gon.

$$n \cdot 180° = 18 \cdot 180°$$
$$= 3240°$$

In Exercises 17–20, find the measures of ∠*X* and ∠*Y*. *(See Example 4.)*

▶ **17.**

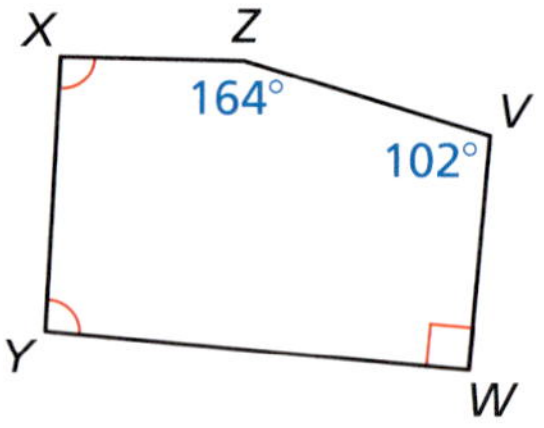

18.

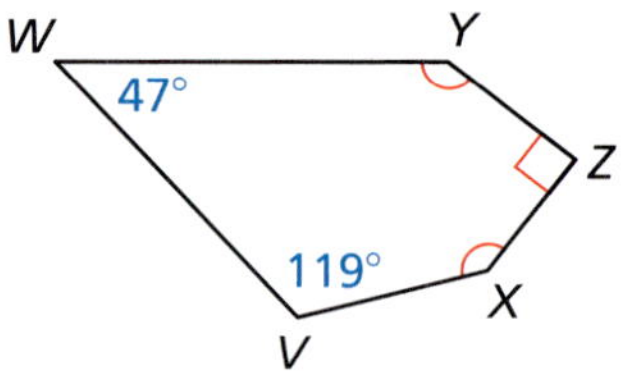

19.

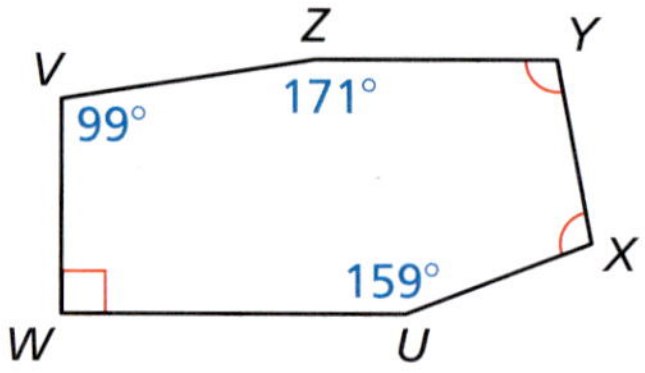

20.

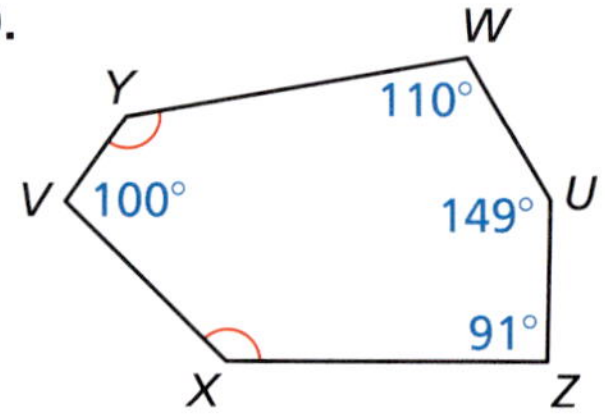

In Exercises 21–24, find the value of *x*. *(See Example 5.)*

▶ **21.**

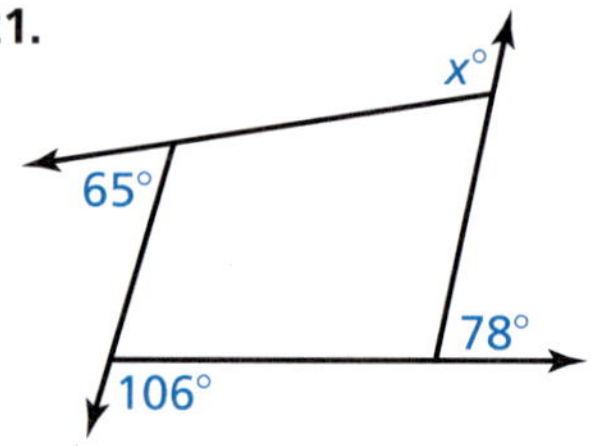

22.

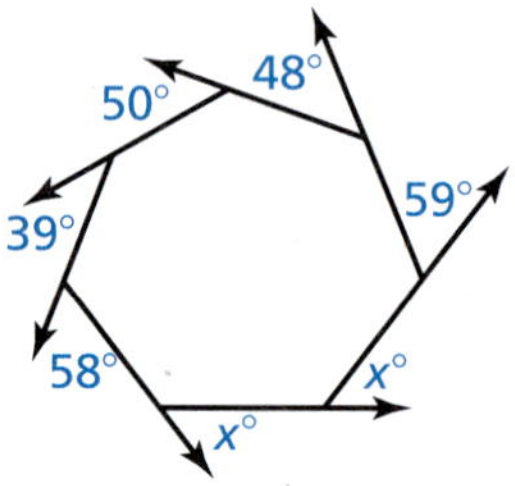

23.

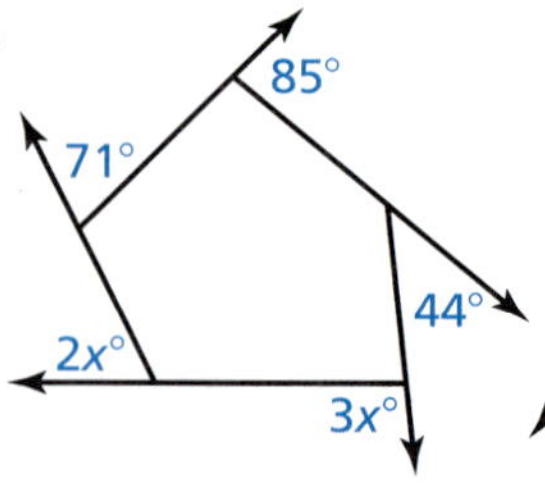

24.

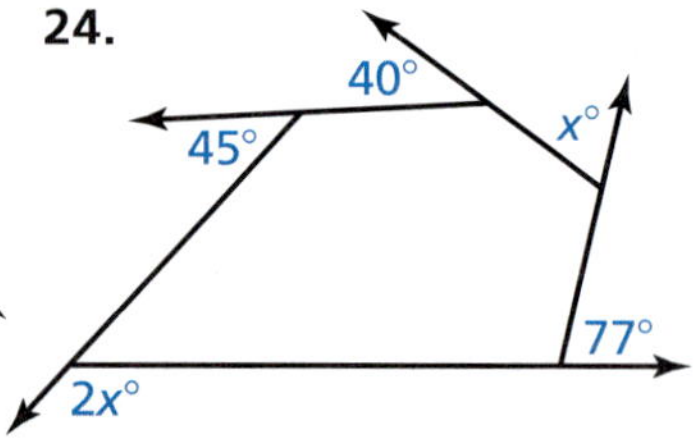

In Exercises 25–28, find the measure of each interior angle and each exterior angle of the indicated regular polygon. *(See Example 6.)*

▶ **25.** pentagon
26. 18-gon
27. 45-gon
28. 90-gon

GO DIGITAL

7 MTR **29.** **MODELING REAL LIFE** Octagon Towers is an octagon-shaped condominium building in Miami, Florida. Assuming the base of Octagon Towers is a regular octagon, find the measure of each interior angle of the base of Octagon Towers.

7 MTR **30.** **MODELING REAL LIFE** NASA's InSight is a robotic lander designed to study the deep interior of Mars. The solar panels on InSight are shaped like regular decagons. Find the measure of each interior angle of the solar panels. Then find the measure of each exterior angle.

31. **WRITING A FORMULA** Write a formula to find the number of sides n in a regular polygon given that the measure of one interior angle is x°.

32. **WRITING A FORMULA** Write a formula to find the number of sides n in a regular polygon given that the measure of one exterior angle is x°.

REASONING In Exercises 33–36, find the number of sides for the regular polygon described.

33. Each interior angle has a measure of 156°.

34. Each interior angle has a measure of 165°.

35. Each exterior angle has a measure of 9°.

36. Each exterior angle has a measure of 6°.

37. **COLLEGE PREP** Which of the following angle measures are possible interior angle measures of a regular polygon? Explain your reasoning. Select all that apply.

Ⓐ 162° Ⓒ 75°

Ⓑ 171° Ⓓ 40°

GO DIGITAL

4 MTR **38.** **MAKING AN ARGUMENT** Your friend claims that to find the interior angle measures of a regular polygon, you do not have to use the Polygon Interior Angles Theorem. Instead you can use the Polygon Exterior Angles Theorem and then the Linear Pair Postulate. Is your friend correct? Explain your reasoning.

39. **PROVING A THEOREM** Write a paragraph proof of the Polygon Interior Angles Theorem for the case when $n = 5$.

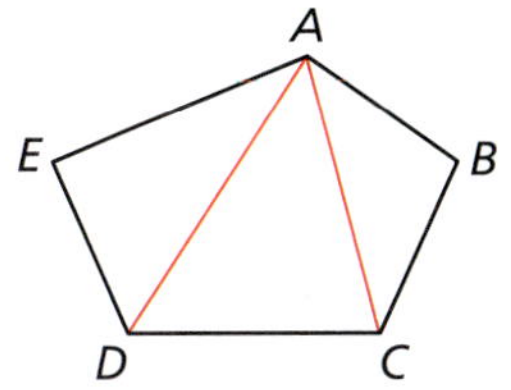

40. **PROVING A COROLLARY** Write a paragraph proof of the Corollary to the Polygon Interior Angles Theorem.

41. **PROBLEM SOLVING** In an equilateral hexagon, four of the exterior angles each have a measure of x°. The other two exterior angles each have a measure of twice the sum of x and 48. Find the measure of each exterior angle.

42. **HOW DO YOU SEE IT?**
Is the hexagon a regular hexagon? Explain your reasoning.

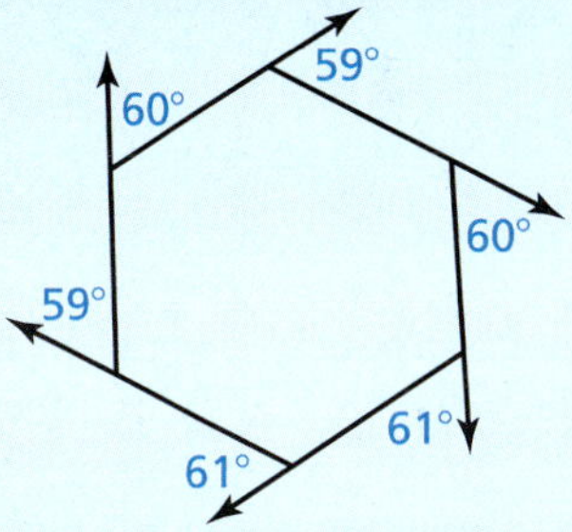

2 MTR **43.** **MULTIPLE REPRESENTATIONS** The formula for the measure of each interior angle in a regular polygon can be written in function notation.

a. Write a function h that represents the measure of any interior angle in a regular polygon with n sides.

b. Find $h(9)$. Then find n when $h(n) = 150^\circ$.

c. Plot the points for $n = 3, 4, 5, 6, 7,$ and 8. What happens to $h(n)$ as n gets larger?

44. **THOUGHT PROVOKING**
Write an expression to find the sum of the measures of the interior angles for a concave polygon. Explain your reasoning.

45. **DIG DEEPER** For a concave polygon, is it true that at least one of the interior angle measures must be greater than 180°? If not, give an example. If so, explain your reasoning.

46. **PROVING A THEOREM** Write a paragraph proof of the Polygon Exterior Angles Theorem.

5 MTR 47. **STRUCTURE** Polygon *ABCDEFGH* is a regular octagon. Suppose sides $\overline{AB}$ and $\overline{CD}$ are extended to meet at a point *P*. Find $m\angle BPC$. Explain your reasoning. Include a diagram with your answer.

48. **REASONING** You are given a convex polygon. You are asked to draw a new polygon by adding additional vertices and increasing the sum of the interior angle measures by 540°. How many more sides does your new polygon have? Explain your reasoning.

REVIEW & REFRESH

WATCH

In Exercises 49 and 50, find the value of *x*.

49.

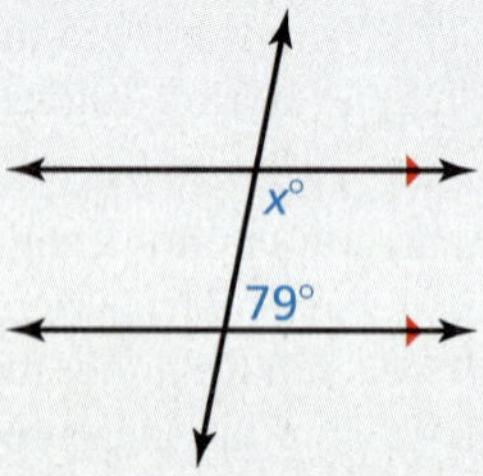

50.

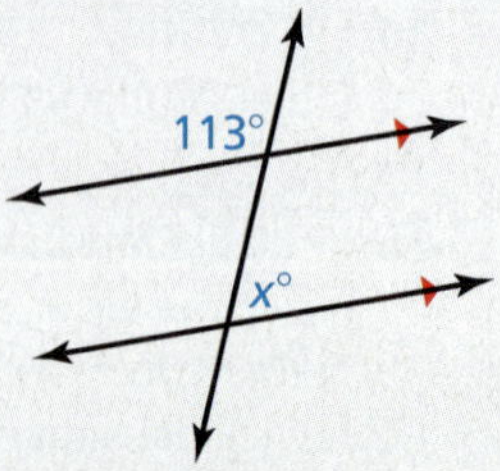

51. Which is greater, $m\angle 1$ or $m\angle 2$? Explain your reasoning.

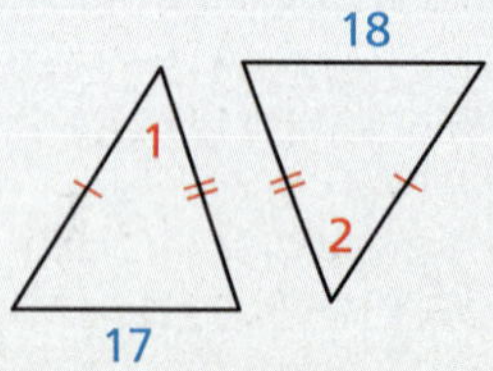

52. Describe the possible lengths of the third side of a triangle with side lengths of 4 inches and 17 inches.

53. Write an equation of the line that passes through $(3, -4)$ and is perpendicular to $y = \frac{1}{2}x + 7$.

54. Determine whether the polygon has line symmetry. If so, draw the line(s) of symmetry and describe any reflections that map the figure onto itself.

55. The sum of the measures of the interior angles of a convex polygon is 1260°. Classify the polygon by the number of sides.

56. $\overline{DE}$ is the midsegment of $\triangle ABC$. Find the value of *x*.

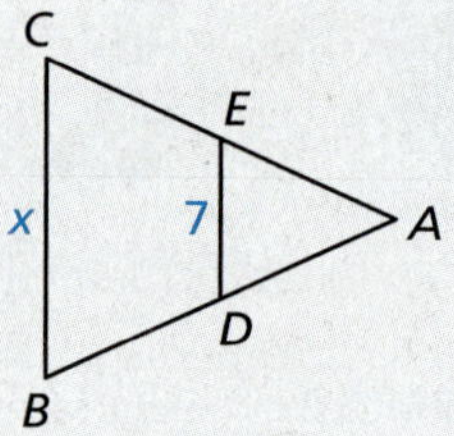

57. Graph $h(x) = 3.5x^2$. Compare the graph to the graph of $f(x) = x^2$.

58. Graph $f(x) = -\frac{1}{4}(8)^x$. Compare the graph to the graph of the parent function. Identify the *y*-intercepts and asymptotes of the graphs. Find the domain and range of *f*.

59. Factor $x^2 + 4x - 21$.

7 MTR 60. **MODELING REAL LIFE** You are designing an obstacle course along a straight path. The distance between obstacle three and obstacle seven is the same as the distance between obstacle four and obstacle nine. Show that the distance between obstacle three and obstacle four is the same as the distance between obstacle seven and obstacle nine.

61. Find the measure of the exterior angle.

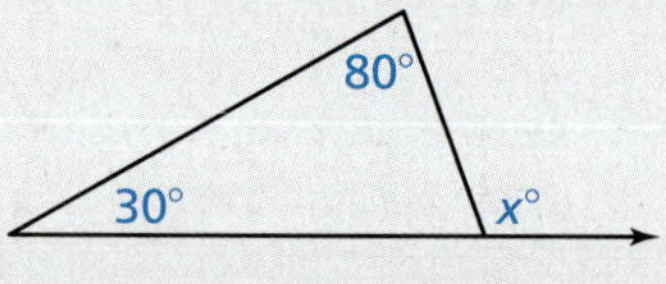

GO DIGITAL

7.2 Properties of Parallelograms

Learning Target: Prove and use properties of parallelograms.

Success Criteria:
- I can prove properties of parallelograms.
- I can use properties of parallelograms.
- I can solve problems involving parallelograms in the coordinate plane.

EXPLORE IT! Discovering Properties of Parallelograms

Work with a partner. A *parallelogram* is a quadrilateral with both pairs of opposite sides parallel.

a. Use technology to construct any parallelogram. Explain your process.

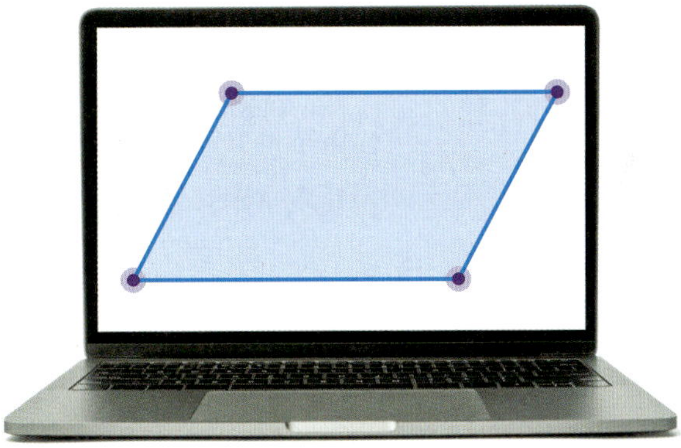

b. Find the angle measures and side lengths of the parallelogram. What do you observe?

c. Repeat parts (a) and (b) for several other parallelograms. Use your results to make conjectures about the angle measures and side lengths of a parallelogram. Prove your conjectures.

d. Use technology to draw the diagonals of any parallelogram. Make a conjecture about any parallelogram that involves its diagonals. Provide examples to support your reasoning.

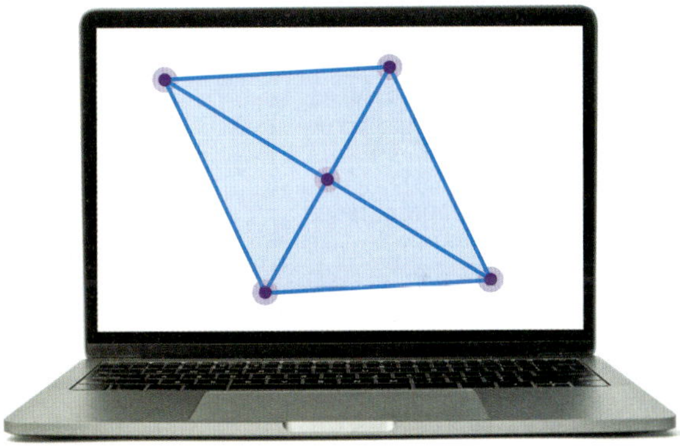

4 MTR CONSTRUCT AN ARGUMENT
Can you construct a parallelogram using its diagonals? Explain your reasoning.

Geometric Reasoning
MA.912.GR.1.4 Prove relationships and theorems about parallelograms. Solve mathematical and real-world problems involving postulates, relationships and theorems of parallelograms.

GO DIGITAL

Using Properties of Parallelograms

Vocabulary AZ VOCAB

parallelogram, *p. 362*

A **parallelogram** is a quadrilateral with both pairs of opposite sides parallel. In $\square PQRS$, $\overline{PQ} \parallel \overline{RS}$ and $\overline{QR} \parallel \overline{PS}$ by definition. The theorems below describe other properties of parallelograms.

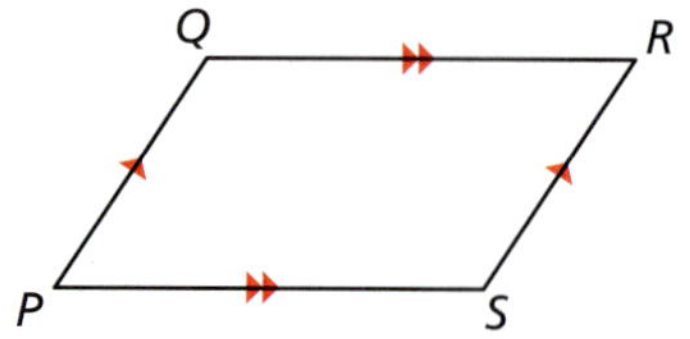

THEOREMS

7.3 Parallelogram Opposite Sides Theorem

If a quadrilateral is a parallelogram, then its opposite sides are congruent.

If $PQRS$ is a parallelogram, then $\overline{PQ} \cong \overline{RS}$ and $\overline{QR} \cong \overline{SP}$.

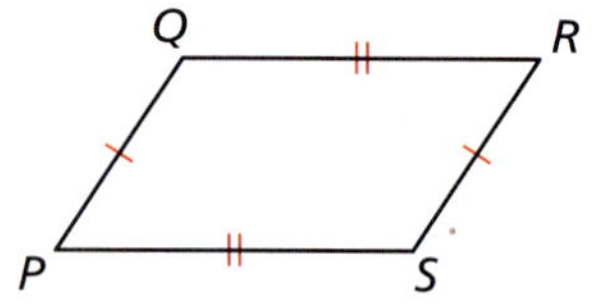

7.4 Parallelogram Opposite Angles Theorem

If a quadrilateral is a parallelogram, then its opposite angles are congruent.

If $PQRS$ is a parallelogram, then $\angle P \cong \angle R$ and $\angle Q \cong \angle S$.

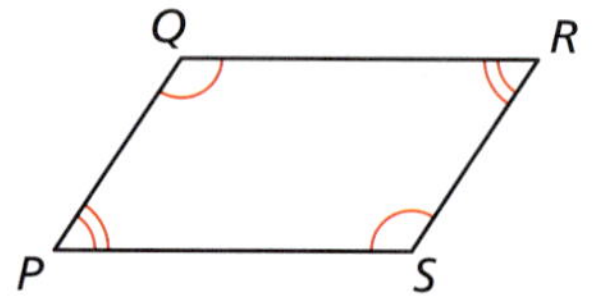

Prove this Theorem Exercise 33, page 367

PROOF Parallelogram Opposite Sides Theorem

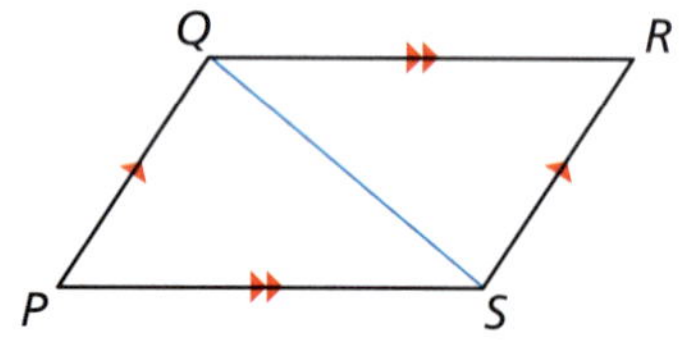

Given $PQRS$ is a parallelogram.

Prove $\overline{PQ} \cong \overline{RS}$, $\overline{QR} \cong \overline{SP}$

Plan for Proof

a. Draw diagonal $\overline{QS}$ to form $\triangle PQS$ and $\triangle RSQ$.

b. Use the ASA Congruence Theorem to show that $\triangle PQS \cong \triangle RSQ$.

c. Use congruent triangles to show that $\overline{PQ} \cong \overline{RS}$ and $\overline{QR} \cong \overline{SP}$.

Plan in Action

	STATEMENTS	REASONS
	1. $PQRS$ is a parallelogram.	**1.** Given
a.	**2.** Draw $\overline{QS}$.	**2.** Through any two points, there exists exactly one line.
	3. $\overline{PQ} \parallel \overline{RS}$, $\overline{QR} \parallel \overline{PS}$	**3.** Definition of parallelogram
b.	**4.** $\angle PQS \cong \angle RSQ$, $\angle PSQ \cong \angle RQS$	**4.** Alternate Interior Angles Theorem
	5. $\overline{QS} \cong \overline{SQ}$	**5.** Reflexive Property of Segment Congruence
	6. $\triangle PQS \cong \triangle RSQ$	**6.** ASA Congruence Theorem
c.	**7.** $\overline{PQ} \cong \overline{RS}$, $\overline{QR} \cong \overline{SP}$	**7.** Corresponding parts of congruent triangles are congruent.

3 MTR REFLECT ON YOUR METHOD

Explain how the two-column proof helps keep track of the steps in the plan.

GO DIGITAL

EXAMPLE 1 Using Properties of Parallelograms

WATCH

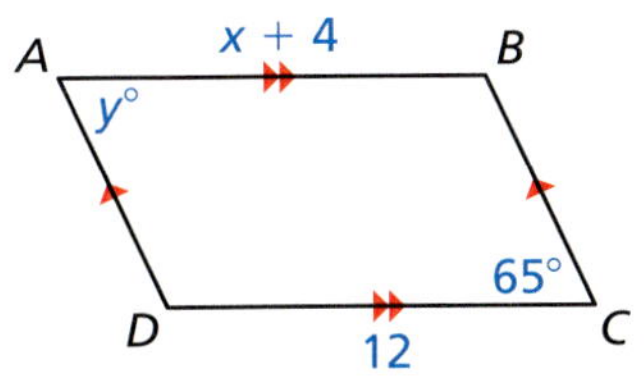

Find the values of x and y.

SOLUTION

$ABCD$ is a parallelogram by the definition of a parallelogram. Use the Parallelogram Opposite Sides Theorem to find the value of x.

$AB = CD$ — Opposite sides of a parallelogram are congruent.

$x + 4 = 12$ — Substitute $x + 4$ for AB and 12 for CD.

$x = 8$ — Subtract 4 from each side.

By the Parallelogram Opposite Angles Theorem, $\angle A \cong \angle C$, or $m\angle A = m\angle C$. So, $y° = 65°$.

▶ In $\square ABCD$, $x = 8$ and $y = 65$.

SELF-ASSESSMENT

1 I don't understand yet. 2 I can do it with help. 3 I can do it on my own. 4 I can teach someone else.

1. Find FG and $m\angle G$.

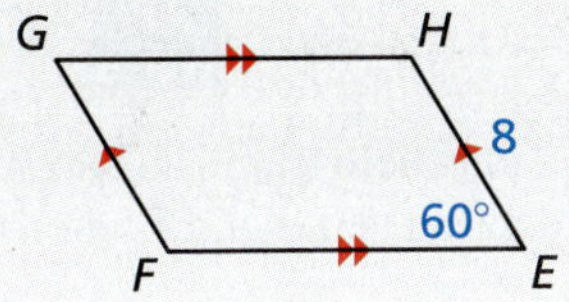

2. Find the values of x and y.

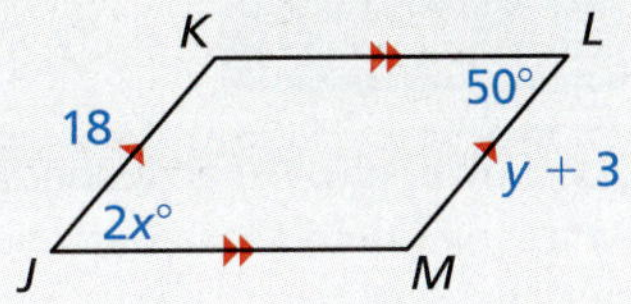

The Consecutive Interior Angles Theorem states that if two parallel lines are cut by a transversal, then the pairs of consecutive interior angles formed are supplementary.

A pair of consecutive angles in a parallelogram is like a pair of consecutive interior angles between parallel lines. This similarity suggests the Parallelogram Consecutive Angles Theorem.

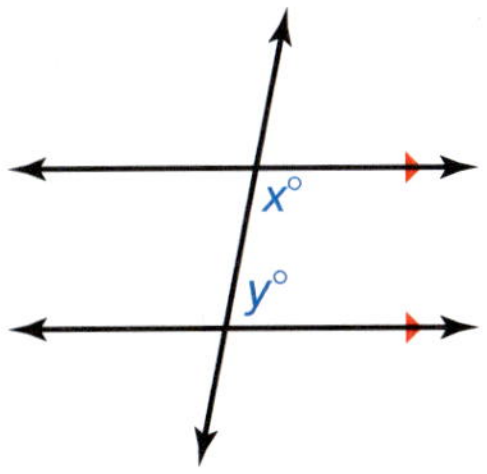

THEOREMS

7.5 Parallelogram Consecutive Angles Theorem

If a quadrilateral is a parallelogram, then its consecutive angles are supplementary.

If $PQRS$ is a parallelogram, then $x° + y° = 180°$.

Prove this Theorem Exercise 34, page 367

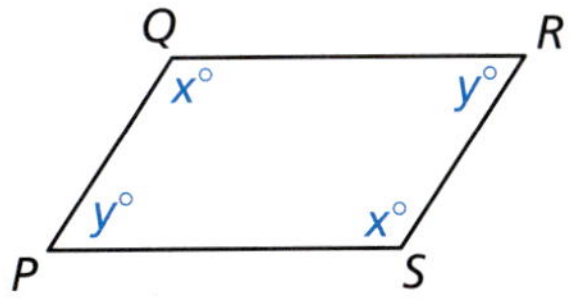

7.6 Parallelogram Diagonals Theorem

If a quadrilateral is a parallelogram, then its diagonals bisect each other.

If $PQRS$ is a parallelogram, then $\overline{QM} \cong \overline{SM}$ and $\overline{PM} \cong \overline{RM}$.

Proof page 364
Prove this Theorem Exercise 35, page 367

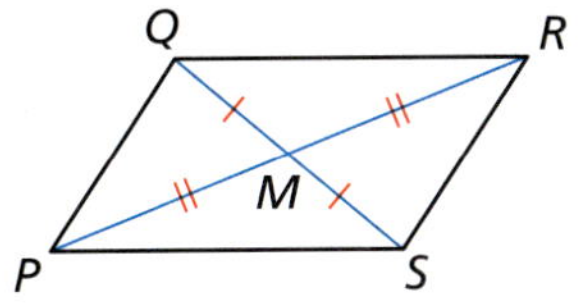

GO DIGITAL

PROOF Parallelogram Diagonals Theorem

Given *PQRS* is a parallelogram. Diagonals $\overline{PR}$ and $\overline{QS}$ intersect at point *M*.

Prove *M* bisects $\overline{QS}$ and $\overline{PR}$.

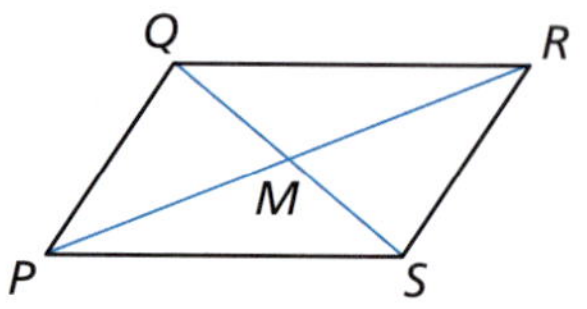

STATEMENTS	REASONS
1. *PQRS* is a parallelogram.	**1.** Given
2. $\overline{PQ} \parallel \overline{RS}$	**2.** Definition of a parallelogram
3. $\angle QPR \cong \angle SRP$, $\angle PQS \cong \angle RSQ$	**3.** Alternate Interior Angles Theorem
4. $\overline{PQ} \cong \overline{RS}$	**4.** Parallelogram Opposite Sides Theorem
5. $\triangle PMQ \cong \triangle RMS$	**5.** ASA Congruence Theorem
6. $\overline{QM} \cong \overline{SM}$, $\overline{PM} \cong \overline{RM}$	**6.** Corresponding parts of congruent triangles are congruent.
7. *M* bisects $\overline{QS}$ and $\overline{PR}$.	**7.** Definition of segment bisector

EXAMPLE 2 Using Properties of a Parallelogram

WATCH

As shown, part of the extending arm of a desk lamp is a parallelogram. The angles of the parallelogram change as the lamp is raised and lowered. Find $m\angle C$ when $m\angle D = 110°$.

SOLUTION

By the Parallelogram Consecutive Angles Theorem, the consecutive angle pairs in ▱*ABCD* are supplementary. So, $m\angle D + m\angle C = 180°$. Because $m\angle D = 110°$, $m\angle C = 180° - 110° = 70°$.

EXAMPLE 3 Writing a Two-Column Proof

WATCH

Write a two-column proof.

Given *ABCD* and *GDEF* are parallelograms.

Prove $\angle B \cong \angle F$

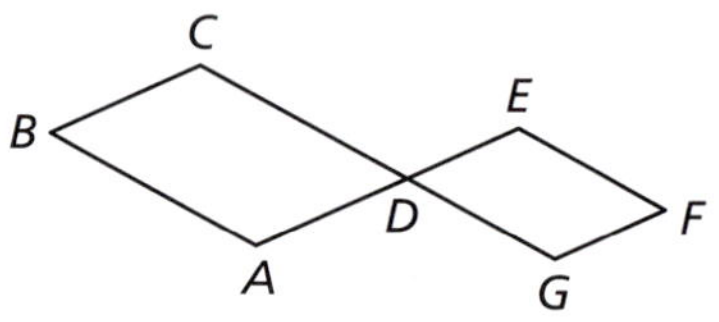

STATEMENTS	REASONS
1. *ABCD* and *GDEF* are parallelograms.	**1.** Given
2. $\angle CDA \cong \angle B$, $\angle EDG \cong \angle F$	**2.** Parallelogram Opposite Angles Theorem
3. $\angle CDA \cong \angle EDG$	**3.** Vertical Angles Congruence Theorem
4. $\angle B \cong \angle F$	**4.** Transitive Property of Angle Congruence

SELF-ASSESSMENT

1 I don't understand yet. 2 I can do it with help. 3 I can do it on my own. 4 I can teach someone else.

3. WHAT IF? In Example 2, find $m\angle C$ when $m\angle D$ is twice the measure of $\angle C$.

4. Using the figure and the given statement in Example 3, prove that $\angle C$ and $\angle F$ are supplementary angles.

GO DIGITAL

Using Parallelograms in the Coordinate Plane

EXAMPLE 4 Using Parallelograms in the Coordinate Plane

Find the coordinates of the intersection of the diagonals of $\square LMNO$ with vertices $L(1, 4)$, $M(7, 4)$, $N(6, 0)$, and $O(0, 0)$.

SOLUTION

By the Parallelogram Diagonals Theorem, the diagonals of a parallelogram bisect each other. So, the coordinates of the intersection are the midpoints of diagonals $\overline{LN}$ and $\overline{OM}$. You can use either diagonal to find the coordinates of the intersection. Use $\overline{OM}$ to simplify the calculation because one endpoint is (0, 0).

$$\text{coordinates of midpoint of } \overline{OM} = \left(\frac{7+0}{2}, \frac{4+0}{2}\right) = \left(\frac{7}{2}, 2\right) \qquad \text{Midpoint Formula}$$

▶ The coordinates of the intersection of the diagonals are $\left(\frac{7}{2}, 2\right)$.

Check Graph $\square LMNO$ and draw the diagonals.

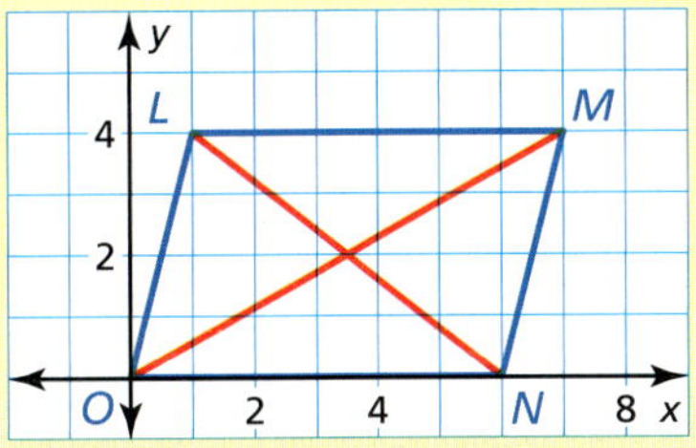

The point of intersection appears to be $\left(\frac{7}{2}, 2\right)$. ✓

EXAMPLE 5 Using Parallelograms in the Coordinate Plane

Three vertices of $\square WXYZ$ are $W(-1, -3)$, $X(-3, 2)$, and $Z(4, -4)$. Find the coordinates of vertex Y.

SOLUTION

REMEMBER
When graphing a polygon in the coordinate plane, the name of the polygon gives the order of the vertices.

Step 1 Graph the vertices W, X, and Z.

Step 2 Find the slope of $\overline{WX}$.

$$\text{slope of } \overline{WX} = \frac{2-(-3)}{-3-(-1)} = \frac{5}{-2}$$

Step 3 Start at $Z(4, -4)$. Use the rise and run from Step 2 to find vertex Y.

A rise of 5 represents a change of 5 units up. A run of -2 represents a change of 2 units left.

So, plot the point that is 5 units up and 2 units left from $Z(4, -4)$. The point is (2, 1). Label it as vertex Y.

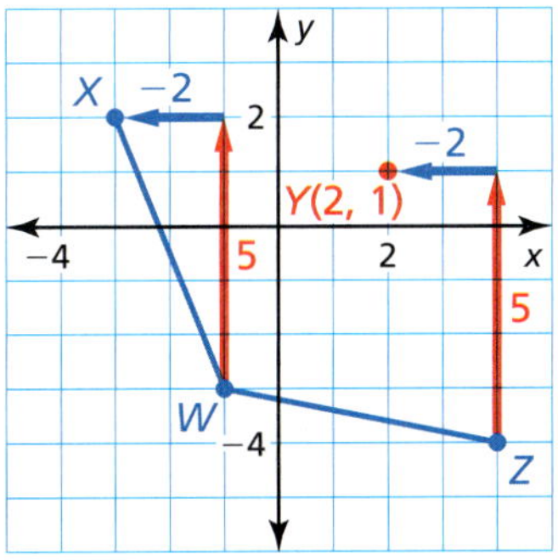

▶ So, the coordinates of vertex Y are (2, 1).

Check Find the slopes of $\overline{XY}$ and $\overline{WZ}$ to verify that they are parallel.

$$\text{slope of } \overline{XY} = \frac{1-2}{2-(-3)} = \frac{-1}{5} = -\frac{1}{5} \qquad \text{slope of } \overline{WZ} = \frac{-4-(-3)}{4-(-1)} = \frac{-1}{5} = -\frac{1}{5}$$

SELF-ASSESSMENT

1 I don't understand yet. 2 I can do it with help. 3 I can do it on my own. 4 I can teach someone else.

5. Find the coordinates of the intersection of the diagonals of $\square STUV$ with vertices $S(-2, 3)$, $T(1, 5)$, $U(6, 3)$, and $V(3, 1)$.

6. Three vertices of $\square ABCD$ are $A(2, 4)$, $B(5, 2)$, and $C(3, -1)$. Find the coordinates of vertex D.

GO DIGITAL

7.2 Practice WITH CalcChat® AND CalcView®

In Exercises 1–4, find the value of each variable in the parallelogram. *(See Example 1.)*

1.

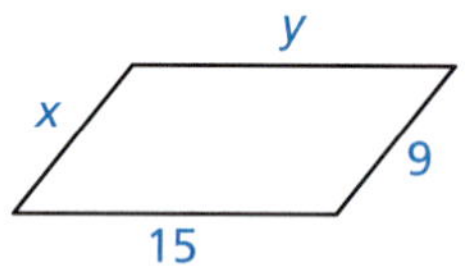

2.

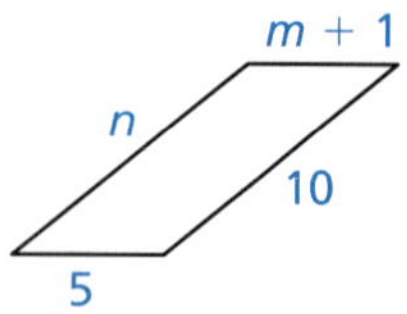

▶ 3.

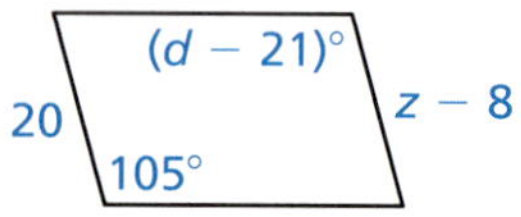

4.

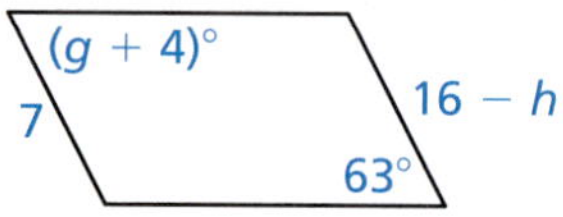

In Exercises 5 and 6, find the measure of the indicated angle in the parallelogram. *(See Example 2.)*

▶ 5. Find $m\angle B$.

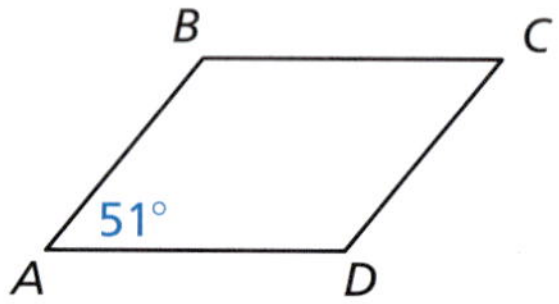

6. Find $m\angle N$.

In Exercises 7–14, find the indicated measure in ▱*LMNQ*. Explain your reasoning.

7. *LM*

8. *LQ*

9. *LP*

10. *MQ*

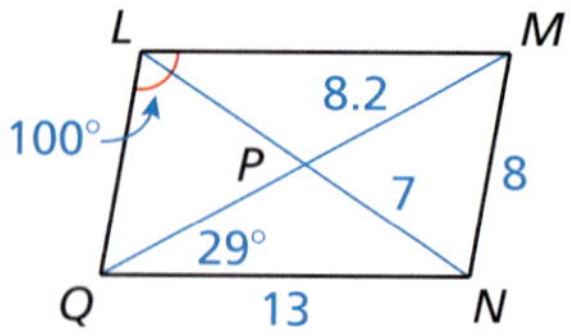

11. $m\angle LMN$

12. $m\angle NQL$

13. $m\angle MNQ$

14. $m\angle LMQ$

In Exercises 15–18, find the value of each variable in the parallelogram.

15.

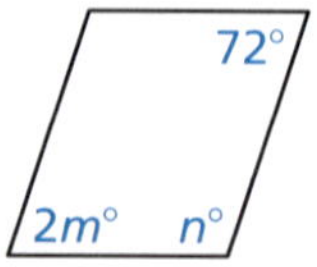

16.

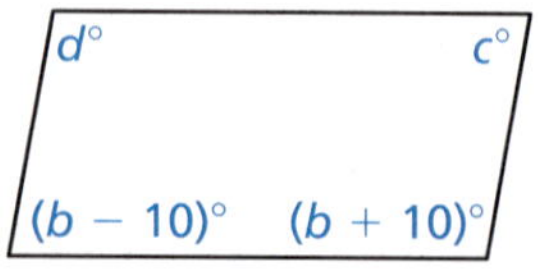

17.

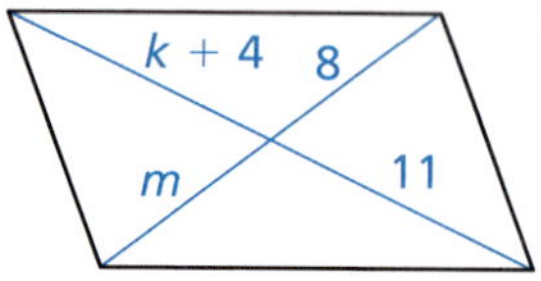

18.

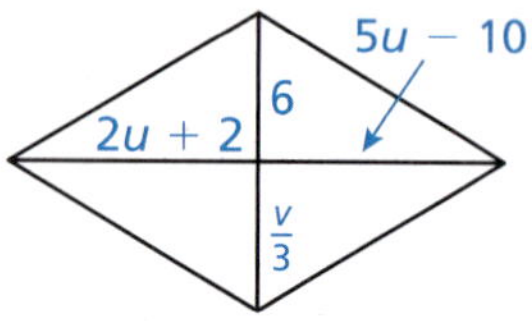

4 MTR ERROR ANALYSIS In Exercises 19 and 20, describe and correct the error in using properties of parallelograms.

19.

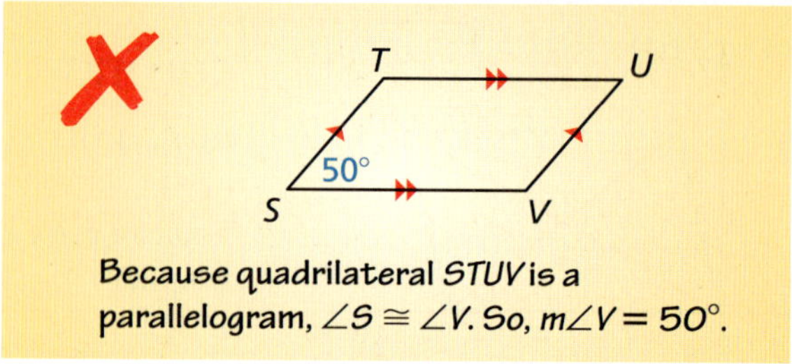

20.

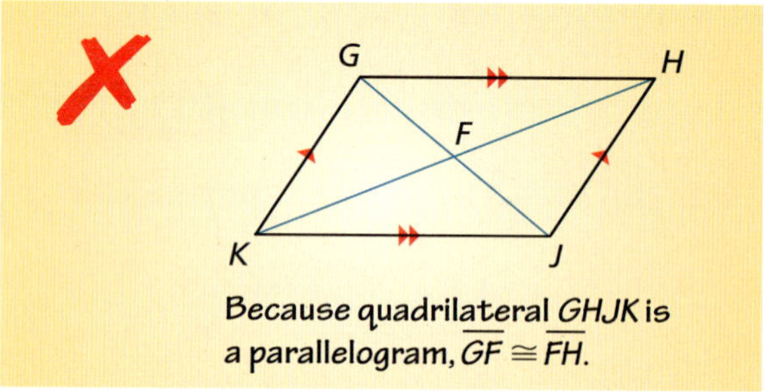

PROOF In Exercises 21 and 22, write a two-column proof. *(See Example 3.)*

▶ 21. Given *ABCD* and *CEFD* are parallelograms.

Prove $\overline{AB} \cong \overline{FE}$

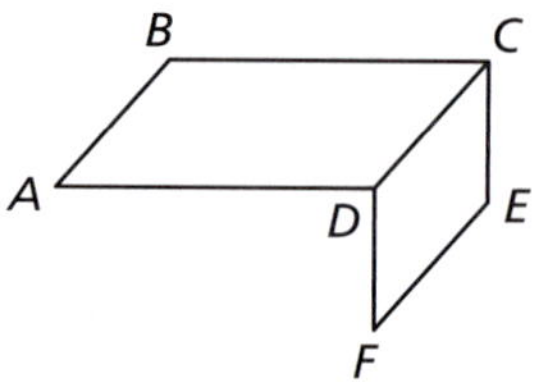

22. Given *ABCD*, *EBGF*, and *HJKD* are parallelograms.

Prove $\angle 2 \cong \angle 3$

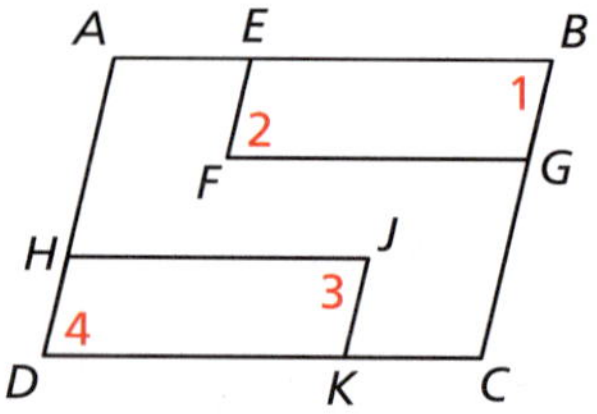

In Exercises 23 and 24, find the coordinates of the intersection of the diagonals of the parallelogram with the given vertices. *(See Example 4.)*

▶ 23. $W(-2, 5)$, $X(2, 5)$, $Y(4, 0)$, $Z(0, 0)$

24. $Q(-1, 3)$, $R(5, 2)$, $S(1, -2)$, $T(-5, -1)$

GO DIGITAL

In Exercises 25–28, three vertices of ▱*DEFG* are given. Find the coordinates of the remaining vertex. *(See Example 5.)*

25. $D(0, 2)$, $E(-1, 5)$, $G(4, 0)$

26. $D(-2, -4)$, $F(0, 7)$, $G(1, 0)$

▶ **27.** $D(-4, -2)$, $E(-3, 1)$, $F(3, 3)$

28. $E(-1, 4)$, $F(5, 6)$, $G(8, 0)$

NUMBER SENSE In Exercises 29 and 30, find the measure of each angle.

29. The measure of one interior angle of a parallelogram is 0.25 times the measure of another angle.

30. The measure of one interior angle of a parallelogram is 50° more than 4 times the measure of another angle.

7 MTR **31. MODELING REAL LIFE** Part of the mount for the pair of astronomy binoculars shown is a parallelogram. The angles of the parallelogram change as the mount is adjusted. Find $m\angle R$, $m\angle S$, and $m\angle T$ when $m\angle Q = 66°$.

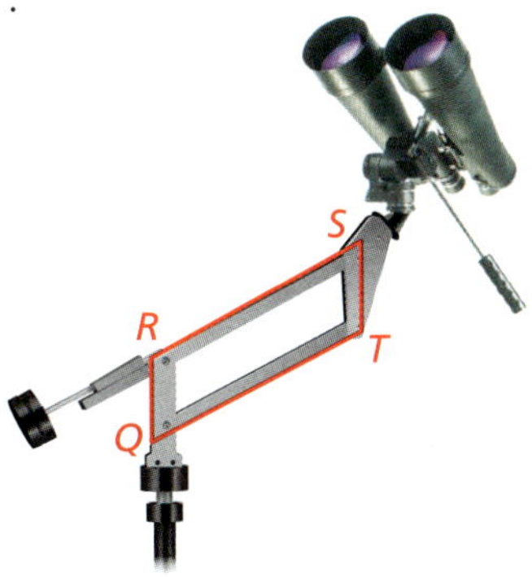

7 MTR **32. MODELING REAL LIFE** The feathers on an arrow form two congruent parallelograms. The parallelograms are reflections of each other over the line that contains their shared side. Show that $m\angle 2 = 2m\angle 1$.

33. PROVING A THEOREM Use the diagram to write a two-column proof of the Parallelogram Opposite Angles Theorem.

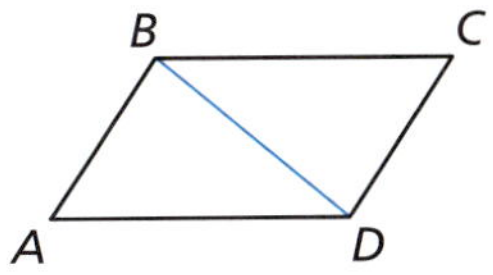

Given *ABCD* is a parallelogram.

Prove $\angle A \cong \angle C$, $\angle B \cong \angle D$

GO DIGITAL

34. PROVING A THEOREM Use the diagram to write a two-column proof of the Parallelogram Consecutive Angles Theorem.

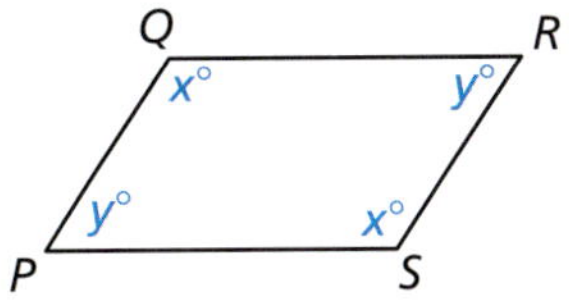

Given *PQRS* is a parallelogram.

Prove $x° + y° = 180°$

35. PROVING A THEOREM *WXYZ* is a parallelogram. Diagonals $\overline{WY}$ and $\overline{XZ}$ intersect at point *V*. Write a paragraph proof of the Parallelogram Diagonals Theorem.

36. CONSTRUCTION Construct any parallelogram and label it *ABCD*. Draw diagonals $\overline{AC}$ and $\overline{BD}$. Explain how to use paper folding to verify the Parallelogram Diagonals Theorem for ▱*ABCD*.

37. PROBLEM SOLVING The sides of ▱*MNPQ* are represented by the expressions below. Find the perimeter of ▱*MNPQ*.

$MQ = -2x + 37$ $\quad$ $QP = y + 14$

$NP = x - 5$ $\quad$ $MN = 4y + 5$

38. PROBLEM SOLVING In ▱*LMNP*, the ratio of *LM* to *MN* is 4 : 3. Find *LM* when the perimeter of ▱*LMNP* is 28.

39. REASONING Points (1, 2), (3, 6), and (6, 4) are three vertices of a parallelogram. How many parallelograms can be created using these three vertices? Find the coordinates of each point that could be the fourth vertex.

40. HOW DO YOU SEE IT? The mirror shown is attached to the wall by an arm that can extend away from the wall. In the figure, points *P*, *Q*, *R*, and *S* are the vertices of a parallelogram. This parallelogram is one of several that change shape as the mirror is extended. What happens to $m\angle P$ as $m\angle Q$ increases? Explain.

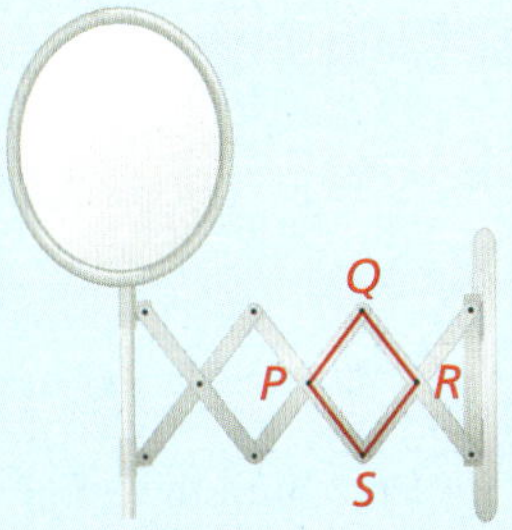

7 MTR **41. MODELING REAL LIFE** Why is a parallelogram structure used in mounts like the one shown in Exercise 31?

42. THOUGHT PROVOKING
Is it possible that any triangle can be partitioned into four congruent triangles that can be rearranged to form a parallelogram? Explain your reasoning.

6 MTR **43. JUSTIFYING STEPS** In $\square STUV$, $m\angle TSU = 32^\circ$, $m\angle USV = (x^2)^\circ$, $m\angle TUV = 12x^\circ$, and $\angle TUV$ is an acute angle. Find $m\angle USV$. Justify your steps.

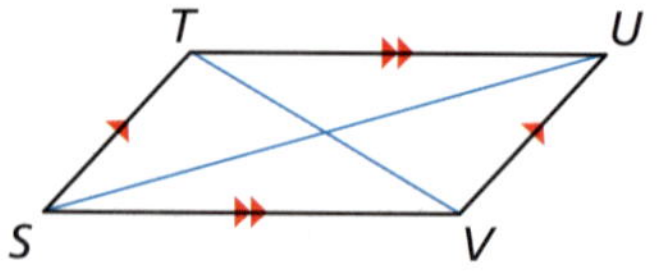

4 MTR **44. DISCUSS MATHEMATICAL THINKING** Let *JKLM* and *PQRS* be parallelograms.

a. Is $\square JKLM$ always congruent to $\square PQRS$ when all corresponding sides are congruent? Explain your reasoning.

b. Is $\square JKLM$ always congruent to $\square PQRS$ when $\overline{KL} \cong \overline{QR}$, $\overline{LM} \cong \overline{RS}$, and $\angle L \cong \angle R$? Explain your reasoning.

45. PROOF In the diagram, $\overline{EK}$ bisects $\angle FEH$, and $\overline{FJ}$ bisects $\angle EFG$. Prove that $\overline{EK} \perp \overline{FJ}$. (*Hint*: Write equations using the angle measures of the triangles and quadrilaterals formed.)

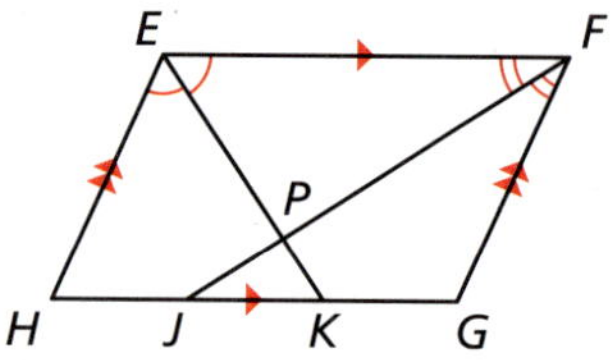

46. PROOF Prove the *Congruent Parts of Parallel Lines Corollary:* If three or more parallel lines cut off congruent segments on one transversal, then they cut off congruent segments on every transversal.

Given $\overleftrightarrow{GH} \parallel \overleftrightarrow{JK} \parallel \overleftrightarrow{LM}$, $\overline{GJ} \cong \overline{JL}$

Prove $\overline{HK} \cong \overline{KM}$

(*Hint:* Draw $\overline{KP}$ and $\overline{MQ}$ such that quadrilateral *GPKJ* and quadrilateral *JQML* are parallelograms.)

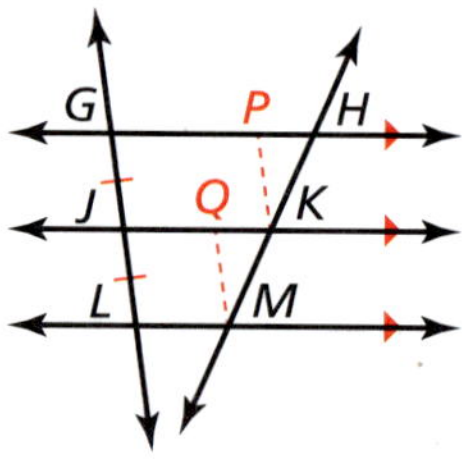

REVIEW & REFRESH

WATCH

47. List the sides of $\triangle DEF$ in order from shortest to longest.

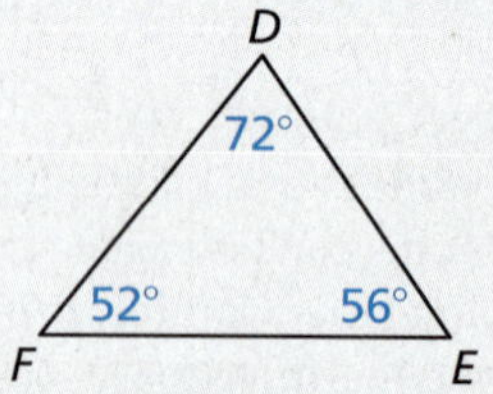

In Exercises 48–50, find the indicated measure in $\square WXYZ$. Explain your reasoning.

48. *YZ*

49. $\angle W$

50. $\angle X$

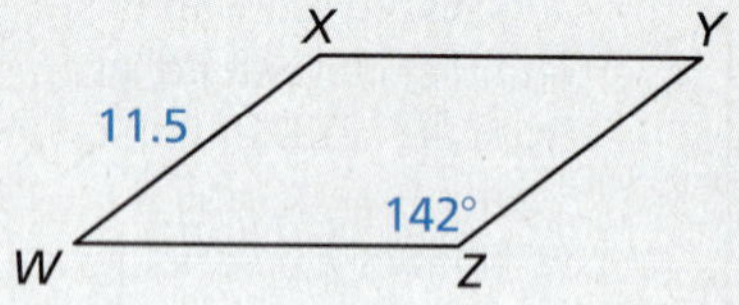

51. Find the value of *x*.

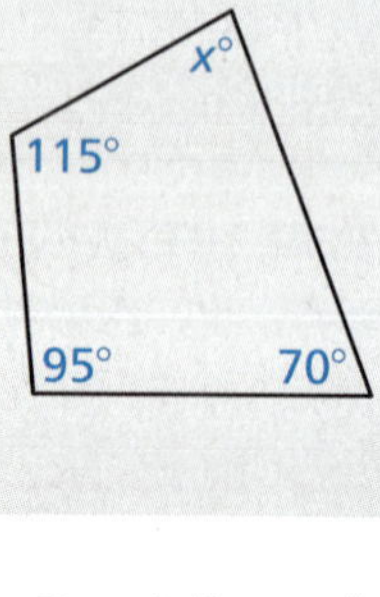

52. The coordinates of a point and its image after a reflection are shown. What is the line of reflection?

$$(7, -5) \rightarrow (-7, -5)$$

53. Decide whether there is enough information to prove that $\ell \parallel m$. If so, state the theorem you can use.

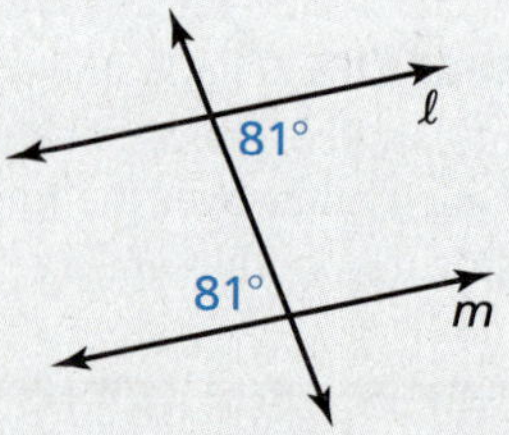

7 MTR **54. MODELING REAL LIFE** The path from *E* to *F* is longer than the path from *E* to *D*. The path from *G* to *D* is the same length as the path from *G* to *F*. What can you conclude about $\angle DGE$ and $\angle EGF$? Explain your reasoning.

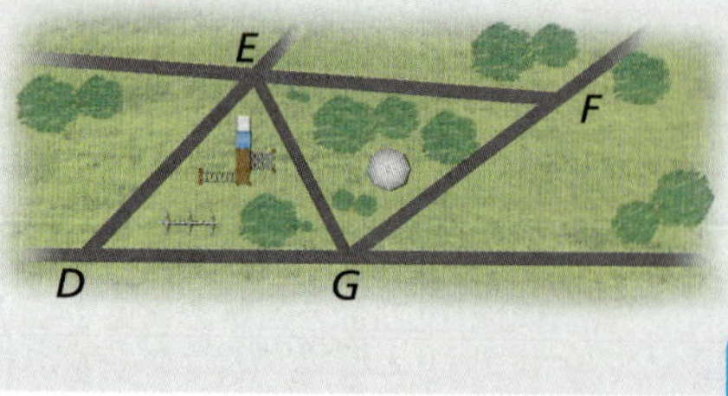

GO DIGITAL

7.3 Proving that a Quadrilateral Is a Parallelogram

Learning Target: Prove that a quadrilateral is a parallelogram.

Success Criteria:
- I can identify features of a parallelogram.
- I can prove that a quadrilateral is a parallelogram.
- I can find missing lengths that make a quadrilateral a parallelogram.
- I can show that a quadrilateral in the coordinate plane is a parallelogram.

EXPLORE IT! Proving that a Quadrilateral Is a Parallelogram

Work with a partner. Use technology.

a. Construct any quadrilateral whose opposite sides are congruent. Is the quadrilateral a parallelogram? Explain how you know.

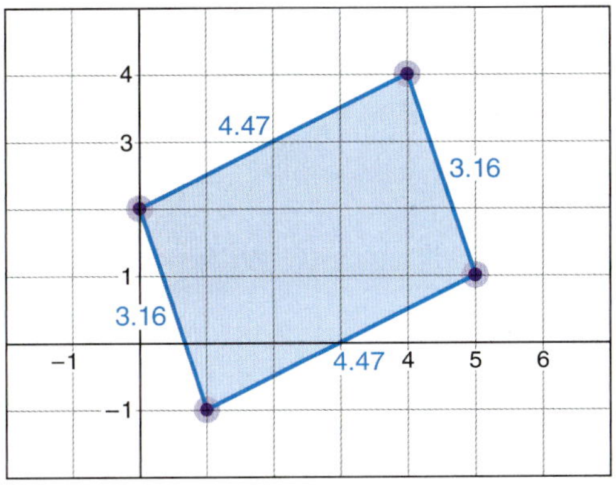

b. Repeat part (a) for several other quadrilaterals. Make a conjecture based on your results.

c. Write the converse of your conjecture. Is the converse true? Explain.

d. Construct any quadrilateral whose opposite angles are congruent. Is the quadrilateral a parallelogram? Explain how you know.

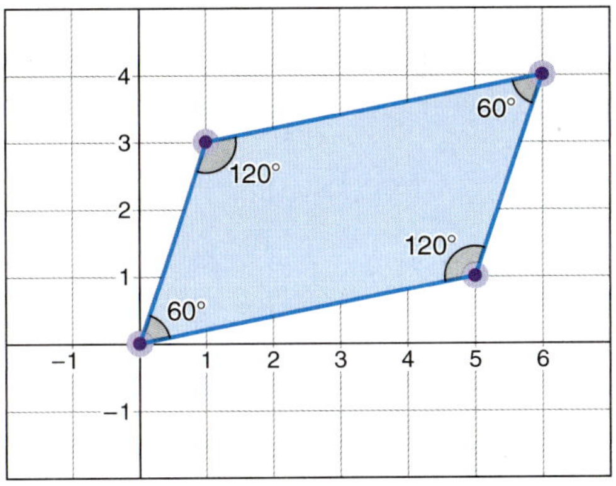

e. Repeat part (d) for several other quadrilaterals. Make a conjecture based on your results.

f. Write the converse of your conjecture in part (e). Is the converse true? Explain.

g. Use congruent triangles to prove that *ABCD* is a parallelogram.

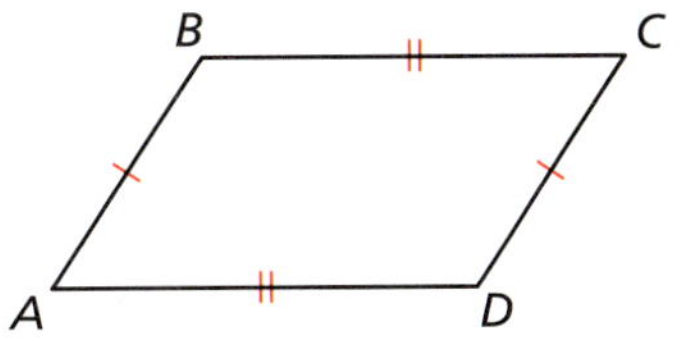

5 MTR MAKE A PLAN

What can you add to quadrilateral *ABCD* that allows you to use congruent triangles in your proof?

Geometric Reasoning

MA.912.GR.1.4 Prove relationships and theorems about parallelograms. Solve mathematical and real-world problems involving postulates, relationships and theorems of parallelograms.

GO DIGITAL

Identifying and Verifying Parallelograms

Given a parallelogram, you can use the Parallelogram Opposite Sides Theorem and the Parallelogram Opposite Angles Theorem to prove statements about the sides and angles of the parallelogram. The converses of the theorems are stated below. You can use these and other theorems in this lesson to prove that a quadrilateral with certain properties is a parallelogram.

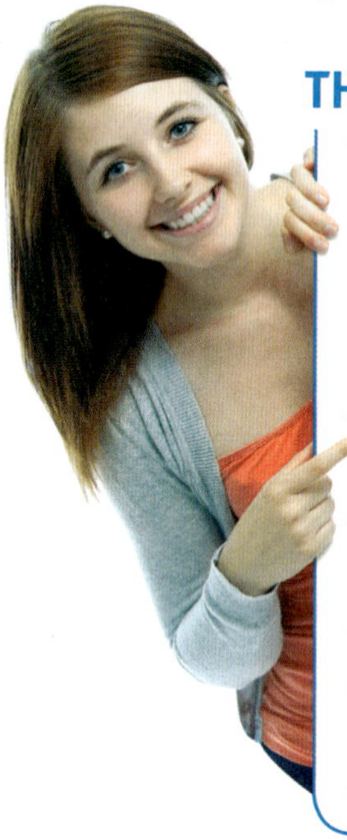

THEOREMS

7.7 Parallelogram Opposite Sides Converse

If both pairs of opposite sides of a quadrilateral are congruent, then the quadrilateral is a parallelogram.

If $\overline{AB} \cong \overline{CD}$ and $\overline{BC} \cong \overline{DA}$, then $ABCD$ is a parallelogram.

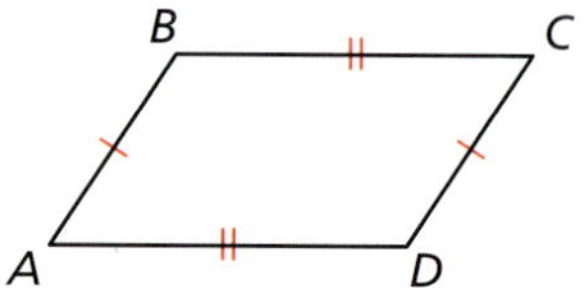

Prove this Theorem Explore It! part (a), page 369

7.8 Parallelogram Opposite Angles Converse

If both pairs of opposite angles of a quadrilateral are congruent, then the quadrilateral is a parallelogram.

If $\angle A \cong \angle C$ and $\angle B \cong \angle D$, then $ABCD$ is a parallelogram.

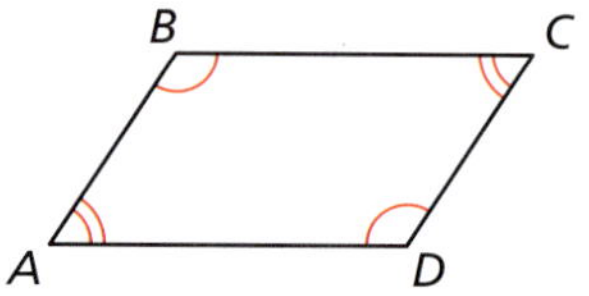

Prove this Theorem Exercise 37, page 377

PROOF Parallelogram Opposite Sides Converse

Given $\overline{AB} \cong \overline{CD}, \overline{BC} \cong \overline{DA}$

Prove $ABCD$ is a parallelogram.

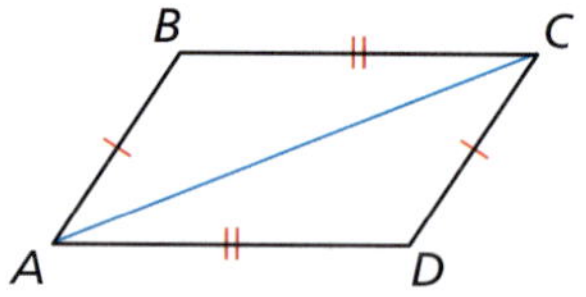

Plan for Proof

a. Draw diagonal $\overline{AC}$ to form $\triangle ABC$ and $\triangle CDA$.

b. Use the SSS Congruence Theorem to show that $\triangle ABC \cong \triangle CDA$.

c. Use the Alternate Interior Angles Converse to show that opposite sides are parallel.

Plan in Action

	STATEMENTS	REASONS
a.	**1.** $\overline{AB} \cong \overline{CD}, \overline{BC} \cong \overline{DA}$	**1.** Given
	2. Draw $\overline{AC}$.	**2.** Through any two points, there exists exactly one line.
	3. $\overline{AC} \cong \overline{CA}$	**3.** Reflexive Property of Segment Congruence
b.	**4.** $\triangle ABC \cong \triangle CDA$	**4.** SSS Congruence Theorem
c.	**5.** $\angle BAC \cong \angle DCA$, $\angle BCA \cong \angle DAC$	**5.** Corresponding parts of congruent triangles are congruent.
	6. $\overline{AB} \parallel \overline{CD}, \overline{BC} \parallel \overline{DA}$	**6.** Alternate Interior Angles Converse
	7. $ABCD$ is a parallelogram.	**7.** Definition of parallelogram

GO DIGITAL

EXAMPLE 1 Identifying a Parallelogram

WATCH

An amusement park ride has a moving platform attached to four swinging arms. The platform swings back and forth, higher and higher, until it goes over the top and around in a circular motion.

In the diagram, $\overline{AD}$ and $\overline{BC}$ represent two of the swinging arms, and $\overline{DC}$ is parallel to the ground (line ℓ). Explain why the moving platform represented by $\overline{AB}$ is always parallel to the ground.

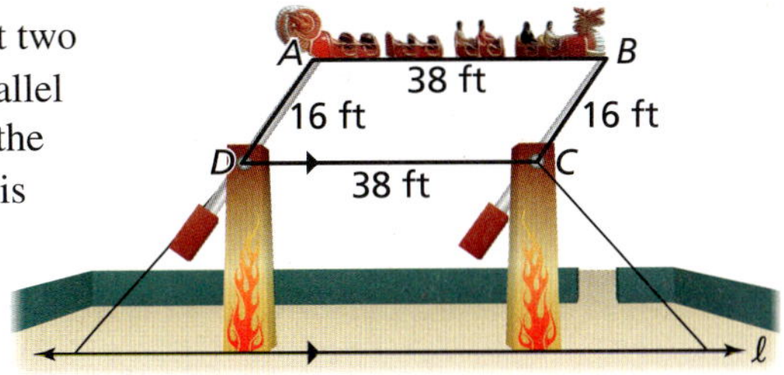

SOLUTION

The shape of quadrilateral *ABCD* changes as the moving platform swings around, but its side lengths do not change. Both pairs of opposite sides are congruent, so *ABCD* is a parallelogram by the Parallelogram Opposite Sides Converse.

By the definition of a parallelogram, $\overline{AB} \parallel \overline{DC}$. Because $\overline{DC}$ is parallel to line ℓ, $\overline{AB}$ is also parallel to line ℓ by the Transitive Property of Parallel Lines. So, the moving platform is parallel to the ground.

EXAMPLE 2 Finding Side Lengths of a Parallelogram

WATCH

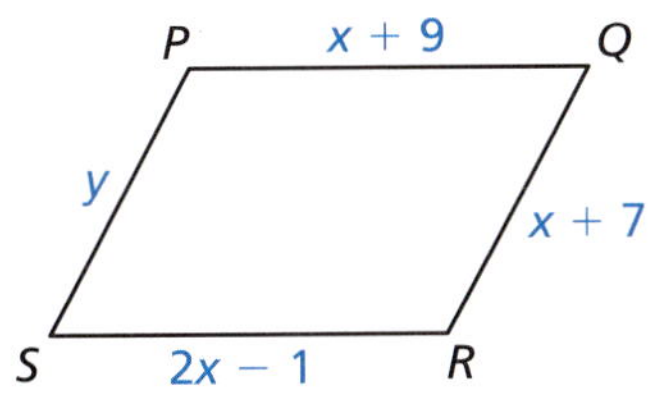

For what values of x and y is quadrilateral *PQRS* a parallelogram?

SOLUTION

By the Parallelogram Opposite Sides Converse, if both pairs of opposite sides of a quadrilateral are congruent, then the quadrilateral is a parallelogram. Find x so that $\overline{PQ} \cong \overline{SR}$.

$PQ = SR$	Set the segment lengths equal.
$x + 9 = 2x - 1$	Substitute $x + 9$ for PQ and $2x - 1$ for SR.
$10 = x$	Solve for x.

When $x = 10$, $PQ = 10 + 9 = 19$ and $SR = 2(10) - 1 = 19$. Find y so that $\overline{PS} \cong \overline{QR}$.

$PS = QR$	Set the segment lengths equal.
$y = x + 7$	Substitute y for PS and $x + 7$ for QR.
$y = 10 + 7$	Substitute 10 for x.
$y = 17$	Add.

When $x = 10$ and $y = 17$, $PS = 17$ and $QR = 10 + 7 = 17$.

▶ Quadrilateral *PQRS* is a parallelogram when $x = 10$ and $y = 17$.

SELF-ASSESSMENT

1 I don't understand yet. 2 I can do it with help. 3 I can do it on my own. 4 I can teach someone else.

1. In quadrilateral *WXYZ*, $m\angle W = 42°$, $m\angle X = 138°$, and $m\angle Y = 42°$. Find $m\angle Z$. Is *WXYZ* a parallelogram? Explain your reasoning.

2. For what values of x and y is quadrilateral *ABCD* a parallelogram? Explain your reasoning.

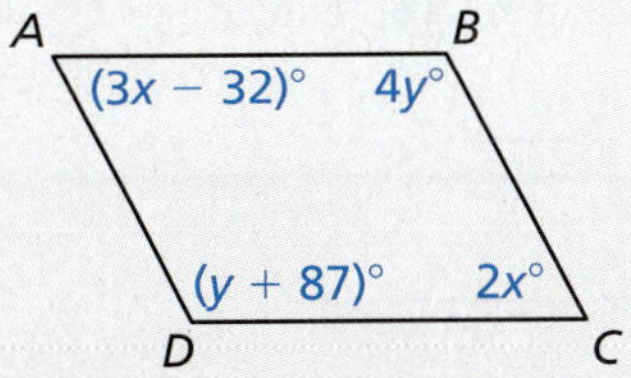

GO DIGITAL

THEOREMS

7.9 Opposite Sides Parallel and Congruent Theorem

If one pair of opposite sides of a quadrilateral are parallel and congruent, then the quadrilateral is a parallelogram.

If $\overline{BC} \parallel \overline{AD}$ and $\overline{BC} \cong \overline{AD}$, then $ABCD$ is a parallelogram.

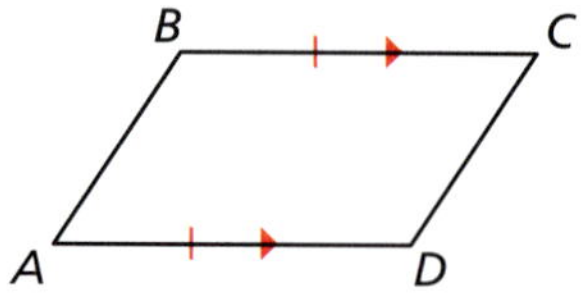

Prove this Theorem Exercise 39, page 377

7.10 Parallelogram Diagonals Converse

If the diagonals of a quadrilateral bisect each other, then the quadrilateral is a parallelogram.

If $\overline{BD}$ and $\overline{AC}$ bisect each other, then $ABCD$ is a parallelogram.

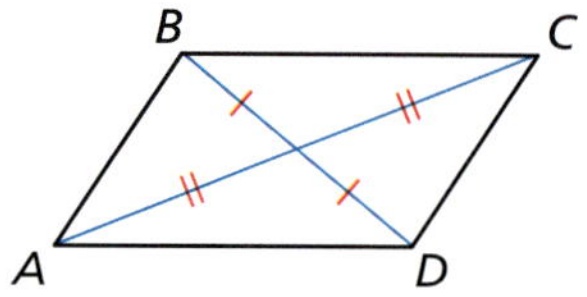

Prove this Theorem Exercise 40, page 377

EXAMPLE 3 Identifying a Parallelogram

The doorway shown is part of a building. Over time, the building has leaned sideways. Explain how you know that $SV = TU$.

SOLUTION

In the photograph, $\overline{ST} \parallel \overline{UV}$ and $\overline{ST} \cong \overline{UV}$. By the Opposite Sides Parallel and Congruent Theorem, quadrilateral $STUV$ is a parallelogram. By the Parallelogram Opposite Sides Theorem, you know that opposite sides of a parallelogram are congruent. So, $SV = TU$.

SELF-ASSESSMENT

1 I don't understand yet. 2 I can do it with help. 3 I can do it on my own. 4 I can teach someone else.

3. WRITING A quadrilateral has four congruent sides. Is the quadrilateral a parallelogram? Justify your answer.

4 MTR **4. DIFFERENT WORDS, SAME QUESTION** Which is different? Find "both" answers.

Construct a quadrilateral with congruent opposite sides.	Construct a quadrilateral with one pair of parallel sides.
Construct a quadrilateral with congruent opposite angles.	Construct a quadrilateral with one pair of opposite sides parallel and congruent.

State the theorem you can use to show that the quadrilateral is a parallelogram.

5.

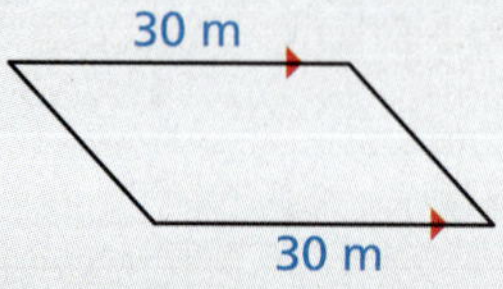

6.

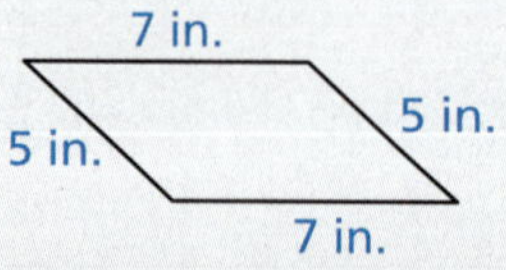

7.

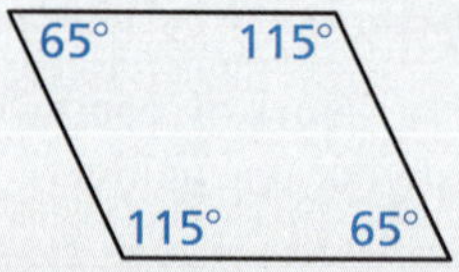

GO DIGITAL

EXAMPLE 4 Finding Diagonal Lengths of a Parallelogram

WATCH

For what value of x is quadrilateral $CDEF$ a parallelogram?

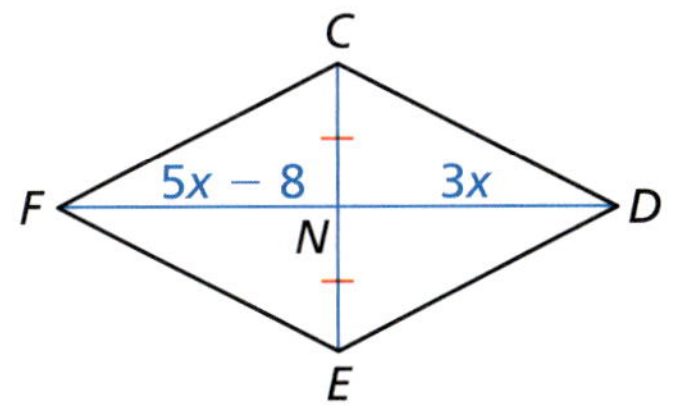

SOLUTION

By the Parallelogram Diagonals Converse, if the diagonals of $CDEF$ bisect each other, then it is a parallelogram. You are given that $\overline{CN} \cong \overline{EN}$. Find x so that $\overline{FN} \cong \overline{DN}$.

$FN = DN$	Set the segment lengths equal.
$5x - 8 = 3x$	Substitute $5x - 8$ for FN and $3x$ for DN.
$-8 = -2x$	Subtract $5x$ from each side.
$4 = x$	Divide each side by -2.

When $x = 4$, $FN = 5(4) - 8 = 12$ and $DN = 3(4) = 12$.

▶ Quadrilateral $CDEF$ is a parallelogram when $x = 4$.

SELF-ASSESSMENT

1 I don't understand yet. 2 I can do it with help. 3 I can do it on my own. 4 I can teach someone else.

8. For what value of x is quadrilateral $MNPQ$ a parallelogram? Explain your reasoning.

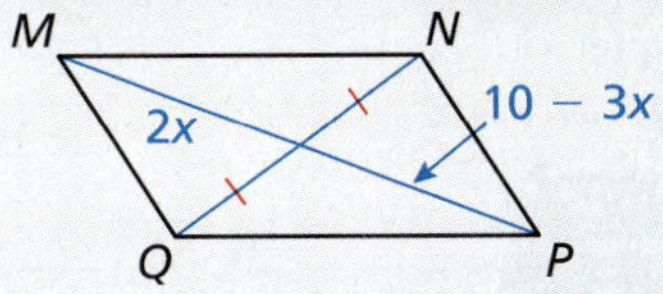

CONCEPT SUMMARY

Ways to Prove a Quadrilateral Is a Parallelogram

1. Show that both pairs of opposite sides are parallel. *(Definition)*	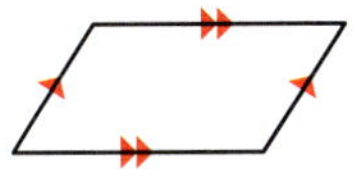
2. Show that both pairs of opposite sides are congruent. *(Parallelogram Opposite Sides Converse)*	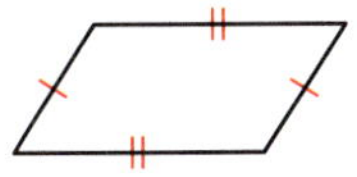
3. Show that both pairs of opposite angles are congruent. *(Parallelogram Opposite Angles Converse)*	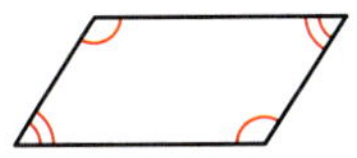
4. Show that one pair of opposite sides are parallel and congruent. *(Opposite Sides Parallel and Congruent Theorem)*	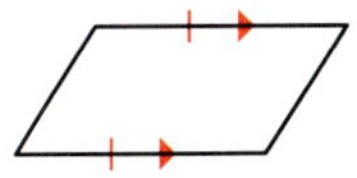
5. Show that the diagonals bisect each other. *(Parallelogram Diagonals Converse)*	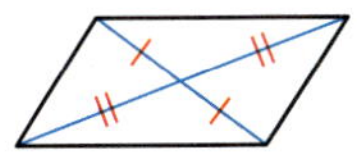

GO DIGITAL

Using Coordinate Geometry

EXAMPLE 5 **Identifying a Parallelogram in the Coordinate Plane**

WATCH

Show that quadrilateral *ABCD* is a parallelogram.

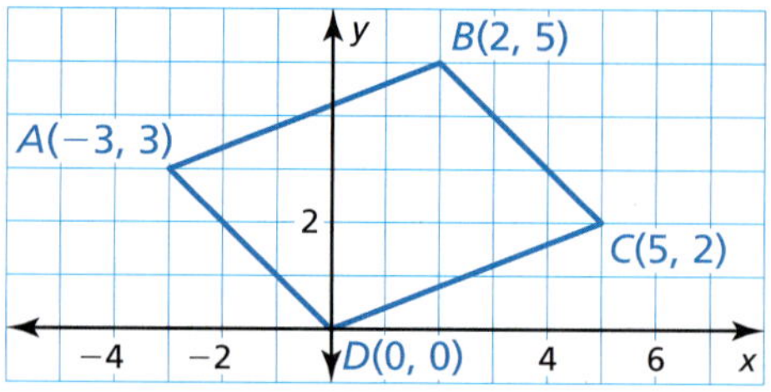

2 MTR USE ANOTHER METHOD
Explain how you can use a transformation to show that $\overline{AB} \cong \overline{CD}$.

SOLUTION

Method 1 Show that a pair of sides are parallel and congruent. Then apply the Opposite Sides Parallel and Congruent Theorem.

First, use the slope formula to show that $\overline{AB} \parallel \overline{CD}$.

$$\text{slope of } \overline{AB} = \frac{5-3}{2-(-3)} = \frac{2}{5} \qquad \text{slope of } \overline{CD} = \frac{2-0}{5-0} = \frac{2}{5}$$

Because $\overline{AB}$ and $\overline{CD}$ have the same slope, they are parallel.

Then use the Distance Formula to show that $\overline{AB}$ and $\overline{CD}$ are congruent.

$$AB = \sqrt{[2-(-3)]^2 + (5-3)^2} = \sqrt{29}$$

$$CD = \sqrt{(5-0)^2 + (2-0)^2} = \sqrt{29}$$

Because $AB = CD = \sqrt{29}$, $\overline{AB} \cong \overline{CD}$.

▶ $\overline{AB}$ and $\overline{CD}$ are parallel and congruent. So, *ABCD* is a parallelogram by the Opposite Sides Parallel and Congruent Theorem.

Method 2 Show that opposite sides are congruent. Then apply the Parallelogram Opposite Sides Converse. In Method 1, you already have shown that because $AB = CD = \sqrt{29}$, $\overline{AB} \cong \overline{CD}$. Now find *AD* and *BC*.

$$AD = \sqrt{(-3-0)^2 + (3-0)^2} = 3\sqrt{2}$$

$$BC = \sqrt{(2-5)^2 + (5-2)^2} = 3\sqrt{2}$$

Because $AD = BC = 3\sqrt{2}$, $\overline{AD} \cong \overline{BC}$.

▶ $\overline{AB} \cong \overline{CD}$ and $\overline{AD} \cong \overline{BC}$. So, *ABCD* is a parallelogram by the Parallelogram Opposite Sides Converse.

SELF-ASSESSMENT

1 I don't understand yet. 2 I can do it with help. 3 I can do it on my own. 4 I can teach someone else.

9. Show that quadrilateral *JKLM* is a parallelogram.

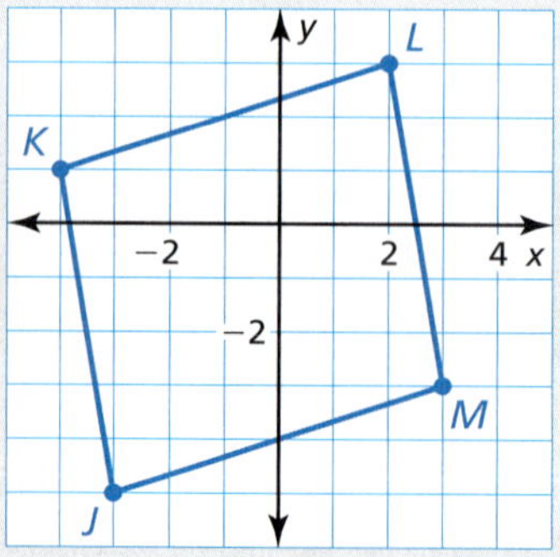

10. Refer to the Concept Summary on page 373. Explain two other methods you can use to show that quadrilateral *ABCD* in Example 5 is a parallelogram.

GO DIGITAL

7.3 Practice WITH CalcChat® AND CalcView®

In Exercises 1–6, state which theorem you can use to show that the quadrilateral is a parallelogram. *(See Examples 1 and 3.)*

1.
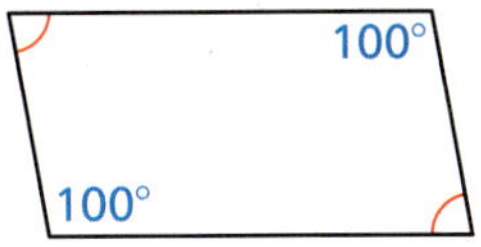

2.
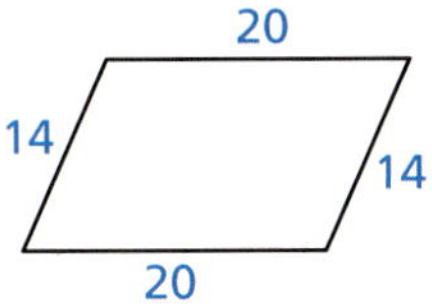

▶ 3.
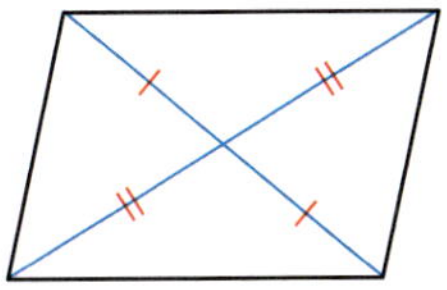

4.
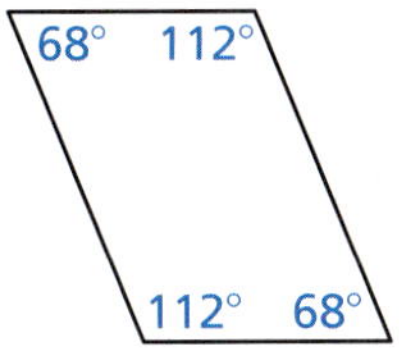

5.
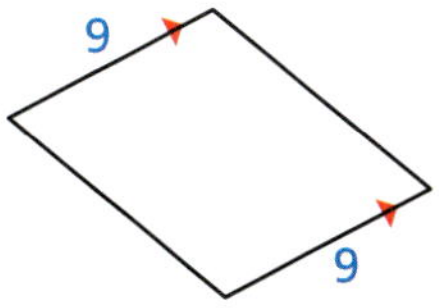

6.
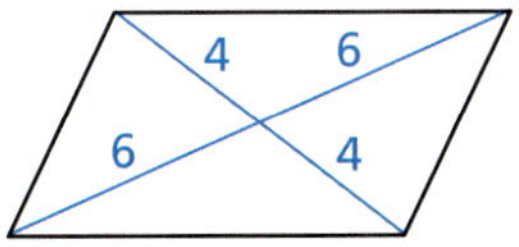

In Exercises 7–10, find the values of *x* and *y* that make the quadrilateral a parallelogram. *(See Example 2.)*

▶ 7.
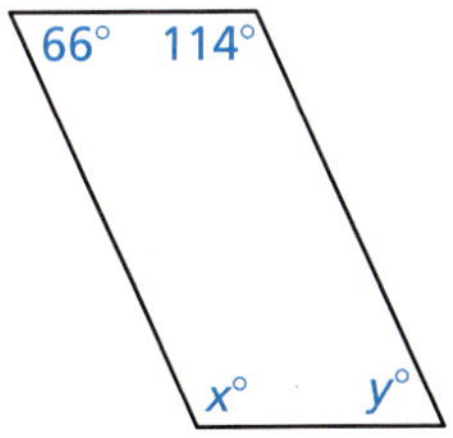

8.
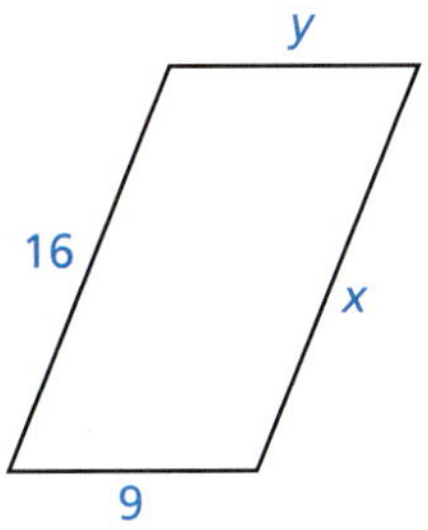

9.
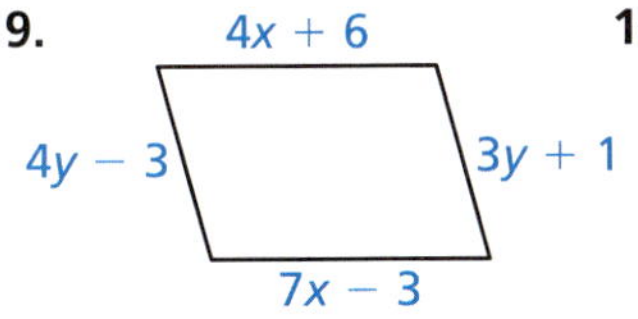

10.
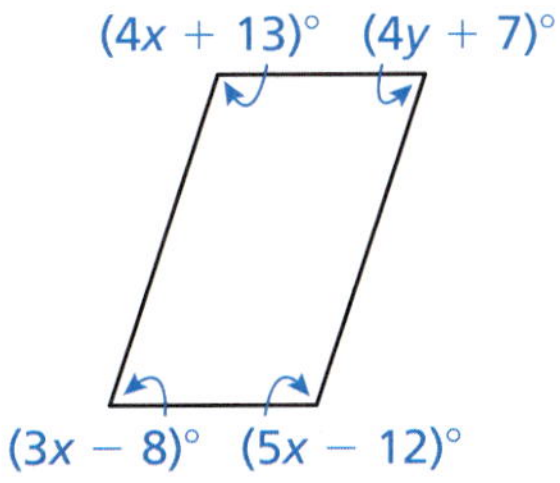

In Exercises 11–14, find the value of *x* that makes the quadrilateral a parallelogram. *(See Example 4.)*

▶ 11.
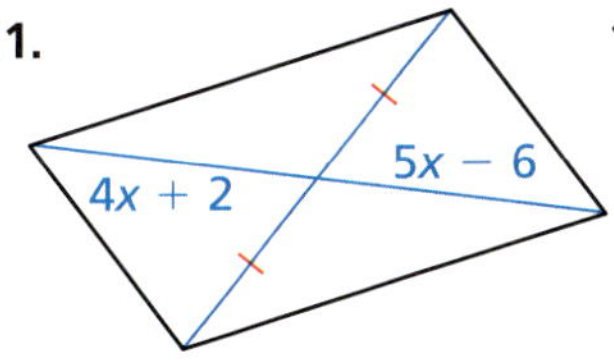

12.
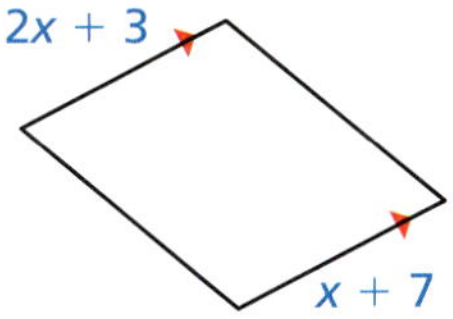

13.
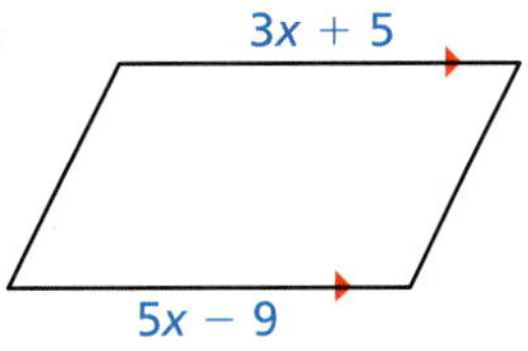

14.
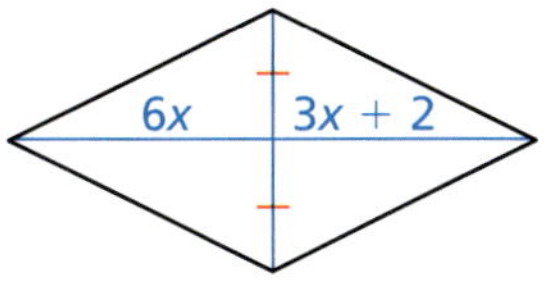

In Exercises 15–18, graph the quadrilateral with the given vertices in a coordinate plane. Then show that the quadrilateral is a parallelogram. *(See Example 5.)*

▶ 15. $A(0, 1)$, $B(4, 4)$, $C(12, 4)$, $D(8, 1)$

16. $E(-3, 0)$, $F(-3, 4)$, $G(3, -1)$, $H(3, -5)$

17. $J(-2, 3)$, $K(-5, 7)$, $L(3, 6)$, $M(6, 2)$

18. $N(-5, 0)$, $P(0, 4)$, $Q(3, 0)$, $R(-2, -4)$

4 MTR **ERROR ANALYSIS In Exercises 19 and 20, describe and correct the error in identifying a parallelogram.**

19.
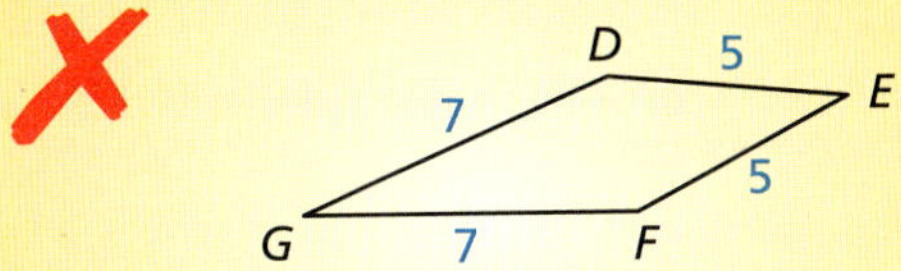

DEFG is a parallelogram by the Parallelogram Opposite Sides Converse.

20.
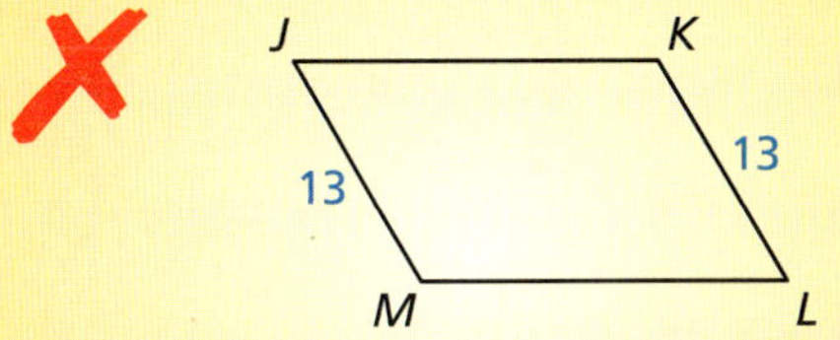

JKLM is a parallelogram by the Opposite Sides Parallel and Congruent Theorem.

21. REASONING Find the value of *x* that makes the quadrilateral a parallelogram.

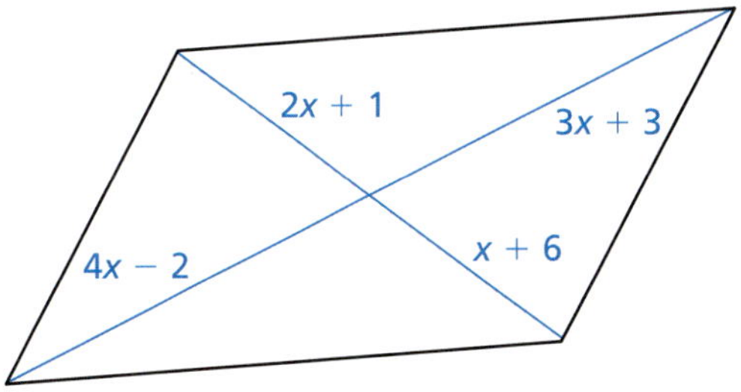

GO DIGITAL

5 MTR **22.** **CONNECTING CONCEPTS** Find the value of x such that the quadrilateral is a parallelogram. Then find the perimeter of the parallelogram. Explain.

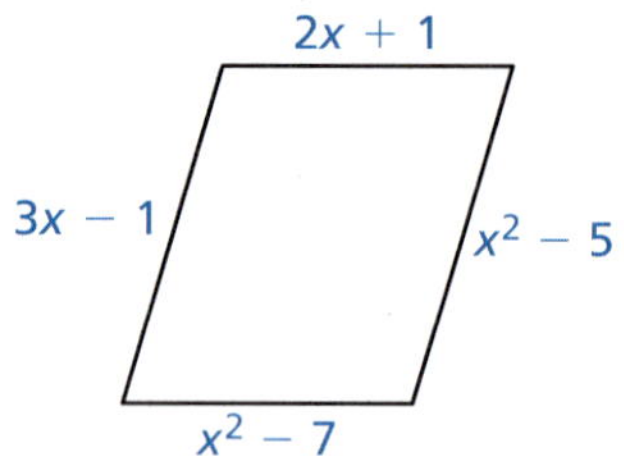

WRITING In Exercises 23–25, write the indicated theorems as a biconditional statement.

23. Parallelogram Opposite Sides Theorem and Parallelogram Opposite Sides Converse

24. Parallelogram Opposite Angles Theorem and Parallelogram Opposite Angles Converse

25. Parallelogram Diagonals Theorem and Parallelogram Diagonals Converse

4 MTR **26.** **MAKING AN ARGUMENT** Can you show that quadrilateral *WXYZ* is a parallelogram by using the Consecutive Interior Angles Converse and the Opposite Sides Parallel and Congruent Theorem? Explain your reasoning.

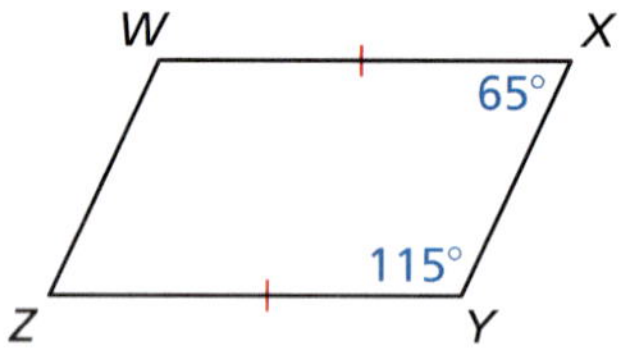

REASONING In Exercises 27 and 28, your classmate incorrectly claims that the marked information can be used to show that the figure is a parallelogram. Draw a quadrilateral with the same marked properties that is clearly *not* a parallelogram.

27.

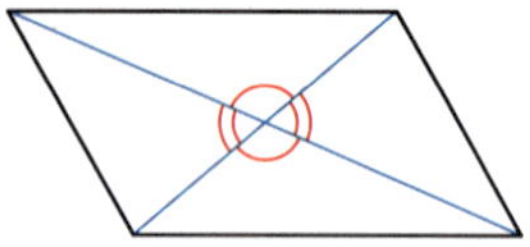

28.

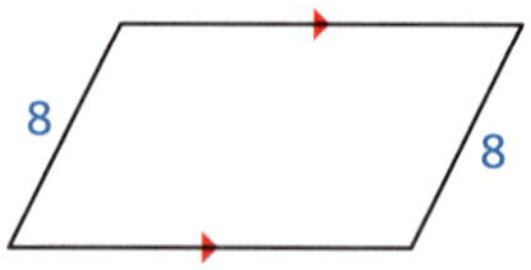

29. **CONSTRUCTION** Describe a method that uses the Opposite Sides Parallel and Congruent Theorem to construct a parallelogram. Then construct a parallelogram using your method.

30. **CONSTRUCTION** Follow the steps below to construct a parallelogram. Use a theorem to explain why this method works.

Step 1 Use a ruler to draw two segments that intersect at their midpoints.

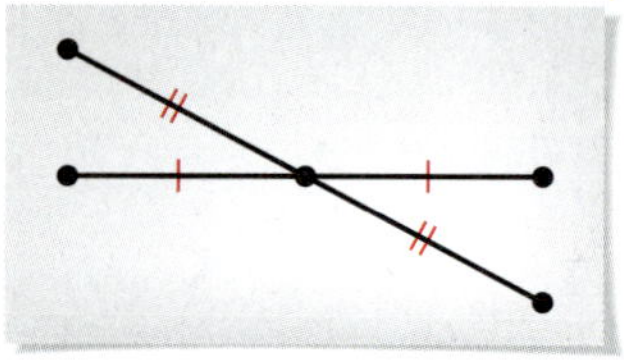

Step 2 Connect the endpoints of the segments to form a parallelogram.

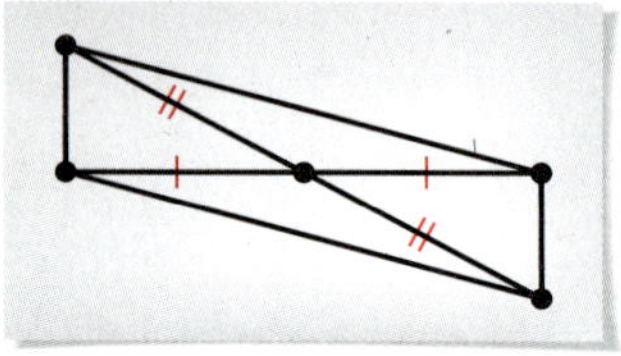

7 MTR **31.** **MODELING REAL LIFE** You shoot a pool ball, and it rolls back to where it started, as shown in the diagram. The ball bounces off each wall at the same angle at which it hits the wall. The ball hits the first wall at an angle of 63°. So, $m\angle AEF = 63°$.

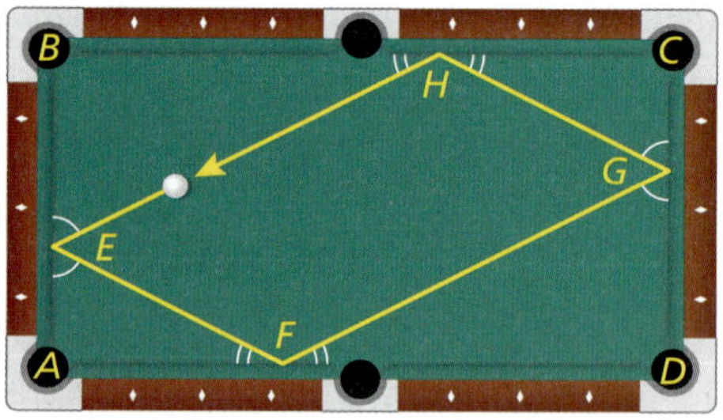

a. Find $m\angle AFE$. Explain your reasoning.

b. Determine whether *EFGH* is a parallelogram. Explain your reasoning.

7 MTR **32.** **MODELING REAL LIFE** In the diagram of the parking lot, $m\angle JKL = 60°$, $JK = LM = 21$ feet, and $KL = JM = 9$ feet.

a. Find $m\angle JML$, $m\angle KJM$, and $m\angle KLM$.

b. $\overline{LM} \parallel \overline{NO}$ and $\overline{NO} \parallel \overline{PQ}$. Which theorem can you use to show that $\overline{JK} \parallel \overline{PQ}$?

GO DIGITAL

REASONING In Exercises 33–36, describe how to prove that *ABCD* is a parallelogram.

33.

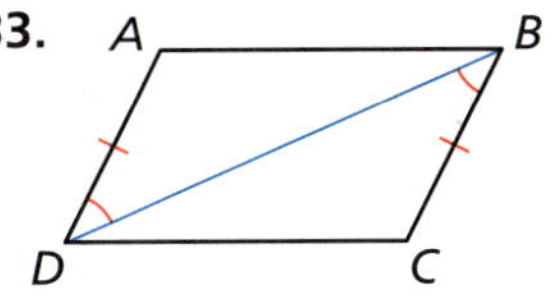

34.

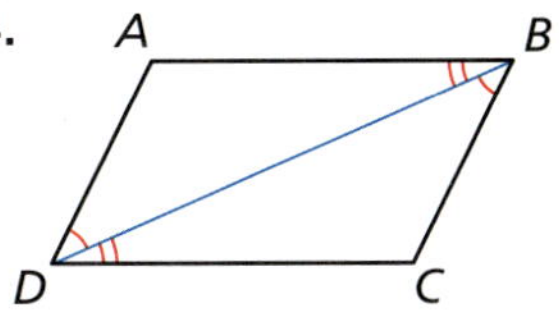

35.

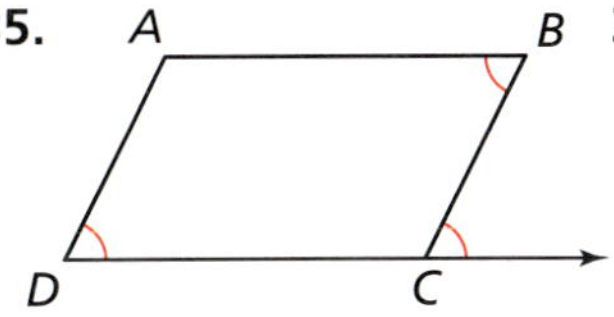

36.

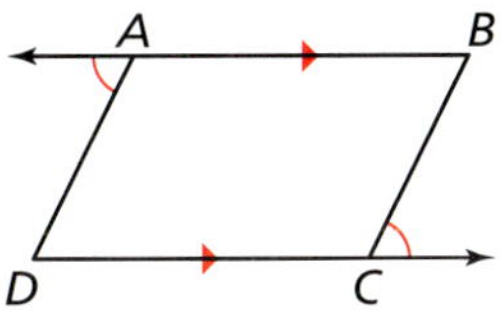

37. PROVING A THEOREM Prove the Parallelogram Opposite Angles Converse. (*Hint*: Let $x°$ represent $m\angle A$ and $m\angle C$. Let $y°$ represent $m\angle B$ and $m\angle D$.)

Given $\angle A \cong \angle C$, $\angle B \cong \angle D$

Prove *ABCD* is a parallelogram.

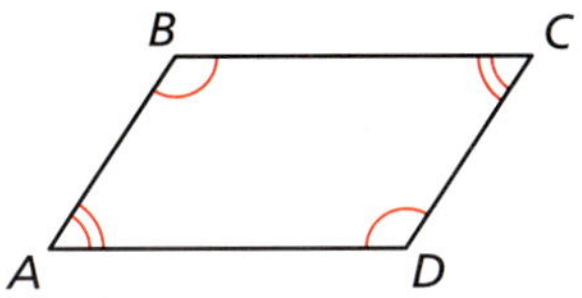

38. REASONING The Parallelogram Consecutive Angles Theorem says that if a quadrilateral is a parallelogram, then its consecutive angles are supplementary. Write the converse of this theorem. Then write a plan for proving the converse. Include a diagram.

39. PROVING A THEOREM Prove the Opposite Sides Parallel and Congruent Theorem.

Given $\overline{QR} \parallel \overline{PS}$, $\overline{QR} \cong \overline{PS}$

Prove *PQRS* is a parallelogram.

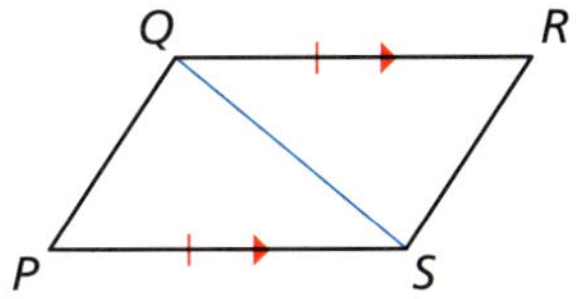

40. PROVING A THEOREM Prove the Parallelogram Diagonals Converse.

Given Diagonals $\overline{JL}$ and $\overline{KM}$ bisect each other.

Prove *JKLM* is a parallelogram.

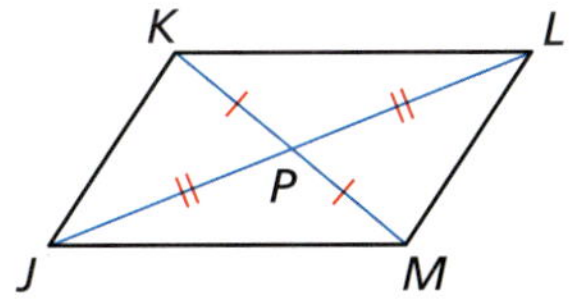

41. PROOF Write a proof.

Given *DEBF* is a parallelogram. $AE = CF$

Prove *ABCD* is a parallelogram.

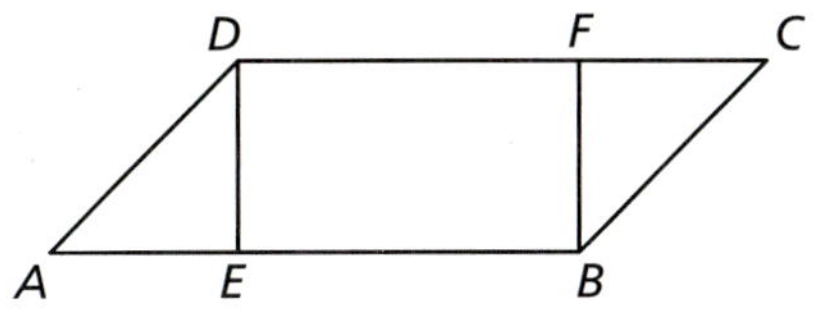

42. HOW DO YOU SEE IT?
A music stand can be folded up, as shown. In the diagrams, *AEFD* and *EBCF* are parallelograms. Which labeled segments remain parallel as the stand is folded?

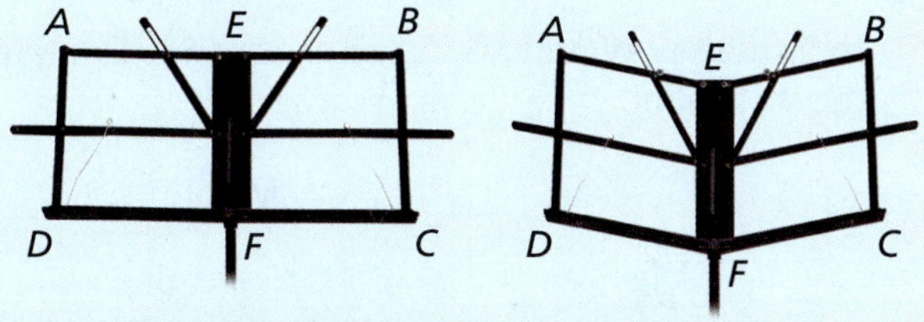

43. REASONING In the diagram, *ABCD* is a parallelogram, $BF = DE = 12$, and $CF = 8$. Find *AE*. Explain your reasoning.

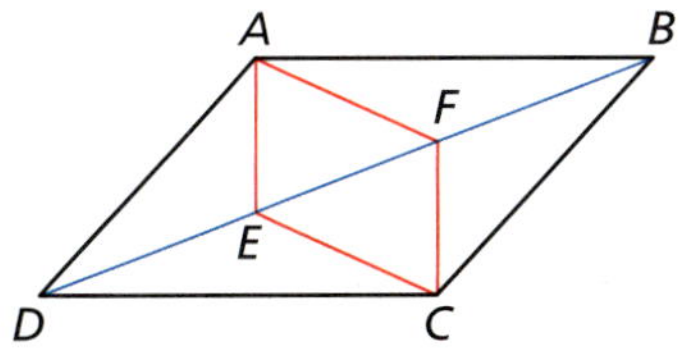

44. THOUGHT PROVOKING
Create a regular hexagon using congruent parallelograms.

45. PROOF Write a proof.

Given *ABCD* is a parallelogram.
$\angle A$ is a right angle.

Prove $\angle B$, $\angle C$, and $\angle D$ are right angles.

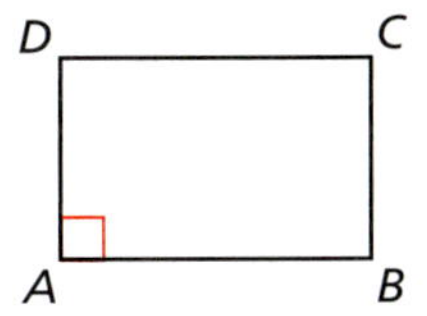

46. REASONING Three interior angle measures of a quadrilateral are 67°, 67°, and 113°. Can you conclude that the quadrilateral is a parallelogram? Explain.

GO DIGITAL

5 MTR **47.** **STRUCTURE** The midpoints of the sides of a quadrilateral have been joined to form what looks like a parallelogram. Show that a quadrilateral formed by connecting the midpoints of the sides of any quadrilateral is always a parallelogram. (*Hint*: Draw a diagram. Include a diagonal of the larger quadrilateral. Show how two sides of the smaller quadrilateral relate to the diagonal.)

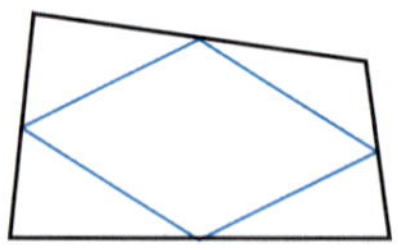

5 MTR **48.** **STRUCTURE** Show that if $ABCD$ is a parallelogram with its diagonals intersecting at E, then you can connect the midpoints F, G, H, and J of $\overline{AE}$, $\overline{BE}$, $\overline{CE}$, and $\overline{DE}$, respectively, to form another parallelogram, $FGHJ$.

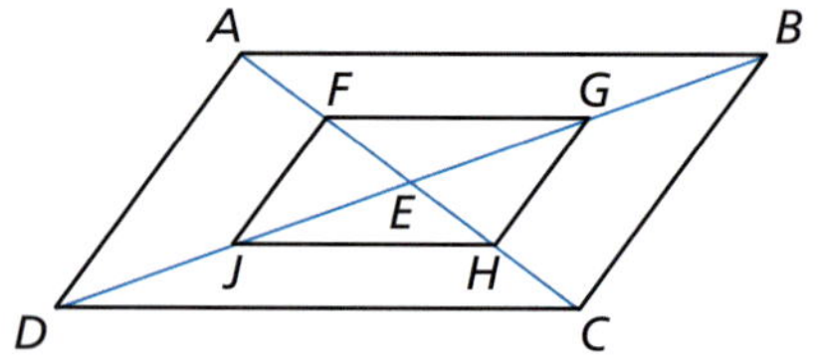

REVIEW & REFRESH

WATCH

In Exercises 49 and 50, $\triangle XYZ \cong \triangle MNL$. Complete the statement.

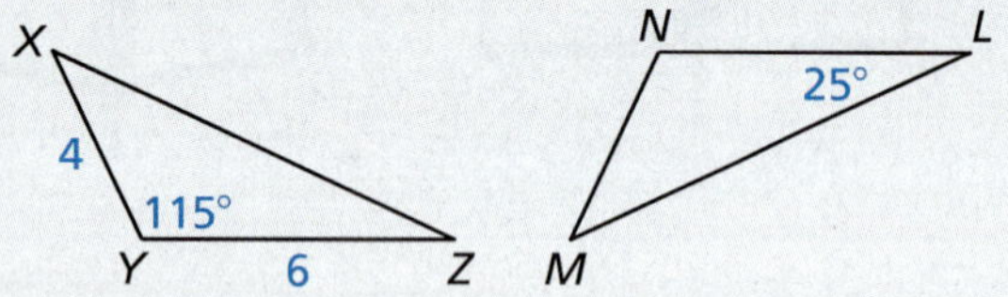

49. $m\angle M =$ ______

50. $MN =$ ______

In Exercises 51 and 52, solve the equation for *y*. Justify each step.

51. $3x + y = 11$

52. $8x + 3 = -9 + 4y$

53. Find the value of x.

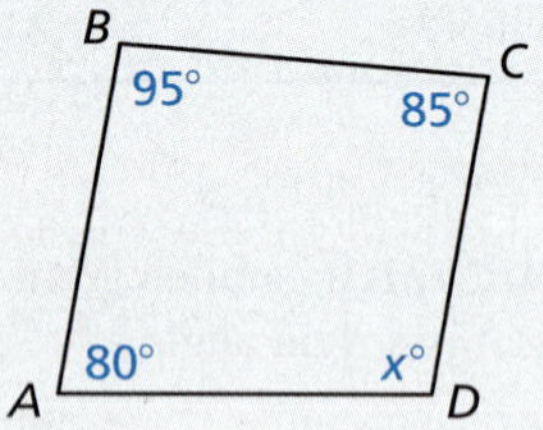

54. Find the distance between $A(4, 2)$ and $B(7, 11)$.

55. Three vertices of ▱$DEFG$ are $D(1, 1)$, $E(5, 2)$, and $F(1, 8)$. Find the coordinates of the remaining vertex.

56. Graph the triangle with vertices $X(2, -2)$, $Y(-4, -6)$, and $Z(0, 8)$ and its image in the coordinate plane after a dilation with scale factor $k = 0.5$.

57. Complete the statement AD ______ CD with < or >. Explain your reasoning.

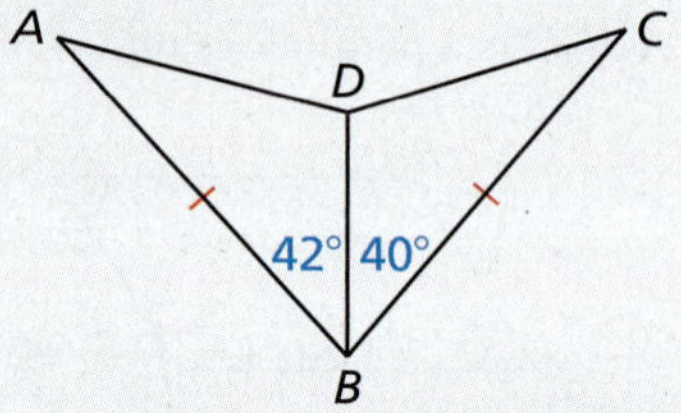

58. State which theorem you can use to show that the quadrilateral is a parallelogram.

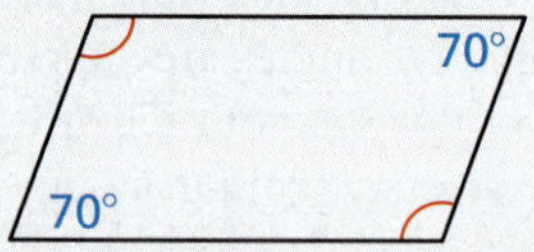

59. Name the vector and write its component form.

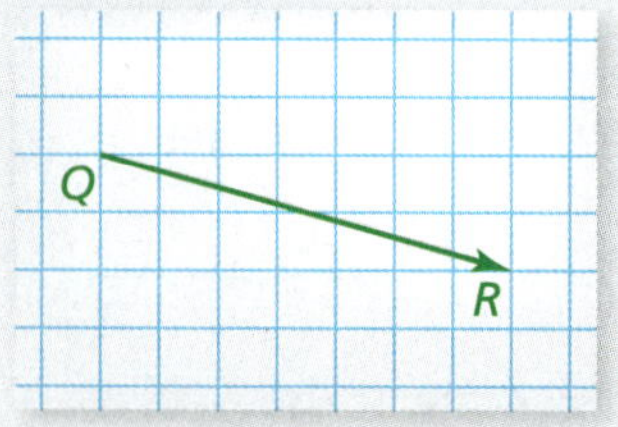

60. Place a rectangle with a length of 4 units and a width of 2 units in the coordinate plane. Find the length of the diagonal.

61. Find the value of x. Show your steps.

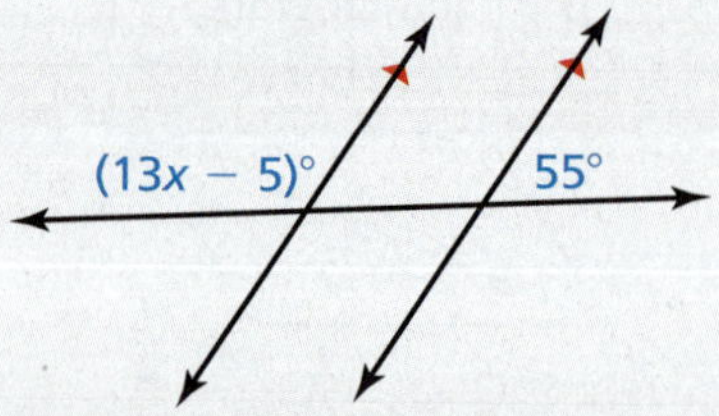

GO DIGITAL

7.4 Properties of Special Parallelograms

Learning Target: Explain the properties of special parallelograms.

Success Criteria:
- I can identify special quadrilaterals.
- I can explain how special parallelograms are related.
- I can find missing measures of special parallelograms.
- I can identify special parallelograms in a coordinate plane.

EXPLORE IT! Analyzing Diagonals of Quadrilaterals

Work with a partner. Recall the three types of parallelograms shown below.

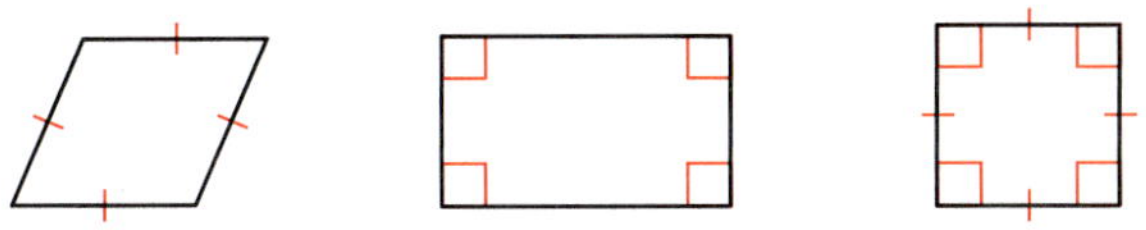

Rhombus   Rectangle   Square

a. Use the diagrams to define each type of quadrilateral.

b. Use technology to construct two congruent line segments that bisect each other. Draw a quadrilateral by connecting the endpoints.

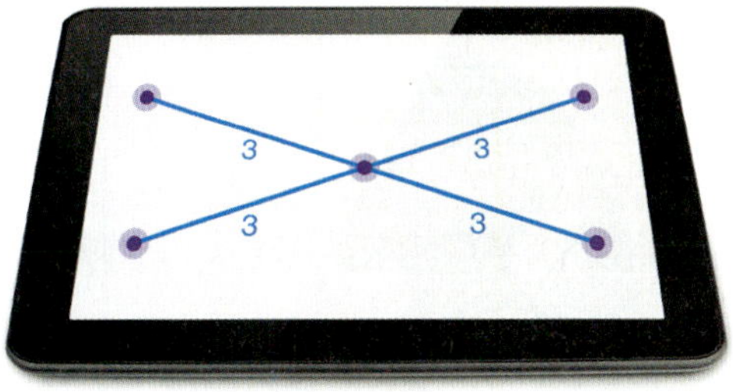

c. Is the quadrilateral you drew a parallelogram? a rectangle? a rhombus? a square? Explain your reasoning.

d. Repeat parts (b) and (c) for several pairs of congruent line segments that bisect each other. Make conjectures based on your results.

e. Use technology to construct two line segments that are perpendicular bisectors of each other. Draw a quadrilateral by connecting the endpoints.

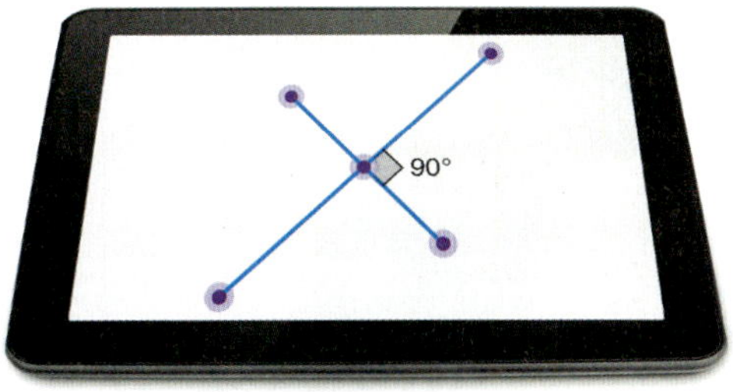

f. Is the quadrilateral you drew a parallelogram? a rectangle? a rhombus? a square? Explain your reasoning.

g. Repeat parts (e) and (f) for several other line segments that are perpendicular bisectors of each other. Make conjectures based on your results.

h. What are some properties of the diagonals of rectangles, rhombuses, and squares?

4 MTR CONSTRUCT AN ARGUMENT
What other quadrilaterals can you form using similar methods? Explain your reasoning.

Geometric Reasoning
MA.912.GR.1.4 Prove relationships and theorems about parallelograms. Solve mathematical and real-world problems involving postulates, relationships and theorems of parallelograms.
MA.912.GR.3.2 Given a mathematical context, use coordinate geometry to classify or justify definitions, properties and theorems involving circles, triangles or quadrilaterals.

GO DIGITAL

Using Properties of Special Parallelograms

Vocabulary

rhombus, *p. 380*
rectangle, *p. 380*
square, *p. 380*

In this lesson, you will learn about corollaries and theorems that correspond to three special types of parallelograms: *rhombuses*, *rectangles*, and *squares*.

KEY IDEAS

Rhombuses, Rectangles, and Squares

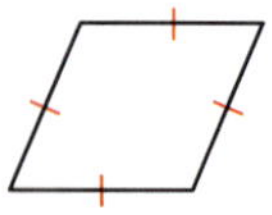

A **rhombus** is a parallelogram with four congruent sides.

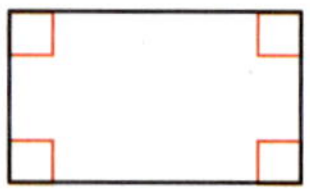

A **rectangle** is a parallelogram with four right angles.

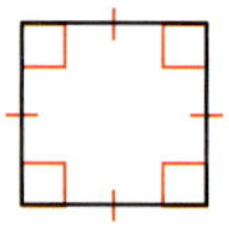

A **square** is a parallelogram with four congruent sides and four right angles.

You can use the corollaries below to prove that a quadrilateral is a rhombus, rectangle, or square, without first proving that the quadrilateral is a parallelogram.

COROLLARIES

7.2 Rhombus Corollary

A quadrilateral is a rhombus if and only if it has four congruent sides.

$ABCD$ is a rhombus if and only if $\overline{AB} \cong \overline{BC} \cong \overline{CD} \cong \overline{AD}$.

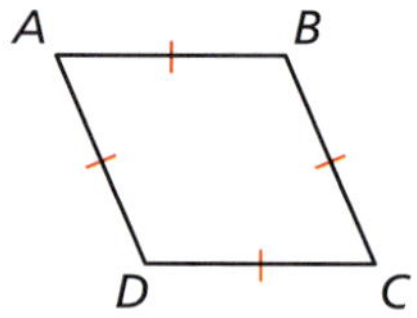

Prove this Corollary Exercise 77, page 387

7.3 Rectangle Corollary

A quadrilateral is a rectangle if and only if it has four right angles.

$ABCD$ is a rectangle if and only if $\angle A$, $\angle B$, $\angle C$, and $\angle D$ are right angles.

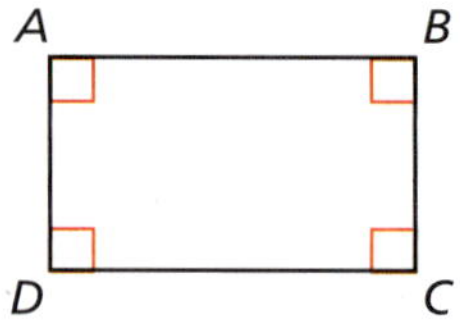

Prove this Corollary Exercise 78, page 387

7.4 Square Corollary

A quadrilateral is a square if and only if it is a rhombus and a rectangle.

$ABCD$ is a square if and only if $\overline{AB} \cong \overline{BC} \cong \overline{CD} \cong \overline{AD}$ and $\angle A$, $\angle B$, $\angle C$, and $\angle D$ are right angles.

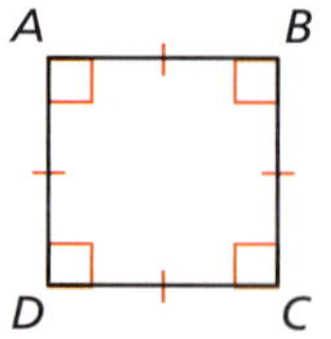

Prove this Corollary Exercise 79, page 387

GO DIGITAL

The Venn diagram below illustrates some important relationships among parallelograms, rhombuses, rectangles, and squares. For example, you can see that a square is a rhombus because it is a parallelogram with four congruent sides. Because it has four right angles, a square is also a rectangle.

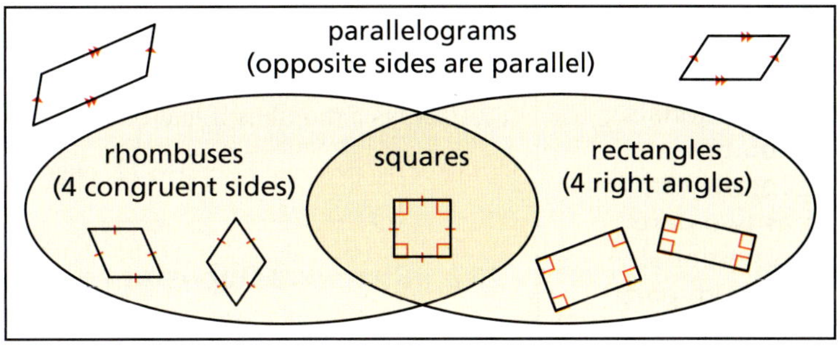

EXAMPLE 1 Using Properties of Special Quadrilaterals

For any rhombus *QRST*, decide whether the statement is *always* or *sometimes* true. Draw a diagram and explain your reasoning.

a. $\angle Q \cong \angle S$ **b.** $\angle Q \cong \angle R$

SOLUTION

a. By definition, a rhombus is a parallelogram with four congruent sides. By the Parallelogram Opposite Angles Theorem, opposite angles of a parallelogram are congruent. So, $\angle Q \cong \angle S$. The statement is *always* true.

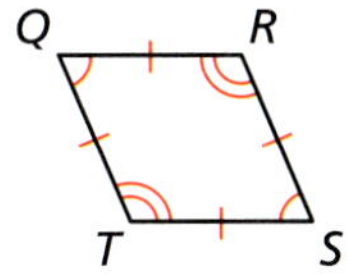

b. If rhombus *QRST* is a square, then all four angles are congruent right angles. So, $\angle Q \cong \angle R$ when *QRST* is a square. Because not all rhombuses are also squares, the statement is *sometimes* true.

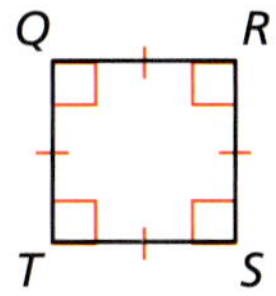

EXAMPLE 2 Classifying Special Quadrilaterals

Classify the special quadrilateral. Explain your reasoning.

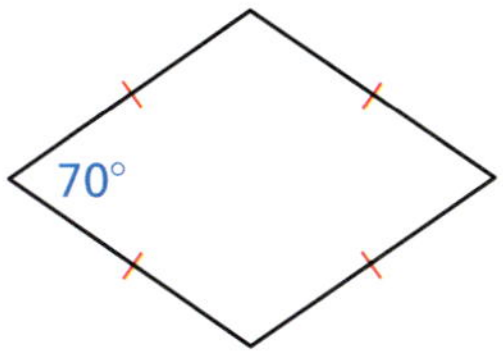

SOLUTION

The quadrilateral has four congruent sides. By the Rhombus Corollary, the quadrilateral is a rhombus. Because one of the angles is not a right angle, the rhombus cannot be a square.

SELF-ASSESSMENT

1 I don't understand yet. 2 I can do it with help. 3 I can do it on my own. 4 I can teach someone else.

1. **VOCABULARY** What is another name for an equilateral rectangle?
2. For any square *JKLM*, is it *always* or *sometimes* true that $\overline{JK} \perp \overline{KL}$? Explain your reasoning.
3. For any rectangle *EFGH*, is it *always* or *sometimes* true that $\overline{FG} \cong \overline{GH}$? Explain your reasoning.
4. A quadrilateral has four congruent sides and three angles that measure 90°. Sketch the quadrilateral and classify it.

GO DIGITAL

Using Properties of Diagonals

THEOREMS

7.11 Rhombus Diagonals Theorem

A parallelogram is a rhombus if and only if its diagonals are perpendicular.

$\square ABCD$ is a rhombus if and only if $\overline{AC} \perp \overline{BD}$.

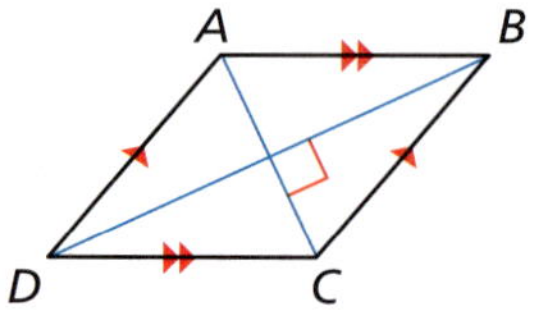

Prove this Theorem Exercise 72, page 387

7.12 Rhombus Opposite Angles Theorem

A parallelogram is a rhombus if and only if each diagonal bisects a pair of opposite angles.

$\square ABCD$ is a rhombus if and only if $\overline{AC}$ bisects $\angle BCD$ and $\angle BAD$, and $\overline{BD}$ bisects $\angle ABC$ and $\angle ADC$.

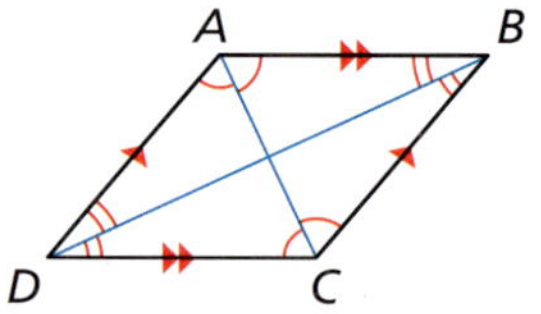

Prove this Theorem Exercises 75 and 76, page 387

READING

Recall that biconditionals, such as the Rhombus Diagonals Theorem, can be rewritten as two parts. To prove a biconditional, you must prove both parts.

PROOF Part of Rhombus Diagonals Theorem

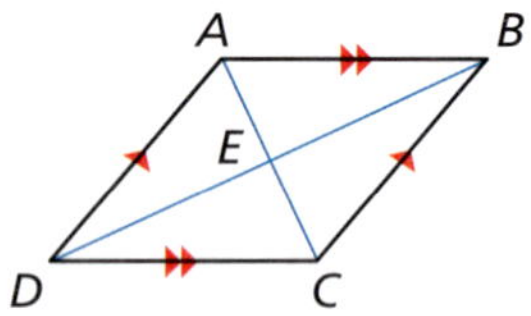

Given $ABCD$ is a rhombus.

Prove $\overline{AC} \perp \overline{BD}$

$ABCD$ is a rhombus. By the definition of a rhombus, $\overline{AB} \cong \overline{BC}$. Because a rhombus is a parallelogram and the diagonals of a parallelogram bisect each other, $\overline{BD}$ bisects $\overline{AC}$ at E. So, $\overline{AE} \cong \overline{EC}$. $\overline{BE} \cong \overline{BE}$ by the Reflexive Property of Segment Congruence. So, $\triangle AEB \cong \triangle CEB$ by the SSS Congruence Theorem. $\angle AEB \cong \angle CEB$ because corresponding parts of congruent triangles are congruent. Then by the Linear Pair Postulate, $\angle AEB$ and $\angle CEB$ are supplementary. Two congruent angles that form a linear pair are right angles, so $m\angle AEB = m\angle CEB = 90°$ by the definition of a right angle. So, $\overline{AC} \perp \overline{BD}$ by the definition of perpendicular lines.

EXAMPLE 3 Finding Angle Measures in a Rhombus

WATCH

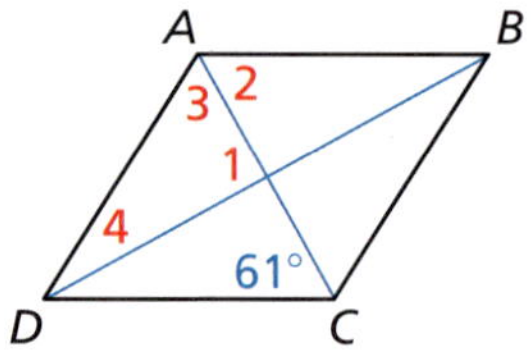

Find the measures of the numbered angles in rhombus $ABCD$.

SOLUTION

Use the Rhombus Diagonals Theorem and the Rhombus Opposite Angles Theorem to find the angle measures.

$m\angle 1 = 90°$	The diagonals of a rhombus are perpendicular.
$m\angle 2 = 61°$	Alternate Interior Angles Theorem
$m\angle 3 = 61°$	Each diagonal of a rhombus bisects a pair of opposite angles, and $m\angle 2 = 61°$.
$m\angle 1 + m\angle 3 + m\angle 4 = 180°$	Triangle Sum Theorem
$90° + 61° + m\angle 4 = 180°$	Substitute 90° for $m\angle 1$ and 61° for $m\angle 3$.
$m\angle 4 = 29°$	Solve for $m\angle 4$.

▶ So, $m\angle 1 = 90°$, $m\angle 2 = 61°$, $m\angle 3 = 61°$, and $m\angle 4 = 29°$.

SELF-ASSESSMENT 1 I don't understand yet. 2 I can do it with help. 3 I can do it on my own. 4 I can teach someone else.

5. In Example 3, find $m\angle ADC$ and $m\angle BCD$.

GO DIGITAL

THEOREM

7.13 Rectangle Diagonals Theorem

A parallelogram is a rectangle if and only if its diagonals are congruent.

$\square ABCD$ is a rectangle if and only if $\overline{AC} \cong \overline{BD}$.

Prove this Theorem Exercises 86 and 87, page 388

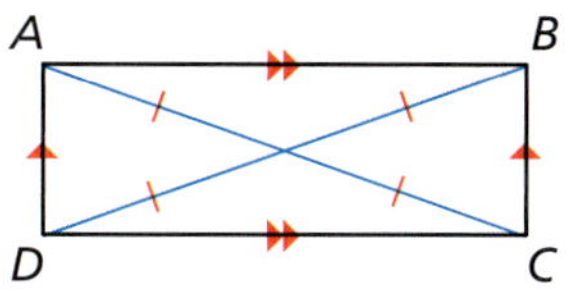

EXAMPLE 4 Finding Diagonal Lengths in a Rectangle

In rectangle $QRST$, $QS = 5x - 31$ and $RT = 2x + 11$. Find the lengths of the diagonals of $QRST$.

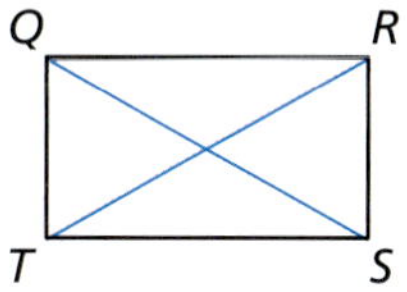

SOLUTION

By the Rectangle Diagonals Theorem, the diagonals of a rectangle are congruent. Find x so that $\overline{QS} \cong \overline{RT}$.

$QS = RT$	Set the diagonal lengths equal.
$5x - 31 = 2x + 11$	Substitute $5x - 31$ for QS and $2x + 11$ for RT.
$x = 14$	Simplify.

When $x = 14$, $QS = 5(14) - 31 = 39$ and $RT = 2(14) + 11 = 39$.

▶ Each diagonal has a length of 39 units.

EXAMPLE 5 Finding Coordinates of Missing Vertices

Square $ABCD$ has a diagonal $\overline{AC}$ with vertices $A(1, 3)$ and $C(5, 6)$. Find the coordinates of the remaining vertices.

SOLUTION

By the Parallelogram Diagonals Theorem, $\overline{AC}$ and $\overline{BD}$ bisect each other. So, they have the same midpoint. Using the Midpoint Formula, the midpoint is $M(3, 4.5)$. By the Rhombus Diagonals Theorem, $\overline{AC} \perp \overline{BD}$. By the Rectangle Diagonals Theorem, $\overline{AC} \cong \overline{BD}$. So, a 90° counterclockwise rotation about $M(3, 4.5)$ maps $\overline{AC}$ to $\overline{BD}$.

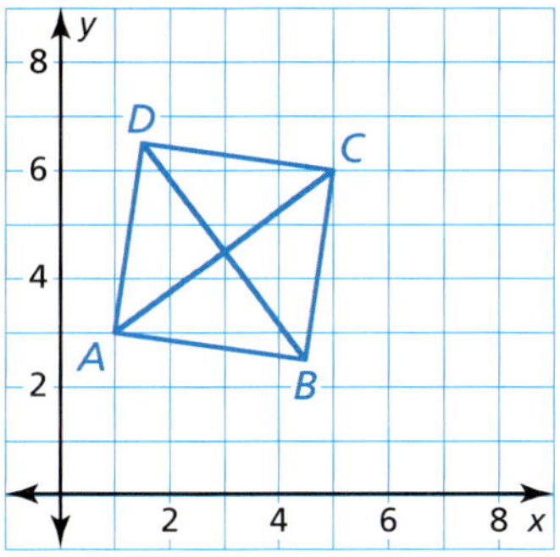

Step 1 Translate $\overline{AC}$ along $\overrightarrow{MO}$.

$$(a, b) \rightarrow (a - 3, b - 4.5)$$

$$A(1, 3) \rightarrow A'(1 - 3, 3 - 4.5) = A'(-2, -1.5)$$

$$C(5, 6) \rightarrow C'(5 - 3, 6 - 4.5) = C'(2, 1.5)$$

Step 2 Use the coordinate rule for a 90° counterclockwise rotation about the origin.

$$(a, b) \rightarrow (-b, a)$$

$$A'(-2, -1.5) \rightarrow A''(1.5, -2)$$

$$C'(2, 1.5) \rightarrow C''(-1.5, 2)$$

Step 3 Translate back along $\overrightarrow{OM}$.

$$(a, b) \rightarrow (a + 3, b + 4.5)$$

$$A''(1.5, -2) \rightarrow A'''(1.5 + 3, -2 + 4.5) = A'''(4.5, 2.5) = B$$

$$C''(-1.5, 2) \rightarrow C'''(-1.5 + 3, 2 + 4.5) = C'''(1.5, 6.5) = D$$

▶ So, the remaining vertices are $B(4.5, 2.5)$ and $D(1.5, 6.5)$.

Check Check that the diagonals are perpendicular.

$$m_{\overline{AC}} = \frac{6 - 3}{5 - 1} = \frac{3}{4}$$

$$m_{\overline{BD}} = \frac{6.5 - 2.5}{1.5 - 4.5} = -\frac{4}{3} \checkmark$$

GO DIGITAL

EXAMPLE 6 Identifying a Parallelogram in the Coordinate Plane

WATCH

Decide whether $\square ABCD$ with vertices $A(-2, 6)$, $B(6, 8)$, $C(4, 0)$, and $D(-4, -2)$ is a *rectangle*, a *rhombus*, or a *square*. Give all names that apply.

SOLUTION

1. Understand the Problem You know the vertices of $\square ABCD$. You need to identify the type of parallelogram.

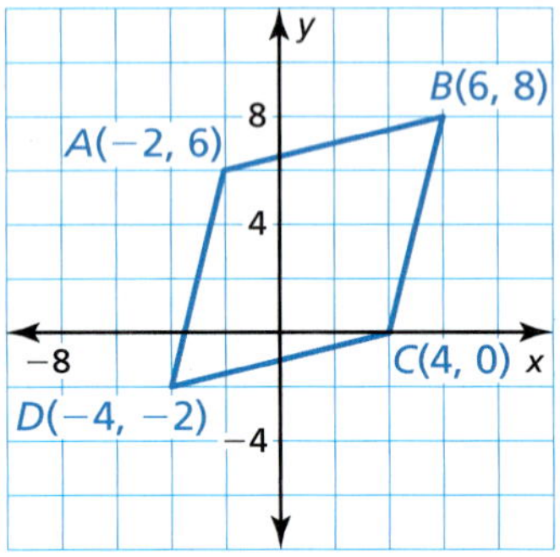

2. Make a Plan Begin by graphing the vertices. From the graph, it appears that all four sides are congruent and there are no right angles.

Check the lengths and slopes of the diagonals of $\square ABCD$. If the diagonals are congruent, then $\square ABCD$ is a rectangle. If the diagonals are perpendicular, then $\square ABCD$ is a rhombus. If they are both congruent and perpendicular, then $\square ABCD$ is a rectangle, a rhombus, and a square.

3. Solve and Check Use the Distance Formula to find AC and BD.

$$AC = \sqrt{(-2-4)^2 + (6-0)^2} = \sqrt{72} = 6\sqrt{2}$$

$$BD = \sqrt{[6-(-4)]^2 + [8-(-2)]^2} = \sqrt{200} = 10\sqrt{2}$$

Because $6\sqrt{2} \neq 10\sqrt{2}$, the diagonals are not congruent. So, $\square ABCD$ is not a rectangle. Because it is not a rectangle, it also cannot be a square.

Use the slope formula to find the slopes of the diagonals $\overline{AC}$ and $\overline{BD}$.

$$\text{slope of } \overline{AC} = \frac{6-0}{-2-4} = \frac{6}{-6} = -1 \qquad \text{slope of } \overline{BD} = \frac{8-(-2)}{6-(-4)} = \frac{10}{10} = 1$$

Because the product of the slopes of the diagonals is -1, the diagonals are perpendicular.

▶ So, $\square ABCD$ is a rhombus.

Check Check the side lengths of $\square ABCD$. Each side has a length of $2\sqrt{17}$ units, so $\square ABCD$ is a rhombus. Check the slopes of two consecutive sides.

$$\text{slope of } \overline{AB} = \frac{8-6}{6-(-2)} = \frac{2}{8} = \frac{1}{4} \qquad \text{slope of } \overline{BC} = \frac{8-0}{6-4} = \frac{8}{2} = 4$$

Because the product of these slopes is not -1, $\overline{AB}$ is not perpendicular to $\overline{BC}$. So, $\angle ABC$ is not a right angle, and $\square ABCD$ cannot be a rectangle or a square. ✓

SELF-ASSESSMENT

1 I don't understand yet. 2 I can do it with help. 3 I can do it on my own. 4 I can teach someone else.

6. In rectangle $WXYZ$, $WY = 4x - 15$ and $XZ = 3x + 8$. Find the lengths of the diagonals of $WXYZ$.

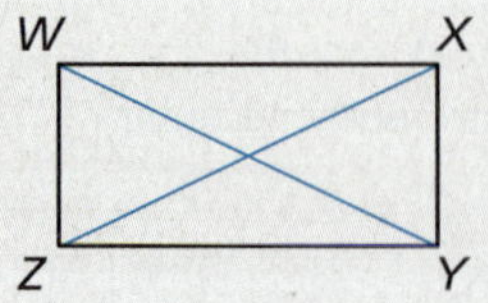

7. Square $ABCD$ has a diagonal $\overline{AC}$ with vertices $A(-2, 1)$ and $C(2, 4)$. Find the coordinates of the remaining vertices.

8. Decide whether $\square PQRS$ with vertices $P(-5, 2)$, $Q(0, 4)$, $R(2, -1)$, and $S(-3, -3)$ is a *rectangle*, a *rhombus*, or a *square*. Give all names that apply.

GO DIGITAL

7.4 Practice WITH CalcChat® AND CalcView®

In Exercises 1–6, for any rhombus *JKLM*, decide whether the statement is *always* or *sometimes* true. Draw a diagram and explain your reasoning. *(See Example 1.)*

1. $\angle L \cong \angle M$
2. $\angle K \cong \angle M$
3. $\overline{JM} \cong \overline{KL}$
4. $\overline{JK} \cong \overline{KL}$
5. $\overline{JL} \cong \overline{KM}$
6. $\angle JKM \cong \angle LKM$

In Exercises 7–10, classify the quadrilateral. Explain your reasoning. *(See Example 2.)*

7. 

8. 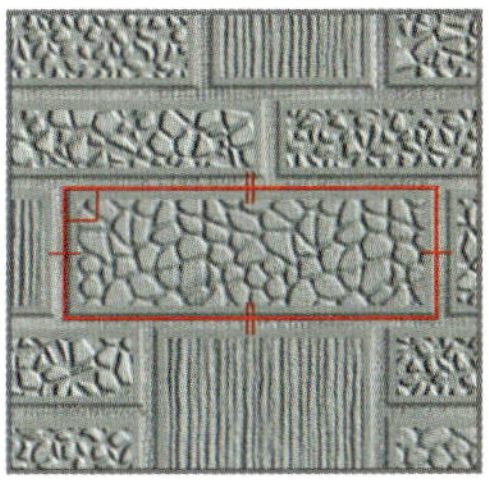

9. 

10. 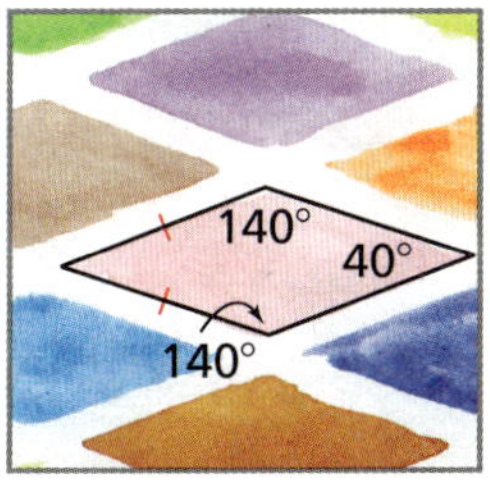

In Exercises 11–14, find the measures of the numbered angles in rhombus *DEFG*. *(See Example 3.)*

11. 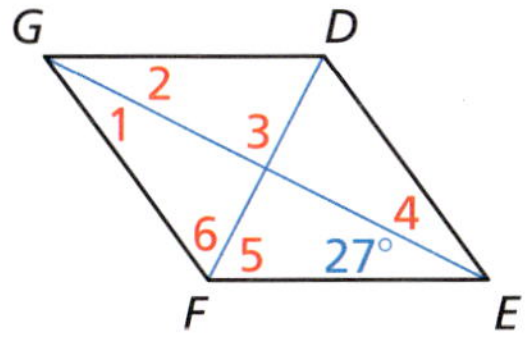

12. 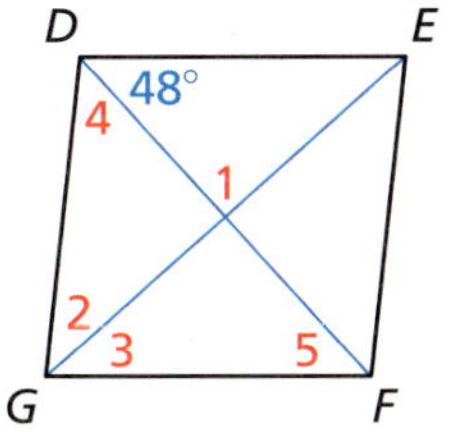

13. 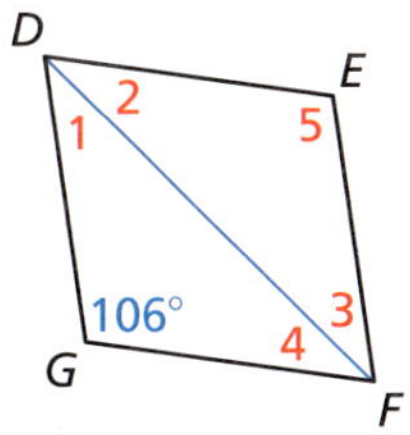

14. 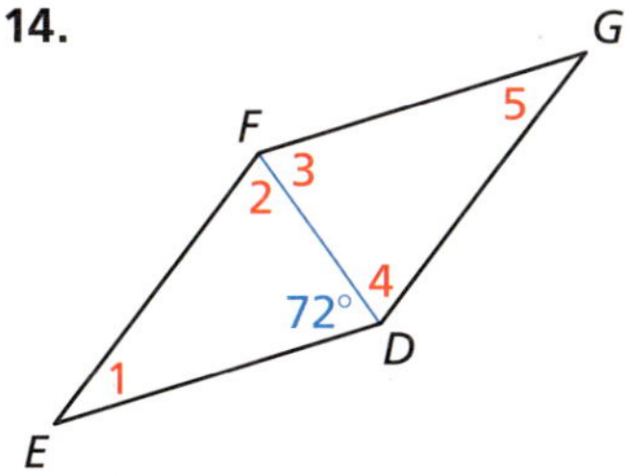

In Exercises 15–20, for any rectangle *WXYZ*, decide whether the statement is *always* or *sometimes* true. Draw a diagram and explain your reasoning.

15. $\angle W \cong \angle X$
16. $\overline{WX} \cong \overline{YZ}$
17. $\overline{WX} \cong \overline{XY}$
18. $\overline{WY} \cong \overline{XZ}$
19. $\overline{WY} \perp \overline{XZ}$
20. $\angle WXZ \cong \angle YXZ$

In Exercises 21–24, find the lengths of the diagonals of rectangle *WXYZ*. *(See Example 4.)*

21. $WY = 6x - 7$
    $XZ = 3x + 2$
22. $WY = 14x + 10$
    $XZ = 11x + 22$
23. $WY = 24x - 8$
    $XZ = -18x + 13$
24. $WY = 16x + 2$
    $XZ = 36x - 6$

4 MTR 25. **ERROR ANALYSIS** Quadrilateral *PQRS* is a rectangle. Describe and correct the error in finding the value of *x*.

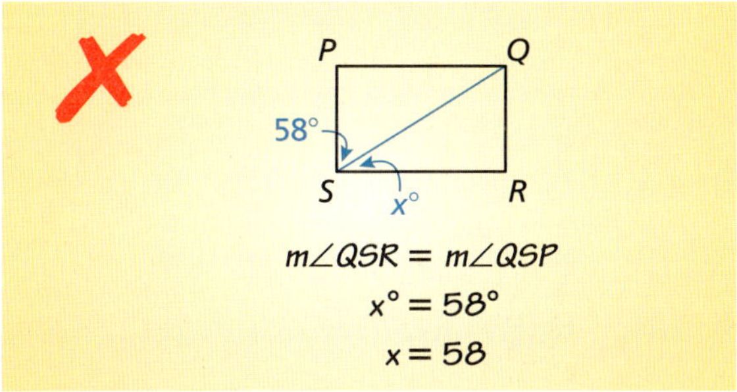

4 MTR 26. **ERROR ANALYSIS** Quadrilateral *PQRS* is a rhombus. Describe and correct the error in finding the value of *x*.

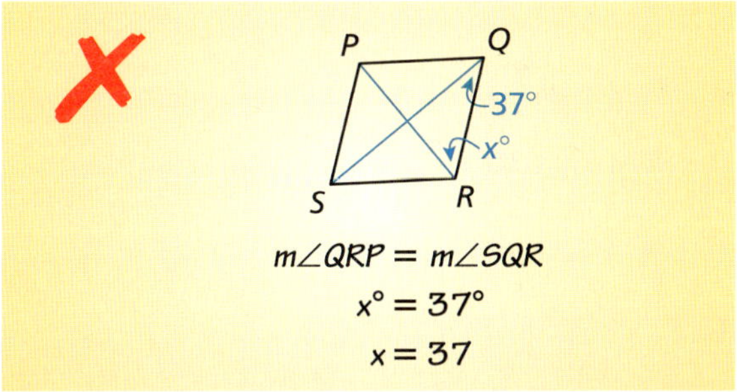

In Exercises 27–30, square *ABCD* has a diagonal $\overline{AC}$ with the given vertices. Find the coordinates of the remaining vertices. *(See Example 5.)*

27. $A(1, 3)$ and $C(6, -2)$
28. $A(-3, 2)$ and $C(0, 5)$
29. $A(-4, -1)$ and $C(2, 5)$
30. $A(2, 2)$ and $C(6, -2)$

GO DIGITAL

In Exercises 31–36, the diagonals of rhombus *ABCD* intersect at *E*. Given that $m\angle BAC = 53°$, $DE = 8$, and $EC = 6$, find the indicated measure.

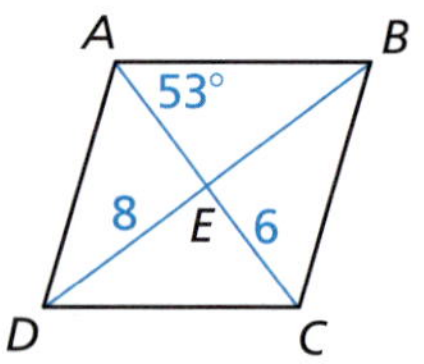

31. $m\angle DAC$

32. $m\angle AED$

33. $m\angle ADC$

34. DB

35. AE

36. AC

In Exercises 37–42, the diagonals of rectangle *QRST* intersect at *P*. Given that $m\angle PTS = 34°$ and $QS = 10$, find the indicated measure.

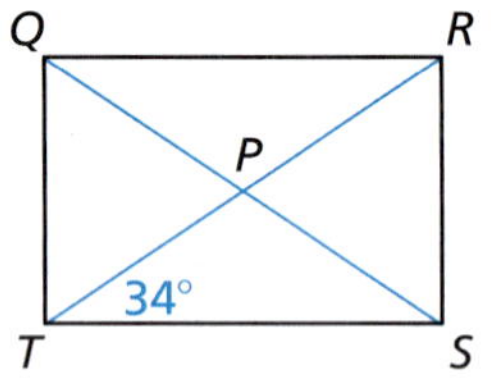

37. $m\angle QTR$

38. $m\angle QRT$

39. $m\angle SRT$

40. QP

41. RT

42. RP

In Exercises 43–48, the diagonals of square *LMNP* intersect at *K*. Given that $LK = 1$, find the indicated measure.

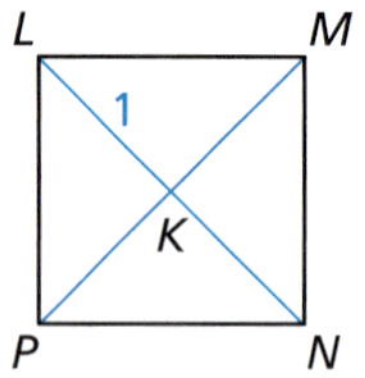

43. $m\angle MKN$

44. $m\angle LMK$

45. $m\angle LPK$

46. KN

47. LN

48. MP

In Exercises 49–54, name each quadrilateral—*parallelogram*, *rectangle*, *rhombus*, or *square*—for which the statement is always true.

49. It is equiangular.

50. It is equiangular and equilateral.

51. The diagonals are perpendicular.

52. Opposite sides are congruent.

53. The diagonals bisect each other.

54. The diagonals bisect opposite angles.

In Exercises 55–60, decide whether ▱*JKLM* is a rectangle, a rhombus, or a square. Give all names that apply. Explain your reasoning. *(See Example 6.)*

▶ **55.** $J(-4, 2), K(0, 3), L(1, -1), M(-3, -2)$

56. $J(-2, 7), K(7, 2), L(-2, -3), M(-11, 2)$

57. $J(3, 1), K(3, -3), L(-2, -3), M(-2, 1)$

58. $J(-1, 4), K(-3, 2), L(2, -3), M(4, -1)$

59. $J(5, 2), K(1, 9), L(-3, 2), M(1, -5)$

60. $J(5, 2), K(2, 5), L(-1, 2), M(2, -1)$

61. **COLLEGE PREP** Which name can be used to classify the quadrilateral? Select all that apply.

Ⓐ parallelogram

Ⓑ rectangle

Ⓒ rhombus

Ⓓ square

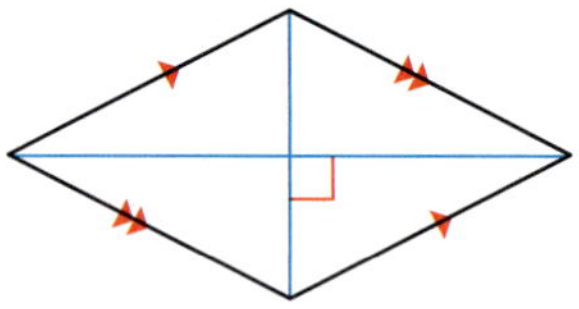

62. **REASONING** Order the terms in a diagram so that each term builds off the previous term(s). Explain why each term is in the location you chose.

quadrilateral

square

rectangle

rhombus

parallelogram

4 MTR **DISCUSS MATHEMATICAL THINKING** **In Exercises 63–68, complete each statement with *always*, *sometimes*, or *never*. Explain your reasoning.**

63. A square is _________ a rhombus.

64. A rectangle is _________ a square.

65. A rectangle _________ has congruent diagonals.

66. The diagonals of a square _________ bisect its angles.

67. A rhombus _________ has four congruent angles.

68. A rectangle _________ has perpendicular diagonals.

69. **REASONING** Which quadrilateral can be called a regular quadrilateral? Explain your reasoning.

GO DIGITAL

70. USING TOOLS You want to mark off a square region for a garden at school. You use a tape measure to mark off a quadrilateral on the ground. Each side of the quadrilateral is 2.5 meters long. Explain how you can use the tape measure to determine whether the quadrilateral is a square.

7 MTR **71. MODELING REAL LIFE** In the window, $\overline{BD} \cong \overline{DF} \cong \overline{BH} \cong \overline{HF}$. Also, $\angle HAB$, $\angle BCD$, $\angle DEF$, and $\angle FGH$ are right angles.

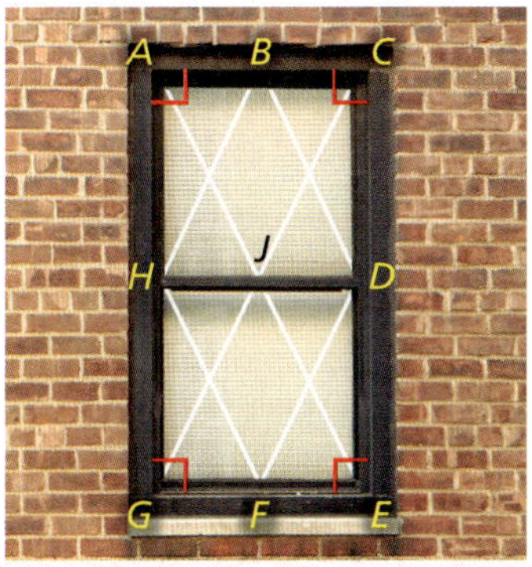

a. Classify *HBDF* and *ACEG*. Explain your reasoning.

b. $BD = 25$ inches and $AE = 50$ inches. What is the total length of the material used to create the white grid on the window? Explain.

72. PROVING A THEOREM Use the plan for proof to write a paragraph proof for one part of the Rhombus Diagonals Theorem.

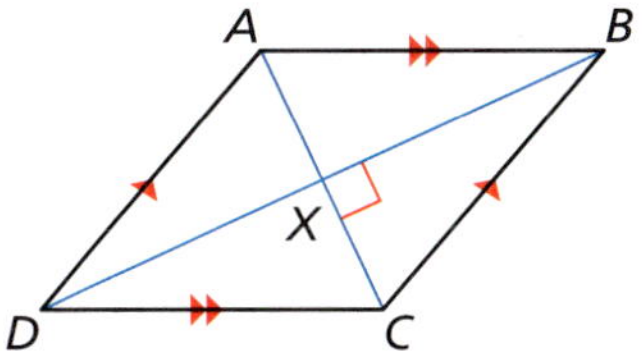

Given *ABCD* is a parallelogram.
$\overline{AC} \perp \overline{BD}$

Prove *ABCD* is a rhombus.

Plan for Proof Because *ABCD* is a parallelogram, its diagonals bisect each other at *X*. Use $\overline{AC} \perp \overline{BD}$ to show that $\triangle BXC \cong \triangle DXC$. Then show that $\overline{BC} \cong \overline{DC}$. Use the properties of a parallelogram to show that *ABCD* is a rhombus.

73. REASONING Determine whether it is possible for a diagonal of the given quadrilateral to divide the quadrilateral into two equilateral triangles. Explain your reasoning.

a. square **b.** rhombus

GO DIGITAL

74. HOW DO YOU SEE IT?
What additional information do you need to determine whether the figure is a rectangle?

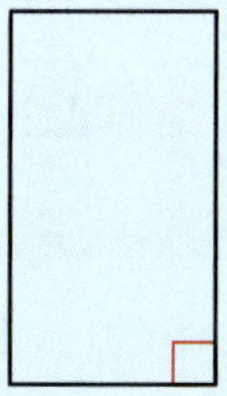

PROVING A THEOREM In Exercises 75 and 76, write a proof for part of the Rhombus Opposite Angles Theorem.

75. Given *PQRS* is a parallelogram.
$\overline{PR}$ bisects $\angle SPQ$ and $\angle QRS$.
$\overline{SQ}$ bisects $\angle PSR$ and $\angle RQP$.

Prove *PQRS* is a rhombus.

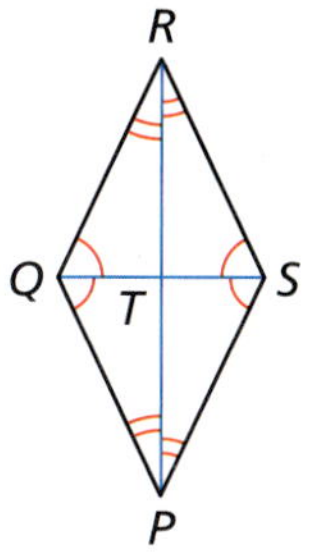

76. Given *WXYZ* is a rhombus.

Prove $\overline{WY}$ bisects $\angle ZWX$ and $\angle XYZ$.
$\overline{ZX}$ bisects $\angle WZY$ and $\angle YXW$.

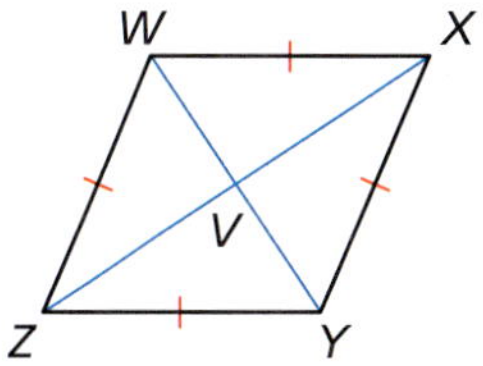

PROVING A COROLLARY In Exercises 77–79, write the corollary as a conditional statement and its converse. Then explain why each statement is true.

77. Rhombus Corollary **78.** Rectangle Corollary

79. Square Corollary

4 MTR **80. MAKING AN ARGUMENT** Is it possible for a rhombus to have congruent diagonals? Explain your reasoning.

4 MTR **81. DISCUSS MATHEMATICAL THINKING** Are all rhombuses similar? Are all squares similar? Explain your reasoning.

82. THOUGHT PROVOKING
Explain why every rhombus has at least two lines of symmetry.

83. PROOF Write a proof.

Given $\triangle XYZ \cong \triangle XWZ$, $\angle XYW \cong \angle ZWY$

Prove *WXYZ* is a rhombus.

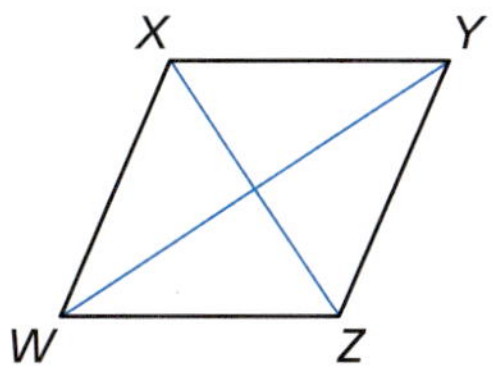

84. PROOF Write a proof.

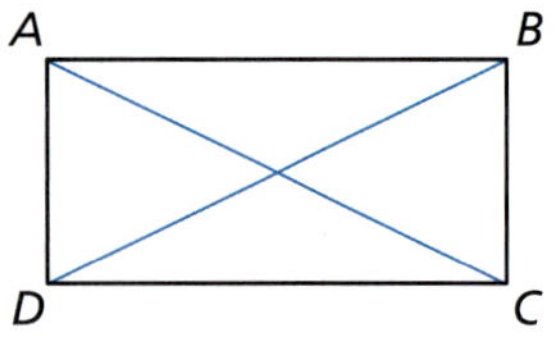

Given $\overline{BC} \cong \overline{AD}$,
$\overline{BC} \perp \overline{DC}$,
$\overline{AD} \perp \overline{DC}$

Prove $ABCD$ is a rectangle.

85. DIG DEEPER The length of one diagonal of a rhombus is 4 times the length of the other diagonal. Write an expression that represents the perimeter of the rhombus.

PROVING A THEOREM In Exercises 86 and 87, write a proof for part of the Rectangle Diagonals Theorem.

86. Given $PQRS$ is a rectangle.

Prove $\overline{PR} \cong \overline{SQ}$

87. Given $PQRS$ is a parallelogram.
$\overline{PR} \cong \overline{SQ}$

Prove $PQRS$ is a rectangle.

REVIEW & REFRESH

WATCH

In Exercises 88 and 89, use the graphs of f and g to describe the transformation from the graph of f to the graph of g.

88. $f(x) = -4x + 7$, $g(x) = f(x - 2)$

89. $f(x) = 6x - 4$, $g(x) = \frac{1}{2}f(x)$

90. $\overline{DE}$ is a midsegment of $\triangle ABC$. Find the values of x and y.

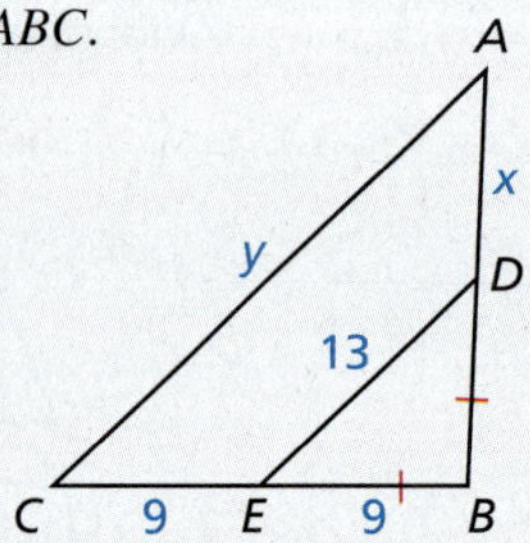

91. Rewrite the definition as a biconditional statement.

Definition A *diagonal* of a polygon is a segment that joins two nonconsecutive vertices.

In Exercises 92 and 93, solve the inequality. Graph the solution, if possible.

92. $|2h - 7| + 10 \leq 6$ **93.** $2(g - 4) > 3(3g + 2)$

94. Find the values of x and y in the parallelogram.

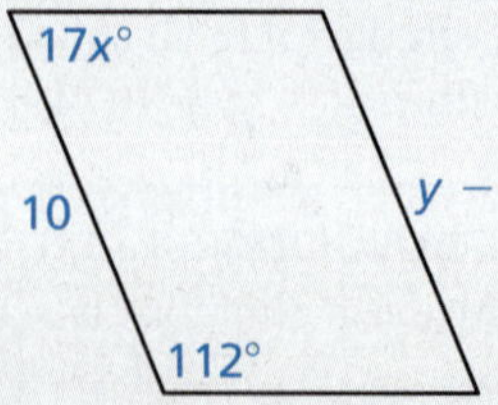

95. Determine whether the relation is a function. Explain.

Input, x	-2	-1	0	1	2
Output, y	7	4	3	4	7

96. Find the measure of each interior angle and each exterior angle of a regular 24-gon.

7 MTR **97. MODELING REAL LIFE** Classify the quadrilateral. Explain your reasoning.

98. Find the perimeter and area of $\triangle PQR$ with vertices $P(3, -2)$, $Q(3, 4)$, and $R(6, 2)$.

In Exercises 99 and 100, decide whether you can use the given information to prove that $\triangle JKL \cong \triangle XYZ$. Explain your reasoning.

99. $\angle K \cong \angle Y$, $\overline{JL} \cong \overline{XZ}$, $\overline{JK} \cong \overline{XY}$

100. $\angle L \cong \angle Z$, $\angle J \cong \angle X$, $\overline{JL} \cong \overline{YZ}$

101. State which theorem you can use to show that the quadrilateral is a parallelogram.

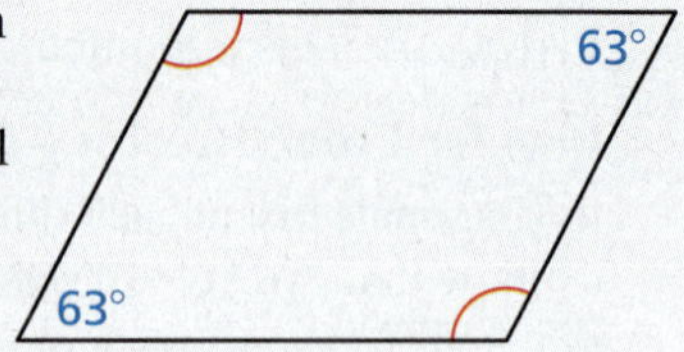

102. Find the length of $\overline{AB}$. Explain your reasoning.

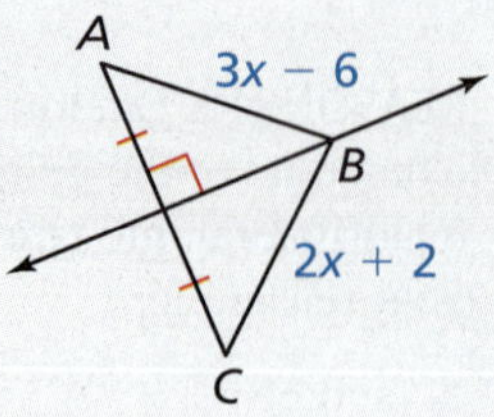

GO DIGITAL

7.5 Properties of Trapezoids and Kites

Learning Target: Use properties of trapezoids and kites to find measures.

Success Criteria:
- I can identify trapezoids and kites.
- I can use properties of trapezoids and kites to solve problems.
- I can find the length of the midsegment of a trapezoid.
- I can explain the hierarchy of quadrilaterals.

EXPLORE IT! Discovering Properties of Trapezoids and Kites

Work with a partner. Consider the types of quadrilaterals shown.

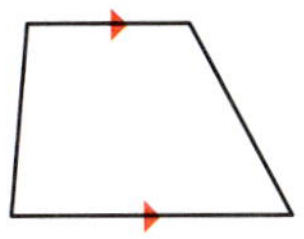
Trapezoid

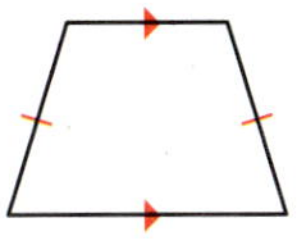
Isosceles trapezoid

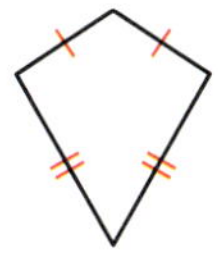
Kite

a. Use the diagrams to define each type of quadrilateral.

b. Use technology to construct an isosceles trapezoid. Explain your method.

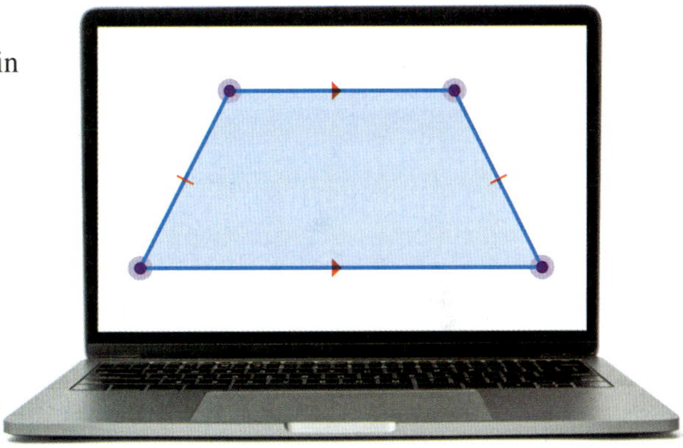

c. Find the angle measures and diagonal lengths of the trapezoid. What do you observe?

d. Repeat parts (b) and (c) for several other isosceles trapezoids. Make conjectures based on your results.

e. Use technology to construct a kite. Explain your method. Make a conjecture about interior angle measures and a conjecture about the diagonals of the kite. Provide examples to support your reasoning.

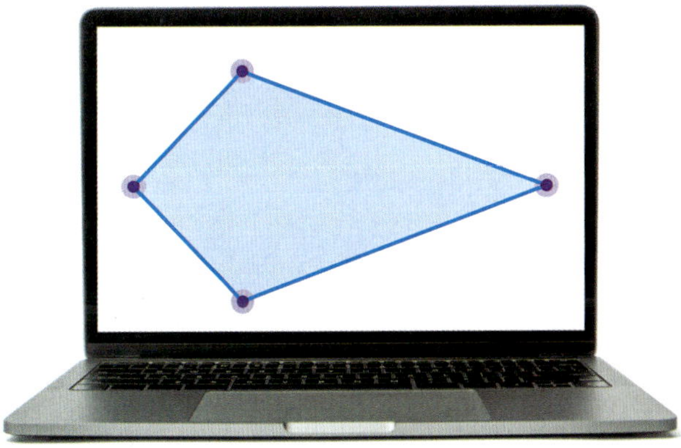

4 MTR CONSTRUCT AN ARGUMENT
Can you construct a kite using its diagonals? Explain your reasoning.

Geometric Reasoning
MA.912.GR.1.5 Prove relationships and theorems about trapezoids. Solve mathematical and real-world problems involving postulates, relationships and theorems of trapezoids.
MA.912.GR.3.2 Given a mathematical context, use coordinate geometry to classify or justify definitions, properties and theorems involving circles, triangles or quadrilaterals.

GO DIGITAL

Using Properties of Trapezoids

Vocabulary

trapezoid, *p. 390*
bases, *p. 390*
legs, *p. 390*
base angles, *p. 390*
isosceles trapezoid, *p. 390*
midsegment of a trapezoid, *p. 392*
kite, *p. 393*

A **trapezoid** is a quadrilateral with at least one pair of parallel sides. The parallel sides are the **bases**. A trapezoid can have one or two pairs of bases. Any nonparallel sides are the **legs**. A *strict trapezoid* has exactly one pair of bases.

Base angles of a trapezoid are two consecutive angles whose common side is a base. A trapezoid has two pairs of base angles. For example, in trapezoid $ABCD$, $\angle A$ and $\angle D$ are one pair of base angles, and $\angle B$ and $\angle C$ are the second pair.

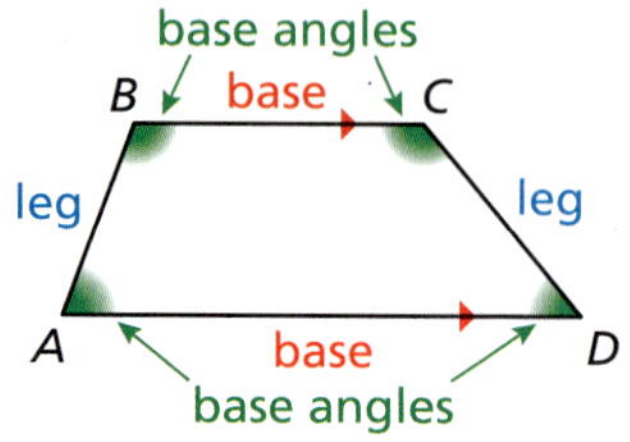

If the legs of a strict trapezoid are congruent, then the trapezoid is an **isosceles trapezoid**.

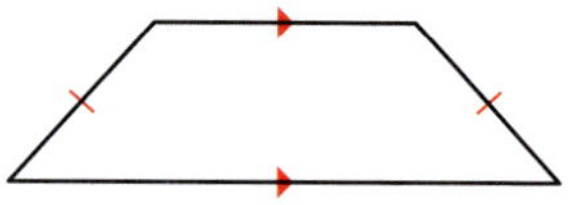

Isosceles trapezoid

EXAMPLE 1 Identifying a Trapezoid in the Coordinate Plane

WATCH

Show that $ORST$ is a trapezoid. Then decide whether it is isosceles.

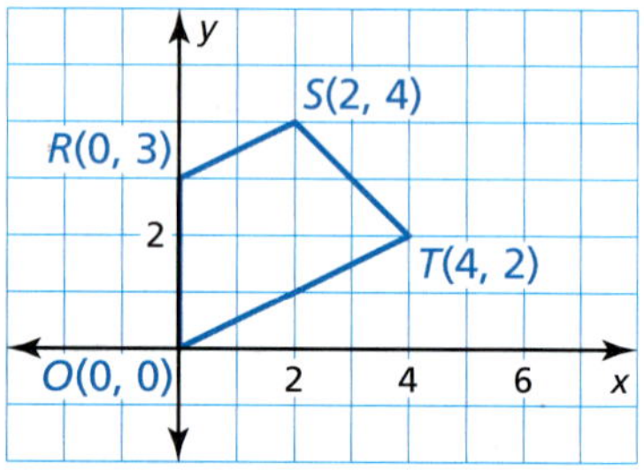

SOLUTION

Step 1 Compare the slopes of opposite sides.

$$\text{slope of } \overline{ST} = \frac{2-4}{4-2} = \frac{-2}{2} = -1$$

$$\text{slope of } \overline{RO} = \frac{3-0}{0-0} = \frac{3}{0} \quad \text{Undefined}$$

The slopes of $\overline{ST}$ and $\overline{RO}$ are not the same, so $\overline{ST}$ is not parallel to $\overline{RO}$.

$$\text{slope of } \overline{RS} = \frac{4-3}{2-0} = \frac{1}{2} \qquad \text{slope of } \overline{OT} = \frac{2-0}{4-0} = \frac{2}{4} = \frac{1}{2}$$

The slopes of $\overline{RS}$ and $\overline{OT}$ are the same, so $\overline{RS} \parallel \overline{OT}$.

▶ Because $ORST$ has a pair of parallel sides, it is a trapezoid.

Step 2 $ORST$ has exactly one pair of parallel sides, so it is a strict trapezoid. Compare the lengths of legs $\overline{RO}$ and $\overline{ST}$ to determine whether it is isosceles.

$$RO = |3 - 0| = 3 \qquad ST = \sqrt{(2-4)^2 + (4-2)^2} = \sqrt{8} \approx 2.83$$

Because $RO \neq ST$, legs $\overline{RO}$ and $\overline{ST}$ are *not* congruent.

▶ So, $ORST$ is not an isosceles trapezoid.

SELF-ASSESSMENT

1 I don't understand yet. 2 I can do it with help. 3 I can do it on my own. 4 I can teach someone else.

In Exercises 1 and 2, use trapezoid $ABCD$.

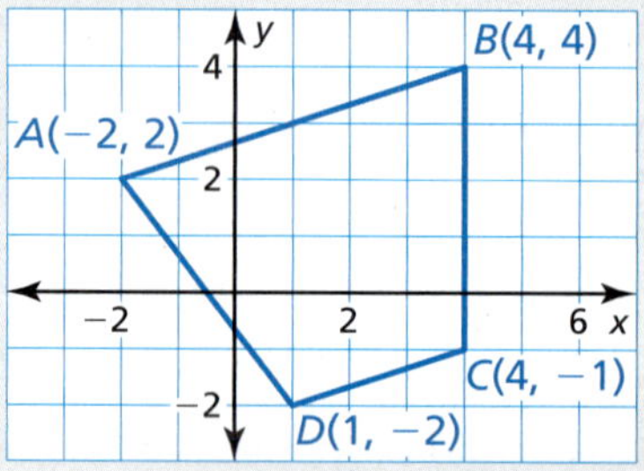

1. Show that $ABCD$ is a trapezoid. Then decide whether it is isosceles.
2. **REASONING** Vertex B moves to $B(1, 3)$. Is $ABCD$ a trapezoid? Explain your reasoning.

GO DIGITAL

THEOREMS

7.14 Isosceles Trapezoid Base Angles Theorem

If a trapezoid is isosceles, then each pair of base angles is congruent.

If trapezoid $ABCD$ is isosceles, then $\angle A \cong \angle D$ and $\angle B \cong \angle C$.

Prove this Theorem Exercise 41, page 397

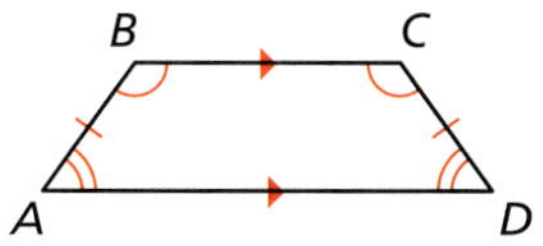

7.15 Isosceles Trapezoid Base Angles Converse

If a trapezoid has a pair of congruent base angles, then it is an isosceles trapezoid.

If $\angle A \cong \angle D$ (or if $\angle B \cong \angle C$), then trapezoid $ABCD$ is isosceles.

Prove this Theorem Exercise 42, page 397

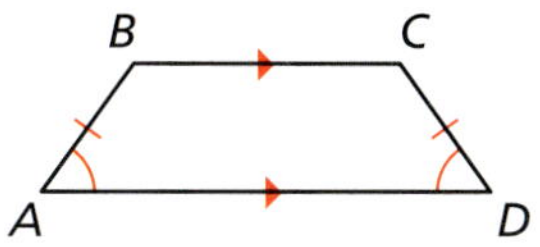

7.16 Isosceles Trapezoid Diagonals Theorem

A trapezoid is isosceles if and only if its diagonals are congruent.

Trapezoid $ABCD$ is isosceles if and only if $\overline{AC} \cong \overline{BD}$.

Prove this Theorem Exercise 51, page 398

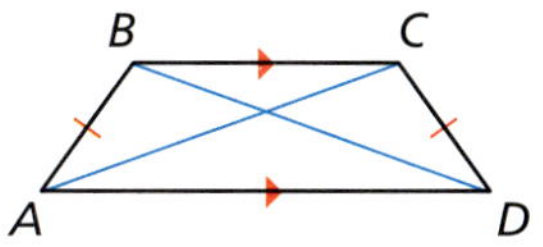

7.17 Isosceles Trapezoid Opposite Angles Theorem

A trapezoid is isosceles if and only if opposite angles are supplementary.

Trapezoid $ABCD$ is isosceles if and only if $m\angle A + m\angle C = 180°$ and $m\angle B + m\angle D = 180°$.

Prove this Theorem Exercise 46, page 397

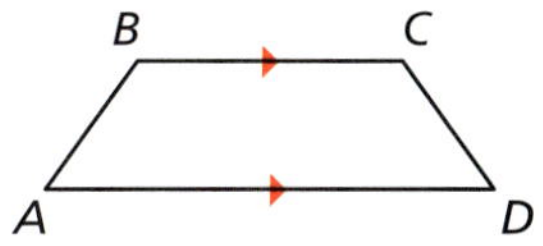

EXAMPLE 2 Using Properties of Isosceles Trapezoids

WATCH

Incan architecture often features trapezoidal doorways and windows. Find $m\angle L$ in the doorway.

SOLUTION

$JKLM$ is an isosceles trapezoid because there is exactly one pair of bases and the legs are congruent. By the Isosceles Trapezoid Opposite Angles Theorem, $\angle J$ and $\angle L$ are supplementary. So, $m\angle L = 180° - 85° = 95°$.

▶ So, $m\angle L = 95°$.

SELF-ASSESSMENT

1 I don't understand yet. 2 I can do it with help. 3 I can do it on my own. 4 I can teach someone else.

In Exercises 3 and 4, use trapezoid *EFGH*.

3. If $EG = FH$, is trapezoid $EFGH$ isosceles? Explain.

4. If $m\angle HEF = 70°$ and $m\angle FGH = 110°$, is trapezoid $EFGH$ isosceles? Explain.

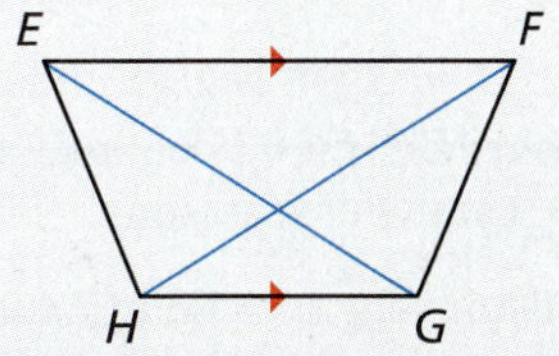

GO DIGITAL

Using the Trapezoid Midsegment Theorem

Recall that a midsegment of a triangle is a segment that connects the midpoints of two sides of the triangle. The **midsegment of a trapezoid** is the segment that connects the midpoints of its legs. The theorem below is similar to the Triangle Midsegment Theorem.

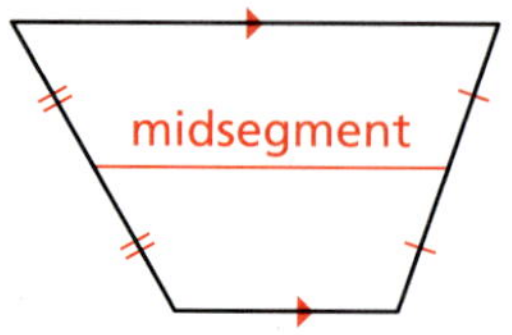

READING

The midsegment of a trapezoid is sometimes called the *median* of the trapezoid.

THEOREM

7.18 Trapezoid Midsegment Theorem

The midsegment of a trapezoid is parallel to each base, and its length is one-half the sum of the lengths of the bases.

If $\overline{MN}$ is the midsegment of trapezoid $ABCD$, then $\overline{MN} \parallel \overline{AB}$, $\overline{MN} \parallel \overline{DC}$, and $MN = \frac{1}{2}(AB + CD)$.

Prove this Theorem Exercise 50, page 397

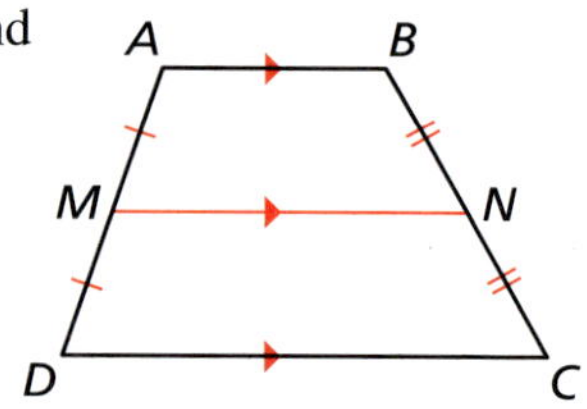

EXAMPLE 3 Using the Midsegment of a Trapezoid

WATCH

In the diagram, $\overline{MN}$ is the midsegment of trapezoid $PQRS$. Find MN.

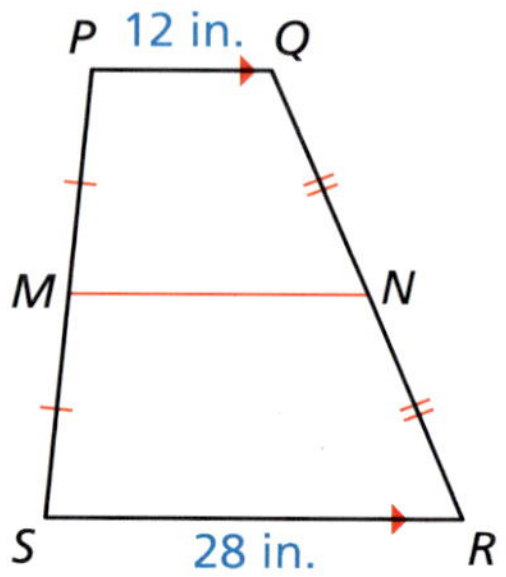

SOLUTION

$MN = \frac{1}{2}(PQ + SR)$ Trapezoid Midsegment Theorem

$= \frac{1}{2}(12 + 28)$ Substitute 12 for PQ and 28 for SR.

$= 20$ Simplify.

▶ So, MN is 20 inches.

EXAMPLE 4 Using a Midsegment in the Coordinate Plane

WATCH

Find the length of midsegment $\overline{YZ}$ in trapezoid $STUV$.

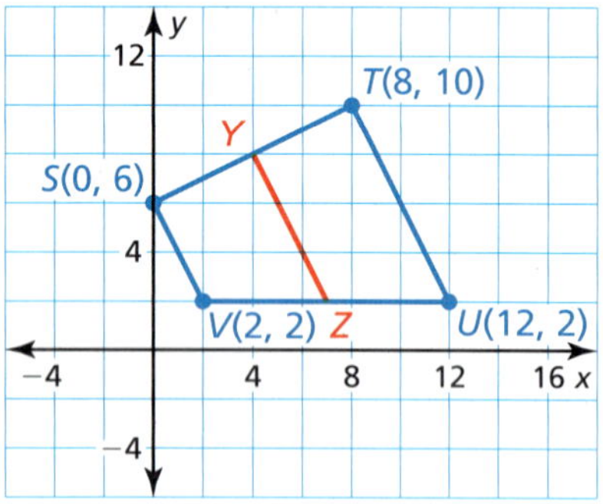

SOLUTION

Step 1 Find the lengths of $\overline{SV}$ and $\overline{TU}$.

$$SV = \sqrt{(0 - 2)^2 + (6 - 2)^2} = \sqrt{20} = 2\sqrt{5}$$

$$TU = \sqrt{(8 - 12)^2 + (10 - 2)^2} = \sqrt{80} = 4\sqrt{5}$$

Step 2 Use the Trapezoid Midsegment Theorem.

$$YZ = \frac{1}{2}(SV + TU) = \frac{1}{2}\left(2\sqrt{5} + 4\sqrt{5}\right) = \frac{1}{2}\left(6\sqrt{5}\right) = 3\sqrt{5}$$

▶ So, the length of $\overline{YZ}$ is $3\sqrt{5}$ units.

SELF-ASSESSMENT

1 I don't understand yet. 2 I can do it with help. 3 I can do it on my own. 4 I can teach someone else.

5. In trapezoid $JKLM$, $\angle J$ and $\angle M$ are right angles, and $JK = 9$ centimeters. The length of midsegment $\overline{NP}$ of trapezoid $JKLM$ is 12 centimeters. Sketch trapezoid $JKLM$ and its midsegment. Find ML. Explain your reasoning.

6. Use a different method to find the length of $\overline{YZ}$ in Example 4.

GO DIGITAL

Using Properties of Kites

A **kite** is a quadrilateral that has two pairs of consecutive congruent sides, but opposite sides are not congruent.

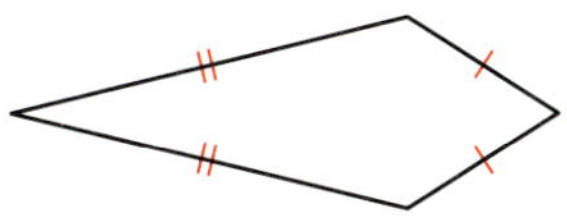

WORDS AND MATH

The most common type of toy kite is shaped like the quadrilateral of the same name. The quadrilateral was named after the toy.

THEOREMS

7.19 Kite Diagonals Theorem

If a quadrilateral is a kite, then its diagonals are perpendicular.

If quadrilateral $ABCD$ is a kite, then $\overline{AC} \perp \overline{BD}$.

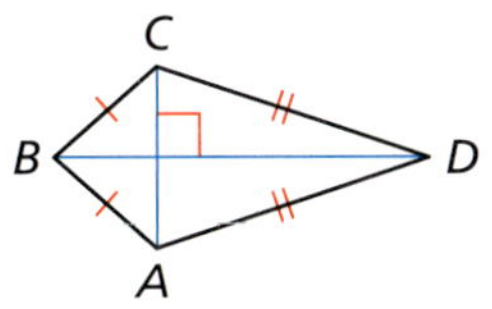

7.20 Kite Opposite Angles Theorem

If a quadrilateral is a kite, then exactly one pair of opposite angles are congruent.

If quadrilateral $ABCD$ is a kite and $\overline{BC} \cong \overline{BA}$, then $\angle A \cong \angle C$ and $\angle B \ncong \angle D$.

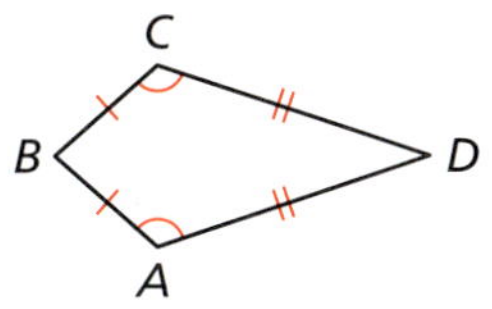

Prove this Theorem Exercise 49, page 397

STUDY TIP

The congruent angles of a kite are formed by the noncongruent adjacent sides.

PROOF Kite Diagonals Theorem

Given $ABCD$ is a kite, $\overline{BC} \cong \overline{BA}$, and $\overline{DC} \cong \overline{DA}$.

Prove $\overline{AC} \perp \overline{BD}$

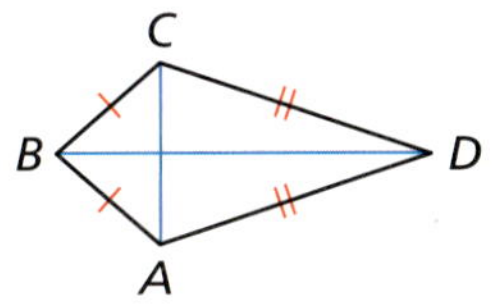

STATEMENTS	REASONS
1. $ABCD$ is a kite with $\overline{BC} \cong \overline{BA}$ and $\overline{DC} \cong \overline{DA}$.	**1.** Given
2. B and D lie on the $\perp$ bisector of $\overline{AC}$.	**2.** Converse of the $\perp$ Bisector Theorem
3. $\overline{BD}$ is the $\perp$ bisector of $\overline{AC}$.	**3.** Through any two points, there exists exactly one line.
4. $\overline{AC} \perp \overline{BD}$	**4.** Definition of $\perp$ bisector

EXAMPLE 5 Finding Angle Measures in a Kite

WATCH

You design the kite shown for the Treasure Island Kite Festival. Find $m\angle D$ in the kite.

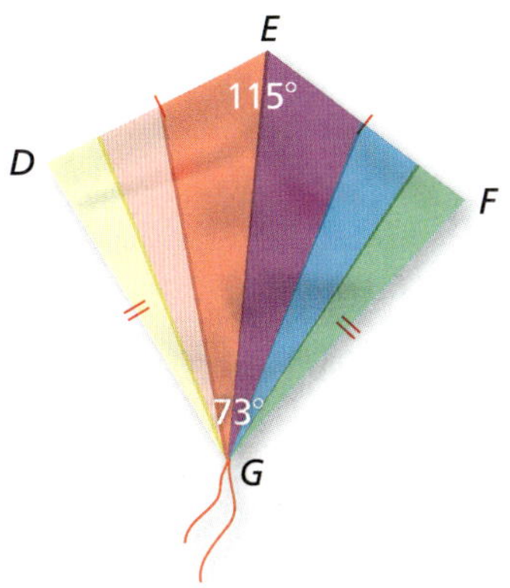

SOLUTION

By the Kite Opposite Angles Theorem, $DEFG$ has exactly one pair of congruent opposite angles. Because $\angle E \ncong \angle G$, $\angle D$ and $\angle F$ must be congruent. So, $m\angle D = m\angle F$. Write and solve an equation to find $m\angle D$.

$m\angle D + m\angle F + 115° + 73° = 360°$	Corollary to the Polygon Interior Angles Theorem
$m\angle D + m\angle D + 115° + 73° = 360°$	Substitute $m\angle D$ for $m\angle F$.
$2(m\angle D) + 188° = 360°$	Combine like terms.
$m\angle D = 86°$	Solve for $m\angle D$.

GO DIGITAL

Identifying Special Quadrilaterals

The diagram shows relationships among the special quadrilaterals you have studied in this chapter. Each shape in the diagram has the properties of the shapes linked above it. For example, a parallelogram has the properties of a trapezoid. So, any theorem you prove about a trapezoid is proven for a parallelogram.

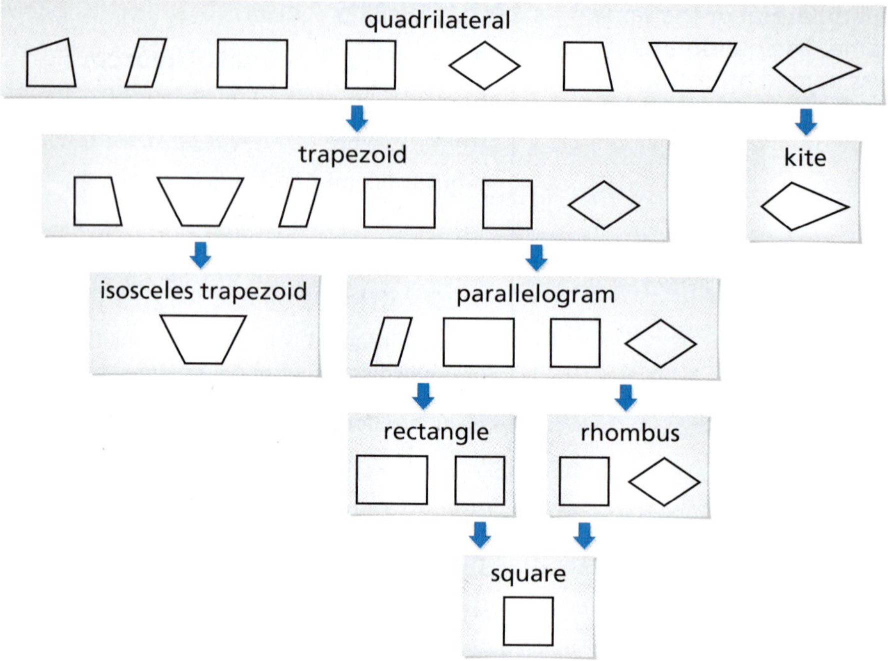

EXAMPLE 6 Identifying a Quadrilateral

WATCH

REMEMBER

In Example 6, *ABCD* looks like a square. But you must rely only on marked information when you interpret a diagram.

What is the most specific name for quadrilateral *ABCD*?

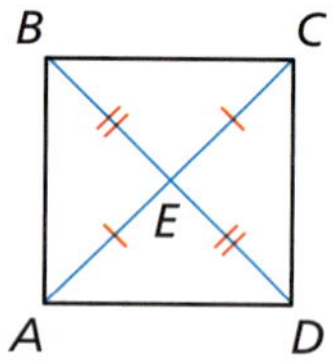

SOLUTION

The diagram shows $\overline{AE} \cong \overline{CE}$ and $\overline{BE} \cong \overline{DE}$. So, the diagonals bisect each other. By the Parallelogram Diagonals Converse, *ABCD* is a parallelogram.

Rectangles, rhombuses, and squares are also parallelograms. However, there is no information given about the side lengths or angle measures of *ABCD*. So, you cannot determine whether it is a rectangle, a rhombus, or a square.

▶ So, the most specific name for *ABCD* is a parallelogram.

SELF-ASSESSMENT

1 I don't understand yet. 2 I can do it with help. 3 I can do it on my own. 4 I can teach someone else.

7. In a kite, the measures of a pair of opposite angles are 50° and 108°. Find the measure of one of the other angles in the kite.

8. Quadrilateral *DEFG* has at least one pair of opposite sides congruent. What types of quadrilaterals meet this condition?

Give the most specific name for the quadrilateral. Explain your reasoning.

9.

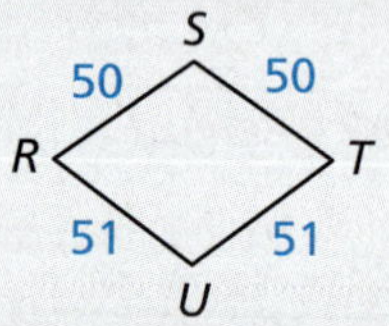

10.

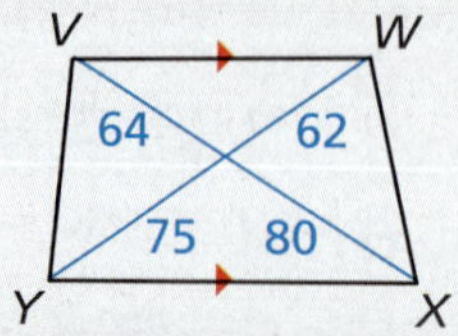

11.

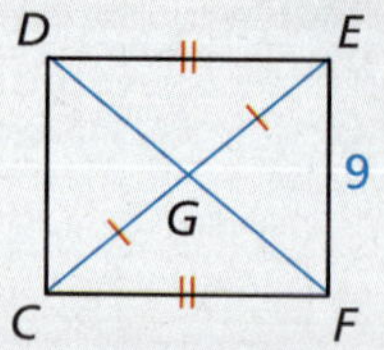

GO DIGITAL

7.5 Practice WITH CalcChat® AND CalcView®

In Exercises 1–4, show that the quadrilateral with the given vertices is a trapezoid. Then decide whether it is isosceles. *(See Example 1.)*

▶ **1.** $W(1, 4), X(1, 8), Y(-3, 9), Z(-3, 3)$

2. $D(-3, 3), E(-1, 1), F(1, -4), G(-3, 0)$

3. $M(-2, 0), N(0, 4), P(5, 4), Q(8, 0)$

4. $H(1, 9), J(4, 2), K(5, 2), L(8, 9)$

In Exercises 5 and 6, find the measure of each angle in the isosceles trapezoid. *(See Example 2.)*

▶ **5.**

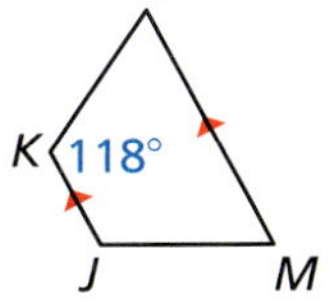

6.

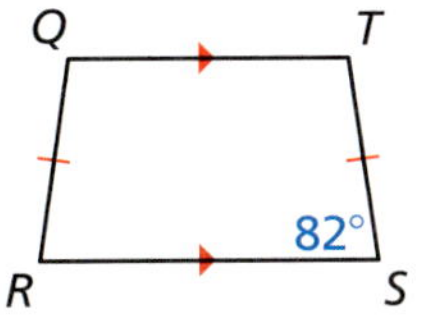

In Exercises 7 and 8, find the length of the midsegment of the trapezoid. *(See Example 3.)*

▶ **7.**

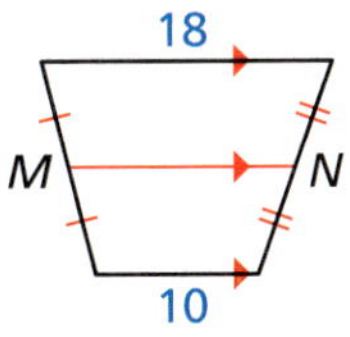

8.

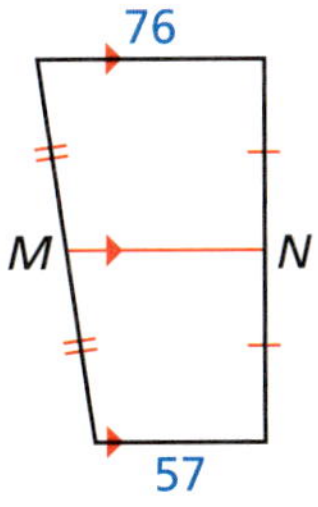

In Exercises 9 and 10, find AB.

9.

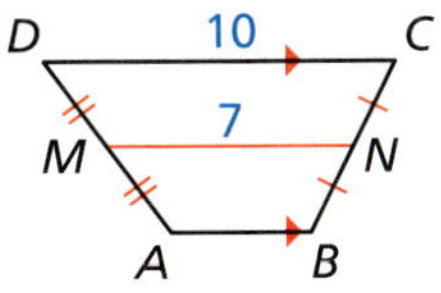

10.

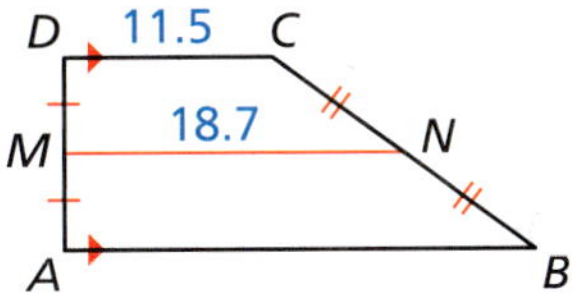

In Exercises 11–14, find the length of the midsegment of the trapezoid with the given vertices. *(See Example 4.)*

▶ **11.** $S(0, 0), T(2, 7), U(6, 10), V(8, 6)$

12. $A(0, 3), B(2, 5), C(6, 4), D(2, 0)$

13. $A(2, 0), B(8, -4), C(12, 2), D(0, 10)$

14. $S(-2, 4), T(-2, -4), U(3, -2), V(13, 10)$

GO DIGITAL

In Exercises 15–18, find $m\angle G$. *(See Example 5.)*

▶ **15.**

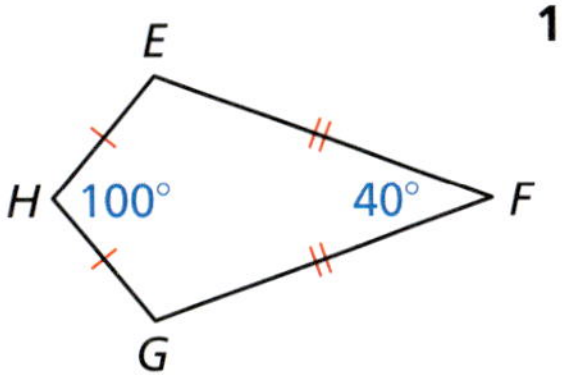

16.

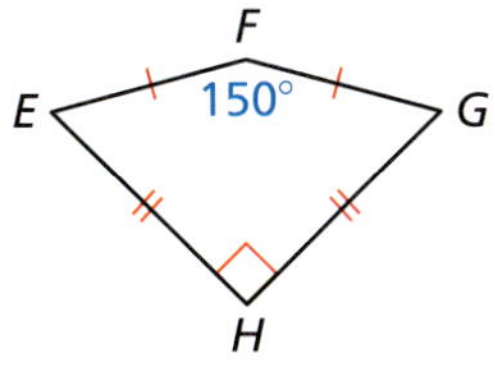

17.

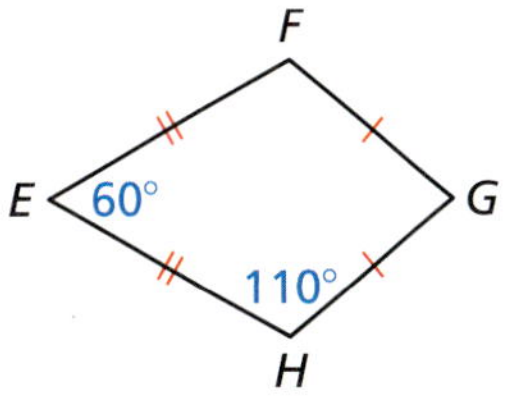

18.

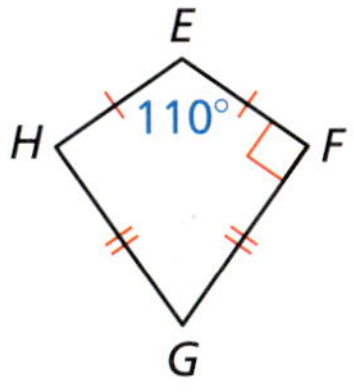

4 MTR **19. ERROR ANALYSIS** Describe and correct the error in finding DC.

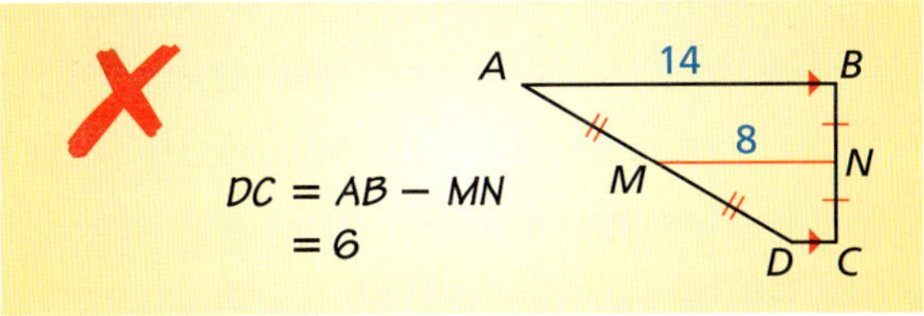

1 MTR **20. HELP A CLASSMATE** Explain to a classmate how to find $m\angle A$.

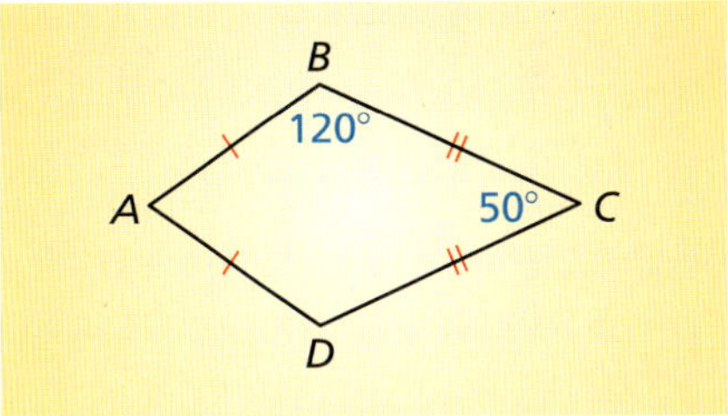

In Exercises 21–24, give the most specific name for the quadrilateral. Explain your reasoning. *(See Example 6.)*

▶ **21.**

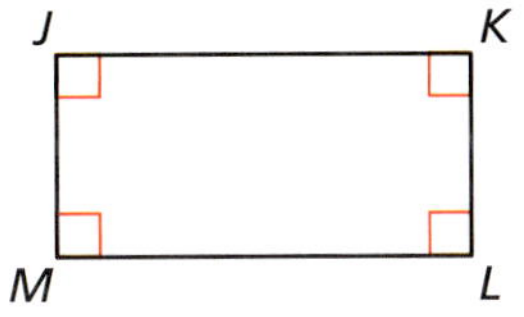

22.

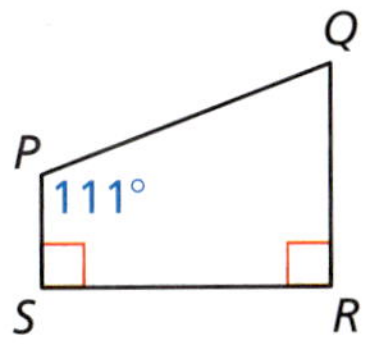

23.

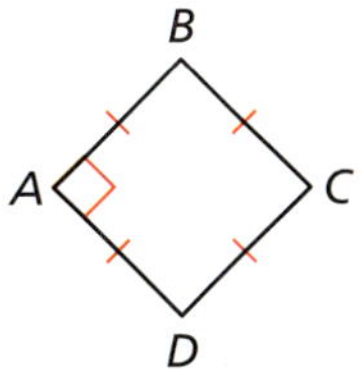

24.

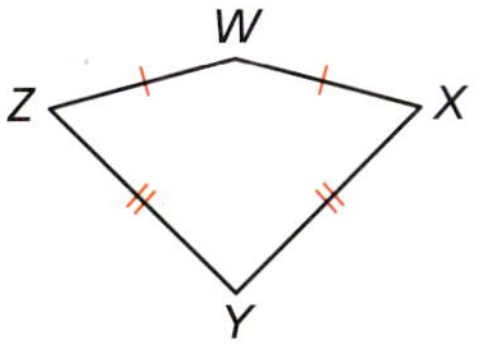

REASONING **In Exercises 25 and 26, tell whether enough information is given in the diagram to classify the quadrilateral by the indicated name. Explain.**

25. rhombus

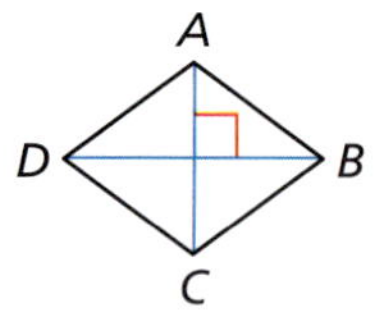

26. square

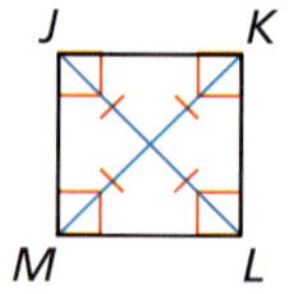

STRUCTURE **In Exercises 27 and 28, find the value of x.**

27.

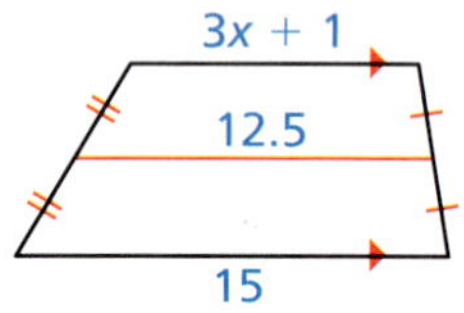

28.

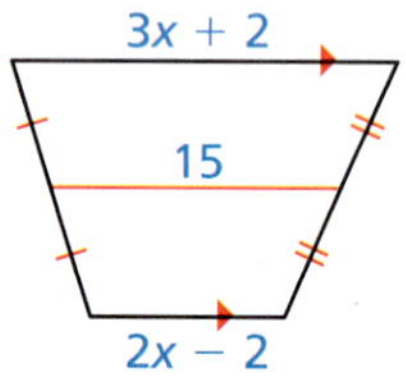

29. **REASONING** In the diagram, $NP = 8$ inches, and $LR = 20$ inches. What is the diameter of the bottom layer of the cake?

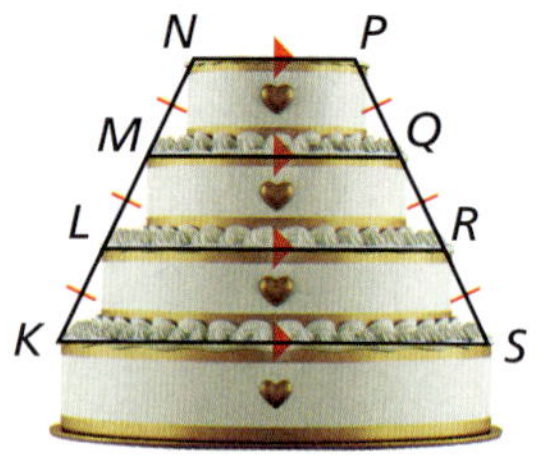

30. **CONNECTING CONCEPTS** You use 94 inches of plastic to frame the perimeter of a kite. One side of the kite has a length of 18 inches. Find the length of each of the three remaining sides.

REASONING **In Exercises 31–34, determine which pairs of segments or angles must be congruent so that you can prove that *ABCD* is the indicated quadrilateral. Explain your reasoning. (There may be more than one correct answer.)**

31. isosceles trapezoid

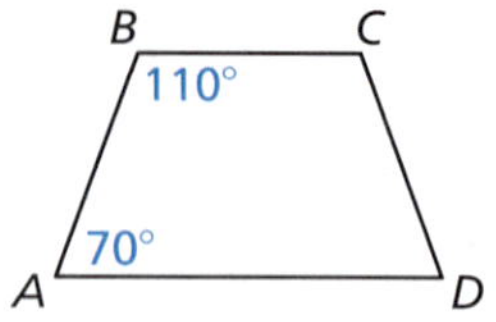

32. kite

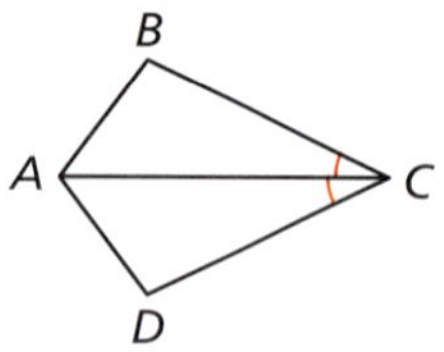

33. parallelogram

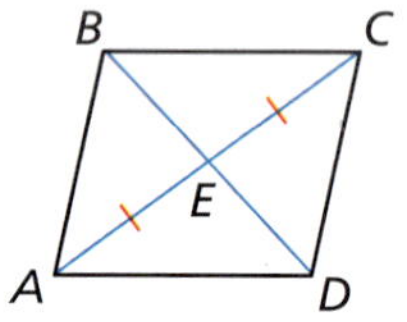

34. square

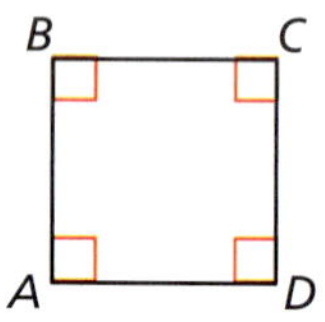

35. **PROOF** Write a proof.

Given $\overline{JL} \cong \overline{LN}$, $\overline{KM}$ is a midsegment of $\triangle JLN$.

Prove Quadrilateral *JKMN* is an isosceles trapezoid.

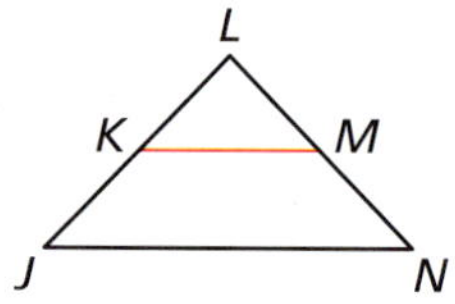

36. **PROOF** Write a proof.

Given *ABCD* is a kite.

$\overline{AB} \cong \overline{CB}$, $\overline{AD} \cong \overline{CD}$

Prove $\overline{CE} \cong \overline{AE}$

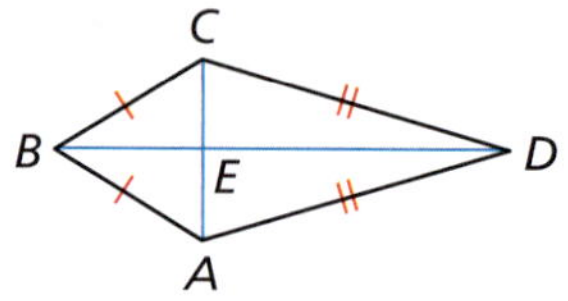

37. **REASONING** Point U lies on the perpendicular bisector of $\overline{RT}$. Describe the set of points S for which *RSTU* is a kite.

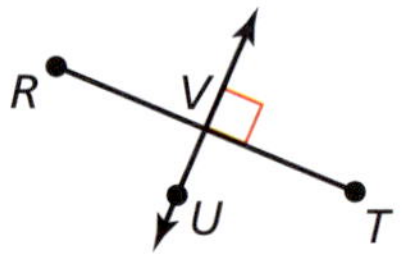

38. **REASONING** Determine whether the points $A(4, 5)$, $B(-3, 3)$, $C(-6, -13)$, and $D(6, -2)$ are the vertices of a kite. Explain your reasoning.

39. **MODELING REAL LIFE** Scientists are researching solar sails, which move spacecraft using radiation pressure from sunlight. What is the area of the solar sail shown? Explain your reasoning.

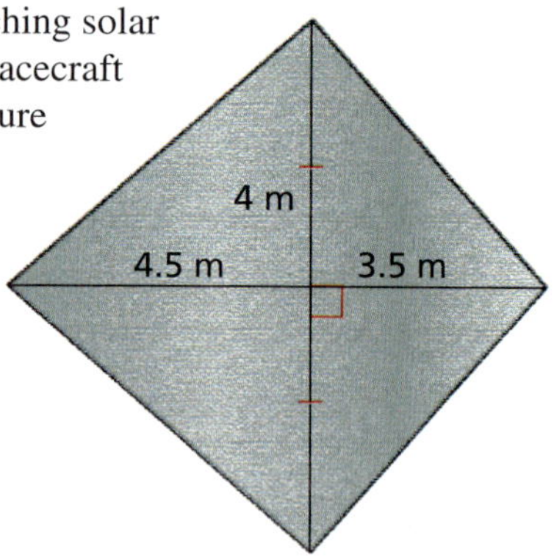

40. **PERFORMANCE TASK** You want to frame a 4-inch-by-6-inch photo. Construct a rectangular frame using four pieces of wood, each in the shape of a trapezoid. Make a diagram showing how to arrange the trapezoids. Include side lengths and angle measures in your diagram.

GO DIGITAL

PROVING A THEOREM **In Exercises 41 and 42, use the diagram to prove the given theorem. In the diagram, $\overline{EC}$ is drawn parallel to $\overline{AB}$.**

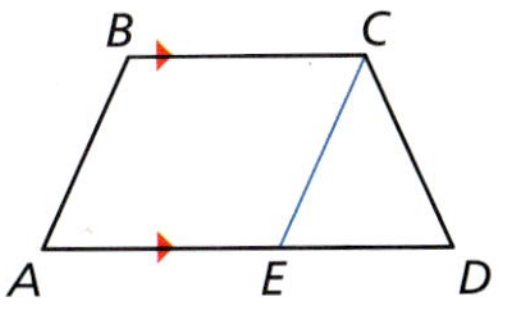

41. Isosceles Trapezoid Base Angles Theorem

Given *ABCD* is an isosceles trapezoid. $\overline{BC} \parallel \overline{AD}$

Prove $\angle A \cong \angle D, \angle B \cong \angle BCD$

42. Isosceles Trapezoid Base Angles Converse

Given *ABCD* is a trapezoid. $\angle A \cong \angle D, \overline{BC} \parallel \overline{AD}$

Prove *ABCD* is an isosceles trapezoid.

5 MTR **43.** **CONNECTING CONCEPTS** The bases of a trapezoid lie on the lines $y = 2x + 7$ and $y = 2x - 5$. Write the equation of the line that contains the midsegment of the trapezoid.

44. **COLLEGE PREP** In trapezoid *PQRS*, $\overline{PQ} \parallel \overline{RS}$ and $\overline{MN}$ is the midsegment of *PQRS*. If $RS = 5 \cdot PQ$, what is the ratio of *MN* to *RS*?

Ⓐ 3 : 5 Ⓒ 1 : 2

Ⓑ 5 : 3 Ⓓ 3 : 1

45. **CONSTRUCTION** $\overline{AC}$ and $\overline{BD}$ bisect each other.

a. Construct quadrilateral *ABCD* so that $\overline{AC}$ and $\overline{BD}$ are congruent but not perpendicular. Classify the quadrilateral. Justify your answer.

b. Construct quadrilateral *ABCD* so that $\overline{AC}$ and $\overline{BD}$ are perpendicular but not congruent. Classify the quadrilateral. Justify your answer.

46. **PROVING A THEOREM** To prove the biconditional statement in the Isosceles Trapezoid Opposite Angles Theorem, you must prove both parts separately.

a. Prove part of the Isosceles Trapezoid Opposite Angles Theorem.

Given *JKLM* is an isosceles trapezoid. $\overline{KL} \parallel \overline{JM}$

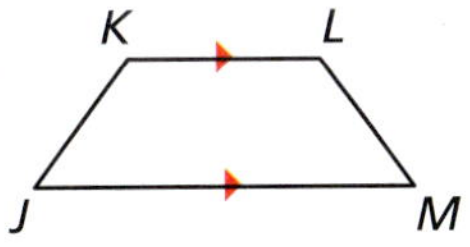

Prove $m\angle J + m\angle L = 180°$ and $m\angle K + m\angle M = 180°$

b. Write the other part of the Isosceles Trapezoid Opposite Angles Theorem as a conditional. Then prove the statement is true.

GO DIGITAL

7 MTR **47.** **MODELING REAL LIFE** A plastic web is made in the shape of a regular dodecagon (12-sided polygon). $\overline{AB} \parallel \overline{PQ}$, and *X* is equidistant from the vertices of the dodecagon.

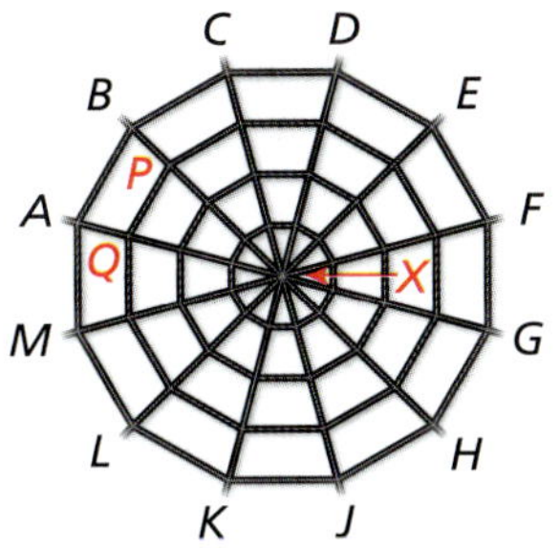

a. Are you given enough information to prove that *ABPQ* is an isosceles trapezoid?

b. What is the measure of each interior angle of *ABPQ*?

48. **HOW DO YOU SEE IT?** One of the earliest shapes used for cut diamonds is called the *table cut*, as shown in the figure. Each face of a cut gem is called a *facet*.

STEM

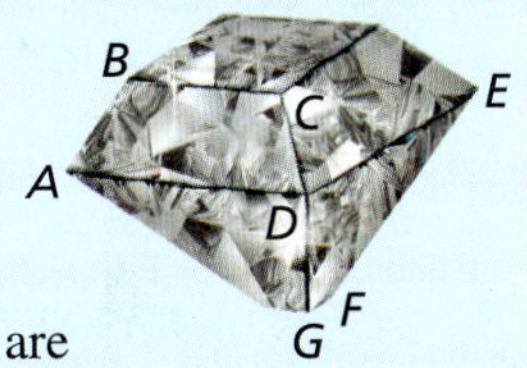

a. $\overline{BC} \parallel \overline{AD}$, and $\overline{AB}$ and $\overline{DC}$ are not parallel. What shape is the facet labeled *ABCD*?

b. $\overline{DE} \parallel \overline{GF}$, and $\overline{DG}$ and $\overline{EF}$ are congruent but not parallel. What shape is the facet labeled *DEFG*?

49. **PROVING A THEOREM** Use the plan for proof below to write a paragraph proof of the Kite Opposite Angles Theorem.

Given *EFGH* is a kite. $\overline{EF} \cong \overline{FG}, \overline{EH} \cong \overline{GH}$

Prove $\angle E \cong \angle G, \angle F \not\cong \angle H$

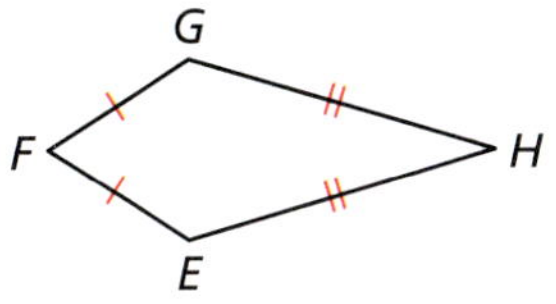

Plan for Proof First show that $\angle E \cong \angle G$. Then use an indirect argument to show that $\angle F \not\cong \angle H$.

50. **PROVING A THEOREM** In the diagram below, $\overline{BG}$ is the midsegment of $\triangle ACD$, and $\overline{GE}$ is the midsegment of $\triangle ADF$. Use the diagram to prove the Trapezoid Midsegment Theorem.

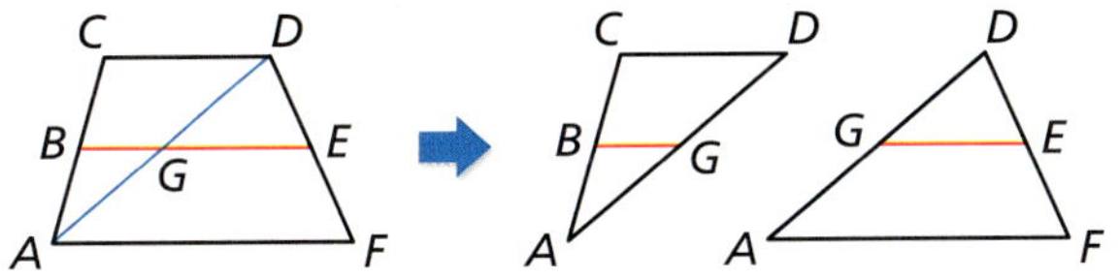

51. PROVING A THEOREM To prove the biconditional statement in the Isosceles Trapezoid Diagonals Theorem, you must prove both parts separately.

a. Prove part of the Isosceles Trapezoid Diagonals Theorem.

Given $JKLM$ is an isosceles trapezoid, $\overline{KL} \parallel \overline{JM}$, and $\overline{JK} \cong \overline{LM}$.

Prove $\overline{JL} \cong \overline{KM}$

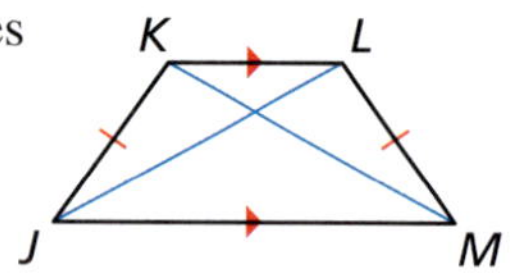

b. Write the other part of the Isosceles Trapezoid Diagonals Theorem as a conditional. Then prove the statement is true.

52. THOUGHT PROVOKING
Is SSSSA a valid congruence theorem for kites? Justify your answer.

53. PROOF What special type of quadrilateral is $EFGH$? Write a proof to show that your answer is correct.

Given In the three-dimensional figure, $\overline{JK} \cong \overline{LM}$. E, F, G, and H are the midpoints of $\overline{JL}$, $\overline{KL}$, $\overline{KM}$, and $\overline{JM}$, respectively.

Prove $EFGH$ is a ________.

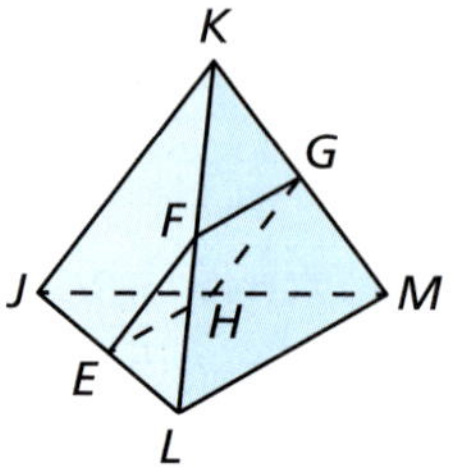

REVIEW & REFRESH

WATCH

54. Decide whether enough information is given to prove that the triangles are congruent using the HL Congruence Theorem.

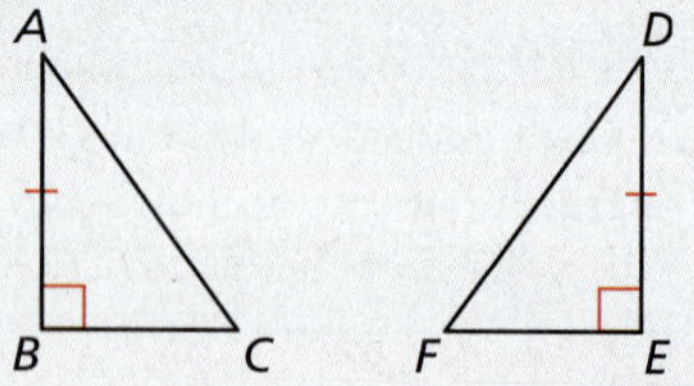

55. Find the distance from $(-4, 7)$ to the line $y = 2x$.

56. Classify the quadrilateral.

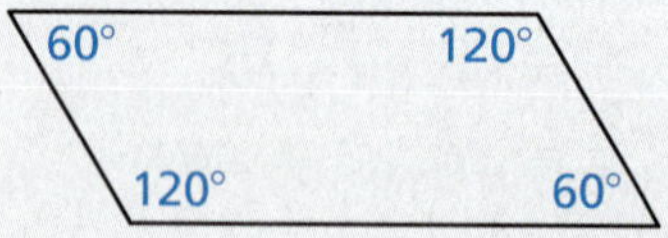

57. Identify the pair(s) of congruent angles in the diagram. Explain how you know they are congruent.

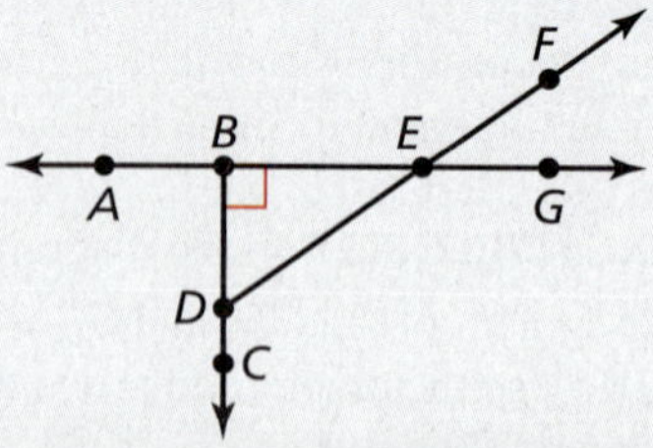

58. Find the measure of LP in ▱$LMNQ$. Explain your reasoning.

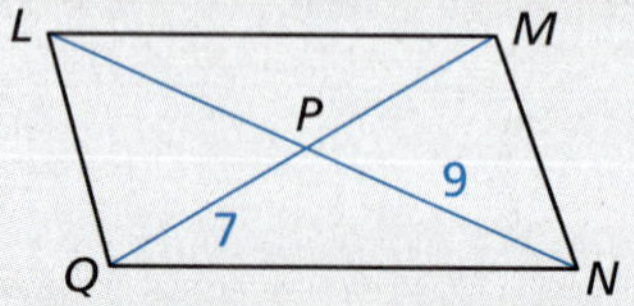

59. Write the conditional statement in if-then form.

The polygon is a triangle because the sum of its interior angle measures is 180°.

60. State which theorem you can use to show that the quadrilateral is a parallelogram.

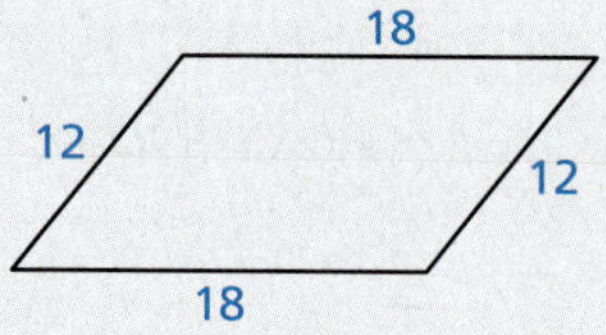

61. Graph $\overline{MN}$ with endpoints $M(1, 3)$ and $N(3, 5)$ and its image after a translation two units right, followed by a rotation 90° clockwise about the origin.

7 MTR **62. MODELING REAL LIFE** Find the perimeter of the kite.

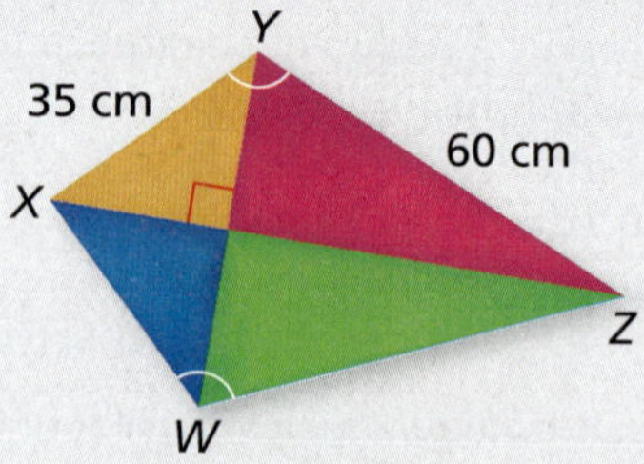

63. Sketch a diagram showing $\overline{AB}$ in plane P and $\overleftrightarrow{CD}$ not in plane P, such that $\overleftrightarrow{CD}$ bisects $\overline{AB}$.

64. Identify all pairs of alternate exterior angles.

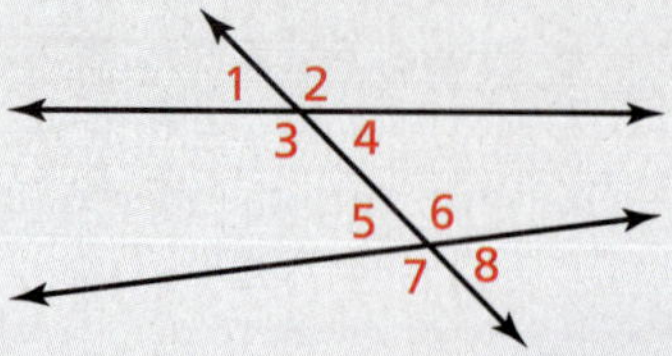

GO DIGITAL

7 Chapter Review WITH CalcChat®

Chapter Learning Target: Understand quadrilaterals and other polygons.

Chapter Success Criteria:
- ◆ I can find angles of a polygon.
- ◆ I can describe properties of parallelograms.
- ■ I can use properties of parallelograms to solve problems.
- ■ I can justify why a figure is a special quadrilateral.

◆ Surface
■ Deep

SELF-ASSESSMENT 1 I don't understand yet. 2 I can do it with help. 3 I can do it on my own. 4 I can teach someone else.

7.1 Angles of Polygons (pp. 353–360) WATCH

Learning Target: Find angle measures of polygons.

Vocabulary AZ VOCAB
diagonal
equilateral polygon
equiangular polygon
regular polygon

1. Find the sum of the measures of the interior angles of a regular 30-gon. Then find the measure of each interior angle and each exterior angle.

Find the value of x.

2.

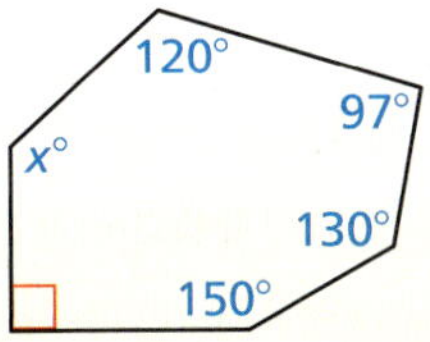

3.

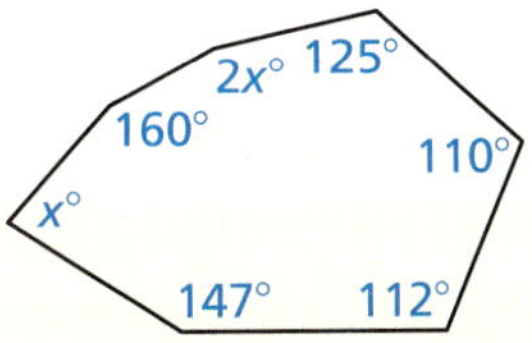

4.

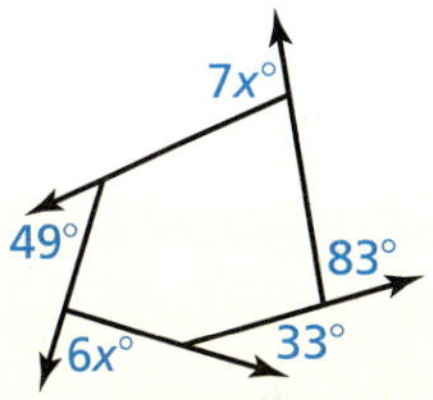

7.2 Properties of Parallelograms (pp. 361–368) WATCH

Learning Target: Prove and use properties of parallelograms.

Vocabulary AZ VOCAB
parallelogram

Find the value of each variable in the parallelogram.

5.

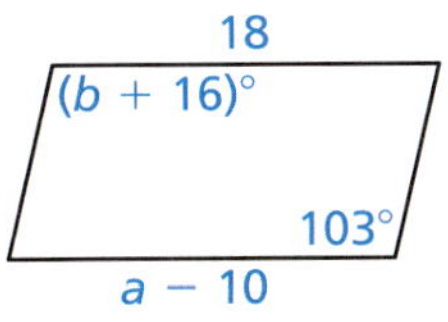

6.

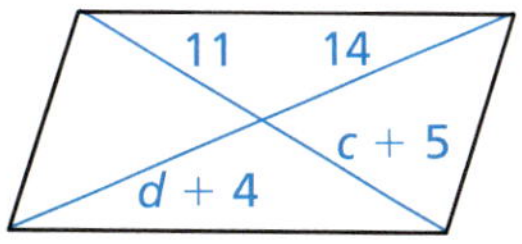

7. Find the coordinates of the intersection of the diagonals of ▱$QRST$ with vertices $Q(-8, 1)$, $R(2, 1)$, $S(4, -3)$, and $T(-6, -3)$.

8. Three vertices of ▱$JKLM$ are $J(1, 4)$, $K(5, 3)$, and $L(6, -3)$. Find the coordinates of vertex M.

9. Three vertices of a parallelogram are located at $(-4, 3)$, $(1, -2)$, and $(-1, 4)$. What are two possible locations of the fourth vertex? Explain your reasoning.

10. The figure shown is composed of two parallelograms. Find x. Justify your answer.

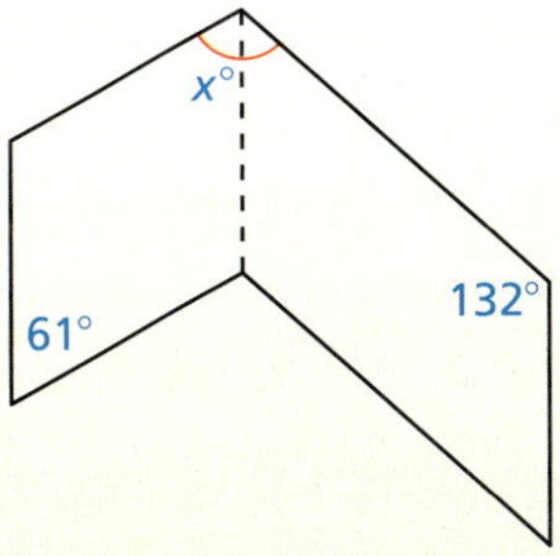

GO DIGITAL

7.3 Proving that a Quadrilateral Is a Parallelogram (pp. 369–378)

WATCH

Learning Target: Prove that a quadrilateral is a parallelogram.

State which theorem you can use to show that the quadrilateral is a parallelogram.

11.

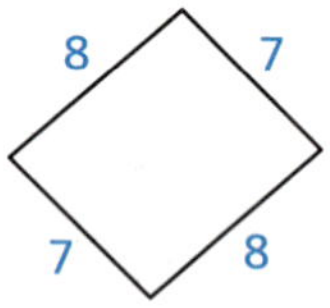

12.

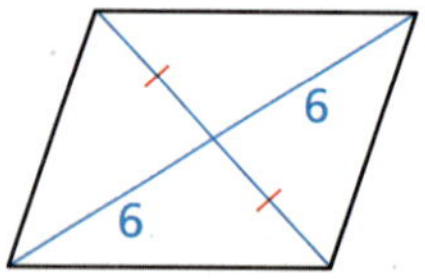

13.

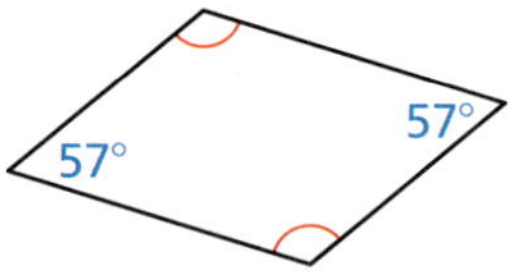

14. Find the values of x and y that make the quadrilateral a parallelogram.

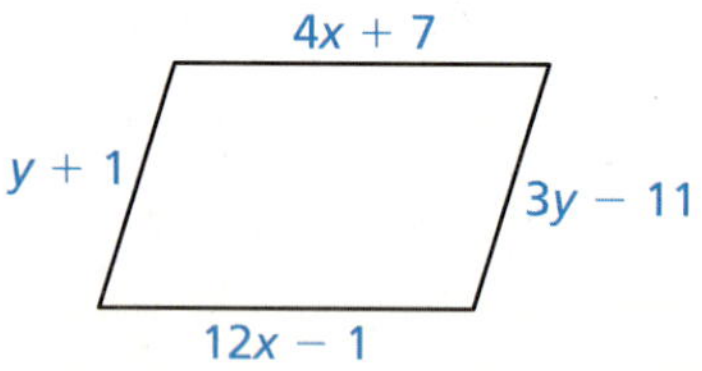

15. Find the value of x that makes the quadrilateral a parallelogram.

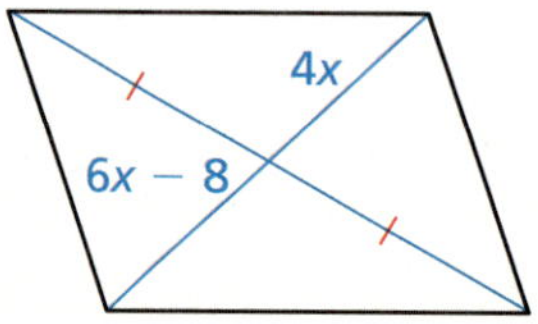

16. In the diagram of the staircase shown, $\overline{QT} \parallel \overline{RS}$, $QT = RS = 9$ feet, $QR = 3$ feet, and $m\angle QRS = 123°$.

a. Which theorem can you use to show that $QRST$ is a parallelogram?

b. Find ST, $m\angle QTS$, $m\angle TQR$, and $m\angle TSR$. Explain your reasoning.

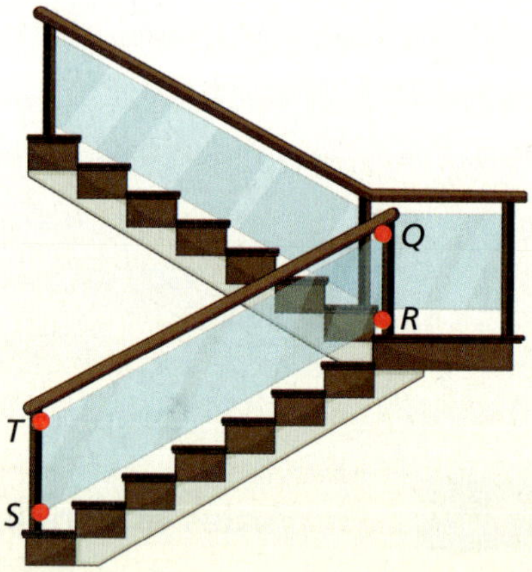

17. Show that quadrilateral $WXYZ$ with vertices $W(-1, 6)$, $X(2, 8)$, $Y(1, 0)$, and $Z(-2, -2)$ is a parallelogram.

7.4 Properties of Special Parallelograms (pp. 379–388)

WATCH

Learning Target: Explain the properties of special parallelograms.

Vocabulary VOCAB

rhombus
rectangle
square

Classify the special quadrilateral. Explain your reasoning.

18.

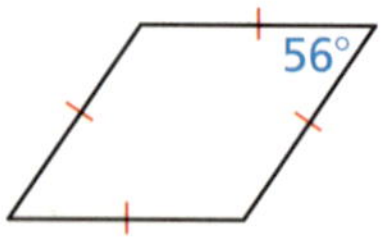

19.

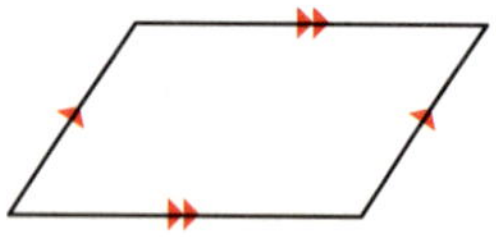

20.

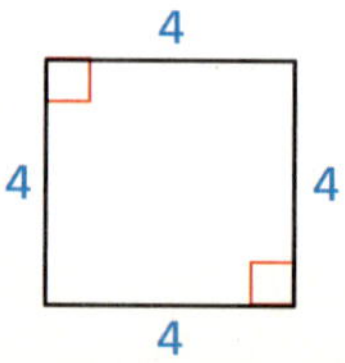

21. Find the lengths of the diagonals of rectangle $WXYZ$ where $WY = -2x + 34$ and $XZ = 3x - 26$.

22. Square $ABCD$ has diagonal $\overline{AC}$ with vertices $A(1, 1)$ and $C(3, -1)$. Find the coordinates of the remaining vertices.

23. Decide whether $\square JKLM$ with vertices $J(5, 8)$, $K(9, 6)$, $L(7, 2)$, and $M(3, 4)$ is a rectangle, a rhombus, or a square. Give all names that apply. Explain.

GO DIGITAL

7.5 Properties of Trapezoids and Kites (pp. 389–398) WATCH

Learning Target: Use properties of trapezoids and kites to find measures.

24. Find the measure of each angle in the isosceles trapezoid *WXYZ*.

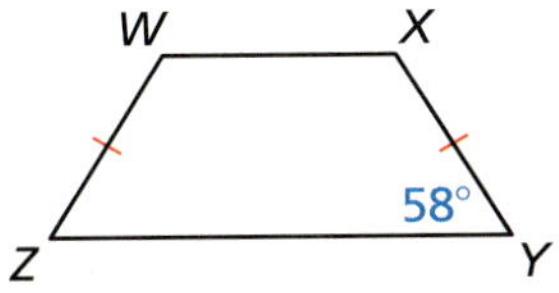

25. Find the length of the midsegment of trapezoid *ABCD*.

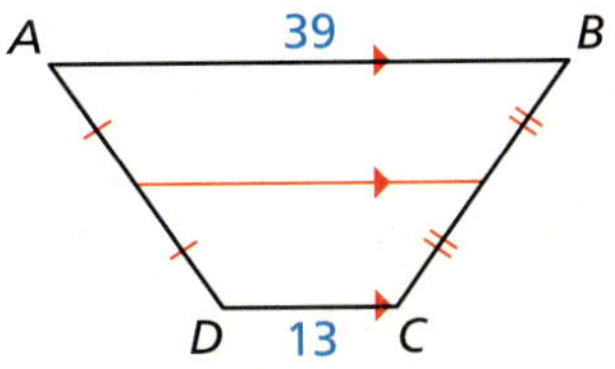

26. Find the length of the midsegment of trapezoid *JKLM* with vertices *J*(6, 10), *K*(10, 6), *L*(8, 2), and *M*(2, 2).

27. A kite has angle measures of $7x°$, 65°, 85°, and 105°. Find the value of x. Which of the angles are opposite angles? Explain your reasoning.

28. Quadrilateral *WXYZ* is a trapezoid with a pair of opposite supplementary angles. Is *WXYZ* an isosceles trapezoid? Explain your reasoning.

Give the most specific name for the quadrilateral. Explain your reasoning.

29.

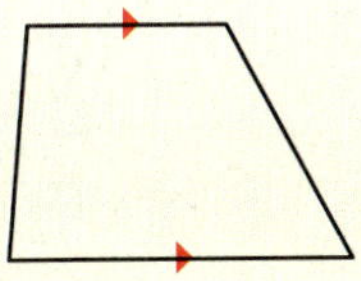

30.

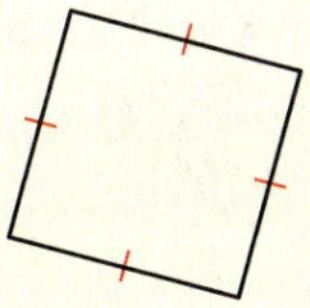

31.

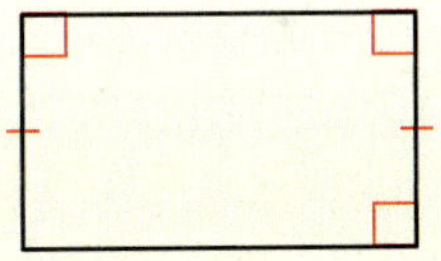

Vocabulary AZ VOCAB

trapezoid
bases
legs
base angles
isosceles trapezoid
midsegment of a trapezoid
kite

Mathematical Thinking and Reasoning

6 MTR ASSESS THE REASONABLENESS OF SOLUTIONS

Mathematicians who assess the reasonableness of solutions verify possible solutions by explaining the methods used.

1. Your friend measures a pair of nonadjacent angles of a quadrilateral and determines that the quadrilateral must be a parallelogram. Is your friend's claim reasonable? If not, explain how your friend could revise their conclusion.

2. In Exercise 38 on page 367, you find that $LM = 16$. Without calculating, determine whether your answer is reasonable.

GO DIGITAL

7 Practice Test WITH CalcChat®

Find the value of each variable in the parallelogram.

1.

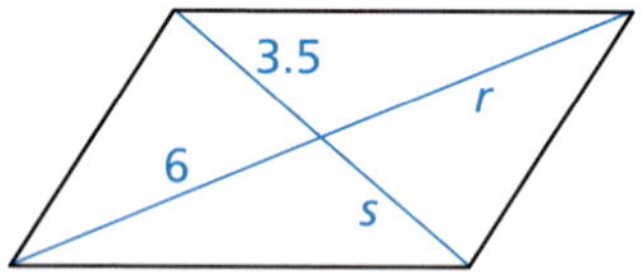

2.

3.

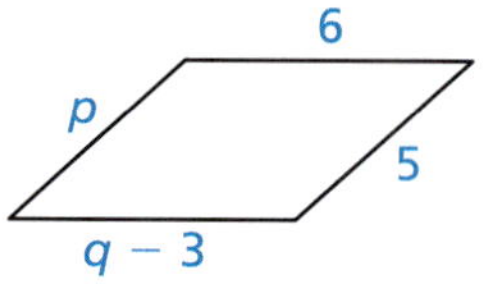

Give the most specific name for the quadrilateral. Explain your reasoning.

4.

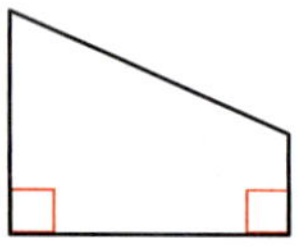

5.

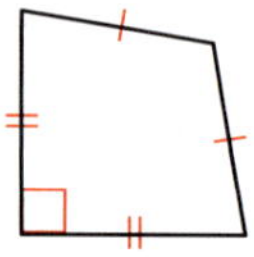

6.

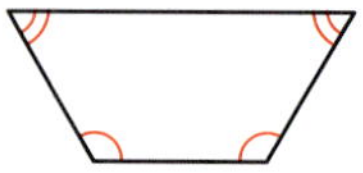

7. In a convex octagon, three of the exterior angles each have a measure of $x°$. The other five exterior angles each have a measure of $(2x + 7)°$. Find the measure of each exterior angle.

8. Quadrilateral $PQRS$ has vertices $P(5, 1)$, $Q(9, 6)$, $R(5, 11)$, and $S(1, 6)$. Classify quadrilateral $PQRS$ using the most specific name.

Determine whether enough information is given to show that the quadrilateral is a parallelogram. Explain your reasoning.

9.

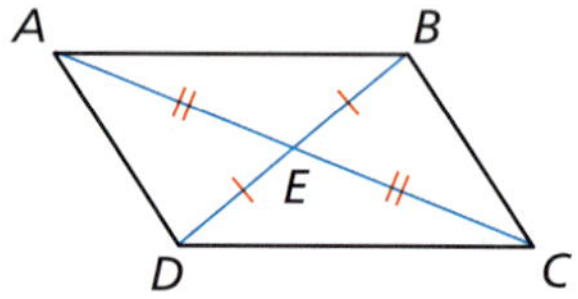

10.

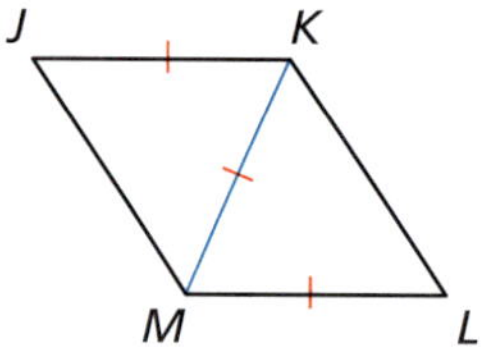

11.

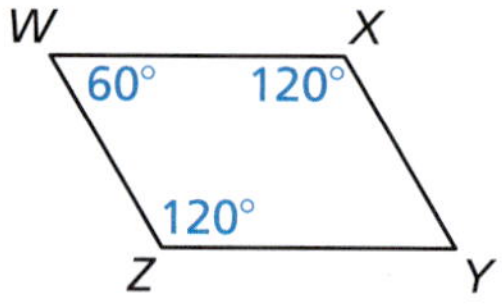

12. Explain why a parallelogram with one right angle must be a rectangle.

13. Summarize the ways you can prove that a quadrilateral is a square.

14. You are building a plant stand with three equally spaced circular shelves. The diagram shows a vertical cross section through the diameters of the shelves. What is the diameter of the middle shelf?

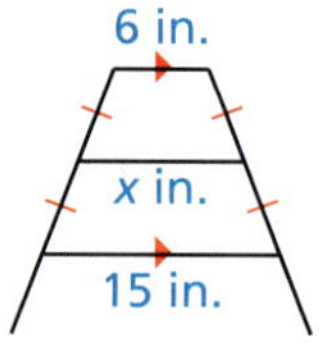

15. Three vertices of $\square JKLM$ are $J(-2, -1)$, $K(0, 2)$, and $L(4, 3)$.

a. Find the coordinates of vertex M.

b. Find the coordinates of the intersection of the diagonals of $\square JKLM$.

c. Find the perimeter of $\square JKLM$.

16. The Pentagon in Washington, DC, is shaped like a regular pentagon. Find the measure of each interior angle.

GO DIGITAL

7 Performance Task
Starstruck

People have identified shapes and patterns in the night sky for thousands of years. Patterns formed by stars are called *constellations*.

There are 88 constellations recognized by the International Astronomical Union.

Although the stars in constellations are different distances from Earth, they appear near each other when seen by an observer looking at the night sky.

REACH FOR THE STARS

Choose one of the constellations shown. Identify any polygons in the drawing and find the sum(s) of the measures of the interior angle measures. Then research the constellation and determine when it was discovered, who discovered it, and the name of its brightest star.

Which of the drawings of the constellations above appear to contain a parallelogram? a kite? Use theorems in this chapter to explain whether each is actually a parallelogram or a kite.

GO DIGITAL

7 Test Prep WITH CalcChat® Cumulative Practice

WATCH Tutorial videos are available for each exercise.

1. In the diagram, $\triangle ABC \cong \triangle DEF$. Find the values of x and y.

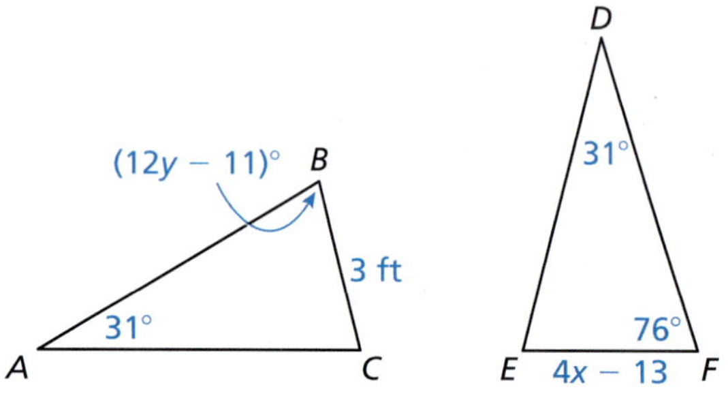

Ⓐ $x = 3, y = 73$

Ⓑ $x = 4, y = 7$

Ⓒ $x = 4, y = \frac{29}{4}$

Ⓓ $x = -4, y = 7$

2. Find the coordinates of point P along the directed line segment AB so that the ratio of AP to PB is 5 to 3.

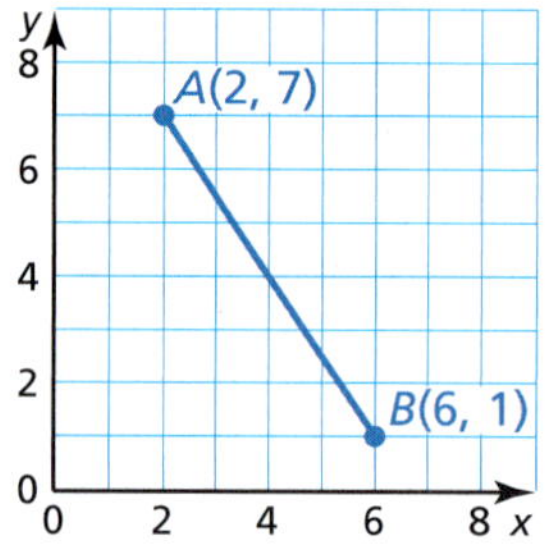

3. Find the value of x that makes the quadrilateral a parallelogram.

Ⓐ $x = -\frac{2}{9}$

Ⓑ $x = \frac{6}{5}$

Ⓒ $x = 2$

Ⓓ $x = 9$

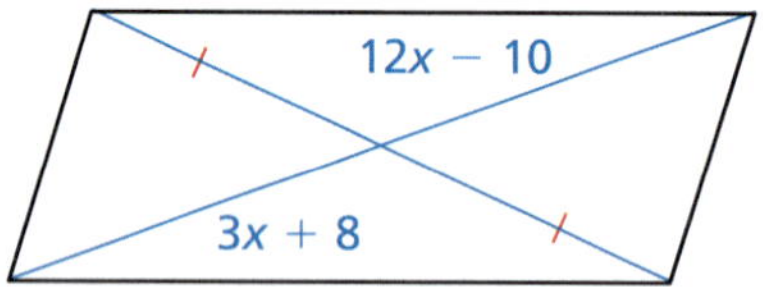

4. Use the Isosceles Trapezoid Base Angles Converse to prove that $ABCD$ is an isosceles trapezoid.

Given $\overline{BC} \parallel \overline{AD}$, $\angle EBC \cong \angle ECB$, $\angle ABE \cong \angle DCE$

Prove $ABCD$ is an isosceles trapezoid.

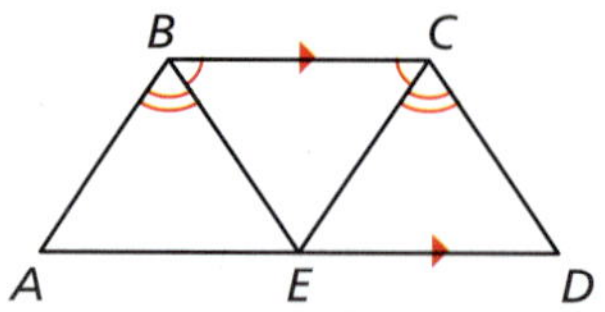

GO DIGITAL

5. One part of the Rectangle Diagonals Theorem says, “If the diagonals of a parallelogram are congruent, then it is a rectangle.” Using the reasons given, there are multiple ways to prove this part of the theorem. Provide a statement for each reason to form one possible proof of this part of the theorem.

Given $QRST$ is a parallelogram.
$\overline{QS} \cong \overline{RT}$

Prove $QRST$ is a rectangle.

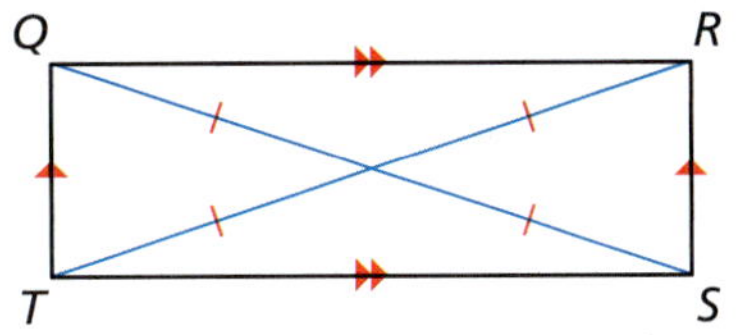

STATEMENTS	REASONS
1. $\overline{QS} \cong \overline{RT}$	1. Given
2. ____________	2. Parallelogram Opposite Sides Theorem
3. ____________	3. SSS Congruence Theorem
4. ____________	4. Corresponding parts of congruent triangles are congruent.
5. ____________	5. Parallelogram Consecutive Angles Theorem
6. ____________	6. Congruent supplementary angles have the same measure.
7. ____________	7. Parallelogram Consecutive Angles Theorem
8. ____________	8. Subtraction Property of Equality
9. ____________	9. Definition of a right angle
10. ____________	10. Definition of a rectangle

6. Graph the quadrilateral and its image after a translation 3 units right and 1 unit up, followed by a reflection in the y-axis.

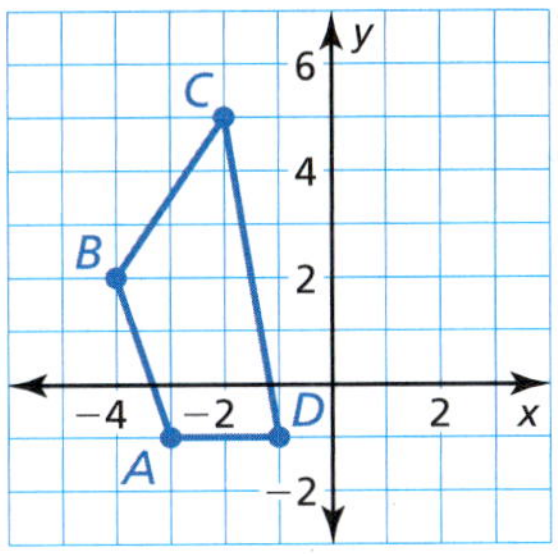

7. Find the perimeter P and area A of $\triangle JKL$ with vertices $J(-5, -7)$, $K(2, 3)$, and $L(-1, -7)$.

Ⓐ $P \approx 26.6$ units, $A = 20$ square units

Ⓑ $P \approx 26.6$ units, $A = 40$ square units

Ⓒ $P = 34$ units, $A = 20$ square units

Ⓓ $P \approx 28.6$ units, $A = 30$ square units

8. Which of the following can you use to prove that two triangles are congruent? Select all that apply.

Ⓐ ASA

Ⓑ SSS

Ⓒ SSA

Ⓓ AAA

GO DIGITAL

8 Similarity

NATIONAL GEOGRAPHIC EXPLORER

Carter Clinton

WATCH

Carter Clinton studies soil taken from the New York African Burial Ground. The burial ground contains the remains of more than 15,000 free and enslaved Africans buried from the middle 1630s to 1795. In 1991, the graves were excavated, and the soil was carefully stored. Carter analyzes bacterial DNA in the soil to learn about the lives of the people buried at the site.

- What is DNA? How can scientists obtain a sample of a person's DNA?
- What is genomics?
- The human genome is composed of about 3 billion *base pairs*. Sequences of base pairs are labeled using the letters A, C, G, and T. What do these letters represent?

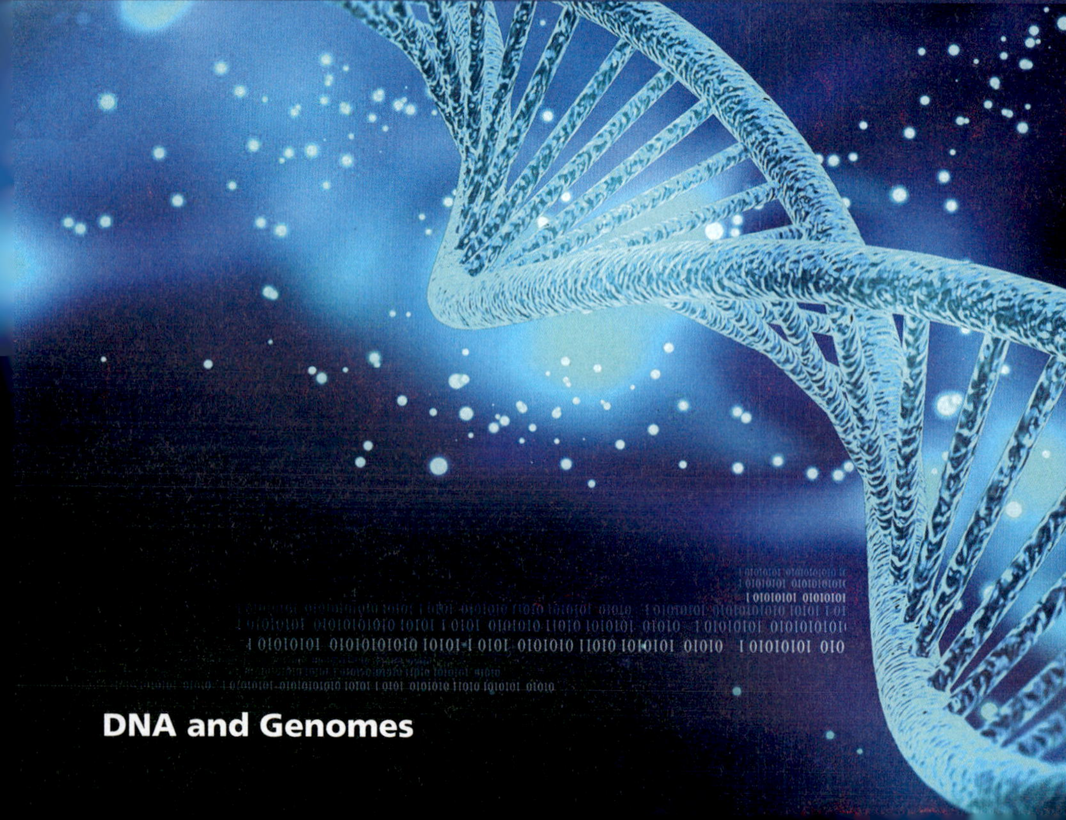

STEM

Artists often create scale drawings of monuments and historical sites for educational materials. In the Performance Task, you will create a brochure for the African Burial Ground National Monument that includes a scale drawing of the site.

DNA and Genomes

GO DIGITAL

Preparing for Chapter 8

Chapter Learning Target: Understand similarity.

Chapter Success Criteria:
- ◆ I can identify corresponding parts of similar polygons.
- ◆ I can find and use scale factors in similar polygons.
- ■ I can prove triangles are similar.
- ■ I can use proportionality theorems to solve problems.

◆ Surface
■ Deep

Chapter Vocabulary

Work with a partner. Discuss each of the terms.

similar figures
similarity transformation
corresponding parts
proportion

Mathematical Thinking and Reasoning

5 MTR

USE PATTERNS AND STRUCTURE TO HELP UNDERSTAND AND CONNECT MATHEMATICAL CONCEPTS

Mathematicians who use patterns and structure to help understand and connect mathematical concepts look for similarities among problems.

Work with a partner.

1. The rectangular photo is dilated by a scale factor k. Complete the table for the perimeter P (in centimeters) and the area A (in square centimeters) of the rectangle for each scale factor. How does the scale factor affect the perimeter and the area?

Scale factor, *k*	$\frac{1}{4}$	$\frac{1}{2}$	1	2	3
Perimeter, *P*					
Area, *A*					

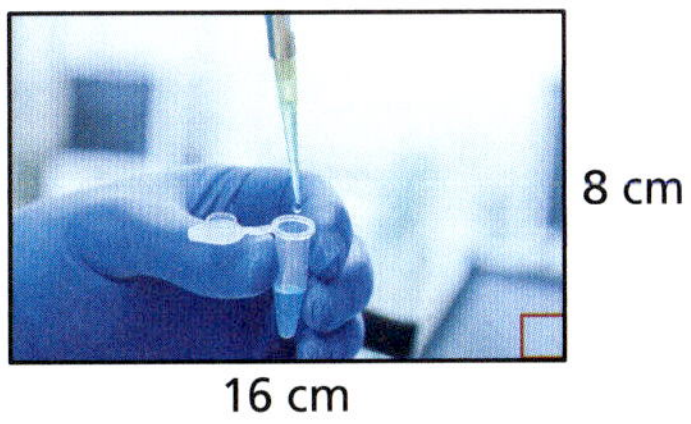

2. Show why the relationships you described in Exercise 1 are true for a rectangle with any length, width, and scale factor.

3. Complete the table in Exercise 1 for the triangle. Compare any patterns in your table with the patterns you described in Exercise 1. Is this true for any triangle with any scale factor? Explain.

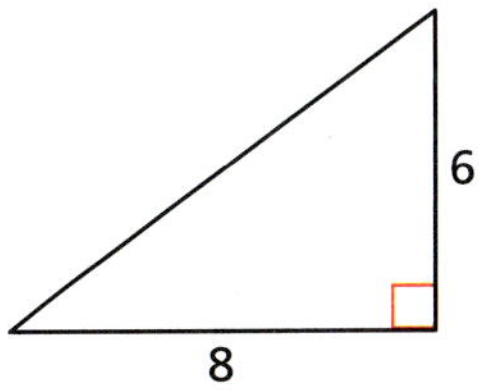

GO DIGITAL

8 Prepare WITH CalcChat®

Determining whether Ratios Form a Proportion

Example 1 **Tell whether $\frac{2}{8}$ and $\frac{3}{12}$ form a proportion.**

Compare the ratios in simplest form.

$$\frac{2}{8} = \frac{2 \div 2}{8 \div 2} = \frac{1}{4}$$
$$\frac{3}{12} = \frac{3 \div 3}{12 \div 3} = \frac{1}{4}$$

The ratios are equivalent.

▶ So, $\frac{2}{8}$ and $\frac{3}{12}$ form a proportion.

Tell whether the ratios form a proportion.

1. $\frac{5}{3}, \frac{35}{21}$

2. $\frac{9}{24}, \frac{24}{64}$

3. $\frac{8}{56}, \frac{6}{28}$

4. $\frac{18}{4}, \frac{27}{9}$

5. $\frac{15}{21}, \frac{55}{77}$

6. $\frac{26}{8}, \frac{39}{12}$

Finding a Scale Factor

Example 2 **Find the scale factor of each dilation.**

a.

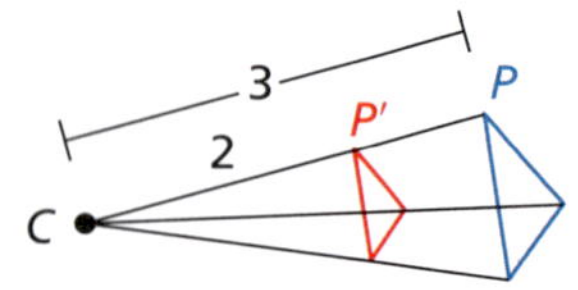

▶ Because $\frac{CP'}{CP} = \frac{2}{3}$, the scale factor is $k = \frac{2}{3}$.

b.

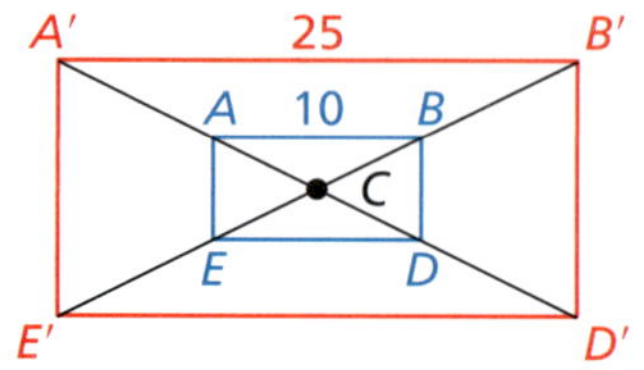

▶ Because $\frac{A'B'}{AB} = \frac{25}{10}$, the scale factor is $k = \frac{25}{10} = \frac{5}{2}$.

Find the scale factor of the dilation.

7.

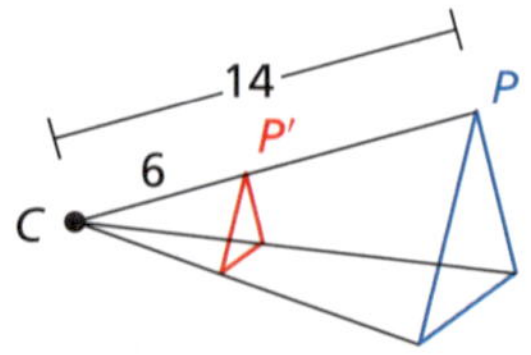

8.

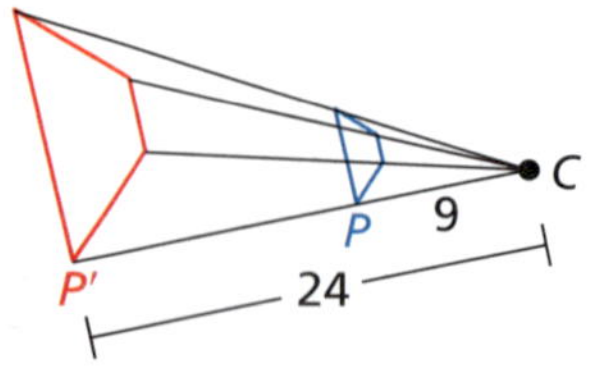

9.

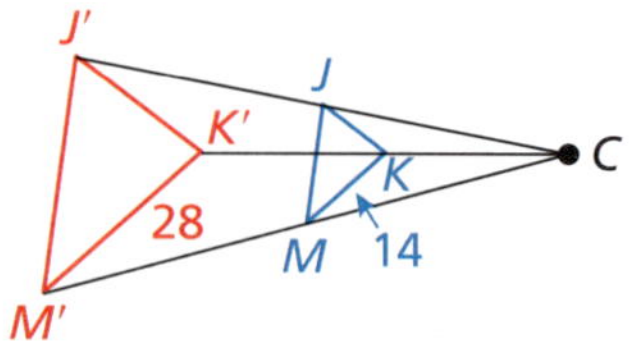

4 MTR **10. DISCUSS MATHEMATICAL THINKING** If ratio X and ratio Y form a proportion and ratio Y and ratio Z form a proportion, do ratio X and ratio Z form a proportion? Explain your reasoning.

GO DIGITAL

8.1 Similar Polygons

Learning Target: Understand the relationship between similar polygons.

Success Criteria:
- I can use similarity statements.
- I can find corresponding lengths in similar polygons.
- I can find perimeters and areas of similar polygons.
- I can decide whether polygons are similar.

EXPLORE IT! Comparing a Figure to Its Dilation

Work with a partner. Use technology.

a. Construct any quadrilateral. Dilate it to form a similar quadrilateral using any scale factor k and any center of dilation.

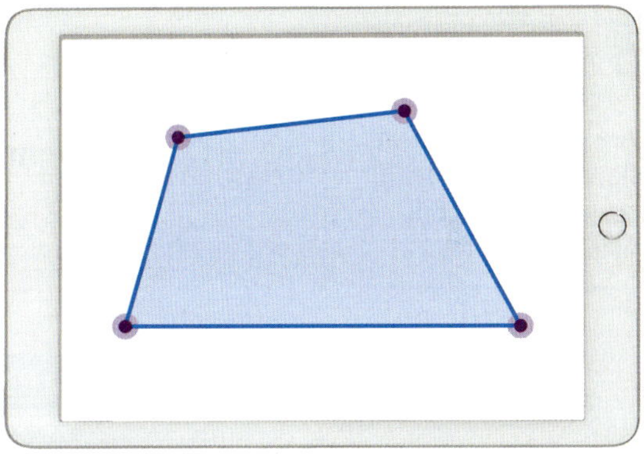

b. Compare the corresponding angles of the original quadrilateral and its image. What do you observe?

c. Find the ratios of the side lengths of the image to the corresponding side lengths of the original quadrilateral. What do you observe?

d. Compare the perimeters and areas of the original quadrilateral and its image. What do you observe?

e. Repeat parts (a)–(d) for several other polygons, scale factors, and centers of dilation. Do you obtain similar results?

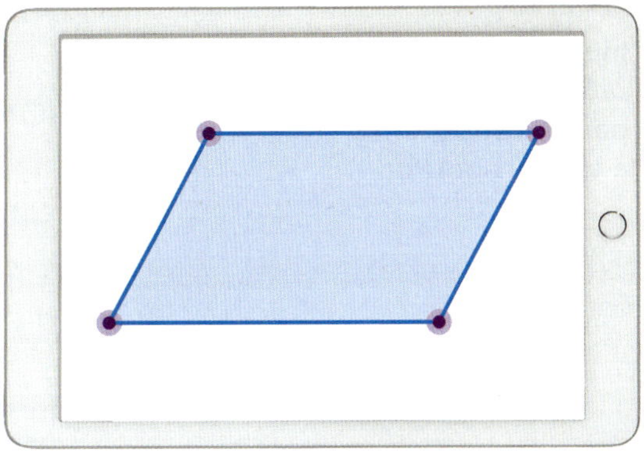

f. Explain, in your own words, how similar polygons are related.

4 MTR CONSTRUCT AN ARGUMENT

Are two regular polygons with the same number of sides always similar? Explain.

Geometric Reasoning

MA.912.GR.1.6 Solve mathematical and real-world problems involving congruence or similarity in two-dimensional figures.

MA.912.GR.4.3 Extend previous understanding of scale drawings and scale factors to determine how dilations affect the area of two-dimensional figures and the surface area or volume of three-dimensional figures.

Also MA.912.GR.2.8, MA.912.GR.4.4

GO DIGITAL

Using Similarity Statements

Recall from Section 4.6 that two geometric figures are similar figures if and only if there is a similarity transformation that maps one figure onto the other.

KEY IDEA

Corresponding Parts of Similar Polygons

In the diagram below, $\triangle ABC$ is similar to $\triangle DEF$. You can write "$\triangle ABC$ is similar to $\triangle DEF$" as $\triangle ABC \sim \triangle DEF$. A similarity transformation preserves angle measure. So, corresponding angles are congruent. A similarity transformation also enlarges or reduces side lengths by a scale factor k. So, corresponding side lengths are proportional.

Corresponding angles

$\angle A \cong \angle D$, $\angle B \cong \angle E$, $\angle C \cong \angle F$

Ratios of corresponding side lengths

$$\frac{DE}{AB} = \frac{EF}{BC} = \frac{FD}{CA} = k$$

5 MTR RELATE CONCEPTS
Are any two congruent figures also similar? If so, what is the scale factor?

EXAMPLE 1 Using Similarity Statements

WATCH

In the diagram, $\triangle RST \sim \triangle XYZ$.

a. Find the scale factor from $\triangle RST$ to $\triangle XYZ$.

b. List all pairs of congruent angles.

c. Write the ratios of the corresponding side lengths in a *statement of proportionality.*

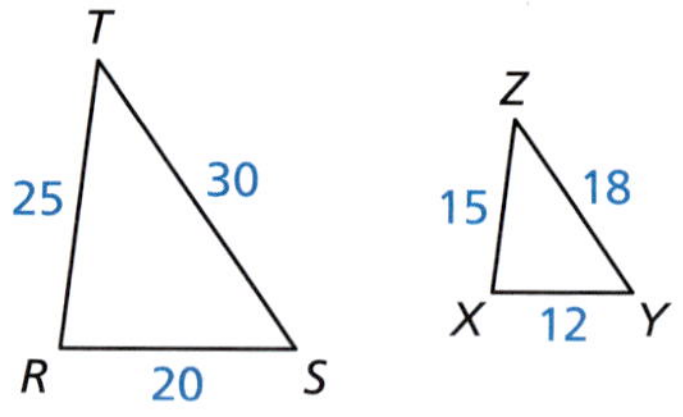

READING
In a *statement of proportionality*, any pair of ratios forms a true proportion.

SOLUTION

a. $\frac{XY}{RS} = \frac{12}{20} = \frac{3}{5}$ $\qquad \frac{YZ}{ST} = \frac{18}{30} = \frac{3}{5}$ $\qquad \frac{ZX}{TR} = \frac{15}{25} = \frac{3}{5}$

So, the scale factor is $\frac{3}{5}$.

b. $\angle R \cong \angle X$, $\angle S \cong \angle Y$, and $\angle T \cong \angle Z$.

c. Because the ratios in part (a) are equal, $\frac{XY}{RS} = \frac{YZ}{ST} = \frac{ZX}{TR}$.

SELF-ASSESSMENT

1 I don't understand yet. 2 I can do it with help. 3 I can do it on my own. 4 I can teach someone else.

1. In the diagram, $\triangle JKL \sim \triangle PQR$.

a. Find the scale factor from $\triangle JKL$ to $\triangle PQR$.

b. List all pairs of congruent angles.

c. Write the ratios of the corresponding side lengths in a statement of proportionality.

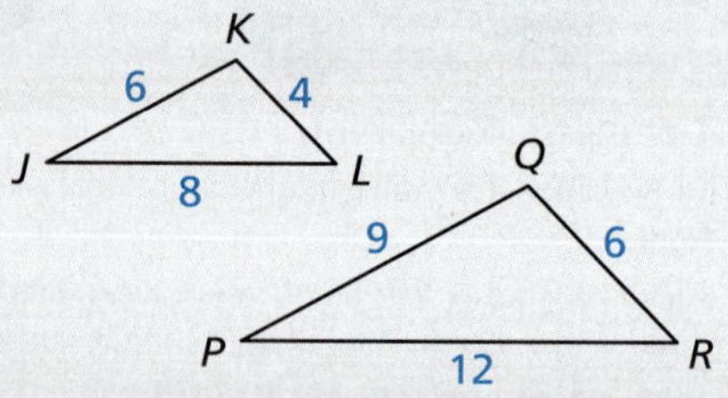

GO DIGITAL

Finding Corresponding Lengths in Similar Polygons

KEY IDEA

Corresponding Lengths in Similar Polygons

If two polygons are similar, then the ratio of any two corresponding lengths in the polygons is equal to the scale factor of the similar polygons.

READING

Corresponding lengths in similar triangles include side lengths, altitudes, medians, and midsegments.

EXAMPLE 2 Finding a Corresponding Length

WATCH

In the diagram, $\triangle DEF \sim \triangle MNP$. Find the value of x.

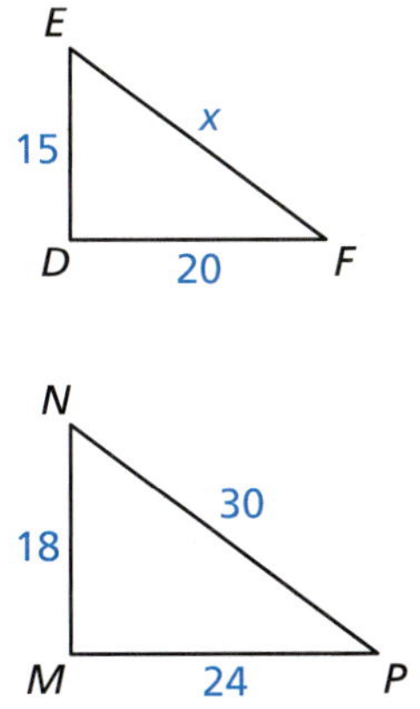

SOLUTION

The triangles are similar, so the corresponding side lengths are proportional.

$$\frac{MN}{DE} = \frac{NP}{EF}$$ Write proportion.

$$\frac{18}{15} = \frac{30}{x}$$ Substitute.

$$18x = 450$$ Cross Products Property

$$x = 25$$ Divide each side by 18.

▶ The value of x is 25.

2 MTR USE ANOTHER METHOD

There are several ways to write the proportion. Write a different proportion that you can use to solve the problem.

EXAMPLE 3 Finding a Corresponding Length

WATCH

In the diagram, $\triangle TPR \sim \triangle XPZ$. Find the length of the altitude $\overline{PS}$.

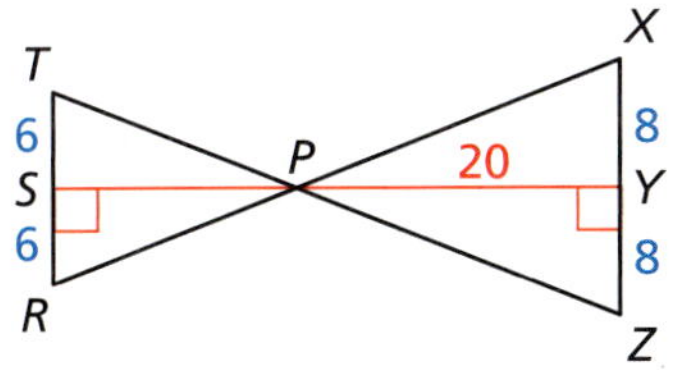

SOLUTION

First, find the scale factor from $\triangle XPZ$ to $\triangle TPR$.

$$\frac{TR}{XZ} = \frac{6+6}{8+8} = \frac{12}{16} = \frac{3}{4}$$

Because the ratio of the lengths of the altitudes in similar triangles is equal to the scale factor, you can write the following proportion.

$$\frac{PS}{PY} = \frac{3}{4}$$ Write proportion.

$$\frac{PS}{20} = \frac{3}{4}$$ Substitute 20 for PY.

$$PS = 15$$ Multiply each side by 20.

▶ The length of the altitude $\overline{PS}$ is 15.

SELF-ASSESSMENT

1 I don't understand yet. 2 I can do it with help. 3 I can do it on my own. 4 I can teach someone else.

2. Find the value of x.

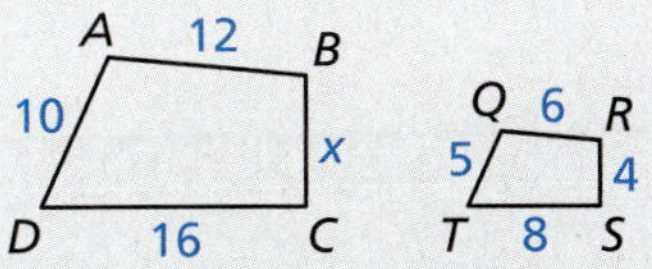

$ABCD \sim QRST$

3. Find KM.

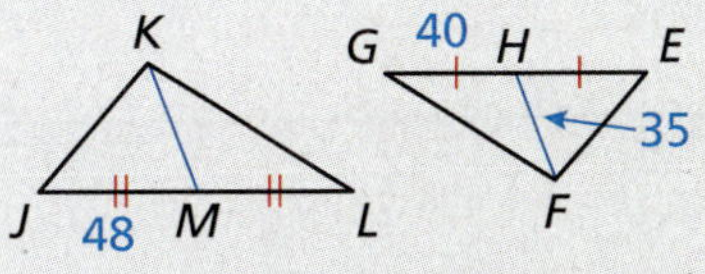

$\triangle JKL \sim \triangle EFG$

GO DIGITAL

Finding Perimeters and Areas of Similar Polygons

5 MTR RELATE CONCEPTS
When two similar polygons have a scale factor of k, what is the ratio of their perimeters?

THEOREM

8.1 Perimeters of Similar Polygons

If two polygons are similar, then the ratio of their perimeters is equal to the ratios of their corresponding side lengths.

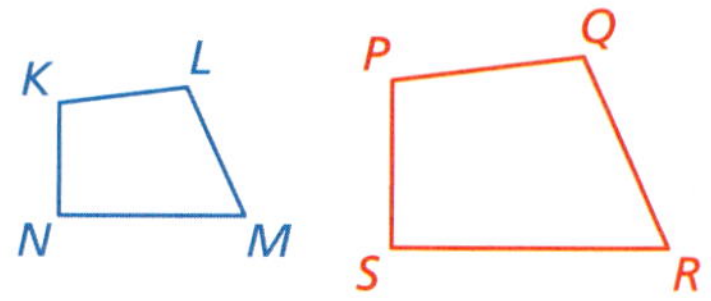

If $KLMN \sim PQRS$, then $\dfrac{PQ + QR + RS + SP}{KL + LM + MN + NK} = \dfrac{PQ}{KL} = \dfrac{QR}{LM} = \dfrac{RS}{MN} = \dfrac{SP}{NK}$.

Proof *BigIdeasMath.com*

Prove this Theorem Exercise 40, page 417

7 MTR **EXAMPLE 4** **Modeling Real Life** WATCH STEM

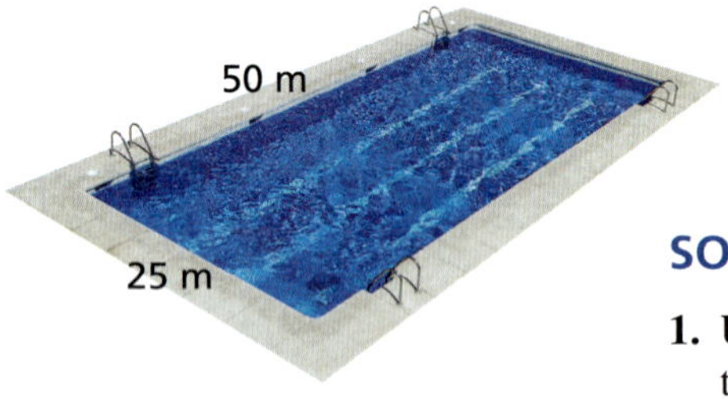

A town plans to build a new swimming pool. An Olympic-sized pool is rectangular with a length of 50 meters and a width of 25 meters. The new pool will be similar in shape to an Olympic-sized pool but will have a length of 40 meters. Find the perimeters of an Olympic-sized pool and the new pool.

SOLUTION

1. **Understand the Problem** You are given the length and width of a rectangle and the length of a similar rectangle. You need to find the perimeters of both rectangles.
2. **Make a Plan** Find the scale factor of the similar rectangles, and find the perimeter of an Olympic-sized pool. Then use the Perimeters of Similar Polygons Theorem to write and solve a proportion to find the perimeter of the new pool.
3. **Solve and Check** Because the new pool will be similar to an Olympic-sized pool, the scale factor is the ratio of the lengths, $\frac{40}{50} = \frac{4}{5}$. The perimeter of an Olympic-sized pool is $2(50) + 2(25) = 150$ meters. Write and solve a proportion to find the perimeter x of the new pool.

$\dfrac{x}{150} = \dfrac{4}{5}$ Perimeters of Similar Polygons Theorem

$x = 120$ Multiply each side by 150.

▶ So, the perimeter of an Olympic-sized pool is 150 meters, and the perimeter of the new pool is 120 meters.

Check Check that the ratio of the perimeters is equal to the scale factor. $\dfrac{120}{150} = \dfrac{4}{5}$ ✓

STUDY TIP
You can also write the scale factor as a decimal. In Example 4, you can write the scale factor as 0.8 and multiply by 150 to get $x = 0.8(150) = 120$.

SELF-ASSESSMENT

1 I don't understand yet. 2 I can do it with help. 3 I can do it on my own. 4 I can teach someone else.

4. The two gazebos shown are similar pentagons. Find the perimeter of Gazebo A.

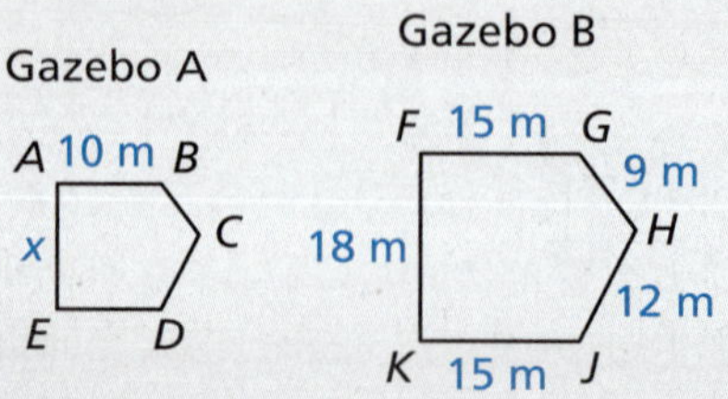

GO DIGITAL

RELATE CONCEPTS
When two similar polygons have a scale factor of k, what is the ratio of their areas?

THEOREM

8.2 Areas of Similar Polygons

If two polygons are similar, then the ratio of their areas is equal to the squares of the ratios of their corresponding side lengths.

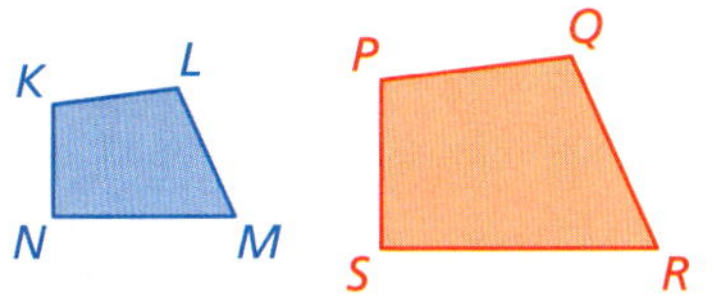

If $KLMN \sim PQRS$, then $\dfrac{\text{Area of } PQRS}{\text{Area of } KLMN} = \left(\dfrac{PQ}{KL}\right)^2 = \left(\dfrac{QR}{LM}\right)^2 = \left(\dfrac{RS}{MN}\right)^2 = \left(\dfrac{SP}{NK}\right)^2.$

Proof *BigIdeasMath.com*

Prove this Theorem Exercise 51, page 418

EXAMPLE 5 Finding Areas of Similar Polygons

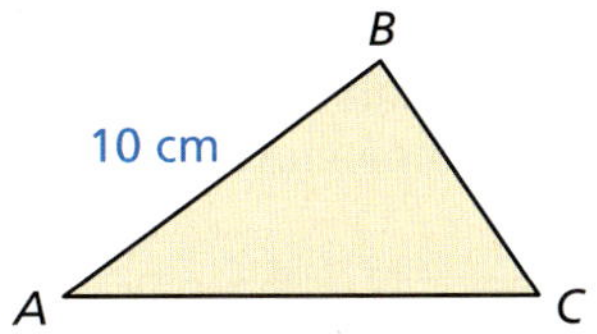

Area of $\triangle ABC = 36 \text{ cm}^2$

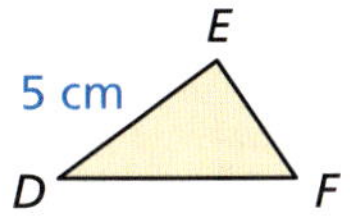

In the diagram, $\triangle ABC \sim \triangle DEF$. Find the area of $\triangle DEF$.

SOLUTION

Because the triangles are similar, the ratio of the area of $\triangle ABC$ to the area of $\triangle DEF$ is equal to the square of the ratio of AB to DE. Write and solve a proportion to find the area of $\triangle DEF$. Let A represent the area of $\triangle DEF$.

$$\frac{\text{Area of } \triangle ABC}{\text{Area of } \triangle DEF} = \left(\frac{AB}{DE}\right)^2 \quad \text{Areas of Similar Polygons Theorem}$$

$$\frac{36}{A} = \left(\frac{10}{5}\right)^2 \quad \text{Substitute.}$$

$$\frac{36}{A} = \frac{100}{25} \quad \text{Square the right side of the equation.}$$

$$900 = 100A \quad \text{Cross Products Property}$$

$$9 = A \quad \text{Divide.}$$

▶ The area of $\triangle DEF$ is 9 square centimeters.

SELF-ASSESSMENT

1 I don't understand yet. 2 I can do it with help. 3 I can do it on my own. 4 I can teach someone else.

4 MTR **5. DIFFERENT WORDS, SAME QUESTION** Which is different? Find "both" answers.

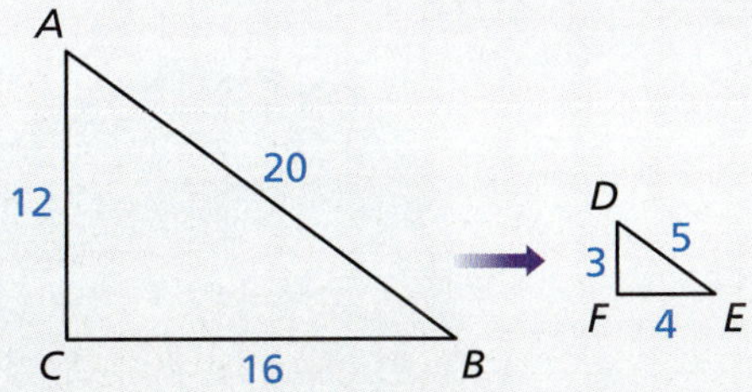

What is the scale factor?

What is the ratio of their areas?

What is the ratio of their corresponding side lengths?

What is the ratio of their perimeters?

6. In the diagram, $GHJK \sim LMNP$. Find the area of $LMNP$.

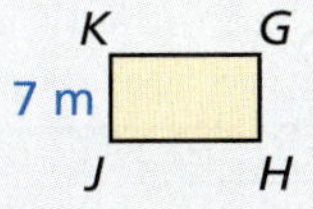

Area of $GHJK = 84 \text{ m}^2$

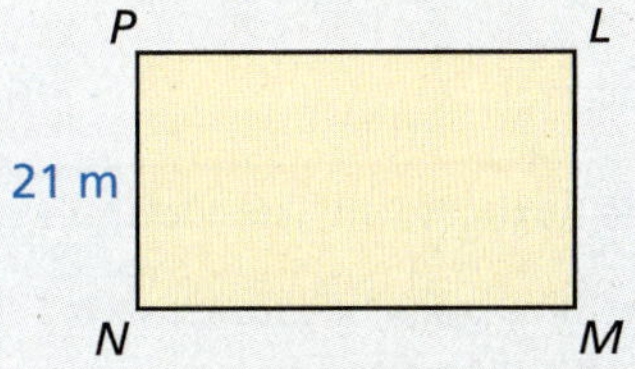

Deciding whether Polygons Are Similar

EXAMPLE 6 **Deciding whether Polygons Are Similar**

Decide whether *ABCDE* and *KLQRP* are similar. Explain your reasoning.

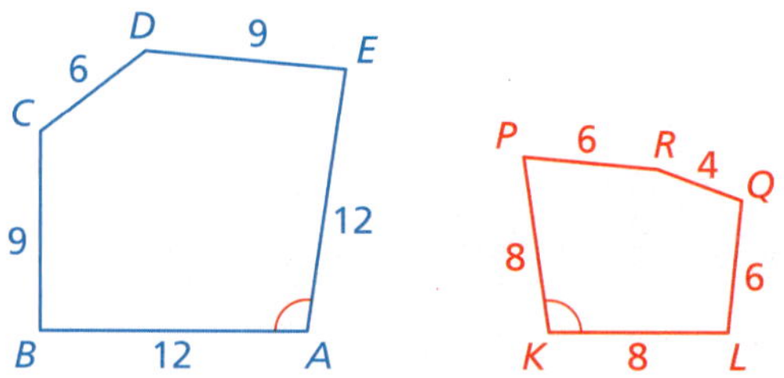

SOLUTION

Corresponding sides of the pentagons are proportional with a scale factor of $\frac{2}{3}$. However, this does not necessarily mean the pentagons are similar. Use a similarity transformation to decide whether the pentagons are similar. A dilation with center *A* and scale factor $\frac{2}{3}$ moves *ABCDE* onto *AFGHJ*. Then a reflection moves *AFGHJ* onto *KLMNP*.

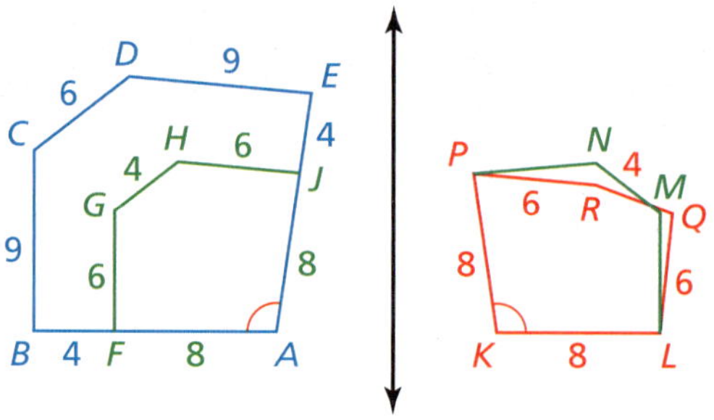

KLMNP does not exactly coincide with *KLQRP*, because not all the corresponding angles are congruent. (Only ∠*A* and ∠*K* are congruent.)

▶ Because angle measure is not preserved, the two pentagons are not similar.

SELF-ASSESSMENT 1 I don't understand yet. 2 I can do it with help. 3 I can do it on my own. 4 I can teach someone else.

Refer to the floor tile designs below. In each design, the red outer shape is a regular hexagon.

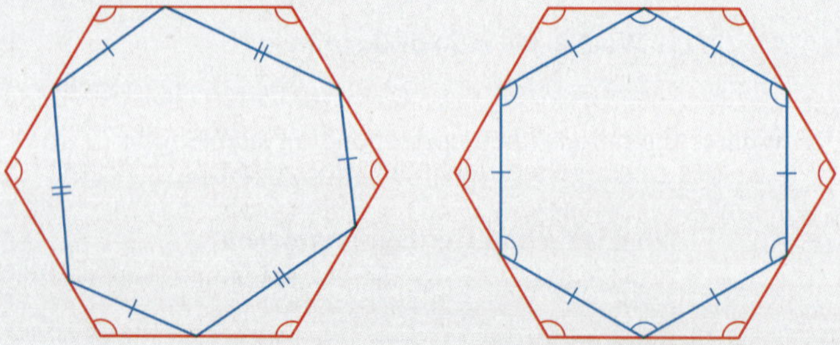

Tile Design 1   Tile Design 2

7. Decide whether the hexagons in Tile Design 1 are similar. Explain.

8. Decide whether the hexagons in Tile Design 2 are similar. Explain.

GO DIGITAL

8.1 Practice WITH CalcChat® AND CalcView®

In Exercises 1 and 2, find the scale factor. Then list all pairs of congruent angles and write the ratios of the corresponding side lengths in a statement of proportionality. *(See Example 1.)*

▶ **1.** $\triangle ABC \sim \triangle LMN$

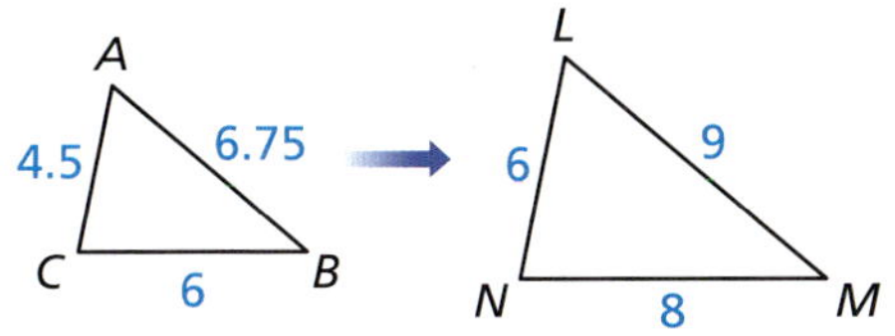

2. $DEFG \sim PQRS$

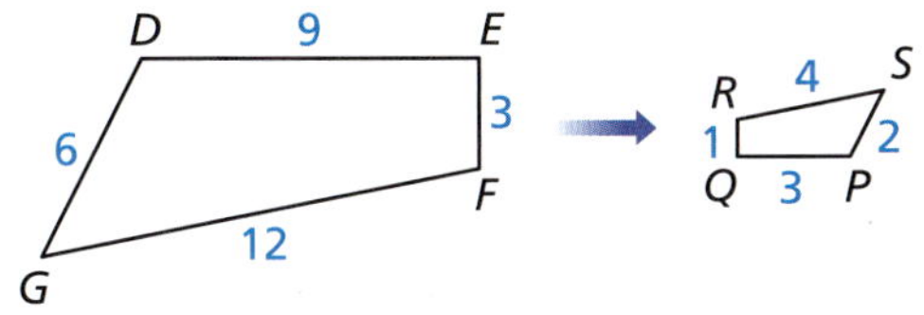

In Exercises 3–6, the polygons are similar. Find the value of x. *(See Example 2.)*

▶ **3.**

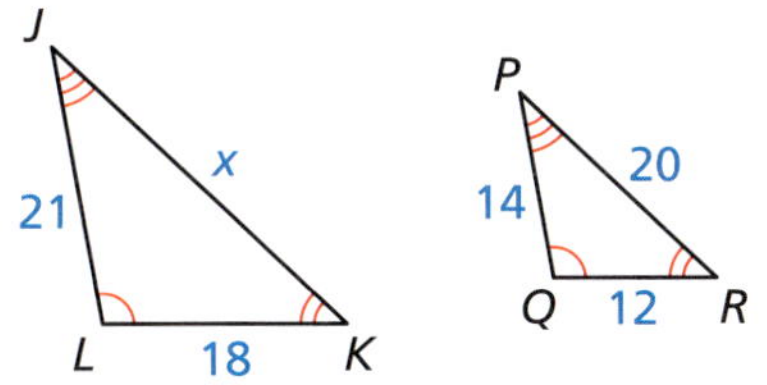

4.

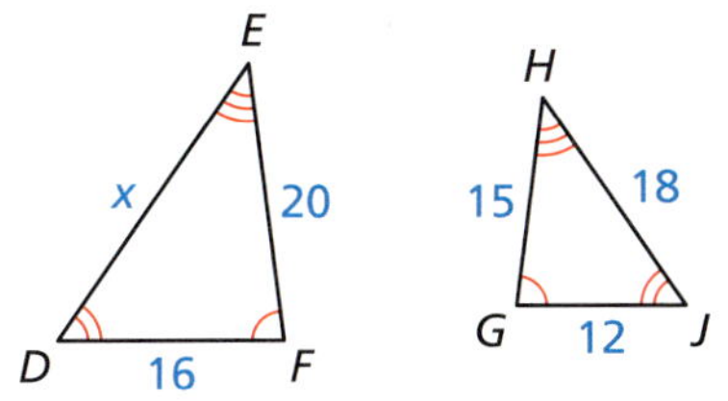

5.

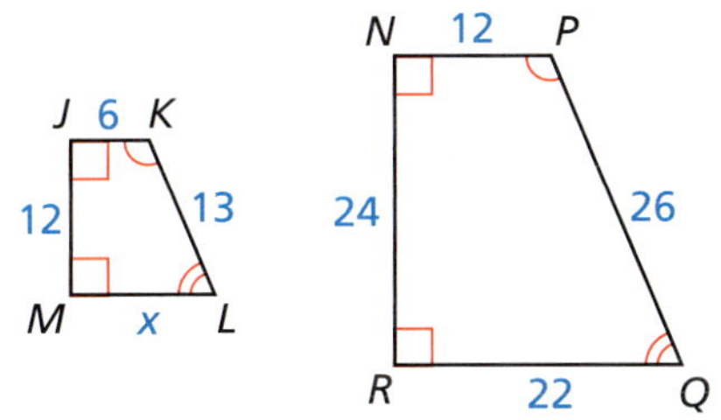

6.

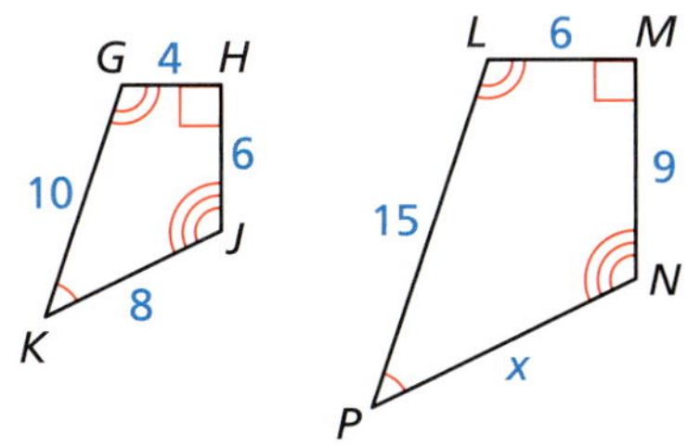

In Exercises 7 and 8, the triangles are similar. Identify the type of segment shown by the dashed line and find the value of the variable. *(See Example 3.)*

▶ **7.**

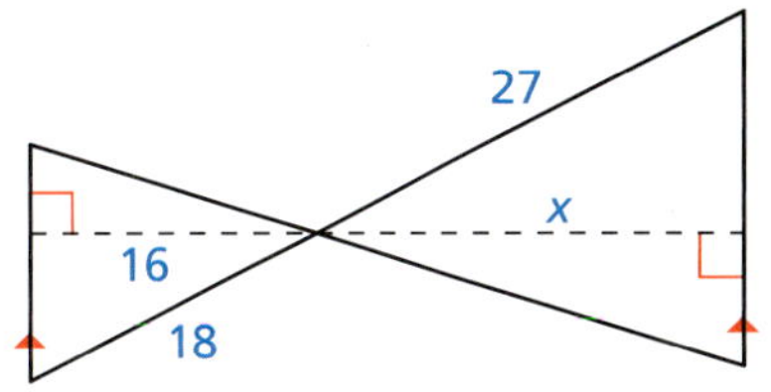

8.

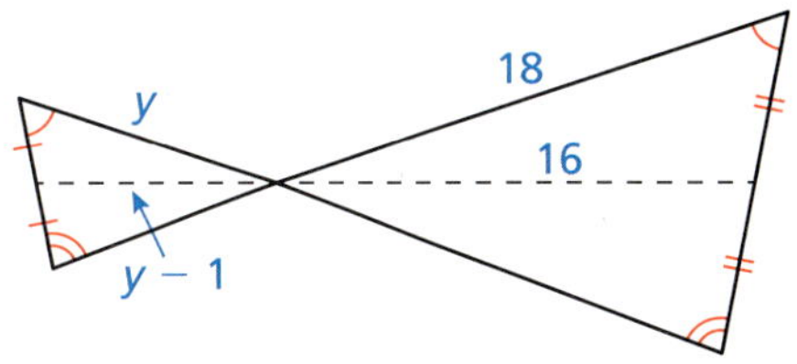

In Exercises 9 and 10, $RSTU \sim ABCD$. Find the ratio of their perimeters.

9.

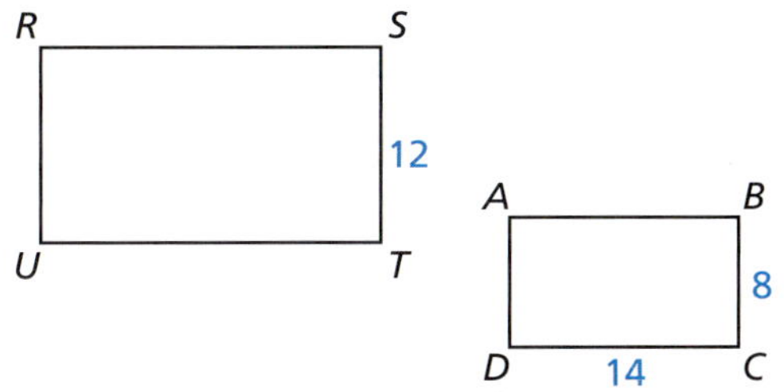

10.

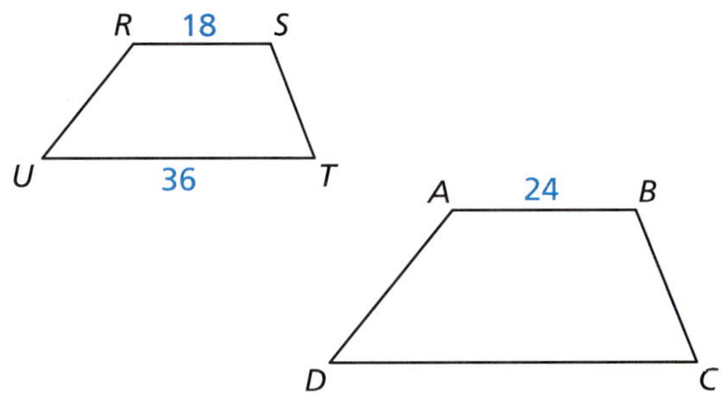

In Exercises 11–14, two polygons are similar. The perimeter of one polygon and the ratio of the corresponding side lengths are given. Find the perimeter of the other polygon.

11. perimeter of smaller polygon: 48 cm; ratio: $\frac{2}{3}$

12. perimeter of smaller polygon: 66 ft; ratio: $\frac{3}{4}$

13. perimeter of larger polygon: 120 yd; ratio: $\frac{1}{6}$

14. perimeter of larger polygon: 85 m; ratio: $\frac{2}{5}$

GO DIGITAL

15. MODELING REAL LIFE Scientists rope off two excavation sites. Site A is rectangular with a length of 60 meters and a width of 40 meters. Site B is similar in shape to Site A but has a length of 45 meters. How much rope is needed for both sites? *(See Example 4.)*

16. MODELING REAL LIFE Your family decides to put a rectangular patio in your backyard, similar to the shape of your backyard. Your backyard has a length of 45 feet and a width of 20 feet. The length of your new patio is 18 feet. Find the perimeter of your backyard and the perimeter of the patio.

In Exercises 17–20, the polygons are similar. The area of one polygon is given. Find the area of the other polygon. *(See Example 5.)*

17.

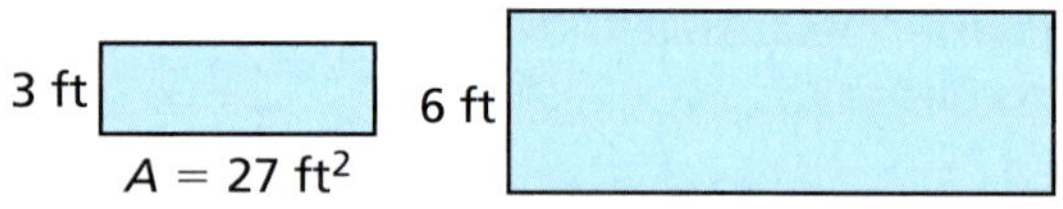

18.

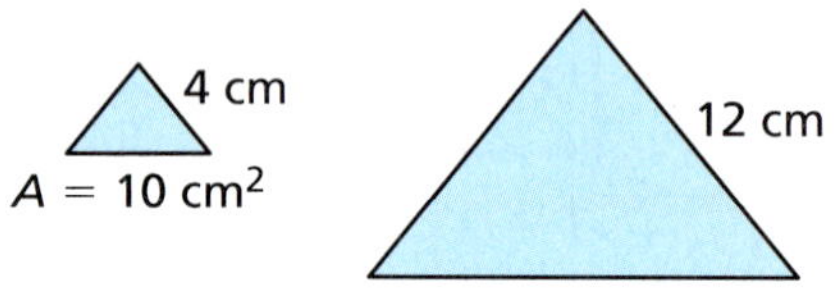

19.

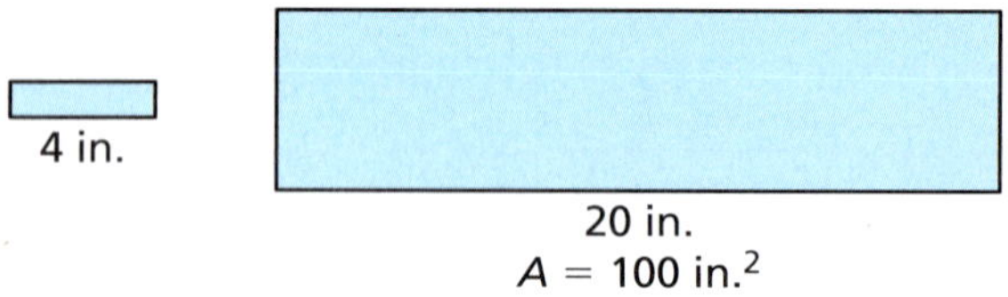

20.

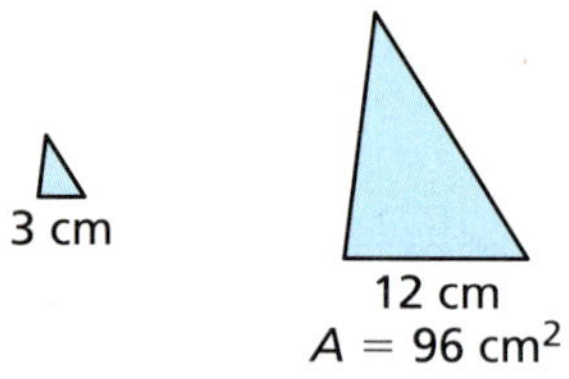

21. ERROR ANALYSIS Describe and correct the error in finding the perimeter of triangle B. The triangles are similar.

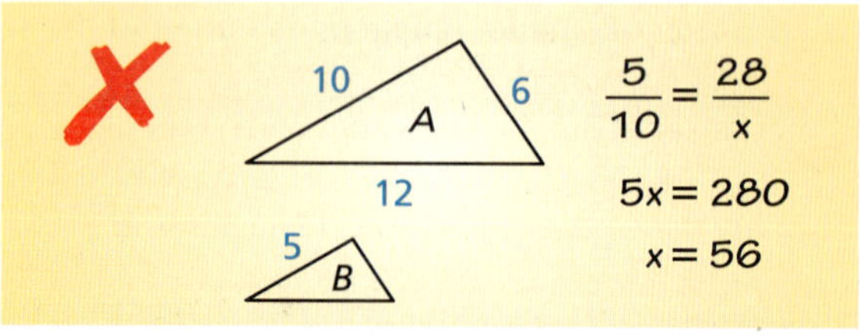

22. HELP A CLASSMATE Your classmate wants to find the area of rectangle B. The rectangles are similar. Explain how your classmate can complete the work.

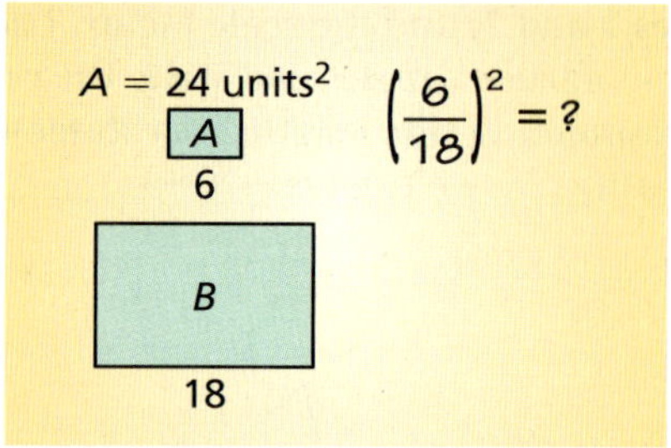

In Exercises 23 and 24, decide whether the red and blue polygons are similar. *(See Example 6.)*

23.

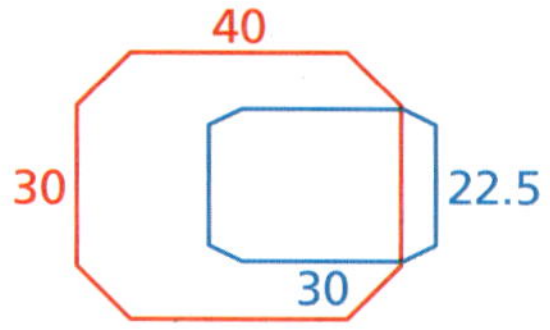

24.

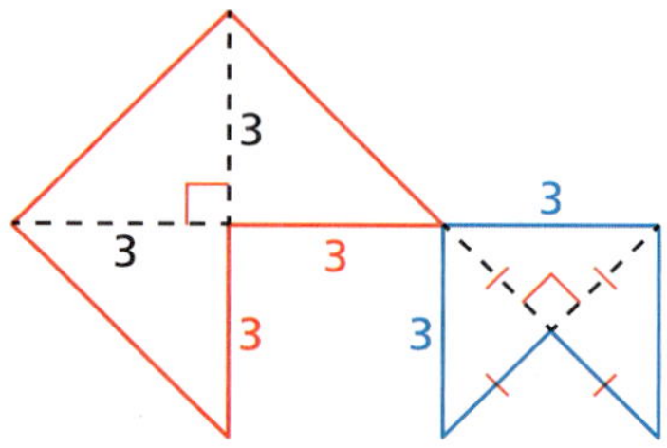

STRUCTURE **In Exercises 25–32, *JKLM* ~ *EFGH*.**

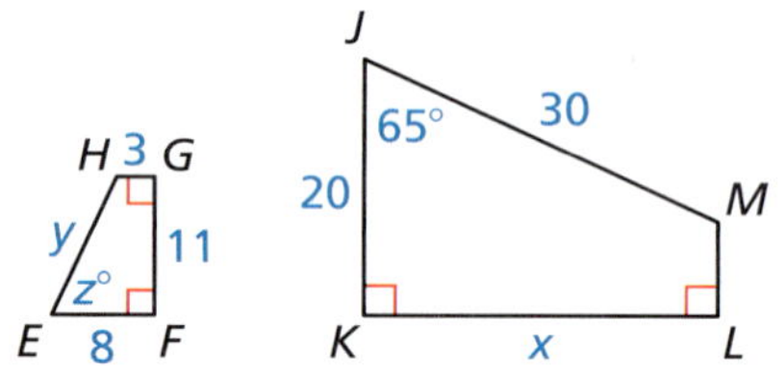

25. Find the scale factor of *JKLM* to *EFGH*.

26. List all pairs of congruent angles.

27. Write the ratios of the corresponding side lengths in a statement of proportionality.

28. Find the values of x, y, and z.

29. Find the perimeter of each polygon.

30. Find the ratio of the perimeters of *JKLM* to *EFGH*.

31. Find the area of each polygon.

32. Find the ratio of the areas of *JKLM* to *EFGH*.

GO DIGITAL

33. REASONING In table tennis, the table is a rectangle 9 feet long and 5 feet wide. A tennis court is a rectangle 78 feet long and 36 feet wide. Are the two surfaces similar? Explain. If so, find the scale factor of the tennis court to the table.

4 MTR **34. MAKING AN ARGUMENT** If the side lengths of two rectangles are proportional, must the two rectangles be similar? Explain your reasoning.

35. REASONING Triangles ABC and DEF are similar. Which statement is correct? Select all that apply.

Ⓐ $\frac{BC}{EF} = \frac{AC}{DF}$ Ⓒ $\frac{AB}{EF} = \frac{BC}{DE}$

Ⓑ $\frac{AB}{DE} = \frac{CA}{FE}$ Ⓓ $\frac{CA}{FD} = \frac{BC}{EF}$

5 MTR **36. STRUCTURE** Rectangle A is similar to rectangle B. Rectangle A has side lengths of 6 and 12. Rectangle B has a side length of 18. What are the possible values for the length of the other side of rectangle B? Select all that apply.

Ⓐ 6 Ⓑ 9 Ⓒ 24 Ⓓ 36

3 MTR **MAINTAIN ACCURACY In Exercises 37–40, the figures are similar. Find the missing corresponding side length.**

37. Figure A has a perimeter of 72 meters, and one of the side lengths is 18 meters. Figure B has a perimeter of 120 meters.

38. Figure A has a perimeter of 24 inches. Figure B has a perimeter of 36 inches, and one of the side lengths is 12 inches.

39. Figure A has an area of 48 square feet, and one of the side lengths is 6 feet. Figure B has an area of 75 square feet.

40. Figure A has an area of 18 square feet. Figure B has an area of 98 square feet, and one of the side lengths is 14 feet.

41. WRITING Use similarity transformations to explain the meaning of similarity for triangles.

GO DIGITAL

42. PROVING A THEOREM Prove the Perimeters of Similar Polygons Theorem for similar rectangles. Include a diagram in your proof.

4 MTR **DISCUSS MATHEMATICAL THINKING In Exercises 43–48, tell whether the polygons are *always, sometimes,* or *never* similar.**

43. two isosceles triangles

44. two isosceles trapezoids

45. two rhombuses

46. two squares

47. two regular polygons

48. a right triangle and an equilateral triangle

7 MTR **49. MODELING REAL LIFE** During a total solar eclipse, the moon passes directly between Earth and the Sun, blocking the Sun's rays. The distance DA between Earth and the Sun is 93,000,000 miles, the distance DE between Earth and the moon is 240,000 miles, and the radius AB of the Sun is 432,500 miles. Use the diagram and the given measurements to estimate the radius EC of the moon.

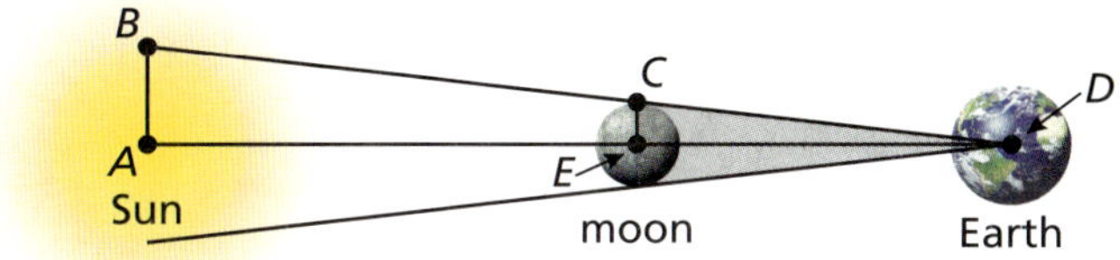

$\triangle DEC \sim \triangle DAB$ *Not drawn to scale*

50. HOW DO YOU SEE IT?
You shine a flashlight directly on an object to project its image onto a parallel screen. Will the object and the image be similar? Explain your reasoning.

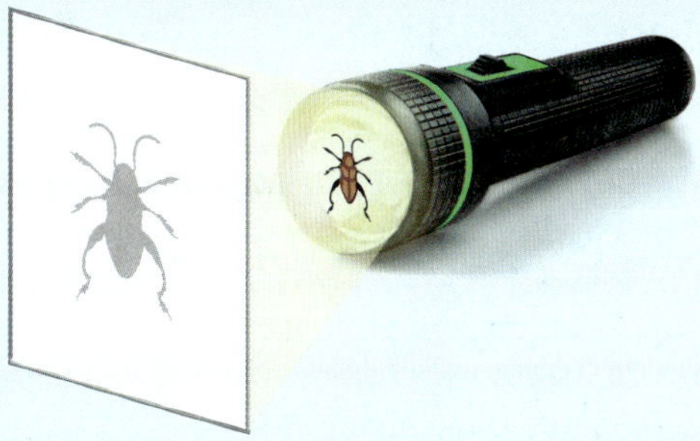

5 MTR **51. CONNECTING CONCEPTS** The equations of the lines shown are $y = \frac{4}{3}x + 4$ and $y = \frac{4}{3}x - 8$. Show that $\triangle AOB \sim \triangle COD$.

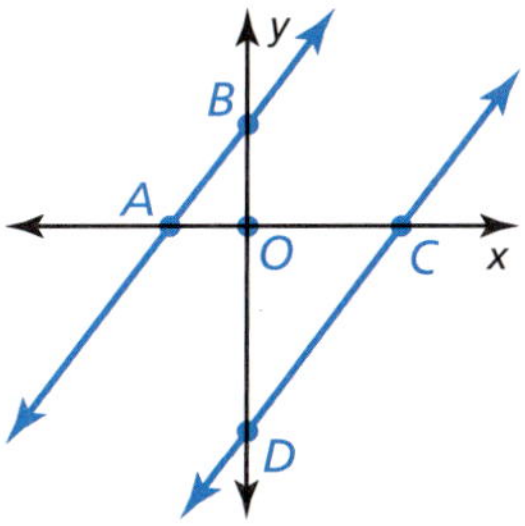

52. THOUGHT PROVOKING
The postulates and theorems in this book represent Euclidean geometry. In spherical geometry, all points are points on the surface of a sphere. A line is a circle on the sphere whose diameter is equal to the diameter of the sphere. A plane is the surface of the sphere. In spherical geometry, is it possible that two triangles are similar but not congruent? Explain your reasoning.

53. PROVING A THEOREM Prove the Areas of Similar Polygons Theorem for similar rectangles. Include a diagram in your proof.

54. DIG DEEPER In the diagram, *PQRS* is a square, and *PLMS* ~ *LMRQ*. Find the exact value of *x*. This value is called the *golden ratio*. Golden rectangles have their length and width in this ratio. Show that the similar rectangles in the diagram are golden rectangles.

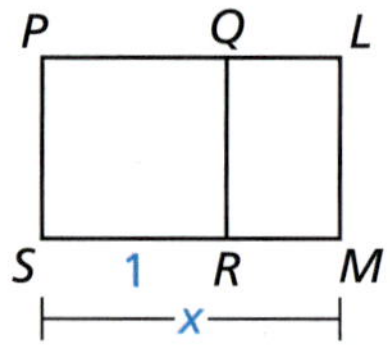

REVIEW & REFRESH

WATCH

In Exercises 55–57, find the value of *x*.

55.

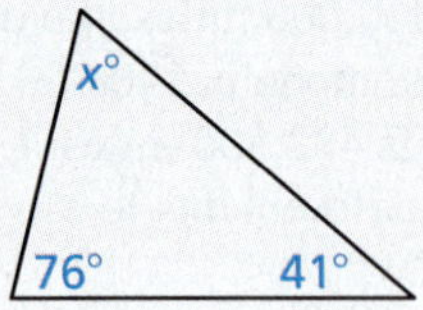

56.

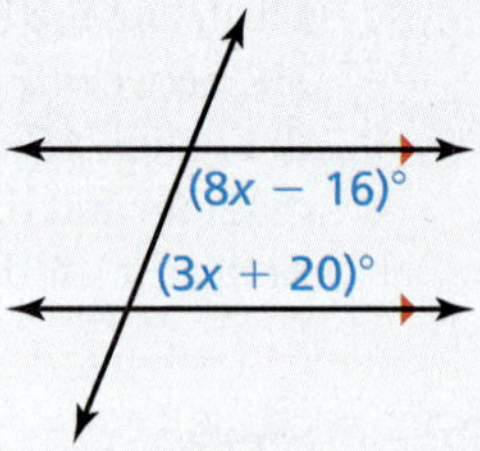

57.

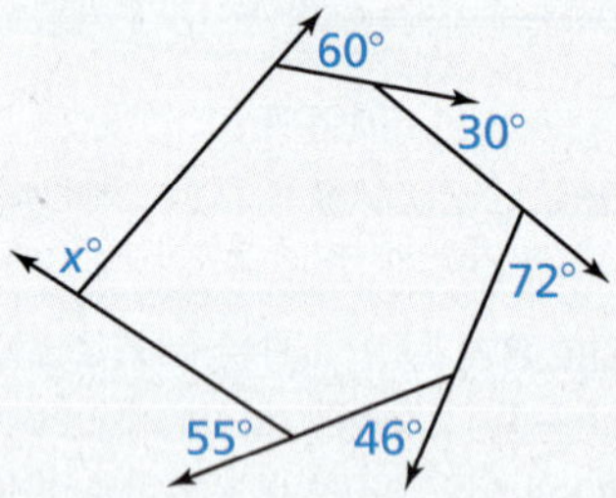

58. In rectangle *ABCD*, $AC = 6x - 15$ and $BD = -2x + 17$. Find the lengths of the diagonals of *ABCD*.

7 MTR **59. MODELING REAL LIFE** You are making a blueprint of a house. You measure the lengths of the walls of a room to be 11 feet by 12 feet. When you draw the room on the blueprint, the lengths of the walls are 8.25 inches by 9 inches. What scale factor dilates the room to the blueprint?

60. Find the values of *x* and *y* that make the quadrilateral a parallelogram.

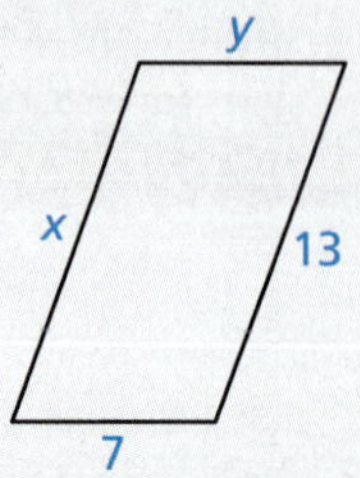

In Exercises 61–63, factor the polynomial.

61. $-3x^2 + 14x - 16$

62. $x^2 + 2x - 24$

63. $x^3 - 3x^2 - 5x + 15$

64. Trapezoid *QRST* has vertices $Q(-1, 4)$, $R(2, 4)$, $S(2, 2)$, and $T(-3, 2)$. Find the length of the midsegment.

65. The two triangles are similar. Find the value of *x*.

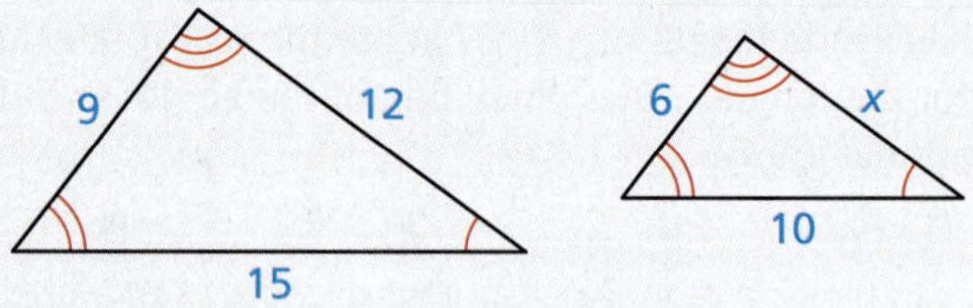

In Exercises 66 and 67, using *f*, graph *g*. Describe the transformation from the graph of *f* to the graph of *g*.

66. $f(x) = x^2 - 4$; $g(x) = f(x + 3)$

67. $f(x) = -x^2 + 3x$; $g(x) = f(x) - 2$

68. Tell whether the table of values represents a *linear*, an *exponential*, or a *quadratic* function.

x	−2	−1	0	1	2
y	2	3	4.5	6.75	10.125

In Exercises 69 and 70, solve the equation.

69. $|2x + 5| = 11$

70. $4x^2 - 16 = 8$

71. The incenter of $\triangle ABC$ is point *N*, $NZ = 4x - 10$, and $NY = 3x - 1$. Find *NW*.

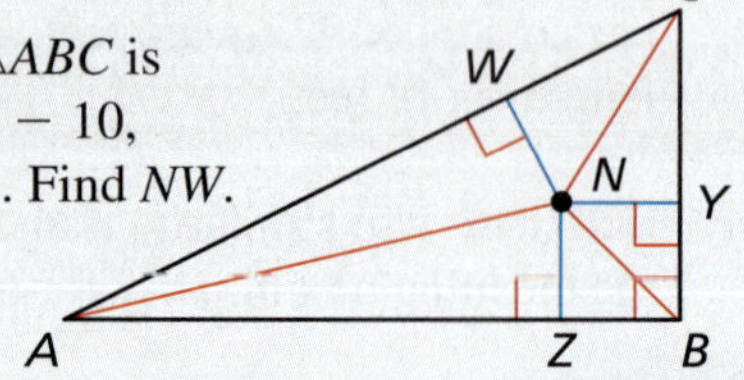

GO DIGITAL

8.2 Proving Triangle Similarity by AA

Learning Target: Understand and use the Angle-Angle Similarity Theorem.

Success Criteria: **H**
- I can use similarity transformations to prove the Angle-Angle Similarity Theorem.
- I can use angle measures of triangles to determine whether triangles are similar.
- I can prove triangle similarity using the Angle-Angle Similarity Theorem.
- I can solve real-life problems using similar triangles.

EXPLORE IT! Comparing Triangles

Work with a partner.

a. Construct any $\triangle ABC$. Find $m\angle A$ and $m\angle B$.

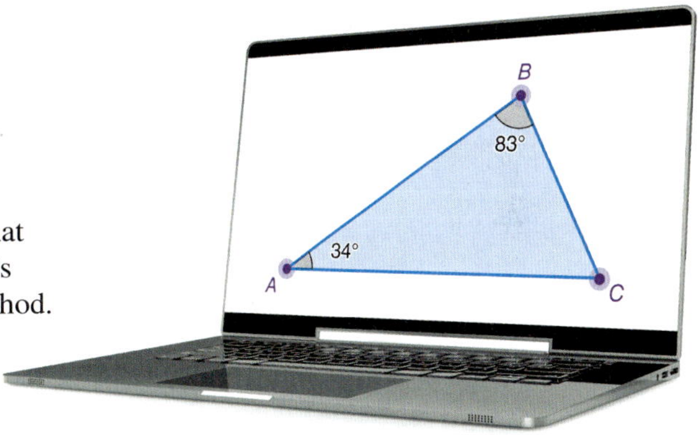

2 MTR USE ANOTHER METHOD
What are some other methods you can use to construct $\triangle DEF$ in part (b)?

b. Construct $\triangle DEF$ so that each of the following is true. Explain your method.

- $m\angle D = m\angle A$
- $m\angle E = m\angle B$
- $\triangle DEF$ is not congruent to $\triangle ABC$.

c. Are the two triangles similar? Explain your reasoning.

d. Repeat parts (a)–(c) for several different triangles. Describe your results. Can you construct two triangles in this way that are *not* similar?

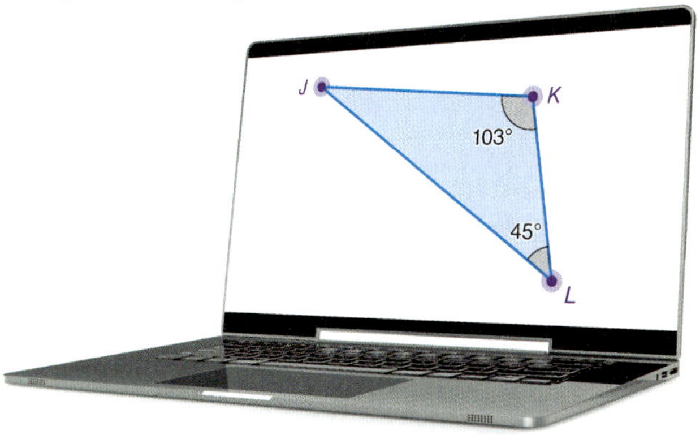

e. Make a conjecture about any two triangles with two pairs of congruent angles. Use transformations to support your answer.

Geometric Reasoning
MA.912.GR.1.2 Prove triangle congruence or similarity using Side-Side-Side, Side-Angle-Side, Angle-Side-Angle, Angle-Angle-Side, Angle-Angle and Hypotenuse-Leg.
MA.912.GR.1.6 Solve mathematical and real-world problems involving congruence or similarity in two-dimensional figures.
H Also MA.912.GR.2.9

GO DIGITAL

Using the Angle-Angle Similarity Theorem

THEOREM

8.3 Angle-Angle (AA) Similarity Theorem

If two angles of one triangle are congruent to two angles of another triangle, then the two triangles are similar.

If $\angle A \cong \angle D$ and $\angle B \cong \angle E$, then $\triangle ABC \sim \triangle DEF$.

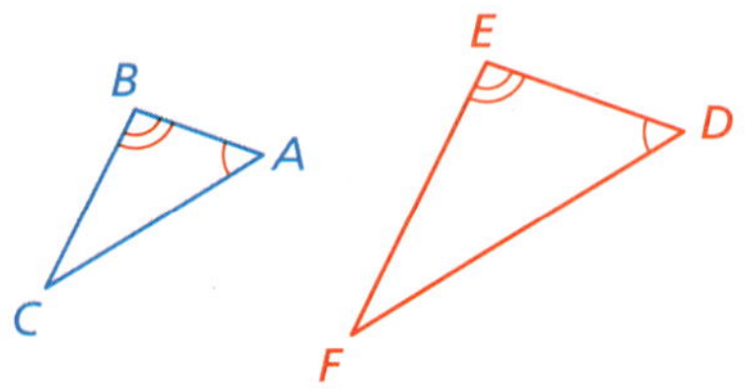

Prove this Theorem Exercise 3, page 421

PROOF Angle-Angle (AA) Similarity Theorem H

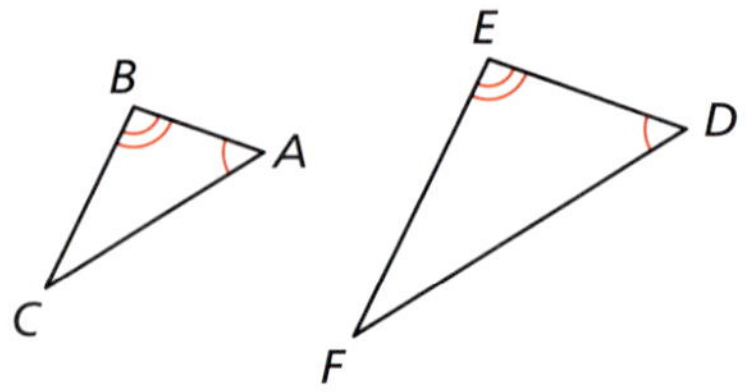

Given $\angle A \cong \angle D$, $\angle B \cong \angle E$

Prove $\triangle ABC \sim \triangle DEF$

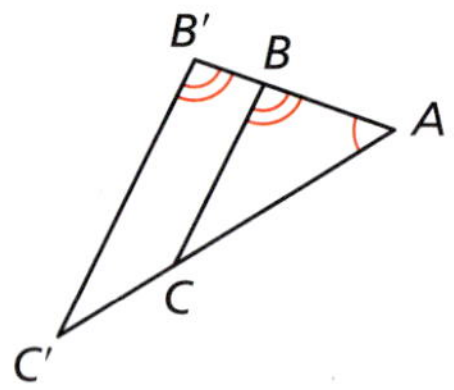

Dilate $\triangle ABC$ using a scale factor of $k = \dfrac{DE}{AB}$ and center A. The image of $\triangle ABC$ is $\triangle AB'C'$.

Because a dilation is a similarity transformation, $\triangle ABC \sim \triangle AB'C'$. Because the ratio of corresponding lengths of similar polygons equals the scale factor, $\dfrac{AB'}{AB} = \dfrac{DE}{AB}$. Multiplying each side by AB yields $AB' = DE$. By the definition of congruent segments, $\overline{AB'} \cong \overline{DE}$.

By the Reflexive Property of Angle Congruence, $\angle A \cong \angle A$. Because corresponding angles of similar polygons are congruent, $\angle B' \cong \angle B$. Because $\angle B' \cong \angle B$ and $\angle B \cong \angle E$, $\angle B' \cong \angle E$ by the Transitive Property of Angle Congruence.

Because $\angle A \cong \angle D$, $\angle B' \cong \angle E$, and $\overline{AB'} \cong \overline{DE}$, $\triangle AB'C' \cong \triangle DEF$ by the ASA Congruence Theorem. So, a composition of rigid motions maps $\triangle AB'C'$ to $\triangle DEF$.

Because a dilation followed by a composition of rigid motions maps $\triangle ABC$ to $\triangle DEF$, $\triangle ABC \sim \triangle DEF$.

READING

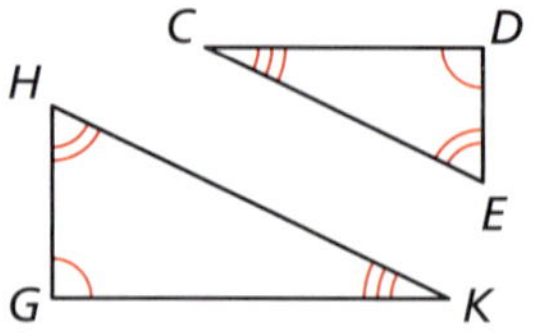

Use multiple arcs to show congruent angles. This will help you write similarity statements.

EXAMPLE 1 Using the AA Similarity Theorem

WATCH

Determine whether the triangles are similar. If they are, write a similarity statement. Explain your reasoning.

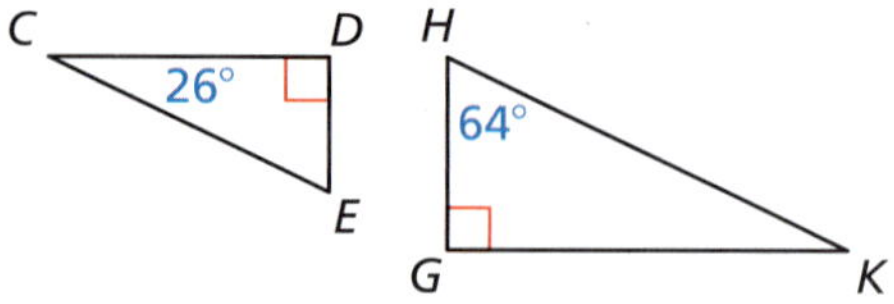

SOLUTION

Because they are both right angles, $\angle D$ and $\angle G$ are congruent.

By the Triangle Sum Theorem, $26° + 90° + m\angle E = 180°$, so $m\angle E = 64°$. So, $\angle E$ and $\angle H$ are congruent.

▶ So, $\triangle CDE \sim \triangle KGH$ by the AA Similarity Theorem.

GO DIGITAL

EXAMPLE 2 Using the AA Similarity Theorem

WATCH

Show that the two triangles are similar and write a similarity statement.

a. $\triangle ABE$ and $\triangle ACD$

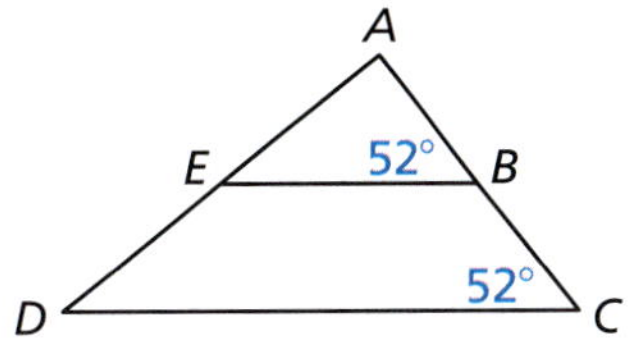

b. $\triangle SVR$ and $\triangle UVT$

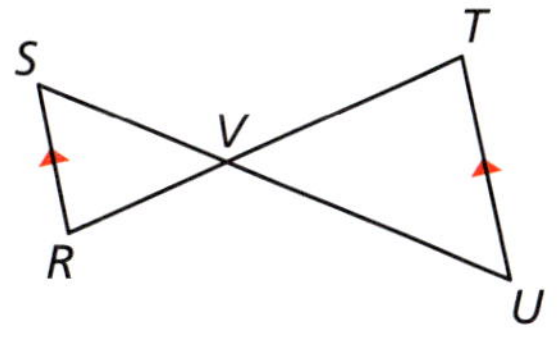

SOLUTION

a. Because $m\angle ABE$ and $m\angle C$ both equal 52°, $\angle ABE \cong \angle C$. By the Reflexive Property of Angle Congruence, $\angle A \cong \angle A$.

▶ So, $\triangle ABE \sim \triangle ACD$ by the AA Similarity Theorem.

b. You know $\angle SVR \cong \angle UVT$ by the Vertical Angles Congruence Theorem. The diagram shows $\overline{RS} \parallel \overline{UT}$, so $\angle S \cong \angle U$ by the Alternate Interior Angles Theorem.

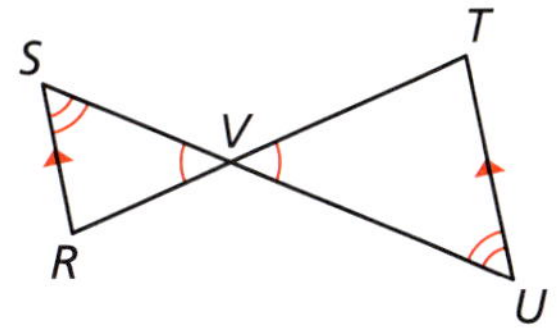

▶ So, $\triangle SVR \sim \triangle UVT$ by the AA Similarity Theorem.

5 MTR DECOMPOSE A PROBLEM

How can you redraw the triangles in Example 2 separately so that the corresponding parts are more easily recognizable?

SELF-ASSESSMENT

1 I don't understand yet. 2 I can do it with help. 3 I can do it on my own. 4 I can teach someone else.

1. WRITING Can you assume that corresponding sides and corresponding angles of any two similar triangles are congruent? Explain.

2. WRITING Explain why all equilateral triangles are similar.

3. Use transformations to prove that $\triangle PQR \sim \triangle STU$.

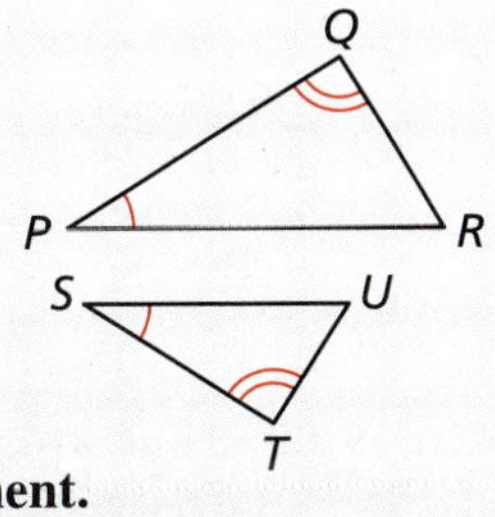

Show that the triangles are similar and write a similarity statement.

4. $\triangle FGH$ and $\triangle RQS$

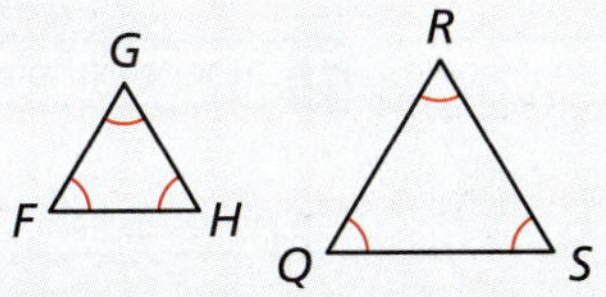

5. $\triangle CDF$ and $\triangle DEF$

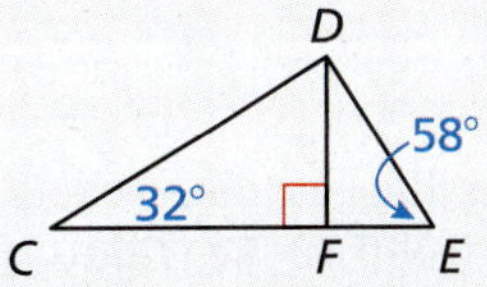

6. WHAT IF? In Example 2(b), $\overline{SR} \nparallel \overline{TU}$. Can the triangles still be similar? Explain.

GO DIGITAL

Solving Real-Life Problems

Previously, you learned a way to use congruent triangles to find measurements indirectly. Another useful way to find measurements indirectly is by using similar triangles.

7 MTR **EXAMPLE 3** **Modeling Real Life** WATCH

A free-fall ride casts a shadow that is 175 feet long. At the same time, a person standing nearby who is 5 feet 5 inches tall casts a shadow that is 34 inches long. How tall is the free-fall ride?

Falcon's Fury at Busch Gardens Tampa Bay is the tallest freestanding drop tower in North America.

SOLUTION

1. **Understand the Problem** You are given the length of the shadow of the ride, the height of the person, and the length of the shadow of the person. You need to find the height of the ride.

2. **Make a Plan** The Sun's rays hit the ride and the person at the same angle. There are two pairs of congruent angles, so the triangles are similar by the AA Similarity Theorem. Use similar triangles to write and solve a proportion.

3. **Solve and Check** The ride and the person form sides of two right triangles with their shadows on the ground.

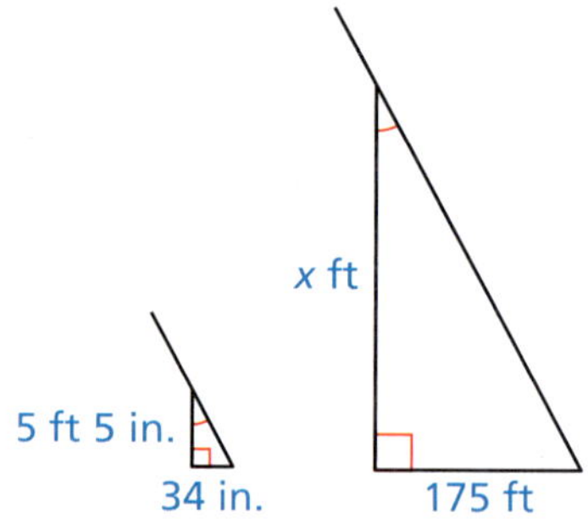

You can use a proportion to find the height x. Write 5 feet 5 inches as 65 inches so that you can form two ratios of feet to inches.

$$\frac{x \text{ ft}}{65 \text{ in.}} = \frac{175 \text{ ft}}{34 \text{ in.}}$$ Write proportion of side lengths.

$$x = \frac{11{,}375}{34}$$ Multiply each side by 65.

$$x \approx 334.56$$ Divide.

▶ So, the free-fall ride is about 335 feet tall.

Check Reasonableness Check that your answer has the correct units. The problem asks for the height of the ride. Because your answer is 335 feet, the units are appropriate.

SELF-ASSESSMENT 1 I don't understand yet. 2 I can do it with help. 3 I can do it on my own. 4 I can teach someone else.

7. **WHAT IF?** A child who is 52 inches tall is standing next to the person in Example 3. Show two different ways to find the length of the child's shadow.

8. You stand outside and measure the lengths of the shadows cast by both you and a tree. Write a proportion showing how you can find the height of the tree.

GO DIGITAL

8.2 Practice WITH CalcChat® AND CalcView®

In Exercises 1–4, determine whether the triangles are similar. If they are, write a similarity statement. Explain your reasoning. *(See Example 1.)*

1.
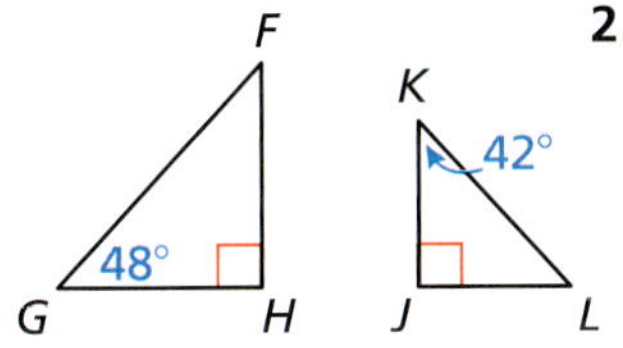

2.
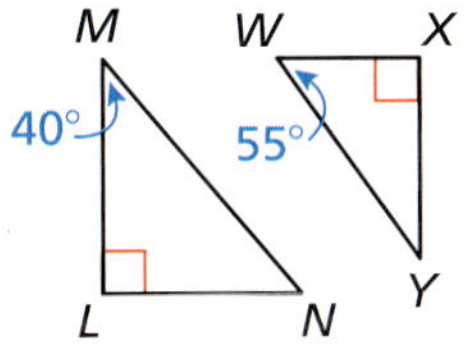

▶ 3.
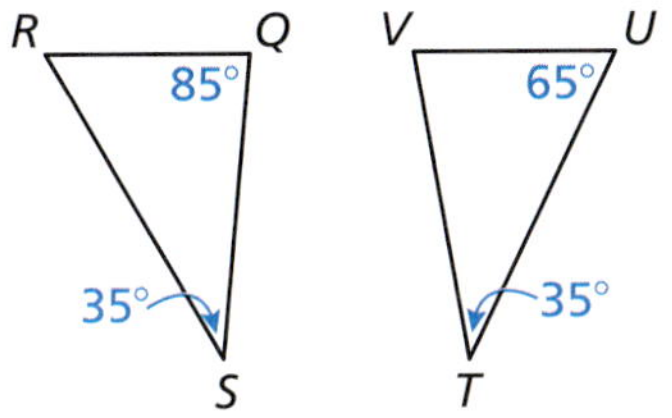

4.
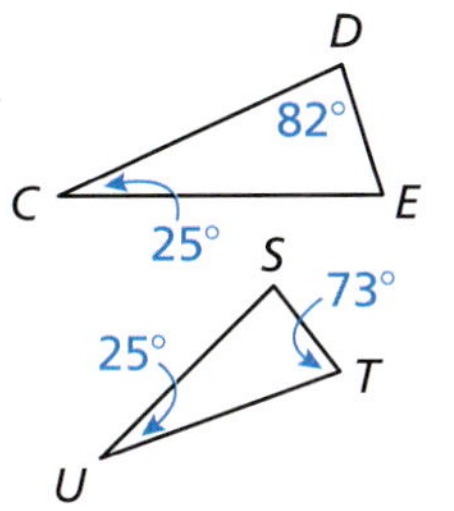

In Exercises 5–8, show that the two triangles are similar and write a similarity statement. *(See Example 2.)*

5.
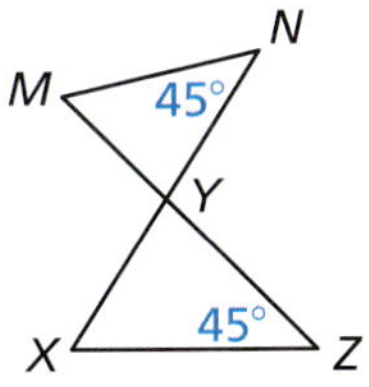

6.
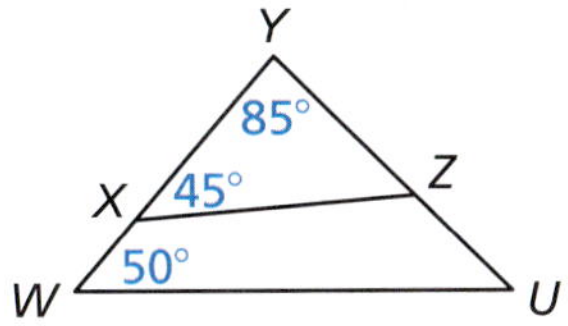

▶ 7.
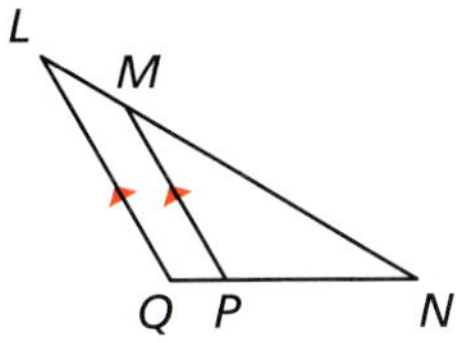

8.
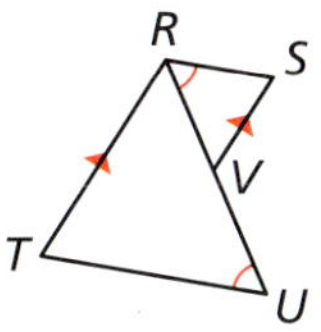

In Exercises 9–16, use the diagram to complete the statement.

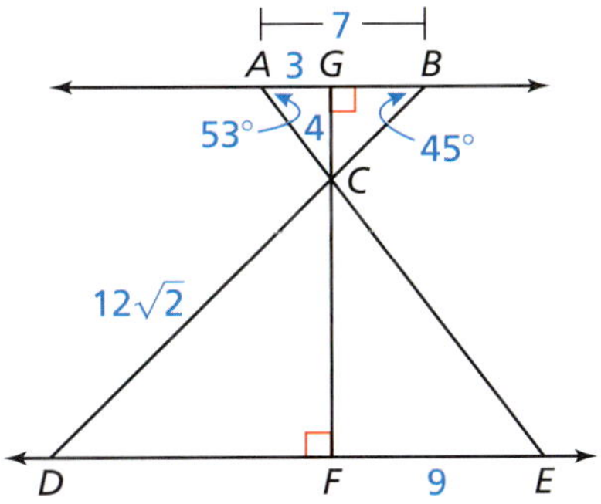

9. $\triangle CAG \sim$ ▭

10. $\triangle DCF \sim$ ▭

11. $\triangle ACB \sim$ ▭

12. $m\angle ECF =$ ▭

13. $m\angle ECD =$ ▭

14. $CF =$ ▭

15. $DE =$ ▭

16. $BC =$ ▭

GO DIGITAL

4 MTR 17. **ERROR ANALYSIS** Describe and correct the error in finding the value of x.

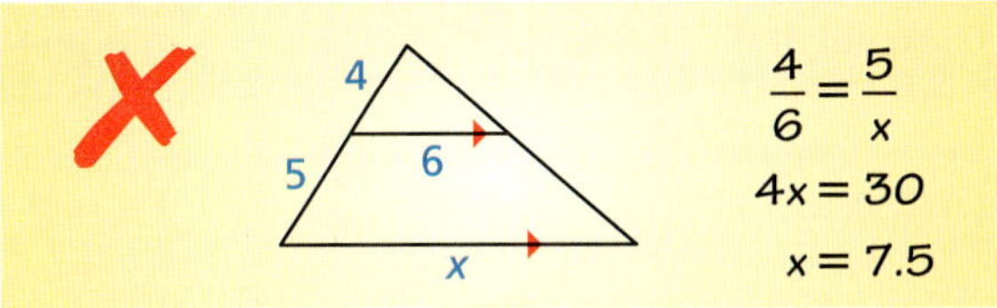

18. **COLLEGE PREP** In one triangle, $m\angle A = 31°$ and $m\angle B = 97°$. In a second triangle, $m\angle D = (x + 6)°$ and $m\angle E = (y - 4)°$. For which of the following values of x and y are the two triangles similar? Select all that apply.

Ⓐ $x = 25, y = 56$

Ⓑ $x = 31, y = 97$

Ⓒ $x = 46, y = 101$

Ⓓ $x = 91, y = 35$

7 MTR ▶ 19. **MODELING REAL LIFE** A *pile* is a post-like support made of timber, steel, or concrete used in foundations that are driven deep into the ground to support structures such as piers. While the tide is out, a pile at Old Pier in Naples, Florida, casts a shadow that is 31 inches long. At the same time, a person standing on the shore who is 6 feet 4 inches tall casts a shadow that is 45 inches long. How much of the pile is above the water? *(See Example 3.)*

7 MTR 20. **MODELING REAL LIFE** You can measure the width of the lake using a surveying technique, as shown in the diagram. Find the width of the lake, WX. Justify your answer.

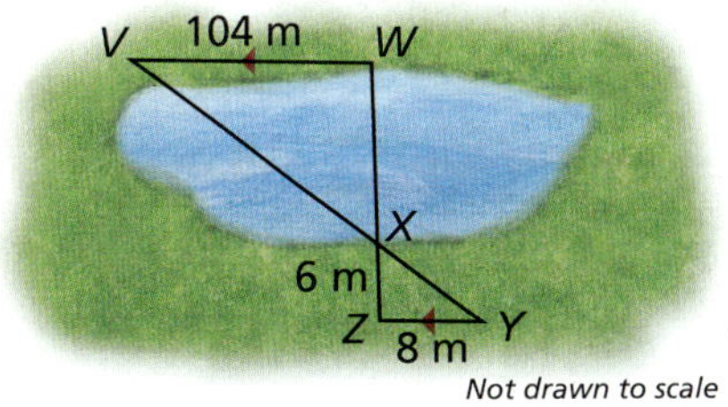

Not drawn to scale

21. **MAKING AN ARGUMENT** To determine the height of the Monticello Opera House, your friend tells you to stand in the building's shadow so that the tip of your shadow coincides with the tip of the building's shadow. Your friend claims to be able to use the distance between the tips of the shadows and you, the distance between you and the Opera House, and your height to estimate the height of the Monticello Opera House. Is this possible? Explain. Include a diagram in your answer.

22. **HOW DO YOU SEE IT?**
In the diagram, which triangles can you use to find the distance x between the shoreline and the buoy, L? Explain your reasoning.

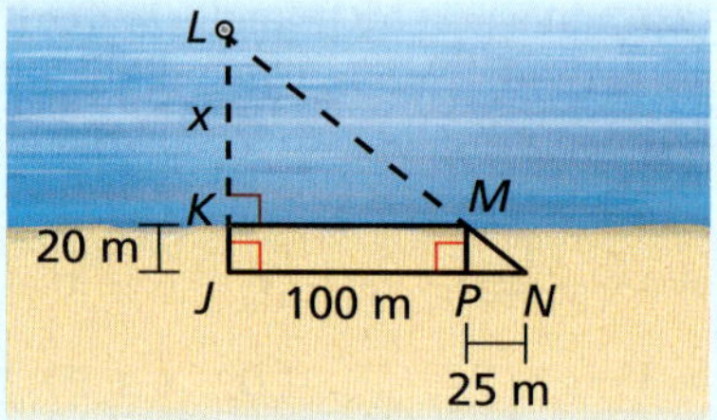

REASONING In Exercises 23–26, is it possible for $\triangle JKL$ and $\triangle XYZ$ to be similar? Explain your reasoning.

23. $m\angle J = 71°$, $m\angle K = 52°$, $m\angle X = 71°$, and $m\angle Z = 57°$

24. $\triangle JKL$ is a right triangle and $m\angle X + m\angle Y = 150°$.

25. $m\angle L = 87°$ and $m\angle Y = 94°$

26. $m\angle J + m\angle K = 85°$ and $m\angle Y + m\angle Z = 80°$

27. **CONNECTING CONCEPTS** Explain how you can use similar triangles to show that any two points on a line can be used to find its slope.

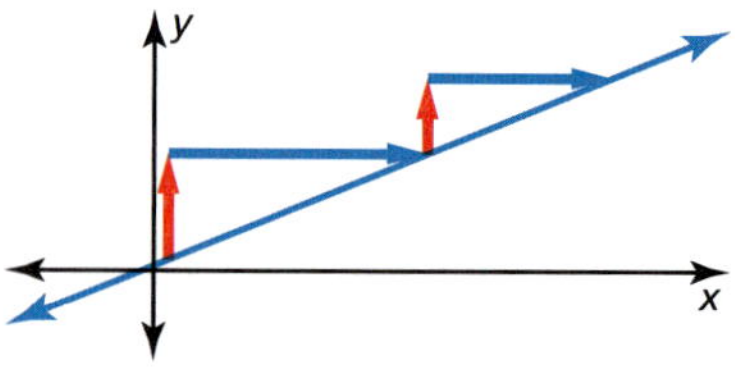

28. **THOUGHT PROVOKING**
Decide whether (a) AAA and (b) AAAA are valid methods of showing that two quadrilaterals are similar. Justify your answer.

29. **PROOF** Prove that if the lengths of two sides of a triangle are a and b, respectively, then the lengths of the corresponding altitudes to those sides are in the ratio $\frac{b}{a}$.

30. **DIG DEEPER** A portion of an amusement park ride is shown. Find EF. Justify your answer.

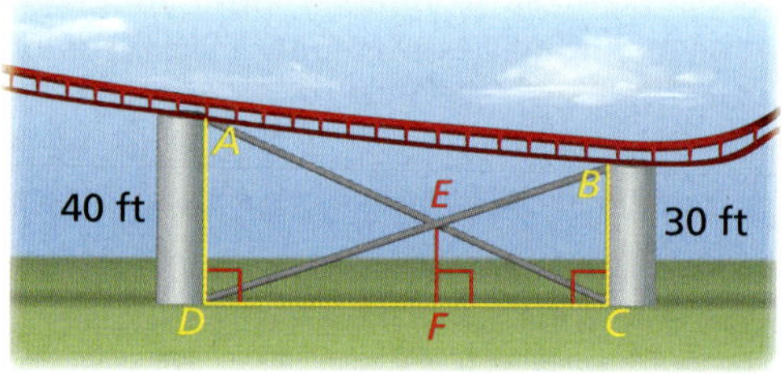

REVIEW & REFRESH

WATCH

31. Decide whether enough information is given to prove that the triangles are congruent. If so, state the theorem you can use.

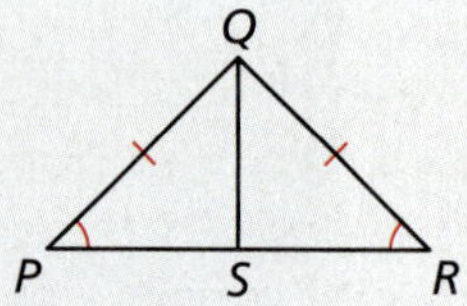

32. Determine whether the triangles are similar. If they are, write a similarity statement. Explain your reasoning.

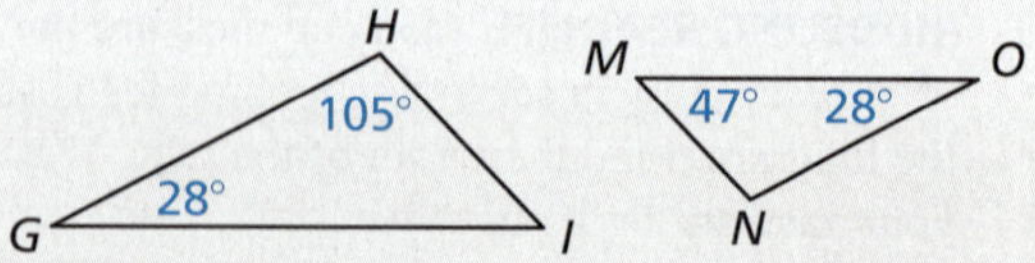

33. **MODELING REAL LIFE** A stop sign is shaped like a regular octagon. Find the measure of each (a) interior angle and (b) exterior angle.

34. Find $m\angle S$.

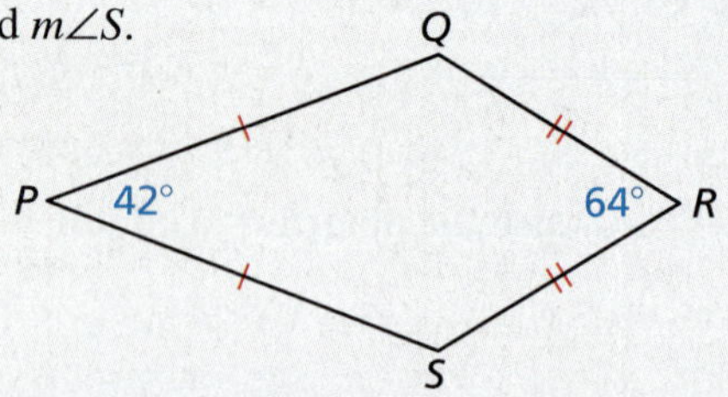

35. In the diagram, $ABCD \sim EFGH$. The area of $ABCD$ is 288 square meters. Find the area of $EFGH$.

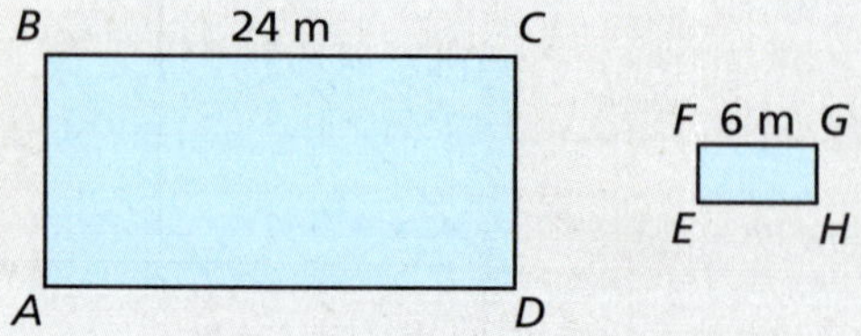

36. Decide whether $\square WXYZ$ with vertices $W(-4, 2)$, $X(-1, 5)$, $Y(1, 3)$, and $Z(-2, 0)$ is a *rectangle*, a *rhombus*, or a *square*. Give all names that apply.

GO DIGITAL

8.3 Proving Triangle Similarity by SSS and SAS

Learning Target: Understand and use additional triangle similarity theorems.

Success Criteria:
- I can use the SSS and SAS Similarity Theorems to determine whether triangles are similar.
- I can use similar triangles to prove theorems about slopes of parallel and perpendicular lines.

EXPLORE IT! Deciding whether Triangles Are Similar

4 MTR JUSTIFY A RESULT
How can you use transformations to justify your conclusion in part (b)?

Work with a partner. Use technology.

a. Construct any $\triangle ABC$. Find AB, AC, and BC.

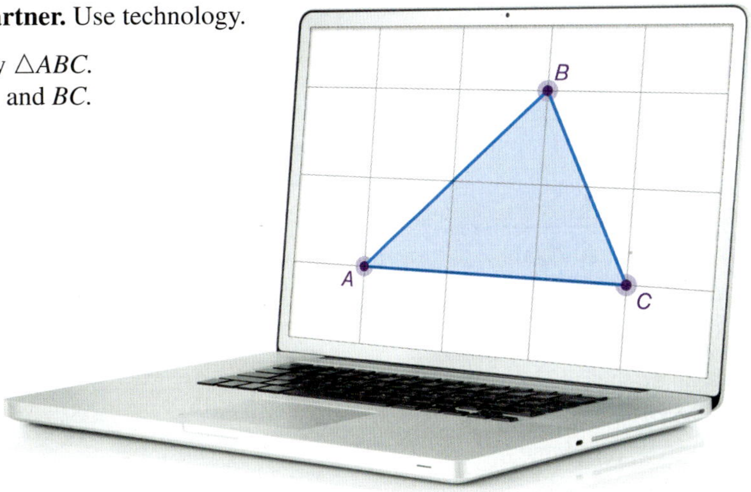

b. Choose any positive rational number k and construct $\triangle DEF$ so that $DE = k \cdot AB$, $DF = k \cdot AC$, and $EF = k \cdot BC$. Is $\triangle DEF$ similar to $\triangle ABC$? Explain your reasoning.

c. Repeat parts (a) and (b) several times by changing $\triangle ABC$ and k. Make a conjecture about the similarity of two triangles based on their corresponding side lengths.

d. If an angle of one triangle is congruent to an angle of a second triangle and the lengths of the sides including these angles are proportional, are the triangles similar? Provide examples to support your reasoning.

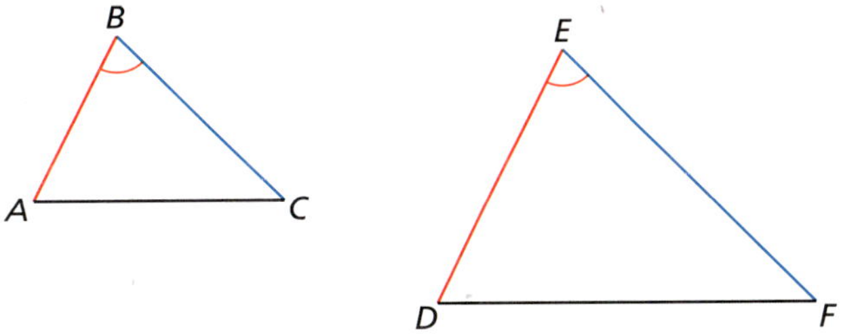

Geometric Reasoning

MA.912.GR.1.2 Prove triangle congruence or similarity using Side-Side-Side, Side-Angle-Side, Angle-Side-Angle, Angle-Angle-Side, Angle-Angle and Hypotenuse-Leg.

MA.912.GR.1.6 Solve mathematical and real-world problems involving congruence or similarity in two-dimensional figures.

GO DIGITAL

Using the Side-Side-Side Similarity Theorem

In addition to using congruent corresponding angles to show that two triangles are similar, you can use proportional corresponding side lengths.

THEOREM

8.4 Side-Side-Side (SSS) Similarity Theorem

If the corresponding side lengths of two triangles are proportional, then the triangles are similar.

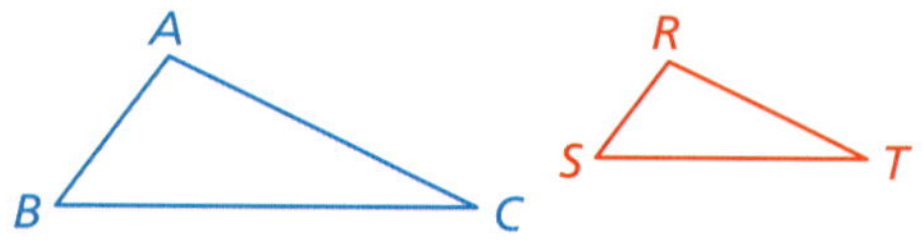

If $\frac{AB}{RS} = \frac{BC}{ST} = \frac{CA}{TR}$, then $\triangle ABC \sim \triangle RST$.

Proof page 427
Prove this Theorem Exercise 32, page 434

EXAMPLE 1 Using the SSS Similarity Theorem WATCH

Is either $\triangle DEF$ or $\triangle GHJ$ similar to $\triangle ABC$? If so, write a similarity statement.

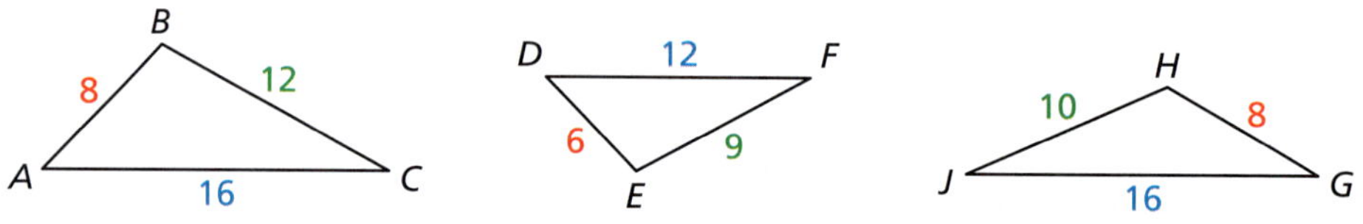

5 MTR MAKE A PLAN
When using the SSS Similarity Theorem, which pairs of side lengths should you compare?

SOLUTION

Compare $\triangle ABC$ and $\triangle DEF$ by finding ratios of corresponding side lengths.

Shortest sides	Longest sides	Remaining sides
$\frac{AB}{DE} = \frac{8}{6} = \frac{4}{3}$	$\frac{CA}{FD} = \frac{16}{12} = \frac{4}{3}$	$\frac{BC}{EF} = \frac{12}{9} = \frac{4}{3}$

▶ All the ratios are equal, so $\triangle ABC \sim \triangle DEF$.

Compare $\triangle ABC$ and $\triangle GHJ$ by finding ratios of corresponding side lengths.

Shortest sides	Longest sides	Remaining sides
$\frac{AB}{GH} = \frac{8}{8} = 1$	$\frac{CA}{JG} = \frac{16}{16} = 1$	$\frac{BC}{HJ} = \frac{12}{10} = \frac{6}{5}$

▶ The ratios are not all equal, so $\triangle ABC$ and $\triangle GHJ$ are not similar.

SELF-ASSESSMENT 1 I don't understand yet. 2 I can do it with help. 3 I can do it on my own. 4 I can teach someone else.

1. Which of the three triangles are similar? Write a similarity statement.

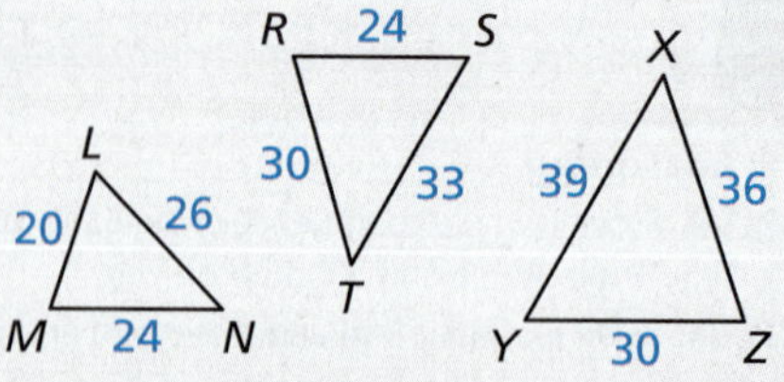

GO DIGITAL

4 MTR JUSTIFY A RESULT

What theorem or postulate allows you to draw an auxiliary line $\overleftrightarrow{PQ}$ in $\triangle RST$ that is parallel to $\overleftrightarrow{RT}$?

PROOF SSS Similarity Theorem

Given $\dfrac{RS}{JK} = \dfrac{ST}{KL} = \dfrac{TR}{LJ}$

Prove $\triangle RST \sim \triangle JKL$

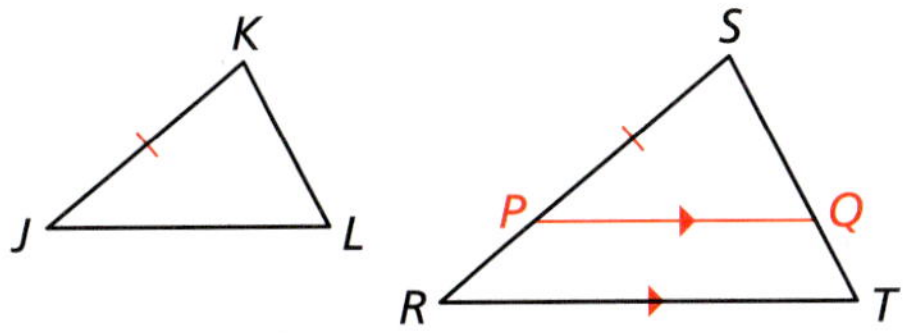

Locate P on $\overline{RS}$ so that $PS = JK$. Draw $\overline{PQ}$ so that $\overline{PQ} \parallel \overline{RT}$. Then $\triangle RST \sim \triangle PSQ$ by the AA Similarity Theorem, and $\dfrac{RS}{PS} = \dfrac{ST}{SQ} = \dfrac{TR}{QP}$. You can use the given proportion and the fact that $PS = JK$ to deduce that $SQ = KL$ and $QP = LJ$. By the SSS Congruence Theorem, it follows that $\triangle PSQ \cong \triangle JKL$. Finally, use the definition of congruent triangles and the AA Similarity Theorem to conclude that $\triangle RST \sim \triangle JKL$.

EXAMPLE 2 Using the SSS Similarity Theorem

WATCH

Find the value of x that makes $\triangle ABC \sim \triangle DEF$.

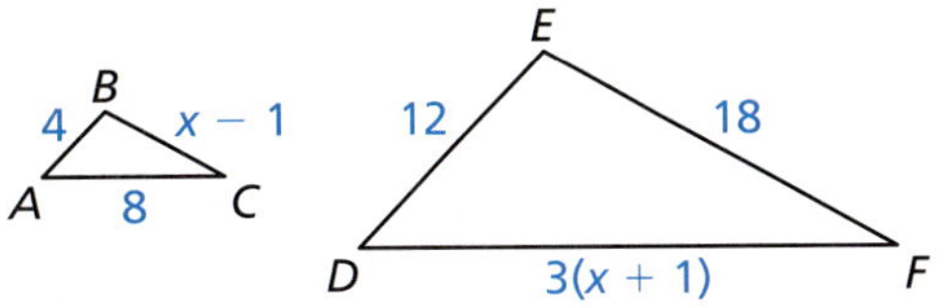

SOLUTION

Step 1 Find the value of x that makes corresponding side lengths proportional.

$\dfrac{AB}{DE} = \dfrac{BC}{EF}$ Write proportion.

$\dfrac{4}{12} = \dfrac{x - 1}{18}$ Substitute.

$6 = x - 1$ Multiply each side by 18.

$7 = x$ Add 1 to each side.

ANOTHER WAY

You can also use $\dfrac{AB}{DE} = \dfrac{AC}{DF}$ in Step 1.

Step 2 Check that the side lengths are proportional when $x = 7$.

$BC = x - 1 = 6$ $\qquad$ $DF = 3(x + 1) = 24$

$\dfrac{AB}{DE} \stackrel{?}{=} \dfrac{BC}{EF} \Rightarrow \dfrac{4}{12} = \dfrac{6}{18}$ ✓ $\qquad$ $\dfrac{AB}{DE} \stackrel{?}{=} \dfrac{AC}{DF} \Rightarrow \dfrac{4}{12} = \dfrac{8}{24}$ ✓

▶ When $x = 7$, the triangles are similar by the SSS Similarity Theorem.

SELF-ASSESSMENT

1 I don't understand yet. 2 I can do it with help. 3 I can do it on my own. 4 I can teach someone else.

2. Find the value of x that makes $\triangle ABC \sim \triangle DEF$.

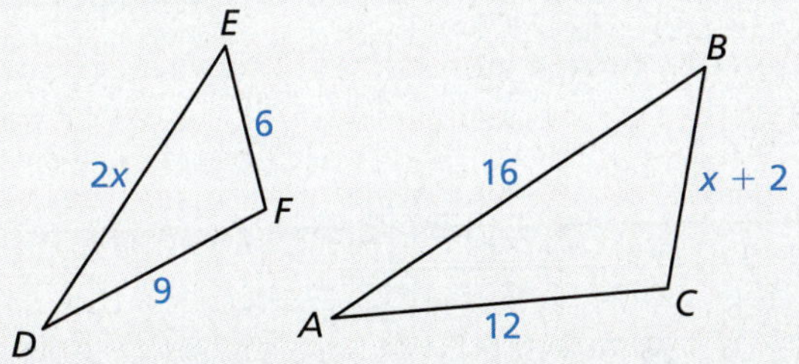

GO DIGITAL

Using the Side-Angle-Side Similarity Theorem

THEOREM

8.5 Side-Angle-Side (SAS) Similarity Theorem

If an angle of one triangle is congruent to an angle of a second triangle and the lengths of the sides including these angles are proportional, then the triangles are similar.

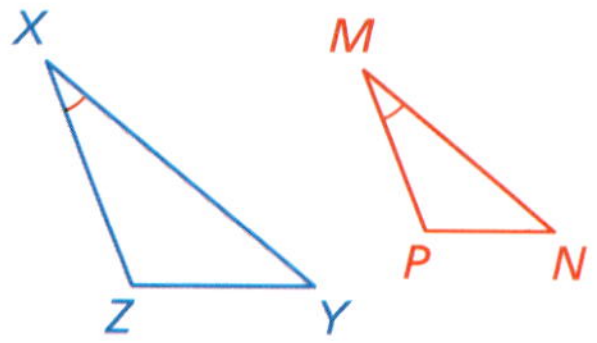

If $\angle X \cong \angle M$ and $\frac{ZX}{PM} = \frac{XY}{MN}$, then $\triangle XYZ \sim \triangle MNP$.

Prove this Theorem Exercise 22, page 432; Exercise 32, page 434

7 MTR **EXAMPLE 3** Modeling Real Life WATCH

You are building a lean-to shelter starting from a tree branch, as shown. Can you construct the right end so it is similar to the left end using the angle measure and lengths shown?

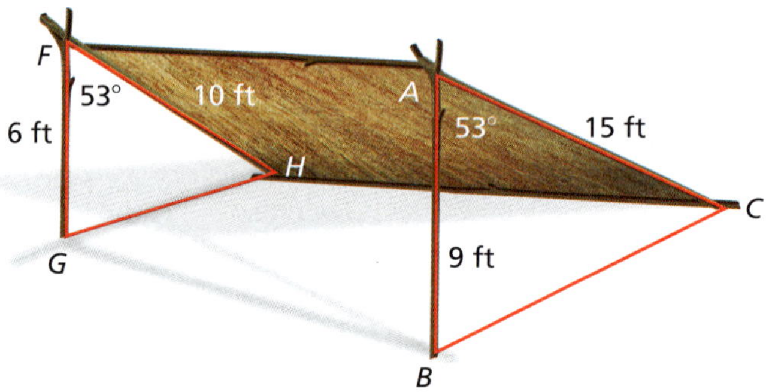

SOLUTION

Both $m\angle A$ and $m\angle F$ equal 53°, so $\angle A \cong \angle F$. Next, compare the ratios of the lengths of the sides that include $\angle A$ and $\angle F$.

Shorter sides

$$\frac{AB}{FG} = \frac{9}{6} = \frac{3}{2}$$

Longer sides

$$\frac{AC}{FH} = \frac{15}{10} = \frac{3}{2}$$

The lengths of the sides that include $\angle A$ and $\angle F$ are proportional. So, by the SAS Similarity Theorem, $\triangle ABC \sim \triangle FGH$.

▶ Yes, you can make the right end similar to the left end of the shelter using the given angle measure and lengths.

SELF-ASSESSMENT 1 I don't understand yet. 2 I can do it with help. 3 I can do it on my own. 4 I can teach someone else.

Explain how to show that the indicated triangles are similar.

3. $\triangle SRT \sim \triangle PNQ$

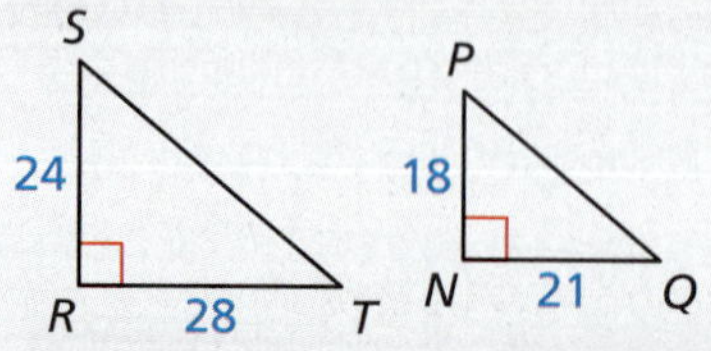

4. $\triangle XZW \sim \triangle YZX$

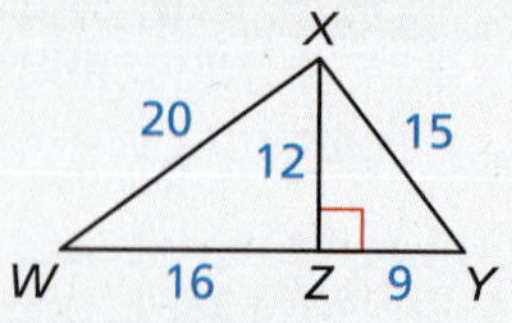

GO DIGITAL

CONCEPT SUMMARY

Triangle Similarity Theorems

AA Similarity Theorem

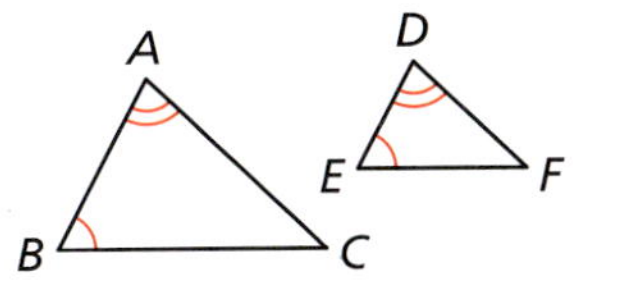

If $\angle A \cong \angle D$ and $\angle B \cong \angle E$, then $\triangle ABC \sim \triangle DEF$.

SSS Similarity Theorem

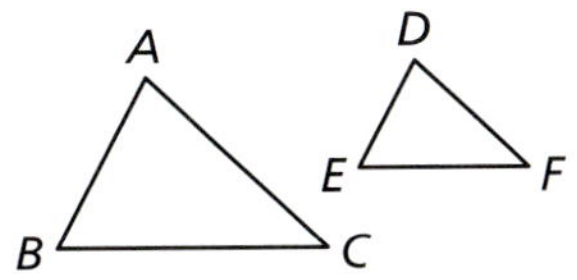

If $\frac{AB}{DE} = \frac{BC}{EF} = \frac{AC}{DF}$, then $\triangle ABC \sim \triangle DEF$.

SAS Similarity Theorem

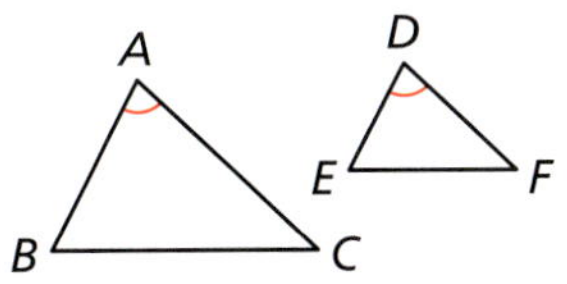

If $\angle A \cong \angle D$ and $\frac{AB}{DE} = \frac{AC}{DF}$, then $\triangle ABC \sim \triangle DEF$.

Proving Slope Criteria Using Similar Triangles

You can use similar triangles to prove the Slopes of Parallel Lines Theorem. Because the theorem is biconditional, you must prove both parts.

1. If two nonvertical lines are parallel, then they have the same slope.
2. If two nonvertical lines have the same slope, then they are parallel.

The first part is proved below. The second part is proved in the exercises.

PROOF **Part of Slopes of Parallel Lines Theorem**

Given $\ell \parallel n$, ℓ and n are nonvertical.

Prove $m_\ell = m_n$

First, consider the case where ℓ and n are horizontal. Because all horizontal lines are parallel and have a slope of 0, the statement is true for horizontal lines.

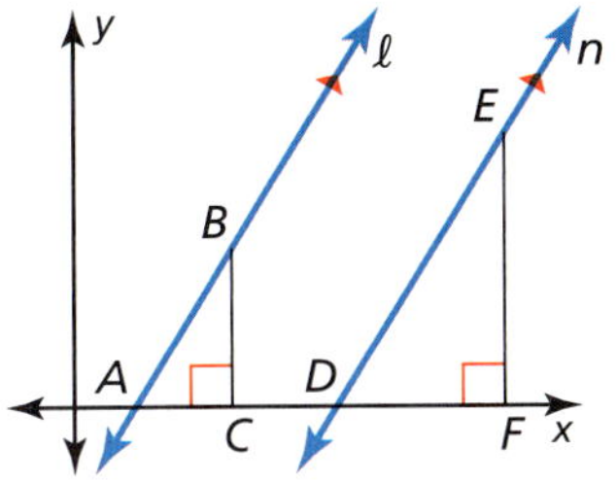

For the case of nonhorizontal, nonvertical lines, draw two such parallel lines, ℓ and n, and label their x-intercepts A and D, respectively. Draw a vertical segment $\overline{BC}$ parallel to the y-axis from point B on line ℓ to point C on the x-axis. Draw a vertical segment $\overline{EF}$ parallel to the y-axis from point E on line n to point F on the x-axis. Because vertical and horizontal lines are perpendicular, $\angle BCA$ and $\angle EFD$ are right angles.

STATEMENTS	REASONS
1. $\ell \parallel n$	**1.** Given
2. $\angle BAC \cong \angle EDF$	**2.** Corresponding Angles Theorem
3. $\angle BCA \cong \angle EFD$	**3.** Right Angles Congruence Theorem
4. $\triangle ABC \sim \triangle DEF$	**4.** AA Similarity Theorem
5. $\frac{BC}{EF} = \frac{AC}{DF}$	**5.** Corresponding sides of similar figures are proportional.
6. $\frac{BC}{AC} = \frac{EF}{DF}$	**6.** Rewrite proportion.
7. $m_\ell = \frac{BC}{AC}$, $m_n = \frac{EF}{DF}$	**7.** Definition of slope
8. $m_n = \frac{BC}{AC}$	**8.** Substitution Property of Equality
9. $m_\ell = m_n$	**9.** Transitive Property of Equality

GO DIGITAL

To prove the Slopes of Perpendicular Lines Theorem, you must prove both parts.

1. If two nonvertical lines are perpendicular, then the product of their slopes is -1.
2. If the product of the slopes of two nonvertical lines is -1, then the lines are perpendicular.

The first part is proved below. The second part is proved in the exercises.

PROOF Part of Slopes of Perpendicular Lines Theorem

Given $\ell \perp n$, ℓ and n are nonvertical.

Prove $m_\ell m_n = -1$

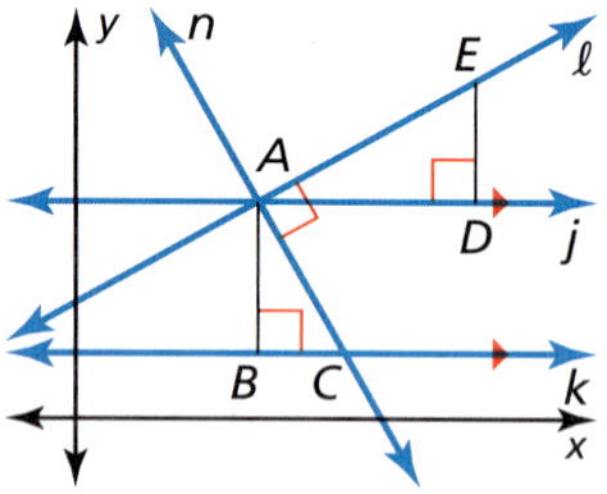

Draw two nonvertical, perpendicular lines, ℓ and n, that intersect at point A. Draw a horizontal line j parallel to the x-axis through point A. Draw a horizontal line k parallel to the x-axis through point C on line n. Because horizontal lines are parallel, $j \parallel k$. Draw a vertical segment $\overline{AB}$ parallel to the y-axis from point A to point B on line k. Draw a vertical segment $\overline{ED}$ parallel to the y-axis from point E on line ℓ to point D on line j. Because horizontal and vertical lines are perpendicular, $\angle ABC$ and $\angle ADE$ are right angles.

STUDY TIP

When a proof has a large number of steps, a two-column proof is often an appropriate proof format. A two-column proof clearly shows the justifications for each statement and how one statement leads to another.

STATEMENTS	REASONS
1. $\ell \perp n$	1. Given
2. $m\angle CAE = 90^\circ$	2. $\ell \perp n$
3. $m\angle CAE = m\angle DAE + m\angle CAD$	3. Angle Addition Postulate
4. $m\angle DAE + m\angle CAD = 90^\circ$	4. Transitive Property of Equality
5. $\angle BCA \cong \angle CAD$	5. Alternate Interior Angles Theorem
6. $m\angle BCA = m\angle CAD$	6. Definition of congruent angles
7. $m\angle DAE + m\angle BCA = 90^\circ$	7. Substitution Property of Equality
8. $m\angle DAE = 90^\circ - m\angle BCA$	8. Solve statement 7 for $m\angle DAE$.
9. $m\angle BCA + m\angle BAC + 90^\circ = 180^\circ$	9. Triangle Sum Theorem
10. $m\angle BAC = 90^\circ - m\angle BCA$	10. Solve statement 9 for $m\angle BAC$.
11. $m\angle DAE = m\angle BAC$	11. Transitive Property of Equality
12. $\angle DAE \cong \angle BAC$	12. Definition of congruent angles
13. $\angle ABC \cong \angle ADE$	13. Right Angles Congruence Theorem
14. $\triangle ABC \sim \triangle ADE$	14. AA Similarity Theorem
15. $\frac{AD}{AB} = \frac{DE}{BC}$	15. Corresponding sides of similar figures are proportional.
16. $\frac{AD}{DE} = \frac{AB}{BC}$	16. Rewrite proportion.
17. $m_\ell = \frac{DE}{AD}, m_n = -\frac{AB}{BC}$	17. Definition of slope
18. $m_\ell m_n = \frac{DE}{AD} \cdot \left(-\frac{AB}{BC}\right)$	18. Multiplication Property of Equality
19. $m_\ell m_n = \frac{DE}{AD} \cdot \left(-\frac{AD}{DE}\right)$	19. Substitution Property of Equality
20. $m_\ell m_n = -1$	20. Simplify.

GO DIGITAL

8.3 Practice WITH CalcChat® AND CalcView®

In Exercises 1 and 2, determine whether $\triangle JKL$ or $\triangle RST$ is similar to $\triangle ABC$. *(See Example 1.)*

1.

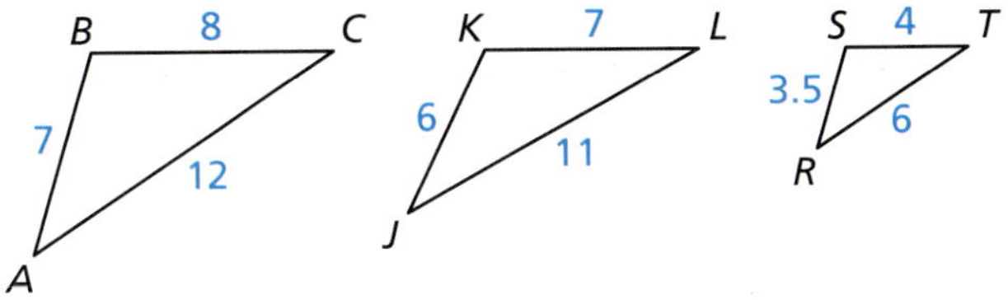

2.

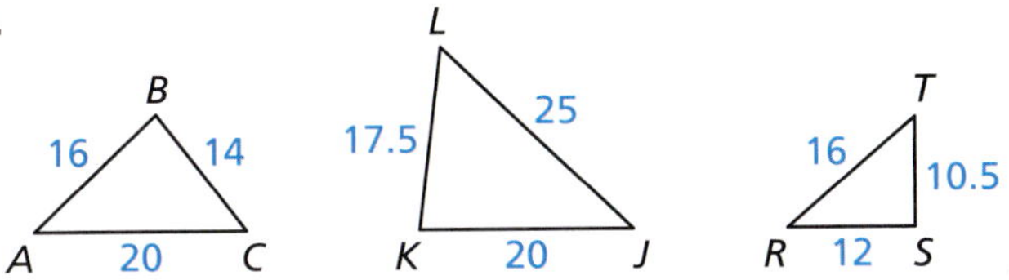

In Exercises 3 and 4, find the value of *x* that makes $\triangle DEF \sim \triangle XYZ$. *(See Example 2.)*

3.

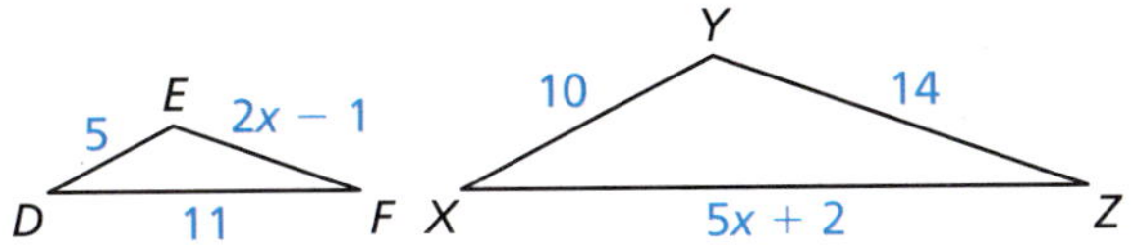

4.

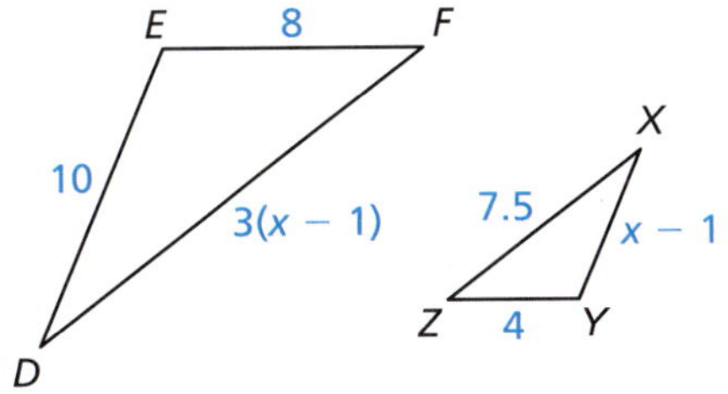

In Exercises 5 and 6, determine whether the two triangles are similar. If they are similar, write a similarity statement and find the scale factor of triangle B to triangle A. *(See Example 3.)*

5.

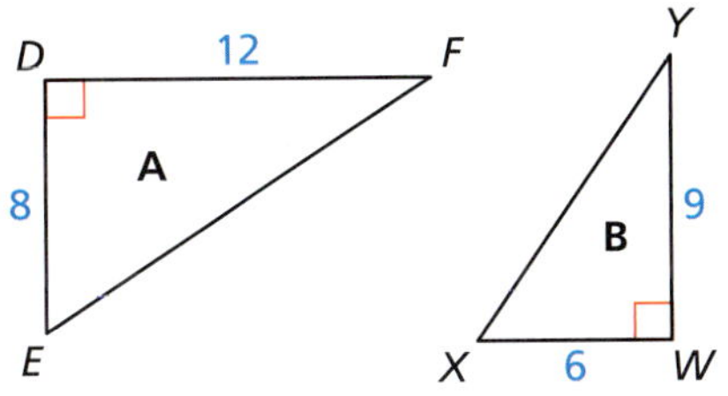

6.

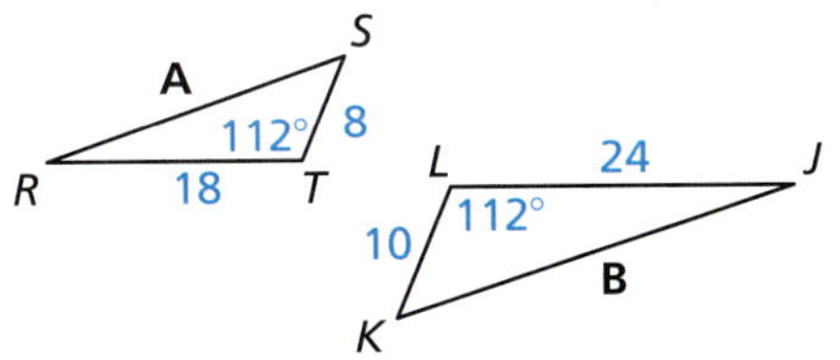

In Exercises 7 and 8, verify that $\triangle ABC \sim \triangle DEF$. Find the scale factor of $\triangle ABC$ to $\triangle DEF$.

7. $\triangle ABC$: $BC = 18$, $AB = 15$, $AC = 12$
   $\triangle DEF$: $EF = 12$, $DE = 10$, $DF = 8$

8. $\triangle ABC$: $AB = 10$, $BC = 16$, $CA = 20$
   $\triangle DEF$: $DE = 25$, $EF = 40$, $FD = 50$

In Exercises 9–12, show that the triangles are similar and write a similarity statement. Explain your reasoning.

9.

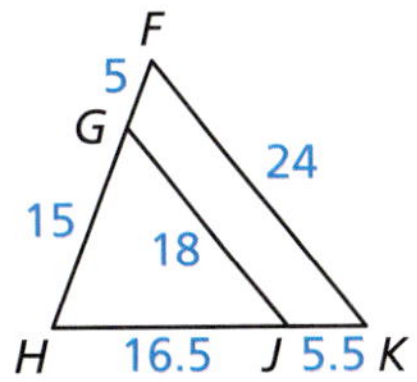

10.

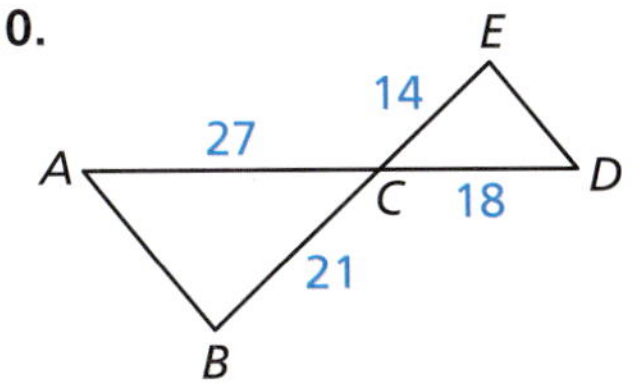

11.

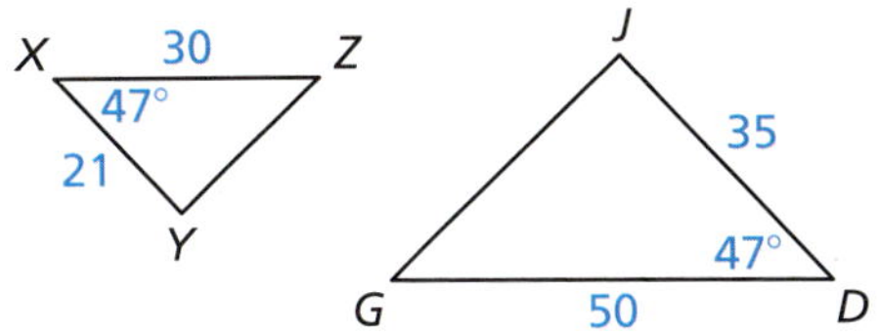

12.

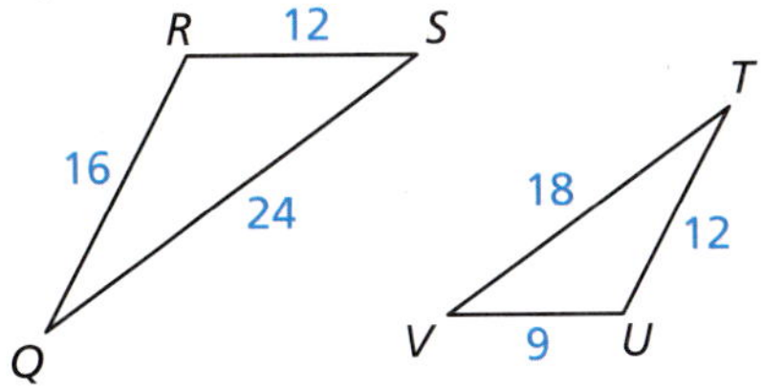

In Exercises 13 and 14, use $\triangle XYZ$.

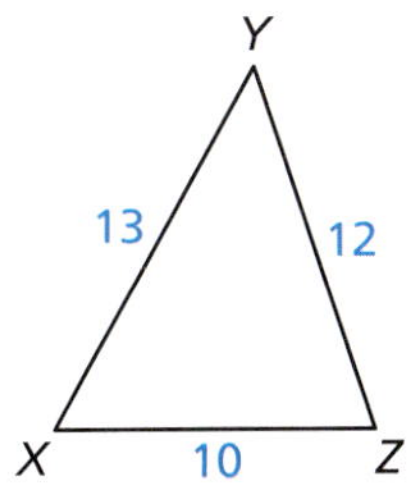

13. The shortest side of a triangle similar to $\triangle XYZ$ is 20 units long. Find the other side lengths of the triangle.

14. The longest side of a triangle similar to $\triangle XYZ$ is 39 units long. Find the other side lengths of the triangle.

5 MTR 15. **STRUCTURE** The side lengths of $\triangle ABC$ are 24, $8x$, and 48. The side lengths of $\triangle DEF$ are 15, 25, and $6x$. Is it possible that the triangles are similar? Explain.

GO DIGITAL

16. **COLLEGE PREP** In the diagram, $\frac{MN}{MR} = \frac{MP}{MQ}$. Which statement(s) must be true? Select all that apply. Explain your reasoning.

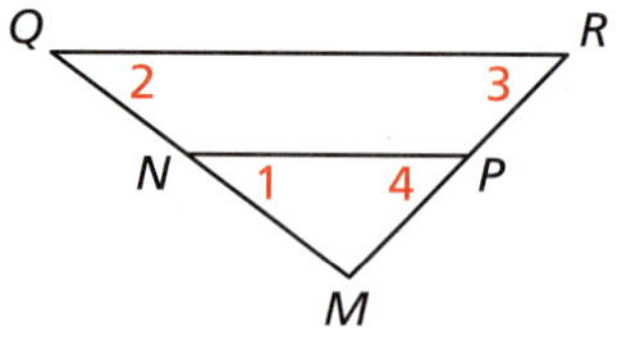

Ⓐ $\angle 1 \cong \angle 2$ Ⓒ $\angle 1 \cong \angle 4$

Ⓑ $\overline{QR} \parallel \overline{NP}$ Ⓓ $\triangle MNP \sim \triangle MRQ$

17. **ERROR ANALYSIS** Describe and correct the error in writing a similarity statement.

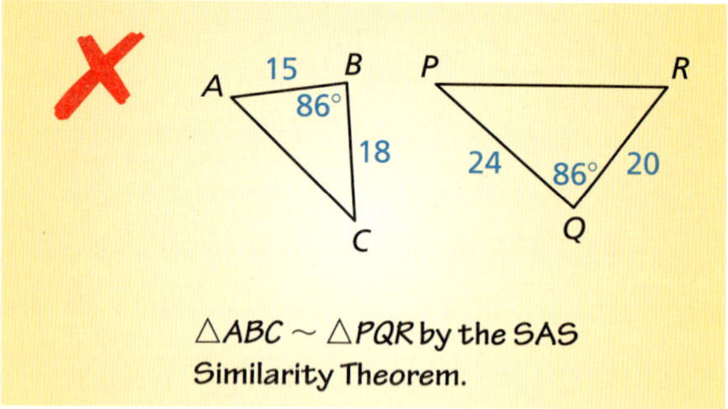

18. **REASONING** You are given two right triangles in which one pair of corresponding legs and the pair of hypotenuses are proportional.

a. The lengths of the given pair of corresponding legs are 6 and 18, and the lengths of the hypotenuses are 10 and 30. Find the lengths of the other pair of corresponding legs.

b. Are these triangles similar? Does this suggest a Hypotenuse-Leg Similarity Theorem for right triangles? Explain.

19. **MODELING REAL LIFE** In the portion of the shuffleboard court shown, $\frac{BC}{AC} = \frac{BD}{AE}$. Determine the additional information you need to show that $\triangle BCD \sim \triangle ACE$ using the given theorem.

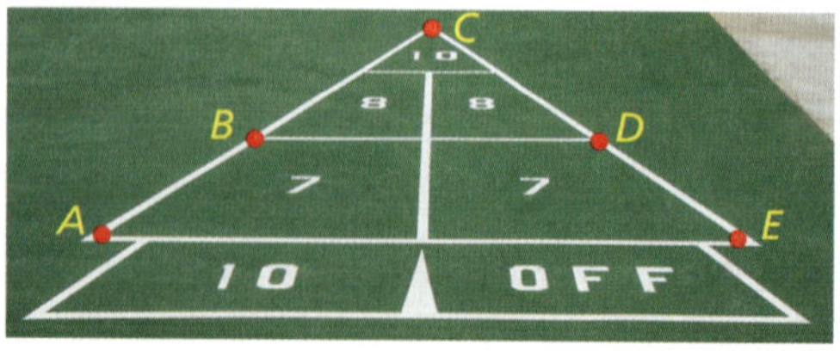

a. SSS Similarity Theorem

b. SAS Similarity Theorem

20. **MODELING REAL LIFE** In the drawing of a DNA molecule, $WX = XY = 0.8$ inch, $VW = YZ = 0.6$ inch, and $WY = 1.0$ inch. Find the width, VZ, of the DNA molecule in the drawing. Justify your answer.

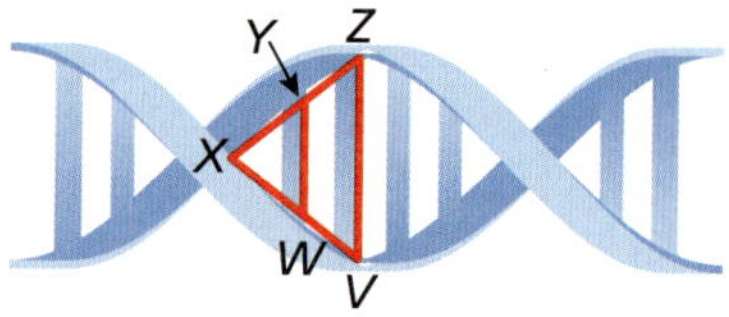

21. **PROOF** Given that $\triangle BAC$ is a right triangle and D, E, and F are midpoints, prove that $m\angle DEF = 90°$.

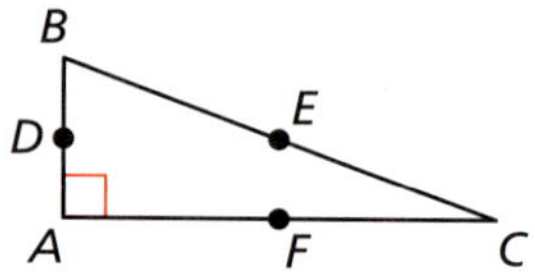

22. **PROVING A THEOREM** Write a two-column proof of the SAS Similarity Theorem.

Given $\angle A \cong \angle D$, $\frac{AB}{DE} = \frac{AC}{DF}$

Prove $\triangle ABC \sim \triangle DEF$

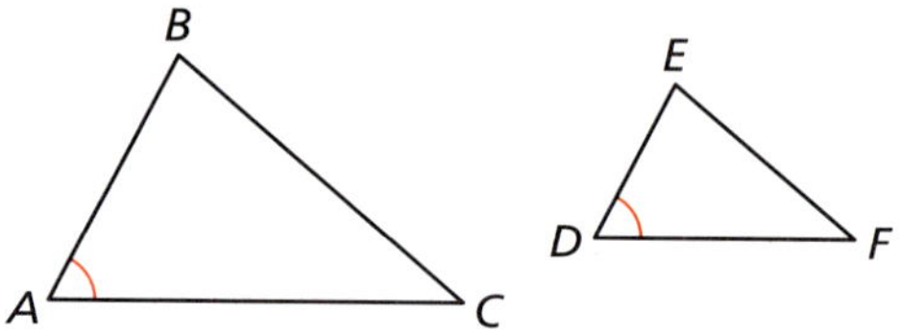

23. **MAKING AN ARGUMENT** For $\triangle JKL$ and $\triangle MNP$, $\frac{JK}{MN} = \frac{KL}{NP}$. Without using the SAS Similarity Theorem, explain why $m\angle K$ must equal $m\angle N$ for the triangles to be similar.

24. **HOW DO YOU SEE IT?**
Which theorem can you use to show that $\triangle OPQ \sim \triangle OMN$ in the portion of the Ferris wheel shown when $PM = QN = 10$ feet and $MO = NO = 20$ feet?

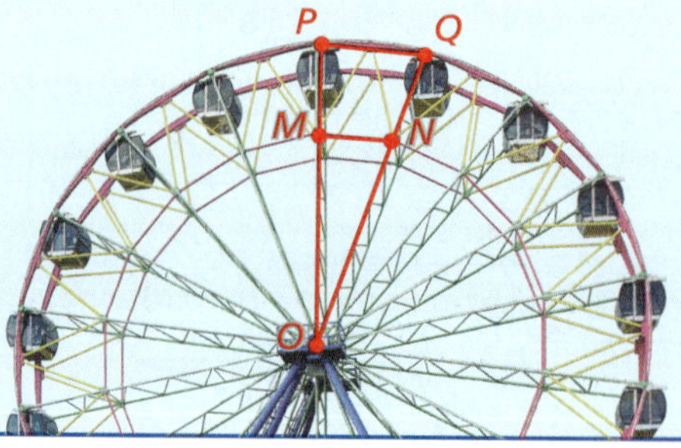

GO DIGITAL

25. REASONING Are any two right triangles similar? Explain your reasoning.

26. PROBLEM SOLVING You want to create a stained-glass window for the opening shown. Show how you can arrange four congruent pieces of glass to create the window. Then show that each piece of stained glass is similar to the opening, and state the side lengths and angle measures of each piece of glass.

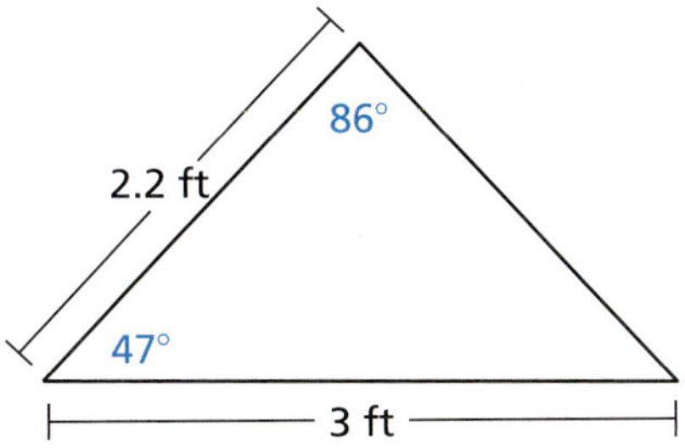

27. CONNECTING CONCEPTS Determine whether $\triangle ABC$ with vertices $A(-5, 4)$, $B(-2, 8)$, and $C(1, 6)$ is similar to $\triangle XYZ$ with vertices $X(-4, -6)$, $Y(2, 2)$, and $Z(8, -2)$. Explain your reasoning.

28. THOUGHT PROVOKING
Decide whether each is a valid method of showing that two quadrilaterals are similar. Justify your answer.

a. SASA **b.** SASAS **c.** SSSS **d.** SSSAS

29. PROVING A THEOREM Prove the second part of the Slopes of Parallel Lines Theorem from page 429.

Given $m_\ell = m_n$, ℓ and n are nonvertical.

Prove $\ell \parallel n$

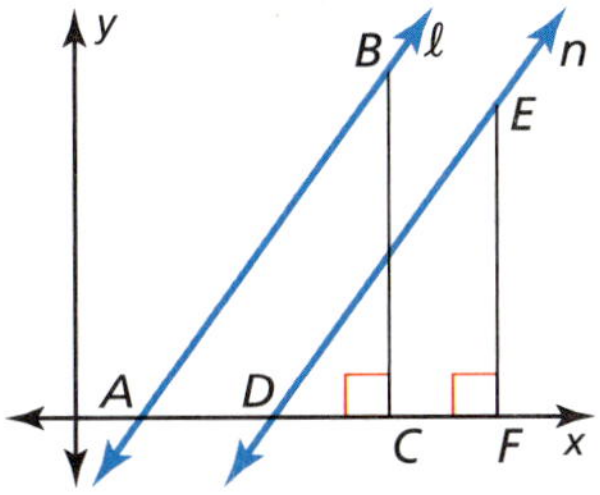

30. PROVING A THEOREM
Complete the two-column proof of the second part of the Slopes of Perpendicular Lines Theorem from page 430.

Given $m_\ell m_n = -1$, ℓ and n are nonvertical.

Prove $\ell \perp n$

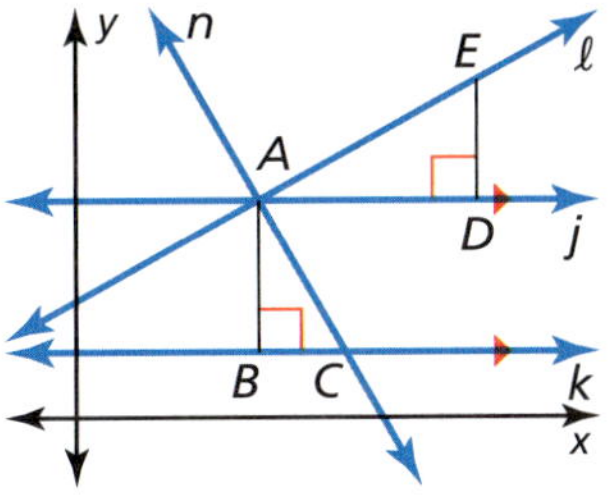

STATEMENTS	REASONS
1. $m_\ell m_n = -1$	1. Given
2. ______	2. Definition of slope
3. $\frac{DE}{AD} \cdot \left(-\frac{AB}{BC}\right) = -1$	3. ______
4. $\frac{DE}{AD} = \frac{BC}{AB}$	4. ______
5. ______	5. Rewrite proportion.
6. ______	6. Right Angles Congruence Theorem
7. ______	7. ______
8. ______	8. Corresponding angles of similar figures are congruent.
9. ______	9. Alternate Interior Angles Theorem
10. $m\angle BAC = m\angle DAE$, $m\angle BCA = m\angle CAD$	10. ______
11. $m\angle BAC + m\angle BCA + 90° = 180°$	11. ______
12. ______	12. Subtraction Property of Equality
13. $m\angle CAD + m\angle DAE = 90°$	13. ______
14. ______	14. Angle Addition Postulate
15. $m\angle CAE = 90°$	15. ______
16. ______	16. ______

GO DIGITAL

31. REASONING Explain why it is not necessary to have an Angle-Side-Angle Similarity Theorem.

32. PROOF Use transformations to prove (a) the SSS Similarity Theorem and (b) the SAS Similarity Theroem.

REVIEW & REFRESH

WATCH

33. In the diagram, ▱*QRST* is similar to ▱*WXYZ*. Find the value of *x*.

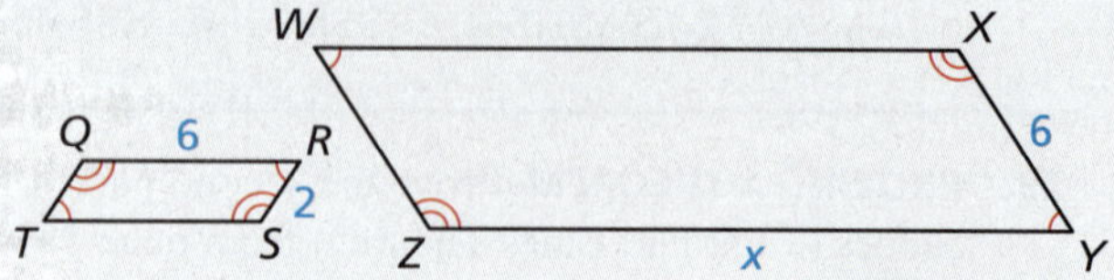

7 MTR **34. MODELING REAL LIFE** The Anthem Veterans Memorial in Anthem, Arizona, consists of five pillars, each of which represents one of the five branches of the United States military. Each pillar is parallel to the pillar directly to the right. Explain why the shortest pillar is parallel to the tallest pillar.

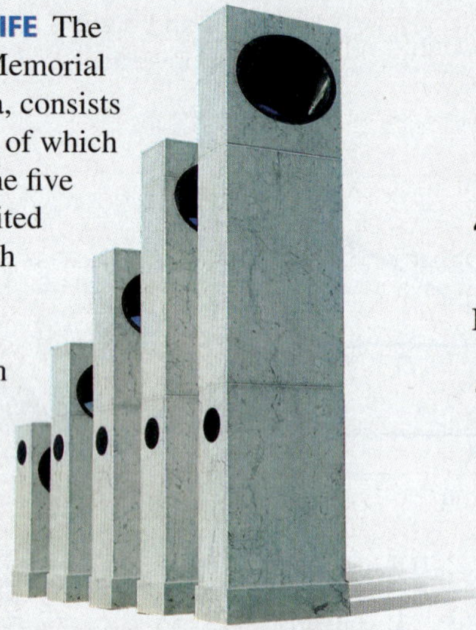

In Exercises 35 and 36, show that the triangles are similar and write a similarity statement. Explain your reasoning.

35.

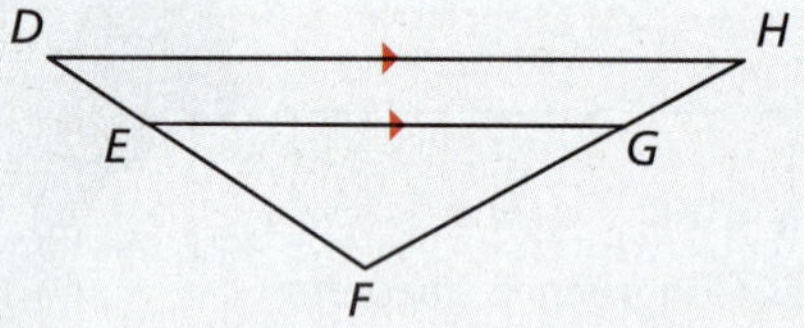

36.

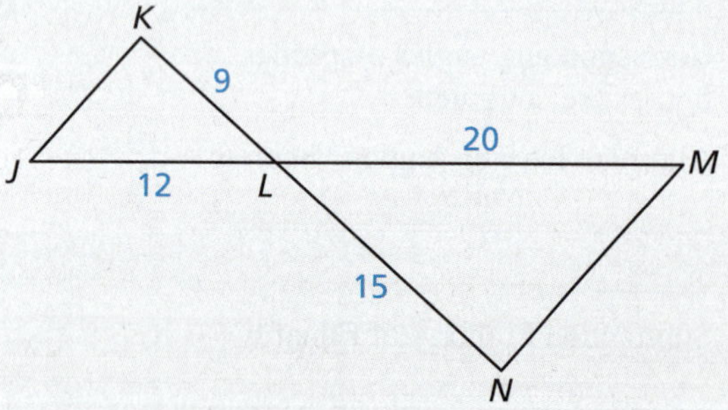

37. Find the coordinates of point *P* along the directed line segment *AB* with vertices *A*(−3, −5) and *B*(9, −1) so that the ratio of *AP* to *PB* is 1 to 3.

38. A triangle has side lengths of 18 inches and 32 inches. Describe the possible lengths of the third side.

39. Determine whether the graph represents a function. Explain.

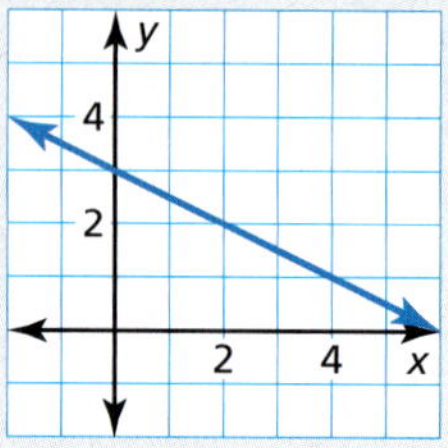

40. Write an equation of the perpendicular bisector of the segment with endpoints *R*(−1, −2) and *S*(5, 2).

In Exercises 41 and 42, find $m\angle C$.

41.

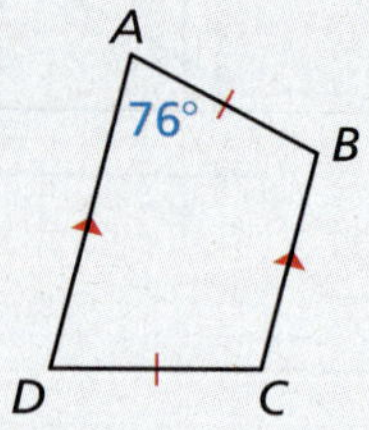

42.

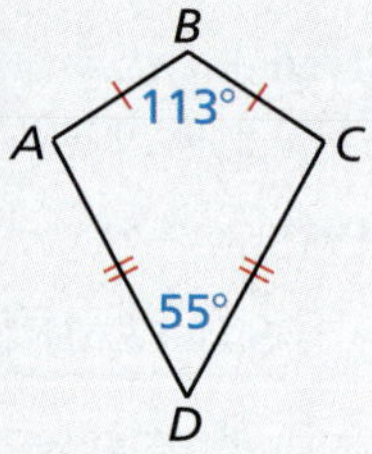

In Exercises 43 and 44, graph △*LMN* with vertices *L*(−3, −2), *M*(−1, 1), and *N*(2, −3) and its image after the composition.

43. Translation: $(x, y) \rightarrow (x - 4, y + 3)$
Rotation: 180° about the origin

44. Rotation: 90° clockwise about the origin
Reflection: in the *y*-axis

In Exercises 45 and 46, decide whether enough information is given to prove that the triangles are congruent. If so, state the theorem you can use.

45. △*JKL*, △*LMJ*

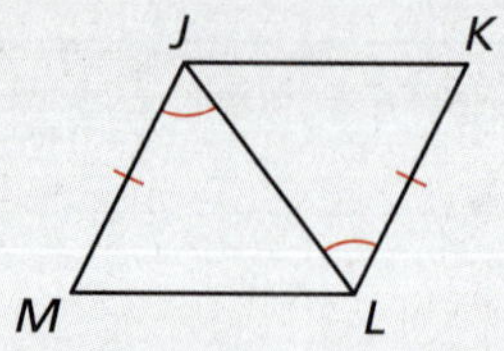

46. △*QRS*, △*QTS*

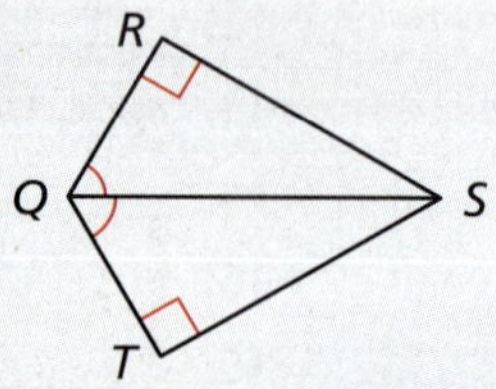

GO DIGITAL

8.4 Proportionality Theorems

Learning Target: Understand and use proportionality theorems.

Success Criteria:
- I can use proportionality theorems to find lengths in triangles.
- I can find lengths when two transversals intersect three parallel lines.
- I can find lengths when a ray bisects an angle of a triangle.

EXPLORE IT! Discovering Proportionality Relationships

Work with a partner. Use technology.

a. Construct any $\triangle ABC$. Construct $\overline{DE}$ parallel to $\overline{BC}$ with endpoints on $\overline{AB}$ and $\overline{AC}$, respectively.

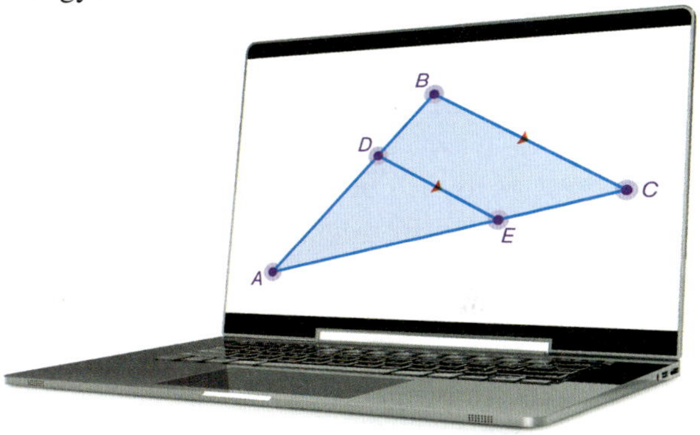

b. Find a relationship among AD, BD, AE, and CE. Is the relationship true when you construct $\overline{DE}$ in other locations parallel to $\overline{BC}$?

c. Change $\triangle ABC$ and repeat parts (a) and (b) several times. Write a conjecture that summarizes your results.

5 MTR MAKE A PLAN
How can you use similarity to prove your conjecture?

d. A line that bisects an angle of a triangle divides the opposite side into two segments. How are the lengths of these segments related to the lengths of the other two sides of the triangle? Provide examples to support your reasoning.

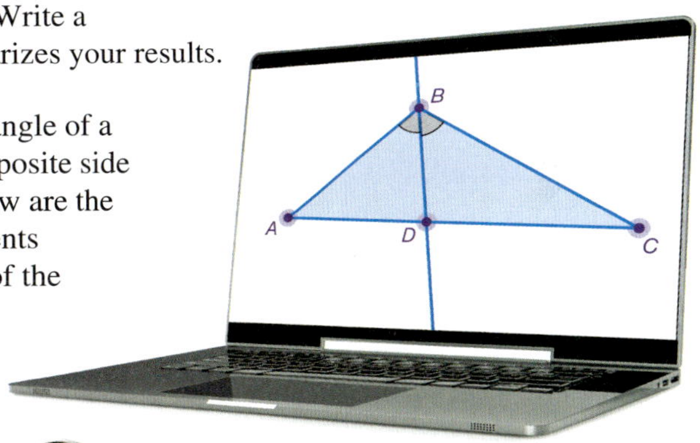

Geometric Reasoning
MA.912.GR.1.6 Solve mathematical and real-world problems involving congruence or similarity in two-dimensional figures.

GO DIGITAL

Using the Triangle Proportionality Theorem

THEOREMS

8.6 Triangle Proportionality Theorem

If a line parallel to one side of a triangle intersects the other two sides, then it divides the two sides proportionally.

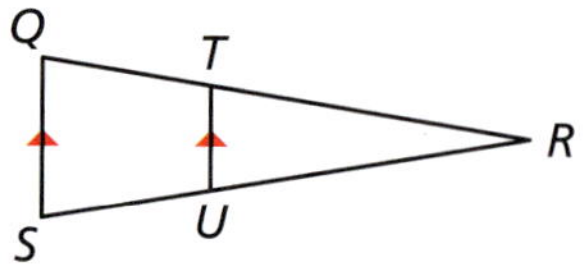

Prove this Theorem Exercise 25, page 441

If $\overline{TU} \parallel \overline{QS}$, then $\frac{RT}{TQ} = \frac{RU}{US}$.

8.7 Converse of the Triangle Proportionality Theorem

If a line divides two sides of a triangle proportionally, then it is parallel to the third side.

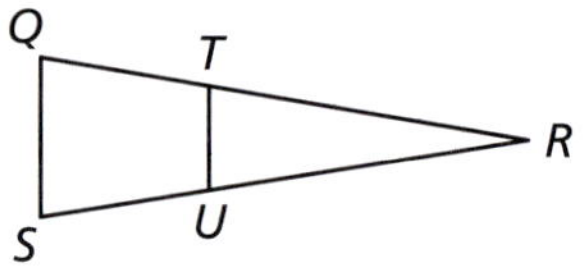

Prove this Theorem Exercise 26, page 441

If $\frac{RT}{TQ} = \frac{RU}{US}$, then $\overline{TU} \parallel \overline{QS}$.

EXAMPLE 1 Finding the Length of a Segment

WATCH

Find the length of $\overline{RQ}$.

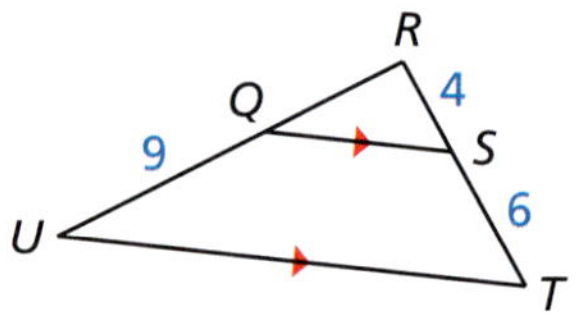

SOLUTION

The diagram shows that $\overline{QS} \parallel \overline{UT}$. So, $\overline{QS}$ divides $\overline{RU}$ and $\overline{RT}$ proportionally. Use the Triangle Proportionality Theorem to write a proportion.

$\frac{RQ}{QU} = \frac{RS}{ST}$ Triangle Proportionality Theorem

$\frac{RQ}{9} = \frac{4}{6}$ Substitute.

$RQ = 6$ Multiply each side by 9.

▶ The length of $\overline{RQ}$ is 6 units.

SELF-ASSESSMENT 1 I don't understand yet. 2 I can do it with help. 3 I can do it on my own. 4 I can teach someone else.

Find the length of $\overline{YZ}$.

1.

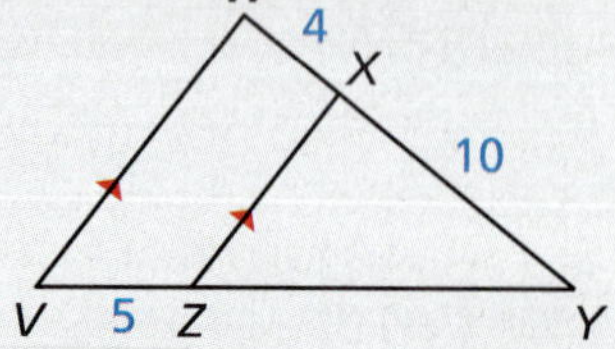

2.

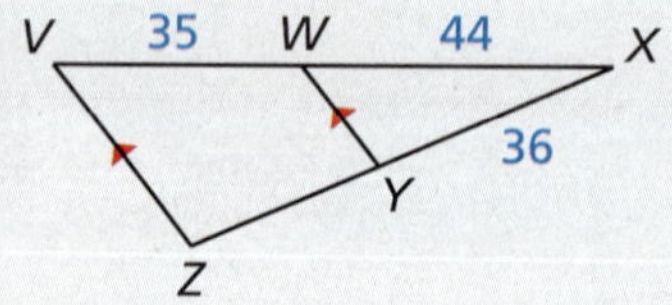

GO DIGITAL

The theorems on the previous page also imply the following.

Contrapositive of the Triangle Proportionality Theorem

If $\frac{RT}{TQ} \neq \frac{RU}{US}$, then $\overline{TU} \nparallel \overline{QS}$.

Inverse of the Triangle Proportionality Theorem

If $\overline{TU} \nparallel \overline{QS}$, then $\frac{RT}{TQ} \neq \frac{RU}{US}$.

7 MTR

EXAMPLE 2 Modeling Real Life

WATCH

On the shoe rack, $BA = 33$ centimeters, $CB = 27$ centimeters, $CD = 44$ centimeters, and $DE = 25$ centimeters. Explain why the shelf is not parallel to the floor.

SOLUTION

Find and simplify the ratios of the lengths.

$$\frac{CD}{DE} = \frac{44}{25} \qquad \frac{CB}{BA} = \frac{27}{33} = \frac{9}{11}$$

▶ Because $\frac{44}{25} \neq \frac{9}{11}$, $\overline{BD}$ is not parallel to $\overline{AE}$. So, the shelf is not parallel to the floor.

4 MTR **CONSTRUCT AN ARGUMENT**

Where can you position point D along $\overline{CE}$ so that the shelf is parallel to the floor?

Recall that you partitioned a directed line segment in the coordinate plane in Section 3.5. You can apply the Triangle Proportionality Theorem to construct a point along a directed line segment that partitions the segment in a given ratio.

CONSTRUCTION Constructing a Point along a Directed Line Segment

WATCH

Construct the point L on $\overline{AB}$ so that the ratio of AL to LB is 3 to 1.

SOLUTION

Step 1

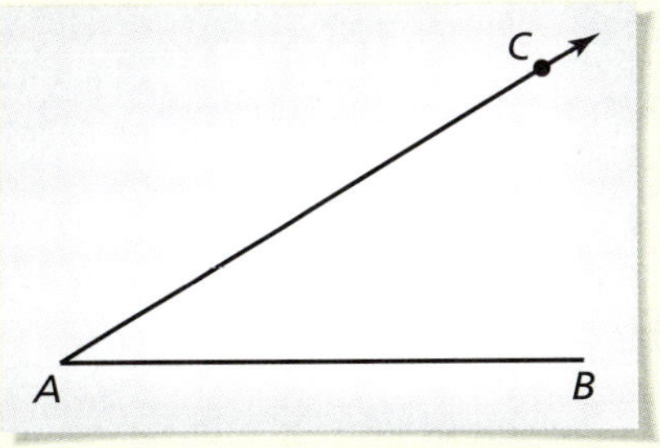

Draw a segment and a ray Draw $\overline{AB}$ of any length. Choose any point C not on $\overleftrightarrow{AB}$. Draw $\overrightarrow{AC}$.

Step 2

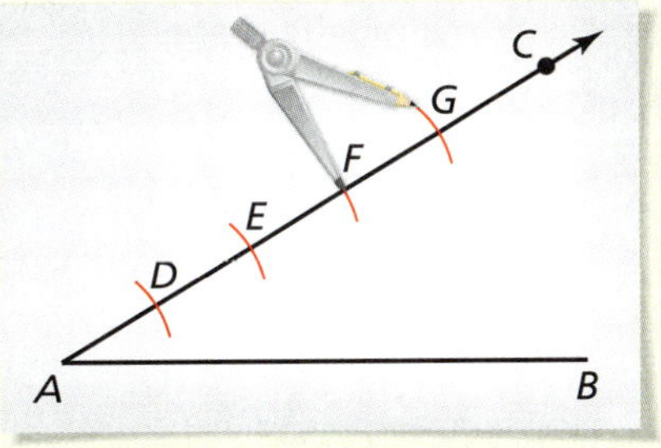

Draw arcs Place the point of a compass at A and make an arc of any radius intersecting $\overrightarrow{AC}$. Label the point of intersection D. Using the same compass setting, make three more arcs on $\overrightarrow{AC}$, as shown. Label the points of intersection E, F, and G and note that $AD = DE = EF = FG$.

Step 3

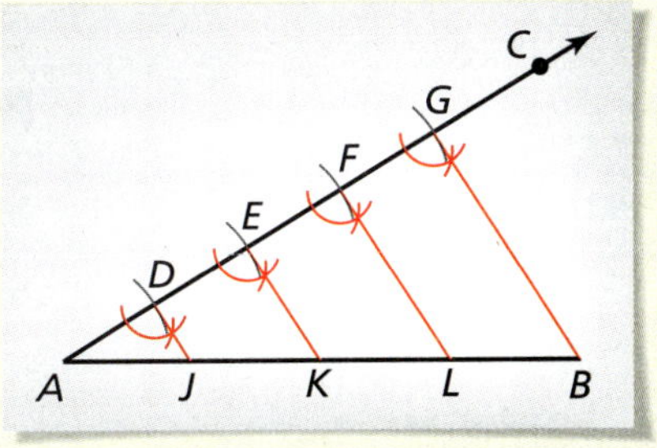

Draw a segment Draw $\overline{GB}$. Copy $\angle AGB$ and construct congruent angles at D, E, and F with sides that intersect $\overline{AB}$ at J, K, and L. Sides $\overline{DJ}$, $\overline{EK}$, and $\overline{FL}$ are all parallel, and they divide $\overline{AB}$ equally. So, $AJ = JK = KL = LB$. Point L divides directed line segment AB in the ratio 3 to 1.

GO DIGITAL

Using Other Proportionality Theorems

THEOREM

8.8 Three Parallel Lines Theorem

If three parallel lines intersect two transversals, then they divide the transversals proportionally.

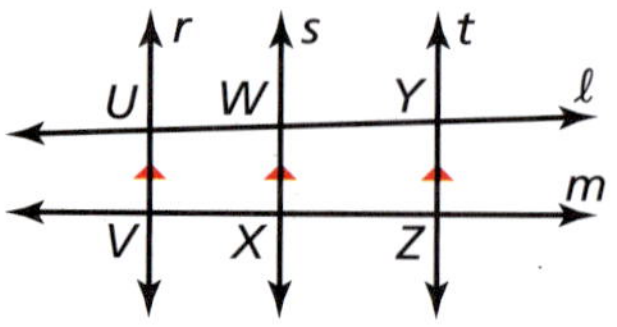

$$\frac{UW}{WY} = \frac{VX}{XZ}$$

Prove this Theorem Exercise 30, page 441

EXAMPLE 3 Using the Three Parallel Lines Theorem

WATCH

Find the distance HF between Main Street and South Main Street.

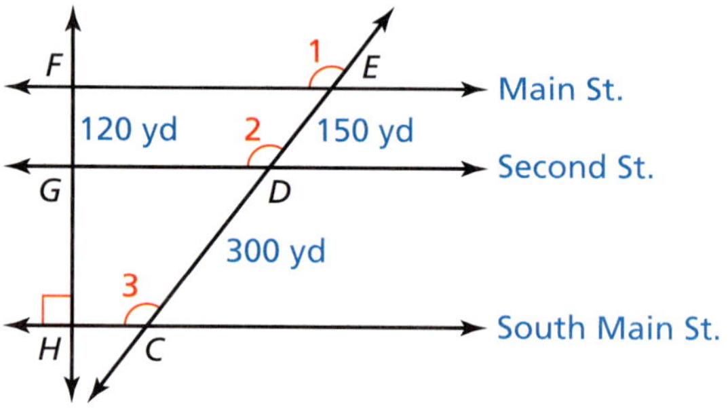

SOLUTION

Because ∠1, ∠2, and ∠3 are all congruent, you can use the Corresponding Angles Converse to determine that $\overleftrightarrow{FE}$, $\overleftrightarrow{GD}$, and $\overleftrightarrow{HC}$ are parallel. These parallel lines divide the transversals proportionally. Find HF.

Method 1 Use the Three Parallel Lines Theorem to write a proportion.

$\frac{HG}{GF} = \frac{CD}{DE}$ — Three Parallel Lines Theorem

$\frac{HG}{120} = \frac{300}{150}$ — Substitute.

$HG = 240$ — Multiply each side by 120.

By the Segment Addition Postulate, $HF = HG + GF = 240 + 120 = 360$.

▶ So, the distance between Main Street and South Main Street is 360 yards.

Method 2 Write a proportion involving total and partial distances.

Step 1 Make a table to compare the distances.

	$\overleftrightarrow{CE}$	$\overleftrightarrow{HF}$
Total distance	$CE = 300 + 150 = 450$	HF
Partial distance	$DE = 150$	$GF = 120$

Step 2 Write and solve a proportion.

$\frac{450}{150} = \frac{HF}{120}$ — Write proportion.

$360 = HF$ — Multiply each side by 120.

▶ So, the distance between Main Street and South Main Street is 360 yards.

GO DIGITAL

THEOREM

8.9 Triangle Angle Bisector Theorem

If a ray bisects an angle of a triangle, then it divides the opposite side into segments whose lengths are proportional to the lengths of the other two sides.

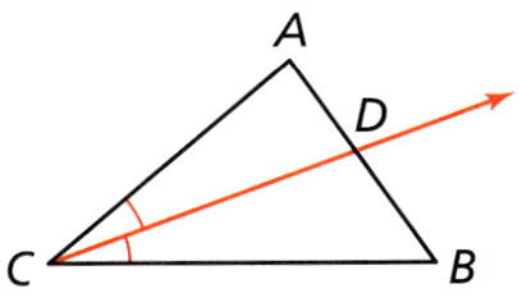

$$\frac{AD}{DB} = \frac{CA}{CB}$$

Prove this Theorem Exercise 33, page 441

EXAMPLE 4 Using the Triangle Angle Bisector Theorem

WATCH

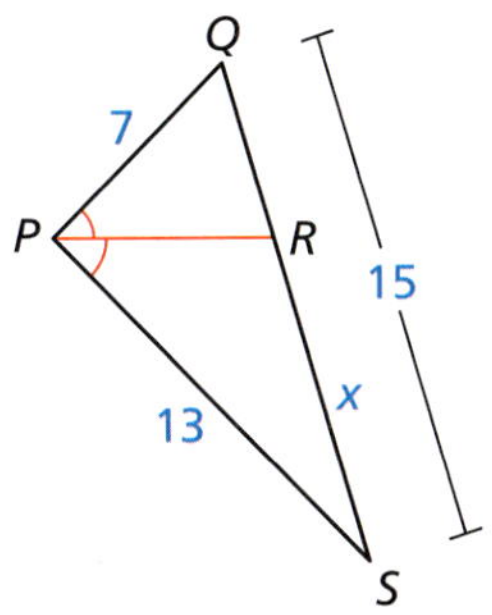

Find the length of $\overline{RS}$.

SOLUTION

Because $\angle QPR \cong \angle SPR$, $\overrightarrow{PR}$ is an angle bisector of $\angle QPS$. So, you can apply the Triangle Angle Bisector Theorem. Use the fact that $RQ = 15 - x$.

$\frac{RQ}{RS} = \frac{PQ}{PS}$	Triangle Angle Bisector Theorem
$\frac{15 - x}{x} = \frac{7}{13}$	Substitute.
$195 - 13x = 7x$	Cross Products Property
$9.75 = x$	Solve for x.

▶ So, the length of $\overline{RS}$ is 9.75 units.

SELF-ASSESSMENT

1 I don't understand yet. 2 I can do it with help. 3 I can do it on my own. 4 I can teach someone else.

3. Determine whether $\overline{PS} \parallel \overline{QR}$.

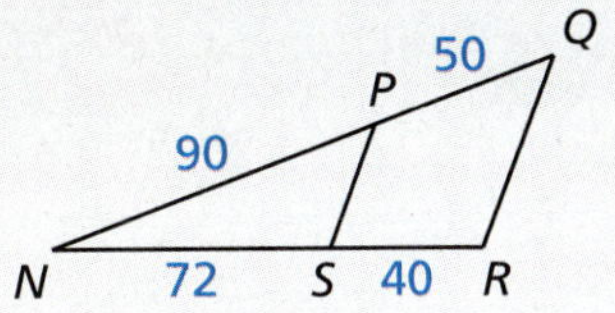

Find the length of the given line segment.

4. $\overline{BD}$

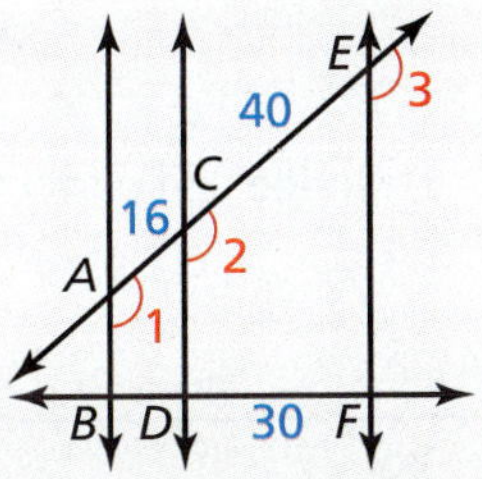

5. $\overline{JM}$

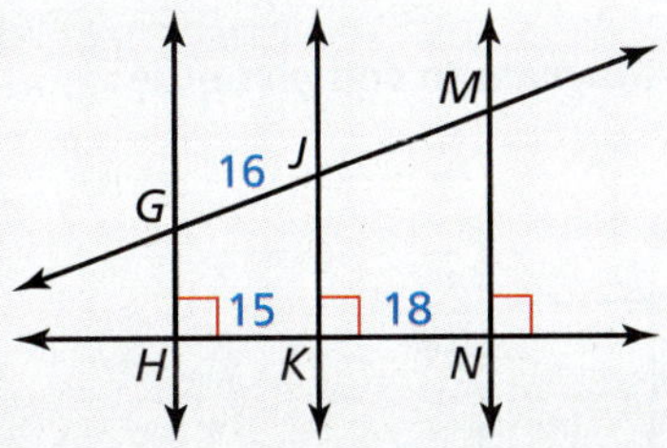

Find the value of the variable.

6.

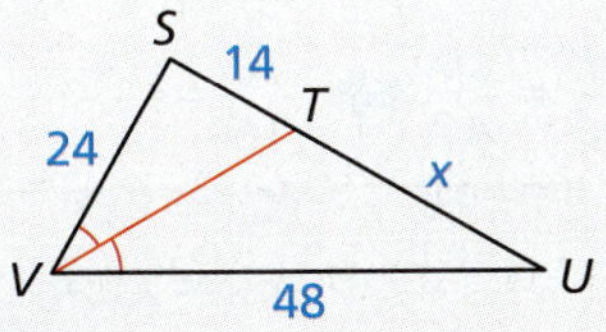

7.

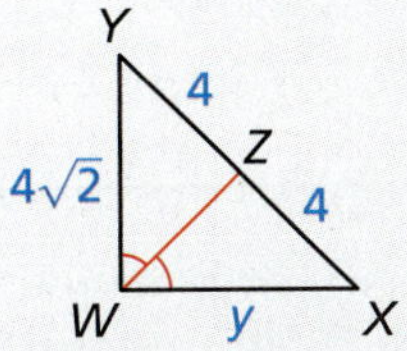

GO DIGITAL

8.4 Practice WITH CalcChat® AND CalcView®

In Exercises 1 and 2, find the length of $\overline{AB}$. *(See Example 1.)*

1.

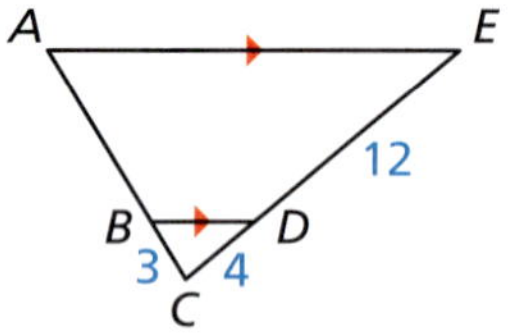

2.

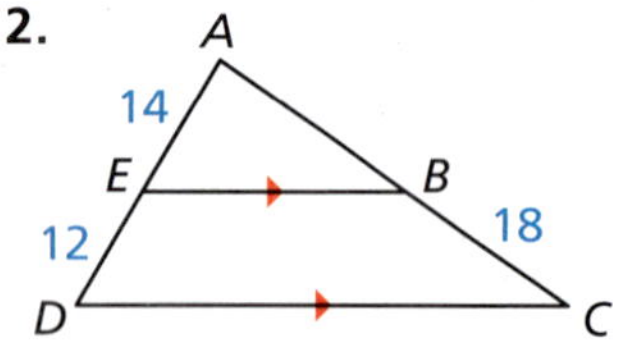

In Exercises 3–6, determine whether $\overline{KM} \parallel \overline{JN}$. *(See Example 2.)*

3.

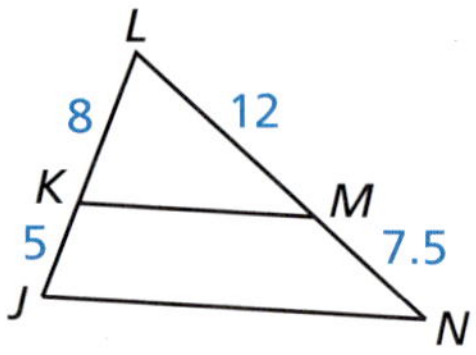

4.

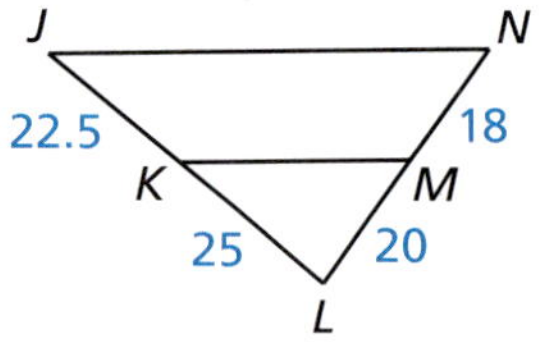

5.

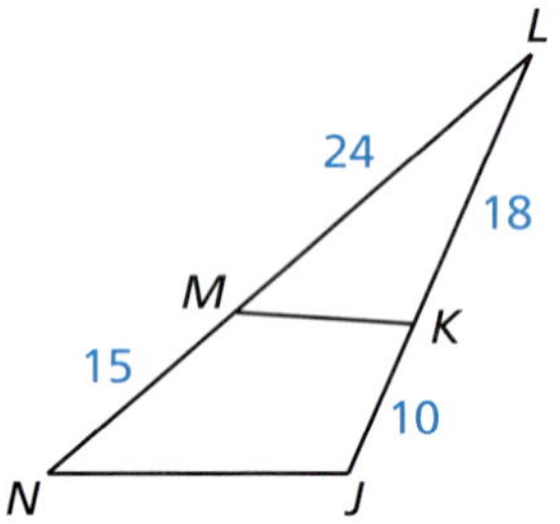

6.

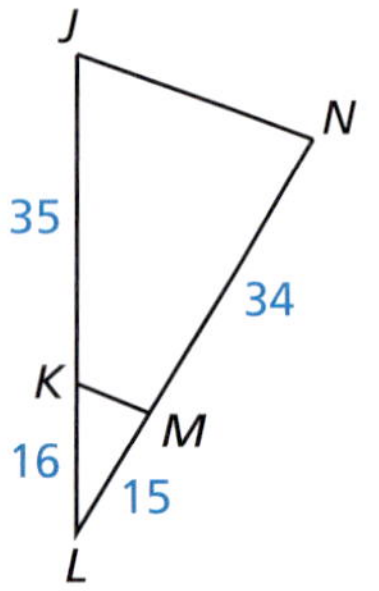

CONSTRUCTION In Exercises 7–10, draw a segment with the given length. Construct the point that divides the segment in the given ratio.

7. 3 in.; 1 to 4

8. 2 in.; 2 to 3

9. 12 cm; 1 to 3

10. 9 cm; 2 to 5

In Exercises 11–14, use the diagram to complete the proportion.

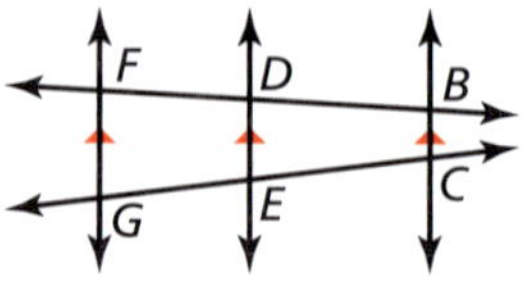

11. $\dfrac{BD}{BF} = \dfrac{\square}{CG}$

12. $\dfrac{CG}{\square} = \dfrac{BF}{DF}$

13. $\dfrac{EG}{CE} = \dfrac{DF}{\square}$

14. $\dfrac{\square}{BD} = \dfrac{CG}{CE}$

In Exercises 15 and 16, find the length of the indicated line segment. *(See Example 3.)*

15. $\overline{VX}$

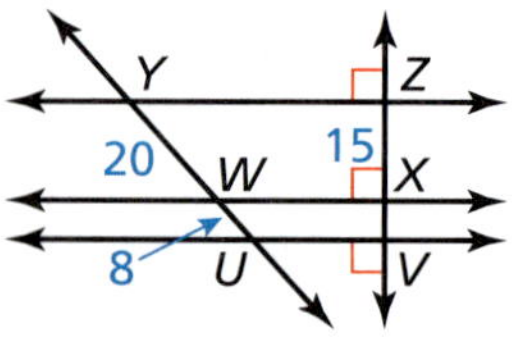

16. $\overline{SU}$

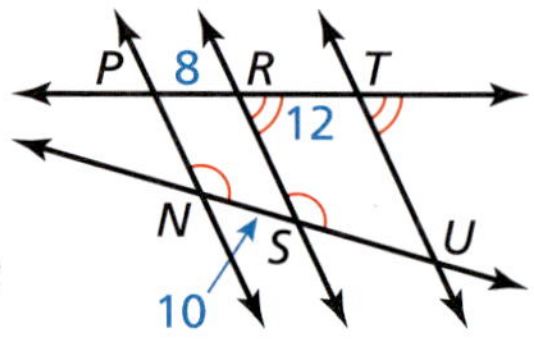

In Exercises 17–20, find the value of the variable. *(See Example 4.)*

17.

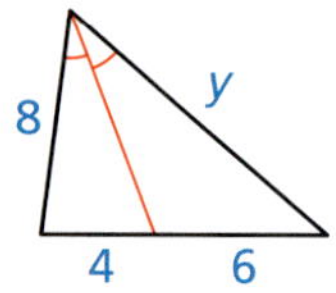

18.

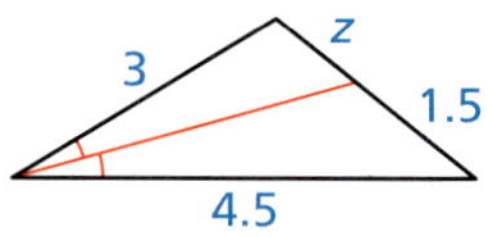

19.

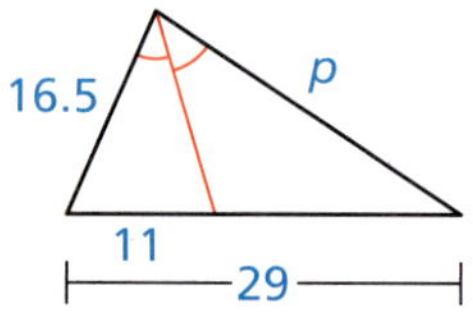

20.

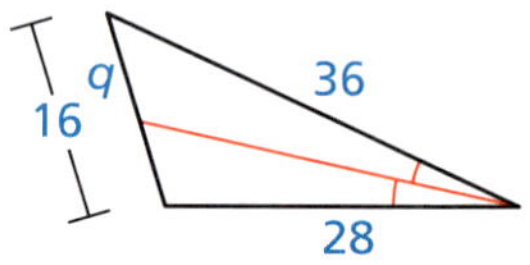

21. ERROR ANALYSIS Describe and correct the error in finding QU.

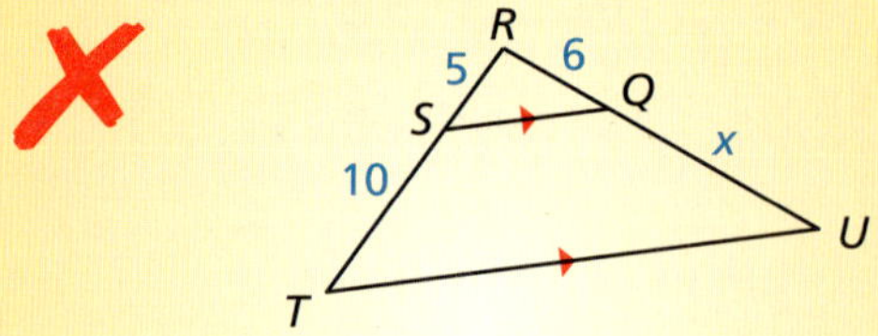

$\dfrac{RS}{QU} = \dfrac{RQ}{ST}$ ➡ $\dfrac{5}{x} = \dfrac{6}{10}$

$6x = 50$

$x = 8\frac{1}{3}$

22. ERROR ANALYSIS Describe and correct the error in finding AD.

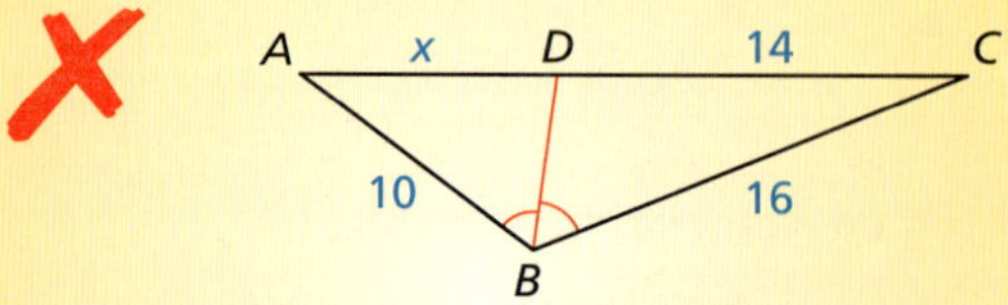

$\dfrac{AB}{BC} = \dfrac{CD}{AD}$ ➡ $\dfrac{10}{16} = \dfrac{14}{x}$

$10x = 224$

$x = 22.4$

GO DIGITAL

5 MTR **STRUCTURE** **In Exercises 23 and 24, find the value of x for which $\overline{PQ} \parallel \overline{RS}$.**

23.

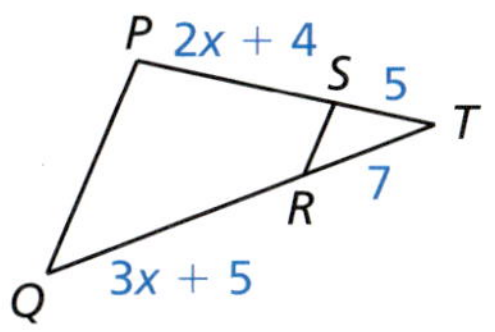

24.

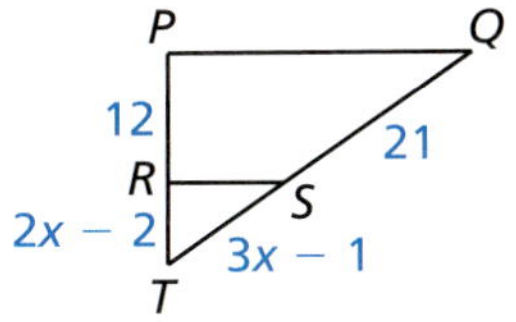

25. PROVING A THEOREM Prove the Triangle Proportionality Theorem.

Given $\overline{QS} \parallel \overline{TU}$

Prove $\frac{QT}{TR} = \frac{SU}{UR}$

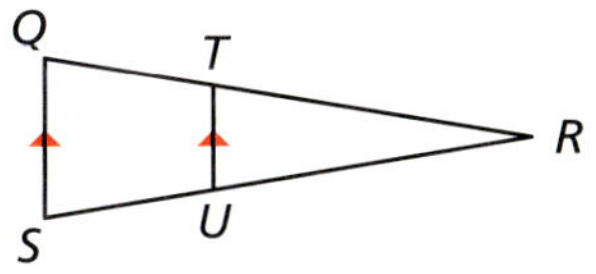

26. PROVING A THEOREM Prove the Converse of the Triangle Proportionality Theorem.

Given $\frac{ZY}{YW} = \frac{ZX}{XV}$

Prove $\overline{YX} \parallel \overline{WV}$

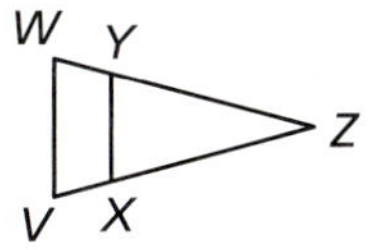

7 MTR **27. MODELING REAL LIFE** The real estate term *lake frontage* refers to the distance along the edge of a piece of property that touches a lake.

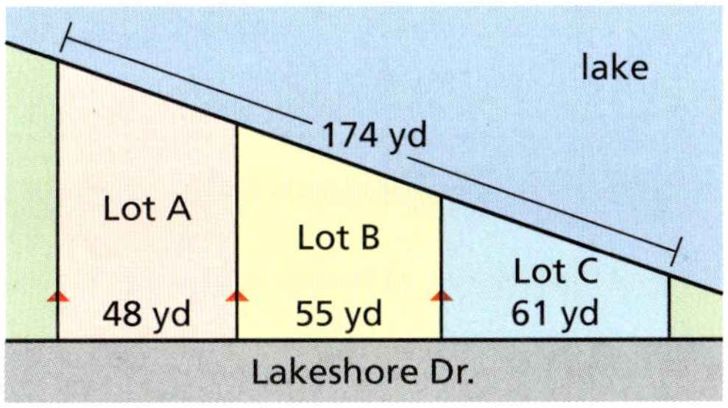

a. Find the lake frontage of each lot shown.

b. In general, the more lake frontage a lot has, the higher its selling price. Which lot(s) should be listed for the highest price?

c. Suppose that lot prices are in the same ratio as lake frontages. If the least expensive lot is $250,000, what are the prices of the other lots? Explain.

6 MTR **28. JUSTIFYING STEPS** Find the values of x and y. Justify your answers.

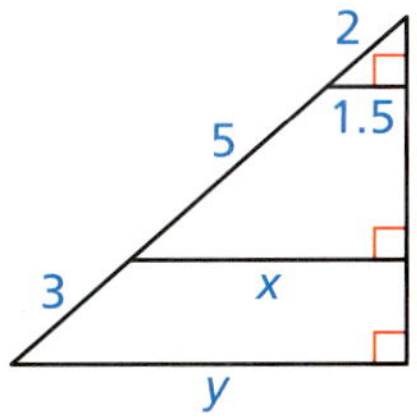

29. REASONING In the construction on page 437, explain why you can apply the Triangle Proportionality Theorem in Step 3.

30. PROVING A THEOREM Use the diagram with the auxiliary line drawn to write a paragraph proof of the Three Parallel Lines Theorem.

Given $k_1 \parallel k_2 \parallel k_3$

Prove $\frac{CB}{BA} = \frac{DE}{EF}$

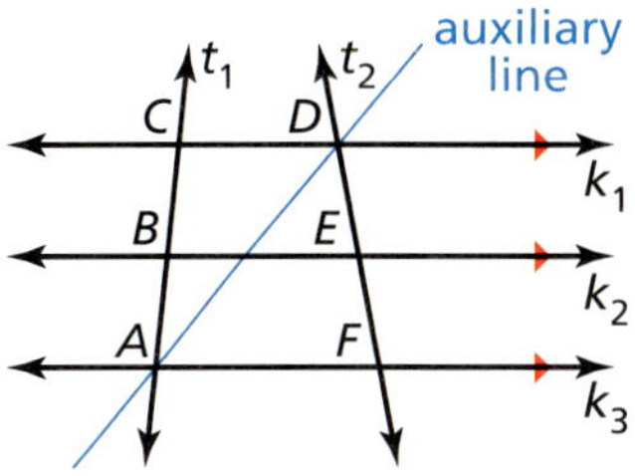

31. REASONING In $\triangle LMN$, the angle bisector of $\angle M$ also bisects $\overline{LN}$. Classify $\triangle LMN$ as specifically as possible. Justify your answer.

32. HOW DO YOU SEE IT?
During a football game, the quarterback throws the ball to the receiver. The receiver is between two defensive players, as shown. If Player 1 is closer to the quarterback when the ball is thrown and both defensive players move at the same speed, which player will reach the receiver first? Explain.

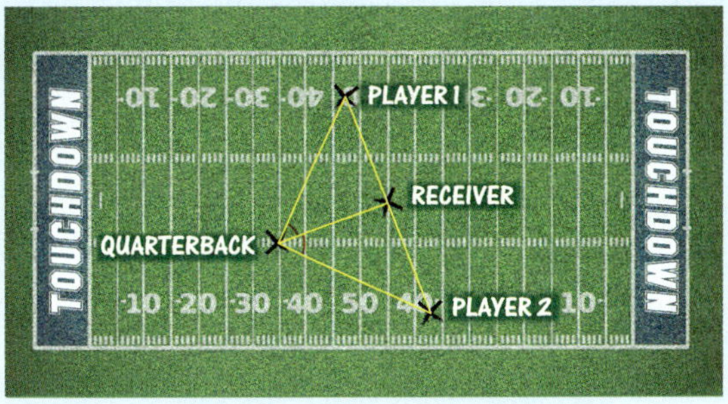

33. PROVING A THEOREM Use the diagram with the auxiliary lines drawn to write a paragraph proof of the Triangle Angle Bisector Theorem.

Given $\angle YXW \cong \angle WXZ$

Prove $\frac{YW}{WZ} = \frac{XY}{XZ}$

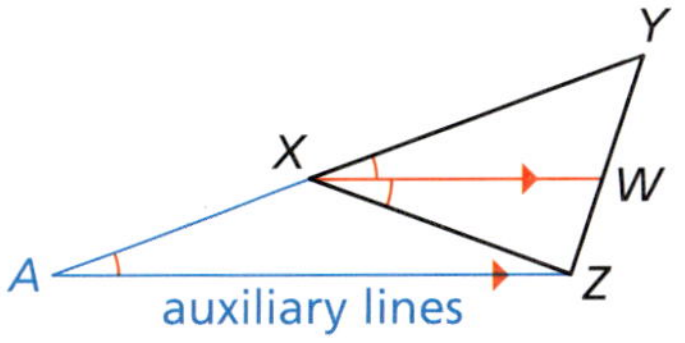

GO DIGITAL

34. THOUGHT PROVOKING Write the converse of the Triangle Angle Bisector Theorem. Is the converse true? Justify your answer.

35. REASONING How is the Triangle Midsegment Theorem related to the Triangle Proportionality Theorem? Explain.

4 MTR **36. MAKING AN ARGUMENT** Two people leave points A and B at the same time. They intend to meet at point C at the same time. The person who leaves point A walks at a speed of 3 miles per hour. To determine how fast the person who leaves point B must walk, do you need to know the length of $\overline{AC}$? Explain your reasoning.

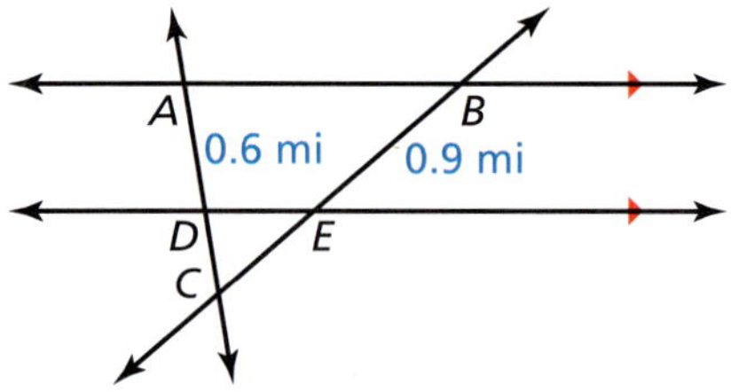

37. CONSTRUCTION Given segments with lengths r, s, and t, construct a segment of length x, such that $\frac{r}{s} = \frac{t}{x}$.

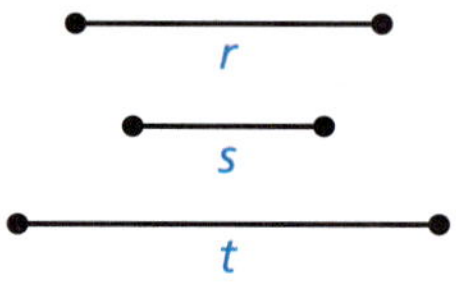

38. PROOF Prove *Ceva's Theorem*: If P is any point inside $\triangle ABC$, then $\frac{AY}{YC} \cdot \frac{CX}{XB} \cdot \frac{BZ}{ZA} = 1$.

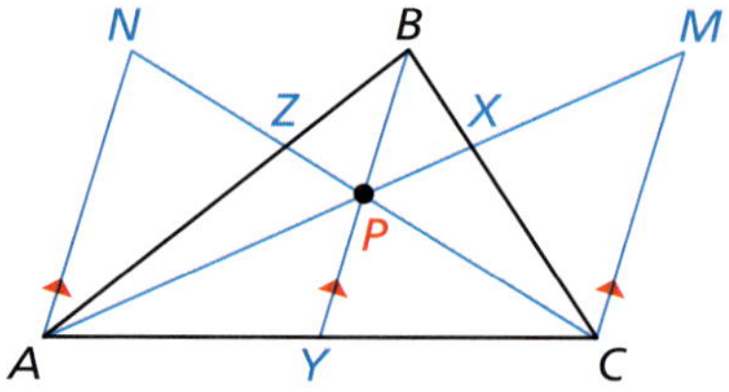

(*Hint*: Draw segments parallel to $\overline{BY}$ through A and C, as shown. Apply the Triangle Proportionality Theorem to $\triangle ACM$.)

REVIEW & REFRESH

WATCH

39. Graph $\triangle FGH$ with vertices $F(-4, -2)$, $G(-2, 0)$, and $H(-4, -4)$ and its image after the similarity transformation.

Translation: $(x, y) \rightarrow (x + 2, y + 4)$

Dilation: $(x, y) \rightarrow \left(\frac{1}{2}x, \frac{1}{2}y\right)$

7 MTR **40. MODELING REAL LIFE** You want to install a public mailbox. The public mailbox should be the same distance from each of the three apartment buildings shown. Use the diagram to determine the location of the public mailbox.

41. Find the value of x when $\triangle DEF \sim \triangle MNP$.

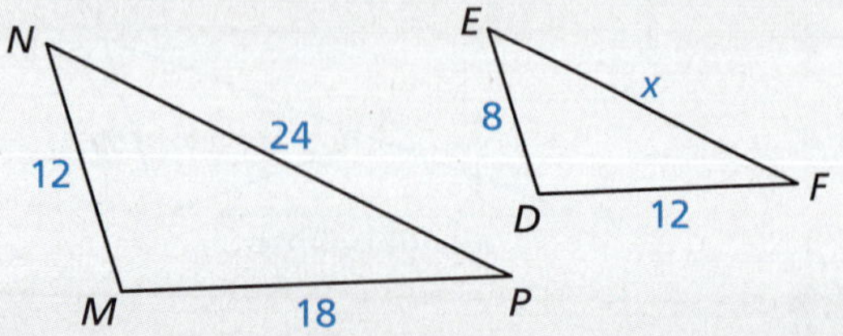

42. Solve the equation $A = \frac{pq}{2}$ for p.

43. Find the length of $\overline{VX}$.

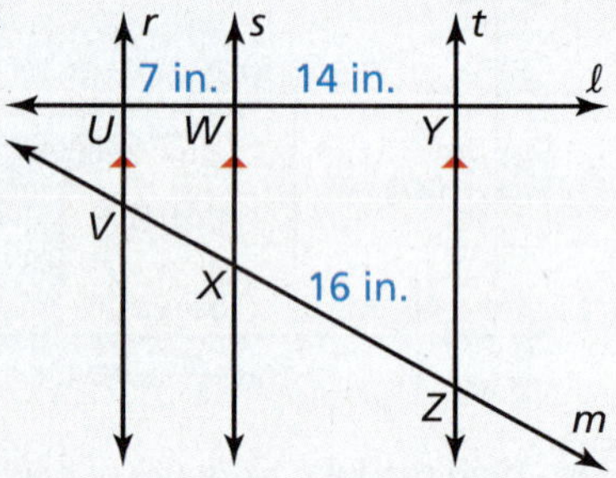

44. Show that the triangles are similar. Write a similarity statement.

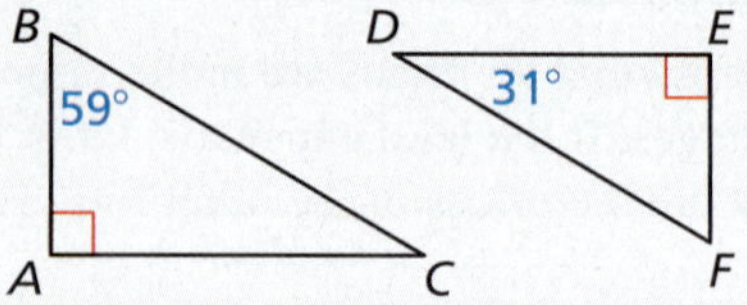

45. The side lengths of $\triangle ABC$ are 10, $6x$, and 20, and the side lengths of $\triangle DEF$ are 25, 30, and 50. Find the value of x that makes $\triangle ABC \sim \triangle DEF$.

46. Find the value of x.

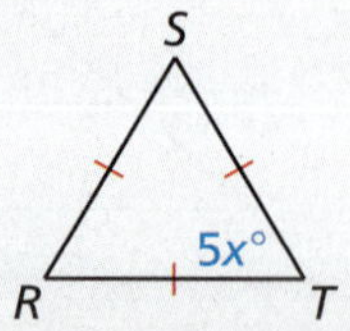

GO DIGITAL

8 Chapter Review WITH CalcChat®

Chapter Learning Target: Understand similarity.

Chapter Success Criteria:
- ◆ I can identify corresponding parts of similar polygons.
- ◆ I can find and use scale factors in similar polygons.
- ■ I can prove triangles are similar.
- ■ I can use proportionality theorems to solve problems.

◆ Surface
■ Deep

SELF-ASSESSMENT 1 I don't understand yet. 2 I can do it with help. 3 I can do it on my own. 4 I can teach someone else.

8.1 Similar Polygons (pp. 409–418) WATCH

Learning Target: Understand the relationship between similar polygons.

Find the scale factor. Then list all pairs of congruent angles and write the ratios of the corresponding side lengths in a statement of proportionality.

1. $ABCD \sim EFGH$

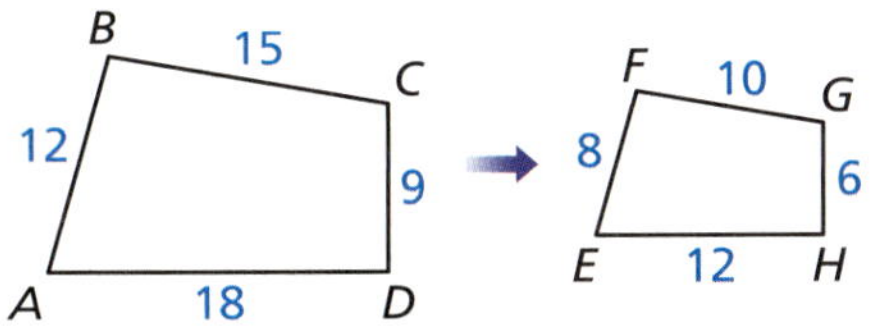

2. $\triangle XYZ \sim \triangle RPQ$

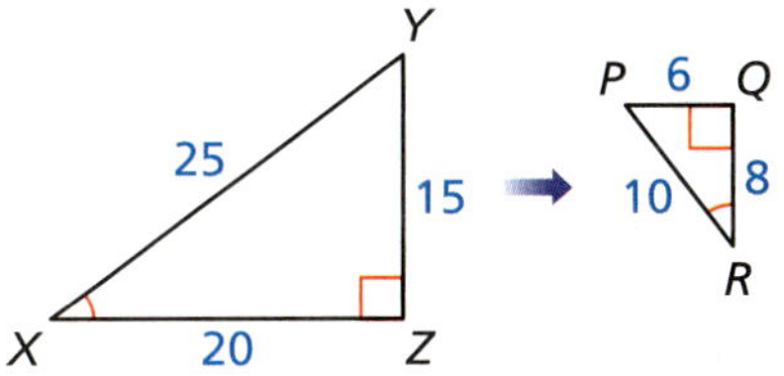

3. The polygons are similar. Find the value of x.

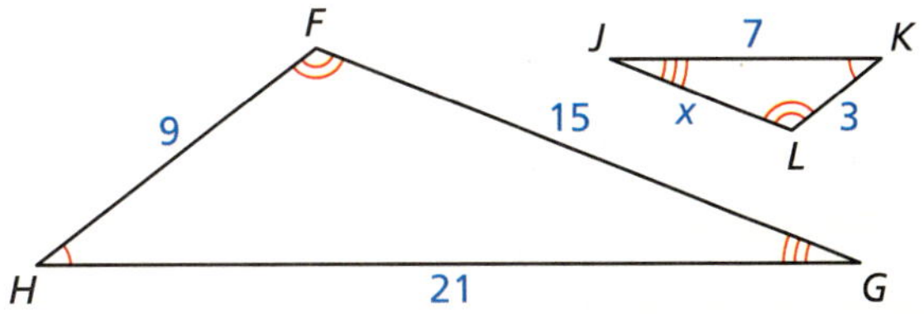

Decide whether the two polygons are similar. Explain your reasoning.

4.

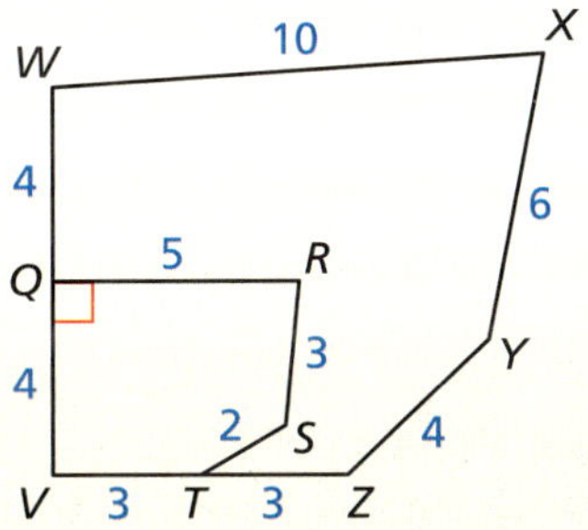

5.

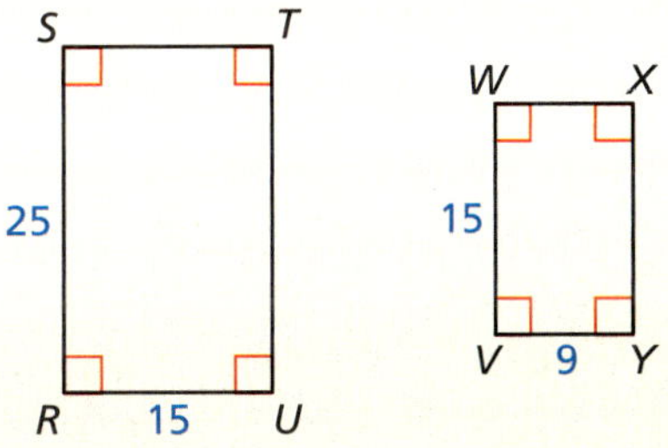

6. A square piece of cloth with an area of 324 square inches is folded in half twice to form the napkin shown. What is the area of the folded napkin? Explain.

GO DIGITAL

8.2 Proving Triangle Similarity by AA (pp. 419–424)

WATCH

Learning Target: Understand and use the Angle-Angle Similarity Theorem.

Determine whether the triangles are similar. If they are, write a similarity statement. Explain your reasoning.

7.

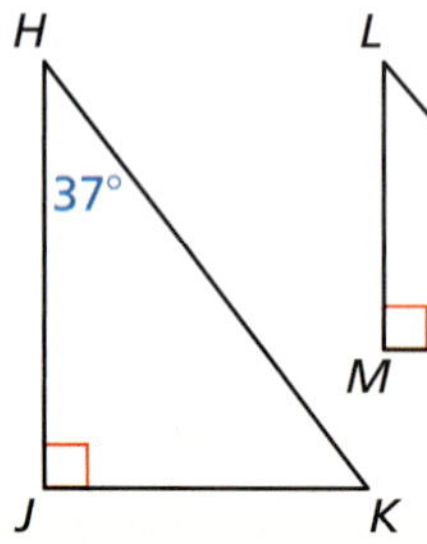

8.

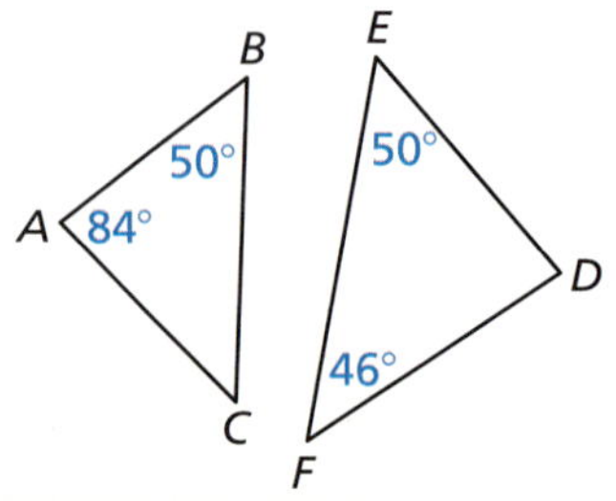

Show that the triangles are similar and write a similarity statement.

9.

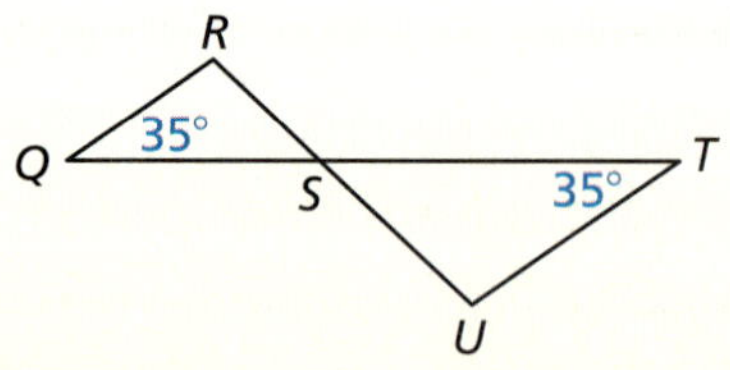

10.

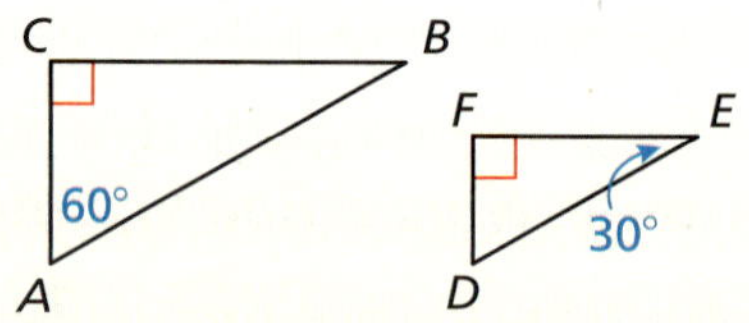

11. A cell tower casts a shadow that is 72 feet long, while a nearby tree that is 27 feet tall casts a shadow that is 6 feet long. How tall is the tower?

8.3 Proving Triangle Similarity by SSS and SAS (pp. 425–434)

WATCH

Learning Target: Understand and use additional triangle similarity theorems.

Use the SSS Similarity Theorem or the SAS Similarity Theorem to show that the triangles are similar and to write a similarity statement. Explain your reasoning.

12.

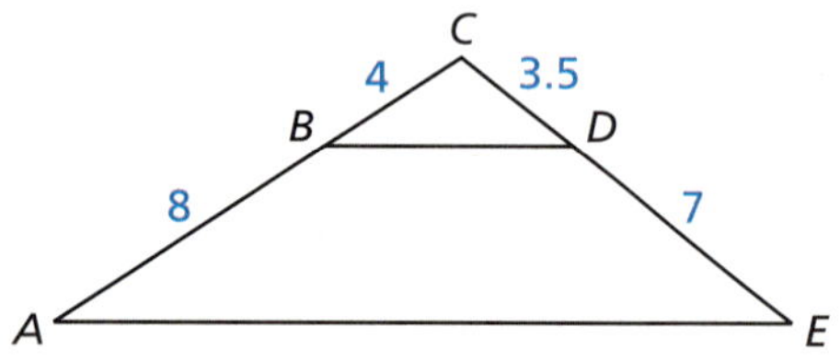

13.

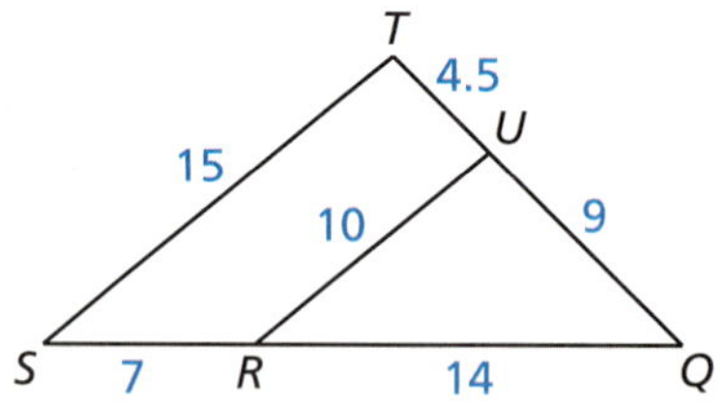

14. Find the value of x that makes $\triangle ABC \sim \triangle DEF$.

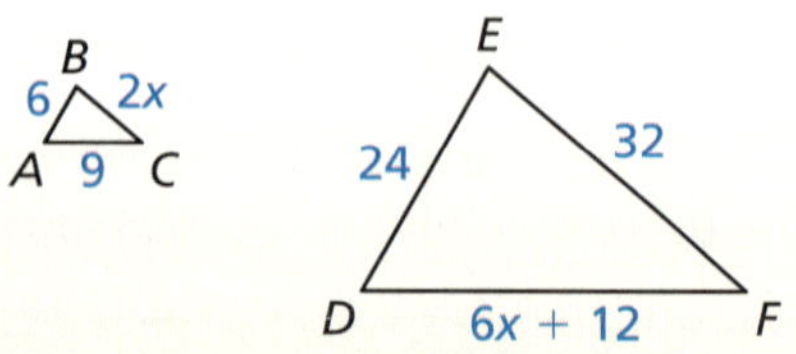

15. The side lengths of $\triangle LMN$ are 20, $8x$, and 60. The side lengths of $\triangle RST$ are 25, 50, and $15x$. Is it possible that the triangles are similar? Explain.

16. $\triangle ABC$ is an isosceles triangle that has an angle with a measure of 50°. $\triangle DEF$ is an isosceles triangle that has an angle with a measure of 40°. Is it possible that the triangles are similar? Explain.

GO DIGITAL

8.4 Proportionality Theorems (pp. 435–442) WATCH

Learning Target: Understand and use proportionality theorems.

Determine whether $\overline{AB} \parallel \overline{CD}$.

17.

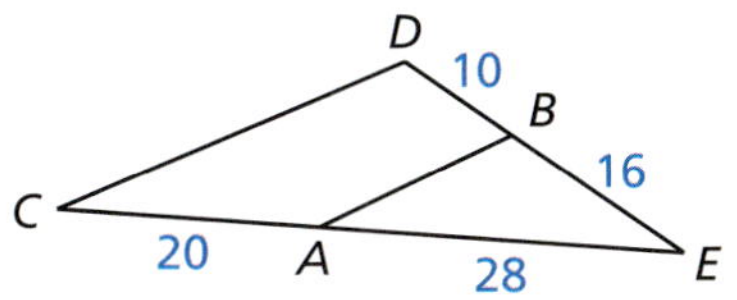

18.

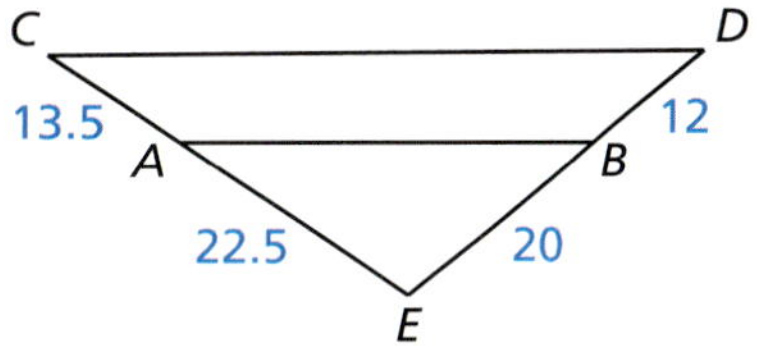

19. Find the length of $\overline{YB}$.

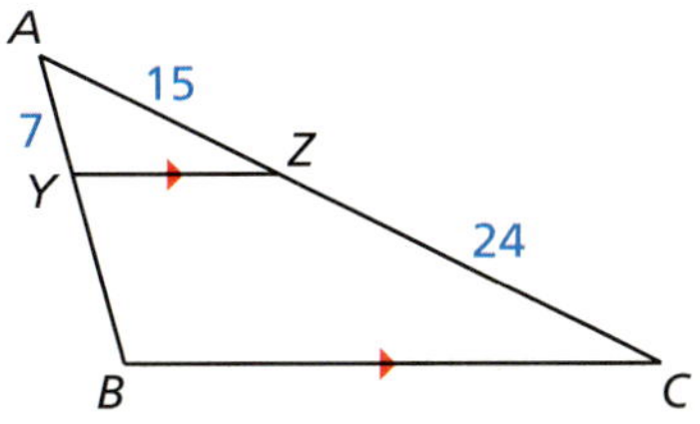

Find the length of $\overline{AB}$.

20.

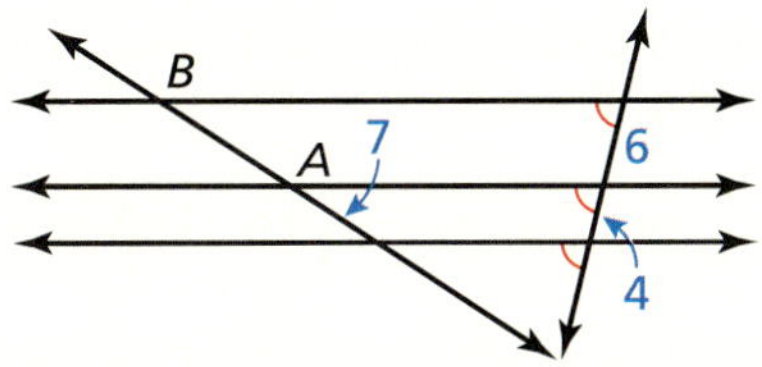

21.

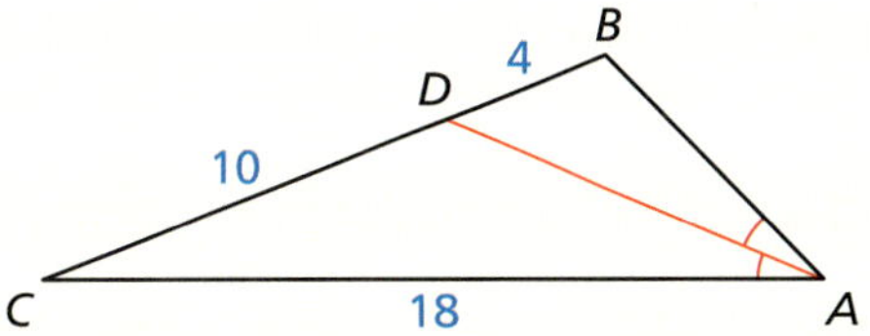

22. Find the distance AC between First Street and Third Street.

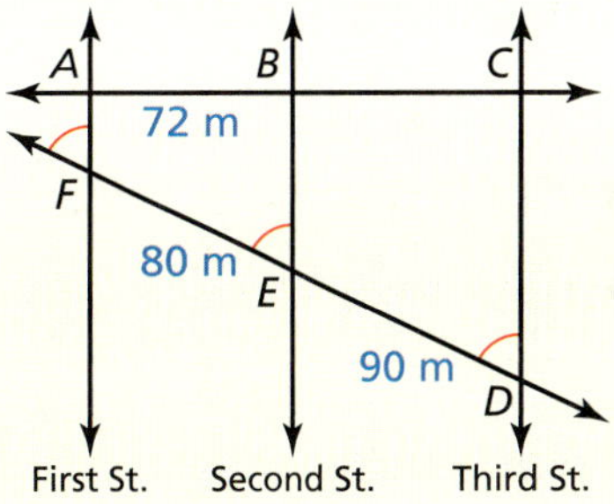

23. Draw $\overline{AB}$ with a length of at least 2 inches. Construct a segment that divides $\overline{AB}$ in the ratio 2 to 3.

Mathematical Thinking and Reasoning

5 MTR **USE PATTERNS AND STRUCTURE TO HELP UNDERSTAND AND CONNECT MATHEMATICAL CONCEPTS**

Mathematicians who use patterns and structure to help understand and connect mathematical concepts look for similarities among problems.

1. In Exercise 27 on page 424, describe the patterns you used to show that any two points on a line can be used to find its slope.
2. Given the side lengths of one triangle, and a side length of a second, similar triangle, explain how to find the perimeter of the second triangle.

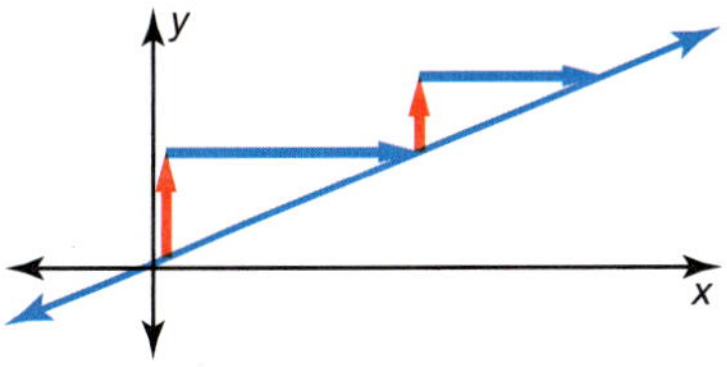

GO DIGITAL

8 Practice Test WITH CalcChat®

Determine whether the triangles are similar. If they are, write a similarity statement. Explain your reasoning.

1.

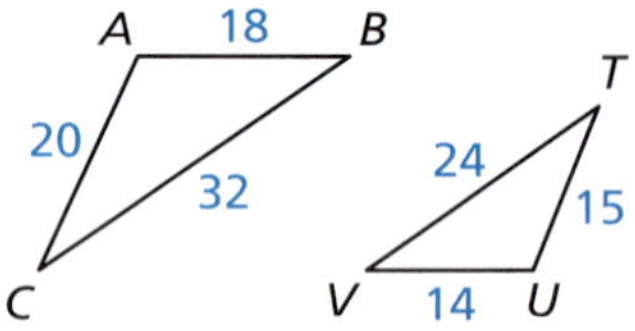

2.

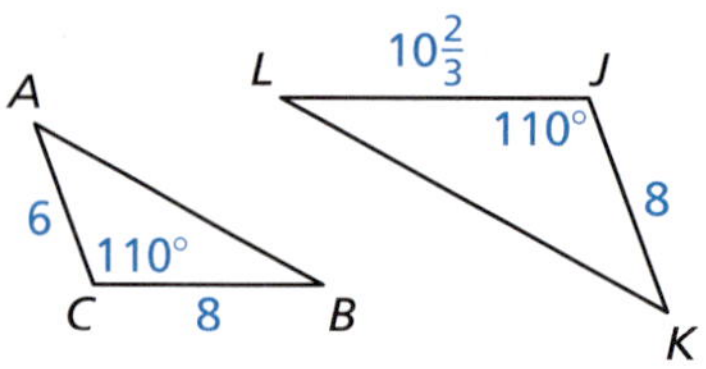

3.

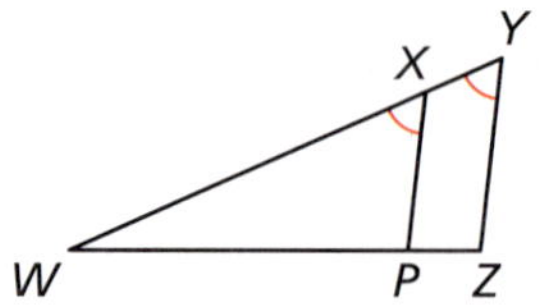

Find the value of the variable.

4.

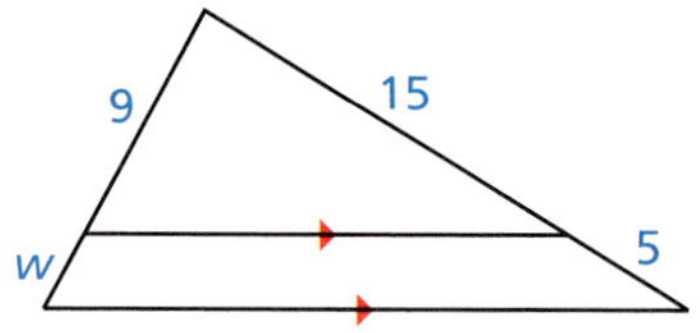

5.

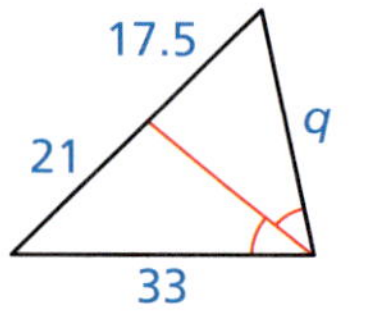

6.

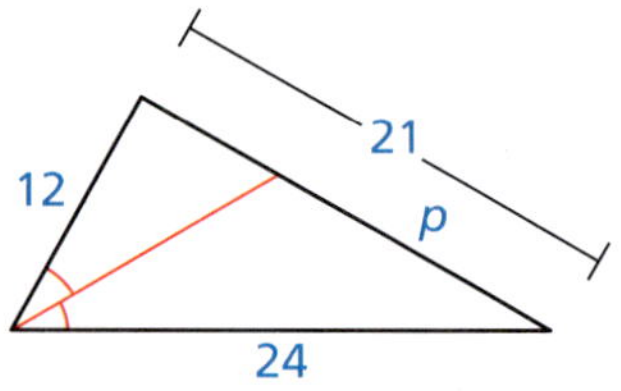

7. Given $\triangle QRS \sim \triangle MNP$, list all pairs of congruent angles. Then write the ratios of the corresponding side lengths in a statement of proportionality.

Use the diagram.

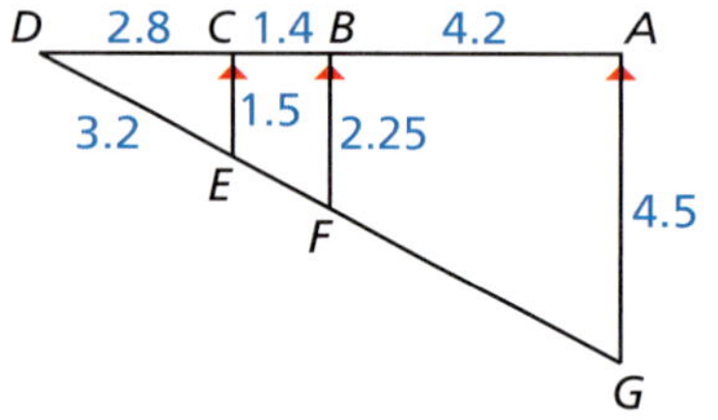

8. Find the length of $\overline{EF}$.

9. Find the length of $\overline{FG}$.

10. Is quadrilateral *FECB* similar to quadrilateral *GFBA*? Explain.

11. Are any two isosceles right triangles similar? Explain.

12. Use the diagram to find the distance *WX* across the water. Justify your answer.

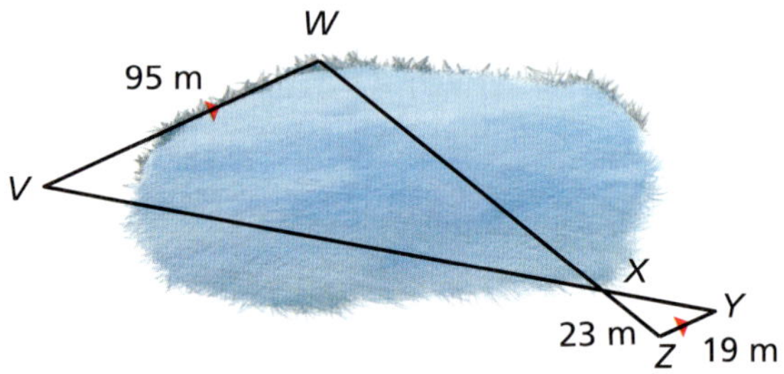

13. You are making a scale model of a rectangular park for a school project. Your model has a length of 2 feet and a width of 1.4 feet. The actual park has a length of 800 yards. What are the perimeter and area of the actual park?

14. You are visiting the Unisphere at Flushing Meadows Corona Park in New York. To estimate the height of the stainless steel model of Earth, you place a mirror on the ground and stand where you can see the top of the model in the mirror. Use the diagram to estimate the height of the model. Explain why this method works.

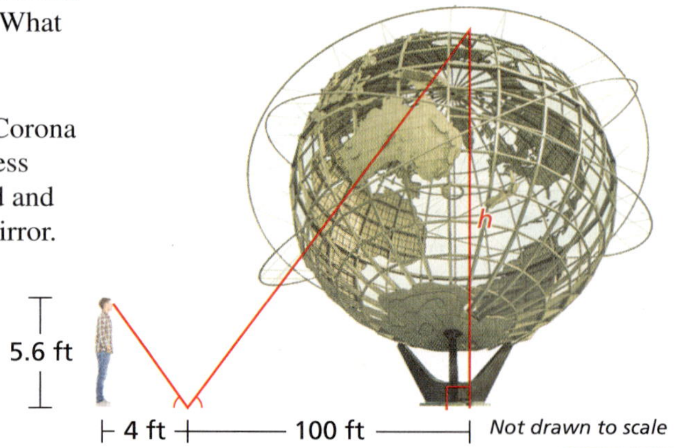

15. Find the value of *a* so that $\triangle ABC$ with vertices $A(-1, 0)$, $B(0, 2)$, and $C(1, 2)$ is similar to $\triangle DBE$ with vertices $D(-3, -4)$, $B(0, 2)$, and $E(a, 2)$.

GO DIGITAL

8 Performance Task

African Burial Ground National Monument

In 1991, archaeologists found human skeletal remains at a site in New York City. A total of 419 bodies were uncovered while excavating the northern portion of the site. Dating from the middle 1630s to 1795, the site is the nation's earliest and largest African burial ground rediscovered in the United States. The site was designated as the nation's 123rd National Monument on February 27, 2006.

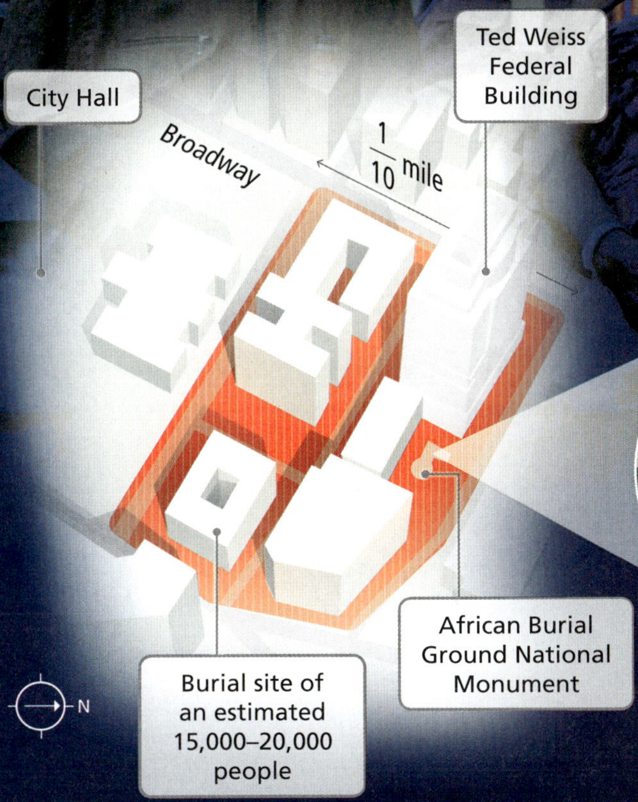

THE MONUMENT CONSISTS OF 7 ELEMENTS:

1. Wall of Remembrance
2. Ancestral Re-interment Grove
3. Memorial Wall
4. Ancestral Chamber
5. Circle of the Diaspora
6. Spiral Processional Ramp
7. Ancestral Libation Court

The Ancestral Chamber extends 24 feet above street level and 6 feet below ground level. The Circle of the Diaspora has an outer radius of 26.5 feet.

CREATING A BROCHURE

An overhead view of the Ancestral Chamber at the African Burial Ground National Monument is shown. What do the two red triangles represent? Explain how you know that the triangles are similar, and approximate the value of x. Then create a brochure for the monument. Your brochure should include information about the seven elements of the monument, the cultural and historical importance of the site, and a scale drawing of the Ancestral Chamber and the Circle of the Diaspora. Indicate the scale factor that you used to create your drawing.

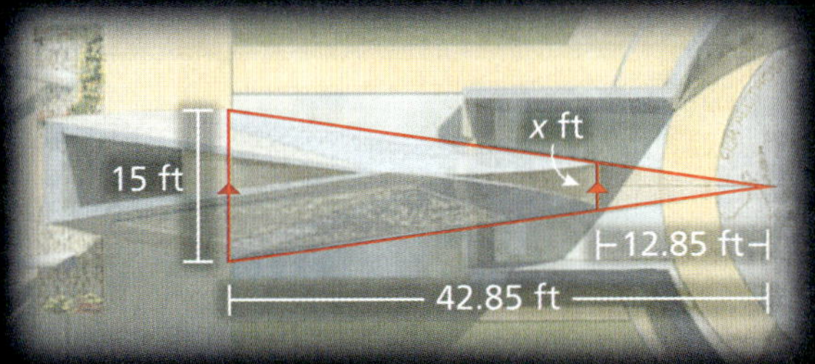

GO DIGITAL

8 Test Prep WITH CalcChat® Cumulative Practice

WATCH Tutorial videos are available for each exercise.

1. Which are possible values for VX and XZ? Select all that apply.

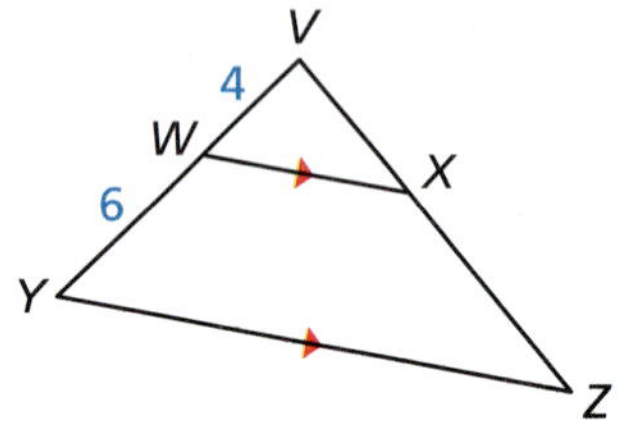

Ⓐ $VX = 5, XZ = 7.5$

Ⓑ $VX = 9, XZ = 6$

Ⓒ $VX = 6, XZ = 8$

Ⓓ $VX = 6, XZ = 9$

2. Classify the quadrilateral using the most specific name.

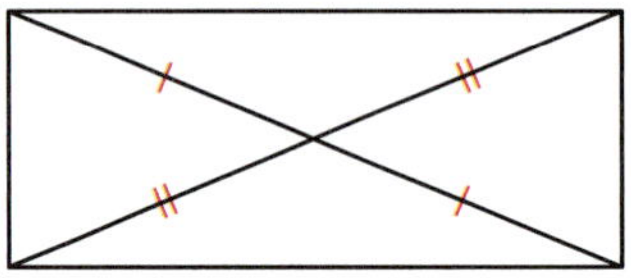

Ⓐ rectangle

Ⓑ square

Ⓒ parallelogram

Ⓓ rhombus

3. A woodworker is cutting the largest wheel possible from a triangular scrap of wood. The wheel touches each side of the triangle, as shown.

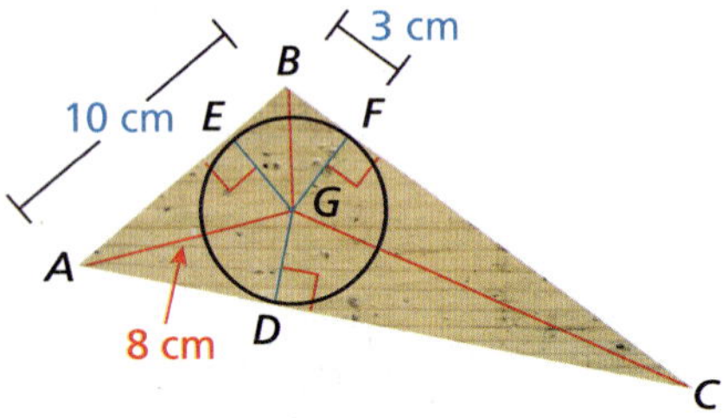

a. Which point of concurrency is the center of the circle?

b. What type of segments are $\overline{BG}$, $\overline{CG}$, and $\overline{AG}$?

c. Find the radius of the wheel. Justify your answer.

4. In the diagram, $ABCD$ is a parallelogram. Which congruence theorem(s) can you use to show that $\triangle AED \cong \triangle CEB$? Select all that apply.

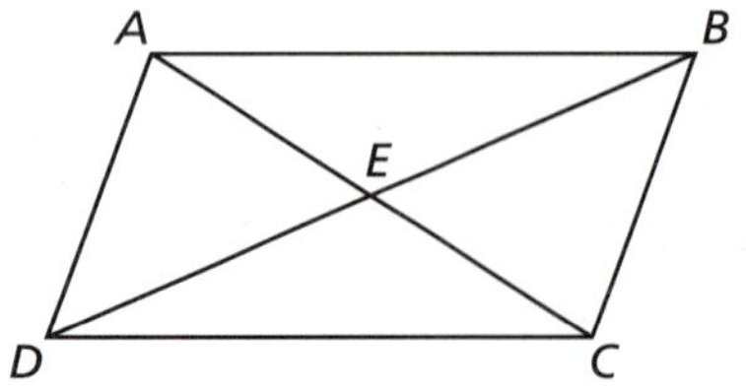

Ⓐ SAS Congruence Theorem

Ⓑ SSS Congruence Theorem

Ⓒ HL Congruence Theorem

Ⓓ SSA Congruence Theorem

Ⓔ AAS Congruence Theorem

GO DIGITAL

5. Complete the two-column proof.

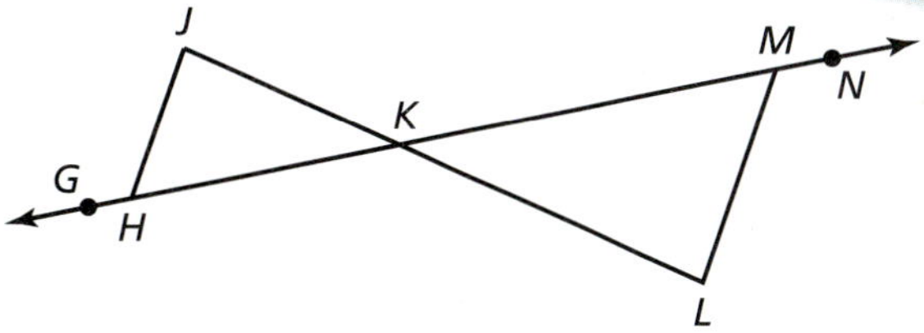

Given $\frac{KJ}{KL} = \frac{KH}{KM}$

Prove $\angle LMN \cong \angle JHG$

STATEMENTS	REASONS
1. $\frac{KJ}{KL} = \frac{KH}{KM}$	1. Given
2. $\angle JKH \cong \angle LKM$	2. ________________
3. $\triangle JKH \sim \triangle LKM$	3. ________________
4. $\angle KHJ \cong \angle KML$	4. ________________
5. ________________	5. Definition of congruent angles
6. $m\angle KHJ + m\angle JHG = 180^\circ$	6. Linear Pair Postulate
7. $m\angle JHG = 180^\circ - m\angle KHJ$	7. ________________
8. $m\angle KML + m\angle LMN = 180^\circ$	8. ________________
9. ________________	9. Subtraction Property of Equality
10. $m\angle LMN = 180^\circ - m\angle KHJ$	10. ________________
11. ________________	11. Transitive Property of Equality
12. $\angle LMN \cong \angle JHG$	12. ________________

6. Find the value of x.

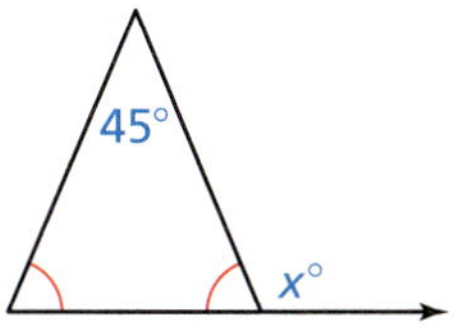

7. Which describes the location of the orthocenter of a right triangle?

Ⓐ inside the triangle

Ⓑ on the triangle

Ⓒ outside the triangle

8. Your friend makes the statement "Quadrilateral *ABCD* is similar to quadrilateral *JKLM*." Describe the relationships between corresponding angles and between corresponding sides that make this statement true.

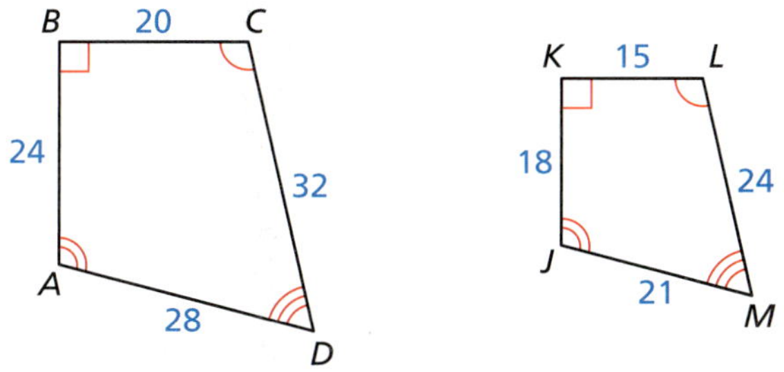

GO DIGITAL

9 Right Triangles and Trigonometry

WATCH

NATIONAL GEOGRAPHIC EXPLORER

Beth Shapiro

Dr. Beth Shapiro researches how populations and species change through time by studying genetic material extracted from fossils and other remains. She wrote *How to Clone a Mammoth*, a book describing the scientific techniques needed and challenges that must be overcome to revive an extinct species.

- What does *de-extinction* mean?
- What does *cloning* mean? Have any animals been cloned?
- Where did woolly mammoths live? When did they become extinct?
- What is a mastodon? How are mammoths and mastodons related to modern-day elephants?

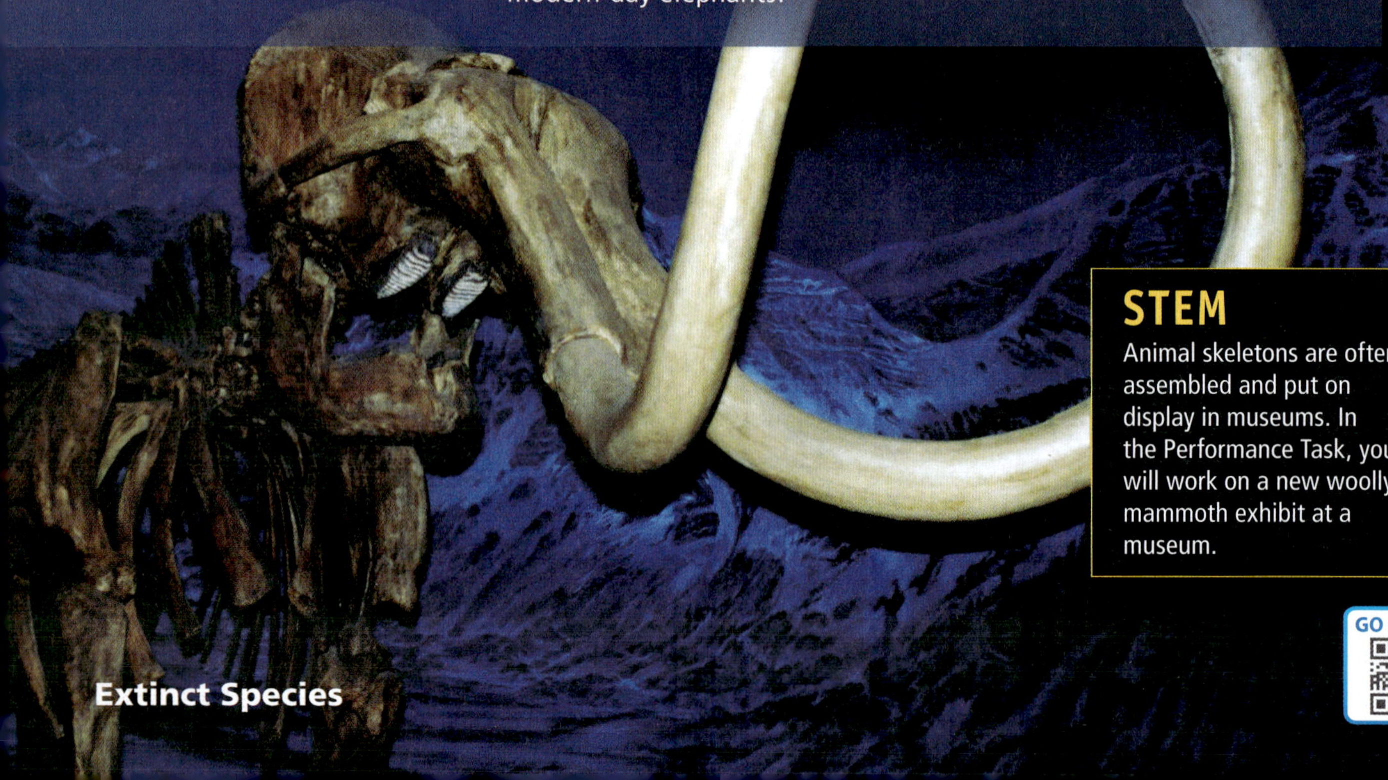

STEM

Animal skeletons are often assembled and put on display in museums. In the Performance Task, you will work on a new woolly mammoth exhibit at a museum.

GO DIGITAL

Extinct Species

Preparing for Chapter 9

Chapter Learning Target: Understand right triangles and trigonometry.

Chapter Success Criteria:
- ◆ I can use the Pythagorean Theorem to solve problems.
- ◆ I can find side lengths in special right triangles.
- ■ I can explain how similar triangles are used with trigonometric ratios.
- ■ I can use trigonometric ratios to solve problems.

◆ Surface
■ Deep

AZ VOCAB

Chapter Vocabulary

Work with a partner. Discuss each of the vocabulary terms.

Pythagorean triple
trigonometric ratio
angle of elevation
angle of depression

Mathematical Thinking and Reasoning

5 MTR

USE PATTERNS AND STRUCTURE TO HELP UNDERSTAND AND CONNECT MATHEMATICAL CONCEPTS

Mathematicians who use patterns and structure to help understand and connect mathematical concepts focus on relevant details in problems.

Work with a partner. You are hiking in a canyon and want to find the height of the canyon wall. You use a cardboard square to line up the top and bottom of the canyon wall with your eye.

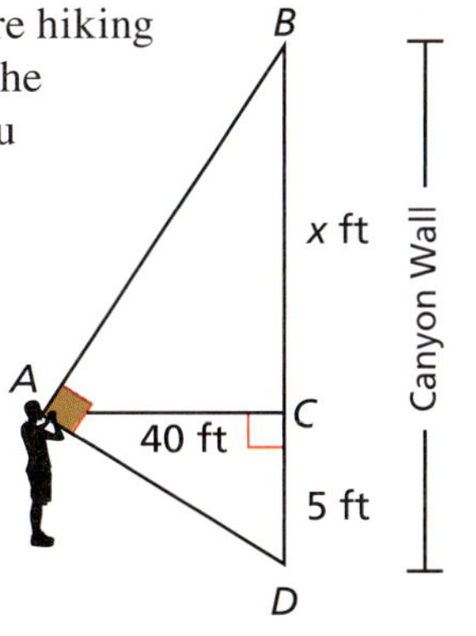

Not drawn to scale

1. Identify any similar triangles in the diagram. Explain how you know that they are similar.
2. Describe the relationship among side lengths in similar triangles.
3. Use your answers in Exercises 1 and 2 to describe the relationships among the lengths in this situation. Then find the height of the canyon wall. Explain your method.

GO DIGITAL

9 Prepare WITH CalcChat®

Solving Proportions

WATCH

Example 1 Solve $\frac{x}{10} = \frac{3}{2}$.

$\frac{x}{10} = \frac{3}{2}$ Write the proportion.

$x \cdot 2 = 10 \cdot 3$ Cross Products Property

$2x = 30$ Multiply.

$\frac{2x}{2} = \frac{30}{2}$ Division Property of Equality

$x = 15$ Simplify.

Solve the proportion.

1. $\frac{x}{12} = \frac{3}{4}$

2. $\frac{x}{3} = \frac{5}{2}$

3. $\frac{4}{x} = \frac{7}{56}$

4. $\frac{10}{23} = \frac{4}{x}$

5. $\frac{x+1}{2} = \frac{21}{14}$

6. $\frac{9}{3x-15} = \frac{3}{12}$

Using Properties of Radicals

WATCH

Example 2 Simplify $\sqrt{128}$.

$\sqrt{128} = \sqrt{64 \cdot 2}$ Factor using the greatest perfect square factor.

$= \sqrt{64} \cdot \sqrt{2}$ Product Property of Square Roots

$= 8\sqrt{2}$ Simplify.

WATCH

Example 3 Simplify $\frac{4}{\sqrt{5}}$.

$\frac{4}{\sqrt{5}} = \frac{4}{\sqrt{5}} \cdot \frac{\sqrt{5}}{\sqrt{5}}$ Multiply by $\frac{\sqrt{5}}{\sqrt{5}}$.

$= \frac{4\sqrt{5}}{\sqrt{25}}$ Product Property of Square Roots

$= \frac{4\sqrt{5}}{5}$ Simplify.

Simplify the expression.

7. $\sqrt{75}$

8. $\sqrt{270}$

9. $\sqrt{135}$

10. $\frac{2}{\sqrt{7}}$

11. $\frac{5}{\sqrt{2}}$

12. $\frac{12}{\sqrt{6}}$

5 MTR **13. STRUCTURE** Are the expressions $3\sqrt{64x^3}$ and $4x\sqrt{36x}$ equivalent? Justify your answer.

GO DIGITAL

9.1 The Pythagorean Theorem

Learning Target: Understand and apply the Pythagorean Theorem.

Success Criteria:
- I can list common Pythagorean triples.
- I can find missing side lengths of right triangles.
- I can classify a triangle as acute, right, or obtuse given its side lengths.

EXPLORE IT! Proving the Pythagorean Theorem

Work with a partner. A **pictorial proof** uses diagrams to demonstrate a logical argument. In this Explore It!, you will perform a pictorial proof of the Pythagorean Theorem.

a. Draw and cut out a right triangle with legs a and b, and hypotenuse c. Make three additional copies of your right triangle. Arrange all four triangles to form a large square, as shown.

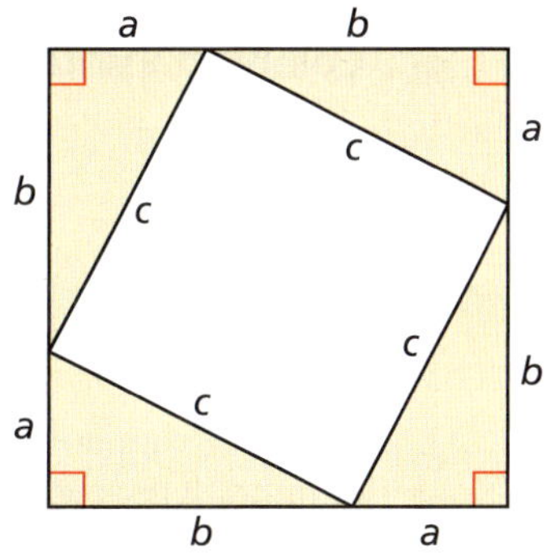

b. Find the area of the large square in terms of a, b, and c. Explain your reasoning.

c. Copy the large square. Divide it into two smaller squares and two congruent rectangles, as shown.

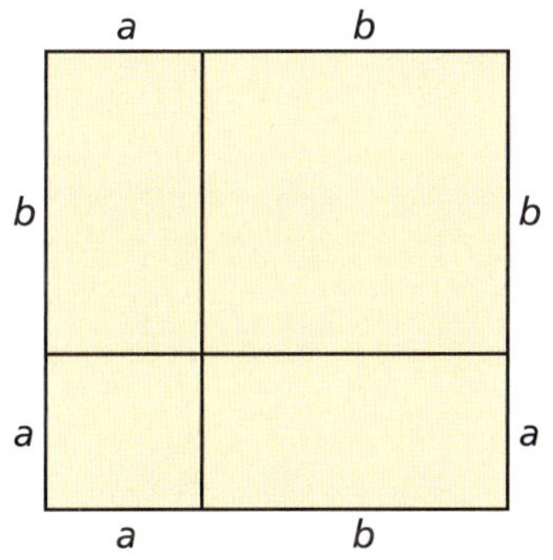

d. Find the area of the large square in terms of a and b. Explain your reasoning.

e. Use your answers in parts (b) and (d) to justify the Pythagorean Theorem.

Pythagorean Theorem:

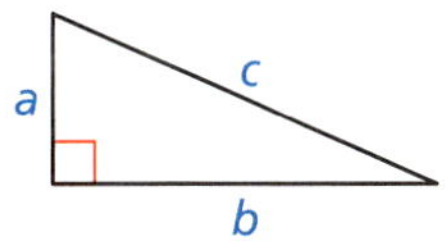

$c^2 = a^2 + b^2$

4 MTR COMMUNICATE CLEARLY
In your own words, explain the meaning of the Pythagorean Theorem.

Geometric Reasoning
MA.912.GR.1.3 Prove relationships and theorems about triangles. Solve mathematical and real-world problems involving postulates, relationships and theorems of triangles.
Trigonometry
MA.912.T.1.2 Solve mathematical and real-world problems involving right triangles using trigonometric ratios and the Pythagorean Theorem.

GO DIGITAL

Using the Pythagorean Theorem

Vocabulary

pictorial proof, *p. 453*
Pythagorean triple, *p. 454*

One of the most famous theorems in mathematics is the Pythagorean Theorem, named for the ancient Greek mathematician Pythagoras. This theorem describes the relationship among the side lengths of a right triangle.

THEOREM

9.1 Pythagorean Theorem

In a right triangle, the square of the length of the hypotenuse is equal to the sum of the squares of the lengths of the legs.

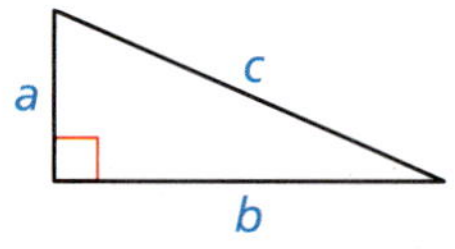

$c^2 = a^2 + b^2$

Prove this Theorem Explore It!, page 453; Exercise 33, page 459; Exercise 35, page 473

A **Pythagorean triple** is a set of three positive integers a, b, and c that satisfy the equation $c^2 = a^2 + b^2$.

KEY IDEA

Common Pythagorean Triples and Some of Their Multiples

3, 4, 5	**5, 12, 13**	**8, 15, 17**	**7, 24, 25**
6, 8, 10	10, 24, 26	16, 30, 34	14, 48, 50
9, 12, 15	15, 36, 39	24, 45, 51	21, 72, 75
$3x, 4x, 5x$	$5x, 12x, 13x$	$8x, 15x, 17x$	$7x, 24x, 25x$

The most common Pythagorean triples are in bold. The other triples are the result of multiplying each integer in a boldfaced triple by the same positive integer.

STUDY TIP

You may find it helpful to memorize the basic Pythagorean triples, shown in **bold**, for standardized tests.

EXAMPLE 1 Using the Pythagorean Theorem WATCH

Find the value of x. Then tell whether the side lengths form a Pythagorean triple.

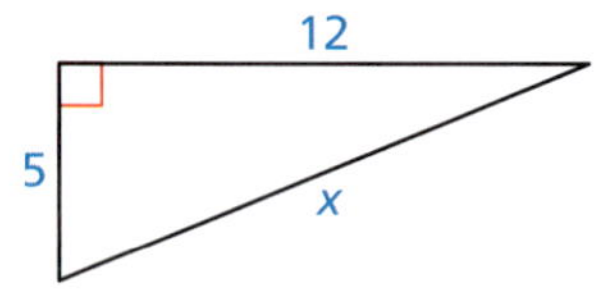

SOLUTION

$c^2 = a^2 + b^2$	Pythagorean Theorem
$x^2 = 5^2 + 12^2$	Substitute.
$x^2 = 25 + 144$	Multiply.
$x^2 = 169$	Add.
$x = 13$	Take the positive square root of each side.

▶ The value of x is 13. Because the side lengths 5, 12, and 13 are integers that satisfy the equation $c^2 = a^2 + b^2$, they form a Pythagorean triple.

GO DIGITAL

EXAMPLE 2 Using the Pythagorean Theorem

WATCH

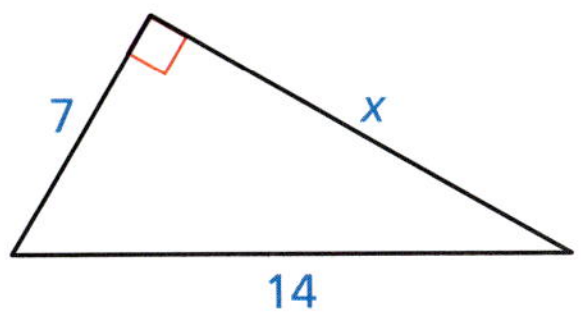

Find the value of x. Then tell whether the side lengths form a Pythagorean triple.

SOLUTION

$c^2 = a^2 + b^2$	Pythagorean Theorem
$14^2 = 7^2 + x^2$	Substitute.
$196 = 49 + x^2$	Multiply.
$147 = x^2$	Subtract 49 from each side.
$\sqrt{147} = x$	Take the positive square root of each side.
$\sqrt{49} \cdot \sqrt{3} = x$	Product Property of Square Roots
$7\sqrt{3} = x$	Simplify.

▶ The value of x is $7\sqrt{3}$. Because $7\sqrt{3}$ is not an integer, the side lengths do not form a Pythagorean triple.

7 MTR

EXAMPLE 3 Modeling Real Life

WATCH

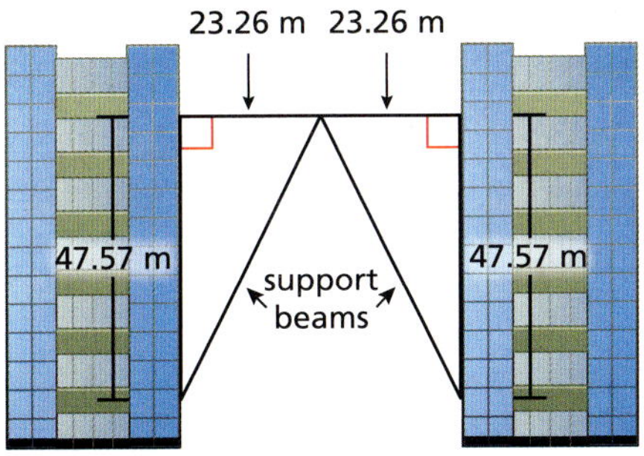

The skyscrapers shown are connected by a skywalk with support beams. Find the length of each support beam.

SOLUTION

Each support beam forms the hypotenuse of a right triangle. Use the Pythagorean Theorem to find the length x (in meters) of each support beam.

$x^2 = (23.26)^2 + (47.57)^2$	Pythagorean Theorem
$x = \sqrt{(23.26)^2 + (47.57)^2}$	Take the positive square root of each side.
$x \approx 52.95$	Use technology.

▶ The length of each support beam is about 52.95 meters.

SELF-ASSESSMENT

1 I don't understand yet. 2 I can do it with help. 3 I can do it on my own. 4 I can teach someone else.

4 MTR **1. DIFFERENT WORDS, SAME QUESTION** Which is different? Find "both" answers.

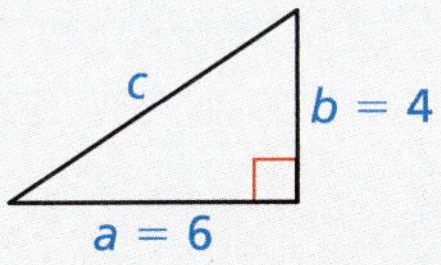

What is the length of the hypotenuse?

What is the value of $a^2 + b^2$?

What is the value of $\sqrt{a^2 + b^2}$?

What is the value of c?

2. Do the side lengths of a right triangle with a hypotenuse of 15 feet and a leg length of 9 feet form a Pythagorean triple? Justify your answer.

3. An anemometer is a device used to measure wind speed. The anemometer shown is attached to the top of a pole. Support wires are attached to the pole 5 feet above the ground. Each support wire is 6 feet long. How far from the base of the pole is each wire attached to the ground?

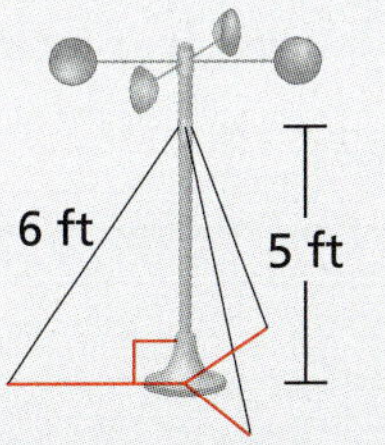

GO DIGITAL

Using the Converse of the Pythagorean Theorem

The converse of the Pythagorean Theorem is also true. You can use it to determine whether a triangle with given side lengths is a right triangle.

THEOREM

9.2 Converse of the Pythagorean Theorem

If the square of the length of the longest side of a triangle is equal to the sum of the squares of the lengths of the other two sides, then the triangle is a right triangle.

If $c^2 = a^2 + b^2$, then $\triangle ABC$ is a right triangle.

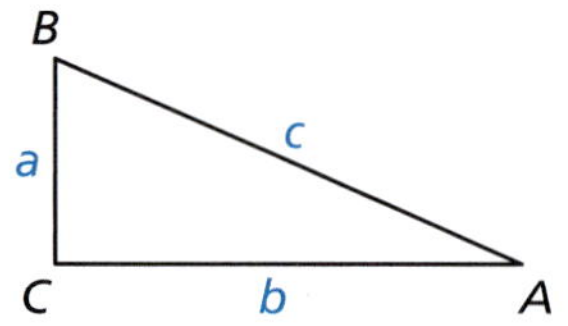

Prove this Theorem Exercise 34, page 459

EXAMPLE 4 Verifying Right Triangles

WATCH

Tell whether each triangle is a right triangle.

a.

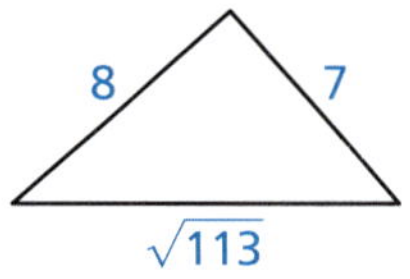

b.

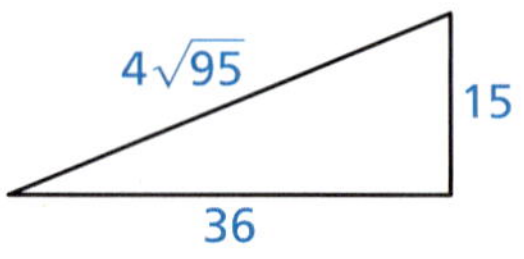

SOLUTION

Let c represent the length of the longest side of the triangle. Check to see whether the side lengths satisfy the equation $c^2 = a^2 + b^2$.

a. $(\sqrt{113})^2 \stackrel{?}{=} 7^2 + 8^2$

$113 \stackrel{?}{=} 49 + 64$

$113 = 113$ ✓

▶ The triangle is a right triangle.

STUDY TIP

Use technology to determine that $\sqrt{113} \approx 10.630$ is the length of the longest side in part (a).

b. $(4\sqrt{95})^2 \stackrel{?}{=} 15^2 + 36^2$

$4^2 \cdot (\sqrt{95})^2 \stackrel{?}{=} 15^2 + 36^2$

$16 \cdot 95 \stackrel{?}{=} 225 + 1296$

$1520 \neq 1521$ ✗

▶ The triangle is *not* a right triangle.

SELF-ASSESSMENT

1 I don't understand yet. 2 I can do it with help. 3 I can do it on my own. 4 I can teach someone else.

Tell whether the triangle is a right triangle.

4.

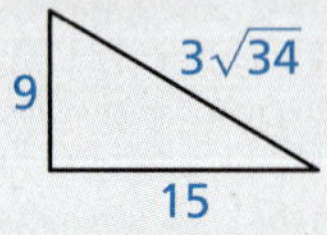

5.

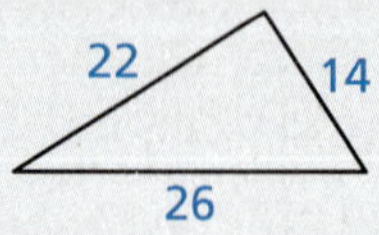

6.

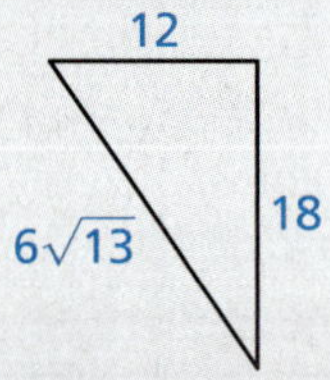

GO DIGITAL

Classifying Triangles

The Converse of the Pythagorean Theorem is used to determine whether a triangle is a right triangle. You can use the theorem below to determine whether a triangle is acute or obtuse.

THEOREM

9.3 Pythagorean Inequalities Theorem

For any $\triangle ABC$, where c is the length of the longest side, the following statements are true.

If $c^2 < a^2 + b^2$, then $\triangle ABC$ is acute.

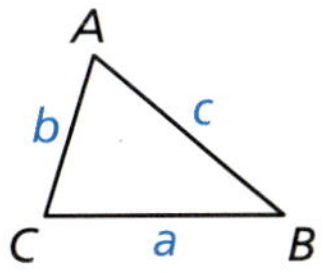

$c^2 < a^2 + b^2$

If $c^2 > a^2 + b^2$, then $\triangle ABC$ is obtuse.

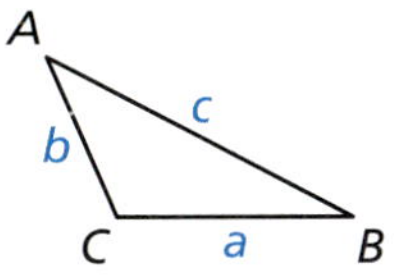

$c^2 > a^2 + b^2$

Prove this Theorem Exercises 35 and 36, page 459

EXAMPLE 5 Classifying Triangles

WATCH

Determine whether segments with lengths of 4.3 feet, 5.2 feet, and 6.1 feet form a triangle. If so, is the triangle *acute*, *right*, or *obtuse*?

SOLUTION

Step 1 Use the Triangle Inequality Theorem to determine whether the segments form a triangle.

$4.3 + 5.2 \overset{?}{>} 6.1$ $\qquad 4.3 + 6.1 \overset{?}{>} 5.2$ $\qquad 5.2 + 6.1 \overset{?}{>} 4.3$

$9.5 > 6.1$ ✓ $\qquad 10.4 > 5.2$ ✓ $\qquad 11.3 > 4.3$ ✓

▶ The segments with lengths of 4.3 feet, 5.2 feet, and 6.1 feet form a triangle.

REMEMBER
The Triangle Inequality Theorem on page 333 states that the sum of the lengths of any two sides of a triangle is greater than the length of the third side.

Step 2 Classify the triangle by comparing the square of the length of the longest side with the sum of the squares of the lengths of the other two sides.

c^2		$a^2 + b^2$	Compare c^2 with $a^2 + b^2$.
6.1^2		$4.3^2 + 5.2^2$	Substitute.
37.21		18.49 + 27.04	Simplify.
37.21	$<$	45.53	Add.

▶ Because $c^2 < a^2 + b^2$, the segments with lengths of 4.3 feet, 5.2 feet, and 6.1 feet form an acute triangle.

SELF-ASSESSMENT 1 I don't understand yet. 2 I can do it with help. 3 I can do it on my own. 4 I can teach someone else.

Determine whether the segment lengths form a triangle. If so, is the triangle *acute*, *right*, or *obtuse*?

7. 3, 4, 6 **8.** 2.1, 2.8, 3.5 **9.** 4.6, 2.8, 7.4

GO DIGITAL

9.1 Practice WITH CalcChat® AND CalcView®

In Exercises 1–4, find the value of x. Then tell whether the side lengths form a Pythagorean triple. *(See Examples 1 and 2.)*

▶ **1.**

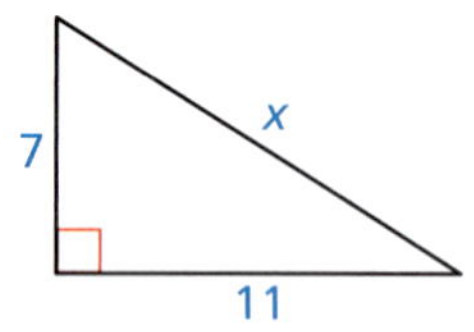

2.

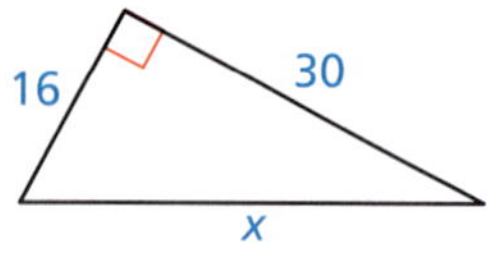

▶ **3.**

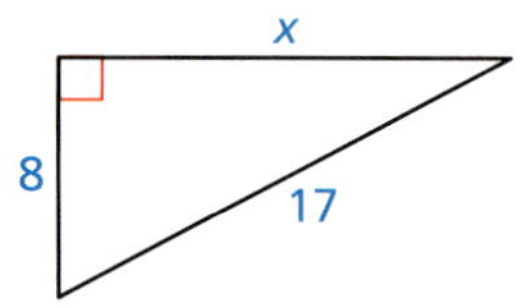

4.

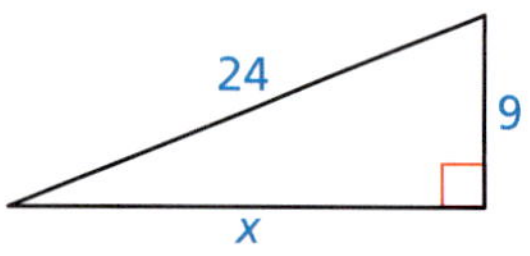

ERROR ANALYSIS **In Exercises 5 and 6, describe and correct the error in using the Pythagorean Theorem.**

5.

✗

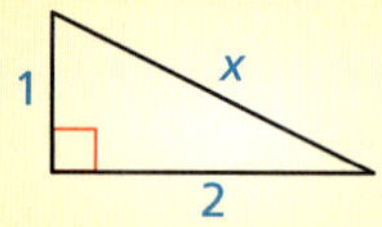

$c = a^2 + b^2$

$x = 1^2 + 2^2$

$x = 1 + 4$

$x = 5$

6.

✗

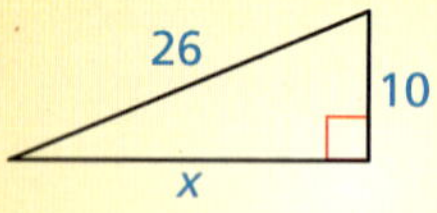

$c^2 = a^2 + b^2$

$x^2 = 10^2 + 26^2$

$x^2 = 100 + 676$

$x^2 = 776$

$x = \sqrt{776}$

$x \approx 27.9$

7. **MODELING REAL LIFE** Find the distance between the two platforms of the fire escape. *(See Example 3.)*

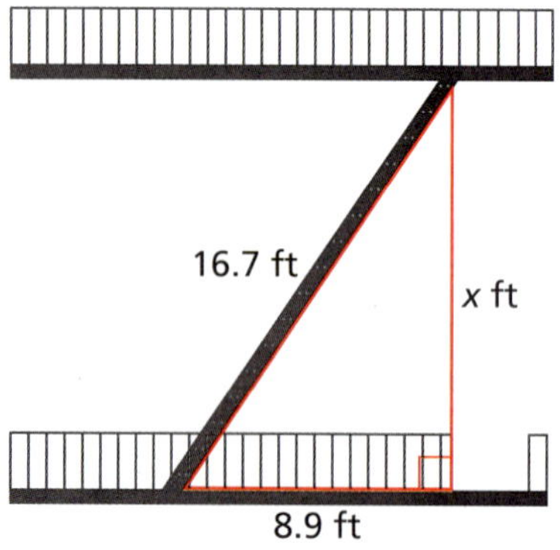

8. **MODELING REAL LIFE** Snowcat Ridge Alpine Snow Park, the first outdoor snow park in Florida, opened in Dade City in 2020. The park features a snow tubing hill shown below. Find the distance x from the top of the hill to the bottom.

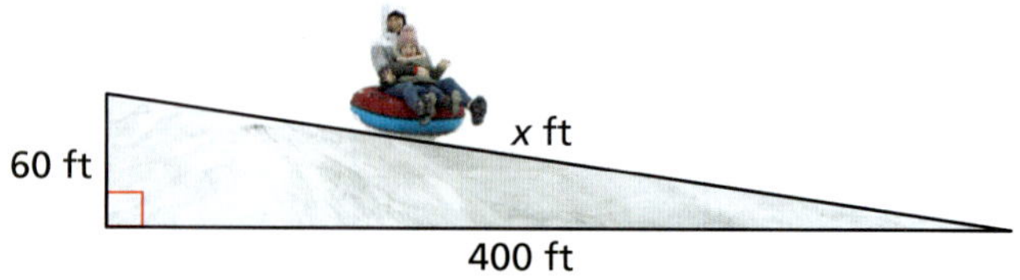

In Exercises 9–14, tell whether the triangle is a right triangle. *(See Example 4.)*

9.

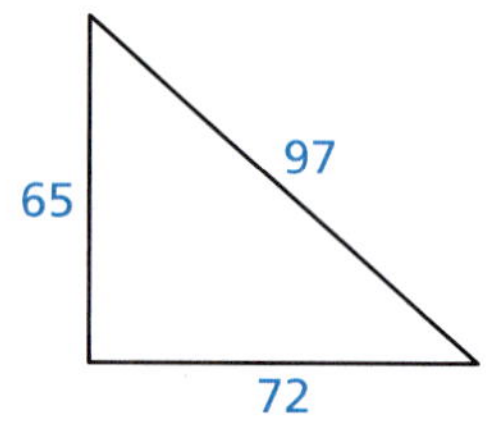

10.

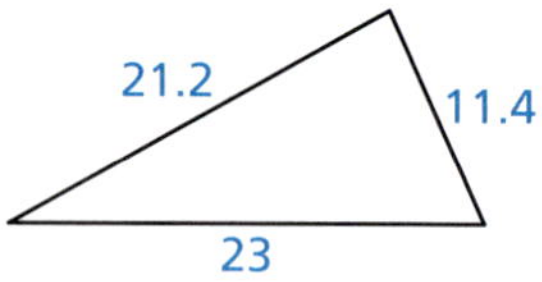

▶ **11.**

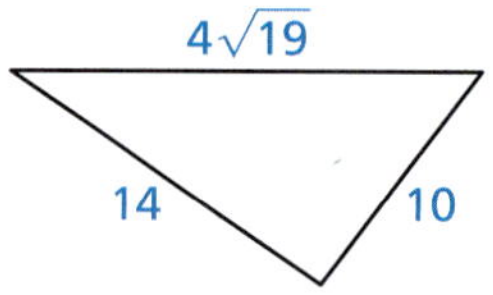

12.

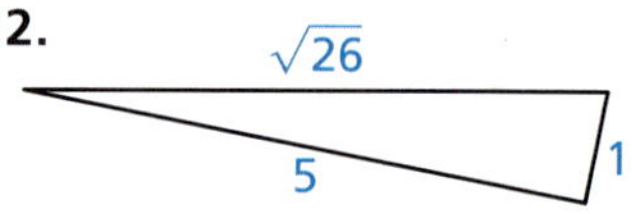

13.

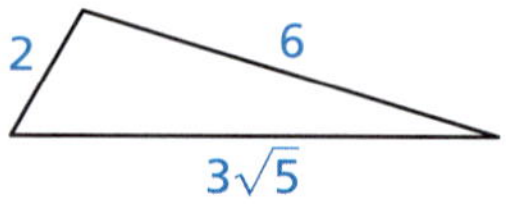

14.

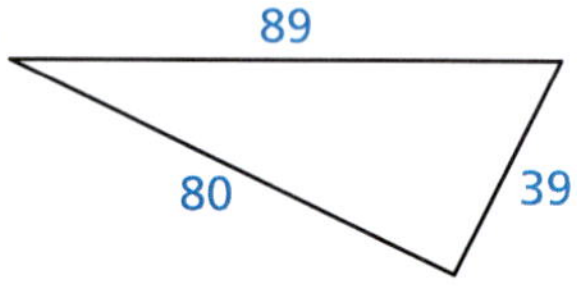

In Exercises 15–22, determine whether the segment lengths form a triangle. If so, is the triangle *acute*, *right*, or *obtuse*? *(See Example 5.)*

▶ **15.** 10, 11, and 14

16. 6, 8, and 14

17. 12, 16, and 20

18. 15, 20, and 26

19. 4.1, 8.2, and 12.4

20. 5.3, 6.7, and 7.8

21. 24, 30, and $6\sqrt{43}$

22. 10, 15, and $5\sqrt{13}$

GO DIGITAL

23. **MODELING REAL LIFE** In baseball, the lengths of the paths between consecutive bases are 90 feet, and the paths form right angles. The catcher throws the ball from home plate to second base. How far does the ball travel?

24. **REASONING** You are making a frame for a painting. The rectangular painting will be 18 inches long and 24 inches wide. Using a yardstick, how can you be certain that the corners of the frame are 90°?

In Exercises 25 and 26, find the area of the isosceles triangle.

25.

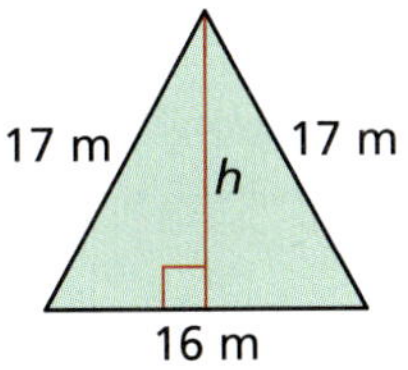

26.

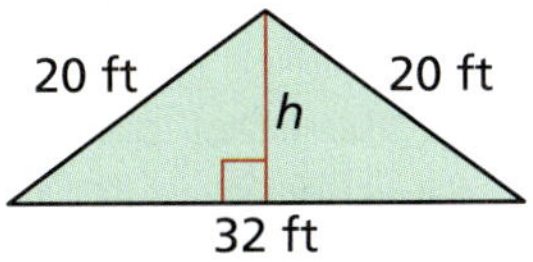

27. **STRUCTURE** Justify the Distance Formula using the Pythagorean Theorem.

28. **HOW DO YOU SEE IT?**
Without performing any calculations, how do you know $\angle C$ is a right angle?

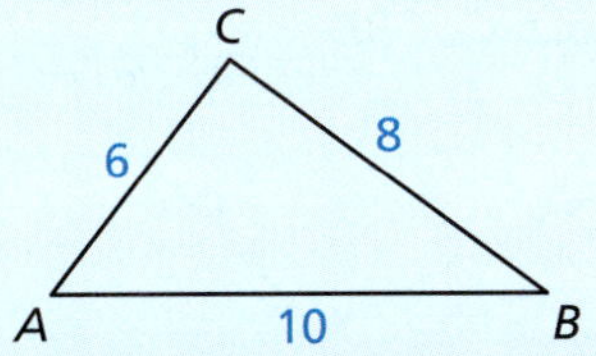

29. **ASSESS REASONABLENESS** You and a friend are making a kite and need to figure out how much binding to buy for the perimeter of the kite. The binding comes in packages of two yards. Your friend claims you need two packages of binding. Is your friend's claim reasonable? Explain.

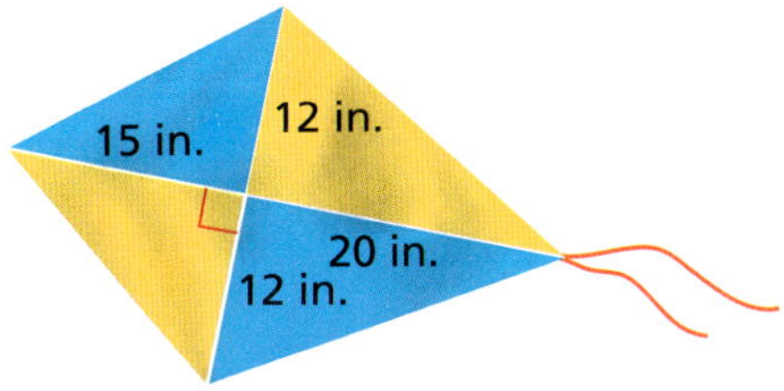

30. **PROVING A THEOREM** Prove the Hypotenuse-Leg (HL) Congruence Theorem.

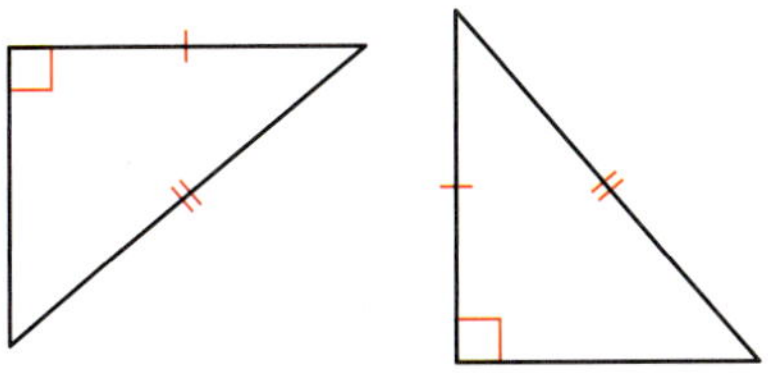

GO DIGITAL

31. **WRITING** Describe two ways, other than measuring, to determine whether $\triangle XYZ$ with vertices $X(1, -2)$, $Y(-1, 6)$, and $Z(5, -1)$ is a right triangle.

32. **REASONING** Can 96 and 100 be part of a Pythagorean triple? Explain your reasoning.

33. **PROVING A THEOREM** Explain why $\triangle ABC$, $\triangle ACD$, and $\triangle CBD$ are similar. Then use the similar triangles to prove the Pythagorean Theorem.

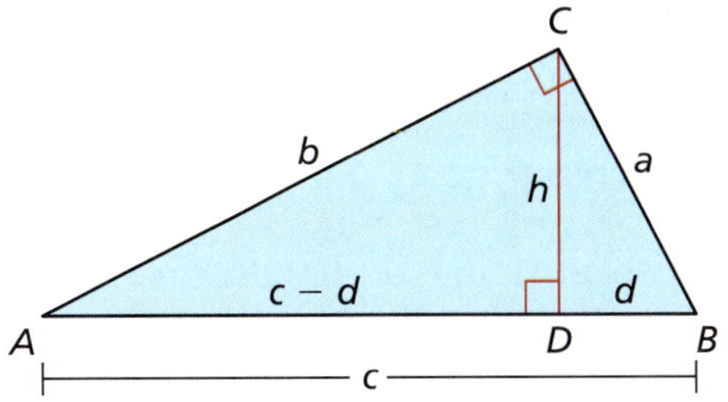

34. **PROVING A THEOREM** Prove the Converse of the Pythagorean Theorem. (*Hint*: Draw $\triangle ABC$ with side lengths a, b, and c, where c is the length of the longest side. Then draw a right triangle with side lengths a, b, and x, where x is the length of the hypotenuse. Compare lengths c and x.)

35. **PROVING A THEOREM** Prove the Pythagorean Inequalities Theorem when $c^2 < a^2 + b^2$.

Given In $\triangle ABC$, $c^2 < a^2 + b^2$, where c is the length of the longest side. $\triangle PQR$ has side lengths a, b, and x, where x is the length of the hypotenuse, and $\angle R$ is a right angle.

Prove $\triangle ABC$ is an acute triangle.

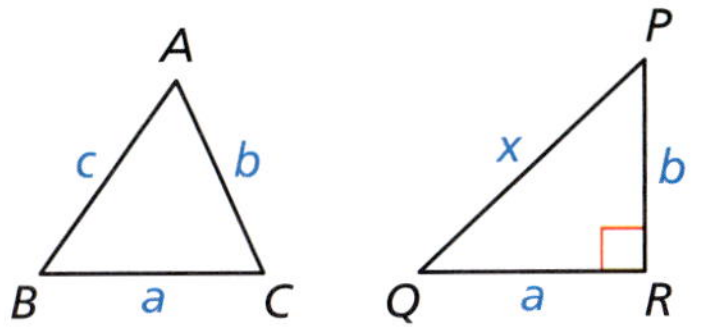

36. **PROVING A THEOREM** Prove the Pythagorean Inequalities Theorem when $c^2 > a^2 + b^2$. *(See Exercise 35.)*

37. **DISCUSS MATHEMATICAL THINKING** The side lengths of a triangle are 9, 12, and x. What values of x make the triangle a right triangle? an acute triangle? an obtuse triangle?

38. **THOUGHT PROVOKING**
Consider two positive integers m and n, where $m > n$. Do the following expressions produce a Pythagorean triple? If so, prove your answer. If not, give a counterexample.

$$2mn, \quad m^2 - n^2, \quad m^2 + n^2$$

39. **CONNECTING CONCEPTS** Let $ABCD$ be a square in the coordinate plane with vertices that have integer coordinates. If possible, draw $ABCD$ so that it has the given area, and show that $ABCD$ is a square. If not possible, explain why not.

a. 5 square units

b. 6 square units

c. 8 square units

d. 13 square units

40. **DIG DEEPER** Without measuring, find $m\angle PTQ$ in the diagram. (*Hint:* Draw square $PQRS$ and its image after a rotation 90° counterclockwise about point P. What do you know about $\triangle T'PT$?)

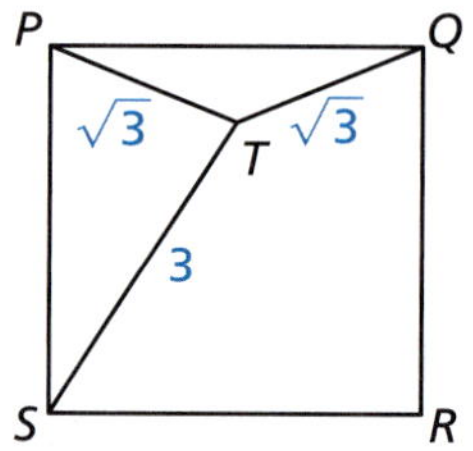

REVIEW & REFRESH

In Exercises 41 and 42, simplify the expression.

41. $\dfrac{8}{\sqrt{2}}$

42. $\dfrac{\sqrt{75}}{\sqrt{3}}$

43. Find the value of x.

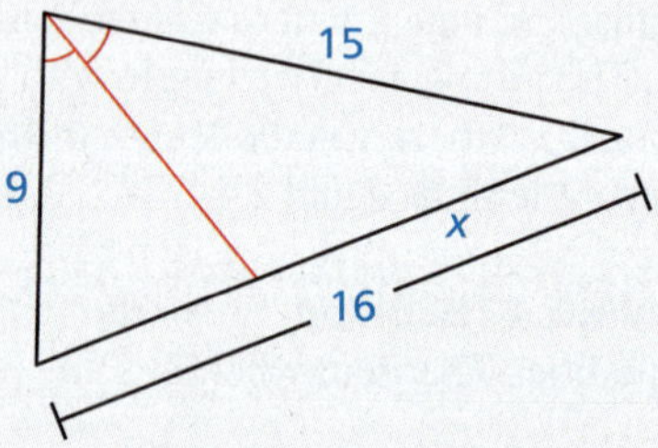

44. $WXYZ$ is an isosceles trapezoid, where $\overline{XY} \parallel \overline{WZ}$, $\overline{WX} \cong \overline{YZ}$, and $m\angle W = 54°$. Find $m\angle X$, $m\angle Y$, and $m\angle Z$.

45. **MODELING REAL LIFE** Two different tent sizes made by the same company are shown. Determine whether the triangular faces of the tents are similar. Explain your reasoning.

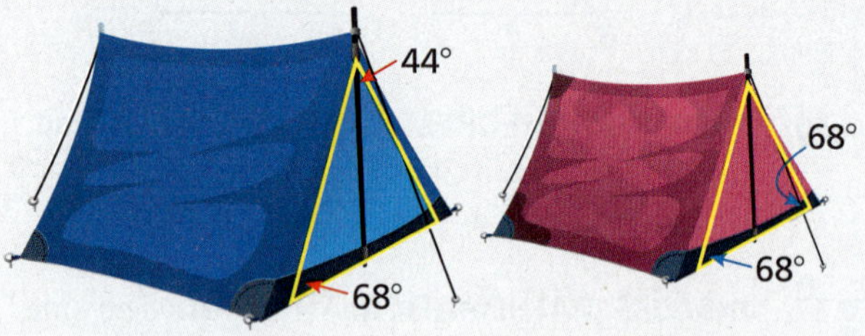

46. Tell whether the triangle is a right triangle.

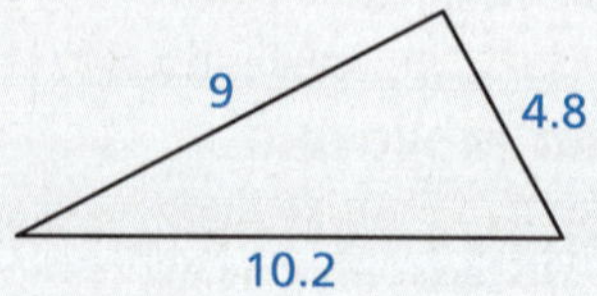

47. Find the coordinates of the centroid of $\triangle JKL$ with vertices $J(-1, 2)$, $K(5, 6)$, and $L(5, -2)$.

48. Show that the triangles are similar and write a similarity statement.

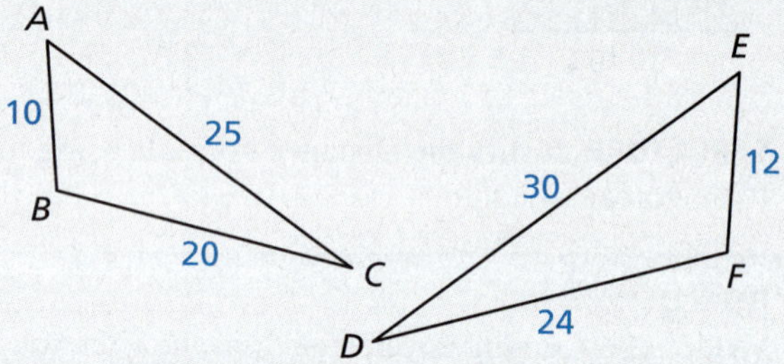

49. Explain how to find $m\angle ABD$ when you are given $m\angle ABC$ and $m\angle CBD$.

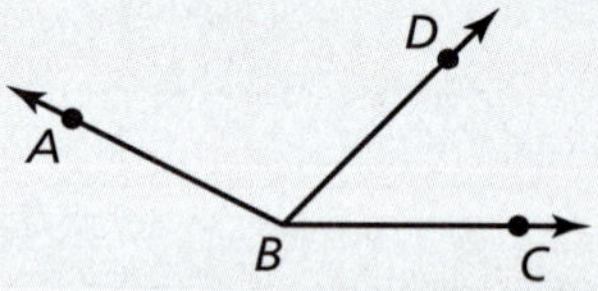

In Exercises 50–52, graph the polygon with the given vertices and its image after the indicated transformation.

50. $A(-4, 1)$, $B(5, -2)$, $C(2, 4)$
Translation: $(x, y) \rightarrow (x - 2, y - 3)$

51. $J(-3, -1)$, $K(1, 2)$, $L(1, -1)$, $M(-3, -4)$
Dilation: scale factor $k = 3$

52. $D(2, 1)$, $E(1, 3)$, $F(6, 3)$, $G(5, 1)$
Rotation: 180° about the origin

53. Find $m\angle 1$, $m\angle 2$, and $m\angle 3$. Explain your reasoning.

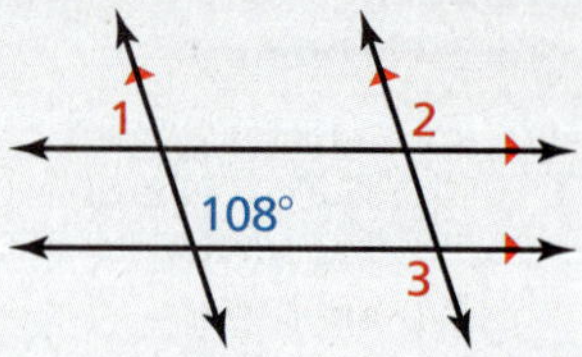

GO DIGITAL

9.2 Special Right Triangles

Learning Target: Understand and use special right triangles.

Success Criteria:
- I can find side lengths in 45°-45°-90° triangles.
- I can find side lengths in 30°-60°-90° triangles.
- I can use special right triangles to solve real-life problems.

EXPLORE IT! Finding Side Ratios of Special Right Triangles

Work with a partner.

a. One type of special right triangle is a 45°-45°-90° triangle.

i. Construct a right triangle with acute angle measures of 45°.

ii. Find the exact ratios of the side lengths.

$\frac{AB}{BC} =$

$\frac{AB}{AC} =$

$\frac{AC}{BC} =$

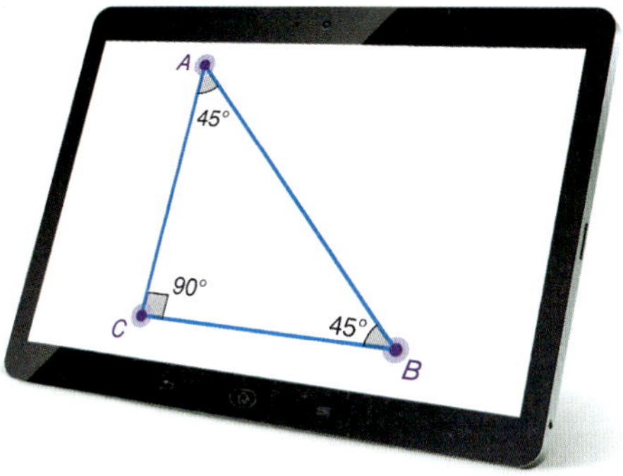

iii. Repeat parts (i) and (ii) for several other 45°-45°-90° triangles. Use your results to write a conjecture about the ratios of the side lengths of 45°-45°-90° triangles.

b. Another type of special right triangle is a 30°-60°-90° triangle.

i. Construct a right triangle with acute angle measures of 30° and 60°.

ii. Find the exact ratios of the side lengths.

$\frac{AB}{BC} =$

$\frac{AB}{AC} =$

$\frac{AC}{BC} =$

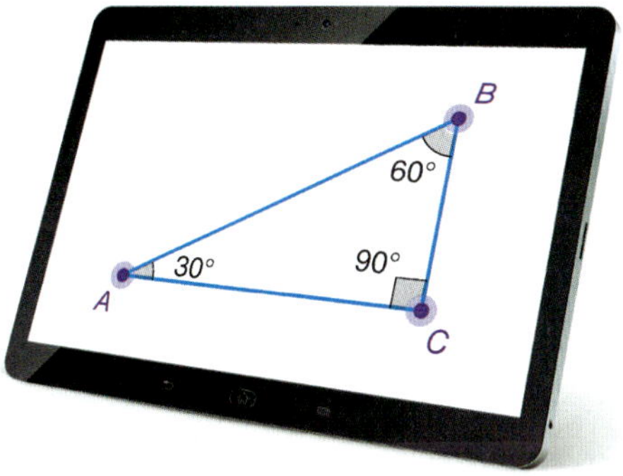

iii. Repeat parts (i) and (ii) for several other 30°-60°-90° triangles. Use your results to write a conjecture about the ratios of the side lengths of 30°-60°-90° triangles.

4 MTR CONSTRUCT AN ARGUMENT

Construct an argument about how to use the length of the hypotenuse to find the leg lengths in a 30°-60°-90° triangle.

Geometric Reasoning

MA.912.GR.1.3 Prove relationships and theorems about triangles. Solve mathematical and real-world problems involving postulates, relationships and theorems of triangles.

Trigonometry

MA.912.T.1.2 Solve mathematical and real-world problems involving right triangles using trigonometric ratios and the Pythagorean Theorem.

GO DIGITAL

Finding Side Lengths in Special Right Triangles

A 45°-45°-90° triangle is an *isosceles right triangle* that can be formed by cutting a square in half diagonally.

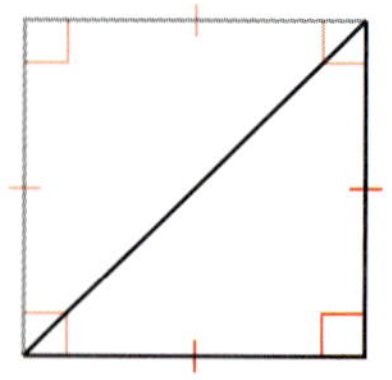

THEOREM

9.4 45°-45°-90° Triangle Theorem

In a 45°-45°-90° triangle, the hypotenuse is $\sqrt{2}$ times as long as each leg.

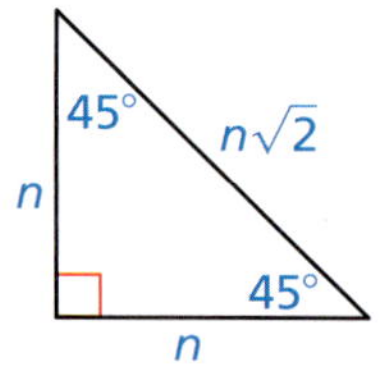

hypotenuse = leg • $\sqrt{2}$

Prove this Theorem Exercise 17, page 465

EXAMPLE 1 **Finding Side Lengths in 45°-45°-90° Triangles** WATCH

Find the value of x. Write your answer in simplest form.

a.

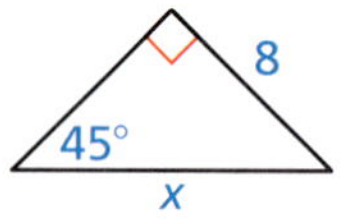

b.

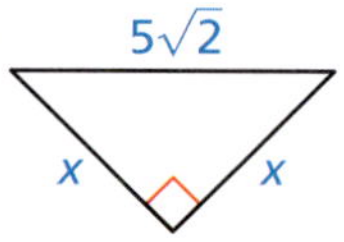

REMEMBER

An expression involving a radical with index 2 is in simplest form when no radicands have perfect squares as factors other than 1, no radicands contain fractions, and no radicals appear in the denominator of a fraction.

SOLUTION

a. By the Triangle Sum Theorem, the measure of the third angle must be 45°, so the triangle is a 45°-45°-90° triangle.

hypotenuse = leg • $\sqrt{2}$	45°-45°-90° Triangle Theorem
$x = 8 \cdot \sqrt{2}$	Substitute.
$x = 8\sqrt{2}$	Simplify.

▶ The value of x is $8\sqrt{2}$.

b. By the Base Angles Theorem and the Corollary to the Triangle Sum Theorem, the triangle is a 45°-45°-90° triangle.

hypotenuse = leg • $\sqrt{2}$	45°-45°-90° Triangle Theorem
$5\sqrt{2} = x \cdot \sqrt{2}$	Substitute.
$\frac{5\sqrt{2}}{\sqrt{2}} = \frac{x\sqrt{2}}{\sqrt{2}}$	Division Property of Equality
$5 = x$	Simplify.

▶ The value of x is 5.

GO DIGITAL

THEOREM

9.5 30°-60°-90° Triangle Theorem

In a 30°-60°-90° triangle, the hypotenuse is twice as long as the shorter leg, and the longer leg is $\sqrt{3}$ times as long as the shorter leg.

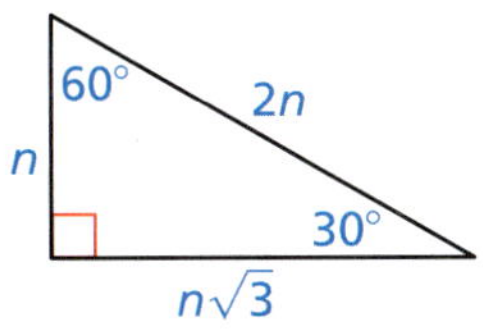

hypotenuse = shorter leg • 2
longer leg = shorter leg • $\sqrt{3}$

Prove this Theorem Exercise 19, page 466

EXAMPLE 2 Finding Side Lengths in a 30°-60°-90° Triangle

WATCH

Find the values of x and y. Write your answers in simplest form.

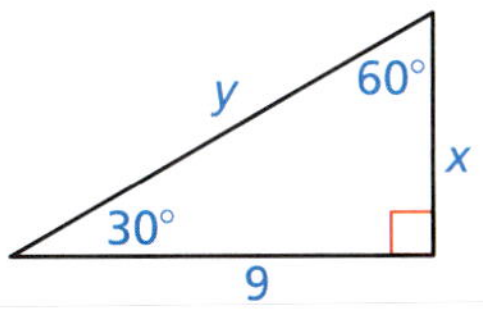

SOLUTION

Step 1 Find the value of x.

longer leg = shorter leg • $\sqrt{3}$	30°-60°-90° Triangle Theorem
$9 = x \cdot \sqrt{3}$	Substitute.
$\dfrac{9}{\sqrt{3}} = x$	Divide each side by $\sqrt{3}$.
$\dfrac{9}{\sqrt{3}} \cdot \dfrac{\sqrt{3}}{\sqrt{3}} = x$	Multiply by $\dfrac{\sqrt{3}}{\sqrt{3}}$.
$\dfrac{9\sqrt{3}}{3} = x$	Multiply fractions.
$3\sqrt{3} = x$	Simplify.

▶ The value of x is $3\sqrt{3}$.

> **REMEMBER**
> Because the angle opposite 9 is larger than the angle opposite x, the leg with length 9 is longer than the leg with length x by the Triangle Larger Angle Theorem.

Step 2 Find the value of y.

hypotenuse = shorter leg • 2	30°-60°-90° Triangle Theorem
$y = 3\sqrt{3} \cdot 2$	Substitute.
$y = 6\sqrt{3}$	Simplify.

▶ The value of y is $6\sqrt{3}$.

SELF-ASSESSMENT

1 I don't understand yet. 2 I can do it with help. 3 I can do it on my own. 4 I can teach someone else.

Find the missing side length(s). Write your answer(s) in simplest form.

1.

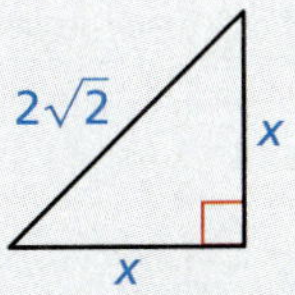

2.

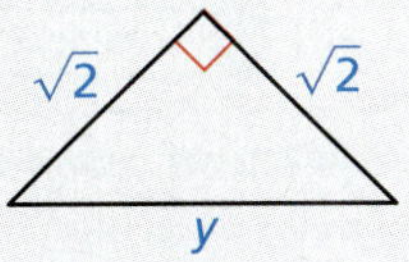

3.

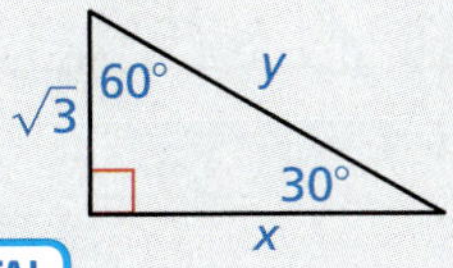

4.

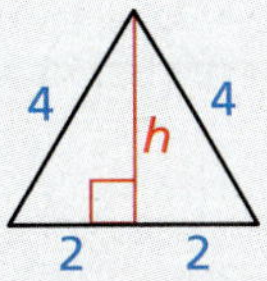

GO DIGITAL

Solving Real-Life Problems

7 MTR **EXAMPLE 3** **Modeling Real Life** WATCH

The biohazard sign is shaped like an equilateral triangle. Estimate the area of the sign.

SOLUTION

First find the height h of the triangle by dividing it into two 30°-60°-90° triangles. The length of the longer leg of one of these triangles is h. The length of the shorter leg is 18 inches.

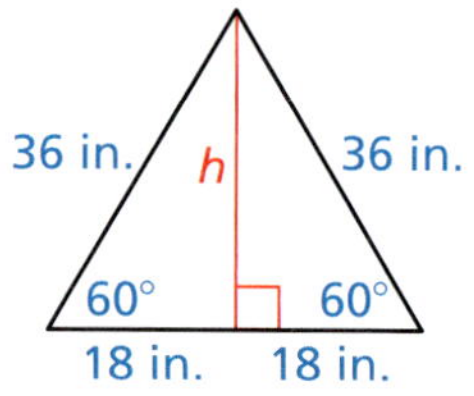

$h = 18 \cdot \sqrt{3} = 18\sqrt{3}$ 30°-60°-90° Triangle Theorem

Use $h = 18\sqrt{3}$ to find the area of the equilateral triangle.

$\text{Area} = \frac{1}{2}bh = \frac{1}{2}(36)(18\sqrt{3}) \approx 561.18$

▶ The area of the sign is about 561 square inches.

7 MTR **EXAMPLE 4** **Modeling Real Life** WATCH

A tipping platform is a ramp used to unload trucks. How high is the end of an 80-foot ramp when the tipping angle is 30°? 45°?

SOLUTION

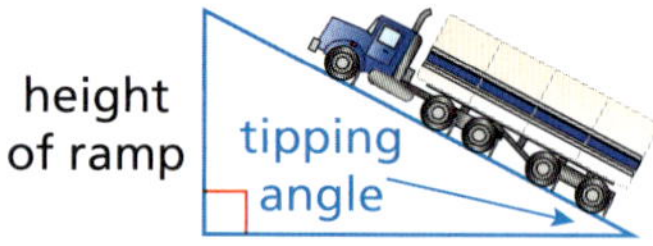

When the tipping angle is 30°, the height h of the ramp is the length of the shorter leg of a 30°-60°-90° triangle. The length of the hypotenuse is 80 feet.

$80 = 2h$ 30°-60°-90° Triangle Theorem

$40 = h$ Divide each side by 2.

When the tipping angle is 45°, the height is the length of a leg of a 45°-45°-90° triangle. The length of the hypotenuse is 80 feet.

$80 = h \cdot \sqrt{2}$ 45°-45°-90° Triangle Theorem

$\frac{80}{\sqrt{2}} = h$ Divide each side by $\sqrt{2}$.

$56.6 \approx h$ Use technology.

▶ When the tipping angle is 30°, the ramp height is 40 feet. When the tipping angle is 45°, the height is about 56 feet 7 inches.

SELF-ASSESSMENT

1 I don't understand yet. 2 I can do it with help. 3 I can do it on my own. 4 I can teach someone else.

5. The logo on a recycling bin resembles an equilateral triangle with side lengths of 6 centimeters. Approximate the area of the logo.

6. The body of a dump truck rests on a frame. The body is raised to empty a load of sand. How far from the frame is the front of the 14-foot-long body when it is tipped upward by a 60° angle?

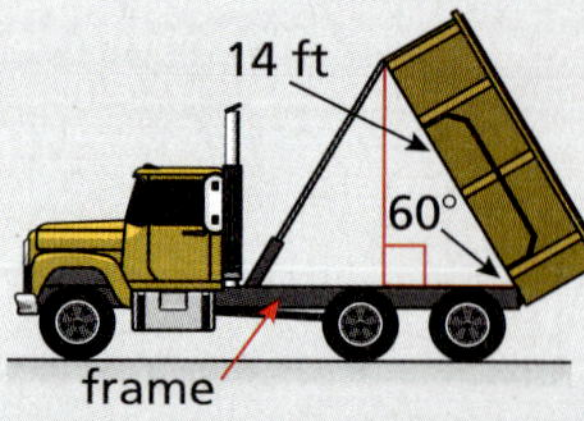

GO DIGITAL

9.2 Practice WITH CalcChat® AND CalcView®

In Exercises 1–4, find the value of x. Write your answer in simplest form. *(See Example 1.)*

1.

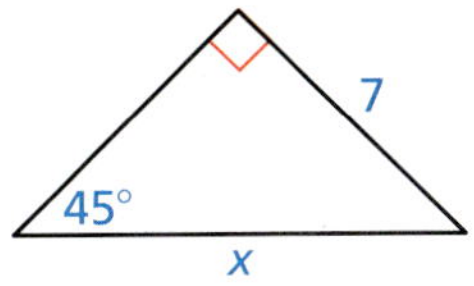

2.

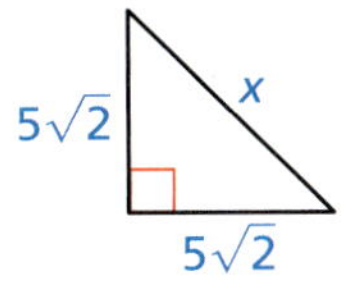

▶ **3.**

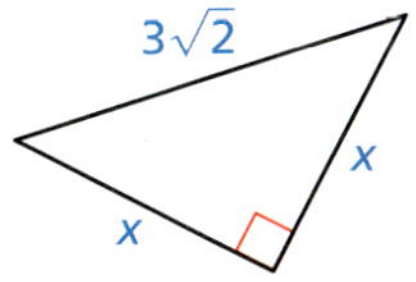

4.

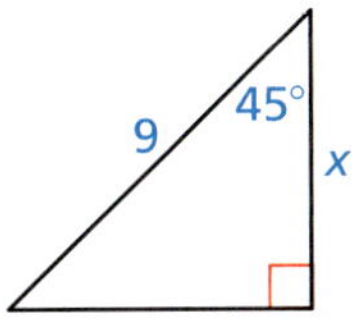

In Exercises 5–8, find the values of x and y. Write your answers in simplest form. *(See Example 2.)*

5.

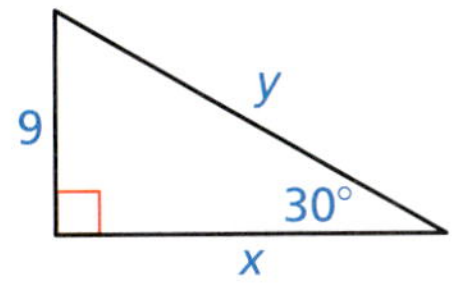

6.

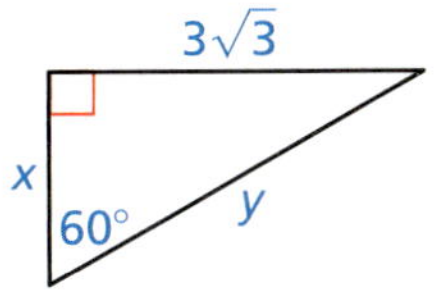

▶ **7.**

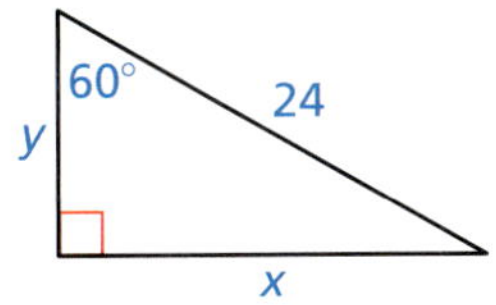

8.

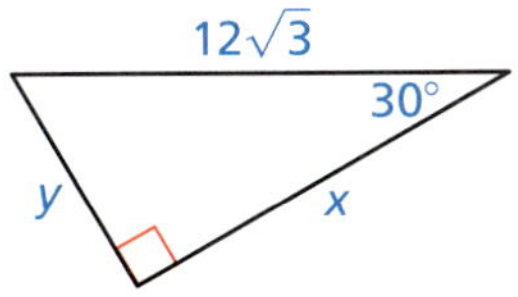

4 MTR **ERROR ANALYSIS** **In Exercises 9 and 10, describe and correct the error in finding the length of the hypotenuse in the special right triangle.**

9.

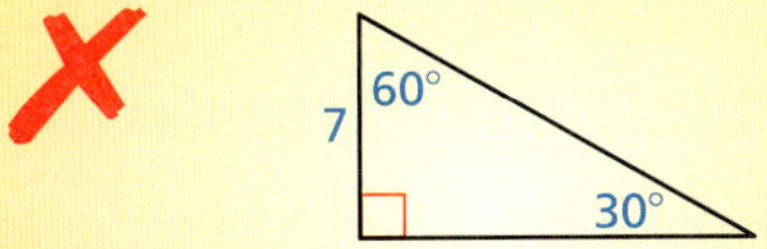

hypotenuse = shorter leg • $\sqrt{3} = 7\sqrt{3}$

So, the length of the hypotenuse is $7\sqrt{3}$ units.

10.

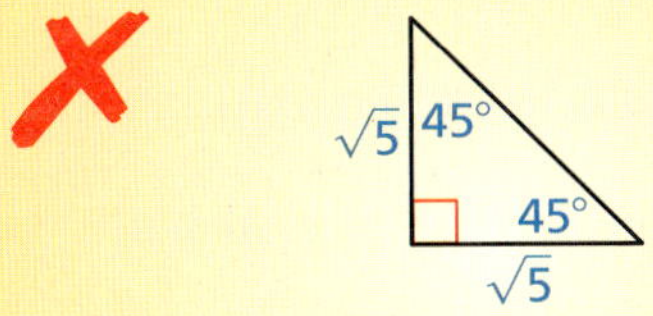

hypotenuse = leg • $2 = 2\sqrt{5}$

So, the length of the hypotenuse is $2\sqrt{5}$ units.

In Exercises 11 and 12, sketch the figure that is described. Find the indicated length. Write your answer in simplest form.

11. The perimeter of a square is 36 inches. Find the length of a diagonal.

12. The side length of an equilateral triangle is 5 centimeters. Find the length of an altitude.

In Exercises 13 and 14, find the area of the figure. Write your answer in simplest form. *(See Example 3.)*

▶ **13.**

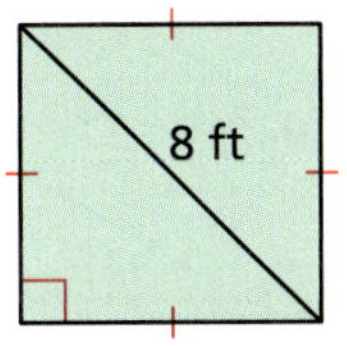

14.

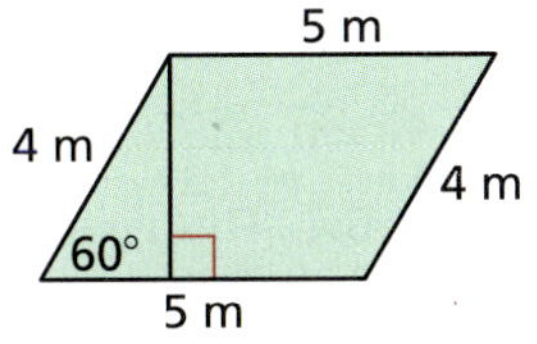

7 MTR **15. MODELING REAL LIFE** Each half of the drawbridge is about 284 feet long. How high does the drawbridge rise when x is 30°? 45°? 60°? *(See Example 4.)*

7 MTR **16. MODELING REAL LIFE** A nut is shaped like a regular hexagon with side lengths of 1 centimeter. Find the value of x.

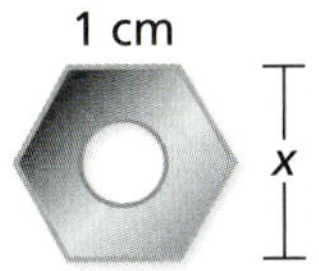

17. PROVING A THEOREM Write a paragraph proof of the 45°-45°-90° Triangle Theorem.

Given $\triangle DEF$ is a 45°-45°-90° triangle.

Prove The hypotenuse is $\sqrt{2}$ times as long as each leg.

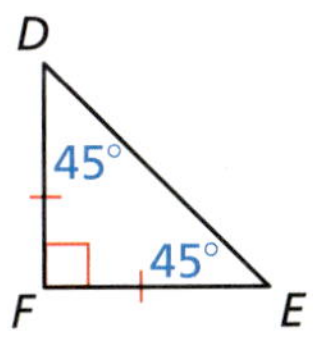

GO DIGITAL

18. HOW DO YOU SEE IT?
The diagram shows part of the *Wheel of Theodorus*.

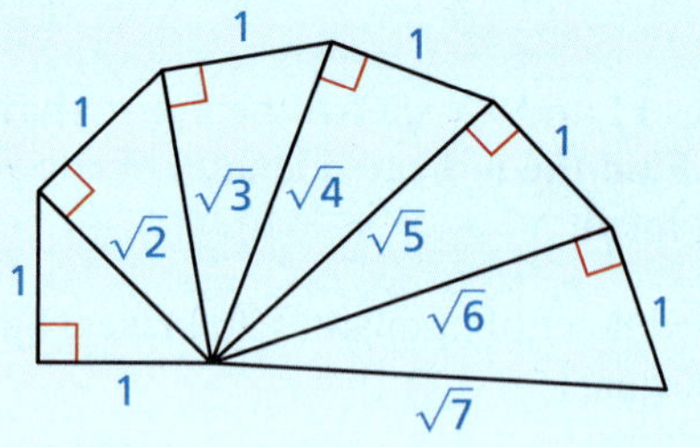

a. Which triangles, if any, are 45°-45°-90° triangles?

b. Which triangles, if any, are 30°-60°-90° triangles?

19. PROVING A THEOREM Write a paragraph proof of the 30°-60°-90° Triangle Theorem. (*Hint*: Construct $\triangle JML$ congruent to $\triangle JKL$.)

Given $\triangle JKL$ is a 30°-60°-90° triangle.

Prove The hypotenuse is twice as long as the shorter leg, and the longer leg is $\sqrt{3}$ times as long as the shorter leg.

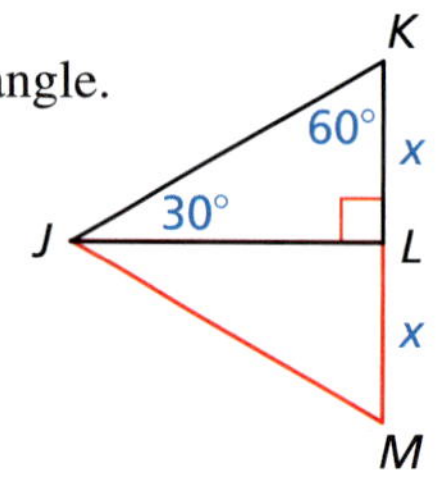

20. THOUGHT PROVOKING
The diagram below is called the *Ailles rectangle*. Each triangle in the diagram has rational angle measures, and each side length contains at most one square root. Label the sides and angles in the diagram. Describe the triangles.

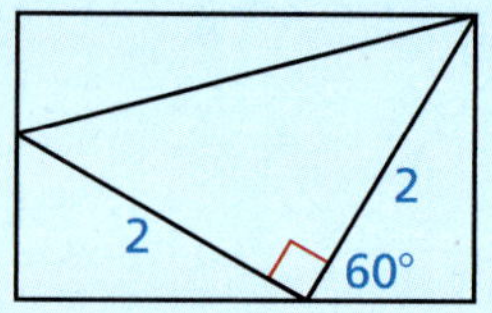

21. WRITING Describe two ways to show that all isosceles right triangles are similar to each other.

22. REASONING The area of an equilateral triangle is $3\sqrt{3}$ square units. Find the side length of the triangle. Justify your answer.

23. DIG DEEPER $\triangle TUV$ is a 30°-60°-90° triangle, where two vertices are $U(3, -1)$ and $V(-3, -1)$, $\overline{UV}$ is the hypotenuse, and point T is in Quadrant I. Find the coordinates of T.

REVIEW & REFRESH

WATCH

24. In the diagram, $\triangle DEF \sim \triangle LMN$. Find the value of x.

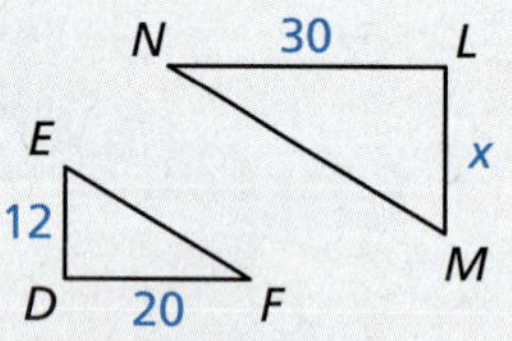

25. Determine whether segments with lengths of 2.6 feet, 4.8 feet, and 6.0 feet form a triangle. If so, is the triangle *acute*, *right*, or *obtuse*?

26. Find the length of $\overline{DC}$.

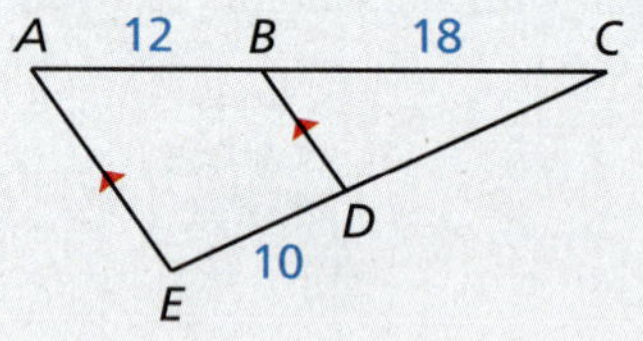

27. Find the values of x and y. Write your answers in simplest form.

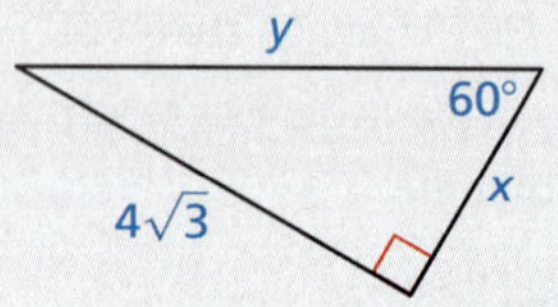

28. The endpoints of $\overline{CD}$ are $C(-2, 9)$ and $D(3, -1)$. Find the coordinates of the midpoint M.

29. Determine whether the polygons with the given vertices are congruent. Use transformations to explain your reasoning.

$D(3, 5)$, $E(8, 0)$, $F(4, -3)$ and $M(-5, 3)$, $N(0, 8)$, $P(3, 4)$

7 MTR **30. MODELING REAL LIFE** Which pieces of stained glass, if any, are similar? Explain.

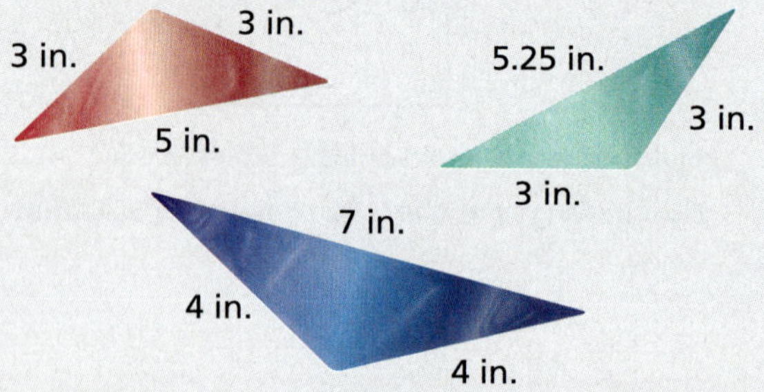

31. Three vertices of $\square JKLM$ are $J(0, 5)$, $K(4, 5)$, and $M(3, 0)$. Find the coordinates of vertex L.

32. Rewrite the definition as a biconditional statement.

Definition A *parallelogram* is a quadrilateral with both pairs of opposite sides parallel.

GO DIGITAL

9.3 Similar Right Triangles

Learning Target: Use proportional relationships in right triangles.

Success Criteria:
- I can explain the Right Triangle Similarity Theorem.
- I can find the geometric mean of two numbers.
- I can find missing dimensions in right triangles.

EXPLORE IT! Analyzing Similarity in Right Triangles

Work with a partner.

a. Construct two congruent right scalene triangles.

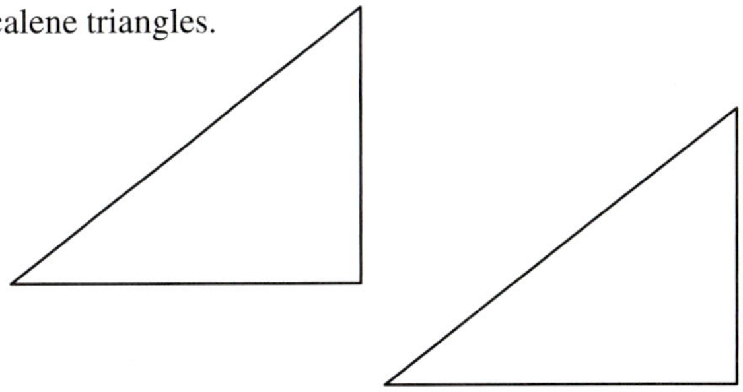

b. Construct an altitude to the hypotenuse in one of the right triangles. Use the altitude to form two smaller triangles. Label the three triangles as shown.

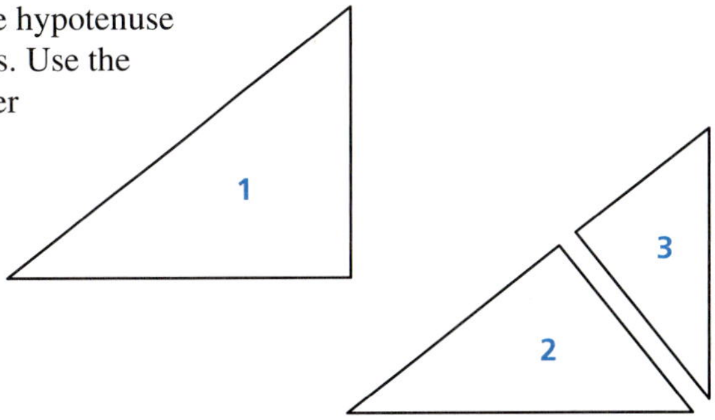

c. Compare the angle measures of Triangle 1 with the angle measures of Triangle 2. What can you conclude about the two triangles?

d. Repeat part (c) for Triangles 1 and 3, and for Triangles 2 and 3.

e. When you draw an altitude to the hypotenuse of a right triangle, what is the relationship between the two smaller triangles that are formed? How are each of the smaller triangles related to the larger triangle?

f. Use your results in part (e) to find x. Justify your answer.

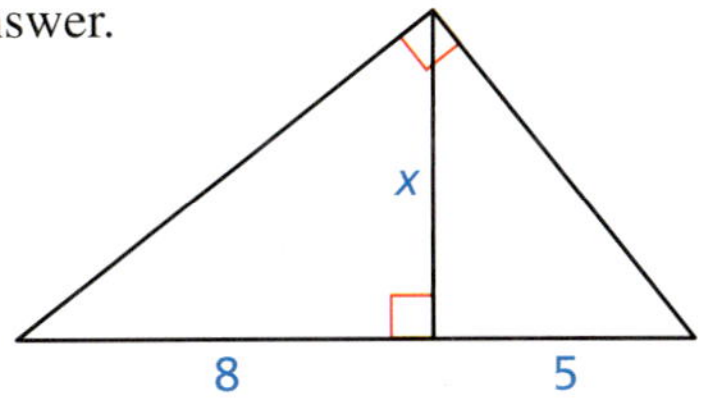

5 MTR MAKE A PLAN
In part (f), explain how you plan to use the relationships among the triangles to find the value of x.

Geometric Reasoning
MA.912.GR.1.3 Prove relationships and theorems about triangles. Solve mathematical and real-world problems involving postulates, relationships and theorems of triangles.
MA.912.GR.1.6 Solve mathematical and real-world problems involving congruence or similarity in two-dimensional figures.

GO DIGITAL

Identifying Similar Triangles

When the altitude is drawn to the hypotenuse of a right triangle, the two smaller triangles are similar to the original triangle and to each other.

Vocabulary AZ VOCAB

geometric mean, *p. 470*

THEOREM

9.6 Right Triangle Similarity Theorem

If the altitude is drawn to the hypotenuse of a right triangle, then the two triangles formed are similar to the original triangle and to each other.

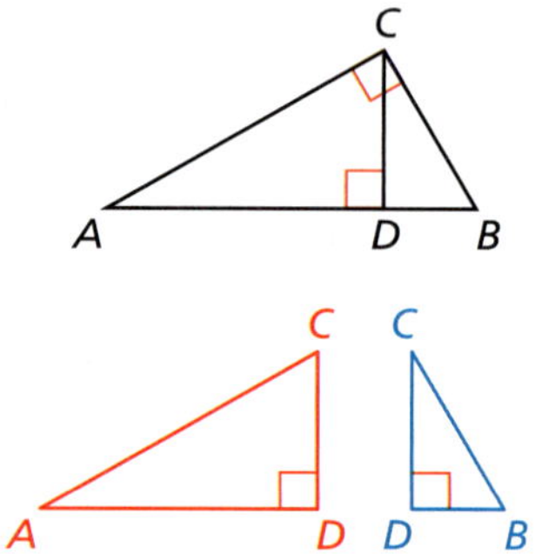

$\triangle CBD \sim \triangle ABC$, $\triangle ACD \sim \triangle ABC$, and $\triangle CBD \sim \triangle ACD$.

Prove this Theorem Exercise 41, page 474

EXAMPLE 1 **Identifying Similar Triangles** WATCH

Identify the similar triangles.

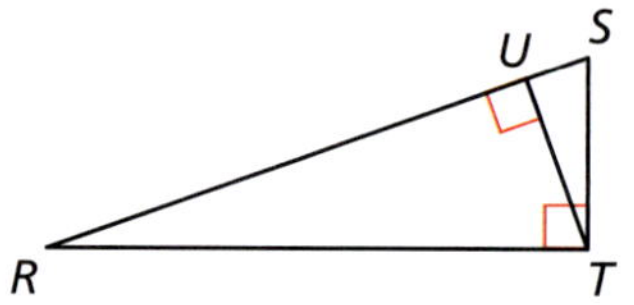

SOLUTION

$\triangle RST$ is a right triangle with altitude $\overline{TU}$ drawn to the hypotenuse. By the Right Triangle Similarity Theorem, the two triangles formed by $\overline{TU}$ are similar to $\triangle RST$ and to each other. Sketch the three similar right triangles so that the corresponding angles and sides have the same orientation.

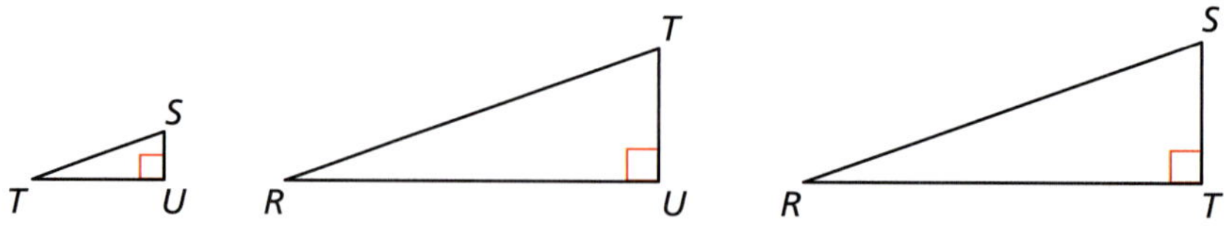

▶ $\triangle TSU \sim \triangle RTU \sim \triangle RST$

SELF-ASSESSMENT 1 I don't understand yet. 2 I can do it with help. 3 I can do it on my own. 4 I can teach someone else.

Identify the similar triangles.

1.

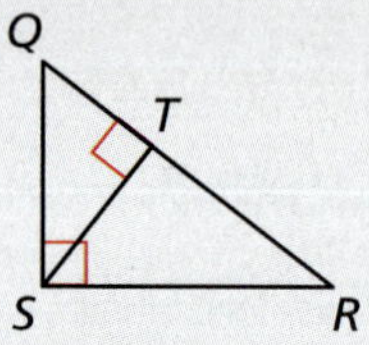

2.

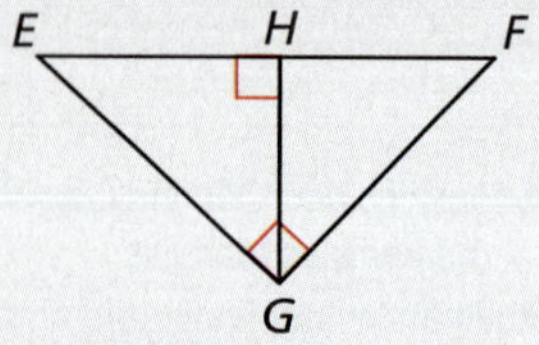

GO DIGITAL

7 MTR

EXAMPLE 2 Modeling Real Life

WATCH

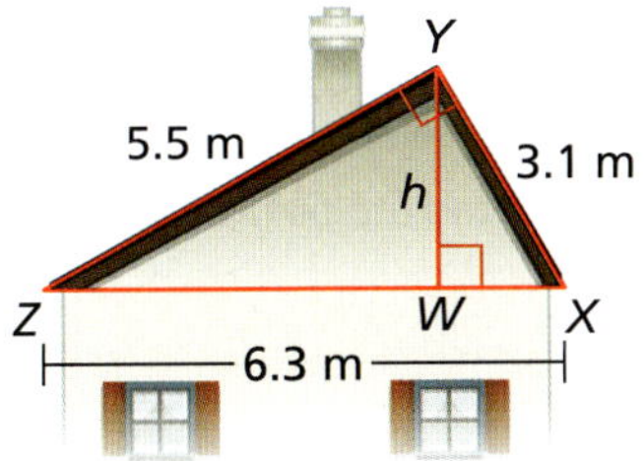

A roof has a cross section that is a right triangle. The diagram shows the approximate dimensions of this cross section. Find the height *h* of the roof.

SOLUTION

1. **Understand the Problem** You are given the side lengths of a right triangle. You need to find the height of the roof, which is the length of the altitude drawn to the hypotenuse.
2. **Make a Plan** Identify any similar triangles. Then use the similar triangles to write and solve a proportion for *h*.
3. **Solve and Check** $\overline{YW}$ is the altitude of $\triangle XZY$ drawn to the hypotenuse. By the Right Triangle Similarity Theorem, the two triangles formed by $\overline{YW}$ are similar to $\triangle XZY$ and to each other. Sketch the three similar right triangles so that the corresponding angles and sides have the same orientation.

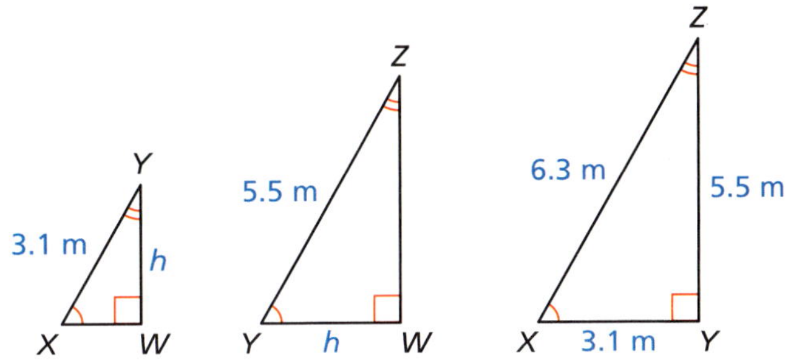

$\triangle XYW \sim \triangle YZW \sim \triangle XZY$

Because $\triangle XYW \sim \triangle XZY$, you can write a proportion.

$$\frac{YW}{ZY} = \frac{XY}{XZ}$$ Corresponding side lengths of similar triangles are proportional.

$$\frac{h}{5.5} = \frac{3.1}{6.3}$$ Substitute.

$$h \approx 2.7$$ Multiply each side by 5.5.

▶ The height of the roof is about 2.7 meters.

5 MTR **STRUCTURE**
Can you solve for *h* by writing a proportion using $\triangle XYW$ and $\triangle YZW$? Explain.

Check Reasonableness Because the height of the roof is a leg length of right $\triangle YZW$ and right $\triangle XYW$, it should be shorter than each of their hypotenuses. The lengths of the two hypotenuses are $YZ = 5.5$ and $XY = 3.1$. Because $2.7 < 3.1$, the answer is reasonable.

SELF-ASSESSMENT

1 I don't understand yet. 2 I can do it with help. 3 I can do it on my own. 4 I can teach someone else.

Find the value of *x*.

3.

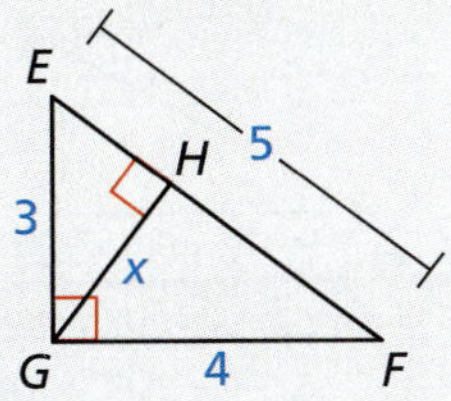

4.

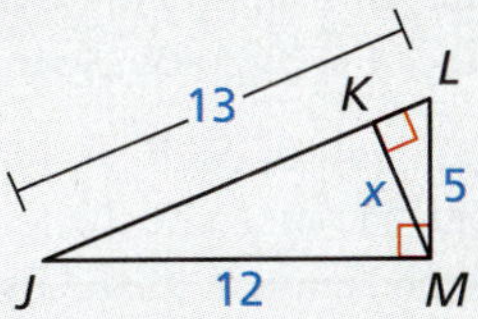

GO DIGITAL

Using a Geometric Mean

KEY IDEA

Geometric Mean

The **geometric mean** of two positive numbers a and b is the positive number x that satisfies $\frac{a}{x} = \frac{x}{b}$. So, $x^2 = ab$ and $x = \sqrt{ab}$.

EXAMPLE 3 **Finding a Geometric Mean** WATCH

Find the geometric mean of 24 and 48.

SOLUTION

$x^2 = ab$	Definition of geometric mean
$x^2 = 24 \cdot 48$	Substitute 24 for a and 48 for b.
$x = \sqrt{24 \cdot 48}$	Take the positive square root of each side.
$x = \sqrt{24 \cdot 24 \cdot 2}$	Factor.
$x = 24\sqrt{2}$	Simplify.

▶ The geometric mean of 24 and 48 is $24\sqrt{2} \approx 33.9$.

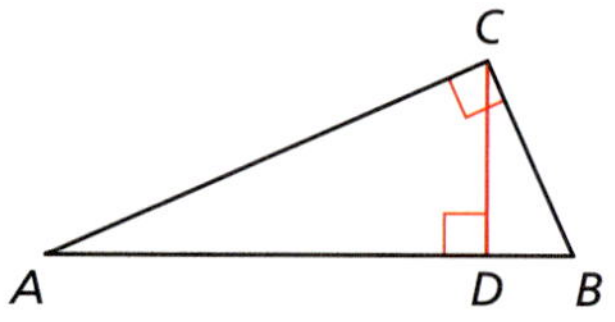

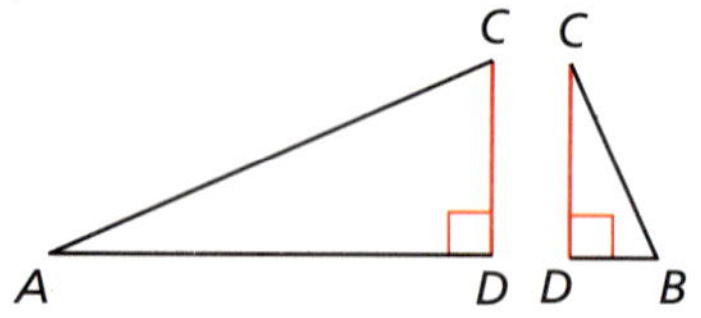

In right $\triangle ABC$, altitude $\overline{CD}$ is drawn to the hypotenuse, forming two smaller right triangles. From the Right Triangle Similarity Theorem, you know that $\triangle CBD \sim \triangle ACD \sim \triangle ABC$. Because the triangles are similar, you can write and simplify the following proportions involving geometric means.

$$\frac{CD}{AD} = \frac{BD}{CD} \qquad \frac{CB}{DB} = \frac{AB}{CB} \qquad \frac{AC}{AD} = \frac{AB}{AC}$$

$$CD^2 = AD \cdot BD \qquad CB^2 = DB \cdot AB \qquad AC^2 = AD \cdot AB$$

THEOREMS

9.7 Geometric Mean (Altitude) Theorem

In a right triangle, the altitude to the hypotenuse divides the hypotenuse into two segments.

The length of the altitude is the geometric mean of the lengths of the two segments of the hypotenuse.

Prove this Theorem Exercise 39, page 474

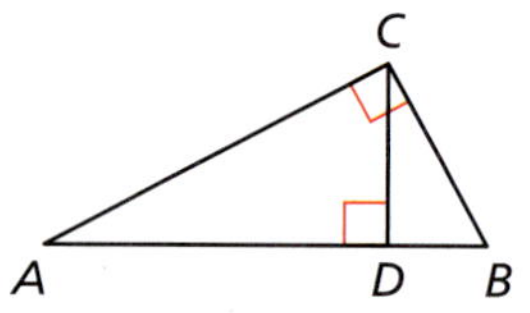

$CD^2 = AD \cdot BD$

9.8 Geometric Mean (Leg) Theorem

In a right triangle, the altitude to the hypotenuse divides the hypotenuse into two segments.

The length of each leg of the right triangle is the geometric mean of the lengths of the hypotenuse and the segment of the hypotenuse that is adjacent to that leg.

Prove this Theorem Exercise 40, page 474

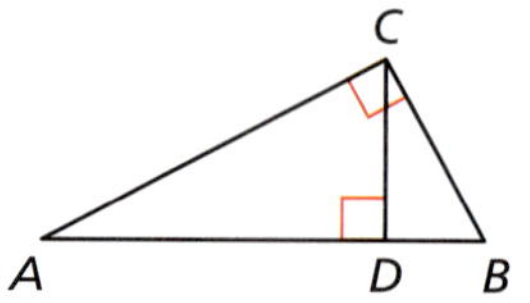

$CB^2 = DB \cdot AB$
$AC^2 = AD \cdot AB$

GO DIGITAL

EXAMPLE 4 Using a Geometric Mean

Find the value of each variable.

a.

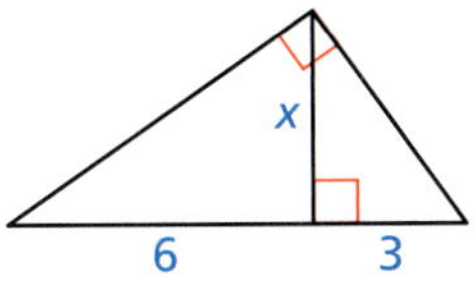

b.

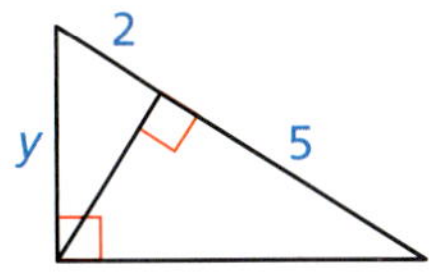

COMMON ERROR

In Example 4(b), the Geometric Mean (Leg) Theorem gives $y^2 = 2 \cdot (5 + 2)$, not $y^2 = 5 \cdot (5 + 2)$, because the side with length y is adjacent to the segment with length 2.

SOLUTION

a. Apply the Geometric Mean (Altitude) Theorem.

$$x^2 = 6 \cdot 3$$
$$x^2 = 18$$
$$x = \sqrt{18}$$
$$x = \sqrt{9} \cdot \sqrt{2}$$
$$x = 3\sqrt{2}$$

▶ The value of x is $3\sqrt{2}$.

b. Apply the Geometric Mean (Leg) Theorem.

$$y^2 = 2 \cdot (5 + 2)$$
$$y^2 = 2 \cdot 7$$
$$y^2 = 14$$
$$y = \sqrt{14}$$

▶ The value of y is $\sqrt{14}$.

EXAMPLE 5 Using Indirect Measurement

To find the cost of installing a rock wall in your school gymnasium, you need to find the height of the gym wall. You use a cardboard square to line up the top and bottom of the gym wall with your eye. Your friend measures the vertical distance from the ground to your eye and the horizontal distance from you to the gym wall. Find the height of the gym wall.

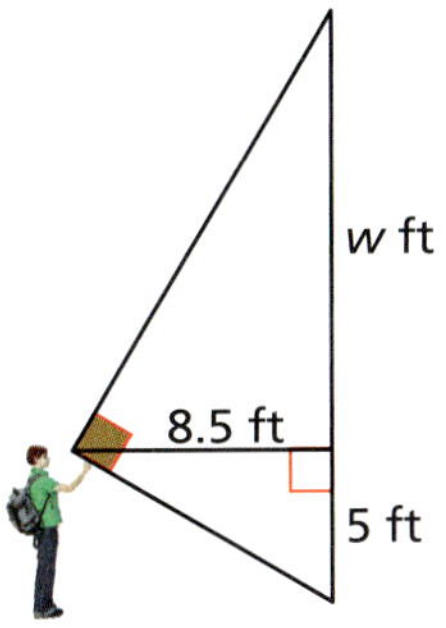

SOLUTION

By the Geometric Mean (Altitude) Theorem, you know that 8.5 is the geometric mean of w and 5.

$8.5^2 = w \cdot 5$	Geometric Mean (Altitude) Theorem
$72.25 = 5w$	Multiply.
$14.45 = w$	Divide each side by 5.

▶ The height of the wall is $5 + w = 5 + 14.45 = 19.45$ feet.

SELF-ASSESSMENT

1 I don't understand yet. 2 I can do it with help. 3 I can do it on my own. 4 I can teach someone else.

Find the geometric mean of the two numbers.

5. 12 and 27

6. 18 and 54

7. 16 and 18

8. Find the value of x in the triangle at the right.

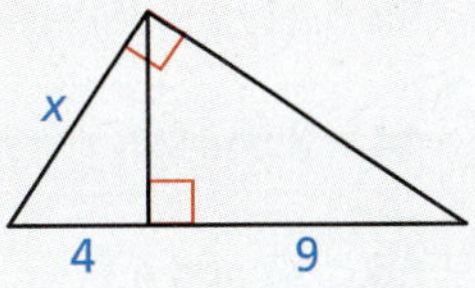

9. **WHAT IF?** In Example 5, the vertical distance from the ground to your eye is 5.5 feet, and the distance from you to the gym wall is 9 feet. Find the height of the gym wall.

GO DIGITAL

9.3 Practice WITH CalcChat® AND CalcView®

In Exercises 1 and 2, identify the similar triangles. *(See Example 1.)*

▶ **1.**

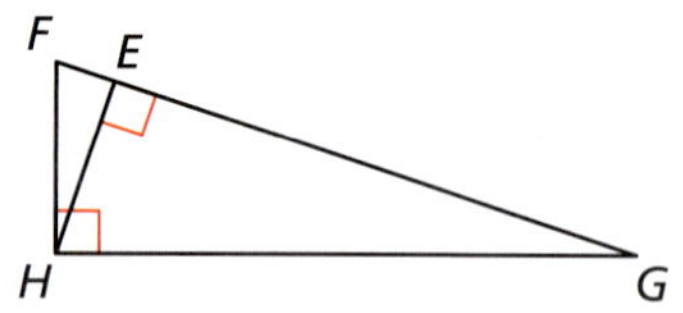

2.

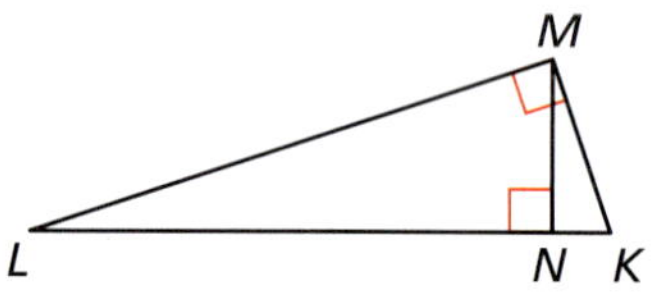

In Exercises 3–8, find the value of *x*. *(See Example 2.)*

▶ **3.**

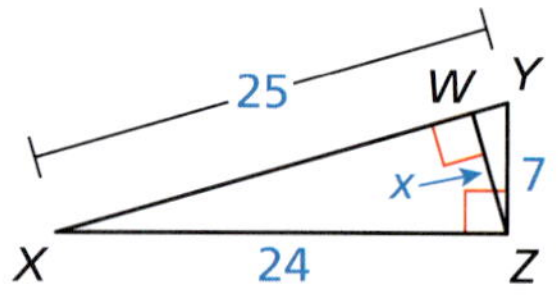

4.

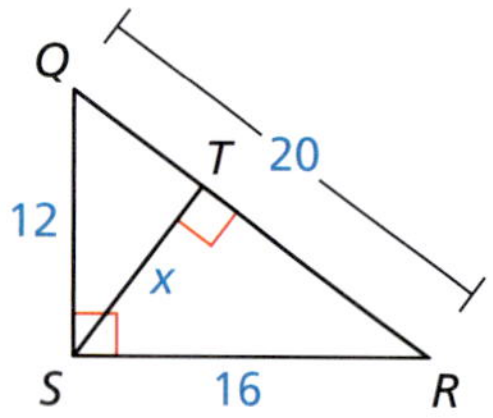

5.

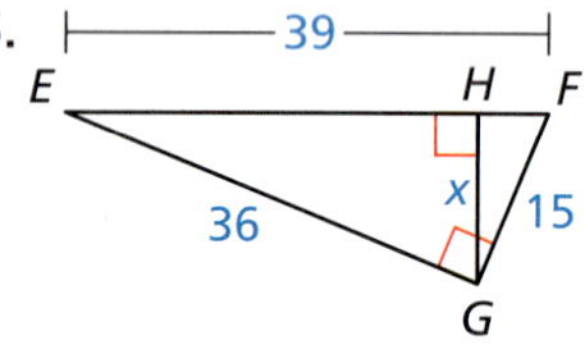

6.

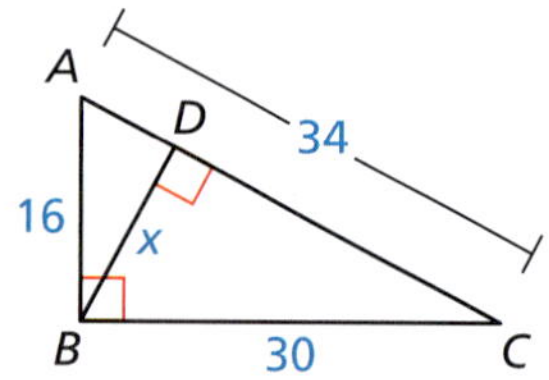

7.

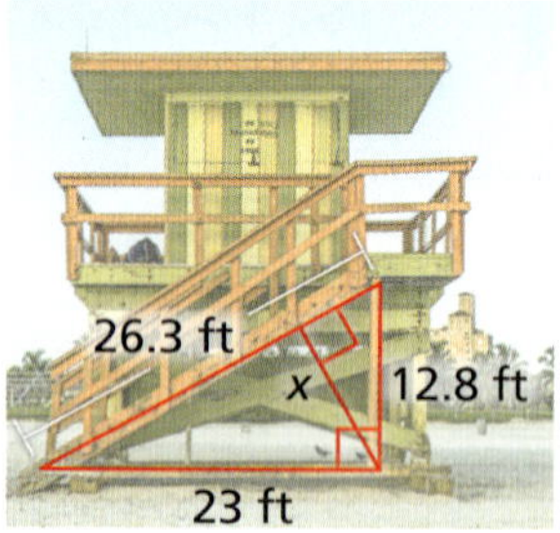

8.

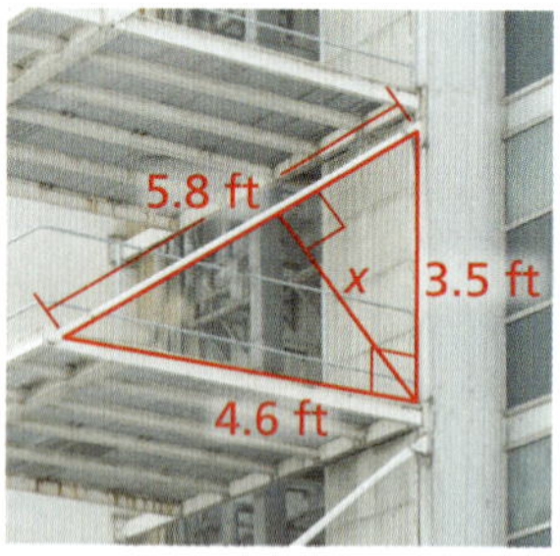

In Exercises 9–16, find the geometric mean of the two numbers. *(See Example 3.)*

▶ **9.** 8 and 32

10. 9 and 16

11. 14 and 20

12. 25 and 35

13. 16 and 25

14. 8 and 28

15. 17 and 36

16. 24 and 45

In Exercises 17–24, find the value of the variable. *(See Example 4.)*

▶ **17.**

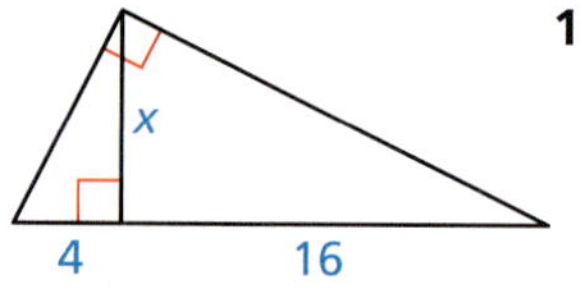

18.

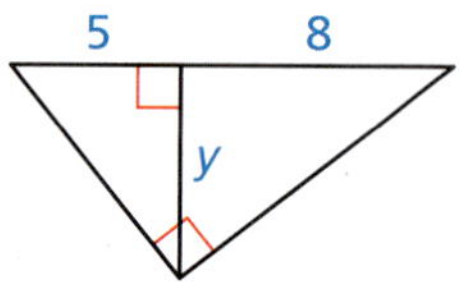

19.

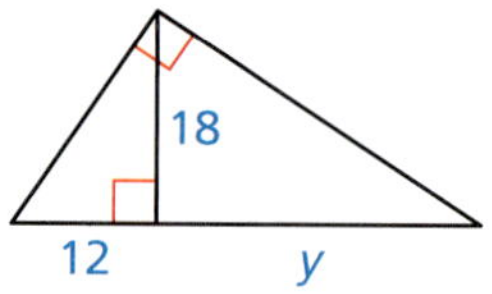

20.

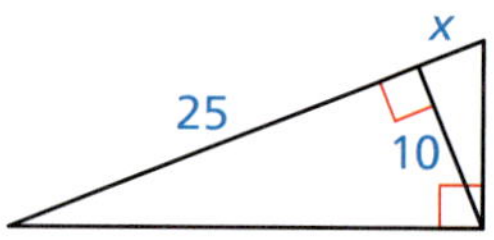

21.

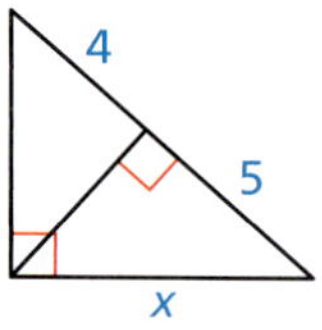

22.

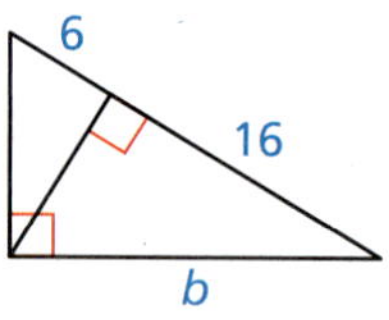

23.

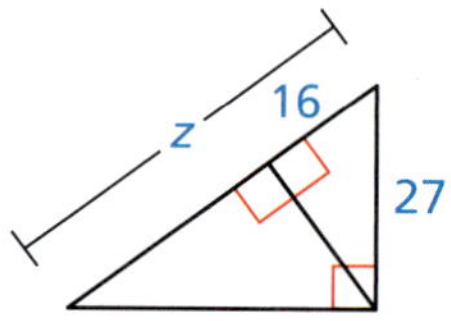

24.

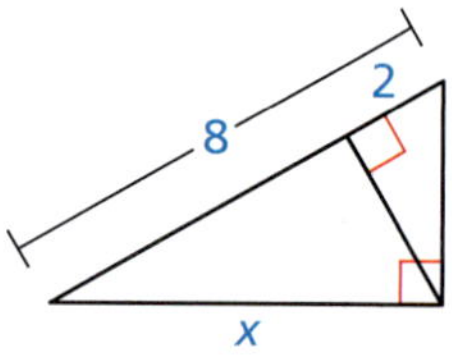

4 MTR **25. ERROR ANALYSIS** Describe and correct the error in applying the Geometric Mean (Leg) Theorem.

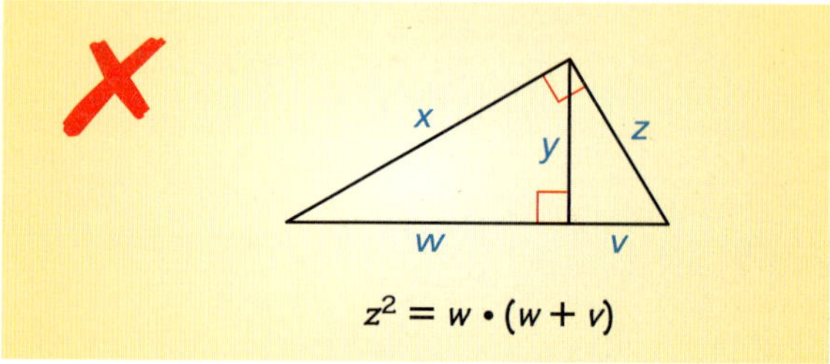

26. COLLEGE PREP Use the diagram. Decide which proportions are true. Select all that apply.

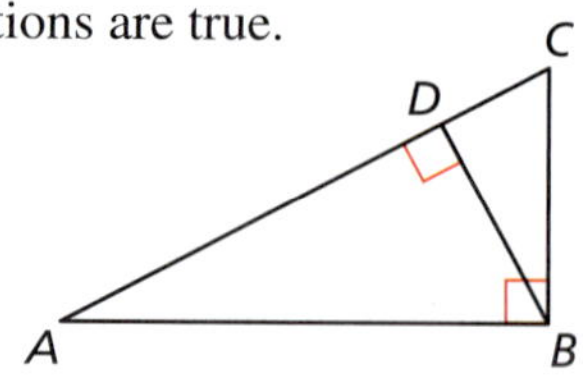

Ⓐ $\dfrac{DB}{DC} = \dfrac{DA}{DB}$

Ⓑ $\dfrac{BA}{CB} = \dfrac{CB}{BD}$

Ⓒ $\dfrac{CA}{BA} = \dfrac{BA}{CA}$

Ⓓ $\dfrac{DB}{BC} = \dfrac{DA}{BA}$

GO DIGITAL

7 MTR **MODELING REAL LIFE** **In Exercises 27 and 28, use the diagram.** *(See Example 5.)*

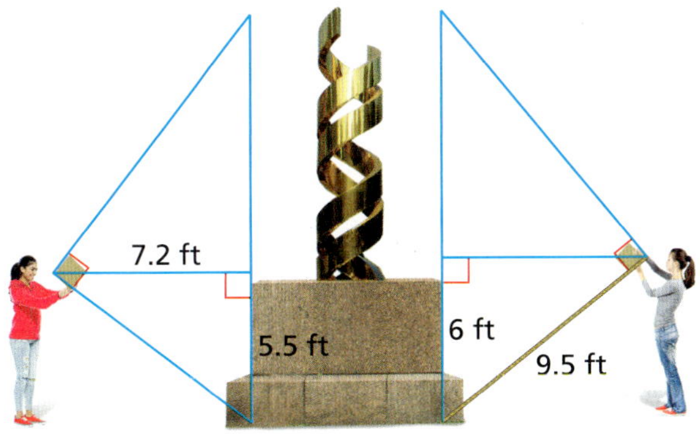

27. You examine a contemporary DNA sculpture. You use a cardboard square to line up the top and bottom of the sculpture with your eye, as shown at the left. Approximate the height of the sculpture.

28. Your classmate is standing on the right side of the sculpture. She has a piece of rope staked at the base of the sculpture. She extends the rope to the cardboard square she is holding lined up to the top and bottom of the sculpture. She uses her measurements to approximate the height of the sculpture. Does she get the same answer as you did in Exercise 27? Explain your reasoning.

5 MTR **STRUCTURE** **In Exercises 29–32, find the value(s) of the variable(s).**

29.

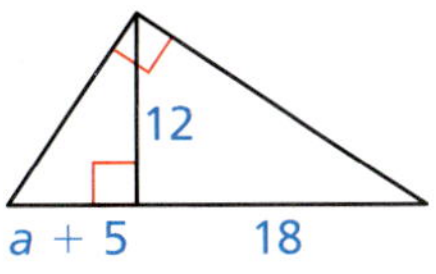

30.

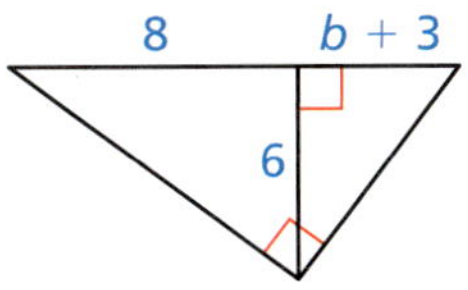

31.

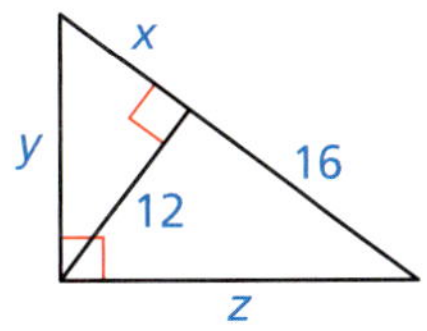

32.

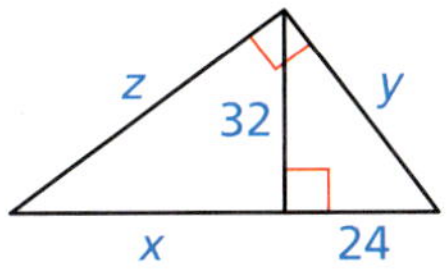

33. PROBLEM SOLVING You are designing a kite. You know that $AD = 44.8$ centimeters, $DC = 72$ centimeters, and $AC = 84.8$ centimeters. You want to use a straight crossbar $\overline{BD}$. About how long should it be? Explain your reasoning.

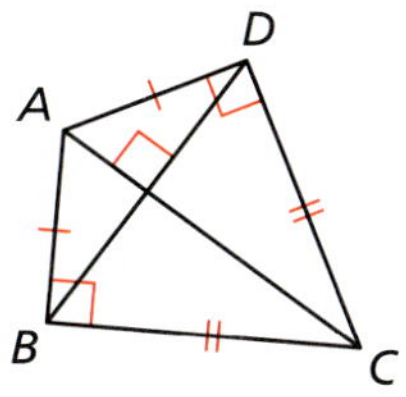

5 MTR **34. STRUCTURE** Use both of the Geometric Mean Theorems to find AC and BD.

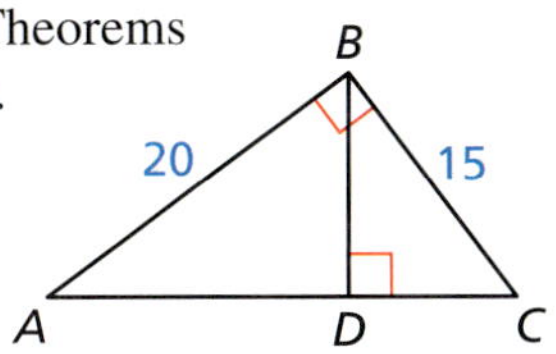

35. PROVING A THEOREM Use the diagram of $\triangle ABC$. Copy and complete the proof of the Pythagorean Theorem.

Given In $\triangle ABC$, $\angle BCA$ is a right angle.

Prove $c^2 = a^2 + b^2$

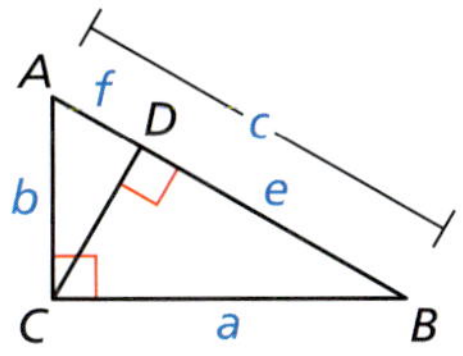

STATEMENTS	REASONS
1. In $\triangle ABC$, $\angle BCA$ is a right angle.	1. ________
2. Draw a perpendicular segment (altitude) from C to $\overline{AB}$.	2. Perpendicular Postulate
3. $ce = a^2$ and $cf = b^2$	3. ________
4. $ce + b^2 =$ ____ $+ b^2$	4. Addition Property of Equality
5. $ce + cf = a^2 + b^2$	5. ________
6. $c(e + f) = a^2 + b^2$	6. ________
7. $e + f =$ ____	7. Segment Addition Postulate
8. $c \cdot c = a^2 + b^2$	8. ________
9. $c^2 = a^2 + b^2$	9. Simplify.

36. HOW DO YOU SEE IT?
In which triangle does the Geometric Mean (Altitude) Theorem apply? Explain.

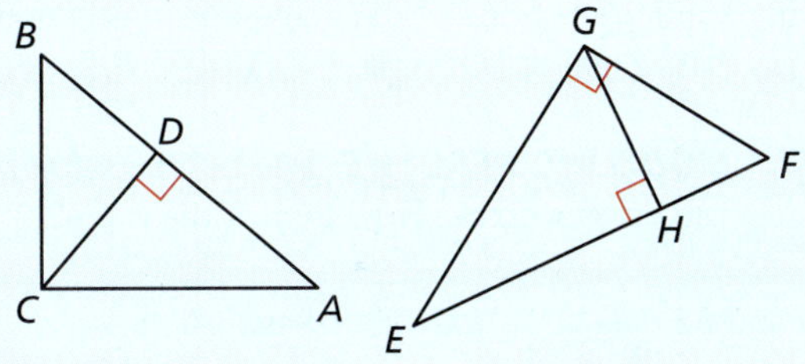

4 MTR **37. MAKING AN ARGUMENT** Your friend claims the geometric mean of 4 and 9 is 6, and then labels the triangle, as shown. Is your friend correct? Explain.

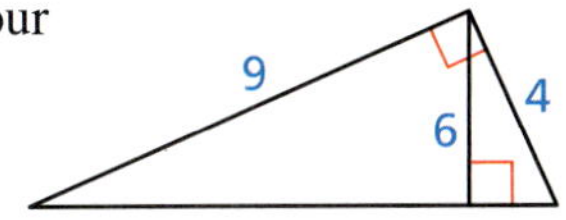

GO DIGITAL

38. MAINTAIN ACCURACY Draw a right isosceles triangle and label the two leg lengths x. Then draw the altitude to the hypotenuse and label its length y. Now use the Right Triangle Similarity Theorem to draw the three similar triangles from the image and label any side length that is equal to either x or y. What can you conclude about the relationship between the two smaller triangles? Explain.

In Exercises 39 and 40, use the given statements to prove the theorem.

Given $\triangle ABC$ is a right triangle.
Altitude $\overline{CD}$ is drawn to hypotenuse $\overline{AB}$.

39. PROVING A THEOREM Prove the Geometric Mean (Altitude) Theorem by showing that $CD^2 = AD \cdot BD$.

40. PROVING A THEOREM Prove the Geometric Mean (Leg) Theorem by showing that $CB^2 = DB \cdot AB$ and $AC^2 = AD \cdot AB$.

41. PROVING A THEOREM Prove the Right Triangle Similarity Theorem.

Given $\triangle ABC$ is a right triangle.
Altitude $\overline{CD}$ is drawn to hypotenuse $\overline{AB}$.

Prove $\triangle CBD \sim \triangle ABC$,
$\triangle ACD \sim \triangle ABC$,
$\triangle CBD \sim \triangle ACD$

42. THOUGHT PROVOKING
The arithmetic mean and geometric mean of two nonnegative numbers x and y are shown.

$$\text{arithmetic mean} = \frac{x + y}{2}$$

$$\text{geometric mean} = \sqrt{xy}$$

Write an inequality that relates these two means. Justify your answer.

REVIEW & REFRESH

WATCH

43. $\overline{DE}$ is a midsegment of $\triangle ABC$. Find the value of x.

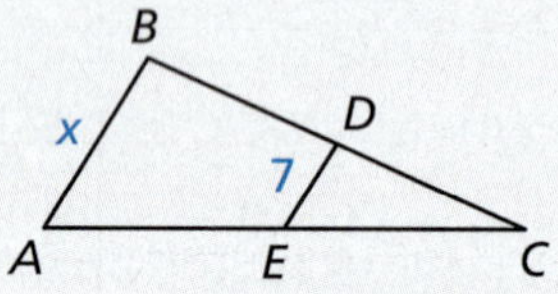

44. Find BC. Explain your reasoning.

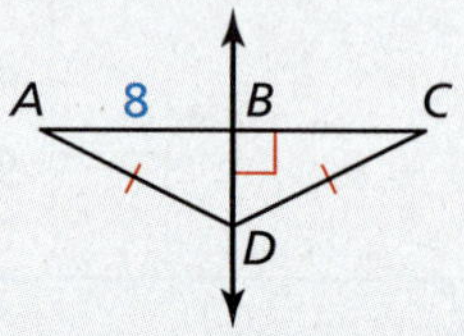

45. Find the geometric mean of 27 and 3.

46. Find the lengths of the diagonals of rectangle $WXYZ$ given $WY = 4x - 1$ and $XZ = 2x + 7$.

47. MODELING REAL LIFE You build a ramp so you can cross a creek. How long is the ramp?

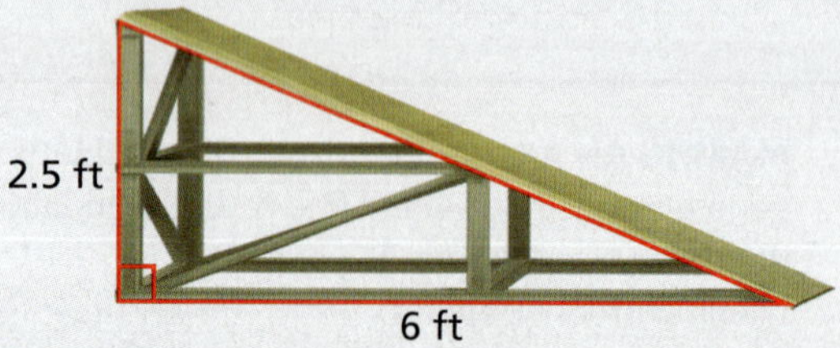

48. Tell whether the triangle is a right triangle.

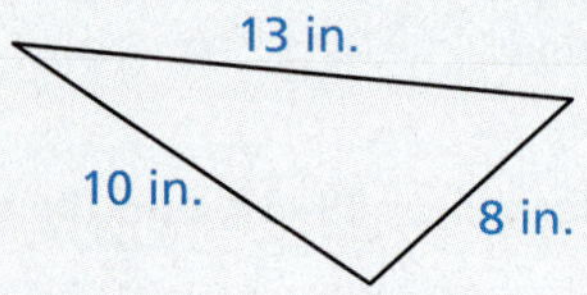

49. Find the value of y.

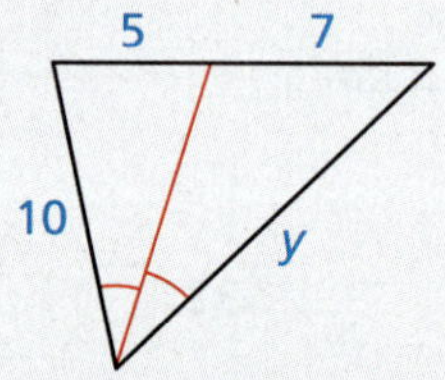

50. Determine whether the triangles are similar. If so, write a similarity statement. Explain your reasoning.

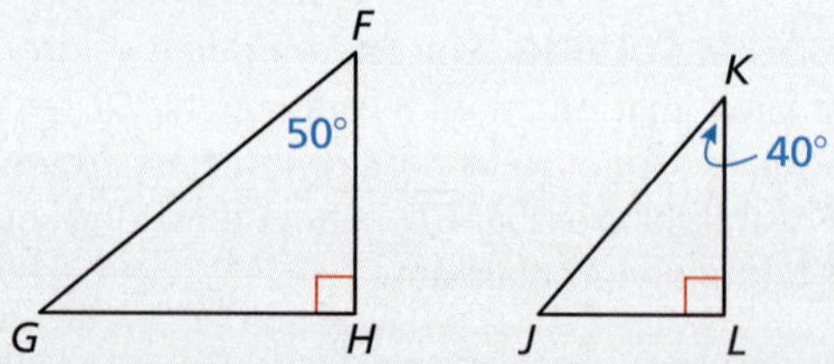

51. Graph $\triangle RST$ with vertices $R(0, 3)$, $S(3, 5)$, and $T(5, 2)$ and its image after a translation 2 units down, followed by a reflection in the y-axis.

GO DIGITAL

9.4 The Tangent Ratio

Learning Target: Understand and use the tangent ratio.

Success Criteria:
- I can explain the tangent ratio.
- I can find tangent ratios.
- I can use tangent ratios to solve real-life problems.

EXPLORE IT! Calculating a Tangent Ratio

Work with a partner.

a. Consider right $\triangle ABC$. Several segments perpendicular to $\overline{AC}$ are shown, forming right triangles that each share vertex A. Explain how the triangles are related.

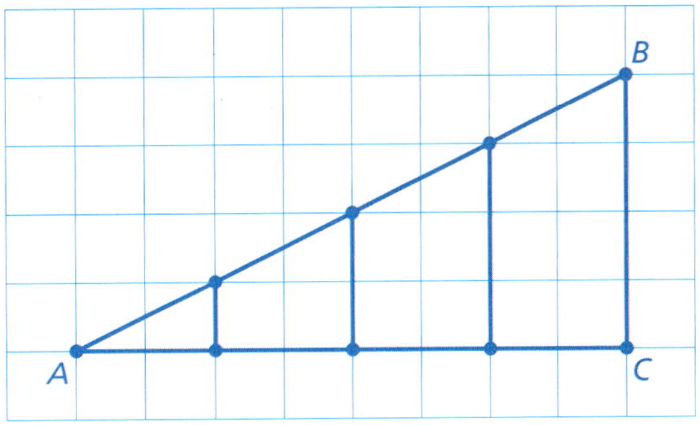

2 MTR USE ANOTHER METHOD

Measure each angle in $\triangle ABC$. Explain how to use technology with a tangent key to calculate tan A and tan B in part (b).

b. Use the definition below to find tan A for each right triangle in part (a). Compare the values. Then find the tangent of the other acute angle in each triangle.

Let $\triangle ABC$ be a right triangle with acute $\angle A$. The *tangent* of $\angle A$ (written as tan A) is defined as follows.

$$\tan A = \frac{\text{length of leg opposite } \angle A}{\text{length of leg adjacent to } \angle A} = \frac{BC}{AC}$$

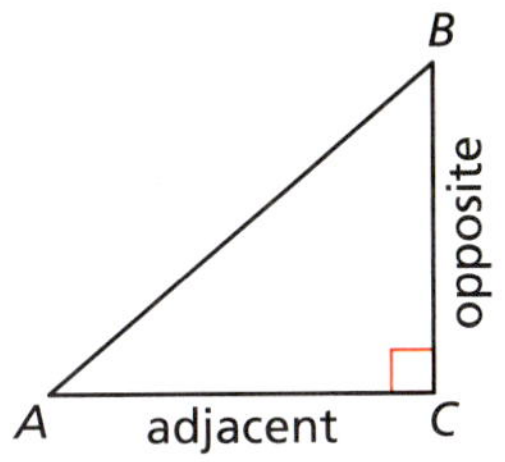

i. Does the size of a right triangle affect the value of the tangent of either acute angle? Justify your conclusion.

ii. Does the measure of an angle affect the value of the tangent of the angle? Explain your reasoning.

c. Summarize what you learned about the tangent ratio in part (b).

Trigonometry

MA.912.T.1.1 Define trigonometric ratios for acute angles in right triangles.

MA.912.T.1.2 Solve mathematical and real-world problems involving right triangles using trigonometric ratios and the Pythagorean Theorem.

GO DIGITAL

Using the Tangent Ratio

Vocabulary

trigonometric ratio, *p. 476*
tangent, *p. 476*
angle of elevation, *p. 478*

A **trigonometric ratio** is a ratio of the lengths of two sides in a right triangle. All right triangles with a given acute angle are similar by the AA Similarity Theorem. So, $\triangle JKL \sim \triangle XYZ$, and you can write $\frac{KL}{YZ} = \frac{JL}{XZ}$. This can be rewritten as $\frac{KL}{JL} = \frac{YZ}{XZ}$, which is a trigonometric ratio. So, trigonometric ratios are constant for a given angle measure.

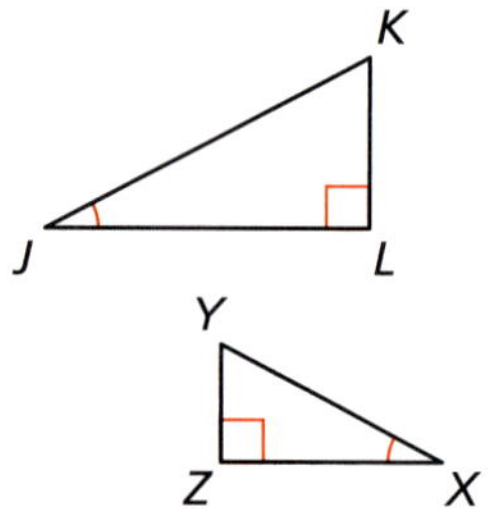

The **tangent** ratio is a trigonometric ratio for acute angles that involves the lengths of the legs of a right triangle.

KEY IDEA

Tangent Ratio

Let $\triangle ABC$ be a right triangle with acute $\angle A$.

The tangent of $\angle A$ (written as tan A) is defined as follows.

$$\tan A = \frac{\text{length of leg opposite } \angle A}{\text{length of leg adjacent to } \angle A} = \frac{BC}{AC}$$

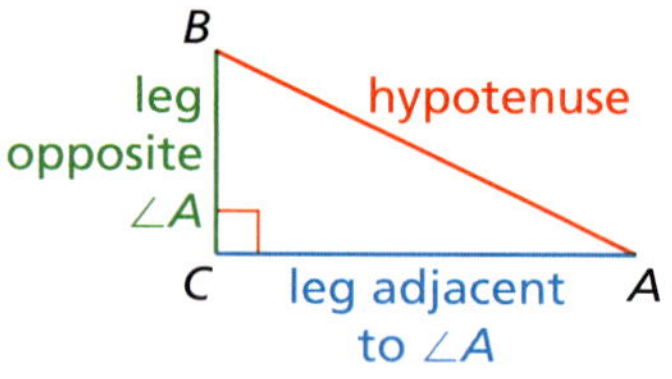

READING

Remember the following abbreviations.

tangent → tan
opposite → opp
adjacent → adj

In the right triangle above, $\angle A$ and $\angle B$ are complementary. So, $\angle B$ is acute. You can use the same diagram to find the tangent of $\angle B$. Notice that the leg adjacent to $\angle A$ is the leg *opposite* $\angle B$ and the leg opposite $\angle A$ is the leg *adjacent* to $\angle B$.

EXAMPLE 1 Finding Tangent Ratios

WATCH

Find tan S and tan R. Write each answer as a fraction and as a decimal.

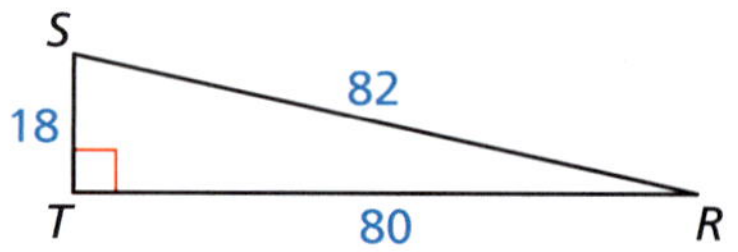

SOLUTION

$$\tan S = \frac{\text{opp } \angle S}{\text{adj to } \angle S} = \frac{RT}{ST} = \frac{80}{18} = \frac{40}{9} \approx 4.4444$$

$$\tan R = \frac{\text{opp } \angle R}{\text{adj to } \angle R} = \frac{ST}{RT} = \frac{18}{80} = \frac{9}{40} = 0.225$$

SELF-ASSESSMENT

1 I don't understand yet. 2 I can do it with help. 3 I can do it on my own. 4 I can teach someone else.

Find tan *J* and tan *K*. Write each answer as a fraction and as a decimal.

1.

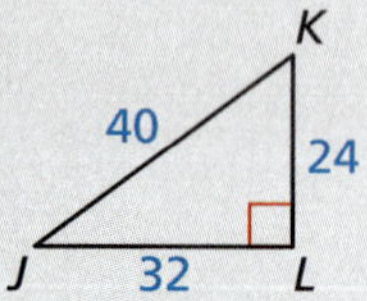

2.

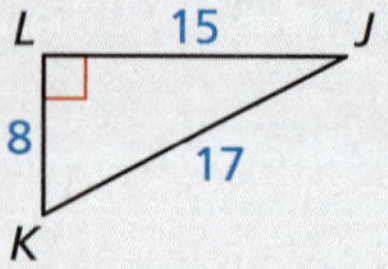

3. WRITING Explain how you know the tangent ratio is constant for a given angle measure.

GO DIGITAL

EXAMPLE 2 Finding a Leg Length

WATCH

Find the value of x.

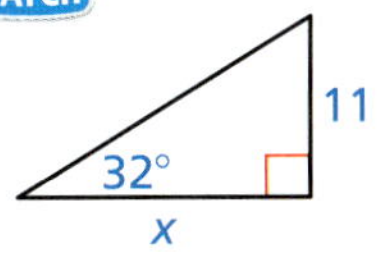

SOLUTION

Use the tangent of an acute angle to find a leg length.

$\tan 32° = \dfrac{\text{opp}}{\text{adj}}$	Write ratio for tangent of 32°.
$\tan 32° = \dfrac{11}{x}$	Substitute.
$x \cdot \tan 32° = 11$	Multiply each side by x.
$x = \dfrac{11}{\tan 32°}$	Divide each side by tan 32°.
$x \approx 17.6$	Use technology.

▶ The value of x is about 17.6.

STUDY TIP

You can also use the Table of Trigonometric Ratios available at *BigIdeasMath.com* to find the decimal approximations of trigonometric ratios.

You can find the tangent of an acute angle measuring 30°, 45°, or 60° by applying what you know about special right triangles.

EXAMPLE 3 Using a Special Right Triangle to Find a Tangent

WATCH

Use a special right triangle to find the tangent of a 60° angle.

SOLUTION

Step 1 Because all 30°-60°-90° triangles are similar, you can simplify your calculations by choosing 1 as the length of the shorter leg. Use the 30°-60°-90° Triangle Theorem to find the length of the longer leg.

longer leg $=$ shorter leg $\cdot \sqrt{3}$	30°-60°-90° Triangle Theorem
$= 1 \cdot \sqrt{3}$	Substitute.
$= \sqrt{3}$	Simplify.

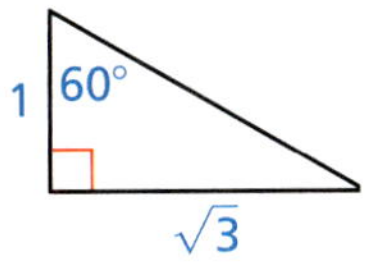

Step 2 Find tan 60°.

$\tan 60° = \dfrac{\text{opp}}{\text{adj}}$	Write ratio for tangent of 60°.
$\tan 60° = \dfrac{\sqrt{3}}{1}$	Substitute.
$\tan 60° = \sqrt{3}$	Simplify.

▶ The tangent of any 60° angle is $\sqrt{3}$, or about 1.7321.

STUDY TIP

The tangents of all 60° angles are the same constant ratio. Any right triangle with a 60° angle can be used to determine this value. A similar statement can be made about any other acute angle.

SELF-ASSESSMENT

1 I don't understand yet. 2 I can do it with help. 3 I can do it on my own. 4 I can teach someone else.

Find the value of x.

4.

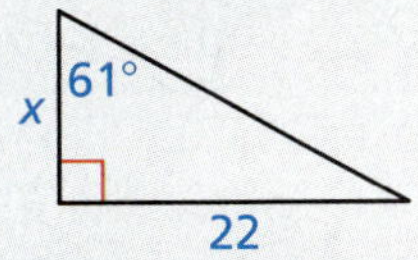

5.

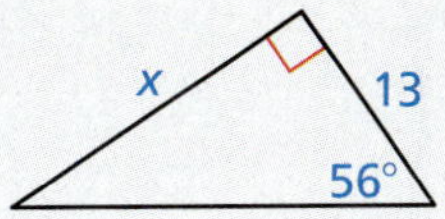

6. **WHAT IF?** In Example 3, the length of the shorter leg is 5 instead of 1. Show that the tangent of 60° is still equal to $\sqrt{3}$.

GO DIGITAL

Solving Real-Life Problems

An **angle of elevation** is an angle formed by a horizontal line and a line of sight *up* to an object.

7 MTR **EXAMPLE 4** **Modeling Real Life** WATCH

You measure your distance from the Ponce de Leon Inlet Lighthouse and the angle of elevation from the ground to the top of the lighthouse. Find the height of the lighthouse.

The Ponce de Leon Inlet Lighthouse is the tallest lighthouse in Florida.

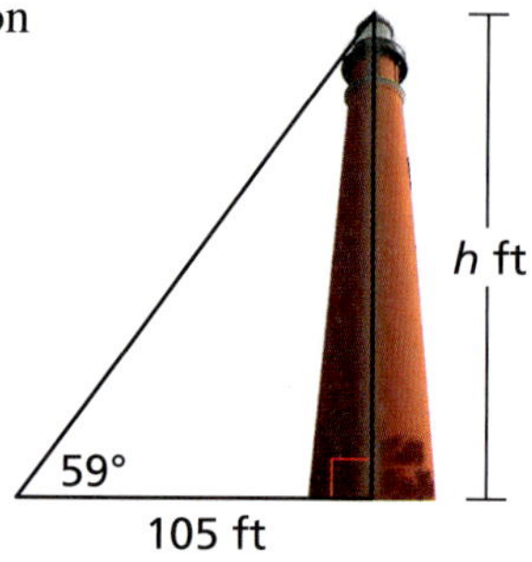

SOLUTION

1. **Understand the Problem** You are given your distance from the lighthouse and the angle of elevation. You need to find the height of the lighthouse.
2. **Make a Plan** Write a trigonometric ratio for the tangent of the angle of elevation involving the height h. Then solve for h.
3. **Solve and Check**

$$\tan 59^\circ = \frac{\text{opp}}{\text{adj}}$$ Write ratio for tangent of 59°.

$$\tan 59^\circ = \frac{h}{105}$$ Substitute.

$$105 \cdot \tan 59^\circ = h$$ Multiply each side by 105.

$$174.7 \approx h$$ Use technology.

▶ The lighthouse is about 175 feet tall.

Check Reasonableness Because 59° is close to 60°, you can use the legs of a 30°-60°-90° triangle to check your answer.

$$\text{longer leg} = \text{shorter leg} \cdot \sqrt{3}$$ 30°-60°-90° Triangle Theorem

$$= 105 \cdot \sqrt{3}$$ Substitute.

$$\approx 181.9$$ Use technology.

The value of 181.9 feet is close to the value of h. ✓

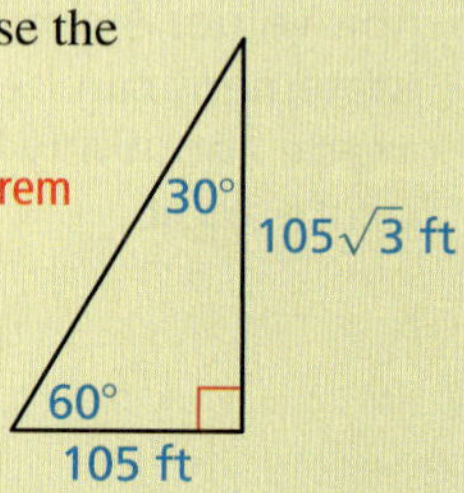

SELF-ASSESSMENT 1 I don't understand yet. 2 I can do it with help. 3 I can do it on my own. 4 I can teach someone else.

7. You measure your distance from a lamppost and the angle of elevation from the ground to the top of the lamppost. Find the height of the lamppost.

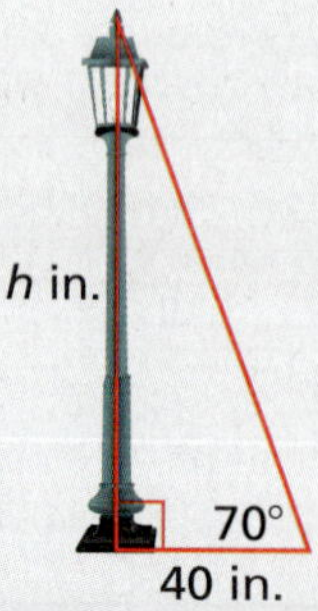

GO DIGITAL

9.4 Practice WITH CalcChat® AND CalcView®

In Exercises 1–4, find the tangents of the acute angles in the right triangle. Write each answer as a fraction and as a decimal. *(See Example 1.)*

1.

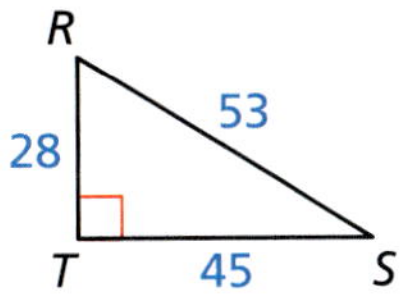

2.

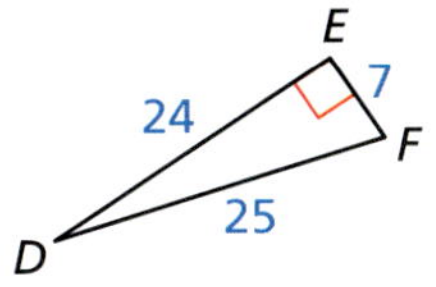

3.

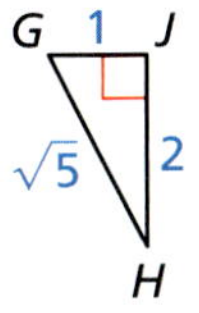

4.

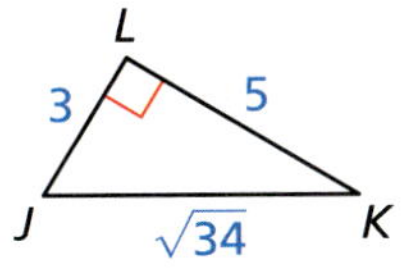

In Exercises 5–8, find the value of x. *(See Example 2.)*

5.

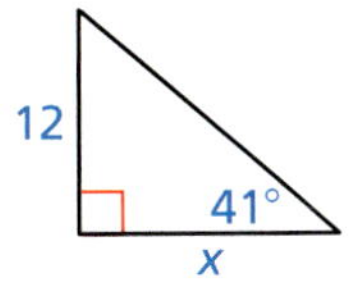

6.

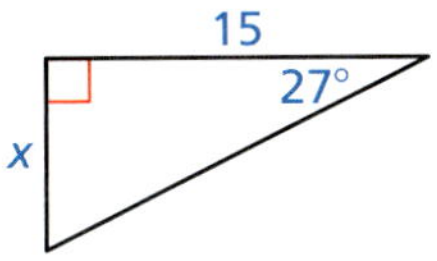

7.

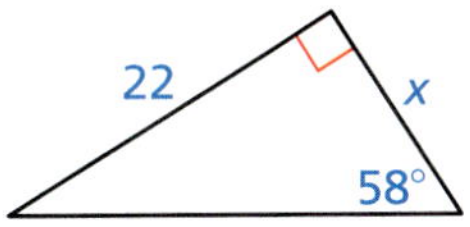

8.

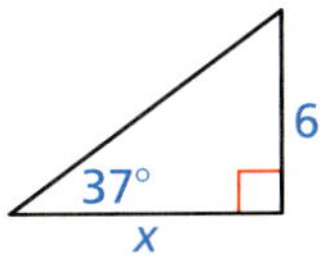

ERROR ANALYSIS In Exercises 9 and 10, describe the error in the statement of the tangent ratio. Correct the error if possible. Otherwise, write not possible.

9.

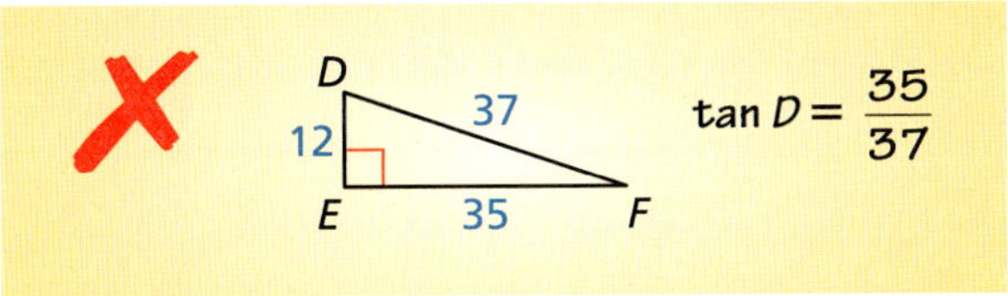

10.

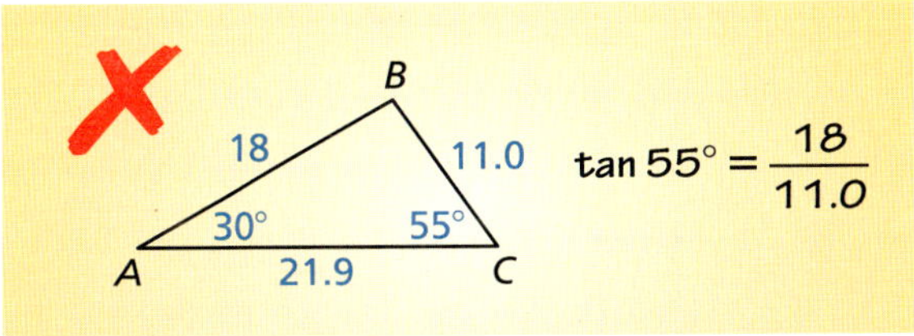

In Exercises 11 and 12, use a special right triangle to find the tangent of the given angle measure. *(See Example 3.)*

11. 45° **12.** 30°

GO DIGITAL

13. MODELING REAL LIFE You measure your distance from the base of the Citrus Tower and the angle of elevation from the ground to the top of the tower. Find the height h of the Citrus Tower. *(See Example 4.)*

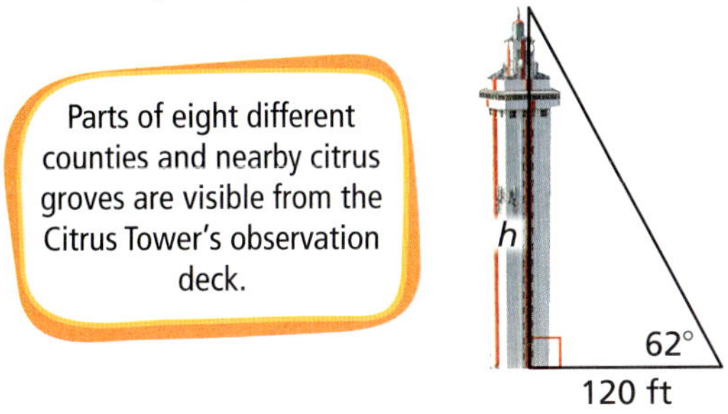

14. MODELING REAL LIFE Scientists can measure the depths of craters on the moon by looking at photos of shadows. Estimate the depth d of the crater.

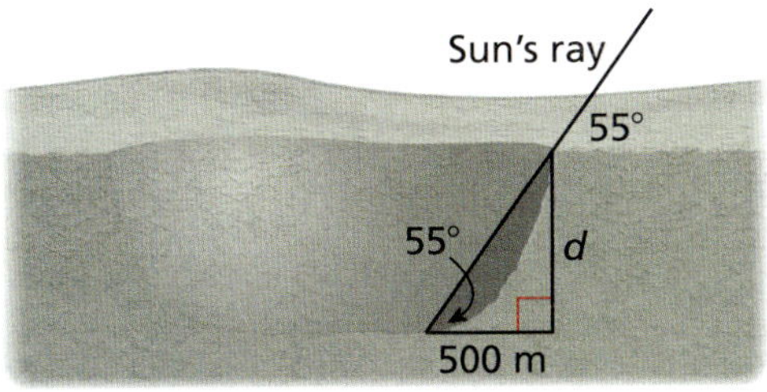

15. STRUCTURE Find the tangent of the smaller acute angle in a right triangle with side lengths 5, 12, and 13.

16. STRUCTURE Find the tangent of the larger acute angle in a right triangle with side lengths 3, 4, and 5.

17. REASONING How does the tangent of an acute angle in a right triangle change as the angle measure increases? Justify your answer.

18. MAKING AN ARGUMENT Your family room has a sliding-glass door. You want to buy an awning for the door that will be just long enough to keep the Sun out when it is at its highest point in the sky. Your friend claims you can determine how far the overhang should extend by multiplying 8 by tan 70°. Is your friend correct? Explain.

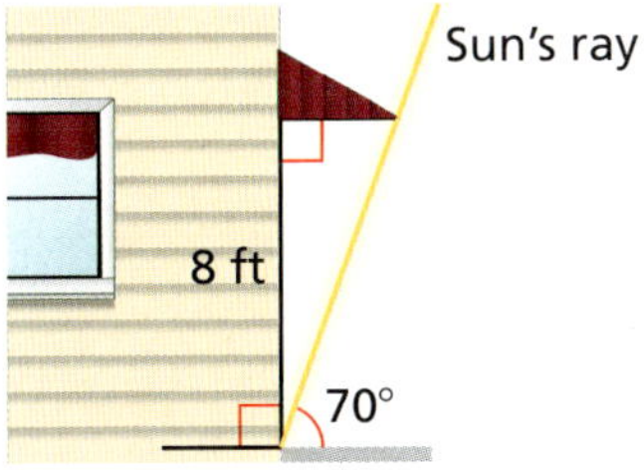

19. PROBLEM SOLVING Your class is having a class picture taken. The photographer is positioned 14 feet away from the center of the class. The photographer turns 50° to look at either end of the class.

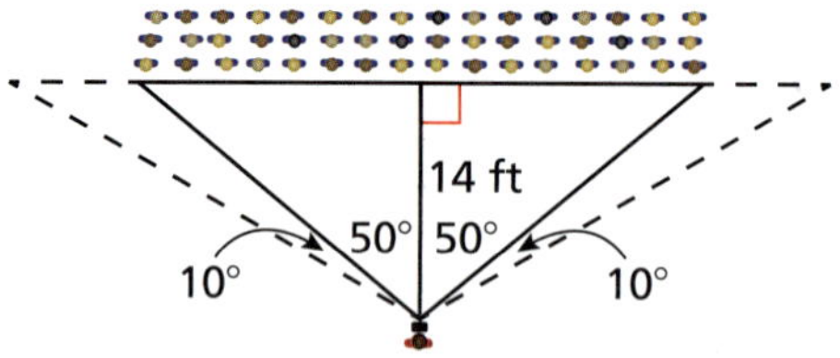

a. Your class banner is 28 feet long. If the class is arranged as shown, can the students hold the banner so that it stretches the full length of the class? Explain.

b. A group of students arrive late, so the photographer turns another 10° either way to see the end of the camera range. If each student needs 2 feet of space, about how many more students can fit at the end of each row? Explain.

20. HOW DO YOU SEE IT?
Write expressions for the tangent of each acute angle in the right triangle. Explain how the tangent of one acute angle is related to the tangent of the other acute angle. What kind of angle pair is $\angle A$ and $\angle B$?

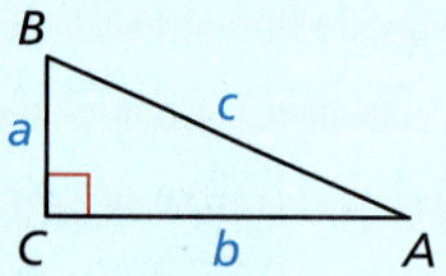

4 MTR **21. DISCUSS MATHEMATICAL THINKING** For what angle measure(s) is the tangent of an acute angle in a right triangle equal to 1? greater than 1? less than 1? Justify your answer.

22. THOUGHT PROVOKING
To create the diagram, you begin with an isosceles right triangle with legs 1 unit long. Then the hypotenuse of the first triangle becomes the leg of a second triangle, whose remaining leg is 1 unit long. The pattern continues so one leg of each new triangle is the hypotenuse of the previous triangle and the other is 1 unit long. Continue the diagram until you have constructed an angle whose tangent is $\frac{1}{\sqrt{6}}$. Approximate the measure of this angle.

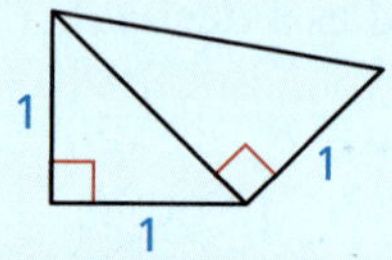

23. DIG DEEPER Find the perimeter of the figure, where $AC = 26$, $AD = BF$, and D is the midpoint of $\overline{AC}$.

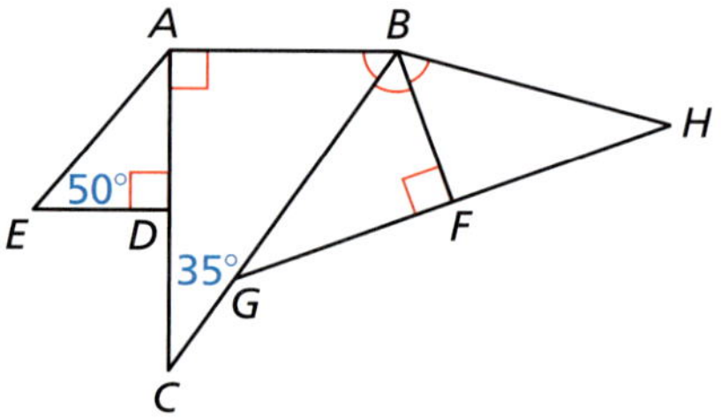

REVIEW & REFRESH

WATCH

24. Find the value of x. Tell whether the side lengths form a Pythagorean triple.

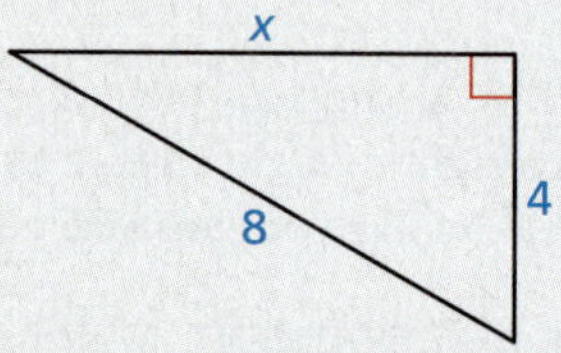

25. Find the geometric mean of 18 and 26.

26. Find the area of the polygon with vertices (1, 1), (2, −2), and (−3, −2).

27. Graph $\overline{XY}$ with endpoints $X(-4, 2)$ and $Y(1, -3)$ and its image after a rotation 180° about the origin, followed by a reflection in the x-axis.

28. Find the coordinates of the circumcenter of the triangle with vertices $A(-1, 2)$, $B(-1, -2)$, and $C(5, -2)$.

In Exercises 29–32, find the value of x.

29.

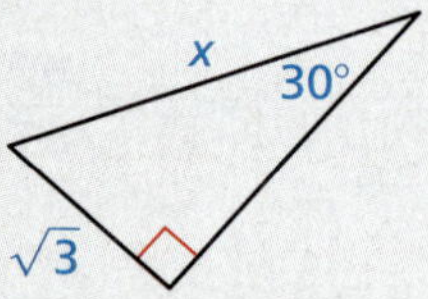

30.

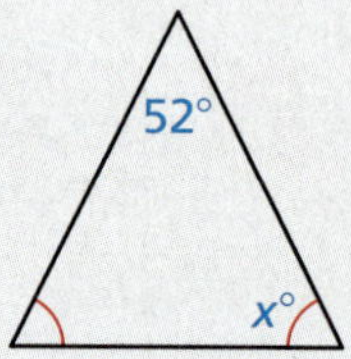

31.

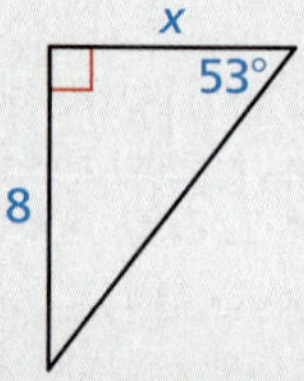

32.

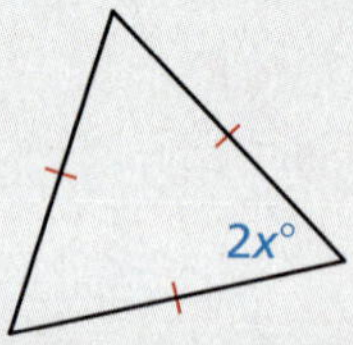

7 MTR **33. MODELING REAL LIFE** The horizontal boards of a fence are parallel. Find $m\angle 2$.

GO DIGITAL

9.5 The Sine and Cosine Ratios

Learning Target: Understand and use the sine and cosine ratios.

Success Criteria:
- I can explain the sine and cosine ratios.
- I can find sine and cosine ratios.
- I can use sine and cosine ratios to solve real-life problems.

EXPLORE IT! Calculating Sine and Cosine Ratios

Work with a partner.

a. Consider right $\triangle ABC$. Several segments perpendicular to $\overline{AC}$ are shown, forming right triangles that each share vertex A. For each triangle, find the length of the hypotenuse.

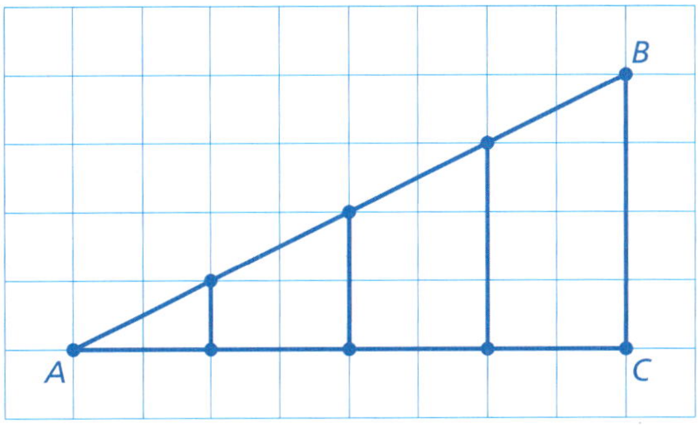

4 MTR CONSTRUCT AN ARGUMENT

For a given acute $\angle A$, describe the possible values of $\cos A$ and $\sin A$. Explain your reasoning.

b. Use the definitions below to find $\sin A$ and $\cos A$ for each right triangle in part (a). Compare the values.

Let $\triangle ABC$ be a right triangle with acute $\angle A$. The *sine* of $\angle A$ and *cosine* of $\angle A$ (written as $\sin A$ and $\cos A$, respectively) are defined as follows.

$$\sin A = \frac{\text{length of leg opposite } \angle A}{\text{length of hypotenuse}} = \frac{BC}{AB}$$

$$\cos A = \frac{\text{length of leg adjacent to } \angle A}{\text{length of hypotenuse}} = \frac{AC}{AB}$$

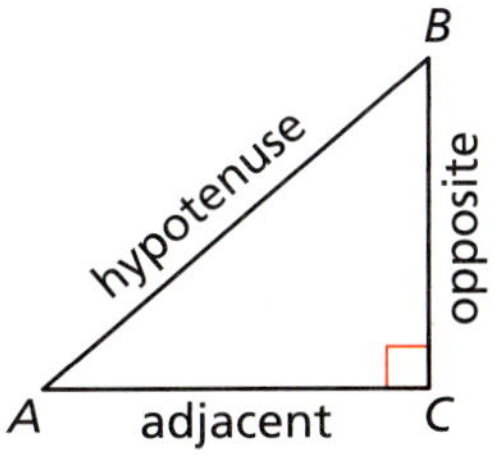

i. Does the size of a right triangle affect the value of the sine and cosine of an acute angle? Justify your conclusion.

ii. Does the measure of an angle affect the value of the sine and cosine of the angle? Explain your reasoning.

c. What is the relationship between the measures of $\angle A$ and $\angle B$? Find $\sin B$ and $\cos B$. How are these values related to $\sin A$ and $\cos A$? Explain why these relationships exist.

d. Summarize what you learned about the sine and cosine ratios in parts (b) and (c).

Trigonometry

MA.912.T.1.1 Define trigonometric ratios for acute angles in right triangles.

MA.912.T.1.2 Solve mathematical and real-world problems involving right triangles using trigonometric ratios and the Pythagorean Theorem.

GO DIGITAL

Using the Sine and Cosine Ratios

The **sine** and **cosine** ratios are trigonometric ratios for acute angles that involve the lengths of a leg and the hypotenuse of a right triangle.

Vocabulary

sine, *p. 482*
cosine, *p. 482*
angle of depression, *p. 485*
unit circle, *p. 486*

READING

Remember the following abbreviations.

sine → sin
cosine → cos
hypotenuse → hyp

KEY IDEA

Sine and Cosine Ratios

Let $\triangle ABC$ be a right triangle with acute $\angle A$. The sine of $\angle A$ and cosine of $\angle A$ (written as $\sin A$ and $\cos A$) are defined as follows.

$$\sin A = \frac{\text{length of leg opposite } \angle A}{\text{length of hypotenuse}} = \frac{BC}{AB}$$

$$\cos A = \frac{\text{length of leg adjacent to } \angle A}{\text{length of hypotenuse}} = \frac{AC}{AB}$$

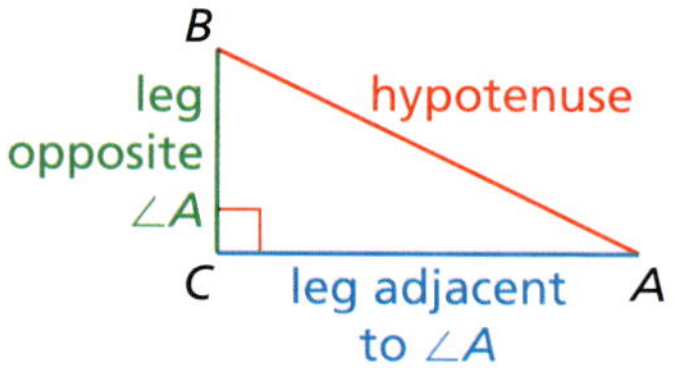

EXAMPLE 1 Finding Sine and Cosine Ratios

WATCH

Find $\sin S$, $\sin R$, $\cos S$, and $\cos R$. Write each answer as a fraction and as a decimal.

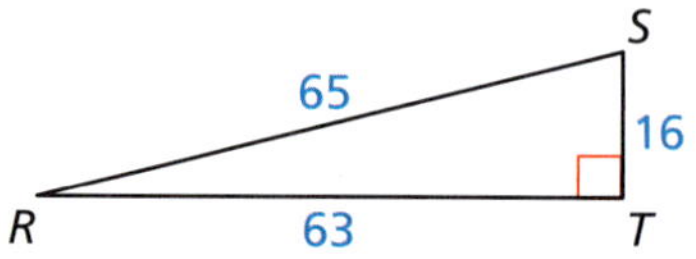

SOLUTION

$$\sin S = \frac{\text{opp } \angle S}{\text{hyp}} = \frac{RT}{SR} = \frac{63}{65} \approx 0.9692 \qquad \sin R = \frac{\text{opp } \angle R}{\text{hyp}} = \frac{ST}{SR} = \frac{16}{65} \approx 0.2462$$

$$\cos S = \frac{\text{adj to } \angle S}{\text{hyp}} = \frac{ST}{SR} = \frac{16}{65} \approx 0.2462 \qquad \cos R = \frac{\text{adj to } \angle R}{\text{hyp}} = \frac{RT}{SR} = \frac{63}{65} \approx 0.9692$$

In Example 1, notice that $\sin S = \cos R$ and $\sin R = \cos S$. This is true because the side opposite $\angle S$ is adjacent to $\angle R$ and the side opposite $\angle R$ is adjacent to $\angle S$. The relationship between the sine and cosine of $\angle S$ and $\angle R$ is true for all complementary angles.

KEY IDEA

Sine and Cosine of Complementary Angles

The sine of an acute angle is equal to the cosine of its complement. The cosine of an acute angle is equal to the sine of its complement.

Let A and B be complementary angles. Then the following statements are true.

$$\sin A = \cos(90° - A) = \cos B \qquad \sin B = \cos(90° - B) = \cos A$$

$$\cos A = \sin(90° - A) = \sin B \qquad \cos B = \sin(90° - B) = \sin A$$

GO DIGITAL

EXAMPLE 2 Rewriting Trigonometric Expressions

WATCH

Write sin 56° in terms of cosine.

SOLUTION

Use the fact that the sine of an acute angle is equal to the cosine of its complement.

$$\sin 56° = \cos(90° - 56°) = \cos 34°$$

▶ The sine of 56° is the same as the cosine of 34°.

EXAMPLE 3 Finding Leg Lengths

WATCH

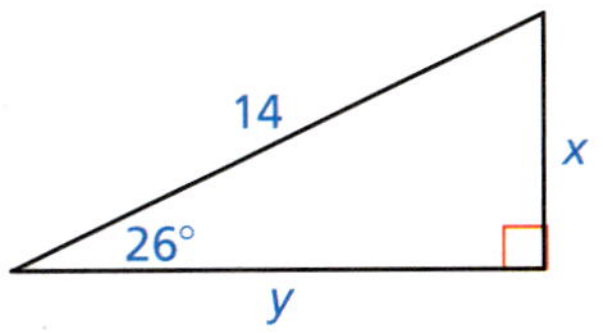

Find the values of x and y using sine and cosine.

SOLUTION

Step 1 Use a sine ratio to find the value of x.

$\sin 26° = \frac{\text{opp}}{\text{hyp}}$ — Write ratio for sine of 26°.

$\sin 26° = \frac{x}{14}$ — Substitute.

$14 \cdot \sin 26° = x$ — Multiply each side by 14.

$6.1 \approx x$ — Use technology.

▶ The value of x is about 6.1.

Step 2 Use a cosine ratio to find the value of y.

$\cos 26° = \frac{\text{adj}}{\text{hyp}}$ — Write ratio for cosine of 26°.

$\cos 26° = \frac{y}{14}$ — Substitute.

$14 \cdot \cos 26° = y$ — Multiply each side by 14.

$12.6 \approx y$ — Use technology.

▶ The value of y is about 12.6.

SELF-ASSESSMENT

1 I don't understand yet. 2 I can do it with help. 3 I can do it on my own. 4 I can teach someone else.

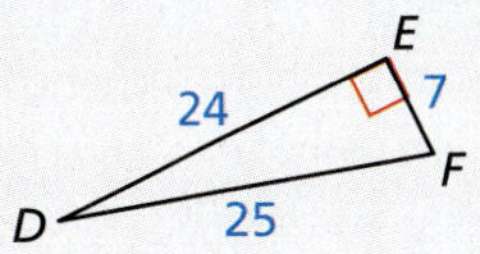

1. Find sin D, sin F, cos D, and cos F. Write each answer as a fraction and as a decimal.
2. Write cos 23° in terms of sine.
3. Find the values of u and t using sine and cosine.

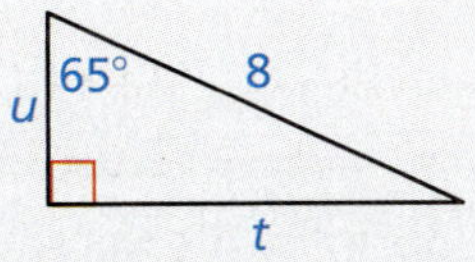

4 MTR

4. **WHICH ONE DOESN'T BELONG?** Which ratio does *not* belong with the other three? Explain your reasoning.

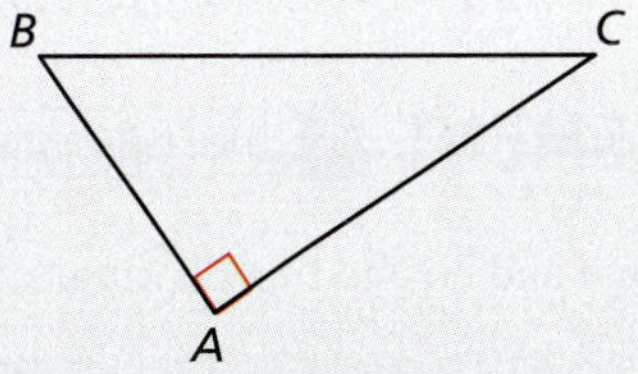

sin B	cos C	tan B	$\frac{AC}{BC}$

GO DIGITAL

Finding Sine and Cosine in Special Right Triangles

EXAMPLE 4 Finding the Sine and Cosine of 45°

WATCH

Find the sine and cosine of a 45° angle.

SOLUTION

Begin by sketching a 45°-45°-90° triangle. Because all such triangles are similar, you can simplify your calculations by choosing 1 as the length of each leg. Using the 45°-45°-90° Triangle Theorem, the length of the hypotenuse is $\sqrt{2}$.

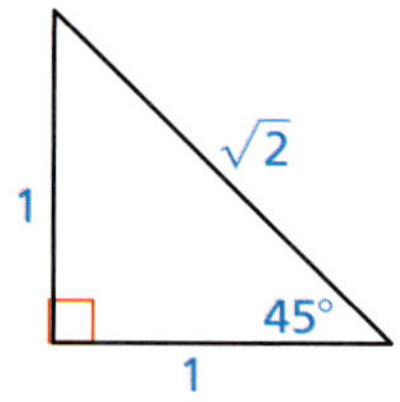

$$\sin 45^\circ = \frac{\text{opp}}{\text{hyp}} = \frac{1}{\sqrt{2}} = \frac{\sqrt{2}}{2} \approx 0.7071$$

$$\cos 45^\circ = \frac{\text{adj}}{\text{hyp}} = \frac{1}{\sqrt{2}} = \frac{\sqrt{2}}{2} \approx 0.7071$$

STUDY TIP

Notice that

$$\sin 45^\circ = \cos(90 - 45)^\circ = \cos 45^\circ.$$

► So, $\sin 45^\circ \approx 0.7071$ and $\cos 45^\circ \approx 0.7071$.

EXAMPLE 5 Finding the Sine and Cosine of 30°

WATCH

Find the sine and cosine of a 30° angle.

SOLUTION

Begin by sketching a 30°-60°-90° triangle. Because all such triangles are similar, you can simplify your calculations by choosing 1 as the length of the shorter leg. Using the 30°-60°-90° Triangle Theorem, the length of the longer leg is $\sqrt{3}$ and the length of the hypotenuse is 2.

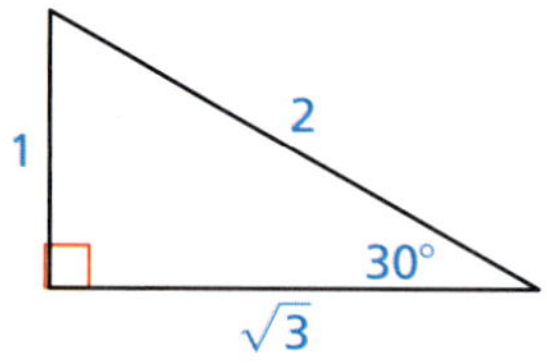

$$\sin 30^\circ = \frac{\text{opp}}{\text{hyp}} = \frac{1}{2} = 0.5$$

$$\cos 30^\circ = \frac{\text{adj}}{\text{hyp}} = \frac{\sqrt{3}}{2} \approx 0.8660$$

► So, $\sin 30^\circ = 0.5$ and $\cos 30^\circ \approx 0.8660$.

SELF-ASSESSMENT 1 I don't understand yet. 2 I can do it with help. 3 I can do it on my own. 4 I can teach someone else.

5. Find the sine and cosine of a 60° angle.

GO DIGITAL

Solving Real-Life Problems

Recall from the previous lesson that an *angle of elevation* is an angle formed by a horizontal line and a line of sight *up* to an object. An **angle of depression** is an angle formed by a horizontal line and a line of sight *down* to an object.

7 MTR

EXAMPLE 6 Modeling Real Life

WATCH

You are skiing on a mountain. Find the distance x from you to the base of the mountain.

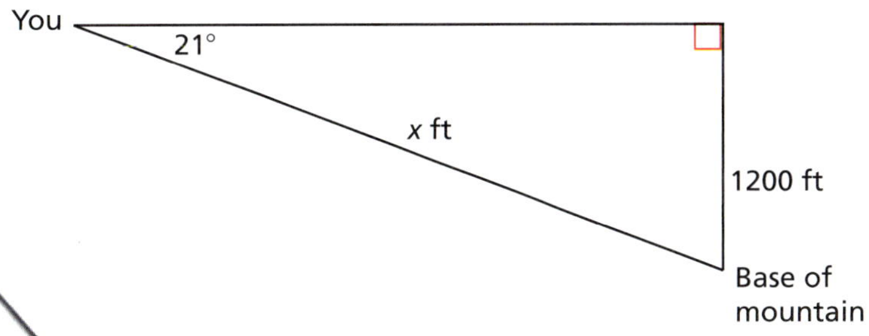

SOLUTION

1. **Understand the Problem** You are given the angle of depression and your altitude above the bottom of the mountain. You need to find the distance x from you to the base of the mountain.
2. **Make a Plan** Write a trigonometric ratio for the sine of the angle of depression involving the distance x. Then solve for x.
3. **Solve and Check**

$$\sin 21^\circ = \frac{\text{opp}}{\text{hyp}}$$ Write ratio for sine of 21°.

$$\sin 21^\circ = \frac{1200}{x}$$ Substitute.

$$x \cdot \sin 21^\circ = 1200$$ Multiply each side by x.

$$x = \frac{1200}{\sin 21^\circ}$$ Divide each side by sin 21°.

$$x \approx 3348.5$$ Use technology.

▶ The distance from you to the base of the mountain is about 3349 feet.

Check The value of sin 21° is about 0.3584. Substitute for x in the sine ratio, and compare the values.

$$\frac{1200}{x} \approx \frac{1200}{3348.5} \approx 0.3584$$

This value is approximately the same as the value of sin 21°. ✓

SELF-ASSESSMENT 1 I don't understand yet. 2 I can do it with help. 3 I can do it on my own. 4 I can teach someone else.

6. **WHAT IF?** In Example 6, the angle of depression is 28°. Find the distance from you to the base of the mountain.

GO DIGITAL

Using the Unit Circle

KEY IDEA

The Unit Circle

The circle with center (0, 0) and radius 1 is called the **unit circle**. Consider an angle A with its vertex at the origin, one side on the x-axis, and one side passing through a point (x, y) on the unit circle in the first quadrant. The values of sin A and cos A are the y-coordinate and x-coordinate, respectively, of the point (x, y). The value of tan A is $\frac{y}{x}$, $x \neq 0$.

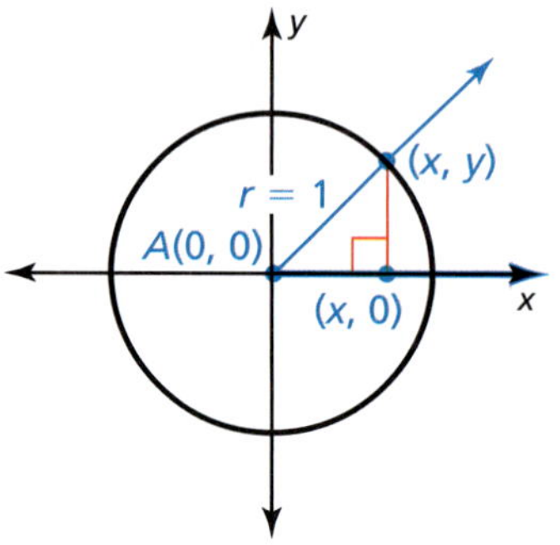

$$\sin A = \frac{\text{opp } \angle A}{\text{hyp}} = \frac{y}{1} = y$$

$$\cos A = \frac{\text{adj } \angle A}{\text{hyp}} = \frac{x}{1} = x$$

$$\tan A = \frac{\text{opp } \angle A}{\text{adj } \angle A} = \frac{y}{x},\ x \neq 0$$

STUDY TIP

Notice that the points (0, 0), (x, y), and $(x, 0)$ form a right triangle with leg lengths x and y, and a hypotenuse with a length of 1.

EXAMPLE 7 Using the Unit Circle

WATCH

Find the sine, cosine, and tangent of $\angle A$.

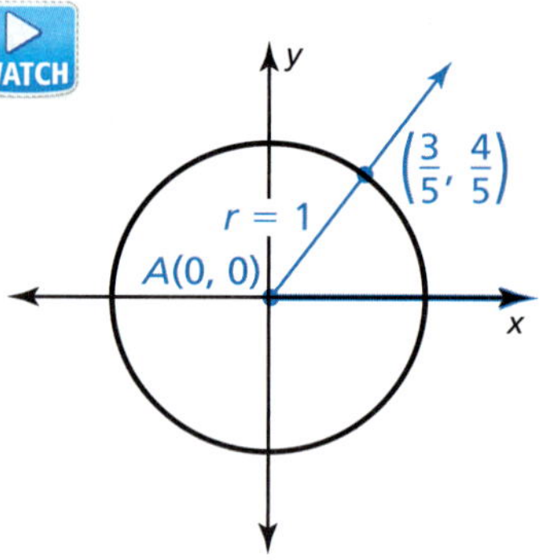

SOLUTION

The vertex of $\angle A$ is at the origin, one side lies on the x-axis, and one side passes through a point on the unit circle. So, you can use the coordinates of the point $\left(\frac{3}{5}, \frac{4}{5}\right)$ to find the values of the trigonometric ratios for $\angle A$.

$$\sin A = y = \frac{4}{5} \qquad \cos A = x = \frac{3}{5} \qquad \tan A = \frac{y}{x} = \frac{\frac{4}{5}}{\frac{3}{5}} = \frac{4}{3}$$

5 MTR RELATE CONCEPTS

Prove that the point $\left(\frac{3}{5}, \frac{4}{5}\right)$ lies on the unit circle.

SELF-ASSESSMENT

1 I don't understand yet. 2 I can do it with help. 3 I can do it on my own. 4 I can teach someone else.

7. Find the sine, cosine, and tangent of $\angle A$.

2 MTR 8. **MAKE A CONNECTION** Use the unit circle to describe what happens to the values of the trigonometric ratios as the measure of an acute angle approaches (a) 0° and (b) 90°.

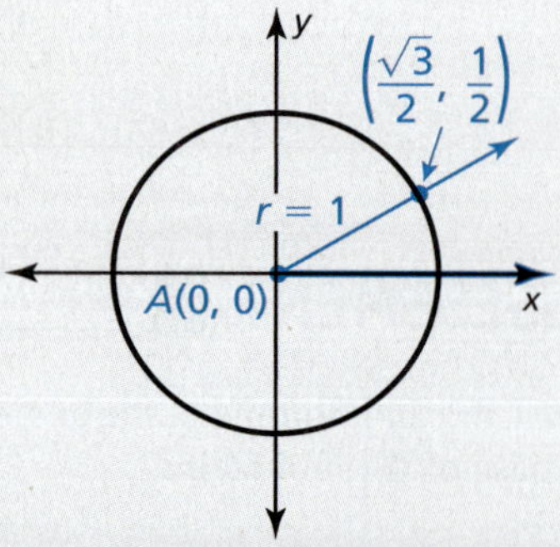

GO DIGITAL

9.5 Practice WITH CalcChat® AND CalcView®

In Exercises 1–6, find sin *D*, sin *E*, cos *D*, and cos *E*. Write each answer as a fraction and as a decimal. *(See Example 1.)*

▶ **1.**

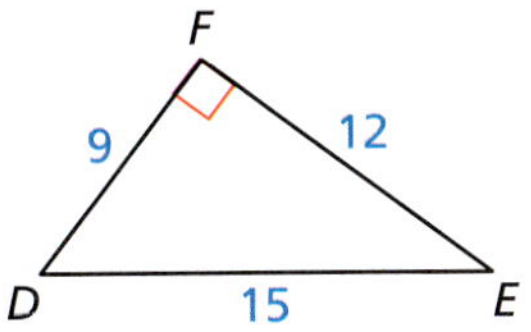

2.

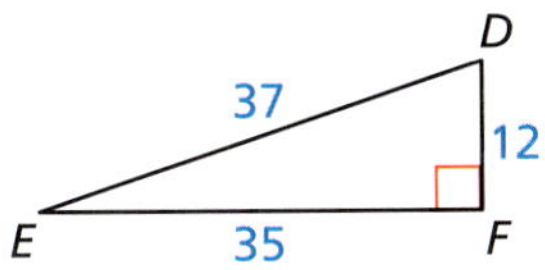

3.

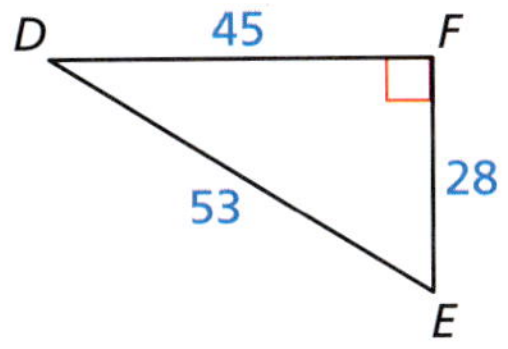

4.

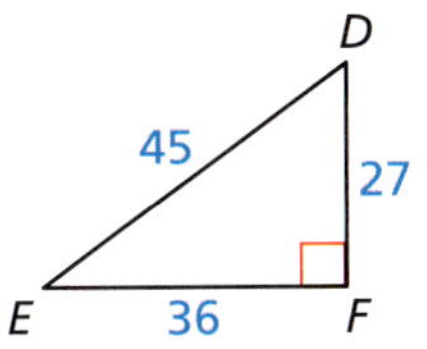

5.

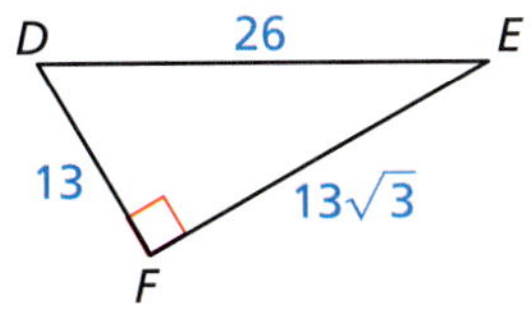

6.

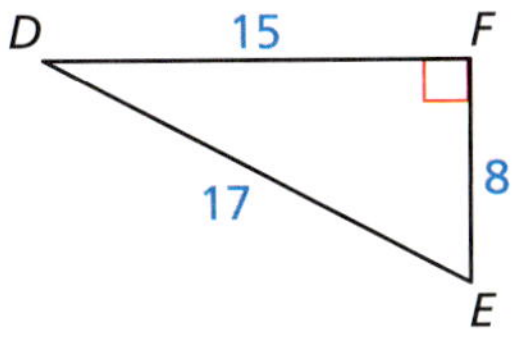

In Exercises 7–10, write the expression in terms of cosine. *(See Example 2.)*

▶ **7.** $\sin 37°$

8. $\sin 81°$

9. $\sin 29°$

10. $\sin 64°$

In Exercises 11–14, write the expression in terms of sine.

▶ **11.** $\cos 59°$

12. $\cos 42°$

13. $\cos 73°$

14. $\cos 18°$

In Exercises 15–20, find the value of each variable using sine and cosine. *(See Example 3.)*

▶ **15.**

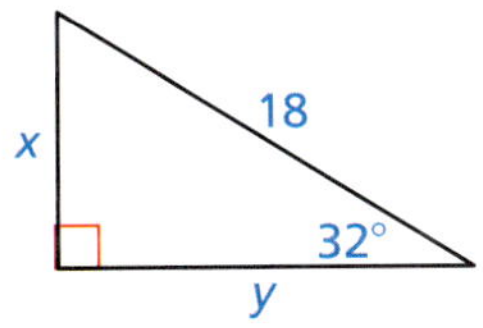

16.

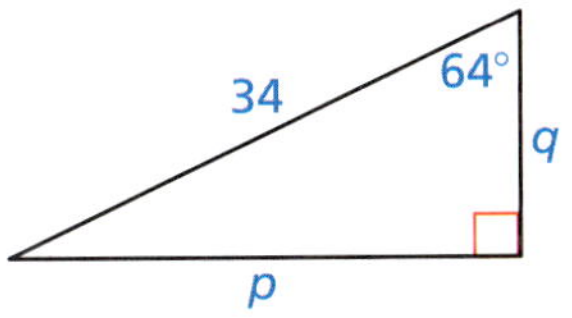

17.

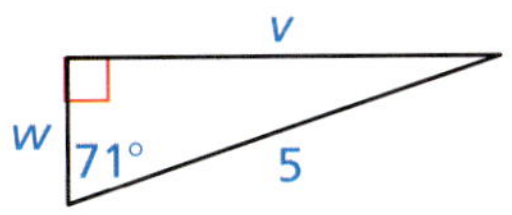

18.

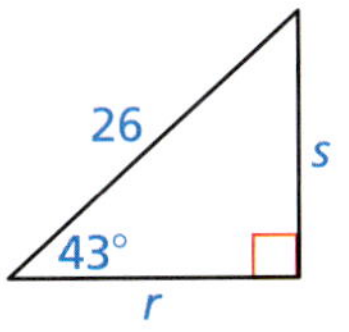

19.

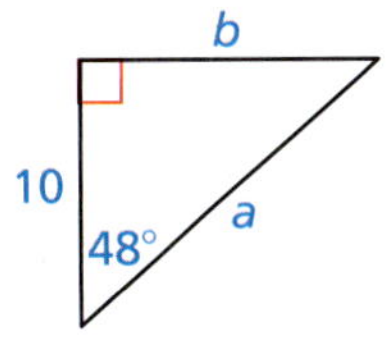

20.

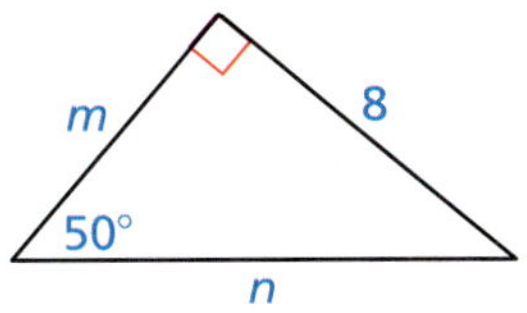

▶ **21. REASONING** Which ratios are equal? Select all that apply. *(See Example 4.)*

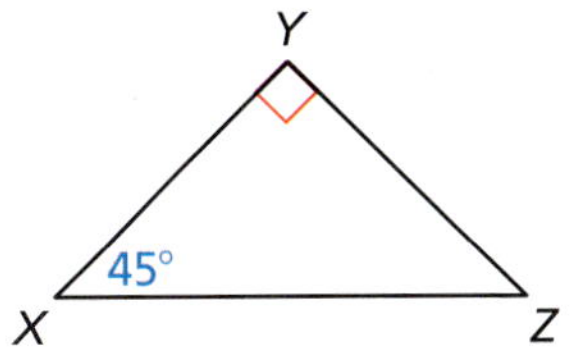

sin *X* | cos *X* | sin *Z* | cos *Z*

22. WRITING Explain how to tell which side of a right triangle is adjacent to an acute angle and which side is the hypotenuse.

▶ **23. REASONING** Which ratios are equal to $\frac{1}{2}$? Select all that apply. *(See Example 5.)*

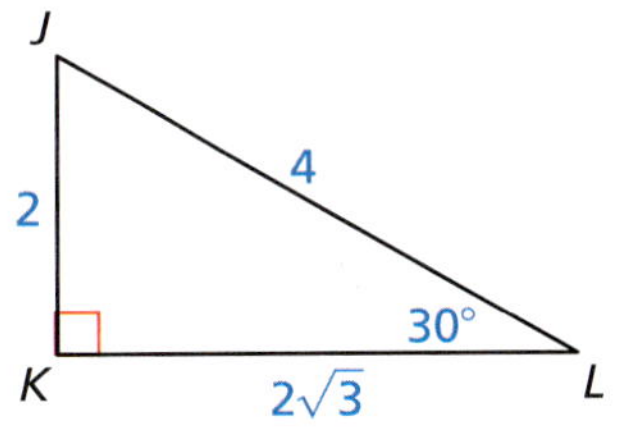

sin *L* | cos *L* | sin *J* | cos *J*

24. WRITING Describe what you must know about a triangle to use the sine ratio or the cosine ratio. Explain.

4 MTR **25. ERROR ANALYSIS** Describe and correct the error in finding sin *A*.

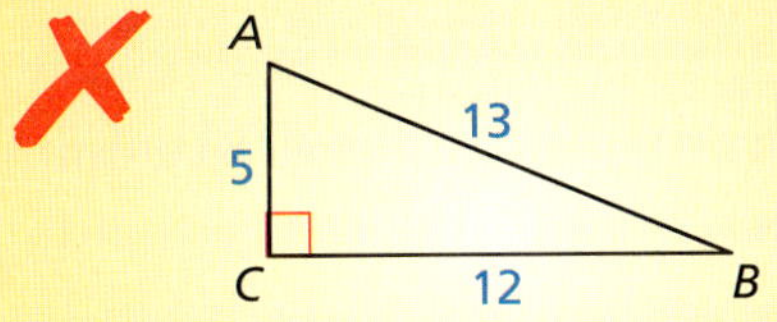

$\sin A = \frac{5}{13}$

GO DIGITAL

26. COLLEGE PREP Which of the following is *not* equal to the ratio of the length of the leg adjacent to $\angle P$ to the length of the hypotenuse?

Ⓐ $\sin R$ Ⓒ $\dfrac{QR}{PR}$

Ⓑ $\cos P$ Ⓓ $\dfrac{PQ}{PR}$

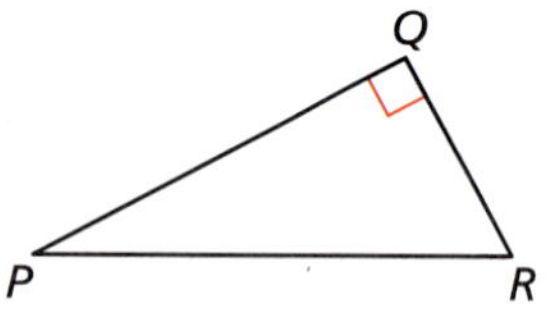

27. MODELING REAL LIFE Find the length of the slide. *(See Example 6.)*

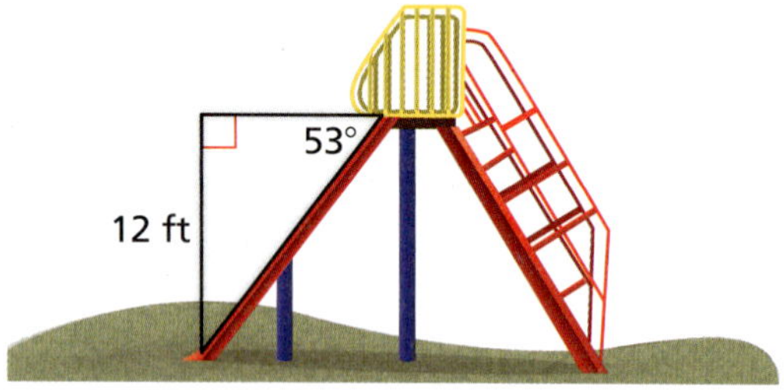

28. MODELING REAL LIFE Find the height of the roller coaster using two different methods. Explain.

29. MULTIPLE REPRESENTATIONS You see a sailboat while standing on a 30-foot-tall cliff next to an ocean.

a. Draw and label a diagram of the situation.

b. Make a table showing the angle of depression and the length of your line of sight. Use the angles 40°, 50°, 60°, 70°, and 80°.

c. Graph the values you found in part (b), with the angle measures on the x-axis.

d. Describe the angle of depression and the length of your line of sight as the boat sails away from you.

30. PROBLEM SOLVING You fly a kite using a 20-foot-long string. The angle of elevation from your hands to the kite is 67°. Your hands are 5 feet above the ground.

a. How far above the ground is the kite?

b. As the angle of elevation increases, does the height of your kite change at a constant rate? Justify your answer.

In Exercises 31–38, find the sine, cosine, and tangent of $\angle A$. *(See Example 7.)*

31.

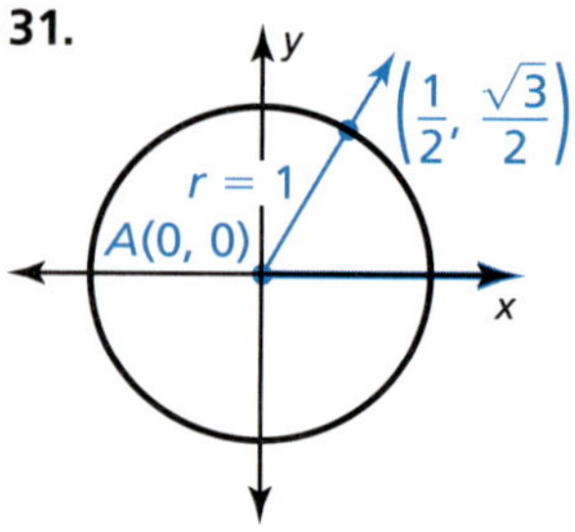

32.

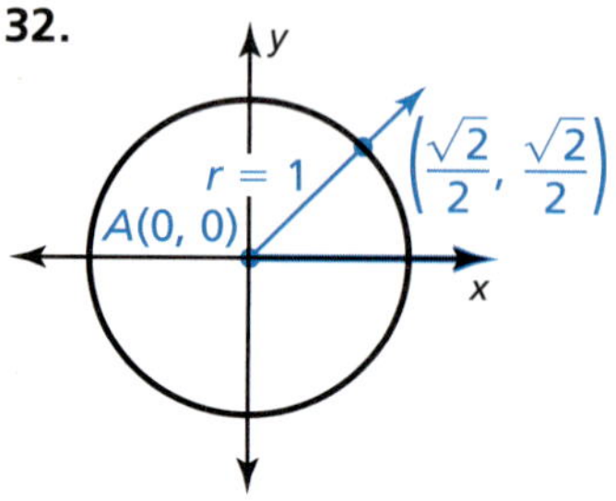

33.

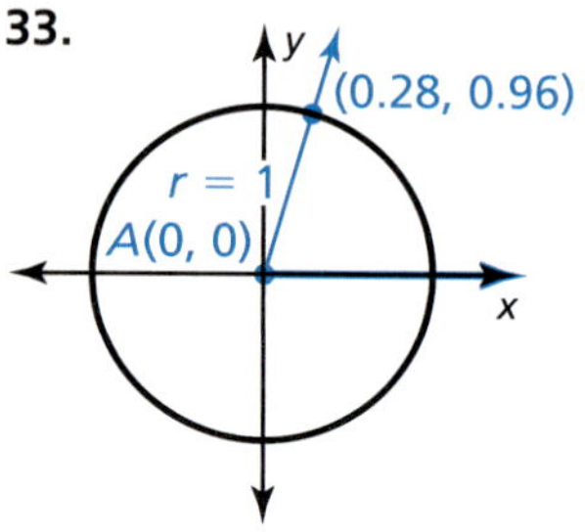

34.

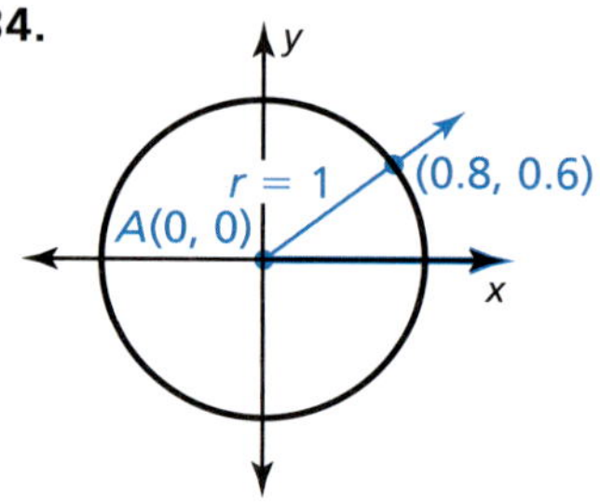

35.

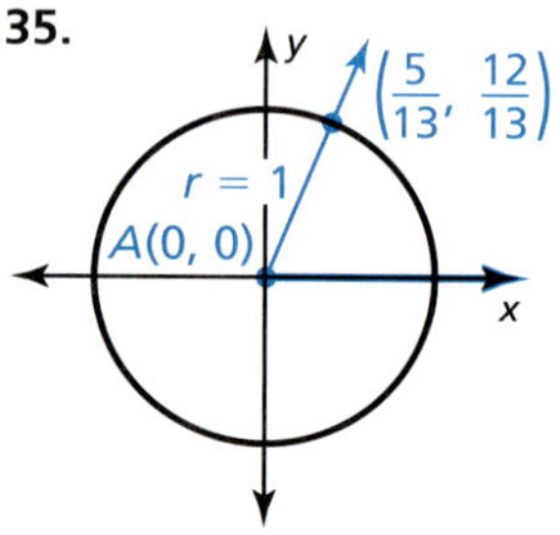

36.

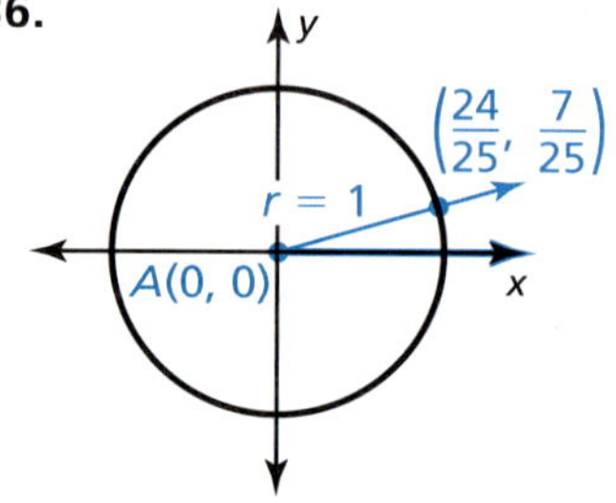

37.

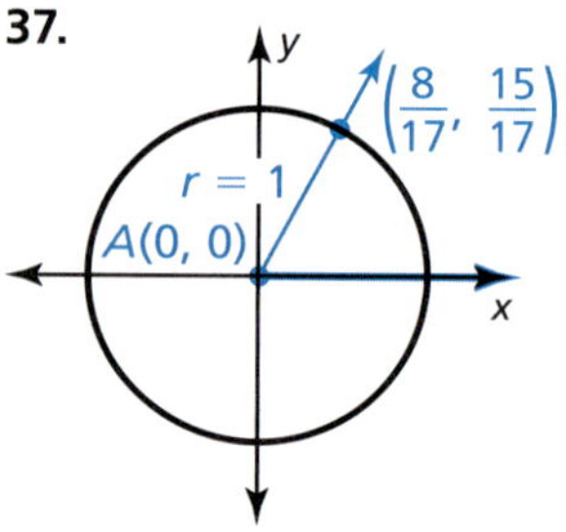

38.

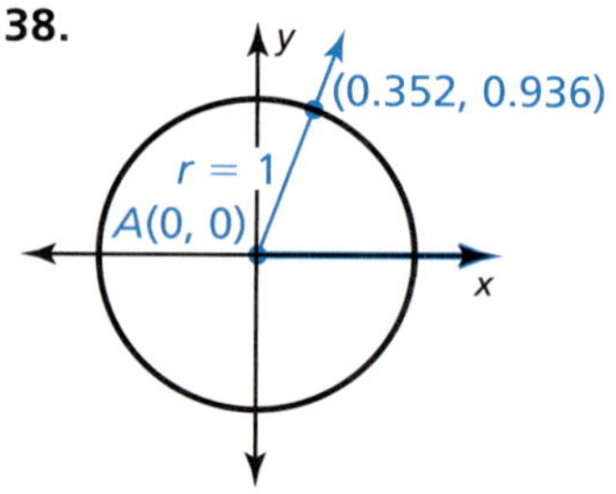

39. REASONING Does the point $\left(\frac{1}{9}, \frac{2}{3}\right)$ lie on the unit circle? Explain how you know.

40. REASONING The vertex of $\angle A$ is at the origin. One side lies on the x-axis, and one side passes through a point on the unit circle in the first quadrant. Are there angle measures for which the values of one or more of the trigonometric ratios of $\angle A$ exceed 1? Explain why or why not.

5 MTR **41. STRUCTURE** A right triangle is placed in the first quadrant of the unit circle. Find the perimeter of the triangle. Explain how you found your answer.

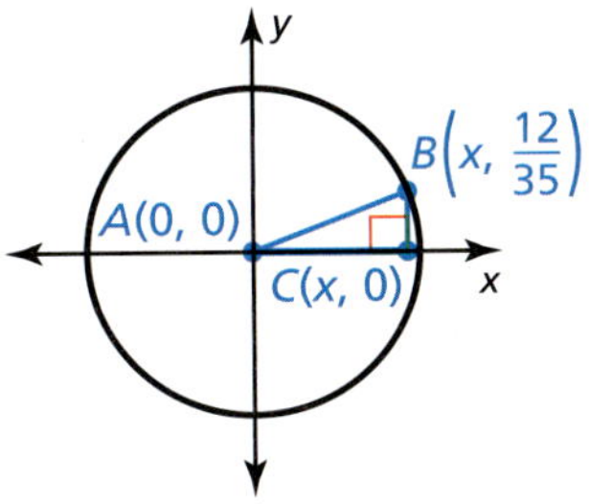

42. HOW DO YOU SEE IT?
Using only the given information, explain whether you would use a sine ratio or a cosine ratio to find the length of the hypotenuse.

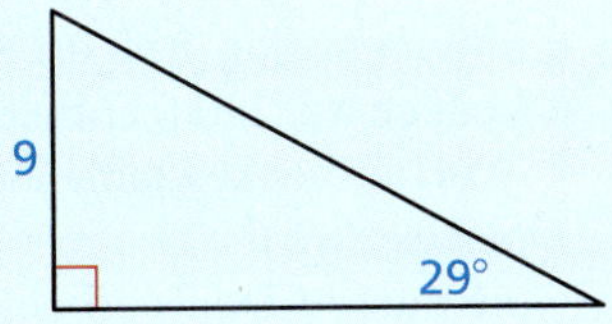

5 MTR **43. CONNECTING CONCEPTS** $\triangle EQU$ is equilateral and $\triangle RGT$ is a right triangle with $RG = 2$, $RT = 1$, and $m\angle T = 90°$. Show that $\sin E = \cos G$.

4 MTR **44. DISCUSS MATHEMATICAL THINKING** Let A be any acute angle of a right triangle. Show that (a) $\tan A = \dfrac{\sin A}{\cos A}$ and (b) $(\sin A)^2 + (\cos A)^2 = 1$.

7 MTR **45. MODELING REAL LIFE** Sonar systems use sound to detect underwater objects. Submarines use sonar systems to detect obstacles.

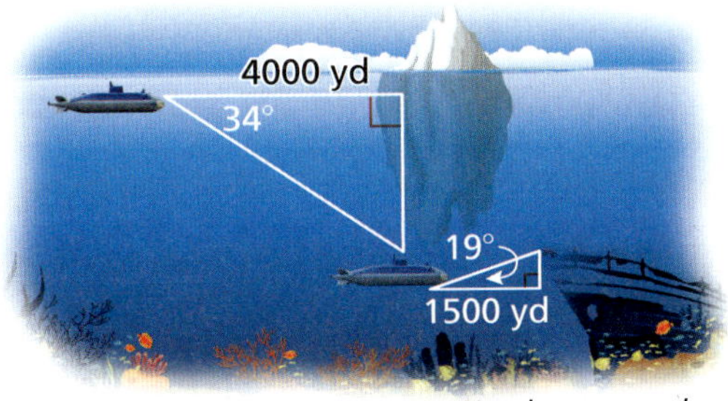

Not drawn to scale

a. The sonar system of a submarine detects an iceberg ahead. How many yards must the submarine descend to pass under the iceberg?

b. The sonar system then detects a sunken ship 1500 yards ahead, with an angle of elevation of 19° to the highest part of the sunken ship. How many yards must the submarine rise to pass over the sunken ship?

46. THOUGHT PROVOKING
Determine which of the following infinite series represents $\sin x$ and which represents $\cos x$ (where x is measured in radians). Then use each series to approximate the sine and cosine of $\frac{\pi}{6}$. Justify your answer. (*Hints*: $\pi = 180°$; $5! = 5 \cdot 4 \cdot 3 \cdot 2 \cdot 1$; Find the values that the sine and cosine ratios approach as the angle measure approaches zero.)

a. $x - \dfrac{x^3}{3!} + \dfrac{x^5}{5!} - \dfrac{x^7}{7!} + \cdots$

b. $1 - \dfrac{x^2}{2!} + \dfrac{x^4}{4!} - \dfrac{x^6}{6!} + \cdots$

47. REASONING The vertex of $\angle 2$ is at the origin. One side lies on the y-axis, and one side passes through a point (x, y). How can you use the values of the trigonometric ratios for $\angle 1$ to find the values of the trigonometric ratios for $\angle 2$? Explain.

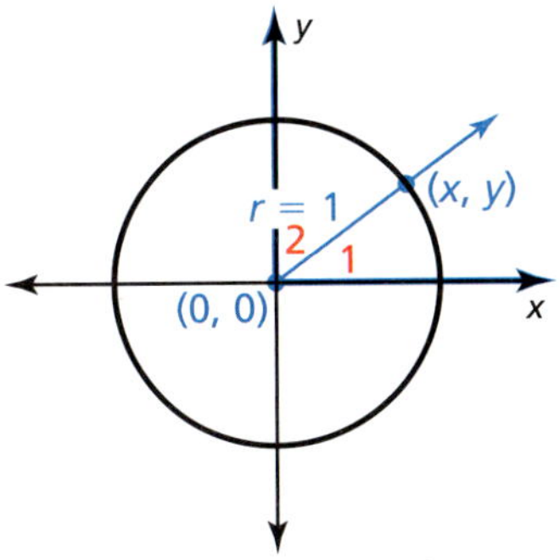

48. REASONING Consider an angle with its vertex at the origin, one side on the x-axis, and one side passing through the point (x, y). The cosine of the angle is $\frac{21}{29}$. Find the sine and tangent of the angle.

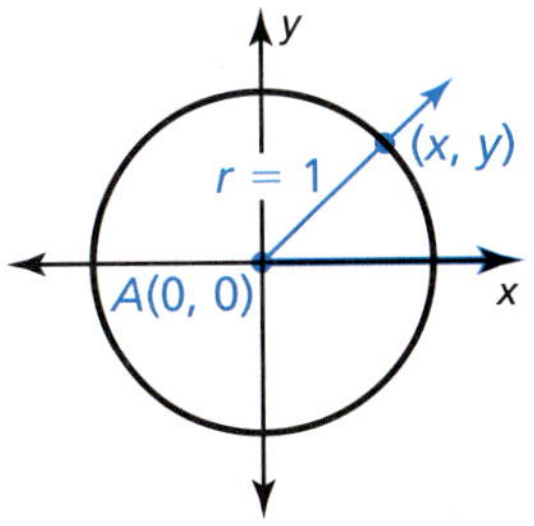

7 MTR **49. PERFORMANCE TASK** You are a pilot for a commercial airliner. In ideal conditions, you want to use a constant angle of depression as you approach your destination and land a plane. Research cruising altitudes and a typical angle of depression. About how far should the airplane be from your destination when you begin to descend? Explain.

GO DIGITAL

50. DIG DEEPER Consider angle A with its vertex at the origin, one side on the x-axis, and one side passing through the point (x, y) in the first quadrant. Find the sine, cosine, and tangent of $\angle A$.

a. (6, 8)

b. (5, 12)

c. (8, 15)

4 MTR **51. DISCUSS MATHEMATICAL THINKING** Explain why the area of $\triangle ABC$ in the diagram can be found using the formula Area $= \frac{1}{2}ab \sin C$. Then calculate the area when $a = 4$, $b = 7$, and $m\angle C = 40°$.

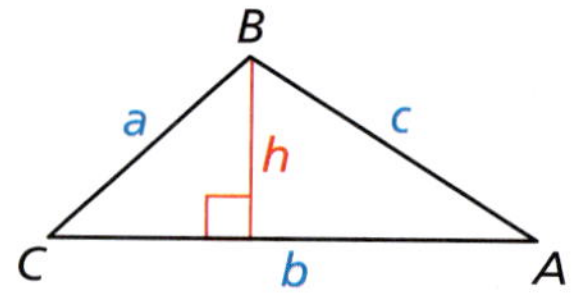

REVIEW & REFRESH

WATCH

In Exercises 52 and 53, find the value of x. Tell whether the side lengths form a Pythagorean triple.

52.

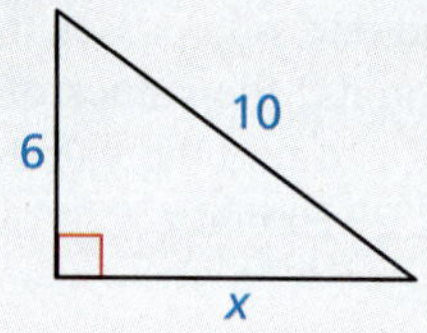

53.

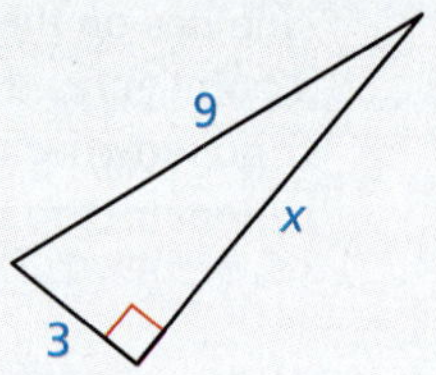

54. Write sin 47° in terms of cosine.

55. Find the value of x.

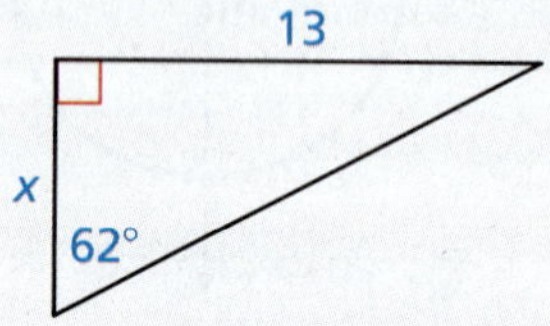

56. Find the measures of each interior angle and each exterior angle of a regular 21-gon.

57. Graph the quadrilateral with vertices $W(-4, -1)$, $X(3, -1)$, $Y(6, -5)$, and $Z(-1, -5)$ in the coordinate plane. Then show that the quadrilateral is a parallelogram.

58. Identify the similar right triangles. Then find the value of x.

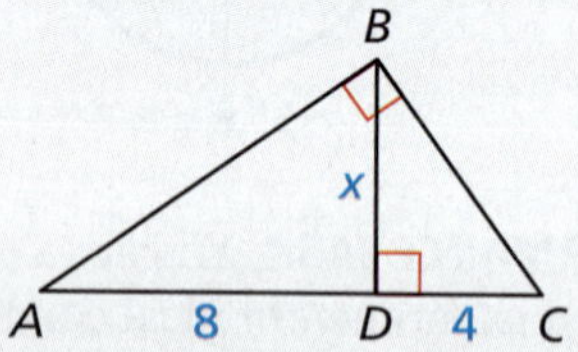

59. Find the values of x and y. Write your answers in simplest form.

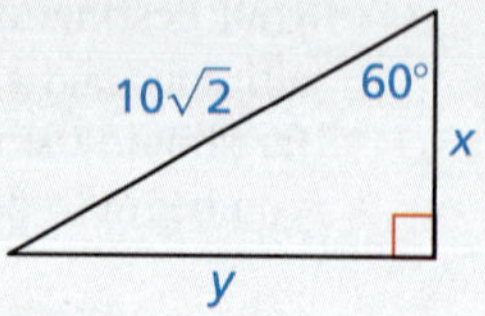

7 MTR **60. MODELING REAL LIFE** The dimensions of an official hockey rink used by the National Hockey League are 200 feet by 85 feet. The dimensions of an air hockey table are 96 inches by 40.8 inches. Are the two rectangles similar? If so, find the ratio of their perimeters and the ratio of their areas. If not, show how you can modify one of the dimensions of the air hockey table so that they are similar.

61. Determine whether AC is *less than*, *greater than*, or *equal to* DF. Explain your reasoning.

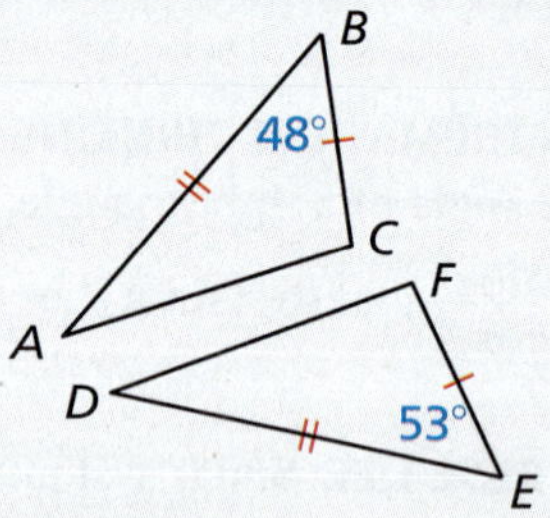

62. The polygons are congruent. Find the values of x and y.

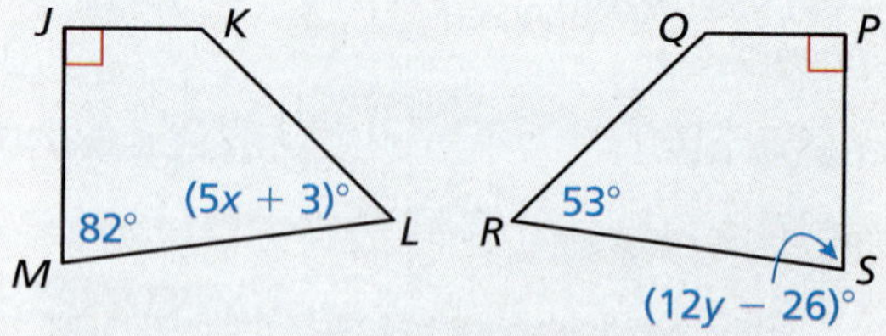

63. Draw a rectangle with a length of 9 units and a width of 3 units in a coordinate plane. Find the length of a diagonal.

64. Given the points $A(-4, 1)$ and $B(8, -7)$, find the coordinates of point P along the directed line segment AB so the ratio of AP to PB is 1 to 3.

GO DIGITAL

9.6 Law of Sines and Law of Cosines (H)

Learning Target: Find unknown side lengths and angle measures of acute and obtuse triangles.

Success Criteria:

- I can find areas of triangles using formulas that involve sine.
- I can use inverse trigonometric ratios to approximate angle measures.
- I can find unknown measures in triangles using the Law of Sines.
- I can find unknown measures in triangles using the Law of Cosines.

EXPLORE IT! Discovering the Law of Sines

Work with a partner.

a. Construct any acute $\triangle ABC$. Construct the altitude from vertex B to $\overline{AC}$ and label it as h. Then write a formula for the area of $\triangle ABC$.

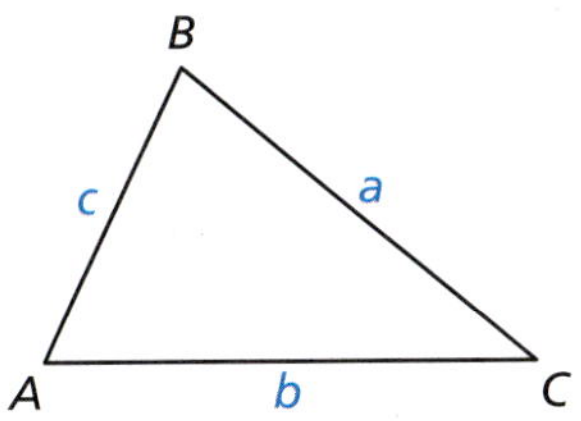

b. Rewrite your area formula for $\triangle ABC$ so that it contains $\sin C$. Explain your reasoning.

c. Rewrite your area formula for $\triangle ABC$ so that it contains $\sin A$. Explain your reasoning.

d. Show that $\dfrac{\sin A}{a} = \dfrac{\sin C}{c}$.

Draw several other acute triangles, and find their side lengths and angle measures. Does $\dfrac{\sin A}{a} = \dfrac{\sin C}{c}$? Write a conjecture about the relationship between the sines of the angles and the lengths of the sides of a triangle.

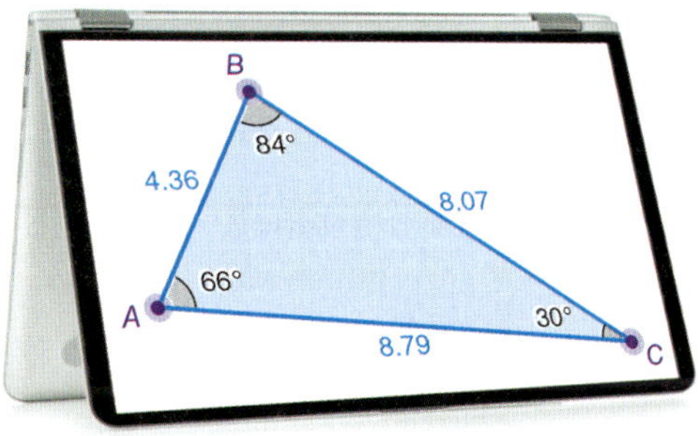

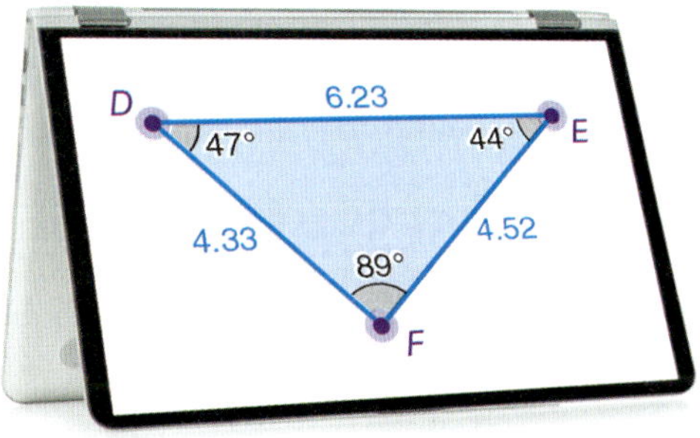

1 MTR ANALYZE A PROBLEM

In part (e), does your conjecture hold true for a triangle with an obtuse angle?

Trigonometry

H **MA.912.T.1.3** Apply the Law of Sines and the Law of Cosines to solve mathematical and real-world problems involving triangles.

H **MA.912.T.1.4** Solve mathematical problems involving finding the area of a triangle given two sides and the included angle.

GO DIGITAL

Finding Areas of Triangles

Vocabulary

inverse tangent, *p. 493*
inverse sine, *p. 493*
inverse cosine, *p. 493*
Law of Sines, *p. 494*
Law of Cosines, *p. 495*

You can also find trigonometric ratios for obtuse angles. For now, use technology to find these trigonometric ratios.

EXAMPLE 1 **Finding Trigonometric Ratios for Obtuse Angles**

WATCH

Use technology to find each trigonometric ratio.

a. $\tan 150^\circ$ **b.** $\sin 120^\circ$ **c.** $\cos 95^\circ$

SOLUTION

a. $\tan 150^\circ \approx -0.5774$ **b.** $\sin 120^\circ \approx 0.8660$ **c.** $\cos 95^\circ \approx -0.0872$

REMEMBER

Round the values of trigonometric ratios to four decimal places, and round lengths and angle measures to the nearest tenth.

KEY IDEA

Area of a Triangle

The area of any triangle is given by one-half the product of the lengths of two sides times the sine of their included angle. For $\triangle ABC$ shown, there are three ways to calculate the area.

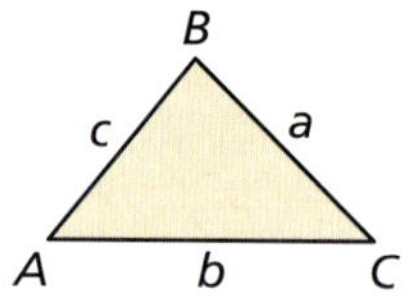

$\text{Area} = \frac{1}{2}bc \sin A$ $\text{Area} = \frac{1}{2}ac \sin B$ $\text{Area} = \frac{1}{2}ab \sin C$

EXAMPLE 2 **Finding the Area of a Triangle**

WATCH

Find the area of the triangle.

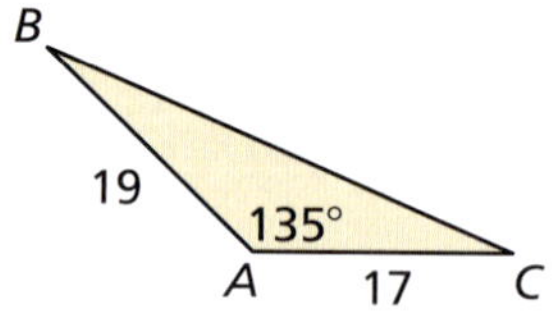

SOLUTION

$\text{Area} = \frac{1}{2}bc \sin A$ Write formula for area.

$= \frac{1}{2}(17)(19) \sin 135^\circ$ Substitute for *b*, *c*, and *A*.

≈ 114.2 Use technology.

▶ The area of the triangle is about 114.2 square units.

SELF-ASSESSMENT 1 I don't understand yet. 2 I can do it with help. 3 I can do it on my own. 4 I can teach someone else.

Use technology to find the trigonometric ratio.

1. $\tan 110^\circ$ **2.** $\sin 97^\circ$ **3.** $\cos 165^\circ$

Find the area of $\triangle ABC$ with the given side lengths and included angle.

4.

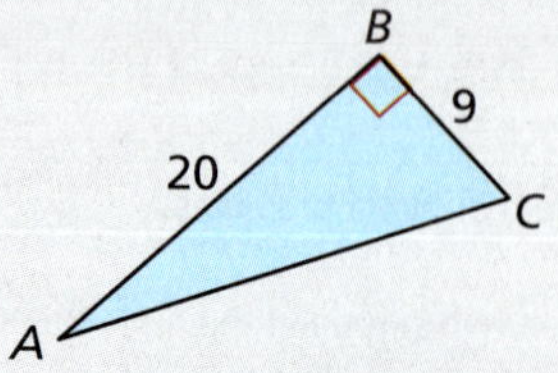

5.

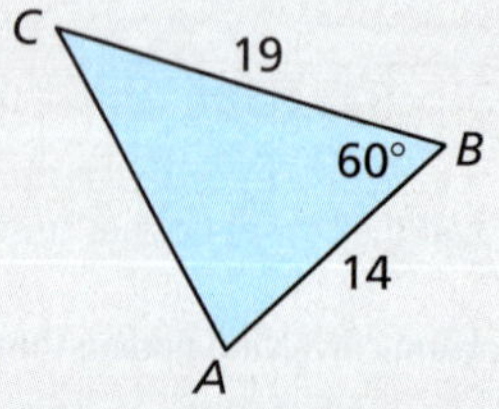

6.

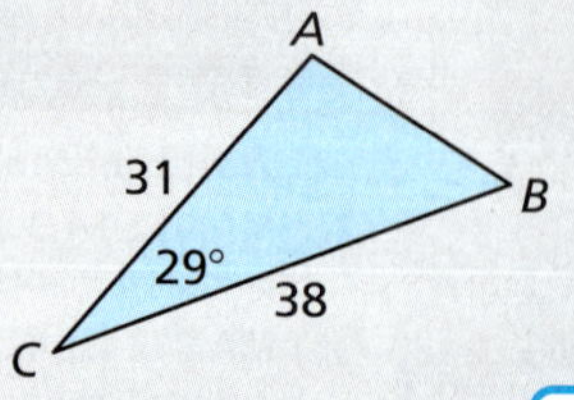

GO DIGITAL

Using Inverse Trigonometric Ratios

EXAMPLE 3 **Identifying Angles from Trigonometric Ratios** WATCH

Determine which of the two acute angles has a cosine of 0.5.

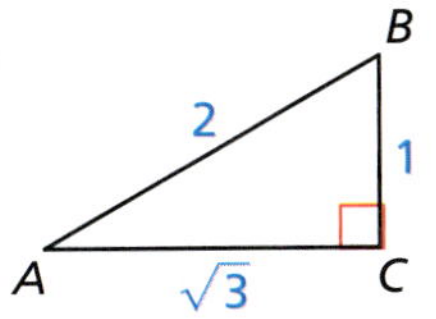

SOLUTION

Find the cosine of each acute angle.

$$\cos A = \frac{\text{adj to } \angle A}{\text{hyp}} = \frac{\sqrt{3}}{2} \approx 0.8660 \qquad \cos B = \frac{\text{adj to } \angle B}{\text{hyp}} = \frac{1}{2} = 0.5$$

▶ The acute angle that has a cosine of 0.5 is $\angle B$.

If the measure of an acute angle is 60°, then its cosine is 0.5. The converse is also true. If the cosine of an acute angle is 0.5, then the measure of the angle is 60°. So, in Example 3, the measure of $\angle B$ must be 60° because its cosine is 0.5.

READING

The expression "$\tan^{-1} x$" is read as "the inverse tangent of x."

KEY IDEA

Inverse Trigonometric Ratios

Let $\angle A$ be an acute angle.

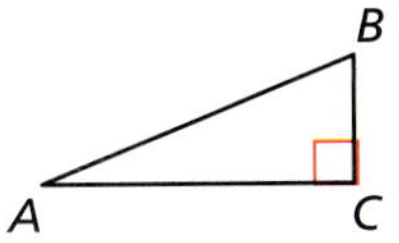

Inverse Tangent If $\tan A = x$, then $\tan^{-1} x = m\angle A$. $\quad \tan^{-1} \frac{BC}{AC} = m\angle A$

Inverse Sine If $\sin A = y$, then $\sin^{-1} y = m\angle A$. $\quad \sin^{-1} \frac{BC}{AB} = m\angle A$

Inverse Cosine If $\cos A = z$, then $\cos^{-1} z = m\angle A$. $\quad \cos^{-1} \frac{AC}{AB} = m\angle A$

2 MTR USE ANOTHER METHOD

Refer to the Table of Trigonometric Ratios available at *BigIdeasMath.com.* How can you use the values in the table to approximate the angle measures?

EXAMPLE 4 **Finding Angle Measures** WATCH

Let $\angle A$, $\angle B$, and $\angle C$ be acute angles. Use technology to approximate the measures of $\angle A$, $\angle B$, and $\angle C$.

a. $\tan A = 0.75$ **b.** $\sin B = 0.87$ **c.** $\cos C = 0.15$

SOLUTION

a. $m\angle A = \tan^{-1} 0.75 \approx 36.9°$

b. $m\angle B = \sin^{-1} 0.87 \approx 60.5°$

c. $m\angle C = \cos^{-1} 0.15 \approx 81.4°$

SELF-ASSESSMENT

1 I don't understand yet. 2 I can do it with help. 3 I can do it on my own. 4 I can teach someone else.

Let $\angle G$, $\angle H$, and $\angle K$ be acute angles. Use technology to approximate the measures of $\angle G$, $\angle H$, and $\angle K$.

7. $\tan G = 0.43$ **8.** $\sin H = 0.68$ **9.** $\cos K = 0.94$

10. Determine which of the two acute angles in $\triangle DEF$ has a sine of $\frac{12}{13}$.

11. Approximate the unknown angle measures of $\triangle DEF$. Explain your reasoning.

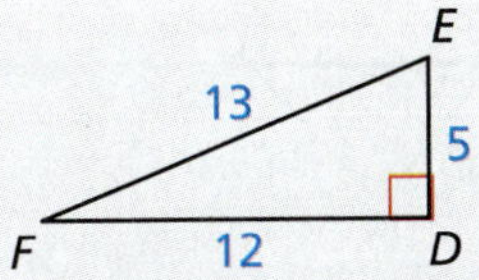

GO DIGITAL

Using the Law of Sines

The trigonometric ratios in the previous sections can be used to find unknown measures only in right triangles. You will learn two laws that can be used to find unknown measures in any triangle.

You can use the **Law of Sines** to find unknown measures in triangles when two angles and the length of any side are known (AAS or ASA cases), or when the lengths of two sides and an angle opposite one of the two sides are known (SSA case).

THEOREM

9.9 Law of Sines

The Law of Sines can be written in either of the following forms for $\triangle ABC$ with sides of length a, b, and c.

$$\frac{\sin A}{a} = \frac{\sin B}{b} = \frac{\sin C}{c} \qquad \frac{a}{\sin A} = \frac{b}{\sin B} = \frac{c}{\sin C}$$

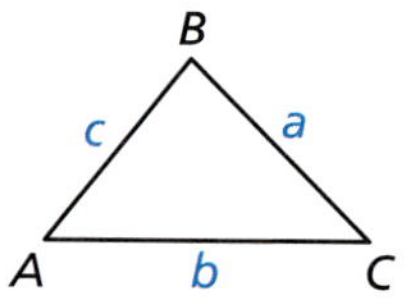

Prove this Theorem Exercise 55, page 499

EXAMPLE 5 Using the Law of Sines

WATCH

Find all unknown measures in the triangle.

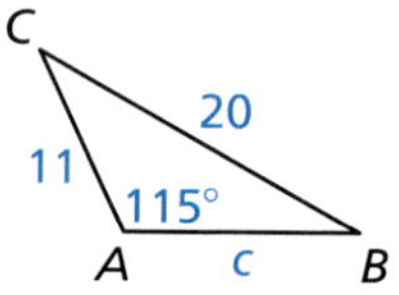

SOLUTION

Use the Law of Sines to find $m\angle B$.

$$\frac{\sin B}{b} = \frac{\sin A}{a}$$ Law of Sines

$$\frac{\sin B}{11} = \frac{\sin 115^\circ}{20}$$ Substitute.

$$\sin B = \frac{11 \sin 115^\circ}{20}$$ Multiply each side by 11.

$$m\angle B \approx 29.9^\circ$$ Use technology.

By the Triangle Sum Theorem, $m\angle C \approx 180^\circ - 115^\circ - 29.9^\circ = 35.1^\circ$.

Use the Law of Sines again to find the remaining side length c of the triangle.

$$\frac{c}{\sin C} = \frac{a}{\sin A}$$ Law of Sines

$$\frac{c}{\sin 35.1^\circ} = \frac{20}{\sin 115^\circ}$$ Substitute.

$$c = \frac{20 \sin 35.1^\circ}{\sin 115^\circ}$$ Multiply each side by sin 35.1°.

$$c \approx 12.7$$ Use technology.

▶ In $\triangle ABC$, $m\angle B \approx 29.9^\circ$, $m\angle C \approx 35.1^\circ$, and $c \approx 12.7$.

SELF-ASSESSMENT

1 I don't understand yet. 2 I can do it with help. 3 I can do it on my own. 4 I can teach someone else.

Find all unknown measures in the triangle.

12.

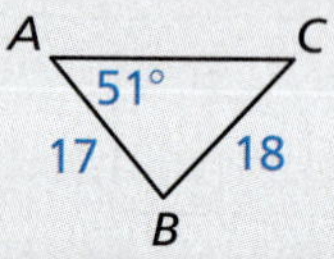

13.

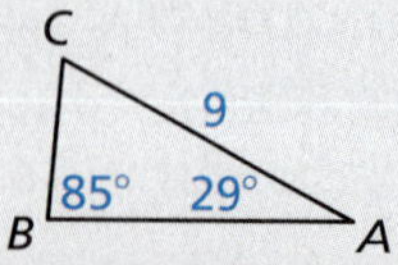

GO DIGITAL

Using the Law of Cosines

You can use the **Law of Cosines** to find unknown measures in triangles when two sides and the included angle are known (SAS case), or when all three sides are known (SSS case).

THEOREM

9.10 Law of Cosines

If $\triangle ABC$ has sides of length a, b, and c, as shown, then the following are true.

$$a^2 = b^2 + c^2 - 2bc \cos A$$
$$b^2 = a^2 + c^2 - 2ac \cos B$$
$$c^2 = a^2 + b^2 - 2ab \cos C$$

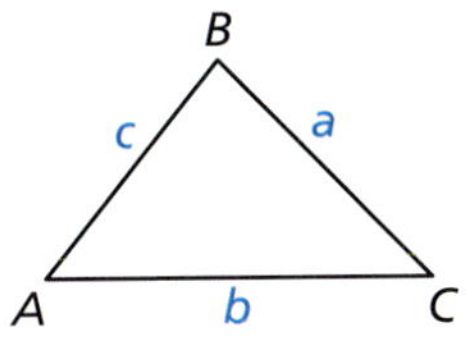

Prove this Theorem Exercise 59, page 500

EXAMPLE 6 Using the Law of Cosines

WATCH

Find all unknown measures in the triangle.

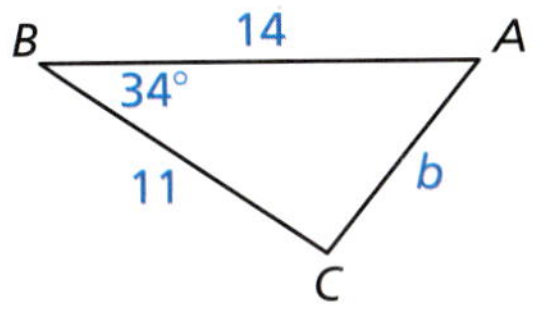

SOLUTION

Use the Law of Cosines to find side length b.

$b^2 = a^2 + c^2 - 2ac \cos B$	Law of Cosines
$b^2 = 11^2 + 14^2 - 2(11)(14) \cos 34°$	Substitute.
$b^2 = 317 - 308 \cos 34°$	Simplify.
$b = \sqrt{317 - 308 \cos 34°}$	Take the positive square root of each side.
$b \approx 7.9$	Use technology.

Use the Law of Sines to find $m\angle A$.

$\dfrac{\sin A}{a} = \dfrac{\sin B}{b}$	Law of Sines
$\dfrac{\sin A}{11} = \dfrac{\sin 34°}{\sqrt{317 - 308 \cos 34°}}$	Substitute.
$\sin A = \dfrac{11 \sin 34°}{\sqrt{317 - 308 \cos 34°}}$	Multiply each side by 11.
$m\angle A \approx 51.6°$	Use technology.

By the Triangle Sum Theorem, $m\angle C \approx 180° - 34° - 51.6° = 94.4°$.

▶ In $\triangle ABC$, $b \approx 7.9$, $m\angle A \approx 51.6°$, and $m\angle C \approx 94.4°$.

COMMON ERROR

In Example 6, the smaller remaining angle is found first because when using technology, the inverse sine feature gives angle measures from only 0° to 90°.

SELF-ASSESSMENT 1 I don't understand yet. 2 I can do it with help. 3 I can do it on my own. 4 I can teach someone else.

Find the unknown measures in the triangle.

14.

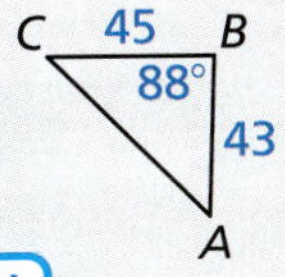

15.

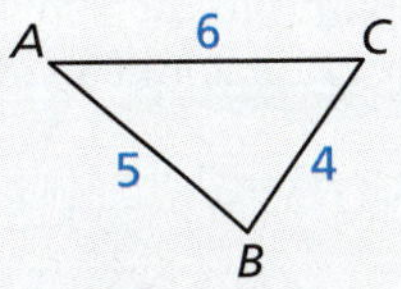

GO DIGITAL

Solving Real-Life Problems

7 MTR **EXAMPLE 7** **Modeling Real Life** WATCH

The distance between consecutive bases on a baseball field is 90 feet. A player catches a ball at point A and wants to throw the ball directly to third base. How far must the player throw the ball?

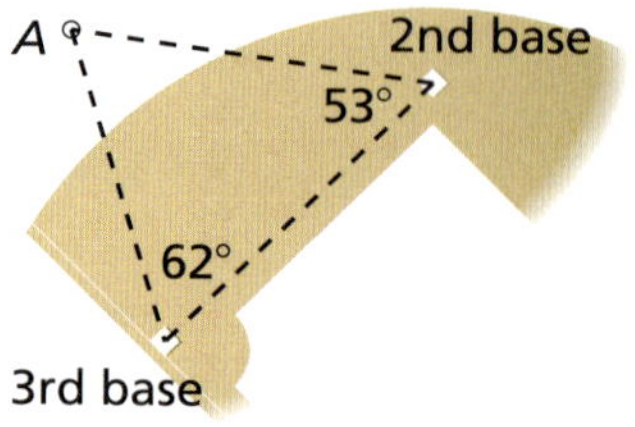

SOLUTION

In the diagram, let b represent the distance from point A to third base. So, b represents how far the player must throw the ball.

By the Triangle Sum Theorem, $m\angle A = 180° - 62° - 53° = 65°$.

Use the Law of Sines to write an equation involving b.

$$\frac{90}{\sin 65°} = \frac{b}{\sin 53°}$$ Write an equation involving b.

$$\frac{90 \sin 53°}{\sin 65°} = b$$ Multiply each side by sin 53°.

$$79.3 \approx b$$ Use technology.

▶ The player must throw the ball about 79 feet.

7 MTR **EXAMPLE 8** **Modeling Real Life** WATCH

An organism's step angle is a measure of walking efficiency. The closer the step angle is to 180°, the more efficiently the organism walks. The diagram shows a set of footprints for a dinosaur. Find the step angle B.

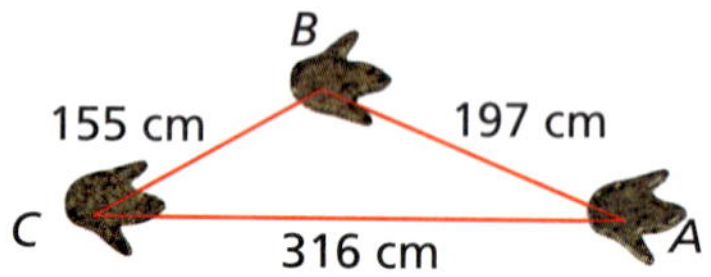

SOLUTION

$$b^2 = a^2 + c^2 - 2ac \cos B$$ Law of Cosines

$$316^2 = 155^2 + 197^2 - 2(155)(197) \cos B$$ Substitute.

$$\frac{316^2 - 155^2 - 197^2}{-2(155)(197)} = \cos B$$ Solve for cos B.

$$127.3° \approx m\angle B$$ Use technology.

▶ The step angle B is about 127.3°.

SELF-ASSESSMENT 1 I don't understand yet. 2 I can do it with help. 3 I can do it on my own. 4 I can teach someone else.

16. WHAT IF? The player in Example 7 decides to throw the ball to second base instead of third base. Which throw is longer and by how much?

17. Determine whether the dinosaur whose footprints are shown at the right walked more efficiently than the dinosaur in Example 8.

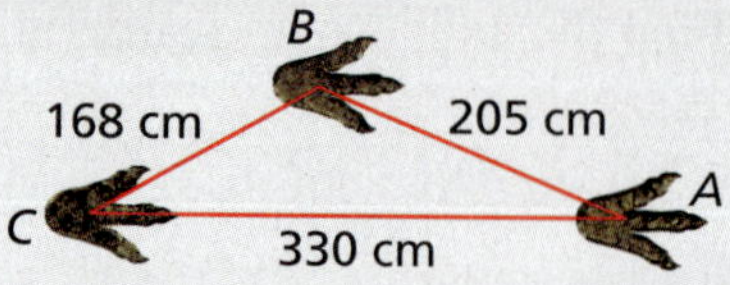

GO DIGITAL

9.6 Practice WITH CalcChat® AND CalcView®

In Exercises 1–6, use technology to find the trigonometric ratio. *(See Example 1.)*

▶ **1.** sin 127°

2. sin 98°

3. cos 139°

4. cos 108°

5. tan 165°

6. tan 116°

In Exercises 7–10, find the area of the triangle. *(See Example 2.)*

▶ **7.**

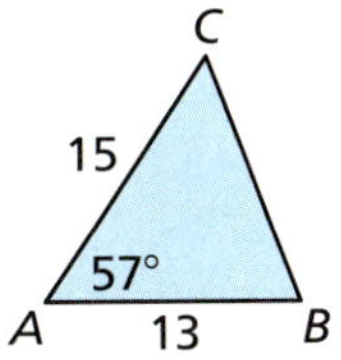

8.

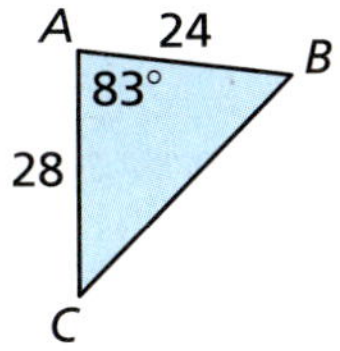

9.

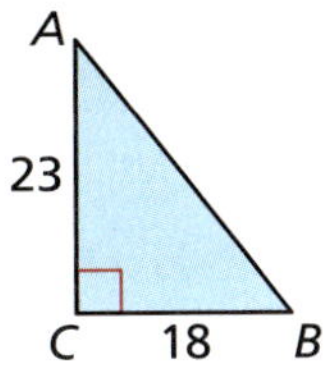

10.

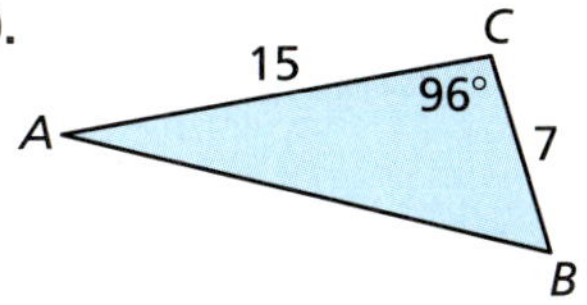

In Exercises 11–14, determine which of the two acute angles has the given trigonometric ratio. *(See Example 3.)*

11. The cosine of the angle is $\frac{4}{5}$.

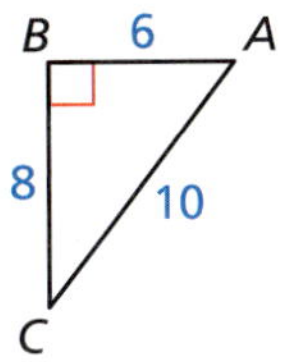

12. The sine of the angle is $\frac{5}{11}$.

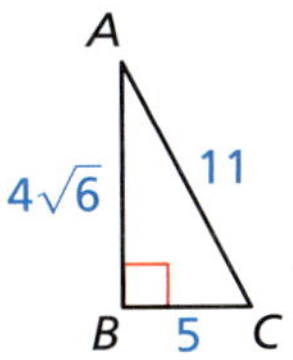

▶ **13.** The sine of the angle is 0.96.

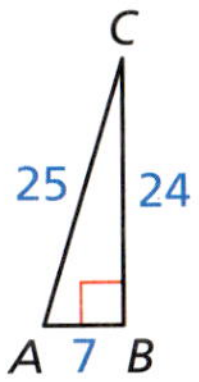

14. The tangent of the angle is 1.5.

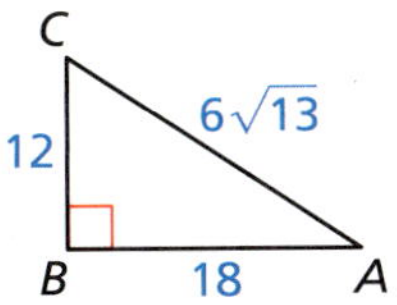

In Exercises 15–20, let $\angle D$ be an acute angle. Use technology to approximate $m\angle D$. *(See Example 4.)*

▶ **15.** $\sin D = 0.75$

16. $\sin D = 0.19$

17. $\cos D = 0.33$

18. $\cos D = 0.64$

19. $\tan D = 0.28$

20. $\tan D = 0.72$

GO DIGITAL

In Exercises 21 and 22, use inverse trigonometric ratios to find all unknown measures in the triangle. Explain your reasoning.

21.

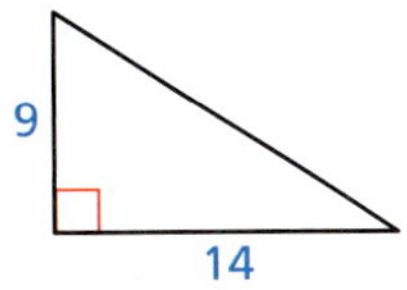

22.

In Exercises 23–28, find all unknown measures in the triangle. *(See Example 5.)*

▶ **23.**

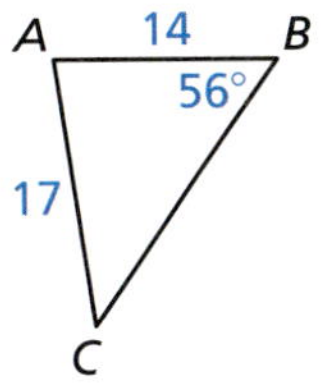

24.

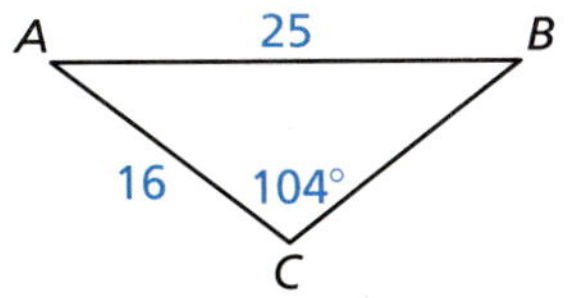

25.

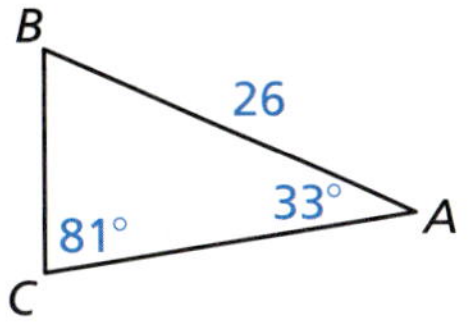

26.

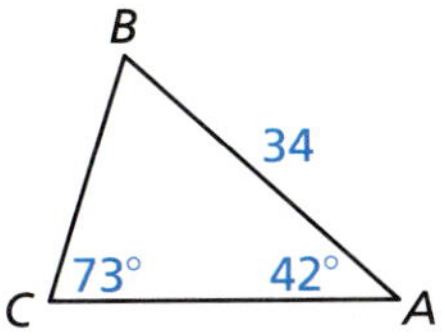

27.

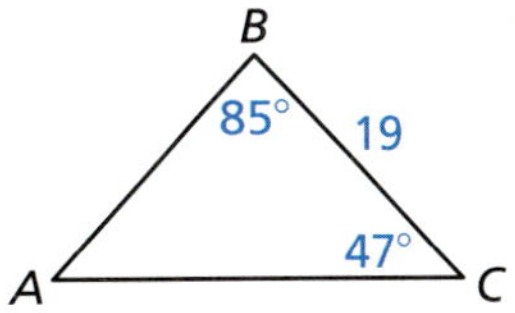

28.

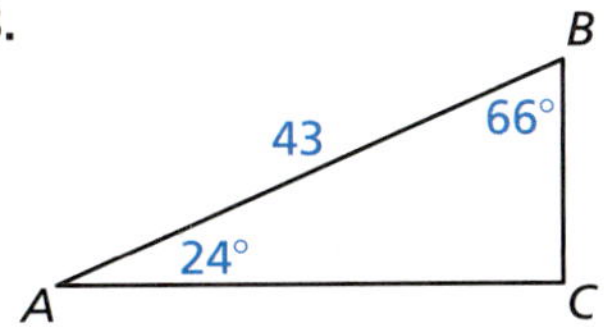

In Exercises 29–32, find all unknown measures in the triangle. *(See Example 6.)*

▶ **29.**

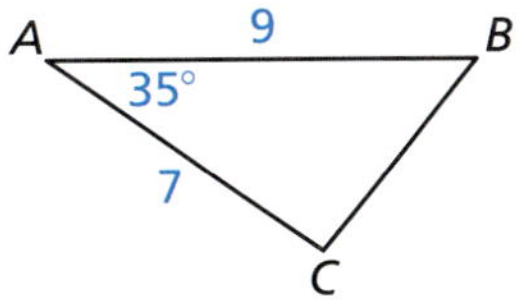

30.

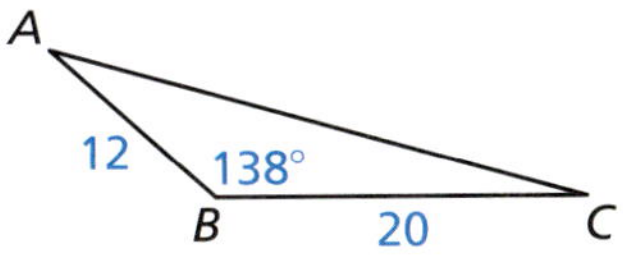

31.

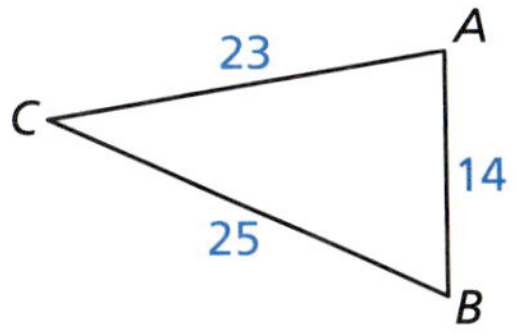

32.

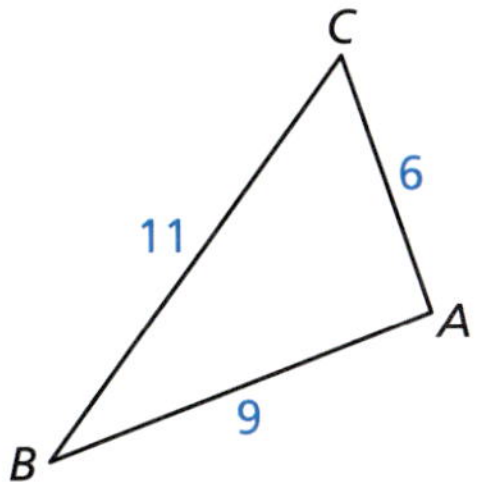

33. **ERROR ANALYSIS** Describe and correct the error in finding $m\angle C$.

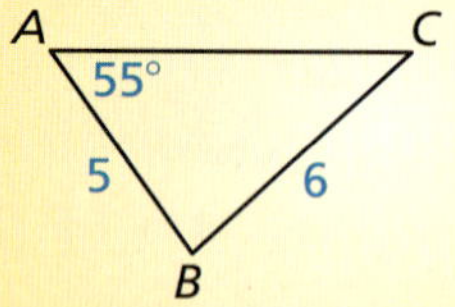

$$\frac{\sin C}{c} = \frac{\sin A}{a}$$

$$\frac{\sin C}{6} = \frac{\sin 55^\circ}{5}$$

$$\sin C = \frac{6 \sin 55^\circ}{5}$$

$$m\angle C \approx 79.4^\circ$$

34. **ERROR ANALYSIS** Describe and correct the error in finding $m\angle A$ in $\triangle ABC$ when $a = 19$, $b = 21$, and $c = 11$.

$$\cos A = \frac{a^2 - b^2 - c^2}{-2(a)(b)}$$

$$\cos A = \frac{19^2 - 21^2 - 11^2}{-2(19)(21)}$$

$$m\angle A \approx 75.4^\circ$$

CHOOSE A METHOD **In Exercises 35–38, tell whether you would use the Law of Sines, the Law of Cosines, or the Pythagorean Theorem and trigonometric ratios to find the unknown measures in the triangle with the given information. Explain your reasoning. Then find all the unknown measures.**

35. $m\angle A = 72^\circ$, $m\angle B = 44^\circ$, $b = 14$

36. $m\angle C = 65^\circ$, $a = 12$, $b = 21$

37. $m\angle B = 90^\circ$, $a = 15$, $c = 6$

38. $a = 34$, $b = 19$, $c = 27$

39. **MODELING REAL LIFE** You build a triangular fence with trees located at each vertex. The distance from Tree A to Tree C is 40 feet. How long is the fence between Tree B and Tree C? *(See Example 7.)*

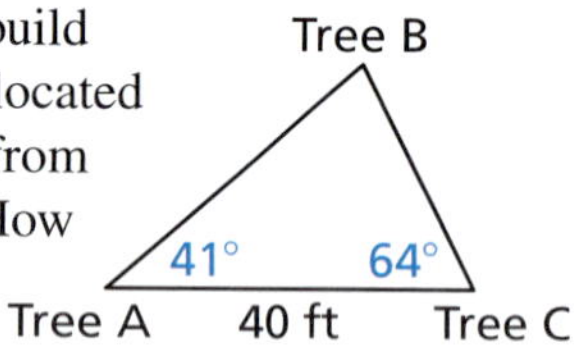

40. **MODELING REAL LIFE** The roof of a birdhouse forms a triangle. How long are the unknown sides of the face?

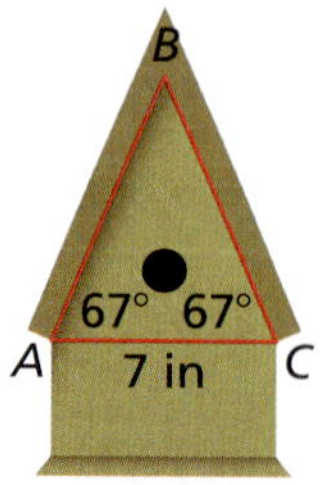

41. **MODELING REAL LIFE** You bounce a basketball to your friend, as shown in the diagram. What is the distance between you and your friend? *(See Example 8.)*

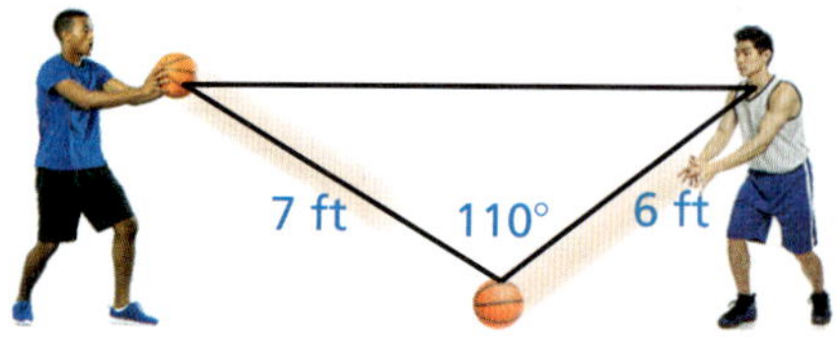

42. **MODELING REAL LIFE** A zip line is constructed across a valley, as shown in the diagram. What is the width w of the valley?

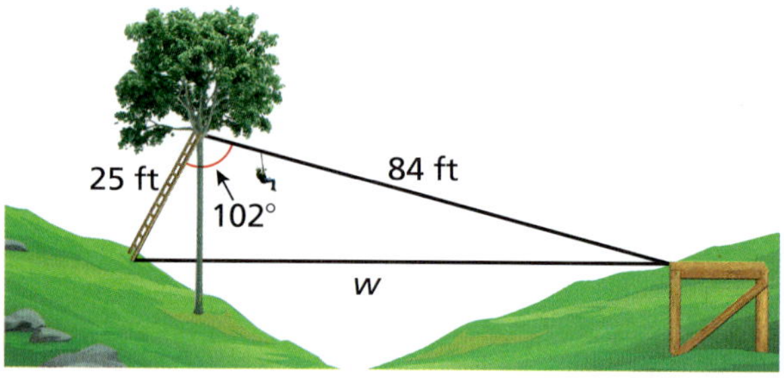

43. **MODELING REAL LIFE** You are on the observation deck of the Empire State Building looking at the Chrysler Building. When you turn 145° clockwise, you see the Statue of Liberty. You know that the Chrysler Building and the Empire State Building are about 0.6 mile apart and that the Chrysler Building and the Statue of Liberty are about 5.6 miles apart. Draw a diagram to represent this situation. Estimate the distance between the Empire State Building and the Statue of Liberty.

44. **MODELING REAL LIFE** The Leaning Tower of Pisa in Italy has a height of 183 feet. In 1990, the tower stood about 5.5° off vertical until it was stabilized a few years later, and now stands 4° off vertical. In terms of horizontal distance, how much farther was the top of the tower off vertical in 1990 when compared with the tower today?

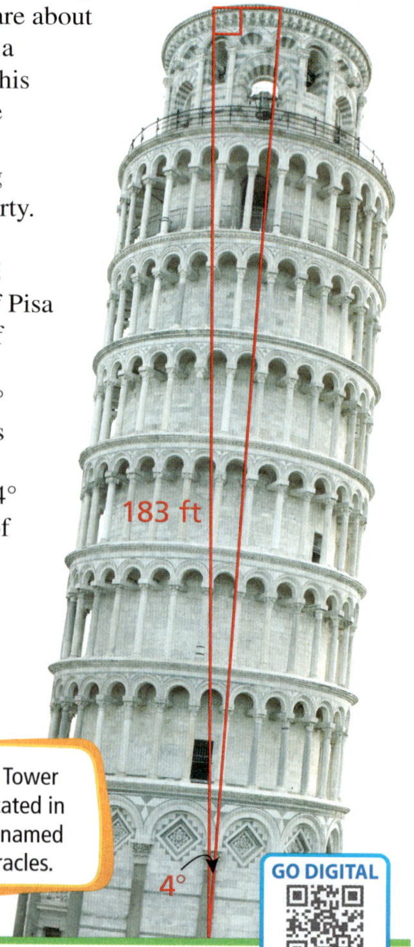

GO DIGITAL

45. MAKING AN ARGUMENT Your friend says that the Law of Sines would be used to find JK. Your cousin says that the Law of Cosines can be used to find JK. Who is correct? Explain your reasoning.

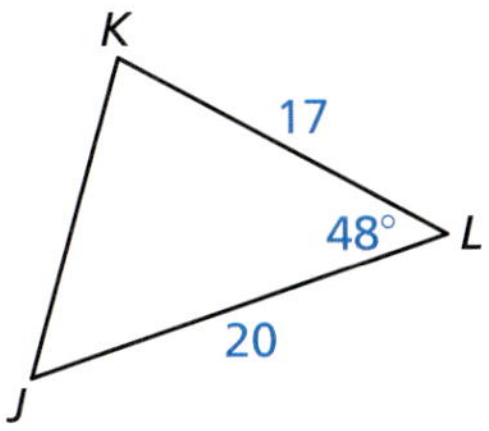

46. REASONING How does finding the area of a triangle using two side lengths and the included angle relate to finding the area of a triangle using base and height?

47. REASONING Use $\triangle XYZ$.

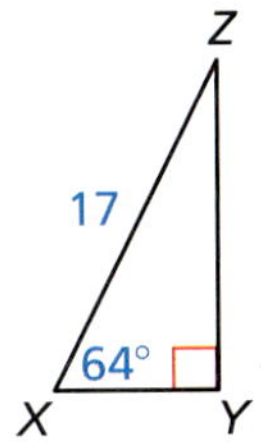

a. Can you use the Law of Sines to find the unknown measures of $\triangle XYZ$? Explain your reasoning.

b. Can you use another method to find the unknown measures of $\triangle XYZ$? Explain your reasoning.

48. MODELING REAL LIFE You are fertilizing a triangular garden. One side of the garden is 62 feet long, and another side is 54 feet long. The angle opposite the 62-foot side is 58°.

a. Draw a diagram to represent this situation.

b. Use the Law of Sines to find the unknown measures of the triangle from part (a).

c. One bag of fertilizer covers an area of 200 square feet. How many bags of fertilizer will you need to cover the entire garden?

49. MODELING REAL LIFE A golfer hits a drive 260 yards on a hole that is 400 yards long. The shot is 15° off target.

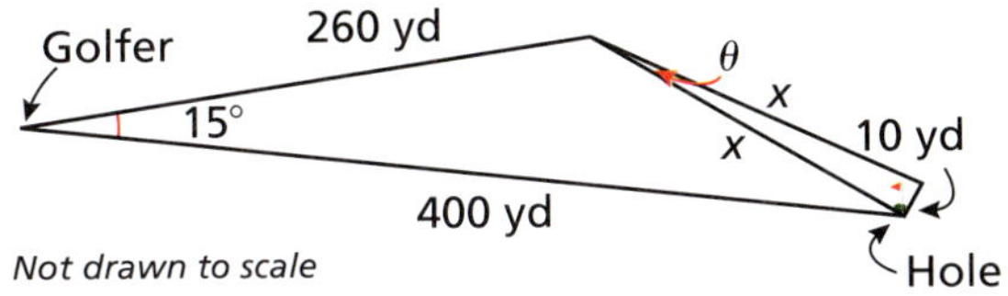

a. What is the distance x from the golfer's ball to the hole?

b. Assume the golfer is able to hit the ball precisely the distance found in part (a). What is the maximum angle θ by which the ball can be off target in order to land no more than 10 yards from the hole?

50. COMPARING METHODS A building is constructed on top of a cliff that is 300 meters high. A person standing on level ground below the cliff observes that the angle of elevation to the top of the building is 72° and the angle of elevation to the top of the cliff is 63°.

a. How far away is the person from the base of the cliff?

b. Describe two different methods you can use to find the height of the building. Use one of these methods to find the building's height.

GO DIGITAL

51. CONNECTING CONCEPTS Find the values of x and y.

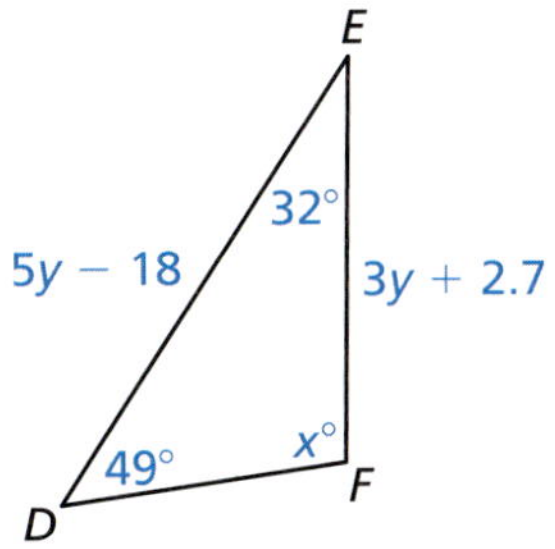

52. STRUCTURE Simplify the Law of Cosines for when the given angle is a right angle.

53. NUMBER SENSE The *ambiguous case* of the Law of Sines occurs when you are given the measure of one acute angle, the length of one adjacent side, and the length of the side opposite that angle, which is less than the length of the adjacent side. This results in two possible triangles. Using the given information, find two possible solutions for $\triangle ABC$. Draw a diagram for each triangle.

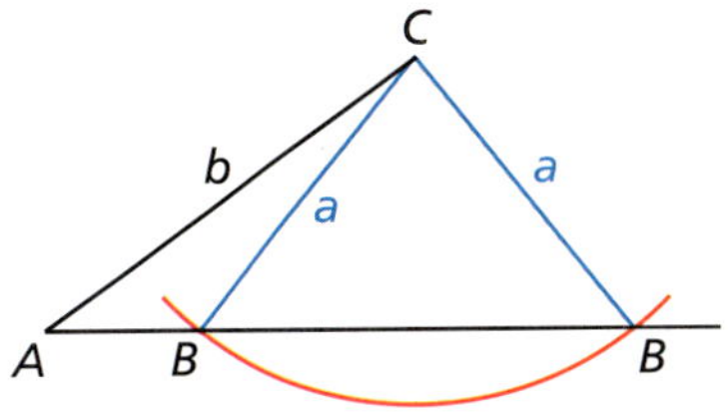

a. $m\angle A = 40°, a = 13, b = 16$

b. $m\angle A = 21°, a = 17, b = 32$

54. HOW DO YOU SEE IT?
Should you use the Law of Sines or the Law of Cosines to solve the triangle? Explain.

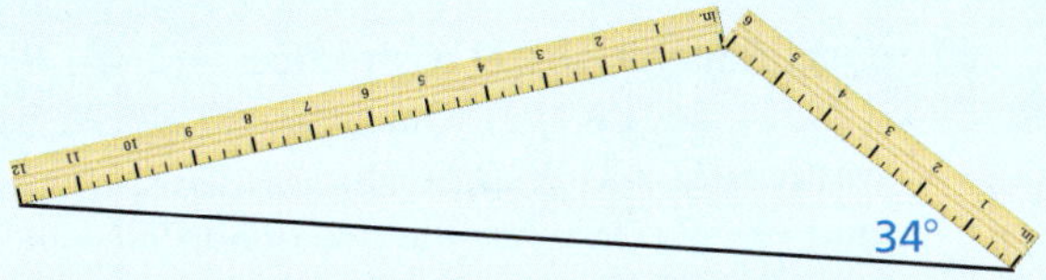

55. PROVING A THEOREM Use the formula for area of a triangle to prove the Law of Sines. Justify each step.

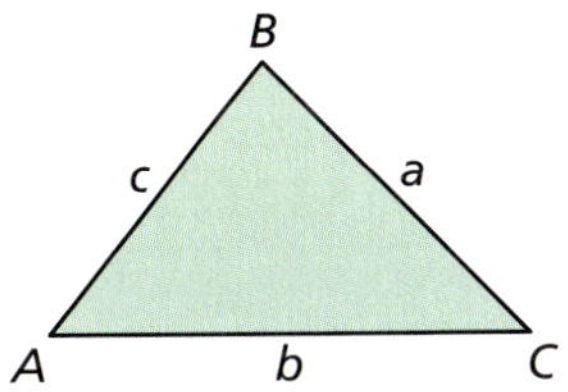

56. CONNECTING CONCEPTS Write an expression that can be used to find the measure of the acute angle formed by each line and the x-axis. Then approximate the angle measure.

a. $y = 3x$

b. $y = \frac{4}{3}x + 4$

57. **REASONING** Explain why the expression $\sin^{-1}(1.2)$ does not make sense.

58. **THOUGHT PROVOKING**
Consider any triangle with side lengths of a, b, and c. Calculate the value of s, which is half the perimeter of the triangle. What measurement of the triangle is represented by $\sqrt{s(s - a)(s - b)(s - c)}$?

5 MTR 59. **STRUCTURE** Use the Law of Cosines to show that the measure of each angle of an equilateral triangle is 60°. Explain your reasoning.

60. **PROBLEM SOLVING** An airplane flies 55° east of north from City A to City B, a distance of 470 miles. Another airplane flies 7° north of east from City A to City C, a distance of 890 miles. What is the distance between Cities B and C?

61. **PROVING A THEOREM**
Use the given information to write a proof of the Law of Cosines.

Given $\overline{BD}$ is an altitude of $\triangle ABC$.

Prove
$a^2 = b^2 + c^2 - 2bc \cos A$

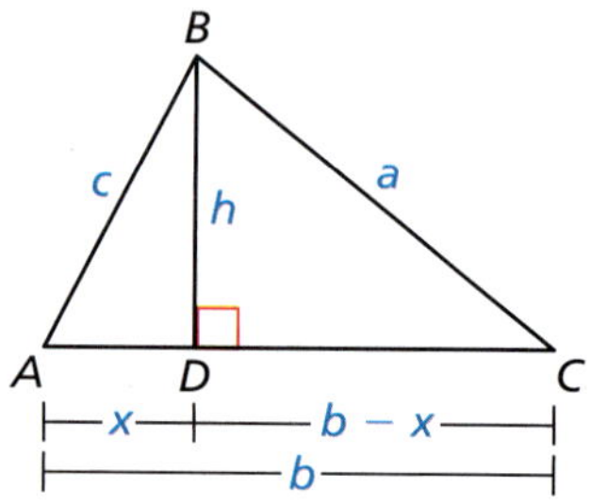

REVIEW & REFRESH

WATCH

In Exercises 62–65, find the value of x.

62.

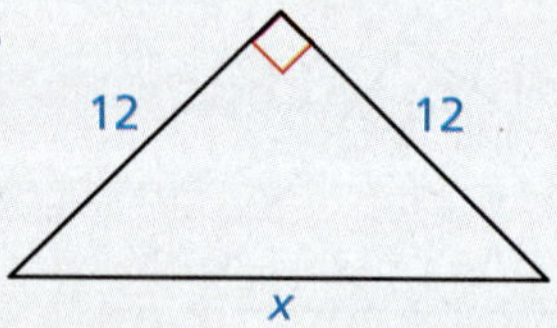

63.

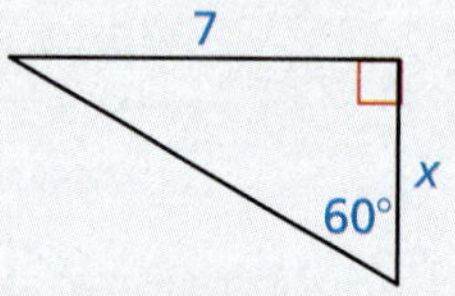

64.

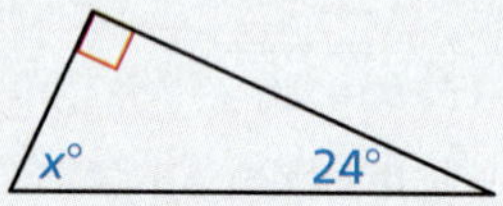

65.

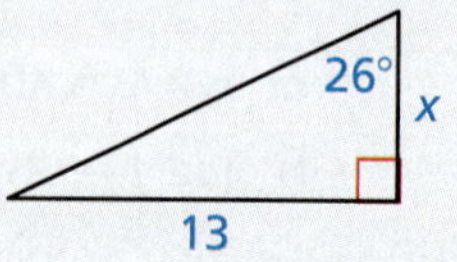

66. A triangle has one side length of 8 inches and another side length of 15 inches. Describe the possible lengths of the third side.

67. Quadrilateral $ABCD$ has vertices $A(2, 4)$, $B(5, 4)$, $C(4, 1)$, and $D(1, 1)$. Quadrilateral $EFGH$ has vertices $E(-3, -2)$, $F(-6, -2)$, $G(-5, -5)$, and $H(-2, -5)$. Are the two quadrilaterals congruent? Use transformations to explain your reasoning.

68. Identify the similar right triangles. Then find the value of y.

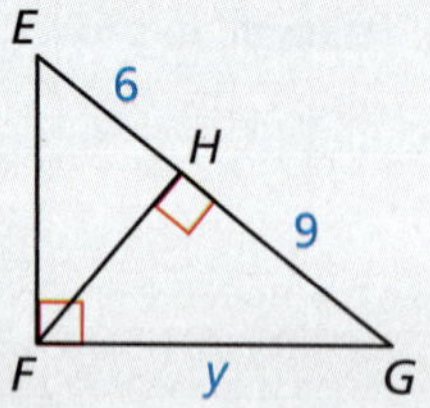

69. Find the values of x and y.

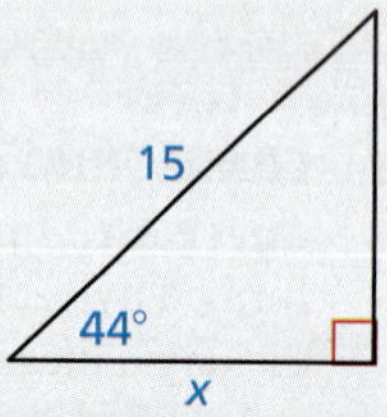

70. State which theorem you can use to show that the quadrilateral is a parallelogram.

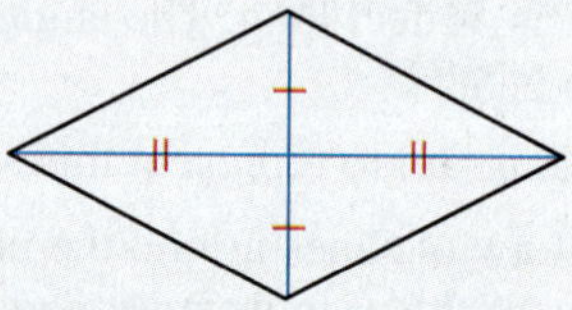

71. $ABCD \cong EFGH$. Find the values of x and y.

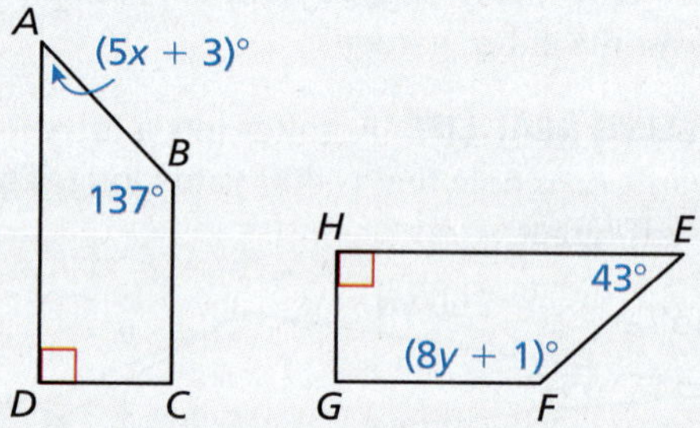

7 MTR 72. **MODELING REAL LIFE** You draw a map of the path from your home to your school on a coordinate plane. The park is exactly halfway between your home and the school. Your home is located at the point (1, 2), and your school is located at the point (5, 6). What point represents the location of the park?

In Exercises 73 and 74, find all the unknown measures in the triangle.

73.

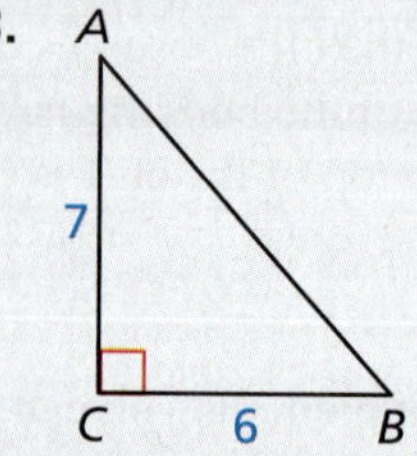

74.

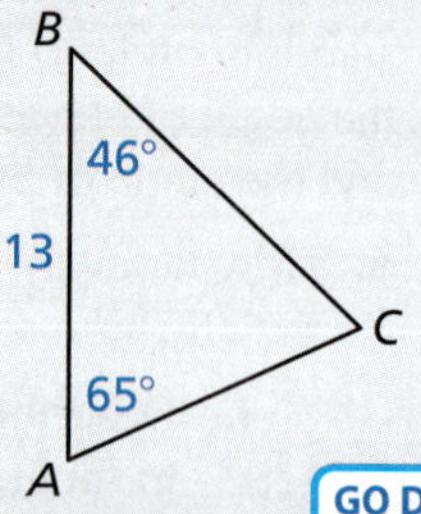

GO DIGITAL

9 Chapter Review WITH CalcChat®

Chapter Learning Target: Understand right triangles and trigonometry.

Chapter Success Criteria:
- ◆ I can use the Pythagorean Theorem to solve problems.
- ◆ I can find side lengths in special right triangles.
- ■ I can explain how similar triangles are used with trigonometric ratios.
- ■ I can use trigonometric ratios to solve problems.

◆ Surface
■ Deep

SELF-ASSESSMENT 1 I don't understand yet. 2 I can do it with help. 3 I can do it on my own. 4 I can teach someone else.

9.1 The Pythagorean Theorem (pp. 453–460) WATCH

Learning Target: Understand and apply the Pythagorean Theorem.

Vocabulary (VOCAB)
pictorial proof
Pythagorean triple

Find the value of x. Then tell whether the side lengths form a Pythagorean triple.

1.
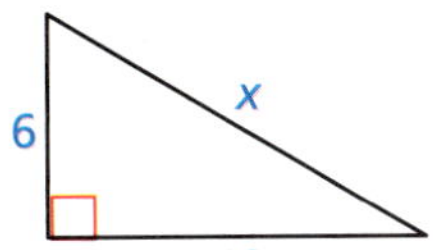

2.
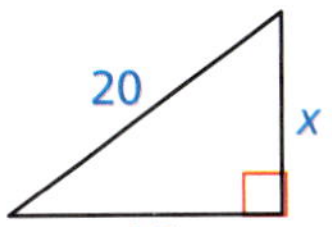

3.
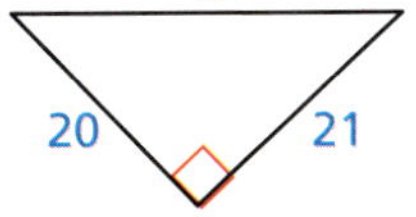

4.
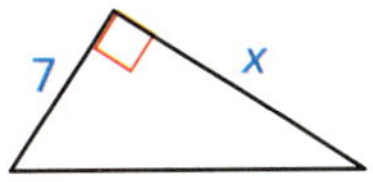

Tell whether the triangle is a right triangle.

5.
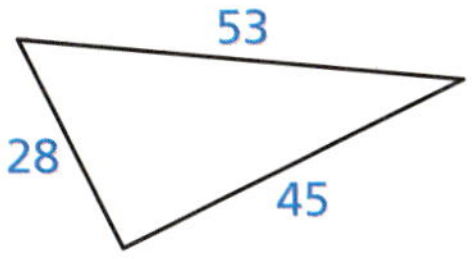

6.
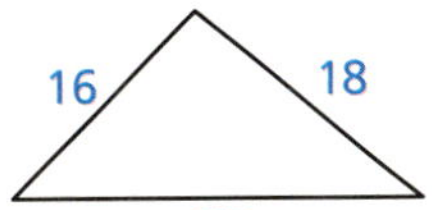

7.
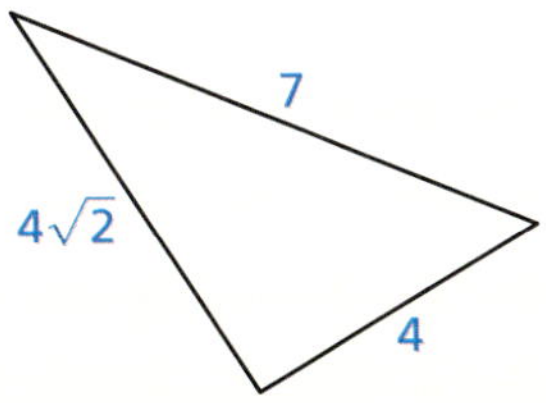

8.
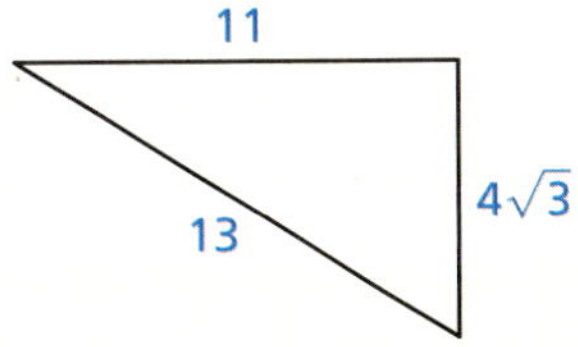

Determine whether the segment lengths form a triangle. If so, is the triangle *acute*, *right*, or *obtuse*?

9. 6, 8, and 9

10. 10, $2\sqrt{2}$, and $6\sqrt{3}$

11. 13, 18, and $3\sqrt{55}$

12. Do the integers 12, 35, and 37 form a Pythagorean triple? If so, use multiples of these integers to write two more Pythagorean triples. If not, explain why not.

13. The integers 9, 40, and 41 form a Pythagorean triple. Without calculating, determine whether a triangle with side lengths of 10 feet, 40 feet, and 41 feet is *acute*, *right*, or *obtuse*. Explain.

GO DIGITAL

9.2 Special Right Triangles (pp. 461–466) WATCH

Learning Target: Understand and use special right triangles.

Find the values of x and y. Write your answer in simplest form.

14.

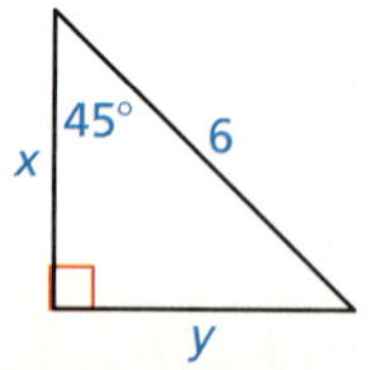

15.

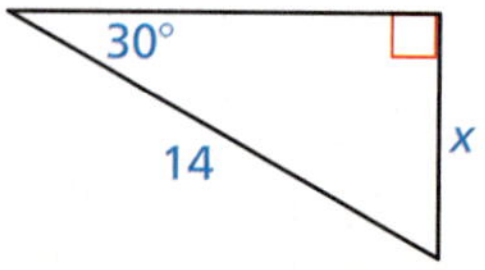

16.

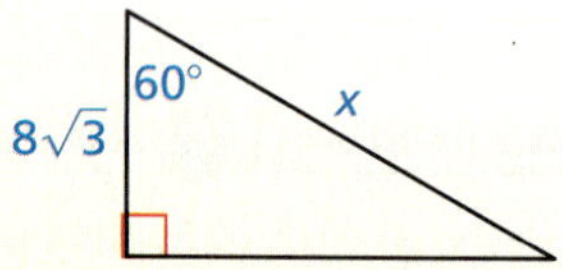

17.

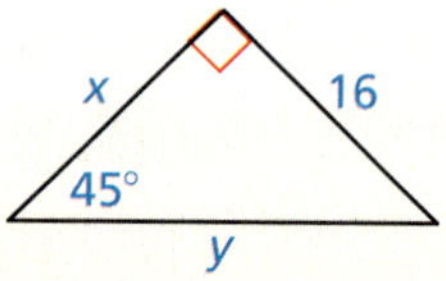

18. Find the area and perimeter of $\triangle ABC$.

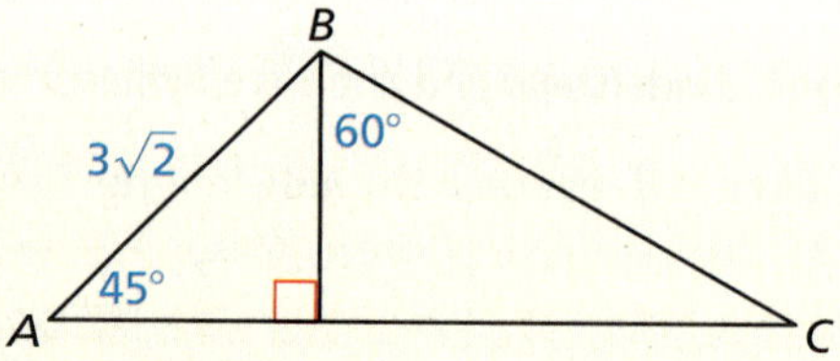

9.3 Similar Right Triangles (pp. 467–474) WATCH

Learning Target: Use proportional relationships in right triangles.

Vocabulary AZ VOCAB

geometric mean

Identify the similar triangles. Then find the value of x.

19.

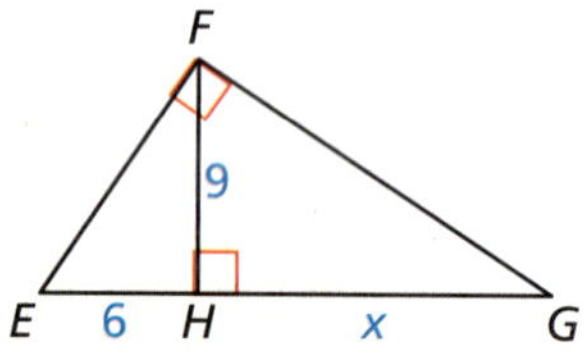

20.

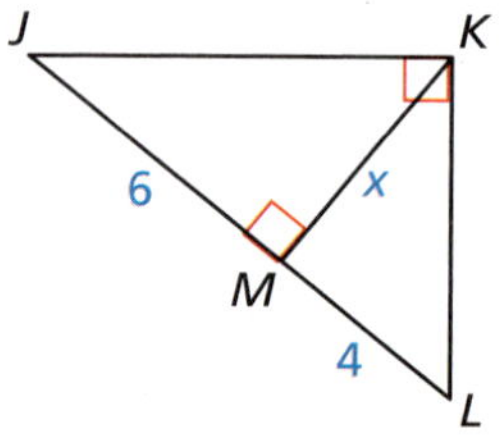

21.

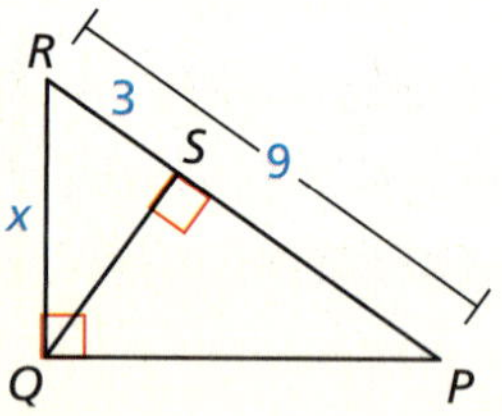

22.

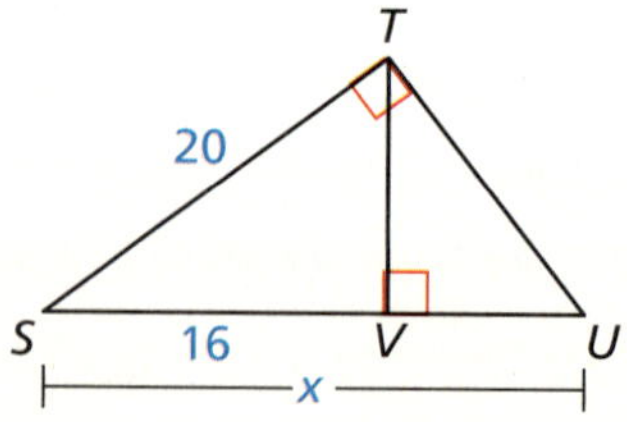

Find the geometric mean of the two numbers.

23. 9 and 25

24. 36 and 48

25. Use the information in the diagram to find the height of the traffic light.

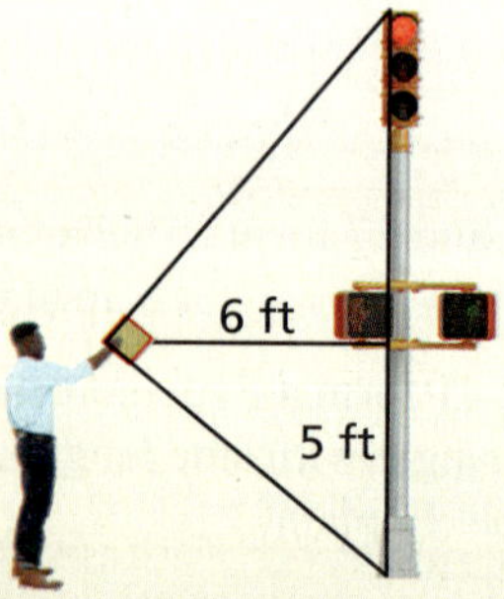

GO DIGITAL

9.4 The Tangent Ratio (pp. 475–480) WATCH

Learning Target: Understand and use the tangent ratio.

Vocabulary
trigonometric ratio
tangent
angle of elevation

Find the tangents of the acute angles in the right triangle. Write each answer as a fraction and as a decimal.

26.

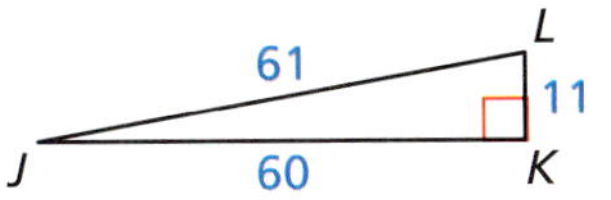

27.

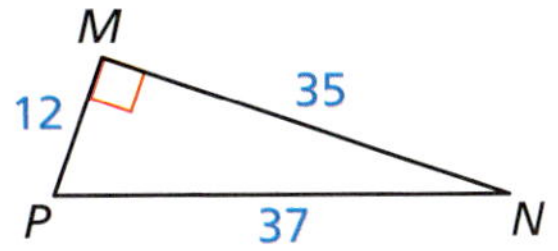

28.

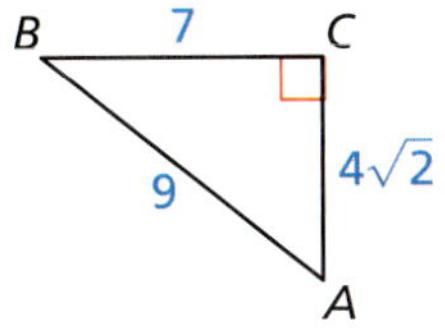

29.

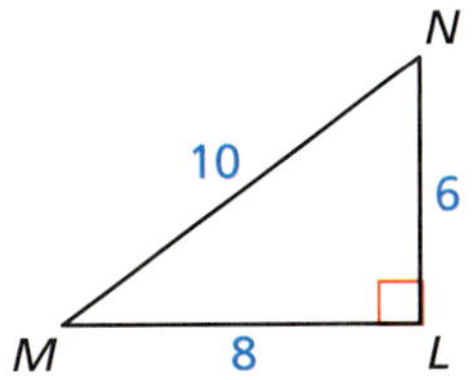

Find the value of *x*.

30.

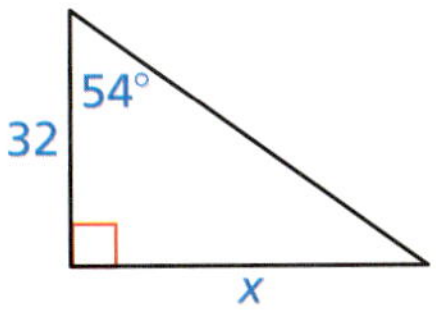

31.

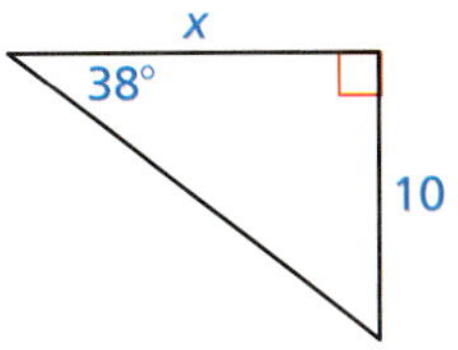

32.

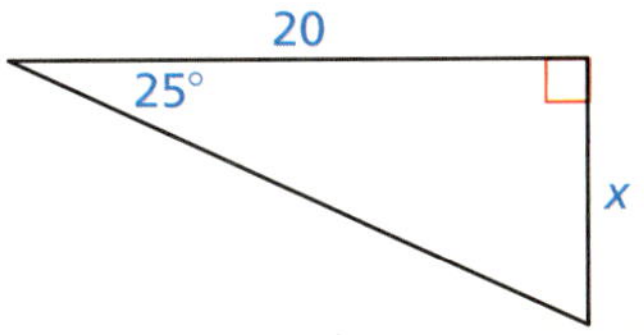

33.

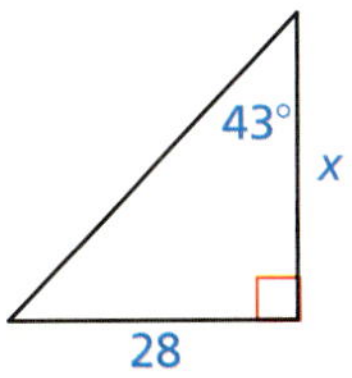

34. The angle of elevation from the bottom of a fence to the top of a tree that is 4 feet from the fence is 75°.

a. How tall is the tree?

b. The angle of elevation from the bottom of the fence to the first limb on the the tree is 62°. How high is the limb?

c. The angle of elevation from the top of the fence to the top of the tree is 70°. How tall is the fence?

35. You stand next to a trampoline. Your eyes are 3 feet above the trampoline, and you are 7 feet from its center, where your friend jumps. The angle of elevation from your eyes to your friend is 35°. How high above the trampoline is your friend?

36. Find the tangent of the smaller acute angle in a right triangle with side lengths 8, 15, and 17.

37. The tangent of an acute angle in a right triangle is $\frac{3}{4}$. Can the sides of the right triangle all have integer lengths? Explain.

GO DIGITAL

9.5 The Sine and Cosine Ratios (pp. 481–490) WATCH

Learning Target: Understand and use the sine and cosine ratios.

Vocabulary VOCAB
sine
cosine
angle of depression
unit circle

Find sin X, sin Z, cos X, and cos Z. Write each answer as a fraction and as a decimal.

38.
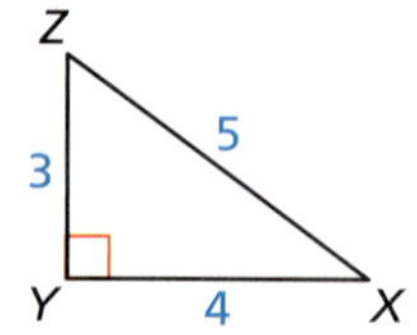

39.
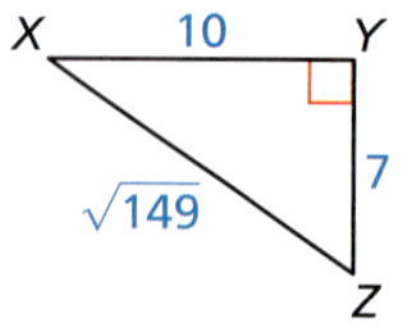

40.
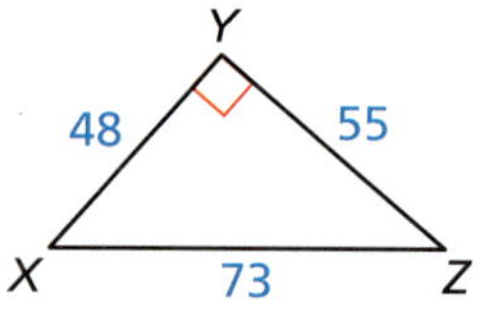

Write the expression in terms of cosine.

41. sin 72°

42. sin 15°

43. sin 45°

Write the expression in terms of sine.

44. cos 29°

45. cos 50°

46. cos 64°

Find the value of each variable using sine and cosine.

47.
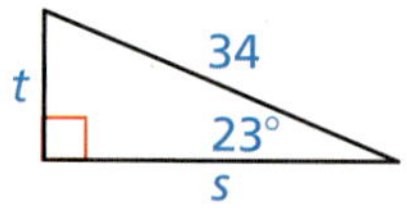

48.
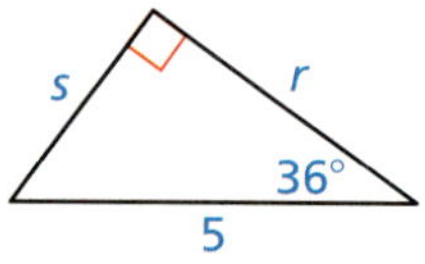

49.
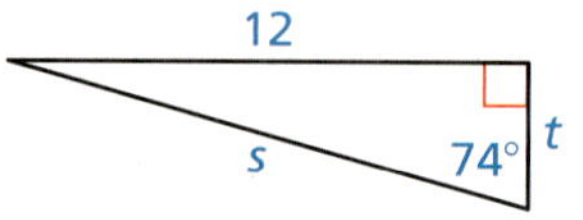

Find the sine, cosine, and tangent of $\angle A$.

50.
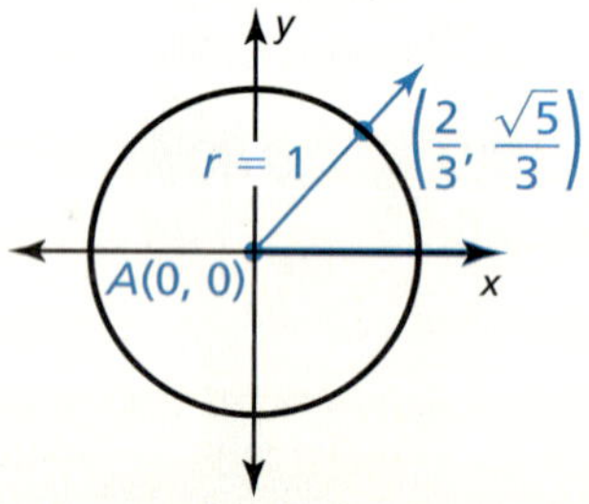

51.
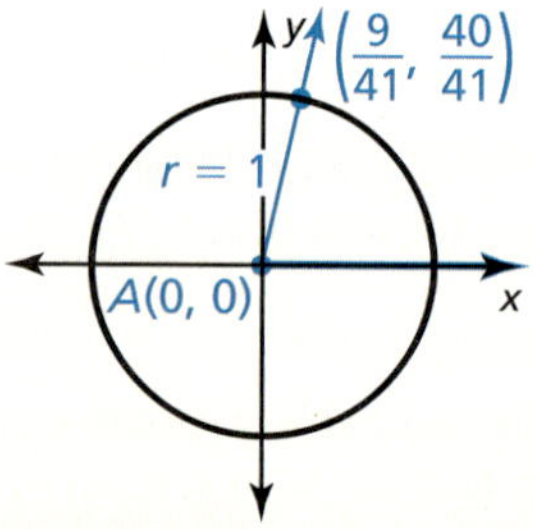

52.
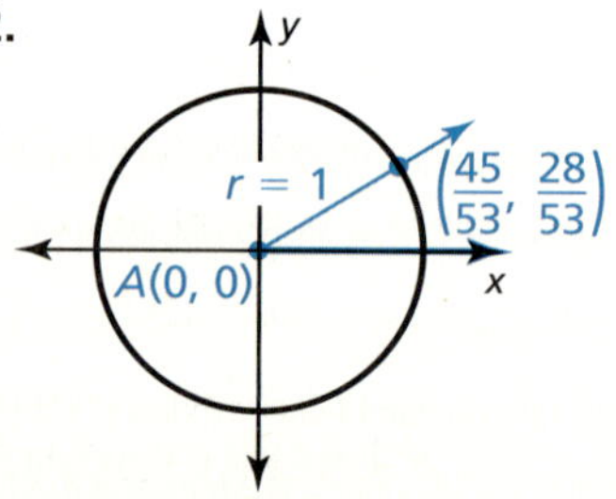

Determine if the point lies on the unit circle. Explain your reasoning.

53. (0.64, 0.36)

54. $\left(\frac{33}{65}, \frac{56}{65}\right)$

55. The Niagara Falls Incline Railway has an angle of elevation of 30° and a total length of 196 feet. How many feet does the Niagara Falls Incline Railway rise vertically?

GO DIGITAL

9.6 Law of Sines and Law of Cosines (pp. 491–500) WATCH

Learning Target: Find unknown side lengths and angle measures of acute and obtuse triangles.

Vocabulary VOCAB

inverse tangent
inverse sine
inverse cosine
Law of Sines
Law of Cosines

Use technology to find the trigonometric ratio.

56. sin 136° **57.** cos 124° **58.** tan 155°

Find the area of $\triangle ABC$ with the given side lengths and included angle.

59.

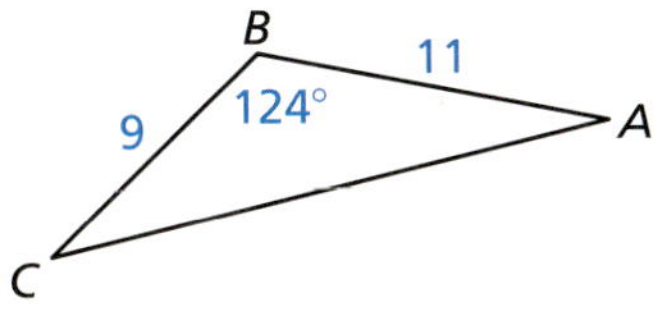

60.

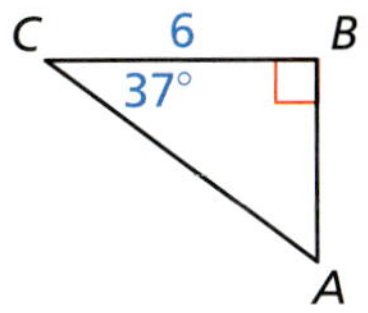

Find all unknown measures in the triangle.

61.

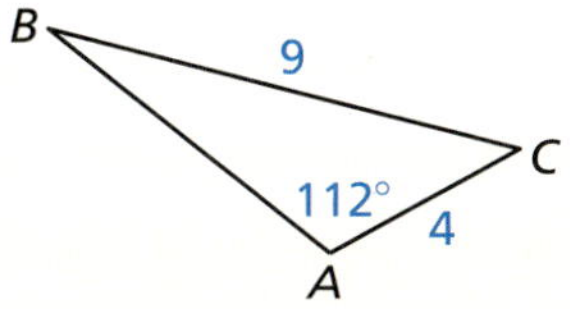

62.

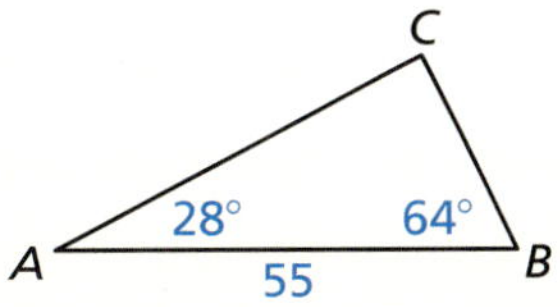

63. $m\angle C = 48°, b = 20, c = 28$ **64.** $m\angle B = 25°, a = 8, c = 3$

65. $m\angle B = 102°, m\angle C = 43°, b = 21$ **66.** $a = 10, b = 3, c = 12$

67. You are sealing a triangular blacktop surface. One side of the triangle is 36 feet long, and another side is 28 feet long. The angle opposite the 28-foot side is 42°.

a. Draw a diagram to represent this situation.

b. Use the Law of Sines to find all the unknown measures in the triangle from part (a).

c. One gallon of sealant covers 80 square feet. How many gallons of sealant do you need to apply two coats to the blacktop surface?

Mathematical Thinking and Reasoning

5 MTR USE PATTERNS AND STRUCTURE TO HELP UNDERSTAND AND CONNECT MATHEMATICAL CONCEPTS

Mathematicians who use patterns and structure to help understand and connect mathematical concepts focus on relevant details in problems.

1. In part (f) of the 9.3 Explore It! on page 467, describe what x represents in each of the Triangles 1, 2, and 3. Then explain how you used the relationships among the triangles to find x.

2. You use a pair of distance-finding binoculars to watch your friend climb a cliff. Using the information shown, describe a way to find your friend's height h above the ground without moving from your position.

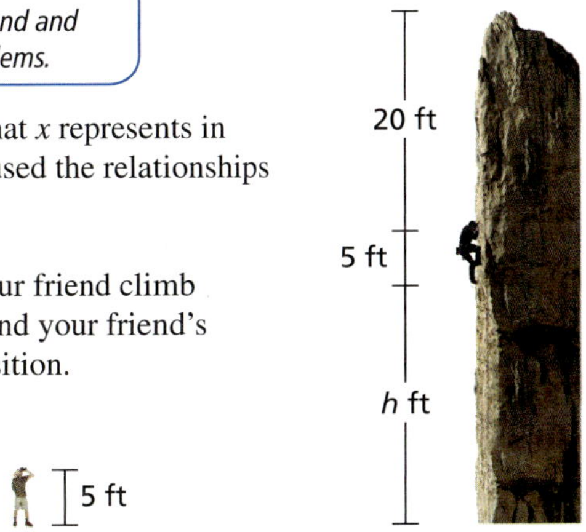

Not drawn to scale

GO DIGITAL

9 Practice Test WITH CalcChat®

Find the value of each variable.

1. 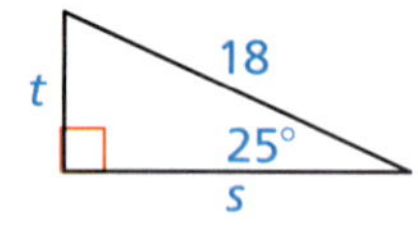

2. 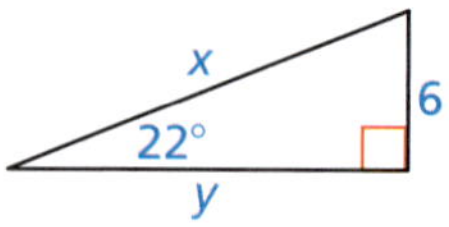

3. 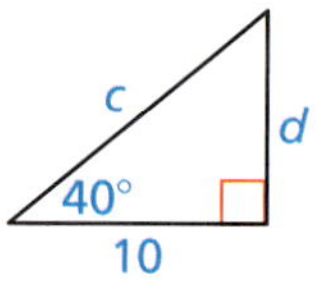

Determine whether the segment lengths form a triangle. If so, is the triangle *acute*, *right*, or *obtuse*?

4. 16, 30, and 34

5. 4, $\sqrt{67}$, and 9

6. $\sqrt{5}$, 5, and 5.5

Find all unknown measures in the triangle.

7. 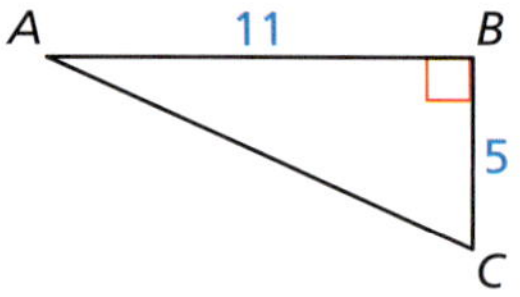

8. 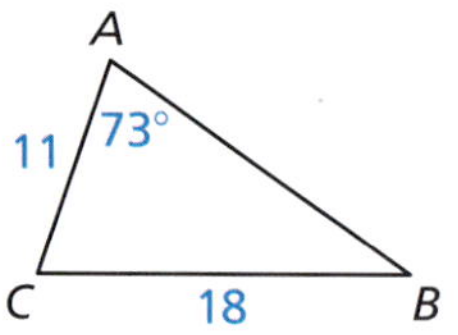

9. 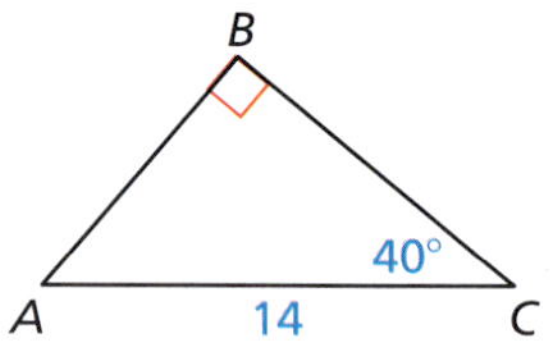

10. 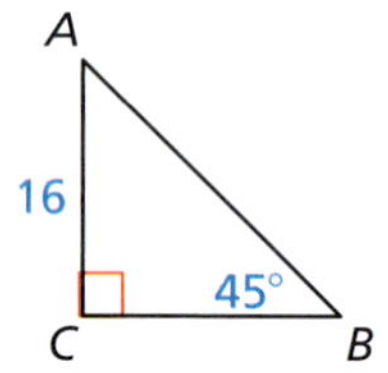

11. 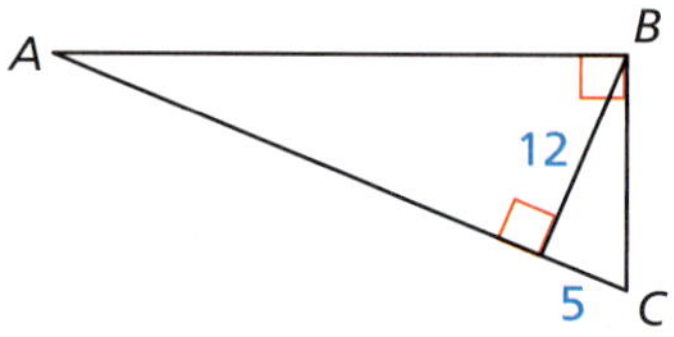

12. 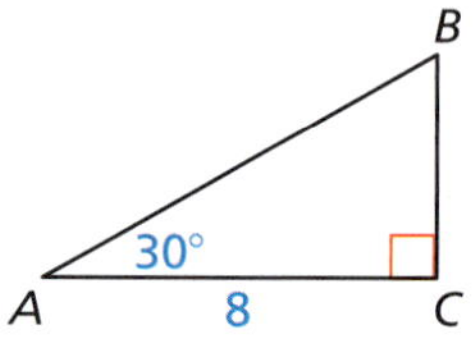

13. $m\angle A = 103°$, $b = 12$, $c = 24$

14. $m\angle A = 26°$, $m\angle C = 35°$, $b = 13$

15. $a = 38$, $b = 31$, $c = 35$

16. Write cos 53° in terms of sine.

17. Find all unknown measures in $\triangle ABC$ with vertices $A(-2, 2)$, $B(1, 6)$, and $C(5, 3)$. Then find the area of $\triangle ABC$.

18. Consider an angle in the first quadrant with its vertex at the origin, one side on the x-axis, and one side passing through the point (x, y). The sine of the angle is 0.352. Can you find the tangent of the angle using this information? Explain your reasoning.

19. You are looking up at a large parade balloon. You are 60 feet from the point on the street directly beneath the balloon. The angle of elevation to the top of the balloon is 53°. The angle of elevation to the bottom of the balloon is 29°. Find the height h of the balloon.

20. You are on Easter Island photographing a 13-foot-tall statue called a *moai*. Your camera is on a tripod that is 5 feet tall. The vertical viewing angle of your camera is 90°. How close to the moai can you place the camera with the entire moai in view?

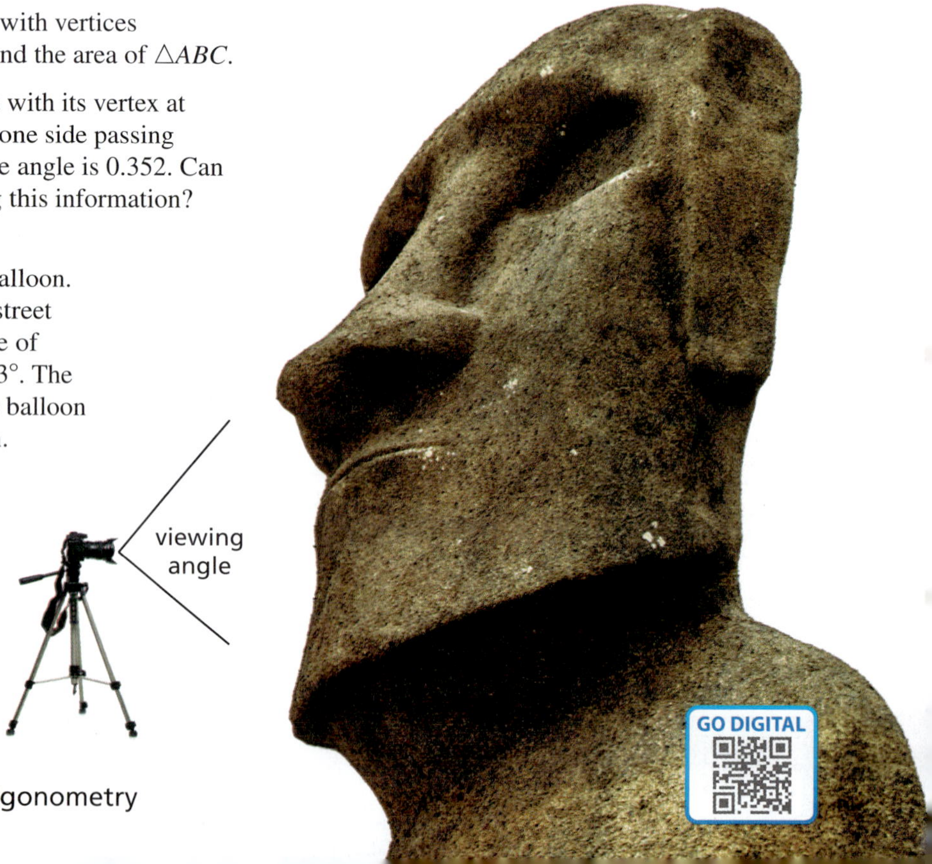

9 Performance Task

Engineering a Mammoth

As Arctic permafrost thaws, large quantities of harmful carbon dioxide and methane gases are released into the atmosphere. Some scientists believe that cloning woolly mammoths could help keep the permafrost frozen. However, because there are no living woolly mammoth cells to clone, scientists instead aim to engineer elephants with mammoth-like characteristics that can thrive in cold environments.

STEP 1:

Extract DNA from mammoth bones.

STEP 2:

Combine several DNA samples and use technology to sequence a mammoth's genome.

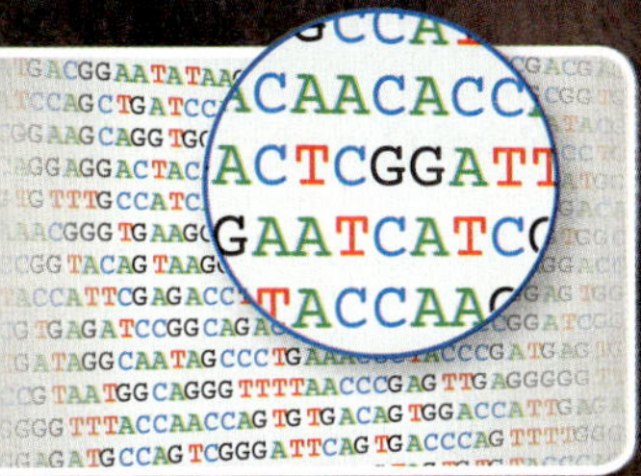

STEP 3:

Compare the mammoth's genome to an elephant's genome and identify key differences.

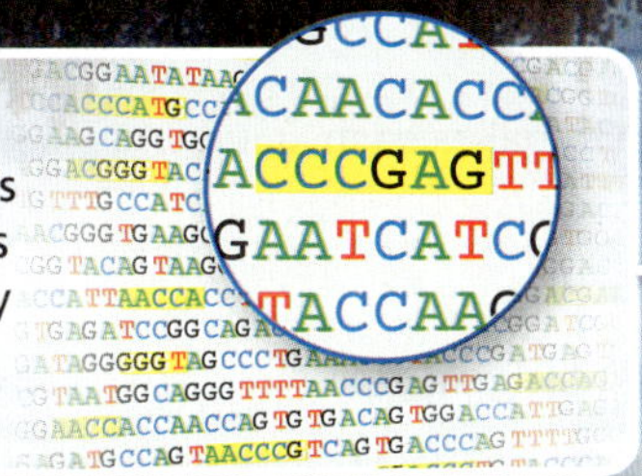

STEP 4:

Retrieve cells from an elephant and use technology to edit the DNA to look more like mammoth DNA in key areas.

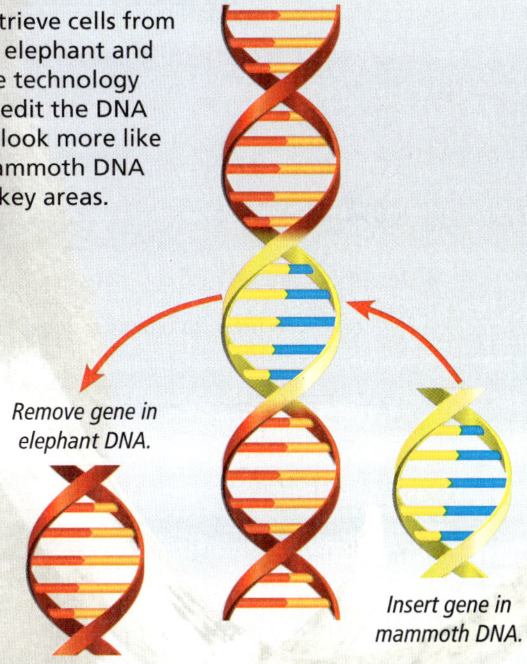

STEP 5:

Perform the cloning process using these new mammoth-like cells.

A MAMMOTH TASK

You work on an exhibit featuring a woolly mammoth skeleton. Standing 15 feet from the assembled skeleton, the angle of elevation from your eyes to the top of the shoulder is 16°. Write your height as a percent of the mammoth's height. Then conduct research to estimate the total length of the mammoth from front to back.

Create an informational display about woolly mammoths that includes the following information:

- typical shoulder heights and weights of woolly mammoths
- the shoulder height of the particular mammoth on display
- habitat and diet
- reasons that the Arctic permafrost is thawing
- how engineering mammoths could help keep the permafrost frozen

GO DIGITAL

9 Test Prep WITH CalcChat®
Cumulative Practice

WATCH Tutorial videos are available for each exercise.

1. In the diagram, *ABCD* is a parallelogram. Which measures are correct? Select all that apply.

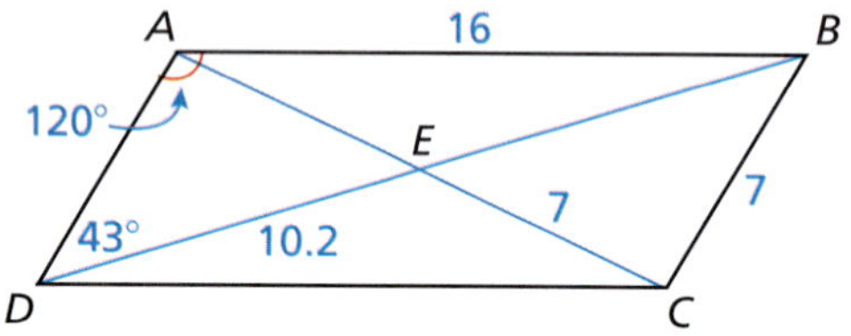

Ⓐ $CD = 16$

Ⓑ $EB = 7$

Ⓒ $BD = 17.2$

Ⓓ $EB = 10.2$

Ⓔ $m\angle ABC = 60°$

Ⓕ $m\angle BCD = 120°$

Ⓖ $m\angle ABD = 43°$

Ⓗ $m\angle BDC = 17°$

2. In $\triangle PQR$ and $\triangle SQT$, *S* is between *P* and *Q*, *T* is between *R* and *Q*, and $\frac{QS}{SP} = \frac{QT}{TR}$. Which statement(s) must be true? Select all that apply.

Ⓐ $\overline{ST} \perp \overline{QP}$

Ⓑ $\overline{ST} \parallel \overline{PR}$

Ⓒ $\frac{ST}{PR} = \frac{QS}{SP}$

Ⓓ $ST = \frac{1}{2}PR$

3. In the diagram, $\triangle JKL \sim \triangle QRS$. Choose the symbol that makes each statement true.

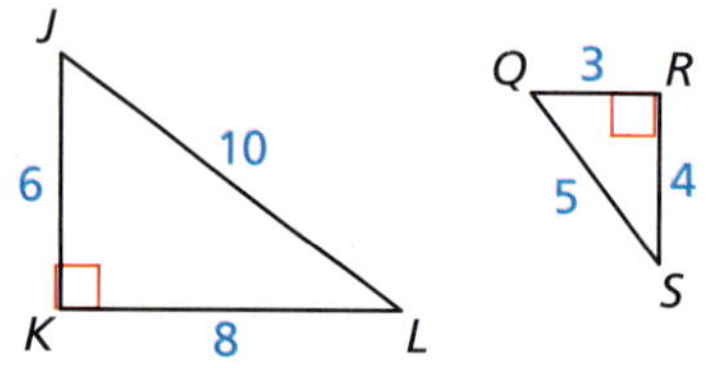

$\sin J$ ☐ $\sin Q$	$\sin L$ ☐ $\cos J$	$\cos L$ ☐ $\tan Q$
$\cos S$ ☐ $\cos J$	$\cos J$ ☐ $\sin S$	$\tan J$ ☐ $\tan Q$
$\tan L$ ☐ $\tan Q$	$\tan S$ ☐ $\cos Q$	$\sin Q$ ☐ $\cos L$

<   =   >

4. A surveyor makes the measurements shown. What is the distance across the river?

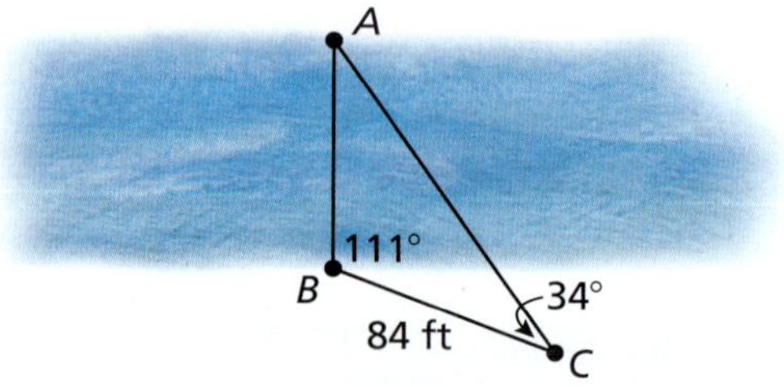

GO DIGITAL

5. Prove that quadrilateral $DEFG$ is a kite.

Given $\overline{HE} \cong \overline{HG}$, $\overline{EG} \perp \overline{DF}$

Prove $\overline{FE} \cong \overline{FG}$, $\overline{DE} \cong \overline{DG}$

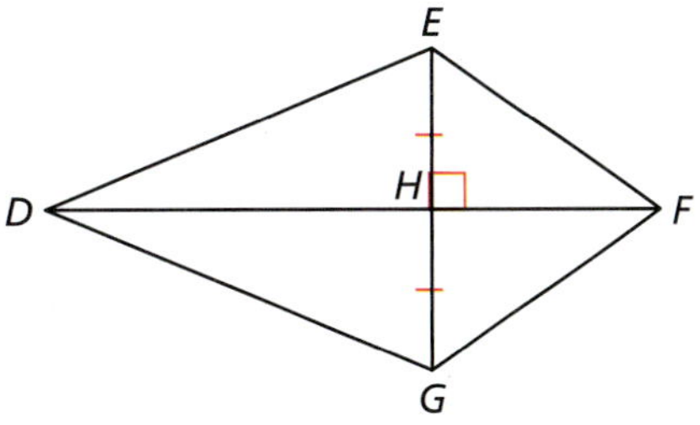

6. The Red Pyramid in Egypt has a square base. Each side of the base measures 722 feet. The height of the pyramid is 343 feet.

a. Use the side length of the base, the height of the pyramid, and the Pythagorean Theorem to find the slant height AB of the pyramid.

b. Find AC.

c. Name three possible ways of finding $m\angle 1$. Then find $m\angle 1$.

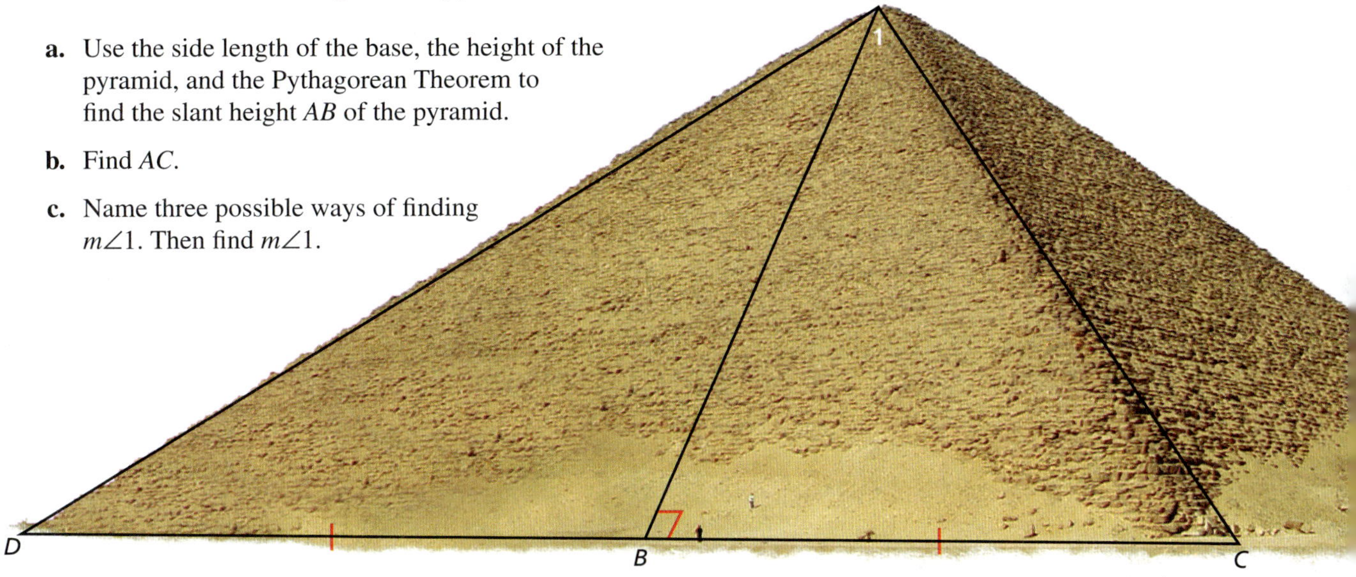

7. Which statements can be proven about the diagram? Select all that apply.

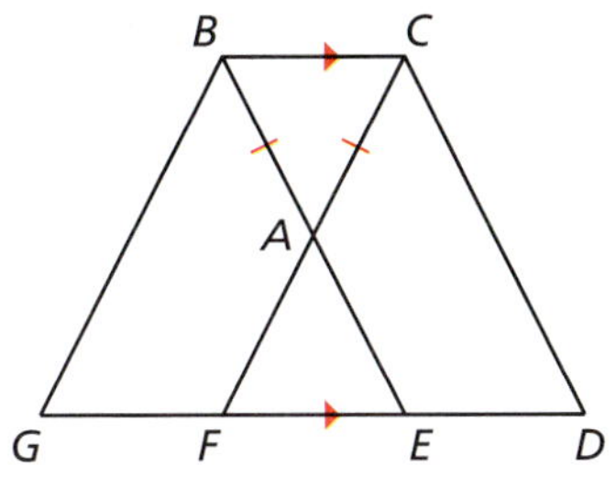

Ⓐ $\triangle ABC \cong \triangle AEF$

Ⓑ $\angle ABC \cong \angle AFE$

Ⓒ $\triangle ABC \sim \triangle AEF$

Ⓓ $\angle CDE \cong \angle AEF$

Ⓔ $\dfrac{AB}{AE} = \dfrac{AC}{AF}$

Ⓕ $\dfrac{CD}{FD} = \dfrac{BG}{GE}$

8. Which is a step in constructing a circle that is inscribed in $\triangle XYZ$?

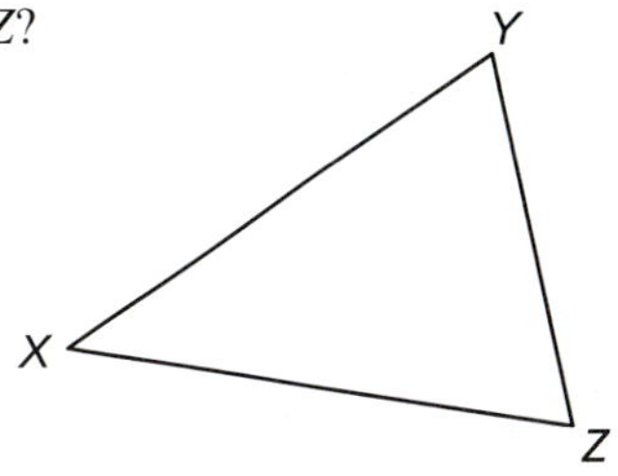

Ⓐ Draw the angle bisector of $\angle X$.

Ⓑ Draw the perpendicular bisector of $\overline{XY}$.

Ⓒ Draw the median from vertex Y to $\overline{XZ}$.

Ⓓ Draw the altitude from vertex Z to $\overline{XY}$.

GO DIGITAL

10 Circles

NATIONAL GEOGRAPHIC EXPLORER

Christine Lee

WATCH

Dr. Christine Lee is a bioarchaeologist, combining biological anthropology and archaeology to study human remains. This field of study looks for physical clues about people's lives, such as what they did for work, what illnesses they had, what traumas they suffered, and their age. In Mongolia, she excavated a royal cemetery for the Xiongnu people, who prompted China to build the Great Wall.

- What other aspects of people's lives can you learn about through bioarchaeology?
- When was the Great Wall of China built? How long is the Great Wall?

STEM

Archaeologists often try to discern the purpose of structures by analyzing their geometric properties. In the Performance Task, you will find geometric relationships in Stonehenge and analyze their possible significance.

GO DIGITAL

Preparing for Chapter 10

Chapter Learning Target: Understand and apply circle relationships.

Chapter Success Criteria:
- ◆ I can identify lines and segments that intersect circles.
- ◆ I can find angle and arc measures in circles.
- ■ I can use circle relationships to solve problems.
- ■ I can use circles to model and solve real-life problems.

◆ Surface
■ Deep

AZ VOCAB

Chapter Vocabulary

Work with a partner. Discuss each of the vocabulary terms.

circle
radius
diameter
concentric circles
central angle
minor arc
major arc
semicircle
congruent circles
inscribed angle
inscribed polygon
circumscribed circle
circumscribed angle

Mathematical Thinking and Reasoning

3 MTR

COMPLETE TASKS WITH MATHEMATICAL FLUENCY

Mathematicians who complete tasks with mathematical fluency select efficient and appropriate methods for solving problems within the given context.

Work with a partner. Two points on a circle define an *arc*. Arcs are measured in degrees.

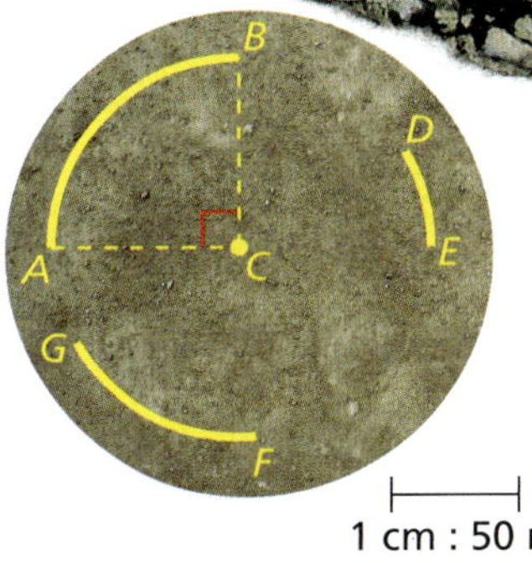

1. Archaeologists uncover partial ruins of the circular stone wall shown, with center C.
   a. Find the measure of arc AB. Explain your method.
   b. What tool(s) can you use to find the measure of the other two arcs? Explain.
   c. Find the measures of arc DE and arc GF.
   d. What percent of the circle has been uncovered?
2. Another portion of the wall is uncovered that has an arc measure of 28°. Use tools to draw this arc. Explain your method.

GO DIGITAL

10 Prepare WITH CalcChat®

Solving Quadratic Equations by Completing the Square

WATCH

Example 1 **Solve $x^2 + 8x - 3 = 0$ by completing the square.**

$x^2 + 8x - 3 = 0$ — Write original equation.

$x^2 + 8x = 3$ — Add 3 to each side.

$x^2 + 8x + 4^2 = 3 + 4^2$ — Complete the square by adding $\left(\frac{8}{2}\right)^2$, or 4^2, to each side.

$(x + 4)^2 = 19$ — Write the left side as a square of a binomial.

$x + 4 = \pm\sqrt{19}$ — Take the square root of each side.

$x = -4 \pm \sqrt{19}$ — Subtract 4 from each side.

▶ The solutions are $x = -4 + \sqrt{19}$ and $x = -4 - \sqrt{19}$.

Solve the equation by completing the square.

1. $x^2 - 2x = 5$

2. $r^2 + 10r = -7$

3. $w^2 - 8w = 9$

4. $p^2 + 10p - 4 = 0$

5. $k^2 - 4k - 7 = 0$

6. $-z^2 + 2z = 1$

Multiplying Binomials

WATCH

Example 2 **Find the product $(x + 3)(2x - 1)$.**

First Outer Inner Last

$(x + 3)(2x - 1) = x(2x) + x(-1) + 3(2x) + (3)(-1)$ — FOIL Method

$= 2x^2 + (-x) + 6x + (-3)$ — Multiply.

$= 2x^2 + 5x - 3$ — Simplify.

▶ The product is $2x^2 + 5x - 3$.

Find the product.

7. $(x + 7)(x + 4)$

8. $(a + 1)(a - 5)$

9. $(q - 9)(3q - 4)$

10. $(2v - 7)(5v + 1)$

11. $(4h + 3)(2 + h)$

12. $(8 - 6b)(5 - 3b)$

4 MTR **13.** **DISCUSS MATHEMATICAL THINKING** Write an expression that represents the product of two consecutive positive odd integers. Explain your reasoning.

GO DIGITAL

10.1 Lines and Segments That Intersect Circles

Learning Target: Identify lines and segments that intersect circles and use them to solve problems.

Success Criteria:
- I can identify special segments and lines that intersect circles.
- I can draw and identify common tangents.
- I can use properties of tangents to solve problems.

EXPLORE IT! Investigating Lines and Segments That Intersect Circles

Work with a partner.

a. Use two pencils and a paper clip to draw a circle by placing a pencil in each end of the paper clip. Anchor one pencil on a piece of paper. Use the other pencil to apply slight pressure to the edge of the paper clip and draw a circle. How would you define a *circle*?

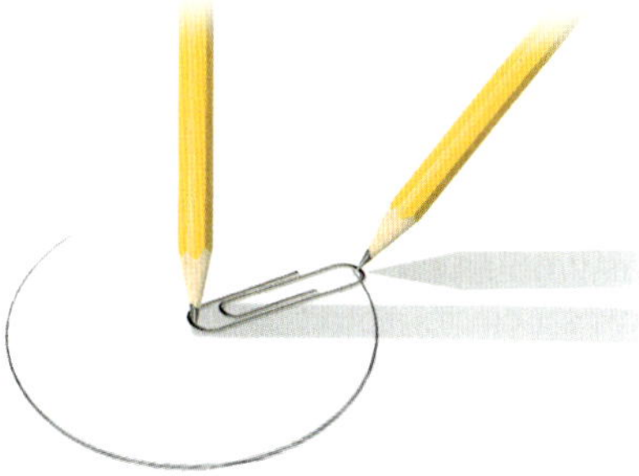

5 MTR USE STRUCTURE
How is the distance between the two pencils in part (a) related to two of the lines or segments in part (b)?

b. The drawing below shows different types of lines or segments that intersect a circle. In your own words, write a definition for each type of line or segment.

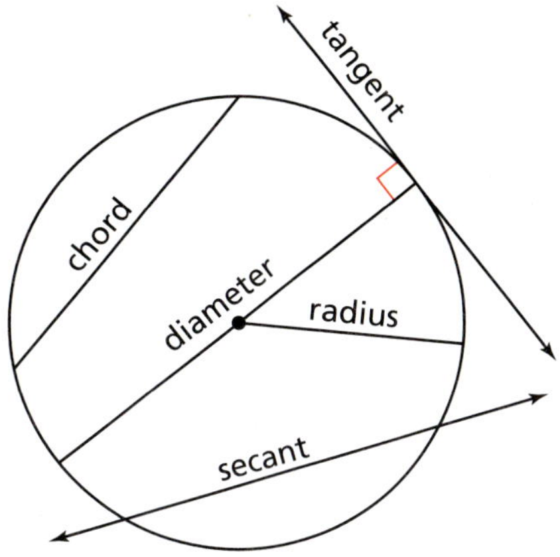

c. Of the five types of lines and segments in part (b), which one is a subset of another? Explain.

d. Draw a circle with center O and a radius $\overline{OA}$. Construct a line m that passes through A and is perpendicular to $\overline{OA}$. What do you notice about line m?

Geometric Reasoning

H MA.912.GR.5.5 Given a point outside a circle, construct a line tangent to the circle that passes through the given point.
MA.912.GR.6.1 Solve mathematical and real-world problems involving the length of a secant, tangent, segment or chord in a given circle.

GO DIGITAL

Vocabulary

AZ VOCAB

circle, *p. 514*
center, *p. 514*
radius, *p. 514*
chord, *p. 514*
diameter, *p. 514*
secant, *p. 514*
tangent, *p. 514*
point of tangency, *p. 514*
tangent circles, *p. 515*
concentric circles, *p. 515*
common tangent, *p. 515*

Identifying Special Segments and Lines

A **circle** is the set of all points in a plane that are equidistant from a given point called the **center** of the circle. A circle with center P is called "circle P" and can be written as $\odot P$.

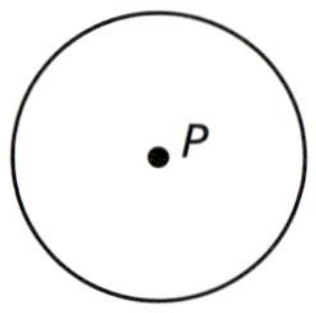

circle *P*, or $\odot P$

REMEMBER

The words *radius* and *diameter* refer to lengths as well as segments. For a given circle, think of *a* radius and *a* diameter as segments and *the* radius and *the* diameter as lengths.

KEY IDEA

Lines and Segments That Intersect Circles

A segment whose endpoints are the center and any point on a circle is a **radius**.

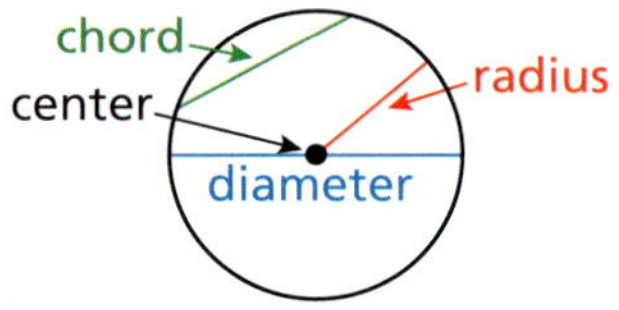

A **chord** is a segment whose endpoints are on a circle. A **diameter** is a chord that contains the center of the circle.

A **secant** is a line that intersects a circle in two points.

A **tangent** is a line in the plane of a circle that intersects the circle in exactly one point, the **point of tangency**. The *tangent ray* $\overrightarrow{AB}$ and the *tangent segment* $\overline{AB}$ are also called tangents.

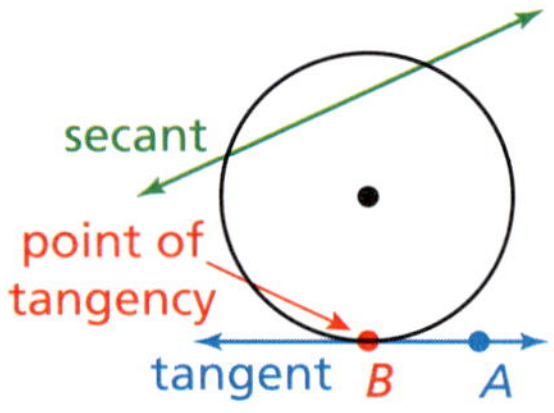

EXAMPLE 1 Identifying Special Segments and Lines

WATCH

Tell whether each line, ray, or segment is best described as a *radius*, *chord*, *diameter*, *secant*, or *tangent* of $\odot C$.

a. $\overline{AC}$ **b.** $\overline{AB}$

c. $\overrightarrow{DE}$ **d.** $\overleftrightarrow{AE}$

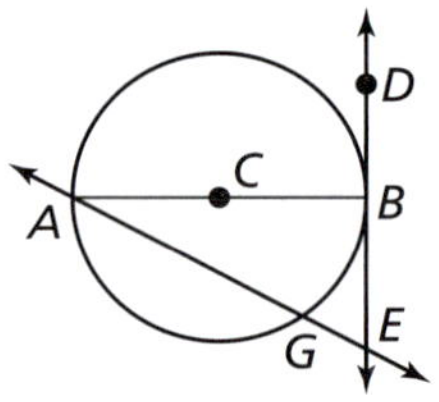

STUDY TIP

In this book, assume that all segments, rays, or lines that appear to be tangent to a circle are tangents.

SOLUTION

a. $\overline{AC}$ is a radius because C is the center and A is a point on the circle.

b. $\overline{AB}$ is a diameter because it is a chord that contains the center C.

c. $\overrightarrow{DE}$ is a tangent ray because it is contained in a line that intersects the circle in exactly one point.

d. $\overleftrightarrow{AE}$ is a secant because it is a line that intersects the circle in two points.

SELF-ASSESSMENT

1 I don't understand yet. 2 I can do it with help. 3 I can do it on my own. 4 I can teach someone else.

1. In Example 1, what word best describes $\overline{AG}$? $\overline{CB}$?

2. In Example 1, name a tangent and a tangent segment.

4 MTR **3.** **WHICH ONE DOESN'T BELONG?** Which type of segment does *not* belong with the other three? Explain your reasoning.

chord radius tangent diameter

GO DIGITAL

Drawing and Identifying Common Tangents

KEY IDEAS

Coplanar Circles and Common Tangents

In a plane, two circles can intersect in two points, one point, or no points. Coplanar circles that intersect in one point are called **tangent circles**. Coplanar circles that have a common center are called **concentric circles**.

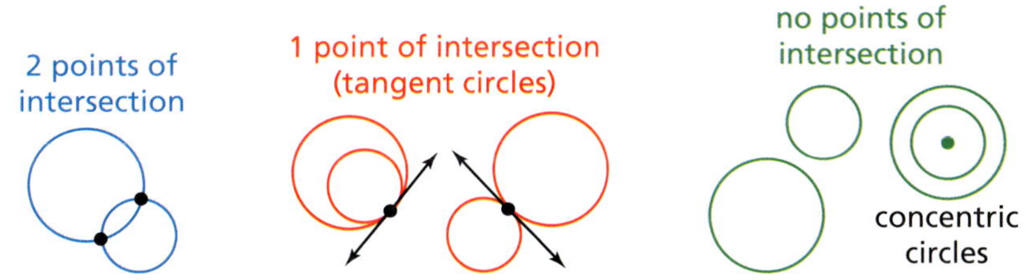

A line or segment that is tangent to two coplanar circles is called a **common tangent**. A *common internal tangent* intersects the segment that joins the centers of the two circles. A *common external tangent* does not intersect the segment that joins the centers of the two circles.

EXAMPLE 2 Drawing and Identifying Common Tangents

Tell how many common tangents the circles have and draw them. State whether the tangents are *external tangents* or *internal tangents*.

a.
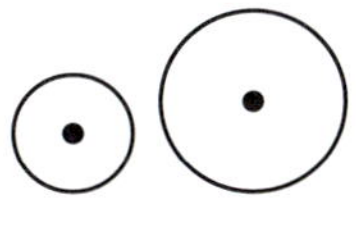

b.
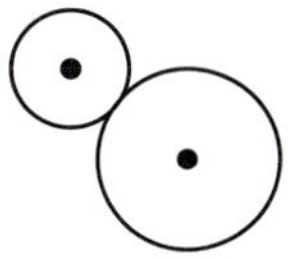

c.
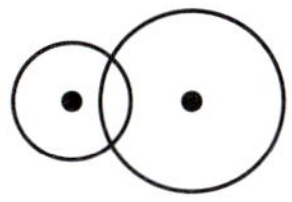

SOLUTION

Draw the segment that joins the centers of the two circles. Then draw the common tangents.

a. 4 common tangents: 2 internal, 2 external

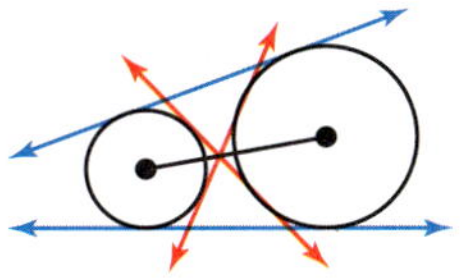

b. 3 common tangents: 1 internal, 2 external

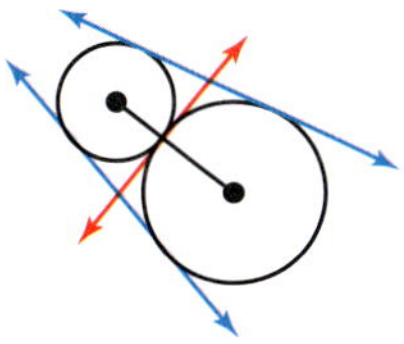

c. 2 common tangents: 2 external

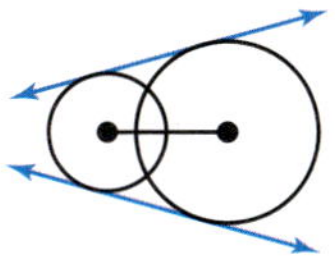

SELF-ASSESSMENT

1 I don't understand yet. 2 I can do it with help. 3 I can do it on my own. 4 I can teach someone else.

Tell how many common tangents the circles have and draw them. State whether the tangents are *external tangents* or *internal tangents*.

4.
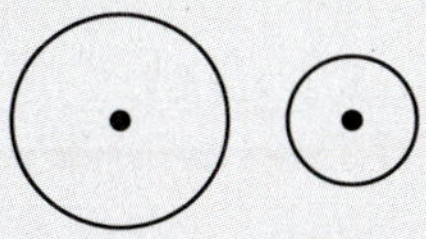

5.
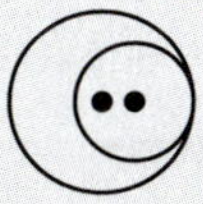

6.
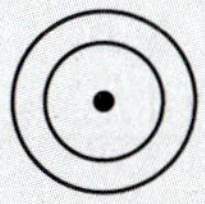

GO DIGITAL

Using Properties of Tangents

THEOREMS

10.1 Tangent Line to Circle Theorem

In a plane, a line is tangent to a circle if and only if the line is perpendicular to a radius of the circle at its endpoint on the circle.

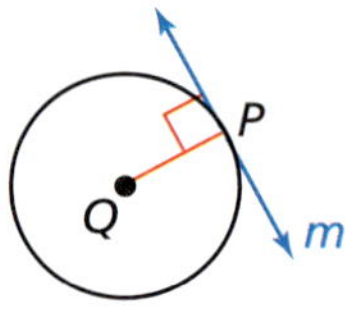

Line m is tangent to $\odot Q$ if and only if $m \perp \overline{QP}$.

Prove this Theorem Exercise 40, page 519

10.2 External Tangent Congruence Theorem

Tangent segments from a common external point are congruent.

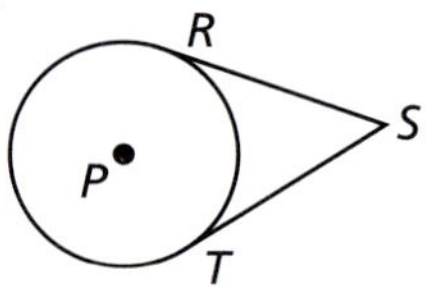

If $\overline{SR}$ and $\overline{ST}$ are tangent segments, then $\overline{SR} \cong \overline{ST}$.

Prove this Theorem Exercise 39, page 519

EXAMPLE 3 Verifying a Tangent to a Circle

WATCH

Is $\overline{ST}$ tangent to $\odot P$?

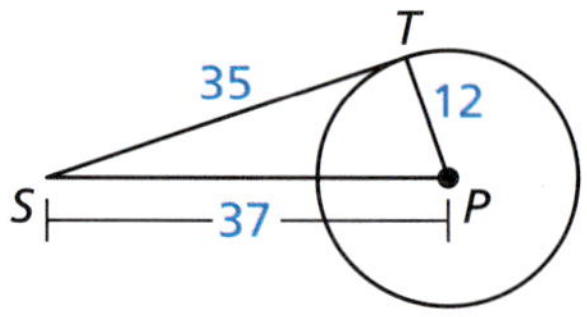

SOLUTION

Use the Converse of the Pythagorean Theorem. Because $12^2 + 35^2 = 37^2$, $\triangle PTS$ is a right triangle and $\overline{ST} \perp \overline{PT}$. So, $\overline{ST}$ is perpendicular to a radius of $\odot P$ at its endpoint on $\odot P$.

▶ By the Tangent Line to Circle Theorem, $\overline{ST}$ is tangent to $\odot P$.

EXAMPLE 4 Finding the Radius of a Circle

WATCH

In the diagram, point B is a point of tangency. Find the radius r of $\odot C$.

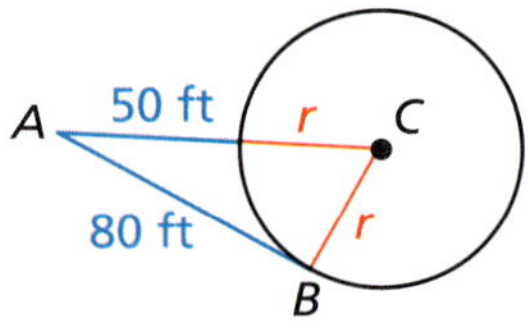

SOLUTION

You know from the Tangent Line to Circle Theorem that $\overline{AB} \perp \overline{BC}$, so $\triangle ABC$ is a right triangle. You can use the Pythagorean Theorem.

$AC^2 = BC^2 + AB^2$	Pythagorean Theorem
$(r + 50)^2 = r^2 + 80^2$	Substitute.
$r^2 + 100r + 2500 = r^2 + 6400$	Multiply.
$100r = 3900$	Subtract r^2 and 2500 from each side.
$r = 39$	Divide each side by 100.

▶ The radius is 39 feet.

GO DIGITAL

4 MTR JUSTIFY A RESULT

Verify that the side lengths of $\triangle ABC$ form a Pythagorean Triple.

CONSTRUCTION Constructing a Tangent to a Circle

WATCH H

Given $\odot C$ and point A, construct a line tangent to $\odot C$ that passes through A. Use a compass and straightedge.

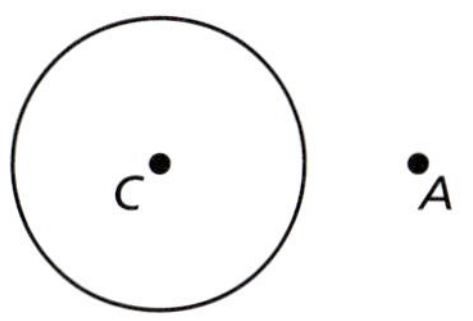

SOLUTION

Step 1

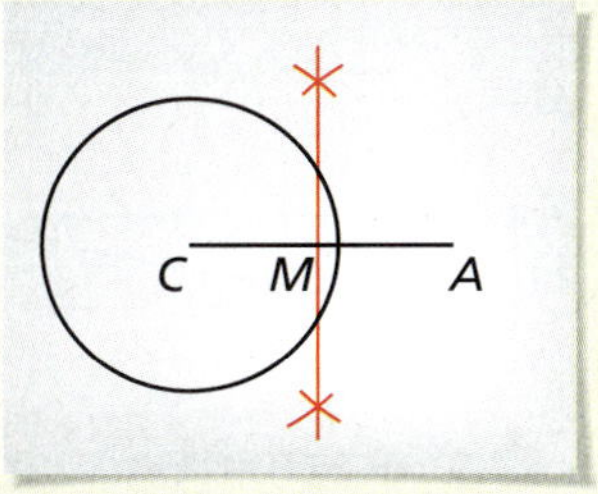

Find a midpoint
Draw $\overline{AC}$. Construct the bisector of the segment and label the midpoint M.

Step 2

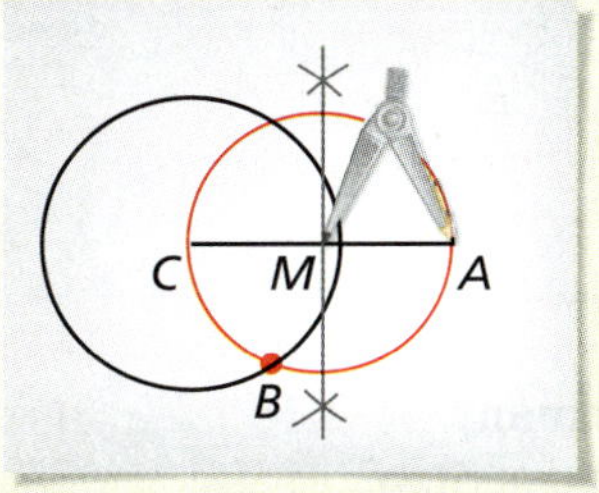

Draw a circle
Construct $\odot M$ with radius MA. Label one of the points where $\odot M$ intersects $\odot C$ as point B.

Step 3

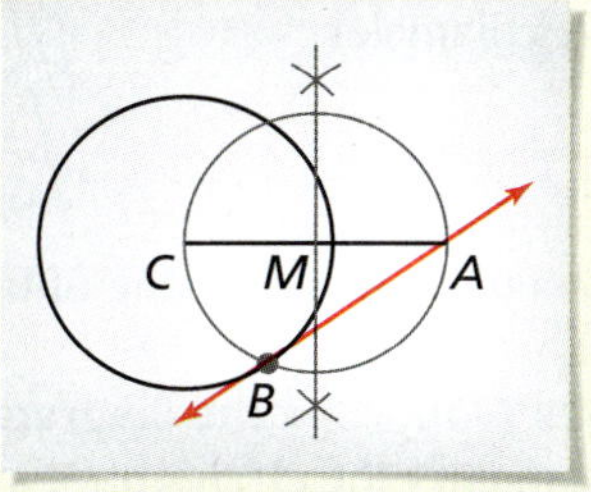

Construct a tangent line
Draw $\overleftrightarrow{AB}$. It is a tangent to $\odot C$ that passes through A.

EXAMPLE 5 Using Properties of Tangents

WATCH

$\overline{RS}$ is tangent to $\odot C$ at S, and $\overline{RT}$ is tangent to $\odot C$ at T. Find the value of x.

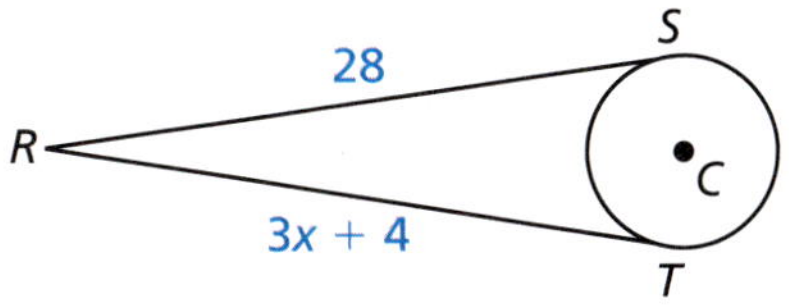

SOLUTION

Because $\overline{RS}$ and $\overline{RT}$ are tangent segments from a common external point, they are congruent by the External Tangent Congruence Theorem.

$RS = RT$	External Tangent Congruence Theorem
$28 = 3x + 4$	Substitute.
$8 = x$	Solve for x.

▶ The value of x is 8.

SELF-ASSESSMENT

1 I don't understand yet. 2 I can do it with help. 3 I can do it on my own. 4 I can teach someone else.

7. Is $\overline{DE}$ tangent to $\odot C$?

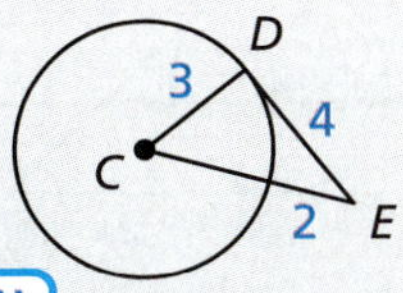

8. $\overline{ST}$ is tangent to $\odot Q$. Find the radius of $\odot Q$.

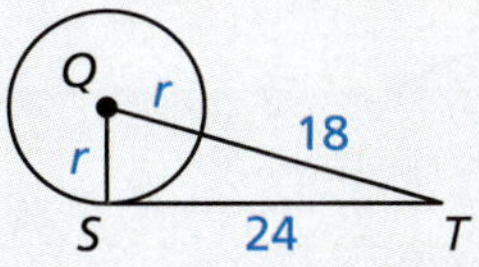

9. Points M and N are points of tangency. Find the value(s) of x.

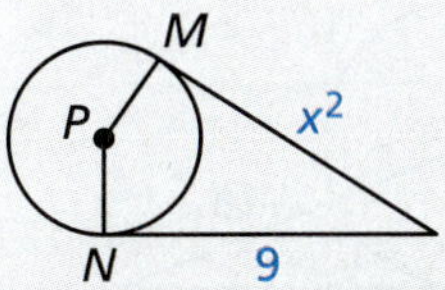

GO DIGITAL

10.1 Practice WITH CalcChat® AND CalcView®

In Exercises 1–6, use the diagram. *(See Example 1.)*

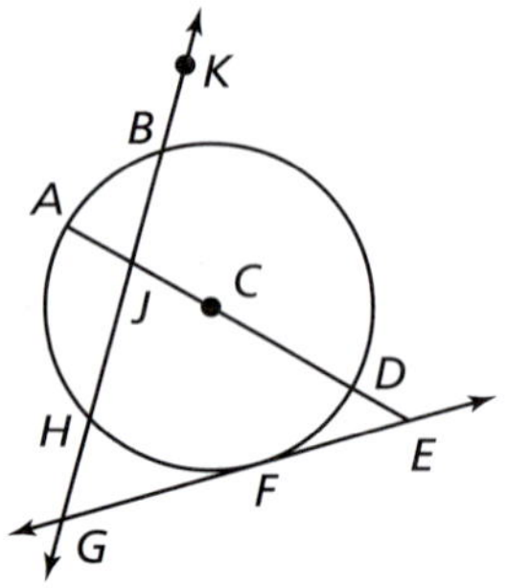

1. Name the circle.
2. Name two radii.
3. Name two chords.
4. Name a diameter.
5. Name a secant.
6. Name a tangent and a point of tangency.

In Exercises 7–10, copy the diagram. Tell how many common tangents the circles have and draw them. State whether the tangents are *external tangents* or *internal tangents*. *(See Example 2.)*

7. 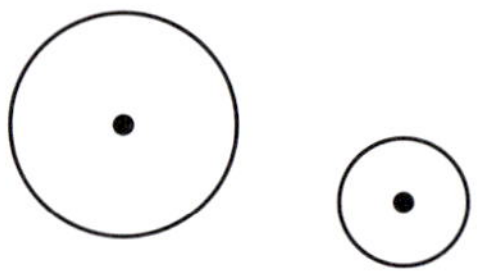

8. 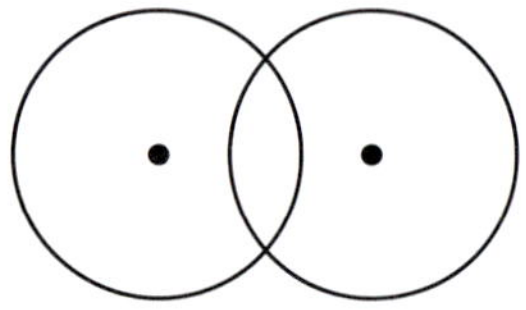

9. 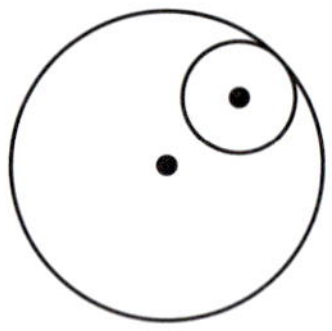

10. 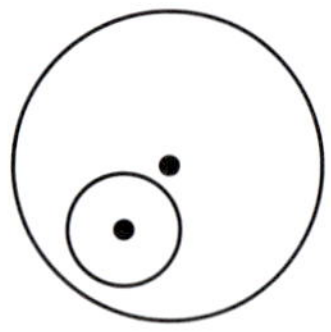

In Exercises 11–14, tell whether $\overline{AB}$ is tangent to $\odot C$. Explain your reasoning. *(See Example 3.)*

11. 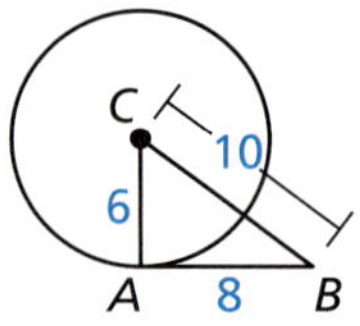

12. 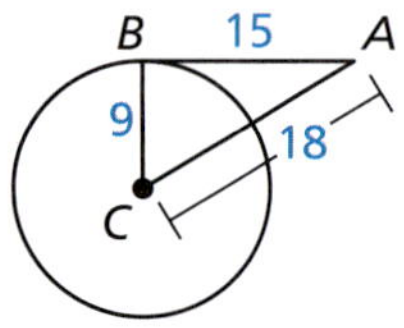

13. 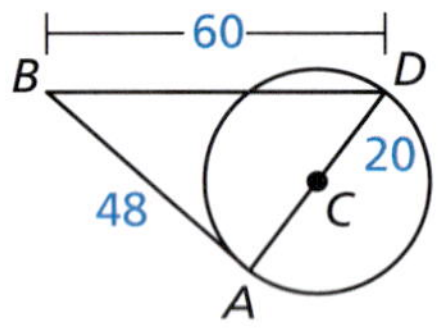

14. 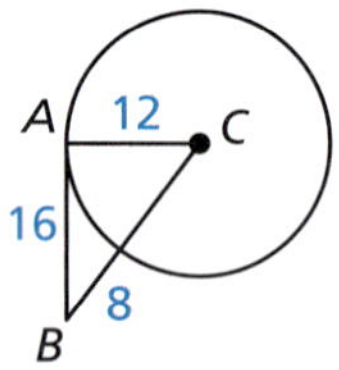

In Exercises 15–18, point *B* is a point of tangency. Find the radius *r* of $\odot C$. *(See Example 4.)*

15. 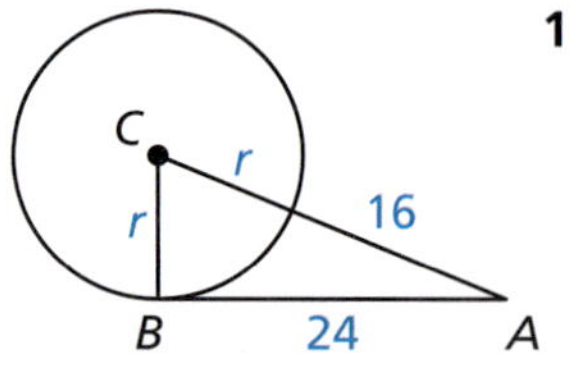

16. 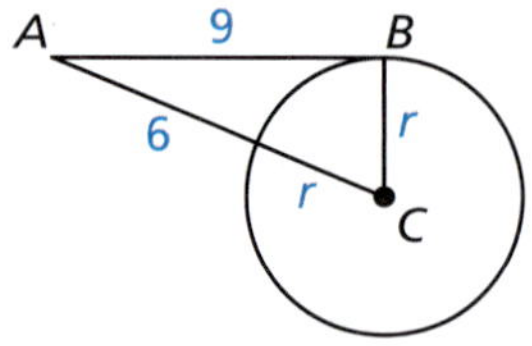

17. 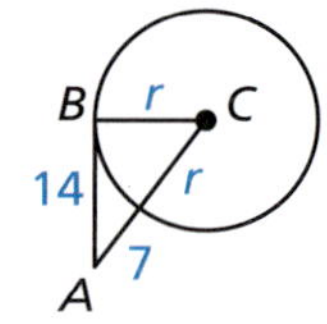

18. 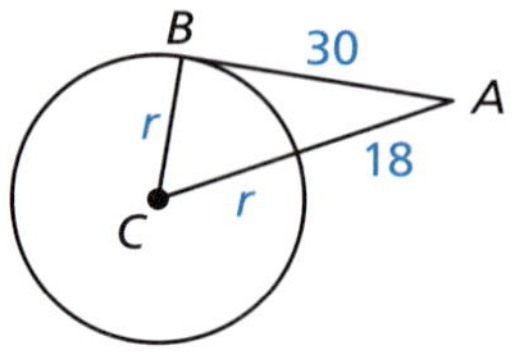

19. **ERROR ANALYSIS** Describe and correct the error in determining whether $\overline{XY}$ is tangent to $\odot Z$.

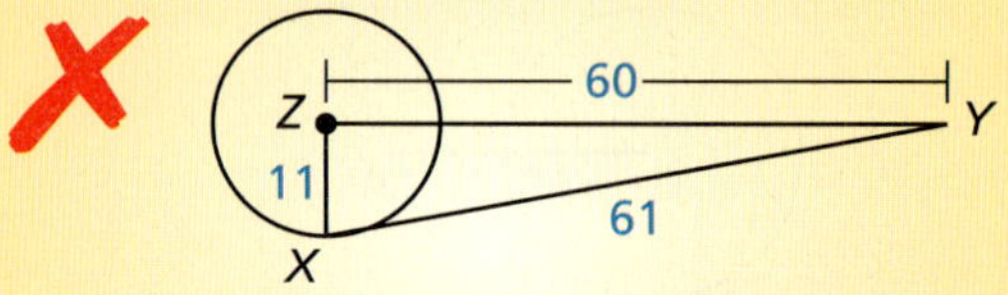

Because $11^2 + 60^2 = 61^2$, $\triangle XYZ$ is a right triangle. So, $\overline{XY}$ is tangent to $\odot Z$.

20. **COLLEGE PREP** Which statement(s) *cannot* be determined from the diagram?

Ⓐ $\overline{CD}$ is tangent to $\odot F$.
Ⓑ $\overline{AE} \cong \overline{ED}$
Ⓒ $\overline{AB} \cong \overline{AE}$
Ⓓ $\overline{FB} \cong \overline{FE}$

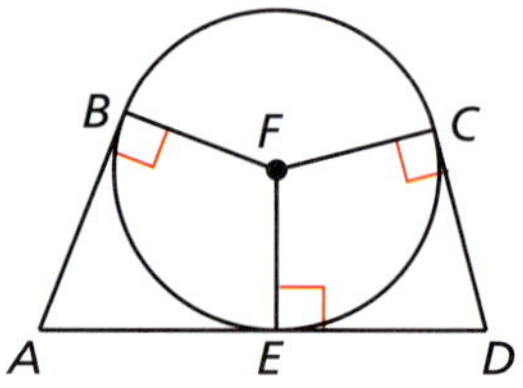

CONSTRUCTION In Exercises 21 and 22, construct $\odot C$ with the given radius and point *A* outside of $\odot C$. Then construct a line tangent to $\odot C$ that passes through *A*.

21. $r = 2$ in.
22. $r = 4.5$ cm

In Exercises 23–28, points *B* and *D* are points of tangency. Find the value(s) of *x*. *(See Example 5.)*

23. 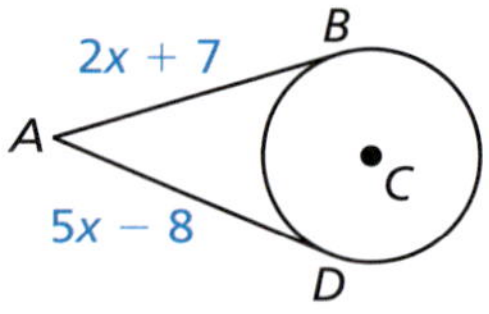

24. 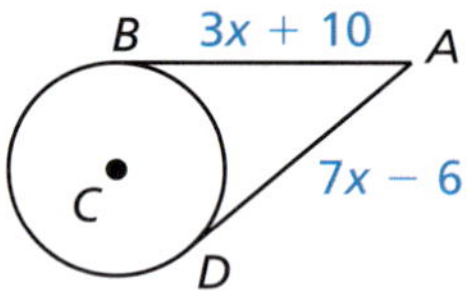

25. 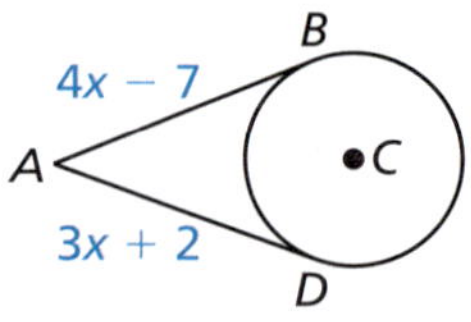

26. 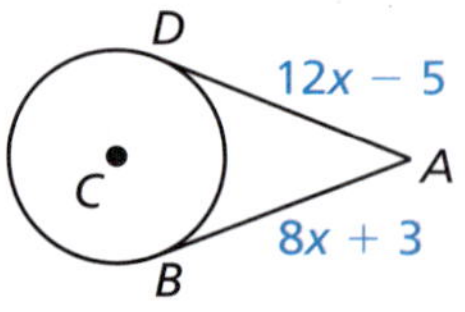

27. 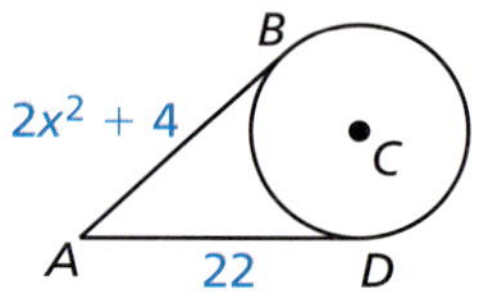

28. 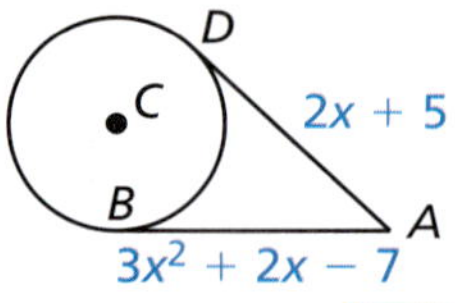

GO DIGITAL

29. REASONING For a point outside of a circle, how many lines exist tangent to the circle that pass through the point? How many such lines exist for a point on the circle? inside the circle? Explain your reasoning.

30. STRUCTURE When will two lines tangent to the same circle not intersect? Justify your answer.

31. WRITING Explain why the diameter of a circle is the longest chord of the circle.

32. HOW DO YOU SEE IT?
In the figure, $\overrightarrow{PA}$ is tangent to $\odot D$, $\overrightarrow{PC}$ is tangent to $\odot E$, and $\overrightarrow{PB}$ is a common internal tangent. How do you know that $\overline{PA} \cong \overline{PB} \cong \overline{PC}$?

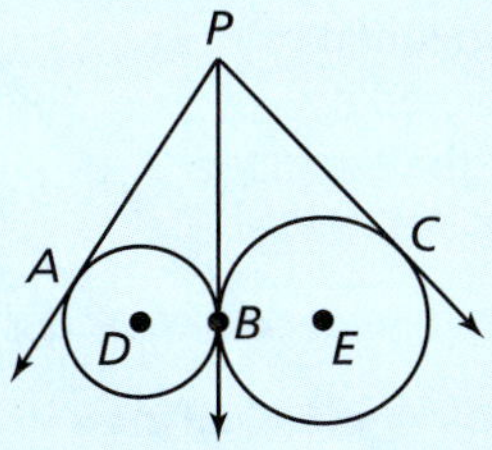

33. MAKING AN ARGUMENT $\overline{PQ}$ and $\overline{PR}$ are tangent to $\odot S$. Can you conclude that $\overline{PS}$ bisects $\angle QPR$? Explain your reasoning.

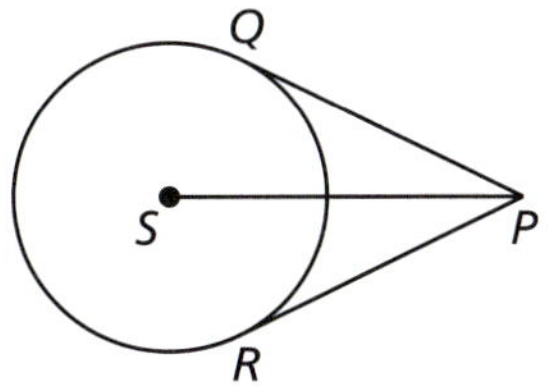

34. PATTERNS In $\odot C$, radii $\overline{CA}$ and $\overline{CB}$ are perpendicular. $\overleftrightarrow{BD}$ and $\overleftrightarrow{AD}$ are tangent to $\odot C$. What type of quadrilateral is $CADB$? Explain your reasoning.

35. MODELING REAL LIFE A bicycle chain is pulled tightly so that $\overline{MN}$ is a common tangent of the gears. Find the distance between the centers of the gears.

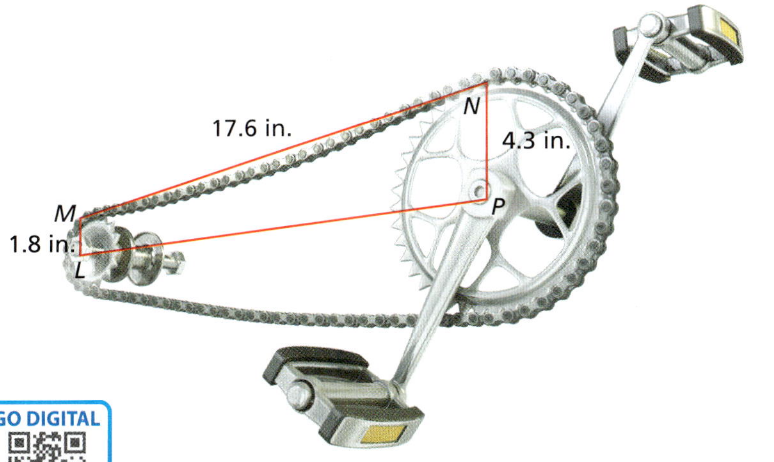

36. PROOF In the diagram, $\overline{RS}$ is a common internal tangent to $\odot A$ and $\odot B$. Prove that $\frac{AC}{BC} = \frac{RC}{SC}$.

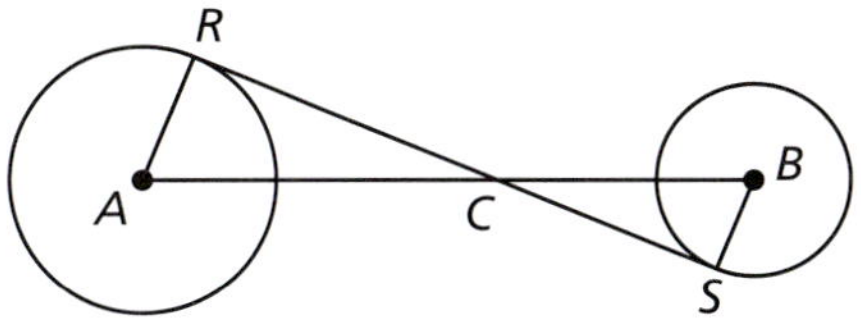

37. REASONING A polygon is *circumscribed* about a circle when every side of the polygon is tangent to the circle. In the diagram, quadrilateral $ABCD$ is circumscribed about $\odot Q$. Is it always true that $AB + CD = AD + BC$? Justify your answer.

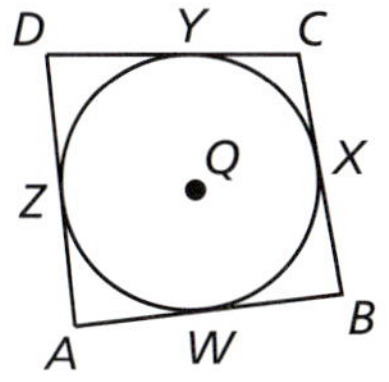

38. CONNECTING CONCEPTS Find the values of x and y.

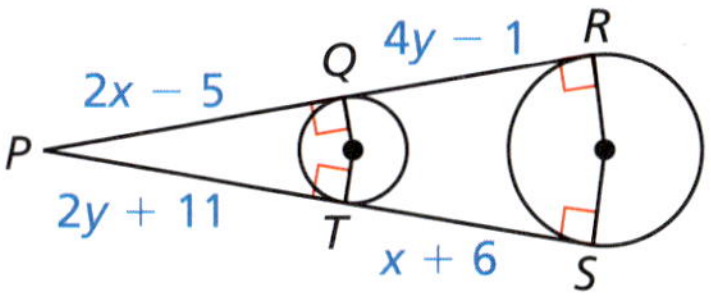

39. PROVING A THEOREM Prove the External Tangent Congruence Theorem.

Given $\overline{SR}$ and $\overline{ST}$ are tangent to $\odot P$.

Prove $\overline{SR} \cong \overline{ST}$

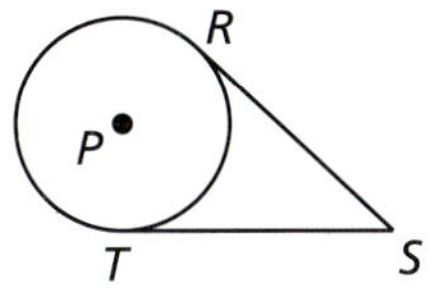

40. PROVING A THEOREM Use the diagram to prove each part of the biconditional in the Tangent Line to Circle Theorem.

a. Prove indirectly that if a line is tangent to a circle, then it is perpendicular to a radius.

Given Line m is tangent to $\odot Q$ at point P.

Prove $m \perp \overline{QP}$

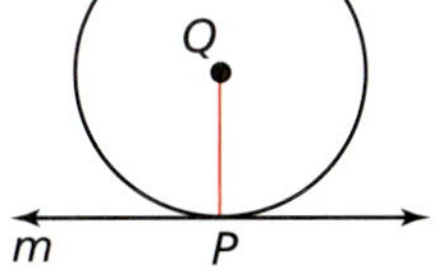

b. Prove indirectly that if a line is perpendicular to a radius at its endpoint, then the line is tangent to the circle.

Given $m \perp \overline{QP}$

Prove Line m is tangent to $\odot Q$.

GO DIGITAL

5 MTR **41.** **CONNECTING CONCEPTS** In the diagram, $\overline{VX}$ is a common external tangent to $\odot Y$ and $\odot Z$. Find VW.

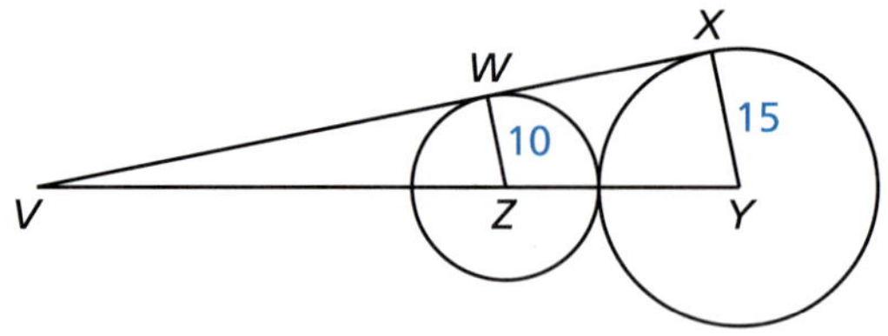

42. **THOUGHT PROVOKING**
In the diagram, $AB = AC = 12$, $BC = 8$, and all three segments are tangent to $\odot P$. What is the radius of $\odot P$? Justify your answer.

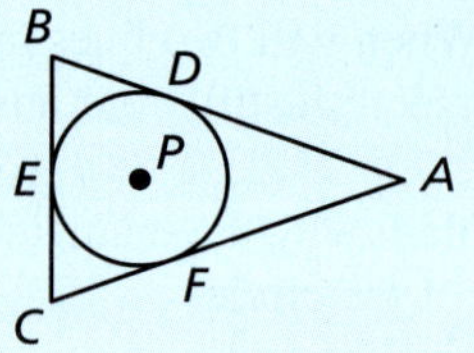

REVIEW & REFRESH

WATCH

In Exercises 43 and 44, find all unknown measures in the triangle.

43.

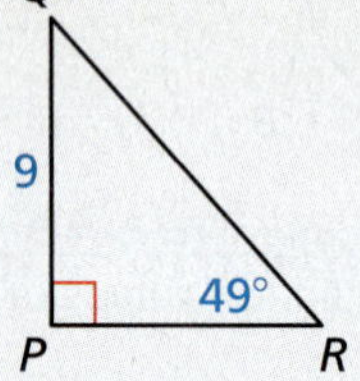

44.

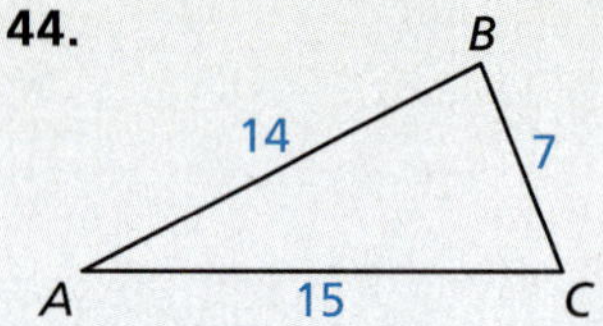

In Exercises 45 and 46, find the indicated measure.

45. $m\angle JKM$

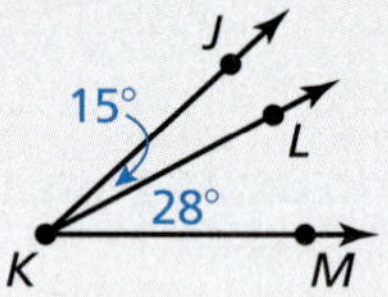

46. AB

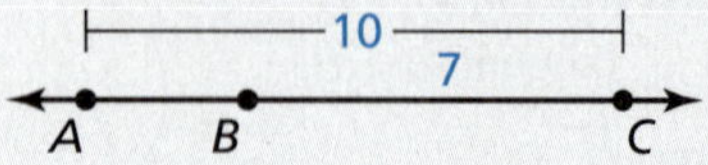

In Exercises 47 and 48, tell whether the lines through the given points are *parallel*, *perpendicular*, or *neither*.

47. **Line 1:** $(-5, -3)$, $(0, 1)$
Line 2: $(4, 2)$, $(8, 7)$

48. **Line 1:** $(-6, -2)$, $(-3, -6)$
Line 2: $(2, 1)$, $(8, -7)$

49. $\angle J$ and $\angle K$ are consecutive angles in a parallelogram, $m\angle J = (3x + 7)°$, and $m\angle K = (5x - 11)°$. Find the measure of each angle.

50. Points J and L are points of tangency. Find the value of x.

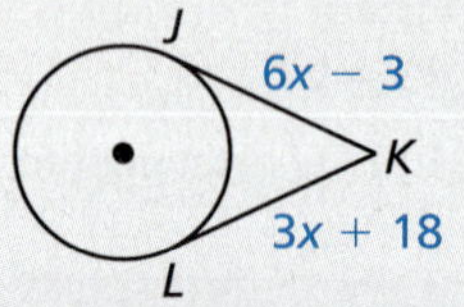

7 MTR **51.** **MODELING REAL LIFE**
Find the horizontal distance covered by the steps of the escalator.

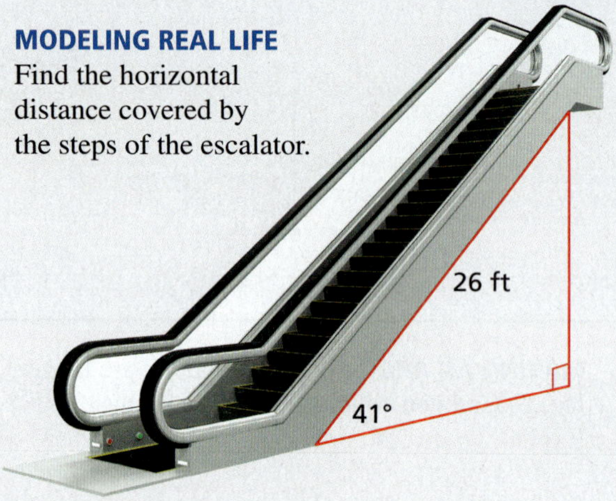

52. Find the value of x.

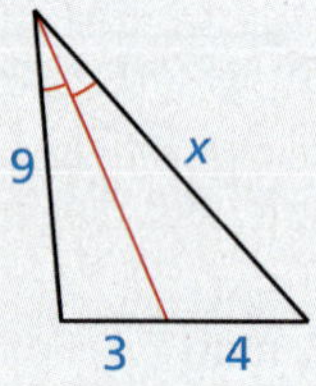

53. Find the length of the midsegment of the trapezoid.

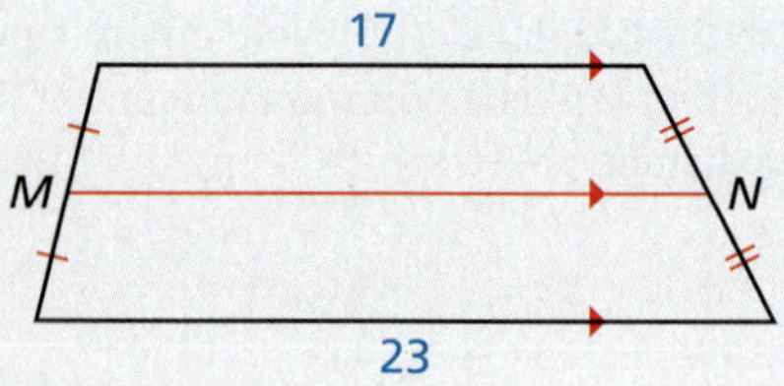

54. Point P is the centroid of $\triangle LMN$. Find LP and PQ when $LQ = 36$.

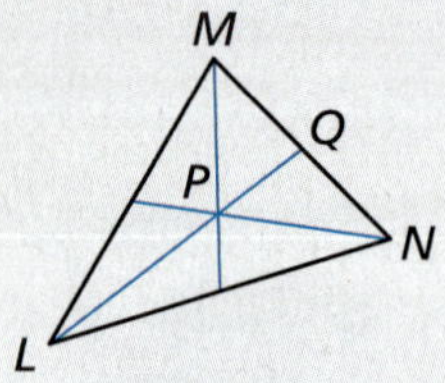

GO DIGITAL

10.2 Finding Arc Measures

Learning Target: Understand arc measures and similar circles.

Success Criteria:
- I can find arc measures.
- I can identify congruent arcs.
- H I can prove that all circles are similar.

A **central angle** of a circle is an angle that is formed by two radii and has a vertex at the center of the circle. A *circular arc* is a portion of a circle that is between two radii. The measure of a circular arc is the measure of its central angle.

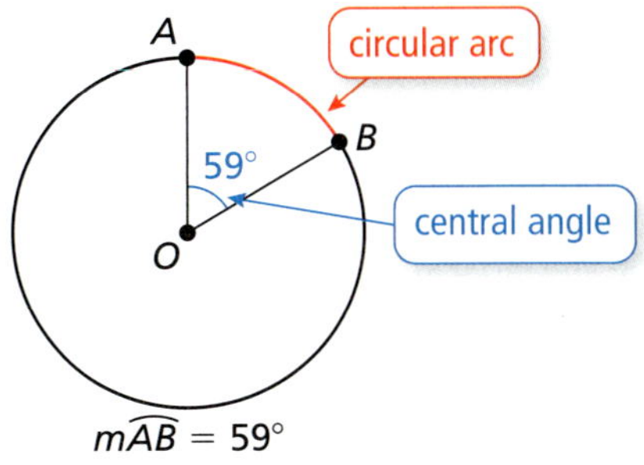

EXPLORE IT! Measuring Circular Arcs

Work with a partner. In each Ferris wheel shown, the passenger cars are equally spaced.

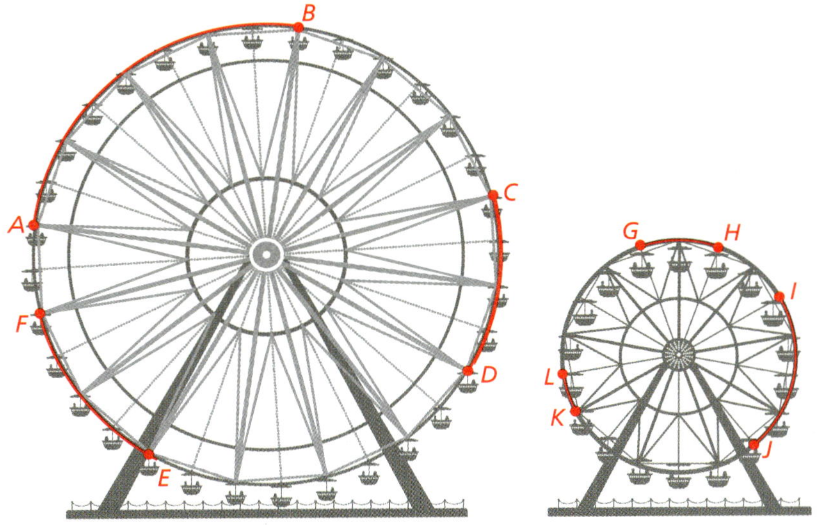

4 MTR CONSTRUCT AN ARGUMENT

Can two circular arcs have the same *measure*, but different *lengths*? Explain.

a. Compare the Ferris wheels. Do they represent *congruent circles*? Explain.

b. Find the measure of each circular arc shown. Explain your reasoning.

c. Are any of the arcs *similar arcs*? *congruent arcs*? Explain.

d. Does the size of the circle affect the measure of a circular arc? Explain your reasoning.

Geometric Reasoning

MA.912.GR.6.2 Solve mathematical and real-world problems involving the measures of arcs and related angles.

H **MA.912.GR.6.5** Apply transformations to prove that all circles are similar.

GO DIGITAL

Finding Arc Measures

Vocabulary

AZ VOCAB

central angle, *p. 522*
minor arc, *p. 522*
major arc, *p. 522*
semicircle, *p. 522*
measure of a minor arc, *p. 522*
measure of a major arc, *p. 522*
adjacent arcs, *p. 523*
congruent circles, *p. 524*
congruent arcs, *p. 524*
similar arcs, *p. 525*

A **central angle** of a circle is an angle that is formed by two radii and has a vertex at the center of the circle. In the diagram, $\angle ACB$ is a central angle of $\odot C$.

If $m\angle ACB$ is less than 180°, then the points on $\odot C$ that lie in the interior of $\angle ACB$ form a **minor arc** with endpoints A and B. The points on $\odot C$ that do not lie on the minor arc AB form a **major arc** with endpoints A and B. A **semicircle** is an arc with endpoints that are the endpoints of a diameter.

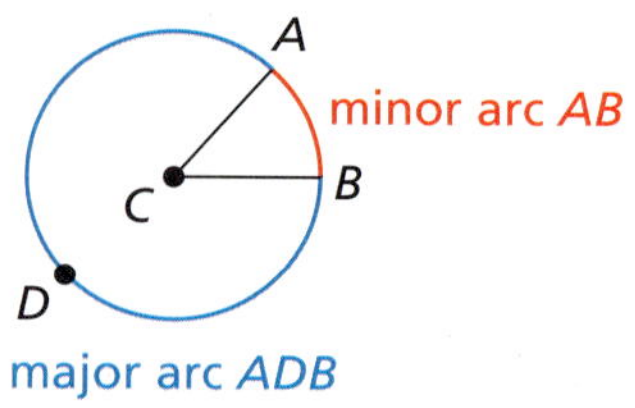

Minor arcs are named by their endpoints. The minor arc associated with $\angle ACB$ is named $\overparen{AB}$. Major arcs and semicircles are named by their endpoints and a point on the arc. The major arc associated with $\angle ACB$ can be named $\overparen{ADB}$.

STUDY TIP

The measure of a minor arc is less than 180°. The measure of a major arc is greater than 180°. The measure of a semicircle is equal to 180°.

KEY IDEA

Measuring Arcs

The **measure of a minor arc** is the measure of its central angle. The expression $m\overparen{AB}$ is read as "the measure of arc AB."

The measure of the entire circle is 360°. The **measure of a major arc** is the difference of 360° and the measure of the related minor arc. The measure of a semicircle is 180°.

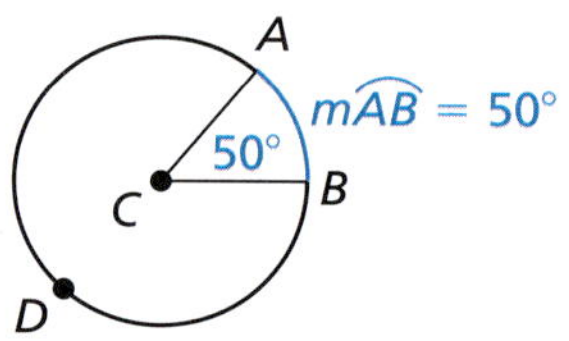

$m\overparen{ADB} = 360° - 50° = 310°$

EXAMPLE 1 Finding Measures of Arcs

WATCH

Find the measure of each arc of $\odot P$, where $\overline{RT}$ is a diameter.

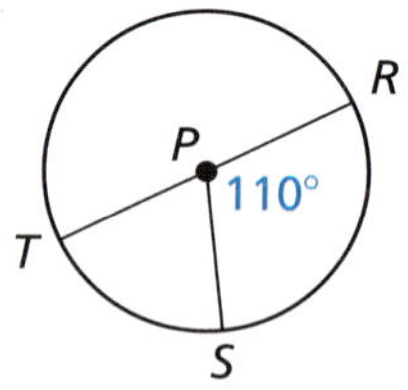

a. $\overparen{RS}$ **b.** $\overparen{RTS}$ **c.** $\overparen{RST}$

SOLUTION

a. $\overparen{RS}$ is a minor arc, so $m\overparen{RS} = m\angle RPS = 110°$.

b. $\overparen{RTS}$ is a major arc, so $m\overparen{RTS} = 360° - m\overparen{RS} = 360° - 110° = 250°$.

c. $\overline{RT}$ is a diameter, so $\overparen{RST}$ is a semicircle, and $m\overparen{RST} = 180°$.

GO DIGITAL

Two arcs of the same circle are **adjacent arcs** when they intersect at exactly one point. You can add the measures of two adjacent arcs.

POSTULATE

10.1 Arc Addition Postulate

The measure of an arc formed by two adjacent arcs is the sum of the measures of the two arcs.

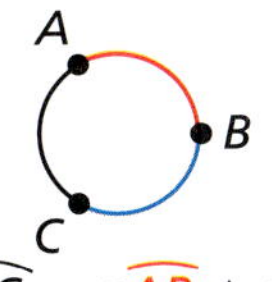

$m\overset{\frown}{ABC} = m\overset{\frown}{AB} + m\overset{\frown}{BC}$

EXAMPLE 2 Using the Arc Addition Postulate

WATCH

Find the measure of each arc.

a. $\overset{\frown}{GE}$ **b.** $\overset{\frown}{GEF}$ **c.** $\overset{\frown}{GF}$

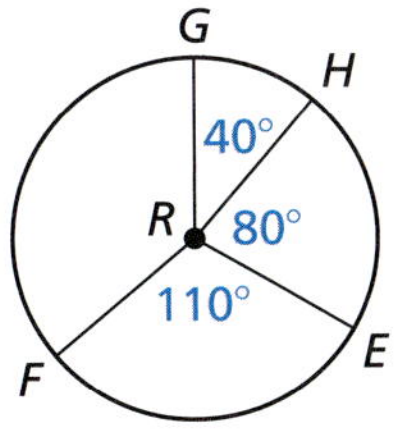

SOLUTION

a. $m\overset{\frown}{GE} = m\overset{\frown}{GH} + m\overset{\frown}{HE} = 40° + 80° = 120°$

b. $m\overset{\frown}{GEF} = m\overset{\frown}{GE} + m\overset{\frown}{EF} = 120° + 110° = 230°$

c. $m\overset{\frown}{GF} = 360° - m\overset{\frown}{GEF} = 360° - 230° = 130°$

EXAMPLE 3 Finding Measures of Arcs

WATCH

The circle graph shows the percents of international visitors to Florida by world region in a recent year. Find the indicated arc measures.

a. $m\overset{\frown}{AC}$

b. $m\overset{\frown}{ACD}$

International Visitors to Florida

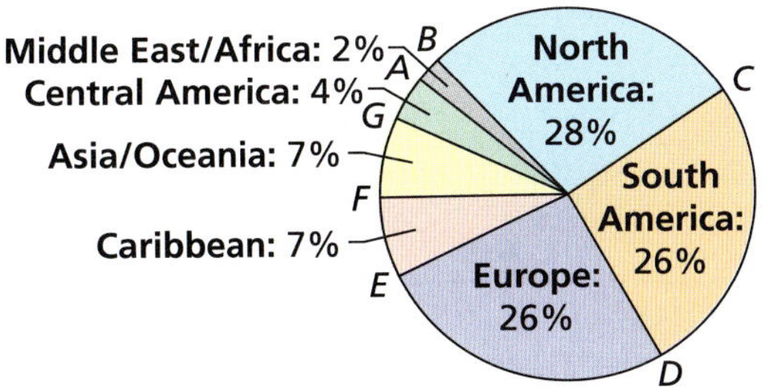

SOLUTION

a. By the Arc Addition Postulate, $m\overset{\frown}{AC} = m\overset{\frown}{AB} + m\overset{\frown}{BC}$.

$\overset{\frown}{AB}$ represents 2% of the circle, so $m\overset{\frown}{AB} = 0.02(360°) = 7.2°$.

$\overset{\frown}{BC}$ represents 28% of the circle, so $m\overset{\frown}{BC} = 0.28(360°) = 100.8°$.

▶ So, $m\overset{\frown}{AC} = 7.2° + 100.8° = 108°$.

b. By the Arc Addition Postulate, $m\overset{\frown}{ACD} = m\overset{\frown}{AC} + m\overset{\frown}{CD}$.

$\overset{\frown}{CD}$ represents 26% of the circle, so $m\overset{\frown}{BC} = 0.26(360°) = 93.6°$.

▶ So, $m\overset{\frown}{ACD} = 108° + 93.6° = 201.6°$.

5 MTR USE STRUCTURE

Show how you can solve part (b) using $m\overset{\frown}{DA}$. Explain your reasoning.

SELF-ASSESSMENT

1 I don't understand yet. 2 I can do it with help. 3 I can do it on my own. 4 I can teach someone else.

Identify the given arc as a *major arc*, *minor arc*, or *semicircle*. Then find the measure of the arc.

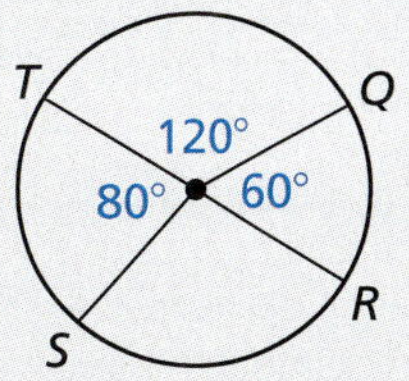

1. $\overset{\frown}{TQ}$ **2.** $\overset{\frown}{QRT}$ **3.** $\overset{\frown}{TQR}$

4. In Example 3, find $m\overset{\frown}{EBD}$.

GO DIGITAL

Identifying Congruent Arcs

Two circles are **congruent circles** if and only if a rigid motion or a composition of rigid motions maps one circle onto the other. This statement is equivalent to the Congruent Circles Theorem below.

Two arcs are **congruent arcs** if and only if they have the same measure and they are arcs of the same circle or of congruent circles.

THEOREMS

10.3 Congruent Circles Theorem

Two circles are congruent circles if and only if they have the same radius.

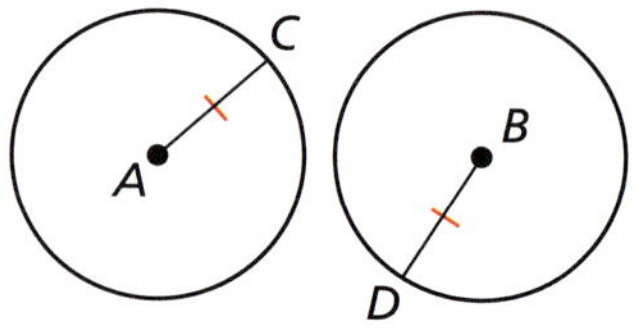

Prove this Theorem Exercise 31, page 528

$\odot A \cong \odot B$ if and only if $\overline{AC} \cong \overline{BD}$.

10.4 Congruent Central Angles Theorem

In the same circle, or in congruent circles, two minor arcs are congruent if and only if their corresponding central angles are congruent.

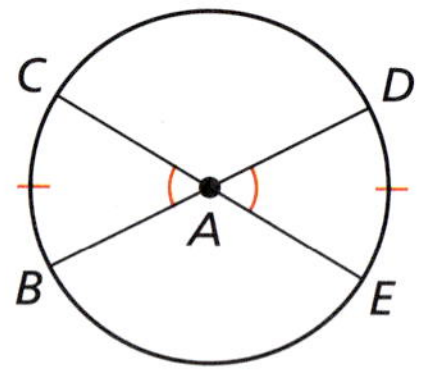

$\overset{\frown}{BC} \cong \overset{\frown}{DE}$ if and only if $\angle BAC \cong \angle DAE$.

Prove this Theorem Exercise 32, page 528

EXAMPLE 4 Identifying Congruent Arcs

WATCH

Tell whether the indicated arcs are congruent. Explain why or why not.

a. $\overset{\frown}{RS}$ and $\overset{\frown}{TU}$

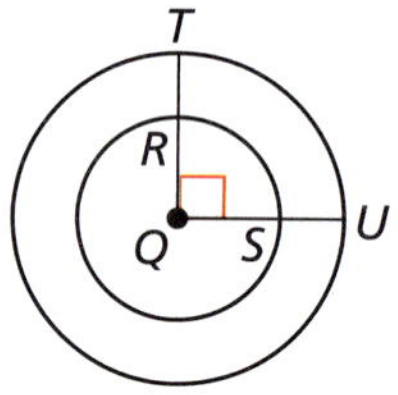

b. $\overset{\frown}{UV}$ and $\overset{\frown}{YZ}$

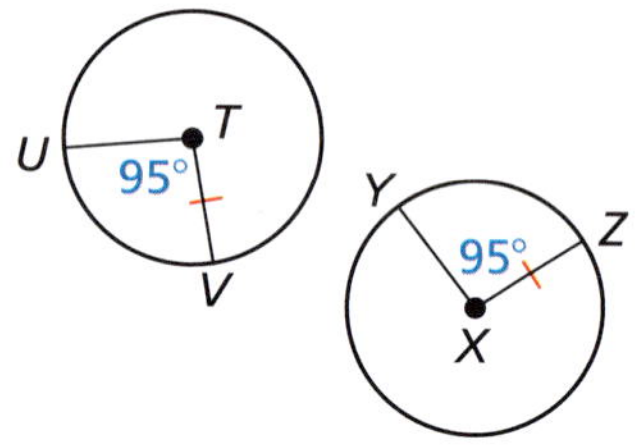

5 MTR USE STRUCTURE
Use a different theorem to show that the arcs in part (b) are congruent. Explain.

SOLUTION

a. $\overset{\frown}{RS}$ and $\overset{\frown}{TU}$ have the same measure, but are not congruent because they are arcs of circles that are not congruent.

b. $\overset{\frown}{UV} \cong \overset{\frown}{YZ}$ by the Congruent Central Angles Theorem because they are arcs of congruent circles and they have congruent central angles, $\angle UTV \cong \angle YXZ$.

SELF-ASSESSMENT

1 I don't understand yet. 2 I can do it with help. 3 I can do it on my own. 4 I can teach someone else.

Tell whether the indicated arcs are congruent. Explain why or why not.

5. $\overset{\frown}{AB}$ and $\overset{\frown}{CD}$

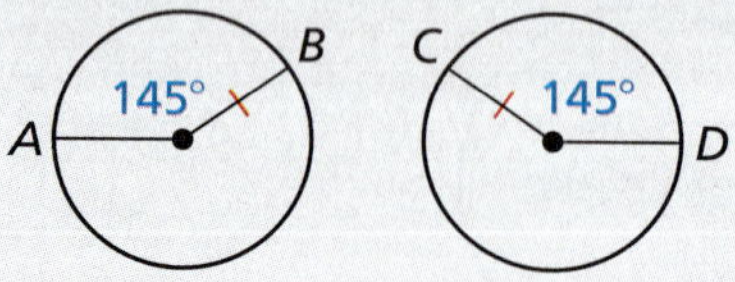

6. $\overset{\frown}{MN}$ and $\overset{\frown}{PQ}$

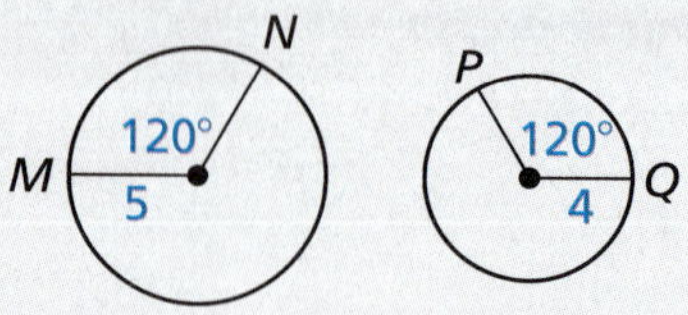

7. $\overset{\frown}{CD}$ and $\overset{\frown}{EF}$

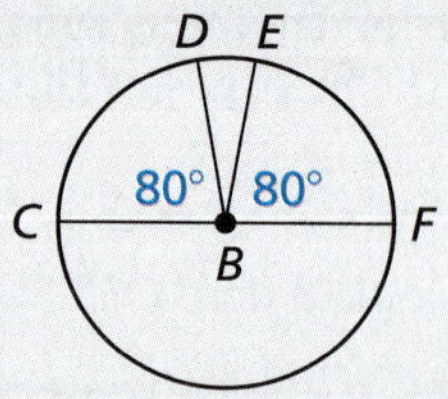

GO DIGITAL

Proving Circles Are Similar

THEOREM

10.5 Similar Circles Theorem

All circles are similar.

Prove this Theorem Exercise 8, page 525

PROOF Similar Circles Theorem H

All circles are similar.

Given $\odot C$ with center C and radius r,
$\odot D$ with center D and radius s

Prove $\odot C \sim \odot D$

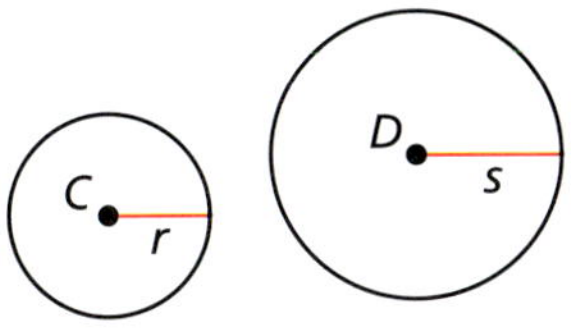

First, translate $\odot C$ so that point C maps to point D. The image of $\odot C$ is $\odot C'$ with center D. So, $\odot C'$ and $\odot D$ are concentric circles.

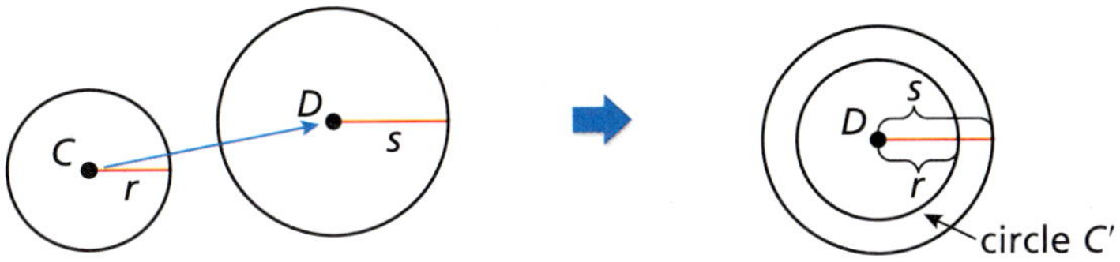

$\odot C'$ is the set of all points that are r units from point D. Dilate $\odot C'$ using center of dilation D and scale factor $\frac{s}{r}$.

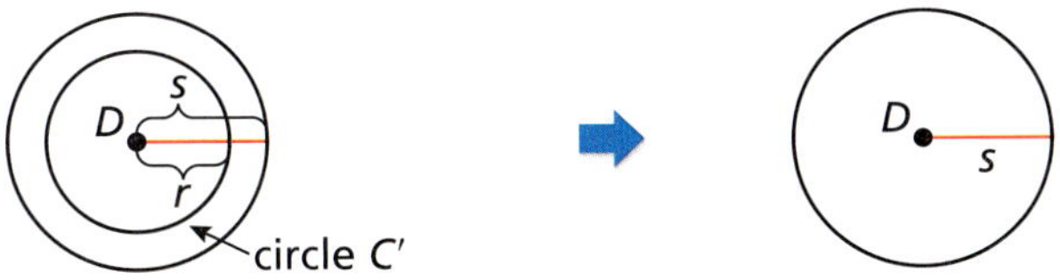

This dilation maps the set of all the points that are r units from point D to the set of all points that are $\frac{s}{r}(r) = s$ units from point D. $\odot D$ is the set of all points that are s units from point D. So, this dilation maps $\odot C'$ to $\odot D$.

Because a similarity transformation maps $\odot C$ to $\odot D$, $\odot C \sim \odot D$.

Two arcs are **similar arcs** if and only if they have the same measure. All congruent arcs are similar, but not all similar arcs are congruent. For instance, in Example 4, the pairs of arcs in parts (a) and (b) are similar, but only the arcs in part (b) are congruent.

SELF-ASSESSMENT

1 I don't understand yet. 2 I can do it with help. 3 I can do it on my own. 4 I can teach someone else.

8. Write a coordinate proof of the Similar Circles Theorem.

Given $\odot O$ with center $O(0, 0)$ and radius r,
$\odot A$ with center $A(a, b)$ and radius s

Prove $\odot O \sim \odot A$

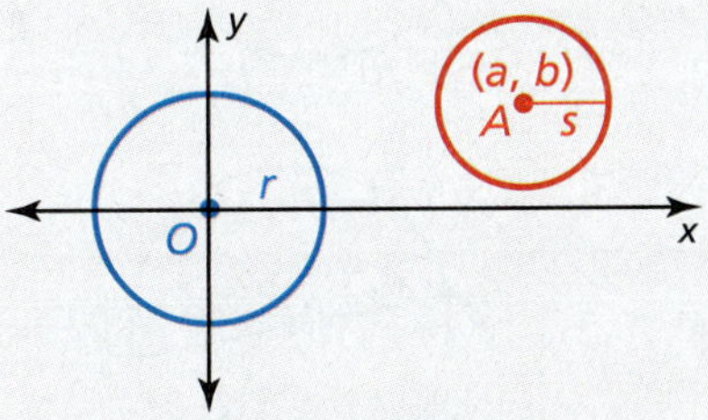

GO DIGITAL

10.2 Practice WITH CalcChat® AND CalcView®

In Exercises 1–6, name the minor arc and find its measure. Then name the major arc and find its measure.

1. 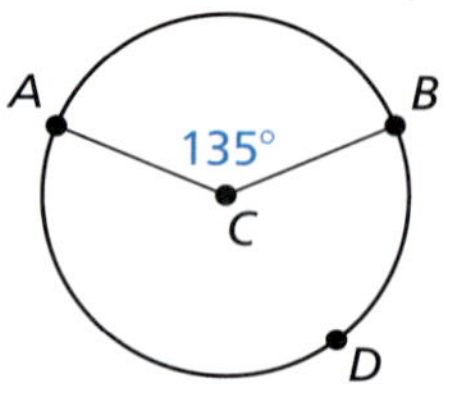

2. 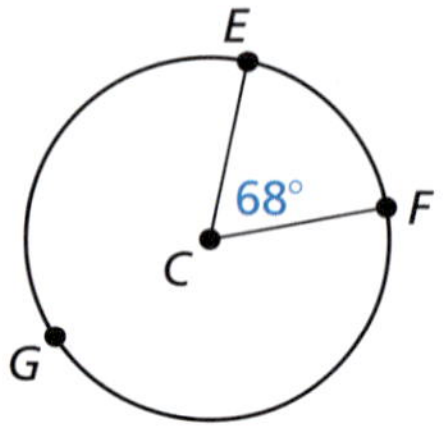

3. 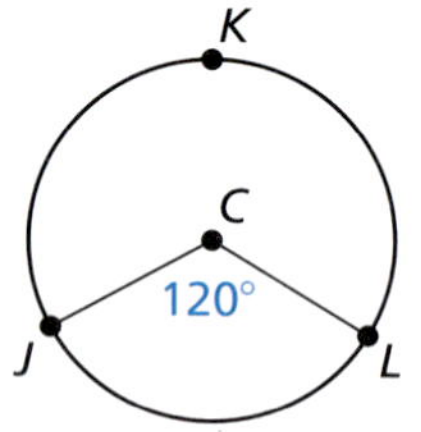

4. 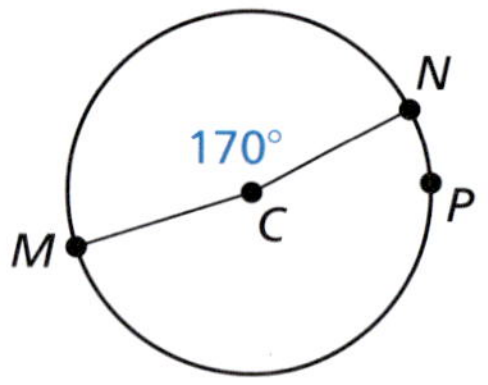

5. 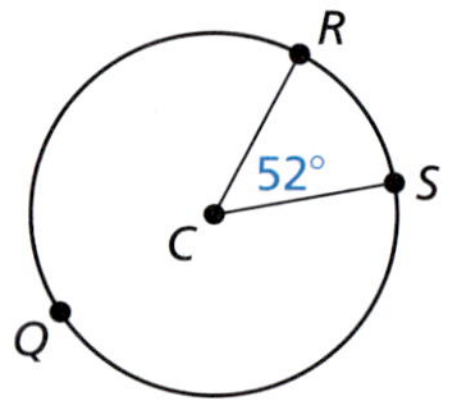

6. 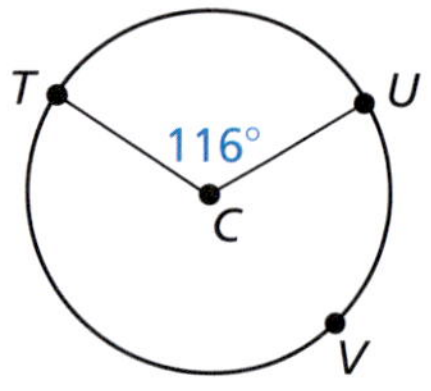

In Exercises 7–10, identify the given arc as a *major arc*, *minor arc*, or *semicircle*. Then find the measure of the arc. *(See Example 1.)*

▶ 7. $\overset{\frown}{AC}$

8. $\overset{\frown}{AB}$

9. $\overset{\frown}{FG}$

10. $\overset{\frown}{EG}$

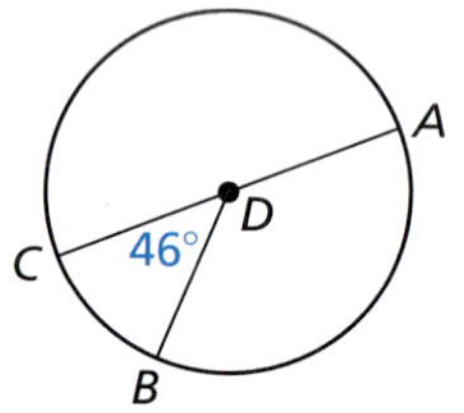

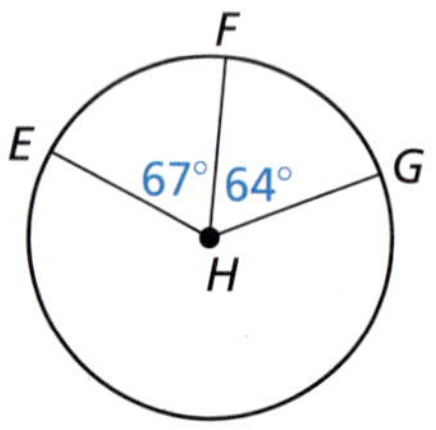

In Exercises 11 and 12, find the measure of each arc. *(See Example 2.)*

▶ 11. a. $\overset{\frown}{JL}$

b. $\overset{\frown}{KM}$

c. $\overset{\frown}{JLM}$

d. $\overset{\frown}{JM}$

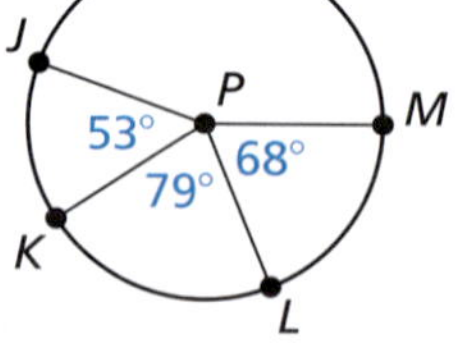

12. a. $\overset{\frown}{RS}$

b. $\overset{\frown}{QRS}$

c. $\overset{\frown}{QST}$

d. $\overset{\frown}{QT}$

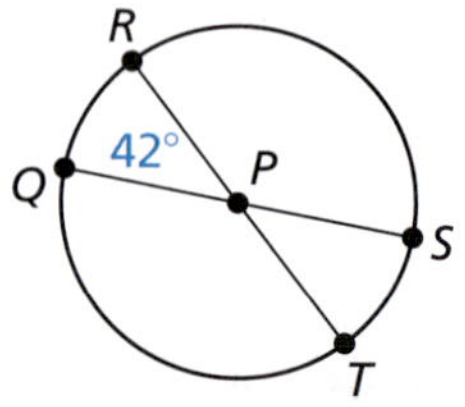

7 MTR 13. **MODELING REAL LIFE** A study investigated land use in Florida. The results are shown in the circle graph. Find each indicated arc measure. *(See Example 3.)*

Florida Land Use

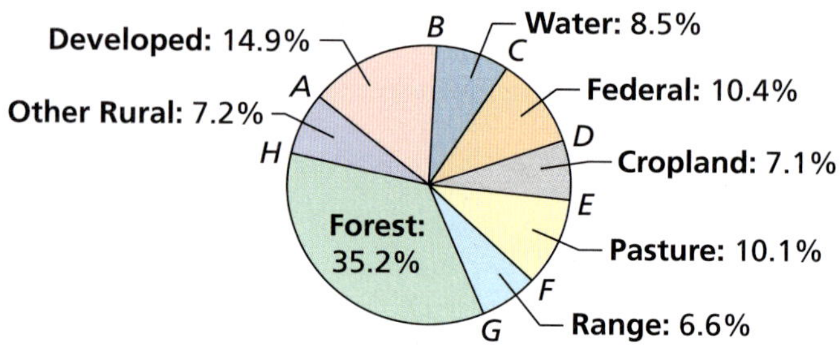

a. $m\overset{\frown}{AC}$ b. $m\overset{\frown}{AE}$ c. $m\overset{\frown}{GDH}$

d. $m\overset{\frown}{BHC}$ e. $m\overset{\frown}{FD}$ f. $m\overset{\frown}{FBD}$

14. **REASONING** The circle graph shows the percentages of students enrolled in fall sports at a high school. Is it possible to find the measure of each minor arc? If so, find the measure of the arc for each category shown. If not, explain why it is not possible.

High School Fall Sports

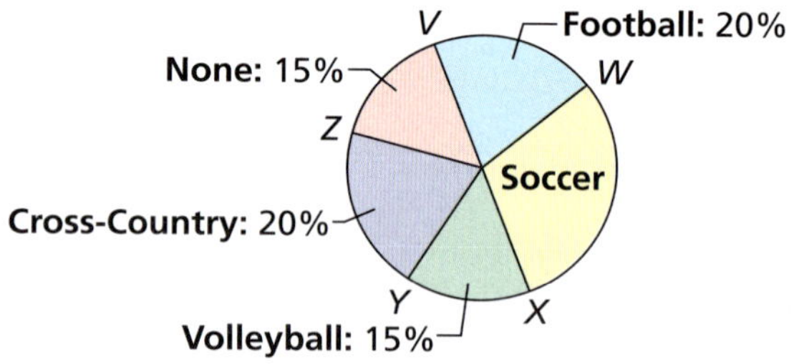

In Exercises 15–18, tell whether the red arcs are congruent. Explain why or why not. *(See Example 4.)*

▶ 15. 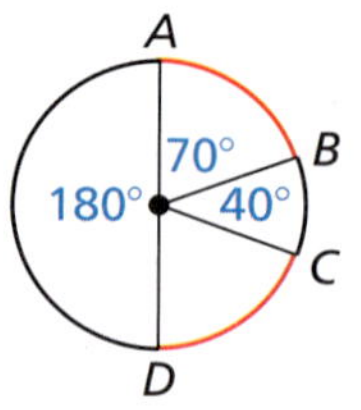

16. 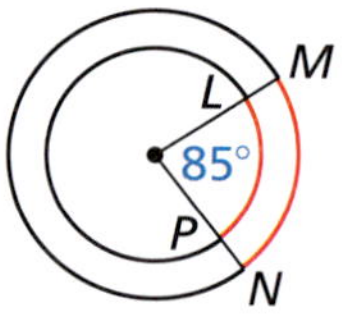

17. 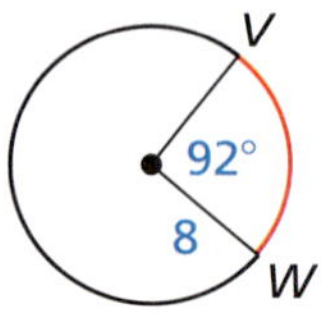

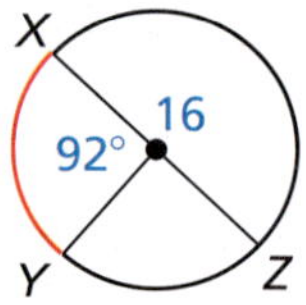

18. 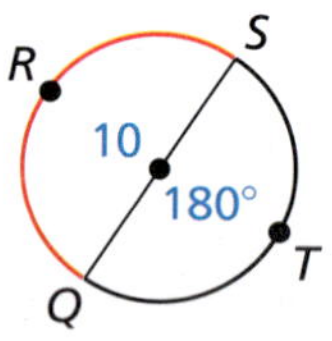

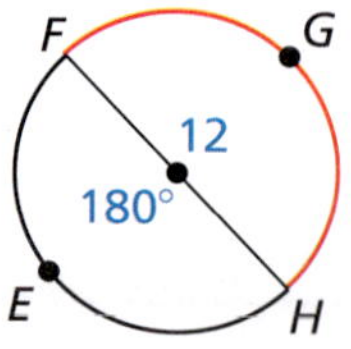

GO DIGITAL

4 MTR **19. ERROR ANALYSIS** Describe and correct the error in naming the red arc.

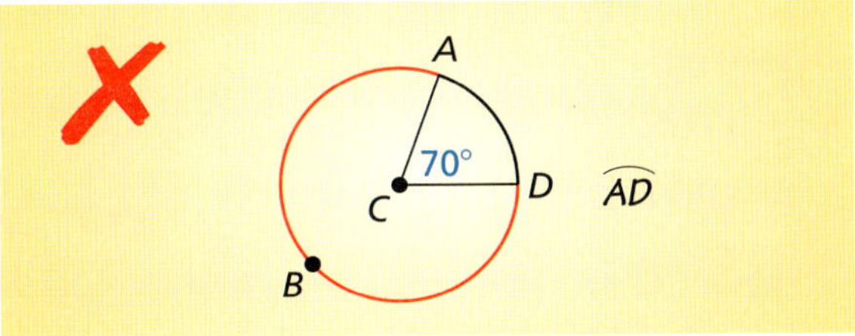

4 MTR **20. ERROR ANALYSIS** Describe and correct the error in naming congruent arcs.

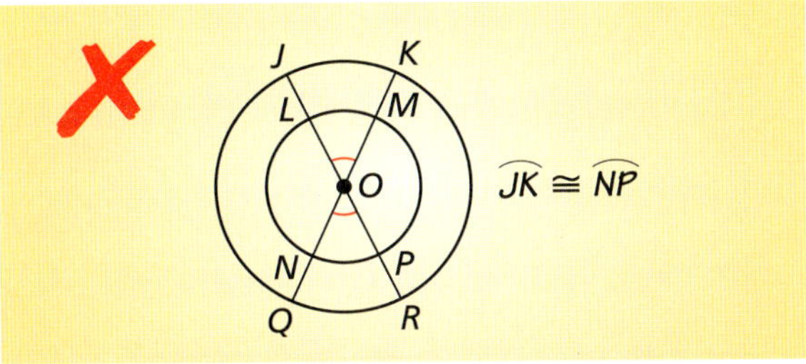

5 MTR **CONNECTING CONCEPTS In Exercises 21 and 22, find the value of *x*. Then find the measure of the red arc.**

21.

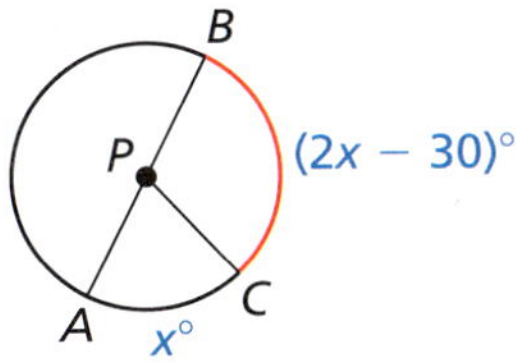

22.

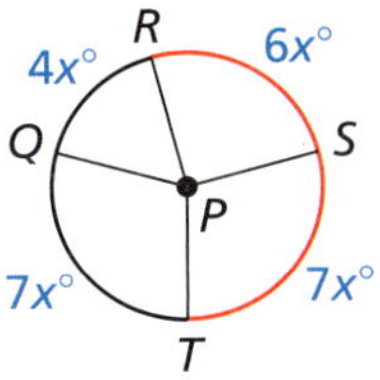

4 MTR **23. MAKING AN ARGUMENT** Your friend says that because $m\widehat{AN}$ is not given, you cannot find the value of x. Is your friend correct? Explain your reasoning.

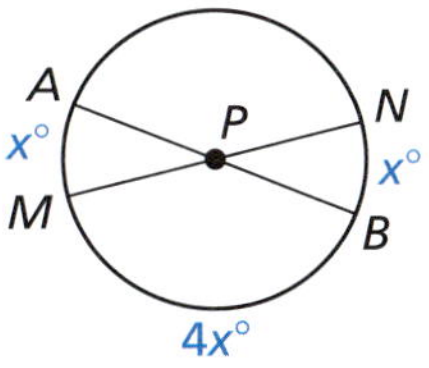

24. HOW DO YOU SEE IT?
Are the circles on the target *similar* or *congruent*? Explain your reasoning.

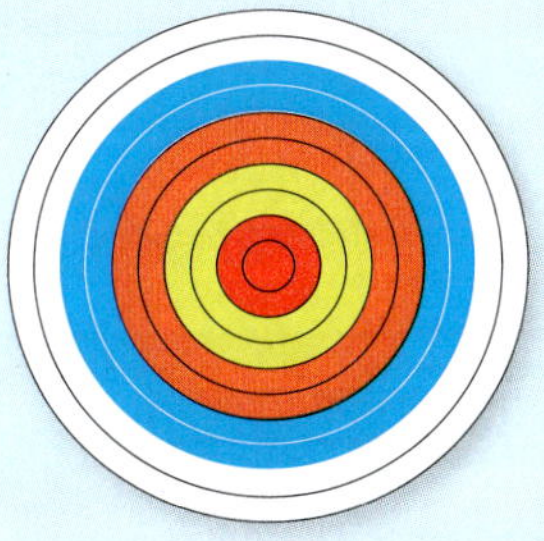

GO DIGITAL

3 MTR **25. MAINTAIN ACCURACY** Two diameters of $\odot P$ are $\overline{AB}$ and $\overline{CD}$. Find $m\widehat{ACD}$ and $m\widehat{AC}$ when $m\widehat{AD} = 20°$.

26. REASONING In $\odot R$, $m\widehat{AB} = 60°$, $m\widehat{BC} = 25°$, $m\widehat{CD} = 70°$, and $m\widehat{DE} = 20°$. Find two possible measures of $\widehat{AE}$.

7 MTR **27. MODELING REAL LIFE**
On a regulation dartboard, the outermost circle is divided into 20 congruent sections. What is the measure of each arc in this circle?

7 MTR **28. MODELING REAL LIFE**
You can use the time zone wheel to find the time in different locations around the world. For example, to find the time in Tokyo when it is 4 P.M. in San Francisco, rotate the small wheel until 4 P.M. and San Francisco line up, as shown. Then look at Tokyo to see that it is 9 A.M. there.

a. What is the arc measure between each time zone on the wheel?

b. What is the measure of the minor arc from the Tokyo zone to the Anchorage zone?

c. If two locations differ by 180° on the wheel, and it is 3 P.M. at one location, what time is it at the other location?

4 MTR **29. DISCUSS MATHEMATICAL THINKING**
Is there enough information to tell whether $\odot C \cong \odot D$? Explain your reasoning.

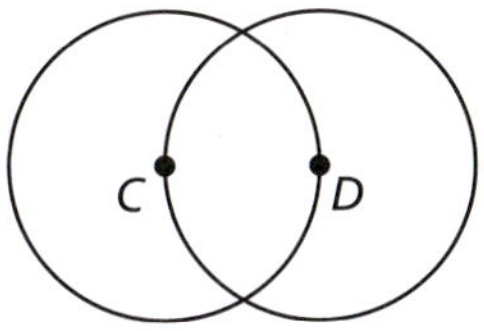

30. THOUGHT PROVOKING Write a formula for the length of a circular arc. Justify your answer.

31. PROVING A THEOREM Use the diagram on page 524 to prove each part of the biconditional in the Congruent Circles Theorem.

a. Given $\overline{AC} \cong \overline{BD}$
Prove $\odot A \cong \odot B$

b. Given $\odot A \cong \odot B$
Prove $\overline{AC} \cong \overline{BD}$

32. PROVING A THEOREM Use the diagram to prove each part of the biconditional in the Congruent Central Angles Theorem.

a. Given $\angle BAC \cong \angle DAE$
Prove $\overparen{BC} \cong \overparen{DE}$

b. Given $\overparen{BC} \cong \overparen{DE}$
Prove $\angle BAC \cong \angle DAE$

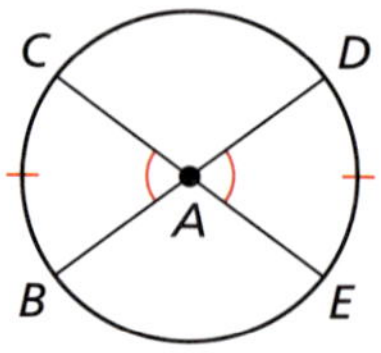

REVIEW & REFRESH

WATCH

33. Points B and D are points of tangency. Find the value(s) of x.

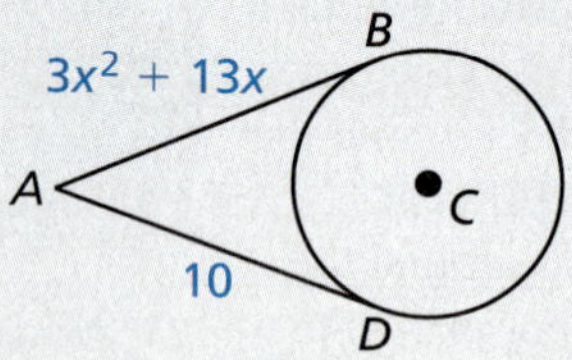

In Exercises 34 and 35, find all unknown measures in the triangle.

34.

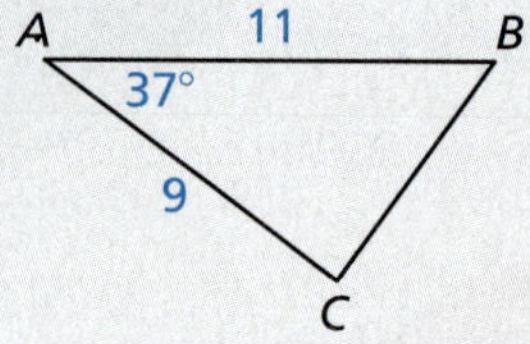

35.

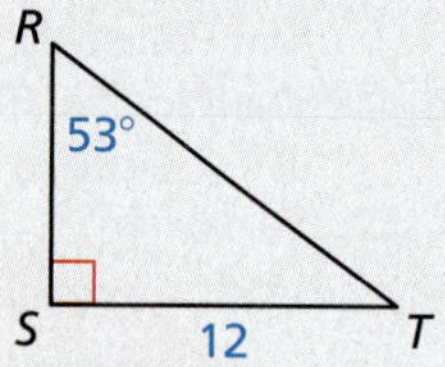

36. Name the minor arc and find its measure. Then name the major arc and find its measure.

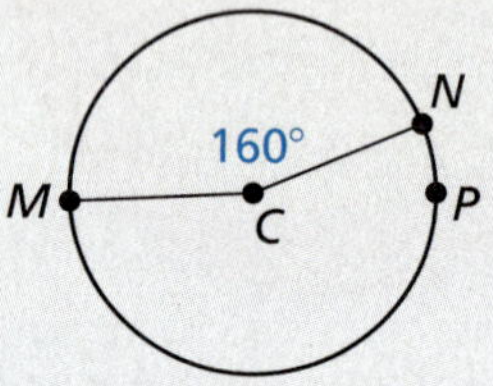

7 MTR **37. MODELING REAL LIFE** The mainsail of a sailboat is shown. Find the length of $\overline{QS}$.

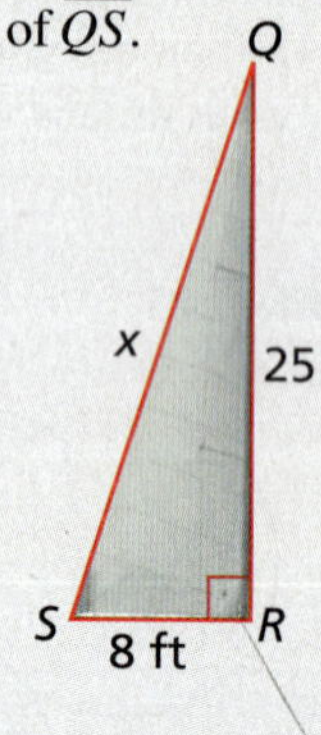

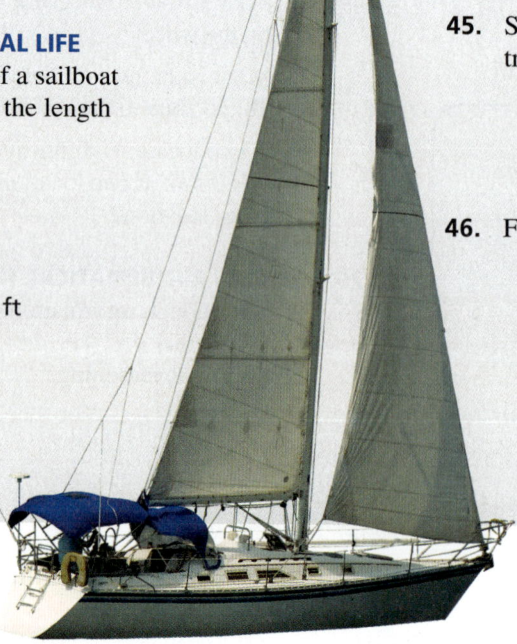

38. Find the geometric mean of 4 and 64.

In Exercises 39–42, find the value of x.

39.

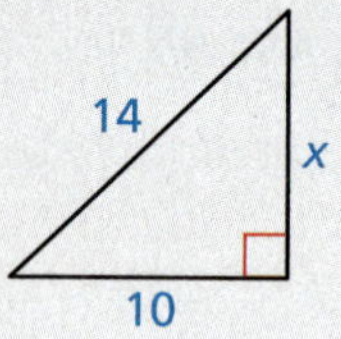

40.

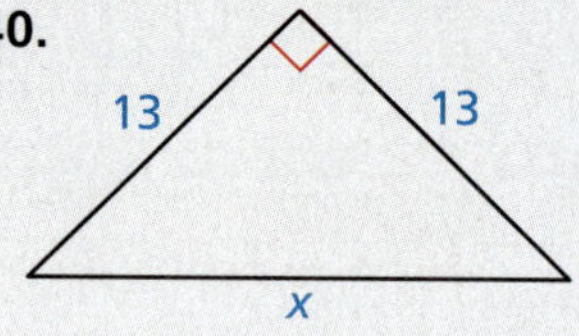

41.

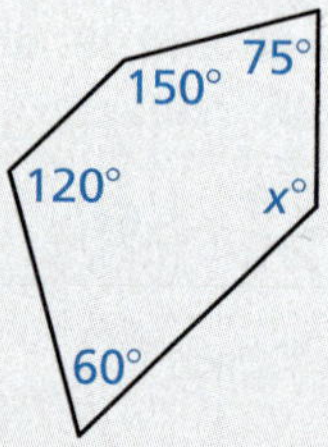

42.

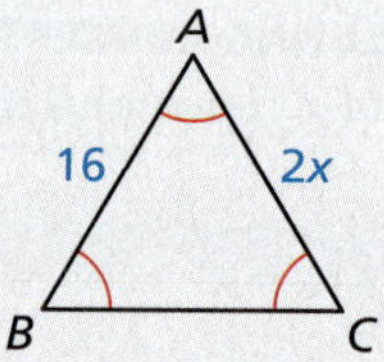

In Exercises 43 and 44, graph the polygon with the given vertices and its image after the indicated transformation.

43. $A(-1, 5)$, $B(-4, 1)$, $C(2, 1)$
Reflection: in the y-axis

44. $E(2, 1)$, $F(2, 5)$, $G(5, 4)$, $H(5, 0)$
Translation: $(x, y) \rightarrow (x + 2, y - 3)$

45. Show that the two triangles are similar.

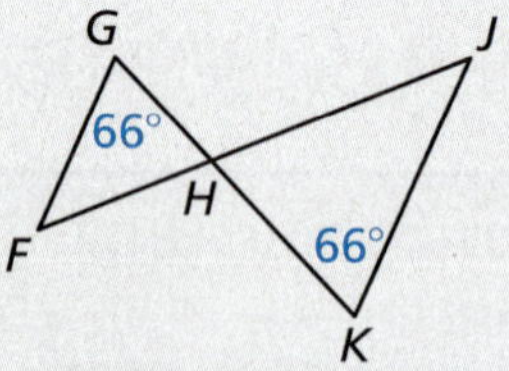

46. Find $m\angle KJL$. Explain your reasoning.

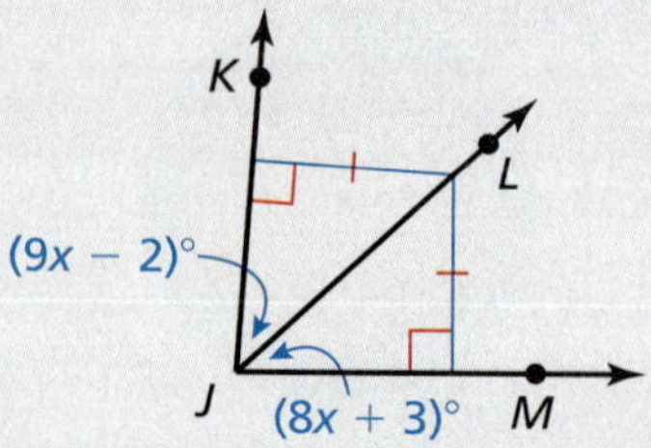

GO DIGITAL

10.3 Using Chords

Learning Target: Understand and apply theorems about chords.

Success Criteria:
- I can use chords of circles to find arc measures.
- I can use chords of circles to find lengths.
- I can describe the relationship between a diameter and a chord perpendicular to a diameter.
- I can find the center of a circle given three points on the circle.

EXPLORE IT! Making Conjectures about Chords

Work with a partner. Use technology.

a. Construct a chord $\overline{BC}$ of a circle A. Then construct a chord on the perpendicular bisector of $\overline{BC}$. What do you notice?

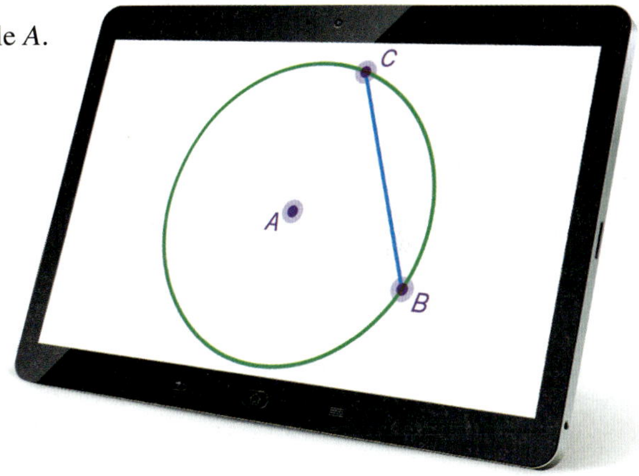

4 MTR COMMUNICATE CLEARLY

Describe two ways to determine whether a chord is also a diameter.

b. In part (a), what happens when you move the endpoints of $\overline{BC}$? What happens when you resize the circle? Make a conjecture about the perpendicular bisector of a chord.

c. Construct a diameter $\overline{BC}$ of a circle A. Then construct a chord $\overline{DE}$ perpendicular to $\overline{BC}$ at a point F. What do you notice?

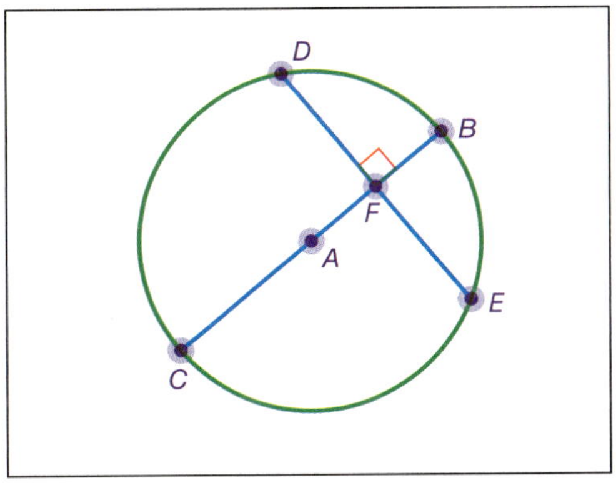

d. In part (c), what happens when you move point F to a different position on $\overline{BC}$? when you resize the circle? Make a conjecture about a chord that is perpendicular to a diameter of a circle.

Geometric Reasoning

MA.912.GR.6.1 Solve mathematical and real-world problems involving the length of a secant, tangent, segment or chord in a given circle.

MA.912.GR.6.2 Solve mathematical and real-world problems involving the measures of arcs and related angles.

GO DIGITAL

Using Chords of Circles

Recall that a *chord* is a segment with endpoints on a circle. Because its endpoints lie on the circle, any chord divides the circle into two arcs. A diameter divides a circle into two semicircles. Any other chord divides a circle into a minor arc and a major arc.

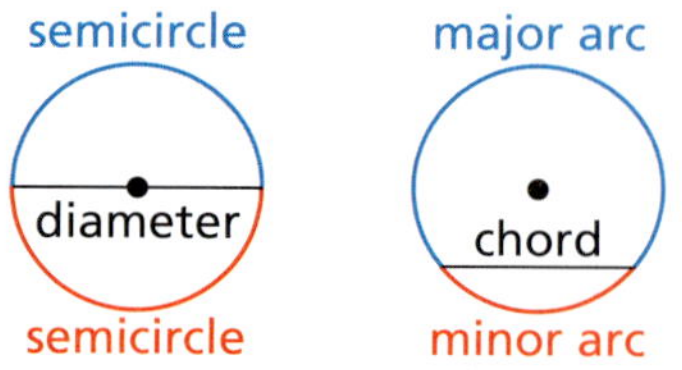

THEOREMS

10.6 Congruent Corresponding Chords Theorem

In the same circle, or in congruent circles, two minor arcs are congruent if and only if their corresponding chords are congruent.

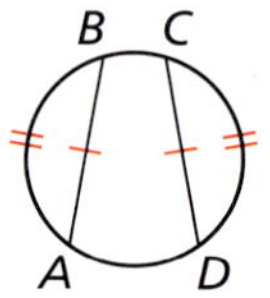

Prove this Theorem Exercise 17, page 533

$\overset{\frown}{AB} \cong \overset{\frown}{CD}$ if and only if $\overline{AB} \cong \overline{CD}$.

10.7 Perpendicular Chord Bisector Theorem

If a diameter of a circle is perpendicular to a chord, then the diameter bisects the chord and its arc.

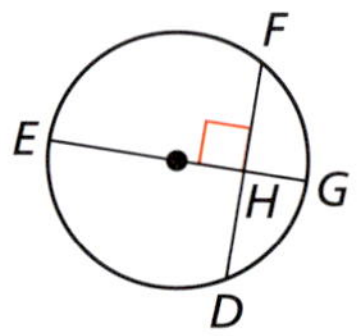

Prove this Theorem Exercise 18, page 534

If $\overline{EG}$ is a diameter and $\overline{EG} \perp \overline{DF}$, then $\overline{HD} \cong \overline{HF}$ and $\overset{\frown}{GD} \cong \overset{\frown}{GF}$.

10.8 Perpendicular Chord Bisector Converse

If one chord of a circle is a perpendicular bisector of another chord, then the first chord is a diameter.

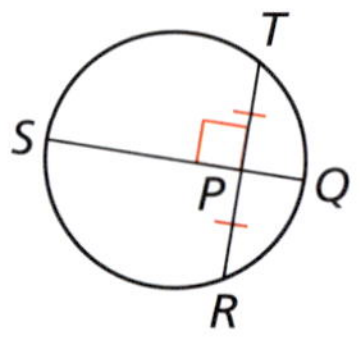

Prove this Theorem Exercise 21, page 534

If $\overline{QS}$ is a perpendicular bisector of $\overline{TR}$, then $\overline{QS}$ is a diameter of the circle.

STUDY TIP

If $\overset{\frown}{GD} \cong \overset{\frown}{GF}$, then the point G, and any line, segment, or ray that contains G, *bisects* $\overset{\frown}{FD}$.

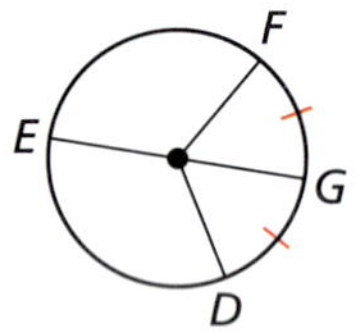

$\overline{EG}$ bisects $\overset{\frown}{FD}$.

EXAMPLE 1 Using Congruent Chords to Find an Arc Measure

WATCH

Find $m\overset{\frown}{FG}$.

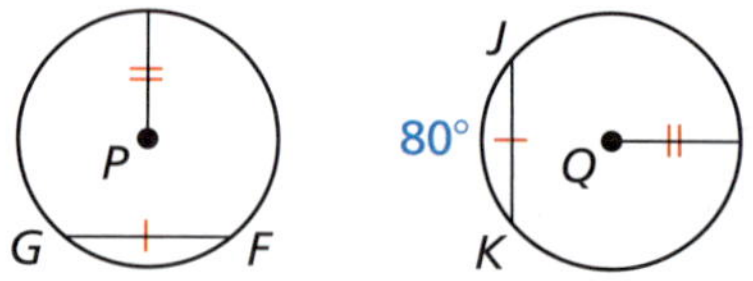

SOLUTION

The circles have the same radius. By the Congruent Circles Theorem, $\odot P \cong \odot Q$. Because $\overline{FG}$ and $\overline{JK}$ are congruent chords in congruent circles, the corresponding minor arcs $\overset{\frown}{FG}$ and $\overset{\frown}{JK}$ are congruent by the Congruent Corresponding Chords Theorem.

▶ So, $m\overset{\frown}{FG} = m\overset{\frown}{JK} = 80°$.

GO DIGITAL

EXAMPLE 2 Using a Diameter

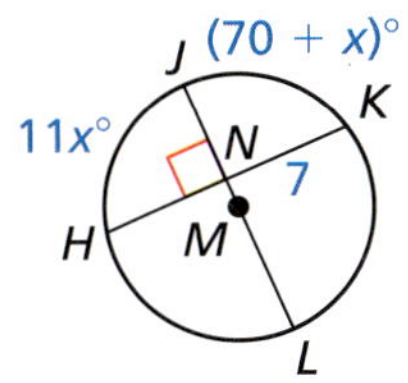

a. Find HK. **b.** Find $m\widehat{HK}$.

SOLUTION

a. Diameter $\overline{JL}$ is perpendicular to $\overline{HK}$. So, by the Perpendicular Chord Bisector Theorem, $\overline{JL}$ bisects $\overline{HK}$, and $HN = NK$.

▶ So, $HK = 2(NK) = 2(7) = 14$.

b. Diameter $\overline{JL}$ is perpendicular to $\overline{HK}$. So, by the Perpendicular Chord Bisector Theorem, $\overline{JL}$ bisects $\widehat{HK}$, and $m\widehat{HJ} = m\widehat{JK}$.

$m\widehat{HJ} = m\widehat{JK}$ Perpendicular Chord Bisector Theorem

$11x^\circ = (70 + x)^\circ$ Substitute.

$10x = 70$ Subtract x from each side.

$x = 7$ Divide each side by 10.

▶ So, $m\widehat{HJ} = m\widehat{JK} = (70 + x)^\circ = (70 + 7)^\circ = 77^\circ$, and $m\widehat{HK} = 2(m\widehat{HJ}) = 2(77^\circ) = 154^\circ$.

EXAMPLE 3 Using Perpendicular Bisectors

Three bushes are arranged in a garden, as shown. Where should you place a sprinkler so that it is the same distance from each bush?

SOLUTION

Place the sprinkler at the center of the circle that passes through the bushes.

Step 1

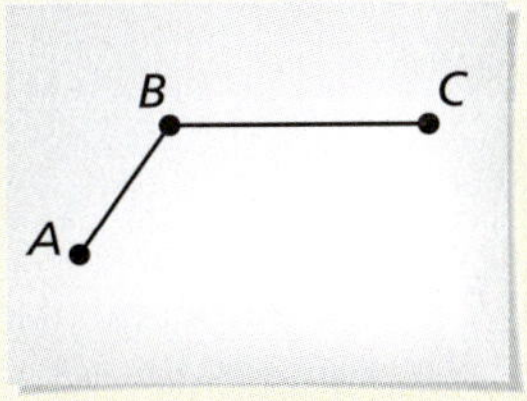

Label the bushes A, B, and C, as shown. Draw $\overline{AB}$ and $\overline{BC}$.

Step 2

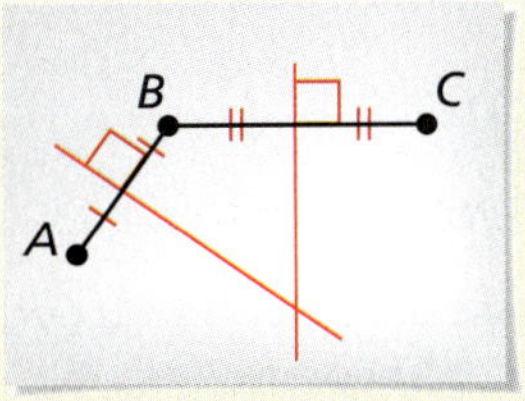

Draw the perpendicular bisectors of $\overline{AB}$ and $\overline{BC}$. By the Perpendicular Chord Bisector Converse, these lie on diameters of the circle containing A, B, and C.

Step 3

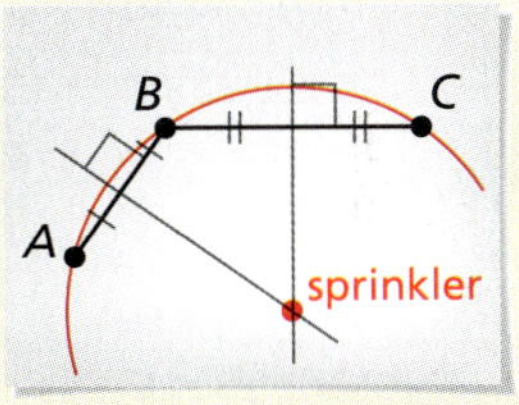

Find the point where the perpendicular bisectors intersect. This is the center of the circle, which is equidistant from points A, B, and C.

SELF-ASSESSMENT

1 I don't understand yet. 2 I can do it with help. 3 I can do it on my own. 4 I can teach someone else.

Find the measure of the red arc or chord.

1.

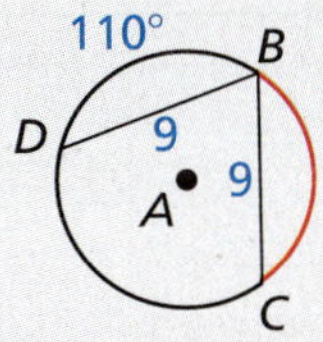

2.

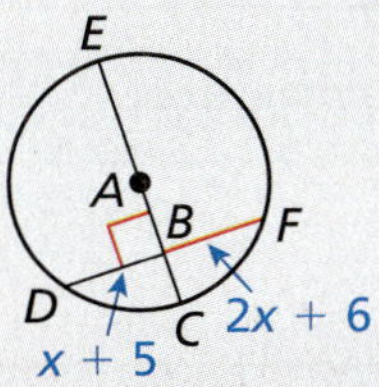

3.

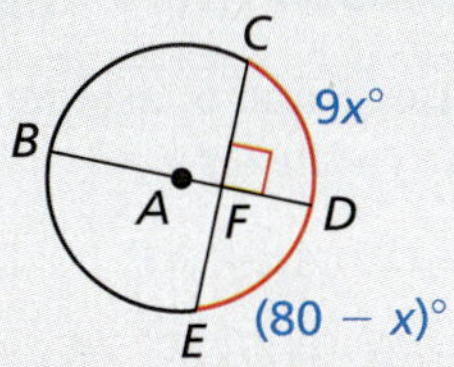

4. In Example 3, you want to plant a fourth bush so that it is equidistant from the sprinkler. Describe the possible locations where you can plant the bush.

GO DIGITAL

THEOREM

10.9 Equidistant Chords Theorem

In the same circle, or in congruent circles, two chords are congruent if and only if they are equidistant from the center.

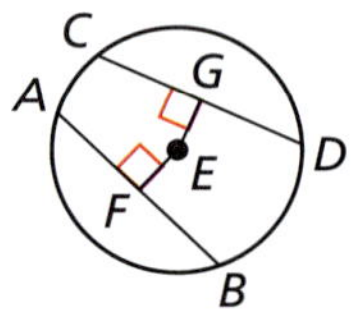

Prove this Theorem Exercise 23, page 534

$\overline{AB} \cong \overline{CD}$ if and only if $EF = EG$.

EXAMPLE 4 Using Congruent Chords to Find a Circle's Radius

WATCH

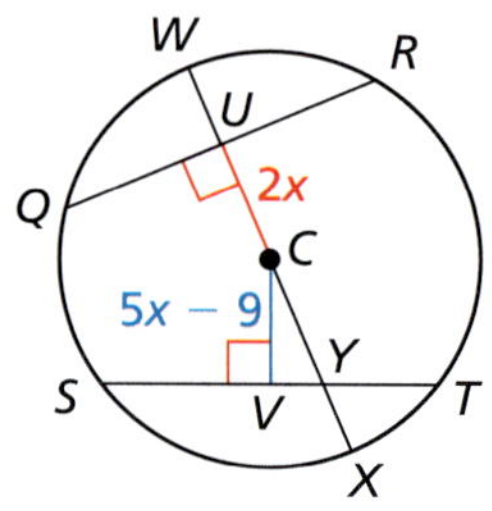

In the diagram, $QR = ST = 16$. Find the radius of $\odot C$.

SOLUTION

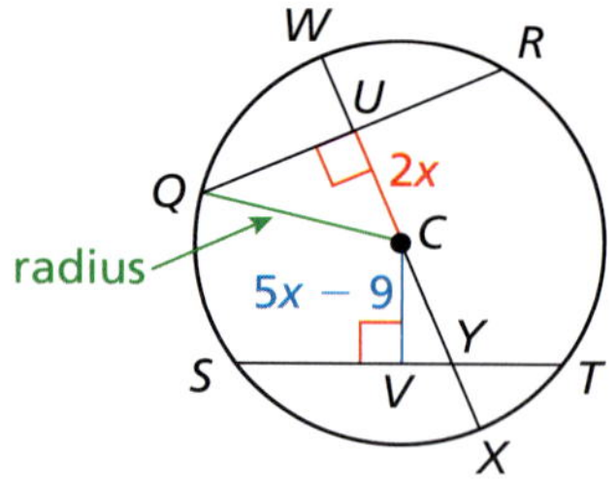

Because $\overline{CQ}$ is a segment whose endpoints are the center and a point on the circle, it is a radius of $\odot C$. Because $\overline{CU} \perp \overline{QR}$, $\triangle QUC$ is a right triangle. Apply properties of chords to find the lengths of the legs of $\triangle QUC$.

Step 1 Find CU.

Because $\overline{QR}$ and $\overline{ST}$ are congruent chords, $\overline{QR}$ and $\overline{ST}$ are equidistant from C by the Equidistant Chords Theorem. So, $CU = CV$.

$CU = CV$	Equidistant Chords Theorem
$2x = 5x - 9$	Substitute.
$x = 3$	Solve for x.

So, $CU = 2x = 2(3) = 6$.

Step 2 Find QU.

Because diameter $\overline{WX} \perp \overline{QR}$, $\overline{WX}$ bisects $\overline{QR}$ by the Perpendicular Chord Bisector Theorem.

So, $QU = \frac{1}{2}QR = \frac{1}{2}(16) = 8$.

Step 3 Find CQ.

The lengths of the legs of right $\triangle QUC$ are $CU = 6$ and $QU = 8$. Because the integers 6, 8, and 10 form a Pythagorean triple, $CQ = 10$.

▶ So, the radius of $\odot C$ is 10 units.

SELF-ASSESSMENT

1 I don't understand yet. 2 I can do it with help. 3 I can do it on my own. 4 I can teach someone else.

In Exercises 5 and 6, use the diagram and the given measures to find the radius of $\odot N$.

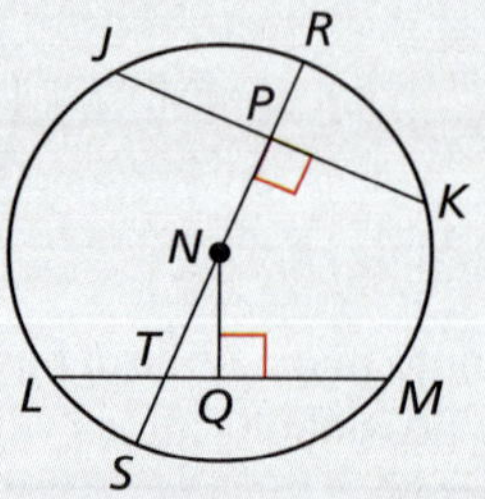

5. $JK = LM = 24$, $NP = 3x$, $NQ = 7x - 12$

6. $NP = NQ = 5$, $JK = 5x - 1$, $LM = 7x - 11$

GO DIGITAL

10.3 Practice WITH CalcChat® AND CalcView®

In Exercises 1–4, find the measure of the red arc or chord in ⊙C. *(See Example 1.)*

▶ 1.

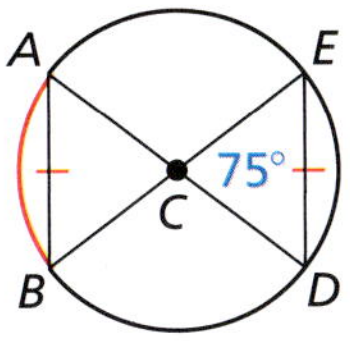

2.

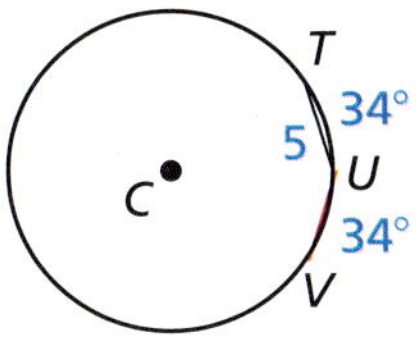

3.

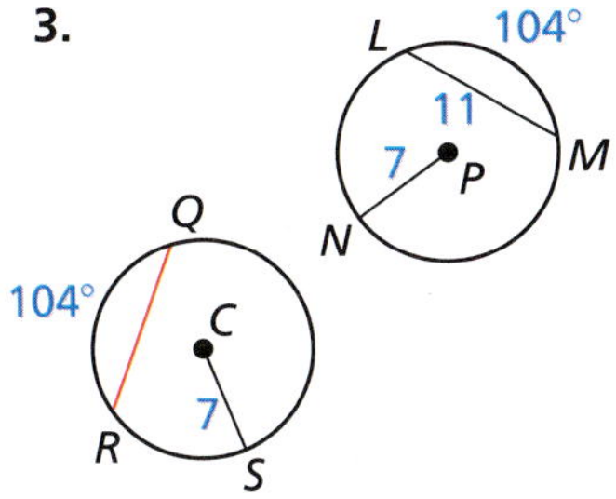

4.

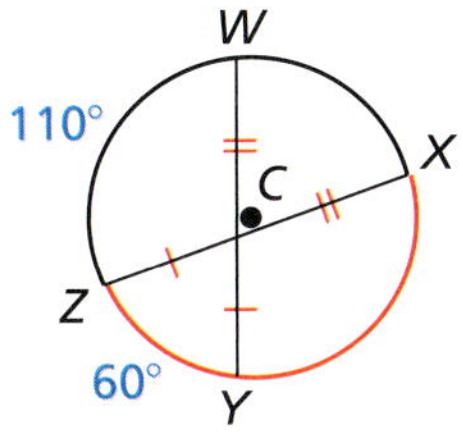

In Exercises 5–8, find the value of x. *(See Example 2.)*

5.

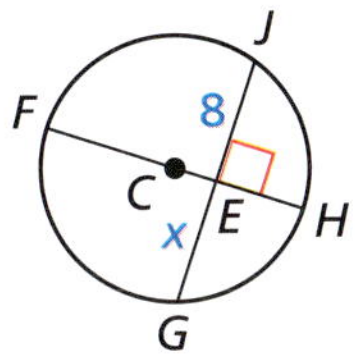

6.

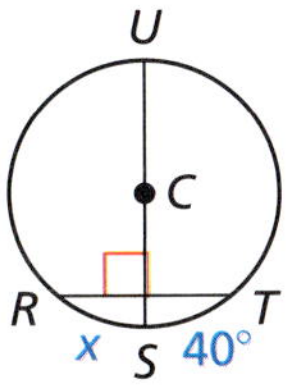

▶ 7.

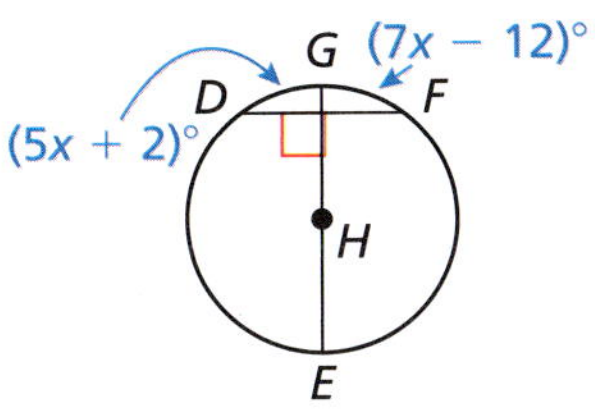

8.

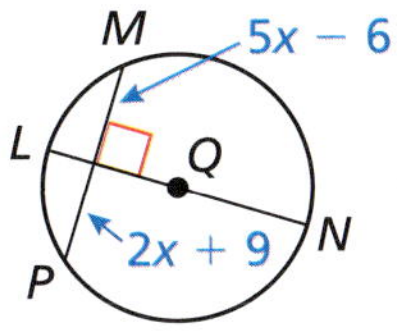

9. **MODELING REAL LIFE** Three tables are arranged on a patio, as shown. Explain how to place a patio heater so that it is the same distance from each table. *(See Example 3.)*

10. **PROBLEM SOLVING** An archaeologist finds part of a circular plate. What was the diameter of the plate? Justify your answer.

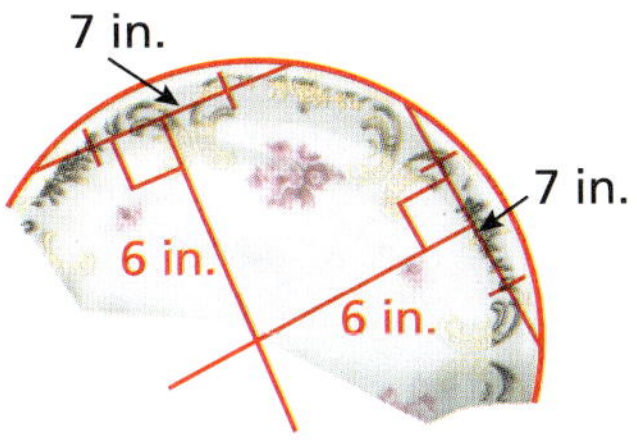

In Exercises 11 and 12, determine whether $\overline{AB}$ is a diameter of the circle. Explain your reasoning.

11.

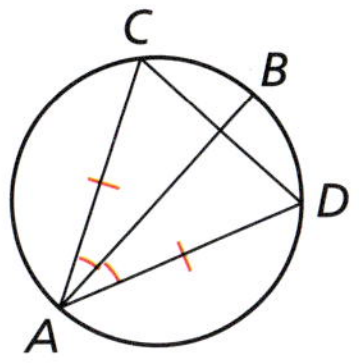

12.

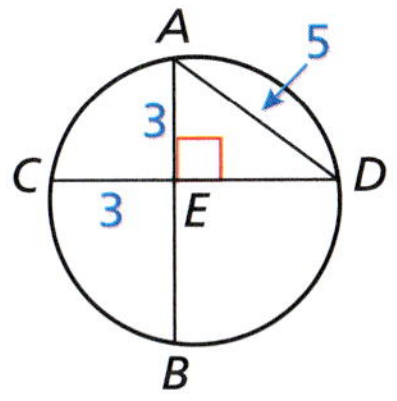

In Exercises 13 and 14, find the radius of ⊙Q. *(See Example 4.)*

▶ 13.

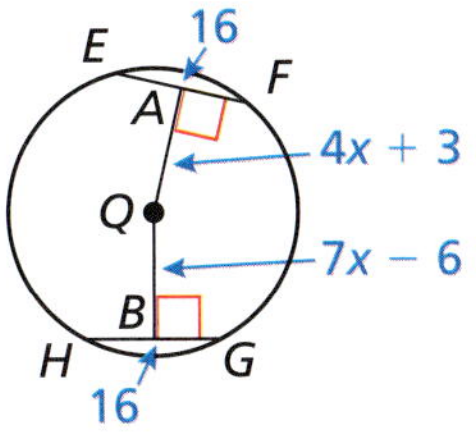

14.

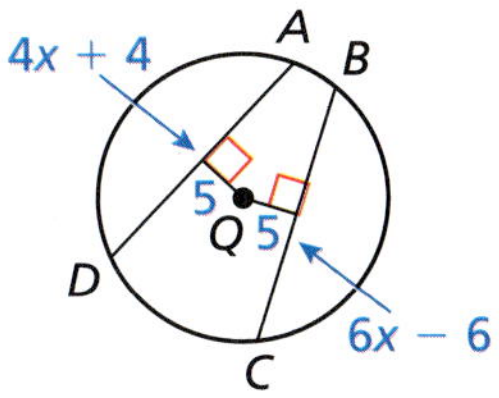

15. **ERROR ANALYSIS** $AB = DE = 24$. Describe and correct the error in finding the radius of ⊙C.

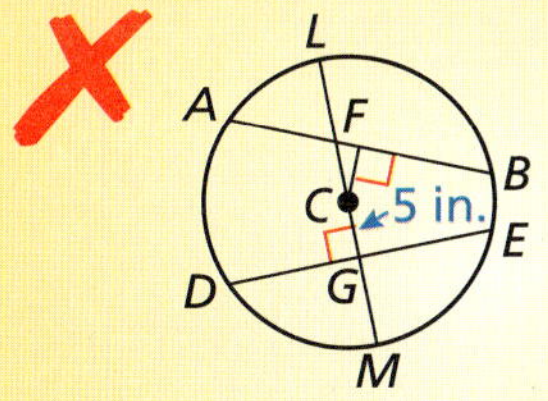

$GE = \frac{1}{2}DE$

$GE = \frac{1}{2}(24)$

$GE = 12$

$5^2 + CE^2 = 12^2$

$CE^2 = 119$

$CE = \sqrt{119}$

So, the radius is $\sqrt{119}$ inches.

16. **HOW DO YOU SEE IT?**
What can you conclude from each diagram? Name a theorem that justifies your answer.

a.

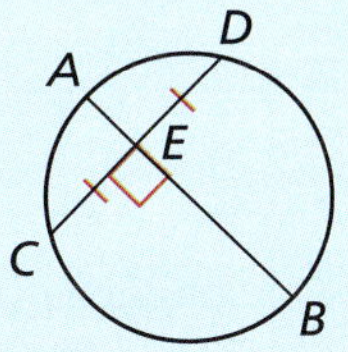

b.

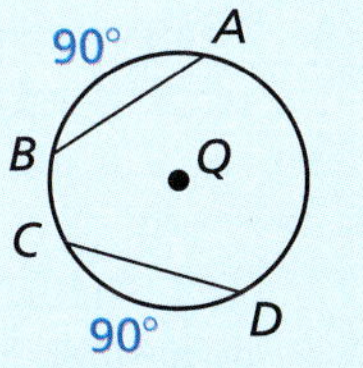

17. **PROVING A THEOREM** Use the diagram to prove each part of the biconditional in the Congruent Corresponding Chords Theorem.

a. **Given** $\overline{AB} \cong \overline{CD}$

Prove $\overset{\frown}{AB} \cong \overset{\frown}{CD}$

b. **Given** $\overset{\frown}{AB} \cong \overset{\frown}{CD}$

Prove $\overline{AB} \cong \overline{CD}$

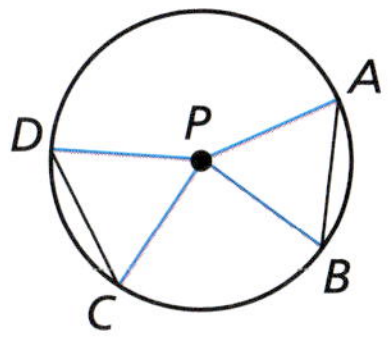

GO DIGITAL

18. PROVING A THEOREM Use congruent triangles to prove the Perpendicular Chord Bisector Theorem.

Given $\overline{EG}$ is a diameter of $\odot L$.
$\overline{EG} \perp \overline{DF}$

Prove $\overline{DC} \cong \overline{FC}$, $\widehat{DG} \cong \widehat{FG}$

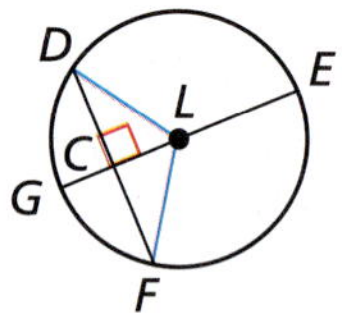

19. REASONING In $\odot P$, all the arcs shown have integer measures. Show that x must be even.

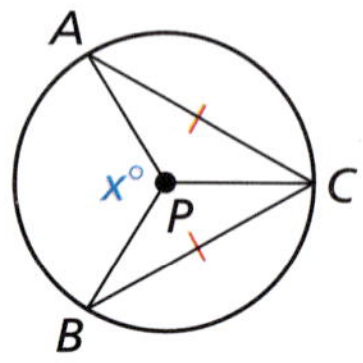

20. MAKING AN ARGUMENT In the diagram, point C is a point of tangency. $\overline{AB}$ is parallel to line n. Does $\overline{CD}$ bisect $\widehat{AB}$? Explain your reasoning.

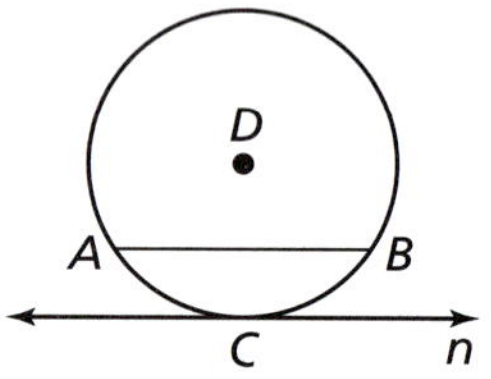

21. PROVING A THEOREM Write a proof of the Perpendicular Chord Bisector Converse.

Given $\overline{QS}$ is the perpendicular bisector of $\overline{RT}$.

Prove $\overline{QS}$ is a diameter of the circle.

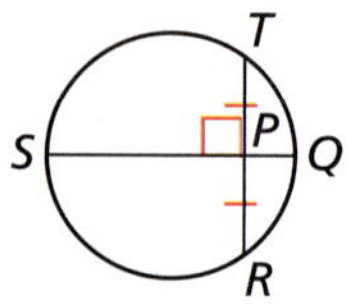

(*Hint:* Let C be the center of the circle. Show that C must lie on $\overline{QS}$.)

22. THOUGHT PROVOKING
In $\odot P$, the lengths of the parallel chords are 20, 16, and 12. Find $m\widehat{AB}$. Explain your reasoning.

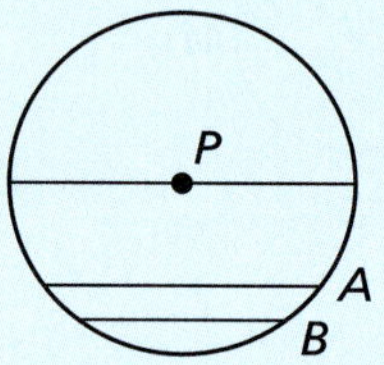

23. PROVING A THEOREM Prove both parts of the biconditional of the Equidistant Chords Theorem.

REVIEW & REFRESH

WATCH

24. In the diagram, point B is a point of tangency. Find the radius x of $\odot C$.

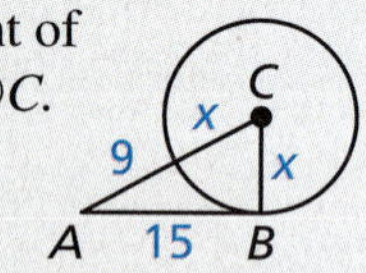

In Exercises 25 and 26, find the missing interior angle measure.

25. Quadrilateral $JKLM$ has angle measures $m\angle J = 32°$, $m\angle K = 125°$, and $m\angle L = 44°$. Find $m\angle M$.

26. Pentagon $PQRST$ has angle measures $m\angle P = 85°$, $m\angle Q = 134°$, $m\angle R = 97°$, and $m\angle S = 102°$. Find $m\angle T$.

27. Tell whether the red arcs are congruent, similar, or neither. Explain.

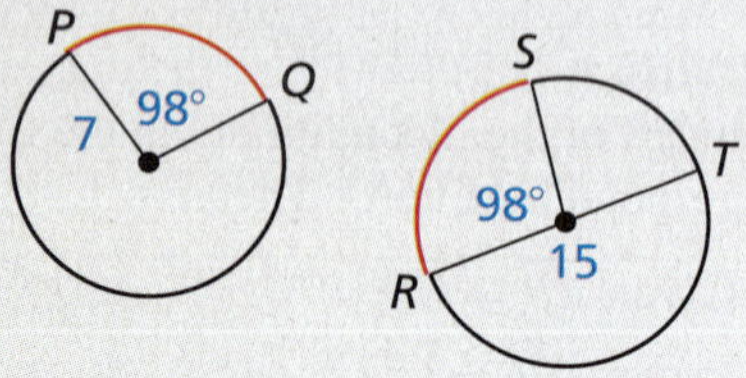

28. MODELING REAL LIFE A surveyor makes the measurements shown to determine the length of a bridge to be built across a small lake from the North Picnic Area to the South Picnic Area. Find the length of the bridge.

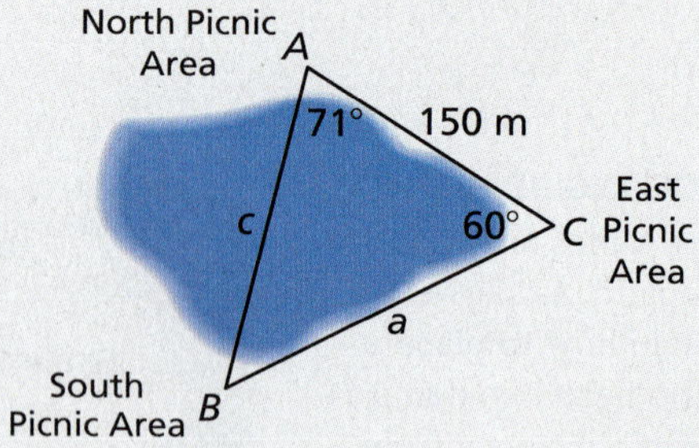

29. Find $m\widehat{EFG}$ in $\odot Q$.

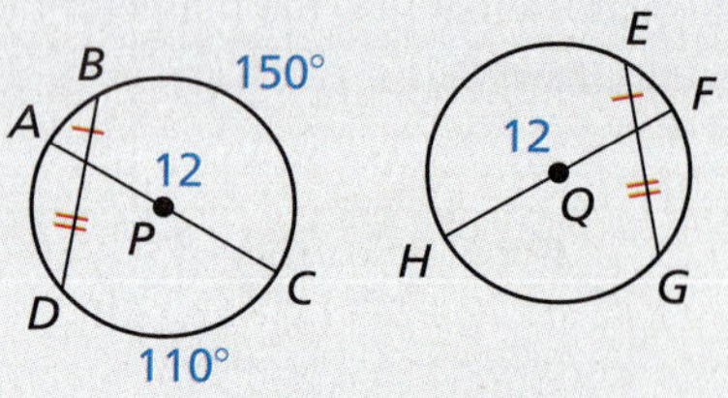

GO DIGITAL

10.4 Inscribed Angles and Polygons

Learning Target: Use properties of inscribed angles and inscribed polygons.

Success Criteria:
- I can find measures of inscribed angles and intercepted arcs.
- I can find angle measures of inscribed polygons.
- (H) I can construct a square inscribed in a circle.

EXPLORE IT! Constructing Inscribed Angles and Central Angles

Work with a partner.

a. Use the diagram to write definitions for an *inscribed angle* and an *intercepted arc*.

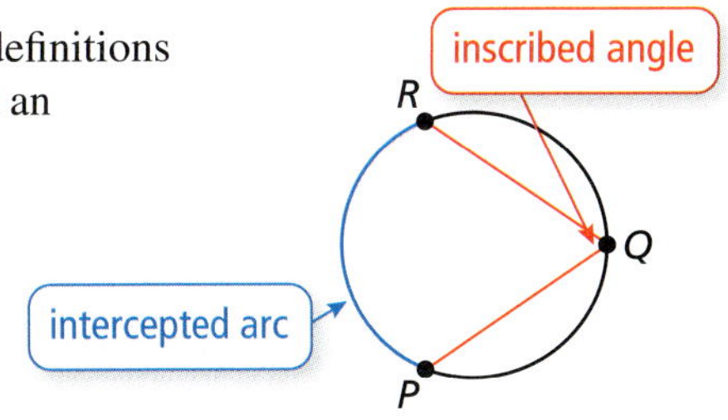

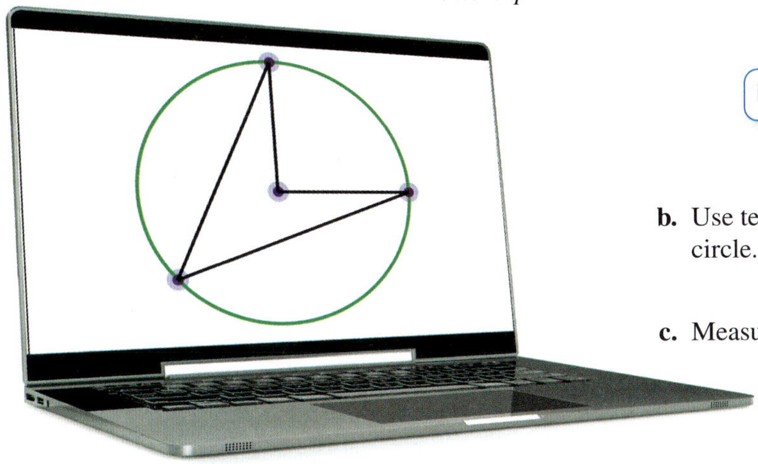

b. Use technology to construct an inscribed angle of a circle. Then construct the corresponding central angle.

c. Measure both angles. What do you notice?

d. What happens to the angle measures when you change the inscribed angle? What happens to the angle measures when you change the size of the circle? Make a conjecture about the relationship between the measure of an inscribed angle and the measure of its intercepted arc.

e. Each vertex of a quadrilateral lies on a circle. Use technology to find a relationship among the measures of the interior angles of the quadrilateral. Make a conjecture that summarizes your results. Provide examples to support your reasoning.

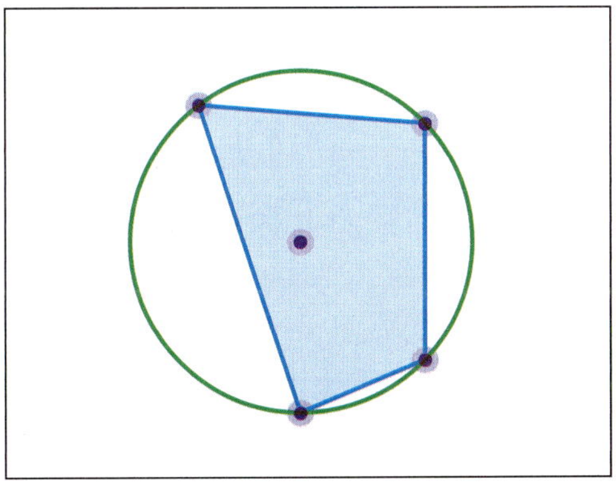

5 MTR USE STRUCTURE
Construct several triangles, each with its vertices on a circle so that one side is a diameter. Classify each triangle by its angle measures. What do you notice?

Geometric Reasoning
MA.912.GR.6.2 Solve mathematical and real-world problems involving the measures of arcs and related angles.
MA.912.GR.6.3 Solve mathematical problems involving triangles and quadrilaterals inscribed in a circle.
(H) **Also MA.912.GR.5.4**

GO DIGITAL

WORDS AND MATH

One meaning of the Latin prefix *in-* is "upon." The vertex of an *inscribed angle* lies upon a circle.

Vocabulary

inscribed angle, *p. 536*
intercepted arc, *p. 536*
subtend, *p. 536*
inscribed polygon, *p. 538*
circumscribed circle, *p. 538*

Using Inscribed Angles

KEY IDEAS

Inscribed Angle and Intercepted Arc

An **inscribed angle** is an angle whose vertex lies on a circle and whose sides contain chords of the circle. An arc that lies between two lines, rays, or segments is called an **intercepted arc**. If the endpoints of a chord or arc lie on the sides of an inscribed angle, then the chord or arc is said to **subtend** the angle.

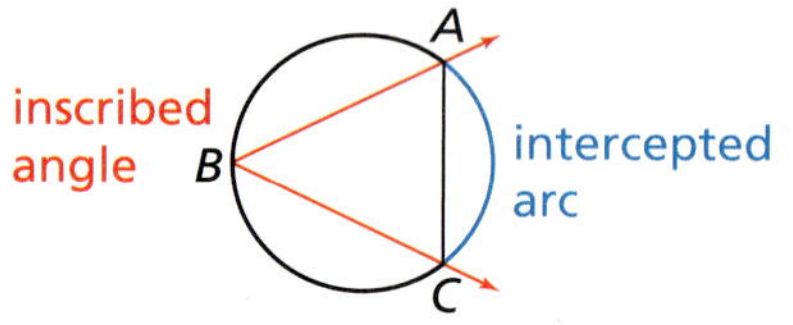

$\angle B$ intercepts $\overset{\frown}{AC}$.
$\overset{\frown}{AC}$ subtends $\angle B$.
$\overline{AC}$ subtends $\angle B$.

THEOREM

10.10 Measure of an Inscribed Angle Theorem

The measure of an inscribed angle is one-half the measure of its intercepted arc.

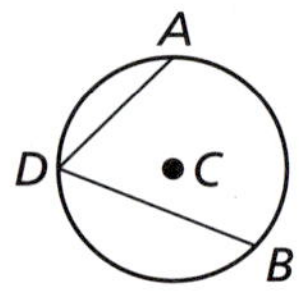

$m\angle ADB = \frac{1}{2}(m\overset{\frown}{AB})$

Prove this Theorem Exercise 23, page 541

The proof of the Measure of an Inscribed Angle Theorem involves three cases.

Case 1 Center C is on a side of the inscribed angle.

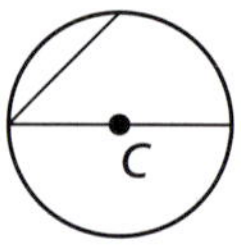

Case 2 Center C is inside the inscribed angle.

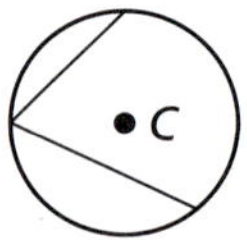

Case 3 Center C is outside the inscribed angle.

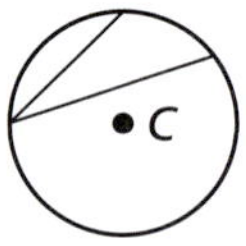

EXAMPLE 1 Using Inscribed Angles

WATCH

Find (a) $m\angle T$ and (b) $m\overset{\frown}{QR}$.

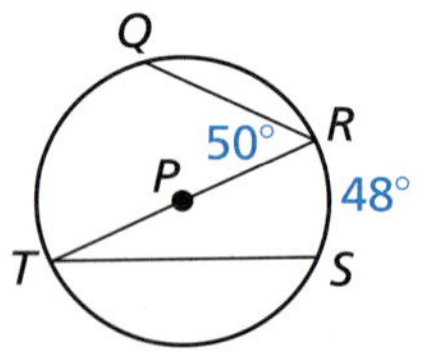

SOLUTION

a. Use the Measure of an Inscribed Angle Theorem to find $m\angle T$.

$$m\angle T = \frac{1}{2}(m\overset{\frown}{RS}) = \frac{1}{2}(48^\circ) = 24^\circ$$

b. Use the Measure of an Inscribed Angle Theorem to find $m\overset{\frown}{TQ}$. Then use the fact that $\overset{\frown}{TQR}$ is a semicircle to find $m\overset{\frown}{QR}$.

Step 1:
$$\frac{1}{2}(m\overset{\frown}{TQ}) = m\angle R$$
$$\frac{1}{2}(m\overset{\frown}{TQ}) = 50^\circ$$
$$m\overset{\frown}{TQ} = 100^\circ$$

Step 2:
$$m\overset{\frown}{QR} = 180^\circ - m\overset{\frown}{TQ}$$
$$= 180^\circ - 100^\circ$$
$$= 80^\circ$$

GO DIGITAL

EXAMPLE 2 Finding the Measure of an Intercepted Arc

WATCH

Find $m\overset{\frown}{RS}$ and $m\angle STR$. What do you notice about $\angle STR$ and $\angle RUS$?

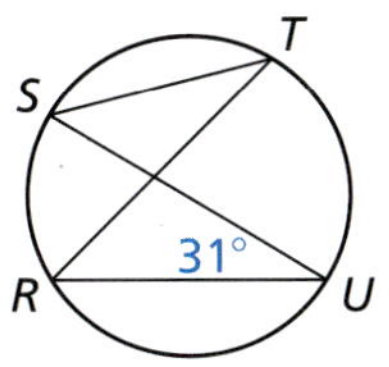

SOLUTION

From the Measure of an Inscribed Angle Theorem, $\frac{1}{2}\left(m\overset{\frown}{RS}\right) = m\angle RUS$.

So, $m\overset{\frown}{RS} = 2(m\angle RUS) = 2(31^\circ) = 62^\circ$.

Also, $m\angle STR = \frac{1}{2}\left(m\overset{\frown}{RS}\right) = \frac{1}{2}(62^\circ) = 31^\circ$.

▶ So, $m\overset{\frown}{RS} = 62^\circ$, $m\angle STR = 31^\circ$, and $\angle STR \cong \angle RUS$.

Example 2 suggests the Inscribed Angles of a Circle Theorem.

THEOREM

10.11 Inscribed Angles of a Circle Theorem

If two inscribed angles of a circle intercept the same arc, then the angles are congruent.

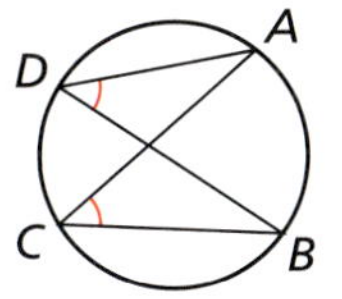

Prove this Theorem Exercise 24, page 541

$\angle ADB \cong \angle ACB$

EXAMPLE 3 Finding the Measure of an Angle

WATCH

Find $m\angle F$.

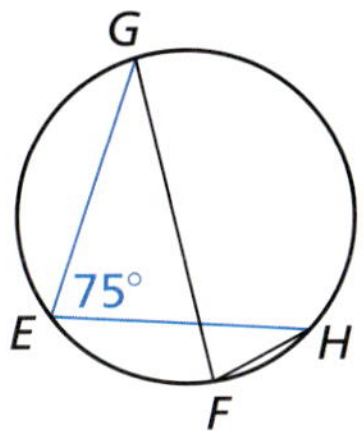

SOLUTION

Both $\angle E$ and $\angle F$ intercept $\overset{\frown}{GH}$. So, $\angle E \cong \angle F$ by the Inscribed Angles of a Circle Theorem.

▶ So, $m\angle F = m\angle E = 75^\circ$.

SELF-ASSESSMENT

1 I don't understand yet. 2 I can do it with help. 3 I can do it on my own. 4 I can teach someone else.

Find the indicated measure.

1. $m\angle G$

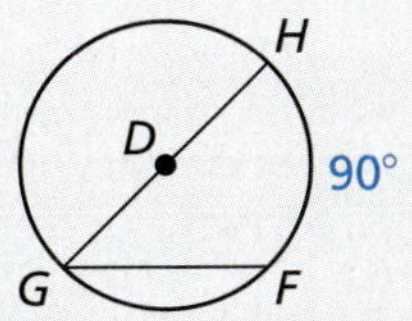

2. $m\overset{\frown}{TV}$

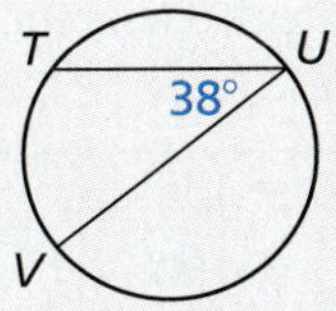

3. $m\angle X$

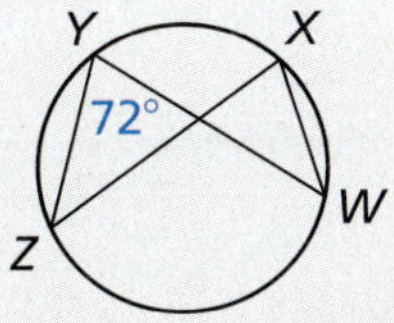

GO DIGITAL

Using Inscribed Polygons

KEY IDEA

Inscribed Polygon

A polygon is an **inscribed polygon** when all its vertices lie on a circle. The circle that contains the vertices is a **circumscribed circle**.

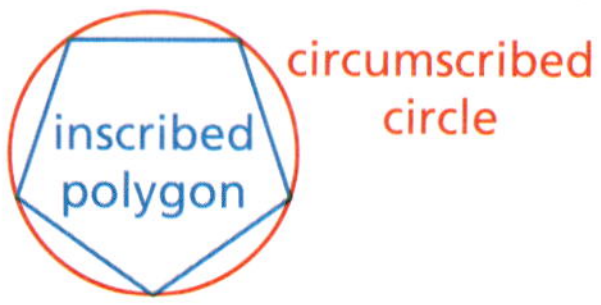

THEOREMS

STUDY TIP

The Inscribed Right Triangle Theorem also applies to right triangles inscribed in semicircles.

10.12 Inscribed Right Triangle Theorem

If a right triangle is inscribed in a circle, then the hypotenuse is a diameter of the circle. Conversely, if one side of an inscribed triangle is a diameter of the circle, then the triangle is a right triangle and the angle opposite the diameter is the right angle.

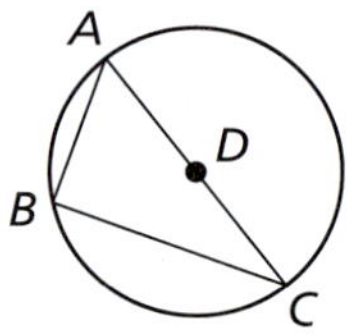

$m\angle ABC = 90°$ if and only if $\overline{AC}$ is a diameter of the circle.

Prove this Theorem Exercise 27, page 541

10.13 Inscribed Quadrilateral Theorem

A quadrilateral can be inscribed in a circle if and only if its opposite angles are supplementary.

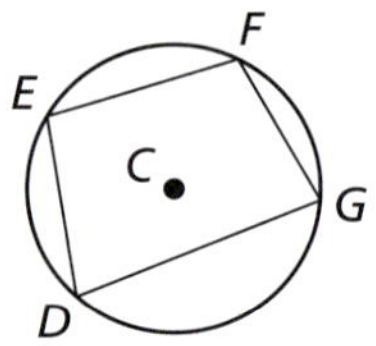

D, E, F, and G lie on $\odot C$ if and only if $m\angle D + m\angle F = m\angle E + m\angle G = 180°$.

Proof *BigIdeasMath.com*

Prove this Theorem Exercise 28, page 541

EXAMPLE 4 Using Inscribed Polygons

WATCH

Find the value of each variable.

a.

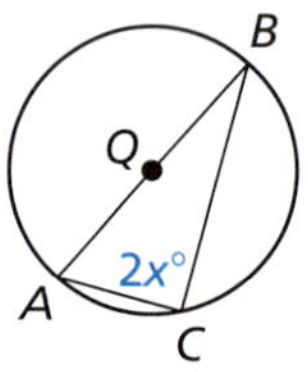

b.

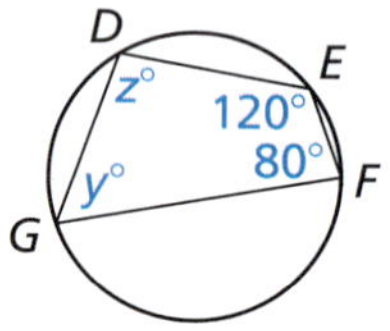

SOLUTION

a. $\overline{AB}$ is a diameter. So, $\angle C$ is a right angle, and $m\angle C = 90°$ by the Inscribed Right Triangle Theorem.

$$2x° = 90°$$
$$x = 45$$

▶ The value of x is 45.

b. $DEFG$ is inscribed in a circle, so opposite angles are supplementary by the Inscribed Quadrilateral Theorem.

$$m\angle D + m\angle F = 180° \qquad m\angle E + m\angle G = 180°$$
$$z + 80 = 180 \qquad 120 + y = 180$$
$$z = 100 \qquad y = 60$$

▶ The value of z is 100 and the value of y is 60.

GO DIGITAL

CONSTRUCTION Constructing a Square Inscribed in a Circle

WATCH H

Given $\odot C$, construct a square inscribed in a circle.

SOLUTION

Step 1

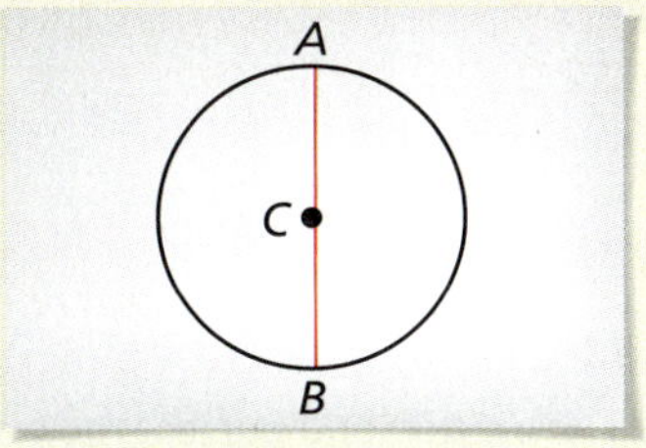

Draw a diameter
Draw any diameter. Label the endpoints A and B.

Step 2

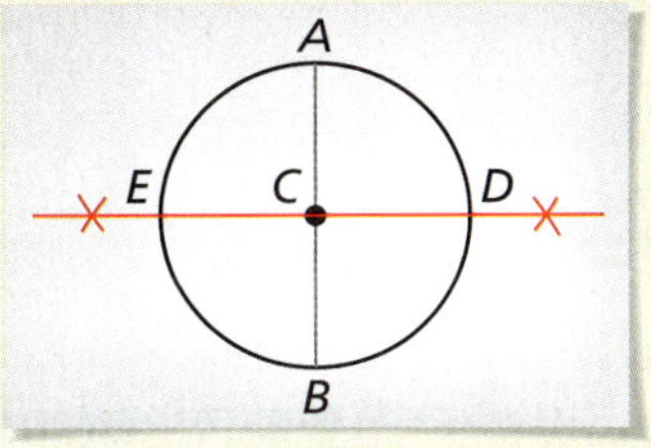

Construct a perpendicular bisector
Construct the perpendicular bisector of the diameter. Label the points where it intersects $\odot C$ as points D and E.

Step 3

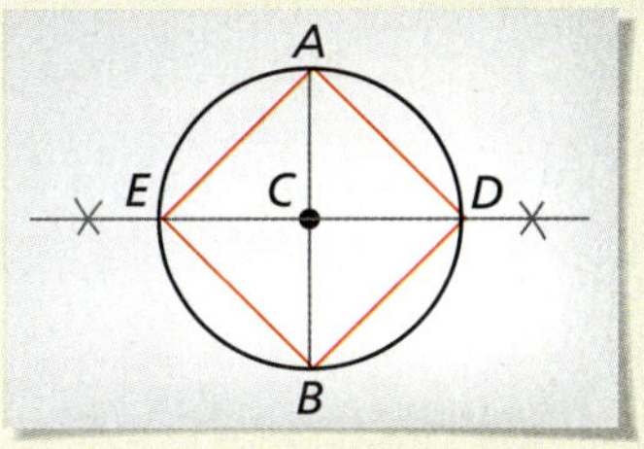

Form a square
Connect points A, D, B, and E to form a square.

EXAMPLE 5 Using a Circumscribed Circle

WATCH

Your camera has a 90° field of vision, and you want to photograph the front of a statue. You stand at a location in which the front of the statue fits perfectly within your camera's field of vision, as shown. You want to change your location. Where else can you stand so that the front of the statue fits perfectly within your camera's field of vision?

SOLUTION

From the Inscribed Right Triangle Theorem, you know that if a right triangle is inscribed in a circle, then the hypotenuse of the triangle is a diameter of the circle. So, draw the circle that has the front of the statue as a diameter.

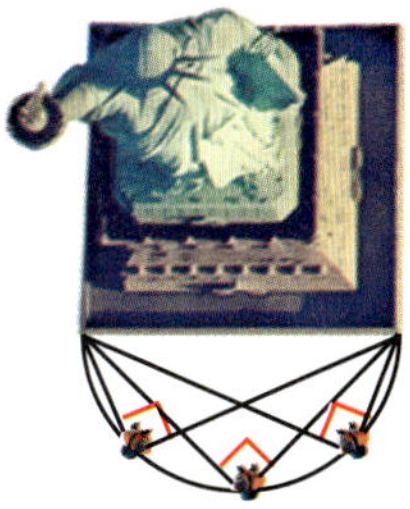

▶ The statue fits perfectly within your camera's 90° field of vision from any point on the semicircle in front of the statue.

SELF-ASSESSMENT

1 I don't understand yet. 2 I can do it with help. 3 I can do it on my own. 4 I can teach someone else.

Find the value of each variable.

4.

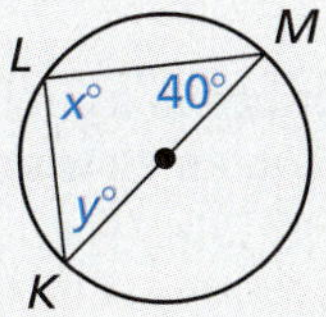

5.

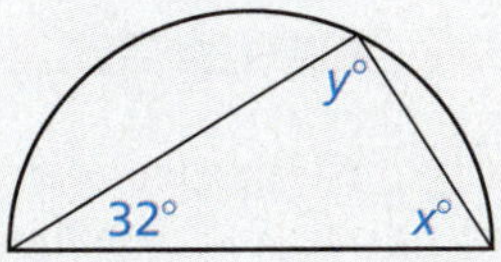

6.

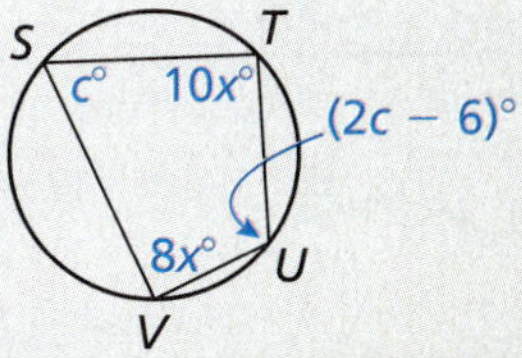

7. In Example 5, explain how to find locations where the left side of the statue fits perfectly within your camera's field of vision.

GO DIGITAL

10.4 Practice WITH CalcChat® AND CalcView®

In Exercises 1–8, find the indicated measure. *(See Examples 1, 2, and 3.)*

▶ **1.** $m\angle A$

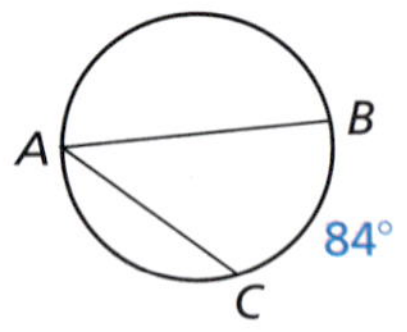

2. $m\angle G$

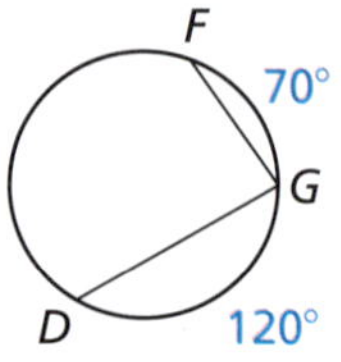

3. $m\angle N$

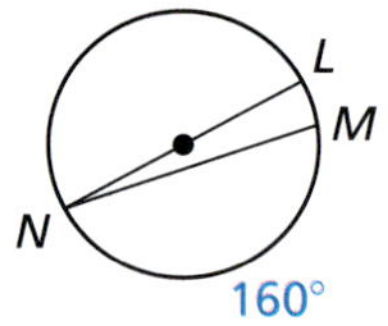

4. $m\overparen{RS}$

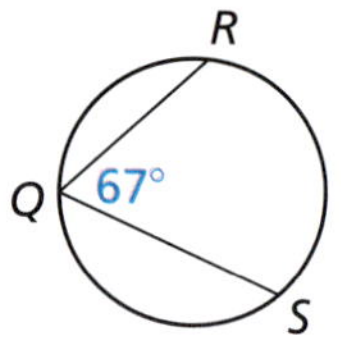

5. $m\overparen{VU}$

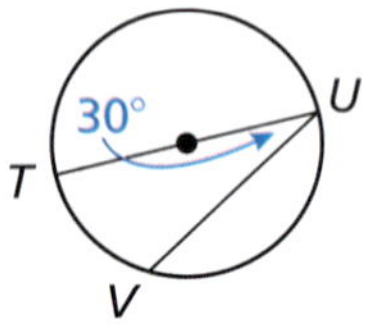

6. $m\overparen{WX}$

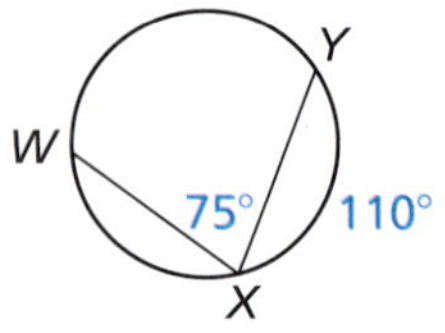

▶ **7.** $m\angle EHF$

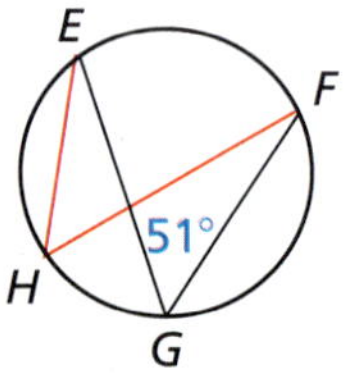

8. $m\overparen{PS}$

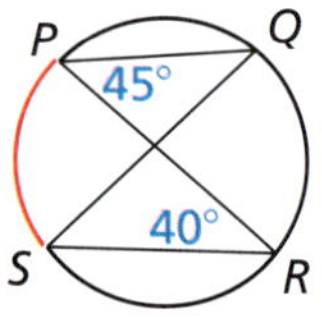

In Exercises 9–12, find the value of each variable. *(See Example 4.)*

▶ **9.**

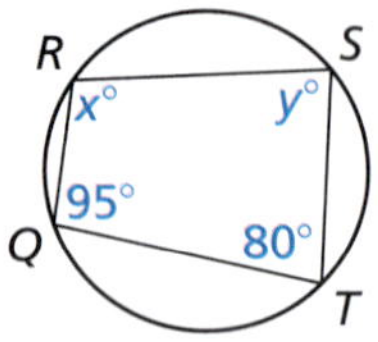

10.

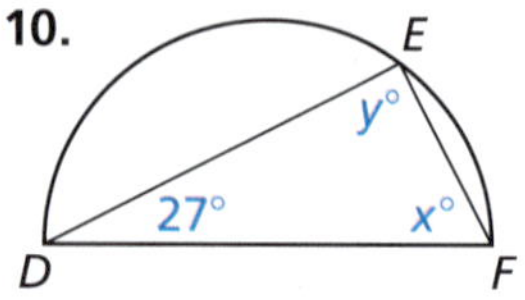

11.

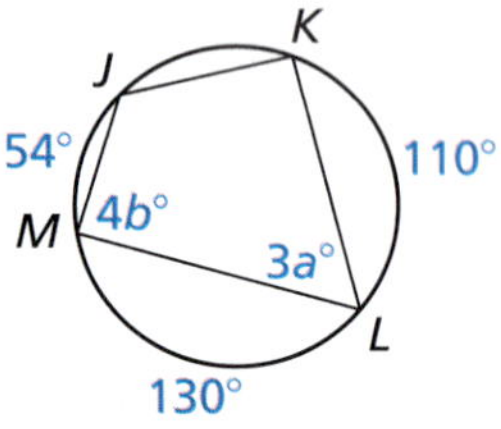

12.

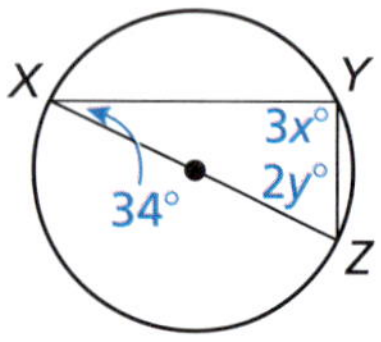

13. **ERROR ANALYSIS** Describe and correct the error in finding $m\overparen{BC}$.

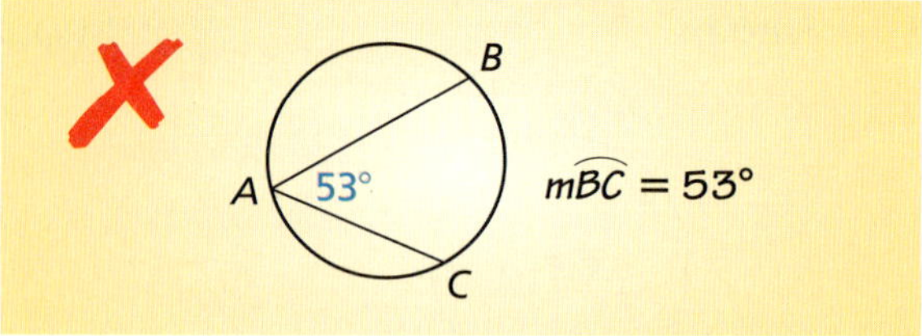

14. **ERROR ANALYSIS** Describe and correct the error in finding the value of x.

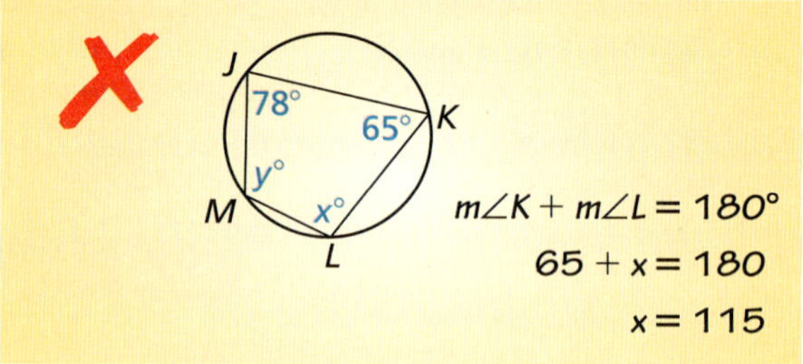

15. **MODELING REAL LIFE** A *carpenter's square* is an L-shaped tool used to draw right angles. You need to cut a circular piece of wood into two semicircles. How can you use the carpenter's square to draw a diameter on the circular piece of wood? *(See Example 5.)*

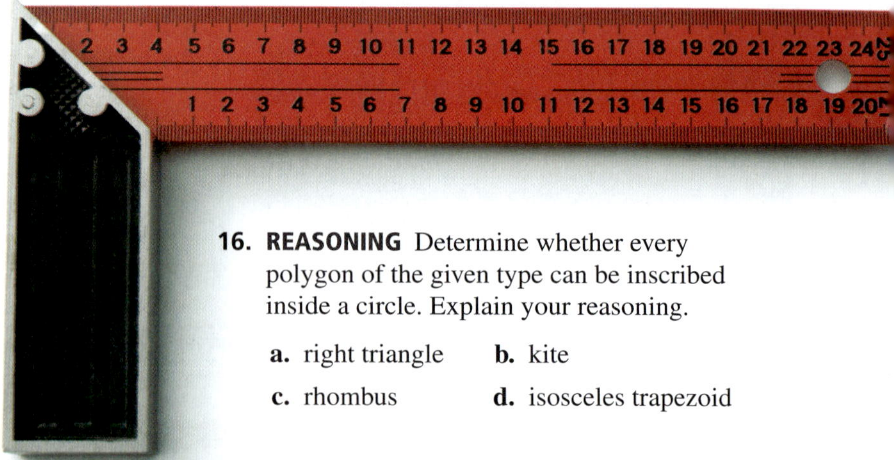

16. **REASONING** Determine whether every polygon of the given type can be inscribed inside a circle. Explain your reasoning.

a. right triangle **b.** kite

c. rhombus **d.** isosceles trapezoid

17. **WRITING** A right triangle is inscribed in a circle, and the radius of the circle is given. Explain how to find the length of the hypotenuse.

18. **WRITING** Explain why the diagonals of a rectangle inscribed in a circle are diameters of the circle.

GO DIGITAL

19. CONSTRUCTION Construct an equilateral triangle inscribed in a circle.

20. CONSTRUCTION The side length of an inscribed regular hexagon is equal to the radius of the circumscribed circle. Use this fact to construct a regular hexagon inscribed in a circle.

5 MTR **CONNECTING CONCEPTS** **In Exercises 21 and 22, find the values of x and y. Then find the measures of the interior angles of the polygon.**

21.

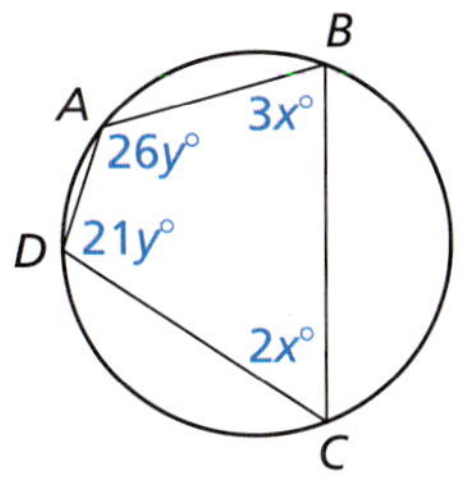

22.

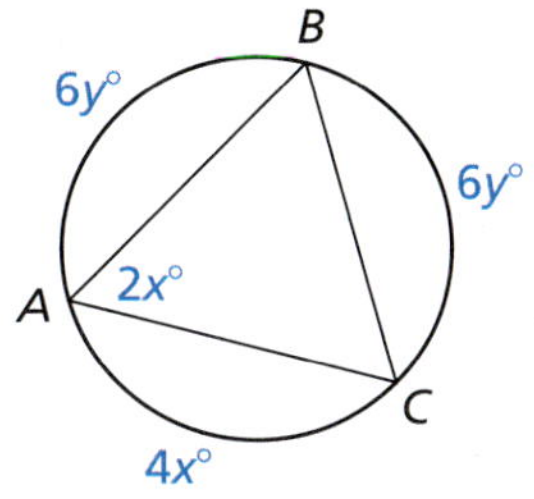

23. PROVING A THEOREM If an angle is inscribed in $\odot Q$, the center Q can be on a side of the inscribed angle, inside the inscribed angle, or outside the inscribed angle. Prove each case of the Measure of an Inscribed Angle Theorem.

a. Case 1

Given $\angle ABC$ is inscribed in $\odot Q$.
Let $m\angle B = x°$.
Center Q lies on $\overline{BC}$.

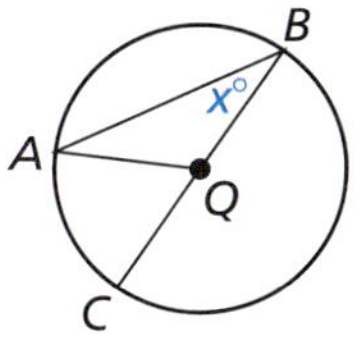

Prove $m\angle ABC = \frac{1}{2} m\overset{\frown}{AC}$

(*Hint*: Show that $\triangle AQB$ is isosceles. Then write $m\overset{\frown}{AC}$ in terms of x.)

b. Case 2 Use the diagram and auxiliary line to write a proof for Case 2.

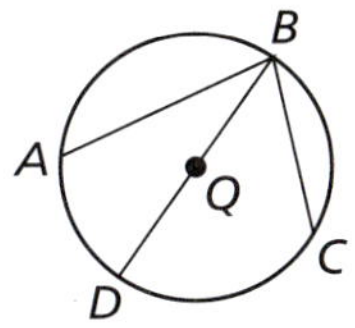

c. Case 3 Use the diagram and auxiliary line to write a proof for Case 3.

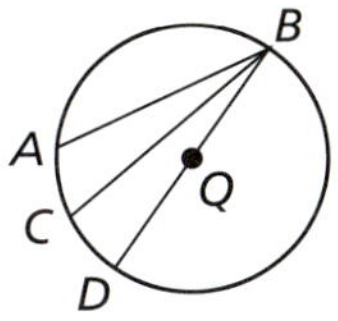

24. PROVING A THEOREM Write a proof of the Inscribed Angles of a Circle Theorem.

GO DIGITAL

4 MTR **25. MAKING AN ARGUMENT** Your friend claims that $\angle PTQ \cong \angle PSQ \cong \angle PRQ$. Is your friend correct? Explain your reasoning.

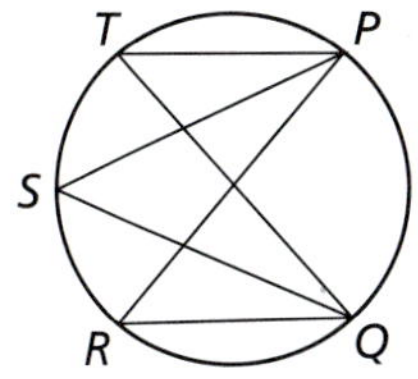

26. HOW DO YOU SEE IT?
Let point Y represent your location on the soccer field below. What type of angle is $\angle AYB$ if you stand anywhere on the circle except at point A or point B?

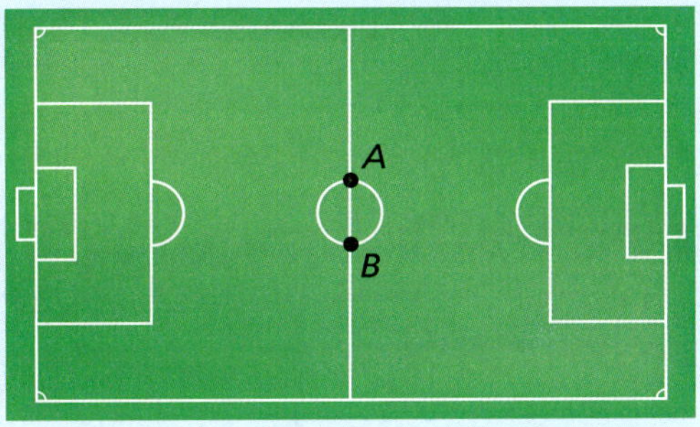

27. PROVING A THEOREM The Inscribed Right Triangle Theorem is written as a conditional statement and its converse. Write a proof for each statement.

28. PROVING A THEOREM Copy and complete the paragraph proof for one part of the Inscribed Quadrilateral Theorem.

Given $\odot C$ with inscribed quadrilateral $DEFG$

Prove $m\angle D + m\angle F = 180°$,
$m\angle E + m\angle G = 180°$

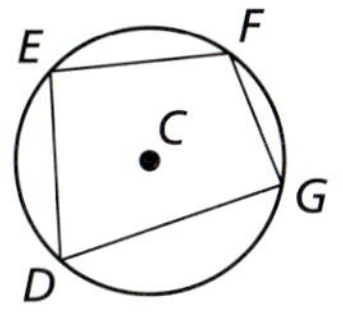

By the Arc Addition Postulate, $m\overset{\frown}{EFG}$ + ____ = 360° and $m\overset{\frown}{FGD} + m\overset{\frown}{DEF} = 360°$. Using the ________ Theorem, $m\overset{\frown}{EDG} = 2(m\angle F)$, $m\overset{\frown}{EFG} = 2(m\angle D)$, $m\overset{\frown}{DEF} = 2(m\angle G)$, and $m\overset{\frown}{FGD} = 2(m\angle E)$. By the Substitution Property of Equality, $2(m\angle D)$ + ___ = 360°, so ___. Similarly, ___.

29. PROBLEM SOLVING Three moons, A, B, and C, are in the same circular orbit 100,000 kilometers above the surface of a planet. The planet is 20,000 kilometers in diameter, and $m\angle ABC = 90°$. The distance between moons B and C is twice the distance between moons A and B. Draw a diagram of the situation. Find the distance between each pair of moons.

30. THOUGHT PROVOKING The figure shows a circle that is circumscribed about $\triangle ABC$. Is it possible to circumscribe a circle about any triangle? Justify your answer.

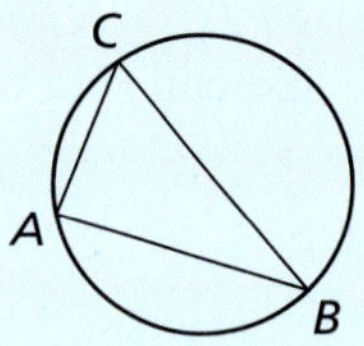

31. DIG DEEPER If you draw the smallest possible circle through C tangent to $\overline{AB}$, the circle will intersect $\overline{AC}$ at J and $\overline{BC}$ at K. Find the exact length of $\overline{JK}$.

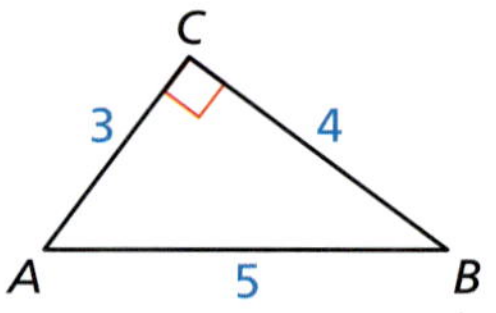

5 MTR **32. CONNECTING CONCEPTS** A cutting board is manufactured by gluing together eight 1-inch boards and cutting out a circle with an 8-inch diameter. $\overline{FH}$ is a diameter of the board shown. What is the length of the seam labeled $\overline{GK}$? Justify your answer.

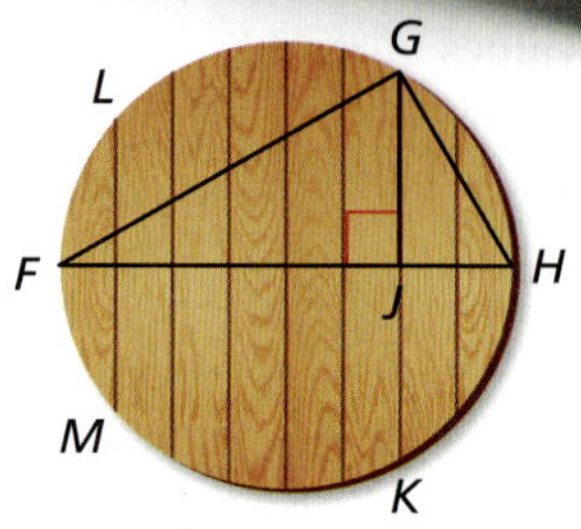

REVIEW & REFRESH

WATCH

33. Describe a congruence transformation that maps $\triangle ABC$ to $\triangle DEF$.

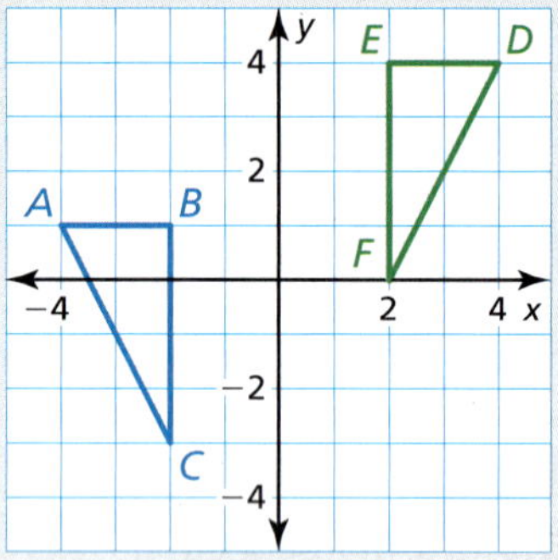

34. Find the radius of $\odot Q$.

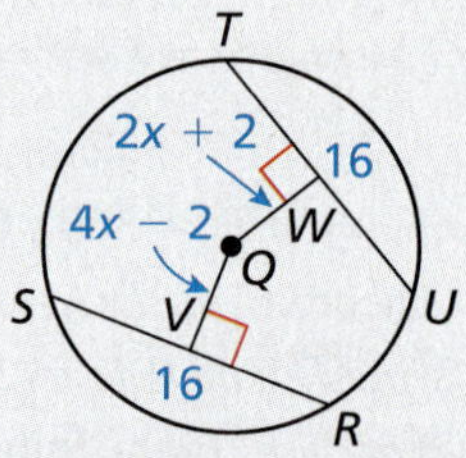

In Exercises 35–38, identify the given arc as a *major arc*, *minor arc*, or *semicircle*. Then find the measure of the arc.

35. $\widehat{BC}$

36. $\widehat{AC}$

37. $\widehat{AB}$

38. $\widehat{ABC}$

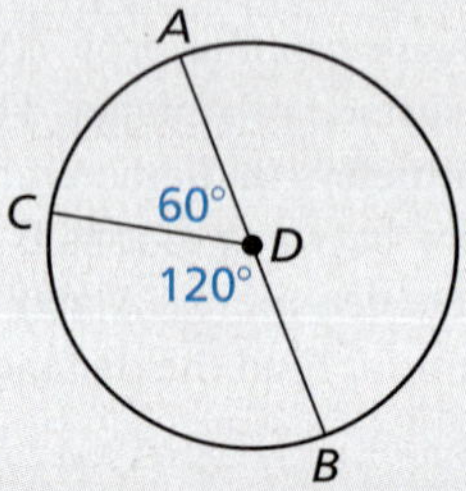

39. Tell whether $\overline{AB}$ is tangent to $\odot C$. Explain your reasoning.

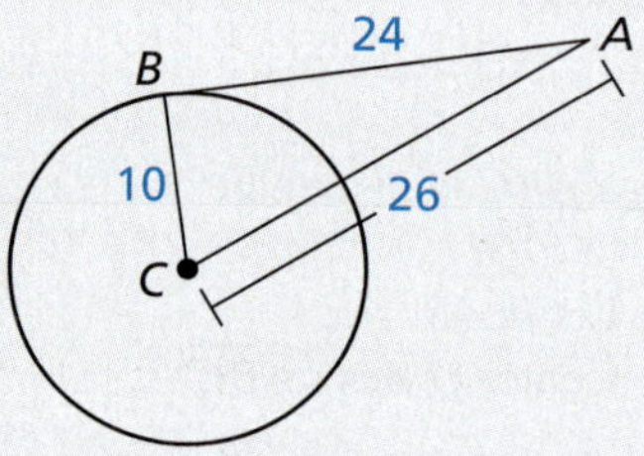

In Exercises 40 and 41, find the indicated measure.

40. $m\angle JKL$

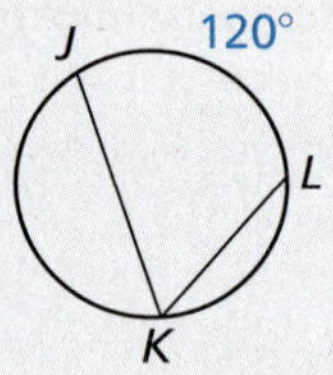

41. $m\angle PQS$

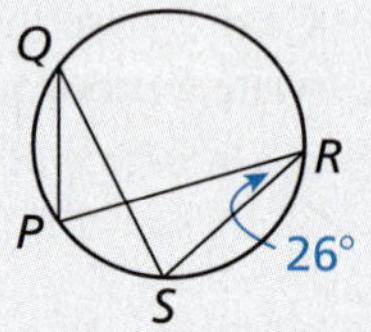

7 MTR **42. MODELING REAL LIFE** You and your friend leave school heading in opposite directions. You each travel 0.3 mile, then change directions and travel 1.1 miles. You start due east and then turn 40° toward south. Your friend starts due west and then turns 25° toward north. Who is farther from the school? Explain your reasoning.

43. Find $m\angle 1$, $m\angle 2$, and $m\angle 3$. Explain your reasoning.

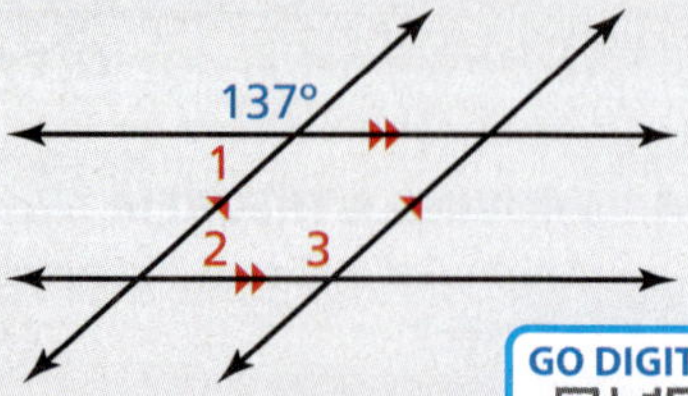

GO DIGITAL

10.5 Angle Relationships in Circles

Learning Target: Understand angles formed by chords, secants, and tangents.

Success Criteria:
- I can identify angles and arcs determined by chords, secants, and tangents.
- I can find angle measures and arc measures involving chords, secants, and tangents.
- I can use circumscribed angles to solve problems.

EXPLORE IT! Investigating Angles in Circles

Work with a partner. Use technology.

a. Construct a chord in a circle. Then construct a tangent line to the circle at one of the endpoints of the chord.

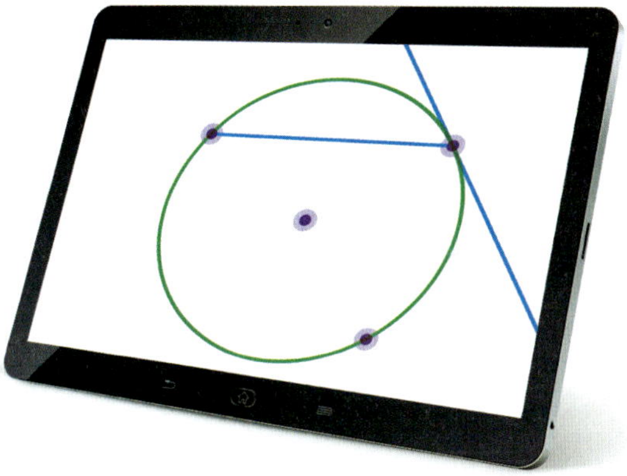

5 MTR RELATE CONCEPTS
Intersecting chords create two pairs of vertical angles. How can this help you find the relationship in part (d)?

b. Find the measures of the angles and arcs determined by the chord and the tangent line. What do you notice?

c. How do the measures of the angles and circular arcs change when you move an endpoint of the chord? How do these measures change when you resize the circle? Use your results to make a conjecture.

d. Construct two chords that intersect inside a circle. Make a conjecture about the measures of the angles and arcs determined by the chords. Provide examples to support your reasoning.

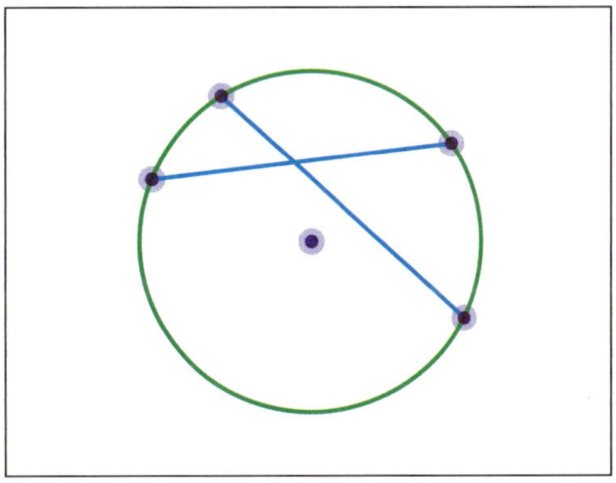

Geometric Reasoning

MA.912.GR.6.1 Solve mathematical and real-world problems involving the length of a secant, tangent, segment or chord in a given circle.

MA.912.GR.6.2 Solve mathematical and real-world problems involving the measures of arcs and related angles.

GO DIGITAL

Finding Angle and Arc Measures

Vocabulary
circumscribed angle, *p. 546*

THEOREM

10.14 Tangent and Intersected Chord Theorem

If a tangent and a chord intersect at a point on a circle, then the measure of each angle formed is one-half the measure of its intercepted arc.

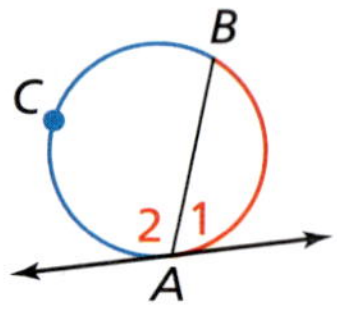

Prove this Theorem Exercise 23, page 549

$m\angle 1 = \frac{1}{2}(m\widehat{AB})$ $\quad m\angle 2 = \frac{1}{2}(m\widehat{BCA})$

EXAMPLE 1 Finding Angle and Arc Measures

Line m is tangent to the circle. Find the indicated measure.

a. $m\angle 1$

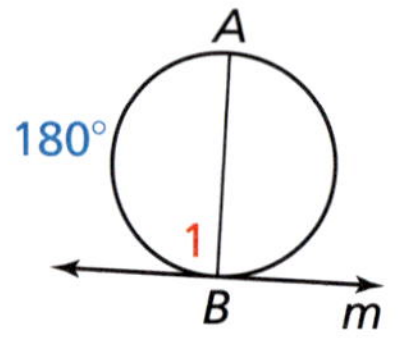

b. $m\widehat{KJL}$

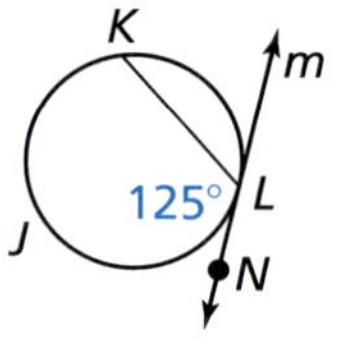

SOLUTION

Use the Tangent and Intersected Chord Theorem to find each measure.

a. $m\angle 1 = \frac{1}{2}(m\widehat{AB})$

▶ So, $m\angle 1 = \frac{1}{2}(180^\circ) = 90^\circ$.

b. $\frac{1}{2}(m\widehat{KJL}) = m\angle KLN$

▶ So, $m\widehat{KJL} = 2(125^\circ) = 250^\circ$.

SELF-ASSESSMENT

1 I don't understand yet. 2 I can do it with help. 3 I can do it on my own. 4 I can teach someone else.

Line *m* is tangent to the circle. Find the indicated measure.

1. $m\angle 1$

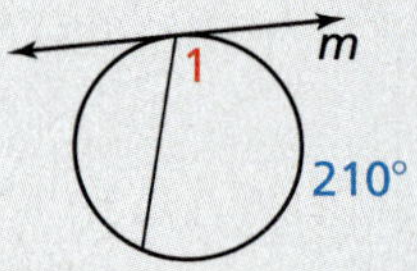

2. $m\widehat{RST}$

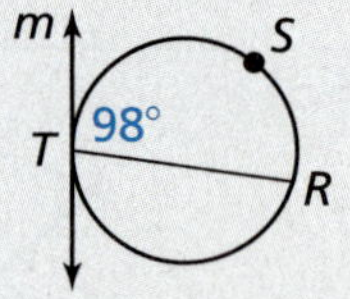

3. $m\widehat{XY}$

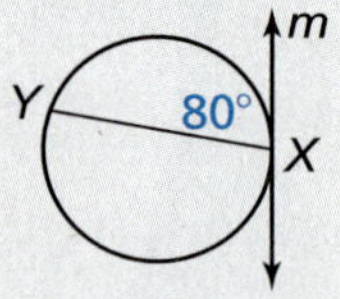

KEY IDEA

Intersecting Lines and Circles

If two nonparallel lines intersect a circle, there are three places where the lines can intersect.

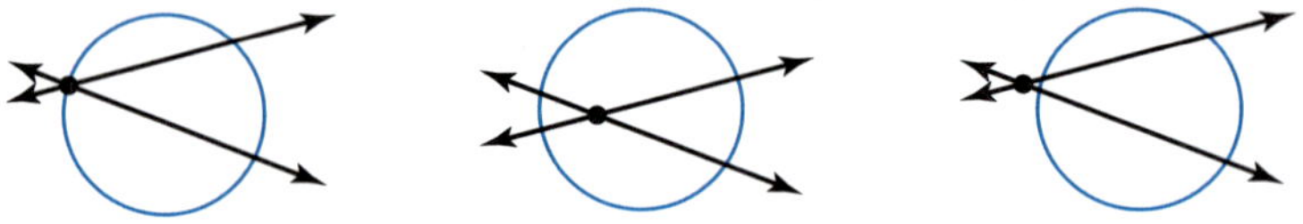

on the circle — inside the circle — outside the circle

GO DIGITAL

THEOREMS

10.15 Angles Inside the Circle Theorem

If two chords intersect *inside* a circle, then the measure of each angle is one-half the *sum* of the measures of the arcs intercepted by the angle and its vertical angle.

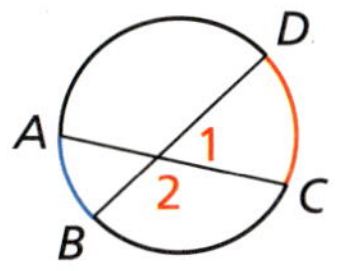

$m\angle 1 = \frac{1}{2}(m\overset{\frown}{DC} + m\overset{\frown}{AB})$

$m\angle 2 = \frac{1}{2}(m\overset{\frown}{AD} + m\overset{\frown}{BC})$

Prove this Theorem Exercise 25, page 549

10.16 Angles Outside the Circle Theorem

If a tangent and a secant, two tangents, or two secants intersect *outside* a circle, then the measure of the angle formed is one-half the *difference* of the measures of the intercepted arcs.

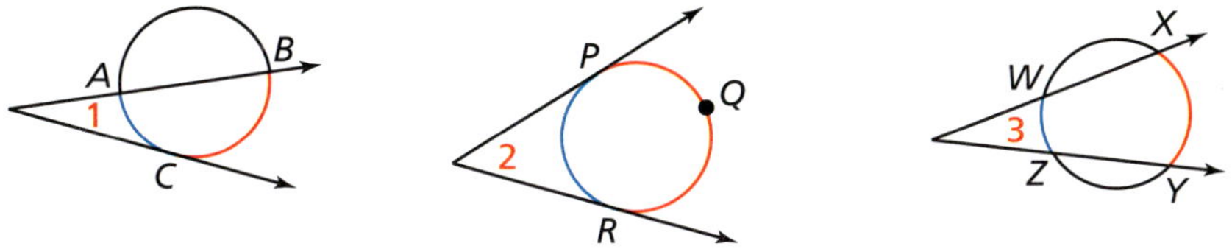

$m\angle 1 = \frac{1}{2}(m\overset{\frown}{BC} - m\overset{\frown}{AC})$ $m\angle 2 = \frac{1}{2}(m\overset{\frown}{PQR} - m\overset{\frown}{PR})$ $m\angle 3 = \frac{1}{2}(m\overset{\frown}{XY} - m\overset{\frown}{WZ})$

Prove this Theorem Exercise 27, page 550

EXAMPLE 2 Finding an Angle Measure

WATCH

Find the value of x.

a.

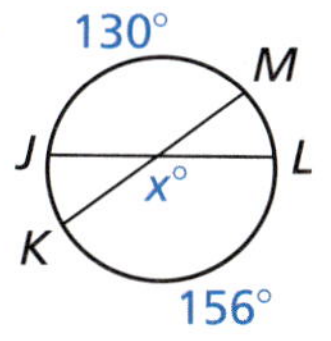

b.

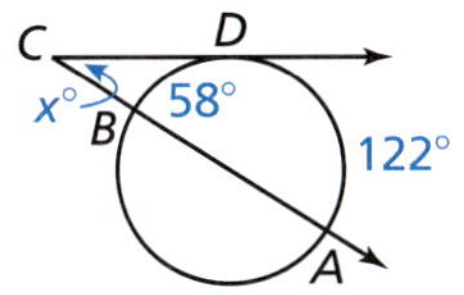

SOLUTION

a. The chords $\overline{JL}$ and $\overline{KM}$ intersect inside the circle. Use the Angles Inside the Circle Theorem.

$$x° = \frac{1}{2}(m\overset{\frown}{JM} + m\overset{\frown}{LK})$$

$$x° = \frac{1}{2}(130° + 156°)$$

$$x = 143$$

b. The tangent $\overleftrightarrow{CD}$ and the secant $\overleftrightarrow{CB}$ intersect outside the circle. Use the Angles Outside the Circle Theorem.

$$m\angle BCD = \frac{1}{2}(m\overset{\frown}{AD} - m\overset{\frown}{BD})$$

$$x° = \frac{1}{2}(122° - 58°)$$

$$x = 32$$

SELF-ASSESSMENT

1 I don't understand yet. 2 I can do it with help. 3 I can do it on my own. 4 I can teach someone else.

Find the value of the variable.

4.

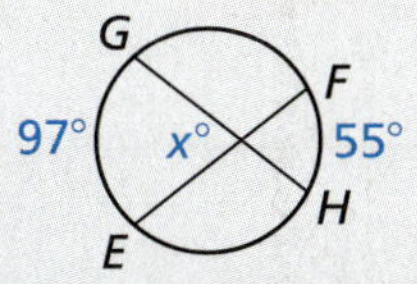

5.

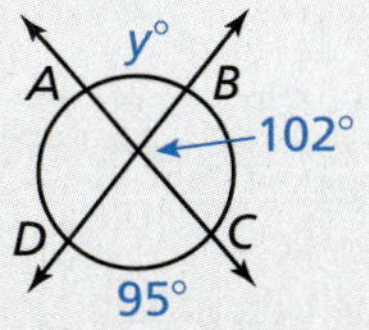

6.

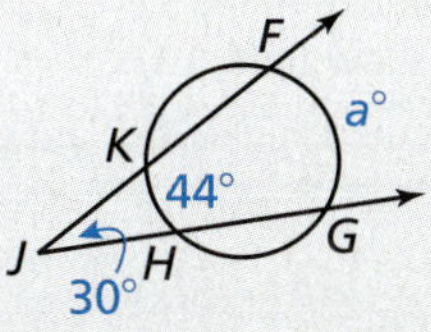

GO DIGITAL

Using Circumscribed Angles

KEY IDEA

Circumscribed Angle

A **circumscribed angle** is an angle whose sides are tangent to a circle.

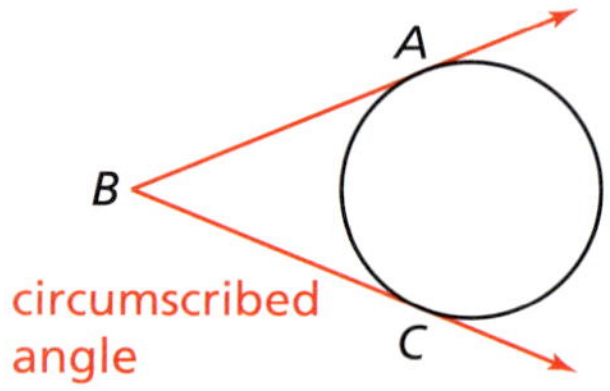

WORDS AND MATH

Circum is a Latin word that means "around." A *circumscribed angle* lies around a circle.

THEOREM

10.17 Circumscribed Angle Theorem

The measure of a circumscribed angle is equal to 180° minus the measure of the central angle that intercepts the same arc.

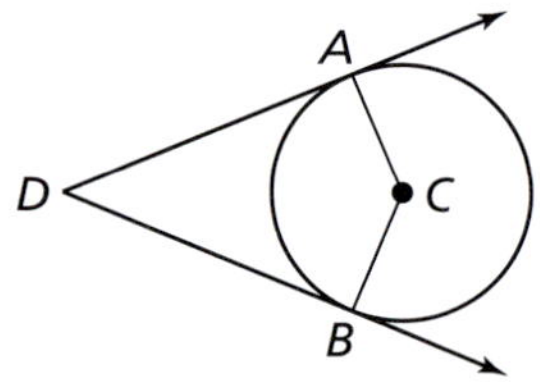

$m\angle ADB = 180° - m\angle ACB$

Prove this Theorem Exercise 28, page 550

EXAMPLE 3 Finding Angle Measures

WATCH

Find the value of x.

a.

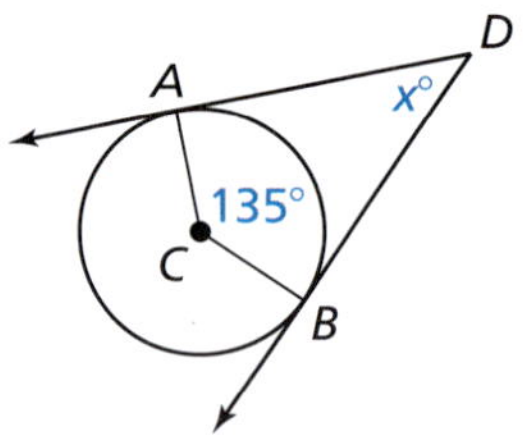

b.

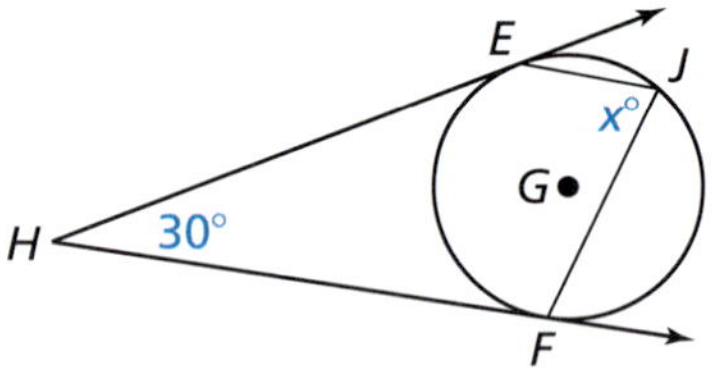

SOLUTION

a. Use the Circumscribed Angle Theorem to find $m\angle ADB$.

$m\angle ADB = 180° - m\angle ACB$	Circumscribed Angle Theorem
$x° = 180° - 135°$	Substitute.
$x = 45$	Subtract.

▶ So, the value of x is 45.

b. Use the Measure of an Inscribed Angle Theorem and the Circumscribed Angle Theorem to find $m\angle EJF$.

$m\angle EJF = \frac{1}{2}(m\overset{\frown}{EF})$	Measure of an Inscribed Angle Theorem
$m\angle EJF = \frac{1}{2}(m\angle EGF)$	Definition of measure of a minor arc
$m\angle EJF = \frac{1}{2}(180° - m\angle EHF)$	Circumscribed Angle Theorem
$x° = \frac{1}{2}(180° - 30°)$	Substitute.
$x = 75$	Simplify.

▶ So, the value of x is 75.

1 MTR HELP A CLASSMATE

Help a classmate understand how to use the Circumscribed Angle Theorem to show that $m\angle EGF = 180° - m\angle EHF$.

GO DIGITAL

7 MTR | WATCH

EXAMPLE 4 Modeling Real Life

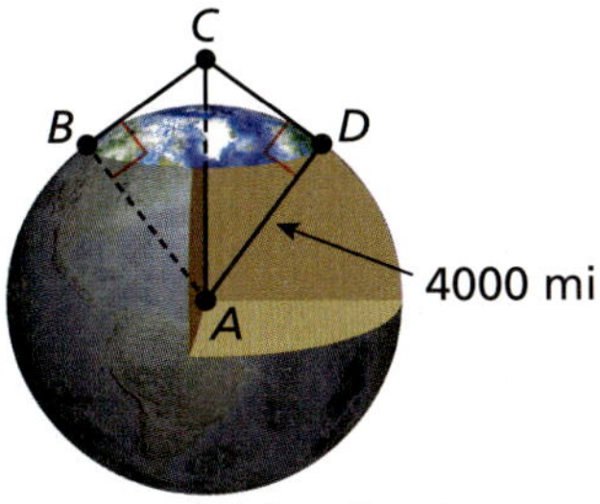

Not drawn to scale

The diagram shows the portion of Earth visible to a satellite in orbit above Earth at point C. Earth's radius is approximately 4000 miles and $m\widehat{BD} = 43°$. How high is the satellite above the surface of Earth?

SOLUTION

1. **Understand the Problem** You are given the radius of Earth and the measure of $\widehat{BD}$. You need to find the height of the satellite above the surface of Earth.
2. **Make a Plan** Use the Circumscribed Angle Theorem and the definition of a minor arc to find $m\angle BCD$ and $m\angle ACB$. Then use trigonometric ratios to find AC.
3. **Solve and Check**

$m\angle BCD = 180° - m\angle BAD$	Circumscribed Angle Theorem
$= 180° - m\widehat{BD}$	Definition of measure of a minor arc
$= 180° - 43°$	Substitute.
$= 137°$	Subtract.

Because $\overline{AB}$ and $\overline{AD}$ are radii of a sphere, $\overline{AB} \cong \overline{AD}$. Also, $\overline{CA} \cong \overline{CA}$ by the Reflexive Property of Segment Congruence. So, $\triangle ABC \cong \triangle ADC$ by the Hypotenuse-Leg Congruence Theorem. Because corresponding parts of congruent triangles are congruent, $\angle ACB \cong \angle ACD$. So, $m\angle ACB = \frac{1}{2}(137°) = 68.5°$. Use the sine ratio to find AC. Let h represent the height (in miles) of the satellite above the surface of Earth.

$\sin \angle ACB = \dfrac{\text{opp}}{\text{hyp}}$	Write ratio for sin $\angle ACB$.
$\sin 68.5° = \dfrac{4000}{4000 + h}$	Substitute.
$(4000 + h)(\sin 68.5°) = 4000$	Multiply each side by $(4000 + h)$.
$4000 + h = \dfrac{4000}{\sin 68.5°}$	Divide each side by sin 68.5°.
$h \approx 299$	Solve for h.

▶ So, the satellite is about 300 miles above the surface of Earth.

Check Because $m\widehat{BD} = 43°$, $m\angle BAC = 21.5°$. To verify that $AC \approx 4300$, compare $\cos \angle BAC \approx \frac{4000}{4300} \approx 0.9302$ and $\cos 21.5° \approx 0.9304$. Because $\frac{4000}{4300} \approx \cos 21.5°$, $AC \approx 4300$. ✓

SELF-ASSESSMENT

1 I don't understand yet. | 2 I can do it with help. | 3 I can do it on my own. | 4 I can teach someone else.

Find the value of x.

7.

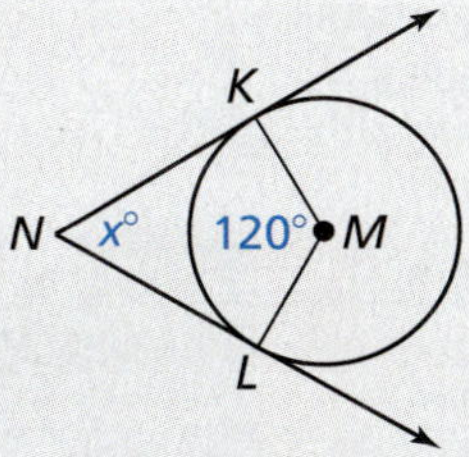

8.

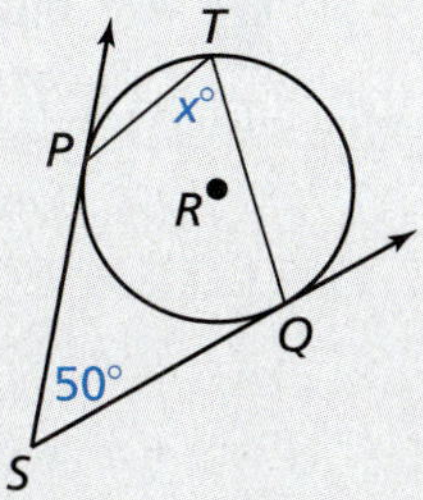

9. The diagram shows the portion of Earth you can see when you are on top of Mount Rainier at point B on a clear day, where $m\widehat{CD} \approx 4°$. What is your height above sea level?

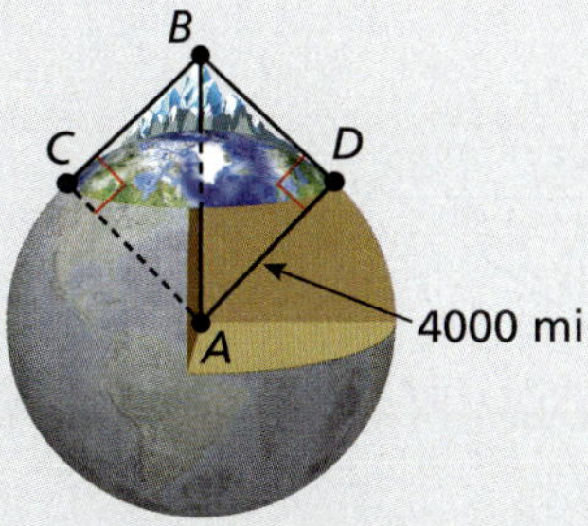

Not drawn to scale

GO DIGITAL

10.5 Practice WITH CalcChat® AND CalcView®

In Exercises 1–4, line t is tangent to the circle. Find the indicated measure. *(See Example 1.)*

1. $m\widehat{AB}$

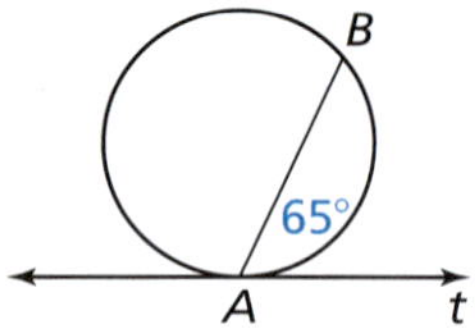

2. $m\widehat{DEF}$

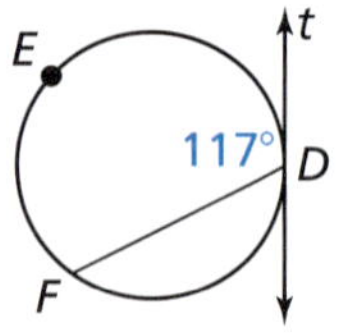

3. $m\angle 1$

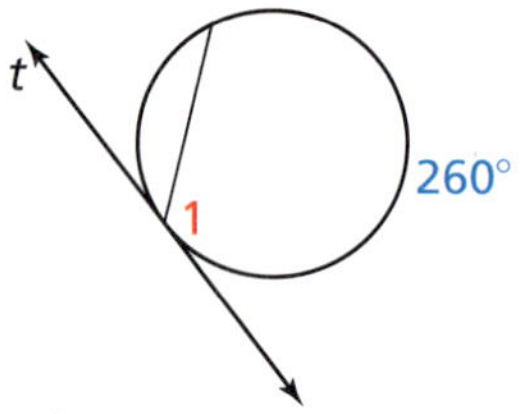

4. $m\angle 3$

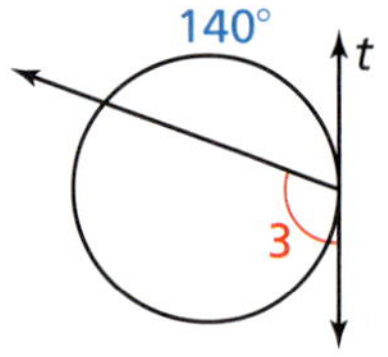

In Exercises 5–12, find the value of x. *(See Examples 2 and 3.)*

5.

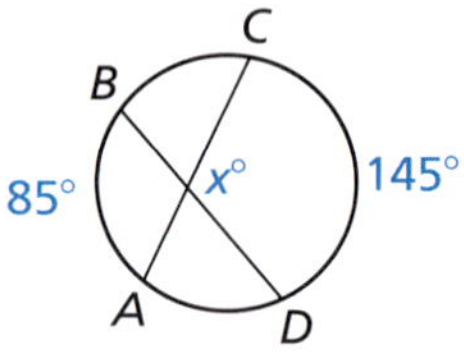

6.

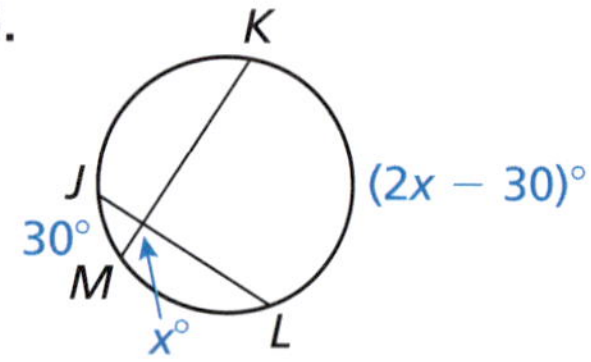

7.

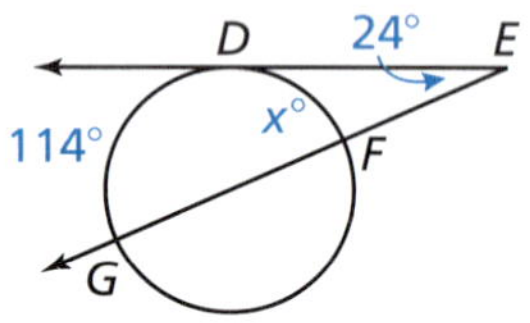

8.

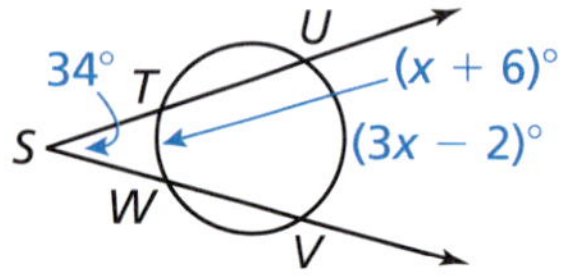

9.

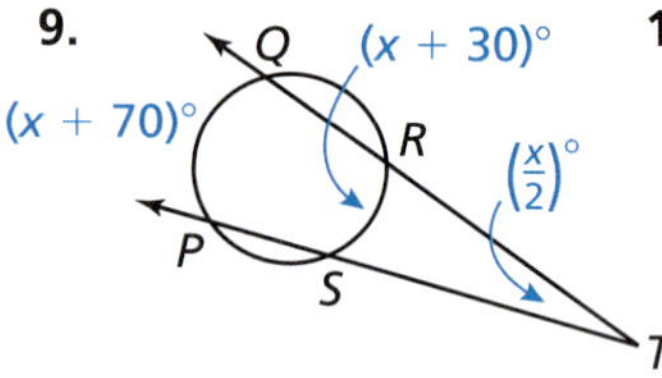

10.

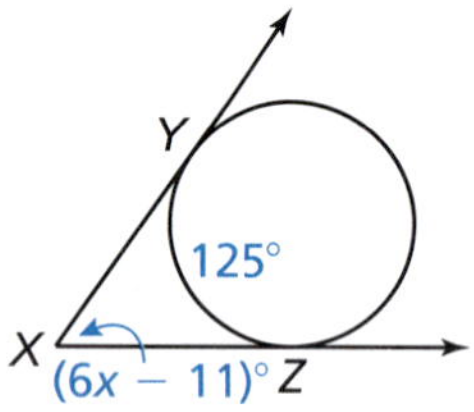

11.

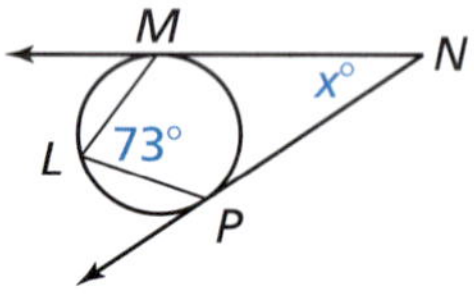

12.

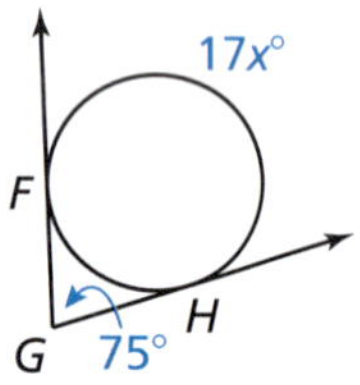

4 MTR ERROR ANALYSIS In Exercises 13 and 14, describe and correct the error in finding the angle measure.

13.

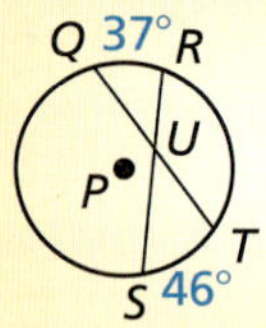

$m\angle SUT = m\widehat{ST} = 46°$

14.

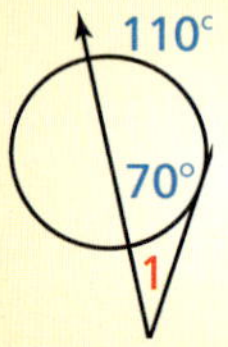

$m\angle 1 = 110° - 70°$
$= 40°$

7 MTR 15. MODELING REAL LIFE The diagram shows the portion of Earth visible from a hot air balloon above the ground at point W. Earth's radius is approximately 4000 miles and $m\widehat{ZX} = 2.8°$. How high is the hot air balloon above the ground? *(See Example 4.)*

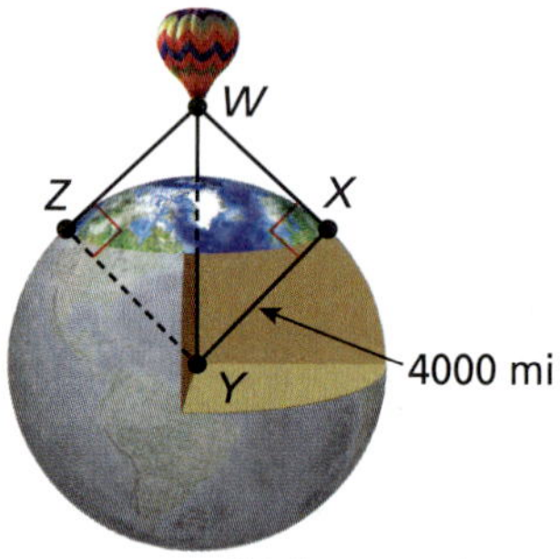

Not drawn to scale

7 MTR 16. MODELING REAL LIFE Fireworks launched from point S reach a maximum height at point F. You are on a ship about 0.01 mile above the water at point E. When the ship reaches point B, you can no longer see the fireworks because of the curvature of Earth. The radius of Earth is about 4000 miles and $m\widehat{SB} = 0.7°$. Find the maximum height of the fireworks.

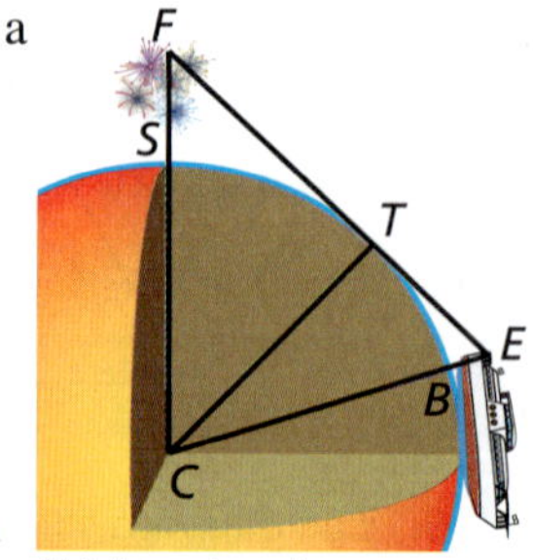

Not drawn to scale

GO DIGITAL

17. CONNECTING CONCEPTS Write an algebraic expression for c in terms of a and b.

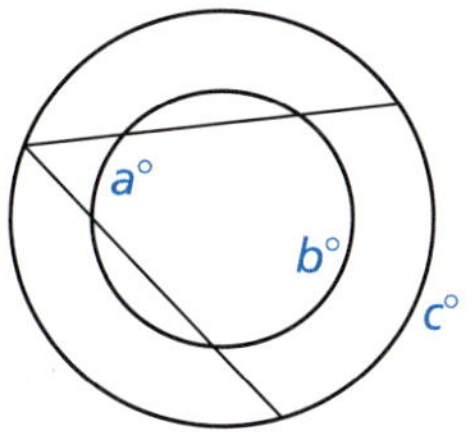

18. DISCUSS MATHEMATICAL THINKING In the diagram, $\overrightarrow{PL}$ is tangent to the circle, and $\overline{KJ}$ is a diameter. What is the range of possible angle measures of $\angle LPJ$? Explain your reasoning.

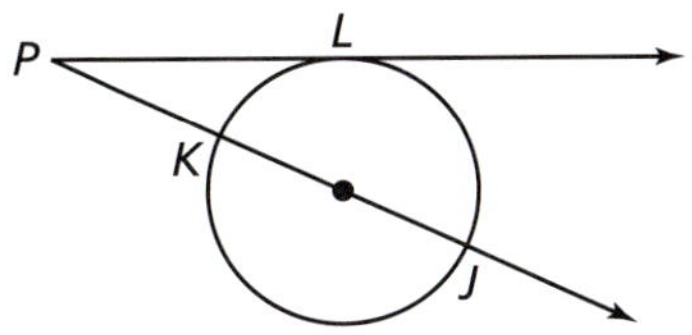

19. PROOF Prove that $m\angle JPN > m\angle JLN$.

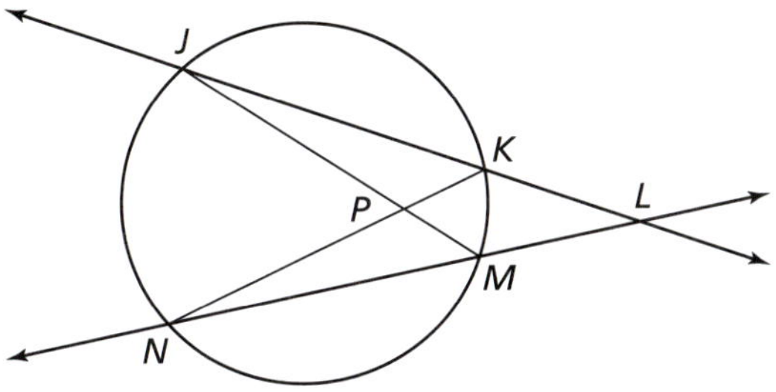

20. MAKING AN ARGUMENT Your friend claims that it is possible for a circumscribed angle to have the same measure as its intercepted arc. Is your friend correct? Explain your reasoning.

21. REASONING Points A and B are on a circle, and t is a tangent line containing A and another point C.

a. Draw two diagrams that illustrate this situation.

b. Write an equation for $m\overparen{AB}$ in terms of $m\angle BAC$ for each diagram.

c. For what measure of $\angle BAC$ can you use either equation to find $m\overparen{AB}$? Explain.

22. REASONING $\triangle XYZ$ is an equilateral triangle inscribed in $\odot P$. $\overline{AB}$ is tangent to $\odot P$ at point X, $\overline{BC}$ is tangent to $\odot P$ at point Y, and $\overline{AC}$ is tangent to $\odot P$ at point Z. Draw a diagram that illustrates this situation. Then classify $\triangle ABC$ by its angles and sides. Justify your answer.

23. PROVING A THEOREM To prove the Tangent and Intersected Chord Theorem, you must prove three cases.

a. The diagram shows the case where $\overline{AB}$ contains the center of the circle. Use the Tangent Line to Circle Theorem to write a proof for this case.

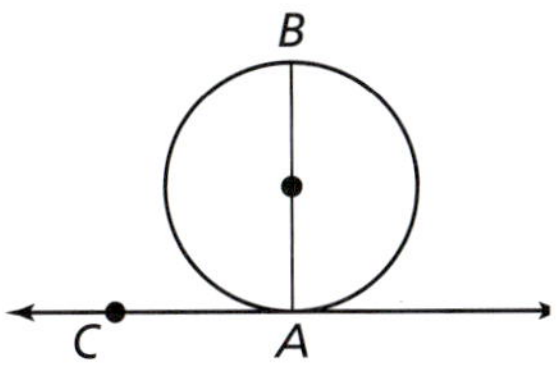

b. Draw a diagram and write a proof for the case where the center of the circle is in the interior of $\angle CAB$.

c. Draw a diagram and write a proof for the case where the center of the circle is in the exterior of $\angle CAB$.

24. HOW DO YOU SEE IT?
In the diagram, television cameras are positioned at A and B to record the news anchors at the desk. The desk is an arc of $\odot A$. You want to move Camera B so that $m\angle B$ is 30°. Should you move the camera closer or farther away? How does this affect $m\overparen{CD}$? Explain your reasoning.

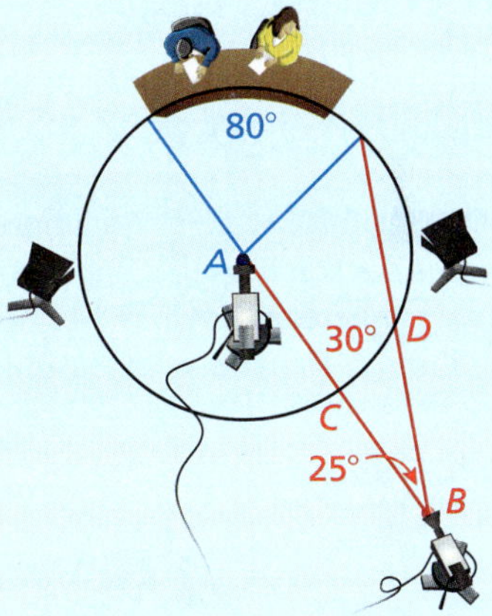

25. PROVING A THEOREM Write a proof of the Angles Inside the Circle Theorem.

Given Chords $\overline{AC}$ and $\overline{BD}$ intersect inside a circle.

Prove $m\angle 1 = \frac{1}{2}(m\overparen{DC} + m\overparen{AB})$

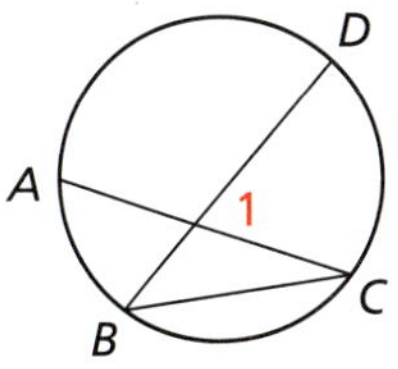

GO DIGITAL

26. THOUGHT PROVOKING In the figure, $\overrightarrow{PB}$ and $\overrightarrow{PC}$ are tangent to the circle. Point A is any point on the major arc formed by the endpoints of the chord $\overline{BC}$. Label all congruent angles in the figure. Justify your reasoning.

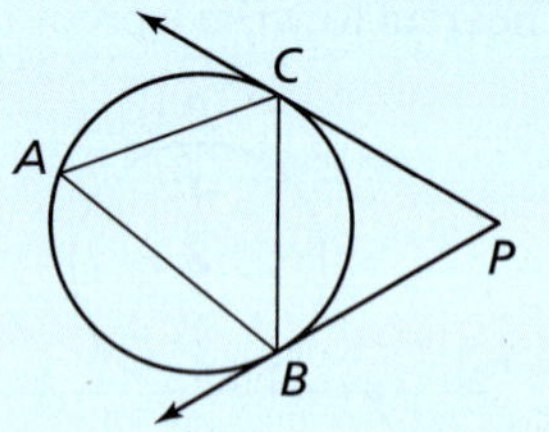

27. PROVING A THEOREM Use the diagram below to prove the Angles Outside the Circle Theorem for the case of a tangent and a secant, where the secant passes through the center of the circle. Then copy the diagrams for the other two cases on page 545 and draw appropriate auxiliary segments. Use your diagrams to prove each case.

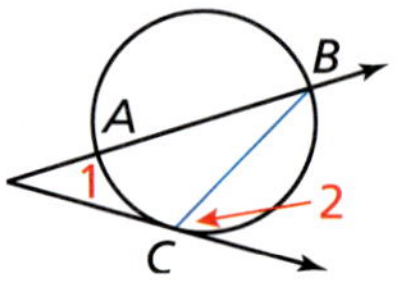

28. PROVING A THEOREM Prove that the Circumscribed Angle Theorem follows from the Angles Outside the Circle Theorem.

DIG DEEPER In Exercises 29 and 30, find the indicated measure(s). Justify your answer.

29. Find $m\angle P$ when $m\widehat{WZY} = 200°$.

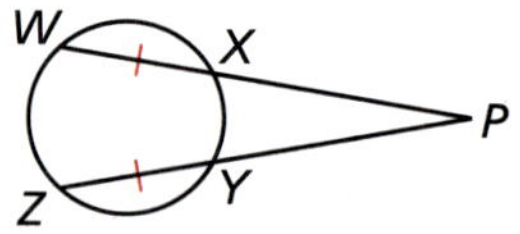

30. Find $m\widehat{AB}$ and $m\widehat{ED}$.

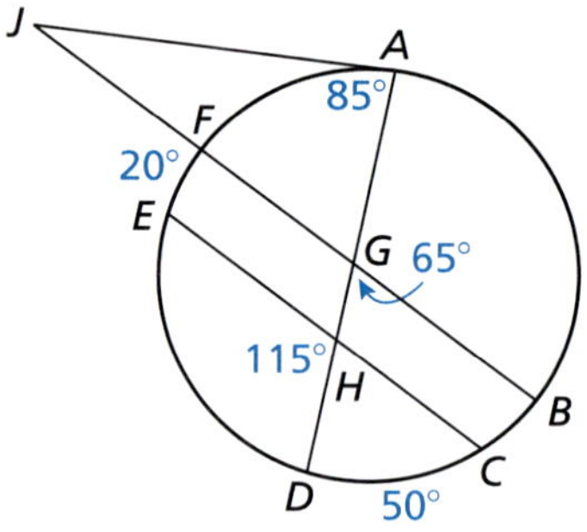

REVIEW & REFRESH

WATCH

31. Find the perimeter and area of the triangle with vertices $P(-3, -7)$, $Q(-3, 8)$, and $R(5, 8)$.

7 MTR **32. MODELING REAL LIFE** An amusement park ride swings back and forth along a circular arc as shown, where $m\angle EAB = 145°$ and $m\angle DAC = 80°$. Find $m\widehat{BC}$. Explain your reasoning.

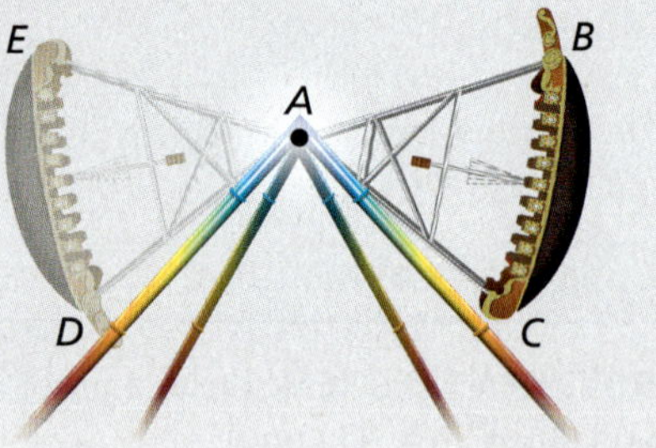

In Exercises 33 and 34, find the value of x.

33.

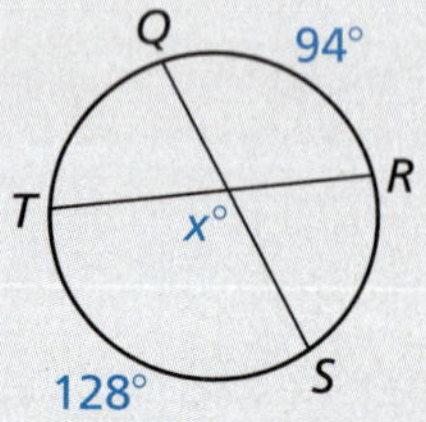

34.

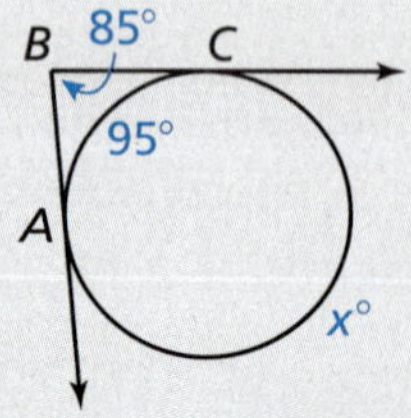

35. A triangle has one side of length 18 and another side of length 11. Describe the possible lengths of the third side.

In Exercises 36 and 37, find the indicated measure.

36. $m\angle B$

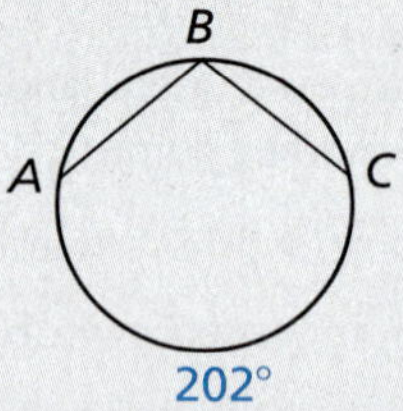

37. $m\widehat{JK}$

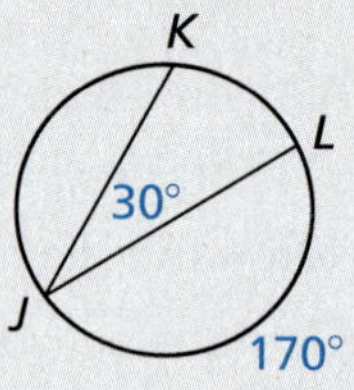

38. Determine whether $\overline{WY}$ is a diameter of the circle. Explain your reasoning.

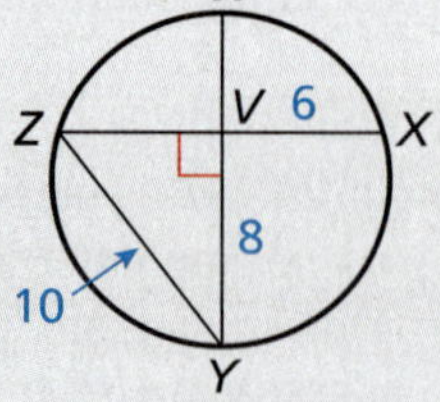

39. Graph $\triangle ABC$ with vertices $A(-8, 6)$, $B(4, 2)$, and $C(-2, -4)$ and its image after a dilation with a scale factor of $\frac{1}{2}$.

GO DIGITAL

10.6 Segment Relationships in Circles

Learning Target: Use theorems about segments of chords, secants, and tangents.

Success Criteria:
- I can find lengths of segments of chords.
- I can identify segments of secants and tangents.
- I can find lengths of segments of secants and tangents.

EXPLORE IT! Investigating Segments of Chords and Secants

Work with a partner. Use technology.

a. Construct two chords $\overline{BC}$ and $\overline{DE}$ that intersect in the interior of a circle at a point F.

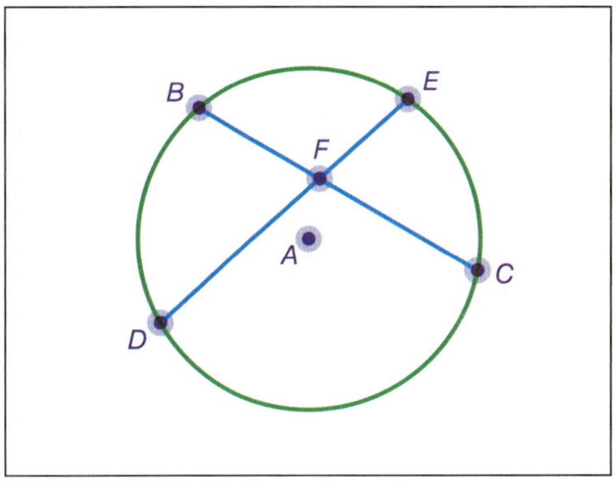

i. Find the segment lengths BF, CF, DF, and EF. How are BF and CF related to DF and EF?

ii. How do the segment lengths change when you move endpoints of the chords? How do the segment lengths change when you resize the circle? Use your results to make a conjecture.

4 MTR CONSTRUCT AN ARGUMENT

For two intersecting chords, is it *sometimes*, *always*, or *never* true that the sum of the lengths of the segments of one chord is equal to the sum of the lengths of the segments of the other chord? Explain.

b. Construct two secants $\overleftrightarrow{BC}$ and $\overleftrightarrow{BD}$ that intersect at a point B outside a circle.

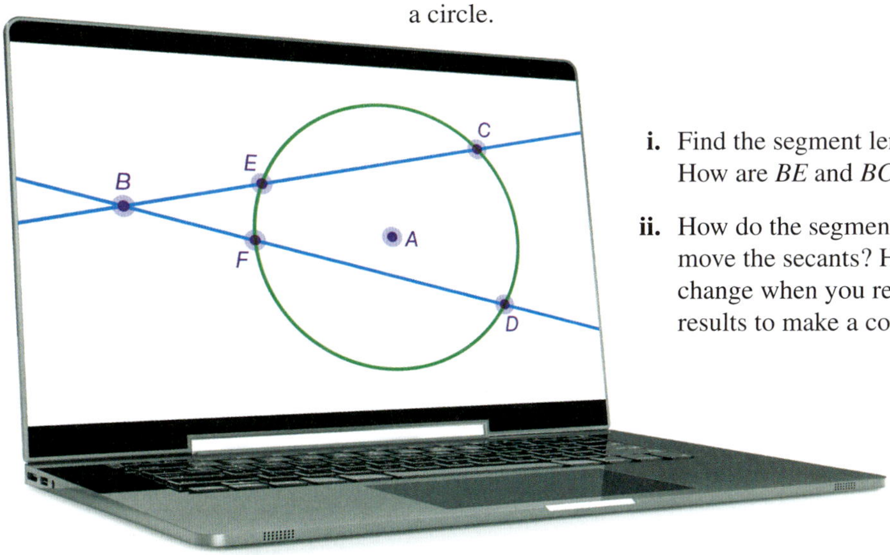

i. Find the segment lengths BE, BC, BF, and BD. How are BE and BC related to BF and BD?

ii. How do the segment lengths change when you move the secants? How do the segment lengths change when you resize the circle? Use your results to make a conjecture.

Geometric Reasoning

MA.912.GR.6.1 Solve mathematical and real-world problems involving the length of a secant, tangent, segment or chord in a given circle.

GO DIGITAL

Using Segments of Chords, Tangents, and Secants

When two chords intersect in the interior of a circle, each chord is divided into two segments that are called **segments of the chord**.

Vocabulary

segments of a chord, *p. 552*
tangent segment, *p. 553*
secant segment, *p. 553*
external segment, *p. 553*

THEOREM

10.18 Segments of Chords Theorem

If two chords intersect in the interior of a circle, then the product of the lengths of the segments of one chord is equal to the product of the lengths of the segments of the other chord.

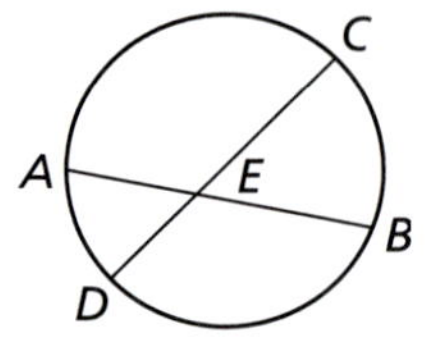

$EA \cdot EB = EC \cdot ED$

Prove this Theorem Exercise 20, page 556

EXAMPLE 1 Using Segments of Chords

Find ML and JK.

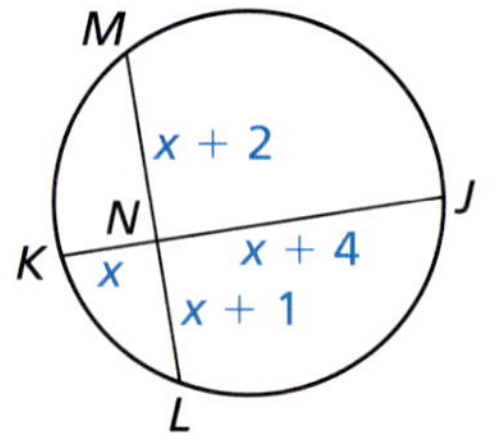

SOLUTION

Use the Segments of Chords Theorem to find the value of x.

$NK \cdot NJ = NL \cdot NM$	Segments of Chords Theorem
$x \cdot (x + 4) = (x + 1) \cdot (x + 2)$	Substitute.
$x^2 + 4x = x^2 + 3x + 2$	Simplify.
$4x = 3x + 2$	Subtract x^2 from each side.
$x = 2$	Subtract $3x$ from each side.

Find ML and JK by substitution.

$$ML = (x + 2) + (x + 1) = 2 + 2 + 2 + 1 = 7$$

$$JK = x + (x + 4) = 2 + 2 + 4 = 8$$

▶ So, $ML = 7$ and $JK = 8$.

SELF-ASSESSMENT

1 I don't understand yet. 2 I can do it with help. 3 I can do it on my own. 4 I can teach someone else.

Find the value of x.

1.

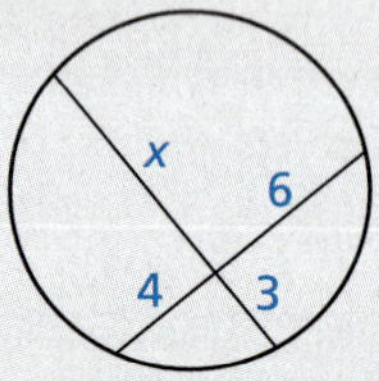

2.

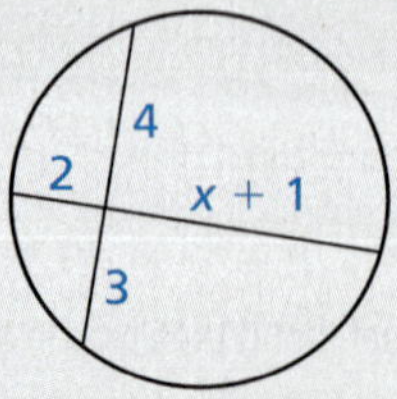

3.

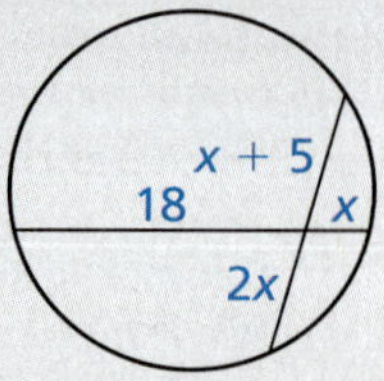

GO DIGITAL

KEY IDEAS

Tangent Segment and Secant Segment

A **tangent segment** is a segment that is tangent to a circle at an endpoint. A **secant segment** is a segment that contains a chord of a circle and has exactly one endpoint outside the circle. The part of a secant segment that is outside the circle is called an **external segment**.

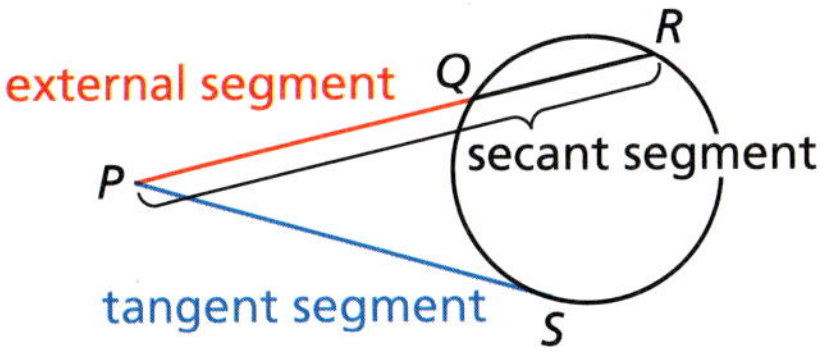

$\overline{PS}$ is a tangent segment.
$\overline{PR}$ is a secant segment.
$\overline{PQ}$ is the external segment of $\overline{PR}$.

THEOREM

10.19 Segments of Secants Theorem

If two secant segments share the same endpoint outside a circle, then the product of the lengths of one secant segment and its external segment equals the product of the lengths of the other secant segment and its external segment.

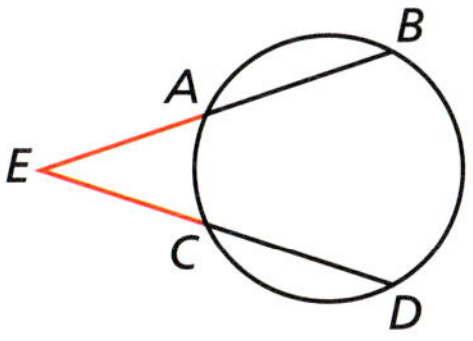

$EA \cdot EB = EC \cdot ED$

Prove this Theorem Exercise 17, page 555

EXAMPLE 2 Using Segments of Secants

WATCH

Find the value of x.

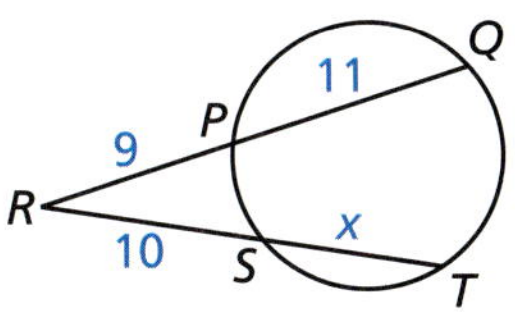

SOLUTION

$RP \cdot RQ = RS \cdot RT$	Segments of Secants Theorem
$9 \cdot (11 + 9) = 10 \cdot (x + 10)$	Substitute.
$180 = 10x + 100$	Simplify.
$80 = 10x$	Subtract 100 from each side.
$8 = x$	Divide each side by 10.

▶ The value of x is 8.

SELF-ASSESSMENT

1 I don't understand yet. 2 I can do it with help. 3 I can do it on my own. 4 I can teach someone else.

Find the value of x.

4.

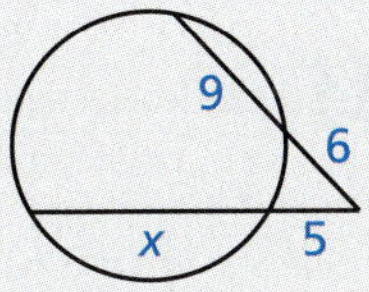

5.

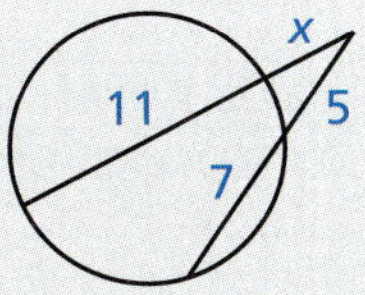

6.

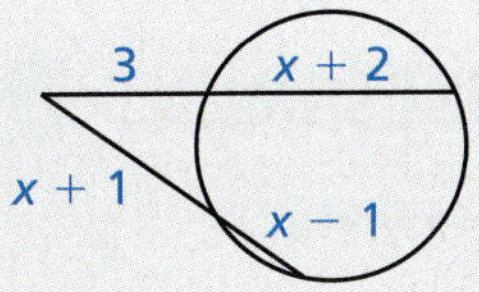

7. WRITING Explain the difference between a tangent segment and a secant segment.

GO DIGITAL

THEOREM

10.20 Segments of Secants and Tangents Theorem

If a secant segment and a tangent segment share an endpoint outside a circle, then the product of the lengths of the secant segment and its external segment equals the square of the length of the tangent segment.

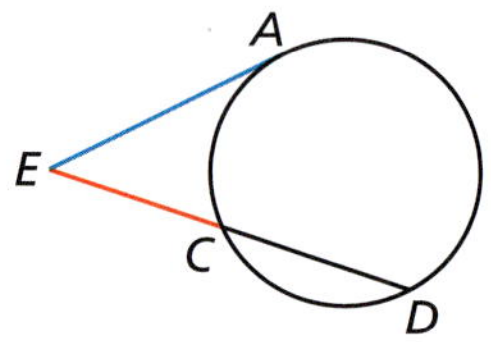

$EA^2 = EC \cdot ED$

Prove this Theorem Exercises 21 and 22, page 556

EXAMPLE 3 Using Segments of Secants and Tangents

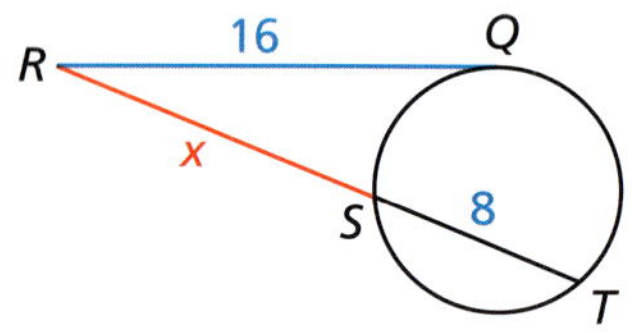

Find RS.

SOLUTION

$RQ^2 = RS \cdot RT$	Segments of Secants and Tangents Theorem
$16^2 = x \cdot (x + 8)$	Substitute.
$256 = x^2 + 8x$	Simplify.
$0 = x^2 + 8x - 256$	Write in standard form.
$x = \dfrac{-8 \pm \sqrt{8^2 - 4(1)(-256)}}{2(1)}$	Use Quadratic Formula.
$x = -4 \pm 4\sqrt{17}$	Simplify.

Use the positive solution because lengths cannot be negative.

▶ So, $RS = -4 + 4\sqrt{17}$, or about 12.49.

EXAMPLE 4 Finding the Radius of a Circle

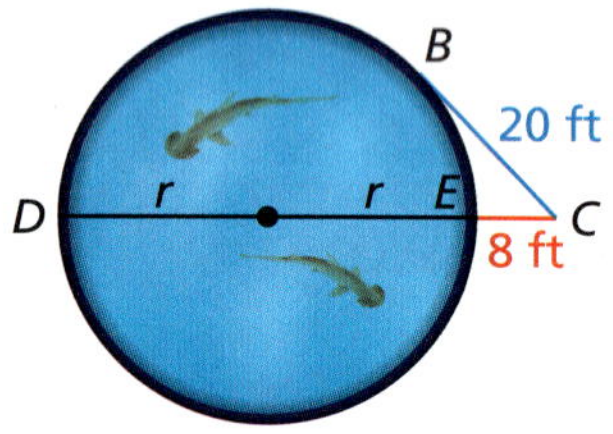

Find the radius of the aquarium tank.

SOLUTION

$CB^2 = CE \cdot CD$	Segments of Secants and Tangents Theorem
$20^2 = 8 \cdot (2r + 8)$	Substitute.
$400 = 16r + 64$	Simplify.
$336 = 16r$	Subtract 64 from each side.
$21 = r$	Divide each side by 16.

▶ So, the radius of the tank is 21 feet.

SELF-ASSESSMENT

1 I don't understand yet. 2 I can do it with help. 3 I can do it on my own. 4 I can teach someone else.

Find the value of x.

8.

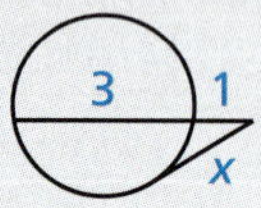

9.

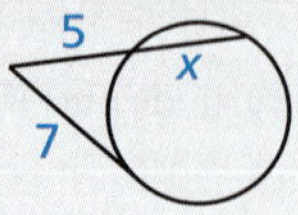

10.

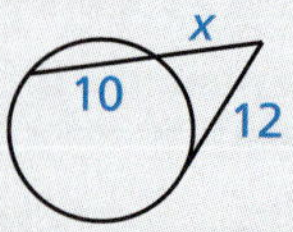

11. WHAT IF? In Example 4, $CB = 35$ feet and $CE = 14$ feet. Find the radius of the tank.

GO DIGITAL

10.6 Practice WITH CalcChat® AND CalcView®

In Exercises 1–4, find the value of *x*. *(See Example 1.)*

▶ **1.**

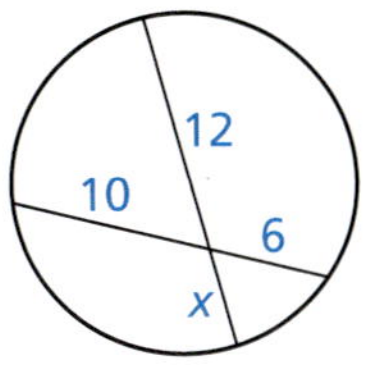

2.

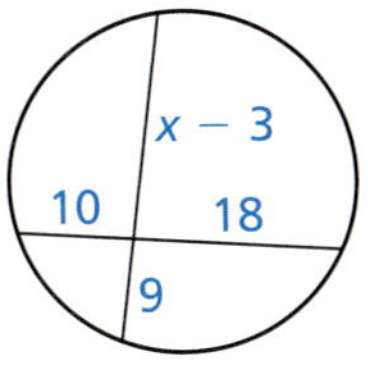

3.

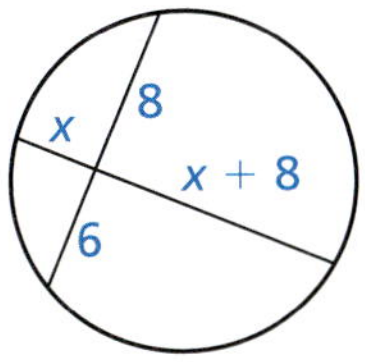

4.

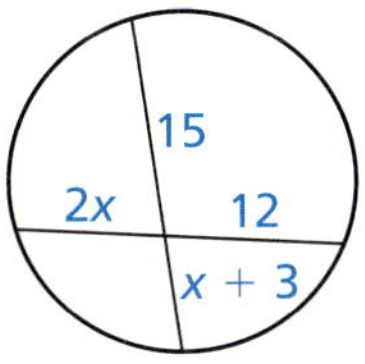

In Exercises 5–8, find the value of *x*. *(See Example 2.)*

▶ **5.**

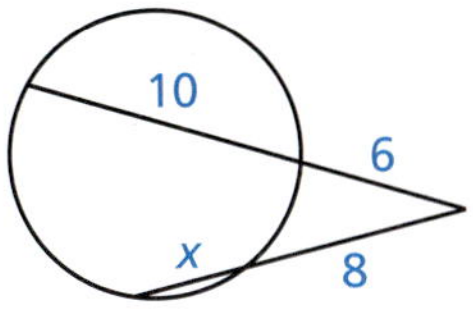

6.

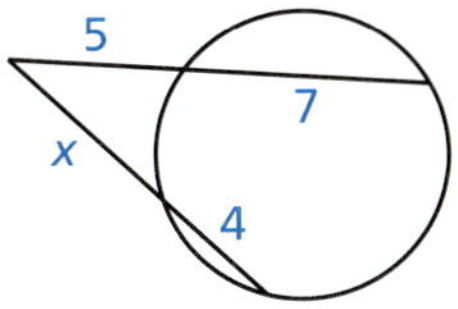

7.

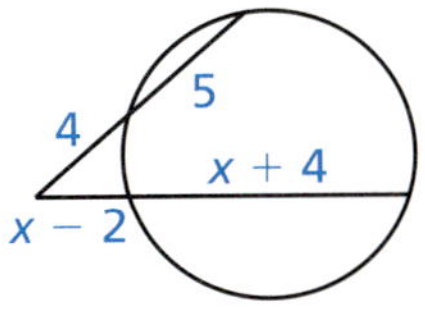

8.

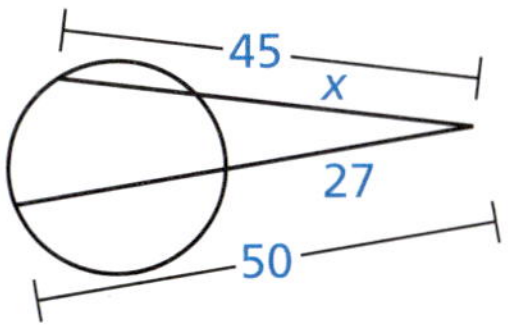

In Exercises 9–12, find the value of *x*. *(See Example 3.)*

▶ **9.**

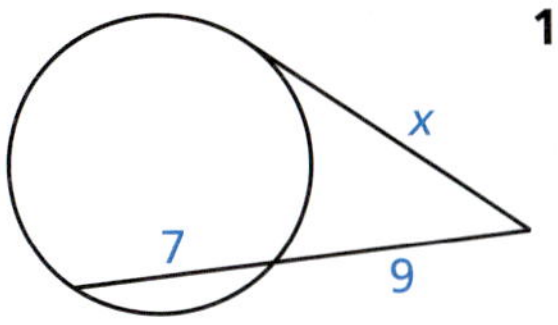

10.

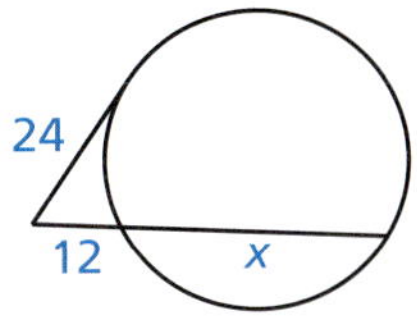

11.

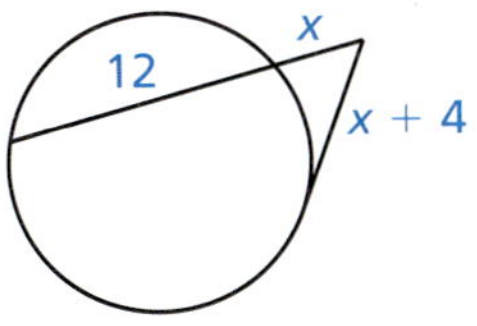

12.

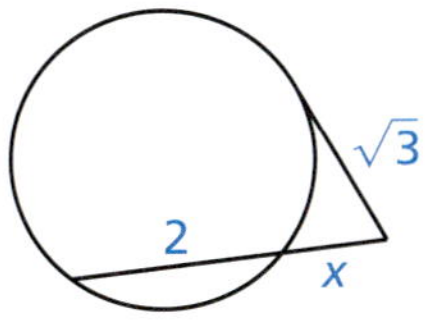

4 MTR **13. ERROR ANALYSIS** Describe and correct the error in finding *CD*.

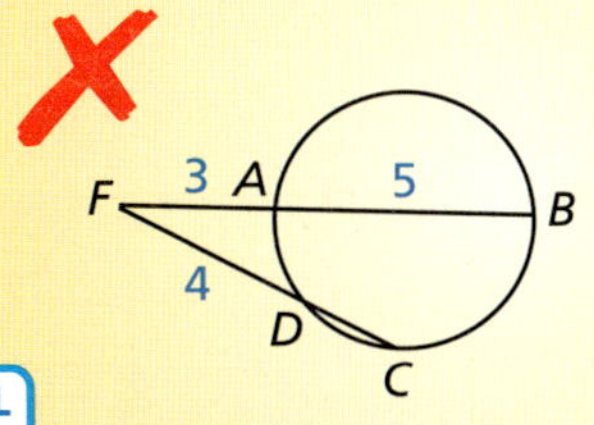

$CD \cdot DF = AB \cdot AF$
$CD \cdot 4 = 5 \cdot 3$
$CD \cdot 4 = 15$
$CD = 3.75$

4 MTR **14. ERROR ANALYSIS** Describe and correct the error in finding *LM*.

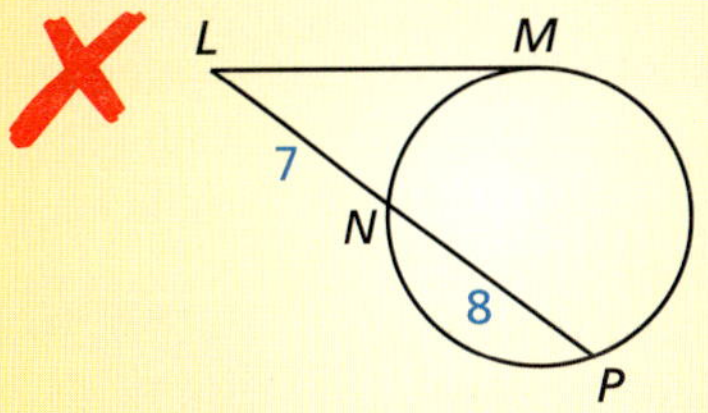

$LM = LN \cdot LP$
$LM = 7 \cdot 15$
$LM = 105$

7 MTR ▶ **15. MODELING REAL LIFE** The *Cassini* spacecraft conducted a series of missions in Saturn's orbit from 2004 to 2017. Three of Saturn's moons, Tethys, Calypso, and Telesto, have nearly circular orbits of radius 295,000 kilometers. The diagram shows the positions of the moons and the spacecraft on one of *Cassini's* missions. Find the distance *DB* from *Cassini* to Tethys when $\overline{AD}$ is tangent to the circular orbit. *(See Example 4.)*

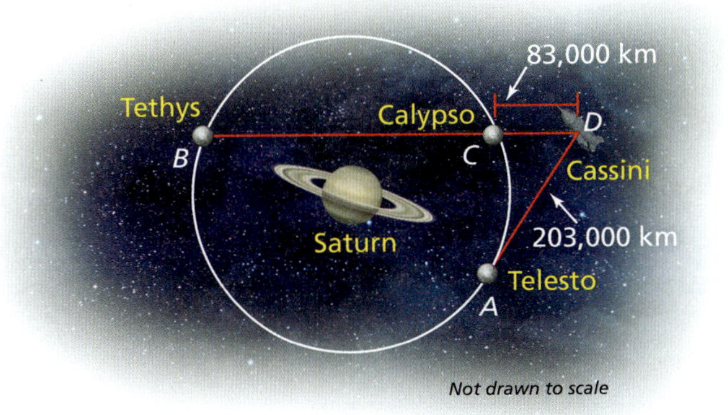

7 MTR **16. MODELING REAL LIFE** Newgrange is a large tomb in Ireland consisting of a circular mound with a diameter of 250 feet. A 62-foot-long passage leads toward the mound's center. Find the perpendicular distance *x* from the end of the passage to either side of the mound.

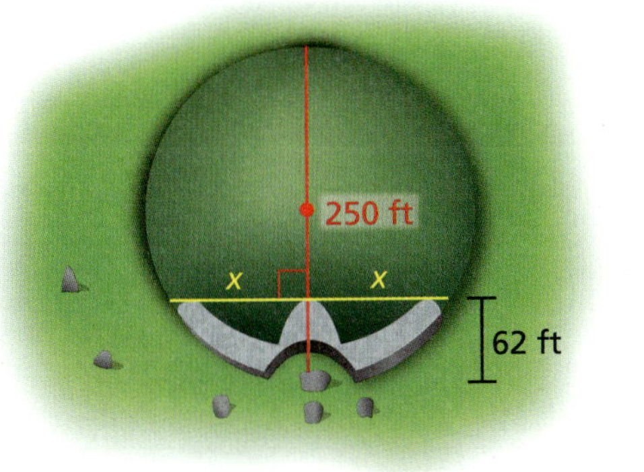

17. PROVING A THEOREM Prove the Segments of Secants Theorem. (*Hint*: Draw a diagram and add auxiliary line segments to form similar triangles.)

GO DIGITAL

18. HOW DO YOU SEE IT?
Which two theorems do you need to use to find PQ? Explain your reasoning.

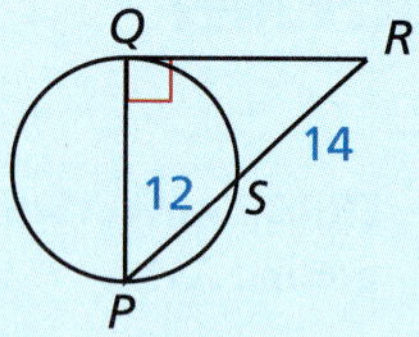

19. MODELING REAL LIFE You are designing an animated logo for your website. Sparkles leave point C and move to the outer circle along the segments shown so that all of the sparkles reach the outer circle at the same time. Sparkles travel from point C to point D at 2 centimeters per second. How fast should sparkles move from point C to point N? Explain.

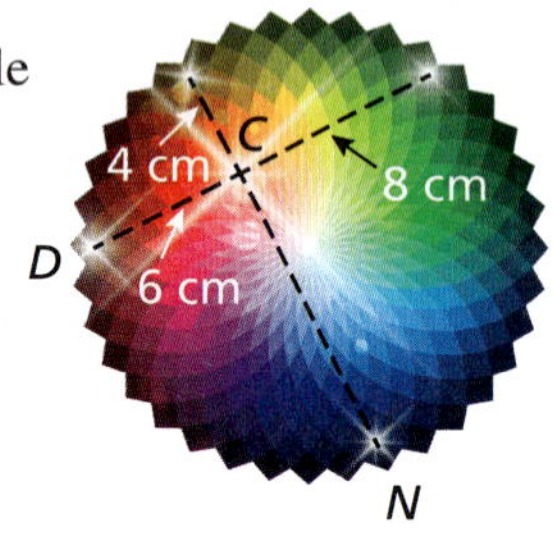

20. PROVING A THEOREM Write a proof of the Segments of Chords Theorem.

Plan for Proof Use the diagram from page 552. Draw $\overline{AC}$ and $\overline{DB}$. Show that $\triangle EAC$ and $\triangle EDB$ are similar. Use the fact that corresponding side lengths in similar triangles are proportional.

21. PROVING A THEOREM Use the Tangent Line to Circle Theorem to prove the Segments of Secants and Tangents Theorem for the special case when the secant segment contains the center of the circle.

22. PROVING A THEOREM Prove the Segments of Secants and Tangents Theorem.

23. CONNECTING CONCEPTS In the figure, $AB = 12$, $BC = 8$, $DE = 6$, $PD = 4$, and A is a point of tangency. Find the radius of $\odot P$.

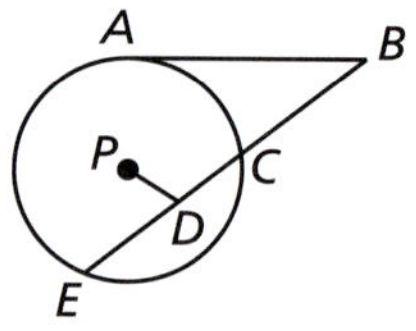

24. THOUGHT PROVOKING
Circumscribe a triangle about a circle. Then, using the points of tangency, inscribe a triangle in the circle. Must it be true that the two triangles are similar? Explain your reasoning.

REVIEW & REFRESH

WATCH

25. In the diagram, $AC = FD = 30$, $PG = x + 5$, and $PJ = 3x - 1$. Find the radius of $\odot P$.

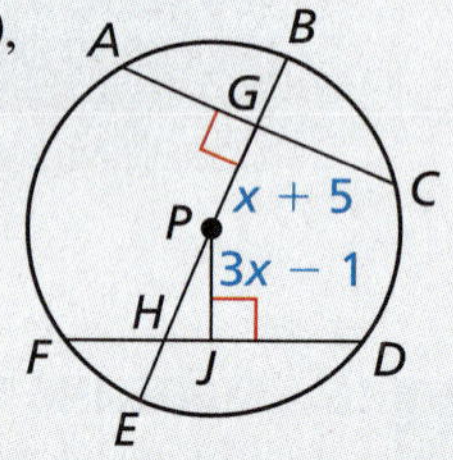

26. MODELING REAL LIFE You are 60 feet from a radio tower and the angle of elevation from the ground to the top of the tower is 69°. Find the height of the radio tower.

In Exercises 27 and 28, find the indicated measure.

27. MP

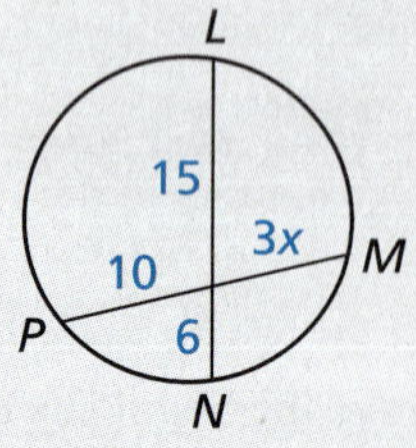

28. $m\widehat{WZY}$

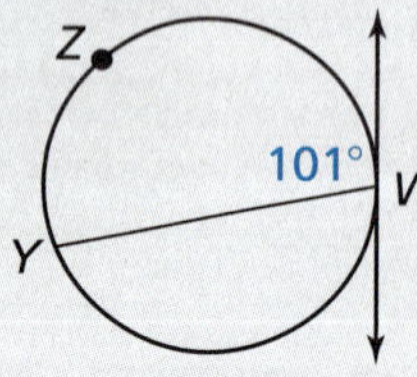

29. Find $m\widehat{WY}$.

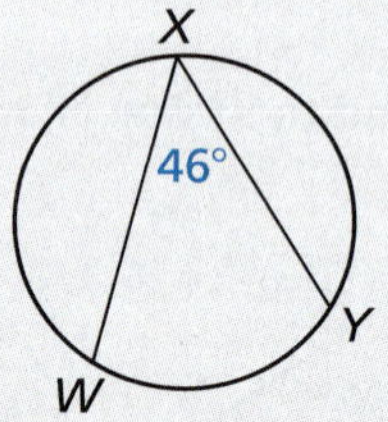

30. Show that the two triangles are similar.

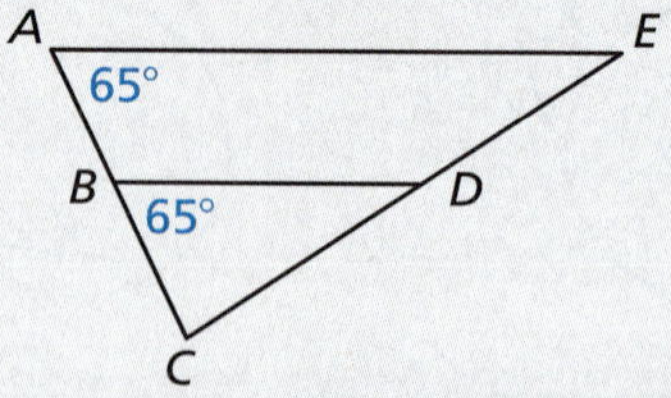

31. Find the value of x that makes $m \parallel n$. Explain your reasoning.

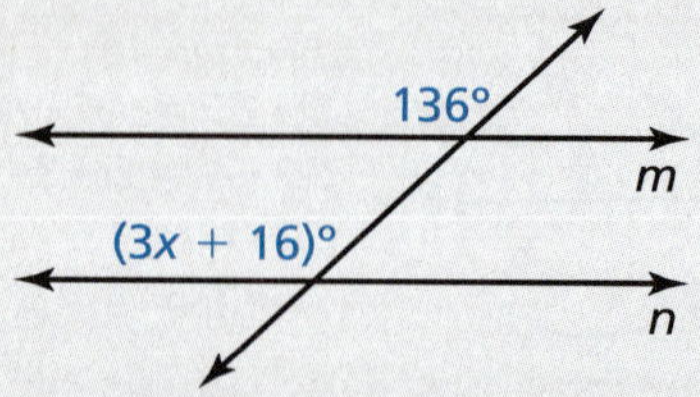

GO DIGITAL

10.7 Circles in the Coordinate Plane

Learning Target: Understand equations of circles.

Success Criteria:
- I can write equations of circles.
- I can find the center and radius of a circle.
- I can graph equations of circles.
- I can write coordinate proofs involving circles.

EXPLORE IT! Deriving Equations of Circles

Work with a partner.

a. Let (0, 0) be the center of a circle with radius 4. Use the Pythagorean Theorem to write an equation that represents the distance between the center of the circle and a point (x, y) on the circle.

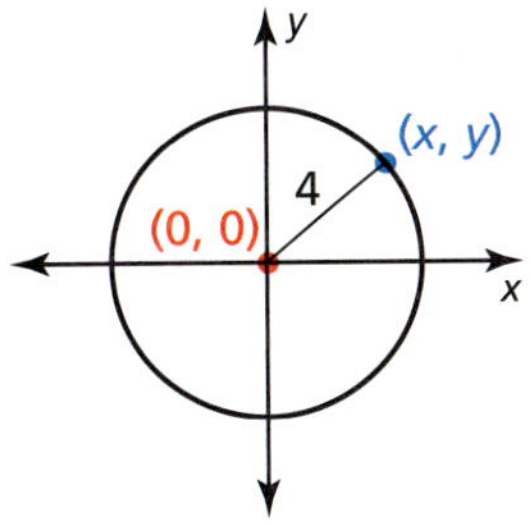

2 MTR MODEL A PROBLEM
What is the equation of a circle with center (0, 0) and radius r in the coordinate plane?

b. How does the equation in part (a) change when the center of the circle is (1, 1)? Write the equation.

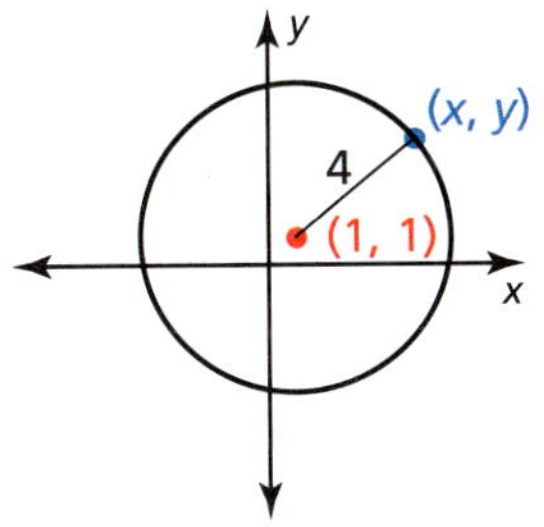

c. What is the equation of a circle with center (h, k) and radius r in the coordinate plane?

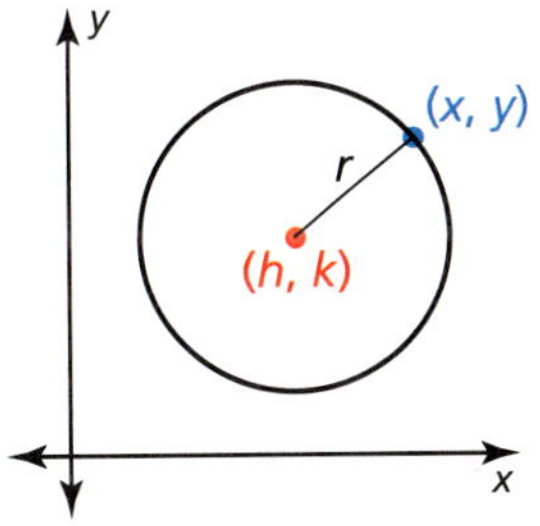

Geometric Reasoning
MA.912.GR.7.2 Given a mathematical or real-world context, derive and create the equation of a circle using key features.
MA.912.GR.7.3 Graph and solve mathematical and real-world problems that are modeled with an equation of a circle. Determine and interpret key features in terms of the context.
Also MA.912.GR.3.2, MA.912.GR.3.3

GO DIGITAL

Writing and Graphing Equations of Circles

Vocabulary AZ VOCAB

standard equation of a circle, *p. 558*

Let (x, y) represent any point on a circle with center at the origin and radius r. By the Pythagorean Theorem,

$$x^2 + y^2 = r^2.$$

This is the equation of a circle with center at the origin and radius r.

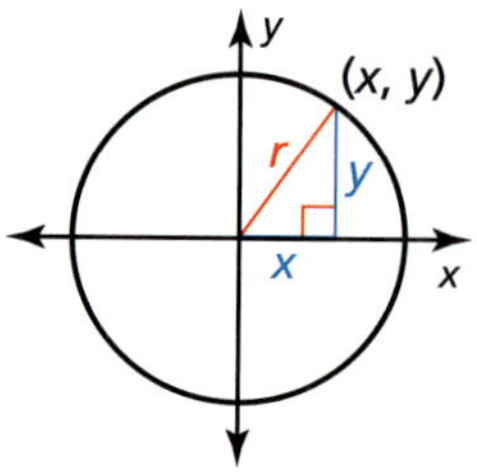

KEY IDEA

Standard Equation of a Circle

Let (x, y) represent any point on a circle with center (h, k) and radius r. By the Pythagorean Theorem,

$$(x - h)^2 + (y - k)^2 = r^2.$$

This is the **standard equation of a circle** with center (h, k) and radius r.

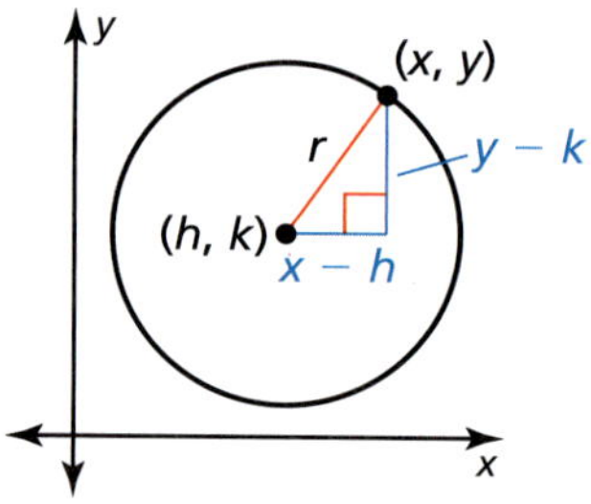

EXAMPLE 1 **Writing the Standard Equation of a Circle** WATCH

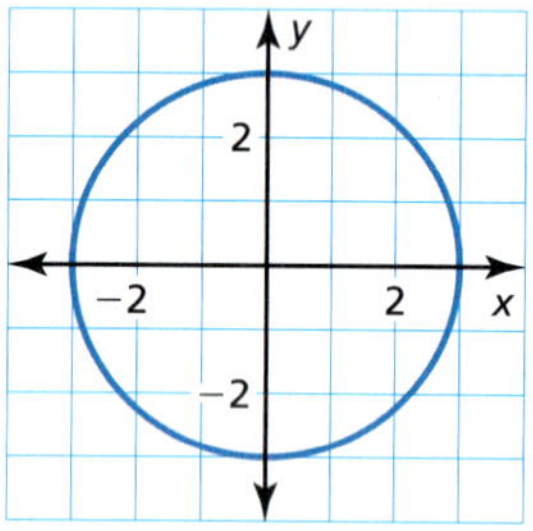

Write the standard equation of each circle.

a. the circle shown at the left

b. the circle with center $(0, -9)$ and diameter 8.4

SOLUTION

a. The radius is 3, and the center is at the origin.

$(x - h)^2 + (y - k)^2 = r^2$	Standard equation of a circle
$(x - 0)^2 + (y - 0)^2 = 3^2$	Substitute $(h, k) = (0, 0)$ and $r = 3$.
$x^2 + y^2 = 9$	Simplify.

▶ The standard equation of the circle is $x^2 + y^2 = 9$.

b. The radius is $8.4 \div 2 = 4.2$, and the center is at $(0, -9)$.

$(x - h)^2 + (y - k)^2 = r^2$	Standard equation of a circle
$(x - 0)^2 + [y - (-9)]^2 = 4.2^2$	Substitute $(h, k) = (0, -9)$ and $r = 4.2$.
$x^2 + (y + 9)^2 = 17.64$	Simplify.

▶ The standard equation of the circle is $x^2 + (y + 9)^2 = 17.64$.

SELF-ASSESSMENT 1 I don't understand yet. 2 I can do it with help. 3 I can do it on my own. 4 I can teach someone else.

Write the standard equation of the circle.

1. the circle with center $(0, 0)$ and radius 2.5

2. the circle with center $(-2, 5)$ and diameter 14

GO DIGITAL

EXAMPLE 2 Writing the Standard Equation of a Circle

WATCH

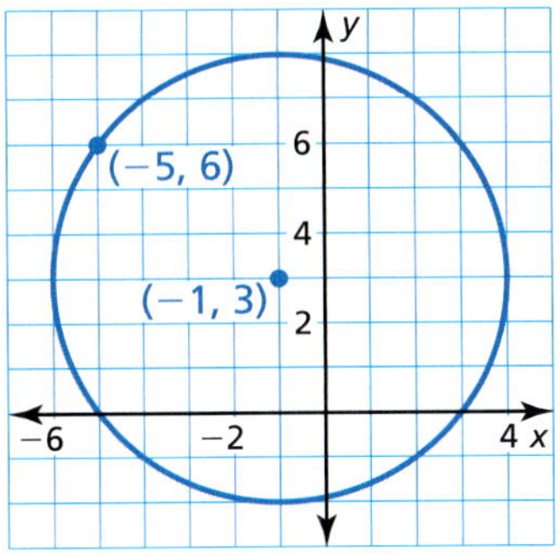

Write the standard equation of the circle shown.

SOLUTION

The center is $(-1, 3)$ and a point on the circle is $(-5, 6)$. To find the radius r, find the distance between $(-1, 3)$ and $(-5, 6)$.

$r = \sqrt{[-5 - (-1)]^2 + (6 - 3)^2}$ Distance Formula

$= \sqrt{(-4)^2 + 3^2}$ Simplify.

$= 5$ Simplify.

Substitute the values of h, k, and r into the standard equation of a circle.

$(x - h)^2 + (y - k)^2 = r^2$ Standard equation of a circle

$[x - (-1)]^2 + (y - 3)^2 = 5^2$ Substitute $(h, k) = (-1, 3)$ and $r = 5$.

$(x + 1)^2 + (y - 3)^2 = 25$ Simplify.

▶ The standard equation of the circle is $(x + 1)^2 + (y - 3)^2 = 25$.

EXAMPLE 3 Graphing a Circle

WATCH

The equation of a circle is $x^2 + y^2 - 8x + 4y - 16 = 0$. Graph the circle. Identify the domain and range.

SOLUTION

You can write the equation in standard form by completing the square on the x-terms and the y-terms.

$x^2 + y^2 - 8x + 4y - 16 = 0$ Equation of circle

$x^2 - 8x + y^2 + 4y = 16$ Isolate constant. Group terms.

$x^2 - 8x + 16 + y^2 + 4y + 4 = 16 + 16 + 4$ Complete the square twice.

$(x - 4)^2 + (y + 2)^2 = 36$ Factor left side. Simplify right side.

$(x - 4)^2 + [y - (-2)]^2 = 6^2$ Rewrite the equation.

REMEMBER

To complete the square for the expression $x^2 + bx$, add the square of half the coefficient of the term bx.

$$x^2 + bx + \left(\frac{b}{2}\right)^2 = \left(x + \frac{b}{2}\right)^2$$

Use the center $(4, -2)$ and the radius 6 to graph the circle.

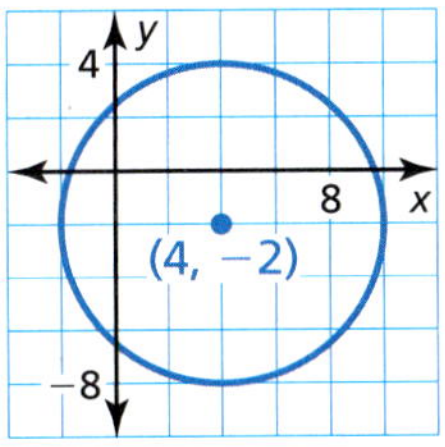

▶ The domain is $-2 \le x \le 10$ and the range is $-8 \le y \le 4$.

SELF-ASSESSMENT

1 I don't understand yet. 2 I can do it with help. 3 I can do it on my own. 4 I can teach someone else.

3. The point $(3, 4)$ is on a circle with center $(1, 4)$. Write the standard equation of the circle.

4. WRITING Explain why knowing the location of the center and one point on a circle is enough to graph the circle.

5. The equation of a circle is $x^2 + y^2 - 8x + 6y + 9 = 0$. Graph the circle. Identify the domain and range.

GO DIGITAL

Finding Tangent Lines

EXAMPLE 4 **Finding a Tangent Line** WATCH

The equation of a circle is $(x + 1)^2 + (y - 2)^2 = 25$. Graph the circle. Then write an equation of the line tangent to the circle at (3, 5).

SOLUTION

Step 1 Use the center and the radius to graph the circle.

The equation is written in standard form. Use the center $(-1, 2)$ and the radius $\sqrt{25} = 5$ to graph the circle.

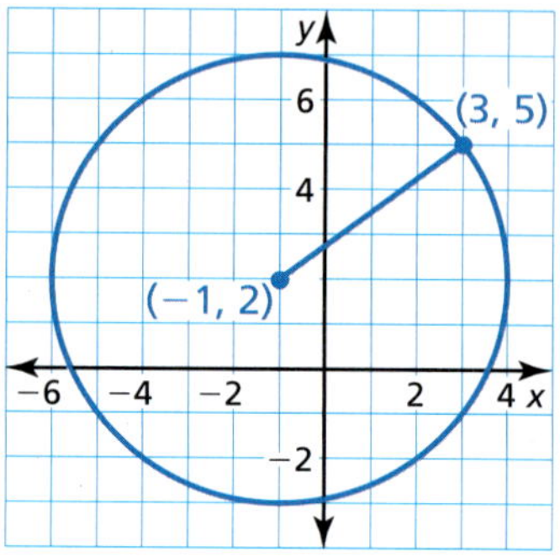

Step 2 Find the slope of the tangent line.

By the Tangent Line to Circle Theorem, the line tangent to the circle at (3, 5) is perpendicular to the radius with endpoints $(-1, 2)$ and (3, 5). Because the slope of this radius is $\frac{5-2}{3-(-1)} = \frac{3}{4}$, the slope of the tangent line is $-\frac{4}{3}$.

Step 3 Use point-slope form to write an equation of the tangent line.

The line passes through (3, 5) with a slope of $-\frac{4}{3}$.

$$y - 5 = -\frac{4}{3}(x - 3), \text{ or } y = -\frac{4}{3}x + 9$$

▶ An equation of the tangent line is $y = -\frac{4}{3}x + 9$.

Check Use technology to verify that the graph of $y = -\frac{4}{3}x + 9$ is tangent to the graph of $(x + 1)^2 + (y - 2)^2 = 25$ at the point (3, 5). ✓

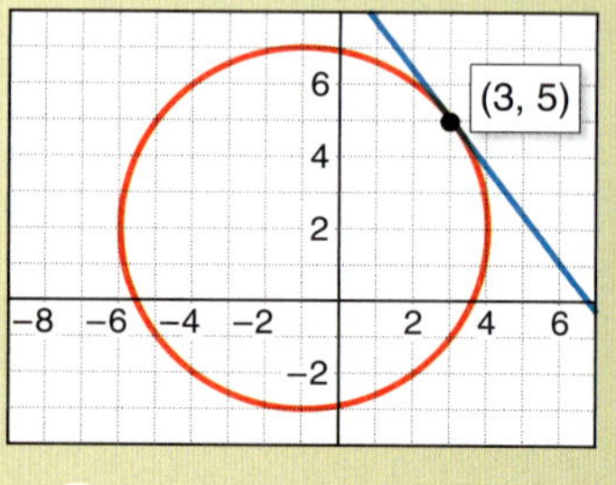

SELF-ASSESSMENT 1 I don't understand yet. 2 I can do it with help. 3 I can do it on my own. 4 I can teach someone else.

6. The equation of a circle is $(x - 5)^2 + (y + 3)^2 = 100$. Graph the circle. Then write an equation of the line tangent to the circle at $(-1, 5)$.

7. DIG DEEPER The graph of $y = \frac{3}{2}x + 15$ is tangent to a circle with center $(-3, 4)$. Find the point of tangency. Then write an equation of a second line that is tangent to the circle and is parallel to the given tangent line.

GO DIGITAL

Writing Coordinate Proofs Involving Circles

EXAMPLE 5 **Writing a Coordinate Proof Involving a Circle** WATCH

The point (2, 0) is on a circle centered at the origin. Prove or disprove that the point $(\sqrt{2}, \sqrt{2})$ is on the circle.

SOLUTION

The circle centered at the origin that contains the point (2, 0) has the following radius.

$$r = \sqrt{(x - h)^2 + (y - k)^2} = \sqrt{(2 - 0)^2 + (0 - 0)^2} = \sqrt{4} = 2$$

So, a point is on the circle if and only if the distance from that point to the origin is 2. The distance from $(\sqrt{2}, \sqrt{2})$ to (0, 0) is

$$d = \sqrt{(\sqrt{2} - 0)^2 + (\sqrt{2} - 0)^2} = \sqrt{2 + 2} = \sqrt{4} = 2.$$

▶ So, the point $(\sqrt{2}, \sqrt{2})$ is on the circle.

Solving Real-Life Problems

7 MTR **EXAMPLE 6** **Modeling Real Life** WATCH STEM

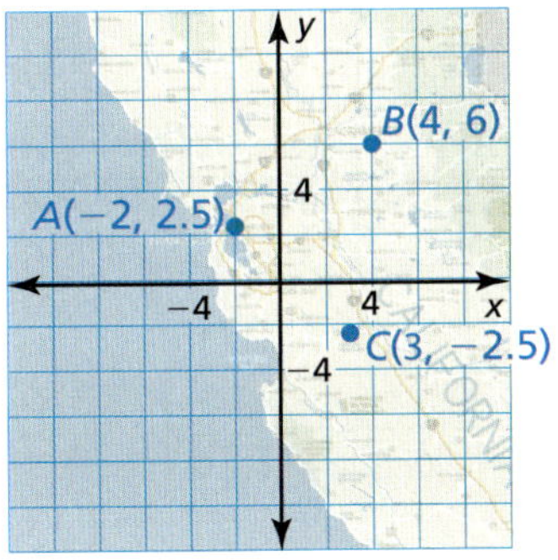

The epicenter of an earthquake is the point on Earth's surface directly above the earthquake's origin. A seismograph can be used to determine the distance to the epicenter of an earthquake. Seismographs are needed in three different places to locate an earthquake's epicenter.

Use the seismograph readings from points A, B, and C to find the epicenter of an earthquake.

- The epicenter is 7 miles away from point A.
- The epicenter is 4 miles away from point B.
- The epicenter is 5 miles away from point C.

SOLUTION

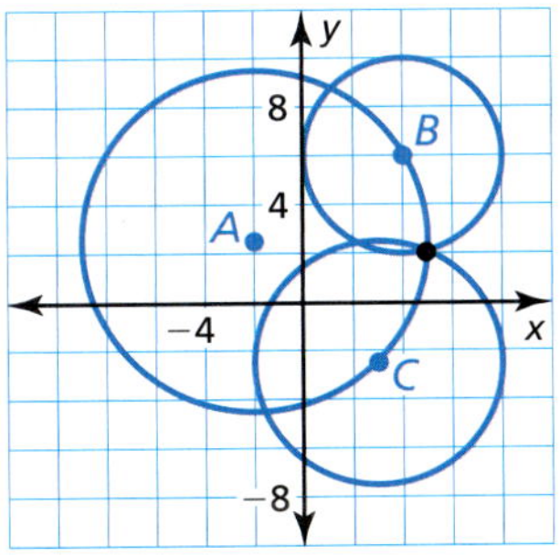

The set of all points equidistant from a given point is a circle, so the epicenter is located on each of the following circles.

$\odot A$ with center (−2, 2.5) and radius 7

$\odot B$ with center (4, 6) and radius 4

$\odot C$ with center (3, −2.5) and radius 5

To find the epicenter, graph the circles on a coordinate plane where each unit corresponds to one mile. Find the point of intersection of the three circles.

▶ The epicenter is at about (5, 2).

SELF-ASSESSMENT 1 I don't understand yet. 2 I can do it with help. 3 I can do it on my own. 4 I can teach someone else.

8. The point (0, 1) is on a circle centered at the origin. Prove or disprove that the point $(1, \sqrt{5})$ is on the circle.

9. REASONING Why are three seismographs needed to locate an earthquake's epicenter?

GO DIGITAL

10.7 Practice WITH CalcChat® AND CalcView®

In Exercises 1–6, write the standard equation of the circle. *(See Example 1.)*

1. 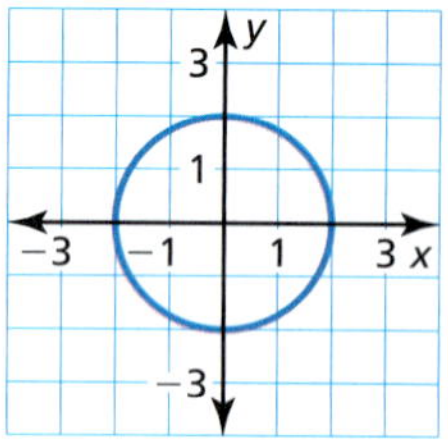

2. 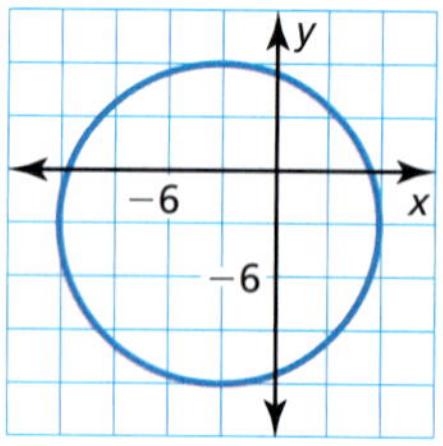

3. the circle with center (0, 0) and radius 7

4. the circle with center (4, 1) and radius 5

5. the circle with center (−3, 4) and diameter 2

6. the circle with center (3, −5) and diameter 22

In Exercises 7–10, use the given information to write the standard equation of the circle. *(See Example 2.)*

7. The center is (0, 0), and a point on the circle is (0, 6).

8. The center is (0, 0), and a point on the circle is (3, −7).

9. The center is (1, 2), and a point on the circle is (4, 2).

10. The center is (−7, −2), and a point on the circle is (1, −8).

In Exercises 11–16, graph the circle. Identify the domain and range. *(See Example 3.)*

11. $x^2 + y^2 = 49$

12. $(x + 5)^2 + (y - 3)^2 = 9$

13. $x^2 + y^2 - 6x = 7$

14. $x^2 + y^2 + 4y = 32$

15. $x^2 + y^2 - 8x - 2y + 16 = 0$

16. $x^2 + y^2 + 4x + 12y + 15 = 0$

In Exercises 17–20, graph the circle. Then write an equation of the line tangent to the circle at the given point. *(See Example 4.)*

17. $(x + 3)^2 + (y + 2)^2 = 4$; $(-3, 0)$

18. $x^2 + y^2 = 100$; $(6, 8)$

19. $x^2 + y^2 + 4x - 2y - 20 = 0$; $(1, -3)$

20. $x^2 + y^2 - 6x - 10y + 16 = 0$; $(0, 8)$

2 MTR **MAKE A CONNECTION** **In Exercises 21–24, the endpoints of a diameter of a circle are given. Write the standard equation of the circle.**

21. (0, −1), (−5, −6)

22. (−3, −6), (5, 4)

23. (11, 5), (−13, −13)

24. (2, 15), (−10, −1)

4 MTR **25. ERROR ANALYSIS** Describe and correct the error in writing the standard equation of a circle.

✗ The standard equation of a circle with center (−3, −5) and radius 3 is $(x - 3)^2 + (y - 5)^2 = 9$.

4 MTR **26. ERROR ANALYSIS** Describe and correct the error in graphing and identifying the domain and range of $(x - 4)^2 + (y + 6)^2 = 36$.

✗ 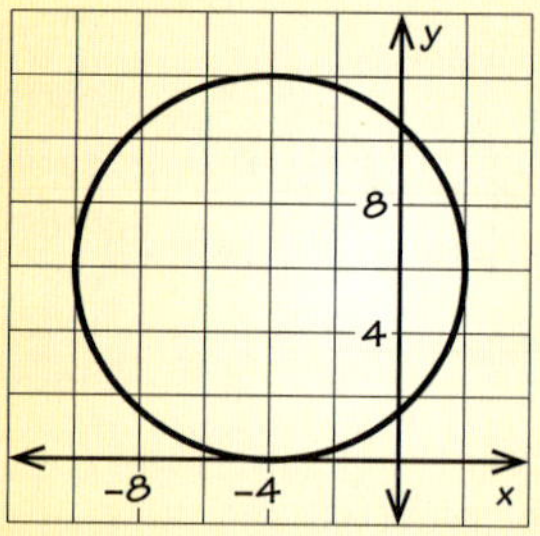

The domain is $-10 \leq x \leq 2$ and the range is $0 \leq y \leq 12$.

In Exercises 27–30, prove or disprove the statement. *(See Example 5.)*

27. The point (2, 3) lies on the circle centered at the origin with radius 8.

28. The point $(2, \sqrt{5})$ lies on the circle centered at the origin with radius 3.

29. The point $(\sqrt{6}, 2)$ lies on the circle centered at the origin and containing the point (3, −1).

30. The point $(\sqrt{7}, 5)$ lies on the circle centered at the origin and containing the point (5, 2).

7 MTR **31. MODELING REAL LIFE** A field is centered at the origin. You search an area of the field represented by $(x - 10)^2 + (y - 13)^2 = 81$, where x and y are measured in feet. Graph the circle. Identify and interpret the domain and range.

GO DIGITAL

6 MTR **32. ASSESS REASONABLENESS** Your friend says the equation of the line tangent to the circle at $(-7, -3)$ is $y = \frac{3}{4}x + \frac{9}{4}$. Is your friend's claim reasonable? Explain.

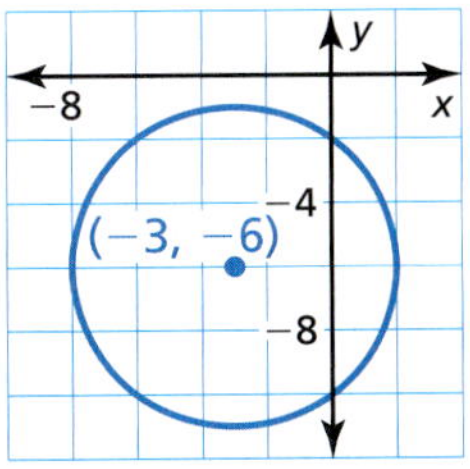

7 MTR **33. MODELING REAL LIFE** A city's commuter system has three zones. Zone 1 serves people living within 3 miles of the city's center. Zone 2 serves those between 3 and 7 miles from the center. Zone 3 serves those more than 7 miles from the center. *(See Example 6.)*

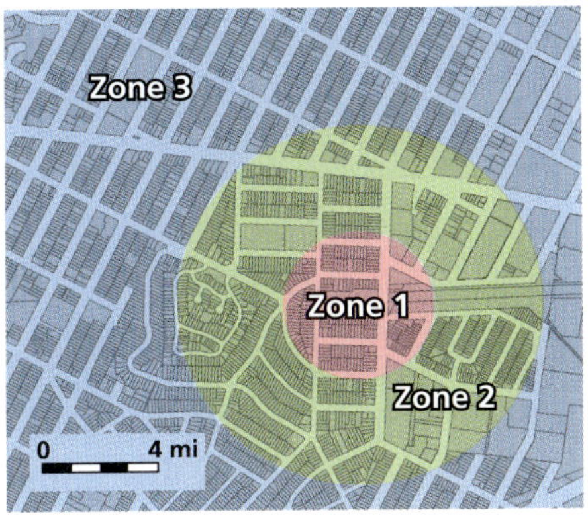

a. Graph this situation in a coordinate plane where each unit corresponds to 1 mile. Locate the city's center at the origin.

b. Determine which zone serves people whose homes are represented by the points $(3, 4)$, $(6, 5)$, $(1, 2)$, $(0, 3)$, and $(1, 6)$.

7 MTR **34. MODELING REAL LIFE** Telecommunication towers can be used to transmit cellular phone calls. A graph shows towers at points $(0, 0)$, $(0, 5)$, and $(6, 3)$ with units measured in kilometers. Each tower has a range of 3 kilometers.

a. Sketch a graph and locate the towers. Are there any locations that may receive calls from more than one tower? Explain your reasoning.

b. The center of City A is located at $(-2, 2.5)$, and the center of City B is located at $(5, 4)$. Each city has a radius of 1.5 kilometers. Which city seems to have better cell phone coverage? Explain your reasoning.

4 MTR **35. MAKING AN ARGUMENT** Your friend claims that the equation of a circle passing through the points $(-1, 0)$ and $(1, 0)$ is $x^2 - 2yk + y^2 = 1$ with center $(0, k)$. Is your friend correct? Explain your reasoning.

36. HOW DO YOU SEE IT?
Match each graph with its equation.

a. b. c. d.

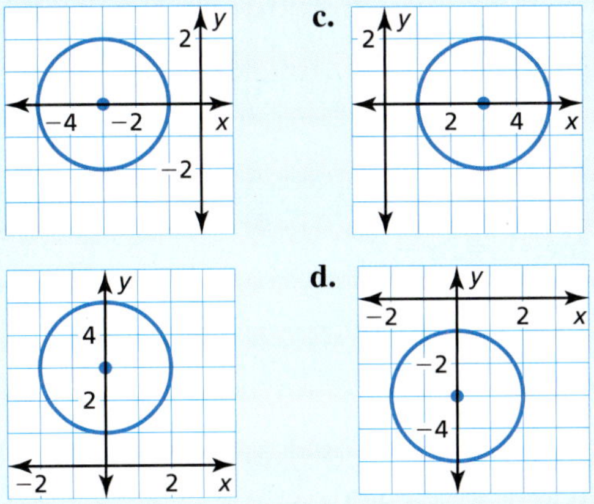

A. $x^2 + (y + 3)^2 = 4$ **C.** $(x + 3)^2 + y^2 = 4$

B. $(x - 3)^2 + y^2 = 4$ **D.** $x^2 + (y - 3)^2 = 4$

37. OPEN-ENDED The Rakotzbrücke Bridge in Saxony, Germany, and its reflection form a circle. Use the image below to write and graph an equation that represents this phenomenon. Find and interpret the domain, range, and radius.

7 MTR **38. PERFORMANCE TASK** Your class plans a field trip to Madame Tussauds Orlando. Your teacher asks you to find a possible lunch location within a three-mile radius of the wax museum.

a. Graph the situation in a coordinate plane. Interpret the domain, range, and center of the graph.

b. Write a list of 10 possible restaurants. Explain why you chose these restaurants.

4 MTR **39. DISCUSS MATHEMATICAL THINKING** Describe when the slope of a line tangent to a circle is (a) positive, (b) negative, (c) zero, and (d) undefined. What do you notice?

7 MTR **40. INVESTIGATE** Describe a career that uses circles in a coordinate plane.

GO DIGITAL

5 MTR **CONNECTING CONCEPTS** **In Exercises 41–44, determine whether the line is a *tangent*, a *secant*, a *secant that contains the diameter*, or *none of these*. Explain your reasoning.**

41. **Circle:** $(x - 4)^2 + (y - 3)^2 = 9$;
Line: $y = 6$

42. **Circle:** $(x + 2)^2 + (y - 2)^2 = 16$;
Line: $y = 2x - 4$

43. **Circle:** $(x - 5)^2 + (y + 1)^2 = 4$;
Line: $y = \frac{1}{5}x - 3$

44. **Circle:** $(x + 3)^2 + (y - 6)^2 = 25$;
Line: $y = -\frac{4}{3}x + 2$

5 MTR **45.** **STRUCTURE** The vertices of $\triangle XYZ$ are $X(4, 5)$, $Y(4, 13)$, and $Z(8, 9)$. Find the equation of the circle circumscribed about $\triangle XYZ$. Justify your answer.

46. **THOUGHT PROVOKING**
Four tangent circles are centered on the x-axis. The radius of $\odot A$ is twice the radius of $\odot O$. The radius of $\odot B$ is three times the radius of $\odot O$. The radius of $\odot C$ is four times the radius of $\odot O$. All circles have integer radii, and the point (63, 16) is on $\odot C$. What is the equation of $\odot A$? Explain your reasoning.

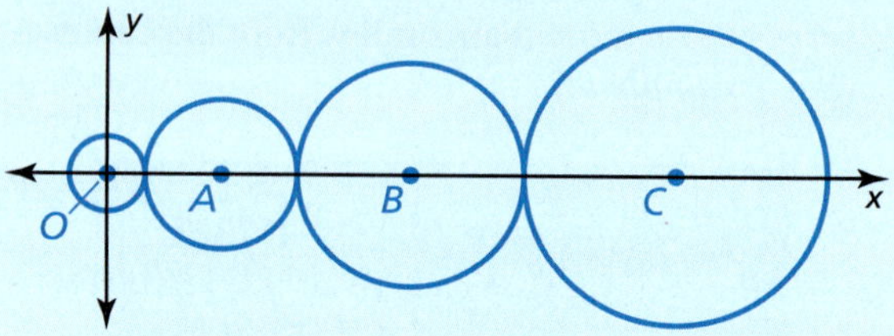

REVIEW & REFRESH

WATCH

47. Write the standard equation for the circle with center (5, 4) that passes through the point (2, 0).

In Exercises 48 and 49, find the value of x.

48.

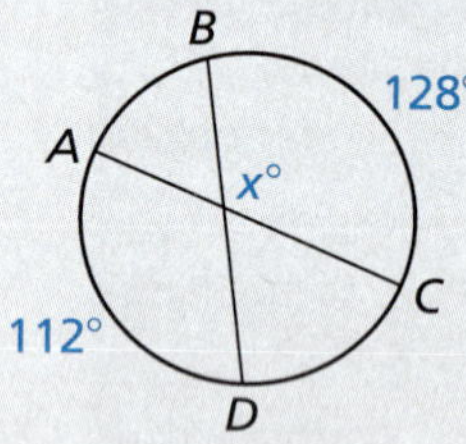

49.

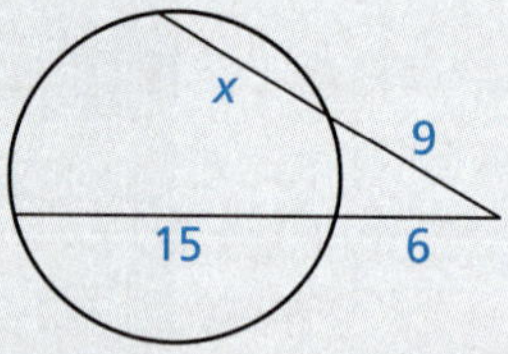

In Exercises 50–54, identify the arc as a *major arc*, *minor arc*, or *semicircle*. Then find the measure of each arc.

50. $\overset{\frown}{RS}$

51. $\overset{\frown}{PR}$

52. $\overset{\frown}{ST}$

53. $\overset{\frown}{PRT}$

54. $\overset{\frown}{RST}$

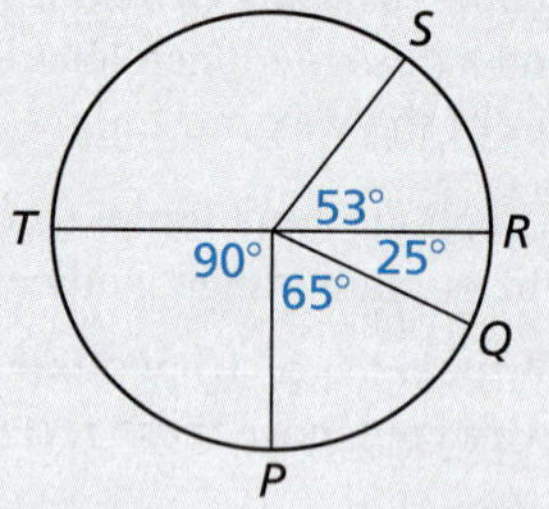

7 MTR **55.** **MODELING REAL LIFE** The vertical supports on a staircase are parallel. Find $m\angle 2$.

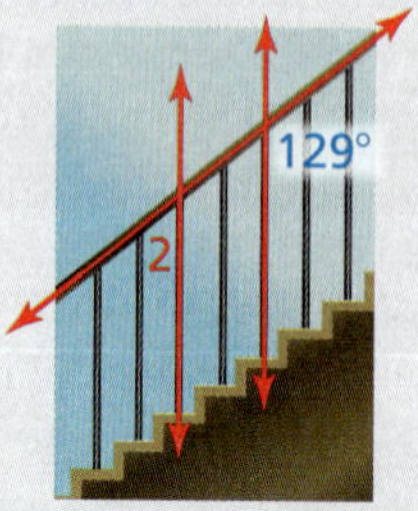

56. Find $m\angle N$.

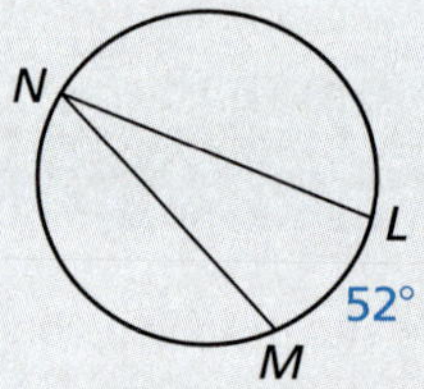

57. Find the value of x that makes $\triangle ABC \sim \triangle XYZ$.

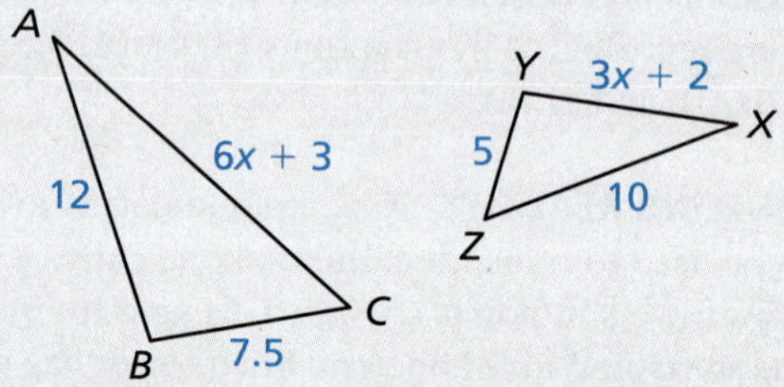

7 MTR **58.** **MODELING REAL LIFE** The stride angle of Horse A is shown. Horse B is the same size as Horse A but has a stride angle of 112°. Which horse has a longer stride? Explain.

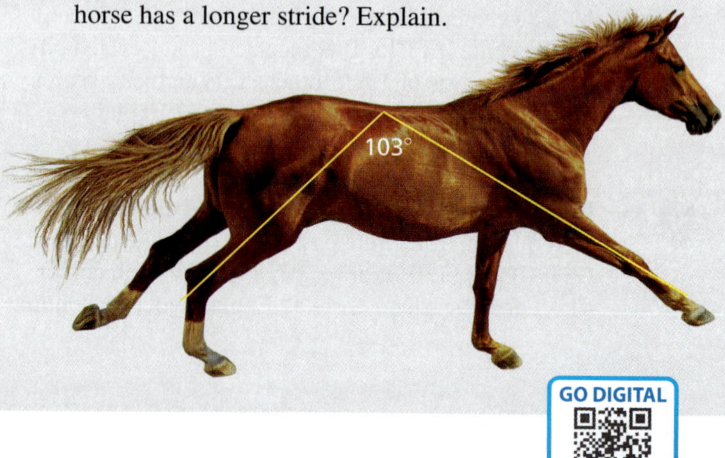

GO DIGITAL

10 Chapter Review WITH CalcChat®

Chapter Learning Target: Understand and apply circle relationships.

Chapter Success Criteria:
- ◆ I can identify lines and segments that intersect circles.
- ◆ I can find angle and arc measures in circles.
- ■ I can use circle relationships to solve problems.
- ■ I can use circles to model and solve real-life problems.

◆ Surface
■ Deep

SELF-ASSESSMENT 1 I don't understand yet. 2 I can do it with help. 3 I can do it on my own. 4 I can teach someone else.

10.1 Lines and Segments That Intersect Circles (pp. 513–520)

WATCH

Learning Target: Identify lines and segments that intersect circles and use them to solve problems.

Vocabulary VOCAB

circle
center
radius
chord
diameter
secant
tangent
point of tangency
tangent circles
concentric circles
common tangent

Tell whether the line, ray, or segment is best described as a *radius*, *chord*, *diameter*, *secant*, or *tangent* of ⊙*P*.

1. $\overline{PK}$
2. $\overline{NM}$
3. $\overrightarrow{JL}$
4. $\overline{KN}$
5. $\overleftrightarrow{NL}$
6. $\overline{PN}$

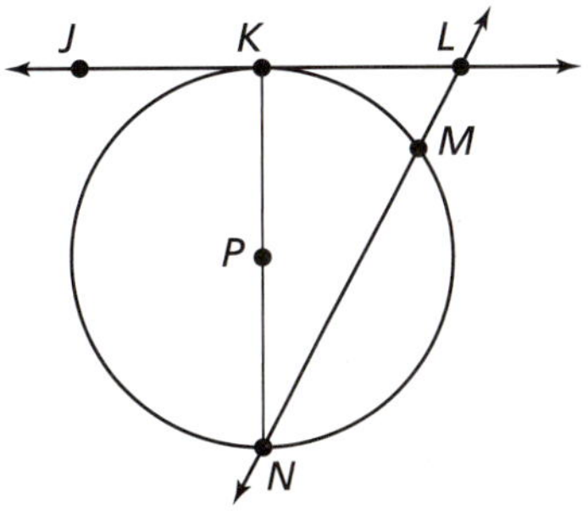

Tell how many common tangents the circles have and draw them. State whether the tangents are *external tangents* or *internal tangents*.

7.

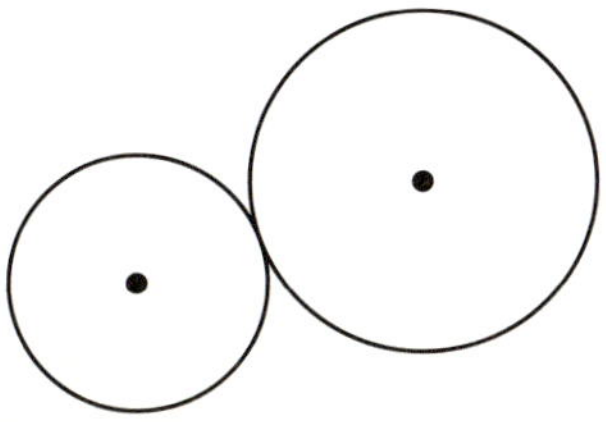

8.

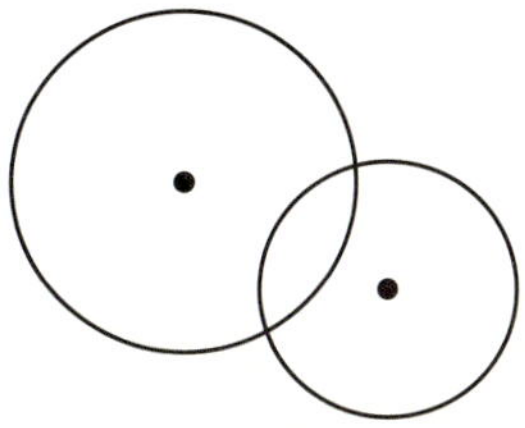

Points *Y* and *Z* are points of tangency. Find the value of the variable.

9.

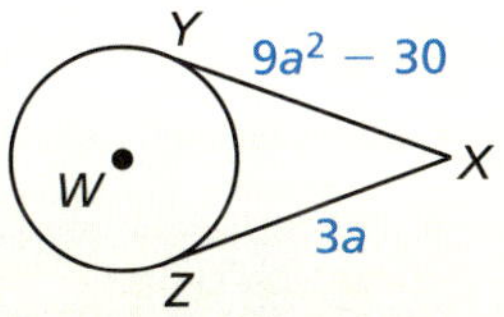

10.

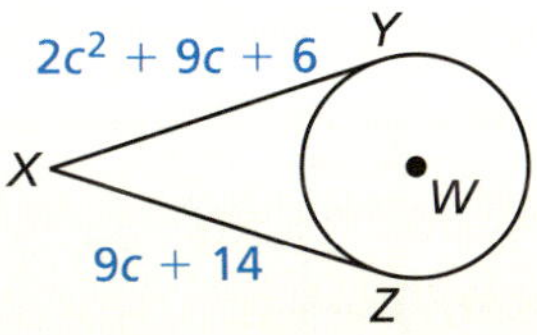

11.

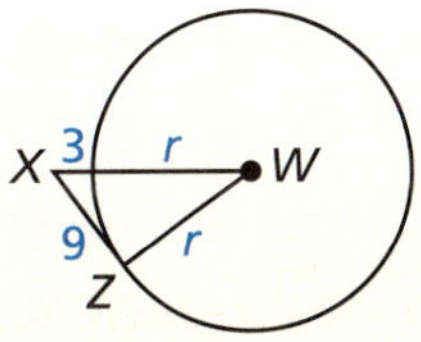

12. Tell whether $\overline{AB}$ is tangent to ⊙*C*. Explain.

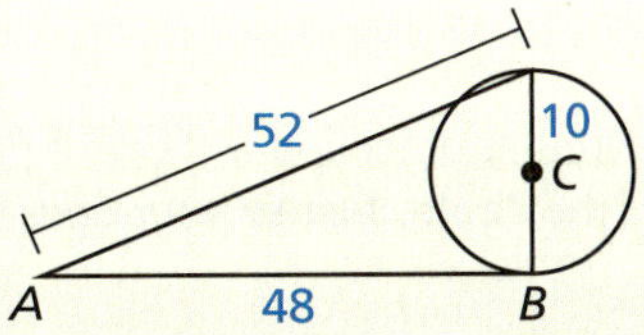

GO DIGITAL

10.2 Finding Arc Measures (pp. 521–528)

Learning Target: Understand arc measures and similar circles.

Find the measure of the indicated arc.

13. $\widehat{KL}$ **14.** $\widehat{LM}$

15. $\widehat{KM}$ **16.** $\widehat{KN}$

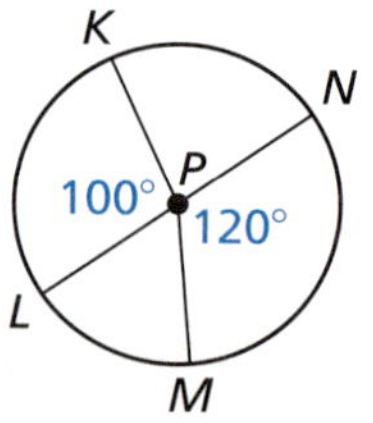

Vocabulary

central angle
minor arc
major arc
semicircle
measure of a minor arc
measure of a major arc
adjacent arcs
congruent circles
congruent arcs
similar arcs

Tell whether the red arcs are congruent. Explain why or why not.

17.

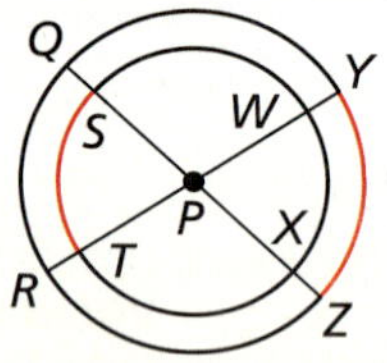

18.

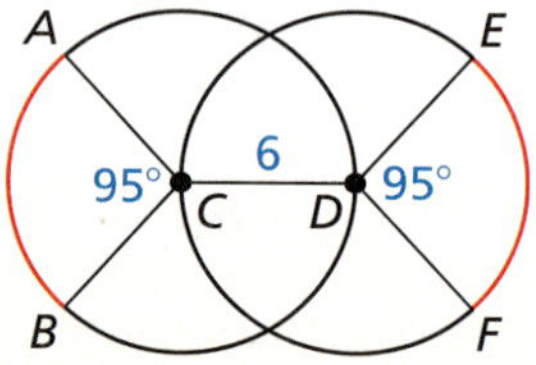

19. A survey asked high school seniors what they intend to do after graduating. The circle graph shows the results. Find each indicated arc measure.

a. $\widehat{AB}$ **b.** $\widehat{AC}$

c. $\widehat{BD}$ **d.** $\widehat{ABF}$

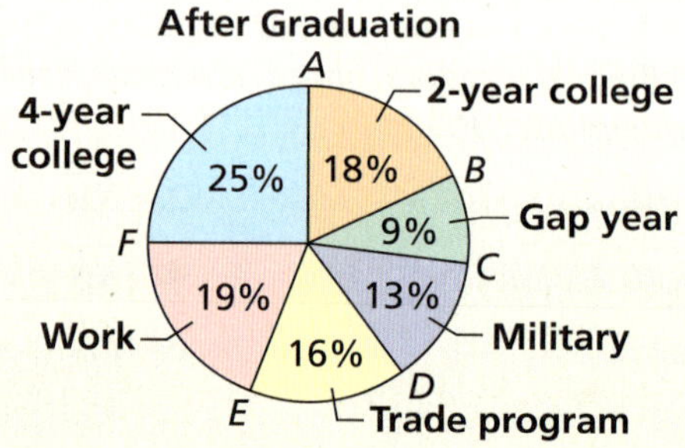

10.3 Using Chords (pp. 529–534)

Learning Target: Understand and apply theorems about chords.

Find the measure of $\widehat{AB}$.

20.

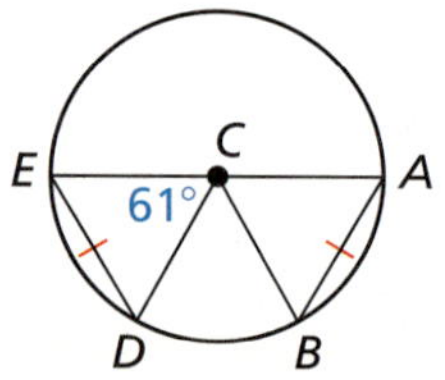

21.

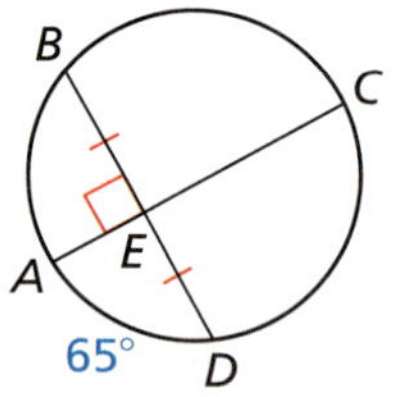

22.

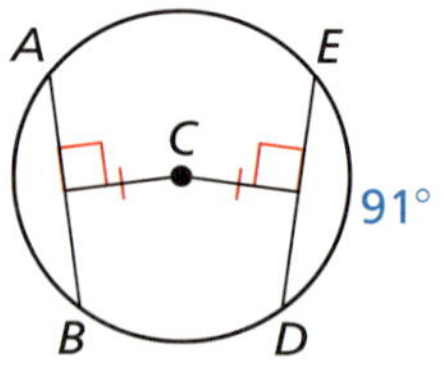

23. In the diagram, $QN = QP = 10$, $JK = 4x$, and $LM = 6x - 24$. Find the radius of $\odot Q$.

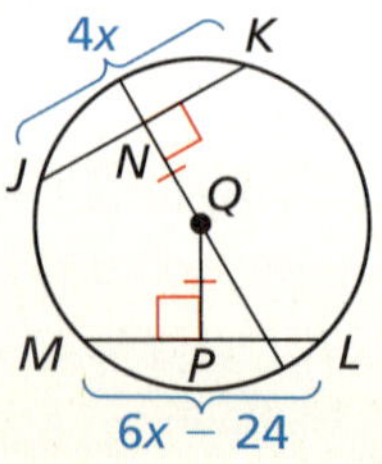

24. Trace the outline of a circular object on a sheet of paper. Construct the center of the circle. Justify your construction.

GO DIGITAL

10.4 Inscribed Angles and Polygons (pp. 535–542) WATCH

Learning Target: Use properties of inscribed angles and inscribed polygons.

Vocabulary (AZ VOCAB)
- inscribed angle
- intercepted arc
- subtend
- inscribed polygon
- circumscribed circle

Find the value(s) of the variable(s).

25.

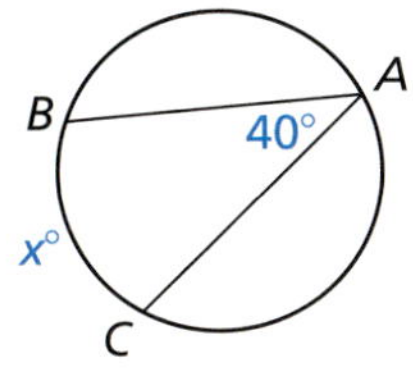

26.

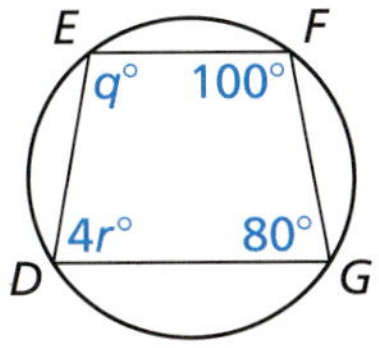

27.

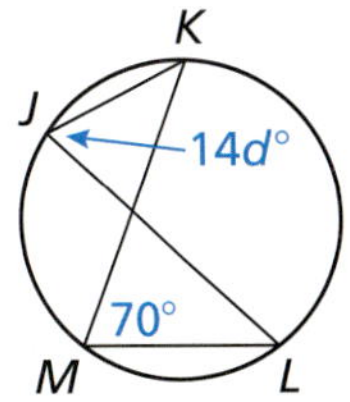

28.

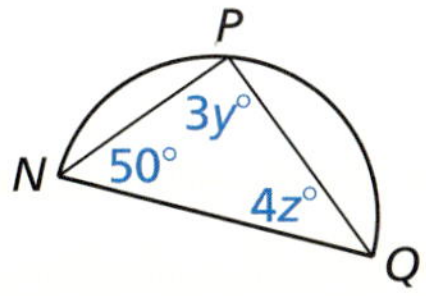

29.

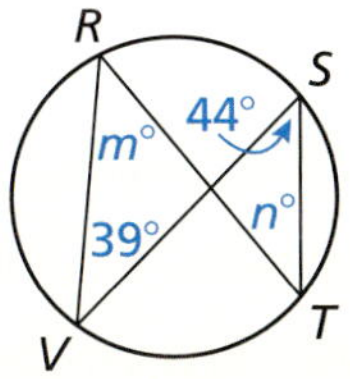

30.

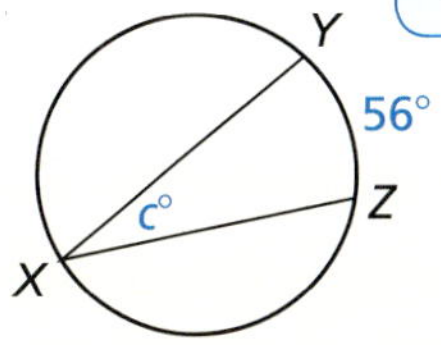

31. List all the congruent angles in the figure at the right.

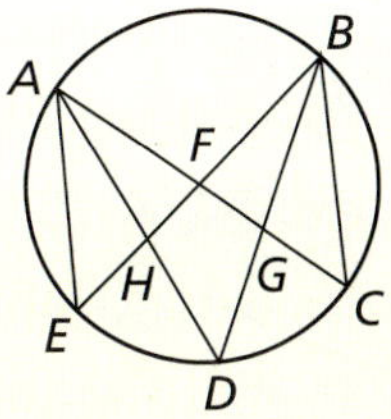

32. Construct a 30°-60°-90° right triangle inscribed in a circle. Justify your construction.

10.5 Angle Relationships in Circles (pp. 543–550) WATCH

Learning Target: Understand angles formed by chords, secants, and tangents.

Vocabulary (AZ VOCAB)
- circumscribed angle

Find the value of x.

33.

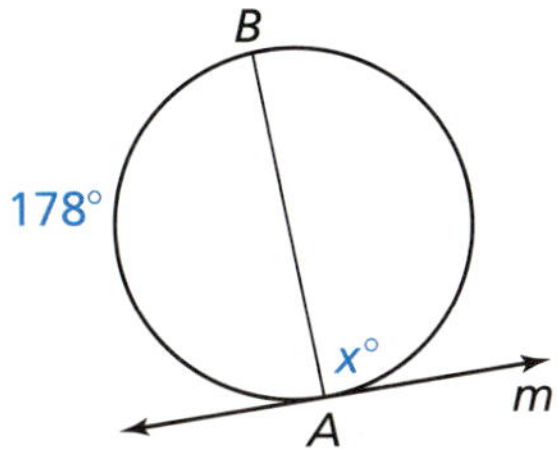

34.

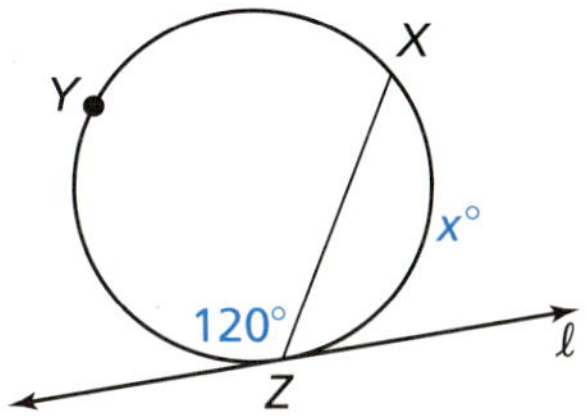

35.

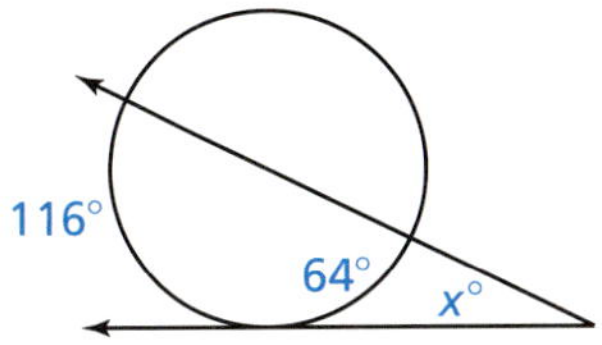

36.

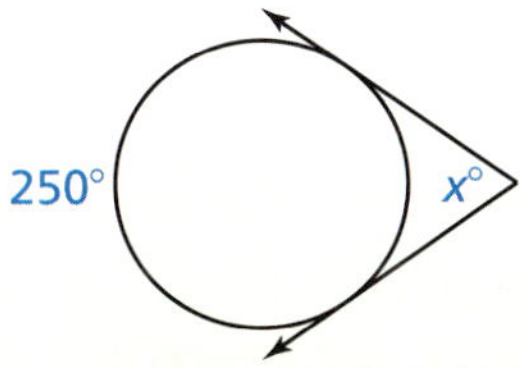

37.

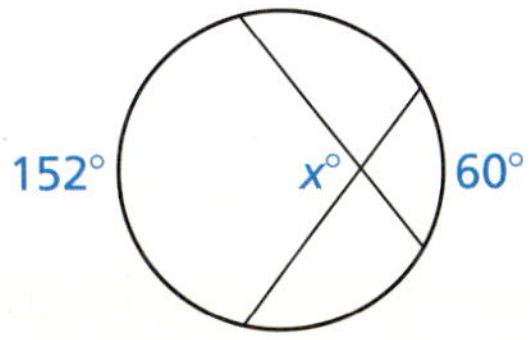

38.

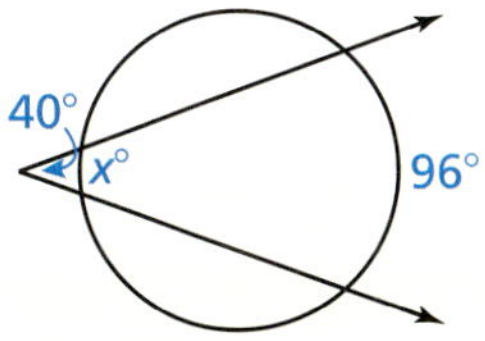

39. In the diagram, $\overrightarrow{AC}$ and $\overrightarrow{AD}$ are tangent to the circle and $m\angle A = \frac{1}{2} m\angle B$. Find $m\angle COD$.

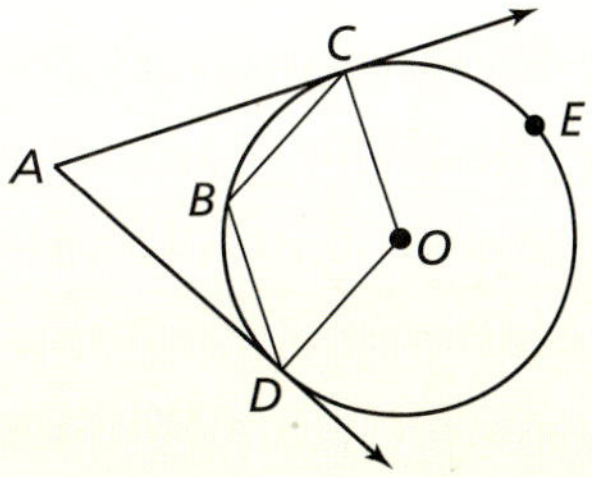

GO DIGITAL

10.6 Segment Relationships in Circles (pp. 551–556)

Learning Target: Use theorems about segments of chords, secants, and tangents.

Vocabulary

segments of a chord
tangent segment
secant segment
external segment

Find the value of x.

40.

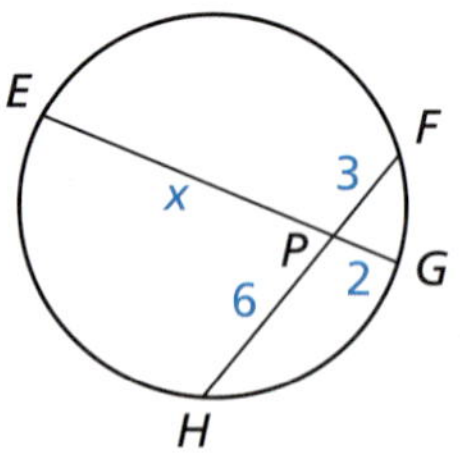

41.

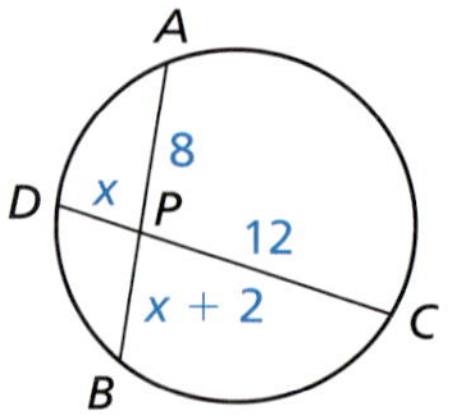

42.

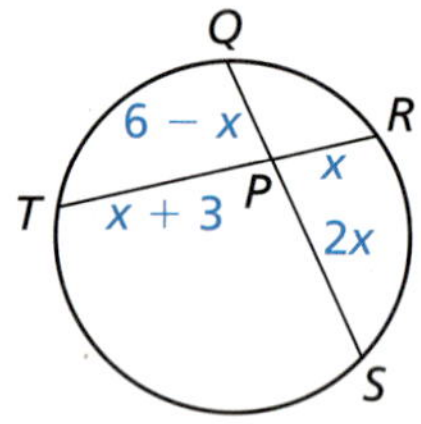

43.

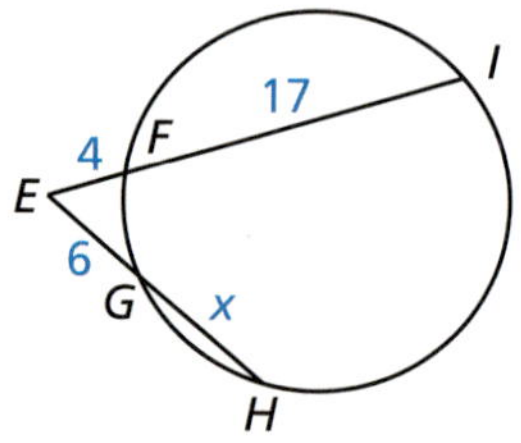

44.

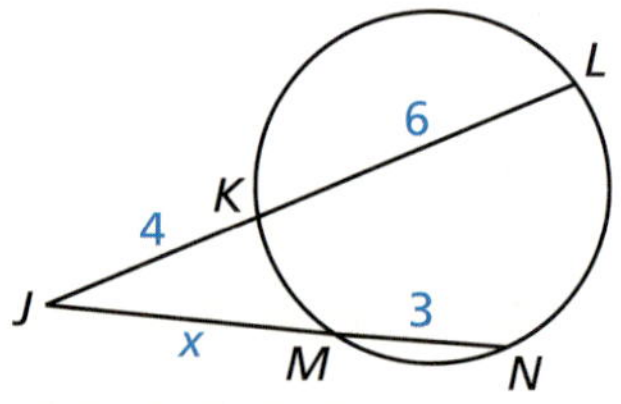

45.

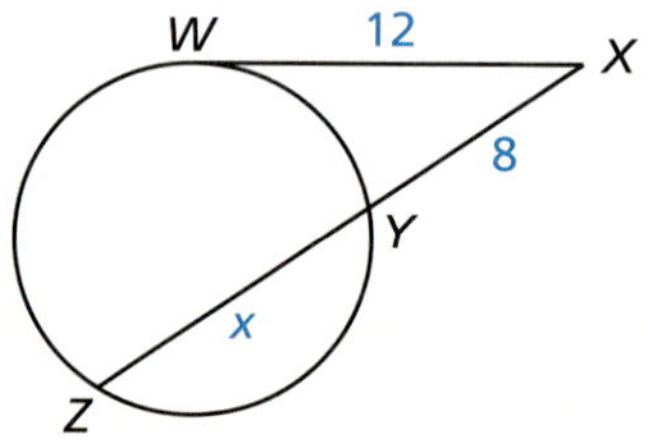

46. Identify the secant segment(s) in Exercise 44. Explain why the segment is a secant segment.

47. Identify the tangent segment(s) in Exercise 45. Explain why the segment is a tangent segment.

48. A local park has a circular ice skating rink. You are standing at point A, about 12 feet from the edge of the rink. The distance from you to a point of tangency on the rink is about 20 feet. Estimate the radius of the rink.

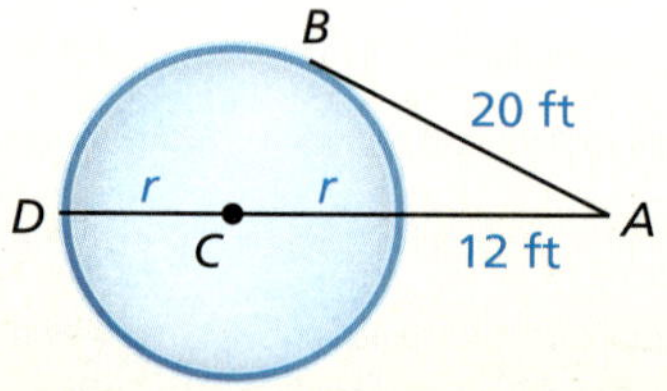

49. You and your friend determine who can swim the fastest by timing each other swimming across a circular pond. You swim from point A to point B, and your friend swims from point C to point D. Is the race fair? Explain why or why not.

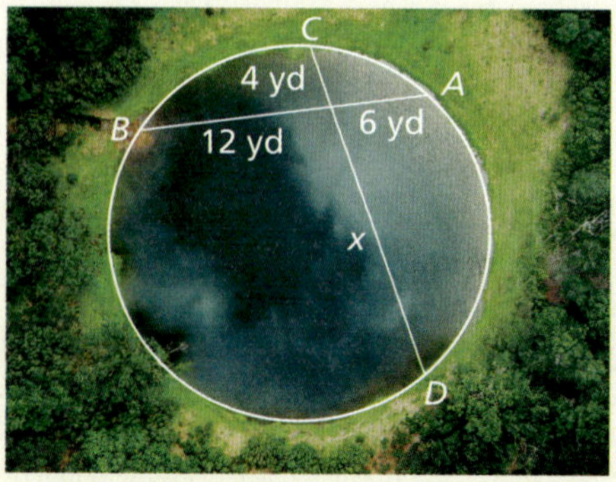

GO DIGITAL

10.7 Circles in the Coordinate Plane *(pp. 557–564)* WATCH

Learning Target: Understand equations of circles.

Vocabulary AZ VOCAB
standard equation of a circle

Write the standard equation of the circle.

50.

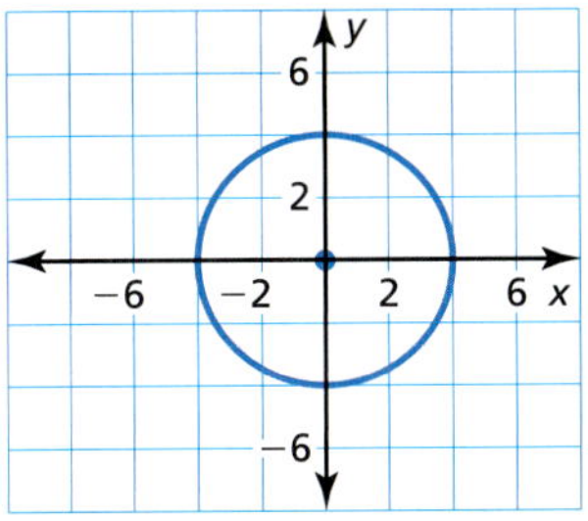

51.

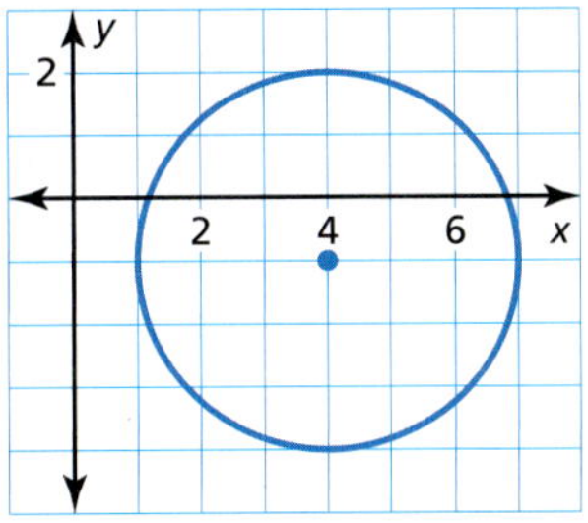

52. the circle with center (0, 0) and radius 9

53. the circle with center (0, 0) and radius 5.2

54. the circle with center (6, 21) and diameter 8

55. the circle with center (10, 7) and diameter 7

56. The center is $(-7, 6)$, and a point on the circle is $(-7, 1)$.

Graph the circle. Identify the domain and range.

57. $x^2 + y^2 = 81$

58. $(x + 3)^2 + (y + 6)^2 = 49$

Graph the circle. Then write an equation of the line tangent to the circle at the given point.

59. $x^2 + y^2 = 25$; (4, 3)

60. $(x - 1)^2 + (y + 3)^2 = 100$; (7, 5)

61. $x^2 + y^2 + 8x - 4y - 5 = 0$; $(0, -1)$

62. The equation of a circle is $x^2 + y^2 - 12x + 8y + 48 = 0$. Find the center and the radius of the circle. Then graph the circle.

63. The point $(-5, 0)$ is on a circle centered at the origin. Prove or disprove that the point $(4, -3)$ is on the circle.

Mathematical Thinking and Reasoning

3 MTR **COMPLETE TASKS WITH MATHEMATICAL FLUENCY**
Mathematicians who complete tasks with mathematical fluency select efficient and appropriate methods for solving problems within the given context.

1. In Exercise 20 on page 541, describe two types of tools that can be used to perform the construction. Describe both construction methods and explain the advantages of each. Tell which type you used and explain why.

2. In 10.5 Explore It! on page 543, how could you complete the exploration without using technology? Why would you avoid using this method?

GO DIGITAL

10 Practice Test WITH CalcChat®

Find the measure of each numbered angle in $\odot P$. Justify your answer.

1.

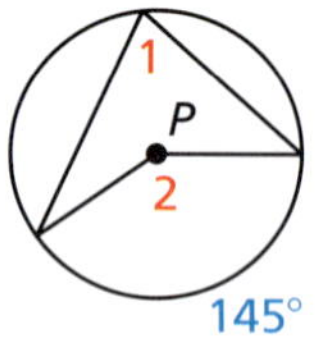

2.

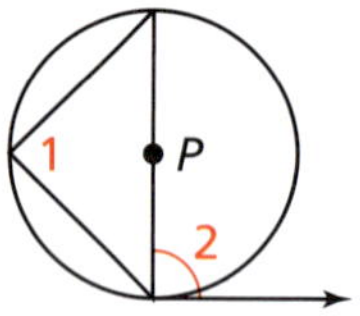

3.

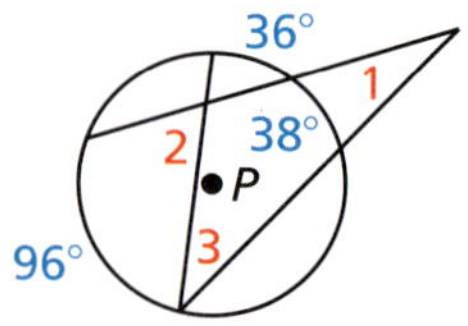

4.

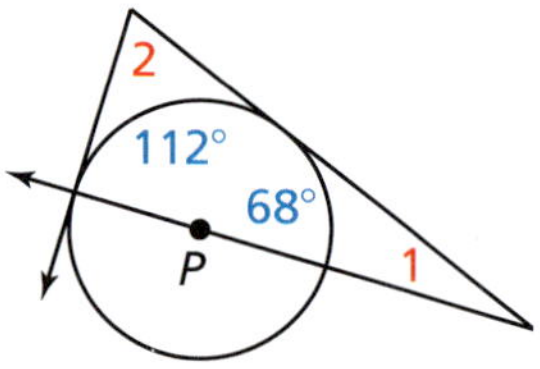

Use the diagram.

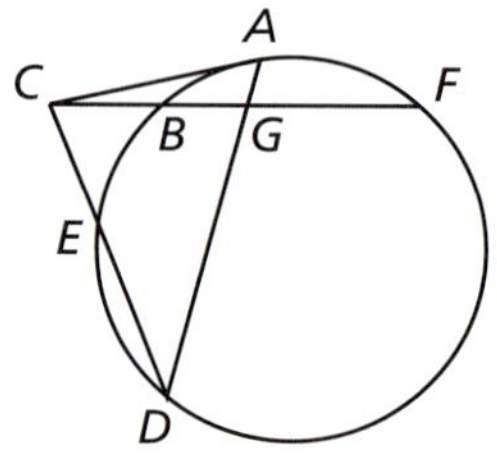

5. $AG = 2$, $GD = 9$, and $BG = 3$. Find GF.

6. $CF = 12$, $CB = 3$, and $CD = 9$. Find CE.

7. $BF = 9$ and $CB = 3$. Find CA.

Find the value of the variable(s). Justify your answer.

8.

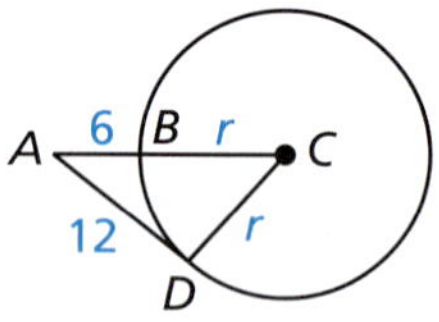

9.

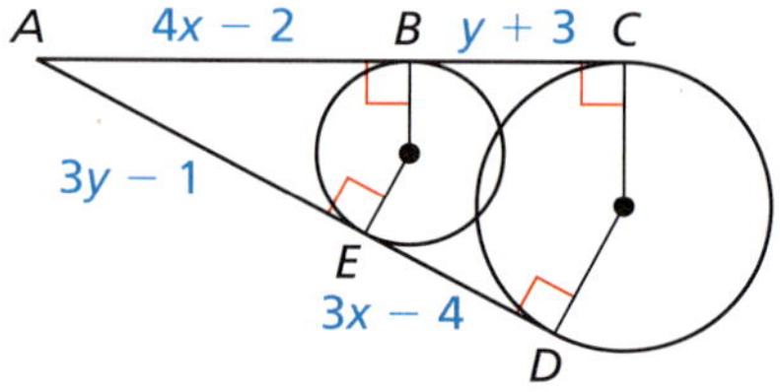

10. The point $(-1, 4)$ is on a circle centered at $(0, 2)$. Prove or disprove that the point $\left(2\sqrt{2}, -1\right)$ is on the circle.

Prove the given statement.

11. $\widehat{ST} \cong \widehat{RQ}$

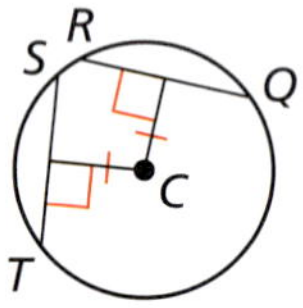

12. $\widehat{JM} \cong \widehat{LM}$

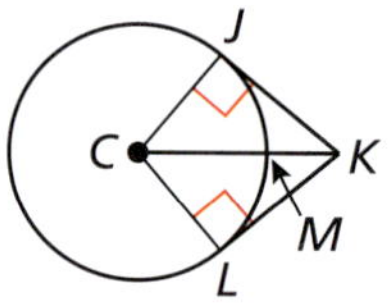

13. $\widehat{DG} \cong \widehat{FG}$

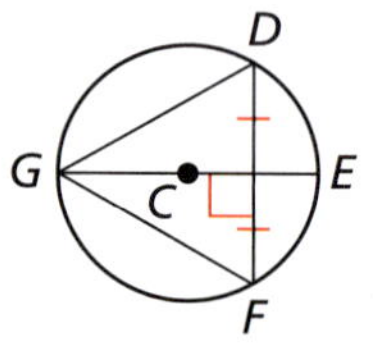

14. A stage scene is modeled in a coordinate plane. The locations of three actors are given by the points $A(11, 4)$, $B(8, 5)$, and $C(15, 5)$, and the region illuminated by an overhead light is given by the equation $(x - 13)^2 + (y - 4)^2 = 16$. Graph the equation. Which actors are illuminated by the light?

15. A car goes around a turn too quickly, leaving tire marks in the shape of a circular arc. The formula $S = 3.87\sqrt{fr}$ is used to estimate the speed of such a car in miles per hour, where f is the *coefficient of friction* and r is the radius of the arc in feet.

a. Accident investigators create the diagram shown. Given that $AB = 320$ feet, find the radius r of the arc. Justify your answer.

b. The coefficient of friction of the road is $f = 0.7$. Estimate the speed of the car.

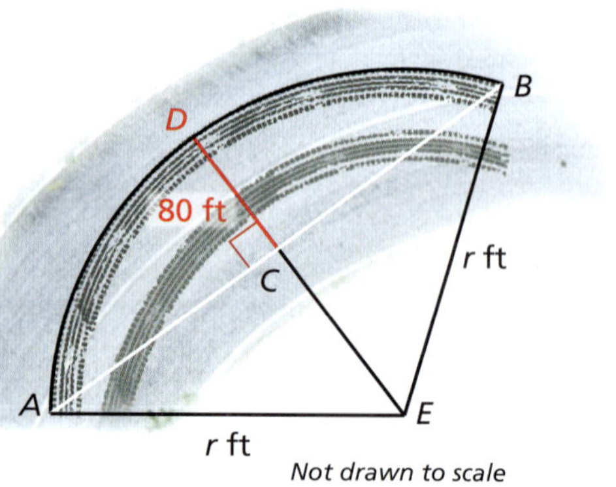

Not drawn to scale

GO DIGITAL

10 Performance Task
Stonehenge

Stonehenge is a prehistoric monument in Wiltshire, England. The first monument at the site was constructed about 5000 years ago.

Possible Functions and Purposes

BURIAL GROUND

Human remains as much as 5000 years old have been found at the site.

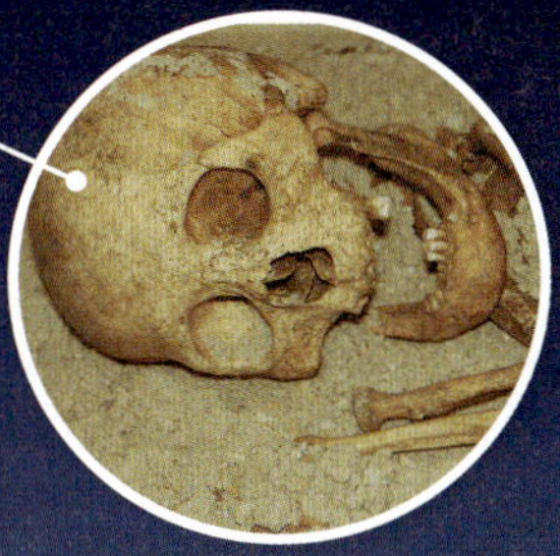

CEREMONIAL SITE

The site may have served as the location of religious rituals and ceremonies.

SCIENTIFIC OBSERVATORY

The geometry of Stonehenge suggests that it may have been built to observe and predict the patterns of the moon and the Sun.

A PLACE OF HEALING

Skeletons at the site show signs of illness or injury. This indicates that ancient peoples may have traveled to the region believing that the stones contained healing powers.

A *megalith* is a large stone used to construct a structure or monument. Some researchers believe that a unit of length called a *megalithic yard* (MY) was used in the construction of Stonehenge.

$$1 \text{ MY} \approx 0.829 \text{ m}$$

THE GEOMETRY OF STONEHENGE

Use the Internet or another resource to research and describe each component of Stonehenge labeled in the diagram.

The four Station Stones approximately form a rectangle with the given dimensions (in megalithic yards). Use these dimensions to approximate the diameter of the Aubrey ring. Then find the measure of each arc subtended by adjacent Station Stones. Determine which regular polygon can be inscribed in the Aubrey ring using the shorter sides of the Station Stone rectangle. Explain your reasoning. Then use the Internet or another resource to research the possible significance of this polygon.

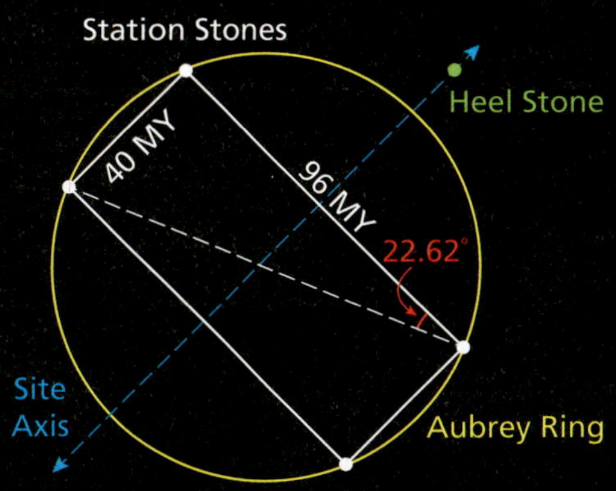

GO DIGITAL

10 Test Prep WITH CalcChat®
Cumulative Practice

WATCH Tutorial videos are available for each exercise.

1. Given that $\triangle ABC \sim \triangle DEF$, find the area of $\triangle DEF$.

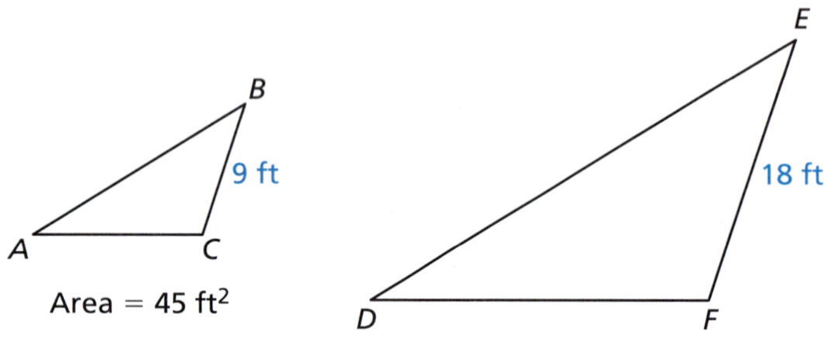

Ⓐ 90 ft²

Ⓑ 126 ft²

Ⓒ 135 ft²

Ⓓ 180 ft²

2. Complete the paragraph proof.

Given Circle C with center (2, 1) and radius 1, Circle D with center (0, 3) and radius 4

Prove Circle C is similar to Circle D.

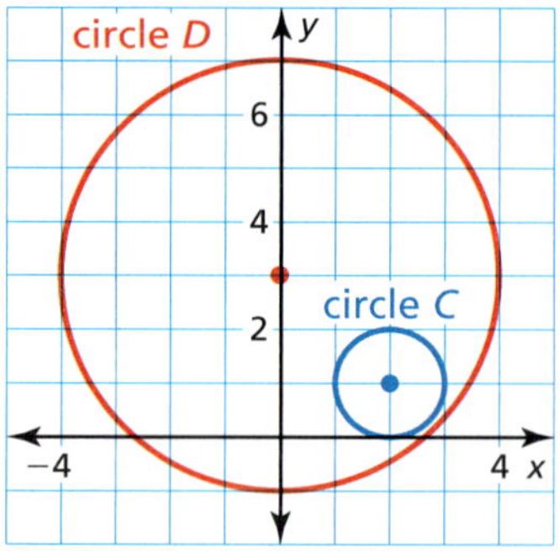

Map Circle C to Circle C' by using the __________ $(x, y) \rightarrow$ ________ so that Circle C' and Circle D have the same center at (__, __). Dilate Circle C' using a center of dilation (__, __) and a scale factor of ___. Because there is a ________ transformation that maps Circle C to Circle D, Circle C is _________ Circle D.

3. Use the diagram to write a proof.

Given $\triangle JPL \cong \triangle NPL$
$\overline{PK}$ is an altitude of $\triangle JPL$.
$\overline{PM}$ is an altitude of $\triangle NPL$.

Prove $\triangle PKL \sim \triangle NMP$

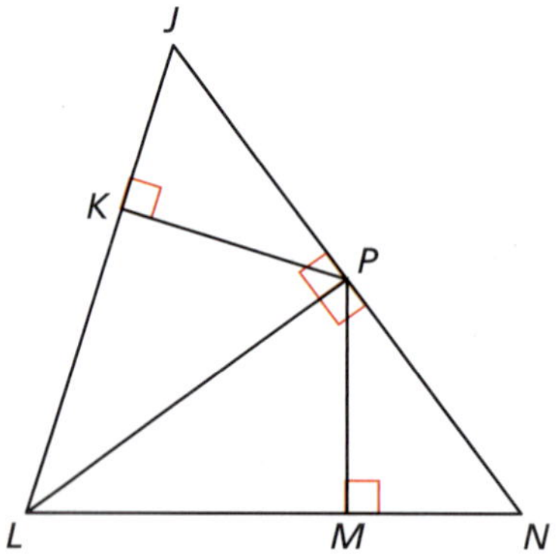

4. The equation of a circle is $x^2 + y^2 + 14x - 16y + 77 = 0$. What are the center and radius of the circle?

Ⓐ center: (14, −16), radius: 8.8

Ⓑ center: (−7, 8), radius: 6

Ⓒ center: (−14, 16), radius: 8.8

Ⓓ center: (7, −8), radius: 6

GO DIGITAL

5. Which describes the location of the circumcenter of a right triangle?

Ⓐ inside the triangle

Ⓑ outside the triangle

Ⓒ at a vertex of the triangle

Ⓓ on the hypotenuse

6. Create as many true equations as possible.

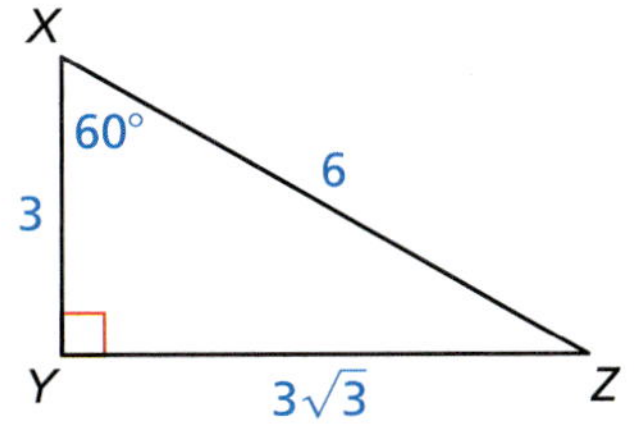

▢ = ▢

sin *X*	cos *X*	tan *X*	$\frac{XY}{XZ}$	$\frac{YZ}{XZ}$
sin *Z*	cos *Z*	tan *Z*	$\frac{XY}{YZ}$	$\frac{YZ}{XY}$

7. Which angles have the same measure as $\angle ACB$? Select all that apply.

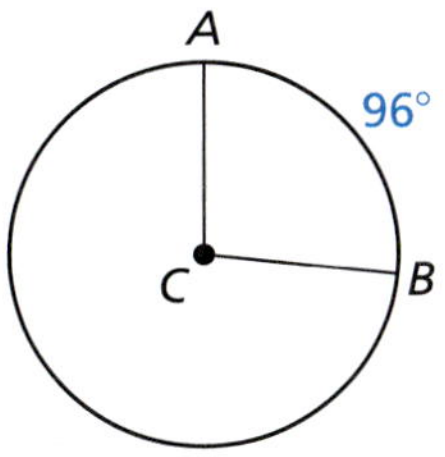

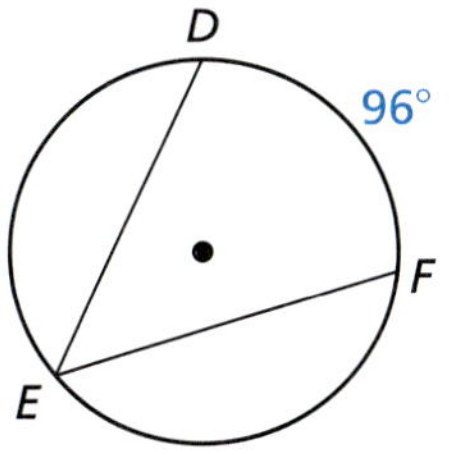

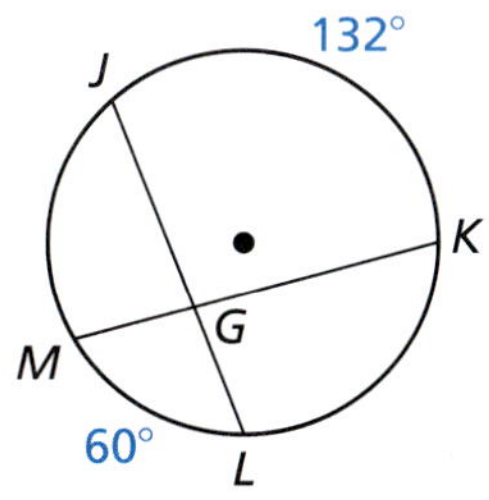

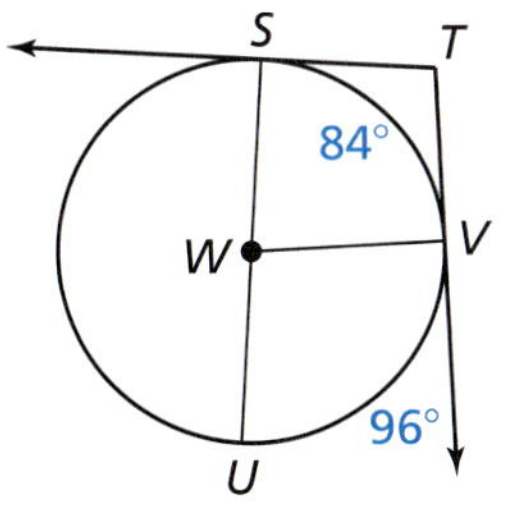

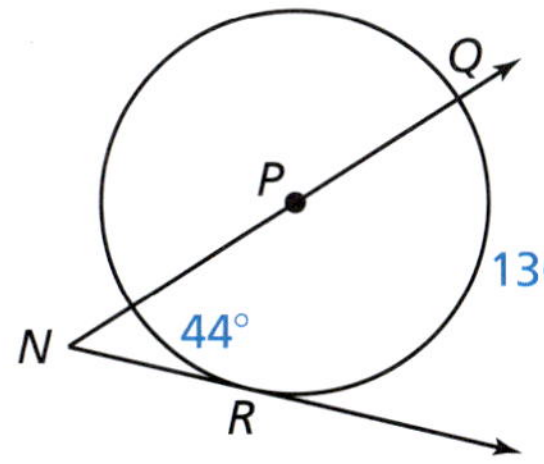

Ⓐ $\angle DEF$

Ⓑ $\angle JGK$

Ⓒ $\angle KGL$

Ⓓ $\angle LGM$

Ⓔ $\angle STV$

Ⓕ $\angle VWU$

Ⓖ $\angle SWV$

Ⓗ $\angle QNR$

8. $\overline{BD}$ is a common side of $\triangle ABD$ and $\triangle BCD$. $m\angle A = m\angle ADB = 45°$ and $m\angle C = 110°$. Which is the longest side of the two triangles?

Ⓐ $\overline{AB}$

Ⓑ $\overline{BC}$

Ⓒ $\overline{AD}$

Ⓓ $\overline{BD}$

GO DIGITAL

11 Circumference and Area

NATIONAL GEOGRAPHIC EXPLORER

Alizé Carrère

WATCH

Alizé Carrère has a bachelor's degree in environmental sciences and international development, and a master's degree in bioresource engineering. She has worked in the Middle East on water resource management and electronic waste between Israel and Palestine. She has also conducted research in Madagascar, studying farmers who adapted to severe deforestation by using erosional gullies as farmland.

- What is bioresource engineering?
- Where is Madagascar? How large is it? What is unique about this island?
- What percent of the species of animals living in Madagascar are not found anywhere else on Earth?

STEM

One focus of bioresource engineering is irrigation. In the 1940s, Frank Zybach invented the center-pivot system to irrigate crops more effectively. In the Performance Task, you will design your own center-pivot irrigation system.

GO DIGITAL

Bioresource Engineering

Preparing for Chapter 11

Chapter Learning Target: Understand circumference and area.

Chapter Success Criteria:
- ◆ I can find circumferences of circles and arc lengths of sectors.
- ◆ I can find areas of circles and sectors.
- ■ I can find areas of polygons.
- ■ I can solve real-life problems involving area.

◆ Surface
■ Deep

AZ VOCAB

Chapter Vocabulary

Work with a partner. Discuss each of the vocabulary terms.

circumference
arc length
radian
sector of a circle
center of a regular polygon
radius of a regular polygon
central angle of a regular polygon
population density

Mathematical Thinking and Reasoning

7 MTR

APPLY MATHEMATICS TO REAL-WORLD CONTEXTS
Mathematicians who apply mathematics to real-world contexts connect mathematical concepts to everyday experiences.

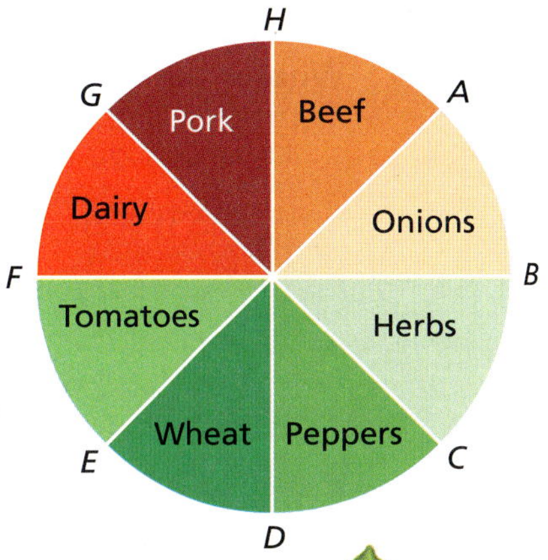

Work with a partner. A *pizza farm* is an educational attraction where ingredients are produced that are used to make pizza.

1. Estimate $m\overset{\frown}{AB}$. Explain your reasoning.
2. The length of the outer row of tomato plants is about 45 feet. Use this information to find the radius of the farm. Justify your answer.
3. A farmer places a sprinkler at the center of the farm. The sprinkler can be set to rotate any number of degrees.
   a. What angle of rotation should the farmer use to water each sector of the farm except the dairy, pork, and beef sectors?
   b. Find the total area watered by the sprinkler in part (a).

GO DIGITAL

11 Prepare WITH CalcChat®

Finding Areas of Triangles

WATCH

Example 1 **A triangle has a base of 8 inches and a height of 7 inches. Find the area of the triangle.**

$A = \frac{1}{2}bh$ — Write formula for area of a triangle.

$= \frac{1}{2}(8)(7)$ — Substitute 8 for *b* and 7 for *h*.

$= 28$ — Multiply.

▶ The area of the triangle is 28 square inches.

Find the area of the triangle.

1.

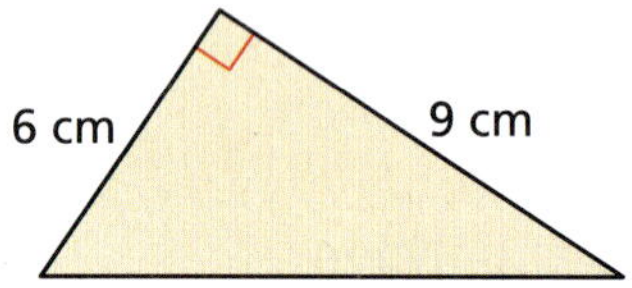

2.

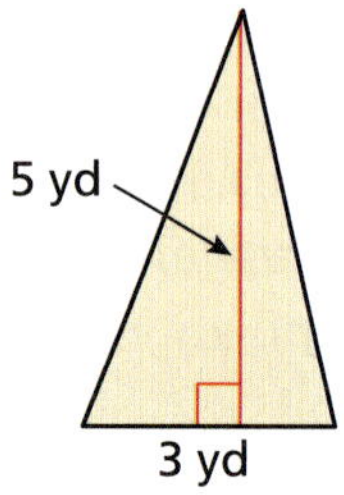

3.

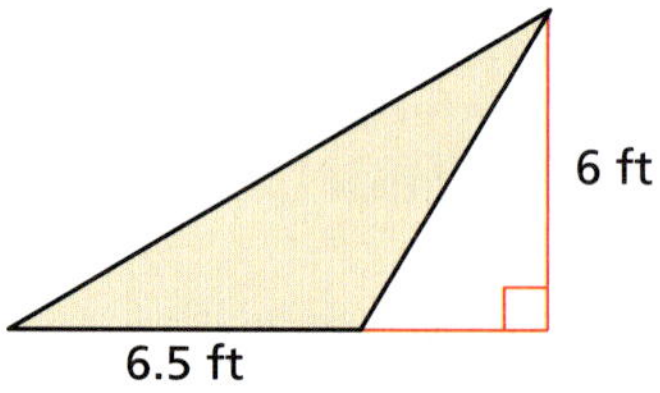

Finding a Missing Dimension

WATCH

Example 2 **A rectangle has a perimeter of 10 meters and a length of 3 meters. What is the width of the rectangle?**

$P = 2\ell + 2w$ — Write formula for perimeter of a rectangle.

$10 = 2(3) + 2w$ — Substitute 10 for *P* and 3 for ℓ.

$10 = 6 + 2w$ — Multiply 2 and 3.

$4 = 2w$ — Subtract 6 from each side.

$2 = w$ — Divide each side by 2.

▶ The width is 2 meters.

Find the missing dimension.

4. A rectangle has a perimeter of 28 inches and a width of 5 inches. What is the length of the rectangle?

5. A triangle has an area of 12 square centimeters and a height of 12 centimeters. What is the base of the triangle?

6. A rectangle has an area of 84 square feet and a width of 7 feet. What is the length of the rectangle?

5 MTR **7.** **STRUCTURE** A rectangle has an area of 64 square meters. The length of the rectangle is 4 times its width. Find the perimeter of the rectangle.

GO DIGITAL

11.1 Circumference and Arc Length

Learning Target: Understand circumference, arc length, and radian measure.

Success Criteria:
- I can use the formula for the circumference of a circle to find measures.
- I can find arc lengths and use arc lengths to find measures.
- I can solve real-life problems involving circumference.
- I can explain radian measure and convert between degree and radian measure.

EXPLORE IT! Finding the Length of a Circular Arc

Work with a partner. A roundabout at a playground has a radius of 10 feet. As it rotates, a person can ride the roundabout at different distances from the center of the circular ride.

a. Find the distance that each person travels along the red circular arc.

i. one full rotation

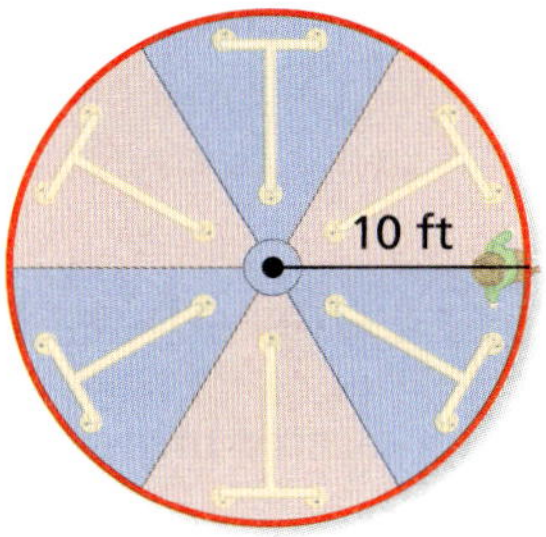

ii. one-fourth of a rotation

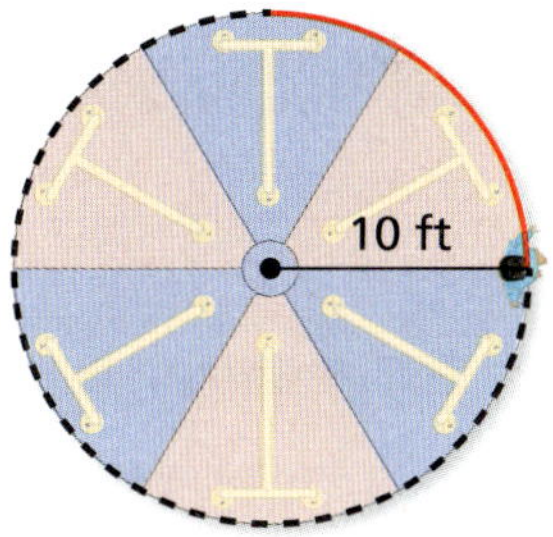

iii. one-third of a rotation

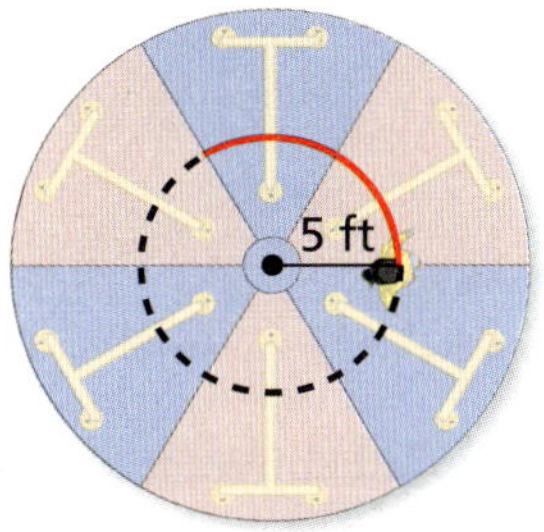

iv. five-eighths of a rotation

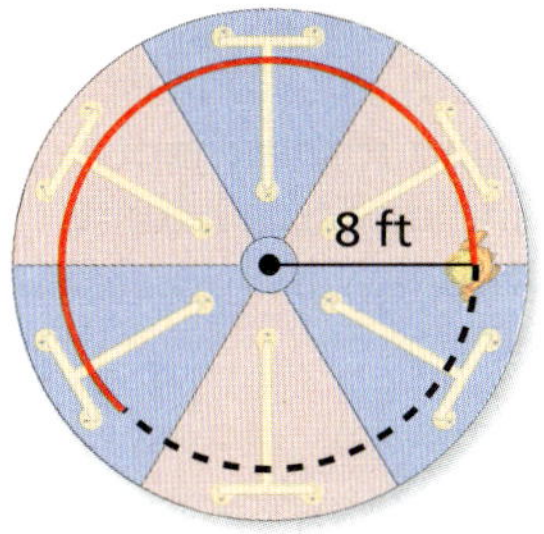

b. A person standing 8 feet from the center of the roundabout travels one-fourth of a rotation. Without performing any calculations, who travels farther, this person or the person in part (ii)? How do you know?

c. For what fraction of a rotation would a person standing 10 feet from the center of the roundabout travel the same distance as the person in part (iv)? Explain your reasoning.

d. Explain how to find the length of any circular arc.

4 MTR COMMUNICATE CLEARLY
Explain how you can use arc measure to find the length of a circular arc. Write a formula to support your answer.

Geometric Reasoning
MA.912.GR.6.4 Solve mathematical and real-world problems involving the arc length and area of a sector in a given circle.

GO DIGITAL

Using the Formula for Circumference

Vocabulary
circumference, *p. 578*
arc length, *p. 579*
radian, *p. 581*

The **circumference** of a circle is the distance around the circle. Consider a regular polygon inscribed in a circle. As the number of sides increases, the polygon approximates the circle, and the ratio of the perimeter of the polygon to the diameter of the circle approaches $\pi \approx 3.14159\ldots$.

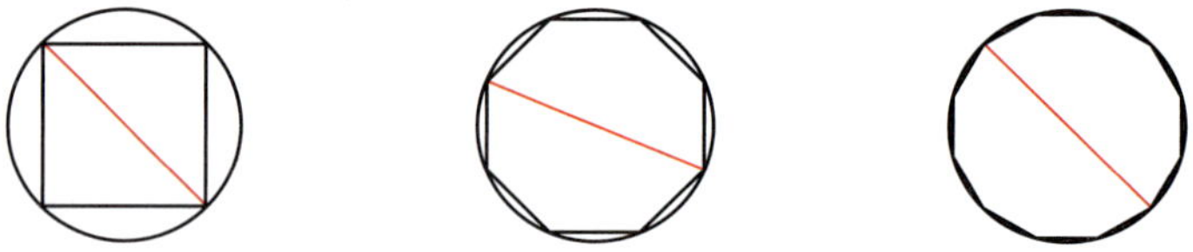

For all circles, the ratio of the circumference C to the diameter d is the same. This ratio is $\frac{C}{d} = \pi$. Solving for C yields the formula for the circumference of a circle, $C = \pi d$. Because $d = 2r$, where r is the radius, you can also write the formula as $C = \pi(2r) = 2\pi r$.

KEY IDEA

Circumference of a Circle

The circumference C of a circle is $C = \pi d$ or $C = 2\pi r$, where d is the diameter of the circle and r is the radius of the circle.

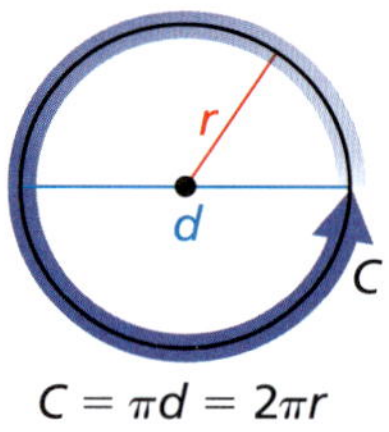

EXAMPLE 1 Using the Formula for Circumference

Find each indicated measure.

a. circumference of a circle with a radius of 9 centimeters

b. radius of a circle with a circumference of 26 meters

STUDY TIP
You have used 3.14 to approximate the value of π. Throughout this chapter, use technology to evaluate expressions involving π.

SOLUTION

a. $C = 2\pi r$ — Write formula for circumference.

$= 2 \cdot \pi \cdot 9$ — Substitute 9 for r.

$= 18\pi$ — Multiply.

≈ 56.55 — Use technology.

▶ The circumference is about 56.55 centimeters.

b. $C = 2\pi r$ — Write formula for circumference.

$26 = 2\pi r$ — Substitute 26 for C.

$\frac{26}{2\pi} = r$ — Divide each side by 2π.

$4.14 \approx r$ — Use technology.

▶ The radius is about 4.14 meters.

SELF-ASSESSMENT 1 I don't understand yet. 2 I can do it with help. 3 I can do it on my own. 4 I can teach someone else.

1. Find the circumference of a circle with a diameter of 5 inches.

2. Find the diameter of a circle with a circumference of 17 feet.

GO DIGITAL

Finding and Using Arc Lengths

> **STUDY TIP**
> Just as the terms *point*, *line*, and *plane* are undefined, the distance around a circular arc is another example of an undefined geometric term.

An **arc length** is a portion of the circumference of a circle. You can use the measure of the arc (in degrees) to find its length (in linear units).

KEY IDEA

Arc Length

In a circle, the ratio of the length of a given arc to the circumference is equal to the ratio of the measure of the arc to 360°.

$$\frac{\text{Arc length of } \widehat{AB}}{2\pi r} = \frac{m\widehat{AB}}{360^\circ}, \text{ or}$$

$$\text{Arc length of } \widehat{AB} = \frac{m\widehat{AB}}{360^\circ} \cdot 2\pi r$$

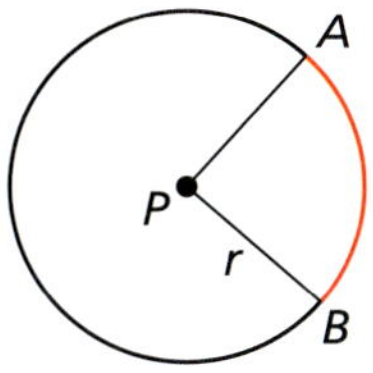

EXAMPLE 2 Finding and Using Arc Lengths

WATCH

Find each indicated measure.

a. arc length of $\widehat{AB}$ **b.** circumference of $\odot Z$ **c.** $m\widehat{RS}$

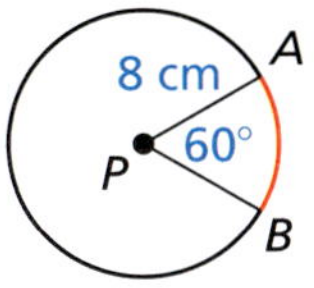

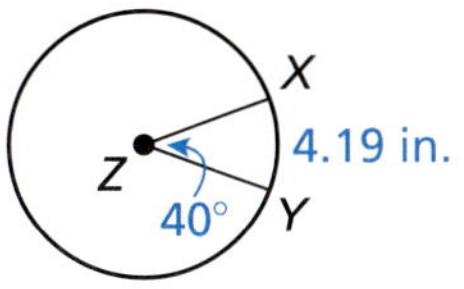

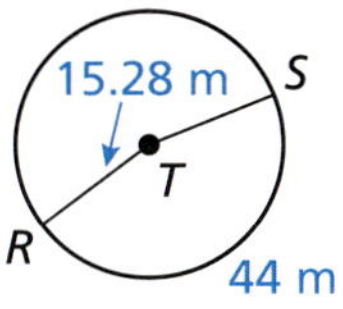

> **5 MTR RELATE CONCEPTS**
> For a given radius, is the length of an arc proportional to the measure of the arc? Explain your reasoning.

SOLUTION

a. $\text{Arc length of } \widehat{AB} = \frac{60^\circ}{360^\circ} \cdot 2\pi(8)$

$\approx 8.38 \text{ cm}$

b. $\frac{\text{Arc length of } \widehat{XY}}{C} = \frac{m\widehat{XY}}{360^\circ}$

$\frac{4.19}{C} = \frac{40^\circ}{360^\circ}$

$\frac{4.19}{C} = \frac{1}{9}$

$37.71 \text{ in.} = C$

c. $\frac{\text{Arc length of } \widehat{RS}}{2\pi r} = \frac{m\widehat{RS}}{360^\circ}$

$\frac{44}{2\pi(15.28)} = \frac{m\widehat{RS}}{360^\circ}$

$360^\circ \cdot \frac{44}{2\pi(15.28)} = \frac{m\widehat{RS}}{360^\circ} \cdot 360^\circ$

$165^\circ \approx m\widehat{RS}$

SELF-ASSESSMENT

1 I don't understand yet. 2 I can do it with help. 3 I can do it on my own. 4 I can teach someone else.

3. WRITING Describe the difference between an arc measure and an arc length.

Find the indicated measure.

4. arc length of $\widehat{PQ}$

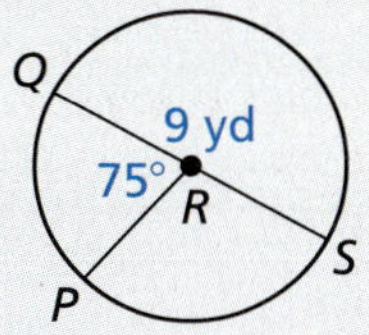

5. circumference of $\odot N$

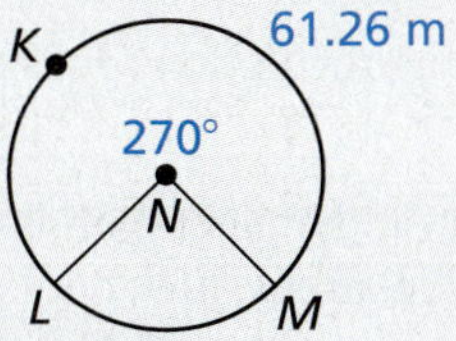

6. radius of $\odot G$

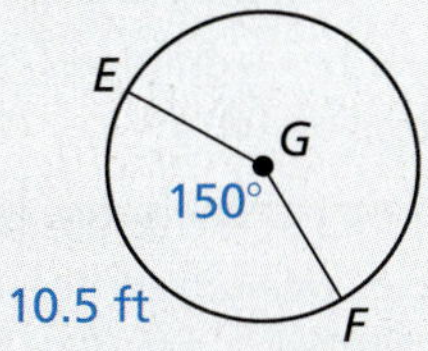

GO DIGITAL

Solving Real-Life Problems

EXAMPLE 3 Using Circumference to Find Distance Traveled

WATCH

The dimensions of a car tire are shown. How many feet does the tire travel when it makes 15 revolutions?

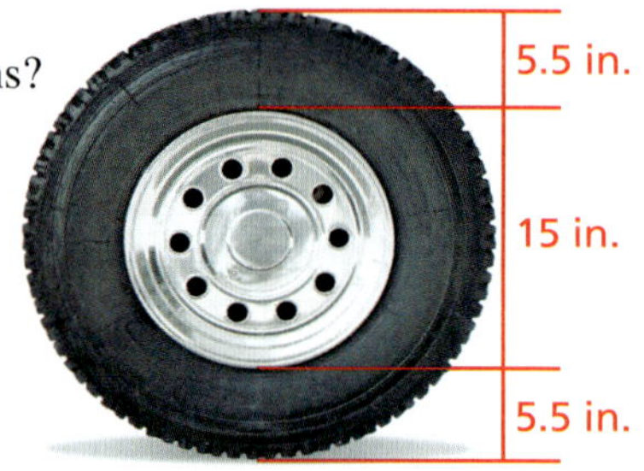

SOLUTION

Step 1 Find the diameter of the tire.

$$d = 15 + 2(5.5) = 26 \text{ in.}$$

Step 2 Find the circumference of the tire.

$$C = \pi d = \pi \cdot 26 = 26\pi \text{ in.}$$

Step 3 Find the distance the tire travels in 15 revolutions. In one revolution, the tire travels a distance equal to its circumference. So, in 15 revolutions, the tire travels a distance equal to 15 times its circumference.

$$\text{Distance traveled} = \text{Number of revolutions} \cdot \text{Circumference}$$

$$= 15 \cdot 26\pi$$

$$\approx 1225.2 \text{ in.}$$

COMMON ERROR

Always pay attention to units. In Example 3, you need to convert units to obtain the correct answer.

Step 4 Use unit analysis. Change 1225.2 inches to feet.

$$1225.2 \text{ in.} \cdot \frac{1 \text{ ft}}{12 \text{ in.}} = 102.1 \text{ ft}$$

▶ The tire travels about 102 feet.

EXAMPLE 4 Using Arc Length to Find Distances

WATCH

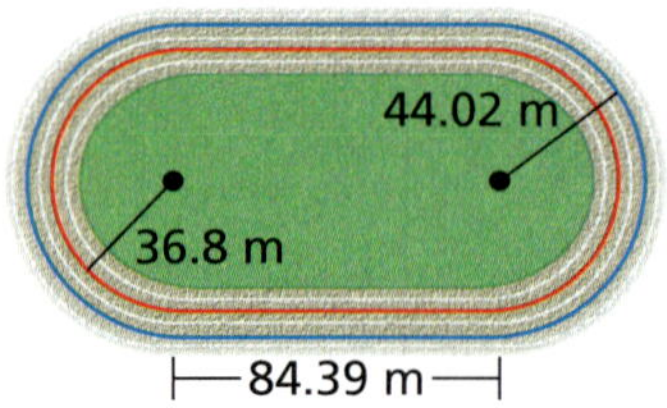

The curves at the ends of the track shown are 180° arcs of circles. The radius of the arc for a runner on the inner path shown in the diagram is 36.8 meters. About how far does the runner travel in one lap?

SOLUTION

The path of the runner is made of two straight sections and two semicircles. To find the total distance, find the sum of the lengths of each part.

$$\text{Distance} = 2 \cdot \text{Length of each straight section} + 2 \cdot \text{Length of each semicircle}$$

$$= 2(84.39) + 2\left(\tfrac{1}{2} \cdot 2\pi \cdot 36.8\right)$$

$$\approx 400.0$$

▶ The runner travels about 400 meters.

SELF-ASSESSMENT 1 I don't understand yet. 2 I can do it with help. 3 I can do it on my own. 4 I can teach someone else.

7. A car tire has a diameter of 28 inches. How many revolutions does the tire make while traveling 500 feet?

8. In Example 4, the radius of the arc for a runner on the outer path is 44.02 meters. The runner completes one lap. Without performing any calculations, how do you know which runner travels farther? Calculate how much farther this runner travels. Explain your reasoning.

GO DIGITAL

Measuring Angles in Radians

In a circle of radius 1, the *radian* measure of a given central angle can be thought of as the length of the arc associated with the angle. The radian measure of a complete circle (360°) is exactly 2π radians, because the circumference of a circle of radius 1 is exactly 2π. You can use this fact to convert from degree measure to radian measure and vice versa.

In a circle, you now know the ratio of the length of a given arc to the circumference is equal to the ratio of the measure of the arc to 360°. To see why, consider the diagram.

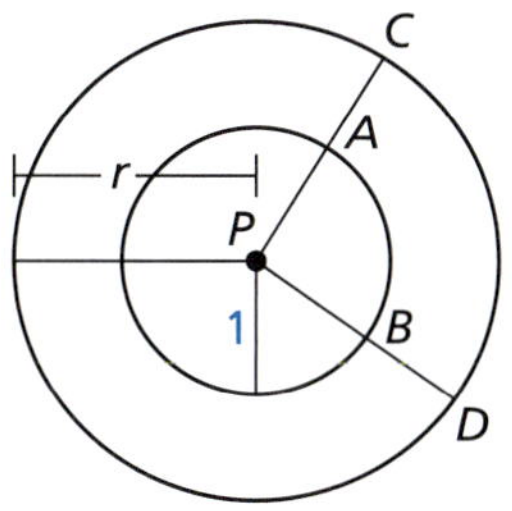

Because a circle of radius 1 has circumference 2π, the arc length of $\widehat{AB}$ is $\dfrac{m\widehat{AB}}{360^\circ} \cdot 2\pi$.

Recall that two arcs are similar arcs if and only if they have the same measure. Because $m\widehat{CD} = m\widehat{AB}$, $\widehat{CD}$ and $\widehat{AB}$ are similar. So, you can write the following proportion.

$$\frac{\text{Arc length of } \widehat{CD}}{\text{Arc length of } \widehat{AB}} = \frac{r}{1}$$

$$\text{Arc length of } \widehat{CD} = r \cdot \text{Arc length of } \widehat{AB}$$

$$\text{Arc length of } \widehat{CD} = r \cdot \frac{m\widehat{AB}}{360^\circ} \cdot 2\pi$$

STUDY TIP

Radian measure is simply another measurement for angles. In some mathematical applications and formulas, calculations are made easier by using radians instead of degrees.

This form of the equation shows that the arc length associated with central angle CPD is *proportional to the radius* of the circle. The constant of proportionality, $\dfrac{m\widehat{AB}}{360^\circ} \cdot 2\pi$, is defined to be the **radian** measure of the central angle associated with the arc.

KEY IDEA

Converting between Degrees and Radians

Degrees to radians Multiply degree measure by $\dfrac{2\pi \text{ radians}}{360^\circ}$, or $\dfrac{\pi \text{ radians}}{180^\circ}$.

Radians to degrees Multiply radian measure by $\dfrac{360^\circ}{2\pi \text{ radians}}$, or $\dfrac{180^\circ}{\pi \text{ radians}}$.

EXAMPLE 5 Converting between Degree and Radian Measure

WATCH

a. Convert 45° to radians.

b. Convert $\dfrac{3\pi}{2}$ radians to degrees.

SOLUTION

a. $45^\circ \cdot \dfrac{\pi \text{ radians}}{180^\circ} = \dfrac{\pi}{4}$ radian

▶ So, $45^\circ = \dfrac{\pi}{4}$ radian.

b. $\dfrac{3\pi}{2}$ radians $\cdot \dfrac{180^\circ}{\pi \text{ radians}} = 270^\circ$

▶ So, $\dfrac{3\pi}{2}$ radians $= 270^\circ$.

SELF-ASSESSMENT 1 I don't understand yet. 2 I can do it with help. 3 I can do it on my own. 4 I can teach someone else.

9. Convert 15° to radians.

10. Convert $\dfrac{4\pi}{3}$ radians to degrees.

GO DIGITAL

11.1 Practice WITH CalcChat® AND CalcView®

In Exercises 1–8, find the indicated measure. *(See Examples 1 and 2.)*

▶ **1.** circumference of a circle with a radius of 6 inches

2. circumference of a circle with a diameter of 5 inches

3. diameter of a circle with a circumference of 63 feet

4. radius of a circle with a circumference of 28π

▶ **5.** arc length of $\widehat{AB}$

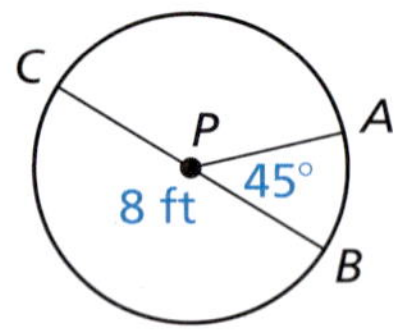

6. $m\widehat{DE}$

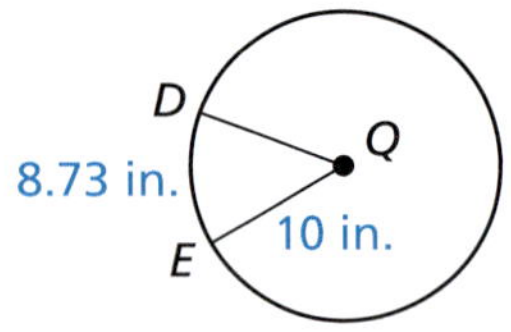

7. circumference of $\odot C$

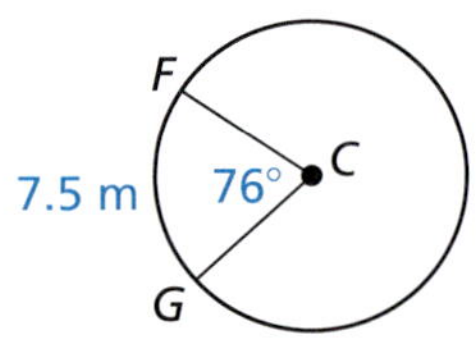

8. radius of $\odot R$

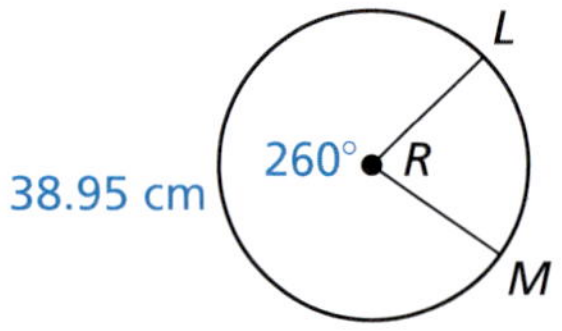

4 MTR **9. ERROR ANALYSIS** Describe and correct the error in finding the circumference of $\odot C$.

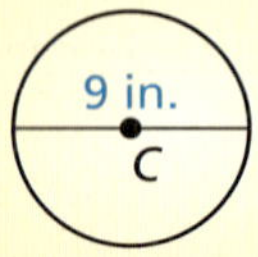

$C = 2\pi r$
$= 2\pi(9)$
$= 18\pi$ in.

4 MTR **10. ERROR ANALYSIS** Describe and correct the error in finding the length of $\widehat{GH}$.

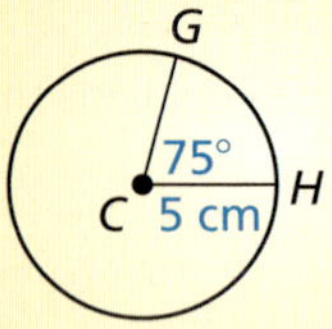

Arc length of $\widehat{GH}$
$= m\widehat{GH} \cdot 2\pi r$
$= 75 \cdot 2\pi(5)$
$= 750\pi$ cm

7 MTR ▶ **11. MODELING REAL LIFE** A measuring wheel is used to calculate the length of a path. The diameter of the wheel is 8 inches. The wheel makes 87 complete revolutions along the length of the path. How many feet long is the path? *(See Example 3.)*

7 MTR **12. MODELING REAL LIFE** You ride your bicycle 40 meters. How many complete revolutions does the front wheel make?

In Exercises 13–16, find the perimeter of the shaded region. *(See Example 4.)*

▶ **13.**

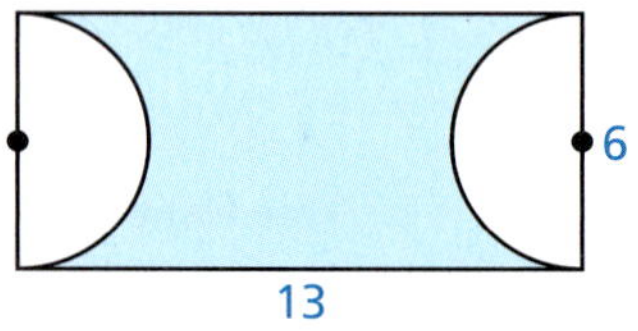

14.

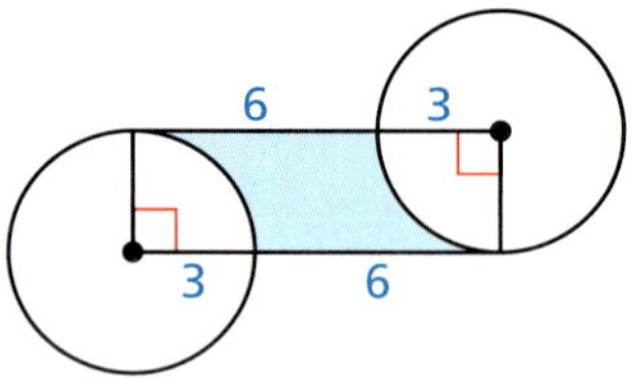

15.

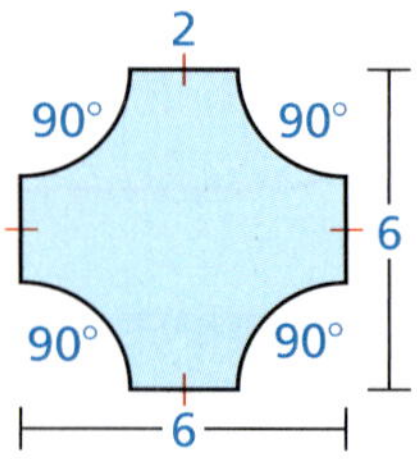

16.

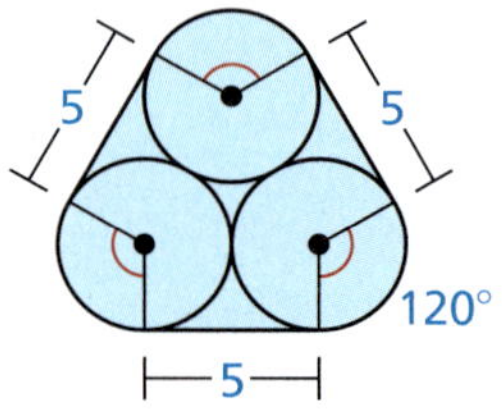

In Exercises 17–20, convert the angle measure. *(See Example 5.)*

▶ **17.** Convert 70° to radians.

18. Convert 300° to radians.

19. Convert $\frac{11\pi}{12}$ radians to degrees.

20. Convert $\frac{\pi}{8}$ radian to degrees.

GO DIGITAL

In Exercises 21 and 22, find the circumference of the circle represented by the given equation. Write the circumference in terms of π.

21. $x^2 + y^2 = 16$

22. $(x + 2)^2 + (y - 3)^2 = 9$

23. **STRUCTURE** A semicircle has endpoints $(-2, 5)$ and $(2, 8)$. Find the arc length of the semicircle.

24. **REASONING** $\widehat{EF}$ is an arc on a circle with radius r. Let $x°$ be the measure of $\widehat{EF}$. Describe the effect on the length of $\widehat{EF}$ if you (a) double the radius of the circle, and (b) double the measure of $\widehat{EF}$.

25. **MODELING REAL LIFE** How many revolutions does the smaller gear complete during a single revolution of the larger gear?

26. **MODELING REAL LIFE** The Orlando Eye in Florida is a Ferris wheel with cars that travel about 0.927 foot per second. How many minutes does it take the Orlando Eye to complete one full revolution?

27. **STRUCTURE** Use the diagram to show that the length of $\widehat{PQ}$ is proportional to the radius r.

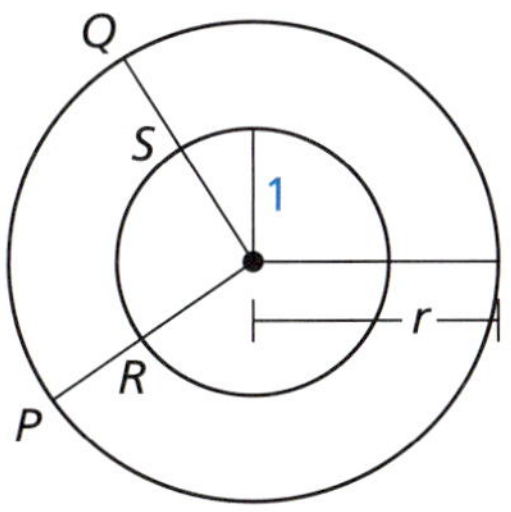

GO DIGITAL

28. **COLLEGE PREP** A 45° arc in $\odot C$ and a 30° arc in $\odot P$ have the same length. What is the ratio of the radius r_1 of $\odot C$ to the radius r_2 of $\odot P$?

Ⓐ 1 to 12 Ⓒ 2 to 3

Ⓑ 1 to 8 Ⓓ 3 to 2

29. **PROBLEM SOLVING** Over 2000 years ago, the Greek scholar Eratosthenes estimated Earth's circumference by assuming that the Sun's rays were parallel. He chose a day when the Sun shone straight down into a well in the city of Syene. At noon, he measured the angle the Sun's rays made with a vertical stick in the city of Alexandria. Eratosthenes assumed that the distance from Syene to Alexandria was about 575 miles. Explain how Eratosthenes was able to use this information to estimate Earth's circumference. Then estimate Earth's circumference.

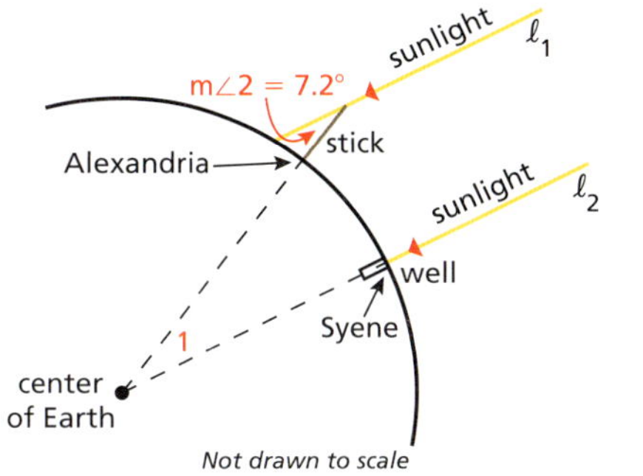

30. **HOW DO YOU SEE IT?**
Compare the circumference of $\odot P$ to the length of $\widehat{DE}$. Explain your reasoning.

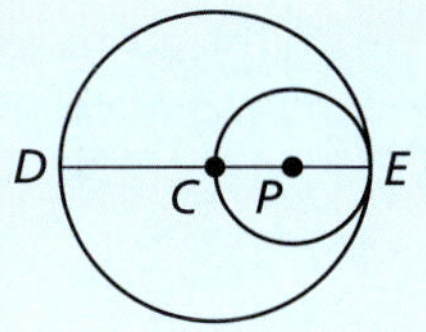

31. **STRUCTURE** Find the circumference of each circle.

a. a circle circumscribed about a square with a side length of 6 centimeters

b. a circle inscribed in an equilateral triangle with a side length of 9 inches

32. **PERFORMANCE TASK** Tire sizes usually involve three numbers. Research and explain the meanings of these numbers. Then choose a tire, and state its size and its warranty. Estimate the number of rotations for which the tire is covered by its warranty.

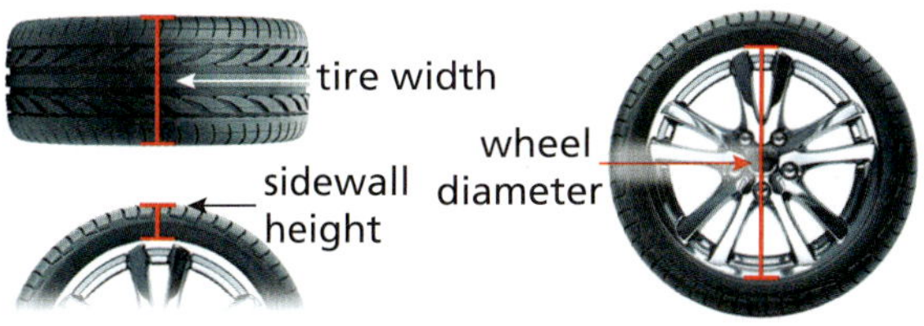

4 MTR **33. MAKING AN ARGUMENT** In the diagram, the measure of the red shaded angle is 30°. The arc length a is 2. Your classmate claims that it is possible to find the circumference of the blue circle without finding the radius of either circle. Is your classmate correct? Explain your reasoning.

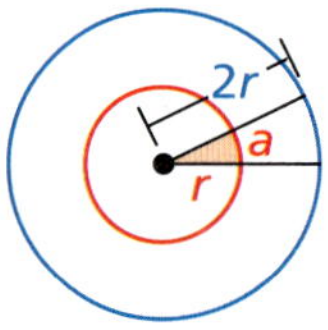

7 MTR **34. MODELING WITH MATHEMATICS** What is the measure (in radians) of the angle formed by the hands of a clock at each time? Explain your reasoning.

a. 1:30 P.M. **b.** 3:15 P.M.

35. DIG DEEPER The sum of the circumferences of circles A, B, and C is 63π. Find AC.

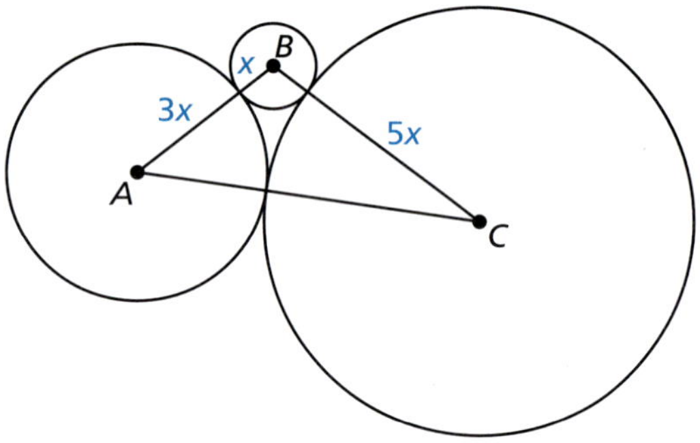

36. THOUGHT PROVOKING
A circle is represented by the equation $(x + 7)^2 + (y - 4)^2 = 169$. The points $A(-19, 9)$ and $B(-2, 16)$ lie on the circle. Find the length of $\widehat{AB}$. Explain your reasoning.

37. PROOF The circles in the diagram are concentric, and $\overline{FG} \cong \overline{GH}$. Prove that $\widehat{JK}$ and $\widehat{NG}$ have the same length.

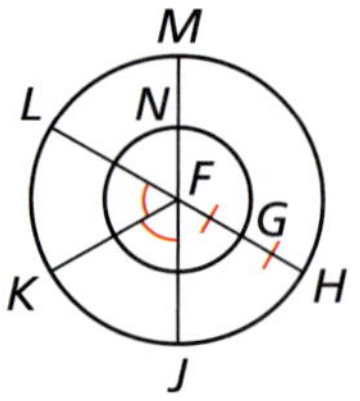

38. REASONING $\overline{AB}$ is divided into congruent segments, each of which is the diameter of a semicircle.

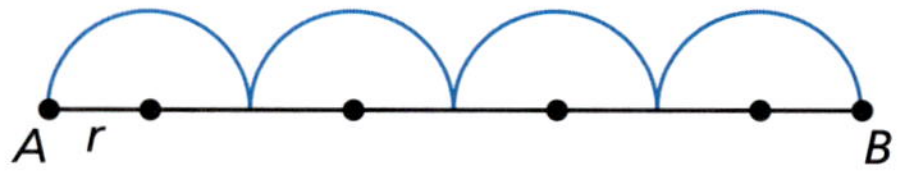

a. What is the sum of the arc lengths?

b. What is the sum of the arc lengths when $\overline{AB}$ is divided into 8 congruent segments? 16 congruent segments? n congruent segments? Explain your reasoning.

REVIEW & REFRESH

WATCH

In Exercises 39 and 40, find the area of the polygon with the given vertices.

39. $X(2, 4)$, $Y(8, -1)$, $Z(2, -1)$

40. $L(-3, 1)$, $M(4, 1)$, $N(4, -5)$, $P(-3, -5)$

In Exercises 41 and 42, find the indicated measure.

41. arc length of $\widehat{AC}$

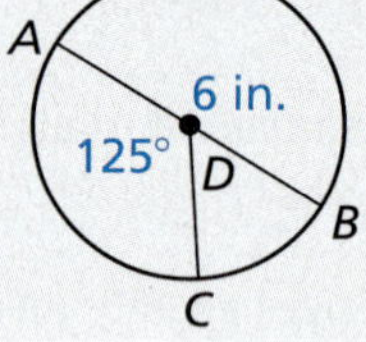

42. $m\widehat{PQ}$

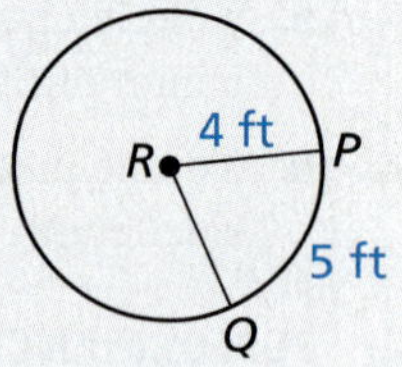

43. Find the length of the midsegment of the trapezoid.

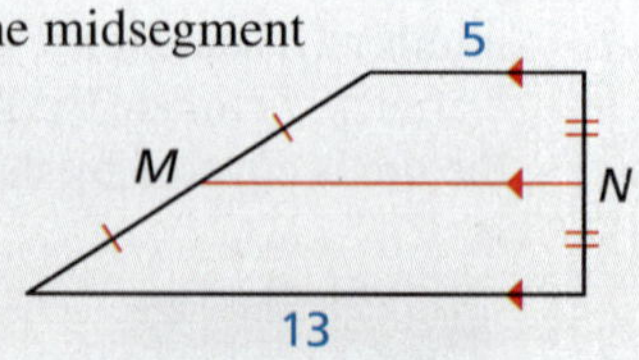

In Exercises 44 and 45, find the center and radius of the circle. Then graph the circle.

44. $x^2 + y^2 = 16$ **45.** $(x - 3)^2 + (y - 7)^2 = 81$

In Exercises 46–48, find the value of x.

46.

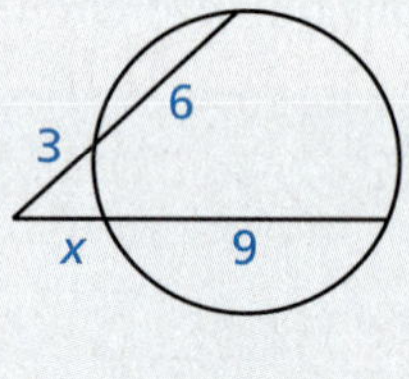

47.

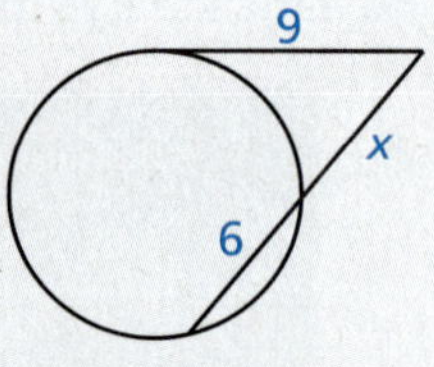

48.

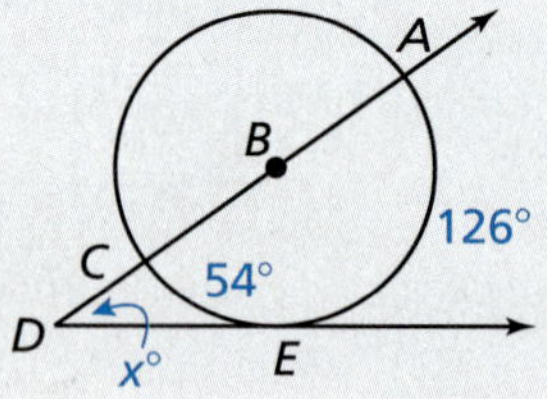

49. Find the value of x that makes $\triangle PQR \sim \triangle XYZ$.

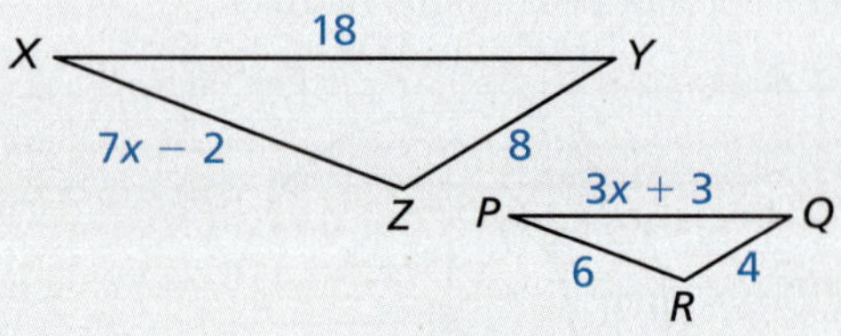

GO DIGITAL

11.2 Areas of Circles and Sectors

Learning Target: Find areas of circles and areas of sectors of circles.

Success Criteria:
- I can use the formula for area of a circle to find measures.
- I can find areas of sectors of circles.
- I can solve problems involving areas of sectors.

EXPLORE IT! Finding the Area of a Sector of a Circle

4 MTR COMMUNICATE CLEARLY

In part (c), explain how you can use the measure of the intercepted arc to find the area of a sector of a circle. Write a formula to support your answer.

Work with a partner. The concentrated beam from the light in a lighthouse can be seen from many miles away. The design of this light was developed at the beginning of the 18th century by French engineer Augustin-Jean Fresnel. Known as the *Fresnel* lens, it has been referred to as "the invention that saved a million ships."

a. A *sector of a circle* is the region bounded by two radii of the circle and their intercepted arc. The figures below show four different lighthouses. Find the area of the shaded circle or sector of a circle that is covered by the light beam from each lighthouse.

i. one full rotation

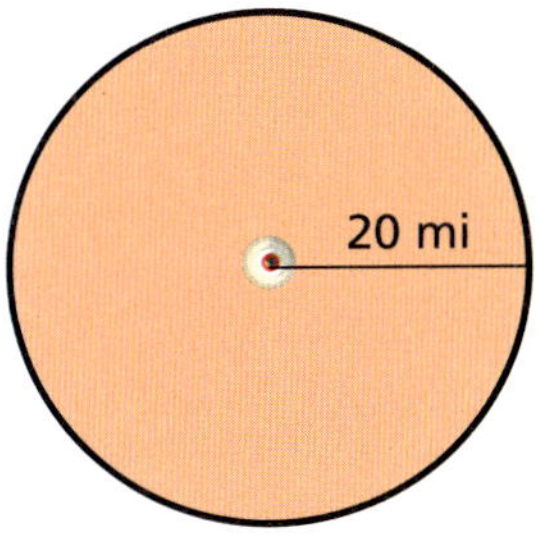

ii. one-fourth of a rotation

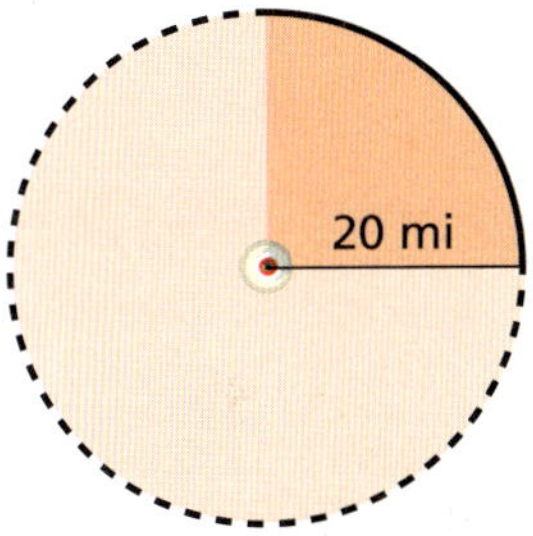

iii. one-third of a rotation

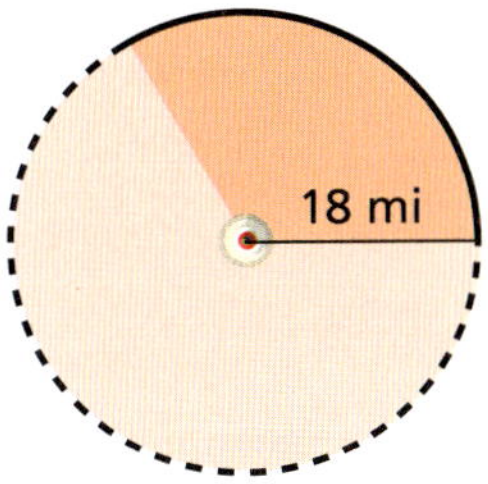

iv. five-eighths of a rotation

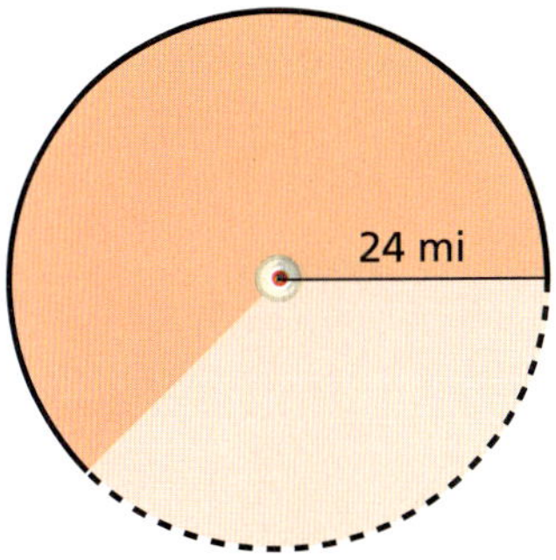

b. For what fraction of a rotation would the lighthouse in part (ii) cover the same area as the lighthouse in part (iv)? Explain your reasoning.

c. Explain, in your own words, how you can find the area of a sector of a circle.

Geometric Reasoning

MA.912.GR.4.4 Solve mathematical and real-world problems involving the area of two-dimensional figures.

MA.912.GR.6.4 Solve mathematical and real-world problems involving the arc length and area of a sector in a given circle.

GO DIGITAL

Vocabulary

sector of a circle, p. 587

Using the Formula for the Area of a Circle

You can divide a circle into congruent sections and rearrange the sections to form a figure that resembles a parallelogram. Increasing the number of congruent sections increases the figure's resemblance to a parallelogram.

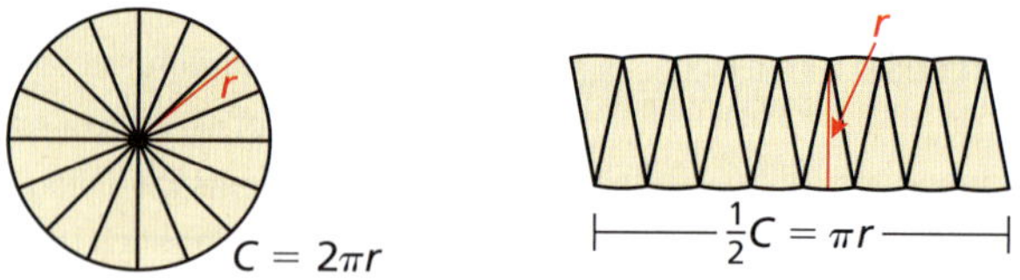

The base of the parallelogram that the figure approaches is half of the circumference, so $b = \frac{1}{2}C = \frac{1}{2}(2\pi r) = \pi r$. The height is the radius, so $h = r$. So, the area of the parallelogram is $A = bh = (\pi r)(r) = \pi r^2$.

KEY IDEA

Area of a Circle

The area of a circle is

$$A = \pi r^2$$

where r is the radius of the circle.

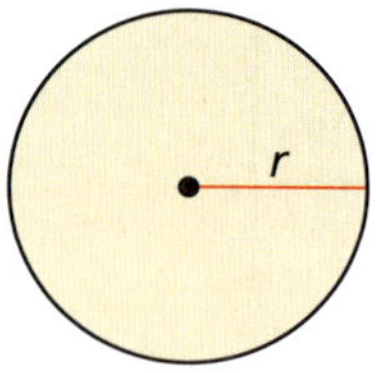

EXAMPLE 1 Using the Formula for the Area of a Circle

Find each indicated measure.

a. area of a circle with a radius of 2.5 centimeters

b. diameter of a circle with an area of 113.1 square centimeters

SOLUTION

a.

$A = \pi r^2$	Write formula for area of a circle.
$= \pi \cdot (2.5)^2$	Substitute 2.5 for r.
$= 6.25\pi$	Simplify.
≈ 19.63	Use technology.

▶ The area of the circle is about 19.63 square centimeters.

b.

$A = \pi r^2$	Write formula for area of a circle.
$113.1 = \pi r^2$	Substitute 113.1 for A.
$\frac{113.1}{\pi} = r^2$	Divide each side by π.
$6 \approx r$	Take the positive square root of each side.

▶ The radius is about 6 centimeters, so the diameter is about 12 centimeters.

SELF-ASSESSMENT

1 I don't understand yet. 2 I can do it with help. 3 I can do it on my own. 4 I can teach someone else.

1. Find the area of a circle with a radius of 4.5 meters.

2. Find the radius of a circle with an area of 176.7 square feet.

GO DIGITAL

Finding Areas of Sectors

A **sector of a circle** is the region bounded by two radii of the circle and their intercepted arc. In the diagram below, sector APB is bounded by $\overline{AP}$, $\overline{BP}$, and $\widehat{AB}$.

KEY IDEA

Area of a Sector

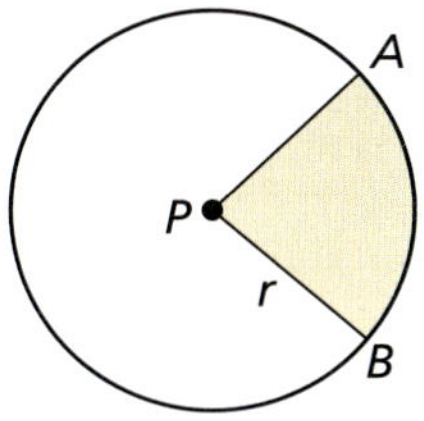

The ratio of the area of a sector of a circle to the area of the whole circle (πr^2) is equal to the ratio of the measure of the intercepted arc to $360°$.

$$\frac{\text{Area of sector } APB}{\pi r^2} = \frac{m\widehat{AB}}{360°}, \quad \text{or} \quad \text{Area of sector } APB = \frac{m\widehat{AB}}{360°} \cdot \pi r^2$$

5 MTR USE STRUCTURE The area of a sector formed by a $45°$ arc is what fraction of the area of the circle?

EXAMPLE 2 Finding Areas of Sectors

WATCH

Find the areas of the sectors formed by $\angle UTV$.

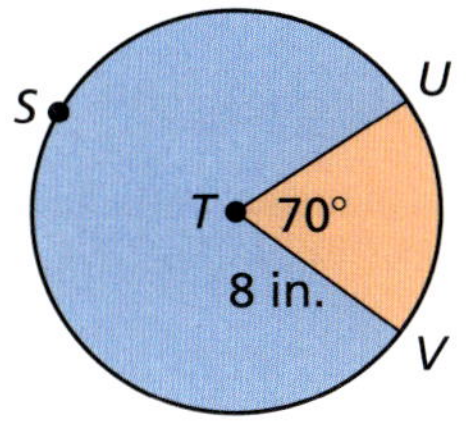

SOLUTION

Step 1 Find the measures of the minor and major arcs.

Because $m\angle UTV = 70°$, $m\widehat{UV} = 70°$ and $m\widehat{USV} = 360° - 70° = 290°$.

Step 2 Find the areas of the small and large sectors.

Area of small sector	$= \dfrac{m\widehat{UV}}{360°} \cdot \pi r^2$	Write formula for area of a sector.
	$= \dfrac{70°}{360°} \cdot \pi \cdot 8^2$	Substitute.
	≈ 39.10	Use technology.
Area of large sector	$= \dfrac{m\widehat{USV}}{360°} \cdot \pi r^2$	Write formula for area of a sector.
	$= \dfrac{290°}{360°} \cdot \pi \cdot 8^2$	Substitute.
	≈ 161.97	Use technology.

▶ The areas of the small and large sectors are about 39.10 square inches and about 161.97 square inches, respectively.

SELF-ASSESSMENT

1 I don't understand yet. 2 I can do it with help. 3 I can do it on my own. 4 I can teach someone else.

Find the areas of the sectors formed by $\angle KJL$.

3.

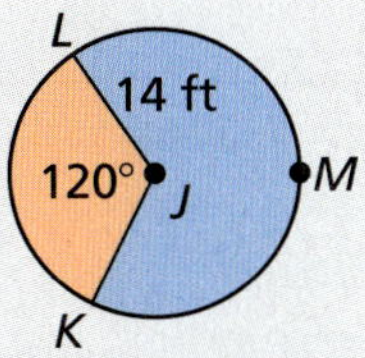

4.

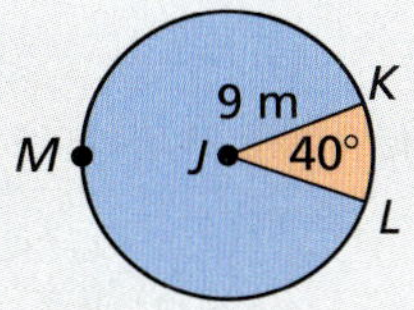

5. WRITING The arc measure of a sector in a given circle is less than $180°$. If the arc measure is doubled, will the area of the sector also be doubled? Explain.

GO DIGITAL

Solving Real-Life Problems

7 MTR **EXAMPLE 3** **Modeling Real Life** WATCH

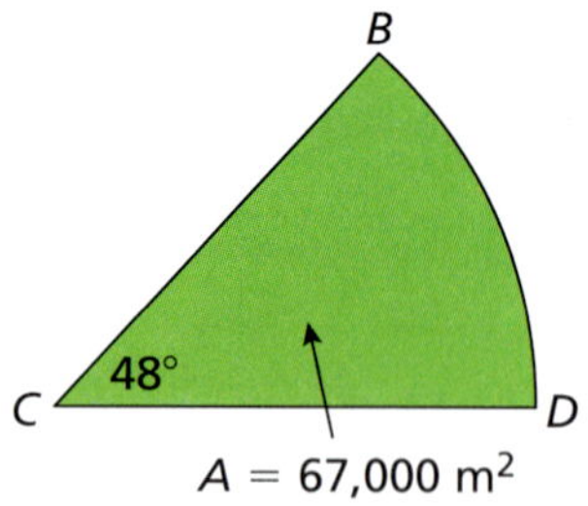

A center-pivot irrigation system consists of sprinkler equipment that rotates around a central pivot point to irrigate a circular region. Find the area of ⊙C.

SOLUTION

To find the area of ⊙C, use the formula for the area of a sector.

$$\text{Area of sector } BCD = \frac{m\overparen{BD}}{360^\circ} \cdot \text{Area of } \odot C \qquad \text{Write formula for area of a sector.}$$

$$67{,}000 = \frac{48^\circ}{360^\circ} \cdot \text{Area of } \odot C \qquad \text{Substitute.}$$

$$502{,}500 = \text{Area of } \odot C \qquad \text{Solve for area of } \odot C.$$

▶ The area of ⊙C is 502,500 square meters.

7 MTR **EXAMPLE 4** **Modeling Real Life** WATCH

For \$15, you can buy one 12-inch pizza or three 6-inch pizzas at a pizza shop. Which option provides more pizza?

SOLUTION

Use the given diameters and the formula for the area of a circle to calculate the amount of pizza in each option.

one 12-inch pizza: $A = \pi r^2 = \pi \cdot 6^2 = 36\pi$ in.²

one 6-inch pizza: $A = \pi r^2 = \pi \cdot 3^2 = 9\pi$ in.²

So, three 6-inch pizzas is $3 \cdot 9\pi = 27\pi$ square inches of pizza.

▶ Because $36\pi > 27\pi$, the 12-inch pizza option provides more pizza.

Another Way A circle with a diameter of 12 inches is a dilation of a circle with a diameter of 6 inches by a scale factor of 2. So, the area of a 12-inch pizza is $2^2 = 4$ times the area of a 6-inch pizza.

Because it takes four 6-inch pizzas to equal the area of a 12-inch pizza, the 12-inch pizza option provides more pizza. ✓

SELF-ASSESSMENT

1 I don't understand yet. 2 I can do it with help. 3 I can do it on my own. 4 I can teach someone else.

6. Find the area of ⊙H.

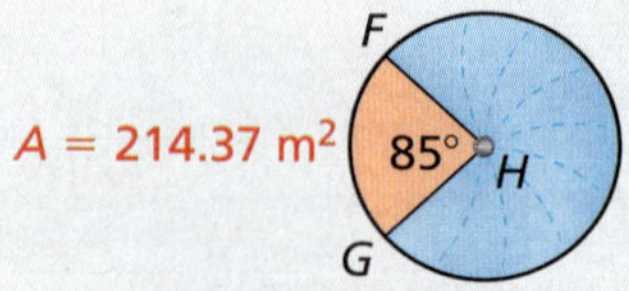

7. **WHAT IF?** You can also buy two 8-inch pizzas for \$15 at the pizza shop in Example 4. Which option provides the most pizza? the least pizza?

3 MTR 8. **ADAPT A PROCEDURE** If you know the area and radius of a sector of a circle, can you find the measure of the intercepted arc? Explain.

GO DIGITAL

11.2 Practice WITH CalcChat® AND CalcView®

In Exercises 1–6, find the indicated measure. *(See Example 1.)*

▶ **1.** area of a circle with a radius of 5 inches

2. area of a circle with a diameter of 16 feet

3. radius of a circle with an area of 89 square feet

4. radius of a circle with an area of 380 square inches

5. diameter of a circle with an area of 12.6 square inches

6. diameter of a circle with an area of 676π square centimeters

In Exercises 7–10, find the areas of the sectors formed by $\angle DFE$. *(See Example 2.)*

▶ **7.**

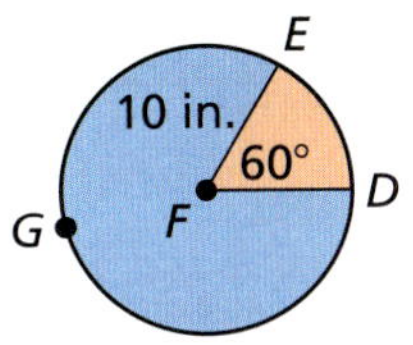

8.

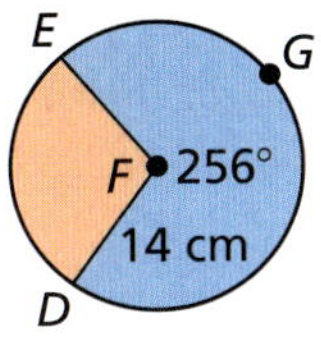

9.

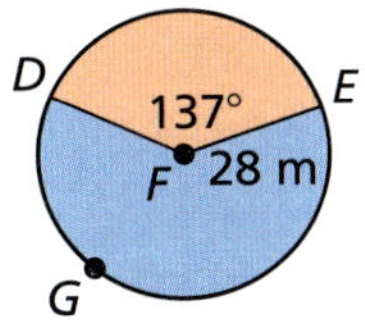

10.

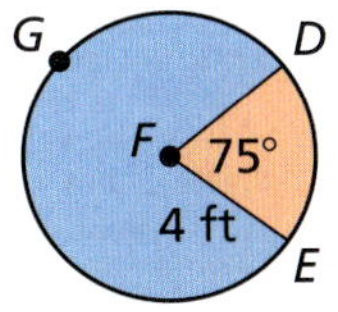

11. ERROR ANALYSIS Describe and correct the error in finding the area of the circle.

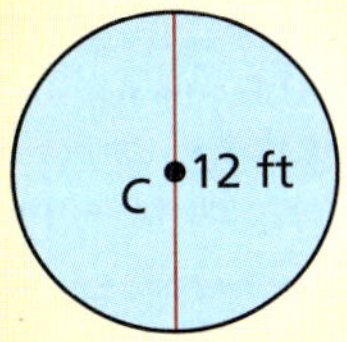

$A = \pi(12)^2$

$= 144\pi$

$\approx 452.39 \text{ ft}^2$

12. ERROR ANALYSIS Describe and correct the error in finding the area of sector XZY when the area of $\odot Z$ is 255 square feet.

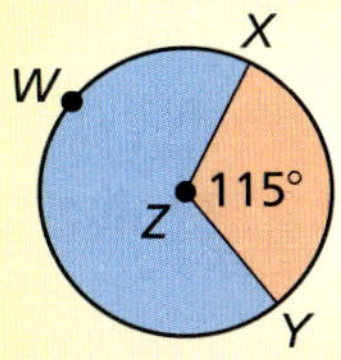

Let n be the area of sector XZY.

$\frac{n}{360} = \frac{115}{255}$

$n \approx 162.35 \text{ ft}^2$

In Exercises 13 and 14, the area of the shaded sector is shown. Find the indicated measure. *(See Example 3.)*

▶ **13.** area of $\odot M$

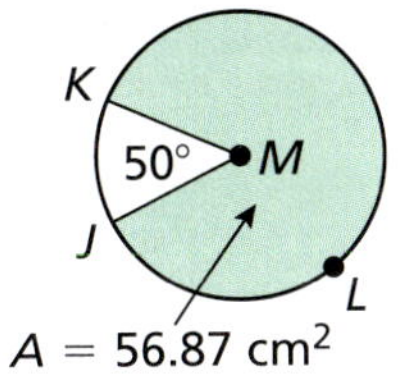

14. radius of $\odot M$

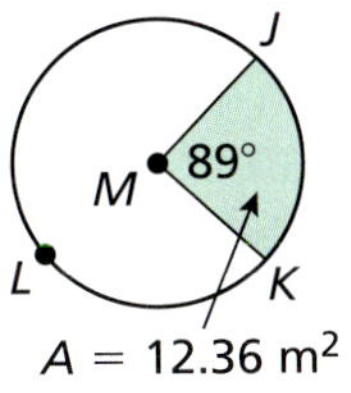

15. MODELING REAL LIFE For \$20, you can buy two 16-inch diameter flatbreads or four 10-inch diameter flatbreads at a restaurant. Which option provides more flatbread? *(See Example 4.)*

16. MODELING REAL LIFE You are painting circles on a mural. You paint two 24-inch diameter circles red and paint four 18-inch diameter circles blue. Do you use more red paint or blue paint?

In Exercises 17 and 18, find the area of the shaded region.

17.

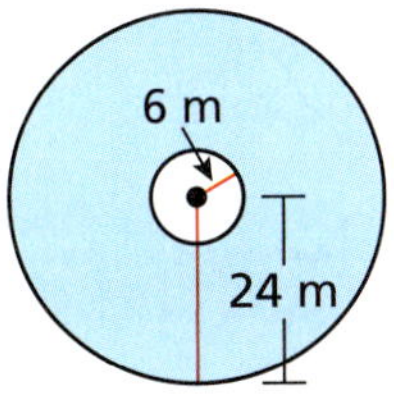

18.

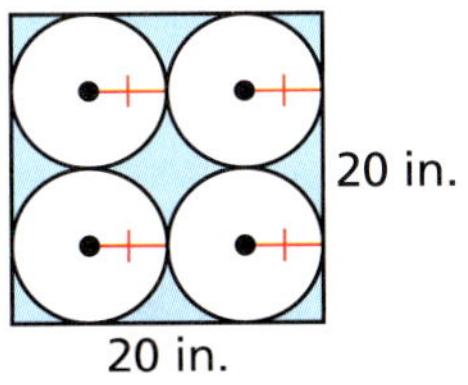

19. PROBLEM SOLVING The diagram shows the shape of a putting green at a miniature-golf course. One part of the green is a sector of a circle. Find the area of the putting green.

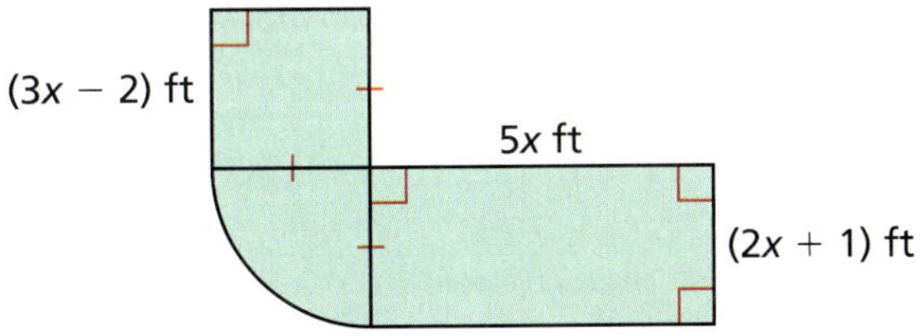

20. HOW DO YOU SEE IT? The outermost edges of the pattern shown form a square. If you know the dimensions of the outer square, is it possible to compute the total colored area? Explain.

GO DIGITAL

7 MTR **21.** **MODELING REAL LIFE** The diagram shows the area of a lawn covered by a water sprinkler.

a. Find the area of the lawn that is covered by the sprinkler.

b. The water pressure is weakened so that the radius is 12 feet. By what percent does the area of the lawn that is covered by the sprinkler decrease?

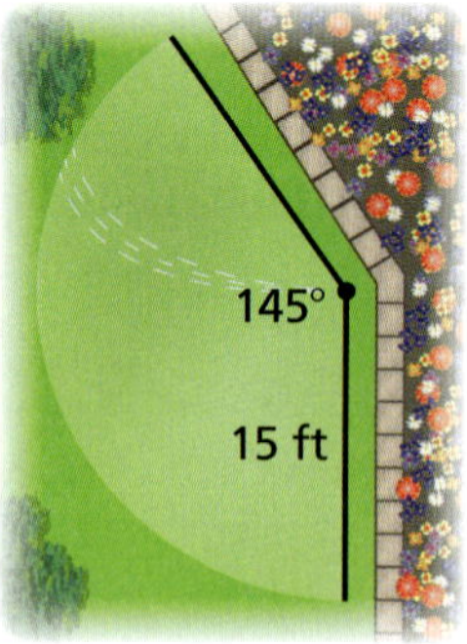

7 MTR **22.** **MODELING REAL LIFE** The diagram shows a projected beam of light from a lighthouse. How much more water is covered than land by the light from the lighthouse? Explain.

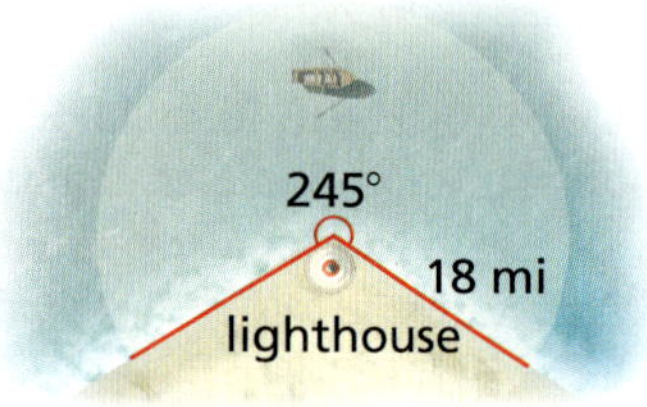

5 MTR **23.** **STRUCTURE** A square is inscribed in a circle. The same square is also circumscribed about a smaller circle. Find the ratio of the area of the larger circle to the area of the smaller circle.

24. **THOUGHT PROVOKING**
You know that the area of a circle is πr^2. Find the formula for the area of an *ellipse*, shown below.

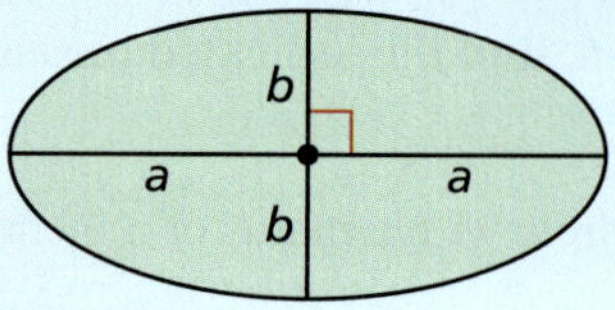

25. **DIG DEEPER** Find the area between the three congruent tangent circles. The radius of each circle is 6 inches.

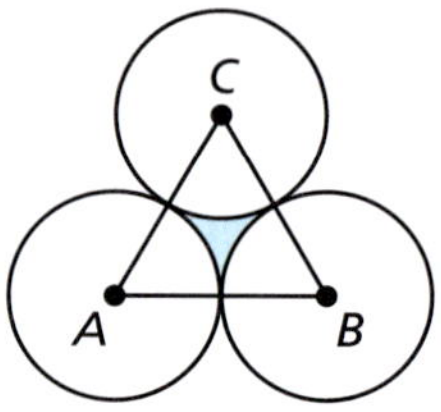

6 MTR **26.** **JUSTIFYING STEPS** The diagram shows semicircles with diameters equal to three sides of a right triangle. Prove that the sum of the areas of the two shaded crescents equals the area of the triangle. Justify your steps.

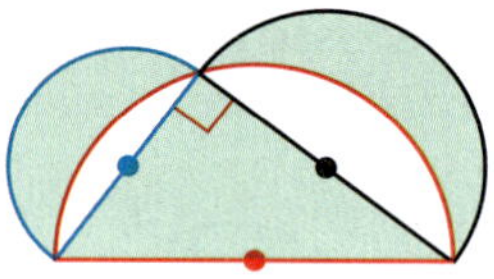

REVIEW & REFRESH

WATCH

27. Find the area of the shaded region.

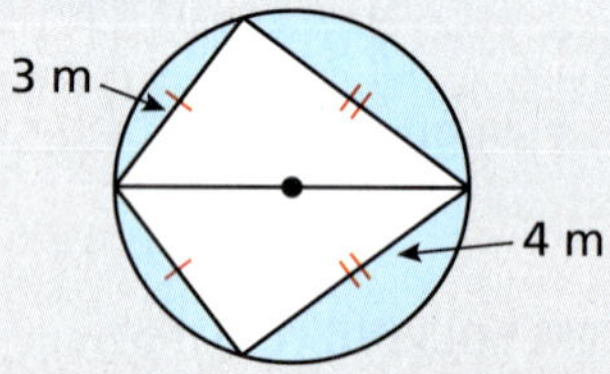

In Exercises 28 and 29, find the indicated measure.

28. circumference of $\odot M$

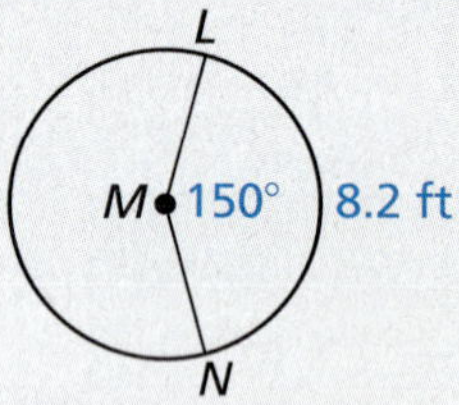

29. $m\angle ADB$

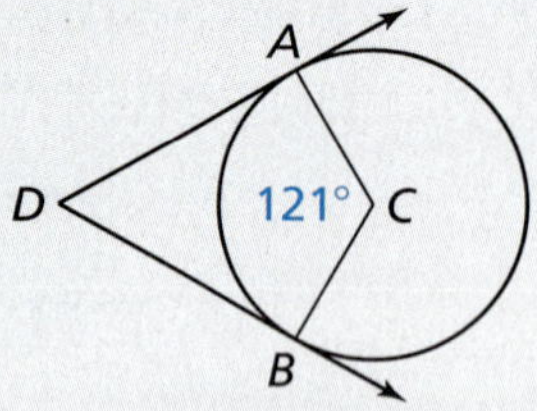

30. Graph $x^2 + y^2 - 16x + 6y + 9 = 0$. State the center and radius of the circle.

31. Find the value of x.

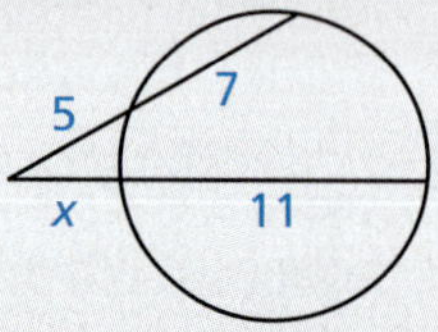

32. Find the distance from the point $(-6, 4)$ to the line $y = 3x - 1$.

33. Point D is the centroid of $\triangle ABC$. Find CD and CE when $DE = 7$.

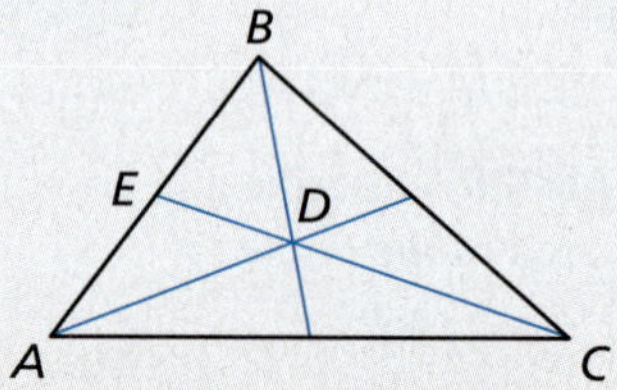

7 MTR **34.** **MODELING REAL LIFE** An agility A-frame is used to test the agility of dogs. Prove that the distance up both sides of the ramp is the same.

35. Find the lengths of the diagonals of rectangle $JKLM$ when $JL = 4x + 2$ and $KM = 3x + 12$.

36. The endpoints of $\overline{QR}$ are $Q(-3, -1)$ and $R(5, 7)$. Write an equation of the perpendicular bisector of $\overline{QR}$.

GO DIGITAL

11.3 Areas of Polygons

Learning Target: Find angle measures and areas of regular polygons.

Success Criteria:

- I can find areas of rhombuses and kites.
- I can find angle measures in regular polygons.
- I can find areas of regular polygons.
- I can explain how the area of a triangle is related to the area formulas for rhombuses, kites, and regular polygons.

The *center of a regular polygon* is the center of its circumscribed circle.

The distance from the center to any side of a regular polygon is called the *apothem of a regular polygon.*

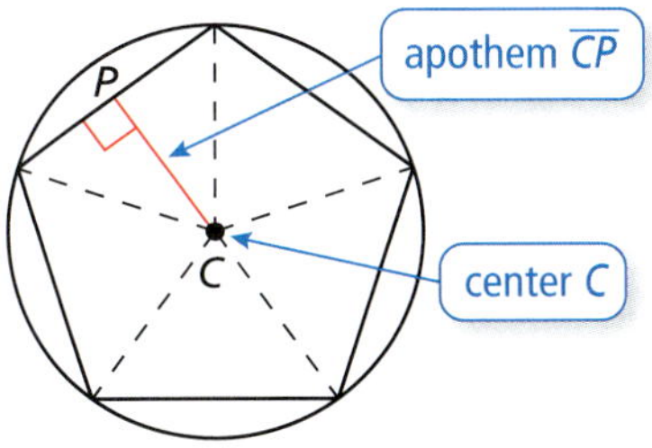

EXPLORE IT! Finding the Area of a Regular Polygon

Work with a partner.

a. Construct an equilateral triangle and find its area. Explain your method.

b. Find the apothem of the triangle you constructed in part (a) and use it to find the area of the triangle. Explain your method. Compare your answer with the area you found in part (a).

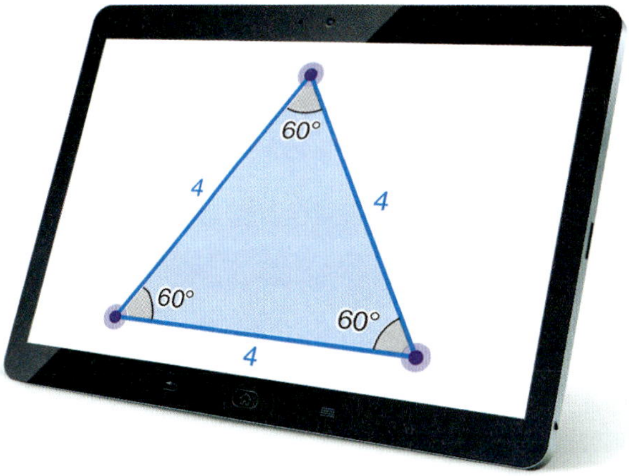

c. Construct each regular polygon using any side length. Then use the apothem of each polygon to find its area. Explain your method.

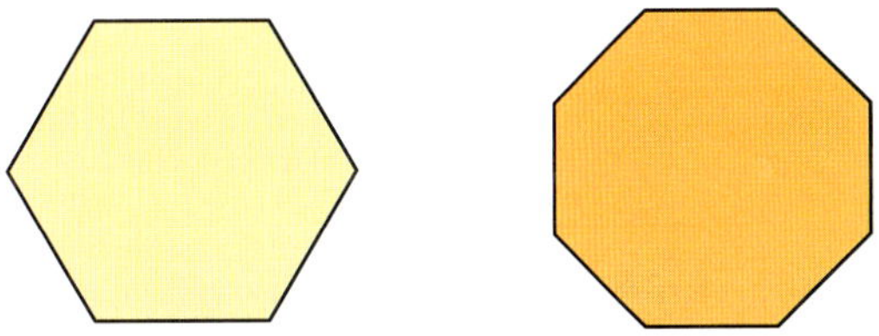

Use your results to develop a strategy to find the area of any regular polygon, regardless of the number of sides.

d. Use your results in part (c) to write a formula for the area of any regular polygon.

2 MTR USE ANOTHER METHOD
What other method(s) can you use to find the area of the polygons in part (c)?

Geometric Reasoning
MA.912.GR.4.4 Solve mathematical and real-world problems involving the area of two-dimensional figures.

GO DIGITAL

Finding Areas of Rhombuses and Kites

Vocabulary

center of a regular polygon, *p. 593*
radius of a regular polygon, *p. 593*
apothem of a regular polygon, *p. 593*
central angle of a regular polygon, *p. 593*

You can decompose a rhombus or kite with diagonals d_1 and d_2 into two congruent triangles with base d_1 and height $\frac{1}{2}d_2$. The area of one of these triangles is $\frac{1}{2}d_1\left(\frac{1}{2}d_2\right) = \frac{1}{4}d_1d_2$. So, the area of a rhombus or kite is $2\left(\frac{1}{4}d_1d_2\right) = \frac{1}{2}d_1d_2$.

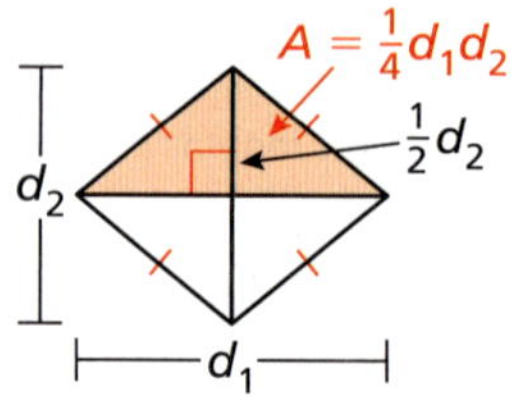

rhombus

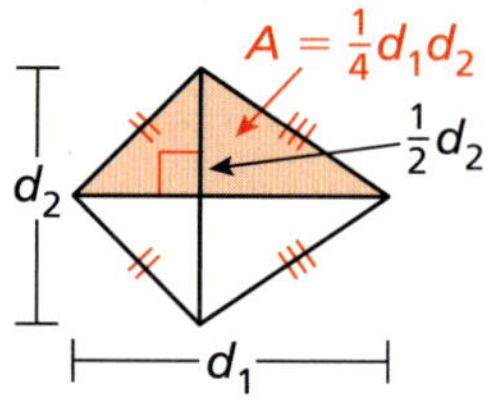

kite

KEY IDEA

Area of a Rhombus or Kite

The area of a rhombus or kite is one-half the product of the diagonals d_1 and d_2.

$A = \frac{1}{2}d_1d_2$

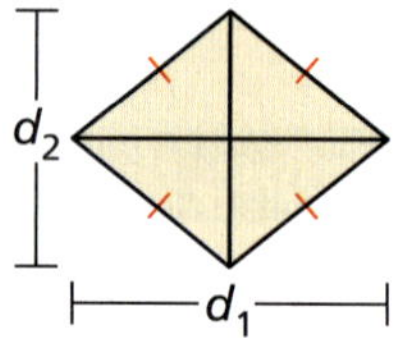

rhombus

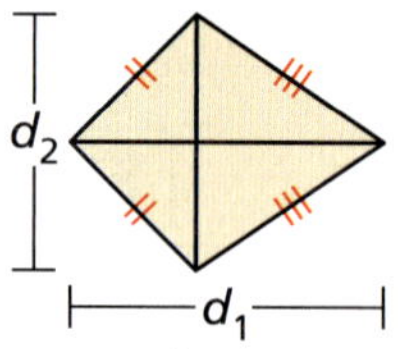

kite

EXAMPLE 1 Finding the Area of a Rhombus or Kite

Find the area of each rhombus or kite.

a.

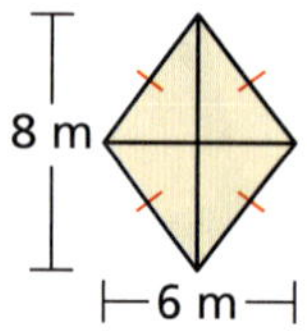

b.

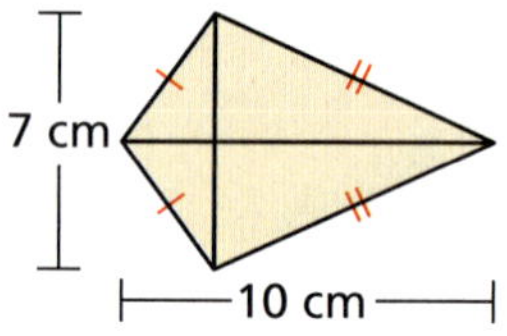

SOLUTION

a. $A = \frac{1}{2}d_1d_2$
$= \frac{1}{2}(6)(8)$
$= 24$

So, the area is 24 square meters.

b. $A = \frac{1}{2}d_1d_2$
$= \frac{1}{2}(10)(7)$
$= 35$

So, the area is 35 square centimeters.

SELF-ASSESSMENT 1 I don't understand yet. 2 I can do it with help. 3 I can do it on my own. 4 I can teach someone else.

1. Find the area of a rhombus with diagonals $d_1 = 4$ feet and $d_2 = 5$ feet.
2. Find the area of a kite with diagonals $d_1 = 12$ inches and $d_2 = 9$ inches.
3. Find the area of the kite.

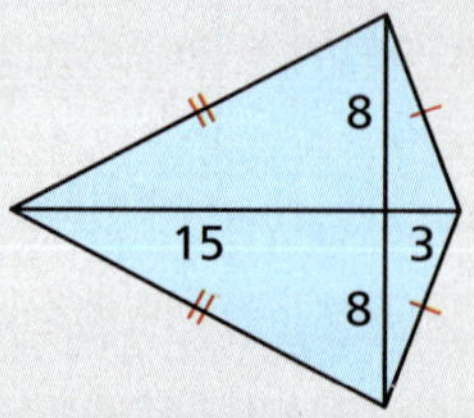

GO DIGITAL

Finding Angle Measures in Regular Polygons

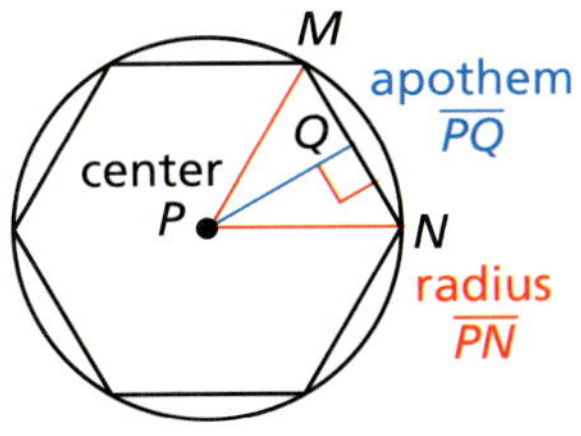

$\angle MPN$ is a central angle.

The diagram shows a regular polygon inscribed in a circle. The **center of a regular polygon** and the **radius of a regular polygon** are the center and the radius of its circumscribed circle.

The distance from the center to any side of a regular polygon is called the **apothem of a regular polygon**. The apothem is the height to the base of an isosceles triangle that has two radii as legs. The word *apothem* refers to a segment as well as a length. For a given regular polygon, think of *an* apothem as a segment and *the* apothem as a length.

A **central angle of a regular polygon** is an angle formed by two radii drawn to consecutive vertices of the polygon. To find the measure of each central angle, divide 360° by the number of sides of the polygon.

EXAMPLE 2 Finding Angle Measures in a Regular Polygon

WATCH

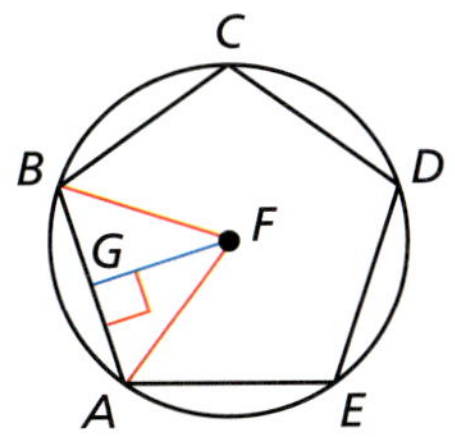

In the diagram, *ABCDE* is a regular pentagon inscribed in $\odot F$. Find each angle measure.

a. $m\angle AFB$ **b.** $m\angle AFG$ **c.** $m\angle GAF$

SOLUTION

a. $\angle AFB$ is a central angle of *ABCDE*, which has 5 sides.

▶ So, $m\angle AFB = \frac{360°}{5} = 72°$.

b. $\overline{FG}$ is an apothem, which makes it an altitude of isosceles $\triangle AFB$. So, $\overline{FG}$ bisects $\angle AFB$ and $m\angle AFG = \frac{1}{2}(m\angle AFB)$.

$$m\angle AFG = \tfrac{1}{2}(m\angle AFB) = \tfrac{1}{2}(72°) = 36°$$

▶ So, $m\angle AFG = 36°$.

c. By the Triangle Sum Theorem, the sum of the angle measures of right $\triangle GAF$ is 180°.

$$m\angle GAF = 180° - 90° - 36° = 54°$$

▶ So, $m\angle GAF = 54°$.

5 MTR RELATE CONCEPTS

$\overline{FG}$ is an altitude of an isosceles triangle. Is $\overline{FG}$ also a median of the isosceles triangle? How do you know?

SELF-ASSESSMENT

1 I don't understand yet. 2 I can do it with help. 3 I can do it on my own. 4 I can teach someone else.

4. **WRITING** Explain how to find the measure of a central angle of a regular 15-gon.

4 MTR **5.** **DIFFERENT WORDS, SAME QUESTION** Which is different? Find "both" answers.

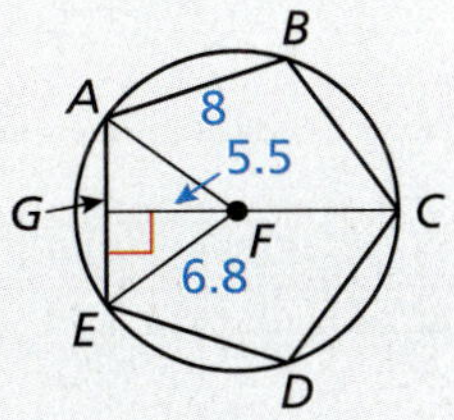

Find the radius of $\odot F$.	Find the apothem of polygon *ABCDE*.
Find *AF*.	Find the radius of polygon *ABCDE*.

In the diagram, *WXYZ* is a square inscribed in $\odot P$.

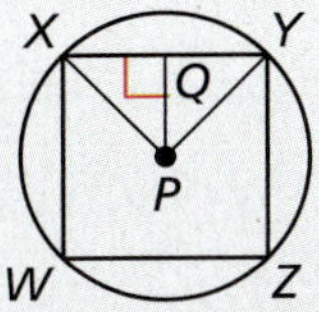

6. Identify the center, a radius, an apothem, and a central angle of *WXYZ*.

7. Find $m\angle XPY$, $m\angle XPQ$, and $m\angle PXQ$.

GO DIGITAL

Finding Areas of Regular Polygons

You can find the area of any regular n-gon by dividing it into congruent triangles.

READING

In this book, a point shown inside a regular polygon marks the center of the circle that can be circumscribed about the polygon.

A = Area of one triangle • Number of triangles

$= \left(\frac{1}{2} \cdot s \cdot a\right) \cdot n$ — Base of triangle is s and height of triangle is a. Number of triangles is n.

$= \frac{1}{2} \cdot a \cdot (n \cdot s)$ — Commutative and Associative Properties of Multiplication

$= \frac{1}{2}a \cdot P$ — There are n congruent sides of length s, so perimeter P is $n \cdot s$.

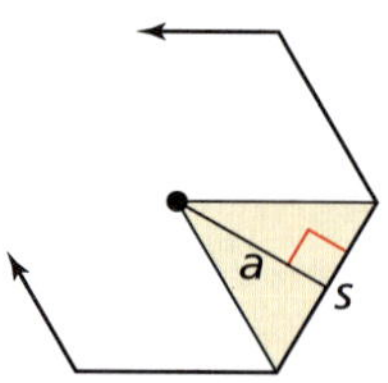

KEY IDEA

Area of a Regular Polygon

The area of a regular n-gon with side length s is one-half the product of the apothem a and the perimeter P.

$$A = \frac{1}{2}aP, \quad \text{or} \quad A = \frac{1}{2}a \cdot ns$$

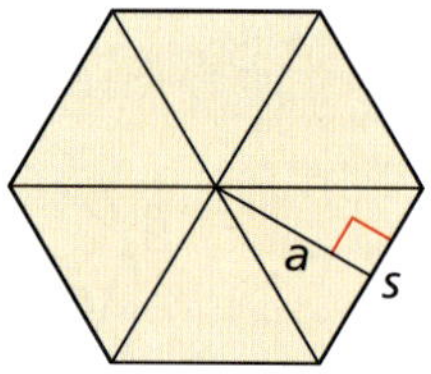

EXAMPLE 3 Finding the Area of a Regular Polygon

WATCH

A regular nonagon is inscribed in a circle with a radius of 4 units. Find the area of the nonagon.

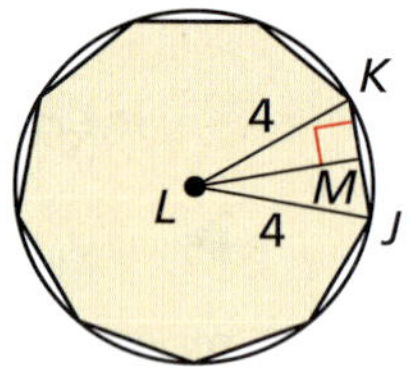

SOLUTION

The measure of central $\angle JLK$ is $\frac{360^\circ}{9}$, or 40°. Apothem $\overline{LM}$ bisects the central angle, so $m\angle KLM$ is 20°. To find the lengths of the legs, use trigonometric ratios for right $\triangle KLM$.

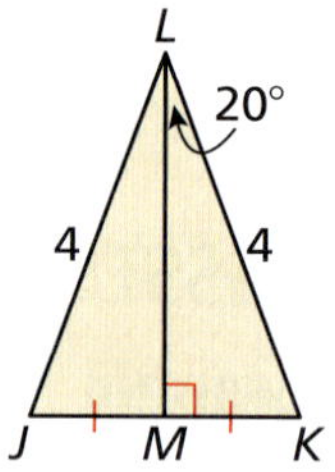

$$\sin 20^\circ = \frac{MK}{LK} \qquad \cos 20^\circ = \frac{LM}{LK}$$

$$\sin 20^\circ = \frac{MK}{4} \qquad \cos 20^\circ = \frac{LM}{4}$$

$$4 \sin 20^\circ = MK \qquad 4 \cos 20^\circ = LM$$

The regular nonagon has side length $s = 2(MK) = 2(4 \sin 20^\circ) = 8 \sin 20^\circ$, and apothem $a = LM = 4 \cos 20^\circ$.

Find the area A of the regular nonagon.

$A = \frac{1}{2}a \cdot ns$ — Write formula for area of a regular polygon.

$= \frac{1}{2}(4 \cos 20^\circ) \cdot (9)(8 \sin 20^\circ)$ — Substitute.

≈ 46.3 — Use technology.

► So, the area of the nonagon is about 46.3 square units.

GO DIGITAL

7 MTR **EXAMPLE 4** **Modeling Real Life** WATCH

You are decorating the top of a table by covering it with small ceramic tiles. The tabletop is a regular octagon with 15-inch sides and a radius of about 19.6 inches. What is the area you are covering?

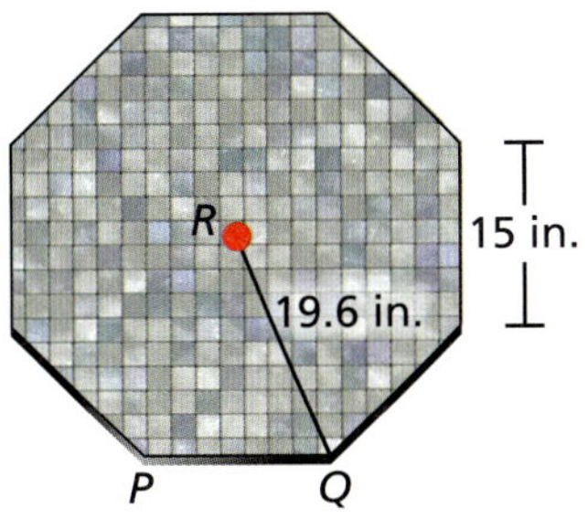

SOLUTION

Step 1 Find the perimeter P of the tabletop. An octagon has 8 sides, so $P = 8(15) = 120$ inches.

Step 2 Find the apothem a. The apothem is height RS of $\triangle PQR$.

Because $\triangle PQR$ is isosceles, altitude $\overline{RS}$ bisects $\overline{QP}$.

So, $QS = \frac{1}{2}(QP) = \frac{1}{2}(15) = 7.5$ inches.

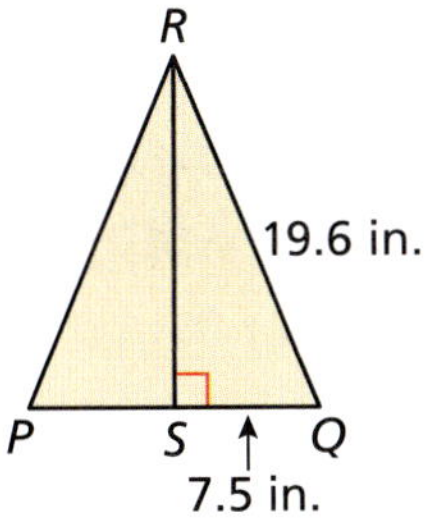

To find RS, use the Pythagorean Theorem for $\triangle RQS$.

$$a = RS = \sqrt{19.6^2 - 7.5^2} = \sqrt{327.91}$$

Step 3 Find the area A of the tabletop.

$A = \frac{1}{2}aP$	Write formula for area of a regular polygon.
$= \frac{1}{2}(\sqrt{327.91})(120)$	Substitute.
≈ 1086.5	Use technology.

▶ The area you are covering with tiles is about 1086.5 square inches.

SELF-ASSESSMENT

1 I don't understand yet. 2 I can do it with help. 3 I can do it on my own. 4 I can teach someone else.

8. A regular 15-gon is inscribed in a circle with a radius of 9 units. Find the area of the 15-gon.

Find the area of the regular polygon.

9.

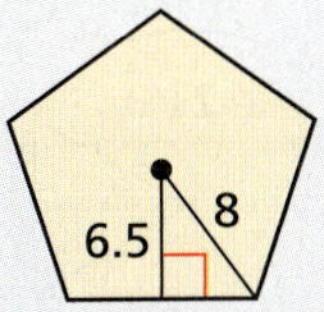

10.

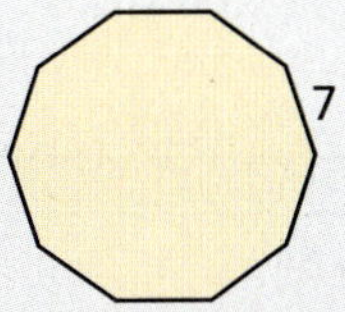

GO DIGITAL

11.3 Practice WITH CalcChat® AND CalcView®

In Exercises 1–4, find the area of the kite or rhombus. *(See Example 1.)*

1. 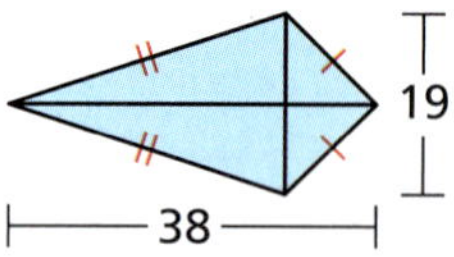

2. 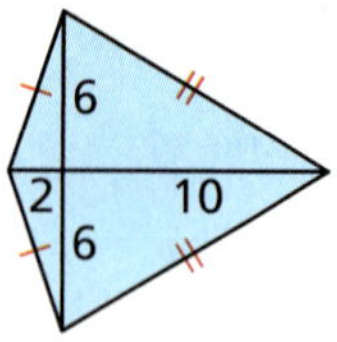

3. 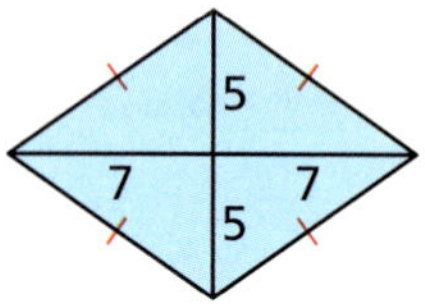

4. 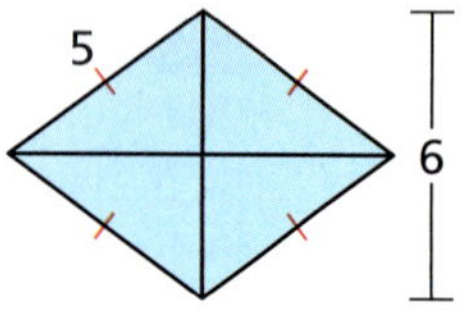

In Exercises 5–8, use the diagram.

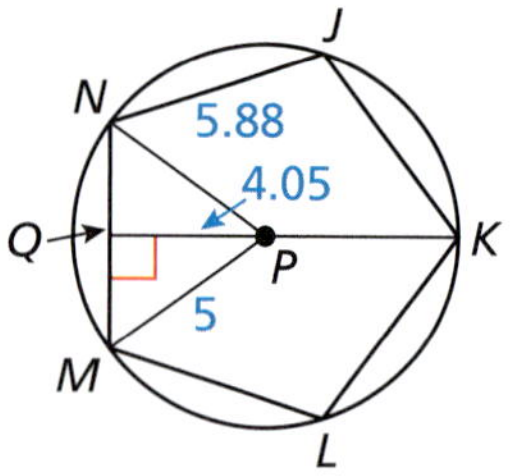

5. Identify the center of polygon *JKLMN*.

6. Identify a central angle of polygon *JKLMN*.

7. What is the radius of polygon *JKLMN*?

8. What is the apothem of polygon *JKLMN*?

In Exercises 9–12, find the measure of a central angle of a regular polygon with the given number of sides.

9. 10 sides
10. 18 sides
11. 24 sides
12. 7 sides

In Exercises 13–16, find the given angle measure for regular octagon *ABCDEFGH*. *(See Example 2.)*

13. $m\angle GJH$
14. $m\angle GJK$
15. $m\angle KGJ$
16. $m\angle EJH$

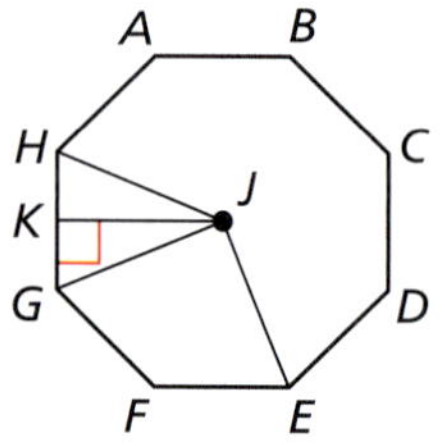

In Exercises 17–24, find the area of the regular polygon. *(See Example 3.)*

17. 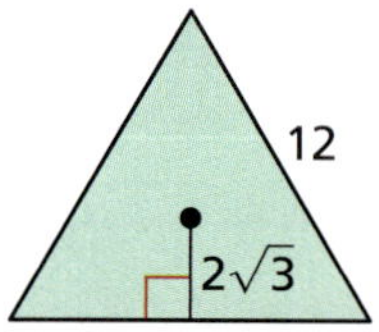

18. 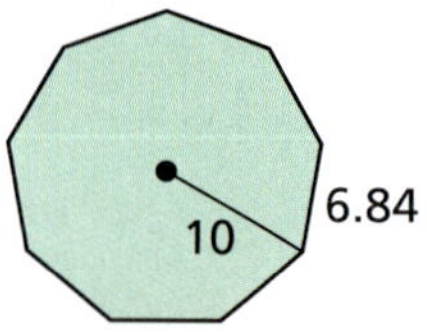

19. 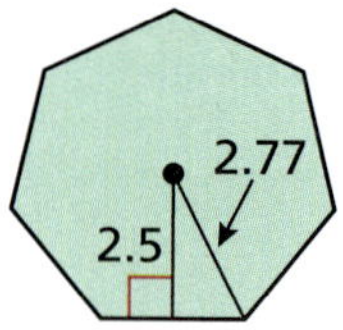

20. 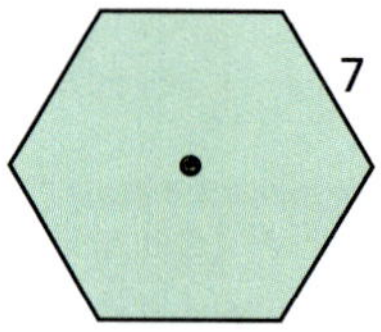

21. an octagon with a radius of 11 units

22. a pentagon with an apothem of 5 units

23. a decagon with an apothem of 6.2 units

24. a dodecagon with a radius of 3.4 units

25. **ERROR ANALYSIS** Describe and correct the error in finding the area of the kite.

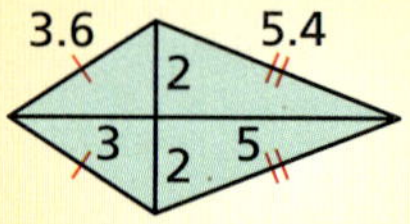

$A = \frac{1}{2}(3.6)(5.4)$

$= 9.72$

So, the area of the kite is 9.72 square units.

26. **ERROR ANALYSIS** Describe and correct the error in finding the area of the regular hexagon.

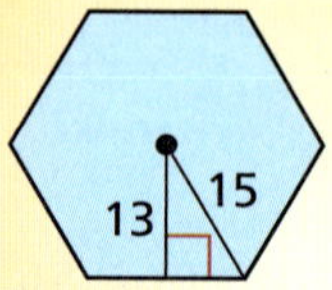

$s = \sqrt{15^2 - 13^2}$

$= \sqrt{56}$

$A = \frac{1}{2}a \cdot ns$

$= \frac{1}{2}(13)(6)(\sqrt{56})$

≈ 292.5

So, the area of the hexagon is about 292.5 square units.

In Exercises 27–30, find the area of the shaded region.

27. 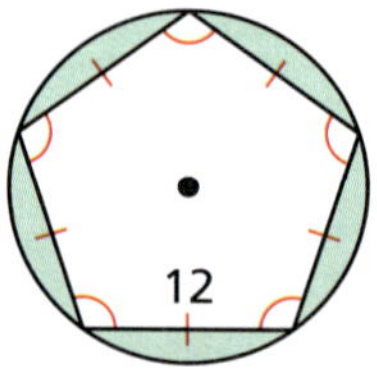

28. 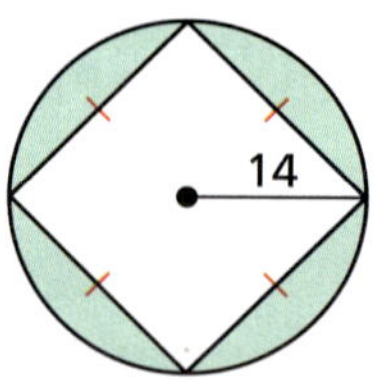

29. 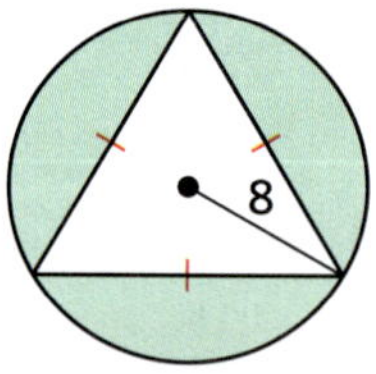

30. 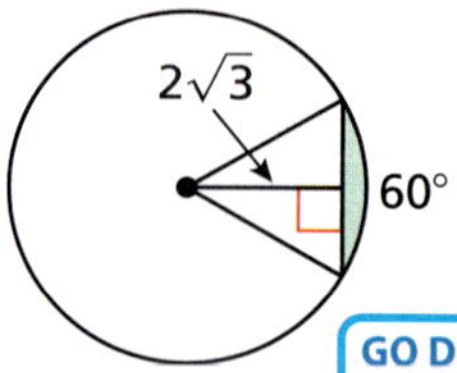

GO DIGITAL

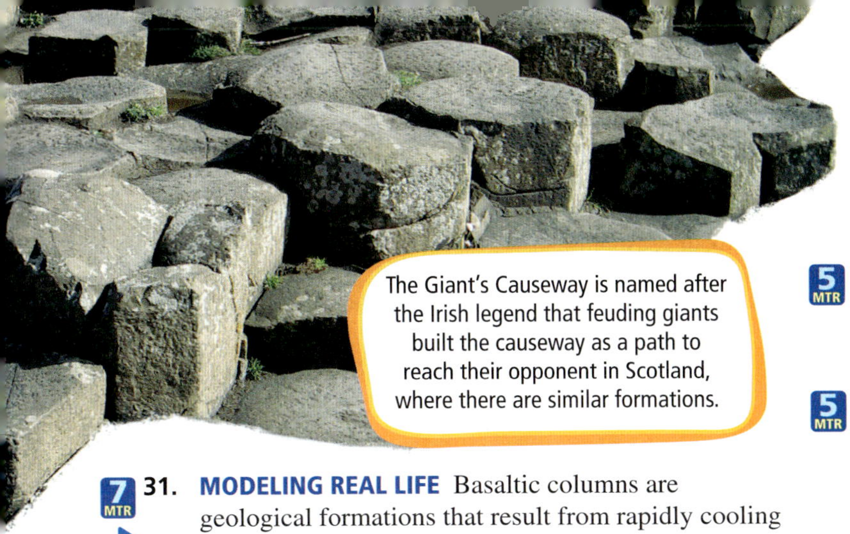

7 MTR **31.** **MODELING REAL LIFE** Basaltic columns are geological formations that result from rapidly cooling lava. Giant's Causeway in Ireland contains many hexagonal basaltic columns. The top of one of the columns is in the shape of a regular hexagon with a radius of 8 inches. Find the area of the top of the column. *(See Example 4.)*

7 MTR **32.** **MODELING REAL LIFE** A watch has a circular surface on a background that is a regular octagon. Find the area of the octagon. Then find the area of the silver border around the circular face.

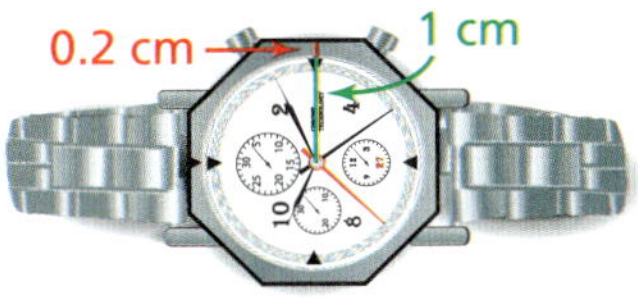

4 MTR **DISCUSS MATHEMATICAL THINKING** **In Exercises 33–35, tell whether the statement is *true* or *false*. Explain your reasoning.**

33. The area of a regular n-gon of a fixed radius r increases as n increases.

34. The apothem of a regular polygon is always less than the radius.

35. The radius of a regular polygon is always less than the side length.

36. **REASONING** Predict which figure has the greatest area and which has the least area. Explain your reasoning. Check by finding the area of each figure.

i.

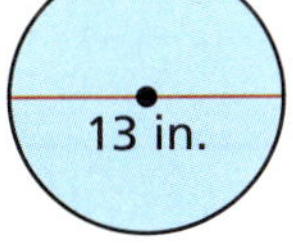

ii.

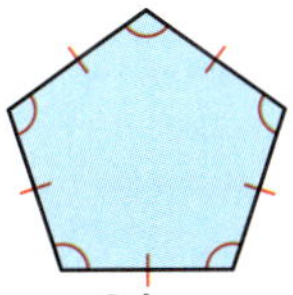

iii.

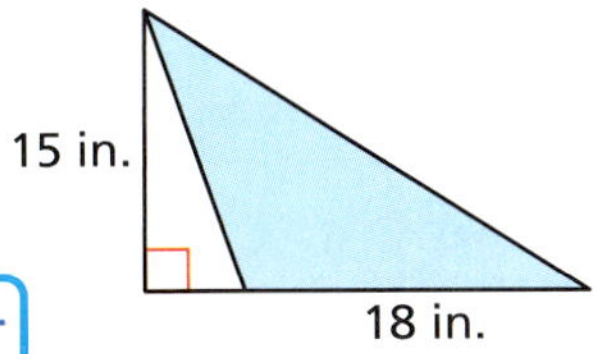

GO DIGITAL

37. **REASONING** What happens to the area of a kite when you (a) double the length of one of the diagonals, (b) double the length of both diagonals, and (c) dilate the kite by a scale factor of 2.5? Explain.

5 MTR **38.** **STRUCTURE** Find the area of a regular pentagon inscribed in a circle whose equation is given by $(x - 4)^2 + (y + 2)^2 = 25$.

5 MTR **39.** **CONNECTING CONCEPTS** One diagonal of a rhombus is four times the length of the other diagonal. The area of the rhombus is 98 square feet. Write and solve an equation to find the length of each diagonal.

40. **REASONING** The perimeter of a regular nonagon, or 9-gon, is 18 inches. Is this enough information to find the area? If so, find the area and explain your reasoning. If not, explain why not.

4 MTR **41.** **MAKING AN ARGUMENT** Your friend claims that it is possible to find the area of any rhombus if you know only the perimeter of the rhombus. Is your friend correct? Explain your reasoning.

42. **PROOF** Prove that the area of any quadrilateral with perpendicular diagonals is $A = \frac{1}{2}d_1d_2$, where d_1 and d_2 are the lengths of the diagonals.

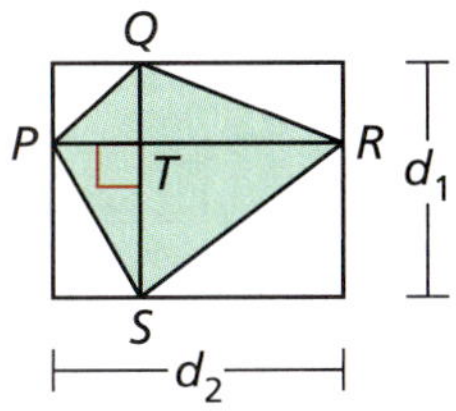

43. **PROBLEM SOLVING** Three vertices of kite $WXYZ$ are $W(1, 1)$, $X(3, 4)$, and $Y(5, 1)$. Find a possible set of coordinates for Z. Then find the perimeter and area of kite $WXYZ$.

44. **HOW DO YOU SEE IT?**
Explain how to find the area of the regular hexagon by dividing the hexagon into equilateral triangles.

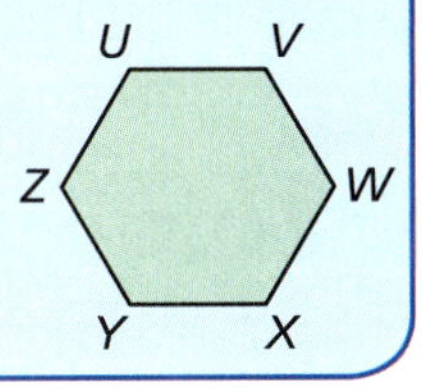

5 MTR **45.** **STRUCTURE** Rewrite the formula for the area of a rhombus for the special case of a square with side length s. Show that this is the same as the formula for the area of a square, $A = s^2$.

5 MTR **46.** **STRUCTURE** Use the formula for the area of a regular polygon to show that the area of an equilateral triangle can be found by using the formula $A = \frac{1}{4}s^2\sqrt{3}$, where s is the side length.

3 MTR **47.** **ADAPT A PROCEDURE** The area of a regular pentagon is 72 square centimeters. Find the length of one side.

3 MTR **48.** **ADAPT A PROCEDURE** The area of a regular dodecagon, or 12-gon, is 140 square inches. Find the apothem of the polygon.

5 MTR **49.** **STRUCTURE** Each polygon in the diagram is regular. Find the approximate area of the entire shaded region.

4 MTR **50.** **COMPARING METHODS** Find the area of regular pentagon $ABCDE$ by using the formula $A = \frac{1}{2}aP$, or $A = \frac{1}{2}a \bullet ns$. Then find the area by adding the areas of smaller polygons. Check that both methods yield the same area. Which method do you prefer? Explain your reasoning.

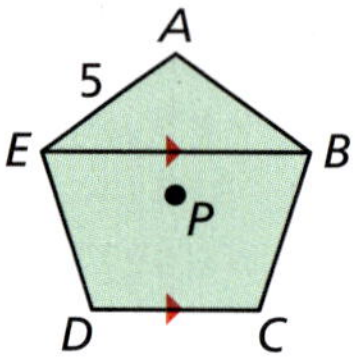

51. **REASONING** The area of a regular n-gon is given by $A = \frac{1}{2}aP$. As n approaches infinity, what does the n-gon approach? What does P approach? What does a approach? What can you conclude? Explain your reasoning.

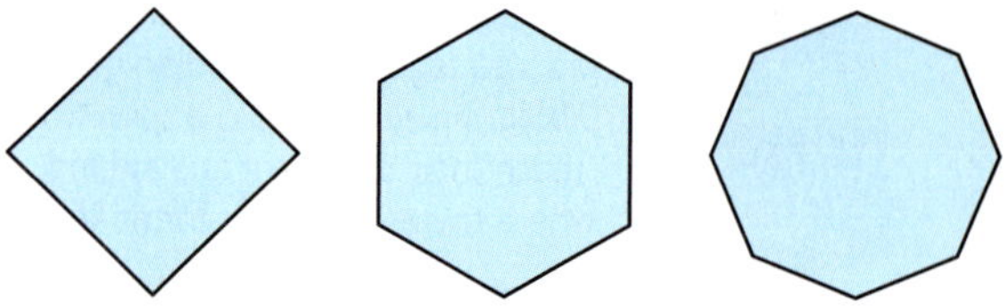

52. **THOUGHT PROVOKING**
Two regular polygons both have n sides. One of the polygons is inscribed in, and the other is circumscribed about, a circle of radius r. Find the area between the two polygons in terms of n and r.

REVIEW & REFRESH

WATCH

In Exercises 53 and 54, find the indicated measure.

53. area of $\odot N$

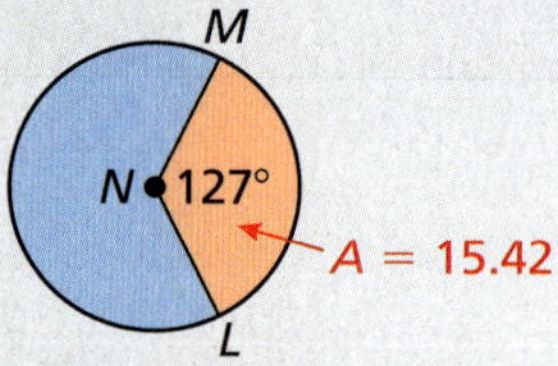

54. arc length of $\overparen{AB}$

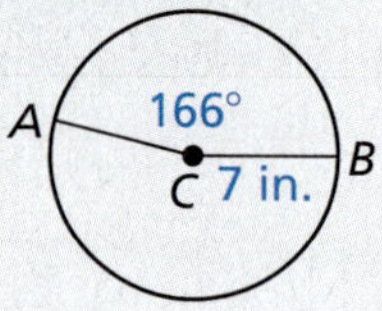

55. Graph the circle $x^2 + y^2 - 10x + 2y = 23$. Identify the domain and range.

56. $\triangle ABC$ has vertices $A(-2, 1)$, $B(0, 3)$, and $C(4, 1)$. $\triangle DEF$ has vertices $D(1, -1)$, $E(3, 1)$, and $F(7, -1)$. Are the triangles similar? Use transformations to explain your reasoning.

In Exercises 57 and 58, find the value of *x*.

57.

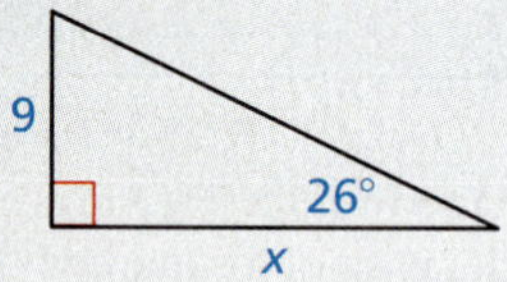

58.

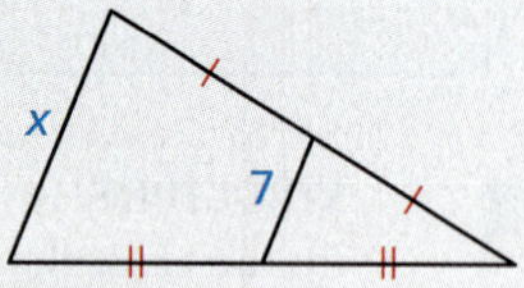

59. Find the area of the regular hexagon.

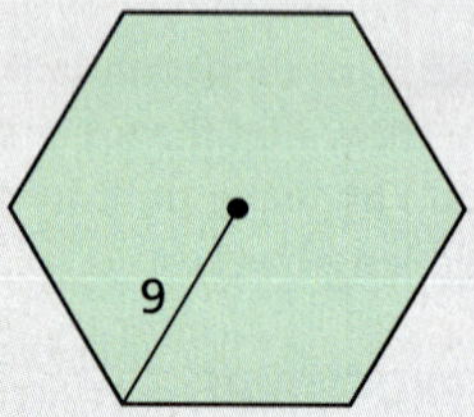

60. Find all the unknown measures in the right triangle. Explain your reasoning.

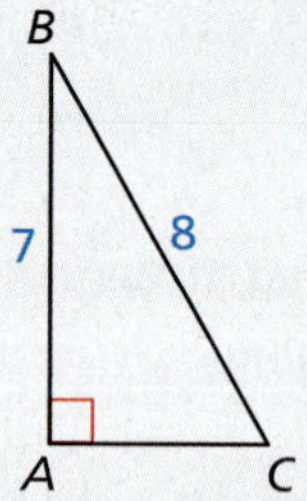

7 MTR **61.** **MODELING REAL LIFE** A wallet is shown below.

a. State which theorem you can use to show that the wallet is in the shape of a parallelogram.

b. Find $m\angle KJM$, $m\angle JML$, and $m\angle KLM$.

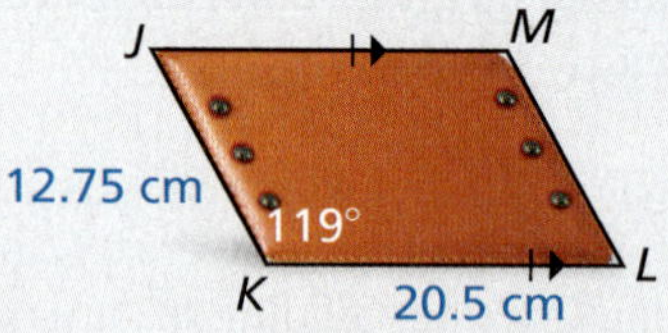

62. Find the value of x that makes $m \parallel n$. Explain your reasoning.

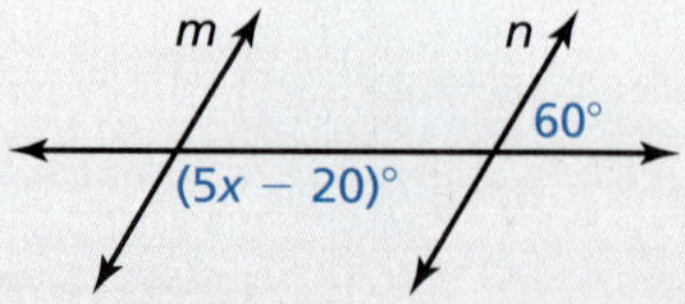

63. Find the measure of the exterior angle.

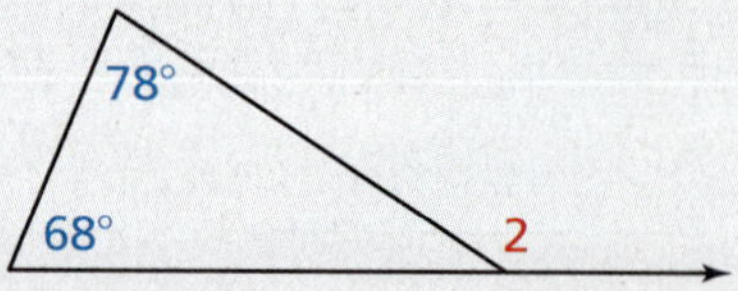

GO DIGITAL

11.4 Modeling with Area

Learning Target: Understand the concept of population density and modeling with area.

Success Criteria:
- I can explain what population density means.
- I can find and use population densities.
- I can use area formulas to solve problems.

EXPLORE IT! Analyzing Population and Area

Work with a partner.

a. Use the Internet to find the population and land area of each county in Florida given below. Then find the number of people per square mile for each county.

i. Indian River County

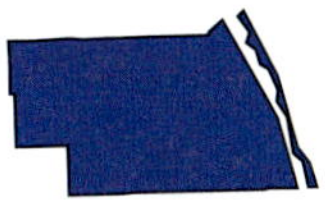

ii. St. Lucie County

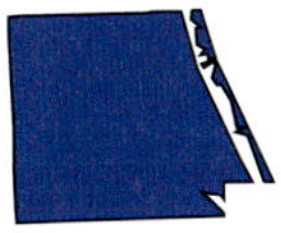

iii. Martin County

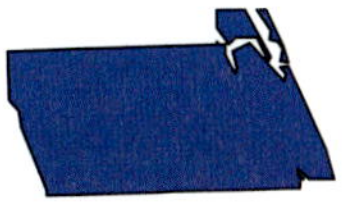

iv. Palm Beach County

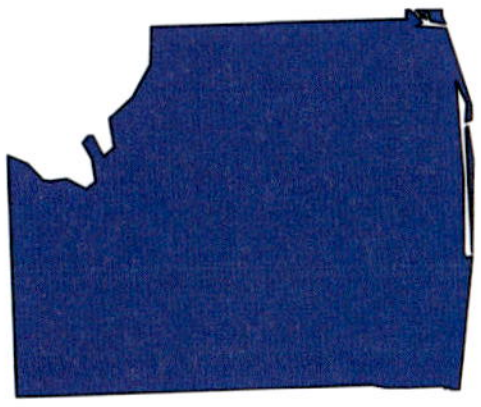

v. Broward County

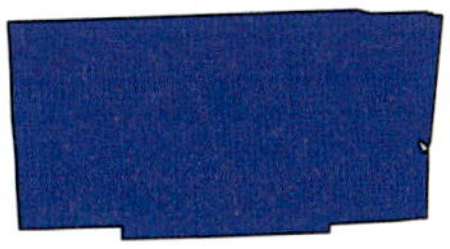

vi. Miami-Dade County

b. Without calculating, how do you expect the number of people per square mile in the entire six-county region to compare to the values for each individual county in part (a)? Justify your answer.

c. Collier County is the largest county in Florida by land area but is not in the top ten largest counties by population. How do you expect the number of people per square mile to change when Collier County is included? Why? Verify your answer.

d. How can you use the population and area of a region to describe how densely the region is populated?

e. Find the population and land area of the county in which you live. How densely populated is your county compared to the counties in part (a)?

7 MTR APPLY MATHEMATICS
Where are the most densely populated and least densely populated places in the world? What factors might influence how densely a region is populated?

Geometric Reasoning
MA.912.GR.4.4 Solve mathematical and real-world problems involving the area of two-dimensional figures.

GO DIGITAL

Using Area Formulas

Vocabulary AZ VOCAB

population density, *p. 600*

The **population density** of a city, county, or state is a measure of how many people live within a given area.

$$\text{Population density} = \frac{\text{number of people}}{\text{area of land}}$$

Population density is usually given in terms of square miles but can be expressed using other units, such as city blocks.

EXAMPLE 1 Finding a Population Density

WATCH

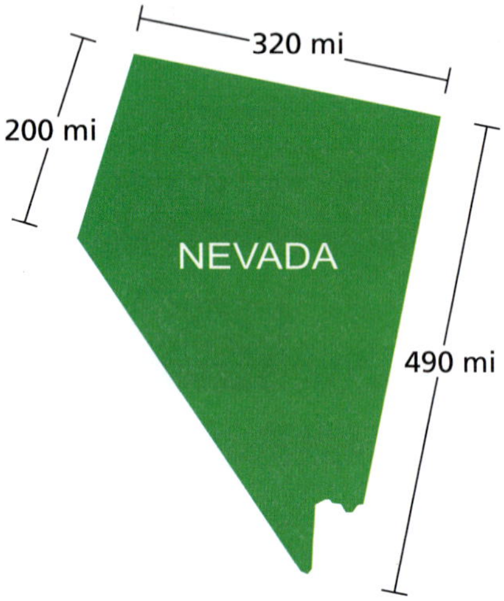

The state of Nevada has a population of about 3.1 million people. Find the population density in people per square mile.

SOLUTION

Step 1 Find the area of Nevada. It is shaped approximately like a trapezoid. Use the formula for the area of a trapezoid to estimate the area of Nevada.

$$A = \frac{1}{2}h(b_1 + b_2) = \frac{1}{2}(320)(200 + 490) = 110{,}400 \text{ mi}^2$$

Step 2 Find the population density.

$$\text{Population density} = \frac{\text{number of people}}{\text{area of land}} = \frac{3{,}100{,}000}{110{,}400} \approx 28$$

▶ The population density is about 28 people per square mile.

EXAMPLE 2 Using the Formula for Population Density

WATCH

A circular region has a population of about 430,000 people and a population density of about 5475 people per square mile. Find the radius of the region.

SOLUTION

Use the formula for population density. Let r represent the radius of the region.

$$\text{Population density} = \frac{\text{number of people}}{\text{area of land}}$$ Formula for population density

$$5475 = \frac{430{,}000}{\pi r^2}$$ Substitute.

$$5475\pi r^2 = 430{,}000$$ Multiply each side by πr^2.

$$r^2 = \frac{430{,}000}{5475\pi}$$ Divide each side by 5475π.

$$r = \sqrt{\frac{430{,}000}{5475\pi}}$$ Take the positive square root of each side.

$$r \approx 5$$ Use technology.

▶ The radius of the region is about 5 miles.

SELF-ASSESSMENT 1 I don't understand yet. 2 I can do it with help. 3 I can do it on my own. 4 I can teach someone else.

1. About 58,000 people live in a circular region with a 2-mile radius. Find the population density in people per square mile.
2. A circular region has a population of about 175,000 people and a population density of about 1318 people per square mile. Find the radius of the region.

GO DIGITAL

7 MTR

EXAMPLE 3 Modeling Real Life

WATCH

You are designing a rectangular corral with an area of 450 square meters. A barn will form one side of the corral. You want to minimize the amount of fencing that you need for the other three sides of the corral. This will include an opening that is 3 meters wide where a gate will be placed. How many meters of fencing do you need?

SOLUTION

Step 1 Use what you know about the area and perimeter of the corral to find an expression that represents the perimeter of the three sides that need fencing.

The area A of a corral of length ℓ and width w is $A = \ell w$.

So, $450 = \ell w$. Solving for ℓ gives $\ell = \frac{450}{w}$.

Let ℓ represent the length of the side of the corral against the barn. So, the expression $2w + \ell$ represents the perimeter of the three sides that need fencing. Using substitution, this expression can be rewritten as $2w + \frac{450}{w}$.

Step 2 Use technology to create tables of values to find the width w that minimizes the value of $2w + \frac{450}{w}$. You may need to decrease the increment for the independent variable, as shown.

Increment of 4

w	$2w + \frac{450}{w}$
4	120.5
8	72.25
12	61.5
16	60.125
20	62.5
24	66.75
28	72.071429

Increment of 0.5

w	$2w + \frac{450}{w}$
13.5	60.333333
14	60.142857
14.5	60.034483
15	60
15.5	60.032258
16	60.125
16.5	60.272727

1 MTR **ANALYZE A PROBLEM**

How can you use the values in the first table to help determine the value of *w* that minimizes the expression?

The width that minimizes the value of $2w + \frac{450}{w}$ is 15 meters.

So, the length of the corral is $\ell = \frac{450}{w} = \frac{450}{15} = 30$ meters.

Step 3 Sketch a diagram of the corral that includes the gate opening, as shown.

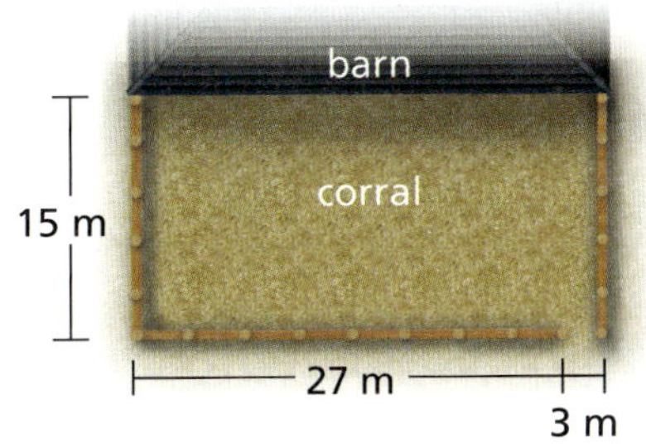

▶ So, you need $2w + \ell - 3 = 2(15) + 30 - 3 = 57$ meters of fencing.

SELF-ASSESSMENT

1 I don't understand yet. 2 I can do it with help. 3 I can do it on my own. 4 I can teach someone else.

3. WHAT IF? You want the corral to have an area of 800 square meters. How many meters of fencing do you need?

GO DIGITAL

7 MTR **EXAMPLE 4** **Modeling Real Life** WATCH

Magnetic computer tape was first used to record and store data in 1951. The first magnetic tape drive recorded data at 768 bits per inch on a $\frac{1}{2}$-inch computer tape.

a. *Areal density* is a measurement of the amount of data that can be stored on a given unit of storage space. The 6-inch segment of magnetic tape stores 4608 bits of data. Find the areal density in bits per square inch.

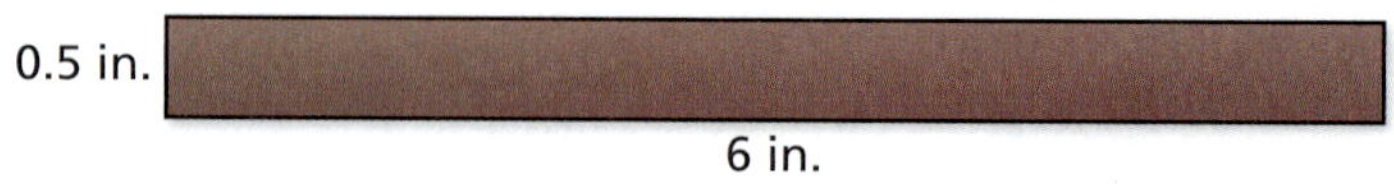

b. Data storage technology has improved dramatically since 1951. Researchers frequently achieve new records in magnetic tape storage. A 4-inch segment of $\frac{1}{2}$-inch magnetic tape can store 402 gigabits of data. About how many times more data per square inch can be stored on this 4-inch segment of tape than on the 6-inch segment in part (a)? (One gigabit contains 1 billion bits.)

SOLUTION

a. Find the area A of the segment of magnetic tape.

$$A = \ell w = (6 \text{ in.})(0.5 \text{ in.}) = 3 \text{ in.}^2$$

Divide to find the areal density in bits per square inch.

$$\frac{4608 \text{ bits}}{3 \text{ in.}^2} = \frac{1536 \text{ bits}}{1 \text{ in.}^2}$$ Divide numerator and denominator by 3.

▶ The areal density is 1536 bits per square inch.

b. Find the areal density of the 4-inch segment of magnetic tape. The area is $(4 \text{ in.})(0.5 \text{ in.}) = 2$ square inches.

$$\frac{402 \text{ gigabits}}{2 \text{ in.}^2} = \frac{201 \text{ gigabits}}{1 \text{ in.}^2}$$ Divide numerator and denominator by 2.

The areal density is 201 gigabits per square inch. Because 1 gigabit contains 1 billion bits, the areal density is 201 billion bits per square inch. Divide to compare the segments of magnetic tape.

$$\frac{201{,}000{,}000{,}000 \text{ bits/in.}^2}{1536 \text{ bits/in.}^2} = 130{,}859{,}375$$

▶ So, about 131 million times more data per square inch can be stored on the 4-inch segment of tape than the 6-inch segment of tape.

The Univac Uniservo was the first magnetic tape drive. Information stored on magnetic tapes was carried to a central computer. Magnetic tape is still used today for data storage because of its low operating costs and reliability.

SELF-ASSESSMENT

1 I don't understand yet. 2 I can do it with help. 3 I can do it on my own. 4 I can teach someone else.

4. The segment of tape in Example 4(a) is a portion of a tape reel that is 1200 feet long. How much data can be stored on the entire reel?

5. **WHAT IF?** A 6-inch segment of $\frac{1}{2}$-inch tape stores 20.1 gigabits of data. Compare the storage of this tape segment with the other segments in Example 4.

GO DIGITAL

11.4 Practice WITH CalcChat® AND CalcView®

In Exercises 1–6, find the indicated measure.
(See Example 1.)

1. The state of Kansas has a population of about 2.85 million people. Find the population density in people per square mile.

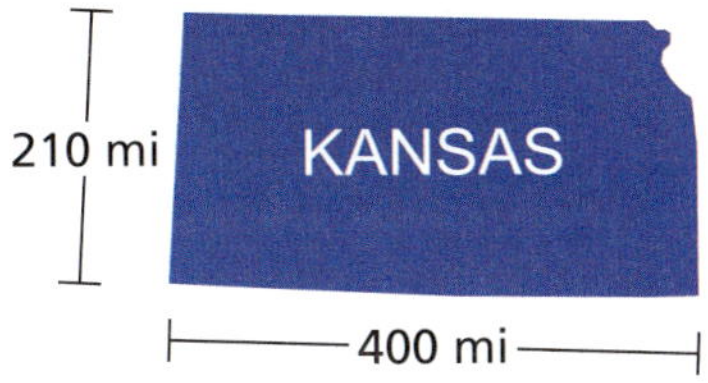

2. Yellowstone National Park has an area of about 2.22 million acres. The table shows the estimated park populations for several animals. Find the population density in animals per acre for each animal.

Animal	Grizzly bear	Elk	Mule deer	Bighorn sheep
Population	728	20,000	1900	345

3. About 210,000 people live in a circular region with a 12-mile radius. Find the population density in people per square mile.

4. About 650,000 people live in a circular region with a 6-mile radius. Find the population density in people per square mile.

5. A town with a population of 80,250 people has 500 city blocks. Each block is one-tenth mile long by one-tenth mile wide. Find the population density of the town in people per square mile.

6. A neighborhood of 1250 people has 30 city blocks. Each block is $\frac{1}{16}$ mile long by $\frac{1}{8}$ mile wide. Find the population density of the neighborhood in people per square mile.

7. PROBLEM SOLVING About 79,000 people live in a circular region with a population density of about 513 people per square mile. Find the radius of the region. *(See Example 2.)*

STEM

8. PROBLEM SOLVING About 1.15 million people live in a circular region with a population density of about 18,075 people per square kilometer. Find the radius of the region.

4 MTR **9. ERROR ANALYSIS** Describe and correct the error in finding the number of people who live in a circular region with a 7.5-mile diameter and a population density of 1550 people per square mile.

✗

$$1550 = \frac{x}{\pi \cdot 7.5^2}$$

$$1550 = \frac{x}{56.25\pi}$$

$$273{,}908 \approx x$$

The number of people who live in the region is about 273,908.

10. HOW DO YOU SEE IT?
The two islands shown below with the given areas have the same population. Which has the greater population density? Explain.

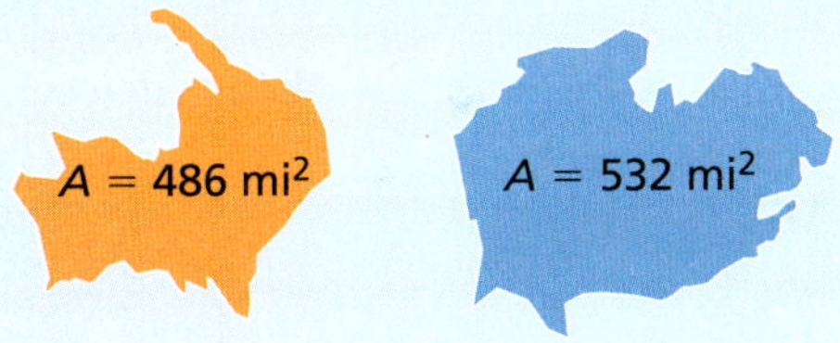

7 MTR **11. MODELING REAL LIFE** The fence surrounding a rectangular giraffe exhibit needs to be replaced. The exhibit covers a total area of 2500 square feet. The zoo wants to change the dimensions of the exhibit to minimize the amount of fencing needed without reducing the exhibit's area. What should the new dimensions be so the exhibit is still rectangular? Explain. *(See Example 3.)*

GO DIGITAL

12. **MODELING REAL LIFE** A field of length ℓ and width w has a perimeter of 320 yards.

a. Write an expression that represents the area of the field in terms of ℓ.

b. Use your expression from part (a) to determine the dimensions of the field that maximize the area. What do you notice?

13. **MODELING REAL LIFE** One measure of rug quality is *knot density*. Higher-quality rugs have a greater knot density. The rug shown is made of 345,600 knots. *(See Example 4.)*

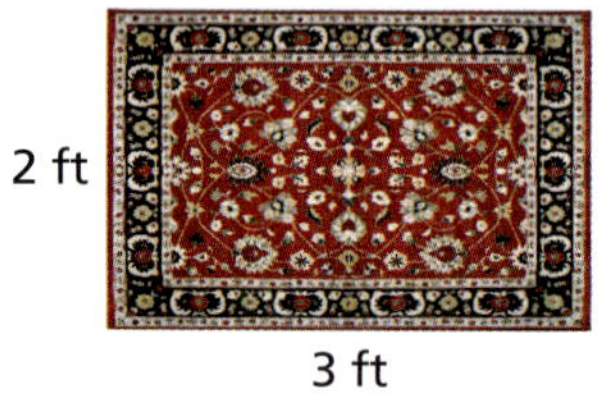

a. Find the knot density per square inch.

b. Another rug is 3 feet by 4 feet and is made of 604,800 knots. Which is the better-quality rug? Explain.

14. **MODELING REAL LIFE** A farmer wants to plant corn so that there are 36,000 plants per acre in the field shown. How many seeds does the farmer need? (1 acre = 43,560 ft^2)

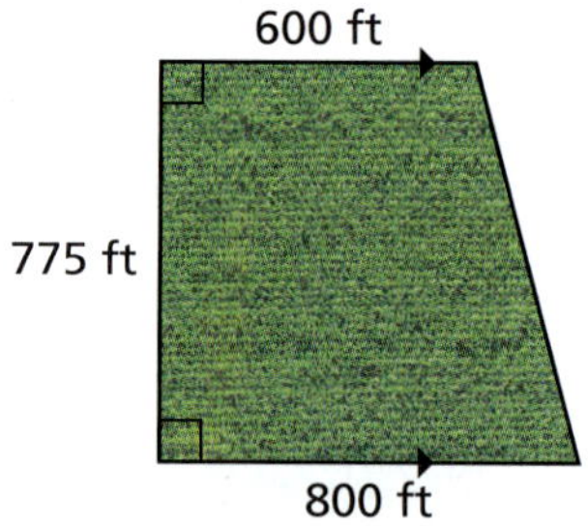

15. **MAKING AN ARGUMENT** You ask your friend which U.S. states have the greatest population densities. Your friend says it must be California and Texas because they have the greatest populations. Do you agree with your friend's reasoning? Explain.

16. **THOUGHT PROVOKING**
Write a real-life problem involving density and area. Show how the area affects the density.

REVIEW & REFRESH

WATCH

In Exercises 17 and 18, find the indicated measure.

17. $m\overparen{EF}$

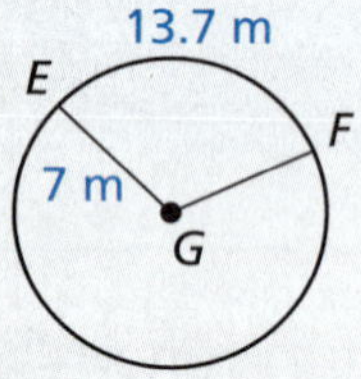

18. area of each sector

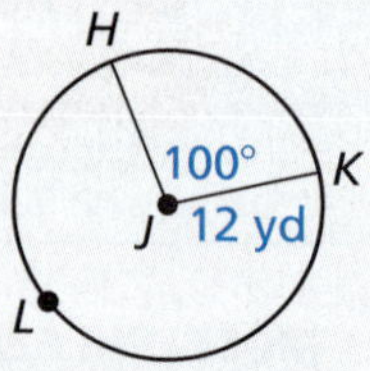

19. In the diagram, *RSTUVWXY* is a regular octagon inscribed in $\odot C$. The radius of the circle is 8 units. Find the area of the octagon.

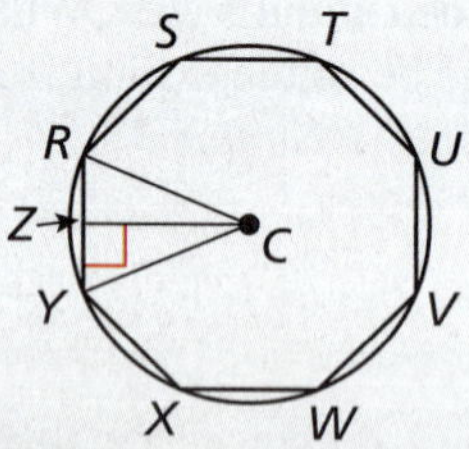

20. **PROBLEM SOLVING** About 82,000 people live in a circular region with a population density of about 275 people per square mile. Find the radius of the region.

21. **MODELING REAL LIFE** A school gymnasium is being remodeled. The basketball court will be similar to an NCAA basketball court, which has a length of 94 feet and a width of 50 feet. The school plans to make the width of the new court 45 feet. Find the perimeters of an NCAA court and the new court.

22. How does *RT* compare to *LN*? Explain your reasoning.

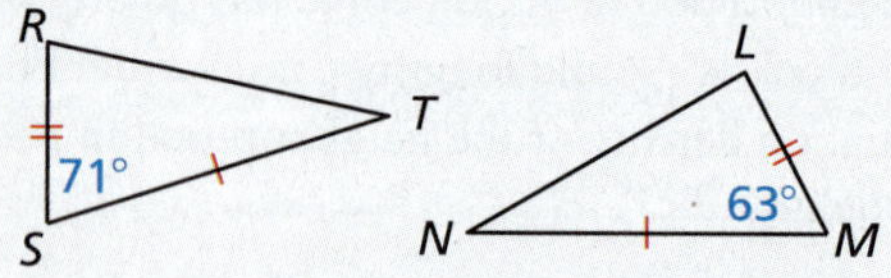

23. Find the value of each variable using sine and cosine.

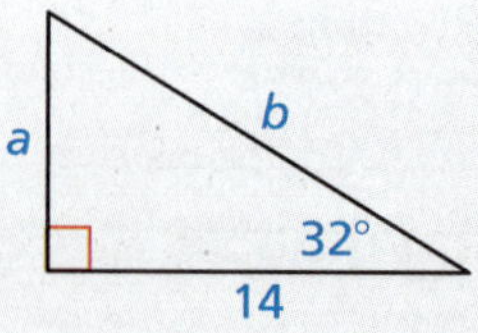

24. Find the sum of the measures of the interior angles of a 17-gon.

25. Find the geometric mean of 4 and 25.

GO DIGITAL

11 Chapter Review WITH CalcChat®

Chapter Learning Target: Understand circumference and area.

Chapter Success Criteria:
- ◆ I can find circumferences of circles and arc lengths of sectors.
- ◆ I can find areas of circles and sectors.
- ■ I can find areas of polygons.
- ■ I can solve real-life problems involving area.

◆ Surface
■ Deep

SELF-ASSESSMENT 1 I don't understand yet. 2 I can do it with help. 3 I can do it on my own. 4 I can teach someone else.

11.1 Circumference and Arc Length (pp. 577–584) WATCH

Learning Target: Understand circumference, arc length, and radian measure.

Vocabulary VOCAB
circumference
arc length
radian

Find the indicated measure.

1. diameter of $\odot P$

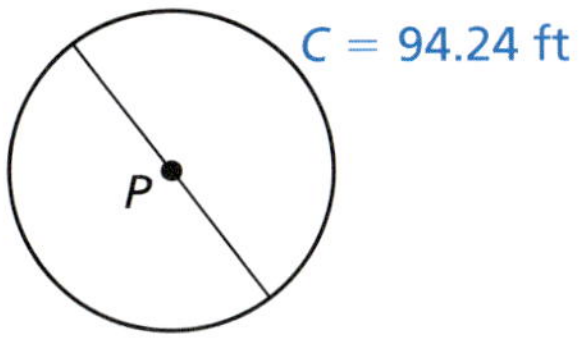

2. circumference of $\odot F$

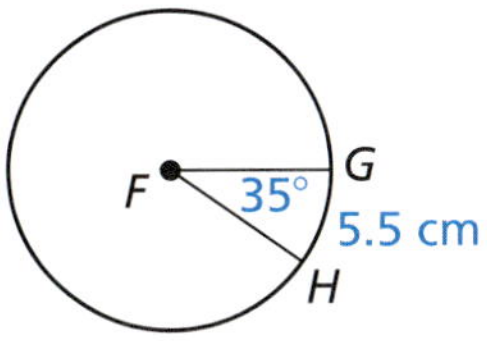

3. arc length of $\widehat{AB}$

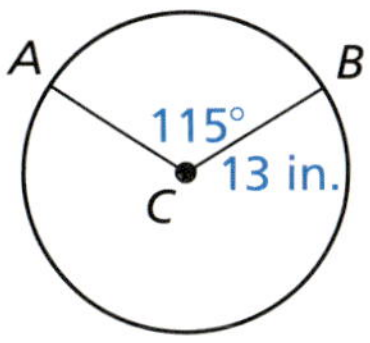

4. $m\widehat{QR}$

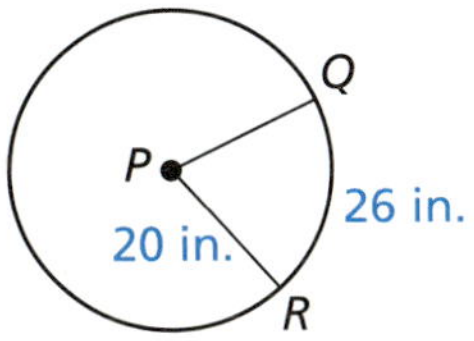

Find the perimeter of the shaded region.

5.

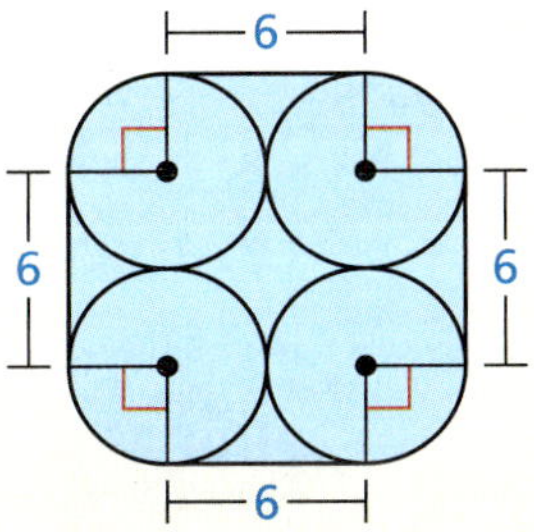

6.

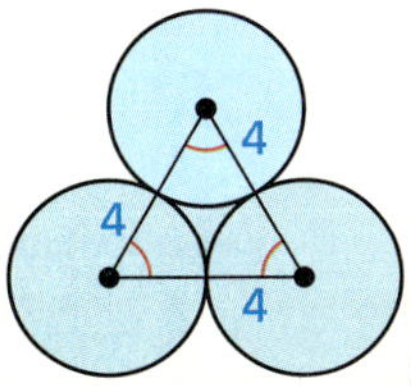

Convert the angle measure.

7. Convert 15° to radians.

8. Convert $\frac{3\pi}{5}$ radians to degrees.

9. A mountain bike tire has a diameter of 26 inches. How far does the tire travel when it makes 32 revolutions?

GO DIGITAL

11.2 Areas of Circles and Sectors (pp. 585–590) WATCH

Learning Target: Find areas of circles and areas of sectors of circles.

Vocabulary AZ VOCAB

sector of a circle

Find the area of the shaded region.

10.

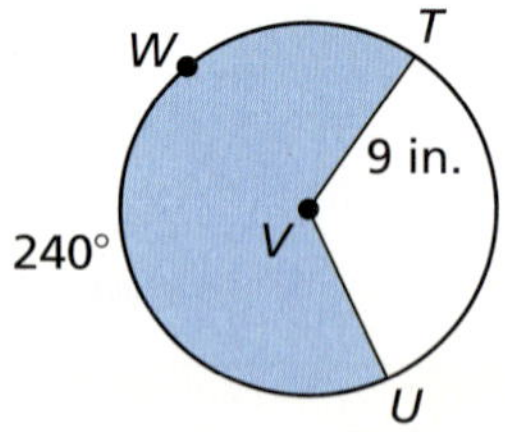

11.

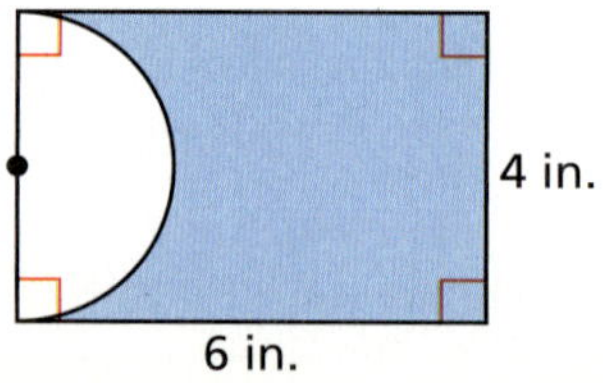

12.

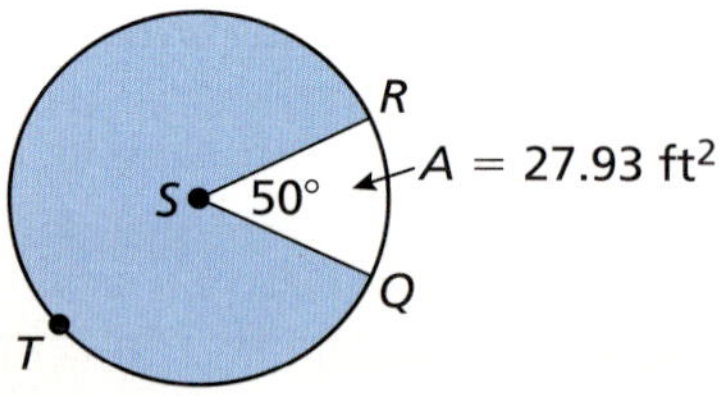

13. A slice of pizza with an area of 38 square inches has been removed from the pizza. What is the diameter of the whole pizza?

14. For \$8, you can buy one 8-inch pie or five 3-inch snack pies at a bakery. Which option provides more pie?

11.3 Areas of Polygons (pp. 591–598) WATCH

Learning Target: Find angle measures and areas of regular polygons.

Vocabulary AZ VOCAB

center of a regular polygon
radius of a regular polygon
apothem of a regular polygon
central angle of a regular polygon

Find the area of the kite or rhombus.

15.

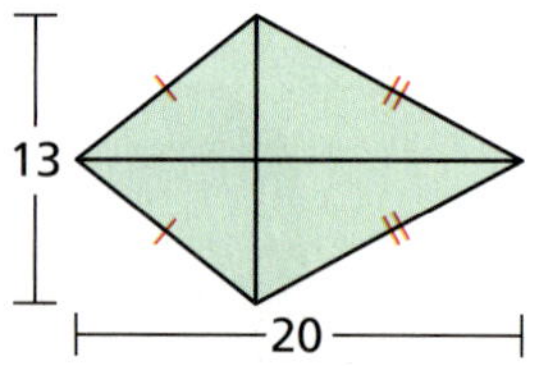

16.

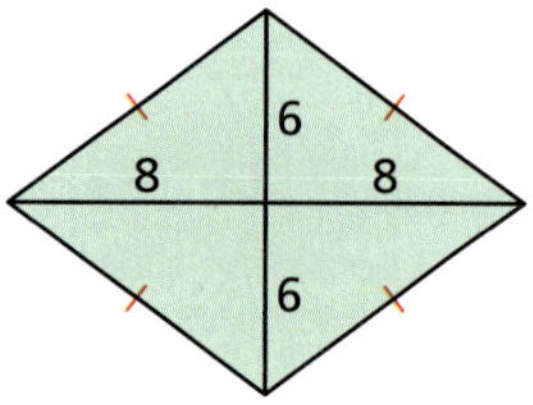

17.

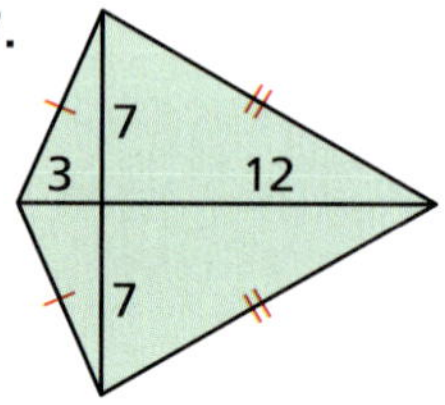

Find the area of the regular polygon.

18.

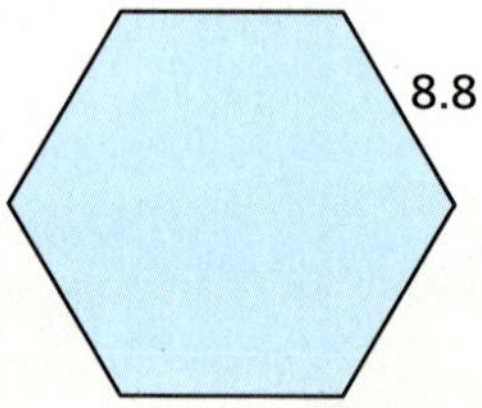

19.

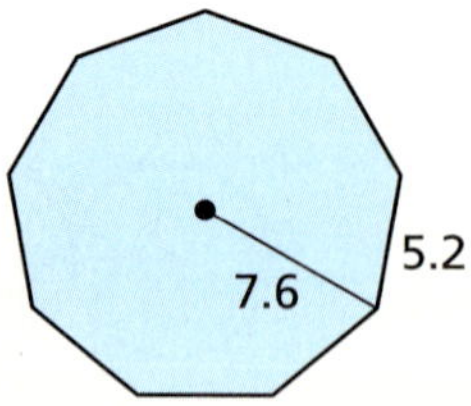

20.

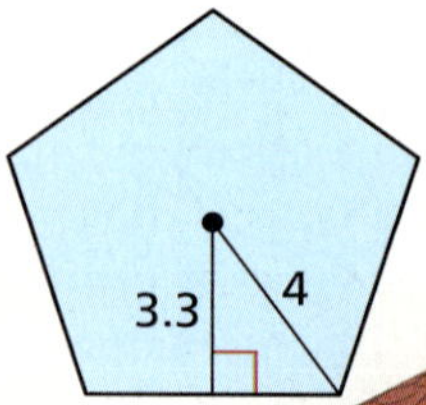

21. A gazebo is in the shape of a regular octagon with a side length of 6 feet. Find the area covered by the gazebo.

22. Find the measure of a central angle of a regular polygon with 11 sides.

23. What regular polygon with a side length of 1 foot can be inscribed in a circle with a diameter of 2 feet? Explain.

GO DIGITAL

11.4 Modeling with Area (pp. 599–604) WATCH

Learning Target: Understand the concept of population density and modeling with area.

Vocabulary
population density

24. About 1.75 million people live in a circular region with a 15-mile diameter. Find the population density in people per square mile.

25. A circular region has a population of about 15,500 people and a population density of about 775 people per square kilometer. Find the radius of the region.

26. You are using 48 feet of portable fencing to enclose a dance area along the front of a stage. What is the greatest rectangular dance area possible with the stage forming one of the sides?

Mathematical Thinking and Reasoning

7 MTR APPLY MATHEMATICS TO REAL-WORLD CONTEXTS
Mathematicians who apply mathematics to real-world contexts connect mathematical concepts to everyday experiences.

1. Kingsley Lake is a circular body of water near Starke, Florida. You need to know the area of the lake for a research project. A source states that its diameter is 10,560 feet. Find the area of the lake in acres.

2. A model for classroom size in high schools suggests that a classroom should have a minimum of 67 square feet of space per student. Write this as the maximum population density of students per 1000 square feet of classroom space. Find the population density for your classroom. Is it within the suggested range? By how many students?

GO DIGITAL

11 Practice Test WITH CalcChat®

1. The smaller circles are congruent. Find the area of the shaded region.

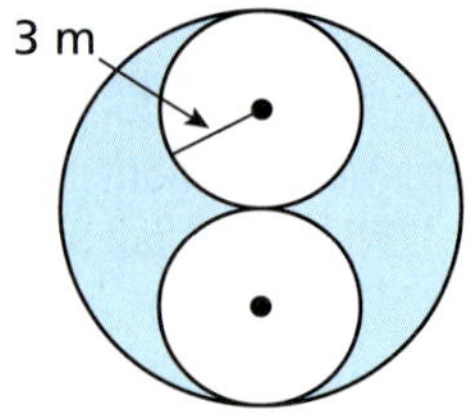

Find the indicated measure.

2. circumference of $\odot F$

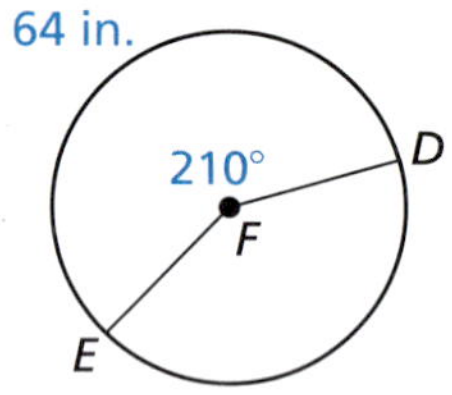

3. $m\widehat{GH}$

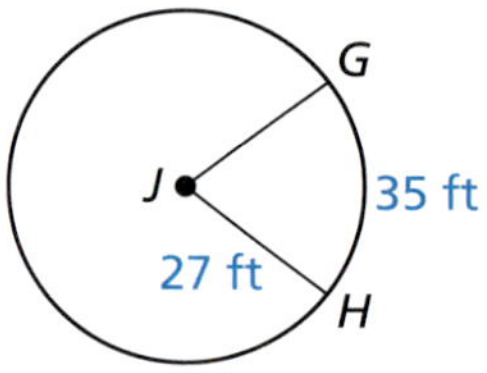

4. area of shaded sector

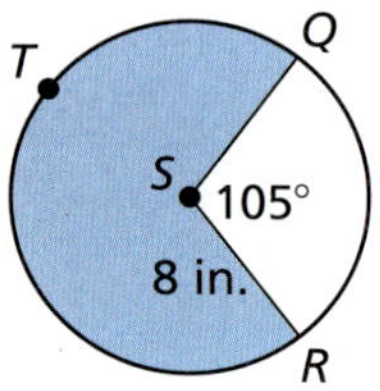

5. The island shown has a population of 12,175 people. Find the population density in people per square kilometer.

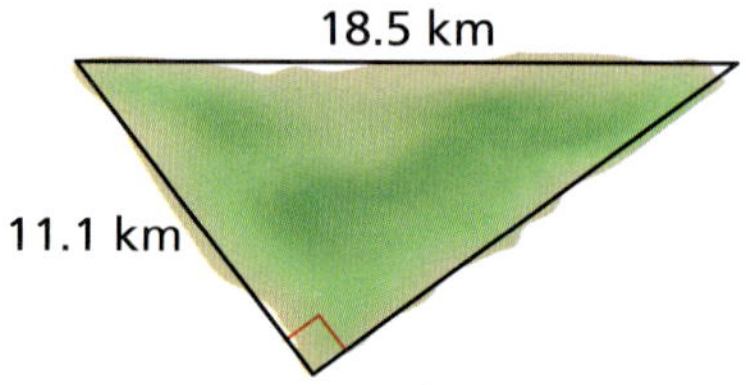

6. Find the area of a regular dodecagon (12-gon) with a side length of 9 inches.

7. Convert 26° to radians and $\frac{5\pi}{9}$ radians to degrees.

8. Find the area of each rhombus-shaped tile. Then find the area of the pattern.

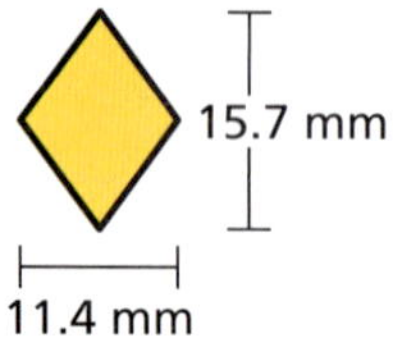

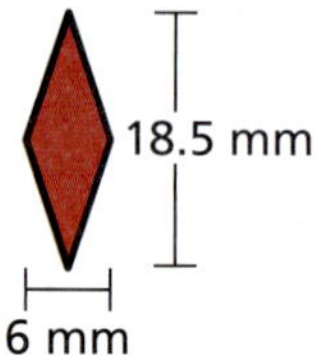

9. A circular region with a 3-mile radius has a population density of about 1000 people per square mile. Find the number of people who live in the region.

10. The area of a kite is 432 square inches. One diagonal is $\frac{3}{2}$ as long as the other diagonal. Find the length of each diagonal.

11. The paper fan shown is shaped like a sector of a circle. Another paper fan is also shaped like a sector of a circle, but with a radius of 6 inches and an intercepted arc of 150°. Which fan will move more air? Explain.

12. There are 414 people per square mile in New York and 404 people per square mile in Florida. There are about 9 million people in New York City and about 1 million people in Jacksonville, Florida. Compare the population density of New York excluding New York City to that of Florida excluding Jacksonville. Explain your reasoning.

GO DIGITAL

11 Performance Task

Center-Pivot Irrigation

A center-pivot irrigation system contains a pipe structure that rotates about a **pivot point**. Water enters the system at the pivot point and then exits through sprinklers along the system. These systems typically complete one full revolution every three days.

Sprinklers are placed along sections of pipe, which are joined together and supported by trusses, and mounted on wheeled towers. Most systems are less than 500 meters long.

The sprinklers in a central-pivot irrigation system must overlap to achieve a more uniform distribution of water. Too much overlap leads to overwatering. Not enough overlap leads to dry spots.

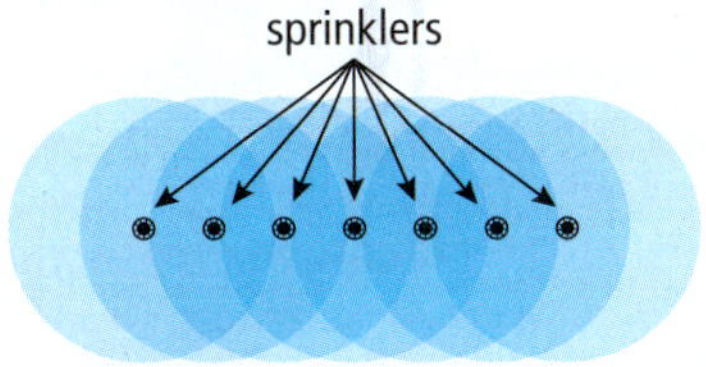

IRRIGATION DESIGN

Design a center-pivot irrigation system. Determine each of the following:

- total radius of the system
- spacing between sprinklers
- number of wheeled towers
- throw radius of each sprinkler
- number of sprinklers
- rotation speed of the system

What happens to the area irrigated by a sprinkler as the distance from the pivot point increases? Provide examples to support your answer. Using this information, explain how you can modify your system to achieve a more uniform distribution of water.

What area can be irrigated by your system in 1 hour? in 1 day?

GO DIGITAL

11 Test Prep WITH CalcChat® Cumulative Practice

WATCH Tutorial videos are available for each exercise.

1. Which statement(s) must be true? Select all that apply.

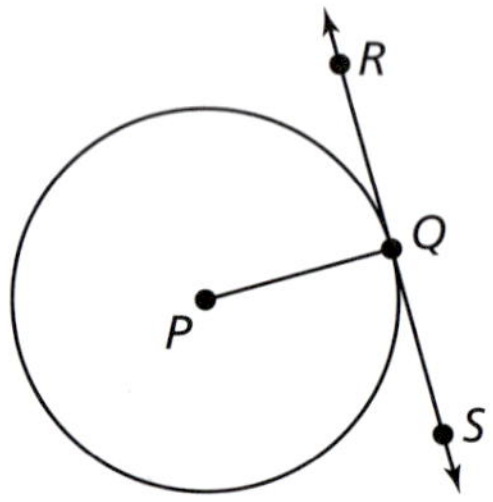

Ⓐ $PQ = \frac{1}{2}RS$

Ⓑ $PQ = RQ$

Ⓒ $\overline{PQ}$ bisects $\overline{RS}$.

Ⓓ $\overline{PQ} \perp \overleftrightarrow{RS}$

2. What is the area of the flat end of the pencil shown?

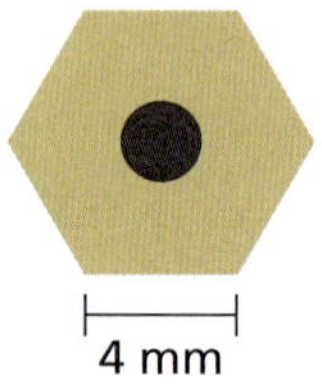

Ⓐ $12\sqrt{3}$ mm^2

Ⓑ $24\sqrt{3}$ mm^2

Ⓒ 48 mm^2

Ⓓ $48\sqrt{3}$ mm^2

3. What is the equation of the line passing through the point (2, 5) that is parallel to the line $x + \frac{1}{2}y = -1$?

Ⓐ $y = -2x + 9$

Ⓑ $y = 2x + 1$

Ⓒ $y = \frac{1}{2}x + 4$

Ⓓ $y = -\frac{1}{2}x + 6$

4. Find the length of each line segment with the given endpoints. Then order the line segments from shortest to longest.

a. $A(1, -5), B(4, 0)$

b. $C(-4, 2), D(1, 4)$

c. $E(-1, 1), F(-2, 7)$

d. $G(-1.5, 0), H(4.5, 0)$

e. $J(-7, -8), K(-3, -5)$

f. $L(10, -2), M(9, 6)$

GO DIGITAL

5. Find the perimeter of the shaded region.

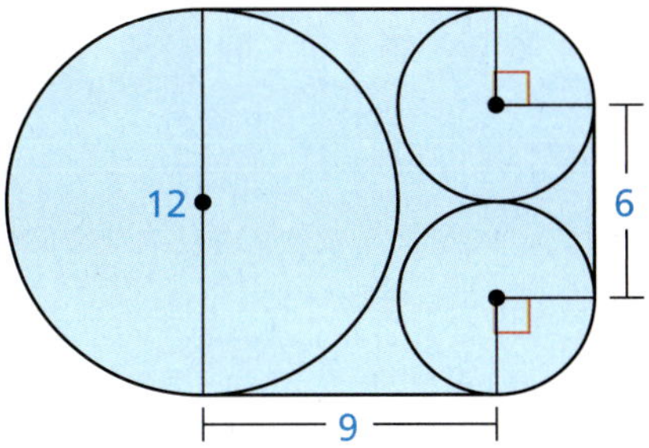

6. What is the arc length of $\widehat{AB}$?

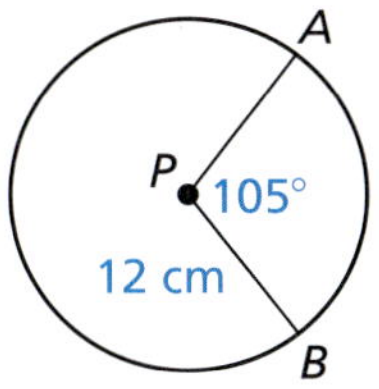

Ⓐ 3.5π cm

Ⓑ 7π cm

Ⓒ 24π cm

Ⓓ 42π cm

7. Each triangle shown below is a right triangle.

a. Are any of the triangles special right triangles? Explain your reasoning.

b. List all similar triangles, if any.

c. Find the lengths of the altitudes of triangles *B* and *C*.

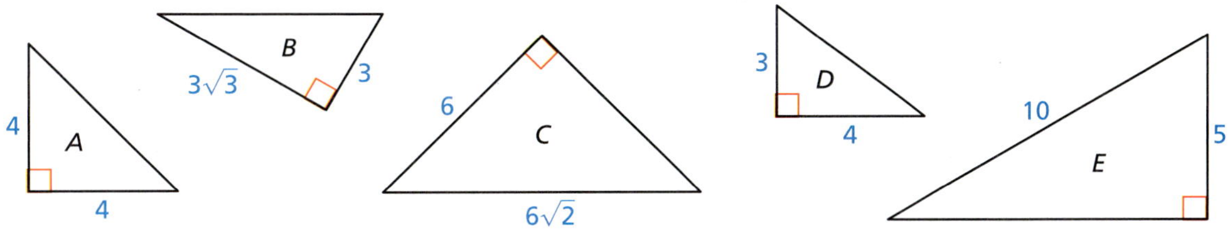

8. The point (0, 2) is on a circle centered at the origin. Prove or disprove that the point $(1, \sqrt{3})$ is on the circle.

9. About 64,660 people live in a circular region with a 1.5-mile radius. Find the population density in people per square mile.

Ⓐ 6861 people per square mile

Ⓑ 9148 people per square mile

Ⓒ 28,738 people per square mile

Ⓓ 43,107 people per square mile

GO DIGITAL

12 Surface Area and Volume

NATIONAL GEOGRAPHIC EXPLORER

Daniela A. Cafaggi

WATCH

Daniela Cafaggi is a biologist who studies the diversity and conservation of bats in the archaeological zones of Yucatán, Mexico. Her team has identified more than 20 different species of bats in the region. She promotes coexistence between humans and bats, and highlights the importance of bats in ancient and current Mayan culture.

- Name several famous archaeological zones of the Yucatán Peninsula in Mexico.
- When did the ancient Mayans build El Castillo at Chichén Itzá?

STEM

Conservationists use artificial bat caves to combat white-nose syndrome. In the Performance Task, you will design an artificial bat cave and estimate the number of hibernating bats that it can accommodate.

GO DIGITAL

Conservation Biology

Preparing for Chapter 12

Chapter Learning Target: Understand surface area and volume.

Chapter Success Criteria:
- ◆ I can describe attributes of solids.
- ◆ I can find surface areas and volumes of solids.
- ■ I can find missing dimensions of solids.
- ■ I can solve real-life problems involving surface area and volume.

◆ Surface
■ Deep

AZ VOCAB

Chapter Vocabulary

Work with a partner. Discuss each of the vocabulary terms.

face
edge
vertex
cross section
lateral face of a cone
volume
chord of a sphere
density
solid of revolution
axis of revolution

Mathematical Thinking and Reasoning

7 MTR

APPLY MATHEMATICS TO REAL-WORLD CONTEXTS

Mathematicians who apply mathematics to real-world contexts use models and methods to understand, represent, and solve problems.

Work with a partner. You want to design and build a bat house.

1. Use the Internet or another resource to research bat houses.
   a. Describe the acceptable sizes for a bat house.
   b. What materials can be used to construct a bat house?
   c. Where should a bat house be located?

2. Sketch a bat house that satisfies the guidelines you described in Exercise 1. Include the dimensions of the bat house in your sketch.

3. Find the volume of your bat house. Explain your method.

4. You want to paint the exterior of your bat house. How many square inches do you need to paint?

GO DIGITAL

12 Prepare WITH CalcChat®

Finding Surface Area

Example 1 **Find the surface area of the prism.**

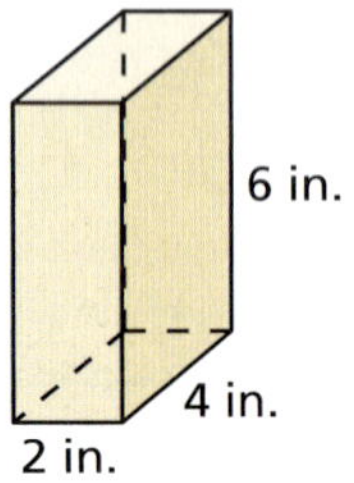

$S = 2\ell w + 2\ell h + 2wh$ — Write formula for surface area of a rectangular prism.

$= 2(2)(4) + 2(2)(6) + 2(4)(6)$ — Substitute 2 for ℓ, 4 for w, and 6 for h.

$= 88$ — Simplify.

▶ The surface area is 88 square inches.

Find the surface area of the prism.

1.

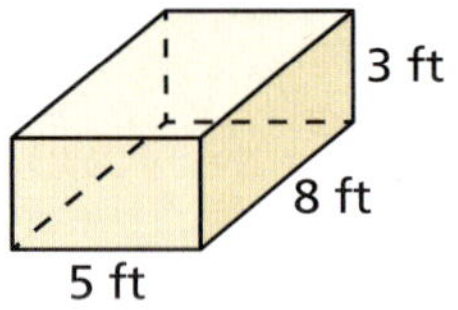

2.

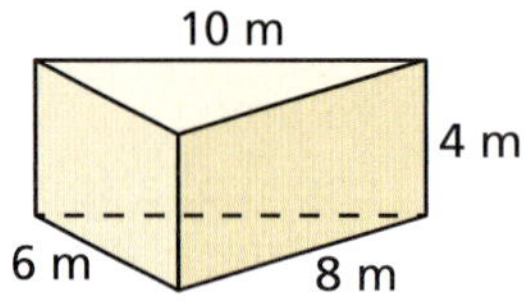

3.

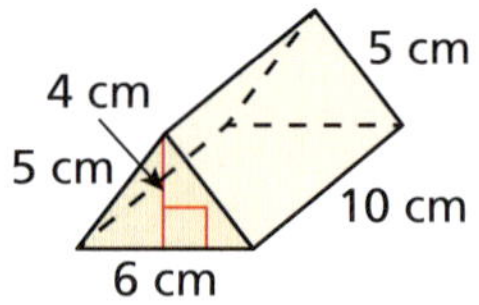

Finding Areas of Similar Polygons

Example 2 **In the diagram, $\triangle JKL \sim \triangle MNP$. Find the area of $\triangle MNP$.**

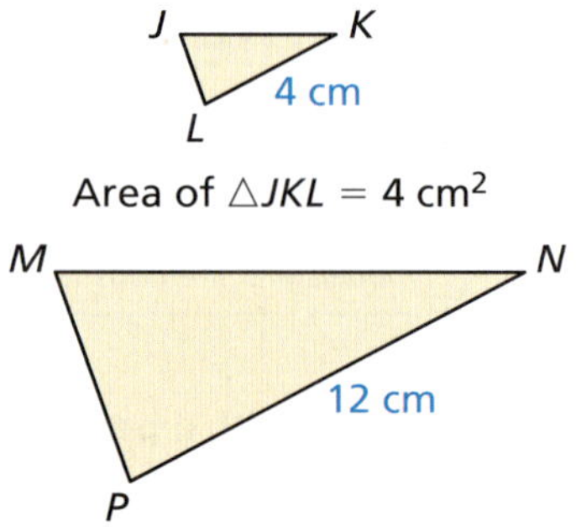

Because the triangles are similar, the ratio of the area of $\triangle JKL$ to the area of $\triangle MNP$ is equal to the square of the ratio of KL to NP. Let A represent the area of $\triangle MNP$.

$$\frac{\text{Area of } \triangle JKL}{\text{Area of } \triangle MNP} = \left(\frac{KL}{NP}\right)^2$$ Areas of Similar Polygons Theorem

$$\frac{4}{A} = \left(\frac{4}{12}\right)^2$$ Substitute.

$$36 = A$$ Simplify.

▶ The area of $\triangle MNP$ is 36 square centimeters.

Find the area of each polygon.

4. In the diagram, $\triangle ABC \sim \triangle DEF$. Find the area of $\triangle DEF$.

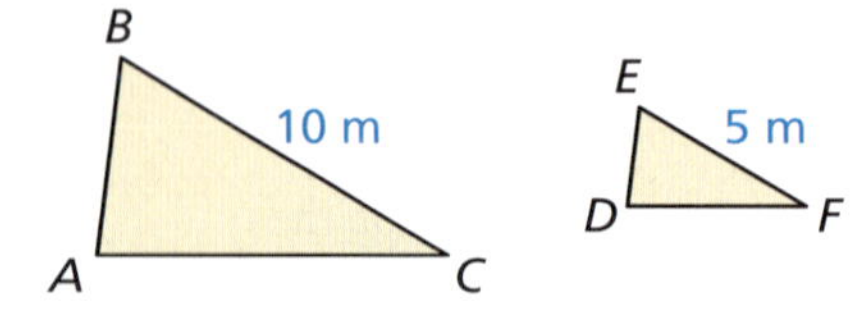

Area of $\triangle ABC$ = 20 m²

5. **PERSEVERE** Rectangle $ABCD$ is similar to rectangle $PQRS$. Find the area of $\triangle PQR$. Justify your answer.

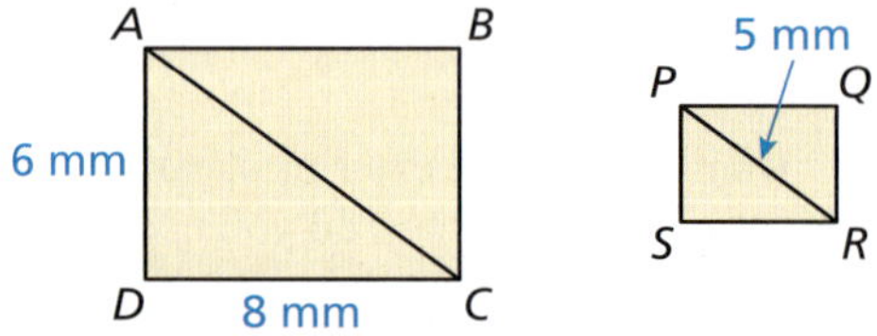

GO DIGITAL

12.1 Cross Sections of Solids

Learning Target: Describe and draw cross sections.

Success Criteria:
- I can describe attributes of solids.
- I can describe and draw cross sections.
- I can solve real-life problems involving cross sections.

Imagine cutting through a piece of food. The intersection formed by the cut you make and the piece of food is called a *cross section*.

EXPLORE IT! Describing Cross Sections

Work with a partner.

a. For each food, describe the shape. If you can cut the food into two congruent pieces, what shape would the cross section be? Compare your results with those of your classmates.

i. wheel of cheese

ii. watermelon

iii. stick of butter

iv. cucumber

5 MTR USE STRUCTURE
When making a cut, what are some things that affect the shape of the cross section?

b. Describe how you can slice the wedge of cheese so that the cross section formed is the given shape.

i. triangle

ii. rectangle

iii. trapezoid

c. Is there more than one way to slice the wedge of cheese in part (b) to form a triangular cross section? Explain. Use drawings to support your answer.

Geometric Reasoning
MA.912.GR.4.1 Identify the shapes of two-dimensional cross-sections of three-dimensional figures.

GO DIGITAL

Classifying Solids

Vocabulary

face, *p. 616*
polyhedron, *p. 616*
edge, *p. 616*
vertex, *p. 616*
cross section, *p. 617*

A three-dimensional figure, or *solid*, is bounded by flat or curved surfaces, called **faces**, that enclose a single region of space. A **polyhedron** is a solid whose faces are all polygons. An **edge** of a solid is a line segment or curve formed by the intersection of two faces. A **vertex** of a polyhedron is a point where three or more edges meet. The plural of polyhedron is *polyhedra* or *polyhedrons*.

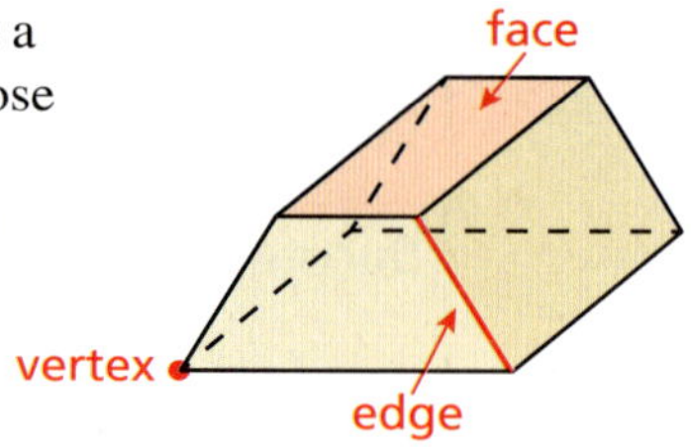

KEY IDEA
Types of Solids

Polyhedra

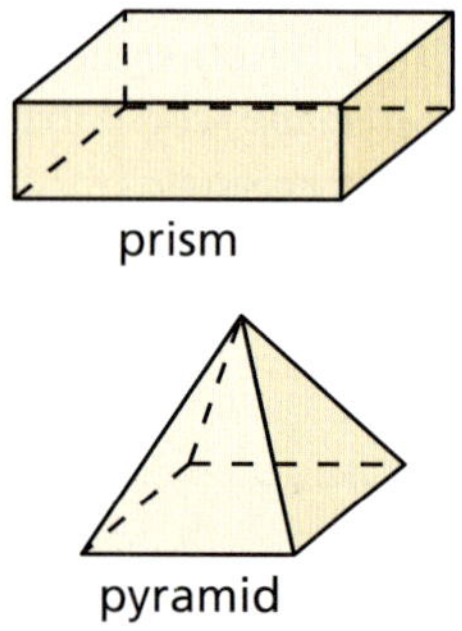

Not Polyhedra

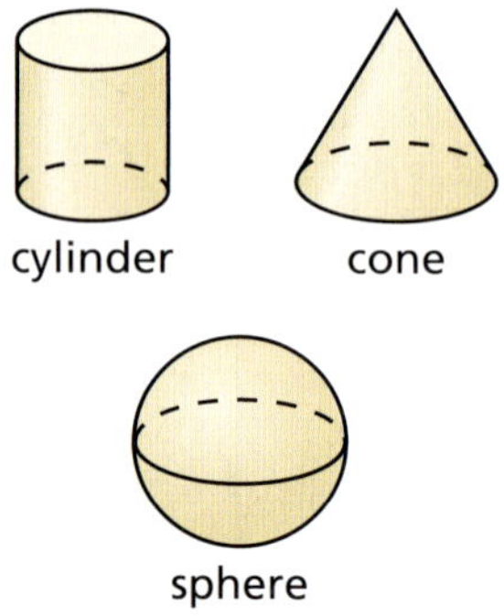

Pentagonal prism

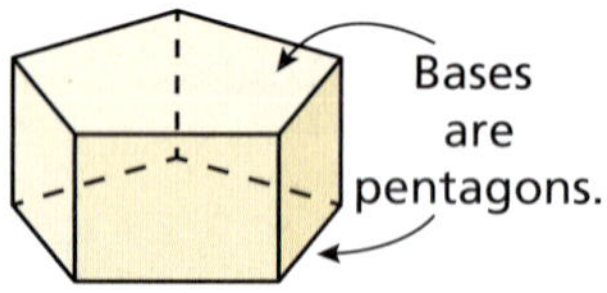

To name a prism or a pyramid, use the shape of the *base.* The two bases of a prism are congruent polygons in parallel planes. For example, the bases of a pentagonal prism are pentagons. The base of a pyramid is a polygon. For example, the base of a triangular pyramid is a triangle.

Triangular pyramid

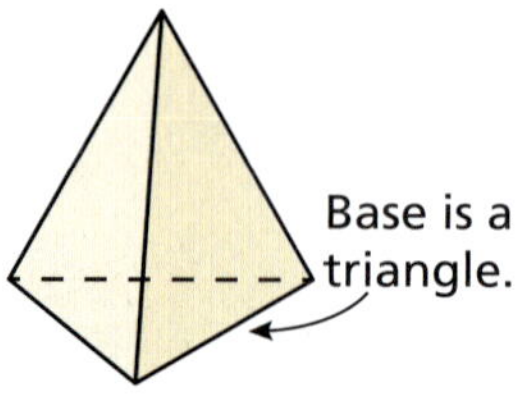

EXAMPLE 1 Classifying Solids

Tell whether each solid is a polyhedron. If it is, name the polyhedron.

a.

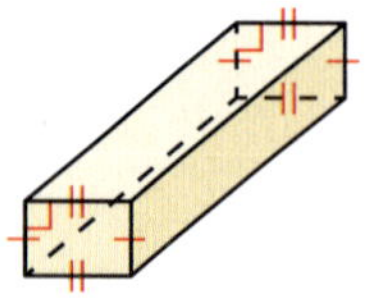

b.

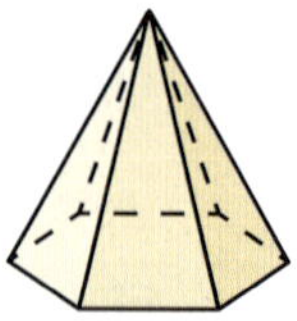

c.

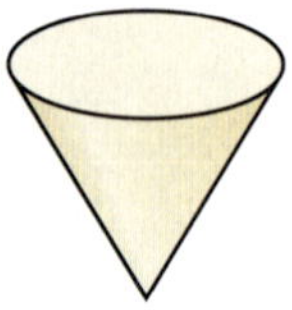

SOLUTION

a. The solid is formed by polygons, so it is a polyhedron. The two bases are congruent rectangles, so it is a rectangular prism.

b. The solid is formed by polygons, so it is a polyhedron. The base is a hexagon, so it is a hexagonal pyramid.

c. The cone has a curved surface, so it is not a polyhedron.

GO DIGITAL

Describing Cross Sections

Imagine a plane slicing through a solid. The intersection of the plane and the solid is called a **cross section**. For example, three different cross sections of a cube are shown below.

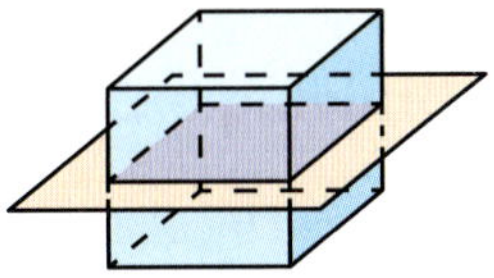
square

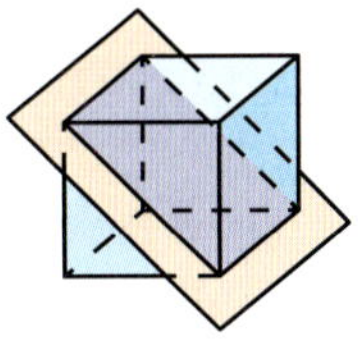
rectangle

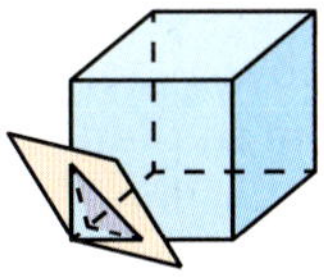
triangle

EXAMPLE 2 Describing Cross Sections WATCH

> **1 MTR ANALYZE A PROBLEM**
> How can rotating a solid help you better visualize a cross section?

Describe the shape formed by the intersection of the plane and the solid.

a.

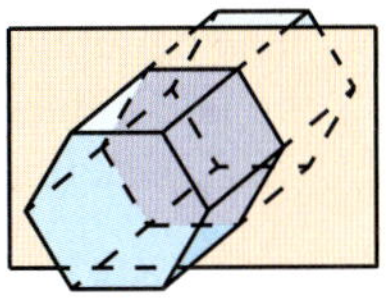

b.

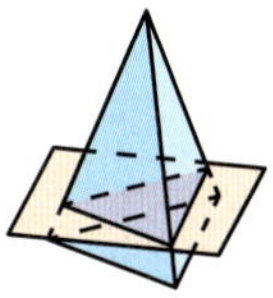

c.

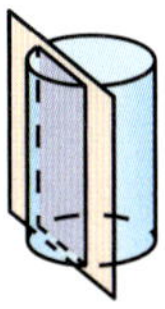

d.

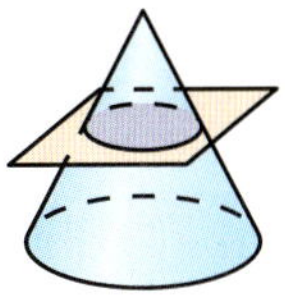

e.

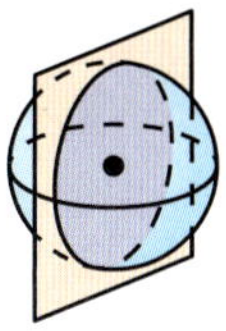

f.

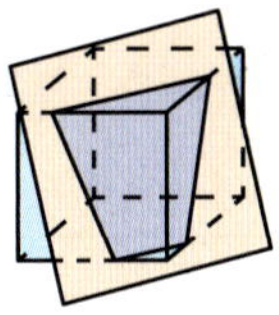

SOLUTION

a. The cross section is a hexagon.

b. The cross section is a triangle.

c. The cross section is a rectangle.

d. The cross section is a circle.

e. The cross section is a circle.

f. The cross section is a trapezoid.

SELF-ASSESSMENT

1 I don't understand yet. 2 I can do it with help. 3 I can do it on my own. 4 I can teach someone else.

Tell whether the solid is a polyhedron. If it is, name the polyhedron.

1.

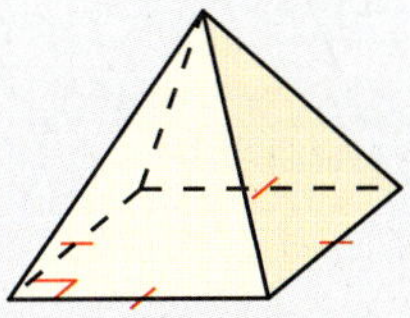

2.

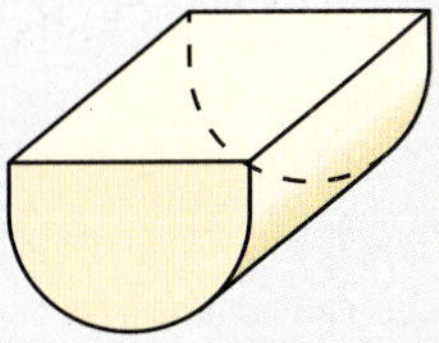

3.

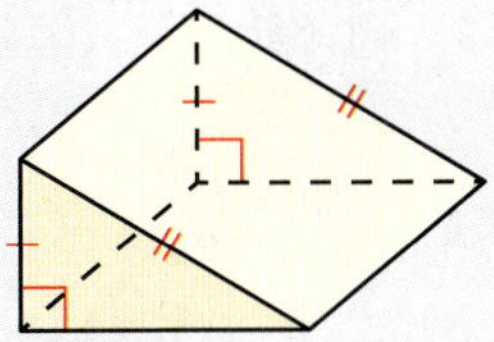

Describe the shape formed by the intersection of the plane and the solid.

4.

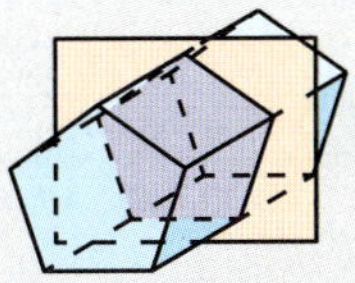

5.

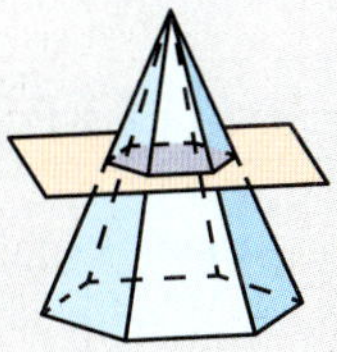

6.

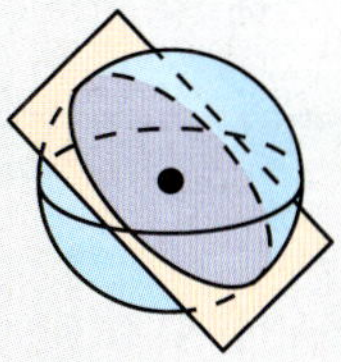

7. WRITING Can a plane intersect a rectangular prism and form a cross section that is a circle? Explain.

GO DIGITAL

Drawing Cross Sections

REMEMBER

An example of the Plane Intersection Postulate is shown below, where the intersection of plane S and plane T is line ℓ.

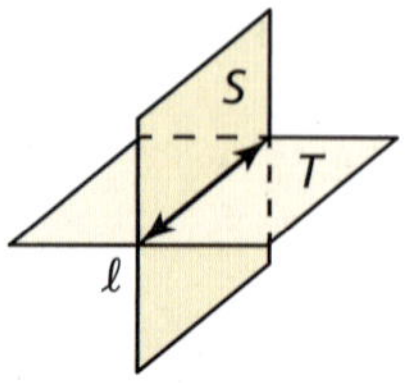

The Plane Intersection Postulate states that if two planes intersect, then their intersection is a line. This postulate can help you when drawing a cross section.

EXAMPLE 3 Drawing a Cross Section

Draw the cross section formed by a plane parallel to the base that contains the red line segment drawn on the square pyramid. What is the shape of the cross section?

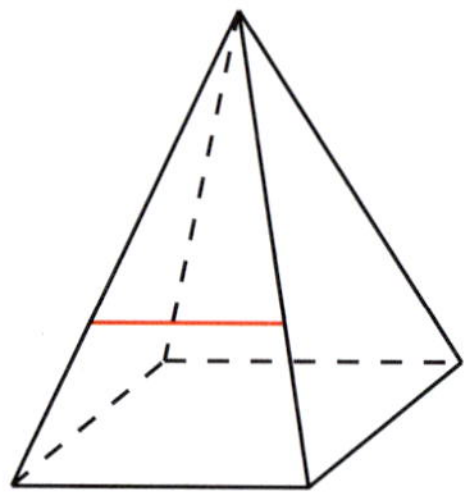

SOLUTION

Step 1 Visualize a horizontal plane parallel to the base that intersects the lateral face and passes through the red line segment.

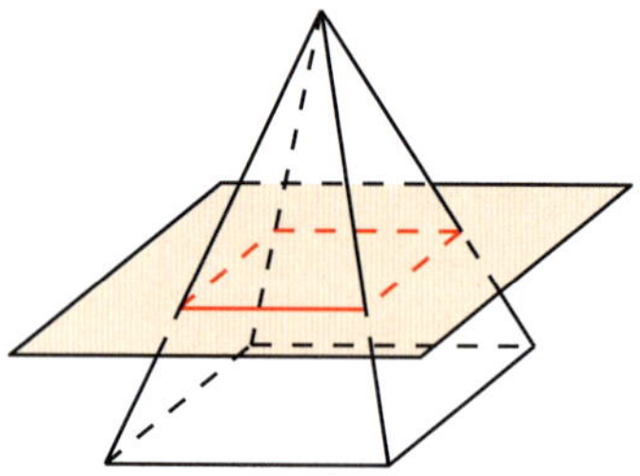

Step 2 The horizontal plane is parallel to the base of the pyramid. So, draw each pair of parallel line segments where the plane intersects the lateral faces of the pyramid.

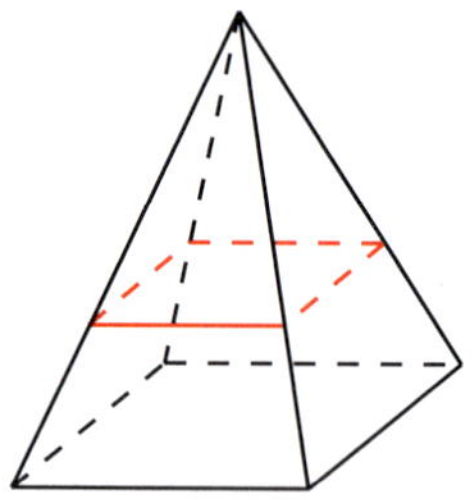

Step 3 Shade the cross section.

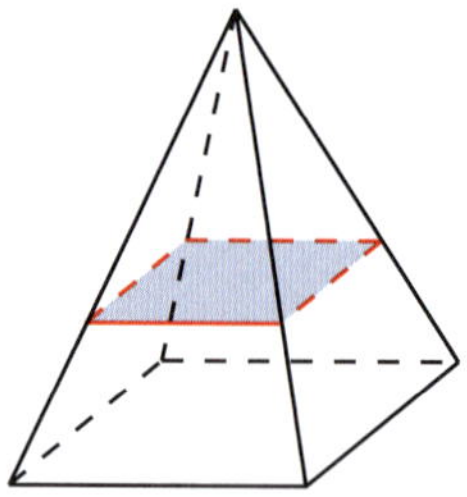

▶ The cross section is a square.

SELF-ASSESSMENT

1 I don't understand yet. 2 I can do it with help. 3 I can do it on my own. 4 I can teach someone else.

8. **WHAT IF?** Draw the cross section formed by a plane perpendicular to the base that contains the vertex of the square pyramid in Example 3. What is the shape of the cross section?

9. **REASONING** Describe how a plane can intersect the pyramid in Example 3 so that it forms a cross section that is (a) a trapezoid and (b) a line segment.

GO DIGITAL

Solving Real-Life Problems

7 MTR **EXAMPLE 4** **Modeling Real Life** WATCH

A machine at a sawmill cuts a 4-inch-by-4-inch piece of wood lengthwise along its diagonal, as shown. Find the perimeter and area of the cross section formed by the cut.

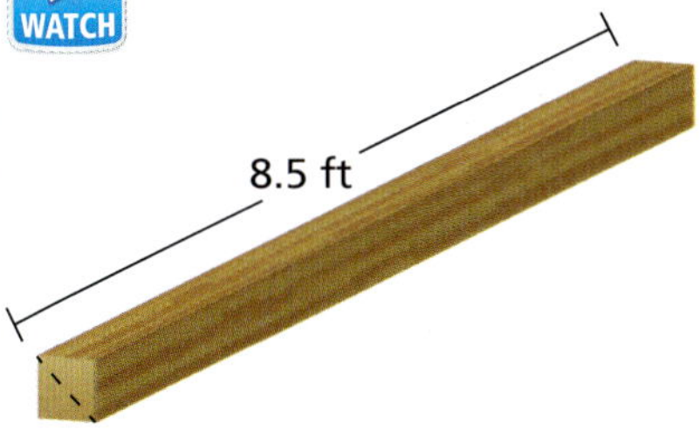

SOLUTION

1. **Understand the Problem** You know that the piece of wood is shaped like a rectangular prism with a length of 8.5 feet and a width and height of 4 inches. You are asked to calculate the perimeter and area of the cross section formed when an 8.5-foot cut is made along its diagonal.

2. **Make a Plan** Determine the shape and the dimensions of the cross section. Then use the dimensions to calculate the perimeter and area of the cross section.

3. **Solve and Check** Draw a diagram of the cross section. It is a rectangle with a length of 8.5 feet, or 102 inches.

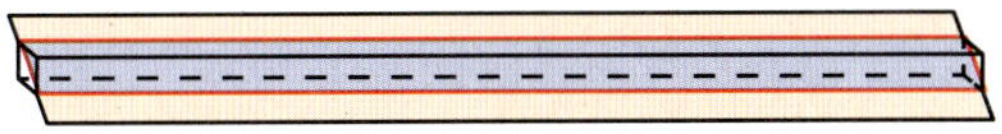

Use the Pythagorean Theorem to find its width. The length and width of the end of the piece of wood is 4 inches.

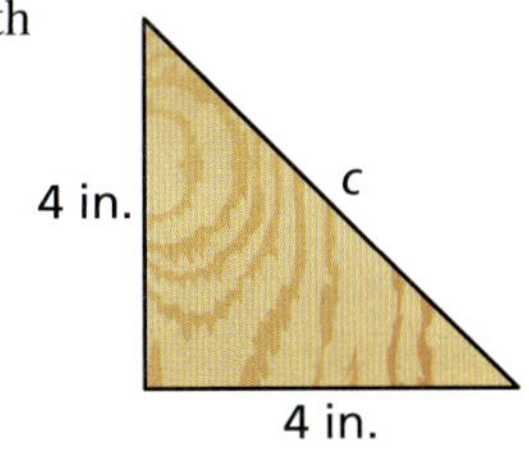

$c^2 = a^2 + b^2$ Pythagorean Theorem

$c^2 = 4^2 + 4^2$ Substitute.

$c^2 = 16 + 16$ Multiply.

$c^2 = 32$ Add.

$c = \sqrt{32}$ Take the positive square root of each side.

$c = 4\sqrt{2}$ Simplify.

The width of the rectangular cross section is $4\sqrt{2}$ inches.

Perimeter of cross section

$$\begin{aligned} P &= 2\ell + 2w \\ &= 2(102) + 2(4\sqrt{2}) \\ &= 204 + 8\sqrt{2} \\ &\approx 215.31 \end{aligned}$$

Area of cross section

$$\begin{aligned} A &= \ell w \\ &= 102 \cdot 4\sqrt{2} \\ &= 408\sqrt{2} \\ &\approx 577 \end{aligned}$$

▶ The perimeter of the cross section is about 215.31 inches, and the area of the cross section is about 577 square inches.

Check Reasonableness Estimate to check that your answer is reasonable. The length of the rectangular cross section is about 9 feet, and its width is about 6 inches, or 0.5 foot.

Perimeter of cross section: $P = 2\ell + 2w = 2(9) + 2(0.5) = 19 \text{ ft} = 228 \text{ in.}$ ✓

Area of cross section: $A = \ell w = 9 \cdot 0.5 = 4.5 \text{ ft}^2 = 648 \text{ in.}^2$ ✓

SELF-ASSESSMENT 1 I don't understand yet. 2 I can do it with help. 3 I can do it on my own. 4 I can teach someone else.

10. A 6-inch-by-6-inch piece of wood that is 10.25 feet long is cut lengthwise along its diagonal. Find the perimeter and area of the cross section formed by the cut.

GO DIGITAL

12.1 Practice WITH CalcChat® AND CalcView®

In Exercises 1–4, match the polyhedron with its name.

1. 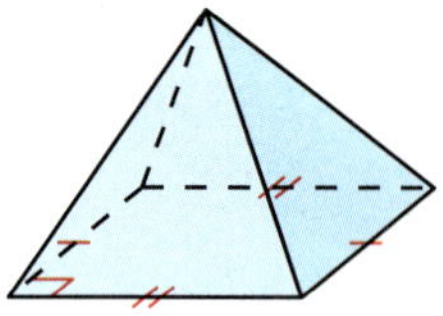

2. 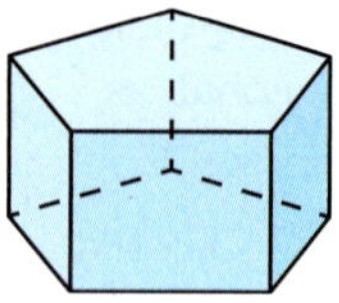

3. 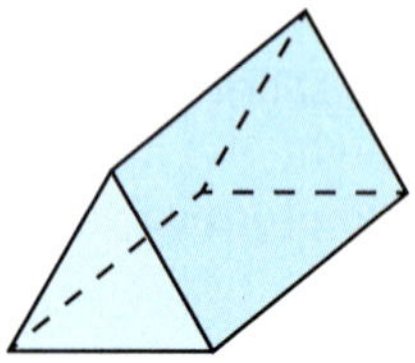

4. 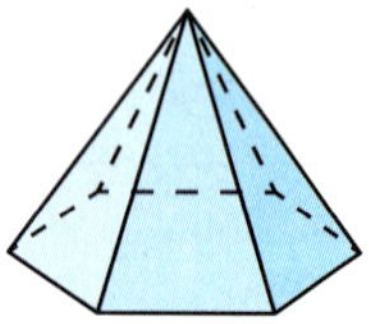

A. triangular prism

B. rectangular pyramid

C. hexagonal pyramid

D. pentagonal prism

In Exercises 5–8, tell whether the solid is a polyhedron. If it is, name the polyhedron. *(See Example 1.)*

▶ 5. 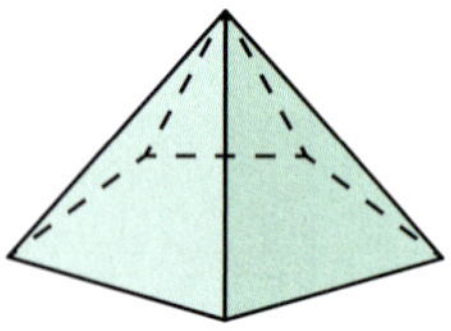

6. 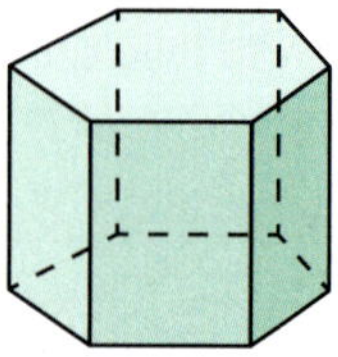

7. 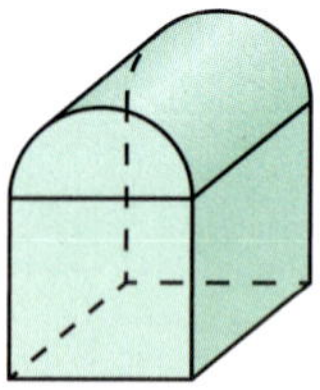

8. 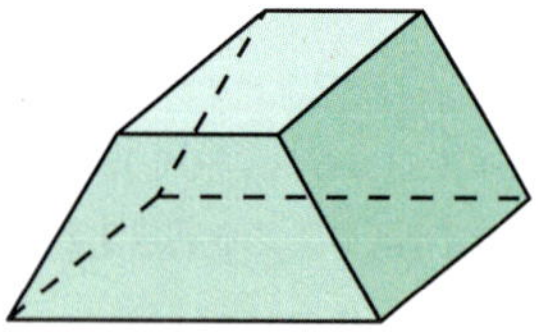

In Exercises 9–14, sketch the polyhedron.

9. triangular prism

10. rectangular prism

11. pentagonal prism

12. hexagonal prism

13. square pyramid

14. pentagonal pyramid

REASONING In Exercises 15–18, name the figure that is described.

15. a pyramid with 6 faces

16. a prism with 10 faces

17. a polyhedron with a rectangular base and 5 total faces

18. a prism with 9 edges

In Exercises 19–22, describe the shape formed by the intersection of the plane and the solid. *(See Example 2.)*

▶ 19. 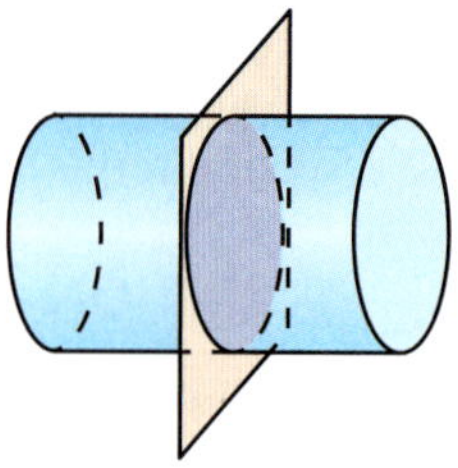

20. 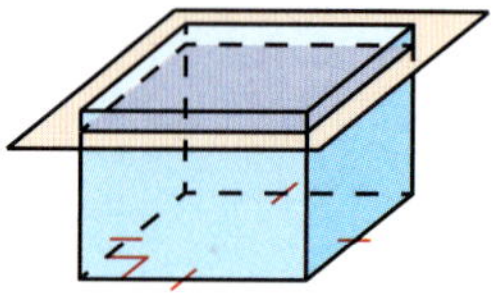

21. 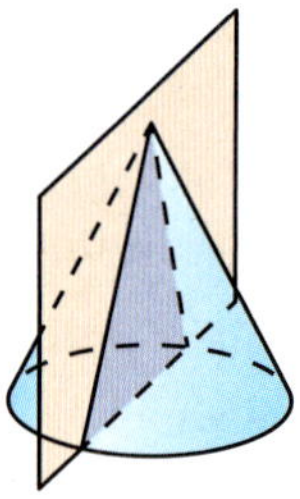

22. 

4 MTR 23. **ERROR ANALYSIS** Describe and correct the error in identifying the solid.

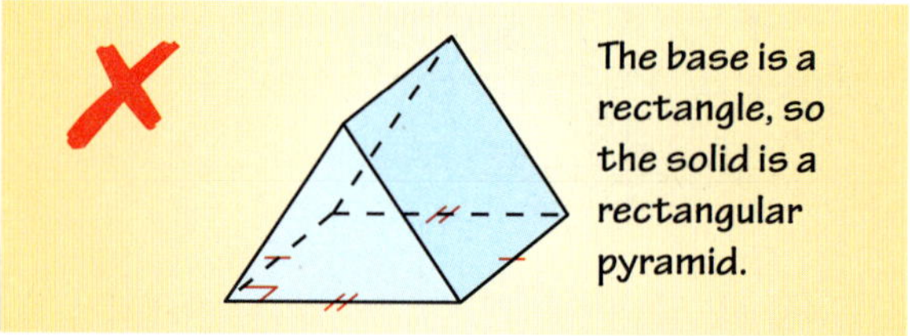

24. **OPEN-ENDED** Give an example of a solid from which a triangular, hexagonal, and trapezoidal cross section can be formed.

In Exercises 25–28, draw the cross section formed by the described plane that contains the red line segment drawn on the solid. What is the shape of the cross section? *(See Example 3.)*

▶ 25. plane is perpendicular to base

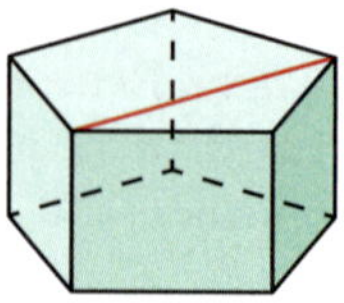

26. plane is parallel to base

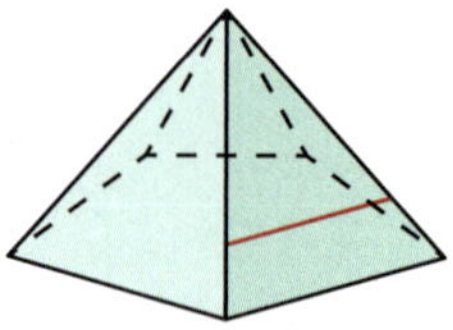

27. plane is parallel to bottom face

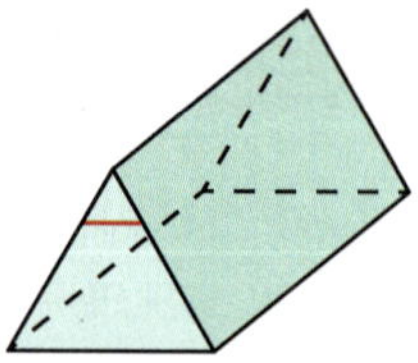

28. plane is perpendicular to bottom face

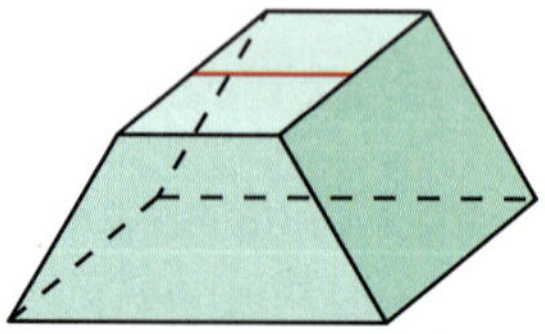

GO DIGITAL

In Exercises 29–34, draw the cross section formed by a vertical plane that divides the solid into two congruent parts. Is there more than one way to use a vertical plane to divide the figure into two congruent parts? If so, does the cross section change? Explain.

29. 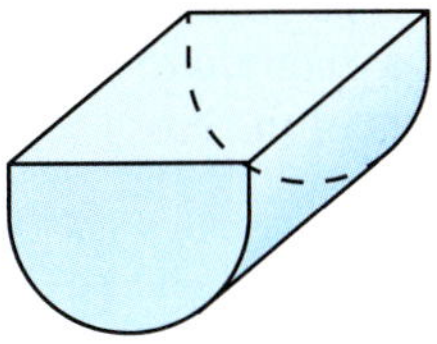

30. 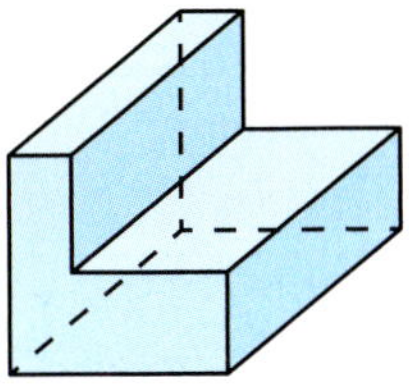

31. 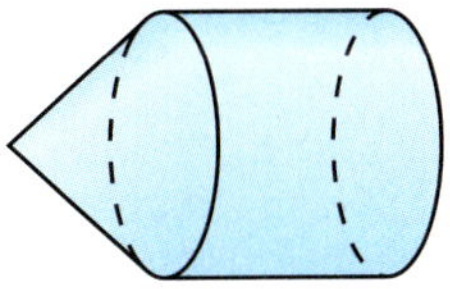

32. 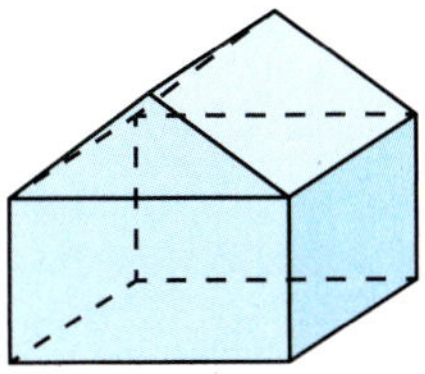

33. 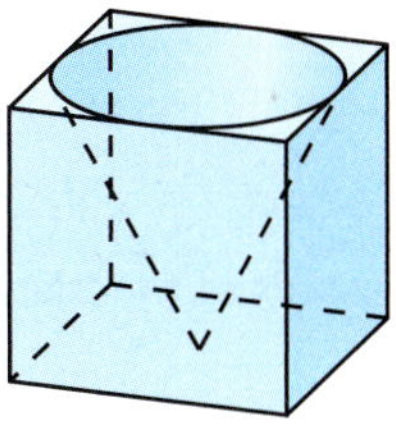

34. 

35. MODELING REAL LIFE You cut the cake vertically into two congruent parts. *(See Example 4.)*

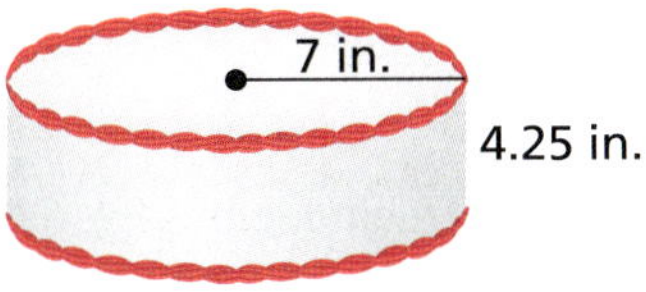

a. Find the perimeter and area of the cross section formed by the cut.

b. Find the surface area of the cake that is not frosted before the cut. How does the unfrosted surface area change after the cut?

c. Can the cake be cut into two congruent parts another way? If so, find the perimeter and area of the cross section formed by the cut.

36. MODELING REAL LIFE A mason uses a concrete saw to cut a block in the shape of a rectangular prism along a diagonal of a base, as shown. Describe the shape of the cross section formed by the cut. Are the two pieces formed by the cut congruent? Explain.

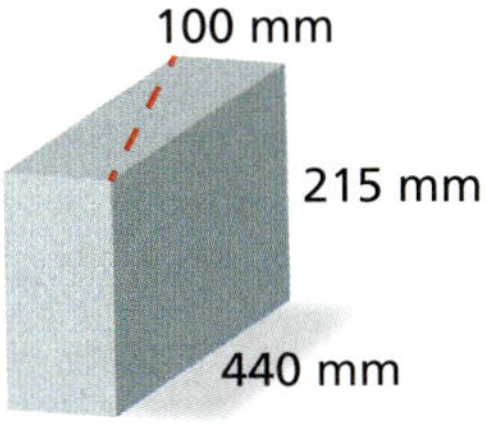

GO DIGITAL

37. REASONING Use the figure shown.

a. One of the hexagonal pipes is cut vertically so that it is divided into two congruent parts. Draw two possible cross sections.

b. How many different ways can a pipe be cut lengthwise to form two congruent parts? Explain.

38. HOW DO YOU SEE IT?
A plane intersects a cube. Determine whether it is possible to form each of the following cross sections. Explain your reasoning.

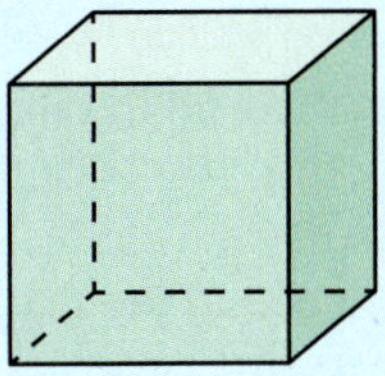

a. triangle
b. circle
c. parallelogram
d. trapezoid
e. hexagon
f. octagon

39. PATTERNS Find the number of faces, edges, and vertices for each solid. Then write an equation that relates F, E, and V. Verify that your equation is true for several other solids in this section.

Solid	Faces, F	Edges, E	Vertices, V
tetrahedron 			
cube 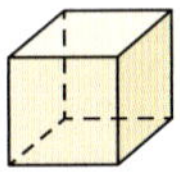			
octahedron 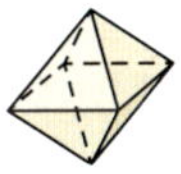			
dodecahedron 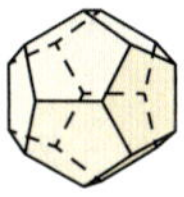			
icosahedron 			

40. THOUGHT PROVOKING Describe a solid that can be intersected by a plane to form the cross section shown. Explain how you form the cross section.

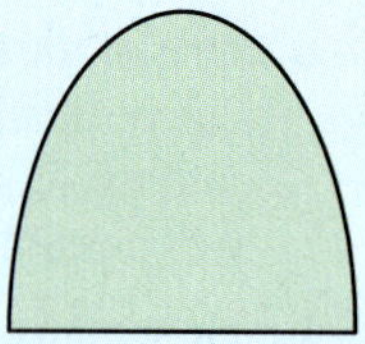

41. MAKING AN ARGUMENT Can every plane that intersects a sphere form a circular cross section? Explain your reasoning.

42. DIG DEEPER A plane that intersects a sphere is 7 meters from the center of the sphere. The radius of the sphere is 25 meters. Draw a diagram to represent this situation. Then find the area of the cross section.

REVIEW & REFRESH

43. Explain how to prove that $\overline{AB} \cong \overline{CB}$.

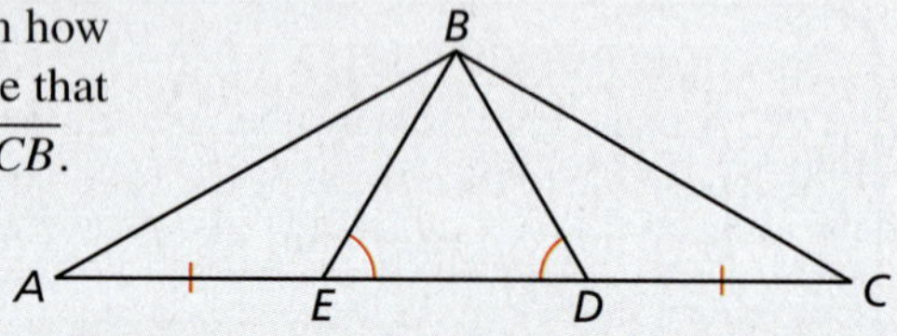

44. MODELING REAL LIFE Tailors want to know the density of fabric when deciding what material to use when making clothing. The piece shown weighs 10 ounces. Find the density of the fabric in ounces per square yard.

In Exercises 45–48, draw the cross section formed by the described plane that contains the red line segment drawn on the solid. What is the shape of the cross section?

45. plane is parallel to base

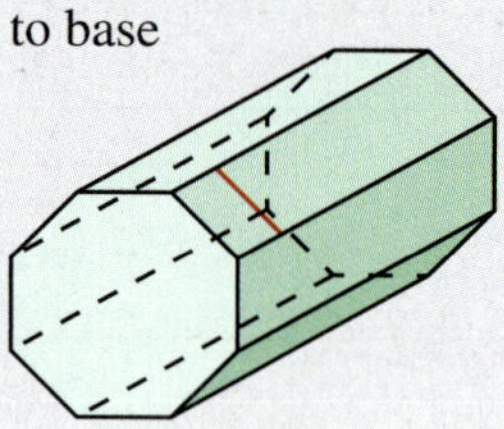

46. plane is parallel to base

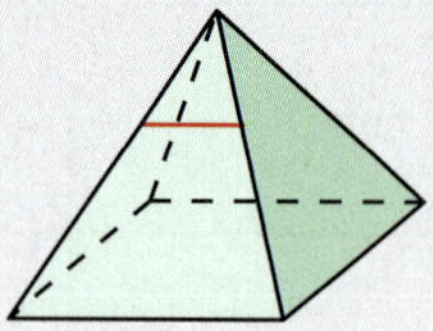

47. plane is perpendicular to base

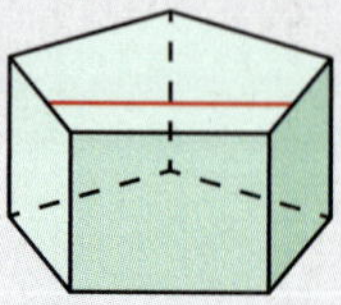

48. plane is perpendicular to base

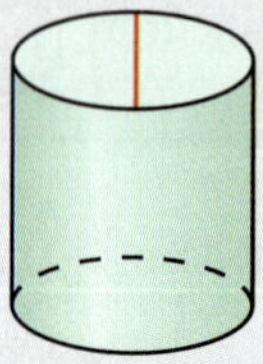

49. Tell whether $\overline{AB}$ is tangent to $\odot C$. Explain your reasoning.

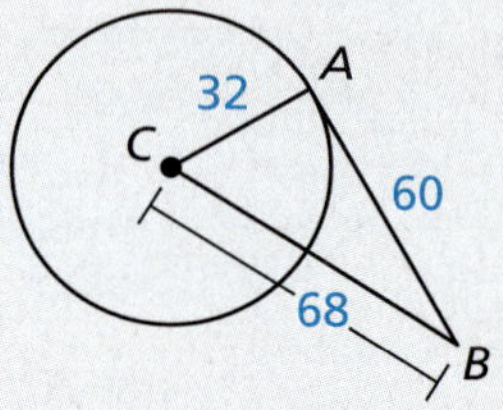

50. Three vertices of $\square MNPQ$ are $M(2, -7)$, $N(-1, 3)$, and $P(5, 9)$. Find the coordinates of the remaining vertex.

51. Approximate the unknown measures of the triangle. Explain your reasoning.

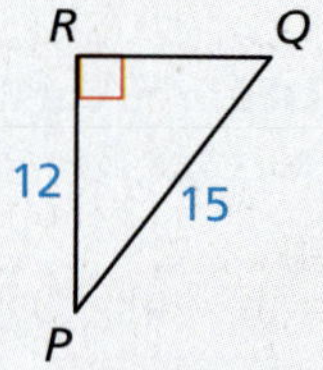

52. Verify that the segment lengths 39, 52, and 64 form a triangle. Is the triangle *acute*, *right*, or *obtuse*?

In Exercises 53 and 54, find the area of the quadrilateral.

53.

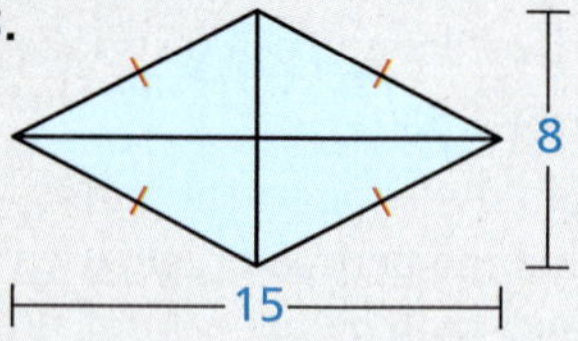

54.

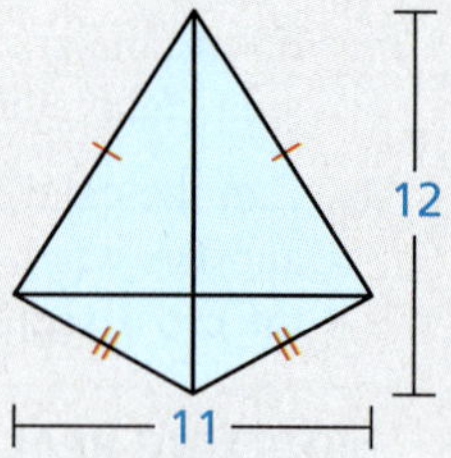

In Exercises 55–58, find the indicated measure.

55. area of a circle with a diameter of 13 centimeters

56. area of a circle with a radius of 4 inches

57. radius of a circle with an area of 67 square meters

58. diameter of a circle with an area of 138 square yards

12.2 Surface Areas of Prisms and Cylinders

Learning Target: Find and use surface areas of prisms and cylinders.

Success Criteria:
- I can find surface areas of prisms and cylinders.
- I can solve real-life problems involving surface areas of prisms and cylinders.
- I can find surface areas of similar prisms and similar cylinders.

EXPLORE IT! Finding Surface Areas of Prisms and Cylinders

Work with a partner. Recall that the *surface area* of a solid is equal to the sum of the area of its faces.

a. Use a net to write an expression for the surface area of the right triangular prism.

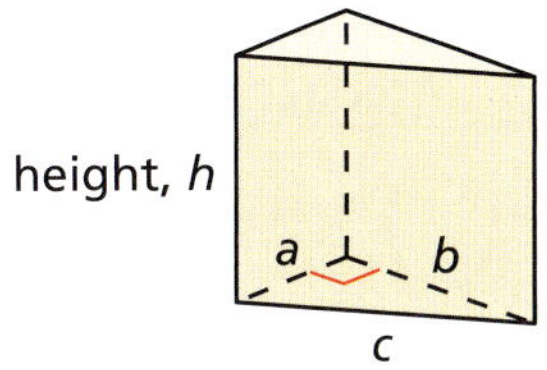

b. Rewrite your expression in part (a) in terms of the perimeter P and area B of the base of the prism. Explain your reasoning.

c. Can you use your expression in part (b) to represent the surface area of a right rectangular prism? a right pentagonal prism? a right hexagonal prism? Explain.

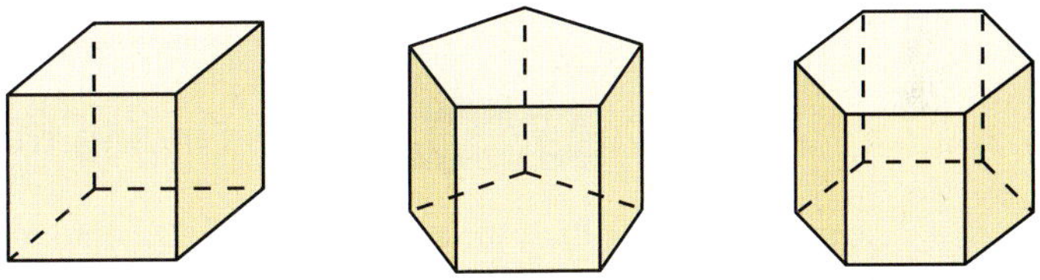

d. Use a net to write an expression for the surface area of the right cylinder.

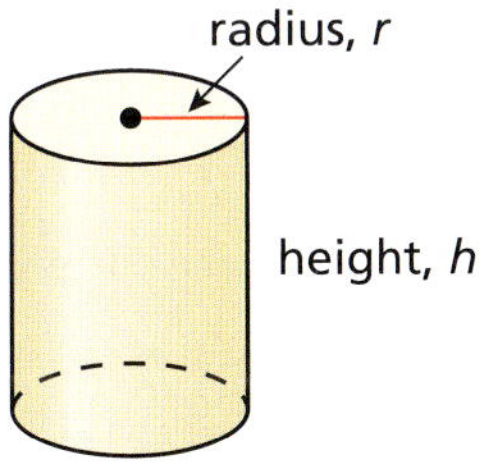

e. Explain how your expression in part (b) is related to your expression in part (d).

6 MTR ASSESS REASONABLENESS

In part (c), what does a prism begin to look like if you continue to increase the number of sides of the base? How does this show that your answer in part (e) is reasonable?

Geometric Reasoning

MA.912.GR.4.3 Extend previous understanding of scale drawings and scale factors to determine how dilations affect the area of two-dimensional figures and the surface area or volume of three-dimensional figures.

MA.912.GR.4.6 Solve mathematical and real-world problems involving the surface area of three-dimensional figures limited to cylinders, pyramids, prisms, cones and spheres.

GO DIGITAL

Vocabulary

surface area, *p. 624*
lateral faces of a polyhedron, *p. 624*
lateral face of a cylinder, *p. 625*
similar solids, *p. 627*

Finding Surface Areas of Right Prisms

The **surface area** of a solid is the sum of the areas of its faces. Recall that a *prism* is a polyhedron with two congruent faces, called *bases*, that lie in parallel planes. The **lateral faces of a polyhedron** are parallelograms formed by connecting the corresponding vertices of the bases. The segments connecting these vertices are *lateral edges*. In a *right prism*, each lateral edge is perpendicular to both bases.

Imagine that you cut some edges of a polyhedron and unfold it to form a net as shown.

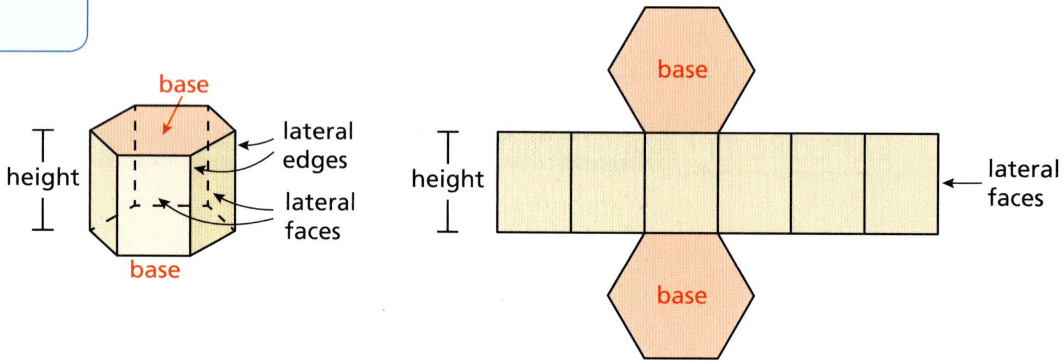

The surface area of a prism is equal to the sum of the base areas and the lateral area.

1 MTR HELP A CLASSMATE

Help a classmate understand why the lateral area of a right prism is equal to the perimeter of a base times the height.

KEY IDEA

Surface Area of a Right Prism

For a right prism with bases that are regular polygons, the surface area S is

$$S = 2B + L = aP + Ph$$

where B is the area of a base, L is the lateral area, a is the apothem of a base, P is the perimeter of a base, and h is the height.

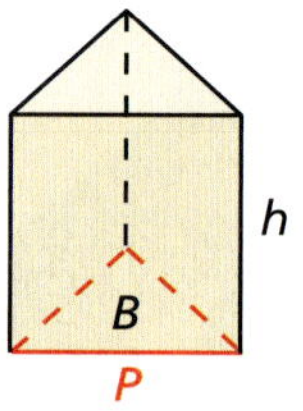

EXAMPLE 1 Finding the Surface Area of a Right Prism

Find the surface area of the regular right prism.

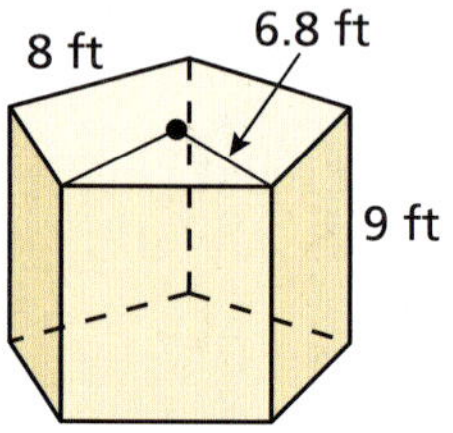

SOLUTION

Find the apothem and perimeter of a base.

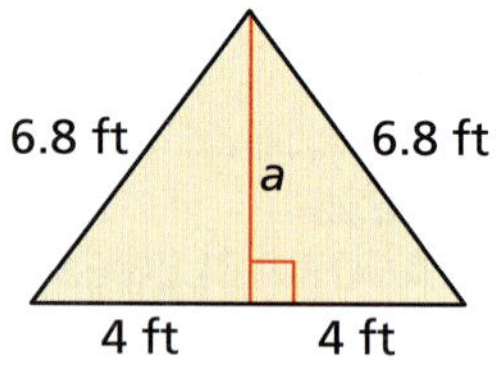

$$a = \sqrt{6.8^2 - 4^2} = \sqrt{30.24}$$

$$P = 5(8) = 40$$

Find the surface area.

$S = aP + Ph$ — Write formula for surface area of a right prism.

$= (\sqrt{30.24})(40) + (40)(9)$ — Substitute.

≈ 580.0 — Use technology.

▶ The surface area is about 580 square feet.

SELF-ASSESSMENT 1 I don't understand yet. 2 I can do it with help. 3 I can do it on my own. 4 I can teach someone else.

1. Find the surface area of a right rectangular prism with a height of 7 inches, a length of 3 inches, and a width of 4 inches.

GO DIGITAL

Finding Surface Areas of Right Cylinders

Recall that a *cylinder* is a solid with congruent circular bases that lie in parallel planes. The height of a cylinder is the perpendicular distance between its bases. The radius of a base is the *radius* of the cylinder. In a *right cylinder*, the segment joining the centers of the bases is perpendicular to the bases.

The **lateral face of a cylinder** is its curved surface. For a right cylinder, the lateral area is equal to the product of the circumference and the height, or $2\pi rh$.

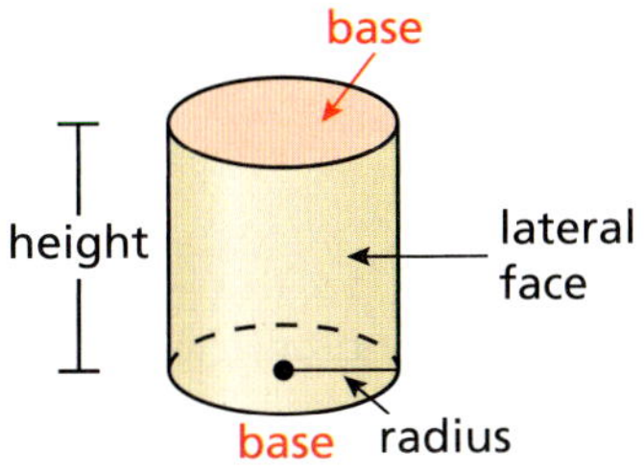

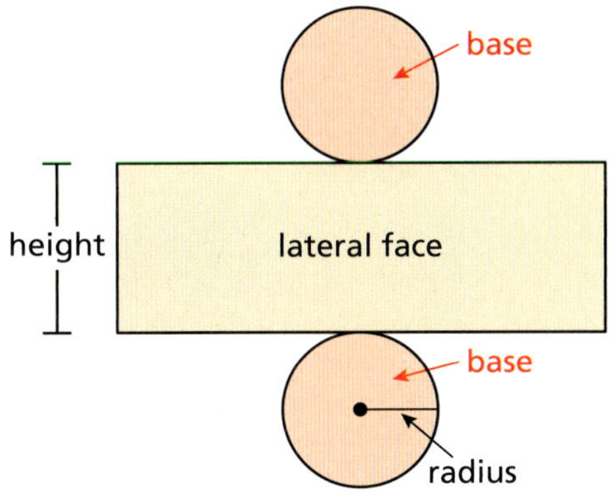

The surface area of a cylinder is equal to the sum of the base areas and the lateral area.

5 MTR USE STRUCTURE
How is calculating the surface area of a right cylinder similar to calculating the surface area of a right prism?

KEY IDEA

Surface Area of a Right Cylinder

The surface area S of a right cylinder is

$$S = 2B + L = 2\pi r^2 + 2\pi rh$$

where B is the area of a base, L is the lateral area, r is the radius, and h is the height.

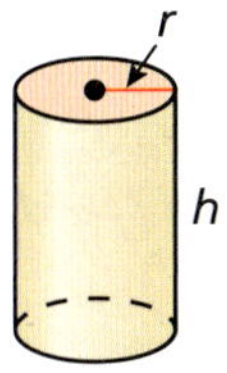

EXAMPLE 2 Finding the Surface Area of a Right Cylinder

WATCH

Find the surface area of the right cylinder.

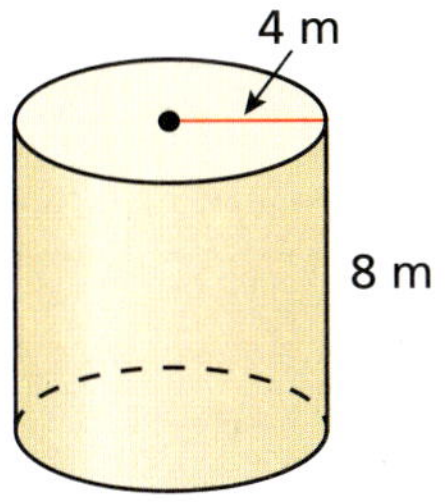

SOLUTION

$S = 2\pi r^2 + 2\pi rh$	Write formula for surface area of a right cylinder.
$= 2\pi(4)^2 + 64\pi$	Substitute.
$= 96\pi$	Simplify.
≈ 301.59	Use technology.

▶ The surface area is about 301.59 square meters.

SELF-ASSESSMENT

1 I don't understand yet. 2 I can do it with help. 3 I can do it on my own. 4 I can teach someone else.

Find the surface area of the right cylinder.

2.

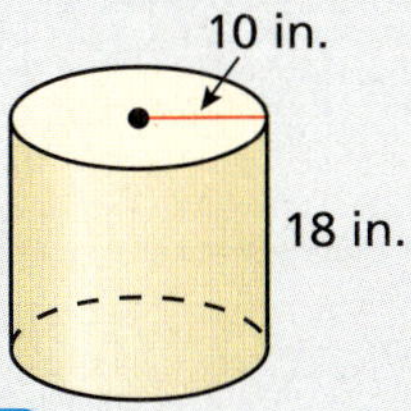

3.

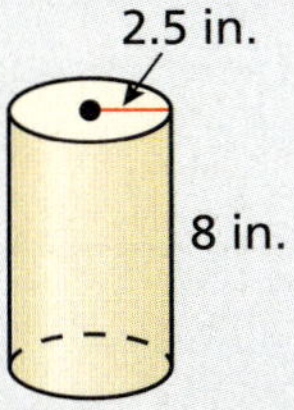

4.

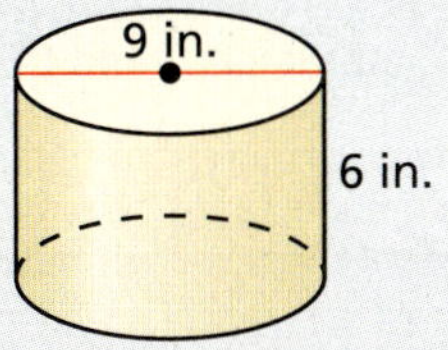

GO DIGITAL

Using Surface Areas of Prisms and Cylinders

7 MTR **EXAMPLE 3** **Modeling Real Life** WATCH

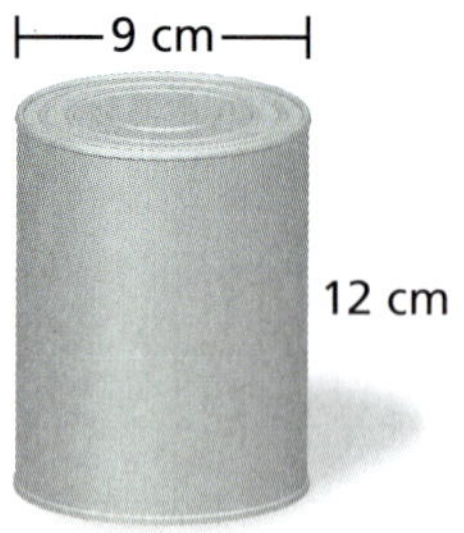

You are designing a label for the cylindrical soup can shown. The label will cover the lateral area of the can. Find the minimum amount of material needed for the label.

SOLUTION

Use the radius $r = \frac{9}{2} = 4.5$ centimeters to find the lateral area.

$L = 2\pi rh$	Definition of lateral area of a cylinder
$= 2\pi(4.5)(12)$	Substitute.
$= 108\pi$	Simplify.
≈ 339.29	Use technology.

▶ The minimum amount of material needed is about 339.29 square centimeters.

EXAMPLE 4 **Finding the Surface Area of a Composite Solid**

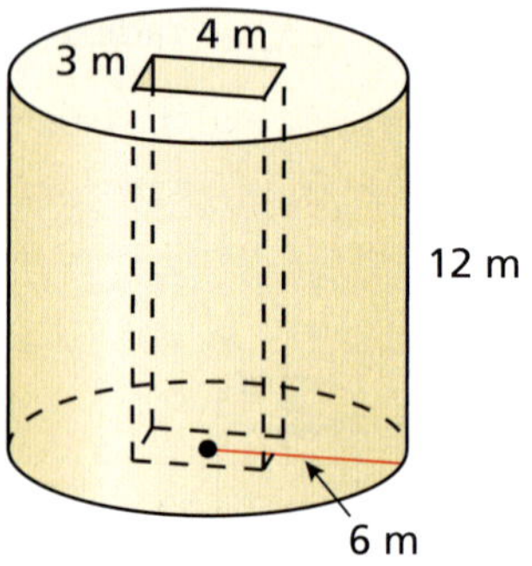

Find the surface area of the *composite solid.* WATCH

SOLUTION

Calculate the lateral area and the base areas of the solid. Then find the sum.

Lateral area of solid = Lateral area of cylinder + Lateral area of prism

$= 2\pi rh + Ph$

$= 2\pi(6)(12) + 14(12)$

$= 144\pi + 168$

Base areas of solids $= 2 \cdot$ (Area of a base of the cylinder − Area of a base of the prism)

$= 2(\pi r^2 - \ell w)$

$= 2[\pi(6)^2 - 4(3)]$

$= 72\pi - 24$

Surface area of solid = Lateral area of solid + Base areas of solid

$= (144\pi + 168) + (72\pi - 24)$

$= 216\pi + 144$

▶ So, the surface area is $216\pi + 144$, or about 822.58 square meters.

SELF-ASSESSMENT 1 I don't understand yet. 2 I can do it with help. 3 I can do it on my own. 4 I can teach someone else.

5. **WHAT IF?** In Example 3, you change the design of the can so that the diameter is 12 centimeters. Find the minimum amount of material needed for the label.

6. Find the surface area of the composite solid.

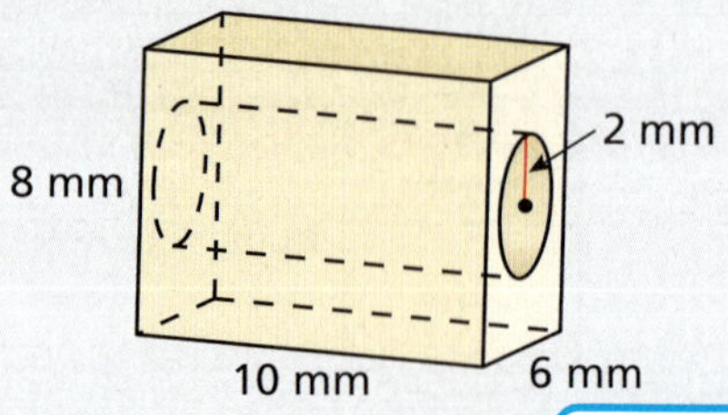

GO DIGITAL

Finding Surface Areas of Similar Solids

5 MTR RELATE CONCEPTS

Explain how your previous understanding of dilations and scale factors in two dimensions is related to finding surface areas of similar solids.

KEY IDEA

Surface Areas of Similar Solids

Two solids of the same type with equal ratios of corresponding linear measures, such as heights or radii, are called **similar solids**. The ratio of the corresponding linear measures of two similar solids is called the *scale factor*.

If two similar solids have a scale factor of k, then the ratio of their surface areas is equal to k^2.

EXAMPLE 5 Finding the Surface Area of a Similar Solid

WATCH

Prism A and prism B are similar. Find the surface area of prism B.

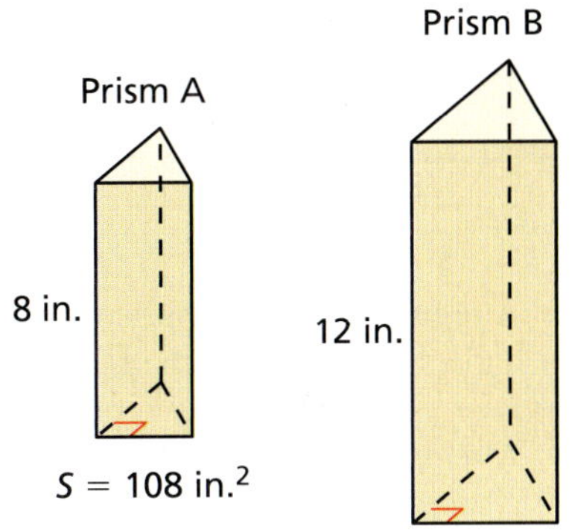

SOLUTION

The scale factor is $k = \dfrac{\text{Height of prism B}}{\text{Height of prism A}}$

$= \dfrac{12}{8}$

$= \dfrac{3}{2}$

Use the scale factor to find the surface area of prism B.

$\dfrac{\text{Surface area of prism B}}{\text{Surface area of prism A}} = k^2$ — The ratio of the surface areas is k^2.

$\dfrac{\text{Surface area of prism B}}{108} = \left(\dfrac{3}{2}\right)^2$ — Substitute.

$\dfrac{\text{Surface area of prism B}}{108} = \dfrac{9}{4}$ — Evaluate power.

$\text{Surface area of prism B} = 243$ — Multiply each side by 108.

▶ The surface area of prism B is 243 square inches.

COMMON ERROR

Be sure to write the ratio of the surface areas in the same order you wrote the ratio of the heights.

SELF-ASSESSMENT

1 I don't understand yet. 2 I can do it with help. 3 I can do it on my own. 4 I can teach someone else.

7. Cylinder C and cylinder D are similar. Find the surface area of cylinder D.

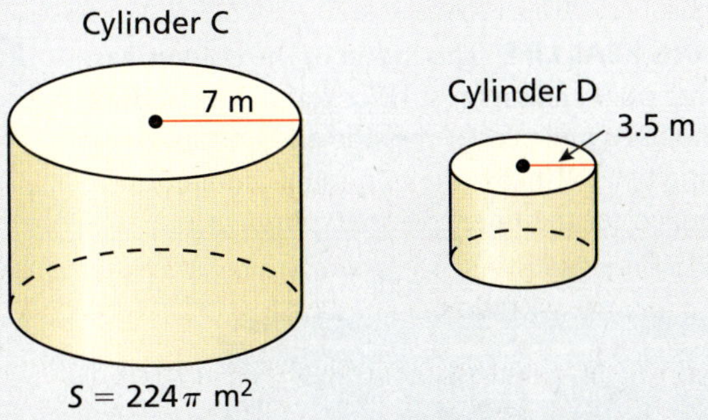

8. **DIG DEEPER** Cylinder M and cylinder N are similar. The radius of cylinder N is equal to its height, and the ratio of the height of cylinder N to the height of cylinder M is 5 : 3. The surface area of cylinder N is 256π square feet greater than the surface area of cylinder M. Find the surface area of each cylinder.

GO DIGITAL

12.2 Practice WITH CalcChat® AND CalcView®

In Exercises 1 and 2, find the surface area of the solid formed by the net.

1.

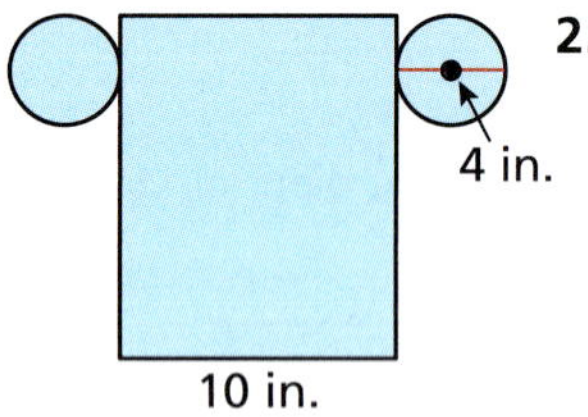

2.

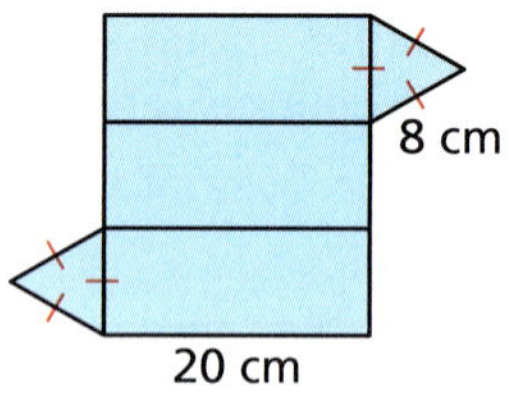

In Exercises 3–6, find the surface area of the right prism. *(See Example 1.)*

3.

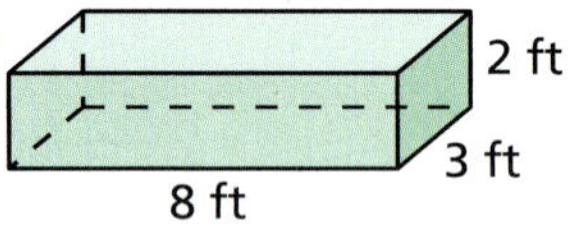

4.

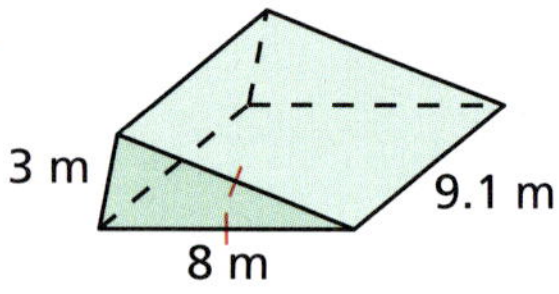

5. A regular pentagonal prism has a height of 3.5 inches and a base edge length of 2 inches.

6. A regular hexagonal prism has a height of 80 feet and a base edge length of 40 feet.

In Exercises 7–10, find the surface area of the right cylinder. *(See Example 2.)*

7.

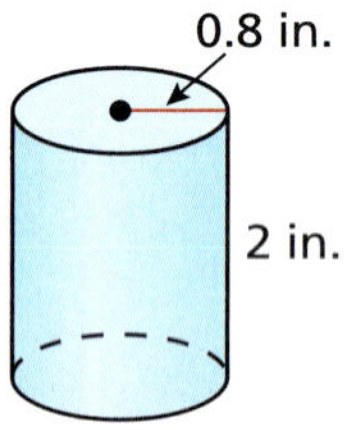

8.

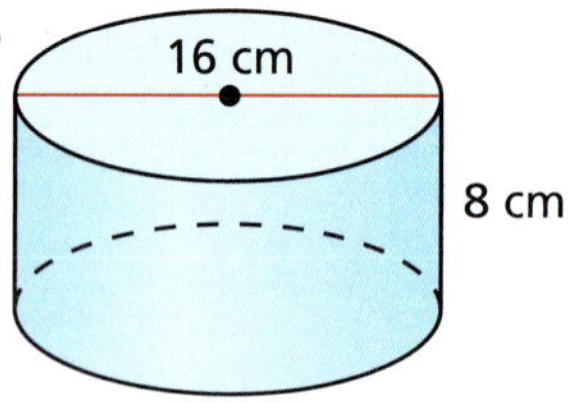

9. A right cylinder has a diameter of 24 millimeters and a height of 40 millimeters.

10. A right cylinder has a radius of 2.5 feet and a height of 7.5 feet.

7 MTR **11.** **MODELING REAL LIFE** The inside of the cylindrical swimming pool shown must be covered with a vinyl liner. The liner must cover the side and bottom of the swimming pool. What is the minimum amount of vinyl needed for the liner? *(See Example 3.)*

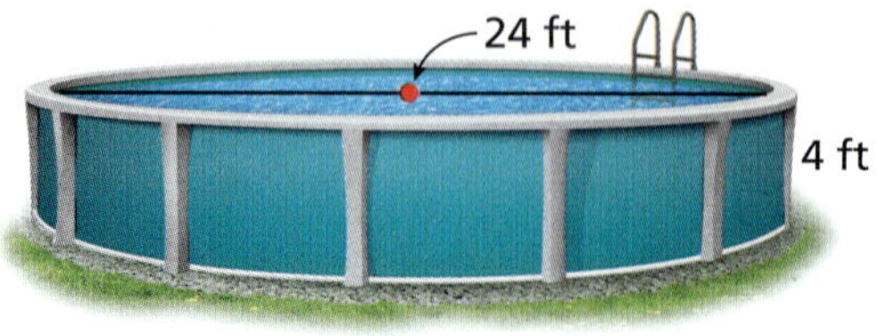

7 MTR **12.** **MODELING REAL LIFE** The tent shown has fabric covering all four sides and the floor. What is the minimum amount of fabric needed to construct the tent?

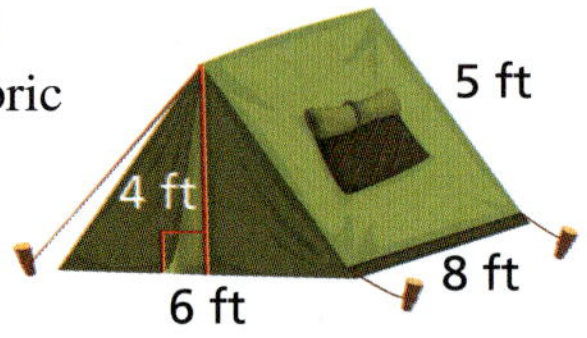

In Exercises 13–16, find the surface area of the composite solid. *(See Example 4.)*

13.

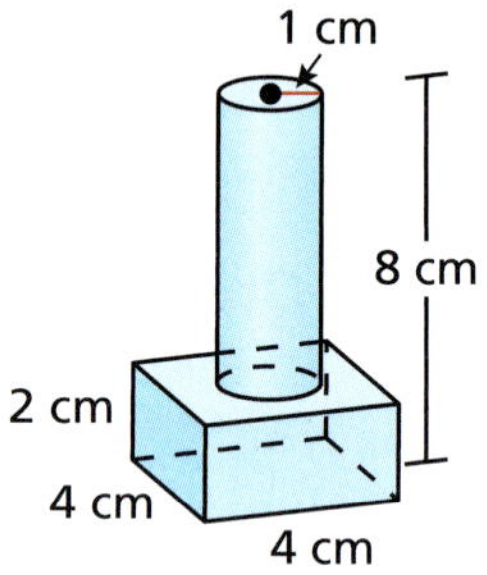

14.

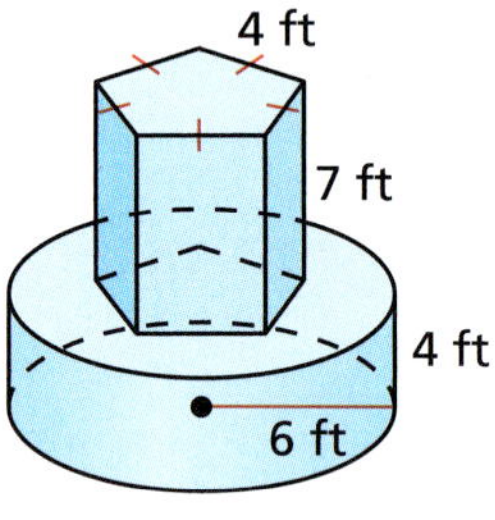

15.

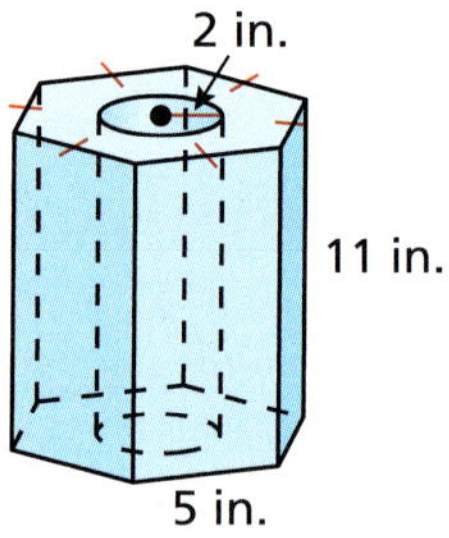

16.

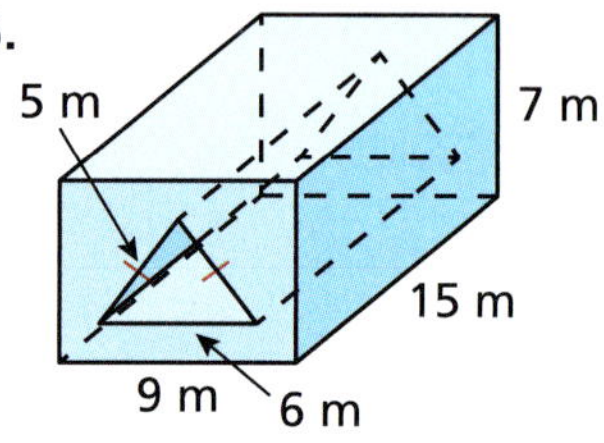

4 MTR **17.** **ERROR ANALYSIS** Describe and correct the error in finding the surface area of the right cylinder.

$S = 2\pi(6)^2 + 2\pi(6)(8)$

$= 168\pi$

$\approx 527.79 \text{ cm}^2$

4 MTR **18.** **ERROR ANALYSIS** Describe and correct the error in finding the surface area of the composite solid.

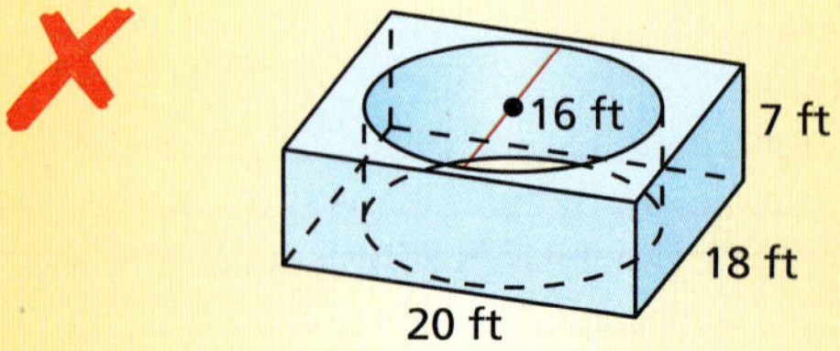

$S = 2(20)(7) + 2(18)(7) + 2\pi(8)(7)$

$+ 2[(18)(20) + \pi(8)^2]$

$\approx 2005.98 \text{ ft}^2$

GO DIGITAL

In Exercises 19 and 20, solids A and B are similar. Use the given information and scale factor *k* from solid A to solid B to find the surface area of solid B.

19. Cylinder A has a surface area of 347π square centimeters and $k = 4$.

20. Prism A has a surface area of 805 square feet and $k = \frac{1}{5}$.

In Exercises 21 and 22, the solids are similar. Find the surface area of solid B. *(See Example 5.)*

21.

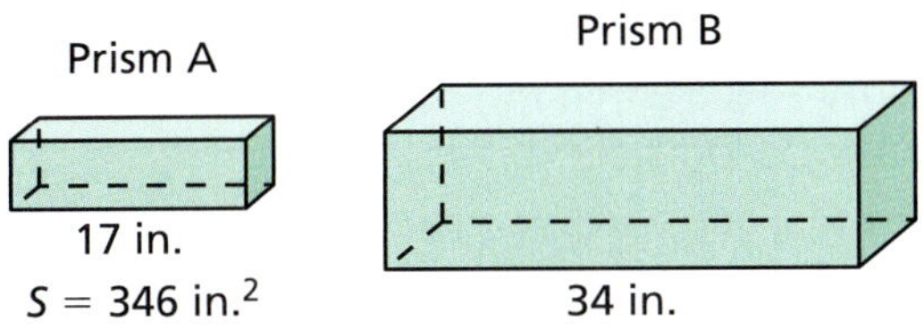

22.

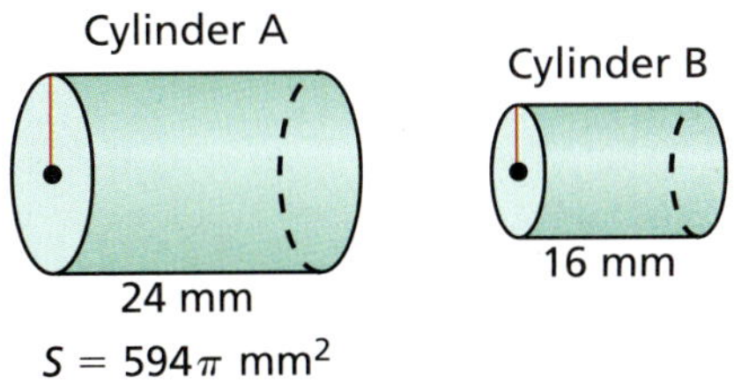

In Exercises 23 and 24, describe how the change affects the surface area of the right prism or right cylinder.

23. tripling the radius

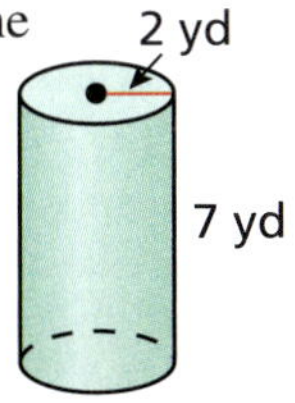

24. multiplying the base edge lengths by $\frac{1}{4}$ and the height by 4

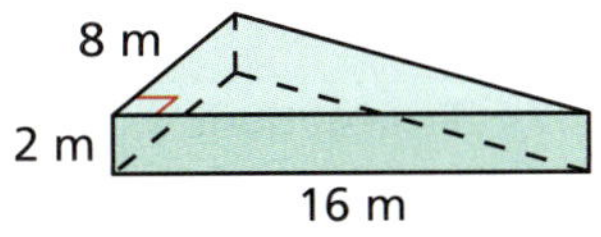

In Exercises 25 and 26, find the height of the right prism or right cylinder.

25. $S = 1097 \text{ m}^2$

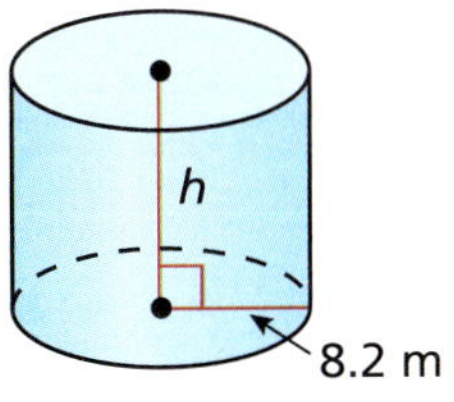

26. $S = 480 \text{ in.}^2$

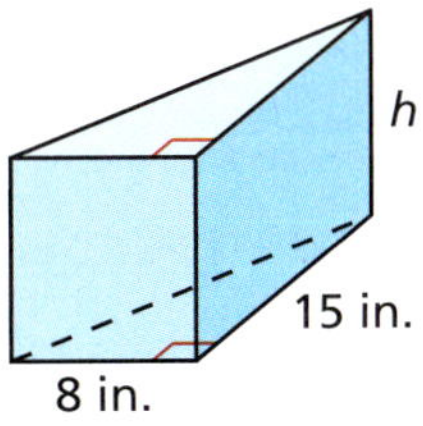

27. **STRUCTURE** A cube has a surface area of 343 square inches. Write and solve an equation to find the length of each edge of the cube.

28. **STRUCTURE** A right cylinder has a surface area of 108 square meters. The radius of the cylinder is twice its height. Write and solve an equation to find the height of the cylinder.

GO DIGITAL

29. **MODELING REAL LIFE** A company makes two types of recycling bins, as shown. Both types of bins have an open top. Which recycling bin requires more material to make? Explain.

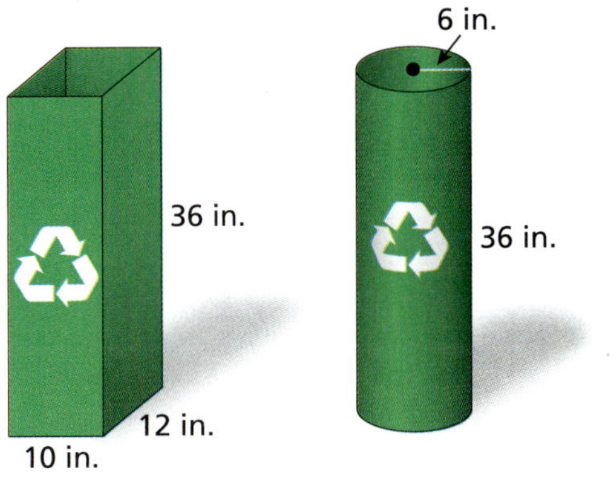

30. **MODELING REAL LIFE** You are painting a rectangular room that is 13 feet long, 9 feet wide, and 8.5 feet high. On one wall is a window that is 2.5 feet wide and 5 feet high. On another wall is a door that is 4 feet wide and 7 feet high. A gallon of paint covers 350 square feet. How many gallons of paint do you need to cover the four walls with one coat of paint, not including the window and door?

31. **OPEN-ENDED** Sketch two solids. Find the surface area of each solid. Use your solids to build a composite solid. How does the surface area of the composite solid differ from the surface area of the individual solids? Justify your answer.

32. **HOW DO YOU SEE IT?**
An open gift box is shown.

a. Why is the area of the net of the box larger than the minimum amount of wrapping paper needed to cover the closed box?

b. When wrapping the box, why would you want to use more than the minimum amount of paper needed?

33. **STRUCTURE** The right triangular prisms shown have the same surface area. Find the height *h* of prism B.

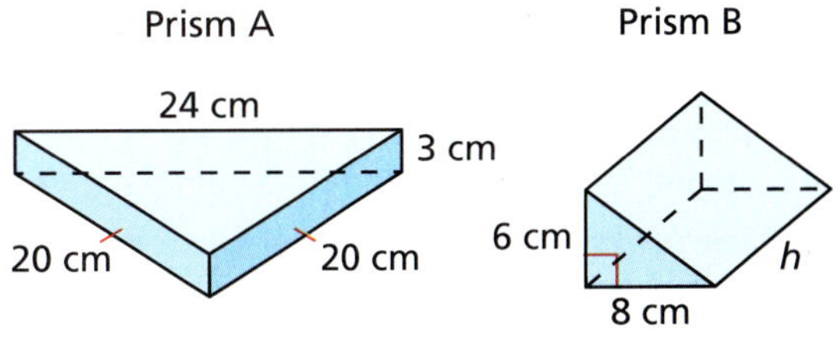

34. REASONING Consider a cube that is built using 27 unit cubes, as shown.

a. Find the surface area of the solid formed when the red unit cubes are removed from the solid shown.

b. Find the surface area of the solid formed when the blue unit cubes are removed from the solid shown.

c. Explain why your answers are different in parts (a) and (b).

35. WRITING Use the diagram to write a formula that can be used to find the surface area S of any cylindrical ring where $0 < r_2 < r_1$.

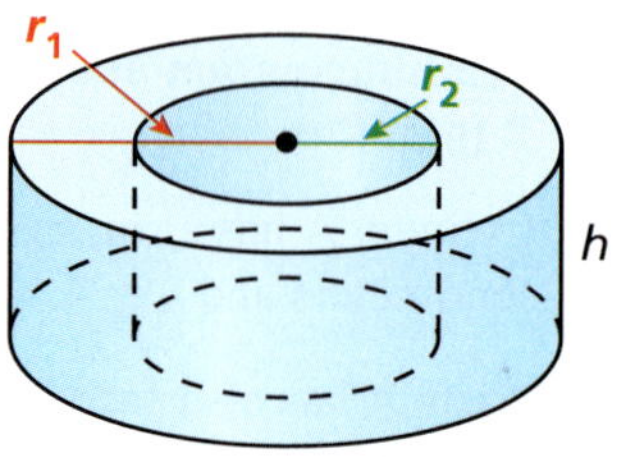

36. THOUGHT PROVOKING
You have 24 cube-shaped building blocks with edge lengths of 1 unit. What arrangement of blocks gives you a rectangular prism with the least surface area? Justify your answer.

REVIEW & REFRESH

WATCH

37. Find the area of the regular polygon.

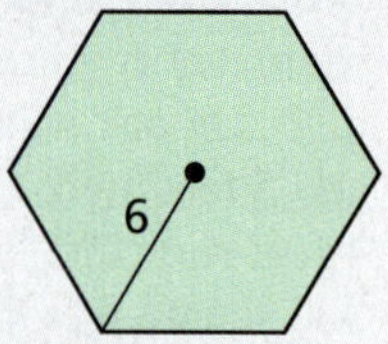

In Exercises 38–41, tell whether the solid is a polyhedron. If it is, name the polyhedron.

38.

39.

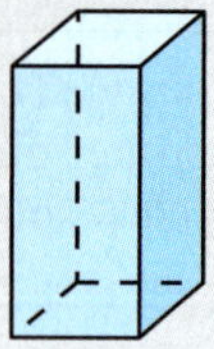

40.

41.

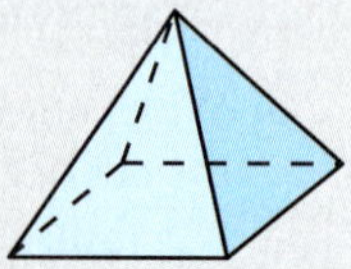

7 MTR **42. MODELING REAL LIFE** Everglades National Park has an area of about 1.51 million acres. The table shows the numbers of several types of animals in the park. Find the population density in animals per acre for each animal.

Animal	Panther	Alligator	Burmese Python	Wood Stork
Population	85	200,000	100,000	2000

In Exercises 43 and 44, sketch the figure described.

43. plane T and line p intersecting at all points on line p

44. plane H and plane U intersecting at $\overleftrightarrow{JC}$

In Exercises 45 and 46, find the surface area of the right prism or cylinder.

45.

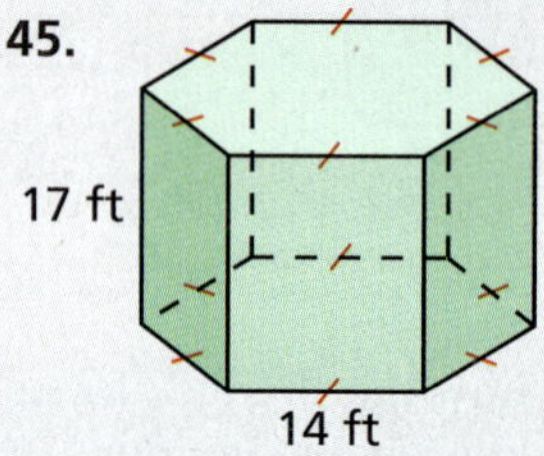

46.

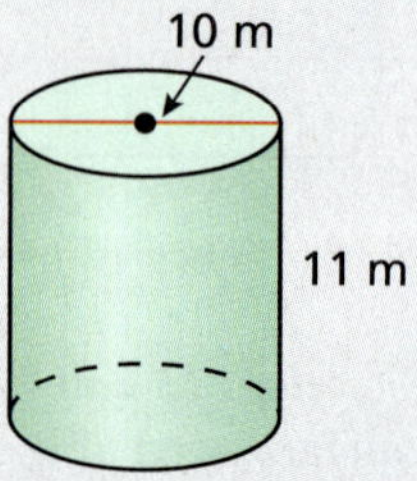

In Exercises 47 and 48, graph quadrilateral *PQRS* with vertices *P*(1, −1), *Q*(4, −2), *R*(3, −4), and *S*(1, −3) and its image after the composition.

47. **Translation:** $(x, y) \rightarrow (x - 3, y + 4)$
Rotation: 90° counterclockwise about the origin

48. **Dilation:** $(x, y) \rightarrow \left(\frac{1}{4}x, \frac{1}{4}y\right)$
Reflection: in the y-axis

49. Find the measure of each angle.

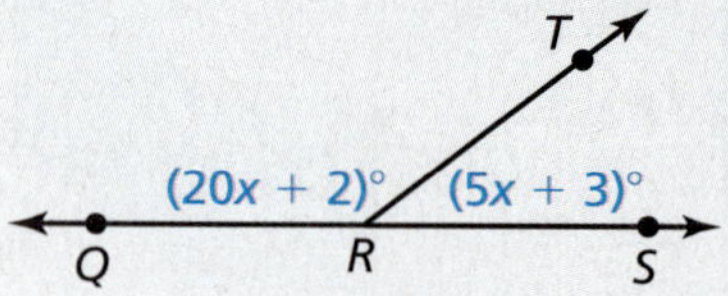

50. Find the geometric mean of 6 and 54.

51. Find the value of x.

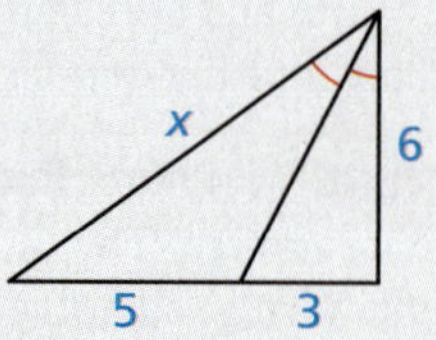

52. Triangle ABC has vertices $A(4, 6)$, $B(8, 8)$, and $C(8, 2)$. Find the circumcenter.

GO DIGITAL

12.3 Surface Areas of Pyramids and Cones

Learning Target: Find and use surface areas of pyramids and cones.

Success Criteria:
- I can find surface areas of pyramids and cones.
- I can find surface areas of similar pyramids and similar cones.
- I can find surface areas of composite figures involving pyramids and cones.

EXPLORE IT! Finding Surface Areas of Pyramids and Cones

Work with a partner. In a *regular pyramid*, the base is a regular polygon and the lateral faces are congruent triangles.

a. Use a net to write an expression for the surface area of the regular triangular pyramid. Let B represent the area of the base.

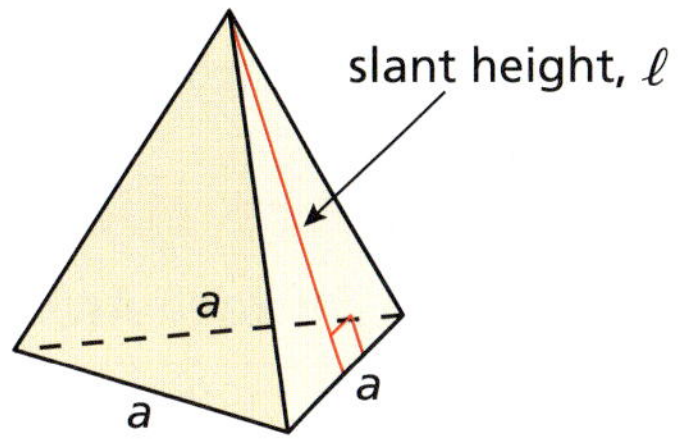

b. Rewrite your expression in part (a) in terms of the perimeter P of the base of the pyramid. Explain your reasoning.

c. Can you use your expression in part (b) to represent the surface area of a square pyramid? a regular pentagonal pyramid? a regular hexagonal pyramid? Explain.

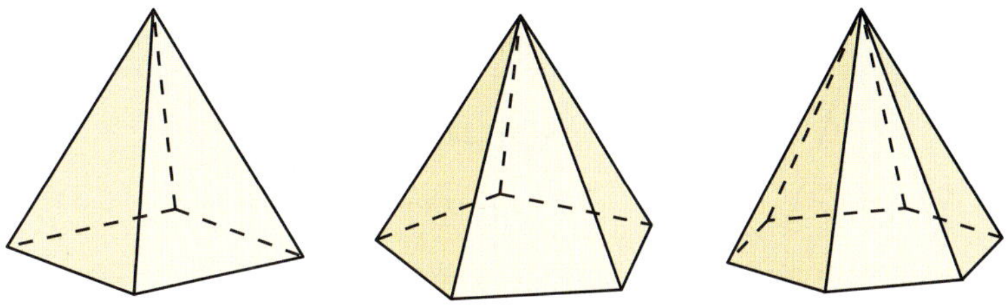

d. Can you use your expression in part (b) to represent the surface area of a right cone? Explain your reasoning.

e. Write an expression for the surface area of a right cone in terms of the radius r and the slant height ℓ. Explain your reasoning.

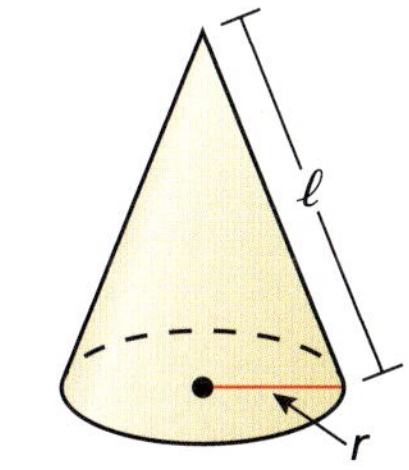

6 MTR ASSESS REASONABLENESS

In part (c), what does a pyramid begin to look like if you continue to increase the number of sides of the base? How does this show that your answer in part (d) is reasonable?

Geometric Reasoning

MA.912.GR.4.3 Extend previous understanding of scale drawings and scale factors to determine how dilations affect the area of two-dimensional figures and the surface area or volume of three-dimensional figures.

MA.912.GR.4.6 Solve mathematical and real-world problems involving the surface area of three-dimensional figures limited to cylinders, pyramids, prisms, cones and spheres.

GO DIGITAL

Finding Surface Areas of Regular Pyramids

Vocabulary

vertex of a pyramid, *p. 632*
slant height of a regular pyramid, *p. 632*
lateral face of a cone, *p. 633*

A *pyramid* is a polyhedron in which the base is a polygon and the lateral faces are triangles with a common vertex, called the **vertex of the pyramid**. The intersection of two lateral faces is a *lateral edge*. The intersection of the base and a lateral face is a *base edge*. The height of a pyramid is the perpendicular distance between the base and the vertex. In a *regular pyramid*, the base is a regular polygon and the segment joining the vertex and the center of the base is perpendicular to the base.

The lateral faces of a regular pyramid are congruent isosceles triangles. The **slant height of a regular pyramid** is the height of each of its lateral faces.

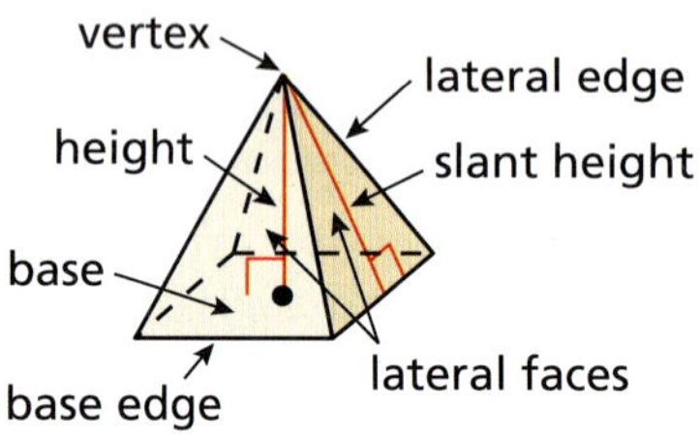

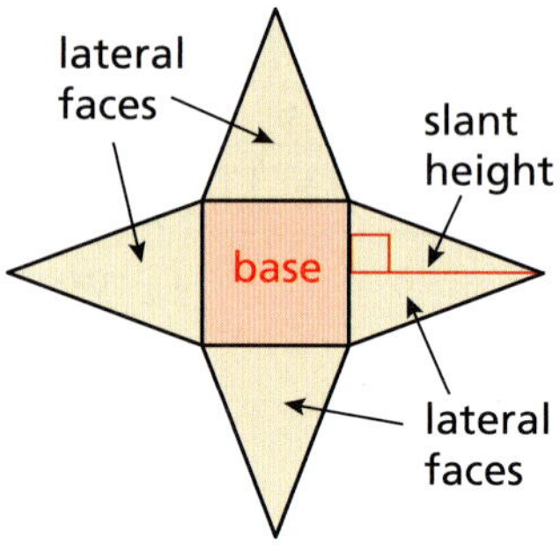

The surface area of a pyramid is equal to the sum of the base area and the lateral area.

1 MTR HELP A CLASSMATE

Help a classmate understand why $L = \frac{1}{2}P\ell$.

KEY IDEA

Surface Area of a Regular Pyramid

The surface area S of a regular pyramid is

$$S = B + L = B + \tfrac{1}{2}P\ell$$

where B is the area of the base, L is the lateral area, P is the perimeter of the base, and ℓ is the slant height.

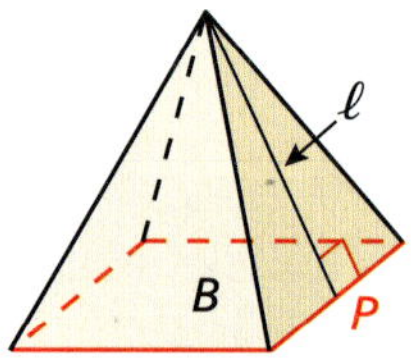

EXAMPLE 1 Finding the Surface Area of a Regular Pyramid

Find the surface area of the regular pentagonal pyramid.

WATCH

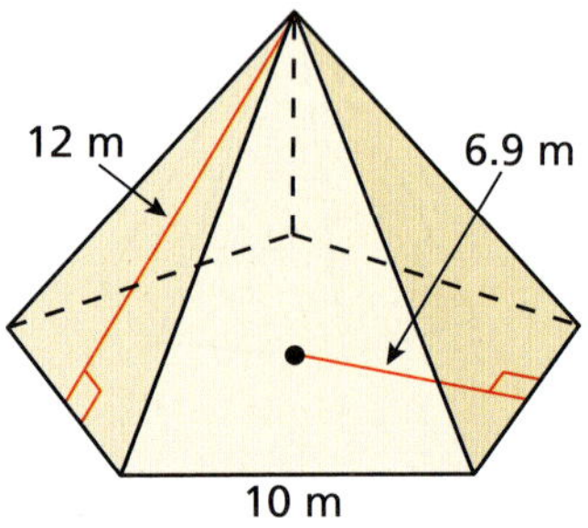

SOLUTION

The perimeter P of the base is $5 \cdot 10 = 50$ meters, and the apothem a is 6.9 meters.

$S = B + \frac{1}{2}P\ell$ — Write formula for surface area of a regular pyramid.

$= \frac{1}{2}(6.9)(50) + \frac{1}{2}(50)(12)$ — Substitute.

$= 172.5 + 300$ — Simplify.

$= 472.5$ — Use technology.

▶ The surface area is 472.5 square meters.

SELF-ASSESSMENT 1 I don't understand yet. 2 I can do it with help. 3 I can do it on my own. 4 I can teach someone else.

1. Find the surface area of the regular hexagonal pyramid.

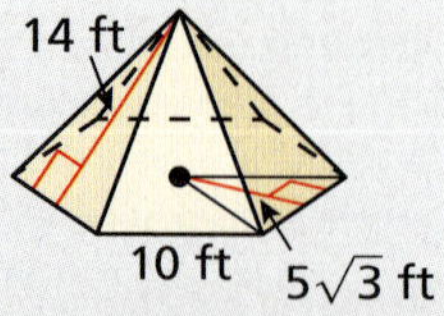

GO DIGITAL

Finding Surface Areas of Right Cones

A *circular cone*, or *cone*, has a circular *base* and a *vertex* that is not in the same plane as the base. The *altitude*, or *height*, is the perpendicular distance between the vertex and the base. In a *right cone*, the height meets the base at its center, and the *slant height* is the distance between the vertex and a point on the base edge.

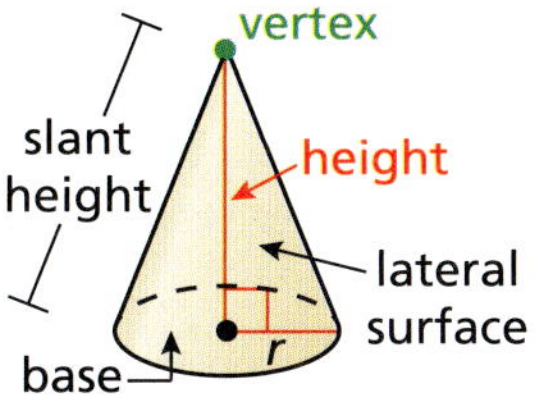

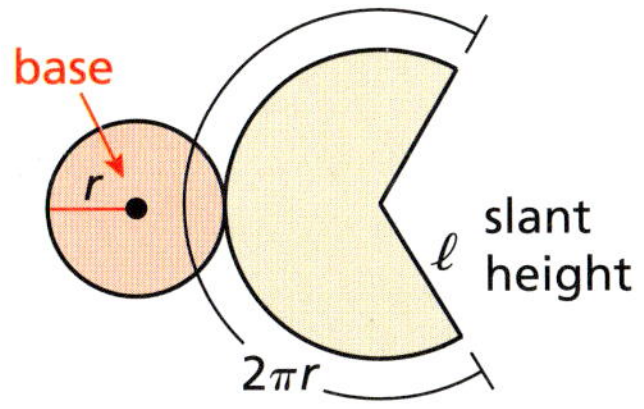

The **lateral face of a cone** consists of all segments that connect the vertex with points on the base edge. When you cut along the slant height and lay the right cone flat, you get the net shown at the left. In the net, the circular base has an area of πr^2 and the lateral face is a sector of a circle. You can find the area of this sector by using a proportion, as shown below.

$$\frac{\text{Area of sector}}{\text{Area of circle}} = \frac{\text{Arc length}}{\text{Circumference of circle}}$$ Set up proportion.

$$\frac{\text{Area of sector}}{\pi \ell^2} = \frac{2\pi r}{2\pi \ell}$$ Substitute.

$$\text{Area of sector} = \pi \ell^2 \cdot \frac{2\pi r}{2\pi \ell}$$ Multiply each side by $\pi \ell^2$.

$$\text{Area of sector} = \pi r \ell$$ Simplify.

The surface area of a right cone is the sum of the base area and the lateral area, $\pi r \ell$.

KEY IDEA

Surface Area of a Right Cone

The surface area S of a right cone is

$$S = B + L = \pi r^2 + \pi r \ell$$

where B is the area of the base, L is the lateral area, r is the radius of the base, and ℓ is the slant height.

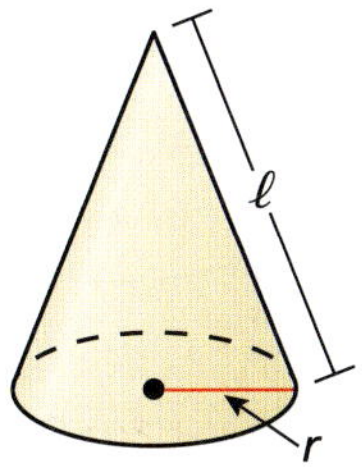

EXAMPLE 2 Finding the Surface Area of a Right Cone

WATCH

Find the surface area of the right cone.

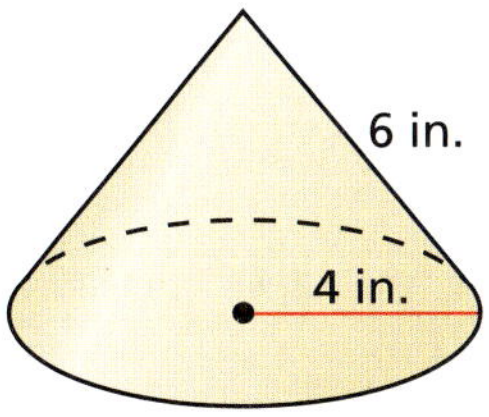

SOLUTION

$S = \pi r^2 + \pi r \ell$	Write formula for surface area of a cone.
$= \pi \cdot 4^2 + \pi(4)(6)$	Substitute.
$= 40\pi$	Simplify.
≈ 125.66	Use technology.

▶ The surface area is about 125.66 square inches.

SELF-ASSESSMENT

1 I don't understand yet. 2 I can do it with help. 3 I can do it on my own. 4 I can teach someone else.

2. Find the surface area of the right cone.

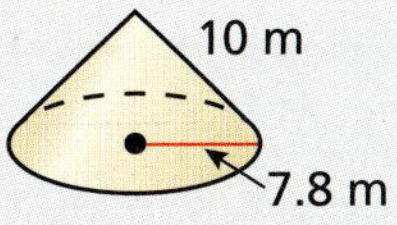

GO DIGITAL

Finding Surface Areas of Similar Solids and Composites

EXAMPLE 3 Finding the Surface Area of a Similar Solid

WATCH

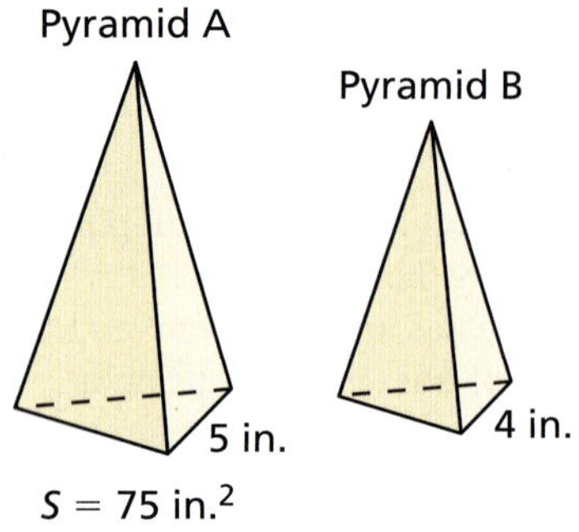

Pyramid A and pyramid B are similar. Find the surface area of pyramid B.

SOLUTION

The scale factor is $k = \dfrac{\text{Edge length of pyramid B}}{\text{Edge length of pyramid A}}$

$= \dfrac{4}{5}$

Use the scale factor to find the surface area of pyramid B.

$\dfrac{\text{Surface area of pyramid B}}{\text{Surface area of pyramid A}} = k^2$ — The ratio of the surface area is k^2.

$\dfrac{\text{Surface area of pyramid B}}{75} = \left(\dfrac{4}{5}\right)^2$ — Substitute.

Surface area of pyramid B $= 48$ — Solve.

▶ The surface area of pyramid B is 48 square inches.

EXAMPLE 4 Finding the Surface Area of a Composite Solid

WATCH

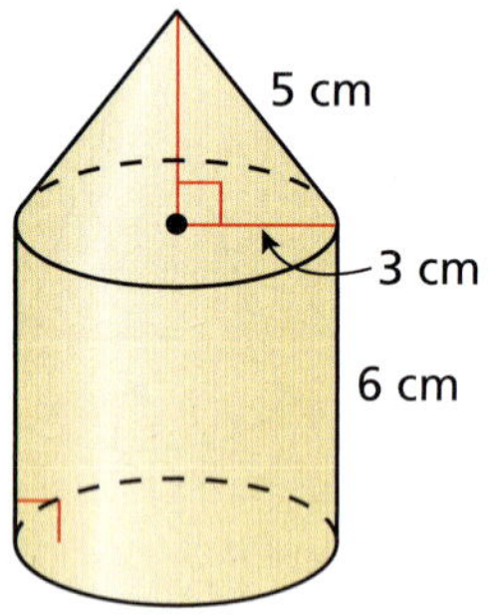

Find the surface area of the composite solid.

SOLUTION

Find the sum of the lateral areas of each solid and one base of the cylinder.

Surface area of solid = Lateral area of cone + Lateral area of cylinder + Area of a base of the cylinder

$= \pi r \ell + 2\pi rh + \pi r^2$

$= \pi(3)(5) + 2\pi(3)(6) + \pi(3)^2$

$= 15\pi + 36\pi + 9\pi$

$= 60\pi$

≈ 188.50

▶ The surface area is about 188.50 square centimeters.

SELF-ASSESSMENT 1 I don't understand yet. 2 I can do it with help. 3 I can do it on my own. 4 I can teach someone else.

3. Cone C and cone D are similar. Find the surface area of cone C.

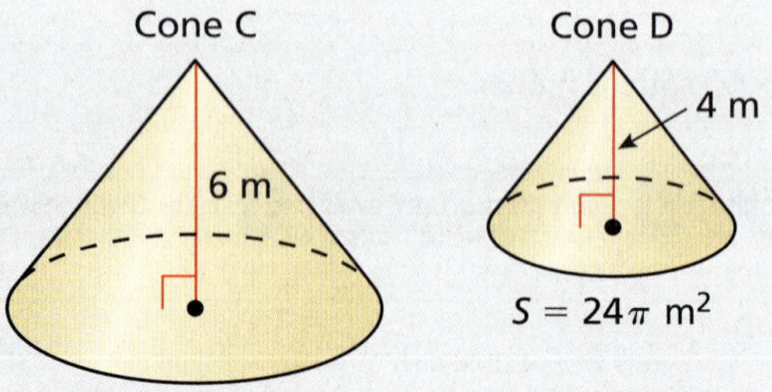

4. Find the surface area of the composite solid.

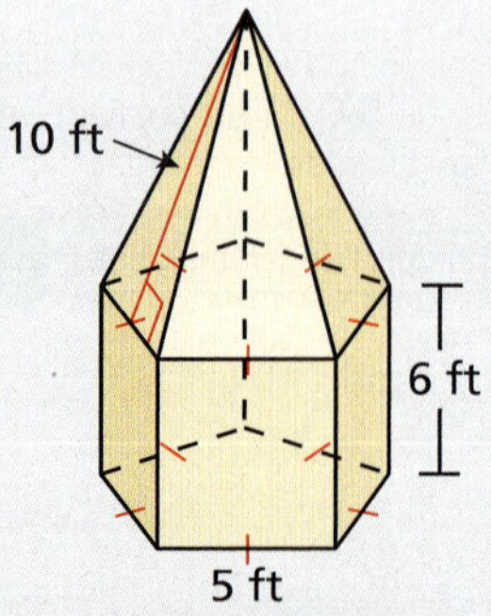

GO DIGITAL

12.3 Practice WITH CalcChat® AND CalcView®

In Exercises 1–4, find the surface area of the regular pyramid. *(See Example 1.)*

1.

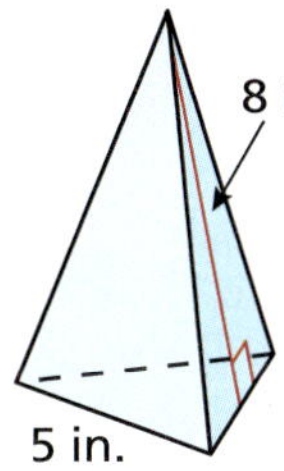

2.

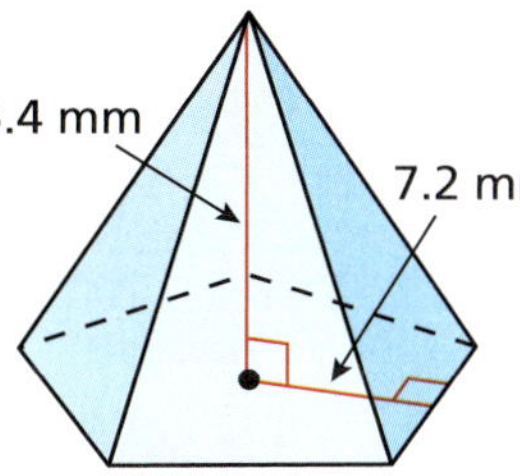

3. A square pyramid has a height of 21 feet and a base edge length of 40 feet.

4. A regular hexagonal pyramid has a slant height of 15 centimeters and a base edge length of 8 centimeters.

In Exercises 5–8, find the surface area of the right cone. *(See Example 2.)*

5.

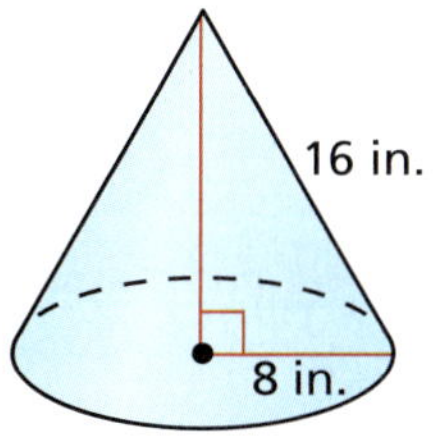

6.

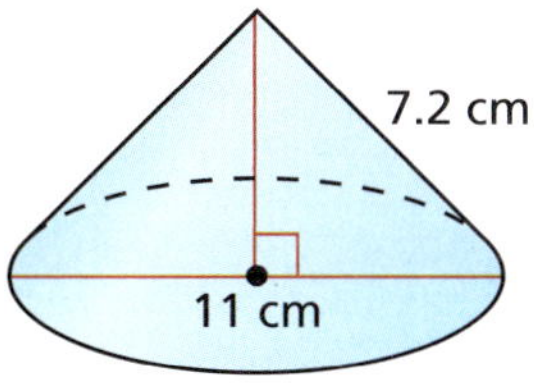

7. A right cone has a radius of 9 inches and a height of 12 inches.

8. A right cone has a diameter of 11.2 feet and a height of 9.2 feet.

9. **ERROR ANALYSIS** Describe and correct the error in finding the surface area of the regular pyramid.

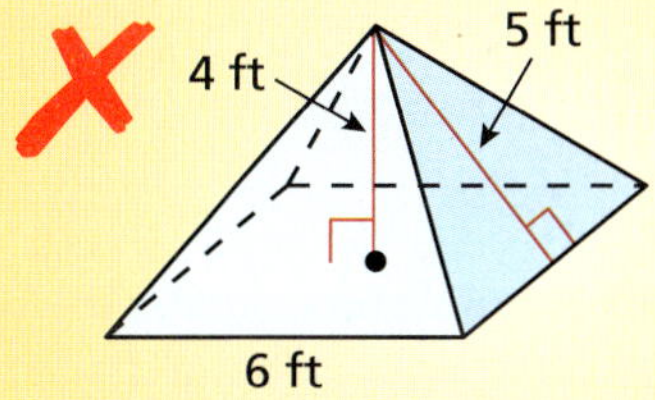

$$S = B + \frac{1}{2}P\ell$$
$$= 6^2 + \frac{1}{2}(24)(4)$$
$$= 84 \text{ ft}^2$$

10. **ERROR ANALYSIS** Describe and correct the error in finding the surface area of the right cone.

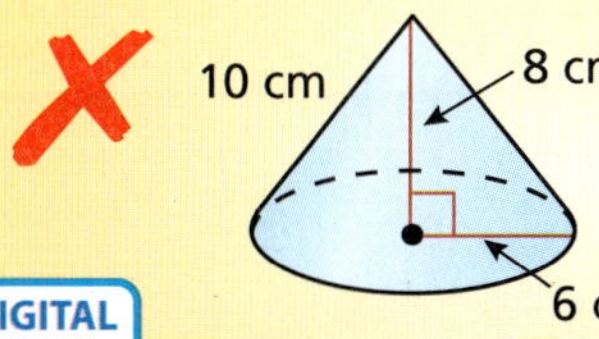

$$S = \pi r^2 + \pi r^2 \ell$$
$$= \pi(6)^2 + \pi(6)^2(10)$$
$$= 396\pi \text{ cm}^2$$

In Exercises 11 and 12, solids A and B are similar. Use the given information and scale factor k from solid A to solid B to find the surface area of solid B.

11. The surface area of pyramid A is 72 square centimeters and $k = \frac{2}{3}$.

12. The surface area of cone A is 150π square meters and $k = \frac{5}{4}$.

In Exercises 13 and 14, the solids are similar. Find the surface area of solid B. *(See Example 3.)*

13.

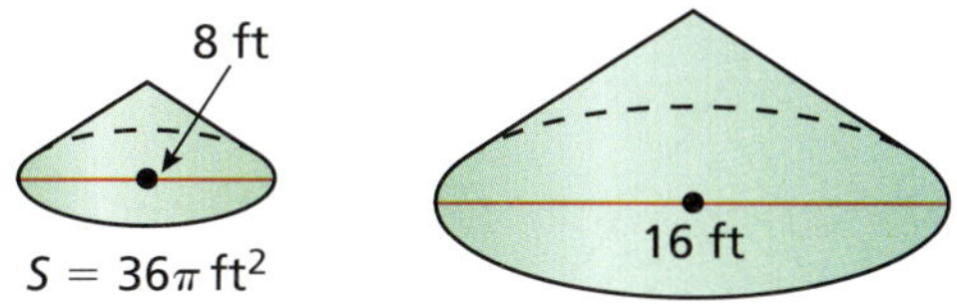

14.

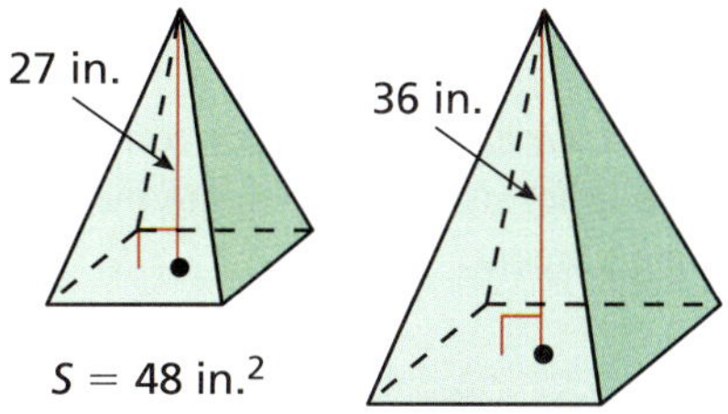

In Exercises 15–18, find the surface area of the composite solid. *(See Example 4.)*

15.

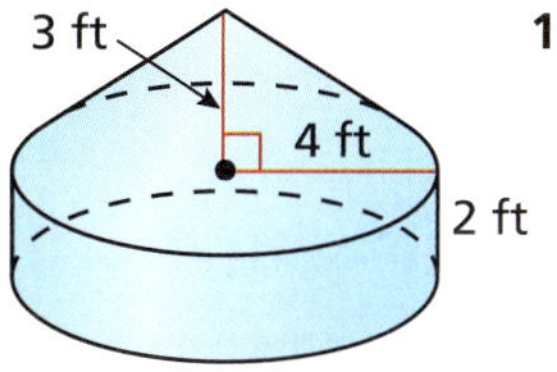

16.

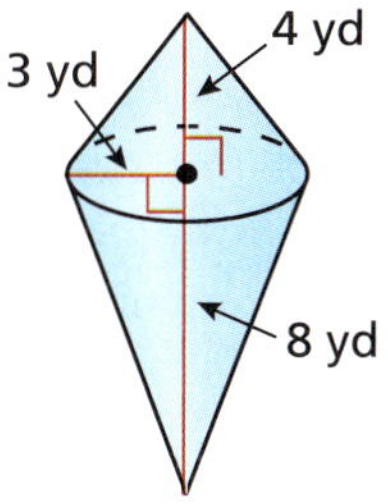

17.

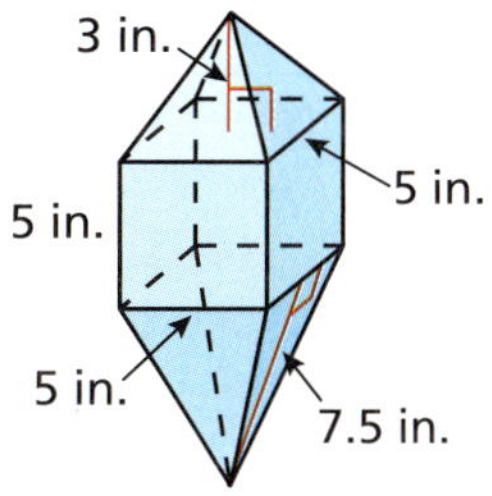

18.

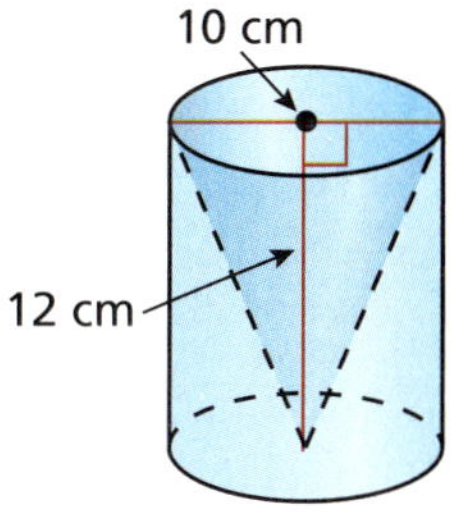

19. **STRUCTURE** A right cone with a radius of 4 inches and a square pyramid both have a slant height of 5 inches. Both solids have the same surface area. Find the length of a base edge of the pyramid.

GO DIGITAL

20. HOW DO YOU SEE IT?
Name the figure that is represented by each net. Justify your answer.

a.

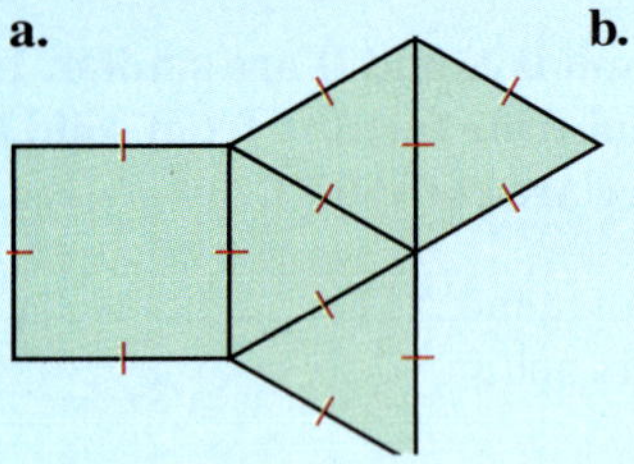

b.

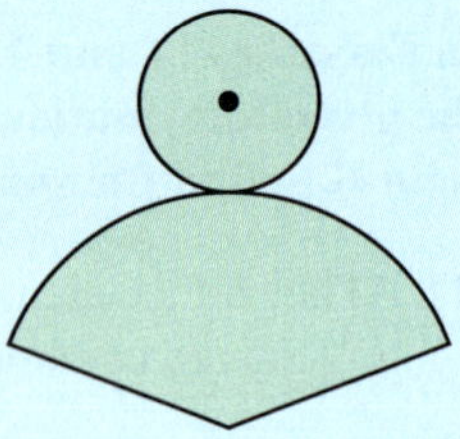

21. STRUCTURE In the figure, $AC = 4$, $AB = 3$, and $DC = 2$.

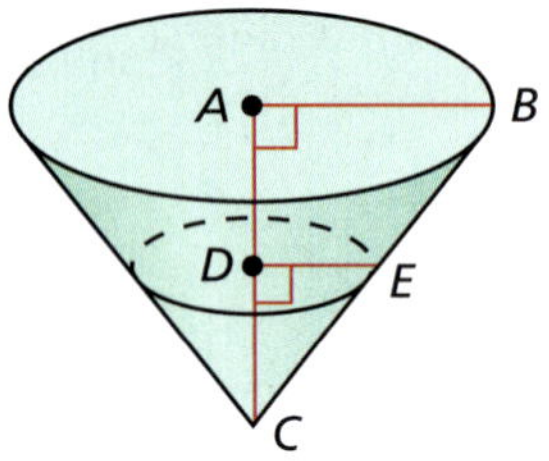

a. Prove $\triangle ABC \sim DEC$.

b. Find BC, DE, and EC.

c. Find the surface areas of the larger cone and the smaller cone in terms of π. Compare the surface areas using a percent.

22. REASONING A regular hexagonal pyramid with a base edge of 9 feet and a height of 12 feet is inscribed in a right cone. Find the lateral area of the cone.

23. REASONING An *Elizabethan collar* is used to prevent an animal from irritating a wound. The angle between the opening with a 16-inch diameter and the side of the collar is 53°. Find the surface area of the collar shown.

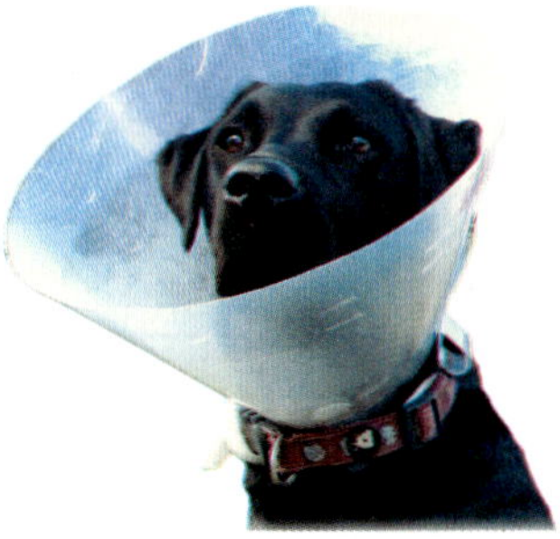

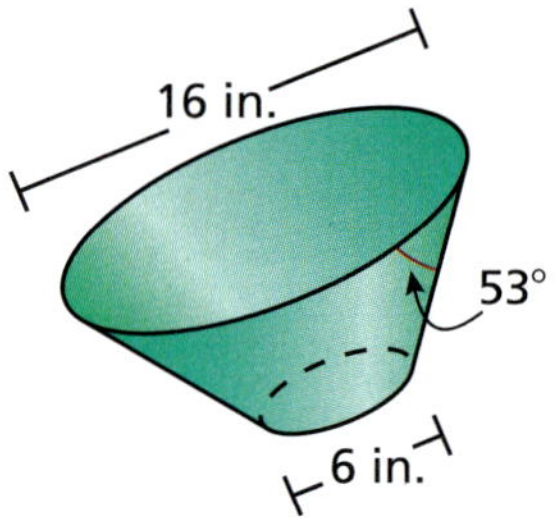

24. THOUGHT PROVOKING
To make a paper drinking cup, follow the given steps. How does the surface area of the cup compare to the original paper circle? Find $m\angle ABC$ where segment AC is the diameter.

REVIEW & REFRESH

WATCH

In Exercises 25–27, find the indicated measure.

25. area of a circle with a radius of 7 feet

26. diameter of a circle with an area of 256π square meters

27. A right cone has a radius of 5 feet and a slant height of 16 feet. Find the surface area of the cone.

28. Find the surface area of the cylinder.

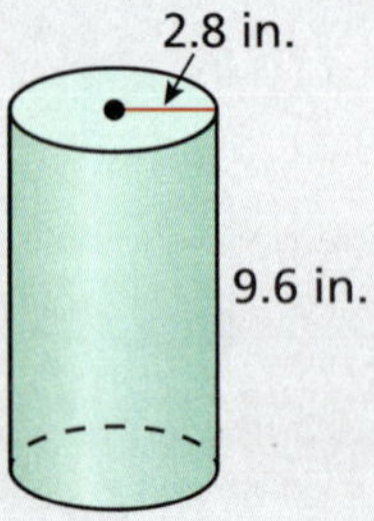

29. Two polygons are similar. The perimeter of one polygon is 54 inches. The ratio of corresponding side lengths is $\frac{2}{3}$. Find two possible perimeters of the other polygon.

30. MODELING REAL LIFE You cut an orange in half. Find the perimeter and area of the cross section formed by the cut.

31. About 350,000 people live in a circular region with a 10-mile radius. Find the population density in people per square mile.

32. Tell whether $\widehat{JM} \cong \widehat{KL}$. Explain why or why not.

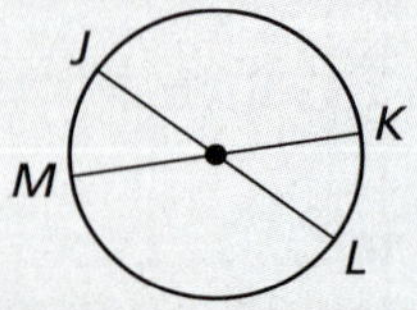

GO DIGITAL

12.4 Volumes of Prisms and Cylinders

Learning Target: Find and use volumes of prisms and cylinders.

Success Criteria:
- I can find volumes of prisms and cylinders.
- I can solve real-life problems involving volumes of prisms and cylinders.
- I can find volumes of similar prisms and similar cylinders.

EXPLORE IT! Finding Volumes of Prisms and Cylinders

Work with a partner. Recall that the *volume* of a right prism or a right cylinder is equal to the product of the area of a base and the height.

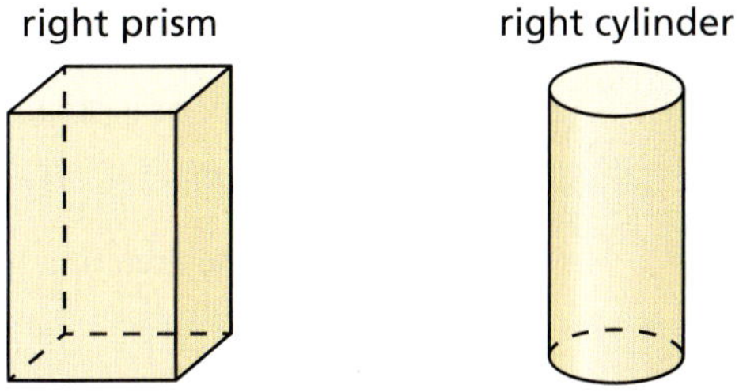

a. What does the volume of a solid represent?

b. Consider a deck of 52 cards. Each card is about $\frac{1}{100}$ inch thick. What is the volume of the deck of cards?

c. A magician performing a trick twists the deck as shown. For various heights, describe the areas of the horizontal cross sections of the deck before and after it is twisted. What do you notice? Use your observations to write a conjecture about the volume of the deck of cards before and after it is twisted.

d. For another trick, the magician uses two stacks of coins. Each coin has a diameter of 22 millimeters and a thickness of 1.78 millimeters. Compare the volumes of the stacks. Explain your reasoning.

4 MTR CONSTRUCT AN ARGUMENT
Does your conjecture in part (c) change if the cards are not rectangular? Explain your reasoning.

e. How can you find the volume of a prism or cylinder that is not a right prism or right cylinder?

Geometric Reasoning

MA.912.GR.4.3 Extend previous understanding of scale drawings and scale factors to determine how dilations affect the area of two-dimensional figures and the surface area or volume of three-dimensional figures.

MA.912.GR.4.5 Solve mathematical and real-world problems involving the volume of three-dimensional figures limited to cylinders, pyramids, prisms, cones and spheres.

GO DIGITAL

Finding Volumes of Prisms and Cylinders

Vocabulary

volume, *p. 638*
Cavalieri's Principle, *p. 638*

The **volume** of a solid is the number of cubic units contained in its interior. **Cavalieri's Principle**, named after Bonaventura Cavalieri (1598–1647), states that if two solids have the same height and the same cross-sectional area at every level, then they have the same volume. The prisms below have equal heights h and equal cross-sectional areas B at every level. By Cavalieri's Principle, the prisms have the same volume.

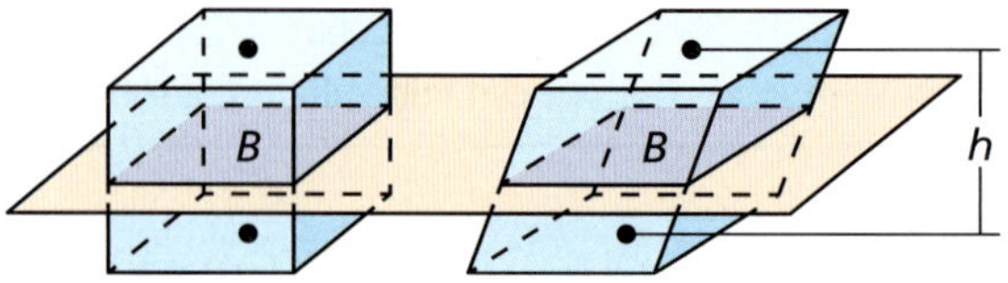

STUDY TIP

The slanted figure is called an *oblique* solid. In an *oblique prism*, the lateral edges are not perpendicular to the bases.

KEY IDEA

Volume of a Prism

The volume V of a prism is

$$V = Bh$$

where B is the area of a base and h is the height.

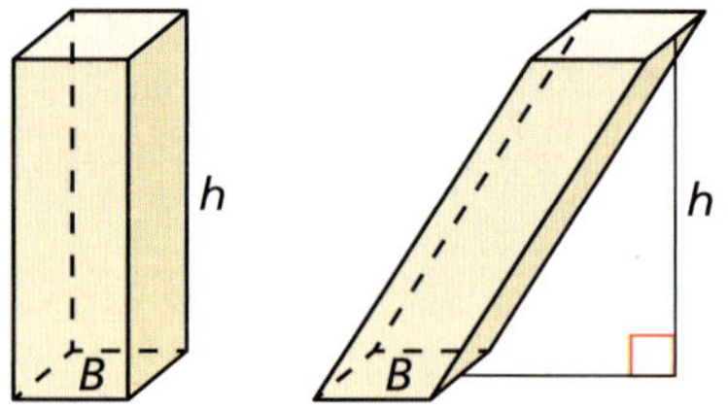

EXAMPLE 1 Finding Volumes of Prisms

Find the volume of each prism.

a.

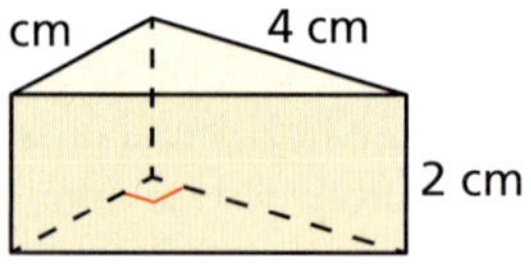

b.

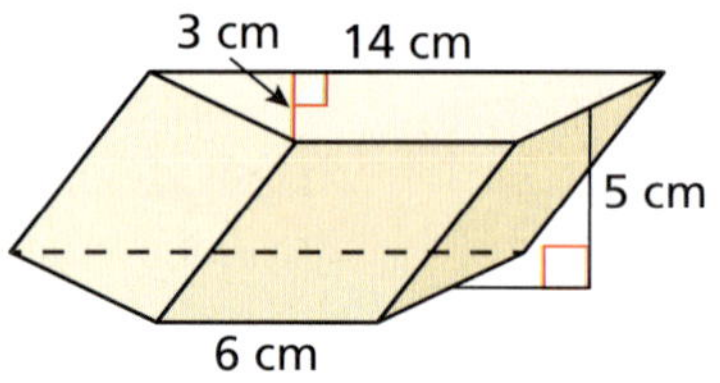

SOLUTION

a. The area of a triangular base is $B = \frac{1}{2}(3)(4) = 6 \text{ cm}^2$ and the height is $h = 2$ cm.

$V = Bh$	Formula for volume of a prism
$= 6(2)$	Substitute.
$= 12$	Simplify.

▶ The volume is 12 cubic centimeters.

b. The area of a trapezoidal base is $B = \frac{1}{2}(3)(6 + 14) = 30 \text{ cm}^2$ and the height is $h = 5$ cm.

$V = Bh$	Formula for volume of a prism
$= 30(5)$	Substitute.
$= 150$	Simplify.

▶ The volume is 150 cubic centimeters.

GO DIGITAL

Consider a cylinder with height h and base radius r, and a rectangular prism with the same height that has a square base with sides of length $r\sqrt{\pi}$.

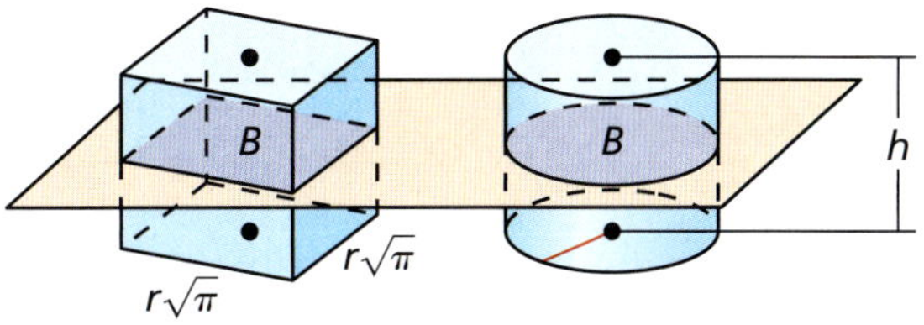

The cylinder and the prism have the same cross-sectional area, πr^2, at every level and the same height. By Cavalieri's Principle, the prism and the cylinder have the same volume. The volume of the prism is $V = Bh = \pi r^2 h$, so the volume of the cylinder is also $V = Bh = \pi r^2 h$.

KEY IDEA

Volume of a Cylinder

The volume V of a cylinder is

$$V = Bh = \pi r^2 h$$

where B is the area of a base, h is the height, and r is the radius of a base.

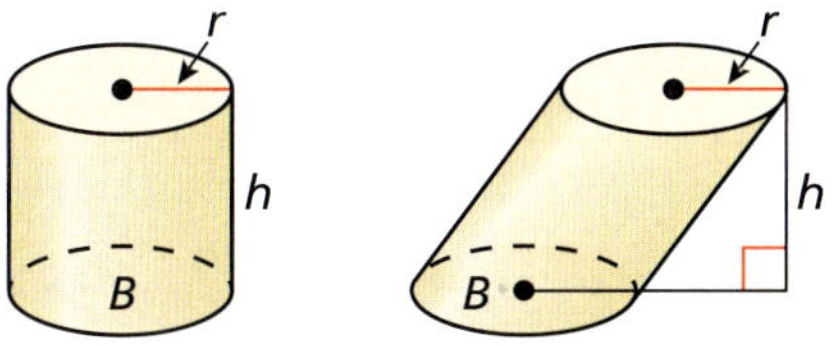

STUDY TIP

In an *oblique cylinder*, the segment that connects the centers of the bases is not perpendicular to the bases.

EXAMPLE 2 Finding Volumes of Cylinders

WATCH

Find the volume of each cylinder.

a.

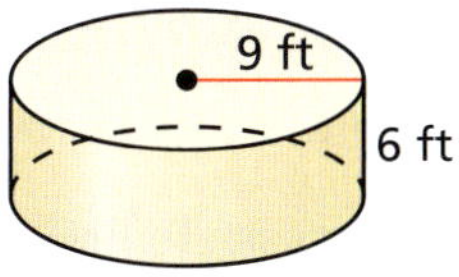

b.

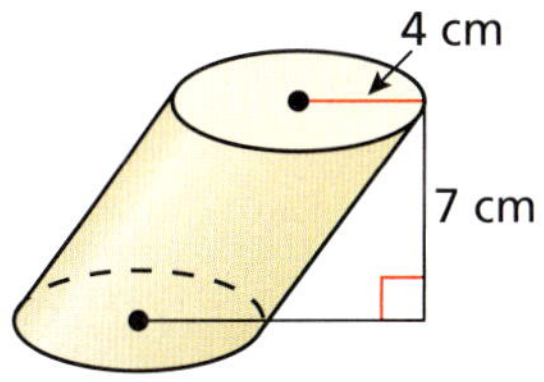

SOLUTION

a. The dimensions of the cylinder are $r = 9$ ft and $h = 6$ ft.

$$V = \pi r^2 h = \pi(9)^2(6) = 486\pi \approx 1526.81$$

▶ The volume is about 1526.81 cubic feet.

b. The dimensions of the cylinder are $r = 4$ cm and $h = 7$ cm.

$$V = \pi r^2 h = \pi(4)^2(7) = 112\pi \approx 351.86$$

▶ The volume is about 351.86 cubic centimeters.

SELF-ASSESSMENT

1 I don't understand yet. 2 I can do it with help. 3 I can do it on my own. 4 I can teach someone else.

Find the volume of the solid.

1.

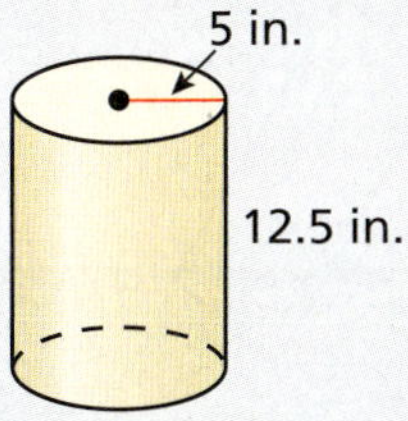

2.

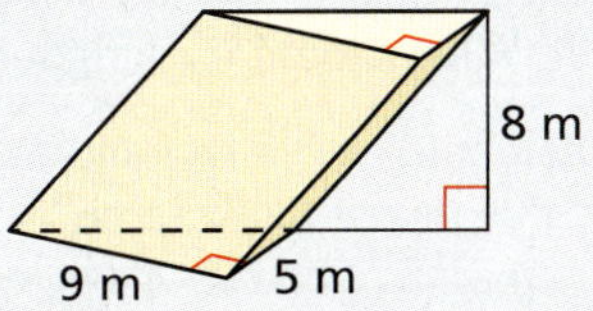

3.

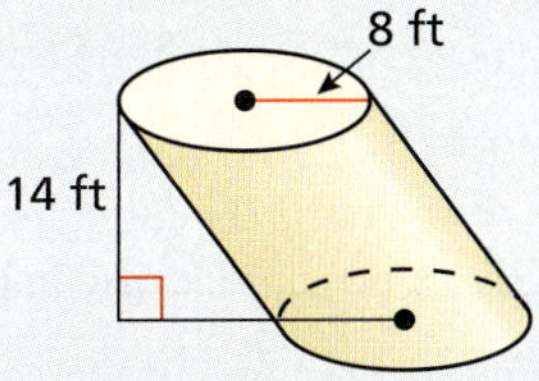

GO DIGITAL

Using Volumes of Prisms

7 MTR **EXAMPLE 3** **Modeling Real Life** WATCH

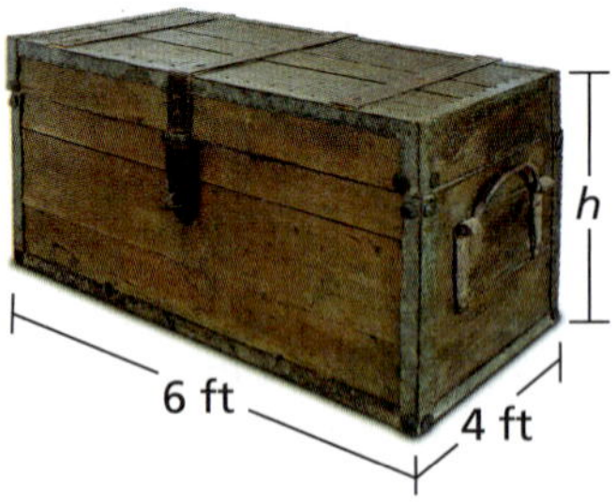

The rectangular chest has a volume of 72 cubic feet. What is the height of the chest?

SOLUTION

1. **Understand the Problem** You know the dimensions of the base of a rectangular prism and the volume. You are asked to find the height.
2. **Make a Plan** Write the formula for the volume of a rectangular prism, substitute known values, and solve for the height h.
3. **Solve and Check** The area of a rectangular base is $B = 6(4) = 24\text{ ft}^2$ and the volume is $V = 72\text{ ft}^3$.

$V = Bh$ — Formula for volume of a prism

$72 = 24h$ — Substitute.

$3 = h$ — Divide each side by 24.

Check $V = Bh$
$= 24(3)$
$= 72$ ✓

▶ The height of the chest is 3 feet.

EXAMPLE 4 **Finding the Volume of a Composite Solid** WATCH

Find the volume of the concrete block.

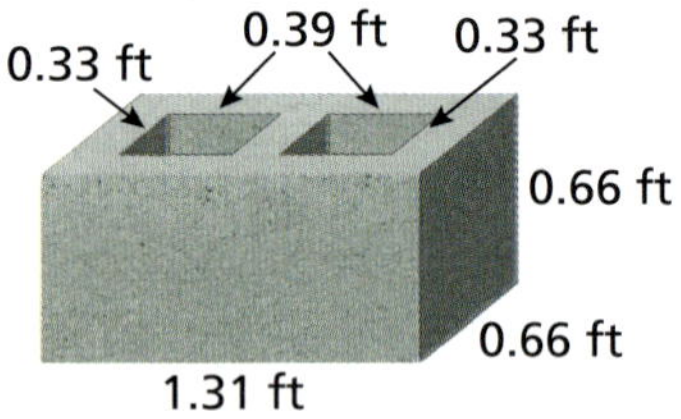

SOLUTION

The small rectangles are congruent. To find the area of the base, subtract two times the area of a small rectangle from the area of the large rectangle.

$B =$ Area of large rectangle $- 2 \cdot$ Area of small rectangle

$= 1.31(0.66) - 2(0.33)(0.39)$

$= 0.6072$

Using the formula for the volume of a prism, the volume is

$V = Bh = 0.6072(0.66) \approx 0.40.$

▶ The volume is about 0.40 cubic foot.

2 MTR **USE ANOTHER METHOD**
Describe another way you can find the volume of the concrete block.

SELF-ASSESSMENT 1 I don't understand yet. 2 I can do it with help. 3 I can do it on my own. 4 I can teach someone else.

4. **WHAT IF?** In Example 3, you want the length to be 5 feet, the width to be 2 feet, and the volume to be 30 cubic feet. What should the height be?
5. **WHAT IF?** In Example 3, you want the height to be 2.5 feet and the volume to be 25 cubic feet. What should the area of the base be? Give a possible length and width.
6. Find the volume of the composite solid.

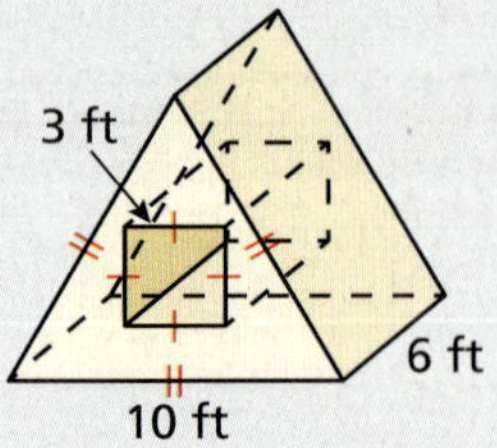

GO DIGITAL

Finding Volumes of Similar Solids

KEY IDEA

Volumes of Similar Solids

If two similar solids have a scale factor of k, then the ratio of their volumes is equal to k^3.

EXAMPLE 5 Finding the Volume of a Similar Solid

WATCH

Cylinder A and cylinder B are similar. Find the volume of cylinder B.

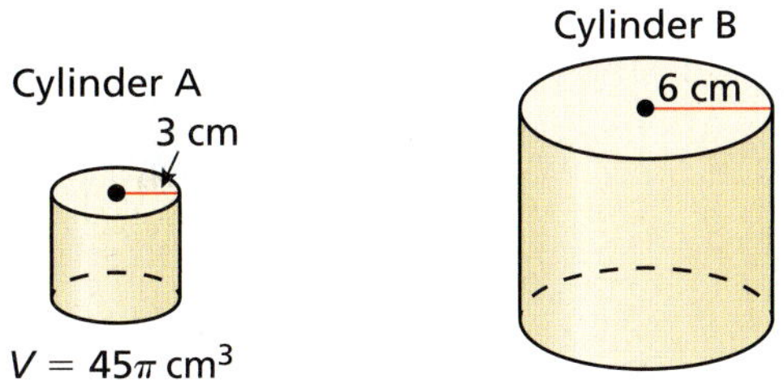

SOLUTION

The scale factor is $k = \dfrac{\text{Radius of cylinder B}}{\text{Radius of cylinder A}}$

$= \dfrac{6}{3} = 2.$

Use the scale factor to find the volume of cylinder B.

$\dfrac{\text{Volume of cylinder B}}{\text{Volume of cylinder A}} = k^3$	The ratio of the volumes is k^3.
$\dfrac{\text{Volume of cylinder B}}{45\pi} = 2^3$	Substitute.
$\dfrac{\text{Volume of cylinder B}}{45\pi} = 8$	Evaluate power.
$\text{Volume of cylinder B} = 360\pi$	Multiply each side by 45π.

▶ The volume of cylinder B is 360π cubic centimeters.

COMMON ERROR

Be sure to write the ratio of the surface areas in the same order you wrote the ratio of the radii.

SELF-ASSESSMENT 1 I don't understand yet. 2 I can do it with help. 3 I can do it on my own. 4 I can teach someone else.

7. Prism C and prism D are similar. Find the volume of prism D.

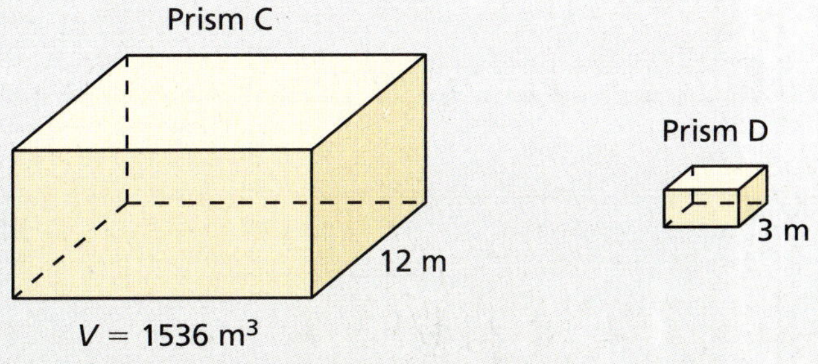

5 MTR **8.** **PATTERNS** Two similar cylinders have a scale factor of $\frac{1}{3}$. The smaller cylinder has a volume of 9 cubic yards. What is the volume of the larger cylinder?

GO DIGITAL

12.4 Practice WITH CalcChat® AND CalcView®

In Exercises 1–4, find the volume of the prism. *(See Example 1.)*

▶ **1.**

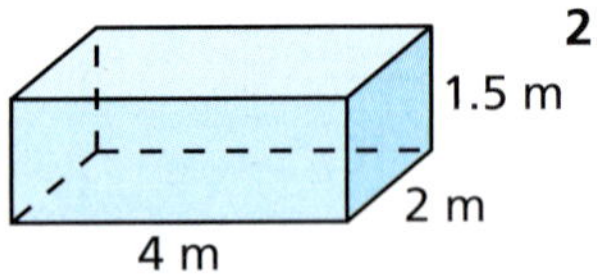

2.

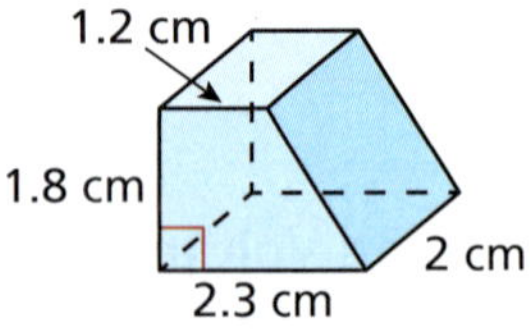

3.

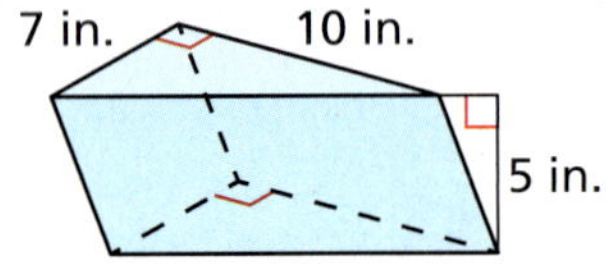

4.

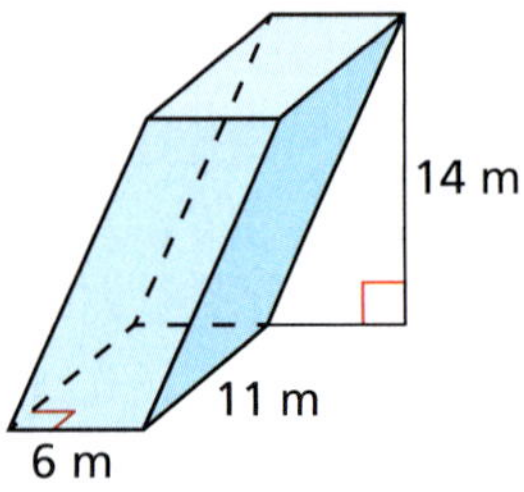

In Exercises 5–8, find the volume of the cylinder. *(See Example 2.)*

▶ **5.**

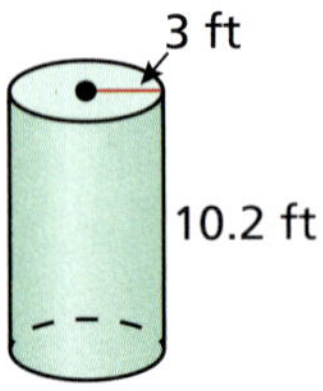

6.

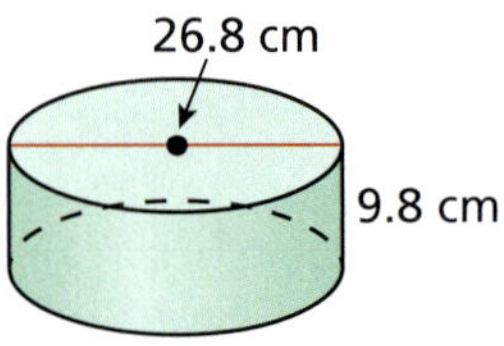

7.

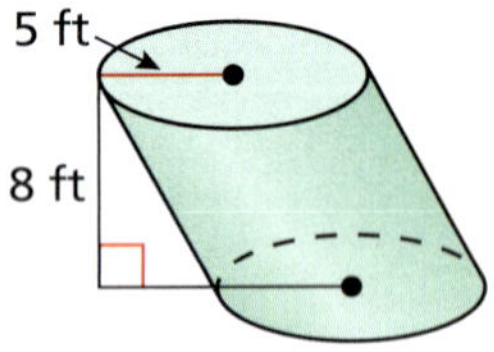

8.

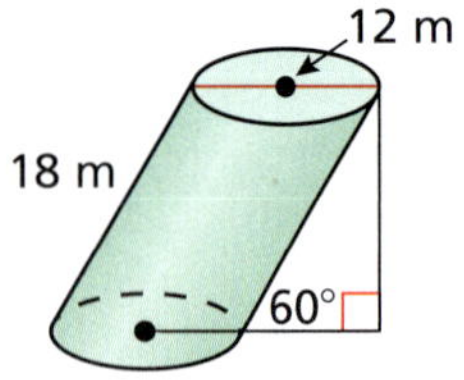

9. **STRUCTURE** Sketch the solid formed by the net. Then find the volume of the solid.

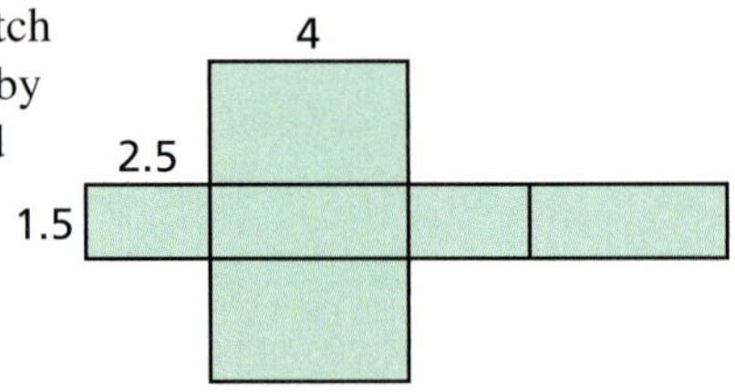

10. **ERROR ANALYSIS** Describe and correct the error in finding the volume of the cylinder.

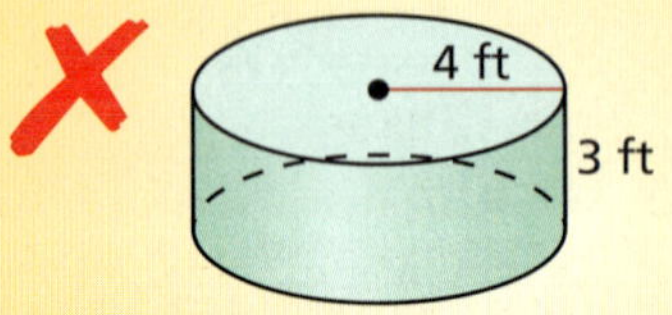

$V = 2\pi rh$

$= 2\pi(4)(3)$

$= 24\pi$

So, the volume of the cylinder is 24π cubic feet.

In Exercises 11 and 12, sketch the solid and find its volume.

11. A prism has a height of 11.2 centimeters and an equilateral triangle for a base, where each base edge is 8 centimeters.

12. A pentagonal prism has a height of 9 feet and each base edge is 3 feet.

In Exercises 13–18, find the missing dimension of the prism or cylinder. *(See Example 3.)*

▶ **13.** Volume = 560 ft³

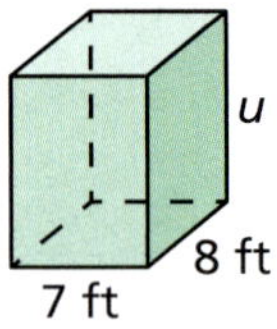

14. Volume = 2700 yd³

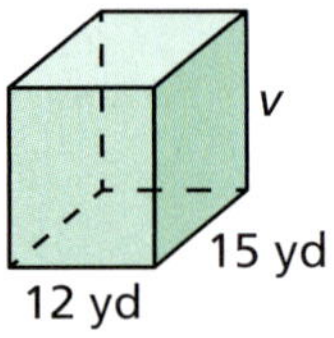

15. Volume = 80 cm³

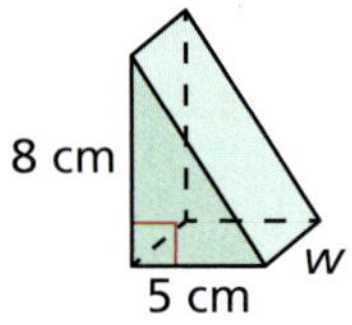

16. Volume = 72.66 in.³

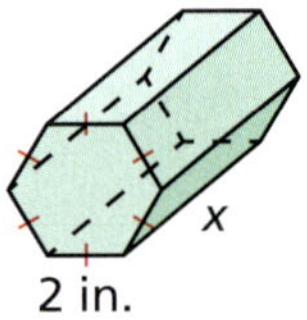

17. Volume = 3000 ft³

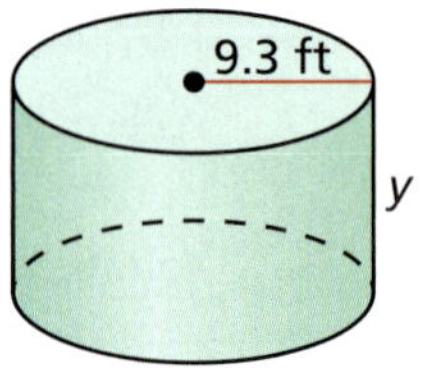

18. Volume = 1696.5 m³

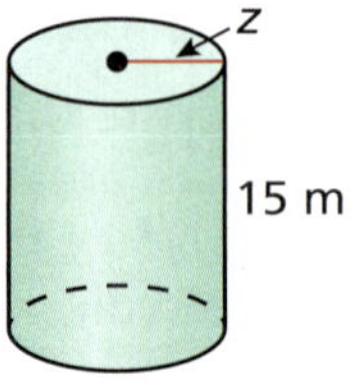

In Exercises 19–22, find the volume of the composite solid. *(See Example 4.)*

▶ **19.**

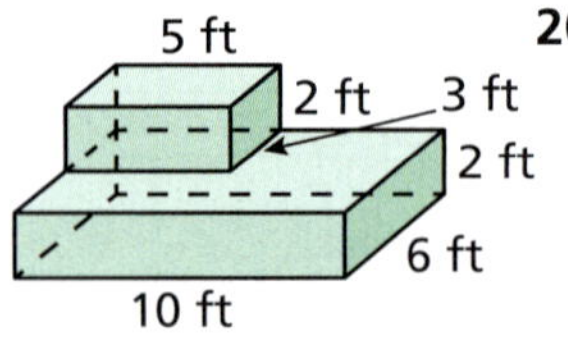

20.

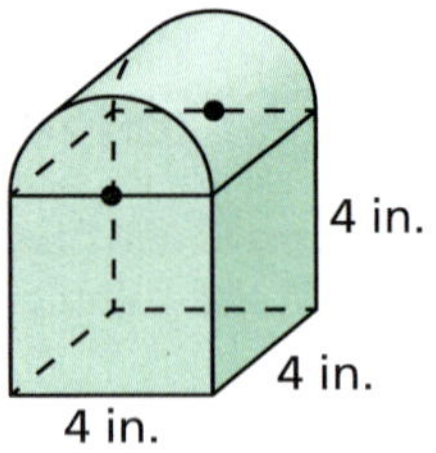

21.

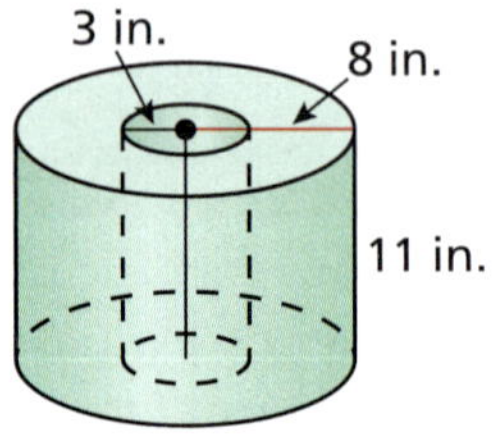

22.

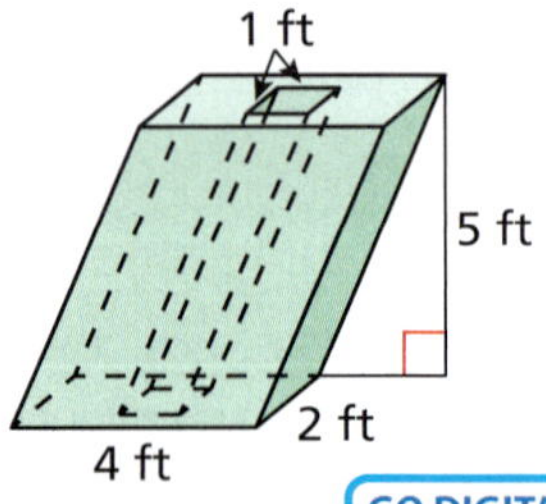

GO DIGITAL

In Exercises 23 and 24, solids A and B are similar. Use the given information and scale factor k from solid A to solid B to find the volume of solid B.

23. Prism A has a volume of 45 cubic inches and $k = 3$.

24. Cylinder A has a volume of 112π cubic meters and $k = \frac{1}{4}$.

In Exercises 25 and 26, the solids are similar. Find the volume of solid B. *(See Example 5.)*

25.

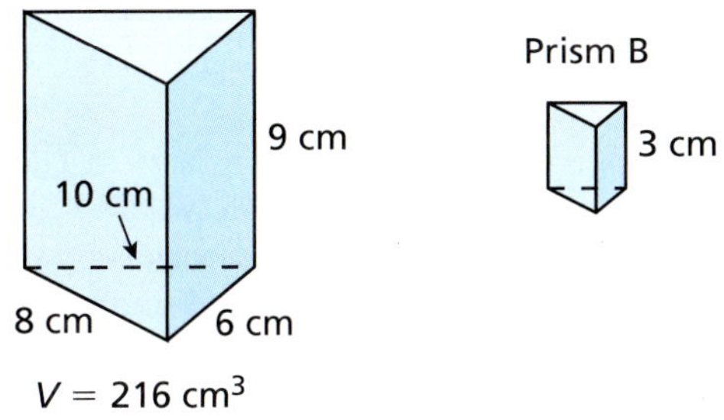

26.

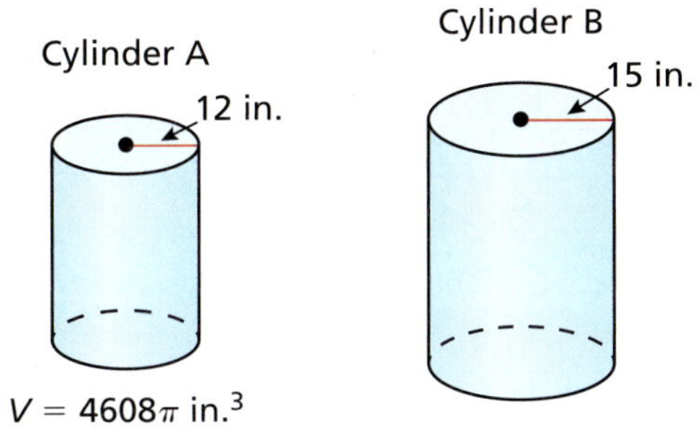

27. MODELING REAL LIFE The Great Blue Hole is a cylindrical trench located off the coast of Belize. It is approximately 1000 feet wide and 400 feet deep. About how many gallons of water does the Great Blue Hole contain? (1 ft^3 $\approx$ 7.48 gallons)

28. MODELING REAL LIFE You melt a rectangular block of wax to make candles. How many candles of the given shape can you make using a block that measures 10 centimeters by 9 centimeters by 20 centimeters?

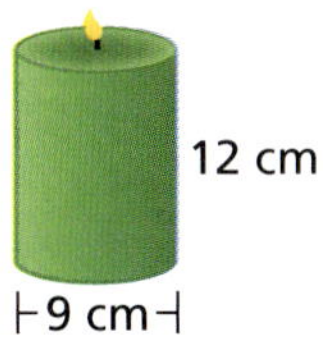

29. PROBLEM SOLVING An aquarium shaped like a rectangular prism has a length of 30 inches, a width of 10 inches, and a height of 20 inches. You fill the aquarium $\frac{3}{4}$ full with water. When you submerge a rock in the aquarium, the water level rises 0.25 inch. Find the volume of the rock. How many rocks of this size can you place in the aquarium before water spills out?

GO DIGITAL

30. PROBLEM SOLVING Which box gives you more cereal for your money? Explain.

31. WRITING Both of the figures shown are made up of the same number of congruent rectangles. Explain how Cavalieri's Principle can be adapted to compare the areas of these figures.

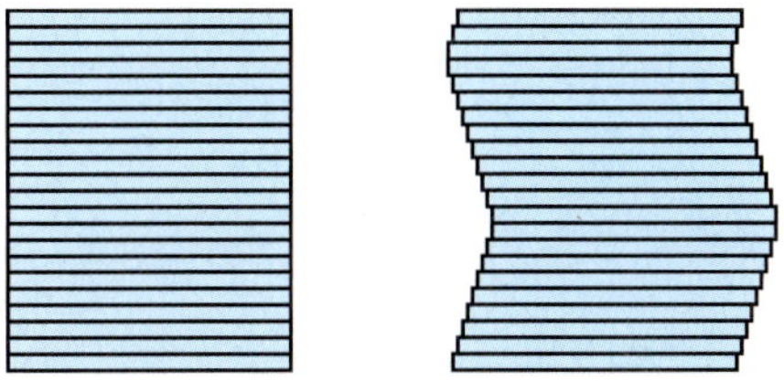

32. HOW DO YOU SEE IT?
Each stack of memo papers contains 500 equally sized sheets of paper. Compare their volumes. Explain your reasoning.

33. OPEN-ENDED Sketch two rectangular prisms that have volumes of 100 cubic inches but different surface areas. Include dimensions in your sketches.

34. CONNECTING CONCEPTS How can you change the edge length of a cube so that the volume is reduced by 40%?

35. DIG DEEPER Estimate the volume of the bag of yellow onions shown. Explain your method.

36. **MAINTAIN ACCURACY** The height of cylinder X is twice the height of cylinder Y. The radius of cylinder X is half the radius of cylinder Y. Compare the volumes of cylinder X and cylinder Y. Justify your answer.

37. **CONNECTING CONCEPTS** Find the volume of the solid shown. The bases are sectors of circles.

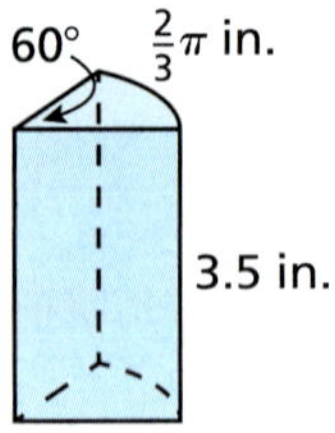

38. **THOUGHT PROVOKING** The two rectangular prisms have the same height, and the bases of prism A are congruent to the bases of prism B. Do the prisms have the same volume? the same surface area? Explain.

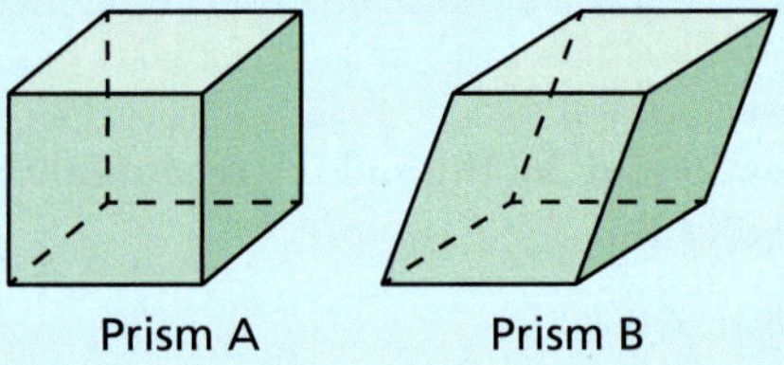

REVIEW & REFRESH

WATCH

39. In the diagram, $JKLM \cong PQRS$. Find the values of x and y.

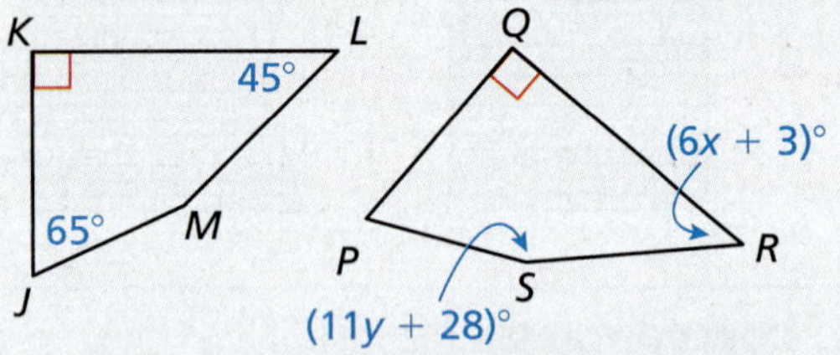

In Exercises 40–43, find the surface area of the solid.

40.

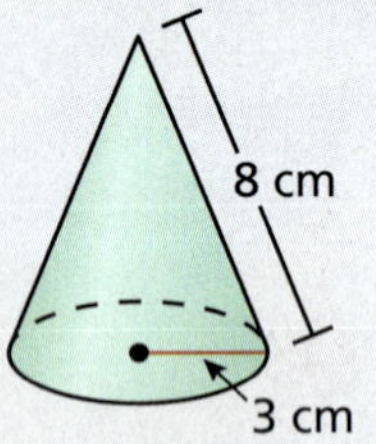

41.

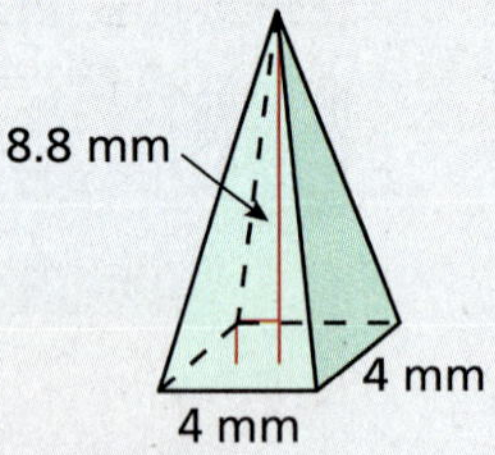

42.

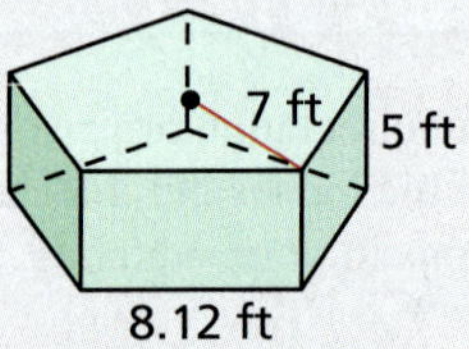

43.

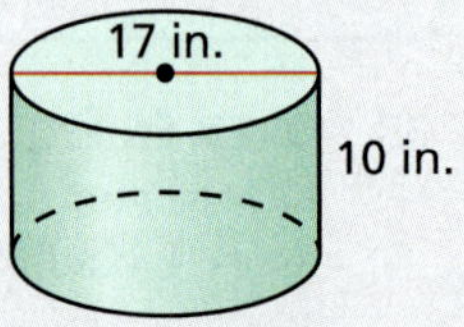

44. In $\triangle ABC$, $m\angle A = 32°$ and $m\angle C = 67°$. In $\triangle DEF$, $m\angle E = 81°$ and $m\angle F = 67°$. Are the triangles similar? Explain your reasoning.

45. Find the value of x.

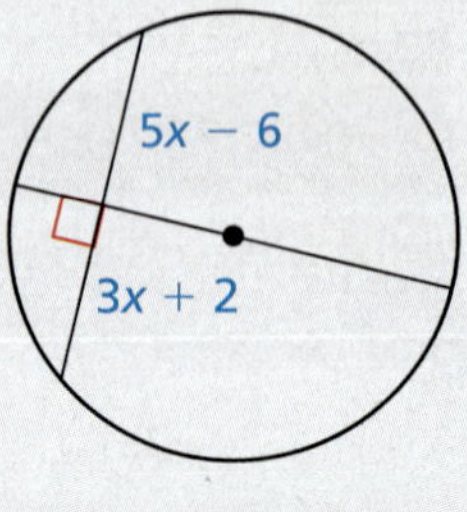

46. **MODELING REAL LIFE** An airplane travels 600 miles west, then turns 60° toward south and travels another 250 miles. How far is the airplane from its starting location?

In Exercises 47–50, find the volume of the prism or cylinder.

47.

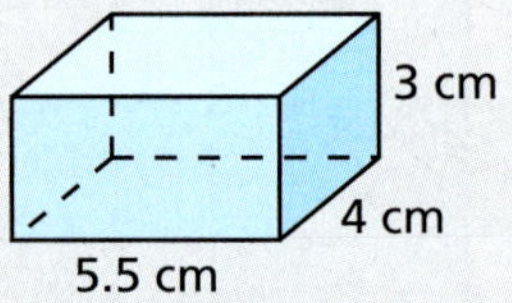

48.

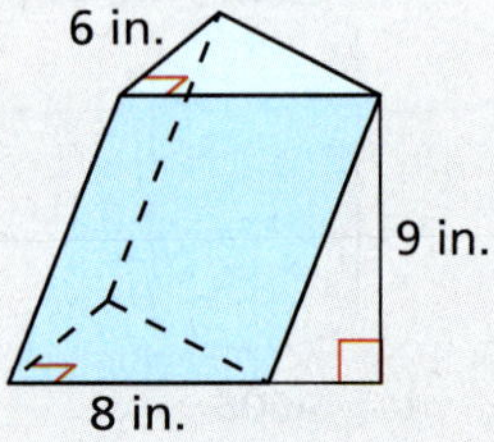

49.

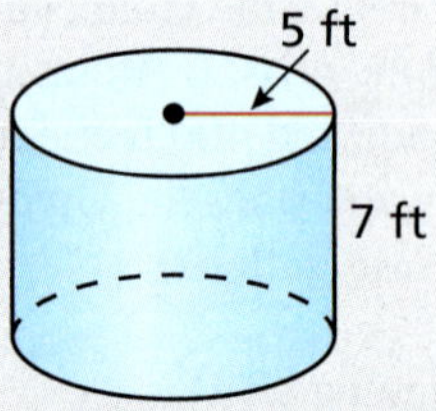

50.

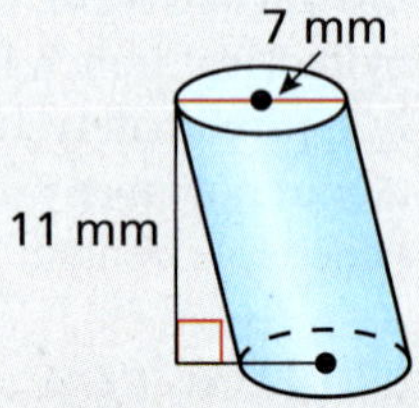

51. The state of Colorado has a population of about 5.76 million people. Find the population density in people per square mile.

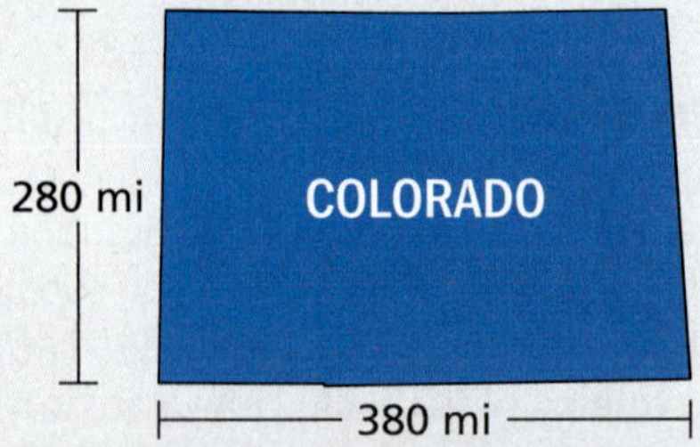

52. Sketch each polyhedron.

a. octagonal prism

b. triangular pyramid

GO DIGITAL

12.5 Volumes of Pyramids and Cones

Learning Target: Find and use volumes of pyramids and cones.

Success Criteria:
- I can find volumes of pyramids and cones.
- I can use volumes of pyramids and cones to find measures.
- I can find volumes of similar pyramids and similar cones.
- I can find volumes of composite solids containing pyramids or cones.

EXPLORE IT! Finding the Volume Formula for Pyramids

Work with a partner.

a. The diagonals divide the cube into congruent pyramids. How many congruent pyramids are formed? Write an expression for the volume of one of the pyramids. Explain your reasoning.

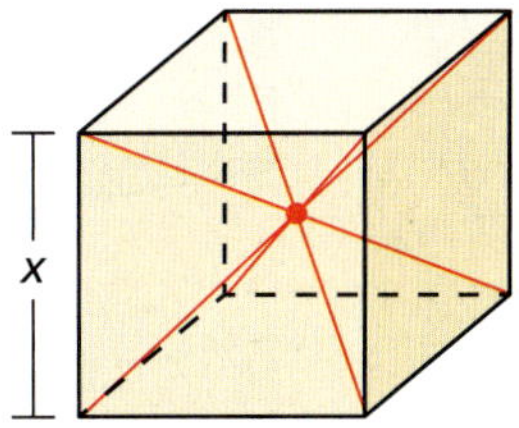

b. The volume of any rectangular pyramid is proportional to its base length ℓ, base width w, and height h. So, you can write

$$V = c(\ell wh)$$

where c is a constant. Using a pyramid from part (a), rewrite each side of this equation in terms of x and solve for c.

c. Given the constant you found in part (b), what is the formula for the volume of any rectangular pyramid? When the pyramid below is filled with sand and poured into the prism, how many pyramids does it take to fill the prism? Explain your reasoning.

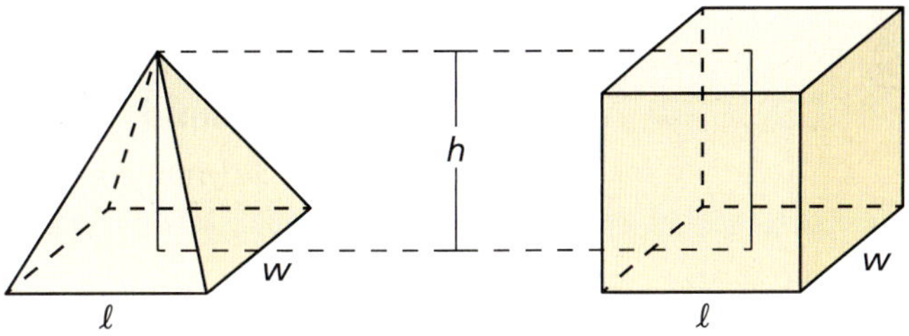

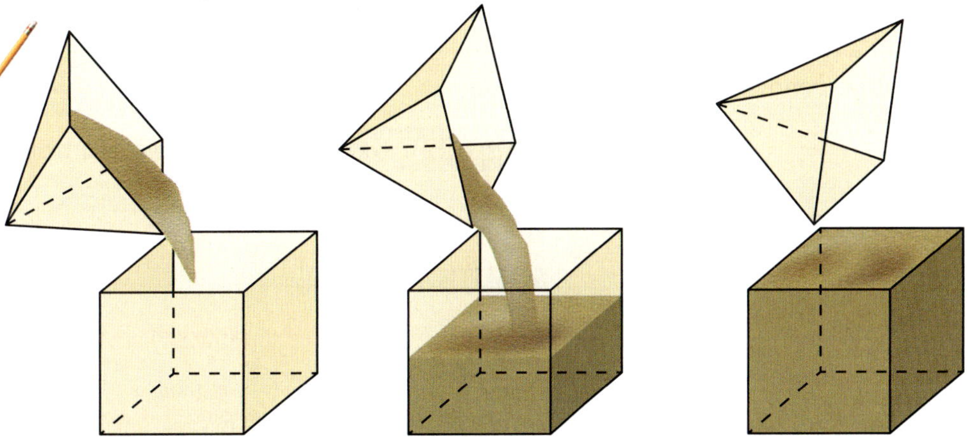

5 MTR MAKE A PLAN

Design an experiment that uses a cone and a cylinder to find a formula for the volume of a cone. Explain how the cone and cylinder you would use are related.

Geometric Reasoning

MA.912.GR.4.5 Solve mathematical and real-world problems involving the volume of three-dimensional figures limited to cylinders, pyramids, prisms, cones and spheres.

GO DIGITAL

Finding Volumes of Pyramids

Consider a triangular prism with parallel, congruent bases $\triangle JKL$ and $\triangle MNP$. You can divide this triangular prism into three triangular pyramids.

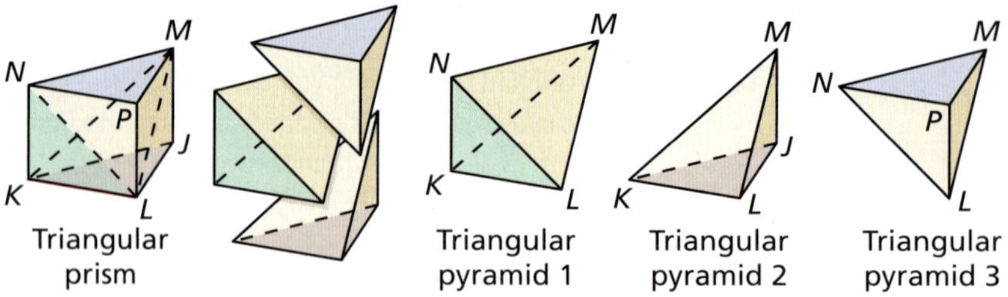

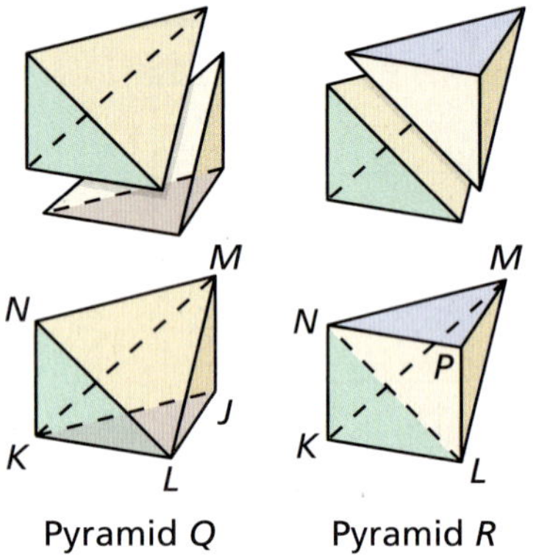

You can combine triangular pyramids 1 and 2 to form a pyramid with a base that is a parallelogram, as shown at the left. Name this pyramid *Q*. Similarly, you can combine triangular pyramids 1 and 3 to form pyramid *R* with a base that is a parallelogram.

In pyramid *Q*, diagonal $\overline{KM}$ divides $\square JKNM$ into two congruent triangles, so the bases of triangular pyramids 1 and 2 are congruent. Similarly, you can divide any cross section parallel to $\square JKNM$ into two congruent triangles that are the cross sections of triangular pyramids 1 and 2.

By Cavalieri's Principle, triangular pyramids 1 and 2 have the same volume. Similarly, using pyramid *R*, you can show that triangular pyramids 1 and 3 have the same volume. By the Transitive Property of Equality, triangular pyramids 2 and 3 have the same volume.

The volume of each pyramid must be one-third the volume of the prism, or $V = \frac{1}{3}Bh$. You can generalize this formula to say that the volume of any pyramid with any base is equal to $\frac{1}{3}$ the volume of a prism with the same base and height because you can divide any polygon into triangles and any pyramid into triangular pyramids.

KEY IDEA

Volume of a Pyramid

The volume V of a pyramid is

$$V = \frac{1}{3}Bh$$

where B is the area of the base and h is the height.

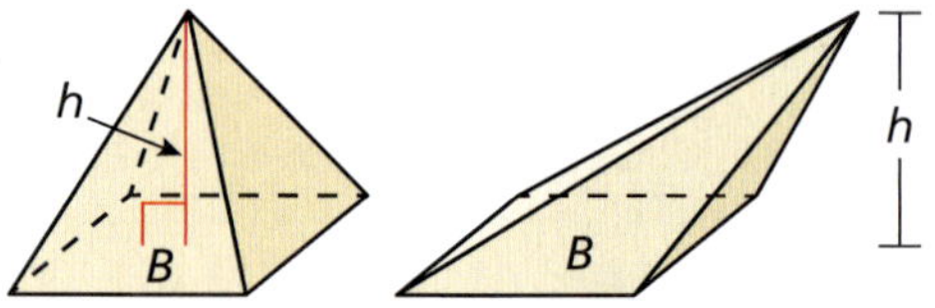

STUDY TIP

In an *oblique pyramid* with a regular base, the segment that connects the vertex and the center of the base is not perpendicular to the base.

EXAMPLE 1 Finding the Volume of a Pyramid

Find the volume of the pyramid.

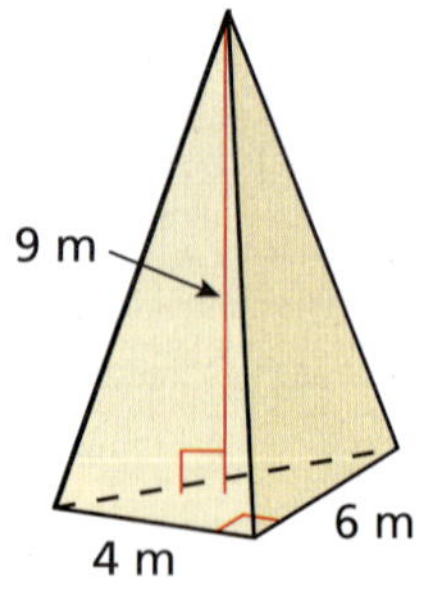

SOLUTION

$V = \frac{1}{3}Bh$ — Write formula for volume of a pyramid.

$= \frac{1}{3}\left(\frac{1}{2} \cdot 4 \cdot 6\right)(9)$ — Substitute.

$= 36$ — Simplify.

▶ The volume is 36 cubic meters.

GO DIGITAL

Finding Volumes of Cones

Consider a cone with a regular polygon inscribed in the base. The pyramid with the same vertex as the cone has volume $V = \frac{1}{3}Bh$. As you increase the number of sides of the polygon, it approaches the base of the cone and the pyramid approaches the cone. The volume approaches $\frac{1}{3}\pi r^2 h$ as the base area B approaches πr^2.

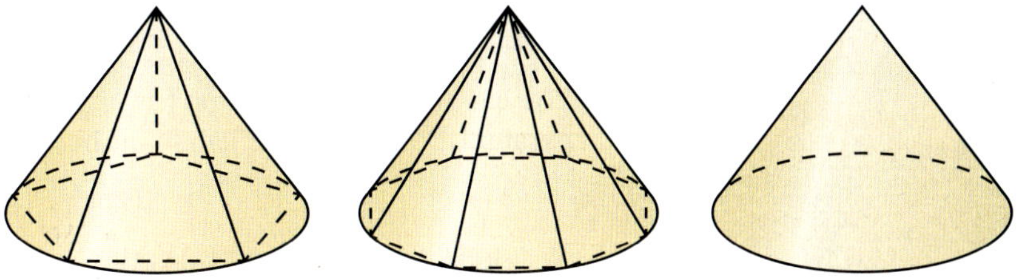

KEY IDEA

Volume of a Cone

The volume V of a cone is

$$V = \frac{1}{3}Bh = \frac{1}{3}\pi r^2 h$$

where B is the area of the base, h is the height, and r is the radius of the base.

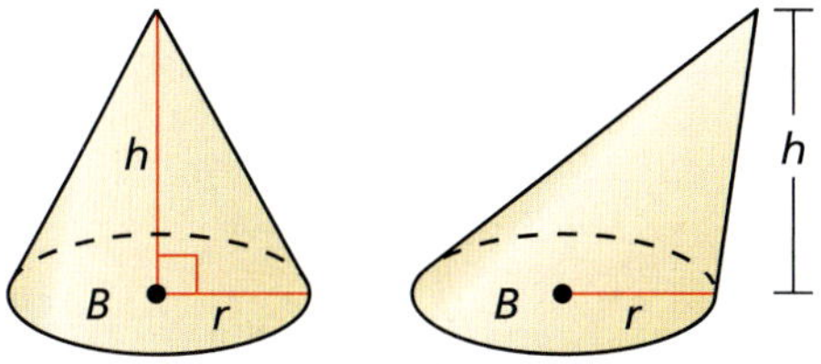

STUDY TIP

In an *oblique cone*, the segment that connects the vertex and the center of the base is not perpendicular to the base.

EXAMPLE 2 Finding the Volume of a Cone

WATCH

Find the volume of the cone.

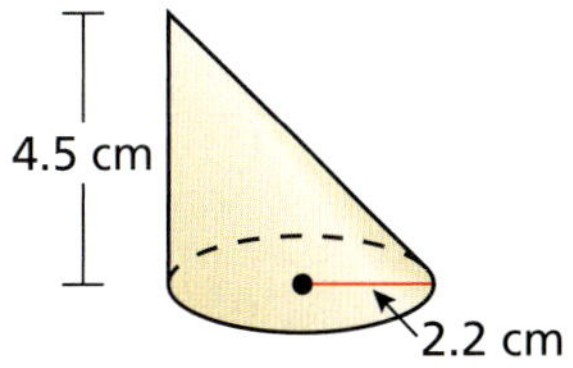

SOLUTION

$V = \frac{1}{3}\pi r^2 h$	Write formula for volume of a cone.
$= \frac{1}{3}\pi \cdot (2.2)^2 \cdot 4.5$	Substitute.
$= 7.26\pi$	Simplify.
≈ 22.81	Use technology.

▶ The volume is about 22.81 cubic centimeters.

SELF-ASSESSMENT 1 I don't understand yet. 2 I can do it with help. 3 I can do it on my own. 4 I can teach someone else.

1. **WRITING** Describe the similarities and differences between pyramids and cones.

Find the volume of the solid.

2. 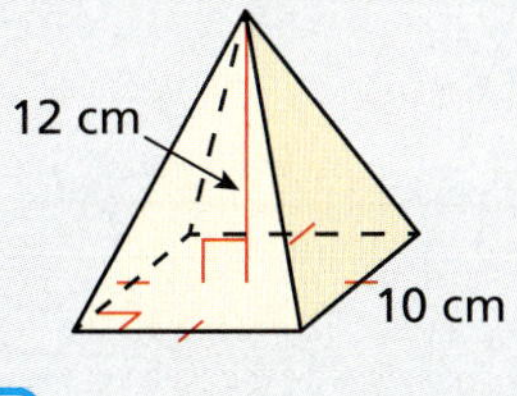

3. 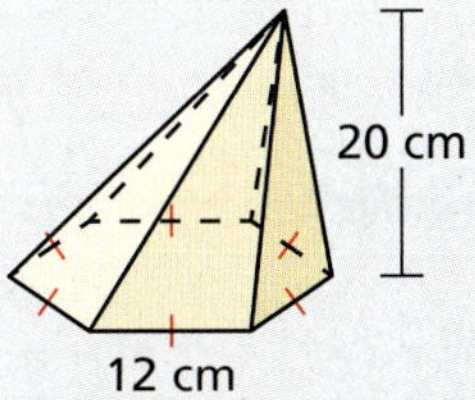

4. 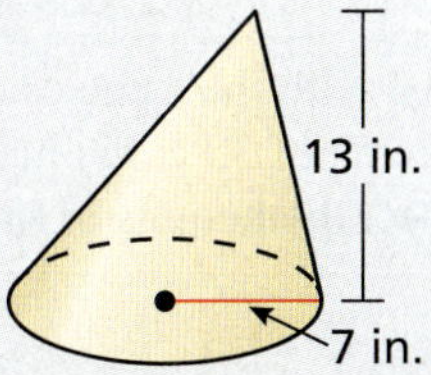

5. 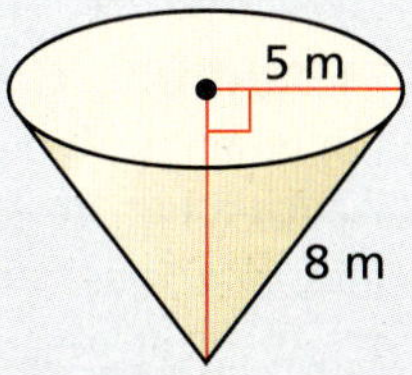

GO DIGITAL

Using Volumes of Pyramids and Cones

EXAMPLE 3 Using the Volume of a Pyramid

WATCH

The Mayan pyramid El Castillo in Mexico has a height of about 30 meters and a volume of about 30,580 cubic meters. Find the side length of the base.

As the Sun sets during the spring and fall equinoxes, El Castillo casts shadows that create the appearance of a snake moving down the stairway of the pyramid.

SOLUTION

You can see in the photo that El Castillo is approximately shaped like a square pyramid. So, let the side length of the base be x. Then the area of the base is x^2.

$V = \frac{1}{3}Bh$ Write formula for volume of a pyramid.

$30{,}580 \approx \frac{1}{3}x^2(30)$ Substitute.

$30{,}580 \approx 10x^2$ Simplify.

$3058 \approx x^2$ Divide each side by 10.

$55.3 \approx x$ Take the positive square root of each side.

▶ The side length of the base is about 55 meters.

EXAMPLE 4 Using the Volume of a Cone

WATCH

Find the height of the cone.

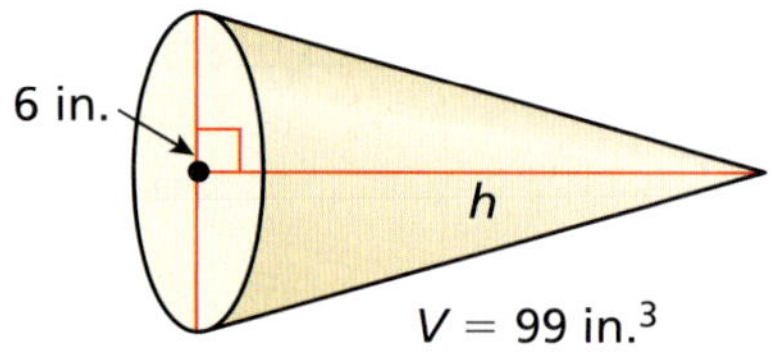

SOLUTION

The diameter is 6 inches, so the radius is $r = \frac{6}{2} = 3$ inches, and the volume is $V = 99$ cubic inches.

$V = \frac{1}{3}\pi r^2 h$ Write formula for volume of a cone.

$99 = \frac{1}{3}\pi \cdot 3^2 \cdot h$ Substitute.

$99 = 3\pi h$ Simplify.

$\frac{99}{3\pi} = h$ Solve for h.

$10.5 \approx h$ Use technology.

▶ The height is about 10.5 inches.

SELF-ASSESSMENT 1 I don't understand yet. 2 I can do it with help. 3 I can do it on my own. 4 I can teach someone else.

6. Find the height of the triangular pyramid at the right.

7. The volume of a cone is 351π cubic yards and the height is 13 yards. Find the radius.

5 MTR 8. **STRUCTURE** Consider a cone with the same diameter as the cone in Example 4, but with a volume of 33 cubic inches. Without calculating, how do you expect the heights of the cones to differ? Explain your reasoning.

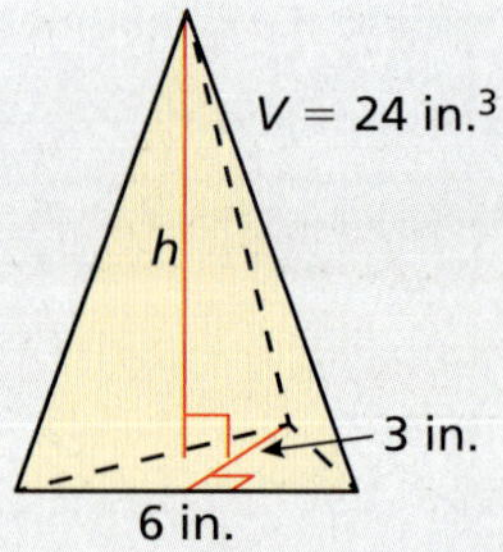

GO DIGITAL

Finding Volumes of Similar Solids and Composites

Pyramid A

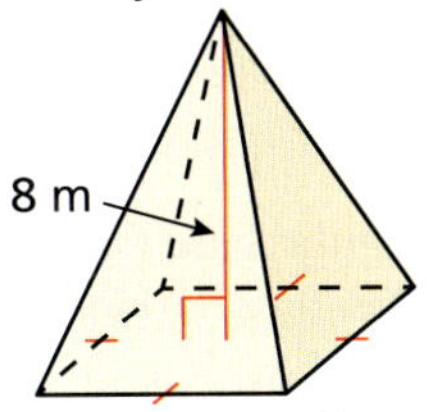

$V = 96\ \text{m}^3$

Pyramid B

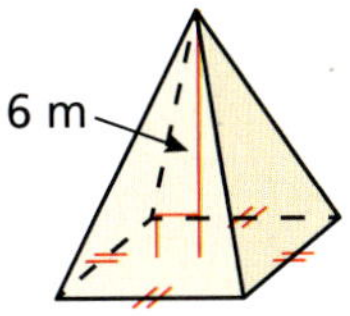

EXAMPLE 5 Finding the Volume of a Similar Solid

WATCH

Pyramid A and pyramid B are similar. Find the volume of pyramid B.

SOLUTION

The scale factor is $k = \dfrac{\text{Height of pyramid B}}{\text{Height of pyramid A}} = \dfrac{6}{8} = \dfrac{3}{4}$.

Use the scale factor to find the volume of pyramid B.

$$\frac{\text{Volume of pyramid B}}{\text{Volume of pyramid A}} = k^3$$ The ratio of the volumes is k^3.

$$\frac{\text{Volume of pyramid B}}{96} = \left(\frac{3}{4}\right)^3$$ Substitute.

$$\text{Volume of pyramid B} = 40.5$$ Multiply each side by 96.

▶ The volume of pyramid B is 40.5 cubic meters.

EXAMPLE 6 Finding the Volume of a Composite Solid

WATCH

Find the volume of the composite solid.

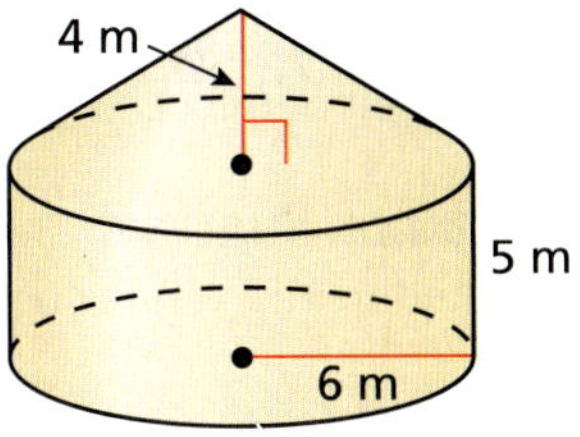

SOLUTION

Let h_1 be the height of the cylinder, and let h_2 be the height of the cone.

Volume of solid	=	Volume of cylinder	+	Volume of cone

$= \pi r^2 h_1 + \frac{1}{3}\pi r^2 h_2$ Write formulas.

$= \pi \cdot 6^2 \cdot 5 + \frac{1}{3}\pi \cdot 6^2 \cdot 4$ Substitute.

$= 180\pi + 48\pi$ Simplify.

$= 228\pi$ Add.

≈ 716.28 Use technology.

▶ The volume is 716.28 cubic meters.

6 MTR ASSESS REASONABLENESS

Use a cylinder with a radius of 6 meters and a height of 9 meters to determine whether your answer is reasonable.

SELF-ASSESSMENT

1 I don't understand yet. 2 I can do it with help. 3 I can do it on my own. 4 I can teach someone else.

9. Cone C and cone D are similar. Find the volume of cone D.

Cone C

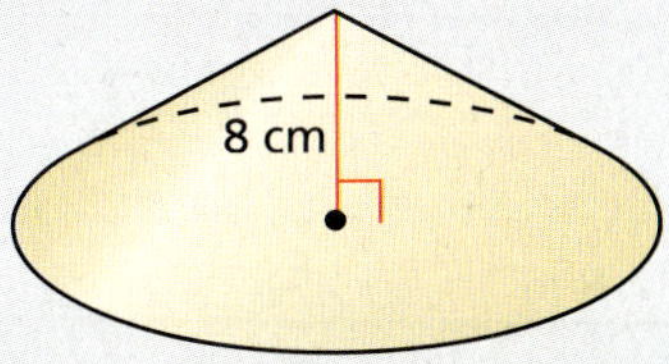

$V = 600\pi\ \text{cm}^3$

Cone D

10. Find the volume of the composite solid.

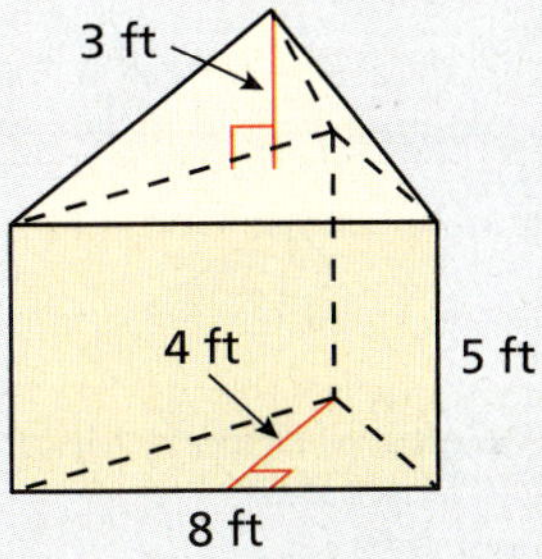

GO DIGITAL

12.5 Practice WITH CalcChat® AND CalcView®

In Exercises 1–6, find the volume of the solid. *(See Examples 1 and 2.)*

1.

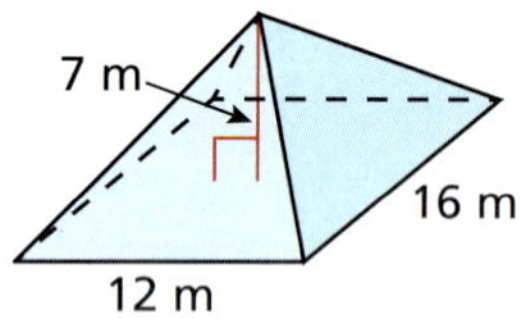

2.

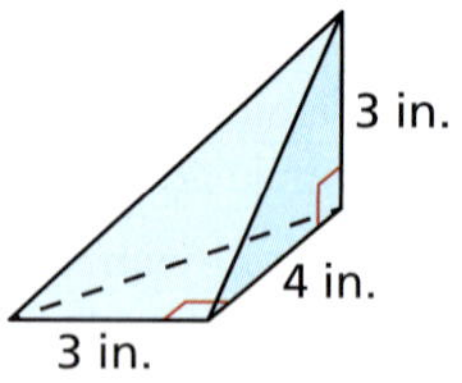

3.

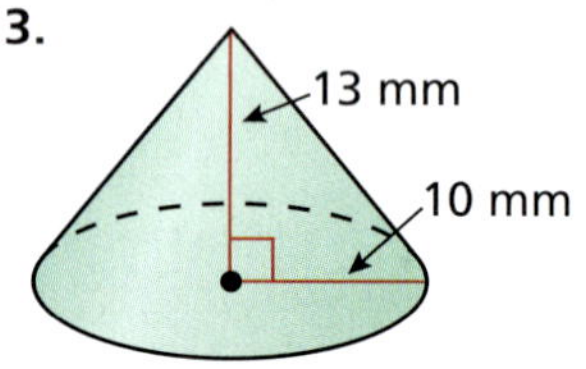

4.

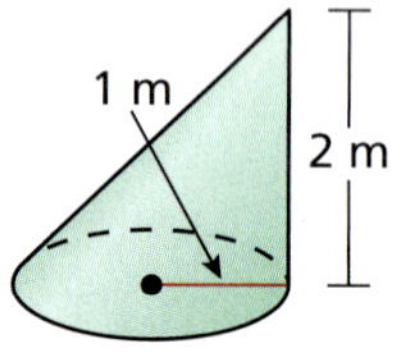

5. A cone has a diameter of 11.5 inches and a height of 15.2 inches.

6. A square pyramid has a base length of 3 feet and a height of 6 feet.

7. ERROR ANALYSIS Describe and correct the error in finding the volume of the pyramid.

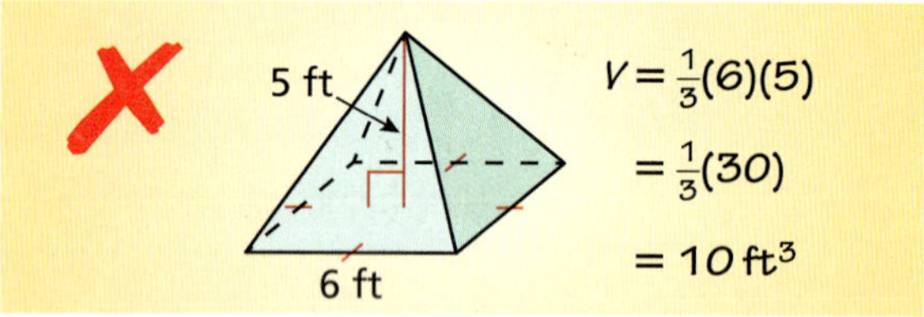

8. ERROR ANALYSIS Describe and correct the error in finding the volume of the cone.

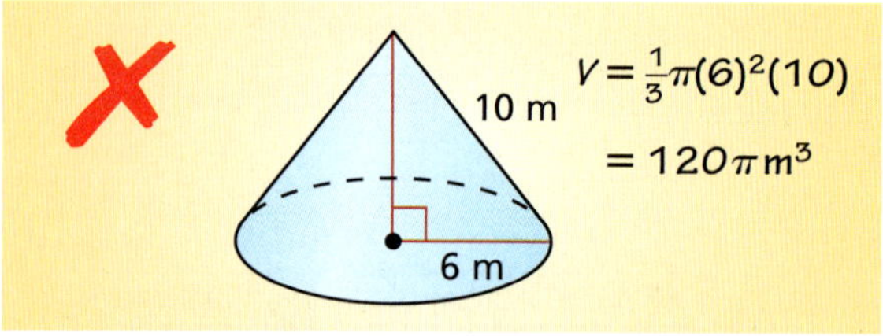

In Exercises 9–12, find the indicated measure. *(See Example 3.)*

9. Volume = 15 ft^3

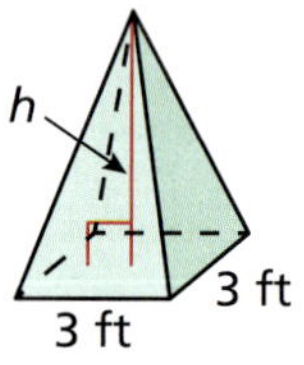

10. Volume = 224 in.^{3}

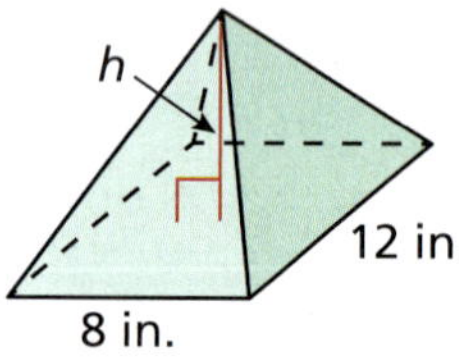

11. A pyramid with a square base has a volume of 120 cubic meters and a height of 10 meters. Find the side length of the base.

12. A pyramid with a rectangular base has a volume of 480 cubic inches and a height of 10 inches. The width of the base is 9 inches. Find the length of the base.

In Exercises 13–16, find the indicated measure. *(See Example 4.)*

13. Volume = 24π m^3

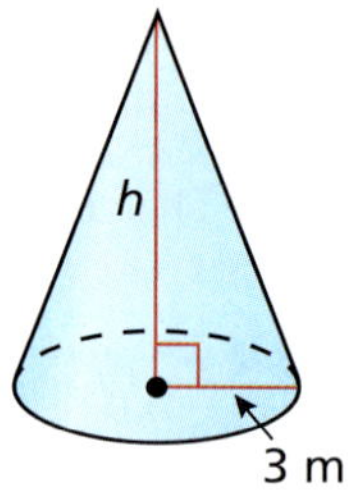

14. Volume = 216π in.^{3}

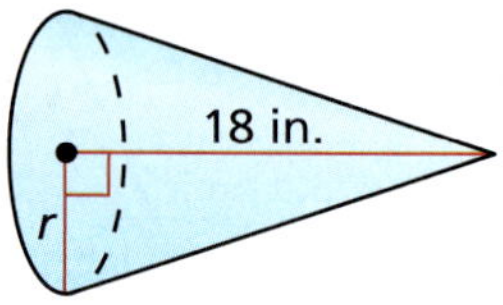

15. A cone has a volume of 48π cubic feet and a height of 9 feet. Find the radius.

16. A cone has a volume of 245π cubic yards and a diameter of 14 yards. Find the height.

In Exercises 17 and 18, solids A and B are similar. Use the given information and scale factor k from solid A to solid B to find the volume of solid B.

17. The volume of cone A is 16π cubic centimeters and $k = \frac{3}{2}$.

18. The volume of cone A is 135π cubic yards and $k = \frac{1}{3}$.

In Exercises 19 and 20, the solids are similar. Find the volume of solid B. *(See Example 5.)*

19.

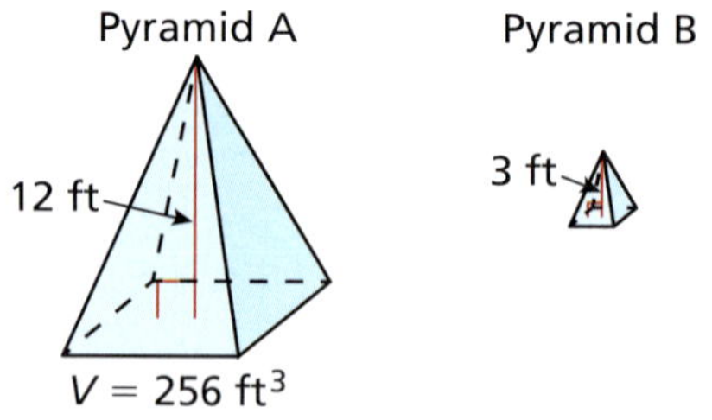

20.

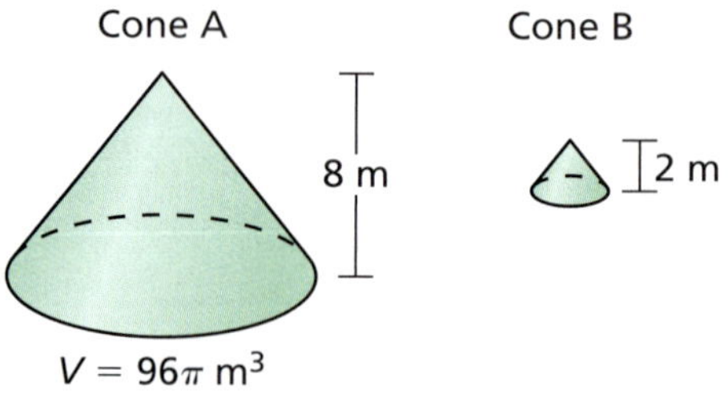

GO DIGITAL

In Exercises 21–24, find the volume of the composite solid. *(See Example 6.)*

21. 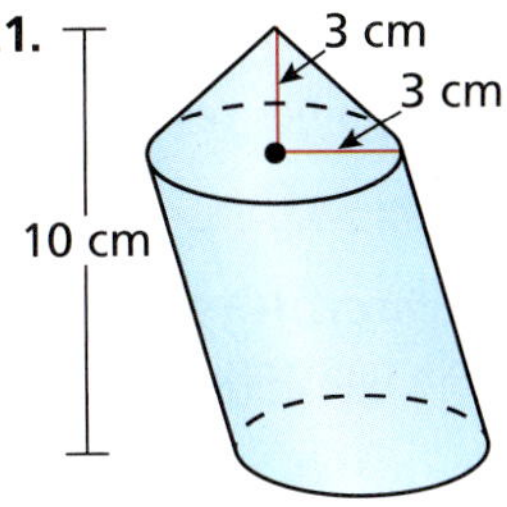

22. 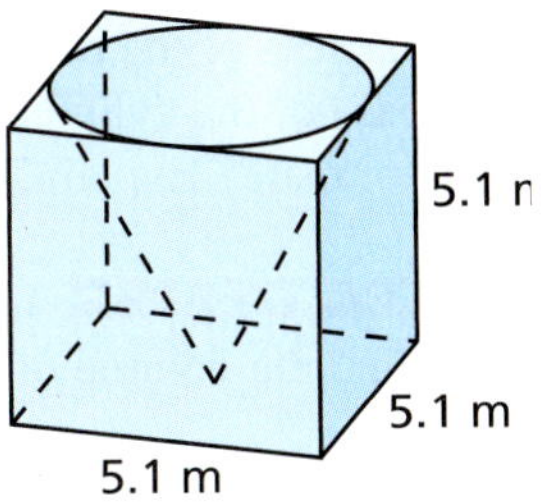

23. 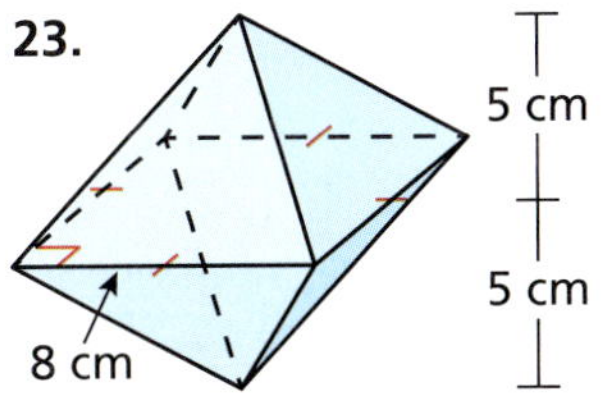

24. 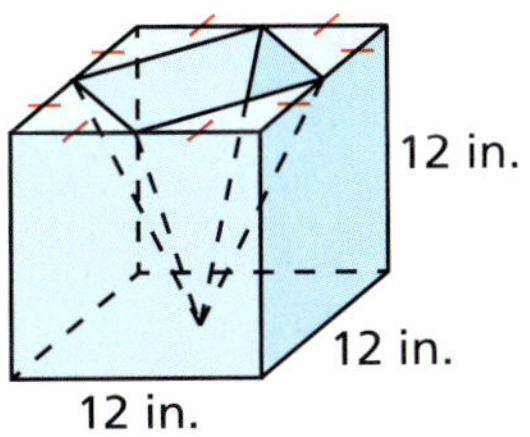

25. **MODELING REAL LIFE** Before the use of electricity, nautical deck prisms were used as a safe way to illuminate decks on ships. This deck prism is composed of three solids: a regular hexagonal prism with an edge length of 3.5 inches and a height of 1.5 inches, a regular hexagonal prism with an edge length of 3.25 inches and a height of 0.25 inch, and a regular hexagonal pyramid with an edge length of 3.25 inches and a height of 3 inches. Find the volume of the deck prism.

26. **MODELING REAL LIFE** A pile of road salt has the dimensions shown. After a winter storm, the linear dimensions of the pile are one-half of the original dimensions.

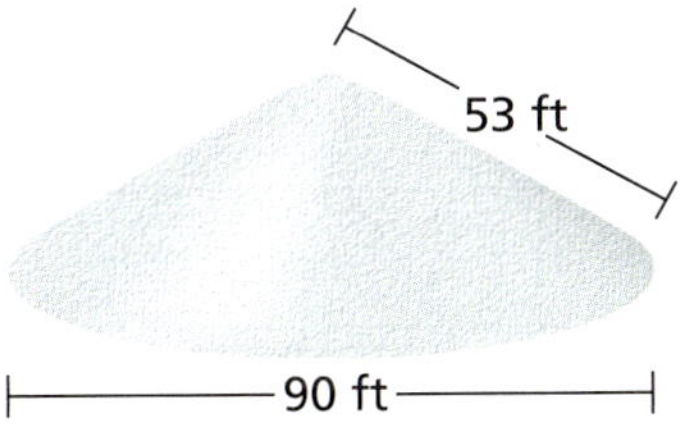

a. How does this change affect the volume of the pile?

b. A *lane mile* is an area of pavement that is one mile long and one lane wide. During the storm, about 400 pounds of road salt were used for every lane mile. Estimate the number of lane miles that were covered with road salt during the storm. A cubic foot of road salt weighs about 80 pounds.

27. **REASONING** A cone has height h and a base with radius r. You want to change the cone so its volume is doubled. What is the new height if you change only the height? What is the new radius if you change only the radius? Explain.

28. **STRUCTURE** A pyramid has a height of 8 feet and a square base with a side length of 6 feet.

a. How does the volume of the pyramid change when the base stays the same and the height is doubled?

b. How does the volume of the pyramid change when the height stays the same and the side length of the base is doubled?

c. Are your answers to parts (a) and (b) true for any square pyramid? Explain your reasoning.

29. **PROBLEM SOLVING** During a chemistry lab, you use a funnel to pour a solvent into a flask. The radius of the base of the funnel is 5 centimeters, and its height is 10 centimeters. You pour the solvent into the funnel at a rate of 80 milliliters per second, and the solvent flows out of the funnel at a rate of 65 milliliters per second. How long will it be before the funnel overflows? ($1\text{ mL} = 1\text{ cm}^3$)

30. **HOW DO YOU SEE IT?**
The rectangular prism shown is formed by three congruent pyramids. How can you use this to verify the formula for the volume of a pyramid?

31. **MAKING AN ARGUMENT** In the figure, the two cylinders are congruent. The combined height of the two smaller cones equals the height of the larger cone. Does this mean the total volume of the two smaller cones is equal to the volume of the larger cone? Justify your answer.

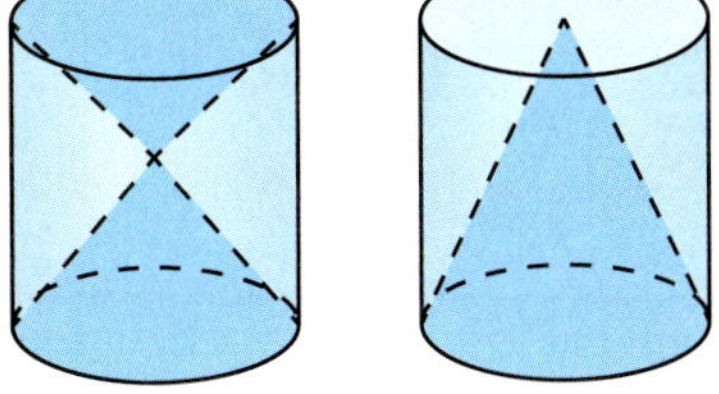

GO DIGITAL

32. THOUGHT PROVOKING
A *frustum* of a pyramid or cone is the part of the solid that lies between the base and a plane parallel to the base, as shown. Write a formula for the volume of the frustum of a square pyramid in terms of a, b, and h. Then write a formula for the volume of the frustum of a cone in terms of s, r, and h.

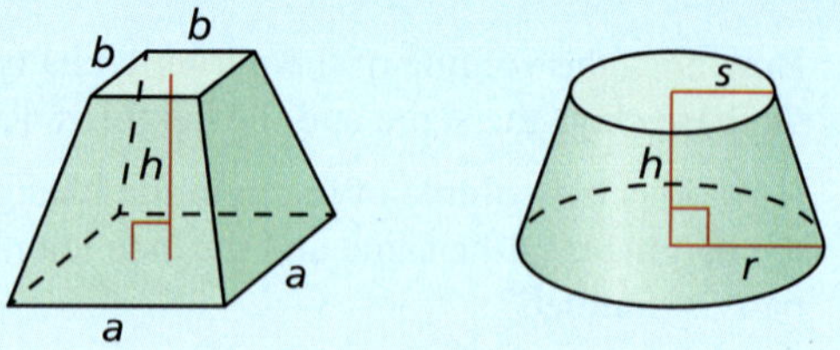

33. REASONING A right cone is inscribed in a cylinder, and a square pyramid is inscribed in the right cone so that all three solids have the same height and the vertices of the cone and pyramid lie on the center of a base of the cylinder. How do the volumes of the solids compare? Explain.

34. DIG DEEPER In the diagram, $m\angle ABC = 35°$. Find the volume of the regular pentagonal pyramid.

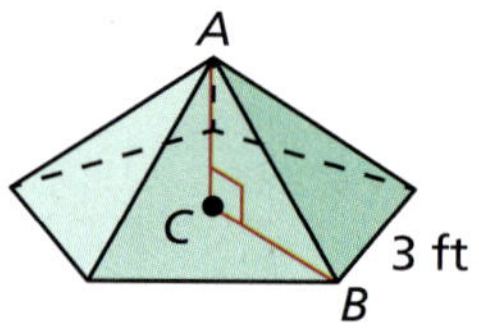

REVIEW & REFRESH

WATCH

In Exercises 35–38, find the value of x.

35.

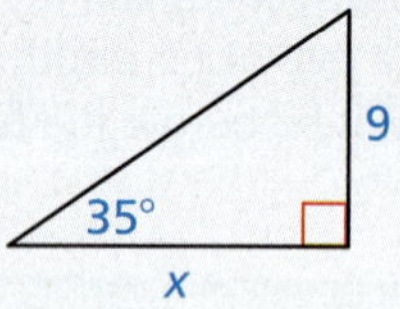

36.

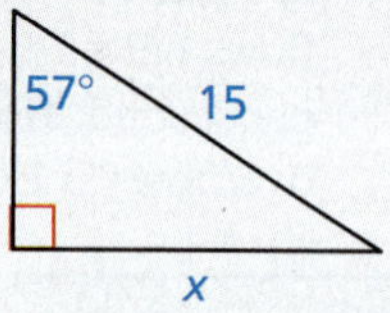

37.

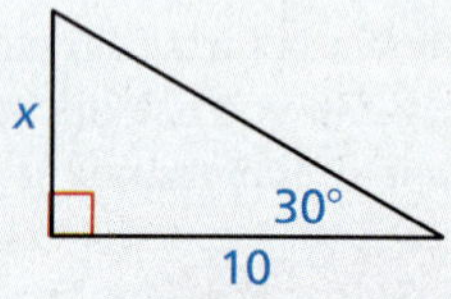

38.

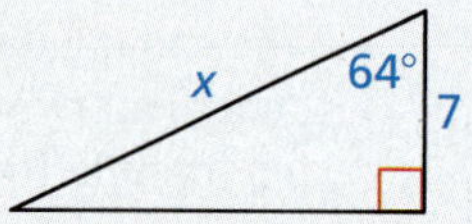

39. A circular region has a population of about 175,000 people and a population density of about 580 people per square mile. Find the radius of the region.

In Exercises 40 and 41, find the surface area of the solid.

40.

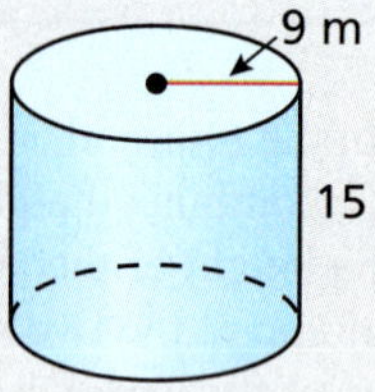

41.

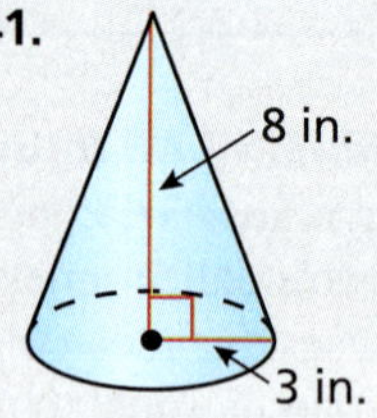

In Exercises 42 and 43, find the volume of the solid.

42.

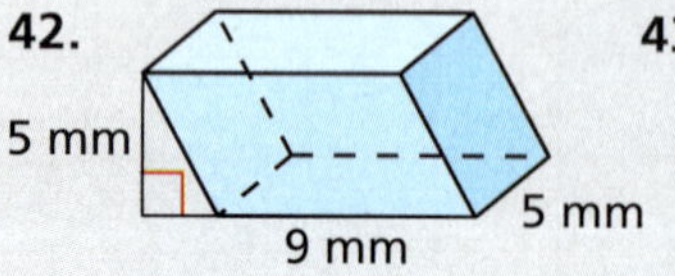

43.

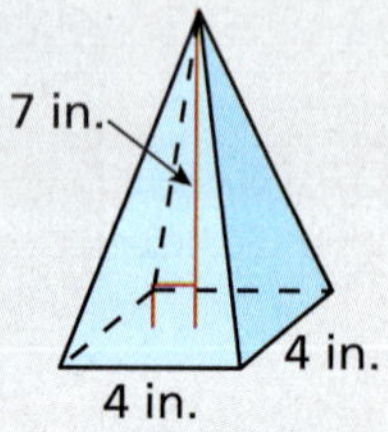

44. Describe the shape formed by the intersection of the plane and the solid.

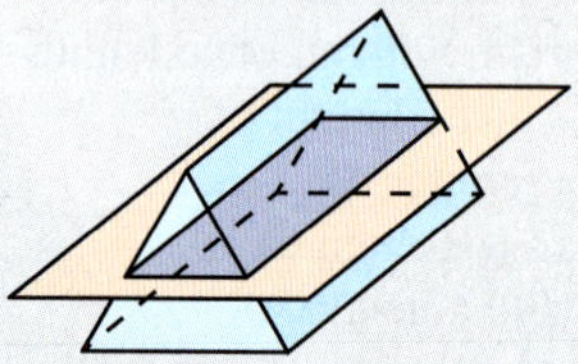

45. The vertices of $\triangle JKL$ are $J(4, 1)$, $K(0, -2)$, and $L(-1, 3)$. Translate $\triangle JKL$ using the vector $\langle -4, 5 \rangle$. Graph $\triangle JKL$ and its image.

7 MTR **46. MODELING REAL LIFE** The diagram shows the location of a campsite and a hiking trail. You want to choose a campsite that is at least 100 feet from the trail. Does this campsite meet your requirement? Explain.

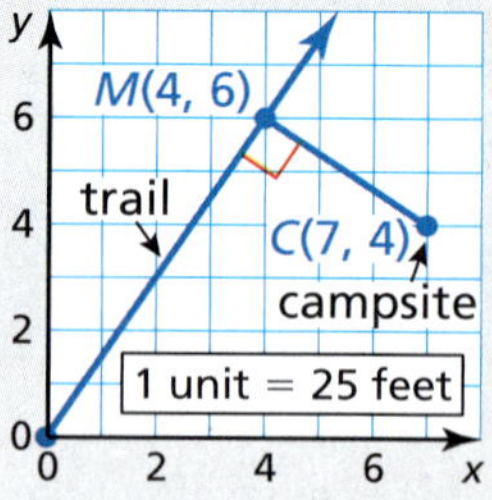

47. Let p be "it is Thanksgiving" and let q be "it is November." Write the conditional statement $p \rightarrow q$ and the contrapositive $\sim q \rightarrow \sim p$ in words. Then decide whether each statement is true or false.

48. Find the distance between $(-3, -9)$ and $(5, -3)$ in the coordinate plane.

GO DIGITAL

12.6 Surface Areas and Volumes of Spheres

Learning Target: Find and use surface areas and volumes of spheres.

Success Criteria:
- I can find surface areas of spheres.
- I can find volumes of spheres.
- I can find the volumes of composite solids.

EXPLORE IT! Finding Surface Area and Volume Formulas for Spheres

Work with a partner.

a. Two identical pieces of material make up the covering of a baseball with radius r. Use the circle with radius r and the covering of the ball to estimate the surface area S of the ball in terms of r.

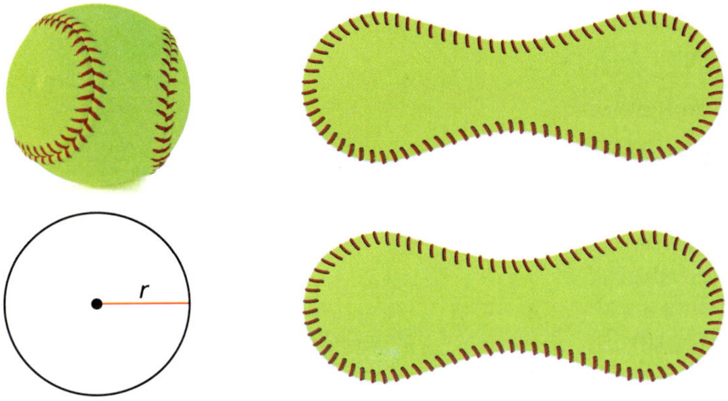

b. A cylinder is circumscribed about a sphere, as shown. Write a formula for the volume V of the cylinder in terms of the radius r.

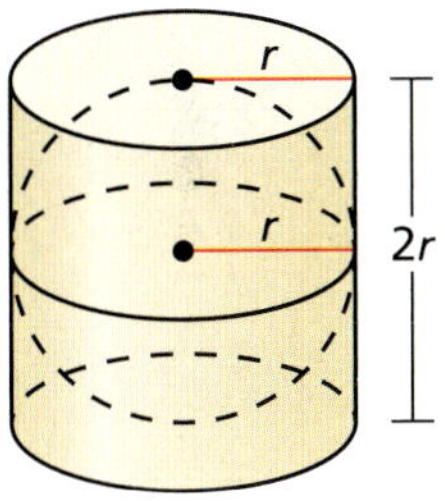

c. When half of the sphere (a *hemisphere*) from part (b) is filled with sand and poured into the cylinder, it takes three hemispheres to fill the cylinder. Use this information to write a formula for the volume V of a sphere in terms of the radius r.

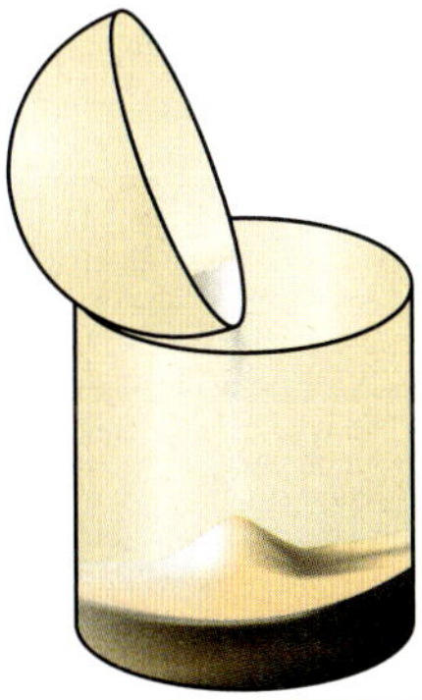

6 MTR ASSESS REASONABLENESS
How can you use a rectangular prism to determine whether your formulas are reasonable?

Geometric Reasoning
MA.912.GR.4.5 Solve mathematical and real-world problems involving the volume of three-dimensional figures limited to cylinders, pyramids, prisms, cones and spheres.
MA.912.GR.4.6 Solve mathematical and real-world problems involving the surface area of three-dimensional figures limited to cylinders, pyramids, prisms, cones and spheres.

GO DIGITAL

Finding Surface Areas of Spheres

Vocabulary

chord of a sphere, *p. 654*
great circle, *p. 654*

A *sphere* is the set of all points in space equidistant from a given point. This point is called the *center* of the sphere. A *radius* of a sphere is a segment from the center to a point on the sphere. A **chord of a sphere** is a segment whose endpoints are on the sphere. A *diameter* of a sphere is a chord that contains the center.

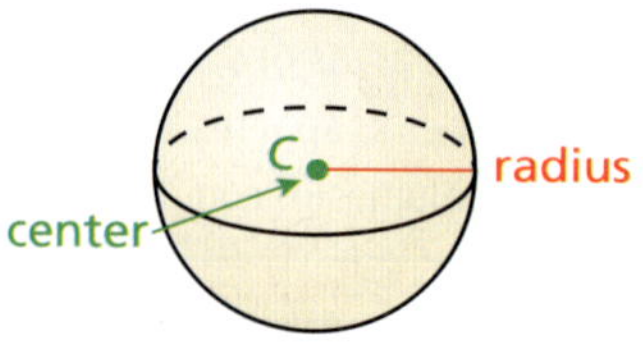

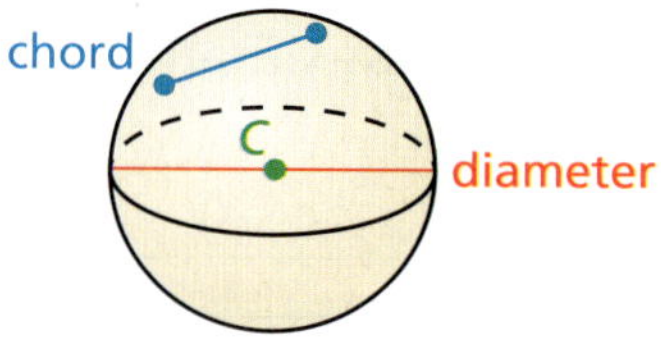

As with circles, the terms *radius* and *diameter* also represent distances, and the diameter is twice the radius.

If a plane intersects a sphere, then the intersection is either a single point or a circle. If the plane contains the center of the sphere, then the intersection is a **great circle** of the sphere. The circumference of a great circle is the circumference of the sphere. Every great circle of a sphere separates the sphere into two congruent halves called *hemispheres*.

WORDS AND MATH

You may be familiar with the word *hemisphere* as it applies to Earth. The Northern Hemisphere and Southern Hemisphere are divided by the equator, which approximates a great circle of Earth.

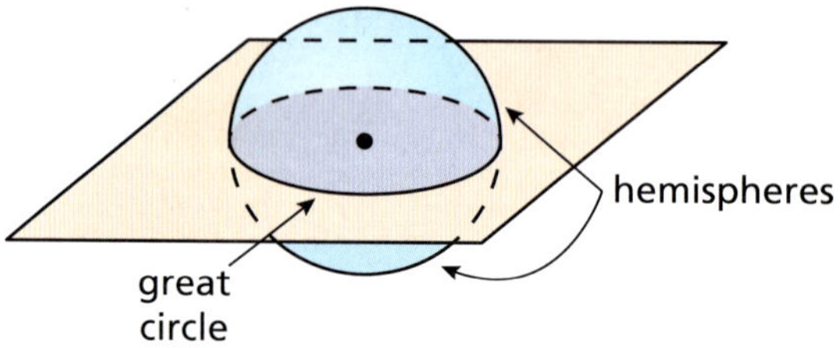

KEY IDEA

Surface Area of a Sphere

The surface area S of a sphere is

$$S = 4\pi r^2$$

where r is the radius of the sphere.

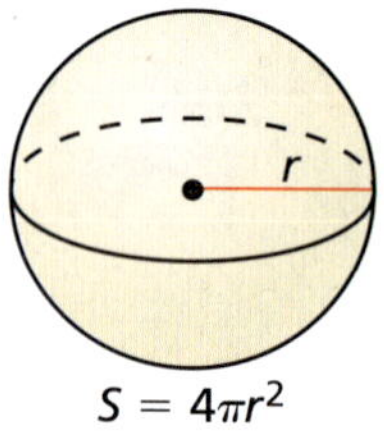

To check whether the surface area formula is reasonable, think of a baseball with radius r.

The covering of the ball consists of two congruent shapes. The area of each shape can be approximated using two great circles of the ball, as shown. The area A of each great circle is $A = \pi r^2$, so it makes sense that the formula for the surface area S of a sphere with radius r is $S = 4\pi r^2$.

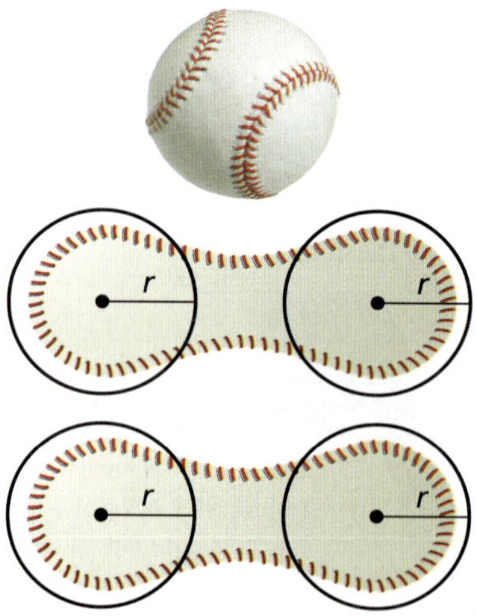

GO DIGITAL

EXAMPLE 1 Finding the Surface Areas of Spheres

Find the surface area of each sphere.

a.

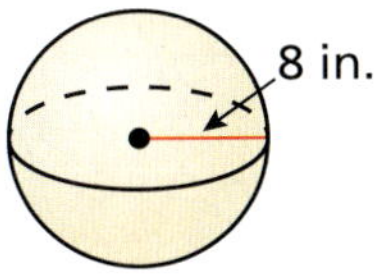

b.

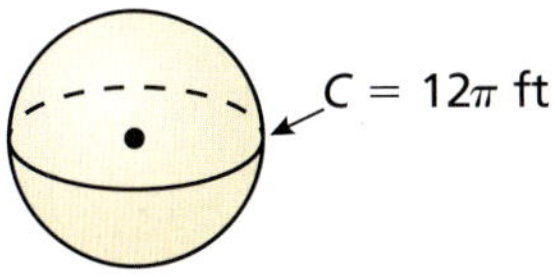

SOLUTION

a. $S = 4\pi r^2$ Formula for surface area of a sphere

$= 4\pi(8)^2$ Substitute 8 for *r*.

$= 256\pi$ Simplify.

≈ 804.25 Use technology.

▶ The surface area is about 804.25 square inches.

b. The circumference is 12π feet, so the radius of the sphere is $\frac{12\pi}{2\pi} = 6$ feet.

$S = 4\pi r^2$ Formula for surface area of a sphere

$= 4\pi(6)^2$ Substitute 6 for *r*.

$= 144\pi$ Simplify.

≈ 452.39 Use technology.

▶ The surface area is about 452.39 square feet.

EXAMPLE 2 Finding a Length in a Sphere

Find the diameter of the sphere.

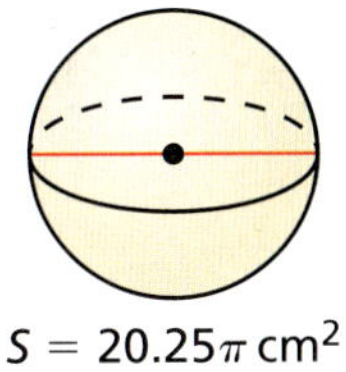

SOLUTION

$S = 4\pi r^2$ Formula for surface area of a sphere

$20.25\pi = 4\pi r^2$ Substitute 20.25π for *S*.

$5.0625 = r^2$ Divide each side by 4π.

$2.25 = r$ Take the positive square root of each side.

▶ The diameter is $2r = 2 \cdot 2.25 = 4.5$ centimeters.

COMMON ERROR

Be sure to multiply the value of *r* by 2 to find the diameter.

SELF-ASSESSMENT

1 I don't understand yet. 2 I can do it with help. 3 I can do it on my own. 4 I can teach someone else.

Find the surface area of the sphere.

1.

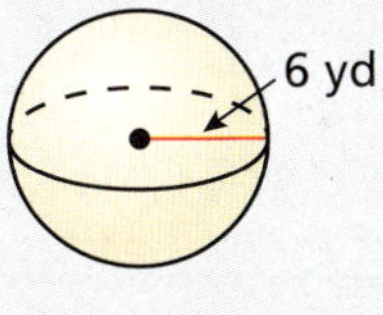

2.

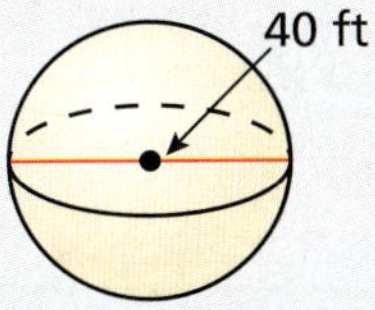

3.

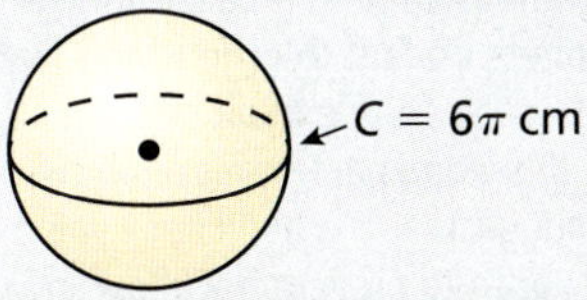

4. A sphere has a surface area of 30π square meters. Find the radius of the sphere.

GO DIGITAL

Finding Volumes of Spheres

The figure shows a hemisphere and a cylinder with a cone removed. A plane parallel to their bases intersects the solids z units above their bases.

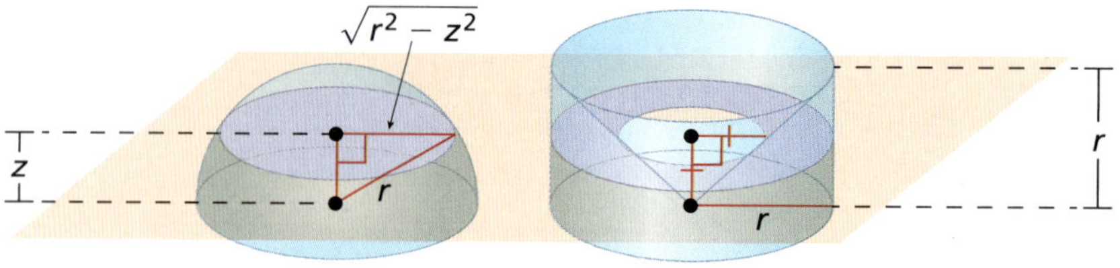

Using the AA Similarity Theorem, you can show that the radius of the cross section of the cone at height z is z. The area of the cross section formed by the plane is $\pi(r^2 - z^2)$ for both solids. Because the solids have the same height and the same cross-sectional area at every level, they have the same volume by Cavalieri's Principle.

$$V_{\text{hemisphere}} = V_{\text{cylinder}} - V_{\text{cone}}$$
$$= \pi r^2(r) - \frac{1}{3}\pi r^2(r)$$
$$= \frac{2}{3}\pi r^3$$

So, the volume of a sphere of radius r is

$$2 \cdot V_{\text{hemisphere}} = 2 \cdot \frac{2}{3}\pi r^3 = \frac{4}{3}\pi r^3.$$

1 MTR HELP A CLASSMATE

Help a classmate understand how to use the AA Similarity Theorem to show that the radius of the cross section of the cone at height z is z.

KEY IDEA

Volume of a Sphere

The volume V of a sphere is

$$V = \frac{4}{3}\pi r^3$$

where r is the radius of the sphere.

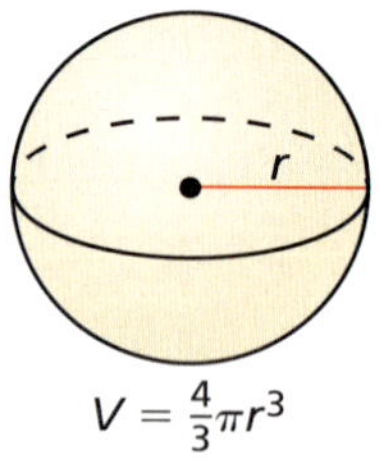

EXAMPLE 3 Finding the Volume of a Sphere

WATCH

Find the volume of the sphere.

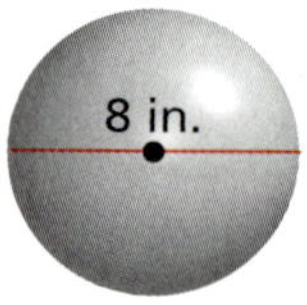

SOLUTION

$V = \frac{4}{3}\pi r^3$	Formula for volume of a sphere
$= \frac{4}{3}\pi(4)^3$	Substitute 4 for r.
$= \frac{256}{3}\pi$	Simplify.
≈ 268.08	Use technology.

▶ The volume of the sphere is about 268.08 cubic inches.

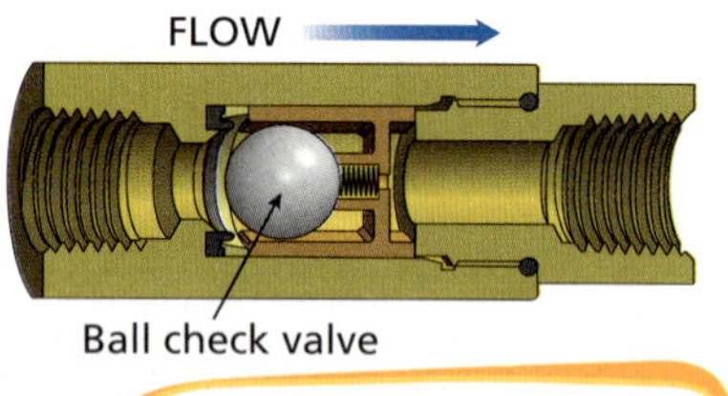

In 1974, a family in Florida famously found an 8-inch metal sphere on Fort George Island that seemed to exhibit strange phenomena. Recent analysis suggests it is a ball check valve used to prevent backflow in water pipes.

GO DIGITAL

EXAMPLE 4 Finding the Volume of a Sphere

WATCH

The surface area of a sphere is 324π square centimeters. Find the volume of the sphere.

SOLUTION

Step 1 Use the surface area to find the radius.

$S = 4\pi r^2$	Formula for surface area of a sphere
$324\pi = 4\pi r^2$	Substitute 324π for S.
$81 = r^2$	Divide each side by 4π.
$9 = r$	Take the positive square root of each side.

The radius is 9 centimeters.

Step 2 Use the radius to find the volume.

$V = \frac{4}{3}\pi r^3$	Formula for volume of a sphere
$= \frac{4}{3}\pi(9)^3$	Substitute 9 for r.
$= 972\pi$	Simplify.
≈ 3053.63	Use technology.

▶ The volume is about 3053.63 cubic centimeters.

EXAMPLE 5 Finding the Volume of a Composite Solid

WATCH

Find the volume of the composite solid.

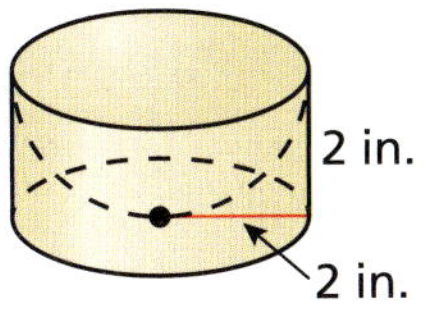

SOLUTION

Volume of solid = Volume of cylinder − Volume of hemisphere

$= \pi r^2 h - \frac{1}{2}\left(\frac{4}{3}\pi r^3\right)$	Write formulas.
$= \pi(2)^2(2) - \frac{1}{2}\left(\frac{4}{3}\pi(2)^3\right)$	Substitute 2 for r and 2 for h.
$= 8\pi - \frac{16}{3}\pi$	Multiply.
$= \frac{24}{3}\pi - \frac{16}{3}\pi$	Rewrite fractions using least common denominator.
$= \frac{8}{3}\pi$	Subtract.
≈ 8.38	Use technology.

▶ The volume is about 8.38 cubic inches.

SELF-ASSESSMENT

1 I don't understand yet. 2 I can do it with help. 3 I can do it on my own. 4 I can teach someone else.

5. The radius of a sphere is 5 yards. Find the volume of the sphere.

6. The diameter of a sphere is 36 inches. Find the volume of the sphere.

7. The surface area of a sphere is 576π square centimeters. Find the volume of the sphere.

8. Find the volume of the composite solid at the right.

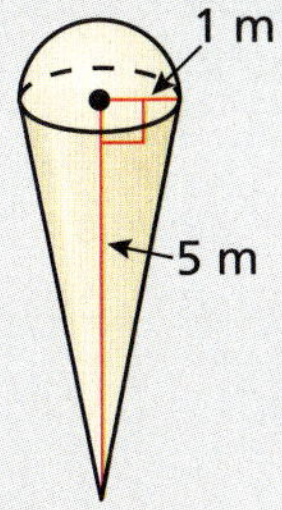

GO DIGITAL

12.6 Practice WITH CalcChat® AND CalcView®

In Exercises 1–4, find the surface area of the sphere. *(See Example 1.)*

▶ **1.**
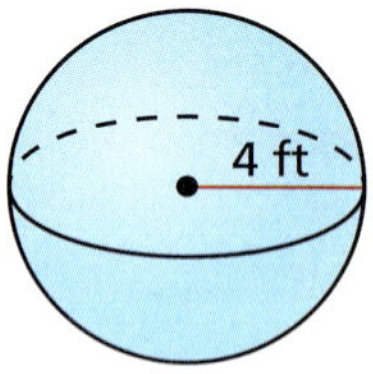

2.
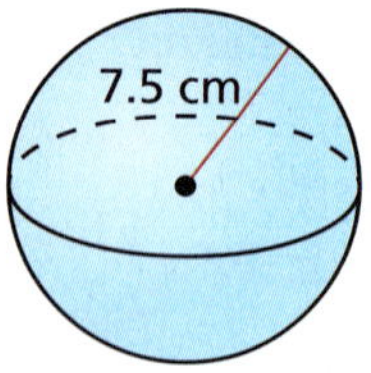

3.
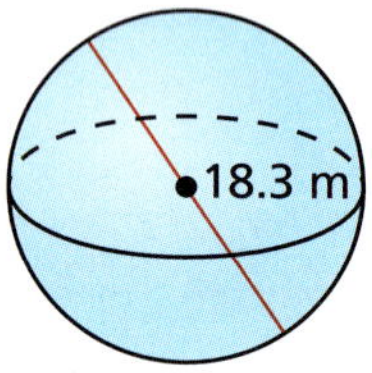

4.
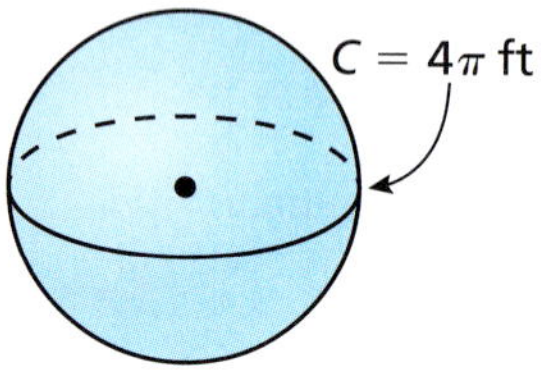

In Exercises 5–8, find the indicated measure. *(See Example 2.)*

5. Find the radius of a sphere with a surface area of 4π square feet.

6. Find the radius of a sphere with a surface area of 1024π square inches.

▶ **7.** Find the diameter of a sphere with a surface area of 900π square meters.

8. Find the diameter of a sphere with a surface area of 196π square centimeters.

In Exercises 9–14, find the volume of the sphere. *(See Example 3.)*

▶ **9.**
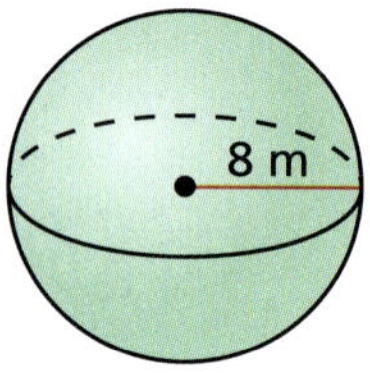

10.
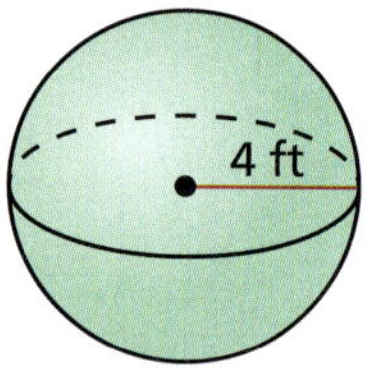

11.
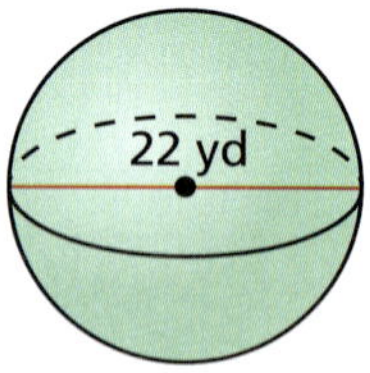

12.
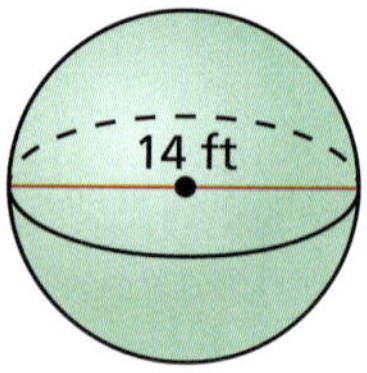

13.
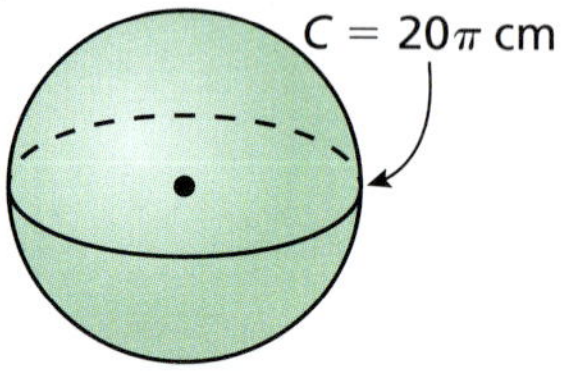

14.
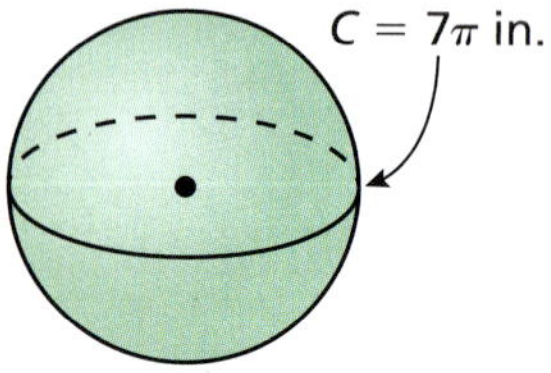

4 MTR ERROR ANALYSIS In Exercises 15 and 16, describe and correct the error in finding the volume of the sphere.

15.
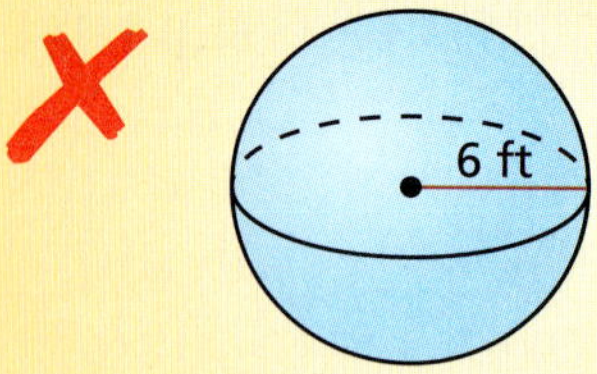

$$V = \frac{4}{3}\pi(6)^2$$
$$= 48\pi$$
$$\approx 150.80 \text{ ft}^3$$

16.
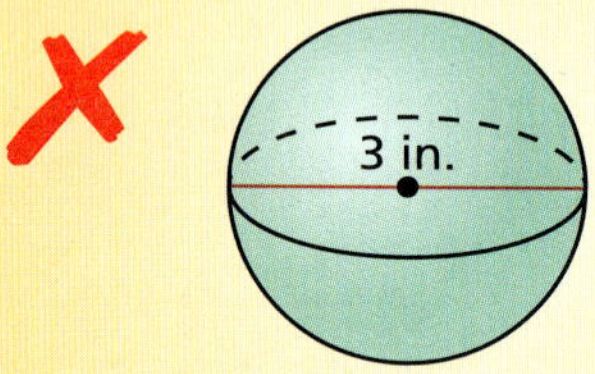

$$V = \frac{4}{3}\pi(3)^3$$
$$= 36\pi$$
$$\approx 113.10 \text{ in.}^3$$

In Exercises 17 and 18, find the surface area and volume of the hemisphere.

17.
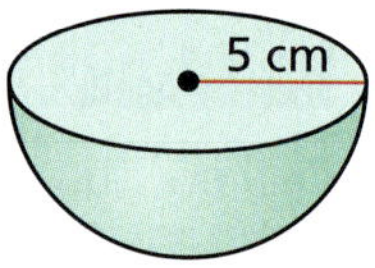

18.
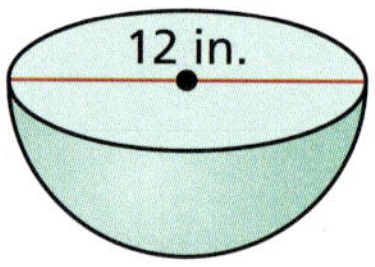

In Exercises 19 and 20, find the volume of the sphere with the given surface area. *(See Example 4.)*

▶ **19.** Surface area $= 16\pi$ ft^2

20. Surface area $= 484\pi$ cm^2

In Exercises 21–24, find the volume of the composite solid. *(See Example 5.)*

▶ **21.**
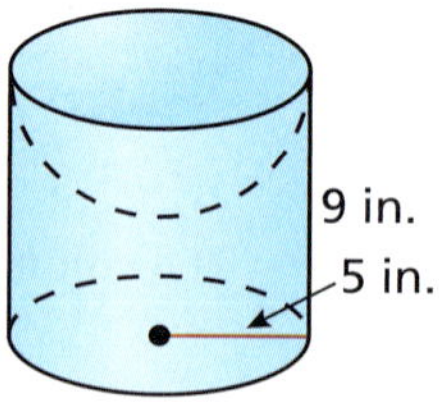

22.
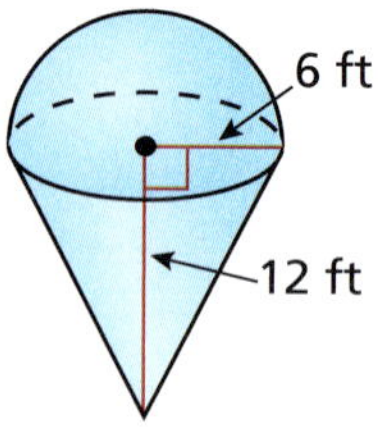

23.
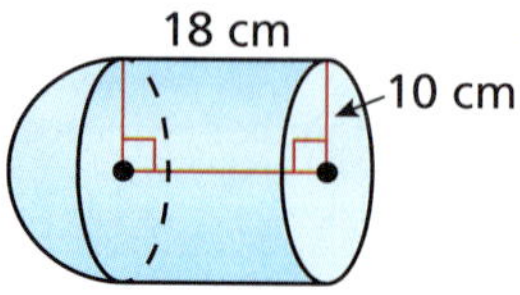

24.
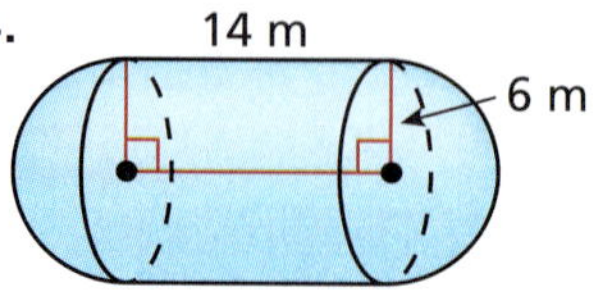

GO DIGITAL

25. **MODELING REAL LIFE** A silo has the dimensions shown. The top of the silo is a hemisphere. Find the volume of the silo.

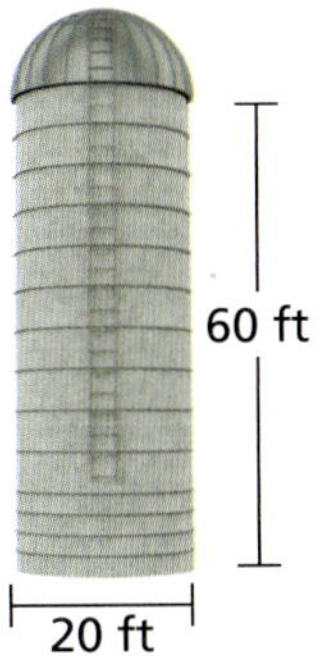

26. **MODELING REAL LIFE** Three tennis balls are stored in a cylindrical container with a height of 8 inches and a radius of 1.43 inches. The circumference of a tennis ball is 8 inches. Find the amount of space within the cylinder not taken up by the tennis balls.

In Exercises 27–30, spheres A and B are similar. Use the given information and scale factor k from sphere A to sphere B to find the indicated measure.

27. The surface area of sphere A is 144π square inches and $k = 4$. Find the surface area of sphere B.

28. The surface area of sphere A is 68π square yards and $k = \frac{2}{3}$. Find the surface area of sphere B.

29. The volume of sphere A is 120π cubic feet and $k = 2$. Find the volume of sphere B.

30. The volume of sphere A is 81π cubic centimeters and $k = \frac{3}{4}$. Find the volume of sphere B.

31. **MODELING REAL LIFE** The largest rubber-band ball ever recorded was measured in 2008 in Lauderhill, Florida. The ball stands 6 feet 7 inches high and weighs 9032 pounds.

a. Assume the ball is a perfect sphere. Find the surface area and volume of the ball.

b. Make a ball out of rubber bands. Is your ball similar to the largest rubber-band ball? Explain why or why not. What conclusion can you make about all spheres?

GO DIGITAL

32. **COLLEGE PREP** A sphere has a diameter of $4(x + 3)$ centimeters and a surface area of 784π square centimeters. What is the value of x?

Ⓐ $x \approx -0.91$
Ⓑ $x = 0.5$
Ⓒ $x \approx 1.19$
Ⓓ $x = 4$

33. **MODELING REAL LIFE** The radius of Earth is about 4000 miles. The radius of the moon is about 1080 miles.

a. Find the surface area of Earth and the moon.

b. Compare the surface areas of Earth and the moon.

c. About 70% of the surface of Earth is water. How many square miles of water are on Earth's surface?

34. **MODELING REAL LIFE** The Torrid Zone on Earth is the area between the Tropic of Cancer and the Tropic of Capricorn.

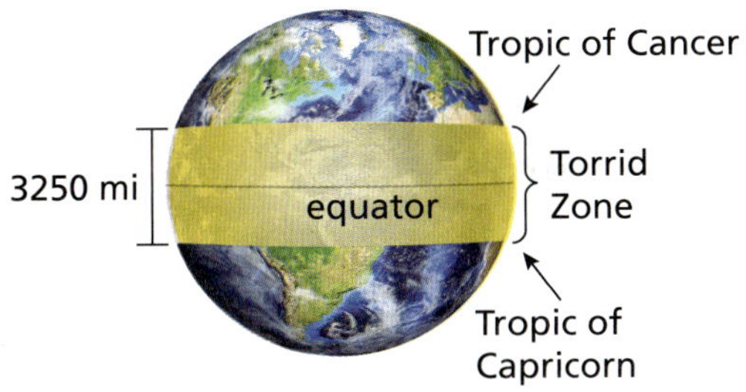

A meteorite is equally likely to hit anywhere on Earth. Estimate the probability that a meteorite will land in the Torrid Zone. (The radius of Earth is about 4000 miles.)

35. **STRUCTURE** Let V be the volume of a sphere, S be the surface area of the sphere, and r be the radius of the sphere. Write an equation for V in terms of r and S. $\left(\textit{Hint}\text{: Start with the ratio } \frac{V}{S}.\right)$

36. **HOW DO YOU SEE IT?**
The formula for the volume of a hemisphere and a cone are shown. If each solid has the same radius and $r = h$, which solid has the greater volume? Explain your reasoning.

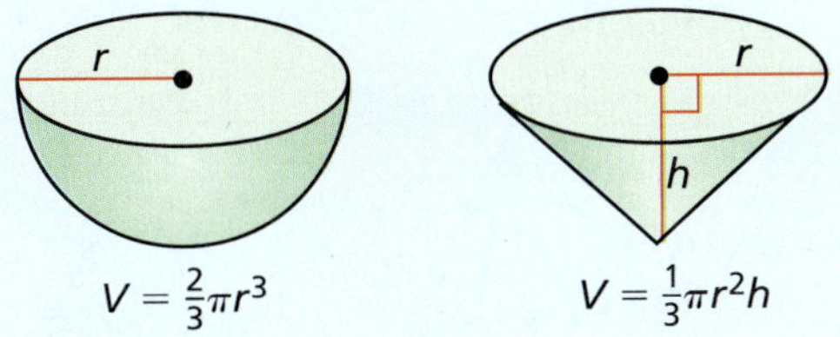

7 MTR **37. PERFORMANCE TASK** The propane tank shown is used as the primary heat source for your residence. Research to estimate the amount of propane that you will use each year. Then estimate your average monthly propane cost. About how often will the tank need to be refilled?

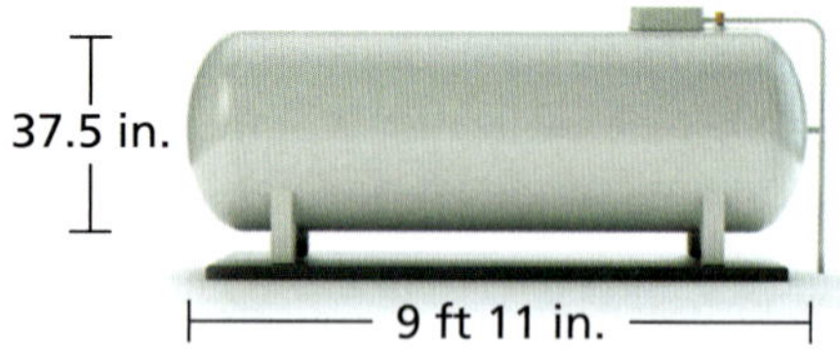

38. THOUGHT PROVOKING
A *spherical lune* is the region between two great circles of a sphere. Find the formula for the surface area of a lune.

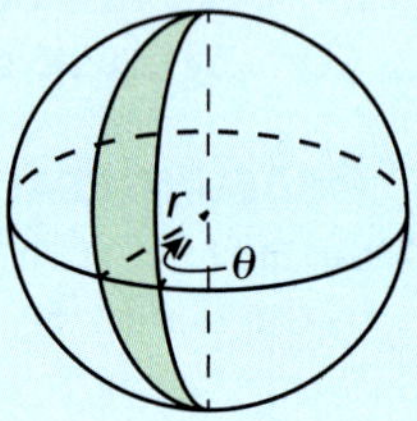

5 MTR **39. CONNECTING CONCEPTS** A *spherical cap* is a portion of a sphere cut off by a plane. The formula for the volume of a spherical cap is $V = \frac{\pi h}{6}(3a^2 + h^2)$, where a is the radius of the base of the cap and h is the height of the cap. Find the volume of a spherical cap when the radius of the sphere is 13 meters and the height of the cap is 8 meters.

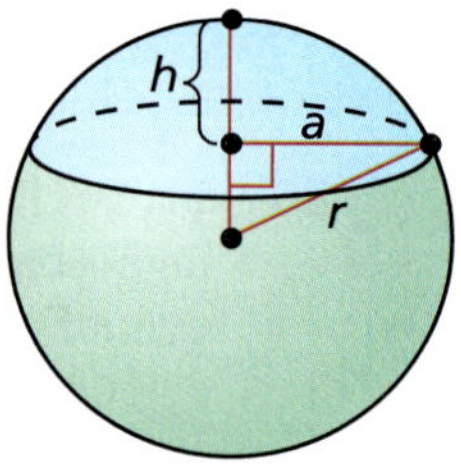

40. DIG DEEPER A sphere with a radius of 2 inches is inscribed in a right cone with a height of 6 inches. Find the surface area and the volume of the cone.

REVIEW & REFRESH

WATCH

41. Find all unknown measures in the triangle.

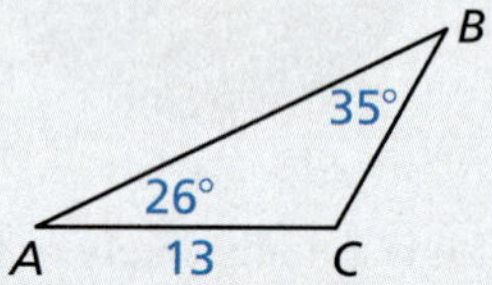

7 MTR **42. MODELING REAL LIFE** The diagram shows the portion of Earth visible from an airplane flying above Earth at point B. Earth's radius is approximately 4000 miles and $m\widehat{AC} = 6.5°$. How high is the airplane above Earth?

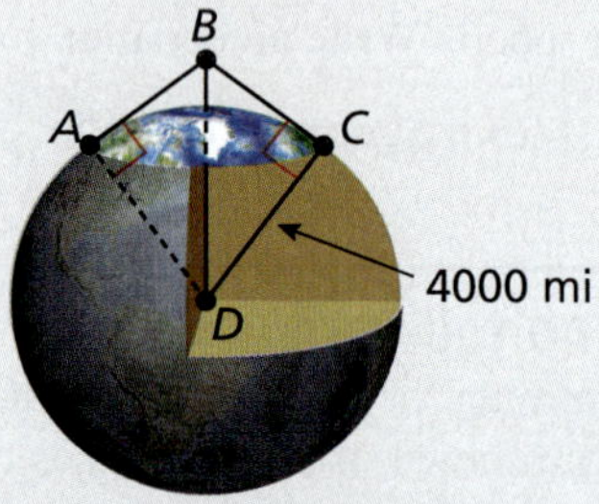

Not drawn to scale

43. The pyramids are similar. Find the volume of pyramid B.

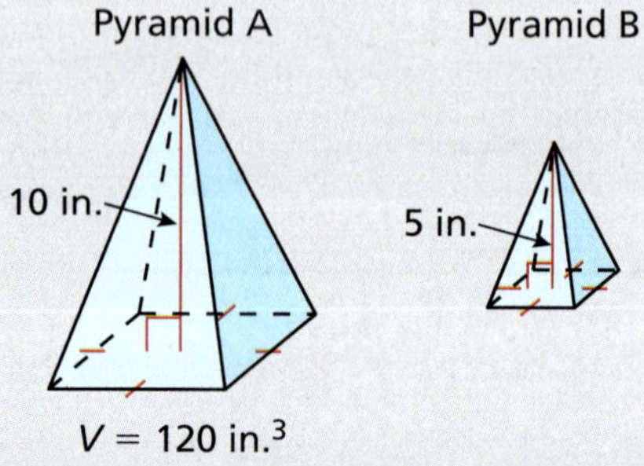

44. Find the missing dimension of the cylinder.

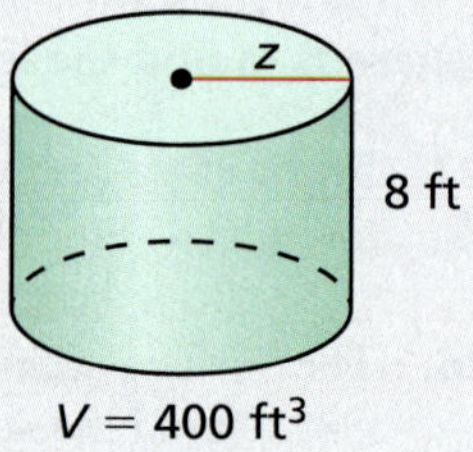

45. In rectangle $WXYZ$, $WY = 4x + 7$ and $XZ = 6x - 3$. Find the lengths of the diagonals of $WXYZ$.

46. Determine whether $\overline{KM} \parallel \overline{JN}$. Explain your reasoning.

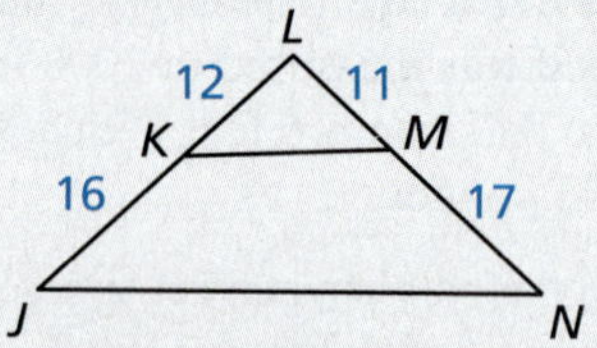

In Exercises 47 and 48, find the surface area and the volume of the solid.

47.

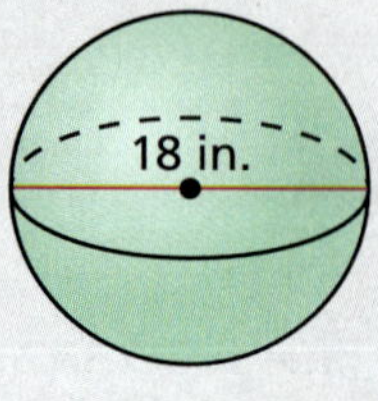

48.

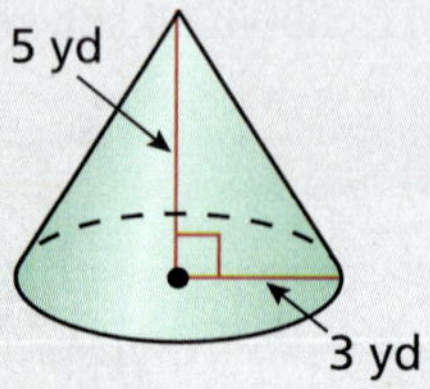

GO DIGITAL

12.7 Modeling with Surface Area and Volume

Learning Target: Understand the concept of density and modeling with volume.

Success Criteria:
- I can explain what density means.
- I can use the formula for density to solve problems.
- I can use geometric shapes to model objects.
- I can solve modeling problems.

EXPLORE IT! Finding Densities

Work with a partner.

a. Approximate the volume of each object with the given mass. Then find the mass per unit of volume, or *density*, of each object.

i. Brick: 2.3 kg

ii. Log: 18.1 kg

iii. Golf ball: 45.9 g

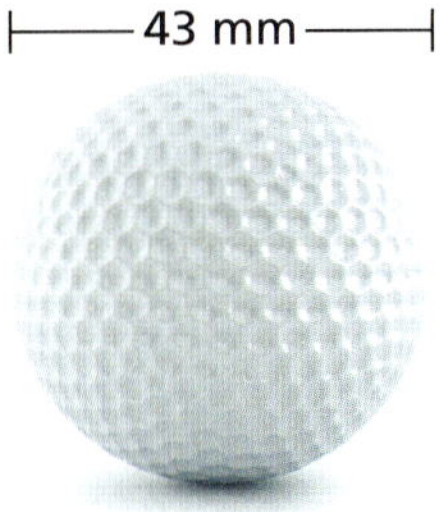

iv. Cork: 2.6 g

b. The objects in part (a) with a density greater than 1 gram per cubic centimeter will sink in water. The objects with a density less than 1 gram per cubic centimeter will float in water. Which of the object(s) sink in water? Which of the object(s) float? Explain your reasoning.

c. Do your answers in part (b) change when each object is cut in half? Explain your reasoning.

d. You dissolve salt in a bucket of water until one of the sunken objects floats to the surface. Which object is it? Why do you think this happens?

1 MTR ANALYZE A PROBLEM

In your own words, explain what the units represent for one of the densities you calculated in part (a). Are any of these densities better represented using different units? Explain.

Geometric Reasoning

MA.912.GR.4.5 Solve mathematical and real-world problems involving the volume of three-dimensional figures limited to cylinders, pyramids, prisms, cones and spheres.

MA.912.GR.4.6 Solve mathematical and real-world problems involving the surface area of three-dimensional figures limited to cylinders, pyramids, prisms, cones and spheres.

GO DIGITAL

Using Volume Formulas

Vocabulary
density, *p. 662*

Density is the amount of matter that an object has in a given unit of volume. The density of an object is calculated by dividing its mass by its volume.

$$\text{Density} = \frac{\text{Mass}}{\text{Volume}}$$

Different materials have different densities, so density can be used to distinguish between materials that look similar. For example, table salt and sugar look alike. However, table salt has a density of 2.16 grams per cubic centimeter, whereas sugar has a density of 1.58 grams per cubic centimeter.

EXAMPLE 1 Using the Formula for Density

The diagram shows the dimensions of a standard gold bar at Fort Knox. Gold has a density of about 19.3 grams per cubic centimeter. Find the mass of the gold bar.

SOLUTION

The units for density are given in grams per cubic centimeter, and the dimensions of the gold bar are given in inches. So, convert the dimensions to centimeters. Then use the formula for density to find the mass.

Step 1 Convert the dimensions to centimeters using 1 inch = 2.54 centimeters.

Length $7\ \cancel{\text{in.}} \cdot \dfrac{2.54\text{ cm}}{1\ \cancel{\text{in.}}} = 17.78\text{ cm}$

Width $3.625\ \cancel{\text{in.}} \cdot \dfrac{2.54\text{ cm}}{1\ \cancel{\text{in.}}} = 9.2075\text{ cm}$

Height $1.75\ \cancel{\text{in.}} \cdot \dfrac{2.54\text{ cm}}{1\ \cancel{\text{in.}}} = 4.445\text{ cm}$

Step 2 Let x represent the mass (in grams) of the gold bar. Use the formula for density to find x. Because the gold bar is shaped like a rectangular prism, use $V = Bh = \ell wh$ for the volume.

$$\text{Density} = \frac{\text{Mass}}{\text{Volume}} \qquad \text{Formula for density}$$

$$19.3 \approx \frac{x}{17.78(9.2075)(4.445)} \qquad \text{Substitute.}$$

$$14{,}044 \approx x \qquad \text{Solve for } x.$$

▶ The mass of the gold bar is about 14,044 grams.

SELF-ASSESSMENT

1 I don't understand yet. 2 I can do it with help. 3 I can do it on my own. 4 I can teach someone else.

1. A concrete cylinder has a radius of 24 inches and a height of 32 inches. The density of concrete is 2.3 grams per cubic centimeter. Find the mass of the concrete cylinder.

2. DIFFERENT WORDS, SAME QUESTION Which is different? Find "both" answers.

What is the mass per unit of volume?	What is the mass in kilograms?
What is the mass divided by the volume?	What is the density?

Tourmaline

Mass = 6.2 g
Volume = 2 cm^3

GO DIGITAL

EXAMPLE 2 Using a Volume Formula

WATCH

A tree harvester estimates the trunk of a sequoia tree to have a height of about 50 meters and a base diameter of about 0.8 meter.

a. The wood of a sequoia tree has a density of about 450 kilograms per cubic meter. Find the mass of the trunk.

b. Each year, the tree trunk forms new cells that arrange themselves in concentric circles called *growth rings*. These rings indicate how much wood the tree produces annually. The harvester estimates that the trunk will put on a growth ring about 1 centimeter thick and its height will increase by about 0.25 meter this year. How many cubic meters of wood does the tree trunk produce this year? If the tree grows at these same rates for the next five years, will it produce the same amount of wood each year? Explain.

SOLUTION

a. Let x represent the mass (in kilograms) of the trunk. Use the formula for density to find x. Because the trunk is approximately cylindrical, use $V = Bh = \pi r^2 h$ for the volume. The radius is $\frac{0.8}{2} = 0.4$ meter.

$$\text{Density} = \frac{\text{Mass}}{\text{Volume}} \qquad \text{Formula for density}$$

$$450 \approx \frac{x}{\pi(0.4)^2(50)} \qquad \text{Substitute.}$$

$$3600\pi \approx x \qquad \text{Solve for } x.$$

$$11{,}310 \approx x \qquad \text{Use technology.}$$

▶ The mass of the trunk is about 11,310 kilograms.

b. Make a table that shows the trunk dimensions and volume for five years.

Year	0	1	2	3	4	5
Height (meters)	50	50.25	50.5	50.75	51	51.25
Base radius (meters)	0.4	0.41	0.42	0.43	0.44	0.45
Volume (cubic meters)	25.13	26.54	27.99	29.48	31.02	32.60

+ 1.41 + 1.45 + 1.49 + 1.54 + 1.58

5 MTR USE A PATTERN
In your table, why do you show the radius increasing by 0.01 meter each year?

▶ The tree will produce about $26.54 - 25.13 = 1.41$ cubic meters of wood in the first year. The tree will not produce the same amount of wood each year for five years because the differences between the volumes from year to year are increasing.

SELF-ASSESSMENT

1 I don't understand yet. 2 I can do it with help. 3 I can do it on my own. 4 I can teach someone else.

3. WHAT IF? The tree harvester makes the same growth estimates for the trunk of a sequoia tree that has a height of about 40 meters and a base diameter of about 0.75 meter. (a) Find the mass of the trunk. (b) How many cubic meters of wood will the trunk gain after four years?

GO DIGITAL

Using Surface Area Formulas

7 MTR

EXAMPLE 3 **Modeling Real Life** WATCH

The density of aluminum is about 2.7 grams per cubic centimeter. The mass of the empty aluminum can is about 14.9 grams. Estimate the thickness of the aluminum.

SOLUTION

Use the density of aluminum and the mass of the can to find the volume of aluminum in the can. Then find the surface area of the can. The product of the surface area and the thickness approximates the volume of aluminum in the can.

> **STUDY TIP**
> You can solve the density formula for the mass and for the volume.
>
> Mass = Density × Volume
>
> $\text{Volume} = \dfrac{\text{Mass}}{\text{Density}}$

Step 1 Estimate the volume of aluminum in the can.

$$\text{Volume} = \frac{\text{Mass}}{\text{Density}} \qquad \text{Write formula.}$$

$$\approx \frac{14.9}{2.7} \qquad \text{Substitute.}$$

$$\approx 5.5 \text{ cm}^3 \qquad \text{Divide.}$$

> 4 MTR **JUSTIFY A RESULT**
> Use unit analysis to show that the units for volume in Example 3 are cubic centimeters.

Step 2 Estimate the surface area of the can. The aluminum can is approximately cylindrical. Assume that the thickness of the aluminum is uniform throughout the can. The radius is $\frac{6.8}{2} = 3.4$ centimeters.

$$S = 2\pi r^2 + 2\pi rh \qquad \text{Formula for surface area of a cylinder}$$

$$= 2\pi(3.4)^2 + 2\pi(3.4)(12.2) \qquad \text{Substitute.}$$

$$= 106.08\pi \qquad \text{Simplify.}$$

$$\approx 333 \text{ cm}^2 \qquad \text{Use technology.}$$

Step 3 Estimate the thickness x of the can.

Surface area × Thickness ≈ Volume of aluminum

$$333x \approx 5.5$$

$$x \approx 0.017$$

▶ So, the thickness of the aluminum is about 0.017 centimeter, or about 0.17 millimeter.

SELF-ASSESSMENT

1 I don't understand yet. 2 I can do it with help. 3 I can do it on my own. 4 I can teach someone else.

4. The mass of the empty aluminum lunch box is about 700 grams. Estimate the thickness of the aluminum.

GO DIGITAL

12.7 Practice WITH CalcChat® AND CalcView®

7 MTR **1. MODELING REAL LIFE** The diagram shows the dimensions of a block of ice. Ice has a density of about 0.92 gram per cubic centimeter. Find the mass of the block of ice. *(See Example 1.)*

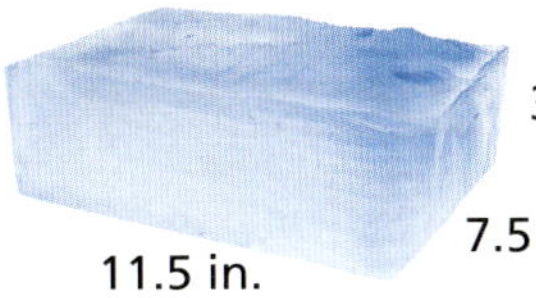

7 MTR **2. MODELING REAL LIFE** The United States has minted one-dollar silver coins called American Eagle Silver Bullion Coins since 1986. Each coin has a diameter of 40.6 millimeters and is 2.98 millimeters thick. The density of silver is 10.5 grams per cubic centimeter. Find the mass of an American Eagle Silver Bullion Coin.

7 MTR **3. MODELING REAL LIFE** An apple growing on a tree has a circumference of 6 inches. *(See Example 2.)*

a. The apple has a density of 0.46 gram per cubic centimeter. Find the mass of the apple.

b. The radius of the apple increases $\frac{1}{8}$ inch per week for the next five weeks. How does the volume change during the five-week period? Explain.

7 MTR **4. MODELING REAL LIFE** The height of a tree trunk is 20 meters and the base diameter is 0.5 meter.

a. The wood has a density of 380 kilograms per cubic meter. Find the mass of the trunk.

b. For each of the next 5 years, the trunk puts on a growth ring 4 millimeters thick. In the first year, the height increases by 0.2 meter. The tree produces the same amount of wood each year. What is the height of the trunk after 5 years?

GO DIGITAL

4 MTR **5. ERROR ANALYSIS** Describe and correct the error in finding the density of an object that has a mass of 24 grams and a volume of 28.3 cubic centimeters.

✗

$$\text{density} = \frac{28.3}{24} \approx 1.18$$

So, the density is about 1.18 cubic centimeters per gram.

4 MTR **6. ERROR ANALYSIS** Describe and correct the error in finding the mass m of an object that has a density of 5.4 grams per cubic centimeter and a volume of 2.5 cubic meters.

✗

$$5.4 = \frac{m}{2.5}$$

$$13.5 = m$$

So, the mass is 13.5 grams.

7 MTR **7. MODELING REAL LIFE** The density of steel is approximately 7.8 grams per cubic centimeter. An empty steel can with a mass of 76 grams has a radius of 6.1 centimeters and a height of 6.3 centimeters. Estimate the thickness of the steel. *(See Example 3.)*

7 MTR **8. MODELING REAL LIFE** The density of the glass in the dish shown is approximately 2.23 grams per cubic centimeter. The mass of the dish is about 0.9 kilogram. Estimate the thickness of the glass.

4 MTR **9. MAKING AN ARGUMENT** As ocean depth increases, water molecules become closer together in part because of decreasing temperatures and increasing pressure. What happens to the density of the water as depth increases? Explain.

10. HOW DO YOU SEE IT?
The two stone blocks shown below with the given densities have the same volume. Which block has a greater mass? Explain.

Granite: 2.7 g/cm³ Sandstone: 2.3 g/cm³

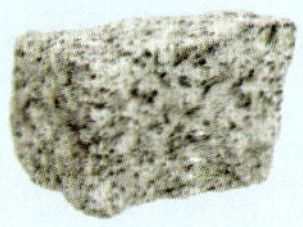

11. **PROBLEM SOLVING** A pool in the shape of a rectangular prism is 6 meters long and 3 meters wide. The water in the pool is 1 meter deep.

  a. The density of water is about 1 gram per cubic centimeter. Find the number of kilograms of water in the pool.

  b. You add 6000 kilograms of water to the pool. What is the depth of the water in the pool?

12. **THOUGHT PROVOKING**
You place two cans of regular soda and two cans of diet soda in a container full of water. The two regular cans sink, but the two diet cans float. Use the Internet to research the contents of regular soda and diet soda. Then make a conjecture about why the diet cans float but the regular cans sink. Include a discussion of *density* and *buoyancy* in your explanation.

13. **REASONING** Links of a chain are made from cylindrical metal rods with a diameter of 6 millimeters. The density of the metal is about 8 grams per cubic centimeter.

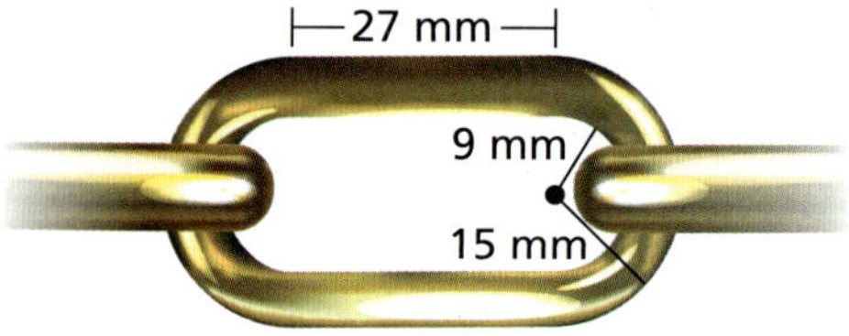

  a. To approximate the length of a rod used to make a link, should you use the perimeter around the inside of the link? the outside? the average of these perimeters? Explain your reasoning. Then approximate the mass of a chain with 100 links.

  b. Approximate the length of a taut chain with 100 links. Explain your procedure.

REVIEW & REFRESH

WATCH

14. Show that a quadrilateral with vertices $P(3, 5)$, $Q(6, 4)$, $R(7, -1)$, and $S(1, 1)$ is a trapezoid. Then decide whether it is isosceles.

7 MTR 15. **MODELING REAL LIFE** Copper has a density of about 8.96 grams per cubic centimeter. The copper rod has a diameter of $\frac{5}{16}$ inch. Find the mass of the copper rod.

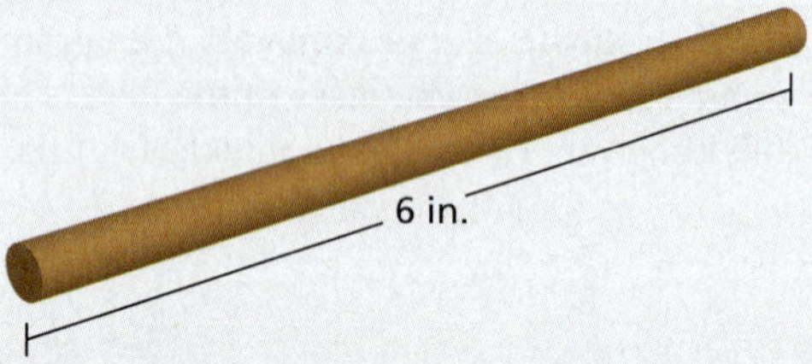

16. Find the surface area and the volume of the sphere.

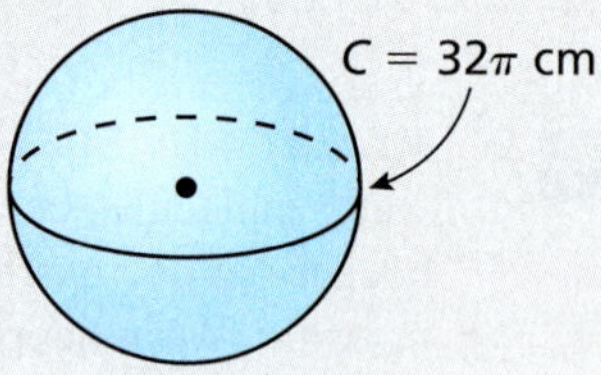

17. Find the volume of the cylinder.

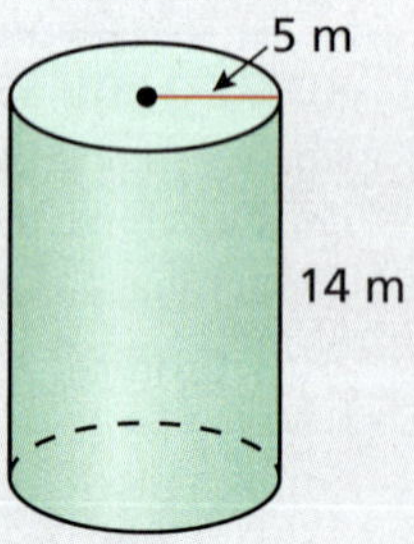

18. The pyramids are similar. Find the volume of pyramid B.

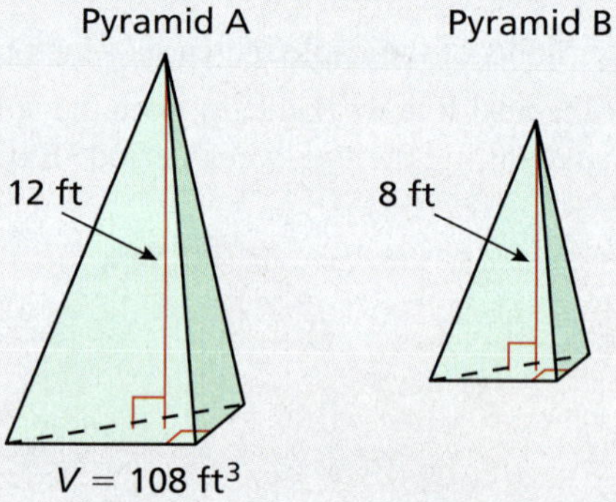

19. Decide whether enough information is given to prove that $\triangle ABC$ and $\triangle DCB$ are congruent. Explain.

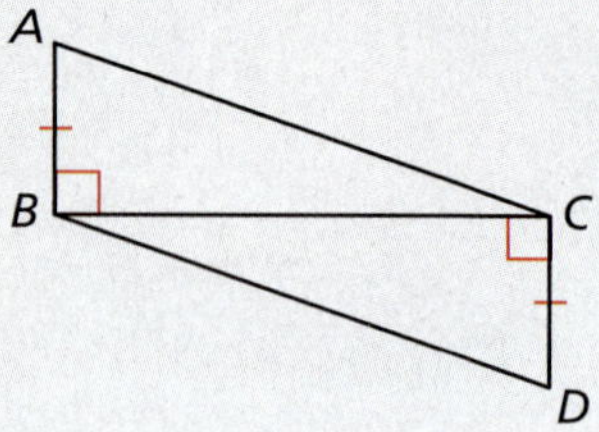

20. A triangle has side lengths of 12 feet, 9 feet, and x feet. Describe the possible values of x. Explain your reasoning.

In Exercises 21 and 22, find the value of x. Write your answer in simplest form.

21.

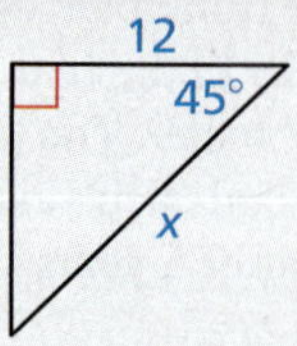

22.

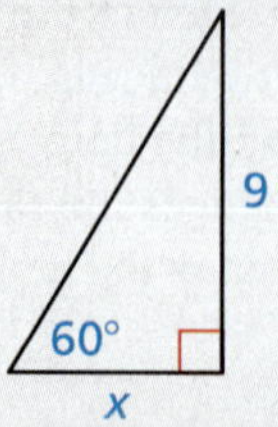

GO DIGITAL

12.8 Solids of Revolution

Learning Target: Sketch and use solids of revolution.

Success Criteria:
- I can sketch and describe solids of revolution.
- I can find surface areas and volumes of solids of revolution.
- I can form solids of revolution in the coordinate plane.

EXPLORE IT! Modeling Solids of Revolution

Work with a partner.

a. Tape 3-inch-by-5-inch index cards to pencils as shown. When you rotate each pencil, describe the solid modeled by the rotating index card.

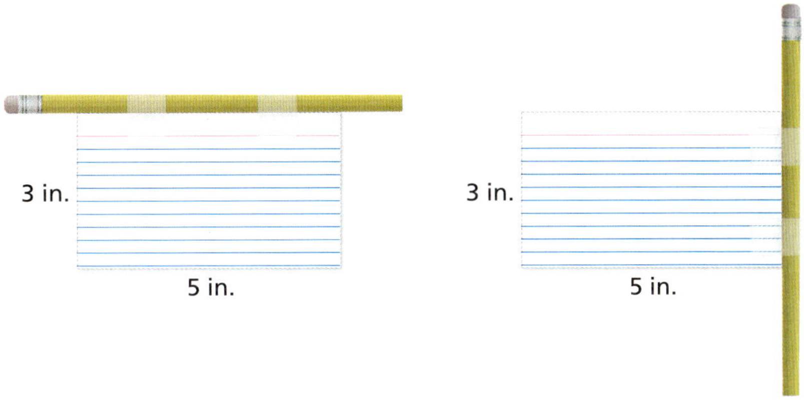

b. Do the solids in part (a) have the same surface area? the same volume? Justify your answers.

c. Tape the straight side of a protractor to a pencil. When you rotate the pencil, describe the solid modeled by the rotating protractor.

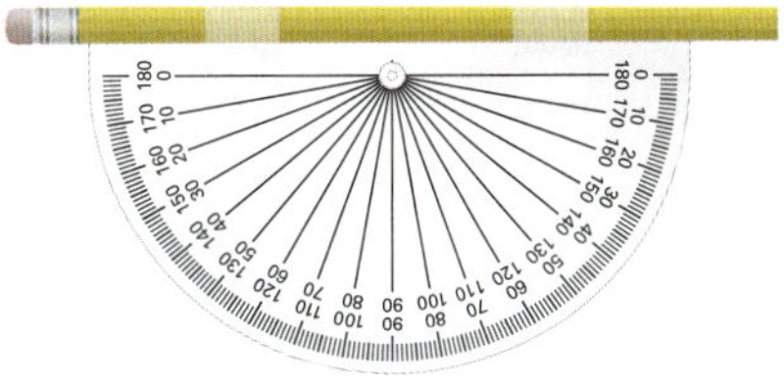

d. Find the surface area and volume of the solid in part (c). Explain how you found your answers.

e. Can you tape an object to a pencil and rotate the pencil to model a cone? a cube? Explain your reasoning.

4 MTR COMMUNICATE CLEARLY
The solids modeled in this Explore It! are called *solids of revolution*. In your own words, define solid of revolution. Give some examples of real-life objects that are solids of revolution.

Geometric Reasoning
MA.912.GR.4.2 Identify three-dimensional objects generated by rotations of two-dimensional figures.

GO DIGITAL

Sketching and Describing Solids of Revolution

Vocabulary

solid of revolution, *p. 668*
axis of revolution, *p. 668*

A **solid of revolution** is a three-dimensional figure that is formed by rotating a two-dimensional shape around an axis. The line around which the shape is rotated is called the **axis of revolution**.

For example, when you rotate a rectangle around a line that contains one of its sides, the solid of revolution that is produced is a cylinder.

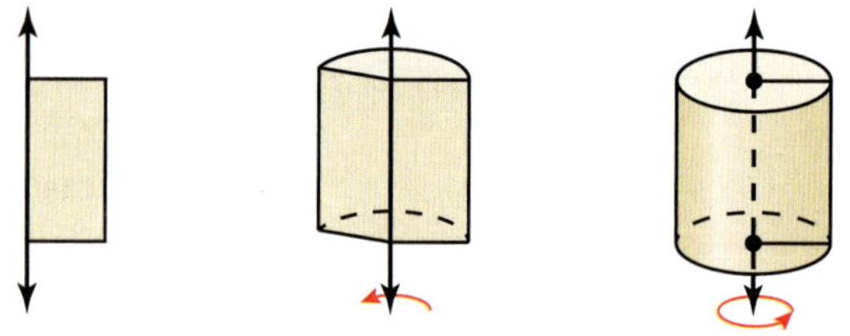

EXAMPLE 1 **Sketching and Describing Solids of Revolution**

Sketch the solid produced by rotating the figure around the given axis. Then identify and describe the solid.

a.

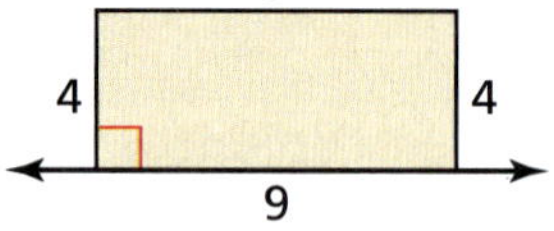

b.

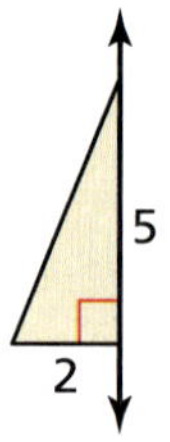

SOLUTION

a.

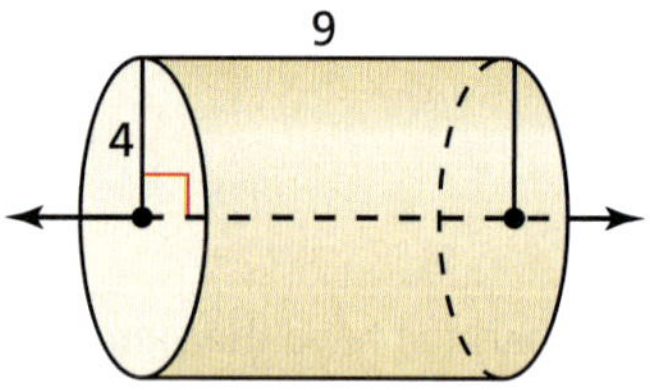

▶ The solid is a cylinder with a height of 9 units and a radius of 4 units.

b.

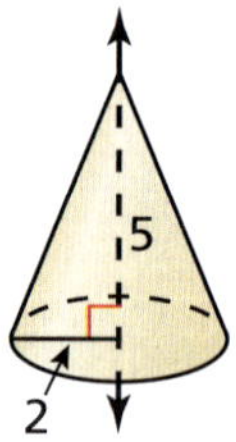

▶ The solid is a cone with a height of 5 units and a radius of 2 units.

SELF-ASSESSMENT

1 I don't understand yet. 2 I can do it with help. 3 I can do it on my own. 4 I can teach someone else.

Sketch the solid produced by rotating the figure around the given axis. Then identify and describe the solid.

1.

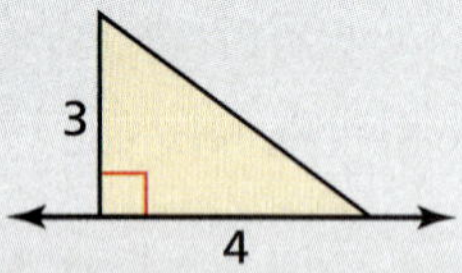

2.

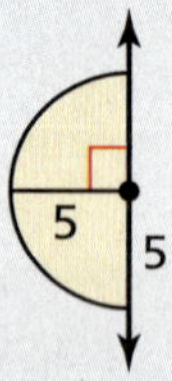

3.

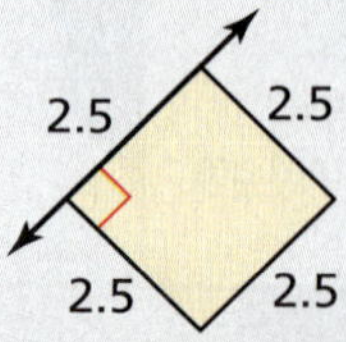

4. **WHICH ONE DOESN'T BELONG?** Which object does *not* belong with the other three? Explain your reasoning.

GO DIGITAL

EXAMPLE 2 Sketching a Two-Dimensional Shape and Axis

Most vases are solids of revolution. Sketch a two-dimensional shape and an axis of revolution that forms the vase shown.

SOLUTION

The two-dimensional shape should match the outline of one side of the vase.

Finding Surface Areas and Volumes of Solids of Revolution

EXAMPLE 3 Finding the Surface Area and Volume of a Solid of Revolution

Sketch and describe the solid produced by rotating the figure around the given axis. Then find its surface area and volume.

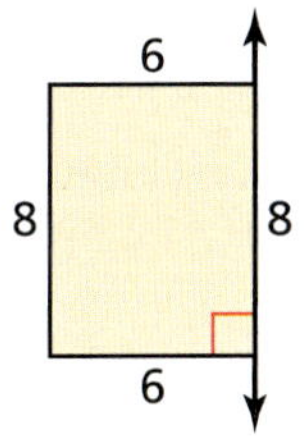

SOLUTION

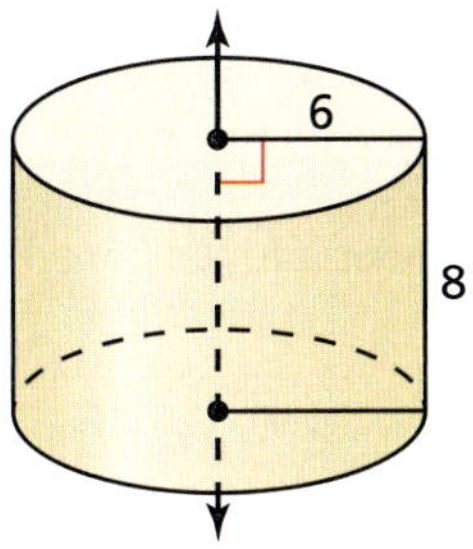

The solid is a cylinder with a height of 8 units and a radius of 6 units.

Surface area: $S = 2\pi r^2 + 2\pi rh = 2\pi(6)^2 + 2\pi(6)(8) = 168\pi \approx 527.79$

Volume: $V = \pi r^2 h = \pi(6)^2(8) = 288\pi \approx 904.78$

▶ The cylinder has a surface area of about 527.79 square units and a volume of about 904.78 cubic units.

SELF-ASSESSMENT

1 I don't understand yet. 2 I can do it with help. 3 I can do it on my own. 4 I can teach someone else.

5. Sketch a two-dimensional shape and an axis of revolution that forms the birdbath shown.

6. Sketch and describe the solid produced by rotating the figure around the given axis. Then find its surface area and volume.

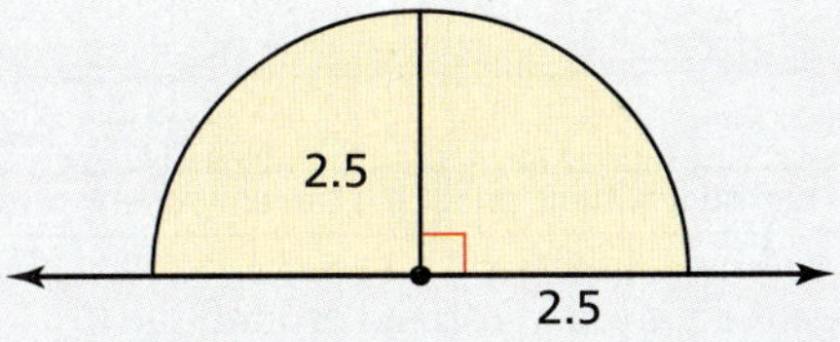

GO DIGITAL

Forming Solids of Revolution in the Coordinate Plane

EXAMPLE 4 **Forming a Solid of Revolution** WATCH

Sketch and describe the solid that is produced when the region enclosed by $y = 0$, $y = x$, and $x = 5$ is rotated around the y-axis. Then find the volume of the solid.

SOLUTION

REMEMBER

When (a, b) is reflected in the y-axis, its image is the point $(-a, b)$.

Step 1 Graph each equation and determine the region that will be rotated around the y-axis.

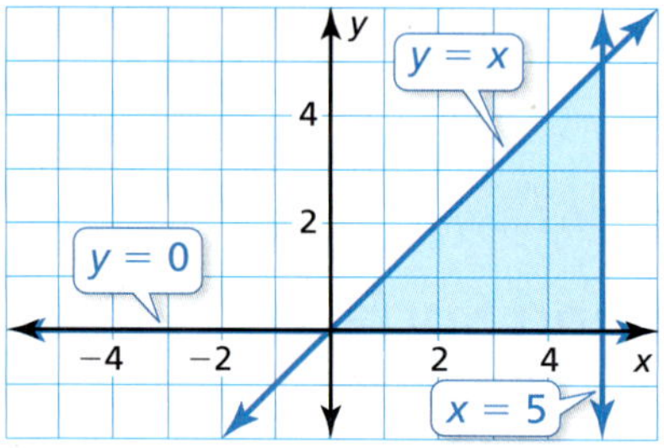

Step 2 Reflect the region in the y-axis.

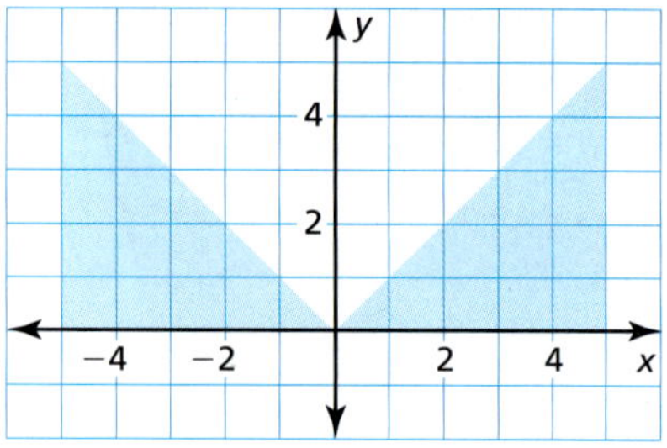

Step 3 Connect the vertices of the triangles using curved lines.

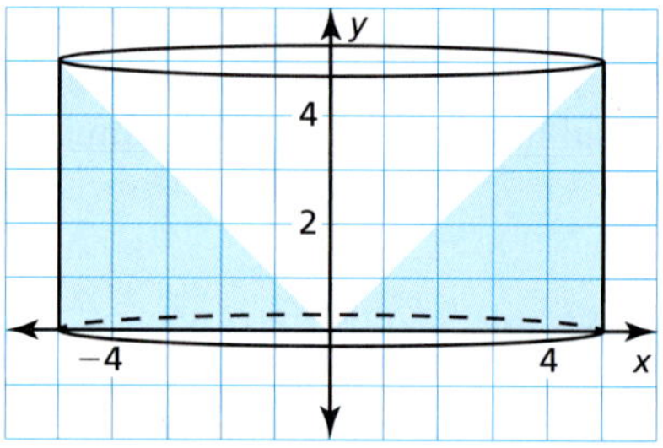

Step 4 The composite solid consists of a cylinder with a cone removed.

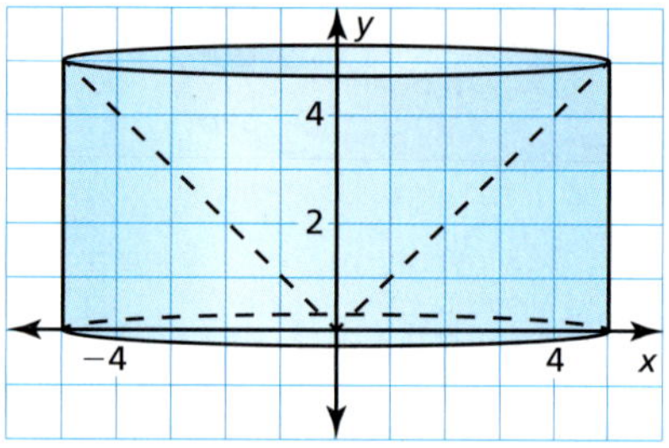

Step 5 Find the volume of the composite solid. The cylinder and the cone both have a height of 5 units and a radius of 5 units.

Volume of solid = Volume of cylinder − Volume of cone

$= \pi r^2 h - \frac{1}{3}\pi r^2 h$ Write formulas.

$= \pi \cdot 5^2 \cdot 5 - \frac{1}{3}\pi \cdot 5^2 \cdot 5$ Substitute 5 for r and 5 for h.

$= 125\pi - \frac{125}{3}\pi$ Simplify.

≈ 261.80 Use technology.

ANOTHER WAY

You can also simplify $\pi r^2 h - \frac{1}{3}\pi r^2 h$ to obtain $\frac{2}{3}\pi r^2 h$. Then substitute 5 for r and h, and evaluate.

▶ The volume of the solid is about 261.80 cubic units.

SELF-ASSESSMENT 1 I don't understand yet. 2 I can do it with help. 3 I can do it on my own. 4 I can teach someone else.

7. **WHAT IF?** In Example 4, does the solid change when the region is rotated around the x-axis? Explain.

8. Sketch and describe the solid that is produced when the region enclosed by $x = 0$, $y = -x$, and $y = -3$ is rotated around the x-axis. Then find the volume of the solid.

GO DIGITAL

12.8 Practice WITH CalcChat® AND CalcView®

In Exercises 1–6, sketch the solid produced by rotating the figure around the given axis. Then identify and describe the solid. *(See Example 1.)*

1.
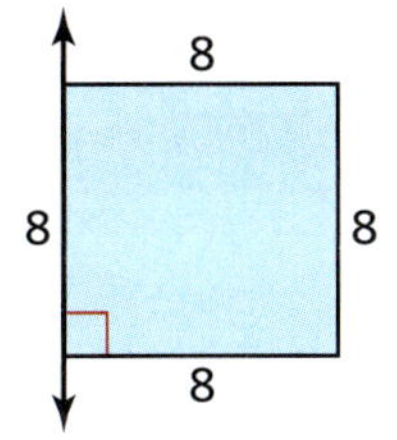

2.
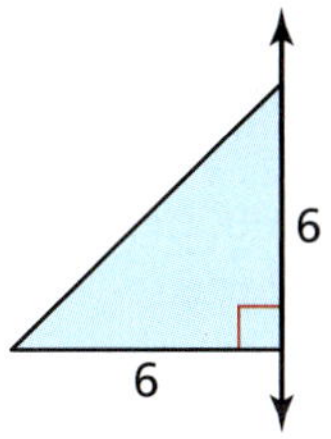

3.
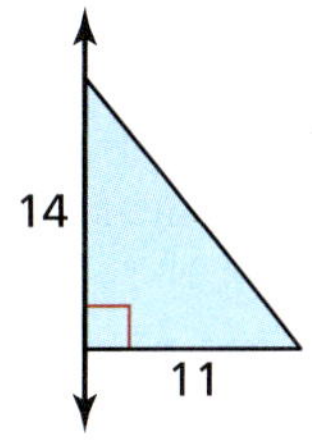

4.
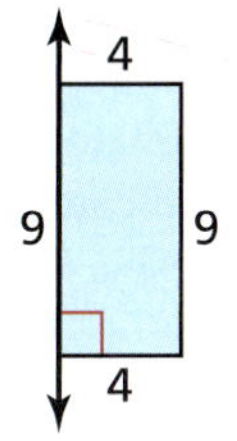

5.
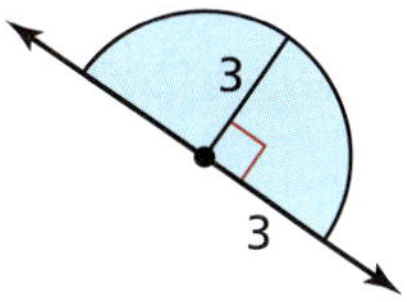

6.
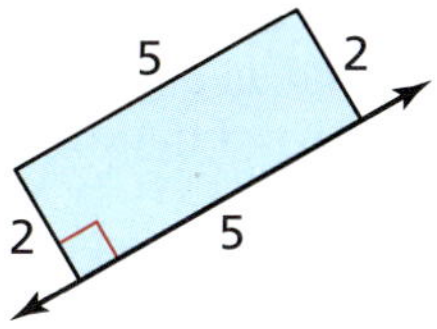

7. ERROR ANALYSIS Describe and correct the error in identifying and describing the solid produced by rotating the figure around the given axis.

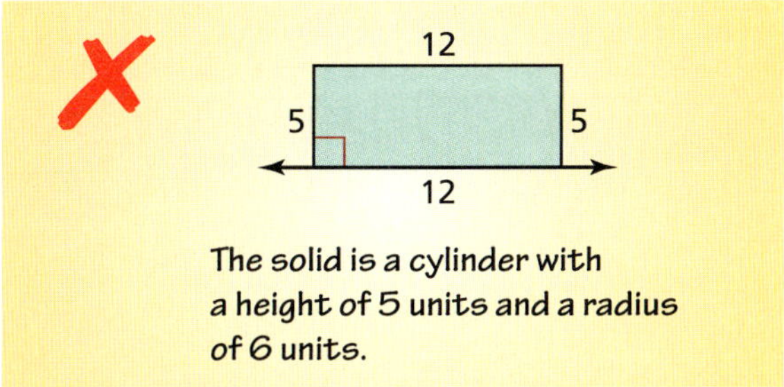

8. REASONING Can you form any solid by rotating a two-dimensional figure around an axis? Explain.

In Exercises 9–12, sketch the solid of revolution. Then identify and describe the solid.

9. a square with side length 4 rotated around one side

10. a rectangle with length 6 and width 3 rotated around one of its shorter sides

11. a right triangle with legs of lengths 6 and 9 rotated around its longer leg

12. a semicircle with radius 10 rotated around its diameter

GO DIGITAL

In Exercises 13–16, sketch a two-dimensional shape and an axis of revolution that forms the object shown. *(See Example 2.)*

13.

14.

15.

16.

In Exercises 17–22, sketch and describe the solid produced by rotating the figure around the given axis. Then find its surface area and volume. *(See Example 3.)*

17.
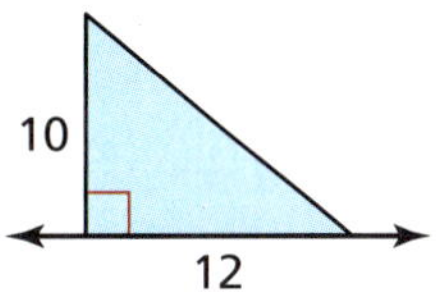

18.
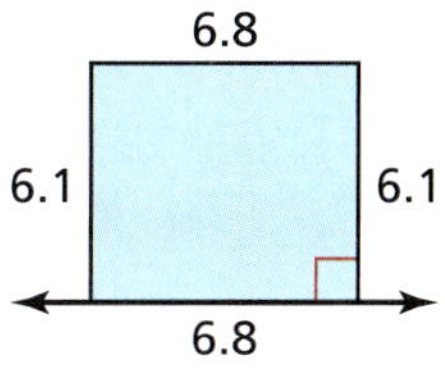

19.
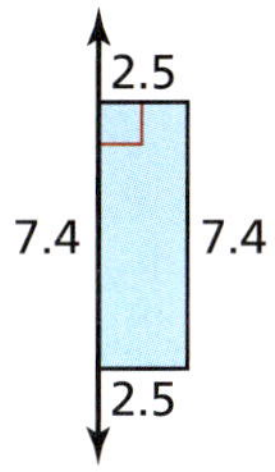

20.
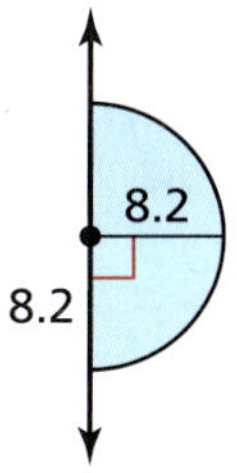

21.
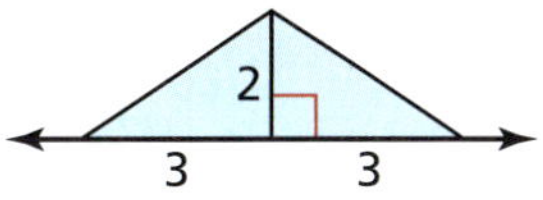

22.
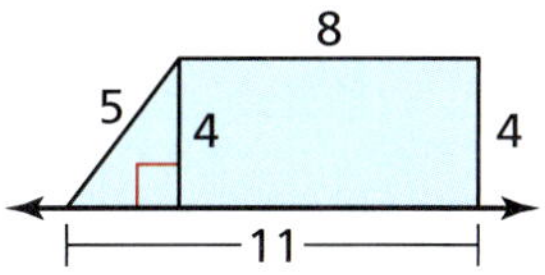

In Exercises 23–26, sketch and describe the solid that is produced when the region enclosed by the given equations is rotated around the given axis. Then find the volume of the solid. *(See Example 4.)*

23. $x = 0, y = 0, y = x + 3;$ x-axis

24. $x = 0, y = 0, y = -2x + 5;$ y-axis

25. $x = 3, y = 0, y = \frac{1}{2}x;$ y-axis

26. $x = -4, y = 0, y = x;$ x-axis

4 MTR **27. MAKING AN ARGUMENT** Your friend says when you rotate the figure shown around either the x-axis or the y-axis, the resulting solid is a sphere. Is your friend correct? Explain.

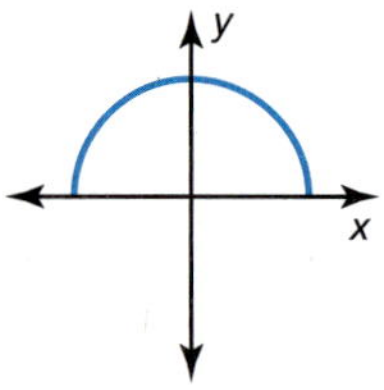

28. HOW DO YOU SEE IT?
The figure shows the graph of a function f where the domain is $\{x \mid a \le x \le b\}$. Sketch the solid produced when the region enclosed by the graph of f and the equations $x = a$, $x = b$, and $y = 0$ is rotated around the x-axis.

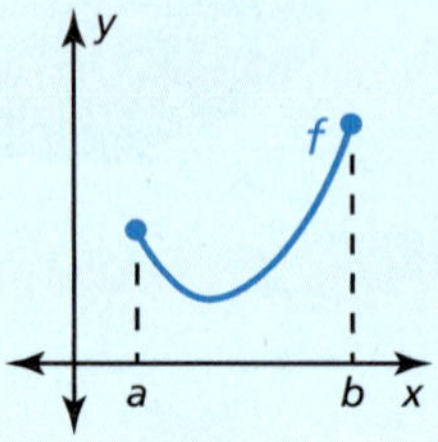

5 MTR **29. STRUCTURE** A right triangle has sides with lengths 15, 20, and 25, as shown. Describe the three solids formed when the triangle is rotated around each of its sides. Then find the volumes of the solids. Give your answers in terms of π.

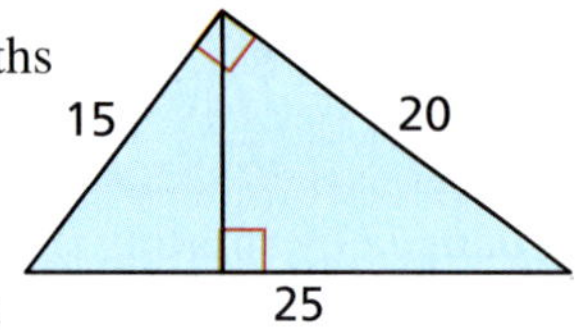

30. THOUGHT PROVOKING
Write a system of equations whose enclosed region, when rotated around the x-axis or y-axis, produces the same solid with the same dimensions.

31. REASONING The solid shown is a type of *torus*.

a. Sketch a two-dimensional shape and an axis of revolution that forms the torus.

b. Which solid can you manipulate to create a torus similar to the one above? Explain, in your own words, how to manipulate the solid to form the torus. You can think of the surface of the solid you choose as being stretchable.

32. DIG DEEPER A 30°-30°-120° isosceles triangle has two legs of length 4 units. When it is rotated around an axis that contains one leg, what is the volume of the solid of revolution?

REVIEW & REFRESH

WATCH

33. PROBLEM SOLVING A circular region has a population of about 2.5 million people and a population density of about 9824 people per square mile. Find the radius of the region.

34. Sketch and describe the solid produced by rotating the figure around the given axis. Then find its surface area and volume.

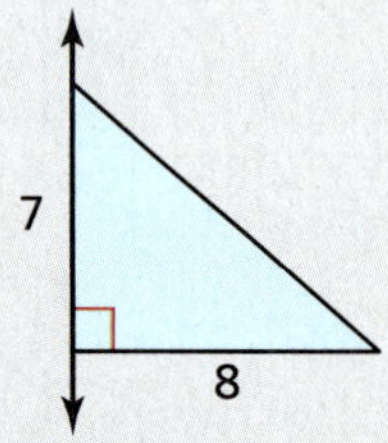

35. The diagram shows the radius of a titanium ball. Titanium has a density of about 4.51 grams per cubic centimeter. Find the mass of the titanium ball.

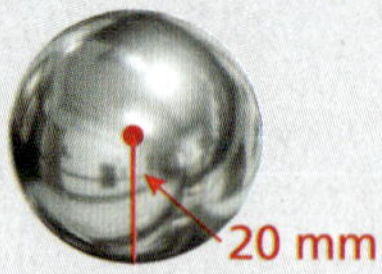

36. Find the value of x. Tell whether the side lengths form a Pythagorean Triple.

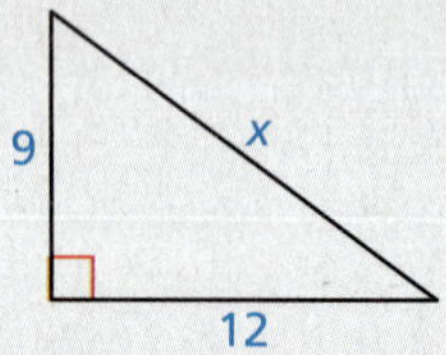

37. Find the surface area of the sphere.

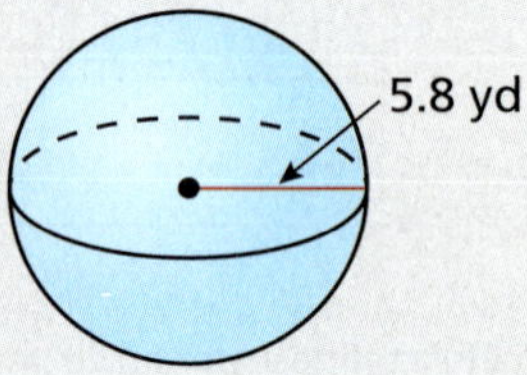

38. Find the volume of the cone.

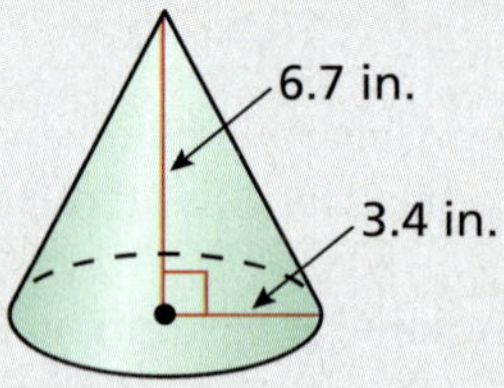

7 MTR **39. MODELING REAL LIFE** You are running on a circular path at a constant rate of 8.8 feet per second. The path is one mile in diameter. How long will it take you to run two complete laps?

40. Classify $\triangle ABC$ with vertices $A(1, 2)$, $B(4, 3)$, and $C(2, 6)$ by its sides. Then determine whether it is a right triangle.

GO DIGITAL

12 Chapter Review WITH CalcChat®

Chapter Learning Target: Understand surface area and volume.

Chapter Success Criteria:
- ◆ I can describe attributes of solids.
- ◆ I can find surface areas and volumes of solids.
- ■ I can find missing dimensions of solids.
- ■ I can solve real-life problems involving surface area and volume.

◆ Surface
■ Deep

SELF-ASSESSMENT 1 I don't understand yet. 2 I can do it with help. 3 I can do it on my own. 4 I can teach someone else.

12.1 Cross Sections of Solids *(pp. 615–622)* WATCH

Learning Target: Describe and draw cross sections.

Vocabulary AZ VOCAB
face
polyhedron
edge
vertex
cross section

Describe the shape formed by the intersection of the plane and the solid.

1.

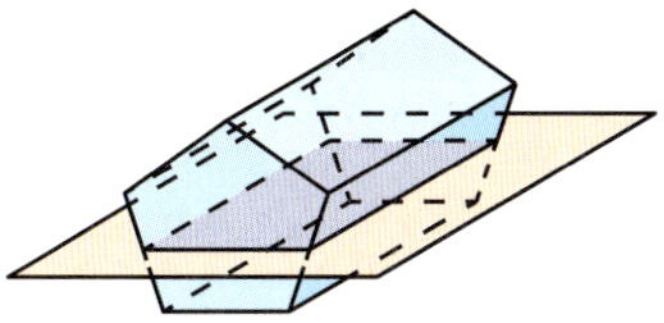

2.

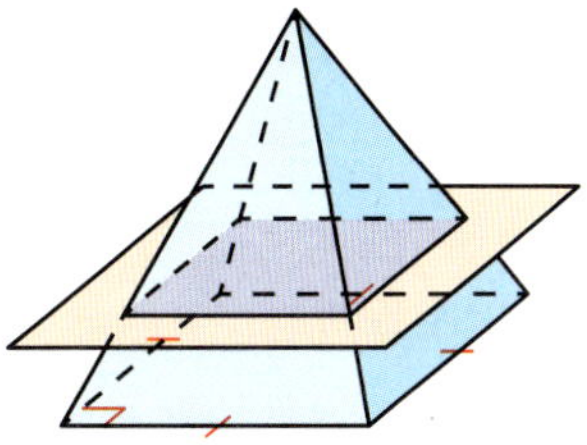

3.

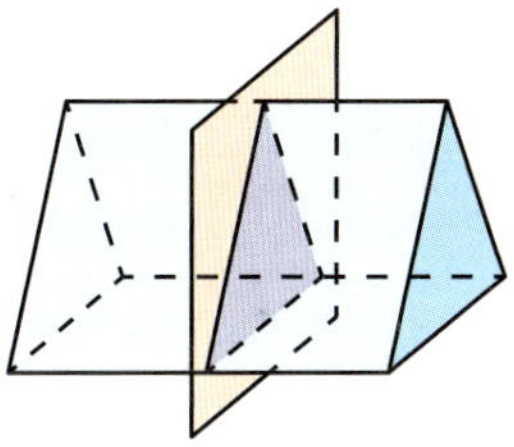

4.

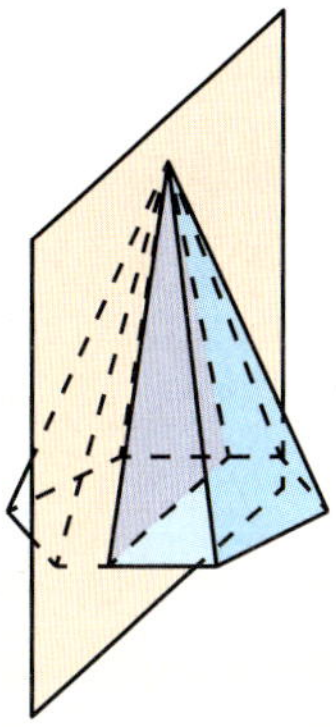

Draw the cross section formed by the described plane that contains the red line segment drawn on the solid. What is the shape of the cross section?

5. plane is parallel to base

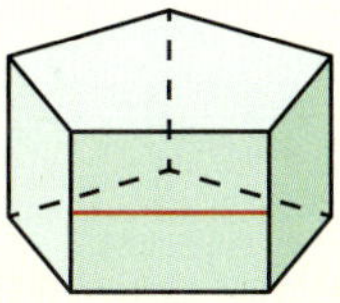

6. plane is parallel to bottom face

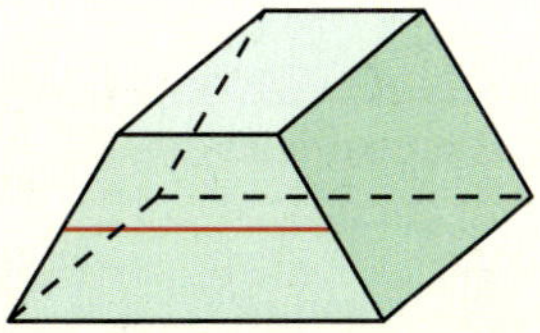

7. Describe and draw two cross sections that can be formed by a plane intersecting the solid in Exercise 5. The shapes of the cross sections should be different than the shape of the cross section in Exercise 5.

GO DIGITAL

12.2 Surface Areas of Prisms and Cylinders (pp. 623–630)

WATCH

Learning Target: Find and use surface areas of prisms and cylinders.

Vocabulary AZ VOCAB

surface area
lateral faces of a polyhedron
lateral face of a cylinder
similar solids

Find the surface area of the right prism or right cylinder.

8.

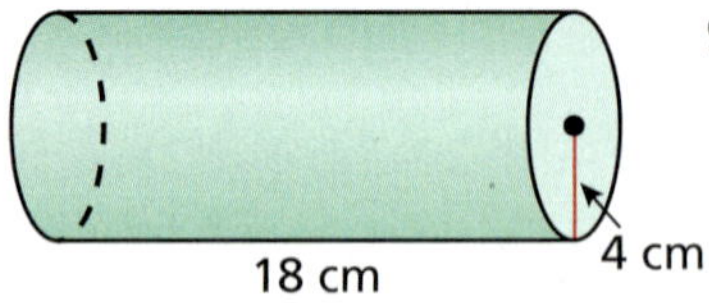

9.

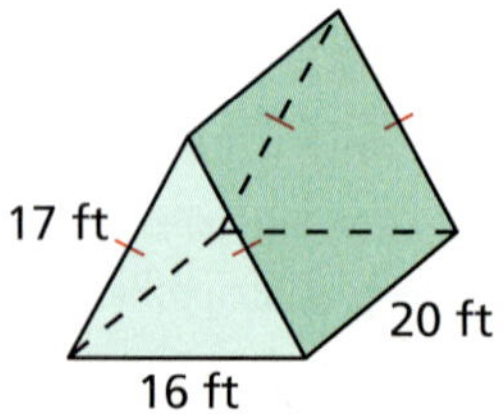

10.

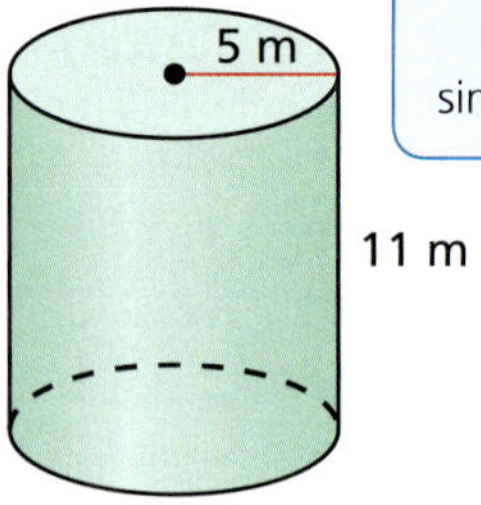

11. Find the surface area of the composite solid.

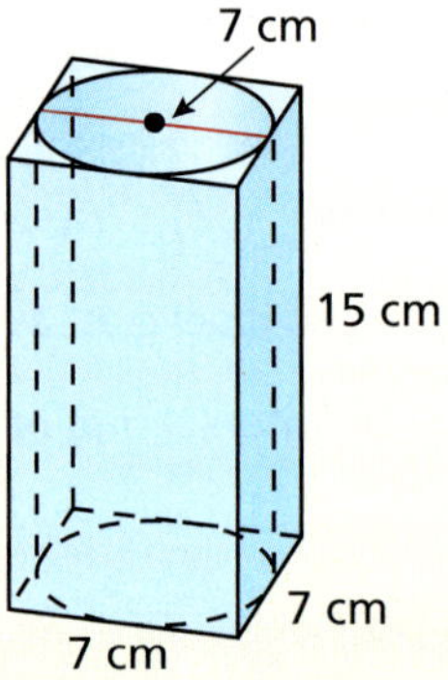

12. A regular pentagonal prism has a height of 10.5 centimeters and a base edge length of 6 centimeters. Another regular pentagonal prism has a height of 6 centimeters and a base edge length of 10.5 centimeters. Which prism has a greater surface area? Explain your reasoning.

12.3 Surface Areas of Pyramids and Cones (pp. 631–636)

WATCH

Learning Target: Find and use surface areas of pyramids and cones.

Vocabulary AZ VOCAB

vertex of a pyramid
slant height of a regular pyramid
lateral face of a cone

Find the surface area of the regular pyramid or right cone.

13.

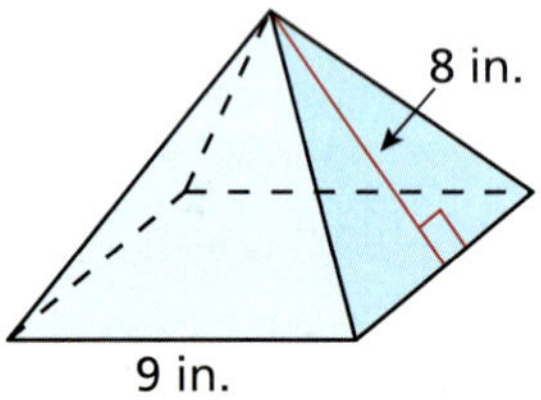

14.

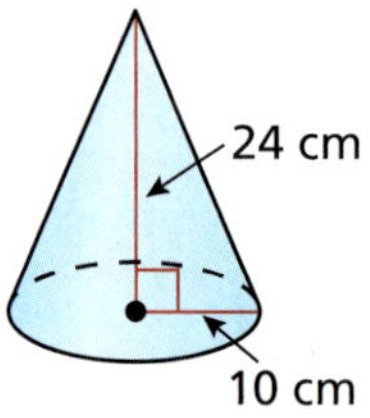

15.

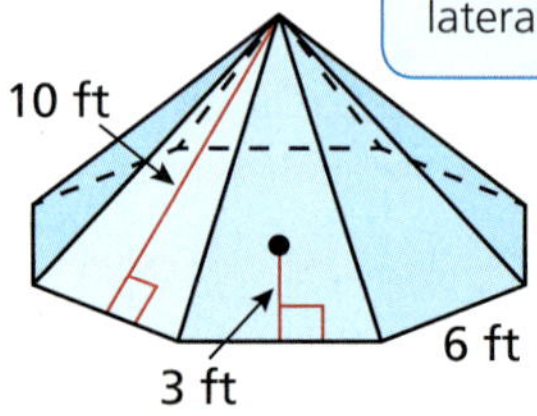

16. Find the surface area of the composite solid.

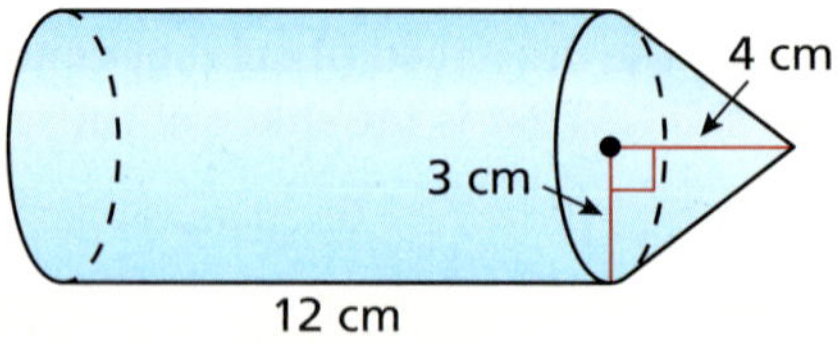

17. Pyramid A and pyramid B are similar. Find the surface area of pyramid B.

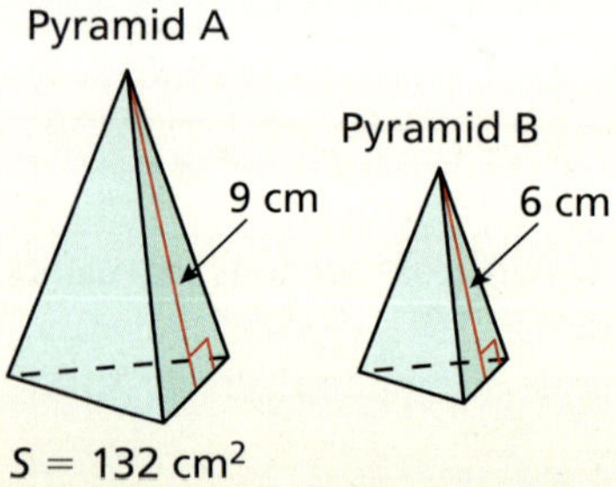

GO DIGITAL

12.4 Volumes of Prisms and Cylinders (pp. 637–644)

WATCH

Learning Target: Find and use volumes of prisms and cylinders.

Vocabulary
volume
Cavalieri's Principle

Find the volume of the prism or cylinder.

18.

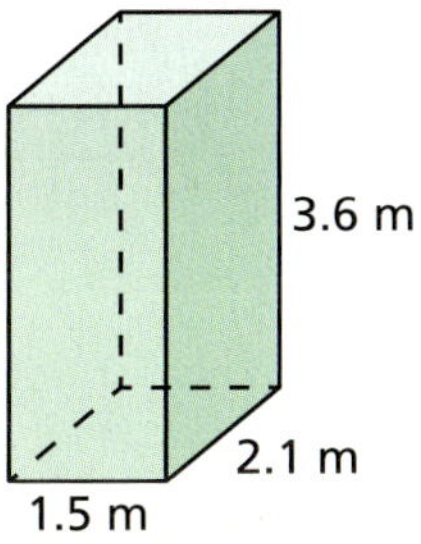

19.

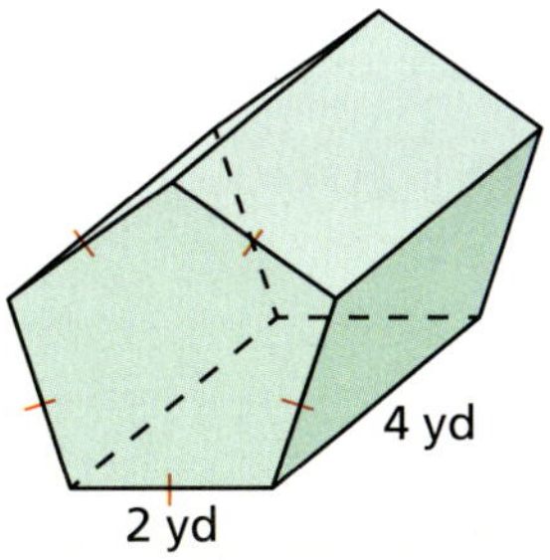

20.

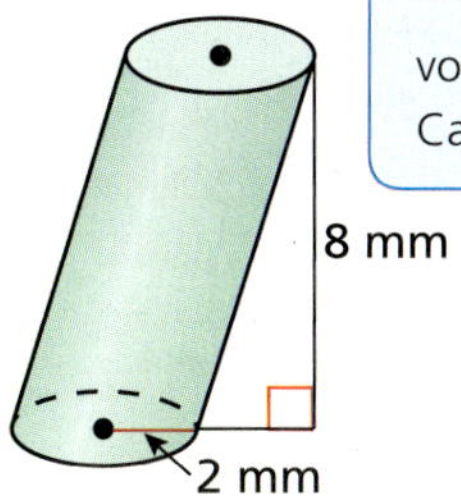

21. Cylinder A and cylinder B are similar. Find the volume of cylinder B.

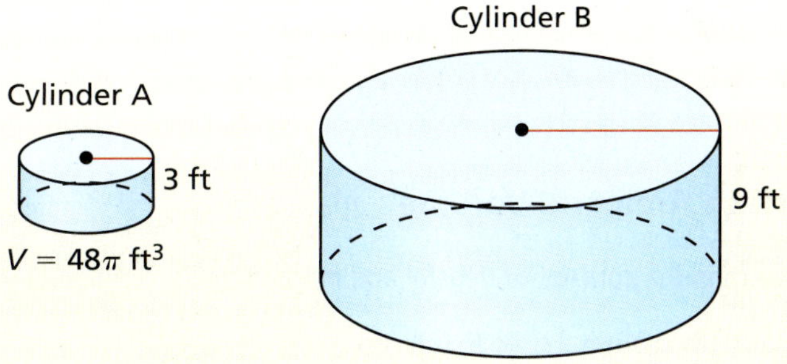

12.5 Volumes of Pyramids and Cones (pp. 645–652)

WATCH

Learning Target: Find and use volumes of pyramids and cones.

Find the volume of the solid.

22.

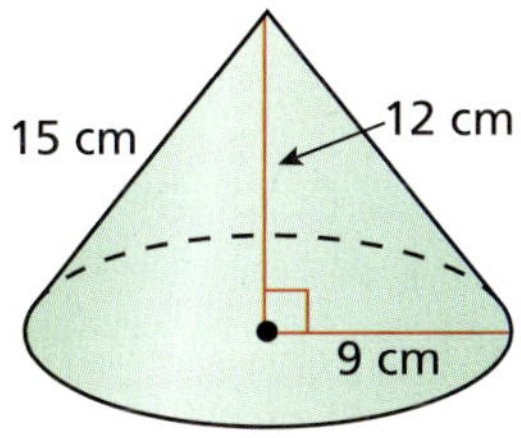

23.

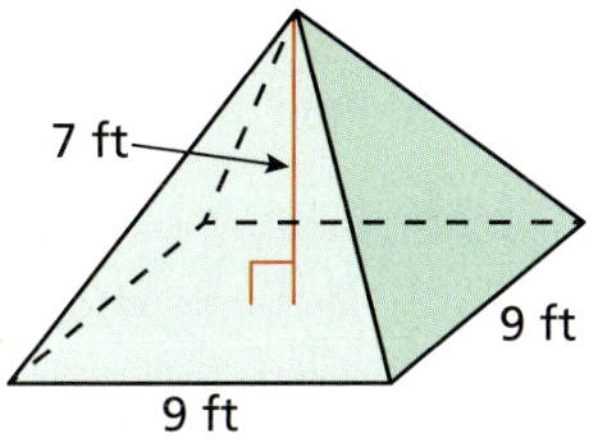

24.

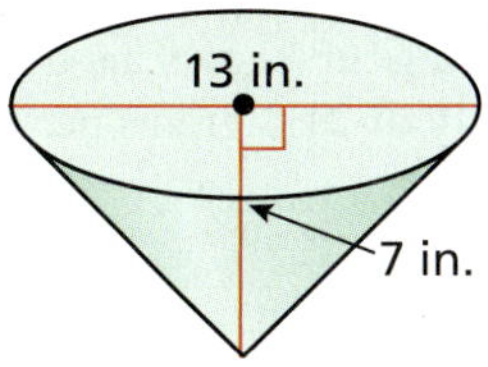

25. The Temple of the Masks is one of five Tikal temples built by the ancient Mayan people in Guatemala. The pyramid has a height of approximately 38 meters and a square base with side lengths of approximately 32 meters. Estimate the volume of the pyramid.

26. A cone with a diameter of 16 centimeters has a volume of 320π cubic centimeters. Find the height of the cone.

Find the volume of the composite solid.

27.

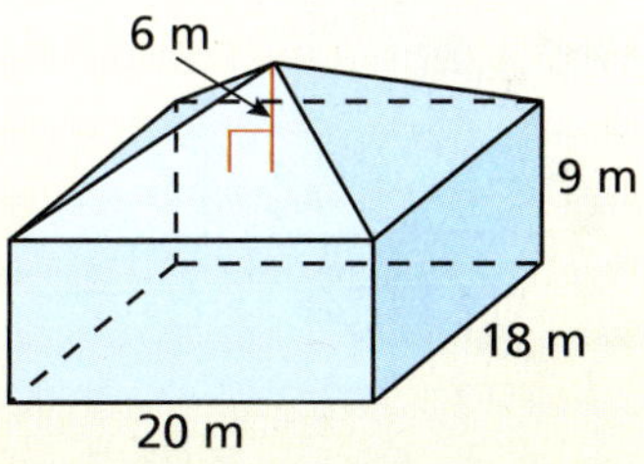

28.

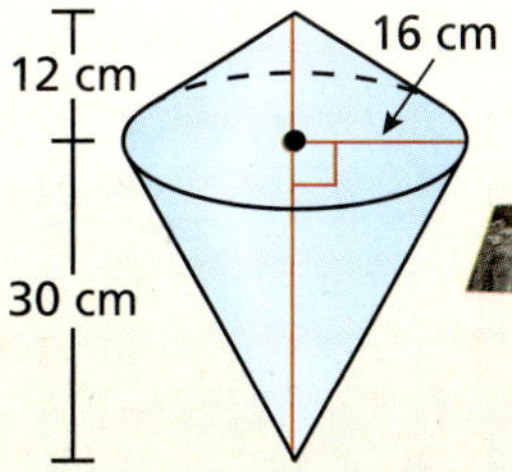

GO DIGITAL

12.6 Surface Areas and Volumes of Spheres *(pp. 653–660)* WATCH

Learning Target: Find and use surface areas and volumes of spheres.

Vocabulary AZ VOCAB
chord of a sphere
great circle

Find the surface area and the volume of the sphere.

29.
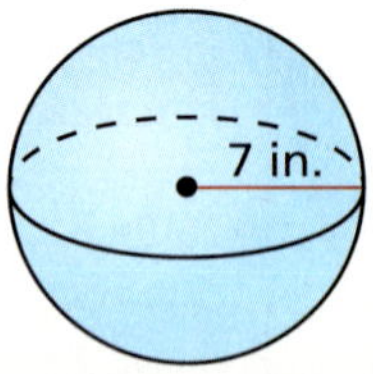

30.
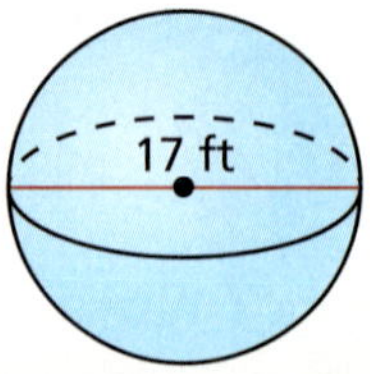

31.
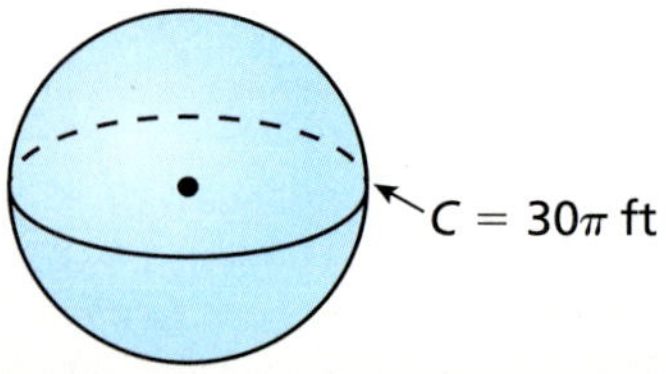

32. The shape of Mercury can be approximated by a sphere with a diameter of 4880 kilometers. Find the surface area and the volume of Mercury.

33. A solid is composed of a cube with a side length of 6 meters and a hemisphere with a diameter of 6 meters. Find the volume of the composite solid.

12.7 Modeling with Surface Area and Volume *(pp. 661–666)* WATCH

Learning Target: Understand the concept of density and modeling with volume.

Vocabulary AZ VOCAB
density

34. A part for a toy train is made by drilling a hole that has a diameter of 0.6 centimeter through a wooden ball that has a diameter of 4 centimeters.

a. Estimate the volume of the wooden ball after the hole is made. Explain your reasoning.

b. Do the surface area and volume of the wood decrease after the hole is made? Explain.

35. The diagram shows the dimensions of a bar of platinum. Platinum has a density of 21.4 grams per cubic centimeter. Find the mass of the bar.

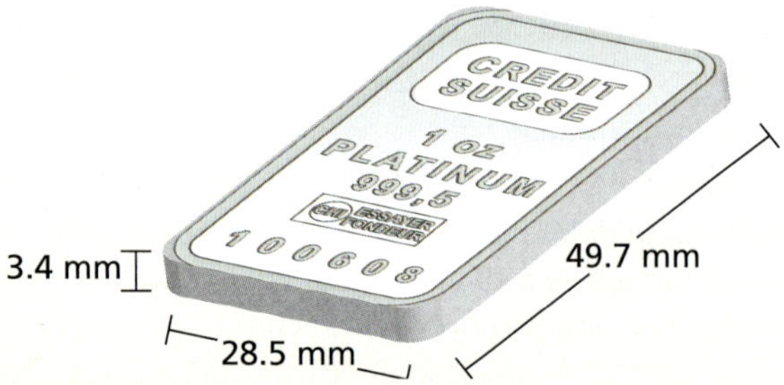

36. The density of porcelain is about 2.4 grams per cubic centimeter. The mass of the mug is about 378 grams. Estimate the thickness of the porcelain.

GO DIGITAL

12.8 Solids of Revolution (pp. 667–672) WATCH

Learning Target: Sketch and use solids of revolution.

Vocabulary
solid of revolution
axis of revolution

Sketch and describe the solid produced by rotating the figure around the given axis. Then find its surface area and volume.

37.

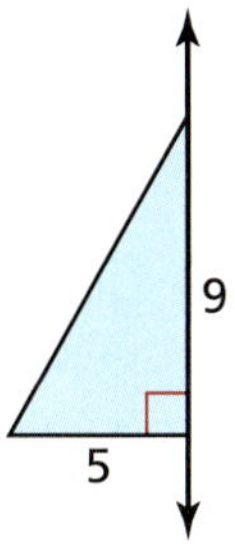

38.

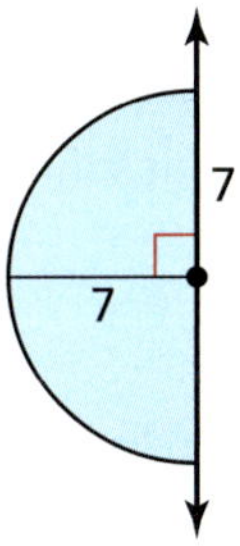

39.

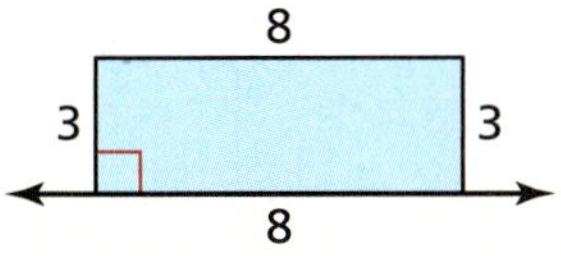

40.

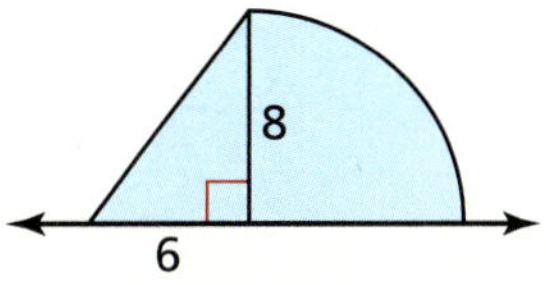

Sketch a two-dimensional shape and an axis of revolution that forms the object shown.

41.

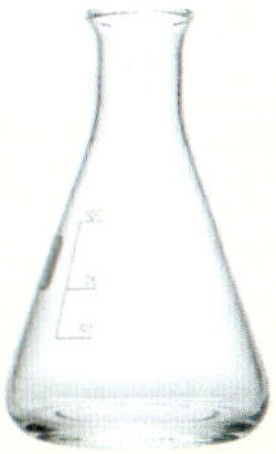

42.

43. Sketch and describe the solid that is produced when the region enclosed by $y = 0$, $y = x$, and $x = 2$ is rotated around the y-axis. Then find the volume of the solid.

Mathematical Thinking and Reasoning

7 MTR **APPLY MATHEMATICS TO REAL-WORLD CONTEXTS**
Mathematicians who apply mathematics to real-world contexts use models and methods to understand, represent, and solve problems.

1. The Leaning Tower of Pisa is a famous landmark in Italy. Research the dimensions of the tower and use them to draw a scale model. Find the volume of the model and use it to approximate the volume of the actual tower. Justify your answer.

2. Mexican Talavera, an ancient form of pottery featuring intricate designs and bright colors, was one of the first forms of pottery to be created using a potter's wheel. Use the Internet or another resource to research different types and shapes of pottery. What technique from this chapter can be used to design different shapes of pottery?

GO DIGITAL

12 Practice Test WITH CalcChat®

1. You slice a bagel in half horizontally to split with your friend. Describe the shape of the cross section you make. What does the cross section look like if you cut the bagel in half vertically?

2. Find the surface area of a right cone with a diameter of 10 feet and a height of 12 feet.

Find the volume of the solid.

3. 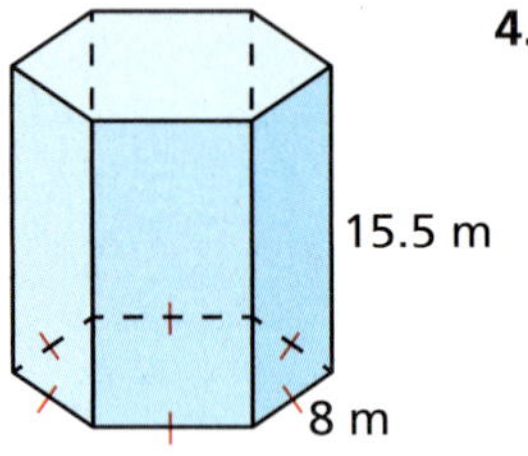

4. 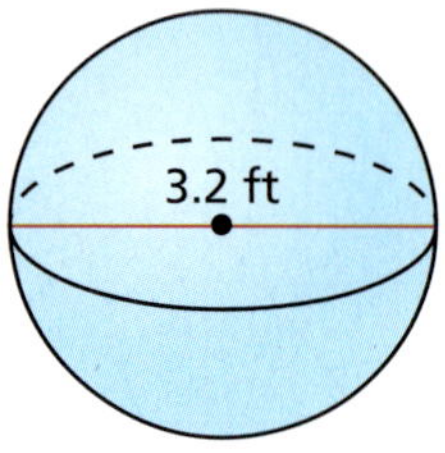

5. 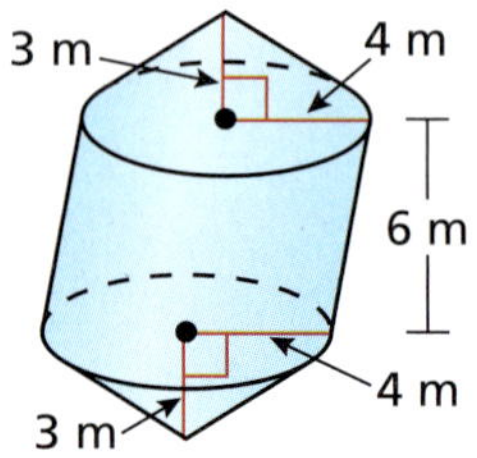

6. 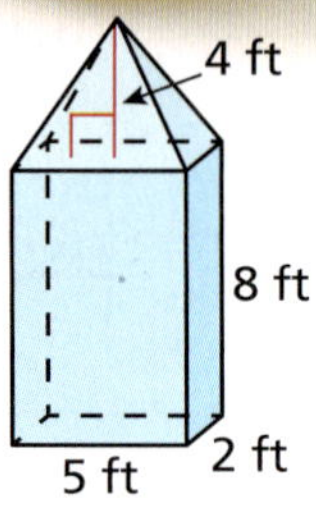

7. Sketch and describe the solid produced by rotating the figure around the given axis. Then find its surface area and volume.

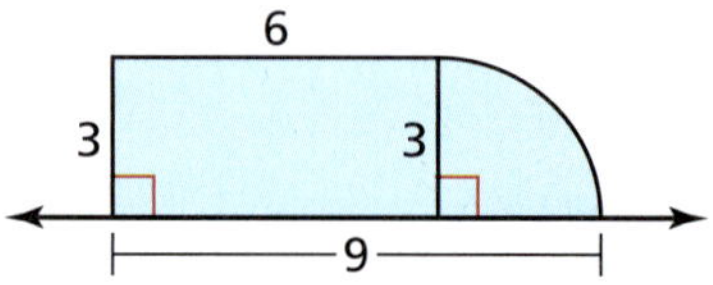

The two solids are similar. Find the surface area of solid B.

8. 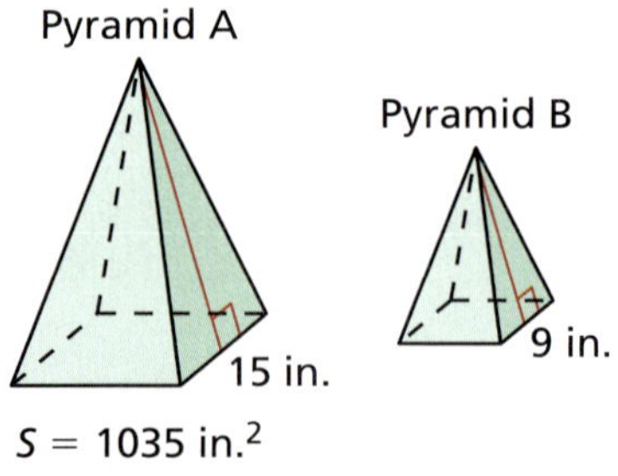

9. 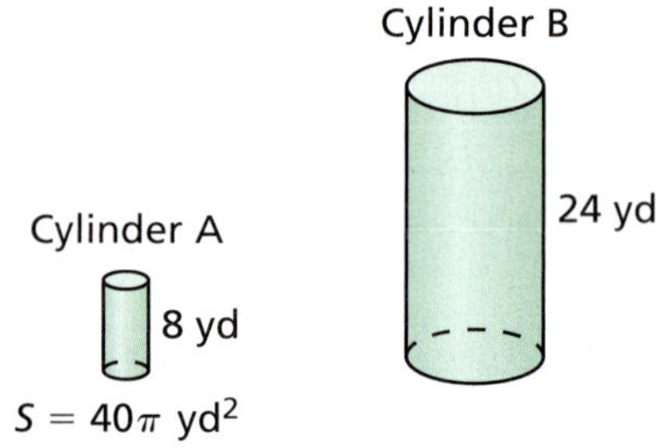

10. The candle has a mass of 1200 grams.

    a. What is the density of the candle wax in grams per cubic centimeter?

    b. After three hours of burning, the height of the candle is about 8.3 inches. How does this change affect the surface area and volume of the candle?

    c. How much mass, to the nearest gram, did the candle lose after 3 hours of burning? If this rate remains constant, estimate how much longer the candle will burn before melting entirely.

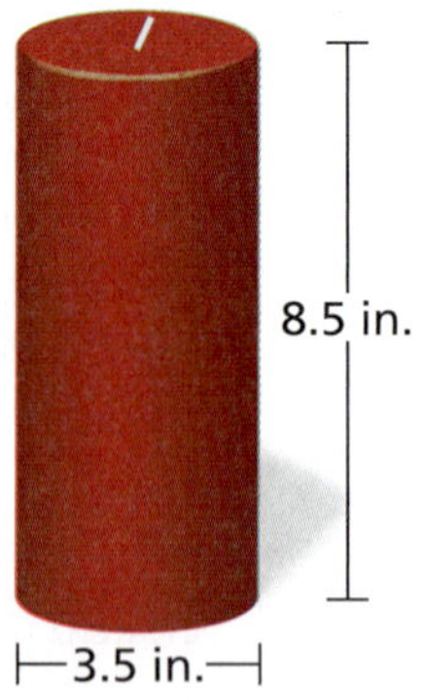

11. A bronze weight weighs 1 gram and has a density of 8.76 grams per cubic centimeter. The bronze is a mixture of tin with a density of 7.29 grams per cubic centimeter and copper with a density of 8.96 grams per cubic centimeter. Find the amount of tin and copper in the 1-gram bronze weight.

12. Which is greater, the volume of the solid produced by rotating the rectangle shown around the *x-axis* or the *y-axis*? Explain.

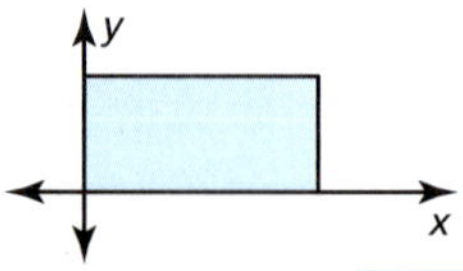

GO DIGITAL

12 Performance Task

White-Nose Syndrome

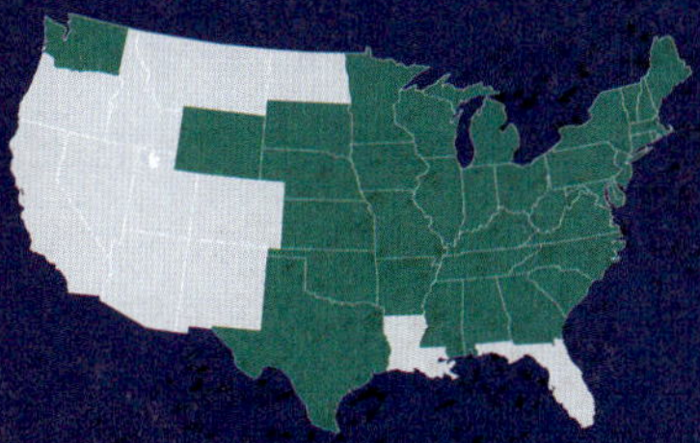

The fungus that causes WNS has spread to at least 36 states in the United States.

What Is White-Nose Syndrome?

White-nose syndrome (WNS) is a fungal disease that affects hibernating bats. The fungus looks like white fuzz on a bat's nose.

WNS causes bats to become overactive and burn up fat needed during the winter. As a result, many bats starve to death before spring.

What Is Being Done?

Some conservationists are building artificial bat caves that can be easily disinfected.

In some artificial caves, mesh materials are used to line the walls so that the bats can grip on to the surface. Wood panels can also be installed to accommodate crevice-dwelling bats.

Why Should We Care?

Bats bring many benefits to their ecosystems:

- Insect Control
- Pollination
- Seed Dispersal
- Fertilization

DESIGN A BAT CAVE

Research the approximate number of bats that roost in Bracken Cave in Comal County, Texas. Then research an appropriate volume for an artificial bat cave. Design an artificial cave that can be used to accommodate the bats in Bracken Cave. Provide a sketch of your cave, and describe the shape of the cave, the thickness of the walls, and the size and shape of the cave opening. Then determine the amount of mesh material that you will need to line the inner walls of your cave.

Finally, conduct research to estimate the maximum number of roosting bats that your cave can accommodate. About how many of these artificial caves are needed to accommodate all of the bats in Bracken Cave?

GO DIGITAL

12 Test Prep WITH CalcChat® Cumulative Practice

WATCH Tutorial videos are available for each exercise.

1. Identify the shape of the cross section formed by the intersection of the plane and the solid.

a. 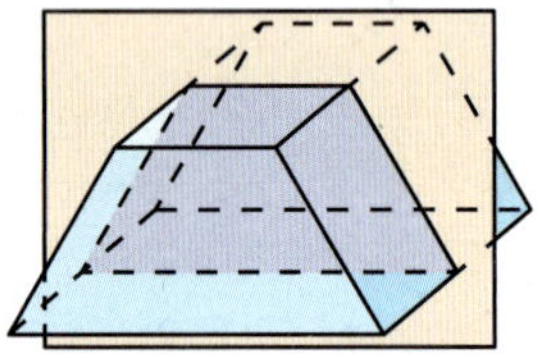

b. 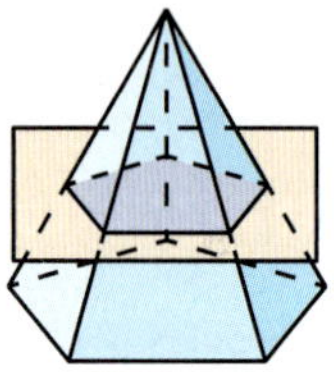

c. 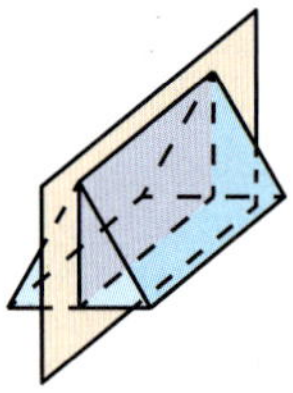

2. $\triangle DEF$ is the image of $\triangle ABC$ after the composition.

Rotation: 270° clockwise about the origin

Translation: $(x, y) \rightarrow (x - 2, y + 3)$

Are $\triangle DEF$ and $\triangle ABC$ congruent? Explain.

3. Which is most helpful as the next step in constructing a line tangent to $\odot P$ that passes through S?

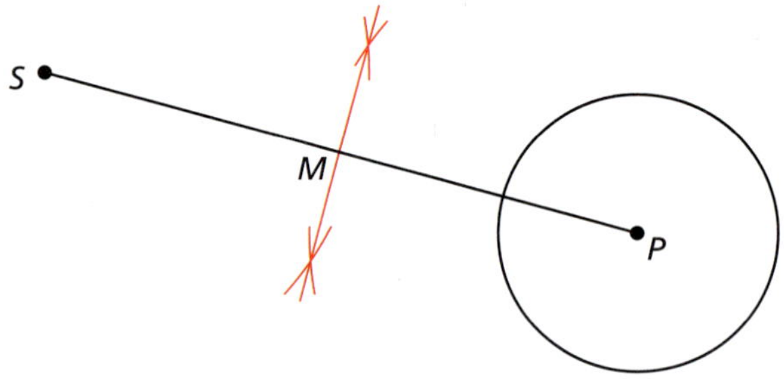

Ⓐ Construct $\odot M$ with radius SM.

Ⓑ Construct $\odot S$ with radius SM.

Ⓒ Construct $\odot M$ with radius SP.

Ⓓ Construct $\odot S$ with radius SP.

4. Which of the following statements are true regarding triangle ABC? Select all that apply.

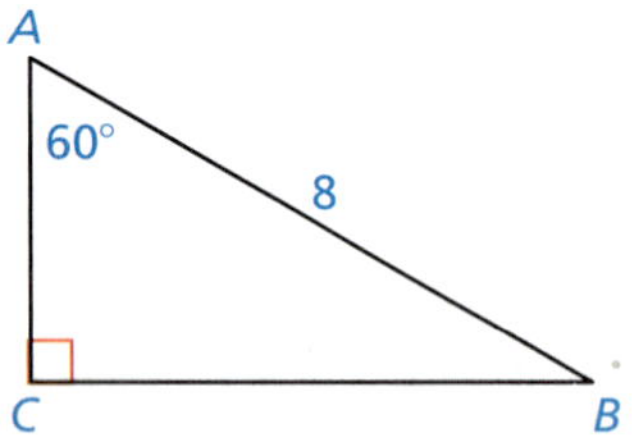

Ⓐ $m\angle B = 30°$

Ⓑ $m\angle B - m\angle A = m\angle C$

Ⓒ $AC = \frac{1}{2}AB$

Ⓓ $AC = 2$

Ⓔ $CB = \sqrt{3}$

Ⓕ $CB = 4\sqrt{3}$

GO DIGITAL

5. The point (4, 3) is on a circle with center $(-2, -5)$. What is the standard equation of the circle?

Ⓐ $(x - 2)^2 + (y - 5)^2 = 10$

Ⓑ $(x + 2)^2 + (y + 5)^2 = 10$

Ⓒ $(x - 2)^2 + (y - 5)^2 = 100$

Ⓓ $(x + 2)^2 + (y + 5)^2 = 100$

6. You have enough wood to build a rectangular deck with an area of 200 square feet along the back of a house. The deck needs a railing around the three sides that are not along the house. This will include an opening that is 4 feet wide for access to and from the yard. What dimensions x and y minimize the length of railing needed?

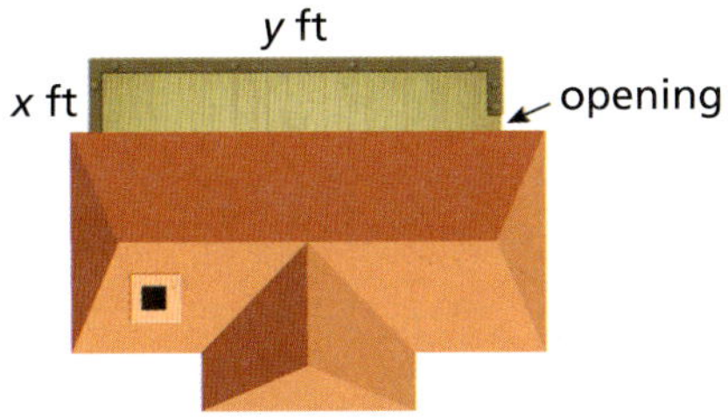

7. You and several friends rent the cottage shown for a weekend. The inside temperature is 55°F when you arrive, so you adjust the heater's thermostat to 72°F. The heater must provide 4 Btu of heat per cubic foot of space to reach this temperature. How many Btu must the heater provide in all?

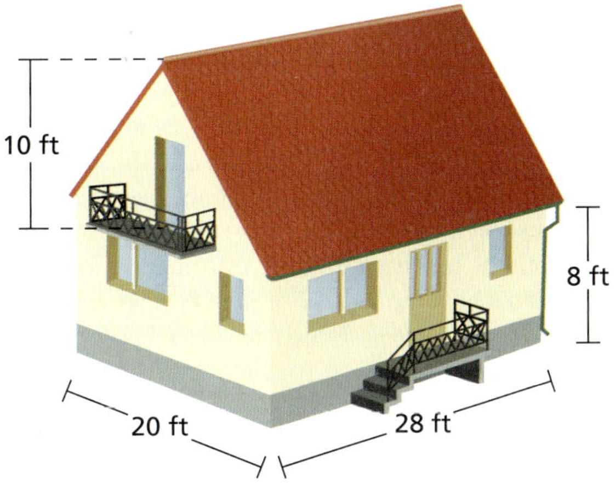

8. The diagram shows a square pyramid and a cone. Both solids have the same height, h, and the base of the cone has radius r. According to Cavalieri's Principle, the solids will have the same volume when the square base of the pyramid has sides of what length?

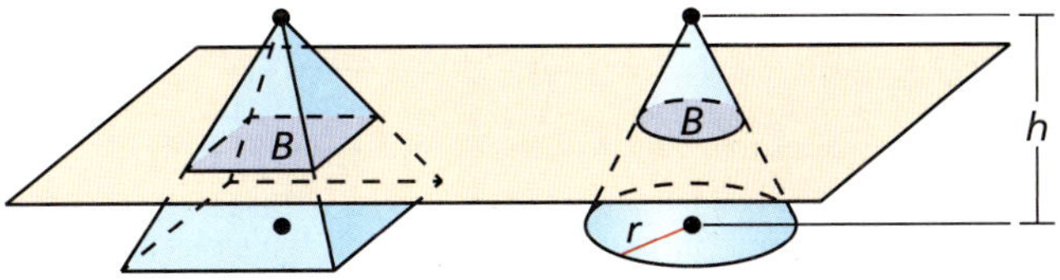

Ⓐ $\frac{\pi r}{2}$

Ⓑ $r\sqrt{\pi}$

Ⓒ $2r$

Ⓓ πr

GO DIGITAL

Geometry Course Project

Introduction

Work with a partner or a group. Choose one of the following projects to complete or create your own. These projects are a fun way to showcase your ability to apply the mathematics learned throughout this course to a real-life situation.

Your project will require you to be creative. It may involve conducting research, incorporating factual information, or imagining and using fictitious ideas and situations. You will work on your project throughout the course, incorporating relevant mathematics from multiple chapters and sections. Possible formats for your project include, but are not limited to: a written report, slide presentation, video, website, or webinar.

The Old Pier

The Old Pier is one of the best-known landmarks of Naples, Florida. Originally constructed in 1888, the pier has had an eventful life. Four hurricanes, a fire, and worms (yes—worms) have all damaged or destroyed the pier over the years, requiring a series of reconstructions and repairs.

Piers are surprisingly complex structures and contain numerous geometric and mathematical features. The length, width, and layout of the pier; the height and shape of the pilings; the dimensions of the building materials used, their number, and how they are assembled; and the depth of the water and contours of the ocean floor can all be described using geometry and other types of mathematics.

Research the Old Pier and its history. Pay particular attention to the specifications of the various reconstructions, repairs, and remodels the pier has undergone. Use what you learn to choose a topic and generate a driving question that you will seek to answer. Use factual information or invent fictitious stories and situations based on real-life information to create a presentation that answers your question and explains how the history of the Old Pier relates to what you are learning in Geometry.

GO DIGITAL

Marine Navigation

Today, sailors at sea typically navigate using the *Global Positioning System* (GPS), a technology first developed in the latter half of the 20th century. Prior to the creation of satellite-based positioning, sailors relied on a variety of tools and techniques to navigate. Celestial navigation allowed sailors to identify their direction long before the invention of the magnetic compass. A method called bearing triangulation used geometry to identify your location by calculating angles between landmarks visible in the distance. Other important techniques and technologies include the sun compass, dead reckoning, latitude and longitude, the sextant, and the marine chronometer. Some of these devices and methods date back to the 6th century B.C.

Research the history of ocean navigation. Focus on the tools and techniques sailors employed prior to the invention of GPS. Use what you learn to choose a topic and generate a driving question that you will seek to answer. Use factual information or invent fictitious stories and situations based on real-life information to create a presentation that answers your question and explains how navigation relates to what you are learning in Geometry.

Port Problems

Florida has the 8th greatest land area of any state but, because it is surrounded almost entirely by water, it has the 2nd longest coastline. Florida is also located at the southeastern-most extreme of the continental United States, placing it closer to Latin America and the Caribbean than any other state. These factors combined have made Florida an important shipping hub.

The State of Florida operates fifteen seaports, massive operations through which tens of billions of dollars of goods pass every year. Mathematics is crucial to ensuring a port runs efficiently. The amount of goods that can be unloaded is dictated by the space available at the port and the volumes of the shipping containers. The geometry of the port's layout must balance the space required for docking massive container ships with the needs of warehouse facilities to store and prepare goods for transport. Costs associated with shipping must also be accounted for—if shipping by sea is expensive, it may be more cost effective to deliver goods by truck or airplane instead.

Research container ships and seaports, and how each are optimized to allow goods to be transported efficiently. Use what you learn to choose a topic and generate a driving question that you will seek to answer. Use factual information or invent fictitious stories and situations based on real-life information to create a presentation that answers your question and explains how shipping relates to what you are learning in Geometry.

Selected Answers

Chapter 1

Prepare with CalcChat

1. 4
2. 11
3. 5
4. 9
5. 8
6. 6
7. 1
8. 5
9. 17
10. 154 m^2
11. 84 yd^2
12. 200 in.^2
13. x and y can be any real number, $x \neq y$; $x = y$; no; Absolute value is never negative.

Section 1.1

1.1 Practice

1. *Sample answer:* A, B, D, E
3. plane S
5. $\overleftrightarrow{QW}$, line g
7. R, Q, S; *Sample answer:* T
9. $\overline{DB}$
11. $\overrightarrow{AC}$
13. $\overrightarrow{EB}$ and $\overrightarrow{ED}$, $\overrightarrow{EA}$ and $\overrightarrow{EC}$
15. $\overrightarrow{AD}$ and $\overrightarrow{AC}$ are not opposite rays because A, C, and D are not collinear; $\overrightarrow{AD}$ and $\overrightarrow{AB}$ are opposite rays because A, B, and D are collinear, and A is between B and D.
17. *Sample answer:*

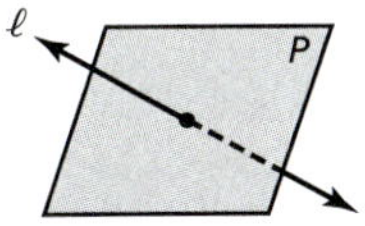

19. *Sample answer:*

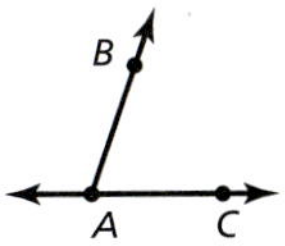

21. *Sample answer:*

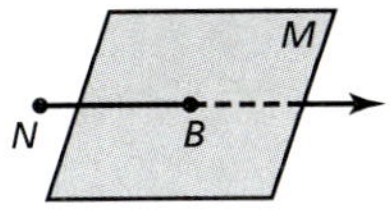

23. *Sample answer:*

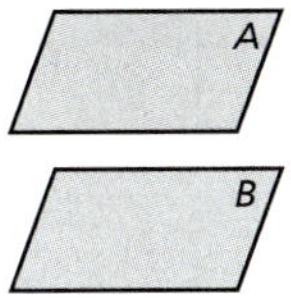

25. J
27. *Sample answer:* D
29. *Sample answer:* C
31. $\overleftrightarrow{AE}$
33. K, N
35. One airplane passes above the other because they are traveling in different planes.
37. point
39. segment
41. P, Q, R, S
43. K, L, M, N
45. no; Two planes can intersect in a line, overlap completely, or not intersect at all.
47. Three legs of the chair will meet on the floor to define a plane, but the point at the bottom of the fourth leg may not be in the same plane. When the chair tips so that this leg is on the floor, the plane defined by this leg and the two legs closest to it now lies in the plane of the floor; no; Three points define a plane, so the legs of the three-legged chair will always meet in the flat plane of the floor.
49. 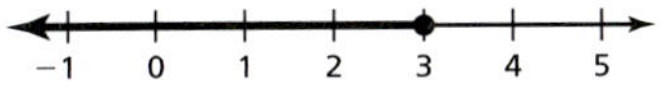

ray
51. never; A line extends without end.
53. sometimes; The point may be in the plane.
55. always; There is exactly one line through any two points.
57. always; There is exactly one plane through any three points not on the same line.
59. 6; The first two lines intersect at one point. The third line could intersect each of the first two lines. The fourth line can be drawn to intersect each of the first 3 lines. Then the total is $1 + 2 + 3 = 6$.

1.1 Review & Refresh

61. Lines a and b are parallel; They have the same slope.
62. Lines a and b are perpendicular; Their slopes are negative reciprocals.
63. $x = 25$
64. $x = 42$
65. $|10t - 30| = 5$; $t = 2.5$ and $t = 3.5$
66. 32
67. 6
68. 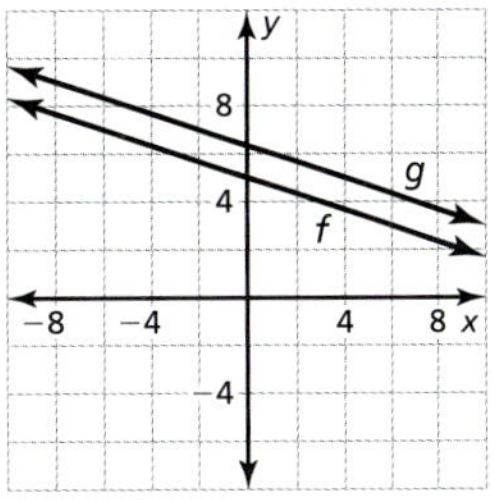

The graph of g is a horizontal translation 4 units right of the graph of f.
69. *Sample answer:* A, C, F, K
70. *Sample answer:* $\overleftrightarrow{FG}$, $\overleftrightarrow{AB}$
71. *Sample answer:* $\overrightarrow{CD}$, $\overrightarrow{LA}$, $\overrightarrow{FH}$
72. *Sample answer:* F, G, H
73. *Sample answer:* C, E, K
74. *Sample answer:* plane ADK, plane CEK
75. *Sample answer:* $\overline{CD}$, $\overline{AB}$, $\overline{GF}$
76. $x = -8, x = 0, x = 5$
77. $x = -4, x = 0, x = 4$
78. $x = -2, x = 0, x = 1$
79. $x = -1, x = 0, x = 7$
80. 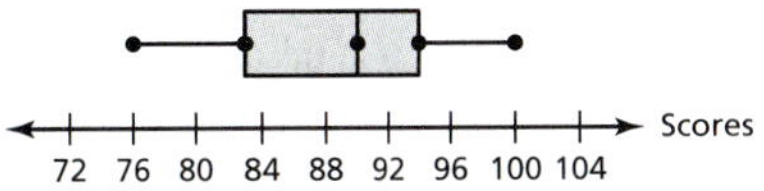

81. 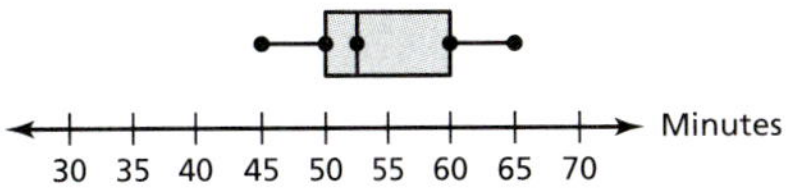

Section 1.2

1.2 Practice

1. 3.5 cm
3. 4.5 cm
5.

7. 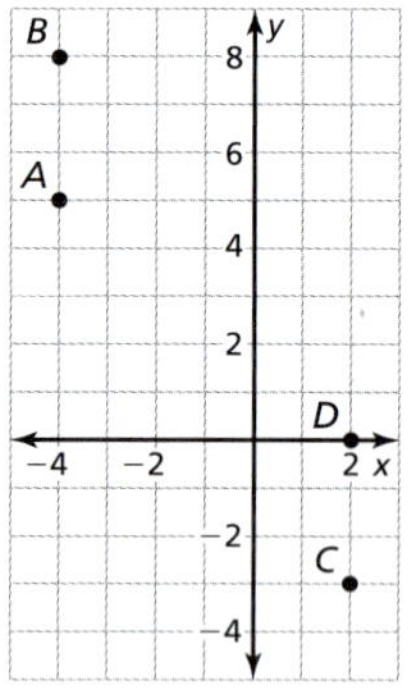

yes

9. 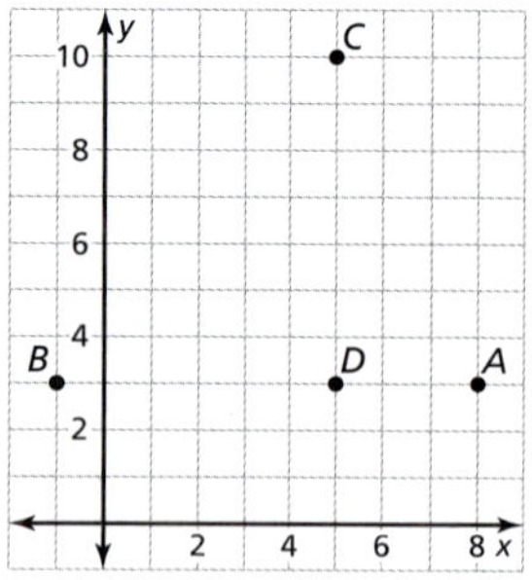

no

11. 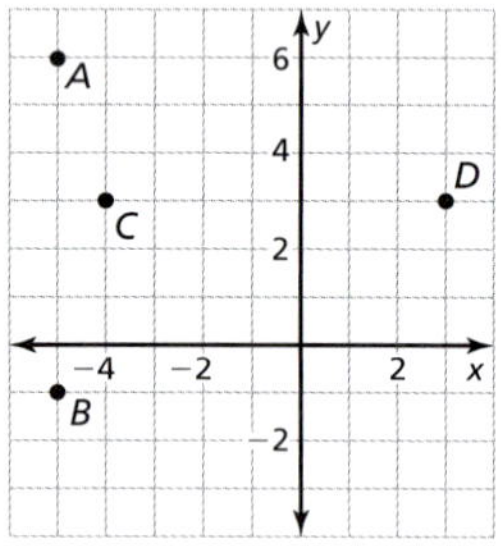

yes

13. 22

15. 23

17. 24

19. 20

21. The absolute value should have been found; $AB = |1 - 4.5| = 3.5$

23. B

25. **a.** 1883 mi
    b. about 50 mi/h

27. **a.** true; B is on $\overleftrightarrow{AC}$ between A and C.
    b. false; B, C, and E are not collinear.
    c. true; D is on $\overleftrightarrow{AH}$ between A and H.
    d. false; C, E, and F are not collinear.

29. no; If you do not line up an object at zero, then take the absolute value of the difference of the measurements at both ends of the object.

31. 593 mi; If the round-trip distance is 647 miles, then the one-way distance is $647 \div 2 = 323.5$ miles; $323.5 - 27 = 296.5$ and $2(296.5) = 593$

33. $AB = 3$, $BC = 3$, $BD = 9$, $AC = 6$, $CD = 6$, $AD = 12$; $\frac{2}{3}$; Two of the segments are 3 units long. The other four are longer than that.

35. yes, no; $FC + CB = FB$, so $FB > CB$. $\overline{AC}$ and $\overline{DB}$ overlap but do not share an endpoint.

1.2 Review & Refresh

37. $y = 9$

38. $x = -2$

39. $x = 6$

40. $x = -13$

41. *Sample answer:*

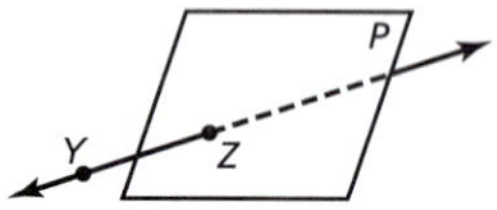

42. $x > 2$

43. 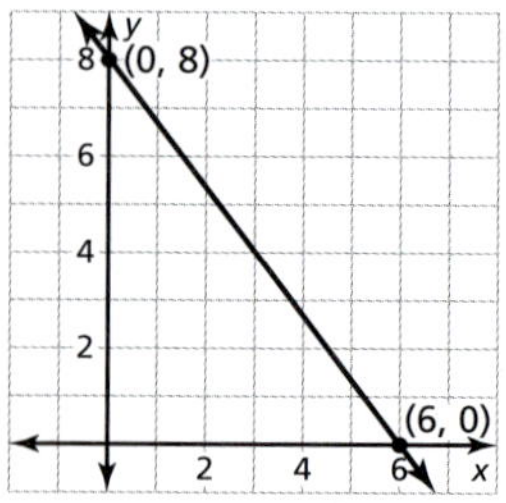

44. 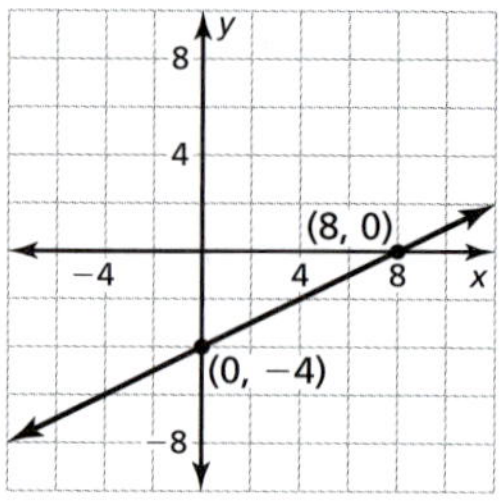

45. not a function; The input -1 is paired with two outputs, -3 and 1.

46. $x \le 19$

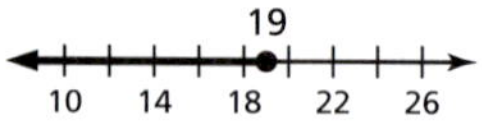

47. $t < -5$

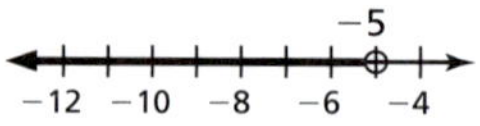

48. $c > -21$

49. $v \le -10$ *or* $v > -2$

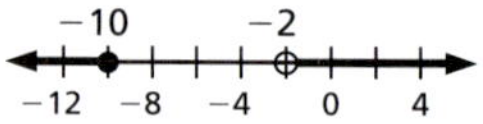

50. 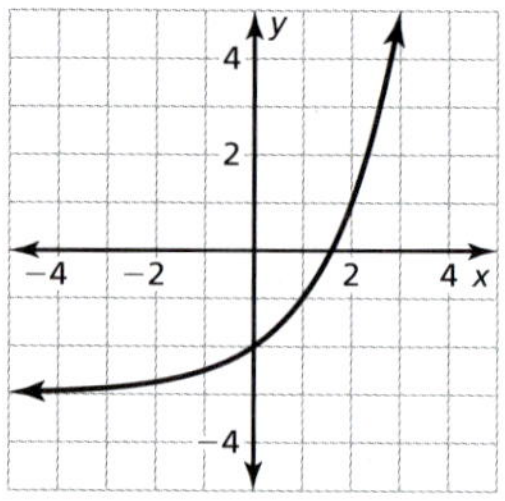

asymptote: $y = -3$; domain: all real numbers, range: $y > -3$

51. 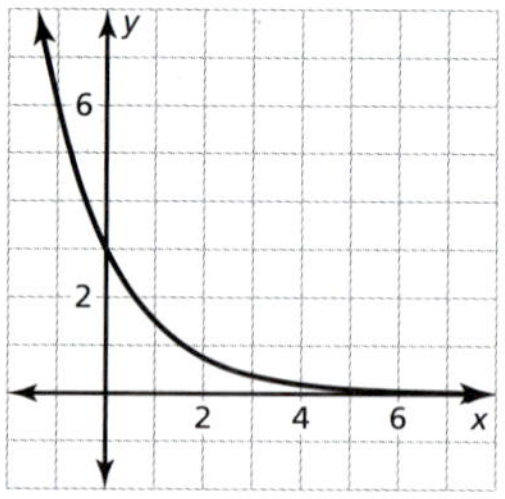

asymptote: $y = 0$; domain: all real numbers, range: $y > 0$

52. yes; yes; As more people attend an amusement park, the lines are likely to get longer.

53. 21

54. 9

55. 5 touchdowns; 2 field goals

56. $y = -6x + 3$

57. $y = \frac{1}{4}x - 6$

Section 1.3

1.3 Practice

1. line k; 34

3. M; 44

5. M; 40

7. $\overrightarrow{MN}$; 32

9. 

11. 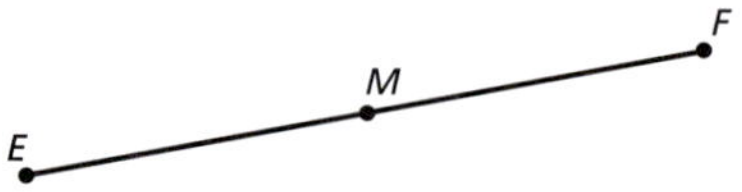

13. 6 **15.** 2

17. 5 **19.** 1

21. (5, 2) **23.** $\left(1, \frac{9}{2}\right)$

25. (0.5, 1.5) **27.** (3, 12)

29. (5, 4) **31.** (18, −9)

33. 10 **35.** $\sqrt{13}$, or about 3.6

37. $\sqrt{97}$, or about 9.8 **39.** 6.5

41. The square root should have been taken. The distance is $\sqrt{61}$.

43. $AB = \sqrt{45}$, $CD = \sqrt{40}$; no; $AB > CD$

45. $WX = \sqrt{61}$, $YZ = 5\sqrt{2}$; no; $WX > YZ$

47. 51 in. **49.** about 8.1 mi

51. no; You have to find the absolute value of the difference.

53. **a.** 86%

b. yes; If your friend scores greater than 97.5% of the final their final grade will be higher.

55.

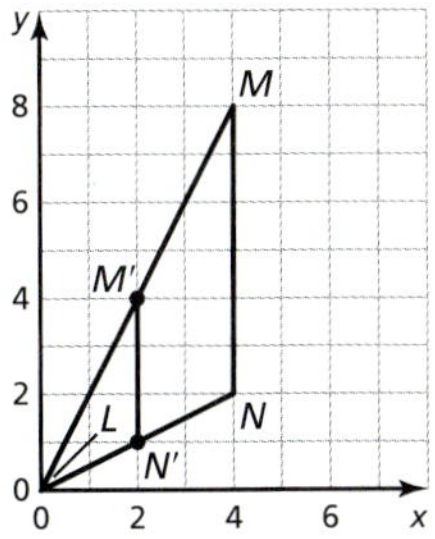

c. yes; The length of each side in $LM'N'$ is half its corresponding side in LMN.

57. about 1.5 min; $MR \approx 40$ yd,

total distance $\approx 40 + 40 + 40 + 70 + 40 = 230$ yd,

$\frac{230}{150} \approx 1.5$ min

59. $\sqrt{613}$

1.3 Review & Refresh

61. 20 cm, 25 cm^2 **62.** 26 ft, 30 ft^2

63. 12 m, 6 m^2 **64.** 36 yd, 60 yd^2

65. $a < -11$

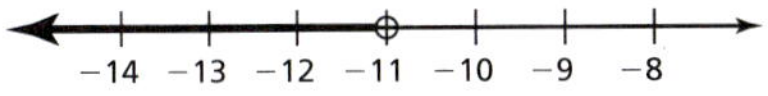

66. $y \geq 13$

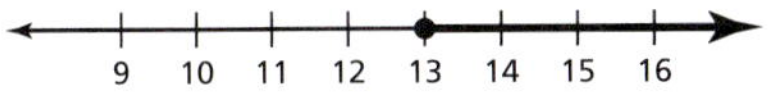

67. $x < -8$

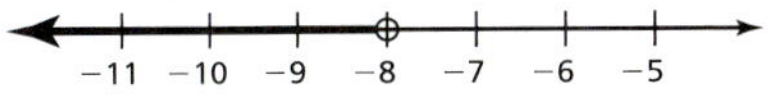

68. $z \leq 48$

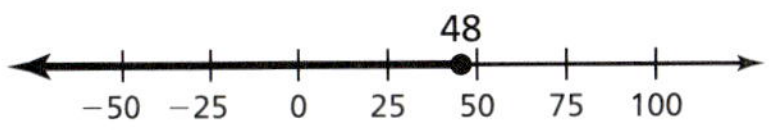

69. $\left(-\frac{1}{2}, 1\right)$; $\sqrt{205}$ **70.** $y = -\frac{1}{3}x - 2$

71. 12 **72.** $3x(x - 12)$

73. $(n - 7)(n + 10)$ **74.** $(11p + 10)(11p - 10)$

75. $(5y - 2)(3y + 2)$ **76.** $\overrightarrow{TP}$ and $\overrightarrow{TQ}$, $\overrightarrow{TR}$ and $\overrightarrow{TS}$

77. $\frac{1}{b^8}$ **78.** $\frac{125t^{12}}{8}$

79.

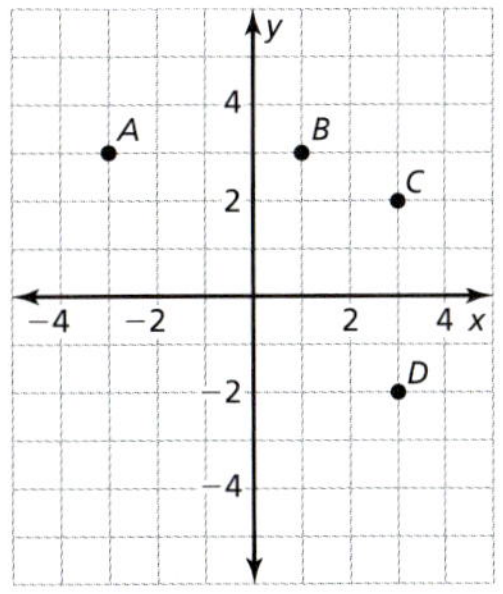

yes

80. **a.** 76 points

b. 6 incorrect answers

81. 2.5 gal

82. The function is negative for all real values of x. The function is increasing for $x < 0$ and decreasing for $x > 0$. $f(x) \to -\infty$ as $x \to -\infty$ and $f(x) \to -\infty$ as $x \to +\infty$

83. $y = \frac{3}{2}x + 2$

Section 1.4

1.4 Practice

1. quadrilateral; concave **3.** pentagon; convex

5. 22 units

7. $13 + \sqrt{89}$, or about 22.43 units

9. $7 + 2\sqrt{10} + \sqrt{13}$, or about 16.93 units

11. 7.5 square units **13.** 9 square units

15. $4 + 4\sqrt{2}$, or about 9.66 units; 4 square units

17. $12\sqrt{2}$, or about 16.97 units; 16 square units

19. The height should be the distance from A to $\overline{BC}$; $h = |3 - 1| = 2$, $A = \frac{1}{2}bh = \frac{1}{2}(4)(2) = 4$; The area is 4 square units.

21. B **23.** 34 ft; 60 ft^2

25. **a.** about 10.47 mi

b. about 17.42 mi

27. **a.** 16 units, 16 square units

b. yes; The sides are all the same length because each one is the hypotenuse of a right triangle with legs that are each 2 units long. Because the slopes of the lines of each side are either 1 or −1, they are perpendicular.

c. $8\sqrt{2}$ units, 8 square units; It is half of the area of the larger square.

29. no; triangle: $P = 12$ units and $A = 6$ square units; A rectangle that is 1×5 will have the same perimeter as the triangle (12 units) but not the same area (5 square units).

31. $x = 2$ **33.** 13 square units

1.4 Review & Refresh

34. linear; As x increases by 1, y increases by 2.

35. $x = 3$ **36.** $x = -1$

37. no solution **38.** $x = 1$

39. $\overline{TR}$ **40.** $\overrightarrow{QV}$ and $\overrightarrow{QS}$, $\overrightarrow{QR}$ and $\overrightarrow{QT}$

41. (3, 0), $2\sqrt{10}$, or about 6.3

42. $\left(\frac{1}{2}, 1\right)$, $\sqrt{145}$, or ab out 12

43. 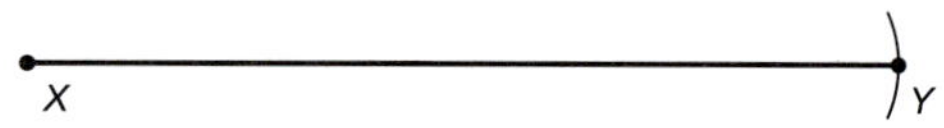

44. $y = 200(1.0125)^{4t}$

45. 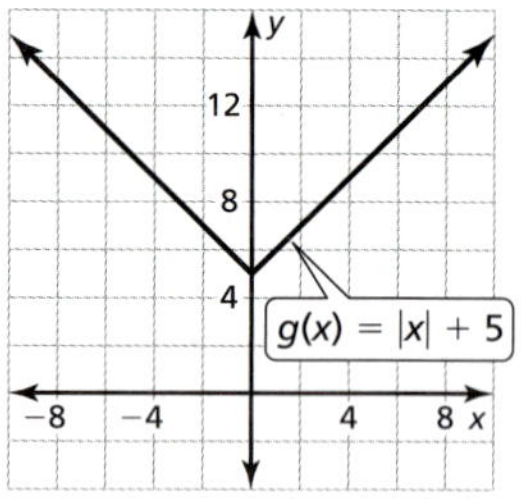

The graph of g is a vertical translation 5 units up of the graph of f.

46. 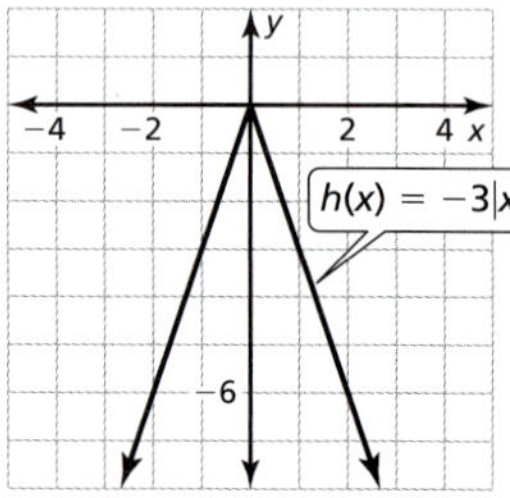

The graph of h is a vertical stretch by a factor of 3 and a reflection in the x-axis of the graph of f.

47. $6 + 4\sqrt{10}$, or about 18.65 units; 18 square units

Section 1.5

1.5 Practice

1. $\angle B$, $\angle ABC$, $\angle CBA$

3. $\angle 1$, $\angle K$, $\angle JKL$ (or $\angle LKJ$)

5. $\angle HMK$, $\angle KMN$, $\angle HMN$

7. 65°; acute

9. 115°; obtuse

11. The outer scale was used, but the inner scale should have been used because $\overrightarrow{OB}$ passes through 0° on the inner scale; 155°

13. 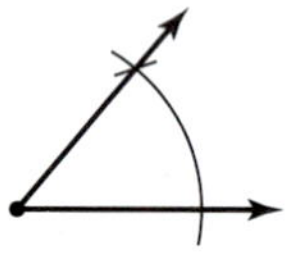

15. $\angle ADE$, $\angle BDC$, $\angle BCD$

17. 34°

19. 58°

21. 42°

23. 37°, 58°

25. 77°, 103°

27. 32°, 58°

29. 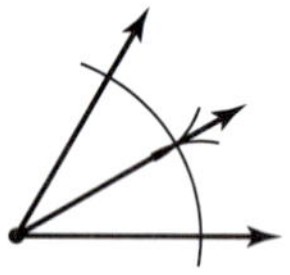

31. 63°, 126°

33. 62°, 62°

35. 44°, 44°, 88°

37. 65°, 65°, 130°

39. 75°

41. **a.** 112°

b. 56°

c. 56°

d. 56°

43. 76°, 16°

45. no; Each obtuse angle has to be greater than 90°, so their combined measure would be greater than 180°.

47. **a.** acute

b. acute

c. acute

d. right

49. A, B, D

51. 68°; $m\angle RSP = 2(m\angle VSP) = 2(17°) = 34°$, $m\angle TSQ = m\angle RSQ = 2(m\angle RSP) = 2(34°) = 68°$

1.5 Review & Refresh

53. $3 + \sqrt{10} + \sqrt{13}$, or about 9.77 units; 4.5 square units

54. $x = 12$

55. $x = -5$

56. infinitely many solutions

57. $4\sqrt{10}$

58. $3\sqrt[3]{5}$

59. $\frac{\sqrt{21}}{10}$

60. $\frac{\sqrt{55}}{5}$

61. **a.** Player A

b. about 18.9 m

62. 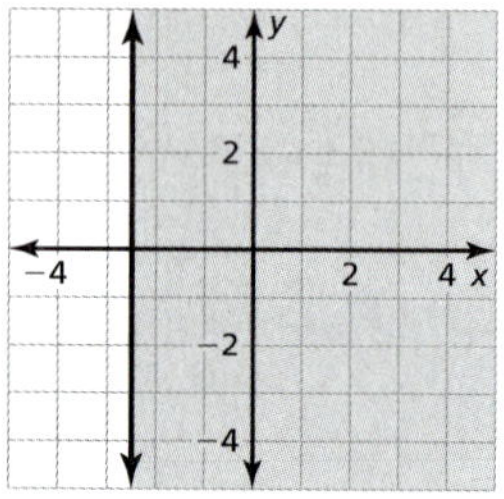

63. 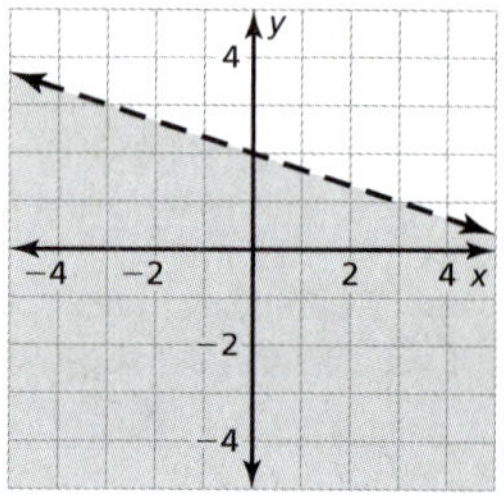

64. 55°; 110°

65. 32°; 44°

66. $(-6, 5)$; Explanations will vary.

67. $(4, -3)$; Explanations will vary.

68. 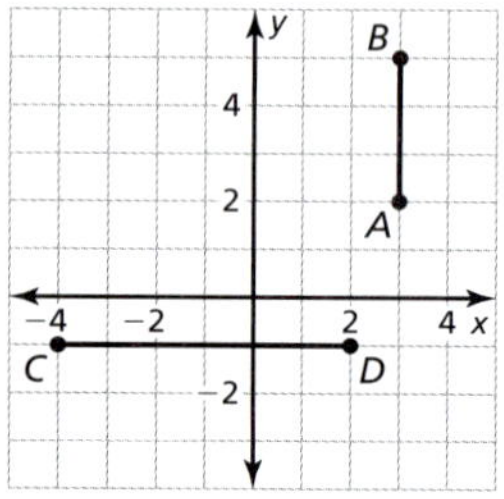

no

69. 35

Section 1.6

1.6 Practice

1. $\angle LJM$, $\angle MJN$

3. $\angle FGH$, $\angle LJM$

5. 67°

7. 102°

9. $m\angle QRT = 47°$, $m\angle TRS = 133°$

11. $m\angle UVW = 12°$, $m\angle XYZ = 78°$

13. $\angle 1$ and $\angle 5$

15. yes; The sides form two pairs of opposite rays.

17. They do not share a common side, so they are not adjacent; $\angle 1$ and $\angle 2$ are adjacent.

19. 60°, 120°

21. 9°, 81°

23. 40°, 50°

25. 122°

27. 63°

29. 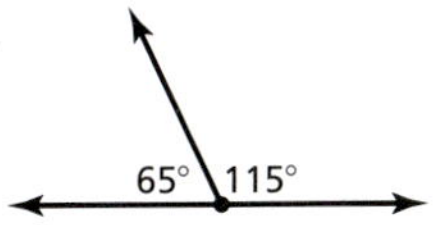

31. $x + (x - 6) = 90$; 48° and 42°

33. $x + \left(\frac{1}{2}x + 3\right) = 180$; 118° and 62°

35. B

37. sometimes; The angles could share a common side and make a right angle.

39. never; Vertical angles are formed by two pairs of opposite rays.

41. never; Its complement will be acute, and its supplement will be obtuse.

43. $\frac{1}{3}$; Because all 4 angles have supplements, the first paper can be any angle. Then there is a 1 in 3 chance of drawing its supplement.

45. yes; The angle of incidence is always congruent to the angle of reflection, so their complements will always be congruent.

47. a. $y°, (180 - y)°, (180 - y)°$
    b. They are always congruent; They are both supplementary to the same angle. So, their measures must be equal.

49. 37°, 53°; If two angles are complementary, then their sum is 90°. If x is one of the angles, then $(90 - x)$ is the complement. Write and solve the equation $90 = (x - (90 - x)) + 74$. The solution is $x = 53$.

1.6 Review & Refresh

50. 24 square units

51. 12.5 square units

52. $(6, -1)$

53. M; 32

54. $t = -8$ and $t = -2$

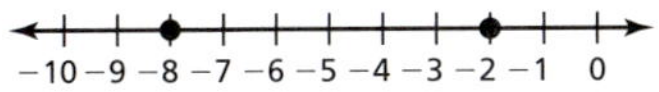

55. $d = -8$ and $d = 16$

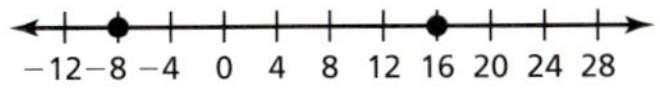

56. no solution

57. $q = \pm 4.5$

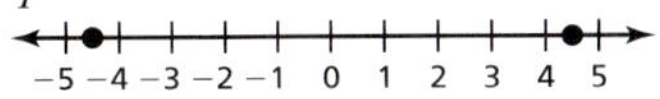

58. $-12x^8 - 32x^7 + 64x^5$

59. $28s^2 - s - 15$

60. The graph of g is a horizontal shrink by a factor of $\frac{1}{2}$ of the graph of f.

61. -5; 2

62. $\frac{3}{2}$; -7

63. 68°, 58°

64. 123°

65. 56°

Chapter 1 Review

1. *Sample answer:* plane XYN

2. *Sample answer:* line g

3. *Sample answer:* line h

4. *Sample answer:* $\overrightarrow{XZ}$, $\overrightarrow{YP}$

5. $\overrightarrow{YX}$ and $\overrightarrow{YZ}$

6. P

7. no; yes; no

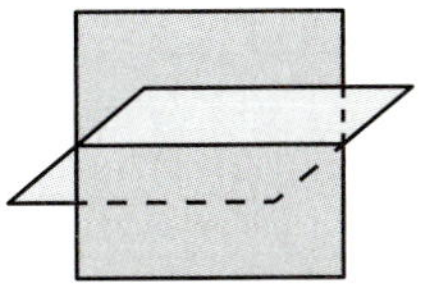

8. 41

9. 11

10. 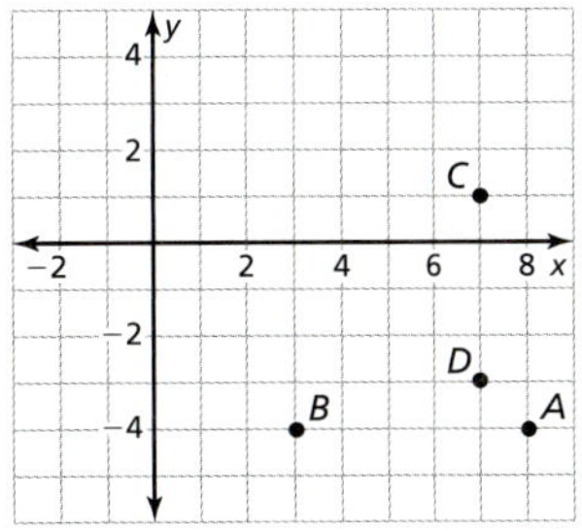

no

11. 126 m; about 19.26 min

12. $\left(\frac{1}{2}, \frac{13}{2}\right)$; $\sqrt{50}$, or about 7.1

13. $\left(\frac{13}{2}, -\frac{5}{2}\right)$; $\sqrt{2}$, or about 1.4

14. $(-2, -3)$

15. 40

16. 5

17. 6

18. about 60.6 ft; $\sqrt{(45 - 0)^2 + (45 - 0)^2} \approx 63.6$ and $63.6 - 3 = 60.6$.

19. hexagon; convex

20. octagon; concave

21. 20 units, 21 square units

22. $14 + 7\sqrt{2}$, or about 23.9 units, 24.5 square units

23. The area of the convex polygon is greater; the perimeter of the concave polygon is greater.

24. $11 + 5\sqrt{2} + \sqrt{17}$, or about 22.2 units

25. 29.5 square units

26. 49°, 28°

27. 88°, 23°

28. 127°

29. 34°, 86°

30. 78°

31. 7°

32. 64°

33. 124°

34. 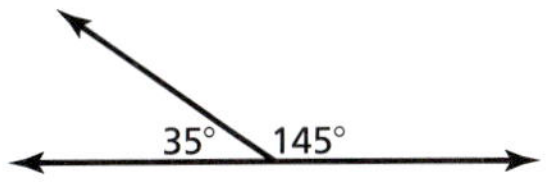

35. 144°, 36°

36. 112°, 68°

Chapter 1 Practice Test

1. 31; Segment Addition Postulate: $QR + RS = QS$, so $QS = 12 + 19 = 31$.

2. 12; Segment Addition Postulate: $QS + SR = QR$, so $QS = QR - SR = 59 - 47 = 12$.

3. $\left(-\frac{5}{2}, -2\right)$; $3\sqrt{17}$, or about 12.4 units

4. $(5, -4)$

5. false

6. true

7. false

8. false

9. $8 + 4\sqrt{10}$, or about 20.6 units; 24 square units

10. $6 + 2\sqrt{34}$, or about 17.66 units, 15 square units

11. 9

12. acute, right, or obtuse

13. Let x represent $m\angle RSU$ and y represent $m\angle TSU$.
$x + y = 48$
$y = 5x - 6$
$x = m\angle RSU = 9°$ and
$y = m\angle TSU = 39°$

14. supplementary: $\angle AFB$ and $\angle BFE$, $\angle AFC$ and $\angle CFE$, $\angle AFD$ and $\angle DFE$; complementary: $\angle AFB$ and $\angle BFC$, $\angle CFD$ and $\angle DFE$; The pairs of supplementary angles each form a linear pair. The pairs of complementary angles each form a right angle; $m\angle DFE = 63°$, $m\angle BFC = 51°$, $m\angle BFE = 141°$

15. *Sample answer:*

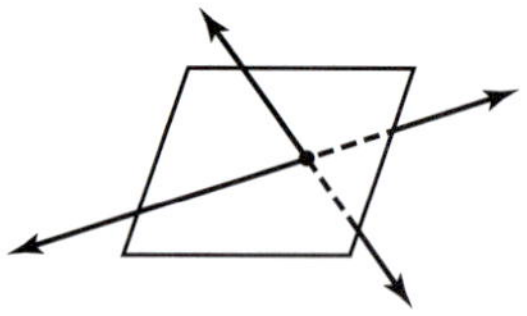

16.

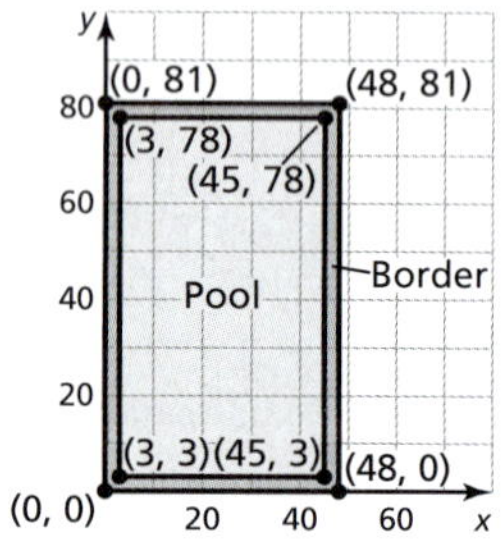

perimeter = 234 ft, area = 3150 ft^2

17. A

Chapter 1 Test Prep

1. C

2. point, segment, ray, line, plane

3. B

4. (4, −7)

5. C

6. **a.** $\angle KJL$

b. $\angle KJL$ and $\angle NJP$, $\angle KJP$ and $\angle LJN$

c. $\angle KJL$ and $\angle LJN$, $\angle KJM$ and $\angle MJN$, $\angle LJM$ and $\angle MJP$, $\angle LJN$ and $\angle NJP$, $\angle NJP$ and $\angle KJP$, $\angle KJL$ and $\angle KJP$

7. In Step 1, a compass is used to draw an arc. The two points (C and B) where this arc intersects the sides of the angle are the same distance from vertex A. In Step 2, the compass is used to draw an arc of points equidistant from B and a separate arc of points equidistant from C. The point where these two arcs intersect is a point that is equidistant from each side of the angle. In Step 3, a straightedge is used to draw the angle bisector, a ray that has an endpoint at the vertex and that passes through the point that is equidistant from each side of the angle.

8. D

9.

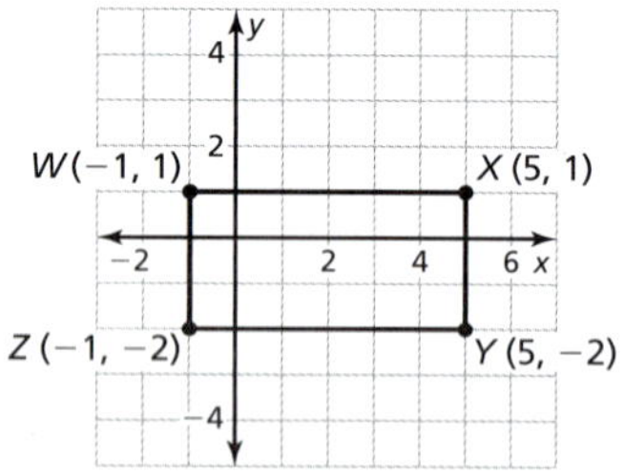

rectangle; yes; If each unit on the coordinate plane represents 15 feet on the basketball court, then each square unit represents $15^2 = 225$ square feet. The rectangle on the coordinate plane that is 3 units by 6 units makes a good model of a basketball court that is 45 feet by 90 feet with the given perimeter and area.

Chapter 2

Prepare with CalcChat

1. $x = -3$

2. $x = 2$

3. $x = 12$

4. $x = \frac{3}{7}$

5. $x = \frac{3}{5}$

6. $x = -2.7$

7. $x = y - 5$

8. $x = -4y + 3$

9. $x = y - 3$

10. $x = \frac{y}{7}$

11. $x = \frac{y - 6}{z + 4}$

12. $x = \frac{z}{6y + 2}$

13. true; equivalent equations have the same solution, and the solution can be obtained by properties of equality, so one equivalent equation can be obtained from the other by using the properties of equality.

Section 2.1

2.1 Practice

1. hypothesis: a polygon is a pentagon; conclusion: it has five sides

3. hypothesis: you run; conclusion: you are fast

5. If $x = 2$, then $9x + 5 = 23$.

7. If a glacier melts, then the sea level rises.

9. If you are allowed to vote, then you are registered.

11. The sky is not blue.

13. The ball is pink.

15. conditional: If two angles are supplementary, then the measures of the angles sum to 180°; true

converse: If the measures of two angles sum to 180°, then they are supplementary; true

inverse: If the two angles are not supplementary, then their measures do not sum to 180°; true

contrapositive: If the measures of two angles do not sum to 180°, then they are not supplementary; true

17. conditional: If you do your math homework, then you will do well on the test; false

converse: If you do well on the test, then you did your math homework; false

inverse: If you do not do your math homework, then you will not do well on the test; false

contrapositive: If you do not do well on the test, then you did not do your math homework; false

19. conditional: If an instrument is a *cuatro*, then it is a guitar; true

converse: If an instrument is a guitar, then it is a *cuatro*; false

inverse: If an instrument is not a *cuatro*, then it is not a guitar; false

contrapositive: If an instrument is not a guitar, then it is not a *cuatro*; true

21. conditional: If $3x - 7 = 20$, then $x = 9$; true

converse: If $x = 9$, then $3x - 7 = 20$; true

inverse: If $3x - 7 \neq 20$, then $x \neq 9$; true

contrapositive: If $x \neq 9$, then $3x - 7 \neq 20$; true

23. true; By definition of right angle, the measure of the right angle shown is 90°.

25. true; If angles form a linear pair, then the sum of the measures of their angles is 180°.
27. A point is the midpoint of a segment if and only if it is the point that divides the segment into two congruent segments.
29. Two angles are adjacent angles if and only if they share a common vertex and side, but have no common interior points.
31. A polygon has three sides if and only if it is a triangle.
33. An angle is a right angle if and only if it measures 90°.
35. The inverse was used instead of the converse; If I bring an umbrella, then it is raining.
37. no; The contrapositive is equivalent to the original conditional statement. In order to write a conditional statement as a true biconditional statement, you must know that the converse (or inverse) is true.
39. no; "If $x^2 - 10 = x + 2$, then $x = 4$" is a false statement because $x = -3$ is also possible. In order for a biconditional statement to be true, both the conditional statement and its converse must be true.
41. A; The conditional and its contrapositive are equivalent statements.
43. *Sample answer:* slogan: "This treadmill is a fat-burning machine!"; conditional statement: If you use this treadmill, then you will burn fat.
45. *Sample answer:* If you are a vegetarian, then you do not eat hamburgers.
47. By definition of a linear pair, $\angle 1$ and $\angle 2$ are supplementary. So, if $m\angle 1 = 90°$, then $m\angle 2 = 90°$. Also, by definition of a linear pair, $\angle 2$ and $\angle 3$ are supplementary. So, if $m\angle 2 = 90°$, then $m\angle 3 = 90°$. Finally, by definition of a linear pair, $\angle 3$ and $\angle 4$ are supplementary. So, if $m\angle 3 = 90°$, then $m\angle 4 = 90°$.

2.1 Review & Refresh

48. no; The input 3 is paired with two outputs, 1 and 5.
49. $x = 1 \pm 2\sqrt{2}$
50. $x = -\frac{5}{2}, x = 1$
51. The graph of g is a translation 5 units up of the graph of f.
52. The graph of g is a vertical stretch by a factor of 3 of the graph of f.
53. 114°, 66°
54. $8 + 4\sqrt{13}$, or about 22.42 units; 24 square units
55. 384,400
56. 49°, 98°
57. $-3x^3 + 21x^2$
58. $z^2 + 7z - 8$
59. $5b^2 - 10b + 5$
60. $-4n^3 + 5n^2 + 5n - 1$
61. $x \le 6$
62. conditional statement: If you play a video game, then you beat the video game; false;

    converse: If you beat a video game, then you play the video game; true;

    inverse: If you do not play a video game, then you do not beat the video game; true;

    contrapositive: If you do not beat a video game, then you do not play the video game; false

Section 2.2

2.2 Practice

1. The absolute value of each number in the list is 1 greater than the absolute value of the previous number in the list, and the signs alternate from positive to negative; −6, 7
3. Each letter is the letter before the previous letter in the alphabet; U, T
5. This is a sequence of regular polygons, each polygon having one more side than the previous polygon.

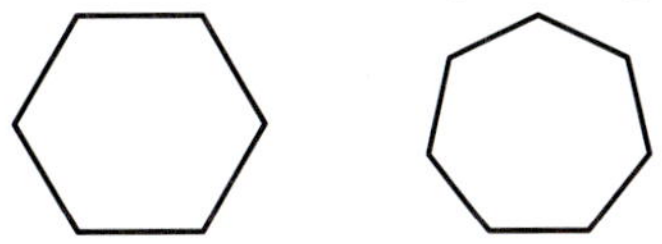

7. The sum of an even integer and an odd integer is an odd integer; *Sample answer:* $2 + 5 = 7$, $-4 + 9 = 5$, $10 + 7 = 17$
9. The quotient of a number and its reciprocal is the square of that number; *Sample answer:*
   $9 \div \frac{1}{9} = 9 \cdot 9 = 9^2$, $\frac{2}{3} \div \frac{3}{2} = \frac{2}{3} \cdot \frac{2}{3} = \left(\frac{2}{3}\right)^2$, $\frac{1}{7} \div 7 = \frac{1}{7} \cdot \frac{1}{7} = \left(\frac{1}{7}\right)^2$
11. *Sample answer:* $1 \cdot 5 = 5$, $5 \not> 5$
13. They could both be right angles. Then, neither are acute.
15. Your device crashes.
17. not possible
19. not possible
21. If a figure is a rhombus, then the figure has two pairs of opposite sides that are parallel.
23. Law of Syllogism
25. Law of Detachment
27. The sum of two odd integers is an even integer; Let m and n be integers. Then $(2m + 1)$ and $(2n + 1)$ are odd integers. $(2m + 1) + (2n + 1) = 2m + 2n + 2 = 2(m + n + 1)$; $2(m + n + 1)$ is divisible by 2 and is therefore an even integer.
29. inductive reasoning; The conjecture is based on the assumption that a pattern, observed in specific cases, will continue.
31. deductive reasoning; The Law of Syllogism was used to draw the conclusion.
33. The Law of Detachment cannot be used because the hypothesis is not true; *Sample answer:* Using the Law of Detachment, because a square is a rectangle, you can conclude that a square has four sides.
35. Using inductive reasoning, you can make a conjecture that male tigers weigh more than female tigers because this was true in all of the specific cases listed in the table.
37. The value of y is 2 more than three times the value of x; $y = 3x + 2$; *Sample answer:* If $x = 10$, then $y = 3(10) + 2 = 32$; If $x = 72$, then $y = 3(72) + 2 = 218$.
39. **a.** Each number in the sequence is the sum of the previous two numbers in the sequence.

    b. 55, 89, 144

    c. *Sample answer:* A spiral can be drawn by connecting the opposite corners of squares with side lengths that follow the Fibonacci sequence. This spiral is similar to the spiral seen on nautilus shells. It is also similar to the golden spiral, which is sometimes found in spiraling galaxies.

41. a. false; When you go camping, you go canoeing, but even though your friend always goes camping when you do, he or she may not choose to go canoeing with you.

b. false; It is known that you and your friend went on a hike, but it is not known where. It is only known that there is a trail near where you are camping.

43. a. Mineral *C* must be Talc. Because it was scratched by all three of the other minerals, it must have the lowest hardness rating. Because Mineral *B* has a higher hardness rating than Mineral *A*, Mineral *A* could be either Gypsum or Calcite, and Mineral *B* could be either Calcite or Fluorite.

b. Check Mineral *B* and Mineral *D*. If Mineral *D* scratches Mineral *B*, then Mineral *D* is Fluorite, Mineral *B* is Calcite, and Mineral *A* is Gypsum. If Mineral *B* scratches Mineral *D*, then Mineral *B* is Fluorite, and you have to check Mineral *D* and Mineral *A*. The one that scratches the other has the higher hardness rating and is therefore Calcite. The one that gets scratched is Gypsum.

2.2 Review & Refresh

45. hypothesis: there is a storm surge; conclusion: erosion of coastline occurs; If there is a storm surge, then the erosion of the coastline will occur.

46. hexagon; convex

47. $y = -2x + 8$

48. 71°

49. nonlinear

50. $x = 2$

51.

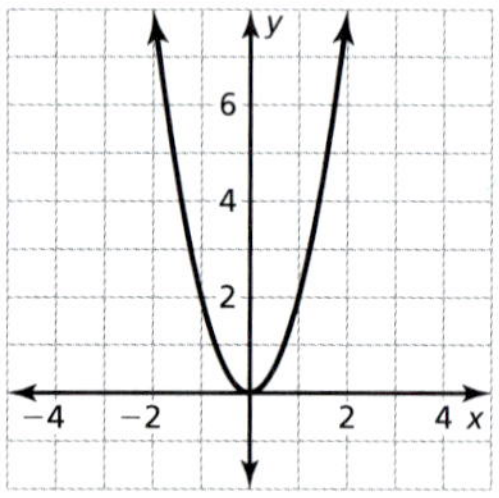

The graph of *g* is a vertical stretch by a factor of 2 of the graph of *f*.

52. 8; 8.4

53. Angle Addition Postulate

54. $-3n^2 - 5n + 2$

55. −4, 5

56. 42 units

Section 2.3

2.3 Practice

1. Two Point Postulate

3. *Sample answer:* Line *q* contains points *J* and *K*.

5. *Sample answer:* Through points *K*, *H*, and *L*, there is exactly one plane, which is plane *M*.

7.

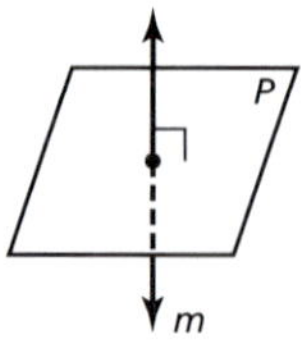

9.

11. yes

13. no

15. yes

17. yes

19. In order to determine that *M* is the midpoint of $\overline{AC}$ or $\overline{BD}$, the segments that would have to be marked as congruent are $\overline{AM}$ and $\overline{MC}$ or $\overline{DM}$ and $\overline{MB}$, respectively; Based on the diagram and markings, you can assume $\overline{AC}$ and $\overline{DB}$ intersect at point *M*, such that $\overline{AM} \cong \overline{MB}$ and $\overline{DM} \cong \overline{MC}$.

21. C, D, F, H

23. Two-Point Postulate

25. If there are two points, then there exists exactly one line that passes through them.

27. no; The postulate says that if two planes intersect, they will intersect in a line. But planes can be parallel and never intersect.

29. Points *E*, *F*, and *G* must be collinear. They must be on the line that intersects plane *P* and plane *Q*; Points *E*, *F*, and *G* can be either collinear or noncollinear.

2.3 Review & Refresh

31. *Sample answer:* trapezoid

32. $t = 2$; Addition Property of Equality, Simplify

33. $x = 7$; Division Property of Equality, Simplify

34. $x = 4$; Subtraction Property of Equality, Simplify

35. $x = 35$; Multiplication Property of Equality, Simplify

36. 153°

37. 60 in., 15 in.

38. $(y - 2) = \frac{3}{4}(x + 8)$

39. You can vote if and only if you are least 18 years old.

40. yes

41. no

42. yes

43. yes

Section 2.4

2.4 Practice

1. Subtraction Property of Equality; Addition Property of Equality; Division Property of Equality

3.

Equation	Explanation and Reason
$5x - 10 = -40$	Write the equation; Given
$5x = -30$	Add 10 to each side; Addition Property of Equality
$x = -6$	Divide each side by 5; Division Property of Equality

5.

Equation	Explanation and Reason
$2x - 8 = 6x - 20$	Write the equation; Given
$-4x - 8 = -20$	Subtract 6*x* from each side; Subtraction Property of Equality
$-4x = -12$	Add 8 to each side; Addition Property of Equality
$x = 3$	Divide each side by −4; Division Property of Equality

7.

Equation	Explanation and Reason
$5(3x - 20) = -10$	Write the equation; Given
$15x - 100 = -10$	Multiply; Distributive Property
$15x = 90$	Add 100 to each side; Addition Property of Equality
$x = 6$	Divide each side by 15; Division Property of Equality

9.

Equation	Explanation and Reason
$2(-x - 5) = 12$	Write the equation; Given
$-2x - 10 = 12$	Multiply; Distributive Property
$-2x = 22$	Add 10 to each side; Addition Property of Equality
$x = -11$	Divide each side by -2; Division Property of Equality

11.

Equation	Explanation and Reason
$4(5x - 9) = -2(x + 7)$	Write the equation; Given
$20x - 36 = -2x - 14$	Multiply on each side; Distributive Property
$22x - 36 = -14$	Add $2x$ to each side; Addition Property of Equality
$22x = 22$	Add 36 to each side; Addition Property of Equality
$x = 1$	Divide each side by 22; Division Property of Equality

13. The Subtraction Property of Equality should be used to subtract x from each side of the equation in order to get the second step.

$7x = x + 24$	Given
$6x = 24$	Subtraction Property of Equality
$x = 4$	Division Property of Equality

15.

Equation	Explanation and Reason
$5x + y = 18$	Write the equation; Given
$y = -5x + 18$	Subtract $5x$ from each side; Subtraction Property of Equality

17.

Equation	Explanation and Reason
$2y + 0.5x = 16$	Write the equation; Given
$2y = -0.5x + 16$	Subtract $0.5x$ from each side; Subtraction Property of Equality
$y = -0.25x + 8$	Divide each side by 2; Division Property of Equality

19.

Equation	Explanation and Reason
$12 - 3y = 30x + 6$	Write the equation; Given
$-3y = 30x - 6$	Subtract 12 from each side; Subtraction Property of Equality
$y = -10x + 2$	Divide each side by -3; Division Property of Equality

21.

Equation	Explanation and Reason
$C = 2\pi r$	Write the equation; Given
$\frac{C}{2\pi} = r$	Divide each side by 2π; Division Property of Equality
$r = \frac{C}{2\pi}$	Rewrite the equation; Symmetric Property of Equality

23.

Equation	Explanation and Reason
$S = 180(n - 2)$	Write the equation; Given
$\frac{S}{180} = n - 2$	Divide each side by 180; Division Property of Equality
$\frac{S}{180} + 2 = n$	Add 2 to each side; Addition Property of Equality
$n = \frac{S}{180} + 2$	Rewrite the equation; Symmetric Property of Equality

25.

Equation	Explanation and Reason
$P = 2\ell + 2w$	Write the equation; Given
$P - 2w = 2\ell$	Subtract $2w$ from each side; Subtraction Property of Equality
$\frac{P - 2w}{2} = \ell$	Divide each side by 2; Division Property of Equality
$\ell = \frac{P - 2w}{2}$	Rewrite the equation; Symmetric Property of Equality

$\ell = 11$ m

27. Multiplication Property of Equality

29. Reflexive Property of Equality

31. Substitution Property of Equality

33. Symmetric Property of Equality

35. $20 + CD$

37. $m\angle 1 = m\angle 3$

39.

Equation	Explanation and Reason
$m\angle ABD = m\angle CBE$	Write the equation; Given
$m\angle ABD = m\angle 1 + m\angle 2$	Add measures of adjacent angles; Angle Addition Postulate
$m\angle CBE = m\angle 2 + m\angle 3$	Add measures of adjacent angles; Angle Addition Postulate
$m\angle ABD = m\angle 2 + m\angle 3$	Substitute $m\angle ABD$ for $m\angle CBE$; Substitution Property of Equality
$m\angle 1 + m\angle 2 = m\angle 2 + m\angle 3$	Substitute $m\angle 1 + m\angle 2$ for $m\angle ABD$; Substitution Property of Equality
$m\angle 1 = m\angle 3$	Subtract $m\angle 2$ from each side; Subtraction Property of Equality

41. Let A represent the start of the race, B represent the first water stop, C represent the second water stop, and D represent the end of the race. A, B, C, and D lie on a line.

Equation	Reason
$AC = BD$	Given
$AC = AB + BC$	Segment Addition Postulate
$BD = BC + CD$	Segment Addition Postulate
$AB + BC = BC + CD$	Substitution Property of Equality
$AB = CD$	Subtraction Property of Equality

So, AB, the distance between the starting line and the first water stop is the same as CD, the distance between the second water stop and the finish line.

43. 1; 2; 3; The definition of the Reflexive Property of Equality only involves one angle or segment. The definition of the Symmetric Property of Equality involves two angles or segments. The definition of the Transitive Property of Equality involves three angles or segments.

45.

Equation	**Explanation and Reason**
$DC = BC$, $AD = AB$	Marked in diagram; Given
$AC = AC$	AC is equal to itself; Reflexive Property of Equality
$AC + AB + BC = AC + AB + BC$	Add $AB + BC$ to each side of $AC = AC$; Addition Property of Equality
$AC + AB + BC = AC + AD + DC$	Substitute AD for AB and DC for BC; Substitution Property of Equality

47. Both properties state basic ideas about equality. The Reflexive Property of Equality states that something is equal to itself. So, both sides of the equal sign are identical. The Symmetric Property of Equality states that you can switch the two sides of an equation. So, two equations are equivalent if they have the same two expressions set equal to each other, but the expressions are on different sides of the equal sign.

49. a.

Equation	**Explanation and Reason**
$C = \frac{5}{9}(F - 32)$	Write the equation; Given
$\frac{9}{5}C = F - 32$	Multiply each side by $\frac{9}{5}$; Multiplication Property of Equality
$\frac{9}{5}C + 32 = F$	Add 32 to each side; Addition Property of Equality
$F = \frac{9}{5}C + 32$	Rewrite the equation; Symmetric Property of Equality

b.

Degrees Celsius (°C)	Degrees Fahrenheit (°F)
0	32
20	68
32	89.6
41	105.8

c.

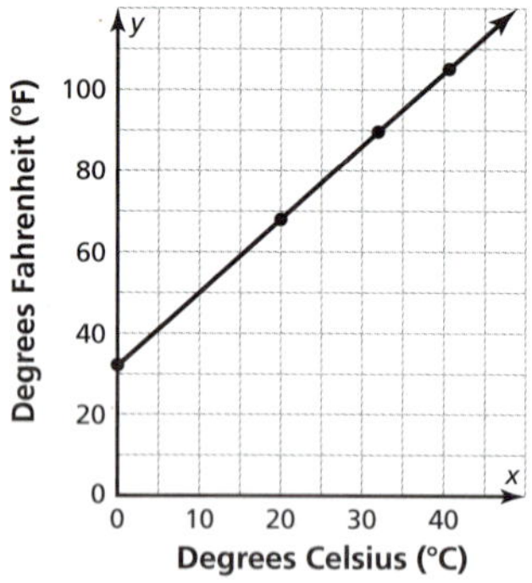

Yes, it is a linear function.

51. A, B, F

2.4 Review & Refresh

53. Segment Addition Postulate

54. angle bisector

55.

$6x - 2y = 12$	Write the equation.
$-2y = -6x + 12$	Subtraction Property of Equality
$y = 3x - 6$	Division Property of Equality

56. $d < -4$ *or* $d > 10$

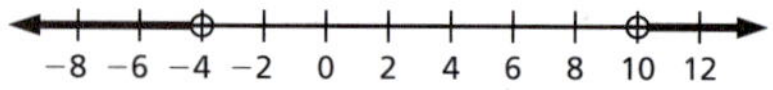

57. $w < -\frac{12}{5}$ *or* $w > \frac{4}{5}$

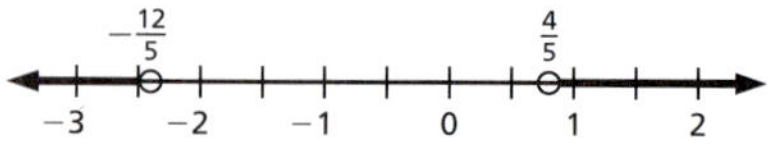

58. $t < -0.4$ *or* $t > 0.9$

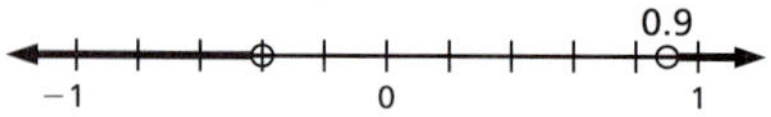

59. *Sample answer:*

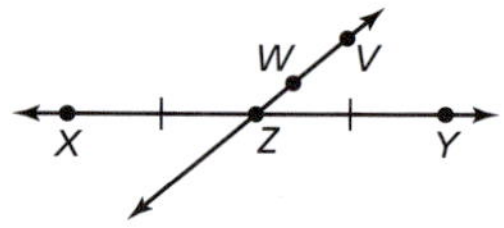

60. If it storms, then soccer practice is cancelled.

61. If a person is taller than 4 feet, then he or she is allowed to ride the roller coaster.

62. $x \geq 20$

$x \leq 40$

$y \geq 16.75x$

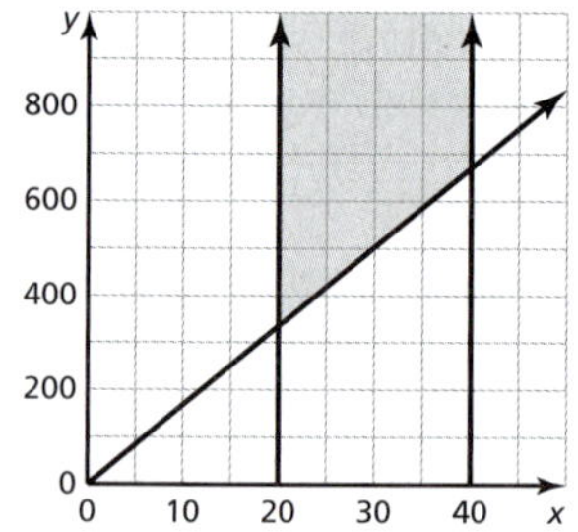

63. The difference of two even integers is an even integer; Let n and m be any integer. Then $2n$ and $2m$ are even integers because they are the product of 2 and an integer. $2n - 2m$ represents the difference of the two even integers. By the Distributive Property, $2n - 2m = 2(n - m)$, and $2(n - m)$ is an even integer because it is the product of 2 and an integer, $(n - m)$.

64. $x^2 + 12x + 32$

65. $y^2 + 7y - 18$

66. The function is positive when $-2.5 < x < -1.5$ and $x > 0.5$ and is negative when $x < -2.5$ and $-1.5 < x < 0.5$. The function is increasing when $x < -2$ and $x > -0.25$ and decreasing when $-2 < x < -0.25$. $y \to -\infty$ as $x \to -\infty$ and $y \to +\infty$ as $x \to +\infty$

Section 2.5

2.5 Practice

1. Given; Addition Property of Equality; $PQ + QR = PR$; Transitive Property of Equality

3. Transitive Property of Segment Congruence

5. Symmetric Property of Angle Congruence

7.

STATEMENTS	REASONS
1. A segment exists with endpoints A and B.	**1.** Given
2. AB equals the length of the segment with endpoints A and B.	**2.** Ruler Postulate
3. $AB = AB$	**3.** Reflexive Property of Equality
4. $\overline{AB} \cong \overline{AB}$	**4.** Definition of congruent segments

9.

STATEMENTS	REASONS
1. $\overline{AB} \cong \overline{CD}, \overline{CD} = \overline{EF}$	**1.** Given
2. $AB = CD; CD = EF$	**2.** Definition of congruent segments
3. $AB = EF$	**3.** Transitive Property of Equality
4. $\overline{AB} \cong \overline{EF}$	**4.** Definition of congruent segments

11.

STATEMENTS	REASONS
1. $\angle GFH \cong \angle GHF$	**1.** Given
2. $m\angle GFH = m\angle GHF$	**2.** Definition of congruent angles
3. $\angle EFG$ and $\angle GFH$ form a linear pair.	**3.** Given (diagram)
4. $\angle EFG$ and $\angle GFH$ are supplementary.	**4.** Definition of linear pair
5. $m\angle EFG + m\angle GFH = 180°$	**5.** Definition of supplementary angles
6. $m\angle EFG + m\angle GHF = 180°$	**6.** Substitution Property of Equality
7. $\angle EFG$ and $\angle GHF$ are supplementary.	**7.** Definition of supplementary angles

13. no; The statements have to have one segment in common in order to use the Transitive Property of Segment Congruence, but in this case, the statements are about four different segments. They may or may not all be congruent to each other.

15. a. Given: $RS = CF$, $SM = MC = FD$; Prove: $RM = CD$

b.

STATEMENTS	REASONS
1. $RS = CF$, $SM = MC = FD$	**1.** Given
2. $RM = RS + SM$	**2.** Segment Addition Postulate
3. $CF + FD = CD$	**3.** Segment Addition Postulate
4. $RS + SM = CD$	**4.** Substitution Property of Equality
5. $RM = CD$	**5.** Substitution Property of Equality

17.

STATEMENTS	REASONS
1. $\overline{QR} \cong \overline{PQ}, \overline{RS} \cong \overline{PQ}$, $QR = 2x + 5$, $RS = 10 - 3x$	**1.** Given
2. $QR = PQ, RS = PQ$	**2.** Definition of congruent segments
3. $QR = RS$	**3.** Transitive Property of Equality

19. a. It is a right angle.

b.

STATEMENTS	REASONS
1. $m\angle 1 + m\angle 1 + m\angle 2 + m\angle 2 = 180°$	**1.** Angle Addition Postulate
2. $2(m\angle 1 + m\angle 2) = 180°$	**2.** Distributive Property
3. $m\angle 1 + m\angle 2 = 90°$	**3.** Division Property of Equality

2.5 Review & Refresh

20. $x = \pm 7$; Explanations will vary.

21. $x = \dfrac{1 \pm \sqrt{22}}{3}$; Explanations will vary.

22. nonlinear; As x increases by a constant amount, y changes by different amounts.

23. 57°

24. 33°

25. The sum of two negative integers is negative; Let a and b be two positive integers. So $-a$ and $-b$ are two negative integers. $-a + (-b) = -(a + b)$. Because the sum of two positive numbers, $a + b$, is positive, $-(a + b)$ is negative, so the sum of two negative numbers, $-a + (-b)$, is negative.

26. *Sample answer:*

Equation	Justification
$-3(6x - 1) = 6x - 9$	Write given equation.
$-18x + 3 = 6x - 9$	Distributive Property
$-18x + 3 - 3 = 6x - 9 - 3$	Subtraction Property of Equality
$-18x = 6x - 12$	Simplify.
$-18x - 6x = 6x - 12 - 6x$	Subtraction Property of Equality
$-24x = -12$	Simplify.
$\frac{-24x}{-24} = \frac{-12}{-24}$	Division Property of Equality
$x = \frac{1}{2}$	Simplify.

27. Reflexive Property of Equality

28. Reflexive Property of Equality

29. \$273.88

30. *Sample answer:*

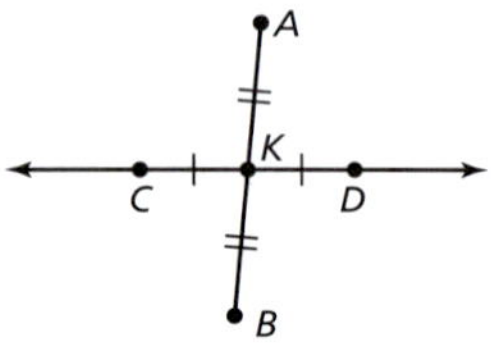

Section 2.6

2.6 Practice

1. $\angle MSN \cong \angle PSQ$ by definition because they have the same measure; $\angle MSP \cong \angle PSR$ by the Right Angles Congruence Theorem. They form a linear pair, which means they are supplementary by the Linear Pair Postulate, and because one is a right angle, so is the other by the Subtraction Property of Equality; $\angle NSP \cong \angle QSR$ by the Congruent Complements Theorem because they are complementary to congruent angles.

3. $\angle GML \cong \angle HMJ$ and $\angle GMH \cong \angle LMJ$ by the Vertical Angles Congruence Theorem; $\angle GMK \cong \angle JMK$ by the Right Angles Congruence Theorem. They form a linear pair, which means they are supplementary by the Linear Pair Postulate, and because one is a right angle, so is the other by the Subtraction Property of Equality.

5. $m\angle 2 = 37°, m\angle 3 = 143°, m\angle 4 = 37°$

7. $m\angle 1 = 146°, m\angle 3 = 146°, m\angle 4 = 34°$

9. $y = 9$

11. $x = 13, y = 20$

13. The expressions should have been set equal to each other because they represent vertical angles;

$(13x + 45)° = (19x + 3)°$

$-6x + 45 = 3$

$-6x = -42$

$x = 7$

15. no; $\angle 1$ and $\angle 4$ are not vertical angles because they do not form two pairs of opposite rays. So, the Vertical Angles Congruence Theorem does not apply.

17. Transitive Property of Angle Congruence; Transitive Property of Angle Congruence

STATEMENTS	REASONS
1. $\angle 1 \cong \angle 3$	1. Given
2. $\angle 1 \cong \angle 2$, $\angle 3 \cong \angle 4$	2. Vertical Angles Congruence Theorem
3. $\angle 2 \cong \angle 3$	3. Transitive Property of Angle Congruence
4. $\angle 2 \cong \angle 4$	4. Transitive Property of Angle Congruence

19. complementary; $m\angle 1 + m\angle 3$; Transitive Property of Equality; $m\angle 2 = m\angle 3$; congruent angles

STATEMENTS	REASONS
1. $\angle 1$ and $\angle 2$ are complementary. $\angle 1$ and $\angle 3$ are complementary.	1. Given
2. $m\angle 1 + m\angle 2 = 90°$, $m\angle 1 + m\angle 3 = 90°$	2. Definition of complementary angles
3. $m\angle 1 + m\angle 2 = m\angle 1 + m\angle 3$	3. Transitive Property of Equality
4. $m\angle 2 = m\angle 3$	4. Subtraction Property of Equality
5. $\angle 2 \cong \angle 3$	5. Definition of congruent angles

21. The purpose of a proof is to ensure the truth of a statement with such certainty that the theorem or rule proved could be used as a justification in proving another statement or theorem. Because inductive reasoning relies on observations about patterns in specific cases, the pattern may not continue or may change. So, the ideas cannot be used to prove ideas for the general case.

23. *Sample answer:* Because $\angle QRS$ and $\angle PSR$ are supplementary, $m\angle QRS + m\angle PSR = 180°$ by the definition of supplementary angles. $\angle QRL$ and $\angle QRS$ form a linear pair and by definition are supplementary, which means that $m\angle QRL + m\angle QRS = 180°$.
So, by the Transitive Property of Equality, $m\angle QRS + m\angle PSR = m\angle QRL + m\angle QRS$, and by the Subtraction Property of Equality, $m\angle PSR = m\angle QRL$.
So, by definition of congruent angles, $\angle PSR \cong \angle QRL$, and by the Symmetric Property of Angle Congruence, $\angle QRL \cong \angle PSR$.

Sample answer: I chose to use paragraph form for my proof because I found it easier to explain my reasoning.

25. *Sample answer:*

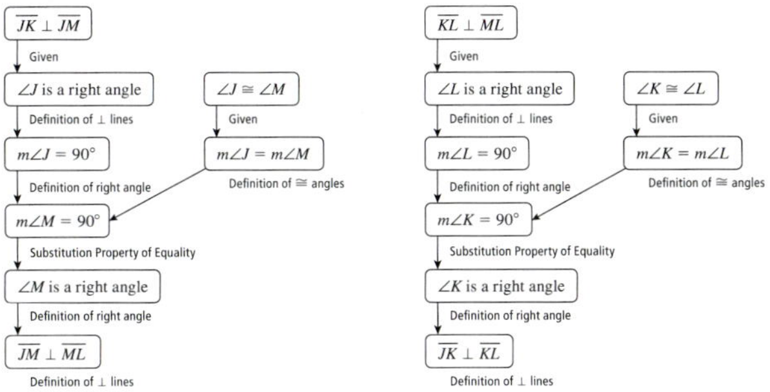

Sample answer: I chose a flow chart proof because I like how I can write my statements and reasons down as I think of them and then add arrows to indicate the flow of the proof.

27. no; The converse would be: "If two angles are supplementary, then they are a linear pair." This is false because angles can be supplementary without being adjacent.

2.6 Review & Refresh

29. $\overline{RS} \cong \overline{VW}$; Transitive Property of Equality

30. *Sample answer:* *B*, *I*, and *C*

31. *Sample answer:* Because *E*, *F*, and *G* are not collinear, there is exactly one plane through points *E*, *F*, and *G*.

32. *Sample answer:* plane *ABC* and plane *BCG*

33. a.

Equation	Justification
$v_f = v_i + at$	Write given equation.
$v_f - v_i = v_i + at - v_i$	Subtraction Property of Equality
$v_f - v_i = at$	Simplify.
$\frac{v_f - v_i}{a} = \frac{at}{a}$	Division Property of Equality
$\frac{v_f - v_i}{a} = t$	Simplify.

b. 6 sec

34. $x^2 - 14x + 49 = (x - 7)^2$

35. Vertical Angles Congruence Theorem; $\angle 1 \cong \angle 3$

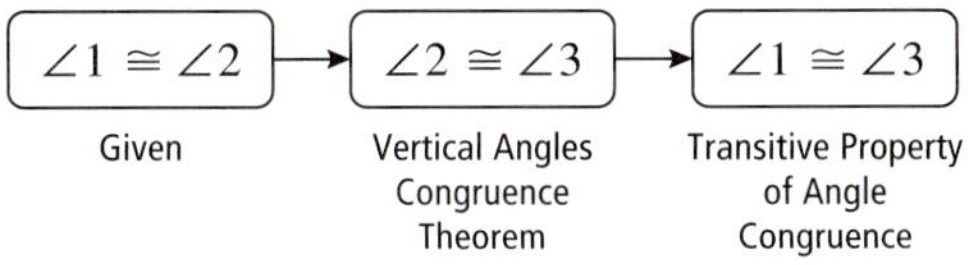

36.

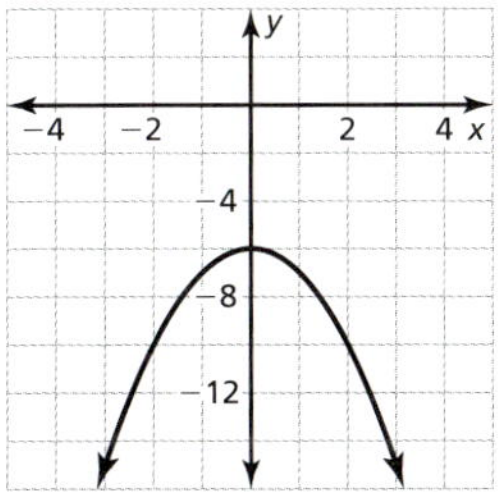

37.

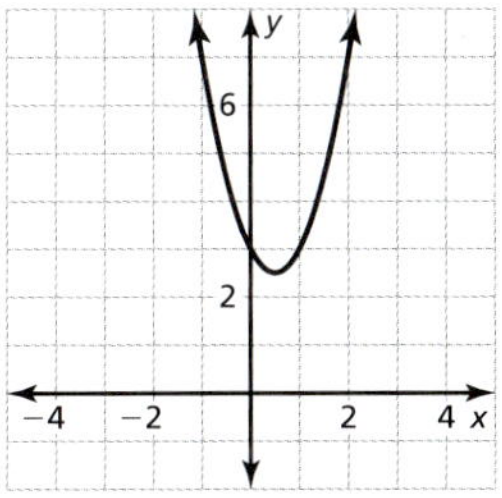

38.

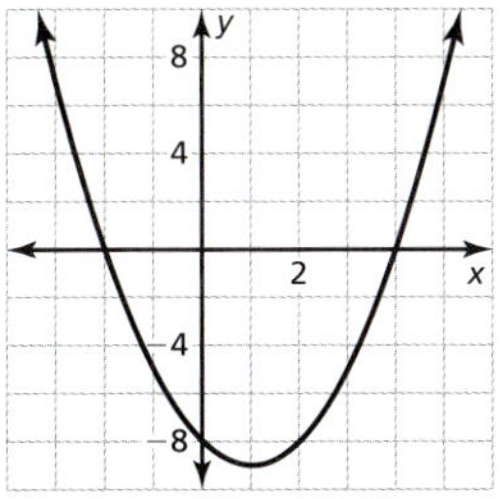

39.

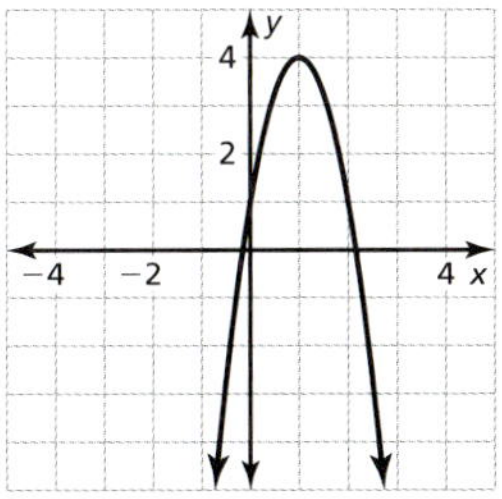

Chapter 2 Review

1. conditional: If two lines intersect, then their intersection is a point.

converse: If two lines intersect in a point, then they are intersecting lines.

inverse: If two lines do not intersect, then they do not intersect in a point.

contrapositive: If two lines do not intersect in a point, then they are not intersecting lines.

biconditional: Two lines intersect if and only if their intersection is a point.

2. conditional: If $4x + 9 = 21$, then $x = 3$.

converse: If $x = 3$, then $4x + 9 = 21$.

inverse: If $4x + 9 \neq 21$, then $x \neq 3$.

contrapositive: If $x \neq 3$, then $4x + 9 \neq 21$.

biconditional: $4x + 9 = 21$ if and only if $x = 3$.

3. conditional: If angles are supplementary, then they sum to 180°.

converse: If angles sum to 180°, then they are supplementary.

inverse: If angles are not supplementary, then they do not sum to 180°.

contrapositive: If angles do not sum to 180°, then they are not supplementary.

biconditional: Angles are supplementary if and only if they sum to 180°.

4. conditional: If an angle is a right angle, then it measures 90°.

converse: If an angle measures 90°, then it is a right angle.

inverse: If an angle is not a right angle, then it does not measure 90°.

contrapositive: If an angle does not measure 90°, then it is not a right angle.

biconditional: An angle is a right angle if and only if it measures 90°.

5. yes; Definition of midpoint

6. no; The diagram does not include tick marks to show the congruence of $\overline{ES}$ and $\overline{ST}$.

7. yes; Definition of segment bisector

8. The difference of any two odd integers is an even integer.

9. The product of an even and an odd integer is an even integer.

10. $m\angle B = 90°$

11. If $4x = 12$, then $2x = 6$.

12. inductive reasoning; The conjecture is based on the assumption that a pattern, observed in specific cases, will continue.

13. deductive reasoning; The conclusion is based on accepted scientific properties and the Law of Detachment.

14. Two Point Postulate

15. *Sample answer:*

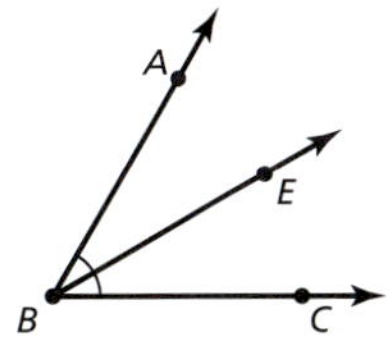

16. *Sample answer:*

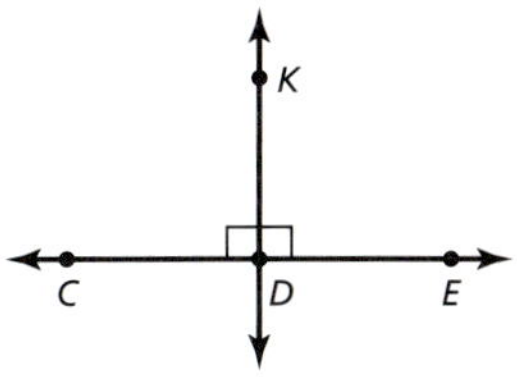

17. *Sample answer:*

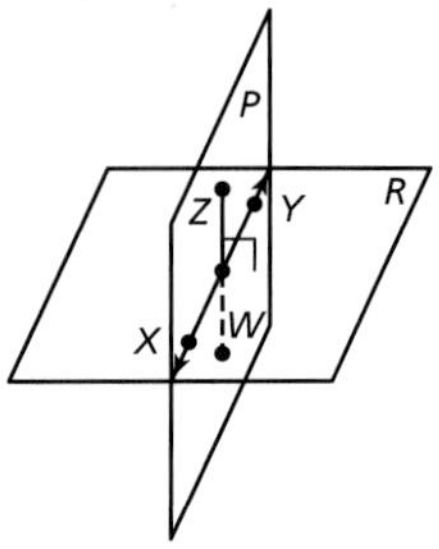

18. yes

19. yes

20. no

21. no

22. *Sample answer:* The intersection of plane R and plane S is $\overleftrightarrow{AB}$.

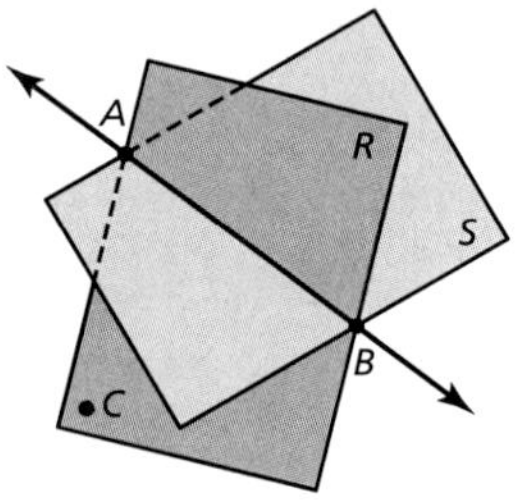

Three Point Postulate, Plane-Line Postulate, Plane Intersection Postulate

23.

Equation	Explanation and Reason
$-9x - 21 = -20x - 87$	Write the equation; Given
$11x - 21 = -87$	Add $20x$ to each side; Addition Property of Equality
$11x = -66$	Add 21 to each side; Addition Property of Equality
$x = -6$	Divide each side by 11; Division Property of Equality

24.

Equation	Explanation and Reason
$15x + 22 = 7x + 62$	Write the equation; Given
$8x + 22 = 62$	Subtract $7x$ from each side; Subtraction Property of Equality
$8x = 40$	Subtract 22 from each side; Subtraction Property of Equality
$x = 5$	Divide each side by 8; Division Property of Equality

25.

Equation	Explanation and Reason
$3(2x + 9) = 30$	Write the equation; Given
$6x + 27 = 30$	Multiply; Distributive Property
$6x = 3$	Subtract 27 from each side; Subtraction Property of Equality
$x = \frac{1}{2}$	Divide each side by 6; Division Property of Equality

26.

Equation	Explanation and Reason
$5x + 2(2x - 23) = -154$	Write the equation; Given
$5x + 4x - 46 = -154$	Multiply; Distributive Property
$9x - 46 = -154$	Combine like terms; Simplify.
$9x = -108$	Add 46 to each side; Addition Property of Equality
$x = -12$	Divide each side by 9; Division Property of Equality

27. Transitive Property of Equality, or Substitution Property of Equality

28. Reflexive Property of Equality

29. Division Property of Equality

30. Symmetric Property of Equality

31.

Equation	Reason
$V = \dfrac{11.25^2\pi h}{231}$	Write given equation.
$231 \cdot V = 231 \cdot \dfrac{11.25^2\pi h}{231}$	Multiplication Property of Equality
$231V = 11.25^2\pi h$	Simplify.
$\dfrac{231V}{11.25^2\pi} = \dfrac{11.25^2\pi h}{11.25^2\pi}$	Division Property of Equality
$\dfrac{231V}{11.25^2\pi} = h$	Simplify.

$h \approx 31.95$ in.

32. Symmetric Property of Angle Congruence

33. Reflexive Property of Angle Congruence

34. Transitive Property of Equality, or Substitution Property of Equality

35.

STATEMENTS	REASONS
1. An angle with vertex A exists.	**1.** Given
2. $m\angle A$ equals the measure of the angle with vertex A.	**2.** Protractor Postulate
3. $m\angle A = m\angle A$	**3.** Reflexive Property of Equality
4. $\angle A \cong \angle A$	**4.** Definition of congruent angles

36.

STATEMENTS	REASONS
1. $\angle BAD \cong \angle CDA$	**1.** Given
2. $m\angle BAD = m\angle CDA$	**2.** Definition of congruent angles
3. $m\angle BAD + m\angle EAB = 180°$, $m\angle CDA + m\angle FDC = 180°$	**3.** Line Pair Postulate
4. $m\angle BAD = 180° - m\angle EAB$, $m\angle CDA = 180° - m\angle FDC$	**4.** Subtraction Property of Equality
5. $180° - m\angle EAB = 180° - m\angle FDC$	**5.** Substitution Property of Equality
6. $m\angle EAB = m\angle FDC$	**6.** Subtraction Property of Equality
7. $\angle EAB \cong \angle FDC$	**7.** Definition of congruent angles

37. Given; Congruent Complements Theorem

STATEMENTS	REASONS
1. $m\angle 1 + m\angle 2 = 90°$	1. Given
2. $\angle 1$ and $\angle 2$ are complementary.	2. Definition of complementary angles
3. $\angle 3$ and $\angle 2$ are complementary.	3. Given
4. $\angle 3 \cong \angle 1$	4. Congruent Complements Theorem

38. $\angle ABD$ and $\angle DBC$ are adjacent angles. By the Angle Addition Postulate, $m\angle ABD + m\angle DBC = m\angle ABC$. From the given values, $m\angle ABD = 24°$ and $m\angle ABC = 48°$. By the Substitution Property of Equality, $24° + m\angle DBC = 48°$. By the Subtraction Property of Equality, $m\angle DBC = 24°$. Because $m\angle ABD = m\angle DBC$, $\angle ABD \cong \angle DBC$ by the definition of congruent angles. So, by the definition of an angle bisector, $\overrightarrow{BD}$ bisects $\angle ABC$.

Sample answer: I chose to use paragraph form for my proof because I found it easier to explain my reasoning.

Chapter 2 Practice Test

1. no; No right angle is marked on $\overleftrightarrow{CD}$.
2. yes; They all lie in plane M.
3. yes; The intersection of two planes is a line by the Plane Intersection Postulate.
4. no; $\overleftrightarrow{DJ}$ is not drawn. So, you cannot be sure about where it intersects $\overleftrightarrow{FG}$.
5.

Equation	Explanation and Reason
$9x + 31 = -23 + 3x$	Write the equation; Given
$6x + 31 = -23$	Subtract $3x$ from each side; Subtraction Property of Equality
$6x = -54$	Subtract 31 from each side; Subtraction Property of Equality
$x = -9$	Divide each side by 6; Division Property of Equality

6.

Equation	Explanation and Reason
$26 + 2(3x + 11) = -18$	Write the equation; Given
$26 + 6x + 22 = -18$	Multiply; Distributive Property
$6x + 48 = -18$	Combine like terms; Simplify.
$6x = -66$	Subtract 48 from each side; Subtraction Property of Equality
$x = -11$	Divide each side by 6; Division Property of Equality

7. conditional: If two planes intersect, then their intersection is a line; true

   converse: If two planes intersect in a line, then they are intersecting planes; true

   inverse: If two planes do not intersect, then they do not intersect in a line; true

   contrapositive: If two planes do not intersect in a line, then they are not intersecting planes; true

   biconditional: Two planes intersect if and only if their intersection is a line; true

8. conditional: If a relation pairs each input with exactly one output, then the relation is a function; true

   converse: If a relation is a function, then each input is paired with exactly one output; true

   inverse: If a relation does not pair each input with exactly one output, then the relation is not a function; true

   contrapositive: If a relation is not a function, then each input is not paired with exactly one output; true

   biconditional: A relation pairs each input with exactly one output if and only if the relation is a function; true

9. The sum of three odd integers is an odd integer; Let m, n, and p be integers. Then $2m + 1$, $2n + 1$, and $2p + 1$ represent three odd integers because they are each 1 more than an even integer. The sum would be $(2m + 1) + (2n + 1) + (2p + 1)$, which equals $2m + 2n + 2p + 3$ when it is simplified using the Associative and Commutative Properties of Equality. According to the Distributive Property, $2m + 2n + 2p + 3 = 2(m + n + p + 1) + 1$. Because this is 1 more than an even integer (the product of 2 and $(m + n + p + 1)$), the sum is an odd integer.

10. no; The Law of Detachment applies when the hypothesis, not the conclusion, of a conditional statement is true.

11.

Equation	Explanation and Reason
$A = \frac{1}{2}bh$	Write the equation; Given
$2A = bh$	Multiply each side by 2; Multiplication Property of Equality
$\frac{2A}{b} = h$	Divide each side by b. Division Property of Equality
$h = 31$ in.	

12. 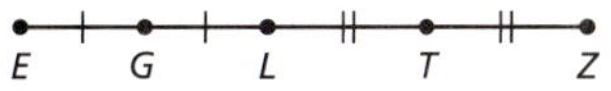

Sample answer:

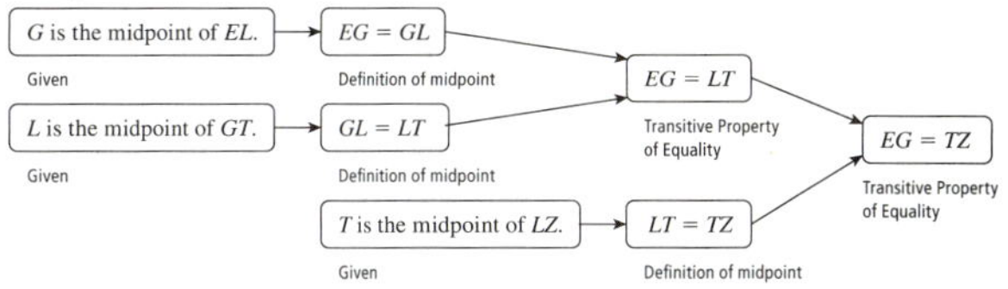

Sample answer: The flowchart proof format shows the flow of the logical arguments.

13. *Sample answer:*

STATEMENTS	REASONS
1. $\angle 2 \cong \angle 3$, $\overrightarrow{TV}$ bisects $\angle UTW$.	1. Given
2. $\angle 1 \cong \angle 2$	2. Definition of angle bisector
3. $\angle 1 \cong \angle 3$	3. Transitive Property of Angle Congruence

Sample answer: The two-column proof format shows the logical order of the deductive reasoning.

Chapter 2 Test Prep

1. a. *Sample answer:* Through points C and D, there exists exactly one line, $\overleftrightarrow{CD}$.
   b. *Sample answer:* The intersection of $\overleftrightarrow{AF}$ and $\overleftrightarrow{BC}$ is point B.
   c. *Sample answer:* Points A and F lie in plane T, so $\overleftrightarrow{AF}$ also lies in plane T.
   d. Plane T and plane S intersect, and their intersection is $\overleftrightarrow{BC}$.
2. Definition of congruent segments; Definition of congruent segments; Segment Addition Postulate; Segment Addition Postulate; Substitution Property of Equality; Substitution Property of Equality; Definition of congruent segments
3. D
4. a. biconditional
   b. inverse
   c. converse
   d. contrapositive
5. C
6. B
7. D
8. B

Chapter 3

Prepare with CalcChat

1. $m = -\frac{3}{4}$
2. $m = 3$
3. $m = 0$
4. $y = -3x + 19$
5. $y = -2x + 2$
6. $y = 4x + 9$
7. $y = \frac{1}{2}x - 5$
8. $y = -\frac{1}{4}x - 7$
9. $y = \frac{2}{3}x + 9$
10. When calculating the slope of a horizontal line, the vertical change is zero. This is the numerator of the fraction, and zero divided by any number is zero. When calculating the slope of a vertical line, the horizontal change is zero. This is the denominator of the fraction, and any number divided by zero is undefined.

Section 3.1

3.1 Practice

1. $\overleftrightarrow{AB}$
3. $\overleftrightarrow{BF}$
5. $\overleftrightarrow{MK}$ and $\overleftrightarrow{LS}$
7. no; They are intersecting lines.
9. $\angle 1$ and $\angle 5$; $\angle 2$ and $\angle 6$; $\angle 3$ and $\angle 7$; $\angle 4$ and $\angle 8$
11. $\angle 1$ and $\angle 8$; $\angle 2$ and $\angle 7$
13. corresponding
15. consecutive interior
17. Lines that do not intersect could also be skew; If two coplanar lines do not intersect, then they are parallel.
19. no; The lines intersect.
21. no; They can both be in a plane that is slanted with respect to the horizontal.
23. *Sample answer:*

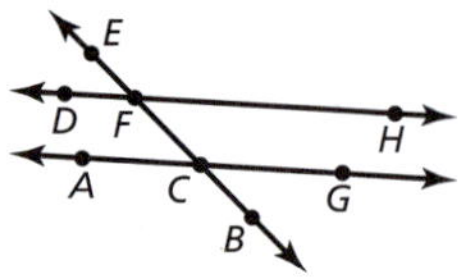

25. *Sample answer:* $a_2 = 1, a_n = a_{n-1} + (n - 1)$

3.1 Review & Refresh

26.

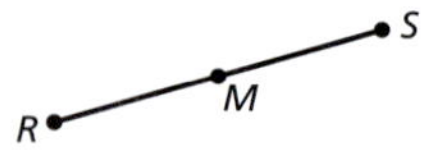

27. $t \geq 4$

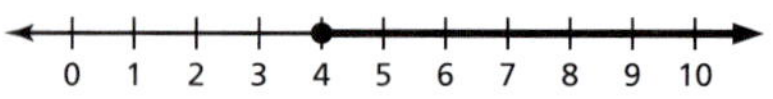

28. Reflexive Property of Segment Congruence
29. alternate exterior angles
30. $(-6, 0)$
31. $(-1, 5)$
32. $m\angle 3 = m\angle 7$
33. *Sample answer:* Because $\angle 1$ and $\angle 3$ are complementary and $\angle 2$ and $\angle 4$ are complementary, $m\angle 1 + m\angle 3 = 90°$ and $m\angle 2 + m\angle 4 = 90°$ by definition of complementary angles. By the Substitution Property of Equality, $m\angle 1 + m\angle 3 = m\angle 2 + m\angle 4$. By the Vertical Angles Congruence Theorem, $m\angle 3 = m\angle 2$. Then, by the Substitution Property of Equality, $m\angle 1 + m\angle 2 = m\angle 2 + m\angle 4$. So, by the Subtraction Property of Equality, $m\angle 1 = m\angle 4$, and $\angle 1 \cong \angle 4$ by the definition of congruent angles.

    Sample explanation: I chose to use paragraph form for my proof because I found it easier to explain my reasoning.
34. $y = -2x + 5$
35. -5
36. about 157.1 in.^{3}

Section 3.2

3.2 Practice

1. $m\angle 1 = 117°$ by the Vertical Angles Congruence Theorem; $m\angle 2 = 117°$ by the Alternate Exterior Angles Theorem
3. $m\angle 1 = 122°$ by the Alternate Interior Angles Theorem; $m\angle 2 = 58°$ by the Consecutive Interior Angles Theorem
5. 64; $2x° = 128°$
   $x = 64$
7. 12; $m\angle 5 = 65°$
   $65° + (11x - 17)° = 180°$
   $11x + 48 = 180$
   $11x = 132$
   $x = 12$
9. 9; $(8x + 9)° + 99° = 180°$
   $8x + 108 = 180$
   $8x = 72$
   $x = 9$
11. $m\angle 1 = 100°, m\angle 2 = 80°, m\angle 3 = 100°$; Because the 80° angle is a consecutive interior angle with both $\angle 1$ and $\angle 3$, they are supplementary by the Consecutive Interior Angles Theorem. Because $\angle 1$ and $\angle 2$ are consecutive interior angles, they are supplementary by the Consecutive Interior Angles Theorem.
13. $m\angle 1 = 80°, m\angle 2 = 80°, m\angle 3 = 100°$; Because $\angle 3$ and the 100° angle are vertical angles, $m\angle 3 = 100°$ by the Vertical Angles Congruence Theorem. $\angle 1$ and $\angle 2$ are corresponding angles with the angle that is supplementary with the 100° angle, so $m\angle 1 = m\angle 2 = 80°$ by the Corresponding Angles Theorem.

15. In order to use the Corresponding Angles Theorem, the angles need to be formed by two parallel lines cut by a transversal, but none of the lines in this diagram appear to be parallel; $\angle 9$ and $\angle 10$ are corresponding angles.

17.

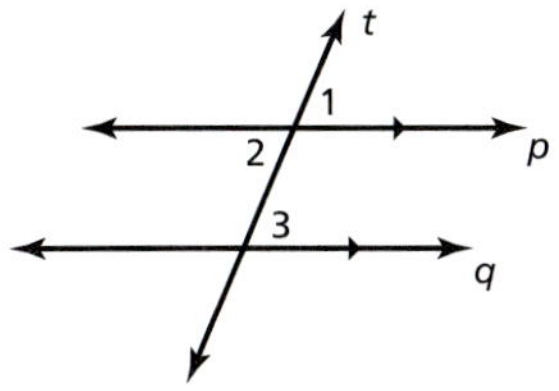

STATEMENTS	REASONS
1. $p \parallel q$	**1.** Given
2. $\angle 3 \cong \angle 1$	**2.** Corresponding Angles Theorem
3. $\angle 1 \cong \angle 2$	**3.** Vertical Angles Congruence Theorem
4. $\angle 3 \cong \angle 2$	**4.** Transitive Property of Angle Congruence

19. $m\angle 2 = 104°$; Because the trees form parallel lines, and the rope is a transversal, the 76° angle and $\angle 2$ are consecutive interior angles. So, they are supplementary by the Consecutive Interior Angles Theorem.

21. yes; If two parallel lines are cut by a perpendicular transversal, then the consecutive interior angles will both be right angles.

23. $19x - 10 = 180$

$14x + 2y - 10 = 180$; $x = 10$, $y = 25$

25. 60°; $\angle 1 \cong \angle 5$ by the Corresponding Angles Theorem, $\angle 2 \cong \angle 4$ by the Alternate Interior Angles Theorem, $\angle 2 \cong \angle 3$ by the definition of angle bisector, and $\angle 4 \cong \angle 5$ is given. So, by the Transitive Property of Congruence, all five of the angles labeled must be congruent to each other. From the diagram, $m\angle 1 + m\angle 2 + m\angle 3 = 180°$, and because they all have the same measure, it must be that they each have a measure of $\frac{180°}{3} = 60°$.

27. *Sample answer:*

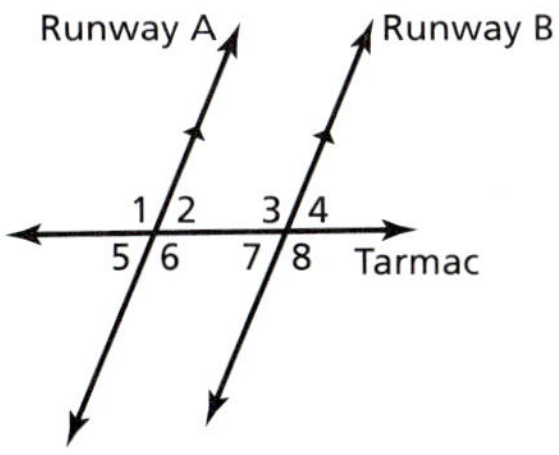

$m\angle 1 = m\angle 3 = m\angle 6 = m\angle 8$, $m\angle 2 = m\angle 4 = m\angle 5 = m\angle 7$

3.2 Review & Refresh

29. *Sample answer:* $\overleftrightarrow{LM}$ and $\overleftrightarrow{QS}$

30. $\overleftrightarrow{LM}$ and $\overleftrightarrow{NP}$

31. $m\angle 1 = m\angle 2 = 114°$; Linear Pair Postulate, Corresponding Angles Theorem

32. Transitive Property of Angle Congruence

33. Symmetric Property of Segment Congruence

34. $(t - 5)(t^2 + 3)$

35. $4x^2(x + 3)(x - 3)$

36. $(4, 0)$, $(0, -7)$

37. $10x - 5$; 115 in.^{2}

38. $x = 9$

Section 3.3

3.3 Practice

1. $x = 40$; Lines m and n are parallel when the marked corresponding angles are congruent.

$3x° = 120°$

$x = 40$

3. $x = 15$; Lines m and n are parallel when the marked consecutive interior angles are supplementary.

$(3x - 15)° + 150° = 180°$

$3x + 135 = 180$

$3x = 45$

$x = 15$

5. $x = 60$; Lines m and n are parallel when the marked consecutive interior angles are supplementary.

$2x° + x° = 180°$

$3x = 180$

$x = 60$

7.

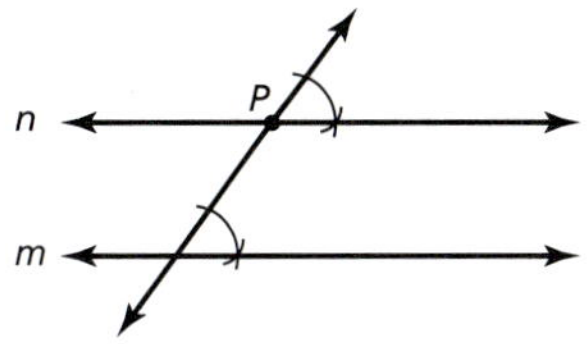

9.

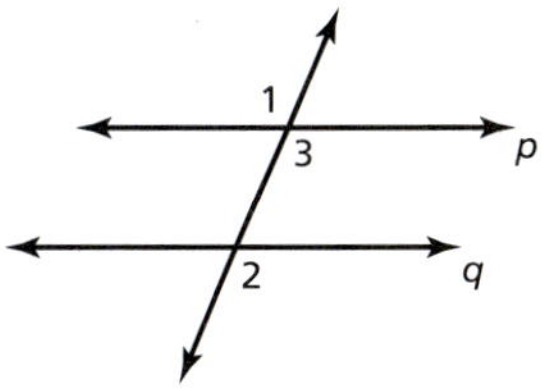

It is given that $\angle 1 \cong \angle 2$. By the Vertical Angles Congruence Theorem, $\angle 1 \cong \angle 3$. Then by the Transitive Property of Congruence, $\angle 2 \cong \angle 3$. So, by the Corresponding Angles Converse, $p \parallel q$.

11. yes; Alternate Interior Angles Converse

13. no

15. no

17. This diagram shows that vertical angles are always congruent. Lines a and b are not parallel unless $x = y$, and you cannot assume that they are equal.

19. yes; $m\angle DEB = 180° - 123° = 57°$ by the Linear Pair Postulate. So, by definition, a pair of corresponding angles are congruent, which means that $\overleftrightarrow{AC} \parallel \overleftrightarrow{DF}$ by the Corresponding Angles Converse.

21. cannot be determined; The marked angles are vertical angles. You do not know anything about the angles formed by the intersection of $\overleftrightarrow{DF}$ and $\overleftrightarrow{BE}$.

23. All of the rungs are parallel to each other by the Transitive Property of Parallel Lines.

25. The two angles marked as 108° are corresponding angles. Because they have the same measure, they are congruent to each other. So, $m \parallel n$ by the Corresponding Angles Converse.

27. $\overrightarrow{EA} \parallel \overrightarrow{HC}$ by the Corresponding Angles Converse. $\angle AEH \cong \angle CHG$ by definition because $m\angle AEH = 62° + 58° = 120°$ and $m\angle CHG = 59° + 61° = 120°$. However, $\overrightarrow{EB}$ is not parallel to $\overrightarrow{HD}$ because corresponding angles $\angle BEH$ and $\angle DHG$ do not have the same measure and are therefore not congruent.

29.

STATEMENTS	REASONS
1. $m\angle 1 = 115°, m\angle 2 = 65°$	**1.** Given
2. $m\angle 1 + m\angle 2 = m\angle 1 + m\angle 2$	**2.** Reflexive Property of Equality
3. $m\angle 1 + m\angle 2 = 115° + 65°$	**3.** Substitution Property of Equality
4. $m\angle 1 + m\angle 2 = 180°$	**4.** Simplify.
5. $\angle 1$ and $\angle 2$ are supplementary.	**5.** Definition of supplementary angles
6. $m \parallel n$	**6.** Consecutive Interior Angles Converse

31.

STATEMENTS	REASONS
1. $\angle 1 \cong \angle 2, \angle 3 \cong \angle 4$	**1.** Given
2. $\angle 2 \cong \angle 3$	**2.** Vertical Angles Congruence Theorem
3. $\angle 1 \cong \angle 3$	**3.** Transitive Property of Congruence
4. $\angle 1 \cong \angle 4$	**4.** Transitive Property of Congruence
5. $\overline{AB} \parallel \overline{CD}$	**5.** Alternate Interior Angles Converse

33. two; *Sample answer:* $\angle 1 \cong \angle 5$, $\angle 2 \cong \angle 7$, $\angle 3 \cong \angle 6$, $\angle 4$ and $\angle 7$ are supplementary.

35. **a.** $x = 54$

b. $y = 47.5$

c. no; If $x = 54$, then $(x + 56)° = 110°$. If $y = 47.5$, then $(y + 7)° = 54.5°$. Because these two angles form a linear pair, their sum should be 180°, but $110° + 54.5° = 164.5°$. So, both pairs of lines cannot be parallel at the same time.

37. no; Based on the diagram, $\overleftrightarrow{AB} \parallel \overleftrightarrow{CD}$ by the Alternate Interior Angles Converse, but you cannot be sure that $\overleftrightarrow{AD} \parallel \overleftrightarrow{BC}$.

3.3 Review & Refresh

39. 13

40. $\sqrt{41}$

41. $x = 19$

42. about 148; The T-shirt reaches a maximum height of about 148 feet after 3 seconds.

43. $x = 27$; When $x = 27$, the angles are congruent. So, by the Alternative Interior Angles Converse, the lines are parallel.

44. *Sample answer:* $\overleftrightarrow{AF}$ and $\overleftrightarrow{CH}$

45. no; There is no information given to prove the lines are parallel.

46. $(-3, -4)$; Substitute $2y + 5$ in for x in the second equation.

47. $(-1, 6)$; Multiply the first equation by 2. Then add the equations.

48. $f(-2) = 11; f(3) = -4; f(5) = -10$

49. *Sample answer:*

STATEMENTS	REASONS
1. $\angle 1 \cong \angle 3$	**1.** Given
2. $\angle 2 \cong \angle 1$	**2.** Vertical Angles Congruence Theorem
3. $\angle 2 \cong \angle 3$	**3.** Transitive Property of Congruence
4. $\angle 3 \cong \angle 4$	**4.** Vertical Angles Congruence Theorem
5. $\angle 2 \cong \angle 4$	**5.** Transitive Property of Congruence

Sample explanation: I chose to use a 2-column proof because I like how it clearly organizes my statements and reasons.

Section 3.4

3.4 Practice

1. $\sqrt{10}$, or about 3.2 units

3.

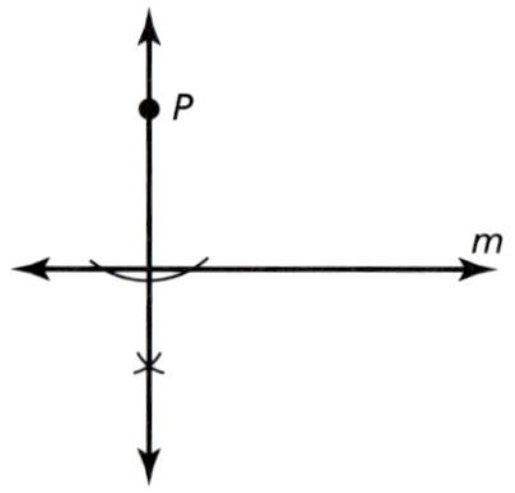

5.

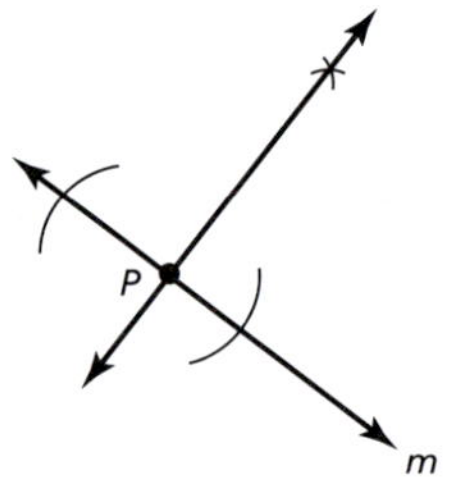

7.

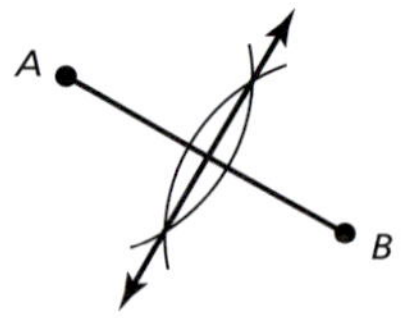

9.

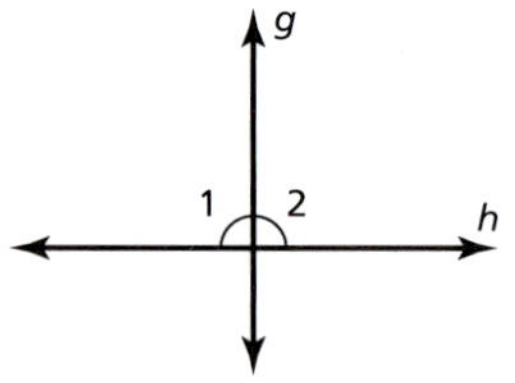

Because $\angle 1 \cong \angle 2$ by definition, $m\angle 1 = m\angle 2$. Also, by the Linear Pair Postulate, $m\angle 1 + m\angle 2 = 180°$. Then, by the Substitution Property of Equality, $m\angle 1 + m\angle 1 = 180°$, and $2(m\angle 1) = 180°$ by the Distributive Property. So, by the Division Property of Equality, $m\angle 1 = 90°$. Finally, $g \perp h$ by the definition of perpendicular lines.

11. none; The only thing that can be concluded in this diagram is that $v \perp y$. In order to say that lines are parallel, you need to know something about both of the intersections between the transversal and the two lines.

13. $m \parallel n$; Because $m \perp q$ and $n \perp q$, lines m and n are parallel by the Lines Perpendicular to a Transversal Theorem. The other lines may or may not be parallel.

15. $n \parallel p$; Because $k \perp n$ and $k \perp p$, lines n and p are parallel by the Lines Perpendicular to a Transversal Theorem.

17. The length of the perpendicular segment should be used; The distance from point C to $\overline{AB}$ is 8 centimeters.

19.

STATEMENTS	REASONS
1. $a \perp b$	**1.** Given
2. $\angle 1$ is a right angle.	**2.** Definition of perpendicular lines
3. $\angle 1 \cong \angle 4$	**3.** Vertical Angles Congruence Theorem
4. $m\angle 1 = 90°$	**4.** Definition of right angle
5. $m\angle 4 = 90°$	**5.** Transitive Property of Equality
6. $\angle 1$ and $\angle 2$ form a linear pair.	**6.** Definition of linear pair
7. $\angle 1$ and $\angle 2$ are supplementary.	**7.** Linear Pair Postulate
8. $m\angle 1 + m\angle 2 = 180°$	**8.** Definition of supplementary angles
9. $90° + m\angle 2 = 180°$	**9.** Substitution Property of Equality
10. $m\angle 2 = 90°$	**10.** Subtraction Property of Equality
11. $\angle 2 \cong \angle 3$	**11.** Vertical Angles Congruence Theorem
12. $m\angle 3 = 90°$	**12.** Transitive Property of Equality
13. $\angle 1$, $\angle 2$, $\angle 3$, and $\angle 4$ are right angles.	**13.** Definition of right angle

21. $m\angle 1 = 90°, m\angle 2 = 60°, m\angle 3 = 30°, m\angle 4 = 20°, m\angle 5 = 90°$;
$m\angle 1 = 90°$, because it is marked as a right angle.
$m\angle 2 = 90° - 30° = 60°$, because it is complementary to the 30° angle.
$m\angle 3 = 30°$, because it is a vertical angle with, and therefore congruent to, the 30° angle.
$m\angle 4 = 90° - (30° + 40°) = 20°$, because it forms a right angle with $\angle 3$ and the 40° angle.
$m\angle 5 = 90°$, because it is a vertical angle with, and therefore congruent to, $\angle 1$.

23. $x = 8$

25. yes; The center of the geyser is $45\sqrt{5}$, or about 100.6 feet from the boardwalk art the closest point.

27.

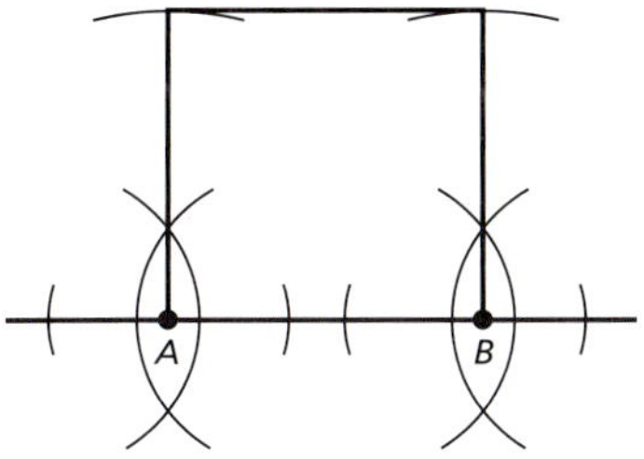

29. no; The area of $\triangle ABC$ is $\frac{1}{2}AB$ times the distance from A to line n.

31. Find the length of the segment that is perpendicular to the plane and that has one endpoint on the given point and one endpoint on the plane; You can find the distance from a line to a plane only if the line is parallel to the plane. Then you can pick any point on the line and find the distance from that point to the plane. If a line is not parallel to a plane, then the distance from the line to the plane is not defined because it would be different for each point on the line.

3.4 Review & Refresh

32. $m = \frac{1}{6}$; $b = -8$

33. $m = 3$; $b = 9$

34. 103°

35. $2\sqrt{2}$, or about 2.8 units

36. $\frac{1}{2}$

37. **a.** 416 yd
b. 1305 yd

38. $(4, -3)$; *Sample answer:* Multiply the first equation by -2 and add the equations.

39. $\overleftrightarrow{HG}$

40. $\overleftrightarrow{FG}$

41. $\overleftrightarrow{CG}$

42. plane BCG

43.

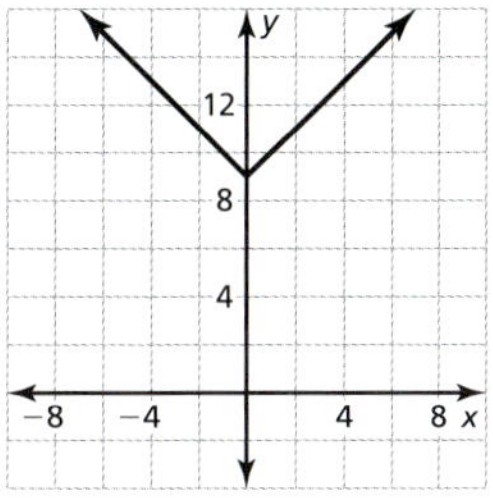

The graph of g is a vertical translation 9 units up of the graph of f; domain: all real numbers, range: $y \geq 9$

44.

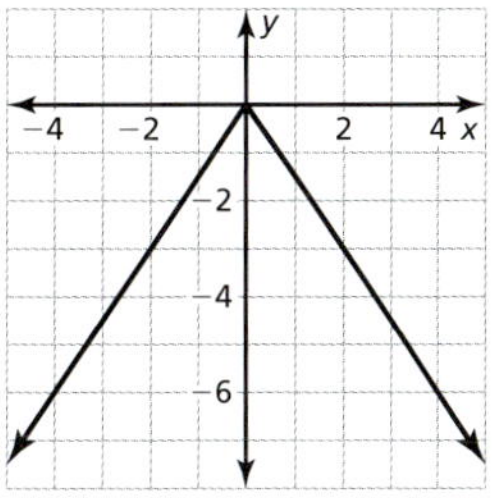

The graph of g is a vertical stretch by a factor of $\frac{3}{2}$ and a reflection in the x-axis of the graph of f; domain: all real numbers, range: $y \leq 0$

45. $m\angle 1 = 55°$; Consecutive Interior Angles Theorem

46. yes; Alternate Interior Angles Converse

Section 3.5

3.5 Practice

1. $P(7, -0.4)$

3. $P(-1.5, -1.5)$

5. $a \parallel c$, $b \perp d$

7. perpendicular; Because $m_1 \cdot m_2 = \left(\frac{2}{3}\right)\left(-\frac{3}{2}\right) = -1$, lines 1 and 2 are perpendicular by the Slopes of Perpendicular Lines Theorem.

9. perpendicular; Because $m_1 \cdot m_2 = 1(-1) = -1$, lines 1 and 2 are perpendicular by the Slopes of Perpendicular Lines Theorem.

11. $y = -2x + 1$

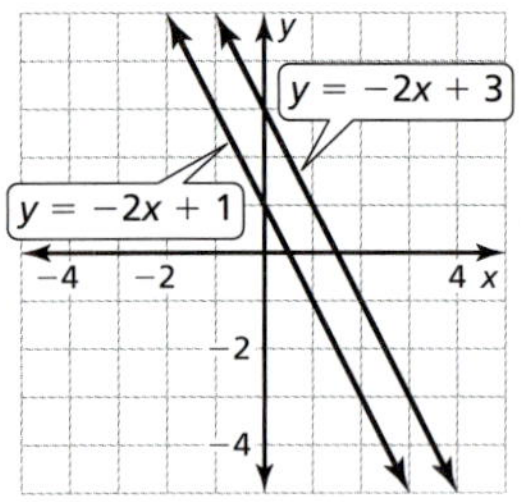

13. $x = -2$

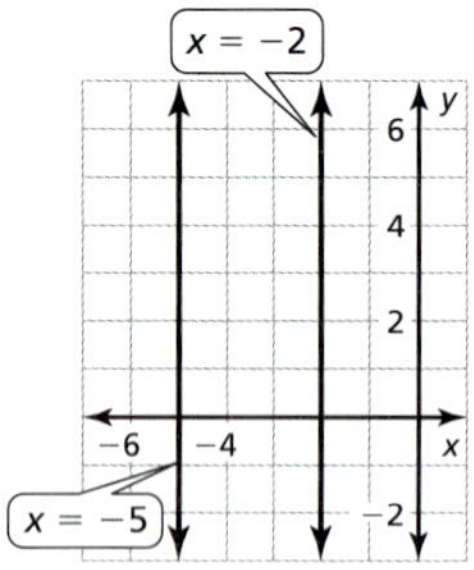

15. $y = \frac{1}{9}x$

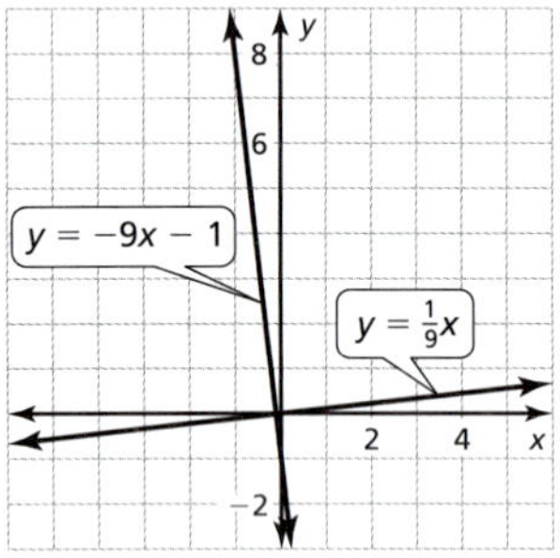

17. $y = \frac{1}{2}x + 2$

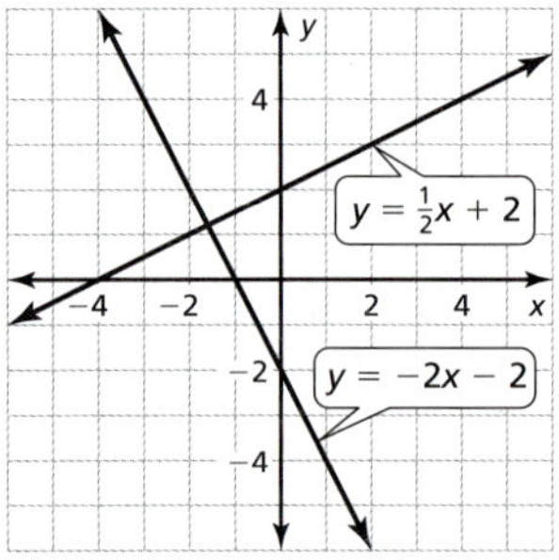

19. $\sqrt{10}$, or about 3.2 units

21. $\sqrt{29}$, or about 5.4 units

23. Because the slopes are opposites but not reciprocals, their product does not equal -1. Lines 1 and 2 are neither parallel nor perpendicular.

25. $(0, 1)$; $y = 2x + 1$

27. $(3, 0)$; $y = \frac{3}{2}x - \frac{9}{2}$

29. $\frac{9\sqrt{5}}{5}$, or about 4 units

31. $\frac{24\sqrt{13}}{13}$, or about 6.7 units

33. $\left(-\frac{11}{5}, -\frac{6}{5}\right)$

35. no; $m\overline{LM} = \frac{2}{5}$, $m\overline{LN} = -\frac{7}{4}$, and $m\overline{MN} = 9$. None of these can pair up to make a product of -1, so none of the segments are perpendicular.

37. about 27.7 ft

39. $m < -1$

41. Answers will vary. Check students' work.

43. **a.** no solution; The lines do not intersect, so they are parallel.

b. $(7, -4)$; The lines intersect at one point.

c. infinitely many solutions; The lines are the same line.

45. $k = 4$

47. If lines x and y are perpendicular to line z, then by the Slopes of Perpendicular Lines Theorem, $m_x \cdot m_z = -1$ and $m_y \cdot m_z = -1$. By the Transitive Property of Equality, $m_x \cdot m_z = m_y \cdot m_z$, and by the Division Property of Equality, $m_x = m_y$. Therefore, by the Slopes of Parallel Lines Theorem, $x \parallel y$.

49. If lines x and y are vertical lines and they are cut by any horizontal transversal z, then $x \perp z$ and $y \perp z$ by the Slopes of Perpendicular Lines Theorem. Therefore, $x \parallel y$ by the Lines Perpendicular to a Transversal Theorem.

51. If two lines are not parallel, then they are not vertical. Proof: Let line x and line y be two non-parallel lines and line x is vertical. Let line z be a transversal that is perpendicular to line x. So, line z is horizontal. Because lines x and y are not parallel, line y is not perpendicular to line z, by the Lines Perpendicular to a Transversal Theorem. Therefore, line y is not vertical, because it is not perpendicular to a horizontal line; If two lines are not parallel, then they are not horizontal. Proof: Let line x and line y be two non-parallel lines and line x is horizontal. So, the slope of line x is 0. Because the lines are not parallel, the lines do not have the same slope by the Slopes of Parallel Lines Theorem. So, the slope of line y is not 0. Therefore, line y is not horizontal.

3.5 Review & Refresh

53. $x = 30$; When $x = 30$, the angles are congruent. So, the lines are parallel by the Corresponding Angles Converse.

54. The product of three consecutive odd numbers is odd; *Sample answer:* $1 \cdot 3 \cdot 5 = 15$; $5 \cdot 7 \cdot 9 = 315$; $9 \cdot 11 \cdot 13 = 1287$

55. $4 + \sqrt{34} + \sqrt{26} \approx 14.93$ units

56. $b \parallel c$; Both are perpendicular to line a, so they are parallel by the Lines Perpendicular to a Transversal Theorem.

57. $(2h + 1)(2h + 3)$

58. $y = 4x - 3$

59. $x = 0$ and $x = 9$

60. $m\angle 2 = 40°$; The measures of consecutive interior angles are supplementary by the Consecutive Interior Angles Theorem.

61. $w < 4$

62. $(1, 6)$; *Sample answer:* Because the coefficients of the y-variable are opposites, the elimination method may be used.

63. $(1, -3)$; *Sample answer:* Because it is easy to solve the first equation for y, the substitution method may be used.

64. $f(x) = -3x + 5$

65.

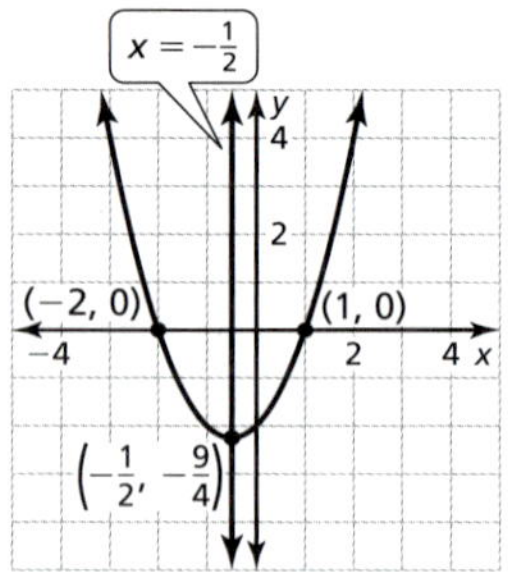

domain: all real numbers; range: $y \geq -\frac{9}{4}$

Chapter 3 Review

1. $\overleftrightarrow{NR}$

2. $\overleftrightarrow{NP}$

3. $\overleftrightarrow{JN}$

4. plane JKN

5. $\angle 3$ and $\angle 5$, $\angle 4$ and $\angle 6$

6. $\angle 3$ and $\angle 6$, $\angle 4$ and $\angle 5$

7. $\angle 1$ and $\angle 5$, $\angle 2$ and $\angle 6$, $\angle 3$ and $\angle 7$, $\angle 4$ and $\angle 8$

8. $\angle 1$ and $\angle 8$, $\angle 2$ and $\angle 7$

9. $x = 145, y = 35;$
$x° + 35° = 180°$
$x = 145$
$y = 35$

10. $x = 13, y = 132;$
$(5x + 17)° = 48°$
$5x = 65$
$x = 13$
$y° + 48° = 180°$
$y = 132$

11. $x = 61, y = 29;$
$2x° + 58° = 180°$
$2x = 122$
$x = 61$
$2y° = 58°$
$y = 29$

12. $x = 14, y = 17;$
$(6x + 32)° = 116°$
$6x = 84$
$x = 14$
$(5y - 21)° + (6x + 32)° = 180°$
$5y - 21 + 6(14) + 32 = 180$
$5y + 95 = 180$
$5y = 85$
$y = 17$

13. $x = 107$; by the Consecutive Interior Angles Converse

14. $x = 133$; by the Alternative Exterior Angles Converse

15. $x = 32$; by the Vertical Angles Congruence Theorem and the Consecutive Interior Angles Converse

16. $x = 23$; by the Corresponding Angles Converse

17. no; $m\angle 1 = 87° \neq 93°$, so the lines are not parallel by the Corresponding Angles Theorem.

18. $x \parallel y$; Because $x \perp z$ and $y \perp z$, lines x and y are parallel by the Lines Perpendicular to a Transversal Theorem.

19. none; The only things that can be concluded in this diagram is that $x \perp z$ and $w \perp y$. In order to say that lines are parallel, you need to know something about *both* of the intersections between the two lines and a transversal.

20. $\ell \parallel m \parallel n$, $a \parallel b$; Because $a \perp n$ and $b \perp n$, lines a and b are parallel by the Lines Perpendicular to a Transversal Theorem. Because $m \perp a$ and $n \perp a$, lines m and n are parallel by the Lines Perpendicular to a Transversal Theorem. Because $\ell \perp b$ and $n \perp b$, lines ℓ and n are parallel by the Lines Perpendicular to a Transversal Theorem. Because $\ell \parallel n$ and $m \parallel n$, lines ℓ and m are parallel by the Transitive Property of Parallel Lines.

21. $a \parallel b$; Because $a \perp n$ and $b \perp n$, lines a and b are parallel by the Lines Perpendicular to a Transversal Theorem.

22.

STATEMENTS	REASONS
1. $\angle 1 \cong \angle 2$	1. Given
2. $g \perp k$	2. Linear Pair Perpendicular Theorem
3. $h \perp k$	3. Given
4. $g \parallel h$	4. Lines Perpendicular to a Transversal Theorem

23. $(-1, 2.75)$

24. $(1.6, -0.2)$

25. $b \parallel c, a \perp b, a \perp c$

26. $y = -x - 1$

27. $y = \frac{1}{2}x + 8$

28. $y = \frac{1}{2}x - 4$

29. $y = -7x - 2$

30. $\frac{3\sqrt{2}}{2}$, or about 2.1 units

31. $\frac{6\sqrt{5}}{5}$, or about 2.7 units

32. $4\sqrt{17}$, or about 16.5 ft

Chapter 3 Practice Test

1. $x = 61, y = 61$; $x° = 61°$ by the Vertical Angles Congruence Theorem. $y° = x° = 61°$ by the Alternate Exterior Angles Theorem.
2. $x = 12, y = 7$; $8x° = 96°$ by the Corresponding Angles Theorem. $(11y + 19)° = 96°$ by the Alternate Interior Angles Theorem.
3. $x = 5, y = 13$; $(8x + 2)° = 42°$ by the Alternate Interior Angles Theorem. $[6(2y - 3)]° + 42° = 180°$ by the Consecutive Interior Angles Theorem.
4. $x = 97$
5. $x = 6$
6. $x = 9$
7. $k \parallel \ell$; Perpendicular Transversal Theorem
8. $3\sqrt{10}$, or about 9.5 units
9. $3y + 2 = 6x - 10$, $(6y + 4) + (4x + 8) = 180$; $x = 12$, $y = 20$
10. **a.** $y = 2x + 12$
    b. $y = -\frac{1}{2}x - \frac{1}{2}$
11. $\left(-\frac{11}{5}, -\frac{6}{5}\right)$
12. **a.** *Sample answer:* $\overleftrightarrow{AB}$ and $\overleftrightarrow{GH}$; non-intersecting, non-coplanar, non-parallel lines
    b. *Sample answer:* $\overleftrightarrow{IJ}$ and $\overleftrightarrow{CD}$; lines intersect at right angle
    c. $\overleftrightarrow{CD}$ and $\overleftrightarrow{EF}$; lines are perpendicular to same transversal
    d. $\angle 1$ and $\angle 3$; angles are corresponding and $\overleftrightarrow{CD} \parallel \overleftrightarrow{EF}$
    e. $\angle 2$ and $\angle 3$; angles are alternate interior and $\overleftrightarrow{CD} \parallel \overleftrightarrow{EF}$

Chapter 3 Test Prep

1. C
2. **a.** $y = -\frac{1}{2}x - 3$
   b. $y = 2x - 5$
3. **a.** supplementary
   b. vertical
   c. complementary
4. B
5. Vertical Angles Congruence Theorem; Transitive Property of Congruence; $\angle 3 \cong \angle 4$; Transitive Property of Congruence
6. D
7. **a.** $\angle 5$
   b. $\angle 6$
   c. $\angle 8$
   d. $\angle 4$
8. B

Chapter 4

Prepare with CalcChat

1. reflection
2. rotation
3. dilation
4. translation
5. 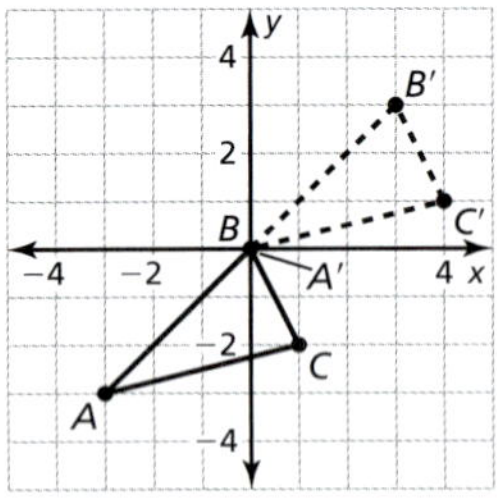

$A'(0, 0), B'(3, 3), C'(4, 1)$
6. 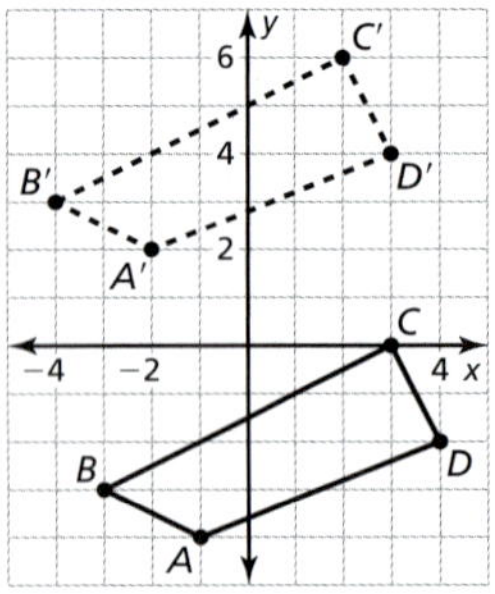

$A'(-2, 2), B'(-4, 3),$
$C'(2, 6), D'(3, 4)$
7. *Sample answer:* Translating a figure 2 units right then 1 unit down results in the same image as translating the figure 1 unit down then 2 units right; The order of the translations does not affect the final position of an image.

Section 4.1

4.1 Practice

1. $\overrightarrow{CD}, \langle 7, -3 \rangle$
3. 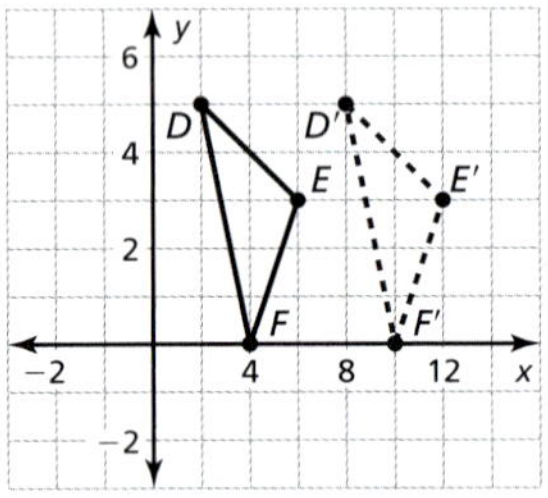

5. 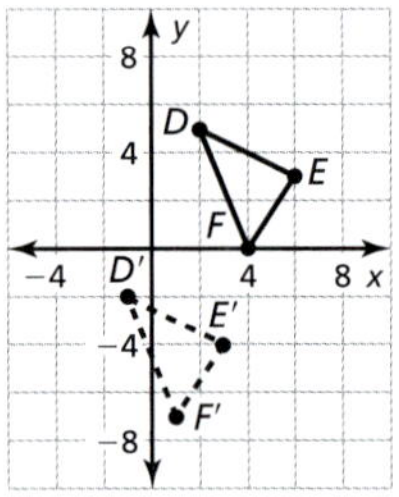

7. $\langle 3, -5 \rangle$
9. $(x, y) \rightarrow (x - 5, y + 2)$
11. $A'(-6, 10)$
13. $C(5, -14)$
15. 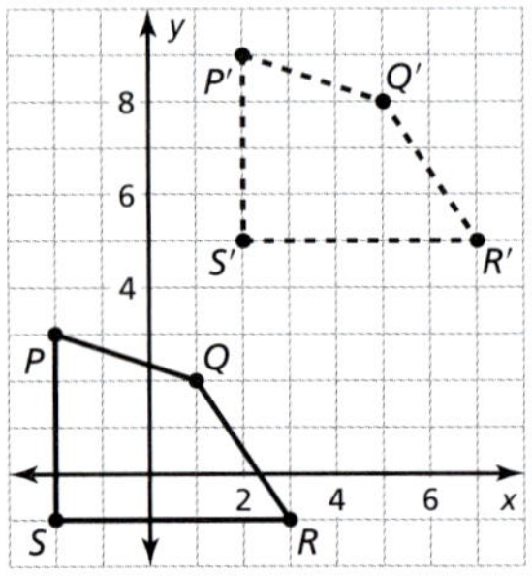

17. 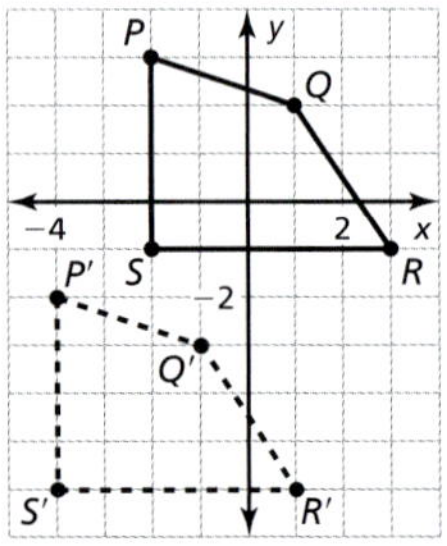

19. The quadrilateral should have been translated left and down;
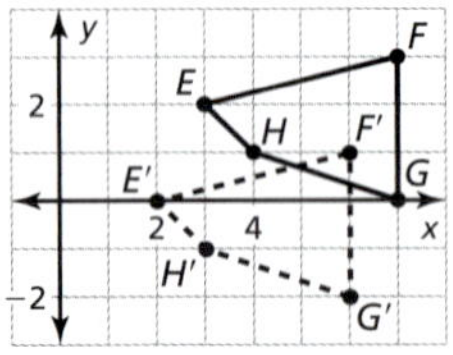

21. 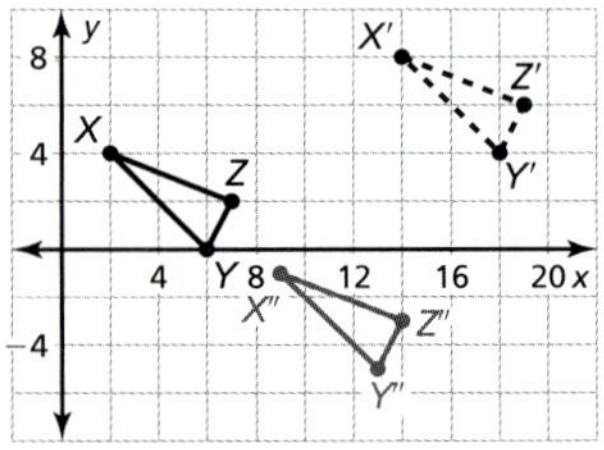

23. translation: $(x, y) \rightarrow (x + 5, y + 1)$,
translation: $(x, y) \rightarrow (x - 5, y - 5)$
25. translation: $(x, y) \rightarrow (x + 2, y - 1)$,
translation: $(x, y) \rightarrow (x + 1, y - 2)$;
$(x, y) \rightarrow (x + 3, y - 3)$
27. $r = 100, s = 8, t = 5, w = 54$
29. **a.** $(x, y) \rightarrow (x + n + s, y + t + m)$
b. $(x, y) \rightarrow (x + s + n, y + m + t)$
c. no; Each image will end up in the same place.
31. $E'(-3, -4), F'(-2, -5), G'(0, -1)$
33. **a.** 21 square units; 21 square units
b. They are equal; The preimage and image of a translation are congruent, so the areas are equal. The translation of Figure A is a rigid motion, so the sides of Figure A' are all the same length and meet at the same angles as the sides of Figure A. So Figure A' is the same size and shape as Figure A and has the same area.
35. **a.** 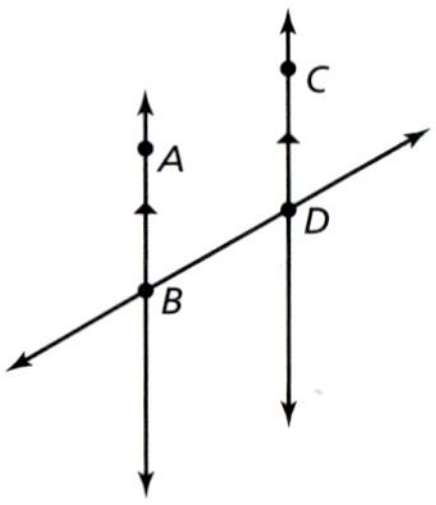

Translate $\overleftrightarrow{AB}$ and $\overleftrightarrow{BD}$ along $\overrightarrow{BD}$ so that point B maps to point D. Then $\overleftrightarrow{BD}$ and its image are the same line. Translations map lines to parallel lines, so $\overleftrightarrow{AB} \parallel \overleftrightarrow{A'B'}$. Because $\overleftrightarrow{A'B'}$ passes through D, by the Parallel Postulate, $\overleftrightarrow{A'B'}$ and $\overleftrightarrow{CD}$ are the same line. Translations are rigid motions which preserve angle measure, so the angles formed by $\overleftrightarrow{AB}$ and $\overleftrightarrow{BD}$ are congruent to the corresponding angles formed by $\overleftrightarrow{CD}$ and $\overleftrightarrow{BD}$.
b. 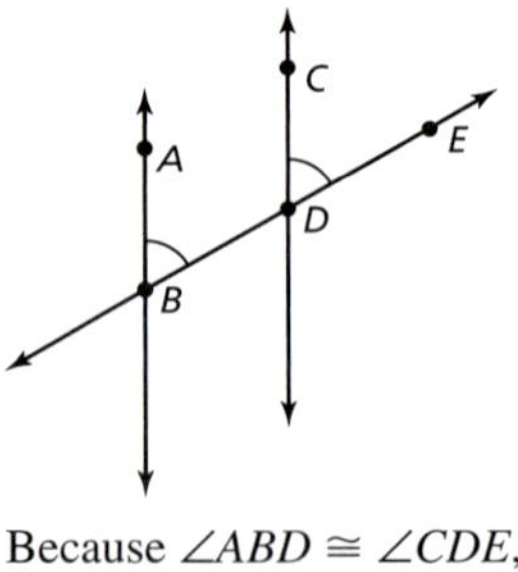

Because $\angle ABD \cong \angle CDE$, a translation along $\overrightarrow{BD}$ maps $\angle ABD$ onto $\angle CDE$. Because translations map lines to parallel lines, $\overleftrightarrow{AB} \parallel \overleftrightarrow{CD}$.
37. yes; According to the definition of translation, the segments connecting corresponding vertices will be congruent and parallel. Also, because a translation is a rigid motion, $\overline{GH} \cong \overline{G'H'}$. So, the resulting figure is a parallelogram.

39. no; Because the value of y changes, you are not adding the same amount to each x-value.

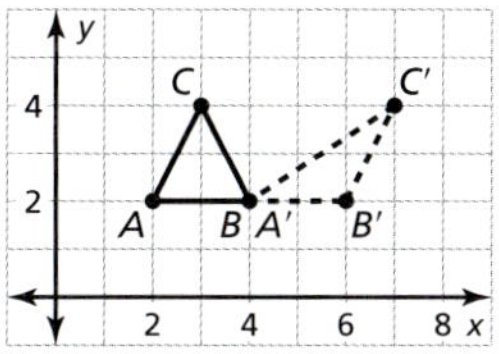

41.

STATEMENTS	REASONS
1. $\overline{MN}$ is perpendicular to line ℓ.	1. Given
2. $\overline{M'N'}$ is the translation of $\overline{MN}$ 2 units left.	2. Given
3. If $M(x_1, y_1)$ and $N(x_2, y_2)$, then $M'(x_1 - 2, y_1)$ and $N'(x_2 - 2, y_2)$.	3. Definition of translation
4. $m_{\overline{MN}} = \dfrac{y_2 - y_1}{x_2 - x_1}$ and $m_{\overline{M'N'}} = \dfrac{y_2 - y_1}{(x_2 - 2) - (x_1 - 2)} = \dfrac{y_2 - y_1}{x_2 - x_1}$	4. Definition of slope
5. $m_{\overline{MN}} = m_{\overline{M'N'}}$	5. Transitive Property of Equality
6. $\overline{MN} \parallel \overline{M'N'}$	6. Slopes of Parallel Lines
7. $\overline{M'N'} \perp \ell$	7. Perpendicular Transversal Theorem

4.1 Review & Refresh

42. yes; Consecutive Interior Angles Converse

43. $y = \frac{1}{4}x - 2$

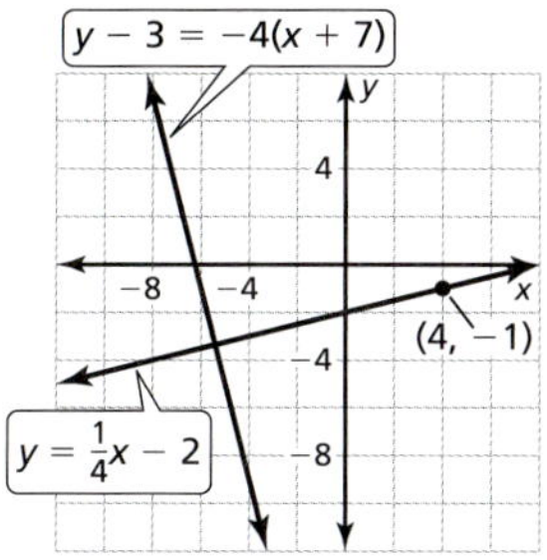

44. $y = -\frac{2}{5}x + \frac{19}{5}$

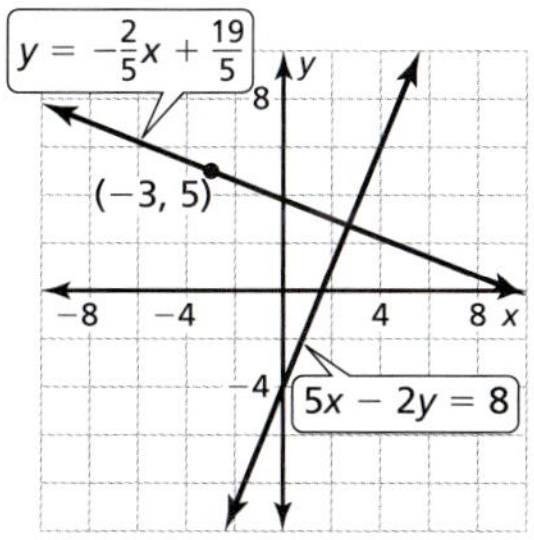

45.

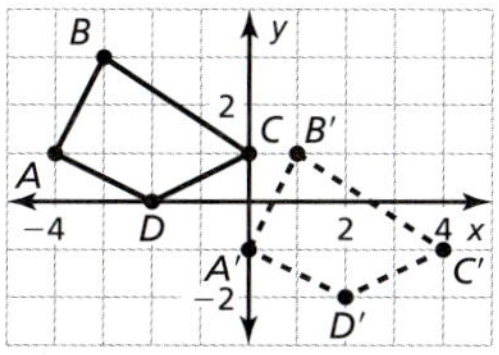

46. $M(-1, 5)$

47. $M(6, 3)$

48. $x = 0$ and $x = 2$

49. $y = -4$ and $y = 0$

50.

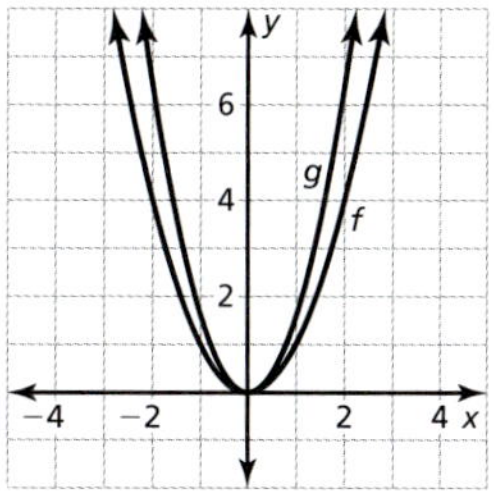

The graph of g is a vertical stretch by a factor of 1.6 of the graph of f.

51.

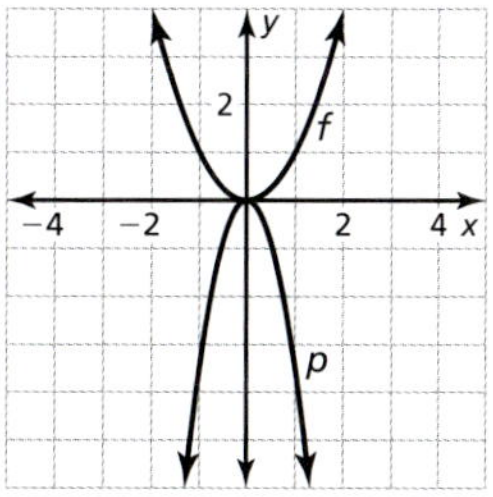

The graph of p is a vertical stretch by a factor of $\frac{7}{2}$ and a reflection in the x-axis of the graph of f.

52. $c \parallel d$; by the Lines Perpendicular to a Transversal Theorem

53. $\frac{3 + \sqrt{3}}{2}$, or about 2.366 sec and $\frac{3 - \sqrt{3}}{2}$, or about 0.634 sec

54. perimeter: $11 + \sqrt{13}$ or about 14.61 units; area: 9 square units

Section 4.2

4.2 Practice

1.

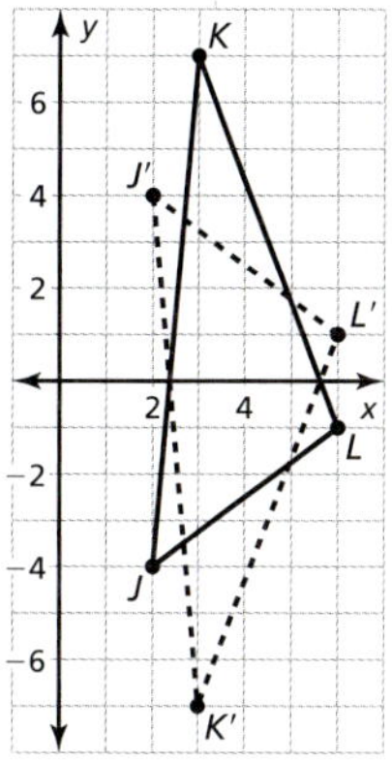

3.

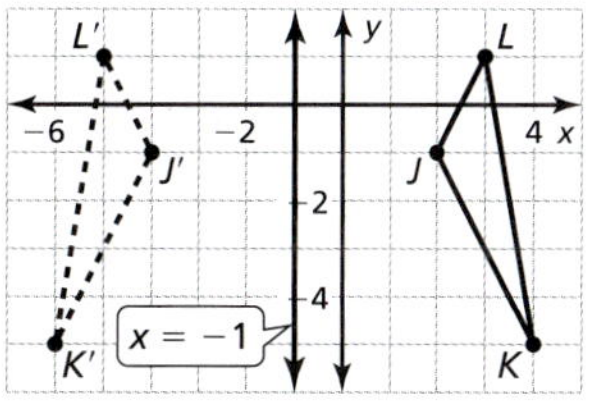

5.

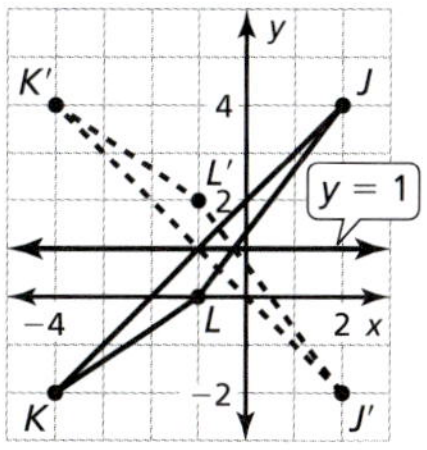

7.

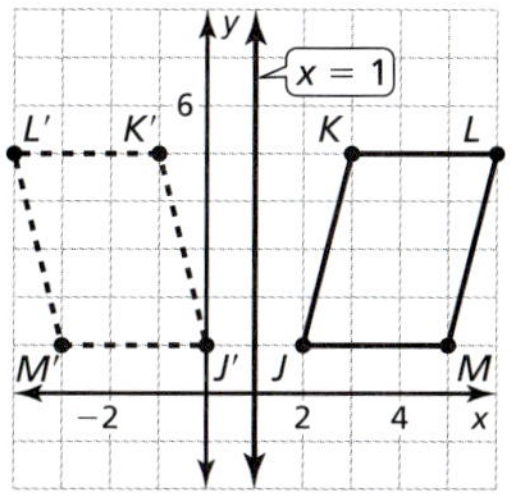

9. $(x, y) \rightarrow (-x, y)$

11. $(x, y) \rightarrow (-x, y)$

13.
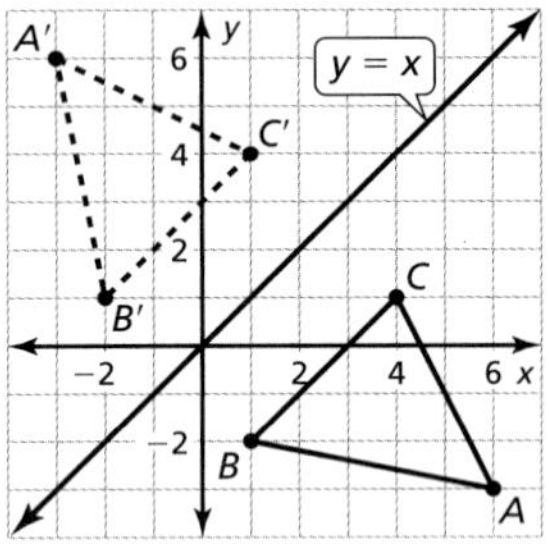

15.
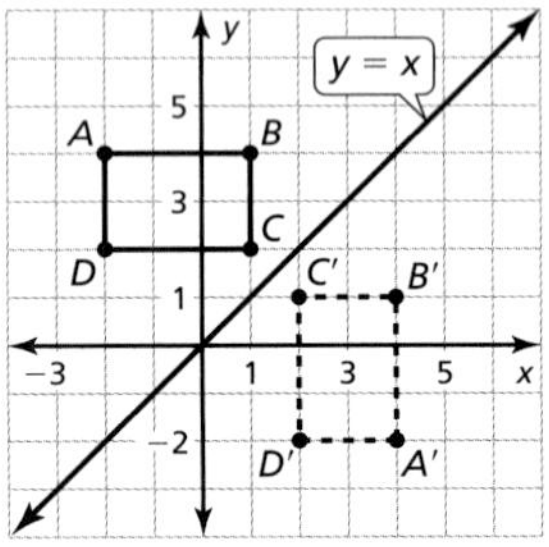

17.
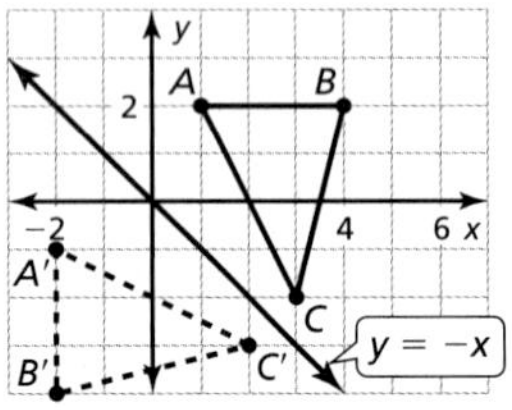

19.
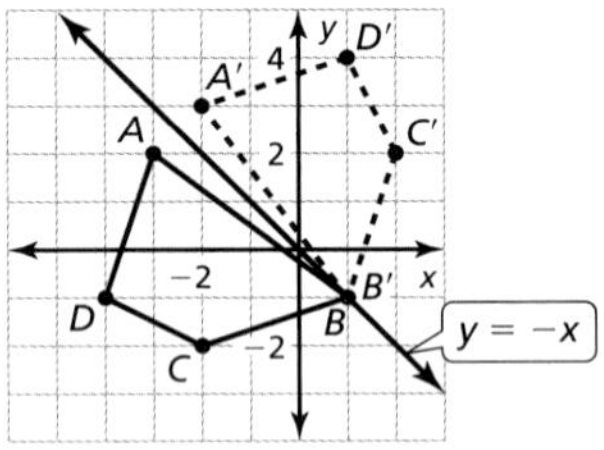

21.
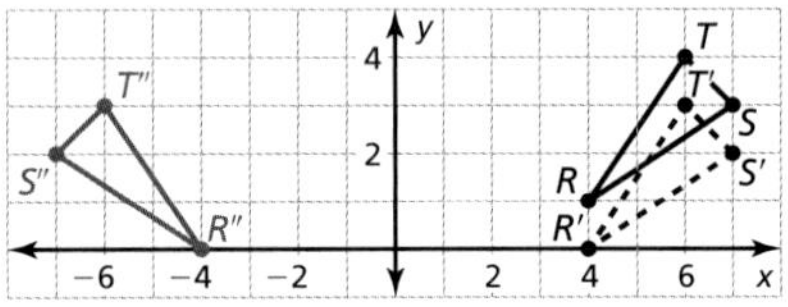

23.
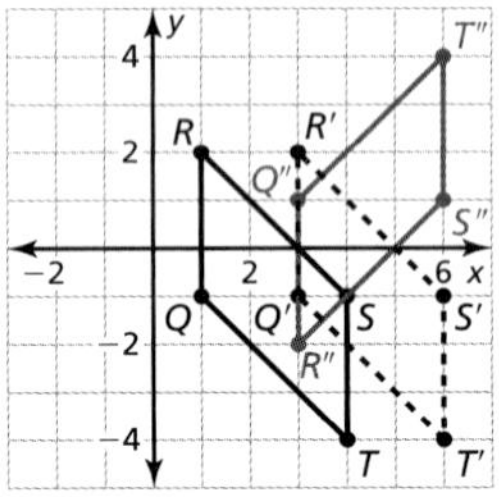

25. yes

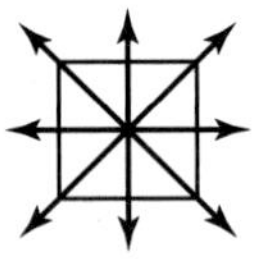

A reflection in the line of symmetry maps the square onto itself.

27. yes

A reflection in the line of symmetry maps the equilateral triangle onto itself.

29. yes

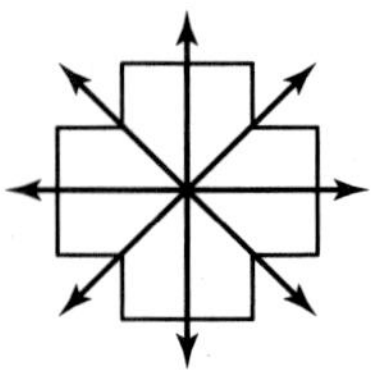

A reflection in the line of symmetry maps the figure onto itself.

31. **a.** none

b. 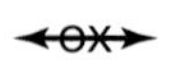

33. $x + 3, y + 3$

35. $y = -3x - 4$

37. $y = 1$

39.
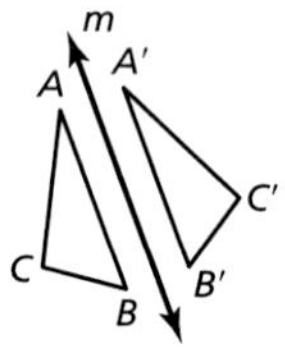

41–45. Check students' work.

47. $y = x + 1$

4.2 Review & Refresh

49. $3\sqrt{2}$, or about 4.2 units

50. $x = -2$ and $x = 6$

51. $x = 3$

52.
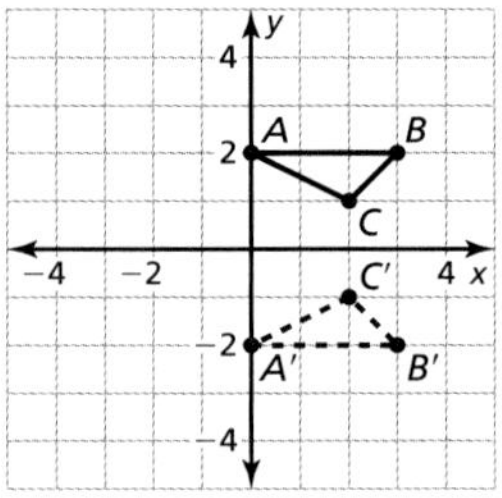

53. Transitive Property of Angle Congruence

54.
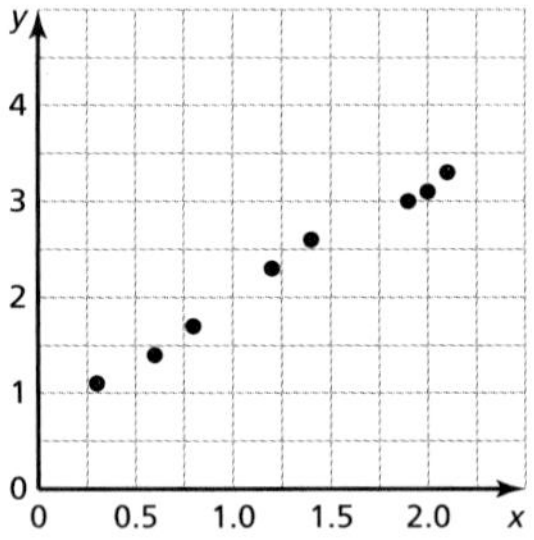

The data show a positive correlation.

55. 44 ft/sec

56. $\sqrt{13}$, or about 3.6 units

57. $h(-1) = 10$

58. $b_2 = \frac{2A}{h} - b_1$

59. $A'(0, 0)$

60. $-(2t - 7)(t - 1)$

61. 5.7×10^6

62. $(-2, 0)$ and $(0, 0)$

63. $x^2 + 7x + 10$

Section 4.3

4.3 Practice

1.

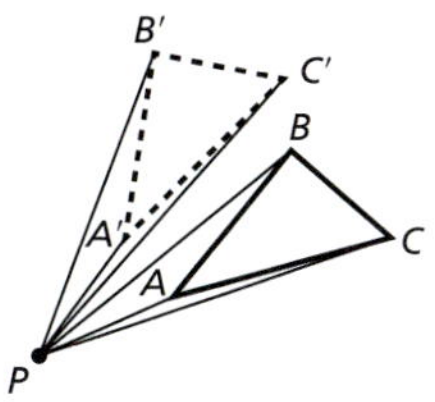

3.

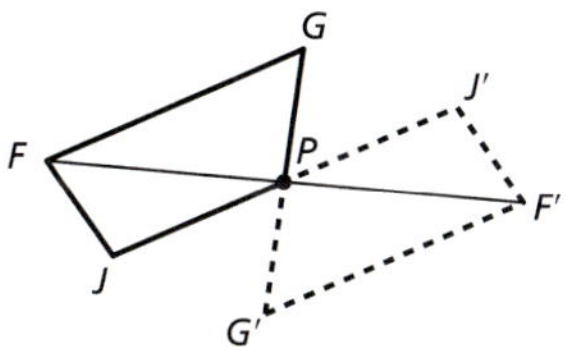

5.

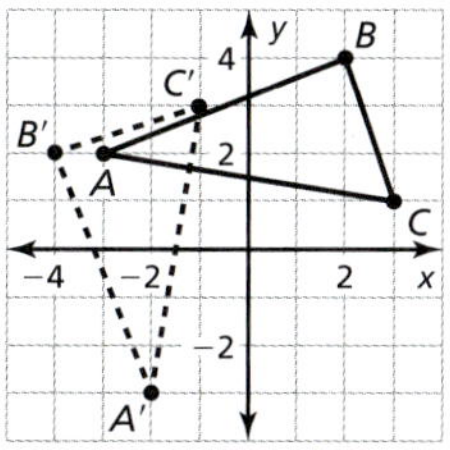

7.

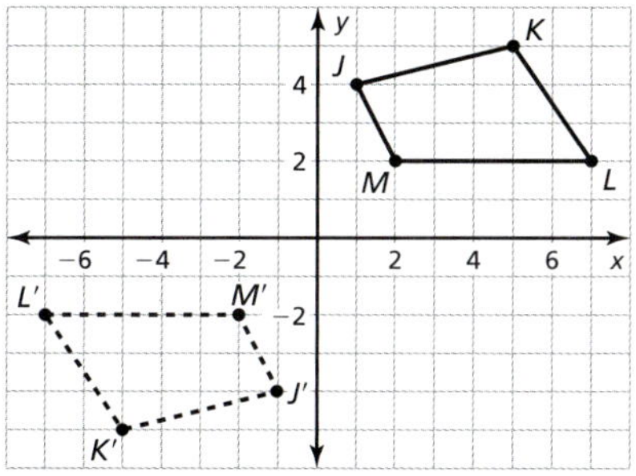

9. **(a)** 90° counterclockwise about the origin
(b) 270° clockwise about the origin; $(x, y) \rightarrow (-y, x)$

11. **(a)** 270° counterclockwise about the origin
(b) 90° clockwise about the origin; $(x, y) \rightarrow (y, -x)$

13. The rule for a rotation 270° counterclockwise about the origin, $(x, y) \rightarrow (y, -x)$, should have been used instead of the rule for a reflection in the x-axis; $C(-1, 1) \rightarrow C'(1, 1)$, $D(2, 3) \rightarrow D'(3, -2)$

15.

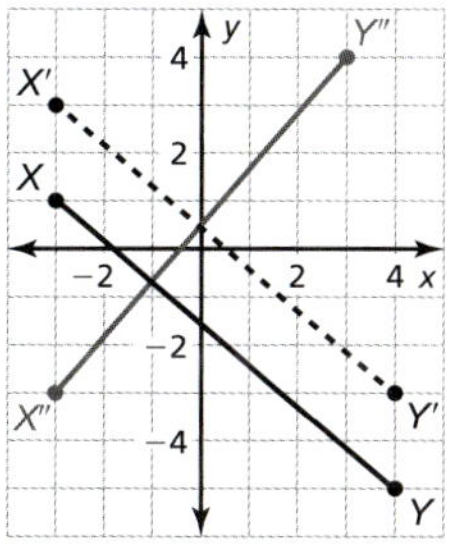

$(x, y) \rightarrow (-y - 2, x)$

17.

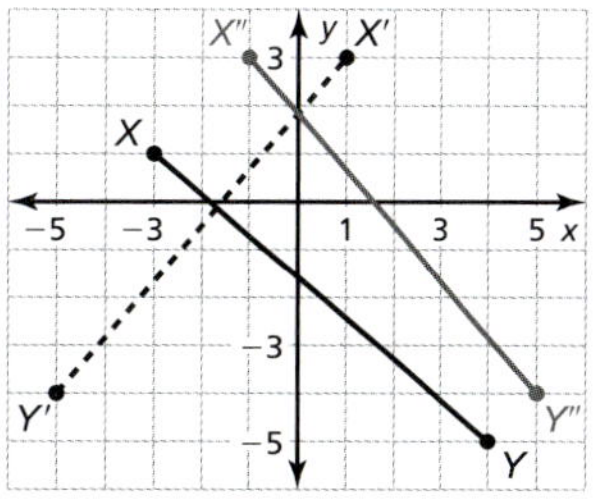

$(x, y) \rightarrow (-y, -x)$

19.

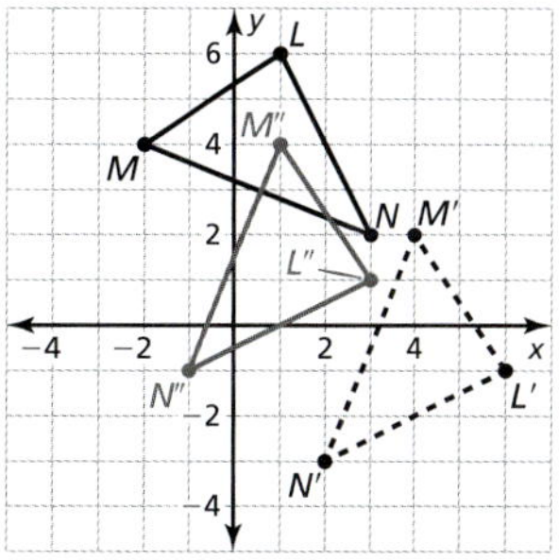

21.

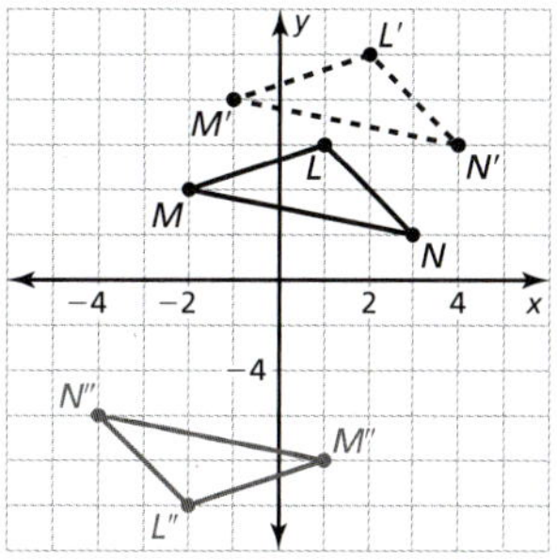

23. $(x, y) \rightarrow (y, -x)$; $(x, y) \rightarrow (-x, -y)$; $(x, y) \rightarrow (-y, x)$

25.

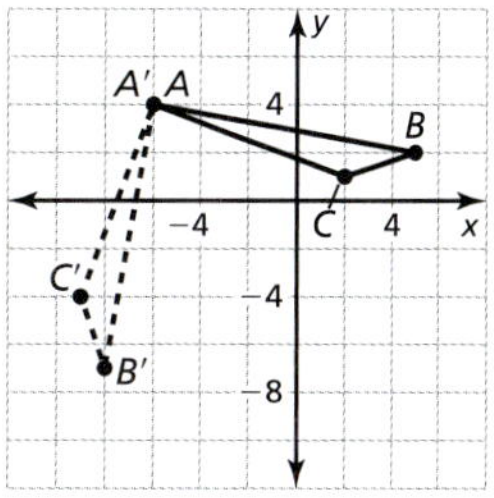

27.

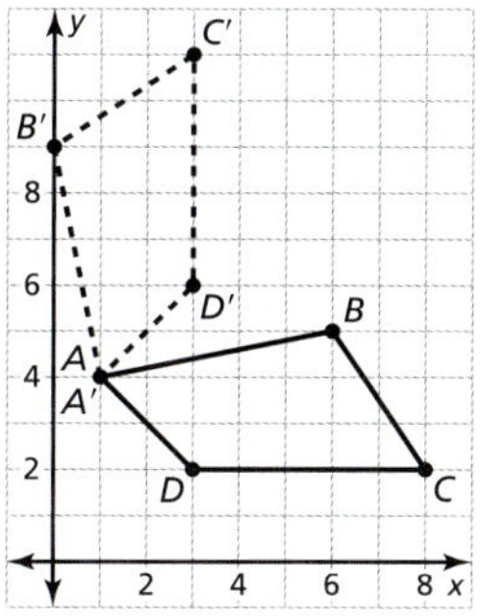

29. yes; Rotations of 90° and 180° about the center map the figure onto itself; the order of rotational symmetry is 4

31. yes; A rotation of 180° about the center maps the rectangle onto itself; the order of rotational symmetry is 2

33. no

35. F

37. D, G

39. B

41.

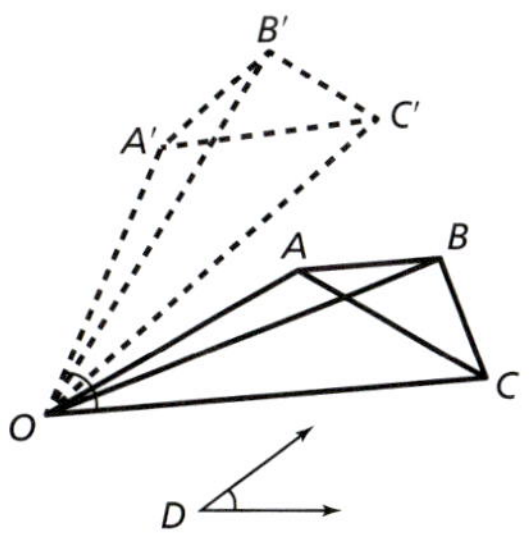

43. **a.** 90°: $y = -\frac{1}{2}x + \frac{3}{2}$, 180°: $y = 2x + 3$, 270°: $y = -\frac{1}{2}x - \frac{3}{2}$; The slope of the line rotated 90° counterclockwise is the opposite reciprocal of the slope of the preimage, and the y-intercept is equal to the x-intercept of the preimage. The slope of the line rotated 180° counterclockwise is equal to the slope of the preimage, and the y-intercepts of the image and preimage are opposites. The slope of the line rotated 270° counterclockwise is the opposite reciprocal of the slope of the preimage, and the y-intercept is the opposite of the x-intercept of the preimage.

45. $J'(-4, 5)$, $K'(-1, 3)$; $J''(5, 4)$, $K''(3, 1)$

47. **a.**

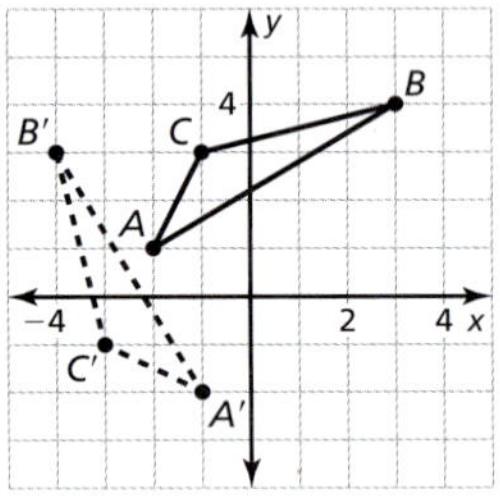

b.

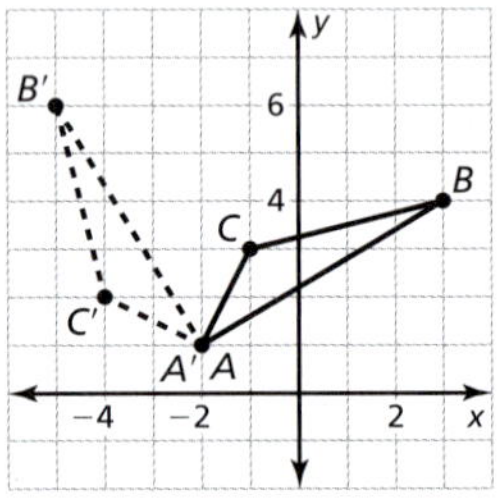

c.

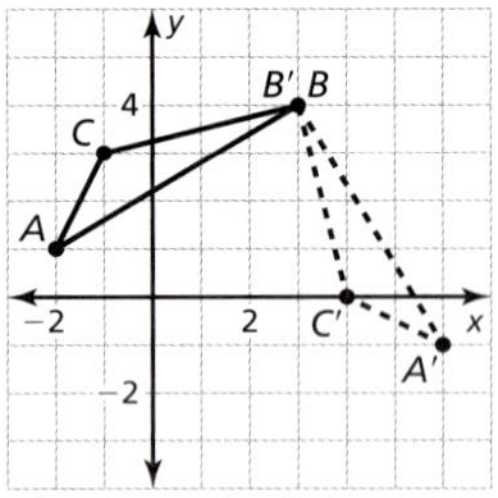

d.

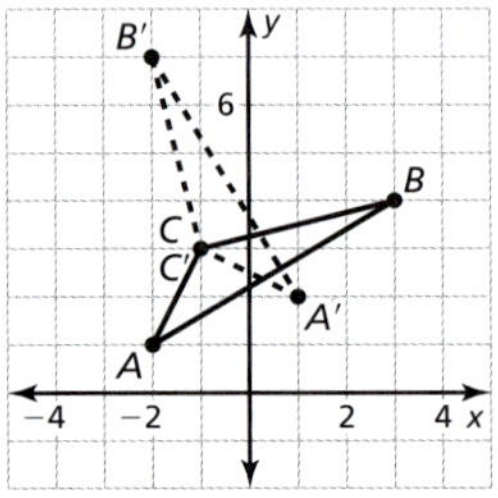

e. (a) $(x, y) \rightarrow (-y, x)$, (b) $(x, y) \rightarrow (-y - 1, x + 3)$, (c) $(x, y) \rightarrow (-y + 7, x + 1)$, (d) $(x, y) \rightarrow (-y + 2, x + 4)$; The rules differ by a horizontal translation and vertical translation because not all of the centers of rotation were the origin.

49. 2

51.

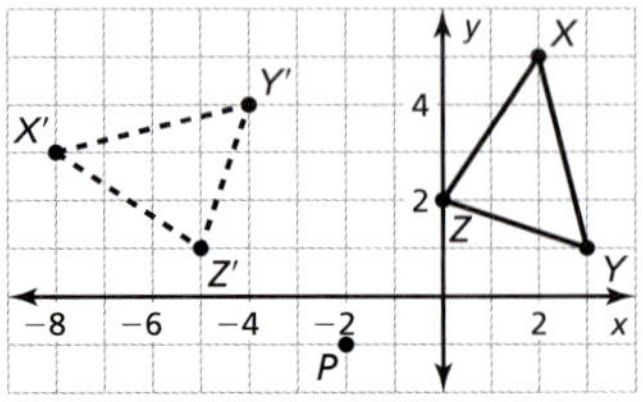

53. $D(-1, 2)$, $E(4, 1)$, $F(3, -2)$, $G(1, -1)$

4.3 Review & Refresh

54. $m\angle EDF = 43°$, $m\angle CDE = 86°$

55. $m\angle CDF = 51°$, $m\angle EDF = 51°$

56. $\angle P$ and $\angle W$, $\angle Q$ and $\angle V$, $\angle R$ and $\angle Z$, $\angle S$ and $\angle Y$, $\angle T$ and $\angle X$; $\overline{PQ}$ and $\overline{WV}$, $\overline{QR}$ and $\overline{VZ}$, $\overline{RS}$ and $\overline{ZY}$, $\overline{ST}$ and $\overline{YX}$, $\overline{TP}$ and $\overline{XW}$

57. $P(4, 3)$

58.

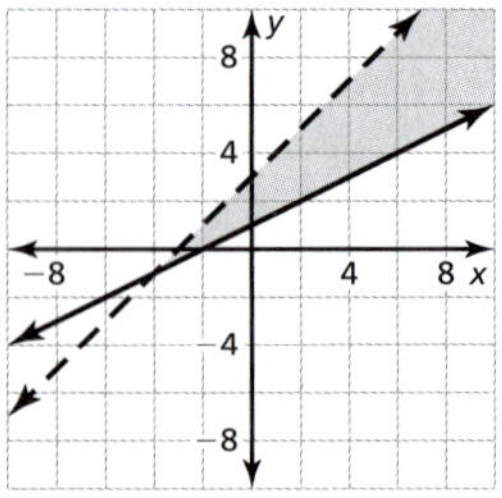

Sample answer: (2, 2)

59. exponential; As x increases by 1, y increases by a constant factor of 4.

60. hexagon; concave

61.

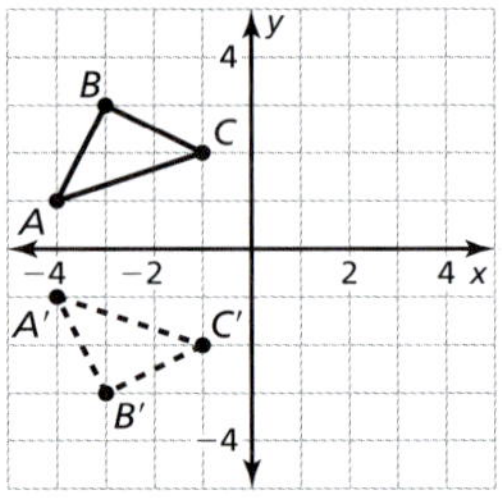

62.

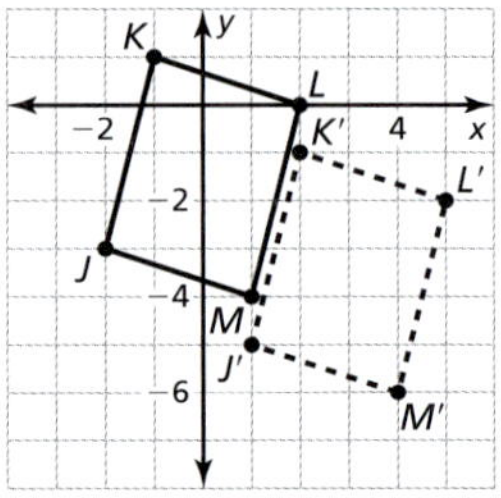

63.

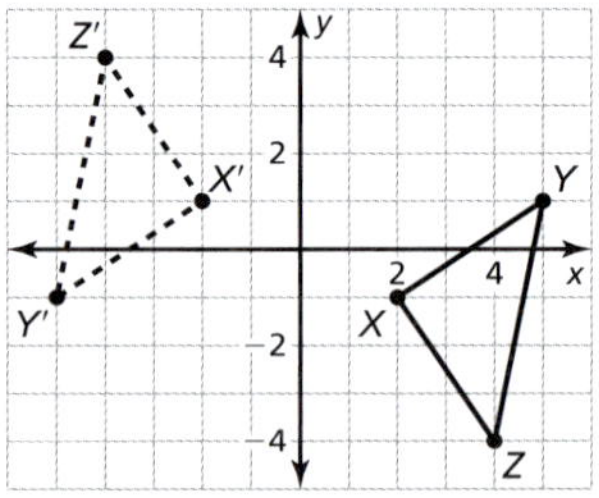

64. inductive reasoning; a pattern is used to reach the conclusion

65. deductive reasoning; facts about animals and the laws of logic are used to reach the conclusion

66. The population in the study is high school students in the United Kingdom and the sample is the 1467 high school students who completed the survey; The sample includes high school students who plan to attend a 4 year college and those who do not.

67. All of the bars are parallel to each other by the Transitive Property of Parallel Lines.

Section 4.4

4.4 Practice

1. $\triangle HJK \cong \triangle QRS$, $\square DEFG \cong \square LMNP$; $\triangle HJK$ is a 90° rotation of $\triangle QRS$. $\square DEFG$ is a translation 7 units right and 3 units down of $\square LMNP$.

3. *Sample answer:* 180° rotation about the origin, followed by a translation 5 units left and 1 unit down

5. yes; $\triangle TUV$ is a translation 4 units right of $\triangle QRS$. So, $\triangle TUV \cong \triangle QRS$.

7. no; M and N are translated 2 units right of their corresponding vertices, L and K, but P is translated only 1 unit right of its corresponding vertex, J. So, this is not a rigid motion.

9. no

11. yes; *Sample answers:*

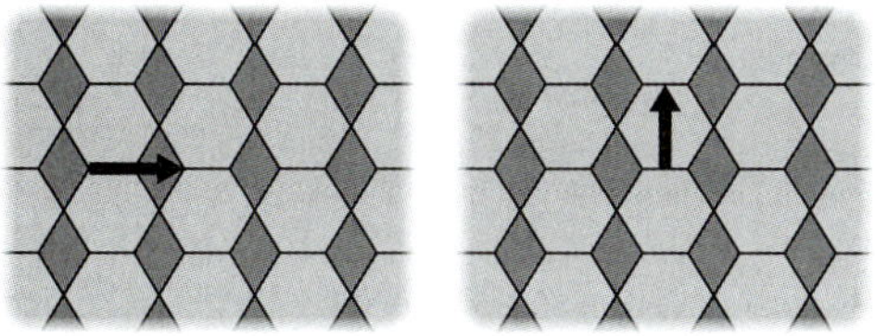

13. yes: *Sample answers:*

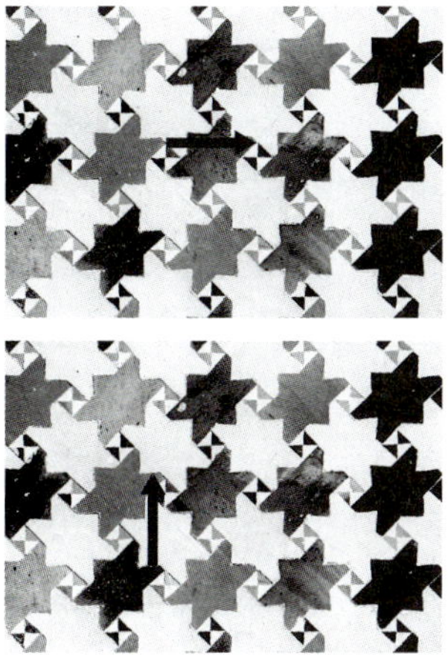

15. $\triangle A''B''C''$

17. 5.2 in.

19. $110°$ rotation about the intersection of lines m and k

21. If $x°$ is the measure of the acute angle formed by the intersecting lines, an angle of $2x°$ should be used to describe the angle of rotation; A $144°$ rotation about point P maps the blue image to the green image.

23. $42°$

25. $90°$

27. Reflect the figure in two parallel lines instead of translating the figure; The third line of reflection is perpendicular to the parallel lines.

29. never; Congruence transformations are rigid motions.

31. sometimes; Reflecting in parallel lines is not a rotation. Reflecting in the y-axis then x-axis is a rotation of $180°$.

33. yes; reflections: $P(1, 3) \rightarrow P'(-1, 3) \rightarrow P''(-1, -3)$ and $Q(3, 2) \rightarrow Q'(-3, 2) \rightarrow Q''(-3, -2)$
$180°$ rotation $P(1, 3) \rightarrow P'(-1, -3)$ and $Q(3, 2) \rightarrow Q'(-3, -2)$

35. no; no; In order to tessellate, you need to be able to arrange interior angles to add up to $360°$ without an overlap or gaps. Polygons such as a heptagon and regular pentagon cannot be used to construct a tessellation.

37.

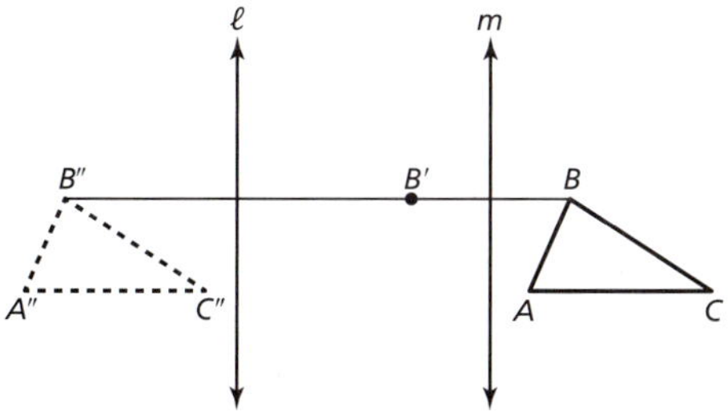

39. *Sample answer:*

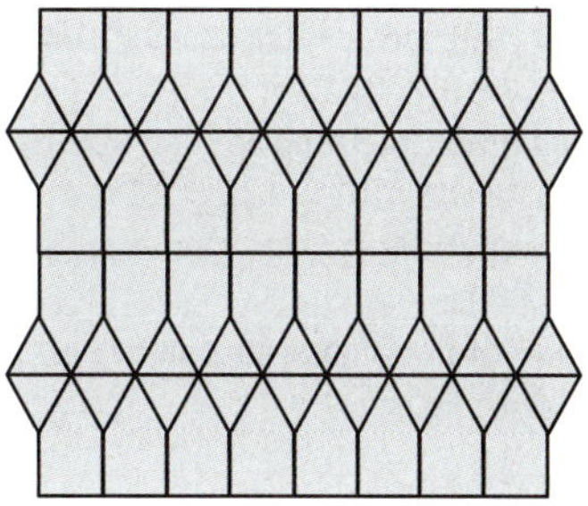

41. yes; *Sample answer:* The distance between the preimage and final image is the same regardless of order, but the final image can be on the opposite side of the preimage if the order of reflections is switched.

43. yes; Any ray's terminal point can be mapped to a different ray's terminal point using translations, and the direction of the ray can be mapped to the other ray's direction using a rotation.

4.4 Review & Refresh

45. $m = -3$

46. $y = 2$

47. $n = -1$

48. 25%

49. *Sample answer:* reflection about the line $y = x$, followed by a reflection about the line $y = -x$

50.

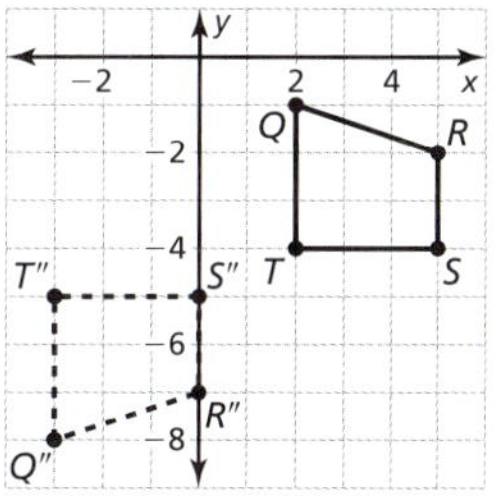

51.

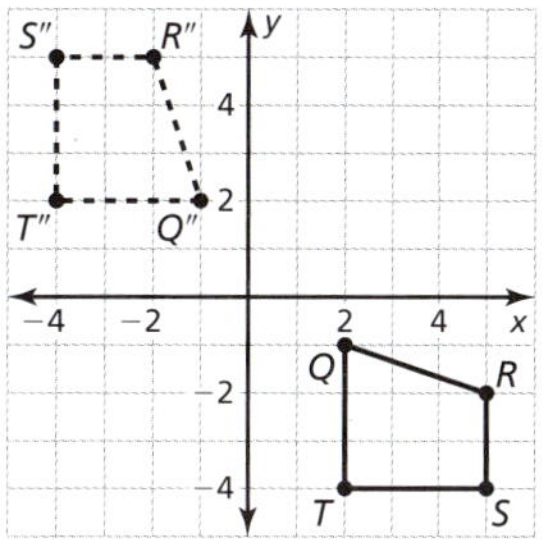

52.

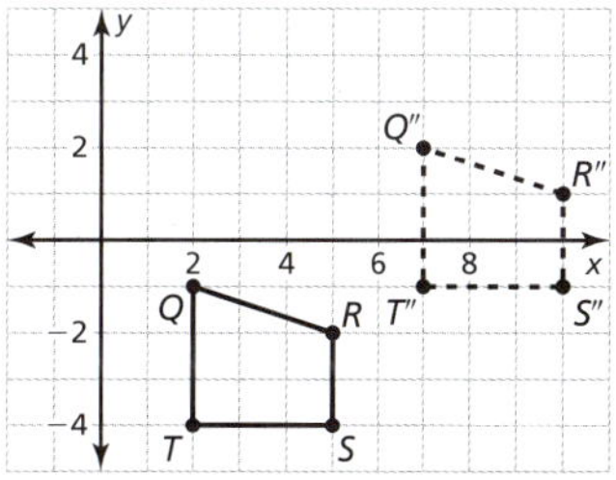

53.

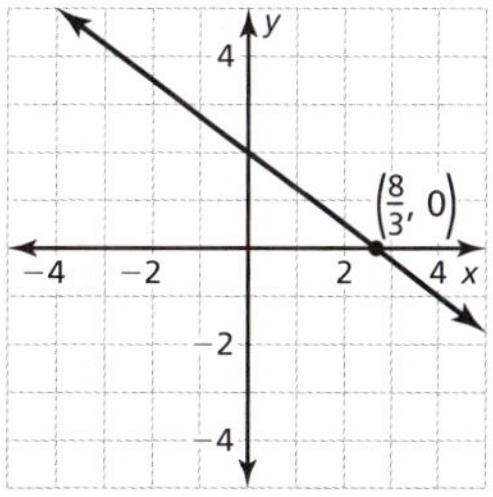

54.

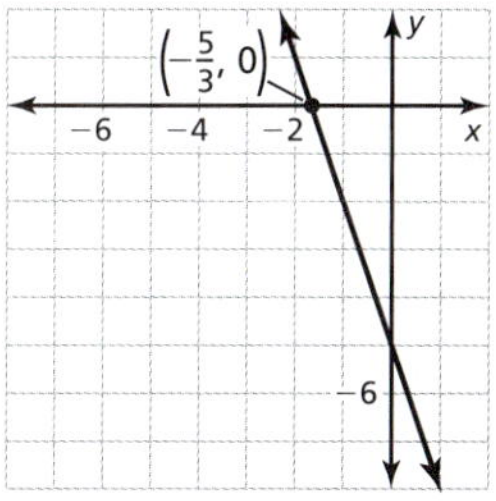

55. $x \geq -\frac{5}{2}$

56. conditional statement: If you ride a roller coaster, then you go to an amusement park; true
converse: If you go to an amusement park, then you ride a roller coaster; false
inverse: If you do not ride a roller coaster, then you do not go to an amusement park; false
contrapositive: If you do not go to an amusement park, then you do not ride a roller coaster; true

57. $6x^\circ = 48^\circ$
$x = 8$

58. $(12x + 30)^\circ + 102^\circ = 180^\circ$
$12x + 132 = 180$
$12x = 48$
$x = 4$

59. 17.5 square units

60. 20 square units

Section 4.5

4.5 Practice

1. $\frac{3}{7}$; reduction

3. $\frac{3}{5}$; reduction

5.

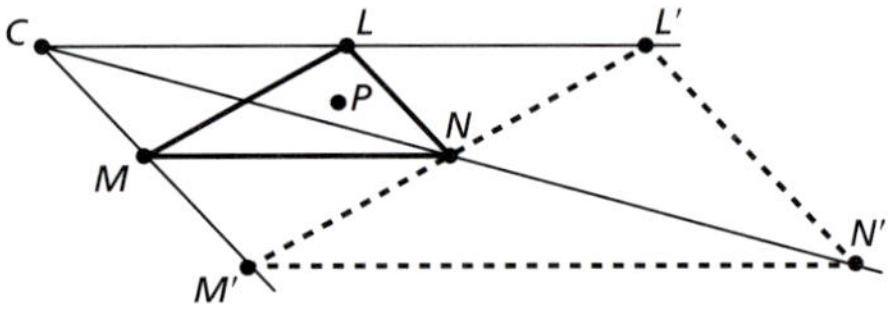

Not drawn to scale.

7.

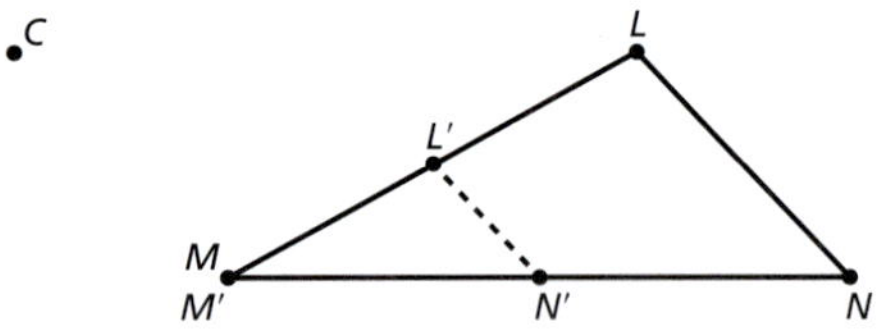

Not drawn to scale.

9.

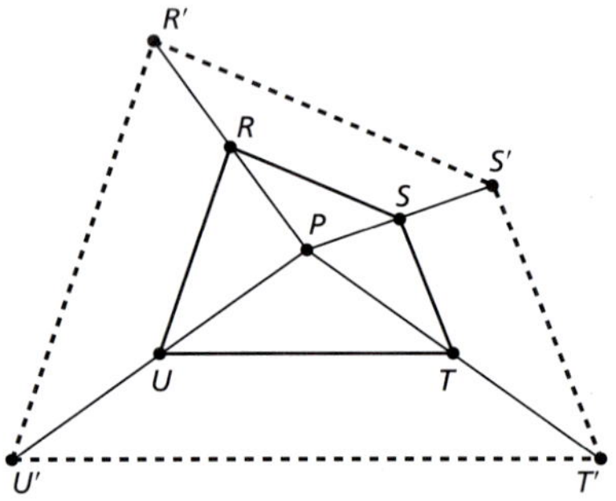

Not drawn to scale.

11.

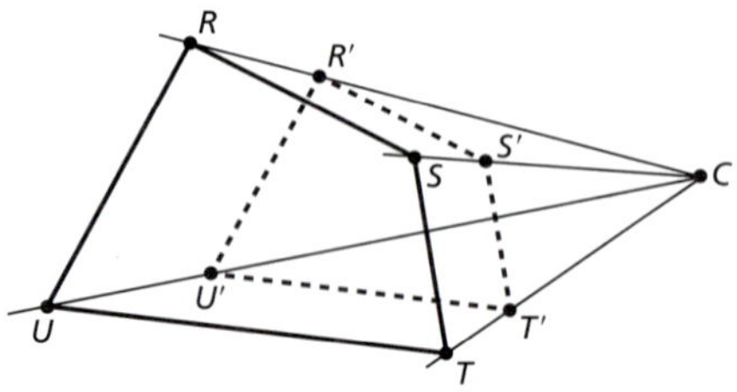

Not drawn to scale.

13.

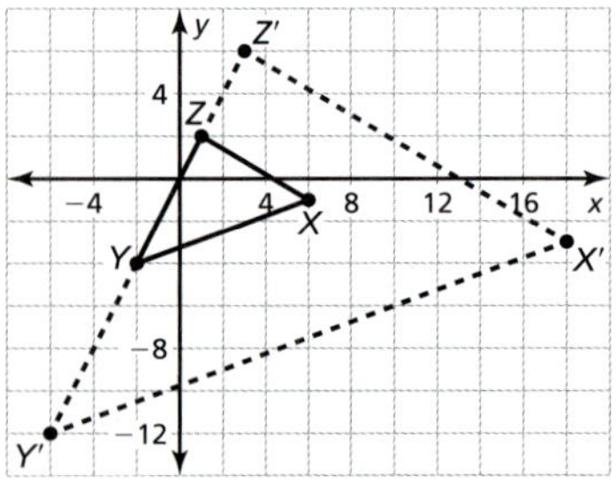

15.

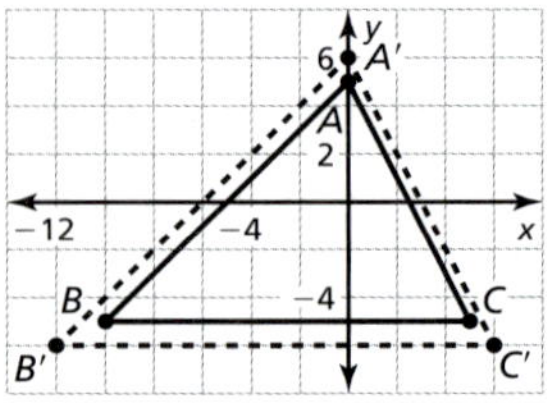

17.

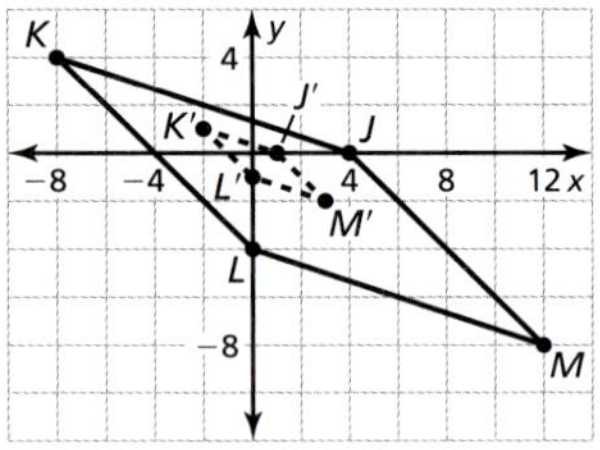

19. $(x, y) \to (3x, 3y)$

21.

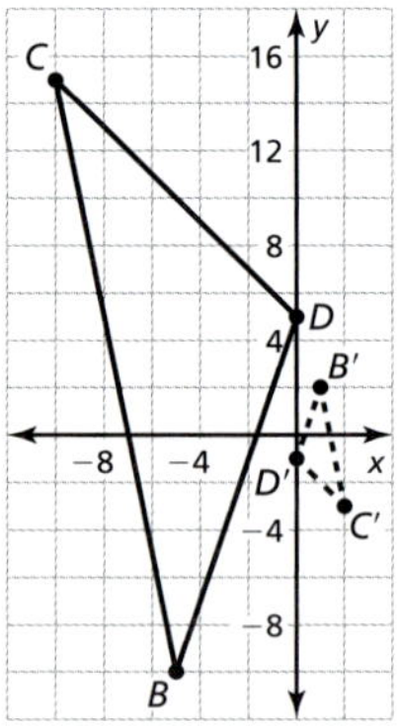

23.

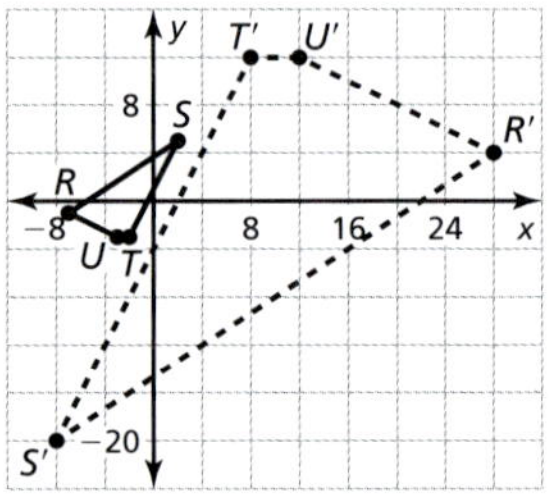

25. $k = 2$

27. The scale factor should be calculated by finding $\frac{CP'}{CP}$, not $\frac{CP}{CP'}$; $k = \frac{3}{12} = \frac{1}{4}$

29. original

31. dilated

33. 300 mm

35. 940 mm

37. grasshopper, honey bee, and monarch butterfly; The scale factor for these three is $k = \frac{15}{2}$. The scale factor for the black beetle is $k = 7$.

39. no; The scale factor for the shorter sides is $\frac{8}{4} = 2$, but the scale factor for the longer sides is $\frac{10}{6} = \frac{5}{3}$. The scale factor for both sides has to be the same or the picture will be distorted.

41.

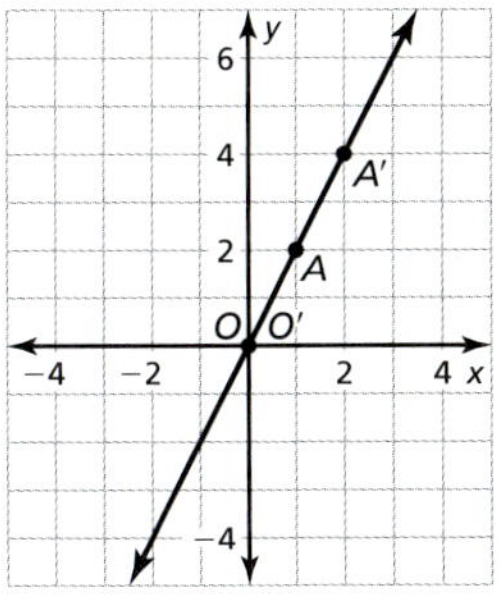

a. $O'A' = 2(OA)$

b. $\overleftrightarrow{O'A'}$ coincides with $\overleftrightarrow{OA}$.

43.

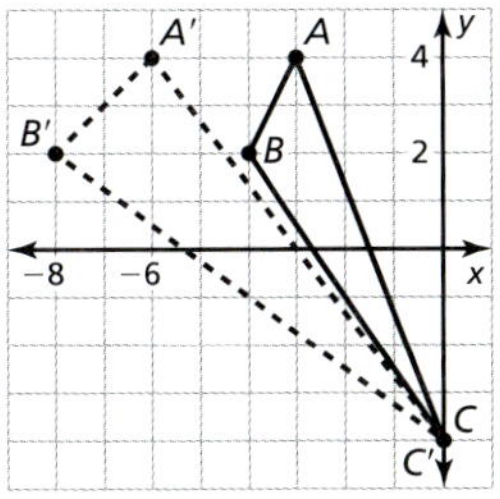

no; The side lengths of the triangle do not all increase by the same factor.

45. The center of dilation must be on the rectangle. So, this point will be in the same place for both the original figure and the reduced figure.

47. a. $P = 24$ units, $A = 32$ square units

b.

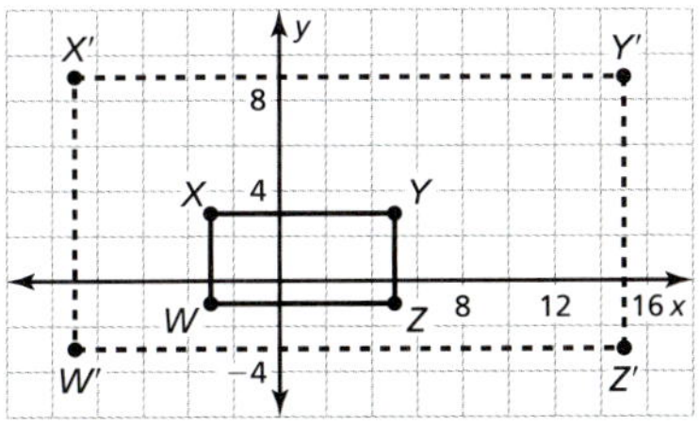

$P = 72$ units, $A = 288$ square units; The perimeter of the dilated rectangle is three times the perimeter of the original rectangle. The area of the dilated rectangle is nine times the area of the original rectangle.

c.

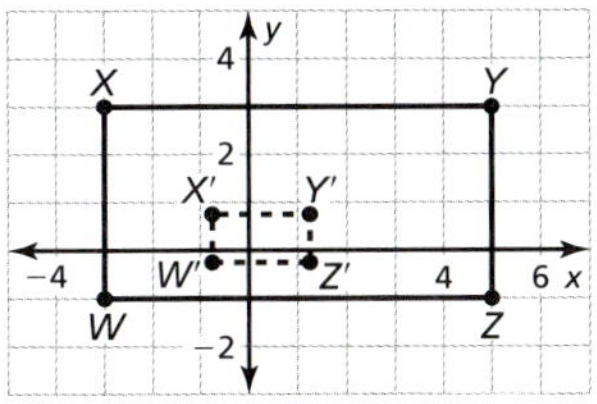

$P = 6$ units, $A = 2$ square units; The perimeter of the dilated rectangle is $\frac{1}{4}$ the perimeter of the original rectangle. The area of the dilated rectangle is $\frac{1}{16}$ the area of the original rectangle.

d. The perimeter changes by a factor of k. The area changes by a factor of k^2.

49. $A'(4, 4)$, $B'(4, 12)$, $C'(10, 4)$

4.5 Review & Refresh

50.

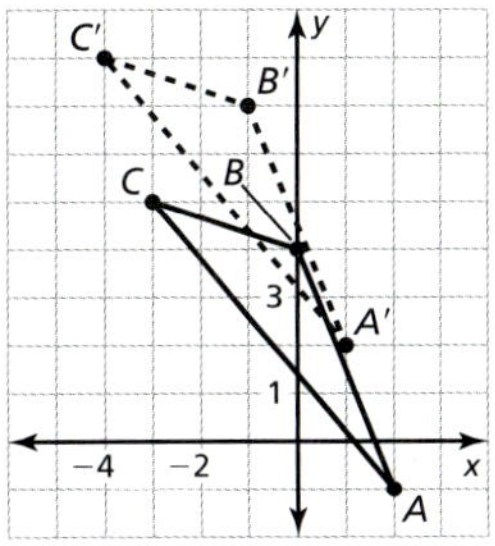

51.

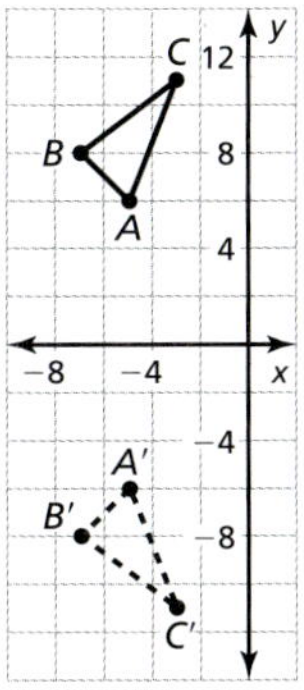

52.

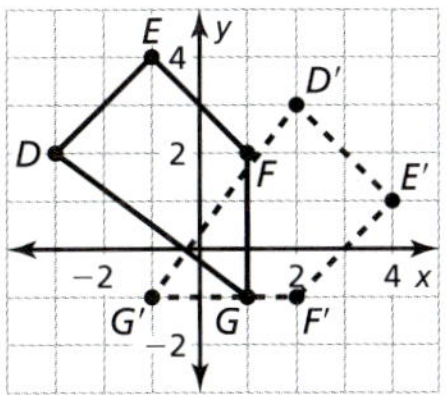

53.

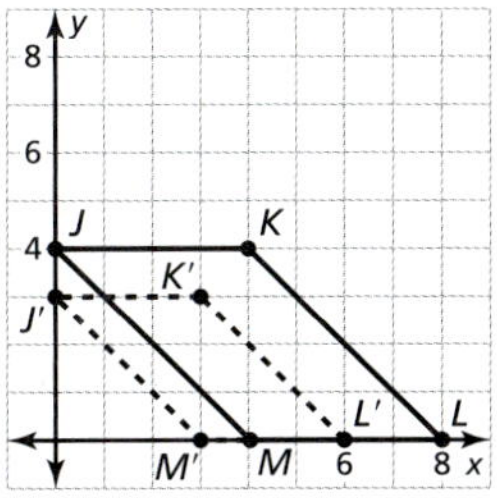

54. $\frac{b}{8}$

55. 999 in.^2

56. *Sample answer:* reflection in the y-axis, translation 5 units down

57.

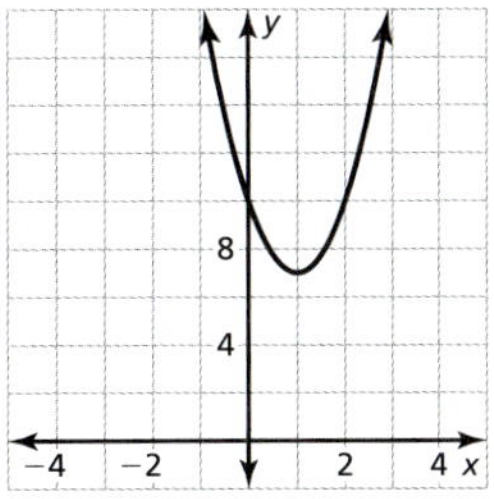

The graph of g is a translation 1 unit right, followed by a vertical stretch by a factor of 3, then a translation 7 units up of the graph of f.

58. $9x^2 - 24x + 16$

59. $2w^2 - 4w - 30$

60. $\left(\frac{1}{2}, \frac{7}{2}\right)$; Substitute $3x + 5$ in for y in the second equation.

61. $\left(-7, \frac{15}{2}\right)$; Solve the second equation for x and substitute the expression into the first equation.

Section 4.6

4.6 Practice

1.

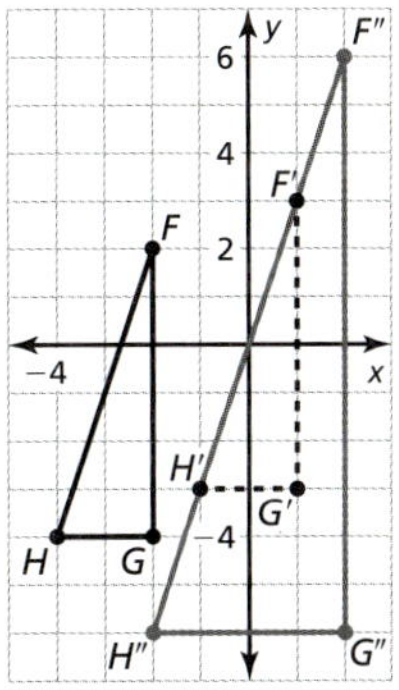

3.

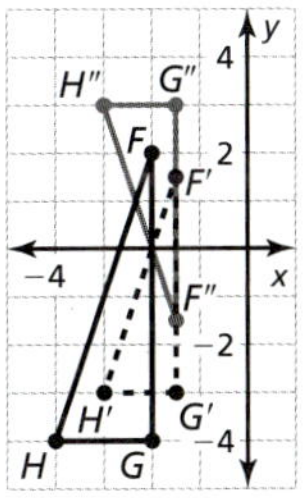

5. *Sample answer:* dilation with a scale factor of 2, followed by a translation 1 unit up

7. *Sample answer:* dilation with a scale factor of $\frac{1}{2}$, followed by a reflection in the y-axis

9. $(x, y) \rightarrow (-2x, -2y)$

11. $(x, y) \rightarrow (2x, 2y + 1)$

13. $(x, y) \rightarrow \left(-\frac{1}{2}x, \frac{1}{2}y\right)$

15. 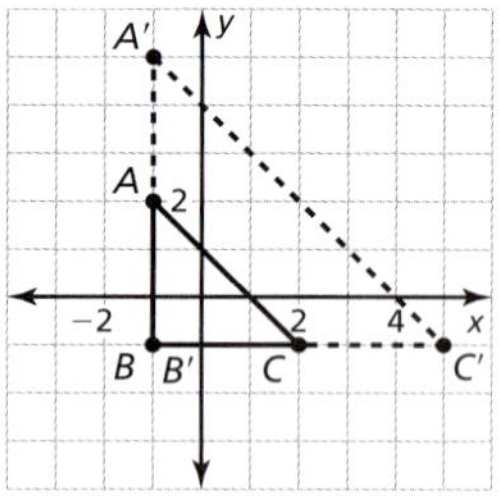

17. 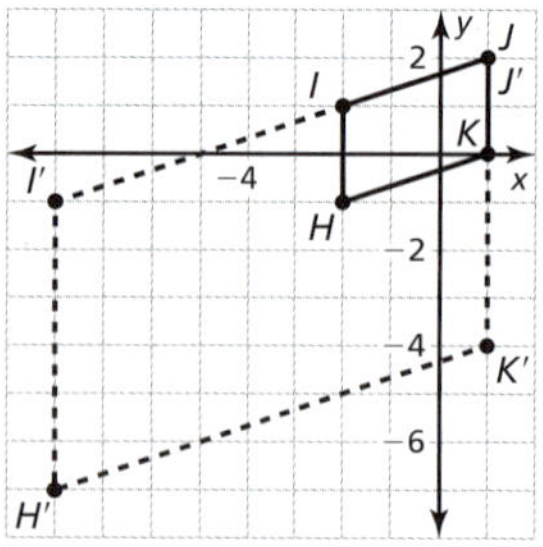

19. yes; $\triangle ABC$ can be mapped to $\triangle DEF$ by a dilation with a scale factor of $\frac{1}{3}$, followed by a translation of 2 units left and 3 units up.

21. no; The scale factor from $\overline{HI}$ to $\overline{JL}$ is $\frac{2}{3}$, but the scale factor from $\overline{GH}$ to $\overline{KL}$ is $\frac{5}{6}$.

23. Reflect $\triangle ABC$ in $\overleftrightarrow{AB}$. Because reflections preserve side lengths and angle measures, the image of $\triangle ABC$, $\triangle ABC'$, is a right isosceles triangle with leg length j. Also because $\overleftrightarrow{AC} \perp \overleftrightarrow{BA}$, point C' is on $\overleftrightarrow{AC}$. So, $\overline{AC'}$ is parallel to $\overline{RT}$.

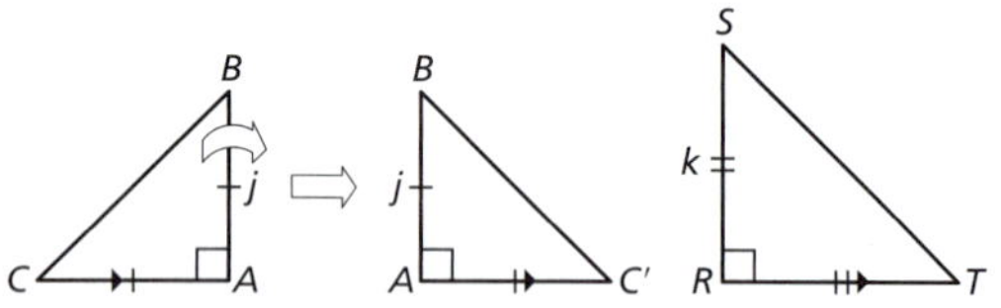

Then translate $\triangle ABC'$ so that point A maps to point R. Because translations map segments to parallel segments and $\overline{AC'} \parallel \overline{RT}$, the image of $\overline{AC'}$ lies on $\overline{RT}$.

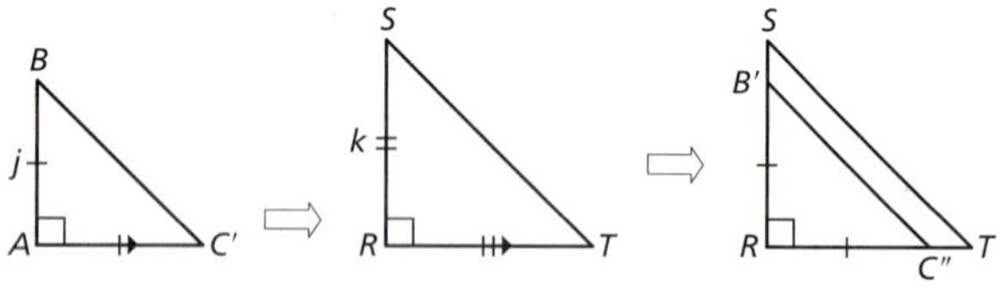

Because translations preserve side lengths and angle measures, the image of $\triangle ABC'$, $\triangle RB'C''$, is a right isosceles triangle with leg length j. Because $\angle B'RC''$ and $\angle SRT$ are right angles, they are congruent. When $\overrightarrow{RC''}$ coincides with $\overrightarrow{RT}$, $\overrightarrow{RB'}$ coincides with $\overrightarrow{RS}$. So, $\overrightarrow{RB'}$ lies on $\overrightarrow{RS}$. Next, dilate $\triangle RB'C''$ using center of dilation R. Choose the scale factor to be the ratio of the side lengths of $\triangle RST$ and $\triangle RB'C''$, which is $\frac{k}{j}$.

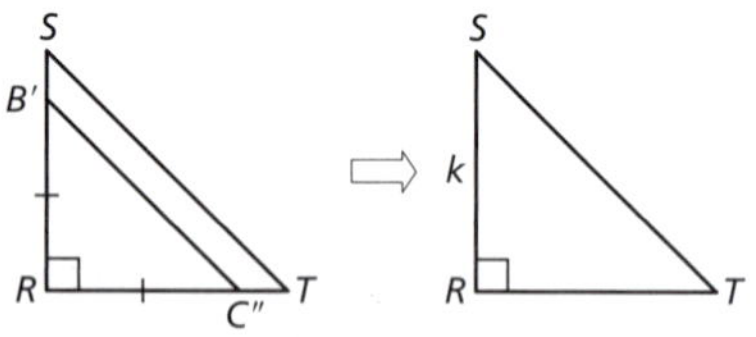

The dilation maps $\overline{RC''}$ to $\overline{RT}$ and $\overline{RB'}$ to $\overline{RS}$ because the images of $\overline{RC''}$ and $\overline{RB'}$ have side length $\frac{k}{j}(j) = k$ and the segments $\overline{RC''}$ and $\overline{RB'}$ lie on lines passing through the center of dilation. So, the dilation maps C'' to T and B' to S. A similarity transformation maps $\triangle ABC$ to $\triangle RST$. So, $\triangle ABC$ is similar to $\triangle RST$.

25. Figure A is not similar to Figure B because the scale factor (A to B) of the shorter legs is $\frac{1}{2}$, and the scale factor (A to B) of the longer legs is $\frac{2}{3}$; Figure A is not similar to Figure B.

27. $J(-8, 0)$, $K(-8, 12)$, $L(-4, 12)$, $M(-4, 0)$; $J''(-9, -4)$, $K''(-9, 14)$, $L''(-3, 14)$, $M''(-3, -4)$; yes; A similarity transformation mapped quadrilateral $JKLM$ to quadrilateral $J''K''L''M''$.

29. dilation by scale factor of 2, then a translation 2 units down and 3 units left; *Sample answer:* Perform opposite transformations in reverse order.

31. **a.** yes; The smaller triangle can be mapped to the larger one by a dilation with $k = -2$, followed by the translation $(x, y) \rightarrow (x + 9, y + 7)$. Because one can be mapped to the other by a similarity transformation, the triangles are similar.

b. The triangle formed when the midpoints of a triangle are connected is always similar to the original triangle.

33. By completing a dilation of rectangle $ABCD$ with center A and scale factor $\sqrt{2}$, $A'D' = 9\sqrt{2}$ and $A'B' = 6\sqrt{2}$, making the area of rectangle $A'B'C'D' = (9\sqrt{2})(6\sqrt{2}) = 108$ square units, which is twice the area of rectangle $ABCD$ which is 54 square units.; To dilate any polygon to produce an image whose area is n times the area of the original polygon, complete a dilation with a scale factor of $\sqrt{n}$.

35. **a.** yes; The smaller triangle can be mapped to the larger one by a dilation with $k = -2$, followed by the translation $(x, y) \rightarrow (x + 9, y + 7)$. Because one can be mapped to the other by a similarity transformation, the triangles are similar.

b. The triangle formed when the midpoints of a triangle are connected is always similar to the original triangle.

4.6 Review & Refresh

36. obtuse

37. straight

38. acute

39. right

40. 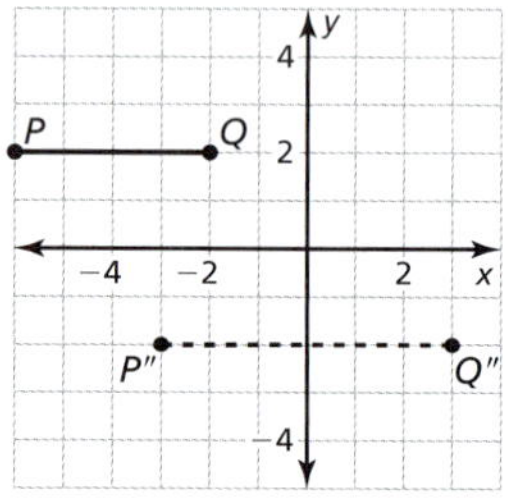

41. 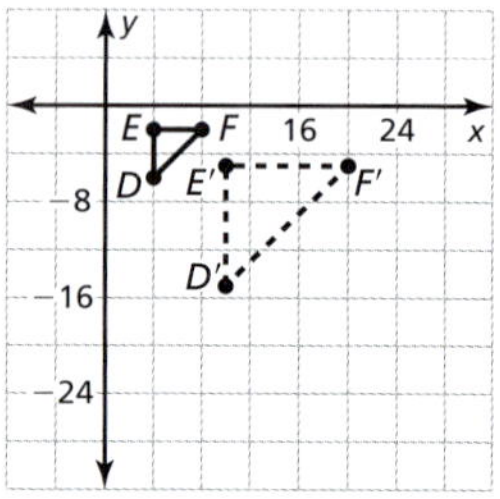

42. 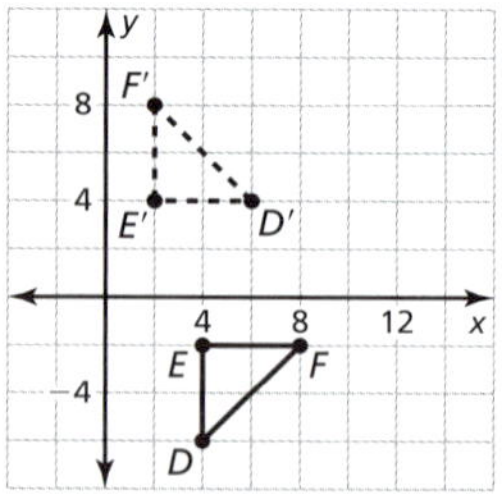

43. $y = 4x - 5$

44.

Equation	**Justification**
$9x - 2 = 13 + 6x$	Write given equation.
$3x - 2 = 13$	Subtraction Property of Equality
$3x = 15$	Addition Property of Equality
$x = 5$	Division Property of Equality

45. a. [0, 1, 2, ..., 10, 11, 12]; discrete; There are only whole numbers of clues.

b. 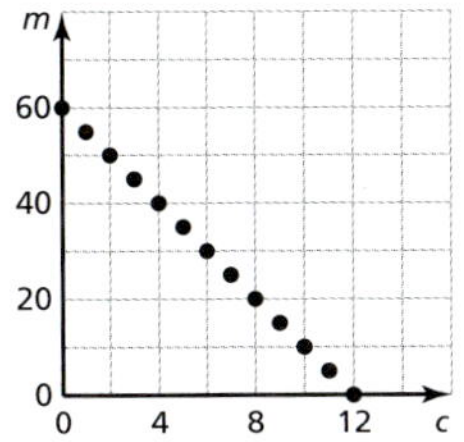

46. $m > 5$

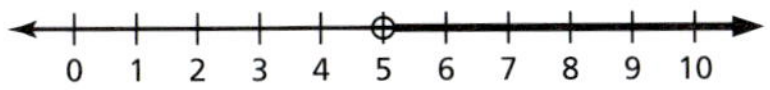

47. $1 < z < 6$

48. yes; *Sample answer:* Quadrilateral *JKLM* can be mapped to quadrilateral *WXYZ* by a reflection in the line $x = 1$, followed by a translation 4 units up.

49. $2x^2 + 3$

50. $3x^2 - 7x - 22$

51. $x^3 + x^2 - 11x + 10$

52. $x^3 - 2x + 3$

53. The graph of g is a vertical translation 3 units up of the graph of f.

54. $(x - 8)(x - 1)$

55. $4y(x + 3)(x + 5)$

Chapter 4 Review

1. 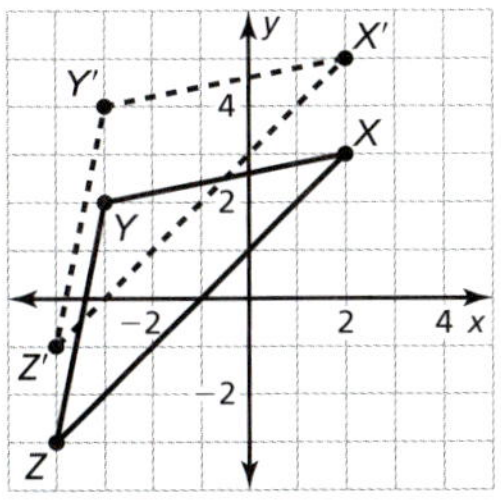

2. 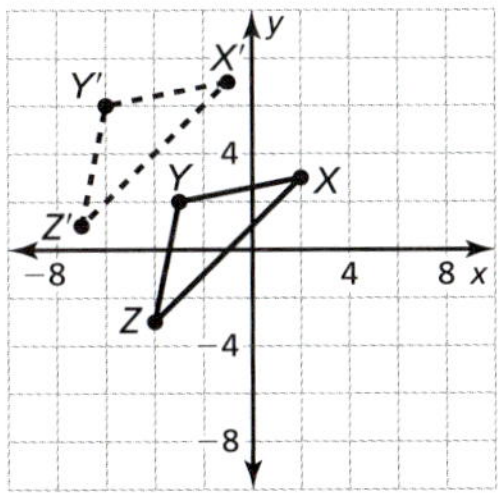

3. 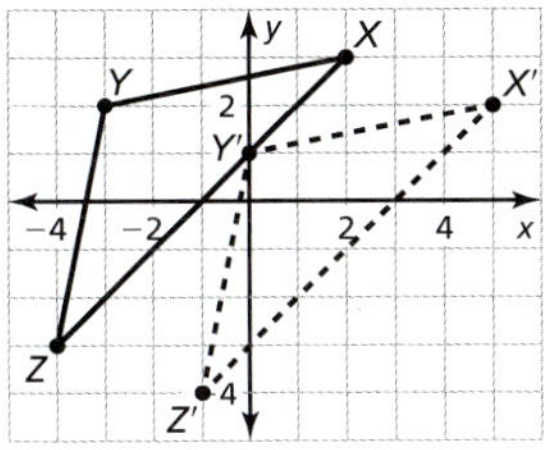

4. 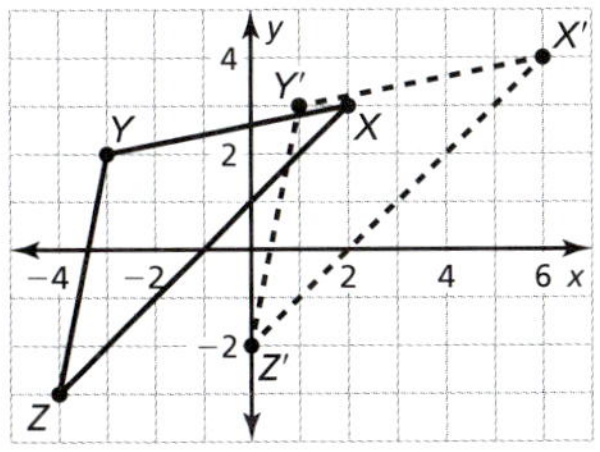

5. 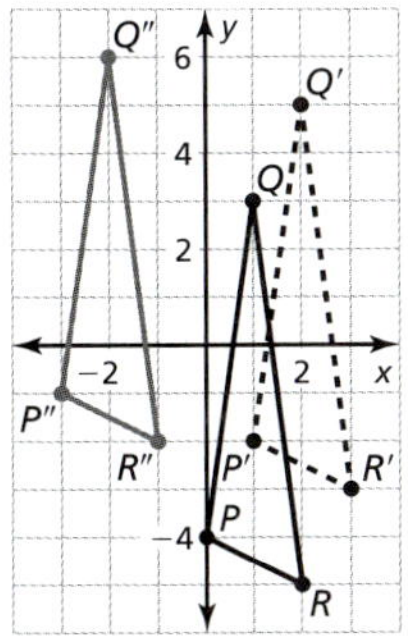

6. 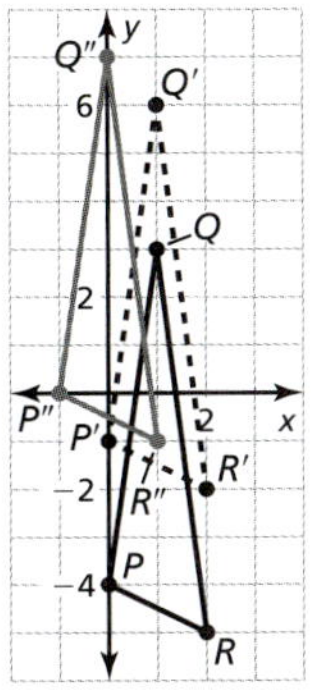

7. $(x, y) \rightarrow (x + 1, y + 3)$

8. 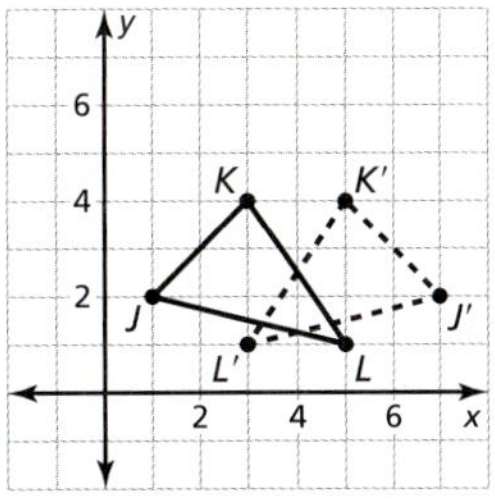

9. $(x, y) \rightarrow (-x, y)$

10. 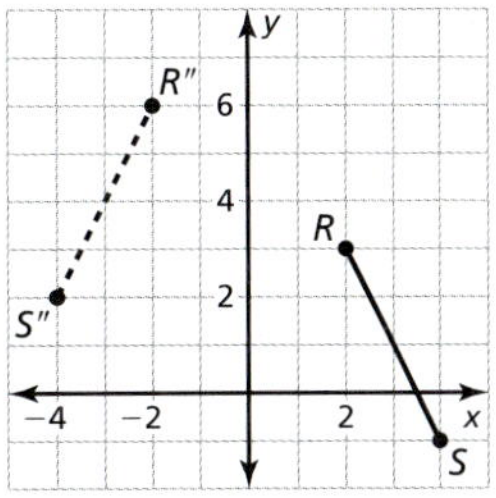

11. 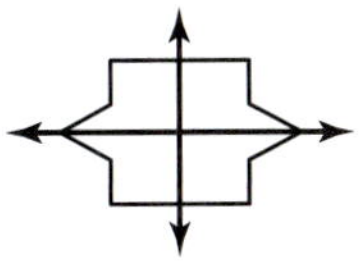

To map the figure onto itself, you can reflect it across the horizontal line, across the vertical line, or across the horiztonal and the vertical lines.

12. 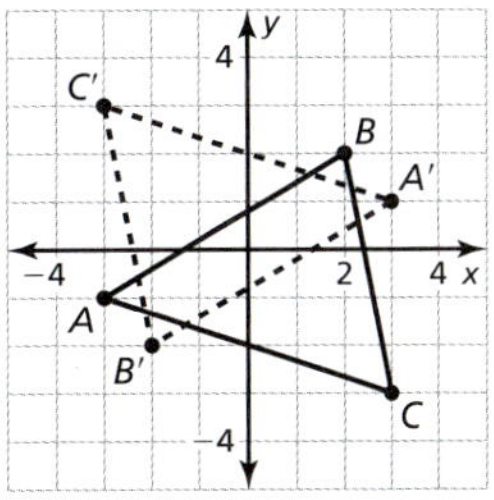

13. 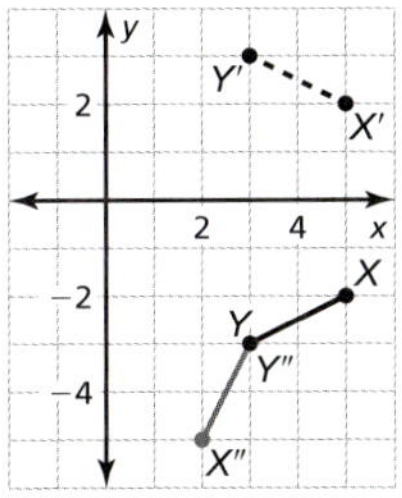

$(x, y) \rightarrow (-y, -x)$

14. yes; Rotations of 60°, 120°, and 180° about the center map the figure onto itself; the order of rotational symmetry is 6

15. yes; Rotations of 72° and 144° about the center map the figure onto itself; the order of rotational symmetry is 5

16. *Sample answer:* orange: rotate 90° around point $(-2, 3)$, translate 4 units right and 5 units down; red: translate 7 units down and 3 units right; purple: rotate 90° counterclockwise around point $(2, 3)$, translate 3 units left and 7 units down

17. no; When a figure can be mapped onto itself by a 270° counterclockwise rotation, it can be mapped onto itself with multiples of a 90° counterclockwise rotation.

18. $\triangle ABC$ can be mapped to $\triangle XYZ$ by a 90° counterclockwise rotation, followed by a reflection in the x-axis. Quadrilateral *MNPQ* can be mapped to quadrilateral *RSTU* by a 270° counterclockwise rotation, followed by a translation 3 units down.

19. yes;

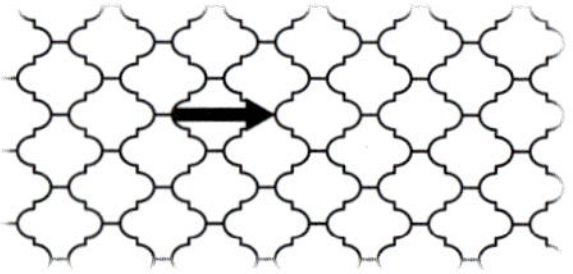

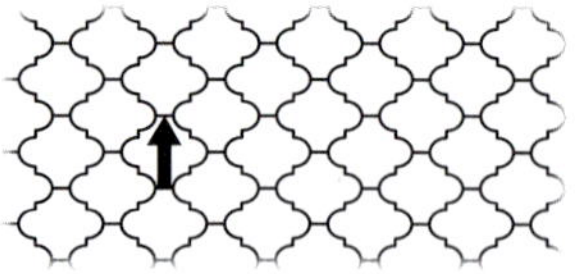

20. **a.** 136 in.

b. reflection in the vertical line through the center of the left marcher, followed by reflection in the vertical line through the center of the right marcher

21. $k = \frac{3}{5}$; reduction

22.

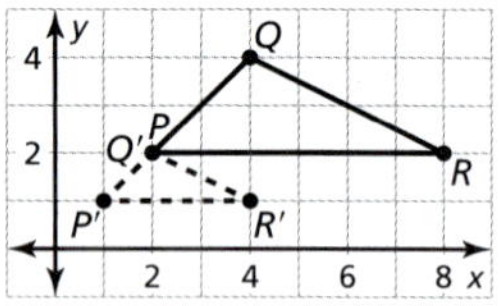

23.

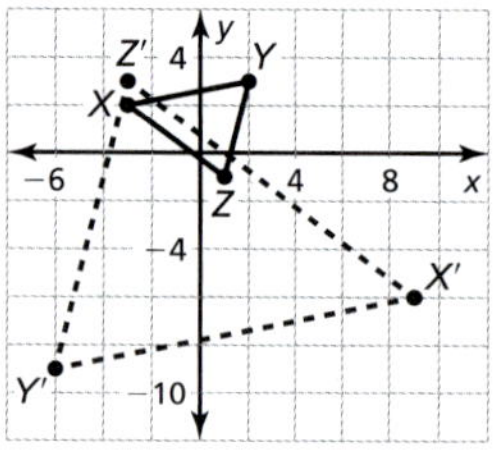

24. $(x, y) \rightarrow (-2x, -2y)$

25. 1.9 cm

26.

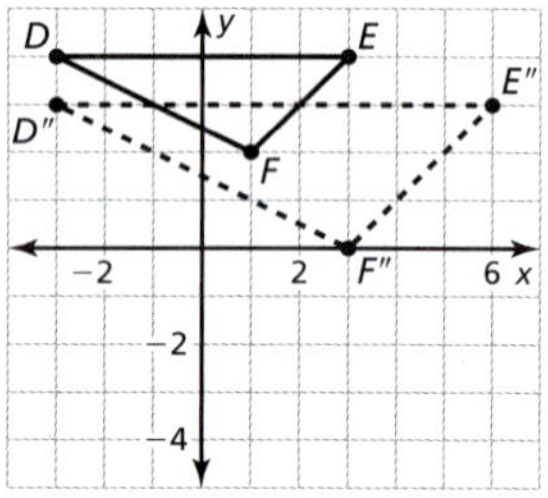

27. *Sample answer:* reflection in the y-axis, followed by a dilation with $k = 3$

28. *Sample answer:* dilation with $k = \frac{1}{2}$, followed by a reflection in the line $y = x$

29. *Sample answer:* 270° counterclockwise rotation about the origin, followed by a dilation with center at the origin and $k = 2$

30. $(x, y) \rightarrow (2y, -2x)$

Chapter 4 Practice Test

1.

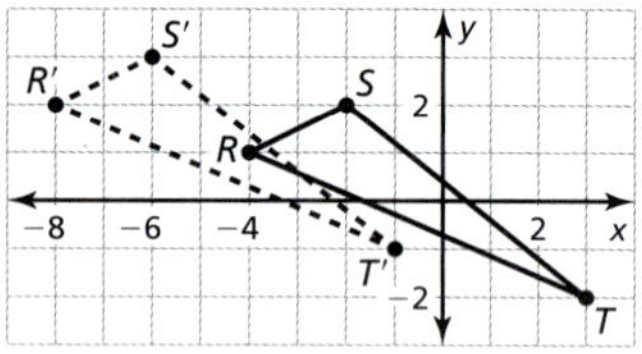

2.

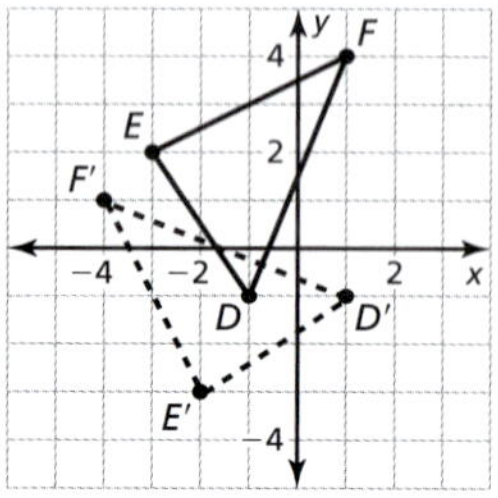

3.

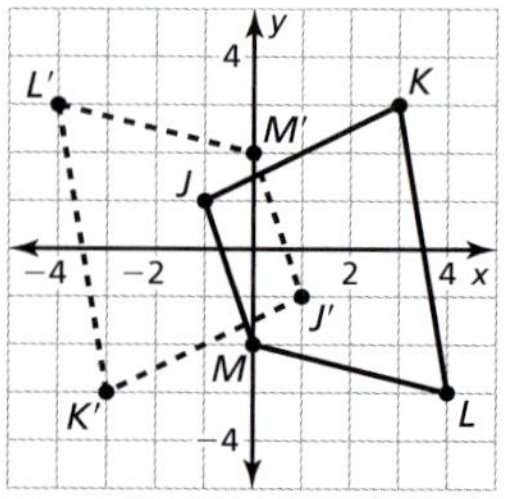

4. similar; Quadrilateral $QRST$ can be mapped to quadrilateral $WXYZ$ by a dilation with the center at the origin and $k = 3$, followed by a reflection in the x-axis. Because this composition has a rigid motion and a dilation, it is a similarity transformation.

5. congruent; $\triangle ABC$ can be mapped to $\triangle DEF$ by a 270° counterclockwise rotation about the origin, followed by a translation 1 unit up and 3 units right. Because this is a composition of two rigid motions, the composition is rigid.

6. yes

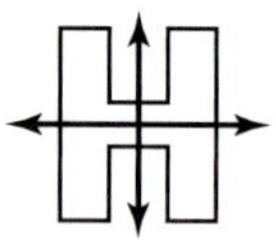

A reflection in the line of symmetry maps the letter onto itself; yes; A rotation of 180° maps the letter onto itself; the order of rotational symmetry is 2

7. yes

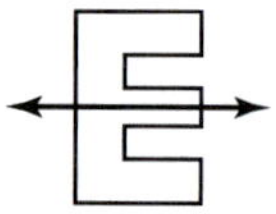

A reflection in the line of symmetry maps the letter onto itself; no rotational symmetry

8. no lines of symmetry; yes; A rotation of 180° maps the letter onto itself; the order of rotational symmetry is 2

9. $k = \frac{1}{3}$

10. *Sample answer:* reflection in the x-axis, followed by the translation $(x, y) \rightarrow (x + 1, y + 2)$; yes; Both reflections and translations are rigid motions. So, according to the Composition Theorem, this composition is a congruence transformation.

11. **a.** *Sample answer:* 270° counterclockwise rotation about the origin, followed by a dilation with $k = \frac{1}{2}$, followed by the translation $(x, y) \rightarrow (x + 2, y - 2)$

b. *Sample answer:* $k = \frac{3}{4}$; A medium slice would be between a small and a large, and $\frac{1}{2} < \frac{3}{4} < 1$.

c. $(x, y) \rightarrow \left(\frac{1}{2}y + 2, -\frac{1}{2}x - 2\right)$

d. no; The scale factor for the shorter sides is $\frac{17}{8}$, but the scale factor for the longer sides is $\frac{11}{6}$. So, the photo would have to be cropped or distorted in order to fit the frame.

12. **a.** reflection, reduction (dilation), and translation

b. 2 in. by 3 in.

c. no; The scale factor for the shorter sides is $\frac{17}{8}$, but the scale factor for the longer sides is $\frac{11}{6}$. So, the photo would have to be cropped or distorted in order to fit the frame.

Chapter 4 Test Prep

1. A, B

2. Step 1. Place the compass at P. Draw an arc that intersects line m in two different places. Label the points of intersection A and B.

Step 2. With the compass at A, draw an arc below line m using a setting greater than $\frac{1}{2}AB$. Using the same compass setting, draw an arc from B that intersects the previous arc. Label the intersection Q.

Step 3. Use a straightedge to draw $\overline{PQ}$.

3. yes; Your friend could find the side lengths and the bottom length by counting units, and then find the angled lengths using the Pythagorean Theorem.

4. C

5. B, C

6. D

7. $\frac{1}{3}, \frac{1}{2}, \frac{3}{4}$

8. **a.** *Sample answer:* $A(2, 0), B(2, 5), C(-2, 5), D(-2, 0)$

b. *Sample answer:* $A(0, 2), B(3, 2), C(3, -2), D(0, -2)$

c. *Sample answer:* $A(2, 0), B(0, 2), C(-2, 0), D(0, -2)$

d. *Sample answer:* $A(2, 2), B(-1, 1), C(-2, -2), D(1, -1)$

9. D

Chapter 5

Prepare with CalcChat

1. $x = -3$

2. $p = 3$

3. $z = -\frac{3}{4}$

4. $M(-2, 4)$; $2\sqrt{13}$, or about 7.2 units

5. $M(6, 2)$; 10 units

6. $M\left(\frac{7}{2}, -1\right)$; $\sqrt{85}$, or about 9.2 units

7. The distance between two points in a coordinate plane can be represented by a right triangle, where the legs of the triangle are the vertical and horizontal components of the distance and the hypotenuse is the actual distance.

Section 5.1

5.1 Practice

1. right isosceles

3. obtuse scalene

5. isosceles; right

7. scalene; not right

9. 71°; acute

11. 52°; right

13. 139°

15. 114°

17. 36°, 54°

19. 37°, 53°

21. 15°, 75°

23. 16.5°, 73.5°

25. The sum of the measures of the angles should be 180°;

$$115° + 39° + m\angle 1 = 180°$$
$$154° + m\angle 1 = 180°$$
$$m\angle 1 = 26°$$

27. 50°

29. 50°

31. 40°

33. 90°

35. *Sample answer:*

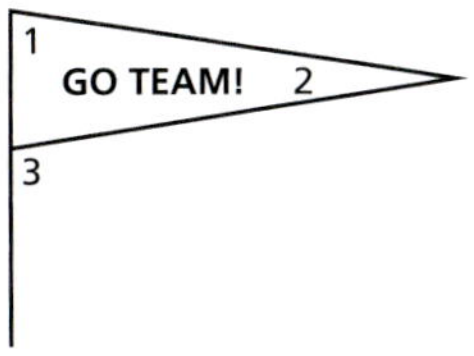

When a triangular pennant is on a stick, the sum of the angle measures of the two angles on the top edge of the pennant ($\angle 1$ and $\angle 2$) is equal to the measure of the angle formed by the stick and the bottom edge of the pennant ($\angle 3$).

37. scalene; right

39. You could make another bend 6 inches from the first bend and leave the last side 8 inches long, or you could make another bend 7 inches from the first bend and then the last side will also be 7 inches long.

41.

STATEMENTS	REASONS
1. $\triangle ABC$ is a right triangle.	1. Given
2. $\angle C$ is a right angle.	2. Given (marked in diagram)
3. $m\angle C = 90°$	3. Definition of a right angle
4. $m\angle A + m\angle B + m\angle C = 180°$	4. Triangle Sum Theorem
5. $m\angle A + m\angle B + 90° = 180°$	5. Substitution Property of Equality
6. $m\angle A + m\angle B = 90°$	6. Subtraction Property of Equality
7. $\angle A$ and $\angle B$ are complementary.	7. Definition of complementary angles

43. yes; no

An obtuse equilateral triangle is not possible, because when two sides form an obtuse angle the third side that connects them must be longer than the other two.

45. **a.** $x = 8, x = 9$

b. one ($x = 4$)

47. $x = 85, y = 65$

49. no; The exterior angle paired with the obtuse angle of an obtuse triangle is less than that interior angle.

5.1 Review & Refresh

51. yes; $\triangle QRS$ is a translation of $\triangle TUV$ 2 units right and 1 unit down.

52. 7.5°; 82.5°

53. $b = -2$ and $b = 7$

54. $x = -2$

55. $k = \frac{2}{5}$; reduction

56. exponential decay; 50%

57. *Sample answer:* $\triangle TUV$ is a dilation with a scale factor of $k = \frac{3}{2}$, followed by a translation 2.5 units right and 4.5 units up of $\triangle DEF$.

58. $\overleftrightarrow{MS}$ and $\overleftrightarrow{LT}$

59. $\overleftrightarrow{NP}$ and $\overleftrightarrow{QR}$

Section 5.2

5.2 Practice

1. *Sample answer:* $\triangle BCA \cong \triangle EFD$; corresponding angles: $\angle A \cong \angle D, \angle B \cong \angle E, \angle C \cong \angle F$; corresponding sides: $\overline{AB} \cong \overline{DE}, \overline{BC} \cong \overline{EF}, \overline{AC} \cong \overline{DF}$

3. 124°

5. 23°

7. $x = 7, y = 8$

9. From the diagram, $\overline{WX} \cong \overline{LM}, \overline{XY} \cong \overline{MN}, \overline{YZ} \cong \overline{NJ}, \overline{VZ} \cong \overline{KJ}$, and $\overline{WV} \cong \overline{LK}$. Also from the diagram, $\angle V \cong \angle K, \angle W \cong \angle L, \angle X \cong \angle M, \angle Y \cong \angle N$, and $\angle Z \cong \angle J$. Because all corresponding parts are congruent, $VWXYZ \cong KLMNJ$.

11. Transitive Property of Triangle Congruence

13. 20°

15.

STATEMENTS	REASONS
1. $\overline{AB} \parallel \overline{DC}$, $\overline{AB} \cong \overline{DC}$, E is the midpoint of $\overline{AC}$ and $\overline{BD}$.	1. Given
2. $\angle AEB \cong \angle CED$	2. Vertical Angles Congruence Theorem
3. $\angle BAE \cong \angle DCE$, $\angle ABE \cong \angle CDE$	3. Alternate Interior Angles Theorem
4. $\overline{AE} \cong \overline{CE}$, $\overline{BE} \cong \overline{DE}$	4. Definition of midpoint
5. $\triangle AEB \cong \triangle CED$	5. All corresponding parts are congruent.

17. From the congruence statement, $\angle R \cong \angle Z$ because they are corresponding angles in congruent triangles. This means means $m\angle R = m\angle Z$. Therefore, $m\angle R = 42°$.

19. Triangles 1 and 4 appear congruent; Triangles 1, 3, and 4 appear similar; *Sample answer:* Triangle 1 appears to be a rotation and a translation of Triangle 4; Triangle 3 appears to be a rotation, a translation, and a dilation of Triangle 1 or Triangle 4: Use a protractor and straightedge to determine the angles and sides lengths.

21.

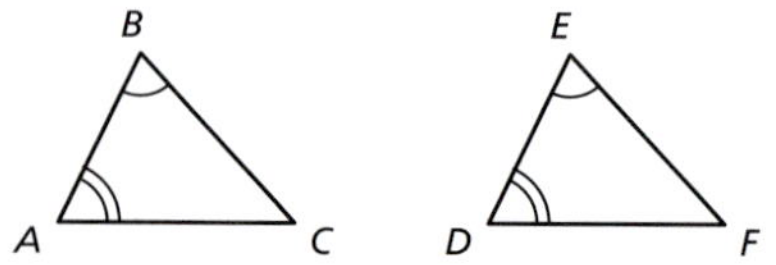

STATEMENTS	REASONS
1. $\angle A \cong \angle D, \angle B \cong \angle E$	1. Given
2. $m\angle A = m\angle D$, $m\angle B = m\angle E$	2. Definition of congruent angles
3. $m\angle A + m\angle B + m\angle C = 180°$, $m\angle D + m\angle E + m\angle F = 180°$	3. Triangle Sum Theorem
4. $m\angle A + m\angle B + m\angle C = m\angle D + m\angle E + m\angle F$	4. Transitive Property of Equality
5. $m\angle A + m\angle B + m\angle C = m\angle A + m\angle B + m\angle F$	5. Substitution Property of Equality
6. $m\angle C = m\angle F$	6. Subtraction Property of Equality
7. $\angle C \cong \angle F$	7. Definition of congruent angles

23. $x = 3, y = 11$

25. A rigid motion maps each part of a figure to a corresponding part of its image. Because rigid motions preserve length and angle measure, corresponding parts of congruent figures are congruent, which means that the corresponding sides and corresponding angles are congruent.

5.2 Review & Refresh

26. 139°

27.

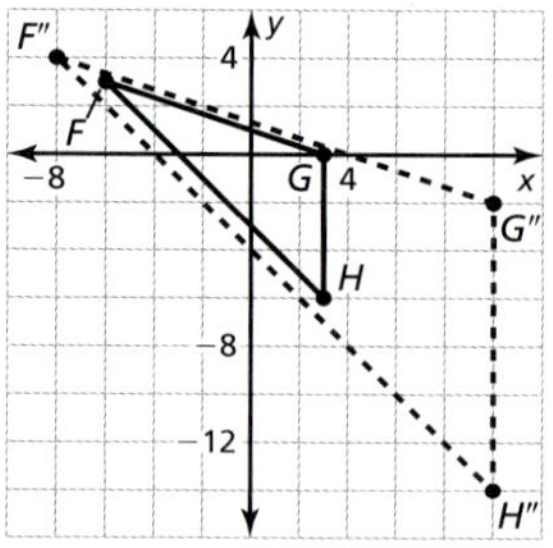

28.

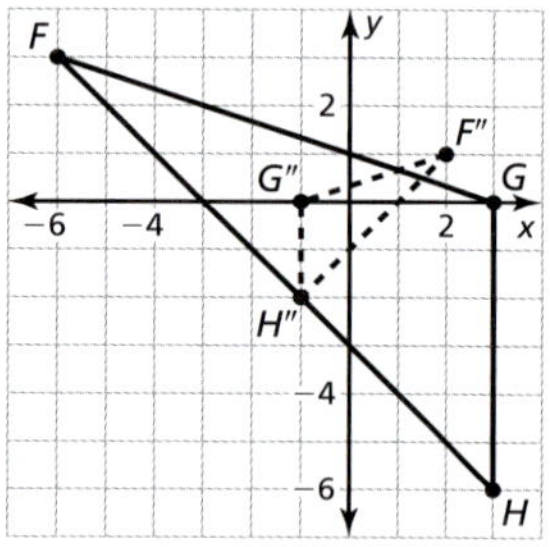

29. *Sample answer:* $RSTQ \cong XYZW$; corresponding sides: $\overline{QR} \cong \overline{WX}, \overline{RS} \cong \overline{XY}, \overline{ST} \cong \overline{YZ}, \overline{QT} \cong \overline{WZ}$; corresponding angles: $\angle Q \cong \angle W, \angle R \cong \angle X, \angle S \cong \angle Y, \angle T \cong \angle Z$

30. $k = \frac{1}{2}$

31. $(t + 5)(t + 2)$

32. $(2x - 3)(x + 4)$

33. 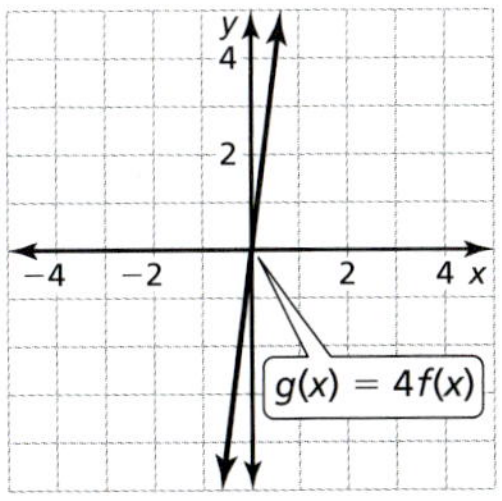

The graph of g is a vertical stretch by a factor of 4.

34. 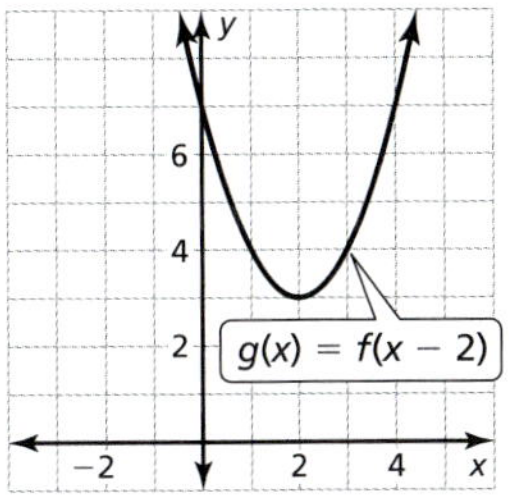

The graph of g is a horizontal translation 2 units right of the graph of f.

35. linear; It can be rewritten in the form $y = mx + b$.

36. nonlinear; It cannot be rewritten in the form $y = mx + b$.

Section 5.3

5.3 Practice

1. no; The congruent angles are not the included angles.

3. yes; Two pairs of sides and the included angles are congruent.

5.

STATEMENTS	REASONS
1. C is the midpoint of $\overline{AE}$ and $\overline{BD}$.	1. Given
2. $\angle ACB \cong \angle ECD$	2. Vertical Angles Congruence Theorem
3. $\overline{AC} \cong \overline{EC}$, $\overline{BC} \cong \overline{DC}$	3. Definition of midpoint
4. $\triangle ABC \cong \triangle EDC$	4. SAS Congruence Theorem

7.

STATEMENTS	REASONS
1. $\overline{SP} \cong \overline{TP}$, $\overrightarrow{PQ}$ bisects $\angle SPT$.	1. Given
2. $\overline{PQ} \cong \overline{PQ}$	2. Reflexive Property of Congruence
3. $\angle SPQ \cong \angle TPQ$	3. Definition of angle bisector
4. $\triangle SPQ \cong \triangle TPQ$	4. SAS Congruence Theorem

9. $\triangle BAD \cong \triangle DCB$; Because the sides of the square are congruent, $\overline{BA} \cong \overline{DC}$ and $\overline{AD} \cong \overline{CB}$. Also, because the angles of the square are congruent, $\angle A \cong \angle C$. So, $\triangle BAD$ and $\triangle DCB$ are congruent by the SAS Congruence Theorem.

11. $\triangle SRT \cong \triangle URT$; $\overline{RT} \cong \overline{RT}$ by the Reflexive Property of Congruence. Also, because all points on a circle are the same distance from the center, $\overline{RS} \cong \overline{RU}$. It is given that $\angle SRT \cong \angle URT$. So, $\triangle SRT$ and $\triangle URT$ are congruent by the SAS Congruence Theorem.

13. 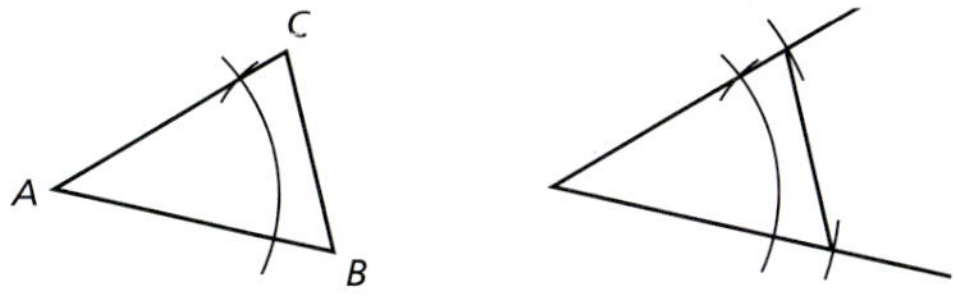

15.

STATEMENTS	REASONS
1. $\triangle AHE$ is equilateral; $\overline{HV}$ bisects $\angle AHE$.	1. Given
2. $\overline{AH} \cong \overline{EH}$	2. Definition of Equilateral Triangle
3. $\overline{VH} \cong \overline{VH}$	3. Reflexive Property of Segment Congruence
4. $\angle AHV \cong \angle EHV$	4. Definition of Angle Bisector
5. $\triangle AHV \cong \triangle EHV$	5. SAS Congruence Theorem

17. no; One of the congruent angles is not the included angle.

19.

STATEMENTS	REASONS
1. $\overline{AC} \cong \overline{DC}$, $\overline{BC} \cong \overline{EC}$	1. Given
2. $\angle ACB \cong \angle DCE$	2. Vertical Angles Congruence Theorem
3. $\triangle ABC \cong \triangle DEC$	3. SAS Congruence Theorem

$x = 4, y = 5$

21. no; The angles between the legs may be different.

5.3 Review & Refresh

23. obtuse isosceles

24. equiangular equilateral

25. 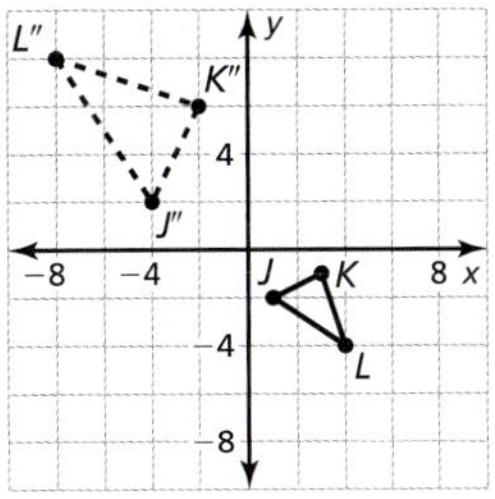

26. no; The congruent angles are not the included angles.

27. $y < -9$

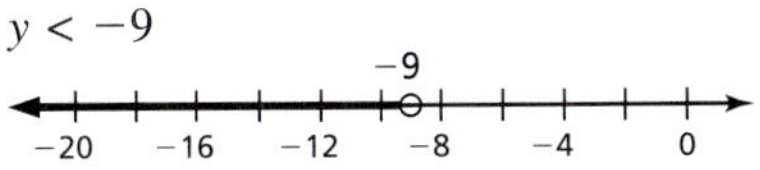

28. $w \geq -24$

29. $d \leq 9$

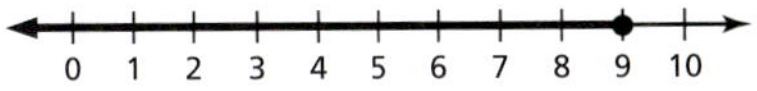

30. $b < -13$ *or* $b > -5$

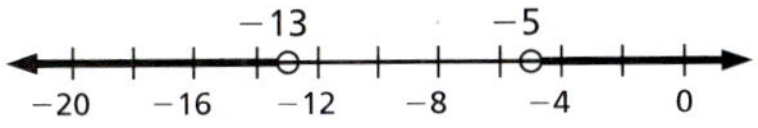

31. $x = 5, y = 13$

SELECTED ANSWERS

32.

		Kickball		
		Like	Don't Like	TOTAL
Frisbee	Like	47	16	63
	Don't Like	23	14	37
	TOTAL	70	30	100

33. $g(-4) = 16$; $g(0) = 14$; $g(8) = 10$

Section 5.4

5.4 Practice

1. *A, D;* Base Angles Theorem

3. $\overline{CD}, \overline{CE}$; Converse of the Base Angles Theorem

5. $x = 12$

7. $x = 52$

9.

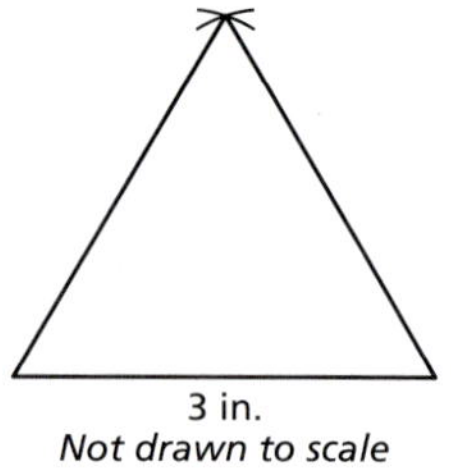

11. $x = 60, y = 60$

13. $x = 30, y = 5$

15. $x = 79, y = 22$

17. When two angles of a triangle are congruent, the sides opposite the angles are congruent; Because $\angle A \cong \angle C$, $\overline{AB} \cong \overline{BC}$. So, $BC = 5$.

19. a. Because $\triangle ABD$ and $\triangle CBD$ are congruent and equilateral, you know that $\overline{AB} \cong \overline{CB}$. So, $\triangle ABC$ is isosceles.

b. Because $\triangle ABC$ is isosceles, $\angle BAE \cong \angle BCE$ by the Base Angles Theorem.

c. By the Reflexive Property of Congruence, $\overline{BE} \cong \overline{BE}$. Because $\triangle ABD$ and $\triangle CBD$ are congruent and equilateral, and also equiangular by the Corollary to the Base Angles Theorem, you can conclude that $\angle ABE \cong \angle CBE$. Also $\overline{AB} = \overline{CB}$ as explained in part (a). So, by the SAS Congruence Theorem, $\triangle ABE \cong \triangle CBE$.

d. 30°

21. 17 in.

23. 6, 8, 10; If $3t = 5t - 12$, then $t = 6$. If $5t - 12 = t + 20$, then $t = 8$. If $3t = t + 20$, then $t = 10$.

25. If the base angles are $x°$, then the vertex angle is $(180 - 2x)°$, or $[2(90 - x)]°$. Because $2(90 - x)$ is divisible by 2, the vertex angle is even when the angles are whole numbers.

27. a. $\angle XVY$, $\angle UXV$; $\angle WUX \cong \angle XVY$ because they are both vertex angles of congruent isosceles triangles. Also, $m\angle UXV + m\angle VXY = m\angle UXY$ by the Angle Addition Postulate, and $m\angle UXY = m\angle WUX + m\angle UWX$ by the Exterior Angle Theorem. So, by the Transitive Property of Equality,
$m\angle UXV + m\angle VXY = m\angle WUX + m\angle UWX$.
Also, $m\angle UWX = m\angle VXY$ because they are base angles of congruent isosceles triangles. By substituting $m\angle UWX$ for $m\angle VXY$, you get
$m\angle UXV + m\angle UWX = m\angle WUX + m\angle UWX$. By the Subtraction Property of Equality, $m\angle UXV = m\angle WUX$, so $\angle UXV \cong \angle WUX$.

b. 8 m

29.

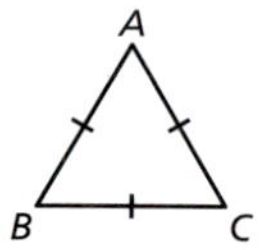

STATEMENTS	REASONS
1. $\triangle ABC$ is equilateral.	**1.** Given
2. $\overline{AB} \cong \overline{AC}$, $\overline{AB} \cong \overline{BC}$, $\overline{AC} \cong \overline{BC}$	**2.** Definition of equilateral triangle
3. $\angle B \cong \angle C$, $\angle A \cong \angle C$, $\angle A \cong \angle B$	**3.** Base Angles Theorem
4. $\triangle ABC$ is equiangular.	**4.** Definition of equiangular triangle

31.

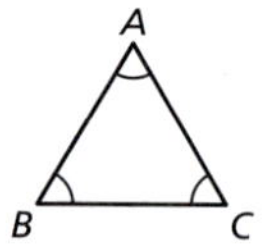

STATEMENTS	REASONS
1. $\triangle ABC$ is equiangular.	**1.** Given
2. $\angle B \cong \angle C$, $\angle A \cong \angle C$, $\angle A \cong \angle B$	**2.** Definition of equiangular triangle
3. $\overline{AB} \cong \overline{AC}$, $\overline{AB} \cong \overline{BC}$, $\overline{AC} \cong \overline{BC}$	**3.** Converse of the Base Angles Theorem
4. $\triangle ABC$ is equilateral.	**4.** Definition of equilateral triangle

33. no; *T*, *U*, and *V* will always be the vertices of an isosceles triangle except when *V* is collinear with *T* and *U*, which happens when the coordinates of *V* are (3, 3).

35.

STATEMENTS	REASONS
1. $\triangle ABC$ is equilateral, $\angle CAD \cong \angle ABE \cong \angle BCF$	**1.** Given
2. $\triangle ABC$ is equiangular.	**2.** Corollary to the Base Angles Theorem
3. $\angle ABC \cong \angle BCA \cong \angle BAC$	**3.** Definition of equiangular triangle
4. $m\angle CAD = m\angle ABE = m\angle BCF$, $m\angle ABC = m\angle BCA = m\angle BAC$	**4.** Definition of congruent angles
5. $m\angle ABC = m\angle ABE + m\angle EBC$, $m\angle BCA = m\angle BCF + m\angle ACF$, $m\angle BAC = m\angle CAD + m\angle BAD$	**5.** Angle Addition Postulate
6. $m\angle ABE + m\angle EBC = m\angle BCF + m\angle ACF = m\angle CAD + m\angle BAD$	**6.** Substitution Property of Equality
7. $m\angle ABE + m\angle EBC = m\angle ABE + m\angle ACF = m\angle ABE + m\angle BAD$	**7.** Substitution Property of Equality
8. $m\angle EBC = m\angle ACF = m\angle BAD$	**8.** Subtraction Property of Equality
9. $\angle EBC \cong \angle ACF \cong \angle BAD$	**9.** Definition of congruent angles
10. $\angle FEB \cong \angle DFC \cong \angle EDA$	**10.** Third Angles Theorem
11. $\angle FEB$ and $\angle FED$ are supplementary, $\angle DFC$ and $\angle EFD$ are supplementary, and $\angle EDA$ and $\angle FDE$ are supplementary.	**11.** Linear Pair Postulate
12. $\angle FED \cong \angle EFD \cong \angle FDE$	**12.** Congruent Supplements Theorem
13. $\triangle DEF$ is equiangular.	**13.** Definition of equiangular triangle
14. $\triangle DEF$ is equilateral.	**14.** Corollary to the Converse of the Base Angles Theorem

5.4 Review & Refresh

36. $\overline{SE}$

37. $\overline{JK}$; $\overline{RS}$

38. $\overline{EF}$; $\overline{UV}$

39. yes; The triangles are congruent by the SAS Congruence Theorem.

40. $m\angle 1 = 53°$

41.

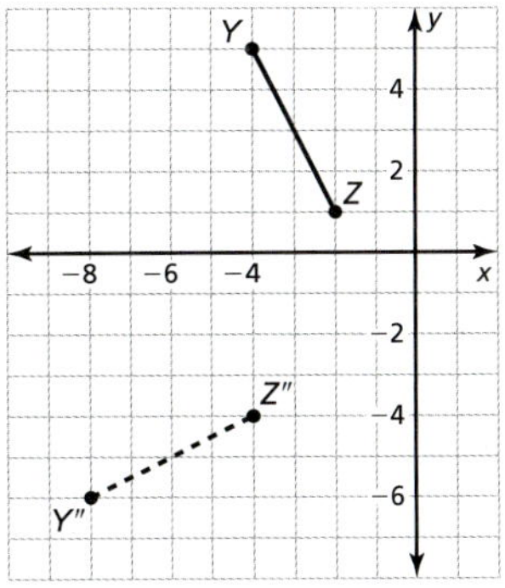

42.

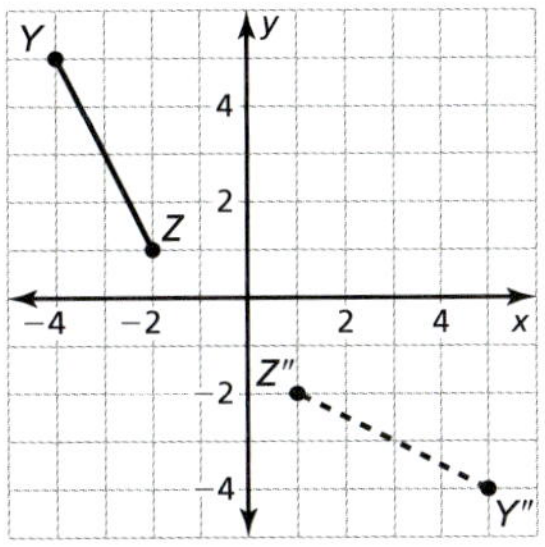

43. $m\angle L = 55°$; $JK = 4$ in.

44. $2\sqrt{5}$, or about 4.5 units

45. $x = 5$, $y = 20$

46. mean = 15, median = 14, mode = 13, range = 8, interquartile range = 4.5

Section 5.5

5.5 Practice

1. yes; $\overline{AB} \cong \overline{DB}$, $\overline{BC} \cong \overline{BE}$, $\overline{AC} \cong \overline{DE}$

3. yes; $\angle B$ and $\angle E$ are right angles, $\overline{AB} \cong \overline{FE}$, $\overline{AC} \cong \overline{FD}$

5. no; You are given that $\overline{RS} \cong \overline{PQ}$, $\overline{ST} \cong \overline{QT}$, and $\overline{RT} \cong \overline{PT}$. So, it should say $\triangle RST \cong \triangle PQT$ by the SSS Congruence Theorem.

7.

STATEMENTS	REASONS
1. $\overline{LM} \cong \overline{JK}$, $\overline{MJ} \cong \overline{KL}$	**1.** Given
2. $\overline{JL} \cong \overline{JL}$	**2.** Reflexive Property of Congruence
3. $\triangle LMJ \cong \triangle JKL$	**3.** SSS Congruence Theorem

9. yes; The diagonal supports in this figure form triangles with fixed side lengths. By the SSS Congruence Theorem, these triangles cannot change shape, so the figure is stable.

11. 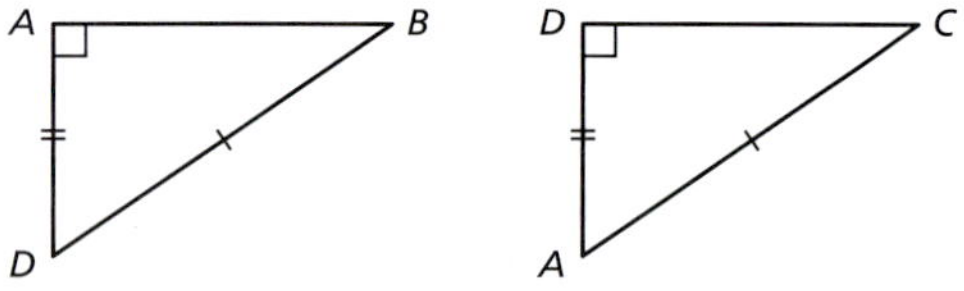

STATEMENTS	REASONS
1. $\overline{AC} \cong \overline{DB}$, $\overline{AB} \perp \overline{AD}$, $\overline{CD} \perp \overline{AD}$	1. Given
2. $\overline{AD} \cong \overline{AD}$	2. Reflexive Property of Congruence
3. $\angle BAD$ and $\angle CDA$ are right angles.	3. Definition of perpendicular lines
4. $\triangle BAD$ and $\triangle CDA$ are right triangles.	4. Definition of a right triangle
5. $\triangle BAD \cong \triangle CDA$	5. HL Congruence Theorem

13. The order of the points in the congruence statement should reflect the corresponding sides and angles; $\triangle TUV \cong \triangle ZYX$ by the SSS Congruence Theorem.

15. 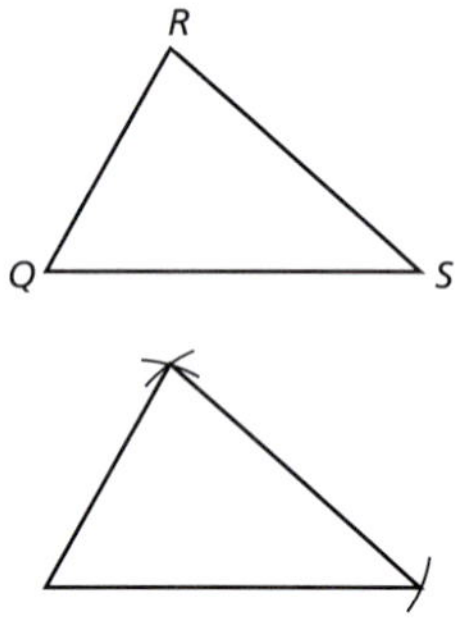

17.

STATEMENTS	REASONS
1. $\overline{HF} \cong \overline{FS} \cong \overline{ST} \cong \overline{TH}$; $\overline{FT} \cong \overline{SH}$; $\angle H$, $\angle F$, $\angle S$, and $\angle T$ are right angles.	1. Given
2. $\overline{SH} \cong \overline{SH}$	2. Reflexive Property of Congruence
3. $\triangle HFS$, $\triangle FST$, and $\triangle STH$ are right triangles.	3. Definition of a right triangle
4. $\triangle HFS \cong \triangle FST \cong \triangle STH$	4. HL Congruence Theorem

19. a. You need to know that the hypotenuses are congruent: $\overline{JL} \cong \overline{ML}$.
    b. SAS Congruence Theorem; By definition of midpoint, $\overline{JK} \cong \overline{MK}$. Also, $\overline{LK} \cong \overline{LK}$, by the Reflexive Property of Congruence, and $\angle JKL \cong \angle MKL$ by the Right Angles Congruence Theorem.

21. congruent

23. Use the string to compare the lengths of the corresponding sides of the two triangles to determine whether SSS Congruence Theorem applies.

25. Use the Pythagorean Theorem to show that $\overline{QP} \cong \overline{RS}$. Then use the *HL* Congruence Theorem to show that $\triangle PTQ \cong \triangle STR$ because a leg and hypotenuse of $\triangle PTQ$ are congruent to a leg and the hypotenuse of $\triangle STR$.

27. *Sample answer:* 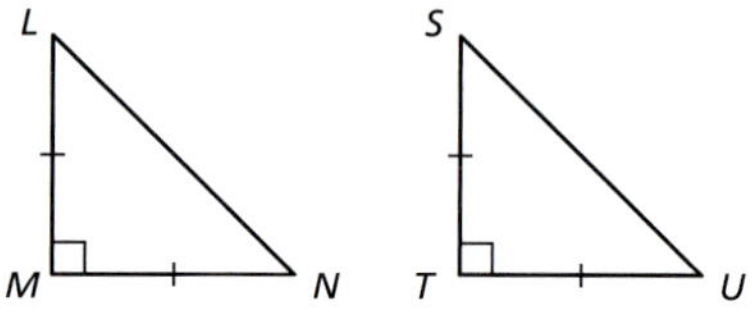

It is given that $\overline{LM} \cong \overline{ST}$ and $\overline{NM} \cong \overline{UT}$. By the definition of perpendicular lines, $m\angle LMN = m\angle STU = 90°$. $\angle LMN \cong \angle STU$ by the Right Angles Congruence Theorem. So, $\triangle LMN \cong \triangle STU$ by the SAS Congruence Theorem.

29. $x = 5$; If $5x = 4x + 3$ and $5x - 2 = 3x + 10$, then $x = 3$ and $x = 6$, which is not possible. If $5x = 3x + 10$ and $5x - 2 = 4x + 3$, then $x = 5$ in both equations. So, when $x = 5$, $AB = CD = 25$ and $AC = BD = 23$, which means that $\overline{AB} \cong \overline{CD}$ and $\overline{AC} \cong \overline{BD}$. Because $\overline{BC} \cong \overline{BC}$, the triangles are congruent by the SSS Congruence Theorem.

5.5 Review & Refresh

31. yes; $m\angle ABE = 139°$ because vertical angles are congruent. $m\angle ABE + m\angle BED = 139° + 41° = 180°$, so $\overleftrightarrow{AC} \parallel \overleftrightarrow{DF}$ by the Consecutive Interior Angles Converse.

32. $m\angle 1 = 35°$

33. $f(x) = 6x - 7$

34. $x = 82$, $y = 16$

35.

STATEMENTS	REASONS
1. $\overline{AE} \cong \overline{DE}$, $\overline{BE} \cong \overline{CE}$	1. Given
2. $\angle AEB \cong \angle DEC$	2. Vertical Angles Congruence Theorem
3. $\triangle AEB \cong \triangle DEC$	3. SAS Congruence Theorem

36. 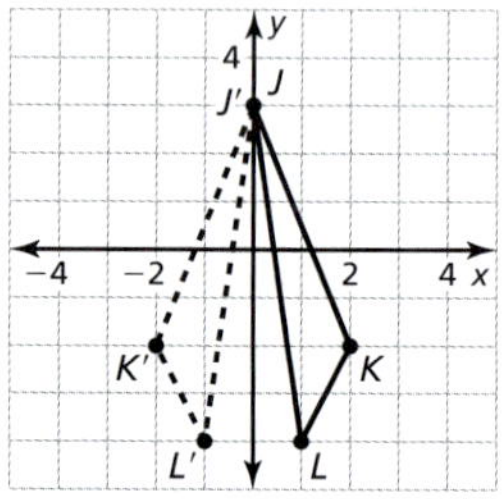

37. 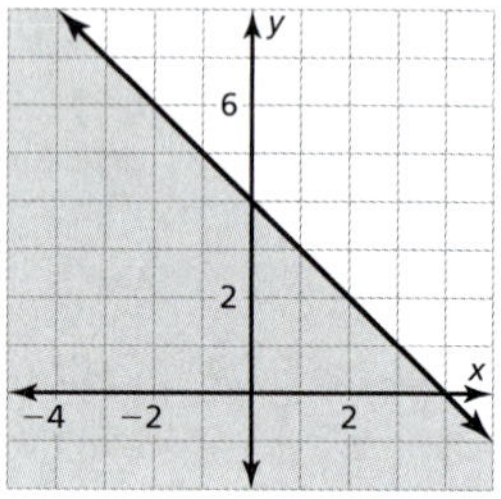

38. no; The corresponding vertices are not written in the same order.

39. quadratic

Section 5.6

5.6 Practice

1. yes; AAS Congruence Theorem

3. no

5. yes; $\triangle ABC \cong \triangle DEF$ by the ASA Congruence Theorem.

7. no; $\overline{AC}$ and $\overline{DE}$ do not correspond.

9. 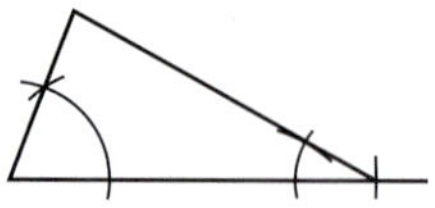

11. The corresponding vertices are not written in the same order; $\triangle JLK \cong \triangle GHF$ by the AAS Congruence Theorem.

13.

STATEMENTS	REASONS
1. M is the midpoint of $\overline{NL}$, $\overline{NL} \perp \overline{NQ}$, $\overline{NL} \perp \overline{MP}$, $\overline{QM} \parallel \overline{PL}$	**1.** Given
2. $\angle QNM$ and $\angle PML$ are right angles.	**2.** Definition of perpendicular lines
3. $\angle QNM \cong \angle PML$	**3.** Right Angles Congruence Theorem
4. $\angle QMN \cong \angle PLM$	**4.** Corresponding Angles Theorem
5. $\overline{NM} \cong \overline{ML}$	**5.** Definition of midpoint
6. $\triangle NQM \cong \triangle MPL$	**6.** ASA Congruence Theorem

15.

STATEMENTS	REASONS
1. $\overline{VW} \cong \overline{UW}$, $\angle X \cong \angle Z$	**1.** Given
2. $\angle W \cong \angle W$	**2.** Reflexive Property of Congruence
3. $\triangle XWV \cong \triangle ZWU$	**3.** AAS Congruence Theorem

17. You are given two right triangles, so the triangles have congruent right angles by the Right Angles Congruence Theorem. Because another pair of angles and a pair of corresponding nonincluded sides (the hypotenuses) are congruent, the triangles are congruent by the AAS Congruence Theorem.

19. You are given two right triangles, so the triangles have congruent right angles by the Right Angles Congruence Theorem. There is also another pair of congruent corresponding angles and a pair of congruent corresponding sides. If the pair of congruent sides is the included side, then the triangles are congruent by the ASA Congruence Theorem. If the pair of congruent sides is a nonincluded pair, then the triangles are congruent by the AAS Congruence Theorem.

21. D

23. yes; When $x = 14$ and $y = 26$, $m\angle ABC = m\angle DBC = m\angle BCA = m\angle BCD = 80°$ and $m\angle CAB = m\angle CDB = 20°$. This satisfies the Triangle Sum Theorem for both triangles. Because $\overline{CB} \cong \overline{CB}$ by the Reflexive Property of Congruence, you can conclude that $\triangle ABC \cong \triangle DBC$ by the ASA Congruence Theorem or the AAS Congruence Theorem.

25. SAS Congruence Theorem: $\overline{EF} \cong \overline{EH}$, $\angle FEG \cong \angle HEG$, and $\overline{EG} \cong \overline{EG}$;
ASA Congruence Theorem: $\angle F \cong \angle H$, $\overline{EH} \cong \overline{EF}$, and $\angle FEG \cong \angle HEG$;
AAS Congruence Theorem: $\angle F \cong \angle H$, $\angle FEG \cong \angle HEG$, and $\overline{EG} \cong \overline{EG}$;
HL Congruence Theorem: $\triangle EFG$ and $\triangle EHG$ are right triangles, $\overline{EG} \cong \overline{EG}$, $\overline{EF} \cong \overline{EH}$;
HA Congruence Theorem: $\triangle EFG$ and $\triangle EHG$ are right triangles, $\overline{EG} \cong \overline{EG}$, and $\angle FEG \cong \angle HEG$;
AL Congruence Theorem: $\triangle EFG$ and $\triangle EHG$ are right triangles, $\angle HEG \cong \angle FEG$, and $\overline{EG} \cong \overline{EG}$

27. yes; By the Triangle Sum Theorem, $m\angle B = 180° - 68° - 59° = 53°$ and $m\angle D = 180° - 53° - 59° = 68°$. So, $\angle A \cong \angle D$ and $\angle B \cong \angle E$ by the definition of congruent angles and $\overline{AB} \cong \overline{DE}$ by the definition of congruent segments. So, $\triangle ABC \cong \triangle DEF$ by the ASA Congruence Theorem.

29. a. $\overline{TU} \cong \overline{XY}, \overline{UV} \cong \overline{YZ}, \overline{TV} \cong \overline{XZ}$;
$\overline{TU} \cong \overline{XY}, \angle U \cong \angle Y, \overline{UV} \cong \overline{YZ}$;
$\overline{UV} \cong \overline{YZ}, \angle V \cong \angle Z, \overline{TV} \cong \overline{XZ}$;
$\overline{TV} \cong \overline{XZ}, \angle T \cong \angle X, \overline{TU} \cong \overline{XY}$;
$\angle T \cong \angle X, \overline{TU} \cong \overline{XY}, \angle U \cong \angle Y$;
$\angle U \cong \angle Y, \overline{UV} \cong \overline{YZ}, \angle V \cong \angle Z$;
$\angle V \cong \angle Z, \overline{TV} \cong \overline{XZ}, \angle T \cong \angle X$;
$\angle T \cong \angle X, \angle U \cong \angle Y, \overline{UV} \cong \overline{YZ}$;
$\angle T \cong \angle X, \angle U \cong \angle Y, \overline{TV} \cong \overline{XZ}$;
$\angle U \cong \angle Y, \angle V \cong \angle Z, \overline{TV} \cong \overline{XZ}$;
$\angle U \cong \angle Y, \angle V \cong \angle Z, \overline{TU} \cong \overline{XY}$;
$\angle V \cong \angle Z, \angle T \cong \angle X, \overline{TU} \cong \overline{XY}$;
$\angle V \cong \angle Z, \angle T \cong \angle X, \overline{UV} \cong \overline{YZ}$

b. $\frac{13}{20}$, or 65%

31. a. no

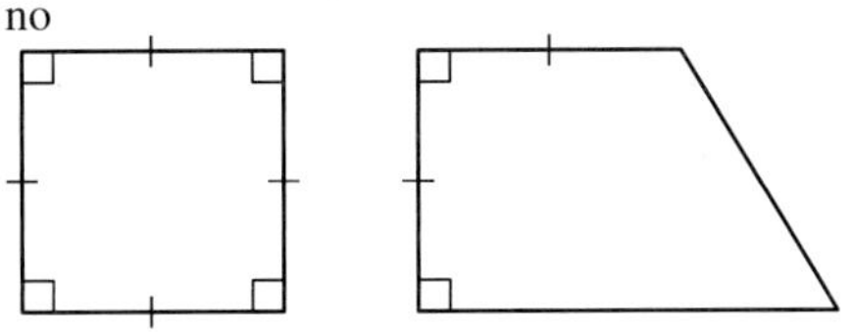

b. yes

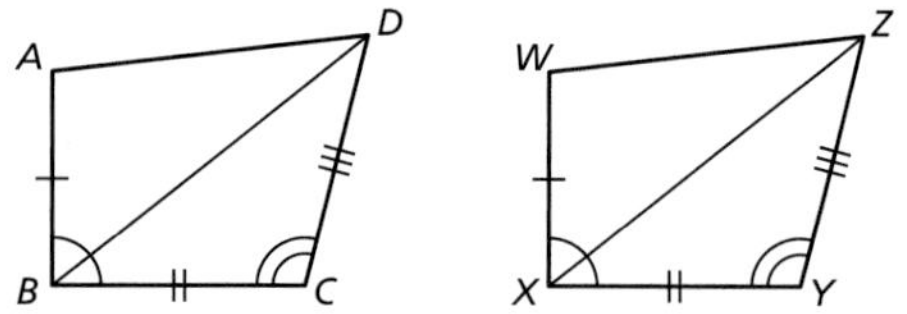

STATEMENTS	REASONS
1. *ABCD* and *WXYZ* with diagonals $\overline{BD}$ and $\overline{XZ}$, $\overline{AB} \cong \overline{WX}$, $\angle ABC \cong \angle WXY$, $\overline{BC} \cong \overline{XY}$, $\angle BCD \cong \angle XYZ$, and $\overline{CD} \cong \overline{YZ}$	**1.** Given
2. $\triangle BCD \cong \triangle XYZ$	**2.** SAS Congruence Theorem
3. $\overline{BD} \cong \overline{XZ}$, $\angle DBC \cong \angle ZXY$, $\angle BDC \cong \angle XZY$	**3.** Corresponding parts of congruent triangles are congruent.
4. $m\angle ABC = m\angle WXY$, $m\angle DBC = m\angle ZXY$	**4.** Definition of congruent angles
5. $m\angle ABC = m\angle ABD + m\angle DBC$, $m\angle WXY = m\angle WXZ + m\angle ZXY$	**5.** Angle Addition Postulate
6. $m\angle ABD + m\angle DBC = m\angle WXZ + m\angle ZXY$	**6.** Transitive Property of Equality
7. $m\angle ABD + m\angle DBC = m\angle WXZ + m\angle DBC$	**7.** Substitution Property of Equality
8. $m\angle ABD = m\angle WXZ$	**8.** Subtraction Property of Equality
9. $\angle ABD \cong \angle WXZ$	**9.** Definition of congruent angles
10. $\triangle ABD \cong \triangle WXZ$	**10.** SAS Congruence Theorem
11. $\overline{AD} \cong \overline{WZ}$, $\angle ADB \cong \angle WZX$	**11.** Corresponding parts of congruent triangles are congruent.
12. $m\angle ADB = m\angle WZX$, $m\angle BDC = m\angle XZY$	**12.** Definition of congruent angles
13. $m\angle ADC = m\angle ADB + m\angle BDC$, $m\angle WZY = m\angle WZX + m\angle ZXY$	**13.** Angle Addition Postulate
14. $m\angle ADC = m\angle WZX + m\angle ZXY$	**14.** Substitution Property of Equality
15. $m\angle ADC = m\angle WZY$	**15.** Transitive Property of Equality
16. $\angle ADC \cong \angle WZY$	**16.** Definition of congruent angles
17. $ABCD \cong WXYZ$	**17.** All corresponding parts are congruent.

c. no

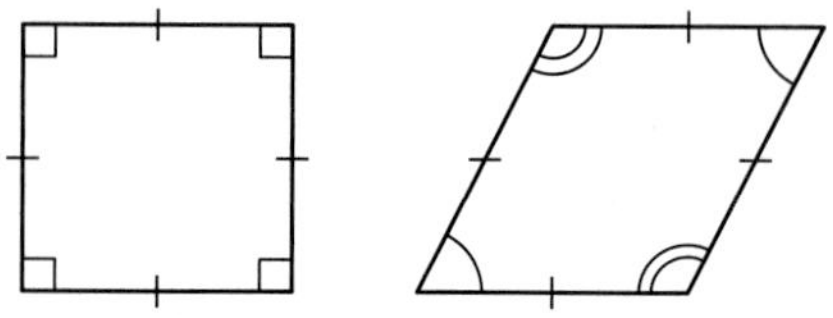

d. no

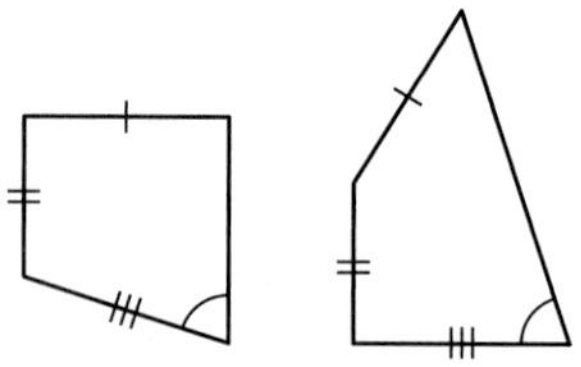

e. no

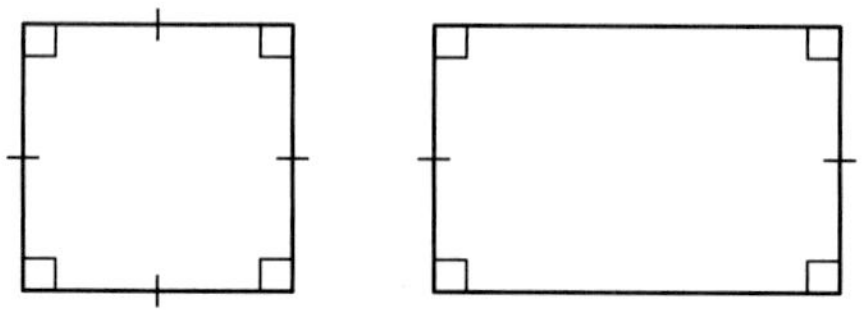

f. no; See counterexample for part (e).

5.6 Review & Refresh

32. (1, 1)

33.

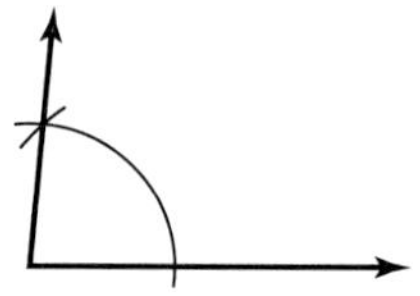

34. one pair of corresponding sides are congruent

35. $\angle WVX$, $\angle WXV$; Base Angles Theorem

36. $\overline{ZX}$, $\overline{ZY}$; Converse of the Base Angles Theorem

37. 60 mm

38. yes; SSS Congruence Theorem

39. no

40. yes; SAS Congruence Theorem

41. yes; AAS Congruence Theorem

Section 5.7

5.7 Practice

1. All three pairs of sides are congruent. So, by the SSS Congruence Theorem, $\triangle ABC \cong \triangle DBC$. Because corresponding parts of congruent triangles are congruent, $\angle A \cong \angle D$.

3. The hypotenuses and one pair of legs of two right triangles are congruent. So, by the HL Congruence Theorem, $\triangle JMK \cong \triangle LMK$. Because corresponding parts of congruent triangles are congruent, $\overline{JM} \cong \overline{LM}$.

5. From the diagram, $\angle JHN \cong \angle KGL$, $\angle N \cong \angle L$, and $\overline{JN} \cong \overline{KL}$. So, by the AAS Congruence Theorem, $\triangle JNH \cong \triangle KLG$. Because corresponding parts of congruent triangles are congruent, $\overline{GK} \cong \overline{HJ}$.

7. Use the AAS Congruence Theorem to prove that $\triangle FHG \cong \triangle GKF$. Then state that $\angle FGK \cong \angle GFH$. Use the Congruent Complements Theorem to prove that $\angle 1 \cong \angle 2$.

9. Use the ASA Congruence Theorem to prove that $\triangle STR \cong \triangle QTP$. Then state that $\overline{PT} \cong \overline{RT}$ because corresponding parts of congruent triangles are congruent. Use the SAS Congruence Theorem to prove that $\triangle STP \cong \triangle QTR$. So, $\angle 1 \cong \angle 2$.

11.

STATEMENTS	REASONS
1. $\overline{AP} \cong \overline{BP}$, $\overline{AQ} \cong \overline{BQ}$	**1.** Given
2. $\overline{PQ} \cong \overline{PQ}$	**2.** Reflexive Property of Congruence
3. $\triangle APQ \cong \triangle BPQ$	**3.** SSS Congruence Theorem
4. $\angle APQ \cong \angle BPQ$	**4.** Corresponding parts of congruent triangles are congruent.
5. $\overline{PM} \cong \overline{PM}$	**5.** Reflexive Property of Congruence
6. $\triangle APM \cong \triangle BPM$	**6.** SAS Congruence Theorem
7. $\angle AMP \cong \angle BMP$	**7.** Corresponding parts of congruent triangles are congruent.
8. $\angle AMP$ and $\angle BMP$ form a linear pair.	**8.** Definition of a linear pair
9. $\overline{MP} \perp \overline{AB}$	**9.** Linear Pair Perpendicular Theorem
10. $\angle AMP$ and $\angle BMP$ are right angles.	**10.** Definition of perpendicular lines

13.

STATEMENTS	REASONS
1. $\overline{FG} \cong \overline{GJ} \cong \overline{HG} \cong \overline{GK}$, $\overline{JM} \cong \overline{LM} \cong \overline{KM} \cong \overline{NM}$	**1.** Given
2. $\angle FGJ \cong \angle HGK$, $\angle JML \cong \angle KMN$	**2.** Vertical Angles Congruence Theorem
3. $\triangle FGJ \cong \triangle HGK$, $\triangle JML \cong \triangle KMN$	**3.** SAS Congruence Theorem
4. $\angle F \cong \angle H$, $\angle L \cong \angle N$	**4.** Corresponding parts of congruent triangles are congruent.
5. $FG = GJ = HG = GK$	**5.** Definition of congruent segments
6. $HJ = HG + GJ$, $FK = FG + GK$	**6.** Segment Addition Postulate
7. $FK = HG + GJ$	**7.** Substitution Property of Equality
8. $FK = HJ$	**8.** Transitive Property of Equality
9. $\overline{FK} \cong \overline{HJ}$	**9.** Definition of congruent segments
10. $\triangle HJN \cong \triangle FKL$	**10.** AAS Congruence Theorem
11. $\overline{FL} \cong \overline{HN}$	**11.** Corresponding parts of congruent triangles are congruent.

15. Because $\overline{AC} \perp \overline{BC}$ and $\overline{ED} \perp \overline{BD}$, $\angle ACB$ and $\angle EDB$ are congruent right angles. Because B is the midpoint of $\overline{CD}$, $\overline{BC} \cong \overline{BD}$. The vertical angles $\angle ABC$ and $\angle EBD$ are congruent. So, $\triangle ABC \cong \triangle EBD$ by the ASA Congruence Theorem. Then, because corresponding parts of congruent triangles are congruent, $\overline{AC} \cong \overline{ED}$. So, you can find the distance ED across the canyon by measuring AC.

17.

STATEMENTS	REASONS
1. $\overline{AD} \parallel \overline{BC}$, E is the midpoint of $\overline{AC}$.	**1.** Given
2. $\overline{AE} \cong \overline{CE}$	**2.** Definition of midpoint
3. $\angle AEB \cong \angle CED$, $\angle AED \cong \angle BEC$	**3.** Vertical Angles Congruence Theorem
4. $\angle DAE \cong \angle BCE$	**4.** Alternate Interior Angles Theorem
5. $\triangle DAE \cong \triangle BCE$	**5.** ASA Congruence Theorem
6. $\overline{DE} \cong \overline{BE}$	**6.** Corresponding parts of congruent triangles are congruent.
7. $\triangle AEB \cong \triangle CED$	**7.** SAS Congruence Theorem

19. yes; You can show that *WXYZ* is a rectangle. This means that the opposite sides are congruent. Because $\triangle WZY$ and $\triangle YXW$ share a hypotenuse, the two triangles have congruent hypotenuses and corresponding legs, which allows you to use the HL Congruence Theorem to prove that the triangles are congruent.

21. $\triangle GHJ$, $\triangle DEF$, $\triangle NPQ$

5.7 Review & Refresh

22. It is given that $\overline{EF} \cong \overline{GH}$ and $\angle EFH \cong \angle GHF$. Use the Reflexive Property of Segment Congruence to show $\overline{HF} \cong \overline{FH}$. Then use the SAS Congruence Theorem to show that $\triangle EFH \cong \triangle GHF$. Because corresponding parts of congruent triangles are congruent, $\overline{EH} \cong \overline{GF}$.

23. 16 units

24. $\sqrt{13} + \sqrt{34} + \sqrt{65}$, or about 17.5 units

25. $x = 30$

26. $2y + 2$

27. yes; ASA Congruence Theorem

28. yes; HL Congruence Theorem

29. $y = 380(1.02)^x$

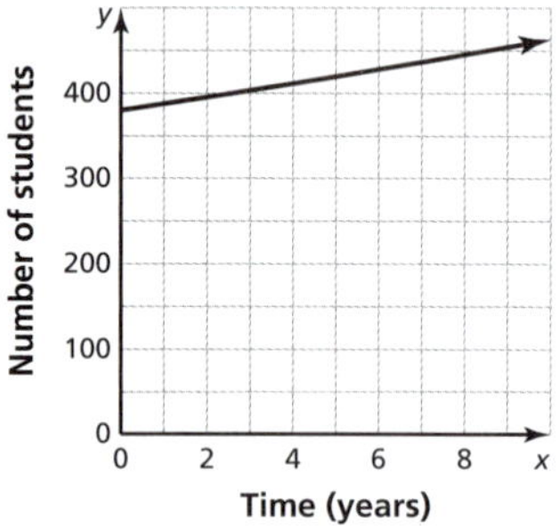

Section 5.8

5.8 Practice

1. *Sample answer:*

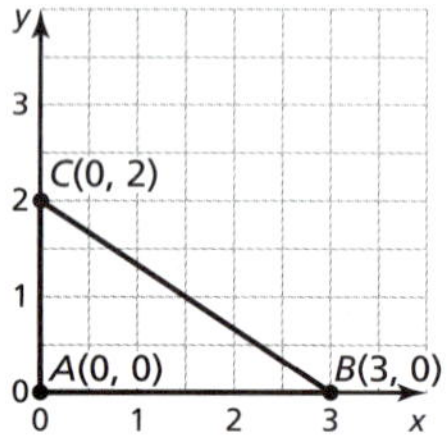

It is easy to find the lengths of horizontal and vertical segments and distances from the origin.

3. *Sample answer:*

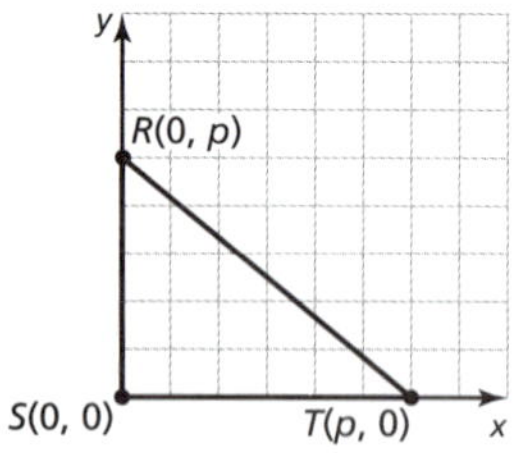

It is easy to find the lengths of horizontal and vertical segments and distances from the origin.

5. *Sample answer:*

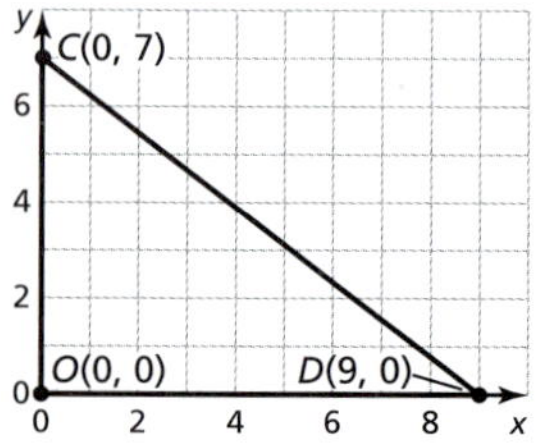

$\sqrt{130}$, or about 11.4 units

7. *Sample answer:*

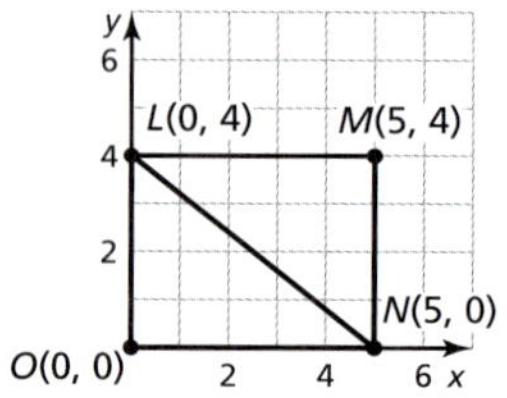

$\sqrt{41}$, or about 6.4 units

9.

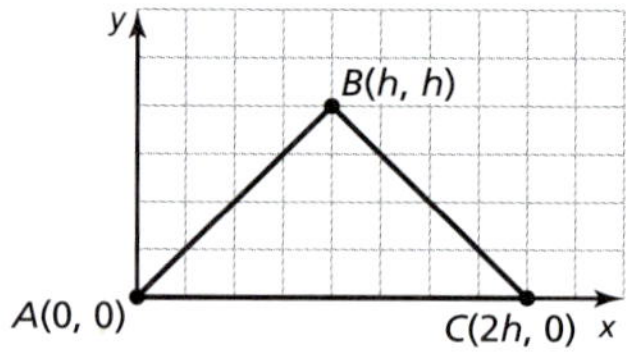

$AB = h\sqrt{2}$, $m_{\overline{AB}} = 1$, $M_{\overline{AB}}\left(\frac{h}{2}, \frac{h}{2}\right)$, $BC = h\sqrt{2}$, $m_{\overline{BC}} = -1$, $M_{\overline{BC}}\left(\frac{3h}{2}, \frac{h}{2}\right)$, $AC = 2h$, $m_{\overline{AC}} = 0$, $M_{\overline{AC}}(h, 0)$; yes; yes; Because $m_{\overline{AB}} \cdot m_{\overline{BC}} = -1$, $\overline{AB} \perp \overline{BC}$ by the Slopes of Perpendicular Lines Theorem. So $\angle ABC$ is a right angle. $\overline{AB} \cong \overline{BC}$ because $AB = BC$. So, $\triangle ABC$ is a right isosceles triangle.

11. $N(h, k)$; $ON = \sqrt{h^2 + k^2}$, $MN = \sqrt{h^2 + k^2}$

13. Find the lengths of $\overline{OP}$, $\overline{PM}$, $\overline{MN}$, and $\overline{NO}$ to show that $\overline{OP} \cong \overline{PM}$ and $\overline{MN} \cong \overline{NO}$.

15. $DC = k$, $BC = k$, $DE = h$, $OB = h$, $EC = \sqrt{h^2 + k^2}$, $OC = \sqrt{h^2 + k^2}$

So, $\overline{DC} \cong \overline{BC}$, $\overline{DE} \cong \overline{OB}$, and $\overline{EC} \cong \overline{OC}$. By the SSS Congruence Theorem, $\triangle DEC \cong \triangle BOC$.

17. Using the Distance Formula,
$AB = \sqrt{(5 - 0)^2 + (12 - 0)^2} = 13$ mm and
$CB = \sqrt{(10 - 5)^2 + (0 - 12)^2} = 13$ mm.
Because $\overline{AB} \cong \overline{CB}$, $\triangle ABC$ is isosceles.

19. yes;
$PQ = 3\sqrt{5}$, $m_{\overline{PQ}} = -2$
$SR = 3\sqrt{5}$, $m_{\overline{SR}} = -2$
$SP = \sqrt{5}$, $m_{\overline{SP}} = \frac{1}{2}$
$RQ = \sqrt{5}$, $m_{\overline{RQ}} = \frac{1}{2}$

So, $\overline{PQ} \cong \overline{SR}$ and $\overline{SP} \cong \overline{RQ}$, which shows that opposite sides are congruent. Also, $m_{\overline{PQ}} \cdot m_{\overline{SP}} = -1$, $m_{\overline{PQ}} \cdot m_{\overline{RQ}} = -1$, $m_{\overline{SR}} \cdot m_{\overline{SP}} = -1$, and $m_{\overline{SR}} \cdot m_{\overline{RQ}} = -1$.

So, $\overline{PQ} \perp \overline{SP}$, $\overline{PQ} \perp \overline{RQ}$, $\overline{SR} \perp \overline{SP}$, and $\overline{SR} \perp \overline{RQ}$ by the Slopes of Perpendicular Lines Theorem. So, by definition of perpendicular lines, $\angle PSR$, $\angle SRQ$, $\angle RQP$, and $\angle QPS$ are right angles. So, the quadrilateral is a rectangle.

21. $(-k, -m)$ and (k, m)

23. Let $A(0, 0)$, $B(\ell, 0)$, $C(\ell, w)$, and $D(0, w)$ represent the vertices of the rectangle. Using the Distance Formula, $AC = \sqrt{(\ell - 0)^2 + (w - 0)^2} = \sqrt{w^2 + \ell^2}$ and $BD = \sqrt{(0 - \ell)^2 + (w - 0)^2} = \sqrt{w^2 + \ell^2}$. Because $AC = BD$, $\overline{AC} \cong \overline{BD}$.

5.8 Review & Refresh

25.

Equation	Justification
$6x + 13 = -5$	Write given equation.
$6x = -18$	Subtraction Property of Equality
$x = -3$	Division Property of Equality

26.

Equation	Justification
$3(x - 1) = -(x + 10)$	Write given equation.
$3x - 3 = -x - 10$	Distributive Property
$3x = -x - 7$	Addition Property of Equality
$4x = -7$	Addition Property of Equality
$x = -\frac{7}{4}$	Division Property of Equality

27. $(7a + 1)(2a + 3)$

28. $\triangle MKJ$ and $\triangle MKL$ are right triangles. $\overline{JK} \cong \overline{LK}$ is given and $\overline{MK} \cong \overline{MK}$ by the Reflexive Property of Congruence. So, by the HL Congruence Theorem, $\triangle MKJ \cong \triangle MKL$. Because corresponding parts of congruent triangles are congruent, $\angle J \cong \angle L$.

29. no

30.

STATEMENTS	REASONS
1. $\overline{XY} \cong \overline{ZY}$, $\overline{WY} \cong \overline{VY}$, $\overline{VW} \cong \overline{ZX}$	1. Given
2. $\overline{VZ} \cong \overline{ZV}$	2. Symmetric Property of Segment Congruence
3. $XY = ZY$, $WY = VY$	3. Definition of congruent segments
4. $WZ = WY + ZY$, $XV = XY + VY$	4. Segment Addition Postulate
5. $WZ = VY + XY$	5. Substitution Property of Equality
6. $WZ = XV$	6. Substitution Property of Equality
7. $\overline{WZ} \cong \overline{XV}$	7. Definition of congruent segments
8. $\triangle VWZ \cong \triangle ZXV$	8. SSS Congruence Theorem

31.

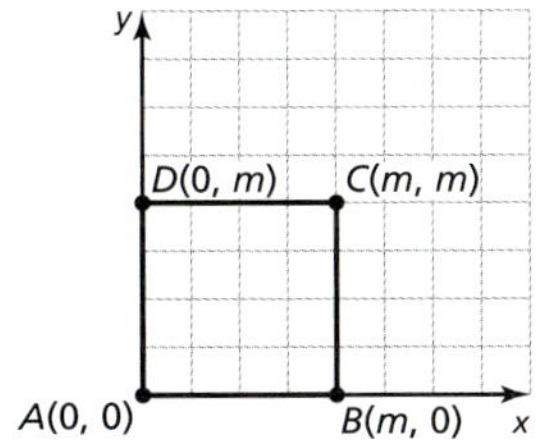

32. $x = 13$

33. $x = 34$

Chapter 5 Review

1. acute isosceles

2. obtuse scalene

3. scalene; not right

4. isosceles; right

5. 132°

6. 90°

7. 42°, 48°

8. 35°, 55°

9. 22°, 68°

10. corresponding sides: $\overline{GH} \cong \overline{LM}$, $\overline{HJ} \cong \overline{MN}$, $\overline{JK} \cong \overline{NP}$, and $\overline{GK} \cong \overline{LP}$; corresponding angles: $\angle G \cong \angle L$, $\angle H \cong \angle M$, $\angle J \cong \angle N$, and $\angle K \cong \angle P$; *Sample answer:* $JHGK \cong NMLP$

11. 16°

12. quadrilateral $ABCF \cong$ quadrilateral $EDCF$; It is given that $\angle BAF \cong \angle DEF$, $\overline{AF} \cong \overline{EF}$, $\angle EFC \cong \angle AFC$, $\overline{BC} \cong \overline{DC}$, and $\overline{AB} \cong \overline{ED}$. Because $\overline{AB} \parallel \overline{FC}$ and $\overline{FC} \parallel \overline{ED}$, $\angle ABC \cong \angle FCD$ and $\angle FCD \cong \angle EDC$ by the Corresponding Angles Theorem. $\overline{FC} \cong \overline{FC}$ by the Reflexive Property of Segment Congruence. Because all corresponding sides are congruent and all corresponding angles are congruent, quadrilateral $ABCF \cong$ quadrilateral $EDCF$.

13. no; There are two pairs of congruent sides and one pair of congruent angles, but the angles are not the included angles.

14. yes;

STATEMENTS	REASONS
1. $\overline{WX} \cong \overline{YZ}$, $\angle WXZ \cong \angle YZX$	1. Given
2. $\overline{XZ} \cong \overline{XZ}$	2. Reflexive Property of Congruence
3. $\triangle WXZ \cong \triangle YZX$	3. SAS Congruence Theorem

15. *Sample answer:*

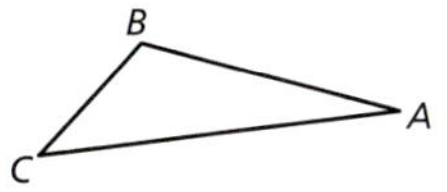

16.

STATEMENTS	REASONS
1. $AD = CD$, $\angle ADB \cong \angle CDB$	1. Given
2. $\overline{AD} \cong \overline{CD}$	2. Definition of Congruent Segments
3. $\overline{DB} \cong \overline{DB}$	3. Reflexive Property of Segment Congruence
4. $\triangle ADB \cong \triangle CDB$	4. SAS Congruence Theorem

17. P; PRQ

18. $\overline{TR}$; $\overline{TV}$

19. RQS; RSQ

20. $\overline{SR}$; $\overline{SV}$

21. $x = 40$, $y = 50$

22. $x = 15$, $y = 5$

23. 45 ft

24. no; There is only enough information to conclude that two pairs of sides are congruent.

25. yes;

STATEMENTS	REASONS
1. $\overline{WX} \cong \overline{YZ}$, $\angle XWZ$ and $\angle ZYX$ are right angles.	1. Given
2. $\overline{XZ} \cong \overline{XZ}$	2. Reflexive Property of Congruence
3. $\triangle WXZ$ and $\triangle YZX$ are right triangles.	3. Definition of a right triangle
4. $\triangle WXZ \cong \triangle YZX$	4. HL Congruence Theorem

26. **a.** $\overline{BD} \perp \overline{AC}$
b. $\overline{AB} \cong \overline{CB}$

27. yes;

STATEMENTS	REASONS
1. $\angle E \cong \angle H$, $\angle F \cong \angle J$, $\overline{FG} \cong \overline{JK}$	1. Given
2. $\triangle EFG \cong \triangle HJK$	2. AAS Congruence Theorem

28. no; There is only enough information to conclude that one pair of angles and one pair of sides are congruent.

29. yes;

STATEMENTS	REASONS
1. $\angle PLN \cong \angle MLN$, $\angle PNL \cong \angle MNL$	1. Given
2. $\overline{LN} \cong \overline{LN}$	2. Reflexive Property of Congruence
3. $\triangle LPN \cong \triangle LMN$	3. ASA Congruence Theorem

30. no; There is only enough information to conclude that one pair of angles and one pair of sides are congruent.

31. By the SAS Congruence Theorem, $\triangle HJK \cong \triangle LMN$. Because corresponding parts of congruent triangles are congruent, $\angle K \cong \angle N$.

32. By the Vertical Angles Congruence Theorem, $\angle AEB \cong \angle CED$. By the AAS Congruence Theorem, $\triangle AEB \cong \triangle CED$. Because corresponding parts of congruent triangles are congruent, $\overline{AE} \cong \overline{CE}$ and $\overline{DE} \cong \overline{BE}$. By the Vertical Angles Congruence Theorem, $\angle AED \cong \angle CEB$. So, $\triangle AED \cong \triangle CEB$ by the SAS Congruence Theorem. Because corresponding parts of congruent triangles are congruent, $\overline{AD} \cong \overline{CB}$.

33. First, state that $\overline{QV} \cong \overline{QV}$. Then use the SSS Congruence Theorem to prove that $\triangle QSV \cong \triangle QTV$. Because corresponding parts of congruent triangles are congruent, $\angle QSV \cong \angle QTV$. $\angle QSV \cong \angle 1$ and $\angle QTV \cong \angle 2$ by the Vertical Angles Congruence Theorem. So, by the Transitive Property of Congruence, $\angle 1 \cong \angle 2$.

34. First, use the ASA Congruence Theorem to prove that $\triangle MFK \cong \triangle JHL$. Because corresponding parts of congruent triangles are congruent, $\angle 2 \cong \angle 1$. By the Symmetric Property of Congruence $\angle 1 \cong \angle 2$.

35. Given: $\overline{AB} \cong \overline{CD}$, $\overline{BC} \cong \overline{DA}$. Construct line segment $\overline{AC}$. By the Symmetric Property of Segment Congruence, $\overline{AC} \cong \overline{CA}$. So, $\triangle CDA \cong \triangle ABC$ by the SAS Congruence Theorem. Because corresponding parts of congruent triangles are congruent, $\angle D \cong \angle B$.

36.

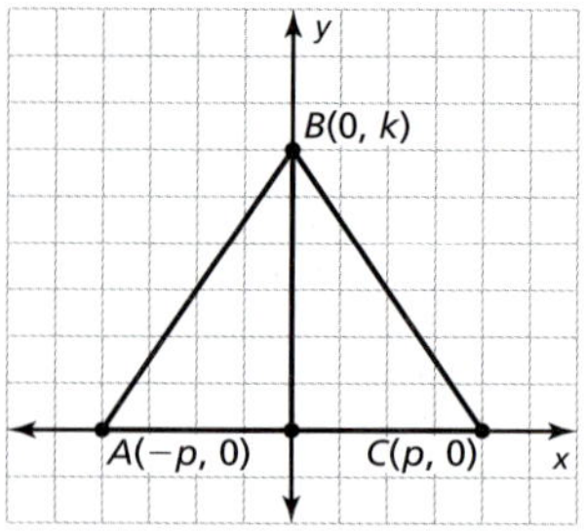

;

It is easy to find the lengths of horizontal segments and distances from the origin. Because the triangle is isosceles, the height from the vertex angle to the base serves as a line of symmetry for the isosceles triangle.

37.

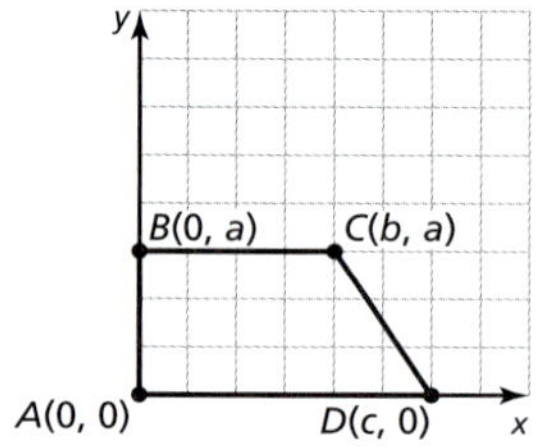

;

It is easy to find the lengths of horizontal and vertical segments and distances from the origin. Horizontal and vertical line segments are perpendicular, forming right angles.

38. $(2k, k)$ **39.** $(0, -k)$

40. Segments $\overline{OD}$ and $\overline{BD}$ have the same length.

$OD = \sqrt{(j-0)^2 + (j-0)^2} = \sqrt{j^2 + j^2} = \sqrt{2j^2} = j\sqrt{2}$

$BD = \sqrt{(j-2j)^2 + (j-0)^2} = \sqrt{(-j)^2 + j^2} = \sqrt{2j^2} = j\sqrt{2}$

Segments $\overline{DB}$ and $\overline{DC}$ have the same length.

$DB = BD = j\sqrt{2}$

$DC = \sqrt{(2j-j)^2 + (2j-j)^2} = \sqrt{j^2 + j^2} = \sqrt{2j^2} = j\sqrt{2}$

Segments $\overline{OB}$ and $\overline{BC}$ have the same length.

$OB = |2j - 0| = 2j$

$BC = |2j - 0| = 2j$

So, you can apply the SSS Congruence Theorem to conclude that $\triangle ODB \cong \triangle BDC$.

41. Using the Distance Formula, $OP = \sqrt{h^2 + k^2}$, $QR = \sqrt{h^2 + k^2}$, $OR = j$, and $QP = j$. So, $\overline{OP} \cong \overline{QR}$ and $\overline{OR} \cong \overline{QP}$. Also, by the Reflexive Property of Congruence, $\overline{QO} \cong \overline{QO}$. So, you can apply the SSS Congruence Theorem to conclude that $\triangle OPQ \cong \triangle QRO$.

Chapter 5 Practice Test

1.

STATEMENTS	REASONS
1. $\overline{CA} \cong \overline{CB} \cong \overline{CD} \cong \overline{CE}$	1. Given
2. $\angle ACB \cong \angle ECD$	2. Vertical Angles Congruence Theorem
3. $\triangle ABC \cong \triangle EDC$	3. SAS Congruence Theorem

2.

STATEMENTS	REASONS
1. $\overline{JK} \parallel \overline{ML}, \overline{MJ} \parallel \overline{KL}$	1. Given
2. $\overline{MK} \cong \overline{KM}$	2. Reflexive Property of Congruence
3. $\angle JKM \cong \angle LMK$, $\angle JMK \cong \angle LKM$	3. Alternate Interior Angles Theorem
4. $\triangle MJK \cong \triangle KLM$	4. ASA Congruence Theorem

3.

STATEMENTS	REASONS
1. $\overline{QR} \cong \overline{RS}, \angle P \cong \angle T$	1. Given
2. $\angle R \cong \angle R$	2. Reflexive Property of Congruence
3. $\triangle SRP \cong \triangle QRT$	3. AAS Congruence Theorem

4. 44°, 46°
5. no; By the Corollary to the Base Angles Theorem, if a triangle is equilateral, then it is also equiangular.
6. no; The Third Angles Theorem can be used to prove that two triangles are equiangular, but AAA is not sufficient to prove that the triangles are congruent. You need to know that at least one pair of corresponding sides are congruent.
7. First, use the HL Congruence Theorem to prove that $\triangle ACD \cong \triangle BED$. Because corresponding parts of congruent triangles are congruent, $\overline{AD} \cong \overline{BD}$. Then use the Base Angles Theorem to prove that $\angle 1 \cong \angle 2$.
8. Use the SSS Congruence Theorem to prove that $\triangle SVX \cong \triangle SZX$. Use the Vertical Angles Congruence Theorem and the SAS Congruence Theorem to prove that $\triangle VXW \cong \triangle ZXY$. Because corresponding parts of congruent triangles are congruent, $\angle SZX \cong \angle SVX$ and $\angle W \cong \angle Y$. Use the Segment Addition Postulate to show that $\overline{VY} \cong \overline{ZW}$. Then use the ASA Congruence Theorem to prove that $\triangle VYT \cong \triangle ZWR$. Because corresponding parts of congruent triangles are congruent, $\angle 1 \cong \angle 2$.
9. HL Congruence Theorem: $\overline{AB} \cong \overline{CD}$ and $\triangle ABD$ and $\triangle CDB$ are right triangles; ASA Congruence Theorem: $\angle ADB \cong \angle CBD$, $\overline{DB} \cong \overline{BD}$, and $\angle ABD \cong \angle CDB$; AAS Congruence Theorem: $\angle ADB \cong \angle CBD$, $\angle ABD \cong \angle CDB$, and $\overline{DB} \cong \overline{BD}$; SAS Congruence Theorem: $\overline{AB} \cong \overline{CD}$, $\angle CDB \cong \angle ABD$, and $\overline{DB} \cong \overline{BD}$.
10. Using the Distance Formula, $PQ = 18$ and $ST = 18$. So, $\overline{PQ} \cong \overline{ST}$. Also, the horizontal segments $\overline{PQ}$ and $\overline{ST}$ each have a slope of 0, which implies that they are parallel. So, $\overline{PS}$ intersects $\overline{PQ}$ and $\overline{ST}$ to form congruent alternate interior angles, $\angle P$ and $\angle S$. By the Vertical Angles Congruence Theorem, $\angle PRQ \cong \angle SRT$. So, by the AAS Congruence Theorem, $\triangle PQR \cong \triangle STR$.
11. $x = 6, y = 7$
12. a. isosceles
    b. $m\angle 1 = m\angle 2 = 70°$; By the Triangle Sum Theorem, $m\angle 1 + m\angle 2 + m\angle 3 = 180°$. Because $m\angle 3 = 40°$, $m\angle 1 + m\angle 2 = 140°$. By the Base Angles Theorem, $\angle 1 \cong \angle 2$, so $m\angle 1 = m\angle 2 = \frac{140°}{2} = 70°$.

Chapter 5 Test Prep

1. By step 1, a line through point P intersects line m in point Q. By steps 2 and 3, $\overline{QA} \cong \overline{QB} \cong \overline{PC} \cong \overline{PD}$ and $\overline{AB} \cong \overline{CD}$ because congruent segments were drawn with the same compass setting. So, in step 4, you see that if $\overline{AB}$ and $\overline{CD}$ were drawn, then $\triangle AQB$ and $\triangle CPD$ would be congruent by the SSS Congruence Theorem. Because corresponding parts of congruent triangles are congruent, $\angle CBD \cong \angle AQB$, which means that $\overleftrightarrow{PD} \parallel m$ by the Corresponding Angles Converse.
2. D
3. C
4. a. The segments $\overline{AB}$ and $\overline{DE}$ have the same measure ($AB = DE = 3$) and are therefore congruent. Also, from the markings in the diagram, $\angle B \cong \angle E$ and $\overline{BC} \cong \overline{EF}$. So, by the SAS Congruence Theorem, $\triangle ABC \cong \triangle DEF$.
   b. a 90° counterclockwise rotation about the origin, followed by a translation 13 units right
5. A, B, D
6. B
7. $AD = 3\sqrt{2}, m_{\overline{AD}} = -1, BC = 3\sqrt{2}, m_{\overline{BC}} = -1, AB = 2\sqrt{2}, m_{\overline{AB}} = 1, DC = 2\sqrt{2}, = m_{\overline{DC}} = 1$
   Sides $\overline{AD}$ and $\overline{BC}$ have the same measure and the same slope, as do $\overline{AB}$ and $\overline{DC}$. So, by the Slopes of Parallel Lines Theorem, $\overline{AD} \parallel \overline{BC}$ and $\overline{AB} \parallel \overline{DC}$. Because the product of their slopes is -1, $\overline{AD} \perp \overline{AB}$, $\overline{AD} \perp \overline{DC}$, $\overline{BC} \perp \overline{AB}$, and $\overline{BC} \perp \overline{DC}$. So, $ABCD$ is a rectangle.
8. Because the same compass settings were used, $\overline{AB} \cong \overline{AC} \cong \overline{BC}$. So, $\triangle ABC$ is equilateral.

Chapter 6

Prepare with CalcChat

1. $y = -3x + 10$
2. $y = x - 7$
3. $y = \frac{1}{4}x - \frac{7}{4}$
4. $-3 \leq w \leq 8$
5. $d < -1$ or $d \geq 5$
6. yes; If the graphs of the two inequalities overlap going in opposite directions and the variable only has to make one or the other true, then every number on the number line makes the compound inequality true.

Section 6.1

6.1 Practice

1. yes; Because point N is equidistant from L and M, point N is on the perpendicular bisector of $\overline{LM}$ by the Converse of the Perpendicular Bisector Theorem. Because only one line can be perpendicular to $\overline{LM}$ at point K, $\overleftrightarrow{NK}$ must be the perpendicular bisector of $\overline{LM}$, and P is on $\overleftrightarrow{NK}$.
3. no; You would need to know that $\overleftrightarrow{PN} \perp \overleftrightarrow{ML}$.
5. 4.6; Because $GK = KJ$ and $\overleftrightarrow{HK} \perp \overleftrightarrow{GJ}$, point H is on the perpendicular bisector of $\overline{GJ}$. So, by the Perpendicular Bisector Theorem, $GH = HJ = 4.6$.
7. 15; Because $\overleftrightarrow{DB} \perp \overleftrightarrow{AC}$ and point D is equidistant from A and C, point D is on the perpendicular bisector of $\overline{AC}$ by the Converse of the Perpendicular Bisector Theorem. By definition of segment bisector, $AB = BC$. So, $5x = 4x + 3$, and the solution is $x = 3$. So, $AB = 5x = 5(3) = 15$.

9. yes; Because H is equidistant from $\overrightarrow{EF}$ and $\overrightarrow{EG}$, $\overrightarrow{EH}$ bisects $\angle FEG$ by the Angle Bisector Theorem.

11. no; Because neither $\overline{BD}$ nor $\overline{DC}$ are marked as perpendicular to $\overrightarrow{AB}$ or $\overrightarrow{AC}$ respectively, you cannot conclude that $DB = DC$.

13. 20°; Because D is equidistant from $\overrightarrow{BC}$ and $\overrightarrow{BA}$, $\overrightarrow{BD}$ bisects $\angle ABC$ by the Converse of the Angle Bisector Theorem. So, $m\angle ABD = m\angle CBD = 20°$.

15. 16; $\overrightarrow{EG}$ is an angle bisector of $\angle FEH$, $\overline{FG} \perp \overrightarrow{EF}$, and $\overline{GH} \perp \overrightarrow{EH}$. So, by the Converse of the Angle Bisector Theorem, $FG = GH$. This means that $x + 11 = 3x + 1$, and the solution is $x = 5$. So, $FG = x + 11 = 5 + 11 = 16$.

17. Because $\overline{DC}$ is not necessarily congruent to $\overline{EC}$, $\overleftrightarrow{AB}$ will not necessarily pass through point C; Because $AD = AE$, and $\overleftrightarrow{AB} \perp \overline{DE}$, $\overleftrightarrow{AB}$ is the perpendicular bisector of $\overline{DE}$.

19. $y = x - 2$

21. $y = -3x + 15$

23. yes; By the Perpendicular Bisector Theorem, $AD = CD$. So, $\overline{AD} \cong \overline{CD}$ by the definition of segment congruence.

25.

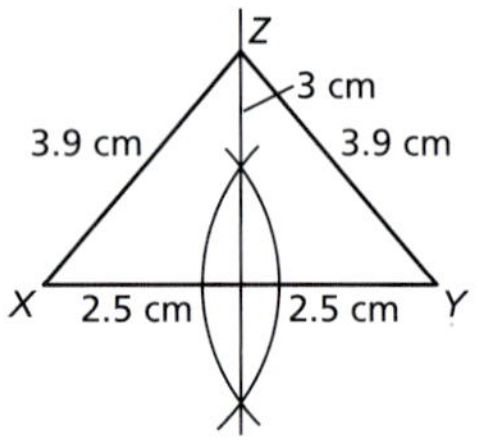

Perpendicular Bisector Theorem

27. B

29. yes; If the triangle is an isosceles triangle, then the angle bisector of the vertex angle will also be the perpendicular bisector of the base.

31. a.

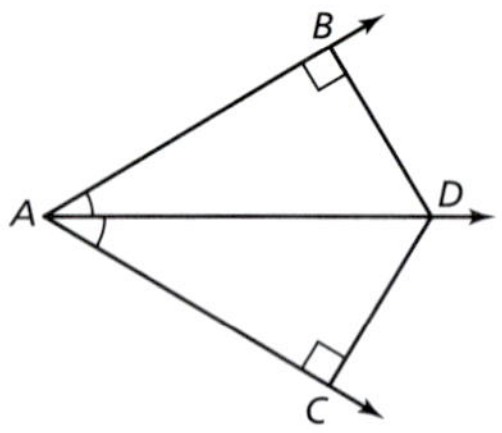

If $\overrightarrow{AD}$ bisects $\angle BAC$, then by definition of angle bisector, $\angle BAD \cong \angle CAD$. Also, because $\overline{DB} \perp \overrightarrow{AB}$ and $\overline{DC} \perp \overrightarrow{AC}$, by definition of perpendicular lines, $\angle ABD$ and $\angle ACD$ are right angles, and congruent to each other by the Right Angles Congruence Theorem. Also, $\overline{AD} \cong \overline{AD}$ by the Reflexive Property of Congruence. So, by the AAS Congruence Theorem, $\triangle ADB \cong \triangle ADC$. Because corresponding parts of congruent triangles are congruent, $DB = DC$. This means that point D is equidistant from each side of $\angle BAC$.

b.

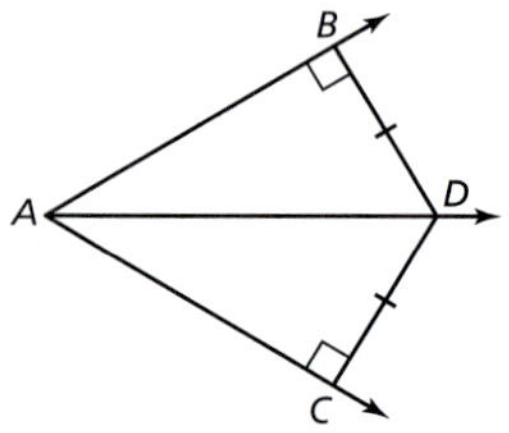

STATEMENTS	REASONS
1. $\overline{DC} \perp \overrightarrow{AC}$, $\overline{DB} \perp \overrightarrow{AB}$, $BD = CD$	1. Given
2. $\angle ABD$ and $\angle ACD$ are right angles.	2. Definition of perpendicular lines
3. $\triangle ABD$ and $\triangle ACD$ are right triangles.	3. Definition of a right triangle
4. $\overline{BD} \cong \overline{CD}$	4. Definition of congruent segments
5. $\overline{AD} \cong \overline{AD}$	5. Reflexive Property of Congruence
6. $\triangle ABD \cong \triangle ACD$	6. HL Congruence Theorem
7. $\angle BAD \cong \angle CAD$	7. Corresponding parts of congruent triangles are congruent.
8. $\overrightarrow{AD}$ bisects $\angle BAC$.	8. Definition of angle bisector

33. a. $y = x$
b. $y = -x$
c. $y = |x|$
d. $y = -|x|$

35. Because $\overline{YW}$ is on plane P, and plane P is a perpendicular bisector of $\overline{XZ}$ at point Y, $\overline{YW}$ is a perpendicular bisector of $\overline{XZ}$ by definition of a plane perpendicular to a line. So, by the Perpendicular Bisector Theorem, $\overline{XW} \cong \overline{ZW}$.

Because $\overline{YV}$ is on plane P, and plane P is a perpendicular bisector of $\overline{XZ}$ at point Y, $\overline{YV}$ is a perpendicular bisector of $\overline{XZ}$ by definition of a plane perpendicular to a line. So, by the Perpendicular Bisector Theorem, $\overline{XV} \cong \overline{ZV}$.

$\overline{WV} \cong \overline{WV}$ by the Reflexive Property of Congruence. Then, because $\overline{XW} \cong \overline{ZW}$ and $\overline{XV} \cong \overline{ZV}$, $\triangle WVX \cong \triangle WVZ$ by the SSS Congruence Theorem. So, $\angle VXW \cong \angle VZW$ because corresponding parts of congruent triangles are congruent.

37. Find the intersection of the three perpendicular bisectors of $\overline{JJ'}$, $\overline{KK'}$, and $\overline{LL'}$.

6.1 Review & Refresh

38. scalene

39. equilateral

40. acute

41. right

42. $m\angle 2 = 35°$; Because the lines are parallel, $\angle 2$ and the 145° angle are supplementary by the Consecutive Interior Angles Theorem.

43. *Sample answer:* Use the SSS Congruence Theorem to prove that $\triangle ADE \cong \triangle CDE$. Because corresponding parts of congruent triangles are congruent, $\angle ADE \cong \angle CDE$. Then use the SAS Congruence Theorem to prove that $\triangle BDA \cong \triangle BDC$.

44. $-6x^6 - 24x^5 + 28x^2$

45. 18; $\overrightarrow{SQ}$ is an angle bisector of $\angle PSR$, $\overline{PQ} \perp \overrightarrow{SP}$, and $\overline{RQ} \perp \overrightarrow{SR}$. So, by the Angle Bisector Theorem, $PQ = RQ$. So, $6x = 3x + 9$, and the solution is $x = 3$, which means that $QP = 6x = 6(3) = 18$.

46. 14.4; $TU = TW$, so $\overline{TU} \cong \overline{TW}$. $\overline{TV}$ is a perpendicular bisector of $\overline{UW}$ by the Converse of the Perpendicular Bisector Theorem. So, $UW = UV + VW = VW + VW = 7.2 + 7.2 = 14.4$.

47. $\angle S \cong \angle Y$; $\angle T \cong \angle Z$

48. 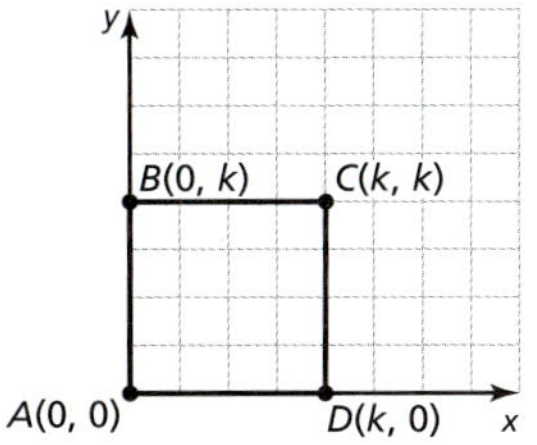

$\left(0, \frac{k}{2}\right), \left(\frac{k}{2}, k\right), \left(k, \frac{k}{2}\right), \left(\frac{k}{2}, 0\right)$

49. 4

Section 6.2

6.2 Practice

1. 9

3. You could copy the positions of the three residences and connect the points to draw a triangle. Then draw the three perpendicular bisectors of the triangle. The point where the perpendicular bisectors meet, the circumcenter, should be the location of the meeting place.

5. 9

7. (3, 4)

9. (−4, 9)

11. 16

13. 6

15. 32

17. *Sample answer:*

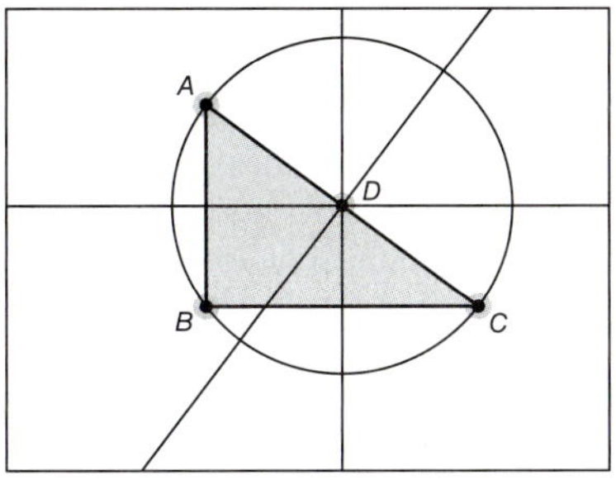

19. *Sample answer:*

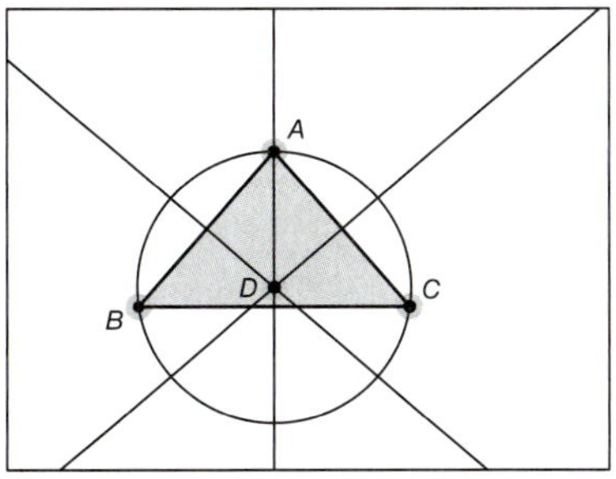

21. 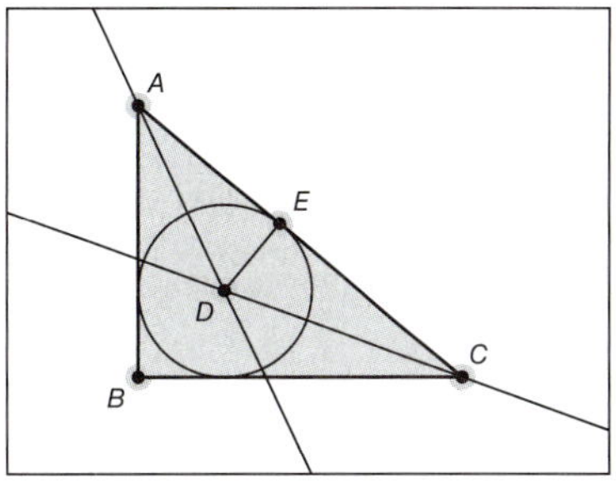

23. 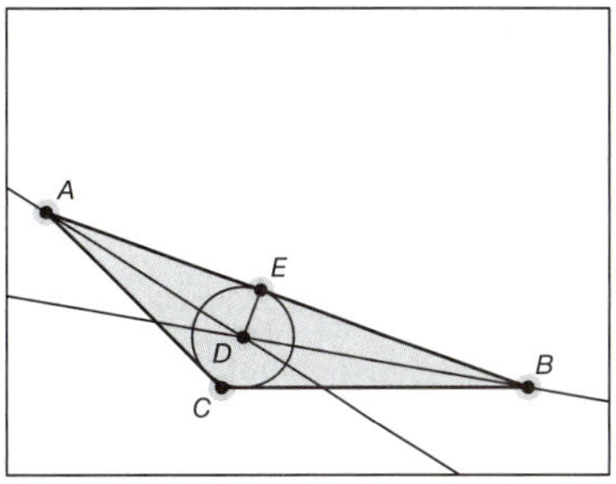

25. Because point G is the intersection of the angle bisectors, it is the incenter. But, because $\overline{GD}$ and $\overline{GF}$ are not necessarily perpendicular to a side of the triangle, there is not sufficient evidence to conclude that $\overline{GD}$ and $\overline{GF}$ are congruent; Point G is equidistant from the sides of the triangle.

27. incenter; The incenter is equidistant from the sides of the triangle.

29. sometimes; If the scalene triangle is obtuse or right, then the circumcenter is outside or on the triangle, respectively. However, if the scalene triangle is acute, then the circumcenter is inside the triangle.

31. sometimes; This only happens when the triangle is equilateral.

33. $\left(\frac{35}{6}, -\frac{11}{6}\right)$

35. $x = 6$

37. The circumcenter of any right triangle is located at the midpoint of the hypotenuse of the triangle. Let $A(0, 2b)$, $B(0, 0)$, and $C(2a, 0)$ represent the vertices of a right triangle where $\angle B$ is the right angle. The midpoint of $\overline{AB}$ is $M_{\overline{AB}}(0, b)$. The midpoint of $\overline{BC}$ is $M_{\overline{BC}}(a, 0)$. The midpoint of $\overline{AC}$ is $M_{\overline{AC}}(a, b)$. Because $\overline{AB}$ is vertical, its perpendicular bisector is horizontal. So, the equation of the horizontal line passing through $M_{\overline{AB}}(0, b)$ is $y = b$. Because $\overline{BC}$ is horizontal, its perpendicular bisector is vertical. So, the equation of the vertical line passing through $M_{\overline{BC}}(a, 0)$ is $x = a$. The circumcenter of $\triangle ABC$ is the intersection of perpendicular bisectors, $y = b$ and $x = a$, which is (a, b). This point is also the midpoint of $\overline{AC}$.

39. **a.** The archaeologists need to locate the circumcenter of the three stones, because that will be the center of the circle that contains all three stones. In order to locate the circumcenter, the archaeologists need to find the point of concurrency of the perpendicular bisectors of the sides of the triangle formed by the three stones.

b. (7, 7)

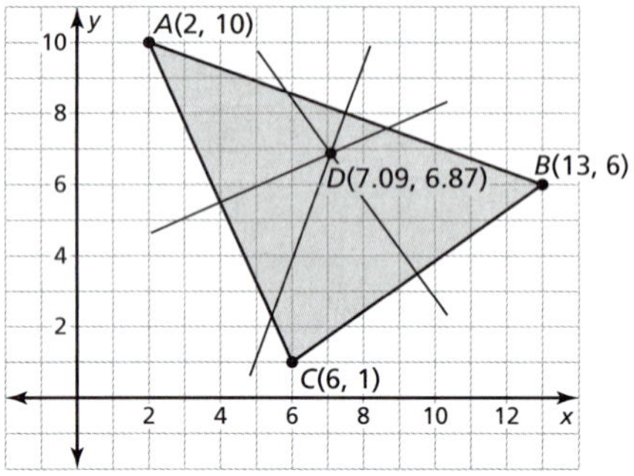

41. yes; In an equilateral triangle, each perpendicular bisector passes through the opposite vertex and divides the triangle into two congruent triangles. So, it is also an angle bisector.

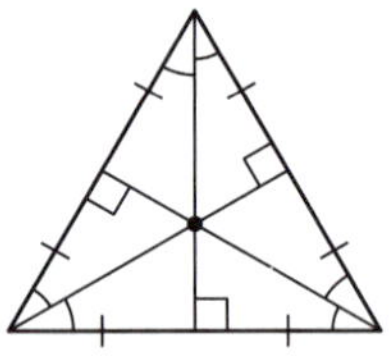

43. If the incenter was outside the triangle, it would be closer to one or more sides of the triangle than the other(s). The Incenter Theorem says that this is not possible.

45. about 3 in.; angle bisectors

47. $x = \dfrac{AB + AC - BC}{2}$ or $x = \dfrac{AB \cdot AC}{AB + AC + BC}$

6.2 Review & Refresh

49. no; Vertices T and V are translations 1 unit down of vertices Q and S, but vertex U is a translation 2 units down of vertex R. So, this is not a rigid motion.

50. *Sample answer:* If a concentrated solar power plant is the largest in the world, then it is the Noor Complex; If a concentrated solar power plant is the Noor Complex, then it cost 3.9 billion dollars to construct.

51. 100°; obtuse

52.

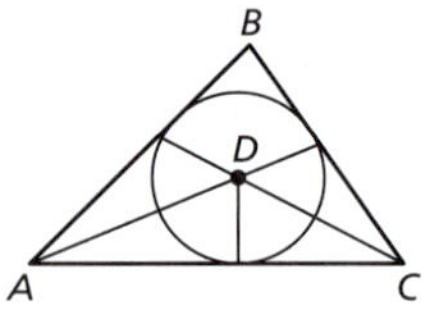

53. Use $\angle ACB \cong \angle DCB$, the Right Angles Congruence Theorem, and the Reflexive Property of Segment Congruence to prove that $\triangle ACB \cong \triangle DCB$ by the ASA Congruence Theorem. Then show that $\angle A \cong \angle D$ because corresponding parts of congruent triangles are congruent.

54. $5n^3(n^2 + 5)$

55. $m\angle ABD = 30°$

56. $\sqrt{10}$, or about 3.2 units

57. Because $AB = BC$, $\triangle ABC$ is isosceles.

58. $M(0, 5)$; $AB = 6$

59. $x = -6$

60. linear; As x increases by 2, y increases by 3.

Section 6.3

6.3 Practice

1. 6, 3

3. 20, 10

5. 10, 15

7. 18, 27

9. 12

11. 10

13. The length of $\overline{DE}$ should be $\frac{1}{3}$ of the length of $\overline{AE}$ because it is the shorter segment from the centroid to the side;

$DE = \frac{1}{3}AE$

$DE = \frac{1}{3}(18)$

$DE = 6$

15. $\left(5, \frac{11}{3}\right)$

17. outside; $(0, -5)$

19. inside; $(-1, 2)$

21.

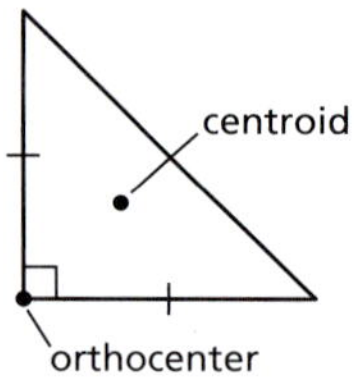

23. *Sample answer:*

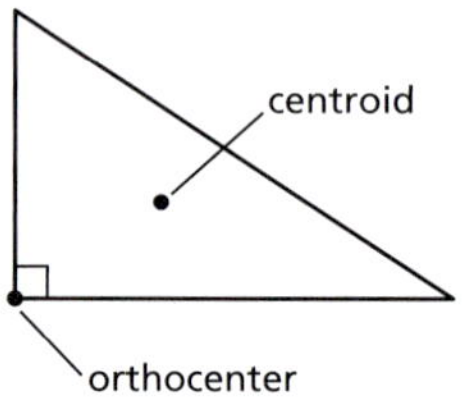

25.

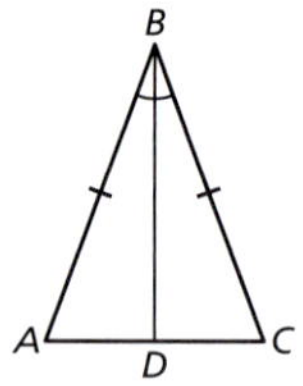

Legs $\overline{AB}$ and $\overline{BC}$ of isosceles $\triangle ABC$ are congruent. $\angle ABD \cong \angle CBD$ because $\overline{BD}$ is an angle bisector of vertex angle ABC. Also, $\overline{BD} \cong \overline{BD}$ by the Reflexive Property of Congruence. So, $\triangle ABD \cong \triangle CBD$ by the SAS Congruence Theorem. $\overline{AD} \cong \overline{CD}$ because corresponding parts of congruent triangles are congruent. So, $\overline{BD}$ is a median.

27. never; Because medians are always inside a triangle, and the centroid is the point of concurrency of the medians, it will always be inside the triangle.

29. sometimes; A median is the same line segment as the perpendicular bisector if the triangle is equilateral or if the segment is connecting the vertex angle to the base of an isosceles triangle. Otherwise, the median and the perpendicular bisectors are not the same segment.

31. sometimes; The centroid and the orthocenter are not the same point unless the triangle is equilateral.

33. Both segments are perpendicular to a side of a triangle, and their point of intersection can fall either inside, on, or outside the triangle. However, the altitude does not necessarily bisect the side, but the perpendicular bisector does. Also, the perpendicular bisector does not necessarily pass through the opposite vertex, but the altitude does.

35. 6.75 in.^{2}; altitude

37. 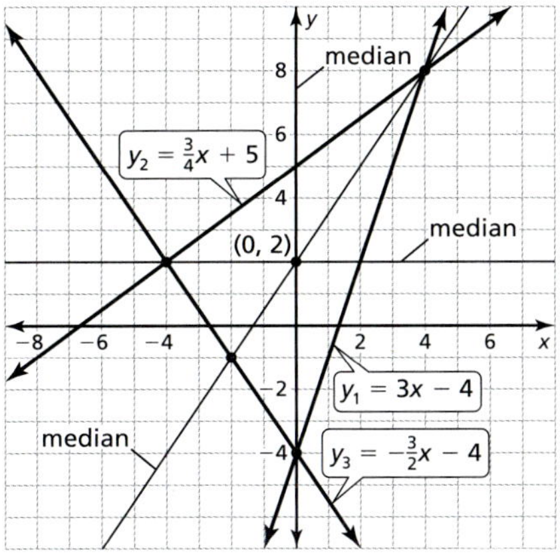

(0, 2)

39. $x = 2.5$

41. $x = 4$

43. $PE = \frac{1}{3}AE, PE = \frac{1}{2}AP, PE = AE - AP$

45. yes; If the triangle is equilateral, then the perpendicular bisectors, angle bisectors, medians, and altitudes will all be the same three segments.

47. 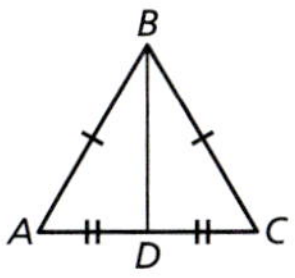

Sides $\overline{AB}$ and $\overline{BC}$ of equilateral $\triangle ABC$ are congruent. $\overline{AD} \cong \overline{CD}$ because $\overline{BD}$ is the median to $\overline{AC}$. Also, $\overline{BD} \cong \overline{BD}$ by the Reflexive Property of Congruence. So, $\triangle ABD \cong \triangle CBD$ by the SSS Congruence Theorem. $\angle ADB \cong \angle CDB$ and $\angle ABD \cong \angle CBD$ because corresponding parts of congruent triangles are congruent. Also, $\angle ADB$ and $\angle CDB$ are a linear pair. Because $\overline{BD}$ and $\overline{AC}$ intersect to form a linear pair of congruent angles, $\overline{BD} \perp \overline{AC}$. So, median $\overline{BD}$ is also an angle bisector, altitude, and perpendicular bisector of $\triangle ABC$.

49. a.

STATEMENTS	REASONS
1. $\overline{LP}$ and $\overline{MQ}$ are medians of scalene $\triangle LMN$; $\overline{LP} \cong \overline{PR}$, $\overline{MQ} \cong \overline{QS}$	**1.** Given
2. $\overline{NP} \cong \overline{MP}$, $\overline{LQ} \cong \overline{NQ}$	**2.** Definition of median
3. $\angle LPM \cong \angle RPN$, $\angle MQL \cong \angle SQN$	**3.** Vertical Angles Congruence Theorem
4. $\triangle LPM \cong \triangle RPN$, $\triangle MQL \cong \triangle SQN$	**4.** SAS Congruence Theorem
5. $\overline{NR} \cong \overline{LM}$, $\overline{NS} \cong \overline{LM}$	**5.** Corresponding parts of congruent triangles are congruent.
6. $\overline{NS} \cong \overline{NR}$	**6.** Transitive Property of Congruence

b. It was shown in part (a) that $\triangle LPM \cong \triangle RPN$ and $\triangle MQL \cong \triangle SQN$. So, $\angle LMP \cong \angle RNP$ and $\angle MLQ \cong \angle SNQ$ because corresponding parts of congruent triangles are congruent. Then, $\overline{NS} \parallel \overline{LM}$ and $\overline{NR} \parallel \overline{LM}$ by the Alternate Interior Angles Converse.

c. Because $\overline{NS}$ and $\overline{NR}$ are both parallel to the same segment, $\overline{LM}$, they would have to be parallel to each other by the Transitive Property of Parallel Lines. However, because they intersect at point N, they cannot be parallel. So, they must be collinear.

51. Let $\overline{AE}$ and $\overline{BF}$ be two medians of $\triangle ABC$ and G be their point of intersection. Draw a line segment through point F and parallel to $\overline{AE}$, and let point Q be the intersection with $\overline{BC}$.

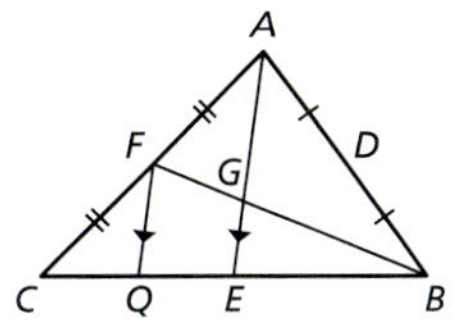

Because $\overline{FQ} \parallel \overline{AE}$, $\angle CFQ \cong \angle CAE$ and $\angle CQF \cong \angle CEA$ by the Corresponding Angles Theorem. $\triangle CFQ$ and $\triangle CAE$ also have $\angle C$ in common, so $\triangle CAE$ is a similarity transformation of $\triangle CFQ$ because the two triangles have the same shape and size. So, $CA = k_1CF$, $CE = k_1CQ$, and $AE = k_1FQ$ for some scale factor k_1. Similarly, note that $\triangle BFQ$ is a similarity transformation of $\triangle BGE$, so $BF = k_2BG$, $BQ = k_2BE$, and $FQ = k_2GE$ for some scale factor k_2. Because F is the midpoint of $\overline{CA}$ and $CA = k_1CF$, $k_1 = 2$. So, $CE = 2CQ$ and Q is the midpoint of $\overline{CE}$. By the Segment Addition Postulate, $QB = QE + EB$. Because $QB = QE + EB = \frac{1}{2}CE + EB = \frac{1}{2}EB + EB = \frac{3}{2}EB$, $k_2 = \frac{3}{2}$. So, $BF = \frac{3}{2}BG$ and the intersection of medians $\overline{AE}$ and $\overline{BF}$, point G, lies on $\overline{BF}$, $\frac{2}{3}$ of the way from B to F.

Let $\overline{CD}$ be the third median of $\triangle ABC$ and H be its point of intersection with $\overline{BF}$. Draw a line segment through point F and parallel to $\overline{CD}$, and let point R be the intersection with $\overline{AB}$.

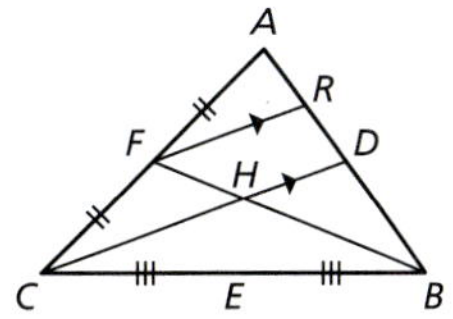

You can use the same reasoning as before to show that $BF = \frac{3}{2}BH$ and the intersection of medians $\overline{CD}$ and $\overline{BF}$, point H, lies on $\overline{BF}$, $\frac{2}{3}$ of the way from B to F.

So, points G and H are the same point, which means medians $\overline{AE}$, $\overline{BF}$, and CD are concurrent at that point.

6.3 Review & Refresh

52. 59°; Because J is equidistant from $\overrightarrow{GH}$ and $\overrightarrow{GK}$, $\overrightarrow{GJ}$ bisects $\angle HGK$ by the Angle Bisector Theorem. This means that $5x - 4 = 4x + 3$, and the solution is $x = 7$. So, $m\angle JGK = 4x + 3 = 4(7) + 3 = 31°$. So, by the Triangle Sum Theorem, $m\angle GJK = 180 - 90 - 31 = 59°$.

53. (7, 5)

54. on; (7, −4)

55. yes

56. no

57.

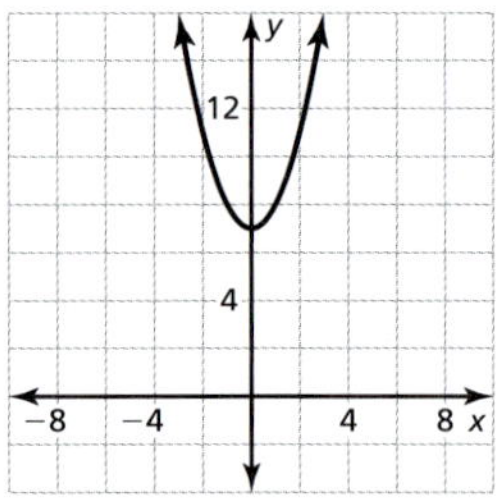

The graph of g is a vertical translation 7 units up of the graph of f.

58. (6, 4)
59. (−2, 3)
60. $x = \pm 4i$
61. $x = 4 \pm 2\sqrt{3}$
62. no; My friend calculated the midpoint location incorrectly. The midpoint is calculated by using $\left(\frac{2+5}{2}, \frac{3+7}{2}\right)$, which results in the location (3.5, 5).
63. *Sample answer:* First find ED and EF using the Distance Formula. Then, $\overline{ED} \cong \overline{EF}$ by the definition of congruent segments. Use the Reflexive Property of Segment Congruence to show that $\overline{EG} \cong \overline{EG}$. $\angle DEG \cong \angle FEG$ by the definition of angle bisector. Finally, prove that $\triangle EGD \cong \triangle EGF$ by the SAS Congruence Theorem.

Section 6.4

6.4 Practice

1. $D(-4, -2)$, $E(-2, 0)$, $F(-1, -4)$
3. Because the slopes of $\overline{EF}$ and $\overline{AC}$ are the same (-4), $\overline{EF} \parallel \overline{AC}$. $EF = \sqrt{17}$ and $AC = 2\sqrt{17}$. Because $\sqrt{17} = \frac{1}{2}\left(2\sqrt{17}\right)$, $EF = \frac{1}{2}AC$.
5. The midpoint of $\overline{OC}$ is $F(p, 0)$. Because the slopes of $\overline{DF}$ and $\overline{BC}$ are the same $\left(-\frac{r}{p-q}\right)$, $\overline{DF} \parallel \overline{BC}$.
$DF = \sqrt{p^2 - 2pq + q^2 + r^2}$ and
$BC = 2\sqrt{p^2 - 2pq + q^2 + r^2}$.
Because $\sqrt{p^2 - 2pq + q^2 + r^2} = \frac{1}{2}\left(2\sqrt{p^2 - 2pq + q^2 + r^2}\right)$,
$DF = \frac{1}{2}BC$.
7. $x = 6$
9. $x = 13$
11. $\overline{JK} \parallel \overline{YZ}$
13. $\overline{XY} \parallel \overline{KL}$
15. $\overline{JL} \cong \overline{XK} \cong \overline{KZ}$
17. 45 ft
19. $\overline{DE}$ is not parallel to $\overline{BC}$. So, $\overline{DE}$ is not a midsegment. So, according to the contrapositive of the Triangle Midsegment Theorem, $\overline{DE}$ does not connect the midpoints of $\overline{AC}$ and $\overline{AB}$.
21. 17
23. **a.** (−1, 2), (9, 8), (5, 0)
b. (2.6, 16.2), (7.4, 13.8), (5.4, 7.8)
Find the slope of each midsegment. Graph the line parallel to each midsegment passing through the opposite vertex. The intersection of these lines will be the vertices of the original triangle. You can check your answer by finding the midsegments of the triangle.

6.4 Review & Refresh

24. *Sample answer:* $-2 - (-7) = 5$, and $5 > -2$
25. 15; Because $\overline{SW} \cong \overline{UW}$ and $\overleftrightarrow{VW} \perp \overline{SU}$, point V is on the perpendicular bisector of $\overline{SU}$. So, by the Perpendicular Bisector Theorem, $SV = UV$. So, $2x + 11 = 8x - 1$, and the solution is $x = 2$, which means that $UV = 8x - 1 = 8(2) - 1 = 15$.
26. (−4, −4)
27. 11
28. $x = 10$, $y = 14$
29.

		Walt Disney World		
		Yes	No	TOTAL
Universal Studios	Yes	23	62	85
	No	71	49	120
	TOTAL	94	111	205

Out of 205 students surveyed, 85 have been to Universal Studios (120 have not) and 94 have been to Walt Disney World (111 have not).

30. nonlinear; It is an absolute value function.
31. linear; The function is of the form $y = mx + b$.

Section 6.5

6.5 Practice

1. Assume temporarily that $WV = 7$ inches.
3. Assume temporarily that $\angle B$ is a right angle.
5. A and C; The angles of an equilateral triangle are always 60°. So, an equilateral triangle cannot have a 90° angle, and cannot be a right triangle.
7. Assume temporarily that an odd number is divisible by 4. Let the odd number be represented by $2y + 1$ where y is a positive integer. Then there must be a positive integer x such that $4x = 2y + 1$. However, when you divide each side of the equation by 4, you get $x = \frac{1}{2}y + \frac{1}{4}$, which is not an integer. So, the assumption must be false, and an odd number is not divisible by 4.
9. *Sample answer:*

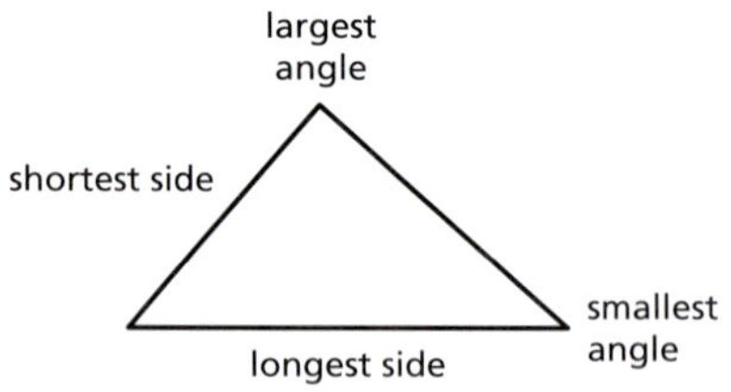

The longest side is across from the largest angle, and the shortest side is across from the smallest angle.

11. $\angle S, \angle R, \angle T$
13. $\angle D, \angle F, \angle E$
15. $\overline{AB}, \overline{BC}, \overline{AC}$
17. $\overline{NP}, \overline{MN}, \overline{MP}$
19. 7 in. $< x <$ 17 in.
21. 16 in. $< x <$ 64 in.
23. yes
25. no; $28 + 17 \not> 46$
27. Because $\sqrt{3} < 2$, $m\angle B < m\angle A$; The correct order is $\angle C$, $\angle B$, $\angle A$.

29. Assume temporarily that the client is guilty. Then the client would have been in Los Angeles, California, at the time of the crime. Because the client was in New York at the time of the crime, the assumption must be false, and the client must be innocent.

31. C

33. The right angle of a right triangle must always be the largest angle because the other two will have a sum of 90°. So, according to the Triangle Larger Angle Theorem, because the right angle is larger than either of the other angles, the side opposite the right angle, which is the hypotenuse, will always have to be longer than either of the legs.

35. **a.** $x > 5$ km, $x < 211$ km

b. The distance between Orlando and Daytona Beach must equal the shortest side of the triangle. $x > 5$ km, $x < 103$ km

37. $2 < x < 15$

39. By the Exterior Angle Theorem, $m\angle 1 = m\angle A + m\angle B$. Then by the Subtraction Property of Equality, $m\angle 1 - m\angle B = m\angle A$. If you assume temporarily that $m\angle 1 \leq m\angle B$, then $m\angle A \leq 0$. Because the measure of any angle in a triangle must be a positive number, the assumption must be false. So, $m\angle 1 > m\angle B$. Similarly, by the Subtraction Property of Equality, $m\angle 1 - m\angle A = m\angle B$. If you assume temporarily that $m\angle 1 \leq m\angle A$, then $m\angle B \leq 0$. Because the measure of any angle in a triangle must be a positive number, the assumption must be false. So, $m\angle 1 > m\angle A$.

41. It is given that $BC > AB$ and $BD = BA$. By the Base Angles Theorem, $m\angle 1 = m\angle 2$. By the Angle Addition Postulate, $m\angle BAC = m\angle 1 + m\angle 3$. So, $m\angle BAC > m\angle 1$. Substituting $m\angle 2$ for $m\angle 1$ produces $m\angle BAC > m\angle 2$. By the Exterior Angle Theorem, $m\angle 2 = m\angle 3 + m\angle C$. So, $m\angle 2 > m\angle C$. Finally, because $m\angle BAC > m\angle 2$ and $m\angle 2 > m\angle C$, you can conclude that $m\angle BAC > m\angle C$.

43. greater than 4 and less than 24; Because of the Triangle Inequality Theorem, FG must be greater than 2 and less than 8, GH must be greater than 1 and less than 7, and FH must be greater than 1 and less than 9. So, the perimeter must be greater than $2 + 1 + 1 = 4$ and less than $8 + 7 + 9 = 24$.

45. Assume temporarily that another segment, $\overline{PA}$, where A is on plane M, is the shortest segment from P to plane M. By definition of the distance between a point and a plane, $\overline{PA} \perp$ plane M. This contradicts the given statement because there cannot be two different segments that share an endpoint and are both perpendicular to the same plane. So, the assumption is false, and because no other segment exists that is the shortest segment from P to plane M, it must be $\overline{PC}$ that is the shortest segment from P to plane M.

6.5 Review & Refresh

47. $(-4, 1), (0, 2), (-1, -1)$

48. $k = \frac{1}{4}$

49. 15

50. inside; (0, 4)

51.

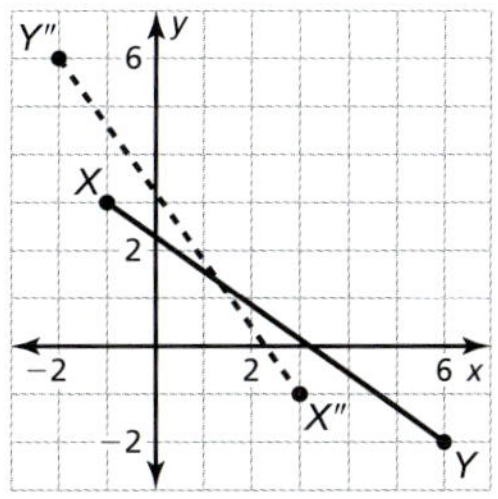

52. yes; HL Congruence Theorem

53. no

54. 6 cm < x < 14 cm

55. $x = 12$

56. *Sample answer:*

Equation	Justification
$\frac{1}{4}x - \frac{3}{2}y = -1$	Write given equation.
$-\frac{3}{2}y = -\frac{1}{4}x - 1$	Subtraction Property of Equality
$-\frac{2}{3}\left(-\frac{3}{2}y\right) = -\frac{2}{3}\left(-\frac{1}{4}x - 1\right)$	Multiplication Property of Equality
$y = \frac{1}{6}x + \frac{2}{3}$	Distributive Property

Section 6.6

6.6 Practice

1. $m\angle 1 > m\angle 2$; By the Converse of the Hinge Theorem, because $\angle 1$ is the included angle in the triangle with the longer third side, its measure is greater than that of $\angle 2$.

3. $m\angle 1 < m\angle 2$; By the Converse of the Hinge Theorem, because $\angle 1$ is the included angle in the triangle with the shorter third side, its measure is less than that of $\angle 2$.

5. $AC > DC$; By the Hinge Theorem, because $\overline{AC}$ is the third side of the triangle with the larger included angle, it is longer than $\overline{DC}$.

7. $TR < UR$; By the Hinge Theorem, because $\overline{TR}$ is the third side of the triangle with the smaller included angle, it is shorter than $\overline{UR}$.

9. $\overline{XY} \cong \overline{YZ}$ and $m\angle WYZ > m\angle WYX$ are given. By the Reflexive Property of Congruence, $\overline{WY} \cong \overline{WY}$. So, by the Hinge Theorem, $WZ > WX$.

11. Flight 1; Because 160° > 150°, the distance the first plane flew is a greater distance than the distance the second plane flew by the Hinge Theorem.

13. The measure of the included angle in $\triangle PSQ$ is greater than the measure of the included angle in $\triangle SQR$; By the Hinge Theorem, $PQ > RS$.

15. $x > \frac{3}{2}$; By the Exterior Angle Theorem, $m\angle ABD = m\angle BDC + m\angle C$. So, $m\angle ABD > m\angle BDC$.

$$AD > BC$$
$$4x - 3 > 2x$$
$$2x > 3$$
$$x > \frac{3}{2}$$

17. Because $\overline{NR}$ is a median, $\overline{PR} \cong \overline{QR}$. $\overline{NR} \cong \overline{NR}$ by the Reflexive Property of Congruence. So, by the Converse of the Hinge Theorem, $\angle NRQ > \angle NRP$. Because $\angle NRQ$ and $\angle NRP$ form a linear pair, they are supplementary. So, $\angle NRQ$ must be obtuse and $\angle NRP$ must be acute.

19. Because $\overline{BC} \cong \overline{EF}$, $\angle CBP \cong \angle FED$ by construction, and $\overline{BP} \cong \overline{ED}$ by construction, you have $\triangle PBC \cong \triangle DEF$ by the SAS Congruence Theorem.

Because $\overrightarrow{BH}$ bisects $\angle PBA$ by construction, $\angle PBH \cong \angle ABH$. By the Transitive Property of Congruence, $\overline{AB} \cong \overline{PB}$. By the Reflexive Property of Congruence, $\overline{BH} \cong \overline{BH}$. So, $\triangle ABH \cong \triangle PBH$ by the SAS Congruence Theorem.

By the Segment Addition Postulate, $AC = AH + HC$. Because corresponding parts of congruent triangle are congruent, $\overline{AH} \cong \overline{PH}$. By the definition of congruent segments, $AH = PH$. By the Addition Property of Equality, $AH + HC = PH + HC$. By the Triangle Inequality Theorem, $PH + HC > PC$. By substitution, $AH + HC > PC$ and $AC > PC$. Because corresponding parts of congruent triangles are congruent, $\overline{PC} \cong \overline{DF}$. By the definition of congruent segments, $PC = DF$. So, by substitution $AC > DF$.

6.6 Review & Refresh

20. $x = 72$

21. $x = 17$

22. 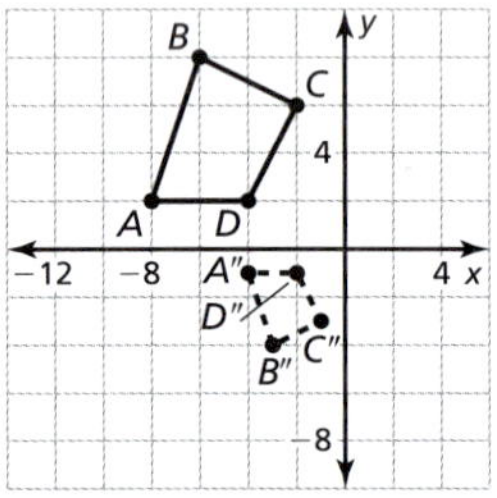

23. $\overline{FG}$; Because $\overline{KL}$ is the third side of the triangle with the smaller included angle, it is shorter than $\overline{FG}$ by the Hinge Theorem.

24. a. centroid

b. no; This point is not equidistant from the three cities. the circumcenter would be equidistant from the cities.

25. no; Because $9 + 11 \ngtr 21$, by the Triangle Inequality Theorem, the triangle is not possible.

Chapter 6 Review

1. 20; Point B is equidistant from A and C, and $\overleftrightarrow{BD} \perp \overline{AC}$. So, by the Converse of the Perpendicular Bisector Theorem, $DC = AD = 20$.

2. 23; $\angle PQS \cong \angle RQS$, $\overline{SR} \perp \overrightarrow{QR}$, and $\overline{SP} \perp \overrightarrow{QP}$. So, by the Angle Bisector Theorem, $SR = SP$. This means that $6x + 5 = 9x - 4$, and the solution is $x = 3$. So, $RS = 9(3) - 4 = 23$.

3. 47°; Point J is equidistant from $\overrightarrow{FG}$ and $\overrightarrow{FH}$. So, by the Converse of the Angle Bisector Theorem, $m\angle JFH = m\angle JFG = 47°$.

4. no; You would need to know $\overline{RQ} \cong \overline{TQ}$, or one pair of corresponding angles are congruent.

5. $(-3, -3)$

6. $(4, 3)$

7. $x = 5$

8. *Sample answer:*

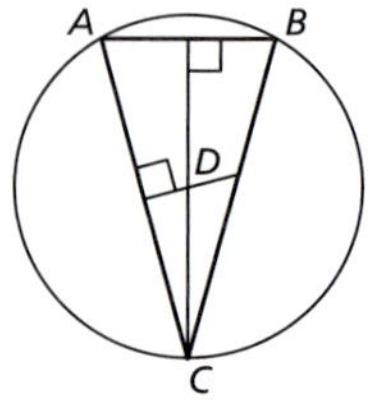

9. *Sample answer:*

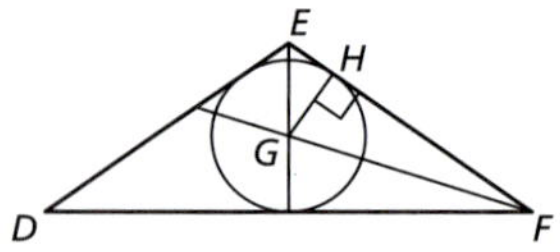

10. $ED = 6$, $DC = 12$

11. $ED = 9$, $DC = 18$

12. $(-6, 3)$

13. $(4, -4)$

14. inside; (3, 5.2)

15. outside; $(-6, -1)$

16. $\triangle ABC$ is isosceles.

17. 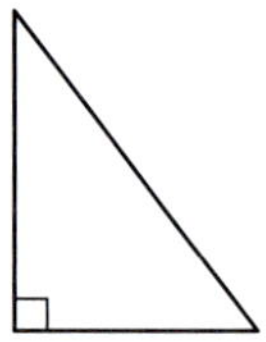

The orthocenter of any right triangle is the vertex containing the right angle because the legs of the triangle are altitudes.

18. $(-6, 6)$, $(-3, 6)$, $(-3, 4)$

19. $(0, 3)$, $(2, 0)$, $(-1, -2)$

20. $x = 50$

21. $x = 14$

22. 1160 m

23. Assume temporarily that $YZ \ngtr 4$. Then it follows that either $YZ < 4$ or $YZ = 4$. If $YZ < 4$, then $XY + YZ < XZ$ because $4 + YZ < 8$ when $YZ < 4$. If $YZ = 4$, then $XY + YZ = XZ$ because $4 + 4 = 8$. Both conclusions contradict the Triangle Inequality Theorem, which says that $XY + YZ > XZ$. So, the temporary assumption that $YZ \ngtr 4$ cannot be true. This proves that in $\triangle XYZ$, if $XY = 4$ and $XZ = 8$, then $YZ > 4$.

24. $\overline{FH}, \overline{FG}, \overline{GH}$

25. $\overline{JK}, \overline{KL}, \overline{JL}$

26. 4 in. $< x <$ 12 in.

27. 3 m $< x <$ 15 m

28. Assume temporarily that $\triangle LRT$ has three congruent angles. $LT = 36$ in. and $LR \leq 24$ in., so $LT > LR$. Because $LT > LR$, $m\angle R > m\angle T$ by the Triangle Longer Side Theorem. So, $m\angle R \neq m\angle T$. So, the assumption must be false, which proves that there can be at most two congruent angles of $\triangle LRT$.

29. $QT > ST$

30. $m\angle QRT > m\angle SRT$

31. second boat; Because $170° > 160°$, the distance the second boat traveled is greater than the distance the first boat traveled by the Hinge Theorem.

Chapter 6 Practice Test

1. $x = 6$

2. $x = 9$

3. $ST = 17$; Because $\overline{RV} \cong \overline{TV}$ and $\overleftrightarrow{SV} \perp \overline{RT}$, point S lies on the perpendicular bisector of $\overline{RT}$. So, by the Perpendicular Bisector Theorem, $RS = TS$. So, $3x + 8 = 7x - 4$. The solution is $x = 3$, so $ST = 7(3) - 4 = 17$.

4. $WY = 32$; Because $\angle WXY \cong \angle WXZ$, point W is on the angle bisector of $\angle ZXY$. So, by the Angle Bisector Theorem, $WY = WZ$. So, $6x + 2 = 9x - 13$. The solution is $x = 5$, so $WY = 6(5) + 2 = 32$.

5. $BW = 20$; Because $m\angle AZW = m\angle BZW$, $m\angle BYW = m\angle CYW$, and $m\angle CXW = m\angle AXW$, point W is the incenter of $\triangle XYZ$. By the Incenter Theorem, $BW = CW$. So, $BW = 20$.

6. $AB > CB$

7. $m\angle 1 < m\angle 2$

8. $m\angle MNP < m\angle NPM$

9. $(2, 2)$; $(0, -2)$; $\left(\frac{4}{3}, \frac{2}{3}\right)$

10. Assume temporarily that $\triangle PQR$ is equilateral and equiangular. Then it follows that $m\angle P \neq m\angle Q$, $m\angle Q \neq m\angle R$, or $m\angle P \neq m\angle R$. By the contrapositive of the Base Angles Theorem, if $m\angle P \neq m\angle Q$, then $PR \neq QR$, if $m\angle Q \neq m\angle R$, then $QP \neq RP$, and if $m\angle P \neq m\angle R$, then $PQ \neq RQ$. All three conclusions contradict the fact that $\triangle PQR$ is equilateral. So, the temporary conclusion must be false. This proves that if $\triangle PQR$ is equilateral, it must also be equiangular.

11. $\frac{ab}{8}$ units^2; $\triangle DGH$ has vertices $D(0, 0)$, $G\left(\frac{a}{2}, \frac{b}{2}\right)$, and $H\left(\frac{a}{2}, 0\right)$. So it has a base of $\frac{a}{2}$ units, a height of $\frac{b}{2}$ units, and $A = \frac{1}{2}bh = \frac{1}{2}\left(\frac{a}{2}\right)\left(\frac{b}{2}\right) = \frac{ab}{8}$ units^2.

12. *Sample answer:* $C(0, 0)$

13. the first hiker; Because $140° > 128°$, the first hiker is a farther distance, because the longer side is opposite the larger angle.

14. Pine Avenue must be longer than 2 miles and shorter than 16 miles.

15. 9 mi; Because the path represents the shortest distance from the beach entrance to Main Street, it must be perpendicular to Main Street, and you ended up at the midpoint between home and the movie theater. So, the trail must be the perpendicular bisector of the portion of Main Street between home and the movie theater. By the Perpendicular Bisector Theorem, the beach entrance must be the same distance from your house and the movie theater. So, Pine Avenue is the same length as the 9-mile portion of Hill Street between home and the beach entrance.

16.

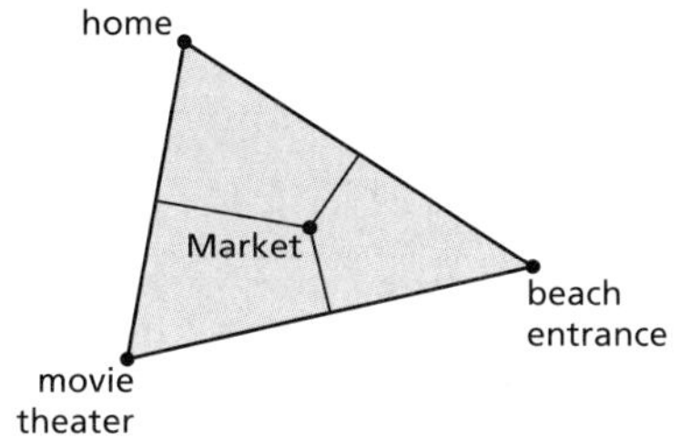

circumcenter

Chapter 6 Test Prep

1.

STATEMENTS	REASONS
1. $\overline{YG}$ is the perpendicular bisector of $\overline{DF}$.	1. Given
2. $\overline{DE} \cong \overline{FE}$, $\overline{YG} \perp \overline{DF}$	2. Definition of perpendicular bisector
3. $\angle DEY$ and $\angle FEY$ are right angles.	3. Definition of perpendicular lines
4. $\angle DEY \cong \angle FEY$	4. Right Angles Congruence Theorem
5. $\overline{YE} \cong \overline{YE}$	5. Reflexive Property of Congruence
6. $\triangle DEY \cong \triangle FEY$	6. SAS Congruence Theorem

2. D

3. C

4. B

5. definition of angle bisector; given; Reflexive Property of Congruence; SAS Congruence Theorem; Corresponding parts of congruent triangles are congruent.

6. a. $T(0, 7)$, $U(2, 4)$, $V(-1, 5)$

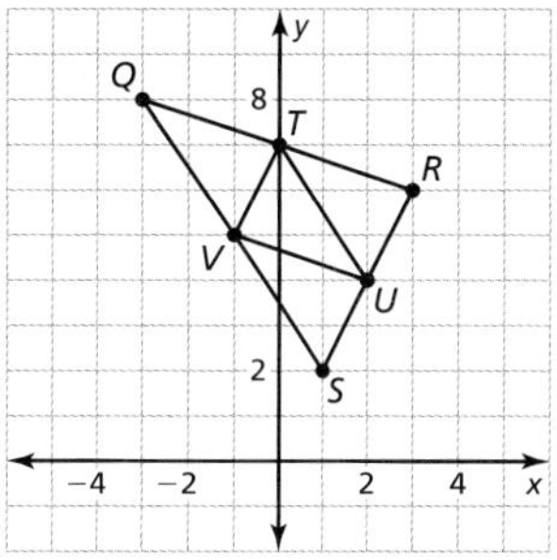

b. slope of $\overline{TV} = 2$, slope of $\overline{SR} = 2$,
Because the slopes are the same, $\overline{TV} \parallel \overline{RS}$.
slope of $\overline{TU} = -\frac{3}{2}$, slope of $\overline{QS} = -\frac{3}{2}$,
Because the slopes are the same, $\overline{TU} \parallel \overline{QS}$.
slope of $\overline{VU} = -\frac{1}{3}$, slope of $\overline{QR} = -\frac{1}{3}$,
Because the slopes are the same, $\overline{VU} \parallel \overline{QR}$.
$TV = \sqrt{5}$, $SR = 2\sqrt{5}$,
Because $\sqrt{5} = \frac{1}{2}(2\sqrt{5})$, $TV = \frac{1}{2}SR$.
$TU = \sqrt{13}$, $QS = 2\sqrt{13}$,
Because $\sqrt{13} = \frac{1}{2}(2\sqrt{13})$, $TU = \frac{1}{2}QS$.
$VU = \sqrt{10}$, $QR = 2\sqrt{10}$,
Because $\sqrt{10} = \frac{1}{2}(2\sqrt{10})$, $VU = \frac{1}{2}QR$.

7. B

8. slope of $\overleftrightarrow{BD} = \frac{2}{3}$, slope of $\overleftrightarrow{B'D'} = \frac{2}{3}$,
Because the slopes are the same, $\overleftrightarrow{BD} \parallel \overleftrightarrow{B'D'}$.

Chapter 7

Prepare with CalcChat

1. $x = 3$
2. $x = 4$
3. $x = 7$
4. $a \parallel b, c \perp d$
5. $a \parallel b, c \parallel d, a \perp c, a \perp d, b \perp c, b \perp d$
6. $b \parallel c, b \perp d, c \perp d$
7. Line a is perpendicular to line e.

Section 7.1

7.1 Practice

1. 1260°
3. 2520°
5. hexagon
7. 16-gon
9. $x = 64$
11. $x = 89$
13. $x = 70$
15. The right angle was not included and the sum of the angle measures should be 720°, not 540°;

$$x° + 121° + 96° + 101° + 162° + 90° = 720°$$
$$x + 570 = 720$$
$$x = 150$$

17. $m\angle X = m\angle Y = 92°$
19. $m\angle X = m\angle Y = 100.5°$
21. $x = 111$
23. $x = 32$
25. 108°, 72°
27. 172°, 8°
29. 135°
31. $n = \dfrac{360}{180 - x}$
33. 15
35. 40
37. A, B; Solving the equation found in Exercise 31 for n yields a positive integer greater than or equal to 3 for A and B, but not for C and D.
39. In a pentagon, when all the diagonals from one vertex are drawn, the polygon is divided into three triangles. Because the sum of the measures of the interior angles of each triangle is 180°, the sum of the measures of the interior angles of the pentagon is $(5 - 2) \cdot 180° = 3 \cdot 180° = 540°$.
41. 21°, 21°, 21°, 21°, 138°, 138°
43. **a.** $h(n) = \dfrac{(n - 2) \cdot 180°}{n}$

    b. $h(9) = 140°$; $n = 12$

    c.

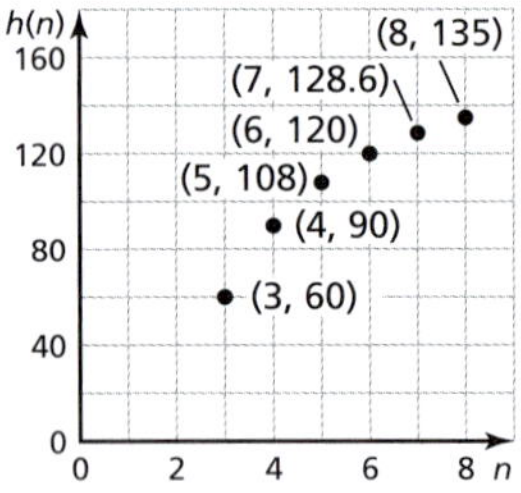

The value of $h(n)$ increases on a curve that gets less steep as n increases.

45. yes; The measure of the angle where the polygon caves in is greater than 180° but less than 360°.
47. 90°; The base angles of $\triangle BPC$ are congruent exterior angles of the regular octagon, each with a measure of 45°. So, $m\angle BPC = 180° - 2(45°) = 90°$.

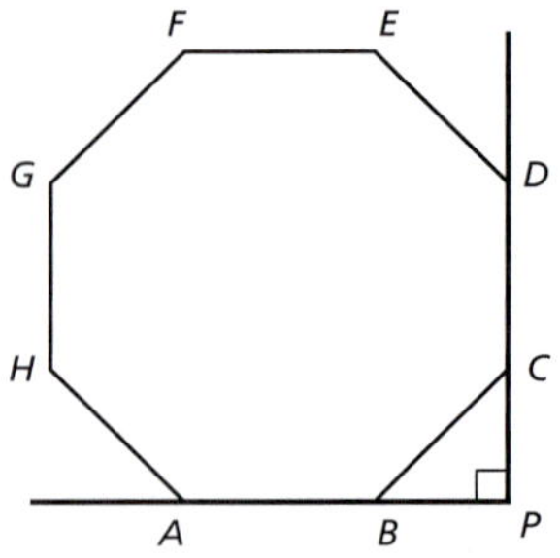

7.1 Review & Refresh

49. $x = 101$
50. $x = 67$
51. $m\angle 2$; By the Converse of the Hinge Theorem, $\angle 1$ is the included angle in the triangle with the shorter third side, so its measure is less than that of $\angle 2$.
52. 13 in. $< x <$ 21 in.
53. $y = -2x + 2$
54. yes; The polygon has 6 lines of symmetry, as shown.

A reflection in the line of symmetry maps the polygon onto itself.

55. nonagon
56. $x = 14$
57.

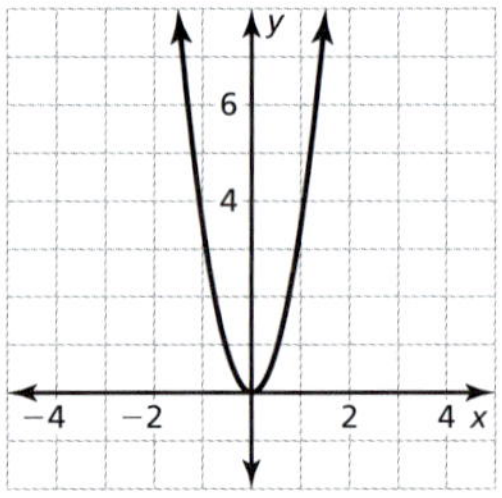

The graph of g is a vertical stretch by a factor of 3.5 of the graph of f.

58.

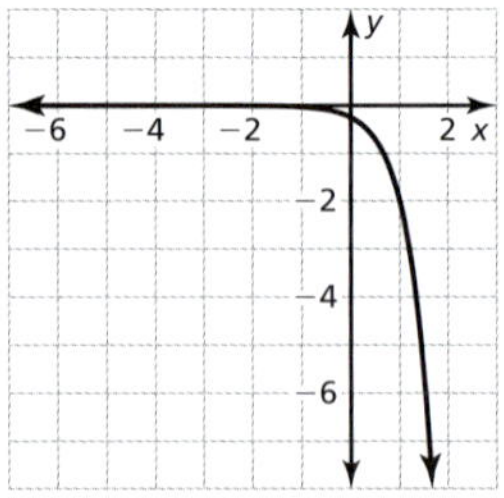

The graph of f is a vertical shrink by a factor of $\frac{1}{4}$, followed by a reflection in the x-axis of the graph of $g(x) = 8^x$; The y-intercept of f is $\left(0, -\frac{1}{4}\right)$ and the y-intercept of g is $(0, 1)$. The asymptotes of f and g are both $y = 0$. The domain of f is all real numbers and the range of f is $y < 0$.

59. $(x - 3)(x + 7)$

60. 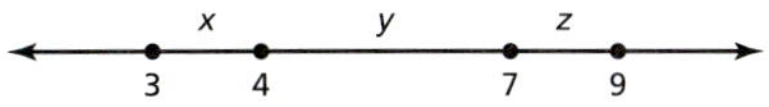

Let x, y, and z represent the distances between obstacles as shown. It is given that $x + y = y + z$. By the Subtraction Property of Equality, $x = z$. So, the distance between obstacles 3 and 4 is the same as the distance between obstacles 7 and 9.

61. 110°

Section 7.2

7.2 Practice

1. $x = 9, y = 15$

3. $d = 126, z = 28$

5. 129°

7. 13; By the Parallelogram Opposite Sides Theorem, $LM = QN$.

9. 7; By the Parallelogram Diagonals Theorem, $LP = PN$.

11. 80°; By the Parallelogram Consecutive Angles Theorem, $\angle QLM$ and $\angle LMN$ are supplementary. So, $m\angle LMN = 180° - 100°$.

13. 100°; By the Parallelogram Opposite Angles Theorem, $m\angle QLM = m\angle MNQ$.

15. $m = 36, n = 108$

17. $k = 7, m = 8$

19. In a parallelogram, consecutive angles are supplementary; Because quadrilateral $STUV$ is a parallelogram, $\angle S$ and $\angle V$ are supplementary. So, $m\angle V = 180° - 50° = 130°$.

21.

STATEMENTS	REASONS
1. $ABCD$ and $CEFD$ are parallelograms.	**1.** Given
2. $\overline{AB} \cong \overline{DC}$, $\overline{DC} \cong \overline{FE}$	**2.** Parallelogram Opposite Sides Theorem
3. $\overline{AB} \cong \overline{FE}$	**3.** Transitive Property of Congruence

23. (1, 2.5)

25. $F(3, 3)$

27. $G(2, 0)$

29. 36°, 144°

31. $m\angle R = 114°, m\angle S = 66°, m\angle T = 114°$

33.

STATEMENTS	REASONS
1. $ABCD$ is a parallelogram.	**1.** Given
2. $\overline{AB} \parallel \overline{DC}$, $\overline{BC} \parallel \overline{AD}$	**2.** Definition of parallelogram
3. $\angle BDA \cong \angle DBC$, $\angle DBA \cong \angle BDC$	**3.** Alternate Interior Angles Theorem
4. $\overline{BD} \cong \overline{BD}$	**4.** Reflexive Property of Congruence
5. $\triangle ABD \cong \triangle CDB$	**5.** ASA Congruence Theorem
6. $\angle A \cong \angle C$	**6.** Corresponding parts of congruent triangles are congruent.
7. $m\angle BDA = m\angle DBC$, $m\angle DBA = m\angle BDC$	**7.** Definition of congruent angles
8. $m\angle B = m\angle DBC + m\angle DBA$, $m\angle D = m\angle BDA + m\angle BDC$	**8.** Angle Addition Postulate
9. $m\angle D = m\angle DBC + m\angle DBA$	**9.** Substitution Property of Equality
10. $m\angle D = m\angle B$	**10.** Transitive Property of Equality
11. $\angle D \cong \angle B$	**11.** Definition of congruent angles

35. By the definition of a parallelogram, $WX \parallel YZ$. So, $\angle WZV \cong \angle YXV$ and $\angle ZWV \cong \angle XYV$ by the Alternate Interior Angles Theorem. By the Parallelogram Opposite Sides Theorem, $\overline{ZW} \cong \overline{XY}$. $\triangle ZWV \cong \triangle XYV$ by the ASA Congruence Theorem. Because corresponding parts of congruent triangles are congruent, $\overline{WV} \cong \overline{YV}$ and $\overline{ZV} \cong \overline{XV}$. Because point V is the midpoint of both $\overline{WY}$ and $\overline{XZ}$ by the definition of midpoint, $\overline{WY}$ and $\overline{XZ}$ are segment bisectors of each other by definition.

37. 52 units

39. 3; (−2, 4), (4, 0), (8, 8)

41. Because the side holding the binoculars is always parallel to the fixed side, the binoculars can move forward and backward to adjust the height without changing the orientation.

43. 16°; *Sample answer:* The opposite angles of a parallelogram are congruent, so $m\angle TSV = m\angle TUV$. By Angle Addition Postulate, $m\angle TSU + m\angle USV = m\angle TSV$, so $m\angle TSU + m\angle USV = m\angle TUV$.

$x^2 + 32 = 12x$	Substituting the angle measures.
$x^2 - 12x + 32 = 0$	Subtract $12x$ from both sides of the equation.
$(x - 4)(x - 8) = 0$	Solve by factoring.
$x = 4$ or $x = 8$	Zero Product Property

Then $m\angle USV = 16°$ or 64°. However, $m\angle USV = 16°$ because $\angle TUV$ is acute.

45.

STATEMENTS	REASONS
1. $\overline{EK}$ bisects $\angle FEH$ and $\overline{FJ}$ bisects $\angle EFG$. $EFGH$ is a parallelogram.	1. Given
2. $m\angle PEH = m\angle PEF$, $m\angle PFE = m\angle PFG$	2. Definition of angle bisector
3. $m\angle HEF = m\angle PEH + m\angle PEF$, $m\angle EFG = m\angle PFE + m\angle PFG$	3. Angle Addition Postulate
4. $m\angle HEF = m\angle PEF + m\angle PEF$, $m\angle EFG = m\angle PFE + m\angle PFE$	4. Substitution Property of Equality
5. $m\angle HEF = 2(m\angle PEF)$, $m\angle EFG = 2(m\angle PFE)$	5. Distributive Property
6. $m\angle HEF + m\angle EFG = 180°$	6. Parallelogram Consecutive Angles Theorem
7. $2(m\angle PEF) + 2(m\angle PFE) = 180°$	7. Substitution Property of Equality
8. $2(m\angle PEF + m\angle PFE) = 180°$	8. Distributive Property
9. $m\angle PEF + m\angle PFE = 90°$	9. Division Property of Equality
10. $m\angle PEF + m\angle PFE + m\angle EPF = 180°$	10. Triangle Sum Theorem
11. $90° + m\angle EPF = 180°$	11. Substitution Property of Equality
12. $m\angle EPF = 90°$	12. Subtraction Property of Equality
13. $\angle EPF$ is a right angle.	13. Definition of right angle
14. $\overline{EK} \perp \overline{FJ}$	14. Definition of perpendicular lines

7.2 Review & Refresh

47. $\overline{DE}, \overline{DF}, \overline{EF}$
48. 11.5; By the Parallelogram Opposite Sides Theorem, $\overline{YZ} \cong \overline{WX}$.
49. 38°; By the Parallelogram Consecutive Angles Theorem, $\angle W$ and $\angle Z$ are supplementary.
50. 142°; By the Parallelogram Opposite Angles Theorem, $\angle X \cong \angle Z$.
51. $x = 80$
52. y-axis
53. yes; Alternate Interior Angles Converse
54. $m\angle EGF > m\angle DGE$ by the Converse of the Hinge Theorem.

Section 7.3

7.3 Practice

1. Parallelogram Opposite Angles Converse
3. Parallelogram Diagonals Converse
5. Opposite Sides Parallel and Congruent Theorem
7. $x = 114, y = 66$
9. $x = 3, y = 4$
11. $x = 8$
13. $x = 7$
15.

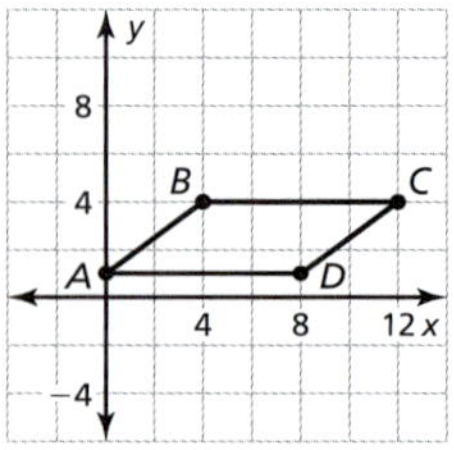

Sample answer: Because $BC = AD = 8$, $\overline{BC} \cong \overline{AD}$. Because both $\overline{BC}$ and $\overline{AD}$ are horizontal lines, their slope is 0, and they are parallel. $\overline{BC}$ and $\overline{AD}$ are opposite sides that are both congruent and parallel. So, $ABCD$ is a parallelogram by the Opposite Sides Parallel and Congruent Theorem.

17.

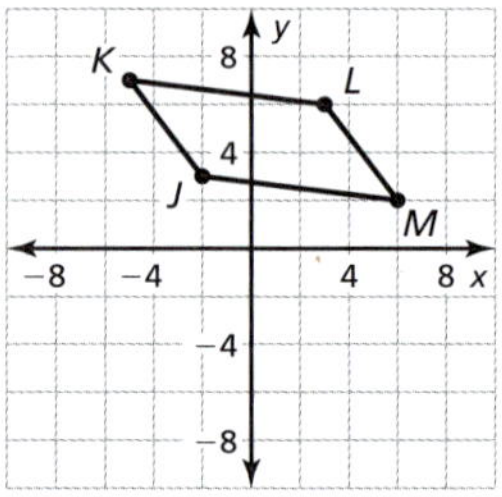

Sample answer: Because $JK = LM = 5$ and $KL = JM = \sqrt{65}$, $\overline{JK} \cong \overline{LM}$ and $\overline{KL} \cong \overline{JM}$. Because both pairs of opposite sides are congruent, quadrilateral $JKLM$ is a parallelogram by the Parallelogram Opposite Sides Converse.

19. In order to be a parallelogram, the quadrilateral must have two pairs of opposite sides that are congruent, not consecutive sides; $DEFG$ is not a parallelogram.
21. $x = 5$
23. A quadrilateral is a parallelogram if and only if both pairs of opposite sides are congruent.
25. A quadrilateral is a parallelogram if and only if the diagonals bisect each other.
27. *Sample answer:*

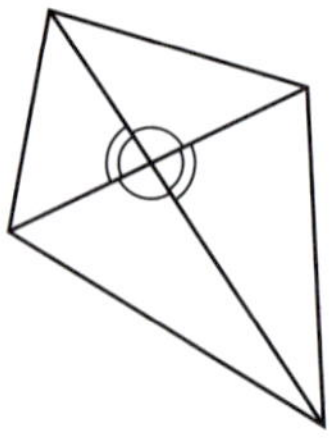

29. *Sample answer:* Draw two horizontal segments that are the same length and connect the endpoints.

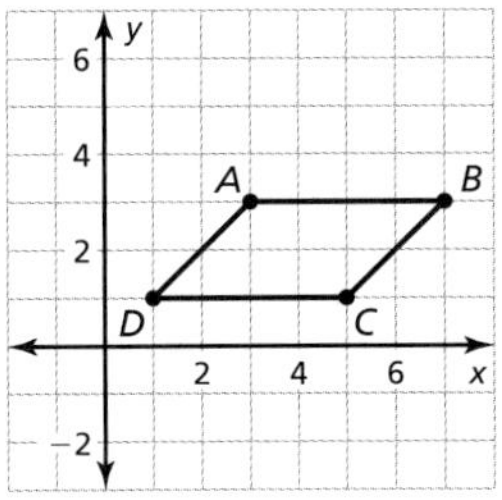

31. **a.** 27°; Because $\angle EAF$ is a right angle, the other two angles of $\triangle EAF$ must be complementary. So, $m\angle AFE = 90° - 63° = 27°$.

b. yes; $\angle HEF \cong \angle HGF$ because they both are adjacent to two congruent angles that together add up to 180°, and $\angle EHG \cong \angle GFE$ for the same reason. So, *EFGH* is a parallelogram by the Parallelogram Opposite Angles Converse.

33. You can use the Alternate Interior Angles Converse to show that $\overline{AD} \parallel \overline{BC}$. Then $\overline{AD}$ and $\overline{BC}$ are both congruent and parallel. So, *ABCD* is a parallelogram by the Opposite Sides Parallel and Congruent Theorem.

35. Use the Corresponding Angles Converse to show that $\overline{AD} \parallel \overline{BC}$ and the Alternate Interior Angles Converse to show that $\overline{AB} \parallel \overline{DC}$.

37.

STATEMENTS	REASONS
1. $\angle A \cong \angle C$, $\angle B \cong \angle D$	**1.** Given
2. Let $m\angle A = m\angle C = x°$ and $m\angle B = m\angle D = y°$.	**2.** Definition of congruent angles
3. $m\angle A + m\angle B + m\angle C + m\angle D = x° + y° + x° + y° = 360°$	**3.** Corollary to the Polygon Interior Angles Theorem
4. $2(x°) + 2(y°) = 360°$	**4.** Simplify.
5. $2(x° + y°) = 360°$	**5.** Distributive Property
6. $x° + y° = 180°$	**6.** Division Property of Equality
7. $m\angle A + m\angle B = 180°$, $m\angle A + m\angle D = 180°$	**7.** Substitution Property of Equality
8. $\angle A$ and $\angle B$ are supplementary. $\angle A$ and $\angle D$ are supplementary.	**8.** Definition of supplementary angles
9. $\overline{BC} \parallel \overline{AD}$, $\overline{AB} \parallel \overline{DC}$	**9.** Consecutive Interior Angles Converse
10. *ABCD* is a parallelogram.	**10.** Definition of parallelogram

39.

STATEMENTS	REASONS
1. $\overline{QR} \parallel \overline{PS}$, $\overline{QR} \cong \overline{PS}$	**1.** Given
2. $\angle SQR \cong \angle QSP$	**2.** Alternate Interior Angles Theorem
3. $\overline{QS} \cong \overline{QS}$	**3.** Reflexive Property of Congruence
4. $\triangle QRS \cong \triangle SPQ$	**4.** SAS Congruence Theorem
5. $\angle QSR \cong \angle SQP$	**5.** Corresponding parts of congruent triangles are congruent.
6. $\overline{QP} \parallel \overline{RS}$	**6.** Alternate Interior Angles Converse
7. *PQRS* is a parallelogram.	**7.** Definition of parallelogram

41.

STATEMENTS	REASONS
1. *DEBF* is a parallelogram. $AE = CF$	**1.** Given
2. $\overline{DE} \cong \overline{BF}, \overline{FD} \cong \overline{EB}$	**2.** Parallelogram Opposite Sides Theorem
3. $\angle DFB \cong \angle DEB$	**3.** Parallelogram Opposite Angles Theorem
4. $\angle AED$ and $\angle DEB$ form a linear pair. $\angle CFB$ and $\angle DFB$ form a linear pair.	**4.** Definition of linear pair
5. $\angle AED$ and $\angle DEB$ are supplementary. $\angle CFB$ and $\angle DFB$ are supplementary.	**5.** Linear Pair Postulate
6. $\angle AED \cong \angle CFB$	**6.** Congruent Supplements Theorem
7. $\overline{AE} \cong \overline{CF}$	**7.** Definition of congruent segments
8. $\triangle AED \cong \triangle CFB$	**8.** SAS Congruence Theorem
9. $\overline{AD} \cong \overline{CB}$	**9.** Corresponding parts of congruent triangles are congruent.
10. $AB = AE + EB$, $DC = CF + FD$	**10.** Segment Addition Postulate
11. $FD = EB$	**11.** Definition of congruent segments
12. $AB = CF + FD$	**12.** Substitution Property of Equality
13. $AB = DC$	**13.** Transitive Property of Equality
14. $\overline{AB} \cong \overline{DC}$	**14.** Definition of congruent segments
15. *ABCD* is a parallelogram.	**15.** Parallelogram Opposite Sides Converse

43. 8; By the Parallelogram Opposite Sides Theorem, $\overline{AB} \cong \overline{CD}$. Also, $\angle ABE$ and $\angle CDF$ are congruent alternate interior angles of parallel segments $\overline{AB}$ and $\overline{CD}$. Then you can use the Segment Addition Postulate, the Substitution Property of Equality, and the Reflexive Property of Congruence to show that $\overline{DF} \cong \overline{BE}$. So, $\triangle ABE \cong \triangle CDF$ by the SAS Congruence Theorem, which means that $AE = CF = 8$ because corresponding parts of congruent triangles are congruent.

45. By the definition of a right angle, $m\angle A = 90°$. Because *ABCD* is a parallelogram, and opposite angles of a parallelogram are congruent, $m\angle A = m\angle C = 90°$. Because consecutive angles of a parallelogram are supplementary, $\angle C$ and $\angle B$ are supplementary, and $\angle C$ and $\angle D$ are supplementary. So, $90° + m\angle B = 180°$ and $90° + m\angle D = 180°$. This gives you $m\angle B = m\angle D = 90°$. So, $\angle B$, $\angle C$, and $\angle D$ are right angles.

47. Given quadrilateral *ABCD* with midpoints *E*, *F*, *G*, and *H* that are joined to form a quadrilateral, you can construct diagonal $\overline{BD}$. Then $\overline{FG}$ is a midsegment of $\triangle BCD$, and $\overline{EH}$ is a midsegment of $\triangle DAB$. So, by the Triangle Midsegment Theorem, $\overline{FG} \parallel \overline{BD}$, $FG = \frac{1}{2}BD$, $\overline{EH} \parallel \overline{BD}$, and $EH = \frac{1}{2}BD$. So, by the Transitive Property of Parallel Lines, $\overline{EH} \parallel \overline{FG}$ and by the Transitive Property of Equality, $EH = FG$. Because one pair of opposite sides is both congruent and parallel, *EFGH* is a parallelogram by the Opposite Sides Parallel and Congruent Theorem.

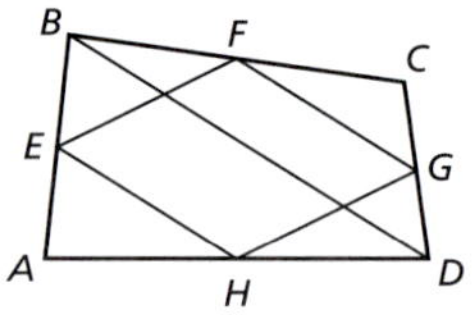

7.3 Review & Refresh

49. 40°

50. 4

51.

Equation	**Justification**
$3x + y = 11$	Write the equation.
$y = -3x + 11$	Subtraction Property of Equality

52.

Equation	**Justification**
$8x + 3 = -9 + 4y$	Write the equation.
$8x + 12 = 4y$	Addition Property of Equality
$2x + 3 = y$	Division Property of Equality

53. $x = 100$

54. $3\sqrt{10}$, or about 9.5 units

55. $(-3, 7)$

56.

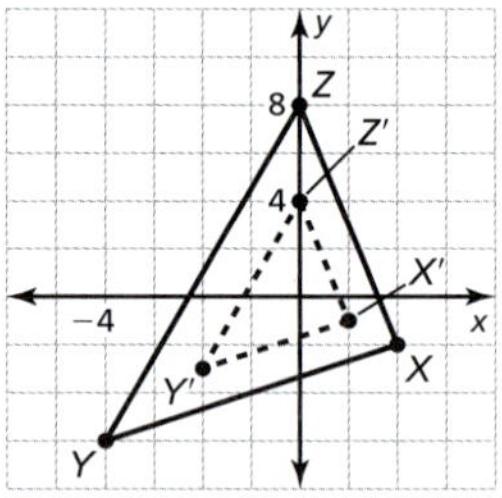

57. $AD > CD$; Because $\overline{AD}$ is the third side of the triangle with the larger included angle, by the Hinge Theorem it is longer than $\overline{CD}$.

58. Parallelogram Opposite Angles Converse

59. $\overrightarrow{QR}$; $\langle 7, -2 \rangle$

60. $2\sqrt{5}$, or about 4.5 units

61. $x = 10$; The angle to the right of the angle labeled $(13x - 5)°$ has a measure of 55° by the Corresponding Angles Theorem.

$(13x - 5)° + 55° = 180°$	Linear Pair Postulate
$13x - 5 = 125$	Subtraction Property of Equality
$13x = 130$	Addition Property of Equality
$x = 10$	Division Property of Equality

Section 7.4

7.4 Practice

1. sometimes; Some rhombuses are squares.

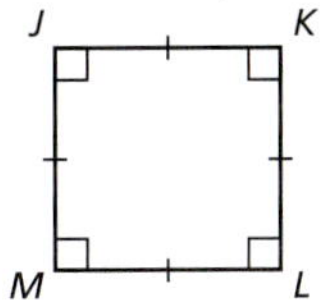

3. always; By definition, a rhombus is a parallelogram, and opposite sides of a parallelogram are congruent.

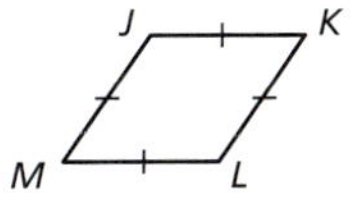

5. sometimes; Some rhombuses are squares.

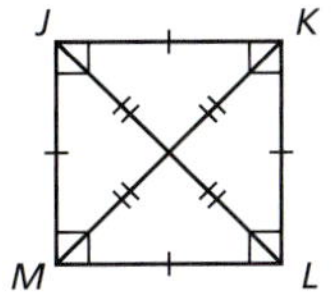

7. square; All of the sides are congruent, and all of the angles are congruent.

9. rectangle; Opposite sides are parallel and the angles are 90°.

11. $m\angle 1 = m\angle 2 = m\angle 4 = 27°$, $m\angle 3 = 90°$, $m\angle 5 = m\angle 6 = 63°$

13. $m\angle 1 = m\angle 2 = m\angle 3 = m\angle 4 = 37°$, $m\angle 5 = 106°$

15. always; All angles of a rectangle are congruent.

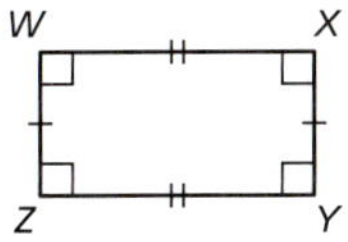

17. sometimes; Some rectangles are squares.

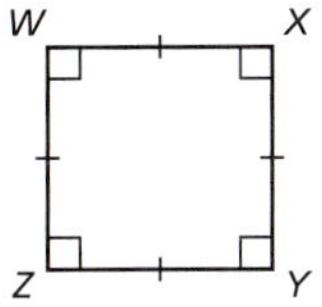

19. sometimes; Some rectangles are squares.

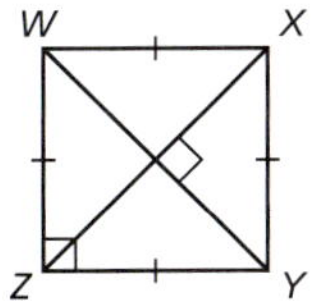

21. 11

23. 4

25. Diagonals do not necessarily bisect opposite angles of a rectangle;

$m\angle QSR = 90° - m\angle QSP$

$x = 32$

27. (6, 3), (1, −2)

29. (−4, 5), (2, −1)

31. 53°

33. 74°

35. 6

37. 56°

39. 56°

41. 10

43. 90°

45. 45°

47. 2

49. rectangle, square

51. rhombus, square

53. parallelogram, rectangle, rhombus, square

55. rectangle, rhombus, square; The diagonals are congruent and perpendicular.

57. rectangle; The sides are perpendicular and not congruent.

59. rhombus; The diagonals are perpendicular and not congruent.

61. A, C

63. always; By the Square Corollary, a square is a rhombus.

65. always; The diagonals of a rectangle are congruent by the Rectangle Diagonals Theorem.

67. sometimes; Some rhombuses are squares.

69. square; A square has four congruent sides and four congruent angles.

71. **a.** rhombus; rectangle; *HBDF* has four congruent sides; *ACEG* has four right angles.

b. 200 in.; $AJ = BH = BD = CJ = HF = JC = JE = DF = 25$ and the product of 25 and 8 is 200.

73. **a.** no; The diagonals of a square always create two right triangles.

b. yes; If the angles of a rhombus are 60°, 120°, 60°, and 120°, the diagonal that bisects the opposite 120° angles will divide the rhombus into two equilateral triangles.

75.

STATEMENTS	REASONS
1. *PQRS* is a parallelogram. $\overline{PR}$ bisects $\angle SPQ$ and $\angle QRS$. $\overline{SQ}$ bisects $\angle PSR$ and $\angle RQP$.	**1.** Given
2. $\angle SRT \cong \angle QRT$, $\angle RQT \cong \angle RST$	**2.** Definition of angle bisector
3. $\overline{TR} \cong \overline{TR}$	**3.** Reflexive Property of Congruence
4. $\triangle QRT \cong \triangle SRT$	**4.** AAS Congruence Theorem
5. $\overline{QR} \cong \overline{SR}$	**5.** Corresponding parts of congruent triangles are congruent.
6. $\overline{QR} \cong \overline{PS}$, $\overline{PQ} \cong \overline{SR}$	**6.** Parallelogram Opposite Sides Theorem
7. $\overline{PS} \cong \overline{QR} \cong \overline{SR} \cong \overline{PQ}$	**7.** Transitive Property of Congruence
8. *PQRS* is a rhombus.	**8.** Definition of rhombus

77. If a quadrilateral is a rhombus, then it has four congruent sides; If a quadrilateral has four congruent sides, then it is a rhombus; The conditional statement is true by the definition of rhombus. The converse is true because if a quadrilateral has four congruent sides, then both pairs of opposite sides are congruent. So, by the Parallelogram Opposite Sides Converse, it is a parallelogram with four congruent sides, which is the definition of a rhombus.

79. If a quadrilateral is a square, then it is a rhombus and a rectangle; If a quadrilateral is a rhombus and a rectangle, then it is a square; The conditional statement is true because if a quadrilateral is a square, then by definition of a square, it has four congruent sides, which makes it a rhombus by the Rhombus Corollary, and it has four right angles, which makes it a rectangle by the Rectangle Corollary; The converse is true because if a quadrilateral is a rhombus and a rectangle, then by the Rhombus Corollary, it has four congruent sides, and by the Rectangle Corollary, it has four right angles. So, by the definition, it is a square.

81. no; yes; Corresponding angles of two rhombuses might not be congruent; Corresponding angles of two squares are congruent.

83.

STATEMENTS	REASONS
1. $\triangle XYZ \cong \triangle XWZ$, $\angle XYW \cong \angle ZWY$	**1.** Given
2. $\angle YXZ \cong \angle WXZ$, $\angle YZX \cong \angle WZX$, $\overline{XY} \cong \overline{XW}$, $\overline{YZ} \cong \overline{WZ}$	**2.** Corresponding parts of congruent triangles are congruent.
3. $\overline{XZ}$ bisects $\angle WXY$ and $\angle WZY$.	**3.** Definition of angle bisector
4. $\angle XWY \cong \angle XYW$, $\angle WYZ \cong \angle ZWY$	**4.** Base Angles Theorem
5. $\angle XYW \cong \angle WYZ$, $\angle XWY \cong \angle ZWY$	**5.** Transitive Property of Congruence
6. $\overline{WY}$ bisects $\angle XWZ$ and $\angle XYZ$.	**6.** Definition of angle bisector
7. $WXYZ$ is a rhombus.	**7.** Rhombus Opposite Angles Theorem

85. $2d\sqrt{17}$, where d is the length of the shorter diagonal

87.

STATEMENTS	REASONS
1. $PQRS$ is a parallelogram. $\overline{PR} \cong \overline{SQ}$	**1.** Given
2. $\overline{PS} \cong \overline{QR}$	**2.** Parallelogram Opposite Sides Theorem
3. $\overline{PQ} \cong \overline{PQ}$	**3.** Reflexive Property of Congruence
4. $\triangle PQR \cong \triangle QPS$	**4.** SSS Congruence Theorem
5. $\angle SPQ \cong \angle RQP$	**5.** Corresponding parts of congruent triangles are congruent.
6. $m\angle SPQ = m\angle RQP$	**6.** Definition of congruent angles
7. $m\angle SPQ + m\angle RQP = 180°$	**7.** Parallelogram Consecutive Angles Theorem
8. $2m\angle SPQ = 180°$ and $2m\angle RQP = 180°$	**8.** Substitution Property of Equality
9. $m\angle SPQ = 90°$ and $m\angle RQP = 90°$	**9.** Division Property of Equality
10. $m\angle RSP = 90°$ and $m\angle QRS = 90°$	**10.** Parallelogram Opposite Angles Theorem
11. $\angle SPQ$, $\angle RQP$, $\angle RSP$, and $\angle QRS$ are right angles.	**11.** Definition of a right angle
12. $PQRS$ is a rectangle.	**12.** Definition of a rectangle

7.4 Review & Refresh

88. The graph of g is a translation 2 units right of the graph of f.

89. The graph of g is a vertical shrink by a factor of $\frac{1}{2}$ of the graph of f.

90. $x = 9$, $y = 26$

91. A segment of a polygon is a diagonal if and only if it joins two nonconsecutive vertices.

92. no solution

93. $g < -2$

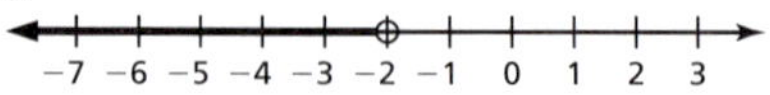

94. $x = 4$, $y = 13$

95. yes; Each input has exactly one output.

96. 165°, 15°

97. square; All of the sides are congruent, and all of the angles are right.

98. $11 + \sqrt{13}$, or about 14.6 units; 9 square units

99. no; There is no SSA Congruence Theorem.

100. no; The congruent segments are not corresponding sides.

101. Parallelogram Opposite Angles Converse

102. $AB = 18$; $\overline{AB} \cong \overline{CB}$ by the Perpendicular Bisector Theorem, so $3x - 6 = 2x + 2$. The solution is $x = 8$, so $AB = 3(8) - 6 = 18$.

Section 7.5

7.5 Practice

1. slope of $\overline{YZ}$ = slope of $\overline{XW}$ and slope of $\overline{XY} \neq$ slope of $\overline{WZ}$; $XY = WZ$, so $WXYZ$ is isosceles.

3. slope of $\overline{MQ}$ = slope of $\overline{NP}$ and slope of $\overline{MN} \neq$ slope of $\overline{PQ}$; $MN \neq PQ$, so $MNPQ$ is not isosceles.

5. $m\angle L = m\angle M = 62°$, $m\angle K = m\angle J = 118°$

7. 14 **9.** 4

11. 7.5 **13.** $3\sqrt{13}$

15. 110° **17.** 80°

19. Because $MN = \frac{1}{2}(AB + DC)$, when you solve for DC, you should get $DC = 2(MN) - AB$; $DC = 2(8) - 14 = 2$.

21. rectangle; $JKLM$ is a quadrilateral with 4 right angles.

23. square; All four sides are congruent and the angles are 90°.

25. no; It could be a kite. **27.** $x = 3$

29. 26 in.

31. *Sample answer:* $\angle B \cong \angle C$; $\overline{BC} \parallel \overline{AD}$, so base angles need to be congruent.

33. *Sample answer:* $\overline{BE} \cong \overline{DE}$; Then the diagonals bisect each other.

35.

STATEMENTS	REASONS
1. $\overline{JL} \cong \overline{LN}$, $\overline{KM}$ is a midsegment of $\triangle JLN$.	1. Given
2. $\overline{KM} \parallel \overline{JN}$	2. Triangle Midsegment Theorem
3. $JKMN$ is a trapezoid.	3. Definition of trapezoid
4. $\angle LJN \cong \angle LNJ$	4. Base Angles Theorem
5. $JKMN$ is an isosceles trapezoid.	5. Isosceles Trapezoid Base Angles Converse

37. any point on $\overleftrightarrow{UV}$ such that $UV \neq SV$ and $SV \neq 0$

39. 32 m^2; The area of the sail is the sum of the areas of the triangles.

41. Given isosceles trapezoid $ABCD$ with $\overline{BC} \parallel \overline{AD}$, construct $\overline{CE}$ parallel to $\overline{BA}$. Then, $ABCE$ is a parallelogram by definition, so $\overline{AB} \cong \overline{EC}$. Because $\overline{AB} \cong \overline{CD}$ by the definition of an isosceles trapezoid, $\overline{CE} \cong \overline{CD}$ by the Transitive Property of Congruence. So, $\angle CED \cong \angle D$ by the Base Angles Theorem and $\angle A \cong \angle CED$ by the Corresponding Angles Theorem. So, $\angle A \cong \angle D$ by the Transitive Property of Congruence. Next, by the Consecutive Interior Angles Theorem, $\angle B$ and $\angle A$ are supplementary and so are $\angle BCD$ and $\angle D$. So, $\angle B \cong \angle BCD$ by the Congruent Supplements Theorem.

43. $y = 2x + 1$

45. a.

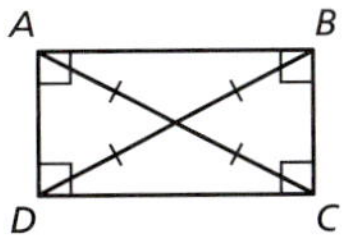

rectangle; The diagonals are congruent, but not perpendicular.

b.

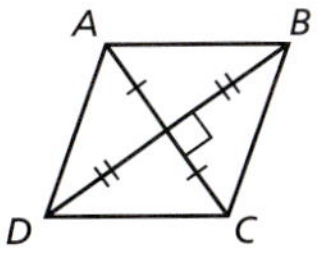

rhombus; The diagonals are perpendicular, but not congruent.

47. a. yes

b. 75°, 75°, 105°, 105°

49. Given kite $EFGH$ with $\overline{EF} \cong \overline{FG}$ and $\overline{EH} \cong \overline{GH}$, construct diagonal $\overline{FH}$, which is congruent to itself by the Reflexive Property of Congruence. So, $\triangle FGH \cong \triangle FEH$ by the SSS Congruence Theorem, and $\angle E \cong \angle G$ because corresponding parts of congruent triangles are congruent. Next, assume temporarily that $\angle F \cong \angle H$. Then $EFGH$ is a parallelogram by the Parallelogram Opposite Angles Converse, and opposite sides are congruent. However, this contradicts the definition of a kite, which says that opposite sides cannot be congruent. So, the assumption cannot be true and $\angle F$ is not congruent to $\angle H$.

51. a.

STATEMENTS	REASONS
1. $JKLM$ is an isosceles trapezoid. $\overline{KL} \parallel \overline{JM}$, $\overline{JK} \cong \overline{LM}$	1. Given
2. $\angle JKL \cong \angle MLK$	2. Isosceles Trapezoid Base Angles Theorem
3. $\overline{KL} \cong \overline{KL}$	3. Reflexive Property of Congruence
4. $\triangle JKL \cong \triangle MLK$	4. SAS Congruence Theorem
5. $\overline{JL} \cong \overline{KM}$	5. Corresponding parts of congruent triangles are congruent.

b. If the diagonals of a trapezoid are congruent, then the trapezoid is isosceles. Let $JKLM$ be a trapezoid, $\overline{KL} \parallel \overline{JM}$ and $\overline{JL} \cong \overline{KM}$. Construct line segments through K and L perpendicular to $\overline{JM}$ as shown below.

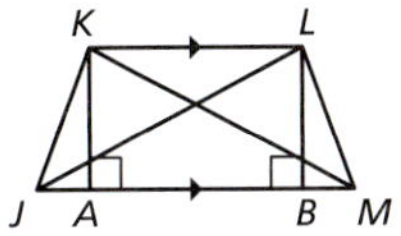

Because $\overline{KL} \parallel \overline{JM}$, $\angle AKL$ and $\angle KLB$ are right angles, so $KLBA$ is a rectangle and $\overline{AK} \cong \overline{BL}$. Then $\triangle JLB \cong \triangle MKA$ by the HL Congruence Theorem. So, $\angle LJB \cong \angle KMA$. $\overline{JM} \cong \overline{JM}$ by the Reflexive Property of Congruence. So, $\triangle KJM \cong \triangle LMJ$ by the SAS Congruence Theorem. Then $\angle KJM \cong \angle LMJ$, and the trapezoid is isosceles by the Isosceles Trapezoid Base Angles Converse.

53. rhombus;

STATEMENTS	REASONS
1. $\overline{JK} \cong \overline{LM}$, E is the midpoint of $\overline{JL}$, F is the midpoint of $\overline{KL}$, G is the midpoint of $\overline{KM}$, H is the midpoint of $\overline{JM}$.	**1.** Given
2. $\overline{EF}$ is a midsegment of $\triangle JKL$, $\overline{FG}$ is a midsegment of $\triangle KML$, $\overline{GH}$ is a midsegment of $\triangle KMJ$, $\overline{EH}$ is a midsegment of $\triangle JML$.	**2.** Definition of midsegment
3. $\overline{EF} \parallel \overline{JK}$, $\overline{FG} \parallel \overline{LM}$, $\overline{GH} \parallel \overline{JK}$, $\overline{EH} \parallel \overline{LM}$	**3.** Triangle Midsegment Theorem
4. $\overline{EF} \parallel \overline{GH}$, $\overline{FG} \parallel \overline{EH}$	**4.** Transitive Property of Parallel Lines
5. $EFGH$ is a parallelogram.	**5.** Definition of parallelogram
6. $EF = \frac{1}{2}JK$, $FG = \frac{1}{2}LM$, $GH = \frac{1}{2}JK$, $EH = \frac{1}{2}LM$	**6.** Triangle Midsegment Theorem
7. $JK = LM$	**7.** Definition of congruent segments
8. $FG = \frac{1}{2}JK$, $EH = \frac{1}{2}JK$	**8.** Substitution Property of Equality
9. $EF = FG = GH = EH$	**9.** Transitive Property of Equality
10. $\overline{EF} \cong \overline{FG} \cong \overline{GH} \cong \overline{EH}$	**10.** Definition of congruent segments
11. $EFGH$ is a rhombus.	**11.** Definition of a rhombus

7.5 Review & Refresh

54. no

55. $3\sqrt{5}$, or about 6.7 units

56. parallelogram

57. $\angle ABD \cong \angle EBD$ by the Right Angles Congruence Theorem; $\angle BED \cong \angle GEF$ by the Vertical Angles Congruence Theorem.

58. 9; Parallelogram Diagonals Theorem

59. If the sum of the interior angle measures of a polygon is 180°, then it is a triangle.

60. Parallelogram Opposite Sides Converse

61.

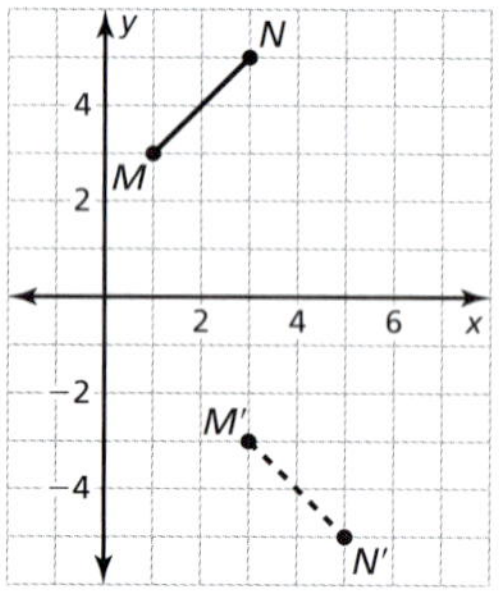

62. 190 cm

63.

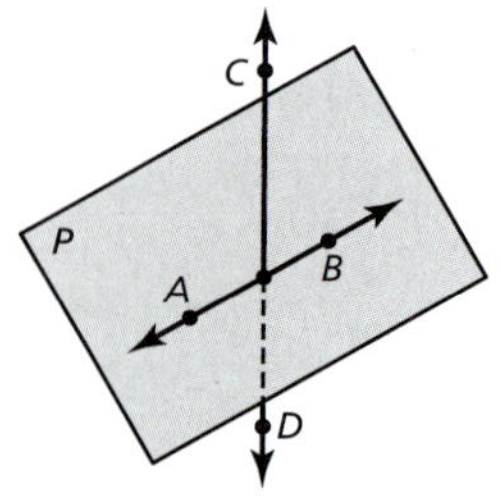

64. $\angle 1$ and $\angle 8$, $\angle 2$ and $\angle 7$

Chapter 7 Review

1. 5040°; 168°; 12°

2. $x = 133$

3. $x = 82$

4. $x = 15$

5. $a = 28$, $b = 87$

6. $c = 6$, $d = 10$

7. $(-2, -1)$

8. $M(2, -2)$

9. *Sample answer:* $(-6, 9)$, $(-2, -3)$; Either vertex results in a quadrilateral with two pairs of opposite parallel sides.

10. $x = 109$; $x° = 61° + (180 - 132)°$ by the Parallelogram Opposite Angles Theorem and by the Parallelogram Consecutive Angles Theorem, respectively.

11. Parallelogram Opposite Sides Converse

12. Parallelogram Diagonals Converse

13. Parallelogram Opposite Angles Converse

14. $x = 1$, $y = 6$

15. $x = 4$

16. **a.** Opposite Sides Parallel and Congruent Theorem

b. 3 ft, 123°, 57°, 57°; $ST = QR$ by the Parallelogram Opposite Sides Theorem, $m\angle QTS = m\angle QRS$ by the Parallelogram Opposite Angles Theorem, $m\angle TQR = m\angle TSR = 180° - 123°$ by the Parallelogram Consecutive Angles Theorem

17. Because $WX = YZ = \sqrt{13}$, $\overline{WX} \cong \overline{YZ}$. Because the slopes of $\overline{WX}$ and $\overline{YZ}$ are both $\frac{2}{3}$, they are parallel. $\overline{WX}$ and $\overline{YZ}$ are opposite sides that are both congruent and parallel. So, $WXYZ$ is a parallelogram by the Opposite Sides Parallel and Congruent Theorem.

18. rhombus; There are four congruent sides.

19. parallelogram; There are two pairs of parallel sides.

20. square; There are four congruent sides and the angles are 90°.

21. 10

22. $(3, 1)$ and $(1, -1)$

23. rectangle, rhombus, square; The diagonals are congruent and perpendicular.

24. $m\angle Z = m\angle Y = 58°$, $m\angle W = m\angle X = 122°$

25. 26

26. $3\sqrt{5}$

27. $x = 15$; Because $360 - (65 + 85 + 105) = 105$, $7x°$ is 105° and it is opposite the other 105° angle because kites have exactly one pair of congruent opposite angles. The 65° angle is opposite the 85° angle because they are the only two angles remaining.

28. yes; By the Consecutive Interior Angles Theorem, the two adjacent angles of a trapezoid with a leg in common are supplementary. By the Congruent Supplements Theorem, two base angles of the trapezoid are congruent. So, the trapezoid is isosceles by the Isosceles Trapezoid Base Angles Converse.

29. trapezoid; There is one pair of parallel sides.

30. rhombus; There are four congruent sides.

31. rectangle; There are four right angles.

Chapter 7 Practice Test

1. $r = 6, s = 3.5$
2. $a = 79, b = 101$
3. $p = 5, q = 9$
4. trapezoid; There is one pair of parallel sides, by the Lines Perpendicular to a Transversal Theorem.
5. kite; There are two pairs of consecutive congruent sides but opposite sides are not congruent.
6. isosceles trapezoid; Base angles are congruent.
7. 25°, 25°, 25°, 57°, 57°, 57°, 57°, 57°
8. rhombus
9. yes; The diagonals bisect each other.
10. no; $\overline{JK}$ and $\overline{ML}$ might not be parallel.
11. yes; Opposite angles are congruent.
12. Consecutive angles are supplementary.
13. Show that a quadrilateral is a parallelogram with four congruent sides and four right angles, or show that a quadrilateral is both a rectangle and a rhombus.
14. 10.5 in.
15. **a.** $M(2, 0)$
    b. $(1, 1)$
    c. $2\sqrt{13} + 2\sqrt{17}$, or about 15.5 units
16. 108°

Chapter 7 Test Prep

1. B
2. $\left(4\frac{1}{2}, 3\frac{1}{4}\right)$
3. C
4.

STATEMENTS	REASONS
1. $\overline{BC} \parallel \overline{AD}$, $\angle EBC \cong \angle ECB$, $\angle ABE \cong \angle DCE$	1. Given
2. $ABCD$ is a trapezoid.	2. Definition of trapezoid
3. $m\angle EBC = m\angle ECB$, $m\angle ABE = m\angle DCE$	3. Definition of congruent angles
4. $m\angle ABE + m\angle EBC = m\angle ABC$, $m\angle DCE + m\angle ECB = m\angle DCB$	4. Angle Addition Postulate
5. $m\angle ABE + m\angle EBC = m\angle ABE + m\angle EBC$	5. Reflexive Property of Equality
6. $m\angle ABE + m\angle EBC = m\angle DCE + m\angle ECB$	6. Substitution Property of Equality
7. $m\angle ABC = m\angle DCB$	7. Transitive Property of Equality
8. $\angle ABC \cong \angle DCB$	8. Definition of congruent angles
9. $ABCD$ is an isosceles trapezoid.	9. Isosceles Trapezoid Base Angles Converse

5. *Sample answer:*
   2. $\overline{QT} \cong \overline{RS}, \overline{QR} \cong \overline{TS}$
   3. $\triangle QRS \cong \triangle RQT$
   4. $\angle QRS \cong \angle RQT$
   5. $m\angle QRS + m\angle RQT = 180°$
   6. $m\angle QRS = m\angle RQT = 90°$
   7. $m\angle TSR + 90° = 180°, m\angle STQ + 90° = 180°$
   8. $m\angle TSR = 90°, m\angle STQ = 90°$
   9. $\angle RQT, \angle QRS, \angle TSR$, and $\angle STQ$ are right angles.
   10. $QRST$ is a rectangle.

6. 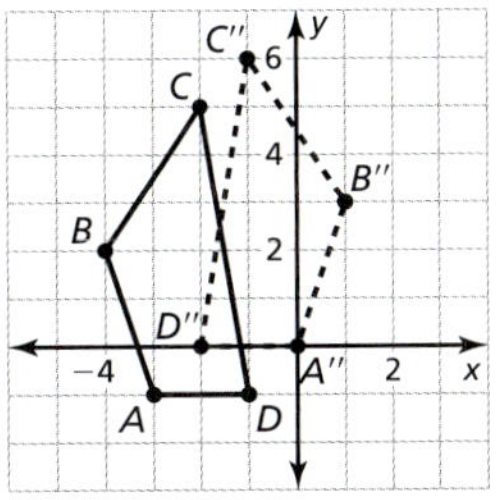

7. A
8. A, B

Chapter 8

Prepare with CalcChat

1. yes
2. yes
3. no
4. no
5. yes
6. yes
7. $k = \frac{3}{7}$
8. $k = \frac{8}{3}$
9. $k = 2$
10. yes; All the ratios are equivalent by the Transitive Property of Equality.

Section 8.1

8.1 Practice

1. $\frac{4}{3}$; $\angle A \cong \angle L, \angle B \cong \angle M, \angle C \cong \angle N$; $\frac{LM}{AB} = \frac{MN}{BC} = \frac{NL}{CA}$
3. $x = 30$
5. $x = 11$
7. altitude; 24
9. $\frac{2}{3}$
11. 72 cm
13. 20 yd
15. 350 m
17. 108 ft^2
19. 4 in.^2
21. Because the first ratio has a side length of B over a side length of A, the second ratio should have the perimeter of B over the perimeter of A;
$$\frac{5}{10} = \frac{x}{28}$$
$$x = 14$$
23. no; Corresponding angles are not congruent.
25.
27. $\frac{JK}{EF} = \frac{KL}{FG} = \frac{LM}{GH} = \frac{MJ}{HE}$
29. 34, 85
31. 60.5, 378.125
33. no; Corresponding side lengths are not proportional.
35. A, D
37. 30 m
39. 7.5 ft

41. Two triangles are similar when there is a similarity transformation that maps one triangle to the other, such that corresponding angles are congruent and corresponding side lengths are proportional.

43. sometimes
45. sometimes

47. sometimes
49. about 1116 mi

51. The coordinates of the points are $A(-3, 0)$, $B(0, 4)$, $C(6, 0)$, $D(0, -8)$, and $O(0, 0)$. The side lengths are $OA = 3$, $OB = 4$, $AB = 5$, $OC = 6$, $OD = 8$, and $CD = 10$. Corresponding side lengths are proportional with a scale factor of 2, so the triangles are proportional.

53.

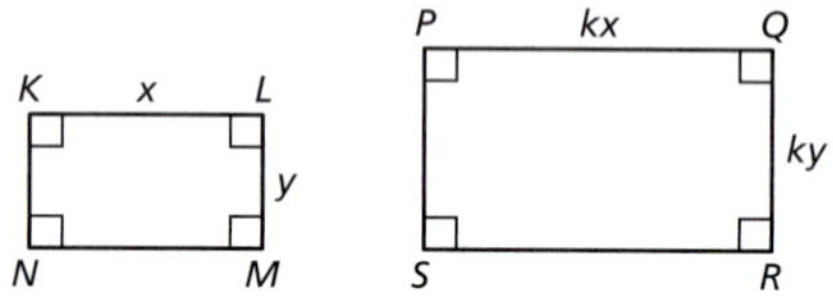

Let *KLMN* and *PQRS* be similar rectangles as shown. The ratio of corresponding side lengths is $\frac{KL}{PQ} = \frac{x}{kx} = \frac{1}{k}$. The area of *KLMN* is xy and the area of *PQRS* is $(kx)(ky) = k^2xy$. So, the ratio of the areas is $\frac{xy}{k^2xy} = \frac{1}{k^2} = \left(\frac{1}{k}\right)^2$. Because the ratio of corresponding side lengths is $\frac{1}{k}$, any pair of corresponding side lengths can be substituted for $\frac{1}{k}$. So,
$\frac{\text{Area of } KLMN}{\text{Area of } PQRS} = \left(\frac{KL}{PQ}\right)^2 = \left(\frac{LM}{QR}\right)^2 = \left(\frac{MN}{RS}\right)^2 = \left(\frac{NK}{SP}\right)^2.$

8.1 Review & Refresh

55. $x = 63$
56. $x = 16$

57. $x = 97$
58. 9 units

59. $k = \frac{1}{16}$
60. $x = 13, y = 7$

61. $-(3x - 8)(x - 2)$
62. $(x - 4)(x + 6)$

63. $(x - 3)(x^2 - 5)$
64. 4 units

65. $x = 8$

66.

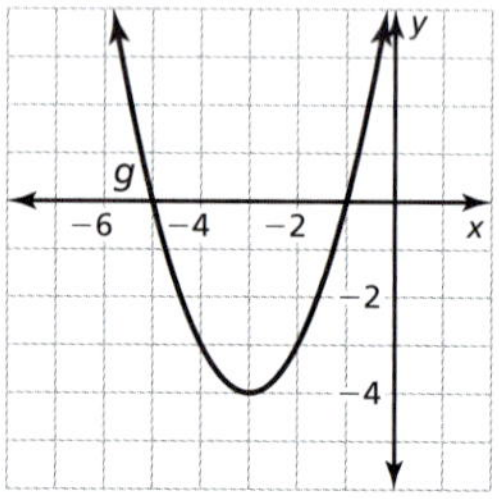

The graph of g is a horizontal translation 3 units left of the graph of f.

67.

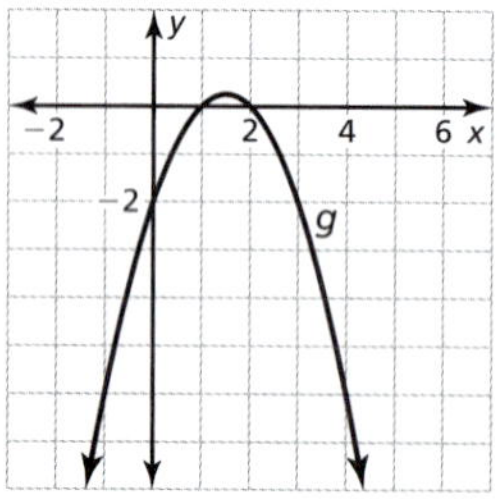

The graph of g is a vertical translation 2 units down of the graph of f.

68. exponential
69. $x = -8$ or 3

70. $x = \pm\sqrt{6}$
71. 26

Section 8.2

8.2 Practice

1. yes; $\triangle FGH \sim \triangle KLJ$; $\angle H \cong \angle J$ and $\angle F \cong \angle K$

3. no; $m\angle R \neq m\angle U$

5. $\angle N \cong \angle Z$ and $\angle MYN \cong \angle XYZ$, so $\triangle MYN \sim \triangle XYZ$.

7. $\angle Q \cong \angle MPN$ and $\angle N \cong \angle N$, so $\triangle LNQ \sim \triangle MNP$.

9. $\triangle CAG \sim \triangle CEF$
11. $\triangle ACB \sim \triangle ECD$

13. $m\angle ECD = 82°$
15. $DE = 21$

17. The side of the larger triangle has a length of 9, not 5.
$\frac{4}{6} = \frac{9}{x}$
$x = 13.5$

19. about 4 ft 4 in.

21. yes;

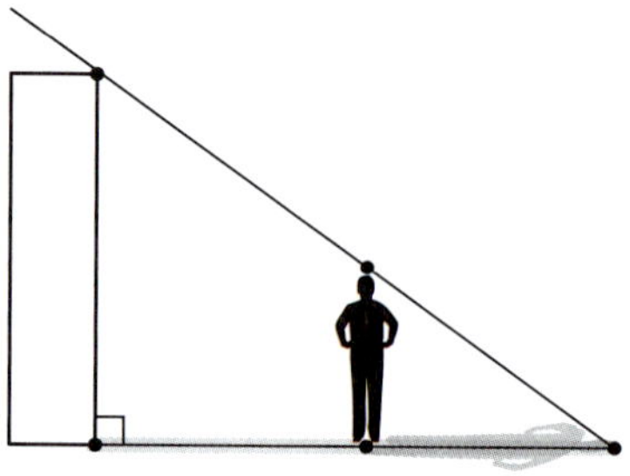

Because two similar triangles are formed, you can use a proportion to estimate the height.

23. yes; Corresponding angles are congruent.

25. no; $94° + 87° > 180°$

27. *Sample answer:* The triangles formed by the distance between the two points, the vertical distance, and the horizontal distance are similar by the AA Similarity Theorem. Because the triangles are similar, the ratios of the vertical sides to the horizontal sides are equal.

29. Let $\triangle XYZ$ be any triangle such that $a = XY$ and $b = XZ$, and the altitudes to $\overline{XY}$ and $\overline{XZ}$ are $\overline{AZ}$ and $\overline{BY}$, respectively. Then, $\angle XAZ$ and $\angle XBY$ are congruent right angles and $\angle YXB \cong \angle ZXA$, so $\triangle BXY \sim \triangle AXZ$ by the AA Similarity Theorem. Because corresponding side lengths of similar figures are proportional, $\frac{AZ}{BY} = \frac{XZ}{XY} = \frac{b}{a}$. Note: When $m\angle X = 90°$, $XY = BY$ and $XZ = AZ$, so the ratio is still $\frac{b}{a}$.

8.2 Review & Refresh

31. no

32. yes; $\angle G \cong \angle O$ and $\angle H \cong \angle N$, so $\triangle GHI \sim \triangle ONM$.

33. **a.** 135°
b. 45°

34. 127°
35. 18 m^2

36. rectangle

Section 8.3

8.3 Practice

1. $\triangle RST$
3. $x = 4$

5. similar; $\triangle DEF \sim \triangle WXY$; $\frac{4}{3}$

7. $\frac{12}{18} = \frac{10}{15} = \frac{8}{12} = \frac{2}{3}$

9. $\frac{HG}{HF} = \frac{HJ}{HK} = \frac{GJ}{FK}$, so $\triangle GHJ \sim \triangle FHK$.

11. $\angle X \cong \angle D$ and $\frac{XY}{DJ} = \frac{XZ}{DG}$, so $\triangle XYZ \sim \triangle DJG$.

13. 24, 26

15. yes; $\frac{24}{15} = \frac{8x}{25} = \frac{48}{6x}$ when $x = 5$.

17. Because $\overline{AB}$ corresponds to $\overline{RQ}$ and $\overline{BC}$ corresponds to $\overline{QP}$, the similarity statement should be $\triangle ABC \sim \triangle RQP$.

19. **a.** $\frac{CD}{CE} = \frac{BC}{AC}$

b. $\angle CBD \cong \angle CAE$

21. $EF = \frac{1}{2}BA$, $DE = \frac{1}{2}AC$, and $DF = \frac{1}{2}BC$ by the Triangle Midsegment Theorem. So, $\frac{EF}{BA} = \frac{DE}{AC} = \frac{DF}{BC} = \frac{1}{2}$, and $\triangle DEF \sim \triangle CAB$. Because corresponding angles of similar figures are congruent, $m\angle CAB = m\angle DEF = 90°$.

23. *Sample answer:* All similarity transformations preserve angle measure, so the corresponding angles of two similar figures must be congruent.

25. not necessarily; The acute angles might not be congruent.

27. yes; Because corresponding side lengths of the two triangles are proportional, $\triangle ABC \sim \triangle XYZ$ by the SSS Similarity Theorem.

29. You are given that $m_\ell = m_n$. By the definition of slope, $m_\ell = \frac{BC}{AC}$ and $m_n = \frac{EF}{DF}$. By the Substitution Property of Equality, $\frac{BC}{AC} = \frac{EF}{DF}$. By the Multiplication Property of Equality, $\frac{BC}{EF} = \frac{AC}{DF}$. By the Right Angles Congruence Theorem, $\angle ACB \cong \angle DFE$. So, $\triangle ABC \sim \triangle DEF$ by the SAS Similarity Theorem. Because corresponding angles of similar triangles are congruent, $\angle BAC \cong \angle EDF$. By the Corresponding Angles Converse, $\ell \parallel n$.

31. If two angles are congruent, then the triangles are similar by the AA Similarity Theorem.

8.3 Review & Refresh

33. $x = 18$

34. By the Transitive Property of Parallel Lines, the shortest pillar is parallel to the third pillar. The same theorem can be applied two more times to show that the shortest pillar is parallel to the tallest pillar.

35. $\angle GEF \cong \angle HDF$ and $\angle F \cong \angle F$, so $\triangle DHF \sim \triangle EGF$.

36. $\angle KLJ \cong \angle NLM$ and $\frac{KL}{NL} = \frac{ML}{JL}$, so $\triangle KLJ \sim \triangle NLM$.

37. $P(0, -4)$

38. $14 < x < 50$

39. yes; Each input x is paired with one output y.

40. $y = -\frac{3}{2}x + 3$

41. $m\angle C = 104°$

42. $m\angle C = 96°$

43.

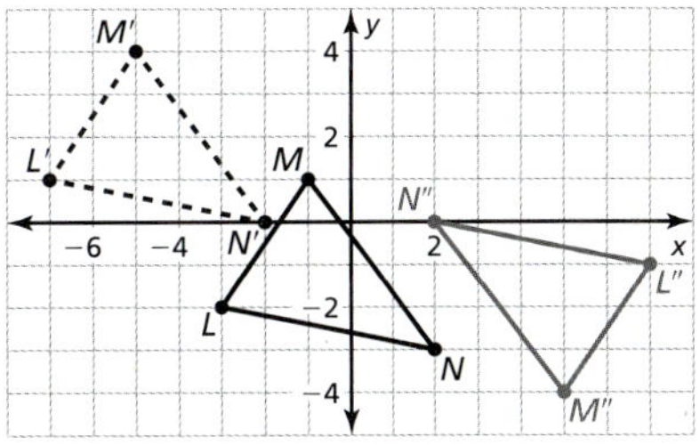

44.

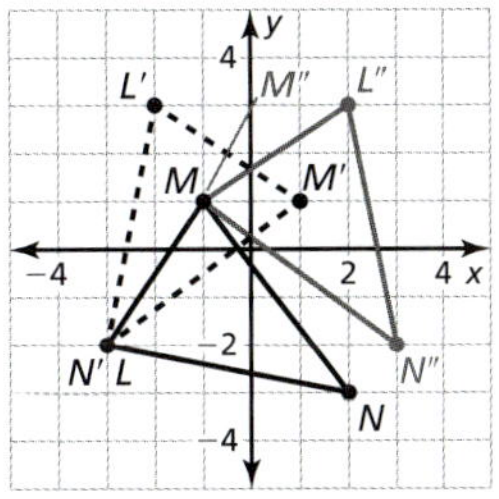

45. yes; SAS Congruence Theorem

46. yes; AAS Congruence Theorem

Section 8.4

8.4 Practice

1. 9

3. yes

5. no

7.

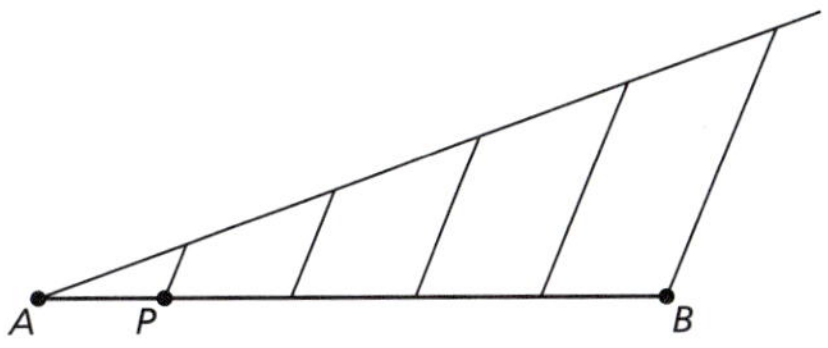

9.

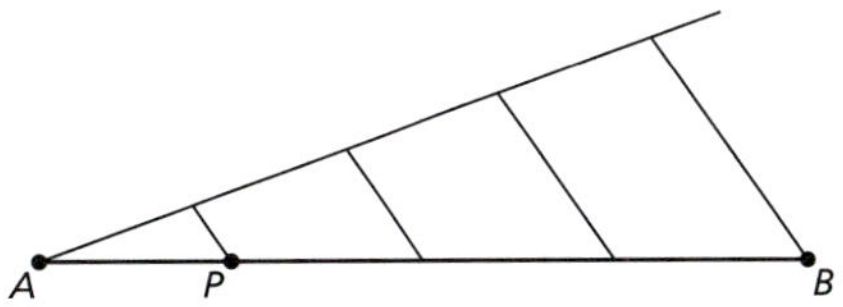

11. CE

13. BD

15. 6

17. $y = 12$

19. $p = 27$

21. The proportion should be $\frac{RS}{ST} = \frac{RQ}{QU}$;

$$\frac{5}{10} = \frac{6}{x}$$
$$5x = 60$$
$$x = 12$$

23. $x = 3$

25.

STATEMENTS	REASONS
1. $\overline{QS} \parallel \overline{TU}$	1. Given
2. $\angle RQS \cong \angle RTU$, $\angle RSQ \cong \angle RUT$	2. Corresponding Angles Theorem
3. $\triangle RQS \sim \triangle RTU$	3. AA Similarity Theorem
4. $\frac{QR}{TR} = \frac{SR}{UR}$	4. Corresponding side lengths of similar figures are proportional.
5. $QR = QT + TR$, $SR = SU + UR$	5. Segment Addition Postulate
6. $\frac{QT + TR}{TR} = \frac{SU + UR}{UR}$	6. Substitution Property of Equality
7. $\frac{QT}{TR} + \frac{TR}{TR} = \frac{SU}{UR} + \frac{UR}{UR}$	7. Rewrite the proportion.
8. $\frac{QT}{TR} + 1 = \frac{SU}{UR} + 1$	8. Simplify.
9. $\frac{QT}{TR} = \frac{SU}{UR}$	9. Subtraction Property of Equality

27. a. about 50.9 yd, about 58.4 yd, about 64.7 yd

b. Lot C

c. about \$287,000, about \$318,000; $\frac{50.9}{250{,}000} \approx \frac{58.4}{287{,}000}$ and $\frac{50.9}{250{,}000} \approx \frac{64.7}{318{,}000}$

29. Because $\overline{DJ}, \overline{EK}, \overline{FL}$, and $\overline{GB}$ are cut by a transversal $\overrightarrow{AC}$, and $\angle ADJ \cong \angle DEK \cong \angle EFL \cong \angle FGB$ by construction, $\overline{DJ} \parallel \overline{EK} \parallel \overline{FL} \parallel \overline{GB}$ by the Corresponding Angles Converse.

31. isosceles; By the Triangle Angle Bisector Theorem, the ratio of the lengths of the segments of $\overline{LN}$ equals the ratio of the other two side lengths. Because $\overline{LN}$ is bisected, the ratio is 1, and $ML = MN$.

33. Because $\overline{WX} \parallel \overline{ZA}$, $\angle XAZ \cong \angle YXW$ by the Corresponding Angles Theorem and $\angle WXZ \cong \angle XZA$ by the Alternate Interior Angles Theorem. So, by the Transitive Property of Congruence, $\angle XAZ \cong \angle XZA$. Then $\overline{XA} \cong \overline{XZ}$ by the Converse of the Base Angles Theorem, and by the Triangle Proportionality Theorem, $\frac{YW}{WZ} = \frac{XY}{XA}$. Because $XA = XZ$, $\frac{YW}{WZ} = \frac{XY}{XZ}$.

35. The Triangle Midsegment Theorem is a specific case of the Triangle Proportionality Theorem when the segment parallel to one side of a triangle that connects the other two sides also happens to pass through the midpoints of those two sides.

37.

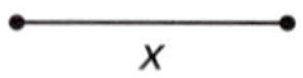

8.4 Review & Refresh

39.

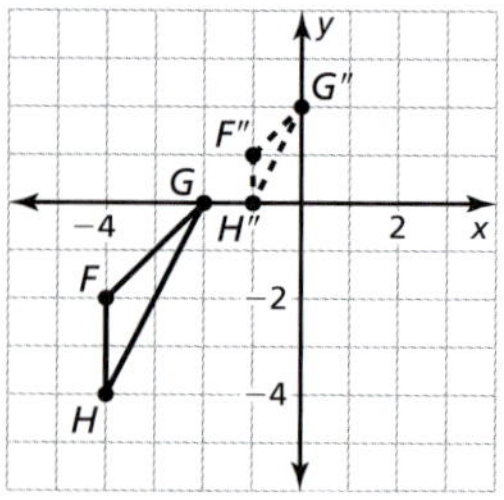

40. The public mailbox is located at point C, which is the circumcenter of the triangle formed by the apartment buildings.

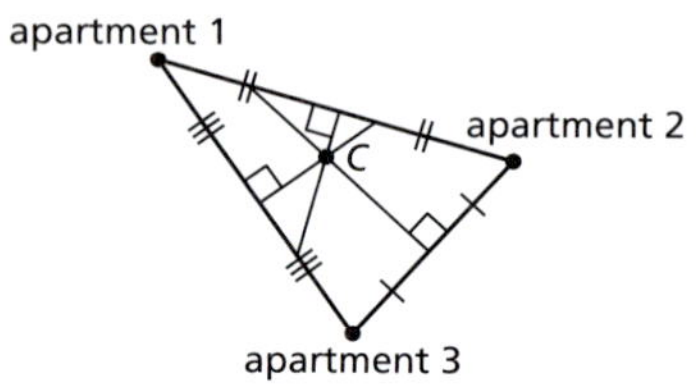

41. $x = 16$ **42.** $p = \frac{2A}{q}$

43. 8 in.

44. $\angle A \cong \angle E$ and $\angle B \cong \angle F$, so $\triangle ABC \sim \triangle EFD$.

45. $x = 2$ **46.** $x = 12$

Chapter 8 Review

1. $\frac{2}{3}$; $\angle A \cong \angle E, \angle B \cong \angle F, \angle C \cong \angle G, \angle D \cong \angle H$; $\frac{AB}{EF} = \frac{BC}{FG} = \frac{CD}{GH} = \frac{DA}{HE}$

2. $\frac{2}{5}$; $\angle X \cong \angle R, \angle Y \cong \angle P, \angle Z \cong \angle Q$; $\frac{RP}{XY} = \frac{PQ}{YZ} = \frac{RQ}{XZ}$

3. $x = 5$

4. no; Corresponding angles are not congruent.

5. yes; Corresponding side lengths are proportional and corresponding angles are congruent.

6. 81 in.^{2} **7.** no; $m\angle H \neq m\angle L$

8. yes; $\triangle ABC \sim \triangle DEF$; $m\angle C = 46°$

9. $\angle Q \cong \angle T$ and $\angle RSQ \cong \angle UST$, so $\triangle RSQ \sim \triangle UST$.

10. $\angle C \cong \angle F$ and $\angle B \cong \angle E$, so $\triangle ABC \sim \triangle DEF$.

11. 324 ft

12. $\angle C \cong \angle C$ and $\frac{CD}{CE} = \frac{CB}{CA}$, so $\triangle CBD \sim \triangle CAE$; SAS Similarity Theorem

13. $\frac{QU}{QT} = \frac{QR}{QS} = \frac{UR}{TS}$, so $\triangle QUR \sim \triangle QTS$; SSS Similarity Theorem

14. $x = 4$

15. yes; $\frac{20}{25} = \frac{8x}{50} = \frac{60}{15x}$ when $x = 5$.

16. no; $\triangle ABC$ either has angle measures 50°, 50°, 80°, or 65°, 65°, 50°.

17. no **18.** yes

19. 11.2 **20.** 10.5

21. 7.2 **22.** 153 m

23. *Sample answer:* Point D on $\overline{CD}$ divides $\overline{AB}$ in the ratio 2 to 3.

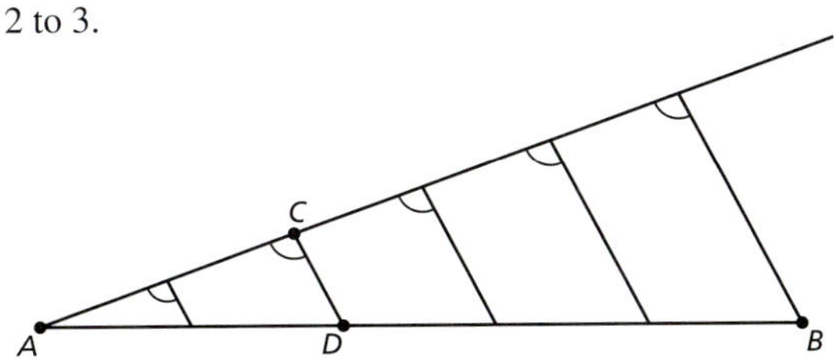

Chapter 8 Practice Test

1. no; Corresponding side lengths are not proportional.

2. yes; $\triangle ABC \sim \triangle KLJ$; $\frac{AC}{KJ} = \frac{BC}{LJ}$ and $\angle C \cong \angle J$.

3. yes; $\triangle WXP \sim \triangle WYZ$; $\angle X \cong \angle Y$ and $\angle W \cong \angle W$

4. $w = 3$

5. $q = 27.5$

6. $p = 14$

7. $\angle Q \cong \angle M$, $\angle R \cong \angle N$, $\angle S \cong \angle P$; $\frac{QR}{MN} = \frac{RS}{NP} = \frac{QS}{MP}$

8. 1.6

9. 4.8

10. no; Corresponding side lengths are not proportional.

11. yes; The angle measures of any right isosceles triangle are 45°, 45°, and 90°.

12. 115 m; $\angle W \cong \angle Z$ and $\angle V \cong \angle Y$ by the Alternate Interior Angles Theorem. So, $\triangle VWX \sim \triangle YZX$ by the AA Similarity Theorem. Because corresponding sides are proportional, $\frac{19}{95} = \frac{23}{WX}$.

13. $P = 2720$ yd; $A = 448{,}000$ yd^2

14. 140 ft; By the AA Similarity Theorem, the triangles are similar.

15. $a = 3$

Chapter 8 Test Prep

1. A, D

2. C

3. **a.** incenter

b. angle bisectors

c. about 3.9 cm; Because $\triangle BGF \cong \triangle BGE$ and corresponding parts of congruent triangles are congruent, $BE = BF = 3$ centimeters. So, $AE = 10 - 3 = 7$ centimeters. Then you can use the Pythagorean Theorem for $\triangle AEG$ to find EG, which is the radius of the wheel.

4. A, B, E

5. Vertical Angles Congruence Theorem; SAS Similarity Theorem; Corresponding angles of similar figures are congruent; $m\angle KHJ = m\angle KML$; Subtraction Property of Equality; Linear Pair Postulate; $m\angle LMN = 180° - m\angle KML$; Substitution Property of Equality; $m\angle LMN = m\angle JHG$; Definition of congruent angles

6. $x = 112.5$

7. B

8. Corresponding angles are congruent and corresponding side lengths are proportional.

Chapter 9

Prepare with CalcChat

1. $x = 9$

2. $x = 7.5$

3. $x = 32$

4. $x = 9.2$

5. $x = 2$

6. $x = 17$

7. $5\sqrt{3}$

8. $3\sqrt{30}$

9. $3\sqrt{15}$

10. $\frac{2\sqrt{7}}{7}$

11. $\frac{5\sqrt{2}}{2}$

12. $2\sqrt{6}$

13. yes; Both expressions have the same domain and have the same value for every value of x in the domain.

Section 9.1

9.1 Practice

1. $x = \sqrt{170} \approx 13.0$; no

3. $x = 15$; yes

5. The formula should be $c^2 = a^2 + b^2$; $x = \sqrt{5} \approx 2.2$

7. $\sqrt{199.68}$, or about 14.1 ft

9. yes

11. no

13. no

15. yes; acute

17. yes; right

19. no

21. yes; obtuse

23. $\sqrt{16{,}200}$, or about 127.3 ft

25. 120 m^2

27. The horizontal distance between any two points is given by $(x_2 - x_1)$, and the vertical distance is given by $(y_2 - y_1)$. The horizontal and vertical segments that represent these distances form a right angle, with the segment between the two points being the hypotenuse. So, you can use the Pythagorean Theorem to say $d^2 = (x_2 - x_1)^2 + (y_2 - y_1)^2$, and when you solve for d, you get the distance formula: $d = \sqrt{(x_2 - x_1)^2 + (y_2 - y_1)^2}$.

29. yes; $2\sqrt{12^2 + 15^2} + 2\sqrt{12^2 + 20^2}$ is approximately 85.07 inches. There are 36 inches in a yard, so you need about 2.36 yards. Each package has 2 yards of binding, so you will need 2 packages.

31. *Sample answer:* Find the side lengths and apply the Converse of the Pythagorean Theorem; Find and compare the slope of each side. Two sides are perpendicular when their slopes are opposite reciprocals.

33. $\triangle CBD \sim \triangle ABC$ by the AA Similarity Theorem because both triangles have a right angle, and both triangles include $\angle B$. $\triangle ACD \sim \triangle ABC$ by the AA Similarity Theorem because both triangles have a right angle, and both triangles include $\angle A$. $\triangle ACD \sim \triangle CBD$ by the AA Similarity Theorem because both triangles have a right angle, and both $\angle B$ and $\angle ACD$ are complementary to $\angle BCD$, so $\angle B \cong \angle ACD$.

STATEMENTS	REASONS
1. $\triangle ABC \sim \triangle ACD \sim \triangle CBD$	**1.** Given
2. $\frac{c}{b} = \frac{b}{c-d}, \frac{c}{a} = \frac{a}{d}$	**2.** Corresponding sides of similar figures are proportional.
3. $c(c-d) = b^2, cd = a^2$	**3.** Cross Products Property
4. $c^2 - cd = b^2$	**4.** Distributive Property
5. $c^2 - a^2 = b^2$	**5.** Substitution Property of Equality
6. $c^2 = a^2 + b^2$	**6.** Addition Property of Equality

35.

STATEMENTS	REASONS
1. In $\triangle ABC$, $c^2 < a^2 + b^2$, where c is the length of the longest side. $\triangle PQR$ has side lengths a, b, and x, where x is the length of the hypotenuse, and $\angle R$ is a right angle.	**1.** Given
2. $a^2 + b^2 = x^2$	**2.** Pythagorean Theorem
3. $c^2 < x^2$	**3.** Substitution Property
4. $c < x$	**4.** Take the positive square root of each side.
5. $m\angle R = 90°$	**5.** Definition of a right angle
6. $m\angle C < m\angle R$	**6.** Converse of the Hinge Theorem
7. $m\angle C < 90°$	**7.** Substitution Property
8. $\angle C$ is an acute angle.	**8.** Definition of acute angle
9. $\triangle ABC$ is an acute triangle.	**9.** Definition of acute triangle

37. $x = 15$ and $x = 3\sqrt{7}$; $3\sqrt{7} < x < 15$; $3 < x < 3\sqrt{7}$ and $15 < x < 21$

39. a. *Sample answer:*

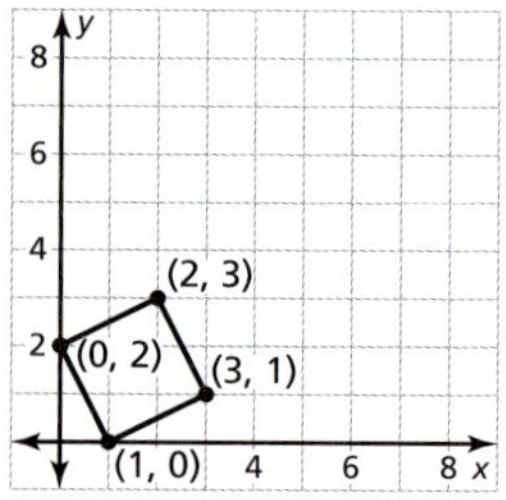

The adjacent sides are perpendicular and each has a length of $\sqrt{5}$.

b. not possible; There are no two integer coordinates that have a distance between them of $\sqrt{6}$.

c. *Sample answer:*

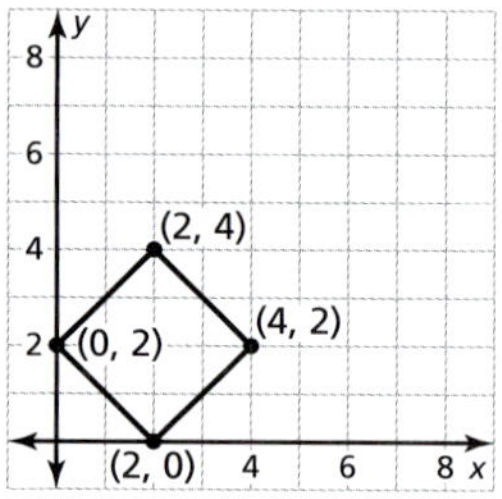

The adjacent sides are perpendicular and each have a length of $2\sqrt{2}$.

d. *Sample answer:*

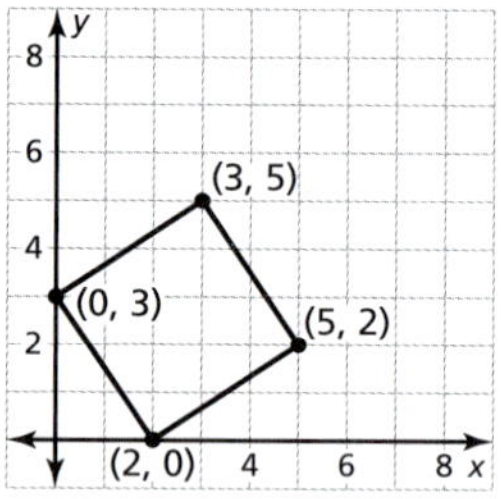

The adjacent sides are perpendicular and each have a length of $\sqrt{13}$.

9.1 Review & Refresh

41. $4\sqrt{2}$

42. 5

43. $x = 10$

44. $m\angle X = m\angle Y = 126°$, $m\angle Z = 54°$

45. yes; Corresponding angles are congruent.

46. yes

47. (3, 2)

48. $\frac{AB}{EF} = \frac{BC}{FD} = \frac{AC}{ED}$, so $\triangle ABC \sim \triangle EFD$.

49. Subtract $m\angle CBD$ from $m\angle ABC$ to find $m\angle ABD$.

50.

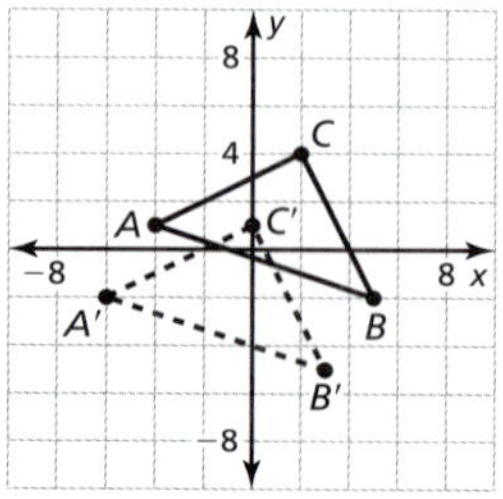

51.

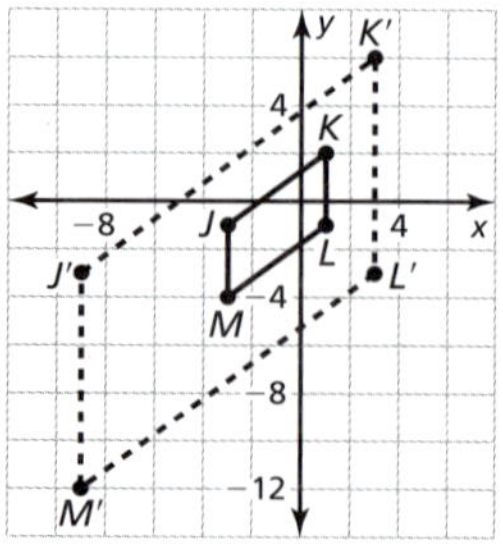

52. 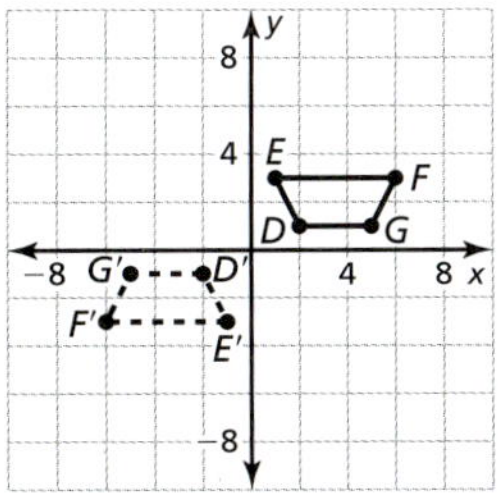

53. $m\angle 1 = 72°$, $m\angle 2 = 108°$, $m\angle 3 = 108°$; $\angle 1$ is supplementary to an angle that is a corresponding angle with the 108° angle. $\angle 3$ and the 108° angle are alternate interior angles, so they are congruent. $\angle 2$ and $\angle 3$ are alternate exterior angles, so they are congruent.

Section 9.2

9.2 Practice

1. $x = 7\sqrt{2}$

3. $x = 3$

5. $x = 9\sqrt{3}$, $y = 18$

7. $x = 12\sqrt{3}$, $y = 12$

9. The hypotenuse of a 30°-60°-90° triangle is equal to the shorter leg times 2; hypotenuse = shorter leg • 2 = 7 • 2 = 14; So, the length of the hypotenuse is 14 units.

11. 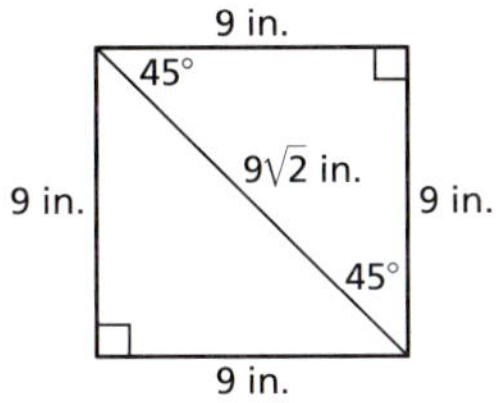

$9\sqrt{2}$ in.

13. 32 ft^2

15. 142 ft; about 200.82 ft; about 245.95 ft

17. Because $\triangle DEF$ is a 45°-45°-90° triangle, by the Converse of the Base Angles Theorem, $\overline{DF} \cong \overline{FE}$. So, let $x = DF = FE$. By the Pythagorean Theorem, $x^2 + x^2 = c^2$, where c is the length of the hypotenuse. So, $2x^2 = c^2$ by the Distributive Property. Take the positive square root of each side to get $x\sqrt{2} = c$. So, the hypotenuse is $\sqrt{2}$ times as long as each leg.

19. Given $\triangle JKL$, which is a 30°-60°-90° triangle, whose shorter leg, $\overline{KL}$, has length x, construct $\triangle JML$, which is congruent and adjacent to $\triangle JKL$. Because corresponding parts of congruent triangles are congruent, $LM = KL = x$, $m\angle M = m\angle K = 60°$, $m\angle MJL = m\angle KJL = 30°$, and $JM = JK$. Also, by the Angle Addition Postulate, $m\angle KJM = m\angle KJL + m\angle MJL$, and by substituting, $m\angle KJM = 30° + 30° = 60°$. So, $\triangle JKM$ has three 60° angles, which means that it is equiangular by definition, and by the Corollary to the Converse of the Base Angles Theorem, it is also equilateral. By the Segment Addition Postulate, $KM = KL + LM$, and by substituting, $KM = x + x = 2x$. So, by the definition of an equilateral triangle, $JM = JK = KM = 2x$. By the Pythagorean Theorem, $(JL)^2 + (KL)^2 = (JK)^2$. By substituting, you get $(JL)^2 + x^2 = (2x)^2$, which is equivalent to $(JL)^2 + x^2 = 4x^2$, when simplified. When the Subtraction Property of Equality is applied, you get $(JL)^2 = 4x^2 - x^2$, which is equivalent to $(JL)^2 = 3x^2$. By taking the positive square root of each side, $JL = x\sqrt{3}$. So, the hypotenuse of the 30°-60°-90° triangle, $\triangle JKL$, is twice as long as the shorter leg, and the longer leg is $\sqrt{3}$ times as long as the shorter leg.

21. *Sample answer:* Because all isosceles right triangles are 45°-45°-90° triangles, they are similar by the AA Similarity Theorem. Because both legs of an isosceles right triangle are congruent, the legs will always be proportional. So, 45°-45°-90° triangles are similar by the SAS Similarity Theorem.

23. $T(1.5, 1.5\sqrt{3} - 1)$

9.2 Review & Refresh

24. $x = 18$

25. yes; obtuse

26. 15

27. $x = 4$, $y = 8$

28. $\left(\frac{1}{2}, 4\right)$

29. yes; $\triangle DEF$ can be mapped to $\triangle MNP$ by a rotation 90° counterclockwise about the origin.

30. the pieces with side lengths of 5.25 inches and 7 inches

31. (7, 0)

32. A quadrilateral is a parallelogram if and only if both pairs of opposite sides are parallel.

Section 9.3

9.3 Practice

1. $\triangle HFE \sim \triangle GHE \sim \triangle GFH$

3. $x = \frac{168}{25} = 6.72$

5. $x = \frac{180}{13} \approx 13.8$

7. about 11.2 ft

9. 16

11. $2\sqrt{70} \approx 16.7$

13. 20

15. $6\sqrt{17} \approx 24.7$

17. $x = 8$

19. $y = 27$

21. $x = 3\sqrt{5} \approx 6.7$

23. $z = \frac{729}{16} \approx 45.6$

25. The length of leg z should be the geometric mean of the length of the hypotenuse, $(w + v)$, and the segment of the hypotenuse that is adjacent to z, which is v, not w; $z^2 = v \cdot (w + v)$

27. about 14.9 ft

29. $a = 3$

31. $x = 9$, $y = 15$, $z = 20$

33. about 76.1 cm; Use the Geometric Mean (Leg) Theorem to find the length of the short segment of $\overline{AC}$. Then use the Pythagorean Theorem to find $\frac{1}{2}$ of BD, and double your answer.

35. given; Geometric Mean (Leg) Theorem; a^2; Substitution Property of Equality; Distributive Property; c; Substitution Property of Equality

37. no; The geometric mean of 4 and 9 is 6, but the labels are incorrect on the triangle. The altitude could be 6, and would be the geometric mean of the two segments that make up the hypotenuse. Or, a leg could be 6, and would be the geometric mean of the length of the hypotenuse and the segment of the hypotenuse that is adjacent to the leg.

39.

STATEMENTS	REASONS
1. Draw $\triangle ABC$, $\angle BCA$ is a right angle.	1. Given
2. Draw a perpendicular segment (altitude) from C to $\overline{AB}$, and label the new point on $\overline{AB}$ as D.	2. Perpendicular Postulate
3. $\triangle ADC \sim \triangle CDB$	3. Right Triangle Similarity Theorem
4. $\frac{BD}{CD} = \frac{CD}{AD}$	4. Corresponding sides of similar figures are proportional.
5. $CD^2 = AD \cdot BD$	5. Cross Products Property

41.

STATEMENTS	REASONS
1. $\triangle ABC$ is a right triangle. Altitude $\overline{CD}$ is drawn to hypotenuse $\overline{AB}$.	1. Given
2. $\angle BCA$ is a right angle.	2. Definition of right triangle
3. $\angle ADC$ and $\angle BDC$ are right angles.	3. Definition of perpendicular lines
4. $\angle BCA \cong \angle ADC \cong \angle BDC$	4. Right Angles Congruence Theorem
5. $\angle A$ and $\angle ACD$ are complementary. $\angle B$ and $\angle BCD$ are complementary.	5. Corollary to the Triangle Sum Theorem
6. $\angle ACD$ and $\angle BCD$ are complementary.	6. Definition of complementary angles
7. $\angle A \cong \angle BCD$, $\angle B \cong \angle ACD$	7. Congruent Complements Theorem
8. $\triangle CBD \sim \triangle ABC$, $\triangle ACD \sim \triangle ABC$, $\triangle CBD \sim \triangle ACD$	8. AA Similarity Theorem

9.3 Review & Refresh

43. $x = 14$

44. 8; By the Converse of the Perpendicular Bisector Theorem, point D is equidistant from the endpoints of $\overline{AC}$, so it lies on the perpendicular bisector of $\overline{AC}$. So, $AB = BC$.

45. 9 **46.** 15

47. 6.5 ft **48.** not a right triangle

49. $y = 14$

50. yes; $\angle H \cong \angle L$, $\angle G \cong \angle K$, $\angle F \cong \angle J$, so $\triangle FGH \sim \triangle JKL$.

51.

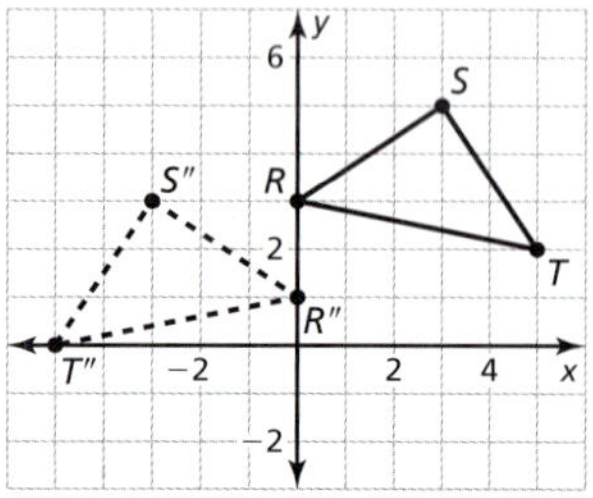

Section 9.4

9.4 Practice

1. $\tan R = \frac{45}{28} \approx 1.6071$, $\tan S = \frac{28}{45} \approx 0.6222$

3. $\tan G = \frac{2}{1} = 2.0000$, $\tan H = \frac{1}{2} = 0.5000$

5. $x \approx 13.8$ **7.** $x \approx 13.7$

9. The tangent ratio should be the length of the leg opposite $\angle D$ to the length of the leg adjacent to $\angle D$, not the length of the hypotenuse; $\tan D = \frac{35}{12}$

11. 1 **13.** about 225.69 ft

15. $\frac{5}{12} \approx 0.4167$

17. it increases; The opposite side gets longer.

19. **a.** no; The distance between the ends of the class is about 33.4 ft.

b. 3 students at each end; The triangle formed by the 60° angle has an opposite leg that is about 7.5 feet longer than the opposite leg of the triangle formed by the 50° angle. Because each student needs 2 feet of space, 3 more students can fit on each end with about 1.5 feet of space left over.

21. 45°; greater than 45°; less than 45°; If the ratio of the legs is equal to 1, then the legs are congruent, and all isosceles right triangles are 45°-45°-90° triangles. If one acute angle measure of a right triangle is less than 45°, then the other acute angle is greater than 45°. So, the side opposite the smaller acute angle will be the shorter leg, which means the ratio will be less than 1, and the leg opposite the larger acute angle will be the longer leg, which means the ratio will be greater than 1.

23. about 128.0 units

9.4 Review & Refresh

24. $x = 4\sqrt{3} \approx 6.9$; no **25.** $6\sqrt{13} \approx 21.6$

26. 7.5 units^2

27.

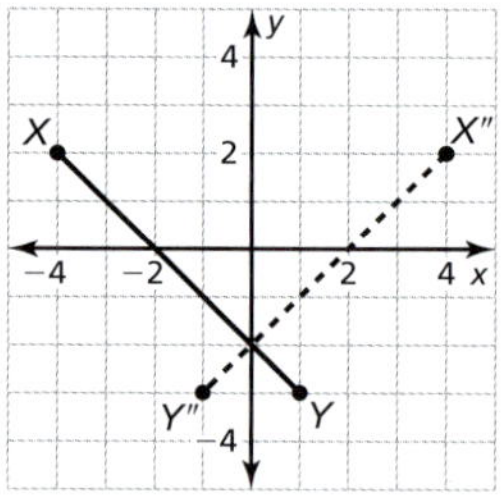

28. (2, 0) **29.** $x = 2\sqrt{3} \approx 3.5$

30. $x = 64$ **31.** $x \approx 6.0$

32. $x = 30$ **33.** $m\angle 2 = 135°$

Section 9.5

9.5 Practice

1. $\sin D = \frac{4}{5} = 0.8000$, $\sin E = \frac{3}{5} = 0.6000$,

$\cos D = \frac{3}{5} = 0.6000$, $\cos E = \frac{4}{5} = 0.8000$

3. $\sin D = \frac{28}{53} \approx 0.5283$, $\sin E = \frac{45}{53} \approx 0.8491$,

$\cos D = \frac{45}{53} \approx 0.8491$, $\cos E = \frac{28}{53} \approx 0.5283$

5. $\sin D = \frac{\sqrt{3}}{2} \approx 0.8660$, $\sin E = \frac{1}{2} = 0.5000$,

$\cos D = \frac{1}{2} = 0.5000$, $\cos E = \frac{\sqrt{3}}{2} \approx 0.8660$

7. cos 53°

9. cos 61°

11. sin 31°

13. sin 17°

15. $x \approx 9.5$, $y \approx 15.3$

17. $v \approx 4.7$, $w \approx 1.6$

19. $a \approx 14.9$, $b \approx 11.1$

21. $\sin X = \cos X = \sin Z = \cos Z$

23. $\sin L$, $\cos J$

25. The sine of $\angle A$ should be equal to the ratio of the length of the leg opposite the angle, to the length of the hypotenuse; $\sin A = \frac{12}{13}$

27. about 15 ft

29. a.

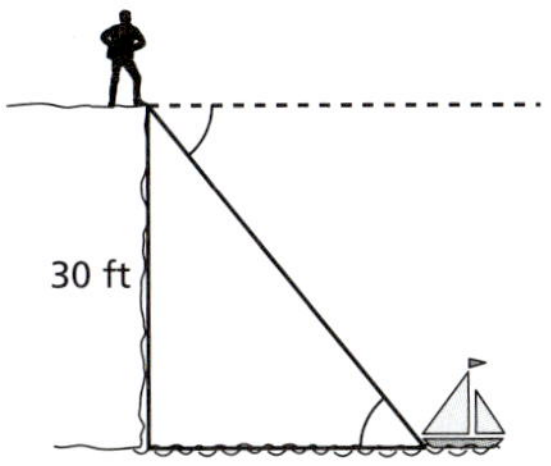

b.

Angle of depression	40°	50°	60°	70°	80°
Approximate length of line of sight (feet)	46.7	39.2	34.6	31.9	30.5

c.

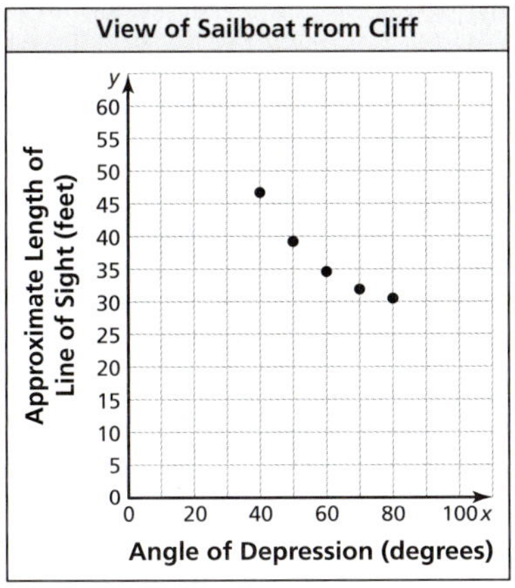

d. As the boat sails away from you, the angle of depression decreases, and the length of your line of sight increases.

31. $\sin A = \frac{\sqrt{3}}{2}$, $\cos A = \frac{1}{2}$, $\tan A = \sqrt{3}$

33. $\sin A = 0.96$, $\cos A = 0.28$, $\tan A = \frac{24}{7}$ or about 3.43

35. $\sin A = \frac{12}{13}$, $\cos A = \frac{5}{13}$, $\tan A = \frac{12}{5}$

37. $\sin A = \frac{15}{17}$, $\cos A = \frac{8}{17}$, $\tan A = \frac{15}{8}$

39. no; $\left(\frac{1}{9}\right)^2 + \left(\frac{2}{3}\right)^2 = \frac{37}{81}$, so $r = \frac{\sqrt{37}}{9}$. The unit circle has a radius of 1.

41. about 2.28 units; Using the Pythagorean Theorem, $x^2 + \left(\frac{12}{35}\right)^2 = 1^2$, making $x = \frac{\sqrt{1081}}{35}$. Add the side lengths to find the perimeter: $1 + \frac{12}{35} + \frac{\sqrt{1081}}{35}$.

43.

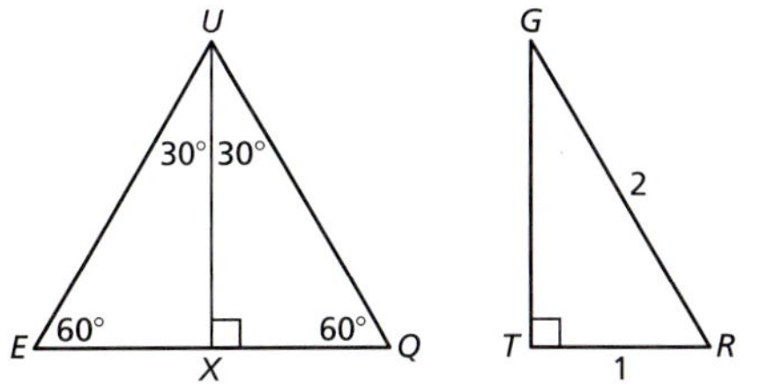

Because $\triangle EQU$ is an equilateral triangle, all three angles have a measure of 60°. When an altitude, $\overline{UX}$, is drawn from U to $\overline{EQ}$ as shown, two congruent 30°-60°-90° triangles are formed, where $m\angle E = 60°$. So, $\sin E = \sin 60° = \frac{\sqrt{3}}{2}$. Also, in $\triangle RGT$, because the hypotenuse is twice as long as one of the legs, it is also a 30°-60°-90° triangle. Because $\angle G$ is across from the shorter leg, it must have a measure of 30°, which means that $\cos G = \cos 30° = \frac{\sqrt{3}}{2}$. So, $\sin E = \cos G$.

45. a. more than 2698.0 yd

b. more than 516.5 yd

47. Angles 1 and 2 are complementary angles. The cosine of angle 2 is equal to the sine of angle 1, or y. The sine of angle 2 is equal to the cosine of angle 1, or x. The tangent of angle 1 is $\frac{y}{x}$, so the tangent of ange 2 is $\frac{x}{y}$.

49. *Sample answer:* about 3.6 miles away for each 1000 feet of altitude at which you are flying; Airplanes usually descend an angle of depression of 3°. So, if the airplane is flying at an elevation of h feet, it should start to descend $\frac{h \text{ ft}}{\tan 3°} \bullet \frac{1 \text{mi}}{5280 \text{ ft}} \approx 0.0036h$ miles away.

51. In $\triangle ABC$, $\sin C = \frac{h}{a}$. So, $h = a \sin C$. When you substitute this into the formula for the area of a triangle, you get $A = \frac{1}{2}bh = \frac{1}{2}b(a \sin C) = \frac{1}{2}ab \sin C$; about 9.0 square units

9.5 Review & Refresh

52. $x = 8$; yes

53. $x = 6\sqrt{2} \approx 8.5$; no

54. cos 43°

55. $x \approx 6.9$

56. about 162.9°, about 17.1°

57.

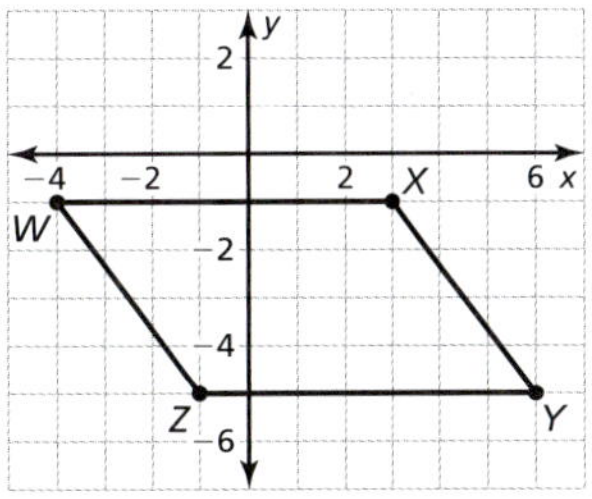

The slopes of $\overline{WX}$ and $\overline{YZ}$ are $\frac{-1-(-1)}{3-(-4)} = \frac{-5-(-5)}{6-(-1)} = 0$, so the sides are parallel. The slopes of $\overline{WZ}$ and $\overline{XY}$ are $\frac{-5-(-1)}{-1-(-4)} = \frac{-5-(-1)}{6-3} = -\frac{4}{3}$, so the sides are parallel. Quadrilateral $WXYZ$ has two pairs of parallel sides, so it is a parallelogram.

58. $\triangle ABC \sim \triangle BDC \sim \triangle ADB$; $x = 4\sqrt{2} \approx 5.7$

59. $x = 5\sqrt{2}$, $y = 5\sqrt{6}$

60. yes; $\frac{25}{1}$; $\frac{625}{1}$

61. $AC < DF$; By the Hinge Theorem, $\overline{AC}$ is the third side of the triangle with the smaller included angle, so it is shorter than $\overline{DF}$.

62. $x = 10, y = 9$

63. *Sample answer:*

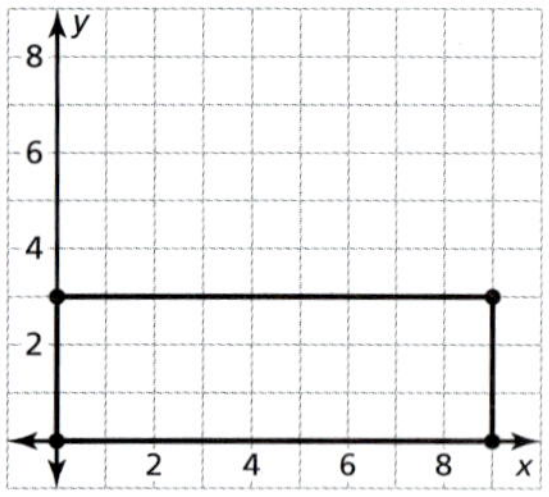

$3\sqrt{10}$, or about 9.5 units

64. $P(-1, -1)$

Section 9.6

9.6 Practice

1. about 0.7986

3. about −0.7547

5. about −0.2679

7. about 81.8 square units

9. 207 square units

11. $\angle C$

13. $\angle A$

15. about 48.6°

17. about 70.7°

19. about 15.6°

21. about 16.6 units, about 57.3°, about 32.7°; $\sqrt{9^2 + 14^2} \approx 16.6$ units; $\tan^{-1} \frac{14}{9} \approx 57.3°$; $\tan^{-1} \frac{9}{14} \approx 32.7°$

23. $m\angle A \approx 80.9°, m\angle C \approx 43.1°, a \approx 20.2$

25. $m\angle B = 66°, a \approx 14.3, b \approx 24.0$

27. $m\angle A = 48°, b \approx 25.5, c \approx 18.7$

29. $a \approx 5.2, m\angle B \approx 50.5°, m\angle C \approx 94.5°$

31. $m\angle A \approx 81.1°, m\angle B \approx 65.3°, m\angle C \approx 33.6°$

33. According to the Law of Sines, the ratio of the sine of an angle's measure to the length of its opposite side should be equal to the ratio of the sine of another angle measure to the length of its opposite side; $\frac{\sin C}{5} = \frac{\sin 55°}{6}$, $\sin C = \frac{5 \sin 55°}{6}$, $m\angle C \approx 43.0°$

35. Law of Sines; given two angle measures and the length of a side; $m\angle C = 64°, a \approx 19.2, c \approx 18.1$

37. Pythagorean Theorem and trigonometric ratios; given the length of the legs of a right triangle; $b \approx 16.2, m\angle A \approx 68.2°, m\angle C \approx 21.8°$

39. about 27.17 ft

41. about 10.7 ft

43.

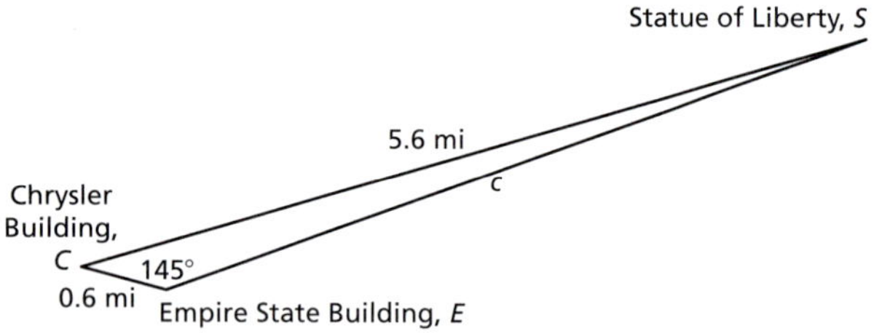

about 5.1 mi

45. your cousin; You are given the lengths of two sides and the measure of their included angle.

47. **a.** yes; You are given the measure of two angles and the length of a side.

b. yes; You can also use the Pythagorean Theorem and trigonometric ratios to solve the triangle, because $\triangle XYZ$ is a right triangle.

49. **a.** about 163.4 yd

b. about 3.5°

51. $x = 99, y \approx 20.1$

53. **a.** $m\angle B \approx 52.3°, m\angle C \approx 87.7°, c \approx 20.2$; $m\angle B \approx 127.7°, m\angle C \approx 12.3°, c \approx 4.3$

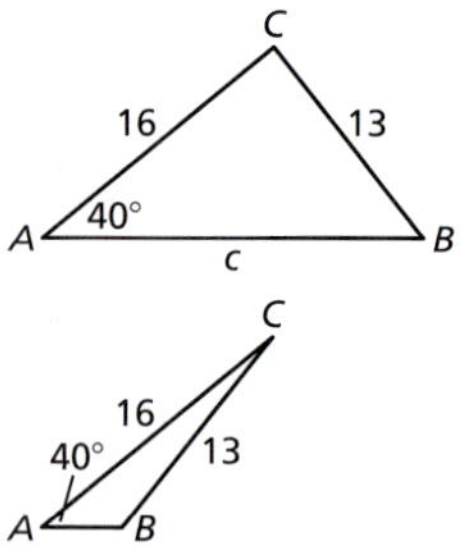

b. $m\angle B \approx 42.4°, m\angle C \approx 116.6°, c \approx 42.4$; $m\angle B \approx 137.6°, m\angle C \approx 21.4°, c \approx 17.3$

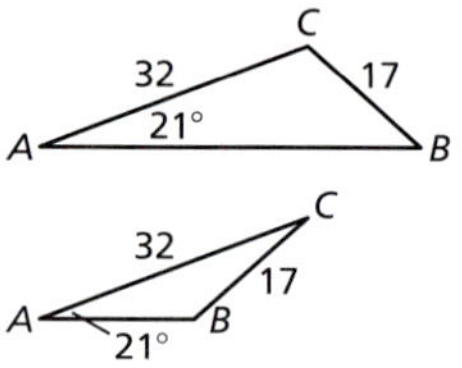

55.

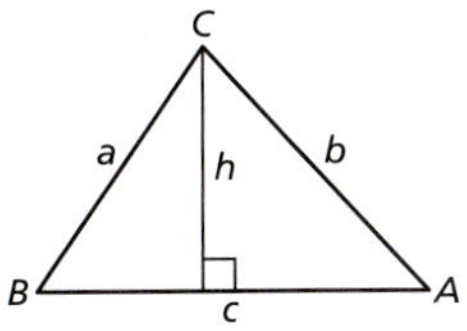

The formula for the area of $\triangle ABC$ with altitude h drawn from C to $\overline{AB}$ as shown is Area $= \frac{1}{2}ch$. Because $\sin A = \frac{h}{b}$, $h = b \sin A$. By substituting, you get

Area $= \frac{1}{2}c(b \sin A) = \frac{1}{2}bc \sin A$.

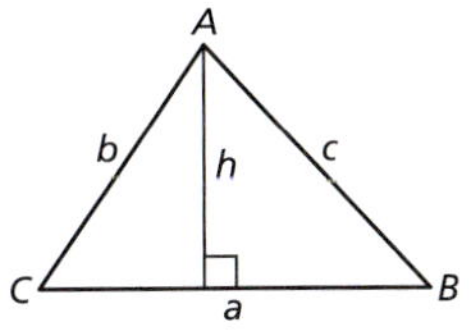

The formula for the area of $\triangle ABC$ with altitude h drawn from A to $\overline{BC}$ as shown is Area $= \frac{1}{2}ah$. Because $\sin B = \frac{h}{c}$, $h = c \sin B$. By substituting, you get

Area $= \frac{1}{2}a(c \sin B) = \frac{1}{2}ac \sin B$.

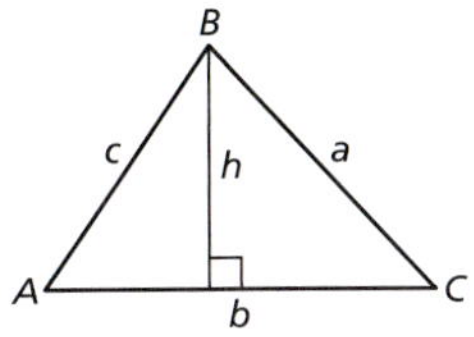

The formula for the area of $\triangle ABC$ with altitude h drawn from B to $\overline{AC}$ is Area $= \frac{1}{2}bh$. Because $\sin C = \frac{h}{a}$, $h = a \sin C$. By substituting, you get Area $= \frac{1}{2}ab \sin C$.

They are all expressions for the area of the same triangle, so they are all equal to each other by the Transitive Property. By the Multiplication Property of Equality, multiply all three expressions by 2 to get $bc \sin A = ac \sin B = ab \sin C$. By the Division Property of Equality, divide all three expressions by abc to get $\frac{\sin A}{a} = \frac{\sin B}{b} = \frac{\sin C}{c}$.

57. Because the sine is the ratio of the length of a leg to the length of the hypotenuse, and the hypotenuse is always longer than either of the legs, the sine cannot have a value greater than 1.

59. Because the triangle is equilateral, you can say $a = b = c = x$. So, the Law of Cosines is $x^2 = x^2 + x^2 - 2(x)(x)\cos X$ for all of the angles. By the Subtraction Property of Equality, $0 = x^2 - 2x^2 \cos X$. Then, by the Addition Property of Equality, $2x^2 \cos X = x^2$. By the Division Property of Equality, $2 \cos X = 1$, or $\cos X = \frac{1}{2}$. So, $X = \cos^{-1} \frac{1}{2} = 60°$. So, the measure of each angle of an equilateral triangle is 60°.

61.

STATEMENTS	REASONS
1. $\overline{BD}$ is an altitude of $\triangle ABC$.	1. Given
2. $\triangle ADB$ and $\triangle CDB$ are right triangles.	2. Definition of altitude
3. $a^2 = (b - x)^2 + h^2$	3. Pythagorean Theorem
4. $a^2 = b^2 - 2bx + x^2 + h^2$	4. Expand binomial.
5. $x^2 + h^2 = c^2$	5. Pythagorean Theorem
6. $a^2 = b^2 - 2bx + c^2$	6. Substitution Property of Equality
7. $\cos A = \frac{x}{c}$	7. Definition of cosine ratio
8. $x = c \cos A$	8. Multiplication Property of Equality
9. $a^2 = b^2 + c^2 - 2bc \cos A$	9. Substitution Property of Equality

9.6 Review & Refresh

62. $x = 12\sqrt{2} \approx 17.0$

63. $x = \frac{7\sqrt{3}}{3} \approx 4.0$

64. $x = 66$

65. $x \approx 26.7$

66. 7 in. $< x <$ 23 in.

67. yes; *Sample answer:* Quadrilateral $ABCD$ can be mapped onto Quadrilateral $GHEF$ by a reflection about the x-axis, followed by a translation 1 unit down and 7 units left.

68. $\triangle EFG \sim \triangle FHG \sim \triangle EHF$; $y = 3\sqrt{15} \approx 11.6$

69. $x \approx 10.8$, $y \approx 10.4$

70. Parallelogram Diagonals Converse

71. $x = 8$, $y = 17$

72. (3, 4)

73. $AB = \sqrt{85} \approx 9.2$, $m\angle A \approx 40.6°$, $m\angle B \approx 49.4°$

74. $a \approx 12.6$, $b \approx 10.0$, $m\angle C = 69°$

Chapter 9 Review

1. $x = 2\sqrt{34} \approx 11.7$; no

2. $x = 12$; yes

3. $x = 29$; yes

4. $x = 2\sqrt{30} \approx 11.0$; no

5. yes

6. no

7. no

8. yes

9. yes; acute

10. yes; right

11. yes; obtuse

12. yes; *Sample answer:* 24, 70, and 74; 36, 105, and 111

13. acute; When the length of one leg of the triangle increases from 9 to 10, the angle opposite that leg increases, so the measure of the right angle decreases.

14. $x = 3\sqrt{2}$, $y = 3\sqrt{2}$

15. $x = 7$

16. $x = 16\sqrt{3}$

17. $x = 16$; $y = 16\sqrt{2}$

18. Area $= \frac{9}{2} + \frac{9\sqrt{3}}{2} \approx 12.3$ square units,

Perimeter $= 9 + 3\sqrt{3} + 3\sqrt{2} \approx 18.4$ units

19. $\triangle GFH \sim \triangle FEH \sim \triangle GEF$; $x = 13.5$

20. $\triangle KLM \sim \triangle JKM \sim \triangle JLK$; $x = 2\sqrt{6} \approx 4.9$

21. $\triangle QRS \sim \triangle PQS \sim \triangle PRQ$; $x = 3\sqrt{3} \approx 5.2$

22. $\triangle TUV \sim \triangle STV \sim \triangle SUT$; $x = 25$

23. 15
24. $24\sqrt{3} \approx 41.6$
25. 12.2 ft
26. $\tan J = \frac{11}{60} \approx 0.1833$, $\tan L = \frac{60}{11} \approx 5.4545$
27. $\tan N = \frac{12}{35} \approx 0.3429$, $\tan P = \frac{35}{12} \approx 2.9167$
28. $\tan A = \frac{7\sqrt{2}}{8} \approx 1.2374$, $\tan B = \frac{4\sqrt{2}}{7} \approx 0.8081$
29. $\tan M = \frac{3}{4} = 0.75$, $\tan N = \frac{4}{3} \approx 1.3333$
30. $x \approx 44.0$
31. $x \approx 12.8$
32. $x \approx 9.3$
33. $x \approx 30.0$
34. **a.** about 14.9 ft
b. about 7.5 ft
c. about 3.9 ft
35. about 7.9 ft
36. $\frac{8}{15}$
37. yes; The side lengths can be 3, 4, and 5, or any multiple of those lengths.
38. $\sin X = \frac{3}{5} = 0.6000$, $\sin Z = \frac{4}{5} = 0.8000$, $\cos X = \frac{4}{5} = 0.8000$, $\cos Z = \frac{3}{5} = 0.6000$
39. $\sin X = \frac{7\sqrt{149}}{149} \approx 0.5735$, $\sin Z = \frac{10\sqrt{149}}{149} \approx 0.8192$, $\cos X = \frac{10\sqrt{149}}{149} \approx 0.8192$, $\cos Z = \frac{7\sqrt{149}}{149} \approx 0.5735$
40. $\sin X = \frac{55}{73} \approx 0.7534$, $\sin Z = \frac{48}{73} \approx 0.6575$, $\cos X = \frac{48}{73} \approx 0.6575$, $\cos Z = \frac{55}{73} \approx 0.7534$
41. $\cos 18°$
42. $\cos 75°$
43. $\cos 45°$
44. $\sin 61°$
45. $\sin 40°$
46. $\sin 26°$
47. $s \approx 31.3$, $t \approx 13.3$
48. $r \approx 4.0$, $s \approx 2.9$
49. $s \approx 12.5$, $t \approx 3.4$
50. $\sin A = \frac{\sqrt{5}}{3}$, $\cos A = \frac{2}{3}$, $\tan A = \frac{\sqrt{5}}{2}$
51. $\sin A = \frac{40}{41}$, $\cos A = \frac{9}{41}$, $\tan A = \frac{40}{9}$
52. $\sin A = \frac{28}{53}$, $\cos A = \frac{45}{53}$, $\tan A = \frac{28}{45}$
53. no; $0.64^2 + 0.36^2 = 0.5392$ so $r = \sqrt{0.5392}$. The unit circle has a radius of 1.
54. yes; $\left(\frac{33}{65}\right)^2 + \left(\frac{56}{65}\right)^2 = 1$ so $r = \sqrt{1} = 1$.
55. 98 ft
56. about 0.6947
57. about -0.5592
58. about -0.4663
59. about 41.0 square units
60. about 13.6 square units
61. $m\angle B \approx 24.3°$, $m\angle C \approx 43.7°$, $c \approx 6.7$
62. $m\angle C = 88°$, $a \approx 25.8$, $b \approx 49.5$
63. $m\angle A \approx 99.9°$, $m\angle B \approx 32.1°$, $a \approx 37.1$
64. $b \approx 5.4$, $m\angle A \approx 141.4°$, $m\angle C \approx 13.6°$
65. $m\angle A = 35°$, $a \approx 12.3$, $c \approx 14.6$
66. $m\angle A \approx 42.6°$, $m\angle B \approx 11.7°$, $m\angle C \approx 125.7°$
67. **a.**

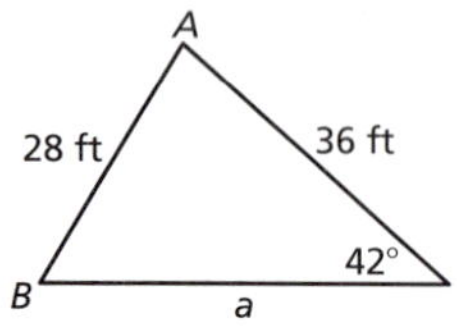

b. $a \approx 41.0$ ft, $m\angle A \approx 78.6°$, $m\angle B \approx 59.4°$
c. 13 gal

Chapter 9 Practice Test

1. $s \approx 16.3$, $t \approx 7.6$
2. $x \approx 16.0$, $y \approx 14.9$
3. $c \approx 13.1$, $d \approx 8.4$
4. yes; right
5. yes; acute
6. yes; obtuse
7. $m\angle A \approx 24.4°$, $m\angle C \approx 65.6°$, $b \approx 12.1$
8. $m\angle B \approx 35.8°$, $m\angle C \approx 71.2°$, $c \approx 17.8$
9. $m\angle A = 50°$, $a \approx 10.7$, $c \approx 9.0$
10. $m\angle A \approx 45°$, $c = 16\sqrt{2}$, $a = 16$
11. $m\angle A \approx 22.6°$, $m\angle C \approx 67.4°$, $a = 13$, $b = 33.8$, $c = 31.2$
12. $m\angle B = 60°$, $a = \frac{8\sqrt{3}}{3}$, $c = \frac{16\sqrt{3}}{3}$
13. $a \approx 29.1$, $m\angle B \approx 23.7°$, $m\angle C \approx 53.3°$
14. $m\angle B = 119°$, $a \approx 6.5$, $c \approx 8.5$
15. $m\angle A \approx 70.0°$, $m\angle B \approx 50.0°$, $m\angle C \approx 60°$
16. $\sin 37°$
17. $a = 5$, $c = 5$, $b = 5\sqrt{2}$, $m\angle A = 45°$, $m\angle C = 45°$, $m\angle B = 90°$; 12.5 square units
18. yes; The sine of the angle is 0.352 or $\frac{44}{125}$. Draw a triangle where the side opposite the angle has a length of 44 and the hypotenuse has a length of 125. Using the Pythagorean Theorem, the remaining leg of the right triangle has a length of 117. As a result, the tangent of the angle is $\frac{44}{117}$.
19. about 46.3 ft
20. about 6.3 ft

Chapter 9 Test Prep

1. A, D, E, F, H
2. B
3. $\sin J = \sin Q$; $\sin L = \cos J$; $\cos L < \tan Q$; $\cos S > \cos J$; $\cos J = \sin S$; $\tan J = \tan Q$; $\tan L < \tan Q$; $\tan S > \cos Q$; $\sin Q = \cos L$
4. about 81.9 ft
5.

STATEMENTS	REASONS
1. $\overline{HE} \cong \overline{HG}$, $\overline{EG} \perp \overline{DF}$	1. Given
2. $\overline{HF} \cong \overline{HF}$, $\overline{DH} \cong \overline{DH}$	2. Reflexive Property of Congruence
3. $\angle EHF$, $\angle GHF$, $\angle GHD$, and $\angle EHD$ are right angles.	3. Definition of perpendicular lines
4. $\angle EHF \cong \angle GHF$, $\angle GHD \cong \angle EHD$	4. Right Angles Congruence Theorem
5. $\triangle EHF \cong \triangle GHF$, $\triangle EHD \cong \triangle GHD$	5. SAS Congruence Theorem
6. $\overline{FE} \cong \overline{FG}$, $\overline{DE} \cong \overline{DG}$	6. Corresponding parts of congruent triangles are congruent.

6. **a.** about 498.0 ft
b. about 615.1 ft
c. Use any of the three trigonometric functions, sine, cosine, or tangent; about 35.9°
7. B, C, E
8. A

Chapter 10

Prepare with CalcChat

1. $x = 1 \pm \sqrt{6}$
2. $r = -5 \pm 3\sqrt{2}$
3. $w = -1, w = 9$
4. $p = -5 \pm \sqrt{29}$
5. $k = 2 \pm \sqrt{11}$
6. $z = 1$
7. $x^2 + 11x + 28$
8. $a^2 - 4a - 5$
9. $3q^2 - 31q + 36$
10. $10v^2 - 33v - 7$
11. $4h^2 + 11h + 6$
12. $18b^2 - 54b + 40$
13. *Sample answer:* $(2n + 1)(2n + 3)$; $2n + 1$ is positive and odd when n is a nonnegative integer. The next positive, odd integer is $2n + 3$.

Section 10.1

10.1 Practice

1. $\odot C$
3. $\overline{BH}, \overline{AD}$
5. $\overleftrightarrow{KG}$
7. 2 internal, 2 external

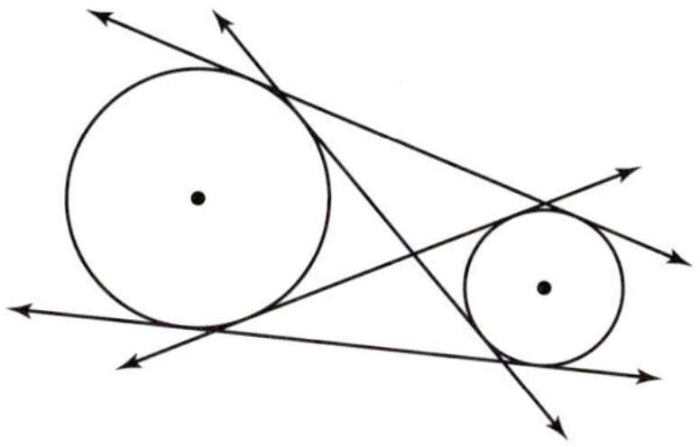

9. 1 external

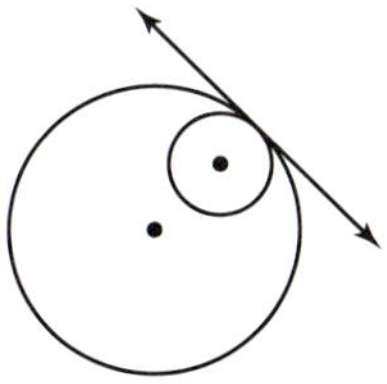

11. yes; $\triangle ABC$ is a right triangle.
13. no; $\triangle ABD$ is not a right triangle.
15. 10
17. 10.5
19. $\angle Z$ is a right angle, not $\angle YXZ$; So, $\overline{XY}$ is not tangent to $\odot Z$.
21. *Sample answer:*

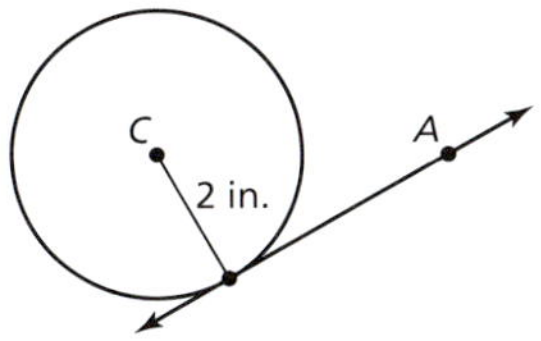

23. $x = 5$
25. $x = 9$
27. $x = \pm 3$
29. 2; 1; 0; *Sample answer:* There are two possible points of tangency from a point outside the circle, one from a point on the circle, and none from a point inside the circle.
31. *Sample answer:* Every point is the same distance from the center, so the farthest two points can be from each other is opposite sides of the center.
33. yes; Draw $\overline{SQ}$ and $\overline{SR}$. $\angle SQR$ and $\angle SRP$ are right angles by the Tangent Line to Circle Theorem. $\overline{SP} \cong \overline{SP}$ by the Reflexive Property. $\triangle SQP \cong \triangle SRP$ by the HL Congruence Theorem. Because corresponding parts of congruent triangles are congruent, $\angle QPS \cong \angle RPS$. By the definition of an angle bisector, $\overline{PS}$ bisects $\angle QPR$.
35. about 17.78 in.
37. yes; $AZ = AW$, $BW = BX$, $CX = CY$, and $DY = DZ$, so, $(AW + WB) + (CY + YD) = (AZ + DZ) + (BX + CX)$.
39. By the Tangent Line to Circles Theorem, $m\angle PRS = m\angle PTS = 90°$. $\overline{PS} \cong \overline{PS}$, so $\triangle PTS \cong \triangle PRS$ by the HL Congruence Theorem. So, $\overline{SR} \cong \overline{ST}$ because corresponding parts of congruent triangles are congruent.
41. $VW = 20\sqrt{6}$

10.1 Review & Refresh

43. $m\angle Q = 41°$, $PR \approx 7.8$, $RQ \approx 11.9$
44. $m\angle A \approx 27.7°$, $m\angle B \approx 84.1°$, $m\angle C \approx 68.2°$
45. 43°
46. 3
47. neither
48. parallel
49. 76°, 104°
50. $x = 7$
51. about 19.6 ft
52. $x = 12$
53. $MN = 20$
54. $LP = 24$, $PQ = 12$

Section 10.2

10.2 Practice

1. $\overset{\frown}{AB}$, 135°; $\overset{\frown}{ADB}$, 225°
3. $\overset{\frown}{JL}$, 120°; $\overset{\frown}{JKL}$, 240°
5. $\overset{\frown}{RS}$, 52°; $\overset{\frown}{RQS}$, 308°
7. semicircle; 180°
9. minor arc; 64°
11. **a.** 132°
b. 147°
c. 200°
d. 160°
13. **a.** 84.24°
b. 147.24°
c. 233.28°
d. 329.4°
e. 61.92°
f. 298.08°
15. congruent; They are arcs of the same circle and they have congruent central angles.
17. congruent; The circles are congruent and they have congruent central angles.
19. $\overset{\frown}{AD}$ is the minor arc; $\overset{\frown}{ABD}$
21. $x = 70$; 110°
23. no; $\overset{\frown}{AMB}$ is a semicircle, so $x + 4x = 180$ and $x = 36$.
25. 340°; 160°
27. 18°
29. yes; Both radii are $\overline{CD}$.
31. **a.** Translate $\odot B$ so that point B maps to point A. The image of $\odot B$ is $\odot B'$ with center A. Because $\overline{AC} \cong \overline{BD}$, this translation maps $\odot B'$ to $\odot A$. A rigid motion maps $\odot B$ to $\odot A$, so $\odot A \cong \odot B$.
b. Because $\odot A \cong \odot B$, the distance from the center of the circle to a point on the circle is the same for each circle. So, $\overline{AC} \cong \overline{BD}$.

10.2 Review & Refresh

33. $x = -5, x = \frac{2}{3}$

34. $a \approx 6.6, m\angle B \approx 55.2^\circ, m\angle C \approx 87.8^\circ$

35. $t \approx 9.0, s \approx 15.0, m\angle T = 37^\circ$

36. $\widehat{MN}$, 160°; $\widehat{MPN}$, 200°

37. $\sqrt{689}$, or about 26.2 ft

38. 16

39. $x = 4\sqrt{6} \approx 9.8$

40. $x = \sqrt{338} \approx 18.4$

41. $x = 135$

42. $x = 8$

43.

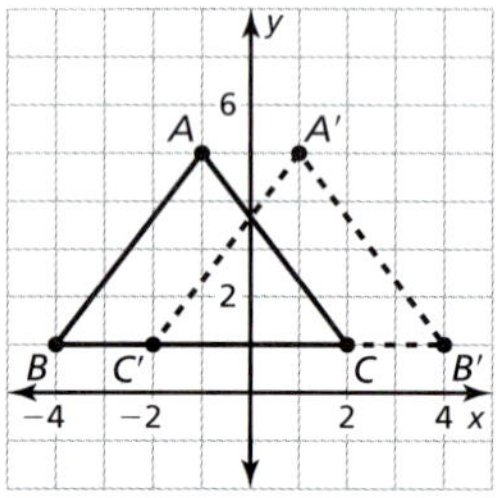

44.

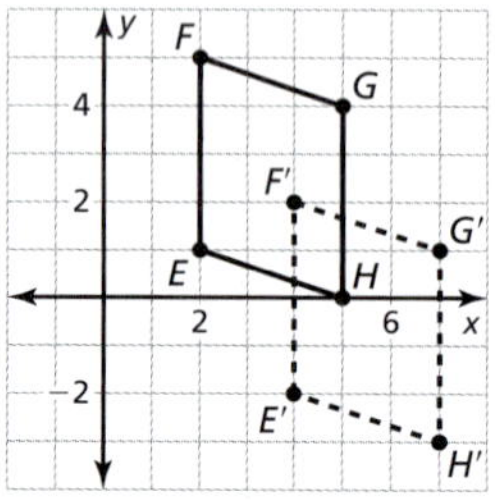

45. $\angle G \cong \angle K$ and $\angle FHG \cong \angle JHK$ because they are vertical angles. So, $\triangle FGH \sim \triangle JKH$ by the AA Similarity Theorem.

46. 43°; By the Converse of the Angle Bisector Theorem, $\overrightarrow{JL}$ bisects $\angle KJM$. When $x = 5$, $9x - 2 = 8x + 3$. So, $m\angle KJL = (9x - 2)^\circ = (9(5) - 2)^\circ = 43^\circ$.

Section 10.3

10.3 Practice

1. 75°

3. 11

5. $x = 8$

7. $x = 7$

9. Draw a segment between each pair of tables. Draw the perpendicular bisectors of two segments. Place the patio heater at the point of intersection.

11. yes; The triangles are congruent, so $\overline{AB}$ is a perpendicular bisector of $\overline{CD}$.

13. 17

15. $\overline{CE}$ is the hypotenuse of $\triangle CGE$.

$CE^2 = 5^2 + 12^2$

$CE^2 = 169$

$CE = 13$

So, the radius is 13 inches.

17. **a.** Because $PA = PB = PC = PD$, $\triangle PDC \cong \triangle PAB$ by the SSS Congruence Theorem. So, $\angle DPC \cong \angle APB$ and $\widehat{AB} \cong \widehat{CD}$.

b. $PA = PB = PC = PD$, and because $\widehat{AB} \cong \widehat{CD}$, $\angle DPC \cong \angle APB$. By the SAS Congruence Theorem, $\triangle PDC \cong \triangle PAB$, so $\overline{AB} \cong \overline{CD}$.

19. *Sample answer:* $\overline{AC} \cong \overline{BC}$, so $\widehat{AC} \cong \widehat{BC}$ and $m\widehat{AB} = 360^\circ - 2m\widehat{AC}$. $m\widehat{AC}$ is an integer, so $2m\widehat{AC}$ is even and $360^\circ - 2m\widehat{AC}$ is even.

21. $\overline{QS}$ is the perpendicular bisector of $\overline{RT}$. Let C be the center of the circle. Because $\overline{CT}$ and $\overline{CR}$ are radii of the same circle, $\overline{CT} \cong \overline{CR}$. By the Converse of the Perpendicular Bisector Theorem, C lies on $\overline{QS}$. By definition, $\overline{QS}$ is a diameter of the circle.

23.

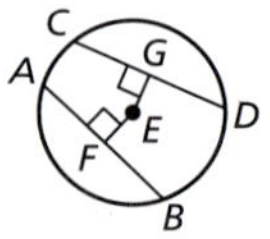

If $\overline{AB} \cong \overline{CD}$, then $\overline{GC} \cong \overline{FA}$. Because $\overline{EC} \cong \overline{EA}$, $\triangle ECG \cong \triangle EAF$ by the HL Congruence Theorem, so $\overline{EF} \cong \overline{EG}$ and $EF = EG$. If $EF = EG$, then because $\overline{EC} \cong \overline{ED} \cong \overline{EA} \cong \overline{EB}$, $\triangle AEF \cong \triangle BEF \cong \triangle DEG \cong \triangle CEG$ by the HL Congruence Theorem. Then $\overline{AF} \cong \overline{BF} \cong \overline{DG} \cong \overline{CG}$, so $\overline{AB} \cong \overline{CD}$.

10.3 Review & Refresh

24. 8

25. 159°

26. 122°

27. similar; The arcs have the same measure but the circles are not congruent.

28. about 172.1 m

29. 100°

Section 10.4

10.4 Practice

1. 42°

3. 10°

5. 120°

7. 51°

9. $x = 100, y = 85$

11. $a = 20, b = 22$

13. The inscribed angle was not doubled; $m\angle BAC = 2(53^\circ) = 106^\circ$

15. Place the right angle of the carpenter's square on the edge of the circle and connect the points where the sides intersect the edge of the circle.

17. When a right triangle is inscribed in a circle, its hypotenuse is the diameter of the circle. So, double the radius to find the length of the hypotenuse.

19.

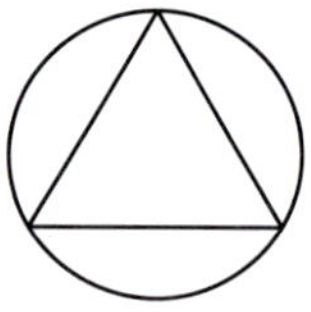

21. $x = 25, y = 5$; 130°, 75°, 50°, 105°

23. **a.** $\overline{QB} \cong \overline{QA}$, so $\triangle ABC$ is isosceles. By the Base Angles Theorem, $\angle QBA \cong \angle QAB$, so $m\angle BAQ = x^\circ$. By the Exterior Angles Theorem, $m\angle AQC = 2x^\circ$. Then $m\widehat{AC} = 2x^\circ$, so $m\angle ABC = x^\circ = \frac{1}{2}(2x)^\circ = \frac{1}{2}m\widehat{AC}$.

b. Given: $\angle ABC$ is inscribed in $\odot Q$. $\overline{DB}$ is a diameter; Prove: $m\angle ABC = \frac{1}{2}m\widehat{AC}$; By Case 1, proved in part (a), $m\angle ABD = \frac{1}{2}m\widehat{AD}$ and $m\angle CBD = \frac{1}{2}m\widehat{CD}$. By the Arc Addition Postulate, $m\widehat{AD} + m\widehat{CD} = m\widehat{AC}$. By the Angle Addition Postulate, $m\angle ABD + m\angle CBD = m\angle ABC$.

Then $m\angle ABC = \frac{1}{2}m\widehat{AD} + \frac{1}{2}m\widehat{CD}$

$= \frac{1}{2}\left(m\widehat{AD} + m\widehat{CD}\right)$

$= \frac{1}{2}m\widehat{AC}$.

c. Given: $\angle ABC$ is inscribed in $\odot Q$. $\overline{DB}$ is a diameter; Prove: $m\angle ABC = \frac{1}{2}m\widehat{AC}$; By Case 1, proved in part (a), $m\angle DBA = \frac{1}{2}m\widehat{AD}$ and $m\angle DBC = \frac{1}{2}m\widehat{CD}$. By the Arc Addition Postulate, $m\widehat{AC} + m\widehat{CD} = m\widehat{AD}$, so $m\widehat{AC} = m\widehat{AD} - m\widehat{CD}$. By the Angle Addition Postulate, $m\angle DBC + m\angle ABC = m\angle DBA$, so $m\angle ABC = m\angle DBA - m\angle DBC$. Then

$$m\angle ABC = \tfrac{1}{2}m\widehat{AD} - \tfrac{1}{2}m\widehat{CD} = \tfrac{1}{2}(m\widehat{AD} - m\widehat{CD}) = \tfrac{1}{2}m\widehat{AC}.$$

25. yes; The angles intercept the same arc.

27. Conditional: Let right $\triangle ABC$ be inscribed in a circle and $\angle C$ be the right angle. By the Measure of an Inscribed Angle Theorem, $m\widehat{AB} = 2(90°) = 180°$. So, $\widehat{AB}$ is a semicircle, and $\overline{AB}$ is a diameter of the circle by the definition of a semicircle.

Converse: Let side $\overline{AB}$ of an inscribed triangle $\triangle ABC$ be a diameter of the circle. Because A and B are endpoints of a diameter, $\widehat{AB}$ is a semicircle by definition. So, $m\widehat{AB} = 180°$. $\angle C$ is an inscribed angle with intercepted arc $\widehat{AB}$. By the Measure of an Inscribed Angle Theorem, $m\angle C = \frac{1}{2}(m\widehat{AB}) = \frac{1}{2}(180°) = 90°$. So, $\angle C$ is a right angle and $\triangle ABC$ is a right triangle.

29.

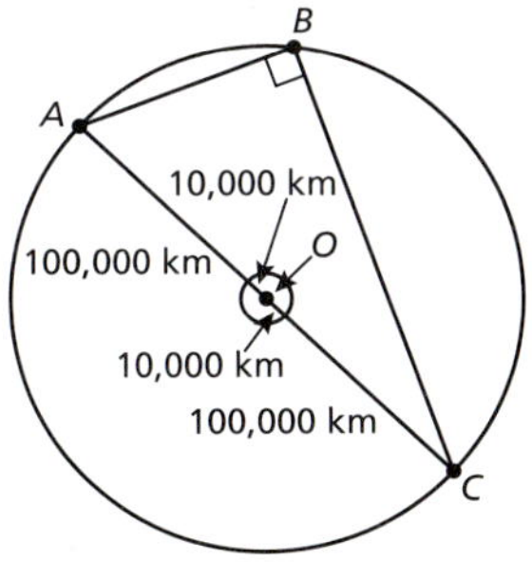

$AC = 220{,}000$ km, $AB = 44{,}000\sqrt{5} \approx 98{,}387$ km, $BC = 88{,}000\sqrt{5} \approx 196{,}774$ km

31. 2.4 units

10.4 Review & Refresh

33. *Sample answer:* Reflect $\triangle ABC$ in the y-axis, then translate 3 units up.

34. 10

35. minor arc; 120°

36. minor arc; 60°

37. semicircle; 180°

38. major arc; 300°

39. yes; $\triangle ABC$ is a right triangle.

40. 60°

41. 26°

42. your friend; Because $155° > 140°$, your friend is farther from school by the Hinge Theorem.

43. $m\angle 1 = 43°$, $m\angle 2 = 43°$, $m\angle 3 = 137°$; $\angle 1$ and the 137° angle are a linear pair, so they are supplementary. $\angle 2$ and $\angle 1$ are congruent by the Alternate Interior Angles Theorem. $\angle 2$ and $\angle 3$ are supplementary by the Consecutive Interior Angles Theorem.

Section 10.5

10.5 Practice

1. 130°

3. 130°

5. 115

7. 66

9. 40

11. 34

13. $\angle SUT$ is not a central angle; $m\angle SUT = \frac{1}{2}(m\widehat{QR} + m\widehat{ST}) = 41.5°$

15. about 1.19 mi

17. $c = b - a$

19. By the Angles Inside a Circle Theorem, $m\angle JPN = \frac{1}{2}(m\widehat{JN} + m\widehat{KM})$. By the Angles Outside the Circle Theorem, $m\angle JLN = \frac{1}{2}(m\widehat{JN} - m\widehat{KM})$. Because the angle measures are positive, $\frac{1}{2}(m\widehat{JN} + m\widehat{KM}) > \frac{1}{2}m\widehat{JN} > \frac{1}{2}(m\widehat{JN} - m\widehat{KM})$, so, $m\angle JPN > m\angle JLN$.

21. a.

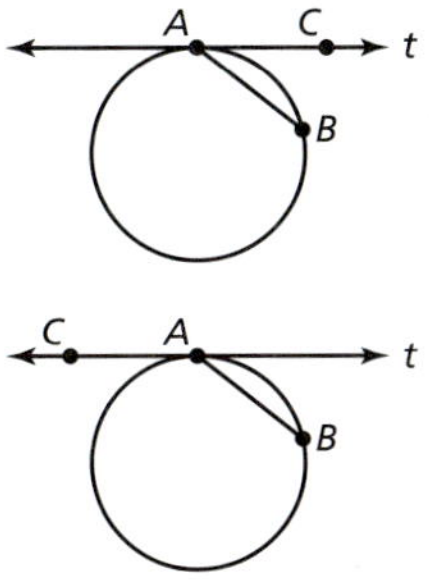

b. $m\widehat{AB} = 2m\angle BAC$, $m\widehat{AB} = 360° - 2m\angle BAC$

c. 90°; $2m\angle BAC = 360° - 2m\angle BAC$ when $m\angle BAC = 90°$.

23. a. By the Tangent Line to Circle Theorem, $m\angle BAC$ is 90°, which is half the measure of the semicircular arc.

b.

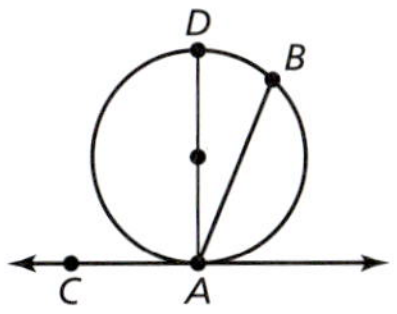

By the Tangent Line to Circle Theorem, $m\angle CAD = 90°$. $m\angle DAB = \frac{1}{2}m\widehat{DB}$ and by part (a), $m\angle CAD = \frac{1}{2}m\widehat{AD}$. By the Angle Addition Postulate, $m\angle BAC = m\angle BAD + m\angle CAD$. So, $m\angle BAC = \frac{1}{2}m\widehat{DB} + \frac{1}{2}m\widehat{AD} = \frac{1}{2}(m\widehat{DB} + m\widehat{AD})$. By the Arc Addition Postulate, $m\widehat{DB} + m\widehat{AD} = m\widehat{ADB}$, so $m\angle BAC = \frac{1}{2}(m\widehat{ADB})$.

25.

STATEMENTS	REASONS
1. Chords $\overline{AC}$ and $\overline{BD}$ intersect.	**1.** Given
2. $m\angle ACB = \frac{1}{2}m\widehat{AB}$ and $m\angle DBC = \frac{1}{2}m\widehat{DC}$	**2.** Measure of an Inscribed Angle Theorem
3. $m\angle 1 = m\angle DBC + m\angle ACB$	**3.** Exterior Angle Theorem
4. $m\angle 1 = \frac{1}{2}m\widehat{DC} + \frac{1}{2}m\widehat{AB}$	**4.** Substitution Property of Equality
5. $m\angle 1 = \frac{1}{2}(m\widehat{DC} + m\widehat{AB})$	**5.** Distributive Property

27. By the Exterior Angle Theorem, $m\angle 2 = m\angle 1 + m\angle ABC$, so $m\angle 1 = m\angle 2 - m\angle ABC$. By the Tangent and Intersected Chord Theorem, $m\angle 2 = \frac{1}{2}m\widehat{BC}$ and by the Measure of an Inscribed Angle Theorem, $m\angle ABC = \frac{1}{2}m\widehat{AC}$. By the Substitution Property,

$m\angle 1 = \frac{1}{2}m\widehat{BC} - \frac{1}{2}m\widehat{AC} = \frac{1}{2}(m\widehat{BC} - m\widehat{AC})$;

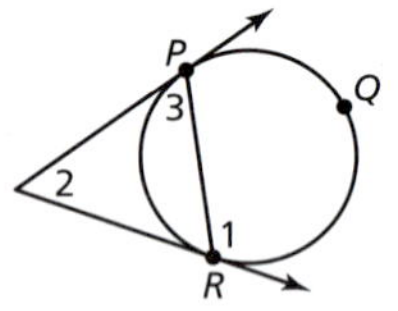

By the Exterior Angle Theorem, $m\angle 1 = m\angle 2 + m\angle 3$, so $m\angle 2 = m\angle 1 - m\angle 3$. By the Tangent and Intersected Chord Theorem, $m\angle 1 = \frac{1}{2}m\widehat{PQR}$ and $m\angle 3 = \frac{1}{2}m\widehat{PR}$. By the Substitution Property,

$m\angle 2 = \frac{1}{2}m\widehat{PQR} - \frac{1}{2}m\widehat{PR} = \frac{1}{2}(m\widehat{PQR} - m\widehat{PR})$;

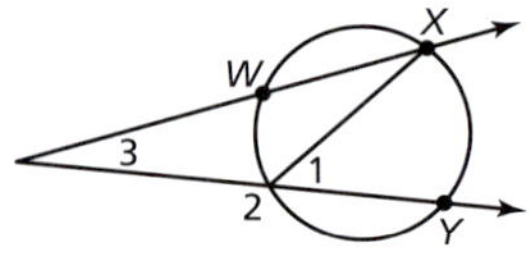

By the Exterior Angle Theorem, $m\angle 1 = m\angle 3 + m\angle WXZ$, so $m\angle 3 = m\angle 1 - m\angle WXZ$. By the Measure of an Inscribed Angle Theorem, $m\angle 1 = \frac{1}{2}m\widehat{XY}$ and $m\angle WXZ = \frac{1}{2}m\widehat{WZ}$. By the Substitution Property,

$m\angle 3 = \frac{1}{2}m\widehat{XY} - \frac{1}{2}m\widehat{WZ} = \frac{1}{2}(m\widehat{XY} - m\widehat{WZ})$.

29. 20°; *Sample answer:* $m\widehat{WXY} = 160°$ and $m\widehat{WX} = m\widehat{ZY}$, so

$$m\angle P = \tfrac{1}{2}(m\widehat{WZ} - m\widehat{XY})$$
$$= \tfrac{1}{2}\left[(200° - m\widehat{ZY}) - (160° - m\widehat{WX})\right]$$
$$= \tfrac{1}{2}(40°) = 20°.$$

10.5 Review & Refresh

31. 40 units; 60 square units

32. 67.5°; Because the sum of the arcs of a circle is 360°,

$$m\widehat{DE} + m\widehat{BC} = 360° - m\widehat{EB} - m\widehat{CD}$$
$$= 360° - 145° - 80°$$
$$= 135°,$$

so $m\widehat{DE} + m\widehat{BC} = 135°$. Because the ride swings the same distance in each direction, $m\widehat{DE} = m\widehat{BC}$, so $m\widehat{BC} = 135° \div 2 = 67.5°$.

33. $x = 111$ **34.** $x = 265$

35. $7 < x < 29$ **36.** 101°

37. 130°

38. yes; By the Pythagorean Theorem, $VZ = 6$, so $\overline{WY}$ bisects $\overline{XZ}$. By the Perpendicular Chord Bisector Converse, $\overline{WY}$ is a diameter.

39.

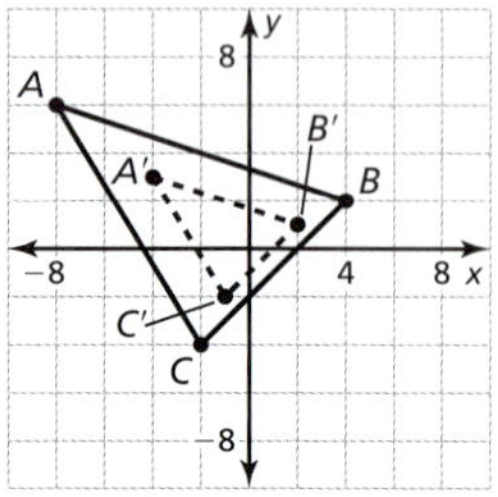

Section 10.6

10.6 Practice

1. $x = 5$ **3.** $x = 4$

5. $x = 4$ **7.** $x = 5$

9. $x = 12$ **11.** $x = 4$

13. The chords were used instead of the secant segments; $CF \cdot DF = BF \cdot AF$; $CD = 2$

15. about 496,494 km

17.

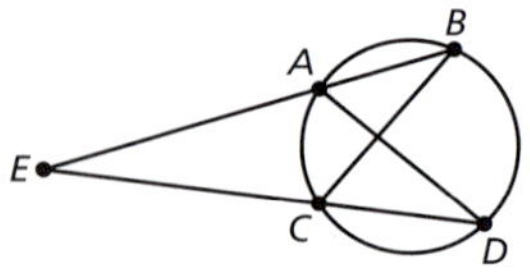

STATEMENTS	REASONS
1. $\overline{EB}$ and $\overline{ED}$ are secant segments.	1. Given
2. $\angle BEC \cong \angle DEA$	2. Reflexive Property of Congruence
3. $\angle ABC \cong \angle ADC$	3. Inscribed Angles of a Circle Theorem
4. $\triangle BCE \sim \triangle DAE$	4. AA Similarity Theorem
5. $\frac{EA}{EC} = \frac{ED}{EB}$	5. Corresponding side lengths of similar triangles are proportional.
6. $EA \cdot EB = EC \cdot ED$	6. Cross Products Property

19. It takes the sparkles 3 seconds to move from point C to point D. Because $CN = 12$ centimeters and the sparkles have 3 seconds to move from point C to point N, the sparkles need to move at a speed of 4 centimeters per second from point C to point N.

21.

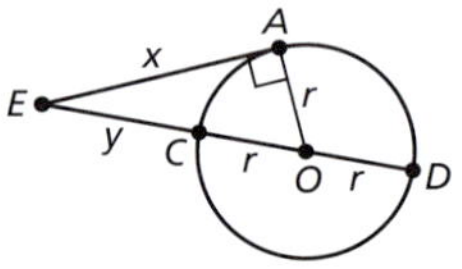

By the Tangent Line to Circle Theorem, $\angle EAO$ is a right angle, which makes $\triangle AEO$ a right triangle. By the Pythagorean Theorem, $(r + y)^2 = r^2 + x^2$. So, $r^2 + 2yr + y^2 = r^2 + x^2$. By the Subtraction Property of Equality, $2yr + y^2 = x^2$. Then $y(2r + y) = x^2$, so $EC \cdot ED = EA^2$.

23. $2\sqrt{10}$

10.6 Review & Refresh

25. 17 **26.** about 156.3 ft

27. 19 **28.** 202°

29. 92°

30. $\angle CBD \cong \angle A$ and $\angle C \cong \angle C$, so $\triangle ACE \sim \triangle BCD$ by the AA Similarity Theorem.

31. $x = 40$; By Corresponding Angles Converse, $m \parallel n$ when the 136° angle is equal to the $(3x + 16)°$ angle. So, $136 = 3x + 16$, and $x = 40$.

Section 10.7

10.7 Practice

1. $x^2 + y^2 = 4$
3. $x^2 + y^2 = 49$
5. $(x + 3)^2 + (y - 4)^2 = 1$
7. $x^2 + y^2 = 36$
9. $(x - 1)^2 + (y - 2)^2 = 9$

11. 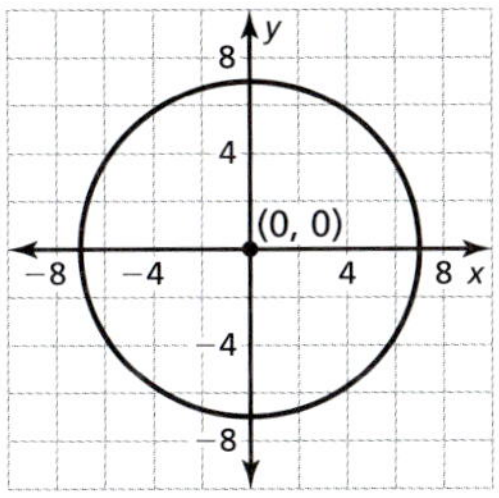

domain: $-7 \le x \le 7$, range: $-7 \le y \le 7$

13. 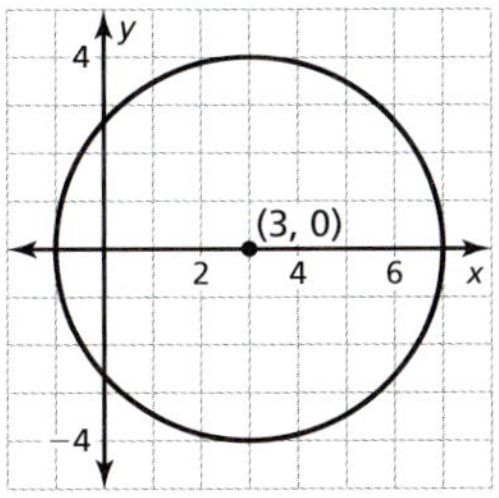

domain: $-1 \le x \le 7$, range: $-4 \le y \le 4$

15. 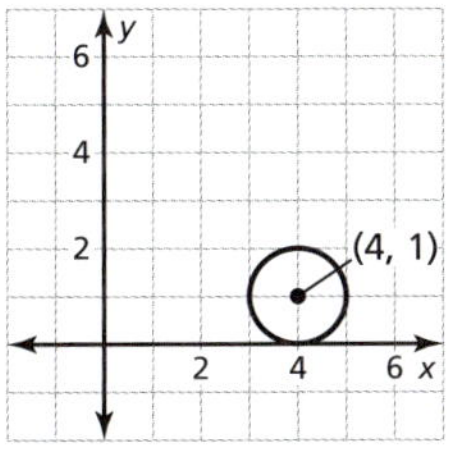

domain: $3 \le x \le 5$, range: $0 \le y \le 2$

17. 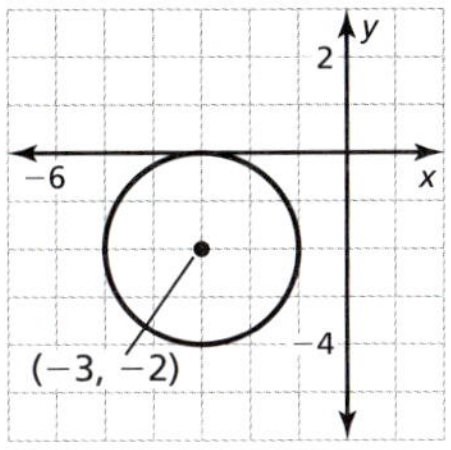

$y = 0$

19. 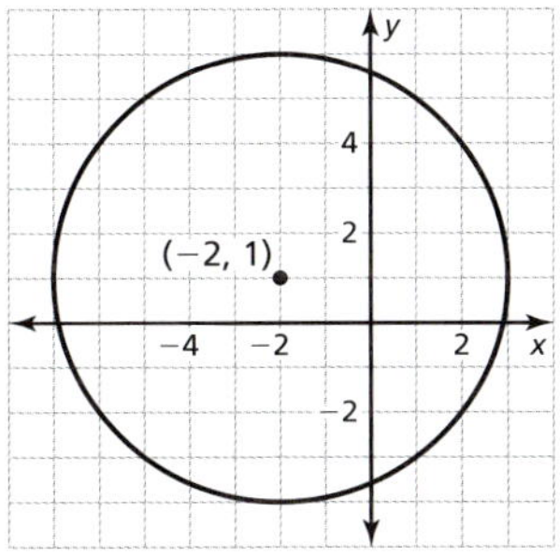

$y = \frac{3}{4}x - \frac{15}{4}$

21. $(x + 2.5)^2 + (y + 3.5)^2 = 12.5$

23. $(x + 1)^2 + (y + 4)^2 = 225$

25. The coordinates of the center should be subtracted; $[x - (-3)]^2 + [y - (-5)]^2 = (x + 3)^2 + (y + 5)^2 = 9$

27. The radius of the circle is 8. $\sqrt{(2 - 0)^2 + (3 - 0)^2} = \sqrt{13} \neq 8$, so (2, 3) does not lie on the circle.

29. The radius of the circle is $\sqrt{10}$. $\sqrt{(\sqrt{6} - 0)^2 + (2 - 0)^2} = \sqrt{10}$, so $(\sqrt{6}, 2)$ does lie on the circle.

31. 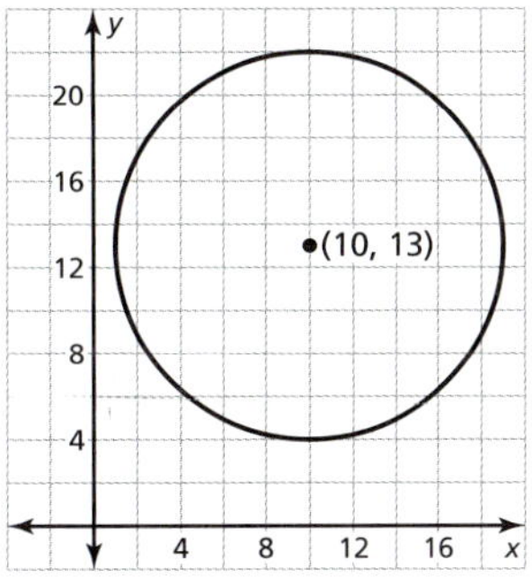

The domain is $1 \le x \le 19$ and represents the horizontal distance searched away from the center of the field. The range is $4 \le y \le 22$ and represents the vertical distance searched away from the center of the field.

33. a. 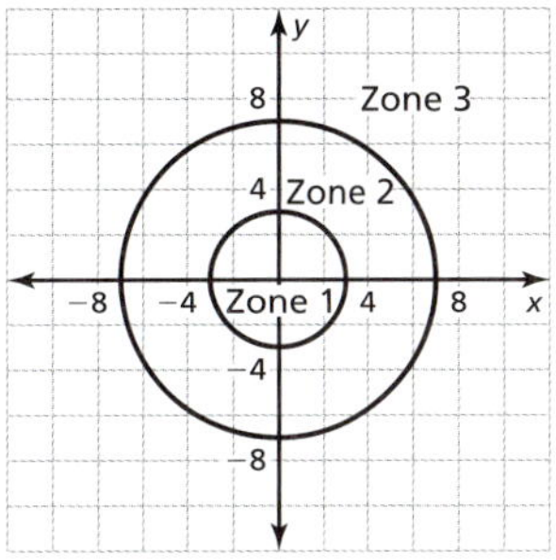

b. zone 2, zone 3, zone 1, zone 1, zone 2

35. yes; The diameter perpendicularly bisects the chord from $(-1, 0)$ to $(1, 0)$, so the center is on the y-axis at $(0, k)$ and the radius is $k^2 + 1$.

37. $(x - 10)^2 + y^2 = 100$; The domain is $0 \le x \le 20$ and represents the horizontal span of the bridge. The range is $-10 \le y \le 10$ and represents the span from the maximum height of the bridge to its reflected point. The radius is 10 units and represents the distance from the center of the circle to a point on the bridge or its reflection.

39. a. When the slope of the radius to the point of tangency is negative.
   b. When the slope of the radius to the point of tangency is positive.
   c. When the slope of the radius to the point of tangency is undefined.
   d. When the slope of the radius to the point of tangency is zero.

   The slope of the line tangent to a circle is the opposite reciprocal of the slope of the radius to the point of tangency.

41. tangent; The system has one solution.

43. secant; The system has two solutions, and $(5, -1)$ is not on the line.

45. $(x - 4)^2 + (y - 9)^2 = 16$; $m\angle Z = 90°$, so $\overline{XY}$ is a diameter by the Inscribed Right Triangle Theorem.

10.7 Review & Refresh

47. $(x - 5)^2 + (y - 4)^2 = 25$

48. $x = 120$

49. $x = 5$

50. minor arc; 53°

51. minor arc; 90°

52. minor arc, 127°

53. major arc; 270°

54. semicircle; 180°

55. 51°

56. 26°

57. $x = 2$

58. Horse B; Because 112° > 103°, the stride of Horse B is longer than the stride of Horse A by the Hinge Theorem.

Chapter 10 Review

1. radius

2. chord

3. tangent

4. diameter

5. secant

6. radius

7. 1 internal, 2 external

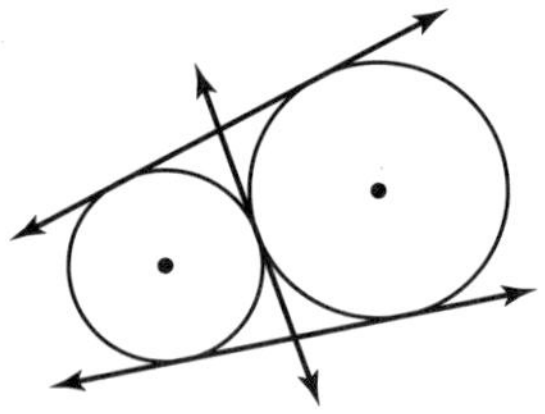

8. 2 external

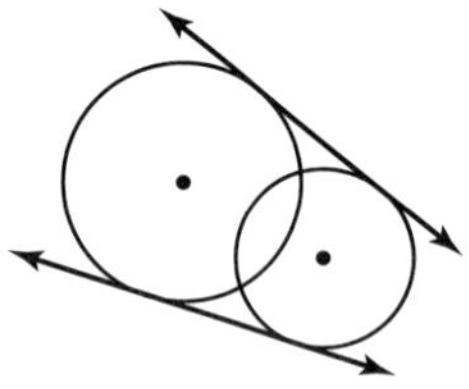

9. $a = 2$

10. $c = 2$

11. $r = 12$

12. yes; $\triangle ABC$ is a right triangle.

13. 100°

14. 60°

15. 160°

16. 80°

17. not congruent; The circles are not congruent.

18. congruent; The circles are congruent and $m\widehat{AB} = m\widehat{EF}$.

19. **a.** 64.8°

b. 97.2°

c. 79.2°

d. 270°

20. 61°

21. 65°

22. 91°

23. 26

24.

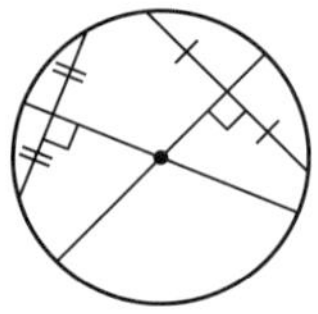

Sample answer: Draw any chord and then construct its perpendicular bisector. By the Perpendicular Chord Bisector Converse, this segment is a diameter of the circle. Repeat this process with another chord and its perpendicular bisector. The center of the circle is at the intersection of the two diameters.

25. $x = 80$

26. $q = 100, r = 20$

27. $d = 5$

28. $y = 30, z = 10$

29. $m = 44, n = 39$

30. $c = 28$

31. $\angle ADB \cong \angle AEB \cong \angle ACB$, $\angle EAC \cong \angle EBC$, $\angle EAD \cong \angle EBD$, $\angle CBD \cong \angle CAD$, $\angle AFE \cong \angle CFB$, $\angle AFB \cong \angle CFE$, $\angle AHE \cong \angle DHB$, $\angle AHB \cong \angle DHE$, $\angle AGD \cong \angle CGB$, $\angle AGB \cong \angle CGD$

32.

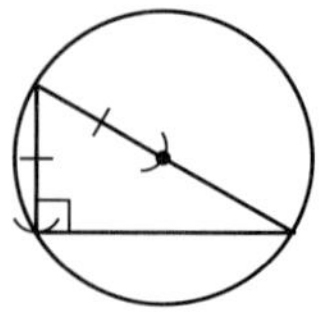

Sample answer: Draw a diameter. By the Inscribed Right Triangle Theorem, the diameter is the hypotenuse of the right triangle. By the 30°-60°-90° Triangle Theorem, the hypotenuse is twice the length of the shorter leg. So, draw a line segment, the length of the radius, from one endpoint of the diameter to a point on the circle. Then connect this point to the other endpoint of the diameter.

33. $x = 91$

34. $x = 120$

35. $x = 26$

36. $x = 70$

37. $x = 106$

38. $x = 16$

39. 120°

40. $x = 9$

41. $x = 4$

42. $x = 3$

43. $x = 8$

44. $x = 5$

45. $x = 10$

46. $\overline{JL}, \overline{JN}$; $\overline{JL}$ is a secent segment because it intersects the circle at two points, K and L; $\overline{JN}$ is a secent segment because it intersects the circle at two points, M and N.

47. $\overline{WX}$; It is a tangent segment because it intersects the circle at exactly one point, W.

48. about 10.7 ft

49. no; Using the Segments of Chords Theorem, $4 \cdot x = 12 \cdot 6$, which results in $x = 18$. $AB = 18$ yd and $CD = 22$ yd. The race is not fair because the distance from A to B is less than the distance from C to D.

50. $x^2 + y^2 = 16$

51. $(x - 4)^2 + (y + 1)^2 = 9$

52. $x^2 + y^2 = 81$

53. $x^2 + y^2 = 27.04$

54. $(x - 6)^2 + (y - 21)^2 = 16$

55. $(x - 10)^2 + (y - 7)^2 = 12.25$

56. $(x + 7)^2 + (y - 6)^2 = 25$

57.

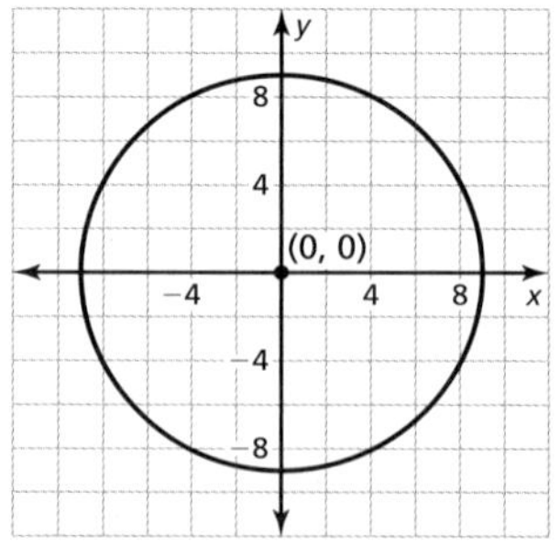

The domain is $-9 \le x \le 9$ and the range is $-9 \le y \le 9$.

58.

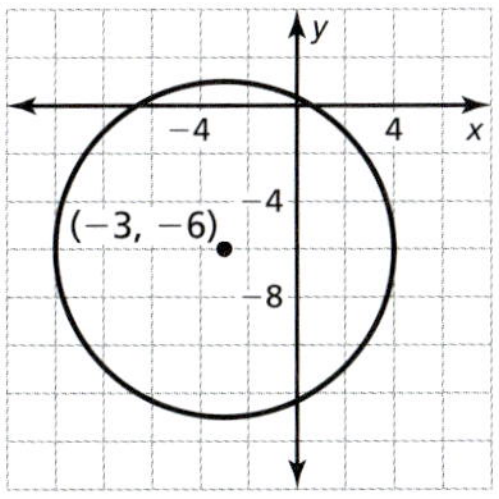

The domain is $-10 \le x \le 4$ and the range is $-13 \le y \le 1$.

59.

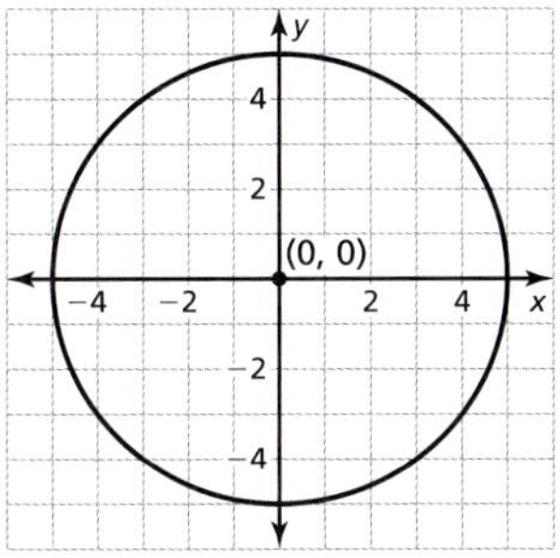

$y = -\frac{4}{3}x + \frac{25}{3}$

60.

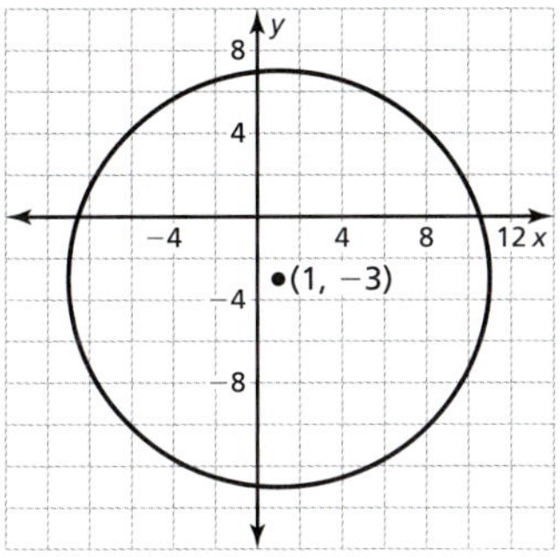

$y = -\frac{3}{4}x + \frac{41}{4}$

61.

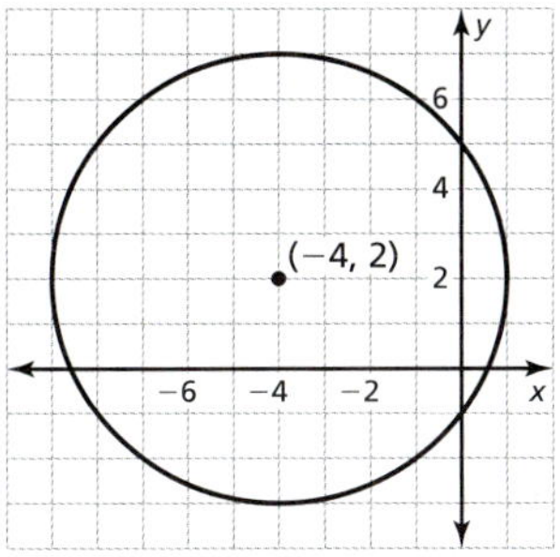

$y = \frac{4}{3}x - 1$

62. center: $(6, -4)$, radius: 2

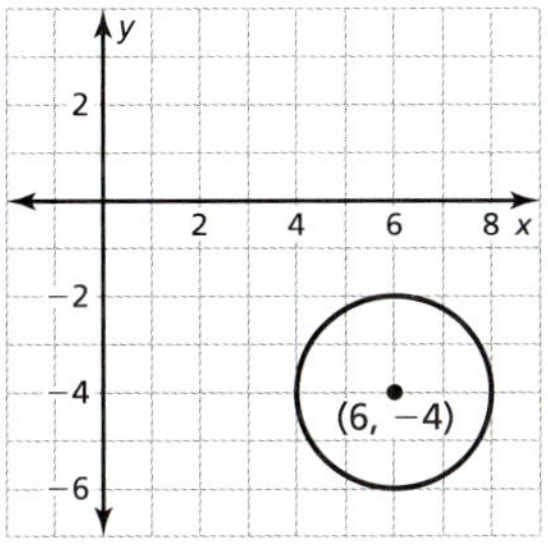

63. The radius of the circle is 5. $d = \sqrt{(0-4)^2 + (0+3)^2} = 5$, so $(4, -3)$ is on the circle.

Chapter 10 Practice Test

1. $m\angle 1 = 72.5°$, $m\angle 2 = 145°$; $\angle 1$ is an inscribed angle and $\angle 2$ is a central angle.

2. $m\angle 1 = 90°$, $m\angle 2 = 90°$; $\angle 1$ and $\angle 2$ intercept a semicircle.

3. $m\angle 1 = 29°$, $m\angle 2 = 66°$, $m\angle 3 = 37°$; $\angle 1$ is outside the circle, $\angle 2$ is inside the circle, and $\angle 3$ is an inscribed angle.

4. $m\angle 1 = 22°$, $m\angle 2 = 68°$; $\angle 1$ and $\angle 2$ are outside the circle.

5. 6

6. 4

7. 6

8. $r = 9$; $m\angle D = 90°$, so $r^2 + 12^2 = (r+6)^2$.

9. $x = 4$, $y = 5$; By the External Tangent Congruence Theorem, $AB = AE$ and $AC = AD$. So, $4x - 2 = 3y - 1$. Because $AB = AE$ and $AC = AD$, $BC = ED$, so $y + 3 = 3x - 4$. Solve the system of equations.

10. The radius of the circle is $\sqrt{5}$.
$d = \sqrt{(2\sqrt{2} - 0)^2 + (-1-2)^2} = \sqrt{17} \neq \sqrt{5}$,
so $(2\sqrt{2}, -1)$ does not lie on the circle.

11. By the Equidistant Chords Theorem, $\overline{ST} \cong \overline{RQ}$. By the Congruent Corresponding Chords Theorem, $\overset{\frown}{ST} \cong \overset{\frown}{RQ}$.

12. $\overline{CJ} \cong \overline{CL}$ and $\overline{CK} \cong \overline{CK}$, so $\triangle CKJ \cong \triangle CKL$ by the HL Congruence Theorem. Then $\angle LCK \cong \angle JCK$, and by the Congruent Central Angles Theorem, $\overset{\frown}{JM} \cong \overset{\frown}{LM}$.

13. $\overline{GE}$ is a perpendicular bisector of $\overline{DF}$, so $DG = FG$ by the Perpendicular Bisector Theorem. By the Congruent Corresponding Chords Theorem, $\overset{\frown}{DG} \cong \overset{\frown}{FG}$.

14.

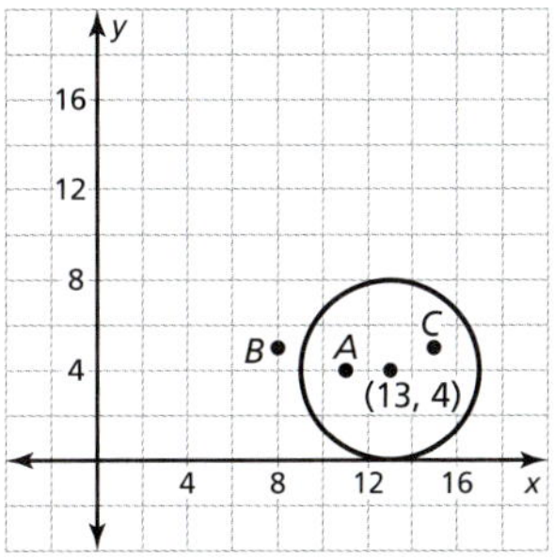

the actors at points A and C

15. **a.** 200 ft; $\overline{DE}$ is a radius, so $CE = (r - 80)$ feet and $\overline{DE}$ lies on the diameter of the circle containing the arc. By the Perpendicular Chord Bisector Theorem, $\overline{AC} \cong \overline{BC}$ and $AC = 160$ feet. So, $160^2 + (r-80)^2 = r^2$.

b. about 46 mi/h

Chapter 10 Test Prep

1. D

2. translation; $(x - 2, y + 2)$; 0; 3; 0; 3; 4; similarity; similar

3. $m\angle JPL = 90°$, $m\angle LPN = 90°$, and $\overline{PM}$ and $\overline{PK}$ are altitudes, so $\triangle JPL \sim \triangle PKL$ and $\triangle NPL \sim \triangle NMP$ by the Right Triangle Similarity Theorem. By the Transitive Property, $\triangle PKL \sim \triangle NMP$.

4. B

5. D

6. $\sin X = \frac{YZ}{XZ}$, $\cos X = \frac{XY}{XZ}$,
$\tan X = \frac{YZ}{XY}$, $\sin Z = \frac{XY}{XZ}$,
$\cos Z = \frac{YZ}{XZ}$, $\tan Z = \frac{XY}{YZ}$,
$\sin X = \cos Z$, $\cos X = \sin Z$

7. B, D, E, F

8. C

Chapter 11

Prepare with CalcChat

1. 27 cm^2
2. 7.5 yd^2
3. 19.5 ft^2
4. 9 in.
5. 2 cm
6. 12 ft
7. 40 m

Section 11.1

11.1 Practice

1. $12\pi \approx 37.70$ in.
3. about 20.05 ft
5. about 3.14 ft
7. about 35.53 m
9. The diameter was used as the radius; $C = \pi d = 9\pi$ in.
11. 182 ft
13. about 44.85 units
15. about 20.57 units
17. $\frac{7\pi}{18}$ radians
19. 165°
21. 8π units
23. about 7.85 units
25. $2\frac{1}{3}$
27. Two arcs are similar when they have the same measure, so $\widehat{RS} \sim \widehat{PQ}$. So, you can write the proportion $\frac{\text{Arc length of } \widehat{PQ}}{\text{Arc length of } \widehat{RS}} = \frac{r}{1}$, or Arc length of $\widehat{PQ} = \left(\text{Arc length of } \widehat{RS}\right) \cdot r$. So the length of $\widehat{PQ}$ is proportional to r.
29. *Sample answer:* Angles 1 and 2 are alternate interior angles, so the arc measure is 7.2°; 28,750 mi
31. **a.** about 26.66 cm
b. about 16.32 in.
33. yes; *Sample answer:* The circumference of the red circle can be found using $2 = \frac{30°}{360°}C$. The circumference of the blue circle is double the circumference of the red circle.
35. 28 units
37. *Sample answer:*

STATEMENTS	REASONS
1. $\overline{FG} \cong \overline{GH}$, $\angle JFK \cong \angle KFL$	**1.** Given
2. $FG = GH$	**2.** Definition of congruent segments
3. $FH = FG + GH$	**3.** Segment Addition Postulate
4. $FH = 2FG$	**4.** Substitution Property of Equality
5. $m\angle JFK = m\angle KFL$	**5.** Definition of congruent angles
6. $m\angle JFL = m\angle JFK + m\angle KFL$	**6.** Angle Addition Postulate
7. $m\angle JFL = 2m\angle JFK$	**7.** Substitution Property of Equality
8. $\angle NFG \cong \angle JFL$	**8.** Vertical Angles Congruence Theorem
9. $m\angle NFG = m\angle JFL$	**9.** Definition of congruent angles
10. $m\angle NFG = 2m\angle JFK$	**10.** Substitution Property of Equality
11. arc length of $\widehat{JK} = \frac{m\angle JFK}{360°} \cdot 2\pi FH$, arc length of $\widehat{NG} = \frac{m\angle NFG}{360°} \cdot 2\pi FG$	**11.** Formula for arc length
12. arc length of $\widehat{JK} = \frac{m\angle JFK}{360°} \cdot 2\pi(2FG)$, arc length of $\widehat{NG} = \frac{2m\angle JFK}{360°} \cdot 2\pi FG$	**12.** Substitution Property of Equality
13. arc length of $\widehat{NG}$ = arc length of $\widehat{JK}$	**13.** Transitive Property of Equality

11.1 Review & Refresh

39. 15 square units
40. 42 square units
41. about 6.5 in.
42. about 71.6°
43. 9
44. center: (0, 0), radius: 4

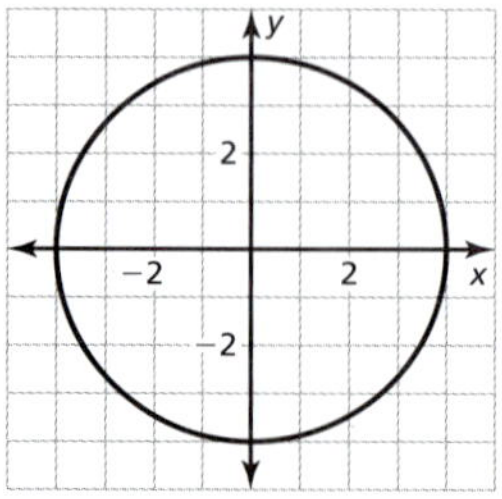

45. center: (3, 7), radius: 9

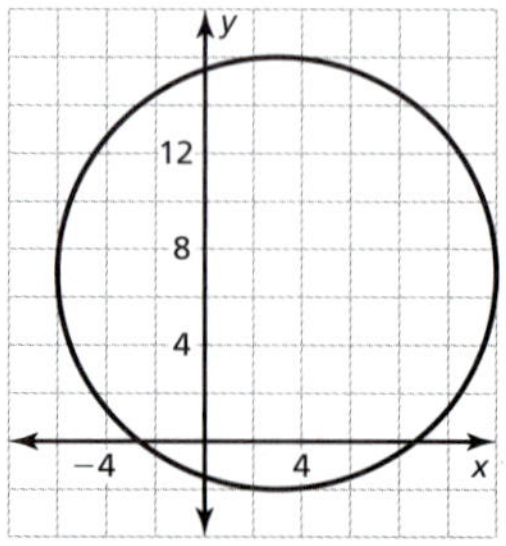

46. $x = \frac{-9 + 3\sqrt{21}}{2} \approx 2.4$

47. $x = -3 + 3\sqrt{10} \approx 6.5$

48. $x = 36$

49. $x = 2$

Section 11.2

11.2 Practice

1. about 78.54 in.^{2}
3. about 5.32 ft
5. about 4.01 in.
7. about 52.36 in.^{2}; about 261.80 in.^{2}
9. about 937.31 m^2; about 1525.70 m^2
11. The diameter was substituted in the formula for area as the radius; $A = \pi(6)^2 \approx 113.10$ ft^2
13. about 66.04 cm^2
15. two 16-inch diameter flatbreads
17. about 1696.46 m^2
19. about 192.48 ft^2
21. **a.** about 285 ft^2
    b. about 36%
23.

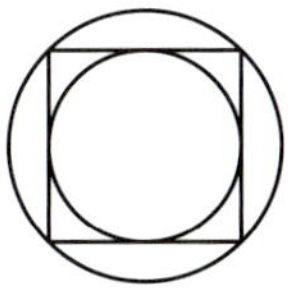

2 to 1

25. about 5.81 in.^{2}

11.2 Review & Refresh

27. about 7.63 m^2
28. about 19.68 ft
29. 59°
30. center: $(8, -3)$, radius: 8

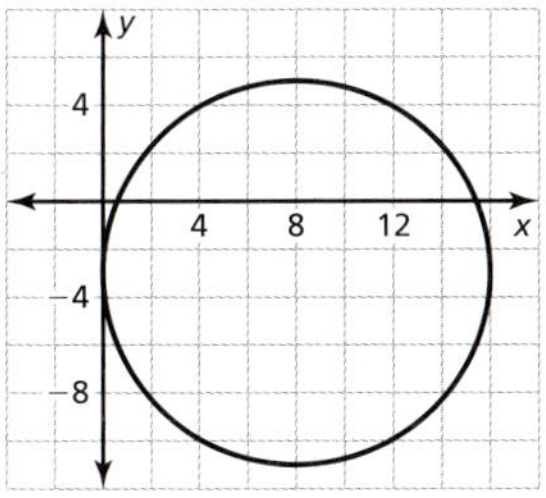

31. $x = 4$
32. $\sqrt{52.9}$, or about 7.3 units
33. $CD = 14$, $CE = 21$
34. By the SAS Congruence Theorem, the right triangles that represent each half of the ramp when looking at it from the sides are congruent. Because corresponding parts of congruent triangles are congruent, the lengths of the ramp are the same.
35. 42
36. $y = -x + 4$

Section 11.3

11.3 Practice

1. 361 square units
3. 70 square units
5. P
7. 5 units
9. 36°
11. 15°
13. 45°
15. 67.5°
17. about 62.35 square units
19. about 20.87 square units
21. about 342.24 square units
23. about 124.90 square units
25. The side lengths were used instead of the diagonals; $A = \frac{1}{2}(8)(4) = 16$
27. about 79.60 square units
29. about 117.92 square units
31. about 166 in.^{2}
33. true; *Sample answer:* As the number of sides increases, the polygon fills more of the circle.
35. false; *Sample answer:* The radius can be less than or greater than the side length.
37. **a.** the area doubles: $A = \frac{1}{2}(2d_1)d_2 = d_1d_2$ which is $2\left(\frac{1}{2}d_1d_2\right)$
    b. the area quadruples: $A = \frac{1}{2}(2d_1)(2d_2) = 2d_1d_2$ which is $4\left(\frac{1}{2}d_1d_2\right)$
    c. the area is 6.25 times the original kite's area: $A = \frac{1}{2}(2.5d_1)(2.5d_2) = 3.125\,d_1d_2$ which is $6.25\left(\frac{1}{2}d_1d_2\right)$
39. $2x^2 = 98$; 7 ft; 28 ft
41. no; You need to know the lengths of the diagonals.
43. (Note: $Z(3, z)$, where $z \neq -2, 4$) *Sample answer:* $(3, 0)$; $\left(2\sqrt{5} + 2\sqrt{13}\right)$ units, 8 square units
45. $A = \frac{1}{2}d^2$; $A = \frac{1}{2}d^2 = \frac{1}{2}(s^2 + s^2) = \frac{1}{2}(2s^2) = s^2$
47. about 6.47 cm
49. about 92 units^2
51. a circle; the circumference; the radius; *Sample answer:* The area of the circle is $\frac{1}{2}rC$; Substitute the expression into the formula for the area of a regular n-gon.

11.3 Review & Refresh

53. about 43.71 square units
54. about 20.28 in.
55.

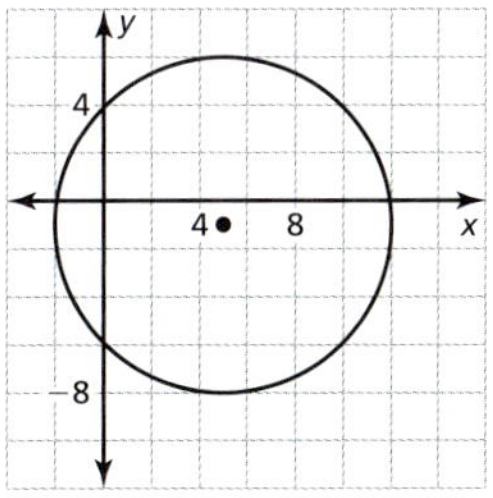

domain: $-2 \leq x \leq 12$, range: $-8 \leq y \leq 6$

56. yes; $\triangle DEF$ is a translation 3 units right and 2 units down of $\triangle ABC$.
57. $x \approx 18.5$
58. $x = 14$
59. about 210.44 square units
60. $AC \approx 3.9$, $m\angle B \approx 29.0°$, $m\angle C \approx 61.0°$; *Sample answer:* Using the Pythagorean Theorem, $AC = \sqrt{8^2 - 7^2} = \sqrt{15} \approx 3.9$. $C = \sin^{-1}\left(\frac{7}{8}\right) \approx 61.0°$. Using the Triangle Sum Theorem, $B \approx 180° - 90° - 61.0° \approx 29.0°$.
61. **a.** Opposite Sides Parallel and Congruent Theorem
    b. 61°, 119°, 61°
62. $x = 28$; By the Consecutive Interior Angles Converse, $m \parallel n$ when the $(5x - 20)°$ angle and the angle vertical to the 60° angle are supplementary. So, $5x - 20 + 60 = 180$ when $x = 28$.
63. 146°

Section 11.4

11.4 Practice

1. about 34 people per mi^2
3. about 464 people per mi^2
5. 16,050 people per mi^2
7. about 7 mi
9. The diameter was substituted into the formula instead of the radius; $1550 = \frac{x}{\pi \cdot 3.75^2}$; $1550 = \frac{x}{14.0625\pi}$; $x \approx 68{,}477$; The number of people who live in the region is about 68,477.
11. 50 feet by 50 feet; These dimensions result in the lowest possible perimeter of a rectangular region with an area of 2500 square feet.
13. a. 400 knots/in.^2

    b. rug in part (a); This rug has a knot density of 350 knots/in.^2, which is less than 400 knots/in.^2
15. no; California and Texas also have two of the greatest areas, so they may not have the greatest population densities.

11.4 Review & Refresh

17. about 112.14°
18. about 125.66 yd^2; about 326.73 yd^2
19. about 181 square units
20. about 9.74 mi
21. 288 ft, 259.2 ft
22. $RT > LN$; Because $\overline{RT}$ is the third side of the triangle with the larger included angle, it is longer than $\overline{LN}$, by the Hinge Theorem.
23. $a \approx 8.7$, $b \approx 16.5$
24. 2700°
25. 10

Chapter 11 Review

1. about 30.00 ft
2. about 56.57 cm
3. about 26.09 in.
4. about 74.48°
5. $24 + 6\pi \approx 42.8$ units
6. $20\pi \approx 62.83$ units
7. $\frac{\pi}{12}$ radian
8. 108°
9. about 218 ft
10. about 169.65 in.^2
11. about 17.72 in.^2
12. 173.166 ft^2
13. about 16 in.
14. one 8-inch pie
15. 130 square units
16. 96 square units
17. 105 square units
18. about 201.20 square units
19. about 167.11 square units
20. about 37.30 square units
21. about 173.8 ft^2
22. about 32.73°
23. hexagon; Because the radius is 1 foot and each side length is 1 foot, the congruent triangles are equilateral, and also equiangular. So, the central angle is 60° and $360 \div 60 = 6$, which is the number of the triangles and the number of sides of the regular polygon.
24. about 9903 people per mi^2
25. about 2.5 km
26. 288 ft^2

Chapter 11 Practice Test

1. $18\pi \approx 56.55\ m^2$
2. about 109.71 in.
3. about 74.27°
4. about 142.42 in.^2
5. about 148 people per km^2
6. about 906.89 in.^2
7. $\frac{13\pi}{90}$ radian; 100°
8. 89.49 mm^2; 55.5 mm^2; 4029.18 mm^2
9. about 28,274 people
10. 24 in., 36 in.
11. the fan shown; The surface area of the fan shown is about 89 square inches and the surface area of the other fan is about 47 square inches.
12. *Sample answer:* the population density of New York excluding New York City is most likely less than the population density of Florida excluding Jacksonville; New York City and Jacksonville have a similar area, but New York City makes up almost half of New York's population whereas Jacksonville only makes up about $\frac{1}{25}$ of Florida's population, so removing New York City from the population density calculation for New York will reduce the density more than removing Jacksonville from the calculation for Florida's population density.

Chapter 11 Test Prep

1. D
2. B
3. A
4. a. $AB \approx 5.83$

   b. $CD \approx 5.39$

   c. $EF \approx 6.08$

   d. $GH = 6$

   e. $JK = 5$

   f. $LM \approx 8.06$

   $\overline{JK}, \overline{CD}, \overline{AB}, \overline{GH}, \overline{EF}, \overline{LM}$
5. $9\pi + 24 \approx 52.27$ units
6. B
7. a. yes; Triangles A and C are 45°-45°-90° triangles, because the ratio of their sides is $x : x : x\sqrt{2}$. Triangles B and E are 30°-60°-90° triangles because the ratio of their side lengths is $x : x\sqrt{3} : 2x$.

   b. $\triangle A \sim \triangle C$, $\triangle B \sim \triangle E$

   c. $\frac{3\sqrt{3}}{2} \approx 2.6$, $3\sqrt{2} \approx 4.2$
8. *Sample answer:* The radius of the circle is 2.

   $d = \sqrt{(0-1)^2 + (0-\sqrt{3})^2} = 2$, so $(1, \sqrt{3})$ is on the circle.
9. B

Chapter 12

Prepare with CalcChat

1. 158 ft^2
2. 144 m^2
3. 184 cm^2
4. 5 m^2
5. 6 mm^2; Because $ABCD \sim PQRS$, $\triangle ABC \sim \triangle PQR$. By the Pythagorean Theorem, $AC = \sqrt{6^2 + 8^2} = 10$. The area of $\triangle ABC$ is $\frac{1}{2}(6)(8) = 24\ mm^2$. So, by the Areas of Similar Polygons Theorem, the area of $\triangle PQR$ is $24\left(\frac{5}{10}\right)^2 = 6\ mm^2$.

Section 12.1

12.1 Practice

1. B
3. A
5. yes; pentagonal pyramid
7. no

9.

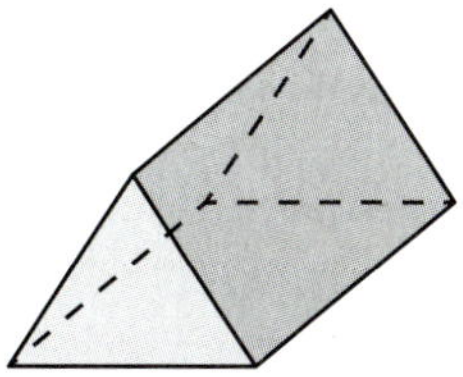

11.

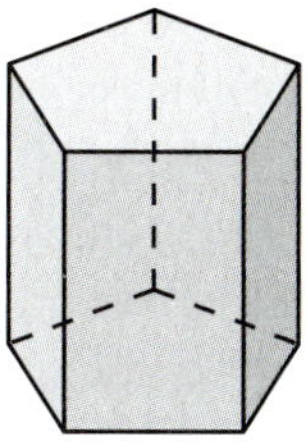

13.

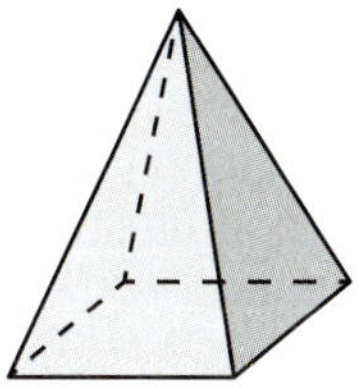

15. pentagonal pyramid

17. rectangular pyramid

19. circle

21. triangle

23. There are two parallel, congruent bases, so it is a prism, not a pyramid; The base is a triangle, so the solid is a triangular prism.

25.

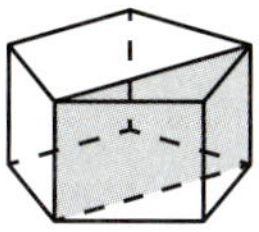

rectangle

27.

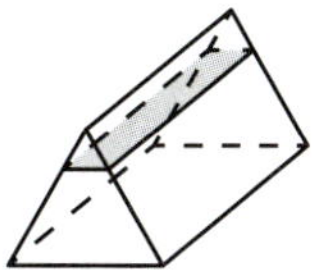

rectangle

29. *Sample answer:*

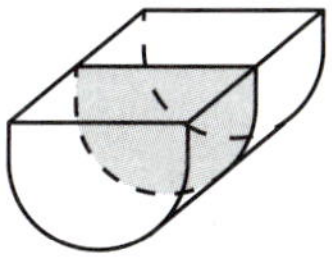

yes; yes; The cross section can also be a rectangle.

31.

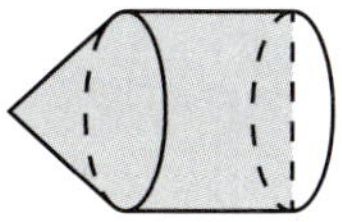

no

33. *Sample answer:*

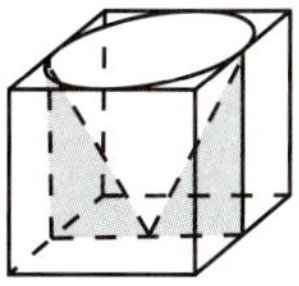

yes; When the vertical plane used in the diagram shown is rotated 90°, the cross section does not change. When the vertical plane is rotated 45°, the cross section is two trapezoids.

35. **a.** 36.5 in., 59.5 in.^{2}

b. about 153.94 in.^{2}; increases by 119 square inches

c. yes; *Sample answer:* about 43.98 in., about 153.94 in.^{2}

37. *Sample answer:*

a.

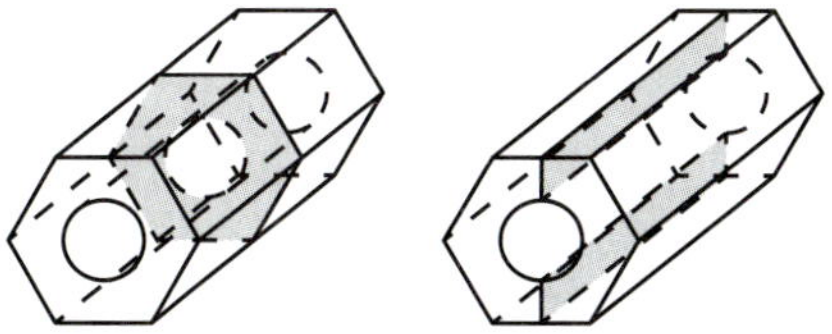

b. infinitely many ways; Any cut made lengthwise through the center of the hexagon will form two congruent parts.

39.

Solid	Faces, *F*	Edges, *E*	Vertices, *V*
tetrahedron	4	6	4
cube	6	12	8
octahedron	8	12	6
dodecahedron	12	30	20
icosahedron	20	30	12

$V - E + F = 2$; Answers will vary.

41. no; The plane can intersect the sphere at a point.

12.1 Review & Refresh

43. First prove that $\overline{BE} \cong \overline{BD}$ using the Converse of the Base Angles Theorem. Then use the Congruent Supplement Theorem to show that $\angle AEB \cong \angle CDB$. $\triangle ABE \cong \triangle CBD$ by the SAS Congruence Theorem. Finally, prove that $\overline{AB} \cong \overline{CB}$ because corresponding parts of congruent triangles are congruent.

44. 12 oz per yd^2

45.

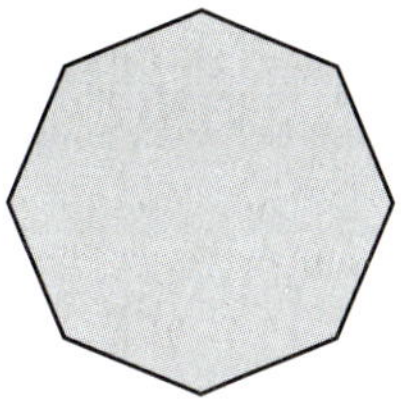

octagon

46.

square

47.

rectangle

48.

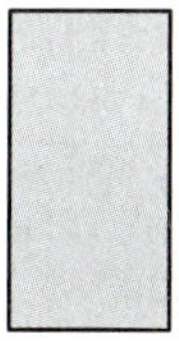

rectangle

49. yes; Because $32^2 + 60^2 = 68^2$, $\triangle ABC$ is a right triangle.

50. $(8, -1)$

51. $RQ = 9$, $m\angle P \approx 36.9°$, $m\angle Q \approx 53.1°$;
Sample answer: Using the Pythagorean Theorem, $RQ = \sqrt{15^2 - 12^2} = \sqrt{81} = 9$. Using trigonometry, $m\angle P = \sin^{-1}\left(\frac{9}{15}\right) \approx 36.9°$. Using the Triangle Sum Theorem, $m\angle Q \approx 180° - 90° - 36.9° \approx 53.1°$.

52. $39 + 52 > 64$, $52 + 64 > 39$, $39 + 64 > 52$; acute

53. 60 square units

54. 66 square units

55. $42.25\pi \approx 132.73$ cm^2

56. $16\pi \approx 50.27$ in.^{2}

57. about 4.62 m

58. about 13.26 yd

Section 12.2

12.2 Practice

1. about 150.80 in.^{2}

3. 92 ft^2

5. about 48.76 in.^{2}

7. about 14.07 in.^{2}

9. about 3920.71 mm^2

11. about 753.98 ft^2

13. about 101.70 cm^2

15. about 573.00 in.^{2}

17. The diameter was used as the radius; $S \approx 207.35$ cm^2

19. 5552π or about 17,442.12 cm^2

21. 1384 in.^{2}

23. The surface area is $4\frac{1}{3}$ times the original surface area.

25. about 13.09 m

27. $6s^2 = 343$; about 7.56 in.

29. the rectangular prism bin; The rectangular prism bin requires 1704 square inches of material and the cylinder bin requires about 1470.27 square inches.

31. Check students' work.

Sample answer: The surface area of the composite figure is the sum of the surface areas of the individual solids minus 2 times the area where the solids touch.

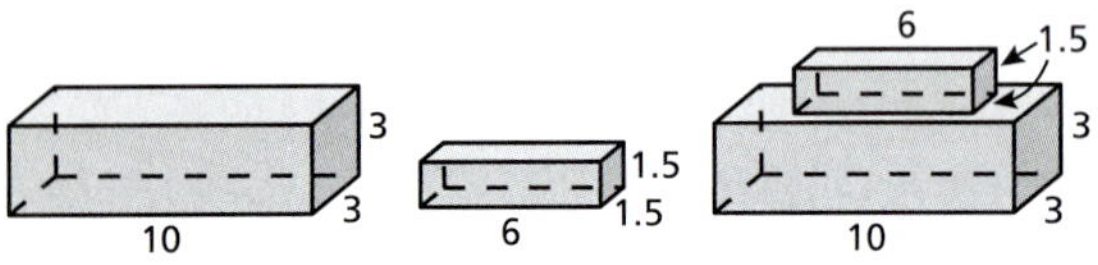

33. 22 cm

35. $S = 2\pi r_1 h + 2\pi r_2 h + 2(\pi r_1^2 - \pi r_2^2)$

12.2 Review & Refresh

37. about 93.53 square units

38. no

39. yes; rectangular prism

40. yes; octagonal prism

41. yes; square pyramid

42. about 0.000056 panther per acre, about 0.13245 alligator per acre, about 0.06623 Burmese python per acre, and about 0.00132 wood stork per acre

43.

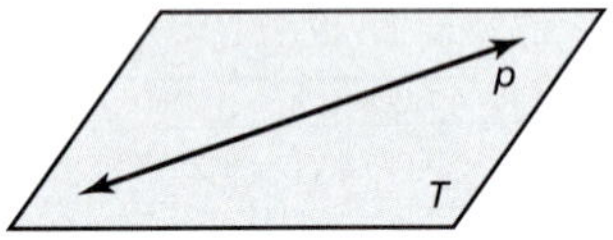

44.

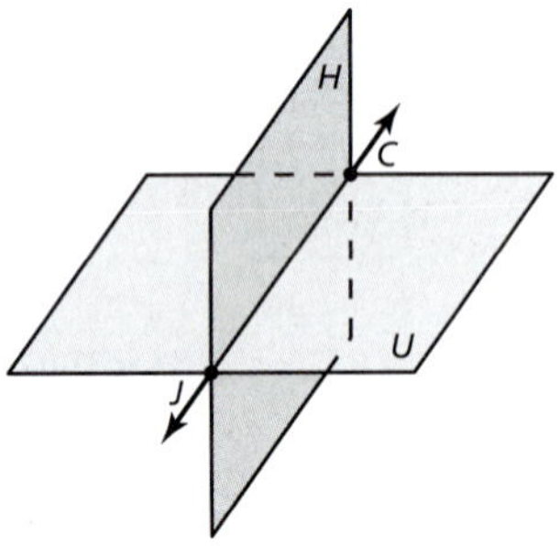

45. about 2446.4 ft^2

46. 160π or about 502.65 m^2

47.

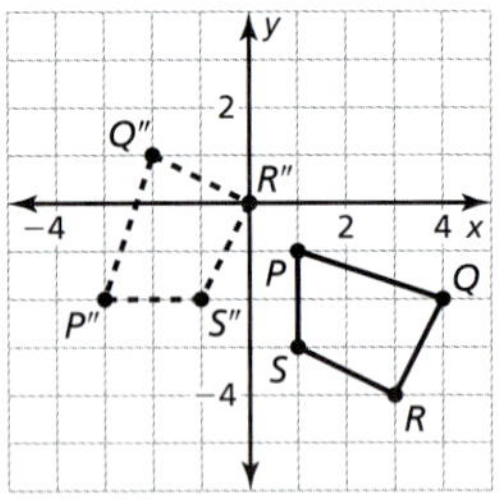

48.

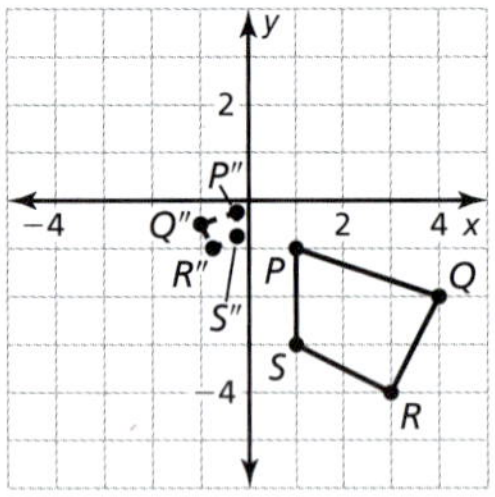

49. $m\angle QRT = 142°$, $m\angle SRT = 38°$

50. 18

51. $x = 10$

52. (7, 5)

Section 12.3

12.3 Practice

1. about 70.83 in.^{2}

3. 3920 ft^2

5. about 603.19 in.^{2}

7. about 678.58 in.^{2}

9. The height of the pyramid was used as the slant height; $S = 6^2 + \frac{1}{2}(24)(5) = 96$ ft^2

11. 32 cm^2

13. 144π ft^2

15. about 163.36 ft^2

17. about 214.05 in.^{2}

19. about 6.75 in.

21. a. $\angle A$ and $\angle D$ are congruent right angles and $\angle C \cong \angle C$ by the Reflexive Property of Congruence (Thm. 2.2), so $\triangle ABC \sim \triangle DEC$ by the AA Similarity Theorem (Thm. 8.3).

b. $BC = 5$, $DE = 1.5$, $EC = 2.5$

c. $24\pi^2$ square units; $6\pi^2$ square units; The surface area of the smaller cone is 25% the surface area of the larger cone.

23. about 574.22 in.^{2}

12.3 Review & Refresh

25. $49\pi \approx 153.94$ ft^2

26. 32 m

27. about 329.87 ft^2

28. about 218.15 in.^{2}

29. 36 in., 81 in.

30. about 9.42 in.; about 7.07 in.^{2}

31. about 1114 people per square mile

32. congruent; They are in the same circle and $m\widehat{JM} = m\widehat{KL}$.

Section 12.4

12.4 Practice

1. 12 m^3

3. 175 in.^{3}

5. $91.8\pi \approx 288.40$ ft^3

7. $200\pi \approx 628.32$ ft^3

9.

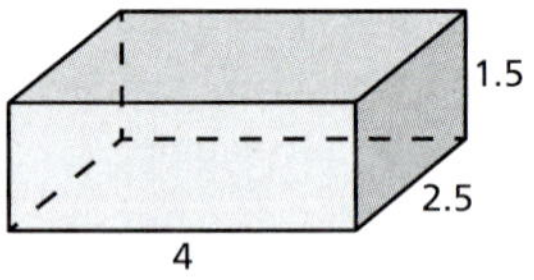

15 cubic units

11.

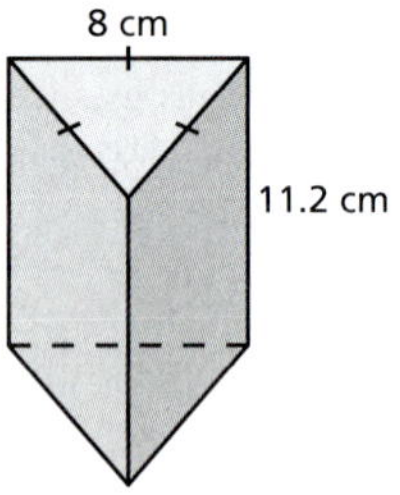

about 310.38 cm^3

13. 10 ft

15. 4 cm

17. about 11.04 ft

19. 150 ft^3

21. about 1900.66 in.^{3}

23. 1215 in.^{3}

25. 8 cm^3

27. about 2,350,000,000 gal

29. 75 in.^{3}; 20

31. *Sample answer:* The stacks have the same height and the rectangles have the same lengths, so the stacks have the same area.

33. *Sample answer:*

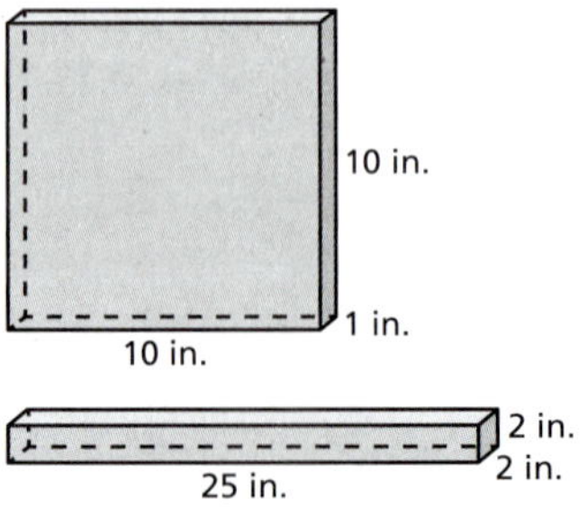

35. *Sample answer:* about 3054 in.^{3}; The bag is approximately cylindrical with a base diameter of about 12 inches, and a height of about 27 inches. So, the volume of the bag is about $V = \pi(6)^2(27) \approx 3054$ in.^{3}

37. about 7.33 in.^{3}

12.4 Review & Refresh

39. $x = 7, y = 12$

40. about 103.67 cm^2

41. about 88.20 mm^2

42. about 434.51 ft^2

43. about 988.03 in.^{2}

44. yes; $m\angle B = 81°$, so $\triangle ABC \sim \triangle DEF$ by the AA Similarity Theorem.

45. $x = 4$

46. about 756.6 mi

47. 66 cm^3

48. 216 in.^{3}

49. about 549.78 ft^3

50. about 423.33 mm^3

51. about 54 people per square mile

52. **a.**

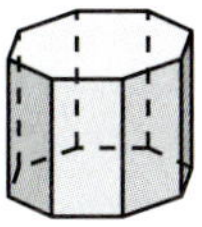

b.

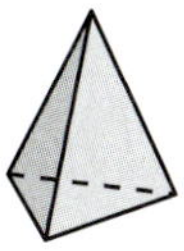

Section 12.5

12.5 Practice

1. 448 m^3

3. about 1361.36 mm^3

5. about 526.27 in.^{3}

7. One side length was used in the formula as the base area; $V = \frac{1}{3}(6^2)(5) = 60$ ft^3

9. 5 ft

11. 6 m

13. 8 m

15. 4 ft

17. 54π cm^3

19. 4 ft^3

21. about 226.19 cm^3

23. about 213.33 cm^3

25. about 82.04 in.^{3}

27. $2h$; $r\sqrt{2}$; *Sample answer:* The original volume is $V = \frac{1}{3}\pi r^2 h$ and the new volume is $V = \frac{2}{3}\pi r^2 h$.

29. about 17.45 sec

31. yes; *Sample answer:* The base areas are the same and the total heights are the same.

33. *Sample answer:* The volume of the cylinder is 3 times the volume of the cone and about 4.7 times the volume of the square pyramid. $V_{\text{cylinder}} = \pi r^2 h$, $V_{\text{cone}} = \frac{1}{3}\pi r^2 h$, and $V_{\text{pyramid}} = \frac{2}{3}r^2 h$.

12.5 Review & Refresh

35. $x \approx 12.9$

36. $x \approx 12.6$

37. $x \approx 5.8$

38. $x \approx 16.0$

39. about 9.80 mi

40. about 1357.17 m^2

41. about 108.80 in.^{2}

42. 225 mm^3

43. about 37.33 in.^{3}

44. rectangle

45.

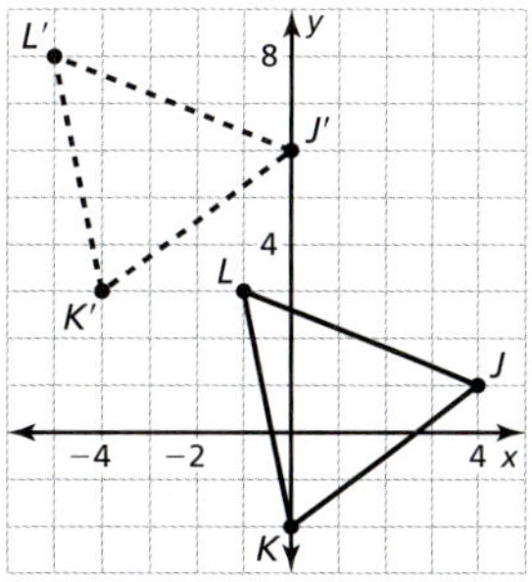

46. no; The campsite is about 90.1 feet away from the trail.

47. If it is Thanksgiving, then it is November; true; If it is not November, then it is not Thanksgiving; true

48. 10 units

Section 12.6

12.6 Practice

1. about 201.06 ft^2

3. about 1052.09 m^2

5. 1 ft

7. 30 m

9. about 2144.66 m^3

11. about 5575.28 yd^3

13. about 4188.79 cm^3

15. The radius was squared instead of cubed; $V = \frac{4}{3}\pi(6)^3 \approx 904.78$ ft^3

17. about 235.62 cm^2; about 261.80 cm^3

19. about 33.51 ft^3

21. about 445.06 in.^{3}

23. about 7749.26 cm^3

25. about 20,944 ft^3

27. 2304π in.^{3}

29. 960π ft^3

31. **a.** about 19,606.68 in.^{2}, about 258,154.62 in.^{3}

b. yes; Answers will vary.; All spheres are similar because they have the same shape.

33. **a.** Earth: about 201.1 million mi^2; moon: about 14.7 million mi^2

b. The surface area of Earth is about 13.7 times greater than the surface area of the moon.

c. about 140.7 million mi^2

35. $V = \frac{1}{3}rS$

37. Answers will vary.

39. about 2077.64 m^3

12.6 Review & Refresh

41. $m\angle C = 119°, a \approx 9.9, c \approx 19.8$

42. about 6.44 mi

43. 15 in.^{3}

44. $z = 3.99$ ft

45. 27 units

46. no; $\frac{JK}{KL} \neq \frac{NM}{ML}$, so $\overline{KM}$ is not parallel to $\overline{JN}$ by the Triangle Proportionality Theorem.

47. about 1017.88 in.^{2}; about 3053.63 in.^{3}

48. about 83.23 yd^2; about 47.12 yd^3

Section 12.7

12.7 Practice

1. about 4551 g

3. **a.** about 27.5 g

b. increases by about 12.9 in.^{3}; The initial volume is about 3.6 in.^{3} and the volume after 5 weeks is about 16.5 in.^{3}.

5. Density is $\frac{\text{mass}}{\text{volume}}$ not $\frac{\text{volume}}{\text{mass}}$; density $= \frac{24}{28.3} \approx 0.85 \text{ g/cm}^3$

7. about 0.21 mm

9. increases; A section of water that is deep in the ocean will have more water molecules, and therefore more mass than a section of water with the same volume that is closer to the surface. So, the density of water deeper in the ocean is greater than the density of water closer to the surface.

11. **a.** 18,000 kg

b. $\frac{4}{3}$ m

13. **a.** *Sample answer:* the average of the outside perimeter and the inside perimeter; Because different metals bend different ways, the average is a good estimate of the length; about 2926.92 g

b. about 451.2 cm; The total length of 100 links is $100(1.5 + 2.7 + 1.5) = 570$ centimeters. However, when the links are connected to form a taut chain, the links overlap 99 times. So, you must subtract $99(0.6 + 0.6) = 118.8$ centimeters from the total length of the 100 links.

12.7 Review & Refresh

14. slope of $\overline{PQ}$: $m = \frac{4-5}{6-3} = -\frac{1}{3}$ and slope of $\overline{RS}$: $m = \frac{-1-1}{7-1} = -\frac{1}{3}$, so $\overline{PQ} \parallel \overline{SR}$. slope of $\overline{PS}$: $m = \frac{5-1}{3-1} = 2$ and slope of $\overline{QR}$: $m = \frac{-1-4}{7-6} = -5$, so $\overline{PS} \nparallel \overline{QR}$. The quadrilateral has a pair of parallel sides, so it is a trapezoid; not isosceles

15. about 67.57 g

16. about 3216.99 cm^2; about 17,157.28 cm^3

17. about 1099.56 m^3

18. 32 ft^3

19. yes; Because $\overline{AB} \cong \overline{CD}$, $\angle ABC \cong \angle DCB$, and $\overline{BC} \cong \overline{BC}$, $\triangle ABC \cong \triangle DCB$ by the SAS Congruence Theorem.

20. $3 < x < 21$; By the Triangle Inequality Theorem, the sum of x and 9 must be greater than 12 and the sum 9 and 12 must be greater than x.

21. $x = 12\sqrt{2}$

22. $x = 3\sqrt{3}$

Section 12.8

12.8 Practice

1.

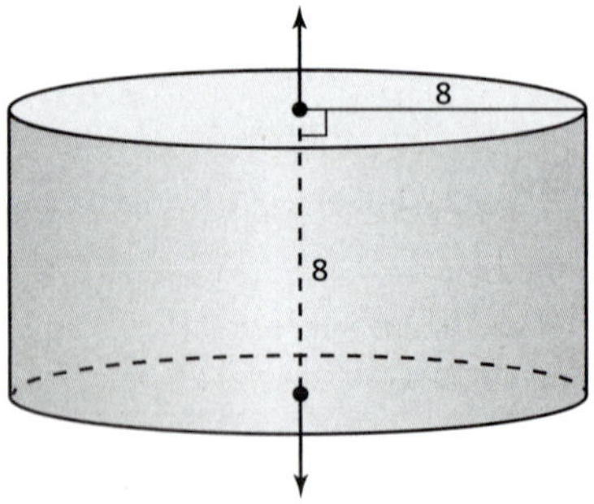

cylinder with a height of 8 units and a radius of 8 units

3.

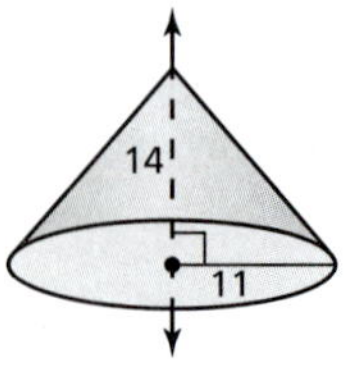

cone with a height of 14 units and a radius of 11 units

5.

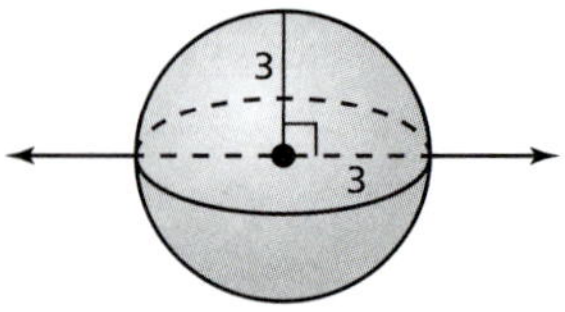

sphere with a radius of 3 units

7. The height and base radius are wrong; The solid is a cylinder with a height of 12 units and a radius of 5 units.

9.

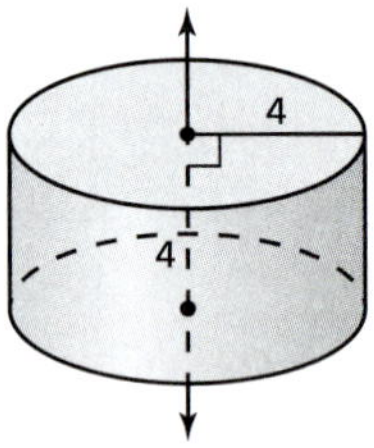

cylinder with a height of 4 units and a radius of 4 units

11.

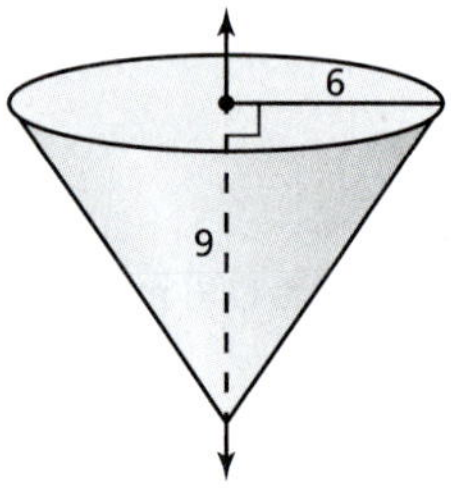

cone with a height of 9 units and a radius of 6 units

13.

15.

17.

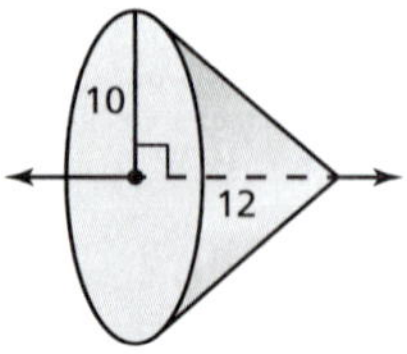

cone with a height of 12 units and a radius of 10 units; about 804.89 square units, about 1256.64 cubic units

19. 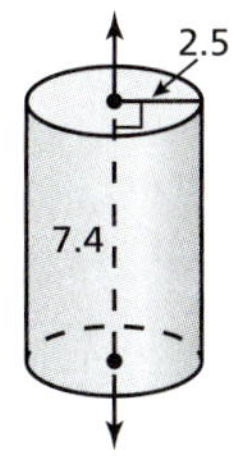

cylinder with a height of 7.4 units and a radius of 2.5 units; about 155.51 square units, about 145.3 cubic units

21. 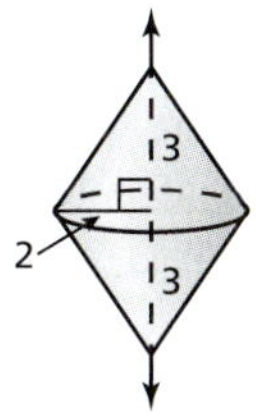

two cones, both with a height of 3 units and a radius of 2 units; about 45.31 square units, about 25.13 cubic units

23. 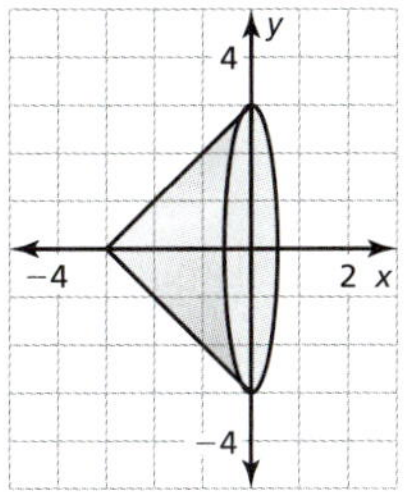

cone with a height of 3 units and a radius of 3 units; about 28.27 cubic units

25. 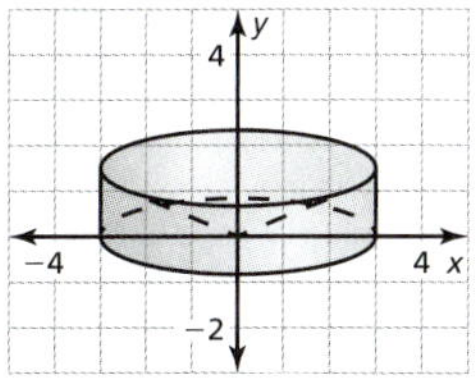

cylinder with a cone removed, both with a height of 1.5 units and a radius of 3 units; about 28.27 cubic units

27. no; The solid produced by rotating the figure around the x-axis is a sphere and the solid produced by rotating the figure around the y-axis is a hemisphere.

29. cone with a height of 15 units and a radius of 20 units, 2000π cubic units; cone with a height of 20 units and a radius of 15 units, 1500π cubic units; two cones, one with a radius of 12 units and a height of 9 units, the other with a radius of 12 units and a height of 16 units, 1200π cubic units

31. **a.** 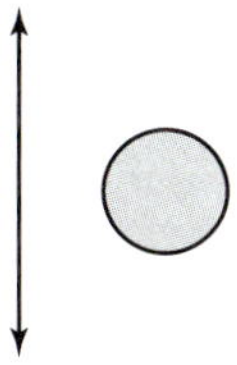

b. *Sample answer:* cylinder; Stretch the cylinder and connect the bases.

12.8 Review & Refresh

33. about 9 mi

34. 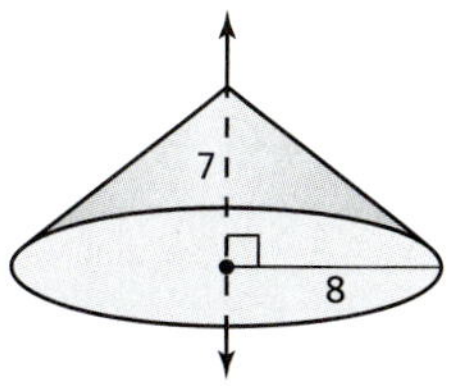

cone with a height of 7 units and a radius of 8 units

35. about 151.13 g

36. about 15; yes

37. about 422.73 yd^2

38. about 81.11 $in.^3$

39. about 62 min 50 sec

40. scalene triangle; no

Chapter 12 Review

1. rectangle

2. square

3. triangle

4. triangle

5. 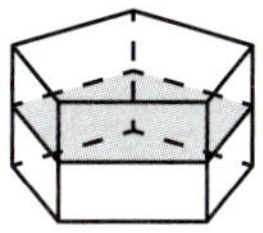

pentagon

6. 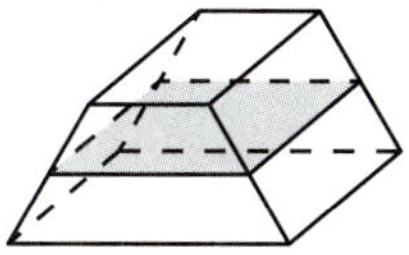

rectangle

7. *Sample answer:* rectangle, triangle

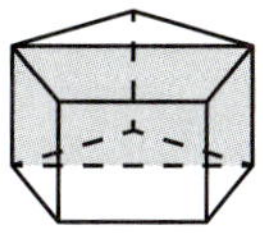

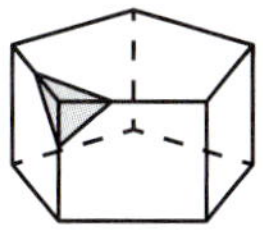

8. about 552.92 cm^2

9. 1240 ft^2

10. 502.65 ft^2

11. about 770.90 cm^2

12. regular pentagonal prism with a base edge length of 10.5 centimeters; *Sample answer:* 694.37 cm^2 > 438.87 cm^2

13. 225 $in.^2$

14. 1130.97 cm^2

15. 312 ft^2

16. about 301.59 cm^2

17. about 58.67 cm^2

18. 11.34 m^3

19. about 27.53 yd^3

20. about 100.53 mm^3

21. 1296π ft^3

22. about 1017.88 cm^3

23. 189 ft^3

24. about 309.71 $in.^3$

25. about 12,970.67 m^3

26. 15 cm

27. 3960 m^3

28. about 11,259.47 cm^3

29. about 615.75 $in.^2$; about 1436.76 $in.^3$

30. about 907.92 ft^2; about 2572.44 ft^3

31. about 2827.43 ft^2; about 14,137.17 ft^3

32. about 74.8 million km^2; about 60.8 billion km^3

33. about 272.55 m^3

34. **a.** about 32.4 cm^3; The volume is approximately equal to the volume of the sphere minus the volume of the cylindrical hole.

b. no, yes; For the surface area, the lateral surface area of the cylinder is added, while the areas of the bases of the cylinder are subtracted. Because the lateral surface area is greater than the total area of the bases, the surface area of the wooden ball increases after the hole is made. The volume decreases because part of the sphere is removed.

35. about 103 g

36. about 0.50 cm

37.

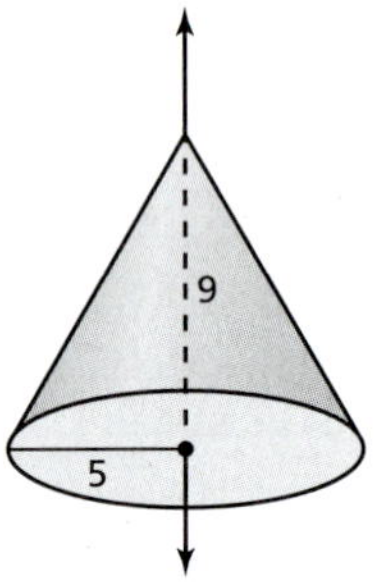

cone with a height of 9 units and a radius of 5 units; about 240.26 square units, about 235.62 cubic units

38.

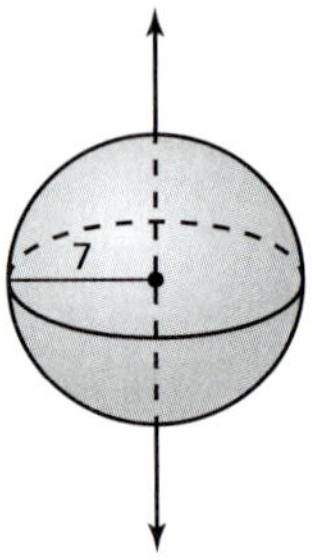

sphere with a radius of 7 units; about 615.75 square units, about 1436.76 cubic units

39.

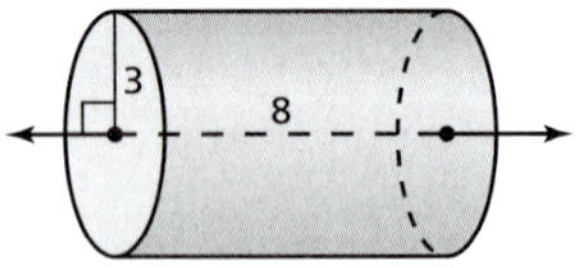

cylinder with a radius of 3 units and a height of 8 units; about 207.35 square units; about 226.19 cubic units

40.

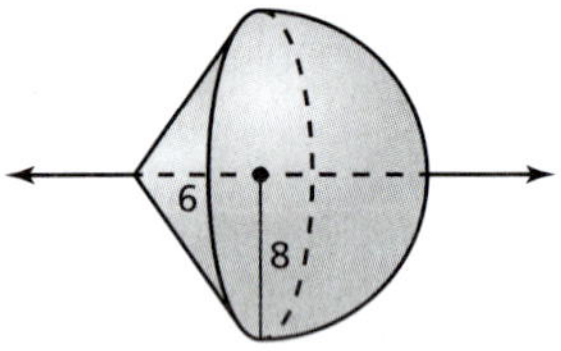

cone with a height of 6 units and a radius of 8 units and a hemisphere with a radius of 8 units; about 653.45 square units, about 1474.45 cubic units

41.

42.

43.

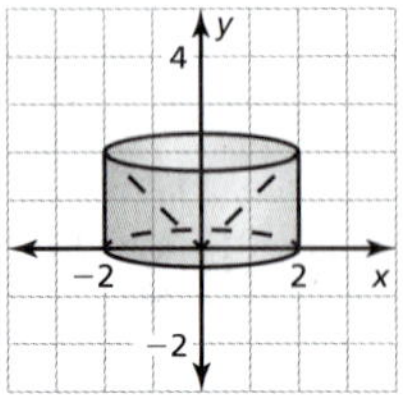

cylinder with a cone removed, both with a height of 2 units and a base radius of 2 units; about 16.76 cubic units

Chapter 12 Practice Test

1. two concentric circles; two separated circles

2. $90\pi \approx 282.74$ ft^2

3. about 2577.29 m^3

4. about 17.16 ft^3

5. about 402.12 m^3

6. $93\frac{1}{3}$ ft^3

7.

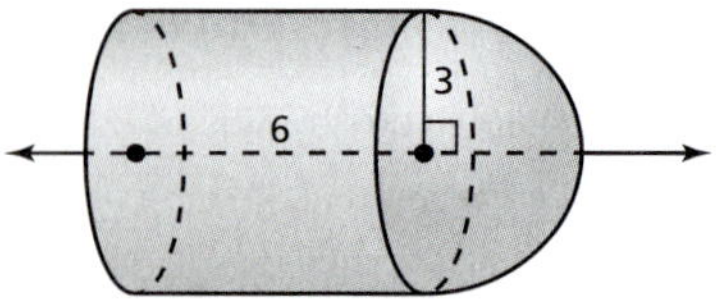

cylinder with height 6 and radius 3, and hemisphere with radius 3; about 197.92 square units; about 226.19 cubic units

8. about 372.6 in.^{2}

9. about 1130.97 yd^2

10. **a.** about 0.9 g/cm^3

b. The surface area is about 0.980 times the original surface area and the volume is about 0.976 times the original volume.

c. about 28.8 g; about 125 h

11. about 0.1 g tin, about 0.9 g copper

12. y-axis; The volume $\pi r^2 h$ is greater when r is greater.

Chapter 12 Test Prep

1. **a.** trapezoid

b. pentagon

c. rectangle

2. yes; There is a composition of rigid motions that maps $\triangle ABC$ to $\triangle DEF$.

3. A

4. A, C, F

5. D

6. $x = 10$, $y = 20$

7. 29,120 Btu

8. B

English-Spanish Glossary

English | Spanish

A

acute angle *(p. 39)* An angle that has a measure greater than 0° and less than 90°

ángulo agudo *(p. 39)* Un ángulo que tiene una medida mayor que 0° y menor que 90°

adjacent angles *(p. 48)* Two angles that share a common vertex and side, but have no common interior points

ángulos adyacentes *(p. 48)* Dos ángulos que comparten un vértice y lado en común, pero que no tienen puntos interiores en común

adjacent arcs *(p. 523)* Arcs of a circle that have exactly one point in common

arcos adyacentes *(p. 523)* Arcos de un círculo que tienen exactamente un punto en común

alternate exterior angles *(p. 124)* Two angles that are formed by two lines and a transversal that are outside the two lines and on opposite sides of the transversal

ángulos exteriores alternos *(p. 124)* Dos ángulos que son formados por dos rectas y una transversal que están fuera de las dos rectas y en lados opuestos de la transversal

alternate interior angles *(p. 124)* Two angles that are formed by two lines and a transversal that are between the two lines and on opposite sides of the transversal

ángulos interiores alternos *(p. 124)* Dos ángulos que son formados por dos rectas y una transversal que están entre las dos rectas y en lados opuestos de la transversal

altitude of a triangle *(p. 317)* The perpendicular segment from a vertex of a triangle to the opposite side or to the line that contains the opposite side

altitud de un triángulo *(p. 317)* El segmento perpendicular desde el vértice de un triángulo al lado opuesto o a la recta que contiene el lado opuesto

angle *(p. 38)* A set of points consisting of two different rays that have the same endpoint

ángulo *(p. 38)* Un conjunto de puntos que consiste en dos rayos distintos que tienen el mismo punto extremo

angle bisector *(p. 42)* A ray that divides an angle into two angles that are congruent

bisectriz de un ángulo *(p. 42)* Un rayo que divide un ángulo en dos ángulos congruentes

angle of depression *(p. 485)* An angle formed by a horizontal line and a line of sight *down* to an object

ángulo de depresión *(p. 485)* El ángulo formado por la recta horizontal y la recta de vista descendente hacia un objeto

angle of elevation *(p. 478)* An angle formed by a horizontal line and a line of sight *up* to an object

ángulo de elevación *(p. 478)* El ángulo formado por la recta horizontal y la recta de vista ascendente hacia un objeto

angle of rotation *(p. 184)* The angle that is formed by rays drawn from the center of rotation to a point and its image

ángulo de rotación *(p. 184)* El ángulo que está formado por rayos dibujados desde el centro de rotación hacia un punto y su imagen

apothem of a regular polygon *(p. 593)* The distance from the center to any side of a regular polygon

apotema de un polígono regular *(p. 593)* La distancia desde el centro a cualquier lado de un polígono regular

arc length *(p. 579)* A portion of the circumference of a circle

longitud de arco *(p. 579)* Una porción de la circunferencia de un círculo

axiom *(p. 12)* A rule that is accepted without proof

axioma *(p. 12)* Una regla que es aceptada sin demostración

axis of revolution *(p. 668)* The line around which a two-dimensional shape is rotated to form a three-dimensional figure

eje de revolución *(p. 668)* La recta alrededor de la cual una forma bidimensional rota para formar una figura tridimensional

B

base angles of an isosceles triangle *(p. 250)* The two angles adjacent to the base of an isosceles triangle

ángulos de la base de un triángulo isósceles *(p. 250)* Los dos ángulos adyacentes a la base de un triángulo isósceles

base angles of a trapezoid *(p. 390)* Either pair of consecutive angles whose common side is a base of a trapezoid

ángulos de la base de un trapecio *(p. 390)* Cualquier par de ángulos consecutivos cuyo lado común es la base de un trapezoide

base of an isosceles triangle *(p. 250)* The side of an isosceles triangle that is not one of the legs

base de un triángulo isósceles *(p. 250)* El lado de un triángulo isósceles que no es uno de los catetos

bases of a trapezoid *(p. 390)* The parallel sides of a trapezoid

bases de un trapecio *(p. 390)* Los lados paralelos de un trapezoide

between *(p. 14)* When three points are collinear, one point is between the other two.

entre *(p. 14)* Cuando tres puntos son colineales, un punto está entre los otros dos.

biconditional statement *(p. 69)* A statement that contains the phrase "if and only if"

enunciado bicondicional *(p. 69)* Un enunciado que contiene la frase "si y sólo si"

C

Cavalieri's Principle *(p. 638)* If two solids have the same height and the same cross-sectional area at every level, then they have the same volume.

Principio de Cavalieri *(p. 638)* Si dos sólidos tienen la misma altura y la misma área transversal en todo nivel, entonces tienen el mismo volumen.

center of a circle *(p. 514)* The point from which all points on a circle are equidistant

centro de un círculo *(p. 514)* El punto desde donde todos los puntos en un círculo son equidistantes

center of dilation *(p. 204)* The fixed point in a dilation

centro de dilatación *(p. 204)* El punto fijo en una dilatación

center of a regular polygon *(p. 593)* The center of a polygon's circumscribed circle

centro de un polígono regular *(p. 593)* El centro del círculo circunscrito de un polígono

center of rotation *(p. 184)* The fixed point in a rotation

centro de rotación *(p. 184)* El punto fijo en una rotación

center of symmetry *(p. 188)* The center of rotation in a figure that has rotational symmetry

centro de simetría *(p. 188)* El centro de rotación en una figura que tiene simetría rotacional

central angle of a circle *(p. 522)* An angle whose vertex is the center of a circle

ángulo central de un círculo *(p. 522)* Un ángulo cuyo vértice es el centro de un círculo

central angle of a regular polygon *(p. 593)* An angle formed by two radii drawn to consecutive vertices of a polygon

ángulo central de un polígono regular *(p. 593)* Un ángulo formado por dos radios extendidos a vértices consecutivos de un polígono

centroid *(p. 316)* The point of concurrency of the three medians of a triangle

centroide *(p. 316)* El punto de concurrencia de las tres medianas de un triángulo

chord of a circle *(p. 514)* A segment whose endpoints are on a circle

cuerda de un círculo *(p. 514)* Un segmento cuyos puntos extremos están en un círculo

chord of a sphere *(p. 654)* A segment whose endpoints are on a sphere

cuerda de una esfera *(p. 654)* Un segmento cuyos puntos extremos están en una esfera

circle *(p. 514)* The set of all points in a plane that are equidistant from a given point

círculo *(p. 514)* El conjunto de todos los puntos en un plano que son equidistantes de un punto dado

circumcenter *(p. 306)* The point of concurrency of the three perpendicular bisectors of a triangle

circuncentro *(p. 306)* El punto de concurrencia de las tres bisectrices perpendiculares de un triángulo

circumference *(p. 578)* The distance around a circle

circunferencia *(p. 578)* La distancia alrededor de un círculo

circumscribed angle *(p. 546)* An angle whose sides are tangent to a circle

ángulo circunscrito *(p. 546)* Un ángulo cuyos lados son tangentes a un círculo

circumscribed circle *(p. 538)* A circle that contains all the vertices of an inscribed polygon

círculo circunscrito *(p. 538)* Un círculo que contiene todos los vértices de un polígono inscrito

collinear points *(p. 4)* Points that lie on the same line

puntos colineales *(p. 4)* Puntos que descansan en la misma recta

common tangent *(p. 515)* A line or segment that is tangent to two coplanar circles

tangente común *(p. 515)* Una recta o segmento que es tangente a dos círculos coplanarios

complementary angles *(p. 48)* Two angles whose measures have a sum of 90°

ángulos complementarios *(p. 48)* Dos ángulos cuyas medidas suman 90°

component form *(p. 168)* A form of a vector that combines the horizontal and vertical components

forma componente *(p. 168)* Una forma de un vector que combina los componentes horizontales y verticales

composition of transformations *(p. 170)* The combination of two or more transformations to form a single transformation

composición de transformaciones *(p. 170)* La combinación de dos o más transformaciones para formar una transformación única

concentric circles *(p. 515)* Coplanar circles that have a common center

círculos concéntricos *(p. 515)* Círculos coplanarios que tienen un centro en común

conclusion *(p. 66)* The "then" part of a conditional statement written in if-then form

conclusión *(p. 66)* La parte después de "entonces" en un enunciado condicional escrito de la forma "si..., entonces..."

concurrent *(p. 306)* Three or more lines, rays, or segments that intersect in the same point

concurrente *(p. 306)* Tres o más rectas, rayos o segmentos que se intersectan en el mismo punto

conditional statement *(p. 66)* A logical statement that has a hypothesis and a conclusion

enunciado condicional *(p. 66)* Un enunciado lógico que tiene una hipótesis y una conclusión

congruence transformation *(p. 195)* A transformation that preserves length and angle measure

See rigid motion.

transformación de congruencia *(p. 195)* Una transformación que preserva la longitud y medida del ángulo

Ver movimiento rígida.

congruent angles *(p. 40)* Two angles that have the same measure

ángulos congruentes *(p. 40)* Dos ángulos que tienen la misma medida

congruent arcs *(p. 524)* Arcs that have the same measure and are of the same circle or of congruent circles

arcos congruentes *(p. 524)* Arcos que tienen la misma medida y que son del mismo círculo o de círculos congruentes

congruent circles *(p. 524)* Circles that can be mapped onto each other by a rigid motion or a composition of rigid motions

círculos congruentes *(p. 524)* Círculos que pueden superponerse sobre sí mismos mediante un movimiento rígido o una composición de movimientos rígidos

congruent figures *(p. 194)* Geometric figures that have the same size and shape

figuras congruentes *(p. 194)* Figuras geométricas que tienen el mismo tamaño y forma

congruent segments *(p. 13)* Line segments that have the same length

segmentos congruentes *(p. 13)* Segmentos de rectas que tienen la misma longitud

conjecture *(p. 74)* An unproven statement that is based on observations

conjetura *(p. 74)* Una afirmación no comprobada que se basa en observaciones

consecutive interior angles *(p. 124)* Two angles that are formed by two lines and a transversal that lie between the two lines and on the same side of the transversal

ángulos interiores consecutivos *(p. 124)* Dos ángulos que son formados por dos rectas y una transversal que descansan entre las dos rectas y en el mismo lado de la transversal

construction *(p. 13)* A geometric drawing that uses a limited set of tools, usually a compass and a straightedge

construcción *(p. 13)* Un dibujo geométrico que usa un conjunto limitado de herramientas, generalmente una regla y compás

contrapositive *(p. 67)* The statement formed by negating both the hypothesis and conclusion of the converse of a conditional statement

contrapositivo *(p. 67)* El enunciado formado por la negación de la hipótesis y conclusión del converso de un enunciado condicional

converse *(p. 67)* The statement formed by exchanging the hypothesis and conclusion of a conditional statement

converso *(p. 67)* El enunciado formado por el intercambio de la hipótesis y conclusión de un enunciado condicional

coordinate *(p. 12)* A real number that corresponds to a point on a line

coordenada *(p. 12)* Un número real que corresponde a un punto en una línea

coordinate proof *(p. 280)* A style of proof that involves placing geometric figures in a coordinate plane

prueba de coordenadas *(p. 280)* Un estilo de prueba que implica colocar figuras geométricas en un plano coordenado

coplanar points *(p. 4)* Points that lie in the same plane

puntos coplanarios *(p. 4)* Puntos que descansan en el mismo plano

corollary to a theorem *(p. 233)* A statement that can be proved easily using the theorem

corolario de un teorema *(p. 233)* Un enunciado que puede comprobarse fácilmente usando el teorema

corresponding angles *(p. 124)* Two angles that are formed by two lines and a transversal that are in corresponding positions

ángulos correspondientes *(p. 124)* Dos ángulos que están formados por dos líneas y una transversal que están en las posiciones correspondientes

corresponding parts *(p. 238)* A pair of sides or angles that have the same relative position in two congruent figures

partes correspondientes *(p. 238)* Un par de lados o ángulos que tienen la misma posición relativa en dos figuras congruentes

cosine *(p. 482)* For an acute angle of a right triangle, the ratio of the length of the leg adjacent to the acute angle to the length of the hypotenuse

coseno *(p. 482)* Para un ángulo agudo de un triángulo rectángulo, la razón de la longitud del cateto adyacente al ángulo agudo a la longitud de la hipotenusa

counterexample *(p. 75)* A specific case for which a conjecture is false

contraejemplo *(p. 75)* Un caso específico para el que una conjetura es falsa

cross section *(p. 617)* The intersection of a plane and a solid

sección transversal *(p. 617)* La intersección de un plano y un sólido

D

deductive reasoning *(p. 76)* A process that uses facts, definitions, accepted properties, and the laws of logic to form a logical argument

razonamiento deductivo *(p. 76)* Un proceso que usa hechos, definiciones, propiedades aceptadas y las leyes de la lógica para formar un argumento lógico

defined terms *(p. 5)* Terms that can be described using known words, such as *point* or *line*

términos definidos *(p. 5)* Términos que pueden describirse usando palabras conocidas, como *punto* o *línea*

density *(p. 662)* The amount of matter that an object has in a given unit of volume

densidad *(p. 662)* La cantidad de materia que tiene un objeto en una unidad de volumen dada

diagonal *(p. 354)* A segment that joins two nonconsecutive vertices of a polygon

diagonal *(p. 354)* Un segmento que une dos vértices no consecutivos de un polígono

diameter *(p. 514)* A chord that contains the center of a circle

diámetro *(p. 514)* Una cuerda que contiene el centro de un círculo

dilation *(p. 204)* A transformation in which a figure is enlarged or reduced with respect to a fixed point

dilatación *(p. 204)* Una transformación en la cual una figura se agranda o reduce con respecto a un punto fijo

directed line segment *(p. 150)* A segment that represents moving from point A to point B is called the directed line segment AB.

segmento de línea dirigido *(p. 150)* Un segmento que representa el moverse del punto A al punto B se llama el segmento de línea dirigido AB.

distance between two points *(p. 12)* The absolute value of the difference of two coordinates on a line

distancia entre dos puntos *(p. 12)* El valor absoluto de la diferencia de dos coordenadas en una recta

distance from a point to a line *(p. 142)* The length of the perpendicular segment from the point to the line

distancia desde un punto a una recta *(p. 142)* La longitud del segmento perpendicular desde el punto a la recta

E

edge *(p. 616)* A line segment or curve formed by the intersection of two faces of a solid

borde *(p. 616)* Un segmento de línea o curva formados por la intersección de dos caras de un sólido

endpoints *(p. 5)* Points that represent the ends of a line segment or ray

puntos extremos *(p. 5)* Puntos que representan los extremos de una semirrecta o un segmento de recta

enlargement *(p. 204)* A dilation in which the scale factor is greater than 1

agrandamiento *(p. 204)* Una dilatación en donde el factor de escala es mayor que 1

equiangular polygon *(p. 356)* A polygon in which all angles are congruent

polígono equiangular *(p. 356)* Un polígono en donde todos los ángulos son congruentes

equidistant *(p. 298)* A point is equidistant from two figures when it is the same distance from each figure.

equidistante *(p. 298)* Un punto es equidistante desde dos figuras cuando está a la misma distancia de cada figura.

equilateral polygon *(p. 356)* A polygon in which all sides are congruent

polígono equilátero *(p. 356)* Un polígono en donde todos los lados son congruentes

equivalent statements *(p. 67)* Two related conditional statements that are both true or both false

enunciados equivalentes *(p. 67)* Dos enunciados condicionales relacionados que son ambos verdaderos, o ambos falsos

exterior of an angle *(p. 38)* The region that contains all the points outside of an angle

exterior de un ángulo *(p. 38)* La región que contiene todos los puntos fuera de un ángulo

exterior angles *(p. 231)* Angles that form linear pairs with the interior angles of a polygon

ángulos exteriores *(p. 231)* Ángulos que forman pares lineales con los ángulos interiores de un polígono

external segment *(p. 553)* The part of a secant segment that is outside the circle

segmento externo *(p. 553)* La parte de un segmento secante que está fuera del círculo

F

face *(p. 616)* A flat or curved surface of a solid

cara *(p. 616)* Una superficie plana o curva de un sólido

flowchart proof (flow proof) *(p. 102)* A type of proof that uses boxes and arrows to show the flow of a logical argument

prueba de organigrama (prueba de flujo) *(p. 102)* Un tipo de prueba que usa casillas y flechas para mostrar el flujo de un argumento lógico

G

geometric mean *(p. 470)* The positive number x that satisfies $\frac{a}{x} = \frac{x}{b}$

So, $x^2 = ab$ and $x = \sqrt{ab}$.

media geométrica *(p. 470)* El número positivo x que satisface $\frac{a}{x} = \frac{x}{b}$

Entonces, $x^2 = ab$ and $x = \sqrt{ab}$.

glide reflection *(p. 178)* A transformation involving a translation followed by a reflection

reflexión por deslizamiento *(p. 178)* Una transformación que implica una traslación seguida de una reflexión

great circle *(p. 654)* The intersection of a plane and a sphere such that the plane contains the center of the sphere

gran círculo *(p. 654)* La intersección de un plano y una esfera, de tal forma que el plano contiene el centro de la esfera

H

horizontal component *(p. 168)* The horizontal change from the starting point of a vector to the ending point

componente horizontal *(p. 168)* El cambio horizontal desde el punto de inicio de un vector hasta el punto final

hypotenuse *(p. 260)* The side opposite the right angle of a right triangle

hipotenusa *(p. 260)* El lado opuesto al ángulo recto de un triángulo recto

hypothesis *(p. 66)* The "if" part of a conditional statement written in if-then form

hipótesis *(p. 66)* La parte después de "si" en un enunciado condicional escrito de la forma "si..., entonces..."

I

if-then form *(p. 66)* A conditional statement in the form "if p, then q"

forma "si..., entonces..." *(p. 66)* Un enunciado condicional en la forma de "si p, entonces q"

image *(p. 168)* A figure that results from the transformation of a geometric figure

imagen *(p. 168)* Una figura que resulta de la transformación de una figura geométrica

incenter *(p. 309)* The point of concurrency of the angle bisectors of a triangle

incentro *(p. 309)* El punto de concurrencia de las bisectrices de los ángulos de un triángulo

inductive reasoning *(p. 74)* A process that includes looking for patterns and making conjectures

razonamiento inductivo *(p. 74)* Un proceso que incluye buscar patrones y hacer conjeturas

initial point *(p. 168)* The starting point of a vector

punto inicial *(p. 168)* El punto de inicio de un vector

inscribed angle *(p. 536)* An angle whose vertex lies on a circle and whose sides contain chords of the circle

ángulo inscrito *(p. 536)* Un ángulo cuyo vértice está en un círculo y cuyos lados contienen cuerdas del círculo

inscribed polygon *(p. 538)* A polygon in which all the vertices lie on a circle

polígono inscrito *(p. 538)* Un polígono en donde todos los vértices descansan sobre un círculo

intercepted arc *(p. 536)* An arc that lies between two lines, rays, or segments

arco interceptado *(p. 536)* Un arco que descansa entre dos rectas, rayos o segmentos

interior of an angle *(p. 38)* The region that contains all the points between the sides of an angle

interior de un ángulo *(p. 38)* La región que contiene todos los puntos entre los lados de un ángulo

interior angles *(p. 231)* Angles of a polygon

ángulos interiores *(p. 231)* Los ángulos de un polígono

intersection *(p. 6)* The set of points two or more geometric figures have in common

intersección *(p. 6)* El conjunto de puntos que dos o más figuras geométricas tienen en común

inverse *(p. 67)* The statement formed by negating both the hypothesis and conclusion of a conditional statement

inverso *(p. 67)* El enunciado formado por la negación de la hipótesis y conclusión de un enunciado condicional

inverse cosine *(p. 493)* An inverse trigonometric ratio, abbreviated as $\cos^{-1}$

For acute angle A, if $\cos A = z$, then $\cos^{-1} z = m\angle A$.

coseno inverso *(p. 493)* Una razón trigonométrica inversa, abreviada como $\cos^{-1}$

Para un ángulo agudo A, si $\cos A = z$, entonces $\cos^{-1} z = m\angle A$.

inverse sine *(p. 493)* An inverse trigonometric ratio, abbreviated as $\sin^{-1}$

For acute angle A, if $\sin A = y$, then $\sin^{-1} y = m\angle A$.

seno inverso *(p. 493)* Una razón trigonométrica inversa, abreviada como $\sin^{-1}$

Para un ángulo agudo A, si $\sin A = y$, entonces $\sin^{-1} y = m\angle A$.

inverse tangent *(p. 493)* An inverse trigonometric ratio, abbreviated as $\tan^{-1}$

For acute angle A, if $\tan A = x$, then $\tan^{-1} x = m\angle A$.

tangente inversa *(p. 493)* Una razón trigonométrica inversa, abreviada como $\tan^{-1}$

Para un ángulo agudo A, si $\tan A = x$, entonces $\tan^{-1} x = m\angle A$.

isosceles trapezoid *(p. 390)* A strict trapezoid with congruent legs

trapecio isósceles *(p. 390)* Un trapecio en sentido estricto con catetos congruentes

K

kite *(p. 393)* A quadrilateral that has two pairs of consecutive congruent sides, but opposite sides are not congruent

papalote *(p. 393)* Un cuadrilátero que tiene dos pares de lados congruentes consecutivos, pero los lados opuestos no son congruentes

L

lateral face of a cone *(p. 633)* Consists of all segments that connect the vertex with points on the base edge of a cone

cara lateral de un cono *(p. 633)* Consiste en todos los segmentos que conectan el vértice con puntos en el borde base de un cono

lateral face of a cylinder *(p. 625)* The curved surface of a cylinder

cara lateral de un cilindro *(p. 625)* La superficie curva de un cilindro

lateral faces of a polyhedron *(p. 624)* The parallelograms formed by connecting the corresponding vertices of the bases

caras laterales de un poliedro *(p. 624)* Los paralelogramos formados al conectar los vértices correspondientes de las bases

Law of Cosines *(p. 495)* For $\triangle ABC$ with side lengths of a, b, and c,

$$a^2 = b^2 + c^2 - 2bc \cos A,$$
$$b^2 = a^2 + c^2 - 2ac \cos B, \text{ and}$$
$$c^2 = a^2 + b^2 - 2ab \cos C.$$

Ley de cosenos *(p. 495)* Para $\triangle ABC$ con longitudes de lados de a, b, y c,

$$a^2 = b^2 + c^2 - 2bc \cos A,$$
$$b^2 = a^2 + c^2 - 2ac \cos B, \text{ y}$$
$$c^2 = a^2 + b^2 - 2ab \cos C.$$

Law of Sines *(p. 494)* For $\triangle ABC$ with side lengths of a, b, and c,

$$\frac{\sin A}{a} = \frac{\sin B}{b} = \frac{\sin C}{c} \text{ and}$$
$$\frac{a}{\sin A} = \frac{b}{\sin B} = \frac{c}{\sin C}.$$

Ley de senos *(p. 494)* Para $\triangle ABC$ con longitudes de lados de a, b, y c,

$$\frac{\sin A}{a} = \frac{\sin B}{b} = \frac{\sin C}{c} \text{ y}$$
$$\frac{a}{\sin A} = \frac{b}{\sin B} = \frac{c}{\sin C}.$$

legs of an isosceles triangle *(p. 250)* The two congruent sides of an isosceles triangle

catetos de un triángulo isósceles *(p. 250)* Los dos lados congruentes de un triángulo isósceles

legs of a right triangle *(p. 260)* The sides adjacent to the right angle of a right triangle

catetos de un triángulo recto *(p. 260)* Los lados adyacentes al ángulo recto de un triángulo recto

legs of a trapezoid *(p. 390)* Any nonparallel sides of a trapezoid

catetos de un trapecio *(p. 390)* Cualquiera de los lados no paralelos de un trapecio

line *(p. 4)* A line has one dimension. It is represented by a line with two arrowheads, but it extends without end.

recta *(p. 4)* Una recta tiene una dimensión. Se representa por una línea con dos flechas, pero se extiende sin fin.

line perpendicular to a plane *(p. 84)* A line that intersects the plane in a point and is perpendicular to every line in the plane that intersects it at that point

recta perpendicular a un plano *(p. 84)* Una recta que intersecta el plano en un punto y es perpendicular a cada recta en el plano que la intersecta en ese punto

line of reflection *(p. 176)* A line that acts as a mirror for a reflection

recta de reflexión *(p. 176)* Una recta que actúa como un espejo para una reflexión

line segment *(p. 5)* A part of a line that consists of two endpoints and all points on the line between the endpoints
See segment.

segmento de recta *(p. 5)* La parte de una recta que consiste en dos puntos extremos y todos los puntos entre ellos
Ver segmento.

line symmetry *(p. 179)* A figure in the plane has line symmetry when the figure can be mapped onto itself by a reflection in a line.

simetría de recta *(p. 179)* Una figura en el plano tiene simetría de recta cuando la figura puede superponerse sobre sí misma por una reflexión en una recta.

line of symmetry *(p. 179)* A line of reflection that maps a figure onto itself

recta de simetría *(p. 179)* Una recta de reflexión que superpone una figura sobre sí misma

linear pair *(p. 50)* Two adjacent angles whose noncommon sides are opposite rays

par lineal *(p. 50)* Dos ángulos adyacentes cuyos lados no comunes son rayos opuestos

M

major arc *(p. 522)* An arc with a measure greater than 180°

arco mayor *(p. 522)* Un arco con una medida mayor de 180°

measure of an angle *(p. 38)* The absolute value of the difference between the real numbers matched with the two rays that form the angle on a protractor

medida de un ángulo *(p. 38)* El valor absoluto de la diferencia entre los números reales asociados con los dos rayos que forman el ángulo en un transportador

measure of a major arc *(p. 522)* The measure of a major arc's central angle

medida de arco mayor *(p. 522)* La medida del ángulo central de un arco mayor

measure of a minor arc *(p. 522)* The measure of a minor arc's central angle

medida de arco menor *(p. 522)* La medida del ángulo central de un arco menor

median of a triangle *(p. 316)* A segment from a vertex of a triangle to the midpoint of the opposite side

mediana de un triángulo *(p. 316)* Un segmento desde el vértice de un triángulo hasta el punto medio del lado opuesto

midpoint *(p. 20)* The point that divides a segment into two congruent segments

punto medio *(p. 20)* El punto que divide un segmento en dos segmentos congruentes

midsegment of a trapezoid *(p. 392)* The segment that connects the midpoints of the legs of a trapezoid

segmento medio de un trapezoide *(p. 392)* El segmento que conecta los puntos medios de los catetos de un trapezoide

midsegment of a triangle *(p. 324)* A segment that connects the midpoints of two sides of a triangle

segmento medio de un triángulo *(p. 324)* Un segmento que conecta los puntos medios de dos lados de un triángulo

minor arc *(p. 522)* An arc with a measure less than 180°

arco menor *(p. 522)* Un arco con una medida menor de 180°

N

narrative proof *(p. 104)* A style of proof that presents the statements and reasons as sentences in a paragraph, using words to explain the logical flow of an argument
See paragraph proof.

prueba en lenguaje narrativo *(p. 104)* Un estilo de prueba que presenta los enunciados y las razones como oraciones en un párrafo, usando palabras para explicar el flujo lógico del argumento; *Ver prueba en forma de párrafo.*

negation *(p. 66)* The opposite of a statement
If a statement is p, then the negation is "not p," written $\sim p$.

negación *(p. 66)* Lo opuesto de un enunciado o afirmación
Si un enunciado es p, entonces la negación es "no p," y se escribe $\sim p$.

O

obtuse angle *(p. 39)* An angle that has a measure greater than 90° and less than 180°

ángulo obtuso *(p. 39)* Un ángulo que tiene una medida mayor que 90° y menor que 180°

opposite rays *(p. 5)* Two rays that have the same endpoint and form a line

rayos opuestos *(p. 5)* Dos rayos tienen el mismo punto extremo y forman una recta

order of rotational symmetry *(p. 188)* The number of times a figure can be mapped onto itself in one 360° rotation about the center of the figure

orden de simetría rotacional *(p. 188)* El número de veces que se puede mapear una figura sobre sí misma en una rotación de 360° alrededor del centro de la figura

orthocenter *(p. 317)* The point of concurrency of the lines containing the altitudes of a triangle

ortocentro *(p. 317)* El punto de concurrencia de las líneas que contienen las alturas de un triángulo

P

paragraph proof *(p. 104)* A style of proof that presents the statements and reasons as sentences in a paragraph, using words to explain the logical flow of an argument *See narrative proof.*

prueba en forma de párrafo *(p. 104)* Un estilo de prueba que presenta los enunciados y las razones como oraciones en un párrafo, usando palabras para explicar el flujo lógico del argumento; *Ver prueba en lenguaje narrativo.*

parallel lines *(p. 122)* Coplanar lines that do not intersect

rectas paralelas *(p. 122)* Rectas coplanarias que no se intersectan

parallel planes *(p. 122)* Planes that do not intersect

planos paralelos *(p. 122)* Planos que no se intersectan

parallelogram *(p. 362)* A quadrilateral with both pairs of opposite sides parallel

paralelogramo *(p. 362)* Un cuadrilátero con ambos pares de lados opuestos paralelos

perpendicular bisector *(p. 143)* A ray, line, line segment, or plane that is perpendicular to a segment at its midpoint

bisectriz perpendicular *(p. 143)* Un rayo, una recta, un segmento de recta o un plano que es perpendicular a un segmento en su punto medio

perpendicular lines *(p. 68)* Two lines that intersect to form a right angle

rectas perpendiculares *(p. 68)* Dos líneas que se intersectan para formar un ángulo recto

plane *(p. 4)* A flat surface made up of points that has two dimensions and extends without end and is represented by a shape that looks like a floor or wall

plano *(p. 4)* Una superficie plana formada por puntos que tiene dos dimensiones y se extiende sin fin y que está representada por una forma que parece un piso o una pared

point *(p. 4)* A location in space that is represented by a dot and has no dimension

punto *(p. 4)* Un lugar en el espacio que está representado por un punto y no tiene dimensión

point of concurrency *(p. 306)* The point of intersection of concurrent lines, rays, or segments

punto de concurrencia *(p. 306)* El punto de intersección de rectas, rayos o segmentos concurrentes

point of tangency *(p. 514)* The point at which a tangent line intersects a circle

punto de tangencia *(p. 514)* El punto en donde una recta tangente intersecta a un círculo

polyhedron *(p. 616)* A solid whose faces are all polygons

poliedro *(p. 616)* Un sólido cuyas caras son todas polígonos

population density *(p. 600)* A measure of how many people live within a given area

densidad de población *(p. 600)* Medición de la cantidad de personas que habitan un área dada

postulate *(p. 12)* A rule that is accepted without proof

postulado *(p. 12)* Una regla que es aceptada sin demostración

preimage *(p. 168)* The original figure before a transformation

preimagen *(p. 168)* La figura original antes de una transformación

proof *(p. 96)* A logical argument that uses deductive reasoning to show that a statement is true

prueba *(p. 96)* Un argumento lógico que usa el razonamiento deductivo para mostrar que un enunciado es verdadero

proof by contradiction *(p. 330)* A style of proof in which you temporarily assume that the desired conclusion is false, then reason logically to a contradiction

This proves that the original statement is true.

prueba por contradicción *(p. 330)* Un estilo de prueba en donde se asume temporalmente que la conclusión deseada es falsa; luego se razona lógicamente hasta llegar a una contradicción

Esto demuestra que el enunciado original es verdadero.

Pythagorean triple *(p. 454)* A set of three positive integers a, b, and c that satisfy the equation $c^2 = a^2 + b^2$

triple pitagórico *(p. 454)* Un conjunto de tres números enteros positivos a, b, y c que satisfacen la ecuación $c^2 = a^2 + b^2$

R

radian *(p. 581)* A unit of measurement for angles

radián *(p. 581)* Una unidad de medida para ángulos

radius of a circle *(p. 514)* A segment whose endpoints are the center and any point on a circle

radio de un círculo *(p. 514)* Un segmento cuyos puntos extremos son el centro y cualquier punto en un círculo

radius of a regular polygon *(p. 593)* The radius of a polygon's circumscribed circle

radio de un polígono regular *(p. 593)* El radio del círculo circunscrito de un polígono

ray *(p. 5)* A part of a line that consists of an endpoint and all points on the line on one side of the endpoint

rayo *(p. 5)* La parte de una recta que consiste en un punto extremo y todos los puntos de la recta de un lado del punto extremo

rectangle *(p. 380)* A parallelogram with four right angles

rectángulo *(p. 380)* Un paralelogramo con cuatro ángulos rectos

reduction *(p. 204)* A dilation in which the scale factor is greater than 0 and less than 1

reducción *(p. 204)* Una dilatación en donde el factor de escala es mayor que 0 y menor que 1

reflection *(p. 176)* A transformation that uses a line like a mirror to reflect a figure

reflexión *(p. 176)* Una transformación que usa una recta como un espejo para reflejar una figura

regular polygon *(p. 356)* A convex polygon that is both equilateral and equiangular

polígono regular *(p. 356)* Un polígono convexo que es tanto equilátero como equiángulo

rhombus *(p. 380)* A parallelogram with four congruent sides

rombo *(p. 380)* Un paralelogramo con cuatro lados congruentes

right angle *(p. 39)* An angle that has a measure of 90°

ángulo recto *(p. 39)* Un ángulo que tiene una medida de 90°

rigid motion *(p. 170)* A transformation that preserves length and angle measure

See congruence transformation.

movimiento rígido *(p. 170)* Una transformación que preserva la longitud y medida del ángulo

Ver transformación de congruencia.

rotation *(p. 184)* A transformation in which a figure is turned about a fixed point

rotación *(p. 184)* Una transformación en la cual una figura gira sobre un punto fijo

rotational symmetry *(p. 188)* A figure has rotational symmetry when the figure can be mapped onto itself by a rotation of 180° or less about the center of the figure.

simetría de rotación *(p. 188)* Una figura tiene simetría de rotación cuando la figura puede superponerse sobre sí misma mediante una rotación de 180° o menos en el centro de la figura.

S

scale factor *(p. 204)* The ratio of the lengths of the corresponding sides of the image and the preimage of a dilation

factor de escala *(p. 204)* La razón de las longitudes de los lados correspondientes de la imagen y la preimagen de una dilatación

secant *(p. 514)* A line that intersects a circle in two points

secante *(p. 514)* Una recta que intersecta a un círculo en dos puntos

secant segment *(p. 553)* A segment that contains a chord of a circle and has exactly one endpoint outside the circle

segmento de secante *(p. 553)* Un segmento que contiene una cuerda de un círculo y que tiene exactamente un punto extremo fuera del círculo

sector of a circle *(p. 587)* The region bounded by two radii of the circle and their intercepted arc

sector de un círculo *(p. 587)* La región encerrada por dos radios del círculo y su arco interceptado

segment *(p. 5)* A part of a line that consists of two endpoints and all points on the line between the endpoints

See line segment.

segmento *(p. 5)* La parte de una línea que consiste en dos puntos extremos y todos los puntos entre ellos

Ver segmento de recta.

segment bisector *(p. 20)* A point, ray, line, line segment, or plane that intersects the segment at its midpoint

bisectriz de segmento *(p. 20)* Un punto, rayo, recta, segmento de recta o plano que intersecta el segmento en su punto medio

segments of a chord *(p. 552)* The segments formed from two chords that intersect in the interior of a circle

segmentos de una cuerda *(p. 552)* Los segmentos formados a partir de dos cuerdas que se intersectan en el interior de un círculo

semicircle *(p. 522)* An arc with endpoints that are the endpoints of a diameter

semicírculo *(p. 522)* Un arco con puntos extremos que son los puntos extremos de un diámetro

sides of an angle *(p. 38)* The rays of an angle

lados de un ángulo *(p. 38)* Los rayos de un ángulo

similar arcs *(p. 525)* Arcs that have the same measure

arcos similares *(p. 525)* Arcos que tienen la misma medida

similar figures *(p. 212)* Geometric figures that have the same shape but not necessarily the same size; Two geometric figures are similar if and only if there is a similarity transformation that maps one of the figures to the other.

figuras similares *(p. 212)* Figuras geométricas que tienen la misma forma pero no necesariamente el mismo tamaño; Dos figuras geométricas son similares, sí y solo sí hay una transformación de similitud que relaciona una de las figuras con la otra

similar solids *(p. 627)* Two solids of the same type with equal ratios of corresponding linear measures

sólidos similares *(p. 627)* Dos sólidos del mismo tipo con razones iguales de medidas lineales correspondientes

similarity transformation *(p. 212)* A dilation or a composition of rigid motions and dilations

transformación de similitud *(p. 212)* Una dilatación o composición de movimientos rígidos y dilataciones

sine *(p. 482)* For an acute angle of a right triangle, the ratio of the length of the leg opposite the acute angle to the length of the hypotenuse

seno *(p. 482)* Para un ángulo agudo de un triángulo rectángulo, la razón de la longitud del cateto enfrente del ángulo agudo a la longitud de la hipotenusa

skew lines *(p. 122)* Lines that do not intersect and are not coplanar

rectas sesgadas *(p. 122)* Rectas que no se intersectan y que no son coplanarias

slant height of a regular pyramid *(p. 632)* The height of a lateral face of a regular pyramid

altura inclinada de una pirámide regular *(p. 632)* La altura de una cara lateral de una pirámide regular

solid of revolution *(p. 668)* A three-dimensional figure that is formed by rotating a two-dimensional shape around an axis

sólido de revolución *(p. 668)* Una figura tridimensional que se forma por la rotación de una forma bidimensional alrededor de un eje

square *(p. 380)* A parallelogram with four congruent sides and four right angles

cuadrado *(p. 380)* Un paralelogramo con cuatro lados congruentes y cuatro ángulos rectos

standard equation of a circle *(p. 558)* $(x - h)^2 + (y - k)^2 = r^2$, where r is the radius and (h, k) is the center

ecuación estándar de un círculo *(p. 558)* $(x - h)^2 + (y - k)^2 = r^2$, donde r es el radio y (h, k) es el centro

straight angle *(p. 39)* An angle that has a measure of 180°

ángulo llano *(p. 39)* Un ángulo que tiene una medida de 180°

subtend *(p. 536)* If the endpoints of a chord or arc lie on the sides of an inscribed angle, the chord or arc is said to subtend the angle.

subtender *(p. 536)* Si los puntos extremos de una cuerda o arco descansan en los lados de un ángulo inscrito, se dice que la cuerda o arco subtiende el ángulo.

supplementary angles *(p. 48)* Two angles whose measures have a sum of 180°

ángulos suplementarios *(p. 48)* Dos ángulos cuyas medidas suman 180°

surface area of a solid *(p. 624)* The sum of the areas of the faces

área superficial de un sólido *(p. 624)* La suma de las áreas de las caras

T

tangent *(p. 476)* For an acute angle of a right triangle, the ratio of the length of the leg opposite the acute angle to the length of the leg adjacent to the acute angle

tangente *(p. 476)* Para un ángulo agudo de un triángulo rectángulo, la razón de la longitud del cateto enfrente del ángulo agudo a la longitud del cateto adyacente al ángulo agudo

tangent of a circle *(p. 514)* A line in the plane of a circle that intersects the circle at exactly one point

tangente de un círculo *(p. 514)* Una recta en el plano de un círculo que intersecta el círculo en exactamente un punto

tangent circles *(p. 515)* Coplanar circles that intersect in one point

círculos tangentes *(p. 515)* Círculos coplanarios que se intersectan en un punto

tangent segment *(p. 553)* A segment that is tangent to a circle at an endpoint

segmento de tangente *(p. 553)* Un segmento que es tangente a un círculo en un punto extremo

terminal point *(p. 168)* The ending point of a vector

punto terminal *(p. 168)* El punto final de un vector

tessellation *(p. 196)* The covering of a plane with one or more congruent figures so that there are no gaps or overlaps

teselación *(p. 196)* La cobertura de un plano con una o más figuras congruentes para que no haya brechas o superposiciones

theorem *(p. 97)* A statement that can be proven

teorema *(p. 97)* Un enunciado que puede comprobarse

transformation *(p. 168)* A function that moves or changes a figure in some way to produce a new figure

transformación *(p. 168)* Una función que mueve o cambia una figura de cierta manera para producir una nueva figura

translation *(p. 168)* A transformation that moves every point of a figure the same distance in the same direction

traslación *(p. 168)* Una transformación que mueve cada punto de una figura la misma distancia en la misma dirección

translational symmetry *(p. 196)* A tessellation has translational symmetry when the tessellation can be mapped onto itself by a translation.

simetría traslacional *(p. 196)* Una teselación tiene simetría traslacional cuando la teselación puede mapearse sobre sí misma mediante una traslación.

transversal *(p. 124)* A line that intersects two or more coplanar lines at different points

transversal *(p. 124)* Una recta que intersecta dos o más rectas coplanarias en puntos distintos

trapezoid *(p. 390)* A quadrilateral with at least one pair of parallel sides

trapecio *(p. 390)* Un cuadrilátero con al menos un par de lados paralelos

trigonometric ratio *(p. 476)* A ratio of the lengths of two sides in a right triangle

razón trigonométrica *(p. 476)* Una razón de las longitudes de dos lados en un triángulo recto

two-column proof *(p. 96)* A type of proof that has numbered statements and corresponding reasons that show an argument in a logical order

prueba de dos columnas *(p. 96)* Un tipo de prueba que tiene enunciados numerados y motivos correspondientes que muestran un argumento en un orden lógico

U

undefined terms *(p. 4)* Words that do not have formal definitions, but there is agreement about what they mean

In geometry, the words *point*, *line*, and *plane* are undefined terms.

términos no definidos *(p. 4)* Palabras que no tienen definiciones formales, pero hay un consenso acerca de lo que significan

En geometría, las palabras *punto*, *línea* y *plano* son términos no definidos.

unit circle *(p. 486)* A circle with center (0, 0) and radius 1; the sine and cosine of an angle with its vertex at the origin, one side on the x-axis, and one side passing through a point on the unit circle in the first quadrant are equal to the y-coordinate and x-coordinate of that point, respectively.

círculo unitario *(p. 486)* Un círculo con centro (0, 0) y radio 1; el seno y el coseno de un ángulo con su vértice en el origen, un lado sobre el eje x y un lado que pasa por un punto sobre el círculo unitario en el primer cuadrante son iguales a la coordenada en y y la coordenada en x de un punto, respectivamente.

V

vector *(p. 168)* A quantity that has both direction and magnitude and is represented in the coordinate plane by an arrow drawn from one point to another

vector *(p. 168)* Una cantidad que tiene tanto dirección como magnitud y que está representada en el plano coordenado por una flecha dibujada de un punto a otro

vertex angle *(p. 250)* The angle formed by the legs of an isosceles triangle

ángulo del vértice *(p. 250)* El ángulo formado por los catetos de un triángulo isósceles

vertex of an angle *(p. 38)* The common endpoint of two rays

vértice de un ángulo *(p. 38)* El punto extremo que dos rayos tienen en común

vertex of a polyhedron *(p. 616)* A point of a polyhedron where three or more edges meet

vértice de un poliedro *(p. 616)* Un punto de un poliedro donde se encuentran tres o más bordes

vertex of a pyramid *(p. 632)* The common point of the lateral faces of a pyramid

vértice de una pirámide *(p. 632)* El punto en común de las caras laterales de una pirámide

vertical angles *(p. 50)* Two angles whose sides form two pairs of opposite rays

ángulos verticales *(p. 50)* Dos ángulos cuyos lados forman dos pares de rayos opuestos

vertical component *(p. 168)* The vertical change from the starting point of a vector to the ending point

componente vertical *(p. 168)* El cambio vertical desde el punto de inicio de un vector hasta el punto final

volume *(p. 638)* The number of cubic units contained in the interior of a solid

volumen *(p. 638)* El número de unidades cúbicas contenidas en el interior de u sólido

W

weighted average *(p. 22)* The average of two points on a number line that are weighted unequally

promedio ponderado *(p. 22)* El promedio de dos puntos en una recta numérica que se ponderan de manera desigual

Index

D

INDEX

S

T

INDEX

U

V

W

Postulates

1.1 Ruler Postulate *(p. 12)*

The points on a line can be matched one to one with the real numbers. The real number that corresponds to a point is the coordinate of the point. The distance between points A and B, written as AB, is the absolute value of the difference of the coordinates of A and B.

1.2 Segment Addition Postulate *(p. 14)*

If B is between A and C, then $AB + BC = AC$.
If $AB + BC = AC$, then B is between A and C.

1.3 Protractor Postulate *(p. 38)*

Consider $\overleftrightarrow{OB}$ and a point A on one side of $\overleftrightarrow{OB}$. The rays of the form $\overrightarrow{OA}$ can be matched one to one with the real numbers from 0 to 180. The measure of $\angle AOB$, which can be written as $m\angle AOB$, is equal to the absolute value of the difference between the real numbers matched with $\overrightarrow{OA}$ and $\overrightarrow{OB}$ on a protractor.

1.4 Angle Addition Postulate *(p. 41)*

If P is in the interior of $\angle RST$, then the measure of $\angle RST$ is equal to the sum of the measures of $\angle RSP$ and $\angle PST$.

2.1 Two Point Postulate *(p. 82)*

Through any two points, there exists exactly one line.

2.2 Line-Point Postulate *(p. 82)*

A line contains at least two points.

2.3 Line Intersection Postulate *(p. 82)*

If two lines intersect, then their intersection is exactly one point.

2.4 Three Point Postulate *(p. 82)*

Through any three noncollinear points, there exists exactly one plane.

2.5 Plane-Point Postulate *(p. 82)*

A plane contains at least three noncollinear points.

2.6 Plane-Line Postulate *(p. 82)*

If two points lie in a plane, then the line containing them lies in the plane.

2.7 Plane Intersection Postulate *(p. 82)*

If two planes intersect, then their intersection is a line.

2.8 Linear Pair Postulate *(p. 104)*

If two angles form a linear pair, then they are supplementary.

3.1 Parallel Postulate *(p. 123)*

If there is a line and a point not on the line, then there is exactly one line through the point parallel to the given line.

3.2 Perpendicular Postulate *(p. 123)*

If there is a line and a point not on the line, then there is exactly one line through the point perpendicular to the given line.

4.1 Translation Postulate *(p. 170)*

A translation is a rigid motion.

4.2 Reflection Postulate *(p. 178)*

A reflection is a rigid motion.

4.3 Rotation Postulate *(p. 186)*

A rotation is a rigid motion.

10.1 Arc Addition Postulate *(p. 523)*

The measure of an arc formed by two adjacent arcs is the sum of the measures of the two arcs.

Theorems

2.1 Properties of Segment Congruence *(p. 97)*

Segment congruence is reflexive, symmetric, and transitive.

Reflexive For any segment AB, $\overline{AB} \cong \overline{AB}$.

Symmetric If $\overline{AB} \cong \overline{CD}$, then $\overline{CD} \cong \overline{AB}$.

Transitive If $\overline{AB} \cong \overline{CD}$ and $\overline{CD} \cong \overline{EF}$, then $\overline{AB} \cong \overline{EF}$.

2.2 Properties of Angle Congruence *(p. 97)*

Angle congruence is reflexive, symmetric, and transitive.

Reflexive For any angle A, $\angle A \cong \angle A$.

Symmetric If $\angle A \cong \angle B$, then $\angle B \cong \angle A$.

Transitive If $\angle A \cong \angle B$ and $\angle B \cong \angle C$, then $\angle A \cong \angle C$.

2.3 Right Angles Congruence Theorem *(p. 102)*

All right angles are congruent.

2.4 Congruent Supplements Theorem *(p. 103)*

If two angles are supplementary to the same angle (or to congruent angles), then they are congruent.

2.5 Congruent Complements Theorem *(p. 103)*

If two angles are complementary to the same angle (or to congruent angles), then they are congruent.

2.6 Vertical Angles Congruence Theorem *(p. 104)*

Vertical angles are congruent.

3.1 Corresponding Angles Theorem *(p. 128)*

If two parallel lines are cut by a transversal, then the pairs of corresponding angles are congruent.

3.2 Alternate Interior Angles Theorem *(p. 128)*

If two parallel lines are cut by a transversal, then the pairs of alternate interior angles are congruent.

3.3 Alternate Exterior Angles Theorem *(p. 128)*

If two parallel lines are cut by a transversal, then the pairs of alternate exterior angles are congruent.

3.4 Consecutive Interior Angles Theorem *(p. 128)*

If two parallel lines are cut by a transversal, then the pairs of consecutive interior angles are supplementary.

3.5 Corresponding Angles Converse *(p. 134)*

If two lines are cut by a transversal so the corresponding angles are congruent, then the lines are parallel.

3.6 Alternate Interior Angles Converse *(p. 135)*

If two lines are cut by a transversal so the alternate interior angles are congruent, then the lines are parallel.

3.7 Alternate Exterior Angles Converse *(p. 135)*

If two lines are cut by a transversal so the alternate exterior angles are congruent, then the lines are parallel.

3.8 Consecutive Interior Angles Converse *(p. 135)*

If two lines are cut by a transversal so the consecutive interior angles are supplementary, then the lines are parallel.

3.9 Transitive Property of Parallel Lines *(p. 137)*

If two lines are parallel to the same line, then they are parallel to each other.

3.10 Linear Pair Perpendicular Theorem *(p. 144)*

If two lines intersect to form a linear pair of congruent angles, then the lines are perpendicular.

3.11 Perpendicular Transversal Theorem *(p. 144)*

In a plane, if a transversal is perpendicular to one of two parallel lines, then it is perpendicular to the other line.

3.12 Lines Perpendicular to a Transversal Theorem *(p. 144)*

In a plane, if two lines are perpendicular to the same line, then they are parallel to each other.

3.13 Slopes of Parallel Lines *(p. 151)*

In a coordinate plane, two distinct nonvertical lines are parallel if and only if they have the same slope. Any two vertical lines are parallel.

3.14 Slopes of Perpendicular Lines *(p. 151)*

In a coordinate plane, two nonvertical lines are perpendicular if and only if the product of their slopes is -1. Horizontal lines are perpendicular to vertical lines.

4.1 Composition Theorem *(p. 170)*

The composition of two (or more) rigid motions is a rigid motion.

4.2 Reflections in Parallel Lines Theorem *(p. 197)*

If lines k and m are parallel, then a reflection in line k followed by a reflection in line m is the same as a translation. If A'' is the image of A, then

1. $\overline{AA''}$ is perpendicular to k and m, and
2. $AA'' = 2d$, where d is the distance between k and m.

4.3 Reflections in Intersecting Lines Theorem *(p. 198)*

If lines k and m intersect at point P, then a reflection in line k followed by a reflection in line m is the same as a rotation about point P. The angle of rotation is $2x°$, where $x°$ is the measure of the acute or right angle formed by lines k and m.

5.1 Triangle Sum Theorem *(p. 231)*

The sum of the measures of the interior angles of a triangle is 180°.

5.2 Exterior Angle Theorem *(p. 232)*

The measure of an exterior angle of a triangle is equal to the sum of the measures of the two nonadjacent interior angles

Corollary 5.1 Corollary to the Triangle Sum Theorem *(p. 233)*

The acute angles of a right triangle are complementary.

5.3 Properties of Triangle Congruence *(p. 239)*

Triangle congruence is reflexive, symmetric, and transitive.

Reflexive For any triangle $\triangle ABC$, $\triangle ABC \cong \triangle ABC$.

Symmetric If $\triangle ABC \cong \triangle DEF$, then $\triangle DEF \cong \triangle ABC$.

Transitive If $\triangle ABC \cong \triangle DEF$ and $\triangle DEF \cong \triangle JKL$, then $\triangle ABC \cong \triangle JKL$.

5.4 Third Angles Theorem *(p. 240)*

If two angles of one triangle are congruent to two angles of another triangle, then the third angles are also congruent.

5.5 Side-Angle-Side (SAS) Congruence Theorem *(p. 244)*

If two sides and the included angle of one triangle are congruent to two sides and the included angle of a second triangle, then the two triangles are congruent.

5.6 Base Angles Theorem *(p. 250)*

If two sides of a triangle are congruent, then the angles opposite them are congruent.

5.7 Converse of the Base Angles Theorem *(p. 250)*

If two angles of a triangle are congruent, then the sides opposite them are congruent.

Corollary 5.2 Corollary to the Base Angles Theorem *(p. 251)*

If a triangle is equilateral, then it is equiangular.

Corollary 5.3 Corollary to the Converse of the Base Angles Theorem *(p. 251)*

If a triangle is equiangular, then it is equilateral.

5.8 Side-Side-Side (SSS) Congruence Theorem *(p. 258)*

If three sides of one triangle are congruent to three sides of a second triangle, then the two triangles are congruent.

5.9 Hypotenuse-Leg (HL) Congruence Theorem *(p. 260)*

If the hypotenuse and a leg of a right triangle are congruent to the hypotenuse and a leg of a second right triangle, then the two triangles are congruent.

5.10 Angle-Side-Angle (ASA) Congruence Theorem *(p. 266)*

If two angles and the included side of one triangle are congruent to two angles and the included side of a second triangle, then the two triangles are congruent.

5.11 Angle-Angle-Side (AAS) Congruence Theorem *(p. 267)*

If two angles and a non-included side of one triangle are congruent to two angles and the corresponding non-included side of a second triangle, then the two triangles are congruent.

6.1 Perpendicular Bisector Theorem *(p. 298)*

In a plane, if a point lies on the perpendicular bisector of a segment, then it is equidistant from the endpoints of the segment.

6.2 Converse of the Perpendicular Bisector Theorem *(p. 298)*

In a plane, if a point is equidistant from the endpoints of a segment, then it lies on the perpendicular bisector of the segment.

6.3 Angle Bisector Theorem *(p. 300)*

If a point lies on the bisector of an angle, then it is equidistant from the two sides of the angle.

6.4 Converse of the Angle Bisector Theorem *(p. 300)*

If a point is in the interior of an angle and is equidistant from the two sides of the angle, then it lies on the bisector of the angle.

6.5 Circumcenter Theorem *(p. 306)*

The circumcenter of a triangle is equidistant from the vertices of the triangle.

6.6 Incenter Theorem *(p. 309)*

The incenter of a triangle is equidistant from the sides of the triangle.

6.7 Centroid Theorem *(p. 316)*

The centroid of a triangle is two-thirds of the distance from each vertex to the midpoint of the opposite side.

6.8 Triangle Midsegment Theorem *(p. 325)*

The segment connecting the midpoints of two sides of a triangle is parallel to the third side and is half as long as that side.

6.9 Triangle Longer Side Theorem *(p. 331)*

If one side of a triangle is longer than another side, then the angle opposite the longer side is larger than the angle opposite the shorter side.

6.10 Triangle Larger Angle Theorem *(p. 331)*

If one angle of a triangle is larger than another angle, then the side opposite the larger angle is longer than the side opposite the smaller angle.

6.11 Triangle Inequality Theorem *(p. 333)*

The sum of the lengths of any two sides of a triangle is greater than the length of the third side.

6.12 Hinge Theorem *(p. 338)*

If two sides of one triangle are congruent to two sides of another triangle, and the included angle of the first is larger than the included angle of the second, then the third side of the first is longer than the third side of the second.

6.13 Converse of the Hinge Theorem *(p. 338)*

If two sides of one triangle are congruent to two sides of another triangle, and the third side of the first is longer than the third side of the second, then the included angle of the first is larger than the included angle of the second.

7.1 Polygon Interior Angles Theorem *(p. 354)*

The sum of the measures of the interior angles of a convex n-gon is $(n - 2) \cdot 180°$.

Corollary 7.1 Corollary to the Polygon Interior Angles Theorem *(p. 355)*

The sum of the measures of the interior angles of a quadrilateral is 360°.

7.2 Polygon Exterior Angles Theorem *(p. 357)*

The sum of the measures of the exterior angles of a convex polygon, one angle at each vertex, is 360°.

7.3 Parallelogram Opposite Sides Theorem *(p. 362)*

If a quadrilateral is a parallelogram, then its opposite sides are congruent.

7.4 Parallelogram Opposite Angles Theorem *(p. 362)*

If a quadrilateral is a parallelogram, then its opposite angles are congruent.

7.5 Parallelogram Consecutive Angles Theorem *(p. 363)*

If a quadrilateral is a parallelogram, then its consecutive angles are supplementary.

7.6 Parallelogram Diagonals Theorem *(p. 363)*

If a quadrilateral is a parallelogram, then its diagonals bisect each other.

7.7 Parallelogram Opposite Sides Converse *(p. 370)*

If both pairs of opposite sides of a quadrilateral are congruent, then the quadrilateral is a parallelogram.

7.8 Parallelogram Opposite Angles Converse *(p. 370)*

If both pairs of opposite angles of a quadrilateral are congruent, then the quadrilateral is a parallelogram.

7.9 Opposite Sides Parallel and Congruent Theorem *(p. 372)*

If one pair of opposite sides of a quadrilateral are parallel and congruent, then the quadrilateral is a parallelogram.

7.10 Parallelogram Diagonals Converse *(p. 372)*

If the diagonals of a quadrilateral bisect each other, then the quadrilateral is a parallelogram.

Corollary 7.2 Rhombus Corollary *(p. 380)*

A quadrilateral is a rhombus if and only if it has four congruent sides.

Corollary 7.3 Rectangle Corollary *(p. 380)*

A quadrilateral is a rectangle if and only if it has four right angles.

Corollary 7.4 Square Corollary *(p. 380)*

A quadrilateral is a square if and only if it is a rhombus and a rectangle.

7.11 Rhombus Diagonals Theorem *(p. 382)*

A parallelogram is a rhombus if and only if its diagonals are perpendicular.

7.12 Rhombus Opposite Angles Theorem *(p. 382)*

A parallelogram is a rhombus if and only if each diagonal bisects a pair of opposite angles.

7.13 Rectangle Diagonals Theorem *(p. 383)*

A parallelogram is a rectangle if and only if its diagonals are congruent.

7.14 Isosceles Trapezoid Base Angles Theorem *(p. 391)*

If a trapezoid is isosceles, then each pair of base angles is congruent.

7.15 Isosceles Trapezoid Base Angles Converse *(p. 391)*

If a trapezoid has a pair of congruent base angles, then it is an isosceles trapezoid.

7.16 Isosceles Trapezoid Diagonals Theorem *(p. 391)*

A trapezoid is isosceles if and only if its diagonals are congruent.

7.17 Isosceles Trapezoid Opposite Angles Theorem *(p. 391)*

A trapezoid is isosceles if and only if opposite angles are supplementary.

7.18 Trapezoid Midsegment Theorem *(p. 392)*

The midsegment of a trapezoid is parallel to each base, and its length is one-half the sum of the lengths of the bases.

7.19 Kite Diagonals Theorem *(p. 393)*

If a quadrilateral is a kite, then its diagonals are perpendicular.

7.20 Kite Opposite Angles Theorem *(p. 393)*

If a quadrilateral is a kite, then exactly one pair of opposite angles are congruent.

8.1 Perimeters of Similar Polygons *(p. 412)*

If two polygons are similar, then the ratio of their perimeters is equal to the ratios of their corresponding side lengths.

8.2 Areas of Similar Polygons *(p. 413)*

If two polygons are similar, then the ratio of their areas is equal to the squares of the ratios of their corresponding side lengths.

8.3 Angle-Angle (AA) Similarity Theorem *(p. 420)*

If two angles of one triangle are congruent to two angles of another triangle, then the two triangles are similar.

8.4 Side-Side-Side (SSS) Similarity Theorem *(p. 426)*

If the corresponding side lengths of two triangles are proportional, then the triangles are similar.

8.5 Side-Angle-Side (SAS) Similarity Theorem *(p. 428)*

If an angle of one triangle is congruent to an angle of a second triangle and the lengths of the sides including these angles are proportional, then the triangles are similar.

8.6 Triangle Proportionality Theorem *(p. 436)*

If a line parallel to one side of a triangle intersects the other two sides, then it divides the two sides proportionally.

8.7 Converse of the Triangle Proportionality Theorem *(p. 436)*

If a line divides two sides of a triangle proportionally, then it is parallel to the third side.

8.8 Three Parallel Lines Theorem *(p. 438)*

If three parallel lines intersect two transversals, then they divide the transversals proportionally.

8.9 Triangle Angle Bisector Theorem *(p. 439)*

If a ray bisects an angle of a triangle, then it divides the opposite side into segments whose lengths are proportional to the lengths of the other two sides.

9.1 Pythagorean Theorem *(p. 454)*

In a right triangle, the square of the length of the hypotenuse is equal to the sum of the squares of the lengths of the legs.

9.2 Converse of the Pythagorean Theorem *(p. 456)*

If the square of the length of the longest side of a triangle is equal to the sum of the squares of the lengths of the other two sides, then the triangle is a right triangle.

9.3 Pythagorean Inequalities Theorem *(p. 457)*

For any $\triangle ABC$, where c is the length of the longest side, the following statements are true.
If $c^2 < a^2 + b^2$, then $\triangle ABC$ is acute.
If $c^2 > a^2 + b^2$, then $\triangle ABC$ is obtuse.

9.4 45°-45°-90° Triangle Theorem *(p. 462)*

In a 45°-45°-90° triangle, the hypotenuse is $\sqrt{2}$ times as long as each leg.

9.5 30°-60°-90° Triangle Theorem *(p. 463)*

In a 30°-60°-90° triangle, the hypotenuse is twice as long as the shorter leg, and the longer leg is $\sqrt{3}$ times as long as the shorter leg.

9.6 Right Triangle Similarity Theorem *(p. 468)*

If the altitude is drawn to the hypotenuse of a right triangle, then the two triangles formed are similar to the original triangle and to each other.

9.7 Geometric Mean (Altitude) Theorem *(p. 470)*

In a right triangle, the altitude to the hypotenuse divides the hypotenuse into two segments. The length of the altitude is the geometric mean of the lengths of the two segments of the hypotenuse.

9.8 Geometric Mean (Leg) Theorem *(p. 470)*

In a right triangle, the altitude to the hypotenuse divides the hypotenuse into two segments. The length of each leg of the right triangle is the geometric mean of the lengths of the hypotenuse and the segment of the hypotenuse that is adjacent to that leg.

9.9 Law of Sines *(p. 494)*

The Law of Sines can be written in either of the following forms for $\triangle ABC$ with sides of length a, b, and c.

$$\frac{\sin A}{a} = \frac{\sin B}{b} = \frac{\sin C}{c}$$

$$\frac{a}{\sin A} = \frac{b}{\sin B} = \frac{c}{\sin C}$$

9.10 Law of Cosines *(p. 495)*

If $\triangle ABC$ has sides of length a, b, and c, then the following are true.
$a^2 = b^2 + c^2 - 2bc \cos A$
$b^2 = a^2 + c^2 - 2ac \cos B$
$c^2 = a^2 + b^2 - 2ab \cos C$

10.1 Tangent Line to Circle Theorem *(p. 516)*

In a plane, a line is tangent to a circle if and only if the line is perpendicular to a radius of the circle at its endpoint on the circle.

10.2 External Tangent Congruence Theorem *(p. 516)*

Tangent segments from a common external point are congruent.

10.3 Congruent Circles Theorem *(p. 524)*

Two circles are congruent circles if and only if they have the same radius.

10.4 Congruent Central Angles Theorem *(p. 524)*

In the same circle, or in congruent circles, two minor arcs are congruent if and only if their corresponding central angles are congruent.

10.5 Similar Circles Theorem *(p. 525)*

All circles are similar.

10.6 Congruent Corresponding Chords Theorem *(p. 530)*

In the same circle, or in congruent circles, two minor arcs are congruent if and only if their corresponding chords are congruent.

10.7 Perpendicular Chord Bisector Theorem *(p. 530)*

If a diameter of a circle is perpendicular to a chord, then the diameter bisects the chord and its arc.

10.8 Perpendicular Chord Bisector Converse *(p. 530)*

If one chord of a circle is a perpendicular bisector of another chord, then the first chord is a diameter.

10.9 Equidistant Chords Theorem *(p. 532)*

In the same circle, or in congruent circles, two chords are congruent if and only if they are equidistant from the center.

10.10 Measure of an Inscribed Angle Theorem *(p. 536)*

The measure of an inscribed angle is one-half the measure of its intercepted arc.

10.11 Inscribed Angles of a Circle Theorem *(p. 537)*

If two inscribed angles of a circle intercept the same arc, then the angles are congruent.

10.12 Inscribed Right Triangle Theorem *(p. 538)*

If a right triangle is inscribed in a circle, then the hypotenuse is a diameter of the circle. Conversely, if one side of an inscribed triangle is a diameter of the circle, then the triangle is a right triangle and the angle opposite the diameter is the right angle.

10.13 Inscribed Quadrilateral Theorem *(p. 538)*

A quadrilateral can be inscribed in a circle if and only if its opposite angles are supplementary.

10.14 Tangent and Intersected Chord Theorem *(p. 544)*

If a tangent and a chord intersect at a point on a circle, then the measure of each angle formed is one-half the measure of its intercepted arc.

10.15 Angles Inside the Circle Theorem *(p. 545)*

If two chords intersect inside a circle, then the measure of each angle is one-half the sum of the measures of the arcs intercepted by the angle and its vertical angle.

10.16 Angles Outside the Circle Theorem *(p. 545)*

If a tangent and a secant, two tangents, or two secants intersect outside a circle, then the measure of the angle formed is one-half the difference of the measures of the intercepted arcs.

10.17 Circumscribed Angle Theorem *(p. 546)*

The measure of a circumscribed angle is equal to 180° minus the measure of the central angle that intercepts the same arc.

10.18 Segments of Chords Theorem *(p. 552)*

If two chords intersect in the interior of a circle, then the product of the lengths of the segments of one chord is equal to the product of the lengths of the segments of the other chord.

10.19 Segments of Secants Theorem *(p. 553)*

If two secant segments share the same endpoint outside a circle, then the product of the lengths of one secant segment and its external segment equals the product of the lengths of the other secant segment and its external segment.

10.20 Segments of Secants and Tangents Theorem *(p. 554)*

If a secant segment and a tangent segment share an endpoint outside a circle, then the product of the lengths of the secant segment and its external segment equals the square of the length of the tangent segment.

Reference

Properties

Properties of Equality

Addition Property of Equality
If $a = b$, then $a + c = b + c$.

Subtraction Property of Equality
If $a = b$, then $a - c = b - c$.

Multiplication Property of Equality
If $a = b$, then $a \cdot c = b \cdot c, c \neq 0$.

Division Property of Equality
If $a = b$, then $\frac{a}{c} = \frac{b}{c}, c \neq 0$.

Reflexive Property of Equality
$a = a$

Symmetric Property of Equality
If $a = b$, then $b = a$.

Transitive Property of Equality
If $a = b$ and $b = c$, then $a = c$.

Substitution Property of Equality
If $a = b$, then a can be substituted for b (or b for a) in any equation or expression.

Properties of Segment and Angle Congruence

Reflexive Property of Congruence
For any segment AB, $\overline{AB} \cong \overline{AB}$.

For any angle A, $\angle A \cong \angle A$.

Symmetric Property of Congruence
If $\overline{AB} \cong \overline{CD}$, then $\overline{CD} \cong \overline{AB}$.

If $\angle A \cong \angle B$, then $\angle B \cong \angle A$.

Transitive Property of Congruence
If $\overline{AB} \cong \overline{CD}$ and $\overline{CD} \cong \overline{EF}$, then $\overline{AB} \cong \overline{EF}$.

If $\angle A \cong \angle B$ and $\angle B \cong \angle C$, then $\angle A \cong \angle C$.

Other Properties

Transitive Property of Parallel Lines
If $p \parallel q$ and $q \parallel r$, then $p \parallel r$.

Distributive Property
Sum
$a(b + c) = ab + ac$
Difference
$a(b - c) = ab - ac$

Triangle Inequalities

Triangle Inequality Theorem

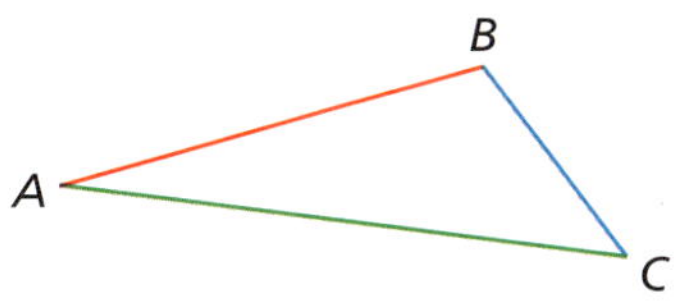

$AB + BC > AC$
$AC + BC > AB$
$AB + AC > BC$

Pythagorean Inequalities Theorem

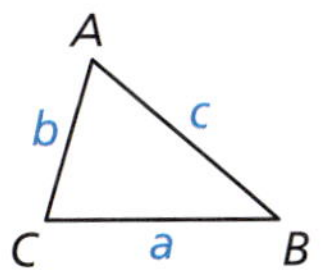

If $c^2 < a^2 + b^2$, then $\triangle ABC$ is acute.

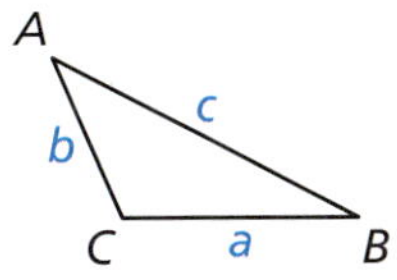

If $c^2 > a^2 + b^2$, then $\triangle ABC$ is obtuse.

Formulas

Coordinate Geometry

Slope

$m = \dfrac{y_2 - y_1}{x_2 - x_1}$

Slope-intercept form

$y = mx + b$

Point-slope form

$y - y_1 = m(x - x_1)$

Standard form of a linear equation

$Ax + By = C$

Standard equation of a circle

$(x - h)^2 + (y - k)^2 = r^2$, with center (h, k) and radius r

Midpoint Formula

$\left(\dfrac{x_1 + x_2}{2}, \dfrac{y_1 + y_2}{2}\right)$

Distance Formula

$d = \sqrt{(x_2 - x_1)^2 + (y_2 - y_1)^2}$

Polygons

Triangle Sum Theorem

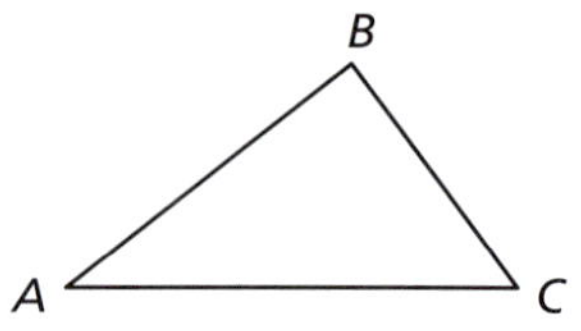

$m\angle A + m\angle B + m\angle C = 180°$

Exterior Angle Theorem

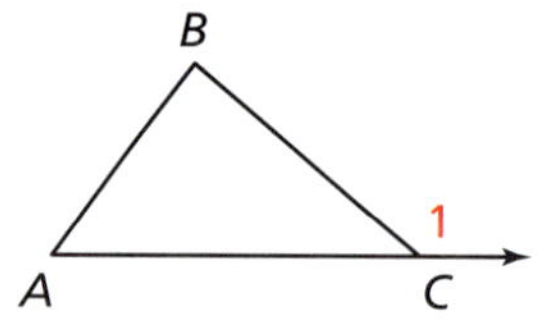

$m\angle 1 = m\angle A + m\angle B$

Triangle Midsegment Theorem

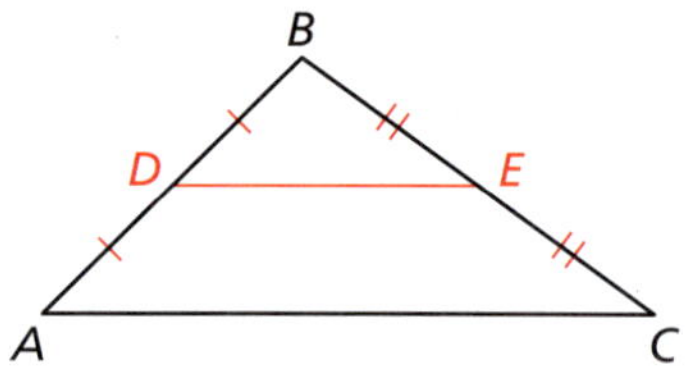

$\overline{DE} \parallel \overline{AC}$, $DE = \frac{1}{2}AC$

Trapezoid Midsegment Theorem

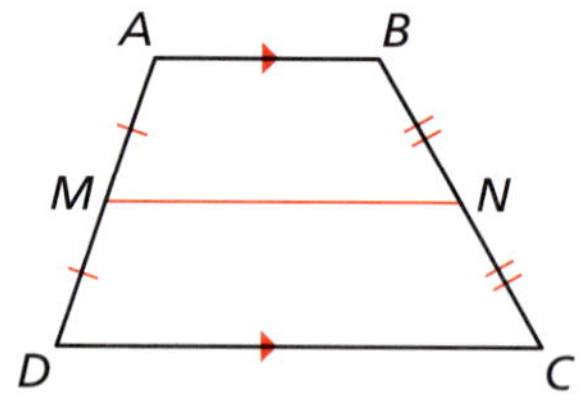

$\overline{MN} \parallel \overline{AB}$, $\overline{MN} \parallel \overline{DC}$, $MN = \frac{1}{2}(AB + CD)$

Polygon Interior Angles Theorem

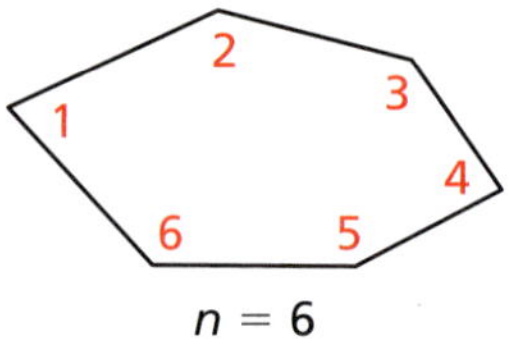

$m\angle 1 + m\angle 2 + \cdots + m\angle n = (n - 2) \cdot 180°$

Polygon Exterior Angles Theorem

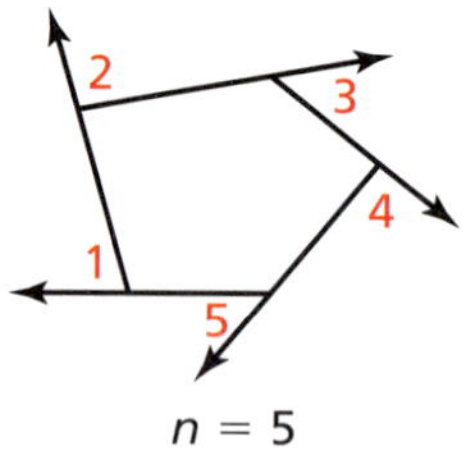

$m\angle 1 + m\angle 2 + \cdots + m\angle n = 360°$

Geometric Mean (Altitude) Theorem

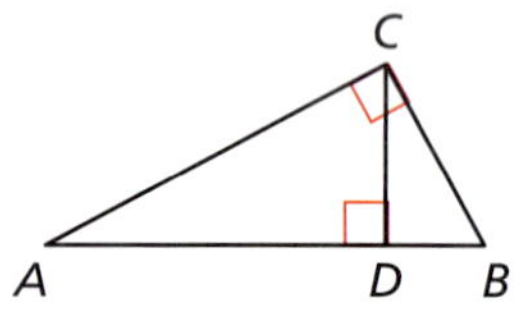

$CD^2 = AD \cdot BD$

Geometric Mean (Leg) Theorem

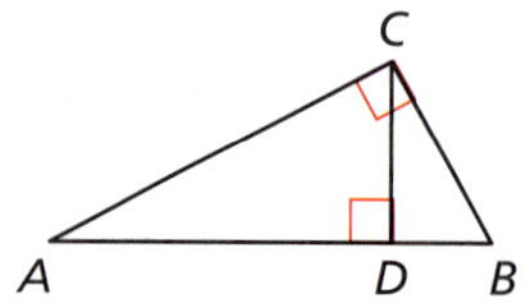

$CB^2 = DB \cdot AB$ $\qquad$ $AC^2 = AD \cdot AB$

Right Triangles

Pythagorean Theorem

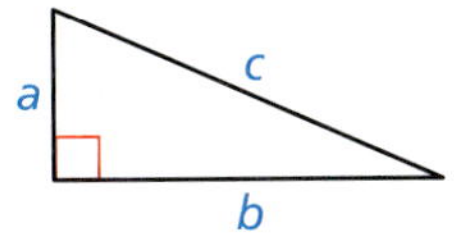

$a^2 + b^2 = c^2$

45°-45°-90° Triangles

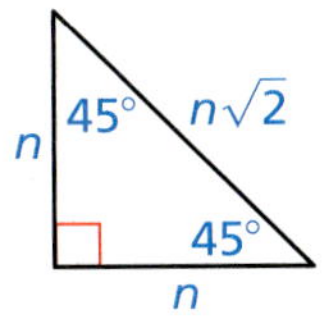

hypotenuse = leg • $\sqrt{2}$

30°-60°-90° Triangles

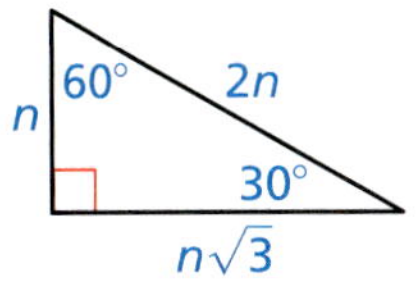

hypotenuse = shorter leg • 2
longer leg = shorter leg • $\sqrt{3}$

Trigonometry

Ratios

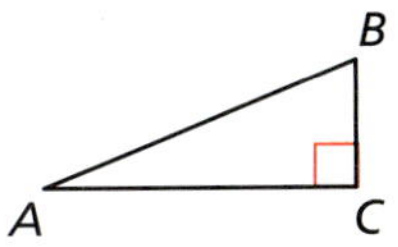

$\sin A = \dfrac{BC}{AB}$ $\qquad \cos A = \dfrac{AC}{AB}$ $\qquad \tan A = \dfrac{BC}{AC}$

$\sin^{-1}\dfrac{BC}{AB} = m\angle A$ $\qquad \cos^{-1}\dfrac{AC}{AB} = m\angle A$ $\qquad \tan^{-1}\dfrac{BC}{AC} = m\angle A$

Conversion between degrees and radians

$180° = \pi$ radians

Sine and cosine of complementary angles

Let A and B be complementary angles. Then the following statements are true.

$\sin A = \cos(90° - A) = \cos B$ $\qquad \sin B = \cos(90° - B) = \cos A$

$\cos A = \sin(90° - A) = \sin B$ $\qquad \cos B = \sin(90° - B) = \sin A$

Any Triangle

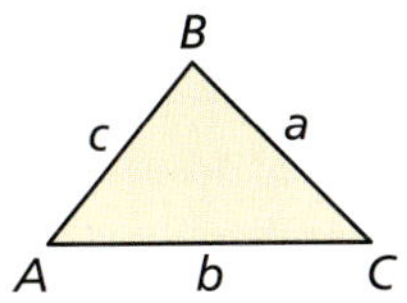

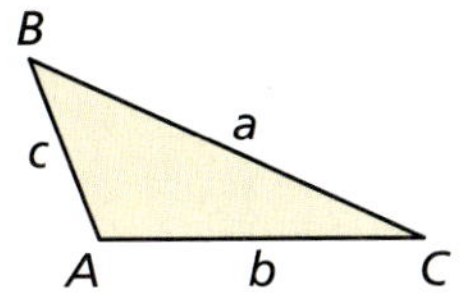

Area

$\text{Area} = \frac{1}{2}bc \sin A$

$\text{Area} = \frac{1}{2}ac \sin B$

$\text{Area} = \frac{1}{2}ab \sin C$

Law of Sines

$\dfrac{\sin A}{a} = \dfrac{\sin B}{b} = \dfrac{\sin C}{c}$

$\dfrac{a}{\sin A} = \dfrac{b}{\sin B} = \dfrac{c}{\sin C}$

Law of Cosines

$a^2 = b^2 + c^2 - 2bc \cos A$

$b^2 = a^2 + c^2 - 2ac \cos B$

$c^2 = a^2 + b^2 - 2ab \cos C$

Circles

Arc length

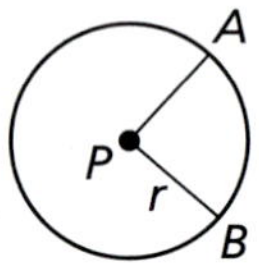

Arc length of $\overset{\frown}{AB} = \frac{m\overset{\frown}{AB}}{360°} \cdot 2\pi r$

Area of a sector

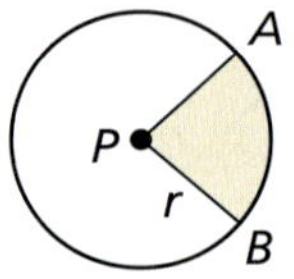

Area of sector $APB = \frac{m\overset{\frown}{AB}}{360°} \cdot \pi r^2$

Central angles

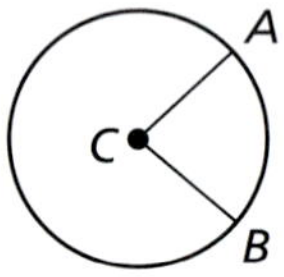

$m\angle ACB = m\overset{\frown}{AB}$

Inscribed angles

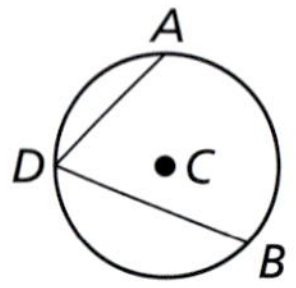

$m\angle ADB = \frac{1}{2}(m\overset{\frown}{AB})$

Tangent and intersected chord

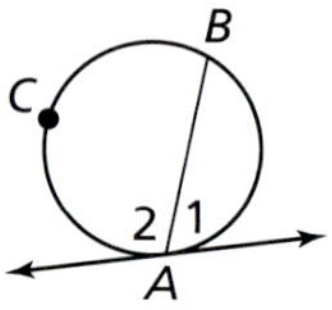

$m\angle 1 = \frac{1}{2}(m\overset{\frown}{AB})$

$m\angle 2 = \frac{1}{2}(m\overset{\frown}{BCA})$

Angles and Segments of Circles

Two chords

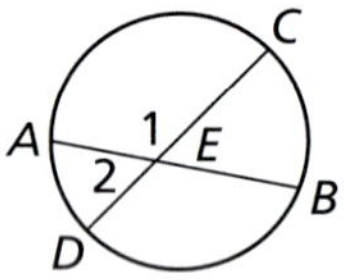

$m\angle 1 = \frac{1}{2}(m\overset{\frown}{AC} + m\overset{\frown}{DB})$

$m\angle 2 = \frac{1}{2}(m\overset{\frown}{AD} + m\overset{\frown}{CB})$

$EA \cdot EB = EC \cdot ED$

Two secants

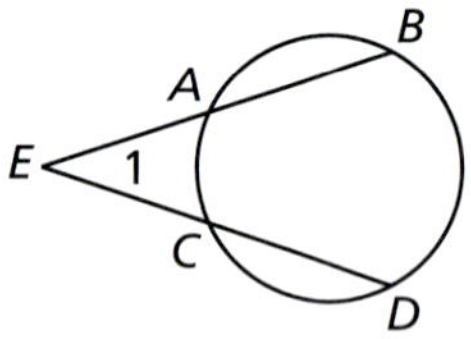

$m\angle 1 = \frac{1}{2}(m\overset{\frown}{BD} - m\overset{\frown}{AC})$

$EA \cdot EB = EC \cdot ED$

Tangent and secant

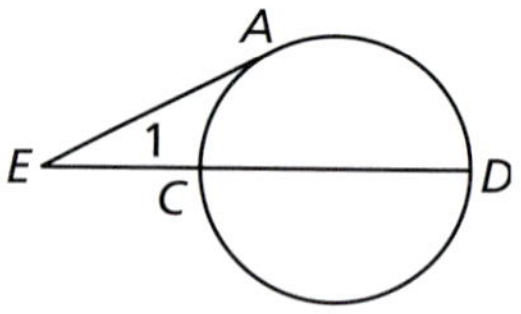

$m\angle 1 = \frac{1}{2}(m\overset{\frown}{AD} - m\overset{\frown}{AC})$

$EA^2 = EC \cdot ED$

Two tangents

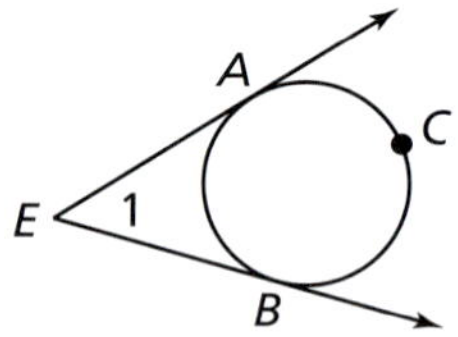

$m\angle 1 = \frac{1}{2}(m\overset{\frown}{ACB} - m\overset{\frown}{AB})$

$EA = EB$

Perimeter, Area, and Volume Formulas

Square

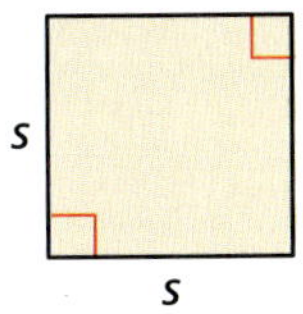

$P = 4s$
$A = s^2$

Rectangle

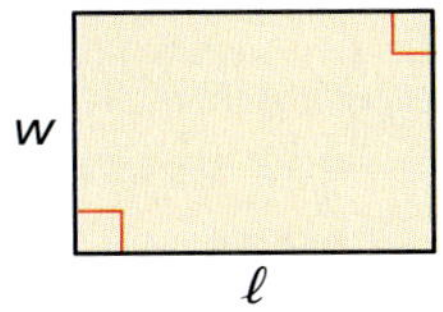

$P = 2\ell + 2w$
$A = \ell w$

Triangle

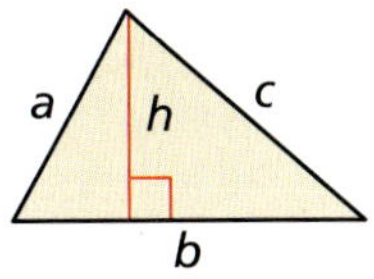

$P = a + b + c$
$A = \frac{1}{2}bh$

Circle

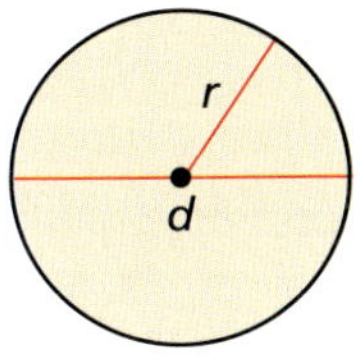

$C = \pi d$ or $C = 2\pi r$
$A = \pi r^2$

Parallelogram

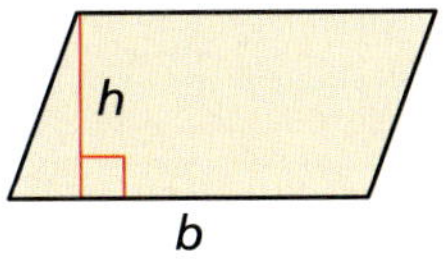

$A = bh$

Trapezoid

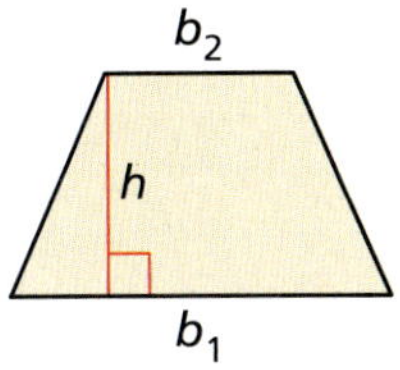

$A = \frac{1}{2}h(b_1 + b_2)$

Rhombus/Kite

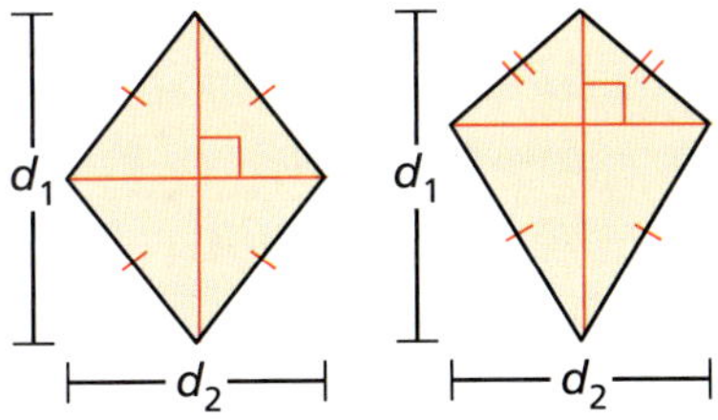

$A = \frac{1}{2}d_1d_2$

Regular *n*-gon

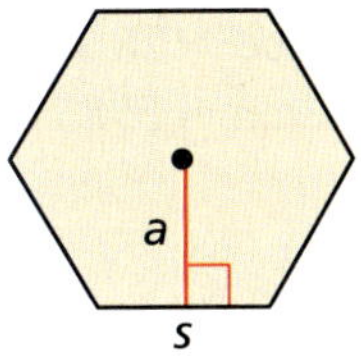

$A = \frac{1}{2}aP$ or $A = \frac{1}{2}a \cdot ns$

Prism

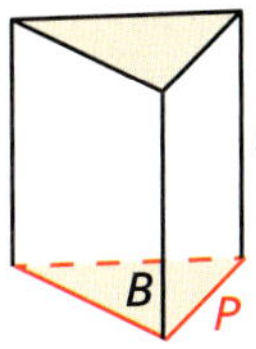

$L = Ph$
$S = 2B + L = aP + Ph$
$V = Bh$

Cylinder

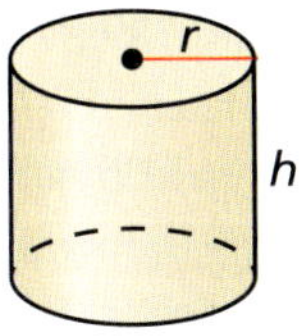

$L = 2\pi rh$
$S = 2B + L = 2\pi r^2 + 2\pi rh$
$V = \pi r^2 h$

Pyramid

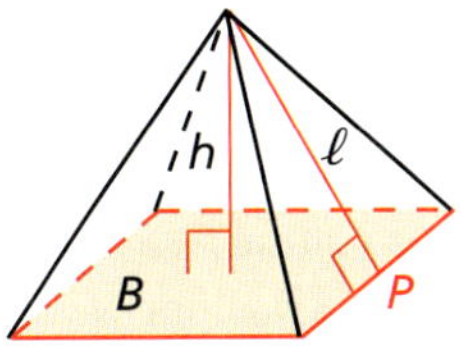

$L = \frac{1}{2}P\ell$
$S = B + L = B + \frac{1}{2}P\ell$
$V = \frac{1}{3}Bh$

Cone

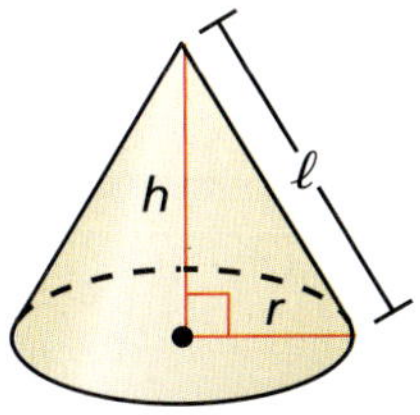

$L = \pi r\ell$
$S = B + L = \pi r^2 + \pi r\ell$
$V = \frac{1}{3}\pi r^2 h$

Sphere

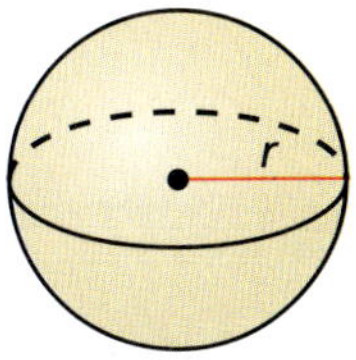

$S = 4\pi r^2$
$V = \frac{4}{3}\pi r^3$

Other Formulas

Geometric mean

$x = \sqrt{a \cdot b}$

Quadratic Formula

$x = \dfrac{-b \pm \sqrt{b^2 - 4ac}}{2a}$,

where $a \neq 0$ and $b^2 - 4ac \geq 0$

Density

$\text{Density} = \dfrac{\text{Mass}}{\text{Volume}}$

Similar polygons or similar solids with scale factor $a : b$

Ratio of perimeters $= a : b$

Ratio of areas $= a^2 : b^2$

Ratio of volumes $= a^3 : b^3$

Population Density

$\text{Population density} = \dfrac{\text{number of people}}{\text{area of land}}$

Conversions

U.S. Customary

1 foot = 12 inches
1 yard = 3 feet
1 mile = 5280 feet
1 mile = 1760 yards
1 acre = 43,560 square feet
1 cup = 8 fluid ounces
1 pint = 2 cups
1 quart = 2 pints
1 gallon = 4 quarts
1 gallon = 231 cubic inches
1 pound = 16 ounces
1 ton = 2000 pounds
1 cubic foot ≈ 7.5 gallons

U.S. Customary to Metric

1 inch = 2.54 centimeters
1 foot ≈ 0.3 meter
1 mile ≈ 1.61 kilometers
1 quart ≈ 0.95 liter
1 gallon ≈ 3.79 liters
1 cup ≈ 237 milliliters
1 pound ≈ 0.45 kilogram
1 ounce ≈ 28.3 grams
1 gallon ≈ 3785 cubic centimeters

Time

1 minute = 60 seconds
1 hour = 60 minutes
1 hour = 3600 seconds
1 year = 52 weeks

Temperature

$C = \frac{5}{9}(F - 32)$

$F = \frac{9}{5}C + 32$

Metric

1 centimeter = 10 millimeters
1 meter = 100 centimeters
1 kilometer = 1000 meters
1 liter = 1000 milliliters
1 kiloliter = 1000 liters
1 milliliter = 1 cubic centimeter
1 liter = 1000 cubic centimeters
1 cubic millimeter = 0.001 milliliter
1 gram = 1000 milligrams
1 kilogram = 1000 grams

Metric to U.S. Customary

1 centimeter ≈ 0.39 inch
1 meter ≈ 3.28 feet
1 meter ≈ 39.37 inches
1 kilometer ≈ 0.62 mile
1 liter ≈ 1.06 quarts
1 liter ≈ 0.26 gallon
1 kilogram ≈ 2.2 pounds
1 gram ≈ 0.035 ounce
1 cubic meter ≈ 264 gallons

Credits

Chapter 1

0 *top* ©Rae Wynn-Grant; *bottom* lassedesignen/Shutterstock.com; **1** Alexey Seafarer/Shutterstock.com; **3** drbimages/E+/Getty Images; **7** *left* BirdShutterB/iStock/Getty Images Plus; *right* Benjah-bmm27; **9** *top left* virusowy/E+/Getty Images; *bottom left* Yayasya/iStock/Getty Images Plus; *Exercise 37* photos777/iStock/Getty Images Plus; *Exercise 38* loops7/DigitalVision Vectors/Getty Images; peepo/E+/Getty Images; *Exercise 39* joppo/Shutterstock.com; *Exercise 40* Nyord/Shutterstock.com; **11** *left* wavebreakmedia/Shutterstock.com; *right* Nirut Punshiri/iStock/Getty Images Plus, venakr/iStock/Getty Images Plus, venakr/iStock/Getty Images Plus; **17** MediaProduction/iStock/Getty Images Plus; **19** Cisco Freeze/Shutterstock.com; **20** Tomwang112/iStock/Getty Images Plus; **23** Atstock Productions/Shutterstock.com; **26** *left* Mark Herreid/Shutterstock.com; *right* **27** stevanovicigor/iStock/Getty Images Plus; **29** diogoppr/Shutterstock.com; **33** aoldman/iStock/Getty Images Plus; **35** *Exercise 23* Nelson_A_Ishikawa/iStock/Getty Images Plus; *Exercise 24* GlobalP/iStock/Getty Images Plus; JordiStock/iStock/Getty Images Plus; **37** PIKSEL/iStock/Getty Images Plus; **40** Ekaterina Romanova/iStock/Getty Images Plus; **43** Mongpro/iStock/Getty Images Plus ; **45** Lewis Calvert Cooper (gonzoshots.com); **47** Jonny White/Alamy Stock Photo; **49** XiXinXing/iStock/Getty Images Plus; **51** Ebtikar/Shutterstock.com; **53** *top* Francisco Javier Alcerreca Gomez/Shutterstock.com; *bottom* Shane Trotter/Shutterstock.com; **54** _LeS_/iStock/Getty Images Plus; **57** TommL/E+/Getty Images; **59** *top* pigphoto/iStock/Getty Images Plus; *bottom* Andyworks/E+/Getty Images

Chapter 2

62 *top* ©Caroline Quanbeck; *bottom* Sharp/Shutterstock.com; **67** Rawpixel/iStock/Getty Images Plus; **68** Tom Wang/Shutterstock.com; **70** Nynke van Holten/iStock/Getty Images Plus; **71** Marat Musabirov/iStock/Getty Images Plus; **74** Dean Drobot/Shutterstock.com; **76** track5/iStock/Getty Images Plus; **79** BMCL/Shutterstock.com; **80** *left* vectorOK/Shutterstock.com; *Exercise 43 in order* Nastya Pirieva/Shutterstock.com; MarcelC/iStock/Getty Images Plus; Dafinchi/iStock/Getty Images Plus; Coldmoon_photo/iStock/Getty Images Plus; **81** Elenathewise/iStock/Getty Images Plus; **83** Rawpixel/iStock/Getty Images Plus; **87** SDI Productions/E+/Getty Images; **89** Pixel-Shot/Shutterstock.com; **90** *left* hjalmeida/iStock/Getty Images Plus; *right* Twin Design/Shutterstock.com; **91** Prostock-Studio/iStock/Getty Images Plus; **93** *left* Rocksweeper/Shutterstock.com; *right* Tiger Images/Shutterstock.com; **96** Twin Design/Shutterstock.com; **100** Oliver Hoffmann/Shutterstock.com; **104** Kamenetskiy Konstantin/Shutterstock.com; **106** caimacanul/Shutterstock.com; **107** Floortje/iStock/Getty Images Plus; **108** m-imagephotography/iStock/Getty Images Plus; **112** Nattstudio/Shutterstock.com; **114** Natykach Nataliia/Shutterstock.com; **115** imaginima/E+/Getty Images, Kativ/E+/Getty Images; **117** Dean Drobot Shutterstock.com

Chapter 3

120 *top* ©Cengage Learning/National Geographic Learning; *bottom* Mateusz Liberra/Shutterstock.com; **121** *top* Nirut Punshiri/iStock/Getty Images Plus, Ivantsov/iStock/Getty Images Plus; *left* paffy/Shutterstock.com; *right* stockbob/Shutterstock.com; **123** AndreyPopov/iStock/Getty Images Plus; **125** *top right* Aldona Griskeviciene/Shutterstock.com; *bottom right* Nastya22/iStock/Getty Images Plus; **126** Iamnee/Shutterstock.com; **127** londoneye/iStock/Getty Images Plus; **130** Hayri Er/E+/Getty Images; **133** Lucian Coman/Shutterstock.com; **137** Sanit Fuangnakhon/Shutterstock.com; **138** koosen/iStock/Getty Images Plus; **139** EpicStockMedia/Shutterstock.com; **141** szefei/iStock/Getty Images Plus; **144** michaeljung/iStock/Getty Images Plus; **145** *top* kokouu/iStock/Getty Images Plus; *bottom* Orbon Alija/E+/Getty Images; **146** *Exercise 15* valio84sl/iStock/Getty Images Plus; *Exercise 16* gowithstock/Shutterstock.com; **149** Ljupco/iStock/Getty Images Plus; **158** okandilek/iStock/Getty Images Plus; **159** bashta/iStock/Getty Images Plus; **160** dnd_project/Shutterstock.com; **161** *background* Johann Ragnarsson/Shutterstock.com; *top* Mopic/Shutterstock.com; **162** Inhabitant/Shutterstock.com

Chapter 4

164 *top* ©Aaron Pomerantz/National Geographic Image Collection; suns07butterfly/Shutterstock.com; *bottom* Anton-Burakov/Shutterstock.com; **165** Hintau Aliaksei/Shutterstock.com; **167** *right* oatintro/iStock/Getty Images Plus; *left* Wavebreakmedia/iStock/Getty Images Plus; **170** 4x6/iStock/Getty Images Plus; **171** SLP_London/iStock/Getty Images Plus; **173** ChrisAt/iStock/Getty Images Plus; **177** ferlistockphoto/iStock/Getty Images Plus; **179** Kokai Zsuzsanna/Shutterstock.com; **183** *left* suhendri/Shutterstock.com; *right* Billion Photos/Shutterstock.com; **190** Biletskiy_Evgeniy/iStock/Getty Images Plus; **191** *top* denisik11/iStock/Getty Images Plus; *bottom* dja65/iStock/Getty Images Plus; *right* Carol Mellema/Shutterstock.com; **193** Hans Kim/Shutterstock.com; **195** VladimirFLoyd/iStock/Getty Images Plus; **197** Dean Mitchell/E+/Getty Images; **198** mocker_bat/iStock/Getty Images Plus; **200** *top* javarman3/iStock/Getty Images Plus; *bottom* Graphics Studio MH. We offer you the best things/iStock/Getty Images Plus; **203** Veronica Louro/Shutterstock.com; **207** *top* martin-dm/E+/Getty Images; *bottom* Henrik Larsson/Shutterstock.com; **208** Kevin Peterson/Photodisc/Getty Images; **209** *In order* Protasov AN/Shutterstock.com; irin-k/Shutterstock.com; photomaster/Shutterstock.com; Eric Isselee/Shutterstock.com; Lightspring/Shutterstock.com; **211** Kaesler Media/Shutterstock.com; **213** My Life Graphic/Shutterstock.com; **222** FatCamera/E+/Getty Images; **223** *background* golden_SUN/iStock/Getty Images Plus; *in order from left to right* Mathee Boonphram/123RF; Thawat Tanhai/123RF; Cosmin Manci/123RF; Geza Farkas/123RF; Geza Farkas/123RF; Vac1/iStock/Getty Images Plus; MarkMirror/iStock/Getty Images Plus; Iuliia Morozova/iStock/Getty Images Plus; Image from the Smithsonian, "Bug Info: Butterflies in the United States" (https://www.si.edu/spotlight/buginfo/butterflyus). Prepared by the Department of Systematic Biology, Entomology Section, National Museum of Natural History; Marco Uliana/123RF; mage from the Smithsonian, "Bug Info: Butterflies in the United States" (https://www.si.edu/spotlight/buginfo/butterflyus). Prepared by the Department of Systematic Biology, Entomology Section, National Museum of Natural History; Onfokus/E+/Getty Images; *bottom* Sari Oneal/Shutterstock.com; Hugo Felix/Shutterstock.com

Chapter 5

226 *top* ©Mark Thiessen/National Geographic Image Collection; *bottom* traffic_analyzer/DigitalVision Vectors/Getty Images; **229** *right* tuulijumala/iStock/Getty Images Plus; *left* ferlistockphoto/iStock/Getty Images Plus; **230** Vinnikava Viktoryia/Shutterstock.com; **232** PIKSEL/iStock/Getty Images Plus; **233** AndreyPopov/iStock/Getty Images Plus; **237** hexvivo/iStock/Getty Images Plus; **240** Dean Drobot/Shutterstock.com; **243** *right* hocus-focus/E+/Getty Images; *left* tetmc/iStock/Getty Images Plus; *top* Arthur Eugene Preston/Shutterstock.com; *bottom left* ritno kurniawan/Shutterstock.com; *bottom right* Krakenimages.com/Shutterstock.com; **249** *top* OvsiankaStudio/iStock/Getty Images Plus; *bottom* Elnur/Shutterstock.com; **253** *left* Ljupco/iStock/Getty Images Plus; *right* Colin D. Young/Shutterstock.com; **254** Leonard Zhukovsky/Shutterstock.com; **257** *right* OvsiankaStudio/iStock/Getty Images Plus; *left* g-stockstudio/iStock/Getty Images Plus; **263** *top left* antpkr/Shutterstock.com; *bottom left* Koksharov Dmitry/Shutterstock.com; *bottom right* tarasov_vl/iStock/Getty Images Plus; **264** style-photography/iStock/Getty Images Plus; **265** *right* tuulijumala/iStock/Getty Images Plus; *left* furtaev/iStock/Getty Images Plus; **271** m-imagephotography/iStock/Getty Images Plus; **272** image_jungle/iStock/Getty Images Plus; **273** Ollyy/Shutterstock.com; **274** DanCardiff/iStock/Getty Images Plus; **275** Kevin Fleming/Corbis Documentary/Getty Images; **276** Ebtikar/Shutterstock.com; **277** *top* pisaphotography/Shutterstock.com; *bottom* abzee/E+/Getty Images; **279** Hogan Imaging/Shutterstock.com; **281** Elenathewise/iStock/Getty Images Plus; **284** zyxeos30/iStock/Getty Images Plus; **286** Photology1971/Shutterstock.com; **287** *top* Glot/Shutterstock; *bottom* Taurus106/Shutterstock.com; **290** Cris Foto/Shutterstock.com; **291** *background* luismmolina/iStock/Getty Images Plus, ktsimage/iStock/Getty Images Plus; *In order from top to bottom* Science Photo Library | KATERYNA KON; Kateryna Kon/Shutterstock.com; Design Cells/iStock/Getty Images Plus; Science Photo Library | KATERYNA KON

Chapter 6

294 *top* ©Amy Gusick/National Geographic Image Collection; *bottom* John Kepchar/Shutterstock.com; **297** *right* Luis Molinero/Shutterstock.com; *left* Flashon Studio/Shutterstock.com; **299** RailROL82/Shutterstock.com; **300** *left* leolintang/iStock/Getty Images Plus, Roman ValieviStock/Getty Images Plus; *right* simonkr/E+/Getty Images; **305** *right* OvsiankaStudio/iStock/Getty Images Plus; *left* michaeljung/Shuterstock.com; **307** *top left* Marcelo_Krelling/Shutterstock.com; *top right* Alex Tuzhikov/Shutterstock.com; *bottom* FG Trade/E+/Getty Images Plus; **312** amwu/iStock/Getty Images Plus; **313** *top* Ozgur Coskun/iStock/Getty Images Plus; *Exercise 42* pedrosala/iStock/Getty Images Plus; *Exercise 46* FrankRamspott/E+/Getty Images; **315** *right* tuulijumala/iStock/Getty Images Plus; *left* Take A Pix Media?shutterstock.com; **317** *right* tuulijumala/iStock/Getty Images Plus; *left* BCFC/Shutterstock.com; **325** Ljupco/iStock/Getty Images Plus; **327** pics721/Shutterstock.com; **329** *right* oatintro/iStock/Getty Images Plus; *left* Dean Drobot/Shutterstock.com; **332** mahiruysal/iStock/Getty Images Plus; **334** Dzm1try/Shutterstock.com; **337** *right* tuulijumala/iStock/Getty Images Plus; *left* Pixel-Shot/Shutterstock.com; **339** 4x6/iStock/Getty Images Plus; **340** Ridofranz/iStock/Getty Images Plus; **343** stocksolutions/Shutterstock.com; **345** Dan Baciu/Shutterstock.com; **347** *background* de_zla/Shutterstock.com; *top left* jonnysek/iStock/Getty Images Plus; *top right* Dewin ' Indew/Shutterstock.com; *bottom* sb-borg/iStock/Getty Images Plus, Comstock/Stockbyte/Getty Images; **349** AlexRaths/iStock/Getty Images Plus

Chapter 7

350 *top* ©Rebecca Drobis/National Geographic Image Collection; *bottom* Sergey Nivens/Shutterstock.com; **351** Jasonfang/E+/Getty Images; **353** iodrakon/iStock/Getty Images Plus; **354** peterspiro/iStock Editorial/Getty Images Plus; **354** Ljupco/iStock/Getty Images Plus; **359** FABIO BISPO/iStock/Getty Images Plus; **361** *top* Roman Samborskyi/Shutterstock.com; *bottom* OvsiankaStudio/iStock/Getty Images Plus; **364** Africa Studio /Shutterstock.com; **369** Sinisa Bobic/Shutterstock.com; **370** AlexRaths/iStock/Getty Images Plus; **372** JackF/iStock/Getty Images Plus; **374** Boarding1Now/iStock/Getty Images Plus; **376** nonnie192/iStock/Getty Images Plus; **377** artisteer/iStock/Getty Images Plus; **379** *right* Roman Samokhin/iStock/Getty Images Plus; *left* Billion Photos/Shutterstock.com; **380** michaeljung/Shutterstock.com; **385** *Exercise 7* Ksenia Palimski/Shutterstock.com; *Exercise 8* jojotextures/iStock/Getty Images Plus; *Exercise 9* mphillips007/iStock/Getty Images Plus; *Exercise 10* donatas1205/Shutterstock.com; **387** Bahadirkar/iStock/Getty Images Plus; **389** *left* Victoria Kisel/Shutterstock.com; *right* OvsiankaStudio/iStock/Getty Images Plus; **391** Roman Tiraspolsky/Shutterstock.com; **396** archideaphoto/iStock/Getty Images Plus; **400** Matthew Cole/Shutterstock.com; **402** gokturk_06/Shutterstock.com; **403** valio84sl/iStock/Getty Images Plus; **404** amanalang/iStock/Getty Images Plus

Chapter 8

406 *top* ©Carter Clinton; *bottom* Creations/Shutterstock.com; **407** Chepko/iStock/Getty Images Plus; **403** *top* 24Novembers/Shutterstock.com; *bottom* feedough/iStock/Getty Images Plus; **409** *bottom* feedough/iStock/Getty Images Plus; *top* 24Novembers/Shutterstock.com; **412** Photitos2016/iStock/Getty Images Plus; **414** Dean Drobot/Shutterstock.com; **416** Kryuchka Yaroslav/Shutterstock.com; **417** Diabluses/iStock/Getty Images Plus; **419** *left* lenetstan/Shutterstock.com; *right* oatintro/iStock/Getty Images Plus; **422** Dolores M. Harvey/Shutterstock.com; **423** LagunaticPhoto/iStock/Getty Images Plus; left AndreyPopov/iStock/Getty Images Plus; **425** *left* AndreyPopov/iStock/Getty Images Plus; *right* CostinT/E+/Getty Images; **428** Asergieiev/iStock/Getty Images Plus; **432** *left* liveslow/iStock/Getty Images Plus; *right* spkphotostock/iStock/Getty Images Plus; *bottom right* KIRYAKOVA ANNA/Shutterstock.com; **433** pathdoc/Shutterstock.com; **434** Used by permission of Anthem Community Council/Renee Palmer-Jones, Designer; **435** *bottom* cc-stock/E+/Getty Images; *top* oatintro/iStock/Getty Images Plus; **436** nilimage/iStock/Getty Images Plus; **437** *left* Andregric/iStock/Getty Images Plus; *right* MaleWitch/iStock/Getty Images Plus; **442** Ryan Herron/iStock/Getty Images Plus; **443** Seregam/Shutterstock.com; **446** littlenySTOCK/Shutterstock.com, Viorel Sima/Shutterstock.com; **447** *In order from top to bottom* Tami Heilemann, Department of the Interior, U.S. National Archives; 'Infographic originally made by and for journalistic platform De Correspondent; African Burial Ground, New York, New York. Photograph in the Carol M. Highsmith Archive, Library of Congress, Prints and Photographs Division.; **449** aldomurillo/iSock/Getty Images Plus

Chapter 9

450 *top* ©Mark Thiessen/National Geographic Image Collection; *bottom* mjones/Shutterstock.com; **451** Africa Studio/Shutterstock.com; **455** lumpynoodles/DigitalVision Vectors/Getty Images; **458** SVPhilon/E+/Getty Images Plus; **461** *left* Digital Genetics/Shutterstock.com; *right* tuulijumala/iStock/Getty Images Plus; **463** sasha2109/Shutterstock.com; **465** DeborahMaxemow/iStock/Getty Images Plus; 467 Samuel Borges PhotographyShutterstock.com; **471** rappensuncle/Shutterstock.com; **472** *left* travelview/Shutterstock.com; *right* et-anan/Shutterstock.com; **473** chinahbzyg/Shutterstock.com, 4x6/iStock/Getty Images Plus, Kamenetskiy Konstantin/Shutterstock.com; **475** Samuel Borges Photography/Shutterstock.com; **478** *top* Kenneth Keifer/Shutterstock.com; *bottom* Aldo Murillo/iStock/Getty Images Plus; **479** Noah Densmore/Shutterstock.com; **480** BerndBrueggemann/iStock/Getty Images Plus; **481** Krakenimages.com/Shutterstock.com; **485** Kesu01/iStock/Getty Images Plus; **488** *Exercise 27* coffeekai/iStock/Getty Images Plus; *Exercise 27* Andyd/E+/Getty Images; *bottom* Cozine/Shutterstock.com; **491** *left* drbimages/E+/Getty Images; *right* Sergey Peterman/iStock/Getty Images; **496** Master1305/Shutterstock.com; **498** *left* Luis Molinero/Shutterstock.com, nickp37/iStock/Getty Images Plus; *right* zoom-zoom/iStock/Getty Images Plus;

502 4x6/iStock/Getty Images Plus, ARTYuSTUDIO/iStock/Getty Images Plus; **503** londoneye/E+/Getty Images; **504** Javen/Shutterstock.com; **505** BettinaRitter/E+/Getty Images; **506** *left* alina_danilova/ Shutterstock.com; *right* Bairachnyi Dmitry/Shutterstock.com; **507** *background* Orla/iStock/Getty Images Plus; *top* Escaflowne/E+/Getty Images; *bottom* Design Cells/iStock/Getty Images Plus; **509** Zhang Baohuan/Shutterstock.com

Chapter 10

510 *top* ©Mark Thiessen/National Geographic Image Collection; *bottom* CW Pix/Shutterstock.com; **511** Eduardo Ibarra/iStock/Getty Images Plus; **513** Asier Romero/Shutterstock.com; **519** ilyarexi/iStock/Getty Images Plus; **520** Nerthuz/iStock/Getty Images Plus; **521** Elr.Sanchez/Shutterstock.com; **527** ruslansemichev/iStock/Getty Images Plus; **528** robophobic/iStock/Getty Images Plus; **529** *left* Dean Drobot/Shutterstock.com; *right* tuulijumala/ iStock/Getty Images Plus; 533 *top* numismarty/iStock/Getty Images Plus; *bottom* M. Unal Ozmen/Shutterstock.com; **535** oatintro/iStock/Getty Images Plus; **537** michaeljung/iStock/Getty Images Plus; **539** studiocasper/E+/Getty Images; **540** Floortje/iStock/Getty Images Plus; **542** Nobilior/iStock/Getty Images Plus; **543** *left* leungchopan/Shutterstock.com; *right* tuulijumala/ iStock/Getty Images Plus; **545** Elnur/Shutterstock.com; **547** leonello/iStock/ Getty Images Plus; **548** ElementalImaging/iStock/Getty Images Plus; **551** oatintro/iStock/Getty Images Plus; **556** *left* Alhovik/Shutterstock.com; *right* Thanwit Intrarak/Shutterstock.com; **557** Francesco83/Shutterstock.com; **563** *left* Timothy Messick/DigitalVision Vectors/Getty Images; *right* DaLiu/ iStock/Getty Images Plus; **564** Maksimovairina/iStock/Getty Images Plus; **566** ramzihachicho/iStock/Getty Images Plus; **568** cicerocastro/iStock/Getty Images Plus; **571** *background* Yas.K/iStock/Getty Images Plus; *In order from left to right (starting at top)* Nicholas Grey/Shutterstock.com; TonyBaggett/ iStock/Getty Images Plus; Gerasimov174/iStock/Getty Images Plus; ipopba/ iStock/Getty Images Plus; duncan1890/DigitalVision Vectors/Getty Images

Chapter 11

574 *top* ©Alize Carrere/National Geographic Image Collection; *bottom* Angela N Perryman/Shutterstock.com; **575** phant/iStock/Getty Images Plus; **577** losw/iStock/Gett Images Plus; **580** Gwoeii/Shutterstock.com; **582** *bottom* ©iStockphoto.com/Prill Mediendesign & Fotografie; *top* ConstantinosZ/iStock/Getty Images Plus; **583** *top* Gearstd/iStock/Getty Images Plus; *middle* Chris Chetters/Shutterstock.com; *bottom* BlackJack3D/ E+/Getty Images; **585** Dovapi/iStock/Getty Images Plus; **586** Ranta Images/ Shutterstock.com; **588** Mizina/iStock/Getty Images Plus, alle12/iStock/ Getty Images Plus; **589** Victoria Kalinina/Shutterstock.com; **590** Agata Buczek/Shutterstock.com; **591** *left* anon_tae/Shutterstock.com; *right* tuulijumala/iStock/Getty Images Plus; **595** *left* ©iStockphoto.com/ Olgertas; *right* keellla/Shutterstock.com; **597** *top* Smoo/iStock/Getty Images Plus; *bottom* Laschon Maximilian/Shutterstock.com; **599** GagliardiPhotography/Shutterstock.com; **601** Edoma/Shutterstock.com; **602** Hugnoi/iStock/Getty images Plus; **603** SeanPavonePhoto'/iStock/Getty Images Plus; **604** inhauscreative/iStock/Getty Images Plus; **605** Ebtikar/ Shutterstock.com; **606** *top* ninikas/iStock/Getty Images Plus; *bottom* mipan/ iStock/Getty Images Plus; **607** *top* Andrii Yalanskyi/Shutterstock.com; *bottom* Ljupco/iStock/Getty Images Plus; **608** ryasick/E+/Getty Images; **609** *background* Jurie Maree/Shutterstock.com; *top right* BrianBrownImages/ iStock/Getty Images Plus; *center right* Songbird839/iStock/Getty Images Plus; **610** mocker_bat/iStock/Getty Images Plus

Chapter 12

612 *top* ©Daniela Cafaggi Lemus; *bottom* Photoongraphy/Shutterstock.com; **613** GlobalP/iStock/Getty Images Plus; **615** *left* drbimages/E+/Getty Images; *In order from left to right* matejmm/iStock/Getty Images Plus; vladimir_karpenyuk/iStock/Getty Images Plus; pamela_d_mcadams/iStock/ Getty Images Plus; ChubarovY/iStock/Getty Images Plus; **618** ImageDB/ iStock/Getty Images Plus; **619** popovaphoto/iStock/Getty Images Plus; **621** penguiiin/iStock/Getty Images Plus; **622** dimarik/iStock/Getty Images Plus; **623** Ljupco/iStock/Getty Images Plus; **631** 4x6/iStock/Getty Images Plus; **636** *top* Louise Addison/iStock/Getty Images Plus; *bottom* Inna Tarasenko/ iStock/Getty Images Plus; **637** *top* Tatiana Popova/ Shutterstock.com; *middle* goodze/iStock/Getty Images Plus; *bottom* Steve Heap/Shutterstock.com; **640** Blade_kostas/iStock/Getty Images Plus; **643** *left* Gunnar Pippel/Shutterstock.com; *top* ronstik/Shutterstock.com, YuriyZhuravov/Shutterstock.com, Petrenko Andriy/Shutterstock.com, Rafa Irusta/Shutterstock.com, *right* nzfhatipoglu/iStock/Getty Images Plus; **645** aldomurillo/E+/Getty Images; **648** koya79/iStock/Getty Images Plus; **651** © Richard Lowthian | Dreamstime.com; **653** *left* PRImageFactory/iStock/ Getty Images Plus; *right* Mark Herreid/Shutterstock.com; **654** Dan Thornberg/Shutterstock.com; **659** *top* Nerthuz/iStock/Getty Images Plus; *Exercise 26* stockcam/E+/Getty Images; *bottom* catalinr/iStock/Getty Images Plus; *Exercise 34* Npeter/Shutterstock.com; **660** KangeStudio/ iStock/Getty Images Plus; **661** *In order from left to right* Mega Pixel/ Shutterstock.com; JIANG HONGYAN/Shutterstock.com; forest_strider/ Shutterstock.com; Mario Savoia/Shutterstock.com; **662** *top* witoldkr1/ iStock/Getty Images Plus; *bottom* Only Fabrizio/Shutterstock.com; **663** Bill Oxford/iStock/Getty Images Plus; **664** *top* karandaev/iStock/Getty Images Plus; *bottom* Suzanne Tucker/Shutterstock.com; **665** *top* paperbees/ Shutterstock.com; *Exercise 3* visual7/E+/Getty Image; *Exercise 8* Андрей Елкин/iStock/Getty Images Plus; *bottom* Gyvafoto/Shutterstock.com, slav/ iStock/Getty Images Plus; **667** Paffy69/iStock/Getty Images Plus; **668** *In order from left to right* GeorgeMPhotography/Shutterstock.com; Ferenz/ Shutterstock.com; bitt24/Shutterstock.com ©iStockphoto.com/kedsanee; **669** Ferenz/Shutterstock.com; **671** *Exercise 13* Carlos E. Santa Maria/ Shutterstock.com; *Exercise 14* ravl/Shutterstock.com; *Exercise 15* ravl/ Shutterstock.com; *Exercise 16* Jorg Hackemann/Shutterstock.com; **672** *left* Mananya Kaewthawee/iStock/Getty Images Plus; *right* Flashon Studio/ Shutterstock.com; **675** mbrand85/iStock/Getty Images Plus; **676** ©iStockphoto.com/mlevy; **677** *left* ThomasVogel/iStock/Getty Images Plus; *right* Chimpinski/iStock/Getty Images Plus; *bottom* Alejandro Morales Lozano/iStock/Getty Images Plus; **678** CreativeI/E+/Getty Images; **679** *background* SAHACHATZ/Shutterstock.com; *top* Ryan von Linden/New York Department of Environmental Conservation; *bottom* Remus86/iStock/ Getty Images Plus

Course Project

682 LagunaticPhoto/iStock/Getty Images Plus; **683** malerapaso/iStock/Getty Images Plus;